18982

Y0-ACF-246

Climate Normals
for the U.S.
(Base: 1951-80)

Climate Normals
for the U.S.
(Base: 1951-80)

First Edition

Data Elements Compiled by
National Climatic Center
Environmental Data and Information Service
National Oceanic and Atmospheric Administration

Gale Research Company
Book Tower • Detroit, Michigan 48226

Bibliographic Note

This book is a compilation of fifty separate pamphlets originally issued by the National Climatic Center during 1983 under the title, *Monthly Normals of Temperature, Precipitation, and Heating and Cooling Degree Days 1951-80 (State Name)*. The fifty pamphlets, each depicting climate normals of an individual state, were issued as part of the on-going series, *Climatography of the United States No. 81 (By State)*. The series is not available as a one-volume unity except in this Gale Research Company edition.

CONTENTS

Introduction 7

Facts about Climate Normals 9

STATE CLIMATE NORMALS

Alabama 11

Alaska 22

Arizona 30

Arkansas 46

California 61

Colorado 97

Connecticut 113

Delaware 119

Florida 125

Georgia 139

Hawaii 153

Idaho 163

Illinois 177

Indiana 193

Iowa 206

Kansas 224

Kentucky 243

Louisiana 253

Maine 263

Maryland 270

Massachusetts 278

Michigan 287

Minnesota 304

Mississippi 320

Missouri 333

Montana 350

Nebraska 373

Nevada 392

New Hampshire 399

New Jersey 405

New Mexico 412

New York 428

North Carolina 444

North Dakota 460

Ohio 477

Oklahoma 493

Oregon 511

Pennsylvania 528

Rhode Island 543

South Carolina 549

South Dakota 560

Tennessee 574

Texas 585

Utah 619

Vermont 633

Virginia 639

Washington 649

West Virginia 666

Wisconsin 676

Wyoming 697

The concept of heating and cooling degree days 711

INTRODUCTION

At the close of each decade, the National Climatic Center recalculates one of the most important and comprehensive of all climate measurements, the thirty-year *climatic normals base.* The base is computed from the most recent 30 years of accumulated reports by 5557 weather stations across the nation, and it becomes the standard of reference for all meteorological and climatic evaluations for the next ten years.

Key elements of the new basic data are then released, not as a single volume, but as a series of fifty reports which describe the "new" climate picture for each of the fifty states.

Climate Normals for the U.S. (Base: 1950-80) puts these frequently consulted paperbound reports into a permanent form suitable for reference use over the next 10 years.

Reports from every weather station in every state are given in this compilation. For nearly all the stations, these four key items of information are given in separate sections, as follows:

> **Temperature normals,** by month, showing for each month for each station: 1) the average of the daily *maximum* temperatures; 2) the average of the daily *minimum* temperatures for each month, and 3) the average of the daily *mean* temperatures for each month.
>
> Precipitation normals, by month.
>
> Heating degree day normals, by month. (The heating degree day is the index which has become so important in tracking fuel consumption and cold weather energy needs.)
>
> Cooling degree day normals, by month. (The cooling degree day is the summer weather index which helps quantify air conditioning requirements.)

A supplemental section provides necessary descriptions of the reporting stations. Each station's elevation, as well as its precise longitude and latitude, are noted. To aid in visualizing the reporting locations, each one is pinpointed on the particular state's map.

Meteorologists, commenting on the importance of updating these data, call attention to the difference between subjective experience of the climate and objective measurement. They say that as individuals we may intuitively feel the climate remains stable from one 30-year period to the next, but 10-year updates tend to show otherwise. For example, the new figures show that a cooling trend which in the 1941-70 period reversed the relatively hot tri-decade of 1931-60 became even more definite in the 1951-80 base period. They call attention to "dust-bowl" years of the nineteen thirties, and warn that engineering calculations which are still integrating that decade in climate measurements could result in plans for too much cooling and not enough heating.

Climate Normals for the U.S. (Base: 1951-80) will be the basic reference for the U.S. climate until approximately 1993.

FACTS ABOUT CLIMATIC NORMALS

The climatic normals presented in this publication are based on records for the 30-year period 1951-80, inclusive. Data are assembled by individual states.

The order of presentation within a state's entry is, temperature, precipitation, and heating and cooling degree days. Units used in this publication are °F for temperature and inches for precipitation. Heating and cooling degree day (base 65 °F) normals are derived from the monthly normal temperatures using the technique developed by Thom (1), (2). Degree day normals also have been computed to other bases and may be obtained at cost from the National Climatic Center, Asheville, NC 28801-2696.

Normals for National Weather Service Offices and Principal Climatological Stations

A normal climatological element is the arithmetical mean computed over a time period spanning three consecutive decades. Homogeneity of instrument exposure and station location is assumed. If no exposure changes have occurred at a station, the normal is estimated by simply averaging the 30 values from the 1951-80 record. Since it is next to impossible to maintain a multiple purpose network of meteorological stations without having exposure changes, it is first necessary to identify and evaluate these changes and to make adjustments for them if required.

After the periods of heterogeneity have been determined, adjustments are applied to remove the heterogeneities introduced into the mean. This is done by comparing the record at the base station, for which the normal is desired, to the records at supplementary stations with homogeneous periods which covers the heterogeneous period at the base station. The differences method is applied to the monthly average maximum and minimum temperature and the ratio method to the monthly total precipitation (3). A weighted average of the various partial means of the adjusted and unadjusted record is then prepared to give the normal.

Normals for Substations

Individual station values (by year-month) of average temperature and total precipitation for the 1951-80 period are available at the National Climatic Center, Asheville, North Carolina, and may be obtained in either microfiche or magnetic tape for the cost of duplication. In addition, monthly extremes of precipitation and temperature are included along with the standard deviation of monthly temperatures. The National Climatic Center also prepares special studies of climatological elements to specifications provided by the requesting agency or person. The cost of providing such services is borne by the requester.

References

1. Thom, H.C.S., "The Rational Relationship Between Heating Degree Days amd Temperature," *Monthly Weather Review,* Vol. 82, No. 1, January 1954.

2. Thom, H.C.S., "Normal Heating Degree Days Above Any Base by the Universal Truncation Coefficient," *Monthly Weather Review,* Vol. 94, No. 7, July 1966.

3. Gerald R. Barger, editor, "Climatology at Work," U.S. Weather Bureau, Washington, D.C., 1960.

Notes

Table Content
Precipitation normals less than .005 are shown as zero.
Monthly normals for February are based on 28-day month.

Station Names
Figures and letters following the station name indicate a rural location, and refer to the distance and direction of the station from the nearest post office. WSO, WSMO, AND WSFO denote a Weather Service Office, a Meteorological Observatory, and a Forecast Office, respectively. Station elevations are in feet above mean sea level. "R" or "6" denotes a recording gage. "//" indicates a wind shield is affixed to the gage.

Pacific Stations
Stations located on islands other than Hawaii generally have short records (i.e., less than 30 years) and did not meet the criteria for computation of normals. Short-term or period averages are given for these stations.

Maps
Maps show the locations of stations for which 1951-80 normals have been prepared.

Degree Day Normals*
The usual arithmetical procedures were not applied to obtain the heating and cooling degree day data. The rational conversion formulae developed by Thom (1) and (2) allow the properly adjusted mean temperature normals to be coverted to degree day normals with uniform consistency. In some cases this procedure will yield a small number of degree day values that are unexpected. These cases occur when the standard deviations are computed from a mixed distribution as frequently occurs during the transition months. The unexpected values are low and unimportant for most applications of degree day data.

* The general concept of heating and cooling degree days is discussed on pages 711-12.

ALABAMA

TEMPERATURE NORMALS (DEG F)

STATION		JAN	FEB	MAR	APR	MAY	JUN	JUL	AUG	SEP	OCT	NOV	DEC	ANN
ANDALUSIA 1 NW	MAX	59.2	62.9	70.5	79.6	85.8	91.0	92.2	91.8	87.6	78.6	68.8	61.5	77.5
	MIN	35.2	36.8	43.6	51.1	58.9	65.2	68.1	67.7	63.9	51.0	41.7	36.2	51.6
	MEAN	47.2	49.9	57.1	65.4	72.4	78.1	80.2	79.8	75.8	64.8	55.3	48.9	64.6
ANNISTON FAA AP	MAX	53.5	57.8	65.6	75.7	82.5	88.5	91.0	90.7	85.3	75.3	64.7	56.8	74.0
	MIN	32.3	34.3	41.2	49.4	57.2	64.7	68.5	67.7	62.5	49.2	39.2	33.9	50.0
	MEAN	43.0	46.1	53.4	62.6	69.9	76.6	79.8	79.2	73.9	62.3	52.0	45.4	62.0
AUBURN AGRONOMY FARM	MAX	55.5	59.1	66.4	75.8	83.0	89.0	90.6	90.5	86.1	76.6	66.5	58.4	74.8
	MIN	33.7	35.3	42.3	50.4	58.4	65.2	68.5	67.9	63.7	51.5	41.8	35.7	51.2
	MEAN	44.6	47.2	54.4	63.1	70.7	77.1	79.6	79.2	74.9	64.1	54.2	47.1	63.0
BAY MINETTE	MAX	61.2	64.9	71.4	79.2	84.9	89.6	90.5	90.4	87.1	79.8	70.1	63.6	77.7
	MIN	40.2	42.0	48.3	56.0	63.3	69.1	71.4	71.0	67.5	56.7	47.2	42.0	56.2
	MEAN	50.7	53.5	59.9	67.7	74.1	79.4	81.0	80.7	77.3	68.2	58.7	52.9	67.0
BELLE MINA 2 N	MAX	50.4	55.0	63.2	74.2	81.3	87.9	90.6	90.5	84.6	74.5	62.8	53.8	72.4
	MIN	30.3	32.5	39.7	48.9	56.6	63.9	67.4	66.0	60.2	47.3	38.2	32.6	48.6
	MEAN	40.4	43.8	51.5	61.6	68.9	75.9	79.0	78.3	72.5	60.9	50.5	43.2	60.5
BIRMINGHAM WSO CITY	MAX	51.7	56.6	64.8	75.0	81.7	88.2	90.6	89.8	84.1	73.5	62.3	54.5	72.7
	MIN	33.0	35.2	42.2	50.5	58.4	66.0	70.0	69.4	64.0	50.6	40.7	35.4	51.3
	MEAN	42.4	45.9	53.5	62.8	70.1	77.1	80.3	79.6	74.1	62.1	51.5	45.0	62.0
BIRMINGHAM WSO R	MAX	52.7	57.3	65.2	75.2	81.6	87.9	90.3	89.7	84.6	74.8	63.7	55.9	73.2
	MIN	33.0	35.2	42.1	50.4	58.3	65.9	69.8	69.1	63.6	50.4	40.5	35.2	51.1
	MEAN	42.9	46.3	53.7	62.8	70.0	77.0	80.1	79.5	74.1	62.6	52.1	45.6	62.2
BREWTON 3 SSE	MAX	62.3	66.3	73.2	81.2	87.3	92.1	93.0	92.5	88.3	79.9	70.1	64.0	79.2
	MIN	36.9	38.6	44.8	51.5	58.2	64.7	68.1	67.4	63.5	50.1	41.5	37.8	51.9
	MEAN	49.6	52.5	59.1	66.4	72.8	78.4	80.6	80.0	75.9	65.0	55.8	50.9	65.6
CHATOM 3 N	MAX	60.9	65.1	72.2	80.1	85.8	91.2	92.4	92.1	88.4	80.2	69.9	63.2	78.5
	MIN	36.5	38.7	44.9	52.2	59.3	65.3	68.1	67.4	63.7	51.5	42.8	38.3	52.4
	MEAN	48.7	51.9	58.6	66.2	72.6	78.3	80.3	79.8	76.0	65.9	56.4	50.8	65.5
CLANTON	MAX	54.0	58.0	65.6	75.4	82.3	88.5	90.7	90.4	85.7	76.1	65.4	57.1	74.1
	MIN	31.4	33.6	40.9	49.5	57.4	64.6	68.3	67.2	62.0	48.3	38.5	33.0	49.6
	MEAN	42.7	45.8	53.3	62.5	69.8	76.6	79.5	78.8	73.9	62.2	52.0	45.1	61.9
DAYTON	MAX	55.9	59.9	67.5	76.5	83.5	89.8	91.9	91.6	87.3	78.1	66.9	59.2	75.7
	MIN	35.4	37.5	44.3	52.8	60.3	66.8	69.9	69.1	64.7	51.9	42.6	37.2	52.7
	MEAN	45.7	48.7	55.9	64.7	71.9	78.3	80.9	80.4	76.1	65.0	54.8	48.2	64.2
DEMOPOLIS LOCK AND DAM	MAX	55.1	59.6	66.9	76.3	83.0	89.2	91.7	91.4	86.7	77.0	66.2	58.3	75.1
	MIN	33.4	35.8	43.3	51.2	59.0	65.7	69.3	68.7	63.7	50.3	40.4	35.2	51.3
	MEAN	44.3	47.7	55.1	63.8	71.0	77.5	80.6	80.0	75.2	63.7	53.3	46.8	63.3
EUFAULA	MAX	59.2	62.5	70.4	78.8	84.9	90.6	92.2	92.0	87.8	79.0	69.2	61.6	77.4
	MIN	35.0	37.0	43.7	51.1	58.9	65.8	69.3	68.6	64.0	50.6	41.1	36.3	51.8
	MEAN	47.1	49.8	57.1	65.0	71.9	78.2	80.8	80.3	75.9	64.8	55.2	49.0	64.6
EVERGREEN	MAX	59.4	63.1	70.4	78.7	84.8	90.2	91.6	91.4	87.5	78.9	68.6	62.0	77.2
	MIN	35.1	37.3	44.0	51.8	59.6	66.4	69.8	68.9	64.8	51.5	41.4	36.7	52.3
	MEAN	47.3	50.2	57.2	65.3	72.3	78.3	80.7	80.2	76.2	65.2	55.0	49.3	64.8
FAIRHOPE 2 NE	MAX	61.8	64.6	71.0	78.2	84.6	89.6	90.5	90.4	87.2	79.6	70.0	63.8	77.6
	MIN	41.5	43.4	49.6	57.0	63.8	69.8	72.3	71.8	68.3	57.2	48.5	43.3	57.2
	MEAN	51.6	54.1	60.3	67.6	74.2	79.7	81.4	81.1	77.8	68.4	59.3	53.5	67.4
FAYETTE	MAX	53.2	57.6	65.9	76.1	82.9	89.6	92.3	91.7	86.6	77.3	65.1	56.1	74.5
	MIN	30.7	33.1	40.2	48.2	55.8	63.0	66.8	65.8	60.2	46.3	37.9	32.6	48.4
	MEAN	41.9	45.4	53.1	62.1	69.4	76.3	79.6	78.8	73.4	61.8	51.5	44.4	61.5
FRISCO CITY	MAX	58.4	62.3	69.7	78.3	84.9	90.6	91.6	91.5	87.6	79.0	68.6	61.3	77.0
	MIN	36.5	38.5	45.7	53.9	61.2	67.0	69.7	69.1	65.2	53.1	43.9	38.5	53.5
	MEAN	47.5	50.4	57.7	66.1	73.1	78.8	80.7	80.3	76.5	66.1	56.3	49.9	65.3

ALABAMA

TEMPERATURE NORMALS (DEG F)

STATION		JAN	FEB	MAR	APR	MAY	JUN	JUL	AUG	SEP	OCT	NOV	DEC	ANN
GADSDEN STEAM PLANT	MAX	51.0	55.5	63.7	74.5	81.6	87.8	90.4	90.1	84.8	74.3	63.0	54.5	72.6
	MIN	31.0	32.8	39.8	48.6	56.1	64.0	67.9	67.0	61.5	47.9	38.7	32.8	49.0
	MEAN	41.0	44.2	51.8	61.6	68.9	75.9	79.1	78.6	73.2	61.1	50.9	43.7	60.8
GREENSBORO	MAX	56.7	61.3	68.7	77.8	84.2	90.3	92.3	92.3	87.7	78.3	67.5	59.8	76.4
	MIN	35.0	37.3	43.7	51.7	59.8	66.7	69.9	69.2	64.6	52.2	42.4	36.8	52.4
	MEAN	45.9	49.3	56.2	64.8	72.0	78.5	81.1	80.8	76.2	65.3	55.0	48.3	64.5
HALEYVILLE	MAX	49.7	54.3	62.2	73.3	79.9	86.5	89.4	89.0	83.5	73.6	62.0	53.3	71.4
	MIN	28.7	30.7	38.0	47.8	55.8	63.2	66.6	65.5	60.1	47.5	37.9	31.5	47.8
	MEAN	39.3	42.5	50.1	60.6	67.9	74.9	78.0	77.3	71.9	60.6	50.0	42.4	59.6
HEADLAND	MAX	58.8	62.3	69.6	78.8	84.8	89.9	90.9	90.5	87.1	78.6	68.8	61.6	76.8
	MIN	37.0	39.3	46.0	54.5	61.8	67.6	70.0	69.6	66.0	54.0	44.4	38.6	54.1
	MEAN	48.0	50.8	57.8	66.7	73.3	78.8	80.5	80.1	76.6	66.3	56.6	50.1	65.5
HIGHLAND HOME	MAX	57.3	60.9	68.5	77.3	83.5	89.0	90.7	90.4	86.6	77.6	67.3	59.8	75.7
	MIN	36.0	37.8	44.6	52.7	60.1	66.3	69.2	68.6	64.6	52.9	43.9	37.7	52.9
	MEAN	46.7	49.4	56.6	65.0	71.8	77.7	79.9	79.6	75.6	65.3	55.7	48.8	64.3
HUNTSVILLE WSO	MAX	49.4	53.9	61.9	73.0	79.9	86.8	89.4	89.2	83.5	73.4	61.6	53.0	71.2
	MIN	31.0	33.2	40.6	50.0	57.6	65.3	69.1	67.9	62.0	49.1	39.4	33.6	49.9
	MEAN	40.2	43.6	51.3	61.5	68.8	76.1	79.3	78.6	72.8	61.3	50.5	43.3	60.6
LAFAYETTE	MAX	55.6	59.9	67.6	76.6	82.8	88.8	90.6	90.2	85.2	75.8	65.6	58.1	74.7
	MIN	33.6	35.7	42.5	50.3	57.9	64.2	67.4	66.4	61.8	50.1	41.3	35.5	50.6
	MEAN	44.6	47.8	55.0	63.5	70.4	76.5	79.0	78.3	73.5	63.0	53.5	46.8	62.7
MARION JUNCTION 2 NE	MAX	56.0	60.4	67.9	76.9	83.4	89.6	91.8	91.4	86.7	77.1	66.6	58.9	75.6
	MIN	34.4	36.6	43.5	51.7	59.2	66.2	69.3	68.8	63.9	50.7	41.0	35.8	51.8
	MEAN	45.2	48.6	55.7	64.3	71.4	77.9	80.6	80.1	75.3	63.9	53.9	47.4	63.7
MOBILE WSO	MAX	60.6	63.9	70.3	78.3	84.9	90.2	91.2	90.7	87.0	79.4	69.3	63.1	77.4
	MIN	40.9	43.2	49.8	57.7	64.8	70.8	73.2	72.9	69.3	57.5	47.9	42.9	57.6
	MEAN	50.8	53.6	60.1	68.0	74.9	80.5	82.2	81.8	78.2	68.5	58.6	53.1	67.5
MONTGOMERY WSO R	MAX	57.0	60.9	68.1	77.0	83.6	89.8	91.5	91.2	86.9	77.5	67.0	59.8	75.9
	MIN	36.4	38.8	45.5	53.3	61.1	68.4	71.8	71.1	66.4	53.1	43.0	37.9	53.9
	MEAN	46.7	49.9	56.8	65.2	72.4	79.1	81.7	81.2	76.7	65.3	55.0	48.9	64.9
MUSCLE SHOALS FAA AP	MAX	49.0	53.7	62.0	73.0	80.4	87.7	90.5	89.9	83.9	73.5	61.4	52.9	71.5
	MIN	30.6	33.1	40.6	50.1	57.7	65.5	69.3	67.9	62.2	48.9	39.4	33.5	49.9
	MEAN	39.8	43.4	51.3	61.6	69.1	76.6	79.9	78.9	73.1	61.3	50.4	43.2	60.7
ONEONTA	MAX	50.7	55.0	63.2	73.8	80.9	87.4	90.3	90.1	84.8	74.5	63.0	54.3	72.3
	MIN	28.9	30.9	37.9	47.1	54.8	62.2	66.4	65.3	59.7	46.0	36.8	31.3	47.3
	MEAN	39.8	43.0	50.6	60.5	67.9	74.9	78.3	77.7	72.3	60.2	49.9	42.8	59.8
OZARK 6 NNW	MAX	60.2	63.8	70.9	79.1	85.3	90.6	91.4	91.6	87.8	79.4	69.5	62.6	77.7
	MIN	38.1	40.2	46.5	54.0	61.0	66.9	69.3	68.9	65.3	54.1	45.3	40.1	54.1
	MEAN	49.2	52.0	58.7	66.6	73.2	78.8	80.4	80.2	76.5	66.8	57.4	51.4	65.9
ROBERTSDALE 1 E	MAX	61.8	64.7	71.0	78.4	84.9	89.9	91.0	90.6	87.6	80.2	70.9	64.2	77.9
	MIN	39.5	41.3	47.8	55.5	62.2	68.1	70.7	70.1	66.7	55.1	46.4	41.0	55.4
	MEAN	50.6	53.0	59.4	67.0	73.6	79.0	80.9	80.4	77.1	67.7	58.7	52.6	66.7
ROCK MILLS	MAX	54.5	58.8	66.5	76.8	83.7	89.8	91.6	91.2	86.0	76.4	65.9	57.5	74.9
	MIN	30.7	32.0	38.5	46.3	54.5	62.1	66.2	65.3	60.3	46.7	37.1	31.8	47.6
	MEAN	42.6	45.4	52.5	61.6	69.1	76.0	78.9	78.3	73.2	61.6	51.5	44.7	61.3
SAINT BERNARD	MAX	49.9	54.3	62.8	73.6	80.5	86.8	89.9	89.6	84.6	74.1	62.1	53.2	71.8
	MIN	27.7	29.6	37.3	46.7	55.1	62.5	66.0	64.8	58.9	45.5	35.8	29.7	46.6
	MEAN	38.9	42.0	50.1	60.2	67.8	74.7	78.0	77.2	71.8	59.8	49.0	41.5	59.3
SAND MT SUBSTA AU	MAX	49.4	53.9	62.0	72.4	78.9	85.2	87.8	87.8	82.6	72.5	61.3	52.7	70.5
	MIN	30.4	32.4	39.6	48.6	56.2	63.1	66.2	65.4	60.2	47.9	38.8	32.8	48.5
	MEAN	40.0	43.2	50.8	60.5	67.6	74.2	77.0	76.6	71.4	60.2	50.1	42.8	59.5

ALABAMA

TEMPERATURE NORMALS (DEG F)

STATION		JAN	FEB	MAR	APR	MAY	JUN	JUL	AUG	SEP	OCT	NOV	DEC	ANN
SCOTTSBORO	MAX	49.8	54.3	62.3	73.5	81.0	87.6	90.5	90.2	84.9	74.4	62.6	53.5	72.1
	MIN	28.3	30.6	37.9	46.7	54.3	61.9	65.8	64.3	58.8	45.3	35.9	30.4	46.7
	MEAN	39.1	42.5	50.2	60.1	67.7	74.8	78.2	77.3	71.9	59.9	49.3	42.0	59.4
SELMA	MAX	58.3	62.7	70.2	78.9	85.5	91.1	92.8	92.6	88.2	79.0	68.1	61.1	77.4
	MIN	37.8	40.2	46.8	54.1	61.8	68.4	71.4	70.8	66.0	53.6	44.0	39.4	54.5
	MEAN	48.1	51.5	58.5	66.5	73.7	79.8	82.1	81.7	77.2	66.3	56.1	50.3	66.0
TALLADEGA	MAX	54.6	59.1	67.0	76.2	82.4	88.5	90.6	90.3	85.2	75.3	65.3	57.5	74.3
	MIN	32.2	34.0	40.7	48.3	55.8	63.1	66.7	65.8	61.1	48.1	38.7	33.6	49.0
	MEAN	43.4	46.6	53.8	62.3	69.1	75.9	78.7	78.1	73.2	61.7	52.0	45.6	61.7
THOMASVILLE	MAX	57.0	61.1	68.9	77.7	83.7	89.7	91.6	91.1	86.9	77.7	67.1	59.8	76.0
	MIN	34.3	36.8	43.7	52.5	60.0	66.6	69.5	68.8	64.3	51.8	42.1	36.2	52.2
	MEAN	45.7	49.0	56.3	65.2	71.9	78.2	80.5	80.0	75.6	64.8	54.6	48.0	64.2
TROY	MAX	58.6	62.1	69.5	77.8	84.2	89.4	90.6	90.7	86.9	78.0	68.0	61.2	76.4
	MIN	38.0	39.8	46.4	54.0	61.2	67.2	69.7	69.5	66.1	55.0	45.4	39.8	54.3
	MEAN	48.3	51.0	57.9	65.9	72.7	78.3	80.2	80.1	76.5	66.5	56.7	50.5	65.4
TUSCALOOSA FAA AP	MAX	54.0	58.9	66.9	76.5	83.3	89.6	91.8	91.6	86.6	77.0	65.3	57.3	74.9
	MIN	33.5	35.8	43.1	51.4	59.4	66.9	70.6	69.7	64.2	50.3	40.4	35.4	51.7
	MEAN	43.8	47.4	55.0	64.0	71.4	78.3	81.3	80.7	75.4	63.7	52.9	46.4	63.4
UNION SPRINGS	MAX	56.9	60.4	67.6	77.3	83.9	89.6	91.4	91.5	87.1	77.6	67.4	60.1	75.9
	MIN	34.8	36.8	43.4	51.7	59.4	66.1	68.8	68.7	64.4	51.8	42.0	35.9	52.0
	MEAN	45.9	48.6	55.5	64.5	71.7	77.9	80.1	80.1	75.8	64.7	54.7	48.0	64.0
VALLEY HEAD	MAX	48.6	53.1	60.7	71.9	78.8	85.1	88.1	88.3	82.9	73.0	61.6	52.4	70.4
	MIN	26.2	27.6	35.3	43.8	52.0	60.2	64.0	63.0	57.7	43.6	33.7	28.1	44.6
	MEAN	37.4	40.4	48.0	57.9	65.4	72.7	76.1	75.7	70.4	58.3	47.7	40.3	57.5

ALABAMA

PRECIPITATION NORMALS (INCHES)

STATION	JAN	FEB	MAR	APR	MAY	JUN	JUL	AUG	SEP	OCT	NOV	DEC	ANN
ALBERTA	4.83	4.90	6.87	5.34	4.25	3.87	4.72	4.01	3.75	2.64	3.38	5.52	54.08
ALEXANDER CITY	5.51	5.03	7.30	5.78	4.08	4.40	5.19	3.99	4.35	2.99	3.55	5.54	57.71
ALICEVILLE	5.19	4.82	6.47	5.92	3.42	2.99	4.42	.3.15	3.83	2.74	3.80	5.40	52.15
ANDALUSIA 1 NW	4.80	4.94	6.18	5.25	4.78	4.89	6.24	5.21	4.78	2.65	3.65	5.60	58.97
ANNISTON FAA AP	5.36	4.82	6.82	5.35	3.99	3.89	4.23	3.80	4.15	2.50	3.35	4.99	53.25
ARLEY 3 S	5.84	5.52	7.03	5.56	4.65	3.49	4.75	3.56	4.35	3.12	4.29	5.76	57.92
AUBURN AGRONOMY FARM	5.14	5.34	6.85	5.31	4.23	3.85	5.74	3.59	4.32	2.83	3.42	5.48	56.10
AUTAUGAVILLE 3 N	4.83	4.62	6.51	5.51	3.98	3.85	4.62	3.83	4.28	2.77	3.28	4.71	52.79
BANKHEAD LOCK AND DAM	5.67	4.93	6.80	5.48	4.42	3.65	4.82	3.45	3.82	2.80	3.94	5.28	55.06
BAY MINETTE	4.92	5.08	6.51	5.26	5.55	5.47	7.47	6.40	6.01	3.04	3.75	5.32	64.78
BEATRICE 1 E	4.80	4.75	6.44	5.23	4.44	4.31	5.84	4.04	4.26	2.25	3.16	5.28	54.80
BELLE MINA 2 N	5.21	4.64	6.50	4.82	4.36	3.38	4.54	3.23	3.71	2.94	4.39	5.37	53.09
BIRMINGHAM WSO CITY	4.97	4.64	6.55	5.30	3.80	3.17	4.19	3.88	4.55	2.77	3.51	4.83	52.15
BIRMINGHAM WSO R	5.23	4.72	6.62	5.00	4.53	3.61	5.39	3.85	4.34	2.64	3.64	4.95	54.52
BOAZ	5.41	4.61	6.34	5.31	4.80	3.65	4.21	3.31	4.45	2.65	4.06	5.21	54.01
BRANTLEY	4.87	5.21	6.22	4.79	3.97	4.54	5.58	4.94	4.54	2.43	3.38	5.28	55.75
BREWTON 3 SSE	4.76	5.57	6.20	5.16	4.74	5.90	6.94	5.51	5.02	2.92	4.05	5.60	62.37
BRUNDIDGE	4.54	4.64	6.01	4.72	3.94	3.93	5.81	4.26	4.09	2.79	3.04	4.98	52.75
CALERA 2 SW	5.56	5.13	7.48	5.91	4.17	4.16	5.02	4.06	4.03	2.61	3.49	4.95	56.57
CARBON HILL	6.03	5.52	6.99	5.68	4.89	4.26	5.21	3.66	4.11	3.18	4.40	5.85	59.78
CHATOM 3 N	4.89	4.92	6.37	5.25	4.84	4.78	6.77	4.56	4.48	2.88	3.64	6.02	59.40
CHILDERSBURG WTR PLANT	5.43	5.00	7.39	5.54	4.43	4.16	4.72	4.22	4.24	2.78	3.39	5.10	56.40
CLANTON	5.40	5.13	7.35	6.07	4.17	3.71	5.12	4.09	4.56	2.79	3.45	5.55	57.39
CLAYTON	4.49	4.78	5.81	4.35	4.15	3.92	5.09	4.32	4.34	2.49	2.96	4.98	51.68
COFFEE SPRINGS 2 NW	4.78	5.21	5.54	4.83	4.43	4.15	5.22	4.82	4.39	2.62	3.57	5.19	54.75
CORDOVA	5.86	5.07	7.13	5.49	4.54	3.54	5.35	3.78	4.15	2.96	4.09	5.71	57.67
CUBA	5.22	4.76	6.48	5.25	4.27	3.80	4.80	3.68	3.17	2.75	3.56	5.47	53.21
DADEVILLE	5.29	5.34	6.81	5.74	3.81	3.96	5.63	4.16	4.28	2.81	3.54	5.08	56.45
DAYTON	5.06	4.72	7.05	5.43	4.31	4.01	4.98	3.49	3.77	2.60	3.35	5.54	54.31
DEMOPOLIS LOCK AND DAM	5.11	4.69	6.50	5.49	4.05	3.49	5.19	3.76	3.42	2.90	3.25	5.01	52.86
ELBA	4.72	4.80	6.38	4.96	4.34	4.16	6.10	4.43	4.71	2.54	3.69	5.44	56.27
ELROD	5.47	4.92	6.68	5.64	4.26	3.28	4.91	3.66	3.67	2.88	3.83	5.30	54.50
EUFAULA	4.80	4.94	5.47	4.56	4.04	4.45	5.65	3.89	3.92	2.20	3.16	5.05	52.13
EVERGREEN	5.24	5.34	6.74	5.44	4.76	5.40	6.82	4.10	5.18	2.34	3.68	5.81	60.85
FAIRHOPE 2 NE	4.64	4.74	5.94	4.91	4.99	5.93	7.81	6.01	7.18	3.13	3.75	5.08	64.11
FALKVILLE	5.32	4.74	6.63	5.11	4.29	3.74	4.89	3.80	4.31	2.92	4.01	4.98	54.74
FAYETTE	5.62	5.13	7.13	6.04	4.51	3.78	4.99	3.13	3.96	2.97	3.89	5.66	56.81
FORT PAYNE	5.31	4.87	6.71	5.18	4.65	3.87	4.75	3.74	4.20	2.71	4.10	5.09	55.18
FRISCO CITY	5.23	5.03	6.28	5.56	4.85	4.88	6.33	4.62	4.96	2.85	3.72	5.36	59.67
GADSDEN STEAM PLANT	5.49	4.81	6.87	5.43	4.25	3.71	4.54	3.44	3.83	2.57	3.82	5.33	54.09
GAINESVILLE	5.29	4.91	6.47	5.88	4.06	3.17	4.48	3.07	3.63	2.47	3.51	5.40	52.34
GARDEN CITY	5.71	4.92	6.40	5.51	4.61	3.51	4.72	3.97	4.56	2.83	4.16	5.45	56.35
GORGAS	5.64	5.00	7.14	5.53	4.43	3.57	5.18	3.60	3.98	2.92	3.92	5.43	56.34
GREENSBORO	5.24	5.07	7.14	5.91	4.03	3.67	5.12	3.52	3.58	2.80	3.34	5.38	54.80
GREENVILLE	5.26	5.24	6.66	5.38	4.26	4.45	5.25	4.43	4.78	2.34	3.67	5.38	57.10
HALEYVILLE	5.82	5.32	7.20	6.04	4.80	3.97	4.77	3.71	4.71	3.30	4.69	6.19	60.52
HAYNEVILLE	4.65	4.76	6.58	4.70	3.79	3.47	4.23	3.15	4.33	2.32	3.06	4.61	49.65
HEADLAND	5.27	4.96	5.44	4.58	4.35	4.62	5.95	4.96	4.08	2.33	3.23	4.88	54.65
HIGHLAND HOME	4.94	5.04	6.74	5.14	4.47	4.18	5.20	3.73	4.29	2.49	3.43	4.84	54.49
HIGHTOWER	5.50	4.90	6.98	5.46	4.36	3.82	4.54	3.65	4.10	3.00	3.80	5.03	55.14
HODGES	5.72	4.98	6.71	5.88	4.96	3.41	4.23	3.45	4.04	3.23	4.58	5.93	57.12
HUNTSVILLE WSO	5.17	4.79	6.78	4.92	4.60	3.74	5.05	3.11	3.99	2.90	4.24	5.43	54.74
JORDAN DAM	5.07	5.20	6.42	5.32	3.73	3.80	5.19	3.56	4.34	2.54	3.49	5.17	53.83
LAFAYETTE	5.36	5.63	7.10	5.45	4.28	3.78	5.43	3.74	4.17	2.70	3.66	5.33	56.63
LAY DAM	5.26	4.56	6.62	5.40	3.92	3.75	4.63	4.02	4.23	2.40	3.19	4.97	52.95

ALABAMA

PRECIPITATION NORMALS (INCHES)

STATION	JAN	FEB	MAR	APR	MAY	JUN	JUL	AUG	SEP	OCT	NOV	DEC	ANN
LEEDS	5.68	5.03	7.21	5.34	4.32	3.56	5.24	3.56	4.52	3.00	4.18	5.41	57.05
LEESBURG	5.67	5.00	7.17	5.81	4.28	3.54	4.40	3.30	4.33	2.85	3.84	5.73	55.92
LIVINGSTON 2 SW	5.16	4.80	6.90	5.74	3.98	3.73	5.89	3.55	3.28	2.75	3.62	5.82	55.22
MARION 6 NE	5.36	4.96	7.14	5.85	3.87	3.50	4.85	3.69	3.75	2.64	3.24	4.98	53.83
MARION JUNCTION 2 NE	5.01	4.60	6.82	5.42	3.93	4.11	4.69	3.53	3.55	2.75	3.26	5.43	53.10
MARTIN DAM	4.85	5.10	6.84	5.28	3.80	4.29	5.16	3.66	4.07	2.47	3.45	5.31	54.28
MELVIN	5.37	5.24	7.13	5.54	4.67	3.74	6.07	4.15	3.84	2.76	3.72	5.78	58.01
MILSTEAD	4.88	5.11	6.42	5.17	3.98	3.87	4.34	3.85	4.24	2.27	3.40	5.14	52.67
MITCHELL DAM	5.17	4.84	6.82	5.52	4.04	3.86	4.74	3.32	4.17	2.52	3.27	5.12	53.39
MOBILE WSO	4.59	4.91	6.48	5.35	5.46	5.07	7.74	6.75	6.56	2.62	3.67	5.44	64.64
MONTEVALLO	5.77	5.33	7.24	5.70	4.55	4.22	5.03	4.16	3.77	2.83	3.63	5.09	57.32
MONTGOMERY WSO R	4.20	4.51	5.92	4.38	4.00	3.45	4.78	3.17	4.72	2.27	2.98	4.78	49.16
MUSCLE SHOALS FAA AP	5.17	4.30	6.22	4.71	4.33	3.52	4.59	3.06	3.82	2.81	3.75	5.30	51.58
NEWTON	4.89	5.07	5.77	4.52	3.95	4.79	5.65	4.30	4.66	2.35	3.44	4.72	54.11
ONEONTA	5.75	5.07	6.86	5.81	4.45	3.71	4.75	3.88	4.47	2.87	4.16	5.23	57.01
OZARK 6 NNW	4.73	4.78	6.03	4.75	3.77	4.76	5.80	4.38	4.52	2.27	3.34	5.45	54.58
PALMERDALE 2 W	5.37	4.75	6.65	5.29	4.20	3.43	4.56	3.64	4.62	2.60	3.93	5.07	54.11
PERRYVILLE	5.14	4.77	7.01	5.70	3.76	3.94	5.16	3.79	3.98	2.68	3.35	5.28	54.56
PINE LEVEL	4.77	4.93	6.33	4.81	3.99	4.13	5.17	3.61	4.48	2.51	3.01	5.05	52.79
PLANTERSVILLE 2 SSE	5.44	5.06	7.38	5.91	3.87	4.19	5.05	3.44	3.75	2.72	3.35	5.51	55.67
RED BAY	5.83	5.03	7.20	5.85	5.15	3.35	4.22	3.27	3.77	2.93	4.53	5.91	57.04
REFORM	5.69	5.06	6.59	5.86	4.55	3.30	5.35	4.08	3.80	2.79	3.91	5.52	56.50
ROBERTSDALE 1 E	4.95	4.86	6.08	4.58	5.02	5.76	7.89	7.16	7.42	3.26	4.04	4.70	65.72
ROCK MILLS	5.59	5.07	6.79	5.05	4.05	3.81	5.09	3.63	4.03	2.67	3.74	5.16	54.68
SAINT BERNARD	5.69	5.24	6.87	5.44	4.79	3.87	4.55	3.43	4.91	3.07	4.31	5.64	57.81
SAND MT SUBSTA AU	5.22	4.89	6.68	5.33	4.51	3.77	3.98	3.23	4.63	2.89	4.08	5.44	54.65
SAYRE	5.68	4.74	6.96	5.40	4.65	3.22	5.13	3.23	4.07	2.81	3.93	5.24	55.06
SCOTTSBORO	5.71	5.02	7.28	4.88	4.52	3.75	4.79	3.21	4.69	2.84	4.24	5.52	56.45
SELMA	4.87	4.84	6.90	5.05	4.01	4.20	4.61	3.47	4.15	2.81	3.13	5.47	53.51
SUTTLE	5.25	4.82	6.89	5.39	3.95	4.02	4.29	3.40	3.57	2.45	3.26	5.38	52.67
TALLADEGA	5.58	4.78	7.11	5.50	4.23	3.85	4.44	3.43	4.07	2.51	3.50	4.96	53.96
THOMASVILLE	4.92	4.96	7.25	5.20	4.48	4.12	5.83	4.36	4.13	3.03	3.73	5.72	57.73
THURLOW DAM	5.01	5.10	6.67	4.99	4.15	3.77	5.05	3.79	4.24	2.31	3.31	5.29	53.68
TROY	4.42	4.67	5.91	4.43	3.96	3.80	6.01	3.97	3.92	2.49	3.23	4.83	51.64
TUSCALOOSA FAA AP	5.34	4.55	6.49	5.48	4.16	3.32	4.89	3.65	3.33	2.89	3.49	5.02	52.61
TUSCALOOSA OLIVER DAM	5.49	4.75	6.80	5.64	4.31	3.40	4.81	3.34	3.50	3.00	3.74	5.11	53.89
UNION SPRINGS	4.81	4.96	5.98	4.59	4.06	4.35	5.89	4.22	4.00	2.63	3.14	5.10	53.73
UNIONTOWN	5.23	4.61	6.92	5.33	3.88	3.66	5.02	3.37	3.31	2.55	3.14	5.15	52.17
VALLEY HEAD	5.49	5.08	6.93	5.43	4.35	4.12	5.37	3.37	4.39	2.90	4.42	5.30	57.15
WADLEY	5.26	4.99	6.93	5.36	4.61	3.92	4.91	3.74	3.87	2.89	3.51	5.43	55.42
WALLACE 2 E	5.06	5.12	6.23	4.98	4.63	5.51	6.70	4.58	4.58	2.42	3.48	5.25	58.54
WALNUT GROVE	5.21	4.66	6.83	5.21	4.19	3.79	4.25	3.49	4.17	2.57	3.95	4.96	53.28
WARRIOR LOCK AND DAM	4.97	4.99	6.39	5.41	3.84	3.41	3.85	3.16	3.15	2.64	3.20	5.06	50.07
WEST BLOCTON	5.99	5.18	7.13	5.84	4.34	4.51	5.13	3.53	4.06	2.89	3.57	5.44	57.61
WETUMPKA	4.93	4.96	6.43	5.35	3.61	3.85	4.35	4.22	4.39	2.40	3.41	5.21	53.11
WHATLEY	5.31	5.40	7.04	5.56	4.75	4.75	6.07	4.51	4.39	2.95	3.66	5.76	60.15
WINFIELD 2 SW	5.75	5.08	6.73	5.41	4.88	3.42	5.08	3.13	3.85	2.93	4.57	5.68	56.51

ALABAMA

HEATING DEGREE DAY NORMALS (BASE 65 DEG F)

STATION	JUL	AUG	SEP	OCT	NOV	DEC	JAN	FEB	MAR	APR	MAY	JUN	ANN
ANDALUSIA 1 NW	0	0	0	112	306	506	565	440	271	87	18	0	2305
ANNISTON FAA AP	0	0	7	133	394	608	682	535	374	116	23	0	2872
AUBURN AGRONOMY FARM	0	0	0	102	331	555	641	505	344	118	15	0	2611
BAY MINETTE	0	0	0	44	212	385	469	343	195	40	0	0	1688
BELLE MINA 2 N	0	0	13	174	435	676	763	601	435	134	42	0	3273
BIRMINGHAM WSO CITY	0	0	7	143	405	620	701	542	372	111	35	7	2943
BIRMINGHAM WSO R	0	0	7	137	387	601	685	532	368	110	36	0	2863
BREWTON 3 SSE	0	0	0	93	288	445	499	371	220	54	0	0	1970
CHATOM 3 N	0	0	0	73	272	447	523	383	232	66	5	0	2001
CLANTON	0	0	0	138	390	617	691	538	377	119	26	0	2896
DAYTON	0	0	0	98	317	527	610	470	307	88	14	0	2431
DEMOPOLIS LOCK AND DAM	0	0	6	121	358	564	649	496	331	97	18	0	2640
EUFAULA	0	0	0	93	301	502	570	441	273	72	14	0	2266
EVERGREEN	0	0	0	96	310	496	564	430	262	76	7	0	2241
FAIRHOPE 2 NE	0	0	0	43	198	370	449	335	185	38	0	0	1618
FAYETTE	0	0	8	145	412	639	716	555	384	141	42	0	3042
FRISCO CITY	0	0	0	79	278	473	557	426	258	65	8	0	2144
GADSDEN STEAM PLANT	0	0	10	153	423	660	744	582	421	136	31	0	3160
GREENSBORO	0	0	0	93	308	518	602	452	301	76	11	0	2361
HALEYVILLE	0	0	21	184	450	701	797	630	475	156	40	0	3454
HEADLAND	0	0	0	62	263	467	549	418	253	56	0	0	2068
HIGHLAND HOME	0	0	0	85	288	507	587	453	291	77	14	0	2302
HUNTSVILLE WSO	0	0	12	166	435	673	769	606	441	136	41	0	3279
LAFAYETTE	0	0	9	134	350	564	643	489	329	99	20	0	2637
MARION JUNCTION 2 NE	0	0	0	112	341	546	625	474	310	94	16	0	2518
MOBILE WSO	0	0	0	50	218	382	469	342	191	43	0	0	1695
MONTGOMERY WSO R	0	0	0	86	307	499	580	439	284	72	10	0	2277
MUSCLE SHOALS FAA AP	0	0	14	164	443	676	781	605	440	150	42	0	3315
ONEONTA	0	0	14	183	453	688	781	616	458	162	51	0	3406
OZARK 6 NNW	0	0	0	51	246	430	513	381	239	59	9	0	1928
ROBERTSDALE 1 E	0	0	0	45	213	395	473	358	210	44	0	0	1738
ROCK MILLS	0	0	7	145	405	629	694	556	400	127	33	0	2996
SAINT BERNARD	0	0	17	196	480	729	809	644	474	169	52	0	3570
SAND MT SUBSTA AU	0	0	21	192	447	688	775	617	453	162	56	0	3411
SCOTTSBORO	0	0	24	196	471	713	803	630	471	170	56	0	3534
SELMA	0	0	0	74	276	462	540	394	243	51	0	0	2040
TALLADEGA	0	0	13	150	390	601	679	523	363	123	36	0	2878
THOMASVILLE	0	0	6	91	321	527	613	463	299	78	15	0	2413
TROY	0	0	0	63	263	455	533	409	252	69	8	0	2052
TUSCALOOSA FAA AP	0	0	5	116	367	577	664	501	334	94	17	0	2675
UNION SPRINGS	0	0	0	98	316	532	608	471	313	82	13	0	2433
VALLEY HEAD	0	0	29	236	519	766	856	689	535	224	91	13	3958

ALABAMA

COOLING DEGREE DAY NORMALS (BASE 65 DEG F)

STATION		JAN	FEB	MAR	APR	MAY	JUN	JUL	AUG	SEP	OCT	NOV	DEC	ANN
ANDALUSIA 1 NW		13	17	27	99	247	393	471	459	328	106	15	6	2181
ANNISTON FAA AP		0	5	15	44	175	348	459	440	274	49	0	0	1809
AUBURN AGRONOMY FARM		9	6	15	61	192	363	453	440	301	74	7	0	1921
BAY MINETTE		26	21	37	121	282	432	496	487	369	143	23	10	2447
BELLE MINA 2 N		0	7	17	32	162	331	434	412	238	47	0	0	1680
BIRMINGHAM WSO CITY		0	7	16	45	193	370	474	453	280	53	0	0	1891
BIRMINGHAM WSO	R	0	8	18	44	191	360	468	450	280	62	0	0	1881
BREWTON 3 SSE		22	21	37	96	245	402	484	465	327	93	12	8	2212
CHATOM 3 N		17	16	34	102	241	399	474	459	330	101	14	6	2193
CLANTON		0	0	14	44	175	348	450	428	271	51	0	0	1781
DAYTON		11	13	25	79	228	399	493	477	337	98	11	6	2177
DEMOPOLIS LOCK AND DAM		8	11	25	61	204	375	484	465	312	81	7	0	2033
EUFAULA		15	15	28	72	228	396	490	474	327	87	7	6	2145
EVERGREEN		16	15	20	85	234	399	487	471	336	102	10	9	2184
FAIRHOPE 2 NE		34	30	39	116	285	441	508	499	384	148	27	13	2524
FAYETTE		0	6	15	54	179	339	453	428	260	45	7	0	1786
FRISCO CITY		14	17	32	98	259	414	487	474	345	114	17	5	2276
GADSDEN STEAM PLANT		0	0	11	34	152	327	437	422	256	32	0	0	1671
GREENSBORO		10	13	28	70	228	405	499	490	336	103	8	0	2190
HALEYVILLE		0	0	13	24	130	301	403	381	228	48	0	0	1528
HEADLAND		22	20	30	107	261	414	481	468	348	102	11	5	2269
HIGHLAND HOME		20	16	30	77	225	381	462	453	318	94	9	5	2090
HUNTSVILLE WSO		0	7	16	31	159	333	443	422	246	51	0	0	1708
LAFAYETTE		11	8	19	54	187	345	434	412	264	72	0	0	1806
MARION JUNCTION 2 NE		11	15	21	73	215	387	484	468	313	78	8	0	2073
MOBILE WSO		29	23	39	133	307	465	533	521	396	158	26	13	2643
MONTGOMERY WSO	R	13	17	30	78	240	423	518	502	351	95	7	0	2274
MUSCLE SHOALS FAA AP		0	0	15	48	169	352	462	431	257	49	5	0	1788
ONEONTA		0	0	12	27	141	301	412	394	233	34	0	0	1554
OZARK 6 NNW		23	17	43	107	263	414	477	471	345	107	18	8	2293
ROBERTSDALE 1 E		27	22	36	104	267	420	493	477	363	129	24	11	2373
ROCK MILLS		0	8	13	25	160	330	431	412	253	40	0	0	1672
SAINT BERNARD		0	0	12	25	139	295	403	378	221	35	0	0	1508
SAND MT SUBSTA AU		0	6	13	27	137	281	372	360	213	43	0	0	1452
SCOTTSBORO		0	0	12	23	140	297	409	381	231	38	0	0	1531
SELMA		17	16	42	96	274	444	530	518	366	115	9	6	2433
TALLADEGA		9	8	16	42	163	327	425	406	259	48	0	0	1703
THOMASVILLE		15	15	29	84	229	396	481	465	324	85	9	0	2132
TROY		15	17	32	96	247	399	471	468	345	110	14	5	2219
TUSCALOOSA FAA AP		7	8	24	64	216	399	505	487	317	75	0	0	2102
UNION SPRINGS		16	11	18	67	221	387	468	468	324	88	7	5	2080
VALLEY HEAD		0	0	8	11	103	244	344	332	191	29	0	0	1262

01 — ALABAMA

STATE-STATION NUMBER	STN TYP	NAME		LATITUDE DEG-MIN	LONGITUDE DEG-MIN	ELEVATION (FT)
1-0140	12	ALBERTA		N 3214	W 08725	175
1-0160	12	ALEXANDER CITY		N 3257	W 08557	660
1-0178	12	ALICEVILLE		N 3307	W 08807	140
1-0252	13	ANDALUSIA 1 NW		N 3119	W 08630	260
1-0272	13	ANNISTON FAA AP		N 3335	W 08551	599
1-0338	12	ARLEY 3 S		N 3404	W 08713	500
1-0430	13	AUBURN AGRONOMY FARM		N 3236	W 08530	652
1-0440	12	AUTAUGAVILLE 3 N		N 3228	W 08641	200
1-0505	12	BANKHEAD LOCK AND DAM		N 3329	W 08720	280
1-0583	13	BAY MINETTE		N 3053	W 08747	268
1-0616	12	BEATRICE 1 E		N 3144	W 08712	178
1-0655	13	BELLE MINA 2 N		N 3442	W 08653	600
1-0829	13	BIRMINGHAM WSO CITY		N 3328	W 08650	744
1-0831	13	BIRMINGHAM WSO	R	N 3334	W 08645	620
1-0957	12	BOAZ		N 3413	W 08610	1070
1-1069	12	BRANTLEY		N 3135	W 08616	274
1-1084	13	BREWTON 3 SSE		N 3104	W 08703	85
1-1178	12	BRUNDIDGE		N 3143	W 08549	510
1-1288	12	CALERA 2 SW		N 3305	W 08646	540
1-1377	12	CARBON HILL		N 3354	W 08732	430
1-1566	13	CHATOM 3 N		N 3130	W 08815	285
1-1620	12	CHILDERSBURG WTR PLANT		N 3317	W 08620	420
1-1694	13	CLANTON		N 3251	W 08638	580
1-1725	12	CLAYTON		N 3153	W 08528	596
1-1807	12	COFFEE SPRINGS 2 NW		N 3111	W 08556	250
1-1940	12	CORDOVA		N 3346	W 08711	334
1-2079	12	CUBA		N 3226	W 08823	215
1-2119	12	DADEVILLE		N 3249	W 08545	760
1-2188	13	DAYTON		N 3221	W 08739	225
1-2245	13	DEMOPOLIS LOCK AND DAM		N 3231	W 08750	100
1-2577	12	ELBA		N 3125	W 08604	195
1-2632	12	ELROD		N 3315	W 08748	252
1-2730	13	EUFAULA		N 3159	W 08506	220
1-2758	13	EVERGREEN		N 3126	W 08656	216
1-2813	13	FAIRHOPE 2 NE		N 3033	W 08753	23
1-2840	12	FALKVILLE		N 3422	W 08654	630
1-2883	13	FAYETTE		N 3341	W 08749	365
1-3043	12	FORT PAYNE		N 3427	W 08543	934
1-3105	13	FRISCO CITY		N 3126	W 08724	410
1-3154	13	GADSDEN STEAM PLANT		N 3402	W 08600	565
1-3160	12	GAINESVILLE		N 3249	W 08809	130
1-3200	12	GARDEN CITY		N 3401	W 08645	500
1-3430	12	GORGAS		N 3339	W 08713	300
1-3511	13	GREENSBORO		N 3242	W 08736	220
1-3519	12	GREENVILLE		N 3150	W 08638	445
1-3620	13	HALEYVILLE		N 3414	W 08737	950
1-3748	12	HAYNEVILLE		N 3211	W 08636	180
1-3761	13	HEADLAND		N 3121	W 08520	370
1-3816	13	HIGHLAND HOME		N 3157	W 08619	594
1-3842	12	HIGHTOWER		N 3332	W 08524	1175

01 — ALABAMA

STATE-STATION NUMBER	STN TYP	NAME		LATITUDE DEG-MIN	LONGITUDE DEG-MIN	ELEVATION (FT)
1-3899	12	HODGES		N 3420	W 08756	840
1-4064	13	HUNTSVILLE WSO		N 3439	W 08646	600
1-4306	12	JORDAN DAM		N 3237	W 08615	290
1-4502	13	LAFAYETTE		N 3254	W 08524	830
1-4603	12	LAY DAM		N 3258	W 08631	420
1-4619	12	LEEDS		N 3333	W 08633	636
1-4627	12	LEESBURG		N 3411	W 08546	589
1-4798	12	LIVINGSTON 2 SW		N 3235	W 08812	160
1-5112	12	MARION 6 NE		N 3241	W 08716	170
1-5121	13	MARION JUNCTION 2 NE		N 3228	W 08713	200
1-5140	12	MARTIN DAM		N 3240	W 08555	340
1-5354	12	MELVIN		N 3156	W 08827	350
1-5439	12	MILSTEAD		N 3227	W 08554	201
1-5465	12	MITCHELL DAM		N 3248	W 08627	350
1-5478	13	MOBILE WSO		N 3041	W 08815	211
1-5537	12	MONTEVALLO		N 3306	W 08652	462
1-5550	13	MONTGOMERY WSO	R	N 3218	W 08624	183
1-5749	13	MUSCLE SHOALS FAA AP		N 3445	W 08737	536
1-5875	12	NEWTON		N 3120	W 08536	210
1-6121	13	ONEONTA		N 3357	W 08629	870
1-6218	13	OZARK 6 NNW		N 3131	W 08541	470
1-6246	12	PALMERDALE 2 W		N 3344	W 08641	798
1-6362	12	PERRYVILLE		N 3236	W 08709	500
1-6468	12	PINE LEVEL		N 3204	W 08604	506
1-6508	12	PLANTERSVILLE 2 SSE		N 3237	W 08654	230
1-6805	12	RED BAY		N 3426	W 08808	680
1-6847	12	REFORM		N 3323	W 08801	228
1-6988	13	ROBERTSDALE 1 E		N 3032	W 08740	155
1-7025	13	ROCK MILLS		N 3309	W 08518	745
1-7157	13	SAINT BERNARD		N 3412	W 08647	802
1-7207	13	SAND MT SUBSTA AU		N 3417	W 08558	1195
1-7282	12	SAYRE		N 3343	W 08658	304
1-7304	13	SCOTTSBORO		N 3441	W 08603	615
1-7366	13	SELMA		N 3225	W 08700	147
1-7963	12	SUTTLE		N 3232	W 08711	145
1-8024	13	TALLADEGA		N 3327	W 08606	565
1-8178	13	THOMASVILLE		N 3155	W 08744	405
1-8215	12	THURLOW DAM		N 3232	W 08554	288
1-8323	13	TROY		N 3149	W 08559	580
1-8380	13	TUSCALOOSA FAA AP		N 3314	W 08737	170
1-8385	12	TUSCALOOSA OLIVER DAM		N 3313	W 08735	152
1-8438	13	UNION SPRINGS		N 3208	W 08543	455
1-8446	12	UNIONTOWN		N 3226	W 08731	265
1-8469	13	VALLEY HEAD		N 3434	W 08537	1040
1-8605	12	WADLEY		N 3308	W 08534	640
1-8637	12	WALLACE 2 E		N 3113	W 08712	205
1-8648	12	WALNUT GROVE		N 3404	W 08619	850
1-8673	12	WARRIOR LOCK AND DAM		N 3247	W 08750	110
1-8809	12	WEST BLOCTON		N 3307	W 08707	500
1-8859	12	WETUMPKA		N 3233	W 08613	190

01 — ALABAMA

STATE-STATION NUMBER	STN TYP	NAME	LATITUDE DEG-MIN	LONGITUDE DEG-MIN	ELEVATION (FT)
1-8867	12	WHATLEY	N 3139	W 08743	170
1-8998	12	WINFIELD 2 SW	N 3355	W 08750	468

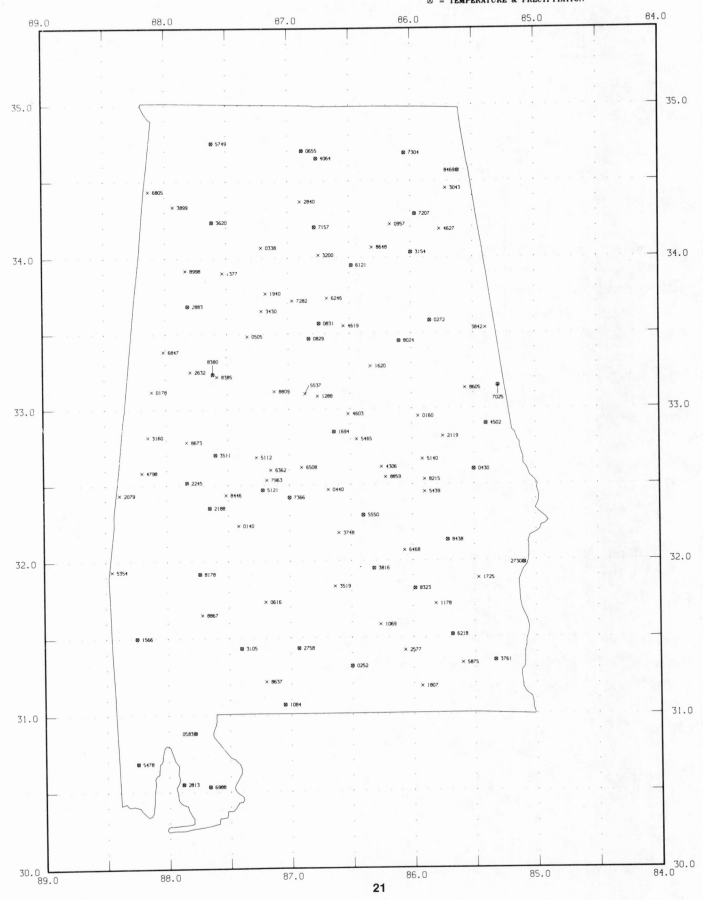

ALASKA

TEMPERATURE NORMALS (DEG F)

STATION			JAN	FEB	MAR	APR	MAY	JUN	JUL	AUG	SEP	OCT	NOV	DEC	ANN
ADAK		MAX	37.1	36.9	38.7	41.4	45.0	48.9	53.6	55.7	52.1	46.6	41.0	37.7	44.6
		MIN	29.3	28.8	30.3	33.1	36.6	40.8	44.6	46.9	43.8	38.3	33.1	30.0	36.3
		MEAN	33.2	32.9	34.6	37.3	40.8	44.9	49.1	51.3	48.0	42.5	37.1	33.9	40.5
ANCHORAGE WSO AP	R	MAX	20.0	25.5	31.7	42.6	54.2	61.8	65.1	63.2	55.2	40.8	27.9	20.4	42.4
		MIN	6.0	10.3	15.7	28.2	38.3	47.0	51.1	49.2	41.1	28.4	15.4	7.1	28.2
		MEAN	13.0	17.9	23.7	35.4	46.3	54.4	58.1	56.2	48.2	34.6	21.7	13.8	35.3
ANNETTE WSO AP	R	MAX	37.3	41.3	43.3	49.1	55.7	60.5	64.1	64.6	60.0	51.8	44.2	39.9	51.0
		MIN	28.1	31.7	32.7	36.6	42.0	47.5	51.5	51.9	48.0	41.9	35.2	31.2	39.9
		MEAN	32.8	36.5	38.0	42.9	48.9	54.0	57.8	58.3	54.1	46.9	39.7	35.6	45.5
ANNEX CREEK		MAX	25.4	32.8	38.6	47.4	54.6	60.9	62.7	60.6	54.8	46.6	36.7	29.5	45.9
		MIN	15.0	22.2	26.7	32.4	37.9	43.9	45.7	44.5	41.5	36.4	28.0	20.8	32.9
		MEAN	20.3	27.5	32.7	39.9	46.3	52.4	54.2	52.6	48.2	41.6	32.4	25.2	39.4
BARROW WSO AP	//R	MAX	-8.0	-13.8	-9.7	5.4	23.6	37.4	44.6	42.4	33.8	18.9	4.6	-7.0	14.4
		MIN	-20.8	-25.5	-22.1	-8.8	13.9	29.3	33.2	33.5	27.3	9.5	-6.6	-18.8	3.7
		MEAN	-14.4	-19.6	-15.9	-1.7	18.8	33.3	38.9	38.0	30.6	14.2	-1.0	-12.9	9.0
BARTER ISLAND WSO AP	R	MAX	-8.5	-14.3	-9.5	7.0	26.1	38.1	45.3	43.7	35.0	20.4	5.7	-6.7	15.2
		MIN	-21.6	-26.7	-23.1	-8.7	15.6	30.1	34.5	34.3	27.7	10.1	-6.1	-19.1	3.9
		MEAN	-15.0	-20.5	-16.3	-.9	20.9	34.1	39.9	39.0	31.4	15.2	-.2	-12.9	9.6
BEAVER FALLS		MAX	34.9	39.5	42.1	48.7	55.6	61.1	64.9	65.0	59.6	50.6	42.5	37.9	50.2
		MIN	25.7	29.5	30.8	34.7	40.3	46.6	51.0	51.7	47.6	40.6	33.5	29.2	38.4
		MEAN	30.3	34.5	36.5	41.7	47.9	53.9	58.0	58.4	53.6	45.6	38.0	33.5	44.3
BETHEL WSO AP	R	MAX	11.8	13.1	19.3	31.7	48.7	58.9	62.1	59.4	52.0	35.6	23.7	11.3	35.6
		MIN	-2.0	-1.6	2.1	15.2	31.9	42.3	47.3	46.2	38.1	23.8	11.2	-1.8	21.1
		MEAN	4.9	5.7	10.7	23.4	40.3	50.6	54.7	52.8	45.0	29.7	17.5	4.8	28.3
BETTLES WSO AP		MAX	-6.5	-.6	12.1	31.6	52.5	67.2	69.0	62.7	48.8	25.5	6.2	-4.7	30.3
		MIN	-22.4	-18.8	-10.6	9.0	33.0	46.3	48.1	43.7	32.1	12.3	-7.9	-20.1	12.1
		MEAN	-14.5	-9.7	.8	20.3	42.8	56.8	58.6	53.2	40.5	18.9	-.9	-12.4	21.2
BIG DELTA WSO AP		MAX	.8	11.0	23.0	40.5	56.5	66.7	69.4	65.0	52.7	31.7	14.6	2.3	36.2
		MIN	-13.6	-7.0	.0	20.8	36.9	47.3	50.6	46.4	35.5	18.1	.6	-11.9	18.6
		MEAN	-6.4	2.0	11.5	30.7	46.7	57.1	60.0	55.7	44.1	24.9	7.6	-4.8	27.4
COLD BAY WSO AP	R	MAX	32.8	32.2	33.6	37.7	44.3	50.0	54.9	55.4	52.0	44.2	38.8	33.9	42.5
		MIN	23.8	22.7	23.6	28.3	34.7	40.8	45.6	46.9	43.0	34.7	29.8	25.0	33.2
		MEAN	28.3	27.5	28.6	33.0	39.5	45.4	50.3	51.2	47.5	39.5	34.3	29.5	37.9
COLLEGE OBSERVATORY		MAX	-1.0	8.3	22.2	40.3	58.4	69.3	71.0	66.3	54.1	32.7	13.2	.2	36.3
		MIN	-15.3	-9.8	.2	19.7	35.6	46.3	49.2	45.3	34.7	17.6	-1.0	-13.7	17.4
		MEAN	-8.2	-.8	11.2	30.0	47.1	57.8	60.1	55.8	44.4	25.2	6.1	-6.8	26.8
CORDOVA FAA AP	//R	MAX	29.5	34.4	37.0	43.8	51.4	57.6	61.3	61.3	56.2	46.6	37.6	31.4	45.7
		MIN	13.0	18.0	20.0	28.4	35.6	42.7	46.6	45.4	40.0	32.6	24.1	16.8	30.3
		MEAN	21.3	26.2	28.5	36.1	43.5	50.2	54.0	53.4	48.1	39.6	30.9	24.1	38.0
EIELSON FIELD		MAX	-4.7	6.5	21.1	40.1	57.7	68.2	70.3	65.7	53.9	32.0	11.2	-2.7	34.9
		MIN	-20.8	-15.0	-5.2	18.8	36.6	47.7	50.5	46.0	34.7	16.4	-4.9	-18.2	15.6
		MEAN	-12.8	-4.2	7.9	29.5	47.2	58.0	60.5	55.9	44.3	24.2	3.1	-10.5	25.3
EKLUTNA PROJECT		MAX	15.2	23.8	32.7	46.6	58.7	66.4	69.2	65.6	55.9	40.0	26.1	16.9	43.1
		MIN	-1.6	5.5	10.5	25.6	35.6	43.9	47.0	44.6	37.3	25.3	11.2	.8	23.8
		MEAN	6.8	14.7	21.6	36.1	47.2	55.2	58.1	55.1	46.6	32.7	18.7	8.9	33.5
ELMENDORF A F BASE		MAX	18.1	24.4	30.7	42.6	54.0	61.0	64.6	62.8	54.8	39.5	26.9	19.2	41.6
		MIN	4.6	10.0	15.3	28.6	39.3	47.8	52.0	50.0	42.0	28.2	15.3	6.8	28.3
		MEAN	11.4	17.2	23.0	35.6	46.7	54.4	58.3	56.4	48.4	33.9	21.2	13.0	35.0
FAIRBANKS WSO AP	R	MAX	-3.9	7.3	21.7	40.8	59.2	70.1	71.8	66.5	54.4	32.6	12.4	-1.7	35.9
		MIN	-21.6	-15.4	-4.8	19.5	37.2	48.5	51.2	46.5	35.4	17.5	-4.6	-18.4	15.9
		MEAN	-12.8	-4.0	8.5	30.2	48.2	59.3	61.5	56.6	44.9	25.0	3.9	-10.1	25.9

ALASKA

TEMPERATURE NORMALS (DEG F)

STATION		JAN	FEB	MAR	APR	MAY	JUN	JUL	AUG	SEP	OCT	NOV	DEC	ANN
FIVE FINGER LIGHT STA	MAX	34.3	37.8	39.7	45.5	51.2	56.4	59.0	58.7	54.3	48.2	42.0	37.7	47.1
	MIN	27.0	30.3	32.5	36.4	41.1	46.4	49.8	48.7	45.3	40.2	34.6	30.2	38.5
	MEAN	30.7	34.1	36.1	41.0	46.2	51.4	54.4	53.7	49.8	44.2	38.3	33.9	42.8
GULKANA WSO	MAX	.2	13.8	27.2	41.8	54.6	64.9	68.4	64.8	54.0	35.7	14.7	1.8	36.8
	MIN	-17.5	-8.0	.2	19.6	32.4	42.3	46.2	42.1	33.3	18.6	-.5	-14.0	16.2
	MEAN	-8.7	2.9	13.7	30.7	43.5	53.7	57.3	53.5	43.6	27.2	7.1	-6.1	26.5
HOMER WSO	MAX	27.0	31.2	34.4	42.1	49.8	56.3	60.5	60.3	54.8	44.0	34.9	27.7	43.6
	MIN	14.4	17.4	19.3	28.1	34.6	41.2	45.1	45.2	39.7	30.6	22.8	15.8	29.5
	MEAN	20.8	24.3	26.9	35.1	42.2	48.8	52.8	52.8	47.3	37.3	28.9	21.8	36.6
JUNEAU WSO AP R	MAX	27.4	33.7	37.4	46.8	54.7	61.1	64.0	62.6	55.9	47.0	37.5	31.5	46.6
	MIN	16.1	21.9	25.0	31.3	38.1	44.2	47.4	46.6	42.3	36.5	28.0	22.1	33.3
	MEAN	21.8	27.8	31.2	39.1	46.5	52.7	55.7	54.6	49.2	41.8	32.7	26.8	40.0
KASILOF	MAX	19.8	26.9	32.3	43.1	53.7	61.0	64.7	63.4	57.0	43.8	29.9	20.6	43.0
	MIN	1.4	6.6	10.5	24.3	32.2	40.2	44.7	43.5	36.8	26.2	14.3	4.4	23.8
	MEAN	10.6	16.8	21.5	33.7	43.0	50.6	54.7	53.5	46.9	35.0	22.1	12.5	33.4
KENAI FAA MUNICIPAL AP	MAX	19.4	26.2	31.4	41.7	52.0	57.7	61.8	61.5	55.4	42.4	29.8	20.4	41.6
	MIN	1.0	5.9	9.7	25.1	34.7	42.2	46.7	45.6	38.6	27.3	13.9	3.6	24.5
	MEAN	10.2	16.1	20.6	33.4	43.4	50.0	54.2	53.6	47.0	34.9	21.9	12.0	33.1
KETCHIKAN	MAX	38.7	42.8	44.4	50.6	57.3	62.1	65.7	66.1	61.2	52.7	45.3	40.9	52.3
	MIN	27.1	31.2	32.0	35.7	40.9	46.6	50.6	51.0	47.0	40.7	34.3	30.6	39.0
	MEAN	32.9	37.0	38.2	43.2	49.1	54.4	58.2	58.6	54.1	46.7	39.8	35.8	45.7
KING SALMON WSO AP //R	MAX	20.0	22.9	27.6	38.7	50.8	58.8	62.9	61.3	54.7	40.5	30.0	19.7	40.7
	MIN	5.0	6.2	10.8	23.5	33.7	41.4	46.3	46.7	39.5	25.8	15.9	4.2	24.9
	MEAN	12.5	14.6	19.2	31.1	42.3	50.1	54.6	54.0	47.1	33.2	23.0	12.0	32.8
KODIAK WSO	MAX	36.6	34.8	38.3	43.3	48.4	56.1	59.3	61.0	55.9	47.1	39.9	35.2	46.3
	MIN	27.2	24.0	27.0	32.7	38.0	43.3	48.1	48.5	43.9	35.2	29.4	24.0	35.1
	MEAN	31.9	29.4	32.7	38.0	43.2	49.7	53.7	54.8	49.9	41.2	34.7	29.6	40.7
KOTZEBUE WSO AP //R	MAX	3.7	1.3	8.0	21.5	38.6	49.8	58.7	56.9	46.9	27.8	13.6	2.2	27.4
	MIN	-9.7	-13.5	-9.3	3.0	24.6	37.7	47.6	46.8	36.3	17.8	2.5	-10.7	14.4
	MEAN	-3.0	-6.1	-.6	12.3	31.6	43.8	53.1	51.9	41.6	22.8	8.1	-4.2	20.9
LITTLE PORT WALTER	MAX	34.8	38.3	40.2	45.5	51.3	56.8	60.7	60.8	55.9	48.8	41.8	37.5	47.7
	MIN	27.1	30.0	30.8	33.8	38.5	44.2	48.4	49.0	45.7	40.1	34.3	30.1	37.7
	MEAN	31.0	34.2	35.5	39.7	44.9	50.5	54.6	54.9	50.8	44.5	38.1	33.8	42.7
MATANUSKA AGR EXP ST1	MAX	19.8	27.2	33.7	46.0	57.8	65.2	67.9	65.3	56.8	42.4	29.0	20.2	44.3
	MIN	2.9	9.6	14.8	27.0	35.8	44.1	47.8	45.8	38.6	26.4	13.5	3.8	25.8
	MEAN	11.4	18.4	24.3	36.5	46.8	54.7	57.9	55.6	47.7	34.4	21.3	12.0	35.1
MCGRATH WSO AP R	MAX	-1.1	9.1	21.5	37.5	54.9	65.5	68.0	63.5	52.7	31.8	13.4	-.8	34.7
	MIN	-19.4	-14.0	-5.2	15.9	34.3	45.0	48.4	45.1	35.2	18.0	-2.5	-17.8	15.3
	MEAN	-10.2	-2.5	8.2	26.7	44.6	55.2	58.2	54.4	44.0	25.0	5.5	-9.4	25.0
MC KINLEY PARK	MAX	8.0	15.4	24.0	38.2	53.0	63.7	66.3	62.1	51.1	32.9	18.5	8.2	36.8
	MIN	-9.9	-5.7	-.5	15.0	29.0	39.0	42.2	39.5	30.2	14.2	.9	-9.1	15.4
	MEAN	-.9	4.8	11.8	26.6	41.0	51.4	54.3	50.8	40.7	23.6	9.7	-.4	26.1
NOME WSO AP R	MAX	13.4	11.8	15.7	25.8	42.1	52.0	56.6	55.7	48.5	33.8	22.7	12.0	32.5
	MIN	-1.9	-5.1	-2.6	9.9	29.3	38.6	44.4	44.0	36.1	22.2	9.7	-3.3	18.4
	MEAN	5.8	3.3	6.6	17.9	35.7	45.4	50.5	49.9	42.3	28.0	16.2	4.4	25.5
NORTHWAY FAA AP	MAX	-13.3	1.5	21.6	40.8	55.7	66.3	69.1	65.0	52.2	29.5	5.7	-9.8	32.0
	MIN	-30.0	-21.1	-9.0	14.8	32.0	44.2	47.9	42.9	31.5	12.6	-10.3	-25.3	10.9
	MEAN	-21.7	-9.8	6.3	27.8	43.9	55.3	58.5	54.0	41.8	21.1	-2.3	-17.6	21.4
PALMER AAES	MAX	19.3	26.1	33.0	45.0	56.6	64.2	66.8	64.1	56.5	41.7	28.1	19.5	43.4
	MIN	2.8	8.4	13.4	26.8	36.2	44.6	48.0	46.1	38.7	26.0	13.1	4.2	25.7
	MEAN	11.0	17.3	23.2	35.9	46.4	54.4	57.4	55.1	47.6	33.9	20.6	11.9	34.6

ALASKA

TEMPERATURE NORMALS (DEG F)

STATION			JAN	FEB	MAR	APR	MAY	JUN	JUL	AUG	SEP	OCT	NOV	DEC	ANN
PETERSBURG		MAX	31.0	36.7	40.4	47.8	55.0	60.5	63.7	62.6	56.9	48.5	40.0	34.5	48.1
		MIN	20.5	25.4	27.6	32.3	38.2	44.5	48.1	47.3	43.2	38.0	30.6	25.4	35.1
		MEAN	25.8	31.1	34.0	40.1	46.6	52.5	55.9	55.0	50.1	43.3	35.3	29.9	41.6
ST PAUL ISLAND WSO AP	R	MAX	30.4	26.4	28.2	32.0	38.6	45.1	49.4	50.8	48.4	41.7	37.2	32.2	38.4
		MIN	22.2	17.3	18.5	23.4	30.9	36.8	41.9	44.1	40.5	33.5	29.4	24.0	30.2
		MEAN	26.3	21.9	23.3	27.7	34.8	40.9	45.7	47.5	44.5	37.6	33.3	28.1	34.3
SEWARD		MAX	28.7	32.6	36.1	44.1	51.7	58.4	62.9	62.2	55.6	44.7	36.0	29.7	45.2
		MIN	18.5	21.6	23.8	31.3	38.0	44.7	49.4	49.0	43.4	34.2	26.5	19.7	33.3
		MEAN	23.7	27.1	30.0	37.7	44.8	51.6	56.2	55.6	49.5	39.5	31.3	24.8	39.3
SHEMYA WSO AP	R	MAX	34.6	33.7	35.3	38.0	41.4	45.1	49.3	51.5	50.6	45.1	39.3	36.0	41.7
		MIN	28.0	26.6	28.0	31.2	35.2	39.4	43.8	46.3	44.4	37.5	31.3	28.7	35.0
		MEAN	31.3	30.2	31.7	34.6	38.3	42.2	46.6	48.9	47.5	41.3	35.3	32.4	38.4
SITKA FAA AIRPORT		MAX	36.7	40.4	41.9	47.6	53.3	57.4	61.2	62.0	58.3	50.5	43.6	38.9	49.3
		MIN	28.3	31.2	32.0	35.6	40.8	46.4	50.8	51.6	47.7	41.5	35.3	31.3	39.4
		MEAN	32.5	35.8	37.0	41.6	47.1	52.0	56.1	56.9	53.0	46.0	39.4	35.2	44.4
SITKA MAGNETIC OBSY		MAX	35.2	40.0	42.2	47.9	53.4	57.3	60.9	62.1	58.6	50.8	42.7	37.5	49.1
		MIN	23.8	27.3	28.1	32.1	37.4	43.6	48.0	48.6	44.1	37.9	31.7	27.5	35.8
		MEAN	29.5	33.7	35.2	40.0	45.4	50.5	54.5	55.4	51.4	44.4	37.2	32.5	42.5
TALKEETNA WSO		MAX	17.9	25.3	32.0	43.7	55.9	64.8	67.8	64.5	55.6	39.8	26.3	17.7	42.6
		MIN	-1.1	3.4	6.5	22.1	33.2	44.4	48.4	45.4	36.2	23.0	9.0	-.4	22.5
		MEAN	8.4	14.4	19.3	32.9	44.6	54.6	58.1	55.0	45.9	31.4	17.7	8.7	32.6
TANANA WSO		MAX	-5.3	2.9	16.9	35.8	56.3	68.6	70.0	64.4	51.2	29.0	8.8	-4.2	32.9
		MIN	-20.2	-16.4	-8.5	12.9	33.5	44.8	47.6	43.9	32.8	15.0	-4.9	-18.6	13.5
		MEAN	-12.8	-6.8	4.2	24.4	44.9	56.7	58.8	54.2	42.0	22.0	2.0	-11.4	23.2
UNIVERSITY EXP STATION		MAX	-.9	10.0	25.0	42.7	60.5	71.4	72.9	67.9	55.9	34.1	13.7	.3	37.8
		MIN	-17.6	-10.9	-.7	18.8	34.4	45.2	48.2	44.5	34.0	17.3	-2.1	-15.9	16.3
		MEAN	-9.2	-.5	12.2	30.8	47.5	58.3	60.6	56.2	45.0	25.7	5.8	-7.8	27.1
VALDEZ WSO		MAX	29.3	30.0	36.2	43.9	51.1	57.5	62.1	61.5	54.1	43.2	33.8	25.1	44.0
		MIN	20.9	19.4	24.1	31.4	38.2	44.5	47.8	46.6	41.3	34.5	25.5	16.8	32.6
		MEAN	25.1	24.7	30.2	37.7	44.7	51.0	55.0	54.1	47.7	38.9	29.7	21.0	38.3
WRANGELL		MAX	32.1	37.6	41.1	48.3	55.7	60.9	63.9	63.1	57.4	49.7	41.2	35.9	48.9
		MIN	21.6	26.8	29.2	34.1	39.5	45.1	48.4	48.4	44.5	38.3	31.0	26.1	36.1
		MEAN	26.9	32.2	35.2	41.2	47.6	53.0	56.2	55.8	51.0	44.0	36.1	31.0	42.5
YAKUTAT WSO AP	R	MAX	30.1	34.8	37.2	43.2	49.7	55.6	59.5	59.7	55.3	47.2	38.4	32.7	45.3
		MIN	16.3	20.6	22.1	28.7	35.8	43.2	47.5	46.3	41.1	34.4	26.5	20.6	31.9
		MEAN	23.2	27.7	29.7	36.0	42.8	49.4	53.5	53.1	48.2	40.8	32.5	26.6	38.6

ALASKA

PRECIPITATION NORMALS (INCHES)

STATION		JAN	FEB	MAR	APR	MAY	JUN	JUL	AUG	SEP	OCT	NOV	DEC	ANN
ADAK		6.20	4.67	6.01	4.66	4.28	3.17	2.98	4.05	5.37	6.86	8.03	7.50	63.78
ANCHORAGE WSO AP	R	.80	.93	.69	.66	.57	1.08	1.97	2.11	2.45	1.73	1.11	1.10	15.20
ANGOON		3.37	3.20	2.54	2.10	1.95	2.04	2.83	3.33	5.37	7.05	4.94	4.46	43.18
ANNETTE WSO AP	R	9.98	10.14	9.11	8.83	6.72	4.97	4.71	7.32	9.92	17.58	13.29	12.90	115.47
ANNEX CREEK		8.07	7.60	7.42	5.78	5.89	5.44	7.29	9.63	13.92	16.73	10.45	9.70	107.92
BARROW WSO AP	//R	.21	.17	.17	.21	.16	.37	.86	.98	.59	.55	.30	.18	4.75
BARTER ISLAND WSO AP	R	.50	.27	.24	.22	.35	.56	1.03	1.08	.80	.81	.40	.23	6.49
BEAVER FALLS		13.42	12.36	11.49	10.36	7.94	6.11	5.66	9.53	14.72	23.59	16.89	16.51	148.58
BETHEL WSO AP	R	.78	.68	.80	.71	.80	1.34	2.11	3.46	2.22	1.29	.96	.98	16.13
BETTLES WSO AP		.76	.68	.71	.60	.50	1.37	1.64	2.34	1.68	1.21	.95	.82	13.26
BIG DELTA WSO AP		.31	.27	.27	.24	.92	2.38	2.37	1.95	1.10	.55	.39	.37	11.12
COLD BAY WSO AP	R	2.70	2.27	2.31	1.95	2.47	2.16	2.50	3.70	3.77	4.29	4.04	2.85	35.01
COLLEGE OBSERVATORY		.61	.48	.47	.28	.60	1.52	2.04	2.22	1.24	.80	.70	.80	11.84
CORDOVA FAA AP	//R	5.20	6.68	5.46	5.75	5.82	4.81	6.63	8.22	12.69	12.94	8.92	7.49	90.61
EIELSON FIELD		.85	.65	.58	.40	.75	1.64	2.41	2.24	1.47	.98	.85	.80	13.62
EKLUTNA PROJECT		1.05	.97	.77	.85	.88	1.81	2.73	2.44	2.64	1.71	1.41	1.36	18.62
ELMENDORF A F BASE		.89	1.02	.86	.68	.57	1.16	2.14	2.20	2.47	1.65	1.13	1.40	16.17
FAIRBANKS WSO AP	R	.53	.42	.40	.27	.57	1.32	1.77	1.86	1.09	.74	.67	.73	10.37
FIVE FINGER LIGHT STA		4.30	4.36	2.91	3.00	3.02	2.83	3.68	5.14	6.06	9.04	6.45	6.21	57.00
GULKANA WSO		.45	.50	.34	.19	.63	1.47	1.81	1.47	1.43	.89	.75	.89	10.82
HOMER WSO		1.65	1.93	1.28	1.31	1.07	1.05	1.47	2.36	2.86	3.28	2.91	2.58	23.75
JUNEAU WSO AP	R	3.69	3.74	3.34	2.92	3.41	2.98	4.13	5.02	6.40	7.71	5.15	4.66	53.15
KASILOF		1.01	1.18	1.00	.76	.83	1.19	1.80	2.25	2.83	2.13	1.63	1.42	18.03
KENAI FAA MUNICIPAL AP		1.09	1.08	1.04	.90	.92	1.29	1.93	2.56	3.30	2.35	1.65	1.35	19.46
KETCHIKAN		13.73	13.57	11.75	12.36	9.62	7.49	7.65	11.31	13.52	24.71	16.93	16.70	159.34
KING SALMON WSO AP	//R	1.04	.88	1.13	1.05	1.18	1.50	2.08	3.13	2.78	1.92	1.40	1.24	19.33
KODIAK WSO		8.29	6.29	4.06	4.84	7.73	3.37	3.91	5.21	7.60	9.99	6.67	6.28	74.24
KOTZEBUE WSO AP	//R	.35	.30	.31	.31	.53	1.46	2.03	1.50	.61	.47	.35	8.53	
LITTLE PORT WALTER		18.63	20.11	16.00	14.24	12.48	8.38	8.88	13.39	22.91	35.48	27.49	25.72	223.71
MATANUSKA AGR EXP ST1		.77	.66	.53	.63	.74	1.59	2.50	2.38	2.33	1.42	1.02	.99	15.56
MCGRATH WSO AP	R	.81	.63	.72	.79	.72	1.44	2.06	2.74	2.04	1.14	1.16	1.05	15.30
MC KINLEY PARK		.74	.59	.57	.38	.82	2.46	3.06	2.38	1.43	.92	.87	.84	15.06
NOME WSO AP	R	.81	.52	.57	.64	.54	1.19	2.20	3.11	2.34	1.26	.94	.65	14.77
NORTHWAY FAA AP		.27	.30	.17	.21	.85	1.81	2.37	1.43	.96	.50	.32	.31	9.50
PALMER AAES		.81	.73	.57	.68	.68	1.40	2.22	2.28	2.32	1.54	1.10	1.03	15.36
PALMER 1 N		.82	.72	.56	.74	.73	1.53	2.46	2.60	2.59	1.47	1.12	.96	16.30
PETERSBURG		8.65	8.27	7.17	6.84	5.89	5.23	5.26	7.08	10.40	16.97	11.46	10.92	104.14
ST PAUL ISLAND WSO AP	R	1.78	1.28	1.26	1.21	1.23	1.24	2.02	3.07	2.52	2.85	2.49	1.76	22.71
SEWARD		4.23	5.48	3.19	3.92	3.95	2.17	2.71	5.01	9.62	10.05	7.28	5.86	63.47
SHEMYA WSO AP	R	2.31	1.85	1.82	1.82	1.73	1.65	2.68	3.64	3.16	4.03	3.96	2.87	31.52
SITKA FAA AIRPORT		6.57	6.10	5.99	5.09	4.70	3.58	4.43	6.68	10.40	13.64	9.36	8.63	85.17
SITKA MAGNETIC OBSY		7.86	7.30	7.09	5.91	4.90	3.75	4.61	6.63	10.96	15.04	10.53	9.96	94.54
TALKEETNA WSO		1.45	1.53	1.49	1.36	1.39	2.50	3.37	4.24	3.97	2.63	1.87	1.41	27.23
TANANA WSO		.60	.51	.56	.39	.63	1.30	1.98	2.68	1.58	.87	.72	.74	12.56
UNIVERSITY EXP STATION		.61	.44	.48	.25	.61	1.53	2.25	2.30	1.28	.81	.74	.78	12.08
VALDEZ WSO		5.05	4.10	3.46	3.13	2.44	2.13	3.95	3.72	8.26	9.12	6.00	5.34	56.70
WALES		.44	.29	.40	.28	.46	.80	1.55	2.76	2.23	1.26	.67	.29	11.43
WRANGELL		6.07	6.07	5.32	4.77	4.25	4.03	4.71	5.78	8.32	13.33	8.76	8.07	79.48
YAKUTAT WSO AP	R	9.39	10.02	9.55	8.62	9.12	5.56	8.26	10.06	15.78	20.12	15.50	12.98	134.96

ALASKA

HEATING DEGREE DAY NORMALS (BASE 65 DEG F)

STATION		JUL	AUG	SEP	OCT	NOV	DEC	JAN	FEB	MAR	APR	MAY	JUN	ANN
ADAK		493	425	510	698	837	964	986	899	942	831	750	603	8938
ANCHORAGE WSO AP	R	214	273	504	942	1299	1587	1612	1319	1280	888	580	318	10816
ANNETTE WSO AP	R	230	213	327	561	759	911	998	798	837	663	499	336	7132
ANNEX CREEK		335	384	504	725	978	1234	1386	1050	1001	753	580	382	9312
BARROW WSO AP	//R	809	837	1032	1575	1980	2415	2461	2369	2508	2001	1432	951	20370
BARTER ISLAND WSO AP	R	778	806	1008	1544	1956	2415	2480	2394	2520	1977	1367	927	20172
BEAVER FALLS		226	218	342	601	810	977	1076	854	884	699	530	340	7557
BETHEL WSO AP	R	319	378	600	1094	1425	1866	1863	1660	1683	1248	766	432	13334
BETTLES WSO AP		217	374	735	1429	1977	2399	2465	2092	1990	1341	688	252	15959
BIG DELTA WSO AP		165	297	627	1243	1722	2164	2213	1764	1659	1029	567	247	13697
COLD BAY WSO AP	R	456	428	525	791	921	1101	1138	1050	1128	960	791	588	9877
COLLEGE OBSERVATORY		166	298	618	1234	1767	2226	2269	1842	1668	1050	555	229	13922
CORDOVA FAA AP	//R	341	360	507	787	1023	1268	1355	1086	1132	867	667	444	9837
EIELSON FIELD		151	289	621	1265	1857	2341	2412	1938	1770	1065	552	224	14485
EKLUTNA PROJECT		214	307	552	1001	1389	1739	1804	1408	1345	867	552	298	11476
ELMENDORF A F BASE		212	267	498	964	1314	1612	1662	1338	1302	882	567	322	10940
FAIRBANKS WSO AP	R	141	270	603	1240	1833	2328	2412	1932	1752	1044	521	198	14274
FIVE FINGER LIGHT STA		329	350	456	645	801	964	1063	865	896	720	583	408	8080
GULKANA WSO		239	357	642	1172	1737	2204	2285	1739	1590	1029	667	343	14004
HOMER WSO		378	378	531	859	1083	1339	1370	1140	1181	897	707	486	10349
JUNEAU WSO AP	R	288	322	474	719	969	1184	1339	1042	1048	777	574	369	9105
KASILOF		319	357	543	930	1287	1628	1686	1350	1349	939	682	432	11502
KENAI FAA MUNICIPAL AP		335	353	540	933	1293	1643	1699	1369	1376	948	670	450	11609
KETCHIKAN		219	209	327	567	756	905	995	784	831	654	493	325	7065
KING SALMON WSO AP	//R	322	341	537	986	1260	1643	1628	1411	1420	1017	704	447	11716
KODIAK WSO		355	316	453	738	909	1097	1026	997	1001	810	676	459	8837
KOTZEBUE WSO AP	//R	375	410	702	1308	1707	2145	2108	1991	2034	1581	1035	636	16032
LITTLE PORT WALTER		322	313	426	636	807	967	1054	862	915	759	623	435	8119
MATANUSKA AGR EXP ST1		220	291	519	949	1311	1643	1662	1305	1262	855	564	309	10890
MCGRATH WSO AP	R	218	333	630	1240	1785	2306	2331	1890	1761	1149	632	299	14574
MC KINLEY PARK		332	440	729	1283	1659	2027	2043	1686	1649	1152	744	408	14152
NOME WSO AP	R	450	468	681	1147	1464	1879	1835	1728	1810	1413	908	588	14371
NORTHWAY FAA AP		206	345	696	1361	2019	2561	2688	2094	1820	1116	654	295	15855
PALMER AAES		236	307	522	964	1332	1646	1674	1336	1296	873	577	318	11081
PETERSBURG		282	310	447	673	891	1088	1215	949	961	747	570	375	8508
ST PAUL ISLAND WSO AP	R	598	543	615	849	951	1144	1200	1207	1293	1119	936	723	11178
SEWARD		273	291	465	791	1011	1246	1280	1061	1085	819	626	402	9350
SHEMYA WSO AP	R	570	499	525	735	891	1011	1045	974	1032	912	828	684	9706
SITKA FAA AIRPORT		276	251	360	589	768	924	1008	818	868	702	555	390	7509
SITKA MAGNETIC OBSY		326	298	408	639	834	1008	1101	876	924	750	608	435	8207
TALKEETNA WSO		214	310	573	1042	1419	1745	1755	1417	1417	963	632	317	11804
TANANA WSO		203	341	690	1333	1890	2368	2412	2010	1885	1218	623	256	15229
UNIVERSITY EXP STATION		150	283	600	1218	1776	2257	2300	1834	1637	1026	543	216	13840
VALDEZ WSO		310	338	519	809	1059	1364	1237	1128	1079	819	629	420	9711
WRANGELL		278	291	420	651	867	1054	1181	918	924	714	539	360	8197
YAKUTAT WSO AP	R	357	369	504	750	975	1190	1296	1044	1094	870	688	468	9605

ALASKA

COOLING DEGREE DAY NORMALS (BASE 65 DEG F)

STATION		JAN	FEB	MAR	APR	MAY	JUN	JUL	AUG	SEP	OCT	NOV	DEC	ANN
ADAK		0	0	0	0	0	0	0	0	0	0	0	0	0
ANCHORAGE WSO AP	R	0	0	0	0	0	0	0	0	0	0	0	0	0
ANNETTE WSO AP	R	0	0	0	0	0	6	7	5	0	0	0	0	18
ANNEX CREEK		0	0	0	0	0	0	0	0	0	0	0	0	0
BARROW WSO AP	//R	0	0	0	0	0	0	0	0	0	0	0	0	0
BARTER ISLAND WSO AP	R	0	0	0	0	0	0	0	0	0	0	0	0	0
BEAVER FALLS		0	0	0	0	0	7	9	13	0	0	0	0	29
BETHEL WSO AP	R	0	0	0	0	0	0	0	0	0	0	0	0	0
BETTLES WSO AP		0	0	0	0	0	6	18	8	0	0	0	0	32
BIG DELTA WSO AP		0	0	0	0	0	10	10	9	0	0	0	0	29
COLD BAY WSO AP	R	0	0	0	0	0	0	0	0	0	0	0	0	0
COLLEGE OBSERVATORY		0	0	0	0	0	13	14	13	0	0	0	0	40
CORDOVA FAA AP	//R	0	0	0	0	0	0	0	0	0	0	0	0	0
EIELSON FIELD		0	0	0	0	0	14	11	7	0	0	0	0	32
EKLUTNA PROJECT		0	0	0	0	0	0	0	0	0	0	0	0	0
ELMENDORF A F BASE		0	0	0	0	0	0	0	0	0	0	0	0	0
FAIRBANKS WSO AP	R	0	0	0	0	0	27	33	10	0	0	0	0	70
FIVE FINGER LIGHT STA		0	0	0	0	0	0	0	0	0	0	0	0	0
GULKANA WSO		0	0	0	0	0	0	0	0	0	0	0	0	0
HOMER WSO		0	0	0	0	0	0	0	0	0	0	0	0	0
JUNEAU WSO AP	R	0	0	0	0	0	0	0	0	0	0	0	0	0
KASILOF		0	0	0	0	0	0	0	0	0	0	0	0	0
KENAI FAA MUNICIPAL AP		0	0	0	0	0	0	0	0	0	0	0	0	0
KETCHIKAN		0	0	0	0	0	7	8	11	0	0	0	0	26
KING SALMON WSO AP	//R	0	0	0	0	0	0	0	0	0	0	0	0	0
KODIAK WSO		0	0	0	0	0	0	0	0	0	0	0	0	0
KOTZEBUE WSO AP	//R	0	0	0	0	0	0	7	0	0	0	0	0	7
LITTLE PORT WALTER		0	0	0	0	0	0	0	0	0	0	0	0	0
MATANUSKA AGR EXP ST1		0	0	0	0	0	0	0	0	0	0	0	0	0
MCGRATH WSO AP	R	0	0	0	0	0	5	7	0	0	0	0	0	12
MC KINLEY PARK		0	0	0	0	0	0	0	0	0	0	0	0	0
NOME WSO AP	R	0	0	0	0	0	0	0	0	0	0	0	0	0
NORTHWAY FAA AP		0	0	0	0	0	0	0	0	0	0	0	0	0
PALMER AAES		0	0	0	0	0	0	0	0	0	0	0	0	0
PETERSBURG		0	0	0	0	0	0	0	0	0	0	0	0	0
ST PAUL ISLAND WSO AP	R	0	0	0	0	0	0	0	0	0	0	0	0	0
SEWARD		0	0	0	0	0	0	0	0	0	0	0	0	0
SHEMYA WSO AP	R	0	0	0	0	0	0	0	0	0	0	0	0	0
SITKA FAA AIRPORT		0	0	0	0	0	0	0	0	0	0	0	0	0
SITKA MAGNETIC OBSY		0	0	0	0	0	0	0	0	0	0	0	0	0
TALKEETNA WSO		0	0	0	0	0	0	0	0	0	0	0	0	0
TANANA WSO		0	0	0	0	0	7	11	6	0	0	0	0	24
UNIVERSITY EXP STATION		0	0	0	0	0	15	14	10	0	0	0	0	39
VALDEZ WSO		0	0	0	0	0	0	0	0	0	0	0	0	0
WRANGELL		0	0	0	0	0	0	5	6	0	0	0	0	11
YAKUTAT WSO AP	R	0	0	0	0	0	0	0	0	0	0	0	0	0

50 — ALASKA

STATE-STATION NUMBER	STN TYP	NAME		LATITUDE DEG-MIN	LONGITUDE DEG-MIN	ELEVATION (FT)
50-0026	13	ADAK		N 5153	W 17639	15
50-0280	13	ANCHORAGE WSO AP	R	N 6110	W 15001	114
50-0310	12	ANGOON		N 5730	W 13435	15
50-0352	13	ANNETTE WSO AP	R	N 5502	W 13134	110
50-0363	13	ANNEX CREEK		N 5819	W 13406	24
50-0546	13	BARROW WSO AP	//R	N 7118	W 15647	31
50-0558	13	BARTER ISLAND WSO AP	R	N 7008	W 14338	39
50-0657	13	BEAVER FALLS		N 5523	W 13128	35
50-0754	13	BETHEL WSO AP	R	N 6047	W 16148	125
50-0761	13	BETTLES WSO AP		N 6655	W 15131	666
50-0770	13	BIG DELTA WSO AP		N 6400	W 14544	1268
50-2102	13	COLD BAY WSO AP	R	N 5512	W 16243	96
50-2107	13	COLLEGE OBSERVATORY		N 6452	W 14750	621
50-2177	13	CORDOVA FAA AP	//R	N 6030	W 14530	41
50-2707	13	EIELSON FIELD		N 6440	W 14706	547
50-2730	13	EKLUTNA PROJECT		N 6128	W 14910	38
50-2820	13	ELMENDORF A F BASE		N 6115	W 14948	192
50-2968	13	FAIRBANKS WSO AP	R	N 6449	W 14752	436
50-3072	13	FIVE FINGER LIGHT STA		N 5716	W 13337	30
50-3465	13	GULKANA WSO		N 6209	W 14527	1570
50-3665	13	HOMER WSO		N 5938	W 15130	67
50-4100	13	JUNEAU WSO AP	R	N 5822	W 13435	12
50-4425	13	KASILOF		N 6019	W 15115	75
50-4546	13	KENAI FAA MUNICIPAL AP		N 6034	W 15115	86
50-4590	13	KETCHIKAN		N 5521	W 13139	15
50-4766	13	KING SALMON WSO AP	//R	N 5841	W 15639	49
50-4988	13	KODIAK WSO		N 5745	W 15230	14
50-5076	13	KOTZEBUE WSO AP	//R	N 6652	W 16238	10
50-5519	13	LITTLE PORT WALTER		N 5623	W 13439	14
50-5733	13	MATANUSKA AGR EXP ST1		N 6134	W 14916	150
50-5769	13	MCGRATH WSO AP	R	N 6258	W 15537	344
50-5778	13	MC KINLEY PARK		N 6343	W 14858	2070
50-6496	13	NOME WSO AP	R	N 6430	W 16526	13
50-6586	13	NORTHWAY FAA AP		N 6257	W 14156	1713
50-6870	13	PALMER AAES		N 6136	W 14906	225
50-6871	12	PALMER 1 N		N 6137	W 14906	220
50-7233	13	PETERSBURG		N 5649	W 13257	50
50-8118	13	ST PAUL ISLAND WSO AP	R	N 5709	W 17013	22
50-8371	13	SEWARD		N 6007	W 14927	70
50-8419	13	SHEMYA WSO AP	R	N 5243	E 17406	122
50-8494	13	SITKA FAA AIRPORT		N 5704	W 13521	15
50-8503	13	SITKA MAGNETIC OBSY		N 5703	W 13520	67
50-8976	13	TALKEETNA WSO		N 6218	W 15006	345
50-9014	13	TANANA WSO		N 6510	W 15206	232
50-9641	13	UNIVERSITY EXP STATION		N 6451	W 14752	475
50-9686	13	VALDEZ WSO		N 6108	W 14621	20
50-9739	12	WALES		N 6537	W 16803	9
50-9919	13	WRANGELL		N 5628	W 13223	37
50-9941	13	YAKUTAT WSO AP	R	N 5931	W 13940	28

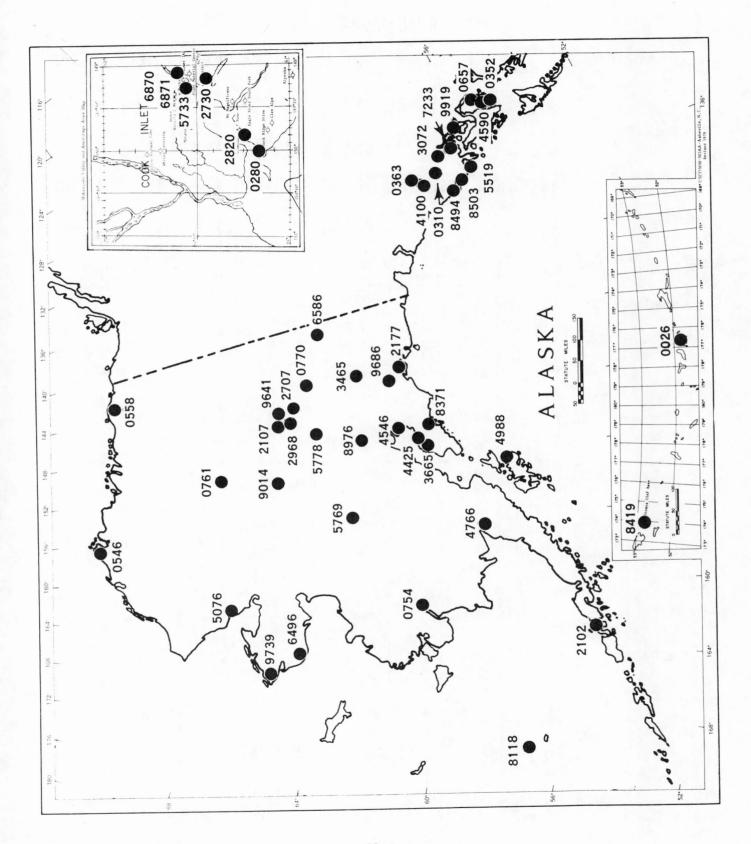

ARIZONA

TEMPERATURE NORMALS (DEG F)

STATION		JAN	FEB	MAR	APR	MAY	JUN	JUL	AUG	SEP	OCT	NOV	DEC	ANN
AJO	MAX	64.3	68.9	73.2	81.3	89.8	99.6	103.2	100.8	97.6	87.8	74.1	66.0	83.9
	MIN	42.0	45.1	49.0	55.5	63.1	72.5	78.7	76.3	72.4	62.2	50.3	43.6	59.2
	MEAN	53.2	57.0	61.1	68.4	76.5	86.1	91.0	88.6	85.0	75.0	62.2	54.8	71.6
ANVIL RANCH	MAX	66.3	69.8	74.3	82.4	91.0	101.0	100.9	98.0	96.0	87.1	74.7	67.0	84.0
	MIN	32.5	34.5	38.7	44.1	51.9	61.8	70.1	67.7	61.2	49.4	38.3	32.6	48.6
	MEAN	49.4	52.2	56.5	63.3	71.5	81.4	85.5	82.8	78.6	68.3	56.5	49.8	66.3
APACHE POWDER COMPANY	MAX	63.3	67.1	71.8	79.9	88.0	97.2	96.6	93.6	91.2	82.8	71.2	64.0	80.6
	MIN	28.5	30.7	35.2	40.6	47.1	56.9	65.5	63.6	56.8	44.7	33.6	28.4	44.3
	MEAN	45.9	48.9	53.5	60.3	67.5	77.1	81.1	78.7	74.1	63.8	52.4	46.2	62.5
BARTLETT DAM	MAX	65.1	69.2	73.0	81.5	91.0	101.5	105.3	102.5	98.5	88.2	74.8	66.8	84.8
	MIN	39.9	41.9	45.0	51.2	59.1	67.8	75.6	73.9	69.4	59.1	47.5	41.4	56.0
	MEAN	52.5	55.6	59.0	66.4	75.1	84.7	90.5	88.2	84.0	73.7	61.2	54.2	70.4
BETATAKIN	MAX	38.9	42.7	48.6	59.1	70.0	81.2	86.4	83.2	76.0	64.0	48.6	40.1	61.6
	MIN	20.2	22.8	26.5	33.2	42.2	52.6	58.5	56.5	50.4	40.4	28.8	21.7	37.8
	MEAN	29.6	32.8	37.6	46.2	56.1	66.9	72.5	69.9	63.2	52.2	38.7	31.0	49.7
BLACK RIVER PUMPS	MAX	48.1	51.5	56.1	65.4	74.6	85.4	87.3	83.7	79.7	70.3	57.9	50.2	67.5
	MIN	20.4	22.8	26.6	32.8	40.5	49.8	57.3	55.5	49.9	39.2	27.5	21.1	37.0
	MEAN	34.3	37.2	41.3	49.1	57.5	67.6	72.4	69.6	64.8	54.8	42.7	35.7	52.3
BOUSE	MAX	65.8	71.7	76.8	85.2	94.1	103.9	108.3	105.9	101.1	89.9	75.2	66.5	87.0
	MIN	33.2	37.0	41.7	48.0	56.8	65.3	76.0	74.3	65.2	52.3	39.7	33.1	51.9
	MEAN	49.5	54.3	59.3	66.6	75.5	84.6	92.2	90.1	83.2	71.1	57.5	49.8	69.5
BUCKEYE	MAX	67.1	72.4	77.3	86.0	94.9	104.9	108.4	105.6	101.2	90.6	76.7	68.4	87.8
	MIN	35.2	38.8	43.0	48.4	56.5	65.0	75.6	73.6	65.8	53.3	41.7	35.3	52.7
	MEAN	51.2	55.6	60.2	67.2	75.8	85.0	92.0	89.6	83.5	72.0	59.2	51.9	70.3
CANELO 1 NW	MAX	57.8	60.7	64.7	72.6	80.5	90.2	88.4	85.2	83.6	76.3	65.3	58.5	73.7
	MIN	26.5	27.5	31.0	36.2	42.8	52.5	60.0	58.1	52.2	41.8	31.9	26.9	40.6
	MEAN	42.2	44.1	47.9	54.4	61.7	71.4	74.3	71.7	67.9	59.1	48.6	42.7	57.2
CASA GRANDE	MAX	66.4	71.4	76.4	85.3	94.4	104.2	106.4	103.6	99.7	89.8	75.8	67.3	86.7
	MIN	35.8	38.8	43.3	48.8	57.4	66.6	76.2	73.9	66.7	54.5	42.7	36.1	53.4
	MEAN	51.1	55.1	59.9	67.1	75.9	85.4	91.4	88.8	83.2	72.2	59.3	51.8	70.1
CASA GRANDE RUINS N M	MAX	67.4	72.1	77.0	85.9	95.3	105.1	107.4	104.4	100.8	90.4	76.7	68.2	87.6
	MIN	34.0	36.4	40.2	45.4	53.7	63.5	74.4	72.4	65.2	52.7	40.7	34.5	51.1
	MEAN	50.7	54.3	58.6	65.7	74.5	84.3	90.9	88.4	83.0	71.6	58.7	51.4	69.3
CHILDS	MAX	59.6	65.5	69.7	78.4	88.0	98.6	102.9	99.2	95.2	84.8	70.5	60.8	81.1
	MIN	32.0	34.5	38.0	43.6	51.3	59.5	68.7	66.3	60.0	49.1	38.2	33.0	47.9
	MEAN	45.9	50.0	53.9	61.0	69.6	79.1	85.8	82.8	77.6	67.0	54.4	46.9	64.5
CHINO VALLEY	MAX	52.4	57.0	61.3	68.7	77.8	88.1	92.7	89.3	85.6	75.7	62.8	54.4	72.2
	MIN	21.2	23.7	27.1	32.8	40.3	48.5	58.3	56.5	48.4	37.8	27.1	21.0	36.9
	MEAN	36.8	40.4	44.3	50.7	59.0	68.3	75.5	72.9	67.1	56.8	45.0	37.7	54.5
CHIRICAHUA NAT MON	MAX	55.8	59.0	63.8	72.3	80.6	89.9	89.2	85.8	82.9	74.7	63.4	56.6	72.8
	MIN	29.6	30.6	33.9	39.4	46.5	55.4	60.1	58.6	54.9	46.0	36.0	30.6	43.5
	MEAN	42.7	44.8	48.8	55.9	63.6	72.7	74.7	72.2	69.0	60.4	49.7	43.6	58.2
CLIFTON	MAX	61.7	66.6	72.4	80.8	89.7	99.4	101.2	98.5	94.2	84.2	71.0	61.9	81.8
	MIN	31.0	35.3	39.8	47.2	55.4	65.3	70.7	69.0	63.9	52.8	39.2	31.6	50.1
	MEAN	46.4	51.0	56.1	64.0	72.5	82.4	86.0	83.8	79.1	68.5	55.1	46.8	66.0
CORDES	MAX	56.7	60.2	63.8	71.9	80.9	91.4	95.6	92.3	88.2	77.8	65.8	58.1	75.2
	MIN	32.0	33.6	36.0	41.1	48.6	57.8	65.8	64.5	58.9	49.4	39.1	33.1	46.7
	MEAN	44.4	46.9	49.9	56.5	64.8	74.6	80.7	78.4	73.6	63.7	52.4	45.6	61.0
DEER VALLEY	MAX	65.4	70.2	74.5	83.0	92.3	102.0	104.9	102.4	98.7	88.2	74.6	66.7	85.2
	MIN	36.9	39.7	43.2	49.0	57.3	66.4	75.7	74.1	67.6	55.5	43.7	37.5	53.9
	MEAN	51.2	54.9	58.9	66.0	74.8	84.2	90.3	88.2	83.2	71.9	59.2	52.1	69.6

ARIZONA

TEMPERATURE NORMALS (DEG F)

STATION		JAN	FEB	MAR	APR	MAY	JUN	JUL	AUG	SEP	OCT	NOV	DEC	ANN
DOUGLAS B D FAA AP	MAX	60.9	64.4	69.2	77.3	85.2	94.2	92.8	90.3	87.8	79.7	68.5	61.5	77.7
	MIN	28.9	30.8	35.8	42.1	49.8	59.6	65.4	63.5	58.5	46.9	35.1	29.0	45.5
	MEAN	44.9	47.6	52.5	59.7	67.5	76.9	79.1	76.9	73.1	63.3	51.9	45.3	61.6
DUNCAN	MAX	58.7	63.4	69.3	77.7	85.8	94.7	95.1	92.4	88.1	78.9	67.0	58.9	77.5
	MIN	23.1	25.7	30.9	36.8	44.9	53.9	64.1	61.9	54.1	41.0	29.0	22.8	40.7
	MEAN	40.9	44.6	50.1	57.2	65.4	74.3	79.6	77.2	71.1	60.0	48.0	40.9	59.1
FLAGSTAFF WSO //R	MAX	41.7	44.5	48.6	57.1	66.7	77.6	81.9	78.9	74.1	63.7	51.0	43.6	60.8
	MIN	14.7	16.9	20.4	25.9	32.9	40.9	50.3	48.7	40.9	30.6	21.5	15.9	30.0
	MEAN	28.2	30.7	34.5	41.6	49.9	59.2	66.1	63.8	57.5	47.2	36.3	29.8	45.4
FLORENCE	MAX	66.9	71.4	76.1	85.0	94.5	104.4	106.5	104.0	100.8	90.0	76.4	68.1	87.0
	MIN	36.3	38.8	42.3	47.4	55.3	64.5	74.6	72.8	66.5	54.7	43.2	37.2	52.8
	MEAN	51.6	55.1	59.2	66.2	74.9	84.5	90.6	88.4	83.7	72.4	59.8	52.7	69.9
FORT VALLEY	MAX	42.4	45.0	48.5	57.3	67.1	78.2	81.8	78.2	74.3	64.6	52.0	44.3	61.1
	MIN	8.4	11.4	15.9	21.1	26.6	33.1	43.9	43.0	34.9	25.3	16.4	9.7	24.1
	MEAN	25.4	28.3	32.2	39.2	46.9	55.6	62.9	60.6	54.6	45.0	34.2	27.0	42.7
GANADO	MAX	41.2	46.2	52.8	62.3	72.7	83.1	87.2	83.4	76.9	66.2	52.0	43.1	63.9
	MIN	14.1	18.8	23.5	29.1	37.1	45.9	55.1	53.3	45.3	34.6	23.0	15.1	32.9
	MEAN	27.7	32.6	38.2	45.7	54.9	64.5	71.2	68.3	61.1	50.4	37.5	29.1	48.4
GILA BEND	MAX	68.1	73.0	78.0	86.7	95.9	105.6	109.3	106.5	102.9	91.8	77.6	69.1	88.7
	MIN	37.9	41.1	45.4	51.7	60.4	68.9	78.7	77.0	69.9	57.8	45.9	38.6	56.1
	MEAN	53.0	57.1	61.8	69.2	78.2	87.3	94.0	91.8	86.4	74.8	61.8	53.9	72.4
HEBER RANGER STATION	MAX	46.3	50.2	55.1	63.4	72.1	82.8	85.3	81.9	77.6	68.2	55.8	47.7	65.5
	MIN	15.6	18.5	22.0	27.0	33.8	42.5	51.7	50.1	43.1	32.4	21.9	16.4	31.3
	MEAN	31.0	34.4	38.6	45.2	53.0	62.7	68.5	66.0	60.4	50.3	38.9	32.1	48.4
HOLBROOK	MAX	47.2	54.2	61.1	70.1	79.5	90.4	94.5	91.2	85.5	73.7	59.3	47.7	71.2
	MIN	18.4	21.9	27.0	33.8	41.2	50.6	59.9	57.8	49.6	37.6	25.7	18.5	36.8
	MEAN	32.8	38.1	44.1	52.0	60.3	70.5	77.2	74.5	67.6	55.6	42.5	33.1	54.0
JEROME	MAX	50.0	53.8	58.1	66.3	75.2	85.5	89.6	86.3	82.3	72.3	59.4	51.7	69.2
	MIN	32.5	34.2	37.6	43.9	52.2	62.1	66.8	64.7	60.6	51.0	40.0	33.7	48.3
	MEAN	41.3	44.0	47.9	55.1	63.7	73.8	78.2	75.5	71.5	61.7	49.7	42.7	58.8
JUNIPINE	MAX	50.8	54.5	58.9	67.2	76.3	86.6	90.2	87.2	83.3	73.0	60.5	52.6	70.1
	MIN	27.2	28.6	31.4	37.0	44.1	51.8	59.1	57.6	52.0	42.6	33.8	28.6	41.2
	MEAN	39.0	41.5	45.2	52.1	60.2	69.2	74.7	72.4	67.6	57.8	47.2	40.6	55.6
KINGMAN NO 2	MAX	54.9	59.8	63.9	71.7	81.4	91.6	97.1	94.6	89.8	79.0	64.9	56.4	75.4
	MIN	31.6	34.2	37.1	43.5	51.6	60.6	68.7	66.8	59.7	49.5	38.6	32.3	47.9
	MEAN	43.2	47.0	50.5	57.6	66.5	76.2	82.9	80.8	74.8	64.2	51.8	44.3	61.7
LAVEEN 3 SSE	MAX	66.3	71.3	75.9	84.3	93.5	103.0	105.6	102.9	99.2	88.9	75.7	67.5	86.2
	MIN	37.6	40.6	44.5	50.8	59.3	68.7	77.6	75.0	68.2	56.0	44.4	38.1	55.1
	MEAN	52.0	56.0	60.2	67.6	76.4	85.9	91.6	89.0	83.7	72.5	60.1	52.8	70.7
LEES FERRY	MAX	48.8	57.6	66.6	77.0	87.3	98.3	103.5	100.2	92.8	78.6	61.0	49.4	76.8
	MIN	27.3	32.8	39.1	47.2	56.2	65.4	72.7	70.4	61.6	49.3	36.4	28.5	48.9
	MEAN	38.0	45.2	52.9	62.1	71.8	81.9	88.1	85.3	77.3	64.0	48.7	39.0	62.9
LEUPP	MAX	46.7	54.9	61.8	70.7	80.4	91.3	95.5	92.6	86.5	74.4	59.0	47.8	71.8
	MIN	16.1	20.4	25.6	32.8	40.3	48.4	58.8	56.8	48.4	35.8	23.7	15.7	35.2
	MEAN	31.4	37.7	43.7	51.7	60.4	69.9	77.2	74.8	67.5	55.1	41.4	31.8	53.6
LITCHFIELD PARK	MAX	67.4	72.3	77.1	85.8	95.0	104.6	107.5	105.0	101.1	90.6	77.0	68.5	87.7
	MIN	36.2	39.3	43.4	49.0	56.9	66.0	75.5	73.4	65.8	53.7	42.5	36.2	53.2
	MEAN	51.8	55.8	60.3	67.4	75.9	85.3	91.5	89.2	83.5	72.2	59.8	52.4	70.4
MC NARY	MAX	44.7	47.0	50.8	60.0	69.1	79.4	81.3	78.3	75.1	66.5	54.0	47.2	62.8
	MIN	16.7	18.4	22.3	27.8	34.1	41.7	49.2	47.8	41.8	32.8	23.9	18.2	31.2
	MEAN	30.7	32.7	36.6	43.9	51.6	60.5	65.2	63.1	58.5	49.7	39.0	32.7	47.0

ARIZONA

TEMPERATURE NORMALS (DEG F)

STATION		JAN	FEB	MAR	APR	MAY	JUN	JUL	AUG	SEP	OCT	NOV	DEC	ANN
MESA EXPERIMENT FARM	MAX	65.2	69.6	73.8	82.5	91.6	101.5	104.6	102.0	98.3	88.3	74.9	67.0	84.9
	MIN	37.0	39.9	44.3	50.3	57.9	66.6	75.5	73.8	67.1	55.9	44.3	37.8	54.2
	MEAN	51.1	54.7	59.1	66.4	74.8	84.1	90.0	87.9	82.7	72.1	59.6	52.4	69.6
MIAMI	MAX	55.6	60.2	64.9	73.8	83.3	93.9	96.9	93.6	89.3	78.8	65.0	56.7	76.0
	MIN	32.5	34.8	39.2	46.1	54.4	64.0	70.3	67.5	62.5	51.6	40.0	33.4	49.7
	MEAN	44.1	47.5	52.1	60.0	68.9	79.0	83.6	80.6	75.9	65.2	52.6	45.1	62.9
MONTEZUMA CASTLE N M	MAX	59.6	65.3	70.2	78.3	87.5	97.9	101.6	98.1	94.0	83.3	69.4	60.1	80.4
	MIN	25.6	28.5	33.0	38.3	45.5	53.5	63.6	61.9	54.0	42.7	31.6	25.3	42.0
	MEAN	42.6	46.9	51.6	58.3	66.6	75.7	82.6	80.0	74.1	63.0	50.5	42.7	61.2
MORMON FLAT	MAX	63.4	68.0	72.3	81.4	90.8	101.3	104.0	101.3	97.4	87.1	73.1	64.6	83.7
	MIN	42.3	43.6	46.9	53.5	62.4	71.8	78.8	76.3	72.0	61.9	50.2	43.4	58.6
	MEAN	52.9	55.8	59.6	67.5	76.6	86.6	91.4	88.8	84.7	74.5	61.7	54.0	71.2
NOGALES	MAX	64.0	66.6	70.2	77.5	85.0	93.9	92.7	90.2	89.1	81.8	71.7	65.0	79.0
	MIN	27.1	28.5	32.5	37.2	43.2	53.5	63.6	61.2	54.1	42.7	32.2	27.3	41.9
	MEAN	45.6	47.5	51.4	57.3	64.1	73.8	78.2	75.7	71.6	62.3	52.0	46.2	60.5
ORACLE 2 SE	MAX	57.9	60.9	64.7	73.2	82.2	92.1	93.1	90.0	87.2	78.3	66.5	58.9	75.4
	MIN	34.3	36.1	38.9	44.6	52.5	62.4	66.7	64.9	60.7	51.5	40.9	35.4	49.1
	MEAN	46.2	48.5	51.9	58.9	67.4	77.3	79.9	77.5	74.0	64.9	53.7	47.2	62.3
ORGAN PIPE CACTUS N M	MAX	67.7	71.7	75.5	83.0	90.9	100.0	103.2	101.2	98.1	88.6	76.2	68.8	85.4
	MIN	38.2	40.2	43.2	48.4	55.2	64.0	73.8	72.0	66.3	55.7	44.6	38.4	53.3
	MEAN	53.0	56.0	59.4	65.7	73.1	82.0	88.5	86.6	82.2	72.2	60.4	53.6	69.4
PARKER	MAX	67.0	72.7	78.1	86.4	95.0	104.4	109.1	106.7	102.1	91.0	76.7	67.7	88.1
	MIN	38.4	42.7	47.2	53.6	62.4	71.0	79.7	78.4	70.9	58.5	45.8	38.6	57.3
	MEAN	52.7	57.7	62.7	70.0	78.7	87.8	94.4	92.6	86.5	74.8	61.3	53.2	72.7
PAYSON	MAX	53.4	57.5	61.4	69.7	78.7	89.3	92.9	89.1	85.1	75.0	62.6	55.0	72.5
	MIN	23.8	25.6	28.6	33.6	40.7	48.8	58.1	56.5	49.1	39.5	29.5	23.8	38.1
	MEAN	38.6	41.5	45.0	51.6	59.7	69.1	75.5	72.8	67.1	57.3	46.1	39.4	55.3
PEARCE	MAX	61.2	64.8	69.3	77.5	85.5	94.9	94.4	91.3	88.8	80.4	69.1	61.5	78.2
	MIN	30.0	31.7	35.8	42.0	49.2	58.8	64.5	62.2	56.7	46.5	35.9	30.4	45.3
	MEAN	45.6	48.3	52.6	59.8	67.4	76.9	79.5	76.8	72.8	63.5	52.5	45.9	61.8
PETRIFIED FOREST NP	MAX	47.2	53.6	60.1	69.0	78.0	88.8	92.4	89.1	83.7	72.4	58.0	47.9	70.0
	MIN	20.8	23.9	28.3	34.9	43.1	52.2	60.5	58.9	52.0	40.3	28.4	20.9	38.7
	MEAN	34.1	38.8	44.2	52.0	60.6	70.5	76.5	74.0	67.9	56.4	43.2	34.4	54.4
PHOENIX WSO R	MAX	65.2	69.7	74.5	83.1	92.4	102.3	105.0	102.3	98.2	87.7	74.3	66.4	85.1
	MIN	39.4	42.5	46.7	53.0	61.5	70.6	79.5	77.5	70.9	59.1	46.9	40.2	57.3
	MEAN	52.3	56.1	60.6	68.0	77.0	86.5	92.3	89.9	84.6	73.4	60.6	53.3	71.2
PRESCOTT	MAX	50.3	54.1	57.7	65.4	74.0	84.7	88.7	85.2	81.5	71.9	59.5	51.8	68.7
	MIN	22.1	24.2	28.0	33.3	40.3	48.8	57.5	55.6	48.6	38.1	28.4	22.5	37.3
	MEAN	36.2	39.2	42.8	49.4	57.2	66.8	73.1	70.4	65.1	55.0	44.0	37.2	53.0
ROOSEVELT 1 WNW	MAX	58.6	64.5	69.6	78.8	88.3	98.7	102.3	99.3	94.9	83.9	68.8	59.7	80.6
	MIN	36.5	39.4	43.9	51.2	59.9	68.9	75.1	72.7	67.4	56.8	44.6	37.7	54.5
	MEAN	47.6	51.9	56.8	65.0	74.1	83.9	88.7	86.0	81.2	70.4	56.8	48.7	67.6
SABINO CANYON	MAX	66.8	69.8	74.2	82.6	91.5	101.4	101.9	99.5	96.8	87.1	75.0	67.9	84.5
	MIN	37.1	39.0	42.8	48.6	56.3	65.9	72.4	70.2	65.7	54.9	43.7	37.8	52.9
	MEAN	52.0	54.5	58.5	65.7	73.9	83.7	87.2	84.9	81.3	71.0	59.4	52.9	68.8
SACATON	MAX	65.9	70.7	75.2	84.5	93.7	103.3	105.9	103.4	99.9	89.3	75.8	67.4	86.3
	MIN	33.7	36.9	41.6	47.1	55.7	65.4	75.5	72.9	65.6	53.1	40.6	33.4	51.8
	MEAN	49.8	53.8	58.4	65.8	74.7	84.4	90.7	88.2	82.8	71.2	58.2	50.4	69.0
SAFFORD EXPERIMENT FRM	MAX	59.4	64.4	69.7	78.8	87.7	97.5	98.3	95.3	91.5	82.0	68.7	60.1	79.5
	MIN	27.9	30.7	35.9	42.1	49.8	59.4	67.5	65.4	58.2	46.5	34.8	28.4	45.6
	MEAN	43.7	47.5	52.9	60.5	68.8	78.5	82.9	80.4	74.9	64.3	51.8	44.2	62.5

ARIZONA

TEMPERATURE NORMALS (DEG F)

STATION		JAN	FEB	MAR	APR	MAY	JUN	JUL	AUG	SEP	OCT	NOV	DEC	ANN
SAINT JOHNS	MAX	48.5	54.6	60.9	70.3	78.6	88.2	90.7	87.9	83.3	73.4	59.7	49.5	70.5
	MIN	18.3	21.8	26.8	33.3	41.2	50.0	58.2	55.6	48.3	36.3	25.1	18.0	36.1
	MEAN	33.4	38.2	43.9	51.8	59.9	69.1	74.5	71.8	65.8	54.9	42.5	33.8	53.3
SAN CARLOS RESERVOIR	MAX	58.4	63.8	68.9	78.0	87.4	97.7	100.1	97.1	93.2	82.8	68.7	59.1	79.6
	MIN	32.4	35.9	40.2	47.1	56.1	65.5	73.2	70.7	64.1	52.7	39.9	33.1	50.9
	MEAN	45.4	49.8	54.6	62.6	71.8	81.7	86.6	83.9	78.7	67.7	54.3	46.1	65.3
SANTA RITA EXP RANGE	MAX	60.4	63.3	67.3	75.4	83.4	93.1	91.8	88.4	86.9	79.3	68.0	61.6	76.6
	MIN	36.9	38.4	41.8	47.2	54.8	64.2	67.2	65.1	62.1	54.7	42.9	37.2	51.0
	MEAN	48.7	50.9	54.6	61.3	69.1	78.7	79.5	76.8	74.5	67.0	55.5	49.4	63.8
SEDONA RANGER STATION	MAX	55.3	59.6	63.8	72.1	81.4	92.2	96.3	93.2	88.7	78.0	64.9	56.5	75.2
	MIN	29.2	31.3	34.4	40.2	47.6	56.2	63.9	62.1	56.9	47.2	35.9	29.7	44.6
	MEAN	42.3	45.4	49.1	56.2	64.6	74.2	80.1	77.7	72.9	62.7	50.4	43.1	59.9
SELIGMAN	MAX	51.5	56.0	60.5	68.4	77.2	87.3	91.6	88.6	84.8	75.0	62.3	53.7	71.4
	MIN	21.5	23.6	26.3	31.6	38.7	46.7	55.3	54.0	47.2	37.4	27.5	21.6	36.0
	MEAN	36.5	39.8	43.4	50.0	58.0	67.0	73.5	71.4	66.0	56.2	44.9	37.7	53.7
SNOWFLAKE	MAX	48.1	54.1	59.5	68.1	76.6	86.7	90.1	87.1	82.5	72.5	58.7	49.2	69.4
	MIN	17.2	19.6	24.3	30.3	38.0	46.0	55.9	53.9	45.5	34.1	23.6	17.2	33.8
	MEAN	32.7	36.9	41.9	49.2	57.3	66.4	73.0	70.5	64.0	53.3	41.2	33.2	51.6
SOUTH PHOENIX	MAX	66.2	71.1	75.9	84.0	92.6	101.6	103.7	101.4	98.2	87.9	74.9	67.1	85.4
	MIN	36.6	39.5	42.8	47.9	55.4	64.1	73.7	71.8	64.9	53.9	42.9	37.4	52.6
	MEAN	51.4	55.3	59.4	66.0	74.0	82.8	88.7	86.6	81.6	70.9	58.9	52.3	69.0
SPRINGERVILLE	MAX	46.9	50.0	55.4	63.2	71.2	80.5	82.4	79.9	76.2	67.9	56.1	48.2	64.8
	MIN	15.4	17.7	22.6	28.0	35.0	43.4	51.8	49.5	42.6	31.5	21.2	15.7	31.2
	MEAN	31.1	33.9	39.0	45.6	53.2	62.0	67.2	64.7	59.4	49.7	38.7	32.0	48.0
TEMPE	MAX	65.5	70.1	74.5	83.1	92.4	102.1	105.0	102.5	98.8	88.6	75.1	67.2	85.4
	MIN	36.2	39.1	43.3	48.5	56.0	64.3	74.4	72.8	65.3	53.4	42.1	36.3	52.6
	MEAN	50.9	54.6	58.9	65.8	74.3	83.2	89.7	87.6	82.1	71.0	58.6	51.8	69.0
TEMPE CITRUS EXP STA	MAX	65.9	71.0	75.9	84.9	93.9	103.5	106.2	103.8	100.0	89.4	75.7	67.5	86.5
	MIN	35.1	37.4	41.3	46.4	53.6	61.7	72.3	70.8	63.7	52.5	41.2	35.4	51.0
	MEAN	50.5	54.2	58.6	65.7	73.8	82.6	89.3	87.3	81.9	71.0	58.5	51.5	68.7
TOMBSTONE	MAX	61.2	64.0	68.4	76.7	85.1	94.5	93.4	90.5	88.4	80.3	68.9	61.4	77.7
	MIN	34.8	36.4	39.7	45.6	53.1	62.5	65.9	64.2	60.8	51.8	41.7	36.0	49.4
	MEAN	48.0	50.2	54.1	61.2	69.1	78.5	79.7	77.4	74.6	66.1	55.3	48.7	63.6
TRUXTON CANYON	MAX	54.9	59.4	63.7	71.8	81.3	91.7	96.9	93.7	88.5	78.3	64.5	56.0	75.1
	MIN	28.4	30.4	33.3	39.1	47.1	56.1	64.0	62.7	55.3	45.1	34.4	28.6	43.7
	MEAN	41.6	44.9	48.5	55.5	64.2	73.9	80.5	78.3	71.9	61.7	49.5	42.3	59.4
TUCSON CAMP AVE EXP FM	MAX	66.4	69.7	74.2	82.6	91.1	100.4	101.0	98.6	96.3	87.0	74.7	67.4	84.1
	MIN	32.8	34.6	38.7	44.0	51.6	61.5	71.0	68.8	62.7	50.7	38.8	33.2	49.0
	MEAN	49.6	52.2	56.5	63.3	71.4	81.0	86.0	83.7	79.5	68.9	56.8	50.3	66.6
TUCSON MAGNETIC OBSY	MAX	65.1	68.3	72.4	81.1	90.3	100.2	100.9	98.2	96.0	86.5	73.9	66.6	83.3
	MIN	33.3	35.5	38.9	44.3	52.2	62.4	71.5	69.8	63.9	51.9	40.0	33.9	49.8
	MEAN	49.2	51.9	55.7	62.7	71.3	81.4	86.2	84.0	79.9	69.2	57.0	50.3	66.6
TUCSON UNIV OF ARIZONA	MAX	67.0	70.4	74.9	82.9	91.4	100.6	101.3	99.0	96.4	87.2	75.2	67.9	84.5
	MIN	38.2	40.2	44.1	50.1	57.8	67.4	74.3	72.6	67.3	55.9	44.8	39.1	54.3
	MEAN	52.6	55.3	59.5	66.5	74.6	84.0	87.8	85.8	81.9	71.6	60.0	53.5	69.4
TUCSON WSO R	MAX	64.1	67.4	71.8	80.1	88.8	98.5	98.5	95.9	93.5	84.1	72.2	65.0	81.7
	MIN	38.1	40.0	43.8	49.7	57.5	67.4	73.8	72.0	67.3	56.7	45.2	39.0	54.2
	MEAN	51.1	53.8	57.8	65.0	73.2	82.9	86.2	84.0	80.4	70.4	58.7	52.0	68.0
TUMACACORI NAT MON	MAX	66.3	69.0	72.9	80.8	89.2	98.8	97.7	94.5	93.5	85.6	74.0	67.3	82.5
	MIN	31.4	32.5	35.9	40.5	47.2	57.3	65.6	63.5	58.1	46.9	37.3	32.6	45.7
	MEAN	48.9	50.8	54.4	60.7	68.2	78.1	81.7	79.0	75.8	66.3	55.6	49.9	64.1

ARIZONA

TEMPERATURE NORMALS (DEG F)

STATION		JAN	FEB	MAR	APR	MAY	JUN	JUL	AUG	SEP	OCT	NOV	DEC	AN..
WALNUT CREEK	MAX	51.1	55.8	60.4	68.2	76.1	86.0	89.7	86.8	83.0	73.4	60.6	52.0	
	MIN	20.2	22.7	24.8	29.2	35.7	43.2	53.3	52.0	43.7	33.5	24.9	19.5	
	MEAN	35.7	39.3	42.6	48.7	56.0	64.6	71.5	69.4	63.4	53.5	42.8	35.8	
WELLTON	MAX	67.6	72.9	77.3	84.3	92.3	101.1	104.9	103.2	99.6	89.9	76.5	68.1	86.5
	MIN	35.2	39.1	44.0	50.0	57.2	66.0	76.5	75.2	67.5	54.9	42.1	35.2	53.6
	MEAN	51.4	56.0	60.7	67.2	74.8	83.5	90.7	89.2	83.6	72.4	59.3	51.7	70.0
WHITERIVER	MAX	52.4	55.7	59.3	67.7	76.8	87.7	90.5	87.0	83.5	74.3	62.0	54.2	70.9
	MIN	22.6	25.2	29.8	35.5	42.6	51.8	59.1	57.1	51.7	41.1	29.9	23.4	3°.2
	MEAN	37.5	40.5	44.6	51.6	59.8	69.8	74.8	72.1	67.6	57.7	46.0	38.8	55.1
WICKENBURG	MAX	63.7	68.1	72.2	80.9	90.5	100.8	104.9	101.6	96.9	86.5	73.4	65.5	83.8
	MIN	31.0	34.0	37.8	43.5	51.1	59.8	70.0	68.0	60.0	48.3	37.4	31.3	47.7
	MEAN	47.4	51.0	55.0	62.2	70.8	80.3	87.5	84.8	78.5	67.4	55.4	48.4	65.7
WILLCOX 3 NNW	MAX	58.9	63.3	68.7	77.2	85.9	95.1	95.3	92.1	89.0	79.8	67.4	59.5	77.7
	MIN	25.3	26.5	31.0	36.0	43.2	53.1	63.2	61.2	53.6	41.2	30.3	25.1	40.8
	MEAN	42.1	45.0	49.8	56.6	64.5	74.1	79.3	76.7	71.3	60.5	48.9	42.3	59.3
WILLIAMS	MAX	44.7	47.0	50.7	58.8	67.6	78.1	82.5	79.2	74.9	65.4	53.4	46.7	62.4
	MIN	22.2	23.3	26.3	32.3	39.9	48.9	55.1	53.5	48.2	38.9	28.9	23.6	36.8
	MEAN	33.5	35.1	38.5	45.6	53.7	63.5	68.8	66.4	61.6	52.2	41.2	35.2	49.6
WINDOW ROCK	MAX	41.9	46.0	51.4	60.8	70.7	81.8	85.5	82.4	77.1	67.1	53.1	44.5	63.5
	MIN	12.8	17.2	22.1	28.1	36.1	45.6	54.1	51.8	43.0	32.1	21.5	13.6	31.5
	MEAN	27.4	31.6	36.8	44.5	53.4	63.7	69.8	67.1	60.0	49.6	37.3	29.1	47.5
WINSLOW WSO //R	MAX	45.0	53.2	60.7	70.0	79.9	91.0	94.5	91.1	85.2	73.1	57.9	46.0	70.6
	MIN	19.0	23.6	29.1	36.0	44.4	53.6	63.0	61.1	52.7	40.1	27.8	19.3	39.1
	MEAN	32.0	38.4	44.9	53.0	62.2	72.3	78.8	76.1	69.0	56.6	42.9	32.7	·54.9
WUPATKI NAT MON	MAX	46.9	55.0	62.5	71.7	81.3	92.2	95.9	92.6	86.7	74.4	58.5	47.2	72.1
	MIN	23.7	28.4	33.3	40.2	48.6	58.7	65.1	62.1	56.0	45.1	32.8	24.3	43.2
	MEAN	35.3	41.7	47.9	56.0	64.9	75.5	80.5	77.4	71.4	59.7	45.7	35.8	57.7
YUMA CITRUS STATION	MAX	67.8	73.3	78.2	85.4	93.1	102.1	106.2	104.9	101.1	90.9	77.3	68.9	87.4
	MIN	37.9	40.4	44.3	49.7	56.6	64.5	74.5	74.0	67.3	55.9	44.6	38.7	54.0
	MEAN	52.9	56.9	61.3	67.6	74.9	83.3	90.4	89.5	84.2	73.4	61.0	53.8	70.8
YUMA WSO R	MAX	68.6	73.9	78.5	85.7	93.6	102.9	106.8	105.3	101.4	90.9	77.4	69.1	87.8
	MIN	43.2	46.1	49.9	55.6	63.0	71.4	80.4	79.5	73.1	61.8	50.2	43.8	59.8
	MEAN	55.9	60.0	64.2	70.7	78.3	87.2	93.6	92.4	87.3	76.4	63.8	56.5	73.9

IZONA

PRECIPITATION NORMALS (INCHES)

STATION	JAN	FEB	MAR	APR	MAY	JUN	JUL	AUG	SEP	OCT	NOV	DEC	ANN
	.66	.51	.79	.24	.10	.09	1.19	2.24	.75	.66	.57	.82	8.62
	1.45	1.07	1.13	.58	.51	.79	3.53	3.98	2.12	2.19	1.17	1.47	19.99
RANCH	.75	.63	.70	.31	.17	.28	2.64	2.20	1.30	.81	.55	.92	11.26
APACHE POWDER COMPANY	.79	.58	.55	.24	.16	.54	3.67	3.29	1.45	.95	.46	.73	13.41
BARTLETT DAM	1.57	1.31	1.62	.55	.23	.33	1.09	1.77	1.14	1.02	1.04	1.46	13.13
BETATAKIN	.95	.85	.89	.62	.46	.45	1.47	1.50	.97	1.09	.98	1.08	11.31
BLACK RIVER PUMPS	1.57	1.31	1.53	.57	.45	.67	3.00	3.40	1.57	1.57	1.00	1.66	18.30
BOUSE	.60	.46	.53	.20	.06	.10	.67	1.00	.51	.40	.44	.47	5.44
BUCKEYE	.80	.68	.74	.23	.08	.07	.62	1.23	.70	.60	.47	.73	6.95
CAMP WOOD	2.44	2.18	2.25	1.05	.65	.51	2.32	3.19	1.38	1.37	1.63	1.79	20.76
CANELO 1 NW	1.11	.85	.90	.35	.13	.67	4.38	3.98	1.63	1.12	.67	1.27	17.06
CASA GRANDE	.81	.66	.85	.29	.11	.20	.90	1.80	.61	.78	.66	.91	8.58
CASA GRANDE RUINS N M	.86	.69	.97	.28	.13	.14	.98	1.20	.63	.89	.69	1.03	8.49
CHANDLER	.86	.68	.87	.24	.12	.11	.76	.92	.57	.68	.54	.88	7.23
CHANDLER HEIGHTS	.92	.76	1.02	.28	.14	.06	.83	1.22	.67	.80	.70	.95	8.35
CHEVELON RS	1.51	1.32	1.43	.81	.60	.40	3.11	3.12	1.61	1.52	1.41	1.67	18.51
CHILDS	2.04	1.58	1.83	.86	.44	.45	1.99	2.70	1.43	1.39	1.34	2.06	18.11
CHINO VALLEY	1.01	.95	1.09	.60	.39	.35	2.04	2.34	1.08	.97	.73	.91	12.46
CHIRICAHUA NAT MON	1.28	.86	1.13	.44	.27	.62	4.08	4.17	1.74	1.09	.79	1.39	17.86
CLIFTON	.98	.78	.78	.28	.22	.37	2.15	2.09	1.50	1.28	.60	1.23	12.26
CORDES	1.41	1.33	1.28	.61	.32	.33	1.79	2.35	1.25	1.09	1.03	1.37	14.16
CROWN KING	3.29	3.02	3.05	1.16	.57	.54	3.58	4.61	1.82	1.26	1.89	3.20	27.99
DEER VALLEY	.95	.70	.94	.31	.16	.14	.76	1.33	.55	.74	.60	.94	8.12
DOUGLAS B D FAA AP	.74	.47	.48	.16	.16	.41	3.51	3.03	1.15	.84	.45	.76	12.16
DUNCAN	.83	.68	.56	.21	.19	.30	2.04	1.90	.99	1.03	.48	.89	10.10
FLAGSTAFF WSO //R	2.10	1.95	2.13	1.35	.75	.57	2.47	2.62	1.47	1.54	1.65	2.26	20.86
FLORENCE	1.06	.82	1.16	.39	.15	.19	1.21	1.37	.73	.95	.78	1.20	10.01
FORT VALLEY	2.04	1.82	2.14	1.43	.81	.70	3.06	3.19	1.51	1.51	1.44	2.01	21.66
FRITZ RANCH	1.14	.94	.92	.45	.25	.41	2.49	2.70	1.49	1.71	.72	1.19	14.41
GANADO	.77	.65	.77	.59	.42	.33	1.59	1.59	.97	1.24	.81	.91	10.64
GILA BEND	.67	.49	.61	.22	.10	.06	.69	1.01	.46	.48	.40	.63	5.82
GISELA	1.87	1.54	1.84	.80	.35	.35	2.24	2.91	1.51	1.57	1.31	1.85	18.14
GRANITE REEF DAM	1.20	.88	1.17	.40	.16	.11	1.01	1.31	.62	.78	.76	1.07	9.47
GRIGGS 3 W	1.00	.69	.87	.21	.09	.06	.63	1.24	.74	.59	.53	.88	7.53
HEBER RANGER STATION	1.48	1.21	1.44	.78	.62	.44	2.88	2.84	1.90	1.71	1.15	1.75	18.20
HOLBROOK	.50	.47	.44	.37	.34	.26	1.08	1.51	.86	.93	.53	.49	7.78
HORSESHOE DAM	1.73	1.43	1.51	.65	.22	.26	1.53	2.23	1.14	1.39	1.12	1.75	14.96
IRVING	2.29	1.86	2.14	1.16	.54	.50	2.20	2.71	1.55	1.48	1.49	2.02	19.94
JEROME	1.62	1.63	1.80	1.07	.58	.45	3.02	3.30	1.22	1.38	1.29	1.50	18.86
JUNIPINE	3.28	2.92	3.04	2.13	1.03	.66	2.38	2.79	1.94	2.11	2.36	3.07	27.71
KELVIN	1.64	1.27	1.53	.55	.22	.20	1.26	2.09	1.11	1.34	1.11	1.64	13.96
KINGMAN NO 2	1.11	.96	1.11	.63	.24	.24	.98	1.36	.75	.67	.71	.89	9.65
LAVEEN 3 SSE	.77	.61	.84	.23	.11	.25	.98	1.34	.68	.69	.53	.82	7.85
LEES FERRY	.43	.34	.51	.38	.33	.23	.83	.86	.50	.60	.53	.46	6.00
LEUPP	.37	.38	.48	.30	.22	.35	1.18	.91	.91	.80	.31	.41	6.62
LITCHFIELD PARK	1.00	.81	.80	.29	.09	.07	.67	1.10	.71	.63	.55	.88	7.60
MC NARY	2.92	2.04	3.08	1.32	.74	.79	3.51	3.78	1.97	2.37	1.99	2.73	27.24
MESA EXPERIMENT FARM	.89	.65	.85	.28	.17	.10	.94	1.11	.68	.72	.57	.89	7.85
MIAMI	2.05	1.53	1.99	.65	.35	.36	2.22	3.29	1.31	1.52	1.39	2.34	19.00
MONTEZUMA CASTLE N M	1.05	.99	1.15	.67	.32	.37	1.38	1.95	1.30	1.03	.87	1.08	12.16
MORMON FLAT	1.51	1.11	1.49	.49	.28	.19	1.44	2.13	1.03	1.34	1.00	1.56	13.57
N LAZY H RANCH	1.21	.89	.95	.35	.17	.26	2.48	2.33	1.21	.96	.72	1.11	12.64
NOGALES	.98	.77	.86	.28	.12	.43	4.68	3.76	1.50	1.24	.69	1.30	16.61
ORACLE 2 SE	1.92	1.72	2.04	.85	.37	.42	3.11	3.22	1.79	1.64	1.36	2.34	20.78
ORGAN PIPE CACTUS N M	.78	.57	.77	.22	.15	.10	1.33	1.75	1.03	.81	.66	1.01	9.18

ARIZONA

PRECIPITATION NORMALS (INCHES)

STATION	JAN	FEB	MAR	APR	MAY	JUN	JUL	AUG	SEP	OCT	NOV	DEC	ANN
PARKER	.60	.43	.47	.17	.05	.02	.26	.49	.40	.38	.35	.46	4.08
PATAGONIA 2	1.19	.87	1.04	.33	.17	.48	4.51	3.79	1.78	1.35	.78	1.37	17.66
PAYSON	2.15	1.85	2.12	1.07	.52	.44	2.81	3.16	1.76	1.85	1.53	2.02	21.28
PEACH SPRINGS	.93	.92	1.16	.62	.39	.28	1.60	1.89	1.11	.73	.75	.94	11.32
PEARCE	.73	.59	.54	.20	.17	.41	2.91	3.13	1.29	.76	.40	.80	11.93
PETRIFIED FOREST NP	.52	.47	.55	.36	.45	.35	1.21	1.64	1.06	1.08	.57	.53	8.79
PHOENIX WSO R	.73	.59	.81	.27	.14	.17	.74	1.02	.64	.63	.54	.83	7.11
PINETOP FISH HATCHERY	2.26	1.65	2.41	1.08	.67	.61	3.46	3.47	1.87	2.07	1.69	2.21	23.45
PLEASANT VALLEY R S	2.08	1.57	2.07	.92	.56	.49	2.68	3.22	1.60	1.80	1.31	1.79	20.09
PORTAL 4 SW	1.34	.93	1.04	.41	.30	.77	4.97	3.99	2.08	1.42	.90	1.76	19.91
PRESCOTT	1.72	1.51	1.53	.76	.50	.53	3.15	3.45	1.49	1.22	1.33	1.65	18.84
REDINGTON	1.10	.89	1.08	.30	.22	.30	2.88	2.32	1.12	1.02	.72	1.30	13.25
ROOSEVELT 1 WNW	1.84	1.27	1.79	.56	.30	.25	1.25	1.85	1.20	1.32	1.08	1.88	14.59
RUBY STAR RANCH	.88	.74	.76	.26	.13	.35	3.51	2.70	1.50	1.04	.77	1.08	13.72
RUCKER CANYON	1.28	.84	1.10	.33	.15	.61	4.32	3.72	1.90	1.33	.73	1.45	17.76
SABINO CANYON	1.26	.91	1.09	.46	.17	.24	2.35	2.31	1.17	1.06	.82	1.26	13.10
SACATON	.89	.68	.83	.29	.10	.12	1.00	1.35	.49	.75	.68	.92	8.10
SAFFORD EXPERIMENT FRM	.63	.55	.56	.18	.12	.24	1.76	1.55	.99	.86	.40	.76	8.60
SAINT JOHNS	.65	.56	.87	.45	.35	.48	1.99	1.99	1.20	1.08	.55	.70	10.87
SAN CARLOS RESERVOIR	1.66	1.22	1.60	.46	.28	.30	1.77	2.22	1.12	1.42	1.06	1.77	14.88
SANTA RITA EXP RANGE	1.60	1.28	1.45	.56	.13	.70	4.75	4.11	2.06	1.46	1.10	1.82	21.02
SASABE 7 NW	1.31	1.05	1.33	.41	.23	.34	3.86	3.56	1.86	1.17	.95	1.90	17.97
SEDONA RANGER STATION	1.88	1.77	1.93	1.15	.62	.46	1.89	2.18	1.57	1.45	1.37	1.48	17.75
SELIGMAN	.96	.81	1.06	.51	.43	.51	1.90	2.15	.80	.80	.68	1.04	11.65
SNOWFLAKE	.69	.58	.79	.43	.53	.40	1.96	2.64	1.30	1.23	.67	.79	12.01
SOUTH PHOENIX	.87	.68	.89	.30	.16	.22	.87	1.39	.63	.70	.57	.81	8.09
SPRINGERVILLE	.55	.44	.55	.32	.34	.45	2.62	3.02	1.39	1.07	.50	.45	11.70
STEPHENS RANCH	.96	.62	.70	.18	.16	.55	3.23	2.85	1.33	1.15	.57	1.11	13.41
STEWART MOUNTAIN	1.51	1.08	1.44	.47	.20	.16	1.29	1.69	.83	1.10	.97	1.48	12.22
SUPERIOR	2.24	1.59	2.13	.77	.30	.29	1.68	2.58	1.28	1.42	1.34	2.22	17.84
TEMPE	.87	.65	.90	.27	.18	.14	.78	1.23	.71	.76	.58	.93	8.00
TEMPE CITRUS EXP STA	.90	.70	.90	.29	.20	.16	.81	1.30	.86	.76	.59	.94	8.41
TOLLESON 1 E	.88	.69	.85	.26	.14	.18	.67	1.17	.68	.63	.51	.84	7.50
TOMBSTONE	.84	.55	.70	.25	.17	.46	3.51	3.23	1.08	.85	.48	.69	12.81
TONTO CREEK FISH HATCH 2	3.99	2.89	3.59	1.64	.92	.70	3.96	4.67	2.56	2.53	2.27	3.58	33.30
TRUXTON CANYON	.95	.99	1.19	.68	.30	.24	1.22	2.06	.97	.73	.66	.89	10.88
TUCSON CAMP AVE EXP FM	.98	.77	.89	.44	.16	.22	2.11	2.16	1.05	.96	.70	1.15	11.59
TUCSON MAGNETIC OBSY	1.17	.81	1.01	.40	.16	.22	2.07	2.17	1.07	1.02	.79	1.09	11.98
TUCSON UNIV OF ARIZONA	.94	.73	.82	.45	.17	.28	2.14	2.03	1.10	.89	.68	1.12	11.35
TUCSON WSO R	.83	.63	.68	.32	.14	.22	2.42	2.13	1.33	.88	.62	.94	11.14
TUMACACORI NAT MON	.95	.69	.77	.26	.11	.48	3.82	3.47	1.29	1.04	.71	1.14	14.73
TUWEEP	1.23	.99	1.26	.67	.56	.47	1.39	1.71	.88	.86	.87	1.24	12.13
WALNUT CANYON NAT MON	1.66	1.41	1.84	.96	.71	.49	2.28	2.17	1.44	1.44	1.26	1.91	17.57
WALNUT CREEK	1.51	1.40	1.36	.65	.56	.48	2.61	2.84	1.26	1.16	1.03	1.21	16.07
WALNUT GROVE	1.68	1.76	1.80	.80	.37	.33	2.05	3.06	1.38	.94	1.24	1.61	17.02
WELLTON	.47	.31	.31	.12	.04	.02	.38	.61	.57	.31	.26	.41	3.81
WHITERIVER	1.65	1.19	1.82	.82	.49	.58	2.82	3.47	1.54	1.59	1.21	1.40	18.58
WICKENBURG	1.10	1.06	1.38	.48	.22	.20	1.27	1.96	1.11	.62	.73	1.12	11.25
WILLCOX 3 NNW	.86	.62	.62	.24	.15	.39	2.74	2.60	1.12	.77	.44	.97	11.52
WILLIAMS	1.85	1.70	2.00	1.20	.82	.58	2.95	3.23	1.62	1.53	1.61	2.01	21.10
WILLOW SPRINGS RANCH	1.52	1.18	1.35	.57	.29	.35	2.42	2.78	1.25	1.32	1.01	1.86	15.90
WINDOW ROCK	.77	.67	.84	.54	.45	.42	1.66	2.17	1.14	1.26	.71	.88	11.51
WINKELMAN 6 S	1.26	1.09	1.13	.47	.26	.33	1.95	2.48	1.49	1.21	.84	1.34	13.85
WINSLOW WSO //R	.43	.46	.51	.32	.30	.35	1.14	1.41	.83	.90	.41	.58	7.64
WUPATKI NAT MON	.36	.48	.59	.24	.35	.37	1.39	1.61	.79	.80	.43	.52	7.93

ARIZONA

PRECIPITATION NORMALS (INCHES)

STATION		JAN	FEB	MAR	APR	MAY	JUN	JUL	AUG	SEP	OCT	NOV	DEC	ANN
Y LIGHTNING RANCH		.84	.57	.57	.16	.09	.48	3.62	3.33	1.22	.93	.51	1.01	13.33
YOUNGTOWN		.94	.78	.91	.30	.11	.11	.68	1.30	.60	.77	.57	.95	8.02
YUMA CITRUS STATION		.44	.28	.28	.12	.02	.02	.11	.54	.30	.40	.20	.34	3.05
YUMA WSO	R	.38	.26	.18	.13	.04	.01	.15	.42	.25	.29	.20	.34	2.65

ARIZONA

HEATING DEGREE DAY NORMALS (BASE 65 DEG F)

STATION	JUL	AUG	SEP	OCT	NOV	DEC	JAN	FEB	MAR	APR	MAY	JUN	ANN
AJO	0	0	0	15	142	333	371	247	183	55	6	0	1352
ANVIL RANCH	0	0	0	36	268	471	484	365	278	107	19	0	2028
APACHE POWDER COMPANY	0	0	0	86	378	583	592	451	362	159	33	0	2644
BARTLETT DAM	0	0	0	13	154	345	392	282	236	82	6	0	1510
BETATAKIN	0	9	108	405	789	1054	1097	902	849	564	289	57	6123
BLACK RIVER PUMPS	0	10	68	323	669	908	952	778	735	477	243	40	5203
BOUSE	0	0	0	23	234	471	481	309	204	64	5	0	1791
BUCKEYE	0	0	0	14	190	406	428	277	188	45	6	0	1554
CANELO 1 NW	0	7	12	194	492	691	707	585	530	318	133	8	3677
CASA GRANDE	0	0	0	16	188	409	431	289	199	51	7	0	1590
CASA GRANDE RUINS N M	0	0	0	15	201	422	443	310	219	70	0	0	1680
CHILDS	0	0	0	88	324	561	592	420	357	170	33	0	2545
CHINO VALLEY	0	0	23	262	600	846	874	689	642	429	205	30	4600
CHIRICAHUA NAT MON	0	0	17	174	459	663	691	566	502	278	92	0	3442
CLIFTON	0	0	0	45	302	564	577	398	295	101	11	0	2293
CORDES	0	0	5	125	378	601	639	507	474	268	94	8	3099
DEER VALLEY	0	0	0	12	194	400	428	297	215	71	8	0	1625
DOUGLAS B D FAA AP	0	0	0	97	393	611	623	487	388	171	26	0	2796
DUNCAN	0	0	0	167	510	747	747	571	462	239	62	0	3505
FLAGSTAFF WSO //R	34	76	229	552	861	1091	1141	960	946	702	468	194	7254
FLORENCE	0	0	0	13	177	385	415	291	211	65	6	0	1563
FORT VALLEY	87	148	312	620	924	1178	1228	1028	1017	774	561	287	8164
GANADO	0	31	132	453	825	1113	1156	907	831	579	317	72	6416
GILA BEND	0	0	0	11	132	351	376	242	150	33	6	0	1301
HEBER RANGER STATION	16	39	147	456	783	1020	1054	857	818	594	372	115	6271
HOLBROOK	0	0	27	298	675	989	998	753	648	396	178	25	4987
JEROME	0	0	12	163	459	691	735	588	537	310	131	8	3634
JUNIPINE	0	0	33	242	534	756	806	658	614	391	179	24	4237
KINGMAN NO 2	0	0	0	117	396	642	676	504	456	252	76	0	3119
LAVEEN 3 SSE	0	0	0	16	170	378	403	266	185	45	5	0	1468
LEES FERRY	0	0	0	98	489	806	837	554	380	134	22	0	3320
LEUPP	0	0	21	307	708	1029	1042	764	660	399	167	19	5116
LITCHFIELD PARK	0	0	0	11	172	391	409	272	183	46	0	0	1484
MC NARY	46	84	197	474	780	1001	1063	904	880	633	415	156	6633
MESA EXPERIMENT FARM	0	0	0	11	180	391	431	302	212	65	6	0	1598
MIAMI	0	0	0	88	372	617	648	490	408	184	39	0	2846
MONTEZUMA CASTLE N M	0	0	0	120	435	691	694	507	420	219	55	0	3141
MORMON FLAT	0	0	0	18	153	352	375	275	220	75	11	0	1479
NOGALES	0	0	0	110	390	583	601	490	422	234	81	0	2911
ORACLE 2 SE	0	0	0	91	350	552	583	462	418	214	76	0	2746
ORGAN PIPE CACTUS N M	0	0	0	11	164	357	372	268	207	74	10	0	1463
PARKER	0	0	0	8	142	366	381	222	133	39	0	0	1291
PAYSON	0	0	35	246	567	794	818	658	620	402	187	20	4347
PEARCE	0	0	0	99	375	592	601	468	390	173	39	0	2737
PETRIFIED FOREST NP	0	0	24	274	654	949	958	734	645	390	165	13	4806
PHOENIX WSO R	0	0	0	13	159	368	394	269	187	52	0	0	1442
PRESCOTT	0	6	54	314	630	862	893	722	688	468	255	57	4949
ROOSEVELT 1 WNW	0	0	0	32	262	505	539	367	279	100	13	0	2097
SABINO CANYON	0	0	0	20	187	375	403	305	230	64	6	0	1590
SACATON	0	0	0	16	215	453	471	324	237	74	8	0	1798
SAFFORD EXPERIMENT FRM	0	0	0	78	396	645	660	490	379	157	22	0	2827
SAINT JOHNS	0	0	35	317	675	967	980	750	654	396	176	17	4967
SAN CARLOS RESERVOIR	0	0	0	57	326	586	608	426	333	127	13	0	2476
SANTA RITA EXP RANGE	0	0	6	71	294	484	505	399	340	148	29	0	2276
SEDONA RANGER STATION	0	0	12	146	438	679	704	549	499	282	108	10	3427

ARIZONA

HEATING DEGREE DAY NORMALS (BASE 65 DEG F)

STATION	JUL	AUG	SEP	OCT	NOV	DEC	JAN	FEB	MAR	APR	MAY	JUN	ANN
SELIGMAN	0	9	55	289	603	846	884	706	670	450	240	52	4804
SNOWFLAKE	0	6	67	363	714	986	1001	787	716	474	246	56	5416
SOUTH PHOENIX	0	0	0	20	194	394	422	287	206	69	14	0	1606
SPRINGERVILLE	23	57	176	474	789	1023	1051	871	806	582	366	119	6337
TEMPE	0	0	0	16	201	409	437	300	216	67	8	0	1654
TEMPE CITRUS EXP STA	0	0	0	14	206	419	450	310	221	65	8	0	1693
TOMBSTONE	0	0	0	71	297	505	527	414	354	152	26	0	2346
TRUXTON CANYON	0	0	8	154	465	704	725	563	512	298	103	6	3538
TUCSON CAMP AVE EXP FM	0	0	0	24	252	456	477	365	277	96	11	0	1958
TUCSON MAGNETIC OBSY	0	0	0	25	247	456	490	374	301	111	13	0	2017
TUCSON UNIV OF ARIZONA	0	0	0	11	167	362	384	287	207	56	0	0	1474
TUCSON WSO R	0	0	0	30	204	403	431	326	246	86	8	0	1734
TUMACACORI NAT MON	0	0	0	63	287	468	499	403	338	148	27	0	2233
WALNUT CREEK	7	9	87	362	666	905	908	720	694	489	287	92	5226
WELLTON	0	0	0	12	183	412	422	263	163	60	9	0	1524
WHITERIVER	0	0	29	239	570	812	853	686	632	402	179	19	4421
WICKENBURG	0	0	0	50	292	515	546	398	330	135	25	0	2291
WILLCOX 3 NNW	0	0	6	164	483	704	710	560	471	262	86	8	3454
WILLIAMS	12	32	130	401	714	924	977	837	822	582	357	104	5892
WINDOW ROCK	20	22	162	477	831	1113	1166	935	874	615	360	101	6676
WINSLOW WSO //R	0	0	12	270	663	1001	1023	745	623	360	132	10	4839
WUPATKI NAT MON	0	0	9	194	579	905	921	652	530	284	101	7	4182
YUMA CITRUS STATION	0	0	0	10	146	347	375	242	154	53	7	0	1334
YUMA WSO R	0	0	0	8	92	276	290	176	104	37	0	0	983

ARIZONA

COOLING DEGREE DAY NORMALS (BASE 65 DEG F)

STATION	JAN	FEB	MAR	APR	MAY	JUN	JUL	AUG	SEP	OCT	NOV	DEC	ANN
AJO	0	23	62	157	362	633	806	732	600	325	58	17	3775
ANVIL RANCH	0	6	15	56	220	492	636	552	408	138	13	0	2536
APACHE POWDER COMPANY	0	0	5	18	111	363	499	425	273	49	0	0	1743
BARTLETT DAM	0	19	50	124	319	591	791	719	570	282	40	10	3515
BETATAKIN	0	0	0	0	14	114	233	161	54	8	0	0	584
BLACK RIVER PUMPS	0	0	0	0	10	118	229	153	62	6	0	0	578
BOUSE	0	9	27	112	331	588	843	778	546	212	9	0	3455
BUCKEYE	0	14	39	111	341	600	837	763	555	231	16	0	3507
CANELO 1 NW	0	0	0	0	30	200	288	214	99	11	0	0	842
CASA GRANDE	0	12	41	114	345	612	818	738	546	239	17	0	3482
CASA GRANDE RUINS N M	0	10	21	91	299	579	803	725	540	220	12	0	3300
CHILDS	0	0	13	50	176	423	645	552	378	150	6	0	2393
CHINO VALLEY	0	0	0	0	19	129	326	248	86	8	0	0	816
CHIRICAHUA NAT MON	0	0	0	0	49	236	301	226	137	31	0	0	980
CLIFTON	0	6	19	71	244	522	651	583	423	153	5	0	2677
CORDES	0	0	6	13	88	296	487	415	263	84	0	0	1652
DEER VALLEY	0	14	26	101	312	576	784	719	546	226	20	0	3324
DOUGLAS B D FAA AP	0	0	0	12	103	357	437	369	243	44	0	0	1565
DUNCAN	0	0	0	0	74	283	453	378	183	12	0	0	1383
FLAGSTAFF WSO //R	0	0	0	0	0	20	68	39	0	0	0	0	127
FLORENCE	0	14	31	101	313	585	794	725	561	242	21	0	3387
FORT VALLEY	0	0	0	0	0	0	22	11	0	0	0	0	33
GANADO	0	0	0	0	0	57	196	133	15	0	0	0	401
GILA BEND	0	20	50	159	415	669	899	831	642	315	36	7	4043
HEBER RANGER STATION	0	0	0	0	0	46	124	70	9	0	0	0	249
HOLBROOK	0	0	0	6	32	190	378	295	105	7	0	0	1013
JEROME	0	0	7	13	90	272	409	326	207	60	0	0	1384
JUNIPINE	0	0	0	0	30	150	301	229	111	18	0	0	839
KINGMAN NO 2	0	0	7	30	123	339	555	490	294	92	0	0	1930
LAVEEN 3 SSE	0	14	37	123	359	627	825	744	561	249	23	0	3562
LEES FERRY	0	0	5	47	233	507	716	629	369	67	0	0	2573
LEUPP	0	0	0	0	24	166	378	304	96	0	0	0	968
LITCHFIELD PARK	0	14	37	118	342	609	822	750	555	235	16	0	3498
MC NARY	0	0	0	0	0	21	52	25	0	0	0	0	98
MESA EXPERIMENT FARM	0	13	29	107	309	573	775	710	531	231	18	0	3296
MIAMI	0	0	8	34	160	420	577	484	327	94	0	0	2104
MONTEZUMA CASTLE N M	0	0	0	18	104	324	546	465	273	58	0	0	1788
MORMON FLAT	0	17	53	150	371	648	818	738	591	312	54	11	3763
NOGALES	0	0	0	0	53	267	409	332	198	26	0	0	1285
ORACLE 2 SE	0	0	12	31	150	373	462	388	270	88	11	0	1785
ORGAN PIPE CACTUS N M	0	16	33	95	261	510	729	670	516	234	26	0	3090
PARKER	0	18	62	189	425	684	911	856	645	312	31	0	4133
PAYSON	0	0	0	0	23	143	326	245	98	7	0	0	842
PEARCE	0	0	6	17	114	357	450	366	234	52	0	0	1596
PETRIFIED FOREST NP	0	0	0	0	29	178	357	279	111	8	0	0	962
PHOENIX WSO R	0	20	51	142	376	645	846	772	588	273	27	6	3746
PRESCOTT	0	0	0	0	13	111	251	174	57	0	0	0	606
ROOSEVELT 1 WNW	0	0	25	100	295	567	735	651	486	199	16	0	3074
SABINO CANYON	0	11	28	85	282	561	688	617	489	206	19	0	2986
SACATON	0	10	32	98	309	582	797	719	534	208	11	0	3300
SAFFORD EXPERIMENT FRM	0	0	0	22	139	405	555	477	297	56	0	0	1951
SAINT JOHNS	0	0	0	0	18	140	295	213	59	0	0	0	725
SAN CARLOS RESERVOIR	0	0	11	55	223	501	670	586	411	141	0	0	2598
SANTA RITA EXP RANGE	0	0	17	37	156	411	450	366	291	133	9	0	1870
SEDONA RANGER STATION	0	0	6	18	96	286	468	394	249	75	0	0	1592

ARIZONA

COOLING DEGREE DAY NORMALS (BASE 65 DEG F)

STATION	JAN	FEB	MAR	APR	MAY	JUN	JUL	AUG	SEP	OCT	NOV	DEC	ANN
SELIGMAN	0	0	0	0	23	112	266	207	85	16	0	0	709
SNOWFLAKE	0	0	0	0	7	98	248	176	37	0	0	0	566
SOUTH PHOENIX	0	16	32	99	293	534	735	670	498	203	11	0	3091
SPRINGERVILLE	0	0	0	0	0	29	91	48	8	0	0	0	176
TEMPE	0	9	27	91	296	546	766	701	513	202	9	0	3160
TEMPE CITRUS EXP STA	0	7	23	86	281	528	753	691	507	200	11	0	3087
TOMBSTONE	0	0	16	38	153	405	456	384	288	105	6	0	1851
TRUXTON CANYON	0	0	0	13	78	273	481	412	215	52	0	0	1524
TUCSON CAMP AVE EXP FM	0	6	14	45	209	480	651	580	435	145	6	0	2571
TUCSON MAGNETIC OBSY	0	7	12	42	208	492	657	589	447	156	7	0	2617
TUCSON UNIV OF ARIZONA	0	15	37	101	302	570	707	645	507	216	17	5	3122
TUCSON WSO R	0	12	22	86	262	537	657	589	462	198	15	0	2840
TUMACACORI NAT MON	0	5	10	19	126	393	518	434	324	103	0	0	1932
WALNUT CREEK	0	0	0	0	8	80	209	146	39	6	0	0	488
WELLTON	0	11	30	126	313	555	797	750	558	241	12	0	3393
WHITERIVER	0	0	0	0	18	163	304	220	107	13	0	0	825
WICKENBURG	0	6	20	51	205	459	698	614	405	125	0	0	2583
WILLCOX 3 NNW	0	0	0	10	71	281	443	363	195	25	0	0	1388
WILLIAMS	0	0	0	0	7	59	130	76	28	0	0	0	300
WINDOW ROCK	0	0	0	0	0	62	168	87	12	0	0	0	329
WINSLOW WSO //R	0	0	0	0	45	229	428	344	132	9	0	0	1187
WUPATKI NAT MON	0	0	0	14	97	322	481	389	201	30	0	0	1534
YUMA CITRUS STATION	0	15	39	131	314	549	787	760	576	271	26	0	3468
YUMA WSO R	8	36	80	208	412	666	887	849	669	361	56	12	4244

STATE-STATION NUMBER	STN TYP	NAME	LATITUDE DEG-MIN	LONGITUDE DEG-MIN	ELEVATION (FT)
2-0080	13	AJO	N 3222	W 11252	1763
2-0159	12	ALPINE	N 3351	W 10908	8050
2-0287	13	ANVIL RANCH	N 3159	W 11123	2750
2-0309	13	APACHE POWDER COMPANY	N 3154	W 11015	3690
2-0632	13	BARTLETT DAM	N 3349	W 11138	1650
2-0750	13	BETATAKIN	N 3641	W 11032	7286
2-0808	13	BLACK RIVER PUMPS	N 3329	W 10946	6040
2-0949	13	BOUSE	N 3357	W 11401	930
2-1026	13	BUCKEYE	N 3322	W 11235	870
2-1216	12	CAMP WOOD	N 3448	W 11252	5708
2-1231	13	CANELO 1 NW	N 3133	W 11032	4985
2-1306	13	CASA GRANDE	N 3253	W 11145	1405
2-1314	13	CASA GRANDE RUINS N M	N 3300	W 11132	1419
2-1511	12	CHANDLER	N 3318	W 11150	1212
2-1514	12	CHANDLER HEIGHTS	N 3313	W 11141	1425
2-1574	12	CHEVELON RS	N 3432	W 11055	7006
2-1614	13	CHILDS	N 3421	W 11142	2650
2-1654	13	CHINO VALLEY	N 3445	W 11227	4750
2-1664	13	CHIRICAHUA NAT MON	N 3200	W 10921	5300
2-1849	13	CLIFTON	N 3303	W 10917	3460
2-2109	13	CORDES	N 3418	W 11210	3773
2-2329	12	CROWN KING	N 3412	W 11220	6000
2-2462	13	DEER VALLEY	N 3335	W 11205	1257
2-2664	13	DOUGLAS B D FAA AP	N 3127	W 10936	4098
2-2754	13	DUNCAN	N 3244	W 10906	3640
2-3010	13	FLAGSTAFF WSO //R	N 3508	W 11140	7006
2-3027	13	FLORENCE	N 3302	W 11123	1505
2-3160	13	FORT VALLEY	N 3516	W 11144	7347
2-3258	12	FRITZ RANCH	N 3320	W 10911	4321
2-3303	13	GANADO	N 3543	W 10934	6340
2-3393	13	GILA BEND	N 3257	W 11243	737
2-3448	12	GISELA	N 3407	W 11117	2900
2-3621	12	GRANITE REEF DAM	N 3331	W 11142	1325
2-3702	12	GRIGGS 3 W	N 3330	W 11229	1160
2-3961	13	HEBER RANGER STATION	N 3424	W 11033	6590
2-4089	13	HOLBROOK	N 3454	W 11010	5069
2-4182	12	HORSESHOE DAM	N 3359	W 11143	2020
2-4391	12	IRVING	N 3424	W 11137	3762
2-4453	13	JEROME	N 3445	W 11207	5245
2-4508	13	JUNIPINE	N 3458	W 11145	5134
2-4594	12	KELVIN	N 3306	W 11058	1850
2-4645	13	KINGMAN NO 2	N 3512	W 11401	3539
2-4829	13	LAVEEN 3 SSE	N 3320	W 11209	1115
2-4849	13	LEES FERRY	N 3652	W 11136	3210
2-4872	13	LEUPP	N 3517	W 11058	4700
2-4977	13	LITCHFIELD PARK	N 3330	W 11222	1030
2-5412	13	MC NARY	N 3404	W 10951	7320
2-5467	13	MESA EXPERIMENT FARM	N 3325	W 11152	1230
2-5512	13	MIAMI	N 3324	W 11053	3560
2-5635	13	MONTEZUMA CASTLE N M	N 3437	W 11150	3180

02 — ARIZONA

STATE-STATION NUMBER	STN TYP	NAME		LATITUDE DEG-MIN	LONGITUDE DEG-MIN	ELEVATION (FT)
2-5700	13	MORMON FLAT		N 3333	W 11127	1715
2-5908	12	N LAZY H RANCH		N 3207	W 11041	3050
2-5921	13	NOGALES		N 3121	W 11055	3808
2-6119	13	ORACLE 2 SE		N 3236	W 11044	4540
2-6132	13	ORGAN PIPE CACTUS N M		N 3156	W 11247	1678
2-6250	13	PARKER		N 3410	W 11417	425
2-6282	12	PATAGONIA 2		N 3133	W 11045	4190
2-6323	13	PAYSON		N 3414	W 11120	4913
2-6328	12	PEACH SPRINGS		N 3533	W 11324	4970
2-6353	13	PEARCE		N 3154	W 10949	4420
2-6468	13	PETRIFIED FOREST NP		N 3448	W 10952	5440
2-6481	13	PHOENIX WSO	R	N 3326	W 11201	1117
2-6601	12	PINETOP FISH HATCHERY		N 3407	W 10955	7200
2-6653	12	PLEASANT VALLEY R S		N 3406	W 11056	5050
2-6716	12	PORTAL 4 SW		N 3153	W 10912	5390
2-6796	13	PRESCOTT		N 3434	W 11228	5510
2-7036	12	REDINGTON		N 3226	W 11029	2869
2-7281	13	ROOSEVELT 1 WNW		N 3340	W 11109	2205
2-7330	12	RUBY STAR RANCH		N 3155	W 11105	3640
2-7334	12	RUCKER CANYON		N 3145	W 10925	5370
2-7355	13	SABINO CANYON		N 3218	W 11049	2640
2-7370	13	SACATON		N 3304	W 11145	1285
2-7390	13	SAFFORD EXPERIMENT FRM		N 3249	W 10941	2954
2-7435	13	SAINT JOHNS		N 3430	W 10922	5730
2-7480	13	SAN CARLOS RESERVOIR		N 3310	W 11031	2532
2-7593	13	SANTA RITA EXP RANGE		N 3146	W 11051	4300
2-7622	12	SASABE 7 NW		N 3135	W 11136	3825
2-7708	13	SEDONA RANGER STATION		N 3452	W 11146	4320
2-7716	13	SELIGMAN		N 3519	W 11253	5250
2-8012	13	SNOWFLAKE		N 3430	W 11005	5642
2-8112	13	SOUTH PHOENIX		N 3323	W 11204	1155
2-8162	13	SPRINGERVILLE		N 3408	W 10917	7060
2-8206	12	STEPHENS RANCH		N 3124	W 10912	4000
2-8214	12	STEWART MOUNTAIN		N 3334	W 11132	1422
2-8348	12	SUPERIOR		N 3318	W 11106	2995
2-8489	13	TEMPE		N 3326	W 11156	1150
2-8499	13	TEMPE CITRUS EXP STA		N 3323	W 11158	1180
2-8598	12	TOLLESON 1 E		N 3327	W 11214	1025
2-8619	13	TOMBSTONE		N 3142	W 11003	4610
2-8650	12	TONTO CREEK FISH HATCH 2		N 3421	W 11108	639
2-8778	13	TRUXTON CANYON		N 3523	W 11340	3820
2-8796	13	TUCSON CAMP AVE EXP FM		N 3217	W 11057	2330
2-8800	13	TUCSON MAGNETIC OBSY		N 3215	W 11050	2526
2-8815	13	TUCSON UNIV OF ARIZONA		N 3215	W 11057	2444
2-8820	13	TUCSON WSO	R	N 3207	W 11056	2584
2-8865	13	TUMACACORI NAT MON		N 3134	W 11103	3267
2-8895	12	TUWEEP		N 3617	W 11304	4775
2-9156	12	WALNUT CANYON NAT MON		N 3510	W 11131	6685
2-9158	13	WALNUT CREEK		N 3456	W 11249	5090
2-9166	12	WALNUT GROVE		N 3418	W 11233	3764

02 — ARIZONA

LEGEND
11 = TEMPERATURE ONLY
12 = PRECIPITATION ONLY
13 = TEMP. & PRECIP.

STATE-STATION NUMBER	STN TYP	NAME		LATITUDE DEG-MIN	LONGITUDE DEG-MIN	ELEVATION (FT)
2-9211	13	WELLTON		N 3240	W 11408	260
2-9271	13	WHITERIVER		N 3350	W 10958	5280
2-9287	13	WICKENBURG		N 3358	W 11244	2095
2-9334	13	WILLCOX 3 NNW		N 3218	W 10951	4190
2-9359	13	WILLIAMS		N 3515	W 11211	6750
2-9382	12	WILLOW SPRINGS RANCH		N 3243	W 11052	3690
2-9410	13	WINDOW ROCK		N 3541	W 10903	6750
2-9420	12	WINKELMAN 6 S		N 3255	W 11043	2075
2-9439	13	WINSLOW WSO	//R	N 3501	W 11044	4895
2-9542	13	WUPATKI NAT MON		N 3531	W 11122	4908
2-9562	12	Y LIGHTNING RANCH		N 3127	W 11012	4550
2-9634	12	YOUNGTOWN		N 3336	W 11218	1135
2-9652	13	YUMA CITRUS STATION		N 3237	W 11439	191
2-9660	13	YUMA WSO	R	N 3240	W 11436	194

02 — ARIZONA

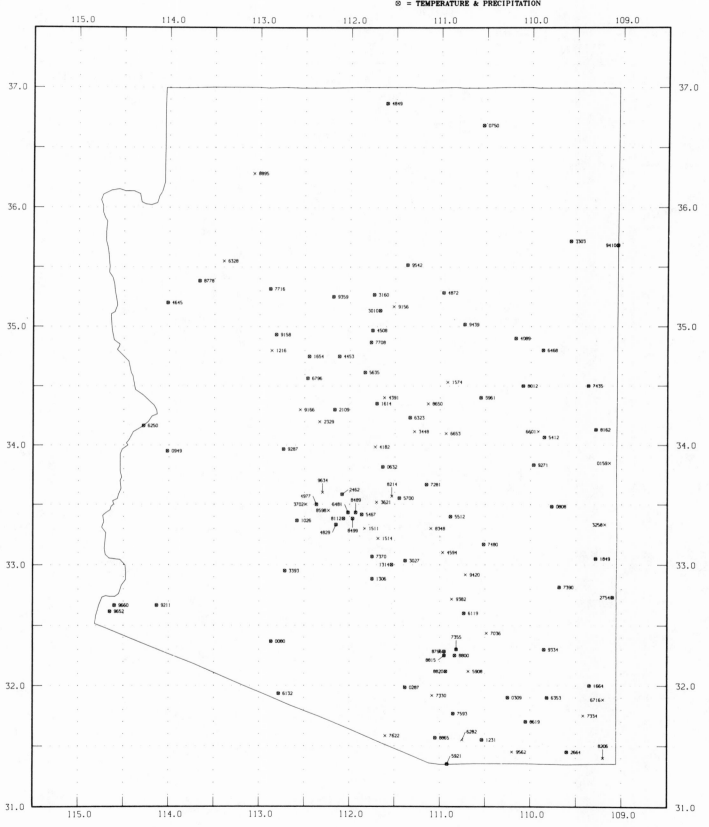

45

ARKANSAS

TEMPERATURE NORMALS (DEG F)

STATION		JAN	FEB	MAR	APR	MAY	JUN	JUL	AUG	SEP	OCT	NOV	DEC	ANN
ALUM FORK	MAX	51.2	56.3	64.3	75.2	81.8	88.5	93.0	92.1	85.7	76.6	63.2	54.4	73.5
	MIN	29.6	32.7	39.7	49.6	57.7	65.1	69.1	67.7	61.8	50.4	39.9	33.2	49.7
	MEAN	40.4	44.5	52.0	62.4	69.7	76.8	81.1	79.9	73.8	63.5	51.6	43.8	61.6
ARKADELPHIA 2 N	MAX	52.8	58.5	66.6	76.3	82.6	89.3	93.1	93.2	86.8	77.3	64.0	55.6	74.7
	MIN	31.1	34.3	41.3	50.6	58.6	66.1	69.9	68.5	62.4	49.9	39.7	33.6	50.5
	MEAN	41.9	46.5	54.0	63.5	70.6	77.7	81.5	80.9	74.6	63.6	51.9	44.6	62.6
BATESVILLE LIVESTOCK	MAX	48.2	53.2	61.5	73.0	80.3	88.0	92.8	91.8	85.0	75.5	61.5	51.6	71.9
	MIN	25.2	28.8	36.8	47.3	54.4	62.4	66.8	64.7	58.3	46.4	36.6	29.2	46.4
	MEAN	36.7	41.0	49.2	60.2	67.4	75.2	79.8	78.3	71.7	61.0	49.1	40.4	59.2
BATESVILLE L AND D 1	MAX	48.9	53.8	62.4	74.2	81.6	89.4	93.4	92.2	85.4	75.6	61.8	52.2	72.6
	MIN	26.2	29.4	37.5	47.6	55.3	63.2	67.4	65.4	58.9	46.3	36.9	30.2	47.0
	MEAN	37.6	41.6	50.0	60.9	68.5	76.3	80.4	78.8	72.2	61.0	49.3	41.2	59.8
BENTON	MAX	52.2	57.3	65.2	75.1	81.8	88.4	92.5	92.1	85.7	76.4	63.8	55.2	73.8
	MIN	28.6	31.9	39.6	49.0	56.7	64.4	68.1	66.4	60.1	47.7	37.8	31.4	48.5
	MEAN	40.4	44.6	52.5	62.1	69.3	76.5	80.3	79.3	73.0	62.1	50.8	43.3	61.2
BENTONVILLE 5 WSW	MAX	45.1	50.3	58.5	70.5	77.3	84.9	90.1	89.5	82.3	72.0	58.2	49.1	69.0
	MIN	23.0	27.4	34.6	45.6	54.0	62.0	66.3	64.0	57.2	45.7	34.9	27.6	45.2
	MEAN	34.1	38.9	46.6	58.1	65.7	73.5	78.2	76.8	69.8	58.9	46.5	38.4	57.1
BLYTHEVILLE	MAX	46.7	51.5	60.6	73.0	82.0	89.8	92.7	90.9	84.5	75.1	60.9	50.4	71.5
	MIN	28.7	32.3	40.3	50.8	60.0	67.7	71.2	68.8	62.1	50.0	40.0	32.8	50.4
	MEAN	37.7	41.9	50.5	61.9	71.1	78.8	82.0	79.9	73.3	62.6	50.5	41.6	61.0
BRINKLEY	MAX	49.3	53.9	62.7	73.3	81.6	89.4	93.0	91.8	85.4	75.8	62.4	52.4	72.6
	MIN	29.0	32.5	39.8	49.9	58.5	66.0	69.3	67.3	60.3	47.4	37.8	32.0	49.2
	MEAN	39.2	43.2	51.3	61.6	70.1	77.7	81.2	79.6	72.9	61.6	50.1	42.2	60.9
CAMDEN 1	MAX	53.9	59.1	67.1	76.7	83.2	90.0	93.6	93.0	87.1	77.8	65.6	57.1	75.4
	MIN	30.9	33.9	41.2	51.1	59.2	66.6	70.2	68.6	62.1	49.4	39.5	33.3	50.5
	MEAN	42.4	46.5	54.2	63.9	71.2	78.3	81.9	80.8	74.6	63.6	52.6	45.2	62.9
CONWAY	MAX	50.8	56.2	64.4	75.3	82.0	89.1	93.6	93.2	86.4	76.6	63.1	53.8	73.7
	MIN	29.1	32.5	40.2	50.2	58.0	65.6	69.6	67.8	61.5	49.2	39.2	32.3	49.6
	MEAN	40.0	44.3	52.3	62.8	70.0	77.4	81.6	80.5	73.9	62.9	51.2	43.1	61.7
CORNING	MAX	45.8	50.7	60.1	72.5	81.0	89.0	92.4	90.9	84.1	74.2	60.0	49.4	70.8
	MIN	26.3	30.0	38.5	49.1	57.2	65.3	68.9	66.6	59.4	47.2	37.5	30.5	48.0
	MEAN	36.1	40.4	49.3	60.8	69.1	77.2	80.7	78.8	71.8	60.8	48.8	40.0	59.5
CROSSETT 7 S	MAX	54.8	60.2	68.0	77.1	83.5	89.9	93.1	92.7	87.3	78.5	65.9	57.8	75.7
	MIN	32.8	35.2	42.0	50.9	58.4	65.7	69.0	67.8	62.3	49.6	40.9	35.0	50.8
	MEAN	43.8	47.7	55.0	64.0	71.0	77.8	81.1	80.3	74.8	64.1	53.5	46.4	63.3
DARDANELLE	MAX	50.0	55.3	63.6	74.7	82.0	88.8	93.2	92.3	86.0	76.6	62.6	53.1	73.2
	MIN	28.3	32.1	39.7	49.9	58.4	66.0	70.1	68.2	61.8	49.2	38.7	31.6	49.5
	MEAN	39.2	43.7	51.7	62.3	70.2	77.5	81.7	80.3	73.9	62.9	50.7	42.4	61.4
DE QUEEN	MAX	53.4	59.0	67.2	76.2	82.8	89.6	93.8	93.7	87.2	77.8	64.8	56.5	75.2
	MIN	30.2	33.4	40.6	50.3	58.7	65.9	69.2	67.8	61.7	49.3	38.9	32.6	49.9
	MEAN	41.8	46.2	53.9	63.3	70.8	77.8	81.5	80.8	74.4	63.6	51.9	44.6	62.6
DES ARC	MAX	49.6	54.6	63.0	74.2	82.0	89.5	93.2	91.9	85.6	76.5	63.1	53.1	73.0
	MIN	30.0	33.3	41.2	51.8	59.9	67.1	70.3	68.0	61.4	49.9	40.3	33.2	50.5
	MEAN	39.8	44.0	52.1	63.0	71.0	78.4	81.8	80.0	73.5	63.2	51.7	43.2	61.8
DUMAS	MAX	52.5	57.7	65.7	75.8	83.5	90.8	93.7	92.6	86.9	77.3	64.6	55.8	74.7
	MIN	32.6	35.8	42.8	52.3	60.5	67.9	70.9	69.2	63.0	50.8	41.7	35.4	51.9
	MEAN	42.6	46.7	54.3	64.1	72.0	79.4	82.3	80.9	75.0	64.1	53.1	45.7	63.4
EL DORADO FAA AIRPORT	MAX	53.7	58.6	66.5	75.8	82.6	89.4	92.6	92.0	86.4	77.1	64.9	57.1	74.7
	MIN	32.4	34.9	42.4	51.5	59.7	67.1	70.8	69.4	63.7	50.7	40.5	34.6	51.5
	MEAN	43.1	46.8	54.5	63.7	71.2	78.3	81.8	80.7	75.1	63.9	52.7	45.9	63.1

ARKANSAS

TEMPERATURE NORMALS (DEG F)

STATION		JAN	FEB	MAR	APR	MAY	JUN	JUL	AUG	SEP	OCT	NOV	DEC	ANN
EUREKA SPRINGS	MAX	46.8	51.6	59.8	71.9	78.3	85.4	90.6	90.0	82.8	73.1	59.1	50.5	70.0
	MIN	25.3	29.5	36.6	47.4	55.0	62.9	67.0	65.0	58.5	48.2	37.4	29.9	46.9
	MEAN	36.0	40.6	48.2	59.7	66.7	74.2	78.8	77.5	70.6	60.6	48.3	40.2	58.5
FAYETTEVILLE FAA AP	MAX	46.2	50.8	58.5	70.1	77.3	84.7	89.9	88.9	82.0	72.2	58.6	50.1	69.1
	MIN	23.3	27.5	35.0	45.8	53.9	62.2	66.6	64.3	57.3	45.0	34.5	27.5	45.2
	MEAN	34.8	39.2	46.8	57.9	65.6	73.5	78.3	76.7	69.7	58.6	46.6	38.8	57.2
FAYETTEVILLE EXP STA	MAX	46.7	51.6	59.2	70.8	77.5	85.0	90.2	89.5	82.6	72.7	59.0	50.7	69.6
	MIN	24.9	29.2	36.5	47.4	55.4	63.6	67.9	65.8	58.9	47.4	36.8	29.3	46.9
	MEAN	35.8	40.4	47.9	59.1	66.5	74.3	79.1	77.7	70.8	60.1	47.9	40.0	58.3
FORT SMITH WSO //R	MAX	48.4	53.8	62.5	73.7	81.0	88.5	93.6	92.9	85.7	75.9	61.9	52.1	72.5
	MIN	26.6	30.9	38.5	49.1	58.2	66.3	70.5	68.9	62.1	49.0	37.7	30.2	49.0
	MEAN	37.5	42.4	50.5	61.4	69.6	77.5	82.1	80.9	73.9	62.5	49.8	41.2	60.8
GILBERT	MAX	50.3	54.9	62.7	74.3	80.6	87.8	92.5	91.8	85.0	76.0	62.6	53.8	72.7
	MIN	23.0	26.7	34.4	45.0	52.9	60.8	64.7	62.6	56.1	43.1	33.5	26.4	44.1
	MEAN	36.7	40.8	48.6	59.7	66.8	74.3	78.6	77.3	70.6	59.6	48.1	40.2	58.4
GRAVETTE	MAX	46.0	51.1	59.3	70.9	77.8	85.4	90.9	90.6	83.5	73.2	58.8	49.9	69.8
	MIN	23.1	27.6	35.2	46.3	54.1	62.3	66.4	64.8	58.0	46.7	35.3	27.6	45.6
	MEAN	34.5	39.3	47.3	58.6	66.0	73.9	78.7	77.7	70.8	60.0	47.1	38.8	57.7
HELENA 5 NW	MAX	49.1	53.8	62.4	73.7	81.5	88.7	91.8	90.7	84.9	75.7	62.3	52.9	72.3
	MIN	30.2	33.2	41.0	51.1	59.6	67.0	70.5	68.5	62.2	49.6	39.9	33.3	50.5
	MEAN	39.7	43.5	51.7	62.4	70.5	77.9	81.1	79.7	73.6	62.6	51.2	43.1	61.4
HOPE 3 NE	MAX	52.1	57.1	64.7	74.7	82.0	88.8	93.0	92.7	86.6	77.3	64.5	55.9	74.1
	MIN	29.8	32.8	40.1	49.8	58.3	66.0	69.5	68.0	61.7	49.2	39.0	32.3	49.7
	MEAN	41.0	45.0	52.4	62.2	70.1	77.4	81.3	80.4	74.2	63.3	51.8	44.2	61.9
HOT SPRINGS 1 NNE	MAX	51.5	56.7	64.8	75.3	82.1	89.1	93.4	93.0	86.5	76.9	63.2	54.5	73.9
	MIN	31.3	34.3	41.4	51.6	59.4	67.0	71.1	69.7	63.4	52.7	41.9	34.7	51.5
	MEAN	41.4	45.5	53.2	63.5	70.8	78.1	82.2	81.4	75.0	64.8	52.6	44.6	62.8
JONESBORO	MAX	46.5	50.9	60.2	72.4	80.8	88.7	92.1	90.7	84.1	74.5	60.5	50.0	71.0
	MIN	28.6	32.1	40.2	50.7	59.3	67.3	70.9	68.7	61.9	49.7	39.9	32.6	50.2
	MEAN	37.6	41.5	50.2	61.6	70.1	78.0	81.5	79.7	73.0	62.2	50.2	41.4	60.6
KEO	MAX	50.2	55.2	63.7	74.4	81.6	88.5	91.7	90.4	84.2	75.4	62.7	53.8	72.7
	MIN	30.5	34.2	41.3	51.1	59.3	66.5	69.7	67.8	61.2	49.5	39.9	33.6	50.4
	MEAN	40.4	44.7	52.5	62.8	70.5	77.5	80.7	79.1	72.7	62.5	51.3	43.7	61.5
LEAD HILL	MAX	47.7	52.7	61.2	72.6	79.9	87.7	93.2	91.7	84.4	74.8	61.1	51.5	71.5
	MIN	23.6	27.9	35.3	46.2	54.4	62.4	67.3	65.0	58.2	45.6	35.2	28.1	45.8
	MEAN	35.7	40.3	48.3	59.4	67.2	75.1	80.3	78.4	71.3	60.2	48.1	39.8	58.7
LITTLE ROCK WSO //R	MAX	49.8	54.5	63.2	73.8	81.7	89.5	92.7	92.3	85.6	75.8	62.4	53.2	72.9
	MIN	29.9	33.6	41.2	50.9	59.2	67.5	71.4	69.6	63.0	50.4	40.0	33.2	50.8
	MEAN	39.9	44.1	52.2	62.4	70.5	78.5	82.1	81.0	74.3	63.1	51.2	43.2	61.9
MAGNOLIA 3 N	MAX	55.2	60.7	68.0	76.7	83.1	89.5	93.3	92.9	86.8	77.9	66.1	58.0	75.7
	MIN	32.8	35.5	42.4	51.1	58.5	65.6	68.7	67.4	62.0	50.2	40.3	35.1	50.8
	MEAN	44.0	48.1	55.2	63.9	70.9	77.6	81.0	80.2	74.4	64.1	53.2	46.5	63.3
MALVERN	MAX	52.6	58.3	66.5	76.8	83.1	89.7	93.4	92.7	86.1	76.7	64.2	55.6	74.6
	MIN	29.2	32.4	39.4	49.1	56.8	64.2	67.8	66.3	60.6	48.1	38.5	32.3	48.7
	MEAN	40.9	45.4	53.0	63.0	70.0	77.0	80.6	79.5	73.4	62.4	51.4	44.0	61.7
MAMMOTH SPRING	MAX	47.2	52.2	61.1	73.2	80.0	87.5	92.0	91.0	84.1	74.8	60.8	50.7	71.2
	MIN	23.1	27.1	35.3	45.8	53.3	61.4	65.4	63.9	56.7	44.1	33.6	26.9	44.7
	MEAN	35.2	39.7	48.2	59.5	66.7	74.5	78.7	77.5	70.4	59.4	47.2	38.8	58.0
MARIANNA 2 S	MAX	49.4	54.4	62.8	73.9	82.2	89.6	92.1	90.7	85.2	76.2	63.0	53.3	72.7
	MIN	30.2	33.8	41.2	51.3	59.7	66.9	69.8	67.6	61.3	49.2	40.0	33.4	50.4
	MEAN	39.8	44.1	52.0	62.6	70.9	78.3	81.0	79.2	73.3	62.7	51.5	43.4	61.6

ARKANSAS

TEMPERATURE NORMALS (DEG F)

STATION		JAN	FEB	MAR	APR	MAY	JUN	JUL	AUG	SEP	OCT	NOV	DEC	ANN
MENA	MAX	50.1	55.4	63.1	73.0	79.7	86.6	91.3	90.7	84.0	74.4	61.4	53.1	71.9
	MIN	29.7	33.0	40.0	50.0	57.8	65.1	68.3	67.1	61.5	50.5	40.0	32.7	49.6
	MEAN	39.9	44.2	51.6	61.5	68.8	75.9	79.8	79.0	72.8	62.4	50.7	42.9	60.8
MONTICELLO 3 SW	MAX	53.0	57.8	65.7	75.2	82.4	89.0	92.2	91.6	85.7	76.9	64.4	56.8	74.2
	MIN	31.7	34.2	41.9	51.0	58.7	66.1	69.0	67.6	62.0	50.1	40.8	34.5	50.6
	MEAN	42.4	46.0	53.8	63.1	70.6	77.5	80.6	79.6	73.9	63.5	52.6	45.7	62.4
MORRILTON	MAX	50.8	55.7	64.1	75.0	82.2	89.6	93.9	93.0	86.5	77.0	63.5	53.9	73.8
	MIN	29.9	33.2	40.8	50.9	58.8	66.3	70.0	68.2	61.8	49.8	40.0	33.0	50.2
	MEAN	40.4	44.5	52.5	63.0	70.5	78.0	82.0	80.6	74.2	63.5	51.8	43.5	62.0
MOUNT IDA 3 SE	MAX	51.0	56.0	63.4	73.8	80.8	87.9	93.0	92.2	85.5	75.9	63.2	54.5	73.1
	MIN	26.3	29.1	36.5	46.6	55.0	63.0	66.8	65.1	58.7	45.6	35.5	29.2	46.5
	MEAN	38.7	42.6	50.0	60.2	67.9	75.5	79.9	78.7	72.1	60.8	49.4	41.9	59.8
MOUNTAIN HOME 1 NNW	MAX	46.3	51.4	59.9	71.9	78.8	86.3	91.2	90.2	83.5	74.0	59.9	50.2	70.3
	MIN	24.2	28.2	36.1	47.0	55.3	63.0	67.4	65.5	58.8	47.0	36.4	28.8	46.5
	MEAN	35.3	39.8	48.0	59.4	67.1	74.7	79.3	77.9	71.1	60.5	48.2	39.5	58.4
NASHVILLE EXP STATION //	MAX	51.0	55.7	63.2	73.2	80.6	87.6	92.2	92.2	85.7	76.3	63.2	54.5	73.0
	MIN	29.6	32.3	39.2	49.2	57.8	65.3	69.0	68.0	62.6	50.4	39.5	32.5	49.6
	MEAN	40.3	44.0	51.3	61.2	69.2	76.5	80.7	80.1	74.2	63.4	51.4	43.5	61.3
NEWPORT	MAX	47.4	52.0	61.1	72.9	81.0	88.9	92.4	90.8	84.5	75.0	61.4	51.0	71.5
	MIN	28.1	31.8	40.2	50.9	59.4	66.7	70.0	67.5	60.5	48.3	39.0	31.9	49.5
	MEAN	37.8	42.0	50.7	61.9	70.2	77.8	81.2	79.2	72.5	61.7	50.2	41.4	60.6
NIMROD DAM	MAX	49.5	54.2	62.2	73.6	80.9	88.2	93.0	92.2	85.5	75.8	62.3	52.8	72.5
	MIN	25.3	29.0	37.0	47.5	55.7	63.5	67.7	65.9	59.5	46.9	36.0	28.6	46.9
	MEAN	37.4	41.6	49.6	60.6	68.3	75.9	80.4	79.1	72.5	61.4	49.2	40.7	59.7
NO. LITTLE ROCK WSFO	MAX	48.6	53.6	62.0	73.1	80.3	87.5	91.6	90.7	84.0	74.5	61.2	51.9	71.6
	MIN	30.3	34.0	42.0	52.5	60.5	68.1	71.8	70.3	64.1	52.2	41.7	34.2	51.8
	MEAN	39.5	43.8	52.0	62.8	70.4	77.8	81.7	80.5	74.1	63.4	51.5	43.1	61.7
OKAY	MAX	55.2	60.4	68.0	77.1	83.6	90.1	94.1	93.9	87.8	79.1	66.5	58.1	76.2
	MIN	32.8	36.2	43.5	53.0	60.7	68.2	71.6	69.8	63.7	51.8	41.8	35.4	52.4
	MEAN	44.0	48.3	55.8	65.0	72.2	79.1	82.9	81.9	75.8	65.5	54.2	46.8	64.3
OZARK	MAX	50.2	55.4	63.3	74.3	81.2	88.6	93.6	93.0	86.2	76.3	62.7	53.3	73.2
	MIN	28.5	32.5	39.7	50.1	58.2	65.9	70.0	68.6	62.6	50.5	39.6	32.0	49.9
	MEAN	39.4	44.0	51.6	62.2	69.7	77.3	81.8	80.8	74.4	63.4	51.2	42.7	61.5
PARAGOULD 1 S	MAX	46.4	51.4	60.6	72.9	81.3	89.2	92.2	90.4	84.2	74.9	60.7	50.1	71.2
	MIN	27.4	31.0	39.0	49.2	57.8	65.6	69.2	67.0	60.1	47.9	38.3	31.4	48.7
	MEAN	36.9	41.2	49.8	61.1	69.6	77.4	80.8	78.7	72.1	61.4	49.5	40.8	59.9
PINE BLUFF	MAX	52.8	57.8	65.8	76.4	83.1	89.9	93.6	92.8	86.6	77.1	64.5	56.0	74.7
	MIN	33.0	36.2	43.2	53.0	61.0	68.4	71.8	70.4	64.0	52.0	42.3	35.7	52.6
	MEAN	42.9	47.0	54.5	64.8	72.1	79.1	82.7	81.6	75.3	64.6	53.4	45.9	63.7
POCAHONTAS 1	MAX	46.4	51.5	60.5	72.6	80.7	88.5	92.1	90.8	84.0	74.3	60.1	49.7	70.9
	MIN	25.7	29.4	37.2	47.6	55.9	64.6	68.6	66.3	59.7	47.3	37.0	29.6	47.4
	MEAN	36.1	40.5	48.9	60.1	68.4	76.6	80.4	78.6	71.9	60.8	48.5	39.7	59.2
PRESCOTT	MAX	53.0	58.7	66.8	76.4	83.3	90.0	94.0	93.6	87.1	77.5	64.4	55.8	75.1
	MIN	31.5	34.6	41.9	51.5	59.8	67.2	70.8	69.8	63.5	51.6	40.9	34.2	51.4
	MEAN	42.2	46.7	54.4	64.0	71.5	78.6	82.4	81.7	75.3	64.6	52.7	45.0	63.3
RUSSELLVILLE 4 N	MAX	50.4	55.8	63.8	75.1	82.2	89.4	93.7	93.0	86.4	76.9	63.0	53.5	73.6
	MIN	27.0	30.8	38.1	48.3	56.8	64.8	69.3	67.5	60.9	47.9	37.8	30.6	48.3
	MEAN	38.7	43.3	51.0	61.7	69.5	77.2	81.5	80.3	73.7	62.4	50.4	42.1	61.0
SAINT CHARLES	MAX	49.6	54.0	62.7	73.6	81.6	89.2	92.6	91.8	85.8	76.4	63.2	53.6	72.8
	MIN	29.4	33.1	41.3	51.5	59.4	67.0	70.4	68.7	62.1	49.4	39.3	32.7	50.4
	MEAN	39.5	43.6	52.0	62.6	70.5	78.1	81.5	80.3	74.0	62.9	51.3	43.2	61.6

ARKANSAS

TEMPERATURE NORMALS (DEG F)

STATION		JAN	FEB	MAR	APR	MAY	JUN	JUL	AUG	SEP	OCT	NOV	DEC	ANN
SEARCY	MAX	49.6	54.8	63.5	74.6	82.1	89.7	93.6	92.6	85.9	76.4	62.7	53.0	73.2
	MIN	28.6	32.2	39.8	50.1	58.1	65.7	69.6	67.7	61.5	48.8	39.0	32.2	49.4
	MEAN	39.1	43.5	51.7	62.4	70.2	77.7	81.7	80.2	73.7	62.6	50.9	42.6	61.4
SILOAM SPRINGS	MAX	46.7	51.6	59.6	70.9	77.9	85.4	91.0	90.5	83.3	73.2	59.3	50.5	70.0
	MIN	24.3	28.5	36.0	46.9	54.8	63.4	67.6	65.7	59.0	47.3	36.2	28.5	46.5
	MEAN	35.6	40.1	47.8	58.9	66.4	74.4	79.3	78.1	71.2	60.3	47.8	39.6	58.3
STUTTGART	MAX	51.3	56.3	64.8	75.6	83.0	89.9	93.4	92.3	86.1	76.9	63.7	54.5	74.0
	MIN	32.1	35.5	42.5	52.7	60.9	68.6	72.0	70.2	63.8	51.9	41.7	35.2	52.3
	MEAN	41.7	45.9	53.7	64.2	72.0	79.3	82.7	81.3	75.0	64.4	52.7	44.9	63.2
STUTTGART 9 ESE	MAX	48.2	52.8	61.3	72.4	80.6	88.2	91.4	90.2	84.4	75.2	62.1	52.3	71.6
	MIN	29.8	33.3	40.9	51.3	60.3	68.0	71.0	69.0	61.9	49.0	39.7	32.9	50.6
	MEAN	39.0	43.0	51.1	61.9	70.5	78.1	81.2	79.6	73.2	62.1	51.0	42.6	61.1
SUBIACO	MAX	50.5	55.9	64.2	75.2	82.1	89.6	94.2	93.3	86.4	76.7	63.1	53.7	73.7
	MIN	28.4	32.6	40.1	50.6	58.5	66.0	69.8	68.0	61.9	50.3	39.7	32.1	49.8
	MEAN	39.4	44.3	52.2	62.9	70.3	77.8	82.0	80.7	74.2	63.6	51.4	42.9	61.8
TEXARKANA FAA AIRPORT	MAX	53.5	58.6	66.0	75.2	82.3	89.1	93.0	92.7	86.6	77.4	65.1	56.9	74.7
	MIN	34.7	37.9	44.8	54.0	62.0	69.3	72.6	71.5	65.6	54.2	44.0	37.6	54.0
	MEAN	44.1	48.3	55.4	64.6	72.2	79.2	82.9	82.1	76.1	65.9	54.6	47.3	64.4
WALDRON	MAX	52.3	57.4	65.3	76.1	83.0	90.3	95.2	94.3	87.6	77.7	64.1	55.8	74.9
	MIN	27.0	30.8	38.4	48.6	56.8	64.3	67.6	65.9	59.4	47.1	36.9	30.1	47.7
	MEAN	39.7	44.1	51.8	62.4	69.9	77.4	81.4	80.1	73.5	62.5	50.5	43.0	61.4
WARREN	MAX	53.6	58.4	66.3	75.7	82.6	89.6	93.4	92.9	87.0	77.7	65.1	56.8	74.9
	MIN	31.8	34.9	41.8	51.2	59.3	66.5	69.9	68.3	62.1	49.9	40.8	34.5	50.9
	MEAN	42.7	46.7	54.1	63.5	71.0	78.1	81.7	80.7	74.6	63.9	52.9	45.7	63.0
WYNNE	MAX	48.4	53.4	62.0	73.8	81.5	89.1	92.5	90.9	85.0	76.2	62.4	52.4	72.3
	MIN	29.2	32.9	40.4	50.6	58.8	66.3	69.7	67.6	61.0	49.1	39.4	32.5	49.8
	MEAN	38.8	43.2	51.3	62.2	70.2	77.7	81.1	79.3	73.0	62.7	50.9	42.5	61.1

ARKANSAS

PRECIPITATION NORMALS (INCHES)

STATION	JAN	FEB	MAR	APR	MAY	JUN	JUL	AUG	SEP	OCT	NOV	DEC	ANN
ABBOTT	2.00	2.49	4.11	4.40	4.92	3.36	3.27	3.34	3.68	2.99	3.30	3.02	40.88
ALICIA	3.54	3.62	5.12	4.55	5.09	3.29	3.67	3.34	4.05	2.57	4.59	3.98	47.41
ALUM FORK	4.06	3.65	5.36	5.25	5.83	4.58	3.91	3.24	4.40	3.28	4.41	4.23	52.20
ALY	3.03	3.36	5.41	4.84	5.69	3.91	4.16	3.09	4.09	3.69	4.03	4.13	49.43
AMITY 3 NE	4.01	4.11	5.26	5.74	6.17	4.57	4.42	3.46	5.00	3.55	5.29	4.60	56.18
ANTOINE	3.67	3.72	4.59	5.48	5.62	4.37	3.99	3.03	4.21	3.27	4.81	4.20	50.96
ARKADELPHIA 2 N	3.99	3.68	4.95	5.62	5.68	4.23	3.87	2.68	4.54	3.37	4.81	4.37	51.79
ARKANSAS CITY	5.11	4.82	5.45	5.43	5.04	3.21	4.34	2.86	3.51	2.59	4.65	4.79	51.80
ATHENS	3.47	3.83	5.73	5.98	6.50	4.89	4.21	3.43	5.05	4.36	5.03	4.33	56.81
AUGUSTA	3.52	3.83	5.12	5.14	5.17	3.83	2.93	3.92	4.21	2.40	4.61	4.44	49.12
BATESVILLE LIVESTOCK	2.90	3.00	4.84	4.51	4.74	3.29	3.60	3.36	4.22	2.80	4.33	3.51	45.10
BATESVILLE L AND D 1	3.21	3.35	4.80	4.72	4.74	3.32	3.95	3.46	4.37	2.74	4.56	3.71	46.93
BEEDEVILLE	4.07	3.92	5.11	4.91	4.81	3.90	3.11	3.29	4.24	2.38	4.23	4.45	48.42
BENTON	3.81	3.85	5.10	5.58	5.53	3.93	4.06	3.20	4.73	3.29	4.76	4.43	52.27
BENTONVILLE 5 WSW	1.86	2.56	3.63	4.29	5.57	5.20	3.57	3.25	4.13	3.49	3.33	2.54	43.42
BIG FORK	3.64	3.87	6.11	5.88	6.37	4.69	4.80	3.34	4.83	4.30	4.52	4.38	56.73
BLACK ROCK	3.46	3.46	4.86	4.30	4.97	3.16	3.56	3.36	4.23	2.52	4.67	3.95	46.50
BLAKELY MOUNTAIN DAM	3.88	3.78	5.18	5.66	6.16	3.91	4.81	3.36	4.38	3.59	5.11	4.19	54.01
BLUFF CITY 3 SW	4.02	3.86	4.79	5.32	5.44	4.15	3.96	3.37	4.47	3.16	4.16	4.52	51.22
BLYTHEVILLE	4.05	4.08	5.37	4.86	5.01	3.88	3.72	3.45	3.92	2.47	4.29	4.33	49.43
BOUGHTON	3.87	3.78	4.66	5.67	5.30	3.97	3.89	2.95	4.67	3.14	4.49	4.32	50.71
BRINKLEY	4.15	4.29	4.92	5.52	5.51	3.65	3.65	3.23	3.90	2.51	4.16	4.59	50.08
BUFFALO TOWER	2.50	3.01	4.62	5.09	5.72	3.97	3.66	3.42	4.67	3.62	4.46	3.39	48.13
CABOT 4 SW	3.61	3.53	4.84	5.20	5.36	3.81	3.35	3.19	3.91	2.75	4.41	4.10	48.06
CALICO ROCK	2.48	2.93	4.31	4.24	4.79	3.99	3.16	3.23	3.93	3.10	4.18	3.27	43.61
CAMDEN 1	4.37	3.94	4.90	5.12	4.74	3.65	4.08	3.08	4.50	2.77	4.57	4.60	50.32
CLINTON	2.93	3.46	5.53	4.90	5.86	4.08	3.86	3.33	4.43	3.26	4.48	4.15	50.27
CONWAY	3.64	3.52	5.11	4.82	5.42	4.06	3.29	2.70	4.74	3.02	4.55	4.12	48.99
CORNING	3.77	3.45	5.27	4.32	5.24	3.16	3.81	3.28	3.92	2.38	4.27	3.90	46.77
COVE	3.14	3.56	5.26	6.01	5.55	4.15	4.51	3.55	4.46	4.25	3.92	4.15	52.51
CROSSETT 7 S	5.21	4.54	5.67	5.91	5.35	3.94	4.72	3.40	4.15	2.60	4.44	4.82	54.75
CRYSTAL VALLEY	4.05	4.05	4.88	5.60	5.89	3.99	3.58	3.03	4.79	3.19	4.79	4.31	52.15
DANVILLE	2.59	2.98	5.07	4.61	5.44	3.59	3.63	3.36	3.97	3.38	4.16	3.75	46.53
DARDANELLE	2.84	3.08	5.20	4.40	5.10	4.17	3.24	3.25	3.67	3.34	4.42	3.76	46.47
DE QUEEN	3.41	3.32	4.91	5.48	6.30	3.58	3.95	3.05	4.57	3.67	4.21	3.94	50.39
DERMOTT 3 NE	5.00	4.84	5.60	5.17	4.50	3.59	4.34	2.82	3.60	2.74	4.73	4.94	51.87
DES ARC	4.08	4.41	5.15	5.75	5.67	3.56	3.97	3.38	3.62	2.56	4.35	4.47	50.97
DUMAS	4.55	4.29	5.56	5.26	5.15	3.24	4.55	3.13	3.72	2.82	4.33	4.28	50.88
EL DORADO FAA AIRPORT	4.74	4.02	4.83	5.37	4.79	3.62	3.91	3.04	3.64	2.96	3.84	4.36	49.12
EUREKA SPRINGS	1.90	2.66	3.69	4.21	5.32	4.84	4.07	3.29	4.04	3.33	3.52	2.67	43.54
EVENING SHADE 1 NE	2.87	3.26	4.86	4.44	4.74	3.40	3.42	3.48	4.10	2.54	4.34	3.52	44.97
FAYETTEVILLE FAA AP	1.96	2.72	3.84	4.58	5.46	4.52	3.71	3.51	4.19	3.32	3.40	2.70	43.91
FAYETTEVILLE EXP STA	1.78	2.45	3.46	4.44	5.17	4.55	3.56	3.47	4.09	3.21	3.23	2.52	41.93
FORDYCE	4.56	3.92	5.06	5.24	4.86	3.23	4.00	3.08	3.95	2.95	4.43	4.76	50.04
FORT SMITH WSO //R	1.86	2.53	3.88	4.20	4.79	3.67	3.15	3.02	3.22	3.24	3.50	2.85	39.91
FORT SMITH WATER PLANT	2.30	2.86	4.45	5.13	6.15	4.02	4.22	3.52	4.49	3.92	3.98	3.32	48.36
GEORGETOWN	3.91	3.79	5.31	5.35	5.22	3.87	3.27	3.33	3.95	2.46	4.35	4.21	49.02
GILBERT	2.15	2.77	3.94	4.38	5.21	3.78	3.30	2.89	3.62	2.96	4.01	3.18	42.19
GLENWOOD 3 ENE	3.52	3.72	5.22	6.01	6.76	4.68	4.73	3.05	4.82	3.74	5.10	4.27	55.62
GRAVELLY 4 E	2.71	3.19	4.95	4.98	5.41	4.05	4.06	2.93	3.93	3.44	3.84	3.87	47.36
GRAVETTE	1.89	2.23	3.82	4.48	5.38	5.18	3.58	3.26	4.11	3.24	3.36	2.58	43.11
GREEN FOREST	2.03	2.53	3.74	4.32	5.06	5.10	3.71	2.66	4.02	3.02	3.42	2.65	42.26
GREENWOOD	2.02	2.80	4.32	4.63	5.03	3.44	3.60	3.09	3.80	3.58	3.84	3.27	43.42
GURDON	3.91	3.78	4.47	5.78	5.52	4.09	3.56	3.47	4.56	3.41	4.43	4.32	51.30
HECTOR	2.75	3.42	4.96	4.57	5.11	4.15	3.61	3.49	4.21	3.60	4.12	3.90	47.89

ARKANSAS

PRECIPITATION NORMALS (INCHES)

STATION	JAN	FEB	MAR	APR	MAY	JUN	JUL	AUG	SEP	OCT	NOV	DEC	ANN
HELENA 5 NW	4.46	4.36	5.31	5.47	5.23	4.03	3.81	2.70	4.06	2.75	4.47	4.87	51.52
HOPE 3 NE	3.84	3.78	4.63	5.55	5.40	4.25	3.67	4.01	4.28	3.29	4.40	4.07	51.17
HOPPER 1 E	3.87	4.00	5.82	5.91	6.27	4.59	4.45	3.32	4.80	4.20	4.90	4.61	56.74
HORATIO	3.12	3.56	4.99	5.53	5.84	3.88	3.61	2.58	4.28	3.78	4.14	3.90	49.21
HOT SPRINGS 1 NNE	3.81	4.08	5.25	5.89	6.43	4.40	5.17	3.35	4.37	3.36	4.80	4.47	55.38
JASPER	2.37	2.94	4.30	4.37	5.52	4.49	3.95	3.46	3.91	3.15	3.99	3.11	45.56
JESSIEVILLE	4.12	4.14	5.79	5.71	5.99	4.55	4.55	3.36	4.94	4.00	5.22	4.49	56.86
JONESBORO	3.86	3.95	5.16	5.11	4.92	3.10	3.44	3.09	3.69	2.66	4.54	4.10	47.62
KEO	4.09	3.95	4.98	4.95	5.04	3.48	3.83	2.45	3.64	2.99	3.97	4.48	47.85
LAKE CITY	4.01	4.01	5.24	5.30	4.85	3.40	3.25	3.76	4.42	2.64	4.34	4.42	49.64
LANGLEY	3.95	4.08	5.87	5.98	6.61	5.07	4.64	3.78	4.85	4.26	5.08	4.71	58.88
LEAD HILL	1.93	2.44	3.90	4.11	4.55	4.52	3.48	3.07	3.77	3.15	3.60	2.65	41.17
LEOLA	4.14	4.05	4.96	5.37	5.37	3.94	4.16	2.99	4.21	3.10	4.42	4.52	51.23
LITTLE ROCK WSO //R	3.91	3.83	4.69	5.41	5.29	3.67	3.63	3.07	4.26	2.84	4.37	4.23	49.20
MADISON	4.20	4.34	4.87	5.17	5.15	3.20	3.55	3.79	3.76	2.73	4.41	4.84	50.01
MAGNOLIA 3 N	4.21	4.07	4.71	5.59	4.75	3.71	4.04	3.46	3.70	2.89	4.47	4.73	50.33
MALVERN	4.19	3.96	5.24	5.49	5.72	4.27	4.60	2.94	4.66	3.35	4.95	4.78	54.15
MAMMOTH SPRING	2.65	3.00	4.50	4.38	4.91	3.49	3.60	3.04	3.84	2.62	3.89	3.32	43.24
MARIANNA 2 S	4.35	4.10	5.36	5.51	5.24	3.37	3.99	2.79	4.02	2.95	4.38	4.83	50.89
MARSHALL	2.17	2.92	4.60	4.34	5.26	3.61	3.36	3.09	4.07	2.95	4.03	3.12	43.52
MELBOURNE 5 WNW	2.72	3.06	4.75	4.40	4.81	3.50	3.70	3.32	3.93	2.66	4.31	3.39	44.55
MENA	3.07	3.45	5.29	5.80	5.67	4.24	4.65	2.85	4.74	4.26	4.23	3.92	52.17
MONTICELLO 3 SW	5.01	4.54	5.63	5.47	4.82	3.43	4.68	3.09	3.99	3.03	4.44	4.53	52.66
MOROBAY LOCK NO 8	4.67	4.27	5.22	5.29	5.05	3.93	4.51	3.04	3.81	2.77	4.04	4.19	50.79
MORRILTON	3.17	3.15	4.89	4.67	5.37	3.99	3.12	2.73	3.96	2.83	4.20	3.91	45.99
MOUNT IDA 3 SE	3.54	3.75	5.73	4.97	6.43	4.29	4.53	3.07	4.70	3.86	4.58	4.26	53.71
MOUNTAIN HOME 1 NNW	2.35	2.86	4.09	4.40	4.90	4.16	3.09	2.62	3.68	2.88	3.85	3.19	42.07
MOUNTAIN VIEW	2.95	3.26	5.19	4.63	5.10	3.78	3.94	3.50	5.00	2.87	4.45	3.51	48.18
MULBERRY 6 NNE	2.27	2.97	4.41	4.79	5.63	3.72	3.96	3.45	4.15	3.85	4.03	3.35	46.58
NASHVILLE EXP STATION //	3.51	3.85	4.86	5.92	5.85	4.81	4.10	3.45	4.36	3.67	4.56	4.13	53.07
NATHAN 4 WNW	3.49	3.40	5.08	5.92	5.71	4.89	3.85	2.88	4.46	3.80	4.62	4.00	52.10
NEWHOPE 3 E	3.64	3.92	5.51	5.96	6.14	4.68	4.48	3.05	4.60	4.00	4.77	4.39	55.14
NEWPORT	3.82	3.57	5.16	4.85	5.22	3.92	3.81	3.91	4.12	2.67	4.40	3.78	49.23
NIMROD DAM	3.09	3.04	5.01	4.85	5.31	3.66	3.34	2.96	3.83	3.00	3.88	3.68	45.65
NO. LITTLE ROCK WSFO	3.70	3.72	4.89	5.29	5.00	3.23	3.12	2.68	3.82	2.72	4.08	4.02	46.26
ODELL 3 N	2.33	2.80	4.08	4.60	5.81	4.51	4.43	3.53	4.80	3.99	3.84	2.98	47.70
ODEN 2 W	3.11	3.71	5.67	5.28	5.65	4.24	4.64	3.01	4.07	3.53	4.30	4.09	51.30
OKAY	3.74	3.37	4.29	5.40	5.54	3.93	3.74	3.39	4.12	3.29	4.52	3.84	49.17
OWENSVILLE	3.74	3.93	5.32	5.71	5.77	4.23	4.60	3.59	4.62	3.53	4.64	4.43	54.11
OZARK	2.25	2.93	4.36	4.35	4.90	3.53	3.57	2.87	4.12	3.32	3.84	3.14	43.18
PARAGOULD 1 S	3.90	3.72	5.29	4.86	4.90	3.42	3.67	3.65	4.14	2.51	4.49	4.24	48.79
PARKS	2.41	2.75	4.73	5.02	5.38	3.68	4.00	2.58	3.78	3.42	3.55	3.41	44.71
PERRY	3.29	3.22	5.12	4.74	5.59	4.29	3.31	3.11	3.99	2.87	4.15	3.74	47.42
PINE BLUFF	4.38	4.34	5.09	5.28	5.63	3.05	3.62	3.08	3.79	3.17	4.13	4.72	50.28
PINE RIDGE	3.18	3.67	5.62	5.43	5.97	4.04	4.58	3.00	4.47	3.78	3.93	4.06	51.73
POCAHONTAS 1	3.88	3.56	5.60	4.54	4.96	3.11	3.88	3.10	3.95	2.49	4.76	3.80	47.63
PRESCOTT	4.18	3.91	5.01	5.91	5.24	4.29	4.30	3.43	4.58	3.60	4.61	4.30	53.36
RATCLIFF	2.08	2.67	3.86	4.03	4.89	3.15	3.50	2.67	3.71	2.98	3.40	3.01	39.95
RUSSELLVILLE 4 N	2.88	3.33	5.00	4.73	5.38	4.31	3.43	3.44	3.83	3.33	4.14	3.83	47.63
SAINT CHARLES	4.42	4.20	5.33	5.61	4.89	3.55	3.93	2.75	3.75	2.96	4.54	4.58	50.51
SAINT FRANCIS	3.73	3.40	5.09	4.55	5.03	3.43	3.51	2.84	4.03	2.27	4.38	4.07	46.33
SEARCY	4.04	3.76	5.53	5.03	5.73	3.73	3.99	3.88	4.34	2.77	4.58	4.38	51.76
SHIRLEY	3.29	3.37	5.30	4.63	5.37	4.11	3.85	3.36	4.70	2.99	4.54	4.03	49.54
SILOAM SPRINGS	1.75	2.29	3.53	4.54	5.56	4.81	3.33	3.90	4.26	3.46	3.39	2.43	43.25
SPARKMAN 3 WSW	4.36	3.74	4.69	5.49	5.24	3.68	3.98	2.75	3.81	3.26	4.17	4.50	49.67

ARKANSAS

PRECIPITATION NORMALS (INCHES)

STATION	JAN	FEB	MAR	APR	MAY	JUN	JUL	AUG	SEP	OCT	NOV	DEC	ANN
STAMPS	4.32	3.97	4.65	5.95	4.82	3.80	4.17	3.45	3.85	2.90	4.36	4.59	50.83
STUTTGART	4.30	4.37	5.57	5.54	5.07	4.13	3.54	2.85	3.80	2.77	4.10	4.85	50.89
STUTTGART 9 ESE	4.04	4.05	5.09	5.25	4.70	3.61	3.44	2.81	4.06	2.71	4.35	4.59	48.70
SUBIACO	2.28	2.88	4.22	4.64	5.37	4.18	3.64	3.33	3.68	3.44	3.65	3.52	44.83
TAYLOR	4.16	3.73	4.36	5.05	4.67	3.49	4.22	3.30	3.45	2.78	4.06	4.24	47.51
TEXARKANA FAA AIRPORT	3.62	3.32	4.16	5.08	4.37	3.92	3.54	3.19	3.57	2.77	3.85	3.86	45.25
WALDRON	2.38	2.82	4.37	4.76	5.24	3.92	3.94	2.87	3.76	3.41	3.67	3.59	44.73
WARREN	4.86	4.50	5.50	5.51	4.94	3.35	4.39	3.06	4.06	2.98	4.49	4.56	52.20
WYNNE	3.96	3.97	4.90	5.54	5.40	3.66	3.15	3.22	4.15	2.57	4.30	4.58	49.40

ARKANSAS

HEATING DEGREE DAY NORMALS (BASE 65 DEG F)

STATION	JUL	AUG	SEP	OCT	NOV	DEC	JAN	FEB	MAR	APR	MAY	JUN	ANN
ALUM FORK	0	0	7	115	406	657	763	574	424	126	22	0	3094
ARKADELPHIA 2 N	0	0	9	127	398	632	716	518	366	97	15	0	2878
BATESVILLE LIVESTOCK	0	0	22	178	477	763	877	672	510	180	59	6	3744
BATESVILLE L AND D 1	0	0	17	167	471	738	849	655	480	157	38	0	3572
BENTON	0	0	12	153	426	673	763	571	407	131	27	0	3163
BENTONVILLE 5 WSW	0	0	31	214	555	825	958	731	580	230	89	10	4223
BLYTHEVILLE	0	0	5	136	435	725	846	647	468	145	25	0	3432
BRINKLEY	0	0	13	169	447	707	800	618	444	136	34	0	3368
CAMDEN 1	0	0	7	127	380	614	701	518	360	104	18	0	2829
CONWAY	0	0	9	134	414	679	775	580	416	111	21	0	3139
CORNING	0	0	17	178	486	775	896	689	502	161	37	0	3741
CROSSETT 7 S	0	0	5	127	352	577	664	494	339	100	14	0	2672
DARDANELLE	0	0	8	129	429	701	800	596	431	130	22	0	3246
DE QUEEN	0	0	8	121	400	632	719	526	363	115	16	0	2900
DES ARC	0	0	11	148	407	676	781	588	427	135	21	0	3194
DUMAS	0	0	0	114	362	598	694	518	360	93	9	0	2748
EL DORADO FAA AIRPORT	0	0	5	118	376	592	679	517	354	101	13	0	2755
EUREKA SPRINGS	0	0	38	181	501	769	899	683	537	192	60	13	3873
FAYETTEVILLE FAA AP	0	0	31	224	552	812	936	722	572	234	83	8	4174
FAYETTEVILLE EXP STA	0	0	24	194	513	775	905	689	545	206	70	8	3929
FORT SMITH WSO //R	0	0	13	143	456	738	853	633	461	147	33	0	3477
GILBERT	0	0	25	210	507	769	877	678	520	201	60	5	3852
GRAVETTE	0	0	31	192	537	812	946	720	561	219	68	10	4096
HELENA 5 NW	0	0	19	151	414	679	784	610	429	122	35	0	3243
HOPE 3 NE	0	0	10	131	401	645	744	560	408	123	23	0	3045
HOT SPRINGS 1 NNE	0	0	10	108	379	632	732	546	388	114	23	0	2932
JONESBORO	0	0	14	164	444	732	849	658	474	147	39	0	3521
KEO	0	0	14	141	411	660	763	568	408	118	17	0	3100
LEAD HILL	0	0	21	195	507	781	908	692	536	204	68	7	3919
LITTLE ROCK WSO //R	0	0	8	132	414	676	778	585	417	124	18	0	3152
MAGNOLIA 3 N	0	0	10	114	354	574	651	481	334	91	15	0	2624
MALVERN	0	0	12	151	408	651	747	549	389	110	21	0	3038
MAMMOTH SPRING	0	0	32	214	534	812	924	708	533	198	65	0	4020
MARIANNA 2 S	0	0	9	143	405	670	781	585	423	127	19	0	3162
MENA	0	0	12	144	433	685	778	582	432	142	34	0	3242
MONTICELLO 3 SW	0	0	10	130	379	598	701	541	372	112	22	0	2865
MORRILTON	0	0	7	123	400	667	763	574	409	114	19	0	3076
MOUNT IDA 3 SE	0	0	13	182	468	716	815	627	475	174	60	0	3530
MOUNTAIN HOME 1 NNW	0	0	21	188	504	791	921	706	542	198	67	6	3944
NASHVILLE EXP STATION //	0	0	8	125	408	667	766	588	436	142	24	0	3164
NEWPORT	0	0	12	166	444	732	843	644	462	143	31	0	3477
NIMROD DAM	0	0	19	174	474	753	856	655	490	165	51	5	3642
NO. LITTLE ROCK WSFO	0	0	9	127	405	679	791	594	423	117	18	0	3163
OKAY	0	0	0	91	333	564	651	473	316	73	9	0	2510
OZARK	0	0	8	125	418	691	794	588	432	130	26	0	3212
PARAGOULD 1 S	0	0	14	165	465	750	871	666	487	158	39	0	3615
PINE BLUFF	0	0	6	108	357	592	692	514	362	85	13	0	2729
POCAHONTAS 1	0	0	18	175	495	784	896	686	513	172	42	0	3781
PRESCOTT	0	0	0	102	377	620	707	512	353	101	10	0	2782
RUSSELLVILLE 4 N	0	0	12	148	438	710	815	608	445	136	34	0	3346
SAINT CHARLES	0	0	8	138	415	676	791	599	422	121	27	0	3197
SEARCY	0	0	8	137	423	694	803	602	431	118	22	0	3238
SILOAM SPRINGS	0	0	24	186	516	787	911	697	545	210	78	11	3965
STUTTGART	0	0	8	114	375	623	722	540	377	91	15	0	2865
STUTTGART 9 ESE	0	0	9	147	420	694	806	616	448	133	24	0	3297

ARKANSAS

HEATING DEGREE DAY NORMALS (BASE 65 DEG F)

STATION	JUL	AUG	SEP	OCT	NOV	DEC	JAN	FEB	MAR	APR	MAY	JUN	ANN
SUBIACO	0	0	11	120	412	685	794	580	419	114	23	0	3158
TEXARKANA FAA AIRPORT	0	0	0	88	324	549	648	473	326	86	7	0	2501
WALDRON	0	0	10	137	435	682	784	585	428	127	25	0	3213
WARREN	0	0	9	125	369	598	691	522	364	105	21	0	2804
WYNNE	0	0	11	153	423	698	812	610	446	133	27	0	3313

ARKANSAS

COOLING DEGREE DAY NORMALS (BASE 65 DEG F)

STATION	JAN	FEB	MAR	APR	MAY	JUN	JUL	AUG	SEP	OCT	NOV	DEC	ANN
ALUM FORK	0	0	21	48	168	354	499	462	271	69	0	0	1892
ARKADELPHIA 2 N	0	0	25	52	189	381	512	493	297	83	5	0	2037
BATESVILLE LIVESTOCK	0	0	20	36	133	312	459	412	223	54	0	0	1649
BATESVILLE L AND D 1	0	0	15	34	146	339	477	428	233	43	0	0	1715
BENTON	0	0	19	44	160	345	474	443	252	64	0	0	1801
BENTONVILLE 5 WSW	0	0	10	23	111	265	409	366	175	25	0	0	1384
BLYTHEVILLE	0	0	18	52	214	414	527	462	254	62	0	0	2003
BRINKLEY	0	7	19	34	192	381	502	453	250	63	0	0	1901
CAMDEN 1	0	0	25	71	210	399	524	490	295	83	8	0	2105
CONWAY	0	0	22	45	176	372	515	481	276	68	0	0	1955
CORNING	0	0	16	35	164	366	487	428	221	48	0	0	1765
CROSSETT 7 S	7	9	29	70	200	384	499	474	299	99	7	0	2077
DARDANELLE	0	0	19	49	183	375	518	474	275	64	0	0	1957
DE QUEEN	0	0	19	64	196	384	512	490	290	78	7	0	2040
DES ARC	0	0	27	75	207	402	521	465	266	92	8	0	2063
DUMAS	0	6	29	66	226	432	536	493	305	86	5	0	2184
EL DORADO FAA AIRPORT	0	7	29	62	205	399	521	487	308	84	7	0	2109
EUREKA SPRINGS	0	0	17	33	112	289	432	393	206	44	0	0	1526
FAYETTEVILLE FAA AP	0	0	7	21	102	263	412	363	172	26	0	0	1366
FAYETTEVILLE EXP STA	0	0	15	29	116	287	437	394	198	42	0	0	1518
FORT SMITH WSO //R	0	0	11	39	175	375	530	493	280	66	0	0	1969
GILBERT	0	0	12	42	115	284	422	381	193	42	0	0	1491
GRAVETTE	0	0	12	27	99	277	425	394	205	37	0	0	1476
HELENA 5 NW	0	8	17	44	206	391	499	456	277	76	0	0	1974
HOPE 3 NE	0	0	17	39	181	372	505	477	286	79	0	0	1956
HOT SPRINGS 1 NNE	0	0	22	69	203	393	533	508	310	102	7	0	2147
JONESBORO	0	0	15	45	197	390	512	456	254	77	0	0	1946
KEO	0	0	21	52	188	375	487	437	245	63	0	0	1868
LEAD HILL	0	0	18	36	136	310	474	415	210	46	0	0	1645
LITTLE ROCK WSO //R	0	0	20	46	188	405	530	496	287	73	0	0	2045
MAGNOLIA 3 N	0	8	30	58	198	378	496	471	292	86	0	0	2017
MALVERN	0	0	17	50	176	360	484	450	264	70	0	0	1871
MAMMOTH SPRING	0	0	12	33	118	289	425	388	194	40	0	0	1499
MARIANNA 2 S	0	0	20	55	202	399	496	440	258	72	0	0	1942
MENA	0	0	16	37	152	327	459	434	246	64	0	0	1735
MONTICELLO 3 SW	0	9	25	55	195	375	484	453	277	84	7	0	1964
MORRILTON	0	0	21	54	189	390	527	484	283	77	0	0	2025
MOUNT IDA 3 SE	0	0	10	30	150	319	462	425	226	52	0	0	1674
MOUNTAIN HOME 1 NNW	0	0	15	30	132	297	443	400	204	48	0	0	1569
NASHVILLE EXP STATION //	0	0	11	28	155	345	487	468	284	76	0	0	1854
NEWPORT	0	0	19	50	192	384	502	440	237	64	0	0	1888
NIMROD DAM	0	0	12	33	153	332	477	437	244	63	0	0	1751
NO. LITTLE ROCK WSFO	0	0	20	51	186	384	518	481	282	77	0	0	1999
OKAY	0	6	31	73	232	423	555	524	324	106	9	0	2283
OZARK	0	0	16	46	172	369	521	490	290	75	0	0	1979
PARAGOULD 1 S	0	0	15	41	182	372	490	425	227	53	0	0	1805
PINE BLUFF	7	10	36	79	233	423	549	515	315	96	9	0	2272
POCAHONTAS 1	0	0	13	25	148	348	477	422	225	45	0	0	1703
PRESCOTT	0	0	25	71	211	408	539	518	313	90	8	0	2183
RUSSELLVILLE 4 N	0	0	11	37	173	366	512	474	273	68	0	0	1914
SAINT CHARLES	0	0	19	49	197	393	512	474	278	73	0	0	1995
SEARCY	0	0	19	40	184	381	518	471	269	63	0	0	1945
SILOAM SPRINGS	0	0	12	27	121	293	443	406	210	40	0	0	1552
STUTTGART	0	5	27	67	232	429	549	505	308	95	6	0	2223
STUTTGART 9 ESE	0	0	17	40	195	393	502	453	255	58	0	0	1913

ARKANSAS

COOLING DEGREE DAY NORMALS (BASE 65 DEG F)

STATION	JAN	FEB	MAR	APR	MAY	JUN	JUL	AUG	SEP	OCT	NOV	DEC	ANN
SUBIACO	0	0	22	51	188	384	527	487	287	77	0	0	2023
TEXARKANA FAA AIRPORT	0	5	28	74	231	426	555	530	337	116	12	0	2314
WALDRON	0	0	19	49	177	372	508	468	265	59	0	0	1917
WARREN	0	9	26	60	207	393	518	487	297	91	6	0	2094
WYNNE	0	0	22	49	188	381	499	443	251	81	0	0	1914

03 — ARKANSAS

LEGEND
11 = TEMPERATURE ONLY
12 = PRECIPITATION ONLY
13 = TEMP. & PRECIP.

STATE-STATION NUMBER	STN TYP	NAME	LATITUDE DEG-MIN	LONGITUDE DEG-MIN	ELEVATION (FT)
3-0006	12	ABBOTT	N 3504	W 09412	624
3-0064	12	ALICIA	N 3554	W 09105	256
3-0130	13	ALUM FORK	N 3448	W 09252	780
3-0136	12	ALY	N 3447	W 09329	854
3-0150	12	AMITY 3 NE	N 3417	W 09325	475
3-0178	12	ANTOINE	N 3402	W 09325	285
3-0220	13	ARKADELPHIA 2 N	N 3409	W 09303	196
3-0234	12	ARKANSAS CITY	N 3337	W 09112	145
3-0300	12	ATHENS	N 3419	W 09358	960
3-0326	12	AUGUSTA	N 3518	W 09122	218
3-0458	13	BATESVILLE LIVESTOCK	N 3549	W 09147	571
3-0460	13	BATESVILLE L AND D 1	N 3545	W 09138	277
3-0536	12	BEEDEVILLE	N 3526	W 09106	221
3-0582	13	BENTON	N 3433	W 09237	285
3-0586	13	BENTONVILLE 5 WSW	N 3621	W 09417	1264
3-0664	12	BIG FORK	N 3429	W 09358	1125
3-0746	12	BLACK ROCK	N 3607	W 09106	269
3-0764	12	BLAKELY MOUNTAIN DAM	N 3436	W 09311	426
3-0800	12	BLUFF CITY 3 SW	N 3341	W 09309	360
3-0806	13	BLYTHEVILLE	N 3555	W 08954	252
3-0848	12	BOUGHTON	N 3352	W 09320	250
3-0936	13	BRINKLEY	N 3453	W 09112	205
3-1010	12	BUFFALO TOWER	N 3552	W 09330	2578
3-1102	12	CABOT 4 SW	N 3457	W 09204	289
3-1132	12	CALICO ROCK	N 3607	W 09208	400
3-1152	13	CAMDEN 1	N 3336	W 09249	116
3-1492	12	CLINTON	N 3535	W 09228	510
3-1596	13	CONWAY	N 3505	W 09228	316
3-1632	13	CORNING	N 3624	W 09035	293
3-1666	12	COVE	N 3426	W 09425	1050
3-1730	13	CROSSETT 7 S	N 3302	W 09156	175
3-1750	12	CRYSTAL VALLEY	N 3442	W 09227	350
3-1834	12	DANVILLE	N 3503	W 09324	370
3-1838	13	DARDANELLE	N 3513	W 09309	330
3-1948	13	DE QUEEN	N 3402	W 09421	420
3-1962	12	DERMOTT 3 NE	N 3333	W 09123	140
3-1968	13	DES ARC	N 3458	W 09130	204
3-2148	13	DUMAS	N 3353	W 09129	160
3-2300	13	EL DORADO FAA AIRPORT	N 3313	W 09248	252
3-2356	13	EUREKA SPRINGS	N 3624	W 09344	1518
3-2366	12	EVENING SHADE 1 NE	N 3605	W 09137	490
3-2443	13	FAYETTEVILLE FAA AP	N 3600	W 09410	1251
3-2444	13	FAYETTEVILLE EXP STA	N 3606	W 09410	1270
3-2540	12	FORDYCE	N 3349	W 09225	264
3-2574	13	FORT SMITH WSO //R	N 3520	W 09422	447
3-2578	12	FORT SMITH WATER PLANT	N 3539	W 09409	793
3-2760	12	GEORGETOWN	N 3508	W 09127	200
3-2794	13	GILBERT	N 3559	W 09243	595
3-2842	12	GLENWOOD 3 ENE	N 3421	W 09330	587
3-2922	12	GRAVELLY 4 E	N 3453	W 09338	451

STATE-STATION NUMBER	STN TYP	NAME	LATITUDE DEG-MIN	LONGITUDE DEG-MIN	ELEVATION (FT)
3-2930	13	GRAVETTE	N 3624	W 09428	1250
3-2946	12	GREEN FOREST	N 3620	W 09326	1360
3-2976	12	GREENWOOD	N 3513	W 09415	518
3-3074	12	GURDON	N 3355	W 09309	210
3-3235	12	HECTOR	N 3528	W 09258	555
3-3242	13	HELENA 5 NW	N 3434	W 09040	220
3-3428	13	HOPE 3 NE	N 3343	W 09333	375
3-3438	12	HOPPER 1 E	N 3422	W 09340	700
3-3442	12	HORATIO	N 3356	W 09422	337
3-3466	13	HOT SPRINGS 1 NNE	N 3431	W 09303	680
3-3600	12	JASPER	N 3601	W 09311	857
3-3704	12	JESSIEVILLE	N 3442	W 09304	722
3-3734	13	JONESBORO	N 3550	W 09042	345
3-3862	13	KEO	N 3436	W 09200	230
3-3998	12	LAKE CITY	N 3549	W 09026	230
3-4060	12	LANGLEY	N 3419	W 09351	796
3-4106	13	LEAD HILL	N 3625	W 09255	810
3-4134	12	LEOLA	N 3410	W 09235	261
3-4248	13	LITTLE ROCK WSO //R	N 3444	W 09214	257
3-4528	12	MADISON	N 3501	W 09043	215
3-4548	13	MAGNOLIA 3 N	N 3319	W 09314	315
3-4562	13	MALVERN	N 3423	W 09249	311
3-4572	13	MAMMOTH SPRING	N 3629	W 09132	690
3-4638	13	MARIANNA 2 S	N 3444	W 09046	234
3-4666	12	MARSHALL	N 3554	W 09238	1050
3-4746	12	MELBOURNE 5 WNW	N 3605	W 09159	600
3-4756	13	MENA	N 3435	W 09415	1207
3-4900	13	MONTICELLO 3 SW	N 3336	W 09149	300
3-4934	12	MOROBAY LOCK NO 8	N 3319	W 09227	85
3-4938	13	MORRILTON	N 3508	W 09244	280
3-4988	13	MOUNT IDA 3 SE	N 3432	W 09336	697
3-5036	13	MOUNTAIN HOME 1 NNW	N 3620	W 09223	800
3-5046	12	MOUNTAIN VIEW	N 3552	W 09207	770
3-5072	12	MULBERRY 6 NNE	N 3534	W 09401	500
3-5112	13	NASHVILLE EXP STATION //	N 3400	W 09356	550
3-5158	12	NATHAN 4 WNW	N 3407	W 09352	550
3-5174	12	NEWHOPE 3 E	N 3414	W 09350	850
3-5186	13	NEWPORT	N 3536	W 09117	220
3-5200	13	NIMROD DAM	N 3457	W 09310	480
3-5320	13	NO. LITTLE ROCK WSFO	N 3450	W 09216	563
3-5354	12	ODELL 3 N	N 3548	W 09424	1500
3-5358	12	ODEN 2 W	N 3438	W 09348	800
3-5376	13	OKAY	N 3346	W 09355	300
3-5498	12	OWENSVILLE	N 3437	W 09249	500
3-5508	13	OZARK	N 3529	W 09350	396
3-5563	13	PARAGOULD 1 S	N 3602	W 09030	270
3-5591	12	PARKS	N 3448	W 09358	668
3-5691	12	PERRY	N 3503	W 09248	330
3-5754	13	PINE BLUFF	N 3413	W 09201	215
3-5760	12	PINE RIDGE	N 3435	W 09354	840

03 — ARKANSAS

STATE-STATION NUMBER	STN TYP	NAME	LATITUDE DEG-MIN	LONGITUDE DEG-MIN	ELEVATION (FT)
3-5820	13	POCAHONTAS 1	N 3616	W 09059	330
3-5908	13	PRESCOTT	N 3348	W 09323	308
3-6008	12	RATCLIFF	N 3518	W 09353	463
3-6352	13	RUSSELLVILLE 4 N	N 3520	W 09309	346
3-6376	13	SAINT CHARLES	N 3423	W 09108	200
3-6380	12	SAINT FRANCIS	N 3627	W 09008	300
3-6506	13	SEARCY	N 3515	W 09145	245
3-6586	12	SHIRLEY	N 3539	W 09219	550
3-6624	13	SILOAM SPRINGS	N 3611	W 09433	1150
3-6768	12	SPARKMAN 3 WSW	N 3354	W 09254	152
3-6804	12	STAMPS	N 3322	W 09329	270
3-6918	13	STUTTGART	N 3429	W 09132	214
3-6920	13	STUTTGART 9 ESE	N 3428	W 09125	198
3-6928	13	SUBIACO	N 3518	W 09339	500
3-7038	12	TAYLOR	N 3306	W 09327	250
3-7048	13	TEXARKANA FAA AIRPORT	N 3327	W 09400	361
3-7488	13	WALDRON	N 3454	W 09406	675
3-7582	13	WARREN	N 3336	W 09204	206
3-8052	13	WYNNE	N 3514	W 09047	260

03 — ARKANSAS

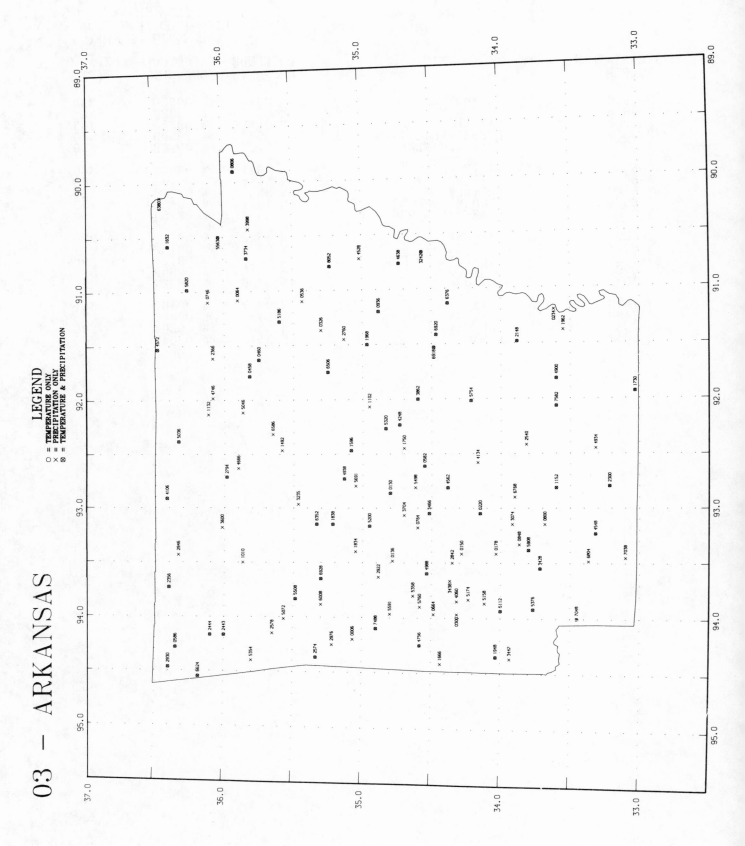

LEGEND

O = TEMPERATURE ONLY
X = PRECIPITATION ONLY
⊗ = TEMPERATURE & PRECIPITATION

CALIFORNIA

TEMPERATURE NORMALS (DEG F)

STATION		JAN	FEB	MAR	APR	MAY	JUN	JUL	AUG	SEP	OCT	NOV	DEC	ANN
ALDERPOINT	MAX	53.5	58.7	61.9	67.0	73.4	81.6	90.6	89.9	86.7	74.6	61.0	53.9	71.1
	MIN	35.7	38.1	38.7	40.7	45.1	49.6	52.1	52.1	48.5	44.1	40.2	36.5	43.5
	MEAN	44.6	48.4	50.3	53.8	59.3	65.7	71.4	71.0	67.6	59.4	50.6	45.2	57.3
ALTURAS RANGER STATION	MAX	41.7	46.9	51.5	59.2	68.3	77.6	88.2	86.5	80.0	67.8	52.9	44.6	63.8
	MIN	16.0	20.9	23.5	27.6	34.4	40.6	43.5	41.4	35.4	28.4	22.9	17.7	29.4
	MEAN	28.9	34.0	37.5	43.4	51.4	59.1	65.9	64.0	57.7	48.1	37.9	31.2	46.6
ASH MOUNTAIN	MAX	57.2	61.1	53.7	69.5	78.6	89.1	97.6	96.0	90.8	80.2	66.8	58.4	75.8
	MIN	36.0	39.0	41.0	45.5	51.9	60.4	68.1	66.7	60.9	52.3	42.8	36.8	50.1
	MEAN	46.6	50.1	52.4	57.5	65.3	74.8	82.9	81.4	75.9	66.3	54.8	47.6	63.0
AUBERRY 1 NW	MAX	54.9	58.3	60.7	67.6	77.4	87.2	95.1	93.2	87.5	76.7	63.5	56.2	73.2
	MIN	33.6	36.0	37.8	42.0	48.8	55.9	63.9	61.9	57.6	48.6	39.3	34.3	46.6
	MEAN	44.3	47.2	49.3	54.8	63.1	71.6	79.5	77.5	72.6	62.7	51.4	45.3	59.9
AUBURN	MAX	53.7	58.6	61.8	67.8	76.2	85.4	93.4	92.0	87.1	76.9	63.5	55.2	72.6
	MIN	35.9	38.6	39.9	43.7	48.9	55.6	61.1	60.1	56.4	49.9	41.9	36.5	47.4
	MEAN	44.8	48.6	50.9	55.8	62.6	70.5	77.2	76.1	71.8	63.4	52.7	45.9	60.0
AVALON PLEASURE PIER	MAX	62.2	62.9	63.4	65.5	67.1	69.5	72.4	74.1	73.4	71.1	67.2	63.4	67.7
	MIN	46.5	47.3	48.1	50.7	53.7	56.7	59.9	61.6	60.0	56.2	50.9	47.2	53.2
	MEAN	54.3	55.1	55.8	58.1	60.4	63.1	66.2	67.9	66.7	63.7	59.1	55.3	60.5
BAKERSFIELD WSO R	MAX	57.4	63.7	68.6	75.1	83.9	92.2	98.8	96.4	90.8	81.0	67.4	57.6	77.7
	MIN	38.9	42.6	45.5	50.1	57.2	64.3	70.1	68.5	63.8	54.9	44.9	38.7	53.3
	MEAN	48.2	53.2	57.1	62.7	70.6	78.3	84.5	82.4	77.3	68.0	56.2	48.2	65.6
BARRETT DAM	MAX	63.7	65.8	66.5	70.8	76.1	85.0	93.7	93.4	89.8	81.4	71.8	66.0	77.0
	MIN	33.1	35.0	38.1	42.3	47.0	51.3	57.9	58.0	53.6	46.1	38.3	33.2	44.5
	MEAN	48.4	50.4	52.3	56.6	61.5	68.2	75.8	75.7	71.8	63.8	55.1	49.6	60.8
BARSTOW	MAX	60.0	65.6	70.0	77.0	85.9	95.7	102.6	100.3	94.2	83.0	68.8	60.4	80.3
	MIN	31.7	35.5	39.6	45.0	52.6	60.4	67.2	65.3	58.5	48.1	37.5	31.0	47.7
	MEAN	45.9	50.5	54.8	61.0	69.2	78.1	84.9	82.8	76.4	65.5	53.1	45.7	64.0
BEAUMONT 1 E	MAX	59.5	62.9	64.9	70.6	77.3	86.8	95.7	94.6	89.9	80.2	68.4	61.4	76.0
	MIN	38.1	38.4	39.1	41.5	46.7	51.9	58.5	58.5	55.2	49.3	42.8	39.3	46.6
	MEAN	48.9	50.7	52.0	56.1	62.0	69.4	77.1	76.6	72.6	64.8	55.6	50.4	61.4
BERKELEY	MAX	56.1	59.5	61.1	63.3	66.4	69.2	69.5	69.6	71.7	69.6	62.9	57.0	64.7
	MIN	43.2	45.8	46.0	47.6	50.3	53.0	53.9	54.7	55.6	52.9	48.3	43.9	49.6
	MEAN	49.7	52.7	53.6	55.5	58.4	61.1	61.7	62.2	63.7	61.3	55.6	50.4	57.2
BISHOP WSO R	MAX	52.9	58.2	63.4	70.9	80.2	90.4	97.5	95.2	88.2	77.1	63.5	55.1	74.4
	MIN	21.4	25.9	29.5	35.8	43.5	50.8	56.3	53.8	46.8	37.4	27.7	22.1	37.6
	MEAN	37.2	42.1	46.5	53.4	61.9	70.7	77.0	74.5	67.5	57.3	45.6	38.7	56.0
BLUE CANYON WSO //R	MAX	43.5	44.9	45.3	51.3	60.3	69.2	77.7	76.5	72.4	62.8	51.4	46.3	58.5
	MIN	30.7	31.3	31.0	35.2	42.8	51.0	58.9	57.3	53.3	45.5	37.1	32.7	42.2
	MEAN	37.1	38.1	38.2	43.3	51.5	60.1	68.3	66.9	62.8	54.2	44.3	39.5	50.4
BLYTHE	MAX	67.6	73.4	78.7	86.5	94.8	104.0	108.7	106.5	102.2	90.7	76.5	67.8	88.1
	MIN	38.2	41.9	46.3	52.1	59.8	67.7	76.8	75.4	67.8	56.1	44.4	38.0	55.4
	MEAN	52.9	57.7	62.6	69.3	77.3	85.9	92.8	91.0	85.0	73.4	60.5	52.9	71.8
BLYTHE FAA AIRPORT	MAX	65.9	71.7	77.2	85.3	94.1	104.2	108.6	106.1	101.2	89.7	75.3	66.5	87.2
	MIN	40.9	45.2	49.5	56.0	64.0	72.8	81.7	80.0	73.3	61.2	48.5	41.2	59.5
	MEAN	53.4	58.5	63.4	70.7	79.1	88.5	95.2	93.1	87.3	75.5	61.9	53.9	73.4
BOCA	MAX	40.2	44.5	47.8	55.4	64.5	73.7	83.1	81.7	76.2	65.9	51.9	43.1	60.7
	MIN	9.1	11.7	16.3	22.9	29.7	34.4	37.7	35.4	30.0	24.1	19.7	12.4	23.6
	MEAN	24.7	28.1	32.0	39.2	47.1	54.1	60.4	58.6	53.1	45.0	35.8	27.7	42.2
BOWMAN DAM	MAX	45.1	47.4	48.8	54.6	62.8	71.7	79.6	78.3	74.1	64.6	52.2	46.9	60.5
	MIN	26.7	27.3	27.5	31.5	38.3	46.7	53.7	52.8	49.2	41.8	33.5	29.0	38.2
	MEAN	36.0	37.4	38.2	43.1	50.6	59.3	66.7	65.6	61.6	53.2	42.9	38.0	49.4

CALIFORNIA

TEMPERATURE NORMALS (DEG F)

STATION		JAN	FEB	MAR	APR	MAY	JUN	JUL	AUG	SEP	OCT	NOV	DEC	ANN
BRAWLEY 2 SW	MAX	68.8	73.7	77.9	84.7	92.8	102.5	107.0	105.5	101.9	91.7	78.6	70.1	87.9
	MIN	38.7	42.5	46.5	52.0	58.8	66.3	75.3	75.1	68.7	58.0	46.0	38.9	55.6
	MEAN	53.8	58.1	62.2	68.3	75.8	84.4	91.2	90.3	85.4	74.9	62.3	54.5	71.8
BROOKS FARNHAM RANCH	MAX	55.9	61.1	65.1	71.9	81.1	90.0	96.6	95.0	90.3	80.4	66.5	57.4	75.9
	MIN	34.0	36.9	38.5	41.3	47.5	54.0	57.6	55.7	52.0	45.7	38.2	33.9	44.6
	MEAN	45.0	49.1	51.8	56.6	64.3	72.0	77.1	75.4	71.2	63.0	52.4	45.7	60.3
BURBANK VALLEY PMP PLT	MAX	66.3	69.0	70.0	73.0	76.3	81.8	89.4	89.0	87.3	81.1	73.2	67.7	77.0
	MIN	41.3	43.5	45.0	48.0	52.8	56.8	60.9	61.3	58.9	53.0	45.9	41.6	50.8
	MEAN	53.8	56.3	57.5	60.6	64.5	69.3	75.1	75.2	73.1	67.1	59.6	54.7	63.9
BURNEY	MAX	44.1	49.8	54.0	60.7	69.6	78.3	86.9	85.6	80.6	69.6	54.0	45.4	64.9
	MIN	19.1	23.3	25.5	29.2	35.5	41.1	43.8	41.6	35.6	29.1	24.7	20.3	30.7
	MEAN	31.6	36.6	39.8	45.0	52.6	59.7	65.4	63.6	58.2	49.3	39.4	32.9	47.8
BUTTONWILLOW	MAX	55.9	62.7	68.3	75.3	83.9	92.3	98.6	96.7	91.5	81.6	67.4	56.6	77.6
	MIN	33.9	38.0	41.7	46.4	53.3	59.8	65.1	63.0	57.4	48.2	38.5	33.2	48.2
	MEAN	44.9	50.4	55.0	60.9	68.6	76.1	81.9	79.9	74.5	64.9	53.0	45.0	62.9
CALAVERAS BIG TREES	MAX	46.2	48.7	50.3	55.5	64.6	74.1	83.0	81.9	76.6	66.9	54.5	48.2	62.5
	MIN	26.5	27.8	28.3	32.0	38.0	45.1	50.7	49.2	45.6	38.7	31.6	27.7	36.8
	MEAN	36.4	38.2	39.4	43.8	51.3	59.6	66.9	65.6	61.1	52.8	43.1	38.0	49.7
CAMPO	MAX	60.9	63.3	64.7	70.0	76.3	85.8	93.9	92.8	88.8	79.6	68.7	62.6	75.6
	MIN	32.9	32.8	33.9	35.8	39.9	44.1	52.5	53.0	48.8	42.0	36.4	33.1	40.4
	MEAN	46.9	48.1	49.3	52.9	58.1	64.9	73.2	73.0	68.8	60.8	52.6	47.9	58.0
CAMP PARDEE	MAX	52.7	58.4	62.6	69.4	78.9	88.2	95.9	93.9	88.5	77.9	63.5	53.8	73.6
	MIN	37.7	40.6	42.0	45.3	50.0	55.6	61.1	60.0	57.4	51.7	44.2	38.7	48.7
	MEAN	45.3	49.5	52.3	57.4	64.5	72.0	78.5	77.0	73.0	64.8	53.9	46.3	61.2
CANOGA PARK PIERCE COL	MAX	67.0	69.9	71.5	75.7	79.7	86.4	94.9	94.0	91.3	83.6	74.4	68.6	79.8
	MIN	39.1	40.4	41.3	44.1	48.6	52.6	57.0	57.5	54.4	48.7	42.6	39.1	47.1
	MEAN	53.1	55.2	56.4	59.9	64.2	69.5	76.0	75.8	72.9	66.2	58.5	53.9	63.5
CANYON DAM	MAX	38.6	43.2	47.8	56.9	66.8	76.1	85.4	83.9	77.2	64.3	48.5	39.9	60.7
	MIN	21.1	23.0	25.0	29.2	35.6	42.0	46.5	44.9	40.3	33.7	27.8	23.0	32.7
	MEAN	29.9	33.1	36.5	43.1	51.2	59.1	66.0	64.4	58.8	49.0	38.2	31.5	46.7
CEDARVILLE	MAX	39.6	45.1	49.8	58.0	67.2	76.6	87.6	85.8	78.5	65.3	50.5	41.8	62.2
	MIN	20.4	24.9	28.0	32.9	40.1	47.4	55.2	52.6	44.9	35.9	27.9	22.2	36.0
	MEAN	30.0	35.1	38.9	45.5	53.7	62.0	71.4	69.2	61.7	50.6	39.2	32.0	49.1
CHICO EXPERIMENT STA	MAX	53.6	59.9	64.6	71.7	80.8	89.2	95.4	93.6	89.3	78.6	63.5	54.4	74.6
	MIN	36.0	39.1	40.5	43.9	50.7	56.8	60.8	59.0	54.8	47.8	40.8	36.5	47.2
	MEAN	44.8	49.5	52.6	57.8	65.8	73.0	78.1	76.4	72.1	63.2	52.2	45.5	60.9
CHULA VISTA	MAX	62.8	63.6	62.9	64.2	65.6	67.5	71.1	73.1	73.5	71.0	67.9	64.2	67.3
	MIN	44.0	45.6	47.8	51.1	55.3	58.7	62.7	64.3	61.9	56.0	49.3	44.4	53.4
	MEAN	53.4	54.6	55.4	57.7	60.4	63.1	66.9	68.7	67.7	63.5	58.6	54.3	60.4
CLAREMONT POMONA COL	MAX	63.6	66.2	67.2	71.1	74.8	81.5	89.7	89.5	86.6	79.5	71.2	65.0	75.5
	MIN	40.7	42.0	43.2	45.9	50.0	54.2	59.1	59.7	57.8	52.3	45.7	41.2	49.3
	MEAN	52.2	54.1	55.2	58.6	62.5	67.9	74.4	74.6	72.3	65.9	58.5	53.1	62.4
CLOVERDALE 3 SSE	MAX	56.3	60.9	64.0	69.5	77.1	84.6	91.0	89.9	86.7	77.8	65.5	57.4	73.4
	MIN	37.3	40.2	41.0	43.7	47.9	52.4	53.2	52.8	51.8	48.4	42.7	38.0	45.8
	MEAN	46.8	50.6	52.5	56.6	62.6	68.5	72.1	71.4	69.3	63.1	54.1	47.7	59.6
COALINGA	MAX	57.0	62.8	67.8	74.6	83.9	92.0	98.8	96.8	91.4	81.1	67.0	57.8	77.6
	MIN	35.3	38.7	40.3	44.6	51.3	58.1	64.0	62.3	57.1	48.7	39.7	35.0	47.9
	MEAN	46.2	50.8	54.1	59.7	67.6	75.1	81.5	79.6	74.3	64.9	53.4	46.4	62.8
COLFAX	MAX	53.8	57.1	59.5	65.9	74.6	83.8	91.9	90.4	85.4	75.0	62.1	55.5	71.3
	MIN	35.1	37.2	38.3	42.1	48.4	56.3	63.2	61.0	57.1	48.8	40.1	35.4	46.9
	MEAN	44.5	47.2	48.9	54.0	61.5	70.1	77.6	75.7	71.3	62.0	51.1	45.4	59.1

CALIFORNIA

TEMPERATURE NORMALS (DEG F)

STATION		JAN	FEB	MAR	APR	MAY	JUN	JUL	AUG	SEP	OCT	NOV	DEC	ANN
COLUSA 1 SSW	MAX	53.5	60.5	65.7	73.1	82.1	90.2	96.2	94.5	89.8	78.5	63.7	54.3	75.2
	MIN	36.3	39.8	40.9	44.2	51.1	56.4	58.7	57.2	53.9	47.4	40.2	36.2	46.9
	MEAN	45.0	50.2	53.3	58.7	66.6	73.3	77.5	75.9	71.8	63.0	52.0	45.3	61.1
CORCORAN IRRIG DIST	MAX	54.5	61.3	67.5	75.3	84.8	93.1	99.7	97.5	91.4	81.1	66.1	55.0	77.3
	MIN	35.6	38.8	41.1	45.3	51.6	57.9	62.7	61.3	56.7	48.2	39.4	34.9	47.8
	MEAN	45.1	50.1	54.3	60.3	68.2	75.5	81.2	79.4	74.1	64.7	52.8	45.0	62.6
CORONA	MAX	66.2	68.8	70.2	74.2	78.6	84.6	91.9	91.4	89.2	82.2	73.6	67.7	78.2
	MIN	40.2	41.4	42.6	45.6	50.3	54.4	58.5	59.0	56.1	50.3	43.8	40.0	48.5
	MEAN	53.2	55.1	56.4	59.9	64.5	69.6	75.2	75.2	72.7	66.3	58.7	53.9	63.4
COVELO	MAX	51.5	57.6	61.3	67.9	75.9	84.7	93.4	92.4	88.1	75.5	60.5	52.1	71.7
	MIN	30.1	32.3	33.5	36.2	40.9	46.6	50.4	49.2	44.3	38.2	33.5	30.5	38.8
	MEAN	40.8	45.0	47.4	52.1	58.4	65.6	72.0	70.8	66.2	56.8	47.1	41.3	55.3
CULVER CITY	MAX	67.5	68.6	68.1	69.3	70.7	73.7	77.7	78.7	78.7	76.6	72.8	69.1	72.6
	MIN	45.5	46.9	48.2	50.6	53.9	57.2	60.7	62.0	60.4	56.5	50.5	46.2	53.2
	MEAN	56.5	57.8	58.1	60.0	62.3	65.5	69.2	70.4	69.6	66.5	61.7	57.7	62.9
CUYAMACA	MAX	50.4	53.0	54.3	60.2	66.9	76.9	85.3	84.5	80.6	70.9	59.3	52.7	66.3
	MIN	28.8	30.2	32.0	35.3	40.7	48.3	55.6	54.0	48.5	39.8	33.4	29.3	39.7
	MEAN	39.6	41.7	43.2	47.8	53.9	62.6	70.5	69.3	64.6	55.4	46.4	41.0	53.0
DAGGETT FAA AP	MAX	60.7	66.0	70.5	78.0	87.1	97.4	104.0	101.4	94.7	83.2	69.4	61.4	81.2
	MIN	35.9	40.3	44.5	50.2	58.2	66.2	73.2	71.6	65.0	54.6	43.1	35.9	53.2
	MEAN	48.3	53.2	57.5	64.1	72.7	81.8	88.6	86.5	79.9	68.9	56.3	48.7	67.2
DAVIS 2 WSW EXP FARM	MAX	53.0	59.6	64.5	71.1	79.6	87.7	93.2	91.6	87.9	78.4	64.1	54.1	73.7
	MIN	37.2	40.3	41.4	44.3	49.2	54.0	55.3	54.5	53.0	48.1	41.5	37.3	46.3
	MEAN	45.1	50.0	53.0	57.7	64.4	70.9	74.3	73.1	70.4	63.2	52.8	45.7	60.1
DEATH VALLEY	MAX	65.3	73.3	80.6	89.5	98.7	108.7	115.4	113.0	106.0	92.9	75.9	65.4	90.4
	MIN	39.6	46.3	54.0	62.1	71.6	81.2	88.4	86.2	76.2	62.9	48.6	39.3	63.0
	MEAN	52.5	59.8	67.3	75.8	85.2	95.0	101.9	99.7	91.1	77.9	62.3	52.4	76.7
DENAIR 3 NNE	MAX	53.7	60.5	66.1	72.6	81.1	88.5	93.9	91.6	87.6	78.3	64.4	54.1	74.4
	MIN	35.5	38.3	39.3	43.0	48.5	53.6	57.1	55.9	52.8	45.9	38.4	35.0	45.3
	MEAN	44.6	49.4	52.7	57.8	64.8	71.1	75.5	73.8	70.2	62.1	51.5	44.5	59.8
DE SABLA	MAX	50.4	54.3	56.8	63.2	71.7	81.1	89.0	87.9	83.2	72.5	58.6	52.3	68.4
	MIN	31.7	33.4	34.2	37.9	43.8	50.5	55.4	53.9	50.7	44.4	36.9	32.6	42.1
	MEAN	41.1	43.8	45.5	50.6	57.8	65.9	72.3	70.9	67.0	58.5	47.8	42.5	55.3
EAGLE MOUNTAIN	MAX	64.1	69.5	73.9	81.5	89.8	99.7	105.2	103.1	98.2	87.5	73.8	65.8	84.3
	MIN	44.1	48.2	52.0	59.2	67.1	76.5	82.8	80.9	75.1	64.6	53.0	45.6	62.4
	MEAN	54.1	58.9	63.0	70.4	78.5	88.1	94.0	92.0	86.7	76.1	63.5	55.7	73.4
EAST PARK RESERVOIR	MAX	54.4	58.5	61.2	67.6	77.1	86.4	93.9	91.9	87.1	76.9	63.5	56.0	72.9
	MIN	31.7	35.1	37.1	40.5	47.2	54.5	59.3	57.2	52.2	44.8	36.9	32.2	44.1
	MEAN	43.1	46.8	49.2	54.0	62.2	70.5	76.6	74.6	69.7	60.9	50.2	44.1	58.5
EL CAPITAN DAM	MAX	68.1	70.1	70.1	73.9	77.9	85.1	92.8	92.9	90.8	84.2	76.1	70.3	79.4
	MIN	41.6	42.9	44.5	47.5	51.1	54.7	58.4	59.8	58.3	53.0	46.7	42.2	50.1
	MEAN	54.8	56.5	57.3	60.7	64.5	69.9	75.7	76.4	74.6	68.6	61.4	56.2	64.7
EL CENTRO 2 SSW	MAX	69.4	74.3	79.0	85.8	94.6	104.0	108.4	106.7	102.7	92.1	78.4	69.9	88.8
	MIN	38.3	42.0	45.5	50.6	57.6	65.8	74.4	74.3	67.7	57.0	45.3	38.2	54.7
	MEAN	53.9	58.2	62.3	68.3	76.1	84.9	91.4	90.5	85.2	74.6	61.9	54.1	71.8
ELECTRA POWER HOUSE	MAX	56.9	61.9	65.1	71.4	80.5	89.9	97.5	95.7	90.7	80.1	66.0	57.2	76.1
	MIN	34.8	37.3	39.1	42.1	47.3	52.4	56.8	55.7	52.5	46.4	39.9	35.2	45.0
	MEAN	45.9	49.6	52.1	56.8	63.9	71.2	77.2	75.7	71.6	63.3	53.0	46.2	60.5
ELSINORE	MAX	66.0	68.3	70.2	75.3	81.4	90.0	98.7	97.8	94.1	84.2	73.9	67.7	80.6
	MIN	35.8	38.0	40.1	43.3	48.4	53.9	59.1	59.4	56.0	47.9	40.4	35.6	46.5
	MEAN	50.9	53.2	55.2	59.3	64.9	72.0	78.9	78.6	75.0	66.1	57.2	51.7	63.6

CALIFORNIA

TEMPERATURE NORMALS (DEG F)

STATION		JAN	FEB	MAR	APR	MAY	JUN	JUL	AUG	SEP	OCT	NOV	DEC	ANN
ESCONDIDO	MAX	65.7	67.4	68.3	71.7	75.3	81.1	87.8	87.7	85.2	79.0	72.3	67.1	75.7
	MIN	38.7	40.3	42.6	46.3	51.0	55.2	58.8	59.7	57.1	50.7	43.1	38.8	48.5
	MEAN	52.2	53.9	55.5	59.0	63.2	68.2	73.3	73.7	71.2	64.9	57.7	53.0	62.2
EUREKA WSO R	MAX	53.4	54.6	54.0	54.7	57.0	59.1	60.3	61.3	62.2	60.3	57.5	54.5	57.4
	MIN	41.3	42.6	42.5	44.0	47.3	50.2	51.9	52.6	51.5	48.3	45.2	42.2	46.6
	MEAN	47.3	48.7	48.3	49.4	52.2	54.7	56.1	57.0	56.8	54.3	51.4	48.3	52.0
FAIRFIELD FIRE STATION	MAX	55.5	61.8	65.6	70.4	77.0	83.4	88.0	87.7	86.2	78.3	65.6	56.2	73.0
	MIN	37.0	40.3	42.4	45.1	49.4	53.1	55.3	55.5	54.2	48.6	41.7	36.9	46.6
	MEAN	46.3	51.1	54.0	57.8	63.2	68.3	71.7	71.6	70.2	63.5	53.7	46.6	59.8
FAIRMONT	MAX	52.8	56.3	58.7	64.3	72.3	81.6	90.3	89.5	85.1	74.9	61.7	54.4	70.2
	MIN	36.0	38.2	39.6	44.1	50.4	58.9	67.4	66.0	60.9	52.7	43.0	36.9	49.5
	MEAN	44.4	47.3	49.1	54.3	61.4	70.3	78.9	77.8	73.1	63.8	52.3	45.7	59.9
FONTANA KAISER	MAX	66.6	69.4	70.4	74.5	79.4	86.7	94.9	94.0	91.1	83.3	74.0	68.4	79.4
	MIN	44.0	45.1	46.3	48.8	52.4	56.7	62.5	63.0	61.1	55.8	48.9	44.7	52.4
	MEAN	55.3	57.3	58.4	61.6	65.9	71.8	78.7	78.5	76.1	69.6	61.5	56.6	65.9
FORT BIDWELL	MAX	39.7	45.9	51.8	60.3	69.2	77.3	86.2	84.8	78.5	67.4	51.7	42.3	62.9
	MIN	19.9	24.5	26.8	30.9	37.5	43.4	48.1	46.0	40.1	33.0	26.4	21.3	33.2
	MEAN	29.9	35.3	39.3	45.6	53.4	60.4	67.2	65.4	59.4	50.2	39.1	31.8	48.1
FORT BRAGG	MAX	55.4	56.8	57.3	58.6	60.9	63.0	63.7	64.2	65.4	63.3	60.0	56.1	60.4
	MIN	39.9	41.1	41.4	42.9	45.8	48.4	49.3	49.9	49.6	47.0	43.9	40.8	45.0
	MEAN	47.6	48.9	49.4	50.8	53.4	55.8	56.6	57.1	57.6	55.2	52.0	48.5	52.7
FORT JONES RANGER STA	MAX	43.6	51.7	57.0	64.1	73.3	82.3	91.5	89.6	83.8	71.0	54.0	43.5	67.1
	MIN	23.7	26.7	28.7	31.5	37.2	43.5	47.9	46.2	39.1	32.5	28.2	25.8	34.3
	MEAN	33.7	39.2	42.9	47.8	55.3	63.0	69.7	67.9	61.5	51.8	41.1	34.7	50.7
FORT ROSS	MAX	56.5	58.2	58.9	60.0	62.2	65.0	65.8	66.5	68.0	65.9	61.1	57.2	62.1
	MIN	41.8	42.9	42.1	42.7	44.7	47.1	48.2	48.9	49.5	47.5	45.1	42.2	45.2
	MEAN	49.2	50.6	50.5	51.3	53.5	56.1	57.0	57.7	58.8	56.7	53.1	49.8	53.7
FRESNO WSO R	MAX	54.2	61.2	66.5	73.7	82.7	91.1	97.9	95.5	90.3	79.9	65.2	54.4	76.1
	MIN	36.8	39.7	42.0	46.5	52.7	58.9	64.1	62.2	57.8	49.7	41.1	36.3	49.0
	MEAN	45.5	50.5	54.3	60.1	67.7	75.0	81.0	78.9	74.1	64.8	53.2	45.3	62.5
FRIANT GOVERNMENT CAMP	MAX	54.8	60.8	65.5	73.2	83.5	92.8	100.6	98.6	92.4	81.5	66.5	55.9	77.2
	MIN	35.7	38.7	39.5	42.7	48.3	54.6	60.4	58.7	55.0	48.5	40.9	35.8	46.6
	MEAN	45.3	49.8	52.6	58.0	65.9	73.7	80.5	78.7	73.7	65.1	53.7	45.8	61.9
GLENNVILLE	MAX	55.8	58.0	58.9	64.6	73.6	83.3	91.7	89.7	84.8	74.6	63.4	57.1	71.3
	MIN	27.5	30.1	31.3	34.3	38.7	43.9	49.8	48.6	44.3	37.5	31.3	27.6	37.1
	MEAN	41.7	44.1	45.1	49.5	56.2	63.6	70.8	69.2	64.6	56.1	47.4	42.4	54.2
GRANT GROVE	MAX	43.0	44.2	44.4	48.9	56.5	66.5	75.1	73.7	68.9	59.9	50.2	45.3	56.4
	MIN	24.7	25.0	25.2	28.9	35.8	43.9	50.7	49.1	45.2	38.7	31.0	26.8	35.4
	MEAN	33.9	34.7	34.9	38.9	46.2	55.2	62.9	61.4	57.1	49.3	40.6	36.1	45.9
GRATON	MAX	55.3	60.5	63.6	68.4	74.4	80.2	83.8	83.2	81.5	74.8	63.9	55.9	70.5
	MIN	35.7	37.9	38.1	39.7	43.4	46.7	47.8	47.9	47.1	43.8	39.3	36.0	42.0
	MEAN	45.5	49.2	50.9	54.1	58.9	63.5	65.8	65.6	64.3	59.3	51.7	46.0	56.2
HAIWEE	MAX	50.7	56.4	61.4	68.6	77.7	88.0	94.9	92.8	86.4	75.5	61.2	52.6	72.2
	MIN	28.9	32.2	35.8	42.5	50.2	59.0	65.5	63.5	57.2	47.4	36.3	29.9	45.7
	MEAN	39.8	44.3	48.6	55.6	64.0	73.5	80.2	78.2	71.8	61.5	48.8	41.2	59.0
HALF MOON BAY	MAX	57.9	59.2	59.3	60.0	61.4	62.7	63.5	64.7	66.9	66.1	62.6	58.8	61.9
	MIN	42.9	43.8	43.5	44.4	47.0	49.7	51.2	52.3	51.7	48.7	45.8	43.5	47.0
	MEAN	50.4	51.5	51.4	52.2	54.2	56.2	57.4	58.5	59.3	57.4	54.2	51.2	54.5
HANFORD	MAX	54.9	61.9	67.5	74.8	83.7	91.1	97.0	95.0	90.3	80.8	66.2	55.2	76.5
	MIN	34.9	38.0	40.7	45.0	51.0	56.9	61.2	59.2	54.6	46.5	38.2	34.1	46.7
	MEAN	44.9	50.0	54.1	59.9	67.4	74.0	79.1	77.2	72.5	63.7	52.2	44.7	61.6

CALIFORNIA

TEMPERATURE NORMALS (DEG F)

STATION		JAN	FEB	MAR	APR	MAY	JUN	JUL	AUG	SEP	OCT	NOV	DEC	ANN
HAPPY CAMP RANGER STA	MAX	46.4	53.9	60.2	68.0	77.7	86.4	95.8	93.8	87.6	72.4	55.9	47.1	70.4
	MIN	30.9	33.2	33.9	36.8	42.4	48.3	52.3	51.4	45.9	40.8	36.5	32.6	40.4
	MEAN	38.7	43.5	47.1	52.4	60.0	67.4	74.0	72.6	66.7	56.6	46.3	39.9	55.4
HAT CREEK PH NO 1	MAX	45.9	51.7	56.3	63.6	73.0	81.8	91.1	89.2	83.3	71.4	55.9	46.8	67.5
	MIN	21.5	24.9	27.3	31.4	37.5	43.1	45.9	43.7	37.9	31.7	26.9	22.3	32.8
	MEAN	33.7	38.3	41.8	47.5	55.3	62.5	68.6	66.5	60.6	51.6	41.4	34.6	50.2
HAYFIELD PUMPING PLANT	MAX	65.0	69.5	73.5	80.8	88.7	98.7	104.3	102.5	98.1	87.7	74.4	66.7	84.2
	MIN	39.0	42.4	45.5	51.6	58.8	66.7	75.8	74.1	67.0	56.5	46.0	39.6	55.3
	MEAN	52.0	56.0	59.5	66.3	73.8	82.7	90.1	88.3	82.6	72.1	60.3	53.2	69.7
HEALDSBURG	MAX	57.1	62.6	66.3	71.4	78.2	84.6	89.3	88.5	86.6	78.5	66.2	57.7	73.9
	MIN	38.2	41.1	41.8	44.1	48.3	52.3	52.5	52.8	52.0	48.2	42.8	38.7	46.1
	MEAN	47.7	51.9	54.1	57.8	63.2	68.5	71.0	70.6	69.3	63.4	54.5	48.3	60.0
HENSHAW DAM	MAX	59.3	61.9	62.9	67.4	73.1	83.1	92.6	92.2	88.1	79.0	67.8	61.4	74.1
	MIN	29.2	30.9	33.5	37.1	42.0	46.2	53.5	53.3	48.3	40.0	33.2	29.1	39.7
	MEAN	44.3	46.5	48.3	52.3	57.6	64.7	73.1	72.8	68.2	59.5	50.5	45.3	56.9
HETCH HETCHY	MAX	47.9	53.3	55.9	61.6	69.5	77.8	86.1	85.4	80.9	71.9	58.1	49.2	66.5
	MIN	28.1	29.8	31.3	36.0	42.4	48.9	55.1	54.0	49.4	41.7	33.7	29.2	40.0
	MEAN	38.0	41.6	43.6	48.9	56.0	63.4	70.6	69.7	65.2	56.9	45.9	39.2	53.3
HUNTINGTON LAKE	MAX	42.9	44.1	44.1	48.4	55.8	65.3	73.5	72.5	67.7	58.9	49.4	44.2	55.6
	MIN	22.0	22.1	22.1	26.2	32.9	40.5	47.5	46.8	42.6	35.9	28.8	24.2	32.6
	MEAN	32.5	33.1	33.2	37.4	44.4	53.0	60.6	59.7	55.2	47.4	39.1	34.3	44.2
IMPERIAL	MAX	69.3	73.9	78.0	84.8	92.8	102.3	106.6	104.9	100.7	90.5	77.9	70.1	87.7
	MIN	41.6	45.6	49.6	55.2	61.8	69.4	77.6	77.3	71.6	61.1	49.4	42.2	58.5
	MEAN	55.5	59.8	63.8	70.0	77.3	85.9	92.1	91.1	86.2	75.8	63.6	56.2	73.1
INDIO U S DATE GARDEN	MAX	71.0	75.8	79.3	85.9	92.9	102.0	107.0	105.4	102.1	92.9	80.4	72.4	88.9
	MIN	38.3	43.4	48.9	56.3	63.7	71.4	77.6	76.4	70.2	59.2	46.2	38.6	57.5
	MEAN	54.7	59.6	64.1	71.1	78.3	86.7	92.3	90.9	86.2	76.1	63.3	55.6	73.2
INYOKERN	MAX	58.8	64.7	69.3	76.6	86.1	96.4	103.1	100.7	94.1	83.1	68.5	59.5	80.1
	MIN	30.0	34.1	37.6	43.2	51.7	60.1	66.0	63.8	57.4	47.6	36.6	29.7	46.5
	MEAN	44.4	49.4	53.5	59.9	68.9	78.2	84.6	82.3	75.8	65.4	52.6	44.6	63.3
IRON MOUNTAIN	MAX	63.9	69.9	75.1	83.2	92.2	102.5	108.3	106.0	100.6	89.0	74.2	65.3	85.9
	MIN	42.3	46.1	50.3	57.5	65.8	75.1	81.8	79.6	72.9	61.8	50.5	43.3	60.6
	MEAN	53.1	58.1	62.7	70.4	79.0	88.8	95.1	92.8	86.8	75.4	62.4	54.3	73.2
JESS VALLEY	MAX	41.4	44.8	47.9	55.5	64.6	73.1	83.1	81.3	75.3	64.6	50.7	43.7	60.5
	MIN	19.3	22.2	23.4	27.7	34.4	40.6	46.2	44.7	39.9	33.0	26.1	21.4	31.6
	MEAN	30.4	33.5	35.7	41.6	49.5	56.9	64.7	63.0	57.6	48.9	38.5	32.6	46.1
JULIAN WYNOLA	MAX	55.1	57.5	58.6	63.7	70.1	80.6	90.0	88.9	84.5	74.6	63.4	57.2	70.4
	MIN	34.0	34.2	34.4	36.4	39.9	45.0	53.1	53.2	48.7	43.3	38.1	35.3	41.3
	MEAN	44.6	45.9	46.5	50.0	55.0	62.9	71.6	71.1	66.6	59.0	50.8	46.3	55.9
KENTFIELD	MAX	55.0	60.3	64.1	68.6	74.2	79.9	83.5	82.5	81.4	74.3	62.9	55.9	70.2
	MIN	38.4	41.1	41.4	43.2	46.3	49.8	50.5	50.7	50.2	47.5	43.1	39.1	45.1
	MEAN	46.7	50.7	52.8	55.9	60.3	64.9	67.0	66.6	65.9	60.9	53.1	47.5	57.7
KERN RIVER PH NO 1	MAX	57.7	63.4	68.0	75.2	84.5	93.2	100.3	97.7	92.0	81.3	67.5	57.8	78.2
	MIN	42.0	46.0	48.1	52.6	59.7	67.3	74.5	73.5	69.1	60.7	50.3	42.7	57.2
	MEAN	49.8	54.7	58.1	63.9	72.1	80.3	87.4	85.7	80.6	71.0	58.9	50.3	67.7
KERN RIVER PH NO 3	MAX	58.8	62.4	65.5	71.7	80.2	89.7	97.4	96.0	91.2	80.6	67.6	60.0	76.8
	MIN	32.4	35.2	37.7	42.8	50.4	58.0	65.2	63.4	58.2	48.0	38.2	32.7	46.9
	MEAN	45.6	48.8	51.6	57.3	65.3	73.9	81.3	79.8	74.7	64.3	53.0	46.4	61.8
KETTLEMAN STATION	MAX	55.5	62.1	67.5	74.9	84.1	92.8	99.3	97.3	91.4	80.8	66.5	56.1	77.4
	MIN	39.0	43.5	45.9	49.9	56.2	63.0	69.4	68.2	64.1	56.7	46.7	39.4	53.5
	MEAN	47.3	52.8	56.7	62.4	70.2	77.9	84.4	82.7	77.8	68.7	56.6	47.8	65.4

CALIFORNIA

TEMPERATURE NORMALS (DEG F)

STATION		JAN	FEB	MAR	APR	MAY	JUN	JUL	AUG	SEP	OCT	NOV	DEC	ANN
KING CITY	MAX	62.7	66.5	68.9	73.4	77.6	82.4	84.5	84.2	84.5	80.1	70.2	63.7	74.9
	MIN	35.1	37.8	38.7	40.7	44.9	48.3	50.5	50.7	48.8	44.2	38.4	35.0	42.8
	MEAN	48.9	52.2	53.8	57.1	61.3	65.4	67.5	67.5	66.7	62.2	54.3	49.3	58.9
KLAMATH R	MAX	53.8	56.3	56.2	58.3	61.7	65.0	66.2	66.7	67.4	64.3	59.0	54.7	60.8
	MIN	37.9	39.5	39.6	41.4	45.2	48.8	51.4	52.0	50.1	46.3	42.4	38.9	44.5
	MEAN	45.9	47.9	47.9	49.9	53.5	56.9	58.8	59.3	58.8	55.3	50.7	46.8	52.6
LAGUNA BEACH	MAX	64.7	65.8	65.7	67.2	69.0	71.3	75.0	76.8	76.8	74.0	69.8	65.8	70.2
	MIN	41.6	42.7	44.0	46.4	51.1	54.8	58.3	59.0	57.3	52.5	46.2	41.9	49.7
	MEAN	53.2	54.2	54.9	56.8	60.1	63.0	66.7	67.9	67.1	63.3	58.0	53.9	59.9
LAKE ARROWHEAD	MAX	45.4	49.3	53.1	58.7	65.7	74.6	81.3	80.1	77.0	66.9	53.2	46.6	62.7
	MIN	29.0	30.1	31.1	34.4	40.5	48.4	56.7	55.8	51.0	42.6	34.2	29.9	40.3
	MEAN	37.2	39.7	42.1	46.5	53.1	61.5	69.0	68.0	64.0	54.8	43.7	38.3	51.5
LAKEPORT	MAX	53.1	57.9	62.0	68.7	77.7	86.8	95.1	93.5	87.7	75.8	61.2	53.8	72.8
	MIN	31.3	34.5	35.4	38.0	42.8	48.8	52.6	51.6	47.7	42.0	36.3	32.6	41.1
	MEAN	42.2	46.2	48.7	53.4	60.3	67.8	73.9	72.6	67.8	58.9	48.8	43.2	57.0
LAKE SPAULDING	MAX	44.7	48.0	49.7	55.2	64.1	73.6	80.9	79.7	75.9	66.0	52.5	46.6	61.4
	MIN	23.7	25.2	25.5	29.0	35.5	42.4	46.8	45.5	42.5	36.2	29.0	25.0	33.9
	MEAN	34.2	36.6	37.7	42.1	49.8	58.0	63.9	62.6	59.2	51.1	40.8	35.8	47.7
LA MESA	MAX	66.7	68.2	68.0	70.3	72.5	76.7	82.6	83.7	83.1	78.7	72.9	68.3	74.3
	MIN	44.0	45.1	46.8	49.6	53.4	56.8	60.9	62.3	60.6	55.4	48.8	44.6	52.4
	MEAN	55.4	56.7	57.4	60.0	63.0	66.8	71.8	73.0	71.8	67.1	60.9	56.5	63.4
LE GRAND	MAX	54.6	61.5	66.8	73.3	81.4	89.0	95.4	93.6	89.3	80.0	65.3	55.0	75.4
	MIN	36.7	39.2	40.5	44.1	50.0	55.9	60.5	59.1	55.3	48.1	40.3	36.0	47.1
	MEAN	45.6	50.4	53.7	58.7	65.7	72.5	77.9	76.4	72.3	64.1	52.9	45.5	61.3
LEMON COVE	MAX	55.7	62.0	67.2	74.4	83.4	92.1	98.5	96.0	90.5	80.1	66.0	55.8	76.8
	MIN	36.8	40.4	42.9	46.9	52.6	58.3	63.4	61.6	57.8	50.6	42.0	36.5	49.2
	MEAN	46.2	51.2	55.0	60.6	68.0	75.2	81.0	78.8	74.1	65.4	54.1	46.2	63.0
LINDSAY	MAX	56.3	62.8	67.9	75.5	84.5	92.6	98.8	96.3	91.0	80.7	66.7	56.4	77.5
	MIN	35.5	38.1	40.1	44.3	50.3	55.9	61.3	59.3	55.1	47.1	39.4	35.1	46.8
	MEAN	45.9	50.5	54.0	59.9	67.4	74.3	80.1	77.8	73.1	63.9	53.0	45.8	62.1
LIVERMORE COUNTY F D	MAX	57.0	61.3	64.6	70.1	76.6	83.6	90.0	89.2	86.8	78.7	66.1	57.7	73.5
	MIN	34.7	37.8	38.8	41.4	45.9	50.5	52.7	52.6	51.2	46.4	39.5	35.2	43.9
	MEAN	45.9	49.6	51.7	55.8	61.3	67.1	71.3	70.9	69.0	62.6	52.8	46.5	58.7
LODI	MAX	53.7	60.5	65.8	72.1	79.9	86.8	91.8	90.3	87.2	78.3	64.3	54.3	73.8
	MIN	37.0	39.6	40.9	44.2	48.8	53.3	56.1	55.1	52.7	46.6	40.3	36.7	45.9
	MEAN	45.4	50.1	53.4	58.2	64.4	70.1	73.9	72.7	69.9	62.4	52.3	45.5	59.9
LOMPOC	MAX	64.2	65.4	65.2	66.4	67.6	69.5	70.9	72.0	74.2	73.8	69.5	65.6	68.7
	MIN	39.5	41.3	42.2	44.2	47.4	50.1	52.2	53.0	51.9	47.8	42.9	39.0	46.0
	MEAN	51.9	53.4	53.7	55.3	57.5	59.8	61.5	62.5	63.1	60.8	56.2	52.3	57.3
LONG BEACH WSO	MAX	66.0	67.3	68.0	70.9	73.4	77.4	83.0	83.8	82.5	78.4	72.7	67.4	74.2
	MIN	44.3	45.9	47.7	50.8	55.2	58.9	62.6	64.0	61.6	56.6	49.6	44.7	53.5
	MEAN	55.2	56.6	57.9	60.9	64.3	68.2	72.8	73.9	72.1	67.5	61.2	56.1	63.9
LOS ANGELES WSO R	MAX	64.6	65.5	65.1	66.7	69.1	72.0	75.3	76.5	76.4	74.0	70.3	66.1	70.1
	MIN	47.3	48.6	49.7	52.2	55.7	59.1	62.6	64.0	62.5	58.5	52.1	47.8	55.0
	MEAN	56.0	57.1	57.4	59.5	62.4	65.6	69.0	70.3	69.5	66.3	61.2	57.0	62.6
LOS ANGELES CIVIC CENTER	MAX	66.6	68.5	68.7	70.9	73.2	77.9	83.8	84.1	83.0	78.5	72.7	68.1	74.7
	MIN	47.7	49.2	50.2	53.0	56.6	60.4	64.3	65.3	63.7	59.2	52.7	48.4	55.9
	MEAN	57.2	58.9	59.5	62.0	64.9	69.2	74.1	74.7	73.4	68.9	62.7	58.3	65.3
LOS BANOS	MAX	54.9	62.2	67.8	74.5	82.3	90.0	96.9	94.9	90.1	80.2	65.8	55.4	76.3
	MIN	36.1	39.3	41.3	45.0	50.6	55.8	59.7	58.6	55.3	48.6	40.6	35.9	47.2
	MEAN	45.5	50.7	54.5	59.8	66.4	72.9	78.3	76.8	72.7	64.4	53.2	45.7	61.7

CALIFORNIA

TEMPERATURE NORMALS (DEG F)

STATION		JAN	FEB	MAR	APR	MAY	JUN	JUL	AUG	SEP	OCT	NOV	DEC	ANN
LOS GATOS	MAX	58.1	62.1	65.5	70.2	75.9	81.9	86.3	85.7	83.7	76.5	65.9	59.0	72.6
	MIN	38.1	40.3	41.0	42.9	46.7	50.5	53.3	53.0	52.2	48.0	42.3	38.2	45.5
	MEAN	48.1	51.2	53.3	56.6	61.3	66.2	69.8	69.4	68.0	62.2	54.1	48.6	59.1
MADERA	MAX	54.4	61.3	67.2	74.6	84.1	92.3	99.1	97.0	91.6	81.5	66.2	55.2	77.0
	MIN	36.2	39.0	41.0	44.8	50.6	56.3	61.1	59.4	55.0	47.6	40.0	35.5	47.2
	MEAN	45.3	50.2	54.1	59.7	67.4	74.3	80.1	78.2	73.3	64.6	53.1	45.4	62.1
MANZANITA LAKE //	MAX	40.4	42.3	43.7	49.9	59.9	69.6	79.1	77.1	71.2	60.4	47.7	42.3	57.0
	MIN	19.7	20.8	21.6	26.9	34.4	41.2	45.7	44.5	40.8	34.5	27.5	22.3	31.7
	MEAN	30.1	31.6	32.7	38.5	47.2	55.4	62.5	60.8	56.0	47.5	37.6	32.3	44.4
MARICOPA	MAX	56.3	62.6	67.7	74.5	83.4	92.0	98.6	96.2	90.6	80.3	66.3	56.8	77.1
	MIN	37.2	41.7	45.1	49.5	56.4	63.7	70.3	68.7	63.7	55.3	44.4	37.3	52.8
	MEAN	46.8	52.2	56.4	62.0	70.0	77.9	84.5	82.5	77.1	67.8	55.4	47.1	65.0
MARYSVILLE	MAX	53.8	60.6	66.0	73.0	81.5	89.8	96.4	94.2	89.7	79.4	64.9	54.9	75.4
	MIN	37.5	41.2	43.2	47.0	52.8	58.0	60.9	59.6	56.5	50.1	42.6	37.7	48.9
	MEAN	45.6	51.0	54.6	60.0	67.2	74.0	78.7	76.9	73.1	64.8	53.7	46.3	62.2
MC CLOUD	MAX	45.0	49.0	52.3	59.9	69.7	78.5	87.4	85.7	80.3	68.8	54.2	47.1	64.8
	MIN	22.9	25.8	27.0	30.4	36.4	42.9	46.6	44.8	39.8	33.7	28.4	24.5	33.6
	MEAN	34.0	37.4	39.7	45.2	53.1	60.7	67.1	65.3	60.1	51.3	41.3	35.8	49.3
MECCA FIRE STATION	MAX	69.2	74.5	79.0	86.3	93.6	101.9	106.1	104.5	101.2	91.3	78.8	70.6	88.1
	MIN	37.7	42.7	47.6	53.6	60.6	67.1	74.9	74.3	67.6	56.9	45.2	37.7	55.5
	MEAN	53.5	58.6	63.4	70.0	77.1	84.5	90.5	89.4	84.5	74.2	62.0	54.2	71.8
MERCED FIRE STATION 2	MAX	55.1	62.2	67.7	74.7	83.3	91.0	97.3	95.3	90.7	80.8	66.3	55.5	76.7
	MIN	36.0	38.6	40.6	44.1	50.1	55.5	60.0	58.2	54.5	47.2	39.6	35.3	46.6
	MEAN	45.6	50.4	54.2	59.4	66.7	73.3	78.7	76.8	72.6	64.0	53.0	45.4	61.7
MINERAL	MAX	40.7	44.0	46.3	53.2	62.6	71.8	81.5	80.2	74.7	63.3	49.0	42.6	59.2
	MIN	21.5	23.5	24.5	27.8	33.2	39.3	43.0	41.8	38.4	33.0	27.6	23.2	31.4
	MEAN	31.1	33.8	35.4	40.5	48.0	55.6	62.3	61.0	56.6	48.2	38.3	33.0	45.3
MODESTO	MAX	53.9	61.0	66.5	73.0	80.9	88.3	94.4	92.2	87.8	78.2	64.3	54.1	74.6
	MIN	37.5	40.6	42.1	45.7	50.6	55.6	59.3	58.3	55.4	49.0	41.5	37.3	47.7
	MEAN	45.7	50.8	54.4	59.4	65.8	72.0	76.8	75.3	71.6	63.6	52.9	45.7	61.2
MONO LAKE	MAX	40.6	44.5	50.6	58.3	67.3	76.6	84.3	82.7	76.5	66.2	51.9	42.5	61.8
	MIN	20.3	21.7	24.8	29.6	36.4	43.3	50.1	49.2	43.1	34.9	27.3	21.6	33.5
	MEAN	30.5	33.1	37.7	43.9	51.9	60.0	67.2	66.0	59.8	50.6	39.6	32.1	47.7
MONTEREY	MAX	59.9	61.7	61.8	62.7	64.3	66.7	67.7	68.8	72.1	70.6	65.7	61.2	65.3
	MIN	42.8	44.3	44.2	45.2	47.6	49.9	51.3	52.2	52.6	50.8	46.9	43.5	47.6
	MEAN	51.4	53.0	53.0	53.9	55.9	58.3	59.5	60.5	62.4	60.7	56.4	52.4	56.5
MOUNT HAMILTON	MAX	47.9	49.1	49.5	55.1	63.0	71.5	79.7	78.2	74.1	65.8	54.9	50.1	61.6
	MIN	35.9	36.4	35.2	38.9	45.3	53.1	62.6	61.2	57.0	50.8	41.8	37.7	46.3
	MEAN	41.9	42.8	42.4	47.0	54.2	62.4	71.2	69.7	65.6	58.3	48.4	44.0	54.0
MOUNT SHASTA WSO //R	MAX	42.1	47.3	50.9	57.9	67.0	75.4	85.1	83.3	77.5	65.4	50.9	43.9	62.2
	MIN	25.5	28.6	29.6	33.2	39.6	46.2	50.7	49.0	44.3	37.4	30.8	26.7	36.8
	MEAN	33.9	38.0	40.2	45.6	53.3	60.9	68.0	66.2	60.9	51.4	40.9	35.3	49.6
MOUNT WILSON FC 338 B //	MAX	51.0	52.0	52.8	57.7	64.7	74.4	81.1	80.1	76.8	68.4	58.2	53.1	64.2
	MIN	35.7	36.0	35.6	39.2	46.2	55.4	63.0	61.6	58.0	50.6	42.1	37.7	46.8
	MEAN	43.4	44.0	44.2	48.5	55.5	64.9	72.1	70.8	67.4	59.5	50.2	45.4	55.5
NAPA STATE HOSPITAL	MAX	57.4	62.6	65.6	69.8	75.1	80.0	82.4	81.9	82.7	77.2	66.2	58.4	71.6
	MIN	37.9	40.8	41.0	42.7	46.8	50.9	52.9	52.8	51.7	47.8	42.3	38.5	45.5
	MEAN	47.6	51.7	53.3	56.3	61.0	65.4	67.7	67.4	67.2	62.5	54.3	48.5	58.6
NEEDLES FAA AIRPORT	MAX	62.9	69.2	74.9	83.4	93.0	103.4	108.7	105.9	100.2	87.9	72.7	63.6	85.5
	MIN	40.9	44.9	49.0	56.3	65.6	75.3	83.0	80.9	73.3	61.1	48.9	41.8	60.1
	MEAN	51.9	57.1	62.0	69.9	79.3	89.4	95.9	93.4	86.8	74.5	60.8	52.8	72.8

CALIFORNIA

TEMPERATURE NORMALS (DEG F)

STATION		JAN	FEB	MAR	APR	MAY	JUN	JUL	AUG	SEP	OCT	NOV	DEC	ANN
NEVADA CITY	MAX	48.7	52.5	55.5	62.2	71.0	80.2	88.9	87.2	80.9	69.5	56.8	50.1	67.0
	MIN	28.8	30.5	32.0	35.4	41.2	46.9	50.6	48.8	44.8	38.6	32.8	29.4	38.3
	MEAN	38.7	41.6	43.8	48.9	56.1	63.6	69.8	68.0	62.9	54.1	44.8	39.7	52.7
NEWPORT BEACH HARBOR	MAX	62.6	63.3	63.4	64.5	66.2	68.8	71.9	73.3	73.1	70.9	67.2	64.0	67.4
	MIN	45.8	47.2	48.7	51.3	54.9	58.3	61.6	63.0	61.2	56.8	50.9	46.6	53.9
	MEAN	54.2	55.3	56.0	57.9	60.6	63.6	66.7	68.2	67.1	63.9	59.1	55.3	60.7
OAKLAND WSO R	MAX	54.5	58.3	60.1	62.6	65.2	68.5	70.6	71.1	72.8	69.4	61.8	55.4	64.2
	MIN	43.4	46.6	47.5	49.4	52.4	55.2	56.8	57.3	57.2	53.8	48.2	43.8	50.9
	MEAN	49.0	52.5	53.8	56.0	58.8	61.9	63.7	64.2	65.0	61.6	55.0	49.6	57.6
OJAI	MAX	66.1	68.8	70.1	73.5	76.9	83.2	91.2	90.7	88.3	82.0	73.8	67.9	77.7
	MIN	36.7	38.8	40.3	43.6	47.7	51.3	55.3	55.6	53.4	47.6	41.1	37.0	45.7
	MEAN	51.5	53.8	55.2	58.6	62.3	67.3	73.3	73.2	70.9	64.9	57.5	52.4	61.7
ORANGE COVE	MAX	54.6	61.0	66.2	73.3	82.5	90.7	97.5	95.8	90.3	80.4	66.3	55.5	76.2
	MIN	36.1	39.0	40.9	44.9	51.2	57.2	62.4	60.9	57.0	49.0	41.2	36.1	48.0
	MEAN	45.4	50.1	53.6	59.1	66.8	74.0	80.0	78.4	73.7	64.8	53.8	45.8	62.1
ORICK PRAIRIE CREEK PK	MAX	52.1	55.7	56.5	58.7	62.5	65.8	68.4	69.6	70.9	65.8	57.5	52.3	61.3
	MIN	36.8	38.3	37.9	39.1	42.7	46.5	48.4	48.9	46.7	43.2	40.6	37.7	42.2
	MEAN	44.5	47.0	47.3	48.9	52.6	56.2	58.4	59.3	58.8	54.5	49.1	45.0	51.8
ORLAND	MAX	53.6	59.7	64.6	71.9	81.1	89.4	95.9	94.0	89.5	78.8	64.0	54.6	74.8
	MIN	35.4	39.0	41.1	45.2	52.1	58.6	61.4	59.4	56.0	49.1	41.2	36.0	47.9
	MEAN	44.6	49.4	52.9	58.6	66.6	74.0	78.7	76.7	72.8	63.9	52.6	45.4	61.4
ORLEANS	MAX	50.0	56.3	61.8	69.1	77.0	85.0	93.9	92.5	87.9	72.9	57.3	50.3	71.2
	MIN	34.7	37.3	37.9	40.0	44.5	49.0	52.9	53.0	48.7	44.5	40.3	35.9	43.2
	MEAN	42.4	46.9	49.8	54.6	60.8	67.0	73.4	72.8	68.3	58.7	48.8	43.1	57.2
OXNARD	MAX	65.6	67.0	66.6	67.8	69.3	71.8	74.7	75.9	76.0	74.9	71.2	67.3	70.7
	MIN	43.5	44.3	45.1	47.1	50.6	53.8	56.7	58.0	56.2	52.3	47.2	44.1	49.9
	MEAN	54.6	55.7	55.9	57.5	60.0	62.8	65.7	66.9	66.1	63.6	59.2	55.7	60.3
PALMDALE	MAX	58.6	63.0	66.6	72.9	80.8	90.1	97.7	96.5	91.4	80.6	67.1	59.4	77.1
	MIN	32.2	35.1	38.0	42.6	49.9	57.4	65.0	63.2	56.9	47.4	37.6	31.9	46.4
	MEAN	45.4	49.1	52.3	57.8	65.4	73.8	81.4	79.9	74.2	64.0	52.4	45.7	61.8
PALM SPRINGS	MAX	69.3	74.7	79.0	86.2	93.9	103.2	109.1	106.8	102.3	92.1	78.5	70.2	88.8
	MIN	40.8	44.5	47.1	52.2	58.8	65.8	74.2	73.2	67.1	57.9	47.3	41.1	55.8
	MEAN	55.1	59.6	63.1	69.2	76.4	84.5	91.7	90.1	84.7	75.0	63.0	55.7	72.3
PALO ALTO JR MUSEUM	MAX	57.1	61.2	63.7	67.7	72.3	76.6	77.9	77.9	77.9	73.2	64.6	57.9	69.0
	MIN	38.5	41.5	42.6	44.6	48.3	52.4	54.7	54.6	52.7	48.1	42.5	38.6	46.6
	MEAN	47.8	51.4	53.2	56.2	60.3	64.5	66.3	66.3	65.3	60.6	53.6	48.3	57.8
PALOMAR MT OBSERVATORY	MAX	52.4	53.9	55.6	61.2	68.4	78.0	84.8	84.1	80.6	71.2	60.5	54.4	67.1
	MIN	33.6	34.1	34.3	38.2	44.3	53.2	61.0	60.4	56.0	47.9	39.3	35.0	44.8
	MEAN	43.0	44.0	45.0	49.7	56.3	65.6	72.9	72.3	68.4	59.6	49.9	44.7	56.0
PARKER RESERVOIR	MAX	63.9	69.7	75.0	83.3	92.5	102.6	107.8	105.7	101.0	89.5	74.3	65.2	85.9
	MIN	42.4	46.6	51.7	58.7	67.0	76.1	83.4	81.8	75.9	64.4	51.3	43.5	61.9
	MEAN	53.2	58.2	63.3	71.1	79.8	89.4	95.6	93.8	88.5	77.0	62.8	54.4	73.9
PASADENA	MAX	66.8	69.0	69.8	72.8	75.7	81.1	88.6	88.7	87.4	81.2	73.6	68.3	76.9
	MIN	43.1	44.8	45.9	48.6	52.5	56.4	60.8	61.5	59.7	54.4	47.9	43.7	51.6
	MEAN	55.0	56.9	57.8	60.7	64.1	68.8	74.7	75.1	73.6	67.8	60.7	56.0	64.3
PASO ROBLES	MAX	61.2	65.1	67.8	73.3	80.1	87.3	93.9	93.2	89.9	82.0	69.7	62.5	77.2
	MIN	32.9	36.2	37.4	39.3	43.2	47.2	50.1	49.7	47.0	41.2	35.2	31.5	40.9
	MEAN	47.1	50.7	52.6	56.3	61.6	67.3	72.0	71.5	68.4	61.6	52.5	47.0	59.1
PASO ROBLES FAA AP	MAX	58.9	62.8	65.8	71.7	79.1	87.4	94.7	93.5	88.7	79.9	67.4	60.0	75.8
	MIN	33.8	36.5	37.8	40.1	45.1	49.8	53.4	52.9	50.3	44.1	37.2	32.8	42.8
	MEAN	46.4	49.7	51.8	55.9	62.1	68.7	74.1	73.2	69.6	62.1	52.3	46.4	59.4

CALIFORNIA

TEMPERATURE NORMALS (DEG F)

STATION		JAN	FEB	MAR	APR	MAY	JUN	JUL	AUG	SEP	OCT	NOV	DEC	ANN
PETALUMA FIRE STA 2	MAX	56.4	61.3	63.8	67.5	72.4	78.1	82.5	82.8	82.5	76.3	65.5	57.0	70.5
	MIN	37.7	40.4	40.7	42.4	46.0	50.0	51.4	51.7	51.4	47.4	41.9	37.9	44.9
	MEAN	47.1	50.8	52.3	55.0	59.2	64.1	67.0	67.3	67.0	61.8	53.7	47.5	57.7
PINNACLES NAT MONUMENT	MAX	60.9	63.9	65.8	71.3	79.3	88.0	96.0	95.0	91.1	81.9	69.7	62.6	77.1
	MIN	33.4	35.6	36.3	38.6	42.5	47.0	50.9	50.3	48.4	43.2	37.2	33.5	41.4
	MEAN	47.2	49.7	51.1	55.0	60.9	67.5	73.5	72.7	69.8	62.6	53.5	48.1	59.3
PISMO BEACH	MAX	62.8	64.4	55.1	66.7	67.7	69.7	69.5	70.4	72.1	71.8	68.1	64.3	67.7
	MIN	42.5	43.9	44.0	45.1	47.1	50.2	51.9	53.1	52.8	50.7	47.0	43.3	47.6
	MEAN	52.7	54.2	54.6	56.0	57.4	59.9	60.7	61.7	62.5	61.3	57.6	53.8	57.7
PLACERVILLE	MAX	51.1	55.5	58.7	64.8	73.5	83.0	91.7	90.4	84.5	73.1	59.5	52.3	69.8
	MIN	31.1	33.4	35.4	38.9	44.4	50.3	55.2	53.9	49.5	42.9	35.8	31.6	41.9
	MEAN	41.1	44.5	47.1	51.9	59.0	66.6	73.5	72.1	67.1	58.0	47.7	42.0	55.9
POMONA CAL POLY	MAX	65.7	68.1	69.0	72.8	76.6	82.9	90.9	90.6	88.4	81.6	73.5	67.8	77.3
	MIN	39.0	40.6	42.2	45.3	49.8	53.9	58.2	59.0	56.7	51.0	43.9	39.1	48.2
	MEAN	52.4	54.4	55.6	59.1	63.2	68.4	74.6	74.9	72.6	66.3	58.7	53.4	62.8
PORTERVILLE	MAX	56.5	63.1	68.3	75.1	83.8	92.2	98.5	96.5	91.5	81.7	67.6	57.0	77.7
	MIN	36.2	39.8	42.5	46.8	52.8	58.9	64.6	62.6	57.5	49.7	41.3	36.1	49.1
	MEAN	46.4	51.5	55.4	61.0	68.3	75.6	81.6	79.5	74.5	65.7	54.4	46.5	63.4
PRIEST VALLEY	MAX	55.8	59.1	61.5	67.9	76.2	86.1	94.3	92.9	87.9	77.9	64.8	58.0	73.5
	MIN	28.2	30.5	31.2	33.5	38.5	43.8	49.4	48.2	44.0	37.1	31.0	27.8	36.9
	MEAN	42.0	44.9	46.4	50.7	57.4	65.0	71.9	70.6	66.0	57.5	47.9	42.9	55.3
QUINCY	MAX	45.5	51.8	56.8	64.5	73.6	82.5	91.0	89.2	84.1	72.5	55.4	46.6	67.8
	MIN	23.1	25.7	27.5	30.4	35.9	40.9	42.9	39.9	35.5	30.3	27.3	23.5	31.9
	MEAN	34.3	38.8	42.2	47.5	54.8	61.7	66.9	64.6	59.8	51.4	41.4	35.0	49.9
RANDSBURG	MAX	53.8	58.6	63.2	71.0	81.1	91.5	98.8	96.5	89.3	77.1	62.9	54.9	74.9
	MIN	35.7	38.5	40.2	45.6	53.1	61.9	68.7	66.8	61.7	53.2	42.6	36.6	50.4
	MEAN	44.7	48.6	51.7	58.4	67.1	76.7	83.8	81.7	75.5	65.2	52.8	45.8	62.7
RED BLUFF WSO R	MAX	53.9	60.0	64.0	71.2	81.2	90.3	98.2	95.7	90.4	78.8	63.6	55.2	75.2
	MIN	37.1	40.7	42.5	46.6	53.9	61.8	66.3	64.4	60.0	51.5	43.0	37.8	50.5
	MEAN	45.5	50.4	53.3	58.9	67.5	76.1	82.3	80.1	75.2	65.2	53.4	46.5	62.9
REDDING FIRE STN 4	MAX	55.7	61.6	66.1	73.3	83.0	91.9	99.5	97.0	91.9	79.8	64.3	56.5	76.7
	MIN	37.3	40.8	42.7	47.1	54.3	61.8	67.4	65.3	60.5	52.6	43.3	38.0	50.9
	MEAN	46.5	51.2	54.4	60.2	68.7	76.9	83.5	81.2	76.2	66.2	53.8	47.3	63.8
REDLANDS	MAX	64.7	67.4	68.7	73.7	78.5	87.0	95.9	94.7	90.8	82.1	72.6	66.4	78.5
	MIN	39.1	41.1	42.8	46.1	50.7	55.1	60.4	60.8	57.6	51.4	43.7	39.1	49.0
	MEAN	51.9	54.3	55.8	59.9	64.6	71.1	78.2	77.8	74.2	66.8	58.2	52.8	63.8
REDWOOD CITY	MAX	58.2	62.3	65.6	70.1	75.3	80.3	82.9	82.5	81.6	74.9	65.7	58.8	71.5
	MIN	39.2	41.6	42.6	44.3	47.8	51.5	53.8	53.8	52.4	48.3	43.2	39.6	46.5
	MEAN	48.7	52.0	54.1	57.2	61.6	65.9	68.4	68.1	67.0	61.6	54.5	49.2	59.0
RICHMOND	MAX	57.4	61.5	63.4	66.0	68.5	70.3	69.7	70.3	73.9	72.0	64.7	58.1	66.3
	MIN	42.0	45.0	46.1	48.3	51.4	54.2	55.1	56.0	56.3	52.9	47.5	42.7	49.8
	MEAN	49.7	53.3	54.8	57.1	60.0	62.3	62.4	63.2	65.1	62.5	56.1	50.5	58.1
RIVERSIDE FIRE STA. #3	MAX	66.0	68.8	70.4	74.7	79.5	86.6	94.2	93.4	90.5	82.5	73.2	67.1	78.9
	MIN	39.5	41.1	42.9	46.3	51.2	55.5	60.2	60.4	57.3	50.6	43.3	39.5	49.0
	MEAN	52.8	55.0	56.7	60.5	65.4	71.1	77.2	77.0	73.9	66.6	58.3	53.3	64.0
SACRAMENTO WSO R	MAX	52.6	59.4	64.1	71.0	79.7	87.4	93.3	91.7	87.6	77.7	63.2	53.2	73.4
	MIN	37.9	41.2	42.4	45.3	50.1	55.1	57.9	57.6	55.8	50.0	42.8	37.9	47.8
	MEAN	45.3	50.3	53.2	58.2	64.9	71.2	75.6	74.7	71.7	63.9	53.0	45.6	60.6
SACRAMENTO CITY WSO R	MAX	53.9	60.6	65.4	71.9	79.7	87.1	93.1	91.5	87.6	78.0	64.1	54.6	74.0
	MIN	40.2	43.7	45.2	48.2	52.8	57.3	60.0	59.6	58.1	52.6	45.3	40.4	50.3
	MEAN	47.1	52.2	55.3	60.1	66.3	72.2	76.6	75.6	72.9	65.3	54.7	47.5	62.2

CALIFORNIA

TEMPERATURE NORMALS (DEG F)

STATION		JAN	FEB	MAR	APR	MAY	JUN	JUL	AUG	SEP	OCT	NOV	DEC	ANN
SAINT HELENA	MAX	56.5	61.7	65.0	70.8	77.6	84.6	89.8	88.8	86.3	77.6	65.5	58.0	73.5
	MIN	36.0	38.7	39.2	41.4	46.2	50.6	52.4	51.9	49.8	45.7	40.4	36.3	44.1
	MEAN	46.3	50.2	52.2	56.1	61.9	67.6	71.1	70.4	68.1	61.7	53.0	47.2	58.8
SAINT MARYS COLLEGE	MAX	53.0	58.2	61.7	66.4	71.7	77.3	82.6	81.7	80.6	72.9	61.7	54.0	68:5
	MIN	35.0	37.8	38.7	41.5	46.5	50.7	53.7	53.9	52.0	45.9	39.2	35.4	44.2
	MEAN	44.1	48.1	50.2	54.0	59.1	64.1	68.2	67.8	66.3	59.4	50.4	44.7	56.4
SALT SPRINGS PWR HOUSE	MAX	52.3	55.8	58.1	63.4	70.9	78.8	87.5	86.7	82.8	73.6	61.0	53.9	68.7
	MIN	33.9	35.2	35.6	39.4	45.9	52.9	60.6	59.4	55.5	48.3	40.0	35.8	45.2
	MEAN	43.1	45.5	46.9	51.4	58.4	65.9	74.1	73.1	69.2	60.9	50.5	44.9	57.0
SAN BERNARDINO CO HOSP	MAX	66.4	68.8	70.0	74.8	80.1	88.7	97.6	96.6	92.9	84.0	74.1	68.0	80.2
	MIN	39.4	41.6	43.3	46.6	51.2	55.8	60.9	61.2	58.0	51.1	43.6	39.4	49.3
	MEAN	52.9	55.3	56.7	60.7	65.7	72.2	79.3	79.0	75.4	67.6	58.8	53.8	64.8
SANDBERG WSO //R	MAX	45.9	49.2	51.0	56.4	65.0	75.1	84.5	83.8	79.3	68.8	55.3	47.8	63.5
	MIN	33.8	35.5	35.9	39.8	45.6	54.5	64.1	63.9	60.3	52.5	42.6	35.5	47.0
	MEAN	39.9	42.4	43.5	48.1	55.3	64.9	74.3	73.9	69.8	60.7	49.0	41.7	55.3
SAN DIEGO WSO R	MAX	65.2	66.4	65.9	67.8	68.6	71.3	75.6	77.6	76.8	74.6	69.9	66.1	70.5
	MIN	48.4	50.3	52.1	54.5	58.2	61.2	64.9	66.8	65.1	60.3	53.6	48.7	57.0
	MEAN	56.8	58.4	59.0	61.2	63.4	66.3	70.3	72.2	71.0	67.5	61.8	57.4	63.8
SAN FRANCISCO WSO R	MAX	55.5	59.0	60.6	63.0	66.3	69.6	71.0	71.8	73.4	70.0	62.7	56.3	64.9
	MIN	41.5	44.1	44.9	46.6	49.3	52.0	53.3	54.2	54.3	51.2	46.3	42.2	48.3
	MEAN	48.5	51.6	52.8	54.8	57.8	60.8	62.2	63.0	63.9	60.6	54.5	49.2	56.6
S F FEDERAL BLDG. WSO R	MAX	56.1	59.4	60.0	61.1	62.5	64.3	64.0	65.0	68.9	68.3	62.9	56.9	62.5
	MIN	46.2	48.4	48.6	49.2	50.7	52.5	53.1	54.2	55.8	54.8	51.5	47.2	51.0
	MEAN	51.2	53.9	54.3	55.2	56.6	58.4	58.5	59.6	62.4	61.6	57.2	52.0	56.7
SAN GABRIEL FIRE DEPT	MAX	68.3	70.4	71.2	74.1	77.0	82.1	89.1	89.3	87.9	82.0	74.9	69.9	78.0
	MIN	41.3	42.9	44.9	47.8	52.2	56.3	60.4	61.1	58.6	53.0	45.9	41.3	50.5
	MEAN	54.8	56.7	58.1	61.0	64.6	69.2	74.8	75.2	73.3	67.5	60.4	55.6	64.3
SAN JOSE R	MAX	57.9	61.8	64.7	68.9	73.7	78.5	81.5	81.2	80.3	74.4	65.0	58.2	70.5
	MIN	41.1	43.7	44.6	46.5	50.3	53.8	56.0	56.0	55.5	51.2	45.5	41.2	48.8
	MEAN	49.5	52.8	54.6	57.7	62.0	66.2	68.8	68.6	67.9	62.8	55.2	49.7	59.7
SAN LUIS OBISPO POLY	MAX	62.4	64.5	64.7	67.0	69.3	73.6	77.4	78.2	78.7	76.3	69.8	64.3	70.5
	MIN	41.7	43.3	43.6	45.2	47.4	50.2	52.1	52.6	52.4	50.3	46.1	42.4	47.3
	MEAN	52.1	54.0	54.2	56.1	58.4	61.9	64.8	65.4	65.6	63.3	58.0	53.4	58.9
SAN RAFAEL	MAX	57.6	62.3	65.3	69.2	74.2	78.9	82.3	81.9	82.0	76.1	65.6	58.3	71.1
	MIN	41.4	43.8	44.2	45.8	48.8	52.4	53.2	53.6	53.6	50.9	46.4	42.2	48.0
	MEAN	49.5	53.1	54.8	57.5	61.5	65.7	67.8	67.8	67.8	63.5	56.0	50.3	59.6
SANTA ANA FIRE STATION	MAX	67.8	69.3	69.7	71.9	74.4	77.8	82.9	83.9	83.7	79.4	73.7	68.8	75.3
	MIN	44.1	45.3	47.0	49.8	54.1	57.8	61.5	62.7	60.8	55.9	49.0	44.5	52.7
	MEAN	55.9	57.4	58.4	60.9	64.3	67.8	72.2	73.3	72.3	67.6	61.4	56.7	64.0
SANTA BARBARA	MAX	65.2	66.4	66.9	68.5	69.7	72.6	76.0	77.3	77.1	74.9	70.8	66.8	71.0
	MIN	42.7	44.4	45.7	48.4	51.0	54.3	57.1	58.6	56.9	52.6	47.3	43.0	50.2
	MEAN	54.0	55.4	56.3	58.5	60.4	63.5	66.6	68.0	67.0	63.8	59.1	54.9	60.6
SANTA BARBARA FAA AP	MAX	63.3	64.4	65.0	66.6	68.4	71.1	73.6	74.8	75.0	72.5	69.0	64.8	69.0
	MIN	40.6	42.9	44.4	47.0	49.9	53.3	56.8	58.0	56.1	51.0	44.5	40.3	48.7
	MEAN	52.0	53.7	54.7	56.8	59.2	62.2	65.2	66.4	65.5	61.8	56.8	52.6	58.9
SANTA CRUZ	MAX	59.5	62.3	64.1	67.0	70.8	74.0	75.0	75.4	77.0	73.7	66.1	60.5	68.8
	MIN	38.4	40.4	40.7	41.9	45.3	48.4	50.5	50.9	49.9	46.5	41.9	38.6	44.5
	MEAN	49.0	51.4	52.4	54.5	58.1	61.2	62.8	63.2	63.4	60.1	54.0	49.6	56.6
SANTA MARIA WSO R	MAX	62.8	64.2	63.9	65.6	67.3	69.9	72.1	72.8	74.2	73.3	68.9	64.6	68.3
	MIN	38.8	40.3	40.9	42.7	46.2	49.6	52.4	53.2	51.8	47.6	42.1	38.3	45.3
	MEAN	50.8	52.3	52.4	54.2	56.8	59.8	62.3	63.1	63.0	60.5	55.5	51.4	56.8

CALIFORNIA

TEMPERATURE NORMALS (DEG F)

STATION		JAN	FEB	MAR	APR	MAY	JUN	JUL	AUG	SEP	OCT	NOV	DEC	ANN
SANTA MONICA PIER	MAX	64.0	63.9	63.2	63.6	64.6	67.7	70.5	72.1	72.1	70.3	68.1	65.5	67.1
	MIN	48.8	49.7	50.2	52.1	54.9	58.0	61.0	62.5	61.2	57.9	53.4	49.9	55.0
	MEAN	56.4	56.8	56.7	57.9	59.8	62.8	65.8	67.3	66.7	64.1	60.8	57.7	61.1
SANTA PAULA	MAX	67.2	69.4	69.8	72.3	73.5	76.8	81.0	81.5	81.5	78.5	73.1	68.4	74.4
	MIN	41.4	42.1	42.4	44.9	48.7	51.7	54.3	55.5	54.2	49.7	44.6	41.9	47.6
	MEAN	54.3	55.8	56.1	58.6	61.1	64.3	67.6	68.6	67.9	64.1	58.9	55.2	61.0
SANTA ROSA	MAX	57.3	62.5	65.5	69.7	75.2	80.9	84.3	84.4	84.0	77.7	66.2	57.9	72.1
	MIN	36.0	38.8	39.0	41.2	45.3	49.2	50.5	50.6	49.7	45.4	39.8	36.7	43.5
	MEAN	46.7	50.7	52.3	55.5	.60.3	65.0	67.4	67.5	66.9	61.6	53.0	47.4	57.9
SCOTIA	MAX	54.2	56.9	57.0	58.6	62.1	65.5	68.2	69.5	70.3	66.5	60.0	55.1	62.0
	MIN	39.2	40.9	41.0	43.0	46.7	50.0	51.9	52.4	50.6	47.3	43.7	40.2	45.6
	MEAN	46.7	48.9	49.0	50.8	54.4	57.8	60.1	60.9	60.4	56.9	51.9	47.6	53.8
SHASTA DAM	MAX	51.7	56.3	60.2	67.5	76.7	85.9	95.1	93.1	86.9	74.6	60.5	53.2	71.8
	MIN	38.4	40.8	42.4	47.3	54.4	62.2	68.3	66.9	61.8	54.3	45.6	39.7	51.8
	MEAN	45.1	48.6	51.3	57.4	65.6	74.1	81.7	80.0	74.4	64.5	53.1	46.5	61.9
SIERRAVILLE RANGER STA//	MAX	42.8	47.2	51.3	58.7	67.2	76.0	84.7	83.7	78.7	68.6	54.1	45.6	63.2
	MIN	15.5	19.6	23.5	28.1	35.0	40.8	44.2	41.5	36.4	29.5	22.9	17.2	29.5
	MEAN	29.2	33.4	37.5	43.4	51.1	58.4	64.5	62.6	57.6	49.1	38.5	31.5	46.4
SONOMA	MAX	56.9	62.7	66.2	71.1	77.5	84.3	89.7	89.1	87.2	78.8	65.7	57.4	73.9
	MIN	36.2	38.8	39.0	40.5	44.2	48.0	49.4	49.2	48.1	44.3	39.7	36.6	42.8
	MEAN	46.6	50.8	52.6	55.8	60.8	66.1	69.6	69.2	67.6	61.6	52.7	47.0	58.4
SONORA RS	MAX	54.3	58.6	61.1	66.9	76.6	86.3	95.3	93.7	88.2	77.8	63.6	55.8	73.2
	MIN	32.4	34.6	36.5	40.2	45.7	51.8	58.3	56.7	52.3	44.7	37.4	32.5	43.6
	MEAN	43.4	46.6	48.9	53.5	61.2	69.1	76.8	75.2	70.3	61.3	50.5	44.2	58.4
SO ENTRANCE YOSEMITE	MAX	47.7	50.0	51.2	56.6	65.8	75.6	84.3	82.8	77.5	67.8	55.8	50.1	63.8
	MIN	25.6	26.5	27.4	30.8	37.2	44.0	49.8	48.5	44.5	37.4	30.4	27.0	35.8
	MEAN	36.7	38.3	39.3	43.8	51.5	59.8	67.0	65.7	61.0	52.7	43.2	38.5	49.8
STOCKTON WSO	MAX	52.8	59.9	65.3	72.4	81.1	89.0	95.0	93.1	88.7	78.6	64.1	53.5	74.5
	MIN	37.5	40.6	42.0	45.6	51.5	57.1	60.9	60.4	57.4	50.5	42.4	37.5	48.6
	MEAN	45.2	50.3	53.7	59.0	66.3	73.1	78.0	76.8	73.1	64.6	53.3	45.5	61.6
STOCKTON FIRE STA NO 4	MAX	53.5	60.1	65.2	71.4	79.4	86.7	92.0	90.5	86.6	77.7	64.2	54.2	73.5
	MIN	36.1	39.3	41.4	44.9	49.5	54.2	57.1	56.3	53.6	47.8	40.6	36.2	46.4
	MEAN	44.8	49.7	53.3	58.2	64.5	70.5	74.6	73.4	70.1	62.8	52.5	45.2	60.0
STONY GORGE RESERVOIR	MAX	54.4	59.4	62.9	69.6	79.9	89.5	96.6	94.5	89.5	78.4	63.8	55.9	74.5
	MIN	33.4	36.8	38.3	41.8	48.6	55.9	60.8	59.4	54.6	46.8	38.5	33.8	45.7
	MEAN	43.9	48.1	50.6	55.8	64.3	72.7	78.7	76.9	72.1	62.6	51.2	44.9	60.2
STRAWBERRY VALLEY	MAX	48.0	50.8	52.8	59.4	68.2	77.1	85.7	84.2	79.9	69.1	55.9	50.2	65.1
	MIN	29.6	30.8	30.9	34.2	40.4	47.5	52.6	51.6	48.4	41.9	34.6	30.8	39.4
	MEAN	38.8	40.8	41.9	46.8	54.4	62.3	69.2	67.9	64.2	55.5	45.3	40.6	52.3
SUSANVILLE AIRPORT	MAX	40.3	46.7	52.9	61.1	70.9	80.5	89.9	87.4	80.1	67.0	51.9	41.8	64.2
	MIN	19.9	24.0	26.9	31.3	37.9	44.5	49.1	46.8	40.8	33.2	26.4	20.6	33.5
	MEAN	30.1	35.4	39.9	46.2	54.4	62.5	69.5	67.1	60.5	50.1	39.2	31.2	48.8
TAHOE CITY	MAX	38.0	40.0	42.4	49.3	58.7	68.0	77.0	75.8	69.0	58.5	46.5	40.5	55.3
	MIN	19.4	20.4	21.9	26.2	32.2	38.6	44.3	43.3	39.0	32.4	26.0	20.9	30.4
	MEAN	28.7	30.3	32.2	37.8	45.5	53.3	60.7	59.6	54.0	45.4	36.3	30.7	42.9
TEHACHAPI	MAX	51.6	54.6	56.2	62.3	70.8	79.0	87.3	85.7	81.2	71.3	59.6	53.2	67.7
	MIN	29.8	31.3	32.8	36.1	43.1	50.6	57.4	54.8	48.6	40.7	34.3	30.6	40.8
	MEAN	40.7	43.0	44.5	49.2	57.0	64.8	72.4	70.3	64.9	56.0	47.0	41.9	54.3
TEJON RANCHO	MAX	56.2	61.1	65.6	72.3	81.3	90.2	96.8	94.6	89.9	79.1	66.1	56.8	75.8
	MIN	36.6	40.4	42.8	47.2	53.7	60.8	67.4	65.4	60.8	52.6	42.7	36.7	50.6
	MEAN	46.4	50.8	54.2	59.8	67.5	75.5	82.1	80.0	75.3	65.8	54.4	46.8	63.2

CALIFORNIA

TEMPERATURE NORMALS (DEG F)

STATION		JAN	FEB	MAR	APR	MAY	JUN	JUL	AUG	SEP	OCT	NOV	DEC	ANN
THERMAL FAA AIRPORT	MAX	69.8	74.6	78.7	85.6	93.0	102.1	106.8	105.0	101.1	91.0	78.7	70.9	88.1
	MIN	38.3	43.1	48.3	55.1	62.7	69.8	76.9	75.6	69.1	57.8	45.4	38.2	56.7
	MEAN	54.1	58.8	63.5	70.4	77.9	85.9	91.9	90.4	85.1	74.4	62.0	54.6	72.4
TIGER CREEK PH	MAX	51.6	57.7	60.5	66.2	74.5	83.7	92.2	91.0	86.5	76.2	61.2	50.9	71.0
	MIN	32.9	34.4	35.5	38.9	44.6	50.5	56.5	55.7	51.9	44.5	37.5	33.5	43.0
	MEAN	42.3	46.1	48.0	52.6	59.6	67.1	74.4	73.3	69.2	60.4	49.4	42.2	57.1
TORRANCE	MAX	66.1	67.3	67.4	69.2	71.4	74.3	78.3	79.4	78.7	76.1	71.8	67.4	72.3
	MIN	44.1	45.4	46.3	48.9	52.8	56.4	59.9	61.1	59.6	55.3	49.0	44.8	52.0
	MEAN	55.1	56.4	56.8	59.1	62.1	65.4	69.1	70.3	69.2	65.7	60.4	56.1	62.1
TRACY CARBONA	MAX	53.6	60.3	66.1	72.8	80.5	88.1	94.2	92.7	88.1	78.6	64.7	54.5	74.5
	MIN	36.6	40.0	42.2	45.3	49.6	54.6	56.8	55.8	53.8	49.0	41.8	36.6	46.8
	MEAN	45.1	50.2	54.2	59.1	65.0	71.3	75.5	74.2	71.0	63.8	53.3	45.6	60.7
TRONA	MAX	58.9	65.4	71.3	78.7	88.2	98.4	105.6	103.4	96.6	85.1	69.6	59.2	81.7
	MIN	31.7	36.7	41.5	48.6	56.9	65.2	72.3	70.7	63.0	51.6	39.1	31.1	50.7
	MEAN	45.3	51.1	56.4	63.7	72.6	81.8	89.0	87.1	79.8	68.4	54.3	45.2	66.2
TRUCKEE RANGER STATION	MAX	38.3	42.0	44.9	52.1	62.1	71.9	81.5	80.1	74.4	63.6	49.1	40.6	58.4
	MIN	13.4	15.8	19.3	24.5	30.4	36.0	40.8	39.1	34.7	28.0	21.5	15.1	26.6
	MEAN	25.8	28.9	32.1	38.3	46.3	54.0	61.2	59.6	54.6	45.8	35.3	27.8	42.5
TULELAKE //R	MAX	40.1	46.6	51.4	59.3	68.7	76.7	85.5	83.6	78.3	66.4	50.8	41.5	62.4
	MIN	19.2	23.0	24.6	28.0	35.2	41.3	45.0	42.6	36.6	29.5	24.4	20.0	30.8
	MEAN	29.7	34.8	38.0	43.7	52.0	59.0	65.3	63.1	57.5	48.0	37.6	30.8	46.6
TUSTIN IRVINE RANCH	MAX	66.9	68.6	69.2	71.8	74.4	78.5	83.4	84.4	84.3	79.4	73.3	68.3	75.2
	MIN	40.3	41.6	43.3	46.5	50.9	54.9	58.2	59.0	56.6	51.3	44.4	40.3	48.9
	MEAN	53.6	55.1	56.3	59.2	62.7	66.7	70.9	71.8	70.5	65.4	58.9	54.3	62.1
TWENTYNINE PALMS	MAX	62.7	68.2	73.3	81.0	90.0	99.9	105.4	103.0	97.2	86.0	71.7	63.3	83.5
	MIN	35.3	38.9	42.0	48.5	56.3	64.5	72.1	70.3	63.5	52.7	41.9	35.6	51.8
	MEAN	49.0	53.6	57.7	64.7	73.1	82.3	88.7	86.7	80.3	69.4	56.8	49.5	67.7
TWIN LAKES	MAX	38.1	39.6	40.3	45.2	52.9	62.0	70.3	69.4	65.4	56.0	45.5	40.5	52.1
	MIN	15.3	15.7	16.2	20.7	28.6	36.3	42.8	41.5	38.2	31.1	23.6	18.0	27.3
	MEAN	26.7	27.7	28.3	33.0	40.8	49.2	56.6	55.5	51.8	43.6	34.6	29.3	39.8
UKIAH	MAX	57.2	62.0	64.9	70.4	77.7	85.2	93.6	92.4	88.9	78.5	64.9	57.9	74.5
	MIN	35.6	38.2	38.7	41.1	45.9	51.2	54.4	53.7	50.4	44.9	39.7	36.1	44.2
	MEAN	46.5	50.1	51.8	55.8	61.8	68.2	74.0	73.1	69.7	61.7	52.3	47.0	59.3
U C L A	MAX	65.3	66.4	66.2	67.4	68.7	71.8	76.7	77.4	77.5	74.8	70.9	66.9	70.8
	MIN	49.7	50.4	50.0	51.9	54.7	57.6	60.6	61.9	61.3	58.4	54.4	51.1	55.2
	MEAN	57.5	58.5	58.1	59.7	61.7	64.7	68.7	69.7	69.4	66.6	62.7	59.0	63.0
UPLAND 3 N	MAX	63.4	66.1	67.4	71.8	76.3	83.4	91.8	90.9	88.3	80.0	70.4	64.8	76.2
	MIN	40.6	41.6	42.0	44.5	48.5	52.5	58.1	58.6	56.8	51.1	45.3	41.1	48.4
	MEAN	52.0	53.9	54.7	58.2	62.4	68.0	75.0	74.8	72.6	65.6	57.9	53.0	62.3
VACAVILLE	MAX	54.3	61.2	65.9	72.3	80.7	88.6	95.1	93.8	89.5	79.5	64.7	54.9	75.0
	MIN	36.1	39.5	41.1	43.6	48.7	53.8	56.1	55.1	52.9	47.5	40.8	36.5	46.0
	MEAN	45.2	50.4	53.5	58.0	64.7	71.2	75.6	74.5	71.2	63.5	52.8	45.7	60.5
VICTORVILLE PUMP PLANT	MAX	57.9	62.1	65.6	72.4	80.7	90.6	97.8	96.3	90.7	80.4	67.1	59.3	76.7
	MIN	29.7	32.8	36.0	40.7	47.3	54.2	61.3	60.2	54.3	44.7	35.2	29.4	43.8
	MEAN	43.8	47.5	50.9	56.5	64.0	72.4	79.6	78.3	72.5	62.6	51.2	44.3	60.3
VISALIA	MAX	55.0	61.9	67.5	74.4	83.1	91.3	97.7	95.8	90.4	80.7	66.4	55.5	76.6
	MIN	36.9	40.6	43.4	47.5	53.4	59.2	64.1	62.0	58.1	50.7	42.2	36.8	49.6
	MEAN	46.0	51.3	55.5	61.0	68.3	75.3	80.9	78.9	74.2	65.7	54.3	46.2	63.1
WASCO	MAX	55.6	62.8	68.7	75.6	84.5	92.9	99.1	97.0	91.5	81.5	67.1	56.2	77.7
	MIN	35.4	39.4	42.9	47.8	54.1	60.6	65.8	63.8	58.8	49.7	40.6	35.2	49.5
	MEAN	45.6	51.1	55.8	61.7	69.4	76.8	82.5	80.4	75.2	65.6	53.9	45.7	63.6

CALIFORNIA

TEMPERATURE NORMALS (DEG F)

STATION		JAN	FEB	MAR	APR	MAY	JUN	JUL	AUG	SEP	OCT	NOV	DEC	ANN
WATSONVILLE WATERWORKS	MAX	59.5	61.9	62.7	64.8	67.0	69.4	70.1	70.3	72.9	71.7	66.2	60.8	66.4
	MIN	38.2	40.5	41.6	43.7	46.9	49.9	51.8	52.3	51.3	47.4	42.2	38.4	45.4
	MEAN	48.9	51.2	52.2	54.3	57.0	59.7	61.0	61.3	62.1	59.6	54.3	49.6	55.9
WEAVERVILLE RANGER STA	MAX	46.7	53.7	58.1	65.5	75.0	83.9	93.4	91.4	86.4	74.0	56.8	46.3	69.3
	MIN	26.7	28.7	29.7	32.8	38.7	44.6	48.6	47.3	41.6	34.8	31.2	28.3	36.1
	MEAN	36.7	41.2	44.0	49.2	56.8	64.3	71.0	69.4	64.0	54.4	44.0	37.3	52.7
WILLOWS	MAX	54.0	60.2	65.0	72.0	80.5	88.4	94.3	92.4	88.8	79.3	64.8	55.0	74.6
	MIN	35.4	38.9	40.6	44.4	51.7	58.1	61.3	59.2	55.8	48.9	40.8	36.0	47.6
	MEAN	44.7	49.5	52.8	58.2	66.1	73.3	77.8	75.9	72.3	64.1	52.8	45.5	61.1
WINTERS	MAX	54.6	61.1	66.0	73.1	81.6	90.1	96.6	94.8	90.1	80.2	65.9	55.8	75.8
	MIN	36.6	40.6	42.8	46.5	52.1	57.3	59.2	58.2	56.1	49.7	41.9	36.8	48.2
	MEAN	45.6	50.9	54.5	59.8	66.9	73.7	78.0	76.5	73.1	65.0	53.9	46.3	62.0
WOODFORDS	MAX	43.7	47.5	51.0	57.9	66.9	76.5	85.4	83.2	76.5	65.7	52.6	45.7	62.7
	MIN	23.0	25.2	27.0	31.6	38.9	46.2	53.1	51.6	45.8	38.1	29.4	23.9	36.2
	MEAN	33.3	36.4	39.0	44.8	52.9	61.4	69.3	67.4	61.1	51.9	41.0	34.8	49.4
WOODLAND 1 WNW	MAX	53.5	60.3	65.8	72.9	81.9	90.2	96.2	94.4	90.0	79.5	64.6	54.4	75.3
	MIN	36.6	40.2	41.7	45.0	50.3	55.1	56.9	55.8	53.9	48.6	41.8	36.9	46.9
	MEAN	45.1	50.3	53.8	59.0	66.1	72.7	76.6	75.1	72.0	64.1	53.2	45.7	61.1
YORBA LINDA	MAX	67.3	69.3	70.4	73.7	76.7	81.8	88.8	89.2	87.4	81.4	74.2	68.9	77.4
	MIN	42.4	43.1	43.6	46.0	50.9	54.5	58.2	59.0	56.7	52.0	46.1	42.3	49.6
	MEAN	54.9	56.2	57.0	59.9	63.8	68.2	73.5	74.1	72.0	66.7	60.1	55.6	63.5
YOSEMITE PARK HEADQTRS	MAX	47.7	54.9	59.1	65.5	73.2	81.9	90.4	90.1	85.5	74.7	58.6	47.5	69.1
	MIN	26.0	28.6	30.5	35.1	41.3	47.6	53.4	51.8	46.7	38.8	30.5	26.3	38.1
	MEAN	36.9	41.8	44.9	50.3	57.3	64.8	71.9	71.0	66.1	56.8	44.6	37.0	53.6
YREKA	MAX	43.4	50.6	55.0	62.4	72.0	80.8	90.7	88.6	82.1	69.2	53.3	44.1	66.0
	MIN	24.6	27.8	30.2	34.5	40.8	48.0	53.3	52.1	45.6	37.1	30.3	26.0	37.5
	MEAN	34.0	39.2	42.6	48.4	56.4	64.4	72.0	70.4	63.9	53.2	41.8	35.1	51.8

CALIFORNIA

PRECIPITATION NORMALS (INCHES)

STATION	JAN	FEB	MAR	APR	MAY	JUN	JUL	AUG	SEP	OCT	NOV	DEC	ANN
ACTON-ESDO CYN FC261F	2.22	1.85	1.64	.85	.28	.03	.03	.17	.25	.29	1.29	1.31	10.21
ADIN RANGER STATION	2.45	1.76	1.52	1.31	1.36	1.08	.27	.44	.60	1.38	1.70	2.19	16.06
ALDERPOINT	11.44	8.22	6.60	3.43	1.51	.40	.03	.48	.84	3.25	7.13	9.67	53.00
ALISO CN OAT MT FC 446	5.60	4.20	3.30	2.00	.50	.09	.01	.18	.27	.48	2.87	3.16	22.66
ALPINE	3.31	2.65	2.98	1.61	.62	.13	.09	.20	.33	.73	1.56	2.18	16.39
ALTADENA	5.08	4.38	3.34	1.82	.44	.13	.01	.12	.37	.39	2.56	2.55	21.19
ALTURAS RANGER STATION	1.67	1.23	1.25	1.00	1.21	1.09	.31	.43	.48	.94	1.31	1.53	12.45
ANGIOLA	1.51	1.37	1.05	.86	.25	.06	.01	.02	.21	.26	.81	1.12	7.53
ANGWIN PAC UNION COL	9.95	6.59	4.76	2.77	.61	.30	.06	.18	.54	2.41	4.99	7.96	41.12
ANZA	2.63	2.34	2.28	1.29	.42	.05	.38	.57	.70	.44	1.29	2.02	14.41
ASH MOUNTAIN	5.00	4.11	3.84	2.86	.91	.25	.06	.11	.50	.86	2.61	4.29	25.40
AUBERRY 1 NW	4.60	4.01	3.57	2.60	.82	.16	.04	.04	.44	.91	2.61	4.04	23.84
AUBURN	7.49	4.99	4.71	2.97	1.07	.31	.15	.15	.46	2.01	4.43	5.72	34.46
AVALON PLEASURE PIER	2.78	2.49	1.85	1.00	.21	.03	.00	.09	.25	.16	1.50	1.79	12.15
BAKERSFIELD WSO R	.98	1.07	.87	.70	.24	.07	.01	.05	.13	.30	.65	.65	5.72
BALCH POWER HOUSE	6.06	4.88	4.20	3.12	1.09	.31	.04	.04	.64	1.21	3.17	5.05	29.81
BARRETT DAM	3.36	2.78	3.00	1.61	.52	.07	.12	.21	.25	.51	1.71	2.32	16.46
BARSTOW	.66	.52	.48	.22	.10	.10	.31	.30	.32	.19	.44	.50	4.14
BEAUMONT 1 E	3.56	3.07	2.90	1.51	.63	.11	.16	.15	.44	.52	1.75	2.20	17.00
BEL AIR FC-10A	4.86	3.90	2.88	1.38	.26	.04	.01	.16	.18	.25	2.25	2.56	18.73
BERKELEY	5.30	3.51	2.97	1.95	.40	.17	.07	.11	.36	1.25	2.93	4.22	23.24
BIG BAR RANGER STATION	8.16	5.84	4.47	2.13	1.19	.50	.11	.45	.68	2.43	5.52	7.48	38.96
BIG DALTON DAM FC223BE	5.92	4.72	4.06	2.33	.74	.18	.02	.14	.49	.58	2.56	3.38	25.12
BIG PINES PARK FC 83B	5.29	5.05	3.76	2.09	.65	.11	.26	.54	.81	.52	2.83	3.68	25.59
BIG SUR STATE PARK	9.03	6.51	5.44	3.50	.77	.19	.07	.11	.52	1.68	4.96	7.70	40.48
BIG TUJUNGA DAM FC46DE	5.88	5.49	4.10	2.30	.61	.03	.02	.11	.39	.41	3.24	3.24	25.82
BISHOP WSO R	1.32	.98	.43	.31	.30	.11	.19	.11	.18	.17	.49	1.02	5.61
BLUE CANYON WSO //R	14.11	9.93	8.96	5.45	2.70	.86	.30	.55	.97	3.93	8.41	11.70	67.87
BLYTHE	.44	.38	.36	.17	.06	.06	.11	.94	.33	.31	.24	.35	3.75
BLYTHE FAA AIRPORT	.44	.31	.27	.22	.01	.03	.21	.72	.31	.33	.17	.35	3.37
BOCA	4.34	2.98	2.49	1.41	1.22	.69	.49	.57	.50	1.16	2.09	3.78	21.72
BORREGO DESERT PARK	1.08	.92	.67	.30	.08	.02	.30	.57	.33	.23	.59	.76	5.85
BOWMAN DAM	13.36	9.55	8.66	5.12	2.98	1.11	.30	.79	.97	4.21	8.37	11.46	66.88
BRAWLEY 2 SW	.41	.25	.26	.11	.03	.00	.08	.40	.29	.21	.19	.31	2.54
BROOKS FARNHAM RANCH	4.42	3.07	2.03	1.20	.33	.18	.04	.07	.24	.99	2.31	3.61	18.49
BRUSH CREEK RS	15.56	10.82	8.36	5.34	2.26	.81	.24	.42	1.07	4.03	8.77	11.65	69.33
BURBANK VALLEY PMP PLT	3.77	3.33	2.52	1.24	.28	.04	.01	.14	.24	.31	1.94	1.96	15.78
BURNEY	5.57	3.93	3.02	1.91	1.26	.88	.12	.37	.73	1.74	3.33	5.20	28.06
BUTTONWILLOW	.97	1.05	.75	.58	.21	.04	.03	.01	.17	.24	.55	.58	5.18
CACHUMA DAM	4.22	4.10	3.07	1.73	.31	.04	.01	.02	.28	.36	2.20	2.89	19.23
CALAVERAS BIG TREES	10.94	8.00	7.42	5.00	1.85	.64	.19	.30	.73	2.38	5.94	9.37	52.76
CALEXICO 2 NE	.39	.23	.14	.08	.02	.01	.13	.34	.32	.28	.20	.34	2.48
CALISTOGA	8.76	6.46	4.35	2.63	.64	.25	.08	.17	.46	2.15	4.62	6.96	37.53
CALLAHAN	3.81	2.76	2.05	1.28	.89	.76	.35	.44	.56	1.58	2.96	4.20	21.64
CAMPO	2.91	2.40	2.51	1.38	.47	.05	.39	.41	.36	.49	1.44	2.04	14.85
CAMP PARDEE	4.01	3.04	2.98	2.15	.59	.18	.10	.12	.28	1.08	2.57	3.42	20.52
CANOGA PARK PIERCE COL	4.03	3.26	2.46	1.19	.25	.02	.00	.15	.16	.25	2.13	2.10	16.00
CANYON DAM	8.54	6.17	4.99	2.66	1.58	.75	.22	.41	.68	2.14	4.57	6.78	39.49
CEDARVILLE	2.02	1.36	1.33	1.02	1.11	.83	.37	.38	.48	1.18	1.61	2.70	14.39
CHALLENGE RANGER STA	14.68	10.50	8.80	5.29	1.91	.61	.17	.28	.87	3.64	8.07	11.51	66.33
CHATSWORTH FC24F	4.13	3.35	2.39	1.42	.24	.03	.01	.19	.17	.23	2.21	2.27	16.64
CHESTER	6.80	5.36	3.77	2.03	1.31	.93	.30	.43	.72	1.92	3.96	5.56	33.09
CHICO EXPERIMENT STA	5.75	3.94	3.07	2.04	.72	.42	.05	.17	.43	1.61	3.55	4.18	25.93
CHITTENDEN PASS	4.30	3.01	2.70	1.83	.37	.12	.07	.11	-.35	.79	2.24	3.71	19.60
CHULA VISTA	1.88	1.21	1.56	.79	.24	.06	.02	.11	.16	.30	1.17	1.17	8.67

CALIFORNIA

PRECIPITATION NORMALS (INCHES)

STATION	JAN	FEB	MAR	APR	MAY	JUN	JUL	AUG	SEP	OCT	NOV	DEC	ANN
CLAREMONT POMONA COL	4.02	3.28	2.96	1.40	.40	.05	.04	.12	.30	.35	1.77	2.44	17.13
CLOVERDALE 3 SSE	10.23	7.25	4.91	3.00	.73	.28	.07	.21	.60	2.52	5.77	8.25	43.82
COALINGA	1.65	1.55	.94	.66	.24	.02	.01	.02	.25	.25	.99	1.25	7.83
COLEMAN FISHERIES STA	5.29	3.84	3.07	2.07	1.26	.55	.16	.31	.63	1.45	3.61	4.60	26.84
COLFAX	10.29	7.10	6.37	3.98	1.62	.51	.18	.24	.63	2.75	5.92	8.25	47.84
COLGATE POWER HOUSE	8.67	6.11	4.97	3.28	1.18	.39	.14	.17	.52	2.47	5.05	6.70	39.65
COLUSA 1 SSW	3.45	2.57	1.73	1.06	.38	.19	.06	.08	.23	.99	2.06	2.61	15.41
CORCORAN IRRIG DIST	1.45	1.34	.93	.74	.19	.05	.01	.00	.18	.22	.80	1.02	6.93
CORNING HOUGHTON RANCH	4.19	3.58	2.35	1.48	.62	.37	.09	.16	.31	.96	2.64	3.36	20.11
CORONA	2.72	2.34	1.75	.94	.21	.03	.04	.12	.29	.19	1.25	1.72	11.60
COVELO	9.82	6.58	5.21	2.43	1.03	.30	.06	.43	.58	2.35	5.45	8.12	42.36
COVINA TEMPLE FC 193 B	4.10	3.30	2.79	1.53	.36	.04	.03	.15	.30	.44	1.84	2.36	17.24
CRESCENT CITY 1 N	11.92	8.90	8.55	4.55	3.05	1.10	.33	.92	1.96	5.08	9.79	11.00	67.15
CULVER CITY	3.57	2.92	2.04	1.09	.14	.04	.00	.10	.14	.20	1.86	1.99	14.09
CUMMINGS	16.20	11.14	9.27	4.80	1.87	.46	.06	.56	1.07	4.50	10.13	13.78	73.84
CUYAMACA	6.12	5.72	5.98	3.37	1.46	.16	.42	.57	.87	1.29	3.36	4.87	34.19
DAGGETT FAA AP	.56	.36	.35	.27	.09	.08	.32	.50	.44	.17	.28	.39	3.81
DAVIS 2 WSW EXP FARM	4.12	2.85	1.97	1.22	.35	.12	.03	.05	.21	.95	2.17	3.10	17.14
DEATH VALLEY	.27	.39	.23	.14	.09	.04	.16	.09	.15	.11	.22	.14	2.03
DEEP SPRINGS COLLEGE	.71	.84	.54	.61	.47	.26	.47	.25	.40	.21	.48	.58	5.82
DELANO	1.37	1.29	1.08	.91	.29	.08	.00	.01	.22	.31	.86	.86	7.28
DENAIR 3 NNE	2.40	2.04	1.77	1.31	.30	.08	.04	.07	.16	.55	1.53	2.01	12.26
DE SABLA	13.78	10.27	8.12	5.20	1.94	.92	.16	.36	.97	3.86	8.77	10.99	65.34
DESCANSO RANGER STA	4.56	3.84	4.29	2.40	1.03	.09	.30	.26	.33	.93	2.18	3.25	23.46
DOWNEY FD FC107C	3.51	3.13	2.17	1.19	.21	.06	.01	.13	.19	.18	1.64	1.96	14.38
DOWNIEVILLE RANGER STA	12.81	9.53	7.77	5.00	2.66	.83	.23	.44	1.04	3.82	7.80	10.77	62.70
DRY CANYON RESERVOIR	2.94	2.47	2.05	1.21	.41	.05	.01	.11	.23	.27	1.58	1.77	13.10
DUNNIGAN	3.77	2.99	1.95	1.20	.29	.15	.04	.05	.23	.96	2.26	2.98	16.87
EAGLE MOUNTAIN	.41	.24	.23	.14	.03	.04	.41	.56	.40	.25	.18	.37	3.26
EAST PARK RESERVOIR	4.55	3.39	1.94	1.14	.40	.31	.03	.17	.26	1.05	2.49	3.50	19.23
EL CAPITAN DAM	3.06	2.59	2.78	1.44	.56	.12	.05	.21	.28	.68	1.56	2.13	15.46
EL CENTRO 2 SSW	.42	.22	.18	.09	.01	.00	.10	.36	.19	.25	.25	.28	2.35
ELECTRA POWER HOUSE	5.81	4.31	4.27	2.99	.97	.29	.17	.15	.36	1.41	3.43	5.02	29.18
ELLERY LAKE //	4.25	2.87	2.53	1.76	.99	.53	.82	.54	.81	1.03	2.86	3.86	22.85
ELLIOTT	3.60	2.62	2.36	1.76	.47	.15	.08	.08	.27	.78	2.10	2.98	17.25
ELSINORE	2.75	2.34	1.89	.76	.20	.02	.03	.17	.32	.22	1.19	1.77	11.66
ESCONDIDO	3.30	2.39	2.55	1.25	.43	.08	.05	.12	.27	.42	1.64	2.03	14.53
EUREKA WSO R	6.99	5.20	5.05	2.91	1.60	.56	.10	.37	.90	2.71	5.90	6.22	38.51
EXCHEQUER RESERVOIR	3.46	2.99	2.87	2.14	.57	.20	.04	.07	.31	.82	2.18	3.26	18.91
FAIRFIELD FIRE STATION	5.11	3.27	2.54	1.43	.36	.15	.04	.10	.27	1.26	2.64	3.77	20.94
FAIRMONT	3.68	3.06	2.24	1.35	.40	.03	.02	.14	.35	.24	2.09	2.21	15.81
FIDDLETOWN LYNCH RANCH	7.17	5.36	4.99	3.58	1.19	.35	.17	.18	.45	1.91	4.31	6.21	35.87
FIVE POINTS 5 SSW	1.42	1.26	.80	.59	.22	.05	.03	.02	.18	.26	.80	.92	6.55
FONTANA KAISER	3.68	2.84	2.59	1.13	.32	.04	.02	.09	.32	.29	1.49	2.12	14.93
FOREST GLEN	13.56	9.63	7.82	3.92	1.70	.50	.11	.50	1.09	3.71	8.43	11.49	62.46
FORESTHILL RANGER STA	10.78	7.65	6.80	4.28	1.72	.50	.19	.30	.69	2.82	6.11	8.98	50.82
FORT BIDWELL	2.55	1.91	1.56	1.07	1.12	.85	.42	.49	.59	1.09	1.97	2.54	16.16
FORT BRAGG	7.80	5.93	5.05	3.05	1.00	.39	.11	.41	.67	2.67	5.43	6.79	39.30
FORT JONES RANGER STA	4.77	2.79	2.00	1.08	.76	.78	.34	.49	.65	1.39	2.94	4.49	22.48
FORT ROSS	8.71	6.01	4.54	2.97	.73	.35	.12	.33	.70	2.66	5.23	6.56	38.91
FRENCH GULCH	7.50	6.02	4.98	3.21	1.44	.70	.16	.42	.84	2.36	5.74	6.60	39.97
FRESNO WSO R	2.05	1.85	1.61	1.15	.31	.08	.01	.02	.16	.43	1.24	1.61	10.52
FRIANT GOVERNMENT CAMP	2.51	2.29	2.04	1.50	.47	.12	.02	.02	.27	.59	1.67	2.23	13.73
GARBERVILLE	12.33	8.96	6.92	3.74	1.46	.34	.04	.39	.96	3.56	7.84	10.55	57.09
GASQUET RANGER STATION	17.92	12.62	12.38	6.26	3.68	1.03	.23	.95	2.29	6.33	14.00	16.53	94.22

CALIFORNIA

PRECIPITATION NORMALS (INCHES)

STATION	JAN	FEB	MAR	APR	MAY	JUN	JUL	AUG	SEP	OCT	NOV	DEC	ANN
GEM LAKE	3.37	2.54	2.13	1.45	.85	.46	.68	.56	.70	.73	2.29	3.13	18.89
GEORGETOWN RANGER STA	11.36	7.72	7.06	4.79	1.77	.57	.23	.28	.68	2.88	6.24	9.35	52.93
GLENDORA WEST FC 185	4.84	3.90	3.28	1.78	.54	.11	.02	.11	.39	.46	2.17	2.65	20.25
GLENNVILLE	3.53	2.74	2.85	2.04	.73	.14	.03	.14	.51	.66	1.80	2.94	18.11
GOLD RUN	12.04	8.42	7.44	4.58	2.03	.66	.25	.32	.80	3.19	7.04	9.91	56.68
GRANT GROVE	8.70	7.09	6.18	4.29	1.47	.42	.08	.11	.89	1.43	4.32	7.16	42.14
GRATON	9.83	6.42	4.54	2.93	.59	.27	.09	.20	.53	2.49	5.42	7.74	41.05
GREENVIEW	4.91	3.10	1.67	1.15	.67	.58	.19	.32	.41	1.42	3.06	4.37	21.85
GREENVILLE RANGER STA	8.47	6.25	4.95	2.72	1.59	.85	.30	.46	.67	2.31	4.64	7.00	40.21
GROVELAND RANGER STA	7.49	6.07	5.07	3.70	1.07	.38	.08	.12	.58	1.24	4.36	6.18	36.34
HAINES CYN UPPER FC367	5.91	5.36	3.99	2.55	.70	.08	.01	.12	.50	.49	3.57	3.41	26.69
HAIWEE	1.19	1.26	.82	.40	.28	.09	.26	.30	.34	.14	.68	.85	6.61
HALF MOON BAY	5.32	3.69	3.50	2.06	.58	.25	.12	.21	.41	1.63	3.02	4.43	25.22
HANFORD	1.55	1.43	1.11	.76	.26	.05	.01	.02	.21	.26	.96	1.18	7.80
HAPPY CAMP RANGER STA	12.18	7.78	6.51	2.78	1.45	.61	.25	.54	1.09	3.67	7.91	10.90	55.67
HARRISON GULCH RS	7.51	5.97	4.06	2.39	1.00	.66	.08	.52	.59	2.20	4.85	6.51	36.34
HARRY ENGLEBRIGHT DAM	7.48	5.45	4.44	2.82	.93	.32	.05	.11	.44	1.96	4.13	5.56	33.69
HAT CREEK PH NO 1	3.24	2.53	2.09	1.22	1.22	.89	.21	.37	.56	1.23	2.09	3.22	18.87
HAYFIELD PUMPING PLANT	.52	.36	.30	.11	.09	.02	.21	.46	.31	.30	.26	.35	3.29
HAYFORK RANGER STATION	7.49	5.25	3.87	2.00	.94	.47	.15	.39	.68	2.16	4.84	6.81	35.05
HEALDSBURG	10.22	6.99	4.48	2.75	.59	.26	.07	.23	.58	2.36	5.84	7.91	42.28
HEALDSBURG NO 2	9.38	6.72	4.42	2.68	.59	.25	.08	.22	.53	2.30	5.47	7.66	40.30
HEMET	2.37	1.91	1.93	1.04	.32	.05	.12	.20	.50	.40	1.18	1.49	11.51
HENSHAW DAM	5.04	4.32	4.43	2.24	.74	.08	.31	.42	.48	.69	2.42	3.43	24.60
HERNANDEZ 2 NW	3.34	3.01	2.37	1.46	.35	.07	.04	.06	.36	.65	1.96	2.91	16.58
HETCH HETCHY	6.33	5.18	4.63	3.58	1.66	.89	.20	.38	.65	1.52	4.04	5.86	34.92
HILTS	4.45	2.72	2.18	1.01	.95	.76	.24	.52	.58	1.64	3.05	4.38	22.48
IDYLLWILD FIRE DEPT	4.76	3.92	4.11	2.19	.94	.13	.71	.77	.93	.80	2.26	3.87	25.39
IMPERIAL	.42	.24	.22	.11	.01	.00	.10	.31	.26	.21	.23	.29	2.40
INDEPENDENCE	1.36	1.01	.32	.27	.22	.08	.17	.13	.23	.10	.47	1.03	5.39
INDIO U S DATE GARDEN	.61	.40	.29	.10	.04	.00	.17	.33	.33	.15	.37	.35	3.14
INYOKERN	.77	.89	.56	.22	.09	.01	.16	.20	.23	.10	.49	.44	4.16
IOWA HILL	10.64	7.33	6.62	4.23	2.06	.53	.22	.29	.71	2.88	6.33	8.74	50.58
IRON MOUNTAIN	.53	.22	.26	.17	.03	.05	.24	.41	.26	.39	.21	.33	3.10
JESS VALLEY	1.99	1.67	1.82	1.80	2.04	1.57	.48	.64	.73	1.38	1.89	1.96	17.97
JULIAN WYNOLA	5.02	4.17	4.34	2.44	1.10	.16	.33	.54	.80	.91	2.52	3.27	25.60
JUNCAL DAM	7.18	6.74	4.22	2.43	.46	.06	.01	.02	.50	.42	3.54	4.33	29.91
KEE RANCH //	2.07	1.70	1.46	.43	.09	.07	.21	.24	.49	.31	1.17	1.26	9.50
KENTFIELD	11.50	7.40	5.50	3.30	.82	.31	.12	.17	.57	2.92	6.01	9.08	47.70
KERLINGER	1.64	1.22	1.09	.77	.28	.07	.05	.14	.18	.38	.99	1.35	8.16
KERN RIVER PH NO 1	1.75	1.79	1.68	1.31	.39	.06	.01	.07	.25	.53	1.15	1.49	10.48
KERN RIVER PH NO 3	3.05	2.29	1.78	1.02	.25	.11	.09	.19	.36	.31	1.16	2.12	12.73
KETTLEMAN STATION	1.41	1.34	.88	.70	.27	.01	.02	.04	.15	.26	.73	.80	6.61
KING CITY	2.50	2.15	1.56	.92	.22	.04	.01	.05	.17	.40	1.28	1.95	11.25
KLAMATH R	15.05	10.65	10.58	5.50	3.32	1.11	.28	.92	2.15	5.86	12.41	13.54	81.37
LA CRESCENTA FC 251 B	5.44	4.70	3.52	2.04	.55	.11	.01	.13	.37	.45	2.75	3.02	23.09
LAGUNA BEACH	2.83	2.33	1.94	1.16	.27	.11	.01	.09	.30	.26	1.42	1.62	12.34
LAKE ARROWHEAD	8.98	7.40	6.52	3.52	1.44	.09	.16	.40	.96	.99	4.53	6.09	41.08
LAKEPORT	7.28	4.69	3.30	1.95	.58	.27	.05	.18	.34	1.67	4.01	5.72	30.04
LAKE SPAULDING	13.84	10.04	9.10	5.64	2.89	.95	.40	.63	.92	4.00	8.16	12.19	68.76
LA MESA	2.80	1.95	2.31	1.09	.43	.08	.04	.12	.24	.46	1.37	1.67	12.56
LEBEC	2.33	2.58	2.14	1.30	.57	.03	.04	.13	.32	.37	1.81	1.71	13.33
LECHUZA PT STA FC 352B	5.78	4.35	3.12	1.72	.30	.06	.01	.09	.27	.27	2.98	3.17	22.12
LE GRAND	2.44	2.08	1.80	1.37	.40	.06	.01	.04	.21	.51	1.61	2.12	12.65
LEMON COVE	2.57	2.18	1.97	1.40	.53	.12	.01	.03	.28	.64	1.53	2.21	13.47

CALIFORNIA

PRECIPITATION NORMALS (INCHES)

STATION	JAN	FEB	MAR	APR	MAY	JUN	JUL	AUG	SEP	OCT	NOV	DEC	ANN
LINDSAY	2.19	1.94	1.78	1.25	.48	.05	.01	.03	.28	.51	1.44	1.77	11.73
LIVERMORE COUNTY F D	3.04	2.19	1.81	1.28	.38	.11	.04	.07	.18	.67	1.77	2.57	14.11
LOCKWOOD 2 N	3.21	2.54	2.04	1.18	.29	.02	.02	.07	.19	.49	1.69	2.38	14.12
LODI	3.45	2.51	2.11	1.53	.42	.13	.07	.07	.26	.72	2.05	2.95	16.27
LOMPOC	2.90	2.74	2.30	1.31	.24	.03	.01	.02	.24	.39	1.61	2.09	13.88
LONG BEACH WSO	2.98	2.50	1.69	.83	.16	.04	.00	.09	.16	.15	1.36	1.58	11.54
LOS ALAMOS	3.12	3.10	2.51	1.46	.33	.02	.00	.03	.35	.41	1.59	2.24	15.16
LOS ANGELES WSO R	3.06	2.49	1.76	.93	.14	.04	.01	.10	.15	.26	1.52	1.62	12.08
LOS ANGELES CIVIC CENTER	3.69	2.96	2.35	1.17	.23	.03	.00	.12	.27	.21	1.85	1.97	14.85
LOS BANOS	1.76	1.62	1.14	.84	.24	.04	.02	.04	.22	.47	1.17	1.44	9.00
LOS BANOS ARBURUA RCH	1.65	1.36	.97	.77	.24	.03	.05	.04	.24	.34	1.02	1.37	8.08
LOS GATOS	6.05	4.55	3.37	1.95	.37	.06	.05	.07	.27	1.08	2.97	4.80	25.59
LOS PRIETOS RANGER STA	5.29	4.61	3.35	2.05	.41	.04	.01	.02	.19	.44	2.80	3.24	22.45
LYTLE CREEK RANGER STA	8.54	6.61	5.88	2.64	.81	.05	.08	.10	.44	.65	3.64	4.36	33.80
MADERA	2.00	1.74	1.45	1.29	.40	.07	.02	.03	.18	.46	1.36	1.72	10.72
MAD RIVER RANGER STA	12.74	9.55	7.37	3.84	1.74	.47	.10	.50	1.06	3.82	8.42	11.75	61.36
MANZANITA LAKE //	6.85	5.46	4.93	3.62	2.65	1.66	.35	.78	1.21	2.95	4.89	5.93	41.28
MARICOPA	1.03	1.20	.94	.64	.25	.02	.00	.03	.25	.30	.70	.66	6.02
MARIPOSA	5.55	4.78	3.99	3.12	1.00	.27	.06	.07	.50	1.25	3.55	5.14	29.28
MARYSVILLE	4.55	3.42	2.40	1.65	.44	.24	.07	.10	.31	1.30	2.70	3.37	20.55
MC CLOUD	10.17	8.01	6.03	3.65	2.04	.91	.27	.56	1.06	2.95	6.96	8.65	51.26
MECCA FIRE STATION	.55	.34	.18	.07	.03	.01	.14	.17	.34	.25	.26	.32	2.66
MENDOTA DAM	1.48	1.33	1.15	.92	.26	.04	.01	.02	.21	.31	.97	1.24	7.94
MERCED FIRE STATION 2	2.41	1.98	1.68	1.33	.34	.07	.03	.03	.16	.52	1.50	2.00	12.05
MIDDLETOWN	11.38	7.62	5.17	3.08	.74	.23	.05	.18	.50	2.42	5.36	8.71	45.44
MINERAL	10.47	7.69	6.10	3.94	2.28	1.37	.20	.74	1.18	3.86	6.74	8.88	53.45
MODESTO	2.33	1.83	1.62	1.21	.29	.07	.06	.08	.17	.57	1.50	1.97	11.70
MOJAVE	1.20	1.07	.76	.39	.12	.03	.06	.09	.27	.16	.74	.73	5.62
MONO LAKE	2.34	1.85	1.20	.65	.65	.33	.60	.43	.58	.52	1.52	2.29	12.96
MONTEREY	4.03	2.81	2.65	1.70	.41	.19	.09	.12	.27	.69	2.21	3.18	18.35
MOUNT DIABLO N GATE	5.26	3.63	2.89	1.90	.57	.16	.05	.08	.26	1.28	2.83	4.19	23.10
MOUNT HAMILTON	4.13	3.54	2.97	2.15	.59	.15	.05	.09	.18	1.08	2.73	3.97	21.63
MOUNT HEBRON RS	1.49	1.19	1.04	.83	.97	.95	.35	.52	.55	.99	1.67	1.89	12.44
MOUNT SHASTA WSO //R	7.21	5.69	4.23	2.75	1.55	.80	.25	.45	.85	2.01	5.19	6.07	37.05
MOUNT WILSON FC 338 B //	7.75	6.95	5.38	2.84	.69	.08	.04	.14	.72	.70	4.08	4.41	33.78
MUIR WOODS	8.42	5.47	4.35	2.58	.70	.39	.09	.19	.53	2.22	4.70	6.72	36.36
NAPA STATE HOSPITAL	5.67	3.72	2.93	1.79	.48	.19	.05	.11	.32	1.45	3.15	4.48	24.34
NEEDLES FAA AIRPORT	.52	.39	.43	.26	.07	.03	.47	.67	.45	.35	.34	.41	4.39
NEVADA CITY	11.82	8.75	7.03	4.41	1.79	.47	.19	.25	.79	2.91	6.82	9.36	54.59
NEWARK	3.17	2.10	1.81	1.18	.33	.11	.04	.08	.19	.70	1.83	2.43	13.97
NEWHALL SOLEDAD FC32CE	4.29	3.72	2.68	1.56	.34	.04	.02	.09	.22	.26	2.35	2.39	17.96
NEWMAN 2 NW	2.32	1.91	1.41	1.08	.23	.04	.02	.04	.17	.43	1.37	1.71	10.73
NEWPORT BEACH HARBOR	2.56	2.07	1.69	1.15	.20	.06	.01	.08	.29	.15	1.32	1.51	11.09
NILAND	.36	.30	.21	.13	.01	.01	.17	.29	.27	.25	.24	.31	2.55
NORTH FORK RANGER STA	6.53	5.29	4.80	3.35	1.22	.37	.06	.07	.54	1.14	3.51	5.60	32.48
OAKLAND WSO R	4.03	2.79	2.32	1.47	.37	.13	.05	.05	.26	1.12	2.26	3.18	18.03
OCCIDENTAL	12.50	8.49	6.06	3.92	1.00	.38	.12	.36	.75	3.70	7.10	9.39	53.77
OJAI	5.14	4.73	3.21	1.82	.40	.04	.01	.03	.39	.33	2.51	3.06	21.67
ORANGE COVE	2.67	2.14	1.76	1.30	.43	.08	.01	.01	.21	.55	1.45	2.22	12.83
ORICK PRAIRIE CREEK PK	12.60	9.11	8.90	4.87	2.97	1.00	.22	.84	1.69	4.70	10.03	11.49	68.42
ORLAND	4.24	3.20	2.11	1.35	.61	.35	.11	.25	.35	1.05	2.85	3.42	19.89
ORLEANS	10.70	7.33	6.59	3.05	1.86	.58	.14	.49	1.10	3.77	8.06	10.17	53.84
OXNARD	3.56	3.06	2.24	1.18	.12	.04	.01	.05	.28	.19	1.78	2.02	14.53
PACIFIC HOUSE	9.80	7.15	6.63	4.69	2.14	.69	.29	.36	.61	2.60	5.50	8.77	49.23
PACOIMA DAM FC 33 A E	4.51	3.54	3.17	1.87	.51	.09	.01	.17	.40	.54	2.33	2.44	19.58

CALIFORNIA

PRECIPITATION NORMALS (INCHES)

STATION		JAN	FEB	MAR	APR	MAY	JUN	JUL	AUG	SEP	OCT	NOV	DEC	ANN
PAICINES OHRWALL RANCH		3.29	2.58	2.17	1.42	.34	.05	.07	.07	.29	.61	1.66	2.96	15.51
PALMDALE		1.73	1.29	1.05	.53	.15	.03	.04	.15	.23	.23	.94	1.01	7.38
PALM SPRINGS		1.26	.75	.55	.17	.08	.03	.23	.23	.36	.16	.62	.76	5.20
PALO ALTO JR MUSEUM		3.41	2.27	1.92	1.17	.33	.07	.04	.06	.18	.65	1.82	2.85	14.77
PALOMA		5.10	4.34	3.27	2.06	.48	.08	.07	.09	.19	.74	2.65	4.03	23.10
PALOMAR MT OBSERVATORY		5.72	4.99	5.46	2.51	.60	.11	.41	.51	.60	.76	2.87	4.56	29.10
PALOS VERDES EST FC43D		2.97	2.37	1.65	.95	.20	.06	.00	.12	.18	.15	1.49	1.58	11.72
PANOCHE		1.82	1.62	1.25	.68	.22	.03	.03	.06	.27	.33	1.06	1.55	8.92
PARKER RESERVOIR		.78	.54	.53	.23	.09	.06	.31	.60	.42	.46	.45	.50	4.97
PASADENA		4.69	3.96	3.11	1.60	.40	.09	.01	.12	.28	.37	2.30	2.36	19.29
PASKENTA RANGER STA		4.90	3.98	2.84	1.78	.68	.49	.13	.20	.34	1.24	3.23	3.93	23.74
PASO ROBLES		3.34	2.70	1.74	1.26	.27	.02	.03	.06	.24	.43	1.65	2.46	14.20
PASO ROBLES FAA AP		2.85	2.44	1.68	1.16	.24	.02	.03	.05	.23	.40	1.39	2.06	12.55
PATTIWAY		1.67	1.81	1.60	1.09	.37	.05	.03	.04	.38	.40	1.08	1.16	9.68
PETALUMA FIRE STA 2		5.95	3.97	2.64	1.71	.31	.17	.05	.08	.25	1.40	3.16	4.33	24.02
PINNACLES NAT MONUMENT		3.31	2.99	2.55	1.52	.36	.06	.03	.07	.22	.66	1.90	2.78	16.45
PISMO BEACH		3.78	2.97	2.28	1.57	.31	.03	.01	.01	.34	.65	1.95	2.67	16.57
PIT RIVER POWER HOUSE		16.30	11.90	8.98	5.51	2.66	1.16	.26	.63	1.97	4.12	10.11	13.42	77.02
PLACERVILLE		7.93	5.34	5.13	3.47	1.30	.36	.19	.17	.46	1.87	4.38	6.39	36.99
PT ARGUELLO LIGHT STA		2.80	2.79	2.04	1.26	.24	.03	.01	.02	.16	.42	1.65	2.00	13.42
POMONA CAL POLY		4.18	3.29	2.82	1.38	.32	.04	.03	.10	.31	.42	1.74	2.39	17.02
PORTERVILLE		2.09	1.78	1.67	1.23	.42	.06	.01	.02	.32	.58	1.33	1.61	11.12
PORTOLA		4.56	3.05	2.65	1.31	1.16	.66	.40	.37	.42	1.26	2.41	3.96	22.21
POTTER VALLEY PH		10.46	7.11	5.30	2.82	1.14	.34	.10	.30	.64	2.60	6.02	8.58	45.41
PRIEST VALLEY		4.35	3.62	2.91	1.80	.49	.07	.05	.07	.39	.72	2.45	3.52	20.44
QUINCY		8.82	6.37	4.82	2.89	1.59	.72	.19	.38	.68	2.57	4.79	7.22	41.04
RANDSBURG		1.23	1.13	.67	.40	.10	.02	.12	.16	.23	.20	.58	.64	5.48
RED BLUFF WSO	R	4.50	3.31	2.39	1.51	.77	.43	.06	.21	.46	1.16	3.10	3.59	21.49
REDDING FIRE STN 4		8.51	6.19	4.96	2.82	1.28	.83	.18	.51	1.05	2.03	5.56	7.03	40.95
REDLANDS		2.79	2.20	2.12	1.17	.49	.07	.10	.15	.41	.37	1.33	1.69	12.89
REDWOOD CITY		4.56	3.13	2.44	1.44	.36	.09	.04	.07	.24	.92	2.25	3.73	19.27
REPRESA		4.54	3.13	2.81	1.87	.68	.22	.10	.12	.30	1.15	2.83	3.28	21.03
RICHMOND		4.96	3.36	2.66	1.75	.31	.13	.07	.07	.31	1.22	2.82	4.17	21.83
RIVERSIDE FIRE STA. #3		2.17	1.77	1.55	.86	.23	.03	.08	.14	.31	.20	1.00	1.30	9.64
SACRAMENTO WSO	R	4.03	2.88	2.06	1.31	.33	.11	.05	.07	.27	.86	2.23	2.90	17.10
SACRAMENTO CITY WSO	R	4.18	2.94	2.18	1.44	.35	.13	.05	.09	.30	.90	2.31	3.00	17.87
SAINT HELENA		8.82	5.91	4.00	2.24	.55	.23	.07	.14	.39	1.96	4.30	6.63	35.24
SAINT MARYS COLLEGE		6.47	4.41	3.61	2.23	.58	.17	.06	.09	.33	1.64	3.51	5.11	28.21
SALINAS FAA AP		2.78	2.08	1.74	1.24	.24	.08	.04	.06	.26	.50	1.51	2.48	13.01
SALINAS DAM		4.73	3.98	3.08	2.01	.36	.03	.01	.04	.34	.53	2.38	3.51	21.00
SALT SPRINGS PWR HOUSE		8.72	6.52	5.96	4.08	2.07	.66	.25	.41	.86	2.08	5.20	7.51	44.32
SAN BERNARDINO CO HOSP		3.49	2.77	2.50	1.32	.54	.08	.04	.13	.49	.52	1.62	2.18	15.68
SAN CLEMENTE DAM		4.73	3.91	3.18	1.83	.38	.08	.05	.06	.24	.67	2.25	3.72	21.10
SANDBERG PTL FC130B		3.08	3.38	2.33	1.52	.47	.04	.03	.13	.41	.29	2.38	1.86	15.92
SANDBERG WSO	//R	2.52	2.81	1.57	1.04	.33	.04	.03	.10	.35	.26	1.99	1.90	12.94
SAN DIEGO WSO	R	2.11	1.43	1.60	.78	.24	.06	.01	.11	.19	.33	1.10	1.36	9.32
SAN DIMAS FIRE WD FC95		4.18	3.46	2.89	1.57	.39	.07	.03	.13	.33	.43	1.87	2.46	17.81
SAN FRANCISCO WSO	R	4.65	3.23	2.64	1.53	.32	.11	.03	.05	.19	1.06	2.35	3.55	19.71
S F FEDERAL BLDG. WSO	R	4.48	2.83	2.58	1.48	.35	.15	.04	.08	.24	1.09	2.49	3.52	19.33
SAN GABRIEL CANYON PH		5.44	4.44	3.52	1.81	.60	.15	.03	.13	.39	.46	2.26	2.86	22.09
SAN GABRIEL DM FC425BE		7.02	5.70	4.30	2.38	.66	.08	.02	.10	.41	.57	3.04	3.93	28.21
SAN GABRIEL FIRE DEPT		4.48	3.74	2.81	1.44	.27	.06	.01	.05	.30	.39	2.00	2.21	17.76
SAN JOSE	R	3.00	2.23	2.03	1.19	.30	.07	.05	.13	.21	.67	1.71	2.27	13.86
SAN LUIS OBISPO POLY		5.46	4.39	3.18	2.02	.39	.04	.04	.05	.33	.78	2.47	3.85	23.00
SAN RAFAEL		9.20	6.14	4.24	2.43	.54	.25	.06	.09	.43	2.10	4.77	7.23	37.48

CALIFORNIA

PRECIPITATION NORMALS (INCHES)

STATION	JAN	FEB	MAR	APR	MAY	JUN	JUL	AUG	SEP	OCT	NOV	DEC	ANN
SANTA ANA FIRE STATION	3.17	2.35	2.05	1.09	.21	.05	.01	.11	.25	.17	1.38	1.76	12.60
SANTA BARBARA	4.28	3.72	2.49	1.50	.30	.07	.01	.03	.23	.26	2.13	2.68	17.70
SANTA BARBARA FAA AP	3.83	3.46	2.43	1.35	.22	.03	.02	.02	.25	.35	1.89	2.33	16.18
SANTA CRUZ	6.51	4.81	3.79	2.47	.49	.19	.15	.14	.39	1.25	3.58	5.21	28.98
SANTA MARGARITA BOOST	7.17	5.64	4.27	2.76	.61	.06	.01	.06	.40	1.03	3.47	5.43	30.91
SANTA MARIA WSO R	2.43	2.63	1.87	1.17	.24	.04	.01	.04	.27	.46	1.37	1.82	12.35
SANTA MONICA PIER	3.33	3.00	2.11	.98	.15	.01	.03	.15	.11	.16	1.73	1.93	13.69
SANTA PAULA	4.36	3.54	2.86	1.38	.22	.03	.00	.05	.26	.26	2.25	2.60	17.81
SANTA ROSA	7.00	4.69	3.42	2.26	.51	.25	.06	.17	.42	1.90	3.71	5.49	29.88
SAUGUS PWR PL NO 1	3.83	3.24	2.74	1.66	.57	.10	.01	.12	.29	.34	2.16	2.46	17.52
SAWYERS BAR RANGER STA	9.23	6.02	5.17	2.43	1.47	.87	.22	.64	.88	3.35	6.79	8.83	45.90
SCOTIA	9.97	7.76	6.33	3.59	1.50	.47	.05	.34	.85	2.93	6.80	8.94	49.53
SHASTA DAM	13.05	9.78	7.43	4.55	1.82	.95	.21	.61	1.32	2.91	8.56	10.73	61.92
SIERRAVILLE RANGER STA//	5.46	3.75	2.90	1.56	1.35	.60	.32	.42	.52	1.97	2.99	4.73	26.57
SNOW CREEK UPPER	2.39	2.25	1.90	.95	.20	.03	.27	.55	.61	.43	1.55	2.16	13.29
SONOMA	6.78	4.56	3.18	1.97	.44	.27	.06	.10	.33	1.63	3.85	4.99	28.16
SONORA RS	6.19	4.89	4.42	3.14	.99	.30	.06	.14	.47	1.44	3.72	5.26	31.02
SO ENTRANCE YOSEMITE	9.07	7.31	6.08	4.29	1.67	.64	.12	.11	.79	1.56	5.19	7.32	44.15
SPRECKELS HWY BRIDGE	2.90	2.16	2.06	1.30	.32	.10	.08	.06	.25	.51	1.57	2.48	13.79
STOCKTON WSO	3.02	2.03	1.81	1.36	.30	.08	.05	.07	.23	.62	1.77	2.43	13.77
STOCKTON FIRE STA NO 4	3.10	2.30	1.83	1.39	.38	.10	.05	.06	.23	.68	1.69	2.59	14.40
STONYFORD RANGER STA	4.98	3.82	2.31	1.24	.41	.33	.04	.17	.23	1.16	2.71	3.78	21.18
STONY GORGE RESERVOIR	4.45	3.41	2.17	1.27	.47	.40	.06	.32	.25	1.05	2.61	3.49	19.95
STRAWBERRY VALLEY	17.80	12.86	10.35	5.95	2.60	.78	.25	.41	1.16	4.54	10.04	14.27	81.01
SUSANVILLE AIRPORT	2.88	1.93	1.38	.64	.75	.67	.30	.22	.36	1.14	1.43	2.59	14.29
SUTTER HILL RANGER STA	5.66	4.19	4.00	2.97	.89	.28	.15	.16	.35	1.38	3.56	4.86	28.45
TAHOE CITY	6.56	4.67	3.65	2.20	1.22	.66	.33	.46	.55	1.82	3.29	6.09	31.50
TEHACHAPI	1.79	1.79	1.55	1.12	.40	.06	.08	.26	.29	.28	1.15	1.63	10.40
TEJON RANCHO	1.72	1.62	1.89	1.49	.49	.07	.05	.09	.26	.50	1.13	1.34	10.65
THERMAL FAA AIRPORT	.52	.37	.23	.07	.06	.01	.18	.30	.32	.14	.30	.32	2.82
TIGER CREEK PH	9.15	6.71	6.14	4.39	1.60	.53	.21	.28	.62	2.21	5.21	8.06	45.11
TOPANGA PATROL STA FC6	6.78	5.02	3.76	1.85	.28	.02	.02	.14	.22	.27	2.99	3.44	24.79
TORRANCE	3.53	2.65	1.82	.96	.18	.03	.00	.11	.19	.13	1.64	1.89	13.13
TRACY CARBONA	1.90	1.41	1.28	.93	.32	.08	.05	.15	.22	.48	1.17	1.56	9.55
TRONA	.87	.71	.51	.19	.10	.09	.18	.18	.22	.15	.40	.35	3.95
TRUCKEE RANGER STATION	6.55	4.67	3.88	2.31	1.39	.69	.41	.46	.52	1.59	3.24	5.86	31.57
TULELAKE //R	1.17	1.04	1.00	.69	1.02	.95	.24	.54	.42	.97	1.29	1.51	10.84
TURLOCK	2.32	1.93	1.58	1.24	.26	.05	.03	.05	.15	.47	1.48	1.83	11.39
TUSTIN IRVINE RANCH	2.92	2.15	1.91	1.15	.22	.04	.01	.08	.27	.20	1.32	1.70	11.97
TWENTYNINE PALMS	.42	.29	.28	.12	.08	.02	.67	.68	.49	.26	.26	.32	3.89
TWIN LAKES	9.35	6.93	6.20	4.38	2.40	1.10	.81	.89	1.02	2.44	5.84	8.42	49.78
UKIAH	8.97	6.19	4.39	2.39	.78	.25	.06	.20	.51	1.93	5.10	7.35	38.12
U C L A	4.39	3.71	2.61	1.30	.24	.04	.01	.16	.19	.22	2.12	2.40	17.39
UPLAND 3 N	4.79	3.77	3.40	1.70	.57	.06	.05	.10	.36	.44	1.96	2.69	19.89
UPPER LAKE 7 W	9.35	7.00	5.19	2.99	1.01	.32	.06	.27	.50	2.18	5.57	7.45	41.89
UPPER MATTOLE	16.73	12.22	9.37	5.53	2.21	.39	.09	.61	1.35	4.83	11.66	13.87	78.86
VACAVILLE	6.10	3.88	2.77	1.68	.40	.12	.06	.04	.35	1.29	2.97	4.63	24.29
VAN NUYS FC 15A	3.96	3.43	2.40	1.37	.28	.02	.01	.16	.20	.19	1.98	1.99	15.99
VENTURA	3.31	3.04	2.42	1.11	.18	.04	.01	.03	.28	.18	1.83	2.11	14.54
VICTORVILLE PUMP PLANT	.97	.83	.68	.33	.15	.05	.11	.18	.36	.22	.55	.57	5.00
VINCENT FIRE STA FC120	1.79	1.49	1.21	.59	.22	.04	.03	.28	.25	.27	.98	1.01	8.16
VINTON	2.39	1.54	1.26	.78	.99	.64	.32	.38	.37	.91	1.33	2.15	13.06
VISALIA	1.94	1.68	1.38	1.07	.36	.05	.01	.02	.19	.49	1.13	1.54	9.86
VOLTA POWER HOUSE	5.99	4.68	3.80	2.91	1.82	.89	.14	.50	.80	2.21	4.61	5.35	33.70
WALNUT PTL STA FC102C	3.90	3.19	2.67	1.37	.30	.06	.02	.11	.28	.30	1.70	2.17	16.07

CALIFORNIA

PRECIPITATION NORMALS (INCHES)

STATION	JAN	FEB	MAR	APR	MAY	JUN	JUL	AUG	SEP	OCT	NOV	DEC	ANN
WASCO	1.20	1.28	1.01	.76	.26	.06	.02	.01	.15	.25	.68	.80	6.48
WATSONVILLE WATERWORKS	4.70	3.35	2.92	1.86	.33	.13	.09	.09	.35	.92	2.61	4.02	21.37
WEAVERVILLE RANGER STA	8.41	5.67	4.12	2.30	1.20	.59	.18	.48	.86	2.42	5.64	7.32	39.19
WESTHAVEN	1.44	1.35	.92	.70	.24	.05	.02	.00	.19	.25	.76	.88	6.80
WHITTIER CTY H FC 106C	3.55	3.07	2.24	1.22	.24	.03	.01	.11	.28	.24	1.63	1.89	14.51
WILLOWS	3.74	2.95	1.76	1.19	.44	.28	.06	.16	.30	.93	2.51	2.96	17.28
WINTERS	5.15	3.66	2.33	1.32	.33	.15	.03	.05	.24	.98	2.51	3.88	20.63
WOFFORD HEIGHTS	2.65	2.00	1.40	.80	.22	.06	.08	.22	.31	.25	1.02	1.98	10.99
WOODFORDS	3.94	2.72	2.12	1.16	.97	.53	.59	.60	.73	1.17	2.47	3.74	20.74
WOODLAND 1 WNW	4.10	3.11	2.12	1.33	.35	.15	.04	.05	.25	.96	2.26	3.17	17.89
WRIGHTS	10.29	7.34	5.65	3.62	.75	.28	.13	.15	.62	2.12	5.36	8.12	44.43
YORBA LINDA	3.58	2.78	2.30	1.15	.27	.05	.02	.15	.29	.23	1.57	2.07	14.46
YOSEMITE PARK HEADQTRS	6.80	5.70	4.74	3.33	1.41	.60	.35	.28	.78	1.47	4.27	6.33	36.06
YREKA	3.68	2.17	1.80	.89	.77	.85	.40	.63	.59	1.25	2.34	3.83	19.20

CALIFORNIA

HEATING DEGREE DAY NORMALS (BASE 65 DEG F)

STATION	JUL	AUG	SEP	OCT	NOV	DEC	JAN	FEB	MAR	APR	MAY	JUN	ANN
ALDERPOINT	6	11	33	184	432	614	632	465	456	343	191	58	3425
ALTURAS RANGER STATION	56	104	232	524	813	1048	1119	868	853	648	427	204	6896
ASH MOUNTAIN	0	0	0	77	312	539	570	417	395	270	105	16	2701
AUBERRY 1 NW	0	0	13	136	408	611	642	498	487	327	150	40	3312
AUBURN	0	7	13	122	369	592	626	459	437	298	133	34	3090
AVALON PLEASURE PIER	68	42	60	111	191	305	337	281	291	224	173	122	2205
BAKERSFIELD WSO R	0	0	0	50	268	521	521	335	255	137	35	6	2128
BARRETT DAM	0	0	19	103	297	477	515	409	394	262	141	40	2657
BARSTOW	0	0	0	84	357	598	592	406	324	179	40	0	2580
BEAUMONT 1 E	0	0	10	102	290	453	499	405	410	279	146	35	2629
BERKELEY	110	98	81	126	282	453	474	344	353	285	209	136	2951
BISHOP WSO R	0	0	31	250	582	815	862	641	574	357	156	20	4288
BLUE CANYON WSO //R	35	76	132	356	621	791	865	753	831	658	432	200	5750
BLYTHE	0	0	0	10	152	375	379	223	135	38	0	0	1312
BLYTHE FAA AIRPORT	0	0	0	11	136	352	366	207	113	36	0	0	1221
BOCA	154	215	357	620	876	1156	1249	1033	1023	774	555	327	8339
BOWMAN DAM	43	84	141	380	663	837	899	773	831	657	452	206	5966
BRAWLEY 2 SW	0	0	0	9	113	329	351	209	134	54	6	0	1205
BROOKS FARNHAM RANCH	0	8	9	111	378	598	620	445	409	267	106	15	2966
BURBANK VALLEY PMP PLT	0	0	7	44	178	325	347	252	246	165	78	37	1679
BURNEY	62	103	214	487	768	995	1035	795	781	600	384	179	6403
BUTTONWILLOW	0	0	0	85	360	620	623	409	315	170	40	0	2622
CALAVERAS BIG TREES	42	77	150	386	657	837	887	750	794	636	431	202	5849
CAMPO	0	13	34	161	372	530	561	473	487	363	222	87	3303
CAMP PARDEE	0	0	0	91	333	580	611	434	394	253	96	21	2813
CANOGA PARK PIERCE COL	0	0	8	62	209	349	374	284	280	186	91	42	1885
CANYON DAM	48	87	208	496	804	1039	1088	893	884	657	433	198	6835
CEDARVILLE	10	47	161	452	774	1023	1085	837	809	585	364	158	6305
CHICO EXPERIMENT STA	0	0	8	109	384	605	626	434	384	239	75	14	2878
CHULA VISTA /	22	11	26	88	198	332	360	291	298	219	147	82	2074
CLAREMONT POMONA COL	0	0	13	73	207	369	397	310	309	216	112	42	2048
CLOVERDALE 3 SSE	0	0	14	110	327	536	564	403	388	273	120	28	2763
COALINGA	0	0	0	78	348	577	583	398	338	196	57	6	2581
COLFAX	0	8	24	151	417	608	636	498	499	357	174	54	3426
COLUSA 1 SSW	0	0	5	104	390	611	620	414	363	219	58	9	2793
CORCORAN IRRIG DIST	0	0	0	84	366	620	617	417	332	184	47	0	2667
CORONA	0	0	6	54	202	344	366	284	273	174	68	24	1795
COVELO	17	16	68	266	537	735	750	560	546	392	218	77	4182
CULVER CITY	19	12	20	53	128	235	268	213	222	166	111	70	1517
CUYAMACA	8	28	79	305	558	744	787	652	676	516	355	139	4847
DAGGETT FAA AP	0	0	0	42	269	505	518	338	255	121	23	0	2071
DAVIS 2 WSW EXP FARM	0	0	13	100	366	598	617	420	372	241	99	17	2843
DEATH VALLEY	0	0	0	0	106	391	388	174	75	14	0	0	1148
DENAIR 3 NNE	0	0	13	132	405	636	632	437	381	237	84	17	2974
DE SABLA	7	24	52	225	516	698	741	594	605	442	245	90	4239
EAGLE MOUNTAIN	0	0	0	9	104	300	343	199	141	42	0	0	1138
EAST PARK RESERVOIR	0	0	20	155	444	648	679	510	490	343	141	27	3457
EL CAPITAN DAM	0	0	6	30	135	282	320	248	249	165	76	23	1534
EL CENTRO 2 SSW	0	0	0	6	125	338	344	208	133	52	7	0	1213
ELECTRA POWER HOUSE	0	0	9	100	360	583	592	431	400	265	102	16	2858
ELSINORE	0	0	7	80	244	412	437	335	311	200	80	21	2127
ESCONDIDO	0	0	0	79	228	372	397	314	298	192	94	32	2006
EUREKA WSO R	276	248	246	332	408	518	549	456	518	468	397	309	4725
FAIRFIELD FIRE STATION	0	6	12	91	339	570	580	389	341	235	94	29	2686
FAIRMONT	0	0	15	139	381	598	639	496	493	342	177	51	3331

81

CALIFORNIA

HEATING DEGREE DAY NORMALS (BASE 65 DEG F)

STATION	JUL	AUG	SEP	OCT	NOV	DEC	JAN	FEB	MAR	APR	MAY	JUN	ANN
FONTANA KAISER	0	5	5	35	153	277	316	239	242	157	77	23	1529
FORT BIDWELL	26	70	190	459	777	1029	1088	832	797	582	366	165	6381
FORT BRAGG	260	245	222	304	390	512	539	451	484	426	360	276	4469
FORT JONES RANGER STA	14	43	147	409	717	939	970	722	685	516	310	117	5589
FORT ROSS	248	226	191	257	357	471	490	403	450	411	357	267	4128
FRESNO WSO R	0	0	0	88	354	611	6 3	406	336	187	52	8	2647
FRIANT GOVERNMENT CAMP	0	0	5	78	339	595	611	426	384	240	82	9	2769
GLENNVILLE	10	38	82	284	528	701	722	585	617	465	283	108	4423
GRANT GROVE	104	155	247	493	732	896	964	848	933	783	583	305	7043
GRATON	33	42	69	184	399	589	605	442	437	327	193	89	3409
HAIWEE	0	0	9	153	486	738	781	580	508	305	128	12	3700
HALF MOON BAY	236	208	175	239	324	428	453	378	422	384	335	264	3846
HANFORD	0	0	0	95	384	629	623	420	343	188	48	6	2736
HAPPY CAMP RANGER STA	0	9	45	266	561	778	815	602	555	389	194	49	4263
HAT CREEK PH NO 1	22	55	158	415	708	942	970	748	719	525	309	119	5690
HAYFIELD PUMPING PLANT	0	0	0	20	161	366	407	268	206	90	13	0	1531
HEALDSBURG	0	10	18	98	315	518	536	367	338	235	103	34	2572
HENSHAW DAM	7	18	48	194	435	611	642	518	518	385	240	95	3711
HETCH HETCHY	11	26	70	269	573	800	837	655	663	493	300	119	4816
HUNTINGTON LAKE	153	183	302	546	777	952	1008	893	986	828	639	367	7634
IMPERIAL	0	0	0	0	93	277	301	172	105	28	0	0	976
INDIO U S DATE GARDEN	0	0	0	0	99	298	325	187	117	33	0	0	1059
INYOKERN	0	0	0	84	372	632	639	437	363	197	48	0	2772
IRON MOUNTAIN	0	0	0	11	124	340	374	219	143	41	0	0	1252
JESS VALLEY	81	124	236	499	795	1004	1073	882	908	702	481	261	7046
JULIAN WYNOLA	28	42	76	224	426	580	632	535	574	450	318	166	4051
KENTFIELD	25	44	50	142	357	543	567	400	378	277	159	67	3009
KERN RIVER PH NO 1	0	0	0	33	200	465	471	293	234	147	35	0	1878
KERN RIVER PH NO 3	0	0	0	106	360	577	601	454	415	266	99	14	2892
KETTLEMAN STATION	0	0	0	36	257	533	549	342	270	155	32	7	2181
KING CITY	23	18	31	119	321	487	499	358	347	248	135	53	2639
KLAMATH R	192	179	190	301	429	564	592	479	530	453	357	243	4509
LAGUNA BEACH	37	29	37	88	214	344	366	302	313	249	159	83	2221
LAKE ARROWHEAD	17	33	92	329	639	828	862	708	710	555	381	156	5310
LAKEPORT	0	15	29	199	486	676	707	526	505	356	177	53	3729
LAKE SPAULDING	80	125	190	436	726	905	955	795	846	687	471	231	6447
LA MESA	0	0	10	45	141	272	302	241	243	166	94	53	1567
LE GRAND	0	0	0	89	363	605	601	409	350	211	60	8	2696
LEMON COVE	0	0	0	77	327	583	583	386	315	188	48	7	2514
LINDSAY	0	0	0	93	360	595	592	406	341	187	53	7	2634
LIVERMORE COUNTY F D	5	11	18	114	366	574	592	431	412	285	155	48	3011
LODI	0	8	15	114	381	605	608	417	360	225	96	32	2861
LOMPOC	120	105	88	148	269	394	406	325	350	291	233	161	2890
LONG BEACH WSO	0	0	6	39	140	281	307	245	225	150	69	23	1485
LOS ANGELES WSO R	17	12	18	55	139	255	286	233	240	180	106	54	1595
LOS ANGELES CIVIC CENTER	0	0	0	27	108	218	252	191	190	129	62	27	1204
LOS BANOS	0	0	0	75	354	598	605	400	329	190	55	10	2616
LOS GATOS	17	13	19	115	327	508	524	386	363	267	142	59	2740
MADERA	0	0	5	83	357	608	611	414	338	195	53	9	2673
MANZANITA LAKE //	124	171	278	543	822	1014	1082	935	1001	795	552	301	7618
MARICOPA	0	0	0	49	291	555	564	358	276	172	32	5	2302
MARYSVILLE	0	0	0	71	339	580	601	392	322	184	53	9	2551
MC CLOUD	41	78	178	425	711	905	961	773	784	594	375	166	5991
MECCA FIRE STATION	0	0	0	9	129	338	357	203	116	34	0	0	1186
MERCED FIRE STATION 2	0	0	0	78	360	608	601	409	335	197	57	8	2653

CALIFORNIA

HEATING DEGREE DAY NORMALS (BASE 65 DEG F)

STATION	JUL	AUG	SEP	OCT	NOV	DEC	JAN	FEB	MAR	APR	MAY	JUN	ANN
MINERAL	118	169	262	521	801	992	1051	874	918	735	527	291	7259
MODESTO	0	0	6	92	363	598	598	398	333	198	70	15	2671
MONO LAKE	27	62	175	446	762	1020	1070	893	846	633	412	173	6519
MONTEREY	174	149	104	145	258	391	422	336	372	333	282	204	3170
MOUNT HAMILTON	30	59	82	254	504	658	716	622	701	552	364	182	4724
MOUNT SHASTA WSO //R	30	67	163	422	723	921	964	756	769	582	369	165	5931
MOUNT WILSON FC 338 B //	9	15	74	239	451	614	670	588	653	509	337	137	4296
NAPA STATE HOSPITAL	13	17	33	104	321	512	539	372	363	270	143	62	2749
NEEDLES FAA AIRPORT	0	0	0	9	151	383	406	242	159	40	0	0	1390
NEVADA CITY	12	35	113	346	606	784	815	655	657	483	287	116	4909
NEWPORT BEACH HARBOR	26	11	28	86	185	301	335	272	279	217	141	73	1954
OAKLAND WSO R	68	67	61	124	300	477	496	350	347	274	196	117	2877
OJAI	0	7	11	78	234	391	419	319	310	215	111	52	2147
ORANGE COVE	0	0	0	90	336	595	608	417	353	212	67	7	2685
ORICK PRAIRIE CREEK PK	205	179	190	326	477	620	636	504	549	483	384	264	4817
ORLAND	0	0	0	97	372	608	632	437	375	225	66	12	2824
ORLEANS	0	9	35	203	486	679	701	507	471	320	158	60	3629
OXNARD	49	34	47	87	190	295	326	267	282	230	162	99	2068
PALMDALE	0	0	7	108	378	598	608	445	399	255	92	18	2908
PALM SPRINGS	0	0	0	7	108	299	317	192	140	46	0	0	1109
PALO ALTO JR MUSEUM	31	28	56	142	342	518	533	381	366	269	159	66	2891
PALOMAR MT OBSERVATORY	6	9	46	225	453	629	682	588	629	472	301	101	4141
PARKER RESERVOIR	0	0	0	6	121	335	370	214	143	35	0	0	1224
PASADENA	0	0	0	38	152	286	314	241	236	166	79	38	1550
PASO ROBLES	0	9	25	138	375	558	555	400	384	270	128	43	2885
PASO ROBLES FAA AP	0	7	19	130	381	577	577	428	409	284	121	40	2973
PETALUMA FIRE STA 2	14	17	23	112	339	543	555	398	394	303	187	75	2960
PINNACLES NAT MONUMENT	5	11	19	117	345	524	552	428	431	313	160	51	2956
PISMO BEACH	143	121	101	137	228	347	381	302	327	273	236	159	2755
PLACERVILLE	6	16	51	232	519	713	741	574	555	399	207	74	4087
POMONA CAL POLY	0	0	7	64	206	365	391	305	299	202	100	33	1972
PORTERVILLE	0	0	0	68	318	574	577	378	303	174	55	9	2456
PRIEST VALLEY	8	20	57	242	513	685	713	563	577	429	247	90	4144
QUINCY	45	105	177	422	708	930	952	734	707	525	325	134	5764
RANDSBURG	0	0	0	108	366	595	629	459	422	249	88	7	2923
RED BLUFF WSO R	0	0	0	90	348	574	605	409	363	226	59	8	2682
REDDING FIRE STN 4	0	0	0	79	340	549	574	386	333	210	63	10	2544
REDLANDS	0	0	7	73	218	382	406	306	297	193	84	26	1992
REDWOOD CITY	11	14	26	122	315	490	505	364	338	242	125	48	2600
RICHMOND	98	85	60	98	267	450	474	328	316	243	160	108	2687
RIVERSIDE FIRE STA. #3	0	0	0	55	210	363	378	286	267	167	70	22	1818
SACRAMENTO WSO R	0	0	7	82	360	601	611	412	366	229	83	21	2772
SACRAMENTO CITY WSO R	0	0	0	61	309	543	555	358	304	186	67	15	2398
SAINT HELENA	0	0	15	126	360	552	580	414	397	281	123	31	2879
SAINT MARYS COLLEGE	16	33	38	183	438	629	648	473	459	336	195	96	3544
SALT SPRINGS PWR HOUSE	5	14	29	190	440	623	679	546	561	430	243	96	3856
SAN BERNARDINO CO HOSP	0	0	0	47	202	353	375	281	268	166	67	18	1777
SANDBERG WSO //R	0	10	32	190	480	722	778	633	667	513	316	94	4435
SAN DIEGO WSO R	5	0	7	32	118	240	258	196	193	124	71	40	1284
SAN FRANCISCO WSO R	103	89	80	148	315	490	512	375	378	306	226	139	3161
S F FEDERAL BLDG. WSO R	202	173	109	124	234	403	428	311	332	294	260	201	3071
SAN GABRIEL FIRE DEPT	0	0	0	39	155	297	320	243	228	151	70	29	1532
SAN JOSE R	7	9	17	94	294	474	481	342	322	231	124	44	2439
SAN LUIS OBISPO POLY	59	44	55	106	221	367	400	312	335	276	208	115	2498
SAN RAFAEL	15	17	32	92	270	456	481	333	316	235	133	59	2439

CALIFORNIA

HEATING DEGREE DAY NORMALS (BASE 65 DEG F)

STATION	JUL	AUG	SEP	OCT	NOV	DEC	JAN	FEB	MAR	APR	MAY	JUN	ANN
SANTA ANA FIRE STATION	0	0	0	39	140	265	289	226	219	150	71	31	1430
SANTA BARBARA	37	16	29	85	185	319	341	273	274	204	147	83	1993
SANTA BARBARA FAA AP	55	40	61	129	249	384	403	316	319	248	180	103	2487
SANTA CRUZ	80	81	83	161	330	477	496	381	391	315	214	127	3136
SANTA MARIA WSO R	97	85	83	152	289	422	440	356	391	324	254	161	3054
SANTA MONICA PIER	66	26	45	80	150	237	271	235	261	220	172	110	1873
SANTA PAULA	34	28	37	92	212	316	337	266	283	205	142	78	2030
SANTA ROSA	14	21	33	127	360	546	567	400	394	289	157	72	2980
SCOTIA	154	134	147	251	393	539	567	451	496	426	329	216	4103
SHASTA DAM	0	0	7	110	361	574	617	459	425	273	100	18	2944
SIERRAVILLE RANGER STA//	71	120	233	493	795	1039	1110	885	853	648	436	212	6895
SONOMA	12	20	34	134	369	558	570	398	384	285	159	75	2998
SONORA RS	0	10	24	165	435	645	670	515	499	363	162	51	3539
SO ENTRANCE YOSEMITE	40	67	146	388	654	822	877	748	797	636	426	189	5790
STOCKTON WSO	0	0	0	76	351	605	614	412	350	206	52	8	2674
STOCKTON FIRE STA NO 4	0	0	13	109	375	614	626	428	363	223	76	19	2846
STONY GORGE RESERVOIR	0	0	9	122	414	623	654	473	446	294	101	12	3148
STRAWBERRY VALLEY	34	67	110	314	591	756	812	678	716	552	339	150	5119
SUSANVILLE AIRPORT	15	57	157	462	774	1048	1082	829	778	564	337	130	6233
TAHOE CITY	146	192	330	608	861	1063	1125	972	1017	816	605	351	8086
TEHACHAPI	0	18	74	288	540	716	753	616	636	474	271	108	4494
TEJON RANCHO	0	0	7	96	323	564	577	398	341	212	68	17	2603
THERMAL FAA AIRPORT	0	0	0	9	122	327	343	199	120	34	0	0	1154
TIGER CREEK PH	0	13	19	179	468	707	704	529	527	384	201	66	3797
TORRANCE	15	12	18	62	157	281	310	250	257	193	110	54	1719
TRACY CARBONA	0	0	6	80	351	601	617	414	335	207	78	15	2704
TRONA	0	0	0	42	321	614	611	389	278	136	24	0	2415
TRUCKEE RANGER STATION	132	186	312	595	891	1153	1215	1011	1020	801	580	334	8230
TULELAKE //R	65	118	238	527	822	1060	1094	846	837	639	408	201	6855
TUSTIN IRVINE RANCH	0	5	9	70	194	335	353	283	273	187	100	45	1854
TWENTYNINE PALMS	0	0	0	42	254	481	496	326	248	115	13	0	1975
TWIN LAKES	264	301	396	663	912	1107	1187	1044	1138	960	750	474	9196
UKIAH	0	0	14	132	381	558	574	417	409	292	142	39	2958
U C L A	19	19	32	65	119	210	245	202	225	181	124	68	1509
UPLAND 3 N	0	5	13	91	231	381	403	319	328	230	124	50	2175
VACAVILLE	0	0	5	102	366	598	614	409	357	237	81	19	2788
VICTORVILLE PUMP PLANT	0	0	8	134	414	642	657	490	442	271	120	14	3192
VISALIA	0	0	0	70	321	583	589	384	298	173	42	0	2460
WASCO	0	0	0	69	333	598	601	389	290	154	32	0	2466
WATSONVILLE WATERWORKS	128	128	110	174	321	477	499	386	397	321	248	164	3353
WEAVERVILLE RANGER STA	12	20	99	332	630	859	877	666	651	479	269	99	4993
WILLOWS	0	0	6	97	366	605	629	434	378	233	74	14	2836
WINTERS	0	0	0	74	333	580	601	395	329	204	65	12	2593
WOODFORDS	17	67	148	415	720	936	983	801	806	606	390	159	6048
WOODLAND 1 WNW	0	0	7	88	354	598	617	412	347	209	70	7	2709
YORBA LINDA	0	0	0	55	172	299	320	253	256	179	84	26	1644
YOSEMITE PARK HEADQTRS	10	20	52	269	612	868	871	650	623	450	259	101	4785
YREKA	8	24	108	366	696	927	961	722	694	498	281	110	5395

CALIFORNIA

COOLING DEGREE DAY NORMALS (BASE 65 DEG F)

STATION		JAN	FEB	MAR	APR	MAY	JUN	JUL	AUG	SEP	OCT	NOV	DEC	ANN
ALDERPOINT		0	0	0	7	14	79	205	197	111	10	0	0	623
ALTURAS RANGER STATION		0	0	0	0	5	27	84	73	13	0	0	0	202
ASH MOUNTAIN		0	0	0	45	115	310	555	508	332	117	6	0	1988
AUBERRY 1 NW		0	0	0	21	92	238	450	392	241	65	0	0	1499
AUBURN		0	0	0	22	59	199	378	351	217	73	0	0	1299
AVALON PLEASURE PIER		5	0	6	17	30	65	105	132	111	70	14	0	555
BAKERSFIELD WSO	R	0	0	10	68	208	405	605	539	369	143	0	0	2347
BARRETT DAM		0	0	0	10	33	136	335	332	223	66	0	0	1135
BARSTOW		0	0	7	59	170	393	617	552	342	99	0	0	2239
BEAUMONT 1 E		0	0	7	12	53	167	375	364	238	96	8	0	1320
BERKELEY		0	0	0	0	0	19	8	11	42	11	0	0	91
BISHOP WSO	R	0	0	0	9	60	191	372	295	106	12	0	0	1045
BLUE CANYON WSO	//R	0	0	0	7	14	53	137	135	66	21	0	0	433
BLYTHE		0	18	60	167	381	627	862	806	600	271	17	0	3809
BLYTHE FAA AIRPORT		6	25	63	207	441	705	936	871	669	337	43	8	4311
BOCA		0	0	0	0	0	0	12	17	0	0	0	0	29
BOWMAN DAM		0	0	0	0	6	35	95	102	39	14	0	0	291
BRAWLEY 2 SW		0	16	47	153	341	582	812	784	612	316	32	0	3695
BROOKS FARNHAM RANCH		0	0	0	15	84	225	375	330	195	49	0	0	1273
BURBANK VALLEY PMP PLT		0	8	13	33	62	166	313	316	250	109	16	6	1292
BURNEY		0	0	0	0	0	20	74	60	10	0	0	0	164
BUTTONWILLOW		0	0	0	47	152	338	524	462	285	82	0	0	1890
CALAVERAS BIG TREES		0	0	0	0	6	40	101	95	33	7	0	0	282
CAMPO		0	0	0	0	8	84	257	261	148	31	0	0	789
CAMP PARDEE		0	0	0	25	81	231	419	372	244	85	0	0	1457
CANOGA PARK PIERCE COL		0	9	14	33	66	177	341	339	245	99	14	0	1337
CANYON DAM		0	0	0	0	0	21	79	69	22	0	0	0	191
CEDARVILLE		0	0	0	0	13	68	208	177	62	5	0	0	533
CHICO EXPERIMENT STA		0	0	0	23	100	254	406	357	221	53	0	0	1414
CHULA VISTA		0	0	0	0	0	25	81	125	107	41	6	0	385
CLAREMONT POMONA COL		0	5	5	24	34	129	291	301	232	101	12	0	1134
CLOVERDALE 3 SSE		0	0	0	21	45	133	223	198	143	51	0	0	814
COALINGA		0	0	0	37	138	309	512	453	279	74	0	0	1802
COLFAX		0	0	0	27	65	207	391	340	213	58	0	0	1301
COLUSA 1 SSW		0	0	0	30	108	258	388	338	209	42	0	0	1373
CORCORAN IRRIG DIST		0	0	0	43	146	319	502	446	273	75	0	0	1804
CORONA		0	7	6	21	52	162	316	316	237	95	13	0	1225
COVELO		0	0	0	0	13	95	234	195	104	12	0	0	653
CULVER CITY		0	11	8	16	27	85	149	179	158	99	29	9	770
CUYAMACA		0	0	0	0	11	67	178	161	67	8	0	0	492
DAGGETT FAA AP		0	7	22	94	262	504	732	667	447	163	8	0	2906
DAVIS 2 WSW EXP FARM		0	0	0	22	81	194	288	255	175	44	0	0	1059
DEATH VALLEY		0	28	147	338	626	900	1144	1076	783	400	25	0	5467
DENAIR 3 NNE		0	0	0	21	78	200	326	277	169	42	0	0	1113
DE SABLA		0	0	0	10	22	117	233	207	112	23	0	0	724
EAGLE MOUNTAIN		6	28	79	204	423	693	899	837	651	353	59	11	4243
EAST PARK RESERVOIR		0	0	0	13	54	192	360	302	161	27	0	0	1109
EL CAPITAN DAM		0	10	10	36	61	170	332	353	294	142	27	9	1444
EL CENTRO 2 SSW		0	17	50	151	351	597	818	791	606	304	32	0	3717
ELECTRA POWER HOUSE		0	0	0	19	68	202	378	332	207	47	0	0	1253
ELSINORE		0	0	7	29	77	231	431	422	307	114	10	0	1628
ESCONDIDO		0	0	0	12	38	128	257	270	190	76	9	0	980
EUREKA WSO	R	0	0	0	0	0	0	0	0	0	0	0	0	0
FAIRFIELD FIRE STATION		0	0	0	19	39	128	212	210	168	45	0	0	821
FAIRMONT		0	0	0	21	65	210	431	401	258	101	0	0	1487

CALIFORNIA

COOLING DEGREE DAY NORMALS (BASE 65 DEG F)

STATION	JAN	FEB	MAR	APR	MAY	JUN	JUL	AUG	SEP	OCT	NOV	DEC	ANN
FONTANA KAISER	15	23	38	55	105	227	425	424	338	178	48	17	1893
FORT BIDWELL	0	0	0	0	6	27	95	82	22	0	0	0	232
FORT BRAGG	0	0	0	0	0	0	0	0	0	0	0	0	0
FORT JONES RANGER STA	0	0	0	0	9	57	160	133	42	0	0	0	401
FORT ROSS	0	0	0	0	0	0	0	0	0	0	0	0	0
FRESNO WSO R	0	0	0	40	135	308	496	431	277	82	0	0	1769
FRIANT GOVERNMENT CAMP	0	0	0	30	110	270	481	425	266	81	0	0	1663
GLENNVILLE	0	0	0	10	66	190	168	70	8	0	0	0	512
GRANT GROVE	0	0	0	0	0	11	39	44	10	6	0	0	110
GRATON	0	0	0	0	0	44	57	61	48	7	0	0	217
HAIWEE	0	0	0	23	97	267	471	409	213	45	0	0	1525
HALF MOON BAY	0	0	0	0	0	0	0	6	0	0	0	0	6
HANFORD	0	0	0	35	123	276	437	378	228	55	0	0	1532
HAPPY CAMP RANGER STA	0	0	0	11	39	121	282	244	96	5	0	0	798
HAT CREEK PH NO 1	0	0	0	0	8	44	133	101	26	0	0	0	312
HAYFIELD PUMPING PLANT	0	16	36	129	286	531	778	722	528	240	20	0	3286
HEALDSBURG	0	0	0	19	47	139	190	183	147	48	0	0	773
HENSHAW DAM	0	0	0	0	11	86	258	260	144	23	0	0	782
HETCH HETCHY	0	0	0	10	21	71	184	172	76	18	0	0	552
HUNTINGTON LAKE	0	0	0	0	0	7	17	19	8	0	0	0	51
IMPERIAL	6	26	68	178	381	627	840	809	636	339	51	0	3961
INDIO U S DATE GARDEN	6	35	89	216	412	651	846	803	636	348	48	6	4096
INYOKERN	0	0	7	44	169	396	608	536	324	97	0	0	2181
IRON MOUNTAIN	6	25	72	203	434	714	933	862	654	333	46	8	4290
JESS VALLEY	0	0	0	0	0	18	71	62	14	0	0	0	165
JULIAN WYNOLA	0	0	0	0	8	103	232	231	124	38	0	0	736
KENTFIELD	0	0	0	0	13	64	87	93	77	15	0	0	349
KERN RIVER PH NO 1	0	0	20	114	255	459	694	642	468	219	17	9	2897
KERN RIVER PH NO 3	0	0	35	108	281	505	459	295	84	0	0	1767
KETTLEMAN STATION	0	0	12	77	193	394	601	549	384	150	5	0	2365
KING CITY	0	0	0	11	21	65	100	95	82	32	0	0	406
KLAMATH R	0	0	0	0	0	0	0	0	0	0	0	0	0
LAGUNA BEACH	0	0	0	0	7	23	90	119	100	36	0	0	375
LAKE ARROWHEAD	0	0	0	0	12	51	141	126	62	13	0	0	405
LAKEPORT	0	0	0	8	31	137	281	251	113	10	0	0	831
LAKE SPAULDING	0	0	0	0	0	21	46	50	16	0	0	0	133
LA MESA	0	8	7	16	32	107	215	251	214	110	18	8	986
LE GRAND	0	0	0	22	82	233	400	353	224	61	0	0	1375
LEMON COVE	0	0	0	56	141	313	496	428	276	90	0	0	1800
LINDSAY	0	0	0	34	127	286	468	397	246	59	0	0	1617
LIVERMORE COUNTY F D	0	0	0	9	41	111	201	194	138	40	0	0	734
LODI	0	0	0	21	77	185	276	247	162	33	0	0	1001
LOMPOC	0	0	0	0	0	0	11	28	31	17	0	0	87
LONG BEACH WSO	0	10	0	27	47	119	242	279	219	117	26	5	1091
LOS ANGELES WSO R	7	12	0	15	25	72	141	176	153	95	25	7	728
LOS ANGELES CIVIC CENTER	10	21	20	39	59	153	282	301	256	148	39	11	1339
LOS BANOS	0	0	0	34	98	247	412	366	234	57	0	0	1448
LOS GATOS	0	0	0	15	28	95	166	150	109	28	0	0	591
MADERA	0	0	0	36	128	288	468	409	254	71	0	0	1654
MANZANITA LAKE //	0	0	0	0	0	13	46	41	8	0	0	0	108
MARICOPA	0	0	9	82	187	392	605	543	363	136	0	0	2317
MARYSVILLE	0	0	0	34	121	279	425	369	246	65	0	0	1539
MC CLOUD	0	0	0	0	6	37	106	87	31	0	0	0	267
MECCA FIRE STATION	0	24	67	184	375	585	791	756	585	294	39	0	3700
MERCED FIRE STATION 2	0	0	0	29	110	257	425	366	231	47	0	0	1465

CALIFORNIA

COOLING DEGREE DAY NORMALS (BASE 65 DEG F)

STATION	JAN	FEB	MAR	APR	MAY	JUN	JUL	AUG	SEP	OCT	NOV	DEC	ANN
MINERAL	0	0	0	0	0	9	34	45	10	0	0	0	98
MODESTO	0	0	0	30	95	225	366	319	204	48	0	0	1287
MONO LAKE	0	0	0	0	6	23	95	93	19	0	0	0	236
MONTEREY	0	0	0	0	0	0	0	10	26	12	0	0	48
MOUNT HAMILTON	0	0	0	12	29	104	222	205	100	46	6	7	731
MOUNT SHASTA WSO //R	0	0	0	0	7	42	123	105	40	0	0	0	317
MOUNT WILSON FC 338 B //	0	0	8	14	42	134	229	195	146	68	7	6	849
NAPA STATE HOSPITAL	0	0	0	9	19	74	97	92	99	26	0	0	416
NEEDLES FAA AIRPORT	0	21	66	187	443	732	958	880	654	304	25	0	4270
NEVADA CITY	0	0	0	0	11	74	161	128	50	8	0	0	432
NEWPORT BEACH HARBOR	0	0	0	0	0	31	79	110	91	52	8	0	371
OAKLAND WSO R	0	0	0	0	0	24	28	43	61	18	0	0	174
OJAI	0	6	6	23	28	121	261	261	188	75	9	0	978
ORANGE COVE	0	0	0	35	123	277	465	415	266	84	0	0	1665
ORICK PRAIRIE CREEK PK	0	0	0	0	0	0	0	0	0	0	0	0	0
ORLAND	0	0	0	33	116	282	425	363	238	63	0	0	1520
ORLEANS	0	0	0	8	28	120	265	251	134	8	0	0	814
OXNARD	0	7	0	0	0	33	71	93	80	43	16	7	357
PALMDALE	0	0	5	39	104	282	508	462	283	77	0	0	1760
PALM SPRINGS	10	41	81	172	358	585	828	778	591	317	48	11	3820
PALO ALTO JR MUSEUM	0	0	0	0	14	51	71	68	65	6	0	0	275
PALOMAR MT OBSERVATORY	0	0	9	13	31	119	251	235	148	58	0	0	864
PARKER RESERVOIR	0	24	90	218	459	732	949	893	705	378	55	7	4510
PASADENA	0	14	13	37	51	152	301	313	263	125	23	7	1299
PASO ROBLES	0	0	0	9	22	112	221	211	127	32	0	0	734
PASO ROBLES FAA AP	0	0	0	11	31	151	282	262	157	40	0	0	934
PETALUMA FIRE STA 2	0	0	0	0	7	48	76	89	83	13	0	0	316
PINNACLES NAT MONUMENT	0	0	0	13	33	126	269	250	163	43	0	0	897
PISMO BEACH	0	0	0	0	0	6	10	19	26	23	6	0	90
PLACERVILLE	0	0	0	6	21	122	269	236	114	15	0	0	783
POMONA CAL POLY	0	8	8	25	44	135	298	311	235	105	17	5	1191
PORTERVILLE	0	0	6	54	157	327	515	450	285	90	0	0	1884
PRIEST VALLEY	0	0	0	0	12	90	222	194	87	10	0	0	615
QUINCY	0	0	0	0	9	35	104	93	21	0	0	0	262
RANDSBURG	0	0	9	51	153	358	583	518	319	114	0	0	2105
RED BLUFF WSO R	0	0	0	43	137	341	536	468	310	96	0	0	1931
REDDING FIRE STN 4	0	0	0	66	177	367	574	502	336	117	0	0	2139
REDLANDS	0	7	12	40	72	209	409	397	283	128	14	0	1571
REDWOOD CITY	0	0	0	8	20	75	116	110	86	17	0	0	432
RICHMOND	0	0	0	6	0	27	17	29	63	20	0	0	162
RIVERSIDE FIRE STA. #3	0	6	9	32	83	205	378	372	272	105	9	0	1471
SACRAMENTO WSO R	0	0	0	25	80	207	329	301	208	48	0	0	1198
SACRAMENTO CITY WSO R	0	0	0	39	107	231	360	329	242	71	0	0	1379
SAINT HELENA	0	0	0	14	26	109	192	171	108	24	0	0	644
SAINT MARYS COLLEGE	0	0	0	6	12	69	115	120	77	9	0	0	408
SALT SPRINGS PWR HOUSE	0	0	0	22	39	123	287	265	155	63	5	0	959
SAN BERNARDINO CO HOSP	0	10	10	37	89	234	443	434	312	127	16	6	1718
SANDBERG WSO //R	0	0	0	6	15	91	291	286	176	57	0	0	922
SAN DIEGO WSO R	0	11	7	10	21	79	170	226	187	109	22	0	842
SAN FRANCISCO WSO R	0	0	0	0	0	13	16	27	47	12	0	0	115
S F FEDERAL BLDG. WSO R	0	0	0	0	0	0	0	6	31	19	0	0	56
SAN GABRIEL FIRE DEPT	0	10	14	31	58	155	304	316	254	116	17	6	1281
SAN JOSE R	0	0	0	12	31	80	125	121	104	25	0	0	498
SAN LUIS OBISPO POLY	0	0	0	9	0	22	53	57	73	53	11	7	285
SAN RAFAEL	0	0	0	10	25	80	102	104	116	46	0	0	483

CALIFORNIA

COOLING DEGREE DAY NORMALS (BASE 65 DEG F)

STATION	JAN	FEB	MAR	APR	MAY	JUN	JUL	AUG	SEP	OCT	NOV	DEC	ANN
SANTA ANA FIRE STATION	7	14	14	27	49	115	223	257	223	120	32	8	1089
SANTA BARBARA	0	0	0	9	0	38	86	109	89	48	8	6	393
SANTA BARBARA FAA AP	0	0	0	0	0	19	61	84	76	29	0	0	269
SANTA CRUZ	0	0	0	0	0	13	12	26	35	9	0	0	95
SANTA MARIA WSO R	0	0	0	0	0	0	14	26	23	13	0	0	76
SANTA MONICA PIER	0	5	0	7	11	44	91	97	96	52	24	11	438
SANTA PAULA	5	9	8	13	21	57	114	140	124	64	29	12	596
SANTA ROSA	0	0	0	0	12	72	88	99	90	21	0	0	382
SCOTIA	0	0	0	0	0	0	0	7	9	0	0	0	16
SHASTA DAM	0	0	0	45	118	291	518	465	289	95	0	0	1821
SIERRAVILLE RANGER STA//	0	0	0	0	5	14	56	45	11	0	0	0	131
SONOMA	0	0	0	9	29	108	155	150	112	28	0	0	591
SONORA RS	0	0	0	18	44	174	366	326	183	50	0	0	1161
SO ENTRANCE YOSEMITE	0	0	0	0	7	33	102	88	26	7	0	0	263
STOCKTON WSO	0	0	0	26	92	251	403	366	247	63	0	0	1448
STOCKTON FIRE STA NO 4	0	0	0	19	61	184	298	263	166	41	0	0	1032
STONY GORGE RESERVOIR	0	0	0	18	79	243	425	369	222	47	0	0	1403
STRAWBERRY VALLEY	0	0	0	6	11	69	164	157	86	20	0	0	513
SUSANVILLE AIRPORT	0	0	0	0	9	55	154	122	22	0	0	0	362
TAHOE CITY	0	0	0	0	0	0	12	24	0	0	0	0	36
TEHACHAPI	0	0	0	0	23	102	234	183	71	9	0	0	622
TEJON RANCHO	0	0	6	56	145	332	530	465	316	121	0	0	1971
THERMAL FAA AIRPORT	5	26	74	196	400	627	834	787	603	300	32	0	3884
TIGER CREEK PH	0	0	0	12	34	129	294	270	145	36	0	0	920
TORRANCE	0	9	0	16	20	66	142	177	144	84	19	0	677
TRACY CARBONA	0	0	0	30	78	204	326	289	186	43	0	0	1156
TRONA	0	0	12	97	260	504	744	685	444	147	0	0	2893
TRUCKEE RANGER STATION	0	0	0	0	0	0	14	19	0	0	0	0	33
TULELAKE //R	0	0	0	0	0	21	75	59	13	0	0	0	168
TUSTIN IRVINE RANCH	0	6	0	13	29	96	186	216	174	82	11	0	813
TWENTYNINE PALMS	0	6	21	106	264	519	735	673	459	178	8	0	2969
TWIN LAKES	0	0	0	0	0	0	0	6	0	0	0	0	6
UKIAH	0	0	0	16	43	135	279	255	155	30	0	0	913
U C L A	12	20	11	22	21	59	134	165	164	114	50	24	796
UPLAND 3 N	0	8	8	26	44	140	310	309	241	110	18	9	1223
VACAVILLE	0	0	0	27	72	205	329	295	191	55	0	0	1174
VICTORVILLE PUMP PLANT	0	0	0	16	89	236	453	412	233	60	0	0	1499
VISALIA	0	0	0	53	144	313	493	431	279	91	0	0	1804
WASCO	0	0	5	55	168	358	543	477	306	87	0	0	1999
WATSONVILLE WATERWORKS	0	0	0	0	0	0	0	13	23	7	0	0	43
WEAVERVILLE RANGER STA	0	0	0	0	15	78	198	156	69	0	0	0	516
WILLOWS	0	0	0	29	108	263	397	338	225	69	0	0	1429
WINTERS	0	0	0	48	124	273	403	357	245	74	0	0	1524
WOODFORDS	0	0	0	0	14	51	150	142	31	9	0	0	397
WOODLAND 1 WNW	0	0	0	29	104	238	360	313	217	60	0	0	1321
YORBA LINDA	7	7	8	26	47	122	264	282	215	108	25	7	1118
YOSEMITE PARK HEADQTRS	0	0	0	9	20	95	224	206	85	15	0	0	654
YREKA	0	0	0	0	14	92	225	191	75	0	0	0	597

04 — CALIFORNIA

STATE-STATION NUMBER	STN TYP	NAME	LATITUDE DEG-MIN	LONGITUDE DEG-MIN	ELEVATION (FT)
4-0014	12	ACTON-ESDO CYN FC261F	N 3430	W 11816	2960
4-0029	12	ADIN RANGER STATION	N 4112	W 12057	4195
4-0088	13	ALDERPOINT	N 4010	W 12337	460
4-0115	12	ALISO CN OAT MT FC 446	N 3419	W 11833	2367
4-0136	12	ALPINE	N 3250	W 11646	1735
4-0144	12	ALTADENA	N 3411	W 11808	1127
4-0161	13	ALTURAS RANGER STATION	N 4130	W 12033	4400
4-0204	12	ANGIOLA	N 3559	W 11929	205
4-0212	12	ANGWIN PAC UNION COL	N 3834	W 12226	1815
4-0235	12	ANZA	N 3333	W 11640	3925
4-0343	13	ASH MOUNTAIN	N 3629	W 11850	1708
4-0379	13	AUBERRY 1 NW	N 3705	W 11930	2140
4-0383	13	AUBURN	N 3854	W 12104	1292
4-0395	13	AVALON PLEASURE PIER	N 3321	W 11819	25
4-0442	13	BAKERSFIELD WSO R	N 3525	W 11903	475
4-0449	12	BALCH POWER HOUSE	N 3655	W 11905	1720
4-0514	13	BARRETT DAM	N 3241	W 11640	1623
4-0519	13	BARSTOW	N 3454	W 11702	2162
4-0609	13	BEAUMONT 1 E	N 3356	W 11658	2605
4-0619	12	BEL AIR FC-10A	N 3405	W 11827	540
4-0693	13	BERKELEY	N 3752	W 12215	345
4-0738	12	BIG BAR RANGER STATION	N 4045	W 12315	1260
4-0758	12	BIG DALTON DAM FC223BE	N 3410	W 11749	1575
4-0779	12	BIG PINES PARK FC 83B	N 3423	W 11741	6862
4-0790	12	BIG SUR STATE PARK	N 3615	W 12147	235
4-0798	12	BIG TUJUNGA DAM FC46DE	N 3418	W 11811	2317
4-0822	13	BISHOP WSO R	N 3722	W 11822	4108
4-0897	13	BLUE CANYON WSO //R	N 3917	W 12042	5280
4-0924	13	BLYTHE	N 3337	W 11436	268
4-0927	13	BLYTHE FAA AIRPORT	N 3337	W 11443	395
4-0931	13	BOCA	N 3923	W 12006	5575
4-0983	12	BORREGO DESERT PARK	N 3316	W 11625	765
4-1018	13	BOWMAN DAM	N 3927	W 12039	5347
4-1048	13	BRAWLEY 2 SW	N 3257	W 11533	-100
4-1112	13	BROOKS FARNHAM RANCH	N 3846	W 12209	294
4-1130	12	BRUSH CREEK RS	N 3941	W 12120	3560
4-1194	13	BURBANK VALLEY PMP PLT	N 3411	W 11821	655
4-1214	13	BURNEY	N 4053	W 12140	3127
4-1244	13	BUTTONWILLOW	N 3524	W 11928	269
4-1253	12	CACHUMA DAM	N 3435	W 11959	781
4-1277	13	CALAVERAS BIG TREES	N 3817	W 12019	4696
4-1288	12	CALEXICO 2 NE	N 3241	W 11528	12
4-1312	12	CALISTOGA	N 3835	W 12235	364
4-1316	12	CALLAHAN	N 4119	W 12248	3185
4-1424	13	CAMPO	N 3237	W 11628	2630
4-1428	13	CAMP PARDEE	N 3815	W 12051	658
4-1484	13	CANOGA PARK PIERCE COL	N 3411	W 11834	790
4-1497	13	CANYON DAM	N 4010	W 12105	4555
4-1614	13	CEDARVILLE	N 4132	W 12010	4670
4-1653	12	CHALLENGE RANGER STA	N 3929	W 12113	2560

STATE-STATION NUMBER	STN TYP	NAME	LATITUDE DEG-MIN	LONGITUDE DEG-MIN	ELEVATION (FT)
4-1680	12	CHATSWORTH FC24F	N 3415	W 11836	948
4-1700	12	CHESTER	N 4018	W 12114	4525
4-1715	13	CHICO EXPERIMENT STA	N 3942	W 12147	205
4-1739	12	CHITTENDEN PASS	N 3654	W 12136	125
4-1758	13	CHULA VISTA	N 3236	W 11706	9
4-1779	13	CLAREMONT POMONA COL	N 3406	W 11743	1201
4-1838	13	CLOVERDALE 3 SSE	N 3846	W 12259	320
4-1864	13	COALINGA	N 3609	W 12021	671
4-1907	12	COLEMAN FISHERIES STA	N 4024	W 12208	420
4-1912	13	COLFAX	N 3906	W 12057	2418
4-1916	12	COLGATE POWER HOUSE	N 3920	W 12111	595
4-1948	13	COLUSA 1 SSW	N 3912	W 12201	60
4-2012	13	CORCORAN IRRIG DIST	N 3606	W 11934	200
4-2027	12	CORNING HOUGHTON RANCH	N 3954	W 12221	487
4-2031	13	CORONA	N 3352	W 11734	710
4-2081	13	COVELO	N 3947	W 12315	1385
4-2090	12	COVINA TEMPLE FC 193 B	N 3405	W 11752	575
4-2147	12	CRESCENT CITY 1 N	N 4146	W 12412	40
4-2214	13	CULVER CITY	N 3401	W 11824	106
4-2218	12	CUMMINGS	N 3950	W 12338	1285
4-2239	13	CUYAMACA	N 3259	W 11635	4650
4-2257	13	DAGGETT FAA AP	N 3452	W 11647	1922
4-2294	13	DAVIS 2 WSW EXP FARM	N 3832	W 12146	60
4-2319	13	DEATH VALLEY	N 3628	W 11652	-194
4-2331	12	DEEP SPRINGS COLLEGE	N 3722	W 11759	5225
4-2346	12	DELANO	N 3547	W 11915	323
4-2389	13	DENAIR 3 NNE	N 3734	W 12047	137
4-2402	13	DE SABLA	N 3952	W 12137	2713
4-2406	12	DESCANSO RANGER STA	N 3251	W 11637	3500
4-2494	12	DOWNEY FD FC107C	N 3356	W 11808	116
4-2500	12	DOWNIEVILLE RANGER STA	N 3934	W 12050	2895
4-2516	12	DRY CANYON RESERVOIR	N 3429	W 11832	1455
4-2568	12	DUNNIGAN	N 3853	W 12158	60
4-2598	13	EAGLE MOUNTAIN	N 3348	W 11527	973
4-2640	13	EAST PARK RESERVOIR	N 3922	W 12231	1205
4-2709	13	EL CAPITAN DAM	N 3253	W 11649	600
4-2713	13	EL CENTRO 2 SSW	N 3246	W 11534	-30
4-2728	13	ELECTRA POWER HOUSE	N 3820	W 12040	715
4-2756	12	ELLERY LAKE	N 3756	W 11914	9645
4-2760	12	ELLIOTT	N 3814	W 12112	92
4-2805	13	ELSINORE	N 3340	W 11720	1285
4-2862	13	ESCONDIDO	N 3307	W 11705	660
4-2910	13	EUREKA WSO R	N 4048	W 12410	43
4-2920	12	EXCHEQUER RESERVOIR	N 3735	W 12016	442
4-2934	13	FAIRFIELD FIRE STATION	N 3816	W 12202	38
4-2941	13	FAIRMONT	N 3442	W 11826	3060
4-3038	12	FIDDLETOWN LYNCH RANCH	N 3832	W 12042	2140
4-3083	12	FIVE POINTS 5 SSW	N 3622	W 12009	285
4-3120	13	FONTANA KAISER	N 3405	W 11730	1090
4-3130	12	FOREST GLEN	N 4023	W 12320	2340

STATE-STATION NUMBER	STN TYP	NAME	LATITUDE DEG-MIN	LONGITUDE DEG-MIN	ELEVATION (FT)
4-3134	12	FORESTHILL RANGER STA	N 3901	W 12051	3015
4-3157	13	FORT BIDWELL	N 4151	W 12008	4498
4-3161	13	FORT BRAGG	N 3927	W 12348	80
4-3182	13	FORT JONES RANGER STA	N 4136	W 12251	2725
4-3191	13	FORT ROSS	N 3831	W 12315	116
4-3242	12	FRENCH GULCH	N 4042	W 12238	1100
4-3257	13	FRESNO WSO R	N 3646	W 11943	328
4-3261	13	FRIANT GOVERNMENT CAMP	N 3659	W 11943	410
4-3320	12	GARBERVILLE	N 4006	W 12348	340
4-3357	12	GASQUET RANGER STATION	N 4152	W 12358	384
4-3369	12	GEM LAKE	N 3745	W 11908	8970
4-3384	12	GEORGETOWN RANGER STA	N 3855	W 12047	3001
4-3452	12	GLENDORA WEST FC 185	N 3408	W 11752	822
4-3463	13	GLENNVILLE	N 3543	W 11842	3140
4-3491	12	GOLD RUN	N 3910	W 12052	3320
4-3551	13	GRANT GROVE	N 3644	W 11858	6600
4-3578	13	GRATON	N 3826	W 12252	200
4-3614	12	GREENVIEW	N 4133	W 12254	2818
4-3621	12	GREENVILLE RANGER STA	N 4008	W 12056	3560
4-3672	12	GROVELAND RANGER STA	N 3749	W 12006	3145
4-3704	12	HAINES CYN UPPER FC367	N 3416	W 11815	3440
4-3710	13	HAIWEE	N 3608	W 11757	3825
4-3714	13	HALF MOON BAY	N 3728	W 12226	60
4-3747	13	HANFORD	N 3620	W 11940	242
4-3761	13	HAPPY CAMP RANGER STA	N 4148	W 12322	1150
4-3791	12	HARRISON GULCH RS	N 4022	W 12258	2750
4-3800	12	HARRY ENGLEBRIGHT DAM	N 3914	W 12116	580
4-3824	13	HAT CREEK PH NO 1	N 4056	W 12133	3015
4-3855	13	HAYFIELD PUMPING PLANT	N 3342	W 11538	1370
4-3859	12	HAYFORK RANGER STATION	N 4033	W 12310	2340
4-3875	13	HEALDSBURG	N 3837	W 12252	102
4-3878	12	HEALDSBURG NO 2	N 3837	W 12252	102
4-3896	12	HEMET	N 3345	W 11657	1655
4-3914	13	HENSHAW DAM	N 3314	W 11646	2700
4-3925	12	HERNANDEZ 2 NW	N 3625	W 12055	2160
4-3939	13	HETCH HETCHY	N 3757	W 11947	3870
4-3987	12	HILTS	N 4200	W 12238	2900
4-4176	11	HUNTINGTON LAKE	N 3714	W 11913	7020
4-4211	12	IDYLLWILD FIRE DEPT	N 3345	W 11643	5397
4-4223	13	IMPERIAL	N 3251	W 11534	-64
4-4232	12	INDEPENDENCE	N 3648	W 11812	3950
4-4259	13	INDIO U S DATE GARDEN	N 3344	W 11615	11
4-4278	13	INYOKERN	N 3539	W 11749	2440
4-4288	12	IOWA HILL	N 3905	W 12050	3056
4-4297	13	IRON MOUNTAIN	N 3408	W 11508	922
4-4374	13	JESS VALLEY	N 4116	W 12018	5300
4-4418	13	JULIAN WYNOLA	N 3306	W 11639	3650
4-4422	12	JUNCAL DAM	N 3429	W 11930	2075
4-4467	12	KEE RANCH	N 3410	W 11632	4325
4-4500	13	KENTFIELD	N 3757	W 12233	120

04 — CALIFORNIA

STATE-STATION NUMBER	STN TYP	NAME	LATITUDE DEG-MIN	LONGITUDE DEG-MIN	ELEVATION (FT)
4-4508	12	KERLINGER	N 3741	W 12126	177
4-4520	13	KERN RIVER PH NO 1	N 3528	W 11847	970
4-4523	13	KERN RIVER PH NO 3	N 3547	W 11826	2703
4-4536	13	KETTLEMAN STATION	N 3604	W 12005	508
4-4555	13	KING CITY	N 3612	W 12108	320
4-4577	13	KLAMATH R	N 4131	W 12402	25
4-4628	12	LA CRESCENTA FC 251 B	N 3413	W 11814	1565
4-4647	13	LAGUNA BEACH	N 3333	W 11747	35
4-4671	13	LAKE ARROWHEAD	N 3415	W 11711	5205
4-4701	13	LAKEPORT	N 3902	W 12255	1347
4-4713	13	LAKE SPAULDING	N 3919	W 12038	5156
4-4735	13	LA MESA	N 3246	W 11701	530
4-4863	12	LEBEC	N 3450	W 11852	3585
4-4867	12	LECHUZA PT STA FC 352B	N 3405	W 11853	1600
4-4884	13	LE GRAND	N 3714	W 12015	255
4-4890	13	LEMON COVE	N 3623	W 11902	513
4-4957	13	LINDSAY	N 3611	W 11904	395
4-4997	13	LIVERMORE COUNTY F D	N 3740	W 12146	490
4-5017	12	LOCKWOOD 2 N	N 3558	W 12105	1104
4-5032	13	LODI	N 3807	W 1211?	40
4-5064	13	LOMPOC	N 3439	W 12027	95
4-5085	13	LONG BEACH WSO	N 3349	W 11809	34
4-5107	12	LOS ALAMOS	N 3445	W 12017	565
4-5114	13	LOS ANGELES WSO R	N 3356	W 11824	105
4-5115	13	LOS ANGELES CIVIC CENTER	N 3403	W 11814	257
4-5118	13	LOS BANOS	N 3703	W 12052	120
4-5119	12	LOS BANOS ARBURUA RCH	N 3653	W 12056	860
4-5123	13	LOS GATOS	N 3714	W 12158	365
4-5147	12	LOS PRIETOS RANGER STA	N 3433	W 11947	1024
4-5218	12	LYTLE CREEK RANGER STA	N 3414	W 11729	2730
4-5233	13	MADERA	N 3658	W 12004	268
4-5244	12	MAD RIVER RANGER STA	N 4027	W 12332	2775
4-5311	13	MANZANITA LAKE //	N 4032	W 12134	5850
4-5338	13	MARICOPA	N 3505	W 11923	675
4-5346	12	MARIPOSA	N 3729	W 11958	2011
4-5385	13	MARYSVILLE	N 3909	W 12135	60
4-5449	13	MC CLOUD	N 4116	W 12208	3300
4-5502	13	MECCA FIRE STATION	N 3334	W 11604	-180
4-5528	12	MENDOTA DAM	N 3647	W 12022	166
4-5532	13	MERCED FIRE STATION 2	N 3718	W 12029	169
4-5598	12	MIDDLETOWN	N 3845	W 12237	1122
4-5679	13	MINERAL	N 4021	W 12136	4911
4-5738	13	MODESTO	N 3739	W 12100	91
4-5756	12	MOJAVE	N 3503	W 11810	2735
4-5779	13	MONO LAKE	N 3800	W 11909	6450
4-5795	13	MONTEREY	N 3636	W 12154	345
4-5915	12	MOUNT DIABLO N GATE	N 3752	W 12156	2100
4-5933	13	MOUNT HAMILTON	N 3720	W 12139	4206
4-5941	12	MOUNT HEBRON RS	N 4147	W 12202	4250
4-5983	13	MOUNT SHASTA WSO //R	N 4119	W 12219	3544

04 — CALIFORNIA

STATE-STATION NUMBER	STN TYP	NAME	LATITUDE DEG-MIN	LONGITUDE DEG-MIN	ELEVATION (FT)
4-6006	13	MOUNT WILSON FC 338 B //	N 3414	W 11804	5709
4-6027	12	MUIR WOODS	N 3754	W 12234	225
4-6074	13	NAPA STATE HOSPITAL	N 3817	W 12216	60
4-6118	13	NEEDLES FAA AIRPORT	N 3446	W 11437	913
4-6136	13	NEVADA CITY	N 3916	W 12101	2520
4-6144	12	NEWARK	N 3731	W 12202	10
4-6162	12	NEWHALL SOLEDAD FC32CE	N 3423	W 11832	1243
4-6168	12	NEWMAN 2 NW	N 3721	W 12103	108
4-6175	13	NEWPORT BEACH HARBOR	N 3336	W 11753	10
4-6197	12	NILAND	N 3317	W 11531	-55
4-6252	12	NORTH FORK RANGER STA	N 3714	W 11930	2630
4-6335	13	OAKLAND WSO R	N 3744	W 12212	6
4-6370	12	OCCIDENTAL	N 3825	W 12258	960
4-6399	13	OJAI	N 3427	W 11915	750
4-6476	13	ORANGE COVE	N 3637	W 11918	431
4-6498	13	ORICK PRAIRIE CREEK PK	N 4122	W 12401	161
4-6506	13	ORLAND	N 3945	W 12212	254
4-6508	13	ORLEANS	N 4118	W 12332	403
4-6569	13	OXNARD	N 3412	W 11911	49
4-6597	12	PACIFIC HOUSE	N 3845	W 12030	3440
4-6602	12	PACOIMA DAM FC 33 A E	N 3420	W 11824	1500
4-6610	12	PAICINES OHRWALL RANCH	N 3644	W 12122	950
4-6624	13	PALMDALE	N 3435	W 11806	2596
4-6635	13	PALM SPRINGS	N 3349	W 11632	411
4-6646	13	PALO ALTO JR MUSEUM	N 3727	W 12208	25
4-6650	12	PALOMA	N 3621	W 12130	1835
4-6657	13	PALOMAR MT OBSERVATORY	N 3321	W 11652	5545
4-6663	12	PALOS VERDES EST FC43D	N 3348	W 11823	216
4-6675	12	PANOCHE	N 3636	W 12050	1265
4-6699	13	PARKER RESERVOIR	N 3417	W 11410	738
4-6719	13	PASADENA	N 3409	W 11809	864
4-6726	12	PASKENTA RANGER STA	N 3953	W 12232	755
4-6730	13	PASO ROBLES	N 3538	W 12041	700
4-6742	13	PASO ROBLES FAA AP	N 3540	W 12038	815
4-6754	12	PATTIWAY	N 3456	W 11923	3868
4-6826	13	PETALUMA FIRE STA 2	N 3814	W 12238	16
4-6926	13	PINNACLES NAT MONUMENT	N 3629	W 12111	1307
4-6943	13	PISMO BEACH	N 3508	W 12038	80
4-6946	12	PIT RIVER POWER HOUSE	N 4059	W 12159	1458
4-6960	13	PLACERVILLE	N 3844	W 12048	1890
4-7016	12	PT ARGUELLO LIGHT STA	N 3434	W 12040	76
4-7050	13	POMONA CAL POLY	N 3404	W 11749	740
4-7077	13	PORTERVILLE	N 3604	W 11901	393
4-7085	12	PORTOLA	N 3948	W 12028	4838
4-7109	12	POTTER VALLEY PH	N 3922	W 12308	1015
4-7150	13	PRIEST VALLEY	N 3611	W 12042	2300
4-7195	13	QUINCY	N 3956	W 12056	3409
4-7253	13	RANDSBURG	N 3522	W 11739	3570
4-7292	13	RED BLUFF WSO R	N 4009	W 12215	342
4-7300	13	REDDING FIRE STN 4	N 4033	W 12223	470

STATE-STATION NUMBER	STN TYP	NAME		LATITUDE DEG-MIN	LONGITUDE DEG-MIN	ELEVATION (FT)
4-7306	13	REDLANDS		N 3403	W 11711	1318
4-7339	13	REDWOOD CITY		N 3729	W 12214	31
4-7370	12	REPRESA		N 3842	W 12110	295
4-7414	13	RICHMOND		N 3756	W 12221	55
4-7470	13	RIVERSIDE FIRE STA. NO.3		N 3357	W 11723	840
4-7630	13	SACRAMENTO WSO	R	N 3831	W 12130	17
4-7633	13	SACRAMENTO CITY WSO	R	N 3835	W 12130	19
4-7643	13	SAINT HELENA		N 3830	W 12228	225
4-7661	13	SAINT MARYS COLLEGE		N 3750	W 12206	623
4-7669	12	SALINAS FAA AP		N 3640	W 12136	75
4-7672	12	SALINAS DAM		N 3520	W 12030	1375
4-7689	13	SALT SPRINGS PWR HOUSE		N 3830	W 12013	3700
4-7723	13	SAN BERNARDINO CO HOSP		N 3408	W 11716	1125
4-7731	12	SAN CLEMENTE DAM		N 3626	W 12142	600
4-7734	12	SANDBERG PTL FC130B		N 3445	W 11843	4025
4-7735	13	SANDBERG WSO	//R	N 3445	W 11844	4517
4-7740	13	SAN DIEGO WSO	R	N 3244	W 11710	13
4-7749	12	SAN DIMAS FIRE WD FC95		N 3406	W 11748	955
4-7769	13	SAN FRANCISCO WSO	R	N 3737	W 12223	8
4-7772	13	S F FEDERAL BLDG. WSO	R	N 3747	W 12225	52
4-7776	12	SAN GABRIEL CANYON PH		N 3409	W 11754	744
4-7779	12	SAN GABRIEL DM FC425BE		N 3412	W 11752	1481
4-7785	13	SAN GABRIEL FIRE DEPT		N 3406	W 11806	450
4-7821	13	SAN JOSE	R	N 3721	W 12154	67
4-7851	13	SAN LUIS OBISPO POLY		N 3518	W 12040	320
4-7880	13	SAN RAFAEL		N 3758	W 12232	40
4-7888	13	SANTA ANA FIRE STATION		N 3345	W 11752	115
4-7902	13	SANTA BARBARA		N 3425	W 11941	5
4-7905	13	SANTA BARBARA FAA AP		N 3426	W 11950	9
4-7916	13	SANTA CRUZ		N 3659	W 12201	125
4-7933	12	SANTA MARGARITA BOOST		N 3522	W 12038	1100
4-7946	13	SANTA MARIA WSO	R	N 3454	W 12027	236
4-7953	13	SANTA MONICA PIER		N 3400	W 11830	15
4-7957	13	SANTA PAULA		N 3421	W 11905	263
4-7965	13	SANTA ROSA		N 3827	W 12242	167
4-8014	12	SAUGUS PWR PL NO 1		N 3435	W 11827	2105
4-8025	12	SAWYERS BAR RANGER STA		N 4118	W 12308	2169
4-8045	13	SCOTIA		N 4029	W 12406	139
4-8135	13	SHASTA DAM		N 4043	W 12225	1076
4-8218	13	SIERRAVILLE RANGER STA//		N 3935	W 12022	4975
4-8317	12	SNOW CREEK UPPER		N 3352	W 11641	1940
4-8351	13	SONOMA		N 3817	W 12227	70
4-8353	13	SONORA RS		N 3759	W 12023	1749
4-8380	13	SO ENTRANCE YOSEMITE		N 3730	W 11938	5120
4-8446	12	SPRECKELS HWY BRIDGE		N 3636	W 12141	60
4-8558	13	STOCKTON WSO		N 3754	W 12115	22
4-8560	13	STOCKTON FIRE STA NO 4		N 3800	W 12119	12
4-8580	12	STONYFORD RANGER STA		N 3923	W 12233	1168
4-8587	13	STONY GORGE RESERVOIR		N 3935	W 12232	791
4-8606	13	STRAWBERRY VALLEY		N 3934	W 12106	3808

04 — CALIFORNIA

STATE-STATION NUMBER	STN TYP	NAME	LATITUDE DEG-MIN	LONGITUDE DEG-MIN	ELEVATION (FT)
4-8702	13	SUSANVILLE AIRPORT	N 4023	W 12034	4148
4-8713	12	SUTTER HILL RANGER STA	N 3823	W 12048	1586
4-8758	13	TAHOE CITY	N 3910	W 12008	6230
4-8826	13	TEHACHAPI	N 3508	W 11827	3975
4-8839	13	TEJON RANCHO	N 3502	W 11845	1425
4-8892	13	THERMAL FAA AIRPORT	N 3338	W 11610	120
4-8928	13	TIGER CREEK PH	N 3827	W 12029	2355
4-8967	12	TOPANGA PATROL STA FC6	N 3405	W 11836	745
4-8973	13	TORRANCE	N 3348	W 11820	110
4-8999	13	TRACY CARBONA	N 3742	W 12125	140
4-9035	13	TRONA	N 3547	W 11723	1695
4-9043	13	TRUCKEE RANGER STATION	N 3920	W 12011	5995
4-9053	13	TULELAKE //R	N 4158	W 12128	4035
4-9073	12	TURLOCK	N 3729	W 12051	115
4-9087	13	TUSTIN IRVINE RANCH	N 3344	W 11747	118
4-9099	13	TWENTYNINE PALMS	N 3408	W 11602	1975
4-9105	13	TWIN LAKES	N 3842	W 12002	7829
4-9122	13	UKIAH	N 3909	W 12312	623
4-9152	13	U C L A	N 3404	W 11827	430
4-9158	13	UPLAND 3 N	N 3408	W 11739	1605
4-9167	12	UPPER LAKE 7 W	N 3911	W 12302	1524
4-9177	12	UPPER MATTOLE	N 4015	W 12411	255
4-9200	13	VACAVILLE	N 3822	W 12157	105
4-9260	12	VAN NUYS FC 15A	N 3411	W 11827	695
4-9285	12	VENTURA	N 3417	W 11917	105
4-9325	13	VICTORVILLE PUMP PLANT	N 3432	W 11718	2858
4-9345	12	VINCENT FIRE STA FC120	N 3429	W 11808	3135
4-9351	12	VINTON	N 3949	W 12011	4950
4-9367	13	VISALIA	N 3620	W 11918	325
4-9390	12	VOLTA POWER HOUSE	N 4028	W 12152	2200
4-9431	12	WALNUT PTL STA FC102C	N 3400	W 11752	488
4-9452	13	WASCO	N 3536	W 11920	333
4-9473	13	WATSONVILLE WATERWORKS	N 3656	W 12146	95
4-9490	13	WEAVERVILLE RANGER STA	N 4044	W 12256	2050
4-9560	12	WESTHAVEN	N 3613	W 11959	285
4-9660	12	WHITTIER CTY H FC 106C	N 3358	W 11802	340
4-9699	13	WILLOWS	N 3932	W 12212	140
4-9742	13	WINTERS	N 3832	W 12158	135
4-9754	12	WOFFORD HEIGHTS	N 3543	W 11827	2730
4-9775	13	WOODFORDS	N 3847	W 11949	5671
4-9781	13	WOODLAND 1 WNW	N 3841	W 12148	69
4-9814	12	WRIGHTS	N 3708	W 12157	1600
4-9847	13	YORBA LINDA	N 3353	W 11749	350
4-9855	13	YOSEMITE PARK HEADQTRS	N 3745	W 11935	3970
4-9866	13	YREKA	N 4143	W 12238	2625

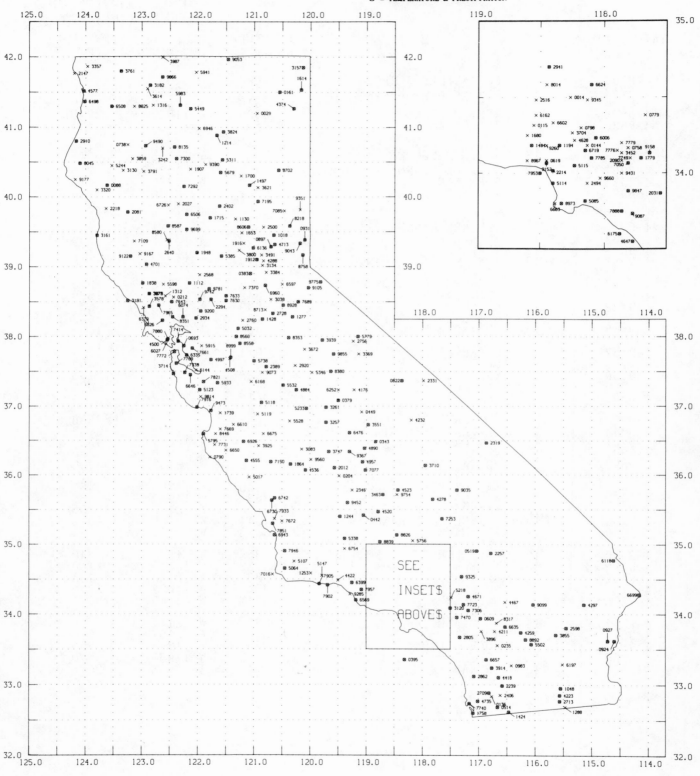

COLORADO

TEMPERATURE NORMALS (DEG F)

STATION		JAN	FEB	MAR	APR	MAY	JUN	JUL	AUG	SEP	OCT	NOV	DEC	ANN
AKRON FAA AIRPORT	MAX	38.1	43.4	48.8	60.1	70.1	81.7	88.2	86.2	77.3	65.7	49.3	41.3	62.5
	MIN	13.0	17.5	22.2	32.1	42.6	52.3	58.7	56.7	47.2	35.9	23.2	16.1	34.8
	MEAN	25.6	30.5	35.5	46.1	56.4	67.0	73.5	71.5	62.2	50.8	36.2	28.8	48.7
ALAMOSA WSO R	MAX	34.2	40.1	48.0	57.8	67.7	78.1	82.0	79.3	73.6	62.9	47.1	36.1	58.9
	MIN	-2.3	5.4	15.1	23.5	33.1	41.4	48.0	45.4	36.1	24.6	11.3	-.3	23.4
	MEAN	16.0	22.8	31.6	40.6	50.4	59.8	65.0	62.4	54.9	43.8	29.2	18.0	41.2
ALLENSPARK	MAX	34.5	36.5	40.1	48.9	58.9	69.6	76.0	72.7	66.1	56.6	43.5	37.3	53.4
	MIN	13.9	15.2	17.2	23.7	32.0	39.4	44.8	43.4	37.2	29.9	20.8	15.9	27.8
	MEAN	24.2	25.9	28.7	36.3	45.5	54.5	60.5	58.1	51.7	43.3	32.2	26.6	40.6
ASPEN	MAX	33.6	37.2	42.5	52.5	63.3	73.9	79.9	77.4	70.3	59.9	44.1	34.9	55.8
	MIN	7.6	9.7	15.5	24.3	32.8	39.1	45.1	43.5	36.5	28.1	17.2	8.7	25.7
	MEAN	20.6	23.5	29.0	38.4	48.1	56.5	62.5	60.5	53.4	44.0	30.7	21.8	40.8
BAILEY	MAX	40.9	43.8	47.2	55.6	65.1	76.3	80.8	78.1	72.9	63.4	48.9	42.7	59.6
	MIN	9.2	10.7	14.5	22.5	30.5	37.2	43.6	41.8	33.5	24.7	15.8	10.7	24.6
	MEAN	25.1	27.3	30.9	39.1	47.9	56.8	62.2	60.0	53.2	44.1	32.4	26.7	42.1
BONNY DAM	MAX	40.6	45.8	50.9	62.8	72.8	84.1	90.8	88.8	79.7	68.9	52.3	44.1	65.1
	MIN	12.3	17.4	22.5	34.1	44.5	54.3	60.9	58.5	48.7	36.1	23.8	16.3	35.8
	MEAN	26.5	31.7	36.7	48.5	58.7	69.2	75.8	73.7	64.2	52.6	38.1	30.2	50.5
BOULDER	MAX	45.0	48.5	52.8	62.2	71.8	82.3	88.1	85.7	78.2	68.0	53.6	47.8	65.3
	MIN	20.1	24.0	27.1	35.5	45.2	54.0	59.8	57.9	49.4	40.0	28.7	23.6	38.8
	MEAN	32.6	36.3	40.0	48.9	58.6	68.2	74.0	71.9	63.8	54.0	41.2	35.7	52.1
BUENA VISTA	MAX	39.9	43.0	47.7	56.1	66.3	77.1	81.9	78.7	73.1	63.7	48.7	41.5	59.8
	MIN	11.4	13.8	18.9	25.9	34.1	41.4	47.5	45.4	37.8	29.1	18.8	12.1	28.0
	MEAN	25.7	28.4	33.3	41.0	50.2	59.3	64.7	62.1	55.5	46.4	33.8	26.8	43.9
BURLINGTON	MAX	42.9	47.9	54.0	64.8	73.9	85.0	90.3	87.9	79.8	69.3	53.5	45.6	66.2
	MIN	15.4	19.8	24.2	34.6	44.8	54.5	60.2	58.2	49.1	37.8	25.6	18.8	36.9
	MEAN	29.2	33.9	39.1	49.7	59.4	69.7	75.3	73.0	64.5	53.6	39.6	32.2	51.6
BYERS 5 NE	MAX	41.9	46.8	52.0	62.2	72.5	83.7	90.1	87.7	79.5	68.1	52.2	45.0	65.1
	MIN	12.0	17.1	21.9	30.9	41.0	49.6	55.9	53.9	45.2	34.2	21.8	15.4	33.2
	MEAN	27.0	32.0	37.0	46.6	56.8	66.7	73.0	70.8	62.4	51.1	37.0	30.2	49.2
CANON CITY	MAX	49.6	52.9	56.2	65.1	74.1	84.6	89.5	87.1	80.1	70.4	57.8	51.7	68.3
	MIN	21.5	24.7	28.2	37.6	46.4	55.4	61.7	59.5	50.7	41.1	29.6	24.7	40.1
	MEAN	35.6	38.8	42.2	51.3	60.3	70.0	75.6	73.3	65.4	55.8	43.7	38.2	54.2
CEDAREDGE	MAX	38.0	44.0	51.6	61.5	71.8	82.0	87.6	84.7	77.5	65.9	49.6	39.7	62.8
	MIN	15.6	20.2	25.7	32.9	41.5	49.7	56.3	54.1	46.2	36.1	25.2	17.5	35.1
	MEAN	26.8	32.1	38.7	47.2	56.7	65.9	72.0	69.4	61.9	51.0	37.5	28.6	49.0
CENTER 4 SSW	MAX	33.8	40.0	48.6	58.4	68.0	77.3	80.8	78.5	73.4	63.3	47.0	36.4	58.8
	MIN	-1.5	5.8	15.7	23.9	33.1	40.0	45.4	43.2	35.7	25.5	12.6	2.0	23.5
	MEAN	16.2	22.9	32.2	41.2	50.6	58.7	63.2	60.8	54.6	44.4	29.8	19.2	41.2
CHEESMAN	MAX	45.5	48.1	51.1	58.8	68.3	79.2	84.4	82.0	76.5	67.1	53.5	47.3	63.5
	MIN	9.1	11.0	17.0	25.7	34.0	42.2	47.5	46.0	38.3	28.4	18.8	13.1	27.6
	MEAN	27.4	29.6	34.1	42.3	51.2	60.7	66.0	64.0	57.4	47.8	36.1	30.2	45.6
CHERRY CREEK DAM	MAX	44.1	48.3	52.6	61.8	71.7	82.7	89.1	86.7	78.9	68.4	53.7	47.2	65.4
	MIN	13.9	18.0	22.5	31.0	40.4	48.9	54.9	53.0	44.3	34.4	23.4	17.5	33.5
	MEAN	29.0	33.2	37.6	46.4	56.1	65.8	72.0	69.8	61.6	51.4	38.6	32.4	49.5
CHEYENNE WELLS	MAX	43.2	48.3	54.2	65.6	75.0	86.2	91.5	89.3	81.1	70.0	53.7	45.5	67.0
	MIN	14.2	18.6	23.4	33.8	44.4	53.9	59.5	57.5	48.4	36.7	24.5	17.4	36.0
	MEAN	28.7	33.4	38.8	49.7	59.7	70.1	75.5	73.4	64.8	53.4	39.2	31.5	51.5
COCHETOPA CREEK	MAX	26.1	31.0	40.4	52.7	64.8	75.5	80.8	78.4	72.0	61.2	43.7	30.1	54.7
	MIN	-6.4	-2.5	9.3	20.1	27.7	34.3	41.7	40.1	31.3	20.9	10.2	-2.4	18.7
	MEAN	9.9	14.3	24.9	36.4	46.3	54.9	61.3	59.3	51.7	41.1	27.0	13.9	36.8

COLORADO

TEMPERATURE NORMALS (DEG F)

STATION		JAN	FEB	MAR	APR	MAY	JUN	JUL	AUG	SEP	OCT	NOV	DEC	ANN
COLORADO NAT MON	MAX	35.8	42.8	51.2	62.7	73.6	85.2	91.6	87.8	79.2	66.0	48.6	38.1	63.6
	MIN	18.0	23.1	28.8	37.1	47.1	57.0	63.7	60.7	52.8	42.2	29.6	20.8	40.1
	MEAN	27.0	33.0	40.0	49.9	60.4	71.1	77.7	74.3	66.0	54.1	39.1	29.4	51.8
COLORADO SPRINGS WSO	MAX	41.4	45.3	49.3	59.5	68.9	79.9	84.9	82.3	74.9	64.6	50.4	43.9	62.1
	MIN	16.2	19.6	23.8	32.9	42.5	51.5	57.4	55.6	47.2	37.0	25.0	18.9	35.6
	MEAN	28.8	32.5	36.6	46.2	55.7	65.7	71.2	69.0	61.1	50.8	37.7	31.4	48.9
CORTEZ	MAX	40.3	45.7	52.5	62.4	72.4	83.5	88.9	86.0	79.1	67.7	52.2	42.1	64.4
	MIN	13.4	18.4	24.0	30.6	39.1	47.0	54.8	53.2	44.7	34.3	23.6	15.0	33.2
	MEAN	26.9	32.1	38.3	46.5	55.8	65.3	71.9	69.6	61.9	51.0	37.9	28.5	48.8
CRESTED BUTTE	MAX	27.4	31.1	37.3	46.7	59.2	70.2	76.0	73.5	67.1	57.0	40.7	29.9	51.3
	MIN	-1.7	.4	7.7	18.0	28.1	33.9	39.4	38.3	31.2	22.9	10.8	.2	19.1
	MEAN	12.9	15.8	22.5	32.4	43.7	52.0	57.7	55.9	49.1	40.0	25.8	15.0	35.2
DEL NORTE	MAX	35.3	40.9	48.6	58.0	67.0	75.3	78.4	76.3	71.6	62.7	47.8	38.0	58.3
	MIN	6.8	11.7	19.0	26.8	35.3	42.2	48.1	46.7	39.5	30.7	18.5	9.2	27.9
	MEAN	21.1	26.3	33.8	42.4	51.2	58.8	63.3	61.5	55.6	46.7	33.2	23.6	43.1
DENVER WSFO //R	MAX	43.1	46.9	51.2	61.0	70.7	81.6	88.0	85.8	77.5	66.8	52.4	46.1	64.3
	MIN	15.9	20.2	24.7	33.7	43.6	52.4	58.7	57.0	47.7	36.9	25.1	18.9	36.2
	MEAN	29.5	33.6	38.0	47.4	57.2	67.0	73.3	71.4	62.6	51.9	38.7	32.6	50.3
DILLON 1 E	MAX	31.8	34.6	38.8	47.8	59.0	69.2	75.0	72.6	66.9	57.1	41.3	34.0	52.3
	MIN	.4	2.2	7.5	17.3	25.6	31.7	37.4	35.9	28.8	20.8	10.8	3.2	18.5
	MEAN	16.1	18.4	23.2	32.6	42.3	50.5	56.2	54.3	47.9	39.0	26.1	18.7	35.4
DURANGO	MAX	40.9	46.6	53.1	62.2	71.7	81.8	86.8	84.2	78.1	67.9	53.4	43.0	64.1
	MIN	10.4	15.0	21.3	27.7	34.7	41.2	49.7	48.1	39.5	30.2	20.7	12.6	29.3
	MEAN	25.7	30.8	37.2	45.0	53.2	61.5	68.3	66.2	58.8	49.1	37.1	27.8	46.7
EADS	MAX	43.9	50.4	56.8	68.3	77.1	88.1	93.1	90.6	82.7	71.8	55.2	47.2	68.8
	MIN	13.6	18.9	24.3	35.0	45.2	55.2	61.1	58.5	49.2	37.0	24.1	17.0	36.6
	MEAN	28.8	34.7	40.6	51.7	61.2	71.7	77.1	74.6	66.0	54.4	39.7	32.1	52.7
EAGLE FAA AP	MAX	33.8	39.7	46.9	57.8	69.3	80.0	86.4	83.2	75.9	64.3	46.9	35.1	59.9
	MIN	2.8	8.3	17.6	25.2	32.9	39.1	45.9	44.2	35.0	25.1	14.6	4.0	24.6
	MEAN	18.3	24.0	32.3	41.5	51.1	59.6	66.2	63.7	55.5	44.8	30.8	19.6	42.3
ESTES PARK	MAX	38.0	40.5	44.2	52.9	62.4	72.8	78.6	76.4	69.9	60.4	46.6	40.1	56.9
	MIN	16.5	17.6	19.7	26.3	33.9	40.9	45.8	44.4	37.3	29.9	22.3	18.3	29.4
	MEAN	27.3	29.1	32.0	39.6	48.2	56.9	62.2	60.4	53.6	45.2	34.4	29.2	43.2
FORT COLLINS	MAX	40.6	45.1	50.2	60.1	69.5	80.0	85.8	83.3	75.4	64.9	50.4	43.7	62.4
	MIN	13.2	18.5	23.3	32.7	42.7	51.2	56.9	54.5	45.2	34.8	23.2	17.1	34.4
	MEAN	26.9	31.8	36.7	46.4	56.1	65.6	71.4	68.9	60.3	49.9	36.8	30.4	48.4
FORT LEWIS	MAX	36.0	39.8	44.9	54.9	65.0	75.8	80.9	77.7	71.5	61.4	47.1	38.3	57.8
	MIN	8.5	11.5	17.7	25.0	32.8	40.2	48.2	46.4	39.1	30.2	19.1	11.0	27.5
	MEAN	22.3	25.6	31.4	40.0	48.9	58.0	64.6	62.1	55.3	45.8	33.1	24.7	42.7
FORT MORGAN	MAX	38.1	44.9	50.8	61.8	72.0	83.3	90.0	87.3	78.4	67.6	50.7	42.0	63.9
	MIN	8.9	16.0	22.7	33.7	44.5	53.8	59.5	56.7	46.4	34.3	21.4	13.3	34.3
	MEAN	23.6	30.4	36.8	47.8	58.3	68.6	74.8	72.0	62.4	51.0	36.1	27.7	49.1
FRUITA	MAX	37.1	46.1	56.1	66.8	77.2	88.0	93.5	90.2	82.4	70.0	52.3	39.8	66.6
	MIN	11.8	18.7	25.1	33.1	42.6	49.7	56.5	54.2	44.3	33.3	22.7	14.3	33.9
	MEAN	24.5	32.4	40.6	50.0	59.9	68.9	75.0	72.2	63.4	51.7	37.5	27.1	50.3
GLENWOOD SPRINGS 1 N	MAX	35.4	42.2	50.3	61.2	71.9	82.9	89.2	85.9	77.9	66.4	48.9	37.3	62.5
	MIN	11.9	16.3	23.1	30.6	38.3	44.4	51.1	49.2	40.9	31.9	21.7	13.5	31.1
	MEAN	23.6	29.3	36.7	45.9	55.1	63.7	70.2	67.6	59.4	49.1	35.3	25.4	46.8
GRAND JUNCTION WSO //R	MAX	35.7	44.5	54.1	65.2	76.2	87.9	94.0	90.3	81.9	68.7	51.0	38.7	65.7
	MIN	15.2	22.4	29.7	38.2	48.0	56.6	63.8	61.5	52.2	41.1	28.2	17.9	39.6
	MEAN	25.5	33.5	41.9	51.7	62.1	72.3	78.9	75.9	67.1	54.9	39.6	28.3	52.7

COLORADO

TEMPERATURE NORMALS (DEG F)

STATION		JAN	FEB	MAR	APR	MAY	JUN	JUL	AUG	SEP	OCT	NOV	DEC	ANN
GRAND LAKE 1 NW	MAX	29.8	33.4	38.4	47.7	58.8	69.3	75.2	72.8	67.1	56.6	40.3	31.7	51.8
	MIN	.8	2.2	8.1	17.5	26.0	31.5	36.6	35.4	28.2	20.7	10.2	2.6	18.3
	MEAN	15.4	17.8	23.3	32.7	42.4	50.4	55.9	54.1	47.7	38.7	25.3	17.2	35.1
GRAND LAKE 6 SSW	MAX	26.3	30.1	36.3	46.8	58.4	68.5	74.4	72.1	65.6	55.1	39.4	29.2	50.2
	MIN	1.0	2.0	8.5	19.9	29.3	35.9	41.8	40.8	33.1	24.9	15.3	5.8	21.5
	MEAN	13.6	16.1	22.5	33.3	43.9	52.2	58.1	56.5	49.4	40.0	27.4	17.5	35.9
GREAT SAND DUNES NM	MAX	34.7	39.2	46.3	56.3	66.2	76.7	80.8	77.8	71.9	61.1	45.5	36.4	57.7
	MIN	9.3	13.1	19.5	27.4	36.2	44.7	50.0	48.1	41.0	31.6	19.7	11.5	29.3
	MEAN	22.0	26.2	32.9	41.9	51.2	60.7	65.4	62.9	56.5	46.4	32.6	24.0	43.6
GREELEY CSC	MAX	39.4	45.5	51.7	62.1	72.3	83.1	89.6	86.8	78.5	67.3	50.6	42.6	64.1
	MIN	10.6	16.9	22.9	33.1	43.3	52.1	57.5	54.7	44.9	33.8	21.9	14.5	33.9
	MEAN	25.0	31.2	37.4	47.6	57.8	67.6	73.6	70.8	61.7	50.5	36.3	28.6	49.0
GREEN MOUNTAIN DAM	MAX	30.6	34.9	42.3	53.7	64.7	75.0	80.7	78.3	71.7	61.0	43.5	32.4	55.7
	MIN	5.6	6.8	13.6	24.2	32.8	39.2	44.8	43.4	36.5	27.8	17.2	7.9	25.0
	MEAN	18.1	20.9	28.0	39.0	48.8	57.1	62.8	60.9	54.1	44.4	30.4	20.2	40.4
GUNNISON	MAX	26.1	31.2	40.9	55.0	66.4	76.6	81.4	78.7	72.8	62.3	44.9	29.8	55.5
	MIN	-6.7	-2.8	9.7	20.6	28.5	35.0	41.9	39.6	30.9	21.3	10.4	-2.5	18.8
	MEAN	9.7	14.2	25.3	37.8	47.5	55.9	61.7	59.2	51.9	41.8	27.7	13.7	37.2
HAYDEN	MAX	29.9	34.8	42.0	56.1	68.2	78.3	85.5	82.5	74.3	62.4	44.9	32.8	57.6
	MIN	4.0	7.4	15.5	26.7	34.6	41.2	47.0	45.5	37.0	27.4	17.3	7.5	25.9
	MEAN	17.0	21.1	28.7	41.4	51.4	59.7	66.3	64.0	55.7	44.9	31.1	20.2	41.8
HERMIT 7 ESE //	MAX	29.9	34.2	38.0	47.6	59.9	70.4	75.2	72.4	67.8	58.9	42.9	32.2	52.5
	MIN	-7.0	-4.0	2.9	15.1	23.6	29.6	36.8	36.0	27.6	18.4	5.9	-4.8	15.0
	MEAN	11.5	15.1	20.5	31.4	41.8	50.0	56.1	54.2	47.7	38.7	24.4	13.7	33.8
HOLLY	MAX	43.9	50.2	57.0	68.5	78.2	89.1	94.5	92.0	83.1	72.6	56.0	47.0	69.3
	MIN	13.0	18.3	24.7	36.9	47.4	58.0	63.3	60.4	50.4	36.8	23.7	16.3	37.4
	MEAN	28.5	34.3	40.9	52.7	62.8	73.6	78.9	76.2	66.8	54.7	39.9	31.7	53.4
HOLYOKE	MAX	40.9	46.5	52.3	64.4	74.0	84.9	91.0	89.3	80.2	69.0	52.2	43.8	65.7
	MIN	13.1	17.6	22.6	32.9	43.7	53.3	59.0	56.6	46.6	35.3	23.2	16.5	35.0
	MEAN	27.0	32.1	37.5	48.7	58.9	69.1	75.0	73.0	63.4	52.1	37.8	30.2	50.4
JOHN MARTIN DAM	MAX	44.3	51.8	58.5	69.8	79.1	90.0	94.9	92.4	84.6	73.7	56.6	47.2	70.2
	MIN	14.6	20.2	26.7	37.9	47.7	57.5	63.3	60.7	51.1	38.2	25.4	18.2	38.5
	MEAN	29.5	36.0	42.6	53.8	63.4	73.8	79.1	76.6	67.9	56.0	41.0	32.7	54.4
KASSLER	MAX	46.4	50.0	53.5	62.7	71.7	82.4	87.6	85.5	78.5	68.4	54.9	48.7	65.9
	MIN	16.4	20.3	24.4	33.7	43.6	52.7	59.1	57.6	48.9	38.4	26.1	19.8	36.8
	MEAN	31.4	35.1	39.0	48.3	57.7	67.6	73.4	71.6	63.7	53.5	40.5	34.3	51.3
KAUFFMAN 4 SSE R	MAX	40.0	44.9	49.5	60.7	70.2	80.8	87.7	86.1	77.4	65.9	50.1	43.1	63.0
	MIN	11.4	15.9	19.3	29.0	39.3	47.8	53.4	51.3	41.9	32.2	20.8	15.4	31.5
	MEAN	25.7	30.4	34.4	44.9	54.8	64.3	70.6	68.7	59.7	49.1	35.5	29.3	47.3
LA JUNTA FAA AIRPORT	MAX	43.4	50.0	56.5	68.1	78.0	89.6	94.3	91.6	83.0	71.8	55.1	46.6	69.0
	MIN	14.9	20.8	26.5	37.6	48.0	58.1	63.9	61.4	52.2	39.6	25.9	18.1	38.9
	MEAN	29.2	35.4	41.5	52.9	63.0	73.9	79.1	76.5	67.6	55.7	40.5	32.4	54.0
LAKEWOOD	MAX	44.2	48.1	51.9	60.9	70.3	81.5	87.5	85.4	77.6	67.5	53.0	47.4	64.6
	MIN	17.5	21.3	25.0	33.6	43.0	51.9	57.6	55.7	46.8	37.3	26.0	21.0	36.4
	MEAN	30.9	34.8	38.5	47.3	56.7	66.7	72.6	70.6	62.2	52.4	39.5	34.2	50.5
LAMAR	MAX	43.8	50.5	57.4	69.0	78.1	88.8	93.7	91.1	82.8	72.1	55.5	46.9	69.1
	MIN	13.5	19.6	25.8	37.2	47.6	57.9	63.6	61.2	51.4	37.8	24.4	16.9	38.1
	MEAN	28.7	35.0	41.7	53.1	62.9	73.4	78.7	76.2	67.1	55.0	40.0	31.9	53.6
LAS ANIMAS	MAX	46.1	52.7	60.0	71.3	80.4	91.1	95.7	93.1	85.4	74.6	57.5	49.2	71.4
	MIN	12.9	18.6	24.8	36.0	46.4	56.2	61.7	59.1	49.6	36.7	23.8	16.3	36.8
	MEAN	29.5	35.7	42.5	53.6	63.4	73.7	78.7	76.1	67.5	55.7	40.7	32.8	54.2

COLORADO

TEMPERATURE NORMALS (DEG F)

STATION		JAN	FEB	MAR	APR	MAY	JUN	JUL	AUG	SEP	OCT	NOV	DEC	ANN
LITTLE HILLS	MAX	37.6	42.4	47.9	58.2	68.6	79.5	86.1	83.0	75.7	64.6	48.7	39.6	61.0
	MIN	5.1	9.1	16.5	24.2	32.1	38.3	45.6	43.9	34.0	23.8	14.7	6.3	24.5
	MEAN	21.3	25.8	32.2	41.2	50.4	58.9	65.9	63.5	54.9	44.2	31.7	23.0	42.8
LONGMONT 2 ESE	MAX	41.2	46.0	51.0	61.3	71.4	82.2	88.6	86.2	77.9	67.3	51.5	44.7	64.1
	MIN	11.6	16.9	22.5	32.3	42.5	50.4	55.6	53.5	44.2	33.5	22.2	15.5	33.4
	MEAN	26.5	31.5	36.8	46.8	57.0	66.3	72.2	69.9	61.1	50.4	36.9	30.1	48.8
MESA VERDE NAT PARK	MAX	39.5	44.0	50.0	60.1	70.7	82.3	87.3	84.3	77.7	65.9	50.1	41.1	62.8
	MIN	18.8	22.0	26.0	33.1	42.3	51.8	57.8	55.8	49.7	39.9	28.1	20.8	37.2
	MEAN	29.2	33.0	38.0	46.6	56.5	67.1	72.6	70.1	63.7	52.9	39.2	31.0	50.0
MONTE VISTA	MAX	33.7	39.6	48.2	58.6	67.9	77.1	80.3	78.2	73.3	63.2	46.8	35.9	58.6
	MIN	.1	6.7	15.8	24.0	33.0	39.9	46.3	44.5	36.1	26.3	13.9	2.7	24.1
	MEAN	17.0	23.2	32.0	41.4	50.4	58.5	63.3	61.4	54.7	44.8	30.4	19.3	41.4
MONTROSE NO 2	MAX	37.9	44.1	52.0	61.8	72.1	83.2	89.1	85.9	78.5	66.7	50.3	40.0	63.5
	MIN	13.3	18.7	25.2	33.4	42.2	50.2	56.4	53.9	45.4	35.0	23.9	15.1	34.4
	MEAN	25.7	31.4	38.6	47.6	57.2	66.7	72.8	69.9	62.0	50.9	37.1	27.6	49.0
NORTHDALE	MAX	36.4	41.2	47.8	58.6	69.2	81.0	86.8	83.7	76.5	64.5	48.4	38.3	61.0
	MIN	9.0	13.7	20.7	27.2	34.6	41.6	50.0	49.1	40.1	30.4	20.1	11.5	29.0
	MEAN	22.7	27.5	34.3	42.9	51.9	61.4	68.5	66.4	58.3	47.4	34.3	24.9	45.0
NORWOOD	MAX	36.3	40.6	46.8	57.1	67.6	78.4	83.7	80.4	73.2	61.9	46.5	37.7	59.2
	MIN	9.0	13.7	20.3	27.0	35.2	43.0	49.0	47.7	40.6	31.2	19.7	11.3	29.0
	MEAN	22.7	27.2	33.6	42.0	51.4	60.7	66.4	64.1	56.9	46.6	33.1	24.5	44.1
OURAY	MAX	38.0	40.3	44.8	54.0	64.0	74.6	79.0	76.4	70.9	60.6	46.4	39.2	57.4
	MIN	14.8	17.0	21.5	29.2	37.5	44.7	51.0	49.8	43.5	34.1	23.3	16.7	31.9
	MEAN	26.4	28.7	33.1	41.6	50.8	59.6	65.0	63.1	57.2	47.4	34.9	27.9	44.6
PAGOSA SPRINGS	MAX	37.8	42.4	48.4	58.8	68.1	78.4	82.6	80.2	74.2	64.3	48.8	39.5	60.3
	MIN	3.0	7.2	15.3	22.4	30.4	36.6	45.4	44.1	35.9	26.5	15.8	6.2	24.1
	MEAN	20.5	24.9	31.9	40.6	49.3	57.5	64.0	62.2	55.1	45.4	32.3	22.9	42.2
PALISADE	MAX	38.8	46.8	55.7	66.2	77.0	88.3	94.8	91.4	83.4	70.9	53.4	41.7	67.4
	MIN	17.6	24.2	31.4	40.1	49.0	57.5	63.8	61.5	53.1	42.4	30.3	20.4	40.9
	MEAN	28.2	35.6	43.6	53.2	63.0	72.9	79.4	76.5	68.3	56.7	41.9	31.1	54.2
PARKER 9 E	MAX	43.3	46.9	50.4	60.0	69.6	80.5	86.4	84.0	76.9	67.0	52.5	46.3	63.7
	MIN	12.4	16.4	20.6	29.7	39.2	48.3	55.0	53.1	44.4	33.7	21.3	15.8	32.5
	MEAN	27.9	31.7	35.5	44.9	54.4	64.4	70.7	68.6	60.7	50.4	36.9	31.1	48.1
PUEBLO WSO //R	MAX	45.2	50.7	55.9	66.5	76.0	87.5	92.2	89.5	81.6	70.7	55.7	48.3	68.3
	MIN	14.3	19.6	25.2	35.7	46.0	55.0	61.5	59.3	50.3	37.5	24.6	17.2	37.2
	MEAN	29.8	35.2	40.6	51.1	61.0	71.3	76.9	74.4	66.0	54.1	40.2	32.8	52.8
RANGELY	MAX	31.7	39.7	50.5	63.2	74.1	85.3	92.4	89.3	80.2	67.2	48.8	35.4	63.2
	MIN	1.3	8.5	20.1	30.9	39.4	47.1	54.2	51.0	40.7	29.9	18.3	5.6	28.9
	MEAN	16.5	24.1	35.3	47.0	56.8	66.3	73.3	70.2	60.5	48.6	33.6	20.5	46.1
RIFLE	MAX	35.8	43.0	52.0	63.0	73.2	83.6	89.7	86.8	79.2	67.7	50.7	38.4	63.6
	MIN	9.0	15.4	22.7	29.9	37.8	44.0	50.7	49.1	40.0	30.0	20.8	11.0	30.0
	MEAN	22.4	29.2	37.4	46.5	55.5	63.8	70.2	68.0	59.6	48.9	35.8	24.7	46.8
ROCKY FORD 2 ESE	MAX	46.1	52.6	59.4	70.0	79.0	89.7	94.0	91.5	84.3	73.4	57.0	49.0	70.5
	MIN	12.2	18.0	24.2	34.9	45.0	54.4	59.7	57.1	47.9	35.3	22.6	15.5	35.6
	MEAN	29.2	35.3	41.8	52.5	62.0	72.1	76.9	74.3	66.1	54.4	39.9	32.3	53.1
RYE	MAX	43.1	45.8	50.6	60.2	69.2	79.5	83.2	80.0	74.0	64.5	51.2	44.8	62.2
	MIN	13.1	15.6	20.5	29.2	37.7	45.2	50.7	49.0	41.6	32.0	21.2	15.2	30.9
	MEAN	28.1	30.7	35.6	44.7	53.5	62.4	67.0	64.5	57.8	48.3	36.2	30.0	46.6
SAGUACHE	MAX	35.8	41.3	49.2	58.9	68.4	77.8	81.8	79.1	73.6	63.6	47.8	37.7	59.6
	MIN	3.5	9.8	16.8	24.1	32.9	40.5	46.7	44.6	36.5	27.2	15.1	6.1	25.3
	MEAN	19.7	25.6	33.0	41.5	50.7	59.2	64.3	61.9	55.0	45.4	31.5	21.9	42.5

COLORADO

TEMPERATURE NORMALS (DEG F)

STATION		JAN	FEB	MAR	APR	MAY	JUN	JUL	AUG	SEP	OCT	NOV	DEC	ANN
SILVERTON	MAX	33.9	37.1	40.1	47.3	58.2	68.7	73.9	71.1	66.1	57.0	43.5	35.5	52.7
	MIN	-.8	.4	7.2	17.9	26.8	32.0	38.0	37.2	30.6	22.6	10.3	1.5	18.6
	MEAN	16.6	18.7	23.7	32.6	42.5	50.4	56.0	54.1	48.4	39.8	26.9	18.6	35.7
STEAMBOAT SPRINGS	MAX	28.5	34.0	41.2	52.6	64.6	75.0	82.6	79.8	72.1	60.9	42.8	30.7	55.4
	MIN	1.9	4.6	12.2	23.5	31.0	35.2	40.8	39.7	31.9	23.7	13.8	4.0	21.9
	MEAN	15.3	19.4	26.8	38.0	47.8	55.1	61.7	59.8	52.0	42.3	28.3	17.4	38.7
STERLING	MAX	37.6	44.5	49.8	61.6	71.7	82.9	89.8	87.4	77.8	66.6	50.0	41.1	63.4
	MIN	10.1	16.5	22.4	33.1	44.0	53.4	58.8	56.0	45.2	33.5	.21.6	13.9	34.0
	MEAN	23.9	30.5	36.1	47.4	57.9	68.2	74.3	71.7	61.5	50.1	35.8	27.5	48.7
STRATTON 3 NE	MAX	42.1	47.2	52.8	64.2	74.0	85.5	90.9	89.0	80.3	69.6	52.6	45.0	66.1
	MIN	14.9	19.2	23.8	33.9	44.3	53.8	59.9	57.9	48.8	37.3	24.8	18.2	36.4
	MEAN	28.5	33.2	38.3	49.1	59.2	69.7	75.4	73.5	64.6	53.5	38.7	31.6	51.3
TAYLOR PARK	MAX	26.6	31.6	37.0	45.4	56.4	67.6	72.1	69.3	63.9	54.4	38.9	27.9	49.3
	MIN	-9.6	-9.3	-.7	13.5	26.5	34.1	40.8	39.8	32.5	23.5	10.1	-6.0	16.3
	MEAN	8.5	11.2	18.2	29.5	41.5	50.8	56.5	54.6	48.2	39.0	24.5	11.0	32.8
TELLURIDE	MAX	37.2	39.6	42.4	51.3	61.9	72.8	77.8	74.9	69.4	60.5	46.4	38.7	56.1
	MIN	6.3	8.1	13.9	22.2	30.1	35.9	41.8	40.7	34.1	.25.8	15.1	7.9	23.5
	MEAN	21.8	23.9	28.2	36.7	46.0	54.4	59.8	57.8	51.8	43.2	30.8	23.3	39.8
TRINIDAD FAA AIRPORT	MAX	46.4	49.9	55.1	64.5	73.9	84.7	88.9	86.6	79.6	69.5	55.5	48.6	66.9
	MIN	15.8	19.5	24.4	34.1	43.9	53.3	59.2	57.2	49.1	37.7	25.2	18.5	36.5
	MEAN	31.1	34.8	39.8	49.4	58.9	69.0	74.0	71.9	64.4	53.6	40.4	33.6	51.7
VALLECITO DAM //	MAX	37.2	41.0	46.3	56.0	65.2	76.2	81.5	78.5	73.0	62.9	48.4	39.8	58.8
	MIN	5.8	7.9	15.6	24.6	32.4	39.7	47.2	45.8	38.9	30.4	20.0	11.4	26.6
	MEAN	21.5	24.5	31.0	40.3	48.8	58.0	64.4	62.2	56.0	46.7	34.2	25.6	42.8
WALDEN	MAX	27.5	31.1	36.9	48.9	60.6	70.9	77.8	74.9	67.6	56.3	39.9	30.5	51.9
	MIN	3.4	5.0	10.8	20.1	27.9	35.7	39.4	36.7	28.9	20.7	11.9	5.4	20.5
	MEAN	15.5	18.1	23.8	34.5	44.3	53.3	58.6	55.8	48.3	38.5	26.0	18.0	36.2
WALSENBURG POWER PLANT	MAX	45.9	49.1	54.0	63.4	72.9	83.4	87.3	84.5	78.6	68.6	54.6	48.0	65.9
	MIN	20.1	22.3	25.6	33.1	42.0	50.5	56.7	55.0	47.5	37.7	27.2	22.3	36.7
	MEAN	33.1	35.7	39.8	48.3	57.5	67.0	72.0	69.8	63.1	53.2	41.0	35.2	51.3
WATERDALE	MAX	42.3	46.2	50.7	60.2	69.8	80.4	87.1	84.5	76.3	65.9	51.3	45.0	63.3
	MIN	13.7	17.8	22.4	31.1	40.4	48.0	53.7	52.2	43.9	34.5	23.3	17.3	33.2
	MEAN	28.0	32.0	36.6	45.7	55.1	64.3	70.4	68.4	60.1	50.3	37.4	31.2	48.3
WESTCLIFFE	MAX	39.5	42.2	46.9	56.1	66.1	76.9	81.7	78.7	72.8	63.0	48.8	41.6	59.5
	MIN	6.5	9.0	16.1	24.4	32.3	39.6	44.4	43.3	35.2	25.2	14.6	7.4	24.8
	MEAN	23.1	25.6	31.6	40.3	49.2	58.3	63.0	61.1	54.0	44.1	31.7	24.6	42.2
WRAY	MAX	43.4	49.3	55.2	66.9	76.3	86.8	92.4	90.8	82.4	71.8	54.8	46.5	68.1
	MIN	12.7	18.2	23.2	34.1	44.8	54.5	60.7	57.7	47.3	34.7	22.5	16.1	35.5
	MEAN	28.1	33.8	39.2	50.5	60.6	70.7	76.6	74.2	64.9	53.3	38.7	31.3	51.8

COLORADO

PRECIPITATION NORMALS (INCHES)

STATION		JAN	FEB	MAR	APR	MAY	JUN	JUL	AUG	SEP	OCT	NOV	DEC	ANN
AKRON FAA AIRPORT		.33	.30	.93	1.27	3.08	2.44	2.75	1.79	1.03	.78	.59	.37	15.66
ALAMOSA WSO	R	.27	.26	.36	.50	.70	.55	1.23	1.13	.74	.68	.35	.36	7.13
ALLENSPARK		1.27	1.01	1.73	2.29	2.68	1.93	2.29	2.30	1.48	1.11	1.21	1.18	20.48
ALTENBERN		1.42	1.25	1.45	1.24	1.33	.95	1.07	1.80	1.15	1.25	1.17	1.24	15.32
ASPEN		1.96	1.61	1.90	1.69	1.52	1.21	1.46	1.85	1.56	1.54	1.55	1.82	19.67
BAILEY		.34	.54	1.05	1.74	2.09	1.34	2.63	2.23	1.34	1.10	.64	.56	15.60
BLANCA		.27	.22	.37	.45	1.00	.66	1.38	1.38	.75	.62	.34	.30	7.74
BONNY DAM		.31	.33	.88	1.45	3.03	2.60	2.49	2.14	1.38	.88	.55	.31	16.35
BOULDER		.63	.75	1.48	2.27	3.28	1.98	1.78	1.51	1.55	1.21	1.01	.67	18.12
BRECKENRIDGE		1.53	1.38	1.58	1.52	1.57	1.51	2.55	2.28	1.45	1.10	1.31	1.36	19.14
BUENA VISTA		.39	.45	.66	.87	1.01	.64	1.59	1.87	.91	.79	.66	.54	10.38
BURLINGTON		.33	.33	.90	1.37	2.72	2.27	2.22	2.07	1.30	.86	.54	.41	15.32
BYERS 5 NE		.42	.39	.92	1.35	2.63	2.00	2.22	1.61	1.40	.79	.64	.39	14.76
CANON CITY		.34	.39	.85	1.35	1.72	1.17	1.80	1.82	1.10	.98	.68	.49	12.69
CEDAREDGE		.92	.85	.96	.87	1.05	.71	.86	1.22	1.03	1.27	.90	.94	11.58
CENTER 4 SSW		.23	.21	.32	.48	.69	.50	1.07	1.27	.72	.68	.43	.32	6.92
CHEESMAN		.44	.61	1.20	1.69	2.06	1.34	2.68	2.34	1.16	1.07	.82	.59	16.00
CHERRY CREEK DAM		.49	.60	1.12	1.81	2.76	2.02	2.05	1.84	1.31	1.09	.91	.55	16.55
CHEYENNE WELLS		.23	.23	.69	1.16	2.78	2.14	2.64	2.10	1.47	.83	.52	.22	15.01
CIMARRON		1.43	.91	1.03	.83	.94	.85	1.21	1.53	1.18	1.19	.84	.92	12.86
CLIMAX 2 NW		2.38	1.91	2.18	2.49	1.89	1.36	2.19	2.21	1.48	1.23	1.74	2.10	23.16
COCHETOPA CREEK		.82	.64	.64	.71	.80	.64	1.54	1.80	.86	.83	.60	.78	10.66
COLORADO NAT MON		.87	.67	1.01	.81	.87	.59	.76	1.36	.81	1.07	.88	.89	10.59
COLORADO SPRINGS WSO		.27	.31	.78	1.35	2.28	2.02	2.85	2.61	1.31	.78	.54	.32	15.42
CORTEZ		1.15	.87	1.09	.93	.99	.40	1.12	1.47	1.14	1.45	.92	1.19	12.72
CRESTED BUTTE		3.26	2.38	2.63	1.80	1.43	1.17	1.87	1.93	1.71	1.54	1.89	2.91	24.52
DEL NORTE		.37	.34	.61	.66	.88	.61	1.57	1.67	.89	.97	.56	.56	9.69
DENVER WSFO	//R	.51	.69	1.21	1.81	2.47	1.58	1.93	1.53	1.23	.98	.82	.55	15.31
DILLON 1 E		1.06	1.08	1.31	1.35	1.40	1.10	1.64	1.63	1.28	.83	.87	1.10	14.65
DOHERTY RANCH		.45	.40	.82	1.11	1.87	1.27	2.16	1.64	1.02	.83	.60	.46	12.63
DOLORES		1.79	1.49	1.74	1.53	1.30	.49	1.33	1.92	1.39	1.84	1.50	1.75	18.07
DURANGO		1.95	1.29	1.52	1.21	1.08	.59	1.70	2.38	1.52	2.07	1.34	1.96	18.61
EADS		.31	.31	.75	1.17	2.48	1.91	2.47	1.78	1.10	.82	.64	.35	14.09
EAGLE FAA AP		.92	.62	.79	.77	.79	.82	1.04	1.03	1.04	.84	.66	.92	10.24
ESTES PARK		.42	.39	.78	1.34	2.13	1.64	2.12	1.97	1.21	.76	.54	.52	13.82
FLAGLER 2 NW		.27	.32	.79	1.23	2.51	2.47	2.94	2.25	1.16	.75	.59	.34	15.62
FLEMING 1 S		.40	.43	1.03	1.61	3.41	3.15	2.63	1.75	.92	.81	.61	.42	17.17
FORT COLLINS		.42	.40	1.07	1.75	2.79	1.75	1.56	1.52	1.09	1.05	.62	.45	14.47
FORT LEWIS		1.74	1.27	1.38	1.12	1.05	.65	1.94	2.12	1.45	2.10	1.23	1.58	17.63
FORT MORGAN		.24	.21	.56	1.20	2.53	1.92	1.91	1.44	1.14	.63	.40	.25	12.43
FOUNTAIN 1 N		.27	.30	.80	1.41	2.21	1.67	2.57	2.42	1.11	.76	.45	.30	14.27
FOWLER 1 SE		.28	.23	.53	.96	1.39	1.26	1.73	1.62	.85	.61	.44	.29	10.19
FRUITA		.67	.53	.75	.69	.74	.43	.58	.97	.68	.84	.67	.65	8.20
GATEWAY 1 SW		.81	.78	.94	.97	.93	.52	.86	1.27	.79	1.22	.94	.78	10.81
GENOA 1 W		.27	.25	.82	1.29	2.50	1.97	2.65	2.38	1.07	.63	.46	.26	14.55
GEORGETOWN		.53	.63	1.19	1.67	1.91	1.34	2.10	2.25	1.38	.95	.73	.76	15.44
GLENWOOD SPRINGS 1 N		1.74	1.17	1.28	1.57	1.31	1.24	1.19	1.43	1.41	1.41	1.07	1.41	16.23
GRAND JUNCTION WSO	//R	.64	.54	.75	.71	.76	.44	.47	.91	.70	.87	.63	.58	8.00
GRAND LAKE 1 NW		1.87	1.39	1.53	1.82	1.87	1.59	2.12	2.18	1.57	1.13	1.27	1.68	20.02
GRAND LAKE 6 SSW		1.12	.83	.97	1.10	1.41	1.27	1.43	1.68	1.17	.85	.83	1.02	13.68
GREAT SAND DUNES NM		.36	.27	.49	.73	1.05	.83	1.83	1.90	1.05	.79	.40	.34	10.04
GREELEY CSC		.36	.28	.82	1.47	2.50	1.76	1.26	1.24	1.20	.86	.56	.37	12.68
GREEN MOUNTAIN DAM		1.16	.97	1.39	1.36	1.49	1.24	1.42	1.69	1.39	1.04	1.04	1.10	15.29
GUFFEY 10 SE		.40	.52	.84	1.25	1.69	1.40	2.80	2.64	1.51	1.00	.61	.51	15.17
GUNNISON		1.03	.84	.75	.66	.65	.52	1.45	1.50	.96	.85	.70	.89	10.80

COLORADO

PRECIPITATION NORMALS (INCHES)

STATION	JAN	FEB	MAR	APR	MAY	JUN	JUL	AUG	SEP	OCT	NOV	DEC	ANN
HAMILTON	1.30	1.28	1.72	1.80	1.73	1.16	1.20	1.58	1.33	1.54	1.38	1.57	17.59
HAYDEN	1.54	1.18	1.19	1.49	1.31	1.20	1.14	1.46	1.15	1.36	1.28	1.60	15.90
HERMIT 7 ESE //	.96	.66	1.20	1.26	1.07	.70	2.24	2.13	1.24	1.66	1.12	1.23	15.47
HOLLY	.33	.29	.71	.97	2.64	2.62	1.94	1.94	1.35	.81	.56	.28	14.44
HOLYOKE	.36	.37	1.06	1.55	3.28	3.39	2.62	1.97	1.31	.85	.51	.36	17.63
IGNACIO 1 N	1.49	.90	1.09	.97	.95	.53	1.50	1.70	1.28	1.66	.94	1.18	14.19
JOHN MARTIN DAM	.21	.21	.54	.95	1.98	1.48	1.99	1.64	.86	.70	.40	.22	11.18
JULESBURG	.40	.41	1.14	1.65	3.48	3.05	2.38	1.66	1.24	.76	.51	.39	17.07
KARVAL	.26	.19	.57	1.01	2.24	1.48	2.66	1.86	.83	.75	.54	.27	12.66
KASSLER	.59	.83	1.47	2.08	2.99	1.62	1.78	1.37	1.42	1.28	1.04	.70	17.17
KAUFFMAN 4 SSE R	.30	.19	.67	1.22	2.49	2.34	2.11	1.44	1.07	.61	.37	.27	13.08
KIT CARSON	.27	.23	.62	1.08	2.35	1.85	2.27	2.05	1.20	.80	.50	.30	13.52
LA JUNTA FAA AIRPORT	.31	.26	.63	1.06	1.80	1.10	2.12	1.42	.84	.71	.50	.27	11.02
LAKE CITY	.91	.76	.92	1.06	.96	.67	1.86	1.84	1.06	1.27	1.03	1.04	13.38
LAKEWOOD	.50	.61	1.12	1.75	2.70	1.74	1.67	1.39	1.40	.98	.82	.47	15.15
LAMAR	.45	.33	.88	1.24	2.54	2.15	2.19	2.01	1.05	.74	.60	.35	14.53
LAS ANIMAS	.32	.30	.61	1.09	2.04	1.49	2.27	1.58	.93	.79	.53	.26	12.21
LEROY 7 WSW	.43	.45	1.14	1.67	3.17	2.83	2.82	1.83	.99	.90	.64	.49	17.36
LITTLE HILLS	.72	.79	1.12	1.37	1.32	1.00	1.05	1.53	1.08	1.09	.97	.95	12.99
LONGMONT 2 ESE	.39	.41	.91	1.70	2.58	1.61	1.09	1.17	1.23	.92	.58	.41	13.00
MANCOS	1.47	1.09	1.37	1.19	1.17	.50	1.78	1.86	1.26	1.75	1.17	1.34	15.95
MESA VERDE NAT PARK	1.91	1.43	1.55	1.22	1.06	.58	1.86	1.96	1.18	1.78	1.27	1.74	17.54
MONTE VISTA	.24	.20	.37	.47	.56	.43	1.23	1.27	.81	.72	.36	.35	7.01
MONTROSE NO 1	.62	.47	.55	.77	.72	.52	.78	1.11	.93	1.05	.68	.68	8.88
MONTROSE NO 2	.60	.47	.53	.78	.75	.53	.83	1.14	1.01	1.05	.70	.68	9.07
NORTHDALE	.84	.80	.74	.88	.96	.37	1.17	1.38	1.08	1.72	1.08	1.02	12.04
NORTH LAKE	.89	1.03	1.70	1.70	2.25	1.31	3.34	3.09	1.46	1.32	1.18	.93	20.20
NORWOOD	1.03	.83	1.01	1.04	1.02	.72	1.68	1.70	1.44	1.54	1.08	1.02	14.11
ORDWAY	.26	.28	.68	.98	1.52	1.20	2.05	1.40	.99	.64	.47	.29	10.76
OTIS 11 NE	.25	.24	.69	1.11	2.72	2.80	2.57	1.71	1.01	.71	.55	.26	14.62
OURAY	1.65	1.57	1.87	1.76	1.60	.99	2.14	2.22	1.70	2.11	1.78	1.55	20.94
PAGOSA SPRINGS	2.08	1.23	1.42	1.19	1.18	.69	1.68	2.45	1.76	1.98	1.50	1.88	19.04
PALISADE	.52	.56	.88	.82	.87	.50	.60	.97	.98	1.06	.73	.57	9.06
PARKER 9 E	.27	.37	.62	1.33	2.33	1.67	1.99	1.76	1.08	.77	.59	.27	13.05
PITKIN	1.76	1.27	1.43	1.31	1.19	1.02	2.01	2.15	1.47	1.02	1.15	1.70	17.48
PLACERVILLE	1.45	1.18	1.52	1.27	1.44	.75	2.01	1.93	1.57	1.40	1.16	1.56	17.24
PUEBLO WSO //R	.25	.27	.68	1.00	1.47	1.16	1.81	1.83	.81	.78	.49	.31	10.87
PYRAMID	1.96	1.77	2.09	1.92	1.48	1.25	1.28	1.76	1.40	1.54	1.66	1.84	19.95
RANGELY	.60	.61	.78	.83	.86	.72	.79	1.06	.96	.96	.61	.59	9.37
RICO	2.59	2.07	2.41	1.93	1.64	1.06	2.77	2.81	2.21	2.24	2.04	2.44	26.21
RIFLE	1.01	.88	.95	.82	.96	.82	.74	1.27	.95	1.18	.84	1.08	11.50
ROCKY FORD 2 ESE	.28	.24	.59	1.14	1.64	1.27	1.90	1.47	1.02	.75	.48	.25	11.03
RYE	.94	1.15	2.01	2.47	3.09	1.49	3.08	3.06	1.51	1.52	1.32	1.06	22.70
SAGUACHE	.30	.24	.39	.62	.78	.55	1.63	1.60	.79	.80	.49	.36	8.55
SHOSHONE	1.99	1.61	1.74	2.11	1.69	1.20	1.17	1.54	1.50	1.74	1.62	1.92	19.83
SILVERTON	1.66	1.49	1.78	1.49	1.42	1.09	2.67	2.88	2.24	2.17	1.50	1.89	22.28
SPICER	1.11	.87	.95	1.14	1.39	1.21	1.52	1.71	1.17	.90	.83	1.08	13.88
SPRINGFIELD	.33	.32	.92	1.35	2.61	2.06	2.54	1.77	1.04	.73	.68	.32	14.67
STEAMBOAT SPRINGS	2.86	2.16	2.07	2.18	1.98	1.44	1.39	1.57	1.54	1.65	1.89	2.57	23.30
STERLING	.31	.21	.70	1.27	3.03	2.59	2.52	1.74	1.02	.84	.46	.33	15.02
STONINGTON	.32	.27	.78	1.20	2.63	2.13	2.44	2.02	1.20	.80	.54	.27	14.60
STRATTON 3 NE	.31	.39	.88	1.36	2.62	2.26	2.63	2.34	1.30	.93	.63	.35	16.00
TACOMA	1.97	1.44	1.62	1.34	1.33	1.01	2.44	2.75	1.92	2.18	1.47	2.00	21.47
TAYLOR PARK	1.48	1.13	1.28	1.19	1.21	.93	1.64	1.83	1.37	1.12	1.10	1.51	15.79
TELLURIDE	1.63	1.50	1.80	1.95	1.70	1.02	2.47	2.53	1.89	2.02	1.48	1.58	21.57

COLORADO

PRECIPITATION NORMALS (INCHES)

STATION	JAN	FEB	MAR	APR	MAY	JUN	JUL	AUG	SEP	OCT	NOV	DEC	ANN
TRINIDAD FAA AIRPORT	.38	.41	.79	1.03	1.82	1.28	1.99	1.77	.87	.84	.59	.53	12.30
TROY 1 SE	.27	.29	.68	1.17	2.31	1.58	2.75	2.10	1.17	.72	.56	.32	13.92
VALLECITO DAM //	2.58	1.80	2.21	1.76	1.51	.93	2.54	2.99	2.15	2.63	1.92	2.62	25.64
VONA	.41	.39	.93	1.30	2.72	2.06	2.65	2.05	1.24	.87	.66	.42	15.70
WALDEN	.58	.45	.57	.70	1.21	.99	1.23	1.36	.97	.76	.56	.62	10.00
WALSENBURG POWER PLANT	.58	.80	1.27	1.72	1.89	1.09	2.11	1.84	.96	1.07	.89	.70	14.92
WATERDALE	.45	.50	1.04	1.92	3.03	1.82	1.69	1.72	1.38	1.12	.68	.48	15.83
WESTCLIFFE	.47	.56	1.16	1.31	1.63	.93	2.26	2.38	1.16	1.19	.82	.71	14.58
WINDSOR 1 SE	.28	.23	.76	1.42	2.36	1.71	1.32	1.27	1.09	.90	.46	.33	12.13
WINTER PARK //	2.68	2.02	2.68	3.00	2.68	2.00	1.99	2.28	1.80	1.74	1.97	2.31	27.15
WRAY	.39	.38	.92	1.52	3.01	3.02	2.66	2.05	1.27	.82	.59	.40	17.03
YAMPA	1.17	.99	1.19	1.31	1.37	1.38	1.78	1.91	1.38	1.16	1.09	1.19	15.92

COLORADO

HEATING DEGREE DAY NORMALS (BASE 65 DEG F)

STATION		JUL	AUG	SEP	OCT	NOV	DEC	JAN	FEB	MAR	APR	MAY	JUN	ANN
AKRON FAA AIRPORT		0	9	145	447	864	1122	1221	966	915	567	279	76	6611
ALAMOSA WSO	R	40	100	303	657	1074	1457	1519	1182	1035	732	453	165	8717
ALLENSPARK		157	217	404	673	984	1190	1265	1095	1125	861	605	323	8899
ASPEN		95	150	348	651	1029	1339	1376	1162	1116	798	524	262	8850
BAILEY		108	165	354	648	978	1187	1237	1056	1057	777	530	256	8353
BONNY DAM		0	7	122	391	807	1079	1194	932	877	495	219	49	6172
BOULDER		0	6	130	357	714	908	1004	804	775	483	220	59	5460
BUENA VISTA		47	116	285	577	936	1184	1218	1025	983	720	459	184	7734
BURLINGTON		6	5	108	364	762	1017	1110	871	803	459	200	38	5743
BYERS 5 NE		0	7	139	438	840	1079	1178	924	868	552	266	78	6369
CANON CITY		0	9	81	301	639	831	911	734	707	411	179	33	4836
CEDAREDGE		0	13	129	434	825	1128	1184	921	815	534	268	70	6321
CENTER 4 SSW		98	150	312	639	1056	1420	1513	1179	1017	714	446	202	8746
CHEESMAN		50	85	242	533	867	1079	1166	991	958	681	428	160	7240
CHERRY CREEK DAM		7	8	154	431	792	1011	1116	890	849	558	284	80	6180
CHEYENNE WELLS		0	7	108	368	774	1039	1125	885	812	459	193	36	5806
COCHETOPA CREEK		120	185	399	741	1140	1584	1708	1420	1243	858	580	307	10285
COLORADO NAT MON		0	6	80	346	777	1104	1178	896	775	453	194	28	5837
COLORADO SPRINGS WSO		8	25	162	440	819	1042	1122	910	880	564	296	78	6346
CORTEZ		0	11	115	434	813	1132	1181	921	828	555	292	68	6350
CRESTED BUTTE		231	286	477	775	1176	1550	1615	1378	1318	978	660	390	10834
DEL NORTE		78	118	282	567	954	1283	1361	1084	967	678	428	192	7992
DENVER WSFO	//R	0	0	135	414	789	1004	1101	879	837	528	253	74	6014
DILLON 1 E		273	332	513	806	1167	1435	1516	1305	1296	972	704	435	10754
DURANGO		9	34	193	493	837	1153	1218	958	862	600	366	125	6848
EADS		0	0	78	339	759	1020	1122	848	756	399	161	30	5512
EAGLE FAA AP		33	80	288	626	1026	1407	1448	1148	1014	705	431	171	8377
ESTES PARK		106	148	342	614	918	1110	1169	1005	1023	762	521	248	7966
FORT COLLINS		5	11	171	468	846	1073	1181	930	877	558	281	82	6483
FORT LEWIS		60	111	294	595	957	1249	1324	1103	1042	750	499	221	8205
FORT MORGAN		0	6	140	438	867	1156	1283	969	874	516	224	47	6520
FRUITA		0	0	101	412	825	1175	1256	913	756	450	176	28	6092
GLENWOOD SPRINGS 1 N		6	34	188	493	891	1228	1283	1000	877	573	312	103	6988
GRAND JUNCTION WSO	//R	0	0	65	325	762	1138	1225	882	716	403	148	19	5683
GRAND LAKE 1 NW		282	338	519	815	1191	1482	1538	1322	1293	969	701	438	10888
GRAND LAKE 6 SSW		214	264	468	775	1128	1473	1593	1369	1318	951	654	384	10591
GREAT SAND DUNES NM		46	97	255	577	972	1271	1333	1086	995	693	428	150	7903
GREELEY CSC		0	0	149	450	861	1128	1240	946	856	522	238	52	6442
GREEN MOUNTAIN DAM		103	141	327	639	1038	1389	1454	1235	1147	780	502	245	9000
GUNNISON		111	188	393	719	1119	1590	1714	1422	1231	816	543	276	10122
HAYDEN		38	77	289	623	1017	1389	1488	1229	1125	708	422	178	8583
HERMIT 7 ESE	//	276	335	519	815	1218	1590	1659	1397	1380	1008	719	450	11366
HOLLY		0	0	90	332	753	1032	1132	860	747	374	137	21	5478
HOLYOKE		0	0	112	409	816	1079	1178	921	853	489	211	44	6112
JOHN MARTIN DAM		0	0	49	294	720	1001	1101	812	694	347	117	13	5148
KASSLER		0	0	113	370	735	952	1042	837	806	501	242	62	5660
KAUFFMAN 4 SSE	R	16	22	188	493	885	1107	1218	969	949	603	320	97	6867
LA JUNTA FAA AIRPORT		0	0	59	306	735	1011	1110	829	729	370	125	15	5289
LAKEWOOD		0	10	145	399	765	955	1057	846	822	531	274	79	5883
LAMAR		0	0	64	319	750	1026	1125	840	722	364	128	12	5350
LAS ANIMAS		0	0	45	296	729	998	1101	820	698	348	102	9	5146
LITTLE HILLS		36	91	306	645	999	1302	1355	1098	1017	714	453	191	8207
LONGMONT 2 ESE		0	6	162	453	843	1082	1194	938	874	546	256	78	6432
MESA VERDE NAT PARK		0	7	99	381	774	1054	1110	896	837	552	282	47	6039
MONTE VISTA		104	130	309	626	1038	1417	1488	1170	1023	708	453	202	8668

COLORADO

HEATING DEGREE DAY NORMALS (BASE 65 DEG F)

STATION	JUL	AUG	SEP	OCT	NOV	DEC	JAN	FEB	MAR	APR	MAY	JUN	ANN
MONTROSE NO 2	0	10	135	437	837	1159	1218	941	818	522	254	69	6400
NORTHDALE	8	38	207	546	921	1243	1311	1050	952	663	406	132	7477
NORWOOD	24	67	250	570	957	1256	1311	1058	973	690	422	159	7737
OURAY	50	90	240	546	903	1150	1197	1016	989	702	440	180	7503
PAGOSA SPRINGS	82	113	297	608	981	1305	1380	1123	1026	732	487	233	8367
PALISADE	0	0	48	272	693	1051	1141	823	663	361	129	14	5195
PARKER 9 E	12	16	174	458	843	1051	1150	932	915	603	333	111	6598
PUEBLO WSO //R	0	0	89	346	744	998	1091	834	756	421	163	23	5465
RANGELY	0	12	175	508	942	1380	1504	1145	921	540	263	58	7448
RIFLE	6	24	177	499	876	1249	1321	1002	856	555	298	82	6945
ROCKY FORD 2 ESE	0	0	59	334	753	1014	1110	832	719	379	127	12	5339
RYE	51	102	243	523	864	1082	1144	960	911	609	362	136	6987
SAGUACHE	68	127	305	608	1005	1336	1404	1103	992	705	443	183	8279
SILVERTON	279	338	498	781	1143	1438	1500	1296	1280	972	698	438	10661
STEAMBOAT SPRINGS	113	169	390	704	1101	1476	1541	1277	1184	810	533	297	9595
STERLING	0	6	157	462	876	1163	1274	966	896	528	235	51	6614
STRATTON 3 NE	0	0	108	367	789	1035	1132	890	828	477	203	40	5869
TAYLOR PARK	264	322	504	806	1215	1674	1752	1506	1451	1065	729	426	11714
TELLURIDE	163	223	396	676	1026	1293	1339	1151	1141	849	589	318	9164
TRINIDAD FAA AIRPORT	0	0	86	359	738	973	1051	846	781	468	207	35	5544
VALLECITO DAM //	60	106	274	567	924	1221	1349	1134	1054	741	502	219	8151
WALDEN	198	285	501	822	1170	1457	1535	1313	1277	915	642	351	10466
WALSENBURG POWER PLANT	0	8	102	370	720	924	989	820	781	501	240	49	5504
WATERDALE	11	14	180	456	828	1048	1147	924	880	579	311	100	6478
WESTCLIFFE	90	132	330	648	999	1252	1299	1103	1035	741	490	208	8327
WRAY	0	0	91	375	789	1045	1144	874	800	435	172	26	5751

COLORADO

COOLING DEGREE DAY NORMALS (BASE 65 DEG F)

STATION		JAN	FEB	MAR	APR	MAY	JUN	JUL	AUG	SEP	OCT	NOV	DEC	ANN
AKRON FAA AIRPORT		0	0	0	0	13	136	268	210	61	7	0	0	695
ALAMOSA WSO	R	0	0	0	0	0	9	40	20	0	0	0	0	69
ALLENSPARK		0	0	0	0	0	8	18	0	0	0	0	0	26
ASPEN		0	0	0	0	0	7	18	11	0	0	0	0	36
BAILEY		0	0	0	0	0	10	21	10	0	0	0	0	41
BONNY DAM		0	0	0	0	24	175	339	276	98	7	0	0	919
BOULDER		0	0	0	0	22	155	283	220	94	16	0	0	790
BUENA VISTA		0	0	0	0	0	13	37	26	0	0	0	0	76
BURLINGTON		0	0	0	0	26	179	325	253	93	11	0	0	887
BYERS 5 NE		0	0	0	0	12	129	252	187	61	7	0	0	648
CANON CITY		0	0	0	0	33	183	329	266	93	15	0	0	919
CEDAREDGE		0	0	0	0	11	97	217	149	36	0	0	0	510
CENTER 4 SSW		0	0	0	0	0	13	43	20	0	0	0	0	76
CHEESMAN		0	0	0	0	0	31	81	54	14	0	0	0	180
CHERRY CREEK DAM		0	0	0	0	8	104	224	157	52	9	0	0	554
CHEYENNE WELLS		0	0	0	0	28	189	326	267	102	9	0	0	921
COCHETOPA CREEK		0	0	0	0	0	0	5	8	0	0	0	0	13
COLORADO NAT MON		0	0	0	0	52	211	394	294	110	8	0	0	1069
COLORADO SPRINGS WSO		0	0	0	0	8	99	200	149	45	0	0	0	501
CORTEZ		0	0	0	0	6	77	214	154	22	0	0	0	473
CRESTED BUTTE		0	0	0	0	0	0	0	0	0	0	0	0	0
DEL NORTE		0	0	0	0	0	6	25	10	0	0	0	0	41
DENVER WSFO	//R	0	0	0	0	11	134	261	203	53	8	0	0	680
DILLON 1 E		0	0	0	0	0	0	0	0	0	0	0	0	0
DURANGO		0	0	0	0	0	20	111	71	7	0	0	0	209
EADS		0	0	0	0	43	231	375	298	108	11	0	0	1066
EAGLE FAA AP		0	0	0	0	0	9	71	39	0	0	0	0	119
ESTES PARK		0	0	0	0	0	5	19	5	0	0	0	0	29
FORT COLLINS		0	0	0	0	5	100	204	132	30	0	0	0	471
FORT LEWIS		0	0	0	0	0	11	47	21	0	0	0	0	79
FORT MORGAN		0	0	0	0	16	155	304	223	62	0	0	0	760
FRUITA		0	0	0	0	18	145	310	228	53	0	0	0	754
GLENWOOD SPRINGS 1 N		0	0	0	0	5	64	167	115	20	0	0	0	371
GRAND JUNCTION WSO	//R	0	0	0	0	58	238	431	338	128	12	0	0	1205
GRAND LAKE 1 NW		0	0	0	0	0	0	0	0	0	0	0	0	0
GRAND LAKE 6 SSW		0	0	0	0	0	0	0	0	0	0	0	0	0
GREAT SAND DUNES NM		0	0	0	0	0	21	58	32	0	0	0	0	111
GREELEY CSC		0	0	0	0	15	130	267	185	50	0	0	0	647
GREEN MOUNTAIN DAM		0	0	0	0	0	8	34	14	0	0	0	0	56
GUNNISON		0	0	0	0	0	0	9	9	0	0	0	0	18
HAYDEN		0	0	0	0	0	19	79	46	10	0	0	0	154
HERMIT 7 ESE	//	0	0	0	0	0	0	0	0	0	0	0	0	0
HOLLY		0	0	0	0	69	279	431	347	144	13	0	0	1283
HOLYOKE		0	0	0	0	22	167	310	253	64	9	0	0	825
JOHN MARTIN DAM		0	0	0	11	68	277	437	360	136	15	0	0	1304
KASSLER		0	0	0	0	16	140	263	207	74	13	0	0	713
KAUFFMAN 4 SSE	R	0	0	0	0	0	76	190	136	29	0	0	0	431
LA JUNTA FAA AIRPORT		0	0	0	7	63	282	437	357	137	18	0	0	1301
LAKEWOOD		0	0	0	0	16	130	240	184	61	8	0	0	639
LAMAR		0	0	0	0	7	63	264	425	347	127	9	0	0	1242
LAS ANIMAS		0	0	0	6	53	270	425	344	120	8	0	0	1226
LITTLE HILLS		0	0	0	0	0	8	64	44	0	0	0	0	116
LONGMONT 2 ESE		0	0	0	0	8	117	227	158	45	0	0	0	555
MESA VERDE NAT PARK		0	0	0	0	19	110	236	165	60	6	0	0	596
MONTE VISTA		0	0	0	0	0	7	52	18	0	0	0	0	77

COLORADO

COOLING DEGREE DAY NORMALS (BASE 65 DEG F)

STATION	JAN	FEB	MAR	APR	MAY	JUN	JUL	AUG	SEP	OCT	NOV	DEC	ANN
MONTROSE NO 2	0	0	0	0	12	120	242	162	45	0	0	0	581
NORTHDALE	0	0	0	0	0	24	116	81	6	0	0	0	227
NORWOOD	0	0	0	0	0	30	67	39	7	0	0	0	143
OURAY	0	0	0	0	0	18	50	31	6	0	0	0	105
PAGOSA SPRINGS	0	0	0	0	0	8	51	27	0	0	0	0	86
PALISADE	0	0	0	7	67	251	446	357	147	15	0	0	1290
PARKER 9 E	0	0	0	0	0	93	189	127	45	6	0	0	460
PUEBLO WSO //R	0	0	0	0	39	212	369	295	119	8	0	0	1042
RANGELY	0	0	0	0	8	97	257	173	40	0	0	0	575
RIFLE	0	0	0	0	0	46	167	117	15	0	0	0	345
ROCKY FORD 2 ESE	0	0	0	0	34	225	369	288	92	0	0	0	1008
RYE	0	0	0	0	5	58	113	87	27	5	0	0	295
SAGUACHE	0	0	0	0	0	9	46	31	0	0	0	0	86
SILVERTON	0	0	0	0	0	0	0	0	0	0	0	0	0
STEAMBOAT SPRINGS	0	0	0	0	0	0	11	7	0	0	0	0	18
STERLING	0	0	0	0	15	147	293	214	52	0	0	0	721
STRATTON 3 NE	0	0	0	0	23	181	327	264	96	10	0	0	901
TAYLOR PARK	0	0	0	0	0	0	0	0	0	0	0	0	0
TELLURIDE	0	0	0	0	0	0	0	0	0	0	0	0	0
TRINIDAD FAA AIRPORT	0	0	0	0	18	155	279	216	68	5	0	0	741
VALLECITO DAM //	0	0	0	0	0		42	19	0	0	0	0	70
WALDEN	0	0	0	0	0	0	0	0	0	0	0	0	0
WALSENBURG POWER PLANT	0	0	0	0	8	109	219	157	45	0	0	0	538
WATERDALE	0	0	0	0	0	79	178	120	33	0	0	0	410
WESTCLIFFE	0	0	0	0	0	7	28	11	0	0	0	0	46
WRAY	0	0	0	0	35	197	360	285	88	12	0	0	977

05 — COLORADO

LEGEND
11 = TEMPERATURE ONLY
12 = PRECIPITATION ONLY
13 = TEMP. & PRECIP.

STATE-STATION NUMBER	STN TYP	NAME		LATITUDE DEG-MIN	LONGITUDE DEG-MIN	ELEVATION (FT)
5-0114	13	AKRON FAA AIRPORT		N 4010	W 10313	4663
5-0130	13	ALAMOSA WSO	R	N 3727	W 10552	7536
5-0183	13	ALLENSPARK		N 4012	W 10532	8500
5-0214	12	ALTENBERN		N 3930	W 10823	5690
5-0370	13	ASPEN		N 3911	W 10650	7928
5-0454	13	BAILEY		N 3924	W 10529	7733
5-0776	12	BLANCA		N 3726	W 10531	7750
5-0834	13	BONNY DAM		N 3938	W 10211	3647
5-0848	13	BOULDER		N 4000	W 10516	5445
5-0909	12	BRECKENRIDGE		N 3929	W 10602	9559
5-1071	13	BUENA VISTA		N 3851	W 10608	7954
5-1121	13	BURLINGTON		N 3919	W 10216	4165
5-1179	13	BYERS 5 NE		N 3945	W 10415	5200
5-1294	13	CANON CITY		N 3826	W 10516	5343
5-1440	13	CEDAREDGE		N 3854	W 10756	6180
5-1458	13	CENTER 4 SSW		N 3744	W 10608	7683
5-1528	13	CHEESMAN		N 3913	W 10517	6875
5-1547	13	CHERRY CREEK DAM		N 3939	W 10451	5647
5-1564	13	CHEYENNE WELLS		N 3849	W 10221	4250
5-1609	12	CIMARRON		N 3827	W 10733	6896
5-1660	12	CLIMAX 2 NW		N 3923	W 10612	11300
5-1713	13	COCHETOPA CREEK		N 3826	W 10646	8000
5-1772	13	COLORADO NAT MON		N 3906	W 10844	5280
5-1778	13	COLORADO SPRINGS WSO		N 3849	W 10443	6145
5-1886	13	CORTEZ		N 3721	W 10834	6177
5-1959	13	CRESTED BUTTE		N 3852	W 10658	8855
5-2184	13	DEL NORTE		N 3740	W 10621	7884
5-2220	13	DENVER WSFO	//R	N 3945	W 10452	5283
5-2281	13	DILLON 1 E		N 3938	W 10602	9065
5-2312	12	DOHERTY RANCH		N 3723	W 10353	5130
5-2326	12	DOLORES		N 3728	W 10830	6950
5-2432	13	DURANGO		N 3717	W 10753	6550
5-2446	13	EADS		N 3829	W 10247	4215
5-2454	13	EAGLE FAA AP		N 3939	W 10655	6500
5-2759	13	ESTES PARK		N 4023	W 10531	7497
5-2932	12	FLAGLER 2 NW		N 3919	W 10305	4975
5-2944	12	FLEMING 1 S		N 4040	W 10250	4250
5-3005	13	FORT COLLINS		N 4035	W 10505	5001
5-3016	13	FORT LEWIS		N 3714	W 10803	7595
5-3038	13	FORT MORGAN		N 4015	W 10348	4321
5-3063	12	FOUNTAIN 1 N		N 3841	W 10442	5570
5-3079	12	FOWLER 1 SE		N 3807	W 10402	4328
5-3146	13	FRUITA		N 3910	W 10844	4507
5-3246	12	GATEWAY 1 SW		N 3840	W 10859	4562
5-3258	12	GENOA 1 W		N 3917	W 10332	5603
5-3261	12	GEORGETOWN		N 3942	W 10542	8500
5-3359	13	GLENWOOD SPRINGS 1 N		N 3934	W 10720	5823
5-3488	13	GRAND JUNCTION WSO	//R	N 3907	W 10832	4855
5-3496	13	GRAND LAKE 1 NW		N 4016	W 10550	8680
5-3500	13	GRAND LAKE 6 SSW		N 4011	W 10552	8288

05 — COLORADO

LEGEND
11 = TEMPERATURE ONLY
12 = PRECIPITATION ONLY
13 = TEMP. & PRECIP.

STATE-STATION NUMBER	STN TYP	NAME		LATITUDE DEG-MIN	LONGITUDE DEG-MIN	ELEVATION (FT)
5-3541	13	GREAT SAND DUNES NM		N 3743	W 10532	8120
5-3553	13	GREELEY CSC		N 4025	W 10442	4653
5-3592	13	GREEN MOUNTAIN DAM		N 3953	W 10620	7740
5-3656	12	GUFFEY 10 SE		N 3841	W 10523	8200
5-3662	13	GUNNISON		N 3832	W 10656	7664
5-3738	12	HAMILTON		N 4022	W 10737	6230
5-3867	13	HAYDEN		N 4029	W 10715	6300
5-3951	13	HERMIT 7 ESE	//	N 3746	W 10708	9001
5-4076	13	HOLLY		N 3803	W 10207	3393
5-4082	13	HOLYOKE		N 4035	W 10218	3746
5-4250	12	IGNACIO 1 N		N 3708	W 10738	6424
5-4388	13	JOHN MARTIN DAM		N 3804	W 10256	3814
5-4413	12	JULESBURG		N 4100	W 10215	3469
5-4444	12	KARVAL		N 3844	W 10332	5075
5-4452	13	KASSLER		N 3930	W 10506	5495
5-4460	13	KAUFFMAN 4 SSE	R	N 4050	W 10356	5250
5-4603	12	KIT CARSON		N 3846	W 10247	4284
5-4720	13	LA JUNTA FAA AIRPORT		N 3803	W 10331	4196
5-4734	12	LAKE CITY		N 3803	W 10719	8880
5-4762	13	LAKEWOOD		N 3945	W 10508	5637
5-4770	13	LAMAR		N 3807	W 10236	3617
5-4834	13	LAS ANIMAS		N 3804	W 10313	3891
5-4945	12	LEROY 7 WSW		N 4029	W 10301	4390
5-5048	13	LITTLE HILLS		N 4000	W 10812	6140
5-5116	13	LONGMONT 2 ESE		N 4010	W 10504	4950
5-5327	12	MANCOS		N 3721	W 10818	7018
5-5531	13	MESA VERDE NAT PARK		N 3712	W 10829	7070
5-5706	13	MONTE VISTA		N 3735	W 10609	7667
5-5717	12	MONTROSE NO 1		N 3829	W 10753	5830
5-5722	13	MONTROSE NO 2		N 3829	W 10753	5830
5-5970	13	NORTHDALE		N 3749	W 10901	6693
5-5990	12	NORTH LAKE		N 3713	W 10503	8800
5-6012	13	NORWOOD		N 3808	W 10817	7017
5-6131	12	ORDWAY		N 3813	W 10345	4310
5-6192	12	OTIS 11 NE		N 4016	W 10251	4263
5-6203	13	OURAY		N 3801	W 10740	7740
5-6258	13	PAGOSA SPRINGS		N 3716	W 10701	7238
5-6266	13	PALISADE		N 3907	W 10821	4780
5-6326	13	PARKER 9 E		N 3931	W 10439	6300
5-6513	12	PITKIN		N 3836	W 10632	9200
5-6524	12	PLACERVILLE		N 3802	W 10803	7322
5-6740	13	PUEBLO WSO	//R	N 3817	W 10431	4639
5-6797	12	PYRAMID		N 4014	W 10705	8009
5-6832	13	RANGELY		N 4005	W 10847	5216
5-7017	12	RICO		N 3741	W 10802	8842
5-7031	13	RIFLE		N 3932	W 10748	5400
5-7167	13	ROCKY FORD 2 ESE		N 3802	W 10342	4178
5-7315	13	RYE		N 3755	W 10456	6790
5-7337	13	SAGUACHE		N 3805	W 10609	7697
5-7618	12	SHOSHONE		N 3934	W 10714	5933

05 — COLORADO

STATE-STATION NUMBER	STN TYP	NAME	LATITUDE DEG-MIN	LONGITUDE DEG-MIN	ELEVATION (FT)
5-7656	13	SILVERTON	N 3748	W 10740	9322
5-7848	12	SPICER	N 4027	W 10628	8379
5-7862	12	SPRINGFIELD	N 3724	W 10237	4405
5-7936	13	STEAMBOAT SPRINGS	N 4030	W 10650	6770
5-7950	13	STERLING	N 4037	W 10312	3939
5-7992	12	STONINGTON	N 3717	W 10211	3801
5-8008	13	STRATTON 3 NE	N 3920	W 10233	4334
5-8154	12	TACOMA	N 3731	W 10747	7300
5-8184	13	TAYLOR PARK	N 3849	W 10637	9206
5-8204	13	TELLURIDE	N 3757	W 10749	8756
5-8434	13	TRINIDAD FAA AIRPORT	N 3715	W 10420	5746
5-8468	12	TROY 1 SE	N 3708	W 10319	5460
5-8582	13	VALLECITO DAM //	N 3722	W 10735	7650
5-8722	12	VONA	N 3918	W 10244	4500
5-8756	13	WALDEN	N 4044	W 10616	8099
5-8781	13	WALSENBURG POWER PLANT	N 3737	W 10448	6221
5-8839	13	WATERDALE	N 4025	W 10512	5260
5-8931	13	WESTCLIFFE	N 3808	W 10529	7860
5-9147	12	WINDSOR 1 SE	N 4029	W 10454	4720
5-9175	12	WINTER PARK //	N 3954	W 10546	9058
5-9243	13	WRAY	N 4004	W 10213	3575
5-9265	12	YAMPA	N 4009	W 10654	7892

05 – COLORADO

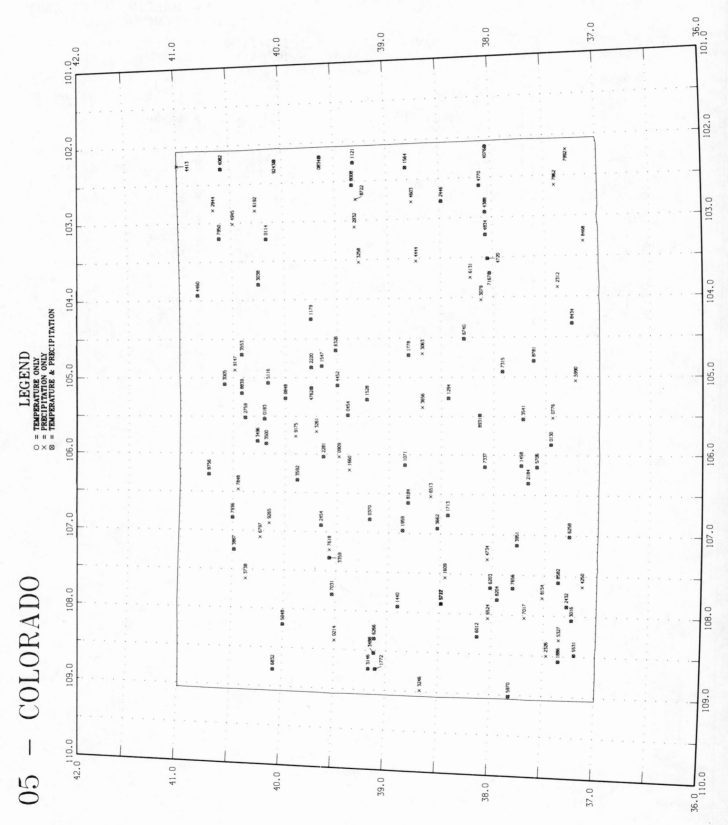

LEGEND

O = TEMPERATURE ONLY
X = PRECIPITATION ONLY
⊗ = TEMPERATURE & PRECIPITATION

CONNECTICUT

TEMPERATURE NORMALS (DEG F)

STATION			JAN	FEB	MAR	APR	MAY	JUN	JUL	AUG	SEP	OCT	NOV	DEC	ANN
BRIDGEPORT WSO	R	MAX	36.5	37.9	45.5	57.2	67.1	76.4	82.1	81.1	74.5	64.5	52.8	41.0	59.7
		MIN	22.5	23.3	30.9	40.0	49.8	59.3	65.9	65.0	57.8	47.4	38.1	27.3	43.9
		MEAN	29.5	30.6	38.2	48.6	58.5	67.9	74.0	73.1	66.2	56.0	45.5	34.2	51.9
DANBURY		MAX	34.8	37.4	46.0	59.4	70.1	78.5	83.2	81.3	73.7	63.3	50.9	38.7	59.8
		MIN	18.3	19.6	27.7	37.0	46.4	55.3	60.2	59.1	51.8	41.3	33.3	22.6	39.4
		MEAN	26.6	28.5	36.9	48.2	58.2	66.9	71.7	70.2	62.8	52.3	42.1	30.7	49.6
FALLS VILLAGE		MAX	33.8	36.3	45.2	59.3	71.6	79.5	83.8	81.6	73.7	63.0	50.0	37.6	59.6
		MIN	12.5	14.4	23.7	32.9	43.0	52.0	56.7	55.4	48.2	37.0	28.9	17.8	35.2
		MEAN	23.2	25.4	34.4	46.1	57.3	65.8	70.3	68.5	61.0	50.0	39.5	27.7	47.4
HARTFORD BRAINARD FLD //R		MAX	35.0	37.5	46.2	59.1	70.1	78.7	83.5	81.6	74.3	63.6	51.3	38.7	60.0
		MIN	17.2	19.0	27.4	36.9	46.6	56.3	61.8	60.0	51.5	40.8	32.6	21.7	39.3
		MEAN	26.2	28.3	36.8	48.0	58.4	67.5	72.7	70.8	62.9	52.2	42.0	30.2	49.7
HARTFORD WSO	R	MAX	33.6	36.3	45.5	60.0	71.4	80.1	84.8	82.6	74.8	63.9	50.6	37.3	60.1
		MIN	16.7	18.8	28.0	37.6	47.3	57.0	61.9	60.0	51.7	40.9	32.5	20.9	39.5
		MEAN	25.2	27.6	36.8	48.8	59.4	68.6	73.4	71.3	63.3	52.4	41.6	29.1	49.8
MIDDLETOWN 4 W		MAX	35.1	37.1	45.5	58.4	69.2	78.1	82.7	80.2	72.4	62.3	51.2	39.1	59.3
		MIN	19.8	20.9	28.9	38.3	47.6	56.9	62.3	61.1	53.6	43.8	35.2	24.5	41.1
		MEAN	27.4	29.1	37.2	48.4	58.5	67.5	72.5	70.7	63.0	53.1	43.2	31.8	50.2
MOUNT CARMEL		MAX	36.2	38.4	46.6	59.0	69.1	77.9	82.9	81.4	74.3	64.3	52.3	40.2	60.2
		MIN	18.6	19.8	28.0	36.9	46.2	55.6	61.2	60.1	52.5	41.9	33.6	22.7	39.8
		MEAN	27.4	29.1	37.3	48.0	57.7	66.7	72.1	70.8	63.4	53.1	43.0	31.5	50.0
NORFOLK 2 SW		MAX	28.0	29.5	38.5	52.6	64.8	73.0	77.5	75.6	67.8	56.5	44.3	31.9	53.3
		MIN	11.6	12.0	21.3	32.4	43.0	52.3	57.3	55.6	48.4	37.8	29.2	17.0	34.8
		MEAN	19.8	20.8	29.9	42.5	53.9	62.7	67.4	65.6	58.1	47.1	36.8	24.5	44.1
NORWALK GAS PLANT		MAX	37.3	39.0	47.1	59.4	69.0	78.2	83.6	82.1	75.3	64.8	53.3	41.3	60.9
		MIN	18.7	19.8	28.4	38.0	46.7	56.1	61.7	60.2	52.7	41.3	33.4	23.2	40.0
		MEAN	28.0	29.5	37.8	48.7	57.9	67.2	72.7	71.2	64.1	53.1	43.4	32.3	50.5
SHEPAUG DAM		MAX	33.3	35.4	43.9	57.4	68.0	75.8	80.0	78.5	71.4	61.8	49.9	37.3	57.7
		MIN	15.0	16.1	24.8	34.9	44.9	53.9	59.0	57.6	50.6	40.1	31.8	20.0	37.4
		MEAN	24.2	25.8	34.3	46.2	56.5	64.9	69.5	68.1	61.0	51.0	40.9	28.7	47.6
STORRS		MAX	33.2	35.0	43.2	56.4	67.2	75.3	79.7	78.1	71.1	61.4	49.5	37.2	57.3
		MIN	17.2	18.4	26.5	36.1	45.4	54.6	60.1	58.6	51.1	41.4	33.2	21.6	38.7
		MEAN	25.2	26.8	34.9	46.3	56.4	65.0	69.9	68.4	61.1	51.5	41.4	29.4	48.0
WEST THOMPSON DAM		MAX	35.2	36.9	45.4	58.5	69.6	77.9	82.8	80.8	73.4	63.4	51.3	38.9	59.5
		MIN	13.2	14.3	24.4	33.5	42.9	52.6	58.0	56.4	48.3	36.8	29.2	17.9	35.6
		MEAN	24.2	25.7	34.9	46.0	56.3	65.2	70.5	68.6	60.9	50.1	40.3	28.4	47.6
WIGWAM RESERVOIR		MAX	34.9	37.1	45.4	59.4	69.6	77.9	82.1	80.2	72.8	63.1	51.3	38.8	59.4
		MIN	15.5	16.3	25.0	34.3	43.6	53.1	58.2	57.0	50.0	39.7	31.2	20.3	37.0
		MEAN	25.2	26.7	35.2	46.9	56.6	65.5	70.2	68.6	61.4	51.4	41.3	29.6	48.2

CONNECTICUT

PRECIPITATION NORMALS (INCHES)

STATION		JAN	FEB	MAR	APR	MAY	JUN	JUL	AUG	SEP	OCT	NOV	DEC	ANN
BARKHAMSTED		3.60	3.20	4.40	4.11	3.58	3.66	3.56	4.46	4.41	3.86	4.14	4.26	47.24
BRIDGEPORT WSO	R	3.25	3.00	3.93	3.74	3.44	2.90	3.46	3.68	3.29	3.33	3.79	3.75	41.56
BULLS BRIDGE DAM		3.16	2.92	3.89	4.16	3.52	3.70	3.99	3.78	4.18	3.73	4.05	4.03	45.11
BURLINGTON		3.87	3.32	4.73	4.43	3.79	3.91	3.87	4.89	4.84	4.15	4.67	4.61	51.08
COCKAPONSET RANGER STA		4.37	3.68	4.81	4.43	4.07	3.19	3.49	4.04	4.31	4.08	4.64	4.88	49.99
DANBURY		3.56	3.20	4.17	4.13	3.99	3.57	3.81	4.62	4.37	4.02	4.49	4.24	48.17
FALLS VILLAGE		3.10	2.56	3.36	3.91	3.38	3.90	3.64	4.50	4.26	3.42	3.83	3.65	43.51
GROTON		4.41	3.96	4.76	4.17	3.84	2.69	3.20	4.01	3.95	3.93	4.62	5.01	48.55
HARTFORD BRAINARD FLD //R		3.54	2.97	4.06	3.79	3.36	3.17	3.68	3.71	3.78	3.46	3.68	3.98	43.18
HARTFORD WSO	R	3.53	3.19	4.15	4.02	3.37	3.38	3.09	4.00	3.94	3.51	4.05	4.16	44.39
MANSFIELD HOLLOW DAM	R	3.75	3.07	3.96	3.83	3.47	2.96	3.91	4.17	3.95	3.64	4.05	4.18	44.94
MIDDLETOWN 4 W		4.00	3.42	4.64	4.37	3.98	3.57	3.88	4.25	4.72	4.16	4.51	4.60	50.10
MOUNT CARMEL		3.93	3.41	4.80	4.34	3.86	3.54	3.88	3.87	4.75	3.96	4.41	4.35	49.10
NORFOLK 2 SW		4.16	3.68	4.83	4.54	3.95	4.24	4.08	4.98	4.81	4.26	4.71	4.84	53.08
NORWALK GAS PLANT		3.54	3.11	4.40	4.29	3.78	3.49	3.77	4.31	4.02	3.85	4.21	4.13	46.90
PUTNAM LAKE		3.71	3.40	4.84	4.52	4.17	3.78	3.86	4.65	4.52	4.08	4.67	4.30	50.50
ROCKY RIVER DAM		3.22	2.90	3.98	4.08	3.58	3.48	3.96	4.23	4.29	3.79	4.14	3.99	45.64
ROUND POND		3.63	3.31	4.53	4.20	4.07	3.73	3.62	4.92	4.28	4.31	4.71	4.29	49.60
SAUGATUCK RESERVOIR		3.77	3.59	5.01	4.37	3.99	3.56	3.65	4.17	4.25	3.92	4.39	4.44	49.11
SHEPAUG DAM		3.63	3.15	4.25	4.19	3.79	3.97	3.84	4.50	4.30	3.94	4.33	4.21	48.10
SHUTTLE MEADOW RESVR		3.96	3.52	4.67	4.52	3.90	3.33	3.78	4.23	4.50	4.03	4.58	4.58	49.60
STEVENSON DAM		4.25	3.71	5.24	4.63	3.96	3.55	3.65	3.94	4.38	4.03	4.82	4.85	51.01
STORRS		3.82	3.13	4.13	3.91	3.78	3.22	4.18	4.45	4.16	3.92	4.25	4.35	47.30
TORRINGTON		3.52	3.17	4.20	4.04	3.40	3.59	3.67	4.67	4.31	3.97	4.27	4.41	47.22
WEST HARTFORD		3.78	3.14	4.54	4.35	3.71	3.70	3.83	4.49	4.49	4.07	4.63	4.34	49.07
WEST THOMPSON DAM		3.93	3.18	4.18	4.00	3.54	3.42	3.72	4.30	4.25	3.79	4.33	4.32	46.96
WIGWAM RESERVOIR		3.57	3.09	4.12	3.97	3.39	3.48	3.84	4.10	4.28	3.85	4.11	4.30	46.10
WOODBURY		3.23	2.82	4.01	3.96	3.67	3.60	4.42	4.26	4.29	3.76	4.12	3.86	46.00

CONNECTICUT

HEATING DEGREE DAY NORMALS (BASE 65 DEG F)

STATION	JUL	AUG	SEP	OCT	NOV	DEC	JAN	FEB	MAR	APR	MAY	JUN	ANN
BRIDGEPORT WSO R	0	0	49	285	585	955	1101	963	831	492	220	20	5501
DANBURY	0	6	107	394	687	1063	1190	1022	871	504	228	28	6100
FALLS VILLAGE	0	21	147	465	765	1156	1296	1109	949	567	252	48	6775
HARTFORD BRAINARD FLD //R	0	10	111	401	690	1079	1203	1028	874	510	220	30	6156
HARTFORD WSO R	0	8	102	391	702	1113	1234	1047	874	486	197	20	6174
MIDDLETOWN 4 W	6	17	117	373	654	1029	1166	1005	862	498	218	31	5976
MOUNT CARMEL	0	6	92	369	660	1039	1166	1005	859	510	239	33	5978
NORFOLK 2 SW	25	52	216	555	846	1256	1401	1238	1088	675	351	98	7801
NORWALK GAS PLANT	0	0	80	369	648	1014	1147	994	843	489	237	26	5847
SHEPAUG DAM	9	21	145	434	723	1125	1265	1098	952	564	274	61	6671
STORRS	7	16	139	419	708	1104	1234	1070	933	561	276	56	6523
WEST THOMPSON DAM	6	26	145	462	741	1135	1265	1100	933	570	277	58	6718
WIGWAM RESERVOIR	7	16	143	422	711	1097	1234	1072	924	543	273	52	6494

CONNECTICUT

COOLING DEGREE DAY NORMALS (BASE 65 DEG F)

STATION	JAN	FEB	MAR	APR	MAY	JUN	JUL	AUG	SEP	OCT	NOV	DEC	ANN
BRIDGEPORT WSO R	0	0	0	0	18	107	279	251	85	6	0	0	746
DANBURY	0	0	0	0	17	85	211	167	41	0	0	0	521
FALLS VILLAGE	0	0	0	0	13	72	169	130	27	0	0	0	411
HARTFORD BRAINARD FLD //R	0	0	0	0	16	105	242	189	48	0	0	0	600
HARTFORD WSO R	0	0	0	0	24	128	260	203	51	0	0	0	666
MIDDLETOWN 4 W	0	0	0	0	16	106	239	193	57	0	0	0	611
MOUNT CARMEL	0	0	0	0	13	84	220	186	44	0	0	0	547
NORFOLK 2 SW	0	0	0	0	7	29	99	70	9	0	0	0	214
NORWALK GAS PLANT	0	0	0	0	17	92	239	197	53	0	0	0	598
SHEPAUG DAM	0	0	0	0	10	58	149	117	25	0	0	0	359
STORRS	0	0	0	0	9	56	158	121	22	0	0	0	366
WEST THOMPSON DAM	0	0	0	0	7	64	176	138	22	0	0	0	407
WIGWAM RESERVOIR	0	0	0	0	13	67	168	128	35	0	0	0	411

06 — CONNECTICUT

11 = TEMPERATURE ONLY
12 = PRECIPITATION ONLY
13 = TEMP. & PRECIP.

STATE-STATION NUMBER	STN TYP	NAME		LATITUDE DEG-MIN	LONGITUDE DEG-MIN	ELEVATION (FT)
6-0299	12	BARKHAMSTED		N 4155	W 07257	660
6-0806	13	BRIDGEPORT WSO	R	N 4110	W 07308	7
6-0961	12	BULLS BRIDGE DAM		N 4139	W 07329	260
6-0973	12	BURLINGTON		N 4148	W 07256	510
6-1488	12	COCKAPONSET RANGER STA		N 4128	W 07231	160
6-1762	13	DANBURY		N 4123	W 07328	510
6-2658	13	FALLS VILLAGE		N 4157	W 07322	580
6-3207	12	GROTON		N 4121	W 07203	39
6-3451	13	HARTFORD BRAINARD FLD //R		N 4144	W 07239	15
6-3456	13	HARTFORD WSO	R	N 4156	W 07241	169
6-4488	12	MANSFIELD HOLLOW DAM	R	N 4145	W 07211	250
6-4767	13	MIDDLETOWN 4 W		N 4133	W 07243	369
6-5077	13	MOUNT CARMEL		N 4124	W 07254	180
6-5445	13	NORFOLK 2 SW		N 4158	W 07313	1337
6-5893	13	NORWALK GAS PLANT		N 4107	W 07325	37
6-6655	12	PUTNAM LAKE		N 4105	W 07338	300
6-6966	12	ROCKY RIVER DAM		N 4135	W 07326	220
6-7002	12	ROUND POND		N 4118	W 07332	800
6-7157	12	SAUGATUCK RESERVOIR		N 4115	W 07321	300
6-7373	13	SHEPAUG DAM		N 4143	W 07318	840
6-7432	12	SHUTTLE MEADOW RESVR		N 4139	W 07249	410
6-8065	12	STEVENSON DAM		N 4123	W 07310	60
6-8138	13	STORRS		N 4148	W 07215	650
6-8436	12	TORRINGTON		N 4148	W 07307	580
6-9162	12	WEST HARTFORD		N 4145	W 07247	275
6-9388	13	WEST THOMPSON DAM		N 4157	W 07154	360
6-9568	13	WIGWAM RESERVOIR		N 4140	W 07308	600
6-9775	12	WOODBURY		N 4133	W 07314	650

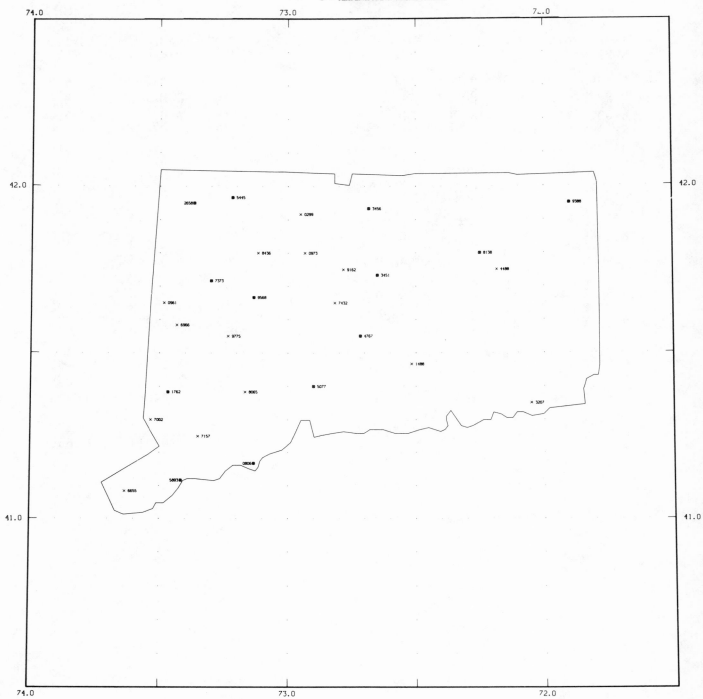

DELAWARE

TEMPERATURE NORMALS (DEG F)

STATION		JAN	FEB	MAR	APR	MAY	JUN	JUL	AUG	SEP	OCT	NOV	DEC	ANN
BRIDGEVILLE 1 NW	MAX	43.3	45.7	54.3	66.1	75.4	83.4	87.1	85.6	79.7	68.8	57.9	47.2	66.2
	MIN	25.3	26.3	33.3	42.2	51.8	60.6	65.3	63.8	57.0	45.9	37.0	28.8	44.8
	MEAN	34.3	36.0	43.8	54.2	63.7	72.0	76.2	74.7	68.4	57.4	47.5	38.1	55.5
DOVER	MAX	43.1	46.1	54.5	66.1	75.7	83.7	87.5	86.1	80.4	69.6	58.0	47.2	66.5
	MIN	25.6	26.6	34.0	43.2	52.7	61.4	66.4	65.4	58.8	47.6	38.3	29.5	45.8
	MEAN	34.4	36.4	44.3	54.7	64.2	72.6	77.0	75.8	69.6	58.6	48.2	38.4	56.2
GEORGETOWN 5 SW	MAX	43.6	45.7	54.2	65.6	74.9	83.0	87.1	85.6	79.7	68.6	58.0	47.4	66.1
	MIN	24.7	25.7	32.8	41.6	51.2	60.0	65.1	64.1	57.0	45.4	36.4	28.1	44.3
	MEAN	34.2	35.8	43.5	53.6	63.1	71.5	76.1	74.9	68.4	57.0	47.2	37.8	55.3
LEWES 1 SW	MAX	43.1	45.0	52.5	63.4	71.8	80.1	84.2	83.2	77.5	67.3	57.5	47.4	64.4
	MIN	25.9	27.1	34.1	42.8	52.3	61.3	66.1	65.3	59.2	47.7	38.4	29.7	45.8
	MEAN	34.6	36.1	43.3	53.1	62.1	70.7	75.2	74.3	68.4	57.5	48.0	38.5	55.2
MILFORD 2 WSW	MAX	43.4	45.5	54.9	66.4	75.7	83.5	87.8	86.2	80.2	69.0	58.2	47.3	66.5
	MIN	25.2	26.0	33.6	42.4	52.0	60.5	65.3	64.1	57.4	46.3	36.9	28.5	44.9
	MEAN	34.3	35.8	44.3	54.4	63.9	72.0	76.5	75.2	68.8	57.7	47.6	37.9	55.7
NEWARK UNIVERSITY FARM	MAX	40.3	43.3	52.5	64.8	75.1	83.1	86.9	85.4	79.4	68.0	55.8	44.2	64.9
	MIN	23.1	24.1	31.7	40.9	50.4	59.4	64.4	63.5	56.4	44.6	35.7	26.9	43.4
	MEAN	31.7	33.7	42.1	52.8	62.7	71.3	75.7	74.5	67.9	56.3	45.8	35.6	54.2
WILMINGTON NCASTLE WSO	MAX	39.2	41.8	50.9	63.0	72.7	81.2	85.6	84.1	77.8	66.7	54.8	43.6	63.5
	MIN	23.2	24.6	32.6	41.8	51.7	61.2	66.3	65.4	58.0	45.9	36.4	27.3	44.5
	MEAN	31.2	33.2	41.8	52.4	62.2	71.2	76.0	74.8	67.9	56.3	45.6	35.5	54.0
WILMINGTON PORTER RES	MAX	38.5	40.8	49.8	61.9	71.7	80.3	84.4	82.9	76.4	65.4	53.5	42.7	62.4
	MIN	23.0	24.4	32.2	41.7	51.3	60.5	65.5	64.4	57.3	46.0	36.8	27.4	44.2
	MEAN	30.8	32.6	41.0	51.8	61.6	70.4	74.9	73.7	66.9	55.7	45.2	35.1	53.3

DELAWARE

PRECIPITATION NORMALS (INCHES)

STATION	JAN	FEB	MAR	APR	MAY	JUN	JUL	AUG	SEP	OCT	NOV	DEC	ANN
BRIDGEVILLE 1 NW	3.41	3.07	3.86	3.31	3.31	3.76	4.36	5.28	3.72	3.42	3.17	3.69	44.36
DOVER	3.24	2.86	3.87	3.21	3.50	3.61	4.27	5.12	4.14	3.56	3.53	3.53	44.44
GEORGETOWN 5 SW	3.35	3.02	3.97	3.20	3.26	3.59	4.02	5.43	3.51	3.48	3.29	3.59	43.71
LEWES 1 SW	3.47	3.33	4.21	3.41	3.54	3.70	4.16	5.36	3.26	3.33	3.49	3.82	45.08
MILFORD 2 WSW	3.39	3.13	4.03	3.41	3.48	3.72	4.38	4.78	3.75	3.47	3.43	3.81	44.78
NEWARK UNIVERSITY FARM	3.10	2.73	3.66	3.50	3.52	3.75	4.32	4.50	3.61	3.18	3.19	3.53	42.59
WILMINGTON NCASTLE WSO	3.11	2.99	3.87	3.39	3.23	3.51	3.90	4.03	3.59	2.89	3.33	3.54	41.38
WILMINGTON PORTER RES	3.29	3.00	4.07	3.88	3.57	4.00	4.13	4.42	3.81	3.23	3.76	3.74	44.90

DELAWARE

HEATING DEGREE DAY NORMALS (BASE 65 DEG F)

STATION	JUL	AUG	SEP	OCT	NOV	DEC	JAN	FEB	MAR	APR	MAY	JUN	ANN
BRIDGEVILLE 1 NW	0	0	24	250	525	834	952	812	657	327	99	0	4480
DOVER	0	0	16	212	504	825	949	801	642	314	93	0	4356
GEORGETOWN 5 SW	0	0	24	264	534	843	955	818	667	345	114	7	4571
LEWES 1 SW	0	0	22	244	510	822	942	809	673	357	120	6	4505
MILFORD 2 WSW	0	0	25	240	522	840	952	818	642	323	95	0	4457
NEWARK UNIVERSITY FARM	0	0	35	278	576	911	1032	876	710	366	122	0	4906
WILMINGTON NCASTLE WSO	0	0	36	282	582	915	1048	890	719	378	130	6	4986
WILMINGTON PORTER RES	0	0	49	297	594	927	1060	907	744	396	144	9	5127

DELAWARE

COOLING DEGREE DAY NORMALS (BASE 65 DEG F)

STATION	JAN	FEB	MAR	APR	MAY	JUN	JUL	AUG	SEP	OCT	NOV	DEC	ANN
BRIDGEVILLE 1 NW	0	0	0	0	59	214	347	301	126	15	0	0	1062
DOVER	0	0	0	5	68	231	372	335	154	14	0	0	1179
GEORGETOWN 5 SW	0	0	0	0	55	202	344	307	126	16	0	0	1050
LEWES 1 SW	0	0	0	0	30	177	316	288	124	11	0	0	946
MILFORD 2 WSW	0	0	0	0	61	215	357	316	139	13	0	0	1101
NEWARK UNIVERSITY FARM	0	0	0	0	51	194	332	295	122	9	0	0	1003
WILMINGTON NCASTLE WSO	0	0	0	0	43	192	341	304	123	12	0	0	1015
WILMINGTON PORTER RES	0	0	0	0	38	171	307	270	106	9	0	0	901

07 — DELAWARE

LEGEND
11 = TEMPERATURE ONLY
12 = PRECIPITATION ONLY
13 = TEMP. & PRECIP.

STATE-STATION NUMBER	STN TYP	NAME	LATITUDE DEG-MIN	LONGITUDE DEG-MIN	ELEVATION (FT)
7-1330	13	BRIDGEVILLE 1 NW	N 3845	W 07537	50
7-2730	13	DOVER	N 3909	W 07531	30
7-3570	13	GEORGETOWN 5 SW	N 3838	W 07527	45
7-5320	13	LEWES 1 SW	N 3846	W 07509	20
7-5915	13	MILFORD 2 WSW	N 3854	W 07528	30
7-6410	13	NEWARK UNIVERSITY FARM	N 3940	W 07544	90
7-9595	13	WILMINGTON NCASTLE WSO	N 3940	W 07536	74
7-9605	13	WILMINGTON PORTER RES	N 3946	W 07532	274

07 – DELAWARE

LEGEND
O = TEMPERATURE ONLY
X = PRECIPITATION ONLY
⊗ = TEMPERATURE & PRECIPITATION

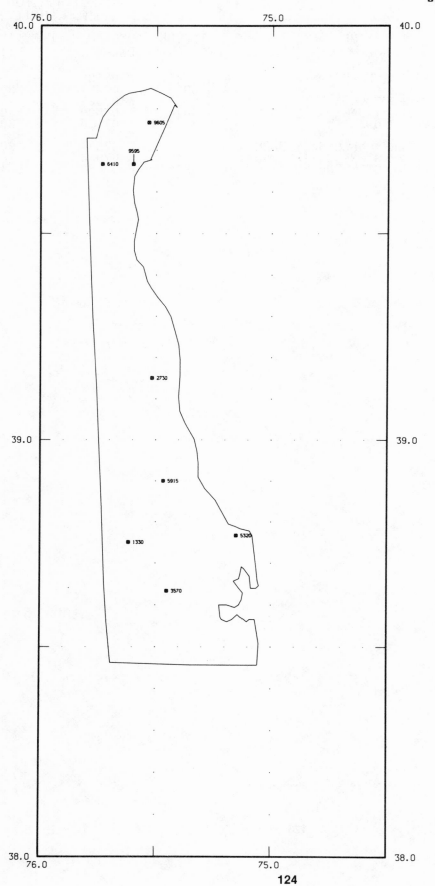

FLORIDA

TEMPERATURE NORMALS (DEG F)

STATION			JAN	FEB	MAR	APR	MAY	JUN	JUL	AUG	SEP	OCT	NOV	DEC	ANN
APALACHICOLA WSO	R	MAX	60.5	62.4	68.0	75.1	81.7	86.6	88.0	88.0	85.3	78.2	69.2	63.0	75.5
		MIN	45.1	46.9	53.4	60.7	67.3	72.9	75.0	74.7	72.3	62.1	52.7	47.0	60.8
		MEAN	52.8	54.7	60.7	67.9	74.5	79.8	81.5	81.4	78.9	70.2	61.0	55.0	68.2
ARCADIA		MAX	74.3	75.7	80.8	85.5	89.8	91.2	92.0	92.1	90.2	85.3	79.6	75.1	84.3
		MIN	48.8	49.3	54.2	57.8	63.2	68.2	70.1	70.9	70.1	63.5	55.6	49.9	60.1
		MEAN	61.6	62.5	67.5	71.7	76.5	79.7	81.1	81.5	80.2	74.4	67.6	62.5	72.2
ARCHBOLD BIOLOGIC STA		MAX	74.2	75.6	80.9	85.8	89.9	91.6	92.6	92.8	90.8	86.1	80.4	75.3	84.7
		MIN	49.0	49.5	54.6	58.5	63.8	68.1	69.8	70.3	70.0	64.0	56.4	50.5	60.4
		MEAN	61.6	62.6	67.8	72.1	76.9	79.9	81.2	81.6	80.4	75.1	68.4	62.9	72.5
AVON PARK		MAX	74.1	75.6	80.7	85.7	90.1	92.0	92.9	92.9	90.9	85.6	79.9	75.0	84.6
		MIN	49.5	50.4	55.4	60.1	65.4	69.9	71.7	72.1	71.1	64.7	56.9	51.2	61.5
		MEAN	61.8	63.0	68.1	72.9	77.8	81.0	82.3	82.5	81.0	75.2	68.4	63.1	73.1
BARTOW		MAX	73.1	74.7	80.1	85.0	89.5	91.5	92.6	92.6	90.5	85.4	78.7	74.0	84.0
		MIN	48.7	49.7	54.8	59.7	65.3	69.8	71.5	71.7	70.5	63.7	55.8	50.0	60.9
		MEAN	60.9	62.2	67.5	72.3	77.4	80.7	82.1	82.2	80.6	74.6	67.3	62.0	72.5
BELLE GLADE EXP STA		MAX	74.1	75.0	79.5	83.0	86.4	88.6	90.2	90.5	88.9	84.6	79.3	75.0	82.9
		MIN	51.0	51.5	55.6	59.5	64.5	69.0	70.6	70.9	70.5	65.3	58.4	52.6	61.6
		MEAN	62.6	63.3	67.6	71.2	75.5	78.9	80.4	80.7	79.7	75.0	68.9	63.8	72.3
BRADENTON 5 ESE		MAX	71.8	72.6	77.1	82.0	86.9	89.5	90.6	90.7	89.4	84.4	78.3	73.2	82.2
		MIN	50.0	50.9	55.7	59.6	65.0	69.9	71.8	72.1	71.5	64.6	56.8	51.1	61.6
		MEAN	60.9	61.8	66.4	70.8	76.0	79.7	81.2	81.5	80.5	74.5	67.5	62.2	71.9
BROOKSVILLE CHIN HILL		MAX	70.3	72.0	77.9	83.1	87.8	90.2	90.5	90.3	88.8	83.4	76.8	71.7	81.9
		MIN	48.4	49.1	54.6	59.7	65.1	70.1	71.6	71.8	70.7	63.5	55.1	49.8	60.8
		MEAN	59.4	60.6	66.3	71.4	76.5	80.1	81.1	81.1	79.8	73.5	66.0	60.8	71.4
BUSHNELL 2 E		MAX	71.2	72.9	78.6	83.8	88.5	90.8	91.5	91.5	89.5	84.0	77.3	72.1	82.6
		MIN	45.8	46.5	52.1	56.9	62.8	68.3	70.5	70.7	69.3	60.7	52.2	46.9	58.6
		MEAN	58.5	59.8	65.3	70.4	75.7	79.5	81.0	81.1	79.4	72.4	64.8	59.6	70.6
CHIPLEY 3 E		MAX	63.0	65.8	72.6	79.8	85.6	90.0	91.1	90.9	88.3	80.3	71.2	65.2	78.7
		MIN	39.7	41.6	48.1	55.1	61.8	68.3	71.0	70.5	67.1	54.7	45.5	40.6	55.3
		MEAN	51.3	53.7	60.4	67.5	73.7	79.2	81.1	80.8	77.8	67.5	58.4	52.9	67.0
CLERMONT 6 SSW		MAX	70.0	72.0	78.3	84.0	88.7	90.8	91.6	91.7	89.8	83.6	76.3	70.8	82.3
		MIN	48.9	49.9	55.2	60.2	65.7	70.4	72.0	72.5	71.5	64.7	56.2	50.4	61.5
		MEAN	59.5	61.0	66.8	72.1	77.2	80.6	81.8	82.1	80.7	74.2	66.3	60.6	71.9
CLEWISTON U S ENG		MAX	73.1	74.4	79.8	83.7	87.1	89.4	91.2	91.2	89.5	84.5	78.8	74.1	83.1
		MIN	53.9	54.2	58.8	63.2	67.3	71.1	72.3	72.9	73.1	69.0	62.1	56.0	64.5
		MEAN	63.5	64.3	69.3	73.5	77.2	80.3	81.8	82.1	81.3	76.8	70.5	65.1	73.8
CROSS CITY 2 WNW		MAX	66.0	67.9	74.5	80.9	86.9	90.4	91.0	90.8	88.8	82.1	74.1	68.1	80.1
		MIN	39.9	41.8	48.0	54.5	60.8	67.3	70.0	70.3	68.3	57.1	47.0	41.0	55.5
		MEAN	53.0	54.9	61.3	67.7	73.9	78.9	80.5	80.6	78.6	69.6	60.6	54.5	67.8
DAYTONA BEACH WSO	R	MAX	68.4	69.3	74.6	80.0	84.8	87.8	89.6	89.0	86.9	81.2	74.8	69.8	79.7
		MIN	47.4	48.2	53.6	59.1	65.3	70.5	72.5	72.8	72.1	65.1	55.5	49.2	60.9
		MEAN	57.9	58.8	64.1	69.6	75.1	79.2	81.1	80.9	79.5	73.2	65.2	59.5	70.3
DE FUNIAK SPRINGS		MAX	62.9	66.3	73.1	80.8	87.2	91.5	92.5	92.2	88.7	81.0	71.3	65.1	79.4
		MIN	39.3	41.1	47.3	54.3	61.1	67.0	69.9	69.3	65.9	53.9	44.7	40.4	54.5
		MEAN	51.1	53.8	60.3	67.6	74.2	79.3	81.2	80.7	77.3	67.5	58.0	52.8	67.0
DELAND 1 SSE		MAX	70.4	71.5	77.0	82.7	87.4	90.1	91.7	91.3	89.0	83.0	76.6	71.4	81.8
		MIN	45.6	46.5	51.6	56.8	63.4	68.9	70.7	71.0	69.7	61.7	52.9	46.6	58.8
		MEAN	58.0	59.0	64.3	69.8	75.4	79.5	81.2	81.2	79.4	72.4	64.8	59.0	70.3
EVERGLADES		MAX	75.3	75.7	79.6	83.7	86.6	88.6	90.3	90.9	89.7	86.2	81.2	76.6	83.7
		MIN	53.8	54.3	59.1	62.8	67.7	72.0	73.5	74.0	73.5	68.5	61.4	55.5	64.7
		MEAN	64.6	65.0	69.4	73.3	77.2	80.3	81.9	82.5	81.6	77.4	71.3	66.1	74.2

125

FLORIDA

TEMPERATURE NORMALS (DEG F)

STATION			JAN	FEB	MAR	APR	MAY	JUN	JUL	AUG	SEP	OCT	NOV	DEC	ANN
FORT LAUDERDALE		MAX	75.7	76.3	79.8	83.0	85.9	88.6	90.1	90.6	88.9	85.2	80.8	76.8	83.5
		MIN	56.6	56.9	61.5	65.8	69.7	72.8	74.6	74.6	74.0	69.9	63.9	58.6	66.6
		MEAN	66.2	66.6	70.7	74.4	77.8	80.7	82.3	82.6	81.5	77.6	72.3	67.7	75.0
FORT MYERS WSO	R	MAX	74.3	75.1	79.8	84.5	88.7	90.1	91.0	91.2	89.6	85.2	80.0	75.6	83.8
		MIN	52.5	53.1	57.8	61.7	67.0	72.0	74.1	74.4	73.8	67.7	59.5	53.7	63.9
		MEAN	63.4	64.1	68.8	73.1	77.9	81.1	82.6	82.8	81.7	76.5	69.8	64.7	73.9
FORT PIERCE		MAX	72.9	73.3	77.8	81.1	84.7	87.5	89.4	89.7	87.9	83.5	78.4	74.3	81.7
		MIN	52.4	52.9	57.8	62.8	67.6	71.2	72.5	73.0	72.7	67.9	60.4	54.2	63.8
		MEAN	62.7	63.1	67.8	71.9	76.2	79.4	81.0	81.4	80.3	75.7	69.4	64.3	72.8
GAINESVILLE 2 WSW		MAX	68.6	70.5	76.7	82.4	87.9	90.7	91.6	91.5	89.1	82.7	75.6	70.0	81.4
		MIN	44.2	45.1	51.1	56.8	63.3	69.0	71.2	71.2	69.6	60.4	51.2	45.5	58.2
		MEAN	56.4	57.8	63.9	69.6	75.6	79.8	81.4	81.4	79.4	71.5	63.4	57.8	69.8
GLEN ST MARY 1 W		MAX	66.4	68.9	75.3	81.8	86.9	90.6	92.2	91.8	88.7	81.4	73.9	67.5	80.5
		MIN	40.1	41.6	46.8	53.1	60.0	66.1	69.0	69.2	67.1	56.6	47.3	40.9	54.8
		MEAN	53.3	55.3	61.1	67.5	73.5	78.4	80.6	80.5	78.0	69.0	60.6	54.2	67.7
HIALEAH		MAX	75.2	75.7	79.4	82.3	85.3	88.0	89.2	89.6	88.1	84.7	80.2	76.3	82.8
		MIN	56.2	56.7	61.9	66.1	69.6	72.6	74.2	74.3	73.7	69.4	63.5	58.1	66.4
		MEAN	65.7	66.2	70.7	74.2	77.5	80.3	81.7	82.0	80.9	77.1	71.9	67.3	74.6
HIGH SPRINGS		MAX	67.5	69.7	76.4	82.6	88.1	90.8	91.6	91.4	89.0	82.5	75.0	69.0	81.1
		MIN	42.3	43.7	49.8	56.2	62.8	69.0	71.1	70.9	68.6	58.5	49.1	43.5	57.1
		MEAN	54.9	56.8	63.1	69.4	75.5	79.9	81.4	81.2	78.8	70.5	62.1	56.3	69.2
HOMESTEAD EXP STA		MAX	76.5	77.3	81.3	84.3	87.0	88.8	90.2	90.7	89.2	85.6	81.2	77.5	84.1
		MIN	53.5	54.0	57.8	61.8	66.1	69.9	71.3	71.7	71.6	67.4	60.6	55.1	63.4
		MEAN	65.0	65.7	69.6	73.1	76.6	79.4	80.8	81.2	80.4	76.5	70.9	66.4	73.8
INVERNESS		MAX	70.3	72.1	77.7	83.5	88.4	91.1	91.9	91.5	89.7	83.5	76.9	71.6	82.4
		MIN	46.5	47.3	53.1	58.9	65.0	70.2	72.3	72.5	71.1	62.8	53.5	47.5	60.1
		MEAN	58.4	59.7	65.4	71.2	76.7	80.7	82.1	82.0	80.4	73.2	65.3	59.6	71.2
JACKSONVILLE WSO	R	MAX	64.6	66.8	73.3	79.7	85.2	88.9	90.7	90.2	86.9	79.7	72.4	66.3	78.7
		MIN	41.7	43.3	49.3	55.7	63.0	69.1	71.8	71.8	69.4	59.2	49.2	43.2	57.2
		MEAN	53.2	55.1	61.3	67.7	74.1	79.0	81.3	81.0	78.2	69.5	60.8	54.8	68.0
JACKSONVILLE BEACH		MAX	64.0	65.6	71.5	77.3	82.9	87.0	89.0	88.3	85.9	79.4	71.9	65.9	77.4
		MIN	45.4	46.4	52.0	58.9	65.5	70.9	72.6	72.9	72.3	64.1	54.4	47.6	60.3
		MEAN	54.7	56.0	61.8	68.1	74.2	79.0	80.8	80.6	79.1	71.8	63.2	56.8	68.8
JASPER		MAX	65.3	68.1	75.2	81.9	87.3	91.1	92.0	92.3	89.0	81.2	73.3	66.9	80.3
		MIN	38.8	40.7	46.8	53.1	59.5	65.9	69.0	68.7	66.4	55.2	45.5	39.8	54.1
		MEAN	52.1	54.4	61.0	67.5	73.4	78.6	80.5	80.5	77.7	68.2	59.4	53.4	67.2
KEY WEST WSO		MAX	71.8	74.8	78.6	82.0	84.9	87.3	88.9	88.9	86.5	84.4	79.6	75.2	81.9
		MIN	65.6	65.3	69.5	73.4	76.2	78.5	80.0	79.6	78.6	75.8	71.4	66.8	73.4
		MEAN	68.7	70.1	74.1	77.7	80.6	82.9	84.5	84.3	82.6	80.1	75.5	71.0	77.7
LAKE ALFRED EXP STA		MAX	71.8	72.9	78.4	83.4	88.0	90.5	91.6	91.8	89.9	84.5	78.1	72.9	82.8
		MIN	47.4	48.6	54.2	59.1	64.8	70.0	71.8	71.9	70.3	63.0	54.7	48.8	60.4
		MEAN	59.6	60.8	66.3	71.3	76.4	80.3	81.7	81.9	80.2	73.8	66.4	60.8	71.6
LAKE CITY 2 E		MAX	65.5	67.8	74.6	81.2	86.8	90.3	91.4	91.4	88.4	81.1	73.4	67.2	79.9
		MIN	41.7	43.2	49.1	55.2	61.7	67.5	70.2	70.0	67.8	58.0	48.9	43.1	56.4
		MEAN	53.6	55.5	61.9	68.2	74.2	78.9	80.8	80.7	78.1	69.6	61.2	55.1	68.2
LAKELAND WSO	R	MAX	71.5	73.3	78.6	83.9	88.6	91.1	92.2	92.0	90.1	84.3	77.8	72.7	83.0
		MIN	50.3	51.4	56.6	61.4	66.5	71.0	72.5	72.9	71.9	65.1	57.1	51.5	62.4
		MEAN	61.0	62.4	67.6	72.7	77.6	81.1	82.4	82.5	81.0	74.8	67.5	62.1	72.7
LOXAHATCHEE		MAX	76.1	76.9	81.0	84.4	87.7	89.8	91.3	91.7	89.8	85.7	81.0	77.2	84.4
		MIN	51.7	51.6	55.6	59.0	64.2	68.6	70.2	70.5	70.3	65.6	58.6	53.1	61.6
		MEAN	63.9	64.3	68.3	71.8	76.0	79.2	80.8	81.1	80.1	75.7	69.8	65.1	73.0

FLORIDA

TEMPERATURE NORMALS (DEG F)

STATION		JAN	FEB	MAR	APR	MAY	JUN	JUL	AUG	SEP	OCT	NOV	DEC	ANN
MADISON	MAX	64.1	67.0	74.2	81.3	87.6	91.2	92.1	91.7	88.4	80.9	72.2	65.7	79.7
	MIN	42.3	43.9	50.1	56.4	63.1	68.6	70.7	70.9	68.0	57.7	49.0	43.3	57.0
	MEAN	53.3	55.5	62.2	68.9	75.4	79.9	81.4	81.3	78.3	69.3	60.7	54.5	68.4
MAYO	MAX	65.4	67.7	74.6	81.3	87.1	90.3	91.3	91.2	88.6	81.6	73.6	67.1	80.0
	MIN	40.8	42.6	48.8	55.1	62.0	68.4	70.9	70.8	68.0	56.5	47.3	41.4	56.1
	MEAN	53.1	55.2	61.7	68.2	74.6	79.4	81.2	81.0	78.3	69.1	60.5	54.2	68.0
MELBOURNE	MAX	71.6	72.2	77.2	81.1	85.1	88.1	90.2	89.9	88.0	82.9	77.2	72.9	81.4
	MIN	51.3	51.3	56.5	61.5	66.7	70.7	72.2	72.9	72.2	66.7	58.9	52.8	62.8
	MEAN	61.5	61.8	66.9	71.3	75.9	79.4	81.3	81.4	80.1	74.8	68.1	62.9	72.1
MIAMI BEACH	MAX	73.9	74.3	77.0	79.7	82.6	85.5	87.0	87.6	86.2	83.0	78.5	75.1	80.9
	MIN	62.2	62.3	66.8	70.6	73.8	76.3	78.1	78.3	77.3	73.6	68.8	64.2	71.0
	MEAN	68.0	68.3	71.9	75.2	78.2	80.9	82.6	83.0	81.8	78.3	73.7	69.7	76.0
MIAMI WSO R	MAX	75.0	75.8	79.3	82.4	85.1	87.3	88.7	89.2	87.8	84.2	79.8	76.2	82.6
	MIN	59.2	59.7	64.1	68.2	71.9	74.6	76.2	76.5	75.7	71.6	65.8	60.8	68.7
	MEAN	67.1	67.8	71.7	75.3	78.5	81.0	82.4	82.8	81.8	77.9	72.8	68.5	75.6
MIAMI 12 SSW	MAX	75.4	76.1	79.6	82.6	85.3	87.3	89.1	89.6	88.3	84.7	80.3	76.6	82.9
	MIN	55.8	56.5	61.7	66.2	69.8	72.7	74.3	74.4	73.8	69.3	63.5	57.5	66.3
	MEAN	65.6	66.3	70.7	74.4	77.6	80.1	81.7	82.1	81.0	77.0	71.9	67.1	74.6
MILTON EXP STATION	MAX	62.1	65.2	71.5	78.9	85.4	90.2	91.3	91.1	87.8	80.4	70.7	64.4	78.3
	MIN	40.2	42.0	48.3	55.6	62.1	67.8	70.4	69.8	66.6	54.8	46.3	41.5	55.5
	MEAN	51.2	53.7	59.9	67.2	73.8	79.0	80.9	80.5	77.2	67.6	58.5	53.0	66.9
MONTICELLO 3 W	MAX	63.2	65.7	72.6	79.7	85.9	90.2	91.2	90.9	87.6	80.2	71.5	65.0	78.6
	MIN	39.2	40.7	47.5	54.1	60.8	66.9	69.5	69.2	66.3	54.7	45.8	39.9	54.6
	MEAN	51.2	53.2	60.1	66.9	73.4	78.6	80.4	80.1	77.0	67.5	58.7	52.5	66.6
MOORE HAVEN LOCK 1	MAX	73.3	74.5	79.3	83.4	87.3	89.6	91.0	90.9	89.1	84.4	78.9	74.3	83.0
	MIN	50.9	51.2	56.5	61.4	66.0	70.3	72.0	72.6	72.3	66.9	59.2	52.8	62.7
	MEAN	62.1	62.9	67.9	72.4	76.7	80.0	81.5	81.8	80.7	75.6	69.1	63.6	72.9
MOUNTAIN LAKE	MAX	73.4	74.7	79.9	84.6	89.1	91.0	92.0	91.7	89.9	84.9	78.9	74.4	83.7
	MIN	48.7	49.5	54.6	59.4	64.7	69.5	71.5	71.8	70.8	63.5	55.7	50.0	60.8
	MEAN	61.1	62.1	67.3	72.0	76.9	80.3	81.8	81.8	80.4	74.3	67.3	62.2	72.3
NAPLES 2 NE	MAX	76.0	76.6	80.6	84.3	87.4	89.5	90.7	91.3	90.1	86.3	81.6	77.3	84.3
	MIN	53.4	53.8	58.4	62.0	66.9	71.3	72.8	73.3	72.8	67.4	60.3	54.7	63.9
	MEAN	64.7	65.2	69.5	73.1	77.1	80.4	81.8	82.3	81.5	76.9	71.0	66.0	74.1
NICEVILLE	MAX	60.8	63.5	69.6	77.3	84.3	89.5	90.8	90.7	87.7	80.2	70.6	63.8	77.4
	MIN	37.7	39.5	46.1	53.7	61.2	67.6	70.7	70.3	66.7	54.1	44.7	39.1	54.3
	MEAN	49.3	51.5	57.9	65.5	72.8	78.6	80.8	80.5	77.3	67.2	57.7	51.5	65.9
OCALA	MAX	70.5	72.5	78.7	84.4	89.3	91.7	92.2	92.3	90.2	84.1	77.1	71.7	82.9
	MIN	46.1	47.0	53.0	58.0	64.3	69.5	71.4	71.4	69.6	61.3	52.8	47.4	59.3
	MEAN	58.3	59.8	65.9	71.2	76.8	80.6	81.8	81.9	79.9	72.7	64.9	59.5	71.1
ORLANDO WSO MCCOY AFB	MAX	71.7	72.9	78.3	83.6	88.3	90.6	91.7	91.6	89.7	84.4	78.2	73.1	82.8
	MIN	49.3	50.0	55.3	60.3	66.2	71.2	73.0	73.4	72.5	65.4	56.8	50.9	62.0
	MEAN	60.5	61.5	66.8	72.0	77.3	80.9	82.4	82.5	81.1	74.9	67.5	62.0	72.4
PALATKA	MAX	68.8	70.9	77.2	83.0	87.8	91.0	92.8	92.3	89.1	82.4	75.5	70.0	81.7
	MIN	46.1	47.2	52.9	58.8	65.2	70.8	72.7	72.9	71.1	62.5	53.6	47.8	60.1
	MEAN	57.5	59.1	65.0	70.9	76.5	80.9	82.8	82.6	80.1	72.5	64.6	58.9	71.0
PENSACOLA WSO	MAX	60.6	63.6	69.2	76.7	83.7	89.0	90.1	89.6	86.6	79.3	69.4	63.2	76.8
	MIN	42.7	44.8	51.4	59.3	66.3	72.1	74.4	73.9	70.9	59.5	49.8	44.4	59.1
	MEAN	51.7	54.2	60.4	68.0	75.0	80.6	82.3	81.8	78.7	69.4	59.7	53.8	68.0
PLANT CITY	MAX	73.7	75.0	79.9	84.8	89.1	90.7	91.5	91.8	90.5	85.6	79.4	74.9	83.9
	MIN	48.6	49.4	54.4	58.7	64.2	69.4	71.4	71.7	70.6	63.1	55.1	49.7	60.5
	MEAN	61.2	62.3	67.2	71.8	76.6	80.1	81.5	81.8	80.6	74.4	67.3	62.3	72.3

FLORIDA

TEMPERATURE NORMALS (DEG F)

STATION		JAN	FEB	MAR	APR	MAY	JUN	JUL	AUG	SEP	OCT	NOV	DEC	ANN
POMPANO BEACH	MAX	76.4	77.3	80.8	83.9	86.5	88.9	90.9	91.2	89.8	85.9	81.3	77.5	84.2
	MIN	57.1	56.9	61.2	65.3	68.7	71.8	73.3	73.6	72.9	69.1	63.8	58.8	66.0
	MEAN	66.8	67.1	71.0	74.6	77.7	80.4	82.1	82.4	81.3	77.5	72.5	68.2	75.1
PUNTA GORDA 4 ENE	MAX	74.4	75.6	80.1	84.8	89.0	90.8	91.8	91.9	90.2	85.7	80.0	75.3	84.1
	MIN	51.7	52.3	56.8	60.9	66.3	70.8	72.5	72.5	72.0	66.3	58.5	52.9	62.8
	MEAN	63.0	64.0	68.5	72.9	77.7	80.9	82.2	82.3	81.2	76.1	69.3	64.1	73.5
QUINCY 3 SSW	MAX	63.3	65.7	72.6	79.6	85.5	89.8	90.7	90.4	87.7	80.2	71.6	65.3	78.5
	MIN	40.4	41.7	48.5	55.0	61.8	67.7	70.0	69.9	67.2	56.1	47.2	41.6	55.6
	MEAN	51.9	53.8	60.6	67.3	73.7	78.8	80.4	80.2	77.5	68.2	59.4	53.5	67.1
SAINT LEO	MAX	71.1	72.6	78.3	83.7	88.5	90.9	91.6	91.7	89.9	84.2	77.6	72.6	82.7
	MIN	48.7	49.7	55.0	59.9	65.4	70.0	71.8	72.1	70.9	63.9	55.7	50.2	61.1
	MEAN	59.9	61.2	66.6	71.8	77.0	80.5	81.7	81.9	80.4	74.0	66.7	61.4	71.9
SAINT MARKS 6 SE	MAX	62.3	64.5	71.0	78.4	85.1	90.0	91.1	91.1	88.5	81.5	71.7	64.4	78.3
	MIN	41.6	43.7	49.9	57.2	63.8	69.6	71.9	71.8	69.1	58.4	49.0	43.0	57.4
	MEAN	52.0	54.2	60.5	67.8	74.5	79.8	81.5	81.5	78.8	70.0	60.4	53.7	67.9
ST PETERSBURG	MAX	70.0	71.1	76.0	81.7	86.7	89.3	90.1	90.0	88.7	83.4	76.8	71.4	81.3
	MIN	53.7	54.7	59.8	65.1	70.4	74.5	76.0	75.9	74.8	68.6	61.0	55.3	65.8
	MEAN	61.9	62.9	67.9	73.4	78.5	81.9	83.1	83.0	81.7	76.0	68.9	63.4	73.6
SANFORD EXP STATION	MAX	71.2	72.2	77.7	83.0	87.9	90.7	92.3	91.8	89.3	83.4	77.2	72.2	82.4
	MIN	47.9	48.7	53.8	58.5	64.3	69.2	71.2	71.7	70.6	64.3	55.8	49.8	60.5
	MEAN	59.6	60.5	65.8	70.8	76.1	80.0	81.8	81.7	80.0	73.9	66.5	61.0	71.5
STUART 1 N	MAX	74.4	75.3	79.3	82.3	85.8	88.6	90.0	90.5	88.8	84.5	79.5	75.4	82.9
	MIN	54.7	54.8	59.5	64.0	68.3	71.6	73.2	73.7	73.4	68.5	62.0	56.5	65.0
	MEAN	64.6	65.0	69.4	73.2	77.1	80.1	81.6	82.1	81.1	76.5	70.7	66.0	74.0
TALLAHASSEE WSO	MAX	63.4	65.9	72.7	80.0	86.0	90.1	90.9	90.6	87.8	80.4	71.5	65.3	78.7
	MIN	39.9	41.2	47.7	54.0	62.0	68.8	71.5	71.6	68.8	56.4	46.0	40.7	55.7
	MEAN	51.6	53.6	60.2	67.1	74.0	79.5	81.2	81.1	78.3	68.4	58.8	53.0	67.2
TAMIAMI TRL 40 MI BEND	MAX	77.1	78.1	82.1	85.6	88.8	90.4	91.5	92.0	90.7	86.8	82.4	78.3	85.3
	MIN	56.0	56.1	59.5	61.4	65.8	70.9	73.3	74.3	74.3	70.3	63.5	57.6	65.3
	MEAN	66.6	67.1	70.8	73.6	77.3	80.7	82.4	83.2	82.5	78.6	73.0	68.0	75.3
TAMPA WSO R	MAX	70.0	71.0	76.2	81.9	87.1	89.5	90.0	90.3	88.9	83.7	76.9	71.6	81.4
	MIN	49.5	50.4	56.1	61.1	67.2	72.3	74.2	74.2	72.8	65.1	56.4	50.9	62.5
	MEAN	59.8	60.8	66.2	71.6	77.1	80.9	82.2	82.2	80.9	74.5	66.7	61.3	72.0
TARPON SPGS SEWAGE PL	MAX	69.5	70.5	75.6	80.6	85.6	88.8	90.1	90.2	89.2	83.6	76.6	71.3	81.0
	MIN	48.9	50.2	55.6	60.9	66.7	71.5	73.2	73.1	71.7	64.3	56.1	50.2	61.9
	MEAN	59.2	60.4	65.6	70.8	76.2	80.2	81.6	81.7	80.4	74.0	66.4	60.8	71.4
TAVERNIER	MAX	75.7	76.2	79.8	82.6	85.4	87.8	89.5	90.1	88.3	84.8	80.4	76.9	83.1
	MIN	62.9	63.2	67.5	70.9	73.8	76.1	77.9	78.0	76.8	73.3	68.9	64.4	71.1
	MEAN	69.3	69.7	73.7	76.8	79.6	82.0	83.7	84.1	82.6	79.1	74.7	70.7	77.2
TITUSVILLE 3 NW	MAX	71.9	72.8	78.3	83.4	87.7	90.5	92.0	91.6	89.0	83.4	77.7	72.9	82.6
	MIN	48.6	49.1	54.1	59.3	64.6	69.2	71.0	71.4	71.0	64.8	56.3	50.2	60.8
	MEAN	60.2	61.0	66.2	71.4	76.2	79.9	81.5	81.5	80.0	74.1	67.0	61.6	71.7
VERO BEACH 4 W	MAX	72.2	72.8	77.3	81.2	85.2	87.9	89.7	89.9	87.9	83.3	77.9	73.4	81.6
	MIN	51.6	52.2	57.1	62.2	67.0	70.9	72.4	72.9	72.4	67.1	59.8	53.3	63.2
	MEAN	61.9	62.6	67.2	71.7	76.2	79.4	81.1	81.4	80.2	75.2	68.9	63.4	72.4
WAUCHULA 2 N	MAX	74.0	75.4	80.4	85.1	89.4	91.2	92.0	92.1	90.4	85.2	79.5	74.8	84.1
	MIN	49.0	49.5	54.4	58.4	64.1	69.0	70.9	71.2	70.4	63.7	55.6	50.0	60.5
	MEAN	61.5	62.5	67.4	71.8	76.8	80.1	81.5	81.7	80.4	74.5	67.6	62.4	72.4
WEST PALM BEACH WSO R	MAX	74.5	75.3	79.3	82.5	85.7	88.1	89.7	90.1	88.4	84.4	79.6	75.7	82.8
	MIN	55.9	56.2	60.8	65.1	69.5	72.7	74.2	74.8	74.3	70.1	63.5	58.2	66.3
	MEAN	65.2	65.8	70.1	73.8	77.6	80.4	82.0	82.5	81.4	77.3	71.6	67.0	74.6

FLORIDA

TEMPERATURE NORMALS (DEG F)

STATION		JAN	FEB	MAR	APR	MAY	JUN	JUL	AUG	SEP	OCT	NOV	DEC	ANN
WINTER HAVEN	MAX	72.9	74.6	79.5	84.1	88.5	90.7	91.9	92.0	90.1	84.9	78.7	73.9	83.5
	MIN	49.6	50.0	55.2	59.9	65.2	69.6	71.2	71.7	71.1	64.8	56.6	50.8	61.3
	MEAN	61.2	62.4	67.4	72.1	76.8	80.2	81.6	81.9	80.6	74.9	67.7	62.4	72.4

FLORIDA

PRECIPITATION NORMALS (INCHES)

STATION		JAN	FEB	MAR	APR	MAY	JUN	JUL	AUG	SEP	OCT	NOV	DEC	ANN
APALACHICOLA WSO	R	3.51	3.64	4.04	3.25	2.94	4.81	7.09	7.53	8.66	3.19	2.82	3.50	54.98
ARCADIA		2.17	2.64	2.69	2.13	4.28	7.76	8.26	7.30	7.35	3.95	1.97	2.19	52.69
ARCHBOLD BIOLOGIC STA		2.05	2.48	2.75	2.33	4.28	8.73	7.55	7.49	6.83	3.96	1.65	1.83	51.93
AVON PARK		2.25	2.96	3.11	2.59	4.20	9.21	7.81	7.06	6.90	3.53	1.78	1.83	53.23
BARTOW		2.52	3.37	3.32	2.59	5.35	7.18	8.33	7.46	6.68	2.76	2.03	2.10	53.69
BELLE GLADE EXP STA		2.22	1.95	2.62	2.57	5.47	8.25	8.37	7.57	8.28	4.84	1.84	1.78	55.76
BLOUNTSTOWN		4.59	4.28	5.07	4.29	4.81	5.68	7.18	5.90	5.92	3.34	3.03	3.87	57.96
BRADENTON 5 ESE		2.77	3.03	2.92	2.02	3.24	7.38	8.82	9.60	8.45	3.10	1.97	2.37	55.67
BROOKSVILLE CHIN HILL		2.87	3.59	4.20	2.26	4.37	6.66	7.69	8.66	6.54	2.39	2.02	2.70	53.95
BUSHNELL 2 E		2.88	3.72	4.12	2.36	4.27	6.15	7.68	7.24	6.10	2.58	1.94	2.37	51.41
CHIPLEY 3 E		4.70	4.78	5.23	4.41	4.04	4.70	6.16	5.40	4.95	3.05	3.18	4.30	54.90
CLERMONT 6 SSW		2.41	3.17	3.55	2.70	3.65	6.98	8.09	7.15	6.64	2.76	1.97	2.15	51.22
CLEWISTON U S ENG		1.87	2.02	2.43	2.28	5.29	7.30	6.66	6.54	6.18	3.78	1.78	1.62	47.75
CRESCENT CITY		2.90	3.60	3.82	2.69	3.85	6.11	7.42	7.10	7.11	3.75	2.32	2.90	53.57
CROSS CITY 2 WNW		3.51	3.80	3.98	3.03	3.84	5.97	9.51	8.61	6.23	2.62	2.28	3.30	56.68
DAYTONA BEACH WSO	R	2.37	3.11	2.99	2.25	3.38	6.41	5.52	6.34	6.68	4.62	2.59	2.20	48.46
DE FUNIAK SPRINGS		4.87	5.14	5.82	4.66	4.51	6.36	8.41	7.24	6.36	3.62	3.46	4.77	65.22
DELAND 1 SSE		2.59	3.53	3.42	2.43	4.17	7.23	7.83	7.71	6.81	4.49	2.32	2.04	54.57
EVERGLADES		1.51	1.82	1.87	2.01	4.97	9.77	8.40	7.14	9.31	4.08	1.33	1.34	53.55
FEDERAL POINT		2.98	3.48	3.51	2.50	4.09	5.91	6.75	7.04	7.19	3.93	2.41	2.94	52.73
FERNANDINA BEACH		2.96	3.67	3.59	2.78	3.88	5.31	5.99	6.49	7.60	3.77	2.83	2.69	51.56
FORT LAUDERDALE		2.15	2.49	2.16	3.44	6.36	9.09	6.04	6.88	8.09	7.66	3.81	2.66	60.83
FORT MYERS WSO	R	1.89	2.06	2.85	1.52	4.11	8.72	8.57	8.58	8.56	3.86	1.35	1.57	53.64
FORT PIERCE		2.11	2.91	2.91	2.97	4.47	6.49	5.93	5.37	7.70	7.05	2.30	2.34	52.55
GAINESVILLE 2 WSW		3.27	3.91	3.67	2.94	4.18	6.63	7.09	7.99	5.60	2.33	2.04	3.19	52.84
GLEN ST MARY 1 W		3.59	3.87	4.12	3.36	5.13	6.60	7.87	7.95	6.53	2.23	2.12	3.29	56.66
HART LAKE		2.25	2.86	3.31	2.24	3.80	7.01	7.56	6.50	6.80	3.26	1.95	1.96	49.50
HIALEAH		2.39	2.06	1.91	3.55	6.67	10.12	6.97	7.86	9.01	7.90	3.06	1.97	63.47
HIGH SPRINGS		3.30	3.73	3.77	3.05	4.34	6.39	7.26	8.03	4.75	2.23	2.07	3.10	52.02
HILLSBOROUGH RVR ST PK		2.63	3.46	3.93	2.23	4.70	6.96	7.91	8.39	7.13	3.05	2.17	2.69	55.25
HOMESTEAD EXP STA		1.71	2.07	2.11	3.03	6.99	10.35	7.55	8.09	8.94	6.88	2.03	1.38	61.13
INVERNESS		3.02	3.53	4.05	2.38	4.12	6.79	8.38	8.99	6.03	2.68	1.98	2.73	54.68
ISLEWORTH		2.41	3.29	3.64	2.35	4.49	7.11	7.49	7.18	6.14	2.92	2.03	1.99	51.04
JACKSONVILLE WSO	R	3.07	3.48	3.72	3.32	4.91	5.37	6.54	7.15	7.26	3.41	1.94	2.59	52.77
JACKSONVILLE BEACH		2.99	3.68	3.41	2.67	4.21	5.68	5.76	5.28	7.19	4.75	1.95	2.78	50.35
JASPER		4.23	4.29	4.43	4.16	4.30	6.53	6.30	6.81	5.16	2.24	2.69	3.70	54.84
KEY WEST WSO		1.74	1.92	1.31	1.49	3.22	5.04	3.68	4.80	6.50	4.76	3.23	1.73	39.42
LA BELLE		1.85	2.26	2.87	2.01	5.15	8.91	8.01	7.20	7.00	3.84	1.51	1.79	52.40
LAKE ALFRED EXP STA		2.36	3.14	3.52	2.20	4.81	7.06	6.98	7.25	6.57	3.02	1.95	1.97	50.83
LAKE CITY 2 E		3.75	3.89	4.24	3.46	4.64	6.71	6.77	6.99	5.68	2.35	2.28	3.48	54.24
LAKELAND WSO	R	2.30	2.82	3.52	2.36	4.18	6.05	7.59	7.47	5.81	2.43	1.88	1.93	48.34
LIVE OAK		3.92	3.81	4.39	3.88	3.96	5.67	7.98	6.63	5.22	2.47	2.41	3.38	53.72
LOXAHATCHEE		2.68	2.46	2.86	3.09	5.59	9.62	8.08	7.00	8.89	6.51	2.59	2.21	61.58
MADISON		4.26	4.12	4.73	3.87	3.88	5.32	6.51	5.76	5.21	2.28	2.77	3.69	52.40
MAYO		3.83	3.87	4.45	3.75	4.17	6.21	8.08	7.52	6.05	2.64	2.51	3.48	56.56
MELBOURNE		2.25	2.86	2.93	2.20	4.10	6.20	5.59	4.90	7.58	4.99	2.66	1.91	48.17
MIAMI BEACH		1.98	1.98	1.45	2.62	5.16	6.76	4.31	4.51	6.82	5.96	2.83	1.67	46.05
MIAMI WSO	R	2.08	2.05	1.89	3.07	6.53	9.15	5.98	7.02	8.07	7.14	2.71	1.86	57.55
MIAMI 12 SSW		2.02	2.33	1.68	2.81	7.11	8.62	5.54	5.89	7.80	7.84	2.83	1.71	56.18
MILTON EXP STATION		5.05	4.97	6.09	4.84	4.34	7.09	7.18	6.44	6.78	3.44	3.87	5.37	65.46
MONTICELLO 3 W		4.52	4.55	4.63	4.40	4.43	5.46	6.88	5.52	5.92	2.67	2.77	4.09	55.84
MOORE HAVEN LOCK 1		1.83	2.12	2.53	2.12	5.00	8.15	7.09	6.30	7.17	4.21	1.28	1.81	49.61
MOUNTAIN LAKE		2.50	3.12	3.31	2.18	4.56	7.22	7.76	7.99	6.09	2.69	1.96	1.95	51.33
MYAKKA RIVER ST PARK		2.55	3.08	2.82	2.16	3.84	8.33	8.43	9.35	8.59	3.37	2.12	2.17	56.81
NAPLES 2 NE		1.91	2.00	2.27	1.67	4.55	7.83	8.07	8.52	9.23	3.96	1.24	1.44	52.69

FLORIDA

PRECIPITATION NORMALS (INCHES)

STATION	JAN	FEB	MAR	APR	MAY	JUN	JUL	AUG	SEP	OCT	NOV	DEC	ANN
NICEVILLE	4.81	5.11	5.10	4.45	4.08	5.81	8.04	7.09	7.05	3.76	3.59	4.79	63.68
OCALA	2.91	3.63	3.67	3.05	4.42	7.01	8.49	7.11	5.92	2.81	2.07	2.77	53.86
ORLANDO WSO MCCOY AFB	2.10	2.83	3.20	2.19	3.96	7.39	7.78	6.32	5.62	2.82	1.78	1.83	47.83
PALATKA	3.05	3.61	3.46	2.67	3.84	5.93	6.64	6.95	6.75	3.89	2.09	2.67	51.55
PENSACOLA WSO	4.47	4.90	5.66	4.45	3.87	5.75	7.18	7.04	6.75	3.52	3.42	4.15	61.16
PLANT CITY	2.52	3.33	3.78	2.11	4.12	7.07	8.10	8.68	6.70	2.83	2.02	2.23	53.49
POMPANO BEACH	2.55	2.60	2.58	3.36	6.14	8.27	5.77	6.48	8.57	9.24	3.40	2.42	61.38
PUNTA GORDA 4 ENE	2.12	2.31	2.38	1.75	4.03	7.79	6.98	7.51	7.53	3.75	1.59	1.79	49.53
QUINCY 3 SSW	4.58	4.58	4.99	4.30	4.67	5.30	7.13	5.26	5.36	2.89	2.67	4.04	55.77
ROYAL PALM RANGER STA	1.70	1.87	1.52	2.84	6.45	10.28	6.98	7.52	8.72	6.12	2.10	1.36	57.46
SAINT LEO	2.60	3.66	4.19	2.75	4.78	6.89	8.05	7.62	6.23	2.52	2.21	2.44	53.94
SAINT MARKS 6 SE	4.16	4.29	4.70	4.37	3.56	5.86	7.46	6.29	6.13	3.06	2.86	3.97	56.71
ST PETERSBURG	2.44	3.13	3.69	2.28	3.32	6.12	8.06	8.73	7.60	3.08	2.10	2.55	53.10
SANFORD EXP STATION	2.42	3.26	3.65	2.42	3.49	6.39	7.25	7.13	6.90	4.02	2.12	2.13	51.18
STUART 1 N	2.58	2.64	3.19	2.68	4.97	6.99	6.46	5.78	7.63	6.78	2.53	2.64	54.87
TALLAHASSEE WSO	4.66	5.00	5.60	4.13	5.16	6.55	8.75	7.30	6.45	3.10	3.31	4.58	64.59
TAMIAMI TRL 40 MI BEND	1.73	1.63	1.91	2.45	5.97	9.51	7.53	7.10	8.26	4.88	1.60	1.28	53.85
TAMPA WSO R	2.17	3.04	3.46	1.82	3.38	5.29	7.35	7.64	6.23	2.34	1.87	2.14	46.73
TARPON SPGS SEWAGE PL	2.65	3.31	4.01	2.11	3.20	5.18	8.07	8.55	7.22	2.46	2.17	2.86	51.79
TAVERNIER	1.94	2.03	1.33	2.17	4.96	7.23	4.33	4.54	6.92	7.01	2.05	1.88	46.39
TITUSVILLE 3 NW	2.20	3.09	3.30	2.16	4.06	6.93	8.40	7.77	8.36	5.45	2.73	2.24	56.69
VERO BEACH 4 W	2.43	2.86	3.05	2.59	4.39	6.52	5.76	5.39	7.96	5.94	2.55	1.97	51.41
WAUCHULA 2 N	2.40	3.01	3.02	2.46	4.94	8.33	8.50	6.87	7.03	2.88	1.76	1.89	53.09
WEST PALM BEACH WSO R	2.71	2.62	2.69	3.21	6.02	7.92	6.06	5.78	9.29	7.77	3.39	2.26	59.72
WINTER HAVEN	2.35	3.06	3.44	2.24	4.73	6.32	7.57	6.99	6.33	2.88	2.13	1.94	49.98

FLORIDA

HEATING DEGREE DAY NORMALS (BASE 65 DEG F)

STATION		JUL	AUG	SEP	OCT	NOV	DEC	JAN	FEB	MAR	APR	MAY	JUN	ANN
APALACHICOLA WSO	R	0	0	0	24	154	320	401	311	168	30	0	0	1408
ARCADIA		0	0	0	0	41	142	179	148	58	0	0	0	568
ARCHBOLD BIOLOGIC STA		0	0	0	0	35	136	175	154	53	0	0	0	553
AVON PARK		0	0	0	0	35	142	189	149	41	0	0	0	556
BARTOW		0	0	0	0	54	150	199	156	56	0	0	0	615
BELLE GLADE EXP STA		0	0	0	0	26	119	148	125	43	0	0	0	461
BRADENTON 5 ESE		0	0	0	0	40	151	194	159	66	6	0	0	616
BROOKSVILLE CHIN HILL		0	0	0	0	72	184	240	189	79	5	0	0	769
BUSHNELL 2 E		0	0	0	7	88	214	260	206	91	8	0	0	874
CHIPLEY 3 E		0	0	0	52	222	387	459	348	186	43	0	0	1697
CLERMONT 6 SSW		0	0	0	0	64	183	228	180	71	0	0	0	726
CLEWISTON U S ENG		0	0	0	0	21	107	140	121	28	0	0	0	417
CROSS CITY 2 WNW		0	0	0	29	171	340	399	313	165	33	0	0	1450
DAYTONA BEACH WSO	R	0	0	0	0	83	209	264	214	116	14	0	0	900
DE FUNIAK SPRINGS		0	0	0	64	231	390	460	347	195	37	7	0	1731
DELAND 1 SSE		0	0	0	6	91	222	263	224	103	16	0	0	925
EVERGLADES		0	0	0	0	14	75	113	118	36	0	0	0	356
FORT LAUDERDALE		0	0	0	0	8	56	84	87	19	0	0	0	254
FORT MYERS WSO	R	0	0	0	0	25	107	150	120	39	0	0	0	441
FORT PIERCE		0	0	0	0	30	119	160	138	53	0	0	0	500
GAINESVILLE 2 WSW		0	0	0	10	115	257	314	243	118	12	0	0	1069
GLEN ST MARY 1 W		0	0	0	30	167	353	390	301	169	24	0	0	1434
HIALEAH		0	0	0	0	13	61	96	88	26	0	0	0	284
HIGH SPRINGS		0	0	0	13	133	290	349	270	128	11	0	0	1194
HOMESTEAD EXP STA		0	0	0	0	10	74	94	85	21	0	0	0	284
INVERNESS		0	0	0	0	78	207	264	215	86	8	0	0	858
JACKSONVILLE WSO	R	0	0	0	21	164	332	396	302	166	21	0	0	1402
JACKSONVILLE BEACH		0	0	0	10	117	278	347	278	153	14	0	0	1197
JASPER		0	0	0	42	194	373	427	322	161	38	0	0	1557
KEY WEST WSO		0	0	0	0	0	22	49	37	6	0	0	0	114
LAKE ALFRED EXP STA		0	0	0	0	67	182	228	184	79	7	0	0	747
LAKE CITY 2 E		0	0	0	20	154	326	386	302	148	25	0	0	1361
LAKELAND WSO	R	0	0	0	0	49	157	198	155	59	0	0	0	618
LOXAHATCHEE		0	0	0	0	18	85	114	107	33	0	0	0	357
MADISON		0	0	0	20	174	343	394	301	157	20	0	0	1409
MAYO		0	0	0	28	169	351	400	310	159	23	0	0	1440
MELBOURNE		0	0	0	0	41	140	186	167	67	6	0	0	607
MIAMI BEACH		0	0	0	0	0	22	53	50	8	0	0	0	133
MIAMI WSO	R	0	0	0	0	5	42	76	62	14	0	0	0	199
MIAMI 12 SSW		0	0	0	0	8	68	95	86	17	0	0	0	274
MILTON EXP STATION		0	0	0	47	218	383	460	343	196	38	0	0	1685
MONTICELLO 3 W		0	0	0	58	217	396	454	362	197	48	7	0	1739
MOORE HAVEN LOCK 1		0	0	0	0	28	124	169	144	45	0	0	0	510
MOUNTAIN LAKE		0	0	0	0	50	152	195	153	62	0	0	0	612
NAPLES 2 NE		0	0	0	0	13	78	108	100	24	0	0	0	323
NICEVILLE		0	0	0	55	240	426	503	392	240	62	0	0	1918
OCALA		0	0	0	8	84	213	266	211	86	13	0	0	881
ORLANDO WSO MCCOY AFB		0	0	0	0	47	157	212	172	68	0	0	0	656
PALATKA		0	0	0	7	87	228	286	220	104	0	0	0	932
PENSACOLA WSO		0	0	0	35	192	359	445	327	184	29	0	0	1571
PLANT CITY		0	0	0	0	48	156	195	151	62	5	0	0	617
POMPANO BEACH		0	0	0	0	11	62	80	72	18	0	0	0	243
PUNTA GORDA 4 ENE		0	0	0	0	29	116	153	128	37	0	0	0	463
QUINCY 3 SSW		0	0	0	28	195	367	437	344	180	42	0	0	1593
SAINT LEO		0	0	0	0	58	166	224	178	76	0	0	0	702

FLORIDA

HEATING DEGREE DAY NORMALS (BASE 65 DEG F)

STATION	JUL	AUG	SEP	OCT	NOV	DEC	JAN	FEB	MAR	APR	MAY	JUN	ANN
SAINT MARKS 6 SE	0	0	0	19	169	359	420	321	170	27	0	0	1485
ST PETERSBURG	0	0	0	0	34	133	179	148	51	0	0	0	545
SANFORD EXP STATION	0	0	0	0	53	177	234	189	76	7	0	0	736
STUART 1 N	0	0	0	0	16	84	117	105	31	0	0	0	353
TALLAHASSEE WSO	0	0	0	38	210	383	441	341	191	48	0	0	1652
TAMIAMI TRL 40 MI BEND	0	0	0	0	0	51	75	66	17	0	0	0	209
TAMPA WSO R	0	0	0	0	65	173	228	186	87	0	0	0	739
TARPON SPGS SEWAGE PL	0	0	0	0	59	187	237	190	95	0	0	0	768
TAVERNIER	0	0	0	0	0	22	34	34	0	0	0	0	90
TITUSVILLE 3 NW	0	0	0	0	56	174	222	182	73	5	0	0	712
VERO BEACH 4 W	0	0	0	0	28	124	166	149	55	0	0	0	522
WAUCHULA 2 N	0	0	0	0	46	148	183	148	57	0	0	0	582
WEST PALM BEACH WSO R	0	0	0	0	9	57	92	86	18	0	0	0	262
WINTER HAVEN	0	0	0	0	47	148	194	156	67	0	0	0	612

FLORIDA

COOLING DEGREE DAY NORMALS (BASE 65 DEG F)

STATION		JAN	FEB	MAR	APR	MAY	JUN	JUL	AUG	SEP	OCT	NOV	DEC	ANN
APALACHICOLA WSO	R	23	23	35	117	295	444	512	508	417	185	34	10	2603
ARCADIA		74	78	136	204	357	441	499	512	456	291	119	64	3231
ARCHBOLD BIOLOGIC STA		70	86	139	217	369	447	502	515	462	313	137	71	3328
AVON PARK		90	93	137	240	397	480	536	543	480	316	137	83	3532
BARTOW		72	78	133	223	384	471	530	533	468	298	123	57	3370
BELLE GLADE EXP STA		74	77	123	189	326	417	477	487	441	310	143	82	3146
BRADENTON 5 ESE		67	70	110	180	341	441	502	512	465	295	115	64	3162
BROOKSVILLE CHIN HILL		67	66	119	197	357	453	499	499	444	267	102	54	3124
BUSHNELL 2 E		58	61	100	170	332	435	496	499	432	237	82	47	2949
CHIPLEY 3 E		34	32	43	118	273	426	499	490	384	129	24	12	2464
CLERMONT 6 SSW		58	68	127	218	378	468	521	530	471	285	103	47	3274
CLEWISTON U S ENG		94	101	161	255	378	459	521	530	489	366	186	110	3650
CROSS CITY 2 WNW		27	31	51	114	279	417	481	484	408	172	39	14	2517
DAYTONA BEACH WSO	R	44	41	88	152	313	426	499	493	435	259	89	39	2878
DE FUNIAK SPRINGS		29	33	49	115	292	429	502	487	369	142	21	11	2479
DELAND 1 SSE		46	56	81	160	322	435	502	502	432	235	85	36	2892
EVERGLADES		101	118	173	249	378	459	524	543	498	384	203	109	3739
FORT LAUDERDALE		121	132	196	282	397	471	536	546	495	391	227	139	3933
FORT MYERS WSO	R	100	94	156	243	400	483	546	552	501	357	169	98	3699
FORT PIERCE		88	85	139	207	347	432	496	508	459	332	162	97	3352
GAINESVILLE 2 WSW		48	41	84	150	329	444	508	508	432	211	67	34	2856
GLEN ST MARY 1 W		27	30	48	99	264	402	484	481	390	154	35	18	2432
HIALEAH		117	122	203	276	388	459	518	527	477	375	220	133	3815
HIGH SPRINGS		36	40	69	143	326	447	508	502	414	183	46	21	2735
HOMESTEAD EXP STA		94	105	163	243	360	432	490	502	462	357	187	118	3513
INVERNESS		60	66	99	194	363	471	530	527	462	259	87	40	3158
JACKSONVILLE WSO	R	30	25	51	102	282	420	505	496	396	160	38	15	2520
JACKSONVILLE BEACH		27	26	54	107	285	420	490	484	423	221	63	23	2623
JASPER		27	25	37	113	260	408	481	481	381	141	26	13	2393
KEY WEST WSO		164	180	288	381	484	537	605	598	528	468	315	208	4756
LAKE ALFRED EXP STA		60	67	119	196	353	459	518	524	456	273	109	52	3186
LAKE CITY 2 E		33	36	52	121	285	417	490	487	393	163	40	19	2536
LAKELAND WSO	R	74	82	139	235	391	483	539	543	480	304	124	67	3461
LOXAHATCHEE		80	87	135	204	341	426	490	499	453	332	162	88	3297
MADISON		32	35	70	137	322	447	508	505	399	153	45	18	2671
MAYO		31	35	56	119	298	432	502	496	399	155	34	17	2574
MELBOURNE		78	77	126	195	338	432	505	508	453	304	134	75	3225
MIAMI BEACH		146	142	222	306	409	477	546	558	504	412	261	167	4150
MIAMI WSO	R	141	140	222	309	419	480	539	552	504	400	239	150	4095
MIAMI 12 SSW		114	122	193	282	391	453	518	530	480	372	215	133	3803
MILTON EXP STATION		32	26	37	104	273	420	493	481	366	128	23	11	2394
MONTICELLO 3 W		26	31	46	105	267	408	477	468	360	135	28	8	2359
MOORE HAVEN LOCK 1		79	85	135	224	363	450	512	521	471	329	151	80	3400
MOUNTAIN LAKE		74	72	133	214	369	459	521	521	462	288	119	65	3297
NAPLES 2 NE		99	105	164	243	375	462	521	536	495	369	193	109	3671
NICEVILLE		16	14	20	77	244	408	490	481	369	124	21	8	2272
OCALA		58	66	114	199	366	468	521	524	447	247	81	43	3134
ORLANDO WSO MCCOY AFB		73	74	124	214	381	477	539	543	483	307	122	64	3401
PALATKA		53	54	104	181	357	477	552	546	453	239	75	39	3130
PENSACOLA WSO		33	24	42	119	310	468	536	521	411	171	33	12	2680
PLANT CITY		77	76	130	209	360	453	512	521	468	291	117	73	3287
POMPANO BEACH		135	131	204	288	394	462	530	539	489	388	236	161	3957
PUNTA GORDA 4 ENE		91	100	145	237	394	477	533	536	486	344	158	88	3589
QUINCY 3 SSW		31	30	43	111	270	414	477	471	375	128	27	11	2388
SAINT LEO		66	72	125	209	372	465	518	524	462	279	109	54	3255

FLORIDA

COOLING DEGREE DAY NORMALS (BASE 65 DEG F)

STATION	JAN	FEB	MAR	APR	MAY	JUN	JUL	AUG	SEP	OCT	NOV	DEC	ANN
SAINT MARKS 6 SE	17	19	31	111	295	444	512	512	414	174	31	9	2569
ST PETERSBURG	83	89	141	252	419	507	561	558	501	341	151	83	3686
SANFORD EXP STATION	67	63	101	181	344	450	521	518	450	276	98	53	3122
STUART 1 N	105	105	168	249	375	453	515	530	483	357	187	115	3642
TALLAHASSEE WSO	25	22	42	111	279	435	502	499	399	143	24	11	2492
TAMIAMI TRL 40 MI BEND	124	125	197	258	381	471	539	564	525	422	245	144	3995
TAMPA WSO R	66	68	124	202	375	477	533	533	477	295	116	58	3324
TARPON SPGS SEWAGE PL	57	61	114	179	347	456	515	518	462	279	101	57	3146
TAVERNIER	168	166	274	354	453	510	580	592	528	437	291	199	4552
TITUSVILLE 3 NW	73	70	110	197	347	447	512	512	450	282	116	68	3184
VERO BEACH 4 W	70	82	123	204	347	432	499	508	456	316	145	75	3257
WAUCHULA 2 N	75	78	131	209	366	453	512	518	462	295	124	67	3290
WEST PALM BEACH WSO R	99	108	176	264	391	462	527	543	492	381	207	119	3769
WINTER HAVEN	76	83	141	217	366	456	515	524	468	307	128	67	3348

08 — FLORIDA

LEGEND
11 = TEMPERATURE ONLY
12 = PRECIPITATION ONLY
13 = TEMP. & PRECIP.

STATE-STATION NUMBER	STN TYP	NAME		LATITUDE DEG-MIN	LONGITUDE DEG-MIN	ELEVATION (FT)
8-0211	13	APALACHICOLA WSO	R	N 2944	W 08502	13
8-0228	13	ARCADIA		N 2714	W 08151	63
8-0236	13	ARCHBOLD BIOLOGIC STA		N 2711	W 08121	140
8-0369	13	AVON PARK		N 2736	W 08130	133
8-0478	13	BARTOW		N 2754	W 08151	125
8-0611	13	BELLE GLADE EXP STA		N 2640	W 08038	16
8-0804	12	BLOUNTSTOWN		N 3027	W 08503	60
8-0945	13	BRADENTON 5 ESE		N 2727	W 08228	20
8-1046	13	BROOKSVILLE CHIN HILL		N 2837	W 08222	240
8-1163	13	BUSHNELL 2 E		N 2840	W 08205	75
8-1544	13	CHIPLEY 3 E		N 3047	W 08529	130
8-1641	13	CLERMONT 6 SSW		N 2829	W 08147	125
8-1654	13	CLEWISTON U S ENG		N 2645	W 08055	20
8-1978	12	CRESCENT CITY		N 2925	W 08130	58
8-2008	13	CROSS CITY 2 WNW		N 2939	W 08310	42
8-2158	13	DAYTONA BEACH WSO	R	N 2911	W 08104	30
8-2220	13	DE FUNIAK SPRINGS		N 3044	W 08607	230
8-2229	13	DELAND 1 SSE		N 2901	W 08118	25
8-2850	13	EVERGLADES		N 2551	W 08123	5
8-2915	12	FEDERAL POINT		N 2945	W 08132	5
8-2944	12	FERNANDINA BEACH		N 3040	W 08127	25
8-3163	13	FORT LAUDERDALE		N 2606	W 08012	16
8-3186	13	FORT MYERS WSO	R	N 2635	W 08152	15
8-3207	13	FORT PIERCE		N 2728	W 08021	25
8-3321	13	GAINESVILLE 2 WSW		N 2938	W 08221	86
8-3470	13	GLEN ST MARY 1 W		N 3016	W 08211	128
8-3840	12	HART LAKE		N 2823	W 08111	60
8-3909	13	HIALEAH		N 2550	W 08017	12
8-3956	13	HIGH SPRINGS		N 2950	W 08236	65
8-3986	12	HILLSBOROUGH RVR ST PK		N 2809	W 08214	53
8-4091	13	HOMESTEAD EXP STA		N 2530	W 08030	11
8-4289	13	INVERNESS		N 2850	W 08220	50
8-4332	12	ISLEWORTH		N 2829	W 08132	115
8-4358	13	JACKSONVILLE WSO	R	N 3030	W 08149	24
8-4366	13	JACKSONVILLE BEACH		N 3017	W 08124	10
8-4394	13	JASPER		N 3031	W 08257	147
8-4570	13	KEY WEST WSO		N 2433	W 08145	4
8-4662	12	LA BELLE		N 2645	W 08126	16
8-4707	13	LAKE ALFRED EXP STA		N 2806	W 08143	145
8-4731	13	LAKE CITY 2 E		N 3011	W 08236	195
8-4797	13	LAKELAND WSO	R	N 2802	W 08157	214
8-5099	12	LIVE OAK		N 3017	W 08258	120
8-5182	13	LOXAHATCHEE		N 2641	W 08016	14
8-5275	13	MADISON		N 3028	W 08325	190
8-5539	13	MAYO		N 3003	W 08310	65
8-5612	13	MELBOURNE		N 2804	W 08037	10
8-5658	13	MIAMI BEACH		N 2547	W 08008	5
8-5663	13	MIAMI WSO	R	N 2548	W 08016	7
8-5678	13	MIAMI 12 SSW		N 2539	W 08018	10
8-5793	13	MILTON EXP STATION		N 3047	W 08708	217

08 — FLORIDA

LEGEND
11 = TEMPERATURE ONLY
12 = PRECIPITATION ONLY
13 = TEMP. & PRECIP.

STATE-STATION NUMBER	STN TYP	NAME		LATITUDE DEG-MIN	LONGITUDE DEG-MIN	ELEVATION (FT)
8-5879	13	MONTICELLO 3 W		N 3032	W 08355	148
8-5895	13	MOORE HAVEN LOCK 1		N 2650	W 08105	22
8-5973	13	MOUNTAIN LAKE		N 2756	W 08136	125
8-6065	12	MYAKKA RIVER ST PARK		N 2714	W 08219	20
8-6078	13	NAPLES 2 NE		N 2610	W 08147	4
8-6240	13	NICEVILLE		N 3031	W 08630	60
8-6414	13	OCALA		N 2911	W 08208	90
8-6628	13	ORLANDO WSO MCCOY AFB		N 2826	W 08120	9
8-6753	13	PALATKA		N 2939	W 08139	20
8-6997	13	PENSACOLA WSO		N 3028	W 08712	112
8-7205	13	PLANT CITY		N 2801	W 08208	121
8-7254	13	POMPANO BEACH		N 2614	W 08009	15
8-7397	13	PUNTA GORDA 4 ENE		N 2658	W 08159	10
8-7429	13	QUINCY 3 SSW		N 3033	W 08436	245
8-7760	12	ROYAL PALM RANGER STA		N 2523	W 08036	10
8-7851	13	SAINT LEO		N 2820	W 08216	190
8-7867	13	SAINT MARKS 6 SE		N 3005	W 08410	17
8-7886	13	ST PETERSBURG		N 2746	W 08238	8
8-7982	13	SANFORD EXP STATION		N 2848	W 08114	17
8-8620	13	STUART 1 N		N 2713	W 08015	10
8-8758	13	TALLAHASSEE WSO		N 3023	W 08422	55
8-8780	13	TAMIAMI TRL 40 MI BEND		N 2545	W 08050	15
8-8788	13	TAMPA WSO	R	N 2758	W 08232	19
8-8824	13	TARPON SPGS SEWAGE PL		N 2809	W 08245	8
8-8841	13	TAVERNIER		N 2501	W 08031	5
8-8942	13	TITUSVILLE 3 NW		N 2837	W 08050	30
8-9219	13	VERO BEACH 4 W		N 2738	W 08027	20
8-9401	13	WAUCHULA 2 N		N 2734	W 08149	119
8-9525	13	WEST PALM BEACH WSO	R	N 2641	W 08006	15
8-9707	13	WINTER HAVEN		N 2801	W 08145	136

08 – FLORIDA

LEGEND
O = TEMPERATURE ONLY
× = PRECIPITATION ONLY
⊗ = TEMPERATURE & PRECIPITATION

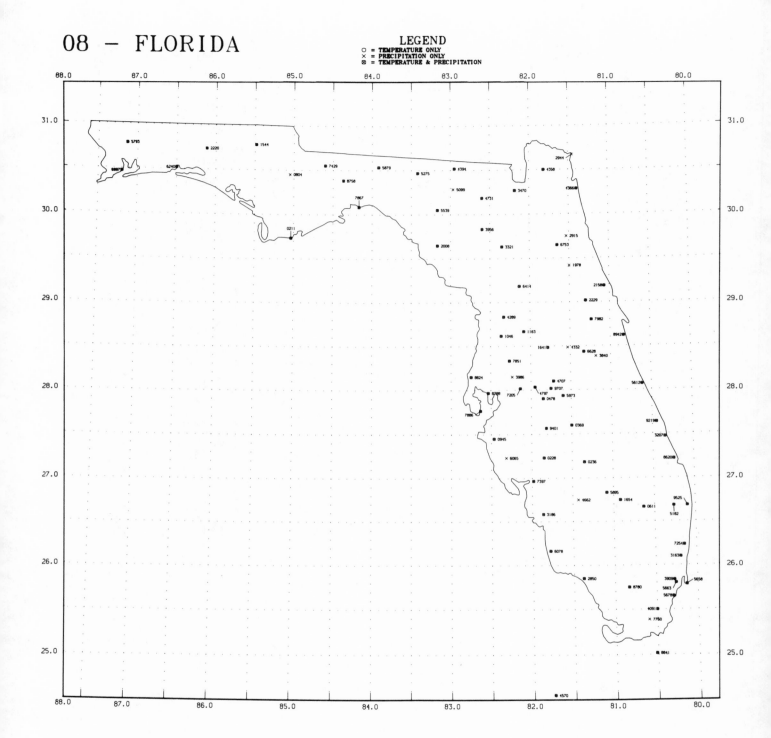

GEORGIA

TEMPERATURE NORMALS (DEG F)

STATION			JAN	FEB	MAR	APR	MAY	JUN	JUL	AUG	SEP	OCT	NOV	DEC	ANN
ALBANY 3 SE		MAX	60.6	63.4	71.0	79.3	85.9	90.6	92.2	92.3	88.5	80.0	70.5	63.0	78.1
		MIN	36.8	38.4	45.3	53.0	60.8	67.5	70.5	70.0	66.0	53.3	43.1	37.6	53.5
		MEAN	48.7	50.9	58.2	66.1	73.3	79.1	81.4	81.2	77.3	66.7	56.8	50.3	65.8
ALMA FAA AIRPORT		MAX	61.6	64.3	71.3	79.2	85.5	89.7	91.7	91.4	86.9	79.1	70.5	63.5	77.9
		MIN	38.4	40.2	46.6	53.5	60.7	67.1	70.1	69.8	66.6	54.9	45.0	39.6	54.4
		MEAN	50.0	52.3	59.0	66.4	73.1	78.4	80.9	80.6	76.8	67.0	57.8	51.6	66.2
AMERICUS 4 ENE		MAX	58.2	62.0	69.4	78.5	84.8	89.6	90.8	91.3	86.8	78.3	68.8	61.1	76.6
		MIN	36.9	38.7	45.4	53.1	60.7	66.9	69.6	69.2	65.3	53.1	43.8	38.4	53.4
		MEAN	47.6	50.4	57.4	65.9	72.8	78.2	80.2	80.2	76.1	65.7	56.3	49.8	65.1
ATHENS WSO		MAX	52.2	55.9	63.6	73.6	80.7	86.8	89.3	88.8	82.9	73.6	63.3	54.7	72.1
		MIN	32.6	34.2	41.0	49.7	58.1	65.2	68.9	68.2	62.9	50.5	40.9	34.7	50.6
		MEAN	42.4	45.1	52.4	61.7	69.4	76.0	79.2	78.5	72.9	62.1	52.1	44.7	61.4
ATLANTA WSO	R	MAX	51.2	55.3	63.2	73.2	79.8	85.6	87.9	87.6	82.3	72.9	62.6	54.1	71.3
		MIN	32.6	34.5	41.7	50.4	58.7	65.9	69.2	68.7	63.6	51.4	41.3	34.8	51.1
		MEAN	41.9	44.9	52.5	61.8	69.3	75.8	78.6	78.2	73.0	62.2	52.0	44.5	61.2
AUGUSTA WSO	R	MAX	56.7	60.1	67.6	76.8	83.7	89.1	91.4	90.9	85.6	76.9	67.5	59.2	75.5
		MIN	33.2	35.0	42.0	49.5	58.3	65.6	69.6	68.9	63.5	50.1	40.3	34.6	50.9
		MEAN	45.0	47.5	54.8	63.2	71.0	77.4	80.6	79.9	74.6	63.5	53.9	46.9	63.2
BLAIRSVILLE EXP STA		MAX	47.9	51.0	58.2	68.7	75.8	81.6	84.4	84.1	79.0	69.9	59.6	51.2	67.6
		MIN	23.6	24.8	32.0	40.6	48.7	56.4	60.6	59.7	54.0	40.5	31.5	25.5	41.5
		MEAN	35.7	37.9	45.1	54.7	62.3	69.0	72.5	71.9	66.5	55.2	45.5	38.4	54.6
BLAKELY		MAX	59.8	63.1	70.5	78.8	85.0	90.0	91.4	91.4	87.8	79.2	69.3	62.1	77.4
		MIN	37.8	39.5	46.0	53.5	60.9	67.4	70.0	69.6	66.2	54.4	44.6	39.2	54.1
		MEAN	48.8	51.3	58.3	66.2	73.0	78.7	80.7	80.5	77.0	66.8	57.0	50.7	65.8
BROOKLET 1 W		MAX	60.3	63.6	70.9	79.5	84.9	88.9	91.1	90.6	86.3	78.3	70.0	62.7	77.3
		MIN	37.8	39.5	46.1	53.7	61.0	66.6	69.4	69.2	65.5	54.4	45.1	39.1	54.0
		MEAN	49.1	51.6	58.5	66.7	73.0	77.7	80.3	79.9	75.9	66.4	57.6	50.9	65.6
BRUNSWICK		MAX	64.1	66.3	72.5	79.1	85.2	89.1	91.6	91.0	87.3	79.9	72.4	65.6	78.7
		MIN	42.2	43.3	50.0	57.5	64.3	70.2	72.6	72.7	70.1	60.0	50.6	44.3	58.2
		MEAN	53.2	54.9	61.3	68.3	74.8	79.7	82.2	81.9	78.7	70.0	61.5	55.0	68.5
BRUNSWICK FAA AP		MAX	60.6	62.3	68.6	76.1	82.4	87.0	89.5	88.7	84.8	77.2	69.2	62.7	75.8
		MIN	41.8	43.5	50.2	58.1	65.8	71.8	74.0	73.9	71.4	60.8	50.7	44.0	58.8
		MEAN	51.2	53.0	59.4	67.1	74.2	79.4	81.8	81.3	78.1	69.0	60.0	53.4	67.3
CAMILLA		MAX	62.5	65.6	73.2	81.1	87.3	91.8	93.1	93.0	89.5	81.0	71.0	64.1	79.4
		MIN	39.4	41.0	47.6	54.4	61.6	67.9	70.6	70.3	66.9	55.1	45.5	40.4	55.1
		MEAN	51.0	53.4	60.5	67.7	74.4	79.9	81.9	81.7	78.2	68.1	58.3	52.3	67.3
CARROLLTON		MAX	53.2	57.5	65.3	75.5	81.6	87.0	89.1	88.8	83.1	74.1	64.0	56.1	72.9
		MIN	30.6	32.3	39.1	46.8	54.7	61.9	65.8	64.8	59.6	46.8	37.8	32.2	47.7
		MEAN	41.9	44.9	52.3	61.2	68.2	74.5	77.4	76.8	71.3	60.5	50.9	44.2	60.3
CARTERSVILLE		MAX	52.3	57.0	64.8	75.4	81.7	87.6	90.2	89.9	84.3	74.6	63.8	55.2	73.1
		MIN	31.3	32.5	39.1	47.4	55.6	63.0	66.7	65.9	60.8	47.7	38.8	33.1	48.5
		MEAN	41.8	44.8	52.0	61.5	68.6	75.4	78.5	77.9	72.6	61.2	51.3	44.2	60.8
CEDARTOWN		MAX	52.6	57.3	65.1	75.4	81.6	87.4	89.9	89.7	84.1	74.4	63.8	55.5	73.1
		MIN	31.0	32.5	39.3	47.5	55.2	62.8	67.1	66.0	60.4	47.0	37.8	32.4	48.3
		MEAN	41.8	44.9	52.2	61.5	68.4	75.1	78.5	77.9	72.3	60.8	50.8	43.9	60.7
CLAYTON 1 W		MAX	51.3	54.4	61.7	71.5	77.3	82.7	85.2	84.8	79.7	71.9	62.0	53.8	69.7
		MIN	28.3	29.2	35.6	43.0	51.2	58.3	62.1	61.5	56.2	44.1	35.3	29.5	44.5
		MEAN	39.8	41.9	48.7	57.3	64.2	70.5	73.7	73.2	68.0	58.0	48.7	41.7	57.1
COLUMBUS WSO	R	MAX	56.9	60.6	68.0	77.4	83.8	89.4	91.1	90.8	86.0	77.0	67.0	59.5	75.6
		MIN	35.4	37.0	43.9	51.9	60.2	67.6	71.0	70.5	65.9	53.1	42.7	37.2	53.0
		MEAN	46.2	48.8	56.0	64.7	72.0	78.5	81.0	80.7	76.0	65.1	54.9	48.4	64.4

GEORGIA

TEMPERATURE NORMALS (DEG F)

STATION		JAN	FEB	MAR	APR	MAY	JUN	JUL	AUG	SEP	OCT	NOV	DEC	ANN
CORNELIA	MAX	49.7	53.5	61.3	71.4	77.6	83.2	86.0	85.5	79.3	70.4	60.2	52.0	69.2
	MIN	29.5	31.1	37.7	45.8	53.3	60.0	63.8	63.4	58.0	46.2	37.3	31.3	46.5
	MEAN	39.6	42.3	49.5	58.6	65.5	71.7	74.9	74.5	68.7	58.3	48.8	41.7	57.8
COVINGTON	MAX	53.4	57.7	65.5	75.2	81.6	87.2	89.5	89.2	83.7	74.2	64.3	55.8	73.1
	MIN	33.0	34.5	41.4	49.4	57.4	64.3	67.8	67.1	62.0	50.0	40.4	34.8	50.2
	MEAN	43.2	46.2	53.5	62.4	69.6	75.8	78.7	78.2	72.8	62.1	52.4	45.3	61.7
CUTHBERT	MAX	60.3	63.8	71.6	80.1	86.3	91.1	92.4	92.4	88.2	79.5	69.8	62.4	78.2
	MIN	38.7	40.2	46.7	54.2	61.9	68.0	70.7	70.4	66.5	55.3	45.9	40.3	54.9
	MEAN	49.5	52.1	59.2	67.2	74.1	79.6	81.6	81.4	77.4	67.4	57.8	51.4	66.6
DAHLONEGA	MAX	48.8	52.3	60.1	71.0	77.9	83.7	86.5	85.7	79.8	70.3	60.1	51.7	69.0
	MIN	29.4	30.6	36.9	45.7	53.6	60.8	64.6	64.2	58.9	46.7	37.6	31.3	46.7
	MEAN	39.2	41.5	48.5	58.4	65.8	72.3	75.5	75.0	69.4	58.5	48.9	41.5	57.9
DALTON	MAX	49.0	53.4	61.6	73.1	80.4	86.3	89.2	89.0	83.7	73.3	61.6	52.3	71.1
	MIN	29.1	31.0	38.2	47.2	55.2	63.0	66.9	65.8	60.3	47.0	37.9	31.5	47.8
	MEAN	39.1	42.2	49.9	60.1	67.8	74.6	78.1	77.4	72.0	60.2	49.8	41.9	59.4
DOUGLAS 2 NNE	MAX	60.5	63.2	70.6	79.0	85.2	89.8	91.8	91.6	87.4	78.9	70.2	62.3	77.5
	MIN	36.6	38.0	45.0	52.6	60.1	66.6	69.7	69.3	65.3	53.2	43.2	37.2	53.1
	MEAN	48.6	50.6	57.8	65.8	72.7	78.2	80.7	80.5	76.4	66.1	56.7	49.8	65.3
DUBLIN 3 S	MAX	58.5	61.9	69.9	79.1	86.0	91.6	93.5	93.4	88.2	79.2	69.6	61.2	77.7
	MIN	34.5	36.0	42.9	50.9	58.9	65.7	68.6	67.6	62.9	50.2	40.8	35.4	51.2
	MEAN	46.5	49.0	56.4	65.1	72.5	78.7	81.0	80.5	75.6	64.8	55.2	48.3	64.5
EASTMAN 1 W	MAX	58.7	62.0	69.6	78.7	85.2	90.4	92.0	92.0	87.3	78.8	69.2	61.4	77.1
	MIN	36.1	37.9	45.0	53.0	60.5	66.9	69.9	69.2	65.0	53.4	44.0	37.8	53.2
	MEAN	47.4	50.0	57.4	65.9	72.9	78.7	81.0	80.7	76.2	66.1	56.6	49.6	65.2
EXPERIMENT	MAX	52.6	56.3	63.9	73.8	80.6	86.6	88.7	88.7	83.5	74.3	64.6	55.8	72.5
	MIN	33.7	35.2	42.2	50.7	58.4	65.1	68.3	67.7	62.7	50.6	41.9	35.6	51.0
	MEAN	43.2	45.8	53.1	62.3	69.5	75.9	78.6	78.2	73.1	62.5	53.3	45.7	61.8
FITZGERALD	MAX	60.9	64.0	71.5	79.9	86.3	90.9	92.8	92.2	87.5	79.0	69.8	62.7	78.1
	MIN	39.8	41.6	48.2	55.8	63.0	69.0	71.5	71.1	67.2	56.1	47.1	41.5	56.0
	MEAN	50.4	52.8	59.9	67.9	74.7	80.0	82.1	81.7	77.4	67.6	58.5	52.1	67.1
FOLKSTON 9 SW	MAX	65.8	68.5	75.2	81.9	87.3	91.1	92.6	92.1	88.6	81.0	73.4	67.0	80.4
	MIN	40.7	42.1	48.0	54.1	60.9	66.7	69.5	69.6	67.1	56.9	47.7	41.8	55.4
	MEAN	53.3	55.3	61.6	68.1	74.1	78.9	81.1	80.9	77.8	69.0	60.6	54.4	67.9
FORT GAINES	MAX	60.5	63.9	71.5	79.0	85.1	89.8	90.9	90.9	87.1	78.6	68.9	62.6	77.4
	MIN	38.0	39.9	46.3	53.8	61.2	67.5	70.4	69.6	66.2	54.6	44.9	39.7	54.3
	MEAN	49.3	51.9	58.9	66.4	73.2	78.7	80.7	80.3	76.7	66.6	56.9	51.2	65.9
FORT STEWART	MAX	62.6	65.7	72.4	79.8	85.8	89.8	92.2	91.3	86.5	79.0	70.9	64.4	78.4
	MIN	39.3	41.1	47.5	54.3	61.8	67.8	70.7	70.6	67.4	56.5	47.0	41.0	55.4
	MEAN	51.0	53.4	60.0	67.1	73.8	78.8	81.5	81.0	77.0	67.8	59.0	52.7	66.9
GAINESVILLE	MAX	50.2	54.3	62.6	73.1	78.7	84.4	87.1	86.8	80.8	71.2	61.3	52.7	70.3
	MIN	30.8	32.2	38.6	47.7	55.6	62.8	66.7	66.2	60.8	48.8	39.7	32.8	48.6
	MEAN	40.5	43.3	50.6	60.4	67.1	73.6	76.9	76.6	70.9	60.0	50.5	42.8	59.4
GLENNVILLE	MAX	61.5	64.6	71.7	79.3	85.7	89.8	91.8	91.3	86.7	78.9	70.4	63.5	77.9
	MIN	39.2	41.0	47.4	54.5	62.0	67.8	70.9	70.6	67.1	56.3	46.8	40.9	55.4
	MEAN	50.4	52.8	59.6	66.9	73.8	78.8	81.4	81.0	76.9	67.6	58.6	52.2	66.7
HARTWELL	MAX	54.4	58.4	66.7	76.8	83.4	89.0	91.8	91.3	85.3	75.8	65.9	56.8	74.6
	MIN	32.7	33.8	40.8	49.5	57.7	64.6	68.2	67.6	62.4	50.1	40.3	34.2	50.2
	MEAN	43.6	46.1	53.8	63.2	70.6	76.9	80.1	79.5	73.9	63.0	53.1	45.6	62.5
HAWKINSVILLE	MAX	59.8	63.0	70.5	79.4	85.8	91.0	92.7	92.9	88.4	80.5	70.8	62.6	78.1
	MIN	34.1	35.9	43.1	51.3	58.9	66.1	68.9	68.1	63.7	51.0	41.1	35.2	51.5
	MEAN	47.0	49.5	56.8	65.4	72.4	78.5	80.8	80.5	76.1	65.8	56.0	48.9	64.8

GEORGIA

TEMPERATURE NORMALS (DEG F)

STATION			JAN	FEB	MAR	APR	MAY	JUN	JUL	AUG	SEP	OCT	NOV	DEC	ANN
JASPER 1 NNW		MAX	49.2	53.4	61.2	71.4	78.1	84.1	86.9	86.4	81.1	71.4	60.5	52.0	69.6
		MIN	30.4	31.9	38.9	47.5	55.0	61.4	64.9	64.5	59.5	47.9	39.1	33.0	47.8
		MEAN	39.8	42.7	50.0	59.5	66.6	72.8	76.0	75.5	70.3	59.7	49.8	42.5	58.8
LA FAYETTE		MAX	49.7	54.4	62.5	73.4	80.7	86.9	89.6	89.3	83.7	73.6	62.1	53.1	71.6
		MIN	27.9	29.5	36.6	45.5	53.5	61.7	65.6	64.6	58.5	45.2	36.6	30.0	46.3
		MEAN	38.8	42.0	49.6	59.5	67.1	74.3	77.6	77.0	71.1	59.4	49.4	41.6	59.0
LA GRANGE		MAX	55.8	60.3	68.4	77.8	83.6	88.6	90.4	89.7	84.8	75.7	66.1	58.3	75.0
		MIN	33.1	34.7	41.3	48.9	56.8	63.8	67.6	66.8	61.7	48.9	39.8	34.5	49.8
		MEAN	44.5	47.5	54.9	63.4	70.3	76.2	79.0	78.3	73.3	62.3	53.0	46.4	62.4
LOUISVILLE 3 S		MAX	58.6	62.3	69.6	78.7	85.0	89.9	92.1	91.7	86.3	77.5	68.5	60.7	76.7
		MIN	35.7	37.4	43.9	51.9	59.7	66.3	69.5	68.8	64.5	52.4	42.5	37.0	52.5
		MEAN	47.2	49.9	56.8	65.3	72.4	78.1	80.8	80.3	75.4	65.0	55.5	48.9	64.6
LUMBER CITY		MAX	60.2	63.1	70.2	79.1	85.6	90.1	92.0	91.9	87.4	79.2	69.9	61.9	77.6
		MIN	34.9	36.7	43.4	51.4	59.0	66.0	69.2	68.6	64.0	51.2	41.2	35.4	51.8
		MEAN	47.6	49.9	56.9	65.2	72.3	78.0	80.6	80.3	75.7	65.2	55.5	48.7	64.7
MACON WSO	R	MAX	57.6	61.1	68.6	78.2	85.0	90.4	92.2	91.9	86.8	78.0	68.1	60.2	76.5
		MIN	35.5	37.4	44.2	52.3	60.3	67.3	70.6	70.0	65.1	52.3	42.5	37.1	52.9
		MEAN	46.6	49.2	56.5	65.3	72.7	78.9	81.4	81.0	76.0	65.2	55.3	48.7	64.7
MILLEDGEVILLE		MAX	55.8	59.5	67.1	76.3	82.7	88.5	90.8	90.6	85.5	76.6	67.2	58.5	74.9
		MIN	31.4	32.5	39.2	47.6	56.3	64.3	68.2	67.5	62.5	49.0	38.8	32.5	49.2
		MEAN	43.7	46.0	53.2	62.0	69.5	76.4	79.5	79.1	74.0	62.8	53.0	45.5	62.1
MILLEN 4 N		MAX	60.0	63.6	70.9	79.2	85.5	89.9	92.2	92.2	87.2	78.8	70.1	62.4	77.7
		MIN	35.3	37.1	43.2	50.6	58.6	65.5	68.7	68.4	64.2	51.5	41.6	36.4	51.8
		MEAN	47.7	50.3	57.1	64.9	72.1	77.7	80.4	80.3	75.7	65.2	55.9	49.4	64.7
MOULTRIE 2 ESE		MAX	62.1	65.1	72.3	79.9	86.0	90.2	91.9	91.6	87.7	79.9	71.0	64.3	78.5
		MIN	39.0	40.8	47.4	54.2	61.3	66.8	69.3	69.0	65.6	54.8	45.5	40.0	54.5
		MEAN	50.5	53.0	59.9	67.1	73.7	78.6	80.6	80.3	76.7	67.4	58.3	52.2	66.5
NEWNAN		MAX	54.2	59.0	66.9	76.4	82.7	88.0	90.2	89.8	84.5	75.2	64.9	56.8	74.1
		MIN	33.0	35.0	41.7	49.9	57.5	63.9	67.0	66.3	61.8	50.1	41.0	35.2	50.2
		MEAN	43.6	47.0	54.3	63.2	70.1	76.0	78.6	78.1	73.2	62.7	53.0	46.0	62.2
ROME		MAX	52.6	57.5	65.7	75.8	81.4	86.9	89.7	89.5	84.6	74.8	64.0	55.3	73.2
		MIN	30.8	32.2	39.3	47.6	55.4	62.8	66.8	66.2	60.6	47.6	38.2	32.2	48.3
		MEAN	41.7	44.9	52.5	61.7	68.4	74.9	78.3	77.9	72.6	61.2	51.1	43.8	60.8
SAVANNAH WSO	R	MAX	60.3	63.1	69.9	77.8	84.2	88.6	90.8	90.1	85.6	77.8	69.5	62.5	76.7
		MIN	37.9	40.0	46.8	54.1	62.3	68.5	71.5	71.4	67.6	55.9	45.5	39.4	55.1
		MEAN	49.2	51.6	58.4	66.0	73.3	78.6	81.2	80.8	76.6	66.9	57.5	51.0	65.9
SILOAM		MAX	55.4	59.4	66.9	76.2	82.7	88.5	90.7	90.6	85.2	76.1	66.4	57.9	74.7
		MIN	34.2	35.5	42.5	50.5	58.3	65.1	68.6	68.1	63.3	51.4	42.1	36.1	51.3
		MEAN	44.8	47.5	54.7	63.4	70.6	76.8	79.7	79.4	74.3	63.8	54.3	47.0	63.0
SWAINSBORO		MAX	59.3	63.0	70.7	79.4	85.6	90.2	92.2	91.9	86.9	78.5	69.4	61.7	77.4
		MIN	36.7	38.3	45.0	52.3	59.9	66.1	69.3	68.9	64.8	53.2	43.5	38.4	53.0
		MEAN	48.0	50.7	57.9	65.9	72.8	78.2	80.8	80.4	75.8	65.9	56.5	50.1	65.3
TALBOTTON		MAX	57.7	61.7	68.9	77.7	83.8	88.8	90.7	90.7	86.2	77.7	68.4	60.2	76.0
		MIN	33.8	35.3	42.2	50.1	57.5	64.5	67.6	67.1	62.4	50.5	40.5	35.2	50.6
		MEAN	45.8	48.5	55.6	63.9	70.7	76.7	79.2	78.9	74.3	64.2	54.5	47.7	63.3
THOMASVILLE 4 SE		MAX	63.7	66.4	73.4	80.8	86.7	90.7	91.8	91.7	88.7	81.2	72.3	65.5	79.4
		MIN	39.0	41.0	47.2	54.2	61.3	67.3	69.9	69.5	66.0	54.3	45.1	39.9	54.6
		MEAN	51.3	53.8	60.4	67.5	74.0	79.0	80.9	80.6	77.4	67.8	58.7	52.7	67.0
TIFTON EXP STATION		MAX	59.6	62.3	69.6	77.7	84.2	89.1	90.7	90.9	87.1	78.9	69.8	62.1	76.8
		MIN	37.4	39.2	46.4	54.4	61.7	67.7	70.3	70.0	66.5	54.6	45.5	39.0	54.4
		MEAN	48.5	50.8	58.0	66.1	73.0	78.4	80.5	80.5	76.8	66.8	57.6	50.6	65.6

GEORGIA

TEMPERATURE NORMALS (DEG F)

STATION		JAN	FEB	MAR	APR	MAY	JUN	JUL	AUG	SEP	OCT	NOV	DEC	ANN
TOCCOA	MAX	52.8	56.6	64.3	74.4	80.7	86.1	88.9	88.2	82.4	73.7	63.3	54.9	72.2
	MIN	32.4	33.6	39.9	48.4	56.3	62.9	66.6	65.9	60.5	49.4	40.4	34.1	49.2
	MEAN	42.6	45.1	52.2	61.5	68.5	74.6	77.8	77.1	71.5	61.5	51.9	44.5	60.7
WARRENTON	MAX	55.9	59.6	67.2	76.7	83.5	88.9	91.4	91.1	84.8	75.5	66.2	58.2	74.9
	MIN	33.8	35.2	41.7	50.0	58.3	65.1	68.7	68.2	63.5	51.0	41.5	35.3	51.0
	MEAN	44.9	47.4	54.4	63.3	70.9	77.0	80.1	79.6	74.2	63.3	53.9	46.8	63.0
WASHINGTON	MAX	54.0	57.8	65.1	74.6	81.6	87.4	90.1	89.8	84.3	74.8	65.5	56.5	73.5
	MIN	30.5	32.0	38.9	47.5	55.9	62.8	66.6	66.1	60.4	47.8	38.2	32.2	48.2
	MEAN	42.3	44.9	52.0	61.1	68.7	75.2	78.4	78.0	72.4	61.3	51.9	44.3	60.9
WAYCROSS 4 NE	MAX	63.0	65.5	72.6	80.4	86.6	90.9	92.8	92.5	88.4	80.6	72.4	64.9	79.2
	MIN	37.1	38.3	44.9	51.8	59.4	66.3	69.3	69.2	65.8	53.7	43.6	37.5	53.1
	MEAN	50.1	51.9	58.8	66.2	73.0	78.6	81.1	80.9	77.1	67.1	58.0	51.2	66.2
WEST POINT	MAX	56.3	60.1	67.7	76.9	83.1	88.9	90.7	90.4	85.5	76.6	66.7	58.8	75.1
	MIN	32.8	34.3	41.1	49.4	57.1	64.7	68.5	67.8	62.5	49.3	39.8	34.2	50.1
	MEAN	44.6	47.2	54.4	63.2	70.1	76.8	79.6	79.1	74.0	63.0	53.3	46.5	62.7

GEORGIA

PRECIPITATION NORMALS (INCHES)

STATION		JAN	FEB	MAR	APR	MAY	JUN	JUL	AUG	SEP	OCT	NOV	DEC	ANN
ABBEVILLE		4.04	4.83	4.95	3.49	3.83	3.88	5.82	4.75	3.90	1.94	2.24	3.34	47.01
ADAIRSVILLE 2 SSE		5.19	4.59	6.70	5.38	4.25	3.71	4.53	3.13	4.06	3.04	3.88	4.73	53.19
AILEY		3.88	4.20	4.73	3.37	4.39	4.08	5.28	5.03	3.76	2.15	2.10	3.56	46.53
ALBANY 3 SE		4.71	4.69	5.03	4.43	4.10	4.66	5.73	4.05	4.17	1.87	2.64	3.68	49.76
ALLATOONA DAM 2		5.04	4.18	6.09	5.24	4.13	3.43	4.64	3.91	3.75	3.02	3.50	4.34	51.27
ALMA FAA AIRPORT		3.63	3.89	4.59	3.23	3.89	4.83	5.62	5.54	3.87	2.08	2.07	3.72	46.96
AMERICUS 4 ENE		4.84	4.67	4.91	3.91	3.73	4.44	5.23	4.20	3.86	1.98	2.75	4.34	48.86
ATHENS WSO		4.85	4.16	5.81	4.04	4.78	4.00	5.18	3.64	3.58	2.70	3.32	4.09	50.15
ATLANTA WSO	R	4.91	4.43	5.91	4.43	4.02	3.41	4.73	3.41	3.17	2.53	3.43	4.23	48.61
AUGUSTA WSO	R	3.99	4.04	4.92	3.31	3.73	3.88	4.40	3.98	3.53	2.02	2.07	3.20	43.07
BALL GROUND		5.69	4.91	7.10	5.65	4.77	3.69	4.71	4.03	3.62	3.13	3.92	5.00	56.22
BEAVERDALE		5.39	4.93	6.43	4.94	4.39	3.83	5.07	4.13	4.35	3.20	4.44	4.98	56.08
BLAIRSVILLE EXP STA		5.29	5.02	6.66	5.10	4.58	4.34	4.87	4.60	4.09	3.36	4.09	4.83	56.83
BLAKELY		4.96	5.02	5.74	4.50	4.76	4.65	5.87	4.44	4.27	2.33	3.12	4.35	54.01
BOWMAN		4.91	4.12	5.80	4.22	3.94	4.04	4.44	3.23	3.08	2.85	3.21	4.08	47.92
BROOKLET 1 W		3.39	3.77	4.30	3.16	4.59	4.62	6.10	4.79	4.31	2.06	2.15	3.17	46.41
BRUNSWICK		3.18	3.41	3.58	2.92	4.21	5.80	6.37	6.54	6.79	2.96	2.14	2.79	50.69
BRUNSWICK FAA AP		3.08	3.48	3.68	2.80	3.92	5.74	6.16	6.42	7.60	3.29	2.26	2.70	51.13
BUTLER		4.56	4.70	5.18	4.10	4.07	4.43	5.40	3.86	3.67	2.20	2.73	4.18	49.08
CAMILLA		4.58	4.74	5.31	4.52	4.41	4.87	5.86	4.39	4.01	2.08	2.78	3.97	51.52
CANTON		5.72	4.97	6.95	5.56	4.29	4.05	4.95	4.10	3.83	3.36	3.77	4.87	56.42
CARNESVILLE 3 N		5.22	4.51	6.30	4.54	4.33	3.94	4.44	3.60	3.72	3.02	3.59	4.36	51.57
CARROLLTON		5.26	4.81	6.50	5.06	4.67	4.00	4.90	3.34	3.94	2.83	3.79	4.88	53.98
CARTERS		5.32	4.81	6.44	5.35	4.68	3.67	4.61	3.82	4.02	3.11	4.11	4.86	54.80
CARTERSVILLE		4.61	4.40	6.18	5.22	4.01	3.34	4.33	3.32	3.63	2.83	3.56	4.66	50.09
CARTERSVILLE 3 SW		5.05	4.34	6.21	5.20	4.01	3.65	4.68	3.54	3.72	3.04	3.48	4.43	51.35
CEDARTOWN		5.09	4.57	6.73	5.58	3.86	4.35	4.73	3.28	4.25	2.79	3.57	4.53	53.33
CHATSWORTH 2		4.95	4.99	5.98	5.06	4.55	3.91	5.22	3.69	4.65	2.85	4.10	4.87	54.82
CHICKAMAUGA PARK		5.25	5.11	6.73	4.74	4.50	3.76	4.70	3.35	4.42	3.28	4.21	5.24	55.29
CLAYTON 1 W		6.77	6.31	8.16	5.95	6.59	5.48	5.72	5.90	5.77	4.76	5.53	6.52	73.46
CLEVELAND		6.22	5.79	8.02	5.82	5.28	4.56	6.01	5.35	4.77	3.74	4.66	5.90	66.12
COLUMBUS WSO	R	4.52	4.52	5.96	4.50	4.44	4.16	5.50	4.02	3.59	2.07	3.06	4.75	51.09
CORDELE		3.97	4.41	5.05	3.41	3.63	3.67	5.51	4.09	3.68	1.82	2.34	3.60	45.18
CORNELIA		6.15	5.42	7.24	5.44	4.83	4.56	5.57	4.42	4.62	3.56	4.26	5.18	61.25
COVINGTON		5.11	4.42	5.90	3.88	4.27	3.71	4.70	3.27	3.18	2.80	3.10	4.20	48.54
CUMMING 2 NNE		5.90	5.06	6.94	5.22	4.43	3.69	4.75	3.63	3.93	3.29	3.86	5.13	55.83
CURRYVILLE 2 W		5.26	4.83	6.79	5.41	4.29	3.82	4.28	3.13	4.04	3.18	3.80	4.93	53.76
CUTHBERT		4.80	4.99	5.55	4.03	4.44	4.03	5.37	3.58	4.41	2.19	2.89	4.73	51.01
DAHLONEGA		6.29	5.74	7.87	5.66	5.19	4.10	5.71	4.97	4.59	3.48	4.78	5.91	64.29
DALLAS 2 SSE		5.36	4.48	6.22	5.15	4.26	4.13	4.51	3.25	3.71	3.19	3.37	4.68	52.31
DALTON		5.36	4.97	6.55	4.92	4.64	4.02	5.19	3.81	4.62	3.19	4.26	4.98	56.51
DAWSON		4.85	5.02	5.64	4.19	4.38	4.28	5.82	3.31	3.72	2.31	2.64	4.39	50.55
DAWSONVILLE		6.15	5.39	7.52	5.75	4.41	3.66	5.21	4.46	4.31	3.33	4.31	5.64	60.14
DOCTORTOWN 1 WSW		3.59	3.45	4.21	3.11	4.75	5.10	7.08	5.53	5.23	2.19	2.06	3.24	49.54
DONALSONVILLE 1 S		4.54	4.83	5.30	4.28	4.50	4.60	5.47	4.55	4.80	2.37	3.08	4.01	52.33
DOUGLAS 2 NNE		4.14	4.28	4.50	3.66	4.37	5.17	6.12	6.05	4.18	2.11	1.98	3.89	50.45
DOUGLASVILLE		5.60	4.78	6.49	5.24	4.67	3.81	4.78	3.05	3.84	3.14	3.54	4.74	53.68
DOVER		3.77	4.02	4.81	3.23	4.34	4.74	5.50	5.23	4.32	2.14	2.05	3.33	47.48
DUBLIN 3 S		4.26	4.63	4.90	3.46	3.73	3.95	4.43	4.34	3.61	2.12	2.44	3.62	45.49
EASTMAN 1 W		4.08	4.42	4.83	3.92	3.72	3.91	5.67	4.66	3.46	2.03	2.39	3.57	46.66
ELBERTON 2 N		5.11	4.27	5.62	4.07	4.32	3.93	4.65	3.49	3.36	2.85	2.99	3.90	48.56
ELLIJAY		6.40	5.85	7.32	5.70	5.07	4.66	5.34	4.46	4.55	3.70	4.67	5.78	63.50
EMBRY		5.55	4.80	6.55	5.36	4.40	4.08	5.02	3.72	4.43	3.11	3.67	4.70	55.39
EXPERIMENT		4.93	4.52	6.04	4.69	4.59	4.49	5.07	4.07	3.64	2.63	3.27	4.59	52.53
FAIRMOUNT		4.83	4.36	6.34	5.25	4.21	3.71	4.48	3.30	3.47	2.92	3.79	4.68	51.34

GEORGIA

PRECIPITATION NORMALS (INCHES)

STATION		JAN	FEB	MAR	APR	MAY	JUN	JUL	AUG	SEP	OCT	NOV	DEC	ANN
FITZGERALD		3.90	4.14	4.55	3.41	3.61	3.98	5.20	4.84	3.88	1.97	2.26	3.49	45.23
FOLKSTON 3 SW		3.08	3.41	3.67	3.35	4.72	5.74	7.18	6.87	5.01	2.57	2.14	2.84	50.58
FOLKSTON 9 SW		3.17	3.48	3.77	3.62	4.34	5.82	7.42	7.29	5.45	2.48	2.14	2.92	51.90
FORT GAINES		4.83	5.06	6.04	4.51	3.80	4.64	5.92	4.24	4.05	2.35	3.33	4.72	53.49
FORT STEWART		3.19	3.25	4.12	2.95	4.95	5.43	6.92	5.49	5.05	1.92	1.92	2.90	48.09
GAINESVILLE		5.81	4.76	6.89	5.06	4.42	4.18	4.69	3.43	4.23	3.43	3.93	4.93	55.76
GLENNVILLE		3.55	3.52	4.04	3.41	4.38	5.04	6.34	5.41	4.50	2.28	2.17	3.34	47.98
HARTWELL		5.23	4.63	6.60	4.30	4.21	4.07	4.72	3.50	4.02	3.13	3.50	4.32	52.23
HAWKINSVILLE		4.47	4.69	4.75	3.71	3.79	3.82	4.92	3.72	3.91	1.78	2.45	3.90	45.91
JASPER 1 NNW		5.72	4.86	7.03	5.83	4.69	3.85	5.25	4.15	4.43	3.37	4.16	5.44	58.78
JONESBORO		5.01	4.71	6.14	4.27	4.38	3.86	4.76	3.38	3.48	2.78	3.22	4.41	50.40
KENSINGTON		5.34	4.84	6.68	4.75	4.58	3.77	4.48	3.39	4.72	3.17	4.16	5.08	54.96
KINGSTON		5.21	4.85	6.75	5.52	4.31	3.56	4.45	3.18	4.10	2.84	3.59	4.72	53.08
LA FAYETTE		5.40	4.84	6.96	4.78	4.44	3.88	4.76	3.43	5.35	3.13	4.41	5.26	56.64
LA GRANGE		5.18	5.24	6.66	5.28	4.03	3.97	5.41	3.76	3.74	2.63	3.52	5.18	54.60
LEXINGTON		4.92	4.02	5.64	3.94	4.12	4.00	4.35	3.47	3.35	2.74	2.99	3.77	47.31
LINCOLNTON		4.93	4.20	5.19	3.91	4.03	4.16	4.62	3.66	3.70	2.37	2.63	3.57	46.97
LOUISVILLE 3 S		4.09	4.23	4.88	3.49	3.96	3.86	4.89	4.33	3.54	2.29	2.28	3.53	45.37
LUMBER CITY		3.81	4.15	4.57	3.36	4.01	4.02	5.86	4.91	3.86	2.05	2.30	3.58	46.48
MACON WSO	R	4.26	4.56	5.18	3.51	3.79	3.83	4.46	3.64	3.29	1.98	2.32	4.04	44.86
MAYSVILLE		5.50	4.54	6.47	4.78	4.68	4.12	4.02	3.67	3.64	3.45	3.81	4.62	53.30
MIDVILLE		4.19	4.36	4.73	3.32	4.44	4.03	5.38	4.90	4.27	2.35	2.30	3.30	47.57
MILLEDGEVILLE		4.60	4.73	5.10	3.56	4.65	3.52	4.47	4.23	3.30	2.07	2.55	3.76	46.54
MILLEN 4 N		3.93	4.28	4.52	3.23	4.35	4.51	4.63	4.56	4.19	2.46	2.28	3.50	46.44
MONTEZUMA		4.68	4.50	4.77	3.72	3.63	3.99	5.42	4.28	3.30	1.97	2.57	3.94	46.77
MONTICELLO		4.77	4.47	6.00	4.04	4.30	3.77	4.82	3.74	3.67	2.41	2.69	4.15	48.83
MOULTRIE 2 ESE		4.23	4.61	4.73	3.97	4.53	4.96	5.74	4.73	4.14	2.30	2.46	4.00	50.40
NEWNAN		5.30	4.64	6.08	4.88	4.44	3.94	4.67	3.70	3.22	2.78	3.57	4.51	51.73
NORCROSS 4 N		5.87	4.42	6.45	5.01	4.16	3.70	4.32	3.44	4.04	2.99	3.76	4.60	52.76
QUITMAN		4.34	4.46	4.47	4.16	4.35	5.09	6.68	5.50	4.43	1.98	2.61	4.10	52.17
RESACA		5.16	4.76	6.54	5.26	4.16	3.49	4.50	2.77	4.23	3.16	3.86	4.97	52.86
RINGGOLD		5.21	4.95	6.57	4.74	4.33	3.77	4.70	3.44	4.58	3.12	4.20	5.12	54.73
ROME		5.34	4.77	6.94	5.42	4.22	3.94	4.44	3.60	4.26	2.74	3.77	4.77	54.21
SAVANNAH WSO	R	3.09	3.17	3.83	3.16	4.62	5.69	7.37	6.65	5.19	2.27	1.89	2.77	49.70
SILOAM		5.24	4.32	5.77	3.79	4.40	3.29	5.11	3.80	3.57	2.52	2.81	3.92	48.54
SPARTA 2 NNW		4.62	4.53	5.43	3.76	4.40	3.56	4.86	3.85	3.16	2.19	2.54	3.61	46.51
SUMMERVILLE 1 SSW		5.24	5.01	6.77	5.28	4.27	3.82	4.31	3.46	4.08	3.16	3.77	5.26	54.43
SWAINSBORO		3.61	4.42	4.62	3.32	3.72	4.15	5.36	4.49	3.61	2.14	2.29	3.54	45.27
TALBOTTON		5.19	5.17	6.72	4.70	4.27	4.41	5.47	3.80	4.13	2.39	3.09	4.91	54.25
TAYLORSVILLE		4.98	4.32	6.18	5.35	3.53	3.73	4.49	2.75	4.12	2.78	3.18	4.39	49.80
THOMASVILLE 4 SE		4.15	4.54	4.79	4.10	4.44	5.57	6.15	4.73	4.64	2.15	2.55	3.95	51.76
TIFTON EXP STATION		4.23	4.28	4.63	3.97	4.11	4.34	5.35	4.56	3.52	2.01	2.18	3.43	46.61
TOCCOA		5.68	5.42	7.41	5.35	4.93	4.46	5.14	4.83	4.35	3.58	4.16	5.28	60.59
WALESKA		5.79	5.04	7.19	5.76	4.23	3.96	4.54	3.74	4.14	3.37	3.93	5.08	56.77
WARRENTON		4.96	4.79	5.81	3.91	4.54	3.65	4.32	3.71	4.08	2.81	2.73	4.20	49.51
WASHINGTON		5.07	4.28	5.46	4.04	4.58	4.23	5.06	3.57	3.63	2.61	2.80	3.91	49.24
WAYCROSS 4 NE		3.85	3.60	4.39	3.50	4.58	4.93	6.45	5.72	4.77	2.38	2.18	3.49	49.84
WAYNESBORO 2 NE		3.97	4.09	4.69	3.12	4.08	4.15	3.94	4.55	3.68	2.75	2.23	3.33	44.58
WEST POINT		4.80	5.25	6.11	5.10	4.04	3.60	5.96	3.11	3.56	2.57	3.14	5.09	52.33
WOODBURY		4.73	4.88	6.38	4.62	4.27	4.08	4.90	3.94	3.93	2.65	3.06	4.71	52.15
WOODSTOCK		5.40	4.58	6.24	4.87	4.14	3.59	4.40	3.40	3.74	3.09	3.44	4.57	51.46

GEORGIA

HEATING DEGREE DAY NORMALS (BASE 65 DEG F)

STATION		JUL	AUG	SEP	OCT	NOV	DEC	JAN	FEB	MAR	APR	MAY	JUN	ANN
ALBANY 3 SE		0	0	0	69	264	464	527	422	247	58	11	0	2062
ALMA FAA AIRPORT		0	0	0	53	236	425	483	368	223	47	0	0	1835
AMERICUS 4 ENE		0	0	0	80	268	478	559	423	267	61	9	0	2145
ATHENS WSO		0	0	0	129	387	629	701	557	402	130	30	0	2965
ATLANTA WSO	R	0	0	7	130	394	636	716	563	400	133	37	5	3021
AUGUSTA WSO	R	0	0	0	107	338	561	626	495	332	92	17	0	2568
BLAIRSVILLE EXP STA		0	0	49	311	585	825	908	759	617	314	138	19	4525
BLAKELY		0	0	0	52	256	449	523	404	237	56	6	0	1983
BROOKLET 1 W		0	0	0	60	240	444	510	391	238	36	5	0	1924
BRUNSWICK		0	0	0	21	142	331	398	310	169	14	0	0	1385
BRUNSWICK FAA AP		0	0	0	26	179	371	447	351	212	25	0	0	1611
CAMILLA		0	0	0	40	222	403	459	349	193	35	0	0	1701
CARROLLTON		0	0	10	165	423	645	716	563	404	138	38	0	3102
CARTERSVILLE		0	0	10	162	411	645	719	566	418	150	52	7	3140
CEDARTOWN		0	0	10	162	426	654	719	563	409	131	36	0	3110
CLAYTON 1 W		0	0	27	229	489	722	781	647	505	238	102	9	3749
COLUMBUS WSO	R	0	0	0	83	313	515	593	460	299	84	9	0	2356
CORNELIA		0	0	30	223	486	722	787	636	486	205	90	14	3679
COVINGTON		0	0	8	129	378	611	676	526	371	112	30	0	2841
CUTHBERT		0	0	0	46	229	427	501	378	217	44	0	0	1842
DAHLONEGA		0	0	24	218	483	729	800	658	518	213	83	9	3735
DALTON		0	0	13	186	456	716	803	638	475	164	49	0	3500
DOUGLAS 2 NNE		0	0	0	70	262	476	530	418	255	57	7	0	2075
DUBLIN 3 S		0	0	0	88	306	523	590	461	288	68	13	0	2337
EASTMAN 1 W		0	0	0	65	264	485	559	435	271	58	11	0	2148
EXPERIMENT		0	0	7	131	356	598	684	543	381	122	33	0	2855
FITZGERALD		0	0	0	40	215	410	473	361	207	26	0	0	1732
FOLKSTON 9 SW		0	0	0	24	172	343	395	299	164	20	0	0	1417
FORT GAINES		0	0	0	52	256	434	503	386	226	59	6	0	1922
FORT STEWART		0	0	0	38	200	392	455	339	202	32	0	0	1658
GAINESVILLE		0	0	18	186	435	688	760	608	454	164	79	12	3404
GLENNVILLE		0	0	0	42	214	407	474	359	212	37	0	0	1745
HARTWELL		0	0	0	119	363	601	663	529	364	101	27	0	2767
HAWKINSVILLE		0	0	0	78	282	499	571	447	277	57	13	0	2224
JASPER 1 NNW		0	0	18	193	456	698	781	624	473	180	69	8	3500
LA FAYETTE		0	0	19	204	468	725	812	650	484	182	67	6	3617
LA GRANGE		0	0	7	128	360	577	645	498	330	98	20	0	2663
LOUISVILLE 3 S		0	0	0	90	291	506	565	433	281	63	10	0	2239
LUMBER CITY		0	0	0	78	295	505	552	433	276	65	7	0	2211
MACON WSO	R	0	0	0	86	299	505	580	452	287	60	10	0	2279
MILLEDGEVILLE		0	0	0	118	365	605	660	532	377	120	29	0	2806
MILLEN 4 N		0	0	0	85	284	490	553	426	274	63	13	0	2188
MOULTRIE 2 ESE		0	0	0	50	216	406	474	359	203	35	0	0	1743
NEWNAN		0	0	7	124	364	589	663	510	348	96	21	0	2722
ROME		0	0	13	161	417	657	722	563	399	137	46	7	3122
SAVANNAH WSO	R	0	0	0	58	240	444	507	387	243	42	0	0	1921
SILOAM		0	0	0	100	328	558	634	496	337	94	22	0	2569
SWAINSBORO		0	0	0	66	268	468	541	413	256	58	8	0	2078
TALBOTTON		0	0	0	102	321	536	607	470	313	89	16	0	2454
THOMASVILLE 4 SE		0	0	0	43	211	394	452	339	193	40	0	0	1672
TIFTON EXP STATION		0	0	0	58	237	454	530	413	246	55	0	0	1993
TOCCOA		0	0	9	145	393	636	694	557	407	134	43	0	3018
WARRENTON		0	0	0	113	338	564	631	500	342	94	19	0	2601
WASHINGTON		0	0	16	162	393	642	704	563	415	147	55	7	3104
WAYCROSS 4 NE		0	0	0	53	234	439	488	390	225	47	0	0	1876

GEORGIA

HEATING DEGREE DAY NORMALS (BASE 65 DEG F)

STATION	JUL	AUG	SEP	OCT	NOV	DEC	JAN	FEB	MAR	APR	MAY	JUN	ANN
WEST POINT	0	0	5	116	357	574	641	507	342	104	28	0	2674

GEORGIA

COOLING DEGREE DAY NORMALS (BASE 65 DEG F)

STATION		JAN	FEB	MAR	APR	MAY	JUN	JUL	AUG	SEP	OCT	NOV	DEC	ANN
ALBANY 3 SE		22	27	36	91	269	423	508	502	369	122	18	9	2396
ALMA FAA AIRPORT		18	13	37	89	254	402	493	484	354	115	20	10	2289
AMERICUS 4 ENE		20	14	32	88	251	396	471	471	333	101	7	7	2191
ATHENS WSO		0	0	12	31	167	330	440	419	242	39	0	0	1680
ATLANTA WSO	R	0	0	12	37	170	329	422	409	247	44	0	0	1670
AUGUSTA WSO	R	6	5	16	38	203	372	484	462	288	61	0	0	1935
BLAIRSVILLE EXP STA		0	0	0	0	55	139	233	214	94	8	0	0	743
BLAKELY		21	21	29	92	254	411	487	481	360	108	16	6	2286
BROOKLET 1 W		17	15	36	87	253	381	474	462	327	103	18	7	2180
BRUNSWICK		32	27	54	113	304	441	533	524	411	176	37	21	2673
BRUNSWICK FAA AP		19	15	39	88	285	432	521	505	393	150	29	11	2487
CAMILLA		25	25	54	116	295	447	524	518	396	136	21	10	2567
CARROLLTON		0	0	10	24	137	290	384	366	199	26	0	0	1436
CARTERSVILLE		0	0	15	45	163	319	419	400	238	44	0	0	1643
CEDARTOWN		0	0	12	26	141	303	419	400	229	32	0	0	1562
CLAYTON 1 W		0	0	0	7	77	174	270	254	117	12	0	0	911
COLUMBUS WSO	R	10	7	20	75	226	405	496	487	330	86	10	0	2152
CORNELIA		0	0	5	13	105	215	311	295	141	16	0	0	1101
COVINGTON		0	0	15	34	173	327	425	409	242	39	0	0	1664
CUTHBERT		20	17	37	110	286	438	515	508	372	121	13	5	2442
DAHLONEGA		0	0	6	15	108	228	326	310	156	16	0	0	1165
DALTON		0	0	6	17	136	291	406	384	223	37	0	0	1500
DOUGLAS 2 NNE		22	15	31	81	246	396	487	481	342	104	13	5	2223
DUBLIN 3 S		16	13	22	71	245	411	496	481	318	82	12	5	2172
EASTMAN 1 W		14	15	36	85	256	411	496	487	336	99	12	8	2255
EXPERIMENT		8	5	12	41	172	327	422	409	250	54	5	0	1705
FITZGERALD		20	20	49	113	301	450	530	518	372	121	20	10	2524
FOLKSTON 9 SW		33	27	58	113	282	417	499	493	384	148	40	15	2509
FORT GAINES		16	19	37	101	260	411	487	474	351	102	13	6	2277
FORT STEWART		21	14	47	95	273	414	512	496	360	125	20	11	2388
GAINESVILLE		0	0	8	26	145	270	369	360	195	31	0	0	1404
GLENNVILLE		21	17	45	94	276	414	508	496	357	122	22	11	2383
HARTWELL		0	0	17	47	200	357	468	450	271	57	6	0	1873
HAWKINSVILLE		13	13	23	69	242	405	490	481	337	103	12	0	2188
JASPER 1 NNW		0	0	8	15	118	242	341	326	177	29	0	0	1256
LA FAYETTE		0	6	7	17	133	285	391	372	202	31	0	0	1444
LA GRANGE		9	8	17	50	184	336	434	412	256	44	0	0	1750
LOUISVILLE 3 S		13	10	26	72	239	393	490	474	312	90	6	7	2132
LUMBER CITY		13	11	24	71	233	390	484	474	321	85	10	0	2116
MACON WSO	R	10	9	24	69	249	417	508	496	335	92	8	0	2217
MILLEDGEVILLE		0	0	11	30	168	342	450	437	274	50	0	0	1762
MILLEN 4 N		17	15	29	60	233	381	477	474	321	91	11	6	2115
MOULTRIE 2 ESE		25	23	44	98	273	408	484	474	351	124	15	9	2328
NEWNAN		0	6	16	42	179	330	422	406	253	52	0	0	1706
ROME		0	0	11	38	151	304	412	400	241	44	0	0	1601
SAVANNAH WSO	R	17	12	38	72	261	408	502	490	348	117	15	10	2290
SILOAM		7	6	18	46	196	354	456	446	283	63	7	0	1882
SWAINSBORO		14	13	36	85	250	396	490	477	324	94	13	6	2198
TALBOTTON		11	8	22	56	193	351	440	431	283	77	6	0	1878
THOMASVILLE 4 SE		27	26	50	115	279	420	493	484	372	129	22	13	2430
TIFTON EXP STATION		19	15	29	88	252	402	481	481	354	113	15	7	2256
TOCCOA		0	0	11	29	152	291	397	375	204	37	0	0	1496
WARRENTON		8	7	13	43	202	360	468	453	280	61	0	0	1895
WASHINGTON		0	0	12	30	170	313	415	403	238	47	0	0	1628
WAYCROSS 4 NE		26	24	32	83	253	408	499	493	363	118	24	11	2334

GEORGIA

COOLING DEGREE DAY NORMALS (BASE 65 DEG F)

STATION	JAN	FEB	MAR	APR	MAY	JUN	JUL	AUG	SEP	OCT	NOV	DEC	ANN
WEST POINT	9	8	13	50	186	354	453	437	275	54	6	0	1845

09 — GEORGIA

STATE-STATION NUMBER	STN TYP	NAME		LATITUDE DEG-MIN	LONGITUDE DEG-MIN	ELEVATION (FT)
9-0010	12	ABBEVILLE		N 3200	W 08318	240
9-0041	12	ADAIRSVILLE 2 SSE		N 3421	W 08455	750
9-0090	12	AILEY		N 3211	W 08234	233
9-0140	13	ALBANY 3 SE		N 3132	W 08408	180
9-0181	12	ALLATOONA DAM 2		N 3410	W 08444	975
9-0211	13	ALMA FAA AIRPORT		N 3132	W 08231	198
9-0253	13	AMERICUS 4 ENE		N 3206	W 08410	476
9-0435	13	ATHENS WSO		N 3357	W 08319	802
9-0451	13	ATLANTA WSO	R	N 3339	W 08426	1010
9-0495	13	AUGUSTA WSO	R	N 3322	W 08158	145
9-0603	12	BALL GROUND		N 3420	W 08423	1100
9-0746	12	BEAVERDALE		N 3455	W 08451	724
9-0969	13	BLAIRSVILLE EXP STA		N 3451	W 08356	1917
9-0979	13	BLAKELY		N 3123	W 08457	300
9-1132	12	BOWMAN		N 3412	W 08302	785
9-1266	13	BROOKLET 1 W		N 3223	W 08141	190
9-1340	13	BRUNSWICK		N 3110	W 08130	13
9-1345	13	BRUNSWICK FAA AP		N 3109	W 08123	20
9-1425	12	BUTLER		N 3233	W 08414	625
9-1500	13	CAMILLA		N 3114	W 08413	175
9-1585	12	CANTON		N 3414	W 08430	870
9-1619	12	CARNESVILLE 3 N		N 3426	W 08314	900
9-1640	13	CARROLLTON		N 3336	W 08505	985
9-1657	12	CARTERS		N 3436	W 08442	695
9-1665	13	CARTERSVILLE		N 3410	W 08448	752
9-1670	12	CARTERSVILLE 3 SW		N 3409	W 08451	700
9-1732	13	CEDARTOWN		N 3401	W 08515	785
9-1863	12	CHATSWORTH 2		N 3446	W 08446	720
9-1906	12	CHICKAMAUGA PARK		N 3455	W 08517	760
9-1982	13	CLAYTON 1 W		N 3453	W 08326	1940
9-2006	12	CLEVELAND		N 3436	W 08346	1570
9-2166	13	COLUMBUS WSO	R	N 3231	W 08457	385
9-2266	12	CORDELE		N 3200	W 08347	308
9-2283	13	CORNELIA		N 3431	W 08332	1470
9-2318	13	COVINGTON		N 3336	W 08352	770
9-2408	12	CUMMING 2 NNE		N 3413	W 08408	1300
9-2429	12	CURRYVILLE 2 W		N 3427	W 08507	650
9-2450	13	CUTHBERT		N 3146	W 08447	461
9-2475	13	DAHLONEGA		N 3432	W 08359	1430
9-2485	12	DALLAS 2 SSE		N 3354	W 08450	1080
9-2493	13	DALTON		N 3446	W 08457	720
9-2570	12	DAWSON		N 3146	W 08426	355
9-2578	12	DAWSONVILLE		N 3425	W 08407	1370
9-2716	12	DOCTORTOWN 1 WSW		N 3139	W 08151	80
9-2736	12	DONALSONVILLE 1 S		N 3101	W 08453	135
9-2783	13	DOUGLAS 2 NNE		N 3131	W 08251	240
9-2791	12	DOUGLASVILLE		N 3345	W 08445	1219
9-2799	12	DOVER		N 3235	W 08143	103
9-2839	13	DUBLIN 3 S		N 3230	W 08254	215
9-2966	13	EASTMAN 1 W		N 3212	W 08312	400

09 — GEORGIA

STATE-STATION NUMBER	STN TYP	NAME		LATITUDE DEG-MIN	LONGITUDE DEG-MIN	ELEVATION (FT)
9-3060	12	ELBERTON 2 N		N 3408	W 08252	540
9-3115	12	ELLIJAY		N 3442	W 08430	1440
9-3147	12	EMBRY		N 3352	W 08459	1200
9-3271	13	EXPERIMENT		N 3316	W 08417	925
9-3295	12	FAIRMOUNT		N 3426	W 08442	740
9-3386	13	FITZGERALD		N 3143	W 08315	371
9-3460	12	FOLKSTON 3 SW		N 3048	W 08202	30
9-3465	13	FOLKSTON 9 SW		N 3044	W 08208	120
9-3516	13	FORT GAINES		N 3136	W 08503	340
9-3538	13	FORT STEWART		N 3152	W 08137	92
9-3621	13	GAINESVILLE		N 3418	W 08351	1170
9-3754	13	GLENNVILLE		N 3156	W 08155	170
9-4133	13	HARTWELL		N 3421	W 08255	690
9-4170	13	HAWKINSVILLE		N 3217	W 08328	245
9-4648	13	JASPER 1 NNW		N 3429	W 08427	1465
9-4700	12	JONESBORO		N 3331	W 08421	930
9-4802	12	KENSINGTON		N 3447	W 08522	850
9-4854	12	KINGSTON		N 3414	W 08456	730
9-4941	13	LA FAYETTE		N 3442	W 08516	797
9-4949	13	LA GRANGE		N 3303	W 08501	715
9-5165	12	LEXINGTON		N 3352	W 08307	729
9-5204	12	LINCOLNTON		N 3347	W 08228	510
9-5314	13	LOUISVILLE 3 S		N 3256	W 08224	325
9-5386	13	LUMBER CITY		N 3155	W 08241	120
9-5443	13	MACON WSO	R	N 3242	W 08339	354
9-5633	12	MAYSVILLE		N 3415	W 08334	920
9-5858	12	MIDVILLE		N 3250	W 08214	180
9-5874	13	MILLEDGEVILLE		N 3306	W 08315	360
9-5882	13	MILLEN 4 N		N 3252	W 08158	195
9-5979	12	MONTEZUMA		N 3217	W 08401	327
9-5988	12	MONTICELLO		N 3318	W 08341	655
9-6087	13	MOULTRIE 2 ESE		N 3110	W 08345	340
9-6335	13	NEWNAN		N 3322	W 08449	990
9-6407	12	NORCROSS 4 N		N 3400	W 08412	950
9-7276	12	QUITMAN		N 3047	W 08334	170
9-7430	12	RESACA		N 3434	W 08457	650
9-7489	12	RINGGOLD		N 3455	W 08507	820
9-7600	13	ROME		N 3415	W 08509	596
9-7847	13	SAVANNAH WSO	R	N 3208	W 08112	46
9-8064	13	SILOAM		N 3332	W 08306	690
9-8223	12	SPARTA 2 NNW		N 3318	W 08259	590
9-8436	12	SUMMERVILLE 1 SSW		N 3427	W 08522	800
9-8496	13	SWAINSBORO		N 3235	W 08222	325
9-8535	13	TALBOTTON		N 3240	W 08432	700
9-8600	12	TAYLORSVILLE		N 3405	W 08459	710
9-8666	13	THOMASVILLE 4 SE		N 3048	W 08354	210
9-8703	13	TIFTON EXP STATION		N 3128	W 08331	360
9-8740	13	TOCCOA		N 3435	W 08319	1019
9-9077	12	WALESKA		N 3419	W 08433	1110
9-9141	13	WARRENTON		N 3325	W 08240	510

09 — GEORGIA

STATE-STATION NUMBER	STN TYP	NAME	LATITUDE DEG-MIN	LONGITUDE DEG-MIN	ELEVATION (FT)
9-9157	13	WASHINGTON	N 3344	W 08244	630
9-9186	13	WAYCROSS 4 NE	N 3115	W 08219	145
9-9194	12	WAYNESBORO 2 NE	N 3307	W 08159	270
9-9291	13	WEST POINT	N 3252	W 08511	575
9-9506	12	WOODBURY	N 3259	W 08435	800
9-9524	12	WOODSTOCK	N 3406	W 08431	992

09 – GEORGIA

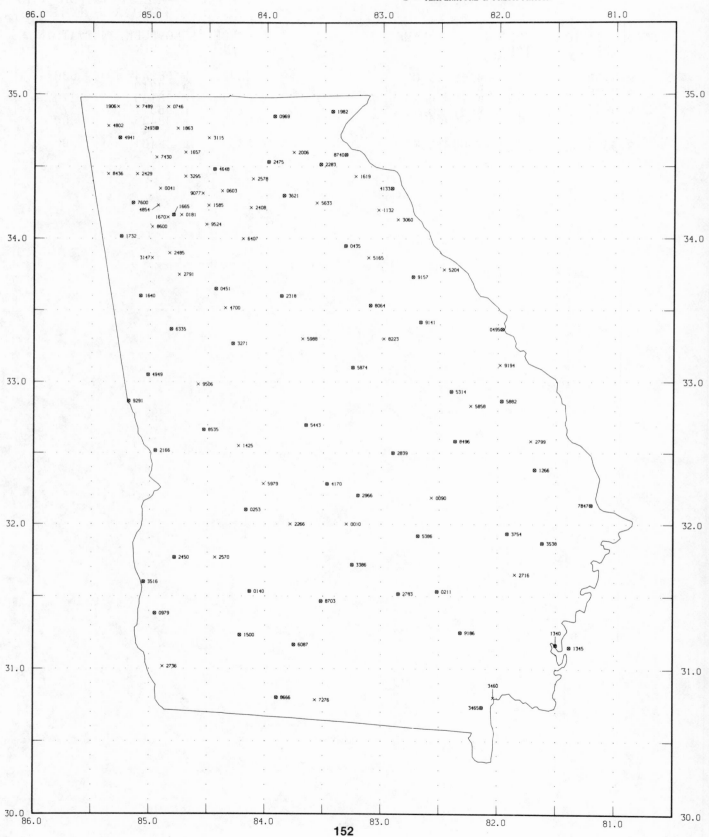

HAWAII

TEMPERATURE NORMALS (DEG F)

STATION			JAN	FEB	MAR	APR	MAY	JUN	JUL	AUG	SEP	OCT	NOV	DEC	ANN
HALEAKALA RS 338		MAX	59.4	58.6	59.0	60.7	62.7	66.1	65.7	66.0	65.2	64.5	62.6	60.5	62.6
		MIN	41.3	40.7	41.1	42.3	43.7	46.3	46.8	46.8	45.5	45.5	44.4	42.9	43.9
		MEAN	50.3	49.7	50.1	51.5	53.2	56.2	56.3	56.4	55.4	55.0	53.5	51.7	53.3
HAWAII VOLCNS NP HQ 54		MAX	66.1	65.5	65.3	66.3	67.7	69.2	70.4	71.3	71.6	70.8	68.3	66.3	68.2
		MIN	49.7	49.7	50.6	51.6	52.6	53.6	54.6	55.2	54.6	54.4	52.9	51.0	52.5
		MEAN	57.9	57.6	58.0	59.0	60.2	61.4	62.6	63.2	63.1	62.6	60.6	58.7	60.4
HILO WSO 87 AP	R	MAX	79.5	79.0	79.0	79.7	81.0	82.5	82.8	83.3	83.6	83.0	80.9	79.5	81.2
		MIN	63.2	63.2	63.9	64.9	66.1	67.1	68.0	68.4	68.0	67.5	66.3	64.3	65.9
		MEAN	71.4	71.2	71.5	72.4	73.6	74.8	75.4	75.9	75.8	75.3	73.6	71.9	73.6
HONOLULU WSFO 703 AP	R	MAX	79.9	80.4	81.4	82.7	84.8	86.2	87.1	88.3	88.2	86.7	83.9	81.4	84.2
		MIN	65.3	65.3	67.3	68.7	70.2	71.9	73.1	73.6	72.9	72.2	69.2	66.5	69.7
		MEAN	72.6	72.9	74.4	75.7	77.5	79.1	80.1	81.0	80.6	79.5	76.6	74.0	77.0
KAHULUI WSO 398 AP	R	MAX	79.5	79.7	81.1	82.2	84.5	85.9	86.5	87.4	87.6	86.4	83.5	81.0	83.8
		MIN	63.4	63.4	64.8	66.2	67.0	68.7	70.4	70.9	69.8	69.1	67.5	65.3	67.2
		MEAN	71.5	71.6	73.0	74.2	75.8	77.3	78.5	79.2	78.7	77.8	75.5	73.2	75.5
KAINALIU 73.2		MAX	76.4	76.6	76.6	76.7	76.9	77.2	78.6	79.6	79.8	80.0	78.8	77.2	77.9
		MIN	58.5	58.5	59.2	60.2	61.6	62.8	63.1	63.7	63.4	63.0	61.5	59.7	61.3
		MEAN	67.5	67.6	67.9	68.5	69.3	70.0	70.8	71.6	71.7	71.5	70.2	68.4	69.6
KAMUELA 192.2		MAX	72.6	72.3	72.4	72.9	73.9	74.3	74.4	75.6	76.8	76.7	74.7	72.6	74.1
		MIN	50.5	50.7	52.0	53.2	54.5	55.5	56.4	57.4	57.1	55.2	54.1	52.2	54.1
		MEAN	61.6	61.5	62.2	63.1	64.2	64.9	65.4	66.5	67.0	66.0	64.4	62.4	64.1
KANEOHE MAUKA 781		MAX	77.1	76.8	77.1	78.2	79.7	81.3	81.9	82.6	83.0	82.1	79.6	78.0	79.8
		MIN	65.1	64.9	65.7	66.9	68.6	70.2	71.2	71.4	70.7	70.2	68.6	66.7	68.4
		MEAN	71.1	70.9	71.4	72.5	74.2	75.8	76.6	77.1	76.9	76.2	74.1	72.4	74.1
KILAUEA POINT 1133		MAX	78.3	77.5	77.3	78.5	80.2	82.2	83.2	83.9	84.8	83.5	80.6	78.6	80.7
		MIN	66.5	65.6	66.3	67.5	69.3	71.2	71.7	72.5	72.6	71.6	69.8	68.0	69.4
		MEAN	72.4	71.6	71.8	73.0	74.8	76.7	77.5	78.2	78.8	77.6	75.3	73.3	75.1
KOHALA MISSION 175.1		MAX	76.0	75.6	75.8	76.4	77.8	79.2	79.9	80.3	81.3	80.5	78.0	76.0	78.1
		MIN	62.9	62.8	63.3	64.2	65.5	66.9	67.9	68.5	68.1	67.6	66.3	64.3	65.7
		MEAN	69.5	69.2	69.6	70.3	71.7	73.1	73.9	74.4	74.7	74.1	72.2	70.2	71.9
KULA SANATORIUM 267		MAX	69.4	69.1	69.8	70.4	71.5	73.0	74.1	75.0	74.7	74.4	72.9	70.8	72.1
		MIN	53.2	52.6	52.8	53.8	55.3	56.1	57.0	57.5	57.6	57.3	56.0	54.2	55.3
		MEAN	61.4	60.9	61.3	62.1	63.4	64.6	65.6	66.3	66.2	65.8	64.4	62.5	63.7
LAHAINA 361		MAX	81.3	81.2	82.0	83.4	84.8	86.2	87.2	87.9	87.9	87.2	84.9	82.7	84.7
		MIN	62.8	62.4	63.2	64.4	65.9	67.1	68.2	68.9	68.6	67.8	65.9	63.9	65.8
		MEAN	72.1	71.8	72.6	73.9	75.4	76.7	77.7	78.5	78.3	77.5	75.5	73.3	75.3
LIHUE WSO 1020.1 AP	//R	MAX	77.8	78.0	78.1	79.2	81.2	83.0	83.8	84.6	84.7	83.2	80.8	78.8	81.1
		MIN	64.6	64.7	65.9	67.6	69.7	71.8	72.9	73.5	72.9	71.4	69.7	66.9	69.3
		MEAN	71.3	71.4	72.0	73.4	75.5	77.4	78.4	79.1	78.9	77.3	75.3	72.9	75.2
MANA 1026		MAX	79.4	80.0	80.9	82.6	84.5	86.4	87.1	87.5	87.4	86.1	83.6	81.0	83.9
		MIN	62.3	61.7	62.4	63.7	65.1	67.3	68.2	68.8	68.1	67.6	65.6	63.7	65.4
		MEAN	70.9	70.9	71.6	73.2	74.8	76.9	77.7	78.2	77.8	76.9	74.7	72.4	74.7
MOUNTAIN VIEW 91		MAX	74.4	73.7	73.3	73.8	75.0	76.4	77.0	77.9	78.5	78.1	75.8	74.5	75.7
		MIN	56.7	56.6	57.4	58.6	59.6	60.5	61.6	62.1	61.3	61.1	60.0	58.2	59.5
		MEAN	65.6	65.2	65.4	66.2	67.3	68.5	69.3	70.0	70.0	69.6	67.9	66.4	67.6
NIU RIDGE 1035		MAX	75.9	76.8	78.0	79.2	80.9	82.8	83.0	84.6	84.1	83.3	80.6	77.6	80.6
		MIN	59.5	58.6	58.9	60.2	61.9	63.2	64.1	64.4	65.0	64.8	63.0	60.9	62.0
		MEAN	67.7	67.7	68.5	69.8	71.4	73.0	73.6	74.5	74.6	74.1	71.8	69.3	71.3
OOKALA 223		MAX	76.8	76.2	76.7	77.4	78.9	80.3	80.7	81.2	81.9	81.1	78.8	77.1	78.9
		MIN	63.6	62.8	63.5	64.3	65.6	67.0	67.8	68.4	68.5	68.0	66.6	64.5	65.9
		MEAN	70.2	69.5	70.1	70.9	72.3	73.7	74.3	74.8	75.2	74.6	72.8	70.8	72.4

HAWAII

TEMPERATURE NORMALS (DEG F)

STATION			JAN	FEB	MAR	APR	MAY	JUN	JUL	AUG	SEP	OCT	NOV	DEC	ANN
OPAEULA 870	A	MAX	75.3	75.2	75.3	76.0	77.7	79.1	79.5	80.6	81.6	81.0	78.4	76.4	78.0
		MIN	59.3	59.0	60.2	61.4	63.0	64.6	65.7	66.2	65.9	65.0	63.8	61.2	62.9
		MEAN	67.4	67.2	67.8	68.7	70.4	71.9	72.6	73.4	73.8	73.0	71.1	68.8	70.5
WAIALUA 847	A	MAX	78.5	78.0	78.6	79.8	82.2	84.2	84.8	85.7	86.0	84.5	81.7	79.3	81.9
		MIN	61.1	60.5	61.4	62.8	64.0	65.8	66.9	67.2	66.5	66.0	64.9	62.8	64.2
		MEAN	69.8	69.3	70.0	71.3	73.1	75.0	75.9	76.5	76.3	75.3	73.3	71.1	73.1

HAWAII

PRECIPITATION NORMALS (INCHES)

STATION		JAN	FEB	MAR	APR	MAY	JUN	JUL	AUG	SEP	OCT	NOV	DEC	ANN
EWA PLANTATION 741		4.61	2.82	2.90	1.50	1.13	.37	.52	.41	.61	1.86	2.98	3.23	22.94
HAINA 214		6.55	7.18	8.51	8.10	3.99	1.93	3.58	4.83	1.95	4.59	7.43	7.27	65.91
HAKALAU 142		9.87	14.03	15.38	13.43	9.89	6.52	9.32	11.07	7.47	10.74	13.85	13.09	134.66
HALEAKALA RS 338		10.18	7.46	6.41	4.50	2.14	.98	2.12	2.64	1.34	2.53	5.65	7.18	53.13
HAMAKUAPOKO 485		5.72	5.25	5.66	4.44	2.81	2.36	3.38	3.46	1.71	3.49	5.40	5.62	49.30
HAWAII VOLCNS NP HQ 54		12.00	11.77	14.39	11.06	7.75	4.45	4.94	6.83	4.24	6.51	12.48	11.19	107.61
HILO WSO 87 AP	R	9.42	13.47	13.55	13.10	9.40	6.13	8.68	10.02	6.63	10.01	14.88	12.86	128.15
HONAUNAU 27		3.73	2.74	3.78	4.62	5.34	5.19	5.38	5.83	5.47	4.74	2.76	2.52	52.10
HONOHINA 137		10.32	14.43	15.91	14.32	10.42	6.47	9.80	12.07	7.32	10.10	14.75	13.71	139.62
HONOLULU WSFO 703 AP	R	3.79	2.72	3.48	1.49	1.21	.49	.54	.60	.62	1.88	3.22	3.43	23.47
HUKIPO 945		4.81	2.76	2.84	1.77	1.31	.30	.51	.49	.84	2.09	2.86	4.32	24.90
KAHULUI WSO 398 AP	R	4.21	3.27	3.00	1.18	.66	.28	.41	.50	.36	.87	2.26	2.85	19.84
KAILUA 446		9.69	11.27	12.69	12.37	8.65	6.79	10.15	10.27	6.48	9.52	12.84	10.61	121.33
KAINALIU 73.2		4.98	2.99	4.75	5.63	6.87	6.68	7.92	7.49	7.33	5.15	3.63	3.31	66.73
KALIHI RES SITE 777	R	9.56	9.71	11.46	9.53	8.62	5.15	7.97	6.65	5.49	8.81	10.51	9.57	103.03
KAMUELA 192.2		4.74	3.76	3.84	3.76	1.93	1.40	2.07	2.40	1.05	2.01	2.91	3.96	33.83
KANEOHE MAUKA 781		8.58	7.52	8.07	7.08	5.20	3.00	4.29	4.01	3.62	6.34	8.79	7.51	74.01
KEAHUA 410		4.41	3.47	2.89	1.61	.67	.19	.40	.51	.28	1.00	2.43	3.31	21.17
KEKAHA 944		4.20	2.19	2.47	1.43	1.10	.32	.35	.45	.78	1.88	2.59	3.91	21.67
KIHEI 311		3.99	2.17	2.07	.67	.29	.11	.12	.14	.14	.54	1.51	2.28	14.03
KILAUEA POINT 1133		6.41	4.64	5.91	5.26	4.17	2.29	3.41	3.15	2.85	4.66	6.77	5.51	55.03
KIPAHULU 258		8.92	7.78	8.13	6.85	6.16	5.71	7.18	6.20	4.77	6.65	7.49	6.98	82.82
KOHALA MISSION 175.1		7.24	5.76	7.27	6.06	4.00	3.42	5.09	5.46	2.53	4.43	6.32	6.16	63.74
KOLO 1033		4.59	2.63	3.08	1.59	1.09	.60	.47	.50	.92	2.04	2.87	3.95	24.33
KUKAIAU 222		10.07	11.87	14.03	13.11	6.91	3.46	5.91	8.83	3.04	6.42	11.19	11.58	106.42
KUKUIHAELE HIC 199		7.46	7.13	9.27	7.94	4.86	2.83	5.01	5.92	2.50	5.14	7.54	7.66	73.26
KULA SANATORIUM 267		6.25	4.09	3.93	3.01	2.26	1.27	1.83	1.89	1.73	1.94	2.57	3.51	34.28
LAHAINA 361		3.79	2.39	2.35	.82	.52	.10	.19	.22	.24	.72	1.95	2.64	15.93
LANAI CITY 672	A	6.31	3.82	4.63	2.76	3.09	1.12	1.92	1.53	2.24	2.55	3.86	4.08	37.91
LIHUE WSO 1020.1 AP	//R	6.24	3.68	4.52	3.29	2.99	1.64	2.03	1.85	2.25	4.52	5.55	5.46	44.02
MANA 1026		4.50	2.42	2.95	1.42	1.14	.52	.65	.45	1.13	2.13	2.70	3.84	23.85
MANOA 712.1		7.23	6.10	7.35	6.30	4.50	3.71	4.53	4.03	3.01	4.79	7.27	7.73	66.55
MANOA TUNNEL 2 716	R	10.54	11.75	14.99	13.73	12.21	9.26	13.17	10.98	7.89	11.77	13.82	14.04	144.15
MOANALUA 770		5.06	3.83	4.83	3.07	2.40	1.45	1.86	1.30	1.21	2.99	4.46	4.93	37.39
MOUNTAIN VIEW 91		13.43	17.65	20.62	20.73	15.47	10.03	13.14	15.70	9.70	13.16	19.71	17.18	186.52
NIU RIDGE 1035		5.16	3.09	3.31	2.06	1.54	.79	.96	.78	1.84	2.90	3.03	4.76	30.22
NUUANU RES 4 783	R	9.79	10.01	13.06	10.60	9.50	6.28	8.61	7.78	6.14	9.40	11.35	10.51	113.03
OLOWALU 296		3.52	1.98	2.12	.46	.41	.09	.07	.15	.22	.61	1.60	2.34	13.57
OOKALA 223		9.60	11.83	13.62	12.69	7.89	4.32	7.06	9.47	4.26	7.26	11.94	11.69	111.63
PAAKEA 350		15.80	19.52	22.94	21.61	16.41	11.50	16.26	17.34	10.18	15.08	20.03	17.34	204.01
PAAUHAU 217		6.76	7.22	8.45	7.92	3.86	1.83	3.43	4.78	1.98	4.37	7.26	7.33	65.19
PAAUILO 221		8.96	10.50	12.07	11.02	5.64	2.65	4.74	6.95	2.35	5.21	9.64	10.19	89.92
PAHALA 21	A	7.69	6.47	6.43	4.08	2.96	1.22	1.33	3.26	2.71	4.21	7.30	4.85	52.51
PAHOA 65		13.21	14.76	15.72	14.28	11.64	8.13	10.21	10.57	8.48	11.60	16.36	15.86	150.82
PAIA 406		4.66	3.89	3.61	2.39	1.40	.85	1.46	1.45	.68	1.79	3.38	3.79	29.35
PALOLO VALLEY 718	R	10.33	11.11	14.10	12.14	10.66	7.55	10.92	9.58	6.62	10.48	12.56	12.92	128.97
PAPAIKOU 144.1		10.19	15.05	15.37	14.73	11.05	6.94	10.28	11.80	8.31	11.48	16.08	14.28	145.56
PAPAIKOU MAUKA 140.1		16.07	19.96	24.40	23.81	17.87	11.13	16.91	19.85	12.12	15.52	22.66	20.73	221.03
PAUOA FLATS 784	R	11.68	13.21	17.52	15.58	14.25	11.72	14.91	13.24	9.16	13.09	15.44	14.92	164.72
PH WAINIHA 1115		12.89	12.84	15.18	13.38	9.43	6.20	9.60	7.55	4.86	8.43	12.00	12.32	124.68
PRINCEVILLE RANCH 1117		8.03	7.14	10.39	8.17	7.22	4.14	6.81	5.70	4.39	6.49	9.75	8.12	86.35
PUEHU RIDGE 1040		6.30	3.69	3.63	2.49	2.02	.73	1.27	1.16	2.24	3.04	3.68	5.39	35.64
PUUNENE 396		4.43	2.85	2.83	1.09	.46	.19	.28	.32	.19	.82	2.08	2.79	18.33
SPRECKELSVILLE 400.2		4.24	3.38	3.16	1.55	.79	.31	.71	.70	.36	1.07	2.56	3.12	21.95
WAHIAWA 930	A	5.55	3.22	3.51	3.07	1.93	1.56	1.84	1.63	1.65	3.12	4.37	4.84	36.29

HAWAII

PRECIPITATION NORMALS (INCHES)

STATION		JAN	FEB	MAR	APR	MAY	JUN	JUL	AUG	SEP	OCT	NOV	DEC	ANN
WAIAHI LOWER 1054		10.55	8.87	10.79	10.66	9.65	7.29	9.81	9.27	7.38	9.94	12.68	11.50	118.39
WAIAHI UPPER 1052		13.06	10.68	13.24	13.63	11.93	9.10	11.89	11.09	8.91	11.61	15.35	13.21	143.70
WAIALUA 847	A	5.54	4.39	4.42	2.56	1.51	.66	.95	1.11	.78	2.46	3.46	4.40	32.24
WAIAWA 943		4.18	2.15	2.44	1.39	1.03	.33	.45	.47	.89	1.89	2.47	3.79	21.48
WAILUKU 386	A	5.83	4.28	3.80	1.89	.89	.43	.61	.68	.46	1.65	2.53	4.07	27.12
WAIMEA 947		4.36	2.52	2.35	1.63	1.13	.35	.45	.40	.74	1.99	2.71	3.99	22.62
WILHELMINA RISE 721	R	5.51	4.95	5.15	4.06	2.81	1.96	2.48	2.13	1.86	3.35	5.05	5.47	44.78

HAWAII

HEATING DEGREE DAY NORMALS (BASE 65 DEG F)

STATION	JUL	AUG	SEP	OCT	NOV	DEC	JAN	FEB	MAR	APR	MAY	JUN	ANN
HALEAKALA RS 338	270	267	288	310	345	412	456	428	462	405	366	264	4273
HAWAII VOLCNS NP HQ 54	81	66	68	81	133	195	220	207	217	180	152	110	1710
HILO WSO 87 AP R	0	0	0	0	0	0	0	0	0	0	0	0	0
HONOLULU WSFO 703 AP R	0	0	0	0	0	0	0	0	0	0	0	0	0
KAHULUI WSO 398 AP R	0	0	0	0	0	0	0	0	0	0	0	0	0
KAINALIU 73.2	0	0	0	0	0	0	6	9	5	0	0	0	20
KAMUELA 192.2	43	36	24	48	55	94	115	104	103	76	75	53	826
KANEOHE MAUKA 781	0	0	0	0	0	0	0	0	0	0	0	0	0
KILAUEA POINT 1133	0	0	0	0	0	0	0	0	0	0	0	0	0
KOHALA MISSION 175.1	0	0	0	0	0	0	0	0	0	0	0	0	0
KULA SANATORIUM 267	26	15	20	22	40	88	122	123	130	96	69	37	788
LAHAINA 361	0	0	0	0	0	0	0	0	0	0	0	0	0
LIHUE WSO 1020.1 AP //R	0	0	0	0	0	0	0	0	0	0	0	0	0
MANA 1026	0	0	0	0	0	0	0	0	0	0	0	0	0
MOUNTAIN VIEW 91	0	0	0	0	0	26	39	42	36	21	12	0	176
NIU RIDGE 1035	0	0	0	0	0	0	11	9	9	0	0	0	29
OOKALA 223	0	0	0	0	0	0	0	0	0	0	0	0	0
OPAEULA 870 A	0	0	0	0	7	22	29	40	36	15	8	0	157
WAIALUA 847 A	0	0	0	0	0	0	0	0	0	0	0	0	0

HAWAII

COOLING DEGREE DAY NORMALS (BASE 65 DEG F)

STATION		JAN	FEB	MAR	APR	MAY	JUN	JUL	AUG	SEP	OCT	NOV	DEC	ANN
HALEAKALA RS 338		0	0	0	0	0	0	0	0	0	0	0	0	0
HAWAII VOLCNS NP HQ 54		0	0	0	0	0	0	7	11	11	7	0	0	36
HILO WSO 87 AP	R	198	176	202	222	267	294	322	338	324	319	258	214	3134
HONOLULU WSFO 703 AP	R	236	221	291	321	388	423	468	496	468	450	348	279	4389
KAHULUI WSO 398 AP	R	202	185	248	276	335	369	419	440	411	397	315	254	3851
KAINALIU 73.2		83	82	95	107	137	153	180	205	201	202	156	108	1709
KAMUELA 192.2		10	6	16	19	50	50	55	83	84	79	37	14	503
KANEOHE MAUKA 781		189	167	198	225	285	324	360	375	357	347	273	229	3329
KILAUEA POINT 1133		229	187	211	240	304	351	388	409	414	391	309	257	3690
KOHALA MISSION 175.1		141	120	143	159	208	243	276	291	291	282	216	161	2531
KULA SANATORIUM 267		11	9	15	9	20	25	45	55	56	47	22	11	325
LAHAINA 361		220	190	236	267	322	351	394	419	399	388	315	257	3758
LIHUE WSO 1020.1 AP	//R	200	184	220	252	326	372	415	437	417	381	309	245	3758
MANA 1026		185	165	205	246	304	357	394	409	384	369	291	229	3538
MOUNTAIN VIEW 91		58	48	48	57	84	108	135	155	150	144	91	70	1148
NIU RIDGE 1035		94	85	118	146	200	240	267	295	288	282	204	135	2354
OOKALA 223		163	128	158	177	226	261	288	304	306	298	234	180	2723
OPAEULA 870	A	104	101	123	126	175	207	236	260	264	252	190	140	2178
WAIALUA 847	A	152	124	155	189	251	300	338	357	339	319	249	189	2962

51 — HAWAII

11 = TEMPERATURE ONLY
12 = PRECIPITATION ONLY
13 = TEMP. & PRECIP.

STATE-STATION NUMBER	STN TYP	NAME		LATITUDE DEG-MIN	LONGITUDE DEG-MIN	ELEVATION (FT)
51-0507	12	EWA PLANTATION 741		N 2121	W 15802	50
51-0840	12	HAINA 214		N 2006	W 15528	460
51-0905	12	HAKALAU 142		N 1954	W 15508	105
51-1004	13	HALEAKALA RS 338		N 2046	W 15615	7030
51-1086	12	HAMAKUAPOKO 485		N 2055	W 15621	320
51-1303	13	HAWAII VOLCNS NP HQ 54		N 1926	W 15516	3971
51-1492	13	HILO WSO 87 AP	R	N 1943	W 15504	27
51-1665	12	HONAUNAU 27		N 1927	W 15553	1089
51-1701	12	HONOHINA 137		N 1956	W 15509	300
51-1919	13	HONOLULU WSFO 703 AP	R	N 2120	W 15755	7
51-2161	12	HUKIPO 945		N 2159	W 15941	800
51-2572	13	KAHULUI WSO 398 AP	R	N 2054	W 15626	48
51-2679	12	KAILUA 446		N 2054	W 15613	700
51-2751	13	KAINALIU 73.2		N 1932	W 15556	1500
51-2960	12	KALIHI RES SITE 777	R	N 2123	W 15750	910
51-3077	13	KAMUELA 192.2		N 2001	W 15540	2670
51-3113	13	KANEOHE MAUKA 781		N 2125	W 15749	225
51-3910	12	KEAHUA 410		N 2052	W 15623	480
51-4272	12	KEKAHA 944		N 2158	W 15943	9
51-4489	12	KIHEI 311		N 2047	W 15627	90
51-4568	13	KILAUEA POINT 1133		N 2214	W 15924	180
51-4634	12	KIPAHULU 258		N 2039	W 15604	260
51-4680	13	KOHALA MISSION 175.1		N 2014	W 15548	537
51-4735	12	KOLO 1033		N 2204	W 15946	36
51-4815	12	KUKAIAU 222		N 2002	W 15521	840
51-4928	12	KUKUIHAELE HIC 199		N 2007	W 15535	980
51-5006	13	KULA SANATORIUM 267		N 2042	W 15622	3004
51-5177	13	LAHAINA 361		N 2053	W 15641	45
51-5286	12	LANAI CITY 672	A	N 2050	W 15655	1620
51-5580	13	LIHUE WSO 1020.1 AP	//R	N 2159	W 15921	103
51-6082	13	MANA 1026		N 2202	W 15946	11
51-6122	12	MANOA 712.1		N 2119	W 15749	220
51-6130	12	MANOA TUNNEL 2 716	R	N 2120	W 15748	650
51-6395	12	MOANALUA 770		N 2121	W 15754	20
51-6552	13	MOUNTAIN VIEW 91		N 1933	W 15507	1530
51-6850	13	NIU RIDGE 1035		N 2202	W 15944	1250
51-6928	12	NUUANU RES 4 783	R	N 2121	W 15749	1048
51-7059	12	OLOWALU 296		N 2049	W 15638	7
51-7131	13	OOKALA 223		N 2001	W 15517	430
51-7150	11	OPAEULA 870	A	N 2134	W 15802	1060
51-7194	12	PAAKEA 350		N 2049	W 15607	1260
51-7204	12	PAAUHAU 217		N 2005	W 15526	415
51-7312	12	PAAUILO 221		N 2003	W 15522	800
51-7421	12	PAHALA 21	A	N 1912	W 15529	870
51-7457	12	PAHOA 65		N 1930	W 15457	670
51-7566	12	PAIA 406		N 2055	W 15623	170
51-7664	12	PALOLO VALLEY 718	R	N 2120	W 15746	995
51-7711	12	PAPAIKOU 144.1		N 1947	W 15506	200
51-7721	12	PAPAIKOU MAUKA 140.1		N 1947	W 15508	1270
51-7810	12	PAUOA FLATS 784	R	N 2121	W 15748	1640

51 – HAWAII

LEGEND
11 = TEMPERATURE ONLY
12 = PRECIPITATION ONLY
13 = TEMP. & PRECIP.

STATE-STATION NUMBER	STN TYP	NAME		LATITUDE DEG-MIN	LONGITUDE DEG-MIN	ELEVATION (FT)
51-8155	12	PH WAINIHA 1115		N 2212	W 15934	101
51-8165	12	PRINCEVILLE RANCH 1117		N 2213	W 15929	295
51-8205	12	PUEHU RIDGE 1040		N 2201	W 15942	1660
51-8543	12	PUUNENE 396		N 2053	W 15627	60
51-8688	12	SPRECKELSVILLE 400.2		N 2054	W 15625	85
51-8941	12	WAHIAWA 930	A	N 2154	W 15934	215
51-8958	12	WAIAHI LOWER 1054		N 2201	W 15927	565
51-8966	12	WAIAHI UPPER 1052		N 2201	W 15928	780
51-9195	13	WAIALUA 847	A	N 2135	W 15807	32
51-9253	12	WAIAWA 943		N 2159	W 15943	10
51-9484	12	WAILUKU 386	A	N 2054	W 15630	180
51-9629	12	WAIMEA 947		N 2158	W 15941	20
51-9980	12	WILHELMINA RISE 721	R	N 2118	W 15747	1100

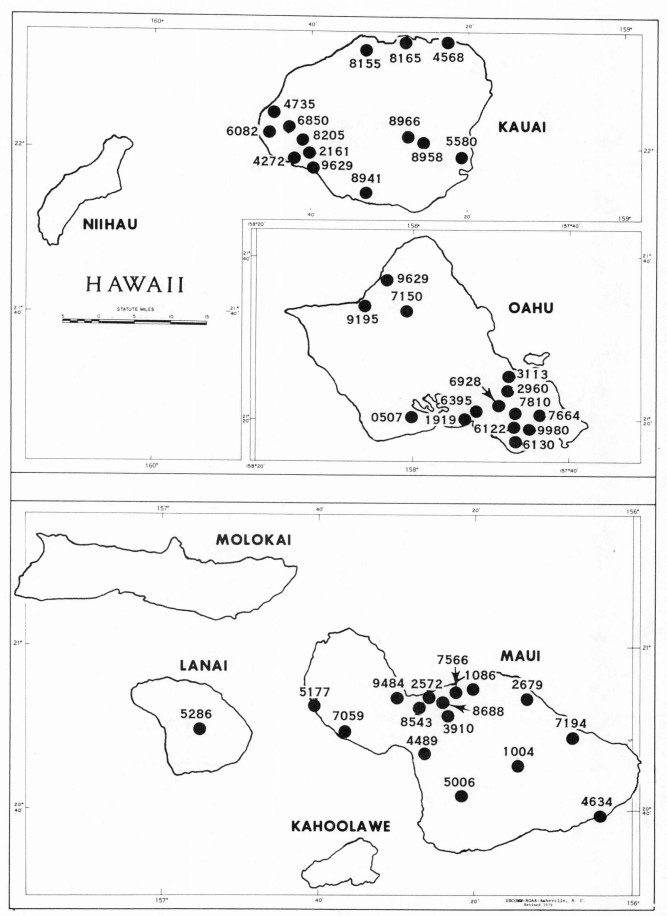

NIIHAU

HAWAII

STATUTE MILES

KAUAI

8155 8165 4568

4735
6850
6082
8205
4272
2161
9629
8941

8966
8958
5580

OAHU

9629
7150
9195

3113
6928
2960
6395
7810
0507
1919
7664
6122
9980
6130

MOLOKAI

LANAI

5286

MAUI

7566

5177
9484 2572 1086
2679
7059
8543
8688
3910
7194
4489
1004
5006
4634

KAHOOLAWE

USCOMM-NOAA-Asheville, N. C.
Revised 1979

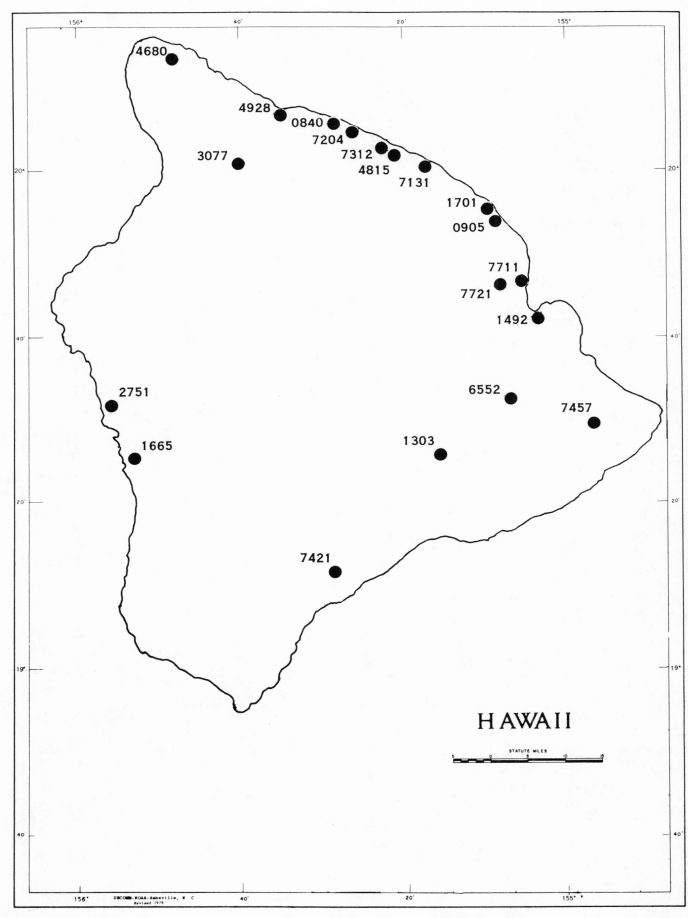

HAWAII

STATUTE MILES

IDAHO

TEMPERATURE NORMALS (DEG F)

STATION		JAN	FEB	MAR	APR	MAY	JUN	JUL	AUG	SEP	OCT	NOV	DEC	ANN
ABERDEEN EXP STATION	MAX	31.0	37.5	45.8	57.6	67.9	77.0	87.1	84.9	75.0	62.8	45.7	34.8	58.9
	MIN	12.2	17.3	22.7	29.5	37.8	44.7	49.8	47.3	38.5	29.2	21.7	15.6	30.5
	MEAN	21.6	27.5	34.2	43.6	52.8	60.9	68.5	66.1	56.8	46.0	33.7	25.2	44.7
AMERICAN FALLS 1 SW	MAX	33.3	39.1	46.9	58.2	68.1	77.9	87.8	85.7	76.3	63.1	46.1	36.2	59.9
	MIN	16.8	21.2	25.9	33.1	40.9	47.7	54.1	52.4	43.7	34.4	26.6	20.4	34.8
	MEAN	25.1	30.2	36.4	45.7	54.5	62.9	71.0	69.1	60.0	48.8	36.4	28.3	47.4
ANDERSON DAM	MAX	34.9	40.2	46.6	58.8	70.6	79.8	91.4	88.6	78.4	65.0	47.5	37.7	61.6
	MIN	19.7	22.1	26.0	34.3	42.3	49.3	56.4	55.0	47.1	38.7	29.9	22.9	37.0
	MEAN	27.3	31.2	36.3	46.6	56.5	64.6	73.9	71.8	62.8	51.9	38.8	30.3	49.3
ARCO 3 SW	MAX	28.5	34.8	42.5	56.0	66.9	76.0	85.3	82.9	73.9	61.7	43.1	31.6	56.9
	MIN	3.2	8.6	17.0	27.3	36.5	42.8	48.3	45.4	37.2	28.5	17.6	7.4	26.7
	MEAN	15.9	21.7	29.8	41.7	51.7	59.4	66.9	64.2	55.6	45.1	30.4	19.6	41.8
ARROWROCK DAM	MAX	33.6	40.6	48.3	58.5	68.9	78.2	90.1	88.0	77.5	63.4	45.7	36.3	60.8
	MIN	20.1	23.6	27.6	34.8	42.4	49.5	56.4	54.4	45.6	36.8	29.1	23.3	37.0
	MEAN	26.9	32.1	37.9	46.7	55.6	63.9	73.3	71.3	61.6	50.1	37.4	29.9	48.9
ASHTON 1 S	MAX	27.8	33.6	39.4	52.7	65.3	73.4	81.9	79.8	71.1	59.3	40.8	30.5	54.6
	MIN	8.9	12.3	16.4	27.5	35.9	42.1	46.4	44.0	36.8	28.7	19.9	11.3	27.5
	MEAN	18.4	23.0	27.9	40.1	50.6	57.8	64.2	61.9	54.0	44.1	30.4	20.9	41.1
AVERY R S NO 2	MAX	32.5	38.7	45.6	56.7	68.5	76.2	86.6	85.2	74.4	58.8	40.9	33.6	58.1
	MIN	21.6	25.1	27.2	32.5	39.0	44.7	47.5	47.0	41.8	35.2	28.9	24.7	34.6
	MEAN	27.1	31.9	36.4	44.6	53.8	60.5	67.1	66.1	58.1	47.0	35.0	29.2	46.4
BAYVIEW MODEL BASIN	MAX	34.2	39.0	44.6	53.8	63.6	71.6	80.0	78.6	68.4	55.6	42.7	36.9	55.8
	MIN	20.1	24.0	26.2	32.0	38.4	45.0	48.6	47.4	40.7	33.3	27.7	23.8	33.9
	MEAN	27.2	31.5	35.4	42.9	51.0	58.3	64.4	63.0	54.6	44.5	35.3	30.4	44.9
BLISS	MAX	37.0	44.4	52.7	63.3	72.9	81.7	92.1	89.5	80.2	67.8	50.4	39.6	64.3
	MIN	20.3	24.5	27.8	33.9	41.6	49.4	55.3	53.0	44.6	35.4	27.9	22.2	36.3
	MEAN	28.7	34.5	40.3	48.6	57.3	65.6	73.8	71.3	62.4	51.7	39.2	30.9	50.4
BOISE WSO //R	MAX	37.1	44.3	51.8	60.8	70.8	79.8	90.6	87.3	77.6	64.6	49.0	39.3	62.8
	MIN	22.6	27.9	30.9	36.4	44.0	51.8	58.5	56.7	48.7	39.1	30.5	24.6	39.3
	MEAN	29.9	36.1	41.4	48.6	57.4	65.8	74.6	72.0	63.2	51.9	39.7	32.0	51.1
BONNERS FERRY 1 SW	MAX	32.1	39.4	47.0	58.8	68.7	75.4	83.9	82.4	72.2	57.0	41.3	34.6	57.7
	MIN	19.0	24.0	26.8	33.3	40.3	46.8	49.8	48.6	41.5	34.0	27.2	23.3	34.6
	MEAN	25.6	31.7	36.9	46.1	54.5	61.1	66.8	65.5	56.9	45.6	34.3	29.0	46.2
BURLEY FAA AIRPORT	MAX	35.5	42.2	49.1	59.0	68.6	77.7	87.6	85.2	76.0	64.3	48.1	38.4	61.0
	MIN	17.8	22.8	26.0	32.3	40.4	47.4	53.5	50.8	42.1	32.6	25.2	20.1	34.3
	MEAN	26.7	32.5	37.6	45.7	54.5	62.6	70.6	68.0	59.1	48.5	36.7	29.3	47.7
CALDWELL	MAX	38.2	46.7	56.2	65.7	75.1	83.7	93.0	90.2	80.3	66.2	50.4	40.3	65.5
	MIN	22.1	26.9	30.4	36.6	44.6	51.5	56.8	54.0	45.2	36.2	28.8	24.2	38.1
	MEAN	30.2	36.9	43.3	51.2	59.9	67.6	74.9	72.2	62.7	51.2	39.6	32.3	51.8
CAMBRIDGE	MAX	32.5	39.6	50.0	62.3	72.2	80.5	91.5	89.1	79.4	65.0	46.9	35.5	62.0
	MIN	14.2	18.9	25.7	32.7	39.5	46.5	52.2	48.9	39.7	31.2	24.4	17.2	32.6
	MEAN	23.4	29.3	37.9	47.5	55.9	63.5	71.9	69.0	59.6	48.1	35.7	26.4	47.4
CASCADE 1 NW	MAX	29.8	35.8	41.3	51.4	62.7	71.5	82.8	80.6	71.1	58.0	40.8	31.7	54.8
	MIN	12.4	15.1	18.3	27.0	34.2	40.7	45.3	43.0	35.8	29.1	22.7	15.8	28.3
	MEAN	21.1	25.5	29.9	39.2	48.5	56.1	64.0	61.8	53.5	43.6	31.8	23.8	41.6
CHALLIS	MAX	30.4	38.1	45.9	57.3	67.2	76.0	86.6	83.7	74.4	61.8	43.6	32.8	58.2
	MIN	10.4	16.2	21.8	30.4	38.5	45.4	50.9	48.7	40.9	31.7	21.7	13.5	30.8
	MEAN	20.4	27.1	33.8	43.9	52.9	60.7	68.8	66.2	57.7	46.7	32.7	23.2	44.5
CHILLY BARTON FLAT	MAX	28.6	34.2	40.1	52.0	62.8	71.7	81.6	79.3	70.8	59.1	41.7	31.0	54.4
	MIN	1.7	6.7	13.8	23.9	31.6	38.3	42.4	39.8	32.1	24.0	14.8	5.3	22.9
	MEAN	15.2	20.4	27.0	38.0	47.2	55.0	62.0	59.6	51.5	41.6	28.3	18.2	38.7

IDAHO

TEMPERATURE NORMALS (DEG F)

STATION		JAN	FEB	MAR	APR	MAY	JUN	JUL	AUG	SEP	OCT	NOV	DEC	ANN
COEUR D'ALENE 1 E	MAX	34.6	41.8	47.6	58.2	68.6	75.9	86.0	84.9	75.6	61.4	44.8	37.4	59.7
	MIN	21.8	25.8	27.7	33.9	41.3	48.0	52.4	51.9	45.0	37.3	30.3	26.3	36.8
	MEAN	28.2	33.8	37.7	46.1	55.0	62.0	69.2	68.5	60.3	49.4	37.6	31.8	48.3
COUNCIL	MAX	33.2	40.4	49.1	61.1	71.6	80.3	91.4	89.3	79.6	65.2	47.5	36.3	62.1
	MIN	17.0	21.0	26.6	34.0	41.3	48.3	55.3	53.3	43.8	34.4	26.6	20.0	35.1
	MEAN	25.1	30.7	37.9	47.6	56.4	64.3	73.4	71.3	61.7	49.8	37.1	28.1	48.6
DEER FLAT DAM	MAX	37.8	45.6	54.4	63.5	72.0	79.7	88.4	86.0	77.6	65.4	49.9	39.9	63.4
	MIN	22.0	27.0	30.8	37.2	44.9	51.7	57.4	55.1	46.8	37.4	29.6	24.5	38.7
	MEAN	29.9	36.4	42.6	50.4	58.5	65.8	72.9	70.6	62.2	51.4	39.8	32.2	51.1
DRIGGS	MAX	29.3	34.6	39.6	51.5	62.7	71.7	81.7	79.9	70.9	59.9	41.6	32.3	54.6
	MIN	6.4	10.0	15.0	25.4	33.9	40.9	47.3	44.6	36.7	27.6	17.3	8.8	26.2
	MEAN	17.9	22.3	27.3	38.5	48.4	56.3	64.5	62.2	53.8	43.8	29.5	20.6	40.4
DUBOIS EXP STATION	MAX	25.7	31.2	37.6	52.4	64.4	73.6	85.0	82.6	72.2	58.2	39.3	29.0	54.3
	MIN	10.4	14.6	19.0	29.2	38.4	45.4	52.7	50.6	42.3	33.1	22.1	13.5	30.9
	MEAN	18.1	23.0	28.3	40.8	51.4	59.5	68.9	66.6	57.3	45.7	30.7	21.3	42.6
EMMETT 2 E	MAX	38.5	46.4	55.1	65.0	74.7	83.3	92.7	89.7	80.6	67.1	50.6	40.5	65.4
	MIN	22.3	26.8	30.3	35.6	42.4	48.8	53.7	51.6	44.2	35.6	28.8	24.2	37.0
	MEAN	30.4	36.6	42.7	50.3	58.6	66.1	73.3	70.7	62.4	51.4	39.7	32.4	51.2
FAIRFIELD RANGER STA	MAX	29.0	34.7	40.3	53.9	66.1	74.7	85.4	83.3	74.9	63.0	44.4	32.7	56.9
	MIN	5.8	9.2	15.3	27.3	34.8	40.6	46.0	43.8	35.6	26.9	18.6	9.4	26.1
	MEAN	17.4	22.0	27.8	40.6	50.5	57.7	65.7	63.6	55.2	45.0	31.5	21.1	41.5
FENN RANGER STATION	MAX	35.5	43.1	49.7	60.6	70.0	77.8	89.3	88.6	76.4	61.0	45.8	37.6	61.3
	MIN	22.8	27.1	29.5	34.9	41.4	47.4	51.2	49.6	44.2	36.8	30.3	26.0	36.8
	MEAN	29.1	35.1	39.7	47.8	55.7	62.7	70.3	69.2	60.3	48.9	38.0	31.8	49.1
FORT HALL INDIAN AGNCY	MAX	33.3	39.9	47.7	58.8	69.3	78.4	88.0	85.7	76.7	64.6	46.7	36.2	60.4
	MIN	14.1	19.3	24.0	30.8	39.4	46.4	52.1	49.7	41.3	31.9	23.5	16.7	32.4
	MEAN	23.7	29.6	35.9	44.9	54.4	62.4	70.1	67.8	59.0	48.3	35.2	26.5	46.5
GARDEN VALLEY RS	MAX	34.0	41.8	49.9	61.0	72.0	80.6	92.2	90.1	80.8	65.6	45.1	34.8	62.3
	MIN	17.5	21.0	24.6	30.4	37.6	44.2	48.1	46.5	39.5	32.1	25.6	19.9	32.3
	MEAN	25.8	31.4	37.3	45.7	54.8	62.4	70.2	68.4	60.2	48.9	35.4	27.4	47.3
GLENNS FERRY	MAX	40.2	48.0	56.1	66.3	76.2	84.9	95.3	92.1	82.2	69.6	52.7	42.2	67.2
	MIN	21.9	25.6	28.6	34.9	42.7	50.3	55.8	52.3	43.6	33.9	27.1	22.6	36.6
	MEAN	31.1	36.8	42.3	50.6	59.5	67.6	75.6	72.2	62.9	51.8	39.9	32.4	51.9
GRACE	MAX	29.7	34.7	41.1	53.7	65.2	73.9	83.4	81.5	73.3	61.2	43.1	32.6	56.1
	MIN	11.0	13.6	18.5	28.0	36.0	42.3	48.0	46.4	38.3	29.9	21.2	13.5	28.9
	MEAN	20.4	24.1	29.8	40.9	50.6	58.1	65.7	64.0	55.8	45.5	32.2	23.1	42.5
GRAND VIEW 2 W	MAX	40.5	49.0	57.6	67.5	76.9	85.6	95.4	92.5	82.8	68.9	52.5	42.2	67.6
	MIN	20.5	24.9	28.2	35.2	43.7	51.2	56.3	53.0	43.5	34.1	26.5	21.7	36.6
	MEAN	30.5	37.0	43.0	51.4	60.4	68.4	75.9	72.7	63.2	51.5	39.5	31.9	52.1
GRANGEVILLE	MAX	36.5	42.3	46.5	54.9	63.5	71.4	82.3	81.5	71.9	59.3	44.8	38.8	57.8
	MIN	20.1	24.5	26.1	31.9	38.4	44.9	50.0	48.7	41.7	34.0	26.8	22.8	34.2
	MEAN	28.3	33.5	36.3	43.4	51.0	58.2	66.2	65.1	56.8	46.7	35.8	30.9	46.0
GROUSE	MAX	28.9	34.6	39.8	50.5	61.5	69.3	79.6	78.3	70.0	58.4	41.7	31.6	53.7
	MIN	-1.9	1.7	9.3	22.3	30.6	36.8	39.9	37.9	30.4	21.8	12.2	1.7	20.2
	MEAN	13.5	18.1	24.6	36.4	46.1	53.1	59.8	58.1	50.2	40.1	27.0	16.7	37.0
HAILEY AIRPORT	MAX	30.8	36.7	42.3	54.2	65.2	74.1	84.9	83.4	74.2	61.8	44.1	33.4	57.1
	MIN	8.3	12.5	18.5	28.4	36.5	42.8	49.1	47.2	39.2	31.0	20.9	11.2	28.8
	MEAN	19.5	24.6	30.5	41.3	50.9	58.5	67.0	65.3	56.7	46.4	32.5	22.3	43.0
HAMER 4 NW	MAX	27.9	34.8	43.9	58.5	69.6	78.5	88.5	85.4	75.7	62.6	43.6	31.4	58.4
	MIN	4.4	9.4	17.3	26.8	36.2	43.2	48.3	45.6	36.7	26.6	16.7	7.4	26.6
	MEAN	16.2	22.1	30.6	42.7	53.0	60.9	68.4	65.5	56.2	44.7	30.1	19.5	42.5

IDAHO

TEMPERATURE NORMALS (DEG F)

STATION		JAN	FEB	MAR	APR	MAY	JUN	JUL	AUG	SEP	OCT	NOV	DEC	ANN
HAZELTON	MAX	35.7	42.7	50.1	60.7	70.0	78.9	88.9	86.5	77.5	65.6	48.4	38.7	62.0
	MIN	18.6	23.2	26.6	32.5	40.8	47.7	54.1	51.6	42.8	33.6	26.1	20.7	34.9
	MEAN	27.2	33.0	38.4	46.6	55.4	63.3	71.5	69.1	60.1	49.6	37.3	29.8	48.4
HILL CITY	MAX	29.2	34.1	39.4	53.4	66.0	74.6	86.6	84.8	75.5	62.7	43.8	32.8	56.9
	MIN	6.5	9.7	15.5	27.4	34.7	39.9	44.9	43.3	35.1	26.7	19.1	10.0	26.1
	MEAN	17.9	21.9	27.5	40.4	50.4	57.3	65.8	64.1	55.4	44.7	31.5	21.4	41.5
HOLLISTER	MAX	37.5	43.4	49.1	59.0	68.0	77.0	87.3	84.6	75.2	63.9	48.5	40.2	61.1
	MIN	19.2	23.7	25.8	31.5	39.1	46.3	54.3	52.4	44.2	35.0	26.3	21.3	34.9
	MEAN	28.3	33.6	37.5	45.3	53.6	61.7	70.8	68.5	59.7	49.5	37.5	30.8	48.1
IDAHO CITY	MAX	34.7	41.5	47.1	57.9	68.2	76.5	87.4	85.3	76.5	63.9	46.1	36.5	60.1
	MIN	13.8	17.0	20.5	27.7	34.5	40.6	45.0	43.2	36.1	29.1	22.8	16.3	28.9
	MEAN	24.3	29.3	33.8	42.8	51.4	58.6	66.2	64.3	56.3	46.6	34.5	26.5	44.6
IDAHO FALLS FAA AP	MAX	27.0	33.6	42.6	55.8	66.9	75.9	86.4	83.8	73.6	60.5	42.8	30.8	56.6
	MIN	10.3	15.4	22.2	30.7	38.9	45.6	51.3	49.1	40.5	31.1	22.1	13.5	30.9
	MEAN	18.7	24.5	32.4	43.2	52.9	60.8	68.9	66.5	57.1	45.8	32.5	22.2	43.8
IDAHO FALLS 46 W	MAX	27.8	34.2	42.3	55.4	66.4	76.1	87.2	84.5	73.8	61.3	42.7	31.2	56.9
	MIN	4.6	9.3	17.8	27.3	36.1	43.7	49.2	46.9	37.3	26.5	17.1	7.6	26.9
	MEAN	16.2	21.8	30.1	41.4	51.3	59.9	68.2	65.7	55.6	43.9	29.9	19.4	41.9
ISLAND PARK DAM	MAX	25.2	31.7	36.8	47.1	59.4	68.7	79.0	77.1	67.7	55.5	37.0	27.2	51.0
	MIN	2.4	3.8	6.5	19.5	30.2	37.6	42.7	40.9	32.6	24.7	14.4	4.3	21.6
	MEAN	13.8	17.8	21.7	33.4	44.8	53.2	60.8	59.0	50.2	40.1	25.7	15.8	36.4
JEROME	MAX	36.1	43.2	51.1	61.7	71.7	81.1	91.3	88.7	78.9	66.4	49.4	39.0	63.2
	MIN	19.0	23.3	26.7	33.2	41.6	49.1	55.9	53.4	44.9	35.8	27.3	21.5	36.0
	MEAN	27.6	33.2	38.9	47.5	56.7	65.2	73.6	71.1	61.9	51.1	38.4	30.3	49.6
KELLOGG	MAX	34.4	41.4	47.0	57.4	67.6	75.3	85.3	83.9	74.2	60.2	44.1	36.6	59.0
	MIN	20.5	25.3	27.7	33.9	40.8	47.0	50.4	48.9	42.9	35.2	28.7	24.2	35.5
	MEAN	27.5	33.3	37.4	45.7	54.2	61.2	67.9	66.4	58.5	47.7	36.4	30.4	47.2
KOOSKIA	MAX	37.4	46.5	54.3	64.0	73.4	81.0	91.4	89.9	80.1	64.2	46.7	39.2	64.0
	MIN	22.6	27.4	29.7	35.2	42.0	48.1	51.3	49.5	43.1	35.1	29.2	25.6	36.6
	MEAN	30.0	37.0	42.0	49.7	57.7	64.6	71.4	69.8	61.6	49.7	38.0	32.4	50.3
KUNA 2 NNE	MAX	37.4	45.6	54.5	63.7	72.0	79.6	88.4	85.7	77.2	65.3	49.8	39.8	63.3
	MIN	21.0	25.6	28.0	33.5	40.9	47.6	52.4	50.7	43.3	35.1	27.8	23.1	35.8
	MEAN	29.2	35.6	41.3	48.6	56.5	63.6	70.4	68.3	60.2	50.2	38.8	31.5	49.5
LEWISTON WSO //R	MAX	38.6	46.3	52.7	61.6	70.7	79.0	89.5	87.3	77.6	63.1	47.6	41.5	63.0
	MIN	25.6	30.6	33.0	38.7	45.9	52.8	58.6	57.4	49.6	40.4	32.6	28.2	41.1
	MEAN	32.1	38.5	42.9	50.2	58.3	65.9	74.1	72.4	63.6	51.8	40.1	34.9	52.1
LIFTON PUMPING STATION	MAX	29.4	32.7	38.4	50.0	62.4	71.9	81.7	79.3	70.0	57.4	41.5	32.2	53.9
	MIN	6.9	7.1	14.0	27.8	38.5	45.8	51.4	47.7	38.3	29.0	19.6	11.1	28.1
	MEAN	18.2	19.9	26.2	38.9	50.5	58.9	66.6	63.5	54.2	43.2	30.6	21.7	41.0
MACKAY RANGER STATION	MAX	28.9	34.8	40.8	53.5	64.3	74.0	84.4	82.0	73.3	60.7	42.4	31.1	55.9
	MIN	5.8	10.9	17.5	27.5	35.7	42.6	48.3	45.9	37.8	29.7	19.1	8.6	27.5
	MEAN	17.3	22.9	29.2	40.5	50.1	58.3	66.3	64.0	55.6	45.2	30.7	19.9	41.7
MALAD	MAX	32.9	38.4	46.4	57.8	68.6	78.1	88.3	85.9	76.7	64.1	46.1	35.3	59.9
	MIN	14.6	18.7	24.0	32.0	39.7	46.3	53.1	51.4	42.4	33.4	24.4	17.2	33.1
	MEAN	23.8	28.6	35.2	44.9	54.2	62.2	70.7	68.7	59.6	48.8	35.3	26.3	46.5
MALAD CITY FAA AP	MAX	32.6	38.5	47.2	58.7	69.9	79.4	90.1	87.6	78.1	65.2	46.8	35.1	60.8
	MIN	10.7	15.3	21.9	29.8	37.4	43.6	49.8	48.2	38.9	29.5	21.6	14.0	30.1
	MEAN	21.6	26.9	34.6	44.3	53.6	61.5	70.0	67.9	58.5	47.4	34.2	24.5	45.4
MAY RANGER STATION	MAX	31.0	39.2	46.6	57.7	67.5	76.1	86.3	84.1	75.1	62.7	44.5	33.4	58.7
	MIN	6.9	12.9	19.3	27.0	34.6	40.9	45.3	43.6	35.9	27.5	18.1	10.3	26.9
	MEAN	19.0	26.1	33.0	42.4	51.1	58.5	65.8	63.9	55.5	45.1	31.3	21.9	42.8

IDAHO

TEMPERATURE NORMALS (DEG F)

STATION		JAN	FEB	MAR	APR	MAY	JUN	JUL	AUG	SEP	OCT	NOV	DEC	ANN
MC CALL	MAX	29.8	35.3	39.8	48.9	60.4	69.7	81.0	79.1	69.7	57.3	40.8	31.7	53.6
	MIN	12.2	14.3	16.8	25.4	33.7	40.3	44.8	42.5	35.9	29.1	22.5	16.0	27.8
	MEAN	21.0	24.9	28.3	37.2	47.1	55.1	62.9	60.8	52.8	43.2	31.7	23.9	40.7
MINIDOKA DAM	MAX	34.3	40.6	48.1	58.5	68.6	78.3	88.4	86.0	76.8	64.0	47.1	37.1	60.7
	MIN	16.7	21.5	26.0	33.1	41.3	48.9	56.3	54.0	45.3	35.8	27.0	20.4	35.5
	MEAN	25.5	31.0	37.1	45.8	55.0	63.6	72.4	70.1	61.0	49.9	37.1	28.8	48.1
MONTPELIER RANGER STA	MAX	29.8	33.6	39.2	51.5	64.1	73.5	84.6	82.7	73.0	60.6	42.4	32.4	55.6
	MIN	6.4	8.0	14.1	25.7	34.5	40.9	47.0	44.7	35.7	26.7	17.3	9.4	25.9
	MEAN	18.1	20.8	26.7	38.6	49.3	57.2	65.8	63.7	54.4	43.7	29.8	20.9	40.8
MOSCOW U OF I	MAX	34.6	41.2	46.8	56.6	65.9	73.7	83.9	82.7	74.2	60.7	44.5	37.2	58.5
	MIN	22.0	27.4	29.8	34.6	40.5	45.6	48.8	48.6	43.7	36.9	30.1	25.6	36.1
	MEAN	28.3	34.3	38.3	45.6	53.2	59.7	66.4	65.7	59.0	48.8	37.3	31.4	47.3
MOUNTAIN HOME	MAX	38.7	45.4	52.9	62.9	73.1	82.7	93.5	91.0	81.2	67.8	51.0	41.0	65.1
	MIN	20.0	24.4	27.2	33.3	41.0	48.8	55.5	52.9	44.0	34.7	26.8	21.9	35.9
	MEAN	29.4	34.9	40.1	48.1	57.1	65.8	74.5	72.0	62.6	51.3	38.9	31.5	50.5
NEW MEADOWS RANGER STA	MAX	30.5	37.5	43.7	54.0	64.8	73.8	84.9	82.5	73.0	59.9	43.2	32.5	56.7
	MIN	8.4	12.1	16.4	26.4	33.0	39.3	42.4	40.0	32.2	24.8	19.6	11.6	25.5
	MEAN	19.5	24.8	30.1	40.2	49.0	56.6	63.7	61.3	52.6	42.4	31.4	22.1	41.1
NEZPERCE	MAX	33.7	40.3	45.2	54.2	63.3	70.7	81.2	80.6	71.2	58.0	43.1	36.4	56.5
	MIN	20.0	25.1	27.1	32.8	39.1	45.1	49.2	48.5	42.5	35.2	27.7	23.5	34.7
	MEAN	26.9	32.7	36.2	43.5	51.3	58.0	65.3	64.6	56.9	46.6	35.4	30.0	45.6
OAKLEY	MAX	38.2	44.2	49.7	58.9	68.0	77.0	85.9	84.1	75.5	64.9	49.3	40.6	61.4
	MIN	19.5	24.1	26.7	32.6	40.2	47.3	54.8	53.0	44.7	35.9	27.6	21.7	35.7
	MEAN	28.9	34.2	38.2	45.7	54.1	62.2	70.4	68.5	60.1	50.4	38.5	31.2	48.5
OLA 4 S	MAX	35.6	43.4	52.0	62.4	72.7	80.9	91.4	89.2	80.3	66.3	48.6	38.3	63.4
	MIN	17.0	21.9	25.7	31.4	38.3	44.9	50.4	47.6	39.7	31.2	24.6	19.9	32.7
	MEAN	26.3	32.7	38.9	46.9	55.5	62.9	70.9	68.4	60.0	48.8	36.6	29.2	48.1
OROFINO	MAX	38.1	47.2	54.6	64.5	73.9	81.8	92.0	90.4	80.3	64.4	47.9	40.1	64.6
	MIN	24.9	29.2	31.4	37.0	43.9	50.0	53.9	52.7	45.7	37.8	31.8	27.9	38.9
	MEAN	31.5	38.3	43.0	50.8	58.9	66.0	73.0	71.6	63.0	51.1	39.9	34.0	51.8
PALISADES DAM	MAX	29.0	34.4	41.1	53.0	65.3	74.4	84.1	81.6	72.8	60.7	42.2	31.6	55.9
	MIN	12.5	14.9	18.9	28.9	37.7	44.6	51.2	49.7	41.6	33.4	24.2	16.6	31.2
	MEAN	20.8	24.7	30.0	41.0	51.6	59.5	67.7	65.6	57.2	47.1	33.3	24.1	43.6
PARMA EXPERIMENT STA	MAX	36.6	44.7	54.8	64.2	73.5	82.1	92.5	89.8	79.9	66.2	49.7	39.1	64.4
	MIN	19.9	24.7	28.3	35.0	42.7	49.1	54.0	51.5	42.8	33.5	26.6	22.4	35.9
	MEAN	28.3	34.7	41.6	49.6	58.1	65.6	73.3	70.7	61.4	49.9	38.2	30.8	50.2
PAUL 1 ENE	MAX	35.4	41.6	48.7	58.3	68.1	77.2	87.5	85.6	76.1	64.3	48.3	38.7	60.8
	MIN	16.8	21.0	24.7	31.4	39.6	46.5	52.5	49.7	40.9	31.9	24.7	19.7	33.3
	MEAN	26.1	31.3	36.7	44.9	53.9	61.9	70.0	67.6	58.6	48.1	36.5	29.2	47.1
PAYETTE	MAX	37.0	45.6	56.1	66.2	75.3	83.6	93.2	90.6	81.0	66.9	50.0	39.1	65.4
	MIN	20.7	25.3	29.6	35.7	43.4	50.4	55.8	53.3	44.5	35.0	27.5	22.7	37.0
	MEAN	28.9	35.5	42.8	51.0	59.4	67.0	74.5	72.0	62.7	51.0	38.8	30.9	51.2
POCATELLO WSO //R	MAX	32.4	38.6	45.8	56.8	67.7	77.6	88.6	86.0	75.7	62.8	45.6	35.3	59.4
	MIN	15.1	20.4	25.2	32.3	40.3	47.3	53.8	51.7	42.7	33.3	24.8	17.9	33.7
	MEAN	23.8	29.5	35.5	44.6	54.0	62.5	71.2	68.9	59.2	48.1	35.2	26.6	46.6
PORTHILL	MAX	31.6	38.6	45.7	58.2	68.5	74.7	82.9	81.4	71.3	56.5	41.3	34.5	57.1
	MIN	16.4	21.4	25.3	32.9	40.0	46.6	49.9	48.1	40.7	33.0	25.9	21.3	33.5
	MEAN	24.0	30.1	35.5	45.5	54.3	60.7	66.4	64.7	56.0	44.8	33.6	27.9	45.3
POTLATCH 3 NNE	MAX	35.4	42.2	47.0	56.4	66.0	73.3	83.2	82.0	73.0	59.9	44.6	38.0	58.4
	MIN	20.6	26.1	27.7	32.5	37.9	42.9	45.1	43.5	38.4	32.7	28.5	24.4	33.4
	MEAN	28.1	34.2	37.4	44.5	52.0	58.1	64.2	62.8	55.7	46.3	36.6	31.2	45.9

IDAHO

TEMPERATURE NORMALS (DEG F)

STATION		JAN	FEB	MAR	APR	MAY	JUN	JUL	AUG	SEP	OCT	NOV	DEC	ANN
PRIEST RIVER EXP STA	MAX	29.9	37.2	44.3	55.7	66.5	73.5	82.4	81.1	71.3	56.3	38.6	31.8	55.7
	MIN	18.4	21.9	24.0	30.1	37.8	44.1	46.7	45.5	39.4	32.8	26.5	22.7	32.5
	MEAN	24.2	29.6	34.2	42.9	52.2	58.8	64.6	63.3	55.4	44.6	32.6	27.3	44.1
RICHFIELD	MAX	30.8	37.3	45.8	58.8	68.6	76.9	86.6	84.1	75.2	62.9	45.3	34.4	58.9
	MIN	13.4	17.4	22.6	29.6	38.1	45.6	52.4	50.0	41.4	32.0	23.7	16.7	31.9
	MEAN	22.1	27.4	34.2	44.3	53.4	61.3	69.5	67.1	58.3	47.5	34.5	25.6	45.4
RIGGINS RANGER STATION	MAX	41.4	49.4	55.9	65.2	73.7	82.1	93.2	92.0	82.0	67.6	51.2	43.0	66.4
	MIN	27.5	31.3	33.7	38.6	45.9	52.6	59.1	58.3	50.8	42.3	34.0	30.0	42.0
	MEAN	34.5	40.4	44.8	51.9	59.8	67.4	76.1	75.2	66.5	55.0	42.6	36.5	54.2
ST ANTHONY 1 WNW	MAX	28.7	34.3	41.3	54.8	66.8	74.5	83.7	81.6	72.6	61.0	42.6	31.6	56.1
	MIN	9.2	12.8	18.0	28.2	36.3	42.7	47.7	45.5	37.7	29.3	20.9	12.0	28.4
	MEAN	19.0	23.6	29.7	41.5	51.6	58.6	65.8	63.6	55.2	45.2	31.8	21.8	42.3
SAINT MARIES	MAX	34.5	42.4	48.4	58.6	68.4	76.2	86.0	84.6	75.8	61.6	44.3	36.8	59.8
	MIN	21.1	25.7	27.4	32.8	40.0	46.1	48.9	47.4	41.2	34.5	28.8	24.8	34.9
	MEAN	27.9	34.1	38.0	45.7	54.2	61.2	67.5	66.0	58.5	48.1	36.5	30.9	47.4
SALMON 1 N	MAX	30.1	38.8	48.2	60.0	69.7	77.8	88.5	85.8	75.4	61.8	43.8	33.2	59.4
	MIN	9.4	16.6	22.6	30.3	38.1	44.8	49.1	47.0	38.8	29.5	21.3	14.1	30.1
	MEAN	19.8	27.7	35.4	45.2	54.0	61.3	68.8	66.4	57.2	45.7	32.6	23.7	44.8
SANDPOINT EXP STATION	MAX	31.5	38.3	44.8	56.0	65.9	73.0	81.3	79.9	69.9	56.1	40.9	34.1	56.0
	MIN	20.2	24.4	26.9	33.4	40.1	46.2	48.5	47.4	41.5	34.1	28.1	24.0	34.6
	MEAN	25.9	31.4	35.9	44.7	53.0	59.6	64.9	63.7	55.7	45.1	34.6	29.1	45.3
SHOSHONE 1 WNW	MAX	33.2	40.3	49.1	61.4	72.4	82.0	92.8	89.6	78.9	65.1	46.7	36.1	62.3
	MIN	16.1	20.5	25.0	32.2	40.8	48.6	56.3	53.9	44.7	34.8	25.7	19.1	34.8
	MEAN	24.7	30.4	37.1	46.8	56.7	65.3	74.6	71.8	61.8	50.0	36.2	27.6	48.6
SWAN FALLS POWER HOUSE	MAX	40.9	49.1	57.7	67.4	77.4	86.2	96.8	93.9	84.1	70.1	52.9	42.6	68.3
	MIN	25.5	30.0	33.8	40.4	48.6	56.2	63.7	61.2	51.9	42.1	33.1	27.2	42.8
	MEAN	33.2	39.6	45.8	53.9	63.0	71.3	80.3	77.6	68.0	56.1	43.0	35.0	55.6
WALLACE WOODLAND PARK	MAX	32.7	39.3	43.9	53.6	63.5	71.0	81.4	80.1	70.6	58.1	42.2	35.1	56.0
	MIN	17.9	22.1	23.7	31.0	37.6	43.7	46.9	46.2	39.8	33.3	26.6	21.9	32.6
	MEAN	25.3	30.7	33.8	42.3	50.6	57.4	64.2	63.2	55.3	45.8	34.5	28.5	44.3
WEISER 2 SE	MAX	36.3	44.8	55.2	64.7	74.3	82.9	92.7	89.9	80.2	66.5	49.8	39.1	64.7
	MIN	20.0	24.5	29.1	35.1	42.5	49.5	54.5	51.5	42.9	33.8	27.4	22.4	36.1
	MEAN	28.2	34.7	42.2	49.9	58.5	66.2	73.6	70.7	61.6	50.2	38.6	30.7	50.4

IDAHO

PRECIPITATION NORMALS (INCHES)

STATION	JAN	FEB	MAR	APR	MAY	JUN	JUL	AUG	SEP	OCT	NOV	DEC	ANN
ABERDEEN EXP STATION	.81	.62	.62	.83	1.15	.97	.36	.56	.59	.79	.66	.85	8.81
AMERICAN FALLS 1 SW	1.07	.78	.79	1.10	1.37	1.00	.47	.64	.62	.91	.89	.78	10.42
ANDERSON DAM	3.80	2.41	1.81	1.29	1.22	1.21	.41	.56	.82	1.17	2.64	3.43	20.77
ARCO 3 SW	1.05	.79	.64	.72	1.22	1.34	.55	.75	.67	.45	.68	1.06	9.92
ARROWROCK DAM	3.26	2.14	1.75	1.44	1.32	1.18	.26	.43	.75	1.18	2.58	2.87	19.16
ASHTON 1 S	2.22	1.85	1.47	1.42	1.91	1.90	.79	1.22	1.14	1.31	1.73	2.15	19.11
AVERY R S NO 2	5.22	3.43	3.02	2.60	2.73	2.41	1.09	1.71	1.98	2.67	3.85	4.63	35.34
BAYVIEW MODEL BASIN	3.30	2.31	1.80	1.58	1.92	1.83	.82	1.23	1.28	2.04	2.72	3.37	24.20
BLACKFOOT 2 SSW —	1.09	.93	.85	.99	1.20	1.19	.45	.56	.70	.80	.93	1.01	10.70
BLISS	1.47	.91	.74	.70	.91	.83	.20	.31	.47	.55	1.14	1.32	9.55
BOISE LUCKY PEAK DAM	1.95	1.27	1.26	1.39	1.32	1.11	.23	.50	.61	.86	1.63	1.68	13.81
BOISE WSO //R	1.64	1.07	1.03	1.19	1.21	.95	.26	.40	.58	.75	1.29	1.34	11.71
BONNERS FERRY 1 SW	3.57	2.06	1.64	1.37	1.63	1.59	.84	1.17	1.40	1.84	3.39	3.59	24.09
BURLEY FAA AIRPORT	1.23	.76	.75	.93	1.18	.95	.37	.53	.52	.59	.88	1.00	9.69
CALDWELL	1.54	1.06	.91	.96	.96	.90	.22	.40	.63	.79	1.19	1.34	10.90
CAMBRIDGE	3.45	2.19	1.81	1.27	1.36	1.29	.29	.55	.76	1.39	2.46	3.21	20.03
CASCADE 1 NW	3.06	1.96	1.88	1.51	1.62	1.83	.39	.80	1.08	1.74	2.39	3.17	21.43
CENTERVILLE ARBAUGH RCH	4.83	3.02	2.66	1.90	1.86	1.83	.45	.83	1.09	1.82	3.23	4.27	27.79
CHALLIS	.55	.36	.40	.58	1.11	1.17	.54	.58	.67	.38	.42	.61	7.37
CHILLY BARTON FLAT	.40	.29	.36	.66	1.15	1.40	.93	.94	.76	.43	.41	.44	8.17
COEUR D'ALENE 1 E	3.88	2.40	2.10	1.65	2.07	1.89	.74	1.24	1.11	1.90	3.06	3.75	25.79
COUNCIL	4.65	2.76	2.26	1.85	1.74	1.61	.40	.71	1.06	1.87	3.19	3.85	25.95
DEER FLAT DAM	1.28	.86	.86	.92	1.03	.95	.21	.47	.57	.72	1.08	1.14	10.09
DIXIE	4.18	2.75	3.03	2.43	2.77	2.81	1.07	1.41	1.61	2.19	2.93	3.78	30.96
DRIGGS	1.51	1.06	1.08	1.23	1.92	2.04	.99	1.27	1.25	1.15	.98	1.46	15.94
DUBOIS EXP STATION	.72	.74	.67	.94	1.65	1.84	.88	.96	.87	.71	.87	.89	11.74
ELK RIVER 1 S	6.06	4.13	3.76	2.95	2.72	2.28	.79	1.28	1.82	2.80	4.38	5.51	38.48
EMMETT 2 E	1.90	1.38	1.07	1.13	1.23	1.04	.15	.38	.79	.92	1.49	1.62	13.10
FAIRFIELD RANGER STA	2.77	1.72	1.21	1.06	1.17	1.14	.41	.54	.66	.77	1.74	2.51	15.70
FENN RANGER STATION	4.90	3.33	3.62	3.63	3.43	2.89	.88	1.50	2.14	2.97	3.89	4.58	37.76
FORT HALL INDIAN AGNCY	.94	.83	.79	1.08	1.34	1.23	.51	.76	.74	1.01	.88	.85	10.96
GARDEN VALLEY RS	4.07	2.46	2.07	1.54	1.50	1.63	.39	.65	.92	1.56	2.80	3.88	23.47
GLENNS FERRY	1.64	1.00	.79	.72	.85	.69	.16	.29	.46	.52	1.21	1.34	9.67
GRACE	1.31	1.16	1.07	1.30	1.63	1.76	.76	1.04	1.07	.95	1.04	1.19	14.28
GRAND VIEW 2 W	.80	.51	.62	.61	.84	.90	.17	.26	.50	.45	.69	.68	7.03
GRANGEVILLE	1.70	1.24	2.07	2.73	3.43	2.90	.96	1.31	1.70	1.90	1.77	1.62	23.33
GROUSE	1.36	.97	.81	.99	1.56	1.49	.62	.96	.76	.57	.93	1.36	12.38
HAILEY AIRPORT	2.67	1.83	1.22	1.07	1.47	1.39	.47	.69	.71	.76	1.44	2.48	16.20
HAMER 4 NW	.57	.52	.53	.67	1.27	1.20	.59	.76	.58	.56	.58	.63	8.46
HAZELTON	1.36	.82	.76	.74	1.09	.85	.27	.40	.55	.62	1.05	1.11	9.62
HILL CITY	2.61	1.65	1.18	.93	1.11	1.16	.34	.46	.66	.83	1.67	2.36	14.96
HOLLISTER	1.01	.66	.87	1.01	1.45	1.29	.57	.59	.66	.71	.89	.94	10.65
HOWE	.72	.54	.48	.72	1.21	1.29	.56	.86	.62	.43	.58	.84	8.85
IDAHO CITY	4.38	2.76	2.39	1.81	1.70	1.64	.34	.69	.98	1.59	2.90	3.88	25.06
IDAHO FALLS FAA AP	.79	.67	.63	.87	1.27	1.27	.48	.76	.69	.76	.75	.83	9.77
IDAHO FALLS 46 W	.78	.63	.56	.76	1.21	1.24	.39	.57	.62	.49	.59	.78	8.62
ISLAND PARK DAM	4.47	3.23	2.93	2.18	2.49	2.73	1.23	1.78	1.68	1.76	2.57	3.88	30.93
JEROME	1.44	.99	.93	.82	.99	.86	.18	.39	.53	.60	1.03	1.36	10.12
KAMIAH	2.35	1.71	2.01	2.48	2.74	2.28	.72	1.12	1.48	2.12	2.18	2.20	23.39
KELLOGG	4.38	2.99	2.57	2.23	2.42	2.24	.92	1.33	1.67	2.22	3.27	4.14	30.38
KOOSKIA	2.26	1.58	2.14	2.84	3.08	2.63	.81	1.32	1.61	2.29	2.15	2.18	24.89
KUNA 2 NNE	1.25	.76	.70	1.02	1.18	.90	.22	.43	.54	.69	1.10	.97	9.76
LEWISTON WSO //R	1.37	.91	1.00	1.13	1.41	1.40	.52	.79	.78	1.01	1.16	1.30	12.78
LIFTON PUMPING STATION	.74	.69	.69	.94	1.14	1.20	.59	.84	.89	.80	.68	.70	9.90
MACKAY RANGER STATION	.98	.52	.52	.71	1.18	1.34	.93	.91	.75	.42	.58	.89	9.73

IDAHO

PRECIPITATION NORMALS (INCHES)

STATION	JAN	FEB	MAR	APR	MAY	JUN	JUL	AUG	SEP	OCT	NOV	DEC	ANN
MALAD	1.66	1.28	1.10	1.21	1.58	1.43	.77	.89	.90	1.00	1.24	1.36	14.42
MALAD CITY FAA AP	1.27	1.01	.95	1.13	1.52	1.41	.79	.88	.85	.88	1.05	1.10	12.84
MAY RANGER STATION	.50	.26	.28	.60	1.26	1.49	.67	.76	.72	.41	.45	.54	7.94
MC CALL	4.29	2.84	2.68	2.08	2.14	2.11	.57	1.09	1.50	2.02	2.89	3.85	28.06
MINIDOKA DAM	1.08	.72	.72	.89	1.16	.96	.31	.49	.54	.66	.80	.90	9.23
MONTPELIER RANGER STA	1.42	1.23	1.20	1.33	1.45	1.55	.65	.94	1.00	.98	1.12	1.30	14.17
MOSCOW U OF I	3.21	2.12	2.04	1.98	1.99	1.65	.71	1.07	1.10	1.83	2.95	3.31	23.96
MOUNTAIN HOME	1.53	.80	.88	.86	.84	.99	.30	.40	.49	.59	1.15	1.34	10.17
NEW MEADOWS RANGER STA	4.06	2.44	2.31	1.88	1.88	1.88	.54	.82	1.32	1.95	2.76	3.82	25.66
NEZPERCE	1.97	1.41	1.76	2.23	2.82	2.34	.92	1.29	1.40	1.88	1.93	1.88	21.83
OAKLEY	.94	.63	.87	1.10	1.69	1.45	.76	.93	.73	.71	.84	.89	11.54
OLA 4 S	2.82	1.91	1.84	1.45	1.41	1.42	.33	.65	.74	1.44	2.17	2.75	18.93
OROFINO	3.27	2.27	2.33	2.21	2.16	1.91	.62	.94	1.30	2.05	2.79	3.54	25.39
PALISADES DAM	2.08	1.67	1.50	1.54	2.05	2.05	1.02	1.43	1.46	1.28	1.50	2.00	19.58
PARMA EXPERIMENT STA	1.66	1.05	.96	.93	1.12	.97	.21	.57	.68	.84	1.23	1.45	11.67
PAUL 1 ENE	1.13	.69	.66	.83	1.26	1.00	.38	.48	.58	.67	.82	.93	9.43
PAYETTE	1.68	1.10	.85	.89	.87	.79	.18	.48	.54	.77	1.30	1.50	10.95
PIERCE //R	5.79	4.02	4.02	3.55	3.49	2.91	.98	1.53	2.15	3.18	4.40	5.60	41.62
POCATELLO WSO //R	1.13	.86	.94	1.16	1.20	1.06	.47	.60	.65	.92	.91	.96	10.86
PORTHILL	2.53	1.72	1.44	1.31	1.68	1.87	.85	1.44	1.34	1.51	2.53	2.67	20.89
POTLATCH 3 NNE	3.43	2.39	2.14	1.96	2.07	1.86	.76	1.07	1.14	1.86	2.76	3.57	25.01
PRIEST RIVER EXP STA	4.58	3.17	2.60	2.02	2.41	2.16	1.05	1.46	1.54	2.44	4.20	4.84	32.47
RICHFIELD	1.87	1.22	.96	.71	.94	.77	.23	.47	.52	.58	1.19	1.63	11.09
RIGGINS RANGER STATION	1.53	1.21	1.68	1.75	2.05	1.83	.73	.94	1.07	1.37	1.41	1.58	17.15
ST ANTHONY 1 WNW	1.52	1.16	1.03	1.17	1.58	1.62	.74	.94	.92	.92	1.26	1.51	14.37
SAINT MARIES	4.63	3.01	2.67	2.19	2.14	2.06	.77	1.32	1.31	2.15	3.56	4.36	30.17
SALMON 1 N	.75	.55	.54	.82	1.34	1.65	.73	.78	.66	.58	.72	.81	9.93
SANDPOINT EXP STATION	4.64	3.32	2.55	2.08	2.30	2.13	1.01	1.57	1.69	2.67	4.34	5.03	33.33
SHOSHONE 1 WNW	1.54	1.05	.93	.67	.92	.71	.19	.38	.54	.56	1.06	1.40	9.95
STREVELL	.67	.63	.72	1.04	1.62	1.43	.84	1.08	.74	.75	.63	.72	10.87
SWAN FALLS POWER HOUSE	.88	.47	.59	.88	1.01	.89	.17	.36	.45	.52	.78	.70	7.70
TETONIA EXP STATION	1.37	1.08	.98	1.15	1.83	1.88	1.07	1.29	1.24	1.25	1.02	1.40	15.56
THREE CREEK	.97	.84	.98	1.34	1.81	1.73	.48	.63	.78	1.03	.92	1.07	12.58
WALLACE WOODLAND PARK	5.70	3.87	3.41	2.63	2.62	2.59	1.13	1.53	2.03	2.96	4.35	5.40	38.22
WEISER 2 SE	1.80	1.18	.92	.93	.97	.80	.15	.39	.53	.80	1.40	1.74	11.61

IDAHO

HEATING DEGREE DAY NORMALS (BASE 65 DEG F)

STATION	JUL	AUG	SEP	OCT	NOV	DEC	JAN	FEB	MAR	APR	MAY	JUN	ANN
ABERDEEN EXP STATION	17	64	261	589	939	1234	1345	1050	955	642	383	154	7633
AMERICAN FALLS 1 SW	0	22	181	502	858	1138	1237	974	887	579	331	120	6829
ANDERSON DAM	0	20	143	406	786	1076	1169	946	890	552	276	98	6362
ARCO 3 SW	34	102	292	617	1038	1407	1522	1212	1091	699	412	191	8617
ARROWROCK DAM	0	26	165	462	828	1088	1181	921	840	549	301	111	6472
ASHTON 1 S	64	126	337	648	1038	1367	1445	1176	1150	747	446	233	8777
AVERY R S NO 2	47	77	244	558	900	1110	1175	927	887	612	356	163	7056
BAYVIEW MODEL BASIN	69	121	317	636	891	1073	1172	938	918	663	434	209	7441
BLISS	0	20	144	412	774	1057	1125	854	766	492	252	78	5974
BOISE WSO //R	0	23	134	406	759	1023	1088	809	732	492	253	83	5802
BONNERS FERRY 1 SW	32	82	258	601	921	1116	1221	932	871	567	326	145	7072
BURLEY FAA AIRPORT	6	44	205	512	849	1107	1187	910	849	579	333	123	6704
CALDWELL	10	28	138	428	762	1014	1079	787	673	414	192	69	5594
CAMBRIDGE	7	30	191	524	879	1197	1290	1000	840	525	288	116	6887
CASCADE 1 NW	77	137	355	663	996	1277	1361	1106	1088	774	512	273	8619
CHALLIS	17	83	235	567	969	1296	1383	1061	967	633	381	162	7754
CHILLY BARTON FLAT	126	197	405	725	1101	1451	1544	1249	1178	810	552	308	9646
COEUR D'ALENE 1 E	17	53	181	484	822	1029	1141	874	846	567	316	131	6461
COUNCIL	0	30	158	471	837	1144	1237	960	840	522	280	110	6589
DEER FLAT DAM	0	16	135	422	756	1017	1088	801	694	438	220	65	5652
DRIGGS	63	121	345	657	1065	1376	1460	1196	1169	795	515	268	9030
DUBOIS EXP STATION	17	59	257	598	1029	1355	1454	1176	1138	726	422	193	8424
EMMETT 2 E	0	26	138	422	759	1011	1073	795	691	441	219	80	5655
FAIRFIELD RANGER STA	47	120	309	620	1005	1361	1476	1204	1153	732	450	229	8706
FENN RANGER STATION	9	28	182	499	810	1029	1113	837	784	516	295	117	6219
FORT HALL INDIAN AGNCY	6	40	203	518	894	1194	1280	991	902	603	336	116	7083
GARDEN VALLEY RS	14	53	189	499	888	1166	1215	941	859	579	320	127	6850
GLENNS FERRY	0	26	128	409	753	1011	1051	790	704	432	198	53	5555
GRACE	35	93	282	605	984	1299	1383	1145	1091	723	446	219	8305
GRAND VIEW 2 W	0	19	121	419	765	1026	1070	784	682	408	176	49	5519
GRANGEVILLE	48	100	271	567	876	1057	1138	882	890	648	434	218	7129
GROUSE	170	225	444	772	1140	1497	1597	1313	1252	858	586	357	10211
HAILEY AIRPORT	37	96	268	577	975	1324	1411	1131	1070	711	437	214	8251
HAMER 4 NW	19	79	279	629	1047	1411	1513	1201	1066	669	376	152	8441
HAZELTON	0	34	182	477	831	1091	1172	896	825	552	308	111	6479
HILL CITY	40	104	298	629	1005	1352	1460	1207	1163	738	453	243	8692
HOLLISTER	6	41	200	481	825	1060	1138	879	853	591	361	149	6584
IDAHO CITY	44	110	276	570	915	1194	1262	1000	967	666	422	204	7630
IDAHO FALLS FAA AP	9	61	257	595	975	1327	1435	1134	1011	654	380	157	7995
IDAHO FALLS 46 W	20	77	292	654	1053	1414	1513	1210	1082	708	425	178	8626
ISLAND PARK DAM	143	207	444	772	1179	1525	1587	1322	1342	948	626	354	10449
JEROME	0	29	151	431	798	1076	1159	890	809	525	276	90	6234
KELLOGG	32	85	219	536	858	1073	1163	888	856	579	342	150	6781
KOOSKIA	0	36	158	474	810	1011	1085	784	713	459	239	87	5856
KUNA 2 NNE	0	38	174	459	786	1039	1110	823	735	492	271	100	6027
LEWISTON WSO //R	0	21	129	409	747	933	1020	742	685	444	225	81	5436
LIFTON PUMPING STATION	17	93	330	676	1032	1342	1451	1263	1203	783	450	195	8835
MACKAY RANGER STATION	32	100	294	614	1029	1398	1479	1179	1110	735	462	221	8653
MALAD	0	32	197	502	891	1200	1277	1019	924	603	341	131	7117
MALAD CITY FAA AP	0	45	211	546	924	1256	1345	1067	942	621	359	139	7455
MAY RANGER STATION	48	104	295	617	1011	1336	1426	1089	992	678	431	209	8236
MC CALL	91	171	371	676	999	1274	1364	1123	1138	834	555	304	8900
MINIDOKA DAM	0	28	165	468	837	1122	1225	952	865	576	319	111	6668
MONTPELIER RANGER STA	40	98	324	660	1056	1367	1454	1238	1187	792	487	245	8948
MOSCOW U OF I	48	92	213	502	831	1042	1138	860	828	582	370	183	6689

IDAHO

HEATING DEGREE DAY NORMALS (BASE 65 DEG F)

STATION	JUL	AUG	SEP	OCT	NOV	DEC	JAN	FEB	MAR	APR	MAY	JUN	ANN
MOUNTAIN HOME	0	23	143	425	783	1039	1104	843	772	507	264	77	5980
NEW MEADOWS RANGER STA	83	155	378	701	1008	1330	1411	1126	1082	744	496	260	8774
NEZPERCE	60	105	267	570	888	1085	1181	904	893	645	425	223	7246
OAKLEY	9	47	182	453	795	1048	1119	862	831	579	348	128	6401
OLA 4 S	6	46	189	502	852	1110	1200	904	809	543	301	115	6577
OROFINO	0	20	123	431	753	961	1039	748	682	426	207	70	5460
PALISADES DAM	21	69	250	555	951	1268	1370	1128	1085	720	415	189	8021
PARMA EXPERIMENT STA	0	26	151	468	804	1060	1138	848	725	462	234	84	6000
PAUL 1 ENE	5	54	216	524	855	1110	1206	944	877	603	350	133	6877
PAYETTE	0	17	122	434	786	1057	1119	826	688	420	195	59	5723
POCATELLO WSO //R	0	32	209	524	894	1190	1277	994	915	612	348	128	7123
PORTHILL	41	100	281	626	942	1150	1271	977	915	585	335	155	7378
POTLATCH 3 NNE	92	135	296	580	852	1048	1144	862	856	615	403	225	7108
PRIEST RIVER EXP STA	72	122	297	632	972	1169	1265	991	955	663	397	199	7734
RICHFIELD	8	63	224	543	915	1221	1330	1053	955	621	365	145	7443
RIGGINS RANGER STATION	0	13	94	319	672	884	946	689	626	398	195	58	4894
ST ANTHONY 1 WNW	41	102	303	614	996	1339	1426	1159	1094	705	415	204	8398
SAINT MARIES	30	91	222	524	855	1057	1150	865	837	579	340	146	6696
SALMON 1 N	15	59	248	598	972	1280	1401	1044	918	594	345	146	7620
SANDPOINT EXP STATION	56	105	286	617	912	1113	1212	941	902	609	372	178	7303
SHOSHONE 1 WNW	0	30	156	465	864	1159	1249	969	865	546	272	94	6669
SWAN FALLS POWER HOUSE	0	0	58	281	660	930	986	711	595	340	125	25	4711
WALLACE WOODLAND PARK	79	129	308	595	915	1132	1231	960	967	681	446	241	7684
WEISER 2 SE	0	19	149	459	792	1063	1141	848	707	453	213	72	5916

IDAHO

COOLING DEGREE DAY NORMALS (BASE 65 DEG F)

STATION	JAN	FEB	MAR	APR	MAY	JUN	JUL	AUG	SEP	OCT	NOV	DEC	ANN
ABERDEEN EXP STATION	0	0	0	0	0	31	125	98	15	0	0	0	269
AMERICAN FALLS 1 SW	0	0	0	0	6	57	191	149	31	0	0	0	434
ANDERSON DAM	0	0	0	0	13	86	276	231	77	0	0	0	683
ARCO 3 SW	0	0	0	0	0	23	93	78	10	0	0	0	204
ARROWROCK DAM	0	0	0	0	10	78	261	221	63	0	0	0	633
ASHTON 1 S	0	0	0	0	0	17	39	30	7	0	0	0	93
AVERY R S NO 2	0	0	0	0	9	28	112	111	37	0	0	0	297
BAYVIEW MODEL BASIN	0	0	0	0	0	8	50	59	5	0	0	0	122
BLISS	0	0	0	0	13	96	273	215	66	0	0	0	663
BOISE WSO //R	0	0	0	0	17	107	298	240	80	0	0	0	742
BONNERS FERRY 1 SW	0	0	0	0	0	28	88	98	15	0	0	0	229
BURLEY FAA AIRPORT	0	0	0	0	8	51	179	137	28	0	0	0	403
CALDWELL	0	0	0	0	34	147	316	251	69	0	0	0	817
CAMBRIDGE	0	0	0	0	6	71	220	154	29	0	0	0	480
CASCADE 1 NW	0	0	0	0	0	6	46	38	10	0	0	0	100
CHALLIS	0	0	0	0	6	33	135	120	16	0	0	0	310
CHILLY BARTON FLAT	0	0	0	0	0	8	33	30	0	0	0	0	71
COEUR D'ALENE 1 E	0	0	0	0	6	41	147	161	40	0	0	0	395
COUNCIL	0	0	0	0	14	89	263	225	59	0	0	0	650
DEER FLAT DAM	0	0	0	0	18	89	245	190	51	0	0	0	593
DRIGGS	0	0	0	0	0	7	47	35	9	0	0	0	98
DUBOIS EXP STATION	0	0	0	0	0	28	138	109	26	0	0	0	301
EMMETT 2 E	0	0	0	0	21	113	260	203	60	0	0	0	657
FAIRFIELD RANGER STA	0	0	0	0	0	10	69	76	15	0	0	0	170
FENN RANGER STATION	0	0	0	0	7	48	173	159	41	0	0	0	428
FORT HALL INDIAN AGNCY	0	0	0	0	7	38	164	126	23	0	0	0	358
GARDEN VALLEY RS	0	0	0	0	0	49	175	158	45	0	0	0	427
GLENNS FERRY	0	0	0	0	28	131	329	249	65	0	0	0	802
GRACE	0	0	0	0	0	12	56	62	6	0	0	0	136
GRAND VIEW 2 W	0	0	0	0	33	151	342	258	67	0	0	0	851
GRANGEVILLE	0	0	0	0	0	14	85	103	25	0	0	0	227
GROUSE	0	0	0	0	0	0	9	11	0	0	0	0	20
HAILEY AIRPORT	0	0	0	0	0	19	99	105	19	0	0	0	242
HAMER 4 NW	0	0	0	0	0	29	124	94	15	0	0	0	262
HAZELTON	0	0	0	0	10	60	204	161	35	0	0	0	470
HILL CITY	0	0	0	0	0	12	65	76	10	0	0	0	163
HOLLISTER	0	0	0	0	7	50	186	149	41	0	0	0	433
IDAHO CITY	0	0	0	0	0	12	81	88	15	0	0	0	196
IDAHO FALLS FAA AP	0	0	0	0	0	31	130	107	20	0	0	0	288
IDAHO FALLS 46 W	0	0	0	0	0	25	119	99	10	0	0	0	253
ISLAND PARK DAM	0	0	0	0	0	0	13	21	0	0	0	0	34
JEROME	0	0	0	0	19	96	267	218	58	0	0	0	658
KELLOGG	0	0	0	0	7	36	122	129	24	0	0	0	318
KOOSKIA	0	0	0	0	12	75	203	185	56	0	0	0	531
KUNA 2 NNE	0	0	0	0	7	58	172	141	30	0	0	0	408
LEWISTON WSO //R	0	0	0	0	17	93	286	251	87	0	0	0	734
LIFTON PUMPING STATION	0	0	0	0	0	12	67	46	6	0	0	0	131
MACKAY RANGER STATION	0	0	0	0	0	20	73	69	12	0	0	0	174
MALAD	0	0	0	0	7	47	181	147	35	0	0	0	417
MALAD CITY FAA AP	0	0	0	0	5	34	158	135	16	0	0	0	348
MAY RANGER STATION	0	0	0	0	0	14	73	70	10	0	0	0	167
MC CALL	0	0	0	0	0	7	26	41	0	0	0	0	74
MINIDOKA DAM	0	0	0	0	9	69	233	186	45	0	0	0	542
MONTPELIER RANGER STA	0	0	0	0	0	11	65	58	6	0	0	0	140
MOSCOW U OF I	0	0	0	0	0	24	92	114	33	0	0	0	263

IDAHO

COOLING DEGREE DAY NORMALS (BASE 65 DEG F)

STATION	JAN	FEB	MAR	APR	MAY	JUN	JUL	AUG	SEP	OCT	NOV	DEC	ANN
MOUNTAIN HOME	0	0	0	0	19	101	295	240	71	0	0	0	726
NEW MEADOWS RANGER STA	0	0	0	0	0	8	43	40	6	0	0	0	97
NEZPERCE	0	0	0	0	0	13	69	92	24	0	0	0	198
OAKLEY	0	0	0	0	10	44	176	156	35	0	0	0	421
OLA 4 S	0	0	0	0	6	52	189	152	39	0	0	0	438
OROFINO	0	0	0	0	18	100	248	225	63	0	0	0	654
PALISADES DAM	0	0	0	0	0	24	104	87	16	0	0	0	231
PARMA EXPERIMENT STA	0	0	0	0	20	102	261	203	43	0	0	0	629
PAUL 1 ENE	0	0	0	0	6	40	160	135	24	0	0	0	365
PAYETTE	0	0	0	0	21	119	295	234	53	0	0	0	722
POCATELLO WSO //R	0	0	0	0	7	53	197	153	35	0	0	0	445
PORTHILL	0	0	0	0	0	26	84	91	11	0	0	0	212
POTLATCH 3 NNE	0	0	0	0	0	18	67	67	17	0	0	0	169
PRIEST RIVER EXP STA	0	0	0	0	0	13	60	69	9	0	0	0	151
RICHFIELD	0	0	0	0	0	34	147	128	23	0	0	0	332
RIGGINS RANGER STATION	0	0	0	5	34	130	344	330	139	9	0	0	991
ST ANTHONY 1 WNW	0	0	0	0	0	12	66	59	9	0	0	0	146
SAINT MARIES	0	0	0	0	0	32	107	122	27	0	0	0	288
SALMON 1 N	0	0	0	0	0	35	132	102	14	0	0	0	283
SANDPOINT EXP STATION	0	0	0	0	0	16	52	65	7	0	0	0	140
SHOSHONE 1 WNW	0	0	0	0	15	103	298	241	60	0	0	0	717
SWAN FALLS POWER HOUSE	0	0	0	7	63	214	474	391	148	5	0	0	1302
WALLACE WOODLAND PARK	0	0	0	0	0	13	54	74	17	0	0	0	158
WEISER 2 SE	0	0	0	0	12	108	267	195	47	0	0	0	629

10 — IDAHO

STATE-STATION NUMBER	STN TYP	NAME	LATITUDE DEG-MIN	LONGITUDE DEG-MIN	ELEVATION (FT)
10-0010	13	ABERDEEN EXP STATION	N 4257	W 11250	4405
10-0227	13	AMERICAN FALLS 1 SW	N 4247	W 11252	4318
10-0282	13	ANDERSON DAM	N 4321	W 11528	3882
10-0375	13	ARCO 3 SW	N 4336	W 11320	5328
10-0448	13	ARROWROCK DAM	N 4336	W 11555	3275
10-0470	13	ASHTON 1 S	N 4404	W 11127	5220
10-0528	13	AVERY R S NO 2	N 4715	W 11555	2390
10-0667	13	BAYVIEW MODEL BASIN	N 4759	W 11633	2075
10-0915	12	BLACKFOOT 2 SSW	N 4310	W 11221	4487
10-1002	13	BLISS	N 4256	W 11457	3265
10-1018	12	BOISE LUCKY PEAK DAM	N 4333	W 11604	2840
10-1022	13	BOISE WSO //R	N 4334	W 11613	2838
10-1079	13	BONNERS FERRY 1 SW	N 4841	W 11619	1860
10-1303	13	BURLEY FAA AIRPORT	N 4232	W 11346	4146
10-1380	13	CALDWELL	N 4340	W 11641	2370
10-1408	13	CAMBRIDGE	N 4434	W 11641	2650
10-1514	13	CASCADE 1 NW	N 4432	W 11603	4896
10-1636	12	CENTERVILLE ARBAUGH RCH	N 4354	W 11551	4300
10-1663	13	CHALLIS	N 4430	W 11414	5175
10-1671	13	CHILLY BARTON FLAT	N 4359	W 11349	6260
10-1956	13	COEUR D'ALENE 1 E	N 4741	W 11645	2158
10-2187	13	COUNCIL	N 4444	W 11626	2950
10-2444	13	DEER FLAT DAM	N 4335	W 11645	2510
10-2575	12	DIXIE	N 4533	W 11528	5610
10-2676	13	DRIGGS	N 4344	W 11107	6116
10-2707	13	DUBOIS EXP STATION	N 4415	W 11212	5452
10-2892	12	ELK RIVER 1 S	N 4646	W 11611	2918
10-2942	13	EMMETT 2 E	N 4352	W 11628	2500
10-3108	13	FAIRFIELD RANGER STA	N 4321	W 11447	5065
10-3143	13	FENN RANGER STATION	N 4606	W 11533	1585
10-3297	13	FORT HALL INDIAN AGNCY	N 4302	W 11226	4460
10-3448	13	GARDEN VALLEY RS	N 4404	W 11555	3212
10-3631	13	GLENNS FERRY	N 4257	W 11518	2570
10-3732	13	GRACE	N 4235	W 11144	5550
10-3760	13	GRAND VIEW 2 W	N 4300	W 11608	2400
10-3771	13	GRANGEVILLE	N 4555	W 11608	3355
10-3882	13	GROUSE	N 4342	W 11337	6100
10-3942	13	HAILEY AIRPORT	N 4331	W 11418	5328
10-3964	13	HAMER 4 NW	N 4358	W 11215	4791
10-4140	13	HAZELTON	N 4236	W 11408	4060
10-4268	13	HILL CITY	N 4318	W 11503	5000
10-4295	13	HOLLISTER	N 4221	W 11434	4525
10-4384	12	HOWE	N 4347	W 11300	4820
10-4442	13	IDAHO CITY	N 4350	W 11550	3965
10-4457	13	IDAHO FALLS FAA AP	N 4331	W 11204	4730
10-4460	13	IDAHO FALLS 46 W	N 4332	W 11257	4938
10-4598	13	ISLAND PARK DAM	N 4425	W 11124	6300
10-4670	13	JEROME	N 4244	W 11431	3740
10-4793	12	KAMIAH	N 4614	W 11601	1212
10-4831	13	KELLOGG	N 4733	W 11610	2312

10 — IDAHO

LEGEND
11 = TEMPERATURE ONLY
12 = PRECIPITATION ONLY
13 = TEMP. & PRECIP.

STATE-STATION NUMBER	STN TYP	NAME	LATITUDE DEG-MIN	LONGITUDE DEG-MIN	ELEVATION (FT)
10-5011	13	KOOSKIA	N 4609	W 11559	1260
10-5038	13	KUNA 2 NNE	N 4331	W 11624	2685
10-5241	13	LEWISTON WSO //R	N 4623	W 11701	1413
10-5275	13	LIFTON PUMPING STATION	N 4207	W 11118	5926
10-5462	13	MACKAY RANGER STATION	N 4355	W 11337	5897
10-5544	13	MALAD	N 4212	W 11215	4600
10-5559	13	MALAD CITY FAA AP	N 4210	W 11219	4476
10-5685	13	MAY RANGER STATION	N 4436	W 11355	5110
10-5708	13	MC CALL	N 4454	W 11607	5025
10-5980	13	MINIDOKA DAM	N 4240	W 11330	4210
10-6053	13	MONTPELIER RANGER STA	N 4219	W 11118	5960
10-6152	13	MOSCOW U OF I	N 4644	W 11658	2660
10-6174	13	MOUNTAIN HOME	N 4309	W 11543	3185
10-6388	13	NEW MEADOWS RANGER STA	N 4458	W 11617	3870
10-6424	13	NEZPERCE	N 4615	W 11615	3145
10-6542	13	OAKLEY	N 4215	W 11353	4600
10-6590	13	OLA 4 S	N 4408	W 11617	2990
10-6681	13	OROFINO	N 4629	W 11615	1027
10-6764	13	PALISADES DAM	N 4321	W 11113	5385
10-6844	13	PARMA EXPERIMENT STA	N 4348	W 11657	2215
10-6877	13	PAUL 1 ENE	N 4237	W 11345	4210
10-6891	13	PAYETTE	N 4405	W 11656	2150
10-7046	12	PIERCE //R	N 4630	W 11548	3190
10-7211	13	POCATELLO WSO //R	N 4255	W 11236	4454
10-7264	13	PORTHILL	N 4900	W 11630	1800
10-7301	13	POTLATCH 3 NNE	N 4658	W 11653	2500
10-7386	13	PRIEST RIVER EXP STA	N 4821	W 11650	2380
10-7673	13	RICHFIELD	N 4304	W 11409	4306
10-7706	13	RIGGINS RANGER STATION	N 4525	W 11619	1801
10-8022	13	ST ANTHONY 1 WNW	N 4358	W 11143	4950
10-8062	13	SAINT MARIES	N 4719	W 11634	2145
10-8080	13	SALMON 1 N	N 4511	W 11354	3970
10-8137	13	SANDPOINT EXP STATION	N 4817	W 11634	2100
10-8380	13	SHOSHONE 1 WNW	N 4258	W 11426	3950
10-8786	12	STREVELL	N 4201	W 11317	5290
10-8928	13	SWAN FALLS POWER HOUSE	N 4315	W 11623	2325
10-9065	12	TETONIA EXP STATION	N 4351	W 11116	6172
10-9119	12	THREE CREEK	N 4205	W 11509	5410
10-9498	13	WALLACE WOODLAND PARK	N 4730	W 11553	2935
10-9638	13	WEISER 2 SE	N 4414	W 11657	2120

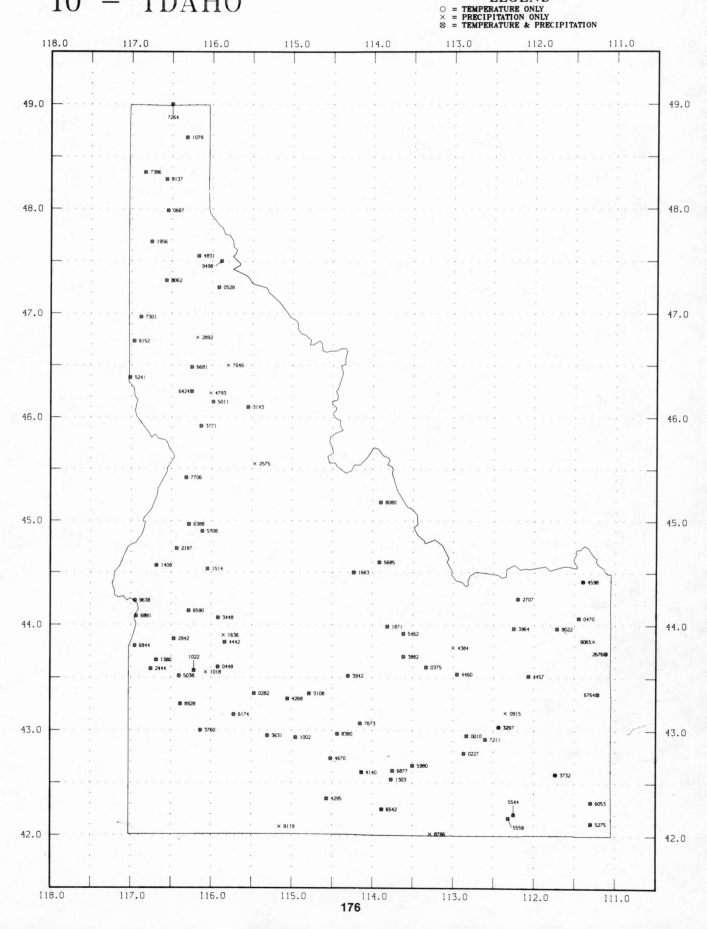

LEGEND
O = TEMPERATURE ONLY
X = PRECIPITATION ONLY
⊗ = TEMPERATURE & PRECIPITATION

ILLINOIS

TEMPERATURE NORMALS (DEG F)

STATION		JAN	FEB	MAR	APR	MAY	JUN	JUL	AUG	SEP	OCT	NOV	DEC	ANN
ALEDO	MAX	29.7	35.7	46.7	63.0	74.4	83.2	86.6	84..8	77.6	66.3	49.0	35.4	61.0
	MIN	11.9	17.3	26.9	39.9	50.3	59.5	63.2	60.8	52.7	42.2	29.9	19.0	39.5
	MEAN	20.9	26.5	36.8	51.5	62.4	71.4	74.9	72.8	65.2	54.3	39.4	27.2	50.3
ALTON DAM 26	MAX	36.4	41.2	51.5	65.2	75.5	84.9	88.7	87.0	80.3	68.3	53.2	41.3	64.5
	MIN	19.0	23.4	32.2	44.9	54.9	64.5	68.7	66.7	58.2	46.6	34.8	25.0	44.9
	MEAN	27.7	32.3	41.8	55.1	65.2	74.7	78.7	76.9	69.3	57.5	44.0	33.1	54.7
ANNA 1 E	MAX	41.3	46.3	55.9	68.5	77.5	86.0	89.2	87.9	81.7	71.1	56.5	45.3	67.3
	MIN	23.4	26.9	35.4	46.8	55.2	63.3	67.2	65.4	58.7	46.9	36.6	28.2	46.2
	MEAN	32.4	36.6	45.7	57.6	66.4	74.7	78.2	76.7	70.2	59.0	46.6	36.8	56.7
ANTIOCH 2 NW	MAX	28.0	32.8	42.5	57.5	69.3	78.8	82.7	80.9	73.7	62.6	46.7	33.5	57.4
	MIN	10.4	14.4	24.0	36.0	45.7	55.5	60.5	59.4	51.7	41.4	29.4	17.6	37.2
	MEAN	19.2	23.6	33.3	46.8	57.5	67.2	71.7	70.2	62.8	52.0	38.1	25.6	47.3
AURORA COLLEGE	MAX	28.9	34.0	44.8	60.1	71.5	80.8	84.2	82.4	76.0	64.4	48.1	34.6	59.2
	MIN	11.3	15.9	26.0	37.9	47.8	57.5	61.8	60.3	52.2	41.0	29.4	18.2	38.3
	MEAN	20.1	25.0	35.4	49.0	59.7	69.2	73.0	71.4	64.1	52.8	38.8	26.4	48.7
BELLEVILLE SO ILL UNIV	MAX	38.7	44.0	54.0	67.3	76.7	85.6	89.1	87.2	81.2	70.2	54.8	43.4	66.0
	MIN	20.8	25.1	33.6	45.0	53.6	62.7	66.3	64.1	56.6	45.1	34.9	26.7	44.5
	MEAN	29.8	34.6	43.8	56.2	65.2	74.1	77.7	75.7	69.0	57.6	44.9	35.0	55.3
CAIRO WSO R	MAX	41.9	46.8	56.4	69.2	78.5	87.0	89.8	87.9	81.4	70.9	56.6	46.3	67.7
	MIN	27.3	31.0	39.5	51.0	60.0	68.2	72.1	70.0	63.1	51.2	40.2	32.1	50.5
	MEAN	34.6	38.9	48.0	60.1	69.3	77.6	81.0	78.9	72.3	61.1	48.5	39.2	59.1
CARBONDALE SEWAGE PL	MAX	41.4	46.4	56.0	69.3	78.2	86.6	89.7	88.6	82.4	71.7	56.9	45.6	67.7
	MIN	22.1	25.4	34.4	45.5	53.7	62.6	66.4	63.9	56.3	43.9	34.4	27.2	44.7
	MEAN	31.7	35.9	45.3	57.4	66.0	74.6	78.1	76.3	69.4	57.8	45.7	36.4	56.2
CARLINVILLE	MAX	36.4	42.1	52.6	66.8	76.4	85.5	89.1	87.1	81.3	69.4	53.5	41.5	65.1
	MIN	17.9	22.5	31.2	43.0	52.1	61.1	64.8	62.9	55.2	44.4	33.1	24.1	42.7
	MEAN	27.2	32.3	41.9	54.9	64.3	73.3	77.0	75.1	68.3	56.9	43.3	32.9	54.0
CARLYLE RESERVOIR	MAX	37.4	42.5	52.4	66.4	76.1	85.3	89.0	87.1	80.8	69.5	54.3	42.6	65.3
	MIN	19.4	23.3	32.7	44.5	54.1	63.0	67.0	64.7	57.1	44.9	34.0	25.3	44.2
	MEAN	28.5	32.9	42.6	55.4	65.1	74.1	78.0	75.9	69.0	57.2	44.2	34.0	54.7
CHARLESTON	MAX	35.0	40.2	50.5	65.1	74.9	83.9	87.2	85.4	80.0	67.8	51.9	39.9	63.5
	MIN	18.4	22.5	31.5	43.4	52.9	61.8	65.5	63.5	56.3	45.1	34.0	24.5	43.3
	MEAN	26.7	31.4	41.0	54.2	64.0	72.8	76.4	74.5	68.2	56.5	42.9	32.3	53.4
CHENOA	MAX	31.3	36.6	47.5	63.0	74.1	83.4	86.0	84.0	78.4	66.3	49.5	36.5	61.4
	MIN	13.7	18.6	28.1	39.7	49.5	59.2	62.8	60.6	53.4	42.4	30.8	20.6	40.0
	MEAN	22.5	27.6	37.9	51.4	61.8	71.3	74.4	72.3	65.9	54.4	40.2	28.6	50.7
CHICAGO O HARE WSO R	MAX	29.2	33.9	44.3	58.8	70.0	79.4	83.3	82.1	75.5	64.1	48.2	35.0	58.7
	MIN	13.6	18.1	27.6	38.8	48.1	57.7	62.7	61.7	53.9	42.9	31.4	20.3	39.7
	MEAN	21.4	26.0	36.0	48.8	59.1	68.6	73.0	71.9	64.7	53.5	39.8	27.7	49.2
CHICAGO UNIVERSITY //R	MAX	30.6	34.8	44.0	57.5	68.5	78.4	82.4	81.3	75.1	63.7	48.4	36.1	58.4
	MIN	17.3	21.8	30.3	41.0	50.4	60.4	66.1	65.3	58.2	47.6	35.1	23.9	43.1
	MEAN	24.0	28.3	37.2	49.3	59.5	69.4	74.3	73.3	66.6	55.7	41.8	30.0	50.8
CHICAGO MIDWAY AP	MAX	29.8	34.5	44.7	59.0	70.7	80.7	84.3	82.8	75.8	64.2	48.4	35.6	59.2
	MIN	15.3	19.8	29.2	40.4	50.2	60.5	65.6	64.5	56.5	45.2	33.0	22.0	41.9
	MEAN	22.6	27.2	37.0	49.7	60.5	70.6	75.0	73.7	66.2	54.7	40.7	28.9	50.6
DANVILLE	MAX	34.4	39.5	50.2	64.8	75.2	84.1	86.9	85.0	79.6	67.7	51.9	39.6	63.2
	MIN	16.8	20.6	29.9	41.1	50.3	59.3	63.4	61.5	54.2	42.8	32.4	23.1	41.3
	MEAN	25.7	30.1	40.1	53.0	62.8	71.7	75.2	73.3	66.9	55.3	42.2	31.4	52.3
DECATUR	MAX	34.5	39.9	50.5	65.3	75.9	84.8	88.0	86.1	80.7	68.6	52.2	39.9	63.9
	MIN	17.1	21.4	30.4	42.4	52.1	61.0	65.1	63.1	55.7	44.3	33.0	23.3	42.4
	MEAN	25.8	30.7	40.5	53.9	64.0	72.9	76.6	74.6	68.2	56.5	42.6	31.6	53.2

ILLINOIS

TEMPERATURE NORMALS (DEG F)

STATION		JAN	FEB	MAR	APR	MAY	JUN	JUL	AUG	SEP	OCT	NOV	DEC	ANN
DE KALB	MAX	27.7	33.2	44.1	59.9	72.5	81.8	84.9	82.8	76.1	64.3	47.2	33.5	59.0
	MIN	10.6	15.7	25.4	37.6	47.9	57.7	61.8	60.1	52.2	41.5	29.3	17.8	38.1
	MEAN	19.2	24.5	34.8	48.8	60.2	69.8	73.4	71.5	64.2	52.9	38.3	25.7	48.6
DIXON	MAX	28.4	33.8	44.9	61.3	73.1	82.2	85.3	83.5	76.5	64.8	48.1	33.7	59.6
	MIN	11.0	16.4	26.3	38.9	49.2	58.8	62.9	60.8	52.8	41.8	29.9	18.2	38.9
	MEAN	19.7	25.1	35.6	50.1	61.2	70.5	74.1	72.2	64.7	53.3	39.0	26.0	49.3
DU QUOIN 1 NNW	MAX	40.2	45.5	55.9	68.7	77.7	86.3	89.9	88.2	82.2	71.3	56.3	44.6	67.2
	MIN	21.8	25.8	34.6	45.7	54.3	62.6	66.4	64.1	56.9	44.9	35.1	27.0	44.9
	MEAN	31.0	35.7	45.2	57.3	66.0	74.5	78.2	76.2	69.6	58.1	45.7	35.8	56.1
EFFINGHAM	MAX	36.2	41.3	51.7	66.0	76.0	85.0	88.5	86.8	80.6	68.8	53.1	41.1	64.6
	MIN	18.3	22.2	31.5	42.9	51.9	61.0	65.1	62.9	55.1	43.5	33.2	24.5	42.7
	MEAN	27.3	31.7	41.6	54.5	64.0	73.0	76.8	74.9	67.9	56.2	43.2	32.8	53.7
FAIRFIELD RADIO WFIW	MAX	39.0	43.9	54.3	67.6	76.7	85.6	88.5	87.0	81.4	70.0	55.0	43.6	66.1
	MIN	21.6	25.2	33.9	44.8	53.5	61.9	65.7	63.7	56.5	45.0	35.3	27.0	44.5
	MEAN	30.3	34.6	44.1	56.2	65.1	73.8	77.1	75.4	69.0	57.5	45.2	35.3	55.3
FLORA 5 NW	MAX	38.3	43.3	53.9	67.5	76.6	85.5	88.9	87.6	81.9	70.3	54.8	43.0	66.0
	MIN	20.2	23.9	33.1	44.1	53.2	61.5	65.2	63.0	55.7	44.2	33.9	25.6	43.6
	MEAN	29.3	33.6	43.5	55.8	64.9	73.5	77.1	75.3	68.8	57.3	44.4	34.4	54.8
FULTON DAM 13	MAX	28.4	33.4	43.9	59.7	71.5	80.6	84.2	82.4	75.3	64.2	47.9	34.0	58.8
	MIN	11.0	15.7	26.0	39.9	51.1	60.6	64.8	62.7	54.5	43.6	30.9	18.8	40.0
	MEAN	19.7	24.6	35.0	49.8	61.3	70.6	74.5	72.6	64.9	53.9	39.4	26.4	49.4
GALESBURG	MAX	29.6	35.3	45.9	61.5	72.7	81.8	85.3	83.4	76.5	65.1	48.8	35.4	60.1
	MIN	12.1	17.6	27.1	40.0	50.5	60.3	64.3	62.4	54.1	42.7	30.3	19.4	40.1
	MEAN	20.9	26.5	36.5	50.8	61.6	71.0	74.8	72.9	65.3	53.9	39.6	27.4	50.1
GALVA	MAX	29.4	34.8	46.0	62.2	73.4	82.3	85.6	83.8	77.2	65.5	48.8	35.2	60.4
	MIN	11.2	16.4	26.0	39.1	49.6	59.0	62.8	60.7	52.8	42.0	29.8	18.4	39.0
	MEAN	20.3	25.6	36.0	50.7	61.5	70.7	74.2	72.3	65.0	53.7	39.3	26.8	49.7
GRIGGSVILLE	MAX	34.3	40.0	50.7	65.5	75.3	83.9	87.9	85.9	79.4	68.1	52.4	39.8	63.6
	MIN	16.4	21.4	30.2	43.0	53.0	61.8	65.6	63.4	55.8	45.2	33.3	22.9	42.7
	MEAN	25.4	30.7	40.5	54.3	64.2	72.9	76.8	74.6	67.6	56.6	42.9	31.4	53.2
HARRISBURG	MAX	42.5	47.4	57.6	70.8	80.0	88.4	91.6	90.2	83.8	72.6	57.7	46.7	69.1
	MIN	23.5	26.9	35.7	46.5	54.7	63.2	67.2	64.9	57.5	45.3	36.1	28.4	45.8
	MEAN	33.0	37.2	46.6	58.7	67.4	75.8	79.4	77.6	70.7	59.0	46.9	37.6	57.5
HILLSBORO 2 SSW	MAX	36.8	42.4	52.9	66.9	76.0	84.9	88.3	86.5	80.7	69.5	53.8	42.0	65.1
	MIN	18.5	22.9	31.5	43.4	52.6	61.4	65.2	62.8	55.5	44.3	33.6	24.5	43.0
	MEAN	27.7	32.7	42.2	55.2	64.4	73.2	76.8	74.7	68.1	57.0	43.7	33.3	54.1
HOOPESTON	MAX	32.0	36.9	47.9	62.6	73.6	82.8	85.6	83.6	78.1	65.9	49.7	37.2	61.3
	MIN	16.2	20.2	29.5	40.8	50.9	60.1	63.7	61.7	54.6	43.6	32.6	22.6	41.4
	MEAN	24.1	28.6	38.7	51.7	62.3	71.5	74.7	72.7	66.4	54.8	41.2	29.9	51.4
JACKSONVILLE	MAX	34.3	39.4	50.3	65.0	74.9	83.9	87.3	85.0	79.4	67.9	52.5	39.6	63.3
	MIN	16.4	20.7	29.8	42.3	52.0	61.0	64.6	62.5	54.5	43.8	32.4	22.7	41.9
	MEAN	25.4	30.1	40.0	53.7	63.5	72.5	75.9	73.7	67.0	55.9	42.4	31.2	52.6
JERSEYVILLE 2 SW	MAX	37.3	42.8	53.3	67.2	76.4	85.0	88.8	86.9	81.2	69.8	54.1	42.2	65.4
	MIN	17.8	22.4	31.0	42.9	52.1	61.2	64.7	62.5	54.7	43.6	32.9	23.9	42.5
	MEAN	27.6	32.6	42.1	55.1	64.3	73.1	76.8	74.7	67.9	56.7	43.5	33.0	54.0
KEWANEE	MAX	29.4	34.7	45.9	62.3	73.7	82.6	85.9	84.1	77.4	65.4	48.9	35.0	60.4
	MIN	12.1	17.0	26.6	39.3	49.8	59.3	63.4	61.2	52.8	42.1	30.1	19.1	39.4
	MEAN	20.7	25.9	36.3	50.8	61.7	71.0	74.7	72.6	65.1	53.8	39.5	27.0	49.9
LA HARPE	MAX	31.9	37.4	48.6	63.9	74.7	83.4	87.2	85.0	78.3	67.2	50.7	37.6	62.2
	MIN	13.1	18.6	28.0	40.7	50.9	60.2	64.0	61.7	53.8	43.1	30.8	20.2	40.4
	MEAN	22.5	28.0	38.4	52.3	62.8	71.8	75.6	73.4	66.1	55.2	40.8	29.0	51.3

ILLINOIS

TEMPERATURE NORMALS (DEG F)

STATION		JAN	FEB	MAR	APR	MAY	JUN	JUL	AUG	SEP	OCT	NOV	DEC	ANN
LINCOLN	MAX	33.2	38.8	49.7	64.9	75.5	84.3	87.4	85.1	79.3	67.6	51.5	38.7	63.0
	MIN	15.9	20.6	29.7	41.7	51.8	60.9	64.7	62.5	55.1	43.6	32.3	22.4	41.8
	MEAN	24.6	29.7	39.7	53.3	63.7	72.6	76.1	73.8	67.2	55.6	41.9	30.6	52.4
MARENGO	MAX	27.6	32.8	43.6	59.7	72.1	81.1	84.8	82.8	75.8	63.9	46.9	33.0	58.7
	MIN	9.8	14.6	24.5	36.1	46.2	56.1	60.4	58.6	50.8	40.3	28.6	17.0	36.9
	MEAN	18.7	23.8	34.1	48.0	59.2	68.6	72.6	70.7	63.3	52.1	37.8	25.1	47.8
MATTOON	MAX	33.8	38.8	49.2	63.6	74.1	83.5	86.5	84.6	78.9	66.7	50.9	39.1	62.5
	MIN	17.4	21.6	30.5	42.2	52.0	61.1	65.1	63.0	55.5	43.7	32.8	23.5	42.4
	MEAN	25.6	30.3	39.9	52.9	63.0	72.4	75.8	73.8	67.2	55.2	41.9	31.3	52.4
MCLEANSBORO 2 E	MAX	39.8	44.7	54.5	68.0	77.4	86.1	89.5	88.3	82.2	70.4	55.7	44.1	66.7
	MIN	21.9	25.3	34.0	45.6	54.0	62.6	66.3	64.0	56.9	45.0	35.4	27.2	44.9
	MEAN	30.9	35.0	44.3	56.8	65.7	74.4	77.9	76.2	69.6	57.7	45.6	35.7	55.8
MINONK 3 NE	MAX	30.2	35.7	47.0	63.2	74.5	83.9	86.8	85.0	79.3	66.9	49.4	35.7	61.5
	MIN	12.9	18.0	27.5	39.4	49.0	58.6	62.0	59.8	52.6	41.6	30.2	19.7	39.3
	MEAN	21.6	26.9	37.3	51.3	61.8	71.3	74.4	72.5	66.0	54.3	39.8	27.7	50.4
MOLINE WSO R	MAX	28.0	33.7	44.8	61.1	72.5	82.1	85.4	83.6	76.2	64.8	48.0	34.3	59.5
	MIN	11.0	16.4	26.5	39.6	50.1	59.9	64.3	62.1	53.2	42.1	30.0	18.4	39.5
	MEAN	19.5	25.1	35.7	50.4	61.3	71.0	74.9	72.9	64.7	53.5	39.0	26.4	49.5
MONMOUTH	MAX	30.7	36.6	47.7	63.4	74.4	82.9	86.5	84.8	77.9	66.4	49.9	36.6	61.5
	MIN	13.4	18.7	27.8	40.5	50.5	59.7	63.6	61.5	53.6	43.1	30.9	20.7	40.3
	MEAN	22.1	27.7	37.8	52.0	62.5	71.3	75.1	73.2	65.8	54.8	40.4	28.7	51.0
MORRISON	MAX	29.2	34.7	45.8	62.1	73.5	82.2	85.6	83.8	77.3	66.2	49.1	34.8	60.4
	MIN	10.3	15.6	25.8	38.5	48.9	58.2	62.1	60.0	51.7	41.1	29.1	17.6	38.2
	MEAN	19.8	25.2	35.8	50.3	61.2	70.2	73.9	71.9	64.6	53.7	39.2	26.2	49.3
MOUNT CARROLL	MAX	27.6	33.3	44.2	60.6	72.2	80.7	84.0	82.0	74.7	63.9	47.2	33.3	58.6
	MIN	9.1	14.2	24.6	37.1	47.6	56.9	60.8	59.0	50.6	39.8	28.1	16.3	37.0
	MEAN	18.4	23.8	34.4	48.9	59.9	68.9	72.4	70.6	62.7	51.9	37.7	24.8	47.9
MOUNT VERNON	MAX	39.2	44.1	54.4	67.6	77.0	85.8	89.3	87.7	81.6	70.3	55.3	43.8	66.3
	MIN	21.0	24.6	33.7	45.2	53.7	62.5	66.5	64.7	57.2	45.2	35.0	26.5	44.7
	MEAN	30.1	34.4	44.0	56.4	65.4	74.2	77.9	76.2	69.5	57.8	45.2	35.2	55.5
NEWTON 2 NE	MAX	37.2	42.4	52.7	66.6	76.6	85.5	89.0	87.4	81.7	69.7	54.0	41.9	65.4
	MIN	19.8	23.4	32.4	43.8	52.6	61.7	65.5	63.4	55.8	44.1	34.2	25.5	43.5
	MEAN	28.5	32.9	42.5	55.2	64.6	73.6	77.3	75.4	68.8	56.9	44.1	33.7	54.5
OLNEY 2 S	MAX	38.3	43.4	54.0	67.4	76.8	85.6	89.0	87.5	82.2	70.6	54.8	43.1	66.1
	MIN	21.0	24.5	33.4	44.5	53.1	61.9	65.6	63.4	56.4	44.9	35.0	26.5	44.2
	MEAN	29.7	34.0	43.7	56.0	65.0	73.8	77.3	75.5	69.3	57.8	44.9	34.8	55.2
OTTAWA	MAX	31.2	36.7	47.6	63.3	74.5	83.7	86.6	85.0	78.7	67.0	50.4	36.7	61.8
	MIN	14.6	19.4	28.9	40.7	50.8	60.6	64.5	62.7	54.9	43.9	32.3	21.3	41.2
	MEAN	22.9	28.1	38.3	52.0	62.7	72.2	75.6	73.9	66.9	55.5	41.4	29.0	51.5
PALESTINE	MAX	37.3	42.4	52.8	66.9	76.5	85.7	88.7	87.1	81.2	69.6	53.9	42.1	65.4
	MIN	20.2	23.9	32.9	44.1	53.0	62.1	65.6	63.4	56.0	44.4	34.3	25.7	43.8
	MEAN	28.8	33.2	42.9	55.5	64.8	73.9	77.2	75.3	68.6	57.0	44.1	33.9	54.6
PANA	MAX	34.6	39.9	50.8	65.4	74.8	83.4	86.7	84.4	78.6	66.8	51.6	39.7	63.1
	MIN	18.5	22.9	31.7	43.3	53.0	61.8	65.7	63.8	56.6	45.3	34.1	24.7	43.5
	MEAN	26.6	31.4	41.2	54.4	63.9	72.7	76.2	74.1	67.6	56.1	42.9	32.3	53.3
PARIS WATERWORKS	MAX	34.9	40.0	50.4	64.7	74.9	84.1	87.3	85.3	79.8	67.6	51.7	39.8	63.4
	MIN	17.7	21.7	31.0	42.7	52.3	61.4	65.1	62.9	55.9	44.4	33.3	23.8	42.7
	MEAN	26.3	30.9	40.7	53.8	63.7	72.8	76.2	74.1	67.9	56.1	42.5	31.8	53.1
PARK FOREST	MAX	29.6	34.3	44.6	59.0	70.5	80.6	83.9	82.2	75.8	64.4	48.2	35.2	59.0
	MIN	13.1	17.3	27.0	38.3	48.2	58.2	62.9	61.4	53.9	42.8	30.7	19.7	39.5
	MEAN	21.4	25.9	35.8	48.7	59.4	69.4	73.4	71.8	64.9	53.6	39.5	27.5	49.3

ILLINOIS

TEMPERATURE NORMALS (DEG F)

STATION			JAN	FEB	MAR	APR	MAY	JUN	JUL	AUG	SEP	OCT	NOV	DEC	ANN
PEORIA WSO	//R	MAX	29.7	35.2	46.5	61.9	72.5	82.1	85.5	83.4	76.7	64.8	48.5	35.4	60.2
		MIN	13.3	18.4	28.1	40.6	50.6	60.2	64.6	62.7	54.5	42.9	30.9	20.2	40.6
		MEAN	21.5	26.8	37.3	51.3	61.6	71.2	75.0	73.1	65.6	53.9	39.8	27.8	50.4
PERU 2 W		MAX	30.3	35.5	46.4	62.6	74.2	83.7	86.7	84.8	78.2	66.3	49.3	35.9	61.2
		MIN	13.0	17.9	27.5	39.4	49.6	59.1	63.3	61.3	53.2	42.3	31.0	20.2	39.8
		MEAN	21.7	26.7	37.0	51.0	61.9	71.4	75.0	73.1	65.7	54.3	40.2	28.1	50.5
PIPER CITY 3 SE		MAX	31.3	36.4	47.6	63.0	74.5	83.7	86.1	84.1	79.1	67.0	49.8	36.8	61.6
		MIN	14.6	19.1	28.6	39.9	49.7	59.2	62.7	60.6	53.5	42.6	31.6	21.1	40.3
		MEAN	23.0	27.7	38.1	51.4	62.2	71.5	74.4	72.4	66.3	54.8	40.7	29.0	51.0
PONTIAC		MAX	31.3	36.6	47.6	63.4	74.5	83.2	85.7	83.7	78.3	66.9	50.0	36.7	61.5
		MIN	15.0	19.8	29.2	41.0	51.2	60.8	64.6	62.6	55.2	44.0	32.3	21.8	41.5
		MEAN	23.2	28.2	38.4	52.2	62.9	72.0	75.2	73.2	66.8	55.4	41.2	29.3	51.5
QUINCY FAA AIRPORT		MAX	31.9	37.6	48.1	63.1	73.4	82.8	87.0	84.6	78.0	66.5	50.7	37.8	61.8
		MIN	15.1	20.4	29.6	42.6	52.7	62.0	66.2	64.1	55.8	44.7	32.4	22.1	42.3
		MEAN	23.6	29.0	38.9	52.9	63.1	72.4	76.7	74.3	66.9	55.6	41.6	29.9	52.1
RANTOUL		MAX	31.7	36.6	47.3	62.6	73.8	83.1	86.1	84.1	78.4	66.2	49.8	37.1	61.4
		MIN	15.3	19.5	29.0	40.9	51.0	60.6	64.4	62.5	54.8	43.3	31.9	22.1	41.3
		MEAN	23.5	28.1	38.2	51.8	62.5	71.9	75.3	73.3	66.6	54.8	40.8	29.6	51.4
ROCKFORD WSO	//R	MAX	26.6	31.8	42.5	58.5	70.6	80.1	83.7	81.9	74.6	63.0	46.3	32.6	57.7
		MIN	9.8	14.9	25.1	37.3	47.6	57.6	62.2	60.6	52.0	41.0	28.7	17.0	37.8
		MEAN	18.3	23.4	33.8	47.9	59.1	68.9	73.0	71.3	63.3	52.1	37.5	24.8	47.8
ROSICLARE		MAX	42.6	47.4	57.3	69.8	77.7	85.4	88.6	87.6	81.8	71.3	57.4	46.5	67.8
		MIN	24.1	27.1	35.8	46.3	54.3	62.5	66.2	64.8	58.1	46.0	36.2	28.6	45.9
		MEAN	33.4	37.2	46.6	58.1	66.0	74.0	77.6	76.3	70.0	58.7	46.8	37.6	56.9
RUSHVILLE		MAX	33.2	39.1	49.6	65.3	75.6	84.6	88.4	86.2	80.0	67.9	51.8	38.6	63.4
		MIN	15.0	19.8	29.2	41.7	51.1	60.7	64.6	62.2	54.2	43.2	31.3	21.3	41.2
		MEAN	24.1	29.5	39.4	53.5	63.4	72.7	76.5	74.3	67.1	55.6	41.6	30.0	52.3
SPARTA		MAX	40.1	45.7	55.7	68.9	78.2	86.6	90.4	88.9	82.8	71.5	56.0	44.5	67.4
		MIN	22.4	26.5	34.9	46.2	54.8	63.6	67.3	65.4	58.1	46.8	36.3	27.9	45.9
		MEAN	31.3	36.1	45.3	57.6	66.5	75.1	78.9	77.1	70.5	59.2	46.2	36.2	56.7
SPRINGFIELD WSO	R	MAX	32.8	38.0	48.9	64.0	74.6	84.1	87.1	84.7	79.3	67.5	51.2	38.4	62.6
		MIN	16.3	20.9	30.3	42.6	52.5	62.0	65.9	63.7	55.8	44.4	32.9	23.0	42.5
		MEAN	24.6	29.5	39.6	53.4	63.6	73.1	76.5	74.2	67.6	55.9	42.1	30.7	52.6
STOCKTON		MAX	25.7	31.3	42.2	59.1	70.8	79.5	83.2	81.1	73.4	62.3	45.3	31.4	57.1
		MIN	9.2	14.1	24.3	37.2	47.9	57.5	61.7	60.0	51.8	41.2	28.5	16.2	37.5
		MEAN	17.4	22.7	33.3	48.2	59.4	68.5	72.5	70.6	62.6	51.8	37.0	23.8	47.3
TUSCOLA		MAX	34.3	39.8	51.0	65.9	76.4	85.3	88.2	86.1	81.1	68.8	52.2	39.7	64.1
		MIN	17.1	21.2	30.4	41.6	51.4	60.7	64.3	62.0	55.0	43.7	32.4	23.4	41.9
		MEAN	25.7	30.6	40.7	53.8	63.9	73.0	76.3	74.1	68.0	56.3	42.3	31.6	53.0
URBANA		MAX	32.3	37.3	47.8	62.6	73.5	82.8	85.5	83.6	78.3	66.2	49.8	37.6	61.4
		MIN	17.0	21.3	30.4	41.9	52.0	60.9	64.8	62.8	55.5	44.4	33.2	23.3	42.3
		MEAN	24.7	29.3	39.1	52.3	62.8	71.9	75.2	73.2	66.9	55.3	41.5	30.5	51.9
VIRDEN 1 N		MAX	34.8	40.1	51.1	65.7	75.8	84.8	88.1	86.1	80.8	68.6	52.6	40.2	64.1
		MIN	17.2	21.9	30.8	42.8	52.4	61.5	64.7	62.6	55.4	44.4	33.0	23.5	42.5
		MEAN	26.0	31.1	41.0	54.3	64.2	73.2	76.4	74.4	68.1	56.5	42.8	31.9	53.3
WALNUT		MAX	28.7	34.5	45.6	62.0	73.9	82.9	86.0	84.1	77.3	65.4	48.2	34.5	60.3
		MIN	11.8	17.0	26.7	39.2	50.0	59.5	63.3	61.4	53.6	42.7	30.3	18.7	39.5
		MEAN	20.3	25.8	36.2	50.6	62.0	71.2	74.7	72.8	65.4	54.1	39.3	26.6	49.9
WATSEKA 2 NW		MAX	31.2	36.1	47.3	62.1	73.2	82.7	85.4	83.2	77.6	65.8	49.5	36.6	60.9
		MIN	15.0	18.8	28.8	40.1	49.7	59.1	62.6	60.4	53.2	42.1	31.5	21.4	40.2
		MEAN	23.1	27.5	38.1	51.1	61.5	70.9	74.0	71.8	65.5	54.0	40.5	29.0	50.6

ILLINOIS

TEMPERATURE NORMALS (DEG F)

STATION		JAN	FEB	MAR	APR	MAY	JUN	JUL	AUG	SEP	OCT	NOV	DEC	ANN
WAUKEGAN 4 WSW	MAX	28.3	32.7	41.9	55.7	67.4	77.1	81.4	80.1	73.3	62.3	47.2	34.1	56.8
	MIN	12.1	16.4	25.5	36.6	46.2	56.1	61.7	60.8	53.4	42.8	30.7	19.1	38.5
	MEAN	20.2	24.6	33.7	46.2	56.8	66.6	71.5	70.5	63.4	52.6	39.0	26.6	47.6
WHEATON 3 SE	MAX	29.2	34.6	45.6	61.3	73.0	82.2	85.4	83.6	77.2	64.9	48.2	35.1	60.0
	MIN	12.6	17.2	26.6	37.8	47.3	57.2	61.6	60.3	52.6	42.3	30.6	19.4	38.8
	MEAN	20.9	25.9	36.1	49.6	60.2	69.7	73.5	71.9	64.9	53.6	39.4	27.3	49.4
WHITE HALL 1 E	MAX	36.2	41.9	52.5	66.8	76.1	84.7	88.1	86.2	80.1	69.3	53.9	41.6	64.8
	MIN	17.4	22.1	31.1	43.3	52.8	61.8	65.6	63.6	56.0	44.9	33.4	23.9	43.0
	MEAN	26.8	32.0	41.8	55.1	64.4	73.3	76.9	74.9	68.1	57.1	43.7	32.7	53.9
WINDSOR	MAX	34.9	40.2	51.0	65.5	75.4	.84.4	87.8	86.0	80.7	68.4	52.2	40.1	63.9
	MIN	17.8	22.1	31.0	42.5	52.3	61.1	64.7	62.6	55.6	44.2	33.5	24.0	42.6
	MEAN	26.4	31.2	41.0	54.0	63.9	72.8	76.3	74.3	68.2	56.3	42.8	32.1	53.3

ILLINOIS

PRECIPITATION NORMALS (INCHES)

STATION	JAN	FEB	MAR	APR	MAY	JUN	JUL	AUG	SEP	OCT	NOV	DEC	ANN
ALBION	2.62	2.84	4.46	4.46	4.53	4.09	3.91	3.45	2.71	2.70	3.53	3.67	42.97
ALEDO	1.28	1.14	2.53	3.98	3.78	4.12	4.32	3.52	3.55	3.15	1.82	1.66	34.85
ALTON DAM 26	1.71	2.12	3.34	3.86	4.15	3.83	3.85	2.95	2.84	2.47	2.76	2.44	36.32
ANNA 1 E	3.05	3.22	5.08	4.49	4.82	3.97	4.11	3.67	3.25	2.37	3.84	3.77	45.64
ANTIOCH 2 NW	1.69	1.27	2.56	3.74	3.20	4.04	4.28	3.71	3.24	2.34	2.21	1.97	34.25
AURORA COLLEGE	1.73	1.42	2.51	3.86	3.64	4.18	4.21	3.68	3.35	2.73	2.08	2.23	35.62
AVON 5 NE	1.52	1.33	2.65	3.84	3.57	4.35	4.29	3.67	3.40	2.90	1.82	1.76	35.10
BEARDSTOWN	1.58	1.67	3.05	4.01	3.96	4.01	3.87	3.79	3.52	3.07	2.16	2.12	36.81
BELLEVILLE SO ILL UNIV	1.91	2.20	3.44	3.74	3.85	3.98	3.36	3.29	3.18	2.37	2.92	2.51	36.75
BENTLEY 1 E	1.41	1.38	2.89	3.84	3.79	3.49	4.05	3.80	3.77	2.92	1.81	1.68	34.83
BENTON FOREST SERVICE	2.72	2.67	4.35	4.14	3.87	3.74	3.44	3.52	2.95	2.57	3.47	3.33	40.77
BLOOMINGTON WATERWORKS	1.65	1.42	2.85	3.86	3.74	3.98	3.88	3.45	3.70	2.45	2.11	2.08	35.17
BLUFFS	1.60	1.68	3.28	3.91	4.13	4.06	3.71	3.59	3.66	2.91	2.21	2.12	36.86
BROOKPORT DAM 52	3.88	3.48	5.26	4.52	4.36	4.14	3.87	3.35	3.58	2.40	4.05	4.13	47.02
CAIRO WSO R	3.47	3.42	4.96	4.44	4.90	4.36	3.96	3.97	3.50	2.54	3.97	4.16	47.65
CANTON 1 ESE	1.69	1.61	2.98	4.04	4.07	4.16	4.06	3.93	3.67	3.04	2.15	2.13	37.53
CARBONDALE SEWAGE PL	2.67	2.94	4.53	4.11	4.29	4.03	3.94	3.78	3.00	2.36	3.71	3.23	42.59
CARLINVILLE	1.63	1.93	3.32	3.96	4.30	3.76	3.68	3.35	3.32	2.45	2.43	2.26	36.39
CARLYLE RESERVOIR	1.97	2.17	3.74	3.84	3.85	4.19	3.54	2.89	2.82	2.40	3.06	2.79	37.26
CARMI 6 NW	2.73	2.67	4.48	3.93	4.20	3.53	3.38	3.13	2.72	2.32	3.27	3.39	39.75
CENTRALIA 2 SW	2.10	2.53	4.01	4.32	4.29	4.05	4.17	3.01	3.11	2.42	3.26	2.91	40.18
CHANNAHON DRESDEN ISL	1.72	1.56	2.83	4.07	3.80	4.13	4.21	3.27	3.45	2.49	2.16	2.25	35.94
CHARLESTON	2.25	1.95	3.37	3.74	4.00	4.27	4.34	3.19	2.96	2.41	2.99	2.74	38.21
CHENOA	1.56	1.28	2.62	4.14	3.42	3.78	4.23	3.47	3.32	2.43	1.94	2.18	34.37
CHESTER	2.17	2.58	4.07	4.10	4.01	3.79	4.26	3.29	2.98	2.16	3.62	3.02	40.05
CHICAGO O HARE WSO R	1.60	1.31	2.59	3.66	3.15	4.08	3.63	3.53	3.35	2.28	2.06	2.10	33.34
CHICAGO UNIVERSITY //R	1.86	1.51	2.81	4.01	3.35	3.91	3.84	3.77	3.05	2.57	2.23	2.37	35.28
CHICAGO MIDWAY AP	1.82	1.51	2.77	4.02	3.24	4.00	4.16	3.66	3.20	2.57	2.14	2.42	35.51
CHILLICOTHE	1.51	1.33	2.74	3.89	3.77	4.35	4.09	3.16	3.60	2.61	2.14	1.97	35.16
CLINTON 1 SSW	1.76	1.80	3.50	4.09	3.98	4.41	3.84	3.85	3.30	2.75	2.37	2.34	37.99
DANVILLE	1.97	1.89	3.00	3.93	3.98	4.48	4.23	3.90	3.08	2.78	2.70	2.60	38.54
DANVILLE SEWAGE PLANT	2.06	1.99	3.09	4.07	3.69	4.26	4.31	3.75	2.71	2.55	2.73	2.73	37.94
DECATUR	2.03	2.03	3.48	4.13	4.09	4.48	4.17	3.69	3.21	2.63	2.56	2.62	39.12
DE KALB	1.66	1.28	2.52	3.65	3.52	4.55	4.62	3.69	3.54	2.86	2.40	2.00	36.29
DIXON	1.52	1.27	2.52	3.81	3.63	4.44	3.86	3.77	3.24	2.72	2.03	1.98	34.79
DU QUOIN 1 NNW	2.43	2.60	4.40	4.00	3.77	3.37	3.75	3.27	2.84	2.31	3.61	3.25	39.60
EDWARDSVILLE	1.90	2.15	3.50	3.84	3.96	4.31	3.95	3.27	3.04	2.56	2.72	2.51	37.71
EFFINGHAM	2.13	2.19	3.62	3.77	3.79	4.53	4.23	2.60	3.20	2.39	3.07	2.93	38.45
ELGIN	1.74	1.34	2.63	3.92	3.42	4.59	4.32	3.34	3.52	2.54	2.29	2.11	35.76
FAIRFIELD RADIO WFIW	2.47	2.60	4.44	4.41	4.64	4.08	4.52	3.33	3.02	2.83	3.66	3.46	43.46
FLORA 5 NW	2.42	2.32	4.10	4.04	4.40	4.42	3.82	3.00	2.94	2.51	3.19	3.29	40.45
FREEPORT SEWAGE PLANT	1.28	1.09	2.30	3.42	3.56	3.99	4.28	3.59	3.51	2.34	2.26	1.73	33.35
FULTON DAM 13	1.32	1.09	2.23	3.53	3.75	4.11	4.06	4.32	3.37	2.69	2.07	1.73	34.27
GALENA 1 N	1.36	1.16	2.42	3.65	3.94	4.08	4.07	4.19	3.84	2.59	2.13	1.73	35.16
GALESBURG	1.59	1.36	2.78	4.03	3.61	4.47	4.23	3.89	3.50	2.73	1.86	1.94	35.99
GALVA	1.56	1.32	2.73	4.06	4.06	4.38	4.12	3.71	3.49	3.02	1.97	2.00	36.42
GENESEO	1.49	1.23	2.62	3.88	4.03	4.13	4.48	3.84	3.44	2.78	1.98	1.83	35.73
GIBSON CITY	1.79	1.52	3.15	3.98	3.77	3.62	4.07	3.32	3.26	2.42	2.08	2.26	35.24
GLADSTONE DAM 18	1.48	1.07	2.52	3.68	3.66	3.96	3.71	3.98	3.61	3.01	1.76	1.65	34.09
GOLCONDA DAM 51	3.55	3.30	5.17	4.50	4.29	4.13	4.46	3.13	3.13	2.43	3.97	4.01	46.07
GOLDEN 1 NW	1.48	1.41	2.92	3.68	3.69	3.85	4.41	3.43	3.78	2.97	1.85	1.69	35.16
GRAFTON	1.66	2.05	3.25	3.70	3.90	3.56	3.69	3.15	3.04	2.42	2.65	2.22	35.29
GRAND TOWER 2 N	2.73	2.92	4.63	4.38	4.66	4.77	4.51	4.10	3.19	2.45	3.88	3.38	45.60
GREENUP	2.15	2.08	3.56	3.97	4.00	3.73	4.36	2.98	2.79	2.28	3.15	2.96	38.01
GREENVILLE 1 E	1.80	2.01	3.32	3.78	4.13	4.18	3.47	2.94	2.87	2.33	2.90	2.67	36.40

ILLINOIS

PRECIPITATION NORMALS (INCHES)

STATION	JAN	FEB	MAR	APR	MAY	JUN	JUL	AUG	SEP	OCT	NOV	DEC	ANN
GRIGGSVILLE	1.56	1.60	3.07	3.82	4.11	3.80	4.04	3.57	3.56	3.12	2.17	2.09	36.51
HARRISBURG	3.04	3.08	4.88	4.33	4.65	3.92	3.66	3.48	2.85	2.44	3.61	3.66	43.60
HARRISBURG DISPOSAL PL	2.93	2.88	4.44	4.02	4.33	3.64	3.32	3.26	2.67	2.29	3.57	3.44	40.79
HAVANA POWER STATION	1.59	1.58	2.95	3.68	3.66	3.72	4.04	3.67	3.48	2.90	2.09	2.15	35.51
HILLSBORO 2 SSW	1.92	2.11	3.53	4.03	4.35	4.29	3.88	3.62	3.06	2.76	2.73	2.71	38.99
HOOPESTON	1.77	1.66	2.90	3.90	3.70	4.11	4.17	3.30	3.32	2.57	2.37	2.20	35.97
HUTSONVILLE POWER PL	2.27	2.12	3.61	3.58	4.17	4.00	3.84	2.89	3.18	2.39	3.04	2.72	37.81
ILLINOIS CITY DAM 16 //	1.47	1.14	2.48	3.72	3.67	4.16	4.04	3.60	3.34	2.66	1.97	1.73	33.98
JACKSONVILLE	1.46	1.61	3.14	3.99	4.47	4.05	3.95	3.66	3.66	2.95	2.21	1.96	37.11
JERSEYVILLE 2 SW	1.58	1.79	3.17	3.91	3.72	3.84	3.62	3.51	3.30	2.58	2.48	1.99	35.49
JOLIET BRANDON RD DAM	1.51	1.48	2.66	3.95	3.75	4.17	4.17	3.49	3.47	2.67	2.10	2.12	35.54
KEITHSBURG 1 NW	1.42	1.17	2.37	3.46	3.70	3.71	3.97	3.26	3.23	2.63	1.62	1.59	32.13
KEWANEE	1.69	1.26	2.59	4.08	3.87	4.54	4.41	3.44	3.23	3.06	2.04	2.23	36.44
LA HARPE	1.60	1.39	2.90	3.94	3.62	4.54	3.81	4.20	4.16	3.02	1.89	1.83	36.90
LAWRENCEVILLE	2.90	2.64	4.33	4.00	4.24	4.30	4.36	3.42	2.82	2.57	3.49	3.34	42.41
LINCOLN	1.81	1.72	3.18	4.01	4.01	4.38	3.85	3.59	3.15	2.62	2.15	2.32	36.79
MACKINAW	1.72	1.62	3.06	4.11	3.78	4.53	3.78	3.42	3.54	2.67	2.21	2.34	36.78
MACOMB	1.47	1.29	2.74	3.76	3.91	4.34	3.92	3.58	3.65	2.85	1.82	1.69	35.02
MARENGO	1.70	1.20	2.52	3.90	3.42	4.53	4.22	3.89	3.58	2.55	2.22	1.98	35.71
MARION 4 NNE	2.80	3.07	4.72	4.10	3.99	3.80	3.61	3.50	3.05	2.19	3.64	3.31	41.78
MARSEILLES LOCK	1.61	1.32	2.59	3.92	3.66	4.07	3.86	3.24	3.29	2.46	1.92	2.07	34.01
MATTOON	1.90	1.96	3.26	3.75	3.71	4.19	4.27	3.14	3.05	2.32	2.73	2.71	36.99
MCLEANSBORO 2 E	2.65	2.63	4.29	4.22	4.08	3.73	3.70	3.14	2.89	2.43	3.54	3.31	40.61
MEDORA	1.71	2.02	3.29	4.01	4.17	3.55	3.71	3.46	3.44	2.75	2.70	2.34	37.15
MINONK 3 NE	1.55	1.41	2.91	3.96	3.76	3.78	4.18	3.33	3.46	2.58	2.04	1.98	34.94
MOLINE WSO R	1.64	1.30	2.77	3.97	4.21	4.32	4.88	3.76	3.74	2.70	1.96	1.92	37.17
MONMOUTH	1.73	1.50	2.86	3.74	3.79	4.06	4.26	3.73	3.53	2.97	1.85	1.97	35.99
MONTICELLO NO 2	1.99	1.78	3.34	4.18	3.65	4.32	4.38	3.87	3.22	2.52	2.49	2.51	38.25
MORRIS	1.37	1.26	2.49	3.81	3.74	4.01	3.85	3.39	3.33	2.46	1.92	1.99	33.62
MORRISON	1.45	1.26	2.66	3.90	4.35	4.11	3.85	4.34	3.30	2.69	2.16	2.09	36.16
MOUNT CARROLL	1.45	1.17	2.40	3.71	4.13	4.16	4.21	3.68	3.45	2.52	2.15	1.89	34.92
MOUNT OLIVE 1 E	1.82	2.11	3.48	3.92	3.93	3.95	3.54	2.96	2.92	2.49	2.66	2.55	36.33
MOUNT PULASKI	1.75	1.84	3.29	4.12	3.54	4.02	4.03	3.19	3.30	2.59	2.15	2.40	36.22
MOUNT STERLING	1.58	1.39	3.09	4.03	4.24	4.03	4.30	3.52	3.49	2.94	2.13	1.96	36.70
MOUNT VERNON	2.26	2.32	3.92	4.23	4.09	3.56	4.13	3.27	3.10	2.33	3.52	3.19	39.92
NASHVILLE 4 NE	1.98	2.45	3.78	4.05	3.88	3.64	3.44	3.15	2.89	2.49	3.11	2.91	37.77
NEW BOSTON DAM 17	1.55	1.15	2.57	3.82	3.68	3.87	4.48	3.67	3.31	2.77	1.67	1.64	34.18
NEWTON 2 NE	2.15	2.33	3.62	3.87	3.62	3.68	4.30	3.40	3.03	2.41	3.12	3.02	38.55
OLNEY 2 S	2.32	2.50	4.09	3.96	4.11	4.08	3.94	3.27	2.57	2.75	3.45	3.28	40.32
OTTAWA	1.63	1.38	2.69	4.00	3.46	4.33	4.07	3.53	3.34	2.39	1.88	2.09	34.79
PALESTINE	2.36	2.49	3.92	3.86	4.18	3.93	4.59	3.21	3.23	2.63	3.24	3.09	40.73
PANA	1.93	2.14	3.63	3.78	3.93	4.40	4.24	3.70	3.16	2.52	2.75	2.82	39.00
PARIS WATERWORKS	2.38	2.21	3.54	3.71	4.09	4.32	4.44	3.41	2.80	2.38	2.92	3.26	39.46
PARK FOREST	1.53	1.42	2.58	4.07	3.66	4.33	4.12	3.57	3.03	2.71	2.18	2.04	35.24
PAW PAW	1.67	1.25	2.53	4.07	3.71	4.20	4.22	3.85	3.55	2.91	2.11	2.04	36.11
PAYSON	1.59	1.63	2.98	3.84	4.31	3.93	4.39	4.05	3.72	3.11	2.00	1.85	37.40
PEORIA WSO //R	1.60	1.41	2.86	3.81	3.84	3.88	3.99	3.39	3.63	2.51	1.96	2.01	34.89
PEOTONE	1.67	1.44	2.64	4.27	3.83	4.74	4.18	3.48	3.11	2.80	2.18	2.28	36.62
PERU 2 W	1.65	1.30	2.78	3.87	3.92	3.96	4.05	3.31	3.38	2.68	1.85	2.08	34.83
PIPER CITY 3 SE	1.65	1.46	2.60	3.67	3.60	3.88	4.00	3.32	3.11	2.23	1.93	2.00	33.45
PONTIAC	1.58	1.28	2.51	3.84	3.66	4.22	4.46	2.83	3.33	2.30	1.83	2.01	33.85
PRAIRIE DU ROCHER 1WSW	2.02	2.34	3.83	3.90	4.00	3.43	4.05	3.67	3.08	2.31	3.26	2.71	38.60
PRINCEVILLE 1 N	1.40	1.18	2.64	3.93	3.81	3.97	4.04	3.25	3.61	2.79	1.84	1.84	34.30
QUINCY FAA AIRPORT	1.36	1.44	3.15	3.84	4.38	4.03	4.32	3.90	4.19	3.26	2.06	1.72	37.65
QUINCY DAM 21 //	1.56	1.49	2.90	3.58	3.94	3.72	3.99	3.51	3.86	2.81	1.77	1.78	34.91

ILLINOIS

PRECIPITATION NORMALS (INCHES)

STATION	JAN	FEB	MAR	APR	MAY	JUN	JUL	AUG	SEP	OCT	NOV	DEC	ANN
QUINCY MEMORIAL BRIDGE	1.66	1.58	3.07	3.75	4.14	3.83	4.13	3.87	3.99	2.92	1.87	1.80	36.61
RANTOUL	1.94	1.64	2.94	3.74	3.45	3.95	4.20	3.40	3.02	2.43	2.40	2.28	35.39
RED BUD	1.87	2.47	3.86	3.67	4.23	3.38	4.03	3.34	2.98	2.31	3.19	2.88	38.21
ROCKFORD WSO //R	1.42	1.18	2.59	4.22	3.75	4.58	4.50	3.71	3.70	2.92	2.30	1.91	36.78
ROSICLARE	3.52	3.40	5.13	4.49	4.49	4.19	3.97	3.40	2.99	2.42	3.89	3.96	45.85
RUSHVILLE	1.57	1.64	3.30	4.13	4.26	4.08	3.81	3.99	3.79	3.18	2.13	2.17	38.05
SAINTE MARIE	2.29	2.32	3.77	3.86	3.90	3.94	4.05	3.40	2.96	2.45	3.21	3.06	39.21
SHAWNEETOWN NEW TOWN	3.14	3.04	4.80	4.02	4.31	4.12	4.01	3.57	2.98	2.35	3.64	3.36	43.34
SIDELL	1.96	1.68	2.81	3.79	3.76	3.97	4.14	3.23	2.75	2.40	2.72	2.49	35.70
SPARTA	1.89	2.51	4.02	4.11	4.21	3.38	4.19	3.51	3.11	2.40	3.37	3.00	39.70
SPRINGFIELD WSO R	1.56	1.78	3.14	3.97	3.34	3.71	3.53	3.20	3.05	2.52	1.93	2.05	33.78
STOCKTON	1.51	1.30	2.50	3.59	3.68	3.85	3.99	3.71	3.65	2.80	2.18	1.79	34.55
STREATOR 2 ENE	1.60	1.48	2.66	3.89	3.57	4.14	4.35	3.21	3.37	2.45	2.01	2.09	34.82
TISKILWA	1.46	1.32	2.72	4.09	3.89	4.18	4.04	3.53	3.38	2.80	1.84	1.94	35.19
TUSCOLA	2.09	2.00	3.45	3.88	3.74	4.05	4.19	3.83	3.09	2.44	2.77	2.75	38.28
URBANA	1.97	1.88	3.32	3.84	3.59	3.92	4.35	3.66	3.02	2.51	2.48	2.50	37.04
UTICA STARVED ROCK DAM	1.63	1.44	2.72	4.06	3.76	4.20	4.20	3.32	3.61	2.52	2.02	2.02	35.50
VIRDEN 1 N	1.78	2.03	3.30	3.83	4.09	3.96	3.67	3.59	3.32	2.50	2.22	2.16	36.45
VIRGINIA	1.51	1.77	3.29	3.74	3.86	4.00	3.77	3.71	3.32	3.00	2.24	2.04	36.25
WALNUT	1.46	1.25	2.80	4.34	3.87	4.22	3.73	3.82	3.67	2.72	1.98	1.91	35.77
WATERLOO 1 WSW	1.89	2.17	3.52	3.58	4.01	3.60	3.85	2.81	3.02	2.46	3.08	2.65	36.64
WATSEKA 2 NW	1.85	1.71	3.28	4.56	4.05	4.57	4.66	3.32	3.27	2.58	2.62	2.54	39.01
WAUKEGAN 4 WSW	1.83	1.30	2.60	3.69	3.09	4.09	3.81	3.48	3.08	2.33	2.21	2.14	33.65
WAYNE CITY	2.44	2.42	4.07	4.20	4.45	3.77	3.66	3.12	2.92	2.46	3.47	3.19	40.17
WHEATON 3 SE	1.84	1.43	2.64	3.94	3.62	3.82	4.21	3.78	3.26	2.49	2.15	2.28	35.46
WHITE HALL 1 E	1.45	1.62	2.98	3.57	4.05	3.78	3.59	3.15	3.10	2.70	2.05	1.96	34.00
WINDSOR	1.99	2.12	3.47	3.80	3.74	4.47	3.98	3.31	3.04	2.47	2.73	2.78	37.90

ILLINOIS

HEATING DEGREE DAY NORMALS (BASE 65 DEG F)

STATION	JUL	AUG	SEP	OCT	NOV	DEC	JAN	FEB	MAR	APR	MAY	JUN	ANN
ALEDO	0	5	81	351	768	1172	1367	1078	874	410	153	15	6274
ALTON DAM 26	0	0	32	265	630	989	1156	916	719	306	108	8	5129
ANNA 1 E	0	0	26	219	552	874	1011	795	607	239	78	5	4406
ANTIOCH 2 NW	9	21	111	414	807	1221	1420	1159	983	546	262	56	7009
AURORA COLLEGE	0	10	82	392	786	1197	1392	1120	918	480	213	28	6618
BELLEVILLE SO ILL UNIV	0	0	40	254	603	930	1091	851	665	276	98	10	4818
CAIRO WSO R	0	0	12	170	495	800	942	731	540	177	46	0	3913
CARBONDALE SEWAGE PL	0	0	36	253	579	887	1032	815	619	246	91	5	4563
CARLINVILLE	0	0	42	276	651	995	1172	916	716	311	120	9	5208
CARLYLE RESERVOIR	0	0	46	275	624	961	1132	899	694	295	114	9	5049
CHARLESTON	0	0	43	290	663	1014	1187	941	744	329	128	11	5350
CHENOA	0	6	69	347	744	1128	1318	1047	840	408	169	18	6094
CHICAGO O HARE WSO R	0	9	75	368	756	1156	1352	1092	899	486	224	38	6455
CHICAGO UNIVERSITY //R	0	0	44	306	696	1085	1271	1028	862	471	213	34	6010
CHICAGO MIDWAY AP	0	5	58	341	729	1119	1314	1058	868	459	198	28	6177
DANVILLE	0	0	59	317	684	1042	1218	977	772	364	150	16	5599
DECATUR	0	0	35	291	672	1035	1215	960	760	339	132	14	5453
DE KALB	5	10	85	386	801	1218	1420	1134	936	486	199	23	6703
DIXON	0	9	79	384	780	1209	1404	1117	911	447	180	22	6542
DU QUOIN 1 NNW	0	0	32	248	579	905	1054	820	623	246	85	6	4598
EFFINGHAM	0	0	70	309	654	998	1169	932	725	325	134	12	5328
FAIRFIELD RADIO WFIW	0	0	41	261	594	921	1076	851	656	275	103	7	4785
FLORA 5 NW	0	0	41	261	618	949	1107	879	667	285	108	8	4923
FULTON DAM 13	0	6	74	361	768	1197	1404	1131	930	456	175	17	6519
GALESBURG	0	0	65	363	762	1166	1367	1078	884	426	171	20	6302
GALVA	0	6	76	369	771	1184	1386	1103	899	429	177	23	6423
GRIGGSVILLE	0	6	52	293	663	1042	1228	960	760	331	133	16	5484
HARRISBURG	0	0	22	215	543	849	992	778	583	211	72	0	4265
HILLSBORO 2 SSW	0	0	54	271	639	983	1156	904	707	302	120	9	5145
HOOPESTON	0	6	62	333	714	1088	1268	1019	815	399	160	16	5880
JACKSONVILLE	0	7	60	310	678	1048	1228	977	775	347	148	17	5595
JERSEYVILLE 2 SW	0	0	53	287	645	992	1159	907	710	308	124	12	5197
KEWANEE	0	7	74	364	765	1178	1373	1095	890	430	175	25	6376
LA HARPE	0	0	56	326	726	1116	1318	1036	825	381	150	13	5947
LINCOLN	0	0	46	309	693	1066	1252	988	784	356	135	11	5640
MARENGO	6	12	98	411	816	1237	1435	1154	958	510	225	39	6901
MATTOON	0	0	57	320	693	1045	1221	972	778	363	153	14	5616
MCLEANSBORO 2 E	0	0	38	258	582	908	1057	840	642	260	103	8	4696
MINONK 3 NE	0	6	62	347	756	1156	1345	1067	859	411	172	17	6198
MOLINE WSO R	0	5	75	370	780	1197	1411	1117	908	438	177	20	6498
MONMOUTH	0	0	59	334	738	1125	1330	1044	843	394	151	15	6033
MORRISON	0	10	78	371	774	1203	1401	1114	905	441	180	22	6499
MOUNT CARROLL	6	13	116	423	819	1246	1445	1154	949	483	201	35	6890
MOUNT VERNON	0	0	36	256	594	924	1082	857	651	271	102	0	4773
NEWTON 2 NE	0	0	48	284	627	970	1132	899	698	305	125	11	5099
OLNEY 2 S	0	0	40	250	603	936	1094	868	660	278	108	6	4843
OTTAWA	0	0	44	318	708	1116	1305	1033	828	390	155	15	5912
PALESTINE	0	0	47	271	627	964	1122	890	685	293	117	11	5027
PANA	0	0	48	295	663	1014	1190	941	738	324	128	11	5352
PARIS WATERWORKS	0	0	48	299	675	1029	1200	955	753	342	136	12	5449
PARK FOREST	0	10	76	368	765	1163	1352	1095	905	489	223	37	6483
PEORIA WSO //R	0	5	64	361	756	1153	1349	1070	859	411	176	22	6226
PERU 2 W	5	12	70	351	744	1144	1342	1072	868	420	182	24	6234
PIPER CITY 3 SE	0	6	62	332	729	1116	1302	1044	834	408	161	17	6011
PONTIAC	0	0	50	319	714	1107	1296	1030	825	389	152	16	5898

ILLINOIS

HEATING DEGREE DAY NORMALS (BASE 65 DEG F)

STATION		JUL	AUG	SEP	OCT	NOV	DEC	JAN	FEB	MAR	APR	MAY	JUN	ANN
QUINCY FAA AIRPORT		0	0	58	314	702	1088	1283	1008	809	368	146	13	5789
RANTOUL		0	0	76	333	726	1097	1287	1033	831	396	160	20	5959
ROCKFORD WSO	//R	5	11	99	412	825	1246	1448	1165	967	513	227	34	6952
ROSICLARE		0	0	35	226	546	849	980	778	581	224	84	9	4312
RUSHVILLE		0	0	56	314	702	1085	1268	994	794	351	148	15	5727
SPARTA		0	0	30	220	564	893	1045	809	620	239	85	6	4511
SPRINGFIELD WSO	R	0	0	48	307	687	1063	1252	994	787	354	149	13	5654
STOCKTON		5	10	108	420	840	1277	1476	1184	983	504	211	37	7055
TUSCOLA		0	0	43	293	681	1035	1218	963	753	341	131	11	5469
URBANA		0	0	57	320	705	1070	1249	1000	803	381	155	18	5758
VIRDEN 1 N		0	0	43	288	666	1026	1209	949	744	328	124	10	5387
WALNUT		0	7	68	356	771	1190	1386	1098	893	432	163	17	6381
WATSEKA 2 NW		0	13	79	359	735	1116	1299	1050	834	417	178	22	6102
WAUKEGAN 4 WSW		7	17	94	395	780	1190	1389	1131	970	564	278	66	6881
WHEATON 3 SE		0	7	72	365	768	1169	1367	1095	896	462	200	26	6427
WHITE HALL 1 E		0	0	38	269	639	1001	1184	924	719	306	117	9	5206
WINDSOR		0	0	46	288	666	1020	1197	946	744	337	130	11	5385

ILLINOIS

COOLING DEGREE DAY NORMALS (BASE 65 DEG F)

STATION	JAN	FEB	MAR	APR	MAY	JUN	JUL	AUG	SEP	OCT	NOV	DEC	ANN
ALEDO	0	0	0	0	72	207	310	247	87	19	0	0	942
ALTON DAM 26	0	0	0	9	114	299	425	369	161	33	0	0	1410
ANNA 1 E	0	0	9	17	121	296	409	363	182	33	0	0	1430
ANTIOCH 2 NW	0	0	0	0	29	122	216	182	45	11	0	0	605
AURORA COLLEGE	0	0	0	0	49	154	252	209	55	14	0	0	733
BELLEVILLE SO ILL UNIV	0	0	8	12	104	283	394	335	160	24	0	0	1320
CAIRO WSO R	0	0	13	30	179	378	496	431	231	49	0	0	1807
CARBONDALE SEWAGE PL	0	0	8	18	122	293	406	350	168	30	0	0	1395
CARLINVILLE	0	0	0	8	98	258	372	313	141	24	0	0	1214
CARLYLE RESERVOIR	0	0	0	7	118	282	403	338	166	33	0	0	1347
CHARLESTON	0	0	0	0	97	245	353	295	139	27	0	0	1156
CHENOA	0	0	0	0	69	207	295	232	96	18	0	0	917
CHICAGO O HARE WSO R	0	0	0	0	41	146	252	223	66	12	0	0	740
CHICAGO UNIVERSITY //R	0	0	0	0	42	166	293	262	92	18	0	0	873
CHICAGO MIDWAY AP	0	0	0	0	59	196	310	275	94	21	0	0	955
DANVILLE	0	0	0	0	82	217	316	261	116	17	0	0	1009
DECATUR	0	0	0	6	101	251	360	298	131	28	0	0	1175
DE KALB	0	0	0	0	50	167	266	212	61	11	0	0	767
DIXON	0	0	0	0	63	187	282	232	70	21	0	0	855
DU QUOIN 1 NNW	0	0	9	15	116	291	409	347	170	34	0	0	1391
EFFINGHAM	0	0	0	10	103	252	366	307	157	36	0	0	1231
FAIRFIELD RADIO WFIW	0	0	8	11	106	271	375	322	161	29	0	0	1283
FLORA 5 NW	0	0	0	9	105	263	375	319	155	22	0	0	1248
FULTON DAM 13	0	0	0	0	60	185	295	241	71	17	0	0	869
GALESBURG	0	0	0	0	66	200	308	250	74	19	0	0	917
GALVA	0	0	0	0	69	194	289	233	76	18	0	0	879
GRIGGSVILLE	0	0	0	10	108	253	371	304	130	32	0	0	1208
HARRISBURG	0	0	13	22	146	327	446	391	193	29	0	0	1567
HILLSBORO 2 SSW	0	0	0	8	102	255	366	301	147	23	0	0	1202
HOOPESTON	0	0	0	0	77	211	304	245	104	17	0	0	958
JACKSONVILLE	0	0	0	8	101	242	342	277	120	28	0	0	1118
JERSEYVILLE 2 SW	0	0	0	11	102	255	366	301	140	29	0	0	1204
KEWANEE	0	0	0	0	72	205	301	243	77	17	0	0	915
LA HARPE	0	0	0	0	82	217	332	264	89	22	0	0	1006
LINCOLN	0	0	0	0	94	239	344	277	112	18	0	0	1084
MARENGO	0	0	0	0	45	147	241	189	47	11	0	0	680
MATTOON	0	0	0	0	91	236	335	273	123	17	0	0	1075
MCLEANSBORO 2 E	0	0	0	14	125	290	400	347	176	31	0	0	1383
MINONK 3 NE	0	0	0	0	73	206	296	239	92	16	0	0	922
MOLINE WSO R	0	0	0	0	62	200	307	250	66	14	0	0	899
MONMOUTH	0	0	0	0	74	204	316	258	83	18	0	0	953
MORRISON	0	0	0	0	62	178	280	224	66	21	0	0	831
MOUNT CARROLL	0	0	0	0	43	152	236	187	47	17	0	0	682
MOUNT VERNON	0	0	0	13	114	280	400	347	171	33	0	0	1358
NEWTON 2 NE	0	0	0	11	112	269	381	322	162	33	0	0	1290
OLNEY 2 S	0	0	0	8	108	270	381	326	169	27	0	0	1289
OTTAWA	0	0	0	0	84	231	329	279	101	23	0	0	1047
PALESTINE	0	0	0	8	111	278	378	319	155	23	0	0	1272
PANA	0	0	0	6	94	242	347	282	126	19	0	0	1116
PARIS WATERWORKS	0	0	0	6	95	246	347	282	135	23	0	0	1134
PARK FOREST	0	0	0	0	49	169	264	221	73	15	0	0	791
PEORIA WSO //R	0	0	0	0	71	208	314	256	82	17	0	0	948
PERU 2 W	0	0	0	0	86	216	315	263	91	19	0	0	990
PIPER CITY 3 SE	0	0	0	0	74	212	295	236	101	16	0	0	934
PONTIAC	0	0	0	0	87	226	316	254	104	21	0	0	1008

ILLINOIS

COOLING DEGREE DAY NORMALS (BASE 65 DEG F)

STATION		JAN	FEB	MAR	APR	MAY	JUN	JUL	AUG	SEP	OCT	NOV	DEC	ANN
QUINCY FAA AIRPORT		0	0	0	0	87	235	367	293	115	23	0	0	1120
RANTOUL		0	0	0	0	82	227	319	260	124	17	0	0	1029
ROCKFORD WSO	//R	0	0	0	0	44	151	253	206	48	12	0	0	714
ROSICLARE		0	0	11	17	115	279	391	354	185	31	0	0	1383
RUSHVILLE		0	0	0	6	99	246	357	288	119	23	0	0	1138
SPARTA		0	0	9	17	132	309	431	375	195	41	0	0	1509
SPRINGFIELD WSO	R	0	0	0	6	106	256	357	289	126	25	0	0	1165
STOCKTON		0	0	0	0	38	142	238	183	36	11	0	0	648
TUSCOLA		0	0	0	5	97	251	350	282	133	23	0	0	1141
URBANA		0	0	0	0	87	225	316	257	114	20	0	0	1019
VIRDEN 1 N		0	0	0	7	99	256	353	291	136	24	0	0	1166
WALNUT		0	0	0	0	70	203	305	249	80	18	0	0	925
WATSEKA 2 NW		0	0	0	0	69	199	282	224	94	18	0	0	886
WAUKEGAN 4 WSW		0	0	0	0	24	114	209	187	46	11	0	0	591
WHEATON 3 SE		0	0	0	0	51	167	267	221	69	12	0	0	787
WHITE HALL 1 E		0	0	0	9	98	258	369	307	131	24	0	0	1196
WINDSOR		0	0	0	7	96	245	350	288	142	19	0	0	1147

STATE-STATION NUMBER	STN TYP	NAME		LATITUDE DEG-MIN	LONGITUDE DEG-MIN	ELEVATION (FT)
11-0055	12	ALBION		N 3823	W 08803	490
11-0072	13	ALEDO		N 4113	W 09045	720
11-0137	13	ALTON DAM 26		N 3853	W 09011	432
11-0187	13	ANNA 1 E		N 3728	W 08914	645
11-0203	13	ANTIOCH 2 NW		N 4230	W 08808	752
11-0338	13	AURORA COLLEGE		N 4145	W 08820	744
11-0356	12	AVON 5 NE		N 4044	W 09022	650
11-0492	12	BEARDSTOWN		N 4001	W 09026	445
11-0510	13	BELLEVILLE SO ILL UNIV		N 3830	W 08951	450
11-0598	12	BENTLEY 1 E		N 4020	W 09106	650
11-0608	12	BENTON FOREST SERVICE		N 3802	W 08855	445
11-0761	12	BLOOMINGTON WATERWORKS		N 4030	W 08901	775
11-0781	12	BLUFFS		N 3945	W 09032	540
11-0993	12	BROOKPORT DAM 52		N 3708	W 08839	330
11-1166	13	CAIRO WSO	R	N 3700	W 08910	314
11-1250	12	CANTON 1 ESE		N 4033	W 09001	628
11-1265	13	CARBONDALE SEWAGE PL		N 3744	W 08912	380
11-1280	13	CARLINVILLE		N 3917	W 08952	625
11-1290	13	CARLYLE RESERVOIR		N 3838	W 08920	501
11-1296	12	CARMI 6 NW		N 3810	W 08812	390
11-1386	12	CENTRALIA 2 SW		N 3830	W 08910	463
11-1420	12	CHANNAHON DRESDEN ISL		N 4124	W 08817	505
11-1436	13	CHARLESTON		N 3929	W 08811	710
11-1475	13	CHENOA		N 4044	W 08844	710
11-1491	12	CHESTER		N 3754	W 08950	460
11-1549	13	CHICAGO O HARE WSO	R	N 4159	W 08754	658
11-1572	13	CHICAGO UNIVERSITY	//R	N 4147	W 08736	594
11-1577	13	CHICAGO MIDWAY AP		N 4147	W 08745	607
11-1627	12	CHILLICOTHE		N 4055	W 08930	510
11-1743	12	CLINTON 1 SSW		N 4008	W 08858	695
11-2140	13	DANVILLE		N 4008	W 08739	558
11-2145	12	DANVILLE SEWAGE PLANT		N 4006	W 08736	527
11-2193	13	DECATUR		N 3951	W 08858	670
11-2223	13	DE KALB		N 4156	W 08846	840
11-2348	13	DIXON		N 4150	W 08929	690
11-2483	13	DU QUOIN 1 NNW		N 3800	W 08915	410
11-2679	12	EDWARDSVILLE		N 3850	W 08957	500
11-2687	13	EFFINGHAM		N 3907	W 08833	595
11-2736	12	ELGIN		N 4202	W 08816	758
11-2931	13	FAIRFIELD RADIO WFIW		N 3823	W 08819	425
11-3109	13	FLORA 5 NW		N 3841	W 08834	500
11-3262	12	FREEPORT SEWAGE PLANT		N 4217	W 08936	750
11-3290	13	FULTON DAM 13		N 4154	W 09009	592
11-3312	12	GALENA 1 N		N 4225	W 09026	770
11-3320	13	GALESBURG		N 4057	W 09023	771
11-3335	13	GALVA		N 4110	W 09002	873
11-3384	12	GENESEO		N 4127	W 09010	639
11-3413	12	GIBSON CITY		N 4028	W 08823	657
11-3455	12	GLADSTONE DAM 18		N 4050	W 09104	538
11-3522	12	GOLCONDA DAM 51		N 3722	W 08829	354

11 — ILLINOIS

STATE-STATION NUMBER	STN TYP	NAME	LATITUDE DEG-MIN	LONGITUDE DEG-MIN	ELEVATION (FT)
11-3530	12	GOLDEN 1 NW	N 4007	W 09102	725
11-3572	12	GRAFTON	N 3858	W 09027	470
11-3595	12	GRAND TOWER 2 N	N 3740	W 08930	367
11-3683	12	GREENUP	N 3915	W 08810	525
11-3693	12	GREENVILLE 1 E	N 3853	W 08924	580
11-3717	13	GRIGGSVILLE	N 3943	W 09044	700
11-3879	13	HARRISBURG	N 3744	W 08832	366
11-3884	12	HARRISBURG DISPOSAL PL	N 3745	W 08832	360
11-3940	12	HAVANA POWER STATION	N 4017	W 09005	464
11-4108	13	HILLSBORO 2 SSW	N 3909	W 08930	720
11-4198	13	HOOPESTON	N 4028	W 08740	710
11-4317	12	HUTSONVILLE POWER PL	N 3908	W 08740	455
11-4355	12	ILLINOIS CITY DAM 16 //	N 4125	W 09101	554
11-4442	13	JACKSONVILLE	N 3944	W 09012	620
11-4489	13	JERSEYVILLE 2 SW	N 3906	W 09021	630
11-4530	12	JOLIET BRANDON RD DAM	N 4130	W 08806	543
11-4655	12	KEITHSBURG 1 NW	N 4106	W 09057	540
11-4710	13	KEWANEE	N 4115	W 08955	820
11-4823	13	LA HARPE	N 4035	W 09058	710
11-4957	12	LAWRENCEVILLE	N 3843	W 08741	450
11-5079	13	LINCOLN	N 4009	W 08922	590
11-5272	12	MACKINAW	N 4032	W 08922	670
11-5280	12	MACOMB	N 4028	W 09040	702
11-5326	13	MARENGO	N 4215	W 08836	825
11-5342	12	MARION 4 NNE	N 3746	W 08854	477
11-5372	12	MARSEILLES LOCK	N 4120	W 08845	486
11-5430	13	MATTOON	N 3928	W 08821	718
11-5515	13	MCLEANSBORO 2 E	N 3805	W 08830	450
11-5539	12	MEDORA	N 3910	W 09008	620
11-5712	13	MINONK 3 NE	N 4056	W 08900	750
11-5751	13	MOLINE WSO R	N 4127	W 09030	582
11-5768	13	MONMOUTH	N 4055	W 09038	770
11-5792	12	MONTICELLO NO 2	N 4002	W 08835	655
11-5820	12	MORRIS	N 4121	W 08826	520
11-5833	13	MORRISON	N 4149	W 08958	603
11-5901	13	MOUNT CARROLL	N 4205	W 08959	837
11-5917	12	MOUNT OLIVE 1 E	N 3904	W 08942	685
11-5927	12	MOUNT PULASKI	N 4001	W 08917	645
11-5935	12	MOUNT STERLING	N 3959	W 09046	711
11-5943	13	MOUNT VERNON	N 3819	W 08855	500
11-6011	12	NASHVILLE 4 NE	N 3823	W 08920	515
11-6080	12	NEW BOSTON DAM 17	N 4111	W 09103	548
11-6159	13	NEWTON 2 NE	N 3900	W 08807	510
11-6446	13	OLNEY 2 S	N 3842	W 08804	480
11-6526	13	OTTAWA	N 4122	W 08850	474
11-6558	13	PALESTINE	N 3900	W 08737	514
11-6579	13	PANA	N 3923	W 08905	696
11-6610	13	PARIS WATERWORKS	N 3938	W 08742	739
11-6616	13	PARK FOREST	N 4130	W 08741	710
11-6661	12	PAW PAW	N 4141	W 08859	895

STATE-STATION NUMBER	STN TYP	NAME		LATITUDE DEG-MIN	LONGITUDE DEG-MIN	ELEVATION (FT)
11-6670	12	PAYSON		N 3949	W 09115	780
11-6711	13	PEORIA WSO	//R	N 4040	W 08941	652
11-6725	12	PEOTONE		N 4120	W 08748	722
11-6753	13	PERU 2 W		N 4120	W 08909	615
11-6819	13	PIPER CITY 3 SE		N 4048	W 08808	690
11-6910	13	PONTIAC		N 4053	W 08837	647
11-6973	12	PRAIRIE DU ROCHER 1WSW		N 3805	W 09007	385
11-7004	12	PRINCEVILLE 1 N		N 4057	W 08946	725
11-7072	13	QUINCY FAA AIRPORT		N 3956	W 09112	763
11-7077	12	QUINCY DAM 21	//	N 3954	W 09126	483
11-7082	12	QUINCY MEMORIAL BRIDGE		N 3956	W 09125	475
11-7150	13	RANTOUL		N 4019	W 08810	740
11-7157	12	RED BUD		N 3813	W 09000	450
11-7382	13	ROCKFORD WSO	//R	N 4212	W 08906	724
11-7487	13	ROSICLARE		N 3725	W 08821	400
11-7551	13	RUSHVILLE		N 4007	W 09034	660
11-7603	12	SAINTE MARIE		N 3856	W 08801	500
11-7859	12	SHAWNEETOWN NEW TOWN		N 3743	W 08811	400
11-7952	12	SIDELL		N 3955	W 08749	685
11-8147	13	SPARTA		N 3808	W 08942	480
11-8179	13	SPRINGFIELD WSO	R	N 3950	W 08941	588
11-8293	13	STOCKTON		N 4221	W 09000	1005
11-8353	12	STREATOR 2 ENE		N 4108	W 08848	650
11-8604	12	TISKILWA		N 4117	W 08930	512
11-8684	13	TUSCOLA		N 3948	W 08817	653
11-8740	13	URBANA		N 4006	W 08814	743
11-8756	12	UTICA STARVED ROCK DAM		N 4119	W 08859	464
11-8860	13	VIRDEN 1 N		N 3931	W 08946	675
11-8870	12	VIRGINIA		N 3957	W 09013	615
11-8916	13	WALNUT		N 4134	W 08935	760
11-9002	12	WATERLOO 1 WSW		N 3820	W 09009	645
11-9021	13	WATSEKA 2 NW		N 4047	W 08746	620
11-9029	13	WAUKEGAN 4 WSW		N 4221	W 08753	700
11-9040	12	WAYNE CITY		N 3821	W 08835	420
11-9221	13	WHEATON 3 SE		N 4149	W 08804	680
11-9241	13	WHITE HALL 1 E		N 3926	W 09023	578
11-9354	13	WINDSOR		N 3926	W 08836	685

11 — ILLINOIS

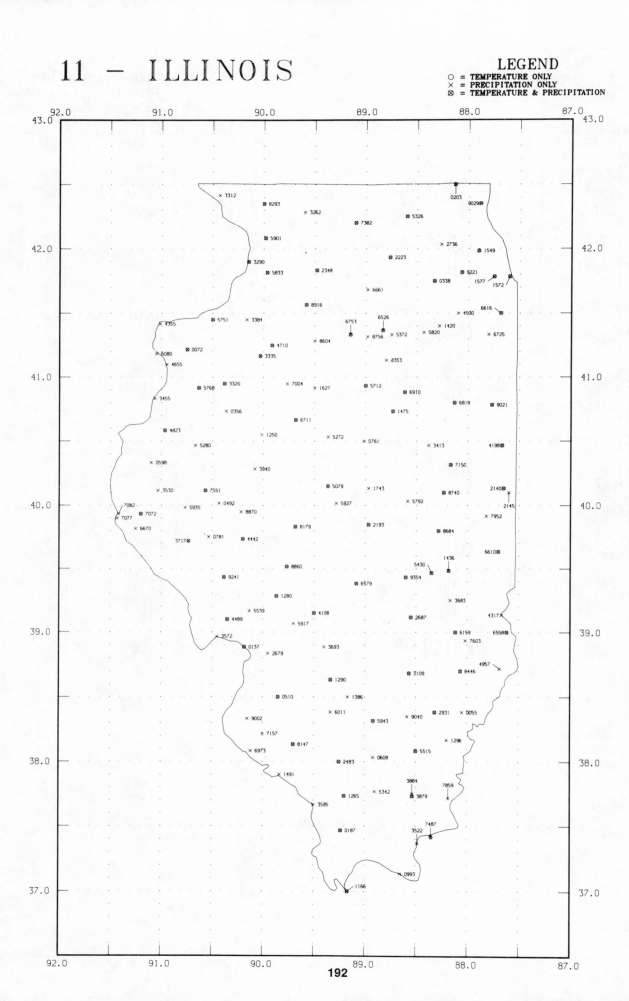

INDIANA

TEMPERATURE NORMALS (DEG F)

STATION		JAN	FEB	MAR	APR	MAY	JUN	JUL	AUG	SEP	OCT	NOV	DEC	ANN
ANDERSON SEWAGE PLANT	MAX	33.4	37.3	48.0	61.5	72.1	81.1	84.5	82.9	77.5	65.4	50.2	38.1	61.0
	MIN	17.7	20.7	30.0	40.6	50.4	59.4	63.1	61.0	54.1	42.7	33.1	23.5	41.4
	MEAN	25.6	29.0	39.0	51.1	61.3	70.3	73.8	72.0	65.8	54.1	41.7	30.8	51.2
BERNE	MAX	33.0	36.8	47.4	61.4	72.6	82.1	85.3	83.7	77.6	65.5	49.7	37.9	61.1
	MIN	17.4	19.8	28.9	39.7	49.6	58.8	62.8	60.7	54.0	42.9	33.0	23.1	40.9
	MEAN	25.2	28.3	38.2	50.6	61.1	70.5	74.0	72.2	65.8	54.2	41.4	30.5	51.0
BROOKVILLE	MAX	36.9	40.4	50.8	64.0	74.5	82.6	86.0	85.0	79.6	67.7	53.2	41.5	63.5
	MIN	16.8	18.7	27.4	37.7	47.4	57.0	61.2	59.2	51.5	38.7	30.1	22.0	39.0
	MEAN	26.9	29.5	39.1	50.9	61.0	69.8	73.6	72.1	65.6	53.2	41.7	31.8	51.3
CAMBRIDGE CITY	MAX	34.1	37.6	48.0	61.7	72.5	81.4	85.2	84.0	78.2	66.2	50.9	39.0	61.6
	MIN	15.1	17.4	27.1	37.9	47.9	56.8	60.3	58.2	50.7	38.6	30.2	21.1	38.4
	MEAN	24.6	27.5	37.6	49.8	60.2	69.1	72.8	71.1	64.5	52.4	40.5	30.1	50.0
COLUMBIA CITY 1 S	MAX	31.4	35.5	46.1	60.6	71.7	81.2	84.5	83.0	77.1	64.9	48.8	36.4	60.1
	MIN	15.3	17.6	26.8	37.8	47.9	57.4	61.1	59.0	51.9	40.8	31.5	21.4	39.0
	MEAN	23.4	26.5	36.4	49.2	59.8	69.3	72.8	71.0	64.5	52.9	40.2	28.9	49.6
COLUMBUS	MAX	37.1	41.0	51.1	64.5	74.4	82.8	86.4	85.2	79.8	68.0	53.3	41.8	63.8
	MIN	18.1	20.5	29.5	40.7	50.1	59.6	63.3	60.8	53.4	41.1	31.8	23.6	41.0
	MEAN	27.6	30.8	40.3	52.6	62.3	71.2	74.9	73.0	66.6	54.6	42.6	32.7	52.4
CRANE NAVAL DEPOT	MAX	39.3	44.4	54.3	67.9	76.9	84.5	87.6	86.6	81.2	70.0	55.0	43.8	66.0
	MIN	21.8	25.0	33.5	45.0	53.8	62.5	66.4	64.8	58.4	47.0	36.2	27.2	45.1
	MEAN	30.6	34.7	43.9	56.5	65.4	73.5	77.0	75.7	69.8	58.5	45.7	35.5	55.6
CRAWFORDSVILLE PWR PL	MAX	33.5	37.7	48.0	62.1	73.3	82.3	86.0	84.2	78.7	66.8	50.9	38.6	61.8
	MIN	15.7	18.3	28.1	39.3	48.9	58.5	62.5	60.3	52.8	40.8	31.1	21.9	39.9
	MEAN	24.6	28.0	38.1	50.7	61.1	70.4	74.3	72.3	65.8	53.9	41.1	30.3	50.9
DELPHI 3 NNE	MAX	34.0	38.4	49.6	64.0	74.4	83.0	86.0	83.9	78.6	67.1	51.3	39.0	62.4
	MIN	16.6	20.0	29.4	40.0	49.5	58.8	62.4	60.3	53.3	42.2	32.3	22.8	40.6
	MEAN	25.3	29.2	39.5	52.0	62.0	70.9	74.2	72.1	66.0	54.6	41.8	30.9	51.5
EDWARDSPORT POWER PL	MAX	37.2	41.3	51.5	65.7	75.7	84.6	87.9	86.4	80.8	69.2	54.0	42.0	64.7
	MIN	18.8	21.8	31.1	42.7	52.0	61.3	65.0	62.8	55.4	42.9	33.2	24.4	42.6
	MEAN	28.0	31.6	41.3	54.3	63.9	73.0	76.5	74.6	68.1	56.0	43.6	33.2	53.7
ELWOOD WATERWORKS	MAX	32.7	36.4	46.8	60.8	72.0	81.3	84.9	83.1	77.6	65.9	50.2	38.0	60.8
	MIN	14.9	17.5	27.2	38.4	48.4	57.6	61.1	58.6	51.4	39.9	30.6	21.0	38.9
	MEAN	23.8	27.0	37.0	49.6	60.2	69.5	73.0	70.9	64.5	52.9	40.4	29.5	49.9
EVANSVILLE	MAX	41.4	46.4	56.7	70.0	79.0	87.4	90.1	88.9	82.9	71.4	56.8	45.7	68.1
	MIN	24.8	28.0	36.4	47.0	55.6	64.2	68.2	66.2	59.3	47.0	37.6	29.5	47.0
	MEAN	33.1	37.2	46.6	58.5	67.3	75.8	79.2	77.6	71.2	59.2	47.2	37.6	57.5
EVANSVILLE WSO //R	MAX	39.3	44.2	54.5	67.6	76.8	85.9	88.8	87.3	81.4	70.0	55.1	44.0	66.2
	MIN	21.9	25.5	34.6	45.4	54.3	63.2	67.4	65.0	57.4	44.5	35.0	27.2	45.1
	MEAN	30.6	34.9	44.6	56.5	65.6	74.6	78.1	76.2	69.4	57.3	45.1	35.6	55.7
FORT WAYNE WSO //R	MAX	30.8	34.5	45.2	59.7	70.9	80.5	84.1	82.3	76.0	63.9	48.2	36.0	59.3
	MIN	15.8	18.3	28.0	38.7	48.9	58.7	62.5	60.5	53.3	42.0	31.9	21.9	40.0
	MEAN	23.3	26.4	36.6	49.2	59.9	69.6	73.3	71.4	64.7	53.0	40.1	29.0	49.7
FRANKFORT DISPOSAL PL	MAX	32.7	36.9	47.5	61.5	72.4	81.6	84.8	82.9	77.1	65.4	49.5	37.7	60.8
	MIN	15.2	18.4	27.9	39.0	49.4	58.5	62.2	59.8	52.8	41.4	31.0	21.5	39.8
	MEAN	24.0	27.6	37.7	50.3	60.9	70.1	73.5	71.4	65.0	53.4	40.3	29.6	50.3
GARY	MAX	30.6	34.9	44.4	58.4	69.8	80.0	83.7	82.0	75.9	64.1	48.8	36.1	59.1
	MIN	15.1	19.2	28.6	39.5	49.5	59.1	64.3	63.0	55.4	44.9	33.5	21.9	41.2
	MEAN	22.9	27.1	36.5	49.0	59.7	69.6	74.0	72.5	65.7	54.5	41.2	29.1	50.2
GOSHEN COLLEGE	MAX	30.8	34.9	45.3	59.9	71.1	80.5	83.9	82.0	75.5	63.7	48.4	35.9	59.3
	MIN	15.9	18.5	27.7	38.2	48.2	57.8	61.6	59.8	52.8	42.2	32.3	22.0	39.8
	MEAN	23.4	26.7	36.6	49.1	59.7	69.2	72.8	70.9	64.2	53.0	40.4	29.0	49.6

INDIANA

TEMPERATURE NORMALS (DEG F)

STATION		JAN	FEB	MAR	APR	MAY	JUN	JUL	AUG	SEP	OCT	NOV	DEC	ANN
GREENCASTLE 1 E	MAX	34.5	39.3	49.8	63.8	74.5	83.6	87.0	85.7	79.8	67.6	51.4	39.3	63.0
	MIN	17.0	20.2	29.9	41.7	51.4	60.4	63.8	62.0	54.9	43.6	33.0	23.0	41.7
	MEAN	25.8	29.8	39.9	52.8	63.0	72.0	75.4	73.9	67.4	55.6	42.2	31.2	52.4
GREENFIELD	MAX	34.5	38.4	49.0	62.8	73.7	82.6	85.9	84.4	78.8	66.6	51.3	39.2	62.3
	MIN	16.8	19.5	29.0	40.5	50.3	59.6	63.4	61.1	54.3	42.3	32.3	22.7	41.0
	MEAN	25.7	29.0	39.0	51.6	62.1	71.1	74.7	72.8	66.6	54.5	41.8	30.9	51.7
GREENSBURG 3 SW	MAX	36.7	40.9	51.3	64.4	73.9	82.2	85.5	84.2	78.7	67.2	52.2	41.2	63.2
	MIN	18.1	21.1	30.2	40.7	49.7	58.3	61.5	59.3	52.9	41.1	31.8	23.6	40.7
	MEAN	27.4	31.0	40.8	52.6	61.8	70.3	73.5	71.8	65.8	54.2	42.0	32.4	52.0
HOBART	MAX	32.2	36.6	46.8	61.0	72.6	81.9	85.2	83.6	78.1	66.9	50.6	37.5	61.1
	MIN	15.4	19.1	28.3	38.8	48.1	57.7	62.3	60.8	53.9	43.6	32.6	22.1	40.2
	MEAN	23.8	27.9	37.6	49.9	60.4	69.8	73.8	72.2	66.0	55.3	41.6	29.8	50.7
HUNTINGTON	MAX	32.6	36.7	47.6	62.1	72.7	82.0	85.2	83.9	77.7	65.7	49.7	37.5	61.1
	MIN	16.8	19.3	28.5	39.4	48.9	58.1	61.8	60.0	53.1	42.2	32.6	22.7	40.3
	MEAN	24.7	28.0	38.1	50.8	60.8	70.0	73.5	72.0	65.5	54.0	41.1	30.1	50.7
INDIANAPOLIS WSO //R	MAX	34.2	38.5	49.3	63.1	73.4	82.3	85.2	83.7	77.9	66.1	50.8	39.2	62.0
	MIN	17.8	21.1	30.7	41.7	51.5	60.9	64.9	62.7	55.3	43.4	32.8	23.7	42.2
	MEAN	26.0	29.9	40.0	52.4	62.5	71.6	75.1	73.2	66.6	54.8	41.8	31.5	52.1
INDIANAPOLIS SE SIDE	MAX	34.6	38.6	48.7	62.5	73.1	82.0	85.5	84.0	78.5	66.6	51.5	39.7	62.1
	MIN	18.1	21.1	30.4	41.8	51.9	61.2	65.2	63.1	55.8	43.7	33.5	23.9	42.5
	MEAN	26.4	29.9	39.6	52.2	62.5	71.6	75.4	73.6	67.1	55.2	42.5	31.8	52.3
JOHNSON EXP FARM	MAX	39.1	43.8	53.9	67.4	77.3	86.0	89.0	87.7	82.2	70.5	55.1	43.7	66.3
	MIN	21.8	25.0	33.9	45.3	54.1	62.9	66.8	64.6	57.9	45.6	35.7	27.3	45.1
	MEAN	30.5	34.4	43.9	56.4	65.7	74.5	77.9	76.1	70.1	58.1	45.5	35.5	55.7
KENTLAND	MAX	32.0	36.6	48.0	63.0	74.4	83.4	86.0	84.2	79.0	66.9	50.0	37.1	61.7
	MIN	15.4	19.4	29.3	40.3	50.1	59.6	63.3	61.0	54.0	43.0	32.1	21.8	40.8
	MEAN	23.8	28.0	38.7	51.7	62.2	71.5	74.6	72.6	66.5	55.0	41.1	29.5	51.3
KOKOMO 7 SE	MAX	32.8	37.0	48.0	62.2	72.8	82.6	85.7	83.6	78.1	66.0	49.9	37.8	61.4
	MIN	16.3	18.9	28.5	39.9	50.0	59.3	62.8	60.3	53.8	42.4	32.4	22.3	40.6
	MEAN	24.6	27.9	38.3	51.0	61.5	71.0	74.2	72.0	65.9	54.2	41.1	30.1	51.0
LA PORTE	MAX	30.7	35.0	45.4	59.9	71.7	81.2	84.5	82.6	76.3	64.3	48.5	35.8	59.7
	MIN	14.9	18.6	27.5	38.3	48.2	57.6	62.2	60.6	53.6	42.6	31.9	21.4	39.8
	MEAN	22.8	26.8	36.5	49.1	60.0	69.4	73.4	71.6	65.0	53.5	40.2	28.6	49.7
MARION 2 N	MAX	32.1	35.7	46.2	60.3	71.7	81.2	84.7	82.8	77.4	65.4	49.5	37.4	60.4
	MIN	15.1	17.1	26.5	37.4	47.5	57.4	61.2	58.8	51.5	40.1	31.0	21.2	38.7
	MEAN	23.6	26.5	36.4	48.9	59.6	69.3	72.9	70.8	64.5	52.7	40.3	29.3	49.6
MARTINSVILLE 2 SW	MAX	36.1	39.9	50.3	64.0	74.2	82.6	86.3	85.0	79.0	67.5	52.6	41.0	63.2
	MIN	17.4	20.0	30.0	41.4	50.6	59.5	63.0	60.4	52.8	40.6	31.5	22.9	40.8
	MEAN	26.8	30.0	40.2	52.7	62.4	71.1	74.7	72.7	65.9	54.1	42.1	32.0	52.1
MOUNT VERNON	MAX	39.5	43.7	53.5	67.0	76.8	85.7	89.2	88.0	82.3	70.8	56.0	44.2	66.4
	MIN	21.7	24.7	33.7	45.4	54.1	63.0	66.7	64.4	57.2	44.6	35.1	26.7	44.8
	MEAN	30.6	34.2	43.6	56.2	65.5	74.3	78.0	76.2	69.8	57.7	45.6	35.5	55.6
NEW CASTLE	MAX	33.8	37.2	47.7	61.3	72.4	81.8	85.5	83.8	77.6	65.4	50.4	38.6	61.3
	MIN	15.6	18.0	27.7	38.2	48.2	57.9	61.2	58.8	51.5	39.9	30.7	21.4	39.1
	MEAN	24.7	27.7	37.7	49.8	60.3	69.9	73.4	71.3	64.6	52.7	40.5	30.0	50.2
NORTH VERNON 2 SW	MAX	39.1	44.1	54.2	67.6	76.3	84.0	87.1	86.1	80.5	69.2	54.6	43.6	65.5
	MIN	20.6	23.7	32.5	42.9	51.3	60.2	64.0	62.6	55.8	44.0	34.2	26.1	43.2
	MEAN	29.9	33.9	43.4	55.2	63.8	72.1	75.6	74.3	68.2	56.6	44.4	34.9	54.4
OAKLANDON GEIST RESVR	MAX	34.5	38.6	48.7	62.9	73.1	82.0	85.4	84.1	78.4	66.8	51.3	39.3	62.1
	MIN	16.9	19.5	28.7	40.0	49.7	58.8	62.7	60.5	53.5	41.8	32.1	22.7	40.6
	MEAN	25.8	29.0	38.8	51.5	61.4	70.4	74.0	72.3	66.0	54.3	41.7	31.0	51.4

INDIANA

TEMPERATURE NORMALS (DEG F)

STATION		JAN	FEB	MAR	APR	MAY	JUN	JUL	AUG	SEP	OCT	NOV	DEC	ANN
OGDEN DUNES	MAX	31.2	35.3	45.1	58.8	70.5	79.8	83.5	82.3	76.4	65.0	49.4	36.6	59.5
	MIN	16.3	20.0	28.7	38.8	48.1	58.1	63.4	62.3	55.8	45.1	33.7	22.9	41.1
	MEAN	23.8	27.7	36.9	48.8	59.3	69.0	73.4	72.3	66.1	55.1	41.6	29.8	50.3
OOLITIC PURDUE EXP FRM	MAX	37.9	42.3	52.4	66.2	75.3	83.3	86.8	85.6	79.8	68.3	53.8	42.4	64.5
	MIN	19.0	21.6	30.8	41.8	50.6	59.6	63.4	61.5	54.3	41.6	32.6	24.1	41.7
	MEAN	28.5	31.9	41.6	54.0	63.0	71.5	75.2	73.6	67.1	55.0	43.2	33.3	53.2
PAOLI RADIO WVAK	MAX	39.3	43.5	53.1	66.5	75.9	83.7	87.3	86.4	80.8	69.5	54.9	43.8	65.4
	MIN	17.9	20.5	29.4	40.9	49.5	58.8	62.6	60.5	53.2	40.1	31.0	23.2	40.6
	MEAN	28.6	32.0	41.3	53.7	62.7	71.3	74.9	73.5	67.0	54.8	43.0	33.5	53.0
PLYMOUTH POWER SUBSTA	MAX	32.0	36.5	47.3	62.3	74.0	83.2	86.3	84.5	78.2	65.7	49.6	37.0	61.4
	MIN	14.8	18.1	27.6	38.5	48.2	57.2	61.0	59.1	51.9	41.3	31.2	21.4	39.2
	MEAN	23.4	27.3	37.5	50.4	61.1	70.2	73.7	71.8	65.1	53.6	40.4	29.2	50.3
PRINCETON 1 W	MAX	39.2	44.5	54.2	67.5	76.8	85.7	88.5	87.2	81.8	70.3	54.8	43.5	66.2
	MIN	21.7	25.2	33.9	44.9	54.0	62.6	66.2	64.1	57.3	45.5	35.5	27.0	44.8
	MEAN	30.4	34.9	44.1	56.2	65.4	74.2	77.3	75.7	69.6	57.9	45.2	35.3	55.5
RICHMOND WTR WKS 2 NNE	MAX	34.4	38.1	48.9	62.2	72.5	81.1	84.3	82.9	76.5	64.7	50.1	39.0	61.2
	MIN	16.9	19.5	28.8	38.8	48.5	57.3	61.2	59.3	52.3	40.3	31.4	22.7	39.8
	MEAN	25.6	28.9	38.9	50.5	60.5	69.2	72.8	71.1	64.4	52.5	40.8	30.9	50.5
ROCHESTER	MAX	30.9	35.1	45.2	59.7	71.4	81.0	84.6	82.7	76.8	64.6	48.9	36.3	59.8
	MIN	13.3	16.2	25.8	37.6	47.6	57.2	60.9	58.3	50.8	39.8	30.0	19.8	38.1
	MEAN	22.1	25.7	35.5	48.7	59.5	69.1	72.8	70.5	63.8	52.2	39.5	28.1	49.0
ROCKVILLE	MAX	35.6	40.7	51.4	65.6	75.5	84.1	87.2	85.4	79.8	68.2	52.6	40.5	63.9
	MIN	17.7	21.5	30.6	42.1	51.4	60.2	64.3	62.2	55.6	44.0	33.6	23.9	42.3
	MEAN	26.7	31.1	41.0	53.9	63.5	72.2	75.8	73.8	67.7	56.1	43.1	32.2	53.1
RUSHVILLE SEWAGE PLANT	MAX	34.2	37.8	48.0	61.8	72.3	81.1	84.4	83.2	77.6	65.7	50.6	39.1	61.3
	MIN	16.2	18.9	28.8	40.1	50.0	59.0	62.6	60.3	53.3	41.1	31.6	22.4	40.4
	MEAN	25.2	28.4	38.4	51.0	61.2	70.0	73.5	71.8	65.5	53.4	41.1	30.8	50.9
SCOTTSBURG	MAX	38.9	43.2	53.2	66.3	76.0	84.3	87.9	86.8	81.2	69.6	54.8	43.6	65.5
	MIN	19.3	21.5	30.8	41.8	51.1	60.4	64.0	61.8	54.3	41.4	32.3	24.7	42.0
	MEAN	29.1	32.4	42.0	54.1	63.6	72.4	76.0	74.3	67.8	55.5	43.6	34.2	53.8
SEYMOUR 2 N	MAX	38.2	42.4	52.5	66.1	75.6	83.5	86.8	86.2	80.8	69.2	54.3	42.7	64.9
	MIN	17.8	20.2	29.4	40.5	49.9	59.0	62.6	59.8	52.4	39.9	31.0	23.1	40.5
	MEAN	28.0	31.3	41.0	53.3	62.8	71.3	74.7	73.0	66.6	54.6	42.7	32.9	52.7
SHELBYVILLE SEWAGE PL	MAX	35.7	39.8	50.2	63.9	74.2	82.8	86.2	84.7	79.1	67.3	52.3	40.5	63.1
	MIN	17.0	19.7	29.8	40.9	50.8	59.8	63.0	61.1	53.9	41.7	31.9	22.7	41.0
	MEAN	26.3	29.7	40.0	52.4	62.5	71.3	74.6	72.9	66.5	54.5	42.1	31.7	52.0
SHOALS HIWAY 50 BRIDGE	MAX	38.9	43.1	53.0	66.9	76.5	84.5	87.7	86.8	80.9	69.6	54.6	43.4	65.5
	MIN	18.9	21.2	30.1	41.4	49.9	59.2	63.2	61.2	53.8	41.2	32.2	24.1	41.4
	MEAN	28.9	32.2	41.6	54.2	63.2	71.9	75.5	74.0	67.4	55.4	43.5	33.8	53.5
SOUTH BEND WSO //R	MAX	30.4	34.1	44.3	58.6	69.9	79.5	82.7	81.0	74.6	63.1	47.8	35.7	58.5
	MIN	15.9	18.6	27.7	38.4	48.1	58.1	62.3	60.8	53.7	43.4	32.8	22.5	40.2
	MEAN	23.2	26.4	36.0	48.5	59.1	68.8	72.5	70.9	64.2	53.2	40.3	29.1	49.4
SPENCER	MAX	36.6	41.1	51.2	65.2	75.3	83.9	87.3	85.9	80.2	68.7	53.2	41.4	64.2
	MIN	17.2	19.6	29.1	40.1	49.5	59.2	63.0	60.7	53.0	40.3	31.1	22.7	40.5
	MEAN	27.0	30.4	40.2	52.6	62.4	71.6	75.1	73.3	66.6	54.5	42.2	32.1	52.3
TELL CITY POWER PLANT	MAX	40.5	44.6	54.2	67.1	77.0	85.1	88.8	87.9	81.7	70.7	56.5	45.2	66.6
	MIN	22.8	25.2	33.6	44.3	53.2	62.4	66.5	64.7	57.8	44.9	35.8	27.7	44.9
	MEAN	31.7	34.9	43.9	55.8	65.1	73.8	77.7	76.3	69.8	57.8	46.2	36.5	55.8
TERRE HAUTE 8 S	MAX	35.0	39.6	49.7	63.7	74.1	83.2	86.8	85.0	79.5	68.0	52.4	40.4	63.1
	MIN	17.3	20.7	30.8	42.5	51.7	60.9	64.4	62.4	55.0	43.3	32.8	23.3	42.1
	MEAN	26.2	30.2	40.3	53.2	62.9	72.1	75.6	73.7	67.3	55.7	42.6	31.9	52.6

INDIANA

TEMPERATURE NORMALS (DEG F)

STATION		JAN	FEB	MAR	APR	MAY	JUN	JUL	AUG	SEP	OCT	NOV	DEC	ANN
VALPARAISO WATERWORKS	MAX	30.8	35.2	45.7	60.2	71.4	80.3	83.4	81.6	75.8	64.8	49.0	36.1	59.5
	MIN	14.4	18.1	27.5	38.3	47.9	57.3	61.5	59.9	53.0	42.9	31.9	21.3	39.5
	MEAN	22.6	26.7	36.6	49.2	59.7	68.8	72.5	70.8	64.4	53.9	40.4	28.7	49.5
VEVAY	MAX	40.4	44.7	54.9	67.8	77.0	85.0	88.4	87.2	81.3	69.6	55.4	44.8	66.4
	MIN	22.0	24.2	32.6	42.3	51.2	60.4	64.4	62.8	56.1	43.9	34.4	26.6	43.4
	MEAN	31.2	34.4	43.7	55.1	64.1	72.7	76.4	75.0	68.7	56.8	44.9	35.7	54.9
VINCENNES 1 NW	MAX	37.5	42.1	52.3	66.2	76.0	85.2	88.5	87.0	81.5	69.8	54.3	42.3	65.2
	MIN	19.3	22.3	31.5	43.3	52.3	61.7	65.5	63.0	55.8	43.1	33.1	24.7	43.0
	MEAN	28.5	32.2	41.9	54.8	64.2	73.5	77.0	75.1	68.7	56.4	43.7	33.5	54.1
WABASH	MAX	31.7	35.6	46.0	60.5	71.7	81.0	84.5	82.6	77.4	65.5	49.6	37.0	60.3
	MIN	13.8	15.9	25.6	36.9	47.0	56.4	60.3	57.7	50.3	38.8	29.9	20.0	37.7
	MEAN	22.8	25.8	35.8	48.7	59.4	68.8	72.4	70.2	63.9	52.2	39.8	28.5	49.0
WASHINGTON	MAX	38.9	43.9	54.3	68.0	76.8	85.3	88.0	86.5	80.7	69.1	54.4	43.1	65.8
	MIN	22.4	25.7	34.5	45.6	54.4	62.9	66.6	64.9	58.2	46.5	36.2	27.8	45.5
	MEAN	30.6	34.8	44.4	56.8	65.6	74.1	77.3	75.7	69.5	57.8	45.3	35.5	55.6
WATERLOO	MAX	31.3	35.2	46.2	60.8	72.0	81.5	84.5	82.9	76.8	64.9	49.0	36.3	60.1
	MIN	14.9	17.1	26.6	37.0	46.8	56.3	60.1	58.0	51.0	40.2	30.8	20.7	38.3
	MEAN	23.1	26.2	36.4	48.9	59.5	68.9	72.3	70.5	63.9	52.6	39.9	28.6	49.2
WEST LAFAYETTE 6 NW	MAX	31.5	35.7	46.3	60.4	71.6	80.8	84.5	82.5	77.1	65.3	49.5	37.0	60.2
	MIN	14.6	17.7	27.9	39.7	49.8	59.0	62.5	60.3	53.1	41.7	31.4	21.0	39.9
	MEAN	23.0	26.7	37.1	50.1	60.7	70.0	73.6	71.4	65.2	53.5	40.5	29.0	50.1
WHEATFIELD 2 NNW	MAX	30.4	34.6	45.3	59.9	72.0	82.0	85.2	83.4	77.7	65.6	49.0	35.7	60.1
	MIN	11.7	14.9	25.7	37.1	47.0	56.4	59.9	57.4	49.8	38.9	28.9	18.5	37.2
	MEAN	21.0	24.8	35.5	48.5	59.6	69.2	72.6	70.4	63.7	52.3	39.0	27.1	48.6
WHITESTOWN	MAX	33.2	37.5	48.3	62.6	73.4	82.3	85.6	83.8	78.0	66.1	50.6	38.1	61.6
	MIN	15.6	18.5	28.1	39.5	49.5	58.5	61.9	59.6	52.4	41.0	31.2	22.0	39.8
	MEAN	24.4	28.0	38.3	51.0	61.5	70.4	73.8	71.7	65.3	53.6	40.9	30.1	50.8
WINAMAC 5 SW	MAX	31.8	36.3	47.4	62.1	73.0	81.4	84.5	82.4	76.2	64.8	49.3	36.8	60.5
	MIN	14.4	17.9	27.3	38.3	48.3	57.4	61.1	59.2	52.1	41.5	30.7	20.8	39.1
	MEAN	23.1	27.1	37.4	50.2	60.7	69.4	72.8	70.8	64.2	53.2	40.0	28.8	49.8
WINCHESTER AIRPORT	MAX	32.4	35.7	46.1	59.9	70.7	80.1	83.6	82.3	76.5	64.5	49.5	37.3	59.9
	MIN	15.3	17.9	27.4	38.7	49.1	58.5	61.9	59.7	52.6	41.3	31.4	21.4	39.6
	MEAN	23.9	26.8	36.8	49.3	59.9	69.3	72.8	71.0	64.6	52.9	40.5	29.4	49.8

INDIANA

PRECIPITATION NORMALS (INCHES)

STATION	JAN	FEB	MAR	APR	MAY	JUN	JUL	AUG	SEP	OCT	NOV	DEC	ANN
ANDERSON SEWAGE PLANT	2.31	2.19	3.29	3.98	3.81	3.76	3.90	3.23	2.79	2.55	2.93	2.80	37.54
ANDERSON WATERWORKS	2.42	2.12	3.19	4.05	3.69	3.85	3.89	3.51	2.85	2.47	2.94	2.83	37.81
BEDFORD 4 SW	2.69	2.52	4.11	3.41	4.26	4.08	4.37	3.72	2.69	2.51	3.23	3.07	40.66
BERNE	2.30	2.06	3.28	3.90	3.56	4.15	3.80	3.19	3.24	2.46	2.74	2.65	37.33
BLOOMINGTON INDIANA U	2.86	2.56	3.83	3.79	4.46	4.33	4.62	3.59	3.18	2.37	3.44	3.46	42.49
BOWLING GREEN 3 NE	2.65	2.53	3.82	3.85	4.43	4.73	4.65	3.48	2.95	2.54	3.29	3.13	42.05
BROOKVILLE	2.85	2.39	3.60	3.68	4.32	4.17	4.73	3.71	2.70	2.58	2.84	2.92	40.49
CAMBRIDGE CITY	2.68	2.30	3.40	4.01	4.21	4.24	4.17	3.37	2.70	2.52	3.14	2.89	39.63
COLUMBIA CITY 1 S	2.16	1.71	2.90	3.70	3.38	3.89	4.02	3.64	3.01	2.54	2.93	2.63	36.51
COLUMBUS	2.94	2.59	3.72	3.95	4.42	3.48	4.64	3.26	2.72	2.29	3.10	3.14	40.25
COVINGTON 3 SW	2.14	1.88	3.11	3.81	4.03	4.35	4.25	3.59	2.80	2.48	2.90	2.78	38.12
CRANE NAVAL DEPOT	2.99	2.74	4.25	3.83	4.46	4.36	4.87	3.76	2.98	2.73	3.51	3.66	44.14
CRAWFORDSVILLE PWR PL	2.40	2.09	3.21	4.23	4.19	4.84	4.42	3.83	2.76	2.56	3.01	2.84	40.38
DECATUR 1 N	2.08	1.70	2.98	3.70	3.79	3.70	3.66	3.04	2.81	2.42	2.71	2.40	34.99
DELPHI 3 NNE	1.95	1.88	2.85	3.79	3.69	4.14	4.49	3.75	2.91	2.46	2.53	2.53	36.97
EDWARDSPORT POWER PL	2.42	2.31	3.68	3.63	4.05	3.95	3.98	3.35	2.95	2.37	3.02	2.98	38.69
ELLISTON	2.51	2.51	3.89	3.81	4.41	4.62	4.64	3.42	2.76	2.67	3.40	3.28	41.92
ELWOOD WATERWORKS	2.37	2.07	3.17	3.92	3.70	4.10	4.23	3.47	3.07	2.64	2.96	2.86	38.56
EVANSVILLE	3.33	3.20	4.79	4.18	4.29	3.82	4.44	3.25	3.02	2.45	3.75	3.55	44.07
EVANSVILLE WSO //R	2.99	3.02	4.58	4.08	4.37	3.50	3.98	3.07	2.67	2.48	3.36	3.45	41.55
FT WAYNE DISPOSAL PL	2.02	1.84	2.92	3.84	3.72	3.79	4.06	3.43	2.79	2.62	2.80	2.44	36.27
FORT WAYNE WSO //R	2.07	1.96	2.94	3.56	3.47	3.62	3.39	3.29	2.53	2.56	2.57	2.44	34.40
FRANKFORT DISPOSAL PL	2.03	2.06	2.98	3.84	3.97	4.41	4.57	3.64	3.01	2.75	2.80	2.72	38.78
GARY	1.71	1.41	2.74	3.78	3.69	3.82	3.78	3.62	3.35	2.91	2.30	2.37	35.48
GOSHEN COLLEGE	1.78	1.58	2.60	3.59	2.97	3.61	3.61	3.66	3.03	2.73	2.32	2.23	33.71
GREENCASTLE 1 E	2.69	2.40	3.69	3.79	4.20	4.70	4.58	4.01	3.35	2.73	3.14	2.98	42.26
GREENFIELD	2.74	2.32	3.23	3.83	4.10	4.17	4.78	3.84	2.97	2.63	3.06	2.82	40.49
GREENSBURG 3 SW	2.57	2.46	3.64	4.09	4.50	4.05	4.53	3.43	2.81	2.46	3.14	3.11	40.79
HOBART	1.71	1.46	2.40	3.76	3.28	3.84	3.84	3.52	3.52	2.88	2.40	2.21	34.82
HUNTINGTON	2.23	2.00	3.24	3.71	3.75	4.19	3.61	3.34	2.85	2.67	2.73	2.72	37.04
INDIANAPOLIS WSO //R	2.65	2.46	3.61	3.68	3.66	3.99	4.32	3.46	2.74	2.51	3.04	3.00	39.12
INDIANAPOLIS SE SIDE	2.37	2.12	3.10	3.67	4.02	4.27	4.72	3.69	2.64	2.60	3.03	2.62	38.85
JOHNSON EXP FARM	2.75	2.66	4.38	4.05	4.49	3.76	4.39	3.48	2.69	2.42	3.32	3.38	41.77
KENTLAND	1.73	1.64	2.88	4.13	3.72	4.82	4.51	3.54	3.07	2.59	2.45	2.33	37.41
KNIGHTSTOWN	2.64	2.17	3.20	4.02	4.14	4.24	4.33	3.55	2.90	2.46	3.38	2.87	39.90
KOKOMO 7 SE	2.23	2.04	3.01	4.03	3.67	3.97	4.40	3.67	2.80	2.90	2.75	2.64	38.11
LA PORTE	2.43	2.16	3.21	4.32	3.21	4.20	4.46	4.07	3.80	3.76	2.81	3.09	41.52
LOGANSPORT CICOT ST BR	2.24	2.01	2.98	4.05	3.82	4.11	4.15	3.45	3.09	2.74	2.77	2.61	38.02
MADISON SEWAGE PLANT	3.31	3.15	4.33	3.92	4.43	3.98	4.01	3.08	3.08	2.72	3.47	3.22	42.70
MARION 2 N	2.17	2.00	3.08	3.92	3.44	3.92	4.20	3.38	2.91	2.35	2.80	2.56	36.73
MARTINSVILLE 2 SW	2.63	2.41	3.41	3.67	4.42	4.32	4.48	3.32	2.87	2.48	3.34	2.93	40.28
MONROEVILLE 3 ENE	2.01	1.67	2.63	3.26	3.21	3.31	3.26	2.92	2.52	2.31	2.49	2.26	31.85
MONTICELLO	2.00	1.80	2.82	4.04	3.54	4.19	3.93	3.67	2.98	2.61	2.55	2.40	36.53
MOUNT VERNON	3.14	3.05	4.92	4.24	4.52	3.69	4.14	3.33	2.69	2.58	3.72	3.40	43.42
MUNCIE	2.40	2.13	3.27	3.99	3.71	4.33	3.69	3.44	2.69	2.44	2.84	2.78	37.71
NEWBURGH DAM 47	3.22	3.01	4.58	4.15	4.11	3.79	3.86	3.02	3.29	2.38	3.57	3.29	42.27
NEW CASTLE	2.45	2.23	3.27	4.19	4.06	4.39	3.84	3.62	2.62	2.66	3.00	2.64	38.97
NEW HARMONY	2.82	2.68	4.54	4.04	4.58	3.75	3.80	3.33	2.84	2.48	3.30	3.15	41.31
NOBLESVILLE	2.33	2.22	3.03	3.85	3.71	4.14	4.29	3.39	2.60	2.42	2.71	2.62	37.31
NORTH VERNON 2 SW	3.66	2.88	4.52	3.92	4.58	4.19	4.69	3.45	2.93	2.61	3.43	3.43	44.29
OAKLANDON GEIST RESVR	2.60	2.37	3.42	4.09	4.21	4.18	4.62	3.74	2.93	2.74	3.09	2.95	40.94
OGDEN DUNES	1.98	1.64	2.80	3.82	3.17	4.03	3.82	3.36	3.42	3.08	2.33	2.41	35.86
OOLITIC PURDUE EXP FRM	3.20	2.76	4.45	4.04	4.68	4.62	4.61	3.69	3.05	2.67	3.48	3.60	44.85
PAOLI RADIO WVAK	3.54	3.13	4.76	4.42	4.51	4.45	4.73	3.72	2.82	2.70	3.74	3.51	46.03
PETERSBURG 61 BRIDGE	2.79	2.54	4.47	3.78	4.19	4.12	4.20	3.44	2.81	2.47	3.35	3.40	41.56

INDIANA

PRECIPITATION NORMALS (INCHES)

STATION	JAN	FEB	MAR	APR	MAY	JUN	JUL	AUG	SEP	OCT	NOV	DEC	ANN
PLYMOUTH POWER SUBSTA	1.91	1.75	2.76	4.09	3.52	4.35	3.97	3.40	3.21	3.09	2.62	2.50	37.17
PORTLAND 1 SW	2.24	1.85	2.91	3.80	3.42	4.07	3.61	3.46	2.91	2.51	2.58	2.54	35.90
PRINCETON 1 W	2.83	2.73	4.56	4.39	4.67	4.22	4.29	3.32	3.06	2.71	3.46	3.61	43.85
RICHMOND WTR WKS 2 NNE	2.63	2.25	3.33	3.74	4.17	4.26	4.03	3.28	2.69	2.48	2.93	2.91	38.70
ROCHESTER	1.80	1.63	2.64	4.07	3.53	4.27	3.99	3.43	3.33	2.62	2.90	2.46	36.67
ROCKVILLE	2.50	2.25	3.55	4.11	3.76	4.50	4.38	3.61	2.74	2.51	3.17	3.30	40.38
RUSHVILLE SEWAGE PLANT	2.82	2.45	3.26	3.78	4.44	4.20	4.45	3.51	2.79	2.47	3.22	2.93	40.32
SCOTTSBURG	3.25	2.99	4.31	3.91	4.11	4.13	4.54	3.46	3.17	2.67	3.20	2.98	42.72
SEYMOUR 2 N	3.35	2.86	4.07	3.78	4.33	4.13	4.88	3.24	2.93	2.56	3.24	3.12	42.49
SHELBYVILLE SEWAGE PL	2.67	2.45	3.42	3.84	4.55	3.84	4.56	3.68	2.79	2.42	3.29	3.01	40.52
SHOALS HIWAY 50 BRIDGE	3.20	2.85	4.55	3.90	4.44	4.08	4.85	3.85	3.15	2.56	3.58	3.36	44.37
SOUTH BEND WSO //R	2.48	1.99	3.05	4.06	2.81	3.94	3.67	3.94	3.22	3.22	2.83	2.95	38.16
SPENCER	2.81	2.55	3.86	3.96	4.45	4.70	4.67	3.95	3.06	2.74	3.55	3.22	43.52
TELL CITY POWER PLANT	3.59	3.21	4.73	4.71	4.49	4.41	4.25	3.59	3.13	2.56	3.89	3.59	46.15
TERRE HAUTE 8 S	2.27	2.06	3.44	3.78	3.98	4.07	4.40	3.12	3.01	2.36	3.02	2.63	38.14
VALPARAISO WATERWORKS	1.95	1.61	2.85	4.32	3.61	4.09	3.95	3.99	3.69	3.39	2.63	2.56	38.64
VEVAY	3.13	3.07	4.34	3.85	3.80	3.88	4.05	3.44	3.26	2.61	3.32	3.14	41.89
VINCENNES 1 NW	2.69	2.49	4.27	4.01	4.24	4.00	4.41	4.00	2.95	2.40	3.55	3.32	42.33
WABASH	2.20	1.80	2.95	3.90	3.63	4.02	4.14	3.85	2.98	2.57	2.64	2.54	37.22
WASHINGTON	2.67	2.59	4.07	3.73	4.28	3.52	4.81	3.76	2.70	2.46	3.33	3.39	41.31
WATERLOO	1.93	1.75	2.87	3.53	3.48	3.47	3.61	3.23	2.95	2.53	2.75	2.71	34.81
WEST LAFAYETTE FAA AP	1.96	1.84	2.91	3.77	3.84	4.29	3.98	3.47	2.84	2.58	2.46	2.36	36.30
WEST LAFAYETTE 6 NW	1.84	1.51	2.72	3.95	3.92	4.44	4.35	3.53	2.96	2.57	2.58	2.23	36.60
WHEATFIELD 2 NNW	1.66	1.47	2.86	4.13	3.35	4.40	4.06	3.60	3.31	2.67	2.36	2.43	36.30
WHITESTOWN	2.46	2.19	3.40	4.05	3.92	4.57	4.41	3.61	2.72	2.89	2.95	2.86	40.03
WILLIAMS	2.95	2.58	4.23	3.88	4.53	4.24	4.58	3.93	2.87	2.60	3.59	3.36	43.34
WINAMAC 5 SW	1.96	1.71	2.78	4.07	3.26	4.51	3.92	3.84	3.07	2.57	2.66	2.61	36.96
WINCHESTER AIRPORT	2.34	1.93	3.23	4.01	3.77	4.25	3.98	3.40	2.73	2.52	2.89	2.72	37.77

INDIANA

HEATING DEGREE DAY NORMALS (BASE 65 DEG F)

STATION	JUL	AUG	SEP	OCT	NOV	DEC	JAN	FEB	MAR	APR	MAY	JUN	ANN
ANDERSON SEWAGE PLANT	6	11	77	352	699	1060	1221	1008	806	421	191	32	5884
BERNE	0	6	72	350	708	1070	1234	1028	831	432	190	14	5935
BROOKVILLE	0	8	77	374	699	1029	1181	994	803	423	185	17	5790
CAMBRIDGE CITY	0	7	90	396	735	1082	1252	1050	849	456	202	22	6141
COLUMBIA CITY 1 S	6	20	105	391	744	1119	1290	1078	887	474	218	39	6371
COLUMBUS	0	5	70	334	672	1001	1159	958	766	372	157	15	5509
CRANE NAVAL DEPOT	0	0	27	236	579	915	1066	848	663	269	107	7	4717
CRAWFORDSVILLE PWR PL	0	7	75	354	717	1076	1252	1036	834	429	189	21	5990
DELPHI 3 NNE	0	7	66	334	696	1057	1231	1002	791	390	165	17	5756
EDWARDSPORT POWER PL	0	0	48	298	642	986	1147	935	735	327	134	9	5261
ELWOOD WATERWORKS	0	11	95	387	738	1101	1277	1064	868	462	204	27	6234
EVANSVILLE	0	0	25	210	534	849	989	778	581	216	73	5	4260
EVANSVILLE WSO //R	0	0	40	259	597	911	1066	843	640	267	100	6	4729
FORT WAYNE WSO //R	0	7	99	384	747	1116	1293	1081	880	474	212	27	6320
FRANKFORT DISPOSAL PL	0	9	82	372	741	1097	1271	1047	846	441	196	21	6123
GARY	6	8	74	340	714	1113	1305	1061	884	480	225	41	6251
GOSHEN COLLEGE	0	13	88	381	738	1116	1290	1072	880	477	213	23	6291
GREENCASTLE 1 E	0	0	54	308	684	1048	1215	986	778	370	152	14	5609
GREENFIELD	0	6	67	340	696	1057	1218	1008	806	402	172	18	5790
GREENSBURG 3 SW	0	10	77	344	690	1011	1166	952	750	377	166	19	5562
HOBART	0	0	62	315	702	1091	1277	1039	849	457	200	29	6021
HUNTINGTON	0	6	71	354	717	1082	1249	1036	834	426	187	19	5981
INDIANAPOLIS WSO //R	0	0	63	330	696	1039	1209	983	775	382	158	15	5650
INDIANAPOLIS SE SIDE	0	0	56	322	675	1029	1197	983	787	384	161	14	5608
JOHNSON EXP FARM	0	0	30	250	585	915	1070	857	662	270	106	7	4752
KENTLAND	0	6	56	325	717	1101	1277	1036	815	399	166	18	5916
KOKOMO 7 SE	9	15	82	358	717	1082	1252	1039	828	426	193	34	6035
LA PORTE	0	10	70	368	744	1128	1308	1070	884	477	213	30	6302
MARION 2 N	0	9	87	390	741	1107	1283	1078	887	483	223	30	6318
MARTINSVILLE 2 SW	0	6	80	352	687	1023	1184	980	769	373	164	18	5636
MOUNT VERNON	0	0	33	257	582	915	1066	862	663	278	101	0	4757
NEW CASTLE	5	9	88	389	735	1085	1249	1044	846	456	205	24	6135
NORTH VERNON 2 SW	0	0	45	274	618	933	1088	871	677	303	128	10	4947
OAKLANDON GEIST RESVR	0	6	71	348	699	1054	1215	1008	812	405	185	21	5824
OGDEN DUNES	0	6	54	324	702	1091	1277	1044	871	486	221	33	6109
OOLITIC PURDUE EXP FRM	0	0	68	332	654	983	1132	927	725	338	154	19	5332
PAOLI RADIO WVAK	0	6	67	327	660	977	1128	924	735	344	145	16	5329
PLYMOUTH POWER SUBSTA	0	11	81	362	738	1110	1290	1056	853	438	185	19	6143
PRINCETON 1 W	0	0	39	252	594	921	1073	843	657	274	105	6	4764
RICHMOND WTR WKS 2 NNE	0	7	100	393	726	1057	1221	1011	809	435	193	21	5973
ROCHESTER	0	17	104	405	765	1144	1330	1100	915	489	225	31	6525
ROCKVILLE	0	6	48	295	657	1017	1187	949	744	342	141	12	5398
RUSHVILLE SEWAGE PLANT	0	5	75	371	717	1060	1234	1025	825	420	183	21	5936
SCOTTSBURG	0	0	49	308	642	955	1113	913	713	332	133	8	5166
SEYMOUR 2 N	0	6	73	338	669	995	1147	944	744	355	150	13	5434
SHELBYVILLE SEWAGE PL	0	6	70	340	687	1032	1200	988	775	382	160	20	5660
SHOALS HIWAY 50 BRIDGE	0	0	56	314	645	967	1119	918	725	331	147	11	5233
SOUTH BEND WSO //R	6	17	88	376	741	1113	1296	1081	899	495	230	35	6377
SPENCER	0	0	65	340	684	1020	1178	969	769	377	160	16	5578
TELL CITY POWER PLANT	0	0	32	255	564	884	1032	843	654	287	108	7	4666
TERRE HAUTE 8 S	0	0	56	307	672	1026	1203	974	766	358	149	10	5521
VALPARAISO WATERWORKS	0	11	80	354	738	1125	1314	1072	880	474	211	32	6291
VEVAY	0	0	38	275	603	908	1048	857	660	306	121	7	4823
VINCENNES 1 NW	0	0	40	288	639	977	1132	918	716	313	125	8	5156
WABASH	5	13	99	405	756	1132	1308	1098	905	489	229	39	6478

INDIANA

HEATING DEGREE DAY NORMALS (BASE 65 DEG F)

STATION	JUL	AUG	SEP	OCT	NOV	DEC	JAN	FEB	MAR	APR	MAY	JUN	ANN
WASHINGTON	0	0	35	248	591	915	1066	846	647	260	100	6	4714
WATERLOO	0	10	87	391	753	1128	1299	1086	887	483	214	30	6368
WEST LAFAYETTE 6 NW	5	9	80	369	735	1116	1302	1072	865	447	196	27	6223
WHEATFIELD 2 NNW	0	15	110	403	780	1175	1364	1126	915	495	217	34	6634
WHITESTOWN	0	5	83	365	723	1082	1259	1036	828	420	184	22	6007
WINAMAC 5 SW	0	10	91	374	750	1122	1299	1061	856	444	197	27	6231
WINCHESTER AIRPORT	6	11	90	384	735	1104	1274	1070	874	471	213	30	6262

INDIANA

COOLING DEGREE DAY NORMALS (BASE 65 DEG F)

STATION	JAN	FEB	MAR	APR	MAY	JUN	JUL	AUG	SEP	OCT	NOV	DEC	ANN
ANDERSON SEWAGE PLANT	0	0	0	0	77	191	279	228	101	14	0	0	890
BERNE	0	0	0	0	70	179	279	229	96	15	0	0	868
BROOKVILLE	0	0	0	0	61	161	267	228	95	8	0	0	820
CAMBRIDGE CITY	0	0	0	0	53	145	244	197	75	5	0	0	719
COLUMBIA CITY 1 S	0	0	0	0	57	168	248	206	90	16	0	0	785
COLUMBUS	0	0	0	0	73	201	307	253	118	12	0	0	964
CRANE NAVAL DEPOT	0	0	9	14	119	262	372	332	171	35	0	0	1314
CRAWFORDSVILLE PWR PL	0	0	0	0	68	183	291	234	99	10	0	0	885
DELPHI 3 NNE	0	0	0	0	72	194	288	227	96	11	0	0	888
EDWARDSPORT POWER PL	0	0	0	6	100	249	357	298	141	19	0	0	1170
ELWOOD WATERWORKS	0	0	0	0	55	162	253	194	80	12	0	0	756
EVANSVILLE	0	0	11	21	144	329	440	391	211	30	0	0	1577
EVANSVILLE WSO //R	0	0	8	12	119	294	406	347	172	20	0	0	1378
FORT WAYNE WSO //R	0	0	0	0	54	165	260	205	90	12	0	0	786
FRANKFORT DISPOSAL PL	0	0	0	0	69	174	267	207	82	13	0	0	812
GARY	0	0	0	0	61	179	285	240	95	15	0	0	875
GOSHEN COLLEGE	0	0	0	0	49	149	245	196	64	9	0	0	712
GREENCASTLE 1 E	0	0	0	0	90	224	322	279	126	17	0	0	1058
GREENFIELD	0	0	0	0	82	201	301	248	115	14	0	0	961
GREENSBURG 3 SW	0	0	0	0	67	178	267	221	101	9	0	0	843
HOBART	0	0	0	0	57	173	276	228	92	15	0	0	841
HUNTINGTON	0	0	0	0	57	169	264	223	86	13	0	0	812
INDIANAPOLIS WSO //R	0	0	0	0	80	213	313	257	111	14	0	0	988
INDIANAPOLIS SE SIDE	0	0	0	0	84	212	322	267	119	19	0	0	1023
JOHNSON EXP FARM	0	0	8	12	128	292	400	344	183	37	0	0	1404
KENTLAND	0	0	0	0	79	213	298	241	101	15	0	0	947
KOKOMO 7 SE	0	0	0	6	84	214	294	232	109	23	0	0	962
LA PORTE	0	0	0	0	58	162	265	214	70	11	0	0	780
MARION 2 N	0	0	0	0	56	159	249	189	72	9	0	0	734
MARTINSVILLE 2 SW	0	0	0	0	83	201	301	245	107	14	0	0	951
MOUNT VERNON	0	0	0	14	117	282	403	347	177	31	0	0	1371
NEW CASTLE	0	0	0	0	59	171	266	204	76	8	0	0	784
NORTH VERNON 2 SW	0	0	7	9	91	223	329	292	141	14	0	0	1106
OAKLANDON GEIST RESVR	0	0	0	0	74	183	282	233	101	17	0	0	890
OGDEN DUNES	0	0	0	0	44	153	264	232	87	17	0	0	797
OOLITIC PURDUE EXP FRM	0	0	0	8	92	214	316	271	131	22	0	0	1054
PAOLI RADIO WVAK	0	0	0	5	74	205	307	270	127	11	0	0	999
PLYMOUTH POWER SUBSTA	0	0	0	0	64	175	273	222	84	9	0	0	827
PRINCETON 1 W	0	0	9	10	117	282	381	332	177	32	0	0	1340
RICHMOND WTR WKS 2 NNE	0	0	0	0	54	147	242	197	82	6	0	0	728
ROCHESTER	0	0	0	0	55	154	246	188	68	9	0	0	720
ROCKVILLE	0	0	0	9	94	228	335	279	129	20	0	0	1094
RUSHVILLE SEWAGE PLANT	0	0	0	0	65	171	264	216	90	11	0	0	817
SCOTTSBURG	0	0	0	0	90	230	341	288	133	14	0	0	1096
SEYMOUR 2 N	0	0	0	0	81	202	301	254	121	16	0	0	975
SHELBYVILLE SEWAGE PL	0	0	0	0	83	209	298	251	115	14	0	0	970
SHOALS HIWAY 50 BRIDGE	0	0	0	7	91	218	326	279	128	16	0	0	1065
SOUTH BEND WSO //R	0	0	0	0	47	149	239	200	64	11	0	0	710
SPENCER	0	0	0	0	79	214	313	260	113	15	0	0	994
TELL CITY POWER PLANT	0	0	0	11	111	271	394	350	176	32	0	0	1345
TERRE HAUTE 8 S	0	0	0	0	84	223	329	270	125	18	0	0	1049
VALPARAISO WATERWORKS	0	0	0	0	47	146	236	190	62	9	0	0	690
VEVAY	0	0	0	9	93	238	353	310	149	21	0	0	1173
VINCENNES 1 NW	0	0	0	7	100	263	372	313	151	21	0	0	1227
WABASH	0	0	0	0	55	153	235	175	66	8	0	0	692

INDIANA

COOLING DEGREE DAY NORMALS (BASE 65 DEG F)

STATION	JAN	FEB	MAR	APR	MAY	JUN	JUL	AUG	SEP	OCT	NOV	DEC	ANN
WASHINGTON	0	0	8	14	119	279	381	332	170	25	0	0	1328
WATERLOO	0	0	0	0	43	147	231	180	54	7	0	0	662
WEST LAFAYETTE 6 NW	0	0	0	0	63	177	272	207	86	13	0	0	818
WHEATFIELD 2 NNW	0	0	0	0	50	160	240	182	71	10	0	0	713
WHITESTOWN	0	0	0	0	75	184	273	213	92	11	0	0	848
WINAMAC 5 SW	0	0	0	0	64	159	247	189	67	8	0	0	734
WINCHESTER AIRPORT	0	0	0	0	55	159	248	197	78	9	0	0	746

STATE-STATION NUMBER	STN TYP	NAME	LATITUDE DEG-MIN	LONGITUDE DEG-MIN	ELEVATION (FT)
12-0177	13	ANDERSON SEWAGE PLANT	N 4006	W 08543	847
12-0182	12	ANDERSON WATERWORKS	N 4006	W 08541	870
12-0550	12	BEDFORD 4 SW	N 3850	W 08632	508
12-0676	13	BERNE	N 4040	W 08457	858
12-0784	12	BLOOMINGTON INDIANA U	N 3910	W 08631	820
12-0877	12	BOWLING GREEN 3 NE	N 3925	W 08658	690
12-1030	13	BROOKVILLE	N 3925	W 08501	670
12-1229	13	CAMBRIDGE CITY	N 3949	W 08510	940
12-1739	13	COLUMBIA CITY 1 S	N 4108	W 08529	850
12-1747	13	COLUMBUS	N 3912	W 08555	621
12-1843	12	COVINGTON 3 SW	N 4007	W 08727	620
12-1869	13	CRANE NAVAL DEPOT	N 3852	W 08650	760
12-1882	13	CRAWFORDSVILLE PWR PL	N 4003	W 08654	679
12-2096	12	DECATUR 1 N	N 4051	W 08456	790
12-2149	13	DELPHI 3 NNE	N 4037	W 08640	560
12-2549	13	EDWARDSPORT POWER PL	N 3848	W 08714	463
12-2605	12	ELLISTON	N 3902	W 08658	550
12-2638	13	ELWOOD WATERWORKS	N 4016	W 08551	853
12-2731	13	EVANSVILLE	N 3758	W 08733	384
12-2738	13	EVANSVILLE WSO //R	N 3803	W 08732	381
12-3027	12	FT WAYNE DISPOSAL PL	N 4106	W 08507	730
12-3037	13	FORT WAYNE WSO //R	N 4100	W 08512	791
12-3082	13	FRANKFORT DISPOSAL PL	N 4019	W 08630	833
12-3213	13	GARY	N 4137	W 08723	597
12-3418	13	GOSHEN COLLEGE	N 4134	W 08550	806
12-3513	13	GREENCASTLE 1 E	N 3939	W 08651	835
12-3527	13	GREENFIELD	N 3947	W 08546	900
12-3547	13	GREENSBURG 3 SW	N 3920	W 08533	875
12-4008	13	HOBART	N 4132	W 08715	600
12-4176	13	HUNTINGTON	N 4053	W 08530	802
12-4259	13	INDIANAPOLIS WSO //R	N 3944	W 08616	792
12-4272	13	INDIANAPOLIS SE SIDE	N 3945	W 08607	750
12-4407	13	JOHNSON EXP FARM	N 3816	W 08745	440
12-4527	13	KENTLAND	N 4046	W 08727	685
12-4642	12	KNIGHTSTOWN	N 3947	W 08532	830
12-4662	13	KOKOMO 7 SE	N 4025	W 08603	855
12-4837	13	LA PORTE	N 4136	W 08643	810
12-5117	12	LOGANSPORT CICOT ST BR	N 4045	W 08623	596
12-5237	12	MADISON SEWAGE PLANT	N 3844	W 08524	455
12-5337	13	MARION 2 N	N 4034	W 08540	791
12-5407	13	MARTINSVILLE 2 SW	N 3924	W 08627	605
12-5815	12	MONROEVILLE 3 ENE	N 4059	W 08449	775
12-5837	12	MONTICELLO	N 4045	W 08646	677
12-6001	13	MOUNT VERNON	N 3757	W 08753	415
12-6023	12	MUNCIE	N 4011	W 08521	957
12-6151	12	NEWBURGH DAM 47	N 3757	W 08724	380
12-6164	13	NEW CASTLE	N 3956	W 08523	973
12-6179	12	NEW HARMONY	N 3808	W 08756	390
12-6338	12	NOBLESVILLE	N 4002	W 08601	780
12-6435	13	NORTH VERNON 2 SW	N 3900	W 08539	790

12 — INDIANA

STATE-STATION NUMBER	STN TYP	NAME	LATITUDE DEG-MIN	LONGITUDE DEG-MIN	ELEVATION (FT)
12-6506	13	OAKLANDON GEIST RESVR	N 3954	W 08559	795
12-6542	13	OGDEN DUNES	N 4137	W 08711	610
12-6580	13	OOLITIC PURDUE EXP FRM	N 3853	W 08633	650
12-6705	13	PAOLI RADIO WVAK	N 3832	W 08629	640
12-6872	12	PETERSBURG 61 BRIDGE	N 3830	W 08717	485
12-7028	13	PLYMOUTH POWER SUBSTA	N 4120	W 08619	785
12-7069	12	PORTLAND 1 SW	N 4025	W 08500	915
12-7125	13	PRINCETON 1 W	N 3821	W 08735	482
12-7370	13	RICHMOND WTR WKS 2 NNE	N 3953	W 08453	1014
12-7482	13	ROCHESTER	N 4104	W 08613	770
12-7522	13	ROCKVILLE	N 3946	W 08714	692
12-7646	13	RUSHVILLE SEWAGE PLANT	N 3936	W 08527	955
12-7875	13	SCOTTSBURG	N 3841	W 08546	550
12-7935	13	SEYMOUR 2 N	N 3859	W 08554	573
12-7999	13	SHELBYVILLE SEWAGE PL	N 3931	W 08547	750
12-8036	13	SHOALS HIWAY 50 BRIDGE	N 3840	W 08648	550
12-8187	13	SOUTH BEND WSO //R	N 4142	W 08619	773
12-8290	13	SPENCER	N 3917	W 08646	550
12-8698	13	TELL CITY POWER PLANT	N 3757	W 08646	394
12-8723	13	TERRE HAUTE 8 S	N 3921	W 08725	555
12-8999	13	VALPARAISO WATERWORKS	N 4131	W 08702	801
12-9080	13	VEVAY	N 3845	W 08504	480
12-9112	13	VINCENNES 1 NW	N 3841	W 08732	420
12-9138	13	WABASH	N 4045	W 08548	800
12-9253	13	WASHINGTON	N 3839	W 08710	510
12-9271	13	WATERLOO	N 4125	W 08502	908
12-9424	12	WEST LAFAYETTE FAA AP	N 4025	W 08656	599
12-9430	13	WEST LAFAYETTE 6 NW	N 4028	W 08700	705
12-9511	13	WHEATFIELD 2 NNW	N 4114	W 08704	658
12-9557	13	WHITESTOWN	N 4000	W 08620	819
12-9605	12	WILLIAMS	N 3848	W 08639	592
12-9670	13	WINAMAC 5 SW	N 4100	W 08639	700
12-9678	13	WINCHESTER AIRPORT	N 4011	W 08455	1109

12 – INDIANA

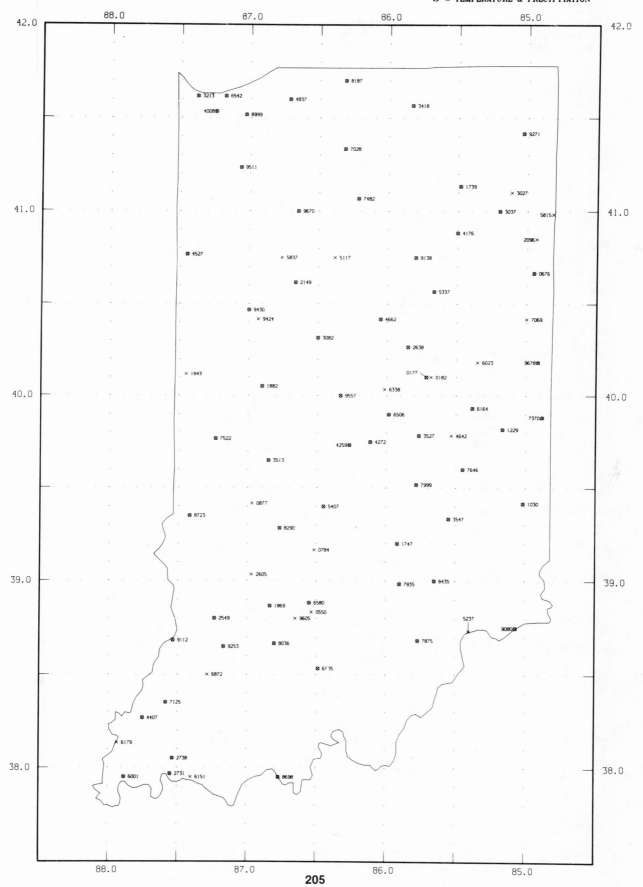

IOWA

TEMPERATURE NORMALS (DEG F)

STATION		JAN	FEB	MAR	APR	MAY	JUN	JUL	AUG	SEP	OCT	NOV	DEC	ANN
ALBIA	MAX	30.4	36.8	47.4	63.1	74.3	83.0	87.6	85.5	77.8	67.2	50.1	36.6	61.7
	MIN	12.3	17.9	27.5	41.0	51.6	60.8	65.4	63.2	54.6	44.1	30.9	19.6	40.7
	MEAN	21.3	27.4	37.5	52.1	63.0	71.9	76.6	74.3	66.2	55.6	40.5	28.1	51.2
ALGONA 3 W	MAX	23.3	30.0	40.3	59.0	72.1	81.5	84.7	82.6	74.1	63.1	44.4	29.7	57.1
	MIN	4.6	10.8	21.6	35.6	47.3	57.1	61.1	58.7	49.7	39.2	25.3	12.5	35.3
	MEAN	14.0	20.4	31.0	47.4	59.7	69.3	72.9	70.7	61.9	51.2	34.8	21.1	46.2
ALLISON	MAX	24.5	31.1	41.5	59.5	72.5	81.6	84.9	83.2	75.1	63.8	45.2	30.6	57.8
	MIN	6.0	12.2	22.9	37.0	48.5	58.1	62.0	60.1	50.9	40.5	26.9	14.2	36.6
	MEAN	15.3	21.7	32.2	48.3	60.5	69.9	73.5	71.7	63.0	52.2	36.1	22.4	47.2
AMES 8 WSW	MAX	26.3	32.6	43.4	61.0	72.8	81.4	85.0	82.9	75.4	64.8	46.7	32.8	58.8
	MIN	7.9	14.2	24.5	38.0	49.3	58.7	63.0	60.4	51.5	40.7	27.3	15.7	37.6
	MEAN	17.1	23.4	34.0	49.5	61.1	70.1	74.0	71.7	63.5	52.8	37.0	24.3	48.2
ANAMOSA 1 WNW	MAX	26.5	32.5	43.4	60.3	72.3	80.9	84.8	82.9	75.1	64.2	46.8	32.7	58.5
	MIN	7.4	13.0	23.6	36.6	47.1	56.5	60.8	58.7	50.1	39.3	27.1	15.4	36.3
	MEAN	17.0	22.7	33.6	48.5	59.7	68.7	72.8	70.8	62.6	51.8	37.0	24.1	47.4
ANKENY 2 SW R	MAX	27.0	33.3	43.8	60.8	72.5	81.1	85.4	83.3	75.5	64.9	47.6	33.5	59.1
	MIN	8.4	14.2	24.9	38.5	49.9	59.4	63.7	61.2	52.0	41.1	28.1	15.9	38.1
	MEAN	17.7	23.8	34.4	49.7	61.2	70.3	74.6	72.3	63.7	53.1	37.9	24.7	48.6
ATLANTIC 1 NE	MAX	28.2	34.7	45.6	62.6	73.5	82.4	86.5	84.2	76.1	65.7	47.9	34.9	60.2
	MIN	7.5	13.6	24.1	37.0	48.6	58.1	62.2	59.7	50.4	39.1	26.3	15.1	36.8
	MEAN	17.9	24.1	34.8	49.8	61.1	70.3	74.4	71.9	63.3	52.4	37.2	25.0	48.5
AUDUBON	MAX	28.0	34.5	45.1	62.3	73.6	82.6	87.0	84.9	76.3	65.7	47.8	34.4	60.2
	MIN	7.6	13.7	23.7	37.4	48.8	58.3	62.8	60.5	51.2	40.1	26.2	15.0	37.1
	MEAN	17.8	24.1	34.4	49.9	61.2	70.5	75.0	72.7	63.8	52.9	37.0	24.7	48.7
BEDFORD	MAX	31.9	38.6	49.2	64.7	75.1	83.5	88.3	86.3	78.5	68.2	51.1	38.0	62.8
	MIN	11.8	17.3	26.9	39.8	50.5	59.5	64.1	61.6	52.7	41.7	29.3	18.8	39.5
	MEAN	21.9	28.0	38.1	52.3	62.8	71.5	76.2	73.9	65.6	55.0	40.2	28.4	51.2
BELLE PLAINE	MAX	27.4	33.5	44.3	60.9	72.7	81.3	85.2	83.3	75.5	64.7	47.2	33.5	59.1
	MIN	8.7	14.3	24.6	38.0	49.0	58.6	62.8	60.5	51.8	40.8	28.1	16.4	37.8
	MEAN	18.1	23.9	34.5	49.5	60.9	70.0	74.0	71.9	63.7	52.8	37.7	25.0	48.5
BELLEVUE L AND D NO 12	MAX	27.8	33.3	43.4	59.5	71.4	80.4	84.1	82.3	74.8	63.9	47.4	33.7	58.5
	MIN	9.7	14.7	25.0	38.2	49.0	58.4	62.7	60.7	52.2	41.5	29.2	17.7	38.3
	MEAN	18.8	24.0	34.2	48.8	60.2	69.4	73.4	71.6	63.5	52.7	38.3	25.7	48.4
BLOOMFIELD	MAX	30.8	36.9	47.7	63.1	74.1	82.9	87.6	85.2	77.6	66.9	50.3	36.8	61.7
	MIN	11.6	17.1	26.9	40.2	50.6	59.7	63.6	61.1	52.9	42.2	29.5	18.8	39.5
	MEAN	21.2	27.0	37.3	51.7	62.4	71.4	75.6	73.2	65.2	54.5	40.0	27.9	50.6
BOONE	MAX	27.0	33.5	43.8	60.9	73.2	82.0	86.4	84.1	76.2	65.4	47.4	33.3	59.4
	MIN	6.6	12.6	23.4	37.5	48.9	58.0	62.2	59.7	50.3	39.2	26.0	14.5	36.6
	MEAN	16.8	23.1	33.6	49.2	61.0	70.0	74.4	71.9	63.3	52.3	36.7	23.9	48.0
BRITT	MAX	23.1	29.9	40.1	58.7	72.3	81.6	84.4	82.2	74.2	63.3	44.3	29.6	57.0
	MIN	4.0	10.3	21.5	36.0	47.8	57.9	61.6	59.3	50.0	39.5	25.4	12.3	35.5
	MEAN	13.6	20.1	30.8	47.4	60.1	69.7	73.0	70.8	62.1	51.5	34.9	21.0	46.3
CARROLL 2 SSW	MAX	26.3	32.8	43.3	61.8	73.6	82.6	86.8	84.6	76.2	64.9	46.8	33.2	59.4
	MIN	6.6	12.4	22.5	35.9	47.4	57.0	61.1	58.7	49.5	38.7	25.4	14.1	35.8
	MEAN	16.4	22.7	32.9	48.9	60.5	69.8	74.0	71.7	62.9	51.8	36.1	23.7	47.6
CASCADE R	MAX	26.3	32.4	43.0	59.5	71.7	80.7	84.5	82.6	74.5	63.5	46.1	32.4	58.1
	MIN	7.1	12.7	23.6	36.8	47.8	57.0	61.0	58.7	49.7	38.9	26.7	14.8	36.2
	MEAN	16.7	22.6	33.3	48.2	59.8	68.9	72.8	70.7	62.1	51.2	36.4	23.6	47.2
CASTANA EXP FARM	MAX	27.0	33.8	43.9	61.3	72.4	81.3	85.5	83.3	75.1	64.8	47.0	33.8	59.1
	MIN	8.0	14.4	23.9	37.5	48.9	58.3	63.0	61.1	51.8	40.9	27.1	15.4	37.5
	MEAN	17.5	24.1	33.9	49.4	60.7	69.8	74.3	72.2	63.4	52.9	37.1	24.6	48.3

IOWA

TEMPERATURE NORMALS (DEG F)

STATION		JAN	FEB	MAR	APR	MAY	JUN	JUL	AUG	SEP	OCT	NOV	DEC	ANN
CEDAR RAPIDS NO 1	MAX	27.4	33.6	44.3	61.0	72.9	81.6	85.3	83.2	75.4	64.5	47.3	33.2	59.1
	MIN	9.7	15.2	25.4	38.6	49.6	59.0	63.2	61.1	52.5	41.9	29.1	17.3	38.6
	MEAN	18.5	24.4	34.9	49.8	61.3	70.3	74.3	72.2	64.0	53.2	38.2	25.3	48.9
CENTERVILLE 4 SSW	MAX	30.6	37.0	47.6	63.1	73.8	82.6	87.4	85.1	77.3	66.8	50.2	37.0	61.5
	MIN	12.2	17.8	27.4	40.5	51.3	60.6	65.1	62.7	54.2	43.5	30.4	19.6	40.4
	MEAN	21.4	27.4	37.5	51.8	62.6	71.6	76.3	73.9	65.8	55.2	40.4	28.3	51.0
CHARITON	MAX	30.8	37.4	48.0	63.9	74.4	82.9	87.5	85.7	78.0	67.6	50.5	37.0	62.0
	MIN	10.2	16.0	26.1	38.9	49.1	58.4	62.8	60.5	51.7	40.9	28.5	17.8	38.4
	MEAN	20.5	26.7	37.0	51.5	61.8	70.7	75.2	73.1	64.9	54.2	39.5	27.4	50.2
CHARLES CITY	MAX	23.4	30.0	40.4	58.4	71.6	80.7	84.3	82.6	74.3	62.9	44.3	29.7	56.9
	MIN	5.2	11.5	22.5	36.7	48.1	57.5	61.6	59.5	50.4	40.1	26.6	13.9	36.1
	MEAN	14.3	20.8	31.5	47.6	59.9	69.1	73.0	71.1	62.4	51.5	35.5	21.8	46.5
CHEROKEE	MAX	25.4	31.9	41.7	59.9	72.4	81.3	85.7	83.5	74.9	64.3	46.3	32.1	58.3
	MIN	3.6	9.6	21.4	35.4	46.6	56.5	61.0	58.5	48.3	36.8	23.8	11.7	34.4
	MEAN	14.5	20.8	31.6	47.7	59.5	68.9	73.4	71.0	61.6	50.6	35.1	21.9	46.4
CLARINDA	MAX	31.1	37.7	48.4	64.4	75.0	83.7	88.3	86.0	77.8	67.0	50.1	37.3	62.2
	MIN	10.8	16.8	26.5	39.5	50.5	59.9	64.4	62.0	52.8	41.6	28.6	17.6	39.3
	MEAN	21.0	27.3	37.4	51.9	62.8	71.9	76.4	74.0	65.3	54.4	39.3	27.5	50.8
CLARION	MAX	23.4	29.6	39.9	58.4	71.7	80.7	83.9	81.6	73.9	62.8	44.6	29.8	56.7
	MIN	4.1	10.5	21.6	35.8	47.4	57.2	61.4	58.8	49.5	38.6	25.1	12.3	35.2
	MEAN	13.8	20.1	30.8	47.2	59.6	69.0	72.7	70.2	61.8	50.7	34.9	21.1	46.0
CLINTON NO 1	MAX	28.1	34.0	45.1	61.5	73.1	81.7	85.1	83.2	76.0	65.0	48.1	34.2	59.6
	MIN	10.9	16.2	26.5	39.4	50.2	59.6	63.8	61.7	53.2	42.3	29.8	18.3	39.3
	MEAN	19.5	25.1	35.9	50.5	61.7	70.7	74.4	72.5	64.6	53.7	39.0	26.3	49.5
COLUMBUS JUNCTION	MAX	29.7	35.6	46.7	63.1	74.4	83.0	86.8	84.9	77.8	67.0	49.8	36.0	61.2
	MIN	11.0	17.1	26.8	40.1	50.3	59.5	63.5	61.3	52.9	42.1	29.7	18.6	39.4
	MEAN	20.4	26.4	36.8	51.6	62.3	71.3	75.2	73.1	65.4	54.6	39.8	27.3	50.4
CORNING	MAX	29.6	35.8	46.0	61.9	72.7	81.5	86.7	84.4	76.4	65.9	49.3	36.2	60.5
	MIN	8.9	14.6	24.2	38.1	49.0	58.1	62.9	60.9	51.5	40.2	27.2	16.2	37.7
	MEAN	19.3	25.2	35.1	50.0	60.9	69.8	74.8	72.6	63.9	53.1	38.3	26.2	49.1
CORYDON	MAX	29.8	36.0	46.6	62.2	73.3	81.8	86.7	84.8	77.0	66.2	49.6	36.2	60.9
	MIN	10.8	16.5	26.1	40.1	50.9	59.9	64.7	62.3	53.7	42.5	29.6	18.4	39.6
	MEAN	20.3	26.3	36.4	51.2	62.1	70.9	75.8	73.6	65.3	54.4	39.6	27.3	50.3
CRESCO	MAX	21.4	27.7	38.3	56.1	69.1	78.0	82.0	79.9	71.2	60.0	42.3	27.7	54.5
	MIN	3.6	9.5	20.4	35.3	46.8	56.4	60.3	58.2	49.0	38.6	25.1	12.1	34.6
	MEAN	12.5	18.7	29.4	45.7	58.0	67.2	71.2	69.1	60.1	49.3	33.7	19.9	44.6
CRESTON 2 SW	MAX	28.7	35.3	46.0	62.5	73.3	82.3	86.8	84.4	76.2	65.3	48.3	35.4	60.4
	MIN	10.6	16.1	25.8	38.6	49.9	59.0	63.7	61.0	52.4	41.8	28.4	17.5	38.7
	MEAN	19.6	25.7	36.0	50.6	61.6	70.7	75.3	72.7	64.3	53.6	38.4	26.5	49.6
DAVENPORT L AND D NO 15	MAX	28.6	34.2	44.5	60.3	72.1	81.5	85.0	83.2	75.8	64.7	47.9	34.7	59.4
	MIN	12.6	18.0	27.5	41.1	52.4	62.1	66.6	64.6	56.3	45.1	31.8	19.9	41.5
	MEAN	20.6	26.1	36.0	50.7	62.3	71.9	75.8	73.9	66.1	54.9	39.9	27.3	50.5
DECORAH 2 N	MAX	23.6	29.9	40.5	57.9	70.7	79.4	83.4	81.6	73.0	61.8	44.0	29.8	56.3
	MIN	4.5	10.1	21.5	35.7	47.3	56.6	60.6	58.6	49.7	39.4	25.9	13.2	35.3
	MEAN	14.0	20.0	31.0	46.8	59.1	68.0	72.0	70.1	61.4	50.6	35.0	21.5	45.8
DENISON	MAX	26.6	33.0	43.2	60.7	72.0	80.8	85.0	82.6	74.5	63.9	46.3	33.2	58.5
	MIN	7.6	14.0	23.8	37.5	49.1	58.8	63.1	61.0	51.6	40.9	26.9	15.1	37.5
	MEAN	17.1	23.5	33.6	49.1	60.6	69.8	74.1	71.8	63.1	52.4	36.6	24.2	48.0
DES MOINES WSO //R	MAX	27.0	33.2	44.2	61.0	72.6	81.8	86.2	84.0	75.7	65.0	47.6	33.7	59.3
	MIN	10.1	15.8	26.0	39.9	51.6	61.4	66.3	63.7	54.4	43.3	29.5	17.6	40.0
	MEAN	18.6	24.5	35.1	50.5	62.1	71.6	76.3	73.9	65.1	54.2	38.6	25.7	49.7

IOWA

TEMPERATURE NORMALS (DEG F)

STATION		JAN	FEB	MAR	APR	MAY	JUN	JUL	AUG	SEP	OCT	NOV	DEC	ANN
DUBUQUE L AND D NO 11	MAX	26.7	32.1	42.2	58.4	70.5	79.4	83.2	81.4	73.5	62.6	46.3	32.6	57.4
	MIN	9.8	14.7	25.4	39.4	51.2	60.8	65.3	63.4	54.9	44.4	30.8	17.8	39.8
	MEAN	18.3	23.4	33.9	48.9	60.9	70.1	74.3	72.4	64.2	53.5	38.6	25.2	48.6
DUBUQUE WSO //R	MAX	23.7	29.6	40.6	57.3	69.0	78.1	82.0	80.0	72.1	61.1	44.2	30.2	55.7
	MIN	7.4	12.9	23.3	37.0	47.8	57.2	61.7	59.8	51.1	40.6	27.3	15.1	36.8
	MEAN	15.6	21.3	32.0	47.2	58.4	67.7	71.9	69.9	61.6	50.9	35.8	22.7	46.3
EMMETSBURG	MAX	23.4	30.0	40.2	58.6	72.0	81.1	84.7	82.4	74.0	63.0	44.2	29.9	57.0
	MIN	4.8	11.0	21.8	36.1	47.8	57.7	61.8	59.2	49.8	39.1	25.3	12.8	35.6
	MEAN	14.1	20.5	31.0	47.4	59.9	69.4	73.3	70.9	61.9	51.1	34.8	21.4	46.3
ESTHERVILLE 2 N	MAX	22.1	28.5	38.5	56.9	70.6	79.9	83.7	81.5	73.0	61.9	43.2	28.4	55.7
	MIN	3.0	9.1	20.0	34.8	46.7	56.9	61.0	58.7	48.6	37.6	24.3	11.2	34.3
	MEAN	12.6	18.8	29.3	45.9	58.7	68.4	72.4	70.1	60.8	49.8	33.8	19.8	45.0
FAIRFIELD	MAX	29.5	35.7	46.4	62.7	74.0	83.3	87.3	84.9	77.1	66.0	49.2	35.9	61.0
	MIN	11.5	17.3	26.9	40.4	51.1	60.6	64.8	62.6	54.0	43.2	30.4	19.3	40.2
	MEAN	20.5	26.5	36.7	51.6	62.6	72.0	76.1	73.8	65.6	54.6	39.8	27.6	50.6
FAYETTE	MAX	23.9	30.3	41.1	58.3	70.7	79.4	83.3	81.3	73.2	62.1	44.2	30.1	56.5
	MIN	4.9	10.4	21.4	35.1	46.7	56.2	60.4	58.2	49.1	38.5	25.2	12.8	34.9
	MEAN	14.4	20.4	31.2	46.8	58.7	67.8	71.9	69.8	61.2	50.3	34.7	21.5	45.7
FOREST CITY	MAX	22.4	29.0	39.5	57.5	71.3	80.5	83.6	81.9	73.2	62.3	43.2	28.8	56.1
	MIN	3.3	9.9	20.9	35.5	47.3	56.9	61.3	58.9	49.6	39.0	25.0	11.9	35.0
	MEAN	12.9	19.5	30.2	46.5	59.3	68.7	72.5	70.4	61.5	50.7	34.1	20.4	45.6
FORT DODGE	MAX	25.2	31.8	42.0	59.6	72.4	81.6	85.2	83.0	75.0	63.9	46.1	31.9	58.1
	MIN	6.3	12.4	23.0	37.0	48.4	58.0	62.5	60.0	50.5	40.0	26.5	14.4	36.6
	MEAN	15.8	22.1	32.5	48.3	60.4	69.8	73.9	71.5	62.7	52.0	36.3	23.1	47.4
GLENWOOD 2 NNW	MAX	30.9	38.0	48.6	64.9	75.3	84.2	88.5	86.4	78.2	68.2	50.6	37.4	62.6
	MIN	10.7	16.9	26.8	39.9	50.7	60.3	64.8	62.7	53.2	41.8	28.7	17.7	39.5
	MEAN	20.8	27.5	37.7	52.4	63.0	72.3	76.7	74.5	65.7	55.0	39.7	27.6	51.1
GREENFIELD	MAX	28.9	35.5	46.0	62.8	73.9	82.4	86.8	84.7	76.8	66.3	48.8	35.5	60.7
	MIN	9.9	15.6	25.3	38.7	49.9	59.4	64.3	61.9	53.1	42.3	28.5	17.1	38.8
	MEAN	19.4	25.6	35.7	50.7	61.9	70.9	75.6	73.3	65.0	54.3	38.7	26.3	49.8
GRINNELL 3 SW	MAX	26.6	32.5	43.4	60.0	71.6	80.7	85.1	83.0	75.2	64.2	47.3	33.2	58.6
	MIN	7.1	12.5	22.8	36.6	47.5	57.3	61.9	59.1	50.1	38.8	26.3	14.9	36.2
	MEAN	16.9	22.6	33.1	48.3	59.5	69.0	73.6	71.1	62.7	51.6	36.8	24.1	47.4
GRUNDY CENTER	MAX	24.3	30.3	41.1	58.6	70.9	79.8	83.4	81.2	73.7	62.9	45.5	30.7	56.9
	MIN	5.4	11.1	22.2	35.9	47.2	57.0	61.0	58.3	49.2	38.4	25.4	13.1	35.4
	MEAN	14.9	20.8	31.7	47.3	59.1	68.4	72.2	69.8	61.5	50.7	35.5	21.9	46.2
GUTTENBERG L AND D 10	MAX	26.5	32.3	42.4	58.9	71.5	80.2	84.1	82.2	74.1	63.1	46.0	32.3	57.8
	MIN	7.8	12.9	24.1	38.4	50.1	59.5	63.8	61.9	53.0	42.4	29.0	16.3	38.3
	MEAN	17.2	22.6	33.3	48.7	60.8	69.9	74.0	72.1	63.6	52.8	37.5	24.3	48.1
HAMPTON 3 NNE	MAX	23.3	29.8	40.1	58.2	71.2	80.3	83.3	81.7	74.0	63.0	44.7	29.7	56.6
	MIN	5.1	11.5	21.9	36.0	47.9	57.6	61.7	59.2	50.0	39.6	26.0	13.2	35.8
	MEAN	14.2	20.6	31.0	47.1	59.5	69.0	72.5	70.5	62.0	51.3	35.4	21.5	46.2
HARLAN	MAX	28.4	35.2	45.6	62.8	73.9	82.7	86.9	84.4	75.9	65.2	48.1	34.9	60.3
	MIN	8.3	14.7	24.7	37.8	49.2	58.8	63.3	61.1	51.8	40.3	27.0	15.9	37.7
	MEAN	18.4	24.9	35.2	50.3	61.6	70.7	75.1	72.8	63.9	52.8	37.6	25.4	49.1
HAWARDEN	MAX	26.0	32.7	43.3	61.3	73.3	82.2	86.6	84.4	75.5	64.7	46.1	32.2	59.0
	MIN	5.0	11.6	22.5	36.5	48.0	58.6	63.2	61.0	50.5	38.8	24.4	12.5	36.1
	MEAN	15.5	22.1	32.9	49.0	60.7	70.4	74.9	72.7	63.0	51.8	35.3	22.4	47.6
HUMBOLDT NO 2	MAX	24.5	31.0	41.3	59.2	72.2	81.4	85.1	82.6	74.6	63.6	45.2	30.8	57.6
	MIN	5.4	11.6	22.6	36.7	48.5	57.7	62.0	59.2	49.7	39.1	25.6	13.3	36.0
	MEAN	15.0	21.3	32.0	48.0	60.4	69.6	73.6	70.9	62.2	51.4	35.4	22.1	46.8

IOWA

TEMPERATURE NORMALS (DEG F)

STATION		JAN	FEB	MAR	APR	MAY	JUN	JUL	AUG	SEP	OCT	NOV	DEC	ANN
IDA GROVE	MAX	26.7	33.3	43.4	61.3	73.1	82.2	86.5	84.3	75.9	64.8	46.7	33.2	59.3
	MIN	6.0	12.3	22.7	36.0	47.6	57.5	61.9	59.6	50.0	38.6	25.0	13.5	35.9
	MEAN	16.4	22.8	33.1	48.6	60.3	69.9	74.2	72.0	63.0	51.7	35.9	23.4	47.6
INDEPENDENCE 2 SW	MAX	24.9	30.7	41.3	58.8	71.4	80.3	84.1	82.0	74.6	63.4	45.9	31.2	57.4
	MIN	5.1	11.0	22.1	36.0	47.3	56.7	60.3	57.7	49.0	38.3	25.7	13.3	35.2
	MEAN	15.0	20.9	31.7	47.4	59.3	68.5	72.2	69.9	61.8	50.9	35.8	22.3	46.3
INDIANOLA 2 SSW	MAX	29.9	36.5	47.0	63.5	74.6	83.3	87.6	85.6	77.8	67.4	49.9	36.3	61.6
	MIN	10.4	16.2	26.1	39.7	50.2	59.4	63.9	61.2	52.3	41.8	28.9	17.8	39.0
	MEAN	20.2	26.4	36.6	51.6	62.4	71.4	75.8	73.4	65.1	54.6	39.4	27.1	50.3
IOWA CITY	MAX	29.0	35.2	46.2	62.8	74.2	83.1	86.8	85.0	77.7	66.9	49.4	35.2	61.0
	MIN	10.5	16.3	26.1	39.3	50.0	59.5	63.9	61.5	53.0	42.0	29.4	18.1	39.1
	MEAN	19.8	25.8	36.2	51.0	62.1	71.3	75.4	73.3	65.4	54.5	39.4	26.7	50.1
IOWA FALLS 1 E	MAX	25.1	31.5	42.0	59.9	72.3	81.3	84.6	82.4	74.9	64.0	45.8	31.5	57.9
	MIN	5.7	11.9	22.6	36.7	48.1	57.5	61.3	58.6	49.6	39.2	25.9	13.8	35.9
	MEAN	15.4	21.7	32.3	48.3	60.2	69.4	73.0	70.5	62.3	51.6	35.9	22.6	46.9
JEFFERSON	MAX	27.6	34.1	44.7	62.2	73.7	82.2	86.5	84.0	76.5	66.1	48.0	34.2	60.0
	MIN	8.4	14.4	24.7	38.1	49.7	59.4	63.7	61.2	51.9	41.1	27.6	16.0	38.0
	MEAN	18.0	24.3	34.7	50.2	61.7	70.8	75.2	72.6	64.2	53.6	37.8	25.1	49.0
KEOKUK LOCK AND DAM 19	MAX	32.7	38.3	48.8	64.0	74.6	83.4	87.5	85.3	78.1	67.4	51.4	38.5	62.5
	MIN	15.7	20.9	30.2	43.5	53.9	63.4	67.7	65.5	57.2	46.0	33.6	22.8	43.4
	MEAN	24.2	29.6	39.6	53.8	64.2	73.4	77.6	75.4	67.7	56.7	42.5	30.7	53.0
KEOSAUQUA	MAX	32.0	38.3	49.1	64.6	75.2	83.6	88.0	86.2	78.7	68.1	51.5	37.9	62.8
	MIN	12.8	18.4	28.0	40.9	50.9	60.3	64.7	62.4	53.8	43.1	31.1	20.4	40.6
	MEAN	22.4	28.4	38.6	52.8	63.1	71.9	76.4	74.3	66.3	55.6	41.3	29.2	51.7
KNOXVILLE	MAX	29.4	36.0	46.7	63.0	74.2	83.0	87.5	85.1	77.3	66.5	49.4	35.9	61.2
	MIN	11.1	17.0	26.6	40.2	51.2	60.2	64.8	62.3	53.7	43.1	29.9	18.4	39.9
	MEAN	20.3	26.5	36.7	51.6	62.7	71.6	76.2	73.7	65.5	54.8	39.7	27.2	50.5
LAKE PARK	MAX	22.2	28.6	38.6	56.9	70.6	79.8	84.2	82.0	73.1	61.7	43.0	28.3	55.8
	MIN	2.4	8.7	19.7	34.8	46.5	56.6	61.1	58.7	48.8	38.0	24.1	10.7	34.2
	MEAN	12.3	18.6	29.2	45.9	58.6	68.2	72.7	70.4	61.0	49.8	33.6	19.5	45.0
LAMONI	MAX	30.3	36.4	46.9	62.1	72.9	81.9	86.8	84.5	76.2	65.6	49.3	36.0	60.7
	MIN	11.5	17.3	26.5	39.7	50.8	59.9	64.5	62.1	53.3	42.5	29.3	18.5	39.7
	MEAN	20.9	26.9	36.7	51.0	61.9	70.9	75.7	73.3	64.8	54.0	39.3	27.3	50.2
LE CLAIRE L AND D 14	MAX	29.3	34.3	44.5	60.3	71.9	80.8	84.4	82.7	75.6	64.6	48.2	34.9	59.3
	MIN	12.2	17.2	27.3	40.9	52.4	62.3	66.6	64.5	56.3	45.4	31.9	19.9	41.4
	MEAN	20.8	25.8	35.9	50.6	62.2	71.5	75.5	73.6	66.0	55.0	40.1	27.4	50.4
LE MARS 2 N	MAX	25.9	33.0	43.6	62.2	74.5	83.6	87.8	85.2	76.7	65.5	46.5	32.7	59.8
	MIN	4.9	11.7	22.7	36.1	48.0	58.2	62.4	60.1	50.1	38.6	24.8	12.9	35.9
	MEAN	15.4	22.4	33.2	49.2	61.2	70.9	75.1	72.7	63.5	52.1	35.7	22.8	47.9
LOGAN	MAX	29.1	36.1	46.9	63.9	74.8	83.4	87.6	85.3	77.1	66.7	49.0	35.8	61.3
	MIN	9.0	15.5	25.5	38.9	50.4	59.9	64.5	62.2	52.8	41.4	27.9	16.3	38.7
	MEAN	19.1	25.8	36.2	51.5	62.6	71.7	76.0	73.8	65.0	54.1	38.4	26.1	50.0
MAPLETON 4 NNW	MAX	27.0	34.0	43.9	61.9	73.1	80.9	84.8	83.1	75.0	65.4	47.4	33.8	59.2
	MIN	6.5	13.3	23.6	37.0	48.9	58.8	63.2	61.3	51.5	39.8	25.9	14.1	37.0
	MEAN	16.8	23.7	33.8	49.5	61.0	69.9	74.0	72.2	63.3	52.6	36.7	24.0	48.1
MAQUOKETA	MAX	27.6	33.4	44.3	61.1	72.4	81.1	84.9	83.0	75.5	64.5	47.5	33.5	59.1
	MIN	9.0	14.2	24.5	38.1	48.3	58.1	62.3	60.2	51.1	40.2	28.0	16.3	37.5
	MEAN	18.3	23.9	34.4	49.6	60.4	69.6	73.6	71.6	63.3	52.4	37.7	24.9	48.3
MARSHALLTOWN	MAX	26.6	32.8	43.2	60.2	72.1	81.4	85.2	83.0	75.0	64.2	46.9	32.8	58.6
	MIN	7.0	12.7	23.7	37.7	48.9	58.5	62.4	59.3	50.1	39.5	26.8	15.0	36.8
	MEAN	16.8	22.8	33.5	49.0	60.5	70.0	73.8	71.2	62.6	51.9	36.9	23.9	47.7

IOWA

TEMPERATURE NORMALS (DEG F)

STATION		JAN	FEB	MAR	APR	MAY	JUN	JUL	AUG	SEP	OCT	NOV	DEC	ANN
MASON CITY 3 N	MAX	22.5	29.2	39.5	57.5	70.9	80.2	83.6	81.6	73.4	62.4	44.0	29.1	56.2
	MIN	3.8	10.2	21.2	35.7	47.3	57.1	61.2	58.8	49.6	39.2	25.2	12.2	35.1
	MEAN	13.2	19.7	30.4	46.6	59.1	68.7	72.4	70.2	61.5	50.8	34.6	20.7	45.7
MASON CITY FAA AP	MAX	21.4	27.6	38.0	56.7	70.1	79.6	83.2	81.2	72.6	61.5	43.2	28.0	55.2
	MIN	3.5	9.6	20.7	35.3	46.7	56.7	61.1	58.7	48.8	38.3	24.5	11.6	34.6
	MEAN	12.5	18.6	29.4	46.0	58.4	68.2	72.2	70.0	60.7	49.9	33.9	19.8	44.9
MILFORD 4 NW	MAX	22.0	29.0	39.2	57.7	71.2	80.6	84.6	82.2	73.3	62.2	43.1	28.6	56.1
	MIN	2.2	9.0	20.0	34.6	46.3	56.3	60.7	58.7	49.1	38.2	23.7	10.6	34.1
	MEAN	12.1	19.0	29.7	46.1	58.8	68.5	72.7	70.5	61.3	50.2	33.4	19.6	45.2
MOUNT PLEASANT	MAX	30.1	36.2	47.3	63.4	74.0	82.9	86.9	84.9	77.6	66.8	49.9	36.2	61.4
	MIN	11.8	17.4	27.1	40.4	50.7	59.9	64.0	61.6	53.4	42.9	30.3	19.2	39.9
	MEAN	21.0	26.8	37.2	51.9	62.4	71.5	75.5	73.2	65.6	54.8	40.2	27.7	50.7
MUSCATINE 4 ENE	MAX	29.3	35.6	46.4	62.3	73.6	82.5	86.1	84.3	77.0	66.1	49.1	35.4	60.6
	MIN	12.1	17.7	27.4	40.6	51.2	60.8	65.1	63.1	54.8	44.1	31.3	19.7	40.7
	MEAN	20.7	26.7	36.9	51.5	62.4	71.7	75.6	73.7	65.9	55.1	40.2	27.6	50.7
NEW HAMPTON	MAX	23.0	29.4	39.4	57.0	69.8	78.5	81.8	80.1	72.1	61.8	43.6	29.4	55.5
	MIN	4.4	10.6	21.2	35.4	47.3	56.5	60.9	58.7	49.7	39.5	25.3	12.5	35.2
	MEAN	13.8	20.0	30.4	46.2	58.6	67.6	71.4	69.4	60.9	50.7	34.5	21.0	45.4
NEWTON 1 E	MAX	27.9	34.6	45.4	62.4	74.0	82.8	86.6	84.5	76.8	66.0	48.3	34.4	60.3
	MIN	9.6	15.6	25.6	39.2	50.5	59.8	64.1	61.8	53.2	42.3	29.2	17.3	39.0
	MEAN	18.8	25.1	35.5	50.8	62.3	71.3	75.4	73.2	65.0	54.2	38.7	25.9	49.7
NORTHWOOD	MAX	22.0	28.8	39.1	57.4	71.2	80.4	83.9	81.9	73.5	62.1	43.0	28.4	56.0
	MIN	3.3	9.6	20.7	35.4	47.2	57.3	61.5	59.0	49.4	39.2	25.0	11.9	35.0
	MEAN	12.6	19.2	29.9	46.5	59.3	68.9	72.7	70.5	61.5	50.6	34.0	20.2	45.5
OAKLAND 2 E	MAX	29.4	36.1	46.8	63.7	74.4	83.0	86.7	84.4	76.7	66.7	49.2	36.1	61.1
	MIN	8.1	14.7	24.8	37.7	49.2	58.6	62.9	60.6	51.3	39.7	26.6	15.6	37.5
	MEAN	18.8	25.4	35.8	50.8	61.9	70.8	74.9	72.5	64.0	53.2	37.9	25.9	49.3
OELWEIN 2 SE	MAX	24.2	30.3	40.8	58.6	70.7	79.4	83.0	81.3	73.5	62.6	45.0	30.7	56.7
	MIN	5.5	11.4	22.3	35.9	47.4	56.6	60.6	58.6	49.8	39.5	25.8	13.8	35.6
	MEAN	14.9	20.9	31.6	47.3	59.0	68.0	71.8	70.0	61.7	51.1	35.4	22.3	46.2
ONAWA	MAX	28.4	35.8	46.1	63.8	75.1	84.0	88.0	85.5	76.9	66.5	48.7	35.2	61.2
	MIN	7.9	14.5	24.7	38.4	49.9	59.6	64.0	61.8	52.1	40.9	26.9	15.4	38.0
	MEAN	18.2	25.2	35.4	51.1	62.5	71.9	76.1	73.7	64.5	53.7	37.8	25.3	49.6
OSAGE	MAX	22.5	28.9	39.2	57.0	70.1	79.0	82.4	80.6	72.2	61.3	43.4	29.0	55.5
	MIN	4.2	10.7	21.8	36.1	47.8	57.4	61.6	59.4	50.1	39.9	25.7	12.8	35.6
	MEAN	13.4	19.8	30.5	46.6	59.0	68.3	72.0	70.0	61.2	50.6	34.6	20.9	45.6
OSCEOLA	MAX	29.9	36.5	47.0	62.9	74.0	82.6	87.2	85.2	77.1	66.5	49.6	36.0	61.2
	MIN	10.6	16.7	26.2	39.8	50.6	59.5	63.8	61.4	53.0	42.4	29.0	17.7	39.2
	MEAN	20.3	26.6	36.6	51.4	62.3	71.0	75.5	73.3	65.1	54.5	39.3	26.9	50.2
OSKALOOSA	MAX	29.6	35.9	46.5	62.8	74.1	82.9	87.2	85.1	77.5	66.7	49.5	35.9	61.1
	MIN	10.9	16.9	26.6	39.8	50.7	60.2	64.4	61.8	53.0	42.6	29.7	18.7	39.6
	MEAN	20.3	26.4	36.6	51.4	62.4	71.6	75.8	73.5	65.3	54.6	39.6	27.3	50.4
OTTUMWA FAA AIRPORT R	MAX	28.5	34.6	45.2	61.2	72.8	82.0	86.2	83.9	76.0	65.1	48.3	34.8	59.9
	MIN	11.2	17.2	27.0	40.9	52.2	61.6	66.1	63.8	54.9	43.9	30.5	19.0	40.7
	MEAN	19.9	25.9	36.1	51.1	62.5	71.9	76.2	73.9	65.5	54.5	39.5	26.9	50.3
PERRY	MAX	26.8	33.1	43.6	60.8	72.7	81.2	85.4	83.1	75.7	65.1	47.6	33.4	59.0
	MIN	7.4	13.3	24.0	37.8	49.0	58.7	62.7	60.0	50.7	39.6	26.8	15.3	37.1
	MEAN	17.1	23.2	33.8	49.3	60.9	70.0	74.1	71.5	63.2	52.4	37.2	24.4	48.1
POCAHONTAS	MAX	24.5	31.1	41.1	59.6	72.6	81.6	85.2	82.8	74.7	63.8	45.5	31.0	57.8
	MIN	5.0	11.3	22.1	36.3	47.8	57.8	62.0	59.4	49.8	38.7	25.3	13.0	35.7
	MEAN	14.7	21.2	31.6	48.0	60.2	69.7	73.6	71.1	62.3	51.3	35.4	22.0	46.8

IOWA

TEMPERATURE NORMALS (DEG F)

STATION		JAN	FEB	MAR	APR	MAY	JUN	JUL	AUG	SEP	OCT	NOV	DEC	ANN
PRIMGHAR	MAX	24.1	31.3	41.3	59.9	72.9	82.2	86.3	83.9	75.4	64.6	45.2	30.5	58.1
	MIN	4.6	11.4	21.7	36.2	48.1	58.0	62.3	60.2	50.5	39.6	25.2	12.6	35.9
	MEAN	14.4	21.4	31.5	48.1	60.5	70.1	74.3	72.1	63.0	52.1	35.2	21.6	47.0
RED OAK	MAX	30.3	37.1	48.2	64.7	75.9	84.9	89.2	86.9	78.4	67.1	49.4	36.4	62.4
	MIN	10.6	16.6	26.6	39.6	50.6	59.9	64.4	62.2	52.8	41.2	28.3	17.4	39.2
	MEAN	20.5	26.9	37.4	52.2	63.3	72.4	76.8	74.6	65.6	54.2	38.9	27.0	50.8
ROCK RAPIDS	MAX	23.7	30.2	40.6	59.2	72.1	81.0	86.0	83.6	74.1	63.2	44.6	30.3	57.4
	MIN	2.4	9.1	20.7	35.1	46.7	57.3	61.8	59.4	48.9	37.1	23.2	10.7	34.4
	MEAN	13.0	19.7	30.7	47.2	59.4	69.2	73.9	71.5	61.5	50.2	34.0	20.5	45.9
ROCKWELL CITY	MAX	25.8	32.5	42.7	60.7	73.1	82.3	86.1	83.7	75.7	64.7	46.3	32.3	58.8
	MIN	6.5	12.9	23.4	37.2	48.8	58.5	62.7	60.2	51.1	40.3	26.4	14.4	36.9
	MEAN	16.2	22.7	33.1	49.0	61.0	70.4	74.4	71.9	63.4	52.5	36.4	23.4	47.9
SAC CITY	MAX	25.6	32.6	43.0	61.1	73.2	82.3	86.5	84.2	75.7	64.9	46.2	32.0	58.9
	MIN	6.9	13.1	23.6	37.3	48.7	58.4	63.1	60.8	51.5	40.7	26.8	14.6	37.1
	MEAN	16.3	22.9	33.3	49.2	61.0	70.4	74.8	72.5	63.7	52.8	36.5	23.3	48.1
SANBORN	MAX	22.5	29.4	39.8	58.5	71.7	80.8	84.6	82.3	73.2	62.0	43.4	29.1	56.4
	MIN	2.9	9.7	20.5	34.7	46.6	56.5	61.0	58.8	49.2	38.5	23.9	11.1	34.5
	MEAN	12.7	19.6	30.2	46.7	59.2	68.7	72.8	70.6	61.2	50.3	33.7	20.1	45.5
SHELDON	MAX	23.2	30.0	40.3	58.8	71.7	80.7	84.7	82.1	73.4	62.4	43.8	29.6	56.7
	MIN	3.5	10.2	21.1	35.0	46.6	56.7	61.1	58.8	49.1	38.1	23.9	11.4	34.6
	MEAN	13.4	20.1	30.7	46.9	59.2	68.7	72.9	70.5	61.3	50.3	33.9	20.6	45.7
SHENANDOAH	MAX	31.7	38.7	49.5	65.5	76.1	84.8	88.9	86.9	78.9	68.5	51.1	38.0	63.2
	MIN	12.4	18.4	27.9	40.8	51.7	61.2	65.8	63.3	54.2	43.0	30.0	19.2	40.7
	MEAN	22.1	28.6	38.7	53.2	63.9	73.1	77.4	75.1	66.6	55.8	40.6	28.6	52.0
SIBLEY	MAX	23.2	30.1	40.5	59.1	71.9	80.9	85.0	82.7	74.4	63.5	44.6	30.0	57.2
	MIN	3.2	9.7	20.6	34.5	45.9	55.7	60.2	57.8	48.3	37.7	23.7	11.3	34.1
	MEAN	13.2	20.0	30.6	46.8	58.9	68.3	72.6	70.3	61.4	50.6	34.1	20.7	45.6
SIDNEY	MAX	32.3	39.4	49.9	65.7	76.0	84.8	89.1	87.1	79.2	69.1	51.4	38.4	63.5
	MIN	12.4	18.5	28.0	41.3	52.3	61.5	66.0	63.6	54.8	43.8	30.5	19.6	41.0
	MEAN	22.4	29.0	39.0	53.5	64.2	73.1	77.6	75.4	67.0	56.5	41.0	29.0	52.3
SIGOURNEY	MAX	29.1	35.5	46.1	62.2	73.5	82.6	86.9	84.9	77.1	66.2	49.5	35.7	60.8
	MIN	10.8	16.3	26.3	39.8	51.1	60.9	65.2	62.6	54.2	43.0	30.0	18.4	39.9
	MEAN	20.0	25.9	36.2	51.0	62.3	71.7	76.1	73.7	65.7	54.6	39.8	27.1	50.3
SIOUX CITY WSO //R	MAX	26.0	33.0	43.5	61.5	73.1	82.0	86.5	84.2	75.4	64.9	46.9	32.7	59.1
	MIN	6.3	13.3	24.0	37.9	49.8	59.7	64.6	62.3	53.6	40.0	25.8	13.9	37.6
	MEAN	16.2	23.2	33.8	49.7	61.5	70.9	75.6	73.3	64.5	52.5	36.4	23.3	48.4
SIOUX RAPIDS	MAX	24.6	31.3	40.7	59.6	72.4	81.7	85.8	83.5	74.9	63.7	45.4	30.9	57.9
	MIN	4.0	10.1	21.0	35.6	47.0	57.2	61.5	58.9	49.2	38.1	24.4	12.0	34.9
	MEAN	14.3	20.7	30.9	47.6	59.7	69.5	73.6	71.2	62.1	50.9	34.9	21.5	46.4
SPENCER 1 N //	MAX	22.9	29.4	39.6	57.7	71.0	80.3	84.3	82.2	73.2	62.3	43.6	29.4	56.3
	MIN	2.5	8.9	20.3	34.3	46.2	56.6	60.8	58.3	47.8	36.9	23.1	10.5	33.9
	MEAN	12.7	19.2	30.0	46.1	58.6	68.5	72.6	70.3	60.6	49.6	33.4	20.0	45.1
STORM LAKE 2 E	MAX	23.5	29.8	39.8	58.0	70.3	79.5	83.9	81.5	73.4	62.6	44.8	30.5	56.5
	MIN	3.4	9.3	19.8	34.3	46.1	56.3	60.9	58.6	48.9	38.0	24.2	11.5	34.3
	MEAN	13.5	19.6	29.8	46.2	58.2	68.0	72.4	70.1	61.2	50.3	34.6	21.0	45.4
TIPTON	MAX	27.6	33.7	45.0	61.4	73.4	82.2	85.8	83.6	76.4	65.5	48.4	33.9	59.7
	MIN	8.6	14.0	24.8	38.0	49.2	58.7	62.6	60.1	51.6	40.7	28.4	16.4	37.8
	MEAN	18.1	23.9	35.0	49.7	61.3	70.5	74.2	71.9	64.0	53.1	38.4	25.2	48.8
TOLEDO	MAX	26.2	32.3	42.6	59.6	71.3	80.2	84.6	82.3	75.1	64.3	47.1	32.6	58.2
	MIN	6.7	12.3	23.1	36.9	47.8	57.6	61.6	59.5	49.9	39.0	26.7	14.7	36.3
	MEAN	16.5	22.3	32.9	48.3	59.6	68.9	73.1	70.9	62.5	51.7	36.9	23.7	47.3

IOWA

TEMPERATURE NORMALS (DEG F)

STATION		JAN	FEB	MAR	APR	MAY	JUN	JUL	AUG	SEP	OCT	NOV	DEC	ANN
TRIPOLI	MAX	24.2	30.8	41.3	58.8	71.5	80.4	84.0	82.0	73.9	62.9	44.8	30.6	57.1
	MIN	5.4	11.8	22.8	36.6	48.0	57.5	61.7	59.5	50.5	40.3	26.7	13.8	36.2
	MEAN	14.9	21.3	32.0	47.7	59.8	69.0	72.9	70.8	62.2	51.6	35.8	22.2	46.7
VINTON	MAX	26.8	33.1	43.8	61.0	73.0	82.1	86.0	83.7	76.0	65.0	47.2	33.1	59.2
	MIN	8.1	13.8	24.6	37.9	48.9	58.2	62.0	60.0	51.0	40.4	27.7	15.9	37.4
	MEAN	17.4	23.5	34.2	49.5	61.0	70.2	74.0	71.9	63.5	52.7	37.4	24.5	48.3
WASHINGTON	MAX	30.1	36.3	47.1	63.5	75.0	83.7	87.5	85.5	78.3	67.3	50.0	36.1	61.7
	MIN	11.8	17.7	27.4	40.7	51.4	60.7	64.7	62.4	54.1	43.3	30.8	19.3	40.4
	MEAN	21.0	27.0	37.3	52.1	63.2	72.2	76.1	74.0	66.2	55.3	40.4	27.7	51.0
WATERLOO WSO	MAX	23.2	29.5	40.5	58.2	70.5	80.1	83.4	81.6	73.4	62.3	44.7	30.1	56.5
	MIN	4.7	10.7	22.2	36.0	47.5	57.6	61.8	59.2	49.8	38.8	25.6	13.0	35.6
	MEAN	14.0	20.1	31.4	47.1	59.0	68.9	72.6	70.4	61.6	50.6	35.2	21.6	46.1
WAUKON	MAX	22.3	28.7	39.4	56.8	69.0	77.1	81.2	79.4	71.3	60.8	43.0	28.6	54.8
	MIN	5.4	11.0	21.8	36.1	47.6	56.8	61.1	59.2	50.4	40.1	25.9	13.3	35.7
	MEAN	13.9	19.9	30.6	46.4	58.3	67.0	71.2	69.3	60.9	50.5	34.5	20.9	45.3
WEBSTER CITY	MAX	25.1	31.6	42.2	60.2	72.6	81.1	84.6	82.4	74.9	64.1	45.7	31.5	58.0
	MIN	6.4	12.5	23.2	36.8	48.3	58.0	62.3	59.9	50.5	39.6	26.1	14.1	36.5
	MEAN	15.8	22.0	32.7	48.5	60.5	69.5	73.5	71.2	62.8	51.9	35.9	22.8	47.3
WILLIAMSBURG	MAX	28.5	34.6	45.6	62.4	73.8	82.3	86.0	84.1	76.7	65.8	48.6	34.7	60.3
	MIN	9.0	14.8	25.0	38.1	49.4	59.0	63.2	60.7	52.0	41.1	28.4	16.8	38.1
	MEAN	18.8	24.7	35.4	50.3	61.6	70.7	74.7	72.4	64.4	53.5	38.5	25.8	49.2
WINTERSET 3 NW	MAX	28.9	35.1	45.5	62.2	73.3	81.9	86.5	84.3	76.1	65.8	48.7	35.4	60.3
	MIN	8.5	14.3	24.1	37.8	48.8	58.1	62.7	60.6	51.4	40.7	27.5	16.0	37.5
	MEAN	18.7	24.7	34.8	50.0	61.1	70.0	74.6	72.4	63.8	53.3	38.1	25.7	48.9

IOWA

PRECIPITATION NORMALS (INCHES)

STATION	JAN	FEB	MAR	APR	MAY	JUN	JUL	AUG	SEP	OCT	NOV	DEC	ANN
AKRON	.52	1.00	1.60	2.26	3.60	4.28	3.44	3.60	2.44	1.71	.91	.77	26.13
ALBIA	1.25	1.13	2.67	3.80	4.26	4.49	4.10	3.94	4.08	2.60	1.98	1.32	35.62
ALGONA 3 W	.72	.97	1.90	2.55	3.71	4.11	3.75	4.00	2.88	1.87	1.35	.84	28.65
ALLISON	.85	1.01	2.21	3.19	4.12	5.12	4.60	4.44	3.40	2.31	1.57	1.16	33.98
ALTON	.53	.87	1.58	2.15	3.47	3.87	3.75	3.59	2.82	1.78	.99	.82	26.22
AMES 8 WSW	.74	.95	2.07	3.40	4.37	5.11	3.45	3.89	3.21	2.31	1.33	.86	31.69
ANAMOSA 1 WNW	1.14	1.16	2.34	3.54	4.10	4.82	4.16	3.82	3.61	2.39	1.88	1.42	34.38
ANKENY 2 SW R	.71	.89	1.96	3.10	4.05	4.75	3.58	3.82	3.18	2.17	1.35	.81	30.37
ATLANTIC 1 NE	.76	.97	2.07	3.00	3.91	4.17	3.37	4.05	3.76	2.17	1.36	.78	30.37
AUDUBON	.95	1.14	2.25	3.31	4.05	4.49	3.90	4.38	3.66	2.12	1.52	.98	32.75
BEDFORD	1.07	1.11	2.24	3.48	4.62	5.21	4.05	4.69	3.75	3.07	1.82	1.04	36.15
BELLE PLAINE	1.13	1.16	2.72	3.97	4.11	4.37	4.17	4.56	3.52	2.35	1.70	1.35	35.11
BELLEVUE L AND D NO 12	1.22	1.07	2.30	3.40	3.58	4.11	3.71	4.09	3.61	2.51	2.03	1.54	33.17
BLOOMFIELD	1.27	1.09	2.66	3.83	3.84	4.43	3.74	4.84	4.04	2.85	1.61	1.46	35.66
BOONE	.98	1.13	2.14	3.32	4.51	5.17	3.82	3.87	3.35	2.31	1.42	1.05	33.07
BRITT	1.09	1.07	2.18	2.86	3.90	4.57	4.09	3.72	3.10	1.93	1.50	1.30	31.31
BURLINGTON RADIO KBUR	1.45	1.27	2.78	3.66	3.63	4.07	3.77	3.98	3.65	2.77	1.67	1.67	34.37
CARROLL 2 SSW	.82	.99	2.13	3.10	4.19	4.76	3.61	3.85	3.05	2.06	1.36	.80	30.72
CASCADE R	1.20	1.07	2.24	3.39	3.94	4.56	3.52	3.85	3.43	2.37	1.95	1.51	33.03
CASTANA EXP FARM	.62	.89	1.99	2.72	4.10	4.36	3.37	3.84	3.10	2.18	1.13	.69	28.99
CEDAR RAPIDS NO 1	1.25	1.06	2.52	3.80	4.39	4.84	4.38	3.97	3.87	2.55	1.88	1.50	36.01
CENTERVILLE 4 SSW	1.13	.95	2.52	3.66	3.56	4.13	3.80	4.30	3.79	2.77	1.37	1.18	33.16
CHARITON	1.05	1.11	2.46	3.64	3.96	4.84	3.84	4.01	4.44	2.59	1.74	1.13	34.81
CHARLES CITY	.84	.92	2.17	3.34	4.10	4.67	4.21	4.13	3.60	2.32	1.44	1.02	32.76
CHEROKEE	.49	.88	1.55	2.25	3.83	4.43	3.66	3.92	2.95	1.88	1.09	.66	27.59
CLARINDA	.93	1.11	2.24	3.30	4.43	4.82	4.46	4.75	4.05	2.57	1.73	.95	35.34
CLARION	.74	.93	1.94	2.78	3.91	4.70	4.07	3.93	3.04	2.19	1.38	.95	30.56
CLINTON NO 1	1.59	1.27	2.68	3.70	3.82	4.27	3.85	4.46	3.33	2.68	2.12	2.01	35.78
CLINTON NO 2	1.51	1.33	2.65	3.72	3.72	4.39	4.02	4.36	3.33	2.65	2.21	1.99	35.88
COLUMBUS JUNCTION	1.37	1.08	2.65	3.80	4.27	4.43	4.35	4.09	3.89	3.10	1.95	1.80	36.78
CORNING	.90	.92	2.14	3.15	4.06	4.55	4.09	4.90	4.28	2.33	1.58	.89	33.79
CORYDON	1.11	1.03	2.62	3.49	3.79	4.31	4.05	3.91	4.33	2.90	1.70	1.17	34.41
CRESCO	.89	.90	2.24	3.20	3.95	4.58	4.51	4.18	3.74	2.44	1.67	1.11	33.41
CRESTON 2 SW	.89	1.18	2.16	3.31	4.26	4.58	4.15	4.42	3.77	2.36	1.78	.85	33.71
DAVENPORT L AND D NO 15	1.46	1.08	2.40	3.52	3.76	4.01	4.36	3.71	3.29	2.49	1.87	1.72	33.67
DECORAH 2 N	.76	.87	1.96	3.40	3.99	4.51	4.63	3.82	3.47	2.04	1.50	1.04	31.99
DENISON	.74	1.01	1.80	2.62	3.96	4.75	3.21	3.75	3.47	2.02	1.16	.82	29.31
DES MOINES WSO //R	1.01	1.12	2.20	3.21	3.96	4.18	3.22	4.11	3.09	2.16	1.52	1.05	30.83
DONNELLSON 4 N	1.65	1.23	2.87	3.74	3.65	4.15	4.13	4.07	4.05	3.11	1.92	1.75	36.32
DORCHESTER	.95	.83	2.12	3.42	3.70	4.83	4.25	3.99	3.22	2.17	1.62	1.36	32.46
DUBUQUE L AND D NO 11	1.07	.94	2.13	3.32	3.50	3.94	3.63	3.91	3.52	2.29	1.79	1.35	31.39
DUBUQUE WSO //R	1.43	1.31	2.92	4.17	4.43	4.17	4.33	4.47	4.13	2.89	2.47	1.87	38.58
DUMONT 3 NNW	.68	.80	1.93	2.98	3.81	4.49	4.56	4.02	3.61	2.29	1.30	.88	31.35
EDDYVILLE	1.20	1.00	2.41	3.51	3.77	3.92	4.01	3.71	3.85	2.33	1.94	1.24	32.89
ELDORA	.80	1.02	2.01	3.10	4.15	4.86	4.35	4.63	3.05	2.29	1.42	.99	32.67
ELKADER	.94	1.14	2.25	3.45	4.18	4.23	4.63	3.97	3.19	2.42	1.63	1.30	33.33
EMMETSBURG	.78	1.15	2.37	2.69	3.70	4.32	3.56	3.95	2.90	1.90	1.30	.95	29.57
ESTHERVILLE 2 N	.79	1.02	1.82	2.68	3.69	4.34	3.37	3.81	3.19	1.79	1.26	.83	28.59
FAIRFIELD	1.24	1.02	2.48	3.49	3.84	3.92	4.82	4.43	3.59	2.56	1.78	1.64	34.81
FAYETTE	.94	1.05	2.26	3.36	4.03	4.53	4.38	3.97	3.19	2.45	1.64	1.19	32.99
FOREST CITY	.91	.96	1.99	2.91	3.97	4.68	4.26	4.59	2.87	1.98	1.32	1.00	31.44
FORT DODGE	.87	1.04	2.24	3.03	3.67	5.06	4.25	4.36	3.34	2.07	1.39	.96	32.28
GLENWOOD 2 NNW	.80	.96	1.99	2.99	4.37	4.42	3.69	4.75	3.73	2.27	1.41	.96	32.34
GREENFIELD	.89	1.21	2.33	3.34	4.11	4.72	3.69	3.96	3.87	2.32	1.47	.87	32.78
GRINNELL 3 SW	1.24	1.15	2.44	3.55	4.12	4.87	4.11	4.19	3.78	2.32	1.89	1.36	35.02

IOWA

PRECIPITATION NORMALS (INCHES)

STATION	JAN	FEB	MAR	APR	MAY	JUN	JUL	AUG	SEP	OCT	NOV	DEC	ANN
GRUNDY CENTER	.82	.98	2.13	3.18	4.12	4.56	4.34	4.04	3.14	2.37	1.43	1.02	32.13
GUTTENBERG L AND D 10	.90	1.01	1.99	3.31	3.71	4.28	4.64	3.71	2.90	2.19	1.71	1.24	31.59
HAMPTON 3 NNE	.83	.97	2.16	3.10	4.14	4.93	4.79	4.18	3.53	2.45	1.43	.98	33.49
HARLAN	.74	1.02	2.10	3.24	4.21	4.39	3.62	4.35	4.02	2.12	1.45	.82	32.08
HAWARDEN	.45	.92	1.45	2.31	3.46	4.11	3.48	3.57	2.75	1.71	.92	.77	25.90
HOLSTEIN	.67	1.03	1.80	2.73	4.10	4.64	3.43	3.84	2.82	2.18	1.19	.80	29.23
HUMBOLDT NO 2	.70	.92	2.01	2.92	3.75	4.54	4.05	3.95	3.14	1.90	1.26	.84	29.98
IDA GROVE	.73	1.00	1.78	2.61	4.13	4.71	3.36	4.27	2.99	2.28	1.05	.79	29.70
INDEPENDENCE 2 SW	.76	.90	1.88	3.07	4.19	4.77	4.12	4.18	3.47	2.27	1.62	1.04	32.27
INDIANOLA 2 SSW	1.04	1.17	2.22	3.61	4.05	4.62	3.46	4.06	3.53	2.52	1.66	1.24	33.18
IOWA CITY	1.08	1.02	2.39	3.76	3.92	4.28	4.88	3.90	3.51	2.61	1.79	1.41	34.55
IOWA FALLS 1 E	.85	1.07	2.06	2.99	3.63	4.60	4.42	4.24	3.33	2.26	1.50	.94	31.89
JEFFERSON	.87	1.02	2.20	3.27	3.99	4.68	3.48	4.24	2.94	2.12	1.37	.89	31.07
KANAWHA	.58	.70	1.67	2.66	3.71	4.70	4.03	3.76	3.10	1.96	1.22	.75	28.84
KEOKUK LOCK AND DAM 19	1.57	1.28	2.83	3.66	4.19	4.02	4.32	3.57	4.15	3.06	1.69	1.56	35.90
KEOSAUQUA	1.49	1.16	2.63	3.80	3.83	4.15	4.23	4.34	4.23	3.09	1.75	1.73	36.43
KNOXVILLE	1.00	1.10	2.24	3.83	3.94	4.37	3.78	3.92	3.61	2.29	1.76	1.15	32.99
LAKE PARK	.60	.86	1.60	2.22	3.30	4.53	3.75	3.69	2.98	1.73	1.14	.86	27.26
LAMONI	.97	1.10	2.42	3.61	4.00	4.24	4.27	4.34	4.11	2.82	1.69	1.16	34.73
LANSING	.97	1.03	2.28	3.29	3.80	4.53	4.21	3.68	3.14	2.05	1.71	1.21	31.90
LE CLAIRE L AND D 14	1.25	1.02	2.16	3.42	3.63	3.96	4.14	3.76	3.13	2.58	1.81	1.59	32.45
LE MARS 2 N	.54	.88	1.71	2.27	3.65	3.91	3.19	3.55	2.79	1.72	.96	.76	25.93
LOGAN	.84	.94	1.94	3.13	4.40	4.36	3.44	4.22	3.20	2.10	1.21	.83	30.61
LORIMOR	1.04	1.16	2.12	3.53	4.16	4.62	3.62	3.99	4.30	2.37	1.80	1.06	33.77
MAPLETON 4 NNW	.56	.95	1.80	2.52	4.09	4.46	3.18	3.63	2.80	2.13	1.02	.68	27.82
MAQUOKETA	1.37	1.16	2.39	3.46	3.75	4.42	3.82	4.04	3.75	2.51	1.96	1.79	34.42
MARSHALLTOWN	.94	1.14	2.34	3.38	4.20	4.71	4.02	4.34	3.17	2.25	1.46	1.08	33.03
MASON CITY 3 N	.71	.83	1.91	3.10	3.98	4.69	4.30	4.67	3.38	2.22	1.46	.89	32.14
MASON CITY FAA AP	.82	.86	1.85	2.76	4.10	4.57	4.21	4.14	3.25	2.02	1.27	.96	30.81
MILFORD 4 NW	.50	.97	1.64	2.48	3.32	4.28	3.74	3.69	3.32	1.85	1.18	.72	27.69
MOUNT AYR	.88	.96	2.24	3.36	3.97	4.55	3.96	4.30	3.95	2.71	1.93	1.11	33.92
MOUNT PLEASANT	1.44	1.07	2.60	3.66	3.51	4.15	4.16	4.40	3.94	2.82	1.74	1.63	35.12
MUSCATINE 4 ENE	1.42	1.15	2.66	3.73	3.89	4.33	4.41	3.53	3.41	2.69	1.90	1.79	34.91
MUSCATINE	1.30	1.02	2.35	3.43	3.61	4.48	3.87	3.62	3.36	2.44	1.99	1.61	33.08
NEW HAMPTON	.80	.82	1.81	3.11	3.96	4.51	4.59	4.08	3.42	2.34	1.45	1.00	31.89
NEWTON 1 E	.93	1.11	2.25	3.34	4.12	4.60	3.65	3.88	3.54	2.35	1.63	1.04	32.44
NORTHWOOD	.88	.82	2.05	2.84	4.04	4.78	4.06	4.01	3.36	1.97	1.31	.96	31.08
OAKLAND 2 E	.79	.99	1.98	3.00	3.91	4.85	3.76	4.47	3.91	2.30	1.33	.84	32.13
OELWEIN 2 SE	.98	1.14	2.02	3.50	4.18	4.78	4.65	3.93	3.20	2.34	1.60	1.26	33.58
ONAWA	.68	1.01	1.98	2.78	4.12	4.41	3.33	3.62	2.79	2.19	1.11	.77	28.79
OSAGE	.88	.92	2.24	2.95	3.91	4.48	4.43	4.53	3.38	2.24	1.38	1.04	32.38
OSCEOLA	.98	1.14	2.38	3.52	4.29	4.77	3.87	4.07	4.19	2.58	1.89	1.12	34.80
OSKALOOSA	1.21	1.09	2.51	3.89	4.05	4.45	3.93	3.58	3.85	2.52	2.01	1.42	34.51
OTTUMWA FAA AIRPORT R	1.15	.99	2.41	3.52	3.80	3.93	4.42	4.01	3.61	2.46	1.66	1.29	33.25
PELLA	1.12	1.12	2.38	3.55	4.10	4.85	3.97	3.66	3.39	2.25	1.86	1.23	33.48
PERRY	.78	1.00	1.99	3.26	4.18	4.68	3.72	4.40	3.16	2.10	1.30	.90	31.47
POCAHONTAS	.64	.89	2.03	2.83	3.68	4.39	3.83	4.11	3.00	2.00	1.24	.80	29.44
PRIMGHAR	.50	.91	1.77	2.53	3.66	4.25	3.52	4.36	2.87	1.98	1.05	.77	28.17
RED OAK	.87	1.11	2.18	3.37	3.95	4.90	3.94	4.72	4.01	2.30	1.47	1.01	33.83
ROCK RAPIDS	.51	.91	1.50	2.30	3.21	4.02	3.19	3.74	2.73	1.65	1.02	.75	25.53
ROCKWELL CITY	.73	.93	2.01	2.79	3.83	4.71	4.09	3.89	3.24	2.03	1.27	.80	30.32
SAC CITY	.70	1.02	2.01	2.65	3.89	4.36	3.42	3.74	3.19	2.03	1.18	.87	29.06
SANBORN	.52	.97	1.70	2.37	3.49	4.02	3.29	4.18	3.07	1.81	1.08	.74	27.24
SHELDON	.54	.96	1.69	2.25	3.56	4.18	3.41	3.89	2.81	1.71	1.01	.82	26.83
SHENANDOAH	.73	.96	2.08	2.94	4.08	4.53	4.18	4.46	3.35	2.48	1.45	.98	32.22

IOWA

PRECIPITATION NORMALS (INCHES)

STATION		JAN	FEB	MAR	APR	MAY	JUN	JUL	AUG	SEP	OCT	NOV	DEC	ANN
SIBLEY		.59	.98	1.72	2.42	3.52	4.45	3.70	4.36	2.92	1.85	1.05	.83	28.39
SIDNEY		.73	.98	2.19	3.03	3.98	4.36	4.46	4.46	3.75	2.80	1.52	.95	33.21
SIGOURNEY		1.05	.89	2.07	3.73	4.04	3.97	4.37	3.89	3.65	2.56	1.97	1.17	33.36
SIOUX CENTER		.53	.89	1.63	2.32	3.50	4.52	3.74	3.42	2.69	1.71	.98	.76	26.69
SIOUX CITY WSO	//R	.60	.94	1.71	2.29	3.43	3.99	3.36	3.14	2.51	1.73	.93	.74	25.37
SIOUX CITY 4 N		.55	.87	1.60	2.16	3.83	4.12	3.45	3.22	2.49	1.82	.88	.69	25.68
SIOUX RAPIDS		.56	.86	1.74	2.57	3.52	4.41	3.57	4.29	3.01	1.91	1.16	.77	28.37
SPENCER 1 N	//	.58	.95	1.79	2.53	3.67	3.92	3.79	3.92	2.88	1.87	1.10	.71	27.71
STORM LAKE 2 E		.54	.84	1.71	2.55	3.80	4.45	3.88	4.20	3.03	1.99	1.04	.65	28.68
TIPTON		1.24	1.33	2.63	4.03	4.66	4.67	4.42	4.24	3.86	2.85	2.31	1.93	38.17
TITONKA 5 NE		.60	.79	1.79	2.66	3.86	4.41	4.02	4.30	2.96	1.94	1.27	.77	29.37
TOLEDO		.98	.98	2.39	3.58	4.42	4.38	4.17	4.69	3.39	2.28	1.68	1.07	34.01
TRIPOLI		.92	1.05	2.38	3.72	4.30	4.26	4.66	4.40	3.27	2.56	1.63	1.20	34.35
VINTON		.98	.94	2.20	3.51	4.16	3.71	3.73	4.07	3.55	2.19	1.63	1.25	31.92
WAPELLO		1.38	1.08	2.57	3.98	3.82	4.11	4.42	3.86	3.51	2.76	1.75	1.67	34.91
WASHINGTON		1.26	1.02	2.49	3.52	3.85	4.27	4.20	3.76	3.46	2.61	1.85	1.61	33.90
WATERLOO WSO		.81	1.02	2.24	3.56	4.15	4.31	4.70	3.69	3.43	2.37	1.67	1.15	33.10
WAUKON		.62	.64	1.63	3.15	3.81	4.15	4.15	3.74	2.96	1.99	1.40	.82	29.06
WEBSTER CITY		.69	.94	1.88	2.83	3.72	4.47	4.18	4.17	2.77	2.07	1.24	.90	29.86
WILLIAMSBURG		1.10	.93	2.24	3.68	4.15	4.53	4.27	4.22	3.68	2.42	1.97	1.29	34.48
WINTERSET 3 NW		.98	1.12	2.07	3.45	3.97	4.75	3.39	4.40	3.69	2.31	1.61	.93	32.67

IOWA

HEATING DEGREE DAY NORMALS (BASE 65 DEG F)

STATION	JUL	AUG	SEP	OCT	NOV	DEC	JAN	FEB	MAR	APR	MAY	JUN	ANN
ALBIA	0	0	61	312	735	1144	1355	1053	853	392	135	17	6057
ALGONA 3 W	11	7	130	436	906	1361	1581	1249	1054	528	205	28	7496
ALLISON	7	8	104	411	867	1321	1541	1212	1017	501	186	25	7200
AMES 8 WSW	7	6	99	390	840	1262	1485	1165	961	465	170	24	6874
ANAMOSA 1 WNW	9	15	116	419	840	1268	1488	1184	973	495	210	27	7044
ANKENY 2 SW R	6	0	97	387	813	1249	1466	1154	949	459	175	26	6781
ATLANTIC 1 NE	9	13	99	403	834	1240	1460	1145	936	456	175	24	6794
AUDUBON	8	0	106	386	840	1249	1463	1145	949	453	165	24	6788
BEDFORD	7	5	79	332	744	1135	1336	1036	834	386	139	20	6053
BELLE PLAINE	6	7	91	392	819	1240	1454	1151	946	465	178	24	6773
BELLEVUE L AND D NO 12	7	10	95	393	801	1218	1432	1148	955	486	197	24	6766
BLOOMFIELD	0	7	79	340	750	1150	1358	1064	859	404	148	20	6179
BOONE	8	8	103	407	849	1274	1494	1173	973	474	169	28	6960
BRITT	10	13	128	430	903	1364	1593	1257	1060	528	201	32	7519
CARROLL 2 SSW	9	11	118	420	867	1280	1507	1184	995	483	182	29	7085
CASCADE R	9	13	112	437	858	1283	1497	1187	983	504	204	29	7116
CASTANA EXP FARM	7	10	111	389	837	1252	1473	1145	964	468	177	27	6860
CEDAR RAPIDS NO 1	6	6	81	381	804	1231	1442	1137	933	456	173	21	6671
CENTERVILLE 4 SSW	0	0	66	328	738	1138	1352	1053	853	400	145	18	6091
CHARITON	0	5	81	355	765	1166	1380	1072	868	405	163	20	6280
CHARLES CITY	9	10	115	430	885	1339	1572	1238	1039	522	201	30	7390
CHEROKEE	9	13	143	456	897	1336	1566	1238	1035	519	210	42	7464
CLARINDA	5	0	72	344	771	1163	1364	1056	856	397	138	14	6180
CLARION	9	13	127	455	903	1361	1587	1257	1060	534	210	34	7550
CLINTON NO 1	0	6	84	365	780	1200	1411	1117	902	435	164	18	6482
COLUMBUS JUNCTION	0	0	68	346	756	1169	1383	1081	874	406	159	19	6261
CORNING	10	7	104	389	801	1203	1417	1114	927	450	178	32	6632
CORYDON	0	6	79	353	762	1169	1386	1084	887	414	167	24	6331
CRESCO	14	26	167	495	939	1398	1628	1296	1104	579	246	52	7944
CRESTON 2 SW	9	6	90	368	798	1194	1407	1100	899	432	161	20	6484
DAVENPORT L AND D NO 15	0	0	54	332	753	1169	1376	1089	899	429	159	14	6274
DECORAH 2 N	12	9	132	456	900	1349	1581	1260	1054	546	222	42	7563
DENISON	8	7	106	400	852	1265	1485	1162	973	477	180	28	6943
DES MOINES WSO //R	0	0	80	357	792	1218	1438	1134	927	435	156	17	6554
DUBUQUE L AND D NO 11	0	8	80	373	792	1234	1448	1165	964	483	179	23	6749
DUBUQUE WSO //R	11	18	124	446	876	1311	1531	1224	1023	534	235	42	7375
EMMETSBURG	8	10	132	439	906	1352	1578	1246	1054	528	200	31	7484
ESTHERVILLE 2 N	12	23	159	482	936	1401	1624	1294	1107	573	230	45	7886
FAIRFIELD	0	0	72	344	756	1159	1380	1078	877	407	157	17	6247
FAYETTE	13	16	141	463	909	1349	1569	1249	1048	546	227	42	7572
FOREST CITY	11	16	134	455	927	1383	1615	1274	1079	555	216	36	7701
FORT DODGE	8	10	126	414	861	1299	1525	1201	1008	501	189	33	7175
GLENWOOD 2 NNW	0	0	73	329	759	1159	1370	1050	846	385	137	19	6127
GREENFIELD	7	0	80	350	789	1200	1414	1103	908	434	155	20	6460
GRINNELL 3 SW	11	24	126	434	846	1268	1491	1187	989	501	227	43	7147
GRUNDY CENTER	9	20	136	452	885	1336	1553	1238	1032	531	215	39	7446
GUTTENBERG L AND D 10	6	6	87	393	825	1262	1482	1187	983	489	182	23	6925
HAMPTON 3 NNE	12	12	126	434	888	1349	1575	1243	1054	537	213	36	7479
HARLAN	6	9	100	390	822	1228	1445	1123	924	441	161	21	6670
HAWARDEN	0	9	117	421	891	1321	1535	1201	995	480	177	30	7177
HUMBOLDT NO 2	7	18	122	433	888	1330	1550	1224	1023	510	195	31	7331
IDA GROVE	8	11	124	419	873	1290	1507	1182	989	492	182	35	7112
INDEPENDENCE 2 SW	10	27	142	450	876	1324	1550	1235	1032	528	218	35	7427
INDIANOLA 2 SSW	5	6	85	346	768	1175	1389	1081	880	407	142	19	6303
IOWA CITY	6	6	68	343	768	1187	1401	1098	893	425	161	20	6376

IOWA

HEATING DEGREE DAY NORMALS (BASE 65 DEG F)

STATION	JUL	AUG	SEP	OCT	NOV	DEC	JAN	FEB	MAR	APR	MAY	JUN	ANN
IOWA FALLS 1 E	8	14	122	425	873	1314	1538	1212	1014	501	194	32	7247
JEFFERSON	8	6	94	369	816	1237	1457	1140	939	444	164	22	6696
KEOKUK LOCK AND DAM 19	0	0	39	284	675	1063	1265	991	787	346	123	14	5587
KEOSAUQUA	0	0	52	315	711	1110	1321	1025	818	371	140	13	5876
KNOXVILLE	0	0	79	341	759	1172	1386	1078	877	406	140	18	6256
LAKE PARK	10	20	154	481	942	1411	1634	1299	1110	573	236	47	7917
LAMONI	0	0	82	360	771	1169	1367	1067	877	420	163	22	6298
LE CLAIRE L AND D 14	0	0	63	335	747	1166	1370	1098	902	436	156	14	6287
LE MARS 2 N	9	12	120	410	879	1308	1538	1193	986	474	173	25	7127
LOGAN	0	0	81	352	798	1206	1423	1098	893	410	142	16	6419
MAPLETON 4 NNW	6	8	110	396	849	1271	1494	1156	967	465	170	30	6922
MAQUOKETA	0	10	91	403	819	1243	1448	1151	949	462	190	26	6792
MARSHALLTOWN	8	12	111	420	843	1274	1494	1182	977	480	186	26	7013
MASON CITY 3 N	14	18	135	451	912	1373	1606	1268	1073	552	224	40	7666
MASON CITY FAA AP	13	19	153	477	933	1401	1628	1299	1104	570	239	45	7881
MILFORD 4 NW	11	16	147	466	948	1407	1640	1288	1094	567	221	39	7844
MOUNT PLEASANT	0	6	69	338	744	1156	1364	1070	862	399	156	17	6181
MUSCATINE 4 ENE	0	0	57	330	744	1159	1373	1072	871	409	152	16	6183
NEW HAMPTON	16	15	148	452	915	1364	1587	1260	1073	564	226	42	7662
NEWTON 1 E	6	9	85	362	789	1212	1432	1117	915	431	151	19	6528
NORTHWOOD	10	19	134	456	930	1389	1624	1282	1088	555	219	35	7741
OAKLAND 2 E	6	7	91	379	813	1212	1432	1109	905	426	157	25	6562
OELWEIN 2 SE	10	14	129	441	888	1324	1553	1235	1035	531	223	36	7419
ONAWA	0	6	98	365	816	1231	1451	1114	918	423	145	18	6585
OSAGE	14	14	139	455	912	1367	1600	1266	1070	552	219	38	7646
OSCEOLA	7	0	82	349	771	1181	1386	1075	880	413	154	23	6321
OSKALOOSA	0	0	66	341	762	1169	1386	1081	880	408	151	17	6261
OTTUMWA FAA AIRPORT R	0	0	67	347	765	1181	1398	1095	896	421	152	17	6339
PERRY	6	9	105	405	834	1259	1485	1170	967	471	179	26	6916
POCAHONTAS	8	14	123	437	888	1333	1559	1226	1035	510	195	29	7357
PRIMGHAR	6	8	122	411	894	1345	1569	1221	1039	507	184	30	7336
RED OAK	0	0	70	353	783	1178	1380	1067	856	389	129	15	6220
ROCK RAPIDS	7	12	145	466	930	1380	1612	1268	1063	534	210	36	7663
ROCKWELL CITY	0	7	103	400	858	1290	1513	1184	989	480	173	26	7023
SAC CITY	5	11	106	394	855	1293	1510	1179	983	474	174	30	7014
SANBORN	9	12	150	464	939	1392	1621	1271	1079	549	213	40	7739
SHELDON	10	14	147	463	933	1376	1600	1257	1063	543	213	39	7658
SHENANDOAH	0	0	54	307	732	1128	1330	1019	815	361	124	14	5884
SIBLEY	8	21	145	454	927	1373	1606	1260	1066	546	218	48	7672
SIDNEY	0	0	61	287	720	1116	1321	1008	806	352	113	17	5801
SIGOURNEY	0	0	69	340	756	1175	1395	1095	893	425	153	24	6325
SIOUX CITY WSO //R	0	8	94	398	858	1293	1513	1170	967	463	159	24	6947
SIOUX RAPIDS	7	14	120	449	903	1349	1572	1240	1057	522	205	30	7468
SPENCER 1 N //	12	13	162	484	948	1395	1621	1282	1085	567	232	39	7840
STORM LAKE 2 E	11	23	155	464	912	1364	1597	1271	1091	564	241	54	7747
TIPTON	6	13	91	388	798	1234	1454	1151	930	459	176	24	6724
TOLEDO	8	18	123	426	843	1280	1504	1196	995	501	209	34	7137
TRIPOLI	9	8	121	426	876	1327	1553	1224	1023	519	202	28	7316
VINTON	7	7	98	396	828	1256	1476	1162	955	465	176	25	6851
WASHINGTON	0	0	57	323	738	1156	1364	1064	859	392	137	15	6105
WATERLOO WSO	8	23	137	457	894	1345	1581	1257	1042	537	222	34	7537
WAUKON	12	13	144	457	915	1367	1584	1263	1066	558	237	44	7660
WEBSTER CITY	10	9	108	417	873	1308	1525	1204	1001	495	185	32	7167
WILLIAMSBURG	0	5	81	372	795	1215	1432	1128	918	441	162	18	6567
WINTERSET 3 NW	7	6	110	381	807	1218	1435	1128	936	450	173	26	6677

IOWA

COOLING DEGREE DAY NORMALS (BASE 65 DEG F)

STATION	JAN	FEB	MAR	APR	MAY	JUN	JUL	AUG	SEP	OCT	NOV	DEC	ANN
ALBIA	0	0	0	5	73	224	360	292	97	21	0	0	1072
ALGONA 3 W	0	0	0	0	41	157	255	184	37	8	0	0	682
ALLISON	0	0	0	0	47	172	270	216	44	14	0	0	763
AMES 8 WSW	0	0	0	0	49	177	286	214	54	12	0	0	792
ANAMOSA 1 WNW	0	0	0	0	45	138	251	195	44	10	0	0	683
ANKENY 2 SW R	0	0	0	0	58	185	304	231	58	18	0	0	854
ATLANTIC 1 NE	0	0	0	0	54	183	301	227	48	12	0	0	825
AUDUBON	0	0	0	0	47	189	318	244	70	10	0	0	878
BEDFORD	0	0	0	5	71	215	354	281	97	22	0	0	1045
BELLE PLAINE	0	0	0	0	51	174	285	221	52	14	0	0	797
BELLEVUE L AND D NO 12	0	0	0	0	48	156	267	215	50	12	0	0	748
BLOOMFIELD	0	0	0	0	67	212	333	261	85	14	0	0	972
BOONE	0	0	0	0	45	178	300	222	52	13	0	0	810
BRITT	0	0	0	0	49	173	258	193	41	11	0	0	725
CARROLL 2 SSW	0	0	0	0	43	173	288	219	55	11	0	0	789
CASCADE R	0	0	0	0	43	146	251	189	25	10	0	0	664
CASTANA EXP FARM	0	0	0	0	44	171	295	233	63	14	0	0	820
CEDAR RAPIDS NO 1	0	0	0	0	59	180	294	229	51	15	0	0	828
CENTERVILLE 4 SSW	0	0	0	0	71	216	350	280	90	24	0	0	1031
CHARITON	0	0	0	0	64	191	320	256	78	20	0	0	929
CHARLES CITY	0	0	0	0	43	153	257	199	37	11	0	0	700
CHEROKEE	0	0	0	0	40	159	270	199	41	10	0	0	719
CLARINDA	0	0	0	0	69	221	358	282	81	16	0	0	1027
CLARION	0	0	0	0	42	154	248	174	31	11	0	0	660
CLINTON NO 1	0	0	0	0	62	189	296	238	72	15	0	0	872
COLUMBUS JUNCTION	0	0	0	0	75	208	316	256	80	24	0	0	959
CORNING	0	0	0	0	51	176	314	243	71	20	0	0	875
CORYDON	0	0	0	0	77	201	339	273	88	25	0	0	1003
CRESCO	0	0	0	0	29	118	206	153	20	8	0	0	534
CRESTON 2 SW	0	0	0	0	56	191	328	245	69	15	0	0	904
DAVENPORT L AND D NO 15	0	0	0	0	75	221	335	280	87	19	0	0	1017
DECORAH 2 N	0	0	0	0	39	132	229	167	24	10	0	0	601
DENISON	0	0	0	0	43	172	290	218	49	9	0	0	781
DES MOINES WSO //R	0	0	0	0	66	215	354	279	83	22	0	0	1019
DUBUQUE L AND D NO 11	0	0	0	0	52	176	293	238	56	16	0	0	831
DUBUQUE WSO //R	0	0	0	0	31	123	225	170	22	9	0	0	580
EMMETSBURG	0	0	0	0	42	163	265	193	39	9	0	0	711
ESTHERVILLE 2 N	0	0	0	0	35	147	241	181	33	11	0	0	648
FAIRFIELD	0	0	0	5	82	227	344	277	90	22	0	0	1047
FAYETTE	0	0	0	0	31	126	227	164	27	7	0	0	582
FOREST CITY	0	0	0	0	39	147	243	183	29	12	0	0	653
FORT DODGE	0	0	0	0	46	177	283	211	57	11	0	0	785
GLENWOOD 2 NNW	0	0	0	7	75	238	368	298	94	19	0	0	1099
GREENFIELD	0	0	0	5	59	197	336	261	80	18	0	0	956
GRINNELL 3 SW	0	0	0	0	56	163	278	214	57	19	0	0	787
GRUNDY CENTER	0	0	0	0	32	141	232	168	31	9	0	0	613
GUTTENBERG L AND D 10	0	0	0	0	52	170	285	226	45	15	0	0	793
HAMPTON 3 NNE	0	0	0	0	42	156	244	182	36	9	0	0	669
HARLAN	0	0	0	0	56	192	319	251	67	11	0	0	896
HAWARDEN	0	0	0	0	44	192	312	248	57	12	0	0	865
HUMBOLDT NO 2	0	0	0	0	52	169	274	201	38	12	0	0	746
IDA GROVE	0	0	0	0	36	182	293	228	64	7	0	0	810
INDEPENDENCE 2 SW	0	0	0	0	41	140	233	179	46	13	0	0	652
INDIANOLA 2 SSW	0	0	0	0	62	211	340	267	88	24	0	0	992
IOWA CITY	0	0	0	0	71	209	328	263	80	17	0	0	968

IOWA

COOLING DEGREE DAY NORMALS (BASE 65 DEG F)

STATION	JAN	FEB	MAR	APR	MAY	JUN	JUL	AUG	SEP	OCT	NOV	DEC	ANN
IOWA FALLS 1 E	0	0	0	0	46	164	256	185	41	10	0	0	702
JEFFERSON	0	0	0	0	62	196	324	242	70	16	0	0	910
KEOKUK LOCK AND DAM 19	0	0	0	10	98	266	391	322	120	27	0	0	1234
KEOSAUQUA	0	0	0	5	81	220	353	288	91	24	0	0	1062
KNOXVILLE	0	0	0	0	69	216	351	274	94	25	0	0	1029
LAKE PARK	0	0	0	0	37	143	249	188	34	10	0	0	661
LAMONI	0	0	0	0	67	199	336	260	76	19	0	0	957
LE CLAIRE L AND D 14	0	0	0	0	69	209	326	271	93	25	0	0	993
LE MARS 2 N	0	0	0	0	56	202	322	251	75	10	0	0	916
LOGAN	0	0	0	5	68	217	346	278	81	15	0	0	1010
MAPLETON 4 NNW	0	0	0	0	46	177	285	232	59	11	0	0	810
MAQUOKETA	0	0	0	0	48	164	270	214	40	12	0	0	748
MARSHALLTOWN	0	0	0	0	47	176	281	204	39	14	0	0	761
MASON CITY 3 N	0	0	0	0	41	151	243	179	30	11	0	0	655
MASON CITY FAA AP	0	0	0	0	34	141	237	174	24	9	0	0	619
MILFORD 4 NW	0	0	0	0	28	144	250	186	36	7	0	0	651
MOUNT PLEASANT	0	0	0	6	75	212	329	261	87	22	0	0	992
MUSCATINE 4 ENE	0	0	0	0	72	217	329	273	84	23	0	0	998
NEW HAMPTON	0	0	0	0	27	120	214	152	25	9	0	0	547
NEWTON 1 E	0	0	0	0	67	208	328	263	85	27	0	0	978
NORTHWOOD	0	0	0	0	42	152	249	189	29	10	0	0	671
OAKLAND 2 E	0	0	0	0	61	199	313	240	61	13	0	0	887
OELWEIN 2 SE	0	0	0	0	37	126	221	169	30	10	0	0	593
ONAWA	0	0	0	6	67	225	348	275	83	15	0	0	1019
OSAGE	0	0	0	0	33	137	231	169	25	9	0	0	604
OSCEOLA	0	0	0	0	70	203	332	260	85	24	0	0	974
OSKALOOSA	0	0	0	0	70	215	338	266	75	19	0	0	983
OTTUMWA FAA AIRPORT R	0	0	0	0	75	224	347	279	82	21	0	0	1028
PERRY	0	0	0	0	52	176	288	210	51	15	0	0	792
POCAHONTAS	0	0	0	0	46	170	274	203	42	12	0	0	747
PRIMGHAR	0	0	0	0	45	183	294	228	62	11	0	0	823
RED OAK	0	0	0	0	77	237	370	298	88	18	0	0	1088
ROCK RAPIDS	0	0	0	0	37	162	283	213	40	8	0	0	743
ROCKWELL CITY	0	0	0	0	49	188	296	221	55	12	0	0	821
SAC CITY	0	0	0	0	50	192	309	243	67	16	0	0	877
SANBORN	0	0	0	0	33	151	250	186	36	8	0	0	664
SHELDON	0	0	0	0	33	150	255	185	36	7	0	0	666
SHENANDOAH	0	0	0	7	90	257	388	313	102	22	0	0	1179
SIBLEY	0	0	0	0	29	147	244	185	37	8	0	0	650
SIDNEY	0	0	0	7	88	260	395	322	121	24	0	0	1217
SIGOURNEY	0	0	0	5	69	225	348	275	90	18	0	0	1030
SIOUX CITY WSO //R	0	0	0	0	51	201	333	266	79	10	0	0	940
SIOUX RAPIDS	0	0	0	0	40	165	274	206	33	12	0	0	730
SPENCER 1 N //	0	0	0	0	34	144	247	177	30	7	0	0	639
STORM LAKE 2 E	0	0	0	0	30	144	240	181	41	8	0	0	644
TIPTON	0	0	0	0	61	189	292	227	61	19	0	0	849
TOLEDO	0	0	0	0	42	151	259	201	48	14	0	0	715
TRIPOLI	0	0	0	0	41	148	254	188	37	11	0	0	679
VINTON	0	0	0	0	52	181	286	221	53	15	0	0	808
WASHINGTON	0	0	0	5	81	231	344	282	93	23	0	0	1059
WATERLOO WSO	0	0	0	0	36	151	244	190	35	11	0	0	667
WAUKON	0	0	0	0	29	104	204	146	21	8	0	0	512
WEBSTER CITY	0	0	0	0	46	167	273	201	42	11	0	0	740
WILLIAMSBURG	0	0	0	0	57	189	305	235	63	15	0	0	864
WINTERSET 3 NW	0	0	0	0	52	176	304	236	74	18	0	0	860

STATE-STATION NUMBER	STN TYP	NAME		LATITUDE DEG-MIN	LONGITUDE DEG-MIN	ELEVATION (FT)
13-0088	12	AKRON		N 4249	W 09633	1168
13-0112	13	ALBIA		N 4102	W 09248	944
13-0133	13	ALGONA 3 W		N 4304	W 09418	1230
13-0157	13	ALLISON		N 4246	W 09247	1060
13-0181	12	ALTON		N 4259	W 09601	1352
13-0200	13	AMES 8 WSW		N 4202	W 09348	1099
13-0213	13	ANAMOSA 1 WNW		N 4207	W 09118	805
13-0241	13	ANKENY 2 SW	R	N 4143	W 09337	950
13-0364	13	ATLANTIC 1 NE		N 4125	W 09500	1195
13-0385	13	AUDUBON		N 4143	W 09456	1190
13-0576	13	BEDFORD		N 4040	W 09443	1120
13-0600	13	BELLE PLAINE		N 4154	W 09216	855
13-0608	13	BELLEVUE L AND D NO 12		N 4216	W 09025	603
13-0753	13	BLOOMFIELD		N 4045	W 09225	830
13-0807	13	BOONE		N 4203	W 09353	1041
13-0923	13	BRITT		N 4304	W 09349	1234
13-1060	12	BURLINGTON RADIO KBUR		N 4049	W 09110	703
13-1233	13	CARROLL 2 SSW		N 4202	W 09453	1250
13-1257	13	CASCADE	R	N 4218	W 09101	845
13-1277	13	CASTANA EXP FARM		N 4204	W 09549	1417
13-1319	13	CEDAR RAPIDS NO 1		N 4202	W 09135	818
13-1354	13	CENTERVILLE 4 SSW		N 4041	W 09254	1017
13-1394	13	CHARITON		N 4100	W 09319	960
13-1402	13	CHARLES CITY		N 4303	W 09240	1013
13-1442	13	CHEROKEE		N 4245	W 09532	1158
13-1533	13	CLARINDA		N 4044	W 09501	1048
13-1541	13	CLARION		N 4244	W 09345	1160
13-1635	13	CLINTON NO 1		N 4150	W 09013	595
13-1640	12	CLINTON NO 2		N 4149	W 09014	635
13-1731	13	COLUMBUS JUNCTION		N 4117	W 09122	610
13-1833	13	CORNING		N 4100	W 09445	1215
13-1848	13	CORYDON		N 4045	W 09319	1075
13-1954	13	CRESCO		N 4323	W 09206	1320
13-1962	13	CRESTON 2 SW		N 4102	W 09424	1320
13-2069	13	DAVENPORT L AND D NO 15		N 4131	W 09034	570
13-2110	13	DECORAH 2 N		N 4319	W 09147	1134
13-2171	13	DENISON		N 4202	W 09520	1401
13-2203	13	DES MOINES WSO	//R	N 4132	W 09339	938
13-2299	12	DONNELLSON 4 N		N 4042	W 09134	715
13-2311	12	DORCHESTER		N 4328	W 09130	778
13-2364	13	DUBUQUE L AND D NO 11		N 4232	W 09039	616
13-2367	13	DUBUQUE WSO	//R	N 4224	W 09042	1056
13-2388	12	DUMONT 3 NNW		N 4247	W 09259	1005
13-2541	12	EDDYVILLE		N 4110	W 09238	680
13-2573	12	ELDORA		N 4221	W 09306	1094
13-2603	12	ELKADER		N 4251	W 09124	782
13-2689	13	EMMETSBURG		N 4306	W 09440	1230
13-2724	13	ESTHERVILLE 2 N		N 4325	W 09450	1302
13-2789	13	FAIRFIELD		N 4102	W 09157	754
13-2864	13	FAYETTE		N 4250	W 09148	1005

STATE-STATION NUMBER	STN TYP	NAME	LATITUDE DEG-MIN	LONGITUDE DEG-MIN	ELEVATION (FT)
13-2977	13	FOREST CITY	N 4316	W 09339	1282
13-2999	13	FORT DODGE	N 4231	W 09410	1111
13-3290	13	GLENWOOD 2 NNW	N 4105	W 09545	1250
13-3438	13	GREENFIELD	N 4118	W 09428	1340
13-3473	13	GRINNELL 3 SW	N 4143	W 09244	910
13-3487	13	GRUNDY CENTER	N 4222	W 09247	1020
13-3517	13	GUTTENBERG L AND D 10	N 4247	W 09106	624
13-3584	13	HAMPTON 3 NNE	N 4246	W 09313	1160
13-3632	13	HARLAN	N 4139	W 09519	1160
13-3718	13	HAWARDEN	N 4300	W 09629	1206
13-3909	12	HOLSTEIN	N 4229	W 09532	1370
13-3985	13	HUMBOLDT NO 2	N 4241	W 09412	1052
13-4038	13	IDA GROVE	N 4220	W 09528	1255
13-4052	13	INDEPENDENCE 2 SW	N 4227	W 09155	940
13-4063	13	INDIANOLA 2 SSW	N 4121	W 09334	840
13-4101	13	IOWA CITY	N 4139	W 09132	640
13-4142	13	IOWA FALLS 1 E	N 4231	W 09315	1168
13-4228	13	JEFFERSON	N 4201	W 09422	1050
13-4308	12	KANAWHA	N 4256	W 09348	1185
13-4381	13	KEOKUK LOCK AND DAM 19	N 4024	W 09122	527
13-4389	13	KEOSAUQUA	N 4044	W 09158	625
13-4502	13	KNOXVILLE	N 4119	W 09308	915
13-4561	13	LAKE PARK	N 4327	W 09519	1465
13-4585	13	LAMONI	N 4037	W 09356	1114
13-4620	12	LANSING	N 4322	W 09113	643
13-4705	13	LE CLAIRE L AND D 14	N 4135	W 09025	577
13-4735	13	LE MARS 2 N	N 4249	W 09610	1165
13-4894	13	LOGAN	N 4138	W 09548	1052
13-4926	12	LORIMOR	N 4107	W 09403	1230
13-5123	13	MAPLETON 4 NNW	N 4214	W 09548	1240
13-5131	13	MAQUOKETA	N 4205	W 09040	640
13-5198	13	MARSHALLTOWN	N 4204	W 09256	898
13-5230	13	MASON CITY 3 N	N 4311	W 09312	1148
13-5235	13	MASON CITY FAA AP	N 4310	W 09320	1194
13-5493	13	MILFORD 4 NW	N 4323	W 09511	1402
13-5769	12	MOUNT AYR	N 4042	W 09415	1206
13-5796	13	MOUNT PLEASANT	N 4057	W 09133	736
13-5837	13	MUSCATINE 4 ENE	N 4127	W 09058	600
13-5842	12	MUSCATINE	N 4124	W 09104	549
13-5952	13	NEW HAMPTON	N 4303	W 09218	1155
13-5992	13	NEWTON 1 E	N 4141	W 09302	936
13-6103	13	NORTHWOOD	N 4327	W 09313	1210
13-6151	13	OAKLAND 2 E	N 4119	W 09523	1150
13-6200	13	OELWEIN 2 SE	N 4240	W 09154	1060
13-6243	13	ONAWA	N 4202	W 09606	1050
13-6305	13	OSAGE	N 4317	W 09248	1170
13-6316	13	OSCEOLA	N 4102	W 09346	1102
13-6327	13	OSKALOOSA	N 4118	W 09239	833
13-6389	13	OTTUMWA FAA AIRPORT R	N 4106	W 09227	842
13-6527	12	PELLA	N 4125	W 09255	870

13 – IOWA

STATE-STATION NUMBER	STN TYP	NAME		LATITUDE DEG-MIN	LONGITUDE DEG-MIN	ELEVATION (FT)
13-6566	13	PERRY		N 4150	W 09407	967
13-6719	13	POCAHONTAS		N 4244	W 09440	1260
13-6800	13	PRIMGHAR		N 4305	W 09538	1520
13-6940	13	RED OAK		N 4100	W 09514	1037
13-7147	13	ROCK RAPIDS		N 4326	W 09610	1350
13-7161	13	ROCKWELL CITY		N 4224	W 09437	1210
13-7312	13	SAC CITY		N 4226	W 09459	1274
13-7386	13	SANBORN		N 4311	W 09541	1551
13-7594	13	SHELDON		N 4311	W 09551	1418
13-7613	13	SHENANDOAH		N 4045	W 09523	1014
13-7664	13	SIBLEY		N 4324	W 09545	1494
13-7669	13	SIDNEY		N 4045	W 09539	1125
13-7678	13	SIGOURNEY		N 4120	W 09212	795
13-7700	12	SIOUX CENTER		N 4304	W 09610	1420
13-7708	13	SIOUX CITY WSO	//R	N 4224	W 09623	1103
13-7713	12	SIOUX CITY 4 N		N 4233	W 09625	1110
13-7726	13	SIOUX RAPIDS		N 4253	W 09508	1275
13-7844	13	SPENCER 1 N	//	N 4310	W 09509	1325
13-7979	13	STORM LAKE 2 E		N 4238	W 09511	1425
13-8266	13	TIPTON		N 4146	W 09109	769
13-8270	12	TITONKA 5 NE		N 4317	W 09359	1170
13-8296	13	TOLEDO		N 4159	W 09235	890
13-8339	13	TRIPOLI		N 4249	W 09215	950
13-8568	13	VINTON		N 4210	W 09201	830
13-8668	12	WAPELLO		N 4111	W 09112	589
13-8688	13	WASHINGTON		N 4117	W 09141	762
13-8706	13	WATERLOO WSO		N 4233	W 09224	868
13-8755	13	WAUKON		N 4316	W 09129	1237
13-8806	13	WEBSTER CITY		N 4228	W 09349	1040
13-9067	13	WILLIAMSBURG		N 4140	W 09201	805
13-9132	13	WINTERSET 3 NW		N 4122	W 09404	1100

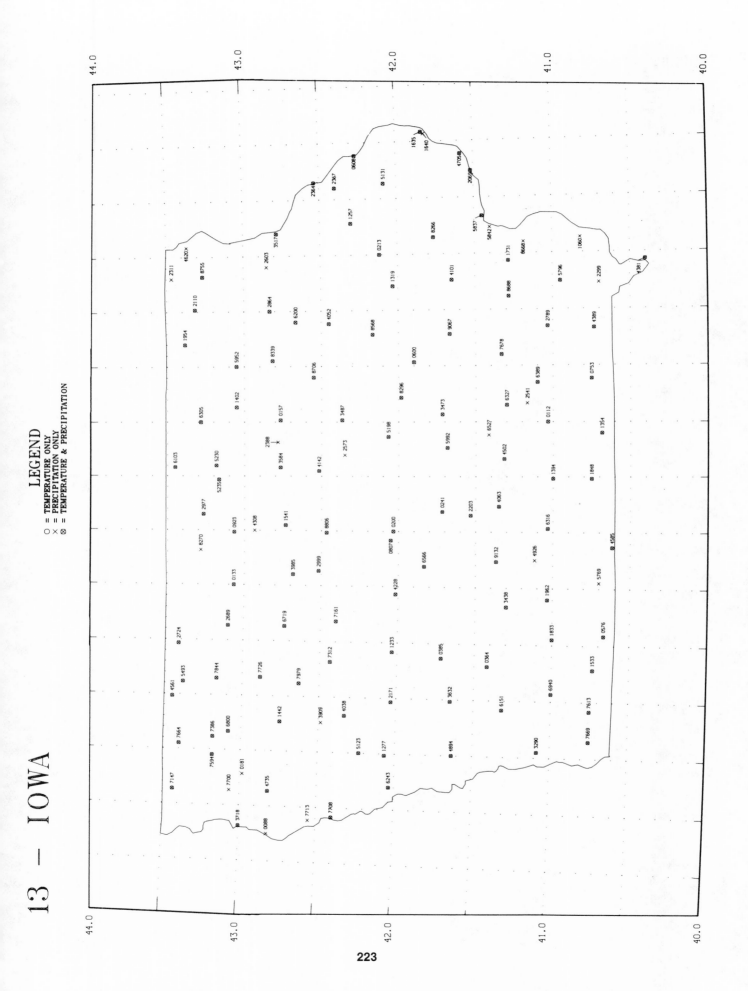

13 — IOWA

LEGEND

O = TEMPERATURE ONLY
X = PRECIPITATION ONLY
⊗ = TEMPERATURE & PRECIPITATION

KANSAS

TEMPERATURE NORMALS (DEG F)

STATION		JAN	FEB	MAR	APR	MAY	JUN	JUL	AUG	SEP	OCT	NOV	DEC	ANN
ABILENE 2 S	MAX	37.9	44.5	54.0	67.0	76.7	87.2	92.8	91.5	82.2	71.6	54.6	43.5	67.0
	MIN	15.1	20.2	29.1	41.6	52.8	62.9	68.0	65.9	56.4	44.2	30.3	20.5	42.3
	MEAN	26.5	32.4	41.6	54.3	64.8	75.1	80.4	78.7	69.3	57.9	42.5	32.0	54.6
ALTON 6 E	MAX	39.4	46.5	55.0	68.3	77.2	87.5	93.0	91.2	82.6	72.5	55.1	44.4	67.7
	MIN	13.8	18.9	27.0	39.7	50.3	60.6	66.0	63.7	53.9	41.7	27.8	18.4	40.2
	MEAN	26.6	32.8	41.0	54.0	63.8	74.1	79.5	77.5	68.3	57.1	41.5	31.5	54.0
ANTHONY	MAX	44.0	50.4	59.4	71.3	80.1	90.6	95.7	94.6	85.1	74.7	57.8	47.6	70.9
	MIN	21.1	25.6	33.2	45.0	54.7	64.4	69.4	67.3	59.4	47.3	34.1	25.3	45.6
	MEAN	32.5	38.0	46.3	58.2	67.5	77.5	82.6	81.0	72.3	61.0	46.0	36.5	58.3
ASHLAND	MAX	45.3	51.5	59.1	71.2	79.4	89.7	95.3	94.0	84.8	74.9	58.8	49.6	71.1
	MIN	16.5	21.0	28.1	40.3	51.1	61.0	66.2	64.3	55.2	41.7	28.6	20.3	41.2
	MEAN	30.9	36.3	43.7	55.7	65.3	75.4	80.8	79.1	70.0	58.3	43.7	35.0	56.2
ATCHISON	MAX	35.6	42.3	52.5	66.7	76.1	84.2	89.5	87.8	79.7	69.4	53.5	41.2	64.9
	MIN	16.2	21.8	30.7	43.4	53.5	62.7	67.2	65.0	56.3	45.7	32.8	22.7	43.2
	MEAN	25.9	32.1	41.6	55.1	64.9	73.5	78.4	76.4	68.0	57.6	43.2	32.0	54.1
ATWOOD	MAX	41.3	47.3	53.8	66.0	75.4	86.3	91.9	90.2	81.6	70.7	53.5	44.6	66.9
	MIN	13.8	18.7	24.8	36.0	46.7	56.7	62.4	59.9	49.5	36.6	24.4	17.3	37.2
	MEAN	27.6	33.0	39.3	51.0	61.1	71.5	77.2	75.1	65.6	53.7	39.0	31.0	52.1
BELLEVILLE	MAX	34.9	41.8	51.6	65.6	75.6	85.5	91.2	89.4	79.9	69.0	52.1	40.7	64.8
	MIN	13.8	19.6	28.0	40.7	51.8	61.7	67.1	65.2	55.3	43.5	29.5	19.7	41.3
	MEAN	24.4	30.7	39.8	53.2	63.7	73.6	79.1	77.3	67.6	56.3	40.8	30.2	53.1
BISON	MAX	41.7	47.8	56.3	68.8	78.4	89.1	94.8	93.3	84.1	73.4	55.8	46.0	69.1
	MIN	16.1	20.7	28.2	40.2	51.0	61.0	66.3	64.2	55.1	42.7	29.4	20.3	41.3
	MEAN	28.9	34.3	42.2	54.5	64.7	75.1	80.6	78.8	69.7	58.1	42.6	33.2	55.2
CENTRALIA	MAX	35.3	42.4	52.2	66.1	75.9	84.9	90.4	88.8	80.3	69.9	53.0	40.9	65.0
	MIN	14.2	19.8	28.8	41.1	51.9	61.7	66.5	64.8	55.6	44.2	30.8	20.7	41.7
	MEAN	24.8	31.1	40.5	53.6	63.9	73.4	78.5	76.8	68.0	57.0	41.9	30.8	53.4
CHANUTE FAA AIRPORT	MAX	39.9	46.3	55.8	68.5	77.0	85.8	91.6	90.3	81.9	71.3	55.8	44.9	67.4
	MIN	19.8	25.0	33.1	45.2	55.1	64.0	68.6	66.6	58.7	47.1	34.4	25.3	45.2
	MEAN	29.9	35.7	44.5	56.9	66.0	75.0	80.1	78.5	70.4	59.2	45.1	35.1	56.4
CIMARRON	MAX	44.5	51.1	58.6	70.3	78.5	89.3	94.0	92.2	84.2	74.1	57.2	48.4	70.2
	MIN	15.9	20.8	27.7	39.2	49.9	59.5	64.3	62.1	53.5	41.0	27.4	19.7	40.1
	MEAN	30.3	36.0	43.2	54.8	64.2	74.4	79.2	77.2	68.8	57.6	42.3	34.1	55.2
CLAY CENTER	MAX	37.5	44.9	55.0	68.7	78.1	87.6	92.9	91.6	82.5	71.8	54.2	42.8	67.3
	MIN	15.9	21.4	30.0	42.4	52.9	62.7	67.7	66.0	56.4	44.6	31.1	21.5	42.7
	MEAN	26.7	33.1	42.5	55.6	65.5	75.2	80.3	78.9	69.5	58.2	42.7	32.2	55.0
COLBY 1 SW	MAX	38.9	44.7	50.5	63.3	73.0	84.3	90.6	88.6	79.4	68.4	51.5	42.8	64.7
	MIN	12.8	17.5	23.2	35.5	46.3	56.6	62.5	59.9	50.1	37.4	24.5	16.6	36.9
	MEAN	25.9	31.1	36.9	49.4	59.7	70.5	76.5	74.3	64.8	52.9	38.0	29.8	50.8
COLDWATER	MAX	44.2	50.9	59.3	71.2	80.1	89.9	95.4	94.0	84.8	74.1	57.4	48.2	70.8
	MIN	19.7	24.6	31.7	43.5	53.5	62.9	67.5	65.5	57.2	45.7	32.4	23.7	44.0
	MEAN	31.9	37.8	45.6	57.4	66.8	76.4	81.5	79.8	71.0	59.9	44.9	36.0	57.4
COLUMBUS 6 NNW	MAX	43.3	49.1	58.2	70.3	78.0	86.2	91.8	91.2	83.7	73.6	58.3	48.0	69.3
	MIN	22.3	27.2	35.0	46.3	55.3	64.0	68.0	65.7	58.7	47.4	35.8	27.3	46.1
	MEAN	32.8	38.2	46.6	58.3	66.7	75.1	79.9	78.5	71.2	60.5	47.0	37.7	57.7
CONCORDIA WSO //R	MAX	35.0	41.7	51.3	64.8	74.8	85.2	90.9	89.2	79.8	69.1	52.0	40.8	64.6
	MIN	14.7	20.1	28.4	40.9	51.7	61.7	67.0	65.4	55.6	44.2	30.2	20.7	41.7
	MEAN	24.9	30.9	39.9	52.9	63.3	73.5	79.0	77.3	67.7	56.7	41.1	30.8	53.2
COTTONWOOD FALLS	MAX	40.5	46.8	56.6	69.2	77.9	86.5	92.7	91.9	83.2	72.9	56.5	45.4	68.3
	MIN	17.7	23.0	31.2	43.8	53.8	63.0	67.7	65.9	57.0	45.7	32.5	23.2	43.7
	MEAN	29.1	34.9	44.0	56.5	65.9	74.8	80.2	78.9	70.1	59.3	44.5	34.3	56.0

KANSAS

TEMPERATURE NORMALS (DEG F)

STATION		JAN	FEB	MAR	APR	MAY	JUN	JUL	AUG	SEP	OCT	NOV	DEC	ANN
COUNCIL GROVE DAM	MAX	36.9	43.1	53.3	66.8	76.0	84.9	91.0	90.0	81.2	70.7	54.5	42.9	65.9
	MIN	15.0	20.2	29.0	42.2	52.7	61.9	66.6	64.6	55.5	43.7	30.9	21.1	42.0
	MEAN	26.0	31.7	41.2	54.5	64.4	73.4	78.8	77.3	68.4	57.2	42.7	32.0	54.0
DODGE CITY WSO //R	MAX	41.1	47.2	55.0	67.4	76.2	87.2	92.5	90.8	81.5	71.0	54.5	45.3	67.5
	MIN	17.9	22.7	29.2	41.1	52.0	62.0	67.4	65.7	56.6	44.4	30.4	22.1	42.6
	MEAN	29.5	35.0	42.1	54.3	64.1	74.6	80.0	78.3	69.1	57.8	42.5	33.7	55.1
EL DORADO	MAX	41.3	47.9	57.4	69.8	78.2	86.7	92.4	91.3	82.6	72.3	56.4	45.8	68.5
	MIN	19.2	24.2	32.4	44.7	54.2	63.6	68.4	66.5	58.1	46.8	33.5	24.3	44.7
	MEAN	30.3	36.1	44.9	57.3	66.3	75.2	80.4	78.9	70.4	59.6	45.0	35.1	56.6
ELKHART	MAX	48.0	53.1	60.1	70.8	79.4	89.4	93.2	91.3	83.5	73.5	58.0	50.9	70.9
	MIN	19.6	23.8	29.4	40.1	50.0	59.8	65.0	63.4	54.7	42.7	29.9	23.3	41.8
	MEAN	33.8	38.5	44.7	55.5	64.7	74.6	79.1	77.4	69.1	58.1	44.0	37.2	56.4
ELLSWORTH	MAX	40.3	46.6	55.9	69.0	78.1	88.4	94.0	92.4	82.8	72.3	55.5	45.1	68.4
	MIN	15.5	20.4	29.0	41.3	52.2	62.4	67.7	65.6	55.9	43.9	29.9	20.5	42.0
	MEAN	27.9	33.5	42.5	55.2	65.2	75.5	80.9	79.0	69.4	58.1	42.7	32.8	55.2
EMPORIA FAA AIRPORT R	MAX	38.1	44.4	54.0	66.9	76.1	84.6	90.4	89.2	80.8	70.1	54.3	43.3	66.0
	MIN	17.4	22.6	30.6	43.2	53.5	62.9	67.6	65.4	56.9	45.3	32.1	23.0	43.4
	MEAN	27.7	33.5	42.3	55.1	64.8	73.8	79.0	77.3	68.9	57.7	43.2	33.2	54.7
ESKRIDGE	MAX	37.5	44.2	54.3	67.6	76.7	85.2	91.1	89.9	81.4	71.0	54.4	42.6	66.3
	MIN	16.7	22.1	30.1	42.9	53.1	62.0	66.5	64.7	56.5	45.8	32.3	22.8	43.0
	MEAN	27.1	33.2	42.2	55.3	64.9	73.6	78.8	77.3	69.0	58.4	43.3	32.7	54.7
EUREKA	MAX	42.4	49.1	58.9	71.2	79.2	87.2	93.3	92.4	83.9	73.7	57.8	47.2	69.7
	MIN	18.6	23.9	31.9	44.2	53.9	62.9	67.9	65.8	57.4	45.5	32.9	24.2	44.1
	MEAN	30.5	36.5	45.4	57.7	66.5	75.1	80.6	79.1	70.7	59.6	45.4	35.7	56.9
FALL RIVER DAM	MAX	41.0	47.3	56.6	69.4	77.2	85.2	91.6	91.2	82.9	72.6	57.3	46.4	68.2
	MIN	18.4	23.3	31.9	44.9	54.6	63.3	67.9	65.8	57.9	45.9	33.3	23.6	44.2
	MEAN	29.7	35.3	44.3	57.2	65.9	74.3	79.8	78.5	70.4	59.3	45.3	35.0	56.3
FLORENCE	MAX	40.1	47.0	56.8	69.5	78.1	87.3	93.3	92.2	83.2	72.7	56.2	45.2	68.5
	MIN	17.5	22.8	31.3	43.7	53.3	62.9	67.8	65.9	57.1	45.5	32.2	22.9	43.6
	MEAN	28.8	35.0	44.1	56.6	65.7	75.1	80.6	79.0	70.2	59.1	44.2	34.1	56.0
FORT SCOTT	MAX	41.4	48.1	58.0	71.2	79.5	87.9	93.6	92.8	84.9	74.1	57.6	46.0	69.6
	MIN	21.1	26.4	34.5	46.9	56.6	65.7	70.4	68.4	60.1	48.6	36.0	26.7	46.8
	MEAN	31.3	37.3	46.3	59.1	68.1	76.8	82.0	80.6	72.5	61.3	46.8	36.4	58.2
FREDONIA 1 E	MAX	42.6	49.4	59.0	71.3	79.4	88.1	93.8	93.2	84.6	73.6	58.3	47.4	70.1
	MIN	20.2	25.4	33.4	45.3	54.7	63.3	67.5	65.8	58.1	46.7	34.5	25.4	45.0
	MEAN	31.4	37.4	46.2	58.3	67.1	75.7	80.7	79.5	71.4	60.2	46.4	36.4	57.6
GARDEN CITY FAA AP	MAX	42.0	48.1	55.0	67.9	76.8	88.2	93.8	91.5	82.2	71.5	54.9	46.0	68.2
	MIN	15.3	20.5	27.5	39.3	50.3	60.6	66.2	64.2	54.6	41.5	27.5	19.1	40.6
	MEAN	28.7	34.3	41.3	53.6	63.6	74.4	80.0	77.8	68.4	56.5	41.2	32.6	54.4
GARNETT	MAX	40.0	46.5	56.6	69.6	78.3	86.2	91.8	90.5	82.3	71.4	55.8	44.6	67.8
	MIN	19.2	24.3	32.5	44.9	54.5	63.4	67.8	65.9	58.0	47.0	34.5	25.1	44.8
	MEAN	29.6	35.4	44.6	57.3	66.4	74.8	79.8	78.3	70.2	59.2	45.1	34.8	56.3
GOODLAND WSO //R	MAX	40.6	45.6	51.1	63.2	72.8	84.4	90.6	88.2	79.2	68.0	51.6	43.4	64.9
	MIN	13.8	18.2	23.2	34.2	45.0	55.2	61.3	59.0	49.3	37.0	24.5	17.1	36.5
	MEAN	27.2	31.9	37.2	48.7	58.9	69.9	76.0	73.7	64.3	52.5	38.1	30.3	50.7
GREAT BEND	MAX	40.8	47.3	55.8	68.7	78.2	88.8	94.0	92.4	83.2	72.8	55.4	45.4	68.6
	MIN	18.2	23.3	31.0	43.1	53.4	63.3	68.6	66.6	57.3	45.9	32.2	23.1	43.8
	MEAN	29.5	35.4	43.4	55.9	65.8	76.1	81.3	79.5	70.3	59.4	43.8	34.3	56.2
GREENSBURG	MAX	42.1	48.3	56.6	69.2	78.1	88.6	94.0	92.6	83.2	72.5	55.9	46.5	69.0
	MIN	17.8	22.5	29.8	41.6	52.1	62.2	67.2	65.2	56.3	44.5	30.7	22.0	42.7
	MEAN	30.0	35.4	43.2	55.4	65.1	75.4	80.6	78.9	69.8	58.5	43.3	34.3	55.8

KANSAS

TEMPERATURE NORMALS (DEG F)

STATION		JAN	FEB	MAR	APR	MAY	JUN	JUL	AUG	SEP	OCT	NOV	DEC	ANN
HAYS 1 S	MAX	38.8	44.9	52.7	66.1	75.2	86.2	92.2	90.8	81.2	70.9	54.0	43.6	66.4
	MIN	13.0	17.8	25.9	39.1	49.9	60.4	65.8	63.5	53.7	40.6	26.8	17.8	39.5
	MEAN	25.9	31.4	39.3	52.6	62.6	73.3	79.0	77.1	67.5	55.8	40.4	30.8	53.0
HEALY	MAX	42.4	48.5	56.1	68.3	77.7	89.1	94.4	92.2	82.7	71.7	54.6	45.9	68.6
	MIN	14.9	19.5	25.6	37.2	48.1	58.5	64.2	62.3	52.8	40.1	26.7	18.8	39.1
	MEAN	28.6	34.0	40.9	52.8	62.9	73.8	79.3	77.3	67.7	55.9	40.7	32.4	53.9
HERINGTON	MAX	38.2	44.9	54.9	67.8	76.9	86.2	91.8	90.8	81.9	70.9	54.2	43.3	66.8
	MIN	16.7	21.9	30.5	42.7	52.9	62.4	67.1	65.5	56.7	45.5	32.1	22.4	43.0
	MEAN	27.5	33.4	42.7	55.3	64.9	74.3	79.5	78.2	69.3	58.2	43.2	32.9	55.0
HILL CITY FAA AIRPORT R	MAX	39.3	45.3	52.5	65.8	75.4	86.5	92.2	90.4	80.9	70.1	53.3	43.6	66.3
	MIN	13.7	18.9	25.8	38.3	49.4	59.6	65.6	63.4	53.0	40.6	26.4	18.3	39.4
	MEAN	26.5	32.1	39.2	52.1	62.4	73.1	78.9	76.9	67.0	55.4	39.8	31.0	52.9
HOLTON 4 NE	MAX	36.7	43.2	53.3	67.4	76.5	84.9	90.2	88.5	80.7	70.3	53.9	42.1	65.6
	MIN	14.7	20.4	28.9	41.5	52.1	61.7	66.3	63.8	55.2	43.8	31.0	21.1	41.7
	MEAN	25.7	31.8	41.1	54.5	64.3	73.3	78.3	76.2	68.0	57.0	42.4	31.6	53.7
HORTON	MAX	35.7	42.4	52.8	67.5	76.9	85.4	90.3	88.6	80.5	70.0	53.4	41.5	65.4
	MIN	13.6	19.7	28.9	41.7	52.1	61.6	66.2	64.0	55.1	43.4	30.5	20.3	41.4
	MEAN	24.7	31.1	40.9	54.6	64.5	73.5	78.3	76.3	67.8	56.7	42.0	30.9	53.4
HOWARD 5 NE	MAX	42.7	49.1	58.5	70.8	78.4	86.5	92.7	92.2	83.8	73.7	58.0	47.2	69.5
	MIN	19.1	24.2	32.4	44.9	54.2	63.0	67.2	65.2	57.3	45.8	33.3	24.3	44.2
	MEAN	30.9	36.7	45.5	57.9	66.3	74.8	79.9	78.7	70.6	59.8	45.7	35.8	56.9
HOXIE	MAX	41.5	47.7	54.9	67.9	77.3	87.8	93.4	91.6	82.6	71.8	54.2	45.0	68.0
	MIN	15.0	19.8	25.7	37.6	48.5	58.4	64.1	62.1	52.4	40.2	26.7	19.3	39.2
	MEAN	28.3	33.8	40.3	52.8	62.9	73.2	78.8	76.9	67.5	56.0	40.5	32.2	53.6
HUDSON	MAX	41.1	47.3	56.3	68.8	78.0	88.6	93.6	91.9	82.7	72.2	55.3	45.5	68.4
	MIN	18.5	23.4	31.0	43.0	53.3	62.8	67.9	66.0	57.4	45.9	32.3	23.2	43.7
	MEAN	29.8	35.4	43.6	55.9	65.7	75.7	80.8	79.0	70.1	59.0	43.8	34.4	56.1
HUGOTON	MAX	46.6	51.6	58.4	69.8	78.5	89.2	93.5	91.2	82.9	73.2	57.6	49.7	70.2
	MIN	17.5	22.5	28.4	40.2	50.6	60.9	66.1	63.9	54.7	42.3	28.6	21.1	41.4
	MEAN	32.0	37.1	43.4	55.0	64.6	75.1	79.8	77.6	68.9	57.8	43.1	35.4	55.8
INDEPENDENCE	MAX	42.9	49.3	58.5	70.7	78.2	86.4	92.2	91.6	83.4	73.2	58.2	47.8	69.4
	MIN	21.2	26.3	34.4	46.4	55.8	64.4	69.0	67.1	59.1	47.5	35.0	26.1	46.0
	MEAN	32.1	37.8	46.5	58.5	67.0	75.4	80.6	79.4	71.2	60.4	46.6	37.0	57.7
JETMORE	MAX	44.0	50.5	58.6	70.9	78.9	89.2	94.3	92.7	84.3	74.1	56.8	47.9	70.2
	MIN	16.9	21.7	28.9	41.1	51.8	61.7	66.9	64.9	55.6	43.4	29.3	20.8	41.9
	MEAN	30.5	36.1	43.8	56.0	65.4	75.4	80.6	78.8	69.9	58.7	43.1	34.4	56.1
JOHN REDMOND DAM	MAX	39.3	45.6	55.9	68.9	77.3	85.9	91.8	90.9	82.7	72.4	56.4	45.0	67.7
	MIN	16.9	22.6	31.0	43.8	53.4	62.6	67.1	65.0	56.5	45.0	32.6	23.5	43.3
	MEAN	28.2	34.1	43.5	56.4	65.4	74.3	79.5	78.0	69.6	58.7	44.5	34.3	55.5
JOHNSON 11 ESE	MAX	45.9	51.7	59.0	70.4	79.4	90.4	94.8	92.1	84.3	74.3	57.4	49.1	70.7
	MIN	15.4	20.1	26.4	37.7	48.3	58.7	63.9	61.6	52.6	39.4	26.2	18.9	39.1
	MEAN	30.6	36.0	42.7	54.1	63.9	74.6	79.3	76.9	68.4	56.9	41.9	34.0	54.9
KANOPOLIS DAM	MAX	37.8	43.8	53.0	66.2	75.7	86.2	92.5	91.2	81.5	71.2	54.2	43.3	66.4
	MIN	14.7	19.6	28.0	41.1	51.8	61.6	66.7	64.4	54.9	43.6	30.4	20.6	41.5
	MEAN	26.3	31.7	40.5	53.6	63.8	73.9	79.6	77.9	68.2	57.4	42.3	32.0	53.9
KINGMAN	MAX	42.5	48.9	58.0	70.3	79.1	89.2	94.7	93.4	84.2	73.7	56.7	46.7	69.8
	MIN	19.4	24.4	32.0	44.2	54.0	63.6	68.5	66.6	58.0	46.2	33.1	23.9	44.5
	MEAN	31.0	36.7	45.0	57.3	66.6	76.4	81.6	80.1	71.1	60.0	44.9	35.3	57.2
KINSLEY	MAX	41.4	47.2	55.1	67.9	77.3	88.1	93.6	92.2	82.9	72.6	55.5	45.8	68.3
	MIN	16.6	21.2	29.0	41.1	51.8	61.8	67.2	65.0	55.8	43.5	30.1	21.2	42.0
	MEAN	29.0	34.3	42.1	54.6	64.6	75.0	80.4	78.6	69.4	58.0	42.8	33.5	55.2

KANSAS

TEMPERATURE NORMALS (DEG F)

STATION		JAN	FEB	MAR	APR	MAY	JUN	JUL	AUG	SEP	OCT	NOV	DEC	ANN
KIRWIN	MAX	37.3	43.9	52.0	65.8	75.7	86.3	92.7	90.9	81.2	70.9	54.0	42.9	66.1
	MIN	11.1	16.7	24.8	38.2	48.9	59.2	64.5	61.8	51.2	38.5	25.5	16.4	38.1
	MEAN	24.2	30.3	38.4	52.0	62.3	72.8	78.6	76.4	66.2	54.8	39.8	29.7	52.1
LAKIN	MAX	43.6	49.4	56.4	68.7	77.6	88.9	94.1	91.5	82.6	72.4	55.9	47.0	69.0
	MIN	15.2	20.3	26.5	38.3	49.0	59.1	64.5	62.1	52.8	40.1	27.2	19.3	39.5
	MEAN	29.4	34.8	41.5	53.6	63.3	74.0	79.3	76.9	67.7	56.3	41.5	33.2	54.3
LARNED	MAX	42.0	48.5	57.3	70.2	78.8	88.9	93.9	92.4	83.2	72.9	55.8	46.3	69.2
	MIN	18.3	23.2	30.4	42.2	52.7	62.7	67.6	65.9	57.0	45.2	31.5	22.7	43.3
	MEAN	30.2	35.9	43.9	56.2	65.8	75.8	80.8	79.2	70.1	59.1	43.7	34.5	56.3
LAWRENCE	MAX	38.1	44.6	54.8	68.4	77.4	85.9	91.3	89.8	81.9	71.3	55.0	43.3	66.8
	MIN	18.6	24.1	32.6	45.3	55.1	64.3	68.8	66.9	58.5	47.8	34.9	25.0	45.2
	MEAN	28.4	34.4	43.7	56.9	66.3	75.1	80.1	78.4	70.3	59.6	44.9	34.2	56.0
LEAVENWORTH 3 S	MAX	36.5	43.3	53.6	67.2	76.7	85.0	90.1	88.5	80.5	70.1	54.0	42.0	65.6
	MIN	16.5	22.2	30.4	43.3	53.5	62.5	67.2	64.8	56.2	44.8	32.3	22.6	43.0
	MEAN	26.6	32.8	42.1	55.3	65.2	73.8	78.7	76.7	68.4	57.5	43.2	32.3	54.4
LEOTI	MAX	42.2	47.9	54.3	66.6	76.1	87.6	92.8	90.0	81.4	71.0	53.8	45.4	67.4
	MIN	14.4	19.4	25.2	36.5	47.1	57.3	62.8	60.4	50.9	38.5	25.7	18.1	38.0
	MEAN	28.4	33.7	39.8	51.5	61.6	72.5	77.8	75.2	66.2	54.8	39.8	31.8	52.8
LIBERAL	MAX	47.7	53.6	60.8	72.6	80.6	91.3	95.6	93.8	85.6	75.5	59.0	50.9	72.3
	MIN	20.0	24.6	30.6	42.0	52.0	62.2	67.3	65.1	56.4	44.4	31.3	23.7	43.3
	MEAN	33.9	39.2	45.7	57.3	66.3	76.8	81.4	79.5	71.0	59.9	45.2	37.3	57.8
LINCOLN 1 ESE	MAX	40.0	46.8	56.4	69.2	78.4	89.3	95.2	93.6	84.1	73.4	55.9	45.2	69.0
	MIN	15.3	20.5	28.9	41.2	51.9	61.9	67.4	65.4	55.8	43.6	29.6	20.2	41.8
	MEAN	27.6	33.7	42.7	55.2	65.2	75.6	81.3	79.6	70.0	58.5	42.7	32.7	55.4
MANHATTAN	MAX	37.7	44.5	54.5	67.9	77.3	86.2	91.7	90.4	81.7	71.1	54.7	43.1	66.7
	MIN	16.5	21.9	30.5	43.1	53.5	63.2	68.0	66.3	56.8	45.2	32.0	22.3	43.3
	MEAN	27.1	33.2	42.5	55.5	65.4	74.8	79.9	78.4	69.3	58.2	43.4	32.7	55.0
MARYSVILLE	MAX	34.2	41.1	50.6	65.4	75.8	85.5	91.0	89.5	80.4	69.6	52.3	40.6	64.7
	MIN	12.3	17.8	27.1	40.0	51.3	61.4	66.2	64.2	54.2	42.2	28.6	18.7	40.3
	MEAN	23.3	29.5	38.9	52.8	63.6	73.5	78.6	76.9	67.3	55.9	40.5	29.7	52.5
MCPHERSON 2 S	MAX	39.4	46.2	55.9	68.4	77.5	88.4	94.1	92.6	82.9	72.3	55.1	44.3	68.1
	MIN	18.3	23.0	30.9	42.9	52.9	62.8	68.1	66.3	57.6	46.1	32.7	23.1	43.7
	MEAN	28.9	34.6	43.5	55.7	65.3	75.6	81.1	79.5	70.3	59.2	43.9	33.7	55.9
MEADE	MAX	46.7	52.7	60.3	72.0	80.0	90.2	95.1	93.6	85.3	75.6	58.9	50.5	71.7
	MIN	18.5	23.2	30.0	41.5	51.8	61.7	66.3	64.5	55.7	43.4	30.3	22.3	42.4
	MEAN	32.6	38.0	45.2	56.8	65.9	76.0	80.7	79.1	70.5	59.5	44.6	36.4	57.1
MEDICINE LODGE	MAX	44.9	51.5	60.0	72.0	80.3	90.2	95.4	94.3	85.1	75.0	58.7	49.0	71.4
	MIN	19.3	24.1	31.8	43.9	53.6	63.2	67.6	65.9	57.5	45.1	32.0	23.3	43.9
	MEAN	32.1	37.8	45.9	57.9	67.0	76.8	81.5	80.2	71.3	60.1	45.4	36.1	57.7
MINNEAPOLIS 2	MAX	38.4	45.6	55.6	68.9	78.5	88.7	94.4	93.0	83.1	71.9	54.6	43.5	68.0
	MIN	17.1	22.2	30.6	42.7	52.9	62.8	68.0	66.3	56.8	45.2	31.7	22.4	43.2
	MEAN	27.8	33.9	43.1	55.9	65.7	75.8	81.2	79.7	70.0	58.6	43.2	33.0	55.7
MOUND CITY	MAX	40.8	47.3	57.0	69.7	78.3	86.8	92.3	91.2	83.2	72.7	56.8	45.7	68.5
	MIN	19.2	24.5	32.7	44.6	54.1	63.0	67.6	65.5	57.2	46.1	34.1	25.1	44.5
	MEAN	30.1	35.9	44.9	57.2	66.3	75.0	80.0	78.4	70.2	59.4	45.5	35.4	56.5
MOUND VALLEY 3 WSW	MAX	44.4	51.0	60.1	71.7	79.3	87.5	93.6	93.1	84.6	74.6	59.0	48.6	70.6
	MIN	21.4	26.2	34.3	46.1	55.5	64.3	68.7	66.6	59.1	47.0	34.8	26.3	45.9
	MEAN	32.9	38.6	47.2	59.0	67.4	76.0	81.2	79.9	71.9	60.8	46.9	37.5	58.3
NESS CITY	MAX	41.8	48.4	56.5	68.9	78.3	89.2	94.1	92.4	83.3	72.9	55.1	45.7	68.9
	MIN	16.5	21.2	28.3	40.2	51.0	61.1	66.6	64.5	54.9	42.4	28.7	20.6	41.3
	MEAN	29.2	34.8	42.4	54.6	64.7	75.2	80.3	78.4	69.1	57.6	41.9	33.2	55.1

227

KANSAS

TEMPERATURE NORMALS (DEG F)

STATION		JAN	FEB	MAR	APR	MAY	JUN	JUL	AUG	SEP	OCT	NOV	DEC	ANN
NEWTON 2 SW	MAX	40.1	46.6	56.4	68.8	77.8	88.1	93.9	92.5	83.2	72.3	55.4	44.8	68.3
	MIN	18.8	23.7	31.6	43.9	53.7	63.3	68.4	66.7	58.2	46.8	33.3	23.8	44.4
	MEAN	29.5	35.2	44.0	56.4	65.8	75.8	81.2	79.6	70.7	59.6	44.4	34.3	56.4
NORTON 8 SSE	MAX	38.8	45.4	53.3	66.5	75.7	86.1	91.9	90.6	81.2	70.5	53.1	42.9	66.3
	MIN	14.5	19.7	26.0	38.2	48.8	58.2	63.7	62.0	52.4	40.9	27.4	19.2	39.3
	MEAN	26.7	32.6	39.7	52.4	62.3	72.2	77.8	76.3	66.8	55.7	40.3	31.1	52.8
OAKLEY	MAX	42.0	48.4	55.2	67.6	76.7	87.4	92.6	90.8	82.5	71.7	54.2	45.3	67.9
	MIN	16.2	20.7	25.9	37.5	48.0	57.9	63.7	61.6	52.4	40.9	27.9	20.3	39.4
	MEAN	29.1	34.6	40.6	52.6	62.4	72.7	78.2	76.2	67.4	56.3	41.1	32.8	53.7
OBERLIN	MAX	41.3	47.7	54.8	67.8	77.0	87.8	93.2	91.6	83.2	72.1	54.2	44.8	68.0
	MIN	13.5	18.7	25.1	37.3	48.2	58.2	64.1	61.5	51.3	38.4	25.1	17.4	38.2
	MEAN	27.4	33.3	40.0	52.5	62.7	73.0	78.7	76.6	67.3	55.3	39.6	31.1	53.1
OLATHE 3 E	MAX	37.4	43.4	53.4	66.6	75.7	83.9	89.3	87.9	80.5	69.8	54.0	42.4	65.4
	MIN	17.8	23.3	31.8	44.1	54.3	63.0	67.3	65.4	57.5	46.8	33.8	24.2	44.1
	MEAN	27.6	33.4	42.6	55.4	65.0	73.5	78.3	76.7	69.0	58.3	43.9	33.3	54.8
OTTAWA	MAX	39.2	45.5	55.3	68.4	77.5	85.8	91.2	90.0	82.1	71.6	56.0	44.4	67.3
	MIN	18.4	24.0	32.5	44.8	54.6	63.8	68.1	65.9	57.5	46.4	34.0	24.6	44.6
	MEAN	28.8	34.8	43.9	56.6	66.1	74.8	79.7	78.0	69.8	59.1	45.0	34.5	55.9
PARSONS	MAX	42.8	49.3	58.6	70.8	78.8	87.2	93.1	92.2	84.0	73.6	57.9	47.3	69.6
	MIN	22.6	27.6	35.5	47.2	56.5	64.9	69.5	67.6	60.3	49.0	36.8	27.8	47.1
	MEAN	32.7	38.5	47.1	59.0	67.7	76.1	81.3	79.9	72.1	61.3	47.4	37.6	58.4
PHILLIPSBURG 1 SSE	MAX	39.1	46.2	54.3	68.0	77.5	88.3	94.1	92.2	82.9	72.0	54.4	43.8	67.7
	MIN	14.1	19.7	27.3	39.7	50.4	60.7	66.1	63.9	53.6	41.4	27.7	18.9	40.3
	MEAN	26.7	32.9	40.8	53.9	64.0	74.5	80.1	78.1	68.3	56.7	41.1	31.3	54.0
PLAINVILLE	MAX	39.0	45.3	53.8	66.8	76.0	87.2	92.8	90.9	81.4	70.9	53.8	43.6	66.8
	MIN	16.0	20.9	28.1	40.0	50.5	60.9	66.3	64.3	54.6	42.5	29.1	20.5	41.1
	MEAN	27.5	33.1	40.9	53.4	63.3	74.0	79.5	77.6	68.0	56.7	41.5	32.1	54.0
PRATT 2 E	MAX	43.4	49.6	58.4	70.5	79.0	89.0	94.1	92.7	83.9	73.7	57.1	47.6	69.9
	MIN	19.0	23.7	31.1	43.0	53.1	62.7	67.5	65.5	57.1	45.4	32.1	23.3	43.6
	MEAN	31.3	36.7	44.8	56.8	66.0	75.9	80.8	79.1	70.5	59.5	44.6	35.5	56.8
QUINTER	MAX	39.6	45.2	52.0	64.8	74.3	85.5	91.4	89.6	80.3	69.9	53.0	43.7	65.8
	MIN	12.9	17.7	23.9	36.4	47.5	58.0	63.8	61.6	51.5	39.0	25.5	17.6	38.0
	MEAN	26.3	31.5	38.0	50.6	61.0	71.8	77.6	75.6	65.9	54.5	39.3	30.7	51.9
RUSSELL FAA AP	MAX	38.3	44.4	52.9	65.9	75.6	87.0	92.5	90.8	81.0	70.4	53.5	43.3	66.3
	MIN	15.1	20.1	28.0	40.4	51.4	61.8	67.4	65.5	55.8	43.5	29.4	20.0	41.5
	MEAN	26.7	32.3	40.5	53.2	63.6	74.5	80.0	78.2	68.4	57.0	41.5	31.7	54.0
SAINT FRANCIS	MAX	43.1	49.0	55.0	67.0	76.0	86.9	92.3	90.5	82.4	71.5	54.3	46.0	67.8
	MIN	15.3	20.2	25.4	36.3	47.1	56.9	63.1	60.7	50.8	38.4	25.9	18.8	38.2
	MEAN	29.2	34.6	40.2	51.7	61.6	71.9	77.7	75.6	66.6	55.0	40.1	32.4	53.1
SALINA FAA AIRPORT	MAX	37.8	44.2	53.8	66.5	76.1	87.0	92.7	91.3	81.4	70.6	53.9	43.0	66.5
	MIN	16.8	21.7	30.3	42.4	53.1	63.3	68.8	67.0	57.4	45.4	31.7	22.2	43.3
	MEAN	27.3	33.0	42.1	54.5	64.6	75.2	80.8	79.2	69.4	58.0	42.8	32.6	55.0
SCOTT CITY	MAX	43.8	49.9	56.9	69.1	77.9	88.6	92.9	90.7	82.7	72.6	55.5	47.2	69.0
	MIN	15.9	20.8	26.7	38.2	48.6	58.7	64.1	61.9	52.5	40.3	27.2	19.9	39.6
	MEAN	29.9	35.4	41.8	53.7	63.3	73.7	78.5	76.3	67.6	56.5	41.4	33.6	54.3
SEDAN	MAX	44.7	51.0	60.1	72.2	79.8	87.9	94.1	93.8	84.9	74.5	59.7	49.2	71.0
	MIN	18.7	23.6	32.1	44.2	53.8	63.1	67.4	65.1	57.2	44.6	32.2	23.4	43.8
	MEAN	31.7	37.4	46.1	58.2	66.8	75.5	80.8	79.5	71.1	59.6	46.0	36.3	57.4
SHARON SPRINGS	MAX	45.9	50.9	57.3	69.2	78.6	89.9	95.1	92.8	84.2	73.6	56.6	48.4	70.2
	MIN	15.5	20.0	25.6	36.7	47.6	57.5	63.4	61.0	51.3	38.8	26.0	19.1	38.5
	MEAN	30.7	35.5	41.5	53.0	63.2	73.7	79.3	76.9	67.8	56.2	41.3	33.8	54.4

KANSAS

TEMPERATURE NORMALS (DEG F)

STATION			JAN	FEB	MAR	APR	MAY	JUN	JUL	AUG	SEP	OCT	NOV	DEC	ANN
SMITH CENTER		MAX	37.4	44.8	53.4	67.3	77.0	87.6	93.6	91.6	82.3	71.1	53.2	42.4	66.8
		MIN	13.7	19.2	27.2	40.1	50.6	60.8	66.1	64.0	54.0	42.3	28.1	18.6	40.4
		MEAN	25.6	32.0	40.3	53.7	63.8	74.2	79.9	77.9	68.2	56.7	40.7	30.5	53.6
SUBLETTE		MAX	45.7	52.0	59.4	70.8	79.0	89.6	94.0	91.9	84.3	74.3	57.2	49.0	70.6
		MIN	17.4	22.3	28.5	39.7	50.2	60.2	65.2	63.2	54.3	42.1	29.0	21.2	41.1
		MEAN	31.6	37.1	44.0	55.3	64.6	75.0	79.7	77.6	69.3	58.2	43.2	35.1	55.9
SYRACUSE 2 W		MAX	44.8	51.2	58.4	70.0	78.9	89.7	94.8	92.1	84.1	73.5	56.5	48.0	70.2
		MIN	13.5	18.8	25.5	37.2	47.8	58.1	63.6	61.3	51.7	37.9	24.5	16.8	38.1
		MEAN	29.2	35.0	42.0	53.6	63.4	73.9	79.2	76.7	67.9	55.7	40.5	32.4	54.1
TOPEKA WSO //R		MAX	36.3	43.0	53.2	66.5	76.0	84.6	89.6	88.5	80.7	69.9	53.8	41.8	65.3
		MIN	15.7	21.9	30.4	42.7	53.2	63.1	67.5	65.6	56.2	44.1	31.5	21.8	42.8
		MEAN	26.1	32.5	41.8	54.6	64.6	73.9	78.6	77.0	68.5	57.0	42.7	31.8	54.1
TORONTO DAM		MAX	40.3	46.7	56.0	68.9	77.1	85.6	91.8	91.2	82.8	72.5	56.8	46.0	68.0
		MIN	17.7	22.7	31.6	44.6	54.3	63.4	68.0	66.0	57.6	45.3	33.2	23.5	44.0
		MEAN	29.0	34.7	43.8	56.8	65.7	74.5	80.0	78.6	70.3	58.9	45.0	34.8	56.0
TRIBUNE 1 W		MAX	43.6	49.3	55.4	66.9	76.0	87.2	92.7	90.0	81.8	71.1	54.5	46.4	67.9
		MIN	14.2	18.9	24.3	35.1	45.6	55.8	61.6	59.5	50.1	37.8	24.9	17.6	37.1
		MEAN	28.9	34.1	39.9	51.0	60.9	71.5	77.2	74.7	66.0	54.5	39.7	32.0	52.5
TROY 3 SW		MAX	34.3	41.1	51.6	66.3	76.0	84.4	88.9	87.2	79.8	69.5	52.8	40.0	64.3
		MIN	15.2	20.9	29.7	42.8	53.5	62.5	66.6	64.4	55.9	45.4	32.1	21.9	42.6
		MEAN	24.8	31.0	40.7	54.6	64.8	73.5	77.8	75.8	67.9	57.5	42.5	31.0	53.5
WAKEENEY		MAX	39.5	45.3	52.4	65.2	74.4	84.9	90.8	89.2	79.8	69.7	53.3	43.8	65.7
		MIN	15.0	19.9	26.7	39.3	49.9	60.0	65.6	63.4	53.5	41.6	28.1	19.6	40.2
		MEAN	27.3	32.6	39.6	52.3	62.2	72.4	78.2	76.3	66.7	55.7	40.7	31.8	53.0
WAMEGO		MAX	38.0	44.9	55.0	68.8	77.9	86.5	91.7	90.5	82.3	71.7	55.2	43.4	67.2
		MIN	16.9	22.5	31.2	43.6	54.1	63.4	67.8	66.1	57.0	45.8	32.5	22.7	43.6
		MEAN	27.5	33.7	43.1	56.2	66.0	75.0	79.8	78.3	69.7	58.7	43.9	33.1	55.4
WASHINGTON		MAX	37.0	44.3	54.1	68.3	78.2	87.5	92.7	91.0	82.2	71.8	54.1	42.5	67.0
		MIN	13.7	19.3	28.4	40.8	51.7	61.8	66.6	64.6	55.1	43.3	29.7	19.9	41.2
		MEAN	25.4	31.8	41.3	54.6	65.0	74.6	79.7	77.8	68.7	57.6	41.9	31.3	54.1
WELLINGTON 2 S		MAX	42.6	48.9	58.2	70.4	78.9	88.7	94.1	93.3	84.1	73.5	57.7	47.1	69.8
		MIN	20.4	25.1	33.0	45.1	54.5	64.3	69.1	67.5	59.1	46.7	34.3	25.1	45.4
		MEAN	31.5	37.1	45.6	57.8	66.8	76.5	81.6	80.4	71.6	60.2	46.0	36.1	57.6
WICHITA WSO //R		MAX	39.8	46.1	55.8	68.1	77.1	87.4	92.9	91.5	82.0	71.2	55.1	44.6	67.6
		MIN	19.4	24.1	32.4	44.5	54.6	64.7	69.8	67.9	59.2	46.9	33.5	24.2	45.1
		MEAN	29.6	35.1	44.1	56.3	65.9	76.1	81.4	79.7	70.6	59.1	44.3	34.4	56.4
WINFIELD		MAX	43.8	50.4	59.8	71.6	79.4	88.5	93.9	93.0	84.7	74.4	58.3	48.0	70.5
		MIN	21.0	25.9	34.0	45.9	55.2	64.6	69.0	67.3	59.6	47.5	35.0	25.9	45.9
		MEAN	32.4	38.2	46.9	58.8	67.4	76.6	81.5	80.2	72.2	61.0	46.7	37.0	58.2

KANSAS

PRECIPITATION NORMALS (INCHES)

STATION		JAN	FEB	MAR	APR	MAY	JUN	JUL	AUG	SEP	OCT	NOV	DEC	ANN
ABILENE 2 S		.64	.89	1.72	2.45	4.04	4.61	3.79	3.01	3.51	2.60	1.31	.87	29.44
AETNA 2 S		.41	.72	1.47	1.84	3.46	3.49	2.71	2.81	2.23	1.77	1.00	.70	22.61
ALEXANDER		.43	.60	1.61	1.77	2.96	3.72	3.21	2.28	2.40	1.42	.89	.52	21.81
ALTA VISTA		.79	.93	2.25	3.29	4.44	4.99	4.52	3.53	4.04	3.00	1.64	1.22	34.64
ALTON 6 E		.59	.79	2.11	2.12	3.76	3.62	3.12	2.96	2.90	1.55	1.08	.56	25.16
ANTHONY		.71	.87	1.97	2.69	3.93	4.51	3.44	2.82	2.80	2.10	1.45	.86	28.15
ARKANSAS CITY		.89	1.28	2.13	3.02	4.40	4.07	3.68	3.54	3.56	2.85	2.07	1.29	32.78
ASHLAND		.40	.57	1.54	1.55	3.59	3.18	2.74	2.61	2.48	1.42	.90	.57	21.55
ATCHISON		.96	1.10	2.43	3.18	4.17	4.89	4.36	4.20	4.64	3.00	1.78	1.23	35.94
ATTICA		.59	.96	1.85	2.32	4.13	4.20	2.78	2.98	3.00	2.21	1.38	.80	27.20
ATWOOD		.47	.57	1.44	1.77	3.09	3.39	3.21	2.19	2.02	1.18	.71	.52	20.56
AUGUSTA		.86	1.25	2.34	2.75	4.42	4.93	4.38	3.42	3.85	2.62	2.01	1.28	34.11
BARNARD		.64	.85	1.91	2.07	3.74	4.05	3.41	3.02	3.24	2.02	1.20	.80	26.95
BELLEVILLE		.63	.91	1.99	2.48	3.95	4.64	3.80	3.66	3.90	2.02	1.15	.77	29.90
BELOIT		.62	.81	1.88	2.25	3.84	4.00	2.97	2.94	2.99	1.77	.93	.65	25.65
BIRD CITY 11 S		.34	.38	.99	1.33	3.01	2.87	2.51	2.24	1.47	.92	.62	.34	17.02
BISON		.51	.74	1.77	2.04	3.16	3.67	3.35	2.83	2.57	1.54	1.04	.59	23.81
BLUE RAPIDS		.72	.82	1.92	2.58	3.86	4.82	3.85	3.94	3.81	2.43	1.36	.82	30.93
BONNER SPRINGS		1.16	1.16	2.38	3.31	4.25	5.40	4.14	3.76	4.41	3.28	1.77	1.27	36.29
BREMEN		.68	.87	2.13	2.56	4.22	4.01	3.41	3.95	3.42	2.30	1.23	.75	29.53
BREWSTER		.34	.41	1.13	1.24	2.93	3.33	2.88	2.26	1.42	1.05	.59	.34	17.92
BUCKLIN 4 SE		.46	.62	1.57	1.82	4.19	3.55	3.02	2.89	2.81	1.83	.96	.60	24.32
BURDETT 6 SSE		.41	.53	1.39	1.71	3.22	3.03	3.36	2.40	2.04	1.57	1.02	.53	21.21
CAWKER CITY		.57	.81	1.68	1.98	3.75	3.86	2.96	3.01	2.82	1.54	.99	.57	24.54
CEDAR BLUFF DAM		.26	.45	1.30	1.78	3.02	3.44	3.26	2.59	2.02	1.24	.73	.33	20.42
CENTRALIA		.95	1.06	2.13	2.79	4.61	5.10	4.14	4.07	4.40	2.69	1.67	.94	34.55
CHANUTE FAA AIRPORT		1.30	1.27	2.80	3.53	4.92	4.90	4.54	3.98	4.54	3.45	2.36	1.51	39.10
CHAPMAN		.72	.99	2.00	2.85	4.33	5.19	4.47	3.07	3.50	2.60	1.39	.95	32.06
CIMARRON		.44	.51	1.54	1.64	3.70	3.53	3.45	2.65	1.95	1.36	.85	.41	22.03
CLAFLIN		.46	.77	1.76	2.06	3.50	4.01	3.31	2.97	2.49	2.03	1.09	.72	25.17
CLAY CENTER		.73	.91	2.00	2.55	4.21	4.80	3.82	3.47	3.47	2.40	1.31	.84	30.51
CLIFTON		.59	.77	1.70	2.38	4.06	4.35	4.10	3.54	3.77	1.98	1.17	.76	29.17
COLBY 1 SW		.37	.42	1.13	1.45	2.95	3.48	3.09	2.02	1.60	1.00	.56	.39	18.46
COLDWATER		.59	.90	1.47	1.77	3.43	3.99	2.67	3.04	2.51	1.89	1.03	.69	23.98
COLLYER 9 S		.31	.53	1.25	1.59	3.12	3.12	2.68	2.91	1.91	1.20	.85	.42	19.89
COLUMBUS 6 NNW		1.33	1.61	2.76	3.52	5.05	5.15	3.46	3.49	5.09	3.32	2.87	1.85	39.50
CONCORDIA WSO	//R	.61	.84	1.86	2.26	3.99	4.25	3.37	3.35	3.00	1.83	1.05	.70	27.11
COTTONWOOD FALLS		.82	.96	2.13	2.90	4.17	4.93	3.96	3.19	4.09	2.74	1.57	1.08	32.54
COUNCIL GROVE DAM		.70	.91	2.09	2.93	4.53	4.73	3.92	3.13	3.86	2.70	1.55	1.09	32.14
COVERT		.52	.83	1.97	1.92	3.62	3.12	3.05	2.95	2.69	1.67	.95	.54	23.83
DENSMORE		.40	.55	1.38	1.93	3.38	3.31	3.39	2.43	2.14	1.39	.74	.43	21.47
DEXTER		.96	1.06	1.89	3.02	4.81	4.89	3.74	3.37	3.38	2.71	2.03	1.10	32.96
DODGE CITY WSO	//R	.45	.57	1.47	1.84	3.28	3.02	3.08	2.54	1.86	1.27	.76	.52	20.66
DRESDEN		.49	.56	1.49	1.74	3.44	3.84	3.14	2.51	2.04	1.24	.83	.49	21.81
EL DORADO		.79	1.00	2.28	2.88	4.79	5.48	3.80	3.36	4.05	2.73	1.85	1.17	34.18
ELGIN		1.19	1.28	2.50	3.46	4.59	4.68	3.75	3.31	3.69	3.20	2.36	1.37	35.38
ELKHART		.42	.45	1.04	1.48	2.65	2.35	2.96	2.36	1.57	.87	.73	.37	17.25
ELLIS		.40	.74	1.67	2.10	3.49	3.94	3.82	2.56	2.28	1.41	.96	.51	23.88
ELLSWORTH		.59	.90	1.96	2.33	4.20	4.00	3.26	3.21	3.33	2.07	1.10	.77	27.72
ELMO 1 NW		.69	.96	2.02	2.87	4.58	5.14	3.78	3.05	4.14	3.00	1.40	.95	32.58
EMMETT		.82	.98	2.18	2.89	4.32	5.37	4.19	3.84	4.48	2.63	1.64	1.12	34.46
EMPORIA FAA AIRPORT	R	.81	.96	2.37	3.02	4.21	5.63	4.22	3.61	4.17	3.08	1.60	1.08	34.76
ENTERPRISE		.83	1.02	2.12	2.99	4.55	5.31	4.13	3.25	3.55	2.87	1.63	1.12	33.37
ESBON 7 N		.59	.84	1.88	2.13	3.50	4.28	3.08	3.21	3.10	1.65	1.04	.66	25.96
ESKRIDGE		.93	1.04	2.31	3.37	4.34	4.81	4.18	3.67	4.33	2.97	1.77	1.36	35.08

KANSAS

PRECIPITATION NORMALS (INCHES)

STATION	JAN	FEB	MAR	APR	MAY	JUN	JUL	AUG	SEP	OCT	NOV	DEC	ANN
EUREKA	1.02	1.21	2.60	3.07	4.70	5.70	3.69	3.49	4.62	2.97	2.15	1.39	36.61
FALL RIVER DAM	.86	.94	2.29	2.91	4.64	4.92	3.67	2.93	4.12	2.88	2.07	1.07	33.30
FLORENCE	.73	.97	2.07	2.92	4.80	4.92	4.01	2.93	3.95	2.89	1.61	1.02	32.82
FORT SCOTT	1.57	1.59	3.05	3.52	4.67	5.36	3.90	3.50	4.60	3.36	2.38	1.80	39.30
FOWLER 3 NNE	.39	.48	1.32	1.36	3.16	2.73	2.98	2.56	2.07	1.22	.80	.43	19.50
FRANKFORT	.78	.98	2.05	2.74	4.28	4.99	3.80	3.75	3.77	2.27	1.47	.93	31.81
FREDONIA 1 E	1.07	1.09	2.46	2.96	4.74	4.90	4.22	3.15	4.10	3.21	2.37	1.37	35.64
GARDEN CITY FAA AP	.38	.46	1.34	1.50	2.92	3.07	2.69	2.41	1.65	1.25	.73	.36	18.76
GARNETT	1.36	1.24	2.82	3.54	4.63	5.26	3.98	4.00	4.78	3.61	2.09	1.52	38.83
GENESEO	.48	.88	1.89	2.48	3.78	4.11	3.41	3.21	3.24	2.23	1.14	.67	27.52
GOODLAND WSO //R	.38	.37	1.04	1.17	2.92	2.72	2.41	1.94	1.44	.91	.60	.41	16.31
GOVE	.33	.50	1.25	1.58	3.33	3.17	2.97	3.10	1.88	1.28	.75	.36	20.50
GREAT BEND	.62	.90	1.78	2.04	3.31	4.00	3.48	2.92	2.60	1.89	1.08	.82	25.44
GREENSBURG	.47	.76	1.50	1.86	3.56	3.88	2.75	2.39	2.40	1.69	.97	.64	22.87
GRENOLA	1.00	1.03	2.14	3.04	4.60	4.79	3.27	3.37	3.68	2.78	2.04	1.17	32.91
GRIDLEY	.91	1.10	2.54	3.11	4.09	5.38	4.51	3.93	4.57	3.01	1.98	1.22	36.35
HADDAM	.59	.86	1.93	2.66	4.16	4.32	3.61	3.65	3.84	1.98	1.29	.73	29.62
HANOVER 4 S	.57	.79	1.88	2.44	4.16	3.95	3.39	3.89	3.40	2.18	1.21	.68	28.54
HARLAN	.46	.69	1.70	1.90	3.61	3.94	3.20	2.90	2.85	1.49	.88	.48	24.10
HARVEYVILLE 3 N	.92	.98	2.17	3.27	4.41	5.19	4.46	3.47	3.87	3.31	1.68	1.36	35.09
HAYS 1 S	.34	.60	1.51	1.82	3.14	3.71	3.43	2.74	2.47	1.34	.84	.43	22.37
HEALY	.44	.55	1.36	1.69	2.92	2.94	2.92	2.59	2.11	1.33	.95	.47	20.27
HERINGTON	.84	1.09	2.27	3.04	4.47	5.83	3.73	3.00	4.16	3.29	1.54	1.13	34.39
HIAWATHA	1.16	1.18	2.47	3.28	4.48	5.19	4.76	4.72	4.79	2.76	1.84	1.21	37.84
HIGHLAND 1 W	.82	1.01	2.13	3.12	4.64	4.93	4.20	4.21	4.59	2.68	1.70	1.09	35.12
HILL CITY FAA AIRPORT R	.47	.60	1.54	1.93	3.70	3.79	3.12	3.03	2.06	1.45	.74	.47	22.90
HOLTON 4 NE	.91	.97	2.24	3.06	4.17	5.42	4.13	4.17	4.65	2.98	1.72	1.26	35.68
HORTON	.98	1.08	2.29	3.32	4.41	5.55	4.46	4.23	4.44	2.96	1.65	1.11	36.48
HOWARD 5 NE	1.02	1.08	2.49	3.26	4.82	4.76	3.85	3.56	4.37	2.97	2.17	1.27	35.62
HOXIE	.38	.52	1.35	1.71	3.33	3.59	3.05	2.71	1.71	1.20	.73	.43	20.71
HOYT	.89	.96	2.02	2.81	4.40	5.12	3.94	3.47	4.28	2.94	1.71	1.11	33.65
HUDSON	.49	.80	1.58	2.16	3.73	3.59	2.93	2.34	2.75	2.11	.95	.78	24.21
HUGOTON	.31	.40	1.06	1.25	3.26	2.55	2.57	2.41	1.56	1.02	.82	.35	17.56
IMPERIAL	.39	.47	1.13	1.55	2.83	2.81	3.00	2.35	1.61	1.16	.77	.38	18.45
INDEPENDENCE	1.17	1.21	2.71	3.43	4.94	5.21	3.52	3.18	4.04	3.44	2.33	1.42	36.60
INMAN	.55	.90	1.96	2.53	4.17	4.21	3.34	3.03	3.64	2.46	1.22	.87	28.88
IOLA 1 W	1.07	1.09	2.55	3.35	4.41	5.41	4.31	3.43	4.71	3.28	2.17	1.29	37.07
JEROME 1 S	.35	.42	1.17	1.54	3.21	3.27	2.99	2.26	2.07	1.19	.70	.35	19.52
JETMORE	.43	.52	1.48	1.65	3.40	3.17	3.16	2.80	1.84	1.42	.75	.49	21.11
JOHN REDMOND DAM	.87	.96	2.49	3.03	4.21	5.22	4.29	3.69	4.56	3.15	1.76	1.08	35.31
JOHNSON 11 ESE	.25	.27	.74	1.29	2.35	2.36	2.04	2.28	1.10	.79	.54	.23	14.24
KANOPOLIS DAM	.42	.64	1.77	2.26	3.80	3.75	3.27	2.86	3.36	2.13	1.12	.70	26.08
KINGMAN	.52	.85	1.79	2.33	3.96	3.82	3.17	2.76	3.51	2.32	1.41	.77	27.21
KINSLEY	.45	.72	1.60	1.87	3.63	3.41	3.19	2.71	2.11	1.67	.99	.64	22.99
KIRWIN	.35	.61	1.51	2.01	3.95	3.67	2.71	2.84	2.62	1.58	.77	.38	23.00
LA CYGNE	1.37	1.21	2.74	3.71	4.30	5.69	4.03	3.72	4.92	3.38	2.10	1.49	38.66
LAKIN	.29	.33	.92	1.32	2.88	2.71	2.35	2.35	1.54	.80	.64	.23	16.36
LARNED	.46	.71	1.53	1.96	3.09	3.47	3.32	2.79	2.73	1.58	1.04	.67	23.35
LAWRENCE	1.11	1.25	2.50	3.27	4.22	5.49	4.45	3.94	3.94	3.22	1.98	1.44	36.81
LEAVENWORTH 3 S	1.09	1.24	2.47	3.28	4.74	4.88	4.68	3.91	4.43	3.56	1.85	1.34	37.47
LEBO	1.00	1.12	2.59	2.97	4.51	5.72	4.04	4.27	4.15	3.23	1.85	1.23	36.68
LECOMPTON	.91	1.04	2.28	3.21	4.18	5.90	4.47	3.77	4.12	3.09	1.91	1.29	36.17
LENORA	.39	.59	1.32	1.73	3.33	3.54	2.95	2.46	2.10	1.42	.76	.46	21.05
LEOTI	.38	.41	1.10	1.27	2.83	2.74	2.58	2.29	1.64	.95	.77	.34	17.30
LE ROY	.95	1.08	2.59	3.36	4.40	5.61	4.34	3.62	4.86	3.09	2.20	1.25	37.35

KANSAS

PRECIPITATION NORMALS (INCHES)

STATION	JAN	FEB	MAR	APR	MAY	JUN	JUL	AUG	SEP	OCT	NOV	DEC	ANN
LIBERAL	.44	.45	1.29	1.23	3.25	2.52	2.92	2.46	1.50	1.11	.76	.40	18.33
LINCOLN 1 ESE	.60	.85	1.95	2.26	4.04	3.75	3.59	3.40	3.26	2.09	1.15	.74	27.68
LINDSBORG	.71	1.01	2.20	2.77	3.98	4.43	3.48	3.09	3.57	2.42	1.30	.98	29.94
LOGAN	.39	.61	1.39	1.90	3.53	3.41	3.27	2.62	2.17	1.39	.80	.43	21.91
LONG ISLAND	.40	.60	1.29	1.93	3.26	3.71	2.90	3.20	2.57	1.46	.68	.44	22.44
LONGTON	1.08	1.12	2.65	3.13	4.72	4.51	3.79	3.31	3.71	3.34	2.19	1.36	34.91
LORETTA	.42	.64	1.58	2.05	3.15	4.06	3.90	2.49	2.79	1.43	.92	.53	23.96
LYNDON 3 ENE	.84	1.00	2.41	3.20	4.47	5.29	4.05	4.03	4.14	3.03	1.89	1.19	35.54
MADISON	1.00	1.18	2.56	2.94	4.35	5.21	4.32	3.70	4.45	2.94	2.05	1.27	35.97
MANHATTAN	.83	.95	2.08	2.79	4.50	5.29	3.96	3.18	4.04	2.89	1.46	.91	32.88
MARION DAM	.64	.91	1.95	2.67	4.37	4.87	4.06	3.02	3.77	2.63	1.52	.99	31.40
MARYSVILLE	.73	.93	1.98	2.69	4.23	4.21	3.97	3.82	3.54	2.40	1.35	.82	30.67
MCCRACKEN	.40	.65	1.66	2.00	2.88	3.64	3.25	2.50	2.24	1.31	.98	.51	22.02
MCFARLAND	.73	.91	2.13	2.94	4.11	5.51	4.14	3.33	4.28	2.59	1.64	1.18	33.49
MCPHERSON 2 S	.60	.87	2.12	2.54	4.33	4.57	3.32	2.75	3.64	2.33	1.27	.88	29.22
MEADE	.52	.65	1.57	1.42	3.38	2.63	2.94	2.81	2.17	1.30	.87	.54	20.80
MEDICINE LODGE	.52	.81	1.74	2.22	3.87	3.57	2.84	2.89	2.47	2.19	1.13	.69	24.94
MILTONVALE	.67	.83	2.05	2.54	4.33	4.67	3.80	3.44	3.75	2.25	1.26	.83	30.42
MINGO 5 E	.45	.50	1.29	1.60	2.98	3.50	2.54	2.40	1.76	1.03	.72	.47	19.24
MINNEAPOLIS 2	.64	.79	1.80	2.29	4.23	4.35	3.61	2.97	3.58	2.23	1.20	.77	28.46
MOUND CITY	1.30	1.28	2.92	3.64	4.54	4.77	3.63	3.97	4.88	3.32	2.13	1.44	37.82
MOUND VALLEY 3 WSW	1.31	1.37	3.08	3.76	5.19	5.29	3.93	3.56	4.59	3.92	2.60	1.67	40.27
MOUNT HOPE	.58	.90	2.01	2.72	4.43	4.45	3.48	2.99	3.73	2.71	1.49	.94	30.43
NATOMA	.60	.89	2.01	2.07	3.68	3.55	3.04	3.17	2.68	1.56	1.01	.57	24.83
NEOSHO RAPIDS	.91	1.12	2.55	3.04	4.32	6.26	4.68	4.13	4.67	3.32	1.88	1.11	37.99
NESS CITY	.42	.59	1.52	1.87	2.88	3.35	3.16	2.45	2.15	1.25	.88	.45	20.97
NEWTON 2 SW	.68	.96	2.13	2.60	4.48	4.69	3.57	3.04	3.69	2.56	1.53	1.03	30.96
NORCATUR 2 N	.45	.61	1.28	1.79	3.41	3.65	3.43	2.54	2.04	1.21	.74	.53	21.68
NORTON 8 SSE	.41	.60	1.33	1.83	3.42	3.68	3.35	2.61	2.16	1.41	.73	.46	21.99
OAKLEY	.41	.51	1.30	1.69	3.19	3.16	2.78	2.38	1.79	1.11	.77	.43	19.52
OBERLIN	.44	.53	1.25	1.72	3.19	3.66	3.16	2.35	2.03	1.06	.65	.41	20.45
OLATHE 3 E	1.07	1.08	2.50	3.47	4.60	5.30	4.55	3.56	4.64	3.29	1.90	1.26	37.22
OSAGE CITY	.95	1.05	2.53	3.17	4.52	4.90	4.25	3.79	4.05	2.93	1.78	1.25	35.17
OSAWATOMIE	1.12	1.12	2.55	3.36	4.56	5.06	3.96	3.70	4.70	3.20	1.98	1.29	36.60
OSWEGO 1 N	1.26	1.57	2.95	3.79	5.12	5.00	3.58	3.50	5.07	3.83	2.62	1.76	40.05
OTTAWA	1.26	1.30	2.76	3.22	4.81	5.40	4.26	4.20	4.65	3.12	1.93	1.36	38.27
OVERBROOK 2 WSW	.93	1.05	2.35	3.22	4.22	5.11	4.35	3.59	3.96	2.92	1.81	1.21	34.72
OXFORD	.78	1.00	1.90	2.87	4.05	4.78	3.66	3.32	3.55	2.69	1.88	1.10	31.58
PAOLA	1.36	1.32	2.57	3.60	4.55	5.61	4.54	3.65	4.78	3.42	1.98	1.40	38.78
PARSONS	1.22	1.34	2.98	3.72	5.18	4.80	3.65	3.43	4.53	3.47	2.54	1.65	38.51
PHILLIPSBURG 1 SSE	.41	.68	1.49	2.16	3.64	3.79	3.21	3.04	2.52	1.52	.80	.42	23.68
PITTSBURG	1.48	1.76	3.07	3.84	5.02	5.37	3.32	3.17	4.57	3.41	3.00	1.97	39.98
PLAINVILLE	.38	.77	1.78	1.92	3.78	3.82	3.09	2.87	2.55	1.38	.88	.47	23.69
PRATT 2 E	.51	.93	1.76	1.97	3.40	3.67	2.89	2.49	2.52	2.16	1.02	.85	24.17
QUINTER	.49	.71	1.68	1.94	3.61	3.61	3.24	3.19	2.13	1.43	.91	.51	23.45
REXFORD 1 SW	.41	.51	1.25	1.53	3.06	3.57	2.78	2.16	1.50	1.09	.59	.39	18.84
RICHFIELD 1 NE	.39	.37	.78	1.32	2.78	2.52	2.26	2.18	1.46	.90	.76	.34	16.06
RICHFIELD 10 WSW	.24	.25	.74	1.30	2.57	2.46	2.29	2.38	1.42	.79	.56	.21	15.21
ROSSVILLE	.91	1.04	2.08	3.01	4.37	5.20	3.93	3.28	4.40	2.83	1.63	1.17	33.85
RUSSELL FAA AP	.56	.79	1.84	2.38	3.87	3.55	3.31	3.40	3.12	1.71	1.00	.69	26.22
RUSSELL SPRINGS	.39	.37	1.22	1.27	2.87	2.98	2.86	2.34	1.60	1.00	.75	.38	18.03
SAINT FRANCIS	.41	.45	1.06	1.39	2.79	2.67	2.64	2.03	1.37	1.01	.63	.43	16.88
SAINT FRANCIS 8 NW	.36	.37	1.09	1.55	2.80	2.59	2.56	2.27	1.36	.94	.52	.36	16.77
SALINA FAA AIRPORT	.71	.85	1.98	2.70	4.15	4.19	3.31	3.03	3.51	2.41	1.22	.89	28.95
SCANDIA	.55	.83	1.83	2.29	3.79	4.66	3.34	3.36	3.47	1.85	1.03	.72	27.72

KANSAS

PRECIPITATION NORMALS (INCHES)

STATION		JAN	FEB	MAR	APR	MAY	JUN	JUL	AUG	SEP	OCT	NOV	DEC	ANN
SCOTT CITY		.53	.58	1.31	1.62	3.09	3.08	2.86	2.35	1.87	1.18	.94	.49	19.90
SEDAN		1.16	1.26	2.73	3.46	5.02	4.88	3.33	3.25	3.92	3.34	2.31	1.38	36.04
SEDGWICK		.63	.92	1.98	2.64	4.45	4.42	3.63	3.31	3.93	2.64	1.55	1.01	31.11
SHARON SPRINGS		.44	.44	1.26	1.33	3.10	3.28	2.75	2.03	1.50	.99	.74	.46	18.32
SMITH CENTER		.39	.59	1.62	1.98	3.56	3.79	2.81	2.92	2.76	1.70	.80	.46	23.38
STILWELL		1.16	1.16	2.52	3.68	4.54	5.41	4.37	3.73	4.70	3.34	1.89	1.39	37.89
STOCKTON		.47	.75	1.68	1.81	3.74	3.40	3.08	2.89	2.11	1.41	.85	.56	22.75
SUBLETTE		.33	.38	1.18	1.30	3.39	2.81	2.54	2.60	1.71	1.14	.76	.30	18.44
SYRACUSE 2 W		.29	.30	.76	1.10	2.62	2.53	2.27	2.08	1.55	.85	.57	.28	15.20
TESCOTT		.53	.75	1.95	2.43	4.40	4.05	3.75	3.05	3.82	2.26	1.13	.70	28.82
THRALL		.79	1.09	2.39	2.96	4.00	5.04	3.63	3.25	3.71	3.00	1.68	1.10	32.64
TOPEKA WSO	//R	.88	1.05	2.18	3.08	3.99	5.14	4.04	3.69	3.45	2.82	1.75	1.31	33.38
TORONTO DAM		.97	1.01	2.43	2.91	4.69	5.16	3.91	3.58	4.20	2.82	2.10	1.15	34.93
TRIBUNE 1 W		.30	.31	.83	1.21	2.57	2.57	2.35	2.31	1.33	.69	.51	.28	15.26
TROUSDALE 1 NE		.48	.72	1.51	1.85	3.53	3.38	2.97	2.67	2.29	1.81	1.03	.70	22.94
TROY 3 SW		.91	1.05	2.44	3.13	4.58	5.07	4.09	3.95	4.96	2.80	1.64	1.16	35.78
ULYSSES		.31	.35	.93	1.39	2.99	2.47	2.18	2.52	1.30	.93	.67	.27	16.31
UTICA		.49	.53	1.43	1.92	3.41	3.40	2.78	2.61	1.90	1.44	.92	.52	21.35
VALLEY FALLS		.90	1.03	2.10	3.08	4.31	5.13	4.59	4.13	4.28	2.91	1.84	1.20	35.50
WAKEENEY		.51	.73	1.68	1.88	3.67	3.41	2.98	3.05	2.21	1.34	1.00	.57	23.03
WAKEFIELD		.62	.74	2.04	2.66	4.56	5.11	4.45	3.31	3.86	2.65	1.34	.80	32.14
WALNUT		1.34	1.37	2.89	3.55	4.87	5.06	4.00	3.47	4.35	3.07	2.27	1.47	37.71
WAMEGO		.82	.96	2.20	2.70	4.30	5.22	4.38	3.20	3.96	2.88	1.54	1.10	33.26
WASHINGTON		.61	.85	2.01	2.75	4.46	4.22	3.69	3.93	3.77	2.10	1.26	.72	30.37
WAVERLY		.90	1.14	2.48	2.91	4.37	5.56	3.79	3.78	4.24	3.07	1.84	1.16	35.24
WELLINGTON 2 S		.82	1.10	2.07	2.93	4.50	4.83	3.82	3.42	3.63	2.49	1.83	1.09	32.53
WICHITA WSO	//R	.68	.85	2.01	2.30	3.91	4.06	3.62	2.80	3.45	2.47	1.47	.99	28.61
WINFIELD		.90	1.11	1.98	2.84	3.97	4.44	3.91	3.16	3.59	2.65	1.82	1.19	31.56
WINONA		.37	.35	1.13	1.24	2.90	3.10	2.62	2.42	1.45	1.04	.63	.39	17.64
WORDEN		1.00	1.13	2.53	3.05	4.22	5.29	4.37	3.80	4.20	3.03	1.94	1.11	35.67
YATES CENTER		1.04	1.13	2.78	3.55	4.46	5.39	4.64	3.79	4.97	3.20	2.37	1.39	38.71

KANSAS

HEATING DEGREE DAY NORMALS (BASE 65 DEG F)

STATION	JUL	AUG	SEP	OCT	NOV	DEC	JAN	FEB	MAR	APR	MAY	JUN	ANN
ABILENE 2 S	5	0	61	259	675	1023	1194	913	725	327	116	9	5307
ALTON 6 E	0	0	54	262	705	1039	1190	902	744	337	135	15	5383
ANTHONY	0	0	22	169	570	884	1008	756	589	225	61	0	4284
ASHLAND	0	0	37	233	639	930	1057	804	660	288	113	9	4770
ATCHISON	0	0	50	261	654	1023	1212	921	725	306	99	10	5261
ATWOOD	0	0	90	358	780	1054	1159	896	797	420	167	27	5748
BELLEVILLE	0	0	67	290	726	1079	1259	960	781	358	130	15	5665
BISON	0	0	35	243	672	986	1119	860	707	323	113	12	5070
CENTRALIA	0	0	50	276	693	1060	1246	949	760	347	124	12	5517
CHANUTE FAA AIRPORT	6	0	40	219	597	927	1088	820	636	258	88	9	4688
CIMARRON	0	0	48	253	681	958	1076	812	684	318	121	15	4966
CLAY CENTER	0	0	40	241	669	1017	1187	893	705	291	101	9	5153
COLBY 1 SW	0	0	117	385	810	1091	1212	949	871	468	202	45	6150
COLDWATER	0	0	31	202	603	899	1026	762	613	246	85	7	4474
COLUMBUS 6 NNW	0	0	25	188	540	846	998	750	578	225	72	7	4229
CONCORDIA WSO //R	0	0	65	277	717	1060	1243	955	778	367	137	16	5615
COTTONWOOD FALLS	6	0	44	214	615	952	1113	843	659	267	88	10	4811
COUNCIL GROVE DAM	8	0	62	269	669	1023	1209	932	738	322	122	17	5371
DODGE CITY WSO //R	0	0	43	251	675	970	1101	840	710	331	124	14	5059
EL DORADO	0	0	41	203	600	927	1076	809	631	248	81	8	4624
ELKHART	0	0	28	238	630	862	967	742	635	294	105	9	4510
ELLSWORTH	0	0	39	241	669	998	1150	882	705	303	105	8	5100
EMPORIA FAA AIRPORT R	7	0	52	255	654	986	1156	882	704	306	107	12	5121
ESKRIDGE	7	0	45	241	651	1001	1175	890	714	300	105	14	5143
EUREKA	6	0	41	207	588	908	1070	798	616	236	76	6	4552
FALL RIVER DAM	5	0	43	220	591	930	1094	832	649	250	90	7	4711
FLORENCE	6	0	46	220	624	958	1122	840	655	267	95	8	4841
FORT SCOTT	0	0	21	182	546	887	1045	776	592	204	58	7	4318
FREDONIA 1 E	0	0	35	194	558	887	1042	773	591	223	74	6	4383
GARDEN CITY FAA AP	0	0	47	284	714	1004	1125	860	735	350	128	14	5261
GARNETT	0	0	36	218	597	936	1097	829	641	246	84	9	4693
GOODLAND WSO //R	0	0	114	394	807	1076	1172	927	862	489	215	43	6099
GREAT BEND	0	0	34	212	636	952	1101	829	680	285	102	8	4839
GREENSBURG	0	0	50	241	651	952	1085	829	676	298	103	8	4893
HAYS 1 S	0	0	67	300	738	1060	1212	941	797	377	147	20	5659
HEALY	0	0	61	298	729	1011	1128	868	747	370	135	20	5367
HERINGTON	0	0	45	238	654	995	1163	885	691	298	108	11	5088
HILL CITY FAA AIRPORT R	0	0	76	314	756	1054	1194	921	800	391	149	24	5679
HOLTON 4 NE	0	0	53	280	678	1035	1218	930	741	320	121	13	5389
HORTON	0	0	45	282	690	1057	1249	949	747	320	113	11	5463
HOWARD 5 NE	0	0	35	203	579	905	1057	792	612	233	80	9	4505
HOXIE	0	0	62	293	735	1017	1138	874	766	372	139	20	5416
HUDSON	0	0	39	221	636	949	1091	829	672	282	94	6	4819
HUGOTON	0	0	40	244	657	918	1023	781	670	310	104	10	4757
INDEPENDENCE	0	0	31	190	552	868	1020	762	582	220	58	5	4288
JETMORE	0	0	31	229	657	949	1070	809	667	283	102	13	4810
JOHN REDMOND DAM	5	0	44	235	615	952	1141	865	667	269	111	13	4917
JOHNSON 11 ESE	0	0	38	267	693	961	1066	812	691	334	117	10	4989
KANOPOLIS DAM	0	0	70	265	681	1023	1200	932	760	347	131	13	5422
KINGMAN	0	0	33	195	603	921	1054	792	628	249	75	7	4557
KINSLEY	0	0	42	250	666	977	1116	860	710	320	120	10	5071
KIRWIN	0	0	88	328	756	1094	1265	972	825	394	153	25	5900
LAKIN	0	0	68	291	705	986	1104	846	729	348	141	18	5236
LARNED	0	0	40	217	639	946	1079	815	663	274	96	8	4777
LAWRENCE	0	0	37	215	603	955	1135	857	668	255	86	8	4819

KANSAS

HEATING DEGREE DAY NORMALS (BASE 65 DEG F)

STATION		JUL	AUG	SEP	OCT	NOV	DEC	JAN	FEB	MAR	APR	MAY	JUN	ANN
LEAVENWORTH 3 S		0	0	42	258	654	1014	1190	902	710	299	103	12	5184
LEOTI		0	0	76	327	756	1029	1135	876	781	409	157	21	5567
LIBERAL		0	0	23	194	594	859	964	722	608	251	100	0	4315
LINCOLN 1 ESE		0	0	36	235	669	1001	1159	876	698	304	109	9	5096
MANHATTAN		0	0	47	246	648	1001	1175	890	705	297	101	9	5119
MARYSVILLE		0	0	65	307	735	1094	1293	994	809	371	137	13	5818
MCPHERSON 2 S		0	0	46	217	633	970	1119	851	674	288	99	9	4906
MEADE		0	0	26	203	612	887	1004	756	623	258	96	7	4472
MEDICINE LODGE		0	0	24	194	588	896	1020	762	599	233	71	0	4387
MINNEAPOLIS 2		0	0	39	229	654	992	1153	871	687	282	102	7	5016
MOUND CITY		6	0	35	211	585	918	1082	815	630	249	84	11	4626
MOUND VALLEY 3 WSW		0	0	22	181	543	853	995	739	560	211	60	0	4164
NESS CITY		0	0	50	259	693	986	1110	846	701	319	112	14	5090
NEWTON 2 SW		0	0	42	209	618	952	1101	834	661	272	93	10	4792
NORTON 8 SSE		0	0	79	306	741	1051	1187	907	784	382	146	23	5606
OAKLEY		0	0	67	288	717	998	1113	851	756	376	145	24	5335
OBERLIN		0	0	69	312	762	1051	1166	888	775	375	135	19	5552
OLATHE 3 E		0	0	49	242	633	983	1159	885	694	296	104	12	5057
OTTAWA		0	0	35	222	600	946	1122	846	654	262	87	9	4783
PARSONS		0	0	24	173	528	849	1001	742	565	209	59	6	4156
PHILLIPSBURG 1 SSE		0	0	57	278	717	1045	1187	899	750	340	128	16	5417
PLAINVILLE		0	0	64	282	705	1020	1163	893	755	353	132	20	5387
PRATT 2 E		0	0	33	207	612	915	1045	792	634	261	91	7	4597
QUINTER		0	0	95	338	771	1063	1200	938	837	432	171	31	5876
RUSSELL FAA AP		0	0	58	272	705	1032	1187	916	760	359	133	16	5438
SAINT FRANCIS		0	0	80	323	747	1011	1110	851	769	399	151	26	5467
SALINA FAA AIRPORT		0	0	49	248	666	1004	1169	896	710	322	116	7	5187
SCOTT CITY		0	0	61	278	708	973	1088	829	719	344	120	17	5137
SEDAN		0	0	40	208	570	890	1032	773	593	223	74	6	4409
SHARON SPRINGS		0	0	48	289	711	967	1063	826	729	365	125	14	5137
SMITH CENTER		0	0	49	276	729	1070	1221	924	766	346	131	16	5528
SUBLETTE		0	0	28	232	654	927	1035	781	658	300	114	10	4739
SYRACUSE 2 W		0	0	56	299	735	1011	1110	840	713	348	120	20	5252
TOPEKA WSO //R		5	0	53	276	669	1029	1206	910	719	321	117	14	5319
TORONTO DAM		5	0	42	226	600	936	1116	848	657	259	90	11	4790
TRIBUNE 1 W		0	0	71	334	759	1023	1119	865	778	420	165	28	5562
TROY 3 SW		0	0	47	266	675	1054	1246	952	753	320	104	11	5428
WAKEENEY		0	0	85	310	729	1029	1169	907	787	385	153	26	5580
WAMEGO		0	0	43	234	633	989	1163	876	686	276	90	8	4998
WASHINGTON		0	0	43	257	693	1045	1228	930	735	318	99	9	5357
WELLINGTON 2 S		0	0	28	195	570	896	1039	781	608	242	88	0	4447
WICHITA WSO //R		0	0	37	219	621	949	1097	837	656	275	89	7	4787
WINFIELD		0	0	27	169	549	868	1011	750	570	207	66	0	4217

KANSAS

COOLING DEGREE DAY NORMALS (BASE 65 DEG F)

STATION	JAN	FEB	MAR	APR	MAY	JUN	JUL	AUG	SEP	OCT	NOV	DEC	ANN
ABILENE 2 S	0	0	0	6	110	312	483	425	190	38	0	0	1564
ALTON 6 E	0	0	0	7	98	288	454	388	153	18	0	0	1406
ANTHONY	0	0	9	21	138	375	546	496	241	45	0	0	1871
ASHLAND	0	0	0	9	122	321	490	437	187	25	0	0	1591
ATCHISON	0	0	0	9	96	265	415	353	140	31	0	0	1309
ATWOOD	0	0	0	0	46	222	378	313	108	8	0	0	1075
BELLEVILLE	0	0	0	0	90	273	441	386	145	20	0	0	1355
BISON	0	0	0	8	104	315	484	428	176	29	0	0	1544
CENTRALIA	0	0	0	0	90	264	419	366	140	28	0	0	1307
CHANUTE FAA AIRPORT	0	0	0	15	119	309	474	419	202	39	0	0	1577
CIMARRON	0	0	8	12	96	297	440	378	162	24	0	0	1417
CLAY CENTER	0	0	8	9	117	315	474	431	175	30	0	0	1559
COLBY 1 SW	0	0	0	0	37	210	361	293	111	10	0	0	1022
COLDWATER	0	0	11	18	141	349	512	459	211	44	0	0	1745
COLUMBUS 6 NNW	0	0	8	24	124	310	462	419	211	49	0	0	1607
CONCORDIA WSO //R	0	0	0	0	84	271	438	381	146	20	0	0	1340
COTTONWOOD FALLS	0	0	8	12	116	304	477	431	197	38	0	0	1583
COUNCIL GROVE DAM	0	0	0	7	104	269	436	385	164	27	0	0	1392
DODGE CITY WSO //R	0	0	0	10	96	302	465	412	166	28	0	0	1479
EL DORADO	0	0	8	17	122	314	477	431	203	36	0	0	1608
ELKHART	0	0	6	9	96	297	437	384	151	25	0	0	1405
ELLSWORTH	0	0	7	9	111	323	493	434	171	28	0	0	1576
EMPORIA FAA AIRPORT R	0	0	0	9	100	276	441	381	169	29	0	0	1405
ESKRIDGE	0	0	7	9	102	272	434	381	165	36	0	0	1406
EUREKA	0	0	8	17	122	309	489	437	212	40	0	0	1634
FALL RIVER DAM	0	0	7	16	118	286	464	419	205	43	0	0	1558
FLORENCE	0	0	7	15	116	311	490	434	202	37	0	0	1612
FORT SCOTT	0	0	12	27	154	361	527	484	246	68	0	0	1879
FREDONIA 1 E	0	0	8	22	139	327	487	450	227	46	0	0	1706
GARDEN CITY FAA AP	0	0	0	8	85	296	465	397	149	20	0	0	1420
GARNETT	0	0	8	15	128	303	459	412	192	38	0	0	1555
GOODLAND WSO //R	0	0	0	0	26	190	345	274	93	6	0	0	934
GREAT BEND	0	0	10	12	126	341	505	450	193	38	0	0	1675
GREENSBURG	0	0	0	10	106	320	484	431	194	39	0	0	1584
HAYS 1 S	0	0	0	0	73	269	439	375	142	14	0	0	1312
HEALY	0	0	0	0	70	284	443	381	142	16	0	0	1336
HERINGTON	0	0	0	7	105	290	454	409	174	27	0	0	1466
HILL CITY FAA AIRPORT R	0	0	0	0	68	267	431	369	136	16	0	0	1287
HOLTON 4 NE	0	0	0	5	99	262	412	347	143	32	0	0	1300
HORTON	0	0	0	8	97	266	412	350	129	25	0	0	1287
HOWARD 5 NE	0	0	8	20	120	303	467	425	203	41	0	0	1587
HOXIE	0	0	0	6	74	266	428	369	137	14	0	0	1294
HUDSON	0	0	9	9	116	327	490	434	192	35	0	0	1612
HUGOTON	0	0	0	10	92	313	459	391	157	21	0	0	1443
INDEPENDENCE	0	0	9	25	120	317	484	446	217	47	0	0	1665
JETMORE	0	0	10	13	115	325	484	428	178	34	0	0	1587
JOHN REDMOND DAM	0	0	0	11	124	292	455	403	182	40	0	0	1507
JOHNSON 11 ESE	0	0	0	7	83	298	443	369	140	15	0	0	1355
KANOPOLIS DAM	0	0	0	5	94	280	457	404	166	29	0	0	1435
KINGMAN	0	0	8	18	124	349	515	468	216	40	0	0	1738
KINSLEY	0	0	0	8	108	310	477	422	174	33	0	0	1532
KIRWIN	0	0	0	0	69	259	422	357	124	12	0	0	1243
LAKIN	0	0	0	6	88	288	443	369	149	21	0	0	1364
LARNED	0	0	9	10	121	332	490	440	193	34	0	0	1629
LAWRENCE	0	0	7	12	126	311	468	415	196	48	0	0	1583

KANSAS

COOLING DEGREE DAY NORMALS (BASE 65 DEG F)

STATION		JAN	FEB	MAR	APR	MAY	JUN	JUL	AUG	SEP	OCT	NOV	DEC	ANN
LEAVENWORTH 3 S		0	0	0	8	109	276	429	363	144	26	0	0	1355
LEOTI		0	0	0	0	52	246	397	316	112	11	0	0	1134
LIBERAL		0	0	10	20	140	358	508	450	203	36	0	0	1725
LINCOLN 1 ESE		0	0	7	10	115	327	505	453	186	33	0	0	1636
MANHATTAN		0	0	7	12	114	303	462	415	176	36	0	0	1525
MARYSVILLE		0	0	0	5	93	268	422	369	134	25	0	0	1316
MCPHERSON 2 S		0	0	8	9	108	327	499	450	205	37	0	0	1643
MEADE		0	0	9	12	123	337	487	437	191	33	0	0	1629
MEDICINE LODGE		0	0	7	20	133	358	512	471	213	42	0	0	1756
MINNEAPOLIS 2		0	0	8	9	123	331	502	456	189	31	0	0	1649
MOUND CITY		0	0	7	15	124	311	471	415	191	37	0	0	1571
MOUND VALLEY 3 WSW		0	0	8	31	135	334	502	462	229	50	0	0	1751
NESS CITY		0	0	0	7	103	320	474	415	173	30	0	0	1522
NEWTON 2 SW		0	0	10	14	118	334	507	453	213	41	0	0	1690
NORTON 8 SSE		0	0	0	0	62	239	401	350	133	18	0	0	1203
OAKLEY		0	0	0	0	64	255	409	347	139	18	0	0	1232
OBERLIN		0	0	0	0	63	259	425	360	138	12	0	0	1257
OLATHE 3 E		0	0	0	8	104	267	417	363	169	35	0	0	1363
OTTAWA		0	0	0	10	121	303	456	403	179	39	0	0	1511
PARSONS		0	0	10	29	143	339	505	462	237	58	0	0	1783
PHILLIPSBURG 1 SSE		0	0	0	7	97	301	468	406	156	21	0	0	1456
PLAINVILLE		0	0	8	5	79	290	450	391	154	24	0	0	1401
PRATT 2 E		0	0	8	15	122	334	490	437	198	37	0	0	1641
QUINTER		0	0	0	0	47	235	391	329	122	13	0	0	1137
RUSSELL FAA AP		0	0	0	0	90	301	465	409	160	24	0	0	1449
SAINT FRANCIS		0	0	0	0	46	233	394	329	128	13	0	0	1143
SALINA FAA AIRPORT		0	0	0	7	104	313	490	440	181	31	0	0	1566
SCOTT CITY		0	0	0	5	67	278	419	350	139	14	0	0	1272
SEDAN		0	0	7	19	129	321	490	450	223	41	0	0	1680
SHARON SPRINGS		0	0	0	5	69	275	443	369	132	16	0	0	1309
SMITH CENTER		0	0	0	7	93	292	462	400	145	19	0	0	1418
SUBLETTE		0	0	7	9	101	310	456	391	157	21	0	0	1452
SYRACUSE 2 W		0	0	0	6	70	287	440	363	143	11	0	0	1320
TOPEKA WSO //R		0	0	0	9	105	281	427	372	158	28	0	0	1380
TORONTO DAM		0	0	0	13	112	296	470	422	201	37	0	0	1551
TRIBUNE 1 W		0	0	0	0	38	223	378	301	101	8	0	0	1049
TROY 3 SW		0	0	0	8	97	266	397	335	134	34	0	0	1271
WAKEENEY		0	0	0	0	66	248	409	350	136	22	0	0	1231
WAMEGO		0	0	7	12	121	308	459	412	184	39	0	0	1542
WASHINGTON		0	0	0	6	99	297	456	397	154	28	0	0	1437
WELLINGTON 2 S		0	0	6	26	144	349	515	477	226	46	0	0	1789
WICHITA WSO //R		0	0	8	14	117	340	508	456	205	36	0	0	1684
WINFIELD		0	0	9	21	140	351	512	471	243	45	0	0	1792

14 — KANSAS

STATE-STATION NUMBER	STN TYP	NAME	LATITUDE DEG-MIN	LONGITUDE DEG-MIN	ELEVATION (FT)
14-0010	13	ABILENE 2 S	N 3853	W 09712	1155
14-0069	12	AETNA 2 S	N 3704	W 09858	1570
14-0135	12	ALEXANDER	N 3828	W 09933	2070
14-0195	12	ALTA VISTA	N 3852	W 09629	1432
14-0201	13	ALTON 6 E	N 3928	W 09850	1591
14-0264	13	ANTHONY	N 3709	W 09801	1340
14-0313	12	ARKANSAS CITY	N 3704	W 09702	1118
14-0365	13	ASHLAND	N 3712	W 09946	1970
14-0405	13	ATCHISON	N 3934	W 09507	945
14-0431	12	ATTICA	N 3715	W 09814	1453
14-0439	13	ATWOOD	N 3949	W 10103	2835
14-0447	12	AUGUSTA	N 3740	W 09659	1257
14-0532	12	BARNARD	N 3912	W 09803	1322
14-0682	13	BELLEVILLE	N 3950	W 09738	1540
14-0693	12	BELOIT	N 3928	W 09807	1392
14-0836	12	BIRD CITY 11 S	N 3936	W 10135	3460
14-0865	13	BISON	N 3831	W 09912	2015
14-0911	12	BLUE RAPIDS	N 3941	W 09640	1150
14-0957	12	BONNER SPRINGS	N 3904	W 09453	860
14-1003	12	BREMEN	N 3954	W 09647	1298
14-1029	12	BREWSTER	N 3921	W 10122	3405
14-1104	12	BUCKLIN 4 SE	N 3732	W 09935	2318
14-1141	12	BURDETT 6 SSE	N 3807	W 09931	2160
14-1371	12	CAWKER CITY	N 3931	W 09826	1483
14-1383	12	CEDAR BLUFF DAM	N 3848	W 09943	2135
14-1408	13	CENTRALIA	N 3943	W 09608	1266
14-1427	13	CHANUTE FAA AIRPORT	N 3740	W 09529	979
14-1435	12	CHAPMAN	N 3858	W 09701	1105
14-1522	13	CIMARRON	N 3748	W 10021	2625
14-1536	12	CLAFLIN	N 3831	W 09832	1795
14-1559	13	CLAY CENTER	N 3923	W 09708	1211
14-1593	12	CLIFTON	N 3934	W 09717	1295
14-1699	13	COLBY 1 SW	N 3923	W 10104	3170
14-1704	13	COLDWATER	N 3716	W 09920	2083
14-1730	12	COLLYER 9 S	N 3854	W 10007	2480
14-1740	13	COLUMBUS 6 NNW	N 3715	W 09452	898
14-1767	13	CONCORDIA WSO //R	N 3933	W 09739	1470
14-1858	13	COTTONWOOD FALLS	N 3822	W 09633	1225
14-1867	13	COUNCIL GROVE DAM	N 3841	W 09631	1320
14-1875	12	COVERT	N 3915	W 09852	1820
14-2086	12	DENSMORE	N 3938	W 09944	2074
14-2126	12	DEXTER	N 3711	W 09643	1175
14-2164	13	DODGE CITY WSO //R	N 3746	W 09958	2582
14-2213	12	DRESDEN	N 3937	W 10025	2731
14-2401	13	EL DORADO	N 3749	W 09650	1282
14-2409	12	ELGIN	N 3700	W 09617	850
14-2432	13	ELKHART	N 3700	W 10154	3585
14-2452	12	ELLIS	N 3856	W 09934	212
14-2459	13	ELLSWORTH	N 3844	W 09814	1530
14-2478	12	ELMO 1 NW	N 3842	W 09714	1300

STATE-STATION NUMBER	STN TYP	NAME	LATITUDE DEG-MIN	LONGITUDE DEG-MIN	ELEVATION (FT)
14-2519	12	EMMETT	N 3918	W 09603	1024
14-2543	13	EMPORIA FAA AIRPORT R	N 3820	W 09612	1209
14-2574	12	ENTERPRISE	N 3854	W 09707	1150
14-2592	12	ESBON 7 N	N 3956	W 09826	1870
14-2602	13	ESKRIDGE	N 3852	W 09606	1417
14-2622	13	EUREKA	N 3749	W 09617	1044
14-2686	13	FALL RIVER DAM	N 3739	W 09605	1020
14-2773	13	FLORENCE	N 3815	W 09656	1294
14-2835	13	FORT SCOTT	N 3751	W 09442	845
14-2855	12	FOWLER 3 NNE	N 3725	W 10011	2480
14-2872	12	FRANKFORT	N 3942	W 09625	1144
14-2894	13	FREDONIA 1 E	N 3732	W 09548	925
14-2975	13	GARDEN CITY FAA AP	N 3756	W 10043	2882
14-3008	13	GARNETT	N 3817	W 09514	1066
14-3037	12	GENESEO	N 3831	W 09809	1755
14-3153	13	GOODLAND WSO //R	N 3922	W 10142	3650
14-3175	12	GOVE	N 3858	W 10029	2638
14-3218	13	GREAT BEND	N 3821	W 09846	1850
14-3239	13	GREENSBURG	N 3737	W 09918	2230
14-3248	12	GRENOLA	N 3721	W 09628	1110
14-3257	12	GRIDLEY	N 3806	W 09553	1110
14-3323	12	HADDAM	N 3951	W 09718	1420
14-3398	12	HANOVER 4 S	N 3950	W 09652	125
14-3432	12	HARLAN	N 3936	W 09846	1590
14-3467	12	HARVEYVILLE 3 N	N 3850	W 09558	1200
14-3527	13	HAYS 1 S	N 3852	W 09920	2000
14-3554	13	HEALY	N 3836	W 10037	2852
14-3594	13	HERINGTON	N 3840	W 09657	1335
14-3634	12	HIAWATHA	N 3951	W 09532	1125
14-3646	12	HIGHLAND 1 W	N 3952	W 09517	990
14-3660	13	HILL CITY FAA AIRPORT R	N 3923	W 09950	2196
14-3759	13	HOLTON 4 NE	N 3930	W 09540	1085
14-3810	13	HORTON	N 3940	W 09531	1029
14-3822	13	HOWARD 5 NE	N 3731	W 09612	1100
14-3837	13	HOXIE	N 3921	W 10027	2690
14-3842	12	HOYT	N 3915	W 09542	1140
14-3847	13	HUDSON	N 3806	W 09839	1867
14-3855	13	HUGOTON	N 3711	W 10121	3100
14-3946	12	IMPERIAL	N 3816	W 10039	2790
14-3954	13	INDEPENDENCE	N 3715	W 09542	780
14-3974	12	INMAN	N 3814	W 09747	1526
14-3984	12	IOLA 1 W	N 3755	W 09526	954
14-4073	12	JEROME 1 S	N 3844	W 10031	2500
14-4081	13	JETMORE	N 3805	W 09954	2270
14-4104	13	JOHN REDMOND DAM	N 3815	W 09545	1090
14-4109	13	JOHNSON 11 ESE	N 3732	W 10134	3172
14-4178	13	KANOPOLIS DAM	N 3836	W 09757	1492
14-4313	13	KINGMAN	N 3740	W 09807	1570
14-4333	13	KINSLEY	N 3756	W 09924	2170
14-4357	13	KIRWIN	N 3940	W 09907	1697

STATE-STATION NUMBER	STN TYP	NAME	LATITUDE DEG-MIN	LONGITUDE DEG-MIN	ELEVATION (FT)
14-4421	12	LA CYGNE	N 3821	W 09446	817
14-4464	13	LAKIN	N 3757	W 10117	3080
14-4530	13	LARNED	N 3811	W 09906	2004
14-4559	13	LAWRENCE	N 3858	W 09516	1000
14-4588	13	LEAVENWORTH 3 S	N 3916	W 09454	880
14-4608	12	LEBO	N 3825	W 09551	1152
14-4613	12	LECOMPTON	N 3903	W 09523	875
14-4642	12	LENORA	N 3936	W 10000	2290
14-4665	13	LEOTI	N 3829	W 10121	3300
14-4675	12	LE ROY	N 3805	W 09538	1005
14-4695	13	LIBERAL	N 3702	W 10055	2838
14-4712	13	LINCOLN 1 ESE	N 3902	W 09808	1400
14-4735	12	LINDSBORG	N 3834	W 09740	1315
14-4775	12	LOGAN	N 3939	W 09935	1940
14-4807	12	LONG ISLAND	N 3957	W 09932	2065
14-4812	12	LONGTON	N 3723	W 09605	940
14-4821	12	LORETTA	N 3840	W 09911	2012
14-4912	12	LYNDON 3 ENE	N 3837	W 09538	1040
14-4937	12	MADISON	N 3808	W 09608	1165
14-4972	13	MANHATTAN	N 3912	W 09635	1065
14-5039	12	MARION DAM	N 3823	W 09705	1369
14-5063	13	MARYSVILLE	N 3950	W 09639	1163
14-5115	12	MCCRACKEN	N 3835	W 09934	2142
14-5132	12	MCFARLAND	N 3903	W 09614	1025
14-5152	13	MCPHERSON 2 S	N 3820	W 09740	1495
14-5171	13	MEADE	N 3717	W 10020	2505
14-5173	13	MEDICINE LODGE	N 3716	W 09835	1450
14-5335	12	MILTONVALE	N 3921	W 09727	1375
14-5355	12	MINGO 5 E	N 3916	W 10052	3025
14-5363	13	MINNEAPOLIS 2	N 3908	W 09742	1315
14-5528	13	MOUND CITY	N 3808	W 09449	865
14-5536	13	MOUND VALLEY 3 WSW	N 3711	W 09527	800
14-5539	12	MOUNT HOPE	N 3752	W 09740	1440
14-5628	12	NATOMA	N 3911	W 09902	1830
14-5680	12	NEOSHO RAPIDS	N 3822	W 09600	1093
14-5692	13	NESS CITY	N 3827	W 09954	2260
14-5744	13	NEWTON 2 SW	N 3802	W 09723	1447
14-5787	12	NORCATUR 2 N	N 3952	W 10012	2578
14-5856	13	NORTON 8 SSE	N 3944	W 09950	2364
14-5888	13	OAKLEY	N 3908	W 10051	3050
14-5906	13	OBERLIN	N 3949	W 10033	2620
14-5972	13	OLATHE 3 E	N 3853	W 09446	1055
14-6076	12	OSAGE CITY	N 3838	W 09550	1080
14-6084	12	OSAWATOMIE	N 3830	W 09458	861
14-6115	12	OSWEGO 1 N	N 3710	W 09506	835
14-6128	13	OTTAWA	N 3838	W 09516	915
14-6154	12	OVERBROOK 2 WSW	N 3846	W 09536	1180
14-6169	12	OXFORD	N 3716	W 09709	1123
14-6209	12	PAOLA	N 3834	W 09452	880
14-6242	13	PARSONS	N 3720	W 09516	908

STATE-STATION NUMBER	STN TYP	NAME		LATITUDE DEG-MIN	LONGITUDE DEG-MIN	ELEVATION (FT)
14-6374	13	PHILLIPSBURG 1 SSE		N 3944	W 09919	1882
14-6414	12	PITTSBURG		N 3721	W 09438	930
14-6435	13	PLAINVILLE		N 3914	W 09918	2145
14-6549	13	PRATT 2 E		N 3738	W 09842	1805
14-6637	13	QUINTER		N 3904	W 10014	2664
14-6787	12	REXFORD 1 SW		N 3928	W 10045	2950
14-6808	12	RICHFIELD 1 NE		N 3717	W 10146	3412
14-6813	12	RICHFIELD 10 WSW		N 3714	W 10157	3525
14-7007	12	ROSSVILLE		N 3908	W 09557	917
14-7046	13	RUSSELL FAA AP		N 3852	W 09849	1864
14-7049	12	RUSSELL SPRINGS		N 3854	W 10111	2960
14-7093	13	SAINT FRANCIS		N 3946	W 10148	3340
14-7095	12	SAINT FRANCIS 8 NW		N 3950	W 10155	3570
14-7160	13	SALINA FAA AIRPORT		N 3848	W 09739	1261
14-7248	12	SCANDIA		N 3947	W 09747	1482
14-7271	13	SCOTT CITY		N 3829	W 10054	2971
14-7305	13	SEDAN		N 3707	W 09610	809
14-7313	12	SEDGWICK		N 3755	W 09726	1377
14-7397	13	SHARON SPRINGS		N 3854	W 10145	3440
14-7542	13	SMITH CENTER		N 3947	W 09847	1790
14-7809	12	STILWELL		N 3846	W 09440	1100
14-7832	12	STOCKTON		N 3926	W 09916	1775
14-7922	13	SUBLETTE		N 3729	W 10051	2916
14-8038	13	SYRACUSE 2 W		N 3759	W 10147	3225
14-8086	12	TESCOTT		N 3900	W 09752	1300
14-8114	12	THRALL		N 3800	W 09619	1350
14-8167	13	TOPEKA WSO	//R	N 3904	W 09538	877
14-8191	13	TORONTO DAM		N 3745	W 09556	961
14-8235	13	TRIBUNE 1 W		N 3828	W 10146	3620
14-8245	12	TROUSDALE 1 NE		N 3749	W 09904	2050
14-8250	13	TROY 3 SW		N 3946	W 09508	1100
14-8287	12	ULYSSES		N 3735	W 10121	3050
14-8323	12	UTICA		N 3839	W 10010	2614
14-8341	12	VALLEY FALLS		N 3921	W 09527	928
14-8495	13	WAKEENEY		N 3901	W 09953	2456
14-8503	12	WAKEFIELD		N 3913	W 09701	1220
14-8549	12	WALNUT		N 3736	W 09505	920
14-8563	13	WAMEGO		N 3912	W 09618	980
14-8578	13	WASHINGTON		N 3949	W 09703	1304
14-8608	12	WAVERLY		N 3824	W 09536	1127
14-8670	13	WELLINGTON 2 S		N 3714	W 09724	1230
14-8830	13	WICHITA WSO	//R	N 3739	W 09726	1321
14-8964	13	WINFIELD		N 3714	W 09659	1135
14-8988	12	WINONA		N 3904	W 10115	3312
14-9040	12	WORDEN		N 3848	W 09522	1100
14-9080	12	YATES CENTER		N 3753	W 09543	1095

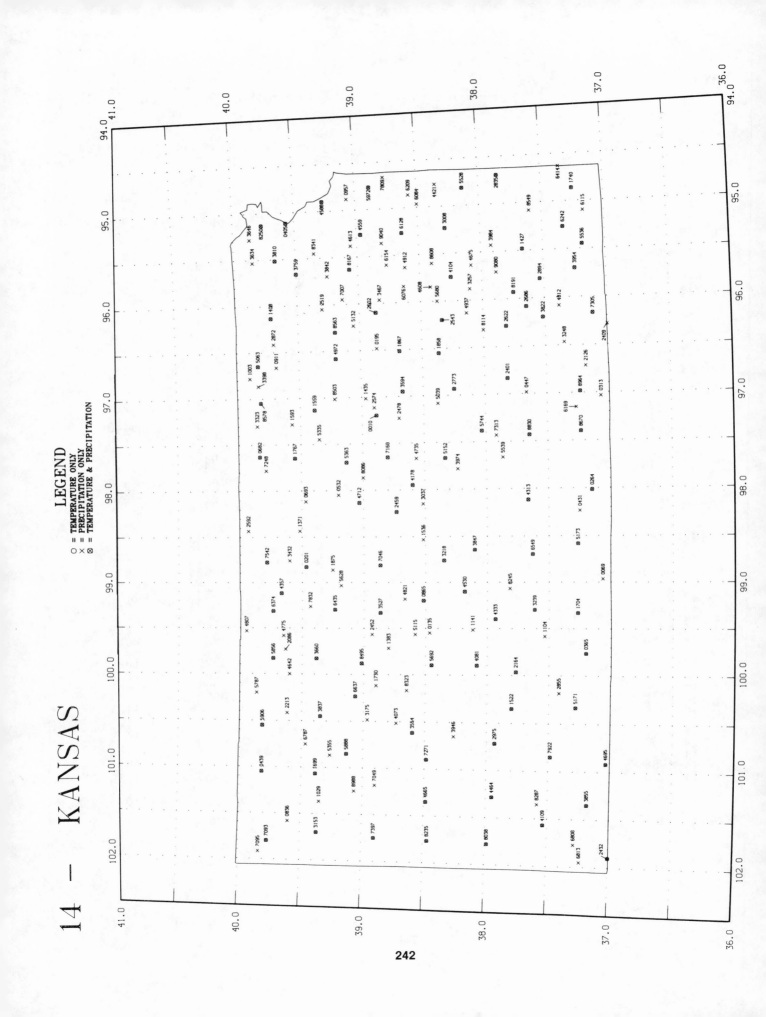

14 — KANSAS

LEGEND
O = TEMPERATURE ONLY
X = PRECIPITATION ONLY
⊗ = TEMPERATURE & PRECIPITATION

KENTUCKY

TEMPERATURE NORMALS (DEG F)

STATION		JAN	FEB	MAR	APR	MAY	JUN	JUL	AUG	SEP	OCT	NOV	DEC	ANN
ANCHORAGE	MAX	41.1	45.4	55.5	67.9	76.0	83.3	86.5	85.5	79.9	69.2	55.4	45.2	65.9
	MIN	22.6	24.9	33.6	43.6	52.3	60.6	64.4	63.1	56.5	44.0	34.5	27.0	43.9
	MEAN	31.9	35.2	44.6	55.8	64.2	72.0	75.5	74.3	68.2	56.6	45.0	36.1	55.0
ASHLAND	MAX	42.3	45.8	55.9	68.4	77.2	83.8	87.0	86.1	80.6	69.5	56.8	46.5	66.7
	MIN	21.6	23.2	31.4	40.6	49.7	58.3	62.7	61.6	55.3	42.2	32.8	25.9	42.1
	MEAN	32.0	34.5	43.7	54.5	63.4	71.1	74.9	73.9	67.9	55.8	44.8	36.2	54.4
BARBOURVILLE	MAX	47.2	50.8	60.0	71.7	79.4	85.5	88.3	87.4	82.6	72.0	60.0	50.7	69.6
	MIN	23.7	25.4	33.1	42.2	50.7	59.2	63.1	62.4	55.8	42.3	33.2	26.8	43.2
	MEAN	35.5	38.1	46.6	57.0	65.1	72.4	75.7	74.9	69.2	57.2	46.6	38.7	56.4
BARDSTOWN	MAX	43.4	47.8	57.7	69.9	78.3	85.6	88.7	87.8	82.4	71.4	57.3	47.6	68.2
	MIN	23.7	26.1	34.8	44.7	52.8	60.8	64.6	63.0	56.5	44.0	34.9	28.1	44.5
	MEAN	33.6	36.9	46.3	57.3	65.6	73.2	76.6	75.4	69.4	57.7	46.1	37.9	56.3
BAXTER	MAX	44.1	47.9	57.1	69.0	77.1	83.3	86.4	85.6	80.4	69.4	57.2	47.5	67.1
	MIN	23.2	24.3	31.8	40.6	49.9	58.3	62.8	62.2	55.8	42.4	32.0	25.6	42.4
	MEAN	33.7	36.1	44.5	54.8	63.5	70.8	74.6	73.9	68.1	55.9	44.6	36.6	54.8
BEAVER DAM	MAX	43.7	48.8	58.4	70.8	79.0	86.1	89.1	88.1	82.4	71.9	58.1	47.9	68.7
	MIN	24.2	27.0	35.5	45.7	53.5	61.8	65.5	63.9	57.2	44.7	35.6	28.5	45.3
	MEAN	34.0	37.9	47.0	58.3	66.3	74.0	77.3	76.0	69.8	58.3	46.9	38.2	57.0
BEREA COLLEGE	MAX	44.0	47.9	57.5	69.2	77.1	84.2	87.3	86.7	81.4	70.2	57.4	48.2	67.6
	MIN	25.3	27.5	35.8	45.7	54.0	61.4	65.3	64.3	58.0	46.4	37.2	29.4	45.9
	MEAN	34.7	37.8	46.7	57.5	65.6	72.8	76.3	75.5	69.7	58.3	47.3	38.8	56.8
BOWLING GREEN FAA AP	MAX	43.1	47.6	57.1	69.3	78.2	86.0	89.2	88.3	82.3	70.8	56.9	47.6	68.0
	MIN	24.6	27.1	35.5	45.5	54.4	62.9	67.1	65.5	58.5	45.2	35.5	28.6	45.9
	MEAN	33.9	37.4	46.3	57.4	66.4	74.5	78.2	77.0	70.4	58.0	46.3	38.1	57.0
CAMPBELLSVILLE 2 SSW	MAX	44.1	48.5	58.1	69.6	77.7	84.9	87.9	87.2	81.9	71.1	57.7	48.3	68.1
	MIN	24.9	27.3	35.7	45.5	53.5	61.2	65.2	63.8	57.4	45.1	36.2	29.4	45.4
	MEAN	34.5	38.0	46.9	57.6	65.6	73.1	76.6	75.6	69.7	58.1	47.0	38.9	56.8
CARROLLTON LOCK 1	MAX	41.6	46.2	56.0	68.5	76.5	83.8	86.8	85.7	80.2	69.9	56.5	46.2	66.5
	MIN	22.9	25.2	33.8	43.7	52.1	60.9	64.9	63.5	56.7	44.4	35.3	27.8	44.3
	MEAN	32.3	35.7	44.9	56.2	64.4	72.4	75.9	74.6	68.5	57.1	45.9	37.0	55.4
COVINGTON WSO R	MAX	37.3	41.2	51.5	64.5	74.2	82.3	85.8	84.8	78.7	66.7	52.6	41.9	63.5
	MIN	20.4	23.0	32.0	42.4	51.7	60.5	64.9	63.3	56.3	43.9	34.1	25.7	43.2
	MEAN	28.9	32.1	41.8	53.5	63.0	71.4	75.4	74.1	67.5	55.3	43.4	33.8	53.4
DANVILLE	MAX	40.9	44.8	54.5	66.8	75.6	82.7	86.5	86.4	81.3	69.6	55.8	45.2	65.8
	MIN	22.2	24.0	32.7	43.2	52.3	60.4	64.2	62.8	56.3	44.0	34.7	26.5	43.6
	MEAN	31.6	34.4	43.6	55.0	64.0	71.6	75.4	74.6	68.9	56.8	45.3	35.9	54.8
FARMERS 1 WNW	MAX	43.4	47.3	57.1	69.3	77.5	84.2	87.2	86.4	80.9	70.1	57.4	47.6	67.4
	MIN	22.4	23.9	32.7	41.8	50.2	58.3	62.5	61.5	54.7	42.2	33.2	26.6	42.5
	MEAN	33.0	35.6	44.9	55.5	63.9	71.3	74.9	74.0	67.8	56.2	45.3	37.1	55.0
FORDS FERRY DAM 50	MAX	42.1	46.3	56.2	69.0	77.8	86.2	89.3	88.0	81.7	70.4	56.7	46.4	67.5
	MIN	24.4	27.7	36.0	47.1	55.6	64.1	68.0	66.6	60.1	47.8	37.4	29.3	47.0
	MEAN	33.3	37.0	46.1	58.1	66.7	75.2	78.7	77.3	70.9	59.1	47.1	37.9	57.3
FRANKFORT LOCK 4	MAX	40.9	44.3	54.1	66.4	75.7	83.5	87.5	86.7	81.3	69.7	56.2	45.7	66.0
	MIN	20.9	22.5	31.0	40.7	50.3	59.4	63.5	62.2	55.3	42.1	33.0	25.4	42.2
	MEAN	30.9	33.4	42.5	53.5	63.0	71.5	75.6	74.5	68.3	55.9	44.6	35.6	54.1
GREENSBURG	MAX	43.4	47.5	57.0	69.4	78.2	85.5	89.1	88.4	83.0	71.7	58.1	47.9	68.3
	MIN	22.4	24.1	32.9	43.0	51.9	60.7	64.8	63.2	56.1	42.3	33.4	26.4	43.4
	MEAN	32.9	35.8	45.0	56.2	65.1	73.1	77.0	75.8	69.6	57.1	45.8	37.2	55.9
HENDERSON 7 SSW	MAX	41.7	46.4	56.2	68.9	77.6	85.8	88.9	88.0	82.2	71.4	56.8	45.9	67.5
	MIN	24.3	27.4	36.2	46.8	55.2	63.3	67.0	64.8	58.3	46.6	37.1	29.1	46.3
	MEAN	33.0	36.9	46.2	57.9	66.4	74.5	77.9	76.4	70.3	59.0	47.0	37.5	56.9

KENTUCKY

TEMPERATURE NORMALS (DEG F)

STATION		JAN	FEB	MAR	APR	MAY	JUN	JUL	AUG	SEP	OCT	NOV	DEC	ANN
HICKMAN 1 E	MAX	44.3	49.5	58.8	70.8	79.5	87.8	90.6	89.7	83.8	73.8	59.3	48.6	69.7
	MIN	25.3	28.5	36.8	47.6	56.1	63.7	67.2	65.1	57.9	45.5	36.1	29.2	46.6
	MEAN	34.8	39.0	47.8	59.3	67.8	75.8	79.0	77.4	70.8	59.7	47.7	38.9	58.2
HOPKINSVILLE	MAX	43.2	47.8	57.3	69.8	78.5	86.5	90.0	89.5	83.6	72.6	58.4	47.5	68.7
	MIN	23.1	25.6	34.4	45.6	54.3	62.6	66.1	64.6	57.8	44.6	34.8	27.5	45.1
	MEAN	33.2	36.7	45.9	57.8	66.4	74.6	78.1	77.0	70.7	58.6	46.6	37.5	56.9
JACKSON WSO AP	MAX	41.4	43.7	52.4	67.2	75.8	82.3	85.2	84.3	78.8	68.9	55.3	44.8	65.0
	MIN	21.3	22.4	28.9	39.2	48.5	55.3	61.3	60.1	52.6	40.4	29.6	22.8	40.2
	MEAN	31.4	33.1	40.7	53.2	62.2	68.8	73.3	72.2	65.7	54.7	42.5	33.8	52.6
LEITCHFIELD 2 N	MAX	42.4	46.8	56.6	68.9	77.3	84.5	87.9	87.1	81.0	70.1	56.5	46.6	67.1
	MIN	23.9	26.2	34.9	45.1	53.5	61.8	65.7	64.4	57.8	45.5	35.6	28.3	45.2
	MEAN	33.2	36.5	45.7	57.0	65.4	73.2	76.8	75.8	69.4	57.8	46.1	37.5	56.2
LEXINGTON WSO R	MAX	39.8	43.7	53.7	65.8	74.9	82.6	85.9	85.0	79.3	67.6	54.1	44.4	64.7
	MIN	23.1	25.4	34.1	44.3	53.6	61.8	65.9	64.8	58.1	45.9	35.7	27.8	45.0
	MEAN	31.5	34.6	43.9	55.1	64.2	72.2	76.0	74.9	68.7	56.8	44.9	36.1	54.9
LOUISVILLE WSO R	MAX	40.8	45.0	54.9	67.5	76.2	84.0	87.6	86.7	80.6	69.2	55.5	45.4	66.1
	MIN	24.1	26.8	35.2	45.6	54.6	63.3	67.5	66.1	59.1	46.2	36.6	28.9	46.2
	MEAN	32.5	35.9	45.1	56.6	65.4	73.7	77.6	76.4	69.9	57.7	46.1	37.2	56.2
LOVELACEVILLE	MAX	44.0	49.0	58.4	70.8	79.0	86.8	89.9	89.0	83.0	73.1	58.7	48.1	69.2
	MIN	24.5	27.7	35.8	46.7	54.8	62.9	66.7	64.8	57.9	45.2	35.7	28.7	46.0
	MEAN	34.3	38.4	47.2	58.8	66.9	74.9	78.4	76.9	70.5	59.2	47.2	38.4	57.6
MADISONVILLE 1 SE	MAX	43.8	48.6	58.2	70.8	78.8	86.2	89.4	88.6	82.8	72.1	58.2	48.1	68.8
	MIN	24.6	27.3	35.8	46.6	54.5	62.4	66.4	64.4	57.6	45.5	36.1	28.9	45.8
	MEAN	34.2	38.0	47.1	58.7	66.7	74.4	77.9	76.5	70.2	58.8	47.2	38.5	57.4
MAMMOTH CAVE PARK	MAX	44.2	49.1	58.8	70.9	78.2	84.7	87.8	87.1	82.0	71.3	57.8	48.4	68.4
	MIN	24.7	26.9	35.1	45.0	52.5	60.1	63.9	62.8	56.7	44.7	35.7	28.8	44.7
	MEAN	34.4	38.0	47.0	58.0	65.4	72.4	75.9	75.0	69.4	58.0	46.8	38.6	56.6
MANCHESTER 4 SE	MAX	46.9	50.6	60.1	71.7	78.4	84.2	87.0	86.1	81.4	71.5	59.7	50.5	69.0
	MIN	24.1	25.7	33.4	41.8	50.6	58.5	63.2	62.2	55.8	42.6	33.4	27.3	43.2
	MEAN	35.5	38.1	46.8	56.8	64.5	71.4	75.1	74.2	68.6	57.1	46.5	38.9	56.1
MAYFIELD RADIO WNGO	MAX	44.5	49.5	59.0	71.0	79.2	86.9	89.6	88.9	83.2	72.9	59.0	48.6	69.4
	MIN	25.6	28.4	36.6	46.8	55.2	63.2	67.2	65.4	58.6	46.4	36.8	29.6	46.7
	MEAN	35.1	39.0	47.8	58.9	67.2	75.1	78.4	77.2	70.9	59.7	47.9	39.1	58.0
MAYSVILLE SEWAGE PLANT	MAX	39.9	43.3	53.3	65.7	75.1	82.8	86.4	85.2	79.9	68.3	55.2	44.4	65.0
	MIN	20.3	21.7	30.2	40.0	48.8	58.1	62.5	61.5	54.7	42.2	32.8	24.8	41.5
	MEAN	30.1	32.5	41.8	52.9	62.0	70.5	74.5	73.4	67.3	55.3	44.0	34.6	53.2
MIDDLESBORO	MAX	45.6	49.5	58.7	70.3	77.8	84.0	87.2	86.5	81.3	70.4	58.4	48.8	68.2
	MIN	24.5	25.8	33.3	42.0	50.3	58.3	62.7	62.2	55.8	42.6	33.1	27.2	43.2
	MEAN	35.1	37.7	46.0	56.2	64.0	71.2	75.0	74.4	68.6	56.5	45.8	38.0	55.7
MOUNT STERLING	MAX	42.7	46.8	56.9	69.1	76.9	83.2	86.2	85.4	80.0	69.6	56.5	46.7	66.7
	MIN	23.9	25.9	34.5	44.1	53.1	60.8	64.7	63.5	57.2	45.2	35.3	28.0	44.7
	MEAN	33.3	36.4	45.8	56.6	65.0	72.0	75.5	74.4	68.6	57.5	45.9	37.4	55.7
MURRAY	MAX	44.3	49.1	58.6	70.7	78.7	86.6	89.6	89.0	82.7	72.1	58.7	48.6	69.1
	MIN	25.9	28.9	37.4	47.9	56.4	64.3	68.0	66.2	59.4	47.1	37.3	30.1	47.4
	MEAN	35.1	39.0	48.0	59.3	67.6	75.5	78.8	77.6	71.1	59.6	48.0	39.4	58.3
OWENSBORO 2 W	MAX	42.0	46.8	56.9	69.5	78.4	86.8	89.8	88.9	83.3	72.3	57.5	46.5	68.2
	MIN	24.4	27.5	35.9	46.4	55.0	63.1	66.7	64.5	58.0	45.6	36.2	29.0	46.0
	MEAN	33.2	37.1	46.4	58.0	66.8	75.0	78.3	76.7	70.7	59.0	46.9	37.8	57.2
PADUCAH SEWAGE PLANT	MAX	43.6	47.9	57.7	70.5	79.0	87.2	90.6	89.2	83.3	72.6	58.6	47.8	69.0
	MIN	24.9	27.7	36.4	47.4	55.6	63.9	68.0	65.8	58.8	46.3	36.7	29.1	46.7
	MEAN	34.3	37.8	47.1	59.0	67.3	75.6	79.3	77.5	71.1	59.5	47.7	38.4	57.9

KENTUCKY

TEMPERATURE NORMALS (DEG F)

STATION		JAN	FEB	MAR	APR	MAY	JUN	JUL	AUG	SEP	OCT	NOV	DEC	ANN
PRINCETON 1 SE	MAX	44.5	49.3	58.9	71.0	78.9	86.6	89.9	89.1	83.2	72.8	59.0	48.7	69.3
	MIN	24.2	27.2	35.5	46.0	54.2	62.5	66.3	64.7	57.8	45.2	35.7	28.6	45.7
	MEAN	34.4	38.3	47.2	58.5	66.6	74.6	78.2	76.9	70.5	59.0	47.4	38.6	57.5
RUSSELLVILLE	MAX	44.1	48.2	57.7	70.1	78.6	86.2	89.4	88.6	82.7	71.4	58.0	48.4	68.6
	MIN	23.9	26.3	35.2	45.6	53.5	61.6	65.4	63.9	57.0	44.0	35.3	28.2	45.0
	MEAN	34.0	37.3	46.5	57.9	66.1	73.9	77.4	76.3	69.8	57.7	46.7	38.3	56.8
SCOTTSVILLE 3 SSW	MAX	44.9	49.7	59.1	70.8	77.8	84.9	87.9	87.5	82.1	71.6	58.3	48.9	68.6
	MIN	26.7	29.3	37.5	47.5	55.4	62.9	66.6	65.4	59.4	47.5	37.7	30.8	47.2
	MEAN	35.8	39.5	48.3	59.2	66.6	73.9	77.3	76.5	70.7	59.6	48.0	39.9	57.9
SHELBYVILLE	MAX	40.1	44.2	54.0	66.3	75.2	82.8	86.6	85.8	80.3	68.8	55.1	44.5	65.3
	MIN	21.1	23.1	32.1	42.6	51.6	60.2	64.0	62.8	56.1	43.4	33.9	25.9	43.1
	MEAN	30.6	33.7	43.1	54.5	63.4	71.5	75.3	74.3	68.3	56.1	44.5	35.2	54.2
SOMERSET 2 N	MAX	44.2	48.3	57.7	69.0	76.8	83.4	86.5	85.9	80.5	69.8	57.1	48.1	67.3
	MIN	24.4	26.4	34.4	43.7	51.9	59.7	63.6	62.3	56.2	43.3	34.5	28.1	44.0
	MEAN	34.3	37.3	46.1	56.3	64.3	71.6	75.0	74.1	68.4	56.6	45.8	38.1	55.7
SUMMER SHADE	MAX	44.1	48.6	57.8	69.6	77.5	84.5	87.7	86.9	81.0	69.8	57.5	48.2	67.8
	MIN	24.8	27.0	35.1	44.9	53.4	61.2	64.9	63.5	57.5	44.6	34.9	28.5	45.0
	MEAN	34.5	37.8	46.5	57.3	65.5	72.9	76.3	75.2	69.2	57.2	46.2	38.4	56.4
VANCEBURG	MAX	41.8	46.0	56.3	69.1	77.4	84.2	87.3	85.8	80.2	69.4	56.3	45.9	66.6
	MIN	22.2	23.8	32.1	41.1	49.8	57.8	61.6	60.8	54.3	42.0	33.1	26.2	42.1
	MEAN	32.0	34.9	44.2	55.1	63.6	71.0	74.5	73.3	67.3	55.7	44.8	36.1	54.4
WEST LIBERTY	MAX	42.1	45.9	55.9	67.9	76.7	83.3	86.6	85.9	80.5	69.8	56.8	46.3	66.5
	MIN	20.2	21.0	29.7	39.2	48.8	57.1	61.7	60.2	52.8	39.1	30.1	23.5	40.3
	MEAN	31.2	33.5	42.8	53.6	62.8	70.2	74.2	73.0	66.7	54.5	43.5	34.9	53.4
WILLIAMSBURG 1 E	MAX	46.4	50.3	59.5	71.4	78.9	84.8	87.6	86.9	81.9	71.0	59.0	49.7	69.0
	MIN	25.1	26.2	34.0	42.9	51.6	59.5	63.7	62.9	56.6	43.7	34.5	28.2	44.1
	MEAN	35.8	38.3	46.8	57.2	65.3	72.2	75.7	74.9	69.3	57.4	46.8	39.0	56.6
WILLIAMSTOWN 3 NW	MAX	39.2	43.2	53.2	65.4	74.6	82.1	85.7	85.0	79.4	68.1	54.2	43.7	64.5
	MIN	20.7	23.1	32.0	42.8	52.2	60.5	64.6	63.1	56.8	45.0	34.7	25.7	43.4
	MEAN	30.0	33.2	42.6	54.1	63.4	71.3	75.2	74.1	68.1	56.6	44.5	34.7	54.0

KENTUCKY

PRECIPITATION NORMALS (INCHES)

STATION	JAN	FEB	MAR	APR	MAY	JUN	JUL	AUG	SEP	OCT	NOV	DEC	ANN
ANCHORAGE	3.60	3.34	4.96	4.30	4.54	4.10	4.97	3.43	3.32	2.82	3.77	3.77	46.92
ASHLAND	3.25	2.85	3.95	3.54	3.98	3.74	4.68	3.77	2.99	2.37	2.81	3.23	41.16
BARBOURVILLE	4.58	4.10	5.25	4.40	4.26	4.66	5.17	3.58	3.51	2.82	3.92	4.37	50.62
BARDSTOWN	3.65	3.34	4.93	4.48	4.65	4.33	5.22	3.41	3.56	2.65	3.55	4.18	47.95
BAXTER	4.57	3.90	5.29	4.17	4.21	4.42	4.95	4.00	3.30	2.85	3.94	4.30	49.90
BEAVER DAM	3.85	3.55	4.61	4.05	4.61	3.60	4.19	3.00	3.49	2.42	3.84	3.85	45.06
BEREA COLLEGE	3.80	3.33	4.61	4.53	4.17	4.28	4.81	4.02	3.74	2.31	3.43	3.95	46.98
BEREA	4.08	3.65	4.92	4.64	4.53	4.37	4.96	4.23	3.84	2.39	3.75	4.18	49.54
BOWLING GREEN	4.64	4.29	5.76	4.26	4.46	4.49	4.43	3.14	3.42	2.79	4.11	4.82	50.61
BOWLING GREEN FAA AP	4.59	3.98	5.52	4.18	4.16	4.53	4.33	3.33	3.18	2.82	3.87	4.53	49.02
CALHOUN LOCK 2	3.53	3.32	4.71	4.58	4.06	3.58	4.09	3.30	2.81	2.28	3.78	3.72	43.76
CAMPBELLSVILLE 2 SSW	4.37	3.70	5.10	4.63	4.41	4.60	5.28	4.21	3.95	2.70	3.76	4.36	51.07
CARROLLTON LOCK 1	3.04	3.09	4.30	3.72	3.90	3.84	3.91	3.17	3.12	2.60	3.16	3.00	40.85
COLLEGE HILL LOCK 11	3.44	3.17	4.35	4.23	4.06	3.95	5.09	3.88	3.48	2.20	3.32	3.67	44.84
COVINGTON WSO R	3.13	2.73	3.95	3.58	3.84	4.09	4.28	2.97	2.91	2.54	3.12	3.00	40.14
DANVILLE	3.91	3.61	4.84	4.48	4.20	4.28	4.99	3.73	3.72	2.41	3.50	4.04	47.71
DUNDEE	3.96	3.71	4.77	4.42	4.50	4.05	4.34	3.35	3.26	2.61	4.10	3.97	47.04
DUNMOR	4.34	3.63	5.22	4.32	4.48	4.05	4.10	3.52	3.52	2.80	4.19	4.12	48.29
FALMOUTH 5 WNW	3.18	2.90	4.43	3.99	4.02	3.96	4.86	3.81	3.64	2.61	3.44	3.23	44.07
FARMERS 1 WNW	3.45	3.24	4.45	3.97	4.32	3.90	5.30	4.20	3.47	2.46	3.09	3.54	45.39
FLEMINGSBURG	3.47	3.13	4.71	3.77	4.36	3.89	4.95	4.10	3.48	2.32	3.14	3.50	44.82
FORD LOCK 10	3.47	3.21	4.34	3.92	4.30	3.96	5.00	3.89	3.54	2.21	3.23	3.73	44.80
FORDS FERRY DAM 50	3.31	3.00	4.56	4.24	4.28	3.81	3.28	3.16	2.78	2.33	3.75	3.67	42.17
FRANKFORT LOCK 4	3.27	2.96	4.51	3.94	4.17	3.76	4.40	3.28	3.56	2.44	3.27	3.37	42.93
FRANKLIN 1 E	4.26	4.00	5.17	4.25	4.34	4.08	4.07	3.15	3.27	2.70	3.92	4.35	47.56
GEORGETOWN WATER WORKS	3.47	3.02	4.67	3.99	4.11	3.78	4.23	3.59	3.28	2.34	3.25	3.57	43.30
GEST LOCK 3	3.05	2.83	4.46	4.02	4.33	3.77	4.69	3.60	3.32	2.49	3.20	3.32	43.08
GLASGOW	4.67	4.17	5.36	4.43	4.48	4.62	4.97	3.52	3.58	2.68	4.34	4.58	51.40
GREENSBURG	4.46	3.96	5.12	4.28	4.81	4.89	4.85	4.33	3.93	2.60	3.88	4.48	51.59
GREENSBURG HIWAY 61	4.19	3.74	5.00	4.15	4.44	4.49	4.60	4.04	3.81	2.46	3.75	4.24	48.91
HARTFORD 6 NW	3.55	3.44	4.73	4.56	4.41	3.64	3.84	3.71	3.16	2.49	3.99	3.76	45.28
HAZARD WATER WORKS	4.27	3.73	4.87	4.20	3.98	4.58	5.30	3.82	3.77	3.04	3.55	3.97	49.08
HEIDELBERG LOCK 14	3.87	3.54	4.76	4.24	4.06	4.04	5.28	3.87	3.38	2.32	3.46	3.78	46.60
HENDERSON 7 SSW	3.38	3.14	4.78	4.23	4.22	3.81	4.14	3.24	3.20	2.44	3.66	3.49	43.73
HICKMAN 1 E	4.44	4.30	5.44	5.20	5.25	3.69	4.07	3.65	3.61	2.48	4.46	4.82	51.41
HIGH BRIDGE LOCK 7	3.76	3.46	4.80	4.38	4.25	4.36	4.85	3.60	3.63	2.27	3.46	3.89	46.71
HOPKINSVILLE	4.41	3.96	5.62	4.49	4.48	3.72	4.22	3.38	3.28	2.64	4.37	4.59	49.16
JACKSON	4.07	3.48	4.68	4.19	3.84	3.70	5.16	3.78	3.34	2.19	3.27	3.90	45.60
JACKSON WSO AP	3.93	3.69	4.62	4.00	3.52	3.78	4.90	3.63	3.10	2.11	3.18	3.53	43.99
JEREMIAH	3.70	3.25	4.61	3.47	3.53	3.57	4.17	3.36	3.02	2.27	3.17	3.44	41.56
KEENE 1 SSW	3.70	3.38	4.91	4.19	4.53	4.28	4.81	3.99	3.52	2.34	3.43	4.05	47.13
LANCASTER	4.08	3.86	5.01	4.47	4.37	4.43	5.10	3.88	3.85	2.36	3.69	4.31	49.41
LEITCHFIELD 2 N	3.86	3.77	4.98	4.25	4.36	4.14	4.57	3.37	3.49	2.80	4.03	4.26	47.88
LEXINGTON WSO R	3.57	3.26	4.83	4.01	4.23	4.25	4.95	3.96	3.28	2.26	3.30	3.78	45.68
LIBERTY	4.78	4.04	5.36	4.30	4.56	4.77	5.03	4.26	4.13	2.74	3.95	4.65	52.57
LITTLE HICKMAN LOCK 8	3.66	3.38	4.59	4.39	4.15	3.93	4.63	3.94	3.56	2.24	3.27	3.79	45.53
LLOYD GREENUP DAM	3.02	2.79	3.95	3.75	3.86	3.90	4.23	3.82	3.34	2.32	2.68	2.98	40.64
LOCKPORT LOCK 2	3.27	3.14	4.86	4.32	4.55	4.02	4.33	3.90	3.50	2.70	3.47	3.49	45.55
LOUISVILLE WSO R	3.38	3.23	4.73	4.11	4.15	3.60	4.10	3.31	3.35	2.63	3.49	3.48	43.56
LOUISVILLE UPPER GAGE	3.33	3.17	4.66	4.05	4.25	3.67	4.45	3.27	3.18	2.66	3.51	3.46	43.66
LOVELACEVILLE	4.00	3.55	5.35	4.87	4.67	4.24	4.15	2.97	3.66	2.71	4.21	4.29	48.67
MADISONVILLE 1 SE	3.79	3.70	4.87	4.67	4.21	3.51	4.34	3.64	3.24	2.52	3.95	3.84	46.28
MAMMOTH CAVE PARK	4.79	4.04	5.60	4.45	4.38	4.73	4.32	3.55	3.74	2.83	4.33	4.82	51.58
MANCHESTER 4 SE	4.46	3.86	4.99	4.45	4.42	4.36	5.45	3.58	3.65	2.79	3.75	4.19	49.95
MAYFIELD RADIO WNGO	4.29	4.24	5.70	5.03	4.69	3.44	4.62	3.49	4.12	2.84	4.48	4.78	51.72

KENTUCKY

PRECIPITATION NORMALS (INCHES)

STATION	JAN	FEB	MAR	APR	MAY	JUN	JUL	AUG	SEP	OCT	NOV	DEC	ANN
MAYSVILLE SEWAGE PLANT	3.54	3.04	4.53	3.94	4.18	3.59	4.56	4.08	3.22	2.49	3.30	3.50	43.97
MIDDLESBORO	4.95	4.16	5.82	4.33	4.35	4.65	4.81	3.85	3.00	3.02	4.31	5.08	52.33
MILLERSTOWN	4.19	3.71	5.04	4.40	4.37	4.29	4.65	3.33	3.72	2.85	4.00	4.45	49.00
MONTICELLO	4.58	4.00	5.33	4.49	4.33	4.27	4.80	3.55	3.90	2.56	3.77	4.38	49.96
MOUNT STERLING	4.07	3.56	4.78	4.07	4.22	4.01	5.11	4.04	3.61	2.26	3.11	3.88	46.72
MUNFORDVILLE	4.65	4.11	5.51	4.52	4.66	4.64	5.40	3.77	3.81	2.74	4.25	4.65	52.71
MURRAY	4.32	4.09	5.59	4.95	4.41	3.89	4.48	3.15	3.87	2.86	4.41	4.45	50.47
OWENSBORO 2 W	3.62	3.40	4.75	4.56	4.49	3.78	3.63	3.45	3.14	2.36	3.90	3.71	44.79
PADUCAH SEWAGE PLANT	3.55	3.13	4.82	4.42	4.54	4.26	3.95	3.30	3.36	2.49	4.06	3.91	45.79
PAINTSVILLE 1 S	3.60	3.34	4.26	3.93	4.13	3.81	5.00	3.94	3.24	2.47	3.00	3.65	44.37
PRINCETON 1 SE	4.53	3.91	5.05	4.61	4.25	3.38	3.90	3.60	3.30	2.45	3.96	4.37	47.31
RAVENNA LOCK 12	3.44	3.09	4.26	4.18	4.29	4.26	5.57	4.07	3.67	2.35	3.48	3.66	46.32
RUSSELLVILLE	4.91	4.31	6.04	4.62	4.57	4.50	4.45	3.38	3.90	2.90	4.46	4.73	52.77
SALVISA LOCK 6	3.92	3.64	5.11	4.36	4.71	4.30	4.90	4.09	3.65	2.58	3.62	4.12	49.00
SCOTTSVILLE 3 SSW	4.68	4.20	5.50	4.55	4.73	4.75	4.38	3.57	3.49	2.69	4.37	4.84	51.75
SEBREE	3.72	3.42	4.82	4.49	4.24	3.59	4.18	3.80	3.06	2.65	3.98	3.71	45.66
SHELBYVILLE	3.43	3.07	4.67	4.07	4.38	3.81	4.78	3.76	3.42	2.70	3.55	3.38	45.02
SHEPHERDSVILLE	3.43	3.15	4.46	4.15	4.09	4.10	4.68	3.61	3.26	2.78	3.59	3.87	45.17
SOMERSET 2 N	4.59	4.03	5.06	4.46	4.37	4.56	4.89	3.76	3.79	2.58	3.71	4.61	50.41
SPRINGFIELD	4.01	3.34	4.53	4.22	4.27	4.47	4.92	3.88	3.57	2.61	3.51	3.80	47.13
SUMMER SHADE	4.62	3.95	5.49	4.44	4.35	4.60	4.82	3.17	3.94	2.44	4.12	4.72	50.66
TAYLORSVILLE	3.45	3.27	4.72	4.23	4.30	4.35	5.06	3.32	3.50	2.64	3.58	3.71	46.13
TYRONE LOCK 5	3.56	3.38	4.91	4.40	4.49	4.23	4.81	4.18	4.10	2.83	3.59	3.75	48.23
VALLEY VIEW LOCK 9	3.47	3.23	4.60	4.28	4.27	3.77	4.93	4.43	3.51	2.28	3.24	3.73	45.74
VANCEBURG	3.29	3.20	4.51	3.88	4.14	3.84	5.01	4.26	3.32	2.40	3.11	3.21	44.17
WAYNESBURG 7 NE	4.47	3.90	5.10	4.58	4.35	4.64	5.20	3.76	3.98	2.43	3.90	4.62	50.93
WEST LIBERTY	3.69	3.32	4.42	4.01	4.12	3.69	5.37	3.67	3.32	2.47	3.18	3.62	44.88
WILLIAMSBURG 1 E	4.44	4.08	5.28	4.33	4.29	4.26	4.70	3.80	3.55	2.83	4.02	4.20	49.78
WILLIAMSTOWN 3 NW	3.01	2.87	4.70	4.01	4.12	3.73	4.59	3.58	3.47	2.62	3.52	3.29	43.51
WILLOW LOCK 13	3.93	3.68	4.69	4.28	4.31	4.03	5.29	4.13	3.54	2.31	3.48	3.91	47.58
WOODBURY LOCK 4	4.17	3.88	4.99	4.15	4.21	4.40	4.19	3.18	3.53	2.74	3.98	4.21	47.63

KENTUCKY

HEATING DEGREE DAY NORMALS (BASE 65 DEG F)

STATION	JUL	AUG	SEP	OCT	NOV	DEC	JAN	FEB	MAR	APR	MAY	JUN	ANN
ANCHORAGE	0	0	48	279	600	896	1026	834	639	287	116	8	4733
ASHLAND	0	6	63	306	606	893	1023	854	660	327	141	21	4900
BARBOURVILLE	0	0	29	264	552	815	915	753	578	252	109	12	4279
BARDSTOWN	0	0	36	249	567	840	973	787	591	247	95	6	4391
BAXTER	0	0	41	299	612	880	970	809	636	310	124	11	4692
BEAVER DAM	0	0	33	231	543	831	961	759	570	218	78	0	4224
BEREA COLLEGE	0	0	31	235	531	812	939	762	580	245	94	5	4234
BOWLING GREEN FAA AP	0	0	28	240	561	834	964	773	589	242	78	0	4309
CAMPBELLSVILLE 2 SSW	0	0	35	240	540	809	946	756	571	237	90	0	4224
CARROLLTON LOCK 1	0	0	39	261	573	868	1014	820	629	273	109	6	4592
COVINGTON WSO R	0	0	52	316	648	967	1119	921	719	350	143	12	5247
DANVILLE	0	0	42	277	591	902	1035	857	663	307	125	13	4812
FARMERS 1 WNW	0	0	49	291	591	865	992	823	630	296	119	11	4667
FORDS FERRY DAM 50	0	0	22	219	537	840	983	784	596	227	80	0	4288
FRANKFORT LOCK 4	0	0	44	298	612	911	1057	885	698	345	137	14	5001
GREENSBURG	0	0	34	267	576	862	995	818	626	273	103	5	4559
HENDERSON 7 SSW	0	0	28	215	540	853	992	787	593	227	78	0	4313
HICKMAN 1 E	0	0	23	193	519	809	936	728	548	197	54	0	4007
HOPKINSVILLE	0	0	23	239	552	853	986	792	601	230	79	6	4361
JACKSON WSO AP	0	0	73	332	675	967	1042	893	753	354	154	26	5269
LEITCHFIELD 2 N	0	0	44	257	567	853	986	798	606	258	100	7	4476
LEXINGTON WSO R	0	0	47	280	603	896	1039	851	661	306	121	10	4814
LOUISVILLE WSO R	0	0	32	250	567	862	1008	815	624	264	98	5	4525
LOVELACEVILLE	0	0	23	204	534	825	952	745	563	205	70	0	4121
MADISONVILLE 1 SE	0	0	26	217	534	822	955	756	568	208	70	0	4156
MAMMOTH CAVE PARK	0	0	30	241	546	818	949	756	572	227	91	10	4240
MANCHESTER 4 SE	0	0	34	262	555	809	915	753	572	255	105	8	4268
MAYFIELD RADIO WNGO	0	0	23	195	513	803	927	728	546	206	63	5	4009
MAYSVILLE SEWAGE PLANT	0	0	57	314	630	942	1082	910	719	363	158	16	5191
MIDDLESBORO	0	0	37	283	576	837	927	764	589	272	125	14	4424
MOUNT STERLING	0	0	41	254	573	856	983	801	603	265	99	6	4481
MURRAY	0	0	22	199	510	794	927	728	543	192	59	0	3974
OWENSBORO 2 W	0	0	20	218	543	843	986	781	587	226	75	0	4279
PADUCAH SEWAGE PLANT	0	0	17	208	519	825	952	762	567	209	71	0	4130
PRINCETON 1 SE	0	0	24	215	528	818	949	748	564	217	69	0	4132
RUSSELLVILLE	0	0	34	249	549	828	961	776	583	229	91	10	4310
SCOTTSVILLE 3 SSW	0	0	23	202	510	778	905	714	534	194	77	6	3943
SHELBYVILLE	0	0	52	302	615	924	1066	876	679	324	140	14	4992
SOMERSET 2 N	0	0	42	275	576	834	952	776	595	269	109	7	4435
SUMMER SHADE	0	0	35	261	564	825	946	762	583	245	95	8	4324
VANCEBURG	0	0	71	306	606	896	1023	843	645	304	135	15	4844
WEST LIBERTY	0	0	63	338	645	933	1048	882	688	346	151	18	5112
WILLIAMSBURG 1 E	0	0	34	267	546	806	905	748	572	249	103	9	4239
WILLIAMSTOWN 3 NW	0	0	47	277	615	939	1085	890	694	333	134	13	5027

KENTUCKY

COOLING DEGREE DAY NORMALS (BASE 65 DEG F)

STATION		JAN	FEB	MAR	APR	MAY	JUN	JUL	AUG	SEP	OCT	NOV	DEC	ANN
ANCHORAGE		0	0	7	11	91	218	326	288	144	19	0	0	1104
ASHLAND		0	0	0	12	91	204	311	282	150	21	0	0	1071
BARBOURVILLE		0	0	8	12	112	234	332	307	155	22	0	0	1182
BARDSTOWN		0	0	11	16	114	252	360	322	168	23	0	0	1266
BAXTER		0	0	0	0	78	185	298	276	134	17	0	0	988
BEAVER DAM		0	0	12	17	118	275	381	341	177	24	0	0	1345
BEREA COLLEGE		0	0	13	20	112	239	350	326	172	28	0	0	1260
BOWLING GREEN FAA AP		0	0	10	14	121	290	409	372	190	23	0	0	1429
CAMPBELLSVILLE 2 SSW		0	0	10	15	108	248	360	329	176	26	0	0	1272
CARROLLTON LOCK 1		0	0	6	9	90	228	338	298	144	16	0	0	1129
COVINGTON WSO	R	0	0	0	5	81	204	322	282	127	16	0	0	1037
DANVILLE		0	0	0	7	94	211	322	298	159	23	0	0	1114
FARMERS 1 WNW		0	0	7	11	85	200	307	279	133	19	0	0	1041
FORDS FERRY DAM 50		0	0	11	20	133	310	425	381	199	36	0	0	1515
FRANKFORT LOCK 4		0	0	0	0	75	209	329	295	143	16	0	0	1067
GREENSBURG		0	0	6	9	106	248	372	335	172	22	0	0	1270
HENDERSON 7 SSW		0	0	10	14	122	289	400	353	187	29	0	0	1404
HICKMAN 1 E		0	0	15	26	141	324	434	384	197	28	0	0	1549
HOPKINSVILLE		0	0	9	14	122	294	406	372	194	40	0	0	1451
JACKSON WSO AP		0	0	0	0	67	140	257	227	94	13	0	0	798
LEITCHFIELD 2 N		0	0	8	18	113	253	366	335	176	33	0	0	1302
LEXINGTON WSO	R	0	0	7	9	96	226	341	307	158	26	0	0	1170
LOUISVILLE WSO	R	0	0	7	12	110	266	391	353	179	24	0	0	1342
LOVELACEVILLE		0	0	11	19	129	302	415	369	188	24	0	0	1457
MADISONVILLE 1 SE		0	0	13	19	123	286	400	357	182	25	0	0	1405
MAMMOTH CAVE PARK		0	0	14	17	103	232	338	310	162	24	0	0	1200
MANCHESTER 4 SE		0	0	8	9	89	200	313	285	142	17	0	0	1063
MAYFIELD RADIO WNGO		0	0	13	23	131	308	415	378	200	31	0	0	1499
MAYSVILLE SEWAGE PLANT		0	0	0	0	65	181	295	264	126	13	0	0	944
MIDDLESBORO		0	0	0	8	94	200	310	291	145	19	0	0	1067
MOUNT STERLING		0	0	8	13	99	216	326	291	149	22	0	0	1124
MURRAY		0	0	16	21	140	319	428	391	205	32	0	0	1552
OWENSBORO 2 W		0	0	11	16	131	304	412	363	191	32	0	0	1460
PADUCAH SEWAGE PLANT		0	0	12	29	142	322	443	388	200	37	0	0	1573
PRINCETON 1 SE		0	0	12	22	118	293	409	369	189	29	0	0	1441
RUSSELLVILLE		0	0	9	16	125	277	384	350	178	23	0	0	1362
SCOTTSVILLE 3 SSW		0	0	16	20	127	273	381	357	194	35	0	0	1403
SHELBYVILLE		0	0	0	9	90	209	319	291	151	26	0	0	1095
SOMERSET 2 N		0	0	9	8	88	205	310	282	144	14	0	0	1060
SUMMER SHADE		0	0	10	14	110	245	350	316	161	19	0	0	1225
VANCEBURG		0	0	0	7	92	195	295	262	140	18	0	0	1009
WEST LIBERTY		0	0	0	0	82	174	285	252	114	13	0	0	920
WILLIAMSBURG 1 E		0	0	8	15	112	225	332	307	163	32	0	0	1194
WILLIAMSTOWN 3 NW		0	0	0	6	85	202	316	282	140	17	0	0	1048

15 – KENTUCKY

STATE-STATION NUMBER	STN TYP	NAME		LATITUDE DEG-MIN	LONGITUDE DEG-MIN	ELEVATION (FT)
15-0155	13	ANCHORAGE		N 3816	W 08532	730
15-0254	13	ASHLAND		N 3827	W 08236	555
15-0381	13	BARBOURVILLE		N 3652	W 08353	980
15-0397	13	BARDSTOWN		N 3748	W 08528	637
15-0450	13	BAXTER		N 3651	W 08320	1164
15-0490	13	BEAVER DAM		N 3725	W 08652	441
15-0619	13	BEREA COLLEGE		N 3734	W 08418	1070
15-0624	12	BEREA		N 3734	W 08419	1000
15-0904	12	BOWLING GREEN		N 3700	W 08626	536
15-0909	13	BOWLING GREEN FAA AP		N 3658	W 08626	535
15-1227	12	CALHOUN LOCK 2		N 3732	W 08716	400
15-1256	13	CAMPBELLSVILLE 2 SSW		N 3719	W 08522	780
15-1345	13	CARROLLTON LOCK 1		N 3841	W 08511	480
15-1704	12	COLLEGE HILL LOCK 11		N 3747	W 08406	611
15-1855	13	COVINGTON WSO	R	N 3904	W 08440	869
15-2040	13	DANVILLE		N 3739	W 08446	955
15-2358	12	DUNDEE		N 3733	W 08646	425
15-2366	12	DUNMOR		N 3705	W 08700	592
15-2775	12	FALMOUTH 5 WNW		N 3843	W 08425	715
15-2791	13	FARMERS 1 WNW		N 3809	W 08333	662
15-2903	12	FLEMINGSBURG		N 3825	W 08344	860
15-2953	12	FORD LOCK 10		N 3754	W 08416	660
15-2961	13	FORDS FERRY DAM 50		N 3728	W 08806	356
15-3028	13	FRANKFORT LOCK 4		N 3814	W 08452	504
15-3036	12	FRANKLIN 1 E		N 3643	W 08634	690
15-3194	12	GEORGETOWN WATER WORKS		N 3812	W 08433	888
15-3203	12	GEST LOCK 3		N 3825	W 08453	490
15-3241	12	GLASGOW		N 3700	W 08555	780
15-3430	13	GREENSBURG		N 3715	W 08530	630
15-3435	12	GREENSBURG HIWAY 61		N 3715	W 08530	562
15-3652	12	HARTFORD 6 NW		N 3732	W 08654	429
15-3714	12	HAZARD WATER WORKS		N 3715	W 08311	881
15-3741	12	HEIDELBERG LOCK 14		N 3733	W 08346	663
15-3762	13	HENDERSON 7 SSW		N 3745	W 08738	430
15-3816	13	HICKMAN 1 E		N 3634	W 08910	375
15-3837	12	HIGH BRIDGE LOCK 7		N 3749	W 08443	544
15-3994	13	HOPKINSVILLE		N 3651	W 08730	540
15-4196	12	JACKSON		N 3733	W 08323	745
15-4202	13	JACKSON WSO AP		N 3726	W 08319	1365
15-4255	12	JEREMIAH		N 3710	W 08256	1160
15-4369	12	KEENE 1 SSW		N 3757	W 08438	941
15-4620	12	LANCASTER		N 3737	W 08435	1032
15-4703	13	LEITCHFIELD 2 N		N 3731	W 08618	620
15-4746	13	LEXINGTON WSO	R	N 3802	W 08436	966
15-4755	12	LIBERTY		N 3721	W 08455	767
15-4825	12	LITTLE HICKMAN LOCK 8		N 3745	W 08435	521
15-4848	12	LLOYD GREENUP DAM		N 3839	W 08252	537
15-4857	12	LOCKPORT LOCK 2		N 3826	W 08458	485
15-4954	13	LOUISVILLE WSO	R	N 3811	W 08544	477
15-4955	12	LOUISVILLE UPPER GAGE		N 3817	W 08548	403

LEGEND
11 = TEMPERATURE ONLY
12 = PRECIPITATION ONLY
13 = TEMP. & PRECIP.

STATE-STATION NUMBER	STN TYP	NAME	LATITUDE DEG-MIN	LONGITUDE DEG-MIN	ELEVATION (FT)
15-4967	13	LOVELACEVILLE	N 3658	W 08850	370
15-5067	13	MADISONVILLE 1 SE	N 3719	W 08729	439
15-5097	13	MAMMOTH CAVE PARK	N 3711	W 08605	800
15-5111	13	MANCHESTER 4 SE	N 3706	W 08343	850
15-5233	13	MAYFIELD RADIO WNGO	N 3647	W 08838	380
15-5243	13	MAYSVILLE SEWAGE PLANT	N 3841	W 08347	515
15-5389	13	MIDDLESBORO	N 3636	W 08344	1175
15-5438	12	MILLERSTOWN	N 3727	W 08603	600
15-5524	12	MONTICELLO	N 3650	W 08450	928
15-5640	13	MOUNT STERLING	N 3804	W 08356	910
15-5684	12	MUNFORDVILLE	N 3716	W 08553	610
15-5694	13	MURRAY	N 3637	W 08819	540
15-6091	13	OWENSBORO 2 W	N 3746	W 08709	420
15-6117	13	PADUCAH SEWAGE PLANT	N 3706	W 08836	325
15-6136	12	PAINTSVILLE 1 S	N 3748	W 08249	620
15-6580	13	PRINCETON 1 SE	N 3707	W 08752	500
15-6679	12	RAVENNA LOCK 12	N 3740	W 08357	634
15-7049	13	RUSSELLVILLE	N 3652	W 08653	655
15-7121	12	SALVISA LOCK 6	N 3756	W 08449	490
15-7215	13	SCOTTSVILLE 3 SSW	N 3644	W 08613	850
15-7234	12	SEBREE	N 3736	W 08732	360
15-7324	13	SHELBYVILLE	N 3813	W 08516	841
15-7334	12	SHEPHERDSVILLE	N 3801	W 08543	450
15-7510	13	SOMERSET 2 N	N 3707	W 08437	1050
15-7604	12	SPRINGFIELD	N 3741	W 08514	738
15-7800	13	SUMMER SHADE	N 3653	W 08543	864
15-7948	12	TAYLORSVILLE	N 3802	W 08521	488
15-8179	12	TYRONE LOCK 5	N 3802	W 08451	524
15-8246	12	VALLEY VIEW LOCK 9	N 3751	W 08427	571
15-8259	13	VANCEBURG	N 3835	W 08320	520
15-8486	12	WAYNESBURG 7 NE	N 3725	W 08436	1215
15-8551	13	WEST LIBERTY	N 3755	W 08315	790
15-8709	13	WILLIAMSBURG 1 E	N 3644	W 08409	1100
15-8714	13	WILLIAMSTOWN 3 NW	N 3839	W 08437	920
15-8724	12	WILLOW LOCK 13	N 3736	W 08350	649
15-8824	12	WOODBURY LOCK 4	N 3711	W 08638	504

15 – KENTUCKY

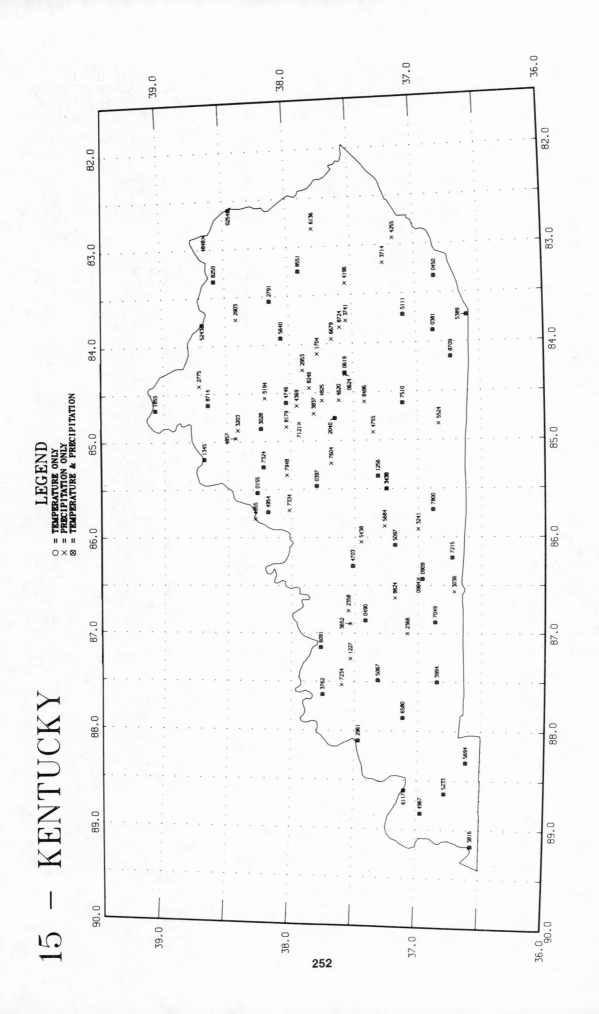

LEGEND

O = TEMPERATURE ONLY
X = PRECIPITATION ONLY
⊗ = TEMPERATURE & PRECIPITATION

LOUISIANA

TEMPERATURE NORMALS (DEG F)

STATION		JAN	FEB	MAR	APR	MAY	JUN	JUL	AUG	SEP	OCT	NOV	DEC	ANN
ALEXANDRIA	MAX	59.1	63.0	70.2	78.1	84.7	90.8	92.9	92.8	88.3	80.4	69.6	62.2	77.7
	MIN	37.7	40.1	47.4	56.1	63.1	69.5	72.0	71.3	66.4	54.1	45.3	39.6	55.2
	MEAN	48.4	51.6	58.8	67.1	73.9	80.1	82.5	82.1	77.4	67.2	57.5	50.9	66.5
AMITE	MAX	60.6	64.4	71.6	79.3	85.7	91.5	92.7	92.2	88.7	80.7	70.2	63.4	78.4
	MIN	38.2	40.4	47.3	55.5	62.2	67.9	70.8	70.4	66.5	54.1	45.5	39.8	54.9
	MEAN	49.4	52.4	59.5	67.4	74.0	79.7	81.8	81.3	77.6	67.4	57.9	51.6	66.7
BASTROP	MAX	56.3	61.1	69.1	78.0	85.1	91.7	94.6	94.0	88.6	79.8	67.8	59.8	77.2
	MIN	35.8	38.6	45.9	54.8	62.2	69.1	72.3	71.0	65.6	53.5	44.5	38.0	54.3
	MEAN	46.1	49.8	57.5	66.4	73.7	80.4	83.4	82.5	77.1	66.7	56.1	48.9	65.7
BATON ROUGE WSO R	MAX	61.1	64.5	71.6	79.2	85.2	90.6	91.4	90.8	87.4	80.1	70.1	63.8	78.0
	MIN	40.5	42.7	49.4	57.5	64.3	70.0	72.8	72.0	68.3	56.3	47.2	42.3	57.0
	MEAN	50.8	53.6	60.5	68.4	74.8	80.3	82.1	81.4	77.9	68.2	58.7	53.1	67.5
BELAH FIRE TOWER	MAX	57.1	61.6	69.2	77.8	84.3	91.0	93.6	93.4	89.0	80.2	68.6	60.7	77.2
	MIN	36.2	38.7	45.7	54.8	62.0	68.0	70.7	69.5	65.5	53.6	44.5	38.3	54.0
	MEAN	46.7	50.2	57.5	66.3	73.2	79.5	82.2	81.5	77.3	66.9	56.6	49.6	65.6
BOGALUSA	MAX	60.7	64.3	71.2	79.3	85.5	91.6	92.6	92.1	88.6	80.6	70.3	63.8	78.4
	MIN	38.1	40.1	47.0	55.6	62.8	69.0	71.6	70.8	66.6	53.7	45.2	39.5	55.0
	MEAN	49.4	52.2	59.1	67.5	74.1	80.3	82.1	81.5	77.6	67.2	57.7	51.7	66.7
BUNKIE	MAX	58.8	62.6	69.9	78.2	85.1	91.1	92.7	92.3	88.2	80.5	69.7	62.3	77.6
	MIN	37.9	40.2	47.2	55.7	62.8	69.1	71.7	70.4	65.4	53.3	45.2	39.5	54.9
	MEAN	48.4	51.4	58.5	67.0	74.0	80.1	82.2	81.4	76.8	66.9	57.4	50.9	66.3
CALHOUN EXP STATION	MAX	55.9	60.7	68.0	76.9	83.7	90.3	93.3	93.3	87.9	78.9	67.3	59.2	76.3
	MIN	34.4	36.9	43.8	52.9	60.6	67.7	71.0	69.5	64.0	51.2	42.4	36.5	52.6
	MEAN	45.2	48.8	55.9	64.9	72.2	79.0	82.2	81.5	76.0	65.1	54.9	47.9	64.5
CARVILLE 2 SW	MAX	62.2	65.3	71.9	79.2	85.5	90.7	91.8	91.5	88.2	80.8	71.0	64.7	78.6
	MIN	41.2	43.0	49.6	57.1	64.1	69.8	72.4	71.9	68.5	57.3	48.9	43.3	57.3
	MEAN	51.7	54.2	60.8	68.2	74.8	80.3	82.1	81.7	78.4	69.1	60.0	54.0	67.9
COTTON VALLEY	MAX	55.5	60.4	67.7	76.5	83.1	89.6	93.4	93.3	87.7	78.3	66.3	58.6	75.9
	MIN	33.8	36.2	42.7	52.1	59.7	67.1	70.6	69.2	63.4	50.4	41.2	35.5	51.8
	MEAN	44.7	48.3	55.2	64.3	71.4	78.4	82.0	81.3	75.6	64.4	53.8	47.1	63.9
COVINGTON 4 NNW	MAX	62.6	66.0	72.2	79.4	85.6	90.9	92.1	91.6	88.2	80.7	70.8	64.9	78.8
	MIN	40.0	41.4	47.9	55.4	61.9	67.6	70.5	70.1	66.7	54.5	46.1	41.4	55.3
	MEAN	51.3	53.7	60.1	67.4	73.8	79.3	81.3	80.9	77.5	67.7	58.5	53.2	67.1
DONALDSONVILLE 3 E	MAX	63.5	66.2	73.2	80.7	86.6	91.7	92.7	92.3	89.0	81.6	72.3	66.0	79.7
	MIN	41.5	43.8	50.2	58.2	64.3	70.2	72.3	72.0	68.6	57.3	48.6	43.6	57.6
	MEAN	52.5	55.0	61.7	69.5	75.5	81.0	82.5	82.2	78.8	69.5	60.5	54.8	68.6
ELIZABETH	MAX	59.9	63.7	70.9	78.6	85.2	91.1	93.5	93.2	88.8	81.3	70.4	63.3	78.3
	MIN	37.5	39.8	46.8	55.2	62.1	67.7	70.4	69.8	65.5	53.4	44.8	39.1	54.3
	MEAN	48.8	51.7	58.8	66.9	73.7	79.5	82.0	81.5	77.2	67.4	57.6	51.3	66.4
FRANKLIN 3 NW	MAX	62.8	65.5	71.6	78.8	84.8	89.6	90.8	90.7	87.8	80.9	71.4	65.7	78.4
	MIN	42.1	44.3	50.5	58.4	64.7	70.4	72.6	72.0	68.4	57.0	48.9	43.8	57.8
	MEAN	52.5	55.0	61.0	68.6	74.8	80.0	81.7	81.4	78.1	69.0	60.2	54.8	68.1
GRAND COTEAU	MAX	61.7	64.9	71.3	78.8	85.2	90.5	91.7	91.6	88.0	81.0	70.8	64.6	78.3
	MIN	41.3	43.5	49.9	57.6	63.8	69.4	72.0	71.3	67.5	56.1	48.0	43.2	57.0
	MEAN	51.5	54.3	60.6	68.2	74.5	80.0	81.9	81.5	77.8	68.6	59.5	54.0	67.7
HACKBERRY 8 SSW	MAX	59.4	62.6	68.7	75.9	82.4	88.0	90.0	89.8	86.7	79.3	69.6	62.9	76.3
	MIN	42.3	45.2	51.9	61.0	67.7	73.2	74.2	73.9	70.8	60.4	50.9	44.8	59.7
	MEAN	50.8	53.9	60.3	68.5	75.1	80.6	82.1	81.8	78.8	69.9	60.3	53.9	68.0
HAMMOND 3 NW	MAX	62.2	65.8	72.6	79.7	85.7	91.4	92.4	91.9	88.4	80.6	70.6	64.2	78.8
	MIN	39.7	41.4	47.7	55.4	61.5	67.1	70.3	69.8	66.3	54.0	45.9	41.1	55.0
	MEAN	51.0	53.6	60.2	67.6	73.6	79.3	81.4	80.9	77.4	67.3	58.3	52.7	66.9

LOUISIANA

TEMPERATURE NORMALS (DEG F)

STATION		JAN	FEB	MAR	APR	MAY	JUN	JUL	AUG	SEP	OCT	NOV	DEC	ANN
HOMER EXP STATION	MAX	55.7	60.1	67.6	76.3	82.9	89.5	92.7	92.7	87.1	78.3	66.7	58.4	75.7
	MIN	34.0	36.6	43.4	52.3	59.8	66.7	70.1	68.8	63.6	51.5	42.4	36.1	52.1
	MEAN	44.9	48.4	55.5	64.4	71.4	78.1	81.4	80.8	75.4	64.9	54.6	47.3	63.9
HOUMA	MAX	64.4	66.7	72.7	79.5	85.1	89.9	90.8	90.6	87.6	80.9	72.4	66.8	79.0
	MIN	44.0	45.7	52.4	59.6	65.4	70.4	72.6	72.2	69.3	58.0	50.3	45.4	58.8
	MEAN	54.2	56.2	62.6	69.6	75.3	80.2	81.7	81.4	78.5	69.4	61.4	56.1	68.9
JENNINGS	MAX	61.2	64.7	71.9	79.1	85.2	90.5	92.1	91.8	88.7	81.5	70.7	64.2	78.5
	MIN	41.5	43.6	50.2	58.5	65.0	70.8	72.9	72.1	68.4	57.5	48.7	43.3	57.7
	MEAN	51.4	54.2	61.1	68.8	75.1	80.7	82.5	82.0	78.6	69.5	59.7	53.8	68.1
LAFAYETTE FAA AIRPORT	MAX	61.4	64.3	71.2	78.8	85.1	90.4	91.2	90.8	87.5	80.3	70.6	64.5	78.0
	MIN	42.0	44.0	50.7	58.5	64.9	70.7	73.0	72.6	68.8	57.2	48.5	43.9	57.9
	MEAN	51.7	54.2	61.0	68.7	75.0	80.6	82.1	81.7	78.2	68.8	59.6	54.2	68.0
LAKE ARTHUR 10 SW	MAX	59.1	62.0	68.7	76.1	82.7	88.6	90.6	90.4	87.3	79.9	69.5	62.3	76.4
	MIN	41.5	44.1	51.3	60.5	67.2	73.0	74.6	73.7	69.5	58.0	49.2	43.6	58.9
	MEAN	50.3	53.1	60.0	68.3	75.0	80.8	82.6	82.1	78.4	69.0	59.4	53.0	67.7
LAKE CHARLES WSO	MAX	60.8	64.0	70.5	77.8	84.1	89.4	91.0	90.8	87.5	80.8	70.5	64.0	77.6
	MIN	42.2	44.5	50.8	58.9	65.6	71.4	73.5	72.8	68.9	57.7	48.9	43.8	58.3
	MEAN	51.5	54.3	60.7	68.4	74.9	80.4	82.3	81.8	78.2	69.3	59.7	53.9	68.0
LAKE PROVIDENCE	MAX	54.3	58.8	66.8	76.2	84.0	90.6	93.0	92.7	87.7	78.6	66.5	57.9	75.6
	MIN	35.1	38.1	45.3	54.4	62.3	69.3	72.1	70.8	65.6	53.3	44.0	37.4	54.0
	MEAN	44.8	48.5	56.1	65.3	73.2	80.0	82.5	81.8	76.7	66.0	55.3	47.7	64.8
LEESVILLE	MAX	59.8	63.9	71.1	78.6	84.7	90.5	93.0	92.8	88.6	80.7	69.7	62.6	78.0
	MIN	37.5	39.3	46.1	54.4	60.9	66.8	70.0	68.9	64.3	51.9	44.1	38.9	53.6
	MEAN	48.7	51.6	58.6	66.5	72.8	78.7	81.5	80.9	76.4	66.3	56.9	50.7	65.8
LOGANSPORT 4 ENE	MAX	58.2	62.8	70.2	78.4	84.7	90.9	94.8	94.7	89.0	80.1	68.2	61.1	77.8
	MIN	35.3	37.5	44.1	52.9	60.2	66.4	69.4	68.2	62.9	50.7	42.2	36.9	52.2
	MEAN	46.8	50.2	57.1	65.7	72.4	78.7	82.1	81.5	75.9	65.4	55.2	49.0	65.0
MELVILLE	MAX	61.3	64.6	71.6	78.7	84.4	89.7	91.1	90.6	86.9	79.5	70.3	64.3	77.8
	MIN	40.1	42.4	49.1	56.8	63.6	69.4	72.1	71.4	67.3	55.1	47.0	41.8	56.3
	MEAN	50.7	53.6	60.4	67.8	74.0	79.6	81.6	81.0	77.1	67.3	58.7	53.1	67.1
MINDEN	MAX	56.0	60.5	67.8	76.8	83.3	90.1	93.2	93.3	87.9	79.0	67.2	59.0	76.2
	MIN	33.9	36.6	43.4	53.0	60.8	67.9	71.2	70.0	64.4	51.7	42.0	35.7	52.6
	MEAN	45.0	48.6	55.6	64.9	72.1	79.0	82.2	81.7	76.2	65.4	54.6	47.4	64.4
MONROE FAA AIRPORT	MAX	54.9	59.8	67.3	76.6	83.9	90.7	92.7	92.0	87.0	78.7	66.5	58.5	75.7
	MIN	35.6	38.7	45.4	54.3	62.5	69.2	72.1	70.3	64.7	51.5	42.6	37.4	53.7
	MEAN	45.3	49.3	56.4	65.4	73.2	80.0	82.4	81.2	75.9	65.1	54.6	48.0	64.7
MORGAN CITY	MAX	63.5	66.0	72.1	79.5	85.2	90.3	91.6	91.3	88.6	81.4	72.1	66.2	79.0
	MIN	43.1	45.0	51.6	59.3	65.4	70.9	72.9	72.4	69.4	58.7	50.0	44.7	58.6
	MEAN	53.3	55.5	61.9	69.4	75.3	80.6	82.3	81.9	79.0	70.1	61.1	55.5	68.8
NATCHITOCHES	MAX	59.8	64.3	71.6	79.8	86.1	92.4	95.0	94.9	89.8	81.3	69.9	62.4	78.9
	MIN	37.4	39.9	46.4	54.9	62.0	68.7	71.6	70.4	65.6	53.4	44.8	39.1	54.5
	MEAN	48.6	52.1	59.0	67.4	74.1	80.5	83.3	82.7	77.7	67.3	57.4	50.8	66.7
NEW IBERIA 5 NW	MAX	62.0	64.6	71.5	78.8	84.9	89.9	91.2	90.7	87.7	80.6	71.3	65.0	78.2
	MIN	41.8	44.0	50.8	58.8	64.9	70.2	72.3	71.8	68.1	56.8	48.5	43.4	57.6
	MEAN	51.9	54.3	61.1	68.8	74.9	80.1	81.8	81.3	77.9	68.8	59.9	54.3	67.9
NEW ORLEANS MOISANT WSO	MAX	61.8	64.6	71.2	78.6	84.5	89.5	90.7	90.2	86.8	79.4	70.1	64.4	77.7
	MIN	43.0	44.8	51.6	58.8	65.3	70.9	73.5	73.1	70.1	59.0	49.9	44.8	58.7
	MEAN	52.4	54.7	61.4	68.7	74.9	80.3	82.1	81.7	78.5	69.2	60.0	54.6	68.2
N O AUDUBON WSO R	MAX	61.8	64.6	71.0	78.3	84.2	89.4	90.6	90.3	87.0	79.5	70.1	64.5	77.6
	MIN	45.3	47.6	54.1	61.2	67.7	73.2	75.3	75.3	72.6	62.1	53.1	47.8	61.3
	MEAN	53.6	56.1	62.6	69.8	76.0	81.3	83.0	82.8	79.8	70.8	61.6	56.2	69.5

LOUISIANA

TEMPERATURE NORMALS (DEG F)

STATION		JAN	FEB	MAR	APR	MAY	JUN	JUL	AUG	SEP	OCT	NOV	DEC	ANN
OLLA 3 SSW	MAX	58.4	62.8	70.5	78.4	84.5	90.9	93.4	93.5	88.7	80.0	68.9	61.4	77.6
	MIN	35.2	37.4	44.6	53.3	60.6	67.0	70.1	69.0	64.0	50.9	42.6	36.8	52.6
	MEAN	46.8	50.1	57.6	65.9	72.6	79.0	81.8	81.3	76.4	65.5	55.8	49.1	65.2
PLAIN DEALING	MAX	54.7	59.6	67.2	76.2	82.7	89.6	93.6	93.1	87.3	78.1	66.1	58.3	75.5
	MIN	32.1	34.5	41.2	50.5	58.8	66.0	69.6	68.7	63.2	50.3	40.5	34.3	50.8
	MEAN	43.4	47.1	54.2	63.4	70.8	77.8	81.6	80.9	75.3	64.2	53.4	46.3	63.2
RESERVE	MAX	61.4	64.3	70.7	78.6	84.6	89.9	91.3	90.9	87.2	80.0	70.4	64.2	77.8
	MIN	40.9	43.0	49.9	58.1	65.2	70.9	73.2	72.7	69.5	57.8	49.0	42.8	57.8
	MEAN	51.1	53.7	60.3	68.3	74.9	80.4	82.2	81.8	78.4	68.9	59.7	53.5	67.8
RUSTON LA POLYTECH INS	MAX	54.9	59.5	67.0	75.9	82.9	89.6	92.7	92.5	87.0	78.1	66.2	58.2	75.4
	MIN	33.8	36.5	43.7	53.0	60.5	67.2	70.3	69.2	64.1	51.6	42.4	35.8	52.3
	MEAN	44.4	48.1	55.4	64.5	71.7	78.4	81.5	80.9	75.6	64.9	54.4	47.0	63.9
SAINT JOSEPH EXP STA	MAX	56.6	60.6	68.0	76.7	83.5	89.8	92.0	91.7	87.4	78.8	68.0	60.1	76.1
	MIN	36.5	38.8	46.0	54.9	62.2	68.8	71.7	70.3	65.2	51.9	43.7	38.4	54.0
	MEAN	46.6	49.8	57.0	65.8	72.8	79.3	81.9	81.0	76.3	65.4	55.9	49.3	65.1
SHREVEPORT WSO R	MAX	55.8	60.6	68.1	76.7	83.5	90.1	93.3	93.2	87.7	78.9	66.8	59.2	76.2
	MIN	36.2	39.0	45.8	54.6	62.4	69.4	72.5	71.5	66.5	54.5	44.5	38.2	54.6
	MEAN	46.0	49.8	57.0	65.7	73.0	79.8	82.9	82.4	77.1	66.7	55.7	48.7	65.4
TALLULAH 2 SW	MAX	55.6	60.1	67.8	76.9	83.6	89.8	92.3	91.8	87.3	78.6	67.5	59.1	75.9
	MIN	35.0	37.6	45.0	53.8	61.2	67.9	70.7	69.0	63.5	50.0	42.3	36.6	52.7
	MEAN	45.3	48.9	56.4	65.4	72.4	78.9	81.5	80.4	75.4	64.3	54.9	47.9	64.3
VERMILION LOCK	MAX	60.1	63.1	69.4	76.5	82.7	88.2	89.8	89.8	86.9	80.0	70.4	63.8	76.7
	MIN	40.2	42.3	49.3	58.5	65.2	70.6	72.3	71.5	67.6	55.8	47.3	42.0	56.9
	MEAN	50.1	52.7	59.4	67.5	74.0	79.4	81.1	80.7	77.3	67.9	58.9	52.9	66.8
WINNFIELD 2 W	MAX	58.1	63.0	70.6	78.5	84.5	90.6	93.1	93.5	88.2	79.6	68.2	60.8	77.4
	MIN	35.9	38.2	44.9	53.5	60.5	67.1	70.0	68.7	63.8	51.1	42.4	37.3	52.8
	MEAN	47.0	50.6	57.8	66.0	72.5	78.9	81.6	81.1	76.1	65.4	55.3	49.0	65.1
WINNSBORO	MAX	56.5	60.6	68.4	77.2	84.2	91.0	93.8	93.6	88.8	80.1	68.2	60.0	76.9
	MIN	35.3	37.8	45.0	54.1	61.4	68.1	71.2	69.5	64.2	51.0	42.8	37.1	53.1
	MEAN	45.9	49.2	56.7	65.6	72.8	79.6	82.5	81.6	76.5	65.5	55.5	48.6	65.0

LOUISIANA

PRECIPITATION NORMALS (INCHES)

STATION		JAN	FEB	MAR	APR	MAY	JUN	JUL	AUG	SEP	OCT	NOV	DEC	ANN
ALEXANDRIA		5.08	4.66	5.26	4.85	5.31	3.89	4.84	3.68	3.92	3.75	4.36	6.30	55.90
AMITE		5.44	5.90	5.66	6.46	5.48	4.51	7.33	4.86	5.18	2.68	4.58	5.82	63.90
ASHLAND 2 S		4.55	4.29	4.46	4.78	5.22	3.57	4.64	2.95	3.90	2.52	3.74	4.62	49.24
BASTROP		4.95	4.66	5.46	5.27	5.27	3.74	4.16	3.00	3.39	2.58	4.51	5.02	52.01
BATON ROUGE WSO	R	4.58	4.97	4.59	5.59	4.82	3.11	7.07	5.05	4.42	2.63	3.95	4.99	55.77
BELAH FIRE TOWER		5.05	4.93	5.94	5.47	5.88	3.61	5.15	3.96	4.26	3.43	4.43	6.04	58.15
BOGALUSA		4.89	5.33	5.57	5.56	5.46	4.66	6.21	4.81	4.70	2.88	4.05	5.99	60.11
BUNKIE		5.36	4.80	5.44	5.63	6.30	4.44	5.22	4.42	4.17	3.58	4.89	6.63	60.88
CALHOUN EXP STATION		4.91	4.27	5.18	4.76	5.53	3.37	4.18	2.85	3.70	2.90	4.05	4.65	50.35
CARVILLE 2 SW		4.81	4.87	4.45	5.00	5.11	4.15	6.95	5.82	5.10	2.65	4.12	5.60	58.63
COTTON VALLEY		4.55	4.17	4.69	4.92	5.06	3.92	4.60	2.85	3.58	2.94	4.16	4.36	49.80
COVINGTON 4 NNW		4.87	5.32	5.71	5.33	5.24	4.64	6.65	5.16	5.39	2.96	4.12	5.89	61.28
CROWLEY EXP STATION		4.93	4.58	3.84	5.28	5.03	4.65	6.09	5.37	4.94	3.12	3.75	5.33	56.91
DE QUINCY 4 N		5.06	4.44	4.33	4.61	5.06	4.15	5.94	4.57	5.45	3.72	4.41	5.62	57.36
DE RIDDER		4.92	4.32	4.34	4.77	5.28	3.93	5.14	3.72	4.33	3.18	5.13	6.33	55.39
DONALDSONVILLE 3 E		4.78	4.93	4.83	5.01	5.29	3.65	7.54	6.18	5.42	2.69	4.19	5.85	60.36
ELIZABETH		5.31	4.73	5.20	5.24	5.98	4.54	5.07	4.29	4.68	3.81	4.64	6.89	60.38
FRANKLIN 3 NW		4.34	4.65	3.84	4.54	5.19	5.27	8.48	8.05	5.56	3.10	4.04	4.81	61.87
GLOSTER 1 W		4.38	3.57	4.06	4.61	5.24	3.38	4.02	2.74	3.86	2.85	3.64	4.34	46.69
GRAND COTEAU		5.00	4.96	4.78	5.35	5.75	4.42	6.13	4.34	4.89	3.59	4.03	5.44	58.68
GRAND ECORE		4.83	4.26	4.57	4.84	5.71	3.92	3.66	3.12	3.86	2.76	3.84	4.98	50.35
HACKBERRY 8 SSW		4.51	3.59	3.32	4.07	4.35	4.39	7.00	7.06	5.42	3.52	4.30	4.79	56.32
HAMMOND 3 NW		4.81	5.65	5.67	6.15	5.68	3.82	7.53	4.82	5.74	2.94	4.44	5.99	63.24
HOMER EXP STATION		4.62	4.25	4.50	5.26	5.48	3.78	4.70	3.05	4.12	2.57	4.20	4.56	51.09
HOSSTON		3.99	3.51	4.11	4.89	4.47	3.53	3.33	3.10	3.13	2.86	4.06	4.18	45.16
HOUMA		4.80	4.46	4.20	4.60	5.43	6.09	7.94	6.68	7.52	3.18	3.96	5.03	63.89
JEANERETTE EXP STA		4.42	4.35	3.88	4.42	4.98	5.10	7.82	6.47	5.12	3.53	3.64	5.24	58.97
JENNINGS		5.00	4.70	4.16	5.44	5.52	4.16	6.26	5.49	5.24	3.38	4.15	5.60	59.10
KEITHVILLE		4.54	3.92	4.23	4.52	5.20	3.68	3.72	2.42	3.78	2.59	3.96	4.20	46.76
KENTWOOD		5.39	5.18	5.78	6.17	6.14	4.49	7.15	5.03	4.74	2.79	4.78	5.91	63.55
KINDER 3 W		5.22	4.81	4.50	4.81	5.66	4.25	5.43	4.81	4.86	3.48	4.21	6.21	58.25
KORAN		4.40	4.01	4.13	4.44	5.50	3.55	3.60	2.68	3.21	2.67	3.99	4.42	46.60
LAFAYETTE FAA AIRPORT		4.72	4.55	4.16	5.10	5.24	4.18	7.19	5.38	5.35	3.20	3.60	5.02	57.69
LAKE ARTHUR 10 SW		4.57	4.53	3.36	4.60	5.40	3.85	6.27	5.96	4.73	3.35	4.22	5.45	56.29
LAKE CHARLES WSO		4.25	3.88	3.05	4.06	5.14	4.19	5.55	5.39	5.21	3.47	3.76	5.08	53.03
LAKE PROVIDENCE		5.42	5.07	6.17	5.27	5.22	3.59	4.57	3.08	3.13	2.64	4.33	5.57	54.06
LEESVILLE		4.66	4.51	4.54	5.02	5.31	4.16	5.47	3.62	3.70	3.05	4.35	5.68	54.07
LOGANSPORT 4 ENE		4.27	3.93	4.27	4.87	6.03	3.56	3.65	3.12	4.12	2.87	3.80	4.49	48.98
LONGVILLE		5.35	4.39	4.44	4.61	5.26	3.86	5.38	4.17	4.83	3.90	4.17	5.74	56.10
MELVILLE		5.14	5.13	4.65	5.25	5.53	3.93	5.18	4.01	4.26	2.89	4.07	5.77	55.81
MERMENTAU		4.99	5.10	3.91	5.63	5.49	4.02	6.64	5.59	5.23	3.68	3.86	5.66	59.80
MINDEN		4.36	4.03	4.38	4.61	5.44	3.76	4.25	2.99	3.28	2.52	4.20	4.44	48.26
MONROE FAA AIRPORT		4.84	4.41	5.21	4.95	5.04	3.30	4.53	2.56	3.49	2.42	3.97	4.84	49.56
MORGAN CITY		4.48	4.60	3.90	4.65	5.28	5.10	8.09	7.17	6.24	3.65	4.30	5.07	62.53
NATCHITOCHES		4.85	4.36	4.53	5.08	5.65	3.69	3.53	2.99	3.83	2.96	3.78	5.07	50.32
NEW IBERIA 5 NW		4.41	4.22	4.06	4.87	4.88	4.98	7.19	5.92	5.16	3.17	3.63	4.76	57.25
NEW ORLEANS MOISANT WSO		4.97	5.23	4.73	4.50	5.07	4.63	6.73	6.02	5.87	2.66	4.06	5.27	59.74
N O AUDUBON WSO	R	4.90	5.19	4.68	4.68	5.06	5.39	7.17	6.67	5.98	2.52	4.01	5.30	61.55
NEW ORLEANS ALGIERS	R	4.76	4.98	5.08	4.76	5.46	5.36	7.81	6.08	5.57	2.88	3.86	5.07	61.68
NEW ORLEANS CITRUS	R	4.89	5.25	4.74	4.68	5.03	4.19	6.54	5.46	5.89	2.50	3.58	4.92	57.67
NEW ORLEANS DUBLIN	R	4.67	4.76	4.31	4.26	4.80	4.92	6.76	6.24	5.65	2.33	3.54	4.55	56.79
NEW ORLEANS JOURDAN	R	4.72	5.37	5.20	4.68	5.47	4.72	7.38	6.13	5.85	2.99	3.79	5.27	61.57
NEW ORLEANS LONDON	R	4.72	5.14	4.88	4.61	4.90	4.73	6.69	5.89	5.76	2.59	3.86	5.03	58.80
OAKNOLIA		5.33	5.39	5.33	5.95	5.64	3.80	6.20	4.77	4.24	3.02	4.27	5.44	59.38
OBERLIN FIRE TOWER		5.49	5.00	4.90	4.98	6.33	4.11	5.74	4.31	5.42	3.85	4.51	6.08	60.72

LOUISIANA

PRECIPITATION NORMALS (INCHES)

STATION	JAN	FEB	MAR	APR	MAY	JUN	JUL	AUG	SEP	OCT	NOV	DEC	ANN
OLLA 3 SSW	5.37	5.02	5.93	5.30	6.23	3.68	4.70	3.32	3.88	2.89	4.32	5.74	56.38
PARADIS 7 S	5.01	5.27	5.05	4.69	5.63	4.74	7.63	6.12	6.59	2.97	4.19	5.17	63.06
PEARL RIVER LOCK NO 1	4.92	5.07	5.85	5.46	5.20	4.16	6.64	5.20	5.53	2.72	4.41	5.78	60.94
PINE GROVE FIRE TOWER	5.88	5.86	6.08	6.52	6.11	4.26	7.78	5.57	4.94	3.20	4.83	6.46	67.49
PLAIN DEALING	4.28	3.86	4.26	5.34	4.74	3.72	4.08	3.33	3.47	2.99	4.24	4.37	48.68
RESERVE	5.13	5.58	5.15	4.51	5.29	4.31	6.50	5.70	5.95	3.01	4.07	5.62	60.82
RODESSA	3.96	3.74	4.45	5.22	4.57	3.69	3.15	2.67	3.32	3.03	4.23	4.20	46.23
RUSTON LA POLYTECH INS	5.07	4.37	4.93	5.03	5.94	3.55	4.59	2.84	4.10	2.72	4.11	4.90	52.15
SAINT JOSEPH EXP STA	5.41	4.69	6.09	5.17	5.18	3.06	3.91	3.16	2.98	2.88	4.07	5.65	52.25
SHERIDAN FIRE TOWER	5.61	5.93	6.21	6.27	6.24	4.86	7.31	6.02	4.98	3.18	4.53	6.59	67.73
SHREVEPORT WSO R	4.02	3.46	3.77	4.71	4.70	3.54	3.56	2.52	3.29	2.63	3.77	3.87	43.84
SIMMESPORT 1 SE	5.33	4.97	5.27	5.70	5.51	3.99	5.30	4.42	4.25	2.99	4.54	6.42	58.69
SPEARSVILLE FIRE TOWER	4.78	4.26	5.12	5.31	5.68	3.61	4.24	3.16	4.20	2.75	4.17	4.70	51.98
STERLINGTON LOCK	4.79	4.34	5.12	4.95	5.30	3.20	4.19	2.40	3.07	2.74	4.16	4.66	48.92
TALLULAH 2 SW	5.16	4.73	6.15	5.61	5.24	3.76	4.29	3.33	2.70	2.59	3.98	5.89	53.43
VERMILION LOCK	4.39	4.23	3.72	4.48	4.11	4.88	8.58	6.93	5.83	3.26	3.83	4.89	59.13
WINNFIELD 2 W	5.16	4.48	5.23	5.47	5.89	3.85	5.26	3.36	4.48	2.76	4.08	5.31	55.33
WINNSBORO	5.25	4.52	5.97	4.91	5.24	3.28	3.86	3.38	3.17	2.60	4.18	5.77	52.13
WINONA FIRE TOWER	5.39	4.79	5.36	5.67	5.91	3.79	5.45	3.94	4.33	3.08	4.43	5.65	57.79
WOODWORTH ST FOREST	5.20	4.92	5.34	5.33	5.74	3.69	5.50	4.11	4.18	3.50	5.11	6.94	59.56

LOUISIANA

HEATING DEGREE DAY NORMALS (BASE 65 DEG F)

STATION	JUL	AUG	SEP	OCT	NOV	DEC	JAN	FEB	MAR	APR	MAY	JUN	ANN
ALEXANDRIA	0	0	0	64	245	442	531	393	239	47	0	0	1961
AMITE	0	0	0	56	237	421	501	373	212	43	0	0	1843
BASTROP	0	0	0	72	279	499	594	438	278	59	0	0	2219
BATON ROUGE WSO R	0	0	0	48	218	380	466	342	187	32	0	0	1673
BELAH FIRE TOWER	0	0	0	67	269	477	574	426	267	64	6	0	2150
BOGALUSA	0	0	0	60	242	423	504	379	221	48	0	0	1877
BUNKIE	0	0	0	66	252	442	527	394	242	56	0	0	1979
CALHOUN EXP STATION	0	0	0	105	311	530	626	468	312	79	9	0	2440
CARVILLE 2 SW	0	0	0	39	188	349	439	328	177	33	0	0	1553
COTTON VALLEY	0	0	5	109	344	555	637	479	328	91	11	0	2559
COVINGTON 4 NNW	0	0	0	47	221	376	449	341	192	41	0	0	1667
DONALDSONVILLE 3 E	0	0	0	44	192	333	424	313	162	24	0	0	1492
ELIZABETH	0	0	0	74	244	430	517	388	231	51	0	0	1935
FRANKLIN 3 NW	0	0	0	40	192	332	421	306	179	29	0	0	1499
GRAND COTEAU	0	0	0	37	200	351	443	319	183	34	0	0	1567
HACKBERRY 8 SSW	0	0	0	37	192	351	462	333	191	29	0	0	1595
HAMMOND 3 NW	0	0	0	55	227	392	460	348	192	37	0	0	1711
HOMER EXP STATION	0	0	0	98	321	549	631	475	324	89	9	0	2496
HOUMA	0	0	0	35	169	295	377	283	136	20	0	0	1315
JENNINGS	0	0	0	50	207	357	448	332	180	28	0	0	1602
LAFAYETTE FAA AIRPORT	0	0	0	42	194	348	437	326	183	30	0	0	1560
LAKE ARTHUR 10 SW	0	0	0	39	210	379	473	351	199	29	0	0	1680
LAKE CHARLES WSO	0	0	0	45	204	351	442	324	184	29	0	0	1579
LAKE PROVIDENCE	0	0	0	86	299	536	634	474	310	69	6	0	2414
LEESVILLE	0	0	0	77	260	443	520	388	231	57	0	0	1976
LOGANSPORT 4 ENE	0	0	0	93	307	496	574	424	273	75	0	0	2242
MELVILLE	0	0	0	63	220	381	467	346	191	37	0	0	1705
MINDEN	0	0	0	91	323	546	629	470	319	78	8	0	2464
MONROE FAA AIRPORT	0	0	0	95	320	527	618	451	305	79	9	0	2404
MORGAN CITY	0	0	0	32	171	311	405	300	157	27	0	0	1403
NATCHITOCHES	0	0	0	65	250	446	526	377	228	44	0	0	1936
NEW IBERIA 5 NW	0	0	0	37	197	340	442	322	187	30	0	0	1555
NEW ORLEANS MOISANT WSO	0	0	0	31	186	336	423	318	171	25	0	0	1490
N O AUDUBON WSO R	0	0	0	20	154	292	391	287	150	18	0	0	1312
OLLA 3 SSW	0	0	0	99	293	498	579	431	265	74	9	0	2248
PLAIN DEALING	0	0	7	121	354	580	670	507	354	110	13	0	2716
RESERVE	0	0	0	44	196	367	455	342	191	30	0	0	1625
RUSTON LA POLYTECH INS	0	0	0	101	328	558	639	480	326	83	7	0	2522
SAINT JOSEPH EXP STA	0	0	0	94	286	487	582	442	278	69	6	0	2244
SHREVEPORT WSO R	0	0	0	76	293	505	597	438	282	69	9	0	2269
TALLULAH 2 SW	0	0	6	117	313	530	621	464	301	75	7	0	2434
VERMILION LOCK	0	0	0	56	221	381	482	361	215	35	0	0	1751
WINNFIELD 2 W	0	0	0	95	302	496	570	417	266	66	6	0	2218
WINNSBORO	0	0	0	100	297	508	604	457	293	69	7	0	2335

LOUISIANA

COOLING DEGREE DAY NORMALS (BASE 65 DEG F)

STATION	JAN	FEB	MAR	APR	MAY	JUN	JUL	AUG	SEP	OCT	NOV	DEC	ANN
ALEXANDRIA	16	18	47	110	279	453	543	530	372	133	20	0	2521
AMITE	17	20	42	115	279	441	521	505	378	130	24	6	2478
BASTROP	8	13	45	101	274	462	570	543	367	125	12	0	2520
BATON ROUGE WSO R	26	23	47	134	304	459	530	508	387	147	29	11	2605
BELAH FIRE TOWER	7	11	35	103	260	435	533	512	369	126	17	0	2408
BOGALUSA	20	21	38	123	282	459	530	512	378	128	23	11	2525
BUNKIE	13	14	40	116	279	453	533	508	354	125	24	0	2459
CALHOUN EXP STATION	12	14	30	76	232	420	533	512	334	108	8	0	2279
CARVILLE 2 SW	26	25	47	129	304	459	530	518	402	166	38	8	2652
COTTON VALLEY	8	11	24	70	209	402	527	505	323	91	8	0	2178
COVINGTON 4 NNW	24	25	41	113	273	429	505	493	375	131	26	10	2445
DONALDSONVILLE 3 E	36	33	59	159	326	480	543	533	414	183	57	17	2840
ELIZABETH	14	16	39	108	272	435	527	512	366	148	22	5	2464
FRANKLIN 3 NW	34	26	55	137	304	450	518	508	393	164	48	15	2652
GRAND COTEAU	24	20	46	130	295	450	524	512	384	149	35	10	2579
HACKBERRY 8 SSW	22	22	45	134	313	468	530	521	414	189	51	7	2716
HAMMOND 3 NW	26	29	43	115	267	429	508	493	372	127	26	11	2446
HOMER EXP STATION	8	10	29	71	207	393	508	490	316	95	9	0	2136
HOUMA	42	37	61	158	319	456	518	508	405	171	61	19	2755
JENNINGS	27	29	59	142	313	471	543	527	408	189	48	9	2765
LAFAYETTE FAA AIRPORT	25	23	59	141	310	468	530	518	396	160	32	13	2675
LAKE ARTHUR 10 SW	18	17	44	128	310	474	546	530	402	163	42	7	2681
LAKE CHARLES WSO	24	24	51	131	307	462	536	521	396	178	45	7	2682
LAKE PROVIDENCE	8	12	34	78	260	450	543	521	355	117	8	0	2386
LEESVILLE	15	13	33	102	242	411	512	493	342	117	17	0	2297
LOGANSPORT 4 ENE	9	10	28	96	233	411	530	512	327	105	13	0	2274
MELVILLE	23	27	48	121	279	438	515	496	363	134	31	12	2487
MINDEN	9	11	28	75	228	420	533	518	340	104	11	0	2277
MONROE FAA AIRPORT	7	11	38	91	263	450	539	502	332	99	8	0	2340
MORGAN CITY	42	34	61	159	319	468	536	524	420	190	54	16	2823
NATCHITOCHES	17	16	42	116	282	465	567	549	381	136	22	5	2598
NEW IBERIA 5 NW	36	23	66	144	307	453	521	505	387	155	44	8	2649
NEW ORLEANS MOISANT WSO	32	30	59	136	307	459	530	518	405	161	36	13	2686
N O AUDUBON WSO R	38	38	76	162	341	489	558	552	444	199	52	19	2968
OLLA 3 SSW	15	14	36	101	245	420	521	505	342	115	17	0	2331
PLAIN DEALING	0	6	19	62	193	384	515	493	316	96	6	0	2090
RESERVE	24	25	45	129	307	462	533	521	402	165	37	11	2661
RUSTON LA POLYTECH INS	0	7	29	68	215	402	512	493	322	97	10	0	2155
SAINT JOSEPH EXP STA	12	16	30	93	248	429	524	496	339	106	13	0	2306
SHREVEPORT WSO R	8	12	34	90	257	444	555	539	363	128	14	0	2444
TALLULAH 2 SW	10	14	34	87	236	417	512	477	318	95	10	0	2210
VERMILION LOCK	20	17	41	110	279	432	499	487	369	146	38	6	2444
WINNFIELD 2 W	12	14	43	96	239	417	515	499	337	108	11	0	2291
WINNSBORO	11	15	36	87	248	438	543	515	350	115	12	0	2370

16 — LOUISIANA

STATE-STATION NUMBER	STN TYP	NAME		LATITUDE DEG-MIN	LONGITUDE DEG-MIN	ELEVATION (FT)
16-0098	13	ALEXANDRIA		N 3119	W 09228	87
16-0205	13	AMITE		N 3043	W 09030	180
16-0349	12	ASHLAND 2 S		N 3207	W 09306	225
16-0537	13	BASTROP		N 3247	W 09154	140
16-0549	13	BATON ROUGE WSO	R	N 3032	W 09108	64
16-0639	13	BELAH FIRE TOWER		N 3138	W 09211	200
16-0945	13	BOGALUSA		N 3047	W 08952	103
16-1287	13	BUNKIE		N 3057	W 09210	80
16-1411	13	CALHOUN EXP STATION		N 3231	W 09220	180
16-1565	13	CARVILLE 2 SW		N 3012	W 09107	26
16-2121	13	COTTON VALLEY		N 3249	W 09325	230
16-2151	13	COVINGTON 4 NNW		N 3032	W 09007	40
16-2212	12	CROWLEY EXP STATION		N 3015	W 09222	25
16-2361	12	DE QUINCY 4 N		N 3031	W 09326	95
16-2367	12	DE RIDDER		N 3050	W 09318	180
16-2534	13	DONALDSONVILLE 3 E		N 3006	W 09056	20
16-2800	13	ELIZABETH		N 3052	W 09248	150
16-3313	13	FRANKLIN 3 NW		N 2949	W 09133	12
16-3657	12	GLOSTER 1 W		N 3212	W 09350	255
16-3800	13	GRAND COTEAU		N 3026	W 09202	55
16-3804	12	GRAND ECORE		N 3148	W 09306	150
16-3979	13	HACKBERRY 8 SSW		N 2953	W 09325	6
16-4034	13	HAMMOND 3 NW		N 3032	W 09029	45
16-4355	13	HOMER EXP STATION		N 3245	W 09304	380
16-4398	12	HOSSTON		N 3253	W 09353	200
16-4407	13	HOUMA		N 2935	W 09044	15
16-4674	12	JEANERETTE EXP STA		N 2957	W 09143	20
16-4700	13	JENNINGS		N 3015	W 09240	30
16-4816	12	KEITHVILLE		N 3221	W 09350	200
16-4859	12	KENTWOOD		N 3056	W 09031	218
16-4884	12	KINDER 3 W		N 3030	W 09254	53
16-4931	12	KORAN		N 3225	W 09328	175
16-5026	13	LAFAYETTE FAA AIRPORT		N 3012	W 09159	38
16-5065	13	LAKE ARTHUR 10 SW		N 3000	W 09248	10
16-5078	13	LAKE CHARLES WSO		N 3007	W 09313	9
16-5090	13	LAKE PROVIDENCE		N 3249	W 09112	105
16-5266	13	LEESVILLE		N 3109	W 09316	240
16-5527	13	LOGANSPORT 4 ENE		N 3159	W 09357	210
16-5584	12	LONGVILLE		N 3036	W 09314	114
16-6117	13	MELVILLE		N 3041	W 09145	30
16-6142	12	MERMENTAU		N 3011	W 09235	16
16-6244	13	MINDEN		N 3236	W 09318	250
16-6303	13	MONROE FAA AIRPORT		N 3231	W 09203	78
16-6394	13	MORGAN CITY		N 2941	W 09111	5
16-6582	13	NATCHITOCHES		N 3146	W 09305	130
16-6657	13	NEW IBERIA 5 NW		N 3003	W 09153	25
16-6660	13	NEW ORLEANS MOISANT WSO		N 2959	W 09015	4
16-6664	13	N O AUDUBON WSO	R	N 2955	W 09008	6
16-6666	12	NEW ORLEANS ALGIERS	R	N 2956	W 09002	2
16-6668	12	NEW ORLEANS CITRUS	R	N 3003	W 08959	0

16 — LOUISIANA

STATE-STATION NUMBER	STN TYP	NAME		LATITUDE DEG-MIN	LONGITUDE DEG-MIN	ELEVATION (FT)
16-6669	12	NEW ORLEANS DUBLIN	R	N 2957	W 09008	22
16-6672	12	NEW ORLEANS JOURDAN	R	N 2959	W 09001	10
16-6675	12	NEW ORLEANS LONDON	R	N 2959	W 09004	12
16-6911	12	OAKNOLIA		N 3044	W 09059	130
16-6938	12	OBERLIN FIRE TOWER		N 3036	W 09247	67
16-6978	13	OLLA 3 SSW		N 3152	W 09216	100
16-7096	12	PARADIS 7 S		N 2947	W 09026	6
16-7161	12	PEARL RIVER LOCK NO 1		N 3027	W 08947	30
16-7304	12	PINE GROVE FIRE TOWER		N 3042	W 09045	190
16-7344	13	PLAIN DEALING		N 3254	W 09341	291
16-7767	13	RESERVE		N 3004	W 09034	13
16-7950	12	RODESSA		N 3258	W 09400	200
16-8067	13	RUSTON LA POLYTECH INS		N 3231	W 09239	275
16-8163	13	SAINT JOSEPH EXP STA		N 3157	W 09114	78
16-8405	12	SHERIDAN FIRE TOWER		N 3051	W 08959	330
16-8440	13	SHREVEPORT WSO	R	N 3228	W 09349	254
16-8507	12	SIMMESPORT 1 SE		N 3059	W 09148	40
16-8669	12	SPEARSVILLE FIRE TOWER		N 3254	W 09234	200
16-8785	12	STERLINGTON LOCK		N 3242	W 09205	55
16-8923	13	TALLULAH 2 SW		N 3224	W 09113	85
16-9319	13	VERMILION LOCK		N 2947	W 09212	14
16-9803	13	WINNFIELD 2 W		N 3156	W 09241	160
16-9806	13	WINNSBORO		N 3209	W 09142	80
16-9809	12	WINONA FIRE TOWER		N 3202	W 09239	220
16-9865	12	WOODWORTH ST FOREST		N 3108	W 09228	90

16 – LOUISIANA

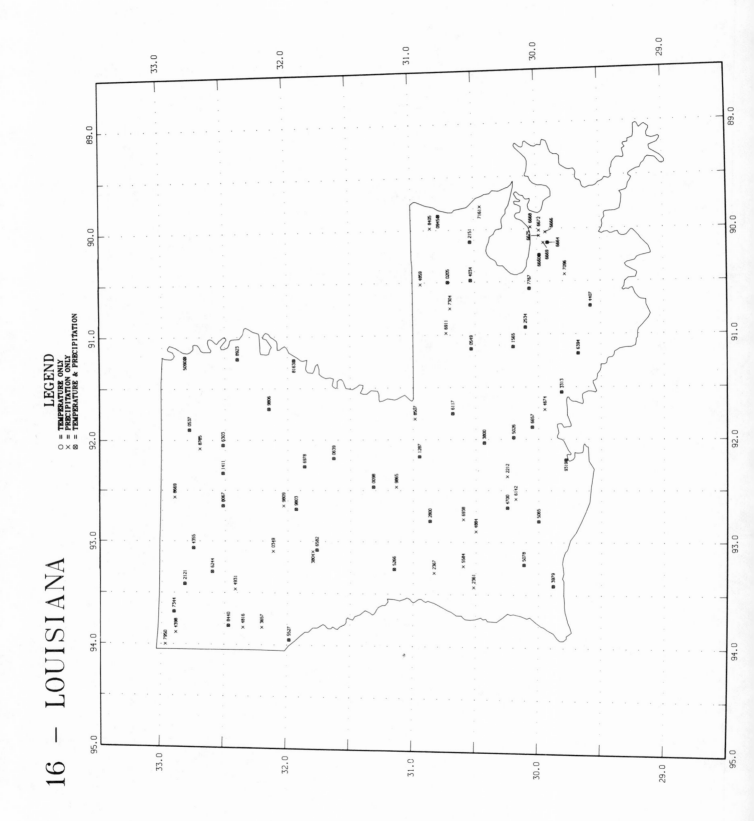

LEGEND

O = TEMPERATURE ONLY
X = PRECIPITATION ONLY
⊗ = TEMPERATURE & PRECIPITATION

MAINE

TEMPERATURE NORMALS (DEG F)

STATION			JAN	FEB	MAR	APR	MAY	JUN	JUL	AUG	SEP	OCT	NOV	DEC	ANN
AUGUSTA FAA AIRPORT		MAX	28.2	30.7	39.5	52.3	65.0	74.1	79.4	77.3	68.7	57.5	44.7	31.8	54.1
		MIN	11.0	12.1	23.4	34.0	44.1	53.5	59.4	57.6	49.3	39.7	30.4	16.2	35.9
		MEAN	19.6	21.5	31.5	43.2	54.6	63.8	69.4	67.5	59.0	48.6	37.6	24.0	45.0
BANGOR FAA AP		MAX	27.1	29.1	38.1	50.7	63.4	72.8	77.9	76.4	67.8	56.6	44.2	31.0	52.9
		MIN	9.3	10.4	21.8	32.8	42.6	52.3	58.1	56.4	48.2	38.7	29.5	14.9	34.6
		MEAN	18.2	19.8	30.0	41.8	53.0	62.6	68.0	66.4	58.0	47.7	36.9	23.0	43.8
BAR HARBOR		MAX	32.1	33.0	40.9	52.4	64.0	72.6	77.6	75.9	67.9	58.0	47.2	36.0	54.8
		MIN	14.9	15.5	24.8	33.7	42.7	51.0	56.6	56.1	49.7	41.5	32.5	19.5	36.5
		MEAN	23.5	24.3	32.8	43.1	53.3	61.8	67.2	66.0	58.8	49.8	39.8	27.8	45.7
BELFAST		MAX	32.4	34.4	42.4	53.7	65.2	74.3	79.7	78.4	70.2	59.7	47.5	35.4	56.1
		MIN	11.7	11.9	22.7	32.3	42.1	51.4	57.1	55.7	48.3	39.1	30.9	17.0	35.0
		MEAN	22.0	23.2	32.6	43.0	53.7	62.9	68.4	67.1	59.3	49.4	39.2	26.3	45.6
CARIBOU WSO	//R	MAX	19.9	22.9	33.5	45.8	60.7	71.0	75.7	73.1	63.9	51.7	38.0	23.9	48.3
		MIN	1.4	3.0	15.1	28.8	39.7	49.6	54.5	51.8	43.2	34.5	24.2	7.5	29.4
		MEAN	10.6	13.0	24.3	37.3	50.2	60.4	65.2	62.5	53.6	43.1	31.1	15.7	38.9
CORINNA		MAX	27.0	30.1	39.5	52.3	66.2	75.3	80.0	77.9	68.9	57.5	43.6	30.1	54.0
		MIN	5.0	6.0	18.2	29.9	40.0	49.8	55.2	52.8	44.8	35.4	26.7	11.1	31.2
		MEAN	16.0	18.1	28.9	41.1	53.1	62.6	67.6	65.4	56.9	46.5	35.1	20.6	42.7
EASTPORT	R	MAX	30.5	31.1	38.3	48.4	58.6	66.9	72.4	72.0	65.0	55.7	45.4	34.4	51.6
		MIN	15.2	15.7	24.5	32.9	40.6	47.5	52.7	53.3	48.5	41.2	33.0	20.4	35.5
		MEAN	22.8	23.4	31.4	40.7	49.6	57.2	62.6	62.7	56.8	48.5	39.2	27.4	43.5
FARMINGTON		MAX	26.7	29.7	38.8	51.3	65.5	74.7	79.8	77.6	69.3	57.7	43.5	30.0	53.7
		MIN	3.5	4.3	17.0	29.1	39.2	48.9	53.9	51.4	42.8	33.5	25.1	9.9	29.9
		MEAN	15.1	17.0	27.9	40.2	52.4	61.8	66.9	64.5	56.1	45.6	34.3	20.0	41.8
GARDINER		MAX	31.1	33.4	41.6	53.9	66.5	75.3	80.5	78.7	70.1	59.6	46.7	34.0	56.0
		MIN	7.8	8.9	20.5	31.0	41.3	50.7	56.4	54.4	46.6	37.0	28.3	13.9	33.1
		MEAN	19.5	21.2	31.1	42.5	53.9	63.0	68.5	66.6	58.4	48.3	37.5	24.0	44.5
HOULTON FAA AP		MAX	23.1	25.9	35.8	48.2	62.8	72.9	78.1	75.4	65.8	54.0	40.4	26.6	50.8
		MIN	2.3	3.2	15.0	27.6	38.5	48.6	54.2	51.5	42.7	34.1	24.2	8.0	29.2
		MEAN	12.7	14.5	25.4	37.9	50.7	60.8	66.2	63.5	54.3	44.1	32.3	17.3	40.0
HOULTON		MAX	22.9	26.0	36.2	48.4	63.4	72.9	77.8	75.2	66.1	54.2	40.2	26.6	50.8
		MIN	4.7	5.9	17.4	29.4	39.8	50.2	55.7	53.2	44.9	35.7	26.0	10.5	31.1
		MEAN	13.8	16.0	26.9	38.9	51.6	61.6	66.8	64.2	55.5	45.0	33.1	18.6	41.0
JONESBORO		MAX	29.8	30.5	38.7	50.0	61.2	69.9	75.8	75.4	67.8	57.2	46.1	33.9	53.0
		MIN	9.0	9.5	20.4	30.6	39.1	47.8	53.4	52.5	45.1	36.3	28.3	14.3	32.2
		MEAN	19.4	20.0	29.6	40.3	50.2	58.9	64.6	64.0	56.5	46.8	37.3	24.1	42.6
LEWISTON		MAX	29.2	31.6	39.8	52.3	65.1	74.7	79.9	78.0	69.6	58.7	45.3	32.7	54.7
		MIN	12.0	13.0	24.0	34.0	44.6	54.5	60.4	58.7	50.8	40.6	31.3	17.9	36.8
		MEAN	20.6	22.3	31.9	43.2	54.9	64.6	70.2	68.4	60.2	49.7	38.3	25.3	45.8
MADISON		MAX	28.0	30.6	39.5	51.9	65.7	74.7	79.8	77.9	69.5	58.1	44.5	31.0	54.3
		MIN	5.4	6.2	18.5	30.2	39.9	49.9	55.4	53.2	45.4	35.7	27.0	11.8	31.6
		MEAN	16.7	18.4	29.0	41.1	52.8	62.3	67.6	65.6	57.5	46.9	35.8	21.4	42.9
PORTLAND WSO	//R	MAX	31.0	33.1	40.5	52.5	63.4	72.8	78.9	77.5	69.6	59.0	47.1	34.9	55.0
		MIN	11.9	12.9	23.7	33.0	42.1	51.4	57.3	55.8	47.7	37.9	29.6	16.7	35.0
		MEAN	21.5	23.0	32.1	42.8	52.8	62.2	68.1	66.6	58.6	48.4	38.4	25.8	45.0
PRESQUE ISLE		MAX	21.5	24.6	34.7	47.0	62.9	73.0	77.4	74.8	65.8	53.4	39.1	25.3	50.0
		MIN	2.7	4.3	15.8	28.4	39.4	49.7	54.7	52.0	44.2	35.3	25.0	8.3	30.0
		MEAN	12.2	14.5	25.3	37.8	51.1	61.4	66.1	63.4	55.0	44.4	32.1	16.8	40.0
RIPOGENUS DAM		MAX	22.3	24.9	35.4	46.4	60.5	71.4	76.3	74.3	66.1	53.9	39.9	26.0	49.8
		MIN	.6	.7	12.1	26.3	37.9	48.8	54.0	52.3	44.1	34.5	24.9	8.1	28.7
		MEAN	11.5	12.8	23.7	36.4	49.2	60.1	65.2	63.3	55.1	44.3	32.4	17.1	39.3

MAINE

TEMPERATURE NORMALS (DEG F)

STATION		JAN	FEB	MAR	APR	MAY	JUN	JUL	AUG	SEP	OCT	NOV	DEC	ANN
RUMFORD 1 SSE	MAX	27.2	29.9	38.8	51.6	65.2	74.3	79.1	76.6	68.4	57.1	43.2	30.2	53.5
	MIN	6.2	7.4	18.9	30.1	40.6	50.1	55.1	53.1	45.3	35.7	27.0	12.2	31.8
	MEAN	16.7	18.7	28.9	40.9	52.9	62.2	67.1	64.9	56.9	46.4	35.1	21.2	42.7
WATERVILLE PUMP STA	MAX	30.3	33.1	42.0	54.8	68.0	76.8	81.7	80.0	71.6	60.2	46.4	33.4	56.5
	MIN	8.5	9.8	21.5	32.0	42.2	52.0	57.1	55.4	47.4	37.7	28.8	14.4	33.9
	MEAN	19.4	21.5	31.8	43.4	55.1	64.4	69.4	67.8	59.5	49.0	37.7	23.9	45.2
WOODLAND	MAX	29.5	30.8	39.9	51.6	65.0	74.6	80.4	78.3	70.0	58.2	45.9	33.2	54.8
	MIN	6.7	6.6	18.8	30.4	40.0	49.7	55.7	53.6	45.0	36.1	27.0	13.0	31.9
	MEAN	18.1	18.7	29.4	41.0	52.5	62.1	68.1	65.9	57.5	47.2	36.5	23.1	43.3

MAINE

PRECIPITATION NORMALS (INCHES)

STATION		JAN	FEB	MAR	APR	MAY	JUN	JUL	AUG	SEP	OCT	NOV	DEC	ANN
AUGUSTA FAA AIRPORT		3.33	3.19	3.53	3.66	3.39	3.32	3.33	3.16	3.23	3.94	4.51	3.91	42.50
BANGOR FAA AP		3.47	3.29	3.24	3.14	3.17	3.02	3.43	3.12	3.50	3.67	4.41	4.16	41.62
BAR HARBOR		4.78	4.24	4.26	4.08	4.06	3.09	2.97	2.77	4.48	4.78	5.79	5.69	50.99
BELFAST		4.18	3.81	4.15	4.02	3.85	3.28	3.33	2.99	3.76	4.67	5.54	5.29	48.87
BRASSUA DAM		2.72	2.49	2.70	3.04	3.00	3.72	3.86	3.72	3.38	3.43	3.75	3.48	39.29
BRUNSWICK		3.93	3.92	4.35	3.94	3.30	3.14	2.80	3.19	3.27	3.77	4.83	4.92	45.36
CARIBOU WSO	//R	2.36	2.14	2.44	2.59	2.88	3.18	4.03	3.97	3.52	3.11	3.22	3.15	36.59
CORINNA		3.30	3.11	3.43	3.55	3.30	3.32	3.46	3.51	3.70	3.84	4.44	3.95	42.91
EASTPORT	R	4.19	3.61	3.41	3.54	3.55	3.08	2.95	3.04	3.34	3.86	4.68	4.57	43.82
ELLSWORTH		4.15	3.85	3.79	3.90	3.48	2.88	3.19	2.82	3.84	4.23	5.27	5.01	46.41
FARMINGTON		3.33	3.47	3.84	3.82	3.59	3.95	3.63	3.69	3.47	4.03	4.61	4.56	45.99
FORT FAIRFIELD		2.75	2.48	2.56	2.88	3.14	3.31	4.30	3.99	3.91	3.33	3.46	3.52	39.63
FORT KENT		2.30	2.02	2.37	2.45	2.87	3.16	3.96	4.13	3.65	3.13	3.04	3.04	36.12
GARDINER		3.52	3.28	3.80	4.02	3.49	3.32	3.30	3.24	3.21	4.09	4.76	4.33	44.36
HOULTON FAA AP		2.74	2.48	2.47	2.78	3.08	3.38	3.91	3.61	3.38	3.49	3.85	3.40	38.57
HOULTON		2.85	2.56	2.62	2.86	2.86	2.96	3.49	3.23	3.08	3.23	3.71	3.75	37.20
JACKMAN		2.51	2.30	2.53	2.85	2.94	3.56	3.82	3.83	3.23	3.15	3.05	3.31	37.08
JONESBORO		4.75	3.95	3.90	4.22	4.09	3.23	3.04	2.97	4.14	4.26	5.14	5.37	49.06
LEWISTON		3.83	3.75	4.14	3.95	3.36	3.39	3.27	3.32	3.21	3.89	4.96	4.69	45.76
LONG FALLS DAM		2.89	2.59	3.10	3.00	3.05	3.41	3.63	3.22	3.09	3.33	3.81	3.67	38.79
MACHIAS		4.67	4.09	3.87	4.21	4.23	3.32	3.40	3.08	4.30	4.38	5.42	5.22	50.19
MADISON		2.97	2.97	3.17	3.22	3.19	3.38	3.41	3.15	3.12	3.55	3.99	3.72	39.84
MIDDLE DAM		2.32	2.17	2.46	2.54	2.94	3.75	3.83	3.82	3.10	3.27	3.16	3.03	36.39
MOOSEHEAD		2.61	2.44	2.65	2.93	2.93	3.88	4.12	3.89	3.40	3.45	3.81	3.54	39.65
PORTLAND WSO	//R	3.78	3.57	3.98	3.90	3.27	3.06	2.83	2.82	3.27	3.83	4.70	4.51	43.52
PRESQUE ISLE		2.36	2.17	2.33	2.32	2.96	3.37	4.15	3.79	3.56	3.17	3.13	2.81	36.12
RIPOGENUS DAM		2.66	2.38	2.46	2.87	3.05	3.66	4.32	3.98	3.32	3.66	3.43	3.38	39.17
RUMFORD 1 SSE		3.03	2.86	3.44	3.53	3.60	3.81	3.84	3.68	3.48	4.18	4.51	3.88	43.84
SQUA PAN DAM		2.57	2.22	2.45	2.83	3.05	3.50	4.12	3.84	3.52	3.34	3.38	3.21	38.03
WATERVILLE PUMP STA		3.07	2.94	3.25	3.48	3.25	3.19	3.26	3.42	3.36	3.75	4.42	4.05	41.44
WOODLAND		4.23	3.67	3.32	3.75	3.61	3.27	3.30	3.13	3.56	4.13	5.13	4.97	46.07

MAINE

HEATING DEGREE DAY NORMALS (BASE 65 DEG F)

STATION		JUL	AUG	SEP	OCT	NOV	DEC	JAN	FEB	MAR	APR	MAY	JUN	ANN
AUGUSTA FAA AIRPORT		25	42	195	508	822	1271	1407	1218	1039	654	329	88	7598
BANGOR FAA AP		32	49	218	536	843	1302	1451	1266	1085	696	372	97	7947
BAR HARBOR		44	45	190	471	756	1153	1287	1140	998	657	363	120	7224
BELFAST		21	28	181	484	774	1200	1333	1170	1004	660	350	94	7299
CARIBOU WSO	//R	86	120	342	679	1017	1528	1686	1456	1262	831	459	150	9616
CORINNA		41	66	249	574	897	1376	1519	1313	1119	717	369	97	8337
EASTPORT	R	100	91	246	512	774	1166	1308	1165	1042	729	477	234	7844
FARMINGTON		51	85	277	601	921	1395	1547	1344	1150	744	397	126	8638
GARDINER		28	45	209	518	825	1271	1411	1226	1051	675	350	90	7699
HOULTON FAA AP		82	99	321	648	981	1479	1621	1414	1228	813	443	145	9274
HOULTON		59	82	285	620	957	1438	1587	1372	1181	783	415	122	8901
JONESBORO		69	74	255	564	831	1268	1414	1260	1097	741	459	187	8219
LEWISTON		13	27	161	474	801	1231	1376	1196	1026	654	321	71	7351
MADISON		34	58	236	561	876	1352	1497	1305	1116	717	382	101	8235
PORTLAND WSO	//R	22	54	201	515	798	1215	1349	1176	1020	666	378	107	7501
PRESQUE ISLE		68	93	300	639	987	1494	1637	1414	1231	816	431	127	9237
RIPOGENUS DAM		80	106	305	642	978	1485	1659	1462	1280	858	490	155	9500
RUMFORD 1 SSE		53	84	250	577	897	1358	1497	1296	1119	723	382	119	8355
WATERVILLE PUMP STA		19	29	182	496	819	1274	1414	1218	1029	648	316	68	7512
WOODLAND		36	50	233	552	855	1299	1454	1296	1104	720	388	112	8099

MAINE

COOLING DEGREE DAY NORMALS (BASE 65 DEG F)

STATION		JAN	FEB	MAR	APR	MAY	JUN	JUL	AUG	SEP	OCT	NOV	DEC	ANN
AUGUSTA FAA AIRPORT		0	0	0	0	6	52	161	119	15	0	0	0	353
BANGOR FAA AP		0	0	0	0	0	25	125	92	8	0	0	0	250
BAR HARBOR		0	0	0	0	0	24	112	76	0	0	0	0	212
BELFAST		0	0	0	0	0	31	126	93	10	0	0	0	260
CARIBOU WSO	//R	0	0	0	0	0	12	92	43	0	0	0	0	147
CORINNA		0	0	0	0	0	25	121	79	6	0	0	0	231
EASTPORT	R	0	0	0	0	0	0	26	20	0	0	0	0	46
FARMINGTON		0	0	0	0	6	30	110	69	10	0	0	0	225
GARDINER		0	0	0	0	6	30	136	94	11	0	0	0	277
HOULTON FAA AP		0	0	0	0	0	19	119	52	0	0	0	0	190
HOULTON		0	0	0	0	0	20	114	58	0	0	0	0	192
JONESBORO		0	0	0	0	0	0	57	43	0	0	0	0	100
LEWISTON		0	0	0	0	8	59	174	132	17	0	0	0	390
MADISON		0	0	0	0	0	20	114	77	11	0	0	0	222
PORTLAND WSO	//R	0	0	0	0	0	23	118	104	9	0	0	0	254
PRESQUE ISLE		0	0	0	0	0	19	102	43	0	0	0	0	164
RIPOGENUS DAM		0	0	0	0	0	8	87	53	8	0	0	0	156
RUMFORD 1 SSE		0	0	0	0	7	35	118	81	7	0	0	0	248
WATERVILLE PUMP STA		0	0	0	0	10	50	155	116	17	0	0	0	348
WOODLAND		0	0	0	0	0	25	132	78	8	0	0	0	243

17 — MAINE

LEGEND
11 = TEMPERATURE ONLY
12 = PRECIPITATION ONLY
13 = TEMP. & PRECIP.

STATE-STATION NUMBER	STN TYP	NAME		LATITUDE DEG-MIN	LONGITUDE DEG-MIN	ELEVATION (FT)
17-0275	13	AUGUSTA FAA AIRPORT		N 4419	W 06948	350
17-0355	13	BANGOR FAA AP		N 4448	W 06849	160
17-0371	13	BAR HARBOR		N 4423	W 06812	30
17-0480	13	BELFAST		N 4424	W 06900	20
17-0814	12	BRASSUA DAM		N 4540	W 06949	1060
17-0934	12	BRUNSWICK		N 4354	W 06956	70
17-1175	13	CARIBOU WSO	//R	N 4652	W 06801	624
17-1628	13	CORINNA		N 4457	W 06913	360
17-2426	13	EASTPORT	R	N 4455	W 06700	79
17-2620	12	ELLSWORTH		N 4432	W 06826	24
17-2765	13	FARMINGTON		N 4441	W 07009	420
17-2868	12	FORT FAIRFIELD		N 4648	W 06746	300
17-2878	12	FORT KENT		N 4715	W 06836	520
17-3046	13	GARDINER		N 4413	W 06947	140
17-3892	13	HOULTON FAA AP		N 4608	W 06747	494
17-3897	13	HOULTON		N 4608	W 06750	410
17-4086	12	JACKMAN		N 4538	W 07016	1180
17-4183	13	JONESBORO		N 4439	W 06739	185
17-4566	13	LEWISTON		N 4406	W 07014	182
17-4781	12	LONG FALLS DAM		N 4513	W 07012	1160
17-4878	12	MACHIAS		N 4443	W 06728	40
17-4927	13	MADISON		N 4448	W 06953	260
17-5261	12	MIDDLE DAM		N 4447	W 07055	1460
17-5460	12	MOOSEHEAD		N 4535	W 06943	1028
17-6905	13	PORTLAND WSO	//R	N 4339	W 07019	43
17-6937	13	PRESQUE ISLE		N 4639	W 06800	599
17-7174	13	RIPOGENUS DAM		N 4553	W 06911	965
17-7325	13	RUMFORD 1 SSE		N 4432	W 07032	630
17-8398	12	SQUA PAN DAM		N 4633	W 06820	610
17-9151	13	WATERVILLE PUMP STA		N 4433	W 06939	89
17-9891	13	WOODLAND		N 4509	W 06724	140

17 — MAINE

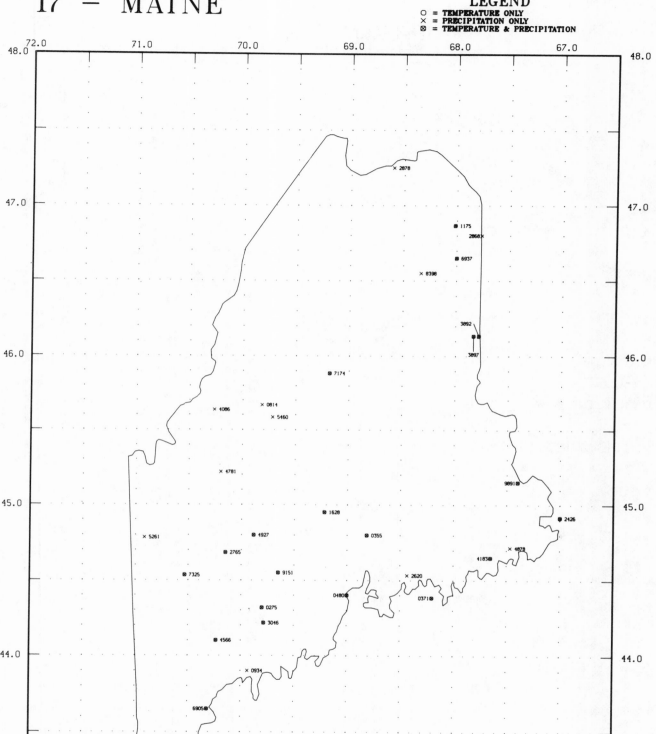

MARYLAND

TEMPERATURE NORMALS (DEG F)

STATION			JAN	FEB	MAR	APR	MAY	JUN	JUL	AUG	SEP	OCT	NOV	DEC	ANN
BALTIMORE WSO	R	MAX	41.0	43.7	53.1	65.1	74.2	82.9	87.1	85.5	79.1	67.7	55.9	45.1	65.0
		MIN	24.3	25.7	33.4	42.9	52.5	61.5	66.5	65.7	58.6	46.1	36.6	27.9	45.1
		MEAN	32.7	34.7	43.3	54.0	63.4	72.2	76.8	75.6	68.9	56.9	46.3	36.5	55.1
BALTIMORE CITY WSO	R	MAX	42.3	44.6	53.6	65.8	75.2	83.8	88.0	86.5	79.9	68.5	56.6	46.1	65.9
		MIN	28.6	29.8	37.4	47.7	57.5	66.7	71.7	70.3	63.5	51.6	41.8	32.6	49.9
		MEAN	35.5	37.3	45.5	56.8	66.3	75.3	79.9	78.4	71.7	60.1	49.2	39.3	57.9
BELTSVILLE		MAX	41.3	43.3	52.8	64.6	73.7	81.9	86.6	85.2	79.1	67.6	56.3	45.1	64.8
		MIN	21.2	22.6	31.0	40.6	49.8	58.7	63.7	62.7	55.3	42.3	33.6	25.5	42.3
		MEAN	31.3	33.0	41.9	52.7	61.8	70.3	75.2	74.0	67.2	55.0	45.0	35.3	53.6
BENSON POLICE BARRACKS		MAX	40.6	43.9	53.3	65.2	74.6	82.1	86.0	84.7	78.9	68.0	55.7	44.4	64.8
		MIN	21.8	23.3	30.9	40.5	50.2	59.0	63.7	62.9	56.1	44.0	34.8	25.8	42.8
		MEAN	31.2	33.6	42.1	52.8	62.4	70.6	74.9	73.8	67.5	56.0	45.3	35.1	53.8
CAMBRIDGE 4 W		MAX	43.1	45.4	54.2	65.6	74.8	83.0	87.0	85.6	79.9	68.9	58.0	47.4	66.1
		MIN	26.2	27.2	34.8	44.1	53.5	62.0	66.8	65.6	59.1	48.0	38.8	30.0	46.3
		MEAN	34.7	36.3	44.5	54.9	64.2	72.5	76.9	75.6	69.5	58.5	48.4	38.7	56.2
CHESTERTOWN		MAX	41.2	43.6	52.7	64.8	74.5	82.8	87.0	85.8	79.6	68.4	56.5	45.2	65.2
		MIN	25.0	26.3	34.0	43.3	52.9	61.8	66.7	65.7	59.0	47.2	37.6	28.7	45.7
		MEAN	33.1	35.0	43.4	54.1	63.7	72.3	76.9	75.8	69.3	57.8	47.0	37.0	55.5
COLLEGE PARK		MAX	43.2	46.3	55.7	67.8	77.0	84.7	88.5	87.2	80.8	69.3	57.5	46.3	67.0
		MIN	24.7	26.1	33.5	43.0	52.4	61.1	65.8	64.7	57.7	45.5	36.4	28.2	44.9
		MEAN	34.0	36.2	44.6	55.4	64.7	72.9	77.2	76.0	69.2	57.4	46.9	37.3	56.0
CONOWINGO DAM		MAX	38.6	41.5	50.8	63.1	73.2	81.8	86.2	84.6	77.8	66.2	53.8	42.4	63.3
		MIN	21.8	23.1	31.3	40.9	50.7	59.3	64.0	63.3	56.6	44.4	34.3	25.8	43.0
		MEAN	30.2	32.3	41.1	52.0	62.0	70.6	75.1	74.0	67.2	55.3	44.1	34.1	53.2
CRISFIELD SOMERS COVE		MAX	43.7	45.6	53.9	65.1	74.2	81.8	86.0	85.1	79.5	68.7	57.9	48.0	65.8
		MIN	29.8	30.4	37.8	47.7	57.1	65.8	71.3	70.5	64.2	52.9	43.0	33.7	50.4
		MEAN	36.8	38.0	45.9	56.4	65.7	73.8	78.7	77.8	71.9	60.8	50.5	40.9	58.1
CUMBERLAND 2		MAX	39.9	43.1	53.3	66.3	76.5	83.7	87.5	86.1	79.8	68.1	54.3	43.0	65.1
		MIN	21.8	23.2	30.8	40.6	49.1	56.9	61.3	60.3	53.4	41.9	33.8	25.6	41.6
		MEAN	30.9	33.2	42.1	53.5	62.8	70.4	74.4	73.2	66.6	55.0	44.1	34.3	53.4
DALECARLIA RESVR D C		MAX	43.4	46.6	56.4	68.2	77.1	84.4	88.0	86.7	80.2	69.3	57.7	46.9	67.1
		MIN	24.4	25.9	33.5	43.1	52.9	61.2	66.3	65.2	58.2	45.8	36.3	27.9	45.1
		MEAN	33.9	36.3	44.9	55.7	65.0	72.8	77.2	75.9	69.2	57.6	47.0	37.4	56.1
DENTON 1 WNW		MAX	43.2	45.8	54.6	67.0	76.3	84.2	88.1	86.7	80.7	69.5	58.3	47.1	66.8
		MIN	24.9	26.3	33.5	42.7	52.2	60.9	65.6	64.5	57.3	45.9	36.8	28.5	44.9
		MEAN	34.1	36.1	44.1	54.9	64.3	72.6	76.9	75.6	69.1	57.7	47.6	37.8	55.9
EASTON POLICE BARRACKS		MAX	43.5	46.2	55.5	66.8	76.1	84.0	87.7	86.3	80.6	69.5	58.1	47.4	66.8
		MIN	26.8	27.8	35.2	44.5	53.8	62.4	67.3	65.9	59.3	48.3	39.2	30.6	46.8
		MEAN	35.2	37.0	45.4	55.7	65.0	73.2	77.5	76.2	70.0	58.9	48.7	39.0	56.8
GLENN DALE BELL STA		MAX	44.0	47.1	56.6	68.2	76.9	84.2	88.1	86.9	80.9	70.1	58.4	47.3	67.4
		MIN	22.4	23.8	31.1	40.3	49.9	58.2	62.9	62.1	55.1	43.0	34.1	25.9	42.4
		MEAN	33.2	35.4	43.9	54.2	63.4	71.2	75.6	74.5	68.0	56.6	46.3	36.7	54.9
HAGERSTOWN		MAX	38.7	41.8	52.2	65.0	75.0	82.8	86.8	85.2	78.6	66.7	53.4	42.3	64.0
		MIN	22.4	23.9	31.9	42.1	51.2	59.1	63.0	61.6	55.0	43.7	35.1	26.1	42.9
		MEAN	30.6	32.9	42.1	53.6	63.1	71.0	74.9	73.4	66.8	55.3	44.3	34.2	53.5
HANCOCK FRUIT LAB		MAX	38.3	40.6	50.7	63.6	73.2	81.2	85.4	83.9	77.4	65.7	53.4	41.8	62.9
		MIN	20.1	21.5	29.2	38.5	47.1	55.8	60.5	59.5	52.3	40.0	32.3	24.0	40.1
		MEAN	29.2	31.1	40.0	51.1	60.2	68.5	72.9	71.7	64.9	52.9	42.8	32.9	51.5
LA PLATA 1 W		MAX	44.4	47.5	57.0	68.5	76.2	83.1	86.9	85.8	80.3	69.8	58.7	47.7	67.2
		MIN	24.6	26.1	33.4	42.5	51.7	60.0	64.8	63.9	57.3	45.3	35.9	27.9	44.5
		MEAN	34.6	36.8	45.2	55.5	64.0	71.6	75.9	74.8	68.8	57.6	47.3	37.9	55.8

MARYLAND

TEMPERATURE NORMALS (DEG F)

STATION		JAN	FEB	MAR	APR	MAY	JUN	JUL	AUG	SEP	OCT	NOV	DEC	ANN
LAUREL 3 W	MAX	41.7	44.5	54.3	66.3	75.4	83.3	87.7	86.4	79.9	68.5	56.2	44.8	65.8
	MIN	25.1	26.8	34.2	44.0	53.3	61.7	66.7	66.0	59.3	47.9	38.7	29.0	46.1
	MEAN	33.4	35.7	44.2	55.2	64.4	72.6	77.2	76.2	69.6	58.2	47.5	36.9	55.9
MILLINGTON	MAX	41.7	44.3	53.8	65.5	74.9	82.9	87.0	85.8	80.1	68.8	56.9	45.6	65.6
	MIN	22.7	23.9	31.8	40.9	50.5	59.3	64.2	63.2	55.9	44.2	35.2	26.4	43.2
	MEAN	32.2	34.1	42.8	53.2	62.7	71.1	75.6	74.5	68.0	56.5	46.1	36.1	54.4
NATIONAL ARBORETUM D C	MAX	43.8	46.5	56.4	68.0	77.0	84.6	88.5	87.4	81.3	69.6	58.2	47.3	67.4
	MIN	25.9	27.2	34.7	44.3	53.3	62.2	67.0	65.9	58.8	46.9	38.0	29.5	46.1
	MEAN	34.8	36.8	45.6	56.2	65.2	73.4	77.7	76.7	70.1	58.3	48.1	38.5	56.8
OAKLAND 1 SE	MAX	36.0	38.4	47.6	59.9	69.2	76.1	79.1	78.4	72.9	62.2	49.9	39.6	59.1
	MIN	16.4	17.5	25.1	34.4	43.5	51.1	55.6	54.9	48.1	36.6	28.5	20.2	36.0
	MEAN	26.2	28.0	36.4	47.2	56.3	63.6	67.4	66.7	60.5	49.5	39.2	29.9	47.6
OWINGS FERRY LANDING	MAX	43.5	45.9	55.1	66.7	75.5	82.9	86.9	85.4	79.3	68.6	57.9	47.2	66.2
	MIN	25.4	26.7	34.1	44.3	53.3	61.5	65.7	64.6	58.3	46.8	37.8	29.0	45.6
	MEAN	34.5	36.3	44.6	55.5	64.4	72.2	76.3	75.0	68.8	57.7	47.9	38.1	55.9
POCOMOKE CITY	MAX	45.6	48.1	56.1	67.1	75.4	82.8	87.1	85.9	80.2	69.6	59.5	49.7	67.3
	MIN	26.8	27.9	34.7	43.2	51.7	60.5	65.6	64.8	57.8	46.7	38.0	30.1	45.7
	MEAN	36.2	38.0	45.5	55.2	63.6	71.7	76.4	75.4	69.1	58.2	48.8	39.9	56.5
PRINCESS ANNE	MAX	45.5	47.7	56.2	67.4	76.3	83.6	87.5	86.3	80.6	69.9	59.6	49.5	67.5
	MIN	25.6	26.8	33.8	42.1	51.2	59.6	64.5	63.7	56.5	45.3	36.4	28.9	44.5
	MEAN	35.6	37.3	45.0	54.8	63.8	71.7	76.0	75.0	68.6	57.6	48.0	39.2	56.1
ROCKVILLE 4 NE	MAX	42.4	45.7	55.6	67.5	76.4	83.5	87.4	85.9	79.6	69.1	57.0	45.8	66.3
	MIN	24.1	25.0	32.5	42.1	51.3	59.4	64.0	62.8	55.8	44.3	35.7	26.8	43.7
	MEAN	33.3	35.4	44.1	54.9	63.9	71.5	75.7	74.4	67.7	56.8	46.3	36.4	55.0
ROYAL OAK	MAX	43.1	45.5	54.7	66.3	75.6	83.4	87.4	86.2	80.3	69.3	57.9	47.1	66.4
	MIN	27.0	27.9	35.6	45.2	54.8	63.5	68.0	66.8	60.3	49.3	39.8	30.8	47.4
	MEAN	35.0	36.7	45.2	55.8	65.2	73.5	77.7	76.6	70.3	59.3	48.9	38.9	56.9
SALISBURY	MAX	45.8	48.1	56.4	67.4	76.1	83.2	86.9	85.9	80.6	70.2	59.9	49.5	67.5
	MIN	27.6	28.6	35.7	44.2	53.5	62.0	67.0	66.1	59.3	47.9	39.1	31.0	46.8
	MEAN	36.7	38.4	46.0	55.8	64.8	72.6	77.0	76.0	69.9	59.1	49.5	40.3	57.2
SALISBURY FAA AIRPORT	MAX	43.9	45.9	53.9	65.4	73.8	81.9	86.1	84.8	78.7	68.0	58.0	47.9	65.7
	MIN	26.3	27.3	34.5	43.1	52.3	61.3	66.8	65.8	58.4	46.6	37.5	29.6	45.8
	MEAN	35.1	36.6	44.2	54.2	63.1	71.6	76.5	75.3	68.6	57.4	47.7	38.8	55.8
SAVAGE RIVER DAM	MAX	35.3	37.5	47.1	59.6	69.7	77.0	80.9	80.1	74.0	62.7	49.9	38.6	59.4
	MIN	17.5	18.1	26.4	37.0	45.8	53.7	58.0	56.9	50.0	38.8	30.7	21.7	37.9
	MEAN	26.4	27.9	36.8	48.4	57.8	65.4	69.5	68.6	62.0	50.8	40.3	30.2	48.7
SNOW HILL 4 N	MAX	45.3	47.3	55.2	66.0	74.5	82.3	86.1	85.3	79.9	69.3	59.1	49.3	66.6
	MIN	26.6	27.6	34.7	43.1	52.4	60.9	65.7	64.8	58.1	47.0	37.7	29.7	45.7
	MEAN	36.0	37.5	45.0	54.6	63.5	71.6	75.9	75.0	69.0	58.1	48.5	39.5	56.2
SOLOMONS	MAX	43.4	45.6	53.8	65.0	74.7	82.3	86.6	85.8	79.7	68.3	57.3	47.4	65.8
	MIN	29.0	29.9	37.2	46.9	56.9	65.5	70.4	70.0	64.4	53.0	42.8	33.3	49.9
	MEAN	36.2	37.8	45.5	56.0	65.8	74.0	78.5	77.9	72.1	60.6	50.1	40.4	57.9
UNIONVILLE	MAX	39.2	42.4	52.4	64.6	73.8	81.7	85.6	84.1	77.7	66.6	54.3	43.1	63.8
	MIN	19.8	21.4	29.3	38.3	48.1	56.9	61.5	60.3	52.9	41.0	32.0	23.6	40.4
	MEAN	29.5	31.9	40.9	51.5	60.9	69.3	73.6	72.2	65.3	53.8	43.2	33.4	52.1
VIENNA	MAX	44.6	46.8	55.7	66.9	76.0	83.8	87.5	86.5	80.6	69.8	59.2	48.6	67.2
	MIN	27.0	28.1	35.4	44.3	53.5	62.2	67.3	66.1	59.1	47.9	38.8	30.4	46.7
	MEAN	35.9	37.5	45.6	55.6	64.8	73.0	77.4	76.3	69.9	58.9	49.0	39.6	57.0
WESTMINSTER POLICE BRK	MAX	37.9	40.8	50.8	63.1	72.6	80.7	84.6	83.2	76.7	65.1	52.8	41.8	62.5
	MIN	22.3	23.5	31.1	41.2	50.4	59.0	63.6	62.6	55.8	44.6	35.7	26.2	43.0
	MEAN	30.1	32.2	41.0	52.2	61.5	69.9	74.2	73.0	66.3	54.9	44.3	34.0	52.8

MARYLAND

TEMPERATURE NORMALS (DEG F)

STATION		JAN	FEB	MAR	APR	MAY	JUN	JUL	AUG	SEP	OCT	NOV	DEC	ANN
WOODSTOCK	MAX	40.7	43.5	53.1	65.4	75.2	82.8	86.9	85.2	78.6	67.8	55.3	44.5	64.9
	MIN	22.0	23.5	31.0	40.2	49.6	57.9	62.8	61.9	55.0	42.6	34.1	25.8	42.2
	MEAN	31.4	33.5	42.1	52.9	62.4	70.3	74.9	73.5	66.8	55.2	44.7	35.2	53.6

MARYLAND

PRECIPITATION NORMALS (INCHES)

STATION	JAN	FEB	MAR	APR	MAY	JUN	JUL	AUG	SEP	OCT	NOV	DEC	ANN
BALTIMORE WSO R	3.00	2.98	3.72	3.35	3.44	3.76	3.89	4.62	3.46	3.11	3.11	3.40	41.84
BALTIMORE CITY WSO R	3.07	3.07	3.93	3.51	3.63	3.71	4.03	4.57	3.66	3.09	3.34	3.78	43.39
BELTSVILLE	2.91	2.50	3.49	3.37	3.81	3.89	4.14	4.71	3.71	3.24	2.97	3.34	42.08
BENSON POLICE BARRACKS	3.43	2.88	4.04	4.00	4.05	4.30	4.35	4.91	4.18	2.98	3.89	3.99	47.00
BRIGHTON DAM	2.96	2.72	3.73	3.47	3.93	3.89	3.54	4.13	3.69	3.12	3.17	3.41	41.76
CAMBRIDGE 4 W	3.49	3.16	4.06	3.30	3.60	3.57	4.31	5.28	3.72	3.28	3.72	3.71	45.20
CHESTERTOWN	3.28	2.95	3.98	3.25	3.62	4.24	3.94	4.61	3.58	3.16	3.46	3.87	43.94
COLLEGE PARK	3.01	2.79	3.65	3.37	3.85	3.85	4.06	5.00	3.56	3.11	3.10	3.38	42.73
CONOWINGO DAM	3.33	2.85	3.99	3.82	3.73	4.29	4.37	4.54	3.89	3.19	3.48	3.87	45.35
CRISFIELD SOMERS COVE	3.23	3.07	4.04	2.97	3.24	3.16	3.77	4.39	3.27	3.21	3.09	3.20	40.64
CUMBERLAND 2	2.53	2.41	3.61	3.36	3.61	3.78	3.19	3.33	2.92	2.71	2.41	2.64	36.50
DALECARLIA RESVR D C	3.07	2.93	3.83	3.60	3.77	3.68	4.04	4.60	3.42	3.30	3.15	3.49	42.88
DENTON 1 WNW	3.32	2.91	3.92	3.32	3.50	3.66	4.34	4.98	3.50	3.37	3.53	3.52	43.87
EASTON POLICE BARRACKS	3.30	3.07	3.77	3.31	3.56	3.14	3.85	4.96	3.52	3.21	3.39	3.61	42.69
EDGEMONT	3.07	2.62	3.71	3.65	3.82	4.19	3.25	3.83	4.16	3.30	3.38	3.02	42.00
FREDERICK 3 E	2.88	2.55	3.52	3.61	3.86	4.02	3.64	3.71	3.55	3.13	3.16	3.23	40.86
GLENN DALE BELL STA	3.06	2.75	3.70	3.52	3.94	3.87	4.31	4.91	3.66	3.30	3.34	3.39	43.75
HAGERSTOWN	2.80	2.26	3.38	3.50	3.74	3.92	3.34	3.59	3.30	3.13	3.02	2.86	38.84
HANCOCK FRUIT LAB	2.61	2.36	3.49	3.24	3.41	3.86	3.56	3.30	3.22	2.93	2.67	2.59	37.24
LA PLATA 1 W	3.14	2.88	3.76	3.17	3.47	3.78	4.03	4.92	3.44	3.41	3.20	3.42	42.62
LAUREL 3 W	2.87	2.69	3.66	3.40	4.06	3.89	3.88	4.17	3.38	3.13	3.40	3.34	41.87
MERRILL	2.91	2.72	3.92	3.67	3.87	3.60	3.67	3.66	2.86	2.88	2.61	3.13	39.50
MILLINGTON	3.16	2.87	3.70	3.18	3.42	3.67	3.87	4.74	3.79	3.05	3.47	3.58	42.50
NATIONAL ARBORETUM D C	3.17	2.85	3.75	3.37	3.93	3.95	4.20	5.10	3.58	3.17	3.08	3.36	43.51
OAKLAND 1 SE	3.79	3.20	4.42	4.51	4.44	4.36	4.68	4.60	3.11	3.17	3.14	3.90	47.32
OWINGS FERRY LANDING	3.14	2.72	3.54	3.24	3.80	3.71	3.66	4.19	3.52	3.27	3.33	3.29	41.41
POCOMOKE CITY	3.44	3.32	4.39	3.19	3.45	3.60	4.01	5.07	3.91	3.64	3.18	3.64	44.84
PRINCESS ANNE	3.33	3.30	3.99	3.32	3.35	3.41	3.82	5.63	3.97	3.68	3.24	3.40	44.44
ROCKVILLE 4 NE	2.95	2.66	3.54	3.24	3.77	3.99	3.78	4.46	3.46	3.00	3.06	3.00	40.91
ROYAL OAK	3.44	3.20	4.07	3.41	3.63	3.43	4.39	5.09	3.72	3.46	3.73	3.74	45.31
SALISBURY	3.52	3.41	4.22	3.31	3.32	3.62	3.76	5.72	3.59	3.63	3.11	3.62	44.83
SALISBURY FAA AIRPORT	3.38	3.33	4.08	3.16	3.38	3.55	4.40	5.72	3.94	3.73	3.06	3.71	45.44
SAVAGE RIVER DAM	2.60	2.35	3.38	3.48	3.67	3.50	3.62	3.61	2.99	2.88	2.35	2.68	37.11
SNOW HILL 4 N	3.63	3.60	4.60	3.32	3.43	3.80	3.99	5.43	3.53	3.80	3.38	3.71	46.22
SOLOMONS	3.50	3.01	3.71	3.05	3.38	3.26	4.67	4.57	3.55	3.02	3.31	3.36	42.39
TOWSON	3.31	3.18	4.04	3.98	4.01	4.04	4.13	4.79	4.15	3.43	3.83	3.82	46.71
UNIONVILLE	2.76	2.41	3.38	3.51	3.67	4.08	3.59	3.76	3.51	2.93	3.09	3.30	39.99
U. S. SOLDIERS HOME DC	3.16	2.85	3.78	3.34	3.85	4.02	4.33	4.84	3.56	3.10	3.01	3.48	43.32
UPPER MARLBORO 3 NNW	3.28	2.73	3.72	3.32	3.60	3.84	4.06	4.66	3.73	3.29	3.16	3.39	42.78
VIENNA	3.52	3.27	3.92	3.34	3.02	3.56	3.94	5.63	3.97	3.44	3.32	3.57	44.50
WESTMINSTER POLICE BRK	3.12	2.82	3.92	3.69	3.87	4.09	4.00	4.00	3.83	3.10	3.21	3.49	43.14
WOODSTOCK	3.14	2.76	3.94	3.62	3.77	4.03	3.56	3.98	4.00	3.12	3.28	3.52	42.72

MARYLAND

HEATING DEGREE DAY NORMALS (BASE 65 DEG F)

STATION	JUL	AUG	SEP	OCT	NOV	DEC	JAN	FEB	MAR	APR	MAY	JUN	ANN
BALTIMORE WSO R	0	0	29	261	561	884	1001	848	673	334	115	0	4706
BALTIMORE CITY WSO R	0	0	12	178	474	797	915	776	605	257	69	0	4083
BELTSVILLE	0	0	40	318	600	921	1045	896	716	369	139	9	5053
BENSON POLICE BARRACKS	0	0	48	292	591	927	1048	879	710	366	129	8	4998
CAMBRIDGE 4 W	0	0	18	220	498	815	939	804	636	307	94	0	4331
CHESTERTOWN	0	0	21	241	540	868	989	840	670	327	104	0	4600
COLLEGE PARK	0	0	24	249	543	859	961	806	632	292	89	0	4455
CONOWINGO DAM	0	0	44	310	627	958	1079	916	741	390	133	8	5206
CRISFIELD SOMERS COVE	0	0	7	163	435	747	874	756	592	265	63	0	3902
CUMBERLAND 2	0	0	57	317	627	952	1057	890	710	349	137	10	5106
DALECARLIA RESVR D C	0	0	25	241	540	856	964	804	623	285	87	0	4425
DENTON 1 WNW	0	0	23	241	522	843	958	809	648	308	94	0	4446
EASTON POLICE BARRACKS	0	0	18	212	489	806	924	784	608	288	82	0	4211
GLENN DALE BELL STA	0	0	28	272	561	877	986	829	654	324	111	7	4649
HAGERSTOWN	0	0	48	307	621	955	1066	899	710	342	127	11	5086
HANCOCK FRUIT LAB	0	0	76	379	666	995	1110	949	775	417	184	22	5573
LA PLATA 1 W	0	0	28	245	531	840	942	790	614	290	98	6	4384
LAUREL 3 W	0	0	21	228	525	871	980	820	645	300	99	0	4489
MILLINGTON	0	0	37	280	567	896	1017	865	688	354	123	10	4837
NATIONAL ARBORETUM D C	0	0	18	231	507	822	936	790	601	274	88	0	4267
OAKLAND 1 SE	25	39	160	481	774	1088	1203	1036	887	534	283	89	6599
OWINGS FERRY LANDING	0	0	23	240	513	834	946	804	632	293	94	5	4384
POCOMOKE CITY	0	0	37	236	486	778	893	756	605	299	102	6	4198
PRINCESS ANNE	0	0	25	245	510	800	911	776	620	311	95	5	4298
ROCKVILLE 4 NE	0	0	51	271	561	887	983	829	648	311	112	10	4663
ROYAL OAK	0	0	13	200	483	809	930	792	614	281	70	0	4192
SALISBURY	0	0	12	203	465	766	877	745	589	282	77	0	4016
SALISBURY FAA AIRPORT	0	0	27	248	519	812	927	795	645	327	104	7	4411
SAVAGE RIVER DAM	11	20	131	440	741	1079	1197	1039	874	498	245	60	6335
SNOW HILL 4 N	0	0	20	228	495	791	899	770	620	316	100	0	4239
SOLOMONS	0	0	6	166	447	763	893	762	605	275	64	0	3981
UNIONVILLE	0	0	75	353	654	980	1101	927	747	405	160	13	5415
VIENNA	0	0	20	216	480	787	902	770	601	286	78	0	4140
WESTMINSTER POLICE BRK	0	0	57	319	621	961	1082	918	744	384	155	9	5250
WOODSTOCK	0	0	50	310	609	924	1042	882	710	363	130	7	5027

MARYLAND

COOLING DEGREE DAY NORMALS (BASE 65 DEG F)

STATION		JAN	FEB	MAR	APR	MAY	JUN	JUL	AUG	SEP	OCT	NOV	DEC	ANN
BALTIMORE WSO	R	0	0	0	0	66	221	366	329	146	10	0	0	1138
BALTIMORE CITY WSO	R	0	0	0	11	109	309	462	415	213	26	0	0	1545
BELTSVILLE		0	0	0	0	40	168	316	279	106	8	0	0	917
BENSON POLICE BARRACKS		0	0	0	0	49	176	307	276	123	13	0	0	944
CAMBRIDGE 4 W		0	0	0	0	69	229	369	329	153	18	0	0	1167
CHESTERTOWN		0	0	0	0	63	223	369	335	150	18	0	0	1158
COLLEGE PARK		0	0	0	0	80	240	378	341	150	14	0	0	1203
CONOWINGO DAM		0	0	0	0	40	176	313	279	110	10	0	0	928
CRISFIELD SOMERS COVE		0	0	0	7	84	264	425	397	214	32	0	0	1423
CUMBERLAND 2		0	0	0	0	69	172	291	258	105	7	0	0	902
DALECARLIA RESVR D C		0	0	0	6	87	237	378	338	151	12	0	0	1209
DENTON 1 WNW		0	0	0	0	72	233	369	329	146	14	0	0	1163
EASTON POLICE BARRACKS		0	0	0	9	82	249	388	347	168	23	0	0	1266
GLENN DALE BELL STA		0	0	0	0	61	193	329	295	118	12	0	0	1008
HAGERSTOWN		0	0	0	0	68	191	307	260	102	7	0	0	935
HANCOCK FRUIT LAB		0	0	0	0	35	127	245	212	73	0	0	0	692
LA PLATA 1 W		0	0	0	0	67	204	338	304	142	16	0	0	1071
LAUREL 3 W		0	0	0	6	81	232	378	347	159	17	0	0	1220
MILLINGTON		0	0	0	0	51	193	329	295	127	16	0	0	1011
NATIONAL ARBORETUM D C		0	0	0	10	94	255	394	363	171	24	0	0	1311
OAKLAND 1 SE		0	0	0	0	13	47	99	92	25	0	0	0	276
OWINGS FERRY LANDING		0	0	0	8	76	221	350	310	137	14	0	0	1116
POCOMOKE CITY		0	0	0	0	58	207	353	322	160	26	0	0	1126
PRINCESS ANNE		0	0	0	0	58	206	341	310	133	15	0	0	1063
ROCKVILLE 4 NE		0	0	0	8	78	205	332	291	132	17	0	0	1063
ROYAL OAK		0	0	0	5	76	255	394	360	172	24	0	0	1286
SALISBURY		0	0	0	6	71	231	372	341	159	20	0	0	1200
SALISBURY FAA AIRPORT		0	0	0	0	45	205	357	319	135	13	0	0	1074
SAVAGE RIVER DAM		0	0	0	0	22	72	150	132	41	0	0	0	417
SNOW HILL 4 N		0	0	0	0	54	202	338	310	140	14	0	0	1058
SOLOMONS		0	0	0	5	89	270	419	400	219	30	0	0	1432
UNIONVILLE		0	0	0	0	33	142	267	227	84	6	0	0	759
VIENNA		0	0	0	0	72	245	384	350	167	27	0	0	1245
WESTMINSTER POLICE BRK		0	0	0	0	47	156	285	251	96	6	0	0	841
WOODSTOCK		0	0	0	0	49	166	307	264	104	6	0	0	896

18 — MARYLAND & D.C.

STATE-STATION NUMBER	STN TYP	NAME		LATITUDE DEG-MIN	LONGITUDE DEG-MIN	ELEVATION (FT)
18-0465	13	BALTIMORE WSO	R	N 3911	W 07640	148
18-0470	13	BALTIMORE CITY WSO	R	N 3917	W 07637	14
18-0700	13	BELTSVILLE		N 3902	W 07653	120
18-0732	13	BENSON POLICE BARRACKS		N 3930	W 07623	365
18-1125	12	BRIGHTON DAM		N 3912	W 07701	330
18-1385	13	CAMBRIDGE 4 W		N 3834	W 07609	5
18-1750	13	CHESTERTOWN		N 3913	W 07604	40
18-1995	13	COLLEGE PARK		N 3859	W 07657	90
18-2060	13	CONOWINGO DAM		N 3939	W 07610	40
18-2215	13	CRISFIELD SOMERS COVE		N 3759	W 07552	8
18-2282	13	CUMBERLAND 2		N 3938	W 07845	700
18-2325	13	DALECARLIA RESVR D C		N 3856	W 07707	146
18-2523	13	DENTON 1 WNW		N 3854	W 07551	40
18-2700	13	EASTON POLICE BARRACKS		N 3845	W 07604	40
18-2770	12	EDGEMONT		N 3940	W 07733	905
18-3355	12	FREDERICK 3 E		N 3924	W 07722	385
18-3675	13	GLENN DALE BELL STA		N 3858	W 07648	150
18-3975	13	HAGERSTOWN		N 3939	W 07744	660
18-4030	13	HANCOCK FRUIT LAB		N 3942	W 07811	428
18-5080	13	LA PLATA 1 W		N 3832	W 07700	140
18-5111	13	LAUREL 3 W		N 3906	W 07654	400
18-5894	12	MERRILL		N 3936	W 07905	1790
18-5985	13	MILLINGTON		N 3916	W 07551	30
18-6350	13	NATIONAL ARBORETUM D C		N 3854	W 07658	50
18-6620	13	OAKLAND 1 SE		N 3924	W 07924	2420
18-6770	13	OWINGS FERRY LANDING		N 3841	W 07640	160
18-7140	13	POCOMOKE CITY		N 3804	W 07533	20
18-7330	13	PRINCESS ANNE		N 3813	W 07541	20
18-7705	13	ROCKVILLE 4 NE		N 3907	W 07706	320
18-7806	13	ROYAL OAK		N 3843	W 07611	10
18-8000	13	SALISBURY		N 3822	W 07535	10
18-8005	13	SALISBURY FAA AIRPORT		N 3820	W 07530	48
18-8065	13	SAVAGE RIVER DAM		N 3931	W 07908	1495
18-8380	13	SNOW HILL 4 N		N 3814	W 07523	28
18-8405	13	SOLOMONS		N 3819	W 07627	12
18-8877	12	TOWSON		N 3923	W 07634	390
18-9030	13	UNIONVILLE		N 3927	W 07711	430
18-9035	12	U. S. SOLDIERS HOME DC		N 3856	W 07701	230
18-9070	12	UPPER MARLBORO 3 NNW		N 3852	W 07647	98
18-9140	13	VIENNA		N 3829	W 07550	12
18-9440	13	WESTMINSTER POLICE BRK		N 3933	W 07658	765
18-9750	13	WOODSTOCK		N 3920	W 07652	415

18 – MARYLAND & D.C.

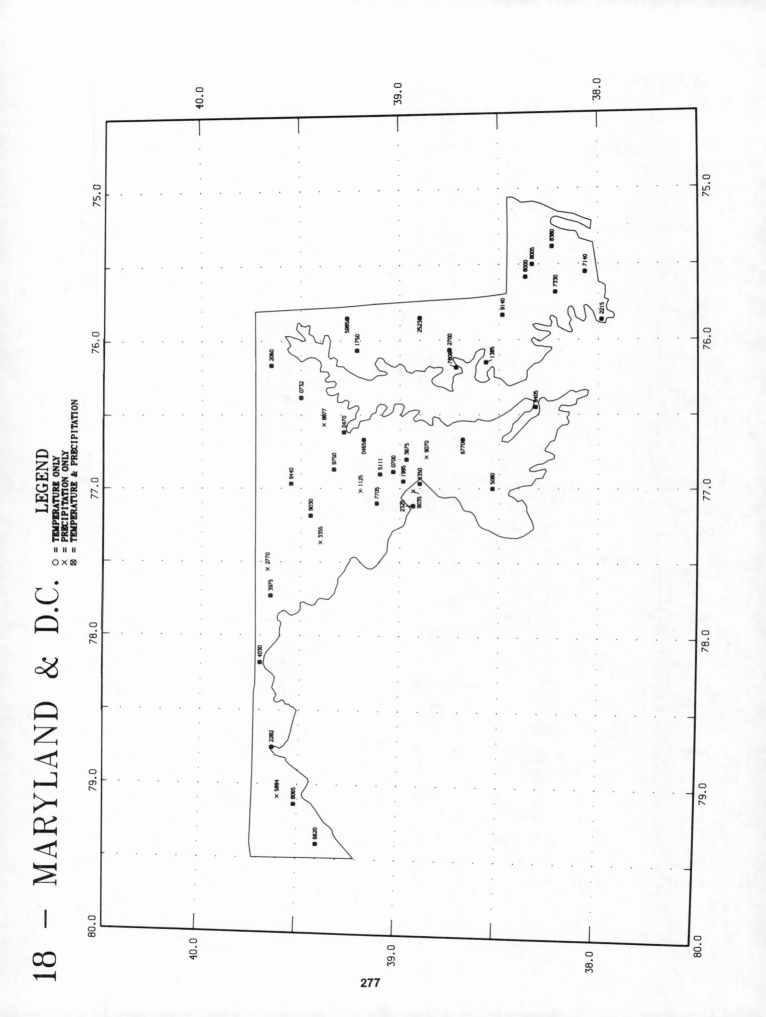

LEGEND

○ = TEMPERATURE ONLY
× = PRECIPITATION ONLY
⊗ = TEMPERATURE & PRECIPITATION

MASSACHUSETTS

TEMPERATURE NORMALS (DEG F)

STATION			JAN	FEB	MAR	APR	MAY	JUN	JUL	AUG	SEP	OCT	NOV	DEC	ANN
AMHERST		MAX	33.1	35.9	45.0	58.8	70.5	78.8	83.3	81.4	73.9	63.4	50.1	36.9	59.3
		MIN	13.4	15.2	25.3	34.9	44.6	54.5	59.4	57.3	49.5	39.1	31.1	18.9	36.9
		MEAN	23.3	25.6	35.2	46.8	57.6	66.7	71.4	69.4	61.7	51.3	40.7	27.9	48.1
BIRCH HILL DAM	R	MAX	31.4	33.7	42.3	56.2	68.5	77.2	81.5	79.3	71.6	60.9	47.6	35.0	57.1
		MIN	9.8	11.2	22.0	32.0	41.2	50.8	55.5	53.3	45.6	35.5	28.1	15.6	33.4
		MEAN	20.6	22.5	32.2	44.1	54.9	64.0	68.5	66.4	58.6	48.2	37.9	25.3	45.3
BLUE HILL WSO	//	MAX	33.5	35.3	43.1	55.7	66.7	75.2	80.7	78.7	70.9	60.9	49.2	37.4	57.3
		MIN	18.3	19.3	27.3	36.7	46.4	55.5	61.6	60.4	53.1	43.5	34.4	22.6	39.9
		MEAN	25.9	27.3	35.3	46.2	56.6	65.4	71.2	69.5	62.0	52.2	41.8	30.0	48.6
BOSTON WSO	//R	MAX	36.4	37.7	45.0	56.6	67.0	76.6	81.8	79.8	72.3	62.5	51.6	40.3	59.0
		MIN	22.8	23.7	31.8	40.8	50.0	59.3	65.1	63.9	56.9	47.1	38.7	27.1	43.9
		MEAN	29.6	30.7	38.4	48.7	58.5	68.0	73.5	71.9	64.6	54.8	45.2	33.7	51.5
CHATHAM WSMO		MAX	37.1	37.0	42.1	50.6	59.4	68.2	74.3	74.1	68.4	60.1	50.8	41.2	55.3
		MIN	25.0	25.1	31.5	38.5	46.2	54.7	60.8	60.8	55.7	47.2	39.5	29.4	42.9
		MEAN	31.1	31.1	36.8	44.6	52.8	61.5	67.6	67.5	62.1	53.7	45.2	35.3	49.1
CHESTNUT HILL		MAX	35.9	37.8	45.6	57.8	68.9	78.0	83.2	81.2	73.9	63.8	51.7	39.3	59.8
		MIN	19.4	20.4	28.5	38.2	47.5	56.7	62.6	61.0	53.4	43.2	35.1	23.7	40.8
		MEAN	27.7	29.1	37.1	48.0	58.2	67.4	72.9	71.1	63.7	53.5	43.4	31.6	50.3
CLINTON		MAX	34.4	35.8	43.7	56.1	67.4	76.2	81.0	79.0	71.6	62.0	50.4	38.1	58.0
		MIN	15.0	15.3	25.2	36.0	46.2	55.9	61.3	59.5	51.1	40.6	32.1	20.4	38.2
		MEAN	24.7	25.5	34.5	46.1	56.8	66.1	71.2	69.2	61.4	51.3	41.3	29.3	48.1
EAST WAREHAM		MAX	36.4	37.4	44.5	55.3	65.7	74.9	80.7	79.3	71.9	62.0	51.3	40.4	58.3
		MIN	18.5	19.0	27.4	35.5	45.5	55.1	61.5	60.2	52.4	42.0	33.3	22.8	39.4
		MEAN	27.4	28.2	36.0	45.4	55.6	65.0	71.1	69.7	62.2	52.0	42.3	31.6	48.9
EDGARTOWN		MAX	38.0	38.2	44.8	53.7	63.3	72.2	78.3	77.9	72.0	63.0	52.8	42.5	58.1
		MIN	20.9	20.8	28.2	36.4	45.2	54.5	60.7	60.2	54.0	44.4	36.2	25.5	40.6
		MEAN	29.5	29.6	36.5	45.1	54.3	63.3	69.5	69.0	63.0	53.8	44.5	34.0	49.3
FITCHBURG 2 S		MAX	34.3	35.8	44.2	57.5	69.7	78.6	84.0	82.0	74.4	63.9	50.6	37.8	59.4
		MIN	13.8	15.1	25.0	35.1	44.9	54.5	59.8	57.6	49.2	38.4	30.4	18.7	36.9
		MEAN	24.1	25.5	34.6	46.4	57.3	66.6	71.9	69.9	61.8	51.2	40.5	28.3	48.2
FRAMINGHAM		MAX	35.3	37.1	45.7	58.5	70.1	79.2	84.3	82.3	74.7	64.2	51.7	39.1	60.2
		MIN	15.8	16.9	26.9	37.2	46.8	56.1	61.8	59.7	51.4	40.4	32.5	20.6	38.8
		MEAN	25.6	27.0	36.4	47.9	58.5	67.7	73.1	71.1	63.0	52.3	42.2	29.9	49.6
HAVERHILL		MAX	34.9	37.5	46.2	59.3	70.8	79.5	84.6	82.4	74.7	64.0	51.0	38.5	60.3
		MIN	17.8	19.2	27.8	37.1	47.1	56.7	62.4	60.6	53.0	42.8	33.9	22.3	40.1
		MEAN	26.4	28.4	37.0	48.3	59.0	68.1	73.5	71.5	63.9	53.4	42.5	30.4	50.2
HYANNIS 2 NNE		MAX	38.3	38.8	44.9	54.6	64.7	73.8	79.6	78.6	72.0	62.9	52.5	42.7	58.6
		MIN	21.3	21.5	28.7	36.5	45.6	55.2	61.8	60.7	53.7	43.7	35.3	25.3	40.8
		MEAN	29.8	30.2	36.8	45.6	55.1	64.5	70.7	69.7	62.8	53.3	43.9	34.0	49.7
KNIGHTVILLE DAM		MAX	32.0	33.9	42.4	56.4	68.5	77.3	81.9	79.8	72.3	61.8	48.7	35.6	57.6
		MIN	9.9	10.9	21.6	32.2	41.4	51.3	56.0	53.6	45.3	34.7	27.7	15.9	33.4
		MEAN	21.0	22.4	32.0	44.3	55.0	64.3	69.0	66.7	58.8	48.3	38.2	25.7	45.5
LAWRENCE		MAX	34.7	36.7	44.5	57.3	68.7	77.7	82.8	80.9	73.1	62.7	50.7	38.2	59.0
		MIN	16.8	17.9	27.4	37.2	47.0	56.4	62.3	60.4	52.4	42.3	33.7	21.6	39.6
		MEAN	25.8	27.3	36.0	47.3	57.9	67.1	72.6	70.7	62.8	52.5	42.2	29.9	49.3
MIDDLETON		MAX	37.0	38.6	45.7	57.4	68.1	76.8	81.6	79.9	72.9	63.4	52.0	40.5	59.5
		MIN	17.9	18.7	27.3	37.0	46.6	56.2	61.8	60.0	52.3	42.5	34.3	22.9	39.8
		MEAN	27.5	28.7	36.5	47.2	57.4	66.5	71.7	70.0	62.6	53.0	43.1	31.7	49.7
NANTUCKET FAA AP		MAX	38.2	37.8	43.0	51.0	59.9	68.5	74.9	74.9	69.5	61.2	52.1	42.9	56.2
		MIN	24.7	24.5	30.3	37.3	45.4	54.5	61.3	61.5	55.5	47.0	39.0	29.1	42.5
		MEAN	31.5	31.2	36.7	44.2	52.7	61.5	68.1	68.2	62.5	54.1	45.6	36.0	49.4

MASSACHUSETTS

TEMPERATURE NORMALS (DEG F)

STATION		JAN	FEB	MAR	APR	MAY	JUN	JUL	AUG	SEP	OCT	NOV	DEC	ANN
NEW BEDFORD	MAX	38.3	39.4	46.2	56.4	66.2	75.2	80.8	80.0	73.2	63.9	53.0	42.3	59.6
	MIN	24.8	25.3	32.5	41.0	50.5	59.3	66.0	65.1	58.1	48.5	39.6	28.8	45.0
	MEAN	31.6	32.4	39.4	48.7	58.4	67.3	73.4	72.6	65.7	56.3	46.4	35.6	52.3
PLYMOUTH	MAX	37.9	39.1	46.1	56.6	67.4	76.9	82.4	80.4	73.2	63.6	52.8	41.8	59.9
	MIN	19.3	19.7	27.7	36.2	45.3	55.1	61.3	60.1	52.6	42.6	34.4	23.7	39.8
	MEAN	28.6	29.4	36.9	46.4	56.4	66.0	71.8	70.3	62.9	53.1	43.6	32.8	49.9
ROCHESTER	MAX	37.2	38.1	45.5	56.2	66.7	75.9	81.4	80.1	73.0	63.4	52.3	41.1	59.2
	MIN	16.6	17.2	26.7	35.5	44.6	54.1	60.1	58.7	50.9	40.1	32.5	21.3	38.2
	MEAN	26.9	27.6	36.1	45.9	55.7	65.0	70.8	69.4	61.9	51.8	42.5	31.2	48.7
ROCKPORT 1 ESE	MAX	34.8	35.6	42.5	53.1	63.1	72.1	77.4	76.0	68.9	59.4	49.2	38.9	55.9
	MIN	20.8	21.4	29.0	37.3	46.3	55.6	61.7	60.8	54.1	44.9	36.1	25.3	41.1
	MEAN	27.8	28.5	35.8	45.2	54.8	63.9	69.6	68.4	61.5	52.2	42.7	32.1	48.5
SPRINGFIELD	MAX	35.0	37.3	46.5	60.3	71.4	79.8	84.6	82.9	75.0	64.5	51.4	38.3	60.6
	MIN	18.1	19.7	28.2	38.1	47.7	56.8	62.5	60.9	52.6	42.8	34.4	22.5	40.4
	MEAN	26.6	28.5	37.4	49.2	59.6	68.3	73.6	71.9	63.8	53.7	42.9	30.5	50.5
STOCKBRIDGE	MAX	31.7	34.0	42.7	56.8	68.5	75.9	80.0	77.8	70.1	60.6	48.0	35.5	56.8
	MIN	11.6	12.9	22.7	32.7	42.2	51.2	55.9	54.6	47.5	37.1	29.3	17.5	34.6
	MEAN	21.6	23.5	32.7	44.8	55.4	63.6	68.0	66.2	58.8	48.8	38.7	26.5	45.7
TAUNTON	MAX	36.8	38.6	46.3	58.3	68.8	77.4	82.5	80.7	73.1	63.2	51.8	40.5	59.8
	MIN	17.2	18.3	26.6	34.9	44.1	53.8	59.9	58.2	50.2	39.5	31.7	21.3	38.0
	MEAN	27.0	28.4	36.5	46.6	56.5	65.6	71.3	69.5	61.7	51.4	41.8	31.0	48.9
TULLY LAKE	MAX	30.5	33.0	42.3	56.1	68.8	77.6	82.3	80.0	71.6	60.1	46.8	33.9	56.9
	MIN	9.3	10.4	22.0	32.5	41.5	51.2	56.0	53.8	45.8	35.5	28.4	15.4	33.5
	MEAN	19.9	21.7	32.2	44.3	55.2	64.4	69.2	66.9	58.7	47.8	37.6	24.7	45.2
WORCESTER WSO	MAX	30.9	32.9	41.1	54.5	65.9	74.4	79.0	77.0	69.4	59.3	46.9	34.7	55.5
	MIN	15.6	16.6	25.2	35.4	45.5	54.8	60.7	59.0	51.3	41.3	32.0	20.1	38.1
	MEAN	23.3	24.8	33.1	45.0	55.7	64.6	69.9	68.0	60.3	50.3	39.5	27.4	46.8

MASSACHUSETTS

PRECIPITATION NORMALS (INCHES)

STATION		JAN	FEB	MAR	APR	MAY	JUN	JUL	AUG	SEP	OCT	NOV	DEC	ANN
AMHERST		3.17	2.67	3.63	3.61	3.41	3.78	3.74	4.10	3.64	3.26	3.61	3.84	42.46
ASHBURNHAM		3.96	3.32	4.35	3.88	3.73	3.58	3.55	3.89	4.12	4.10	4.64	4.36	47.48
BEECHWOOD		4.36	3.77	4.37	4.09	3.96	2.99	3.27	4.24	4.26	4.42	4.99	4.99	49.71
BELCHERTOWN		3.63	3.02	4.01	3.86	3.62	3.73	4.07	4.69	3.96	3.55	3.94	3.94	46.02
BIRCH HILL DAM	R	3.22	2.65	3.37	3.22	3.12	3.39	3.22	3.55	3.19	3.07	3.72	3.58	39.30
BLUE HILL WSO	//	4.57	4.20	4.73	3.97	3.68	2.99	2.94	4.29	4.03	4.05	4.67	5.02	49.14
BOSTON WSO	//R	3.99	3.70	4.13	3.73	3.52	2.92	2.68	3.68	3.41	3.36	4.21	4.48	43.81
BOYLSTON		4.08	3.44	4.23	3.83	3.62	3.49	3.34	4.21	3.99	3.87	4.42	4.21	46.73
BROCKTON		3.92	3.88	4.19	3.99	3.46	2.95	2.95	4.23	3.99	3.92	4.43	4.41	46.32
CHATHAM WSMO		4.34	4.16	4.04	3.90	3.72	2.74	2.94	4.07	3.74	3.71	4.46	4.96	46.77
CHESTERFIELD		3.77	3.41	4.42	4.01	4.00	3.64	3.86	4.35	4.16	4.02	4.53	4.34	48.51
CHESTNUT HILL		4.19	3.55	4.12	3.72	3.53	3.00	3.09	3.78	3.84	3.55	4.44	4.62	45.43
CLINTON		4.09	3.62	4.37	3.79	3.57	3.43	3.26	4.30	3.84	3.79	4.53	4.38	46.97
EAST WAREHAM		4.23	3.94	4.39	4.15	3.83	2.86	2.94	4.43	3.80	3.66	4.48	4.82	47.53
EDGARTOWN		3.93	3.86	4.11	4.02	4.10	2.62	2.65	4.18	3.65	3.56	4.38	4.47	45.53
FITCHBURG 2 S		3.97	3.30	4.30	3.89	3.53	3.43	3.48	3.55	3.76	3.84	4.46	4.27	45.78
FRAMINGHAM		4.11	3.43	4.13	3.74	3.38	3.04	3.40	3.82	3.81	3.60	4.43	4.45	45.34
FRANKLIN		4.24	3.55	4.12	4.06	3.60	2.98	3.46	4.26	4.43	3.90	4.77	4.65	48.02
GARDNER		3.65	3.14	3.98	3.54	3.41	3.42	3.38	3.46	3.65	3.49	4.11	4.08	43.31
HARDWICK		3.40	2.94	3.93	3.73	3.58	3.81	3.95	4.45	4.00	3.60	4.02	3.90	45.31
HATCHVILLE		4.04	3.79	4.15	3.98	3.78	3.03	2.94	4.00	3.65	3.86	4.31	4.44	45.97
HAVERHILL		3.74	3.32	3.77	3.49	3.44	3.00	3.17	3.22	3.57	3.40	4.30	4.27	42.69
HEATH		3.89	3.36	4.25	4.11	4.42	4.27	4.04	4.25	4.29	4.41	4.69	4.46	50.44
HOLYOKE		3.30	2.94	3.97	3.91	3.26	3.48	3.80	4.00	3.96	3.36	3.84	3.94	43.76
HUBBARDSTON		3.23	2.69	3.56	3.50	3.50	3.58	3.61	3.67	3.74	3.45	3.88	3.67	42.08
HYANNIS 2 NNE		3.84	3.72	3.72	3.95	3.84	2.74	2.65	4.09	3.50	3.76	4.48	4.34	44.63
IPSWICH		4.37	3.81	4.34	3.98	3.68	3.05	3.05	3.32	3.82	3.83	4.85	5.03	47.13
KNIGHTVILLE DAM		3.50	3.02	4.10	3.72	3.63	3.49	3.59	4.03	3.74	3.76	4.15	4.05	44.78
LAWRENCE		3.59	3.10	3.67	3.64	3.39	3.19	3.31	3.28	3.69	3.51	4.30	3.89	42.56
MANSFIELD		4.15	3.76	4.32	4.18	3.71	2.94	2.90	4.51	4.11	4.14	4.63	4.39	47.74
MIDDLEBORO		4.19	3.90	4.38	3.94	3.66	2.82	3.19	4.36	3.86	3.68	4.46	4.61	47.05
MIDDLETON		3.82	3.41	3.77	3.49	3.52	2.89	2.98	3.30	3.53	3.71	4.23	4.23	42.88
MILFORD		4.01	3.39	4.04	3.86	3.46	3.09	3.28	4.04	4.04	3.69	4.37	4.28	45.55
NANTUCKET FAA AP		4.04	3.71	3.93	3.57	3.31	2.27	2.77	3.43	3.38	3.41	4.03	4.36	42.21
NEW BEDFORD		4.06	3.84	4.20	3.76	3.35	2.73	2.37	4.26	3.35	3.20	4.16	4.66	43.94
NEWBURYPORT		3.88	3.42	3.94	3.84	3.66	2.98	3.13	3.41	3.70	3.74	4.79	4.46	44.95
NEW SALEM		3.77	3.13	4.07	4.19	3.77	4.55	4.22	4.45	4.09	4.16	4.31	4.42	49.13
NORTHBRIDGE 2		4.00	3.61	4.21	3.79	3.34	3.11	3.22	4.05	4.10	3.64	4.16	4.23	45.46
PELHAM		3.36	2.79	3.93	4.05	3.79	4.09	3.93	4.33	4.08	3.84	4.02	4.02	46.23
PEMBROKE		4.45	4.22	4.36	4.21	3.77	2.71	3.07	4.48	4.08	4.03	4.75	4.90	49.03
PETERSHAM 3 N	//R	3.35	2.72	3.78	3.63	3.27	3.87	3.41	3.79	3.51	3.49	3.89	3.75	42.46
PLAINFIELD		3.51	3.20	4.08	3.89	4.20	3.73	3.96	4.49	4.21	3.95	4.28	4.24	47.74
PLYMOUTH		4.23	3.90	4.12	4.00	3.65	2.70	2.97	4.49	4.16	4.00	4.74	4.75	47.71
QUABBIN INTAKE		3.57	3.07	4.09	3.92	3.65	3.79	3.77	4.27	3.83	3.55	3.89	4.06	45.46
ROCHESTER		4.52	4.09	4.71	4.27	3.84	2.92	3.23	4.53	3.83	3.67	4.68	5.07	49.36
ROCKPORT 1 ESE		4.52	4.04	4.16	3.76	3.64	3.08	2.76	3.29	3.54	3.61	4.28	4.76	45.44
SEGREGANSET		4.15	3.69	4.11	4.05	3.86	2.83	3.18	4.51	3.97	3.68	4.36	4.36	46.75
SOUTHBRIDGE 3 SW		4.28	3.68	4.61	4.11	3.69	3.70	3.66	4.34	4.28	4.25	4.51	4.64	49.75
SPRINGFIELD		3.32	2.87	3.98	4.07	3.56	3.87	3.61	3.97	3.94	3.49	4.02	4.17	44.87
STERLING		4.16	3.50	4.48	4.23	3.82	3.68	3.53	4.03	3.96	4.15	4.83	4.59	48.96
STOCKBRIDGE		3.17	2.71	3.51	3.94	3.77	3.69	3.89	4.27	3.95	3.40	3.81	3.62	43.73
TAUNTON		3.69	3.57	3.94	3.80	3.69	2.93	3.32	4.57	3.84	3.78	4.28	4.36	45.77
TULLY LAKE		3.39	2.79	3.54	3.63	3.50	3.72	3.53	3.93	3.62	3.48	3.80	3.81	42.74
WARE		3.49	2.84	3.70	3.66	3.57	3.55	3.91	4.32	3.86	3.51	3.77	3.92	44.10
WESTFIELD		3.39	2.89	4.03	4.09	3.45	3.64	3.54	4.21	4.05	3.80	4.00	4.19	45.28

MASSACHUSETTS

PRECIPITATION NORMALS (INCHES)

STATION	JAN	FEB	MAR	APR	MAY	JUN	JUL	AUG	SEP	OCT	NOV	DEC	ANN
WEST OTIS	3.21	2.76	3.53	3.41	3.43	3.59	3.81	4.67	4.12	3.57	4.01	3.86	43.97
WORCESTER WSO	3.82	3.29	4.16	3.90	3.86	3.46	3.58	4.42	4.25	4.21	4.43	4.22	47.60

MASSACHUSETTS

HEATING DEGREE DAY NORMALS (BASE 65 DEG F)

STATION		JUL	AUG	SEP	OCT	NOV	DEC	JAN	FEB	MAR	APR	MAY	JUN	ANN
AMHERST		6	15	130	425	729	1150	1293	1103	924	546	247	44	6612
BIRCH HILL DAM	R	21	47	208	521	813	1231	1376	1190	1017	627	320	80	7451
BLUE HILL WSO	//	0	9	119	397	696	1085	1212	1056	921	564	268	55	6382
BOSTON WSO	//R	0	6	80	329	594	970	1097	960	825	489	218	25	5593
CHATHAM WSMO		20	30	121	350	594	921	1051	949	874	612	378	117	6017
CHESTNUT HILL		0	0	96	366	648	1035	1156	1005	865	510	226	33	5940
CLINTON		6	14	134	425	711	1107	1249	1106	946	567	270	59	6594
EAST WAREHAM		6	13	122	403	681	1035	1166	1030	899	588	295	57	6295
EDGARTOWN		8	15	96	347	615	961	1101	991	884	597	332	77	6024
FITCHBURG 2 S		7	12	130	428	735	1138	1268	1106	942	558	254	41	6619
FRAMINGHAM		0	7	101	394	684	1088	1221	1064	887	513	219	27	6205
HAVERHILL		0	6	81	360	675	1073	1197	1025	868	501	210	28	6024
HYANNIS 2 NNE		0	14	108	363	633	961	1091	974	874	582	307	58	5965
KNIGHTVILLE DAM		17	42	200	518	804	1218	1364	1193	1023	621	318	79	7397
LAWRENCE		0	7	98	388	684	1088	1215	1056	899	531	234	32	6232
MIDDLETON		0	15	104	372	657	1032	1163	1016	884	534	247	33	6057
NANTUCKET FAA AP		17	21	111	341	582	899	1039	946	877	624	381	122	5960
NEW BEDFORD		0	11	65	278	558	911	1035	913	794	489	216	35	5305
PLYMOUTH		0	8	108	374	642	998	1128	997	871	558	277	52	6013
ROCHESTER		6	15	124	409	675	1048	1181	1047	896	573	294	54	6322
ROCKPORT 1 ESE		9	14	123	397	669	1020	1153	1022	905	594	316	72	6294
SPRINGFIELD		0	5	100	355	663	1070	1190	1022	856	474	192	26	5953
STOCKBRIDGE		21	60	198	502	789	1194	1345	1162	1001	606	307	84	7269
TAUNTON		6	12	131	422	696	1054	1178	1025	884	552	271	45	6276
TULLY LAKE		16	42	202	533	822	1249	1398	1212	1017	621	313	68	7493
WORCESTER WSO		10	22	159	456	765	1166	1293	1126	989	600	296	68	6950

MASSACHUSETTS

COOLING DEGREE DAY NORMALS (BASE 65 DEG F)

STATION		JAN	FEB	MAR	APR	MAY	JUN	JUL	AUG	SEP	OCT	NOV	DEC	ANN
AMHERST		0	0	0	0	18	95	204	152	31	0	0	0	500
BIRCH HILL DAM	R	0	0	0	0	7	50	130	91	16	0	0	0	294
BLUE HILL WSO	//	0	0	0	0	7	67	197	149	29	0	0	0	449
BOSTON WSO	//R	0	0	0	0	17	115	266	220	68	13	0	0	699
CHATHAM WSMO		0	0	0	0	0	12	100	108	34	0	0	0	254
CHESTNUT HILL		0	0	0	0	16	105	248	194	57	10	0	0	630
CLINTON		0	0	0	0	16	92	198	144	26	0	0	0	476
EAST WAREHAM		0	0	0	0	0	57	195	159	38	0	0	0	449
EDGARTOWN		0	0	0	0	0	26	147	139	36	0	0	0	348
FITCHBURG 2 S		0	0	0	0	15	89	221	164	34	0	0	0	523
FRAMINGHAM		0	0	0	0	17	108	251	196	41	0	0	0	613
HAVERHILL		0	0	0	0	24	121	266	208	48	0	0	0	667
HYANNIS 2 NNE		0	0	0	0	0	43	181	160	42	0	0	0	426
KNIGHTVILLE DAM		0	0	0	0	8	58	141	95	14	0	0	0	316
LAWRENCE		0	0	0	0	14	95	238	184	32	0	0	0	563
MIDDLETON		0	0	0	0	12	78	212	170	32	0	0	0	504
NANTUCKET FAA AP		0	0	0	0	0	17	113	120	36	0	0	0	286
NEW BEDFORD		0	0	0	0	11	104	265	247	86	8	0	0	721
PLYMOUTH		0	0	0	0	10	82	215	172	45	5	0	0	529
ROCHESTER		0	0	0	0	6	54	186	152	31	0	0	0	429
ROCKPORT 1 ESE		0	0	0	0	0	39	151	120	18	0	0	0	328
SPRINGFIELD		0	0	0	0	24	125	267	219	64	0	0	0	699
STOCKBRIDGE		0	0	0	0	9	42	114	97	12	0	0	0	274
TAUNTON		0	0	0	0	7	63	202	151	32	0	0	0	455
TULLY LAKE		0	0	0	0	9	50	146	101	13	0	0	0	319
WORCESTER WSO		0	0	0	0	8	56	162	115	18	0	0	0	359

19 — MASSACHUSETTS

STATE-STATION NUMBER	STN TYP	NAME		LATITUDE DEG-MIN	LONGITUDE DEG-MIN	ELEVATION (FT)
19-0120	13	AMHERST		N 4223	W 07232	150
19-0190	12	ASHBURNHAM		N 4239	W 07153	1190
19-0551	12	BEECHWOOD		N 4213	W 07049	40
19-0562	12	BELCHERTOWN		N 4217	W 07221	556
19-0666	13	BIRCH HILL DAM	R	N 4238	W 07207	840
19-0736	13	BLUE HILL WSO	///R	N 4213	W 07107	629
19-0770	13	BOSTON WSO		N 4222	W 07102	15
19-0801	12	BOYLSTON		N 4221	W 07143	630
19-0860	12	BROCKTON		N 4203	W 07100	80
19-1386	13	CHATHAM WSMO		N 4140	W 06958	50
19-1436	12	CHESTERFIELD		N 4224	W 07251	1425
19-1447	13	CHESTNUT HILL		N 4220	W 07109	124
19-1561	13	CLINTON		N 4224	W 07141	398
19-2451	13	EAST WAREHAM		N 4146	W 07040	20
19-2501	13	EDGARTOWN		N 4123	W 07031	20
19-2806	13	FITCHBURG 2 S		N 4234	W 07146	400
19-2975	13	FRAMINGHAM		N 4217	W 07125	170
19-2997	12	FRANKLIN		N 4205	W 07125	240
19-3052	12	GARDNER		N 4235	W 07159	1110
19-3401	12	HARDWICK		N 4222	W 07211	1000
19-3471	12	HATCHVILLE		N 4137	W 07032	70
19-3505	13	HAVERHILL		N 4246	W 07104	60
19-3549	12	HEATH		N 4240	W 07249	1590
19-3702	12	HOLYOKE		N 4212	W 07236	98
19-3772	12	HUBBARDSTON		N 4229	W 07200	980
19-3821	13	HYANNIS 2 NNE		N 4141	W 07016	35
19-3876	12	IPSWICH		N 4240	W 07052	80
19-3985	13	KNIGHTVILLE DAM		N 4217	W 07252	630
19-4105	13	LAWRENCE		N 4242	W 07110	57
19-4449	12	MANSFIELD		N 4203	W 07112	140
19-4711	12	MIDDLEBORO		N 4153	W 07055	60
19-4744	13	MIDDLETON		N 4236	W 07101	100
19-4760	12	MILFORD		N 4210	W 07131	280
19-5159	13	NANTUCKET FAA AP		N 4115	W 07004	40
19-5246	13	NEW BEDFORD		N 4138	W 07056	70
19-5285	12	NEWBURYPORT		N 4250	W 07055	20
19-5306	12	NEW SALEM		N 4227	W 07220	900
19-5524	12	NORTHBRIDGE 2		N 4207	W 07141	315
19-6251	12	PELHAM		N 4224	W 07224	1100
19-6262	12	PEMBROKE		N 4201	W 07049	74
19-6322	12	PETERSHAM 3 N	//R	N 4232	W 07211	1090
19-6425	12	PLAINFIELD		N 4231	W 07255	1620
19-6486	13	PLYMOUTH		N 4157	W 07040	90
19-6699	12	QUABBIN INTAKE		N 4222	W 07217	530
19-6938	13	ROCHESTER		N 4147	W 07055	60
19-6977	13	ROCKPORT 1 ESE		N 4239	W 07036	80
19-7293	12	SEGREGANSET		N 4150	W 07107	40
19-7627	12	SOUTHBRIDGE 3 SW		N 4203	W 07205	720
19-8046	13	SPRINGFIELD		N 4207	W 07235	190
19-8154	12	STERLING		N 4227	W 07149	490

19 – MASSACHUSETTS

STATE-STATION NUMBER	STN TYP	NAME	LATITUDE DEG-MIN	LONGITUDE DEG-MIN	ELEVATION (FT)
19-8181	13	STOCKBRIDGE	N 4218	W 07320	820
19-8367	13	TAUNTON	N 4154	W 07104	20
19-8573	13	TULLY LAKE	N 4238	W 07213	685
19-8793	12	WARE	N 4216	W 07215	410
19-9191	12	WESTFIELD	N 4207	W 07242	220
19-9371	12	WEST OTIS	N 4210	W 07309	1370
19-9923	13	WORCESTER WSO	N 4216	W 07152	986

19 — MASSACHUSETTS

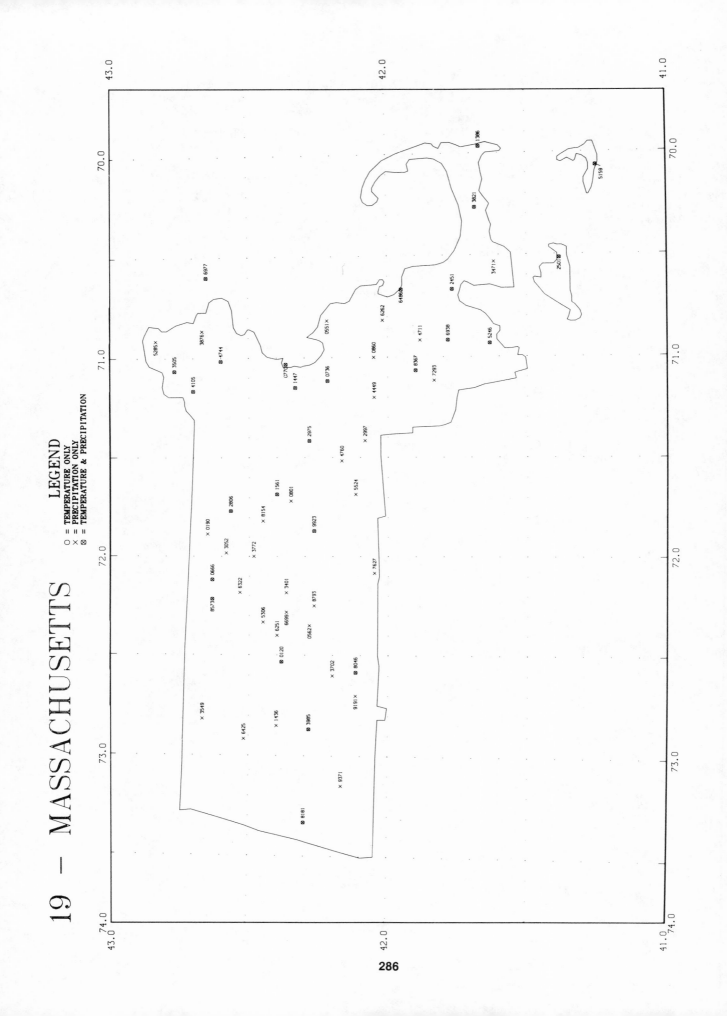

LEGEND

O = TEMPERATURE ONLY
X = PRECIPITATION ONLY
⊗ = TEMPERATURE & PRECIPITATION

MICHIGAN

TEMPERATURE NORMALS (DEG F)

STATION		JAN	FEB	MAR	APR	MAY	JUN	JUL	AUG	SEP	OCT	NOV	DEC	ANN
ADRIAN 2 NNE	MAX	31.1	34.1	44.1	58.6	70.7	80.3	83.8	82.1	75.0	63.2	47.8	35.9	58.9
	MIN	14.9	16.7	25.6	36.0	46.1	55.6	59.5	57.7	50.4	39.5	30.2	20.3	37.7
	MEAN	23.0	25.4	34.9	47.4	58.4	68.0	71.7	69.9	62.7	51.4	39.0	28.1	48.3
ALLEGAN SEWAGE PLANT	MAX	30.8	34.2	43.7	58.6	70.5	79.6	83.4	81.5	74.7	62.8	48.1	35.8	58.6
	MIN	15.8	16.2	25.3	36.3	46.2	55.5	59.6	57.9	50.6	40.1	31.4	21.4	38.0
	MEAN	23.3	25.2	34.5	47.5	58.4	67.5	71.5	69.7	62.7	51.5	39.8	28.6	48.4
ALMA	MAX	28.9	32.2	41.9	57.6	70.2	79.7	83.8	81.5	73.6	61.7	46.3	33.8	57.6
	MIN	13.5	14.1	23.7	35.4	45.5	55.4	59.3	57.5	50.3	40.6	30.8	19.9	37.2
	MEAN	21.2	23.2	32.8	46.5	57.9	67.6	71.6	69.5	62.0	51.2	38.6	26.8	47.4
ALPENA WSO //R	MAX	26.2	28.1	36.9	51.1	64.4	74.2	79.0	76.8	68.1	57.5	42.8	30.9	53.0
	MIN	8.5	7.8	16.7	29.4	38.6	48.1	53.0	52.1	45.0	36.6	27.3	16.0	31.6
	MEAN	17.4	18.0	26.8	40.3	51.5	61.2	66.0	64.5	56.6	47.1	35.1	23.5	42.3
ALPENA SEWAGE PLANT	MAX	26.8	28.2	35.7	48.6	60.8	70.9	76.5	74.9	67.1	56.2	42.9	31.6	51.7
	MIN	12.7	12.2	20.7	32.6	42.5	52.3	58.0	57.1	49.8	40.8	30.7	19.5	35.7
	MEAN	19.7	20.3	28.3	40.6	51.7	61.6	67.3	66.0	58.5	48.6	36.8	25.6	43.8
ANN ARBOR UNIV OF MICH	MAX	30.7	33.9	44.3	58.9	70.9	80.1	83.6	81.6	74.8	62.9	47.9	35.5	58.8
	MIN	16.8	18.3	26.8	38.0	48.5	58.1	62.2	60.7	53.7	43.3	32.8	22.3	40.1
	MEAN	23.8	26.1	35.6	48.5	59.7	69.1	72.9	71.2	64.2	53.1	40.4	28.9	49.5
BAD AXE	MAX	27.9	30.3	39.7	54.7	66.9	76.8	81.3	79.4	71.9	60.5	45.7	33.0	55.7
	MIN	13.3	13.7	22.5	33.8	43.3	53.0	57.4	56.1	49.8	40.6	31.2	20.2	36.2
	MEAN	20.6	22.0	31.1	44.3	55.1	64.9	69.3	67.8	60.9	50.6	38.5	26.6	46.0
BALDWIN	MAX	28.9	32.3	42.0	57.2	70.4	79.2	83.1	80.9	72.3	61.1	45.8	33.5	57.2
	MIN	10.9	9.0	18.5	31.2	41.9	51.0	54.6	53.2	46.0	36.5	27.3	16.5	33.1
	MEAN	19.9	20.7	30.3	44.2	56.2	65.1	68.9	67.1	59.2	48.8	36.6	25.0	45.2
BATTLE CREEK	MAX	29.7	33.0	43.2	58.0	69.9	79.2	82.9	81.1	73.8	61.9	46.7	34.6	57.8
	MIN	15.1	16.5	25.2	36.4	46.5	56.3	60.4	58.8	51.6	41.2	31.2	20.7	38.3
	MEAN	22.4	24.8	34.3	47.2	58.2	67.7	71.7	70.0	62.7	51.5	39.0	27.7	48.1
BEECHWOOD 7 WNW	MAX	20.7	26.2	36.5	51.7	65.8	73.4	77.8	75.5	66.1	56.0	37.8	25.3	51.1
	MIN	1.4	3.5	13.1	27.8	38.8	48.4	53.3	51.9	44.2	35.2	22.2	8.6	29.0
	MEAN	11.1	14.9	24.8	39.8	52.3	60.9	65.6	63.7	55.1	45.6	30.0	17.0	40.1
BENTON HARBOR AIRPORT	MAX	31.4	34.6	44.1	57.9	69.0	78.7	82.1	80.6	74.6	63.4	48.8	36.6	58.5
	MIN	18.0	19.7	27.7	37.8	47.1	57.0	61.4	59.6	53.2	43.2	33.3	23.6	40.1
	MEAN	24.7	27.2	35.9	47.9	58.1	67.9	71.8	70.1	63.9	53.4	41.1	30.1	49.3
BERGLAND DAM	MAX	20.4	25.1	35.9	50.1	64.5	73.7	78.4	75.2	66.0	56.1	38.0	25.6	50.8
	MIN	-.6	-.2	9.3	25.6	37.1	47.0	52.0	50.0	42.6	34.3	21.9	7.5	27.2
	MEAN	9.9	12.5	22.6	37.9	50.8	60.4	65.2	62.6	54.3	45.2	30.0	16.6	39.0
BIG RAPIDS WATERWORKS	MAX	28.0	31.0	40.3	55.6	68.3	77.7	81.7	79.4	70.9	59.4	44.5	32.7	55.8
	MIN	11.2	11.3	20.7	32.6	42.7	51.9	56.0	54.6	46.8	37.1	28.3	17.5	34.2
	MEAN	19.6	21.2	30.5	44.1	55.5	64.8	68.9	67.0	58.9	48.3	36.5	25.1	45.0
CADILLAC	MAX	24.8	27.3	36.7	52.1	65.6	74.6	78.8	76.6	68.1	56.8	41.8	29.8	52.8
	MIN	9.5	7.9	17.0	30.3	40.2	49.9	53.6	52.4	45.5	36.8	27.0	15.4	32.1
	MEAN	17.2	17.6	26.9	41.3	52.9	62.3	66.2	64.5	56.8	46.8	34.4	22.6	42.5
CARO	MAX	29.1	32.1	42.4	58.1	70.3	80.1	84.3	82.0	74.2	62.4	47.2	34.3	58.0
	MIN	12.6	13.3	22.9	33.8	43.1	52.7	56.6	55.1	48.6	39.1	30.3	19.5	35.6
	MEAN	20.9	22.7	32.7	46.0	56.7	66.4	70.5	68.6	61.4	50.8	38.8	26.9	46.9
CHAMPION VAN RIPER PK	MAX	21.2	25.9	36.5	50.8	65.3	74.0	78.3	75.9	66.0	55.2	37.8	25.5	51.1
	MIN	.5	.4	9.3	23.9	35.4	45.0	50.1	48.7	41.8	33.1	21.5	8.0	26.5
	MEAN	10.9	13.2	22.9	37.4	50.4	59.5	64.2	62.3	53.9	44.1	29.7	16.8	38.8
CHARLOTTE	MAX	30.1	33.4	43.5	58.5	70.7	79.9	84.0	82.1	75.0	63.0	47.4	34.9	58.5
	MIN	13.3	13.7	23.4	34.8	44.4	53.6	56.9	54.9	48.4	38.4	29.2	18.9	35.8
	MEAN	21.7	23.6	33.5	46.7	57.6	66.8	70.5	68.5	61.8	50.7	38.3	26.9	47.2

MICHIGAN

TEMPERATURE NORMALS (DEG F)

STATION		JAN	FEB	MAR	APR	MAY	JUN	JUL	AUG	SEP	OCT	NOV	DEC	ANN
CHATHAM EXP FARM	MAX	24.5	27.7	36.4	50.3	64.0	73.3	78.3	76.6	67.2	56.8	41.1	29.3	52.1
	MIN	7.3	7.5	14.8	28.2	37.9	46.3	52.4	52.0	45.5	37.1	26.1	14.1	30.8
	MEAN	15.9	17.6	25.6	39.3	51.0	59.9	65.4	64.3	56.4	47.0	33.6	21.7	41.5
CHEBOYGAN RR LIGHT STA	MAX	26.7	28.5	37.0	50.5	63.6	73.2	79.0	77.3	68.9	58.2	43.6	31.4	53.2
	MIN	11.3	9.7	18.2	30.8	40.7	50.5	57.0	56.3	49.5	40.3	30.2	18.4	34.4
	MEAN	19.0	19.1	27.7	40.7	52.2	61.9	68.0	66.8	59.2	49.3	36.9	25.0	43.8
COLDWATER STATE SCHOOL	MAX	30.1	33.7	43.7	57.8	69.4	78.9	82.4	80.8	74.1	62.3	47.2	35.2	58.0
	MIN	14.5	16.3	25.2	36.1	46.2	55.6	59.3	57.6	50.7	40.4	30.7	20.2	37.7
	MEAN	22.4	25.0	34.5	47.0	57.8	67.3	70.9	69.2	62.4	51.4	39.0	27.7	47.9
DETROIT METROPOLITAN AP	MAX	30.6	33.5	43.4	57.7	69.4	79.0	83.1	81.5	74.4	62.5	47.6	35.4	58.2
	MIN	16.1	18.0	26.5	36.9	46.7	56.3	60.7	59.4	52.2	41.2	31.4	21.6	38.9
	MEAN	23.4	25.8	35.0	47.4	58.1	67.7	71.9	70.5	63.3	51.9	39.5	28.5	48.6
DUNBAR FOREST EXP STA	MAX	23.1	25.8	35.1	49.2	63.2	71.4	75.8	73.8	65.0	55.0	40.6	28.2	50.5
	MIN	4.8	3.7	13.9	28.5	38.2	47.5	53.2	53.4	46.8	37.7	27.2	13.2	30.7
	MEAN	14.0	14.7	24.5	38.9	50.7	59.4	64.5	63.6	55.9	46.4	33.9	20.7	40.6
EAST JORDAN	MAX	27.7	29.8	39.1	54.1	67.4	76.3	80.2	78.2	70.2	59.7	44.4	32.5	55.0
	MIN	12.6	9.4	18.2	31.1	40.5	49.9	54.6	53.6	47.4	38.9	29.7	19.1	33.8
	MEAN	20.2	19.7	28.7	42.6	54.0	63.1	67.4	65.9	58.8	49.4	37.1	25.8	44.4
EAST TAWAS	MAX	28.9	31.2	39.2	53.2	65.3	74.9	79.8	78.1	70.5	59.3	45.5	33.6	55.0
	MIN	11.2	11.1	20.1	32.0	41.3	50.9	55.9	55.1	47.8	38.4	29.0	17.9	34.2
	MEAN	20.0	21.2	29.7	42.6	53.3	62.9	67.9	66.6	59.2	48.9	37.3	25.8	44.6
EAU CLAIRE 4 NE	MAX	30.3	33.7	44.2	58.9	70.2	79.7	83.2	81.4	74.8	63.0	47.8	35.3	58.5
	MIN	16.5	19.0	27.4	38.3	48.1	57.8	62.0	60.6	54.1	43.7	32.8	22.6	40.2
	MEAN	23.4	26.4	35.8	48.6	59.2	68.8	72.6	71.1	64.4	53.4	40.3	29.0	49.4
ESCANABA //R	MAX	23.7	26.5	34.5	46.6	58.5	68.7	75.0	73.2	64.8	54.4	41.0	29.3	49.7
	MIN	8.5	9.1	18.3	31.4	42.1	52.2	57.9	56.6	49.1	40.4	28.8	16.3	34.2
	MEAN	16.1	17.8	26.4	39.0	50.3	60.5	66.5	64.9	57.0	47.4	34.9	22.8	42.0
EVART	MAX	27.2	30.3	39.9	55.8	68.8	77.9	81.6	79.7	71.1	59.6	44.3	31.9	55.7
	MIN	9.5	9.1	18.8	31.0	41.1	50.1	54.0	52.8	45.5	35.9	27.1	16.1	32.6
	MEAN	18.4	19.8	29.4	43.4	55.0	64.0	67.8	66.2	58.3	47.8	35.7	24.0	44.2
FAYETTE	MAX	24.8	27.2	35.6	47.8	59.7	68.7	75.0	73.6	65.6	55.4	42.1	30.3	50.5
	MIN	10.2	10.7	19.4	30.9	40.7	49.9	56.6	56.6	50.2	41.3	30.2	18.0	34.6
	MEAN	17.5	19.0	27.5	39.3	50.2	59.3	65.8	65.2	57.9	48.3	36.2	24.2	42.5
FLINT WSO	MAX	28.6	31.3	39.6	55.9	67.6	77.0	81.2	79.4	72.0	60.6	46.2	33.9	56.1
	MIN	14.0	15.3	24.5	35.7	45.2	54.8	58.9	57.5	50.4	40.6	31.3	20.4	37.4
	MEAN	21.3	23.3	32.1	45.8	56.4	65.9	70.1	68.5	61.2	50.6	38.8	27.2	46.8
GAYLORD CON DEP	MAX	25.4	28.1	38.0	53.5	67.4	76.2	80.3	78.1	69.2	58.4	42.3	29.9	53.9
	MIN	9.6	8.5	17.4	30.8	40.7	50.1	54.4	53.3	46.3	37.5	27.3	15.6	32.6
	MEAN	17.5	18.3	27.7	42.1	54.0	63.2	67.4	65.7	57.8	48.0	34.8	22.8	43.3
GLADWIN	MAX	27.8	30.9	40.4	56.1	68.6	78.4	82.3	80.1	71.6	60.3	44.8	32.6	56.2
	MIN	10.1	10.9	20.5	32.4	42.1	51.3	55.8	54.1	46.1	36.7	27.4	16.8	33.7
	MEAN	19.0	20.9	30.5	44.3	55.4	64.9	69.1	67.1	58.9	48.5	36.1	24.7	45.0
GRAND HAVEN FIRE DEPT	MAX	30.7	32.8	41.9	55.1	66.4	75.4	79.0	77.7	71.6	60.8	47.0	35.7	56.2
	MIN	18.8	19.3	26.8	37.3	46.7	56.1	61.7	60.8	54.0	44.6	34.4	24.1	40.4
	MEAN	24.7	26.1	34.4	46.2	56.6	65.8	70.3	69.3	62.8	52.7	40.7	29.9	48.3
GRAND MARAIS 1 SSE	MAX	24.2	25.8	34.0	47.1	60.4	69.5	74.5	73.4	65.2	55.1	40.5	29.3	49.9
	MIN	10.7	9.4	15.7	28.5	37.1	45.5	51.1	51.5	46.1	37.8	27.7	17.1	31.5
	MEAN	17.5	17.7	24.9	37.8	48.8	57.5	62.8	62.5	55.7	46.5	34.1	23.2	40.8
GRAND RAPIDS KENT WSO //R	MAX	29.0	31.7	41.6	56.9	69.4	78.9	83.0	81.1	73.4	61.4	46.0	33.8	57.2
	MIN	14.9	15.6	24.5	35.6	45.5	55.3	59.8	58.1	50.8	40.4	30.9	20.7	37.7
	MEAN	22.0	23.7	33.1	46.3	57.5	67.1	71.4	69.6	62.1	50.9	38.5	27.3	47.5

MICHIGAN

TEMPERATURE NORMALS (DEG F)

STATION		JAN	FEB	MAR	APR	MAY	JUN	JUL	AUG	SEP	OCT	NOV	DEC	ANN
GRAYLING	MAX	25.9	28.4	38.2	54.1	68.0	76.8	80.9	78.6	69.7	58.4	42.8	30.3	54.3
	MIN	9.3	7.6	16.5	30.3	40.5	50.0	54.1	52.7	45.9	37.1	27.4	15.3	32.2
	MEAN	17.6	18.0	27.4	42.2	54.3	63.4	67.5	65.7	57.8	47.8	35.2	22.9	43.3
GREENVILLE 2 NNE	MAX	28.9	32.1	42.3	58.1	70.5	79.9	83.8	81.8	73.9	62.0	46.3	33.6	57.8
	MIN	13.6	14.4	23.3	34.9	45.1	54.4	58.3	56.8	49.5	39.6	29.8	19.3	36.6
	MEAN	21.3	23.3	32.8	46.5	57.8	67.2	71.1	69.3	61.7	50.8	38.1	26.5	47.2
GROSSE POINTE FARMS	MAX	32.0	34.6	43.9	58.1	69.9	79.9	83.8	81.6	75.1	63.3	49.2	36.9	59.0
	MIN	18.1	19.1	27.0	37.8	48.2	57.9	62.7	61.6	55.1	44.5	34.2	23.8	40.8
	MEAN	25.0	26.9	35.4	48.0	59.1	68.9	73.3	71.6	65.1	53.9	41.7	30.4	49.9
GULL LAKE BIOL. STA.	MAX	30.3	33.8	43.9	58.8	70.9	79.8	83.9	82.2	75.2	63.4	47.7	35.2	58.8
	MIN	14.8	15.9	24.8	36.1	46.5	56.1	60.3	58.9	52.0	41.8	31.7	21.0	38.3
	MEAN	22.6	24.9	34.4	47.5	58.7	68.0	72.2	70.6	63.6	52.6	39.7	28.1	48.6
HALE LOUD DAM	MAX	27.3	30.1	39.6	54.0	66.9	76.1	80.3	78.1	69.9	59.4	44.2	32.0	54.8
	MIN	7.9	7.8	17.3	30.2	40.3	50.0	54.8	53.5	46.2	36.9	27.7	15.8	32.4
	MEAN	17.6	19.0	28.4	42.1	53.6	63.1	67.6	65.8	58.1	48.2	36.0	23.9	43.6
HARBOR BEACH 3NW	MAX	27.7	29.7	38.0	51.9	63.2	73.4	78.0	77.1	70.4	59.3	45.5	33.2	54.0
	MIN	14.7	15.2	23.6	34.2	43.2	52.9	58.7	58.4	52.0	42.4	32.2	21.2	37.4
	MEAN	21.2	22.5	30.8	43.1	53.2	63.2	68.4	67.8	61.2	50.9	38.8	27.2	45.7
HART	MAX	29.3	31.8	41.0	55.8	68.0	77.1	81.4	79.5	72.0	60.7	46.0	34.2	56.4
	MIN	15.6	15.4	23.2	34.6	44.0	53.4	58.1	57.3	50.5	41.5	31.4	21.2	37.2
	MEAN	22.5	23.6	32.1	45.2	56.0	65.3	69.8	68.4	61.3	51.1	38.7	27.8	46.8
HASTINGS	MAX	30.0	33.4	43.6	58.8	70.9	80.0	83.8	82.0	74.8	63.0	47.4	34.7	58.5
	MIN	14.5	14.9	24.1	35.6	45.6	54.9	58.7	57.1	50.1	39.9	30.7	20.1	37.2
	MEAN	22.3	24.2	33.9	47.2	58.3	67.5	71.2	69.6	62.5	51.5	39.1	27.4	47.9
HESPERIA 4 E	MAX	28.6	31.4	41.2	56.4	69.0	78.2	82.1	80.1	72.2	60.7	45.7	33.4	56.6
	MIN	12.7	12.4	21.3	32.9	42.3	51.8	56.2	54.7	47.6	38.3	29.0	18.5	34.8
	MEAN	20.7	21.9	31.3	44.7	55.7	65.1	69.2	67.4	59.9	49.5	37.4	26.0	45.7
HILLSDALE	MAX	30.1	33.7	43.7	58.4	69.7	79.0	82.1	80.6	73.9	62.3	47.2	34.9	58.0
	MIN	14.5	15.8	24.9	35.8	45.6	54.7	58.4	56.9	50.0	39.8	30.2	19.9	37.2
	MEAN	22.3	24.8	34.3	47.1	57.7	66.9	70.3	68.8	62.0	51.1	38.7	27.4	47.6
HOLLAND WJBL	MAX	30.6	33.6	43.1	57.4	69.3	78.6	82.3	80.8	74.0	62.4	47.6	35.5	57.9
	MIN	17.6	18.0	26.0	36.7	46.2	55.4	59.7	58.2	52.1	42.5	32.8	22.6	39.0
	MEAN	24.1	25.8	34.6	47.1	57.8	67.0	71.0	69.5	63.1	52.5	40.3	29.1	48.5
HOUGHTON FAA AP	MAX	19.8	22.1	30.9	45.0	59.5	69.5	74.9	72.5	62.8	52.4	36.7	25.3	47.6
	MIN	7.0	6.7	15.3	29.1	39.2	48.3	54.4	53.6	45.8	37.5	25.6	13.8	31.4
	MEAN	13.4	14.4	23.2	37.1	49.4	58.9	64.7	63.1	54.3	45.0	31.1	19.6	39.5
HOUGHTON LAKE 3 NW	MAX	26.2	29.3	38.8	53.9	67.4	76.3	80.4	78.3	69.6	58.5	43.1	30.6	54.4
	MIN	9.6	8.4	17.9	31.7	41.7	50.9	54.6	53.1	46.0	37.2	27.9	16.2	32.9
	MEAN	17.9	18.9	28.4	42.8	54.6	63.6	67.5	65.7	57.8	47.8	35.5	23.4	43.7
HOUGHTON LAKE WSO	MAX	25.2	28.0	37.3	52.6	65.8	74.8	78.9	76.5	68.0	56.7	41.8	29.9	53.0
	MIN	8.7	8.0	17.5	31.2	41.5	50.8	55.0	53.8	46.5	37.5	27.9	15.8	32.8
	MEAN	17.0	18.0	27.4	41.9	53.7	62.8	67.0	65.2	57.3	47.1	34.9	22.9	42.9
IONIA 5 NW	MAX	30.1	33.2	43.5	59.1	71.3	80.4	84.2	82.3	74.8	63.0	47.3	34.6	58.7
	MIN	14.1	14.6	23.8	35.1	44.9	54.3	58.0	56.3	49.3	39.4	30.4	20.0	36.7
	MEAN	22.1	23.9	33.7	47.1	58.1	67.3	71.1	69.3	62.0	51.2	38.9	27.3	47.7
IRON MOUNTAIN WTR WKS	MAX	23.0	27.8	37.6	53.5	67.1	75.3	79.6	76.9	67.3	57.0	40.3	27.6	52.8
	MIN	3.2	5.3	15.5	29.2	40.2	49.8	54.7	53.0	45.1	36.2	24.3	11.0	30.6
	MEAN	13.1	16.6	26.5	41.4	53.7	62.6	67.2	65.0	56.2	46.6	32.3	19.3	41.7
IRONWOOD	MAX	19.8	25.3	35.6	51.6	65.5	74.3	78.8	76.0	66.5	56.1	38.0	25.0	51.0
	MIN	.8	3.1	13.3	29.4	40.4	49.7	54.9	53.1	45.1	36.2	22.7	8.5	29.8
	MEAN	10.3	14.3	24.5	40.5	53.0	62.1	66.8	64.6	55.8	46.2	30.4	16.8	40.4

MICHIGAN

TEMPERATURE NORMALS (DEG F)

STATION		JAN	FEB	MAR	APR	MAY	JUN	JUL	AUG	SEP	OCT	NOV	DEC	ANN
ISHPEMING	MAX	22.0	26.1	35.6	50.3	64.3	73.3	77.9	75.7	66.2	55.7	38.6	26.4	51.0
	MIN	5.4	6.8	15.0	28.6	39.4	49.1	54.4	52.8	45.5	36.8	24.6	12.3	30.9
	MEAN	13.7	16.5	25.3	39.5	51.9	61.3	66.2	64.3	55.9	46.3	31.6	19.4	41.0
JACKSON FAA AIRPORT	MAX	28.7	31.9	42.1	57.1	69.1	78.9	82.9	81.0	73.7	61.7	46.4	33.9	57.3
	MIN	14.6	16.1	25.2	36.6	46.7	56.4	60.8	59.0	51.6	41.1	31.1	20.5	38.3
	MEAN	21.6	24.0	33.7	46.9	57.9	67.7	71.9	70.0	62.7	51.4	38.8	27.2	47.8
KALAMAZOO STATE HOSP	MAX	30.9	34.7	45.2	60.3	72.3	81.1	84.9	83.1	76.1	64.2	48.2	35.8	59.7
	MIN	16.5	18.3	26.6	37.6	48.0	57.3	61.4	60.0	53.0	42.7	32.3	22.2	39.7
	MEAN	23.7	26.5	35.9	49.0	60.2	69.2	73.2	71.6	64.6	53.5	40.3	29.0	49.7
LAKE CITY EXP FARM	MAX	25.8	28.8	37.9	54.0	67.2	76.6	80.9	78.5	69.8	58.1	42.8	30.3	54.2
	MIN	9.1	7.8	16.9	30.8	40.5	49.4	53.5	52.2	45.3	36.5	26.9	15.0	32.0
	MEAN	17.5	18.3	27.4	42.4	53.9	63.0	67.2	65.4	57.6	47.3	34.9	22.7	43.1
LANSING WSO //R	MAX	29.0	31.7	41.7	56.8	69.1	78.5	82.6	80.9	73.2	61.3	46.3	34.0	57.1
	MIN	14.1	14.9	24.2	35.6	45.3	55.1	58.9	57.3	50.0	40.0	30.6	20.1	37.2
	MEAN	21.6	23.3	33.0	46.3	57.2	66.8	70.8	69.2	61.7	50.7	38.5	27.0	47.2
LAPEER	MAX	29.3	32.6	42.7	58.0	69.6	79.1	83.1	80.9	73.7	62.1	47.0	34.4	57.7
	MIN	13.4	14.4	23.3	34.6	44.3	53.7	57.5	55.7	49.2	39.4	30.4	19.6	36.3
	MEAN	21.4	23.5	33.0	46.3	56.9	66.4	70.3	68.4	61.5	50.8	38.7	27.0	47.0
LUDINGTON 4 SE	MAX	28.7	31.1	40.1	54.2	65.6	74.9	79.5	77.8	70.3	59.0	45.3	33.8	55.0
	MIN	15.6	15.3	23.0	34.3	43.1	52.6	57.6	56.8	50.7	41.3	31.2	20.9	36.9
	MEAN	22.2	23.2	31.6	44.3	54.3	63.8	68.6	67.3	60.5	50.2	38.3	27.4	46.0
LUPTON 1 S	MAX	27.4	30.8	40.3	56.1	69.3	78.3	82.2	79.8	71.2	60.1	44.3	31.6	56.0
	MIN	6.4	6.4	16.1	28.8	37.8	46.9	50.5	49.1	42.5	33.4	25.4	13.7	29.8
	MEAN	16.9	18.6	28.2	42.4	53.6	62.6	66.3	64.5	56.9	46.8	34.9	22.7	42.9
MANISTEE	MAX	28.9	31.3	40.0	53.8	65.9	75.4	80.0	78.2	71.1	60.1	45.7	34.1	55.4
	MIN	16.4	16.0	23.7	34.9	44.1	52.9	58.7	58.1	51.7	42.9	32.6	22.4	37.9
	MEAN	22.7	23.7	31.9	44.4	55.0	64.2	69.4	68.2	61.4	51.5	39.1	28.2	46.6
MARQUETTE WSO	MAX	23.5	26.1	34.2	47.3	59.6	69.4	75.2	73.7	65.0	55.3	40.1	28.8	49.9
	MIN	10.8	12.1	20.4	32.0	41.1	50.4	57.4	57.0	49.3	40.7	28.9	17.7	34.8
	MEAN	17.2	19.1	27.3	39.7	50.4	59.9	66.3	65.3	57.2	48.0	34.5	23.3	42.4
MARQUETTE FAA AP	MAX	20.1	23.8	33.2	47.4	61.5	70.4	75.3	73.3	63.9	53.4	36.5	24.6	48.6
	MIN	4.1	4.8	13.2	27.4	38.9	48.8	53.8	51.8	44.1	35.2	23.6	11.2	29.7
	MEAN	12.1	14.3	23.2	37.4	50.2	59.6	64.6	62.6	54.0	44.3	30.1	17.9	39.2
MIDLAND	MAX	29.0	31.7	41.2	56.9	69.4	79.1	82.9	80.6	72.9	61.5	46.4	34.0	57.1
	MIN	14.9	15.8	24.6	36.2	46.3	55.8	60.1	58.3	51.0	41.3	31.8	20.8	38.1
	MEAN	22.0	23.8	32.9	46.6	57.9	67.5	71.5	69.4	62.0	51.4	39.1	27.4	47.6
MILFORD GM PROVING GRN	MAX	28.0	30.7	40.9	55.5	67.7	77.1	81.1	79.3	71.6	59.8	45.2	33.1	55.8
	MIN	13.8	15.1	24.2	35.9	46.2	56.3	60.5	59.0	51.8	41.4	30.9	20.1	37.9
	MEAN	20.9	22.9	32.6	45.7	56.9	66.7	70.8	69.2	61.7	50.6	38.1	26.6	46.9
MIO HYDRO PLANT	MAX	27.1	29.8	39.4	54.4	67.4	76.4	81.0	78.5	69.7	58.9	44.1	31.7	54.9
	MIN	7.8	7.2	16.5	29.5	39.4	48.9	53.7	52.3	45.1	36.5	27.5	15.0	31.6
	MEAN	17.5	18.6	28.0	42.0	53.5	62.7	67.4	65.4	57.4	47.7	35.8	23.3	43.3
MONROE SEWAGE PLANT	MAX	31.9	34.8	44.0	58.1	69.7	79.8	83.9	82.3	75.6	63.4	48.8	36.7	59.1
	MIN	17.3	19.0	27.4	38.5	49.0	58.9	63.1	61.5	54.7	43.3	33.2	22.8	40.7
	MEAN	24.6	26.9	35.7	48.3	59.4	69.4	73.5	71.9	65.1	53.4	41.1	29.8	49.9
MONTAGUE	MAX	30.5	33.3	42.4	56.2	67.8	76.6	80.4	79.1	71.9	61.2	46.8	35.4	56.8
	MIN	16.0	15.6	23.2	33.6	42.8	51.6	56.5	55.6	49.0	40.2	31.4	21.6	36.4
	MEAN	23.3	24.5	32.8	44.9	55.3	64.1	68.5	67.4	60.5	50.7	39.2	28.5	46.6
MT PLEASANT UNIVERSITY	MAX	28.5	31.6	41.2	57.1	69.7	79.1	83.2	81.2	73.2	61.5	46.3	33.5	57.2
	MIN	13.1	13.6	22.6	34.5	44.4	54.1	58.3	56.8	49.4	39.7	30.1	19.4	36.3
	MEAN	20.8	22.6	31.9	45.8	57.1	66.6	70.8	69.0	61.3	50.6	38.2	26.5	46.8

MICHIGAN

TEMPERATURE NORMALS (DEG F)

STATION			JAN	FEB	MAR	APR	MAY	JUN	JUL	AUG	SEP	OCT	NOV	DEC	ANN
MUSKEGON WSO	//R	MAX	28.9	30.9	40.4	54.6	66.6	76.1	80.3	78.7	71.2	59.7	45.6	34.0	55.6
		MIN	17.2	17.3	25.2	35.8	45.4	54.8	59.9	59.0	51.7	42.2	32.7	22.6	38.7
		MEAN	23.1	24.1	32.8	45.2	56.0	65.5	70.1	68.9	61.4	51.0	39.2	28.3	47.1
NEWBERRY STATE HOSP		MAX	23.1	25.7	34.5	48.7	62.0	71.2	76.4	74.1	64.6	54.0	39.4	28.0	50.1
		MIN	7.8	7.7	16.4	29.1	38.5	47.3	52.7	52.4	45.2	36.8	26.1	14.3	31.2
		MEAN	15.5	16.7	25.4	38.9	50.3	59.3	64.6	63.3	54.9	45.4	32.8	21.2	40.7
ONAWAY STATE PARK		MAX	26.6	29.1	39.0	53.5	67.6	76.6	81.2	79.0	70.1	59.5	43.6	31.6	54.8
		MIN	9.6	8.6	17.4	30.7	40.8	49.9	55.0	53.8	47.2	38.7	28.9	17.0	33.1
		MEAN	18.2	18.9	28.2	42.1	54.2	63.3	68.1	66.4	58.7	49.1	36.3	24.3	44.0
OWOSSO WASTEWATER PL		MAX	29.2	32.2	42.1	57.4	69.1	78.6	82.4	80.4	73.0	61.7	46.8	34.3	57.3
		MIN	14.1	14.9	24.2	35.5	45.4	54.9	58.9	57.2	50.4	40.6	30.9	20.0	37.3
		MEAN	21.7	23.6	33.2	46.5	57.3	66.8	70.7	68.8	61.7	51.2	38.9	27.1	47.3
PELLSTON FAA AIRPORT		MAX	24.8	26.6	35.7	50.6	64.4	74.1	78.5	76.3	67.2	56.6	41.7	29.9	52.2
		MIN	7.1	4.4	14.2	28.7	38.2	47.4	52.3	51.3	44.1	36.1	27.2	14.6	30.5
		MEAN	16.0	15.5	25.0	39.7	51.4	60.8	65.5	63.8	55.7	46.4	34.5	22.3	41.4
PONTIAC STATE HOSPITAL		MAX	29.5	32.8	43.2	58.2	70.2	79.5	83.6	81.8	74.7	62.6	47.2	34.5	58.2
		MIN	15.3	16.5	25.0	36.1	46.6	56.2	60.6	59.2	52.3	42.0	31.7	21.0	38.5
		MEAN	22.4	24.7	34.1	47.2	58.4	67.9	72.1	70.5	63.5	52.3	39.5	27.8	48.4
PORT HURON WASTEWTR PL		MAX	30.2	32.8	41.9	55.9	67.2	77.7	82.2	80.8	73.9	62.2	47.6	35.3	57.3
		MIN	16.9	17.8	26.0	36.0	45.8	56.0	61.7	60.8	53.8	43.6	33.5	22.7	39.6
		MEAN	23.6	25.3	34.0	46.0	56.5	66.8	72.0	70.8	63.9	52.9	40.6	29.0	48.5
SAGINAW FAA AP		MAX	27.1	29.5	39.3	55.0	67.5	77.4	81.7	79.5	71.5	60.0	45.1	32.3	55.5
		MIN	14.2	15.4	24.9	36.6	46.3	56.0	60.2	58.4	51.1	41.2	31.6	20.5	38.0
		MEAN	20.7	22.5	32.1	45.8	56.9	66.7	71.0	69.0	61.3	50.6	38.4	26.4	46.8
SAINT CHARLES		MAX	29.6	32.8	43.0	58.7	71.2	80.8	84.9	82.6	74.9	62.8	47.2	34.7	58.6
		MIN	14.7	15.7	24.8	36.2	46.1	55.4	59.4	57.6	50.7	40.8	31.4	20.8	37.8
		MEAN	22.2	24.3	33.9	47.5	58.7	68.2	72.2	70.1	62.8	51.8	39.3	27.7	48.2
SAINT JOHNS		MAX	29.6	32.9	43.2	58.6	70.6	79.8	83.9	81.9	74.7	63.0	47.4	34.7	58.4
		MIN	14.3	15.2	24.3	35.5	45.5	55.4	59.2	57.5	50.6	40.7	30.8	19.9	37.4
		MEAN	22.0	24.0	33.7	47.1	58.1	67.7	71.6	69.7	62.7	51.9	39.1	27.3	47.9
SANDUSKY		MAX	28.6	31.2	40.8	55.9	68.0	77.8	82.0	80.2	73.1	61.2	46.4	33.6	56.6
		MIN	14.0	14.7	23.3	34.4	43.9	53.8	58.2	57.0	50.2	40.3	31.1	20.2	36.8
		MEAN	21.3	23.0	32.1	45.1	56.0	65.8	70.1	68.7	61.7	50.8	38.8	26.9	46.7
SAULT STE MARIE WSO	//R	MAX	21.2	23.1	32.3	47.1	61.0	70.1	75.1	73.4	64.2	53.6	39.0	26.6	48.9
		MIN	5.4	5.3	15.4	29.0	38.3	46.7	51.9	52.4	45.3	36.9	26.4	12.7	30.5
		MEAN	13.3	14.3	23.9	38.1	49.7	58.4	63.5	62.9	54.8	45.3	32.8	19.7	39.7
SENEY NATL WLR		MAX	24.8	27.6	36.6	50.7	64.9	74.1	79.3	77.0	67.4	56.5	41.2	29.3	52.5
		MIN	7.2	6.1	14.5	29.5	39.6	48.3	53.3	52.8	46.4	37.8	27.0	14.3	31.4
		MEAN	16.0	16.8	25.6	40.1	52.3	61.2	66.3	64.9	57.0	47.2	34.1	21.8	41.9
SOUTH HAVEN EXP FARM		MAX	31.5	34.1	43.3	55.5	66.2	75.8	79.3	78.5	73.4	62.7	48.5	36.5	57.1
		MIN	18.1	19.7	27.2	37.0	46.3	55.8	61.0	59.9	53.4	44.0	34.0	23.6	40.0
		MEAN	24.8	26.9	35.3	46.3	56.3	65.8	70.2	69.2	63.4	53.4	41.3	30.1	48.6
STAMBAUGH 1 S		MAX	21.8	27.1	37.1	52.7	66.3	74.4	78.7	76.2	66.9	56.8	39.0	26.4	52.0
		MIN	1.0	2.6	13.4	27.8	38.4	47.4	52.1	50.3	42.7	34.6	22.3	8.6	28.4
		MEAN	11.4	14.9	25.2	40.3	52.4	60.9	65.4	63.3	54.8	45.7	30.7	17.6	40.2
STANDISH 2 S		MAX	27.8	30.3	39.8	55.0	67.4	77.4	81.6	79.5	71.4	59.9	45.0	32.6	55.6
		MIN	10.4	11.0	20.8	32.6	42.0	51.8	55.9	54.1	47.1	37.5	28.4	17.4	34.1
		MEAN	19.1	20.7	30.3	43.8	54.7	64.6	68.8	66.8	59.3	48.7	36.7	25.1	44.9
STEPHENSON 5 W		MAX	24.2	28.7	38.2	54.0	67.0	75.9	80.5	78.2	68.9	58.3	41.8	29.1	53.7
		MIN	4.3	6.8	17.1	30.6	40.3	49.5	54.3	52.8	45.2	36.5	25.2	12.4	31.3
		MEAN	14.3	17.7	27.7	42.3	53.7	62.7	67.4	65.5	57.1	47.4	33.5	20.8	42.5

MICHIGAN

TEMPERATURE NORMALS (DEG F)

STATION		JAN	FEB	MAR	APR	MAY	JUN	JUL	AUG	SEP	OCT	NOV	DEC	ANN
THREE RIVERS	MAX	30.6	34.5	45.1	60.0	71.6	80.8	84.1	82.4	75.9	63.9	48.2	35.7	59.4
	MIN	15.3	17.3	26.4	37.2	46.8	56.1	59.7	57.9	50.8	40.5	31.3	21.3	38.4
	MEAN	23.0	25.9	35.7	48.6	59.2	68.5	72.0	70.2	63.4	52.2	39.8	28.6	48.9
TRAVERSE CITY FAA AP	MAX	26.3	28.1	37.4	52.9	66.0	76.2	80.7	78.5	69.7	58.5	43.6	31.6	54.1
	MIN	13.1	11.1	19.4	31.8	40.9	51.2	56.9	56.2	49.0	39.8	30.1	19.5	34.9
	MEAN	19.7	19.6	28.4	42.4	53.5	63.7	68.8	67.3	59.4	49.2	36.9	25.6	44.5
VANDERBILT TROUT STA	MAX	25.8	28.2	38.3	53.8	68.1	76.7	80.8	77.8	68.6	57.7	41.9	30.1	54.0
	MIN	5.9	3.4	13.0	26.6	36.0	45.8	49.3	48.4	42.1	33.6	25.1	12.5	28.5
	MEAN	15.9	15.8	25.7	40.3	52.1	61.3	65.0	63.1	55.4	45.7	33.5	21.3	41.3
WATERSMEET	MAX	21.1	25.7	36.2	51.1	65.3	74.5	78.8	76.2	66.9	56.4	38.5	26.0	51.4
	MIN	1.2	2.0	12.3	27.3	38.0	47.0	52.5	50.4	43.1	35.2	22.6	8.9	28.4
	MEAN	11.2	13.9	24.3	39.2	51.7	60.8	65.7	63.3	55.0	45.8	30.5	17.5	39.9
WEST BRANCH 2 N	MAX	26.3	29.4	38.8	54.4	67.6	76.6	80.6	78.1	69.3	58.5	43.4	31.0	54.5
	MIN	9.3	9.9	18.8	31.2	41.0	50.6	55.0	53.6	46.4	36.8	27.2	15.7	33.0
	MEAN	17.8	19.7	28.8	42.8	54.3	63.6	67.8	65.9	57.9	47.7	35.3	23.4	43.8
WILLIS 5 SSW	MAX	30.1	33.2	43.4	57.9	69.5	78.8	82.7	81.1	74.4	62.9	47.7	35.0	58.1
	MIN	13.9	15.7	25.0	35.3	45.4	54.8	58.2	56.3	48.9	38.3	29.5	19.6	36.7
	MEAN	22.0	24.5	34.2	46.7	57.5	66.8	70.4	68.7	61.7	50.6	38.6	27.4	47.4
YPSILANTI EAST MICH U	MAX	30.4	33.5	43.8	58.4	70.3	79.9	83.8	81.7	74.5	62.6	47.5	35.2	58.5
	MIN	16.6	18.3	26.7	37.4	47.4	57.2	61.8	60.4	52.9	42.3	32.3	22.1	39.6
	MEAN	23.5	25.9	35.2	47.9	58.9	68.6	72.8	71.1	63.7	52.5	39.9	28.7	49.1

MICHIGAN

PRECIPITATION NORMALS (INCHES)

STATION	JAN	FEB	MAR	APR	MAY	JUN	JUL	AUG	SEP	OCT	NOV	DEC	ANN
ADRIAN 2 NNE	1.74	1.70	2.65	3.39	3.04	3.30	3.73	3.20	2.62	2.43	2.49	2.50	32.79
ALBION	1.78	1.65	2.47	3.46	3.26	3.39	3.65	3.16	2.52	2.54	2.38	2.34	32.60
ALLEGAN SEWAGE PLANT	2.59	1.69	2.70	3.37	2.93	3.90	3.21	3.31	3.17	2.82	3.01	2.86	35.56
ALMA	1.46	1.19	2.06	2.98	2.79	2.98	2.62	3.66	3.04	2.47	2.31	1.99	29.55
ALPENA WSO //R	1.65	1.34	1.93	2.50	2.83	3.16	3.11	3.17	2.92	1.99	2.21	1.95	28.78
ALPENA SEWAGE PLANT	1.40	1.14	1.65	2.39	2.84	2.95	2.97	3.16	2.97	2.08	2.20	1.85	27.60
ANN ARBOR UNIV OF MICH	1.74	1.54	2.39	3.21	2.78	3.44	2.93	2.84	2.36	2.21	2.38	2.40	30.22
BAD AXE	1.79	1.56	2.19	2.62	2.58	2.88	2.93	3.01	2.67	2.49	2.38	2.18	29.28
BALDWIN	2.29	1.68	2.18	3.19	2.93	3.27	2.88	3.62	3.29	3.00	3.17	2.44	33.94
BATTLE CREEK	1.85	1.58	2.48	3.44	3.15	3.86	3.48	3.28	2.79	2.83	2.66	2.50	33.90
BEECHWOOD 7 WNW	1.29	1.20	1.84	2.56	3.76	4.20	3.89	4.37	3.74	2.85	2.26	1.59	33.55
BENTON HARBOR AIRPORT	2.99	1.87	2.59	3.76	3.00	3.40	3.06	3.17	3.35	3.20	2.90	3.12	36.41
BERGLAND DAM	2.33	1.76	2.34	2.74	3.84	4.10	3.87	4.39	3.97	3.00	3.47	2.56	38.37
BIG RAPIDS WATERWORKS	1.94	1.50	2.12	3.13	2.88	3.22	2.58	3.43	3.30	2.84	2.77	2.22	31.93
BLOOMINGDALE	2.73	1.89	2.80	3.77	3.10	4.04	3.51	3.35	3.48	3.03	3.28	3.27	38.25
CADILLAC	1.62	1.37	2.03	3.03	2.54	3.07	3.20	3.06	3.48	2.88	2.69	1.84	30.81
CARO	1.50	1.18	2.10	2.52	2.55	3.09	2.92	2.96	2.98	2.31	2.27	1.87	28.25
CHAMPION VAN RIPER PK	1.49	1.36	2.13	2.45	3.43	3.86	3.80	3.54	4.27	2.89	2.52	1.91	33.65
CHARLEVOIX	2.03	1.40	2.03	2.49	2.70	2.88	3.06	3.29	3.80	2.56	2.76	2.10	31.10
CHARLOTTE	1.90	1.53	2.39	3.37	2.98	3.84	3.20	3.24	3.03	2.56	2.62	2.30	32.96
CHATHAM EXP FARM	1.97	1.65	1.95	2.46	3.15	3.61	3.56	3.56	4.16	3.24	3.10	2.37	34.78
CHEBOYGAN RR LIGHT STA	1.41	1.18	1.63	2.44	2.54	2.68	2.99	2.90	3.77	2.09	2.38	1.86	27.87
COLDWATER STATE SCHOOL	1.72	1.56	2.36	3.48	3.03	3.67	4.01	3.40	3.03	2.60	2.38	2.19	33.43
CRYSTAL FALLS 6 NE	1.18	1.01	1.73	2.36	3.24	3.96	3.84	3.50	3.46	2.17	1.91	1.44	29.80
DE TOUR 1 N	1.60	1.32	1.95	2.54	2.65	2.94	2.94	2.94	4.06	2.33	2.61	2.03	29.91
DETROIT METROPOLITAN AP	1.86	1.69	2.54	3.15	2.77	3.43	3.10	3.21	2.25	2.12	2.33	2.52	30.97
DUNBAR FOREST EXP STA	1.63	1.39	1.89	2.40	2.91	3.20	3.15	3.50	4.09	2.64	2.98	2.16	31.94
EAST JORDAN	1.99	1.27	1.56	2.55	2.66	2.95	3.19	3.16	4.19	2.91	3.05	2.27	31.75
EAST TAWAS	1.61	1.28	2.06	2.61	2.85	3.21	2.94	3.05	2.98	2.30	2.41	2.22	29.52
EATON RAPIDS	1.66	1.29	2.24	3.16	2.91	3.68	2.89	3.22	2.77	2.30	2.11	1.89	30.12
EAU CLAIRE 4 NE	2.32	1.58	2.45	3.70	3.18	3.46	3.26	3.14	3.36	3.18	2.76	2.71	35.10
ESCANABA //R	1.37	1.07	1.78	2.42	3.01	3.37	3.58	3.32	3.16	1.99	1.99	1.61	28.67
EVART	1.77	1.36	1.95	3.03	2.68	3.34	2.96	3.62	3.28	2.46	2.43	1.89	30.77
FAYETTE	1.64	1.29	1.92	2.56	3.13	2.99	3.28	3.36	3.39	2.12	2.16	1.90	29.74
FLINT WSO	1.59	1.46	2.14	3.05	2.78	3.23	2.81	3.38	2.35	2.13	2.29	2.00	29.21
GAYLORD CON DEP	2.25	1.64	2.06	2.67	2.92	2.99	3.42	3.26	3.93	2.83	3.11	2.52	33.60
GLADWIN	1.70	1.31	2.08	2.87	2.82	3.49	3.21	3.57	3.12	2.73	2.37	2.21	31.48
GRAND HAVEN FIRE DEPT	2.06	1.29	2.20	2.98	2.91	3.08	2.61	3.01	3.27	2.83	2.67	2.36	31.27
GRAND LEDGE 1 NW	1.60	1.41	2.36	3.11	2.84	3.87	3.11	3.03	2.83	2.42	2.44	2.18	31.20
GRAND MARAIS 1 SSE	2.28	1.54	1.77	2.07	2.88	3.14	2.84	3.13	3.60	2.77	2.90	2.41	31.33
GRAND RAPIDS KENT WSO //R	1.91	1.53	2.48	3.56	3.03	3.86	3.02	3.45	3.14	2.89	2.93	2.55	34.36
GRAYLING	1.70	1.36	1.85	2.71	3.03	3.38	3.65	3.48	3.61	2.68	2.67	1.89	32.01
GREENVILLE 2 NNE	1.87	1.56	2.52	3.28	2.88	3.43	2.51	3.84	3.12	2.73	2.80	2.54	33.08
GROSSE POINTE FARMS	1.87	1.61	2.46	3.12	2.78	3.49	3.22	3.44	2.56	2.24	2.47	2.39	31.65
GULL LAKE BIOL. STA.	1.80	1.45	2.03	3.51	3.16	4.20	3.40	3.54	2.98	2.89	2.71	2.28	33.95
HALE LOUD DAM	1.52	1.22	1.84	2.53	2.74	2.76	3.12	3.06	2.95	2.25	2.26	1.79	28.04
HARBOR BEACH 3NW	2.65	2.21	2.43	2.90	2.68	3.13	3.09	3.37	3.04	2.73	2.91	3.14	34.28
HARRISON	1.64	1.37	1.91	2.84	2.82	3.17	3.47	3.24	2.99	2.63	2.41	1.95	30.44
HART	2.55	1.79	2.26	3.25	2.63	3.28	2.91	3.39	3.36	3.30	3.01	2.60	34.33
HASTINGS	1.72	1.38	2.08	3.20	2.74	3.95	2.81	3.14	3.10	2.68	2.34	2.08	31.22
HESPERIA 4 E	2.40	1.64	2.30	3.34	2.70	3.05	2.81	3.79	3.22	3.05	2.78	2.50	33.58
HILLSDALE	2.27	2.01	3.11	3.73	3.45	4.21	4.13	3.27	2.92	2.79	2.94	2.93	37.76
HOLLAND WJBL	2.28	1.48	2.45	3.59	2.80	3.92	3.17	3.32	3.39	3.13	3.13	3.03	35.69
HOUGHTON FAA AP	3.70	2.06	2.22	1.88	3.07	3.03	2.90	3.37	3.49	2.38	2.95	3.08	34.13
HOUGHTON LAKE 3 NW	1.44	1.19	1.68	2.55	2.76	3.09	3.26	2.94	3.14	2.81	2.33	1.97	29.16

MICHIGAN

PRECIPITATION NORMALS (INCHES)

STATION	JAN	FEB	MAR	APR	MAY	JUN	JUL	AUG	SEP	OCT	NOV	DEC	ANN
HOUGHTON LAKE WSO	1.49	1.30	1.88	2.58	2.59	3.10	2.89	2.96	2.77	2.28	2.26	1.89	27.99
HOWELL SEWAGE PLANT	1.65	1.49	2.31	2.90	2.63	3.37	2.96	3.14	2.52	2.27	2.22	2.06	29.52
IONIA 5 NW	1.75	1.42	2.15	3.28	2.90	3.81	2.44	3.68	3.04	2.40	2.42	2.34	31.63
IRON MOUNTAIN WTR WKS	1.13	.96	1.72	2.62	3.35	3.85	3.65	4.01	3.54	2.28	1.78	1.47	30.36
IRONWOOD	1.77	1.31	1.94	2.27	3.71	4.20	3.95	4.49	3.46	2.75	2.92	1.98	34.75
ISHPEMING	1.44	1.25	1.91	2.45	3.26	3.79	3.75	3.45	3.78	2.82	2.27	1.72	31.89
JACKSON FAA AIRPORT	1.60	1.54	2.14	2.96	2.88	3.43	3.12	2.98	2.31	2.11	2.14	1.90	29.11
KALAMAZOO STATE HOSP	2.08	1.67	2.39	3.56	3.14	3.83	3.62	3.17	3.15	3.00	2.67	2.55	34.83
KALKASKA 1 SW	1.64	1.39	1.51	2.85	2.59	3.24	3.13	3.02	3.86	3.18	2.89	2.10	31.40
KENT CITY 2 SW	2.07	1.55	2.39	3.31	2.93	3.17	2.37	3.76	3.30	2.79	2.85	2.48	32.97
KENTON U S FOREST	1.09	.93	1.46	2.15	3.45	3.88	3.70	3.87	3.62	2.63	2.13	1.33	30.24
LAKE CITY EXP FARM	1.26	1.15	1.70	2.88	2.67	3.09	3.26	3.01	3.25	2.65	2.33	1.61	28.86
LANSING WSO //R	1.74	1.56	2.30	2.88	2.57	3.50	2.78	3.04	2.54	2.13	2.33	2.21	29.58
LAPEER	1.44	1.24	1.84	2.92	2.75	3.34	2.46	3.34	2.34	2.25	2.15	1.83	27.90
LOWELL 6 NW	2.13	1.54	2.62	3.24	2.86	3.71	2.82	3.57	3.29	2.63	2.72	2.54	33.67
LUDINGTON 4 SE	2.37	1.77	1.99	2.89	2.47	2.93	2.18	3.79	3.25	2.95	2.77	2.48	31.84
LUPTON 1 S	1.46	1.28	1.93	2.57	2.99	2.82	3.32	3.29	3.12	2.31	2.38	1.90	29.37
MANISTEE	2.07	1.59	2.02	2.86	2.39	2.95	2.81	3.19	3.28	2.96	2.60	2.21	30.93
MARQUETTE WSO	1.73	1.54	2.05	2.54	3.02	3.26	2.83	2.87	3.70	2.75	2.53	2.03	30.85
MARQUETTE FAA AP	2.00	1.87	2.83	3.63	3.96	3.85	3.21	3.25	3.92	3.25	2.92	2.44	37.16
MIDLAND	1.46	1.31	2.18	2.86	2.47	2.92	2.56	3.37	2.83	2.57	2.19	2.03	28.75
MILFORD GM PROVING GRN	1.88	1.76	2.54	3.26	2.81	3.37	2.83	3.44	2.50	2.36	2.43	2.42	31.60
MILLINGTON 3 SW	1.40	1.26	2.05	2.52	2.89	3.11	2.70	3.07	2.85	2.25	2.22	1.84	28.16
MIO HYDRO PLANT	1.50	1.22	1.77	2.38	2.52	2.84	3.18	3.02	2.94	2.03	2.07	1.69	27.16
MONROE SEWAGE PLANT	1.82	1.59	2.57	3.14	2.99	3.60	2.88	3.00	2.58	2.19	2.23	2.35	30.94
MONTAGUE	2.62	1.63	2.33	3.40	2.63	2.75	2.90	3.59	3.14	3.14	2.99	2.54	33.66
MT PLEASANT UNIVERSITY	1.37	1.12	1.99	3.19	2.84	3.20	3.22	3.57	2.95	2.60	2.33	1.86	30.24
MUSKEGON WSO //R	2.37	1.65	2.54	3.16	2.54	2.52	2.42	3.13	2.92	2.78	2.87	2.60	31.50
NEWBERRY STATE HOSP	2.08	1.68	2.12	2.39	3.04	3.41	3.07	3.67	3.67	2.78	2.74	2.29	32.94
NILES	2.51	2.00	2.92	4.14	3.16	4.05	3.64	3.77	3.26	3.57	2.98	3.18	39.18
ONAWAY STATE PARK	1.76	1.40	1.96	2.65	2.86	3.08	3.30	3.21	3.75	2.38	2.55	2.13	31.03
OWOSSO WASTEWATER PL	1.68	1.40	2.04	2.83	2.58	3.32	2.70	3.21	2.68	2.02	2.27	2.06	28.79
PELLSTON FAA AIRPORT	2.28	1.53	2.05	2.70	2.95	2.91	2.93	3.30	4.00	2.60	3.01	2.58	32.84
PONTIAC STATE HOSPITAL	1.55	1.36	2.25	2.87	2.75	3.52	2.82	3.05	2.36	2.31	2.31	2.15	29.30
PORT HURON WASTEWTR PL	1.76	1.43	2.26	3.17	2.79	3.47	2.91	3.02	2.37	2.50	2.60	2.34	30.62
ROCK	1.51	1.33	1.97	2.70	3.48	4.37	3.84	4.11	4.21	2.73	2.45	1.88	34.58
SAGINAW FAA AP	1.79	1.52	2.31	3.00	2.45	2.77	2.60	3.04	2.98	2.66	2.34	2.31	29.77
SAINT CHARLES	1.62	1.34	2.13	2.43	2.49	3.09	2.83	3.29	2.76	2.24	2.17	1.91	28.30
SAINT JOHNS	1.53	1.30	2.05	3.01	2.84	3.39	2.63	3.82	2.79	2.44	2.31	2.01	30.12
SANDUSKY	1.54	1.30	2.05	2.61	2.69	3.15	3.06	2.60	2.37	2.37	2.25	2.02	28.01
SAULT STE MARIE WSO //R	2.20	1.69	2.03	2.38	2.90	3.26	3.00	3.46	3.90	2.89	3.20	2.57	33.48
SEBEWAING MICH SGR CO	1.27	1.10	1.72	2.22	2.47	2.71	2.94	2.76	2.81	2.31	2.07	1.64	26.02
SENEY NATL WLR	2.03	1.72	2.10	2.42	3.12	3.54	3.01	3.33	3.73	2.84	2.95	2.37	33.16
SOUTH HAVEN EXP FARM	2.25	1.64	2.50	3.47	2.85	3.81	3.30	3.47	3.37	3.02	2.69	2.66	35.03
SPALDING	1.31	1.06	1.86	2.74	3.36	3.54	3.70	3.83	3.50	2.32	1.97	1.62	30.81
STAMBAUGH 1 S	1.20	1.03	1.77	2.46	3.69	4.17	3.78	4.28	3.49	2.40	1.98	1.57	31.82
STANDISH 2 S	1.30	1.15	1.85	2.50	2.69	3.15	2.92	2.89	2.99	2.53	2.11	1.73	27.81
STEPHENSON 5 W	1.72	1.33	2.14	2.67	3.57	3.72	3.63	3.86	3.60	2.29	2.05	2.05	32.63
STEUBEN 2 WNW	2.07	1.77	2.24	2.59	2.88	3.36	3.00	3.45	3.56	2.80	2.88	2.30	32.90
THOMPSONVILLE	2.30	1.54	1.92	2.80	2.56	2.87	3.05	3.38	3.65	3.01	2.95	2.38	32.41
THREE RIVERS	1.84	1.49	2.44	3.35	3.12	3.95	3.79	3.16	3.01	2.71	2.38	2.32	33.56
TRAVERSE CITY FAA AP	1.86	1.41	1.78	2.51	2.48	3.15	2.88	2.93	3.60	2.59	2.51	2.01	29.71
VANDERBILT TROUT STA	1.90	1.40	1.88	2.56	2.84	2.89	3.23	3.15	3.59	2.44	2.53	2.13	30.54
WELLSTON TIPPY DAM	2.08	1.52	1.83	2.99	2.57	3.38	3.14	3.29	3.49	2.88	2.62	2.12	31.91
WEST BRANCH 2 N	1.43	1.32	1.88	2.44	2.78	2.80	3.25	3.10	3.04	2.48	2.45	1.86	28.83

MICHIGAN

PRECIPITATION NORMALS (INCHES)

STATION	JAN	FEB	MAR	APR	MAY	JUN	JUL	AUG	SEP	OCT	NOV	DEC	ANN
WILLIAMSTON	1.68	1.50	2.46	2.93	2.88	3.80	3.23	3.29	2.62	2.34	2.41	2.14	31.28
WILLIS 5 SSW	1.77	1.67	2.61	3.25	2.88	3.56	2.73	3.30	2.44	2.17	2.44	2.47	31.29
YALE 1 NNE	1.68	1.38	2.01	2.82	2.77	3.19	2.80	2.92	2.42	2.40	2.29	2.14	28.82
YPSILANTI EAST MICH U	1.73	1.58	2.47	3.10	2.69	3.13	2.91	2.90	2.43	2.07	2.32	2.41	29.74

MICHIGAN

HEATING DEGREE DAY NORMALS (BASE 65 DEG F)

STATION	JUL	AUG	SEP	OCT	NOV	DEC	JAN	FEB	MAR	APR	MAY	JUN	ANN
ADRIAN 2 NNE	5	18	124	429	780	1144	1302	1109	933	528	239	37	6648
ALLEGAN SEWAGE PLANT	9	21	116	430	756	1128	1293	1114	946	525	245	57	6640
ALMA	0	24	120	436	792	1184	1358	1170	998	555	255	52	6944
ALPENA WSO //R	51	91	256	555	897	1287	1476	1316	1184	741	419	137	8410
ALPENA SEWAGE PLANT	34	67	202	508	846	1221	1404	1252	1138	732	412	128	7944
ANN ARBOR UNIV OF MICH	0	10	87	382	738	1119	1277	1089	911	495	205	33	6346
BAD AXE	24	37	149	456	795	1190	1376	1204	1051	621	328	87	7318
BALDWIN	23	54	191	509	852	1240	1398	1240	1076	624	301	93	7601
BATTLE CREEK	6	20	114	427	780	1156	1321	1126	952	534	246	41	6723
BEECHWOOD 7 WNW	80	106	301	601	1050	1488	1671	1403	1246	756	410	157	9269
BENTON HARBOR AIRPORT	6	17	100	372	717	1082	1249	1058	902	513	251	46	6313
BERGLAND DAM	76	132	325	614	1050	1500	1708	1470	1314	813	451	168	9621
BIG RAPIDS WATERWORKS	24	62	201	518	855	1237	1407	1226	1070	627	320	83	7630
CADILLAC	53	89	251	564	918	1314	1482	1327	1181	711	391	131	8412
CARO	8	25	131	446	786	1181	1367	1184	1001	570	283	56	7038
CHAMPION VAN RIPER PK	95	128	333	648	1059	1494	1677	1450	1305	828	464	184	9665
CHARLOTTE	11	34	133	452	801	1181	1342	1159	977	549	259	58	6956
CHATHAM EXP FARM	75	94	263	558	942	1342	1522	1327	1221	771	442	169	8726
CHEBOYGAN RR LIGHT STA	39	60	188	494	843	1240	1426	1285	1156	729	397	120	7977
COLDWATER STATE SCHOOL	7	23	122	431	780	1156	1321	1120	946	540	254	47	6747
DETROIT METROPOLITAN AP	5	12	106	414	765	1132	1290	1098	930	528	247	36	6563
DUNBAR FOREST EXP STA	82	105	278	577	933	1373	1581	1408	1256	783	448	180	9004
EAST JORDAN	55	83	197	484	837	1215	1389	1268	1125	672	358	116	7799
EAST TAWAS	27	54	190	499	831	1215	1395	1226	1094	672	369	110	7682
EAU CLAIRE 4 NE	0	10	82	372	741	1116	1290	1081	905	492	223	35	6347
ESCANABA //R	44	80	244	546	903	1308	1516	1322	1197	780	456	151	8547
EVART	30	68	213	533	879	1271	1445	1266	1104	648	327	97	7881
FAYETTE	58	70	218	518	864	1265	1473	1288	1163	771	459	180	8327
FLINT WSO	9	28	144	455	786	1172	1355	1168	1020	576	288	67	7068
GAYLORD CON DEP	41	77	223	527	906	1308	1473	1308	1156	687	359	113	8178
GLADWIN	23	45	192	512	867	1249	1426	1235	1070	621	314	82	7636
GRAND HAVEN FIRE DEPT	13	26	104	387	729	1088	1249	1089	949	564	281	63	6542
GRAND MARAIS 1 SSE	119	131	288	574	927	1296	1473	1324	1243	816	508	236	8935
GRAND RAPIDS KENT WSO //R	12	23	130	443	795	1169	1333	1156	989	561	262	54	6927
GRAYLING	42	80	228	540	894	1305	1469	1316	1166	684	350	115	8189
GREENVILLE 2 NNE	7	23	127	447	807	1194	1355	1168	998	555	254	47	6982
GROSSE POINTE FARMS	0	7	69	355	699	1073	1240	1067	918	510	215	25	6178
GULL LAKE BIOL. STA.	0	14	90	393	759	1144	1314	1123	949	525	233	37	6581
HALE LOUD DAM	27	68	216	521	870	1274	1469	1288	1135	687	364	104	8023
HARBOR BEACH 3NW	27	38	138	442	786	1172	1358	1190	1060	657	373	104	7345
HART	15	31	133	439	789	1153	1318	1159	1020	594	302	75	7028
HASTINGS	7	20	127	431	777	1166	1324	1142	964	534	243	53	6788
HESPERIA 4 E	22	52	169	488	828	1209	1373	1207	1045	609	310	86	7398
HILLSDALE	9	22	123	438	789	1166	1324	1126	952	537	257	49	6792
HOLLAND WJBL	17	38	107	396	741	1113	1268	1098	942	537	256	56	6569
HOUGHTON FAA AP	86	116	321	620	1017	1407	1600	1417	1296	837	490	194	9401
HOUGHTON LAKE 3 NW	36	82	230	541	885	1290	1460	1291	1135	666	339	118	8073
HOUGHTON LAKE WSO	42	89	242	561	903	1305	1488	1316	1166	693	364	129	8298
IONIA 5 NW	13	24	128	434	783	1169	1330	1151	970	537	250	54	6843
IRON MOUNTAIN WTR WKS	45	66	267	570	981	1417	1609	1355	1194	708	367	113	8692
IRONWOOD	53	98	286	589	1038	1494	1696	1420	1256	735	389	136	9190
ISHPEMING	61	93	278	580	1002	1414	1590	1358	1231	765	422	144	8938
JACKSON FAA AIRPORT	9	20	118	428	786	1172	1345	1148	970	543	255	39	6833
KALAMAZOO STATE HOSP	0	7	78	367	741	1116	1280	1078	902	480	202	30	6281
LAKE CITY EXP FARM	34	75	229	549	903	1311	1473	1308	1166	678	360	112	8198

MICHIGAN

HEATING DEGREE DAY NORMALS (BASE 65 DEG F)

STATION	JUL	AUG	SEP	OCT	NOV	DEC	JAN	FEB	MAR	APR	MAY	JUN	ANN
LANSING WSO //R	12	25	132	452	795	1178	1345	1168	992	561	269	58	6987
LAPEER	8	31	137	450	789	1178	1352	1162	992	561	272	61	6993
LUDINGTON 4 SE	25	43	149	464	801	1166	1327	1170	1035	621	344	104	7249
LUPTON 1 S	49	87	250	564	903	1311	1491	1299	1141	678	366	112	8251
MANISTEE	19	33	135	425	777	1141	1311	1156	1026	618	326	100	7067
MARQUETTE WSO	60	79	241	532	915	1293	1482	1285	1169	759	459	171	8445
MARQUETTE FAA AP	86	123	330	642	1047	1460	1640	1420	1296	828	469	179	9520
MIDLAND	0	22	122	430	777	1166	1333	1154	995	552	249	47	6847
MILFORD GM PROVING GRN	8	21	137	456	807	1190	1367	1179	1004	579	270	59	7077
MIO HYDRO PLANT	39	79	234	536	876	1293	1473	1299	1147	690	371	118	8155
MONROE SEWAGE PLANT	0	0	74	371	717	1091	1252	1067	908	501	212	24	6217
MONTAGUE	23	42	153	449	774	1132	1293	1134	998	603	316	100	7017
MT PLEASANT UNIVERSITY	7	26	145	455	804	1194	1370	1187	1026	576	271	56	7117
MUSKEGON WSO //R	10	22	132	440	774	1138	1299	1145	998	594	297	76	6925
NEWBERRY STATE HOSP	83	115	306	608	966	1358	1535	1352	1228	783	464	188	8986
ONAWAY STATE PARK	37	69	200	493	861	1262	1451	1291	1141	687	349	98	7939
OWOSSO WASTEWATER PL	7	24	130	434	783	1175	1342	1159	986	555	267	47	6909
PELLSTON FAA AIRPORT	74	104	283	577	915	1324	1519	1386	1240	759	432	152	8765
PONTIAC STATE HOSPITAL	0	12	103	404	765	1153	1321	1128	958	534	243	37	6658
PORT HURON WASTEWTR PL	0	15	97	389	732	1116	1283	1112	961	570	283	53	6611
SAGINAW FAA AP	6	23	136	453	798	1197	1373	1190	1020	576	276	55	7103
SAINT CHARLES	0	14	108	416	771	1156	1327	1140	964	525	236	32	6689
SAINT JOHNS	0	14	113	417	777	1169	1333	1148	970	537	249	39	6766
SANDUSKY	12	29	129	448	786	1181	1355	1176	1020	597	298	75	7106
SAULT STE MARIE WSO //R	101	123	306	611	966	1404	1603	1420	1274	807	480	210	9305
SENEY NATL WLR	50	76	244	552	927	1339	1519	1350	1221	747	405	142	8572
SOUTH HAVEN EXP FARM	16	25	101	369	711	1082	1246	1067	921	561	291	75	6465
STAMBAUGH 1 S	68	107	306	598	1029	1469	1662	1403	1234	741	404	152	9173
STANDISH 2 S	19	50	190	505	849	1237	1423	1240	1076	636	331	80	7636
STEPHENSON 5 W	43	64	241	546	945	1370	1572	1324	1156	681	363	111	8416
THREE RIVERS	0	15	99	404	756	1128	1302	1095	908	492	222	35	6456
TRAVERSE CITY FAA AP	33	63	178	490	843	1221	1404	1271	1135	678	369	110	7795
VANDERBILT TROUT STA	83	126	299	598	945	1355	1522	1378	1218	741	414	141	8820
WATERSMEET	75	111	304	601	1035	1473	1668	1431	1262	774	428	158	9320
WEST BRANCH 2 N	33	65	219	542	891	1290	1463	1268	1122	666	345	101	8005
WILLIS 5 SSW	9	22	136	451	792	1166	1333	1134	955	549	255	47	6849
YPSILANTI EAST MICH U	5	16	102	397	753	1125	1287	1095	924	513	234	28	6479

MICHIGAN

COOLING DEGREE DAY NORMALS (BASE 65 DEG F)

STATION	JAN	FEB	MAR	APR	MAY	JUN	JUL	AUG	SEP	OCT	NOV	DEC	ANN
ADRIAN 2 NNE	0	0	0	0	34	127	213	170	55	7	0	0	606
ALLEGAN SEWAGE PLANT	0	0	0	0	41	132	210	167	47	12	0	0	609
ALMA	0	0	0	0	35	130	210	164	30	8	0	0	577
ALPENA WSO //R	0	0	0	0	0	23	82	75	0	0	0	0	180
ALPENA SEWAGE PLANT	0	0	0	0	0	26	105	98	7	0	0	0	236
ANN ARBOR UNIV OF MICH	0	0	0	0	41	156	248	202	63	13	0	0	723
BAD AXE	0	0	0	0	21	84	157	124	26	9	0	0	421
BALDWIN	0	0	0	0	28	96	144	119	17	7	0	0	411
BATTLE CREEK	0	0	0	0	35	122	213	175	45	9	0	0	599
BEECHWOOD 7 WNW	0	0	0	0	17	34	98	66	0	0	0	0	215
BENTON HARBOR AIRPORT	0	0	0	0	37	133	217	175	67	12	0	0	641
BERGLAND DAM	0	0	0	0	11	30	82	58	0	0	0	0	181
BIG RAPIDS WATERWORKS	0	0	0	0	26	77	145	124	18	0	0	0	390
CADILLAC	0	0	0	0	15	50	90	74	0	0	0	0	229
CARO	0	0	0	0	26	98	179	137	23	6	0	0	469
CHAMPION VAN RIPER PK	0	0	0	0	11	19	70	45	0	0	0	0	145
CHARLOTTE	0	0	0	0	30	112	182	142	37	9	0	0	512
CHATHAM EXP FARM	0	0	0	0	8	16	88	72	0	0	0	0	184
CHEBOYGAN RR LIGHT STA	0	0	0	0	0	27	132	115	14	7	0	0	295
COLDWATER STATE SCHOOL	0	0	0	0	31	116	190	153	44	9	0	0	543
DETROIT METROPOLITAN AP	0	0	0	0	33	117	219	183	55	8	0	0	615
DUNBAR FOREST EXP STA	0	0	0	0	0	12	66	62	0	0	0	0	140
EAST JORDAN	0	0	0	0	17	59	129	111	11	0	0	0	327
EAST TAWAS	0	0	0	0	6	47	117	104	16	0	0	0	290
EAU CLAIRE 4 NE	0	0	0	0	43	149	239	199	64	12	0	0	706
ESCANABA //R	0	0	0	0	0	16	90	77	0	0	0	0	183
EVART	0	0	0	0	17	67	117	106	12	0	0	0	319
FAYETTE	0	0	0	0	0	9	83	76	0	0	0	0	168
FLINT WSO	0	0	0	0	21	94	167	136	30	8	0	0	456
GAYLORD CON DEP	0	0	0	0	18	59	115	98	7	0	0	0	297
GLADWIN	0	0	0	0	17	79	150	110	9	0	0	0	365
GRAND HAVEN FIRE DEPT	0	0	0	0	20	87	177	159	38	6	0	0	487
GRAND MARAIS 1 SSE	0	0	0	0	6	11	50	54	9	0	0	0	130
GRAND RAPIDS KENT WSO //R	0	0	0	0	29	117	210	165	43	6	0	0	570
GRAYLING	0	0	0	0	18	67	119	101	12	7	0	0	324
GREENVILLE 2 NNE	0	0	0	0	31	113	197	157	28	7	0	0	533
GROSSE POINTE FARMS	0	0	0	0	32	142	261	212	72	11	0	0	730
GULL LAKE BIOL. STA.	0	0	0	0	38	127	227	187	48	9	0	0	636
HALE LOUD DAM	0	0	0	0	10	47	108	93	9	0	0	0	267
HARBOR BEACH 3NW	0	0	0	0	7	50	132	125	24	5	0	0	343
HART	0	0	0	0	23	84	163	137	22	8	0	0	437
HASTINGS	0	0	0	0	35	128	199	162	52	12	0	0	588
HESPERIA 4 E	0	0	0	0	22	89	152	126	16	7	0	0	412
HILLSDALE	0	0	0	0	31	106	173	140	33	7	0	0	490
HOLLAND WJBL	0	0	0	0	33	116	203	177	50	8	0	0	587
HOUGHTON FAA AP	0	0	0	0	7	11	76	57	0	0	0	0	151
HOUGHTON LAKE 3 NW	0	0	0	0	17	76	114	104	14	7	0	0	332
HOUGHTON LAKE WSO	0	0	0	0	14	63	104	95	11	6	0	0	293
IONIA 5 NW	0	0	0	0	36	123	202	158	38	6	0	0	563
IRON MOUNTAIN WTR WKS	0	0	0	0	16	41	113	66	0	0	0	0	236
IRONWOOD	0	0	0	0	17	49	109	85	10	6	0	0	276
ISHPEMING	0	0	0	0	15	33	98	71	0	0	0	0	217
JACKSON FAA AIRPORT	0	0	0	0	35	120	223	175	49	7	0	0	609
KALAMAZOO STATE HOSP	0	0	0	0	53	156	254	212	66	10	0	0	751
LAKE CITY EXP FARM	0	0	0	0	16	52	103	87	7	0	0	0	265

MICHIGAN

COOLING DEGREE DAY NORMALS (BASE 65 DEG F)

STATION	JAN	FEB	MAR	APR	MAY	JUN	JUL	AUG	SEP	OCT	NOV	DEC	ANN
LANSING WSO //R	0	0	0	0	28	112	192	156	33	9	0	0	530
LAPEER	0	0	0	0	21	103	172	137	32	10	0	0	475
LUDINGTON 4 SE	0	0	0	0	12	68	137	115	14	6	0	0	352
LUPTON 1 S	0	0	0	0	12	40	89	71	7	0	0	0	219
MANISTEE	0	0	0	0	16	76	156	132	27	7	0	0	414
MARQUETTE WSO	0	0	0	0	7	18	100	8	7	5	0	0	225
MARQUETTE FAA AP	0	0	0	0	10	17	73	48	0	0	0	0	148
MIDLAND	0	0	0	0	29	122	206	158	32	8	0	0	555
MILFORD GM PROVING GRN	0	0	0	0	19	110	188	151	38	10	0	0	516
MIO HYDRO PLANT	0	0	0	0	15	49	114	92	6	0	0	0	276
MONROE SEWAGE PLANT	0	0	0	0	39	156	264	219	77	12	0	0	767
MONTAGUE	0	0	0	0	16	73	131	117	18	5	0	0	360
MT PLEASANT UNIVERSITY	0	0	0	0	26	104	187	150	34	8	0	0	509
MUSKEGON WSO //R	0	0	0	0	18	91	169	143	24	6	0	0	451
NEWBERRY STATE HOSP	0	0	0	0	8	17	70	63	0	0	0	0	158
ONAWAY STATE PARK	0	0	0	0	14	47	133	113	11	0	0	0	318
OWOSSO WASTEWATER PL	0	0	0	0	28	101	184	142	31	7	0	0	493
PELLSTON FAA AIRPORT	0	0	0	0	11	26	90	66	0	0	0	0	193
PONTIAC STATE HOSPITAL	0	0	0	0	38	124	224	182	58	10	0	0	636
PORT HURON WASTEWTR PL	0	0	0	0	19	107	222	195	64	14	0	0	621
SAGINAW FAA AP	0	0	0	0	25	106	192	147	25	6	0	0	501
SAINT CHARLES	0	0	0	0	41	128	227	172	42	7	0	0	617
SAINT JOHNS	0	0	0	0	35	120	208	160	44	11	0	0	578
SANDUSKY	0	0	0	0	19	99	170	143	30	8	0	0	469
SAULT STE MARIE WSO //R	0	0	0	0	6	12	55	58	0	0	0	0	131
SENEY NATL WLR	0	0	0	0	11	28	90	73	0	0	0	0	202
SOUTH HAVEN EXP FARM	0	0	0	0	22	99	178	155	53	10	0	0	517
STAMBAUGH 1 S	0	0	0	0	14	29	80	54	0	0	0	0	177
STANDISH 2 S	0	0	0	0	12	68	137	105	19	0	0	0	341
STEPHENSON 5 W	0	0	0	0	13	42	118	80	0	0	0	0	253
THREE RIVERS	0	0	0	0	42	140	221	176	51	7	0	0	637
TRAVERSE CITY FAA AP	0	0	0	0	13	71	151	135	10	0	0	0	380
VANDERBILT TROUT STA	0	0	0	0	14	30	83	67	11	0	0	0	205
WATERSMEET	0	0	0	0	16	32	97	58	0	6	0	0	209
WEST BRANCH 2 N	0	0	0	0	14	59	120	93	6	6	0	0	298
WILLIS 5 SSW	0	0	0	0	22	101	177	137	37	0	0	0	474
YPSILANTI EAST MICH U	0	0	0	0	45	136	247	205	63	10	0	0	706

20 — MICHIGAN

STATE-STATION NUMBER	STN TYP	NAME	LATITUDE DEG-MIN	LONGITUDE DEG-MIN	ELEVATION (FT)
20-0032	13	ADRIAN 2 NNE	N 4155	W 08401	754
20-0094	12	ALBION	N 4215	W 08446	940
20-0128	13	ALLEGAN SEWAGE PLANT	N 4232	W 08551	629
20-0146	13	ALMA	N 4323	W 08440	740
20-0164	13	ALPENA WSO //R	N 4504	W 08334	689
20-0169	13	ALPENA SEWAGE PLANT	N 4504	W 08326	586
20-0230	13	ANN ARBOR UNIV OF MICH	N 4217	W 08344	871
20-0417	13	BAD AXE	N 4349	W 08300	715
20-0446	13	BALDWIN	N 4354	W 08550	825
20-0552	13	BATTLE CREEK	N 4220	W 08511	870
20-0647	13	BEECHWOOD 7 WNW	N 4611	W 08853	1660
20-0710	13	BENTON HARBOR AIRPORT	N 4208	W 08626	649
20-0718	13	BERGLAND DAM	N 4635	W 08933	1300
20-0779	13	BIG RAPIDS WATERWORKS	N 4342	W 08529	930
20-0864	12	BLOOMINGDALE	N 4223	W 08558	725
20-1176	13	CADILLAC	N 4416	W 08524	1295
20-1299	13	CARO	N 4327	W 08324	670
20-1439	13	CHAMPION VAN RIPER PK	N 4631	W 08759	1565
20-1468	12	CHARLEVOIX	N 4519	W 08515	607
20-1476	13	CHARLOTTE	N 4233	W 08450	897
20-1484	13	CHATHAM EXP FARM	N 4621	W 08656	875
20-1492	13	CHEBOYGAN RR LIGHT STA	N 4539	W 08428	586
20-1675	13	COLDWATER STATE SCHOOL	N 4157	W 08500	984
20-1922	12	CRYSTAL FALLS 6 NE	N 4610	W 08814	1360
20-2094	12	DE TOUR 1 N	N 4601	W 08355	598
20-2103	13	DETROIT METROPOLITAN AP	N 4214	W 08320	633
20-2298	13	DUNBAR FOREST EXP STA	N 4619	W 08414	600
20-2381	13	EAST JORDAN	N 4509	W 08508	590
20-2423	13	EAST TAWAS	N 4417	W 08330	586
20-2437	12	EATON RAPIDS	N 4231	W 08439	870
20-2445	13	EAU CLAIRE 4 NE	N 4201	W 08615	870
20-2626	13	ESCANABA //R	N 4545	W 08703	594
20-2671	13	EVART	N 4354	W 08516	997
20-2737	13	FAYETTE	N 4541	W 08642	788
20-2846	13	FLINT WSO	N 4258	W 08345	770
20-3096	13	GAYLORD CON DEP	N 4502	W 08441	1349
20-3170	13	GLADWIN	N 4359	W 08430	770
20-3290	13	GRAND HAVEN FIRE DEPT	N 4304	W 08613	622
20-3306	12	GRAND LEDGE 1 NW	N 4246	W 08446	790
20-3319	13	GRAND MARAIS 1 SSE	N 4640	W 08559	755
20-3333	13	GRAND RAPIDS KENT WSO //R	N 4253	W 08531	784
20-3391	13	GRAYLING	N 4440	W 08442	1140
20-3429	13	GREENVILLE 2 NNE	N 4312	W 08515	882
20-3477	13	GROSSE POINTE FARMS	N 4223	W 08254	613
20-3504	13	GULL LAKE BIOL. STA.	N 4224	W 08524	890
20-3529	13	HALE LOUD DAM	N 4428	W 08343	800
20-3585	13	HARBOR BEACH 3NW	N 4352	W 08241	620
20-3616	12	HARRISON	N 4402	W 08448	1156
20-3632	13	HART	N 4342	W 08622	675
20-3661	13	HASTINGS	N 4239	W 08518	780

STATE-STATION NUMBER	STN TYP	NAME	LATITUDE DEG-MIN	LONGITUDE DEG-MIN	ELEVATION (FT)
20-3769	13	HESPERIA 4 E	N 4334	W 08558	895
20-3823	13	HILLSDALE	N 4155	W 08438	1100
20-3858	13	HOLLAND WJBL	N 4242	W 08606	678
20-3908	13	HOUGHTON FAA AP	N 4710	W 08830	1081
20-3932	13	HOUGHTON LAKE 3 NW	N 4420	W 08449	1136
20-3936	13	HOUGHTON LAKE WSO	N 4422	W 08441	1149
20-3947	12	HOWELL SEWAGE PLANT	N 4238	W 08355	915
20-4078	13	IONIA 5 NW	N 4302	W 08509	850
20-4090	13	IRON MOUNTAIN WTR WKS	N 4550	W 08804	1155
20-4104	13	IRONWOOD	N 4627	W 09010	1520
20-4127	13	ISHPEMING	N 4629	W 08739	1436
20-4150	13	JACKSON FAA AIRPORT	N 4216	W 08428	998
20-4244	13	KALAMAZOO STATE HOSP	N 4217	W 08536	945
20-4257	12	KALKASKA 1 SW	N 4444	W 08512	1032
20-4320	12	KENT CITY 2 SW	N 4312	W 08546	840
20-4328	12	KENTON U S FOREST	N 4629	W 08853	1167
20-4502	13	LAKE CITY EXP FARM	N 4418	W 08512	1230
20-4641	13	LANSING WSO //R	N 4246	W 08436	841
20-4655	13	LAPEER	N 4302	W 08320	865
20-4944	12	LOWELL 6 NW	N 4259	W 08525	815
20-4954	13	LUDINGTON 4 SE	N 4354	W 08624	690
20-4967	13	LUPTON 1 S	N 4425	W 08401	900
20-5065	13	MANISTEE	N 4414	W 08618	585
20-5178	13	MARQUETTE WSO	N 4633	W 08723	665
20-5184	13	MARQUETTE FAA AP	N 4632	W 08733	1415
20-5434	13	MIDLAND	N 4337	W 08413	642
20-5452	13	MILFORD GM PROVING GRN	N 4233	W 08341	1188
20-5488	12	MILLINGTON 3 SW	N 4314	W 08334	757
20-5531	13	MIO HYDRO PLANT	N 4440	W 08408	963
20-5558	13	MONROE SEWAGE PLANT	N 4154	W 08322	582
20-5567	13	MONTAGUE	N 4325	W 08622	600
20-5662	13	MT PLEASANT UNIVERSITY	N 4335	W 08446	796
20-5712	13	MUSKEGON WSO //R	N 4310	W 08614	627
20-5816	13	NEWBERRY STATE HOSP	N 4620	W 08530	886
20-5892	12	NILES	N 4151	W 08616	650
20-6184	13	ONAWAY STATE PARK	N 4526	W 08414	690
20-6300	13	OWOSSO WASTEWATER PL	N 4301	W 08411	738
20-6438	13	PELLSTON FAA AIRPORT	N 4534	W 08448	710
20-6658	13	PONTIAC STATE HOSPITAL	N 4239	W 08318	974
20-6680	13	PORT HURON WASTEWTR PL	N 4259	W 08225	586
20-7068	12	ROCK	N 4604	W 08710	961
20-7227	13	SAGINAW FAA AP	N 4332	W 08405	660
20-7253	13	SAINT CHARLES	N 4318	W 08408	596
20-7280	13	SAINT JOHNS	N 4301	W 08433	755
20-7350	13	SANDUSKY	N 4325	W 08250	774
20-7366	13	SAULT STE MARIE WSO //R	N 4628	W 08422	721
20-7419	12	SEBEWAING MICH SGR CO	N 4344	W 08327	600
20-7515	13	SENEY NATL WLR	N 4617	W 08557	710
20-7690	13	SOUTH HAVEN EXP FARM	N 4224	W 08617	626
20-7742	12	SPALDING	N 4541	W 08730	860

20 — MICHIGAN

LEGEND
11 = TEMPERATURE ONLY
12 = PRECIPITATION ONLY
13 = TEMP. & PRECIP.

STATE-STATION NUMBER	STN TYP	NAME	LATITUDE DEG-MIN	LONGITUDE DEG-MIN	ELEVATION (FT)
20-7812	13	STAMBAUGH 1 S	N 4604	W 08838	1485
20-7820	13	STANDISH 2 S	N 4357	W 08358	616
20-7867	13	STEPHENSON 5 W	N 4524	W 08743	715
20-7880	12	STEUBEN 2 WNW	N 4612	W 08630	645
20-8167	12	THOMPSONVILLE	N 4431	W 08556	795
20-8184	13	THREE RIVERS	N 4156	W 08538	810
20-8251	13	TRAVERSE CITY FAA AP	N 4444	W 08535	618
20-8417	13	VANDERBILT TROUT STA	N 4510	W 08427	925
20-8680	11	WATERSMEET	N 4616	W 08911	1630
20-8772	12	WELLSTON TIPPY DAM	N 4415	W 08557	650
20-8800	13	WEST BRANCH 2 N	N 4418	W 08414	1064
20-9006	12	WILLIAMSTON	N 4241	W 08418	860
20-9014	13	WILLIS 5 SSW	N 4205	W 08335	660
20-9188	12	YALE 1 NNE	N 4309	W 08248	819
20-9218	13	YPSILANTI EAST MICH U	N 4215	W 08337	779

302

20 − MICHIGAN

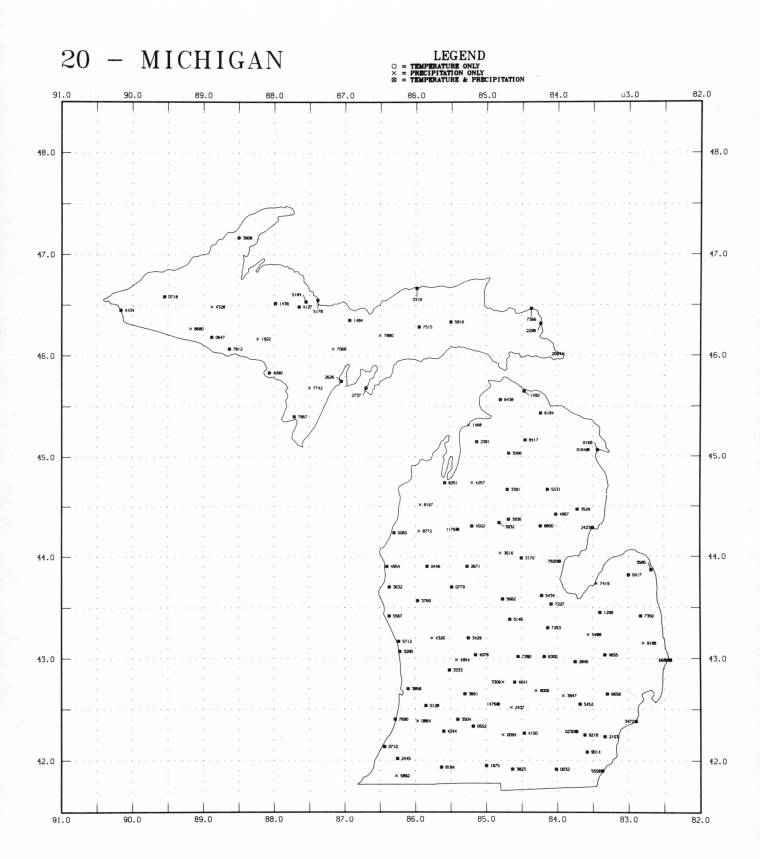

303

MINNESOTA

TEMPERATURE NORMALS (DEG F)

STATION		JAN	FEB	MAR	APR	MAY	JUN	JUL	AUG	SEP	OCT	NOV	DEC	ANN
ADA	MAX	14.2	22.3	35.3	54.6	70.2	78.1	83.5	82.4	71.2	58.9	37.7	22.4	52.6
	MIN	-6.0	1.0	14.5	31.1	42.5	52.9	57.6	55.6	46.0	35.5	19.8	3.8	29.5
	MEAN	4.1	11.7	24.9	42.9	56.4	65.5	70.5	69.0	58.7	47.2	28.8	13.1	41.1
ALBERT LEA	MAX	20.9	27.3	37.4	55.7	69.6	79.3	83.0	80.9	71.8	60.6	42.2	27.4	54.7
	MIN	3.0	8.7	20.4	35.3	47.2	57.3	61.9	59.4	49.6	38.9	24.7	11.1	34.8
	MEAN	12.0	18.0	28.9	45.6	58.4	68.3	72.5	70.2	60.8	49.8	33.4	19.3	44.8
ALEXANDRIA FAA AIRPORT//	MAX	15.3	22.5	33.8	51.8	66.4	75.7	81.4	79.0	68.4	56.8	37.2	22.7	50.9
	MIN	-4.3	2.0	14.6	31.7	44.4	54.4	59.7	57.7	47.1	36.3	21.0	5.6	30.9
	MEAN	5.5	12.2	24.2	41.7	55.4	65.1	70.5	68.4	57.8	46.6	29.1	14.2	40.9
ARGYLE 4 E	MAX	9.9	17.3	30.2	49.9	67.1	75.8	81.2	80.1	68.4	56.1	34.6	18.9	49.1
	MIN	-11.0	-5.4	8.8	28.5	40.3	50.6	54.8	52.4	42.6	32.3	16.5	-.4	25.8
	MEAN	-.5	5.9	19.5	39.2	53.7	63.2	68.0	66.2	55.5	44.2	25.6	9.2	37.5
ARTICHOKE LAKE	MAX	18.5	25.3	36.3	55.0	69.6	78.8	84.3	82.3	72.4	60.5	40.4	25.6	54.1
	MIN	-1.9	4.5	16.9	33.1	45.6	55.7	60.2	57.9	47.9	37.3	21.8	7.4	32.2
	MEAN	8.3	14.9	26.7	44.1	57.6	67.2	72.2	70.1	60.2	48.9	31.1	16.5	43.2
AUSTIN 3 S	MAX	21.5	28.1	38.8	57.3	70.8	79.7	83.3	81.4	73.2	61.8	42.8	28.0	55.6
	MIN	1.7	7.9	19.8	34.3	45.5	55.1	59.1	56.6	47.6	37.5	24.0	10.9	33.3
	MEAN	11.6	18.0	29.3	45.9	58.2	67.4	71.2	69.0	60.5	49.7	33.4	19.4	44.5
BABBITT 2 SE	MAX	14.7	22.3	33.9	49.8	65.0	73.1	77.3	74.7	63.7	53.2	34.0	20.4	48.5
	MIN	-4.0	1.4	13.4	28.8	40.7	50.5	55.5	53.5	44.8	35.3	20.1	4.1	28.7
	MEAN	5.4	11.9	23.7	39.3	52.8	61.8	66.4	64.1	54.3	44.3	27.0	12.3	38.6
BAUDETTE	MAX	13.0	21.4	33.8	51.1	66.1	74.7	79.6	77.3	66.8	55.7	35.6	20.3	49.6
	MIN	-10.8	-5.8	7.8	27.2	39.7	49.9	54.8	52.3	43.1	34.1	18.5	-.4	25.9
	MEAN	1.1	7.8	20.8	39.2	53.0	62.3	67.2	64.9	55.0	44.9	27.0	10.0	37.8
BEMIDJI AIRPORT	MAX	13.1	21.4	33.0	50.1	64.8	74.2	79.4	76.7	65.4	54.4	34.6	20.1	48.9
	MIN	-9.6	-3.9	9.3	27.6	39.7	50.4	55.6	53.1	43.1	33.3	17.4	.1	26.3
	MEAN	1.8	8.8	21.2	38.9	52.3	62.3	67.5	64.9	54.3	43.8	26.1	10.1	37.7
BIG FALLS	MAX	15.1	24.3	36.8	53.3	67.7	76.2	80.7	78.2	67.2	56.2	35.8	21.4	51.1
	MIN	-9.7	-3.8	9.6	27.0	38.3	48.2	52.9	50.7	41.7	32.9	17.6	.3	25.5
	MEAN	2.7	10.3	23.2	40.1	53.0	62.2	66.8	64.5	54.5	44.6	26.7	10.9	38.3
CAMBRIDGE ST HOSPITAL	MAX	17.8	24.9	36.5	54.4	67.9	76.7	81.5	78.9	68.5	57.7	39.0	24.9	52.4
	MIN	-2.1	3.8	16.7	32.6	44.4	54.1	58.8	56.3	46.6	36.6	21.8	7.3	31.4
	MEAN	7.8	14.4	26.6	43.5	56.2	65.4	70.2	67.6	57.6	47.1	30.5	16.1	41.9
CAMPBELL	MAX	15.3	21.8	34.2	52.9	68.2	77.3	83.3	81.9	71.0	59.0	38.5	23.1	52.2
	MIN	-5.7	.5	14.2	31.2	42.4	52.9	57.0	55.0	44.5	33.5	18.9	3.9	29.0
	MEAN	4.8	11.2	24.2	42.0	55.3	65.2	70.2	68.5	57.8	46.3	28.7	13.6	40.7
CANBY	MAX	21.9	28.4	39.0	57.4	71.6	81.5	86.7	84.7	74.6	62.7	42.7	28.4	56.6
	MIN	1.7	8.1	19.3	33.6	45.7	56.0	60.8	58.9	48.9	38.2	23.2	9.7	33.7
	MEAN	11.8	18.3	29.2	45.5	58.7	68.8	73.8	71.9	61.7	50.5	33.0	19.1	45.2
CLOQUET FOR RES CENTER	MAX	17.0	24.2	35.0	51.4	66.1	74.7	79.9	76.9	66.3	55.1	36.4	22.9	50.5
	MIN	-3.5	1.5	13.4	27.9	37.7	47.1	53.4	52.0	44.0	34.6	20.8	5.5	27.9
	MEAN	6.8	12.8	24.2	39.7	51.9	60.9	66.7	64.4	55.1	44.9	28.6	14.2	39.2
COLLEGEVILLE ST JOHN U	MAX	17.7	24.8	35.9	53.9	68.6	77.2	81.9	79.4	69.4	58.4	39.2	24.8	52.6
	MIN	-.9	5.4	17.4	33.6	45.9	55.6	60.5	58.5	49.1	38.8	23.6	8.4	33.0
	MEAN	8.4	15.1	26.7	43.8	57.2	66.4	71.2	69.0	59.3	48.6	31.4	16.7	42.8
CROOKSTON NW EXP STA	MAX	12.6	20.4	32.9	52.5	69.0	77.3	82.5	81.4	70.1	57.8	36.1	20.8	51.1
	MIN	-7.2	-.8	12.9	30.7	41.9	52.5	57.1	55.3	45.6	35.4	19.1	2.7	28.8
	MEAN	2.7	9.8	22.9	41.7	55.5	64.9	69.8	68.4	57.9	46.6	27.6	11.8	40.0
DETROIT LAKES 1 NNE	MAX	14.0	21.6	33.6	52.1	66.7	74.8	79.9	78.3	67.7	56.5	36.4	21.7	50.3
	MIN	-8.3	-2.8	10.8	28.8	40.3	50.6	55.1	53.2	43.1	33.2	18.1	2.2	27.0
	MEAN	2.8	9.5	22.2	40.5	53.6	62.7	67.5	65.8	55.4	44.9	27.3	12.0	38.7

MINNESOTA

TEMPERATURE NORMALS (DEG F)

STATION			JAN	FEB	MAR	APR	MAY	JUN	JUL	AUG	SEP	OCT	NOV	DEC	ANN
DULUTH WSO	//R	MAX	15.5	21.7	31.9	47.6	61.3	70.5	76.4	73.6	63.6	53.0	35.2	21.8	47.7
		MIN	-2.9	2.2	13.9	28.9	39.3	48.2	54.3	52.8	44.3	35.4	21.2	5.8	28.6
		MEAN	6.3	12.0	22.9	38.3	50.3	59.4	65.3	63.2	54.0	44.2	28.2	13.8	38.2
FAIRMONT		MAX	21.2	27.8	38.1	56.1	70.3	79.9	83.6	81.4	72.3	60.9	42.2	27.7	55.1
		MIN	2.8	9.0	20.4	36.0	48.1	58.0	62.0	59.7	50.3	39.7	25.0	11.1	35.2
		MEAN	12.0	18.5	29.3	46.1	59.2	69.0	72.9	70.6	61.3	50.3	33.6	19.4	45.2
FARIBAULT		MAX	21.4	28.0	38.7	57.1	70.7	79.8	84.0	81.4	72.5	61.5	42.4	27.9	55.5
		MIN	1.4	7.2	19.4	34.6	46.2	56.1	60.4	58.3	49.1	38.7	24.6	10.8	33.9
		MEAN	11.4	17.6	29.1	45.9	58.5	68.0	72.3	69.9	60.9	50.2	33.5	19.4	44.7
FARMINGTON 3 NW		MAX	19.9	26.6	37.7	56.5	70.2	78.8	82.9	80.5	71.2	60.3	41.3	26.7	54.4
		MIN	1.1	7.2	19.5	35.0	47.0	56.5	60.6	58.2	49.0	38.8	24.5	10.1	34.0
		MEAN	10.5	16.9	28.6	45.8	58.6	67.7	71.8	69.4	60.2	49.6	32.9	18.4	44.2
FERGUS FALLS		MAX	15.5	23.3	35.3	54.1	69.0	77.7	83.1	81.3	70.5	58.3	37.8	23.1	52.4
		MIN	-3.8	2.8	15.6	31.9	44.2	54.0	58.9	57.0	46.7	36.0	21.0	5.7	30.8
		MEAN	5.9	13.1	25.5	43.0	56.7	65.9	71.0	69.2	58.6	47.1	29.4	14.4	41.7
FOSSTON		MAX	14.2	22.1	34.4	53.0	68.1	76.4	81.4	80.0	68.9	57.5	36.3	21.9	51.2
		MIN	-6.5	-.2	12.9	30.4	42.3	52.2	56.7	54.8	45.1	35.4	19.5	3.3	28.8
		MEAN	3.8	11.0	23.7	41.7	55.3	64.4	69.1	67.4	57.0	46.4	27.9	12.6	40.0
GRAND MARAIS		MAX	20.9	24.9	33.6	46.3	56.3	64.0	70.6	69.7	61.4	52.1	37.8	26.6	47.0
		MIN	2.2	6.1	17.4	30.1	38.1	44.9	52.0	53.4	46.3	37.4	24.6	10.8	30.3
		MEAN	11.6	15.5	25.5	38.2	47.2	54.5	61.3	61.5	53.9	44.8	31.2	18.7	38.7
GRAND MEADOW		MAX	20.0	26.0	36.1	53.7	67.5	76.6	81.0	78.8	70.2	59.3	41.5	26.7	53.1
		MIN	1.1	6.4	18.1	34.0	45.7	55.6	59.8	57.0	47.9	37.4	23.6	10.2	33.1
		MEAN	10.6	16.2	27.1	43.9	56.6	66.1	70.4	68.0	59.1	48.4	32.6	18.5	43.1
GRAND RAPIDS NC SCHOOL		MAX	16.5	24.8	36.2	52.8	67.1	75.6	79.9	77.4	66.8	56.0	36.4	22.5	51.0
		MIN	-5.6	.1	12.6	28.7	40.0	49.4	54.4	52.1	43.6	34.6	20.3	3.8	27.8
		MEAN	5.5	12.5	24.4	40.8	53.6	62.5	67.2	64.8	55.2	45.3	28.4	13.2	39.5
GULL LAKE DAM		MAX	17.4	25.2	36.4	53.5	68.2	76.8	81.3	78.9	68.8	58.1	38.5	23.7	52.2
		MIN	-3.4	2.4	14.5	30.1	42.6	53.1	58.4	56.2	46.8	36.6	21.7	6.2	30.4
		MEAN	7.0	13.8	25.4	41.8	55.4	64.9	69.9	67.6	57.8	47.3	30.1	15.0	41.3
HALLOCK		MAX	9.5	17.0	29.9	50.0	66.7	75.8	81.4	79.7	67.8	55.5	34.3	17.9	48.8
		MIN	-10.7	-4.5	9.5	29.2	41.1	51.8	56.0	53.5	43.6	32.9	16.8	-.4	26.6
		MEAN	-.6	6.2	19.7	39.6	53.9	63.8	68.7	66.6	55.7	44.2	25.6	8.7	37.7
HINCKLEY		MAX	17.9	25.3	37.0	54.4	67.9	76.2	81.4	79.3	68.8	58.1	39.4	24.6	52.5
		MIN	-4.0	1.7	14.9	30.6	41.2	50.4	55.1	53.0	43.8	34.2	21.1	5.8	29.0
		MEAN	7.0	13.5	26.0	42.5	54.6	63.3	68.3	66.1	56.3	46.2	30.2	15.2	40.8
INTNL FALLS WSO	//R	MAX	11.1	19.5	32.0	49.1	63.9	73.3	78.5	75.4	64.1	52.8	32.9	17.8	47.5
		MIN	-11.0	-4.9	8.9	27.1	38.6	49.0	53.7	50.9	41.6	32.8	16.9	-1.4	25.2
		MEAN	.1	7.4	20.5	38.2	51.3	61.2	66.1	63.2	52.8	42.8	24.9	8.2	36.4
ITASCA STATE PARK SCH		MAX	16.1	24.9	36.3	52.7	67.3	75.8	80.8	78.6	67.8	56.8	36.5	22.5	51.3
		MIN	-8.2	-2.4	10.3	26.8	38.7	49.2	54.0	52.0	42.9	33.4	18.3	1.7	26.4
		MEAN	4.0	11.3	23.3	39.8	53.0	62.6	67.4	65.3	55.3	45.1	27.4	12.1	38.9
JORDAN 1 S		MAX	20.8	27.0	37.6	56.2	69.6	78.2	82.4	79.8	70.7	60.0	41.5	27.3	54.3
		MIN	1.2	7.1	19.2	34.6	45.9	55.3	59.4	57.2	48.0	38.3	24.3	10.5	33.4
		MEAN	11.0	17.1	28.4	45.4	57.7	66.8	70.9	68.5	59.4	49.2	32.9	18.9	43.9
LEECH LAKE DAM		MAX	16.6	24.9	36.4	52.7	67.1	75.7	80.3	78.1	67.7	57.0	37.1	22.9	51.4
		MIN	-5.6	.1	12.6	28.8	40.8	51.2	56.5	53.9	45.0	35.2	20.7	3.9	28.6
		MEAN	5.5	12.5	24.5	40.8	53.9	63.5	68.4	66.0	56.4	46.1	28.9	13.4	40.0
LITCHFIELD		MAX	19.6	26.6	37.6	56.3	71.0	79.8	84.3	81.9	72.5	61.0	40.9	26.5	54.8
		MIN	-.1	6.0	18.1	33.9	46.0	56.0	60.7	58.7	48.7	38.3	23.4	9.1	33.2
		MEAN	9.8	16.3	27.9	45.2	58.5	67.9	72.5	70.3	60.6	49.7	32.2	17.9	44.1

MINNESOTA

TEMPERATURE NORMALS (DEG F)

STATION		JAN	FEB	MAR	APR	MAY	JUN	JUL	AUG	SEP	OCT	NOV	DEC	ANN
LITTLE FALLS 1 N	MAX	18.2	25.9	37.2	55.5	69.9	78.2	83.1	80.7	70.6	59.4	39.5	24.9	53.6
	MIN	-4.0	1.7	14.6	31.1	43.1	53.3	58.1	55.7	46.2	35.9	21.4	5.6	30.2
	MEAN	7.1	13.8	25.9	43.3	56.5	65.7	70.7	68.2	58.5	47.7	30.5	15.3	41.9
LONG PRAIRIE	MAX	16.9	24.5	35.6	54.0	68.4	77.1	82.5	80.0	70.0	58.9	38.8	24.0	52.6
	MIN	-4.3	1.3	14.4	31.4	43.1	53.1	57.7	55.4	45.8	35.6	21.0	5.7	30.0
	MEAN	6.3	12.9	25.0	42.7	55.8	65.1	70.1	67.7	57.9	47.3	29.9	14.9	41.3
MADISON SEWAGE PLANT	MAX	21.2	28.1	39.0	57.5	71.7	80.7	86.0	84.0	74.4	63.3	43.5	28.5	56.5
	MIN	.3	6.7	18.5	33.5	45.5	55.5	60.2	57.9	47.3	36.9	22.3	8.9	32.8
	MEAN	10.8	17.4	28.8	45.5	58.6	68.2	73.1	70.9	60.9	50.1	32.9	18.7	44.7
MAPLE PLAIN	MAX	20.0	26.7	38.0	56.2	69.9	78.3	82.9	80.5	70.9	60.0	41.3	26.8	54.3
	MIN	-.2	5.7	18.1	33.7	45.7	55.3	59.7	57.5	48.3	37.9	23.6	8.9	32.9
	MEAN	9.9	16.3	28.0	45.0	57.8	66.8	71.3	69.0	59.6	49.0	32.5	17.9	43.6
MARSHALL	MAX	20.7	26.9	37.5	55.6	69.9	79.3	84.3	82.3	72.5	61.2	42.2	27.5	55.0
	MIN	1.2	7.2	18.9	34.4	46.3	57.0	61.5	59.1	49.0	38.0	23.2	9.7	33.8
	MEAN	11.0	17.1	28.2	45.1	58.1	68.2	72.9	70.7	60.8	49.6	32.7	18.6	44.4
MEADOWLANDS	MAX	17.2	24.8	36.1	52.3	66.7	75.1	79.9	77.2	66.7	55.8	36.7	23.1	51.0
	MIN	-6.7	-1.2	11.3	27.6	38.1	47.6	52.9	50.9	42.4	33.2	19.2	2.6	26.5
	MEAN	5.2	11.9	23.7	39.9	52.4	61.4	66.4	64.1	54.6	44.5	28.0	12.9	38.8
MILACA	MAX	18.3	25.7	36.8	54.9	68.7	77.2	82.3	79.9	69.7	58.5	39.4	25.0	53.0
	MIN	-2.1	3.8	16.1	31.4	42.9	52.5	57.5	55.0	45.8	35.4	21.6	7.2	30.6
	MEAN	8.1	14.7	26.4	43.2	55.8	64.8	69.9	67.5	57.7	47.0	30.6	16.1	41.8
MILAN	MAX	19.1	26.1	37.5	55.7	70.0	79.1	84.0	81.6	71.8	60.2	40.6	26.1	54.3
	MIN	-1.8	4.6	17.2	32.9	44.8	55.2	59.5	57.5	47.1	36.3	21.3	7.3	31.8
	MEAN	8.7	15.3	27.4	44.3	57.4	67.2	71.8	69.6	59.5	48.3	31.0	16.7	43.1
MINN-ST PAUL WSO //R	MAX	19.9	26.4	37.5	56.0	69.4	78.5	83.4	80.9	71.0	59.7	41.1	26.7	54.2
	MIN	2.4	8.5	20.8	36.0	47.6	57.7	62.7	60.3	50.2	39.4	25.3	11.7	35.2
	MEAN	11.2	17.5	29.2	46.0	58.5	68.1	73.1	70.6	60.6	49.6	33.2	19.2	44.7
MONTEVIDEO 1 SW	MAX	19.4	26.4	37.6	56.2	70.6	79.4	84.2	82.2	72.8	61.7	41.7	26.8	54.9
	MIN	-.8	5.5	18.3	34.0	46.0	55.8	60.0	57.7	47.3	37.2	22.4	8.7	32.7
	MEAN	9.3	16.0	28.0	45.2	58.3	67.6	72.1	70.0	60.1	49.5	32.1	17.8	43.8
MOOSE LAKE 1 SSE	MAX	18.4	25.7	36.8	53.4	67.5	75.9	81.3	78.6	68.6	57.8	38.6	24.3	52.2
	MIN	-5.3	-.2	12.7	27.7	37.8	47.2	53.6	51.9	43.8	33.8	20.3	4.4	27.3
	MEAN	6.6	12.8	24.8	40.6	52.7	61.6	67.5	65.3	56.2	45.8	29.4	14.3	39.8
MORA	MAX	19.0	26.5	37.7	55.6	69.5	77.8	82.8	80.5	70.2	59.0	39.6	25.4	53.6
	MIN	-2.9	2.8	15.7	31.5	42.9	52.3	57.5	55.1	45.9	35.8	21.8	6.8	30.4
	MEAN	8.1	14.7	26.7	43.6	56.2	65.0	70.2	67.8	58.1	47.4	30.7	16.1	42.1
MORRIS W C SCHOOL //	MAX	15.9	22.5	34.0	52.5	67.6	76.7	82.1	80.4	70.2	58.8	39.1	23.6	52.0
	MIN	-3.5	2.6	15.9	32.7	44.6	55.0	59.3	57.0	46.5	35.5	20.8	6.0	31.0
	MEAN	6.2	12.6	25.0	42.6	56.1	65.9	70.7	68.8	58.4	47.2	30.0	14.8	41.5
NEW ULM 2 SE	MAX	22.3	28.9	39.6	58.6	72.5	81.6	85.5	83.3	74.5	63.5	43.5	28.8	56.9
	MIN	2.0	8.1	20.3	35.5	46.9	56.8	61.3	58.9	49.3	38.9	24.8	11.3	34.5
	MEAN	12.2	18.5	30.0	47.1	59.7	69.2	73.4	71.1	62.0	51.2	34.2	20.1	45.7
PARK RAPIDS	MAX	15.6	23.9	35.4	52.6	67.1	75.9	81.0	79.1	68.1	56.9	36.6	22.4	51.2
	MIN	-7.7	-1.4	11.7	28.8	41.1	51.3	56.1	54.0	44.0	33.8	18.5	2.3	27.7
	MEAN	3.9	11.3	23.6	40.7	54.2	63.6	68.5	66.6	56.1	45.4	27.6	12.3	39.5
PINE RIVER DAM	MAX	17.7	26.0	37.3	54.0	68.5	76.6	81.4	78.8	68.7	58.1	38.5	24.1	52.5
	MIN	-5.6	.5	12.9	28.6	40.8	50.9	56.3	53.7	45.0	35.1	21.3	4.6	28.7
	MEAN	6.1	13.3	25.1	41.3	54.7	63.8	68.9	66.3	56.8	46.7	29.9	14.4	40.6
PIPESTONE	MAX	19.8	26.4	37.7	55.9	69.6	79.2	84.5	82.7	72.6	61.1	41.8	27.3	54.9
	MIN	-1.2	5.3	16.9	31.7	43.4	53.6	58.5	56.2	45.7	34.5	20.3	7.5	31.0
	MEAN	9.4	15.9	27.3	43.9	56.5	66.4	71.5	69.5	59.2	47.8	31.1	17.4	43.0

MINNESOTA

TEMPERATURE NORMALS (DEG F)

STATION		JAN	FEB	MAR	APR	MAY	JUN	JUL	AUG	SEP	OCT	NOV	DEC	ANN
POKEGAMA DAM	MAX	16.7	24.9	36.4	53.3	67.3	75.6	80.2	77.7	67.2	56.7	37.1	22.9	51.3
	MIN	-6.2	-.5	12.2	28.6	40.3	50.2	55.4	53.4	44.5	35.0	20.5	3.5	28.1
	MEAN	5.3	12.2	24.4	41.0	53.8	62.9	67.8	65.6	55.9	45.9	28.9	13.2	39.7
RED LAKE FALLS	MAX	12.7	20.5	33.0	52.4	68.0	76.4	81.5	79.6	68.9	57.0	35.7	20.7	50.5
	MIN	-8.7	-2.0	11.8	30.4	42.0	52.0	56.5	54.5	45.0	34.7	18.5	1.8	28.0
	MEAN	2.0	9.3	22.5	41.4	55.0	64.2	69.0	67.1	56.9	45.9	27.1	11.3	39.3
RED LAKE INDIAN AGENCY	MAX	12.2	19.6	31.6	48.3	63.4	72.9	78.2	76.1	65.2	54.2	35.1	19.9	48.1
	MIN	-9.7	-4.5	9.5	28.1	41.2	51.8	57.0	54.5	44.8	34.9	18.9	.6	27.3
	MEAN	1.2	7.6	20.6	38.2	52.3	62.4	67.6	65.3	55.0	44.5	27.0	10.3	37.7
ROCHESTER WSO //R	MAX	19.7	26.2	36.7	54.9	68.2	77.6	81.4	79.1	70.3	59.2	41.1	26.3	53.4
	MIN	1.9	7.7	19.2	34.3	45.6	55.5	59.9	57.6	48.1	38.1	24.1	10.7	33.6
	MEAN	10.8	17.0	28.0	44.6	56.9	66.6	70.7	68.4	59.2	48.7	32.6	18.5	43.5
ROSEAU 1 E	MAX	10.5	18.9	31.3	50.6	66.5	74.9	79.6	77.6	66.8	55.1	33.7	18.2	48.6
	MIN	-11.2	-5.3	7.5	27.7	39.2	49.3	53.9	51.3	42.0	32.5	16.5	-1.1	25.2
	MEAN	-.4	6.8	19.4	39.2	52.9	62.2	66.8	64.4	54.4	43.8	25.1	8.6	36.9
ROSEMOUNT AGRI EXP STA//	MAX	20.0	26.7	37.7	56.5	70.2	79.1	83.4	81.1	72.0	60.9	41.5	26.7	54.7
	MIN	.0	6.0	18.6	33.6	45.1	55.1	59.5	57.3	48.0	37.7	23.6	9.6	32.8
	MEAN	10.0	16.4	28.1	45.0	57.7	67.1	71.5	69.2	60.0	49.3	32.6	18.2	43.8
ST CLOUD WSO //R	MAX	17.4	24.5	35.8	54.1	68.3	76.9	81.8	79.2	69.0	58.0	38.8	24.2	52.3
	MIN	-3.3	2.8	15.5	31.8	43.3	53.1	57.7	55.4	45.7	35.5	20.8	6.2	30.4
	MEAN	7.0	13.7	25.7	43.0	55.8	65.0	69.8	67.3	57.4	46.8	29.8	15.2	41.4
ST PETER 2 SW	MAX	21.5	28.1	38.8	57.0	70.9	80.1	84.5	81.9	73.2	62.3	43.0	28.2	55.8
	MIN	.4	6.8	19.4	35.0	46.7	56.7	61.1	58.5	48.6	37.8	24.3	10.4	33.8
	MEAN	11.0	17.5	29.1	46.0	58.8	68.4	72.8	70.3	60.9	50.1	33.7	19.3	44.8
SANDY LAKE DAM LIBBY	MAX	17.9	25.6	36.6	53.1	67.1	75.3	79.9	77.3	67.4	57.1	38.0	23.7	51.6
	MIN	-5.0	.6	12.8	29.3	41.4	51.0	56.1	54.1	45.1	35.6	21.1	4.8	28.9
	MEAN	6.5	13.1	24.8	41.2	54.2	63.2	68.0	65.7	56.3	46.4	29.6	14.3	40.3
SPRINGFIELD 1 NW	MAX	21.0	27.4	37.9	56.7	71.4	80.6	84.3	81.7	73.3	61.9	42.6	27.6	55.5
	MIN	1.5	7.6	19.7	34.8	46.3	56.7	60.3	57.7	48.0	37.8	23.7	10.1	33.7
	MEAN	11.3	17.5	28.9	45.8	58.9	68.7	72.3	69.7	60.7	49.9	33.2	18.9	44.7
TRACY	MAX	21.4	27.3	37.9	56.2	70.4	80.1	85.0	83.1	73.4	61.9	42.7	27.9	55.6
	MIN	.8	7.1	18.5	33.7	45.9	56.3	60.9	58.5	48.4	37.6	22.9	9.6	33.4
	MEAN	11.1	17.2	28.2	45.0	58.2	68.2	73.0	70.8	60.9	49.8	32.8	18.7	44.5
TWO HARBORS	MAX	21.5	26.2	35.0	48.0	59.2	67.8	75.3	73.9	64.9	54.9	39.5	27.3	49.5
	MIN	2.3	6.3	17.5	29.9	38.2	44.6	52.7	54.1	47.0	37.8	24.9	10.7	30.5
	MEAN	12.0	16.3	26.3	39.0	48.7	56.2	64.0	64.0	56.0	46.3	32.3	19.0	40.0
VIRGINIA	MAX	15.9	23.7	35.3	51.7	66.0	74.9	79.3	76.6	65.7	54.8	35.5	21.6	50.1
	MIN	-6.1	-.4	12.0	27.8	39.4	48.9	53.9	51.6	43.3	33.8	18.8	2.5	27.1
	MEAN	4.9	11.7	23.7	39.8	52.7	61.9	66.7	64.1	54.5	44.3	27.2	12.1	38.6
WADENA 3 S	MAX	15.7	23.4	34.7	52.6	67.2	75.8	81.0	79.1	68.2	57.1	37.5	22.8	51.3
	MIN	-5.6	.4	13.4	30.3	42.6	52.7	57.5	55.2	45.0	34.6	19.8	4.1	29.2
	MEAN	5.1	11.9	24.1	41.5	54.9	64.3	69.3	67.2	56.7	45.9	28.7	13.5	40.3
WARROAD	MAX	11.9	20.0	32.1	49.3	64.3	73.6	79.0	76.8	66.4	55.2	34.6	19.4	48.6
	MIN	-10.4	-4.9	8.5	26.9	40.1	51.2	56.0	53.5	44.0	34.0	17.6	-.3	26.4
	MEAN	.8	7.6	20.3	38.1	52.2	62.4	67.5	65.2	55.2	44.6	26.1	9.6	37.5
WASECA EXP. STATION	MAX	19.9	26.4	36.6	55.3	69.6	78.5	82.4	80.0	71.7	60.4	41.8	26.7	54.1
	MIN	.0	6.4	18.2	34.2	45.9	55.7	59.9	57.5	47.9	37.2	23.3	9.2	33.0
	MEAN	10.0	16.4	27.5	44.8	57.8	67.1	71.2	68.8	59.8	48.8	32.6	18.0	43.6
WHEATON	MAX	19.3	26.3	37.8	56.4	71.4	80.0	85.8	84.4	74.5	62.7	41.6	26.4	55.6
	MIN	-.3	6.0	18.3	33.6	45.5	55.8	60.1	58.4	48.4	37.5	22.6	8.2	32.8
	MEAN	9.5	16.2	28.0	45.1	58.4	68.0	73.0	71.4	61.4	50.1	32.1	17.3	44.2

MINNESOTA

TEMPERATURE NORMALS (DEG F)

STATION		JAN	FEB	MAR	APR	MAY	JUN	JUL	AUG	SEP	OCT	NOV	DEC	ANN
WILLMAR STATE HOSPITAL	MAX	18.4	25.6	36.4	55.3	69.8	.78.6	83.1	81.0	71.9	60.4	40.4	25.6	53.9
	MIN	-.4	5.8	18.1	34.2	46.4	56.5	61.0	58.9	49.0	38.6	23.4	8.8	33.4
	MEAN	9.0	15.7	27.3	44.8	58.1	67.6	72.1	69.9	60.5	49.5	31.9	17.2	43.6
WINDOM	MAX	22.2	29.0	39.4	57.7	71.6	81.3	85.4	82.9	74.1	63.0	43.8	28.7	56.6
	MIN	2.3	8.6	19.9	34.5	46.4	56.7	61.0	58.7	48.7	37.9	23.9	10.7	34.1
	MEAN	12.3	18.8	29.7	46.1	59.0	69.0	73.2	70.8	61.5	50.4	33.8	19.7	45.4
WINNEBAGO	MAX	19.9	26.4	36.9	55.2	69.3	78.8	82.9	80.6	71.7	60.4	41.9	26.8	54.2
	MIN	1.1	6.9	18.2	33.7	45.7	56.2	60.8	58.2	48.5	37.5	23.6	9.9	33.4
	MEAN	10.5	16.7	27.5	44.5	57.5	67.5	71.9	69.5	60.1	49.0	32.8	18.4	43.8
WINNIBIGOSHISH DAM	MAX	15.5	23.8	35.6	52.7	67.3	76.2	80.7	78.0	67.0	55.8	36.2	22.1	50.9
	MIN	-6.9	-1.3	11.3	28.0	40.8	51.3	56.7	54.3	45.1	35.4	20.4	3.1	28.2
	MEAN	4.3	11.3	23.4	40.4	54.1	63.8	68.7	66.2	56.1	45.6	28.3	12.6	39.6
WINONA	MAX	23.4	29.9	40.6	57.4	70-.6	79.7	84.5	82.1	73.3	62.1	44.8	29.8	56.5
	MIN	2.0	6.6	18.6	34.4	46.4	56.1	61.0	58.3	48.7	38.4	25.1	11.3	33.9
	MEAN	12.7	18.3	29.6	45.9	58.5	67.9	72.8	70.2	61.0	50.3	34.9	20.6	45.2
ZUMBROTA	MAX	22.1	28.8	39.4	57.9	71.1	79.9	84.5	82.3	73.2	62.4	43.1	28.6	56.1
	MIN	1.1	6.3	18.4	33.5	44.7	54.7	59.1	56.8	47.7	37.3	23.8	10.4	32.8
	MEAN	11.6	17.6	28.9	45.7	57.9	67.3	71.8	69.6	60.5	49.9	33.5	19.5	44.5

MINNESOTA

PRECIPITATION NORMALS (INCHES)

STATION	JAN	FEB	MAR	APR	MAY	JUN	JUL	AUG	SEP	OCT	NOV	DEC	ANN
ADA	.73	.52	.93	2.07	2.63	3.77	3.37	3.12	2.28	1.49	1.12	.82	22.85
AITKIN 2 S	.80	.64	1.45	2.27	3.39	3.83	4.79	4.19	2.59	2.04	1.44	.93	28.36
ALBERT LEA	.80	.84	1.78	2.81	3.97	4.79	4.22	3.81	3.04	2.14	1.47	.90	30.57
ALEXANDRIA FAA AIRPORT//	.84	.69	1.20	2.18	2.92	4.07	3.04	3.60	2.27	1.89	1.13	.76	24.59
ARGYLE 4 E	.80	.51	.83	1.44	2.03	3.07	3.06	2.47	2.24	1.11	.78	.67	19.01
ARTICHOKE LAKE	.63	.68	1.14	2.32	2.98	4.10	3.21	3.09	1.86	1.93	1.02	.64	23.60
AUSTIN 3 S	1.17	.98	1.98	2.78	3.90	4.43	3.95	3.92	3.09	2.13	1.59	1.20	31.12
BABBITT 2 SE	.91	.68	1.17	1.90	2.99	4.06	3.69	4.02	3.53	2.39	1.65	1.01	28.00
BAUDETTE	.58	.44	.75	1.53	2.46	3.66	3.59	3.38	2.80	1.84	.92	.64	22.59
BEAVER	.91	.78	1.96	2.92	3.95	4.53	4.62	4.42	3.55	2.45	1.72	1.04	32.85
BEMIDJI AIRPORT	.58	.45	.79	1.88	2.78	3.65	3.52	3.43	2.52	1.63	.75	.74	22.72
BENSON	.91	.92	1.54	2.34	3.13	4.25	3.40	3.42	2.26	2.01	1.26	.94	26.38
BIG FALLS	.90	.63	1.09	1.85	2.75	3.82	4.03	3.56	3.02	1.98	1.31	1.01	25.95
BLANCHARD POWER STA	.75	.69	1.23	2.25	2.96	4.17	3.71	4.01	2.45	1.87	1.09	.79	25.97
BLUE EARTH	.70	.85	1.68	2.53	3.99	4.90	4.12	3.91	3.02	1.82	1.28	.91	29.71
BRAINERD RANGER STA	.71	.63	1.25	2.14	3.27	3.94	4.38	3.99	2.66	1.90	1.28	.80	26.95
BRICELYN	.82	.85	1.96	2.71	3.98	4.77	4.45	4.49	3.13	2.17	1.39	.86	31.58
CALEDONIA	1.03	1.02	2.41	3.25	4.14	4.63	4.65	3.95	3.69	2.36	1.84	1.28	34.25
CAMBRIDGE ST HOSPITAL	.72	.70	1.37	2.17	3.43	4.57	3.83	4.36	3.11	2.11	1.34	.88	28.59
CAMPBELL	.67	.62	1.03	2.34	2.56	4.05	3.06	2.72	1.93	1.39	.84	.62	21.83
CANBY	.70	.92	1.54	2.49	3.03	3.94	3.50	3.03	2.09	1.58	1.15	.89	24.86
CARIBOU 2 S	.61	.41	.62	1.42	2.50	3.08	2.95	2.55	2.36	1.37	.88	.63	19.38
CASS LAKE	.84	.54	1.05	2.19	2.79	3.93	3.88	3.28	2.68	1.87	1.13	.97	25.15
CHASKA 1 NE	.64	.73	1.53	2.13	3.91	4.73	4.34	3.75	2.85	1.86	1.36	.77	28.60
CLOQUET FOR RES CENTER	1.16	.81	1.71	2.08	3.41	4.10	4.37	4.04	3.31	2.20	1.56	1.21	29.96
COKATO	.89	.94	1.58	2.14	3.44	4.41	4.07	4.08	2.68	1.91	1.35	.89	28.38
COLLEGEVILLE ST JOHN U	.95	.89	1.61	2.22	3.53	4.57	3.13	4.42	2.86	2.05	1.28	.86	28.37
CROOKSTON NW EXP STA	.54	.43	.65	1.39	2.20	3.61	3.17	3.04	2.33	1.39	.76	.52	20.03
DAWSON	.76	.95	1.51	2.38	3.08	4.01	2.84	3.41	1.99	1.60	1.10	.82	24.45
DETROIT LAKES 1 NNE	.67	.58	.87	1.99	2.59	4.13	3.82	3.82	2.30	1.49	.80	.72	23.78
DULUTH WSO //R	1.20	.90	1.78	2.16	3.15	3.96	3.96	4.12	3.26	2.21	1.69	1.29	29.68
ELBOW LAKE	.78	.57	1.09	2.27	2.94	4.19	2.98	3.29	2.22	1.63	.98	.65	23.59
ELGIN	.88	.73	1.87	2.75	3.76	4.66	4.75	3.99	3.28	2.32	1.58	1.08	31.65
ELK RIVER	.78	.78	1.39	2.08	3.64	4.41	3.70	4.38	2.89	1.96	1.36	.83	28.20
FAIRMONT	.72	.93	1.75	2.71	3.71	4.27	3.88	4.20	2.98	1.91	1.24	.89	29.19
FARIBAULT	.86	.84	1.81	2.68	3.79	4.61	4.26	4.19	3.55	2.15	1.31	.98	31.03
FARMINGTON 3 NW	.78	.91	1.83	2.35	3.52	4.69	4.01	4.20	2.96	2.07	1.41	1.01	29.74
FERGUS FALLS	.84	.62	1.05	2.37	2.74	4.36	3.21	3.01	2.09	1.45	.96	.82	23.52
FORT RIPLEY	.75	.74	1.34	2.07	3.08	3.75	3.78	3.97	2.40	1.86	1.16	.79	25.69
FOSSTON	.65	.44	.96	1.78	2.73	3.74	3.21	3.27	2.61	1.67	.91	.74	22.71
GONVICK	.71	.48	.82	1.58	2.54	4.02	3.61	3.23	2.66	1.73	1.05	.73	23.16
GRAND MARAIS	1.02	.77	1.50	1.90	3.05	3.27	3.18	3.26	3.51	2.20	1.59	1.15	26.40
GRAND MEADOW	.95	.91	2.08	2.77	4.14	4.61	4.10	4.07	3.27	2.24	1.51	.99	31.64
GRAND RAPIDS NC SCHOOL	.84	.62	1.26	1.99	3.16	3.79	4.12	3.38	2.99	2.02	1.23	.96	26.36
GULL LAKE DAM	.79	.66	1.37	2.09	3.15	3.95	4.13	3.75	2.49	1.88	1.20	.83	26.29
HALLOCK	.69	.48	.74	1.52	2.55	2.94	2.82	2.61	2.51	1.15	.71	.58	19.30
HARMONY	.84	.98	2.05	3.04	3.82	4.37	4.03	3.81	3.36	2.19	1.53	1.15	31.17
HASTINGS DAM 2	.63	.60	1.49	2.17	3.30	4.61	4.15	3.71	2.91	1.92	1.23	.76	27.48
HIBBING PWR SUBSTATION	1.07	.78	1.45	2.22	3.30	4.20	4.71	3.61	3.44	2.35	1.49	1.15	29.77
HINCKLEY	.95	.78	1.59	2.24	3.33	4.55	3.82	4.35	2.92	2.22	1.48	1.04	29.27
INDUS	.90	.69	.99	1.76	2.45	4.10	4.24	3.47	3.32	2.06	1.46	1.10	26.54
INTNL FALLS WSO //R	.89	.70	1.11	1.60	2.44	3.66	3.85	2.95	3.18	1.78	1.26	.93	24.35
ISLAND LAKE RESERVOIR	.91	.68	1.43	1.96	3.13	3.97	4.24	3.82	3.36	2.03	1.38	1.05	27.96
ITASCA STATE PARK SCH	.93	.64	1.29	2.45	2.80	4.33	3.34	3.47	2.71	1.90	1.28	1.09	26.23
JORDAN 1 S	.67	.72	1.54	2.13	3.68	4.76	4.09	4.01	2.67	1.92	1.17	.77	28.13

MINNESOTA

PRECIPITATION NORMALS (INCHES)

STATION	JAN	FEB	MAR	APR	MAY	JUN	JUL	AUG	SEP	OCT	NOV	DEC	ANN
KETTLE FALLS	1.06	.74	1.28	2.06	2.80	4.02	4.21	3.28	3.49	2.24	1.56	1.06	27.80
LA CRESCENT DAM 7	.88	.84	2.02	3.03	3.98	4.50	4.10	3.97	3.63	2.18	1.64	.98	31.75
LAKE CITY	1.00	.71	1.70	2.59	3.29	4.22	4.11	3.51	3.06	2.24	1.42	.99	28.84
LE CENTER 1 NE	.79	.87	1.70	2.36	3.71	4.38	4.20	4.09	3.09	2.02	1.28	.98	29.47
LEECH LAKE DAM	.72	.49	.99	1.94	2.82	3.81	4.11	3.33	2.64	1.89	1.07	.84	24.65
LITCHFIELD	.76	.79	1.46	2.35	3.30	4.52	3.65	3.85	2.47	1.82	1.23	.76	26.96
LITTLE FALLS 1 N	.71	.67	1.28	2.08	2.99	4.30	3.95	4.10	2.33	2.00	1.20	.77	26.38
LONG PRAIRIE	.92	.79	1.53	2.18	3.05	4.00	3.87	3.89	2.47	1.87	1.24	.92	26.73
MADISON SEWAGE PLANT	.58	.78	1.40	2.57	3.11	4.41	3.14	3.30	2.18	1.75	1.11	.69	25.02
MAHNOMEN 1 W	.86	.61	1.08	1.80	2.52	3.98	3.04	3.24	2.51	1.61	1.04	.89	23.18
MAPLE PLAIN	.84	.78	1.52	2.35	3.93	4.83	4.65	4.09	2.83	2.06	1.46	.86	30.20
MARSHALL	.54	.86	1.47	2.48	3.03	4.09	3.76	2.96	2.44	1.75	1.14	.81	25.33
MEADOWLANDS	.85	.60	1.30	2.18	3.04	3.74	4.17	4.05	3.38	1.99	1.43	.89	27.62
MILACA	.89	.85	1.50	2.45	3.48	4.85	3.93	4.46	2.94	2.03	1.39	.99	29.76
MILAN	.69	.83	1.38	2.49	3.08	4.12	3.28	3.53	2.07	1.82	1.12	.77	25.18
MINN-ST PAUL WSO //R	.82	.85	1.71	2.05	3.20	4.07	3.51	3.64	2.50	1.85	1.29	.87	26.36
MINNEOTA	.57	.70	1.43	2.45	2.99	3.91	3.50	3.17	2.20	1.67	.98	.74	24.31
MINNESOTA CITY DAM 5	.75	.65	1.78	2.78	3.84	4.23	4.14	3.90	3.29	2.19	1.56	.92	30.03
MONTEVIDEO 1 SW	.86	1.05	1.49	2.40	3.29	4.53	3.14	3.80	2.44	1.69	1.21	.88	26.78
MONTGOMERY	.78	.77	1.66	2.35	3.85	4.40	4.24	4.13	3.21	2.01	1.25	.91	29.56
MOOSE LAKE 1 SSE	.87	.67	1.44	2.06	3.19	4.64	4.21	3.84	2.92	2.02	1.45	.97	28.28
MORA	.86	.80	1.50	2.23	3.54	4.47	3.94	4.07	2.84	2.11	1.42	1.00	28.78
MORRIS W C SCHOOL //	.69	.72	1.15	2.45	2.91	3.91	3.29	3.13	1.91	1.85	1.13	.74	23.88
NEW ULM 2 SE	.76	.94	1.72	2.30	3.62	4.37	3.87	3.65	2.61	1.96	1.35	.87	28.02
NORTH MANKATO	.82	.92	1.86	2.51	3.52	4.09	3.91	3.91	2.79	1.89	1.33	.98	28.53
OKLEE	.66	.42	.82	1.64	2.41	3.75	3.82	3.30	2.58	1.50	.84	.70	22.44
ORTONVILLE	.51	.61	1.11	2.30	2.62	3.83	2.89	2.97	1.57	1.60	.92	.57	21.50
OTTERTAIL	.91	.69	1.25	2.45	2.98	4.43	3.75	3.57	2.31	1.78	.94	.95	26.01
PARK RAPIDS	.73	.53	1.12	2.44	2.84	4.33	3.54	3.85	2.61	1.98	1.12	.85	25.94
PINE RIVER DAM	.92	.64	1.41	2.33	3.13	3.93	4.03	4.01	2.78	1.92	1.32	.89	27.31
PIPESTONE	.46	.77	1.30	2.22	3.45	4.13	2.97	3.39	2.85	1.80	.87	.69	24.90
POKEGAMA DAM	.81	.63	1.29	2.12	3.13	3.86	4.26	3.34	3.05	2.04	1.30	.98	26.81
PRESTON	.86	.80	1.99	2.82	3.80	4.67	4.13	4.05	3.26	2.26	1.42	1.04	31.10
RED LAKE FALLS	.55	.37	.73	1.46	2.30	3.65	3.41	3.45	2.43	1.38	.74	.60	21.07
RED LAKE INDIAN AGENCY	.60	.40	.77	1.59	2.43	3.64	3.81	3.25	2.75	1.57	.90	.67	22.38
RED WING	.82	.71	1.67	2.46	3.73	4.81	4.49	3.97	3.39	2.17	1.44	1.02	30.68
RED WING DAM 3	.78	.68	1.59	2.62	3.51	4.36	4.30	3.69	3.14	2.04	1.31	.90	28.92
REDWOOD FALLS FAA AP	.60	.70	1.24	2.30	3.11	3.79	3.84	3.39	2.25	1.84	1.18	.67	24.91
ROCHESTER WSO //R	.74	.69	1.73	2.50	3.42	4.12	3.82	3.85	3.07	2.08	1.39	.84	28.26
ROSEAU 1 E	.59	.39	.65	1.31	2.27	3.31	3.48	3.11	2.59	1.25	.79	.65	20.39
ROSEMOUNT AGRI EXP STA//	.92	.83	1.81	2.46	3.69	4.77	3.99	4.29	3.17	2.22	1.48	1.03	30.66
ST CLOUD WSO //R	.83	.79	1.43	2.26	3.25	4.51	3.35	4.30	2.78	2.06	1.29	.87	27.72
ST JAMES FILTRATION PL	.50	.61	1.45	2.48	3.45	4.12	3.78	3.63	2.97	1.69	1.18	.67	26.53
ST PETER 2 SW	.69	.76	1.57	2.26	3.63	4.67	3.99	3.90	2.74	1.75	1.28	.87	28.11
SANDY LAKE DAM LIBBY	.85	.65	1.36	2.03	3.34	4.09	4.42	4.06	3.01	1.97	1.31	.85	27.94
SPRINGFIELD 1 NW	.57	.82	1.50	2.32	3.24	3.53	3.49	3.54	2.54	1.93	1.18	.72	25.38
SPRING GROVE 1 NW	.62	.67	1.73	3.09	4.02	4.40	4.44	3.98	3.52	2.22	1.41	.83	30.93
STILLWATER 1 SE	.78	.77	1.71	2.44	3.37	5.04	3.78	4.41	3.23	2.12	1.37	.90	29.92
THEILMAN	.84	.70	1.82	2.72	3.52	4.51	4.42	4.03	3.43	2.46	1.58	.89	30.92
THORHULT	.67	.55	.89	1.78	2.36	3.72	3.99	3.48	2.90	1.57	.96	.75	23.62
TOWER	1.06	.83	1.37	2.16	3.28	4.06	3.99	3.89	3.53	2.42	1.60	1.02	29.21
TRACY	.58	.80	1.56	2.44	3.25	3.72	3.24	2.92	2.69	1.86	1.21	.76	25.03
TWO HARBORS	1.02	.63	1.63	2.08	3.36	3.88	3.66	3.83	3.54	2.19	1.56	1.12	28.50
TYLER	.42	.58	1.24	2.19	2.98	4.22	3.14	3.05	2.54	1.64	.91	.66	23.57
VESTA	.64	.84	1.32	2.48	3.25	4.04	3.37	3.39	2.18	1.95	1.09	.74	25.29

MINNESOTA

PRECIPITATION NORMALS (INCHES)

STATION	JAN	FEB	MAR	APR	MAY	JUN	JUL	AUG	SEP	OCT	NOV	DEC	ANN
VIRGINIA	.90	.59	1.23	1.99	2.89	4.19	3.86	3.75	3.20	2.23	1.38	.88	27.09
WABASHA	.87	.68	1.80	2.80	3.88	4.45	4.21	3.94	3.33	2.47	1.61	1.06	31.10
WADENA 3 S	.83	.61	1.38	2.48	3.01	4.40	3.98	3.44	2.48	1.80	1.21	.89	26.51
WARROAD	.56	.45	.70	1.43	2.13	3.41	3.24	2.69	2.58	1.41	.95	.60	20.15
WASECA EXP. STATION	.84	.99	1.99	2.64	3.76	4.48	4.02	4.00	3.36	2.09	1.43	1.02	30.62
WELLS 1 NW	.64	.65	1.61	2.43	3.62	4.53	4.32	3.51	2.88	1.73	1.15	.77	27.84
WHEATON	.61	.59	1.06	2.25	2.86	4.11	2.71	2.42	1.71	1.37	1.03	.60	21.32
WHITEFACE RESERVOIR	.93	.70	1.37	2.00	2.93	3.83	4.16	3.86	3.10	2.01	1.43	.97	27.29
WILLMAR STATE HOSPITAL	.76	.84	1.38	2.46	3.33	4.96	3.41	3.94	2.65	1.97	1.22	.79	27.71
WINDOM	.55	.72	1.51	2.49	3.54	3.97	3.56	3.58	3.11	1.83	1.21	.73	26.80
WINNEBAGO	.89	1.01	1.82	2.47	3.92	4.95	4.23	3.74	3.08	1.91	1.24	1.03	30.29
WINNIBIGOSHISH DAM	.90	.63	1.17	1.90	2.82	4.00	4.06	3.12	2.82	1.98	1.21	1.00	25.61
WINONA	1.14	.93	2.07	2.89	4.17	4.62	4.25	4.25	3.44	2.15	1.63	1.17	32.71
WINONA DAM 5 A	.74	.64	1.74	2.70	4.14	4.12	3.91	3.77	3.26	2.01	1.44	.85	29.32
WINSTED	.70	.74	1.29	2.31	3.58	4.16	4.02	3.67	2.63	1.88	1.44	.71	27.13
ZUMBROTA	.82	.63	1.55	2.52	3.61	4.58	4.07	3.63	3.48	2.30	1.34	.96	29.49

MINNESOTA

HEATING DEGREE DAY NORMALS (BASE 65 DEG F)

STATION	JUL	AUG	SEP	OCT	NOV	DEC	JAN	FEB	MAR	APR	MAY	JUN	ANN
ADA	24	39	212	552	1086	1609	1888	1492	1243	663	296	81	9185
ALBERT LEA	11	20	150	485	948	1417	1643	1316	1119	582	233	39	7963
ALEXANDRIA FAA AIRPORT//	23	33	233	570	1077	1575	1845	1478	1265	699	317	86	9201
ARGYLE 4 E	39	78	294	645	1182	1730	2031	1655	1411	774	371	115	10325
ARTICHOKE LAKE	10	20	173	504	1017	1504	1758	1403	1187	627	254	54	8511
AUSTIN 3 S	20	16	157	481	948	1414	1655	1316	1107	573	238	55	7980
BABBITT 2 SE	58	109	325	642	1140	1634	1848	1487	1280	771	397	136	9827
BAUDETTE	53	88	308	623	1140	1705	1981	1602	1370	774	392	124	10160
BEMIDJI AIRPORT	52	82	326	657	1167	1702	1959	1574	1358	783	409	134	10203
BIG FALLS	63	95	320	632	1149	1677	1931	1532	1296	747	388	130	9960
CAMBRIDGE ST HOSPITAL	23	48	231	555	1035	1516	1773	1417	1190	645	290	72	8795
CAMPBELL	20	39	235	580	1089	1593	1866	1506	1265	690	323	80	9286
CANBY	11	12	147	460	960	1423	1649	1308	1110	585	231	40	7936
CLOQUET FOR RES CENTER	55	88	297	623	1092	1575	1804	1462	1265	759	412	145	9577
COLLEGEVILLE ST JOHN U	17	21	193	514	1008	1497	1755	1397	1187	636	270	58	8553
CROOKSTON NW EXP STA	18	45	233	570	1122	1649	1931	1546	1305	699	319	87	9524
DETROIT LAKES 1 NNE	48	79	294	623	1131	1643	1928	1554	1327	735	368	123	9853
DULUTH WSO //R	71	115	334	645	1104	1587	1820	1484	1305	801	456	179	9901
FAIRMONT	10	17	146	466	942	1414	1643	1302	1107	567	218	33	7865
FARIBAULT	15	23	158	468	945	1414	1662	1327	1113	573	239	46	7983
FARMINGTON 3 NW	12	18	168	487	963	1445	1690	1347	1128	576	231	39	8104
FERGUS FALLS	21	39	213	561	1068	1569	1832	1453	1225	660	279	77	8997
FOSSTON	27	61	254	577	1113	1624	1897	1512	1280	699	326	92	9462
GRAND MARAIS	147	141	333	626	1014	1435	1655	1386	1225	804	552	315	9633
GRAND MEADOW	18	28	191	521	972	1442	1686	1366	1175	633	286	67	8385
GRAND RAPIDS NC SCHOOL	48	82	298	611	1098	1606	1845	1470	1259	726	369	119	9531
GULL LAKE DAM	19	43	228	549	1047	1550	1798	1434	1228	696	320	79	8991
HALLOCK	36	78	289	645	1182	1745	2034	1646	1404	762	372	112	10305
HINCKLEY	35	62	267	583	1044	1544	1798	1442	1209	675	337	103	9099
INTNL FALLS WSO //R	67	116	366	688	1203	1761	2012	1613	1380	804	439	155	10604
ITASCA STATE PARK SCH	45	86	299	617	1128	1640	1891	1504	1293	756	388	124	9771
JORDAN 1 S	15	23	185	498	963	1429	1674	1341	1135	588	253	53	8157
LEECH LAKE DAM	34	66	265	586	1083	1600	1845	1470	1256	726	361	101	9393
LITCHFIELD	12	16	161	482	984	1460	1711	1364	1150	594	235	50	8219
LITTLE FALLS 1 N	17	32	211	536	1035	1541	1795	1434	1212	651	292	67	8823
LONG PRAIRIE	21	43	224	549	1053	1553	1820	1459	1240	669	306	80	9017
MADISON SEWAGE PLANT	10	20	156	471	963	1435	1680	1333	1122	585	228	41	8044
MAPLE PLAIN	18	33	190	508	975	1460	1708	1364	1147	600	254	67	8324
MARSHALL	10	18	162	488	969	1438	1674	1341	1141	597	240	43	8121
MEADOWLANDS	58	104	315	636	1110	1615	1854	1487	1280	753	400	136	9748
MILACA	23	44	226	558	1032	1516	1764	1408	1197	654	306	86	8814
MILAN	16	26	190	518	1020	1497	1745	1392	1166	621	263	55	8509
MINN-ST PAUL WSO //R	12	16	160	488	954	1420	1668	1330	1110	570	238	41	8007
MONTEVIDEO 1 SW	18	24	179	487	987	1463	1727	1372	1147	594	238	55	8291
MOOSE LAKE 1 SSE	47	83	270	595	1068	1572	1810	1462	1246	732	389	128	9402
MORA	23	47	217	546	1029	1516	1764	1408	1187	642	293	78	8750
MORRIS W C SCHOOL //	17	35	217	552	1050	1556	1823	1467	1240	672	300	73	9002
NEW ULM 2 SE	7	10	127	437	924	1392	1637	1302	1085	537	206	31	7695
PARK RAPIDS	33	61	277	608	1122	1634	1894	1504	1283	729	353	99	9597
PINE RIVER DAM	26	61	253	567	1053	1569	1826	1448	1237	711	337	99	9187
PIPESTONE	19	27	202	540	1017	1476	1724	1375	1169	633	289	69	8540
POKEGAMA DAM	35	73	279	592	1083	1606	1851	1478	1259	720	364	118	9458
RED LAKE FALLS	21	63	257	592	1137	1665	1953	1560	1318	708	332	103	9709
RED LAKE INDIAN AGENCY	42	84	304	636	1140	1696	1978	1607	1376	804	413	132	10212
ROCHESTER WSO //R	18	26	188	512	972	1442	1680	1344	1147	612	277	59	8277

MINNESOTA

HEATING DEGREE DAY NORMALS (BASE 65 DEG F)

STATION	JUL	AUG	SEP	OCT	NOV	DEC	JAN	FEB	MAR	APR	MAY	JUN	ANN
ROSEAU 1 E	51	115	325	657	1197	1748	2027	1630	1414	774	392	132	10462
ROSEMOUNT AGRI EXP STA//	18	18	170	494	972	1451	1705	1361	1144	600	257	45	8235
ST CLOUD WSO //R	26	40	240	564	1056	1544	1798	1436	1218	660	305	78	8965
ST PETER 2 SW	11	21	148	472	939	1417	1674	1330	1113	570	233	39	7967
SANDY LAKE DAM LIBBY	42	73	267	577	1062	1572	1814	1453	1246	714	349	106	9275
SPRINGFIELD 1 NW	10	18	159	477	954	1429	1665	1330	1119	576	230	38	8005
TRACY	13	26	167	482	966	1435	1671	1338	1141	600	243	45	8127
TWO HARBORS	105	108	274	580	981	1426	1643	1364	1200	780	505	264	9230
VIRGINIA	61	119	319	642	1134	1640	1863	1492	1280	756	396	141	9843
WADENA 3 S	28	61	262	592	1089	1597	1857	1487	1268	705	332	94	9372
WARROAD	46	87	301	632	1167	1717	1990	1607	1386	807	413	126	10279
WASECA EXP. STATION	13	35	181	511	972	1457	1705	1361	1163	606	258	57	8319
WHEATON	11	18	156	469	987	1479	1721	1366	1147	597	240	43	8234
WILLMAR STATE HOSPITAL	11	15	161	487	993	1482	1736	1380	1169	606	243	45	8328
WINDOM	10	13	141	460	936	1404	1634	1294	1094	567	220	43	7816
WINNEBAGO	14	22	174	502	966	1445	1690	1352	1163	615	268	50	8261
WINNIBIGOSHISH DAM	31	64	273	601	1101	1624	1882	1504	1290	738	355	98	9561
WINONA	11	14	148	467	903	1376	1621	1308	1097	573	249	52	7819
ZUMBROTA	15	19	156	478	945	1411	1655	1327	1119	579	247	49	8000

MINNESOTA

COOLING DEGREE DAY NORMALS (BASE 65 DEG F)

STATION	JAN	FEB	MAR	APR	MAY	JUN	JUL	AUG	SEP	OCT	NOV	DEC	ANN
ADA	0	0	0	0	30	96	195	163	23	0	0	0	507
ALBERT LEA	0	0	0	0	29	138	244	182	24	14	0	0	631
ALEXANDRIA FAA AIRPORT//	0	0	0	0	20	89	193	138	17	0	0	0	457
ARGYLE 4 E	0	0	0	0	21	61	132	115	9	0	0	0	338
ARTICHOKE LAKE	0	0	0	0	24	120	233	178	29	0	0	0	584
AUSTIN 3 S	0	0	0	0	28	127	212	140	22	7	0	0	536
BABBITT 2 SE	0	0	0	0	19	40	101	81	0	0	0	0	241
BAUDETTE	0	0	0	0	20	43	121	85	8	0	0	0	277
BEMIDJI AIRPORT	0	0	0	0	16	53	130	79	0	0	0	0	278
BIG FALLS	0	0	0	0	16	46	118	79	0	0	0	0	259
CAMBRIDGE ST HOSPITAL	0	0	0	0	17	84	184	129	9	0	0	0	423
CAMPBELL	0	0	0	0	23	86	181	148	19	0	0	0	457
CANBY	0	0	0	0	35	154	284	226	48	11	0	0	758
CLOQUET FOR RES CENTER	0	0	0	0	5	22	108	69	0	0	0	0	204
COLLEGEVILLE ST JOHN U	0	0	0	0	28	100	209	145	22	5	0	0	509
CROOKSTON NW EXP STA	0	0	0	0	24	84	167	151	20	0	0	0	446
DETROIT LAKES 1 NNE	0	0	0	0	15	54	125	104	6	0	0	0	304
DULUTH WSO //R	0	0	0	0	0	11	80	59	0	0	0	0	150
FAIRMONT	0	0	0	0	38	153	255	191	35	11	0	0	683
FARIBAULT	0	0	0	0	37	136	241	175	35	9	0	0	633
FARMINGTON 3 NW	0	0	0	0	32	120	223	154	24	10	0	0	563
FERGUS FALLS	0	0	0	0	22	104	207	169	21	6	0	0	529
FOSSTON	0	0	0	0	25	74	154	135	14	0	0	0	402
GRAND MARAIS	0	0	0	0	0	0	32	32	0	0	0	0	64
GRAND MEADOW	0	0	0	0	26	100	186	121	14	6	0	0	453
GRAND RAPIDS NC SCHOOL	0	0	0	0	16	44	117	75	0	0	0	0	252
GULL LAKE DAM	0	0	0	0	22	76	171	123	12	0	0	0	404
HALLOCK	0	0	0	0	28	76	150	127	10	0	0	0	391
HINCKLEY	0	0	0	0	15	52	137	96	6	0	0	0	306
INTNL FALLS WSO //R	0	0	0	0	14	41	101	60	0	0	0	0	216
ITASCA STATE PARK SCH	0	0	0	0	16	52	119	95	8	0	0	0	290
JORDAN 1 S	0	0	0	0	26	107	198	131	17	8	0	0	487
LEECH LAKE DAM	0	0	0	0	16	56	139	97	7	0	0	0	315
LITCHFIELD	0	0	0	0	34	137	244	180	29	8	0	0	632
LITTLE FALLS 1 N	0	0	0	0	28	88	193	131	16	0	0	0	456
LONG PRAIRIE	0	0	0	0	21	83	180	127	11	0	0	0	422
MADISON SEWAGE PLANT	0	0	0	0	30	137	262	203	33	9	0	0	674
MAPLE PLAIN	0	0	0	0	31	121	214	157	28	12	0	0	563
MARSHALL	0	0	0	0	26	139	255	195	36	10	0	0	661
MEADOWLANDS	0	0	0	0	9	28	101	76	0	0	0	0	214
MILACA	0	0	0	0	20	80	175	121	7	0	0	0	403
MILAN	0	0	0	0	28	121	227	168	25	0	0	0	569
MINN-ST PAUL WSO //R	0	0	0	0	36	134	263	190	28	11	0	0	662
MONTEVIDEO 1 SW	0	0	0	0	30	133	239	179	32	7	0	0	620
MOOSE LAKE 1 SSE	0	0	0	0	7	26	125	93	6	0	0	0	257
MORA	0	0	0	0	20	78	185	133	10	0	0	0	426
MORRIS W C SCHOOL //	0	0	0	0	24	100	194	153	19	0	0	0	490
NEW ULM 2 SE	0	0	0	0	42	157	268	199	37	9	0	0	712
PARK RAPIDS	0	0	0	0	18	57	142	110	10	0	0	0	337
PINE RIVER DAM	0	0	0	0	18	63	147	101	7	0	0	0	336
PIPESTONE	0	0	0	0	25	111	220	166	28	7	0	0	557
POKEGAMA DAM	0	0	0	0	16	55	122	92	6	0	0	0	291
RED LAKE FALLS	0	0	0	0	22	79	145	128	14	0	0	0	388
RED LAKE INDIAN AGENCY	0	0	0	0	19	54	122	93	0	0	0	0	288
ROCHESTER WSO //R	0	0	0	0	26	107	194	131	14	7	0	0	479

MINNESOTA

COOLING DEGREE DAY NORMALS (BASE 65 DEG F)

STATION	JAN	FEB	MAR	APR	MAY	JUN	JUL	AUG	SEP	OCT	NOV	DEC	ANN
ROSEAU 1 E	0	0	0	0	17	48	106	97	7	0	0	0	275
ROSEMOUNT AGRI EXP STA//	0	0	0	0	31	108	220	148	20	8	0	0	535
ST CLOUD WSO //R	0	0	0	0	20	78	175	112	12	0	0	0	397
ST PETER 2 SW	0	0	0	0	41	141	252	185	25	10	0	0	654
SANDY LAKE DAM LIBBY	0	0	0	0	14	52	135	95	6	0	0	0	302
SPRINGFIELD 1 NW	0	0	0	0	41	149	237	164	30	9	0	0	630
TRACY	0	0	0	0	32	141	261	206	44	11	0	0	695
TWO HARBORS	0	0	0	0	0	0	74	77	0	0	0	0	151
VIRGINIA	0	0	0	0	15	48	113	91	0	0	0	0	267
WADENA 3 S	0	0	0	0	19	73	162	129	13	0	0	0	396
WARROAD	0	0	0	0	16	48	123	93	7	0	0	0	287
WASECA EXP. STATION	0	0	0	0	34	120	205	153	25	9	0	0	546
WHEATON	0	0	0	0	35	133	259	216	48	7	0	0	698
WILLMAR STATE HOSPITAL	0	0	0	0	30	123	231	167	26	7	0	0	584
WINDOM	0	0	0	0	34	163	264	193	36	8	0	0	698
WINNEBAGO	0	0	0	0	36	125	228	162	27	6	0	0	584
WINNIBIGOSHISH DAM	0	0	0	0	17	62	145	101	6	0	0	0	331
WINONA	0	0	0	0	47	139	253	175	28	11	0	0	653
ZUMBROTA	0	0	0	0	27	118	226	162	21	10	0	0	564

21 – MINNESOTA

LEGEND
11 = TEMPERATURE ONLY
12 = PRECIPITATION ONLY
13 = TEMP. & PRECIP.

STATE-STATION NUMBER	STN TYP	NAME	LATITUDE DEG-MIN	LONGITUDE DEG-MIN	ELEVATION (FT)
21-0018	13	ADA	N 4718	W 09631	906
21-0059	12	AITKIN 2 S	N 4631	W 09342	1240
21-0075	13	ALBERT LEA	N 4339	W 09321	1220
21-0112	13	ALEXANDRIA FAA AIRPORT//	N 4552	W 09523	1421
21-0252	13	ARGYLE 4 E	N 4820	W 09644	870
21-0287	13	ARTICHOKE LAKE	N 4522	W 09608	1075
21-0355	13	AUSTIN 3 S	N 4337	W 09300	1215
21-0390	13	BABBITT 2 SE	N 4741	W 09155	1615
21-0515	13	BAUDETTE	N 4843	W 09437	1075
21-0559	12	BEAVER	N 4409	W 09201	720
21-0643	13	BEMIDJI AIRPORT	N 4730	W 09456	1392
21-0667	12	BENSON	N 4519	W 09536	1040
21-0746	13	BIG FALLS	N 4812	W 09348	1220
21-0826	12	BLANCHARD POWER STA	N 4552	W 09421	1100
21-0852	12	BLUE EARTH	N 4339	W 09406	1095
21-0939	12	BRAINERD RANGER STA	N 4622	W 09412	1214
21-0981	12	BRICELYN	N 4334	W 09348	1190
21-1198	12	CALEDONIA	N 4338	W 09129	1170
21-1227	13	CAMBRIDGE ST HOSPITAL	N 4534	W 09314	1000
21-1245	13	CAMPBELL	N 4606	W 09625	975
21-1263	13	CANBY	N 4443	W 09617	1243
21-1303	12	CARIBOU 2 S	N 4858	W 09627	1020
21-1374	12	CASS LAKE	N 4723	W 09437	1296
21-1465	12	CHASKA 1 NE	N 4448	W 09335	726
21-1630	13	CLOQUET FOR RES CENTER	N 4642	W 09231	1265
21-1669	12	COKATO	N 4505	W 09412	1060
21-1691	13	COLLEGEVILLE ST JOHN U	N 4535	W 09424	1225
21-1891	13	CROOKSTON NW EXP STA	N 4748	W 09637	883
21-2038	12	DAWSON	N 4456	W 09603	1056
21-2142	13	DETROIT LAKES 1 NNE	N 4650	W 09551	1375
21-2248	13	DULUTH WSO //R	N 4650	W 09211	1428
21-2476	12	ELBOW LAKE	N 4559	W 09558	1195
21-2486	12	ELGIN	N 4408	W 09215	1070
21-2500	12	ELK RIVER	N 4518	W 09335	910
21-2698	13	FAIRMONT	N 4338	W 09428	1187
21-2721	13	FARIBAULT	N 4418	W 09316	940
21-2737	13	FARMINGTON 3 NW	N 4440	W 09311	980
21-2768	13	FERGUS FALLS	N 4617	W 09604	1210
21-2904	12	FORT RIPLEY	N 4611	W 09422	1160
21-2916	13	FOSSTON	N 4735	W 09545	1299
21-3206	12	GONVICK	N 4744	W 09530	1454
21-3282	13	GRAND MARAIS	N 4745	W 09020	688
21-3290	13	GRAND MEADOW	N 4342	W 09234	1348
21-3303	13	GRAND RAPIDS NC SCHOOL	N 4714	W 09330	1310
21-3411	13	GULL LAKE DAM	N 4625	W 09421	1215
21-3455	13	HALLOCK	N 4846	W 09657	813
21-3520	12	HARMONY	N 4333	W 09200	1340
21-3567	12	HASTINGS DAM 2	N 4446	W 09252	695
21-3727	12	HIBBING PWR SUBSTATION	N 4726	W 09258	1534
21-3793	13	HINCKLEY	N 4601	W 09256	1035

STATE-STATION NUMBER	STN TYP	NAME	LATITUDE DEG-MIN	LONGITUDE DEG-MIN	ELEVATION (FT)
21-4008	12	INDUS	N 4837	W 09350	1105
21-4026	13	INTNL FALLS WSO //R	N 4834	W 09323	1179
21-4096	12	ISLAND LAKE RESERVOIR	N 4659	W 09214	1372
21-4106	13	ITASCA STATE PARK SCH	N 4713	W 09512	1490
21-4176	13	JORDAN 1 S	N 4439	W 09337	755
21-4306	12	KETTLE FALLS	N 4830	W 09239	1122
21-4418	12	LA CRESCENT DAM 7	N 4352	W 09118	647
21-4438	12	LAKE CITY	N 4427	W 09216	680
21-4641	12	LE CENTER 1 NE	N 4422	W 09343	1090
21-4652	13	LEECH LAKE DAM	N 4715	W 09413	1302
21-4778	13	LITCHFIELD	N 4507	W 09432	1132
21-4793	13	LITTLE FALLS 1 N	N 4559	W 09421	1120
21-4861	13	LONG PRAIRIE	N 4559	W 09451	1290
21-4994	13	MADISON SEWAGE PLANT	N 4500	W 09610	1080
21-5012	12	MAHNOMEN 1 W	N 4719	W 09559	1203
21-5136	13	MAPLE PLAIN	N 4501	W 09339	1030
21-5204	13	MARSHALL	N 4427	W 09547	1165
21-5298	13	MEADOWLANDS	N 4704	W 09244	1280
21-5392	13	MILACA	N 4545	W 09339	1080
21-5400	13	MILAN	N 4507	W 09556	1005
21-5435	13	MINN-ST PAUL WSO //R	N 4453	W 09313	834
21-5482	12	MINNEOTA	N 4434	W 09559	1165
21-5488	12	MINNESOTA CITY DAM 5	N 4410	W 09149	670
21-5563	13	MONTEVIDEO 1 SW	N 4456	W 09545	985
21-5571	12	MONTGOMERY	N 4426	W 09335	1100
21-5598	13	MOOSE LAKE 1 SSE	N 4627	W 09245	1110
21-5615	13	MORA	N 4553	W 09318	990
21-5638	13	MORRIS W C SCHOOL //	N 4535	W 09555	1130
21-5887	13	NEW ULM 2 SE	N 4417	W 09425	826
21-6007	12	NORTH MANKATO	N 4410	W 09402	785
21-6148	12	OKLEE	N 4750	W 09551	1150
21-6224	12	ORTONVILLE	N 4518	W 09627	976
21-6276	12	OTTERTAIL	N 4625	W 09534	980
21-6360	13	PARK RAPIDS	N 4655	W 09504	1434
21-6547	13	PINE RIVER DAM	N 4640	W 09407	1251
21-6565	13	PIPESTONE	N 4401	W 09619	1705
21-6612	13	POKEGAMA DAM	N 4715	W 09335	1280
21-6654	12	PRESTON	N 4340	W 09205	930
21-6787	13	RED LAKE FALLS	N 4753	W 09617	1035
21-6795	13	RED LAKE INDIAN AGENCY	N 4752	W 09502	1216
21-6817	12	RED WING	N 4434	W 09232	688
21-6822	12	RED WING DAM 3	N 4437	W 09237	677
21-6835	12	REDWOOD FALLS FAA AP	N 4433	W 09505	1025
21-7004	13	ROCHESTER WSO //R	N 4355	W 09230	1297
21-7087	13	ROSEAU 1 E	N 4851	W 09545	1047
21-7107	13	ROSEMOUNT AGRI EXP STA//	N 4443	W 09306	950
21-7294	13	ST CLOUD WSO //R	N 4533	W 09404	1034
21-7326	12	ST JAMES FILTRATION PL	N 4359	W 09437	1090
21-7405	13	ST PETER 2 SW	N 4418	W 09358	825
21-7460	13	SANDY LAKE DAM LIBBY	N 4648	W 09319	1234

21 — MINNESOTA

STATE-STATION NUMBER	STN TYP	NAME	LATITUDE DEG-MIN	LONGITUDE DEG-MIN	ELEVATION (FT)
21-7907	13	SPRINGFIELD 1 NW	N 4415	W 09459	1066
21-7915	12	SPRING GROVE 1 NW	N 4334	W 09140	1324
21-8037	12	STILLWATER 1 SE	N 4502	W 09247	710
21-8227	12	THEILMAN	N 4418	W 09212	737
21-8254	12	THORHULT	N 4814	W 09515	1185
21-8311	12	TOWER	N 4748	W 09216	1390
21-8323	13	TRACY	N 4414	W 09537	1403
21-8419	13	TWO HARBORS	N 4701	W 09140	625
21-8429	12	TYLER	N 4417	W 09608	1745
21-8520	12	VESTA	N 4430	W 09525	1080
21-8543	13	VIRGINIA	N 4730	W 09233	1435
21-8552	12	WABASHA	N 4423	W 09202	700
21-8579	13	WADENA 3 S	N 4624	W 09509	1350
21-8679	13	WARROAD	N 4855	W 09519	1069
21-8692	13	WASECA EXP. STATION	N 4404	W 09331	1153
21-8808	12	WELLS 1 NW	N 4345	W 09344	1197
21-8907	13	WHEATON	N 4548	W 09629	1018
21-8939	12	WHITEFACE RESERVOIR	N 4717	W 09211	1492
21-9004	13	WILLMAR STATE HOSPITAL	N 4508	W 09501	1133
21-9033	13	WINDOM	N 4352	W 09507	1370
21-9046	13	WINNEBAGO	N 4346	W 09410	1110
21-9059	13	WINNIBIGOSHISH DAM	N 4726	W 09403	1315
21-9067	13	WINONA	N 4403	W 09138	652
21-9072	12	WINONA DAM 5 A	N 4405	W 09141	663
21-9085	12	WINSTED	N 4458	W 09403	1025
21-9249	13	ZUMBROTA	N 4418	W 09240	985

21 – MINNESOTA

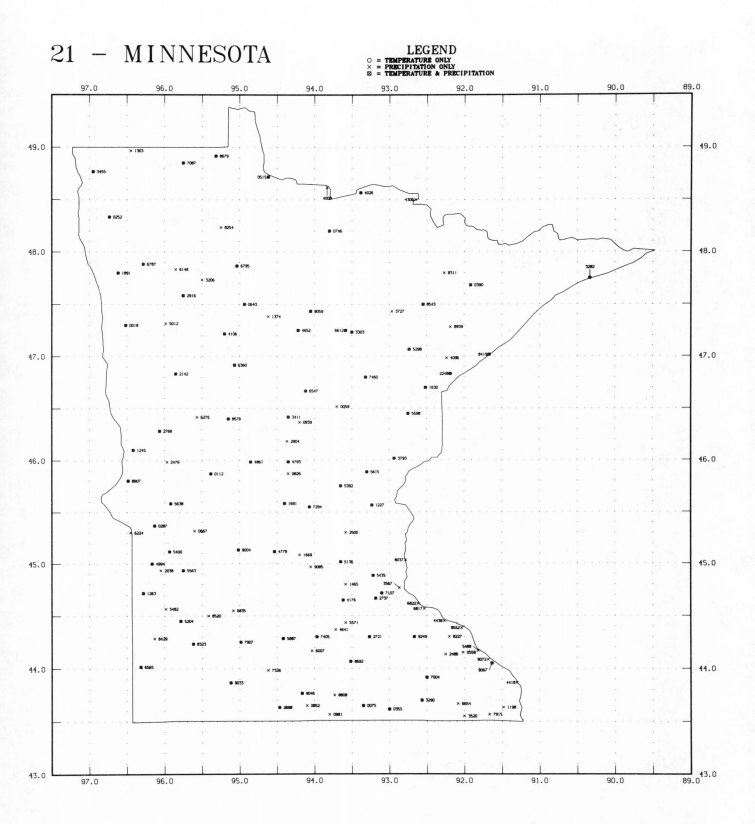

319

MISSISSIPPI

TEMPERATURE NORMALS (DEG F)

STATION		JAN	FEB	MAR	APR	MAY	JUN	JUL	AUG	SEP	OCT	NOV	DEC	ANN
ABERDEEN	MAX	53.8	58.9	66.7	76.7	82.9	89.5	92.2	91.9	86.3	76.7	65.6	57.5	74.9
	MIN	33.0	35.6	42.4	51.3	58.8	65.9	69.5	68.4	62.6	49.6	40.6	35.1	51.1
	MEAN	43.4	47.3	54.6	64.0	70.9	77.7	80.9	80.2	74.5	63.1	53.1	46.3	63.0
BATESVILLE 2 SW	MAX	51.0	55.5	63.7	74.5	81.8	88.9	92.4	91.8	86.5	76.6	64.4	55.0	73.5
	MIN	29.7	32.3	39.8	49.8	57.7	65.2	68.3	66.1	60.0	46.6	37.9	32.2	48.8
	MEAN	40.4	43.9	51.8	62.2	69.8	77.1	80.3	79.0	73.2	61.6	51.2	43.7	61.2
BAY SAINT LOUIS	MAX	58.7	61.8	67.8	75.7	82.6	88.5	90.0	89.8	86.3	78.7	68.2	61.5	75.8
	MIN	41.2	43.5	50.2	59.8	66.3	72.0	73.9	73.4	69.8	58.1	49.2	42.8	58.4
	MEAN	50.0	52.7	59.0	67.8	74.5	80.3	82.0	81.6	78.1	68.4	58.7	52.2	67.1
BAY SPRINGS	MAX	57.3	61.5	68.9	77.3	83.7	90.1	91.8	91.4	87.0	78.1	67.5	60.4	76.3
	MIN	35.8	38.2	44.7	53.2	60.5	67.1	70.2	69.3	64.9	52.5	43.3	37.9	53.1
	MEAN	46.6	49.9	56.8	65.3	72.1	78.6	81.0	80.4	76.0	65.3	55.4	49.2	64.7
BELZONI	MAX	52.8	57.6	65.7	75.8	83.6	90.6	93.1	92.5	87.4	77.9	65.5	57.1	75.0
	MIN	33.3	36.1	43.6	52.9	60.9	68.0	70.9	68.9	62.8	50.1	41.4	35.7	52.1
	MEAN	43.1	46.9	54.7	64.4	72.3	79.4	82.0	80.8	75.1	64.0	53.5	46.4	63.6
BILOXI CITY	MAX	60.6	63.3	69.1	76.5	83.1	88.6	90.2	89.9	86.7	79.6	69.8	63.6	76.8
	MIN	43.2	45.4	51.7	60.0	66.6	72.4	74.3	73.8	70.2	59.2	50.5	45.2	59.4
	MEAN	52.0	54.4	60.4	68.2	74.9	80.5	82.3	81.9	78.5	69.4	60.2	54.4	68.1
BOONEVILLE	MAX	48.4	53.2	61.9	72.9	80.3	87.3	90.8	90.5	84.9	74.4	61.8	52.5	71.6
	MIN	28.7	31.7	39.6	49.5	57.7	65.2	68.7	67.3	61.2	48.1	38.7	31.7	49.0
	MEAN	38.6	42.5	50.8	61.2	69.1	76.3	79.8	78.9	73.1	61.3	50.2	42.1	60.3
BROOKHAVEN	MAX	58.6	62.8	70.1	77.9	83.8	89.7	91.3	91.1	86.7	78.5	68.0	61.4	76.7
	MIN	37.6	39.5	46.4	54.3	61.4	67.4	70.2	69.5	65.5	53.4	45.0	39.6	54.2
	MEAN	48.1	51.2	58.3	66.1	72.6	78.6	80.8	80.3	76.1	66.0	56.5	50.5	65.4
CANTON	MAX	55.8	60.2	67.7	76.9	83.9	90.5	93.3	92.9	88.3	79.2	67.6	59.5	76.3
	MIN	33.4	35.7	43.0	51.8	59.7	66.4	69.4	67.9	62.7	49.3	40.6	35.1	51.3
	MEAN	44.6	48.0	55.4	64.4	71.8	78.5	81.4	80.5	75.6	64.3	54.1	47.3	63.8
CHARLESTON	MAX	51.4	56.1	64.2	75.4	83.1	90.2	93.4	92.7	87.1	76.9	64.6	55.2	74.2
	MIN	30.0	33.4	41.2	50.7	58.7	65.9	69.2	67.4	61.2	48.3	39.3	32.4	49.8
	MEAN	40.7	44.8	52.7	63.1	70.9	78.1	81.3	80.1	74.2	62.6	51.9	43.8	62.0
CLARKSDALE	MAX	50.0	54.5	63.2	74.7	83.2	90.5	93.3	92.0	86.3	76.4	63.4	53.7	73.4
	MIN	32.3	35.3	42.8	53.2	61.6	69.1	72.1	70.4	64.2	52.1	42.0	35.4	52.5
	MEAN	41.1	44.9	53.0	64.0	72.4	79.8	82.7	81.2	75.3	64.3	52.7	44.6	63.0
CLEVELAND	MAX	51.4	56.0	64.2	74.8	83.0	90.1	92.8	91.8	86.4	76.7	63.5	54.7	73.8
	MIN	32.4	35.3	42.3	52.0	60.4	67.9	71.2	69.5	63.5	51.1	41.3	35.0	51.8
	MEAN	41.9	45.7	53.3	63.4	71.7	79.0	82.0	80.7	74.9	63.9	52.4	44.9	62.8
COLLINS	MAX	59.1	63.5	70.8	78.6	84.1	89.9	91.5	91.5	87.6	79.1	68.8	62.3	77.2
	MIN	35.8	38.1	44.5	52.4	59.7	66.1	69.4	68.3	64.1	51.0	42.3	37.4	52.4
	MEAN	47.5	50.8	57.7	65.5	71.9	78.0	80.5	79.9	75.9	65.0	55.6	49.9	64.9
COLUMBIA	MAX	61.0	65.1	72.1	79.8	85.5	91.7	93.0	92.5	88.3	80.1	70.1	63.6	78.6
	MIN	37.9	40.1	46.6	54.6	61.4	67.6	70.7	69.7	65.5	52.7	44.2	39.5	54.2
	MEAN	49.5	52.6	59.4	67.2	73.5	79.6	81.8	81.1	76.9	66.4	57.2	51.6	66.4
COLUMBUS	MAX	53.2	57.9	65.9	76.2	83.1	89.8	92.7	92.5	87.2	77.4	65.9	57.2	74.9
	MIN	31.6	34.1	41.7	50.5	58.3	65.6	68.9	67.8	62.2	48.3	39.3	33.7	50.2
	MEAN	42.4	46.0	53.8	63.4	70.7	77.8	80.9	80.2	74.7	62.9	52.6	45.4	62.6
CORINTH 5 WSW	MAX	50.4	55.6	63.8	75.2	82.5	89.2	92.4	91.5	85.9	75.9	63.3	54.5	73.4
	MIN	29.8	32.4	39.8	49.5	57.7	65.2	68.6	67.0	60.8	47.6	38.5	32.5	49.1
	MEAN	40.1	44.0	51.8	62.3	70.1	77.2	80.5	79.3	73.4	61.8	51.0	43.5	61.3
EUPORA 1 E	MAX	54.1	59.1	66.8	76.4	83.0	89.4	92.1	91.9	86.7	77.3	65.9	57.6	75.0
	MIN	32.1	34.5	41.5	50.2	57.9	64.8	67.9	66.4	61.0	48.2	39.3	34.1	49.8
	MEAN	43.1	46.9	54.2	63.3	70.5	77.2	80.0	79.1	73.8	62.8	52.6	45.9	62.5

MISSISSIPPI

TEMPERATURE NORMALS (DEG F)

STATION		JAN	FEB	MAR	APR	MAY	JUN	JUL	AUG	SEP	OCT	NOV	DEC	ANN
FOREST 3 S	MAX	56.0	60.5	68.1	76.9	83.4	89.6	91.6	91.1	86.9	78.3	66.9	59.5	75.7
	MIN	33.6	35.3	42.2	50.5	58.4	65.6	68.8	67.4	63.0	49.8	40.8	35.2	50.9
	MEAN	44.8	47.9	55.2	63.7	70.9	77.6	80.2	79.3	75.0	64.1	53.9	47.4	63.3
GREENVILLE	MAX	53.1	58.6	66.5	76.6	83.9	90.4	92.7	91.8	86.9	78.1	65.6	57.1	75.1
	MIN	33.0	36.0	43.3	53.0	61.3	68.7	71.6	70.0	64.0	51.8	42.2	35.7	52.6
	MEAN	43.0	47.3	54.9	64.8	72.6	79.6	82.2	81.0	75.5	65.0	53.9	46.4	63.9
GREENWOOD FAA AP	MAX	52.3	56.9	64.9	75.1	82.9	89.8	92.2	91.3	86.2	76.5	64.1	55.6	74.0
	MIN	34.7	37.3	44.5	53.5	61.8	68.8	71.7	70.3	64.6	51.6	42.5	36.9	53.2
	MEAN	43.5	47.1	54.7	64.3	72.4	79.3	82.0	80.8	75.4	64.1	53.3	46.3	63.6
GULFPORT NAVAL CENTER	MAX	61.0	64.1	70.0	77.5	84.2	89.8	91.2	90.9	87.6	80.0	70.0	63.6	77.5
	MIN	42.2	44.2	50.8	59.1	65.6	71.3	73.2	72.7	69.1	58.0	49.1	44.2	58.3
	MEAN	51.6	54.2	60.4	68.3	74.9	80.6	82.2	81.8	78.4	69.1	59.6	53.9	67.9
HATTIESBURG	MAX	59.8	63.9	70.8	78.9	85.0	91.1	92.4	92.2	88.2	79.6	69.6	62.6	77.8
	MIN	36.9	39.3	46.0	54.3	61.0	67.4	70.4	69.5	65.3	52.0	43.7	38.7	53.7
	MEAN	48.4	51.6	58.4	66.6	73.0	79.3	81.4	80.9	76.8	65.8	56.6	50.7	65.8
HERNANDO	MAX	48.8	53.6	61.9	73.2	80.9	88.0	91.7	90.9	85.1	75.2	62.2	53.0	72.0
	MIN	29.8	33.3	41.3	51.7	59.9	67.4	70.6	69.0	63.2	51.4	41.0	33.5	51.0
	MEAN	39.3	43.4	51.6	62.5	70.4	77.7	81.2	80.0	74.1	63.4	51.6	43.3	61.5
HOLLY SPRINGS 4 N	MAX	48.4	52.6	61.0	72.3	79.5	86.6	90.4	89.7	83.9	74.0	61.6	52.4	71.0
	MIN	28.7	31.4	38.9	49.0	56.9	64.5	68.2	66.4	60.3	47.1	38.1	31.8	48.4
	MEAN	38.6	42.0	50.0	60.6	68.3	75.6	79.3	78.1	72.1	60.5	49.9	42.1	59.8
HOUSTON 2 NE	MAX	51.1	55.6	63.6	74.3	81.2	88.0	91.1	90.7	85.0	75.2	63.8	55.0	72.9
	MIN	29.4	31.8	39.4	48.7	56.8	64.0	67.6	65.8	60.2	46.0	37.0	31.0	48.1
	MEAN	40.3	43.7	51.5	61.5	69.0	76.1	79.4	78.3	72.6	60.6	50.4	43.1	60.5
JACKSON WSO	MAX	56.5	60.9	68.4	77.3	84.1	90.5	92.5	92.1	87.6	78.6	67.5	60.0	76.3
	MIN	34.9	37.2	44.2	52.9	60.8	67.9	71.3	70.2	65.1	51.4	42.3	37.1	52.9
	MEAN	45.7	49.1	56.3	65.1	72.5	79.2	81.9	81.2	76.4	65.0	54.9	48.6	64.6
KIPLING	MAX	56.0	60.7	68.0	76.8	83.1	89.1	91.4	91.1	86.4	77.3	66.5	59.0	75.5
	MIN	33.2	35.6	42.1	50.1	57.6	64.6	68.3	67.2	62.3	48.6	39.9	35.1	50.4
	MEAN	44.6	48.2	55.0	63.5	70.4	76.9	79.8	79.2	74.4	63.0	53.2	47.1	62.9
KOSCIUSKO	MAX	53.6	58.3	66.5	75.8	82.1	88.9	91.3	91.0	85.9	76.4	65.2	57.2	74.4
	MIN	31.5	34.0	41.1	50.3	58.2	65.6	68.9	67.6	62.2	48.8	39.4	33.8	50.1
	MEAN	42.6	46.2	53.8	63.1	70.2	77.2	80.1	79.3	74.1	62.6	52.3	45.5	62.3
LAUREL	MAX	57.7	61.6	69.1	77.8	84.1	90.4	92.2	91.8	87.4	78.6	67.9	60.6	76.6
	MIN	35.5	37.7	44.6	53.3	61.0	67.7	70.8	70.0	65.4	52.2	42.7	37.1	53.2
	MEAN	46.6	49.7	56.8	65.6	72.6	79.1	81.6	80.9	76.4	65.4	55.3	48.8	64.9
LIBERTY 1 W	MAX	58.5	62.5	70.0	77.9	84.2	90.4	91.7	91.6	87.8	79.7	68.8	61.5	77.1
	MIN	35.5	37.6	44.5	53.2	60.1	66.2	69.5	68.3	64.4	51.0	42.6	37.1	52.5
	MEAN	47.1	50.1	57.3	65.6	72.2	78.3	80.6	80.0	76.1	65.4	55.7	49.3	64.8
MERIDIAN WSO R	MAX	56.7	61.1	68.6	77.7	84.2	90.4	92.5	92.1	87.2	77.8	67.3	60.0	76.3
	MIN	34.2	36.5	43.1	51.4	59.3	66.4	70.0	69.1	64.2	50.0	40.8	36.0	51.8
	MEAN	45.5	48.8	55.9	64.6	71.7	78.4	81.3	80.6	75.7	63.9	54.1	48.0	64.0
MONTICELLO	MAX	58.8	62.6	70.1	78.4	84.8	91.3	92.9	92.7	88.4	79.3	68.9	61.7	77.5
	MIN	34.3	36.4	43.6	52.3	59.3	65.6	68.9	67.9	63.5	49.6	40.9	35.7	51.5
	MEAN	46.6	49.5	56.9	65.4	72.1	78.5	80.9	80.3	76.0	64.5	54.9	48.7	64.5
MOORHEAD	MAX	53.4	58.3	66.3	76.6	84.3	91.1	93.4	92.7	87.4	77.9	65.3	57.0	75.3
	MIN	33.6	36.6	43.9	53.3	60.7	67.7	70.7	68.8	63.0	50.7	41.9	36.0	52.2
	MEAN	43.5	47.5	55.1	65.0	72.5	79.4	82.1	80.7	75.2	64.3	53.7	46.5	63.8
NATCHEZ	MAX	59.3	63.4	70.7	78.4	84.5	90.3	92.3	92.0	87.7	79.9	68.9	62.5	77.5
	MIN	38.2	40.6	47.3	55.5	62.4	69.0	71.9	70.9	66.3	54.2	45.9	40.5	55.2
	MEAN	48.8	52.0	59.0	67.0	73.5	79.7	82.1	81.4	77.0	67.1	57.4	51.5	66.4

MISSISSIPPI

TEMPERATURE NORMALS (DEG F)

STATION		JAN	FEB	MAR	APR	MAY	JUN	JUL	AUG	SEP	OCT	NOV	DEC	ANN
NEWTON EXP STATION	MAX	57.0	61.5	68.8	77.1	83.5	89.5	92.0	91.4	86.8	77.9	67.5	59.9	76.1
	MIN	34.4	36.6	43.2	51.1	58.5	65.0	68.5	67.4	62.6	48.9	40.8	36.1	51.1
	MEAN	45.7	49.1	56.0	64.1	71.0	77.3	80.3	79.4	74.8	63.4	54.1	48.0	63.6
PHILADELPHIA 1 WSW	MAX	54.8	59.2	67.0	76.1	82.4	88.9	91.3	91.1	86.4	77.2	66.1	58.0	74.9
	MIN	32.2	35.0	42.0	50.8	58.5	65.4	68.8	67.4	62.3	48.4	39.8	34.2	50.4
	MEAN	43.5	47.1	54.5	63.5	70.5	77.2	80.1	79.3	74.4	62.8	53.0	46.1	62.7
PICAYUNE	MAX	63.3	66.7	73.1	79.9	85.8	91.2	92.5	92.1	88.5	81.3	71.5	65.8	79.3
	MIN	39.7	42.1	48.6	55.8	62.0	67.7	70.6	70.3	66.6	54.2	46.2	41.5	55.4
	MEAN	51.5	54.4	60.9	67.9	73.9	79.5	81.6	81.2	77.6	67.8	58.9	53.7	67.4
PICKENS	MAX	55.6	60.4	68.0	76.8	83.3	89.6	91.8	91.3	86.6	77.4	66.1	58.7	75.5
	MIN	34.2	36.8	43.8	52.0	59.0	65.4	68.7	67.4	62.3	49.0	40.7	35.7	51.3
	MEAN	44.9	48.6	55.9	64.5	71.2	77.5	80.3	79.3	74.5	63.2	53.4	47.3	63.4
POPLARVILLE EXP STA	MAX	61.3	64.9	71.3	78.5	84.9	90.7	91.4	90.9	87.1	79.9	69.8	63.9	77.9
	MIN	40.8	42.7	48.9	56.7	63.3	69.0	71.3	70.9	67.7	56.8	48.1	42.9	56.6
	MEAN	51.1	53.8	60.1	67.6	74.1	79.9	81.4	81.0	77.4	68.4	59.0	53.4	67.3
PORT GIBSON	MAX	56.6	61.2	68.8	77.3	83.9	89.7	91.9	91.6	86.8	77.9	66.9	60.0	76.1
	MIN	33.2	35.4	42.7	51.7	59.3	66.1	69.4	68.2	63.2	49.4	40.5	34.8	51.2
	MEAN	44.9	48.3	55.8	64.5	71.6	77.9	80.7	79.9	75.0	63.6	53.8	47.4	63.6
ROSEDALE	MAX	50.8	55.4	63.6	73.9	81.7	88.7	91.6	90.9	85.6	76.2	63.9	54.9	73.1
	MIN	31.6	34.8	42.7	52.0	60.5	68.0	71.3	69.3	62.5	49.4	40.4	34.3	51.4
	MEAN	41.2	45.1	53.1	63.0	71.1	78.4	81.5	80.1	74.1	62.8	52.2	44.7	62.3
RUSSELL 2 WNW	MAX	55.9	60.4	67.7	76.5	82.9	89.1	91.1	90.8	86.4	77.5	66.6	59.0	75.3
	MIN	35.1	37.7	44.9	53.2	60.4	67.2	70.1	69.2	64.6	52.3	43.2	37.5	53.0
	MEAN	45.6	49.1	56.3	64.9	71.7	78.2	80.6	80.0	75.6	64.9	54.9	48.3	64.2
STATE UNIVERSITY	MAX	52.6	57.3	65.1	75.3	82.2	88.8	91.6	91.3	86.0	76.2	64.8	56.2	74.0
	MIN	33.0	36.0	43.3	52.6	60.4	67.6	70.6	69.3	64.0	51.8	42.4	35.8	52.2
	MEAN	42.8	46.6	54.3	64.0	71.3	78.2	81.1	80.4	75.1	64.0	53.6	46.0	63.1
STONEVILLE EXP STA	MAX	51.7	56.3	64.4	74.9	82.8	89.9	92.4	91.5	86.3	77.1	64.4	55.4	73.9
	MIN	33.2	36.1	43.6	53.4	61.4	68.7	71.4	69.5	63.5	51.1	42.3	35.9	52.5
	MEAN	42.5	46.2	54.0	64.2	72.2	79.3	81.9	80.5	74.9	64.1	53.4	45.7	63.2
TUPELO 2 WNW	MAX	51.1	56.4	64.6	75.4	82.7	89.7	92.5	92.0	86.0	76.0	63.1	54.8	73.7
	MIN	31.2	33.5	40.6	49.8	58.0	65.6	69.4	68.2	62.2	48.5	39.1	33.3	50.0
	MEAN	41.2	44.9	52.6	62.6	70.4	77.7	80.9	80.1	74.1	62.3	51.1	44.1	61.8
TYLERTOWN 2 WNW	MAX	60.5	64.8	71.5	78.8	84.8	90.9	92.0	91.6	87.7	80.0	69.5	63.0	77.9
	MIN	37.9	40.0	46.5	54.1	60.6	66.6	69.6	68.9	65.0	52.6	44.7	39.8	53.9
	MEAN	49.2	52.4	59.0	66.5	72.7	78.8	80.9	80.3	76.4	66.3	57.1	51.4	65.9
UNIVERSITY	MAX	49.7	54.3	62.5	73.5	80.4	87.4	90.9	90.2	84.7	74.7	62.5	53.7	72.0
	MIN	29.3	31.9	39.8	49.8	57.2	64.8	68.4	66.8	61.2	47.7	38.8	32.1	49.0
	MEAN	39.5	43.1	51.2	61.7	68.8	76.1	79.7	78.6	73.0	61.2	50.7	42.9	60.5
VICKSBURG MILITARY PK	MAX	56.3	60.5	68.1	76.6	82.9	88.9	91.4	91.0	86.2	77.6	66.6	59.7	75.5
	MIN	37.0	39.4	46.3	55.0	62.5	68.9	71.9	71.1	66.4	54.9	45.3	39.6	54.9
	MEAN	46.7	50.0	57.2	65.8	72.7	79.0	81.7	81.1	76.3	66.3	56.0	49.7	65.2
WATER VALLEY 1 NNE	MAX	51.7	56.0	63.9	74.3	80.8	87.8	91.2	91.1	85.4	75.6	63.9	55.3	73.1
	MIN	30.1	32.6	40.2	49.9	57.5	65.3	69.1	67.6	61.7	48.4	39.1	32.6	49.5
	MEAN	40.9	44.4	52.1	62.2	69.2	76.6	80.2	79.4	73.6	62.0	51.6	44.0	61.4
WIGGINS 4 SE	MAX	61.3	65.3	71.7	79.7	85.8	91.9	93.1	92.6	88.9	80.9	71.0	64.2	78.9
	MIN	37.0	39.0	45.7	53.6	60.7	66.9	69.8	69.3	65.2	53.0	43.9	38.7	53.6
	MEAN	49.2	52.2	58.7	66.7	73.3	79.4	81.5	81.0	77.1	67.0	57.5	51.5	66.3
WOODVILLE 4 ESE	MAX	59.8	63.7	70.9	78.5	84.4	90.0	91.5	91.2	87.5	79.8	69.2	62.9	77.5
	MIN	38.2	39.8	46.7	54.5	61.3	67.1	69.9	69.3	65.2	53.9	45.4	40.0	54.3
	MEAN	49.0	51.8	58.8	66.6	72.9	78.5	80.7	80.3	76.4	66.8	57.3	51.5	65.9

MISSISSIPPI

TEMPERATURE NORMALS (DEG F)

STATION		JAN	FEB	MAR	APR	MAY	JUN	JUL	AUG	SEP	OCT	NOV	DEC	ANN
YAZOO CITY 5 NNE	MAX	55.0	59.5	67.4	76.6	83.8	90.6	92.9	92.6	87.7	78.4	66.5	58.6	75.8
	MIN	34.6	37.2	44.4	53.6	61.5	68.3	71.1	69.9	64.4	51.7	42.4	36.8	53.0
	MEAN	44.8	48.4	55.9	65.2	72.7	79.5	82.1	81.3	76.1	65.1	54.5	47.7	64.4

MÍSSISSIPPI

PRECIPITATION NORMALS (INCHES)

STATION	JAN	FEB	MAR	APR	MAY	JUN	JUL	AUG	SEP	OCT	NOV	DEC	ANN
ABBEVILLE	5.46	4.93	6.38	5.84	5.35	3.53	4.00	2.89	4.06	2.49	5.03	5.19	55.15
ABERDEEN	5.46	4.86	6.54	5.78	4.63	3.81	4.46	2.92	3.55	3.06	4.31	5.47	54.85
ACKERMAN	5.62	4.94	6.49	6.13	4.44	3.49	5.37	2.87	3.94	3.01	4.38	5.52	56.20
ARKABUTLA DAM	4.67	4.66	5.26	5.99	5.21	3.80	3.59	2.96	3.39	2.50	4.32	4.94	51.29
ASHLAND	5.47	5.20	5.92	5.74	5.23	3.74	4.21	3.54	3.93	2.82	4.96	5.41	56.17
BALDWYN	5.39	4.72	6.68	5.48	5.05	3.73	4.04	2.53	3.51	2.96	4.54	5.40	54.03
BATESVILLE 2 SW	5.06	4.80	6.01	5.87	5.40	3.82	4.27	2.85	3.80	2.48	5.12	5.30	54.78
BAY SAINT LOUIS	5.19	4.73	5.16	5.33	5.00	3.81	5.75	6.07	5.94	2.41	3.79	5.11	58.29
BAY SPRINGS	5.27	5.06	6.52	5.87	4.91	3.87	5.53	3.78	3.97	3.04	3.95	5.89	57.66
BELZONI	5.55	4.60	6.19	5.86	4.68	3.37	4.98	2.92	3.14	2.69	4.62	5.59	54.19
BILOXI CITY	5.07	4.79	5.44	4.97	4.87	4.74	6.40	6.24	6.58	3.00	3.64	5.26	61.00
BLACK HAWK	5.32	4.83	6.69	5.81	4.34	3.50	4.94	3.14	3.26	2.90	4.65	5.60	54.98
BLUFF LAKE	5.52	4.74	6.28	5.97	4.08	3.00	5.30	3.02	3.85	2.90	3.94	5.33	53.93
BOONEVILLE	5.54	4.72	6.62	5.62	4.91	3.52	4.25	2.75	3.67	2.99	4.69	5.43	54.71
BROOKHAVEN	5.16	5.02	6.42	5.79	5.06	4.12	5.30	4.36	3.28	2.73	4.46	6.24	57.94
BROOKSVILLE EXP STA	5.37	4.45	6.13	5.89	3.68	2.94	4.66	2.89	3.31	2.69	3.79	5.18	50.98
BRUCE	5.71	5.11	6.99	6.18	4.80	4.40	3.99	2.40	4.19	2.71	4.89	5.61	56.98
BUCKATUNNA	5.81	5.28	7.23	5.58	4.47	4.26	6.52	4.23	4.67	2.83	4.28	5.98	61.14
CALHOUN CITY 2 N	5.62	4.79	5.99	5.77	4.41	3.59	4.51	2.73	3.48	2.57	4.69	5.25	53.40
CANTON	5.42	4.49	5.96	5.74	5.16	2.86	4.24	3.69	2.90	2.71	4.28	5.64	53.09
CARROLLTON 1 SSW	5.82	5.36	6.57	5.73	5.03	3.58	5.07	2.98	3.31	2.92	5.11	5.48	56.96
CHARLESTON	5.40	4.85	6.33	6.08	5.00	4.15	4.41	2.73	3.56	2.35	5.21	4.77	54.84
CLARKSDALE	4.62	4.67	5.35	5.41	5.18	3.75	4.22	2.43	3.67	2.31	4.68	4.90	51.19
CLEVELAND	5.24	5.12	6.53	6.24	5.66	3.90	4.87	2.72	3.91	2.71	4.81	5.33	57.04
COFFEEVILLE	5.23	4.93	6.96	6.08	5.07	3.61	4.18	2.61	4.15	2.71	4.88	5.40	55.81
COLLINS	4.75	4.70	5.82	5.41	4.79	3.49	5.26	4.07	4.38	2.94	3.69	5.63	54.93
COLUMBIA	5.44	5.66	5.96	5.73	5.07	4.18	5.32	5.10	4.09	2.86	4.45	6.30	60.16
COLUMBUS	5.85	5.06	6.70	6.19	4.50	3.06	5.45	3.61	3.52	3.01	4.26	5.54	56.75
CORINTH 5 WSW	5.19	4.91	5.91	5.43	4.62	3.50	3.80	2.93	3.46	2.69	4.72	5.12	52.28
CRANDALL 12 N	5.76	5.45	6.63	5.62	4.74	3.97	5.81	3.81	4.65	2.91	4.13	6.25	59.73
CRAWFORD 5 WSW	5.42	4.56	6.25	5.97	3.90	2.81	4.70	2.86	3.20	2.87	3.91	5.02	51.47
DANCY	5.26	4.77	6.44	5.89	4.60	3.46	4.35	2.97	3.53	2.93	4.21	5.35	53.76
DLO	5.34	4.97	6.32	5.84	4.90	3.75	4.91	4.12	3.69	2.74	4.24	6.08	56.90
EDINBURG	5.47	4.91	6.31	5.81	4.48	3.35	5.36	3.74	3.21	3.01	4.18	5.54	55.37
ELLIOTT	5.30	4.96	6.28	5.60	4.85	3.70	4.85	2.96	3.48	2.65	4.89	5.14	54.66
ENID DAM	4.92	4.61	5.80	5.97	5.03	3.41	4.31	2.70	3.73	2.28	4.88	4.82	52.46
ENTERPRISE	5.19	4.83	6.58	5.44	4.61	3.75	5.80	3.95	3.94	2.57	3.84	5.78	56.28
EUPORA 1 E	5.77	5.09	6.63	6.04	4.43	3.66	4.90	3.21	3.45	3.00	4.15	5.57	55.90
FOREST 3 S	5.53	4.81	6.28	5.42	4.75	4.04	5.78	4.32	3.95	3.22	4.06	5.87	58.03
FULTON 3 W	5.55	4.84	6.86	5.84	5.15	3.57	4.67	3.21	3.97	2.91	4.49	5.65	56.71
GERMANIA	5.20	5.01	6.10	5.29	5.07	3.40	3.67	2.90	2.86	2.65	3.91	5.92	51.98
GHOLSON 8 W	5.19	4.43	5.88	6.02	3.90	2.86	5.33	2.97	3.56	2.87	3.63	4.96	51.60
GREENVILLE	5.17	4.92	5.70	5.69	5.46	3.51	4.33	2.51	3.45	2.55	4.93	5.04	53.26
GREENVILLE 8 SW	4.90	4.55	5.26	5.94	5.17	3.30	4.22	2.73	3.15	2.58	4.77	4.80	51.37
GREENWOOD	5.48	4.91	6.19	5.66	4.90	3.35	4.75	2.71	3.23	3.01	4.82	5.49	54.50
GREENWOOD FAA AP	5.15	4.54	6.22	5.54	4.84	3.35	4.53	2.77	3.31	2.81	4.55	5.40	53.01
GRENADA	5.54	5.10	6.55	5.88	4.98	3.60	4.98	2.58	3.83	2.54	5.14	5.35	56.07
GULFPORT NAVAL CENTER	5.23	4.98	5.41	5.33	4.95	4.64	7.13	5.77	7.23	2.98	3.81	5.39	62.85
HATTIESBURG	5.13	5.47	6.35	5.43	5.32	4.20	5.94	4.81	4.27	2.96	4.03	5.94	59.85
HERNANDO	4.65	4.60	5.49	6.07	5.23	3.85	3.27	3.34	3.62	2.46	4.62	5.04	52.24
HICKORY	5.43	5.07	6.82	5.65	4.44	3.55	4.63	3.70	3.66	2.78	3.78	5.86	55.37
HICKORY FLAT 1 W	5.24	4.75	6.41	5.75	4.85	3.77	4.83	2.86	3.87	2.62	5.21	5.02	55.18
HOLLY SPRINGS 4 N	5.09	4.99	5.85	5.66	5.47	3.27	4.28	3.54	3.77	2.73	4.93	5.17	54.75
HOUSTON 2 NE	5.39	4.93	6.54	5.71	4.64	3.58	4.15	3.08	3.78	2.89	4.53	5.37	54.59
JACKSON WSO	5.00	4.48	5.86	5.85	4.83	2.94	4.40	3.71	3.55	2.62	4.18	5.40	52.82

MISSISSIPPI

PRECIPITATION NORMALS (INCHES)

STATION	JAN	FEB	MAR	APR	MAY	JUN	JUL	AUG	SEP	OCT	NOV	DEC	ANN
KIPLING	5.16	4.68	6.41	5.98	4.15	3.47	6.11	3.43	3.28	2.50	3.56	5.74	54.47
KOSCIUSKO	5.77	4.96	6.51	6.25	4.67	2.94	5.16	3.54	3.72	2.94	4.57	5.66	56.69
LAFAYETTE SPRINGS	5.26	4.88	6.75	6.07	5.31	3.89	4.66	2.80	4.14	2.83	4.97	5.68	57.24
LAKE CORMORANT 1 W	4.50	4.26	5.21	5.38	4.70	3.70	3.28	3.15	3.26	2.50	4.33	4.66	48.93
LAMBERT 5 E	5.13	4.92	5.93	6.05	5.17	3.99	4.74	2.65	3.80	2.27	4.81	4.99	54.45
LAUREL	5.06	4.84	6.21	5.27	4.99	3.54	5.39	4.11	4.52	2.72	3.80	5.84	56.29
LEAKESVILLE	4.68	4.95	6.88	5.44	4.90	4.76	6.68	4.86	5.16	3.13	3.77	5.71	60.92
LEXINGTON 2 NNW	5.22	5.06	5.75	5.84	4.74	3.62	4.36	3.28	3.01	2.63	4.27	4.94	52.72
LIBERTY 1 W	5.04	4.81	5.93	5.38	5.31	4.53	5.85	4.43	4.29	2.48	4.38	6.04	58.47
LOUISVILLE	5.79	4.88	6.47	6.36	4.31	3.16	5.79	3.29	3.77	2.75	4.05	5.39	56.01
MERIDIAN WSO R	4.99	4.58	6.65	5.41	4.20	3.49	5.32	3.36	3.57	2.59	3.48	5.66	53.30
MERRILL	4.89	5.26	6.91	5.19	5.15	4.87	7.52	5.11	6.15	3.24	4.11	5.51	63.91
MIZE	4.98	4.96	6.24	5.98	4.63	3.95	4.60	3.93	3.61	2.82	3.85	5.59	55.14
MONTICELLO	5.39	5.04	6.42	5.76	5.23	4.02	4.71	3.86	4.06	2.68	4.13	5.94	57.24
MOORHEAD	5.42	4.58	6.05	5.98	4.45	3.43	5.34	2.37	3.37	2.90	4.67	5.34	53.90
MOUNT PLEASANT	5.08	4.73	5.57	5.46	4.87	3.55	4.03	3.35	3.77	2.49	4.68	5.06	52.64
NATCHEZ	5.60	4.72	6.02	6.01	5.89	3.41	4.44	4.01	3.70	2.72	4.47	6.44	57.43
NEW ALBANY	5.53	4.74	6.62	5.55	4.79	3.87	4.44	2.85	3.61	2.76	4.99	5.31	55.06
NEWTON EXP STATION	5.00	4.96	6.35	5.58	4.24	3.40	4.93	3.88	3.65	2.59	3.77	5.85	54.20
OAKLEY EXP STATION	5.30	4.52	5.71	5.47	5.19	3.59	4.32	3.46	2.95	2.51	3.88	5.33	52.23
OFAHOMA	5.42	4.78	5.88	5.79	4.59	3.10	4.93	3.11	3.13	2.70	4.05	5.63	53.11
OKOLONA	5.39	4.76	6.81	5.75	4.51	3.59	4.70	3.05	3.52	2.83	4.39	5.58	54.88
PASCAGOULA 2 ENE	5.01	4.67	6.11	5.04	4.91	5.16	6.98	6.13	7.78	3.65	3.75	4.96	64.15
PAULDING	5.15	4.72	6.28	6.02	4.49	3.55	5.62	3.74	3.66	2.59	3.94	5.66	55.42
PELAHATCHIE	5.57	4.89	6.13	5.52	4.82	3.33	5.67	3.69	3.58	2.63	4.29	5.74	55.86
PHILADELPHIA 1 WSW	5.58	4.73	6.48	6.13	4.44	3.67	5.32	3.37	3.67	2.97	4.11	5.36	55.83
PICAYUNE	5.02	5.05	6.02	5.44	5.08	4.64	6.88	5.28	5.48	2.53	4.00	5.56	60.98
PICKENS	5.72	4.59	6.41	5.82	5.34	3.29	5.11	3.15	3.13	2.81	4.40	5.71	55.48
PLEASANT HILL	5.05	4.86	5.61	5.98	5.06	3.71	3.72	3.43	3.76	2.51	4.50	5.19	53.38
PONTOTOC	5.59	4.88	7.05	6.22	4.93	3.83	5.08	2.63	3.95	2.93	4.74	5.81	57.64
POPLARVILLE EXP STA	4.82	5.22	5.74	5.28	5.15	4.51	6.71	5.65	5.05	3.03	4.03	6.17	61.36
PORT GIBSON	5.79	4.83	6.30	5.65	5.23	3.90	4.21	3.39	3.12	2.70	4.32	5.82	55.26
PRENTISS 2 NNE	5.51	4.93	6.27	5.53	4.80	3.87	4.60	3.42	3.75	2.56	3.87	5.82	54.93
RIPLEY	5.30	4.87	6.26	5.62	4.63	3.45	4.75	2.95	3.80	2.75	5.04	5.27	54.69
ROCKPORT	5.60	4.91	6.20	5.80	5.32	3.63	4.81	3.97	3.15	2.90	4.36	6.63	57.28
ROLLING FORK	5.36	4.68	5.81	5.18	4.59	3.77	4.18	2.68	3.20	2.58	4.11	5.50	51.64
ROSEDALE	4.72	4.94	5.57	5.94	5.05	3.32	4.20	2.42	3.82	2.61	4.65	4.74	51.98
RUSSELL 2 WNW	5.04	4.79	6.81	5.23	4.15	3.65	5.59	3.70	3.41	2.87	3.44	5.70	54.38
SARAH	4.90	4.92	5.58	5.39	6.02	3.87	3.84	2.97	3.64	2.59	4.58	5.18	53.48
SARDIS DAM	5.01	4.71	5.80	5.49	5.15	3.52	4.05	2.82	3.47	2.58	4.94	4.89	52.43
SENATOBIA	4.86	4.75	5.57	5.46	5.53	3.59	3.53	2.98	3.69	2.27	4.81	5.20	52.24
SHUBUTA	5.10	5.18	7.37	5.45	4.29	4.14	6.70	4.08	4.18	2.61	3.62	5.73	58.45
SHUQUALAK	5.30	4.46	6.35	6.17	3.77	3.52	5.37	3.32	3.82	2.75	3.58	5.07	53.48
SLEDGE	4.71	4.90	5.44	5.62	5.45	4.31	3.81	2.40	3.63	2.30	4.78	4.82	52.17
STANDARD	5.48	5.11	5.86	5.39	5.44	5.33	7.33	6.01	6.94	2.93	3.92	5.42	65.16
STATE UNIVERSITY	5.42	4.60	6.32	5.72	4.29	3.50	4.86	3.13	3.67	2.79	3.87	5.44	53.61
STONEVILLE EXP STA	5.09	4.78	5.67	5.48	5.09	3.72	4.05	2.37	3.54	2.46	4.82	4.96	52.03
SUMRALL	5.17	5.43	6.04	5.67	4.67	3.99	5.54	4.43	4.30	2.94	4.12	6.28	58.58
SWAN LAKE	5.20	4.63	6.12	5.89	4.71	3.91	4.73	2.58	3.56	2.33	4.92	4.91	53.49
TUPELO 2 WNW	5.65	4.63	6.94	5.66	5.22	3.72	4.59	2.84	3.64	2.99	4.63	5.61	56.12
TYLERTOWN 2 WNW	5.00	5.23	5.48	5.53	5.56	4.17	5.78	4.45	3.60	2.67	4.16	5.93	57.56
UNION	5.51	4.84	6.45	5.89	4.27	3.67	5.36	3.64	3.41	2.64	3.91	5.71	55.30
UNIVERSITY	5.26	4.78	6.06	5.69	5.41	3.58	3.87	2.95	4.06	2.78	5.17	4.94	54.55
VAIDEN 1 SSW	5.60	5.22	6.50	6.45	4.79	3.60	4.72	3.40	3.44	2.90	4.63	5.63	56.88
VANCE	4.98	4.83	5.73	5.97	4.68	3.68	4.06	2.19	3.60	2.38	4.61	4.65	51.36

MISSISSIPPI

PRECIPITATION NORMALS (INCHES)

STATION	JAN	FEB	MAR	APR	MAY	JUN	JUL	AUG	SEP	OCT	NOV	DEC	ANN
VAN VLEET	5.51	4.93	6.49	5.69	5.05	3.50	4.46	2.76	3.79	2.80	4.61	5.16	54.75
VICKSBURG MILITARY PK	5.47	4.86	5.98	5.76	4.81	3.16	3.45	3.06	3.22	2.84	3.86	5.95	52.42
WALNUT GROVE	5.45	4.66	5.94	5.94	4.48	3.08	4.88	3.40	3.53	2.72	3.97	5.34	53.39
WATER VALLEY 1 NNE	4.99	4.73	6.46	5.99	5.01	3.87	3.94	2.70	4.09	2.68	5.03	5.24	54.73
WHITE OAK	5.39	4.96	6.37	5.71	4.90	3.45	4.94	4.23	3.89	2.88	3.92	6.13	56.77
WIGGINS 4 SE	5.14	5.42	6.43	5.15	4.69	4.77	6.51	5.34	5.41	3.11	4.16	5.78	61.91
WOODVILLE 4 ESE	5.41	5.55	6.48	5.80	5.71	4.62	5.83	4.15	4.19	2.67	4.81	6.56	61.78
YAZOO CITY 5 NNE	5.30	5.12	6.18	5.59	5.05	3.52	4.71	3.09	3.20	2.88	4.15	5.97	54.76

MISSISSIPPI

HEATING DEGREE DAY NORMALS (BASE 65 DEG F)

STATION	JUL	AUG	SEP	OCT	NOV	DEC	JAN	FEB	MAR	APR	MAY	JUN	ANN
ABERDEEN	0	0	5	128	362	580	677	505	350	89	15	0	2711
BATESVILLE 2 SW	0	0	11	161	418	660	763	599	426	127	40	0	3205
BAY SAINT LOUIS	0	0	0	47	214	403	483	365	220	36	0	0	1768
BAY SPRINGS	0	0	0	92	302	490	584	441	284	73	13	0	2279
BELZONI	0	0	5	121	356	577	679	518	348	90	13	0	2707
BILOXI CITY	0	0	0	31	178	339	424	318	178	30	0	0	1498
BOONEVILLE	0	0	14	174	444	710	818	630	457	148	37	0	3432
BROOKHAVEN	0	0	0	83	269	456	538	407	245	67	0	0	2065
CANTON	0	0	5	118	339	549	643	490	326	87	13	0	2570
CHARLESTON	0	0	7	148	398	657	753	572	401	115	20	0	3071
CLARKSDALE	0	0	6	125	373	632	741	572	398	96	20	0	2963
CLEVELAND	0	0	6	128	382	623	716	547	390	110	14	0	2916
COLLINS	0	0	0	95	292	468	555	413	261	68	7	0	2159
COLUMBIA	0	0	0	74	253	422	499	371	218	44	0	0	1881
COLUMBUS	0	0	6	135	377	608	701	542	368	105	18	0	2860
CORINTH 5 WSW	0	0	11	165	420	667	772	594	432	120	28	0	3209
EUPORA 1 E	0	0	9	131	372	592	686	517	360	99	17	0	2783
FOREST 3 S	0	0	0	113	343	546	637	490	334	108	16	0	2587
GREENVILLE	0	0	5	96	338	577	682	503	347	79	8	0	2635
GREENWOOD FAA AP	0	0	7	129	359	580	675	509	345	98	14	0	2716
GULFPORT NAVAL CENTER	0	0	0	30	193	353	435	324	175	29	0	0	1539
HATTIESBURG	0	0	0	80	266	448	531	398	238	60	6	0	2027
HERNANDO	0	0	8	128	407	673	797	605	440	121	27	0	3206
HOLLY SPRINGS 4 N	0	0	20	189	453	710	818	651	480	164	57	6	3548
HOUSTON 2 NE	0	0	11	179	438	679	766	596	433	137	29	0	3268
JACKSON WSO	0	0	0	98	316	513	611	462	303	77	9	0	2389
KIPLING	0	0	6	133	360	555	640	480	333	98	15	0	2620
KOSCIUSKO	0	0	13	148	385	605	702	536	369	108	27	0	2893
LAUREL	0	0	0	91	302	502	582	443	282	79	10	0	2291
LIBERTY 1 W	0	0	0	97	293	492	571	436	268	74	8	0	2239
MERIDIAN WSO R	0	0	5	118	335	527	616	464	312	85	17	0	2479
MONTICELLO	0	0	0	112	314	511	588	454	288	70	8	0	2345
MOORHEAD	0	0	0	114	345	574	673	500	341	78	9	0	2634
NATCHEZ	0	0	0	65	252	430	522	383	239	50	0	0	1941
NEWTON EXP STATION	0	0	6	124	337	527	611	459	305	97	10	0	2476
PHILADELPHIA 1 WSW	0	0	7	138	367	586	674	509	348	101	19	0	2749
PICAYUNE	0	0	0	53	211	364	446	324	173	40	0	0	1611
PICKENS	0	0	6	144	355	549	634	474	313	91	18	0	2584
POPLARVILLE EXP STA	0	0	0	38	207	369	454	338	194	39	0	0	1639
PORT GIBSON	0	0	0	119	344	546	632	480	312	87	12	0	2532
ROSEDALE	0	0	8	132	384	629	738	564	391	108	18	0	2972
RUSSELL 2 WNW	0	0	0	92	313	523	614	459	304	79	12	0	2396
STATE UNIVERSITY	0	0	7	117	349	589	688	529	356	89	17	0	2741
STONEVILLE EXP STA	0	0	0	114	353	598	698	535	368	88	12	0	2766
TUPELO 2 WNW	0	0	8	149	421	648	738	563	403	125	33	0	3088
TYLERTOWN 2 WNW	0	0	0	74	254	428	508	372	226	54	0	0	1916
UNIVERSITY	0	0	19	183	429	685	791	623	448	147	52	0	3377
VICKSBURG MILITARY PK	0	0	0	75	284	480	578	432	279	68	5	0	2201
WATER VALLEY 1 NNE	0	0	9	151	402	651	747	585	419	129	43	0	3136
WIGGINS 4 SE	0	0	0	56	251	427	504	387	240	52	6	0	1923
WOODVILLE 4 ESE	0	0	0	66	250	425	511	387	231	49	0	0	1919
YAZOO CITY 5 NNE	0	0	0	103	326	536	633	474	317	80	11	0	2480

MISSISSIPPI

COOLING DEGREE DAY NORMALS (BASE 65 DEG F)

STATION	JAN	FEB	MAR	APR	MAY	JUN	JUL	AUG	SEP	OCT	NOV	DEC	ANN
ABERDEEN	7	9	27	59	198	381	493	471	290	69	0	0	2004
BATESVILLE 2 SW	0	8	17	43	188	363	474	434	257	56	0	0	1840
BAY SAINT LOUIS	18	20	34	120	295	459	527	515	393	152	25	6	2564
BAY SPRINGS	14	18	30	82	233	408	496	477	334	101	14	0	2207
BELZONI	0	11	29	72	239	432	527	490	308	90	11	0	2209
BILOXI CITY	21	21	36	126	307	465	536	524	405	167	34	10	2652
BOONEVILLE	0	0	17	34	164	339	459	431	257	59	0	0	1760
BROOKHAVEN	14	21	37	100	240	408	490	474	333	114	14	7	2252
CANTON	11	14	29	69	224	405	508	481	323	96	12	0	2172
CHARLESTON	0	6	19	58	202	393	505	468	283	73	0	0	2007
CLARKSDALE	0	9	26	66	250	444	549	502	315	103	0	0	2264
CLEVELAND	0	6	27	62	222	420	527	487	303	94	0	0	2148
COLLINS	13	15	35	83	221	390	481	462	327	95	10	0	2132
COLUMBIA	19	24	45	110	267	438	521	499	357	117	19	6	2422
COLUMBUS	0	10	21	57	195	384	493	471	297	70	5	0	2003
CORINTH 5 WSW	0	6	23	39	186	366	481	443	263	65	0	0	1872
EUPORA 1 E	7	10	26	48	188	366	465	437	273	63	0	0	1883
FOREST 3 S	11	11	30	69	199	378	471	443	300	85	10	0	2007
GREENVILLE	0	8	34	73	243	438	533	496	320	96	5	0	2246
GREENWOOD FAA AP	8	8	26	77	244	429	527	490	319	101	8	0	2237
GULFPORT NAVAL CENTER	20	22	33	128	307	468	533	521	402	157	31	9	2631
HATTIESBURG	17	23	33	108	254	429	508	493	354	105	14	0	2338
HERNANDO	0	0	25	46	195	381	502	465	281	79	5	0	1979
HOLLY SPRINGS 4 N	0	7	15	32	159	324	443	406	233	50	0	0	1669
HOUSTON 2 NE	0	0	14	32	153	333	446	412	239	43	0	0	1672
JACKSON WSO	13	17	34	80	241	426	524	502	342	98	13	0	2290
KIPLING	8	10	23	53	183	357	459	440	288	71	6	0	1898
KOSCIUSKO	7	10	22	51	188	366	468	443	286	74	0	0	1915
LAUREL	12	15	28	97	246	423	515	493	342	103	11	0	2285
LIBERTY 1 W	16	18	29	92	231	399	484	465	333	109	14	6	2196
MERIDIAN WSO R	12	10	30	73	224	402	505	484	326	84	8	0	2158
MONTICELLO	18	20	37	82	229	405	493	474	330	96	11	6	2201
MOORHEAD	6	10	34	78	241	432	530	487	311	93	6	0	2228
NATCHEZ	20	19	53	110	267	441	530	508	360	130	24	12	2474
NEWTON EXP STATION	13	14	26	70	196	369	474	446	300	74	10	0	1992
PHILADELPHIA 1 WSW	7	8	23	56	189	366	468	443	289	70	7	0	1926
PICAYUNE	27	27	46	127	276	435	515	502	378	139	28	14	2514
PICKENS	11	15	31	76	210	375	474	443	291	88	7	0	2021
POPLARVILLE EXP STA	23	24	42	117	282	447	508	496	372	144	27	9	2491
PORT GIBSON	9	12	27	72	216	387	487	462	300	76	8	0	2056
ROSEDALE	0	7	22	48	207	402	512	468	281	63	0	0	2010
RUSSELL 2 WNW	12	14	34	76	219	396	484	465	322	89	10	5	2126
STATE UNIVERSITY	0	14	24	59	212	396	499	477	310	86	7	0	2084
STONEVILLE EXP STA	0	9	27	64	235	429	524	481	302	86	0	0	2157
TUPELO 2 WNW	0	0	19	53	201	381	493	468	281	65	0	0	1961
TYLERTOWN 2 WNW	18	19	40	99	241	414	493	474	342	115	17	6	2278
UNIVERSITY	0	10	21	48	170	337	456	422	259	65	0	0	1788
VICKSBURG MILITARY PK	11	12	37	92	244	420	518	499	339	115	14	5	2306
WATER VALLEY 1 NNE	0	8	19	45	173	348	471	446	267	58	0	0	1835
WIGGINS 4 SE	14	28	45	103	263	432	512	496	363	118	26	9	2409
WOODVILLE 4 ESE	15	18	39	97	248	405	487	474	342	122	19	6	2272
YAZOO CITY 5 NNE	7	9	35	86	250	435	530	505	337	106	11	0	2311

22 – MISSISSIPPI

STATE-STATION NUMBER	STN TYP	NAME	LATITUDE DEG-MIN	LONGITUDE DEG-MIN	ELEVATION (FT)
22-0008	12	ABBEVILLE	N 3430	W 08930	390
22-0021	13	ABERDEEN	N 3350	W 08833	207
22-0039	12	ACKERMAN	N 3318	W 08910	528
22-0237	12	ARKABUTLA DAM	N 3445	W 09008	250
22-0290	12	ASHLAND	N 3450	W 08910	610
22-0378	12	BALDWYN	N 3431	W 08838	395
22-0488	13	BATESVILLE 2 SW	N 3418	W 08959	215
22-0519	13	BAY SAINT LOUIS	N 3018	W 08920	20
22-0523	13	BAY SPRINGS	N 3158	W 08916	445
22-0660	13	BELZONI	N 3312	W 09029	110
22-0792	13	BILOXI CITY	N 3024	W 08854	15
22-0841	12	BLACK HAWK	N 3320	W 09001	340
22-0891	12	BLUFF LAKE	N 3317	W 08848	230
22-0955	13	BOONEVILLE	N 3439	W 08835	510
22-1094	13	BROOKHAVEN	N 3135	W 09028	495
22-1111	12	BROOKSVILLE EXP STA	N 3315	W 08834	292
22-1152	12	BRUCE	N 3400	W 08921	240
22-1174	12	BUCKATUNNA	N 3132	W 08832	148
22-1314	12	CALHOUN CITY 2 N	N 3353	W 08919	275
22-1389	13	CANTON	N 3236	W 09002	228
22-1460	12	CARROLLTON 1 SSW	N 3330	W 08956	336
22-1606	13	CHARLESTON	N 3401	W 09003	214
22-1707	13	CLARKSDALE	N 3412	W 09034	178
22-1738	13	CLEVELAND	N 3344	W 09044	140
22-1804	12	COFFEEVILLE	N 3359	W 08940	241
22-1852	13	COLLINS	N 3138	W 08934	290
22-1865	13	COLUMBIA	N 3117	W 08950	160
22-1870	13	COLUMBUS	N 3329	W 08825	184
22-1962	13	CORINTH 5 WSW	N 3455	W 08836	420
22-2034	12	CRANDALL 12 N	N 3205	W 08829	540
22-2046	12	CRAWFORD 5 WSW	N 3317	W 08842	273
22-2160	12	DANCY	N 3340	W 08903	290
22-2385	12	DLO	N 3159	W 08954	292
22-2658	12	EDINBURG	N 3248	W 08920	377
22-2722	12	ELLIOTT	N 3341	W 08945	230
22-2773	12	ENID DAM	N 3409	W 08955	300
22-2795	12	ENTERPRISE	N 3211	W 08849	248
22-2896	13	EUPORA 1 E	N 3332	W 08915	430
22-3107	13	FOREST 3 S	N 3219	W 08929	480
22-3208	12	FULTON 3 W	N 3416	W 08827	350
22-3331	12	GERMANIA	N 3238	W 09036	120
22-3340	12	GHOLSON 8 W	N 3255	W 08852	500
22-3605	13	GREENVILLE	N 3323	W 09101	132
22-3611	12	GREENVILLE 8 SW	N 3318	W 09108	130
22-3614	12	GREENWOOD	N 3332	W 09012	133
22-3627	13	GREENWOOD FAA AP	N 3330	W 09012	128
22-3645	12	GRENADA	N 3347	W 08948	194
22-3671	13	GULFPORT NAVAL CENTER	N 3023	W 08908	35
22-3887	13	HATTIESBURG	N 3118	W 08917	220
22-3975	13	HERNANDO	N 3450	W 09000	386

STATE-STATION NUMBER	STN TYP	NAME	LATITUDE DEG-MIN	LONGITUDE DEG-MIN	ELEVATION (FT)
22-3997	12	HICKORY	N 3219	W 08902	325
22-4001	12	HICKORY FLAT 1 W	N 3437	W 08912	411
22-4173	13	HOLLY SPRINGS 4 N	N 3449	W 08926	480
22-4265	13	HOUSTON 2 NE	N 3355	W 08858	300
22-4472	13	JACKSON WSO	N 3219	W 09005	31
22-4702	13	KIPLING	N 3241	W 08838	316
22-4776	13	KOSCIUSKO	N 3304	W 08936	455
22-4816	12	LAFAYETTE SPRINGS	N 3419	W 08916	450
22-4842	12	LAKE CORMORANT 1 W	N 3454	W 09014	207
22-4869	12	LAMBERT 5 E	N 3411	W 09012	160
22-4939	13	LAUREL	N 3141	W 08907	225
22-4966	12	LEAKESVILLE	N 3110	W 08833	105
22-5062	12	LEXINGTON 2 NNW	N 3308	W 09004	315
22-5070	13	LIBERTY 1 W	N 3110	W 09049	396
22-5247	12	LOUISVILLE	N 3308	W 08904	581
22-5776	13	MERIDIAN WSO R	N 3220	W 08845	290
22-5789	12	MERRILL	N 3059	W 08843	52
22-5943	12	MIZE	N 3151	W 08933	395
22-5987	13	MONTICELLO	N 3133	W 09006	220
22-6009	13	MOORHEAD	N 3327	W 09031	117
22-6084	12	MOUNT PLEASANT	N 3457	W 08931	446
22-6177	13	NATCHEZ	N 3133	W 09123	195
22-6256	12	NEW ALBANY	N 3429	W 08901	340
22-6308	13	NEWTON EXP STATION	N 3220	W 08905	346
22-6476	12	OAKLEY EXP STATION	N 3212	W 09031	205
22-6493	12	OFAHOMA	N 3243	W 08942	348
22-6515	12	OKOLONA	N 3400	W 08845	320
22-6718	12	PASCAGOULA 2 ENE	N 3023	W 08830	16
22-6750	12	PAULDING	N 3202	W 08902	500
22-6811	12	PELAHATCHIE	N 3219	W 08947	355
22-6894	13	PHILADELPHIA 1 WSW	N 3246	W 08908	413
22-6921	13	PICAYUNE	N 3031	W 08941	50
22-6926	13	PICKENS	N 3253	W 08959	222
22-7066	12	PLEASANT HILL	N 3455	W 08953	380
22-7106	12	PONTOTOC	N 3415	W 08900	513
22-7128	13	POPLARVILLE EXP STA	N 3051	W 08933	313
22-7132	13	PORT GIBSON	N 3157	W 09059	180
22-7172	12	PRENTISS 2 NNE	N 3138	W 08951	400
22-7467	12	RIPLEY	N 3444	W 08857	520
22-7537	12	ROCKPORT	N 3148	W 09009	200
22-7560	12	ROLLING FORK	N 3254	W 09053	104
22-7582	13	ROSEDALE	N 3351	W 09101	150
22-7701	13	RUSSELL 2 WNW	N 3225	W 08837	460
22-7807	12	SARAH	N 3434	W 09013	310
22-7815	12	SARDIS DAM	N 3424	W 08948	234
22-7921	12	SENATOBIA	N 3437	W 08957	270
22-8053	12	SHUBUTA	N 3152	W 08842	197
22-8062	12	SHUQUALAK	N 3259	W 08834	217
22-8145	12	SLEDGE	N 3426	W 09013	165
22-8352	12	STANDARD	N 3032	W 08922	140

22 – MISSISSIPPI

STATE-STATION NUMBER	STN TYP	NAME	LATITUDE DEG-MIN	LONGITUDE DEG-MIN	ELEVATION (FT)
22-8374	13	STATE UNIVERSITY	N 3328	W 08848	280
22-8445	13	STONEVILLE EXP STA	N 3325	W 09055	127
22-8556	12	SUMRALL	N 3125	W 08932	290
22-8591	12	SWAN LAKE	N 3353	W 09017	147
22-9003	13	TUPELO 2 WNW	N 3416	W 08844	250
22-9048	13	TYLERTOWN 2 WNW	N 3107	W 09011	440
22-9070	12	UNION	N 3234	W 08907	465
22-9079	13	UNIVERSITY	N 3423	W 08932	380
22-9114	12	VAIDEN 1 SSW	N 3319	W 08945	389
22-9154	12	VANCE	N 3404	W 09021	145
22-9159	12	VAN VLEET	N 3358	W 08854	336
22-9216	13	VICKSBURG MILITARY PK	N 3221	W 09051	260
22-9326	12	WALNUT GROVE	N 3236	W 08928	417
22-9400	13	WATER VALLEY 1 NNE	N 3410	W 08938	380
22-9597	12	WHITE OAK	N 3204	W 08941	420
22-9639	13	WIGGINS 4 SE	N 3048	W 08906	200
22-9793	13	WOODVILLE 4 ESE	N 3106	W 09114	400
22-9860	13	YAZOO CITY 5 NNE	N 3254	W 09023	107

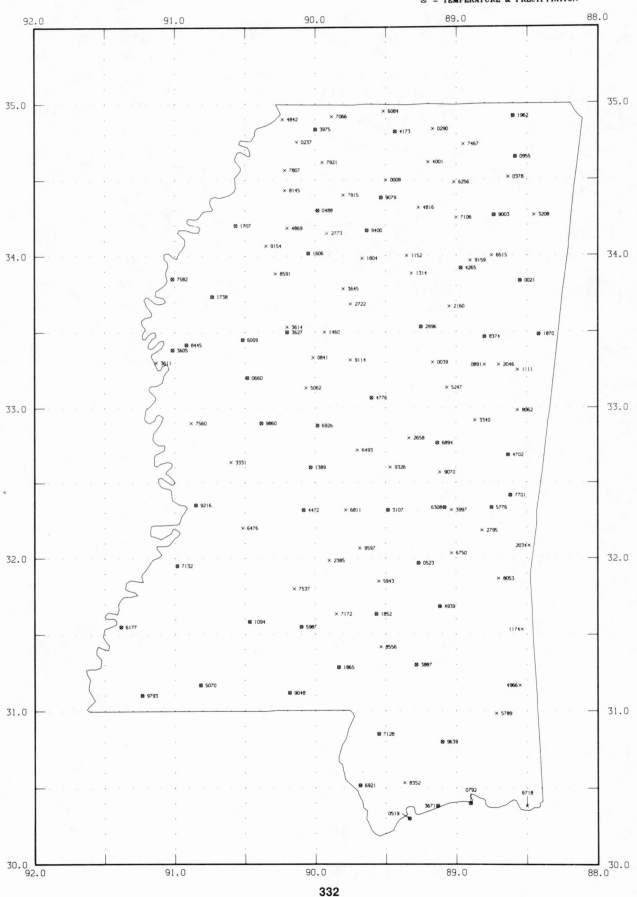

22 — MISSISSIPPI

× 4842　　× 7066　　× 6084　　⊗ 1962
⊗ 3975　　× 4173　× 0290
× 0237　　　　　　　× 7467
× 7921　　　⊗ 0955
× 7807　　　× 4001
× 8145　× 0008　× 6256　× 0378
× 7815　⊗ 9079
⊗ 0488　　× 4816　⊗ 9003　× 3208
⊗ 1707　× 4869　× 2773　⊗ 9400　× 7106
× 9154
⊗ 1606　× 1804　× 1152　× 6515
× 8591　　　× 1314　⊗ 9159
⊗ 7582　　　　　　⊗ 4265　⊗ 0021
⊗ 1738　× 3645
× 2722　× 2160
× 3614　　⊗ 2896
⊗ 3627　× 1460　　⊗ 8374　⊗ 1870
⊗ 8445　⊗ 6009
⊗ 3605　× 0841　× 9114　× 0039　0891×　× 2046
× 3611　　　　　　× 1111
⊗ 0660
× 5062　× 5247
⊗ 4776
⊗ 8062
× 7560　⊗ 9860　⊗ 6926　× 3340
× 2658　⊗ 6894
× 6493　⊗ 4702
× 3331　⊗ 1389　× 9326　× 9070
⊗ 7701
⊗ 9216　⊗ 4472　× 6811　⊗ 3107　6308⊗ × 3997　⊗ 5776
× 6476　× 2795
× 9597　2034×
⊗ 7132　× 2385　× 6750
⊗ 0523　× 8053
× 5943
× 7537
× 7172　⊗ 1852　⊗ 4939
⊗ 1094　　　1174×
⊗ 6177　⊗ 5987
× 8556
⊗ 1865　⊗ 3887
⊗ 5070　　　1966×
⊗ 9793　⊗ 9048　× 5789
⊗ 7128
⊗ 9639
× 8352　　6718
⊗ 6921　　0792　×
0519　3671⊗

MISSOURI

TEMPERATURE NORMALS (DEG F)

STATION		JAN	FEB	MAR	APR	MAY	JUN	JUL	AUG	SEP	OCT	NOV	DEC	ANN
ANDERSON	MAX	45.4	50.5	59.1	71.0	77.3	84.6	90.0	89.0	81.6	72.3	58.6	49.3	69.1
	MIN	21.0	25.4	33.2	44.5	52.8	61.2	65.4	63.3	56.3	44.2	33.1	25.2	43.8
	MEAN	33.2	38.0	46.2	57.8	65.1	73.0	77.7	76.2	69.0	58.3	45.8	37.3	56.5
APPLETON CITY	MAX	40.8	46.9	56.8	69.8	78.4	86.6	91.9	91.3	83.5	72.5	56.9	45.5	68.4
	MIN	19.5	24.7	32.8	44.6	54.0	62.6	66.6	64.7	57.3	45.9	34.0	25.0	44.3
	MEAN	30.2	35.8	44.8	57.2	66.2	74.6	79.2	78.1	70.4	59.2	45.5	35.3	56.4
ARCADIA	MAX	42.8	48.6	58.0	71.1	78.6	85.5	90.0	88.5	81.9	71.9	57.4	47.0	68.4
	MIN	21.5	25.5	33.3	44.2	52.0	60.3	64.5	62.7	55.1	43.4	34.0	26.5	43.6
	MEAN	32.2	37.1	45.6	57.6	65.3	72.9	77.3	75.6	68.6	57.7	45.7	36.8	56.0
BETHANY	MAX	33.5	40.2	51.0	66.0	76.3	84.8	89.7	87.6	79.7	68.9	52.3	39.2	64.1
	MIN	13.0	18.9	28.1	40.9	51.6	60.7	64.9	62.5	53.6	42.6	30.5	20.4	40.6
	MEAN	23.3	29.6	39.5	53.5	64.0	72.8	77.3	75.1	66.7	55.8	41.5	29.8	52.4
BOLIVAR 1NE	MAX	42.1	47.4	56.6	69.3	77.0	84.7	90.0	89.6	82.2	71.6	56.7	46.6	67.8
	MIN	20.6	25.3	33.3	44.9	53.6	62.3	66.6	64.8	57.7	45.6	34.4	25.9	44.6
	MEAN	31.4	36.3	45.0	57.1	65.3	73.5	78.3	77.2	70.0	58.6	45.6	36.3	56.2
BOONVILLE WATERWORKS	MAX	35.8	41.0	51.1	65.1	74.8	83.4	88.9	87.4	80.2	68.8	53.7	41.1	64.3
	MIN	17.0	21.7	30.6	43.8	53.9	63.0	67.5	65.2	57.1	45.6	33.7	23.7	43.6
	MEAN	26.4	31.4	40.9	54.5	64.3	73.3	78.3	76.3	68.7	57.3	43.7	32.5	54.0
BROOKFIELD	MAX	35.6	42.4	53.1	67.5	77.2	85.7	90.5	88.8	81.4	70.6	54.3	41.2	65.7
	MIN	16.1	21.6	30.5	43.1	53.2	62.2	66.7	64.3	56.2	45.2	32.9	22.8	42.9
	MEAN	25.9	32.0	41.8	55.3	65.2	74.0	78.6	76.6	68.8	57.9	43.6	32.0	54.3
BRUNSWICK	MAX	36.3	42.4	53.1	67.4	76.7	84.9	89.8	88.0	81.0	69.9	54.6	41.7	65.5
	MIN	16.7	22.0	30.7	43.6	53.8	63.1	67.1	64.8	56.8	45.7	33.4	23.2	43.4
	MEAN	26.5	32.2	41.9	55.5	65.3	74.0	78.5	76.4	68.9	57.8	44.0	32.5	54.5
CAMDENTON	MAX	43.2	48.8	58.0	71.8	78.9	86.0	91.2	90.4	83.2	72.7	57.8	47.5	69.1
	MIN	20.8	25.6	33.3	45.6	54.3	62.8	67.1	65.6	58.2	46.9	35.3	26.4	45.2
	MEAN	32.0	37.2	45.7	58.7	66.6	74.4	79.2	78.0	70.7	59.8	46.6	36.9	57.2
CANTON LOCK AND DAM 20//	MAX	34.2	39.6	49.7	64.4	74.7	83.7	88.3	85.8	78.9	68.2	52.8	40.0	63.4
	MIN	15.1	20.1	29.4	42.7	53.4	63.0	67.1	64.4	56.2	44.9	32.6	22.3	42.6
	MEAN	24.7	29.9	39.6	53.5	64.0	73.4	77.7	75.1	67.6	56.6	42.7	31.2	53.0
CARROLLTON	MAX	35.7	41.8	53.0	67.3	76.9	85.4	90.3	88.1	80.3	68.6	52.9	41.0	65.1
	MIN	16.6	22.1	30.8	43.1	53.0	62.0	66.1	63.7	55.7	44.7	32.7	23.1	42.8
	MEAN	26.1	32.0	41.9	55.2	65.0	73.7	78.2	75.9	68.0	56.7	42.8	32.1	54.0
CARTHAGE	MAX	44.1	49.3	58.3	70.8	77.9	85.6	91.1	90.6	83.0	72.6	58.5	48.5	69.2
	MIN	22.3	26.8	34.4	45.9	54.7	63.5	68.1	66.0	58.3	46.8	35.4	27.3	45.8
	MEAN	33.2	38.1	46.4	58.4	66.3	74.6	79.6	78.3	70.7	59.7	47.0	37.9	57.5
CARUTHERSVILLE	MAX	44.9	49.6	58.9	71.1	79.8	87.5	90.6	89.2	83.2	73.5	59.8	49.1	69.8
	MIN	27.1	30.4	38.4	49.3	58.5	66.7	70.4	68.6	62.0	49.3	39.1	31.4	49.3
	MEAN	36.0	40.0	48.7	60.2	69.2	77.1	80.5	78.9	72.6	61.4	49.5	40.3	59.5
CHARLESTON	MAX	43.6	48.6	58.2	71.0	80.0	88.1	90.8	89.3	83.0	72.9	58.8	47.5	69.3
	MIN	25.0	28.5	36.9	47.9	56.8	64.9	68.5	66.2	59.0	46.4	37.1	29.4	47.2
	MEAN	34.3	38.6	47.6	59.5	68.4	76.5	79.7	77.8	71.0	59.7	48.0	38.5	58.3
CHILLICOTHE RADIO KCHI	MAX	35.1	41.7	52.6	67.3	77.0	85.6	90.4	88.5	80.9	70.0	53.6	40.8	65.3
	MIN	15.7	21.1	30.0	42.6	52.7	61.6	66.2	63.7	55.3	44.1	32.1	22.0	42.3
	MEAN	25.4	31.4	41.3	55.0	64.9	73.6	78.3	76.1	68.1	57.1	42.9	31.4	53.8
CLEARWATER DAM	MAX	43.7	49.0	58.7	72.1	79.9	87.0	91.6	89.9	82.6	72.5	58.3	47.4	69.4
	MIN	21.2	24.7	33.1	44.4	52.5	60.8	65.2	63.4	56.3	43.7	33.6	25.5	43.7
	MEAN	32.5	36.9	45.9	58.3	66.2	73.9	78.4	76.6	69.5	58.1	46.0	36.5	56.6
CLINTON 3 NW	MAX	38.4	44.0	53.2	67.1	76.3	84.5	90.2	89.3	81.8	71.0	55.3	43.7	66.2
	MIN	16.4	21.3	30.2	42.4	52.6	61.8	66.0	63.5	55.4	43.4	31.4	22.2	42.2
	MEAN	27.4	32.7	41.7	54.8	64.5	73.2	78.1	76.4	68.6	57.2	43.4	33.0	54.3

MISSOURI

TEMPERATURE NORMALS (DEG F)

STATION		JAN	FEB	MAR	APR	MAY	JUN	JUL	AUG	SEP	OCT	NOV	DEC	ANN
COLUMBIA REGION WSO	MAX	36.3	41.6	51.6	65.4	74.5	83.3	88.6	87.2	79.7	68.4	53.1	41.3	64.3
	MIN	18.6	23.4	31.8	44.2	53.7	62.5	66.9	64.8	57.0	45.7	33.9	24.5	43.9
	MEAN	27.5	32.5	41.7	54.8	64.1	72.9	77.8	76.0	68.4	57.1	43.5	32.9	54.1
CONCEPTION	MAX	32.3	38.9	49.6	64.5	74.7	83.8	88.9	87.1	79.2	68.1	51.7	38.7	63.1
	MIN	12.5	18.2	27.4	40.8	51.6	61.0	65.6	63.2	54.4	43.4	30.2	19.4	40.6
	MEAN	22.4	28.5	38.6	52.7	63.2	72.4	77.3	75.2	66.8	55.7	41.0	29.0	51.9
DONIPHAN	MAX	45.6	50.5	59.6	72.1	80.3	87.8	92.2	90.9	83.8	74.0	60.0	48.9	70.5
	MIN	22.2	25.8	34.0	44.8	52.9	61.2	65.8	63.6	56.3	43.0	33.2	25.9	44.1
	MEAN	33.9	38.2	46.8	58.5	66.6	74.6	79.1	77.3	70.1	58.5	46.6	37.4	57.3
EDGERTON	MAX	37.1	43.8	54.1	68.0	77.2	85.3	90.2	88.5	81.1	70.9	55.1	42.9	66.2
	MIN	13.9	19.6	28.6	41.2	51.6	60.9	65.0	62.7	53.9	42.2	30.0	20.4	40.8
	MEAN	25.5	31.7	41.4	54.6	64.5	73.1	77.6	75.6	67.5	56.6	42.6	31.7	53.5
ELDON	MAX	41.1	46.7	56.5	69.7	77.8	85.3	90.7	89.8	82.2	71.2	56.3	45.5	67.7
	MIN	20.8	25.2	33.2	45.4	54.6	63.0	67.8	65.6	57.9	47.0	35.2	25.8	45.1
	MEAN	30.9	36.0	44.9	57.6	66.2	74.2	79.3	77.7	70.1	59.2	45.8	35.7	56.5
ELSBERRY 1 S	MAX	36.4	41.7	52.3	66.3	76.3	85.3	89.4	87.6	80.9	69.7	54.1	41.7	65.1
	MIN	16.0	21.0	30.4	43.0	51.8	61.1	65.2	63.2	55.0	43.0	32.0	22.5	42.0
	MEAN	26.2	31.4	41.4	54.7	64.1	73.2	77.3	75.4	68.0	56.4	43.1	32.1	53.6
FARMINGTON	MAX	41.1	46.3	55.4	68.9	77.0	85.1	89.4	88.2	81.4	70.7	56.2	45.0	67.1
	MIN	19.9	24.0	32.3	43.8	51.6	60.2	64.0	61.9	54.3	42.3	33.1	25.1	42.7
	MEAN	30.5	35.2	43.9	56.4	64.3	72.7	76.8	75.1	67.8	56.5	44.7	35.1	54.9
FAYETTE	MAX	37.2	43.0	53.6	67.3	75.6	83.9	89.0	87.7	80.8	69.9	54.5	42.5	65.4
	MIN	17.4	22.5	31.1	43.6	53.1	61.9	66.2	64.1	56.2	45.2	33.2	23.7	43.2
	MEAN	27.4	32.8	42.4	55.5	64.4	72.9	77.7	75.9	68.5	57.6	43.9	33.1	54.3
FOUNTAIN GROVE WL	MAX	35.7	42.1	53.1	66.8	76.3	84.8	89.8	87.6	80.4	69.9	53.8	41.3	65.1
	MIN	14.7	20.8	30.1	42.7	52.8	61.9	65.7	63.0	54.6	43.7	31.7	21.6	41.9
	MEAN	25.2	31.5	41.6	54.8	64.6	73.4	77.8	75.3	67.5	56.8	42.7	31.5	53.6
FULTON 4 SW	MAX	36.4	42.0	51.8	65.7	75.2	83.9	89.4	88.2	81.1	69.3	54.2	41.8	64.9
	MIN	16.3	20.7	29.4	42.5	52.0	61.2	65.7	63.7	55.6	43.8	32.0	22.5	42.1
	MEAN	26.4	31.4	40.6	54.1	63.6	72.6	77.6	75.9	68.4	56.6	43.1	32.2	53.5
GRANT CITY	MAX	31.7	38.7	49.2	64.6	74.9	83.8	88.9	86.6	78.0	67.4	50.8	37.7	62.7
	MIN	13.0	18.7	27.9	41.2	52.0	60.8	65.6	63.2	54.8	44.1	30.8	20.0	41.0
	MEAN	22.3	28.7	38.6	52.9	63.5	72.3	77.2	74.9	66.4	55.8	40.8	28.9	51.9
GREENVILLE 4 NNW	MAX	44.5	50.2	58.9	71.8	79.7	87.1	91.3	89.9	83.2	73.1	59.2	47.9	69.7
	MIN	20.5	24.1	32.1	43.0	51.0	59.5	63.6	61.4	54.0	40.7	31.9	24.5	42.2
	MEAN	32.5	37.1	45.5	57.4	65.4	73.3	77.5	75.7	68.6	56.9	45.5	36.2	56.0
HAMILTON 2 W	MAX	33.6	39.9	50.6	64.9	74.9	83.8	88.9	87.0	79.5	68.6	52.5	39.6	63.7
	MIN	13.5	19.2	28.7	41.8	51.9	60.9	65.3	63.1	54.7	42.9	30.8	20.4	41.1
	MEAN	23.6	29.6	39.7	53.4	63.4	72.4	77.1	75.1	67.1	55.8	41.7	30.0	52.4
HANNIBAL WATERWORKS	MAX	33.4	39.1	49.7	64.3	74.2	82.9	87.2	85.1	78.3	67.3	51.8	38.8	62.7
	MIN	15.6	20.8	29.8	43.1	53.0	61.9	66.1	64.0	56.0	44.9	32.8	22.1	42.5
	MEAN	24.5	30.0	39.8	53.7	63.6	72.4	76.7	74.6	67.2	56.1	42.3	30.5	52.6
HARRISONVILLE	MAX	38.4	44.4	54.9	68.2	77.1	85.2	90.8	89.5	82.1	71.2	55.7	43.6	66.8
	MIN	18.1	23.4	31.7	44.5	54.1	62.9	67.5	65.4	57.8	46.4	33.7	24.2	44.1
	MEAN	28.3	33.9	43.3	56.4	65.6	74.1	79.2	77.5	70.0	58.8	44.7	33.9	55.5
JACKSON	MAX	42.3	47.9	57.9	70.8	79.8	88.0	91.2	89.9	83.5	72.8	57.7	46.4	69.0
	MIN	23.3	26.9	35.4	46.3	54.6	63.4	67.4	65.4	58.2	45.7	35.7	28.1	45.9
	MEAN	32.8	37.4	46.6	58.5	67.2	75.7	79.3	77.7	70.9	59.3	46.7	37.2	57.4
JEFFERSON CITY L U	MAX	40.9	46.4	56.4	69.9	77.9	85.8	90.9	89.8	82.7	71.6	56.7	45.4	67.9
	MIN	18.0	22.4	30.7	42.7	51.7	60.2	65.0	63.1	55.1	43.3	32.4	23.3	42.3
	MEAN	29.5	34.4	43.6	56.3	64.8	73.0	78.0	76.5	68.9	57.5	44.5	34.4	55.1

MISSOURI

TEMPERATURE NORMALS (DEG F)

STATION		JAN	FEB	MAR	APR	MAY	JUN	JUL	AUG	SEP	OCT	NOV	DEC	ANN
JOPLIN FAA AIRPORT	MAX	41.9	47.3	56.4	69.0	76.7	85.0	90.3	89.1	81.5	71.2	56.4	46.6	67.6
	MIN	23.3	28.0	35.8	47.4	56.4	65.4	69.9	68.0	60.3	48.9	36.8	28.3	47.4
	MEAN	32.6	37.7	46.1	58.2	66.6	75.2	80.1	78.5	70.9	60.1	46.7	37.5	57.5
KANSAS CITY WSO	MAX	34.5	41.1	51.3	65.1	74.6	83.3	88.5	86.8	78.6	67.9	52.1	40.1	63.7
	MIN	17.2	23.0	31.7	44.4	54.6	63.8	68.5	66.5	58.1	47.0	34.0	23.7	44.4
	MEAN	25.9	32.1	41.5	54.8	64.6	73.6	78.5	76.7	68.4	57.5	43.1	31.9	54.1
KANSAS CITY FAA AP	MAX	37.4	43.5	53.6	67.4	76.8	85.3	90.5	89.1	81.4	70.6	54.8	42.8	66.1
	MIN	19.4	25.2	33.6	46.4	57.0	66.4	71.3	69.4	60.5	48.6	34.4	25.6	46.5
	MEAN	28.4	34.4	43.6	56.9	66.9	75.9	80.9	79.3	71.0	59.6	44.6	34.2	56.3
KIRKSVILLE RADIO KIRX	MAX	32.3	38.5	49.5	64.1	74.2	82.7	87.7	85.5	77.8	67.2	51.3	38.2	62.4
	MIN	13.8	19.3	28.2	41.2	51.2	60.3	64.6	62.5	54.1	43.7	31.4	20.6	40.9
	MEAN	23.1	28.9	38.9	52.7	62.7	71.5	76.2	74.0	66.0	55.5	41.3	29.4	51.7
LAKESIDE	MAX	40.7	45.8	54.7	68.6	77.4	85.6	90.9	90.0	82.7	72.1	57.4	46.0	67.7
	MIN	17.9	21.9	30.3	42.1	51.3	60.4	64.7	62.6	55.1	43.5	33.2	24.2	42.3
	MEAN	29.3	33.9	42.5	55.4	64.3	73.0	77.8	76.3	68.9	57.8	45.3	35.1	55.0
LAMAR	MAX	40.3	45.8	54.8	68.0	75.9	83.8	89.6	88.7	80.8	70.6	56.1	45.4	66.7
	MIN	19.7	24.4	32.6	45.0	54.7	63.6	68.0	65.7	57.9	46.0	33.6	25.0	44.7
	MEAN	30.0	35.1	43.7	56.5	65.3	73.7	78.8	77.2	69.4	58.3	44.9	35.3	55.7
LEBANON	MAX	42.0	47.2	56.4	69.0	76.7	84.5	89.3	88.5	81.3	70.9	56.2	46.2	67.4
	MIN	21.4	25.7	33.4	45.0	54.0	62.5	66.9	65.4	58.1	46.8	35.3	26.7	45.1
	MEAN	31.7	36.5	44.9	57.1	65.4	73.5	78.1	76.9	69.7	58.9	45.8	36.5	56.3
LEXINGTON	MAX	34.5	40.7	51.2	66.0	76.2	84.9	90.3	88.5	80.5	68.9	52.9	40.3	64.6
	MIN	16.5	21.8	30.6	44.5	55.0	64.2	68.6	66.1	57.8	46.4	33.7	23.4	44.1
	MEAN	25.5	31.3	40.9	55.2	65.6	74.6	79.5	77.3	69.2	57.7	43.4	31.8	54.3
LICKING 4 N	MAX	40.2	45.1	54.1	67.8	76.1	83.8	88.9	87.8	80.5	69.9	55.4	44.6	66.2
	MIN	17.8	22.4	31.3	43.0	52.0	60.7	65.0	63.1	55.6	43.1	32.2	23.5	42.5
	MEAN	29.0	33.8	42.7	55.4	64.0	72.3	77.0	75.5	68.1	56.5	43.8	34.1	54.4
LOCKWOOD	MAX	43.1	48.6	57.7	70.1	78.0	86.1	91.3	90.9	83.0	72.5	57.4	47.5	68.9
	MIN	22.2	27.0	34.5	45.9	55.0	63.5	67.8	66.0	58.7	47.7	35.8	27.4	46.0
	MEAN	32.7	37.8	46.1	58.0	66.5	74.8	79.6	78.5	70.9	60.1	46.6	37.5	57.4
LOUISIANA STARKS NUR	MAX	37.3	43.3	53.7	67.6	77.0	85.7	90.1	88.5	81.6	70.5	54.9	42.6	66.1
	MIN	16.7	21.7	30.3	42.4	51.2	60.2	64.3	62.0	54.1	43.2	32.4	23.2	41.8
	MEAN	27.0	32.5	42.0	55.0	64.1	73.0	77.2	75.3	67.9	56.9	43.6	32.9	54.0
MACON	MAX	33.3	39.4	49.9	64.4	74.8	83.3	88.7	86.8	79.3	68.1	52.1	39.1	63.3
	MIN	13.6	18.7	27.9	41.3	51.9	61.0	65.3	62.8	54.4	43.1	30.9	20.4	40.9
	MEAN	23.5	29.1	38.9	52.9	63.4	72.2	77.0	74.8	66.9	55.6	41.5	29.8	52.1
MARBLE HILL	MAX	43.4	48.9	58.4	70.9	79.2	87.2	90.9	89.5	82.8	72.8	58.4	47.1	69.1
	MIN	21.0	24.8	33.3	44.4	52.4	60.8	65.0	62.8	55.4	42.7	33.1	25.4	43.4
	MEAN	32.2	36.9	45.9	57.7	65.8	74.0	78.0	76.2	69.1	57.8	45.8	36.3	56.3
MARSHFIELD	MAX	41.7	46.7	55.5	68.2	76.4	84.3	89.3	88.3	80.9	70.1	55.4	45.6	66.9
	MIN	21.3	25.6	33.4	45.1	54.3	62.9	67.3	65.8	58.7	47.2	35.0	26.5	45.3
	MEAN	31.5	36.2	44.5	56.6	65.4	73.6	78.3	77.0	69.8	58.6	45.2	36.1	56.1
MARYVILLE 2 E	MAX	31.2	37.5	47.8	63.3	74.1	82.8	87.7	85.6	77.8	67.0	50.6	37.5	61.9
	MIN	10.6	15.8	25.1	38.5	50.1	59.5	64.0	61.2	52.1	40.4	27.9	17.3	38.5
	MEAN	20.9	26.7	36.5	50.9	62.1	71.2	75.9	73.5	65.0	53.7	39.3	27.5	50.3
MEXICO	MAX	35.2	40.5	50.7	64.9	75.1	84.2	89.4	87.8	80.6	68.7	53.3	40.5	64.2
	MIN	15.1	19.6	28.6	41.9	51.8	61.1	65.3	63.0	54.6	43.2	31.4	21.4	41.4
	MEAN	25.1	30.1	39.7	53.4	63.5	72.7	77.4	75.4	67.6	56.0	42.4	31.0	52.9
MOBERLY RADIO KWIX	MAX	35.8	41.6	52.2	66.4	75.8	84.2	89.2	87.5	80.3	69.2	53.6	41.3	64.8
	MIN	16.7	21.9	30.3	43.1	53.0	61.7	65.9	63.5	55.8	45.1	32.7	23.1	42.7
	MEAN	26.3	31.8	41.3	54.8	64.4	73.0	77.6	75.6	68.1	57.2	43.2	32.2	53.8

MISSOURI

TEMPERATURE NORMALS (DEG F)

STATION		JAN	FEB	MAR	APR	MAY	JUN	JUL	AUG	SEP	OCT	NOV	DEC	ANN
MOUNTAIN GROVE 2 N	MAX	42.8	48.0	56.7	69.2	76.6	84.0	89.1	88.2	81.2	71.1	56.4	46.5	67.5
	MIN	22.2	26.3	33.9	45.4	53.9	61.9	66.3	64.6	57.9	46.8	35.5	27.1	45.2
	MEAN	32.5	37.2	45.4	57.4	65.3	73.0	77.7	76.4	69.6	59.0	46.0	36.8	56.4
MOUNT VERNON MU FARM	MAX	42.2	47.2	56.0	68.2	76.0	84.1	89.6	89.2	81.6	70.8	56.4	46.6	67.3
	MIN	21.1	25.8	33.5	44.8	53.4	62.2	66.6	64.9	57.4	46.0	34.5	26.3	44.7
	MEAN	31.7	36.5	44.8	56.5	64.8	73.2	78.2	77.1	69.5	58.4	45.5	36.5	56.1
NEOSHO	MAX	46.0	51.6	60.3	72.4	78.9	86.2	91.3	90.7	83.6	73.8	59.1	49.9	70.3
	MIN	22.7	27.2	34.7	45.8	54.3	62.7	66.8	64.7	57.6	46.3	35.2	27.5	45.5
	MEAN	34.4	39.4	47.5	59.1	66.7	74.5	79.1	77.7	70.6	60.1	47.2	38.7	57.9
NEVADA SEWAGE PLANT	MAX	42.5	48.6	58.3	71.2	78.8	86.8	92.4	91.7	83.9	73.8	58.3	47.4	69.5
	MIN	19.6	25.0	33.3	45.0	54.3	63.0	67.0	64.7	58.0	46.1	34.0	25.1	44.6
	MEAN	31.1	36.8	45.8	58.1	66.6	75.0	79.7	78.2	70.9	60.0	46.2	36.3	57.1
NEW FLORENCE	MAX	37.4	42.9	53.3	67.3	76.4	84.9	89.5	88.1	81.3	69.7	53.9	42.4	65.6
	MIN	18.6	23.3	31.9	44.2	53.6	62.3	66.5	64.6	56.9	46.1	34.2	24.6	43.9
	MEAN	28.0	33.1	42.6	55.8	65.0	73.6	78.0	76.3	69.1	57.9	44.1	33.5	54.8
OREGON	MAX	33.1	39.8	50.3	65.1	75.0	83.6	88.3	86.9	78.7	68.3	52.0	38.9	63.3
	MIN	13.6	19.4	28.2	41.5	52.5	61.7	66.1	63.9	55.1	44.2	31.2	20.5	41.5
	MEAN	23.4	29.7	39.2	53.4	63.8	72.6	77.2	75.4	66.9	56.3	41.6	29.7	52.4
OZARK BEACH	MAX	45.5	50.4	58.7	71.1	79.0	86.9	92.2	91.1	83.8	74.0	60.0	49.7	70.2
	MIN	19.8	23.4	31.6	42.6	51.5	60.3	64.5	62.2	55.2	42.7	32.3	24.7	42.6
	MEAN	32.7	36.9	45.2	56.9	65.3	73.6	78.4	76.7	69.5	58.4	46.2	37.2	56.4
POPLAR BLUFF R S	MAX	44.1	49.0	58.5	70.9	79.4	87.3	91.2	89.9	83.2	73.3	59.1	48.2	69.5
	MIN	24.1	27.9	36.4	47.4	55.6	63.9	68.1	65.8	58.4	45.5	35.7	28.4	46.4
	MEAN	34.1	38.5	47.4	59.2	67.5	75.6	79.7	77.9	70.8	59.4	47.4	38.3	58.0
SAINT CHARLES	MAX	38.5	43.8	54.1	68.0	77.3	85.9	89.8	88.2	81.7	70.3	55.4	43.6	66.4
	MIN	19.1	23.5	32.4	44.5	53.5	62.9	66.8	64.6	56.5	44.9	34.1	24.7	44.0
	MEAN	28.8	33.7	43.3	56.3	65.4	74.4	78.4	76.4	69.1	57.6	44.8	34.2	55.2
ST. JOSEPH WBAP	MAX	34.7	41.4	51.8	66.3	76.3	85.3	89.5	87.3	79.5	69.6	53.0	40.3	64.6
	MIN	15.0	20.8	30.0	42.9	53.6	63.2	67.4	64.8	55.8	44.0	31.6	21.6	42.6
	MEAN	24.9	31.1	41.0	54.6	65.0	74.3	78.5	76.1	67.7	56.8	42.3	31.0	53.6
SAINT LOUIS WSO R	MAX	37.6	43.1	53.4	67.1	76.4	85.2	89.0	87.4	80.7	69.1	54.0	42.6	65.5
	MIN	19.9	24.5	33.0	45.1	54.7	64.3	68.8	66.6	58.6	46.7	35.1	25.7	45.3
	MEAN	28.8	33.8	43.2	56.1	65.6	74.8	78.9	77.0	69.7	57.9	44.6	34.2	55.4
SALEM	MAX	40.9	46.1	55.0	68.0	76.5	84.1	89.0	87.4	80.7	70.5	56.1	45.3	66.6
	MIN	19.7	24.0	32.3	44.1	52.7	61.2	65.5	63.4	56.0	44.7	33.7	25.3	43.6
	MEAN	30.3	35.1	43.7	56.1	64.6	72.7	77.2	75.4	68.4	57.6	44.9	35.3	55.1
SALISBURY	MAX	36.2	42.3	53.1	66.8	76.2	84.7	89.6	88.0	80.8	69.6	53.9	41.5	65.2
	MIN	17.1	22.5	31.2	43.6	53.3	62.2	66.4	64.1	56.2	45.3	33.4	23.6	43.2
	MEAN	26.7	32.4	42.2	55.2	64.8	73.5	78.1	76.1	68.5	57.5	43.7	32.6	54.3
SAVERTON L AND D 22 //	MAX	36.0	41.1	51.4	65.5	75.6	84.9	89.2	86.8	79.7	68.9	53.9	41.7	64.6
	MIN	17.6	22.5	31.4	44.5	54.7	63.6	67.9	65.7	57.9	47.0	34.7	24.2	44.3
	MEAN	26.8	31.8	41.5	55.0	65.2	74.3	78.5	76.3	68.8	58.0	44.3	32.9	54.5
SEDALIA	MAX	38.8	44.1	54.5	68.1	76.8	85.1	90.4	89.1	81.8	70.5	55.2	43.7	66.5
	MIN	18.5	23.4	32.2	44.3	53.9	62.8	67.0	64.8	57.1	45.8	33.7	24.9	44.0
	MEAN	28.7	33.8	43.4	56.2	65.4	74.0	78.7	77.0	69.5	58.2	44.5	34.3	55.3
SELIGMAN	MAX	44.3	49.0	56.9	69.1	75.4	82.5	87.6	87.3	80.5	70.8	57.3	48.5	67.4
	MIN	22.9	27.0	34.2	46.0	54.6	62.6	66.8	65.1	58.3	47.5	36.3	27.6	45.7
	MEAN	33.6	38.1	45.6	57.6	65.0	72.6	77.2	76.3	69.4	59.2	46.8	38.1	56.6
SHELBINA	MAX	35.8	41.7	52.4	66.8	76.0	84.6	89.5	87.6	81.0	70.2	54.3	41.7	65.1
	MIN	15.4	20.6	29.6	41.8	51.6	60.9	65.0	62.6	54.3	43.3	31.6	22.1	41.6
	MEAN	25.6	31.2	41.0	54.3	63.8	72.8	77.3	75.1	67.7	56.8	43.0	31.9	53.4

MISSOURI

TEMPERATURE NORMALS (DEG F)

STATION		JAN	FEB	MAR	APR	MAY	JUN	JUL	AUG	SEP	OCT	NOV	DEC	ANN
SIKESTON PWR PLT	MAX	42.4	47.3	56.8	69.6	78.9	87.5	90.5	88.7	82.6	72.3	57.8	46.7	68.4
	MIN	24.4	28.0	36.2	47.1	56.4	64.7	68.4	66.1	58.9	46.4	36.6	29.1	46.9
	MEAN	33.4	37.7	46.5	58.4	67.7	76.1	79.5	77.4	70.7	59.4	47.2	37.9	57.7
SPRINGFIELD WSO //R	MAX	42.2	47.1	56.1	68.3	76.5	84.9	89.8	89.3	81.6	70.8	56.2	46.4	67.4
	MIN	20.8	25.3	33.0	44.0	53.1	61.9	66.2	64.7	57.3	45.5	33.9	25.9	44.3
	MEAN	31.5	36.2	44.6	56.2	64.8	73.4	78.0	77.0	69.5	58.2	45.1	36.2	55.9
STEFFENVILLE 1 E	MAX	33.9	39.6	50.5	65.0	74.8	83.1	88.0	86.0	79.2	68.2	52.7	39.7	63.4
	MIN	14.6	20.1	29.2	42.1	51.9	61.1	65.3	62.9	54.7	43.7	32.0	21.7	41.6
	MEAN	24.2	29.9	39.9	53.6	63.4	72.1	76.7	74.5	67.0	56.0	42.4	30.7	52.5
SWEET SPRINGS	MAX	38.1	44.3	54.8	68.4	77.6	86.0	91.3	89.6	82.1	70.9	55.1	43.0	66.8
	MIN	17.2	22.7	31.2	43.2	52.9	61.7	65.9	63.5	55.5	44.2	32.5	23.4	42.8
	MEAN	27.7	33.5	43.0	55.8	65.3	73.9	78.6	76.6	68.9	57.6	43.8	33.2	54.8
TARKIO	MAX	33.0	39.7	50.6	66.1	76.3	84.9	89.3	87.4	79.4	69.2	51.9	39.1	63.9
	MIN	12.5	18.4	27.9	40.9	51.9	61.1	65.5	62.9	53.8	42.6	29.6	19.3	40.5
	MEAN	22.8	29.1	39.2	53.6	64.1	73.0	77.4	75.2	66.6	55.9	40.8	29.2	52.2
TRENTON	MAX	33.8	40.3	51.2	65.8	75.6	84.1	88.6	86.6	79.1	68.5	52.5	39.7	63.8
	MIN	14.3	20.0	29.2	42.3	52.4	61.5	65.6	63.2	54.8	44.1	31.7	21.1	41.7
	MEAN	24.1	30.2	40.2	54.1	64.0	72.8	77.1	74.9	67.0	56.3	42.1	30.5	52.8
UNION	MAX	40.8	46.2	56.2	69.5	77.9	86.1	90.4	89.1	82.3	71.4	56.3	45.4	67.6
	MIN	19.3	23.6	32.0	43.5	51.8	60.6	65.1	62.9	55.1	43.4	32.9	25.0	42.9
	MEAN	30.1	35.0	44.1	56.5	64.9	73.4	77.8	76.0	68.7	57.4	44.6	35.2	55.3
UNIONVILLE	MAX	31.4	37.2	48.1	63.2	73.8	82.6	87.6	85.4	77.5	66.5	50.3	37.0	61.7
	MIN	13.0	18.3	27.5	41.2	51.7	60.8	65.2	63.1	54.9	44.1	31.0	20.2	40.9
	MEAN	22.2	27.8	37.8	52.2	62.8	71.7	76.4	74.3	66.2	55.3	40.7	28.6	51.3
VICHY FAA AIRPORT	MAX	39.3	44.4	53.5	66.9	75.4	83.6	88.2	86.9	79.6	68.9	54.3	43.8	65.4
	MIN	20.2	24.7	32.8	45.3	54.5	63.2	67.7	65.9	58.6	47.4	35.2	25.9	45.1
	MEAN	29.8	34.6	43.2	56.1	65.0	73.4	78.0	76.4	69.1	58.1	44.8	34.9	55.3
WAPPAPELLO DAM	MAX	42.2	46.9	56.3	69.6	78.2	86.6	90.8	89.1	82.2	71.7	57.7	46.3	68.1
	MIN	22.7	26.7	35.7	47.4	55.9	64.3	68.2	66.3	59.4	46.9	36.3	27.6	46.5
	MEAN	32.5	36.8	46.1	58.5	67.1	75.5	79.5	77.7	70.8	59.3	47.0	36.9	57.3
WARRENSBURG	MAX	38.6	44.6	54.8	68.5	77.1	85.1	90.6	89.4	82.1	71.0	55.3	43.4	66.7
	MIN	19.3	24.4	32.6	45.1	54.7	63.5	67.7	65.9	58.0	47.1	34.9	25.2	44.9
	MEAN	29.0	34.5	43.7	56.8	65.9	74.3	79.2	77.7	70.1	59.1	45.1	34.4	55.8
WARSAW NO 1	MAX	42.4	48.4	58.1	71.1	78.3	86.2	91.6	90.2	82.8	72.5	57.7	47.0	68.9
	MIN	18.9	23.9	32.2	43.8	52.9	62.1	66.5	64.6	56.8	44.9	33.4	24.7	43.7
	MEAN	30.7	36.2	45.2	57.4	65.6	74.2	79.1	77.4	69.8	58.7	45.6	35.9	56.3
WASOLA	MAX	44.3	49.2	58.0	70.4	77.5	85.1	90.1	88.9	81.8	72.3	58.1	48.2	68.7
	MIN	22.6	26.1	34.0	45.6	54.0	61.9	66.0	64.4	58.0	46.8	35.8	27.0	45.2
	MEAN	33.5	37.7	46.0	58.1	65.8	73.6	78.1	76.7	69.9	59.6	46.9	37.6	57.0
WAYNESVILLE 2 W	MAX	45.6	50.6	59.5	71.8	78.7	85.4	90.6	89.5	83.1	73.6	59.7	49.7	69.8
	MIN	18.9	23.1	31.0	42.3	50.8	59.4	63.7	61.8	54.1	42.3	31.7	24.0	41.9
	MEAN	32.2	36.9	45.3	57.1	64.7	72.4	77.2	75.6	68.6	58.0	45.7	36.8	55.9
WEST PLAINS	MAX	43.6	48.5	57.1	69.5	77.5	85.1	89.8	88.6	81.6	71.9	57.4	47.4	68.2
	MIN	21.1	25.1	33.0	43.7	52.5	61.1	65.7	63.7	56.3	43.4	32.6	25.5	43.6
	MEAN	32.4	36.8	45.1	56.6	65.0	73.1	77.8	76.2	69.0	57.7	45.0	36.5	55.9
WILLOW SPRINGS	MAX	43.2	48.3	57.1	69.6	77.1	84.6	89.4	88.6	81.4	71.4	57.0	46.9	67.9
	MIN	20.1	24.4	32.1	43.0	51.5	60.0	64.4	62.4	55.5	43.2	32.4	24.7	42.8
	MEAN	31.7	36.4	44.6	56.3	64.3	72.3	77.0	75.6	68.4	57.3	44.7	35.8	55.4
WINDSOR	MAX	38.4	44.0	54.1	67.6	76.1	84.2	89.7	88.6	81.3	70.3	55.2	43.6	66.1
	MIN	17.9	23.0	31.4	43.6	53.3	62.2	66.3	64.3	56.6	45.0	33.2	23.9	43.4
	MEAN	28.2	33.5	42.8	55.6	64.7	73.2	78.0	76.5	69.0	57.7	44.2	33.8	54.8

MISSOURI

PRECIPITATION NORMALS (INCHES)

STATION	JAN	FEB	MAR	APR	MAY	JUN	JUL	AUG	SEP	OCT	NOV	DEC	ANN
ADVANCE	2.86	2.98	4.92	4.43	4.91	3.58	3.79	3.65	3.89	2.42	3.71	3.43	44.57
ALTON	2.61	3.06	4.58	4.54	5.01	3.26	3.48	3.13	4.00	2.76	4.37	3.37	44.17
ANDERSON	1.63	1.98	3.57	4.13	4.76	4.67	3.52	3.59	4.27	3.29	3.34	2.35	41.10
APPLETON CITY	1.46	1.49	2.98	3.47	4.69	4.73	3.75	3.63	4.58	3.53	2.20	1.78	38.29
ARCADIA	2.31	2.55	4.38	4.60	4.42	3.99	3.91	3.63	3.58	2.49	3.80	3.27	42.93
AUXVASSE	1.48	1.62	2.92	3.61	3.97	3.94	3.90	3.09	3.44	3.27	2.07	1.80	35.11
BELLEVIEW	1.71	2.36	3.94	3.94	3.82	3.36	3.67	3.34	3.11	2.42	3.34	2.76	37.77
BERNIE	3.72	3.40	5.17	4.30	4.87	4.02	3.40	3.05	3.78	2.47	4.36	4.19	46.73
BETHANY	1.03	1.06	2.43	3.39	4.22	4.36	3.79	4.50	4.76	3.34	1.69	1.18	35.75
BLOOMFIELD	3.53	3.52	5.44	4.70	5.28	3.74	3.66	3.32	3.70	2.51	4.31	3.93	47.64
BOLIVAR 1NE	1.53	1.96	3.18	3.74	4.40	4.41	3.93	2.89	4.35	3.52	2.66	2.27	38.84
BOONVILLE WATERWORKS	1.53	1.69	2.93	3.95	4.65	4.20	4.07	3.71	4.24	3.70	2.08	1.92	38.67
BOWLING GREEN 2 NE	1.49	1.51	3.12	4.00	4.23	3.84	3.65	3.29	3.42	3.01	2.33	2.02	35.91
BROOKFIELD	1.59	1.35	2.80	3.85	4.62	4.40	4.17	3.82	4.62	3.55	1.70	1.71	38.18
BRUNSWICK	1.55	1.45	2.89	3.71	4.48	4.31	4.30	3.53	4.45	3.71	2.07	1.69	38.14
BUFFALO 3 S	1.58	2.03	3.21	3.67	4.73	4.51	3.48	3.12	4.46	3.65	2.72	2.32	39.48
BUNKER	2.01	2.55	4.14	4.58	4.53	3.50	3.97	3.50	3.75	2.73	3.75	3.01	42.02
BURLINGTON JUNCTION	1.00	1.08	2.06	3.34	4.28	4.36	3.83	4.30	3.97	2.75	1.68	.99	33.64
BUTLER	1.60	1.39	3.00	3.71	4.54	4.83	3.87	3.76	4.55	3.28	2.31	1.78	38.62
CAMDENTON	1.59	2.23	3.52	3.88	4.72	4.23	3.80	3.72	4.43	3.94	2.63	2.40	41.09
CANTON LOCK AND DAM 20//	1.70	1.43	2.93	3.59	3.89	3.91	4.25	3.84	3.97	2.97	1.82	1.72	36.02
CAPE GIRARDEAU FAA	2.90	3.05	4.99	4.48	4.77	3.51	3.74	3.52	3.77	2.39	3.69	3.75	44.56
CAPLINGER MILLS	1.39	1.83	3.11	3.83	4.79	4.38	3.13	3.39	4.73	3.59	2.42	2.18	38.77
CARROLLTON	1.48	1.55	2.93	3.62	4.54	4.39	4.28	4.28	4.24	3.63	1.93	1.65	38.52
CARTHAGE	1.45	1.86	3.17	3.82	4.60	5.23	3.50	3.05	4.66	3.42	2.83	2.11	39.70
CARUTHERSVILLE	3.62	3.76	4.99	4.67	4.80	3.83	3.58	3.16	3.30	2.47	4.25	4.27	46.70
CASSVILLE RANGER STA	1.76	2.35	3.83	4.37	5.08	4.87	3.63	3.32	3.98	3.23	3.30	2.63	42.35
CENTERVILLE	2.49	2.96	4.69	4.54	4.70	3.73	3.82	3.16	3.81	2.86	3.94	3.39	44.09
CENTRALIA	1.48	1.39	2.77	3.64	3.99	3.84	3.72	3.24	3.59	3.19	1.92	1.67	34.44
CHARLESTON	3.46	3.52	5.63	4.58	5.18	3.71	3.84	3.38	3.44	2.59	4.31	4.23	47.87
CHILLICOTHE	1.20	1.01	2.43	3.46	4.06	4.05	4.32	3.84	4.59	3.07	1.63	1.30	34.96
CHILLICOTHE RADIO KCHI	1.45	1.17	2.70	3.59	4.23	4.15	4.44	4.00	4.65	3.07	1.67	1.55	36.67
CLEARWATER DAM	2.63	2.85	4.62	4.14	4.71	3.29	3.67	3.51	3.39	2.28	3.81	3.35	42.25
CLIFTON CITY	1.38	1.64	2.87	3.59	4.35	4.39	4.05	3.35	3.74	3.60	1.88	1.81	36.65
CLINTON 3 NW	1.34	1.51	2.98	3.47	4.99	4.55	4.10	3.68	4.33	3.80	2.17	1.74	38.66
COLOMA	1.22	1.13	2.53	3.47	4.29	4.35	3.96	4.20	3.93	3.47	1.74	1.35	35.64
COLUMBIA REGION WSO	1.57	1.86	3.19	3.83	4.47	3.76	3.51	2.93	3.64	3.34	2.02	1.95	36.06
CONCEPTION	.97	1.15	2.38	3.64	4.37	4.68	4.09	4.40	4.48	2.84	1.69	1.15	35.84
CONCORDIA	1.51	1.53	2.99	3.89	4.40	4.47	4.16	3.88	4.62	3.43	2.07	1.66	38.61
COOK STATION	1.98	2.34	3.60	3.85	4.14	3.43	3.74	3.52	3.29	2.86	3.04	2.48	38.27
DE SOTO	2.10	2.42	3.84	3.99	4.46	3.71	3.42	3.39	2.91	2.32	3.07	2.68	38.31
DEXTER	3.68	3.42	5.30	4.51	5.17	3.75	3.73	3.59	3.71	2.54	4.32	4.03	47.75
DONIPHAN	3.10	3.36	5.27	4.82	5.22	3.68	3.73	3.55	4.02	2.35	4.40	3.58	47.08
EDGERTON	1.13	1.30	2.41	3.16	4.83	4.85	5.01	4.12	4.03	2.94	1.65	1.38	36.81
EDINA	1.70	1.32	2.92	3.56	3.98	4.05	4.09	4.07	4.40	3.49	1.75	1.75	37.08
ELDON	1.48	1.93	3.19	3.86	4.63	4.57	3.38	3.27	4.45	3.28	2.44	2.14	38.62
ELLINGTON	2.31	2.80	4.31	4.31	4.78	3.61	3.65	3.59	3.90	2.59	3.78	3.57	43.20
ELSBERRY 1 S	1.53	1.81	3.15	3.77	3.85	3.59	3.91	3.67	3.50	2.95	2.17	2.20	36.10
FAIRFAX	.93	1.06	2.13	3.18	4.22	4.78	3.91	3.95	4.01	2.44	1.52	.86	32.99
FARMINGTON	2.05	2.56	4.16	4.33	4.09	3.53	3.86	3.59	3.25	2.30	3.41	2.91	40.04
FAYETTE	1.23	1.48	2.84	4.02	4.41	4.14	3.57	3.17	3.52	3.36	1.96	1.63	35.33
FOUNTAIN GROVE WL	1.45	1.27	2.64	3.65	4.59	3.93	4.30	3.79	4.69	3.35	1.66	1.52	36.84
FREDERICKTOWN	2.38	2.62	4.25	4.49	4.29	3.78	3.91	3.85	3.37	2.28	3.66	3.20	42.08
FULTON 4 SW	1.43	1.68	2.88	3.52	4.23	3.65	3.54	2.88	3.55	3.23	2.19	2.00	34.78
GALENA	1.62	1.98	3.56	4.32	4.68	4.69	3.68	3.05	4.07	3.33	3.37	2.57	40.92

MISSOURI

PRECIPITATION NORMALS (INCHES)

STATION	JAN	FEB	MAR	APR	MAY	JUN	JUL	AUG	SEP	OCT	NOV	DEC	ANN
GRANT CITY	.87	1.07	2.20	3.30	4.22	4.71	3.85	4.27	4.22	2.96	1.77	1.11	34.55
GREENVILLE 4 NNW	2.73	2.96	4.78	4.12	4.79	3.63	4.14	3.47	3.74	2.32	4.11	3.44	44.23
GREGORY LANDING	1.67	1.42	2.86	3.46	4.00	3.84	4.51	3.63	4.33	3.09	1.78	1.63	36.22
GROVESPRING	1.73	2.02	3.50	3.92	4.62	4.18	4.31	3.34	3.90	3.33	2.92	2.47	40.24
HAMILTON 2 W	1.18	1.10	2.48	3.43	4.32	4.23	4.01	3.55	4.46	3.30	1.67	1.30	35.03
HANNIBAL WATERWORKS	1.64	1.70	3.18	3.81	4.23	3.78	4.02	3.67	3.62	3.09	1.99	1.96	36.69
HARRISONVILLE	1.31	1.33	2.81	3.76	4.29	4.66	3.59	3.54	4.41	3.37	1.97	1.39	36.43
HERMANN	1.66	1.93	3.04	3.69	4.22	3.97	3.79	3.02	3.60	2.91	2.43	2.05	36.31
HOUSTON	1.94	2.33	3.90	4.12	4.50	3.61	3.89	3.56	3.81	3.18	3.31	2.77	40.92
JACKSON	2.87	2.91	4.79	4.29	4.95	3.99	3.95	3.84	3.76	2.50	3.85	3.63	45.33
JEFFERSON CITY L U	1.41	1.62	2.82	3.37	4.66	4.53	3.20	2.98	3.99	3.47	2.10	1.93	36.08
JEROME	1.82	2.28	3.55	3.40	4.85	3.87	3.89	3.18	3.36	3.18	2.76	2.50	38.64
JOPLIN FAA AIRPORT	1.40	1.92	3.01	3.83	4.68	4.90	3.34	3.50	4.91	3.31	2.57	2.17	39.54
KANSAS CITY WSO	1.08	1.19	2.41	3.23	4.42	4.66	4.35	3.57	4.14	3.10	1.63	1.38	35.17
KANSAS CITY FAA AP	.98	1.05	2.12	2.68	3.42	4.13	3.49	3.16	3.33	2.54	1.23	1.14	29.27
KIRKSVILLE RADIO KIRX	1.37	1.09	2.66	3.55	4.22	3.90	4.00	3.85	4.29	3.10	1.59	1.79	35.41
LA BELLE	1.49	1.24	2.68	3.45	3.72	3.96	4.06	4.22	3.73	3.27	1.69	1.57	35.08
LAKESIDE	1.50	1.91	3.07	3.49	4.51	4.09	3.45	3.35	3.90	3.42	2.45	2.09	37.23
LAMAR	1.51	1.88	3.38	3.78	4.86	4.71	3.17	3.24	4.61	3.53	2.77	2.22	39.66
LEBANON	1.66	2.16	3.51	3.72	4.56	3.71	3.81	3.40	3.90	3.60	2.84	2.36	39.23
LEXINGTON	1.25	1.38	2.67	3.45	4.31	4.42	4.61	3.85	4.09	3.67	1.95	1.42	37.07
LICKING 4 N	1.81	2.16	3.48	3.78	4.37	3.77	3.86	3.16	3.59	2.87	3.05	2.47	38.37
LOCKWOOD	1.40	2.03	3.20	3.83	4.78	4.79	3.28	3.15	4.33	3.53	2.71	2.29	39.32
LOUISIANA STARKS NUR	1.55	1.74	3.12	3.85	4.11	3.64	3.67	3.24	3.64	2.95	2.14	2.05	35.70
LOUISIANA	1.80	1.79	3.33	4.01	4.28	3.82	3.86	3.28	3.52	3.00	2.32	2.23	37.24
LUCERNE	1.03	.95	2.26	3.43	3.89	4.12	4.40	3.81	4.42	2.68	1.58	1.18	33.75
MACON	1.40	1.41	2.97	3.92	4.58	4.73	4.37	3.91	4.25	3.44	1.97	1.71	38.66
MADISON	1.54	1.36	2.99	3.85	3.97	4.01	4.35	3.63	3.77	3.09	1.92	1.60	36.08
MANSFIELD	1.79	2.23	3.63	4.32	4.97	4.26	4.11	3.13	4.28	3.50	3.32	2.65	42.19
MARBLE HILL	2.74	3.03	4.88	4.15	4.63	3.83	3.79	3.63	3.65	2.51	3.91	3.62	44.37
MARSHALL	1.33	1.46	2.89	3.71	4.09	4.14	3.92	3.24	4.01	3.18	1.97	1.50	35.44
MARSHFIELD	1.72	2.10	3.55	3.63	4.71	4.37	3.74	3.01	4.16	3.44	2.87	2.56	39.86
MARTINSBURG	1.66	1.80	3.00	3.61	4.21	3.87	4.25	3.44	3.26	3.03	2.26	1.84	36.23
MARYVILLE 2 E	.74	1.04	2.23	3.55	4.44	4.63	4.03	4.17	4.17	2.78	1.66	1.11	34.55
MEMPHIS	1.55	1.15	2.71	3.84	3.65	3.92	3.95	4.06	4.09	2.97	1.71	1.71	35.31
MEXICO	1.57	1.72	3.30	4.06	4.45	4.28	4.52	2.99	3.82	3.39	2.06	1.95	38.11
MILAN	1.35	1.11	2.61	3.56	4.17	4.11	4.07	4.21	4.23	3.07	1.63	1.47	35.59
MOBERLY RADIO KWIX	1.46	1.57	3.03	3.98	4.24	4.00	4.20	3.65	3.78	3.32	1.98	1.68	36.89
MONROE CITY	1.71	1.58	3.46	4.00	4.44	4.43	4.08	3.89	4.00	3.14	1.95	1.89	38.57
MOUNTAIN GROVE 2 N	1.84	2.29	3.75	4.05	4.42	4.05	4.07	3.38	3.80	2.91	3.34	2.66	40.56
MOUNT VERNON MU FARM	1.51	2.07	3.50	3.89	4.76	5.02	3.16	3.13	4.53	3.44	2.96	2.33	40.30
NEOSHO	1.51	2.12	3.42	4.14	4.65	4.82	3.46	3.30	4.45	3.74	2.99	2.27	40.87
NEVADA SEWAGE PLANT	1.48	1.73	3.23	3.57	5.14	5.12	3.22	3.49	4.31	3.62	2.55	1.90	39.36
NEW FLORENCE	1.63	1.88	3.05	3.68	4.00	3.74	3.77	3.05	3.69	3.07	2.08	2.16	35.80
ODESSA	1.45	1.34	2.73	3.63	4.23	4.63	4.95	3.49	4.18	3.75	1.91	1.44	37.73
OREGON	.90	1.07	2.30	3.17	4.40	4.68	4.23	4.17	4.56	2.53	1.86	1.10	34.97
OWENSVILLE	1.78	2.12	3.57	3.73	4.80	4.20	3.54	3.06	3.47	2.81	2.81	2.31	38.20
OZARK	1.71	2.07	3.46	4.08	5.04	4.34	3.63	2.59	4.29	3.42	3.10	2.54	40.27
OZARK BEACH	1.90	2.39	3.64	4.06	4.67	4.41	3.59	3.07	4.00	3.01	3.40	2.71	40.85
PACIFIC	2.03	2.20	3.26	3.64	4.42	3.94	3.75	3.28	3.04	2.74	2.82	2.30	37.42
PALMYRA	1.49	1.47	3.02	3.78	4.13	3.95	4.04	3.19	3.63	3.04	1.81	1.85	35.40
PARIS 1 SW	1.50	1.51	3.07	3.80	4.00	4.41	4.65	3.91	3.74	3.22	1.95	1.78	37.54
PARMA	3.59	3.35	5.15	4.32	4.79	3.93	3.22	3.19	3.76	2.22	4.16	3.88	45.56
PERRYVILLE WATER PLANT	2.40	2.64	4.30	4.10	4.33	3.63	3.94	3.79	3.17	2.32	3.81	3.03	41.46
PIERCE CITY	1.49	2.08	3.53	4.00	5.08	5.09	3.52	2.84	4.61	3.49	3.10	2.39	41.22

MISSOURI

PRECIPITATION NORMALS (INCHES)

STATION	JAN	FEB	MAR	APR	MAY	JUN	JUL	AUG	SEP	OCT	NOV	DEC	ANN
POLO	1.18	1.16	2.51	3.75	4.37	4.81	4.11	4.19	4.27	3.35	1.71	1.44	36.85
POPLAR BLUFF R S	3.16	3.27	5.31	4.39	5.06	3.74	3.59	3.27	3.73	2.41	4.25	3.62	45.80
PRINCETON	1.22	1.12	2.50	3.45	4.03	4.03	4.32	4.12	4.40	3.05	1.61	1.42	35.27
PUXICO	3.28	3.29	5.51	4.58	5.25	3.91	3.73	3.61	3.90	2.60	4.26	3.70	47.39
QULIN	3.67	3.19	4.97	4.09	4.88	3.69	3.28	2.75	3.90	2.37	4.10	3.77	44.66
REYNOLDS	2.18	2.71	4.06	4.50	4.34	3.63	3.65	3.51	3.46	2.64	3.81	2.85	41.34
RICHWOODS 2 NE	1.93	2.25	3.67	3.63	4.33	3.96	3.12	3.01	2.78	2.46	2.87	2.54	36.55
ROLLA, UNIV OF MO.	1.62	2.12	3.25	3.59	4.70	3.94	3.82	3.29	3.07	3.01	2.70	2.34	37.45
ROUND SPRING 3 NNW	2.08	2.51	4.16	4.34	4.68	3.66	3.91	3.77	3.73	2.44	3.67	2.96	41.91
SAINT CHARLES	1.79	2.06	3.09	3.62	3.84	3.68	4.22	2.60	2.80	2.50	2.55	2.24	34.99
ST. JOSEPH WBAP	.99	1.01	2.35	3.06	4.52	5.10	3.83	4.08	4.13	2.71	1.63	1.10	34.51
SAINT LOUIS WSO R	1.72	2.14	3.28	3.55	3.54	3.73	3.63	2.55	2.70	2.32	2.53	2.22	33.91
SALEM	1.90	2.33	3.81	4.02	4.33	3.24	3.84	3.78	3.63	2.75	3.21	2.75	39.59
SALISBURY	1.58	1.53	3.04	3.70	4.58	4.49	4.08	3.68	4.19	3.53	1.95	1.80	38.15
SAVERTON L AND D 22 //	1.37	1.43	2.88	4.05	4.27	3.61	3.74	3.62	3.36	3.03	2.11	1.75	35.22
SEDALIA	1.47	1.63	2.86	3.91	4.42	4.57	3.98	3.59	3.90	3.75	2.06	1.73	37.87
SELIGMAN	1.72	2.28	3.65	4.20	4.86	4.63	3.69	3.28	4.18	3.35	3.23	2.42	41.49
SHELBINA	1.49	1.39	3.10	3.79	4.10	4.25	4.21	4.08	4.08	3.22	1.88	1.77	37.36
SIKESTON PWR PLT	3.50	3.40	5.16	4.49	4.84	3.96	3.45	3.25	3.92	2.60	4.21	4.05	46.83
SILOAM SPRINGS	2.22	2.73	3.76	4.58	4.39	3.61	3.37	3.08	3.29	2.76	3.51	3.00	40.30
SPRINGFIELD WSO //R	1.60	2.13	3.44	4.03	4.32	4.66	3.58	2.83	4.24	3.20	2.89	2.55	39.47
STEFFENVILLE 1 E	1.45	1.28	2.81	3.43	4.10	3.72	4.36	3.84	3.72	3.24	1.74	1.55	35.24
STOVER	1.55	1.83	3.17	3.62	4.49	4.29	3.69	3.57	3.96	3.83	2.22	2.03	38.25
SULLIVAN	1.97	2.16	3.57	3.67	4.53	4.08	3.88	3.62	3.15	2.77	2.90	2.50	38.80
SUMMERSVILLE	1.95	2.49	4.03	4.36	4.80	3.46	4.12	3.22	3.99	2.78	3.58	2.90	41.68
SUMNER 2 SW	1.35	1.30	2.62	3.52	4.23	4.00	4.29	3.73	4.20	3.53	1.74	1.48	35.99
SWEET SPRINGS	1.33	1.51	3.09	3.99	4.37	4.15	3.93	3.71	4.00	3.26	2.11	1.77	37.22
TARKIO	.98	1.06	2.27	3.37	4.55	4.75	3.75	4.31	4.05	2.62	1.73	1.06	34.50
TECUMSEH 2 N	2.26	2.56	4.07	4.30	4.65	4.12	4.27	3.27	3.66	2.71	3.58	2.86	42.31
TRENTON	1.11	1.05	2.59	3.22	4.22	3.79	4.12	4.14	4.25	2.87	1.45	1.37	34.18
TROY	1.71	1.86	3.13	3.83	3.93	3.91	3.97	3.11	3.33	2.76	2.28	2.15	35.97
UNION	1.76	2.09	3.17	3.63	4.09	4.14	3.74	3.05	3.06	2.70	2.52	2.25	36.20
UNIONVILLE	1.12	1.01	2.47	3.70	4.23	4.13	4.41	4.01	4.34	2.79	1.53	1.33	35.07
VALLEY PARK	1.82	2.18	3.50	3.68	3.98	4.27	3.72	2.96	2.93	2.67	3.00	2.50	37.21
VERSAILLES	1.63	1.94	3.32	3.73	4.50	4.25	3.75	3.56	4.10	3.80	2.37	2.21	39.16
VICHY FAA AIRPORT	1.57	1.94	3.31	3.61	4.71	3.98	3.62	3.50	3.30	3.01	2.66	2.26	37.47
WACO	1.35	1.77	2.85	3.91	4.72	4.82	3.02	3.43	4.88	3.24	2.68	1.99	38.66
WAPPAPELLO DAM	3.10	3.07	5.22	4.35	4.96	3.48	3.93	3.25	3.64	2.48	4.32	3.47	45.27
WARRENSBURG	1.42	1.47	2.93	3.93	4.39	4.83	4.07	3.69	4.13	3.65	1.96	1.68	38.15
WARRENTON 1 N	1.52	1.89	3.13	3.77	3.70	3.92	3.63	2.90	3.55	2.91	2.41	2.15	35.48
WARSAW NO 1	1.59	1.85	3.20	3.86	4.56	4.58	3.87	3.64	4.55	3.62	2.41	1.97	39.70
WASOLA	1.69	2.12	3.63	3.97	4.61	3.55	3.86	3.05	3.82	3.01	3.18	2.61	39.10
WAVERLY	1.32	1.34	2.59	3.42	3.88	3.84	3.58	3.69	4.03	3.34	1.83	1.41	34.27
WAYNESVILLE 2 W	1.93	2.37	3.67	3.67	4.78	3.80	3.82	3.33	3.57	3.39	2.83	2.68	39.84
WEST PLAINS	2.32	2.84	4.34	4.64	4.58	3.92	3.70	3.41	3.63	2.77	3.79	3.33	43.27
WILLIAMSVILLE	2.82	2.91	4.78	4.40	4.83	3.51	4.15	3.37	3.33	2.45	4.40	3.65	44.60
WILLOW SPRINGS	2.13	2.66	4.20	4.71	4.86	3.91	4.07	3.28	3.92	2.85	3.68	3.05	43.32
WINDSOR	1.52	1.54	2.99	3.72	4.43	4.86	4.14	3.66	4.27	3.66	2.22	1.67	38.68
ZALMA 4 E	2.87	2.91	5.10	4.34	4.97	3.96	4.26	3.76	3.76	2.34	4.05	3.44	45.76

MISSOURI

HEATING DEGREE DAY NORMALS (BASE 65 DEG F)

STATION	JUL	AUG	SEP	OCT	NOV	DEC	JAN	FEB	MAR	APR	MAY	JUN	ANN
ANDERSON	0	0	40	231	576	859	986	756	591	238	93	10	4380
APPLETON CITY	0	0	34	220	585	921	1079	818	633	251	84	8	4633
ARCADIA	0	0	40	251	579	874	1017	781	610	242	84	9	4487
BETHANY	0	0	63	310	705	1091	1293	991	791	351	125	14	5734
BOLIVAR 1NE	0	0	42	240	582	890	1042	804	628	255	102	16	4601
BOONVILLE WATERWORKS	0	0	42	271	639	1008	1197	941	747	323	125	11	5304
BROOKFIELD	0	0	42	256	642	1023	1212	924	719	301	110	11	5240
BRUNSWICK	0	0	37	257	630	1008	1194	918	716	300	107	8	5175
CAMDENTON	0	0	28	202	552	871	1023	778	608	216	81	8	4367
CANTON LOCK AND DAM 20//	0	0	39	285	669	1048	1249	983	787	353	129	9	5551
CARROLLTON	0	0	43	279	666	1020	1206	924	716	302	106	9	5271
CARTHAGE	0	0	28	204	540	840	986	753	586	227	78	11	4253
CARUTHERSVILLE	0	0	13	169	465	766	899	700	518	177	46	0	3753
CHARLESTON	0	0	25	210	510	822	952	739	550	199	67	0	4074
CHILLICOTHE RADIO KCHI	0	0	37	277	663	1042	1228	941	735	311	102	10	5346
CLEARWATER DAM	0	0	25	238	570	884	1008	787	604	226	79	8	4429
CLINTON 3 NW	5	0	45	272	648	992	1166	904	722	316	120	13	5203
COLUMBIA REGION WSO	0	0	47	275	645	995	1163	910	722	316	120	13	5206
CONCEPTION	0	0	69	323	720	1116	1321	1022	818	374	146	16	5925
DONIPHAN	0	0	31	235	552	856	964	750	573	216	72	5	4254
EDGERTON	0	0	53	282	672	1032	1225	932	732	321	109	15	5373
ELDON	0	0	33	215	576	908	1057	812	633	242	75	9	4560
ELSBERRY 1 S	0	0	47	294	657	1020	1203	941	732	319	128	10	5351
FARMINGTON	0	0	53	287	609	927	1070	834	662	272	114	15	4843
FAYETTE	0	0	38	258	633	989	1166	902	701	293	114	12	5106
FOUNTAIN GROVE WL	0	0	44	286	669	1039	1234	938	725	316	114	8	5373
FULTON 4 SW	0	0	47	287	657	1017	1197	941	756	335	129	15	5381
GRANT CITY	5	0	66	309	726	1119	1324	1016	818	369	125	16	5893
GREENVILLE 4 NNW	0	0	49	276	585	893	1008	781	612	248	99	15	4566
HAMILTON 2 W	0	0	60	315	699	108?	1283	991	784	356	133	17	5723
HANNIBAL WATERWORKS	0	0	50	302	681	1070	1256	980	781	347	135	11	5613
HARRISONVILLE	0	0	34	237	609	964	1138	871	673	273	98	10	4907
JACKSON	0	0	20	207	549	862	998	773	581	215	68	0	4273
JEFFERSON CITY L U	0	0	46	261	615	949	1101	857	670	276	108	14	4897
JOPLIN FAA AIRPORT	0	0	39	197	549	853	1004	764	594	230	79	12	4321
KANSAS CITY WSO	0	0	42	258	657	1026	1212	921	729	314	112	12	5283
KANSAS CITY FAA AP	0	0	25	213	612	955	1135	857	670	260	78	7	4812
KIRKSVILLE RADIO KIRX	0	0	61	316	711	1104	1299	1011	809	375	148	14	5848
LAKESIDE	0	0	33	250	591	927	1107	871	698	295	115	9	4896
LAMAR	5	0	44	249	603	921	1085	837	660	270	98	12	4784
LEBANON	0	0	39	222	576	884	1032	798	633	257	93	12	4546
LEXINGTON	0	0	34	263	648	1029	1225	944	747	305	101	10	5306
LICKING 4 N	0	0	61	289	636	958	1116	874	698	301	119	19	5071
LOCKWOOD	0	0	32	197	552	853	1001	762	595	233	79	7	4311
LOUISIANA STARKS NUR	0	0	46	274	642	995	1178	910	713	308	123	12	5201
MACON	0	0	48	311	705	1091	1287	1005	809	371	142	11	5780
MARBLE HILL	0	0	33	242	576	890	1017	787	600	235	78	8	4466
MARSHFIELD	0	0	40	233	594	896	1039	806	643	267	97	12	4627
MARYVILLE 2 E	6	9	99	373	771	1163	1367	1072	884	423	159	23	6349
MEXICO	0	0	57	305	678	1054	1237	977	784	354	141	15	5602
MOBERLY RADIO KWIX	0	0	45	268	654	1017	1200	930	735	315	114	10	5288
MOUNTAIN GROVE 2 N	0	0	41	219	570	874	1008	778	618	246	83	13	4450
MOUNT VERNON MU FARM	0	0	44	238	585	884	1032	798	633	272	108	17	4611
NEOSHO	0	0	26	192	534	815	949	717	555	209	71	6	4074
NEVADA SEWAGE PLANT	0	0	37	201	564	890	1051	790	604	228	78	11	4454

MISSOURI

HEATING DEGREE DAY NORMALS (BASE 65 DEG F)

STATION	JUL	AUG	SEP	OCT	NOV	DEC	JAN	FEB	MAR	APR	MAY	JUN	ANN
NEW FLORENCE	0	0	37	252	627	977	1147	893	702	291	101	10	5037
OREGON	0	0	66	305	702	1094	1290	988	800	355	135	16	5751
OZARK BEACH	0	0	30	225	564	862	1001	787	614	257	99	14	4453
POPLAR BLUFF R S	0	0	21	211	528	828	958	742	553	196	64	0	4101
SAINT CHARLES	0	0	46	267	606	955	1122	876	673	277	114	13	4949
ST. JOSEPH WBAP	0	0	56	281	681	1054	1243	949	744	325	109	11	5453
SAINT LOUIS WSO R	0	0	40	258	612	955	1122	874	676	279	110	12	4938
SALEM	0	0	45	253	603	921	1076	837	671	279	101	13	4799
SALISBURY	0	0	39	260	639	1004	1187	913	707	302	110	10	5171
SAVERTON L AND D 22 //	0	0	32	248	621	995	1184	930	729	311	103	8	5161
SEDALIA	0	0	42	260	615	952	1125	874	678	282	111	12	4951
SELIGMAN	0	0	47	220	546	834	973	753	611	245	91	16	4336
SHELBINA	0	0	45	277	660	1026	1221	946	744	327	133	10	5389
SIKESTON PWR PLT	0	0	22	207	534	840	980	764	584	216	62	0	4209
SPRINGFIELD WSO //R	0	0	43	242	597	893	1039	806	640	279	107	14	4660
STEFFENVILLE 1 E	0	0	54	309	678	1063	1265	983	778	349	138	13	5630
SWEET SPRINGS	0	0	42	256	636	986	1156	882	682	286	100	10	5036
TARKIO	0	0	60	306	726	1110	1308	1005	800	347	120	12	5794
TRENTON	0	0	63	301	687	1070	1268	974	769	334	121	13	5600
UNION	0	0	42	258	612	924	1082	840	655	268	104	11	4796
UNIONVILLE	0	0	58	323	729	1128	1327	1042	843	391	145	13	5999
VICHY FAA AIRPORT	0	0	39	245	606	933	1091	851	683	282	100	13	4843
WAPPAPELLO DAM	0	0	22	216	540	871	1008	790	595	214	64	0	4320
WARRENSBURG	0	0	44	239	597	949	1116	854	669	265	102	14	4849
WARSAW NO 1	0	0	41	225	582	902	1063	806	622	241	102	13	4597
WASOLA	0	0	33	212	543	849	977	764	599	231	81	16	4305
WAYNESVILLE 2 W	0	0	48	246	579	874	1017	787	622	260	104	13	4550
WEST PLAINS	0	0	38	244	600	884	1011	790	624	267	92	11	4561
WILLOW SPRINGS	0	0	44	255	609	905	1032	801	641	277	110	17	4691
WINDSOR	0	0	46	265	624	967	1141	882	688	291	119	14	5037

MISSOURI

COOLING DEGREE DAY NORMALS (BASE 65 DEG F)

STATION	JAN	FEB	MAR	APR	MAY	JUN	JUL	AUG	SEP	OCT	NOV	DEC	ANN
ANDERSON	0	0	8	22	97	250	394	347	160	24	0	0	1302
APPLETON CITY	0	0	7	17	121	296	440	406	196	40	0	0	1523
ARCADIA	0	0	9	20	93	246	381	329	148	24	0	0	1250
BETHANY	0	0	0	6	94	248	386	313	114	24	0	0	1185
BOLIVAR 1NE	0	0	8	18	111	271	412	378	192	42	0	0	1432
BOONVILLE WATERWORKS	0	0	0	8	104	260	412	350	153	32	0	0	1319
BROOKFIELD	0	0	0	10	116	281	426	360	156	36	0	0	1385
BRUNSWICK	0	0	0	15	116	278	419	353	154	34	0	0	1369
CAMDENTON	0	0	10	27	130	290	440	403	199	41	0	0	1540
CANTON LOCK AND DAM 20//	0	0	0	8	98	261	394	316	117	25	0	0	1219
CARROLLTON	0	0	0	8	106	270	409	338	133	21	0	0	1285
CARTHAGE	0	0	10	29	118	299	453	412	199	40	0	0	1560
CARUTHERSVILLE	0	0	13	33	176	363	481	431	241	58	0	0	1796
CHARLESTON	0	0	11	34	172	349	456	397	205	46	0	0	1670
CHILLICOTHE RADIO KCHI	0	0	0	11	99	268	412	344	130	32	0	0	1296
CLEARWATER DAM	0	0	12	25	117	275	415	360	160	24	0	0	1388
CLINTON 3 NW	0	0	0	10	105	259	411	353	153	30	0	0	1321
COLUMBIA REGION WSO	0	0	0	10	92	250	397	341	149	30	0	0	1269
CONCEPTION	0	0	0	0	90	238	385	319	123	35	0	0	1190
DONIPHAN	0	0	9	21	122	293	437	381	184	33	0	0	1480
EDGERTON	0	0	0	9	94	258	395	329	128	21	0	0	1234
ELDON	0	0	10	20	113	285	443	394	186	35	0	0	1486
ELSBERRY 1 S	0	0	0	10	100	256	381	322	137	27	0	0	1233
FARMINGTON	0	0	8	14	92	246	366	313	137	23	0	0	1199
FAYETTE	0	0	0	8	95	249	394	338	143	29	0	0	1256
FOUNTAIN GROVE WL	0	0	0	10	102	260	397	319	119	32	0	0	1239
FULTON 4 SW	0	0	0	8	86	243	395	338	149	27	0	0	1246
GRANT CITY	0	0	0	6	79	235	384	307	108	24	0	0	1143
GREENVILLE 4 NNW	0	0	7	20	111	264	388	332	157	25	0	0	1304
HAMILTON 2 W	0	0	0	8	84	239	379	313	123	29	0	0	1175
HANNIBAL WATERWORKS	0	0	0	8	91	233	363	298	116	26	0	0	1135
HARRISONVILLE	0	0	0	15	116	283	440	388	184	44	0	0	1470
JACKSON	0	0	11	20	136	325	443	394	197	30	0	0	1556
JEFFERSON CITY L U	0	0	7	15	102	254	403	357	163	29	0	0	1330
JOPLIN FAA AIRPORT	0	0	8	26	128	318	468	419	216	45	0	0	1628
KANSAS CITY WSO	0	0	0	8	99	270	423	363	144	26	0	0	1333
KANSAS CITY FAA AP	0	0	7	17	137	334	493	443	205	45	0	0	1681
KIRKSVILLE RADIO KIRX	0	0	0	6	76	209	347	279	91	22	0	0	1030
LAKESIDE	0	0	0	7	94	249	397	350	150	26	0	0	1273
LAMAR	0	0	0	15	107	273	433	378	176	41	0	0	1423
LEBANON	0	0	10	20	105	267	406	369	180	33	0	0	1390
LEXINGTON	0	0	0	11	120	298	450	381	160	37	0	0	1457
LICKING 4 N	0	0	7	13	88	238	372	329	154	26	0	0	1227
LOCKWOOD	0	0	9	23	125	301	453	419	209	45	0	0	1584
LOUISIANA STARKS NUR	0	0	0	8	96	252	378	319	133	23	0	0	1209
MACON	0	0	0	8	92	227	376	304	105	20	0	0	1132
MARBLE HILL	0	0	8	16	103	278	403	347	156	19	0	0	1330
MARSHFIELD	0	0	8	15	110	270	412	372	184	35	0	0	1406
MARYVILLE 2 E	0	0	0	0	69	209	344	273	99	22	0	0	1016
MEXICO	0	0	0	6	94	246	388	327	135	26	0	0	1222
MOBERLY RADIO KWIX	0	0	0	9	95	250	391	329	138	26	0	0	1238
MOUNTAIN GROVE 2 N	0	0	10	18	92	253	394	353	179	33	0	0	1332
MOUNT VERNON MU FARM	0	0	7	17	101	263	414	380	179	33	0	0	1394
NEOSHO	0	0	12	32	124	291	437	394	194	40	0	0	1524
NEVADA SEWAGE PLANT	0	0	9	21	127	311	456	409	214	46	0	0	1593

MISSOURI

COOLING DEGREE DAY NORMALS (BASE 65 DEG F)

STATION		JAN	FEB	MAR	APR	MAY	JUN	JUL	AUG	SEP	OCT	NOV	DEC	ANN
NEW FLORENCE		0	0	8	15	101	268	403	350	160	32	0	0	1337
OREGON		0	0	0	7	97	244	382	322	123	35	0	0	1210
OZARK BEACH		0	0	0	14	108	272	415	363	165	21	0	0	1358
POPLAR BLUFF R S		0	0	7	22	141	318	456	400	195	38	0	0	1577
SAINT CHARLES		0	0	0	16	127	295	415	353	169	37	0	0	1412
ST. JOSEPH WBAP		0	0	0	13	109	290	423	348	137	27	0	0	1347
SAINT LOUIS WSO	R	0	0	0	12	128	306	431	372	181	38	0	0	1468
SALEM		0	0	11	12	88	244	378	322	147	24	0	0	1226
SALISBURY		0	0	0	8	104	265	406	344	144	27	0	0	1298
SAVERTON L AND D 22	//	0	0	0	11	109	287	419	350	146	31	0	0	1353
SEDALIA		0	0	8	18	124	282	425	372	177	49	0	0	1455
SELIGMAN		0	0	9	23	91	244	382	350	179	40	0	0	1318
SHELBINA		0	0	0	6	95	244	381	313	126	22	0	0	1187
SIKESTON PWR PLT		0	0	10	18	146	336	450	384	193	34	0	0	1571
SPRINGFIELD WSO	//R	0	0	7	15	101	266	403	372	178	32	0	0	1374
STEFFENVILLE 1 E		0	0	0	7	89	226	363	295	114	30	0	0	1124
SWEET SPRINGS		0	0	0	10	109	277	422	360	159	26	0	0	1363
TARKIO		0	0	0	0	92	252	384	316	108	24	0	0	1176
TRENTON		0	0	0	7	90	247	380	311	123	31	0	0	1189
UNION		0	0	7	13	101	263	397	341	153	23	0	0	1298
UNIONVILLE		0	0	0	7	77	214	353	288	94	22	0	0	1055
VICHY FAA AIRPORT		0	0	7	15	100	265	403	353	162	32	0	0	1337
WAPPAPELLO DAM		0	0	9	19	129	319	450	394	196	39	0	0	1555
WARRENSBURG		0	0	9	19	129	293	440	394	197	56	0	0	1537
WARSAW NO 1		0	0	8	13	120	289	437	384	185	30	0	0	1466
WASOLA		0	0	10	24	105	274	406	363	180	45	0	0	1407
WAYNESVILLE 2 W		0	0	11	23	95	235	378	329	156	29	0	0	1256
WEST PLAINS		0	0	7	15	92	254	397	347	158	18	0	0	1288
WILLOW SPRINGS		0	0	8	16	88	236	372	329	146	16	0	0	1211
WINDSOR		0	0	0	9	110	260	407	357	166	39	0	0	1348

STATE-STATION NUMBER	STN TYP	NAME	LATITUDE DEG-MIN	LONGITUDE DEG-MIN	ELEVATION (FT)
23-0022	12	ADVANCE	N 3706	W 08954	350
23-0127	12	ALTON	N 3642	W 09124	860
23-0164	13	ANDERSON	N 3639	W 09426	1050
23-0204	13	APPLETON CITY	N 3812	W 09402	800
23-0224	13	ARCADIA	N 3735	W 09037	926
23-0357	12	AUXVASSE	N 3901	W 09154	873
23-0539	12	BELLEVIEW	N 3741	W 09044	1040
23-0595	12	BERNIE	N 3640	W 08958	300
23-0608	13	BETHANY	N 4016	W 09402	900
23-0735	12	BLOOMFIELD	N 3653	W 08956	440
23-0789	13	BOLIVAR 1NE	N 3736	W 09325	1080
23-0817	13	BOONVILLE WATERWORKS	N 3858	W 09245	760
23-0856	12	BOWLING GREEN 2 NE	N 3922	W 09111	710
23-0980	13	BROOKFIELD	N 3948	W 09305	790
23-1037	13	BRUNSWICK	N 3925	W 09308	652
23-1087	12	BUFFALO 3 S	N 3736	W 09306	1150
23-1101	12	BUNKER	N 3727	W 09113	1380
23-1141	12	BURLINGTON JUNCTION	N 4027	W 09504	960
23-1145	12	BUTLER	N 3815	W 09419	790
23-1212	13	CAMDENTON	N 3800	W 09245	1040
23-1275	13	CANTON LOCK AND DAM 20//	N 4009	W 09131	488
23-1289	12	CAPE GIRARDEAU FAA	N 3714	W 08935	340
23-1304	12	CAPLINGER MILLS	N 3748	W 09348	750
23-1340	13	CARROLLTON	N 3922	W 09330	750
23-1356	13	CARTHAGE	N 3709	W 09419	1040
23-1364	13	CARUTHERSVILLE	N 3611	W 08939	270
23-1383	12	CASSVILLE RANGER STA	N 3641	W 09352	1340
23-1467	12	CENTERVILLE	N 3726	W 09058	840
23-1482	12	CENTRALIA	N 3913	W 09208	870
23-1540	13	CHARLESTON	N 3656	W 08921	325
23-1580	12	CHILLICOTHE	N 3946	W 09333	700
23-1585	13	CHILLICOTHE RADIO KCHI	N 3948	W 09333	785
23-1674	13	CLEARWATER DAM	N 3707	W 09047	648
23-1704	12	CLIFTON CITY	N 3845	W 09301	720
23-1711	13	CLINTON 3 NW	N 3823	W 09349	770
23-1773	12	COLOMA	N 3932	W 09332	780
23-1791	13	COLUMBIA REGION WSO	N 3849	W 09213	887
23-1822	13	CONCEPTION	N 4015	W 09441	1108
23-1837	12	CONCORDIA	N 3859	W 09334	780
23-1870	12	COOK STATION	N 3749	W 09126	990
23-2220	12	DE SOTO	N 3809	W 09033	553
23-2235	12	DEXTER	N 3648	W 08958	375
23-2289	13	DONIPHAN	N 3637	W 09050	344
23-2474	13	EDGERTON	N 3930	W 09437	840
23-2482	12	EDINA	N 4010	W 09210	780
23-2503	13	ELDON	N 3821	W 09235	930
23-2547	12	ELLINGTON	N 3714	W 09058	695
23-2591	13	ELSBERRY 1 S	N 3909	W 09047	449
23-2729	12	FAIRFAX	N 4020	W 09524	960
23-2809	13	FARMINGTON	N 3747	W 09023	930

23 — MISSOURI

LEGEND
11 = TEMPERATURE ONLY
12 = PRECIPITATION ONLY
13 = TEMP. & PRECIP.

STATE-STATION NUMBER	STN TYP	NAME	LATITUDE DEG-MIN	LONGITUDE DEG-MIN	ELEVATION (FT)
23-2823	13	FAYETTE	N 3909	W 09241	760
23-2995	13	FOUNTAIN GROVE WL	N 3942	W 09318	670
23-3038	12	FREDERICKTOWN	N 3734	W 09018	700
23-3079	13	FULTON 4 SW	N 3848	W 09200	860
23-3094	12	GALENA	N 3648	W 09328	1025
23-3369	13	GRANT CITY	N 4029	W 09425	1134
23-3451	13	GREENVILLE 4 NNW	N 3711	W 09028	410
23-3463	12	GREGORY LANDING	N 4017	W 09130	510
23-3483	12	GROVESPRING	N 3724	W 09237	1320
23-3568	13	HAMILTON 2 W	N 3945	W 09402	900
23-3601	13	HANNIBAL WATERWORKS	N 3943	W 09122	712
23-3649	13	HARRISONVILLE	N 3839	W 09420	904
23-3793	12	HERMANN	N 3842	W 09126	600
23-4019	12	HOUSTON	N 3720	W 09158	1190
23-4226	13	JACKSON	N 3723	W 08940	435
23-4271	13	JEFFERSON CITY L U	N 3834	W 09211	640
23-4291	12	JEROME	N 3755	W 09159	710
23-4315	13	JOPLIN FAA AIRPORT	N 3710	W 09430	982
23-4358	13	KANSAS CITY WSO	N 3919	W 09443	1014
23-4359	13	KANSAS CITY FAA AP	N 3907	W 09436	742
23-4544	13	KIRKSVILLE RADIO KIRX	N 4013	W 09235	970
23-4637	12	LA BELLE	N 4007	W 09155	770
23-4694	13	LAKESIDE	N 3812	W 09237	592
23-4705	13	LAMAR	N 3730	W 09417	960
23-4825	13	LEBANON	N 3742	W 09241	1200
23-4904	13	LEXINGTON	N 3911	W 09353	836
23-4919	13	LICKING 4 N	N 3733	W 09154	1277
23-5027	13	LOCKWOOD	N 3723	W 09357	1078
23-5093	13	LOUISIANA STARKS NUR	N 3926	W 09104	469
23-5098	12	LOUISIANA	N 3927	W 09103	482
23-5121	12	LUCERNE	N 4028	W 09318	950
23-5175	13	MACON	N 3945	W 09228	870
23-5183	12	MADISON	N 3928	W 09213	974
23-5227	12	MANSFIELD	N 3707	W 09235	1470
23-5253	13	MARBLE HILL	N 3718	W 08958	390
23-5298	12	MARSHALL	N 3907	W 09311	790
23-5307	13	MARSHFIELD	N 3720	W 09254	1490
23-5319	12	MARTINSBURG	N 3906	W 09139	807
23-5340	13	MARYVILLE 2 E	N 4021	W 09450	985
23-5492	12	MEMPHIS	N 4028	W 09210	760
23-5541	13	MEXICO	N 3911	W 09154	775
23-5578	12	MILAN	N 4012	W 09307	830
23-5671	13	MOBERLY RADIO KWIX	N 3924	W 09226	850
23-5708	12	MONROE CITY	N 3939	W 09144	720
23-5834	13	MOUNTAIN GROVE 2 N	N 3709	W 09216	1463
23-5862	13	MOUNT VERNON MU FARM	N 3704	W 09353	1220
23-5976	13	NEOSHO	N 3652	W 09422	1011
23-5987	13	NEVADA SEWAGE PLANT	N 3751	W 09422	780
23-6007	13	NEW FLORENCE	N 3855	W 09127	880
23-6269	12	ODESSA	N 3900	W 09358	898

STATE-STATION NUMBER	STN TYP	NAME	LATITUDE DEG-MIN	LONGITUDE DEG-MIN	ELEVATION (FT)
23-6357	13	OREGON	N 3959	W 09508	1090
23-6438	12	OWENSVILLE	N 3821	W 09130	940
23-6452	12	OZARK	N 3701	W 09313	1150
23-6460	13	OZARK BEACH	N 3640	W 09307	700
23-6468	12	PACIFIC	N 3829	W 09044	432
23-6493	12	PALMYRA	N 3948	W 09130	580
23-6509	12	PARIS 1 SW	N 3928	W 09201	760
23-6532	12	PARMA	N 3637	W 08949	280
23-6641	12	PERRYVILLE WATER PLANT	N 3744	W 08955	480
23-6678	12	PIERCE CITY	N 3657	W 09400	1230
23-6775	12	POLO	N 3933	W 09402	1018
23-6791	13	POPLAR BLUFF R S	N 3646	W 09025	380
23-6866	12	PRINCETON	N 4024	W 09335	960
23-6934	12	PUXICO	N 3657	W 09010	360
23-6970	12	QULIN	N 3636	W 09015	315
23-7094	12	REYNOLDS	N 3724	W 09105	1240
23-7122	12	RICHWOODS 2 NE	N 3810	W 09049	910
23-7263	12	ROLLA, UNIV OF MO.	N 3757	W 09146	1202
23-7309	12	ROUND SPRING 3 NNW	N 3719	W 09126	750
23-7397	13	SAINT CHARLES	N 3847	W 09030	470
23-7435	13	ST. JOSEPH WBAP	N 3946	W 09455	811
23-7455	13	SAINT LOUIS WSO R	N 3845	W 09023	535
23-7506	13	SALEM	N 3738	W 09132	1200
23-7514	13	SALISBURY	N 3926	W 09248	723
23-7578	13	SAVERTON L AND D 22 //	N 3938	W 09115	472
23-7632	13	SEDALIA	N 3840	W 09313	780
23-7645	13	SELIGMAN	N 3631	W 09356	1542
23-7720	13	SHELBINA	N 3941	W 09203	775
23-7772	13	SIKESTON PWR PLT	N 3653	W 08936	325
23-7780	12	SILOAM SPRINGS	N 3649	W 09204	1260
23-7976	13	SPRINGFIELD WSO //R	N 3714	W 09323	1268
23-8051	13	STEFFENVILLE 1 E	N 3958	W 09153	650
23-8112	12	STOVER	N 3827	W 09259	1110
23-8171	12	SULLIVAN	N 3812	W 09110	950
23-8184	12	SUMMERSVILLE	N 3711	W 09140	1180
23-8188	12	SUMNER 2 SW	N 3938	W 09317	693
23-8223	13	SWEET SPRINGS	N 3858	W 09325	680
23-8289	13	TARKIO	N 4027	W 09523	1015
23-8313	12	TECUMSEH 2 N	N 3637	W 09217	600
23-8444	13	TRENTON	N 4005	W 09337	820
23-8456	12	TROY	N 3858	W 09059	580
23-8515	13	UNION	N 3827	W 09100	540
23-8523	13	UNIONVILLE	N 4029	W 09300	1062
23-8561	12	VALLEY PARK	N 3833	W 09029	420
23-8603	12	VERSAILLES	N 3826	W 09251	1032
23-8614	13	VICHY FAA AIRPORT	N 3808	W 09146	1125
23-8664	12	WACO	N 3715	W 09436	900
23-8700	13	WAPPAPELLO DAM	N 3656	W 09017	330
23-8712	13	WARRENSBURG	N 3846	W 09344	840
23-8725	12	WARRENTON 1 N	N 3849	W 09108	845

23 – MISSOURI

LEGEND
11 = TEMPERATURE ONLY
12 = PRECIPITATION ONLY
13 = TEMP. & PRECIP.

STATE-STATION NUMBER	STN TYP	NAME	LATITUDE DEG-MIN	LONGITUDE DEG-MIN	ELEVATION (FT)
23-8733	13	WARSAW NO 1	N 3815	W 09322	705
23-8754	13	WASOLA	N 3648	W 09234	1294
23-8768	12	WAVERLY	N 3912	W 09331	800
23-8777	13	WAYNESVILLE 2 W	N 3749	W 09214	885
23-8880	13	WEST PLAINS	N 3644	W 09151	1006
23-8984	12	WILLIAMSVILLE	N 3658	W 09033	520
23-8995	13	WILLOW SPRINGS	N 3659	W 09158	1220
23-9032	13	WINDSOR	N 3832	W 09331	900
23-9178	12	ZALMA 4 E	N 3708	W 09000	440

23 — MISSOURI

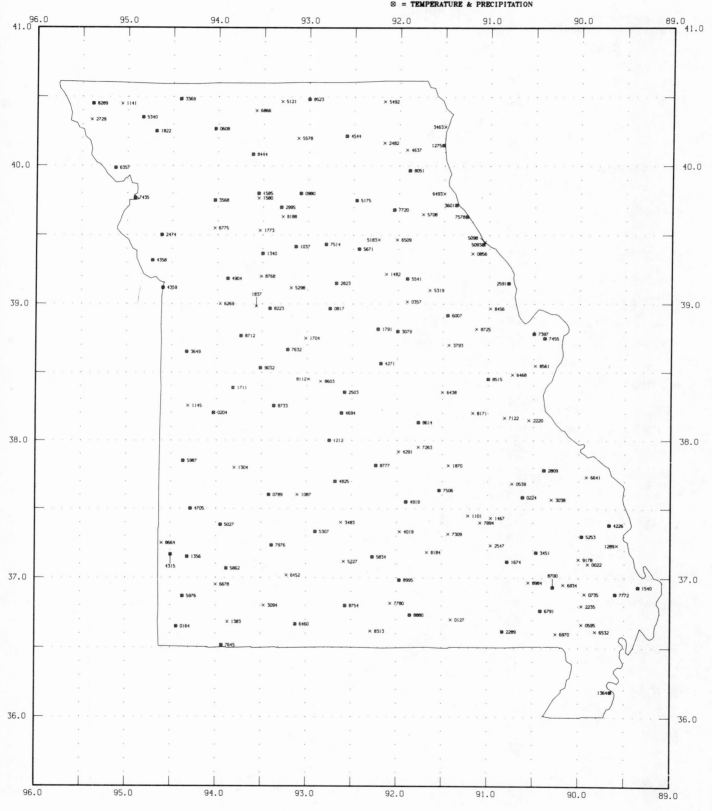

349

MONTANA

TEMPERATURE NORMALS (DEG F)

STATION		JAN	FEB	MAR	APR	MAY	JUN	JUL	AUG	SEP	OCT	NOV	DEC	ANN
AUGUSTA	MAX	32.7	40.8	45.0	55.8	65.7	73.5	83.2	81.7	72.4	62.4	46.4	38.2	58.2
	MIN	8.5	15.8	18.6	27.8	37.5	45.0	49.2	47.7	39.4	32.2	21.5	14.9	29.8
	MEAN	20.6	28.3	31.8	41.8	51.6	59.3	66.2	64.7	55.9	47.3	34.0	26.5	44.0
AUSTIN 1 W	MAX	29.3	35.8	40.3	51.7	61.9	69.8	80.3	78.2	67.8	56.5	40.6	33.3	53.8
	MIN	9.7	16.8	18.5	27.4	36.0	42.8	48.4	46.9	39.2	31.9	21.2	15.0	29.5
	MEAN	19.5	26.3	29.4	39.5	49.0	56.4	64.4	62.6	53.5	44.2	30.9	24.2	41.7
BALLANTINE	MAX	31.5	40.2	48.1	60.8	71.2	80.5	90.6	88.4	77.0	66.0	47.5	37.6	61.6
	MIN	6.7	14.3	20.0	30.6	41.0	49.3	54.7	52.1	42.3	32.7	20.9	13.1	31.5
	MEAN	19.1	27.3	34.1	45.7	56.1	64.9	72.6	70.3	59.7	49.4	34.2	25.4	46.6
BARBER	MAX	33.4	40.7	45.3	57.1	67.0	75.1	84.2	81.7	71.6	62.4	46.2	38.6	58.6
	MIN	7.4	14.7	18.3	28.6	38.3	46.0	51.2	49.0	39.8	31.0	19.4	12.7	29.7
	MEAN	20.5	27.7	31.8	42.9	52.7	60.6	67.7	65.4	55.7	46.7	32.8	25.7	44.2
BELGRADE AP	MAX	27.9	34.8	40.0	53.0	64.0	73.2	84.5	82.3	70.4	58.6	41.8	33.2	55.3
	MIN	4.4	12.2	16.6	28.3	37.1	44.2	49.2	47.7	38.9	29.7	18.0	10.2	28.0
	MEAN	16.2	23.5	28.3	40.7	50.6	58.7	66.9	65.0	54.6	44.2	29.9	21.7	41.7
BIGFORK 13 S	MAX	31.9	37.9	44.0	54.8	64.4	71.9	81.7	79.8	68.4	55.4	41.3	35.7	55.6
	MIN	19.7	24.2	26.4	33.3	40.2	46.6	51.3	49.9	42.9	36.0	28.9	24.5	35.3
	MEAN	25.8	31.0	35.3	44.1	52.3	59.3	66.5	64.9	55.7	45.7	35.1	30.1	45.5
BIG SANDY	MAX	25.2	34.5	43.6	58.3	70.0	78.4	87.8	86.5	74.7	63.0	43.8	32.3	58.2
	MIN	1.6	9.8	18.0	30.4	40.5	48.2	52.5	50.6	40.8	31.2	18.1	8.7	29.2
	MEAN	13.4	22.2	30.8	44.4	55.3	63.3	70.2	68.6	57.8	47.1	31.0	20.5	43.7
BIG TIMBER	MAX	35.2	41.8	47.1	57.7	68.0	77.1	87.0	85.3	74.3	63.2	46.9	39.9	60.3
	MIN	14.8	20.9	22.7	31.0	39.8	47.2	52.5	50.5	41.9	35.2	25.5	20.0	33.5
	MEAN	25.0	31.4	34.9	44.4	53.9	62.2	69.8	67.9	58.1	49.3	36.2	30.0	46.9
BILLINGS WATER PLANT	MAX	34.6	43.0	49.8	61.4	71.2	79.9	89.0	87.3	76.8	66.2	48.8	40.0	62.3
	MIN	9.8	17.1	22.0	31.7	41.6	49.1	54.8	52.3	42.7	34.1	22.9	15.8	32.8
	MEAN	22.2	30.0	35.9	46.6	56.5	64.5	71.9	69.8	59.8	50.2	35.9	27.9	47.6
BILLINGS WSO	R MAX	29.9	37.9	44.0	55.9	66.4	76.3	86.6	84.3	72.3	61.0	44.4	36.0	57.9
	MIN	11.8	18.8	23.6	33.2	43.3	51.6	58.0	56.2	46.5	37.5	25.5	18.2	35.4
	MEAN	20.9	28.4	33.8	44.6	54.9	64.0	72.3	70.3	59.4	49.3	35.0	27.1	46.7
BOULDER STATE SCHOOL	MAX	31.7	38.4	42.9	53.9	63.9	72.4	83.0	81.2	70.9	59.9	43.2	35.8	56.4
	MIN	7.5	14.1	17.4	26.5	35.0	42.4	47.7	45.8	36.6	28.4	18.2	12.4	27.7
	MEAN	19.6	26.3	30.2	40.3	49.5	57.5	65.4	63.5	53.8	44.2	30.7	24.1	42.1
BOZEMAN MONT ST UNIV	MAX	31.1	37.3	41.7	53.1	63.4	72.3	82.1	80.5	70.2	59.0	42.4	35.4	55.7
	MIN	11.3	17.5	20.3	30.0	38.7	45.8	51.8	50.1	41.6	33.2	22.2	16.1	31.6
	MEAN	21.2	27.4	31.0	41.6	51.1	59.1	67.0	65.3	55.9	46.1	32.4	25.8	43.7
BOZEMAN 12 NE	MAX	30.2	35.3	38.4	47.1	57.4	66.0	75.7	74.5	64.4	54.2	39.9	33.8	51.4
	MIN	7.8	12.0	13.5	22.6	30.8	37.2	40.5	39.2	33.0	26.9	17.5	11.7	24.4
	MEAN	19.0	23.7	25.9	34.9	44.1	51.6	58.2	56.9	48.7	40.5	28.7	22.8	37.9
BRADY AZNOE	MAX	28.3	37.7	43.3	56.0	66.9	74.8	84.0	82.5	72.2	61.6	44.9	35.1	57.3
	MIN	3.9	11.5	16.2	26.2	36.6	44.1	47.8	46.9	38.4	30.1	18.2	10.8	27.6
	MEAN	16.1	24.6	29.8	41.1	51.8	59.5	65.9	64.8	55.3	45.9	31.6	23.0	42.5
BREDETTE	MAX	17.7	25.4	36.2	53.6	67.2	76.2	84.0	82.9	70.7	58.4	38.5	25.5	53.0
	MIN	-3.5	4.1	14.2	28.8	40.2	48.7	53.6	51.5	41.5	31.2	16.5	5.2	27.7
	MEAN	7.1	14.8	25.2	41.2	53.8	62.4	68.8	67.2	56.1	44.8	27.5	15.3	40.4
BRIDGER	MAX	32.7	40.9	47.8	58.8	69.5	78.3	87.6	85.8	74.9	63.9	46.5	38.0	60.4
	MIN	10.3	17.9	22.0	31.4	40.8	48.0	53.4	51.2	42.6	34.6	23.7	16.6	32.7
	MEAN	21.5	29.4	34.9	45.1	55.2	63.2	70.5	68.5	58.8	49.3	35.1	27.3	46.6
BROADUS	MAX	30.0	37.3	44.7	57.2	68.3	77.8	87.0	85.5	74.2	62.4	45.1	35.6	58.8
	MIN	3.6	11.6	18.9	30.2	40.9	50.0	55.3	52.4	41.2	30.6	18.5	9.8	30.3
	MEAN	16.8	24.5	31.8	43.7	54.6	63.9	71.2	69.0	57.7	46.5	31.8	22.7	44.5

MONTANA

TEMPERATURE NORMALS (DEG F)

STATION		JAN	FEB	MAR	APR	MAY	JUN	JUL	AUG	SEP	OCT	NOV	DEC	ANN
BROWNING	MAX	26.5	33.4	37.4	49.0	59.8	67.7	77.2	75.6	66.4	55.9	39.7	31.8	51.7
	MIN	7.1	13.9	17.5	27.5	36.2	42.8	47.3	46.1	39.3	32.8	21.0	14.2	28.8
	MEAN	16.8	23.7	27.5	38.2	48.0	55.3	62.3	60.9	52.9	44.4	30.4	23.0	40.3
BUTTE FAA AP	MAX	27.1	33.2	38.0	49.2	60.2	69.2	80.1	77.9	67.0	55.5	39.3	31.0	52.3
	MIN	3.7	9.8	14.5	25.5	34.3	41.4	45.8	43.6	35.2	26.4	15.7	8.1	25.3
	MEAN	15.5	21.5	26.3	37.4	47.3	55.3	63.0	60.8	51.1	41.0	27.5	19.6	38.8
CASCADE 5 S	MAX	32.2	40.0	44.6	55.8	66.1	74.2	84.2	82.6	72.1	61.9	46.2	38.3	58.2
	MIN	10.9	18.2	21.0	31.0	39.6	47.1	50.5	49.0	41.6	35.6	25.1	18.3	32.3
	MEAN	21.6	29.1	32.8	43.4	52.9	60.6	67.4	65.8	56.9	48.7	35.7	28.3	45.3
CHESTER	MAX	24.4	33.8	41.5	56.4	67.7	75.6	84.2	82.9	72.2	60.7	42.6	31.1	56.1
	MIN	-1.2	6.7	13.9	26.3	37.5	45.0	49.4	47.4	37.9	27.5	14.8	5.4	25.9
	MEAN	11.6	20.3	27.7	41.4	52.6	60.3	66.8	65.2	55.1	44.1	28.7	18.3	41.0
CHINOOK	MAX	23.3	32.0	41.9	58.0	70.2	78.0	86.3	84.3	73.4	62.2	42.9	30.9	57.0
	MIN	.2	8.1	16.7	29.8	41.0	48.8	53.2	51.0	41.0	31.5	17.9	7.8	28.9
	MEAN	11.8	20.0	29.3	43.9	55.6	63.5	69.8	67.7	57.2	46.9	30.4	19.4	43.0
CHOTEAU AIRPORT	MAX	31.3	38.9	43.5	54.6	65.4	73.3	82.4	80.6	70.6	60.9	44.8	36.9	56.9
	MIN	8.5	15.8	19.6	29.2	38.9	46.7	51.2	49.5	41.4	34.0	22.5	15.7	31.1
	MEAN	19.9	27.4	31.6	41.9	52.2	60.0	66.8	65.1	56.0	47.5	33.7	26.3	44.0
CIRCLE	MAX	23.2	31.1	40.0	55.8	68.3	77.7	86.1	84.9	72.5	60.8	42.0	31.0	56.1
	MIN	-.4	7.3	16.0	29.0	39.8	49.1	53.7	51.5	41.0	30.6	17.9	7.9	28.6
	MEAN	11.4	19.2	28.0	42.4	54.1	63.4	69.9	68.2	56.7	45.7	30.0	19.5	42.4
COLSTRIP	MAX	33.1	40.6	47.1	58.9	70.0	79.4	89.2	87.3	76.2	64.4	46.9	38.4	61.0
	MIN	6.8	14.3	20.0	30.3	40.3	48.9	54.4	52.3	42.3	33.1	20.9	13.5	31.4
	MEAN	20.0	27.5	33.6	44.6	55.2	64.2	71.8	69.8	59.3	48.8	33.9	26.0	46.2
COLUMBUS	MAX	34.1	42.6	48.1	59.2	69.5	78.2	88.0	86.0	75.3	64.8	47.6	39.0	61.0
	MIN	8.0	15.3	20.4	29.7	39.4	46.9	52.0	49.7	40.9	31.8	20.6	13.2	30.7
	MEAN	21.1	29.0	34.3	44.5	54.5	62.6	70.0	67.9	58.1	48.3	34.1	26.1	45.9
CONRAD AIRPORT	MAX	29.1	37.5	43.3	55.5	66.6	73.8	82.5	81.2	71.5	61.7	44.8	35.1	56.9
	MIN	4.8	12.7	17.6	27.9	38.0	45.9	49.8	48.3	39.7	30.8	18.5	11.7	28.8
	MEAN	17.0	25.1	30.4	41.7	52.3	59.9	66.1	64.8	55.7	46.3	31.7	23.4	42.9
COOKE CITY	MAX	23.5	29.8	34.8	44.4	55.0	64.9	74.3	71.5	62.6	50.9	33.3	24.8	47.5
	MIN	3.5	6.6	8.4	17.8	26.8	32.8	37.4	35.9	29.3	22.4	12.1	5.0	19.8
	MEAN	13.5	18.2	21.6	31.1	40.9	48.8	55.9	53.7	46.0	36.7	22.8	14.9	33.7
CRESTON	MAX	29.4	36.4	42.9	54.9	65.1	72.1	80.8	79.2	68.6	55.6	40.2	33.1	54.9
	MIN	14.1	19.7	23.1	31.1	38.1	44.8	47.7	46.4	38.8	31.4	24.6	19.6	31.6
	MEAN	21.8	28.1	33.0	43.0	51.6	58.5	64.3	62.8	53.7	43.5	32.4	26.4	43.3
CROW AGENCY	MAX	32.6	40.7	48.6	61.0	71.4	80.5	90.2	88.7	77.3	66.4	48.4	38.6	62.0
	MIN	4.9	12.8	19.7	30.4	40.5	48.4	53.2	50.8	41.0	31.1	19.4	10.8	30.3
	MEAN	18.8	26.8	34.1	45.7	56.0	64.5	71.7	69.7	59.2	48.8	33.9	24.7	46.2
CULBERTSON	MAX	18.8	27.4	38.8	56.3	69.3	78.2	85.9	84.8	72.3	60.3	40.0	26.8	54.9
	MIN	-4.6	3.4	13.5	27.9	39.3	48.4	52.8	50.2	39.9	29.5	15.3	3.8	26.6
	MEAN	7.1	15.4	26.2	42.1	54.3	63.3	69.4	67.5	56.1	44.9	27.7	15.4	40.8
CUT BANK FAA AP	MAX	24.7	32.8	38.2	50.6	61.8	69.8	78.9	77.2	67.1	56.5	39.9	30.9	52.4
	MIN	4.2	12.0	16.5	26.9	36.9	44.6	49.0	47.4	39.0	30.9	18.6	11.2	28.1
	MEAN	14.5	22.4	27.4	38.8	49.4	57.2	64.0	62.3	53.1	43.7	29.3	21.1	40.3
DARBY	MAX	34.9	42.2	47.5	57.1	66.4	74.1	83.9	82.1	72.0	60.8	45.1	38.1	58.7
	MIN	16.9	22.0	24.1	30.5	37.7	43.7	47.0	45.9	39.5	32.5	24.5	20.3	32.1
	MEAN	25.9	32.1	35.8	43.8	52.1	58.9	65.5	64.0	55.8	46.7	34.8	29.2	45.4
DILLON FAA AP	MAX	30.8	37.3	42.1	53.3	64.0	73.0	83.3	81.1	70.4	58.8	42.4	34.4	55.9
	MIN	9.4	14.8	18.2	27.7	36.4	43.8	49.1	47.5	39.1	30.8	19.9	13.5	29.2
	MEAN	20.1	26.1	30.2	40.5	50.3	58.4	66.2	64.3	54.8	44.8	31.2	24.0	42.6

MONTANA

TEMPERATURE NORMALS (DEG F)

STATION		JAN	FEB	MAR	APR	MAY	JUN	JUL	AUG	SEP	OCT	NOV	DEC	ANN
DILLON WMCE	MAX	33.7	40.3	45.1	55.8	65.9	74.3	84.0	81.7	72.3	61.8	44.8	36.9	58.1
	MIN	11.1	16.5	19.4	27.7	35.7	42.6	46.8	45.1	37.9	31.0	21.0	15.1	29.2
	MEAN	22.4	28.4	32.2	41.8	50.8	58.5	65.4	63.4	55.1	46.4	32.9	26.0	43.6
DIVIDE 2 NW	MAX	29.6	36.0	40.6	50.9	60.9	69.2	79.4	77.0	67.3	56.7	40.7	32.4	53.4
	MIN	8.5	13.9	17.4	26.7	35.0	41.9	47.3	45.8	37.8	29.7	19.2	12.5	28.0
	MEAN	19.1	25.0	29.0	38.8	48.0	55.6	63.3	61.4	52.6	43.2	30.0	22.5	40.7
EAST ANACONDA	MAX	30.0	35.2	38.6	48.7	59.4	68.1	78.7	76.8	66.2	55.0	39.8	33.4	52.5
	MIN	14.5	19.3	20.8	28.9	37.6	44.9	51.6	50.2	42.3	34.3	23.7	18.7	32.2
	MEAN	22.3	27.3	29.7	38.8	48.6	56.5	65.2	63.6	54.3	44.7	31.8	26.1	42.4
EKALAKA	MAX	28.0	34.3	41.5	55.7	67.5	76.8	85.7	84.6	72.5	60.5	42.9	33.9	57.0
	MIN	4.4	11.2	18.0	29.4	40.5	49.3	55.3	53.1	42.4	32.3	19.0	11.3	30.5
	MEAN	16.3	22.8	29.7	42.6	54.0	63.1	70.5	68.9	57.4	46.4	31.0	22.6	43.8
ENNIS	MAX	32.3	38.2	43.2	54.2	65.0	73.5	82.9	81.0	70.7	60.0	43.5	35.6	56.7
	MIN	13.2	18.1	20.7	28.6	36.3	43.0	47.3	45.5	38.1	31.4	22.9	17.5	30.2
	MEAN	22.7	28.2	32.0	41.4	50.6	58.2	65.2	63.3	54.4	45.7	33.2	26.6	43.5
FAIRFIELD	MAX	30.1	38.6	44.0	55.2	65.6	72.9	81.7	80.3	71.5	61.2	44.4	35.7	56.8
	MIN	9.2	16.6	19.9	29.4	38.7	46.1	51.0	49.8	41.9	34.8	23.3	16.1	31.4
	MEAN	19.7	27.6	32.0	42.3	52.2	59.5	66.3	65.1	56.7	48.0	33.9	25.9	44.1
FLATWILLOW 4 ENE	MAX	31.1	38.8	45.0	57.5	68.4	77.5	87.4	85.8	74.1	63.4	46.6	37.7	59.4
	MIN	6.3	13.9	19.4	30.1	39.7	47.7	53.2	50.9	42.3	33.3	21.1	13.2	30.9
	MEAN	18.7	26.4	32.2	43.8	54.1	62.6	70.3	68.4	58.2	48.4	33.9	25.5	45.2
FORKS 4 NNE	MAX	17.6	25.4	35.3	53.1	66.0	75.0	83.3	82.1	70.2	58.0	38.6	25.8	52.5
	MIN	-3.4	3.8	12.8	27.8	38.9	47.1	52.4	50.9	40.7	30.9	16.0	5.1	26.9
	MEAN	7.1	14.6	24.1	40.5	52.5	61.1	67.9	66.5	55.5	44.5	27.3	15.5	39.8
FORT ASSINNIBOINE	MAX	23.3	31.6	41.0	56.5	68.7	77.2	85.8	84.7	72.8	61.0	42.8	31.0	56.4
	MIN	1.2	8.9	16.5	29.3	40.2	48.2	53.0	51.1	41.5	32.1	18.5	8.6	29.1
	MEAN	12.3	20.2	28.8	42.9	54.5	62.7	69.4	67.9	57.2	46.6	30.7	19.8	42.8
FORT BENTON	MAX	29.7	38.8	46.1	59.0	69.5	77.5	86.2	84.4	73.8	63.3	46.4	36.4	59.3
	MIN	6.1	13.7	19.7	30.5	40.7	48.1	52.4	50.5	41.0	32.0	20.2	12.7	30.6
	MEAN	17.9	26.3	33.0	44.8	55.1	62.8	69.3	67.4	57.4	47.7	33.4	24.6	45.0
FORTINE 1 N	MAX	28.1	37.7	44.7	55.7	66.0	72.9	82.2	80.3	70.4	56.5	39.0	31.5	55.4
	MIN	11.5	18.2	22.0	29.0	35.8	42.5	45.3	43.8	37.1	29.9	22.4	17.0	29.5
	MEAN	19.8	28.0	33.4	42.4	50.9	57.7	63.8	62.1	53.8	43.2	30.7	24.3	42.5
FORT PECK POWER PLANT	MAX	21.6	28.7	38.8	55.3	68.1	77.6	85.8	84.5	72.4	60.6	42.5	30.2	55.5
	MIN	1.1	8.1	17.7	32.1	43.5	52.5	57.7	55.7	45.8	36.7	22.3	10.4	32.0
	MEAN	11.4	18.4	28.3	43.8	55.8	65.1	71.8	70.1	59.1	48.7	32.4	20.4	43.8
GERALDINE	MAX	30.9	38.8	44.5	56.6	67.3	75.3	84.8	83.7	72.8	62.4	46.1	37.3	58.4
	MIN	7.0	14.5	19.7	29.9	39.6	47.0	51.3	50.1	41.2	32.9	20.7	13.4	30.6
	MEAN	19.0	26.7	32.2	43.3	53.5	61.2	68.1	66.9	57.0	47.6	33.4	25.4	44.5
GIBSON DAM	MAX	31.1	38.1	40.7	50.3	60.1	67.9	77.9	76.4	67.2	57.2	42.3	35.6	53.7
	MIN	11.2	17.0	18.9	27.4	35.2	42.1	46.1	45.0	38.3	32.5	23.3	17.3	29.5
	MEAN	21.2	27.6	29.8	38.9	47.7	55.0	62.0	60.7	52.8	44.9	32.9	26.5	41.7
GLASGOW WSO //R	MAX	17.7	25.5	36.6	54.2	67.0	75.8	84.1	82.7	70.5	58.7	39.5	26.4	53.2
	MIN	-1.3	6.3	15.7	30.7	42.1	51.0	56.7	54.9	44.2	33.6	18.8	7.3	30.0
	MEAN	8.2	15.9	26.2	42.5	54.6	63.4	70.4	68.8	57.4	46.2	29.2	16.9	41.6
GLENDIVE	MAX	24.9	33.5	44.0	60.0	72.3	81.1	89.4	88.5	76.5	64.5	44.3	32.2	59.3
	MIN	2.1	10.0	19.0	33.0	44.7	53.9	59.3	56.2	45.2	35.0	21.7	10.6	32.6
	MEAN	13.5	21.8	31.5	46.5	58.5	67.5	74.3	72.4	60.9	49.8	33.0	21.4	45.9
GOLDBUTTE 7 N	MAX	27.3	35.2	40.5	53.1	64.1	71.6	79.8	78.5	68.4	58.2	42.4	33.8	54.4
	MIN	4.5	13.0	17.4	28.4	38.0	45.2	49.0	47.5	39.9	32.4	20.0	12.4	29.0
	MEAN	15.9	24.1	29.0	40.8	51.1	58.4	64.5	63.1	54.2	45.3	31.3	23.1	41.7

MONTANA

TEMPERATURE NORMALS (DEG F)

STATION		JAN	FEB	MAR	APR	MAY	JUN	JUL	AUG	SEP	OCT	NOV	DEC	ANN
GRASSRANGE	MAX	33.3	40.7	46.0	57.2	67.4	76.3	85.7	84.3	73.4	63.6	47.7	39.3	59.6
	MIN	6.6	13.5	18.4	28.3	38.0	45.3	50.5	48.6	40.0	31.6	19.9	12.9	29.5
	MEAN	20.0	27.1	32.2	42.8	52.7	60.8	68.1	66.5	56.8	47.6	33.8	26.1	44.5
GREAT FALLS WSO //R	MAX	28.2	36.5	41.7	54.0	65.3	74.3	84.2	82.0	70.5	59.5	43.5	34.7	56.2
	MIN	9.2	16.8	21.1	31.3	41.1	49.4	54.4	53.0	44.2	36.2	24.5	16.6	33.2
	MEAN	18.7	26.7	31.4	42.7	53.2	61.9	69.3	67.5	57.4	47.9	34.0	25.7	44.7
HAMILTON	MAX	33.7	41.3	47.3	57.6	66.7	74.4	84.1	82.0	71.7	59.6	44.0	36.8	58.3
	MIN	16.3	22.2	25.3	32.2	39.4	45.8	49.8	48.3	40.8	32.1	24.1	19.9	33.0
	MEAN	25.0	31.8	36.3	44.9	53.1	60.1	67.0	65.1	56.3	45.9	34.1	28.3	45.7
HARDIN	MAX	30.8	38.7	46.4	59.4	70.6	79.4	89.7	87.9	75.8	64.5	47.0	37.3	60.6
	MIN	4.4	12.6	19.3	30.4	41.1	49.3	54.2	51.8	41.1	30.6	19.1	10.7	30.4
	MEAN	17.6	25.7	32.9	44.9	55.8	64.3	72.0	69.9	58.5	47.5	33.1	24.0	45.5
HARLEM	MAX	21.8	31.0	41.1	57.6	69.7	77.9	86.0	84.6	73.6	61.6	42.2	29.0	56.3
	MIN	-1.4	6.3	15.2	29.1	39.8	47.8	52.3	50.4	40.1	30.3	16.1	5.6	27.6
	MEAN	10.2	18.7	28.2	43.4	54.8	62.8	69.2	67.5	56.9	46.0	29.2	17.3	42.0
HARLOWTON	MAX	31.9	39.1	43.7	55.0	65.4	73.8	83.4	81.6	71.1	60.5	44.6	36.9	57.3
	MIN	9.2	15.6	18.7	27.8	37.0	44.5	49.5	47.6	38.9	31.8	21.0	15.0	29.7
	MEAN	20.6	27.4	31.2	41.4	51.2	59.2	66.5	64.7	55.0	46.2	32.8	26.0	43.5
HAUGAN 3 E	MAX	31.3	39.9	45.8	56.5	67.1	74.5	84.5	83.1	73.3	58.8	40.9	33.4	57.4
	MIN	13.5	18.8	20.0	27.0	33.6	40.1	41.5	40.1	35.1	29.7	23.5	18.3	28.4
	MEAN	22.5	29.4	32.9	41.8	50.4	57.3	63.1	61.7	54.2	44.3	32.2	25.9	43.0
HAVRE WSO R	MAX	21.5	30.3	39.8	55.3	67.8	76.4	85.4	83.6	71.6	60.0	41.5	29.6	55.3
	MIN	.4	8.8	16.9	30.1	41.2	49.1	53.8	52.0	42.0	31.8	17.9	8.0	29.3
	MEAN	11.0	19.6	28.4	42.7	54.5	62.8	69.6	67.8	56.8	45.9	29.7	18.8	42.3
HAXBY 18 SW	MAX	25.1	34.0	42.5	57.2	68.9	77.5	86.1	84.8	73.1	61.9	43.5	33.0	57.3
	MIN	-.4	7.5	15.5	28.2	39.2	48.6	53.4	51.5	41.3	31.2	18.1	7.7	28.5
	MEAN	12.4	20.7	29.1	42.7	54.1	63.0	69.8	68.2	57.2	46.6	30.9	20.4	42.9
HEBGEN DAM	MAX	21.1	27.9	34.4	45.4	58.8	68.2	78.7	76.4	66.4	52.7	33.7	23.3	48.9
	MIN	2.1	5.4	9.5	21.4	30.9	37.9	43.1	41.8	34.9	27.5	16.9	5.5	23.1
	MEAN	11.6	16.7	22.0	33.4	44.8	53.1	60.9	59.2	50.7	40.1	25.3	14.4	36.0
HELENA WSO //R	MAX	28.1	36.2	42.5	54.7	64.9	73.1	83.6	81.3	70.3	58.6	42.3	33.3	55.7
	MIN	8.1	15.7	20.6	29.8	39.5	47.0	52.2	50.3	40.8	31.5	20.4	13.5	30.8
	MEAN	18.1	26.0	31.6	42.3	52.2	60.1	67.9	65.9	55.6	45.1	31.4	23.5	43.3
HERON 2 NW	MAX	30.7	37.7	43.9	56.2	66.6	73.0	82.0	79.9	69.1	55.2	40.0	33.1	55.6
	MIN	18.2	22.2	24.3	31.0	37.4	43.5	45.5	44.4	39.6	33.2	27.2	22.7	32.4
	MEAN	24.4	30.0	34.1	43.6	52.0	58.2	63.8	62.2	54.4	44.3	33.7	27.9	44.1
HOLTER DAM	MAX	33.0	39.7	44.3	55.2	66.0	74.5	84.7	82.9	71.3	60.2	45.6	38.2	58.0
	MIN	16.5	22.5	25.2	33.5	42.2	49.9	55.1	53.6	46.3	40.8	30.4	23.5	36.6
	MEAN	24.8	31.1	34.8	44.4	54.1	62.2	69.9	68.2	58.9	50.5	38.0	30.9	47.3
HUNGRY HORSE DAM	MAX	28.3	35.4	41.4	51.4	63.1	70.9	80.2	78.4	66.4	52.6	37.9	32.3	53.2
	MIN	14.6	19.5	22.1	30.9	39.1	45.7	50.3	49.2	41.3	33.4	25.5	20.3	32.7
	MEAN	21.5	27.5	31.8	41.2	51.1	58.3	65.3	63.8	53.9	43.0	31.7	26.3	43.0
HUNTLEY EXP STATION	MAX	30.8	39.1	46.6	59.3	69.7	78.2	87.5	85.9	74.7	63.9	46.5	37.0	59.9
	MIN	6.1	13.9	20.1	30.1	40.5	48.3	53.3	51.1	41.6	32.0	20.4	12.1	30.8
	MEAN	18.5	26.6	33.4	44.7	55.1	63.2	70.4	68.5	58.2	48.0	33.5	24.6	45.4
HYSHAM	MAX	30.9	39.5	48.2	61.4	72.1	81.0	90.3	88.6	77.2	66.2	47.4	37.3	61.7
	MIN	5.4	13.0	19.6	30.8	41.1	49.2	55.0	52.3	42.1	32.5	20.6	12.3	31.2
	MEAN	18.2	26.3	33.9	46.1	56.7	65.1	72.6	70.5	59.7	49.4	34.0	24.8	46.4
JOPLIN 1 N	MAX	22.6	31.0	38.2	52.7	64.7	72.9	82.2	80.4	69.1	57.8	40.3	29.6	53.5
	MIN	1.8	9.3	15.8	27.6	38.2	46.0	50.9	49.3	40.3	30.9	17.7	8.7	28.0
	MEAN	12.2	20.2	27.0	40.2	51.5	59.4	66.6	64.9	54.7	44.4	29.0	19.1	40.8

MONTANA

TEMPERATURE NORMALS (DEG F)

STATION		JAN	FEB	MAR	APR	MAY	JUN	JUL	AUG	SEP	OCT	NOV	DEC	ANN
JORDAN	MAX	26.5	35.4	44.3	58.7	70.5	80.2	89.2	87.9	75.7	63.6	44.8	33.8	59.2
	MIN	.8	8.9	17.0	29.4	40.1	49.3	54.4	51.6	41.0	30.7	17.9	8.1	29.1
	MEAN	13.7	22.2	30.7	44.1	55.3	64.8	71.8	69.7	58.4	47.2	31.4	21.0	44.2
KALISPELL WSO R	MAX	27.4	35.0	42.1	54.6	64.8	72.1	82.1	80.3	69.2	55.3	39.0	31.5	54.4
	MIN	11.2	17.5	21.6	30.5	38.1	44.5	47.9	46.7	38.6	29.6	22.7	16.9	30.5
	MEAN	19.3	26.3	31.9	42.6	51.5	58.3	65.0	63.5	53.9	42.5	30.9	24.2	42.5
KALISPELL	MAX	29.6	37.2	44.5	56.0	66.4	73.9	82.7	80.6	69.7	56.2	40.3	32.8	55.8
	MIN	14.7	20.5	23.9	32.1	39.9	46.1	50.3	49.1	41.2	33.2	25.2	19.8	33.0
	MEAN	22.2	28.9	34.2	44.1	53.2	60.0	66.5	64.9	55.5	44.7	32.8	26.3	44.4
LAKEVIEW	MAX	22.6	28.0	33.3	45.1	57.8	66.3	76.5	74.8	65.6	53.5	35.3	25.8	48.7
	MIN	.5	4.3	8.1	20.9	30.4	36.4	41.0	39.9	32.1	24.1	13.2	3.9	21.2
	MEAN	11.6	16.2	20.7	33.0	44.1	51.4	58.8	57.4	48.9	38.9	24.3	14.9	35.0
LEWISTOWN FAA AP	MAX	28.9	35.0	39.1	51.4	62.3	70.7	80.8	79.7	68.6	58.5	43.2	35.1	54.4
	MIN	7.3	13.5	17.3	27.8	37.1	44.5	49.9	48.9	40.3	32.1	20.5	13.5	29.4
	MEAN	18.1	24.2	28.2	39.6	49.7	57.6	65.4	64.4	54.5	45.3	31.9	24.3	41.9
LIBBY 1 NE RANGER STA	MAX	30.7	40.8	48.9	60.4	70.9	78.3	87.9	86.5	75.0	58.4	40.8	32.5	59.3
	MIN	14.7	21.2	23.5	29.6	36.5	43.0	45.5	44.3	38.4	32.3	25.4	20.1	31.2
	MEAN	22.7	31.0	36.2	45.0	53.7	60.7	66.7	65.4	56.8	45.4	33.1	26.3	45.3
LIMA	MAX	27.9	33.5	39.7	51.3	62.0	70.8	81.6	79.2	69.6	58.5	40.2	30.6	53.7
	MIN	5.3	10.0	14.0	23.8	31.8	38.9	43.6	42.1	34.9	26.9	16.2	9.0	24.7
	MEAN	16.6	21.8	26.8	37.6	46.9	54.9	62.6	60.7	52.3	42.7	28.2	19.8	39.2
LINCOLN RANGER STATION	MAX	28.9	36.2	42.1	52.5	63.3	71.1	81.6	80.4	69.5	56.2	38.9	31.2	54.3
	MIN	9.4	14.7	16.8	25.2	32.5	38.9	41.6	39.6	32.9	26.8	19.0	13.7	25.9
	MEAN	19.1	25.5	29.4	38.8	47.9	55.0	61.6	60.0	51.2	41.5	28.9	22.5	40.1
LIVINGSTON	MAX	34.5	40.6	44.8	55.3	65.5	74.3	84.2	82.2	71.8	61.4	45.4	38.2	58.2
	MIN	15.7	21.3	22.9	30.9	39.1	46.5	51.7	50.4	42.6	35.9	26.1	20.5	33.6
	MEAN	25.1	31.0	33.9	43.2	52.3	60.4	68.0	66.3	57.3	48.7	35.7	29.4	45.9
LIVINGSTON FAA AIRPORT	MAX	32.5	38.2	42.1	52.8	63.3	72.5	83.5	81.8	70.3	59.3	43.4	36.8	56.4
	MIN	15.0	20.2	21.7	30.2	38.6	46.1	51.5	49.9	41.5	34.9	25.4	20.3	32.9
	MEAN	23.7	29.2	31.9	41.5	50.9	59.3	67.5	65.8	55.9	47.1	34.4	28.6	44.7
LOMA 1 WNW	MAX	27.8	37.4	45.9	58.5	70.4	78.9	88.8	87.0	75.1	64.2	45.6	35.1	59.6
	MIN	1.5	10.2	18.3	29.5	39.8	47.7	52.4	49.7	40.4	30.7	17.9	8.4	28.9
	MEAN	14.7	23.9	32.1	44.0	55.1	63.3	70.6	68.4	57.8	47.5	31.8	21.8	44.3
LONESOME LAKE	MAX	24.6	33.5	41.6	56.8	68.0	76.4	86.1	84.7	73.1	62.4	43.7	31.8	56.9
	MIN	-.9	7.5	15.6	28.5	38.7	46.2	50.1	48.5	39.1	29.7	15.9	5.9	27.1
	MEAN	11.9	20.5	28.6	42.7	53.5	61.3	68.1	66.6	56.1	46.1	29.8	18.9	42.0
LUSTRE 4 NNW	MAX	17.1	24.6	34.3	51.9	65.4	74.2	82.5	81.1	68.7	56.7	37.5	25.4	51.6
	MIN	-1.2	6.0	14.4	28.8	40.0	49.0	54.2	52.7	42.8	33.3	18.3	7.1	28.8
	MEAN	7.9	15.4	24.4	40.4	52.7	61.7	68.4	66.9	55.8	45.0	27.9	16.3	40.2
MARTINSDALE 3 NNW	MAX	31.9	38.5	42.9	54.0	64.0	71.7	80.2	79.4	70.1	60.2	44.2	36.6	56.1
	MIN	10.2	16.2	17.8	26.4	34.6	41.7	46.0	43.6	36.5	30.3	21.0	15.7	28.3
	MEAN	21.1	27.4	30.4	40.2	49.4	56.7	63.1	61.6	53.3	45.3	32.6	26.2	42.3
MEDICINE LAKE 3 SE	MAX	17.8	26.2	37.4	55.3	68.8	77.6	85.1	83.9	72.2	59.9	39.1	25.8	54.1
	MIN	-3.9	3.8	13.6	29.2	41.3	50.3	54.6	52.5	42.6	32.2	17.1	4.9	28.2
	MEAN	7.0	15.1	25.5	42.3	55.1	64.0	69.9	68.3	57.4	46.1	28.1	15.4	41.2
MELSTONE	MAX	31.4	39.7	46.7	59.1	69.8	79.3	89.0	87.3	75.7	64.7	47.3	37.9	60.7
	MIN	7.3	14.7	20.5	30.8	40.8	49.4	54.9	52.7	43.0	33.4	21.7	13.5	31.9
	MEAN	19.4	27.2	33.6	45.0	55.3	64.4	72.0	70.0	59.4	49.0	34.5	25.7	46.3
MILES CITY	MAX	26.9	35.9	46.1	60.9	72.6	81.9	91.3	88.8	76.3	64.1	45.1	33.9	60.3
	MIN	3.6	11.5	20.1	32.4	43.8	52.9	58.4	55.4	44.5	34.0	21.0	10.7	32.4
	MEAN	15.2	23.7	33.1	46.7	58.2	67.4	74.9	72.1	60.4	49.1	33.1	22.3	46.4

MONTANA

TEMPERATURE NORMALS (DEG F)

STATION		JAN	FEB	MAR	APR	MAY	JUN	JUL	AUG	SEP	OCT	NOV	DEC	ANN
MILES CITY FAA AP	MAX	24.3	32.6	42.6	57.1	69.2	79.2	88.9	86.6	73.8	61.3	42.7	31.5	57.5
	MIN	3.9	11.6	20.0	32.7	44.2	53.5	60.3	57.9	46.4	35.5	21.3	11.4	33.2
	MEAN	14.1	22.1	31.3	44.9	56.8	66.4	74.6	72.3	60.1	48.4	32.0	21.5	45.4
MISSOULA WSO //R	MAX	28.8	36.4	44.4	56.5	65.8	74.0	84.8	82.7	71.3	57.0	40.4	31.8	56.2
	MIN	13.7	19.8	23.8	31.3	38.4	45.1	49.5	48.3	40.1	31.2	23.2	17.9	31.9
	MEAN	21.3	28.1	34.1	43.9	52.1	59.6	67.2	65.5	55.7	44.1	31.8	24.9	44.1
MOCCASIN EXP STATION	MAX	31.0	36.9	39.9	51.2	62.1	70.3	80.6	79.8	68.9	58.7	44.4	37.1	55.1
	MIN	7.3	14.2	18.0	28.4	37.6	45.2	50.4	49.5	41.1	32.5	21.0	13.7	29.9
	MEAN	19.2	25.6	29.0	39.8	49.9	57.8	65.6	64.7	55.0	45.6	32.7	25.4	42.5
MYSTIC LAKE	MAX	31.9	36.1	38.9	47.5	57.6	66.3	74.8	73.5	64.7	55.6	40.9	34.6	51.9
	MIN	14.5	17.9	18.7	26.4	35.7	43.7	50.3	49.6	41.1	34.4	23.7	19.0	31.3
	MEAN	23.3	27.0	28.8	36.9	46.7	55.0	62.6	61.5	52.9	45.0	32.3	26.8	41.6
NOHLY 3 WNW	MAX	18.8	27.1	37.4	55.1	68.3	76.9	84.7	83.7	71.5	59.7	39.5	26.9	54.1
	MIN	-3.6	5.1	15.1	29.6	40.7	49.3	53.8	51.6	41.6	31.2	17.0	5.0	28.0
	MEAN	7.6	16.1	26.3	42.3	54.5	63.1	69.3	67.7	56.6	45.4	28.3	16.0	41.1
NORRIS MADISON PH	MAX	33.9	39.8	44.2	54.6	65.7	74.6	85.4	83.5	72.6	61.0	44.3	37.4	58.1
	MIN	17.6	22.7	24.4	32.7	41.2	48.4	54.6	53.5	45.2	37.6	27.7	22.1	35.6
	MEAN	25.8	31.3	34.3	43.7	53.4	61.5	70.0	68.5	58.9	49.3	36.1	29.8	46.9
OPHEIM 12 SSE	MAX	15.7	23.5	33.4	50.3	64.2	72.9	81.1	79.5	67.3	55.6	36.5	23.8	50.3
	MIN	-5.9	1.9	10.8	25.8	36.5	45.2	49.8	47.6	37.7	27.9	13.5	2.2	24.4
	MEAN	4.9	12.7	22.1	38.1	50.4	59.1	65.5	63.5	52.6	41.8	25.0	13.0	37.4
PHILIPSBURG RANGER STA	MAX	31.0	37.0	41.5	51.9	61.6	69.9	80.5	79.0	69.4	58.4	41.9	35.0	54.8
	MIN	11.5	15.9	17.8	25.3	32.3	38.8	42.0	40.5	34.0	27.7	19.9	15.5	26.8
	MEAN	21.3	26.5	29.7	38.6	47.0	54.4	61.3	59.8	51.7	43.0	30.9	25.3	40.8
PLEVNA	MAX	26.4	32.9	41.2	56.3	68.4	77.8	87.2	86.5	74.7	62.4	43.3	33.0	57.5
	MIN	.0	7.5	16.1	29.0	39.5	49.0	54.6	51.8	40.9	30.0	17.1	7.9	28.6
	MEAN	13.2	20.2	28.7	42.7	54.0	63.5	70.9	69.2	57.8	46.2	30.2	20.5	43.1
POLEBRIDGE	MAX	27.3	36.3	41.9	52.2	63.4	71.0	80.2	78.9	68.3	54.8	38.0	30.4	53.6
	MIN	6.8	12.7	16.1	25.1	32.1	38.4	40.7	39.2	32.9	25.8	18.9	12.3	25.1
	MEAN	17.1	24.5	29.0	38.7	47.8	54.7	60.5	59.1	50.7	40.3	28.5	21.4	39.4
POLSON KERR DAM	MAX	31.1	38.2	44.8	56.0	64.5	72.1	82.3	81.3	70.3	57.4	41.7	34.8	56.2
	MIN	17.9	22.8	25.7	33.3	40.8	47.5	51.9	51.3	43.2	34.8	27.3	22.5	34.9
	MEAN	24.5	30.6	35.3	44.7	52.7	59.9	67.1	66.3	56.8	46.1	34.5	28.7	45.6
POPLAR	MAX	19.2	28.2	39.9	57.4	70.6	79.1	86.8	85.6	74.0	61.8	41.2	27.6	56.0
	MIN	-2.6	5.3	15.2	30.0	41.7	50.7	55.5	53.2	42.6	32.2	18.0	5.9	29.0
	MEAN	8.3	16.8	27.6	43.7	56.1	64.9	71.2	69.4	58.3	47.0	29.6	16.8	42.5
RAPELJE 4 S	MAX	32.4	39.6	44.3	55.8	66.4	75.6	86.3	85.1	73.4	62.3	45.1	37.8	58.7
	MIN	9.2	15.4	19.0	28.5	37.7	45.7	51.5	50.1	41.1	32.3	21.1	14.9	30.5
	MEAN	20.8	27.5	31.7	42.2	52.1	60.7	69.0	67.7	57.3	47.3	33.1	26.4	44.7
RAYMOND BORDER STATION	MAX	15.9	23.9	34.1	52.9	66.6	75.1	82.8	81.3	70.1	58.3	37.3	24.2	51.9
	MIN	-4.8	3.1	12.7	27.8	39.1	48.1	52.9	50.8	40.9	31.2	15.5	3.7	26.8
	MEAN	5.6	13.5	23.4	40.4	52.9	61.6	67.9	66.1	55.5	44.8	26.5	14.0	39.4
RED LODGE	MAX	32.3	37.3	41.0	50.8	61.4	70.3	79.3	77.5	67.1	57.0	42.1	36.1	54.4
	MIN	11.2	16.2	18.5	27.6	37.0	44.0	50.4	49.1	40.5	33.2	21.8	16.4	30.5
	MEAN	21.8	26.8	29.8	39.2	49.2	57.2	64.9	63.3	53.9	45.2	32.0	26.3	42.5
REDSTONE	MAX	18.3	26.8	36.7	54.4	68.2	77.1	84.8	83.4	71.2	59.0	39.3	26.4	53.8
	MIN	-5.5	2.3	12.1	26.9	38.1	47.1	51.1	48.0	38.1	28.5	14.3	2.8	25.3
	MEAN	6.4	14.6	24.5	40.7	53.2	62.1	68.0	65.7	54.7	43.8	26.8	14.6	39.6
ROCK SPRINGS	MAX	24.8	31.7	40.7	55.4	67.1	76.5	86.2	84.9	72.6	61.0	42.7	31.3	56.2
	MIN	2.5	9.7	17.5	29.6	40.5	49.1	54.6	52.8	42.2	32.4	19.1	8.7	29.9
	MEAN	13.7	20.7	29.2	42.5	53.8	62.8	70.4	68.9	57.4	46.7	30.9	20.0	43.1

MONTANA

TEMPERATURE NORMALS (DEG F)

STATION		JAN	FEB	MAR	APR	MAY	JUN	JUL	AUG	SEP	OCT	NOV	DEC	ANN
ROUNDUP	MAX	34.2	42.2	48.0	59.7	70.1	79.3	89.0	87.5	76.1	65.4	48.1	39.7	61.6
	MIN	9.2	16.5	20.9	31.2	40.9	49.1	54.5	51.8	42.2	33.4	22.1	15.3	32.3
	MEAN	21.7	29.4	34.4	45.4	55.5	64.2	71.8	69.7	59.2	49.4	35.1	27.5	46.9
ROY 8 NE	MAX	28.0	36.0	42.7	55.7	66.8	75.8	85.4	84.0	72.6	61.9	44.8	34.9	57.4
	MIN	4.5	11.8	18.4	29.9	40.4	48.2	53.5	51.6	41.9	32.7	19.9	11.1	30.3
	MEAN	16.2	23.9	30.6	42.8	53.6	62.0	69.5	67.9	57.3	47.4	32.4	23.0	43.9
SAINT IGNATIUS	MAX	32.3	39.5	46.6	58.0	67.5	75.1	84.8	83.1	71.8	58.1	42.5	36.1	58.0
	MIN	16.7	22.4	25.0	32.2	39.3	45.9	49.4	48.4	41.1	33.1	25.8	21.2	33.4
	MEAN	24.5	30.9	35.8	45.1	53.4	60.5	67.1	65.8	56.5	45.6	34.2	28.7	45.7
SAVAGE	MAX	21.8	30.3	40.7	57.0	69.6	78.1	86.1	85.1	73.1	61.0	41.6	29.5	56.2
	MIN	-.5	7.7	16.7	30.3	42.0	51.3	56.0	53.4	43.1	33.3	19.4	8.2	30.1
	MEAN	10.7	19.0	28.7	43.7	55.8	64.7	71.1	69.3	58.1	47.2	30.5	18.9	43.1
SCOBEY	MAX	18.5	26.8	37.0	55.1	69.0	77.7	85.4	84.0	72.2	59.8	39.3	26.2	54.3
	MIN	-2.4	5.0	14.4	29.3	40.4	49.3	54.3	51.9	41.7	32.1	17.7	5.8	28.3
	MEAN	8.1	15.9	25.7	42.3	54.7	63.5	69.8	68.0	57.0	46.0	28.5	16.0	41.3
SEELEY LAKE RANGER STA	MAX	30.0	38.6	43.4	53.3	63.4	71.4	82.0	80.9	70.7	58.3	40.5	32.0	55.4
	MIN	8.9	14.2	17.1	26.2	34.1	40.7	42.8	41.4	35.4	29.6	21.9	14.2	27.2
	MEAN	19.4	26.4	30.3	39.8	48.8	56.1	62.4	61.1	53.1	44.0	31.2	23.1	41.3
SIDNEY	MAX	20.7	29.1	39.3	56.1	68.7	77.2	83.9	82.6	71.0	59.5	40.3	28.3	54.7
	MIN	-1.6	6.5	15.6	29.4	40.9	50.0	54.2	51.7	41.8	31.6	18.2	6.8	28.8
	MEAN	9.6	17.8	27.5	42.8	54.8	63.6	69.1	67.2	56.4	45.6	29.3	17.6	41.8
SIMPSON 6 NW	MAX	19.9	28.9	38.2	55.0	68.2	75.9	84.7	83.3	71.6	59.8	40.9	28.0	54.5
	MIN	-3.1	5.4	13.6	27.5	37.9	45.9	50.4	48.3	38.6	28.7	15.0	4.6	26.1
	MEAN	8.5	17.2	26.0	41.3	53.1	61.0	67.5	65.8	55.2	44.3	28.0	16.3	40.4
STANFORD 1 WNW	MAX	32.5	39.3	42.7	53.9	64.0	71.7	81.2	80.1	70.2	60.2	45.6	38.5	56.7
	MIN	8.4	15.1	18.3	27.7	37.0	44.0	49.1	48.0	39.9	31.7	20.8	14.8	29.6
	MEAN	20.5	27.3	30.6	40.8	50.5	57.9	65.2	64.1	55.1	46.0	33.2	26.7	43.2
STEVENSVILLE	MAX	31.9	40.0	47.5	58.8	67.8	75.0	84.6	82.6	72.2	58.6	42.7	35.0	58.1
	MIN	15.1	20.9	23.8	30.0	37.3	44.1	46.8	45.0	38.0	30.4	23.1	18.5	31.1
	MEAN	23.5	30.5	35.7	44.4	52.5	59.6	65.7	63.8	55.1	44.5	32.9	26.8	44.6
SUPERIOR	MAX	32.8	41.7	49.0	59.5	68.9	76.3	86.4	84.5	75.0	60.1	42.8	34.8	59.3
	MIN	17.8	23.0	25.5	31.6	38.6	45.3	48.5	47.2	40.3	33.0	26.2	21.8	33.2
	MEAN	25.4	32.4	37.3	45.6	53.8	60.8	67.5	65.8	57.7	46.6	34.5	28.3	46.3
TERRY	MAX	24.1	31.9	41.8	56.8	69.1	78.4	87.5	86.5	74.0	62.1	43.2	32.0	57.3
	MIN	-1.1	6.7	16.4	30.3	42.0	51.3	56.4	53.4	42.2	30.6	17.2	6.9	29.4
	MEAN	11.5	19.3	29.1	43.6	55.6	64.9	72.0	70.0	58.1	46.4	30.3	19.5	43.4
THOMPSON FALLS PH	MAX	33.4	42.1	49.6	60.9	70.5	77.3	87.2	85.6	75.6	61.0	43.4	35.7	60.2
	MIN	19.3	24.2	26.2	32.4	39.2	45.7	49.2	48.5	42.0	34.5	27.8	23.7	34.4
	MEAN	26.4	33.2	37.9	46.7	54.9	61.6	68.2	67.0	58.8	47.8	35.6	29.7	47.3
TOWNSEND	MAX	30.8	38.7	44.8	56.6	66.5	74.3	83.6	81.9	70.9	59.9	44.4	36.3	57.4
	MIN	8.5	15.6	19.7	29.1	38.1	45.3	49.9	47.4	38.8	30.3	20.3	14.2	29.8
	MEAN	19.7	27.2	32.3	42.9	52.3	59.8	66.8	64.7	54.9	45.1	32.4	25.3	43.6
TRIDENT	MAX	32.4	40.4	46.0	58.0	68.3	77.2	87.2	85.2	74.1	62.2	45.4	37.7	59.5
	MIN	10.2	17.5	21.5	30.7	39.7	46.8	51.6	49.6	40.6	32.4	22.2	15.4	31.5
	MEAN	21.3	29.0	33.8	44.3	54.0	62.0	69.4	67.4	57.4	47.3	33.8	26.6	45.5
TROUT CREEK RANGER STA	MAX	31.8	40.1	47.0	58.4	68.6	75.9	85.9	85.2	75.3	59.6	41.6	34.4	58.7
	MIN	16.2	21.1	23.4	29.4	35.3	41.4	44.0	43.0	37.6	31.3	25.5	21.1	30.8
	MEAN	24.1	30.6	35.2	43.9	52.0	58.7	65.0	64.2	56.5	45.4	33.6	27.8	44.8
TURNER	MAX	20.5	28.2	37.0	53.7	66.6	75.3	83.2	82.4	70.7	58.3	39.6	28.1	53.6
	MIN	-.7	6.8	14.3	28.2	38.8	46.7	51.6	49.8	40.2	31.1	17.1	7.1	27.6
	MEAN	9.9	17.6	25.7	41.0	52.7	61.0	67.4	66.1	55.5	44.7	28.4	17.7	40.6

MONTANA

TEMPERATURE NORMALS (DEG F)

STATION		JAN	FEB	MAR	APR	MAY	JUN	JUL	AUG	SEP	OCT	NOV	DEC	ANN
VALIER	MAX	28.1	36.1	41.1	53.3	64.1	71.9	80.6	79.3	69.7	59.5	42.8	34.2	55.1
	MIN	6.5	13.8	17.9	28.0	38.5	45.9	50.3	49.1	41.2	33.4	21.3	13.5	30.0
	MEAN	17.3	25.0	29.5	40.7	51.3	58.9	65.5	64.2	55.5	46.5	32.1	23.9	42.5
VIDA	MAX	21.2	29.1	38.8	55.3	68.2	77.0	85.1	83.9	72.0	60.2	41.4	29.7	55.2
	MIN	.6	8.2	16.7	30.2	41.5	50.3	55.6	53.7	43.4	33.8	19.7	9.3	30.3
	MEAN	10.9	18.7	27.8	42.8	54.9	63.6	70.4	68.8	57.7	47.0	30.6	19.5	42.7
VIRGINIA CITY	MAX	32.1	37.8	41.5	52.1	62.0	71.1	81.3	79.1	69.0	58.6	42.4	35.2	55.2
	MIN	10.6	15.6	18.1	27.1	35.9	42.6	49.1	47.2	38.6	30.6	20.3	14.2	29.2
	MEAN	21.4	26.7	29.8	39.6	49.0	56.9	65.2	63.2	53.9	44.7	31.4	24.7	42.2
WESTBY	MAX	15.7	24.1	34.9	53.8	68.0	76.8	84.7	82.7	70.6	58.0	37.3	23.9	52.5
	MIN	-5.5	2.0	12.8	28.7	40.7	49.9	54.6	52.1	41.8	31.3	16.3	3.1	27.3
	MEAN	5.1	13.1	23.9	41.3	54.4	63.4	69.7	67.4	56.2	44.7	26.8	13.5	40.0
WEST GLACIER	MAX	27.7	34.7	41.0	52.2	63.7	71.0	79.5	77.6	66.3	52.9	37.4	30.8	52.9
	MIN	13.5	18.9	21.6	29.1	36.5	43.2	46.6	45.7	38.8	31.6	24.4	18.9	30.7
	MEAN	20.6	26.8	31.3	40.6	50.1	57.1	63.1	61.7	52.6	42.3	30.9	24.9	41.8
WEST YELLOWSTONE	MAX	23.4	30.8	36.8	46.4	58.5	68.3	79.4	76.2	65.6	52.8	34.4	25.0	49.8
	MIN	-.1	3.2	7.3	19.3	29.1	36.1	40.8	38.6	30.2	22.1	10.6	1.9	19.9
	MEAN	11.7	17.0	22.1	32.9	43.8	52.2	60.1	57.4	47.9	37.5	22.5	13.5	34.9
WIBAUX 2 E	MAX	22.4	29.9	39.0	54.8	67.8	76.4	85.2	84.5	72.1	60.0	40.5	29.3	55.2
	MIN	-1.0	6.5	15.4	27.9	38.9	48.1	52.7	50.6	40.0	29.9	16.8	7.0	27.7
	MEAN	10.7	18.2	27.2	41.4	53.3	62.3	68.9	67.6	56.1	45.0	28.7	18.2	41.5
WINIFRED	MAX	27.1	35.6	42.5	55.6	66.6	74.9	84.7	83.3	71.8	61.0	43.8	34.2	56.8
	MIN	4.8	12.3	18.5	29.0	38.8	46.3	51.3	49.0	39.8	31.3	19.2	11.2	29.3
	MEAN	16.0	24.0	30.5	42.3	52.7	60.6	68.1	66.2	55.9	46.1	31.5	22.7	43.1
WISDOM	MAX	25.4	31.1	36.0	46.7	58.6	67.3	78.2	76.0	65.8	54.3	37.4	28.4	50.4
	MIN	1.3	4.9	8.3	20.4	28.8	35.8	37.8	34.6	27.6	20.8	12.4	5.0	19.8
	MEAN	13.4	18.0	22.2	33.6	43.7	51.6	58.0	55.3	46.7	37.6	24.9	16.7	35.1
WYOLA	MAX	34.8	41.3	46.8	58.8	69.1	78.1	88.0	86.9	75.7	64.7	47.9	40.0	61.0
	MIN	6.8	14.0	19.2	29.3	38.6	45.7	50.5	48.8	39.3	31.0	19.6	12.3	29.6
	MEAN	20.9	27.7	33.1	44.1	53.9	61.9	69.3	67.9	57.6	47.9	33.8	26.1	45.4

MONTANA

PRECIPITATION NORMALS (INCHES)

STATION	JAN	FEB	MAR	APR	MAY	JUN	JUL	AUG	SEP	OCT	NOV	DEC	ANN
ALBION 1 N	.29	.36	.43	1.40	2.46	2.91	1.92	1.47	1.05	.92	.39	.36	13.96
AUGUSTA	.60	.48	.60	1.29	2.32	2.93	1.39	1.34	.98	.61	.56	.54	13.64
AUSTIN 1 W	1.12	.72	.88	1.33	2.22	2.35	1.16	1.22	1.20	.91	.96	1.21	15.28
BABB 6 NE	.94	.86	.86	1.52	2.77	3.71	1.59	2.10	1.88	.87	.74	.90	18.74
BAKER	.39	.45	.46	1.37	1.88	3.37	1.80	1.28	1.47	.83	.47	.40	14.17
BALLANTINE	.59	.45	.67	1.57	2.40	2`.36	.83	1.22	1.34	.98	.67	.58	13.66
BARBER	.54	.36	.44	.92	2.22	2.61	1.21	1.32	1.13	.86	.46	.42	12.49
BELGRADE AP	.66	.47	.89	1.14	2.23	2.45	1.04	1.18	1.34	1.16	.70	.57	13.83
BELLTOWER	.51	.63	.66	1.10	2.37	2.76	1.52	1.25	.95	.76	.50	.51	13.52
BIDDLE	.40	.28	.43	1.09	2.27	2.75	1.36	.98	1.00	.88	.42	.37	12.23
BIGFORK 13 S	2.20	1.37	1.24	1.46	2.38	2.98	1.32	1.83	1.77	1.60	1.78	2.15	22.08
BIG SANDY	.66	.46	.41	1.02	2.47	2.83	1.29	1.39	1.13	.72	.54	.56	13.48
BIG TIMBER	.66	.44	.80	1.44	2.97	2.65	1.05	1.24	1.42	1.28	.74	.51	15.20
BILLINGS WATER PLANT	.67	.48	.66	1.61	2.37	2.27	.78	1.20	1.28	1.11	.65	.54	13.62
BILLINGS WSO R	.97	.71	1.05	1.93	2.39	2.07	.85	1.05	1.26	1.16	.85	.80	15.09
BLACKLEAF	.63	.63	.66	1.40	2.50	2.87	1.37	1.51	1.08	.52	.57	.56	14.30
BOULDER STATE SCHOOL	.63	.35	.53	.82	1.63	2.06	1.19	1.21	1.08	.57	.51	.54	11.12
BOZEMAN MONT ST UNIV	.91	.63	1.39	1.82	2.94	2.90	1.28	1.44	1.82	1.63	1.11	.76	18.63
BOZEMAN 12 NE	2.86	2.18	2.70	3.12	4.34	4.61	1.90	2.32	2.96	2.72	2.51	2.59	34.81
BRADY AZNOE	.33	.23	.38	1.00	2.26	2.60	1.25	1.34	.80	.51	.32	.26	11.28
BREDETTE	.35	.28	.46	.89	1.81	2.75	1.60	1.72	1.23	.59	.27	.34	12.29
BRIDGER	.73	.44	.93	1.72	1.97	1.90	.60	.86	1.28	1.04	.69	.51	12.67
BROADUS	.50	.54	.67	1.44	2.38	2.72	1.28	1.05	1.14	1.00	.62	.50	13.84
BROADVIEW	.58	.48	.69	1.55	2.62	2.46	1.03	1.33	1.23	.98	.64	.45	14.04
BROWNING	1.13	.84	.81	1.38	2.02	2.87	1.48	1.55	1.14	.71	.79	.80	15.52
BRUSETT 5 NW	.50	.40	.54	1.26	2.11	2.53	1.56	1.49	1.16	.77	.35	.40	13.07
BUSBY	.70	.59	.65	1.59	2.42	2.48	1.12	1.25	1.34	1.10	.65	.67	14.56
BUTTE FAA AP	.55	.38	.65	.98	1.73	2.37	1.11	1.27	1.12	.68	.45	.44	11.73
BYNUM 4 SSE	.55	.47	.54	1.31	2.28	2.59	1.37	1.42	.89	.52	.51	.54	12.99
CANYON CREEK	.55	.39	.39	.96	1.84	2.25	1.20	1.12	.85	.65	.54	.57	11.31
CASCADE 5 S	.62	.53	.84	1.54	2.94	2.96	1.22	1.31	1.22	.94	.58	.57	15.27
CHESTER	.45	.30	.33	.87	1.65	2.50	1.30	1.31	.73	.41	.36	.44	10.65
CHINOOK	.60	.34	.41	1.25	1.81	2.30	1.40	1.22	1.07	.56	.45	.52	11.93
CHOTEAU AIRPORT	.34	.30	.36	.93	2.16	2.63	1.39	1.25	.84	.44	.39	.33	11.36
CIRCLE	.47	.33	.42	1.17	2.02	2.72	1.87	1.34	1.14	.72	.40	.39	12.99
COLSTRIP	.63	.62	.67	1.86	2.65	2.73	1.14	1.39	1.24	1.29	.66	.64	15.52
COLUMBUS	.62	.40	.76	1.63	2.72	2.12	.95	1.02	1.36	1.15	.65	.49	13.87
CONRAD AIRPORT	.51	.37	.46	1.05	2.02	2.72	1.32	1.41	.80	.48	.46	.50	12.10
COOKE CITY	2.79	2.10	2.08	1.93	2.51	2.92	1.94	2.13	2.16	1.66	2.10	2.47	26.79
CRESTON	1.55	1.13	.99	1.36	2.26	2.82	1.35	1.75	1.52	1.39	1.41	1.68	19.21
CROW AGENCY	.77	.67	.90	2.03	2.35	2.36	.77	1.08	1.50	1.35	.85	.68	15.31
CULBERTSON	.45	.37	.44	1.30	2.18	3.01	1.87	1.57	1.47	.83	.39	.41	14.29
CUT BANK FAA AP	.40	.33	.39	.93	1.98	2.76	1.53	1.69	.92	.40	.31	.35	11.99
DARBY	1.72	1.07	.96	1.11	1.93	1.80	.78	.96	1.22	1.12	1.49	1.63	15.79
DENTON 1 NNE	.71	.52	.55	1.06	2.84	2.89	1.60	1.56	1.12	.86	.62	.62	14.95
DILLON FAA AP	.30	.25	.52	.96	1.59	1.97	.89	.84	.93	.58	.43	.27	9.53
DILLON WMCE	.40	.28	.66	1.22	1.87	2.09	1.07	1.04	1.02	.66	.44	.35	11.10
DIVIDE 2 NW	.61	.41	.68	1.12	1.69	2.45	1.11	1.22	1.15	.77	.54	.64	12.39
EAST ANACONDA	1.06	.65	.81	1.04	1.86	2.32	1.27	1.15	1.18	.93	.71	.86	13.84
EKALAKA	.47	.52	.61	1.51	2.63	3.86	1.71	1.29	1.54	.98	.61	.48	16.21
ENNIS	.34	.34	.65	1.01	1.91	2.33	1.14	1.17	1.31	.90	.52	.39	12.01
ETHRIDGE	.34	.31	.29	.97	1.91	2.68	1.34	1.51	.81	.41	.27	.30	11.14
FAIRFIELD	.50	.37	.53	1.16	2.34	2.59	1.36	1.37	.98	.55	.41	.35	12.51
FLATWILLOW 4 ENE	.57	.36	.53	1.25	2.61	2.63	1.30	1.29	1.08	.81	.43	.47	13.33
FORKS 4 NNE	.40	.35	.50	1.03	1.88	2.63	1.89	1.32	.95	.55	.33	.37	12.20

MONTANA

PRECIPITATION NORMALS (INCHES)

STATION	JAN	FEB	MAR	APR	MAY	JUN	JUL	AUG	SEP	OCT	NOV	DEC	ANN
FORT ASSINNIBOINE	.53	.38	.43	1.20	1.77	2.20	1.30	1.34	1.02	.64	.43	.49	11.73
FORT BENTON	.84	.62	.74	1.34	2.46	2.97	1.17	1.25	1.05	.83	.65	.68	14.60
FORTINE 1 N	1.55	1.02	.97	1.16	2.12	2.28	1.23	1.57	1.24	1.20	1.36	1.55	17.25
FORT LOGAN 3 ESE	.41	.27	.45	.66	1.89	2.20	1.12	1.33	1.03	.64	.49	.38	10.87
FORT PECK POWER PLANT	.36	.29	.32	.94	1.97	2.37	1.57	1.47	1.08	.63	.26	.22	11.48
GALATA 16 SSW	.55	.39	.50	1.16	2.18	2.68	1.41	1.43	.87	.64	.44	.48	12.73
GALLATIN GATEWAY 9 SSW	1.07	.91	1.75	2.49	3.54	3.23	1.49	1.53	2.13	1.97	1.26	.98	22.35
GERALDINE	.81	.57	.66	1.29	3.09	3.05	1.37	1.37	1.11	.92	.66	.65	15.55
GIBSON 3 SE	.40	.37	.65	1.34	2.85	2.59	1.04	1.27	1.43	1.05	.63	.40	14.02
GIBSON DAM	1.17	.77	.98	1.73	3.11	3.36	1.40	1.58	1.29	.98	1.12	1.03	18.52
GLASGOW 15 NW	.41	.29	.32	.76	1.77	2.46	1.80	1.33	.85	.53	.29	.36	11.17
GLASGOW WSO //R	.46	.38	.40	.87	1.76	2.47	1.65	1.43	.89	.56	.31	.37	11.55
GLENDIVE	.45	.41	.49	1.17	2.04	3.13	1.72	1.34	1.15	.68	.46	.41	13.45
GOLDBUTTE 7 N	.35	.34	.39	1.03	2.02	3.13	1.44	1.60	1.32	.60	.36	.39	12.97
GRASSRANGE	.81	.45	.76	1.56	3.10	3.15	1.66	1.61	1.40	.92	.61	.53	16.56
GREAT FALLS WSO //R	1.00	.75	.93	1.49	2.52	2.75	1.10	1.31	1.03	.82	.74	.80	15.24
HAMILTON	1.31	.72	.71	.92	1.68	1.74	.78	.93	1.06	.97	1.10	1.19	13.11
HARDIN	.66	.47	.58	1.52	1.88	2.27	.87	.99	1.21	1.01	.62	.53	12.61
HARLEM	.57	.46	.39	1.03	1.61	2.21	1.68	1.20	1.04	.60	.42	.49	11.70
HARLOWTON	.55	.42	.56	.87	2.23	2.88	1.33	1.27	1.09	.74	.47	.47	12.88
HAUGAN 3 E	4.88	3.34	2.73	1.93	1.89	1.93	.83	1.35	1.64	2.34	3.57	4.34	30.77
HAVRE WSO R	.59	.41	.51	1.09	1.61	2.12	1.25	1.15	.92	.57	.43	.52	11.16
HAXBY 18 SW	.47	.37	.45	1.25	2.13	2.78	1.68	1.39	1.32	.86	.42	.40	13.52
HEBGEN DAM	3.47	2.54	2.48	1.95	2.81	3.31	1.58	1.99	1.85	1.69	2.40	3.30	29.37
HELENA 6 N	.47	.26	.41	.75	1.63	2.00	1.07	1.15	.79	.51	.34	.47	9.85
HELENA WSO //R	.66	.44	.69	1.01	1.72	2.01	1.04	1.18	.83	.65	.54	.60	11.37
HERON 2 NW	5.11	3.62	2.57	1.95	2.37	2.70	.93	1.77	1.86	2.52	3.98	4.83	34.21
HIGHWOOD 7 NE	.75	.56	.77	1.48	3.07	3.68	1.36	1.69	1.45	1.02	.74	.72	17.29
HINGHAM	.41	.28	.32	.87	1.74	2.44	1.27	1.26	.88	.48	.32	.40	10.67
HOBSON	.72	.44	.63	.92	2.64	2.88	1.35	1.34	1.05	.71	.62	.59	13.89
HOGELAND 7 WSW	.48	.35	.39	.94	1.62	2.98	1.75	1.34	1.04	.54	.36	.41	12.20
HOLTER DAM	.53	.35	.50	1.20	2.22	2.29	1.23	1.14	.99	.63	.48	.46	12.02
HUNGRY HORSE DAM	3.83	2.66	2.20	2.13	2.74	3.19	1.61	2.20	2.56	3.00	3.56	3.82	33.50
HUNTLEY EXP STATION	.60	.47	.60	1.54	2.26	2.36	.74	1.23	1.32	.98	.65	.61	13.36
HYSHAM	.49	.40	.60	1.46	2.19	2.33	1.05	1.03	1.15	1.02	.53	.52	12.77
JOLIET	.72	.50	1.03	2.09	2.91	2.24	.92	1.14	1.49	1.24	.77	.59	15.64
JOPLIN 1 N	.35	.27	.30	.92	1.92	2.48	1.29	1.23	.83	.41	.27	.31	10.58
JORDAN	.46	.35	.46	1.21	2.02	2.51	1.67	1.25	1.08	.75	.36	.42	12.54
JUDITH GAP	.91	.66	.76	1.12	2.71	3.09	1.56	1.55	1.19	.82	.60	.62	15.59
KALISPELL WSO R	1.62	1.06	.84	1.06	1.76	2.24	.94	1.44	1.11	.98	1.29	1.59	15.93
KALISPELL	1.38	.99	.80	1.01	1.77	2.01	1.01	1.64	1.09	1.05	1.20	1.41	15.36
KNOBS	.29	.32	.42	1.09	2.28	3.15	1.84	1.30	1.20	.74	.47	.26	13.36
KREMLIN	.39	.30	.34	.99	1.94	2.22	1.33	1.22	.99	.56	.32	.37	10.97
LAKEVIEW	1.71	1.11	1.76	1.53	2.36	3.22	1.37	1.65	1.67	1.30	1.26	1.58	20.52
LAME DEER	.66	.71	.78	1.67	2.75	2.74	1.04	1.20	1.17	1.29	.86	.67	15.54
LEWISTOWN 10 S	1.25	.99	1.38	2.13	3.76	4.09	1.82	2.12	2.15	1.47	1.20	1.11	23.47
LEWISTOWN FAA AP	.93	.66	.80	1.21	3.20	3.40	1.60	1.87	1.47	1.03	.72	.77	17.66
LIBBY 1 NE RANGER STA	2.58	1.57	1.22	1.07	1.50	1.55	.79	1.07	1.19	1.48	2.17	2.47	18.66
LIMA	.32	.25	.49	.96	1.80	1.99	1.19	1.24	1.15	.63	.38	.34	10.74
LINCOLN RANGER STATION	2.22	1.41	1.23	1.46	2.22	2.31	1.07	1.16	1.21	1.12	1.51	2.01	18.93
LINDSAY	.53	.49	.57	1.10	1.97	2.56	1.95	1.49	1.17	.70	.58	.39	13.50
LIVINGSTON	.78	.53	1.04	1.40	2.48	2.28	1.31	1.27	1.44	1.33	.86	.56	15.28
LIVINGSTON 12 S	.75	.52	.99	1.39	2.76	2.56	1.38	1.55	1.71	1.32	.82	.67	16.42
LIVINGSTON FAA AIRPORT	.61	.44	.86	1.27	2.73	2.45	1.30	1.38	1.54	1.30	.75	.49	15.12
LOMA 1 WNW	.73	.47	.51	1.13	2.12	2.60	1.11	1.35	.89	.71	.55	.58	12.75

MONTANA

PRECIPITATION NORMALS (INCHES)

STATION	JAN	FEB	MAR	APR	MAY	JUN	JUL	AUG	SEP	OCT	NOV	DEC	ANN
LONESOME LAKE	.67	.41	.39	.93	1.85	2.47	1.24	1.33	.86	.63	.48	.57	11.83
LUSTRE 4 NNW	.34	.25	.38	1.09	1.78	2.80	1.62	1.44	1.13	.62	.27	.31	12.03
MAC KENZIE	.38	.36	.50	1.41	2.16	3.01	1.60	1.40	1.45	.93	.49	.40	14.09
MARTINSDALE 3 NNW	.62	.40	.56	.86	2.54	2.31	1.43	1.46	1.12	.74	.50	.50	13.04
MEDICINE LAKE 3 SE	.42	.40	.46	1.32	2.01	3.07	1.99	1.73	1.32	.77	.40	.41	14.30
MELSTONE	.63	.46	.55	1.42	2.47	2.68	1.27	1.34	1.34	.96	.58	.59	14.29
MILES CITY	.46	.40	.54	1.26	2.04	2.65	1.47	1.25	1.09	.88	.55	.45	13.04
MILES CITY FAA AP	.57	.56	.59	1.37	2.31	2.75	1.52	1.26	1.08	.90	.60	.60	14.11
MISSOULA WSO //R	1.41	.81	.83	1.01	1.62	1.85	.85	.95	1.02	.85	.88	1.21	13.29
MOCCASIN EXP STATION	.66	.47	.66	1.14	2.93	3.34	1.53	1.54	1.19	.77	.59	.51	15.33
MYSTIC LAKE	1.51	1.23	2.07	3.15	3.63	2.99	2.02	1.82	2.16	1.82	1.59	1.27	25.26
NOHLY 3 WNW	.34	.37	.47	1.23	2.27	2.75	1.87	1.73	1.46	.70	.44	.35	13.98
NORRIS MADISON PH	.71	.59	1.26	1.89	3.10	2.95	1.41	1.43	1.70	1.46	.92	.60	18.02
OPHEIM 12 SSE	.23	.22	.36	.89	1.77	2.74	1.82	1.82	1.24	.56	.21	.24	12.10
PENDROY 2 NNW	.66	.45	.62	1.51	2.56	2.81	1.48	1.64	1.16	.64	.56	.60	14.69
PHILIPSBURG RANGER STA	.77	.47	.74	1.24	2.17	2.55	1.15	1.38	1.28	1.03	.74	.70	14.22
PLENTYWOOD	.39	.35	.42	.99	1.80	2.86	1.99	1.66	1.38	.64	.35	.36	13.19
PLEVNA	.39	.42	.45	1.27	1.92	2.69	1.79	1.31	1.36	.77	.42	.42	13.21
POLEBRIDGE	3.03	2.04	1.47	1.41	1.93	2.38	1.21	1.49	1.42	1.72	2.30	2.94	23.34
POLSON KERR DAM	1.29	.89	.75	1.16	2.07	2.18	1.06	1.16	1.22	.97	1.04	1.19	14.98
POPLAR	.40	.32	.42	1.17	1.95	3.04	2.04	1.64	1.20	.68	.30	.35	13.51
PRYOR	.74	.58	.94	2.28	2.79	2.44	.86	1.23	1.51	1.38	.83	.62	16.20
RAPELJE 4 S	.70	.61	.87	1.47	2.59	2.32	1.14	1.26	1.28	1.17	.79	.51	14.71
RAYMOND BORDER STATION	.33	.30	.39	.98	1.98	2.83	1.81	2.00	1.60	.65	.31	.36	13.54
RED LODGE	1.48	1.15	2.53	3.83	3.51	2.93	1.31	1.53	2.23	1.65	1.69	1.18	25.02
REDSTONE	.31	.29	.45	1.00	1.89	2.64	1.78	1.66	1.31	.57	.36	.31	12.57
ROBERTS 1 NNW	.70	.52	1.04	2.04	2.95	2.48	.92	1.17	1.66	1.25	.80	.51	16.04
ROCK SPRINGS	.36	.31	.38	1.14	1.96	2.56	1.33	1.06	1.08	.67	.36	.35	11.56
ROCKY BOY	.79	.62	.62	1.42	2.77	3.45	1.84	1.85	1.46	.99	.72	.72	17.25
ROUNDUP	.44	.32	.39	1.04	2.27	2.36	1.16	1.34	1.14	.84	.33	.38	12.01
ROY 8 NE	.57	.44	.60	1.31	2.69	2.73	1.60	1.50	1.14	.74	.47	.47	14.26
SAINT IGNATIUS	1.34	.78	.97	1.43	2.34	2.53	1.04	1.23	1.34	1.13	.96	1.09	16.18
SAVAGE	.37	.31	.43	1.29	2.07	3.24	1.96	1.52	1.28	.70	.36	.33	13.86
SCOBEY	.54	.48	.61	1.02	1.85	2.78	1.63	1.63	1.33	.58	.45	.59	13.49
SEELEY LAKE RANGER STA	3.25	1.84	1.67	1.38	1.91	2.19	.96	1.17	1.34	1.31	2.10	2.99	22.11
SIDNEY	.40	.40	.47	1.20	2.10	2.78	1.76	1.63	1.29	.78	.47	.42	13.70
SILVER LAKE	1.20	.93	1.47	1.90	2.55	3.00	1.33	1.27	1.53	1.30	.92	.94	18.34
SIMPSON 6 NW	.34	.29	.27	.83	1.38	2.42	1.25	1.22	.85	.43	.32	.30	9.90
SPRINGDALE	.50	.36	.65	1.21	2.55	2.35	1.07	1.03	1.42	1.22	.64	.45	13.45
STANFORD 1 WNW	.73	.52	.60	1.15	3.01	3.07	1.58	1.54	1.16	.82	.58	.58	15.34
STEVENSVILLE	1.45	.86	.75	.78	1.51	1.65	.78	.79	.89	.81	.97	1.25	12.49
SUNBURST 8 E	.36	.32	.29	.91	1.98	2.72	1.50	1.46	1.08	.51	.41	.40	11.94
SUN RIVER 5 SW	.53	.39	.57	1.20	2.35	2.56	1.34	1.18	.91	.66	.47	.45	12.61
SUPERIOR	2.17	1.23	1.35	1.27	1.80	1.86	.85	1.20	1.16	1.27	1.50	1.80	17.46
TERRY	.28	.25	.31	1.00	1.77	2.63	1.56	1.33	1.03	.70	.37	.26	11.49
THOMPSON FALLS PH	3.05	2.03	1.83	1.57	1.86	1.91	.86	1.30	1.33	1.74	2.43	2.86	22.77
TOWNSEND	.51	.27	.60	.80	1.81	2.17	1.13	1.19	1.07	.69	.46	.41	11.11
TRIDENT	.42	.30	.61	.96	2.00	2.21	1.03	1.14	1.25	.80	.46	.28	11.46
TROUT CREEK RANGER STA	4.95	3.32	2.51	1.84	2.02	2.03	.96	1.40	1.48	2.07	3.57	4.34	30.49
TURNER	.35	.31	.36	.93	1.67	2.61	1.94	1.31	1.12	.58	.29	.33	11.80
TWIN BRIDGES	.26	.23	.45	.80	1.56	1.86	.90	.95	.94	.52	.37	.29	9.13
VALIER	.38	.31	.33	1.05	2.14	2.85	1.51	1.54	.96	.47	.38	.36	12.28
VIDA	.73	.55	.68	1.57	2.36	3.07	1.99	1.56	1.38	.83	.48	.50	15.70
VIRGINIA CITY	.74	.53	1.01	1.45	2.50	2.73	1.64	1.43	1.54	1.13	.90	.66	16.26
VOLBORG	.44	.40	.45	1.54	2.46	3.05	1.24	1.26	1.05	.92	.61	.48	13.90

MONTANA

PRECIPITATION NORMALS (INCHES)

STATION	JAN	FEB	MAR	APR	MAY	JUN	JUL	AUG	SEP	OCT	NOV	DEC	ANN
WESTBY	.46	.45	.44	1.15	1.93	2.89	1.92	2.07	1.42	.69	.34	.43	14.19
WEST GLACIER	3.61	2.61	1.78	1.79	2.57	3.24	1.61	1.87	2.14	2.18	2.98	3.54	29.92
WEST YELLOWSTONE	2.39	1.67	1.74	1.51	2.17	2.62	1.38	1.63	1.54	1.42	1.90	2.37	22.34
WHITEFISH 5 NW	2.44	1.81	1.33	1.56	2.48	3.09	1.36	1.78	1.53	1.46	2.05	2.41	23.30
WHITE SULPHUR SP 10 N	.99	.70	.89	.92	2.57	2.73	1.57	1.46	1.27	.84	.82	.94	15.70
WHITEWATER	.37	.33	.34	.84	1.56	2.23	1.77	1.23	.79	.48	.23	.27	10.44
WIBAUX 2 E	.29	.30	.44	1.26	2.22	2.95	1.89	1.51	1.49	.79	.43	.27	13.84
WINIFRED	.73	.50	.60	1.26	2.61	2.85	1.33	1.55	1.05	.84	.60	.63	14.55
WISDOM	.72	.51	.67	.84	1.54	1.89	.95	1.05	1.02	.75	.70	.80	11.44
WISE RIVER 3 WNW	.51	.38	.48	.85	1.68	2.25	1.22	1.12	1.04	.78	.62	.71	11.64
WYOLA	.84	.67	.94	2.23	2.62	2.45	.95	.98	1.42	1.33	.92	.74	16.09

MONTANA

HEATING DEGREE DAY NORMALS (BASE 65 DEG F)

STATION		JUL	AUG	SEP	OCT	NOV	DEC	JAN	FEB	MAR	APR	MAY	JUN	ANN
AUGUSTA		51	97	295	549	930	1194	1376	1028	1029	696	415	200	7860
AUSTIN 1 W		81	144	362	645	1023	1265	1411	1084	1104	765	496	270	8650
BALLANTINE		8	31	214	484	924	1228	1423	1056	958	579	288	106	7299
BARBER		33	83	296	567	966	1218	1380	1044	1029	663	381	160	7820
BELGRADE AP		46	86	328	645	1053	1342	1513	1162	1138	729	446	207	8695
BIGFORK 13 S		46	108	300	598	897	1082	1215	952	921	627	394	191	7331
BIG SANDY		17	57	250	555	1020	1380	1600	1198	1060	618	308	122	8185
BIG TIMBER		21	50	243	493	864	1085	1240	941	933	618	348	130	6966
BILLINGS WATER PLANT		10	30	206	459	873	1150	1327	980	902	552	272	103	6864
BILLINGS WSO	R	9	27	214	487	900	1175	1367	1025	967	612	318	111	7212
BOULDER STATE SCHOOL		66	117	351	645	1029	1268	1407	1084	1079	741	481	240	8508
BOZEMAN MONT ST UNIV		38	83	299	586	978	1215	1358	1053	1054	702	431	200	7997
BOZEMAN 12 NE		219	259	489	760	1089	1308	1426	1156	1212	903	648	402	9871
BRADY AZNOE		58	103	311	592	1002	1302	1516	1131	1091	717	409	189	8421
BREDETTE		38	82	294	626	1125	1541	1795	1406	1234	714	359	136	9350
BRIDGER		18	35	226	487	897	1169	1349	997	933	597	310	113	7131
BROADUS		27	40	251	574	996	1311	1494	1134	1029	639	328	115	7938
BROWNING		120	180	384	639	1038	1302	1494	1156	1163	804	527	300	9107
BUTTE FAA AP		103	175	430	744	1125	1407	1535	1218	1200	828	549	299	9613
CASCADE 5 S		37	86	271	505	879	1138	1345	1005	998	648	375	171	7458
CHESTER		44	92	318	648	1089	1448	1655	1252	1156	708	389	169	8968
CHINOOK		16	65	261	561	1038	1414	1649	1260	1107	633	297	115	8416
CHOTEAU AIRPORT		47	98	301	543	939	1200	1398	1053	1035	693	397	184	7888
CIRCLE		30	61	279	598	1050	1411	1662	1282	1147	678	345	121	8664
COLSTRIP		15	38	220	502	933	1209	1395	1050	973	612	310	109	7366
COLUMBUS		16	48	237	518	927	1206	1361	1008	952	615	329	130	7347
CONRAD AIRPORT		71	113	307	580	999	1290	1488	1117	1073	699	394	184	8315
COOKE CITY		282	350	570	877	1266	1553	1597	1310	1345	1017	747	486	11400
CRESTON		71	124	347	667	978	1197	1339	1033	992	660	415	210	8033
CROW AGENCY		11	31	214	502	933	1249	1432	1070	958	579	285	99	7363
CULBERTSON		35	68	286	623	1119	1538	1795	1389	1203	687	339	120	9202
CUT BANK FAA AP		89	149	374	660	1071	1361	1566	1193	1166	786	484	250	9149
DARBY		51	106	287	567	906	1110	1212	921	905	636	400	202	7303
DILLON FAA AP		63	108	323	626	1014	1271	1392	1089	1079	735	456	214	8370
DILLON WMCE		47	108	308	577	963	1209	1321	1025	1017	696	440	210	7921
DIVIDE 2 NW		92	152	381	676	1050	1318	1423	1120	1116	786	527	289	8930
EAST ANACONDA		79	134	349	629	996	1206	1324	1056	1094	786	508	273	8434
EKALAKA		26	57	272	577	1020	1314	1510	1182	1094	672	349	126	8199
ENNIS		57	96	329	598	954	1190	1311	1030	1023	708	446	215	7957
FAIRFIELD		49	106	286	533	933	1212	1404	1047	1023	681	397	192	7863
FLATWILLOW 4 ENE		23	52	243	515	933	1225	1435	1081	1017	636	342	135	7637
FORKS 4 NNE		43	88	311	636	1131	1535	1795	1411	1268	735	393	157	9503
FORT ASSINNIBOINE		22	63	269	570	1029	1401	1634	1254	1122	663	333	133	8493
FORT BENTON		17	61	253	536	948	1252	1460	1084	992	606	315	125	7649
FORTINE 1 N		88	153	344	676	1029	1262	1401	1036	980	678	437	232	8316
FORT PECK POWER PLANT		15	35	219	505	978	1383	1662	1305	1138	636	298	91	8265
GERALDINE		41	69	270	539	948	1228	1426	1072	1017	651	361	153	7775
GIBSON DAM		117	168	378	623	963	1194	1358	1047	1091	783	536	306	8564
GLASGOW WSO	//R	23	50	260	583	1074	1491	1761	1375	1203	675	332	113	8940
GLENDIVE		8	19	191	471	960	1352	1597	1210	1039	555	226	61	7689
GOLDBUTTE 7 N		77	139	346	611	1011	1299	1522	1145	1116	726	431	219	8642
GRASSRANGE		35	72	277	539	936	1206	1395	1061	1017	666	386	166	7756
GREAT FALLS WSO	//R	20	66	268	536	930	1218	1435	1072	1042	669	369	141	7766
HAMILTON		33	82	278	592	927	1138	1240	930	890	603	373	167	7253
HARDIN		21	36	233	543	957	1271	1469	1100	995	603	296	113	7637

MONTANA

HEATING DEGREE DAY NORMALS (BASE 65 DEG F)

STATION	JUL	AUG	SEP	OCT	NOV	DEC	JAN	FEB	MAR	APR	MAY	JUN	ANN
HARLEM	25	75	263	589	1074	1479	1699	1296	1141	648	324	129	8742
HARLOWTON	47	101	318	583	966	1209	1376	1053	1048	708	428	204	8041
HAUGAN 3 E	90	137	330	642	984	1212	1318	997	995	696	453	239	8093
HAVRE WSO R	21	64	279	592	1059	1432	1674	1271	1135	669	333	131	8660
HAXBY 18 SW	26	50	264	570	1023	1383	1631	1240	1113	669	345	124	8438
HEBGEN DAM	139	191	433	772	1191	1569	1655	1352	1333	948	626	357	10566
HELENA WSO //R	41	77	308	617	1008	1287	1454	1092	1035	681	397	179	8176
HERON 2 NW	81	141	325	642	939	1150	1259	980	958	642	403	212	7732
HOLTER DAM	16	44	223	455	810	1057	1246	949	936	618	342	138	6834
HUNGRY HORSE DAM	71	127	347	682	999	1200	1349	1050	1029	714	431	218	8217
HUNTLEY EXP STATION	17	40	231	527	945	1252	1442	1075	980	609	311	115	7544
HYSHAM	10	24	207	484	930	1246	1451	1084	964	567	270	91	7328
JOPLIN 1 N	49	117	331	639	1080	1423	1637	1254	1178	744	419	194	9065
JORDAN	13	35	235	552	1008	1364	1590	1198	1063	627	309	96	8090
KALISPELL WSO R	66	128	345	698	1023	1265	1417	1084	1026	672	419	218	8361
KALISPELL	46	103	301	629	966	1200	1327	1011	955	627	370	176	7711
LAKEVIEW	197	243	483	809	1221	1553	1655	1366	1373	960	648	408	10916
LEWISTOWN FAA AP	76	123	337	611	993	1262	1454	1142	1141	762	474	238	8613
LIBBY 1 NE RANGER STA	36	84	262	608	957	1200	1311	952	893	600	350	154	7407
LIMA	99	159	386	691	1104	1401	1500	1210	1184	822	561	307	9424
LINCOLN RANGER STATION	126	179	419	729	1083	1318	1423	1106	1104	786	530	304	9107
LIVINGSTON	29	80	266	511	879	1104	1237	952	964	654	398	168	7242
LIVINGSTON FAA AIRPORT	43	72	305	555	918	1128	1280	1002	1026	705	437	194	7665
LOMA 1 WNW	17	69	250	543	996	1339	1559	1151	1020	630	317	128	8019
LONESOME LAKE	28	76	290	586	1056	1429	1646	1246	1128	669	357	153	8664
LUSTRE 4 NNW	40	83	305	620	1113	1510	1770	1389	1259	738	388	142	9357
MARTINSDALE 3 NNW	107	148	361	611	972	1203	1361	1053	1073	744	484	259	8376
MEDICINE LAKE 3 SE	22	55	255	586	1107	1538	1798	1397	1225	681	317	101	9082
MELSTONE	12	35	211	496	915	1218	1414	1058	973	600	309	104	7345
MILES CITY	12	19	202	493	957	1324	1544	1156	989	549	234	65	7544
MILES CITY FAA AP	10	28	215	515	990	1349	1578	1201	1045	603	271	86	7891
MISSOULA WSO //R	29	75	289	648	996	1243	1355	1033	958	633	400	180	7839
MOCCASIN EXP STATION	68	117	326	601	969	1228	1420	1103	1116	756	468	237	8409
MYSTIC LAKE	119	156	382	620	981	1184	1293	1064	1122	843	567	310	8641
NOHLY 3 WNW	37	68	279	608	1101	1519	1779	1369	1200	681	335	118	9094
NORRIS MADISON PH	19	42	234	487	867	1091	1215	944	952	639	364	144	6998
OPHEIM 12 SSE	87	132	384	719	1200	1612	1863	1464	1330	807	453	205	10256
PHILIPSBURG RANGER STA	133	182	405	682	1023	1231	1355	1078	1094	792	558	321	8854
PLEVNA	30	43	255	583	1044	1380	1606	1254	1125	669	348	117	8454
POLEBRIDGE	152	206	434	766	1095	1352	1485	1134	1116	789	533	309	9371
POLSON KERR DAM	40	71	269	586	915	1125	1256	963	921	609	381	173	7309
POPLAR	22	44	237	558	1062	1494	1758	1350	1159	639	296	100	8719
RAPELJE 4 S	33	57	271	549	957	1197	1370	1050	1032	684	400	182	7782
RAYMOND BORDER STATION	53	94	311	626	1155	1581	1841	1442	1290	738	383	153	9667
RED LODGE	81	125	353	614	990	1200	1339	1070	1091	774	490	253	8380
REDSTONE	41	96	324	657	1146	1562	1817	1411	1256	729	371	133	9543
ROCK SPRINGS	35	59	265	567	1023	1395	1590	1240	1110	675	355	128	8442
ROUNDUP	13	34	219	484	897	1163	1342	997	949	588	303	112	7101
ROY 8 NE	24	71	262	546	978	1302	1513	1151	1066	666	358	143	8080
SAINT IGNATIUS	33	75	271	601	924	1125	1256	955	905	597	360	160	7262
SAVAGE	22	37	244	552	1035	1429	1683	1288	1125	639	297	96	8447
SCOBEY	26	81	270	589	1095	1519	1764	1375	1218	681	333	115	9066
SEELEY LAKE RANGER STA	105	149	363	651	1014	1299	1414	1081	1076	756	502	273	8683
SIDNEY	29	54	280	601	1071	1469	1717	1322	1163	666	325	107	8804
SIMPSON 6 NW	33	91	314	642	1110	1510	1752	1338	1209	711	369	166	9245

MONTANA

HEATING DEGREE DAY NORMALS (BASE 65 DEG F)

STATION	JUL	AUG	SEP	OCT	NOV	DEC	JAN	FEB	MAR	APR	MAY	JUN	ANN
STANFORD 1 WNW	74	123	322	589	954	1187	1380	1056	1066	726	450	230	8157
STEVENSVILLE	48	90	304	636	963	1184	1287	966	908	618	388	178	7570
SUPERIOR	26	83	239	570	915	1138	1228	913	859	582	352	160	7065
TERRY	18	40	244	577	1041	1411	1659	1280	1113	642	300	92	8417
THOMPSON FALLS PH	20	65	221	533	882	1094	1197	890	840	549	320	131	6742
TOWNSEND	43	90	316	617	978	1231	1404	1058	1014	663	394	182	7990
TRIDENT	17	48	256	549	936	1190	1355	1008	967	621	346	129	7422
TROUT CREEK RANGER STA	61	107	275	608	942	1153	1268	963	924	633	403	200	7537
TURNER	37	94	309	629	1098	1466	1708	1327	1218	720	386	156	9148
VALIER	63	116	311	574	987	1274	1479	1120	1101	729	429	205	8388
VIDA	23	69	258	558	1032	1411	1677	1296	1153	666	328	116	8587
VIRGINIA CITY	77	129	348	629	1008	1249	1352	1072	1091	762	496	252	8465
WESTBY	30	73	287	629	1146	1597	1857	1453	1274	711	341	120	9518
WEST GLACIER	87	150	381	704	1023	1243	1376	1070	1045	732	462	243	8516
WEST YELLOWSTONE	163	247	513	853	1275	1597	1652	1344	1330	963	657	384	10978
WIBAUX 2 E	39	66	293	620	1089	1451	1683	1310	1172	708	368	139	8938
WINIFRED	36	82	299	586	1005	1311	1519	1148	1070	681	381	176	8294
WISDOM	221	301	549	849	1203	1497	1600	1316	1327	942	660	402	10867
WYOLA	31	54	250	530	936	1206	1367	1044	989	627	347	136	7517

MONTANA

COOLING DEGREE DAY NORMALS (BASE 65 DEG F)

STATION		JAN	FEB	MAR	APR	MAY	JUN	JUL	AUG	SEP	OCT	NOV	DEC	ANN
AUGUSTA		0	0	0	0	0	29	88	88	22	0	0	0	227
AUSTIN 1 W		0	0	0	0	0	12	62	70	17	0	0	0	161
BALLANTINE		0	0	0	0	12	103	244	195	55	0	0	0	609
BARBER		0	0	0	0	0	28	116	96	17	0	0	0	257
BELGRADE AP		0	0	0	0	0	18	105	86	16	0	0	0	225
BIGFORK 13 S		0	0	0	0	0	20	92	105	21	0	0	0	238
BIG SANDY		0	0	0	0	8	71	178	169	34	0	0	0	460
BIG TIMBER		0	0	0	0	0	46	170	140	36	7	0	0	399
BILLINGS WATER PLANT		0	0	0	0	8	88	223	179	50	0	0	0	548
BILLINGS WSO	R	0	0	0	0	0	81	235	191	46	0	0	0	553
BOULDER STATE SCHOOL		0	0	0	0	0	15	78	70	15	0	0	0	178
BOZEMAN MONT ST UNIV		0	0	0	0	0	23	100	92	26	0	0	0	241
BOZEMAN 12 NE		0	0	0	0	0	0	8	8	0	0	0	0	16
BRADY AZNOE		0	0	0	0	0	24	86	96	20	0	0	0	226
BREDETTE		0	0	0	0	12	58	156	150	27	0	0	0	403
BRIDGER		0	0	0	0	7	59	188	144	40	0	0	0	438
BROADUS		0	0	0	0	6	82	219	164	32	0	0	0	503
BROWNING		0	0	0	0	0	9	36	53	21	0	0	0	119
BUTTE FAA AP		0	0	0	0	0	8	41	45	13	0	0	0	107
CASCADE 5 S		0	0	0	0	0	39	111	111	28	0	0	0	289
CHESTER		0	0	0	0	0	28	100	98	21	0	0	0	247
CHINOOK		0	0	0	0	6	70	165	149	27	0	0	0	417
CHOTEAU AIRPORT		0	0	0	0	0	34	103	101	31	0	0	0	269
CIRCLE		0	0	0	0	7	73	182	160	30	0	0	0	452
COLSTRIP		0	0	0	0	6	85	226	187	49	0	0	0	553
COLUMBUS		0	0	0	0	0	58	171	137	30	0	0	0	396
CONRAD AIRPORT		0	0	0	0	0	31	105	107	28	0	0	0	271
COOKE CITY		0	0	0	0	0	0	0	0	0	0	0	0	0
CRESTON		0	0	0	0	0	15	49	56	8	0	0	0	128
CROW AGENCY		0	0	0	0	6	84	218	176	40	0	0	0	524
CULBERTSON		0	0	0	0	7	69	171	146	19	0	0	0	412
CUT BANK FAA AP		0	0	0	0	0	16	58	65	17	0	0	0	156
DARBY		0	0	0	0	0	19	67	75	11	0	0	0	172
DILLON FAA AP		0	0	0	0	0	16	100	86	17	0	0	0	219
DILLON WMCE		0	0	0	0	0	15	60	59	11	0	0	0	145
DIVIDE 2 NW		0	0	0	0	0	7	39	41	9	0	0	0	96
EAST ANACONDA		0	0	0	0	0	18	86	91	28	0	0	0	223
EKALAKA		0	0	0	0	8	69	196	178	44	0	0	0	495
ENNIS		0	0	0	0	0	11	63	43	11	0	0	0	128
FAIRFIELD		0	0	0	0	0	27	89	109	37	6	0	0	268
FLATWILLOW 4 ENE		0	0	0	0	0	63	187	157	39	0	0	0	446
FORKS 4 NNE		0	0	0	0	6	40	133	134	26	0	0	0	339
FORT ASSINNIBOINE		0	0	0	0	8	64	158	153	35	0	0	0	418
FORT BENTON		0	0	0	0	8	59	151	135	25	0	0	0	378
FORTINE 1 N		0	0	0	0	0	13	51	63	8	0	0	0	135
FORT PECK POWER PLANT		0	0	0	0	13	94	226	193	42	0	0	0	568
GERALDINE		0	0	0	0	0	39	137	128	30	0	0	0	334
GIBSON DAM		0	0	0	0	0	6	24	35	12	0	0	0	77
GLASGOW WSO	//R	0	0	0	0	10	65	190	168	32	0	0	0	465
GLENDIVE		0	0	0	0	24	136	296	249	68	0	0	0	773
GOLDBUTTE 7 N		0	0	0	0	0	21	62	80	22	0	0	0	185
GRASSRANGE		0	0	0	0	0	40	131	119	31	0	0	0	321
GREAT FALLS WSO	//R	0	0	0	0	0	48	153	144	40	6	0	0	391
HAMILTON		0	0	0	0	0	20	95	85	17	0	0	0	217
HARDIN		0	0	0	0	11	92	238	188	38	0	0	0	567

MONTANA

COOLING DEGREE DAY NORMALS (BASE 65 DEG F)

STATION		JAN	FEB	MAR	APR	MAY	JUN	JUL	AUG	SEP	OCT	NOV	DEC	ANN
HARLEM		0	0	0	0	8	63	155	153	20	0	0	0	399
HARLOWTON		0	0	0	0	0	30	93	92	18	0	0	0	233
HAUGAN 3 E		0	0	0	0	0	8	31	35	6	0	0	0	80
HAVRE WSO	R	0	0	0	0	8	65	163	151	33	0	0	0	420
HAXBY 18 SW		0	0	0	0	7	64	175	149	30	0	0	0	425
HEBGEN DAM		0	0	0	0	0	0	12	11	0	0	0	0	23
HELENA WSO	//R	0	0	0	0	0	32	131	105	26	0	0	0	294
HERON 2 NW		0	0	0	0	0	8	44	54	7	0	0	0	113
HOLTER DAM		0	0	0	0	0	54	168	143	40	5	0	0	410
HUNGRY HORSE DAM		0	0	0	0	0	17	80	90	14	0	0	0	201
HUNTLEY EXP STATION		0	0	0	0	0	61	185	148	27	0	0	0	421
HYSHAM		0	0	0	0	13	94	246	195	48	0	0	0	596
JOPLIN 1 N		0	0	0	0	0	26	98	114	22	0	0	0	260
JORDAN		0	0	0	0	8	90	224	181	37	0	0	0	540
KALISPELL WSO	R	0	0	0	0	0	17	66	82	12	0	0	0	177
KALISPELL		0	0	0	0	0	26	92	100	16	0	0	0	234
LAKEVIEW		0	0	0	0	0	0	0	8	0	0	0	0	8
LEWISTOWN FAA AP		0	0	0	0	0	16	88	104	22	0	0	0	230
LIBBY 1 NE RANGER STA		0	0	0	0	0	25	88	96	16	0	0	0	225
LIMA		0	0	0	0	0	0	25	26	5	0	0	0	56
LINCOLN RANGER STATION		0	0	0	0	0	0	21	24	0	0	0	0	45
LIVINGSTON		0	0	0	0	0	30	122	120	35	5	0	0	312
LIVINGSTON FAA AIRPORT		0	0	0	0	0	23	120	97	32	0	0	0	272
LOMA 1 WNW		0	0	0	0	10	77	190	174	34	0	0	0	485
LONESOME LAKE		0	0	0	0	0	42	124	126	23	0	0	0	315
LUSTRE 4 NNW		0	0	0	0	7	43	145	142	29	0	0	0	366
MARTINSDALE 3 NNW		0	0	0	0	0	10	48	43	10	0	0	0	111
MEDICINE LAKE 3 SE		0	0	0	0	10	71	174	157	27	0	0	0	439
MELSTONE		0	0	0	0	8	86	229	190	43	0	0	0	556
MILES CITY		0	0	0	0	23	137	319	239	64	0	0	0	782
MILES CITY FAA AP		0	0	0	0	17	128	308	255	68	0	0	0	776
MISSOULA WSO	//R	0	0	0	0	0	18	97	91	10	0	0	0	216
MOCCASIN EXP STATION		0	0	0	0	0	21	86	108	26	0	0	0	241
MYSTIC LAKE		0	0	0	0	0	10	45	48	19	0	0	0	122
NOHLY 3 WNW		0	0	0	0	9	61	170	151	27	0	0	0	418
NORRIS MADISON PH		0	0	0	0	0	39	174	151	51	0	0	0	415
OPHEIM 12 SSE		0	0	0	0	0	28	102	85	12	0	0	0	227
PHILIPSBURG RANGER STA		0	0	0	0	0	0	18	21	6	0	0	0	45
PLEVNA		0	0	0	0	7	72	213	173	39	0	0	0	504
POLEBRIDGE		0	0	0	0	0	0	12	23	0	0	0	0	35
POLSON KERR DAM		0	0	0	0	0	20	105	111	23	0	0	0	259
POPLAR		0	0	0	0	20	97	214	180	36	0	0	0	547
RAPELJE 4 S		0	0	0	0	0	53	157	141	40	0	0	0	391
RAYMOND BORDER STATION		0	0	0	0	8	51	142	128	26	0	0	0	355
RED LODGE		0	0	0	0	0	19	78	73	20	0	0	0	190
REDSTONE		0	0	0	0	5	46	134	118	15	0	0	0	318
ROCK SPRINGS		0	0	0	0	8	62	203	180	37	0	0	0	490
ROUNDUP		0	0	0	0	8	88	224	180	45	0	0	0	545
ROY 8 NE		0	0	0	0	0	53	164	161	31	0	0	0	409
SAINT IGNATIUS		0	0	0	0	0	25	99	99	16	0	0	0	239
SAVAGE		0	0	0	0	12	87	211	171	37	0	0	0	518
SCOBEY		0	0	0	0	14	70	174	174	30	0	0	0	462
SEELEY LAKE RANGER STA		0	0	0	0	0	6	25	28	6	0	0	0	65
SIDNEY		0	0	0	0	9	65	156	123	22	0	0	0	375
SIMPSON 6 NW		0	0	0	0	0	46	110	116	20	0	0	0	292

MONTANA

COOLING DEGREE DAY NORMALS (BASE 65 DEG F)

STATION	JAN	FEB	MAR	APR	MAY	JUN	JUL	AUG	SEP	OCT	NOV	DEC	ANN
STANFORD 1 WNW	0	0	0	0	0	17	80	95	25	0	0	0	217
STEVENSVILLE	0	0	0	0	0	16	70	53	7	0	0	0	146
SUPERIOR	0	0	0	0	0	34	104	107	20	0	0	0	265
TERRY	0	0	0	0	9	89	235	195	37	0	0	0	565
THOMPSON FALLS PH	0	0	0	0	7	29	119	127	35	0	0	0	317
TOWNSEND	0	0	0	0	0	26	99	81	13	0	0	0	219
TRIDENT	0	0	0	0	0	39	153	123	28	0	0	0	343
TROUT CREEK RANGER STA	0	0	0	0	0	11	61	82	20	0	0	0	174
TURNER	0	0	0	0	0	36	112	128	24	0	0	0	300
VALIER	0	0	0	0	0	22	78	91	26	0	0	0	217
VIDA	0	0	0	0	15	74	191	186	39	0	0	0	505
VIRGINIA CITY	0	0	0	0	0	9	84	73	15	0	0	0	181
WESTBY	0	0	0	0	12	72	175	147	23	0	0	0	429
WEST GLACIER	0	0	0	0	0	6	28	48	9	0	0	0	91
WEST YELLOWSTONE	0	0	0	0	0	0	11	12	0	0	0	0	23
WIBAUX 2 E	0	0	0	0	0	58	160	146	26	0	0	0	390
WINIFRED	0	0	0	0	0	44	132	119	26	0	0	0	321
WISDOM	0	0	0	0	0	0	0	0	0	0	0	0	0
WYOLA	0	0	0	0	0	43	164	144	28	0	0	0	379

24 – MONTANA

STATE-STATION NUMBER	STN TYP	NAME		LATITUDE DEG-MIN	LONGITUDE DEG-MIN	ELEVATION (FT)
24-0088	12	ALBION 1 N		N 4512	W 10417	3312
24-0364	13	AUGUSTA		N 4729	W 11223	4070
24-0375	13	AUSTIN 1 W		N 4639	W 11216	5000
24-0392	12	BABB 6 NE		N 4856	W 11322	4300
24-0412	12	BAKER		N 4622	W 10416	2929
24-0432	13	BALLANTINE		N 4557	W 10808	3000
24-0466	13	BARBER		N 4619	W 10922	3730
24-0622	13	BELGRADE AP		N 4547	W 11109	4451
24-0636	12	BELLTOWER		N 4539	W 10423	3300
24-0739	12	BIDDLE		N 4506	W 10520	3339
24-0755	13	BIGFORK 13 S		N 4753	W 11402	3010
24-0770	13	BIG SANDY		N 4810	W 11007	2700
24-0780	13	BIG TIMBER		N 4550	W 10957	4100
24-0802	13	BILLINGS WATER PLANT		N 4546	W 10829	3097
24-0807	13	BILLINGS WSO	R	N 4548	W 10832	3567
24-0877	12	BLACKLEAF		N 4801	W 11229	4323
24-1008	13	BOULDER STATE SCHOOL		N 4614	W 11207	4904
24-1044	13	BOZEMAN MONT ST UNIV		N 4540	W 11103	4856
24-1050	13	BOZEMAN 12 NE		N 4549	W 11053	5950
24-1080	13	BRADY AZNOE		N 4757	W 11120	3329
24-1088	13	BREDETTE		N 4833	W 10516	2687
24-1102	13	BRIDGER		N 4518	W 10855	3680
24-1127	13	BROADUS		N 4526	W 10524	3030
24-1149	12	BROADVIEW		N 4606	W 10853	3880
24-1202	13	BROWNING		N 4834	W 11301	4355
24-1231	12	BRUSETT 5 NW		N 4728	W 10718	3125
24-1297	12	BUSBY		N 4532	W 10657	3440
24-1318	13	BUTTE FAA AP		N 4557	W 11230	5540
24-1342	12	BYNUM 4 SSE		N 4756	W 11218	4015
24-1450	12	CANYON CREEK		N 4649	W 11215	4310
24-1552	13	CASCADE 5 S		N 4713	W 11143	3390
24-1692	13	CHESTER		N 4831	W 11057	3140
24-1722	13	CHINOOK		N 4836	W 10914	2350
24-1737	13	CHOTEAU AIRPORT		N 4749	W 11210	3945
24-1758	13	CIRCLE		N 4726	W 10535	2437
24-1905	13	COLSTRIP		N 4553	W 10636	3221
24-1938	13	COLUMBUS		N 4538	W 10914	3585
24-1974	13	CONRAD AIRPORT		N 4810	W 11158	3537
24-1995	13	COOKE CITY		N 4501	W 10956	7553
24-2104	13	CRESTON		N 4811	W 11408	2940
24-2112	13	CROW AGENCY		N 4536	W 10727	3030
24-2122	13	CULBERTSON		N 4809	W 10431	1920
24-2173	13	CUT BANK FAA AP		N 4836	W 11222	3838
24-2221	13	DARBY		N 4601	W 11410	3880
24-2347	12	DENTON 1 NNE		N 4720	W 10957	3620
24-2404	13	DILLON FAA AP		N 4515	W 11233	5216
24-2409	13	DILLON WMCE		N 4512	W 11238	5228
24-2421	13	DIVIDE 2 NW		N 4546	W 11247	5406
24-2604	13	EAST ANACONDA		N 4606	W 11255	5511
24-2689	13	EKALAKA		N 4553	W 10432	3425

STATE-STATION NUMBER	STN TYP	NAME		LATITUDE DEG-MIN	LONGITUDE DEG-MIN	ELEVATION (FT)
24-2793	13	ENNIS		N 4521	W 11143	4953
24-2820	12	ETHRIDGE		N 4834	W 11208	3544
24-2857	13	FAIRFIELD		N 4737	W 11159	3983
24-3013	13	FLATWILLOW 4 ENE		N 4651	W 10819	3138
24-3089	13	FORKS 4 NNE		N 4847	W 10728	2600
24-3110	13	FORT ASSINNIBOINE		N 4830	W 10948	2613
24-3113	13	FORT BENTON		N 4749	W 11040	2636
24-3139	13	FORTINE 1 N		N 4847	W 11454	3000
24-3157	12	FORT LOGAN 3 ESE		N 4640	W 11107	4700
24-3176	13	FORT PECK POWER PLANT		N 4801	W 10624	2070
24-3346	12	GALATA 16 SSW		N 4815	W 11125	3100
24-3366	12	GALLATIN GATEWAY 9 SSW		N 4528	W 11115	5425
24-3445	13	GERALDINE		N 4736	W 11016	3130
24-3486	12	GIBSON 3 SE		N 4600	W 10928	4410
24-3489	13	GIBSON DAM		N 4736	W 11246	4590
24-3554	12	GLASGOW 15 NW		N 4823	W 10650	2240
24-3558	13	GLASGOW WSO	//R	N 4813	W 10637	2284
24-3581	13	GLENDIVE		N 4706	W 10443	2076
24-3617	13	GOLDBUTTE 7 N		N 4859	W 11124	3499
24-3727	13	GRASSRANGE		N 4702	W 10848	3480
24-3751	13	GREAT FALLS WSO	//R	N 4729	W 11122	3662
24-3885	13	HAMILTON		N 4615	W 11409	3529
24-3915	13	HARDIN		N 4543	W 10736	2905
24-3929	13	HARLEM		N 4832	W 10847	2371
24-3939	13	HARLOWTON		N 4626	W 10950	4160
24-3984	13	HAUGAN 3 E		N 4723	W 11521	3100
24-3996	13	HAVRE WSO	R	N 4833	W 10946	2584
24-4007	13	HAXBY 18 SW		N 4734	W 10642	2650
24-4038	13	HEBGEN DAM		N 4452	W 11120	6489
24-4050	12	HELENA 6 N		N 4640	W 11203	3784
24-4055	13	HELENA WSO	//R	N 4636	W 11200	3828
24-4084	13	HERON 2 NW		N 4805	W 11600	2240
24-4133	12	HIGHWOOD 7 NE		N 4739	W 11040	3600
24-4172	12	HINGHAM		N 4833	W 11026	3030
24-4193	12	HOBSON		N 4700	W 10952	4085
24-4217	12	HOGELAND 7 WSW		N 4849	W 10848	3350
24-4241	13	HOLTER DAM		N 4700	W 11201	3487
24-4328	13	HUNGRY HORSE DAM		N 4821	W 11400	3160
24-4345	13	HUNTLEY EXP STATION		N 4555	W 10815	2989
24-4358	13	HYSHAM		N 4618	W 10714	2660
24-4506	12	JOLIET		N 4529	W 10858	3700
24-4512	13	JOPLIN 1 N		N 4835	W 11047	3360
24-4522	13	JORDAN		N 4719	W 10654	2590
24-4538	12	JUDITH GAP		N 4641	W 10945	4690
24-4558	13	KALISPELL WSO	R	N 4818	W 11416	2965
24-4563	13	KALISPELL		N 4812	W 11418	2971
24-4715	12	KNOBS		N 4555	W 10405	3000
24-4766	12	KREMLIN		N 4835	W 11005	2832
24-4820	13	LAKEVIEW		N 4436	W 11148	6710
24-4839	12	LAME DEER		N 4538	W 10639	3390

STATE-STATION NUMBER	STN TYP	NAME	LATITUDE DEG-MIN	LONGITUDE DEG-MIN	ELEVATION (FT)
24-4978	12	LEWISTOWN 10 S	N 4655	W 10925	4900
24-4985	13	LEWISTON FAA AP	N 4703	W 10927	4145
24-5015	13	LIBBY 1 NE RANGER STA	N 4824	W 11532	2080
24-5030	13	LIMA	N 4439	W 11235	6275
24-5040	13	LINCOLN RANGER STATION	N 4657	W 11239	4540
24-5045	12	LINDSAY	N 4714	W 10509	2681
24-5076	13	LIVINGSTON	N 4540	W 11033	4490
24-5080	12	LIVINGSTON 12 S	N 4529	W 11034	4870
24-5086	13	LIVINGSTON FAA AIRPORT	N 4542	W 11026	4653
24-5153	13	LOMA 1 WNW	N 4756	W 11031	2580
24-5177	13	LONESOME LAKE	N 4815	W 11012	2760
24-5285	13	LUSTRE 4 NNW	N 4827	W 10556	2920
24-5303	12	MAC KENZIE	N 4609	W 10444	2810
24-5387	13	MARTINSDALE 3 NNW	N 4630	W 11020	4800
24-5572	13	MEDICINE LAKE 3 SE	N 4829	W 10427	1952
24-5596	13	MELSTONE	N 4636	W 10752	2890
24-5685	13	MILES CITY	N 4624	W 10549	2360
24-5690	13	MILES CITY FAA AP	N 4626	W 10552	2628
24-5745	13	MISSOULA WSO //R	N 4655	W 11405	3190
24-5761	13	MOCCASIN EXP STATION	N 4703	W 10957	4300
24-5961	13	MYSTIC LAKE	N 4514	W 10945	6558
24-6138	13	NOHLY 3 WNW	N 4801	W 10408	1920
24-6157	13	NORRIS MADISON PH	N 4529	W 11138	4745
24-6238	13	OPHEIM 12 SSE	N 4842	W 10619	2950
24-6426	12	PENDROY 2 NNW	N 4807	W 11220	4365
24-6472	13	PHILIPSBURG RANGER STA	N 4619	W 11318	5270
24-6586	12	PLENTYWOOD	N 4847	W 10433	2040
24-6601	13	PLEVNA	N 4625	W 10430	2765
24-6615	13	POLEBRIDGE	N 4847	W 11416	3690
24-6640	13	POLSON KERR DAM	N 4741	W 11414	2730
24-6660	13	POPLAR	N 4807	W 10512	1995
24-6747	12	PRYOR	N 4526	W 10832	4000
24-6862	13	RAPELJE 4 S	N 4555	W 10915	4125
24-6893	13	RAYMOND BORDER STATION	N 4900	W 10435	2350
24-6918	13	RED LODGE	N 4511	W 10915	5575
24-6927	13	REDSTONE	N 4850	W 10457	2107
24-7128	12	ROBERTS 1 NNW	N 4522	W 10911	4670
24-7136	13	ROCK SPRINGS	N 4649	W 10614	3024
24-7148	12	ROCKY BOY	N 4815	W 10947	3690
24-7214	13	ROUNDUP	N 4627	W 10832	3227
24-7228	13	ROY 8 NE	N 4726	W 10850	3445
24-7286	13	SAINT IGNATIUS	N 4719	W 11406	2900
24-7382	13	SAVAGE	N 4727	W 10421	1985
24-7424	13	SCOBEY	N 4847	W 10525	2458
24-7448	13	SEELEY LAKE RANGER STA	N 4713	W 11331	4100
24-7560	13	SIDNEY	N 4744	W 10409	1920
24-7605	12	SILVER LAKE	N 4610	W 11313	6480
24-7620	13	SIMPSON 6 NW	N 4859	W 11019	2740
24-7800	12	SPRINGDALE	N 4544	W 11014	4221
24-7864	13	STANFORD 1 WNW	N 4710	W 11015	4308

24 — MONTANA

STATE-STATION NUMBER	STN TYP	NAME	LATITUDE DEG-MIN	LONGITUDE DEG-MIN	ELEVATION (FT)
24-7894	13	STEVENSVILLE	N 4631	W 11406	3370
24-7996	12	SUNBURST 8 E	N 4854	W 11144	3610
24-8021	12	SUN RIVER 5 SW	N 4728	W 11146	3560
24-8043	13	SUPERIOR	N 4712	W 11453	2710
24-8165	13	TERRY	N 4648	W 10518	2248
24-8211	13	THOMPSON FALLS PH	N 4736	W 11522	2380
24-8324	13	TOWNSEND	N 4619	W 11131	3833
24-8363	13	TRIDENT	N 4557	W 11129	4036
24-8380	13	TROUT CREEK RANGER STA	N 4752	W 11537	2356
24-8413	13	TURNER	N 4851	W 10824	3045
24-8430	12	TWIN BRIDGES	N 4533	W 11219	4625
24-8501	13	VALIER	N 4819	W 11215	3805
24-8569	13	VIDA	N 4750	W 10529	2409
24-8597	13	VIRGINIA CITY	N 4518	W 11157	5776
24-8607	12	VOLBORG	N 4550	W 10540	3030
24-8777	13	WESTBY	N 4852	W 10403	2105
24-8809	13	WEST GLACIER	N 4830	W 11359	3154
24-8857	13	WEST YELLOWSTONE	N 4439	W 11106	6662
24-8902	12	WHITEFISH 5 NW	N 4829	W 11423	3080
24-8933	12	WHITE SULPHUR SP 10 N	N 4641	W 11052	5440
24-8939	12	WHITEWATER	N 4846	W 10738	2355
24-8957	13	WIBAUX 2 E	N 4659	W 10409	2670
24-9033	13	WINIFRED	N 4733	W 10923	3243
24-9067	13	WISDOM	N 4537	W 11327	6060
24-9082	12	WISE RIVER 3 WNW	N 4548	W 11300	5730
24-9175	13	WYOLA	N 4508	W 10723	3705

24 — MONTANA

LEGEND

O = TEMPERATURE ONLY
X = PRECIPITATION ONLY
⊗ = TEMPERATURE & PRECIPITATION

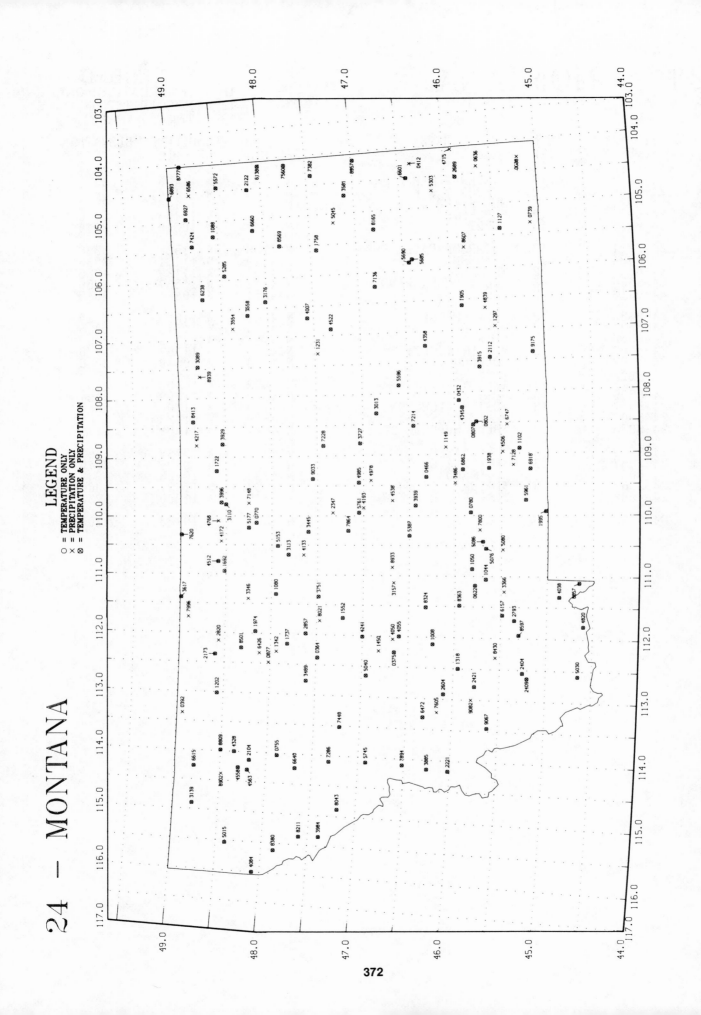

NEBRASKA

TEMPERATURE NORMALS (DEG F)

STATION		JAN	FEB	MAR	APR	MAY	JUN	JUL	AUG	SEP	OCT	NOV	DEC	ANN
AINSWORTH	MAX	32.5	38.1	45.7	60.9	72.0	81.8	88.7	86.9	77.2	66.0	48.1	37.3	61.3
	MIN	10.4	15.5	22.6	35.1	46.4	56.3	61.9	60.0	50.1	39.0	25.2	16.0	36.5
	MEAN	21.5	26.9	34.2	48.0	59.3	69.1	75.3	73.5	63.7	52.5	36.7	26.7	49.0
ALBION	MAX	30.6	37.2	45.4	61.0	71.9	81.6	87.6	85.8	76.6	66.9	49.4	37.2	60.9
	MIN	6.5	12.5	21.7	35.2	46.6	56.7	62.0	59.8	49.3	36.3	23.2	12.8	35.2
	MEAN	18.6	24.9	33.6	48.2	59.2	69.2	74.8	72.8	63.0	51.6	36.3	25.0	48.1
ALLIANCE	MAX	36.0	41.5	46.4	57.9	68.6	79.7	87.1	85.1	75.2	63.6	47.6	39.8	60.7
	MIN	10.5	15.8	21.4	31.8	42.6	52.3	58.2	55.9	45.0	33.7	21.4	14.4	33.6
	MEAN	23.3	28.7	33.9	44.9	55.6	66.1	72.6	70.5	60.1	48.7	34.5	27.2	47.2
ARTHUR	MAX	35.5	41.5	47.9	61.0	70.9	80.6	87.3	85.6	76.8	65.8	49.1	39.8	61.8
	MIN	8.7	13.9	19.8	31.5	42.8	52.7	58.4	56.3	46.3	34.3	20.6	13.1	33.2
	MEAN	22.1	27.8	33.9	46.3	56.9	66.7	72.9	71.0	61.6	50.1	34.9	26.5	47.6
ASHLAND 3 NE	MAX	31.1	38.2	48.0	64.1	74.6	84.0	89.3	87.3	78.7	68.1	51.0	37.8	62.7
	MIN	8.3	14.7	24.7	38.8	49.9	60.2	65.2	62.9	52.5	40.3	27.0	15.9	38.4
	MEAN	19.7	26.4	36.4	51.5	62.3	72.1	77.3	75.2	65.7	54.2	39.1	26.9	50.6
ATKINSON	MAX	30.2	36.4	44.9	61.1	71.9	80.8	86.6	85.3	76.2	65.5	47.7	35.6	60.2
	MIN	8.7	14.3	22.2	35.3	46.7	56.6	61.9	59.8	49.7	38.3	24.5	14.7	36.1
	MEAN	19.5	25.4	33.6	48.2	59.3	68.7	74.3	72.6	63.0	51.9	36.1	25.1	48.1
AUBURN	MAX	33.6	40.4	51.1	66.4	76.2	85.2	89.6	87.7	79.8	69.4	52.3	39.8	64.3
	MIN	12.8	18.8	27.8	41.0	51.7	61.1	65.5	63.3	54.2	42.8	29.9	19.6	40.7
	MEAN	23.2	29.6	39.5	53.7	63.9	73.2	77.6	75.5	67.0	56.2	41.1	29.7	52.5
BEATRICE NO 1	MAX	33.6	40.9	51.1	66.3	76.1	85.4	90.6	89.1	80.4	69.3	51.9	40.0	64.6
	MIN	13.2	18.8	28.0	40.9	51.7	61.4	66.6	64.5	54.7	43.2	29.8	19.7	41.0
	MEAN	23.4	29.9	39.5	53.6	63.9	73.5	78.6	76.8	67.6	56.3	40.9	29.9	52.8
BEAVER CITY	MAX	39.1	46.3	54.6	68.4	77.7	88.5	94.2	92.4	83.3	72.3	54.3	43.7	67.9
	MIN	11.5	17.7	24.8	37.2	47.9	58.1	63.5	61.0	50.3	37.5	24.3	16.0	37.5
	MEAN	25.3	32.0	39.7	52.9	62.9	73.3	78.9	76.7	66.8	54.9	39.3	29.9	52.7
BENKELMAN	MAX	40.3	46.5	52.5	65.2	74.7	85.4	92.0	90.2	81.2	70.5	53.3	44.4	66.4
	MIN	11.9	17.3	23.6	35.4	46.2	56.3	62.2	59.5	48.9	35.6	23.2	15.5	36.3
	MEAN	26.1	31.9	38.1	50.3	60.5	70.9	77.1	74.9	65.0	53.1	38.3	30.0	51.4
BIG SPRINGS	MAX	38.8	44.8	50.9	63.3	72.9	83.3	89.7	87.9	78.8	67.3	51.7	42.4	64.3
	MIN	11.3	16.2	22.8	33.7	44.7	54.3	60.2	57.9	47.1	35.1	22.6	14.8	35.1
	MEAN	25.1	30.5	36.8	48.5	58.8	68.8	75.0	72.9	63.0	51.2	37.2	28.7	49.7
BLAIR	MAX	28.7	35.5	45.4	62.0	73.1	82.0	86.5	84.4	75.6	65.7	49.1	35.9	60.3
	MIN	9.1	15.3	25.4	39.3	50.2	60.3	64.9	62.5	52.9	41.3	27.8	16.6	38.8
	MEAN	18.9	25.4	35.4	50.7	61.6	71.2	75.8	73.5	64.3	53.6	38.5	26.3	49.6
BOX BUTTE EXP STATION	MAX	34.7	40.2	45.1	57.2	68.0	79.0	87.3	85.5	75.7	64.3	47.6	39.2	60.3
	MIN	9.0	14.5	20.2	30.7	41.5	51.1	56.8	54.6	44.0	32.9	20.7	13.3	32.4
	MEAN	21.9	27.4	32.7	44.0	54.8	65.1	72.0	70.1	59.9	48.6	34.1	26.2	46.4
BRIDGEPORT	MAX	39.5	46.3	51.9	63.8	73.6	83.9	91.1	89.2	80.4	69.3	51.9	42.8	65.3
	MIN	10.6	16.2	22.0	32.3	43.1	52.7	58.5	55.8	45.1	32.9	21.1	14.3	33.7
	MEAN	25.1	31.3	37.0	48.1	58.4	68.3	74.8	72.6	62.8	51.1	36.5	28.5	49.5
BROKEN BOW 2 W	MAX	35.0	40.9	48.2	62.6	72.5	82.7	88.3	86.8	78.0	67.8	50.5	40.1	62.8
	MIN	8.8	14.6	21.7	34.1	45.1	55.2	60.8	59.0	48.1	35.5	21.8	13.9	34.9
	MEAN	21.9	27.8	35.0	48.3	58.8	69.0	74.6	72.9	63.1	51.7	36.2	27.0	48.9
BUTTE	MAX	29.8	36.4	45.3	61.7	72.9	82.5	89.1	87.3	77.3	65.9	47.6	35.5	60.9
	MIN	7.4	13.5	21.9	35.3	46.6	56.7	62.4	60.5	49.9	38.3	24.4	14.0	35.9
	MEAN	18.6	25.0	33.6	48.5	59.8	69.7	75.8	73.9	63.6	52.1	36.0	24.8	48.5
CAMBRIDGE	MAX	38.9	45.9	53.4	67.0	75.8	85.9	91.3	90.0	81.9	71.2	53.5	43.1	66.5
	MIN	12.0	17.8	25.0	37.0	48.0	58.1	63.5	61.3	50.6	37.2	24.1	16.1	37.6
	MEAN	25.5	31.9	39.2	52.0	61.9	72.1	77.4	75.7	66.3	54.2	38.8	29.6	52.1

NEBRASKA

TEMPERATURE NORMALS (DEG F)

STATION			JAN	FEB	MAR	APR	MAY	JUN	JUL	AUG	SEP	OCT	NOV	DEC	ANN
CENTRAL CITY		MAX	32.6	39.8	48.9	64.7	75.1	84.9	89.5	87.7	78.8	68.5	50.8	38.5	63.3
		MIN	10.8	17.1	25.7	38.5	49.6	59.7	64.4	62.6	52.4	40.7	27.0	16.9	38.8
		MEAN	21.7	28.4	37.3	51.6	62.3	72.3	77.0	75.2	65.6	54.6	38.9	27.7	51.1
CHADRON FAA AIRPORT	R	MAX	34.6	40.8	47.3	59.1	69.8	80.9	89.4	88.0	77.7	65.2	48.4	38.9	61.7
		MIN	9.1	14.6	21.2	32.2	43.0	52.7	59.0	57.0	45.5	34.0	21.3	13.6	33.6
		MEAN	21.9	27.7	34.3	45.7	56.4	66.8	74.2	72.6	61.7	49.6	34.9	26.3	47.7
CLARKSON		MAX	30.1	36.8	46.7	63.1	74.4	83.8	88.1	85.8	76.8	66.6	49.2	36.5	61.5
		MIN	8.1	14.3	23.8	36.8	48.5	58.4	63.3	61.0	50.9	39.1	25.6	14.9	37.1
		MEAN	19.1	25.6	35.2	50.0	61.5	71.1	75.7	73.4	63.9	52.9	37.4	25.7	49.3
COLUMBUS 3 NE		MAX	30.0	37.1	46.8	62.4	73.6	83.7	88.5	86.3	77.0	66.2	48.9	36.3	61.4
		MIN	9.6	15.9	25.2	38.5	50.0	60.1	65.0	63.0	52.3	40.4	26.7	16.2	38.6
		MEAN	19.8	26.6	36.0	50.5	61.9	72.0	76.8	74.6	64.7	53.4	37.8	26.3	50.0
CRESCENT LAKE NATL WLR		MAX	37.2	43.2	48.6	61.3	70.7	80.3	87.0	86.0	77.3	66.7	50.0	41.0	62.4
		MIN	9.5	14.7	20.7	31.9	42.5	51.5	57.0	54.9	44.2	32.6	20.5	13.4	32.8
		MEAN	23.4	29.0	34.7	46.6	56.6	65.9	72.0	70.5	60.8	49.7	35.3	27.2	47.6
CRETE		MAX	32.5	39.9	49.8	64.9	75.2	84.6	89.5	87.7	79.0	68.6	50.9	38.8	63.5
		MIN	11.9	18.1	27.3	40.1	51.0	60.7	65.5	63.6	54.0	42.7	29.0	18.5	40.2
		MEAN	22.2	29.0	38.5	52.5	63.2	72.7	77.5	75.6	66.5	55.7	40.0	28.7	51.8
CULBERTSON		MAX	37.8	44.6	51.6	64.6	74.5	85.1	91.2	89.6	80.7	69.6	52.0	41.9	65.3
		MIN	10.0	15.6	22.5	34.6	45.7	56.0	61.6	59.3	48.4	34.9	21.9	14.3	35.4
		MEAN	23.9	30.1	37.1	49.6	60.1	70.6	76.4	74.5	64.6	52.3	37.0	28.1	50.4
CURTIS		MAX	38.7	45.4	53.0	66.4	75.5	85.9	91.8	90.7	81.5	70.5	52.4	42.9	66.2
		MIN	9.6	15.6	22.4	34.3	45.4	55.2	60.6	58.2	47.3	34.3	21.4	13.6	34.8
		MEAN	24.1	30.5	37.7	50.4	60.5	70.6	76.2	74.5	64.4	52.4	36.9	28.3	50.5
DAVID CITY		MAX	30.5	37.3	46.9	63.1	74.3	84.0	89.2	86.9	77.7	66.9	49.5	36.7	61.9
		MIN	9.8	15.8	25.0	38.6	49.8	59.7	64.6	62.6	53.0	41.3	27.4	16.4	38.7
		MEAN	20.1	26.5	36.0	50.8	62.0	71.9	77.0	74.8	65.3	54.2	38.4	26.6	50.3
EWING		MAX	30.7	37.2	46.3	62.2	73.0	82.2	87.6	85.9	76.9	66.5	48.5	36.4	61.1
		MIN	7.2	13.5	22.5	36.0	47.3	57.2	62.4	60.2	49.6	37.5	23.7	13.4	35.9
		MEAN	19.0	25.4	34.4	49.1	60.2	69.7	75.0	73.1	63.3	52.0	36.2	24.9	48.5
FAIRBURY 2 SSE		MAX	32.8	39.6	49.2	64.2	74.4	83.8	89.9	88.3	79.1	68.5	51.2	39.2	63.4
		MIN	10.7	16.2	25.3	38.4	49.6	59.7	64.9	62.9	52.8	40.5	27.0	17.2	38.8
		MEAN	21.8	27.9	37.2	51.3	62.0	71.7	77.4	75.6	66.0	54.6	39.1	28.2	51.1
FAIRMONT		MAX	33.8	41.3	50.3	65.4	75.5	85.4	90.6	88.6	80.1	69.8	52.2	40.3	64.4
		MIN	11.2	17.4	26.0	38.7	49.7	59.7	64.6	62.9	52.8	41.7	28.0	17.7	39.2
		MEAN	22.5	29.4	38.2	52.0	62.6	72.6	77.7	75.8	66.5	55.8	40.1	29.0	51.9
FALLS CITY		MAX	34.8	41.5	51.9	66.7	76.2	84.7	89.2	87.4	79.4	69.4	53.1	40.9	64.6
		MIN	14.6	20.4	29.5	42.1	52.9	62.2	66.6	64.4	55.6	44.8	31.7	21.4	42.2
		MEAN	24.8	30.9	40.7	54.5	64.6	73.5	78.0	75.9	67.5	57.1	42.4	31.2	53.4
FRANKLIN		MAX	36.7	43.8	52.1	66.2	76.5	86.7	92.5	90.4	81.0	70.2	52.9	42.1	65.9
		MIN	11.7	17.7	25.6	38.5	49.6	60.0	65.0	62.9	52.5	39.8	25.9	16.8	38.8
		MEAN	24.2	30.8	38.9	52.4	63.0	73.4	78.8	76.6	66.8	55.0	39.4	29.5	52.4
FREMONT		MAX	31.0	38.2	48.5	64.9	75.7	84.6	89.3	87.2	78.7	67.9	50.2	37.5	62.8
		MIN	10.5	16.9	26.3	39.5	50.6	59.9	64.6	62.3	52.7	41.3	28.5	17.6	39.2
		MEAN	20.8	27.5	37.4	52.2	63.2	72.3	77.0	74.7	65.7	54.6	39.4	27.6	51.0
GENEVA		MAX	32.9	39.9	49.6	64.5	74.9	84.3	89.4	87.4	78.4	67.6	50.6	38.9	63.2
		MIN	11.6	17.8	26.4	39.3	50.3	60.2	65.2	63.4	53.8	42.2	28.2	18.1	39.7
		MEAN	22.3	28.9	38.0	51.9	62.6	72.3	77.3	75.4	66.1	55.0	39.4	28.5	51.5
GENOA 2 W		MAX	31.7	38.7	48.2	64.0	74.5	84.1	88.7	86.7	78.1	67.9	50.2	37.5	62.5
		MIN	9.1	15.3	24.5	37.2	48.5	58.4	63.2	61.3	51.1	39.1	25.5	15.3	37.4
		MEAN	20.4	27.0	36.4	50.6	61.5	71.3	76.0	74.0	64.6	53.5	37.9	26.4	50.0

NEBRASKA

TEMPERATURE NORMALS (DEG F)

STATION			JAN	FEB	MAR	APR	MAY	JUN	JUL	AUG	SEP	OCT	NOV	DEC	ANN
GOTHENBURG		MAX	36.1	42.7	50.6	64.4	74.4	84.5	90.0	88.3	79.1	67.8	50.6	40.8	64.1
		MIN	11.6	17.1	24.2	36.5	47.4	57.3	62.7	60.5	50.0	38.0	24.3	16.1	37.1
		MEAN	23.9	29.9	37.4	50.5	60.9	70.9	76.4	74.5	64.6	52.9	37.5	28.5	50.7
GRAND ISLAND WSO	//R	MAX	31.2	38.1	47.1	62.3	73.0	83.6	88.8	86.9	77.1	66.7	49.4	37.3	61.8
		MIN	9.9	16.2	24.7	37.8	49.1	59.2	64.4	62.3	51.5	39.4	25.7	16.0	38.0
		MEAN	20.6	27.2	35.9	50.1	61.1	71.4	76.6	74.6	64.3	53.1	37.6	26.7	49.9
HALSEY 2 W		MAX	35.2	41.5	49.1	63.3	74.3	84.3	90.5	88.7	79.3	68.2	49.8	39.5	63.6
		MIN	8.7	14.2	21.2	33.5	44.5	54.1	59.6	57.5	47.2	34.8	22.2	13.8	34.3
		MEAN	22.0	27.9	35.2	48.4	59.5	69.2	75.1	73.1	63.3	51.6	36.0	26.7	49.0
HARLAN COUNTY DAM		MAX	35.0	41.5	49.5	63.4	73.0	83.3	89.6	88.1	78.9	68.5	51.3	40.3	63.5
		MIN	9.6	15.7	23.7	37.0	47.9	58.1	63.8	61.4	51.2	38.8	24.6	15.3	37.3
		MEAN	22.3	28.6	36.6	50.2	60.5	70.8	76.7	74.7	65.1	53.7	38.0	27.8	50.4
HARRISON		MAX	32.1	37.5	42.5	54.1	65.0	76.3	85.0	83.7	73.8	61.7	44.5	36.4	57.7
		MIN	8.2	13.5	18.4	28.2	38.9	48.5	54.6	52.4	42.0	31.7	19.6	12.9	30.7
		MEAN	20.2	25.5	30.5	41.2	51.9	62.4	69.8	68.1	57.9	46.7	32.1	24.7	44.3
HARTINGTON		MAX	28.3	35.2	45.0	62.5	74.5	84.1	89.5	87.5	78.1	66.8	48.1	34.5	61.2
		MIN	6.6	13.2	22.8	36.8	48.3	58.3	63.6	61.7	51.3	39.8	25.3	13.9	36.8
		MEAN	17.5	24.2	34.0	49.7	61.4	71.2	76.6	74.6	64.7	53.3	36.7	24.2	49.0
HASTINGS		MAX	34.0	40.8	49.2	64.1	74.7	84.8	90.0	87.9	78.6	67.7	50.7	39.3	63.5
		MIN	12.0	17.9	25.5	38.6	49.5	59.1	63.7	61.5	52.0	40.8	27.2	17.8	38.8
		MEAN	23.0	29.4	37.4	51.4	62.1	72.0	76.9	74.7	65.4	54.3	39.0	28.6	51.2
HAYES CENTER		MAX	36.9	42.6	49.1	62.2	71.9	82.3	89.1	87.7	78.1	67.6	50.7	41.4	63.3
		MIN	12.0	17.1	23.4	35.2	46.1	55.7	61.7	59.6	49.6	37.7	24.6	16.9	36.6
		MEAN	24.5	29.9	36.3	48.7	59.0	69.0	75.4	73.7	63.9	52.7	37.7	29.2	50.0
HAY SPRINGS		MAX	34.5	40.4	46.4	58.5	69.7	80.3	87.8	86.7	77.0	64.8	47.5	38.6	61.0
		MIN	9.3	14.8	20.6	31.0	41.7	51.4	57.1	55.3	44.4	33.5	21.3	13.8	32.9
		MEAN	21.9	27.6	33.5	44.8	55.7	65.9	72.5	71.0	60.7	49.2	34.4	26.2	47.0
HEBRON		MAX	32.6	40.0	49.6	64.4	74.6	84.6	90.5	89.0	79.5	69.1	51.0	39.1	63.7
		MIN	11.3	17.2	26.5	40.0	51.1	61.0	66.2	64.1	53.6	41.5	27.8	17.8	39.8
		MEAN	22.0	28.6	38.1	52.3	62.8	72.9	78.4	76.6	66.6	55.4	39.4	28.5	51.8
HOLDREGE 3 SW		MAX	34.5	41.8	50.3	64.7	74.6	85.1	90.2	88.4	79.5	67.7	50.3	39.5	63.9
		MIN	12.8	18.4	25.2	37.5	48.6	58.4	63.6	61.6	51.9	40.7	26.6	17.8	38.6
		MEAN	23.7	30.1	37.7	51.2	61.6	71.8	76.9	75.0	65.7	54.2	38.5	28.7	51.3
IMPERIAL		MAX	39.4	45.2	51.5	64.0	73.8	84.4	90.4	88.6	79.7	68.7	51.6	43.0	65.0
		MIN	13.4	18.4	24.1	35.1	46.1	56.1	61.9	59.7	49.6	37.2	24.5	17.3	37.0
		MEAN	26.4	31.8	37.8	49.6	60.0	70.3	76.2	74.2	64.7	52.9	38.1	30.1	51.0
KEARNEY		MAX	33.0	39.6	47.2	61.7	72.3	82.8	88.6	87.0	77.6	67.0	49.8	38.9	62.1
		MIN	9.2	15.0	23.1	37.0	48.5	58.6	63.7	61.2	50.9	38.5	24.2	15.0	37.1
		MEAN	21.1	27.3	35.2	49.4	60.4	70.7	76.2	74.1	64.3	52.8	37.1	27.0	49.6
KIMBALL		MAX	38.5	43.5	48.0	58.9	68.7	79.8	86.9	84.9	75.9	64.3	49.0	41.9	61.7
		MIN	12.6	17.0	21.2	31.0	41.1	50.3	56.5	54.1	44.0	33.4	22.1	16.1	33.3
		MEAN	25.6	30.3	34.7	45.0	54.9	65.1	71.7	69.5	59.9	48.9	35.6	29.0	47.5
KINGSLEY DAM		MAX	37.0	43.4	49.8	62.5	72.3	82.7	89.5	87.8	79.1	67.9	50.7	41.0	63.6
		MIN	13.9	18.1	24.1	35.4	46.4	56.0	61.9	59.9	50.2	39.4	27.1	19.4	37.7
		MEAN	25.5	30.7	37.0	49.0	59.4	69.4	75.7	73.9	64.7	53.7	38.9	30.2	50.7
LINCOLN WSO AP		MAX	30.4	37.5	47.7	63.6	74.4	84.3	89.5	87.2	78.0	67.4	50.0	37.2	62.3
		MIN	8.9	15.4	25.1	38.6	49.9	60.2	65.6	63.4	52.9	40.6	26.9	16.0	38.6
		MEAN	19.7	26.5	36.4	51.1	62.2	72.3	77.6	75.3	65.5	54.0	38.5	26.6	50.5
LINCOLN WSO CITY	R	MAX	31.0	37.7	47.3	62.9	73.8	83.6	88.6	86.6	77.5	67.0	50.1	37.5	62.0
		MIN	12.7	18.8	28.1	41.7	53.2	63.4	68.4	66.4	56.2	44.4	30.6	19.6	41.9
		MEAN	21.9	28.3	37.7	52.3	63.5	73.5	78.5	76.5	66.9	55.7	40.4	28.6	52.0

NEBRASKA

TEMPERATURE NORMALS (DEG F)

STATION			JAN	FEB	MAR	APR	MAY	JUN	JUL	AUG	SEP	OCT	NOV	DEC	ANN
LODGEPOLE		MAX	40.8	46.4	51.7	63.6	73.0	84.2	91.4	89.4	80.6	69.2	52.3	43.7	65.5
		MIN	12.5	16.9	22.0	32.3	42.8	52.3	58.6	56.3	45.8	34.5	22.6	15.9	34.4
		MEAN	26.6	31.7	36.9	48.0	57.9	68.3	75.0	72.9	63.2	51.9	37.5	29.8	50.0
LOUP CITY		MAX	32.9	39.3	47.7	63.1	73.4	83.2	88.9	.87.1	77.7	67.3	50.2	38.6	62.5
		MIN	8.8	14.8	23.0	35.9	47.5	57.6	62.9	60.5	49.4	36.9	23.6	14.1	36.3
		MEAN	20.8	27.1	35.4	49.5	60.4	70.5	75.9	73.8	63.6	52.1	36.9	26.4	49.4
MADISON		MAX	30.7	37.4	47.1	63.7	74.6	84.5	89.2	86.8	78.1	67.4	49.5	36.9	62.2
		MIN	7.4	13.8	23.5	36.6	48.1	58.3	63.1	60.7	50.2	38.0	24.7	14.0	36.5
		MEAN	19.0	25.6	35.3	50.2	61.4	71.4	76.2	73.8	64.2	52.7	37.1	25.5	49.4
MADRID		MAX	38.1	44.3	51.1	64.3	74.5	85.9	92.2	90.2	81.0	69.3	51.1	41.8	65.3
		MIN	12.2	17.3	23.1	34.3	45.3	55.4	61.3	59.2	48.7	36.4	23.5	16.1	36.1
		MEAN	25.2	30.8	37.2	49.3	59.9	70.7	76.8	74.7	64.9	52.8	37.4	29.0	50.7
MCCOOK		MAX	39.5	45.8	52.9	66.3	75.6	85.9	91.8	89.8	81.5	70.6	53.4	43.6	66.4
		MIN	13.3	18.4	25.0	37.1	47.9	57.9	63.8	61.6	51.2	38.6	25.5	17.7	38.2
		MEAN	26.4	32.1	39.0	51.7	61.8	71.9	77.8	75.7	66.4	54.6	39.5	30.6	52.3
MEDICINE CREEK DAM		MAX	36.0	42.3	49.3	63.1	72.9	83.3	89.7	88.3	79.0	68.4	51.2	41.2	63.7
		MIN	8.8	14.9	22.6	35.6	46.8	57.0	62.7	60.2	49.5	36.5	22.8	14.4	36.0
		MEAN	22.4	28.6	36.0	49.4	59.9	70.2	76.2	74.3	64.3	52.5	37.0	27.9	49.9
MERRIMAN		MAX	34.1	40.1	46.3	59.9	71.3	81.0	88.6	87.4	78.1	66.5	48.5	38.4	61.7
		MIN	8.7	14.4	20.8	32.1	43.4	53.0	58.8	56.4	46.1	35.3	22.1	14.0	33.8
		MEAN	21.4	27.3	33.6	46.0	57.4	67.0	73.7	71.9	62.1	50.9	35.3	26.2	47.7
MINDEN		MAX	34.6	41.8	50.2	64.3	74.2	84.5	89.8	88.0	79.3	68.6	51.0	39.8	63.8
		MIN	12.1	18.1	25.6	37.8	48.9	59.0	64.3	62.4	52.4	40.7	26.4	17.5	38.8
		MEAN	23.4	29.9	37.9	51.1	61.6	71.7	77.1	75.2	65.9	54.7	38.7	28.7	51.3
MITCHELL 5 E		MAX	36.1	42.3	47.3	59.0	69.6	80.4	86.9	84.2	75.8	65.5	49.1	40.5	61.4
		MIN	10.8	15.9	21.6	31.6	42.3	51.8	57.3	54.3	43.7	33.0	21.5	14.3	33.2
		MEAN	23.4	29.1	34.5	45.3	56.0	66.1	72.1	69.3	59.8	49.3	35.3	27.4	47.3
MULLEN		MAX	35.7	41.4	48.1	61.9	72.7	82.9	89.1	87.0	77.8	67.0	49.5	40.1	62.8
		MIN	11.0	15.8	22.0	33.7	44.6	54.4	60.4	58.2	47.9	36.3	23.8	15.8	35.3
		MEAN	23.3	28.6	35.0	47.8	58.7	68.7	74.8	72.6	62.9	51.7	36.7	28.0	49.1
NENZEL 20 S		MAX	35.1	40.7	47.2	61.1	72.1	82.3	88.8	87.1	78.1	67.1	49.4	39.7	62.4
		MIN	8.9	14.1	20.3	31.8	42.4	52.2	57.6	55.6	45.2	34.4	21.7	14.3	33.2
		MEAN	22.0	27.4	33.7	46.4	57.3	67.3	73.2	71.4	61.7	50.7	35.6	27.0	47.8
NIOBRARA		MAX	29.5	36.4	45.6	62.8	74.1	84.1	89.6	87.5	78.1	67.0	48.1	35.6	61.5
		MIN	6.9	13.6	22.9	36.4	47.5	57.6	62.8	60.9	50.5	38.3	24.7	14.0	36.3
		MEAN	18.2	25.0	34.3	49.6	60.8	70.9	76.2	74.3	64.3	52.7	36.4	24.8	49.0
NORFOLK WSO	//R	MAX	27.8	34.2	43.8	60.5	72.2	82.4	87.4	85.0	75.5	64.7	47.1	34.1	59.6
		MIN	6.9	13.3	23.2	36.9	48.8	58.9	64.2	61.9	51.0	38.8	24.8	13.7	36.9
		MEAN	17.4	23.8	33.5	48.7	60.6	70.7	75.8	73.4	63.3	51.8	36.0	23.9	48.2
NORTH LOUP		MAX	33.0	39.7	48.2	63.5	73.6	83.2	87.8	86.0	77.3	67.3	50.1	38.6	62.4
		MIN	8.8	14.8	23.3	35.8	47.3	57.5	62.7	60.5	50.2	37.6	24.0	14.3	36.4
		MEAN	20.9	27.3	35.8	49.7	60.5	70.4	75.3	73.3	63.7	52.5	37.1	26.5	49.4
NORTH PLATTE WSO	//R	MAX	34.2	40.5	47.8	61.5	71.5	81.6	87.8	86.4	77.3	66.7	49.4	39.3	62.0
		MIN	8.3	14.1	21.5	33.6	44.8	55.0	60.6	58.3	46.5	33.7	20.5	12.5	34.1
		MEAN	21.3	27.3	34.7	47.6	58.2	68.3	74.2	72.4	61.9	50.2	35.0	25.9	48.1
OAKDALE		MAX	30.3	36.4	45.4	61.6	72.8	82.4	88.0	85.5	76.3	65.9	48.6	36.2	60.8
		MIN	6.7	12.9	22.3	36.2	47.2	57.2	62.4	60.0	49.3	36.9	24.0	13.1	35.7
		MEAN	18.5	24.6	33.9	48.9	60.0	69.8	75.3	72.8	62.8	51.4	36.3	24.7	48.3
OGALLALA 3 W		MAX	36.3	42.5	48.6	61.9	71.8	82.5	90.0	88.3	78.2	67.2	50.1	40.7	63.2
		MIN	10.2	15.5	22.2	33.7	44.8	54.8	60.3	57.7	46.6	34.0	21.7	14.1	34.6
		MEAN	23.3	29.0	35.5	47.8	58.3	68.7	75.2	73.0	62.4	50.6	35.9	27.4	48.9

NEBRASKA

TEMPERATURE NORMALS (DEG F)

STATION		JAN	FEB	MAR	APR	MAY	JUN	JUL	AUG	SEP	OCT	NOV	DEC	ANN
OMAHA WSO EPPLEY //R	MAX	30.2	37.3	47.7	64.0	74.7	84.2	88.5	86.2	77.5	67.0	50.3	36.9	62.0
	MIN	10.2	17.1	26.9	40.3	51.8	61.7	66.8	64.2	54.0	42.0	28.6	17.4	40.1
	MEAN	20.2	27.2	37.3	52.2	63.3	73.0	77.7	75.2	65.8	54.5	39.5	27.2	51.1
OMAHA NORTH OMAHA WSO	MAX	27.8	34.5	44.8	61.4	72.4	81.6	85.7	83.7	75.0	64.6	47.6	34.5	59.5
	MIN	9.6	16.0	25.6	39.3	51.0	60.8	65.6	63.3	53.8	42.5	28.3	16.9	39.4
	MEAN	18.7	25.3	35.2	50.4	61.7	71.2	75.7	73.5	64.4	53.6	38.0	25.7	49.5
O NEILL	MAX	29.6	36.0	44.9	61.3	72.5	82.3	88.5	86.7	76.7	65.3	47.3	35.2	60.5
	MIN	7.8	13.9	21.8	35.1	46.5	56.2	61.6	59.7	49.4	37.7	24.0	13.7	35.6
	MEAN	18.7	24.9	33.4	48.2	59.5	69.2	75.1	73.2	63.1	51.6	35.7	24.5	48.1
OSCEOLA	MAX	31.2	38.4	48.0	63.8	74.6	84.3	89.2	87.3	78.6	68.1	50.2	37.7	62.6
	MIN	9.2	15.5	24.7	37.7	49.3	59.4	64.2	62.0	52.1	40.2	26.5	16.0	38.1
	MEAN	20.2	27.0	36.4	50.8	62.0	71.9	76.7	74.7	65.4	54.2	38.4	26.9	50.4
OSHKOSH	MAX	37.8	44.0	50.1	63.1	72.7	82.6	89.0	87.2	78.9	67.9	50.9	41.4	63.8
	MIN	10.0	15.4	21.6	32.4	43.5	52.9	58.5	55.7	45.1	33.0	21.0	13.7	33.6
	MEAN	23.9	29.7	35.9	47.8	58.1	67.8	73.8	71.5	62.0	50.5	35.9	27.6	48.7
OSMOND	MAX	28.8	35.6	45.2	62.6	74.2	84.1	88.9	86.6	77.5	66.4	48.4	35.4	61.1
	MIN	6.0	12.7	22.6	36.2	47.9	57.8	62.7	60.5	50.2	38.1	23.9	12.8	36.0
	MEAN	17.4	24.2	34.0	49.4	61.1	71.0	75.8	73.6	63.9	52.3	36.2	24.1	48.6
PAWNEE CITY	MAX	35.3	42.7	52.7	67.7	77.7	86.7	91.8	90.0	81.4	71.0	53.4	41.1	66.0
	MIN	14.1	20.1	29.1	42.0	52.7	61.9	66.6	64.5	55.0	43.6	30.9	20.7	41.8
	MEAN	24.8	31.4	41.0	54.9	65.3	74.3	79.2	77.3	68.3	57.4	42.2	31.0	53.9
PURDUM	MAX	34.6	40.5	47.9	62.4	73.0	82.9	89.0	87.0	78.2	67.4	49.6	39.4	62.7
	MIN	8.9	14.3	21.4	33.7	44.6	54.5	60.0	57.8	47.6	35.7	22.4	13.8	34.6
	MEAN	21.8	27.4	34.7	48.1	58.9	68.7	74.5	72.4	63.0	51.6	36.0	26.6	48.6
RAVENNA	MAX	34.1	41.2	49.7	64.9	75.1	85.4	90.3	88.6	79.7	68.7	50.8	39.3	64.0
	MIN	9.4	15.6	23.8	36.6	47.5	57.6	62.9	60.7	50.1	37.7	23.9	14.5	36.7
	MEAN	21.7	28.5	36.8	50.8	61.3	71.5	76.7	74.7	64.9	53.2	37.3	27.0	50.4
RED CLOUD	MAX	35.9	43.4	52.2	66.5	76.3	86.3	91.9	90.1	80.8	70.0	52.8	41.6	65.7
	MIN	10.5	16.7	25.1	38.2	49.0	58.8	63.9	61.5	51.0	38.7	25.1	15.7	37.9
	MEAN	23.2	30.1	38.7	52.4	62.7	72.5	77.9	75.9	65.9	54.4	39.0	28.7	51.8
SAINT PAUL	MAX	32.3	38.8	47.2	62.4	72.5	82.2	87.2	85.3	76.2	66.0	49.5	37.7	61.4
	MIN	9.5	15.5	24.4	37.5	49.1	59.0	63.9	61.9	51.5	39.2	25.2	15.2	37.7
	MEAN	20.9	27.2	35.8	50.0	60.8	70.6	75.6	73.6	63.9	52.6	37.4	26.5	49.6
SCOTTSBLUFF WSO //R	MAX	37.2	43.4	48.8	60.3	70.8	81.7	89.2	86.7	77.5	66.0	49.8	40.8	62.7
	MIN	11.2	16.7	22.3	32.5	43.6	53.3	59.2	56.5	45.6	34.3	22.0	14.8	34.3
	MEAN	24.2	30.1	35.6	46.4	57.2	67.5	74.2	71.6	61.6	50.2	35.9	27.8	48.5
SEWARD	MAX	31.3	38.4	48.3	64.3	74.9	84.8	89.7	86.6	77.2	66.7	49.6	37.6	62.5
	MIN	11.7	17.8	26.8	39.8	50.7	60.5	65.5	63.4	53.9	42.6	28.8	18.4	40.0
	MEAN	21.5	28.1	37.6	52.1	62.9	72.7	77.6	75.1	65.6	54.7	39.2	28.1	51.3
SIDNEY 6 NNW	MAX	38.8	44.4	49.3	60.5	70.0	80.8	88.0	85.9	77.0	66.1	50.0	42.2	62.8
	MIN	11.3	16.2	21.3	31.1	42.0	51.4	57.7	55.4	45.1	33.5	21.4	15.1	33.5
	MEAN	25.0	30.3	35.3	45.8	56.0	66.2	72.8	70.7	61.0	49.8	35.7	28.7	48.1
SIDNEY FAA AIRPORT X R	MAX	37.6	42.6	47.1	58.7	68.7	80.0	87.6	85.7	76.4	64.8	48.8	40.9	61.6
	MIN	10.9	15.6	19.8	30.0	40.8	50.4	56.8	54.8	44.3	32.7	20.6	14.2	32.6
	MEAN	24.2	29.1	33.5	44.4	54.8	65.3	72.3	70.3	60.4	48.8	34.7	27.6	47.1
SPRINGVIEW	MAX	31.0	36.4	44.0	59.5	71.0	81.2	88.8	87.1	77.0	65.6	47.6	36.2	60.5
	MIN	8.4	13.8	21.4	34.8	46.0	56.1	62.1	60.1	49.8	38.4	24.6	14.7	35.9
	MEAN	19.7	25.1	32.7	47.1	58.6	68.7	75.5	73.6	63.4	52.0	36.1	25.4	48.2
STANTON	MAX	29.7	36.7	46.7	63.5	74.3	83.4	88.2	85.9	76.9	66.4	48.9	35.8	61.4
	MIN	7.9	14.2	23.8	36.9	48.1	57.9	62.8	60.9	50.9	39.4	25.7	14.7	36.9
	MEAN	18.8	25.4	35.3	50.2	61.2	70.7	75.5	73.4	63.9	52.9	37.3	25.3	49.2

NEBRASKA

TEMPERATURE NORMALS (DEG F)

STATION		JAN	FEB	MAR	APR	MAY	JUN	JUL	AUG	SEP	OCT	NOV	DEC	ANN
STAPLETON 5 SSE	MAX	34.4	40.7	47.9	62.1	72.7	82.4	89.4	88.2	78.4	67.1	49.3	39.3	62.7
	MIN	9.4	14.9	21.5	33.2	44.3	54.3	59.9	58.1	47.5	35.6	22.4	14.1	34.6
	MEAN	21.9	27.8	34.7	47.7	58.6	68.3	74.7	73.1	63.0	51.4	35.9	26.7	48.7
SYRACUSE	MAX	31.8	38.7	49.1	64.8	75.6	85.1	89.9	87.9	79.3	68.4	51.2	38.3	63.3
	MIN	10.0	16.2	26.1	39.6	50.8	60.5	65.2	62.8	53.0	40.8	27.7	17.1	39.2
	MEAN	20.9	27.5	37.6	52.2	63.2	72.9	77.5	75.4	66.1	54.6	39.5	27.7	51.3
TECUMSEH	MAX	33.0	40.0	50.0	65.0	75.4	84.6	89.6	87.8	79.3	69.1	52.0	39.5	63.8
	MIN	10.5	16.8	26.5	39.3	49.9	59.9	64.6	62.2	52.4	40.4	27.9	17.6	39.0
	MEAN	21.8	28.4	38.2	52.2	62.7	72.3	77.1	75.0	65.9	54.8	40.0	28.6	51.4
TEKAMAH	MAX	30.1	37.1	47.5	64.5	76.1	85.2	89.3	87.0	78.6	68.1	50.1	36.9	62.5
	MIN	8.9	15.4	25.3	38.7	50.0	60.0	64.5	62.1	52.3	40.9	27.9	16.3	38.5
	MEAN	19.5	26.3	36.4	51.6	63.1	72.6	76.9	74.6	65.5	54.5	39 0	26.6	50.6
TRENTON DAM	MAX	38.4	44.4	51.0	64.2	73.5	84.2	91.0	89.9	80.6	69.6	52.1	42.6	65.1
	MIN	11.0	16.1	22.9	35.4	46.0	56.5	62.5	60.0	49.4	36.4	23.7	15.4	36.3
	MEAN	24.7	30.3	37.0	49.8	59.8	70.4	76.7	75.0	65.1	53.0	37.9	29.0	50.7
VALENTINE LKS GAME REF	MAX	34.3	39.9	46.1	60.1	71.8	81.7	88.8	86.6	77.4	66.0	48.9	38.9	61.7
	MIN	10.7	15.6	22.3	34.7	45.8	55.4	61.3	59.1	48.8	38.1	24.8	16.5	36.1
	MEAN	22.5	27.8	34.2	47.4	58.8	68.6	75.1	72.9	63.1	52.0	36.9	27.7	48.9
VALENTINE WSO R	MAX	31.6	36.9	44.2	59.0	70.6	81.3	88.7	86.7	76.4	64.7	47.3	36.5	60.3
	MIN	5.8	11.4	19.4	32.6	44.0	54.4	60.3	57.8	46.5	34.3	20.7	11.4	33.2
	MEAN	18.7	24.2	31.8	45.8	57.3	67.9	74.5	72.3	61.5	49.5	34.0	24.0	46.8
WAKEFIELD	MAX	28.2	35.2	45.3	62.7	74.0	83.3	87.7	85.5	76.8	66.2	48.2	34.6	60.6
	MIN	5.5	12.3	22.6	35.8	47.4	57.5	62.0	59.5	49.5	37.8	24.1	12.7	35.6
	MEAN	16.9	23.8	34.0	49.3	60.7	70.4	74.9	72.6	63.2	52.0	36.2	23.7	48.1
WEEPING WATER	MAX	32.1	39.3	49.2	65.1	75.7	84.8	89.4	87.2	78.6	68.4	51.2	38.5	63.3
	MIN	10.2	16.7	26.3	39.3	50.2	60.1	64.9	62.6	52.9	41.3	28.0	17.2	39.1
	MEAN	21.2	28.0	37.8	52.2	63.0	72.5	77.2	74.9	65.8	54.9	39.7	27.9	51.3
WEST POINT	MAX	28.9	35.9	45.9	62.9	74.5	84.0	89.0	86.5	77.4	66.6	49.1	35.5	61.4
	MIN	7.5	13.5	23.6	37.5	48.9	59.0	64.0	61.4	51.3	39.4	25.9	14.5	37.2
	MEAN	18.2	24.7	34.8	50.2	61.8	71.5	76.5	74.0	64.4	53.0	37.5	25.1	49.3
YORK	MAX	32.4	39.2	48.4	63.7	74.7	85.0	89.8	87.6	78.3	67.9	50.5	38.4	63.0
	MIN	11.0	17.1	25.9	39.1	50.5	60.5	65.4	63.4	53.2	41.5	27.6	17.5	39.4
	MEAN	21.7	28.2	37.2	51.4	62.6	72.7	77.6	75.6	65.8	54.7	39.1	27.9	51.2

NEBRASKA

PRECIPITATION NORMALS (INCHES)

STATION	JAN	FEB	MAR	APR	MAY	JUN	JUL	AUG	SEP	OCT	NOV	DEC	ANN
AGATE 3 E	.38	.40	.78	1.44	2.86	2.84	1.96	1.58	1.14	.63	.34	.38	14.73
AINSWORTH	.44	.73	1.16	2.21	3.51	3.96	2.99	2.44	2.02	1.13	.76	.52	21.87
ALBION	.43	.83	1.76	2.89	4.26	4.31	3.41	3.76	2.41	1.44	.78	.61	26.89
ALLIANCE	.40	.41	.79	1.75	3.29	3.22	2.24	1.80	1.34	.73	.46	.40	16.83
ANSELMO	.43	.61	1.37	2.24	3.45	3.69	3.57	2.84	2.04	1.15	.73	.47	22.59
ARCADIA	.41	.63	1.33	2.30	3.39	3.68	3.03	2.52	2.24	1.39	.67	.61	22.20
ARTHUR	.31	.31	.76	1.71	3.38	3.46	3.29	2.00	1.62	.87	.49	.33	18.53
ASHLAND 3 NE	.64	.35	1.77	2.92	4.12	4.24	3.45	4.21	3.27	1.98	1.22	.66	29.33
ATKINSON	.39	.71	1.16	2.35	3.36	3.65	3.12	2.26	2.16	1.37	.75	.55	21.83
AUBURN	.90	1.15	2.32	3.04	4.40	4.52	4.39	4.36	3.84	2.63	1.52	.98	34.05
AURORA	.63	.84	1.73	2.82	3.67	4.41	2.88	2.89	2.54	1.52	.91	.73	25.57
BARNESTON	.61	.79	1.73	2.41	4.03	4.36	3.83	4.09	3.69	2.28	1.24	.71	29.77
BARTLETT 7 NNE	.37	.55	1.30	2.59	3.48	3.78	3.06	3.20	2.27	1.35	.64	.51	23.10
BEATRICE NO 1	.70	.93	1.97	2.56	4.02	4.27	3.97	3.89	3.59	2.38	1.19	.69	30.16
BEAVER CITY	.55	.74	1.47	1.85	3.50	4.00	3.32	2.51	2.07	1.37	.75	.57	22.70
BEEMER 5 N	.65	.98	1.58	2.46	3.98	4.11	3.26	3.47	2.50	1.92	.91	.69	26.51
BENKELMAN	.36	.41	1.19	1.53	2.78	2.82	2.83	2.28	1.64	.94	.60	.38	17.76
BERTRAND	.39	.57	1.37	2.07	3.60	3.74	2.97	2.74	2.04	1.32	.60	.45	21.86
BIG SPRINGS	.33	.42	1.08	1.75	3.40	3.22	2.38	1.83	1.18	.83	.48	.39	17.29
BLAIR	.68	.97	2.06	2.76	3.83	4.33	3.13	3.84	3.19	2.12	1.11	.77	28.79
BLOOMFIELD	.66	1.00	1.74	2.67	3.78	4.18	3.05	3.24	2.41	1.44	1.15	.85	26.17
BOX BUTTE EXP STATION	.40	.29	.68	1.52	3.01	2.86	2.21	1.83	1.34	.73	.31	.25	15.43
BRADSHAW	.60	.84	1.66	2.98	3.78	4.04	2.90	3.18	2.95	1.45	.97	.72	26.07
BREWSTER	.43	.77	1.15	2.00	3.44	3.60	3.00	2.74	1.74	1.06	.61	.51	21.05
BRIDGEPORT	.34	.29	.74	1.49	2.88	2.93	2.41	1.37	1.43	.74	.39	.36	15.37
BROKEN BOW 2 W	.45	.56	1.23	2.17	3.20	3.66	3.32	2.82	1.94	1.09	.64	.49	21.57
BRUNING	.57	.95	1.90	2.72	4.02	4.04	3.29	3.28	3.37	1.83	1.11	.71	27.79
BUTTE	.41	.69	1.33	2.37	3.21	3.90	3.15	2.68	2.11	1.35	.87	.62	22.69
CAMBRIDGE	.43	.62	1.34	1.89	3.62	3.76	3.11	2.61	1.73	1.09	.75	.45	21.40
CENTRAL CITY	.49	.83	1.50	2.62	3.90	3.97	3.18	2.49	2.74	1.42	.82	.65	24.61
CHADRON FAA AIRPORT R	.34	.41	.72	1.66	2.79	2.75	2.22	1.29	1.25	.72	.39	.37	14.91
CLARKSON	.69	.98	1.80	2.58	4.28	4.24	3.44	3.49	2.60	1.96	.99	.76	27.81
CLAY CENTER	.54	.80	1.73	2.78	3.99	4.11	3.60	3.03	2.99	1.53	.83	.66	26.59
COLUMBUS 3 NE	.49	.86	1.45	2.58	3.81	4.08	3.03	3.42	2.67	1.73	.88	.72	25.72
COMSTOCK	.44	.61	1.32	2.25	3.21	3.46	2.79	2.58	1.85	1.34	.74	.58	21.17
CREIGHTON	.43	.68	1.45	2.53	3.60	4.00	3.30	3.05	2.30	1.29	.85	.63	24.11
CRESCENT LAKE NATL WLR	.32	.38	.83	1.63	3.30	3.20	2.48	2.00	1.51	.79	.46	.31	17.21
CRETE	.69	1.02	2.01	2.73	4.10	4.73	3.45	3.62	3.40	2.02	1.21	.77	29.75
CRETE 2 S	.66	1.08	2.09	2.80	3.85	4.67	3.51	3.63	3.38	2.16	1.25	.76	29.84
CULBERTSON	.44	.54	1.23	1.74	3.28	3.29	3.12	2.49	1.74	1.14	.74	.45	20.20
CURTIS	.36	.53	1.07	1.98	3.39	3.77	3.08	2.13	1.82	1.05	.51	.40	20.09
DALTON	.53	.56	1.27	2.03	3.23	3.06	2.41	1.80	1.51	1.05	.73	.56	18.74
DAVID CITY	.74	1.08	2.01	2.83	4.40	5.02	2.74	3.45	3.10	1.87	1.10	.84	29.18
DODGE	.60	1.00	1.69	2.74	4.59	4.75	3.13	3.72	2.59	1.79	.98	.72	28.30
ELGIN 9 WSW	.63	.86	1.64	2.48	3.38	3.97	3.08	3.11	2.27	1.34	.70	.71	24.17
ELLSMERE 9 ENE	.56	.83	1.35	2.45	3.70	3.67	3.61	2.81	1.92	1.27	.83	.64	23.64
ELLSWORTH	.25	.34	.62	1.45	2.69	3.21	2.78	1.89	1.49	.76	.36	.28	16.12
ELWOOD 9 SSW	.39	.58	1.22	1.89	3.47	3.59	3.18	2.70	1.95	1.08	.61	.47	21.13
EMERALD	.64	.91	1.79	2.75	3.80	4.12	3.18	3.61	2.96	2.01	1.22	.70	27.69
EMERSON	.63	1.13	2.00	2.68	4.18	4.38	3.52	3.22	2.87	1.98	1.03	.77	28.39
ERICSON 6 WNW	.47	.69	1.28	2.25	3.24	3.59	2.92	2.87	1.98	1.36	.75	.58	21.98
EUSTIS 2NW	.37	.51	1.26	1.99	3.36	3.86	2.69	2.52	1.81	1.09	.56	.38	20.40
EWING	.42	.72	1.31	2.31	3.25	4.47	3.19	2.93	2.24	1.40	.76	.63	23.63
EWING 12 S	.43	.68	1.37	2.37	3.25	3.93	2.95	2.89	2.24	1.41	.64	.58	22.74
FAIRBURY 2 SSE	.64	.93	1.98	2.57	3.97	4.47	3.79	3.91	3.45	2.07	1.14	.71	29.63

NEBRASKA

PRECIPITATION NORMALS (INCHES)

STATION	JAN	FEB	MAR	APR	MAY	JUN	JUL	AUG	SEP	OCT	NOV	DEC	ANN
FAIRMONT	.46	.77	1.68	2.67	3.66	3.84	3.02	3.37	3.06	1.67	.92	.59	25.71
FALLS CITY	1.00	1.12	2.11	2.97	4.35	4.45	4.33	4.46	4.53	2.68	1.71	1.01	34.72
FRANKLIN	.40	.74	1.59	2.12	3.30	4.09	3.17	2.89	2.61	1.35	.75	.46	23.47
FREMONT	.86	.99	2.06	2.90	4.10	4.67	2.98	4.47	3.20	2.14	1.11	.87	30.35
FRIEND	.54	.81	1.77	2.66	3.96	4.47	3.23	3.57	3.36	1.85	.99	.72	27.93
FULLERTON	.47	.92	1.52	2.67	3.94	4.05	3.16	2.88	2.39	1.42	.77	.67	24.86
GENEVA	.66	1.03	2.04	2.88	3.85	4.13	3.27	3.42	3.03	1.86	1.10	.66	27.93
GENOA 2 W	.48	.97	1.55	2.52	4.05	4.23	3.26	3.22	2.43	1.44	.80	.78	25.73
GOTHENBURG	.45	.55	1.35	2.18	3.43	3.67	2.92	2.49	1.69	1.11	.63	.48	20.95
GRAND ISLAND WSO //R	.52	.81	1.55	2.64	3.70	3.72	2.71	2.59	2.51	1.09	.80	.67	23.31
GREELEY	.43	.72	1.48	2.65	3.70	3.80	3.25	3.02	2.27	1.26	.71	.58	23.87
GRESHAM 3 SSW	.55	.87	1.62	2.86	3.64	4.34	3.11	3.20	2.79	1.72	.98	.65	26.33
GUIDE ROCK	.49	.77	1.63	2.29	3.67	4.53	3.29	3.16	3.30	1.59	1.03	.62	26.37
HAIGLER	.34	.41	1.02	1.58	2.66	2.88	2.51	1.91	1.28	.93	.59	.37	16.48
HALSEY 2 W	.49	.76	1.27	2.48	3.43	3.48	3.00	2.61	1.72	1.01	.73	.54	21.52
HARDY	.56	.94	1.94	2.30	3.68	4.53	2.88	3.39	3.89	1.83	1.09	.69	27.72
HARLAN COUNTY DAM	.28	.50	1.18	1.99	3.38	3.95	3.25	2.86	2.46	1.30	.60	.35	22.10
HARRISBURG 10 NW	.39	.30	.84	1.46	2.70	2.36	1.96	1.23	.98	.74	.48	.36	13.80
HARRISON	.42	.57	1.09	1.96	3.12	2.96	2.31	1.20	1.42	.86	.63	.48	17.02
HARTINGTON	.45	.82	1.73	2.39	3.71	4.31	2.89	3.21	2.23	1.54	1.03	.71	25.02
HASTINGS	.64	.92	1.76	2.73	4.26	4.33	3.35	3.38	2.91	1.47	.79	.76	27.30
HAYES CENTER	.33	.51	1.13	1.94	3.14	3.50	3.17	2.03	1.71	.99	.58	.35	19.38
HAY SPRINGS	.52	.68	1.30	2.40	3.52	3.59	3.03	1.80	1.45	.98	.67	.61	20.55
HAY SPRINGS 12 S	.25	.31	.54	1.62	3.00	3.37	2.39	1.72	1.23	.61	.28	.19	15.51
HEBRON	.61	.92	1.87	2.44	3.95	4.21	3.57	3.16	3.34	1.86	1.04	.72	27.69
HERMAN	.64	.88	1.80	2.73	4.04	4.32	2.96	3.80	3.22	2.02	1.00	.67	28.08
HERMAN 6W	.62	.97	1.95	2.89	4.14	4.50	3.10	4.03	3.18	2.07	1.10	.70	29.25
HERSHEY	.45	.56	1.15	1.89	3.62	3.59	2.98	2.05	1.52	.96	.59	.48	19.84
HICKMAN	.69	1.05	2.16	2.98	4.00	4.76	4.20	3.96	3.40	2.54	1.27	.78	31.79
HOLDREGE 3 SW	.53	.78	1.54	2.26	4.01	4.01	3.10	3.00	2.36	1.47	.77	.57	24.40
IMPERIAL	.40	.48	.97	1.67	3.29	3.42	2.89	2.17	1.51	.94	.56	.37	18.67
KEARNEY	.49	.76	1.62	2.49	4.03	3.91	3.27	2.68	2.42	1.49	.74	.63	24.53
KIMBALL	.45	.40	1.19	1.61	3.06	2.86	2.69	1.71	1.23	.90	.55	.48	17.13
KINGSLEY DAM	.46	.54	1.21	1.77	3.19	3.04	2.77	1.88	1.37	.83	.58	.46	18.10
LAMAR	.22	.27	.74	1.47	3.10	3.22	2.60	1.93	1.37	.74	.38	.23	16.27
LAUREL	.52	.87	1.67	2.32	3.79	4.04	3.07	3.25	2.30	1.56	.89	.71	24.99
LINCOLN WSO AP	.64	1.01	1.94	2.81	3.84	3.84	3.20	3.42	2.93	1.68	.96	.65	26.92
LINCOLN WSO CITY R	.66	1.00	1.86	2.72	3.77	4.27	3.57	3.41	3.24	1.99	1.16	.71	28.36
LODGEPOLE	.40	.40	1.14	1.97	3.17	3.23	2.46	2.10	1.28	.94	.63	.42	18.14
LOUP CITY	.45	.72	1.56	2.46	3.73	4.02	3.17	2.57	2.37	1.44	.71	.61	23.81
LYMAN	.41	.39	.87	1.78	3.04	2.54	1.83	1.05	.90	.86	.55	.49	14.71
LYNCH	.41	.81	1.44	2.40	3.14	3.97	3.46	2.69	2.42	1.37	.86	.71	23.68
LYONS	.57	.94	1.73	2.62	4.38	4.41	3.07	3.47	2.77	1.92	.90	.62	27.40
MADISON	.49	.82	1.55	2.48	4.06	4.56	3.25	3.11	2.15	1.50	.76	.61	25.34
MADRID	.41	.57	1.19	1.80	3.38	3.29	2.84	1.95	1.55	.86	.60	.45	18.89
MALCOLM	.64	1.00	1.97	2.93	3.91	4.38	3.21	3.72	2.93	2.02	1.26	.78	28.75
MASON CITY	.47	.61	1.42	2.23	3.77	4.32	3.07	2.72	2.10	1.47	.69	.53	23.40
MCCOOK	.39	.52	1.23	1.92	2.95	3.17	3.24	2.46	1.77	1.07	.72	.41	19.85
MC COOL JUNCTION	.47	.73	1.67	2.75	3.61	4.00	3.58	3.16	2.89	1.57	.96	.71	26.10
MEDICINE CREEK DAM	.32	.39	1.10	1.88	3.39	3.53	3.17	2.32	2.01	1.06	.67	.32	20.16
MERRIMAN	.40	.56	.90	1.98	2.69	3.47	2.43	1.83	1.27	.85	.41	.39	17.18
MILLER	.40	.52	1.34	2.04	3.57	4.10	3.11	2.52	1.92	1.25	.62	.49	21.88
MINDEN	.38	.74	1.63	2.16	3.76	4.46	3.12	3.06	2.55	1.27	.61	.47	24.21
MITCHELL 5 E	.27	.23	.62	1.26	2.62	3.15	1.71	1.10	1.06	.66	.34	.28	13.30
MOOREFIELD	.35	.45	1.15	2.02	3.43	3.86	3.08	2.21	1.76	1.08	.48	.28	20.15

NEBRASKA

PRECIPITATION NORMALS (INCHES)

STATION		JAN	FEB	MAR	APR	MAY	JUN	JUL	AUG	SEP	OCT	NOV	DEC	ANN
MULLEN		.47	.63	1.05	2.27	3.48	3.51	3.22	2.47	1.61	.92	.62	.46	20.71
MULLEN 21 NW		.56	.83	1.17	2.18	3.45	3.81	3.16	2.51	1.60	.95	.78	.56	21.56
NAPONEE		.55	.74	1.51	1.96	3.37	4.13	3.26	3.16	2.66	1.31	.82	.58	24.05
NEBRASKA CITY		.92	1.09	2.29	3.00	3.88	4.49	4.41	4.41	3.55	2.71	1.51	.96	33.22
NELIGH		.52	.78	1.58	2.40	3.57	4.07	2.95	2.57	2.12	1.28	.78	.64	23.26
NELSON		.63	.91	2.00	2.47	3.61	4.38	3.03	3.09	3.41	1.72	1.13	.74	27.12
NENZEL 20 S		.34	.55	1.14	2.19	3.09	3.81	2.61	2.55	1.53	.96	.51	.43	19.71
NEWCASTLE		.38	.84	1.55	2.33	3.67	4.10	3.33	3.35	2.26	1.60	.96	.63	25.00
NEWPORT		.44	.74	1.30	2.32	3.59	3.70	3.13	1.97	2.09	1.29	.88	.62	22.07
NIOBRARA		.35	.58	1.20	2.19	3.28	3.92	3.18	2.75	2.39	1.26	.71	.56	22.37
NORFOLK WSO	//R	.52	.80	1.54	2.21	3.71	4.35	3.21	2.65	2.09	1.36	.72	.63	23.79
NORTH LOUP		.43	.72	1.34	2.31	3.50	3.65	3.17	3.24	2.29	1.29	.63	.56	23.13
NORTH PLATTE WSO	//R	.40	.55	1.12	1.85	3.36	3.72	2.98	1.92	1.67	.91	.56	.43	19.47
OAKDALE		.48	.76	1.49	2.52	3.78	4.16	3.14	2.81	2.21	1.38	.75	.62	24.10
OGALLALA 3 W		.41	.47	1.08	1.79	3.34	3.09	2.58	1.76	1.45	.76	.50	.46	17.69
OMAHA WSO EPPLEY	//R	.77	.91	1.91	2.94	4.33	4.08	3.62	4.10	3.50	2.09	1.32	.77	30.34
OMAHA NORTH OMAHA WSO		.70	.95	2.00	2.74	4.26	4.21	3.50	4.19	3.36	2.11	1.16	.76	29.93
O NEILL		.41	.68	1.28	2.41	3.32	4.00	3.38	2.54	2.09	1.32	.78	.62	22.83
ORLEANS 2 W		.35	.62	1.18	1.76	3.36	3.99	2.91	2.89	2.38	1.39	.61	.43	21.87
OSCEOLA		.62	.97	1.66	2.79	4.10	4.26	3.01	3.23	2.68	1.61	.99	.77	26.69
OSHKOSH		.31	.38	.91	1.84	3.14	3.11	2.49	1.68	1.39	.84	.58	.37	17.04
OSHKOSH 8 SW		.38	.47	1.02	2.11	3.36	3.39	2.68	1.61	1.36	.98	.69	.41	18.46
OSMOND		.51	.95	1.66	2.59	3.75	3.93	3.31	3.11	2.27	1.37	.97	.72	25.14
PALISADE		.39	.54	1.15	1.88	3.07	3.76	3.08	1.94	1.83	1.01	.59	.41	19.65
PAWNEE CITY		.83	1.05	2.15	2.83	4.15	5.13	4.01	4.36	4.07	2.49	1.46	.89	33.42
PILGER		.53	.84	1.53	2.22	4.14	4.43	3.16	3.16	2.25	1.59	.83	.64	25.32
PLATTSMOUTH		.77	.91	2.04	2.73	3.60	4.05	3.36	4.21	3.16	2.15	1.24	.67	28.89
POLK		.67	.94	1.69	2.80	4.00	4.23	3.16	3.02	2.44	1.43	.88	.70	25.96
POTTER		.43	.34	1.02	1.55	2.89	2.69	2.74	1.79	1.18	.87	.51	.37	16.38
PURDUM		.43	.74	1.17	2.27	3.38	3.34	3.01	2.46	1.76	.97	.64	.51	20.68
RAVENNA		.33	.69	1.50	2.37	3.70	4.36	3.19	2.92	2.32	1.52	.77	.54	24.21
RAYMOND		.54	.82	1.79	2.96	3.87	4.40	3.38	3.78	2.97	1.98	1.03	.64	28.16
RED CLOUD		.47	.76	1.76	2.15	3.73	4.57	3.43	3.06	3.13	1.46	.99	.55	26.06
RUSHVILLE		.31	.45	.75	1.79	3.37	3.89	2.71	1.56	1.42	.76	.34	.38	17.73
SAINT PAUL		.35	.70	1.25	2.50	3.97	3.86	2.86	2.71	2.25	1.22	.60	.58	22.85
SCHUYLER		.62	.97	1.65	2.52	4.12	4.57	2.87	3.97	2.72	1.69	.96	.71	27.37
SCOTTSBLUFF WSO	//R	.44	.37	.97	1.43	2.66	2.93	1.96	.97	1.08	.75	.52	.51	14.59
SEWARD		.56	.87	1.67	2.65	3.82	4.18	3.01	3.47	3.05	1.86	1.05	.67	26.86
SIDNEY 6 NNW		.31	.32	.91	1.67	3.03	3.08	2.68	1.82	1.06	.84	.49	.37	16.58
SIDNEY FAA AIRPORT	R	.42	.38	1.12	1.81	3.40	3.31	2.76	1.97	1.20	1.00	.50	.42	18.29
SPENCER 5 SSE		.34	.62	1.13	2.29	3.05	3.73	3.23	2.43	2.24	1.32	.72	.49	21.59
SPRINGVIEW		.37	.61	1.14	2.24	3.35	3.47	3.01	2.00	1.96	1.10	.73	.49	20.47
STANTON		.61	.95	1.76	2.37	4.20	4.31	3.08	2.88	2.24	1.53	.91	.72	25.56
STAPLEHURST		.58	1.04	1.85	2.72	3.75	4.03	3.22	3.61	2.81	1.76	1.03	.77	27.17
STAPLETON 5 SSE		.53	.64	1.20	2.15	3.47	3.62	2.83	2.19	1.87	1.10	.74	.53	20.87
STOCKVILLE		.42	.56	1.13	1.87	3.32	3.70	2.93	2.53	1.74	1.01	.56	.46	20.23
STRATTON		.53	.54	1.33	1.76	3.29	3.30	3.25	2.05	1.66	1.24	.71	.50	20.16
SUPERIOR		.59	.90	1.84	2.14	3.65	4.16	3.20	3.32	3.76	1.75	1.13	.72	27.16
SYRACUSE		.73	.96	2.13	2.84	3.69	4.29	3.78	4.27	3.43	2.36	1.29	.78	30.55
TABLE ROCK 5 N		.89	1.06	2.29	2.86	3.89	4.72	4.47	4.47	3.72	2.46	1.44	1.00	33.27
TAYLOR		.45	.64	1.24	2.41	3.29	4.06	3.03	2.87	1.97	1.14	.81	.56	22.47
TECUMSEH		.92	1.09	2.33	2.90	4.15	4.47	4.55	4.30	3.68	2.59	1.45	.97	33.40
TEKAMAH		.67	.97	2.11	2.86	4.09	4.20	3.30	4.01	2.93	2.02	.97	.79	28.92
TRENTON DAM		.31	.42	1.13	1.61	3.17	3.21	3.03	2.01	1.65	1.06	.65	.38	18.63
ULYSSES 3 NNE		.66	.94	1.75	2.65	3.99	4.23	3.07	3.26	2.84	1.71	.98	.68	26.76

NEBRASKA

PRECIPITATION NORMALS (INCHES)

STATION	JAN	FEB	MAR	APR	MAY	JUN	JUL	AUG	SEP	OCT	NOV	DEC	ANN
UPLAND	.48	.76	1.62	2.31	3.73	3.82	3.04	3.11	2.67	1.35	.76	.55	24.20
UTICA	.41	.69	1.38	2.50	3.69	4.17	3.41	3.51	2.77	1.60	.94	.59	25.66
VALENTINE LKS GAME REF	.34	.53	1.01	2.31	3.32	3.53	3.03	2.37	1.65	1.07	.48	.39	20.03
VALENTINE WSO R	.28	.52	.83	1.82	2.86	2.97	2.42	2.42	1.42	.83	.41	.33	17.11
VIRGINIA	.80	.97	2.14	2.73	3.95	4.55	4.49	4.26	3.60	2.44	1.39	.86	32.18
WAHOO	.67	1.12	2.25	3.34	4.70	4.90	3.44	4.26	3.79	2.44	1.29	.96	33.16
WAKEFIELD	.58	.95	1.81	2.34	3.98	4.11	3.28	3.08	2.53	1.74	.98	.73	26.11
WALLACE 1 ENE	.33	.45	.94	1.55	3.14	3.09	2.66	1.81	1.72	.83	.50	.38	17.40
WAUNETA	.40	.53	1.20	1.70	3.16	3.49	3.07	2.16	1.66	.97	.66	.40	19.40
WAYNE	.54	1.00	1.69	2.27	4.07	4.44	3.09	2.99	2.43	1.55	.90	.65	25.62
WEEPING WATER	.82	1.08	2.10	2.94	4.34	4.07	3.75	4.54	3.55	2.41	1.31	.86	31.77
WELLFLEET	.38	.61	1.17	2.12	3.60	3.46	2.83	2.31	1.82	1.06	.62	.42	20.40
WESTERN	.52	.99	1.84	2.67	4.23	4.29	3.45	3.85	3.18	1.87	1.04	.64	28.57
WEST POINT	.68	1.09	1.76	2.61	4.34	4.47	3.06	3.54	2.64	1.91	1.02	.79	27.91
WHITMAN 24 N	.27	.53	.88	1.90	3.09	3.60	2.41	2.45	1.47	.88	.38	.37	18.23
WINSIDE	.62	.95	1.74	2.29	4.27	4.26	3.29	2.91	2.36	1.59	.83	.72	25.83
YORK	.65	.94	1.67	2.87	3.62	4.19	3.26	3.22	2.74	1.58	1.06	.80	26.60

NEBRASKA

HEATING DEGREE DAY NORMALS (BASE 65 DEG F)

STATION	JUL	AUG	SEP	OCT	NOV	DEC	JAN	FEB	MAR	APR	MAY	JUN	ANN
AINSWORTH	7	6	122	396	849	1187	1349	1067	955	510	201	39	6688
ALBION	8	10	133	423	861	1240	1438	1123	973	504	214	42	6969
ALLIANCE	12	20	192	505	915	1172	1293	1016	964	603	299	84	7075
ARTHUR	10	13	157	467	903	1194	1330	1042	964	561	268	79	6988
ASHLAND 3 NE	7	6	96	352	777	1181	1404	1081	887	409	163	24	6387
ATKINSON	8	9	124	418	867	1237	1411	1109	973	504	204	39	6903
AUBURN	0	0	59	294	717	1094	1296	991	791	345	124	16	5727
BEATRICE NO 1	0	0	58	292	723	1088	1290	983	791	350	129	12	5716
BEAVER CITY	0	0	63	324	771	1088	1231	924	784	368	137	17	5707
BENKELMAN	0	0	121	379	801	1085	1206	927	834	441	179	38	6011
BIG SPRINGS	0	10	128	432	834	1125	1237	966	874	495	213	46	6360
BLAIR	5	6	95	367	795	1200	1429	1109	918	433	162	22	6541
BOX BUTTE EXP STATION	11	18	198	508	927	1203	1336	1053	1001	630	322	97	7304
BRIDGEPORT	0	5	128	431	855	1132	1237	944	868	507	218	53	6378
BROKEN BOW 2 W	10	7	130	418	864	1178	1336	1042	930	501	218	40	6674
BUTTE	6	7	120	408	870	1246	1438	1120	973	495	193	38	6914
CAMBRIDGE	0	0	90	341	786	1097	1225	927	800	394	148	24	5832
CENTRAL CITY	0	0	84	335	783	1156	1342	1025	859	406	146	19	6155
CHADRON FAA AIRPORT R	13	11	167	477	903	1200	1336	1044	952	579	275	74	7031
CLARKSON	5	6	110	389	828	1218	1423	1103	924	450	162	24	6642
COLUMBUS 3 NE	5	5	114	372	816	1200	1401	1075	899	435	160	21	6503
CRESCENT LAKE NATL WLR	9	14	162	474	891	1172	1290	1008	939	552	268	78	6857
CRETE	6	0	66	305	750	1125	1327	1008	822	380	136	13	5938
CULBERTSON	0	0	118	401	840	1144	1274	977	865	462	198	43	6322
CURTIS	5	0	102	396	843	1138	1268	966	846	438	175	32	6209
DAVID CITY	0	0	84	352	798	1190	1392	1078	899	426	153	19	6391
EWING	7	8	121	412	864	1243	1426	1109	949	477	188	34	6838
FAIRBURY 2 SSE	6	7	86	335	777	1141	1339	1039	862	411	159	24	6186
FAIRMONT	0	0	72	301	747	1116	1318	997	831	395	149	17	5943
FALLS CITY	0	0	55	273	678	1048	1246	955	753	323	108	14	5453
FRANKLIN	0	0	70	325	768	1101	1265	958	809	383	137	22	5838
FREMONT	0	0	84	337	768	1159	1370	1050	856	389	129	14	6156
GENEVA	5	0	79	324	768	1132	1324	1011	837	393	142	16	6031
GENOA 2 W	6	7	104	372	813	1197	1383	1064	887	432	163	24	6452
GOTHENBURG	5	0	102	382	825	1132	1274	983	856	435	166	30	6190
GRAND ISLAND WSO //R	7	6	104	377	822	1187	1376	1058	902	447	169	27	6482
HALSEY 2 W	10	7	129	423	870	1187	1333	1039	924	498	203	40	6663
HARLAN COUNTY DAM	8	8	111	359	810	1153	1324	1019	880	444	191	41	6348
HARRISON	31	38	250	567	987	1249	1389	1106	1070	714	406	145	7952
HARTINGTON	0	8	103	378	849	1265	1473	1142	961	459	162	24	6824
HASTINGS	0	5	105	341	780	1128	1302	997	856	408	152	23	6097
HAYES CENTER	9	10	133	389	819	1110	1256	983	890	489	216	59	6363
HAY SPRINGS	13	15	181	490	918	1203	1336	1047	977	606	295	88	7169
HEBRON	0	0	73	314	768	1132	1333	1019	834	386	145	15	6019
HOLDREGE 3 SW	5	0	99	348	795	1125	1280	977	846	414	161	28	6078
IMPERIAL	0	0	102	381	807	1082	1197	930	843	462	181	35	6020
KEARNEY	5	7	118	389	837	1178	1361	1056	924	468	187	37	6567
KIMBALL	9	13	188	499	882	1116	1221	972	939	600	321	91	6851
KINGSLEY DAM	0	0	102	360	783	1079	1225	960	868	480	202	39	6098
LINCOLN WSO AP	5	0	79	353	795	1190	1404	1078	887	417	151	16	6375
LINCOLN WSO CITY R	0	0	60	306	738	1128	1336	1028	846	386	128	11	5967
LODGEPOLE	0	7	125	410	825	1091	1190	932	871	510	235	60	6256
LOUP CITY	6	7	120	408	843	1197	1370	1061	918	465	190	36	6621
MADISON	0	5	105	391	837	1225	1426	1103	921	449	161	23	6646
MADRID	0	0	100	384	828	1116	1234	958	862	471	187	32	6172

NEBRASKA

HEATING DEGREE DAY NORMALS (BASE 65 DEG F)

STATION	JUL	AUG	SEP	OCT	NOV	DEC	JAN	FEB	MAR	APR	MAY	JUN	ANN
MCCOOK	0	0	87	339	765	1066	1197	921	806	404	172	28	5785
MEDICINE CREEK DAM	6	0	118	393	840	1150	1321	1019	899	468	194	36	6444
MERRIMAN	10	14	150	444	891	1203	1352	1056	973	570	252	66	6981
MINDEN	0	0	91	332	789	1125	1290	983	840	417	159	25	6051
MITCHELL 5 E	7	17	193	487	891	1166	1290	1005	946	591	288	87	6968
MULLEN	7	6	136	419	849	1147	1293	1019	930	516	211	48	6581
NENZEL 20 S	10	17	154	449	882	1178	1333	1053	970	558	251	61	6916
NIOBRARA	0	0	112	395	858	1246	1451	1120	952	462	169	34	6799
NORFOLK WSO //R	6	9	125	417	870	1274	1476	1154	977	489	181	27	7005
NORTH LOUP	5	8	113	395	837	1194	1367	1056	905	459	180	32	6551
NORTH PLATTE WSO //R	8	9	151	463	900	1212	1355	1056	939	522	235	59	6909
OAKDALE	8	11	136	434	861	1249	1442	1131	964	483	197	32	6948
OGALLALA 3 W	6	6	148	446	873	1166	1293	1008	915	516	226	58	6661
OMAHA WSO EPPLEY //R	0	0	73	342	765	1172	1389	1058	859	390	130	16	6194
OMAHA NORTH OMAHA WSO	6	6	94	366	810	1218	1435	1112	924	438	158	25	6592
O NEILL	7	10	132	422	879	1256	1435	1123	980	504	201	39	6988
OSCEOLA	5	0	89	350	798	1181	1389	1064	887	426	151	20	6360
OSHKOSH	0	8	143	450	873	1159	1274	988	902	516	232	59	6604
OSMOND	0	7	114	402	864	1268	1476	1142	961	468	176	26	6904
PAWNEE CITY	0	0	41	264	684	1054	1246	941	744	314	103	10	5401
PURDUM	6	0	124	420	870	1190	1339	1053	939	507	210	42	6700
RAVENNA	0	0	96	375	831	1178	1342	1022	874	430	165	26	6339
RED CLOUD	0	0	72	348	780	1125	1296	977	815	383	149	19	5964
SAINT PAUL	7	8	116	392	828	1194	1367	1058	905	450	175	31	6531
SCOTTSBLUFF WSO //R	6	5	172	459	873	1153	1265	977	911	558	254	69	6702
SEWARD	6	0	78	334	774	1144	1349	1033	849	391	137	15	6110
SIDNEY 6 NNW	8	15	167	471	879	1125	1240	972	921	576	296	83	6753
SIDNEY FAA AIRPORT R	8	13	184	502	909	1159	1265	1005	977	618	322	94	7056
SPRINGVIEW	9	10	133	419	867	1228	1404	1117	1001	537	226	55	7006
STANTON	8	9	115	388	831	1231	1432	1109	921	444	166	25	6679
STAPLETON 5 SSE	11	7	138	427	873	1187	1336	1042	939	519	224	52	6755
SYRACUSE	0	0	76	340	765	1156	1367	1050	849	388	134	19	6144
TECUMSEH	0	0	77	333	750	1128	1339	1025	831	389	144	19	6035
TEKAMAH	0	0	75	340	780	1190	1411	1084	887	407	140	15	6329
TRENTON DAM	6	0	109	381	813	1116	1249	972	868	456	199	38	6207
VALENTINE LKS GAME REF	5	5	133	410	843	1156	1318	1042	955	528	213	52	6660
VALENTINE WSO R	7	10	161	486	930	1271	1435	1142	1029	576	253	62	7362
WAKEFIELD	5	9	114	411	864	1280	1491	1154	961	471	173	28	6961
WEEPING WATER	0	0	72	327	759	1150	1358	1036	843	389	138	16	6088
WEST POINT	5	10	102	388	825	1237	1451	1128	936	444	155	25	6706
YORK	0	0	84	333	777	1150	1342	1030	862	408	144	16	6146

NEBRASKA

COOLING DEGREE DAY NORMALS (BASE 65 DEG F)

STATION	JAN	FEB	MAR	APR	MAY	JUN	JUL	AUG	SEP	OCT	NOV	DEC	ANN
AINSWORTH	0	0	0	0	24	162	326	270	83	9	0	0	874
ALBION	0	0	0	0	34	168	312	252	73	7	0	0	846
ALLIANCE	0	0	0	0	7	117	247	190	45	0	0	0	606
ARTHUR	0	0	0	0	17	130	255	199	55	5	0	0	661
ASHLAND 3 NE	0	0	0	0	79	237	389	322	117	18	0	0	1162
ATKINSON	0	0	0	0	28	150	296	245	64	12	0	0	795
AUBURN	0	0	0	6	90	262	395	326	119	22	0	0	1220
BEATRICE NO 1	0	0	0	8	95	267	426	366	136	22	0	0	1320
BEAVER CITY	0	0	0	5	72	266	431	363	117	10	0	0	1264
BENKELMAN	0	0	0	0	40	215	379	311	121	10	0	0	1076
BIG SPRINGS	0	0	0	0	21	160	314	255	68	0	0	0	818
BLAIR	0	0	0	0	57	208	340	269	74	14	0	0	962
BOX BUTTE EXP STATION	0	0	0	0	6	100	228	176	45	0	0	0	555
BRIDGEPORT	0	0	0	0	14	152	304	241	62	0	0	0	773
BROKEN BOW 2 W	0	0	0	0	26	160	307	252	73	5	0	0	823
BUTTE	0	0	0	0	32	179	341	283	78	8	0	0	921
CAMBRIDGE	0	0	0	0	52	237	389	332	129	6	0	0	1145
CENTRAL CITY	0	0	0	0	62	238	372	316	102	13	0	0	1103
CHADRON FAA AIRPORT R	0	0	0	0	8	128	298	247	68	0	0	0	749
CLARKSON	0	0	0	0	54	207	337	266	77	14	0	0	955
COLUMBUS 3 NE	0	0	0	0	64	231	371	303	105	13	0	0	1087
CRESCENT LAKE NATL WLR	0	0	0	0	7	105	226	185	36	0	0	0	559
CRETE	0	0	0	5	80	244	393	329	111	17	0	0	1179
CULBERTSON	0	0	0	0	46	211	358	299	106	8	0	0	1028
CURTIS	0	0	0	0	35	200	353	298	84	5	0	0	975
DAVID CITY	0	0	0	0	60	226	372	307	93	17	0	0	1075
EWING	0	0	0	0	39	175	317	259	70	9	0	0	869
FAIRBURY 2 SSE	0	0	0	0	66	225	390	335	116	13	0	0	1145
FAIRMONT	0	0	0	0	74	245	398	338	117	16	0	0	1188
FALLS CITY	0	0	0	8	96	269	407	338	130	28	0	0	1276
FRANKLIN	0	0	0	5	75	274	428	360	124	15	0	0	1281
FREMONT	0	0	0	5	73	233	377	305	105	15	0	0	1113
GENEVA	0	0	0	0	68	235	386	326	112	14	0	0	1141
GENOA 2 W	0	0	0	0	55	213	347	286	92	16	0	0	1009
GOTHENBURG	0	0	0	0	39	207	359	295	90	7	0	0	997
GRAND ISLAND WSO //R	0	0	0	0	49	219	366	303	83	8	0	0	1028
HALSEY 2 W	0	0	0	0	33	166	323	258	78	7	0	0	865
HARLAN COUNTY DAM	0	0	0	0	51	215	371	308	114	9	0	0	1068
HARRISON	0	0	0	0	0	67	180	134	37	0	0	0	418
HARTINGTON	0	0	0	0	51	210	360	305	94	15	0	0	1035
HASTINGS	0	0	0	0	63	233	369	306	117	9	0	0	1097
HAYES CENTER	0	0	0	0	30	179	331	280	100	7	0	0	927
HAY SPRINGS	0	0	0	0	7	115	245	201	52	0	0	0	620
HEBRON	0	0	0	0	77	252	415	360	121	16	0	0	1241
HOLDREGE 3 SW	0	0	0	0	55	232	374	314	120	13	0	0	1108
IMPERIAL	0	0	0	0	26	194	347	288	93	6	0	0	954
KEARNEY	0	0	0	0	45	208	353	289	97	10	0	0	1002
KIMBALL	0	0	0	0	8	94	216	153	35	0	0	0	506
KINGSLEY DAM	0	0	0	0	28	171	335	280	93	10	0	0	917
LINCOLN WSO AP	0	0	0	0	64	235	396	323	94	12	0	0	1124
LINCOLN WSO CITY R	0	0	0	0	82	266	423	357	117	18	0	0	1263
LODGEPOLE	0	0	0	0	15	159	310	252	71	0	0	0	807
LOUP CITY	0	0	0	0	47	201	344	280	78	8	0	0	958
MADISON	0	0	0	0	50	215	352	278	81	9	0	0	985
MADRID	0	0	0	0	28	203	366	304	97	6	0	0	1004

NEBRASKA

COOLING DEGREE DAY NORMALS (BASE 65 DEG F)

STATION	JAN	FEB	MAR	APR	MAY	JUN	JUL	AUG	SEP	OCT	NOV	DEC	ANN
MCCOOK	0	0	0	5	73	235	397	332	129	16	0	0	1187
MEDICINE CREEK DAM	0	0	0	0	36	192	353	293	97	6	0	0	977
MERRIMAN	0	0	0	0	17	126	280	228	63	7	0	0	721
MINDEN	0	0	0	0	54	226	379	316	118	13	0	0	1106
MITCHELL 5 E	0	0	0	0	9	120	227	151	37	0	0	0	544
MULLEN	0	0	0	0	15	159	310	242	73	6	0	0	805
NENZEL 20 S	0	0	0	0	13	130	264	216	55	5	0	0	683
NIOBRARA	0	0	0	0	39	211	351	293	91	14	0	0	999
NORFOLK WSO //R	0	0	0	0	45	198	341	269	74	8	0	0	935
NORTH LOUP	0	0	0	0	41	194	325	265	74	8	0	0	907
NORTH PLATTE WSO //R	0	0	0	0	24	158	294	239	58	0	0	0	773
OAKDALE	0	0	0	0	42	176	327	253	70	12	0	0	880
OGALLALA 3 W	0	0	0	0	18	169	322	254	70	0	0	0	833
OMAHA WSO EPPLEY //R	0	0	0	6	77	256	394	320	97	16	0	0	1166
OMAHA NORTH OMAHA WSO	0	0	0	0	56	211	338	270	76	13	0	0	964
O NEILL	0	0	0	0	31	165	320	264	75	7	0	0	862
OSCEOLA	0	0	0	0	58	227	368	304	101	15	0	0	1073
OSHKOSH	0	0	0	0	18	143	277	210	53	0	0	0	701
OSMOND	0	0	0	0	55	206	339	274	81	8	0	0	963
PAWNEE CITY	0	0	0	11	112	289	440	381	140	28	0	0	1401
PURDUM	0	0	0	0	21	153	300	234	64	0	0	0	772
RAVENNA	0	0	0	0	50	221	367	305	93	10	0	0	1046
RED CLOUD	0	0	0	0	77	244	404	342	99	19	0	0	1185
SAINT PAUL	0	0	0	0	45	199	335	274	83	8	0	0	944
SCOTTSBLUFF WSO //R	0	0	0	0	13	144	291	210	70	0	0	0	728
SEWARD	0	0	0	0	72	246	396	317	96	15	0	0	1142
SIDNEY 6 NNW	0	0	0	0	17	119	249	192	47	0	0	0	624
SIDNEY FAA AIRPORT R	0	0	0	0	6	103	235	178	46	0	0	0	568
SPRINGVIEW	0	0	0	0	27	166	334	277	85	16	0	0	905
STANTON	0	0	0	0	48	196	333	270	82	12	0	0	941
STAPLETON 5 SSE	0	0	0	0	26	151	312	258	78	6	0	0	831
SYRACUSE	0	0	0	0	78	256	391	326	109	17	0	0	1177
TECUMSEH	0	0	0	0	73	238	380	310	104	17	0	0	1122
TEKAMAH	0	0	0	5	81	243	373	301	90	14	0	0	1107
TRENTON DAM	0	0	0	0	38	200	368	314	112	9	0	0	1041
VALENTINE LKS GAME REF	0	0	0	0	21	160	318	250	76	7	0	0	832
VALENTINE WSO R	0	0	0	0	14	149	301	236	56	5	0	0	761
WAKEFIELD	0	0	0	0	40	190	312	245	60	8	0	0	855
WEEPING WATER	0	0	0	5	76	241	383	310	96	14	0	0	1125
WEST POINT	0	0	0	0	56	220	362	289	84	16	0	0	1027
YORK	0	0	0	0	70	247	391	329	108	14	0	0	1159

STATE-STATION NUMBER	STN TYP	NAME	LATITUDE DEG-MIN	LONGITUDE DEG-MIN	ELEVATION (FT)
25-0030	12	AGATE 3 E	N 4227	W 10350	4670
25-0050	13	AINSWORTH	N 4233	W 09952	2510
25-0070	13	ALBION	N 4141	W 09800	1760
25-0130	13	ALLIANCE	N 4206	W 10252	3971
25-0245	12	ANSELMO	N 4136	W 09950	2590
25-0320	12	ARCADIA	N 4125	W 09908	2186
25-0365	13	ARTHUR	N 4134	W 10141	3500
25-0375	13	ASHLAND 3 NE	N 4104	W 09620	1067
25-0420	13	ATKINSON	N 4233	W 09858	2125
25-0435	13	AUBURN	N 4023	W 09551	1065
25-0445	12	AURORA	N 4052	W 09801	1792
25-0520	12	BARNESTON	N 4003	W 09635	1200
25-0525	12	BARTLETT 7 NNE	N 4159	W 09832	1990
25-0620	13	BEATRICE NO 1	N 4016	W 09645	1235
25-0640	13	BEAVER CITY	N 4008	W 09950	2160
25-0680	12	BEEMER 5 N	N 4201	W 09649	1395
25-0760	13	BENKELMAN	N 4003	W 10132	2970
25-0810	12	BERTRAND	N 4031	W 09938	2518
25-0865	13	BIG SPRINGS	N 4104	W 10205	3369
25-0930	13	BLAIR	N 4133	W 09608	1115
25-0945	12	BLOOMFIELD	N 4236	W 09738	1715
25-1045	13	BOX BUTTE EXP STATION	N 4208	W 10257	4020
25-1065	12	BRADSHAW	N 4053	W 09745	1717
25-1130	12	BREWSTER	N 4157	W 09952	2500
25-1145	13	BRIDGEPORT	N 4140	W 10306	3666
25-1200	13	BROKEN BOW 2 W	N 4125	W 09941	2530
25-1240	12	BRUNING	N 4020	W 09734	1583
25-1365	13	BUTTE	N 4255	W 09851	1810
25-1415	13	CAMBRIDGE	N 4016	W 10010	2258
25-1560	13	CENTRAL CITY	N 4107	W 09800	1695
25-1575	13	CHADRON FAA AIRPORT R	N 4250	W 10305	3300
25-1660	13	CLARKSON	N 4143	W 09707	1490
25-1680	12	CLAY CENTER	N 4032	W 09803	1778
25-1825	13	COLUMBUS 3 NE	N 4128	W 09720	1450
25-1835	12	COMSTOCK	N 4134	W 09915	2335
25-1990	12	CREIGHTON	N 4227	W 09754	1660
25-2000	13	CRESCENT LAKE NATL WLR	N 4145	W 10226	3820
25-2020	13	CRETE	N 4037	W 09657	1435
25-2025	12	CRETE 2 S	N 4036	W 09658	1345
25-2065	13	CULBERTSON	N 4013	W 10050	2605
25-2100	13	CURTIS	N 4038	W 10031	2570
25-2145	12	DALTON	N 4125	W 10258	4273
25-2205	13	DAVID CITY	N 4115	W 09708	1615
25-2380	12	DODGE	N 4143	W 09653	1395
25-2595	12	ELGIN 9 WSW	N 4158	W 09816	2040
25-2629	12	ELLSMERE 9 ENE	N 4210	W 10001	2654
25-2645	12	ELLSWORTH	N 4203	W 10217	3905
25-2690	12	ELWOOD 9 SSW	N 4029	W 09956	2450
25-2706	12	EMERALD	N 4049	W 09651	1200
25-2715	12	EMERSON	N 4217	W 09644	1455

25 — NEBRASKA

LEGEND
11 = TEMPERATURE ONLY
12 = PRECIPITATION ONLY
13 = TEMP. & PRECIP.

STATE-STATION NUMBER	STN TYP	NAME		LATITUDE DEG-MIN	LONGITUDE DEG-MIN	ELEVATION (FT)
25-2770	12	ERICSON 6 WNW		N 4148	W 09847	2095
25-2790	12	EUSTIS 2NW		N 4041	W 10005	2690
25-2805	13	EWING		N 4215	W 09821	1888
25-2806	12	EWING 12 S		N 4205	W 09823	1950
25-2820	13	FAIRBURY 2 SSE		N 4007	W 09710	1460
25-2840	13	FAIRMONT		N 4038	W 09735	1641
25-2850	13	FALLS CITY		N 4004	W 09536	990
25-3035	13	FRANKLIN		N 4006	W 09857	1883
25-3050	13	FREMONT		N 4126	W 09629	1200
25-3065	12	FRIEND		N 4039	W 09717	1545
25-3075	12	FULLERTON		N 4122	W 09758	1630
25-3175	13	GENEVA		N 4032	W 09736	1633
25-3185	13	GENOA 2 W		N 4127	W 09746	1590
25-3365	13	GOTHENBURG		N 4056	W 10010	2565
25-3395	13	GRAND ISLAND WSO	//R	N 4058	W 09819	1841
25-3425	12	GREELEY		N 4133	W 09832	2021
25-3461	12	GRESHAM 3 SSW		N 4059	W 09725	1651
25-3485	12	GUIDE ROCK		N 4004	W 09820	1635
25-3515	12	HAIGLER		N 4001	W 10156	3280
25-3540	13	HALSEY 2 W		N 4154	W 10019	2705
25-3589	12	HARDY		N 4000	W 09755	1520
25-3595	13	HARLAN COUNTY DAM		N 4005	W 09912	2000
25-3605	12	HARRISBURG 10 NW		N 4139	W 10353	4460
25-3615	13	HARRISON		N 4241	W 10353	4850
25-3630	13	HARTINGTON		N 4237	W 09716	1382
25-3660	13	HASTINGS		N 4035	W 09823	1932
25-3690	13	HAYES CENTER		N 4031	W 10101	3044
25-3710	13	HAY SPRINGS		N 4241	W 10241	3855
25-3715	12	HAY SPRINGS 12 S		N 4230	W 10242	3805
25-3735	13	HEBRON		N 4010	W 09735	1458
25-3800	12	HERMAN		N 4140	W 09613	1095
25-3801	12	HERMAN 6W		N 4141	W 09619	1335
25-3810	12	HERSHEY		N 4109	W 10100	2905
25-3825	12	HICKMAN		N 4037	W 09638	1275
25-3910	13	HOLDREGE 3 SW		N 4025	W 09921	2330
25-4110	13	IMPERIAL		N 4031	W 10138	3278
25-4335	13	KEARNEY		N 4042	W 09905	2146
25-4440	13	KIMBALL		N 4114	W 10340	4725
25-4455	13	KINGSLEY DAM		N 4113	W 10139	3300
25-4604	12	LAMAR		N 4034	W 10159	3490
25-4655	12	LAUREL		N 4225	W 09705	1495
25-4795	13	LINCOLN WSO AP		N 4051	W 09645	1180
25-4815	13	LINCOLN WSO CITY	R	N 4049	W 09642	1150
25-4900	13	LODGEPOLE		N 4109	W 10238	3832
25-4985	13	LOUP CITY		N 4117	W 09858	2065
25-5020	12	LYMAN		N 4155	W 10402	4053
25-5040	12	LYNCH		N 4250	W 09827	1411
25-5050	12	LYONS		N 4156	W 09628	1302
25-5080	13	MADISON		N 4150	W 09727	1580
25-5090	13	MADRID		N 4051	W 10133	3200

STATE-STATION NUMBER	STN TYP	NAME	LATITUDE DEG-MIN	LONGITUDE DEG-MIN	ELEVATION (FT)
25-5105	12	MALCOLM	N 4055	W 09652	1350
25-5250	12	MASON CITY	N 4113	W 09918	2270
25-5310	13	MCCOOK	N 4013	W 10035	2586
25-5320	12	MC COOL JUNCTION	N 4045	W 09735	1550
25-5388	13	MEDICINE CREEK DAM	N 4023	W 10013	2387
25-5470	13	MERRIMAN	N 4255	W 10141	3260
25-5525	12	MILLER	N 4056	W 09924	2305
25-5565	13	MINDEN	N 4030	W 09857	2170
25-5590	13	MITCHELL 5 E	N 4157	W 10341	4080
25-5655	12	MOOREFIELD	N 4041	W 10024	2826
25-5700	13	MULLEN	N 4203	W 10103	3220
25-5702	12	MULLEN 21 NW	N 4216	W 10120	3450
25-5780	12	NAPONEE	N 4004	W 09908	1877
25-5810	12	NEBRASKA CITY	N 4041	W 09550	954
25-5830	12	NELIGH	N 4207	W 09801	1746
25-5840	12	NELSON	N 4012	W 09804	1740
25-5860	13	NENZEL 20 S	N 4239	W 10110	3077
25-5895	12	NEWCASTLE	N 4239	W 09653	1355
25-5925	12	NEWPORT	N 4236	W 09920	2230
25-5960	13	NIOBRARA	N 4245	W 09803	1235
25-5995	13	NORFOLK WSO //R	N 4159	W 09726	1544
25-6040	13	NORTH LOUP	N 4130	W 09846	1960
25-6065	13	NORTH PLATTE WSO //R	N 4108	W 10041	2775
25-6135	13	OAKDALE	N 4204	W 09758	1705
25-6200	13	OGALLALA 3 W	N 4108	W 10146	3250
25-6255	13	OMAHA WSO EPPLEY //R	N 4118	W 09554	977
25-6260	13	OMAHA NORTH OMAHA WSO	N 4122	W 09601	1323
25-6290	13	O NEILL	N 4227	W 09838	1975
25-6365	12	ORLEANS 2 W	N 4008	W 09930	1972
25-6375	13	OSCEOLA	N 4111	W 09733	1640
25-6385	13	OSHKOSH	N 4124	W 10221	3393
25-6390	12	OSHKOSH 8 SW	N 4118	W 10226	3825
25-6395	13	OSMOND	N 4221	W 09736	1650
25-6480	12	PALISADE	N 4021	W 10107	2765
25-6570	13	PAWNEE CITY	N 4006	W 09609	1175
25-6735	12	PILGER	N 4201	W 09703	1407
25-6795	12	PLATTSMOUTH	N 4101	W 09553	986
25-6837	12	POLK	N 4105	W 09747	1738
25-6880	12	POTTER	N 4113	W 10319	4425
25-6970	13	PURDUM	N 4204	W 10015	2690
25-7040	13	RAVENNA	N 4102	W 09855	1995
25-7055	12	RAYMOND	N 4057	W 09647	1260
25-7070	13	RED CLOUD	N 4006	W 09831	1720
25-7415	12	RUSHVILLE	N 4243	W 10228	3740
25-7515	13	SAINT PAUL	N 4112	W 09827	1796
25-7640	12	SCHUYLER	N 4126	W 09704	1350
25-7665	13	SCOTTSBLUFF WSO //R	N 4152	W 10336	3957
25-7715	13	SEWARD	N 4054	W 09705	1442
25-7830	13	SIDNEY 6 NNW	N 4114	W 10300	4320
25-7835	13	SIDNEY FAA AIRPORT R	N 4106	W 10259	4292

25 – NEBRASKA

STATE-STATION NUMBER	STN TYP	NAME	LATITUDE DEG-MIN	LONGITUDE DEG-MIN	ELEVATION (FT)
25-8040	12	SPENCER 5 SSE	N 4249	W 09839	1546
25-8090	13	SPRINGVIEW	N 4249	W 09944	2440
25-8110	13	STANTON	N 4157	W 09714	1472
25-8120	12	STAPLEHURST	N 4058	W 09711	1505
25-8130	13	STAPLETON 5 SSE	N 4125	W 10028	3020
25-8215	12	STOCKVILLE	N 4032	W 10023	2504
25-8255	12	STRATTON	N 4009	W 10113	2796
25-8320	12	SUPERIOR	N 4001	W 09804	1578
25-8395	13	SYRACUSE	N 4039	W 09611	1045
25-8410	12	TABLE ROCK 5 N	N 4015	W 09605	1050
25-8455	12	TAYLOR	N 4146	W 09923	2270
25-8465	13	TECUMSEH	N 4022	W 09611	1130
25-8480	13	TEKAMAH	N 4147	W 09613	1052
25-8628	13	TRENTON DAM	N 4010	W 10104	2810
25-8682	12	ULYSSES 3 NNE	N 4106	W 09712	1523
25-8735	12	UPLAND	N 4019	W 09854	2158
25-8745	12	UTICA	N 4054	W 09721	1582
25-8755	13	VALENTINE LKS GAME REF	N 4235	W 10041	2929
25-8760	13	VALENTINE WSO R	N 4252	W 10033	2587
25-8875	12	VIRGINIA	N 4015	W 09630	1545
25-8905	12	WAHOO	N 4112	W 09638	1221
25-8915	13	WAKEFIELD	N 4216	W 09652	1413
25-8920	12	WALLACE 1 ENE	N 4050	W 10110	3080
25-9020	12	WAUNETA	N 4025	W 10122	2940
25-9045	12	WAYNE	N 4214	W 09701	1460
25-9090	13	WEEPING WATER	N 4052	W 09607	1110
25-9115	12	WELLFLEET	N 4045	W 10044	2810
25-9150	12	WESTERN	N 4024	W 09712	1460
25-9200	13	WEST POINT	N 4150	W 09643	1310
25-9266	12	WHITMAN 24 N	N 4224	W 10126	3440
25-9355	12	WINSIDE	N 4210	W 09710	1560
25-9510	13	YORK	N 4052	W 09736	1633

25 — NEBRASKA

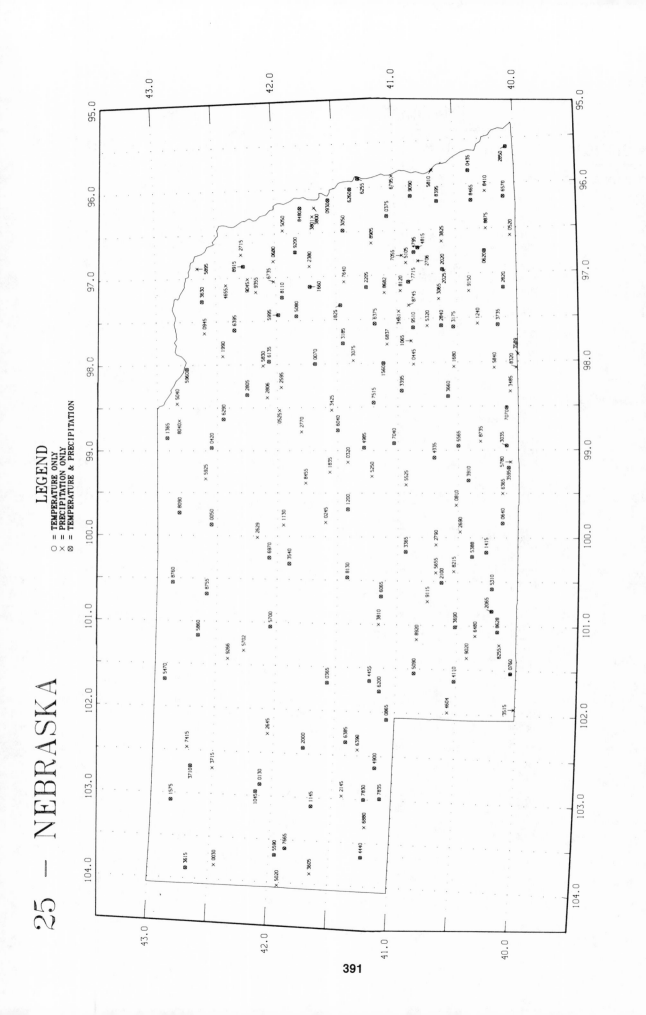

LEGEND

O = TEMPERATURE ONLY
X = PRECIPITATION ONLY
⊗ = TEMPERATURE & PRECIPITATION

391

NEVADA

TEMPERATURE NORMALS (DEG F)

STATION			JAN	FEB	MAR	APR	MAY	JUN	JUL	AUG	SEP	OCT	NOV	DEC	ANN
BATTLE MOUNTAIN AP	R	MAX	41.1	47.7	53.4	62.1	72.3	82.2	93.2	90.4	81.4	68.8	52.3	42.4	65.6
		MIN	16.2	21.9	24.2	29.5	38.0	45.4	51.7	48.1	39.2	29.8	21.9	15.9	31.8
		MEAN	28.7	34.9	38.8	45.8	55.2	63.8	72.5	69.3	60.3	49.3	37.2	29.2	48.8
BOULDER CITY		MAX	54.4	60.5	66.9	75.6	85.3	95.7	101.5	99.1	92.7	80.2	64.6	55.6	77.7
		MIN	38.2	42.6	45.8	52.5	60.8	69.6	76.1	74.2	68.4	58.1	46.2	39.3	56.0
		MEAN	46.3	51.6	56.4	64.1	73.1	82.7	88.8	86.7	80.6	69.2	55.4	47.5	66.9
CALIENTE		MAX	47.5	54.2	60.1	68.6	78.5	89.2	95.9	93.1	86.2	74.7	59.4	49.3	71.4
		MIN	17.3	23.0	26.7	33.2	41.3	48.8	55.9	54.8	45.2	34.7	24.9	18.5	35.4
		MEAN	32.5	38.6	43.4	50.9	59.9	69.1	75.9	74.0	65.7	54.8	42.2	33.9	53.4
CARSON CITY		MAX	46.4	51.6	55.6	62.3	71.0	80.2	89.1	87.1	81.0	70.3	56.4	48.1	66.6
		MIN	20.5	24.3	27.4	31.5	38.6	45.4	50.3	48.0	40.9	32.1	25.0	20.3	33.7
		MEAN	33.5	38.0	41.5	46.9	54.8	62.8	69.7	67.6	61.0	51.2	40.7	34.2	50.2
CONTACT		MAX	39.4	44.1	48.9	58.8	68.7	78.5	89.3	86.1	77.8	66.4	49.5	40.9	62.4
		MIN	15.2	20.1	22.7	27.8	35.3	42.7	48.3	45.9	37.4	28.5	21.5	15.6	30.1
		MEAN	27.3	32.2	35.8	43.3	52.1	60.6	68.8	66.0	57.6	47.5	35.5	28.3	46.3
DESERT NATL WL RANGE		MAX	56.9	62.6	67.8	76.3	85.5	96.1	102.1	99.6	92.8	81.0	66.3	58.3	78.8
		MIN	28.8	32.7	36.8	43.5	51.3	59.7	66.7	65.3	57.6	47.3	36.2	29.6	46.3
		MEAN	42.9	47.7	52.3	59.9	68.5	77.9	84.4	82.5	75.3	64.2	51.3	43.9	62.6
DYER 4 SE		MAX	46.3	52.7	58.7	66.5	75.8	86.4	93.9	91.3	84.1	72.1	56.9	47.6	69.4
		MIN	15.1	21.3	24.3	30.4	38.4	45.9	52.5	50.2	41.8	32.0	21.9	15.5	32.4
		MEAN	30.7	37.0	41.5	48.5	57.1	66.2	73.3	70.8	63.0	52.1	39.4	31.6	50.9
ELKO WSO	R	MAX	36.6	42.6	48.9	58.2	68.5	79.2	90.4	87.8	78.8	66.3	49.4	38.3	62.1
		MIN	13.2	19.4	23.0	28.6	36.1	43.3	49.8	47.3	38.0	28.7	21.2	13.9	30.2
		MEAN	25.0	31.0	36.0	43.4	52.4	61.2	70.1	67.6	58.4	47.5	35.3	26.1	46.2
ELY WSO	//R	MAX	39.0	42.6	47.3	56.2	66.5	77.5	86.8	84.2	76.0	64.0	49.2	40.9	60.9
		MIN	9.7	15.0	19.4	25.7	33.6	40.4	48.1	46.6	37.3	28.0	18.5	11.0	27.8
		MEAN	24.4	28.8	33.4	40.9	50.1	59.0	67.5	65.4	56.7	46.0	33.9	26.0	44.3
FALLON EXPERIMENT STA		MAX	46.3	52.7	58.5	65.5	74.2	83.6	92.3	89.6	81.9	70.3	55.9	47.0	68.2
		MIN	19.2	23.8	27.1	32.7	41.0	48.1	54.0	51.2	43.5	33.9	25.6	19.2	34.9
		MEAN	32.7	38.3	42.8	49.1	57.6	65.9	73.2	70.4	62.7	52.1	40.7	33.1	51.6
GLENBROOK		MAX	40.8	43.2	46.4	53.5	62.6	72.4	81.3	79.7	73.5	61.9	48.9	42.3	58.9
		MIN	24.0	25.3	26.0	29.0	35.2	42.0	49.0	48.5	43.9	36.2	29.4	25.3	34.5
		MEAN	32.4	34.3	36.2	41.3	48.9	57.3	65.2	64.1	58.7	49.1	39.2	33.8	46.7
LAS VEGAS WSO	R	MAX	56.0	62.4	68.3	77.2	87.4	98.6	104.5	101.9	94.7	81.5	66.0	57.1	79.6
		MIN	33.0	37.7	42.3	49.8	59.0	68.6	75.9	73.9	65.6	53.5	41.2	33.6	52.8
		MEAN	44.6	50.1	55.3	63.5	73.3	83.6	90.3	88.0	80.1	67.6	53.6	45.4	66.3
LEHMAN CAVES NAT MON		MAX	40.5	43.7	47.7	56.0	66.0	76.9	85.7	82.9	74.9	62.4	48.6	41.8	60.6
		MIN	18.2	21.2	23.8	30.8	39.5	48.5	57.5	55.5	47.2	37.1	26.1	20.3	35.5
		MEAN	29.4	32.5	35.8	43.4	52.8	62.7	71.6	69.2	61.1	49.8	37.4	31.1	48.1
LOVELOCK FAA AP		MAX	43.7	50.9	56.6	64.7	74.5	84.4	94.5	91.4	82.8	70.5	54.2	44.3	67.7
		MIN	17.0	22.1	25.4	32.0	41.6	49.6	55.6	52.1	43.6	33.3	22.8	16.5	34.3
		MEAN	30.4	36.5	41.0	48.4	58.1	67.0	75.0	71.8	63.3	51.9	38.5	30.4	51.0
MC GILL		MAX	38.9	42.5	47.0	55.5	65.9	76.5	86.1	83.4	75.5	63.9	49.1	41.1	60.5
		MIN	15.8	19.6	22.5	29.4	37.6	46.2	54.5	51.9	43.2	33.5	23.3	17.2	32.9
		MEAN	27.4	31.1	34.8	42.5	51.7	61.4	70.4	67.7	59.4	48.7	36.2	29.2	46.7
MINA		MAX	46.5	52.8	58.1	65.4	75.6	86.3	95.4	93.1	84.8	72.5	56.7	47.7	69.6
		MIN	20.6	25.2	28.4	34.8	44.1	53.1	60.8	57.7	47.8	37.5	28.0	21.3	38.3
		MEAN	33.6	39.0	43.3	50.1	59.8	69.7	78.1	75.4	66.3	55.0	42.4	34.5	53.9
MINDEN		MAX	45.2	50.8	55.2	62.3	71.1	80.7	89.7	88.1	81.6	70.9	56.2	47.6	66.6
		MIN	16.8	21.1	23.8	28.9	36.4	42.9	47.9	45.5	38.5	29.5	22.5	17.2	30.9
		MEAN	31.0	36.0	39.5	45.6	53.8	61.8	68.8	66.8	60.1	50.2	39.4	32.4	48.8

NEVADA

TEMPERATURE NORMALS (DEG F)

STATION		JAN	FEB	MAR	APR	MAY	JUN	JUL	AUG	SEP	OCT	NOV	DEC	ANN
OWYHEE	MAX	37.4	41.9	45.6	54.1	64.1	73.9	85.2	82.8	74.3	62.9	47.8	40.4	59.2
	MIN	17.4	22.0	23.6	29.7	37.6	44.4	51.4	49.4	40.5	31.8	25.2	19.4	32.7
	MEAN	27.4	32.0	34.6	41.9	50.9	59.2	68.3	66.1	57.4	47.4	36.5	29.9	46.0
PIOCHE	MAX	41.2	45.6	50.4	59.2	69.6	81.0	88.5	85.3	77.7	66.0	51.2	43.2	63.2
	MIN	21.2	24.2	27.2	33.9	42.6	51.5	58.4	57.1	49.9	40.0	29.1	22.8	38.2
	MEAN	31.2	34.9	38.8	46.6	56.1	66.3	73.5	71.2	63.8	53.0	40.1	33.0	50.7
RENO WSO //R	MAX	44.8	51.1	55.8	63.3	72.3	81.8	91.3	88.7	81.4	70.0	55.6	46.2	66.9
	MIN	19.5	23.5	25.4	29.4	36.9	43.0	47.7	45.2	38.9	30.5	23.8	18.9	31.9
	MEAN	32.2	37.4	40.6	46.4	54.6	62.4	69.5	66.9	60.2	50.3	39.7	32.5	49.4
RYE PATCH DAM	MAX	44.0	51.1	57.0	65.1	74.9	84.5	94.4	91.7	83.6	71.5	55.3	45.7	68.2
	MIN	18.2	23.0	25.1	30.4	38.9	46.3	52.1	49.3	41.1	31.5	23.7	18.4	33.2
	MEAN	31.2	37.1	41.1	47.8	56.9	65.4	73.3	70.5	62.4	51.5	39.5	32.1	50.7
SEARCHLIGHT	MAX	53.3	58.7	64.0	72.5	81.9	92.4	97.9	95.3	89.3	77.7	63.1	54.5	75.1
	MIN	34.8	37.6	40.2	46.8	55.0	64.6	71.5	69.7	63.4	54.1	42.3	36.0	51.3
	MEAN	44.1	48.2	52.1	59.7	68.5	78.5	84.7	82.5	76.4	65.9	52.7	45.3	63.2
SMOKEY VALLEY	MAX	44.1	49.9	55.4	63.5	73.0	82.9	91.0	88.8	81.5	70.5	54.6	45.1	66.7
	MIN	15.4	20.0	23.6	28.6	37.2	45.1	52.4	50.1	41.3	32.0	22.0	15.4	31.9
	MEAN	29.8	35.0	39.5	46.0	55.2	64.0	71.7	69.4	61.4	51.3	38.4	30.3	49.3
SUNRISE MANR LAS VEGAS	MAX	60.2	66.7	72.6	81.4	91.3	101.9	108.0	105.3	98.4	85.7	70.3	61.2	83.6
	MIN	26.5	31.2	36.3	43.8	51.5	59.3	66.4	65.1	55.9	44.6	33.3	26.5	45.0
	MEAN	43.4	49.0	54.4	62.6	71.4	80.6	87.2	85.2	77.2	65.2	51.9	43.9	64.3
TONOPAH AP	MAX	43.4	48.8	54.3	62.5	72.4	83.1	91.0	88.3	80.5	68.7	53.9	45.2	66.0
	MIN	18.3	23.0	26.5	32.5	41.5	49.9	56.3	54.2	46.7	36.8	26.2	19.4	35.9
	MEAN	30.9	35.9	40.4	47.5	57.0	66.6	73.7	71.3	63.6	52.8	40.1	32.3	51.0
WELLS	MAX	35.7	40.6	46.5	56.7	67.2	77.5	88.7	85.9	76.9	64.1	47.4	37.5	60.4
	MIN	11.2	16.7	21.1	26.7	34.2	41.1	47.5	44.8	35.7	26.8	19.3	11.7	28.1
	MEAN	23.5	28.6	33.8	41.7	50.7	59.3	68.1	65.3	56.4	45.5	33.4	24.6	44.2
WINNEMUCCA WSO //R	MAX	42.3	48.7	53.6	61.8	72.0	81.8	92.7	89.7	80.8	68.5	53.2	43.9	65.8
	MIN	17.2	22.6	23.8	28.8	37.4	45.1	51.2	47.6	38.3	28.9	22.2	17.0	31.7
	MEAN	29.8	35.7	38.7	45.3	54.7	63.5	72.0	68.7	59.6	48.7	37.7	30.4	48.7
YERINGTON	MAX	47.4	53.5	59.1	66.3	75.1	84.1	92.4	90.4	83.3	71.6	56.8	47.9	69.0
	MIN	18.6	22.5	26.0	31.4	39.2	46.5	51.6	48.8	41.4	32.2	23.3	17.5	33.3
	MEAN	33.0	38.0	42.6	48.9	57.2	65.3	72.0	69.7	62.4	51.9	40.1	32.7	51.2

NEVADA

PRECIPITATION NORMALS (INCHES)

STATION	JAN	FEB	MAR	APR	MAY	JUN	JUL	AUG	SEP	OCT	NOV	DEC	ANN
ADAVEN	1.40	1.37	1.36	1.04	.75	.59	1.01	1.09	.85	.78	1.07	1.26	12.57
ARTHUR 4 NW	1.65	1.37	1.44	1.32	1.60	1.25	.59	.64	.79	.91	1.28	1.45	14.29
AUSTIN	1.05	1.16	1.40	1.62	1.44	1.24	.63	.65	.69	.85	.92	1.11	12.76
BATTLE MOUNTAIN AP R	.67	.58	.63	.77	.85	1.01	.30	.36	.46	.56	.57	.73	7.49
BEOWAWE	.75	.63	.63	.79	.93	.96	.36	.47	.50	.47	.65	.77	7.91
BOULDER CITY	.58	.53	.72	.36	.25	.10	.49	.80	.50	.41	.54	.44	5.72
CALIENTE	.84	.78	.97	.75	.64	.30	.92	.99	.63	.85	.77	.61	9.05
CARSON CITY	2.14	1.48	1.00	.52	.62	.41	.28	.29	.38	.50	1.05	2.12	10.79
CONTACT	.78	.53	.68	.78	1.72	1.37	.53	.77	.67	.73	.75	.83	10.14
DEETH 2 SW	1.05	.78	.96	1.12	1.71	1.24	.41	.44	.57	.69	.96	.93	10.86
DESERT NATL WL RANGE	.40	.44	.47	.34	.18	.13	.34	.49	.34	.34	.38	.33	4.18
DYER 4 SE	.28	.43	.35	.46	.51	.33	.51	.38	.46	.34	.47	.37	4.89
ELKO WSO R	1.16	.81	.85	.79	1.03	.91	.33	.58	.47	.56	.83	.98	9.30
ELY WSO //R	.72	.68	.91	.92	1.08	.80	.65	.62	.70	.59	.60	.75	9.02
FALLON EXPERIMENT STA	.45	.50	.40	.42	.67	.57	.21	.30	.30	.34	.36	.36	4.88
GLENBROOK	3.42	2.29	2.10	1.30	1.14	.50	.42	.43	.54	.89	1.88	3.19	18.10
IMLAY	.77	.66	.63	.70	.84	.94	.27	.41	.37	.52	.64	.80	7.55
LAHONTAN DAM	.50	.42	.35	.33	.53	.44	.32	.44	.29	.24	.44	.41	4.71
LAS VEGAS WSO R	.50	.46	.41	.22	.20	.09	.45	.54	.32	.25	.43	.32	4.19
LEHMAN CAVES NAT MON	.92	.99	1.46	1.17	1.23	.75	.86	1.05	.94	1.07	1.07	.98	12.49
LOVELOCK	.56	.55	.42	.54	.51	.56	.26	.32	.35	.33	.45	.63	5.48
LOVELOCK FAA AP	.51	.43	.36	.42	.52	.49	.21	.28	.37	.30	.42	.49	4.80
MC GILL	.46	.47	.61	.86	.99	.93	.77	.74	.63	.56	.49	.59	8.10
MINA	.33	.41	.34	.45	.60	.40	.45	.42	.37	.41	.32	.37	4.87
MINDEN	1.50	.98	.82	.38	.60	.35	.33	.36	.29	.37	.81	1.41	8.20
MONTELLO	.56	.38	.40	.61	1.09	1.12	.61	.75	.46	.48	.57	.54	7.57
OROVADA	1.23	.92	1.02	1.09	1.36	1.22	.30	.47	.49	.75	1.06	1.01	10.92
OWYHEE	1.37	1.09	1.41	1.40	1.80	1.71	.36	.50	.73	1.08	1.19	1.40	14.04
PARADISE VALLEY 1 NW	1.38	1.00	.76	.55	.70	.88	.24	.38	.44	.61	1.02	1.20	9.16
PIOCHE	1.58	1.39	1.44	1.21	.95	.32	1.09	1.22	.81	.87	1.06	1.24	13.18
RENO WSO //R	1.24	.95	.74	.46	.74	.34	.30	.27	.30	.34	.60	1.21	7.49
RUBY LAKE	1.45	1.22	1.16	1.16	1.36	.88	.61	.63	.66	.83	1.22	1.43	12.61
RYE PATCH DAM	.72	.62	.61	.79	.84	.87	.34	.45	.37	.51	.60	.70	7.42
SEARCHLIGHT	.74	.73	.78	.47	.25	.12	.88	.98	.69	.46	.58	.59	7.27
SMOKEY VALLEY	.58	.66	.60	.47	.53	.53	.59	.64	.47	.38	.59	.54	6.58
SUNRISE MANR LAS VEGAS	.50	.46	.39	.28	.18	.10	.36	.50	.32	.29	.51	.33	4.22
TONOPAH AP	.31	.47	.39	.39	.52	.32	.57	.46	.48	.36	.36	.25	4.88
WELLS	.94	.75	.81	.85	1.35	1.12	.46	.54	.63	.72	.83	.89	9.89
WINNEMUCCA WSO //R	.89	.67	.67	.81	.80	.92	.18	.39	.34	.59	.75	.86	7.87
YERINGTON	.60	.57	.37	.36	.77	.58	.32	.34	.30	.26	.45	.60	5.52

NEVADA

HEATING DEGREE DAY NORMALS (BASE 65 DEG F)

STATION	JUL	AUG	SEP	OCT	NOV	DEC	JAN	FEB	MAR	APR	MAY	JUN	ANN
BATTLE MOUNTAIN AP R	0	34	177	487	834	1110	1125	843	812	576	320	102	6420
BOULDER CITY	0	0	0	42	293	543	580	382	299	129	23	0	2291
CALIENTE	0	0	64	323	684	964	1008	739	670	423	197	25	5097
CARSON CITY	8	43	150	428	729	955	977	756	729	543	324	124	5766
CONTACT	14	60	239	543	885	1138	1169	918	905	651	404	170	7096
DESERT NATL WL RANGE	0	0	0	102	411	654	685	484	399	186	48	0	2969
DYER 4 SE	0	17	100	400	768	1035	1063	784	729	495	261	71	5723
ELKO WSO R	19	58	230	543	891	1206	1240	952	899	648	396	166	7248
ELY WSO //R	10	64	261	589	933	1209	1259	1014	980	723	462	196	7700
FALLON EXPERIMENT STA	0	31	125	400	729	989	1001	748	688	477	252	84	5524
GLENBROOK	71	111	210	493	774	967	1011	860	893	711	499	252	6852
LAS VEGAS WSO R	0	0	0	63	346	608	632	417	313	131	22	0	2532
LEHMAN CAVES NAT MON	0	21	163	471	828	1051	1104	910	905	648	390	140	6631
LOVELOCK FAA AP	0	20	116	406	795	1073	1073	798	744	498	246	67	5836
MC GILL	0	41	199	505	864	1110	1166	949	936	675	423	159	7027
MINA	0	6	60	317	678	946	973	728	673	452	211	42	5086
MINDEN	14	48	172	459	768	1011	1054	812	791	582	354	132	6197
OWYHEE	23	77	252	546	855	1088	1166	924	942	693	444	195	7205
PIOCHE	0	7	106	380	747	992	1048	843	812	552	298	62	5847
RENO WSO //R	16	59	171	456	759	1008	1017	773	756	558	333	124	6030
RYE PATCH DAM	0	25	131	419	765	1020	1048	781	741	516	271	89	5806
SEARCHLIGHT	0	0	0	97	375	611	648	470	411	200	57	0	2869
SMOKEY VALLEY	0	25	143	431	798	1076	1091	840	791	570	320	95	6180
SUNRISE MANR LAS VEGAS	0	0	0	79	393	654	670	448	336	126	21	0	2727
TONOPAH AP	0	14	101	384	747	1014	1057	815	763	525	269	64	5753
WELLS	18	81	273	605	948	1252	1287	1019	967	699	443	197	7789
WINNEMUCCA WSO //R	0	42	193	505	819	1073	1091	820	815	591	334	126	6409
YERINGTON	8	39	119	406	747	1001	992	756	694	483	265	82	5592

NEVADA

COOLING DEGREE DAY NORMALS (BASE 65 DEG F)

STATION	JAN	FEB	MAR	APR	MAY	JUN	JUL	AUG	SEP	OCT	NOV	DEC	ANN
BATTLE MOUNTAIN AP R	0	0	0	0	16	66	233	168	36	0	0	0	519
BOULDER CITY	0	7	32	102	274	531	738	673	468	172	5	0	3002
CALIENTE	0	0	0	0	39	148	338	284	85	7	0	0	901
CARSON CITY	0	0	0	0	8	58	154	123	30	0	0	0	373
CONTACT	0	0	0	0	0	38	132	91	17	0	0	0	278
DESERT NATL WL RANGE	0	0	6	33	156	387	601	543	309	77	0	0	2112
DYER 4 SE	0	0	0	0	16	107	257	196	40	0	0	0	616
ELKO WSO R	0	0	0	0	6	52	178	138	32	0	0	0	406
ELY WSO //R	0	0	0	0	0	16	88	76	12	0	0	0	192
FALLON EXPERIMENT STA	0	0	0	0	23	111	258	198	56	0	0	0	646
GLENBROOK	0	0	0	0	0	21	78	83	21	0	0	0	203
LAS VEGAS WSO R	0	0	12	86	279	558	784	713	453	144	0	0	3029
LEHMAN CAVES NAT MON	0	0	0	0	12	71	205	152	46	0	0	0	486
LOVELOCK FAA AP	0	0	0	0	32	127	310	231	65	0	0	0	765
MC GILL	0	0	0	0	10	51	171	124	31	0	0	0	387
MINA	0	0	0	0	50	183	406	329	99	7	0	0	1074
MINDEN	0	0	0	0	7	36	131	104	25	0	0	0	303
OWYHEE	0	0	0	0	7	21	125	111	24	0	0	0	288
PIOCHE	0	0	0	0	23	101	264	199	70	8	0	0	665
RENO WSO //R	0	0	0	0	11	46	156	117	27	0	0	0	357
RYE PATCH DAM	0	0	0	0	19	101	257	196	53	0	0	0	626
SEARCHLIGHT	0	0	11	41	166	405	611	543	342	125	6	0	2250
SMOKEY VALLEY	0	0	0	0	16	65	213	161	35	6	0	0	496
SUNRISE MANR LAS VEGAS	0	0	8	54	220	468	688	626	366	85	0	0	2515
TONOPAH AP	0	0	0	0	21	112	270	209	59	5	0	0	676
WELLS	0	0	0	0	0	26	114	90	15	0	0	0	245
WINNEMUCCA WSO //R	0	0	0	0	14	81	222	157	31	0	0	0	505
YERINGTON	0	0	0	0	23	91	225	185	41	0	0	0	565

STATE-STATION NUMBER	STN TYP	NAME	LATITUDE DEG-MIN	LONGITUDE DEG-MIN	ELEVATION (FT)
26-0046	12	ADAVEN	N 3807	W 11535	6250
26-0438	12	ARTHUR 4 NW	N 4047	W 11511	6280
26-0507	12	AUSTIN	N 3930	W 11705	6605
26-0691	13	BATTLE MOUNTAIN AP R	N 4037	W 11652	4530
26-0795	12	BEOWAWE	N 4036	W 11629	4695
26-1071	13	BOULDER CITY	N 3559	W 11451	2525
26-1358	13	CALIENTE	N 3737	W 11431	4402
26-1485	13	CARSON CITY	N 3909	W 11946	4651
26-1905	13	CONTACT	N 4147	W 11445	5365
26-2189	12	DEETH 2 SW	N 4104	W 11516	5335
26-2243	13	DESERT NATL WL RANGE	N 3626	W 11522	2920
26-2431	13	DYER 4 SE	N 3737	W 11801	4975
26-2573	13	ELKO WSO R	N 4050	W 11547	5075
26-2631	13	ELY WSO //R	N 3917	W 11451	6253
26-2780	13	FALLON EXPERIMENT STA	N 3927	W 11847	3965
26-3205	13	GLENBROOK	N 3905	W 11956	6400
26-3957	12	IMLAY	N 4039	W 11809	4260
26-4349	12	LAHONTAN DAM	N 3928	W 11904	4158
26-4436	13	LAS VEGAS WSO R	N 3605	W 11510	2162
26-4514	13	LEHMAN CAVES NAT MON	N 3900	W 11413	6825
26-4698	12	LOVELOCK	N 4011	W 11828	3975
26-4700	13	LOVELOCK FAA AP	N 4004	W 11833	3900
26-4950	13	MC GILL	N 3924	W 11446	6340
26-5168	13	MINA	N 3823	W 11806	4552
26-5191	13	MINDEN	N 3857	W 11946	4720
26-5352	12	MONTELLO	N 4116	W 11412	4877
26-5818	12	OROVADA	N 4134	W 11747	4310
26-5869	13	OWYHEE	N 4157	W 11606	5396
26-6005	12	PARADISE VALLEY 1 NW	N 4130	W 11732	4675
26-6252	13	PIOCHE	N 3756	W 11427	6165
26-6779	13	RENO WSO //R	N 3930	W 11947	4404
26-7123	12	RUBY LAKE	N 4012	W 11530	6012
26-7192	13	RYE PATCH DAM	N 4028	W 11818	4135
26-7369	13	SEARCHLIGHT	N 3528	W 11455	3540
26-7620	13	SMOKEY VALLEY	N 3847	W 11710	5625
26-7925	13	SUNRISE MANR LAS VEGAS	N 3612	W 11505	1820
26-8170	13	TONOPAH AP	N 3804	W 11705	5426
26-8988	13	WELLS	N 4107	W 11458	5650
26-9171	13	WINNEMUCCA WSO //R	N 4054	W 11748	4301
26-9229	13	YERINGTON	N 3859	W 11910	4375

26 – NEVADA

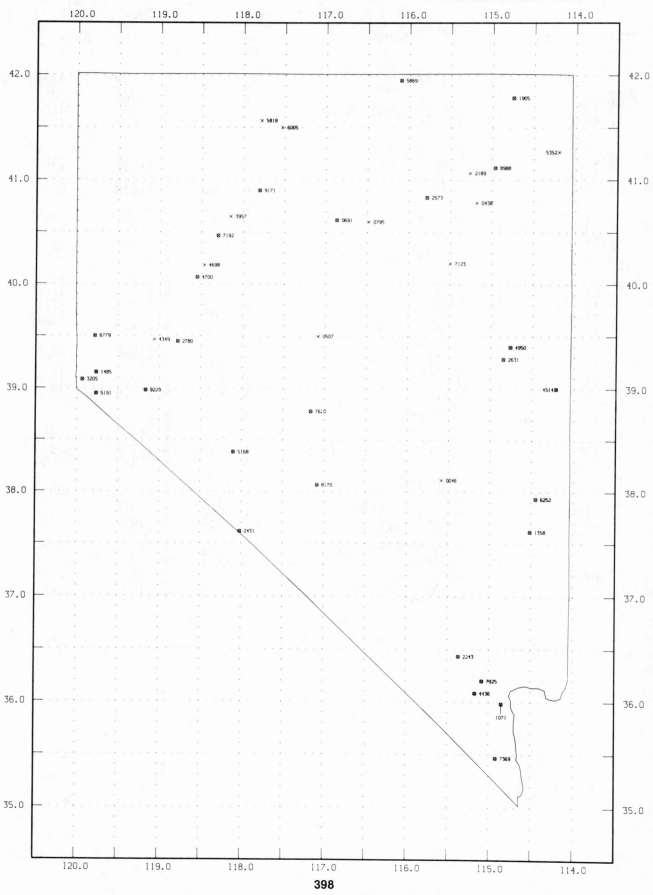

NEW HAMPSHIRE

TEMPERATURE NORMALS (DEG F)

STATION			JAN	FEB	MAR	APR	MAY	JUN	JUL	AUG	SEP	OCT	NOV	DEC	ANN
BETHLEHEM		MAX	25.5	28.5	38.0	51.2	65.3	73.9	77.7	75.2	67.3	56.2	42.2	29.1	52.5
		MIN	5.6	6.9	18.0	29.7	40.6	50.0	54.4	52.5	45.1	35.3	25.7	11.1	31.2
		MEAN	15.6	17.7	28.0	40.5	53.0	61.9	66.1	63.9	56.2	45.8	34.0	20.1	41.9
BLACKWATER DAM	R	MAX	30.0	32.5	41.1	54.6	67.1	76.4	81.3	78.8	70.9	60.1	46.4	33.3	56.0
		MIN	8.6	9.7	21.1	31.8	41.5	51.0	55.8	53.6	45.4	34.9	27.5	14.5	33.0
		MEAN	19.3	21.1	31.1	43.2	54.3	63.7	68.6	66.2	58.2	47.5	37.0	23.9	44.5
CONCORD WSO	R	MAX	30.8	33.2	41.9	56.5	68.9	77.7	82.6	80.1	71.9	61.0	47.2	34.4	57.2
		MIN	9.0	11.0	22.2	31.6	41.4	51.6	56.4	54.5	46.2	35.5	27.3	14.5	33.4
		MEAN	19.9	22.2	32.1	44.1	55.2	64.7	69.5	67.3	59.1	48.2	37.3	24.5	45.3
DURHAM		MAX	33.8	36.4	44.5	57.5	68.9	78.0	83.3	81.3	73.7	62.9	49.4	36.9	58.9
		MIN	12.3	14.3	23.7	32.8	42.1	51.6	56.8	54.9	47.2	37.2	29.3	17.5	35.0
		MEAN	23.1	25.4	34.1	45.2	55.5	64.8	70.1	68.1	60.5	50.1	39.4	27.3	47.0
FIRST CONN LAKE		MAX	21.0	23.1	33.1	45.4	59.9	69.6	73.9	71.5	63.8	52.6	38.7	25.1	48.1
		MIN	-2.7	-3.1	9.1	24.2	36.0	46.4	50.9	48.6	41.0	31.6	22.2	5.6	25.8
		MEAN	9.1	10.0	21.1	34.9	48.0	58.0	62.4	60.1	52.4	42.1	30.5	15.4	37.0
HANOVER		MAX	28.1	31.6	41.1	55.1	68.3	77.0	81.5	79.2	70.4	58.9	44.9	31.7	55.7
		MIN	8.1	10.1	21.4	31.9	42.6	52.6	57.4	55.6	48.3	37.4	28.6	14.5	34.0
		MEAN	18.2	20.9	31.3	43.5	55.5	64.8	69.5	67.5	59.4	48.2	36.8	23.1	44.9
KEENE		MAX	32.3	35.4	44.0	58.3	70.6	78.9	83.5	81.3	73.3	62.5	48.3	35.4	58.7
		MIN	10.9	12.7	23.0	32.8	42.7	52.4	56.8	55.3	47.5	36.8	29.1	16.6	34.7
		MEAN	21.6	24.1	33.5	45.6	56.7	65.7	70.2	68.3	60.5	49.7	38.7	26.0	46.7
LAKEPORT		MAX	28.7	31.4	40.5	54.5	67.6	76.2	80.8	78.2	69.3	57.8	44.7	32.2	55.2
		MIN	10.8	11.7	22.2	32.7	44.1	54.1	59.4	57.6	49.8	39.6	30.3	17.1	35.8
		MEAN	19.8	21.6	31.4	43.6	55.9	65.2	70.1	68.0	59.6	48.8	37.5	24.7	45.5
LEBANON FAA AIRPORT		MAX	28.5	31.7	40.7	54.5	67.6	76.5	81.1	78.7	70.2	59.0	45.0	32.1	55.5
		MIN	5.9	8.0	20.1	30.9	41.1	51.0	55.7	53.9	46.3	35.6	27.3	13.0	32.4
		MEAN	17.2	19.9	30.4	42.7	54.4	63.8	68.4	66.3	58.3	47.4	36.1	22.5	44.0
MASSABESIC LAKE		MAX	32.8	35.3	43.4	56.5	68.3	77.1	81.7	79.7	72.1	61.8	49.0	36.1	57.8
		MIN	10.9	12.8	22.9	32.9	42.8	52.3	57.3	55.7	47.6	37.4	29.6	16.9	34.9
		MEAN	21.9	24.0	33.2	44.7	55.6	64.7	69.5	67.7	59.9	49.6	39.3	26.5	46.4
MT. WASHINGTON WSO		MAX	13.4	13.1	19.1	28.9	40.7	50.7	54.4	52.7	46.3	36.5	26.9	17.0	33.3
		MIN	-3.3	-3.4	4.7	15.9	28.1	38.4	42.9	41.5	34.9	24.5	13.8	1.2	19.9
		MEAN	5.1	4.8	12.0	22.4	34.4	44.6	48.7	47.1	40.6	30.5	20.3	9.1	26.6
NASHUA 2 NNW		MAX	33.7	35.9	44.1	57.4	68.9	77.8	82.7	80.4	72.5	62.0	49.2	36.8	58.5
		MIN	11.4	13.0	23.0	32.3	42.0	51.7	56.7	54.8	46.5	36.2	28.1	16.3	34.3
		MEAN	22.6	24.5	33.6	44.9	55.5	64.8	69.8	67.7	59.5	49.1	38.7	26.6	46.4
PETERBORO 2 S		MAX	31.0	33.1	41.7	55.6	67.6	75.4	79.6	77.4	70.0	60.1	46.8	34.3	56.1
		MIN	12.2	13.0	22.2	32.0	41.7	50.8	55.9	54.4	46.8	37.2	28.8	16.7	34.3
		MEAN	21.6	23.1	32.0	43.8	54.7	63.1	67.8	65.9	58.4	48.7	37.8	25.5	45.2
PINKHAM NOTCH		MAX	25.6	27.2	35.3	46.9	60.9	69.4	73.4	71.2	63.8	54.0	40.8	28.7	49.8
		MIN	5.3	6.2	16.1	27.8	38.6	48.2	52.6	50.5	43.0	33.4	24.1	10.3	29.7
		MEAN	15.5	16.7	25.7	37.3	49.8	58.8	63.1	60.9	53.4	43.7	32.5	19.5	39.7
SURRY MOUNTAIN DAM	R	MAX	30.1	32.3	41.1	54.8	67.4	76.2	80.9	78.6	70.7	60.2	46.9	33.6	56.1
		MIN	7.1	8.3	20.6	31.5	41.1	51.0	55.4	53.5	45.3	34.3	27.3	13.6	32.4
		MEAN	18.6	20.3	30.9	43.2	54.2	63.6	68.2	66.1	58.0	47.2	37.1	23.6	44.3
WOODSTOCK		MAX	28.9	31.8	40.4	54.1	67.7	76.0	80.3	78.0	69.6	59.0	44.7	32.2	55.2
		MIN	8.6	10.0	20.6	30.6	40.8	49.9	54.5	52.7	45.3	35.4	27.3	14.0	32.5
		MEAN	18.8	20.9	30.5	42.4	54.2	62.9	67.4	65.4	57.5	47.2	36.0	23.1	43.9

NEW HAMPSHIRE

PRECIPITATION NORMALS (INCHES)

STATION		JAN	FEB	MAR	APR	MAY	JUN	JUL	AUG	SEP	OCT	NOV	DEC	ANN
BERLIN		2.65	2.36	2.72	2.91	3.09	3.97	3.65	3.59	3.20	3.46	3.44	3.36	38.40
BETHLEHEM		2.30	2.06	2.39	2.87	3.31	4.10	3.94	4.24	3.34	3.14	3.28	3.15	38.12
BLACKWATER DAM	R	3.28	2.95	3.54	3.40	3.36	3.23	3.43	3.27	3.41	3.48	4.05	4.14	41.54
BRADFORD		3.42	3.10	3.76	3.74	3.64	3.37	3.20	3.84	3.65	3.75	4.42	4.37	44.26
CONCORD WSO	R	2.78	2.47	2.93	3.01	2.93	2.91	2.93	3.26	3.12	3.10	3.66	3.43	36.53
DIXVILLE NOTCH		3.09	2.98	3.09	3.46	3.79	4.83	4.52	4.80	3.95	4.05	3.76	3.74	46.06
DURHAM		3.51	3.12	3.66	3.80	3.57	3.00	3.00	3.31	3.37	3.91	4.70	4.28	43.23
ERROL		2.67	2.53	2.86	2.99	3.12	3.76	3.50	3.71	3.05	3.27	3.33	3.32	38.11
FIRST CONN LAKE		2.80	2.55	2.80	3.23	3.60	4.65	4.65	4.39	3.93	3.73	3.72	3.70	43.75
FITZWILLIAM		3.44	2.88	3.91	3.74	3.67	3.91	3.63	3.88	3.66	3.65	4.08	3.97	44.42
FRANKLIN FALLS DAM	R	2.99	2.73	3.21	3.31	3.35	3.18	3.35	3.29	3.17	3.35	3.78	3.74	39.45
HANOVER		2.74	2.45	2.81	3.00	3.29	3.01	3.30	3.26	3.31	3.03	3.26	3.21	36.67
KEENE		3.19	2.66	3.35	3.31	3.48	3.59	3.38	3.60	3.44	3.14	3.59	3.58	40.31
LAKEPORT		3.07	2.80	3.36	3.36	3.42	3.33	3.58	3.41	3.74	3.46	4.08	3.88	41.49
LAKEPORT 2		2.76	2.17	2.63	3.14	3.20	3.14	3.48	3.22	3.44	3.15	3.64	3.22	37.19
LEBANON FAA AIRPORT		2.56	2.17	2.51	2.87	3.28	2.85	3.01	3.24	3.16	2.93	3.24	2.98	34.80
MAC DOWELL DAM	R	4.05	3.42	4.27	3.78	3.56	3.45	3.43	3.75	3.90	3.76	4.50	4.53	46.40
MARLOW		2.75	2.45	2.81	2.95	3.39	3.44	3.05	3.69	3.36	3.24	3.60	3.16	37.89
MASSABESIC LAKE		3.09	2.50	3.14	3.18	3.38	3.22	3.43	3.56	3.32	3.45	3.99	3.61	39.87
MILAN 7 NNW		2.79	2.55	2.69	2.93	3.20	3.77	3.56	3.75	3.02	3.10	3.38	3.37	38.11
MILFORD		3.79	3.27	3.95	3.72	3.64	3.54	3.13	3.68	3.53	3.82	4.49	4.44	45.00
MT. WASHINGTON WSO		7.31	8.01	8.19	7.03	6.46	7.06	6.90	7.60	7.15	6.73	8.54	8.94	89.92
NASHUA 2 NNW		3.56	3.09	3.89	3.55	3.49	3.14	3.18	3.48	3.65	3.67	4.47	4.10	43.27
NEWPORT		2.78	2.62	3.06	3.21	3.28	3.24	3.50	3.47	3.14	3.23	3.42	3.43	38.38
PETERBORO 2 S		3.61	3.31	3.89	3.46	3.50	3.54	3.43	4.02	3.45	3.63	4.19	4.07	44.10
PINKHAM NOTCH		4.41	4.34	5.05	4.37	4.41	4.83	4.63	4.50	4.61	5.23	6.10	5.42	57.90
PLYMOUTH 1 E		3.34	2.90	3.44	3.44	3.92	3.32	4.00	3.38	3.32	3.73	4.13	4.08	43.00
SOUTH DANBURY		3.02	2.85	3.30	3.33	3.66	3.50	3.82	3.52	3.57	3.69	4.08	3.85	42.19
SOUTH LYNDEBORO		3.67	3.38	4.12	3.60	3.72	3.44	3.23	3.55	3.48	4.01	4.45	4.37	45.02
SURRY MOUNTAIN DAM	R	2.96	2.43	3.00	2.99	3.19	3.22	3.06	3.66	3.28	3.06	3.40	3.30	37.55
WEST RUMNEY		3.21	2.99	3.56	3.47	3.81	3.65	3.95	3.57	3.57	3.82	4.15	4.12	43.87
WOODSTOCK		3.14	2.92	3.41	3.60	3.92	3.77	4.32	3.95	3.93	4.01	4.12	4.08	45.17

NEW HAMPSHIRE

HEATING DEGREE DAY NORMALS (BASE 65 DEG F)

STATION		JUL	AUG	SEP	OCT	NOV	DEC	JAN	FEB	MAR	APR	MAY	JUN	ANN
BETHLEHEM		70	95	269	595	930	1392	1531	1324	1147	735	380	117	8585
BLACKWATER DAM	R	22	53	215	543	840	1274	1417	1229	1051	654	339	85	7722
CONCORD WSO	R	20	39	191	521	831	1256	1398	1198	1020	627	314	67	7482
DURHAM		16	26	155	462	768	1169	1299	1109	958	594	303	60	6919
FIRST CONN LAKE		122	171	378	710	1035	1538	1733	1540	1361	903	527	215	10233
HANOVER		16	32	186	521	846	1299	1451	1235	1045	645	309	64	7649
KEENE		10	26	158	474	789	1209	1345	1145	977	582	272	48	7035
LAKEPORT		13	23	176	502	825	1249	1401	1215	1042	642	295	61	7444
LEBANON FAA AIRPORT		29	55	215	546	867	1318	1482	1263	1073	669	340	85	7942
MASSABESIC LAKE		17	37	166	477	771	1194	1336	1148	986	609	299	62	7102
MT. WASHINGTON WSO		505	555	732	1070	1341	1733	1857	1686	1643	1278	949	612	13961
NASHUA 2 NNW		11	29	181	493	789	1190	1314	1134	973	603	302	62	7081
PETERBORO 2 S		33	53	208	505	816	1225	1345	1173	1023	636	329	88	7434
PINKHAM NOTCH		101	145	348	660	975	1411	1535	1352	1218	831	471	191	9238
SURRY MOUNTAIN DAM	R	22	51	221	552	837	1283	1438	1252	1057	654	342	81	7790
WOODSTOCK		38	69	236	552	870	1299	1432	1235	1070	678	345	97	7921

NEW HAMPSHIRE

COOLING DEGREE DAY NORMALS (BASE 65 DEG F)

STATION		JAN	FEB	MAR	APR	MAY	JUN	JUL	AUG	SEP	OCT	NOV	DEC	ANN
BETHLEHEM		0	0	0	0	8	24	104	61	5	0	0	0	202
BLACKWATER DAM	R	0	0	0	0	7	46	133	90	11	0	0	0	287
CONCORD WSO	R	0	0	0	0	10	58	160	111	14	0	0	0	353
DURHAM		0	0	0	0	8	54	174	123	20	0	0	0	379
FIRST CONN LAKE		0	0	0	0	0	5	41	19	0	0	0	0	65
HANOVER		0	0	0	0	14	58	156	109	18	0	0	0	355
KEENE		0	0	0	0	15	69	171	128	23	0	0	0	406
LAKEPORT		0	0	0	0	13	67	171	116	14	0	0	0	381
LEBANON FAA AIRPORT		0	0	0	0	12	49	134	95	14	0	0	0	304
MASSABESIC LAKE		0	0	0	0	7	53	156	121	13	0	0	0	350
MT. WASHINGTON WSO		0	0	0	0	0	0	0	0	0	0	0	0	0
NASHUA 2 NNW		0	0	0	0	7	56	160	112	16	0	0	0	351
PETERBORO 2 S		0	0	0	0	9	31	120	81	10	0	0	0	251
PINKHAM NOTCH		0	0	0	0	0	5	42	18	0	0	0	0	65
SURRY MOUNTAIN DAM	R	0	0	0	0	7	39	121	85	11	0	0	0	263
WOODSTOCK		0	0	0	0	10	34	113	81	11	0	0	0	249

27 – NEW HAMPSHIRE

LEGEND
11 = TEMPERATURE ONLY
12 = PRECIPITATION ONLY
13 = TEMP. & PRECIP.

STATE-STATION NUMBER	STN TYP	NAME		LATITUDE DEG-MIN	LONGITUDE DEG-MIN	ELEVATION (FT)
27-0690	12	BERLIN		N 4429	W 07111	1100
27-0703	13	BETHLEHEM		N 4417	W 07141	1380
27-0741	13	BLACKWATER DAM	R	N 4319	W 07143	480
27-0910	12	BRADFORD		N 4315	W 07158	970
27-1683	13	CONCORD WSO	R	N 4312	W 07130	346
27-2023	12	DIXVILLE NOTCH		N 4452	W 07120	1580
27-2174	13	DURHAM		N 4308	W 07056	70
27-2842	12	ERROL		N 4447	W 07108	1280
27-2999	13	FIRST CONN LAKE		N 4505	W 07117	1660
27-3024	12	FITZWILLIAM		N 4246	W 07209	1060
27-3182	12	FRANKLIN FALLS DAM	R	N 4328	W 07139	430
27-3850	13	HANOVER		N 4342	W 07217	603
27-4399	13	KEENE		N 4255	W 07217	490
27-4475	13	LAKEPORT		N 4333	W 07128	560
27-4480	12	LAKEPORT 2		N 4333	W 07128	501
27-4656	13	LEBANON FAA AIRPORT		N 4338	W 07219	562
27-5013	12	MAC DOWELL DAM	R	N 4254	W 07159	965
27-5150	12	MARLOW		N 4307	W 07212	1180
27-5211	13	MASSABESIC LAKE		N 4259	W 07124	250
27-5400	12	MILAN 7 NNW		N 4440	W 07113	1180
27-5412	12	MILFORD		N 4249	W 07139	300
27-5639	13	MT. WASHINGTON WSO		N 4416	W 07118	6262
27-5712	13	NASHUA 2 NNW		N 4247	W 07129	140
27-5868	12	NEWPORT		N 4323	W 07211	786
27-6697	13	PETERBORO 2 S		N 4251	W 07157	1020
27-6818	13	PINKHAM NOTCH		N 4416	W 07115	2029
27-6945	12	PLYMOUTH 1 E		N 4346	W 07140	560
27-7967	12	SOUTH DANBURY		N 4330	W 07154	930
27-8081	12	SOUTH LYNDEBORO		N 4253	W 07147	646
27-8539	13	SURRY MOUNTAIN DAM	R	N 4300	W 07219	550
27-9474	12	WEST RUMNEY		N 4348	W 07151	560
27-9940	13	WOODSTOCK		N 4359	W 07141	720

27 – NEW HAMPSHIRE

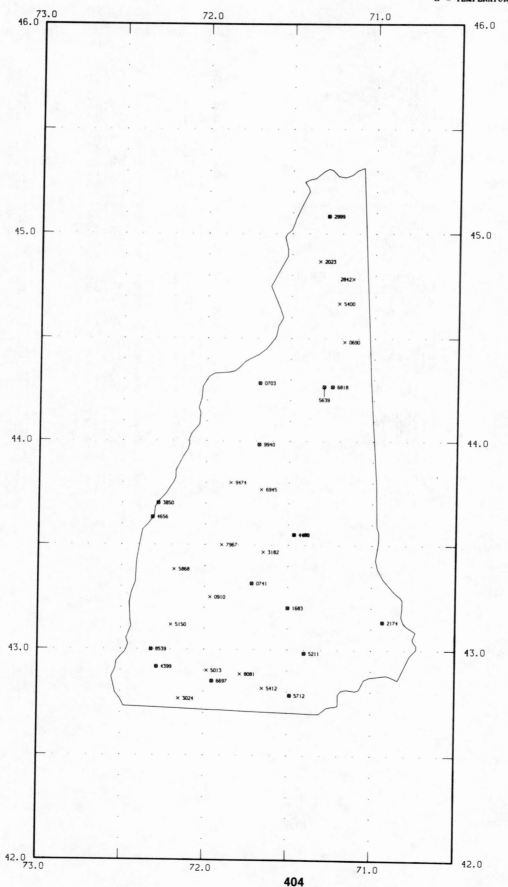

NEW JERSEY

TEMPERATURE NORMALS (DEG F)

STATION		JAN	FEB	MAR	APR	MAY	JUN	JUL	AUG	SEP	OCT	NOV	DEC	ANN
ATLANTIC CITY WSO	MAX	40.6	42.4	50.3	61.6	71.0	79.6	84.0	82.5	76.7	66.1	55.4	45.0	62.9
	MIN	22.9	23.9	31.6	40.4	49.9	58.8	64.8	63.5	56.4	44.8	35.8	26.6	43.3
	MEAN	31.8	33.2	41.0	51.0	60.5	69.2	74.4	73.0	66.6	55.5	45.6	35.8	53.1
ATLANTIC CITY MARINA	MAX	40.3	41.8	47.7	57.7	65.7	74.5	79.9	79.5	73.8	64.4	54.3	44.8	60.4
	MIN	27.9	28.9	35.6	44.2	53.3	62.3	68.2	68.0	61.8	50.7	41.1	31.9	47.8
	MEAN	34.1	35.4	41.7	51.0	59.5	68.4	74.1	73.8	67.8	57.6	47.7	38.4	54.1
BELLEPLAIN ST FOREST	MAX	42.9	45.2	53.6	65.1	74.9	82.3	86.5	85.2	79.2	68.8	57.8	47.0	65.7
	MIN	22.9	23.8	31.3	39.8	49.4	57.9	63.6	62.6	55.8	44.0	35.3	26.8	42.8
	MEAN	32.9	34.5	42.5	52.4	62.2	70.2	75.1	73.9	67.5	56.5	46.6	36.9	54.3
BELVIDERE	MAX	36.2	39.3	49.1	62.1	72.1	79.7	84.1	82.1	75.5	65.3	52.6	40.0	61.5
	MIN	18.3	19.3	27.8	37.2	46.9	56.1	60.4	59.5	52.2	40.9	32.0	22.3	39.4
	MEAN	27.3	29.3	38.5	49.7	59.5	67.9	72.3	70.8	63.9	53.1	42.3	31.2	50.5
BOONTON 1 SE	MAX	36.1	37.9	46.9	59.7	70.0	78.7	83.5	82.0	75.2	64.5	52.4	40.0	60.6
	MIN	18.6	19.4	28.7	38.8	47.4	56.4	61.3	59.8	51.9	40.8	33.2	23.4	40.0
	MEAN	27.4	28.7	37.8	49.3	58.7	67.6	72.4	70.9	63.6	52.7	42.8	31.7	50.3
CANOE BROOK	MAX	37.5	39.4	48.5	61.0	71.2	80.1	85.2	83.3	76.5	65.8	53.9	41.6	62.0
	MIN	17.0	18.2	27.5	37.1	46.3	56.0	61.1	59.9	52.4	40.5	31.9	21.8	39.1
	MEAN	27.3	28.9	38.0	49.1	58.8	68.1	73.2	71.7	64.5	53.2	42.9	31.7	50.6
CAPE MAY 1 NW	MAX	41.3	42.4	49.5	60.0	69.0	78.0	83.3	82.8	77.3	67.1	56.3	46.2	62.8
	MIN	27.6	28.0	34.9	43.6	52.6	61.5	67.1	67.1	61.6	51.4	41.8	31.9	47.4
	MEAN	34.5	35.2	42.2	51.8	60.8	69.8	75.2	74.9	69.4	59.3	49.1	39.1	55.1
CHARLOTTEBURG	MAX	34.9	36.5	45.1	58.2	68.5	76.9	81.8	80.3	73.5	63.1	50.9	38.7	59.0
	MIN	16.3	16.4	25.4	35.7	44.4	53.5	58.2	56.5	49.1	38.6	31.1	20.9	37.2
	MEAN	25.6	26.5	35.3	47.0	56.5	65.2	70.0	68.4	61.4	50.9	41.1	29.8	48.1
ESSEX FELLS SERV BLDG	MAX	36.3	38.7	47.8	60.9	71.5	79.9	84.8	83.0	76.0	65.5	53.3	40.5	61.5
	MIN	18.4	19.8	28.4	38.1	47.3	56.9	61.6	60.0	52.4	41.6	33.3	22.9	40.1
	MEAN	27.4	29.3	38.1	49.5	59.4	68.4	73.2	71.5	64.2	53.6	43.3	31.7	50.8
FLEMINGTON 1 NE	MAX	37.7	39.9	49.5	62.2	72.5	81.8	86.6	84.7	77.7	66.5	53.9	41.6	62.9
	MIN	18.0	19.2	27.6	36.8	45.8	55.4	60.7	59.6	51.7	40.2	31.8	22.5	39.1
	MEAN	27.9	29.6	38.6	49.5	59.2	68.6	73.7	72.2	64.7	53.4	42.9	32.1	51.0
FREEHOLD	MAX	39.0	40.9	49.6	61.9	72.2	80.6	85.2	83.5	77.0	66.2	54.8	43.0	62.8
	MIN	22.0	23.1	30.6	39.7	49.1	58.3	63.2	62.3	55.3	44.5	35.9	26.2	42.5
	MEAN	30.5	32.0	40.1	50.8	60.6	69.5	74.2	72.9	66.2	55.4	45.4	34.6	52.7
GLASSBORO	MAX	39.8	41.8	51.0	63.2	73.0	81.6	86.1	84.6	78.1	66.8	55.3	43.9	63.8
	MIN	23.2	24.4	32.3	41.6	50.8	60.0	65.1	64.2	57.2	45.3	36.7	27.2	44.0
	MEAN	31.6	33.1	41.7	52.4	61.9	70.8	75.6	74.4	67.7	56.1	46.0	35.6	53.9
HAMMONTON 2 NNE	MAX	40.9	42.8	52.0	63.7	73.9	82.8	87.3	86.1	79.6	68.4	57.0	45.2	65.0
	MIN	21.9	23.1	31.5	40.6	49.6	59.0	64.3	62.9	55.4	43.5	35.4	26.0	42.8
	MEAN	31.5	33.0	41.8	52.2	61.8	70.9	75.8	74.5	67.5	56.0	46.2	35.6	53.9
HIGHTSTOWN 1 N	MAX	38.5	40.9	49.8	62.3	72.4	81.2	85.6	83.8	77.0	66.0	54.3	42.5	62.9
	MIN	22.3	23.4	31.1	40.0	49.1	58.3	63.2	62.3	55.2	44.3	35.9	26.7	42.7
	MEAN	30.4	32.2	40.5	51.2	60.8	69.8	74.4	73.1	66.1	55.2	45.1	34.6	52.8
INDIAN MILLS 2 W	MAX	40.3	42.8	51.3	63.6	73.7	82.1	86.1	84.6	78.2	67.3	55.7	44.3	64.2
	MIN	21.4	22.6	30.3	38.7	48.1	57.1	62.1	61.2	53.8	42.4	34.0	25.1	41.4
	MEAN	30.9	32.7	40.8	51.2	60.9	69.6	74.2	72.9	66.0	54.9	44.8	34.7	52.8
JERSEY CITY	MAX	36.7	38.5	46.6	58.1	68.1	77.1	82.1	80.9	73.8	63.3	52.0	40.5	59.8
	MIN	24.5	25.5	33.2	42.9	52.6	61.7	67.1	65.7	58.7	48.4	39.0	28.5	45.7
	MEAN	30.6	32.1	39.9	50.5	60.4	69.4	74.6	73.3	66.3	55.8	45.5	34.5	52.7
LAMBERTVILLE	MAX	39.0	41.7	51.2	64.0	74.7	83.2	87.3	85.5	79.0	67.7	54.8	42.8	64.2
	MIN	21.5	22.7	30.3	39.2	48.7	58.2	63.0	62.2	54.9	43.4	34.7	25.5	42.0
	MEAN	30.3	32.3	40.8	51.6	61.7	70.7	75.2	73.9	67.0	55.6	44.8	34.2	53.2

NEW JERSEY

TEMPERATURE NORMALS (DEG F)

STATION		JAN	FEB	MAR	APR	MAY	JUN	JUL	AUG	SEP	OCT	NOV	DEC	ANN
LITTLE FALLS	MAX	38.4	40.3	49.2	61.8	72.2	81.2	86.3	84.4	77.4	66.5	54.5	42.3	62.9
	MIN	20.7	21.9	30.3	40.0	49.0	58.3	63.2	62.0	54.4	43.1	35.1	25.0	41.9
	MEAN	29.5	31.1	39.8	50.9	60.6	69.8	74.8	73.2	65.9	54.8	44.8	33.7	52.4
LONG BRANCH 2 S	MAX	39.3	40.7	48.0	59.2	68.4	77.6	82.8	81.6	75.3	64.9	54.3	43.6	61.3
	MIN	23.9	24.9	32.2	40.9	50.3	59.7	65.2	64.3	57.9	47.1	37.8	28.1	44.4
	MEAN	31.6	32.9	40.1	50.1	59.4	68.7	74.0	73.0	66.6	56.0	46.1	35.9	52.9
LONG VALLEY	MAX	35.8	38.0	47.2	60.2	70.0	78.2	83.0	80.9	74.0	64.0	51.7	39.6	60.2
	MIN	17.5	18.3	26.1	35.8	44.8	53.7	58.4	57.3	50.1	38.8	30.8	21.3	37.7
	MEAN	26.6	28.2	36.7	48.0	57.4	66.0	70.7	69.2	62.1	51.4	41.3	30.5	49.0
MILLVILLE FAA AIRPORT	MAX	40.5	42.7	51.2	63.0	72.7	81.2	85.6	84.3	78.1	66.9	55.7	44.8	63.9
	MIN	23.1	24.3	32.2	41.2	51.1	60.3	65.9	64.9	57.6	45.6	36.3	27.1	44.1
	MEAN	31.8	33.5	41.7	52.1	61.9	70.8	75.8	74.6	67.9	56.2	46.0	36.0	54.0
MOORESTOWN	MAX	39.0	41.7	50.9	63.0	73.2	81.5	85.8	84.1	77.7	66.2	54.1	42.9	63.3
	MIN	22.8	24.2	31.8	40.7	49.9	58.8	63.9	63.0	55.5	44.3	35.4	26.5	43.1
	MEAN	30.9	32.9	41.4	51.9	61.5	70.2	74.9	73.6	66.6	55.3	44.8	34.8	53.2
MORRIS PLAINS 1 W	MAX	36.4	38.6	47.7	60.6	70.7	79.5	84.4	82.6	75.5	64.9	52.7	40.5	61.2
	MIN	18.2	19.3	27.9	37.5	46.5	55.4	60.4	59.2	51.9	40.8	32.5	22.5	39.3
	MEAN	27.4	29.0	37.8	49.1	58.6	67.5	72.4	70.9	63.7	52.9	42.6	31.6	50.3
NEWARK WSO R	MAX	38.2	40.3	49.1	61.3	71.6	80.6	85.6	84.0	76.9	66.0	54.0	42.3	62.5
	MIN	24.2	25.3	33.3	42.9	53.0	62.4	67.9	67.0	59.4	48.3	39.0	28.6	45.9
	MEAN	31.3	32.8	41.2	52.1	62.3	71.5	76.8	75.5	68.2	57.2	46.5	35.5	54.2
NEW BRUNSWICK	MAX	38.4	40.7	49.8	62.0	72.1	80.6	85.2	83.6	77.1	66.3	54.2	42.4	62.7
	MIN	22.0	23.3	31.0	40.5	49.7	58.8	64.1	63.0	55.6	44.6	36.1	26.2	42.9
	MEAN	30.2	32.0	40.4	51.3	60.9	69.7	74.7	73.3	66.4	55.4	45.2	34.3	52.8
NEWTON	MAX	34.0	36.3	46.0	59.5	70.2	79.0	83.8	81.9	74.6	63.3	50.6	38.0	59.8
	MIN	15.0	16.1	25.9	36.0	44.8	53.9	58.4	56.6	48.8	37.6	30.2	20.0	36.9
	MEAN	24.5	26.2	36.0	47.8	57.5	66.5	71.1	69.3	61.7	50.5	40.4	29.1	48.4
PEMBERTON 3 E	MAX	40.8	43.3	52.2	64.1	74.2	82.6	86.5	85.1	79.1	68.3	56.6	44.7	64.8
	MIN	22.1	23.1	30.5	38.8	48.0	57.0	62.0	61.4	54.3	43.2	35.0	26.1	41.8
	MEAN	31.4	33.2	41.4	51.5	61.1	69.8	74.3	73.3	66.7	55.8	45.8	35.4	53.3
PLAINFIELD	MAX	38.0	41.0	50.3	62.7	72.9	81.4	86.1	84.4	77.6	66.7	54.0	41.8	63.1
	MIN	22.2	23.3	30.8	40.0	49.3	58.4	63.4	62.5	55.4	44.0	35.4	26.4	42.6
	MEAN	30.1	32.2	40.6	51.4	61.1	69.9	74.8	73.5	66.5	55.4	44.7	34.1	52.9
SHILOH	MAX	40.8	43.2	52.0	63.6	73.5	81.4	85.6	84.2	78.6	67.8	55.9	44.9	64.3
	MIN	24.4	25.4	33.0	41.8	51.3	60.4	65.7	64.6	57.9	46.8	37.7	28.6	44.8
	MEAN	32.6	34.3	42.5	52.8	62.4	70.9	75.7	74.5	68.3	57.3	46.8	36.8	54.6
SOMERVILLE 3 NW	MAX	37.5	39.8	49.4	62.3	72.7	81.5	86.4	84.4	77.4	66.0	53.5	41.4	62.7
	MIN	19.7	20.8	28.8	38.1	47.5	56.8	61.7	60.8	53.3	41.9	33.4	23.9	40.6
	MEAN	28.6	30.3	39.1	50.2	60.1	69.2	74.1	72.6	65.4	54.0	43.5	32.7	51.7
SUSSEX 1 SE	MAX	34.0	36.5	45.7	59.5	70.1	78.6	83.1	81.3	74.2	63.5	50.8	37.8	59.6
	MIN	14.1	15.1	24.9	35.1	44.3	53.6	58.2	56.8	48.9	38.1	30.3	19.4	36.6
	MEAN	24.1	25.9	35.3	47.3	57.2	66.1	70.7	69.1	61.6	50.8	40.6	28.7	48.1
TRENTON WSO R	MAX	38.3	40.5	49.2	61.7	71.5	80.5	84.9	83.2	76.1	65.2	53.6	42.4	62.3
	MIN	24.9	26.1	33.6	42.8	52.4	61.7	66.8	65.8	58.5	47.8	39.0	29.3	45.7
	MEAN	31.6	33.3	41.4	52.3	62.0	71.1	75.9	74.5	67.4	56.5	46.3	35.9	54.0

NEW JERSEY

PRECIPITATION NORMALS (INCHES)

STATION	JAN	FEB	MAR	APR	MAY	JUN	JUL	AUG	SEP	OCT	NOV	DEC	ANN
ATLANTIC CITY WSO	3.47	3.34	4.04	3.20	3.07	2.78	4.02	4.72	2.89	3.06	3.73	3.61	41.93
ATLANTIC CITY MARINA	3.25	3.22	3.71	3.12	2.91	2.88	3.89	4.54	2.73	2.76	3.54	3.51	40.06
AUDUBON	3.31	2.99	4.07	3.74	3.69	4.20	4.10	4.90	3.68	3.21	3.75	3.71	45.35
BELLEPLAIN ST FOREST	3.34	3.33	4.21	3.46	3.47	3.45	4.45	4.99	3.30	3.64	3.62	3.96	45.22
BELVIDERE	3.33	2.94	3.76	3.99	3.59	3.87	4.32	4.83	4.06	3.44	3.77	3.69	45.59
BOONTON 1 SE	3.39	3.08	4.22	4.15	4.12	3.98	4.09	4.54	4.53	3.74	4.08	3.93	47.85
BRANCHVILLE	3.25	2.75	3.68	4.07	3.52	3.80	3.96	4.62	3.75	3.41	3.70	3.58	44.09
CANOE BROOK	3.54	3.26	4.50	4.07	4.01	3.79	4.28	4.91	4.36	3.69	4.13	4.08	48.62
CAPE MAY 1 NW	3.28	3.19	3.92	3.24	3.25	3.14	3.58	4.37	3.38	3.14	3.49	3.82	41.80
CHARLOTTEBURG	3.75	3.37	4.70	4.34	3.95	4.19	4.23	5.00	4.39	4.15	4.66	4.23	50.96
ESSEX FELLS SERV BLDG	3.61	3.19	4.37	4.07	3.93	3.82	4.49	4.46	4.34	3.74	4.16	4.14	48.32
FLEMINGTON 1 NE	3.66	3.22	4.18	4.04	3.68	3.75	4.36	4.60	3.93	3.45	3.84	3.95	46.66
FREEHOLD	3.55	3.28	4.44	3.66	3.75	3.47	4.04	4.64	3.67	3.52	3.96	3.91	45.89
GLASSBORO	3.48	3.18	4.18	3.60	3.59	3.60	4.11	4.62	3.75	3.42	3.59	3.72	44.84
GREENWOOD LAKE	3.93	3.55	4.78	4.61	4.01	4.27	4.13	4.86	4.59	4.17	4.74	4.46	52.10
HAMMONTON 2 NNE	3.31	3.23	3.99	3.57	3.53	3.50	4.58	4.60	3.71	3.26	3.73	3.90	44.91
HIGHTSTOWN 1 N	3.31	2.96	3.96	3.65	3.60	3.25	4.33	4.72	3.99	3.42	3.52	3.68	44.39
INDIAN MILLS 2 W	3.55	3.28	4.17	3.60	3.36	3.33	4.19	4.98	3.50	3.32	3.66	4.07	45.01
JERSEY CITY	3.28	2.86	4.24	3.81	3.67	3.26	3.81	4.30	3.82	3.36	3.67	3.69	43.77
LAMBERTVILLE	3.40	2.87	4.10	3.73	3.71	3.34	4.16	4.70	3.73	3.11	3.66	3.62	44.13
LITTLE FALLS	3.53	3.25	4.52	4.11	3.99	3.91	4.24	4.77	4.70	3.92	4.31	4.15	49.40
LONG BRANCH 2 S	3.60	3.59	4.63	3.78	3.85	3.12	3.88	5.11	3.65	3.59	3.88	4.24	46.92
LONG VALLEY	3.86	3.30	4.43	4.33	4.07	3.91	4.68	5.22	4.22	3.92	4.38	4.26	50.58
MAYS LANDING 1 W	3.58	3.40	4.28	3.72	3.35	3.11	4.63	4.78	3.42	3.55	3.68	3.99	45.49
MIDLAND PARK	3.39	3.20	4.51	4.14	4.11	4.20	4.59	5.25	4.51	4.02	4.44	4.17	50.53
MILLVILLE FAA AIRPORT	3.22	3.24	3.99	3.40	3.20	3.49	4.00	4.58	3.21	3.31	3.59	3.72	42.95
MOORESTOWN	3.24	2.99	3.90	3.65	3.63	3.55	4.33	4.91	3.61	3.38	3.50	3.69	44.38
MORRIS PLAINS 1 W	3.65	3.12	4.46	4.40	4.20	4.02	4.32	5.05	4.47	3.79	4.34	4.10	49.92
NEWARK WSO R	3.13	3.05	4.15	3.57	3.59	2.94	3.85	4.30	3.66	3.09	3.59	3.42	42.34
NEW BRUNSWICK	3.45	2.96	4.04	3.77	3.90	3.26	4.39	4.90	3.93	3.33	3.82	3.75	45.50
NEW MILFORD	3.10	2.92	4.04	3.60	3.58	3.40	3.74	4.29	3.77	3.23	4.05	3.62	43.34
NEWTON	3.12	2.63	3.51	4.01	3.50	3.93	4.14	4.78	3.92	3.39	3.64	3.46	44.03
OAK RIDGE RESERVOIR	3.85	3.44	4.43	4.21	3.80	4.48	4.24	5.44	4.58	4.01	4.53	4.27	51.28
PEMBERTON 3 E	3.46	3.05	4.22	3.62	3.44	3.57	4.56	5.28	3.67	3.29	3.61	4.02	45.79
PLAINFIELD	3.55	3.30	4.56	3.98	4.11	3.42	4.76	5.37	4.09	3.66	3.91	4.05	48.76
PRINCETON WATER WORKS	3.35	3.11	4.11	3.57	3.50	3.36	4.78	4.89	3.97	3.32	3.59	3.75	45.30
RAHWAY	2.93	2.84	3.86	3.67	3.43	3.28	4.56	4.52	3.70	3.16	3.56	3.41	42.92
RINGWOOD	3.51	3.14	4.27	4.14	3.66	3.77	3.82	4.50	4.07	3.77	4.15	4.00	46.80
SHILOH	3.08	2.64	3.47	3.11	3.23	3.48	4.23	4.21	3.41	3.28	3.51	3.36	41.01
SOMERVILLE 3 NW	3.33	2.93	4.04	3.78	3.66	3.49	4.54	4.92	3.88	3.42	3.63	3.74	45.36
SPLIT ROCK POND	3.52	3.06	4.23	4.21	4.11	4.41	4.38	4.84	4.57	4.14	4.57	4.10	50.14
SUSSEX 1 SE	3.33	2.68	3.55	4.07	3.58	4.14	4.26	4.97	3.83	3.55	3.85	3.58	45.39
TOMS RIVER	3.55	3.42	4.28	3.95	3.61	3.41	4.65	4.98	3.78	3.91	3.92	4.22	47.68
TRENTON WSO R	3.09	2.92	4.03	3.35	3.37	3.21	4.60	4.47	3.66	2.96	3.25	3.49	42.40
WANAQUE RAYMOND DAM	3.55	3.18	4.37	4.03	3.66	3.92	3.93	4.41	4.24	3.71	4.07	4.07	47.14
WOODCLIFF LAKE	3.46	3.22	4.49	4.06	3.68	3.53	4.26	4.77	4.29	3.55	4.34	3.92	47.57
WOODSTOWN	3.31	2.99	3.77	3.60	3.18	3.37	4.03	4.18	3.63	3.27	3.74	3.74	42.81

NEW JERSEY

HEATING DEGREE DAY NORMALS (BASE 65 DEG F)

STATION		JUL	AUG	SEP	OCT	NOV	DEC	JAN	FEB	MAR	APR	MAY	JUN	ANN
ATLANTIC CITY WSO		0	0	27	298	582	905	1029	890	744	420	165	26	5086
ATLANTIC CITY MARINA		0	0	32	246	519	825	958	829	722	420	189	23	4763
BELLEPLAIN ST FOREST		0	0	36	276	552	871	995	854	698	378	122	10	4792
BELVIDERE		12	12	102	378	681	1048	1169	1000	822	459	204	44	5931
BOONTON 1 SE		0	8	96	381	666	1032	1166	1016	843	471	215	23	5917
CANOE BROOK		0	0	83	370	663	1032	1169	1011	837	477	211	24	5877
CAPE MAY 1 NW		0	0	16	202	477	803	946	834	707	396	150	10	4541
CHARLOTTEBURG		7	16	137	437	717	1091	1221	1078	921	540	276	60	6501
ESSEX FELLS SERV BLDG		0	6	89	359	651	1032	1166	1000	834	465	200	19	5821
FLEMINGTON 1 NE		0	0	81	365	663	1020	1150	991	818	465	209	21	5783
FREEHOLD		0	0	53	306	588	942	1070	924	772	426	166	11	5258
GLASSBORO		0	0	37	286	570	911	1035	893	722	378	138	6	4976
HAMMONTON 2 NNE		0	0	38	291	564	911	1039	896	719	384	140	9	4991
HIGHTSTOWN 1 N		0	0	54	311	597	942	1073	918	760	414	157	11	5237
INDIAN MILLS 2 W		0	0	58	321	606	939	1057	904	750	414	160	7	5216
JERSEY CITY		0	0	59	301	585	946	1066	921	778	435	174	20	5285
LAMBERTVILLE		0	0	44	300	606	955	1076	916	750	402	144	6	5199
LITTLE FALLS		0	0	63	323	606	970	1101	949	781	423	171	11	5398
LONG BRANCH 2 S		0	0	47	286	567	902	1035	899	772	447	186	17	5158
LONG VALLEY		12	12	125	426	711	1070	1190	1030	877	510	253	61	6277
MILLVILLE FAA AIRPORT		0	0	32	284	570	899	1029	882	722	387	134	6	4945
MOORESTOWN		0	0	65	311	606	936	1057	899	732	393	162	10	5171
MORRIS PLAINS 1 W		0	0	95	379	672	1035	1166	1008	843	477	221	33	5929
NEWARK WSO R		0	0	36	254	555	915	1045	902	738	387	140	0	4972
NEW BRUNSWICK		0	0	45	304	594	952	1079	924	763	411	156	11	5239
NEWTON		8	12	134	450	738	1113	1256	1086	899	516	246	46	6504
PEMBERTON 3 E		0	0	45	294	576	918	1042	890	732	405	154	8	5064
PLAINFIELD		0	0	49	305	609	958	1082	918	756	408	159	9	5253
SHILOH		0	0	27	250	546	874	1004	860	698	366	126	6	4757
SOMERVILLE 3 NW		0	0	72	349	645	1001	1128	972	803	444	178	19	5611
SUSSEX 1 SE		11	16	139	446	732	1125	1268	1095	921	531	264	50	6598
TRENTON WSO R		0	0	39	272	561	902	1035	888	732	381	135	5	4950

NEW JERSEY

COOLING DEGREE DAY NORMALS (BASE 65 DEG F)

STATION		JAN	FEB	MAR	APR	MAY	JUN	JUL	AUG	SEP	OCT	NOV	DEC	ANN
ATLANTIC CITY WSO		0	0	0	0	26	152	291	248	75	0	0	0	792
ATLANTIC CITY MARINA		0	0	0	0	19	125	282	273	116	17	0	0	832
BELLEPLAIN ST FOREST		0	0	0	0	35	166	313	276	111	12	0	0	913
BELVIDERE		0	0	0	0	33	131	238	192	69	9	0	0	672
BOONTON 1 SE		0	0	0	0	19	101	232	191	54	0	0	0	597
CANOE BROOK		0	0	0	0	19	117	254	211	68	0	0	0	669
CAPE MAY 1 NW		0	0	0	0	20	154	316	307	148	26	0	0	971
CHARLOTTEBURG		0	0	0	0	13	66	162	122	29	0	0	0	392
ESSEX FELLS SERV BLDG		0	0	0	0	26	121	254	207	65	6	0	0	679
FLEMINGTON 1 NE		0	0	0	0	29	129	272	226	72	5	0	0	733
FREEHOLD		0	0	0	0	30	146	285	245	89	8	0	0	803
GLASSBORO		0	0	0	0	42	180	329	291	118	11	0	0	971
HAMMONTON 2 NNE		0	0	0	0	41	186	335	295	113	12	0	0	982
HIGHTSTOWN 1 N		0	0	0	0	27	155	291	251	87	7	0	0	818
INDIAN MILLS 2 W		0	0	0	0	33	145	285	248	88	7	0	0	806
JERSEY CITY		0	0	0	0	31	152	298	260	98	15	0	0	854
LAMBERTVILLE		0	0	0	0	42	177	316	276	104	9	0	0	924
LITTLE FALLS		0	0	0	0	35	155	304	254	90	7	0	0	845
LONG BRANCH 2 S		0	0	0	0	13	128	279	248	95	7	0	0	770
LONG VALLEY		0	0	0	0	17	91	189	143	38	0	0	0	478
MILLVILLE FAA AIRPORT		0	0	0	0	38	180	335	298	119	11	0	0	981
MOORESTOWN		0	0	0	0	54	166	307	270	113	11	0	0	921
MORRIS PLAINS 1 W		0	0	0	0	22	108	233	187	56	0	0	0	606
NEWARK WSO	R	0	0	0	0	56	199	366	326	132	12	0	0	1091
NEW BRUNSWICK		0	0	0	0	28	152	301	257	87	7	0	0	832
NEWTON		0	0	0	0	14	91	197	145	35	0	0	0	482
PEMBERTON 3 E		0	0	0	0	33	152	288	257	96	9	0	0	835
PLAINFIELD		0	0	0	0	38	156	304	264	94	7	0	0	863
SHILOH		0	0	0	0	46	183	332	295	126	12	0	0	994
SOMERVILLE 3 NW		0	0	0	0	26	145	282	236	84	8	0	0	781
SUSSEX 1 SE		0	0	0	0	22	83	188	143	37	6	0	0	479
TRENTON WSO	R	0	0	0	0	42	188	338	295	111	9	0	0	983

STATE–STATION NUMBER	STN TYP	NAME		LATITUDE DEG–MIN	LONGITUDE DEG–MIN	ELEVATION (FT)
28-0311	13	ATLANTIC CITY WSO		N 3927	W 07434	64
28-0325	13	ATLANTIC CITY MARINA		N 3923	W 07426	11
28-0346	12	AUDUBON		N 3953	W 07505	40
28-0690	13	BELLEPLAIN ST FOREST		N 3915	W 07452	30
28-0729	13	BELVIDERE		N 4050	W 07505	275
28-0907	13	BOONTON 1 SE		N 4054	W 07424	280
28-0978	12	BRANCHVILLE		N 4109	W 07445	580
28-1335	13	CANOE BROOK		N 4045	W 07421	180
28-1351	13	CAPE MAY 1 NW		N 3857	W 07456	17
28-1582	13	CHARLOTTEBURG		N 4102	W 07426	760
28-2768	13	ESSEX FELLS SERV BLDG		N 4050	W 07417	340
28-3029	13	FLEMINGTON 1 NE		N 4031	W 07451	140
28-3181	13	FREEHOLD		N 4016	W 07415	194
28-3291	13	GLASSBORO		N 3942	W 07507	135
28-3516	12	GREENWOOD LAKE		N 4108	W 07420	470
28-3662	13	HAMMONTON 2 NNE		N 3939	W 07448	85
28-3951	13	HIGHTSTOWN 1 N		N 4017	W 07431	100
28-4229	13	INDIAN MILLS 2 W		N 3948	W 07447	100
28-4339	13	JERSEY CITY		N 4044	W 07403	135
28-4635	13	LAMBERTVILLE		N 4022	W 07457	60
28-4887	13	LITTLE FALLS		N 4053	W 07414	150
28-4987	13	LONG BRANCH 2 S		N 4019	W 07401	15
28-5003	13	LONG VALLEY		N 4047	W 07447	550
28-5346	12	MAYS LANDING 1 W		N 3927	W 07445	20
28-5503	12	MIDLAND PARK		N 4059	W 07409	210
28-5581	13	MILLVILLE FAA AIRPORT		N 3922	W 07504	68
28-5728	13	MOORESTOWN		N 3958	W 07458	55
28-5769	13	MORRIS PLAINS 1 W		N 4050	W 07430	400
28-6026	13	NEWARK WSO	R	N 4042	W 07410	11
28-6055	13	NEW BRUNSWICK		N 4029	W 07426	125
28-6146	12	NEW MILFORD		N 4057	W 07402	12
28-6177	13	NEWTON		N 4103	W 07445	565
28-6460	12	OAK RIDGE RESERVOIR		N 4102	W 07430	880
28-6843	13	PEMBERTON 3 E		N 3958	W 07438	80
28-7079	13	PLAINFIELD		N 4036	W 07424	90
28-7328	12	PRINCETON WATER WORKS		N 4020	W 07440	60
28-7393	12	RAHWAY		N 4036	W 07416	20
28-7587	12	RINGWOOD		N 4108	W 07416	305
28-8051	13	SHILOH		N 3928	W 07518	120
28-8194	13	SOMERVILLE 3 NW		N 4036	W 07438	160
28-8402	12	SPLIT ROCK POND		N 4058	W 07428	800
28-8644	13	SUSSEX 1 SE		N 4112	W 07436	390
28-8816	12	TOMS RIVER		N 3957	W 07413	10
28-8883	13	TRENTON WSO	R	N 4013	W 07446	56
28-9187	12	WANAQUE RAYMOND DAM		N 4103	W 07418	320
28-9832	12	WOODCLIFF LAKE		N 4101	W 07403	103
28-9910	12	WOODSTOWN		N 3939	W 07519	50

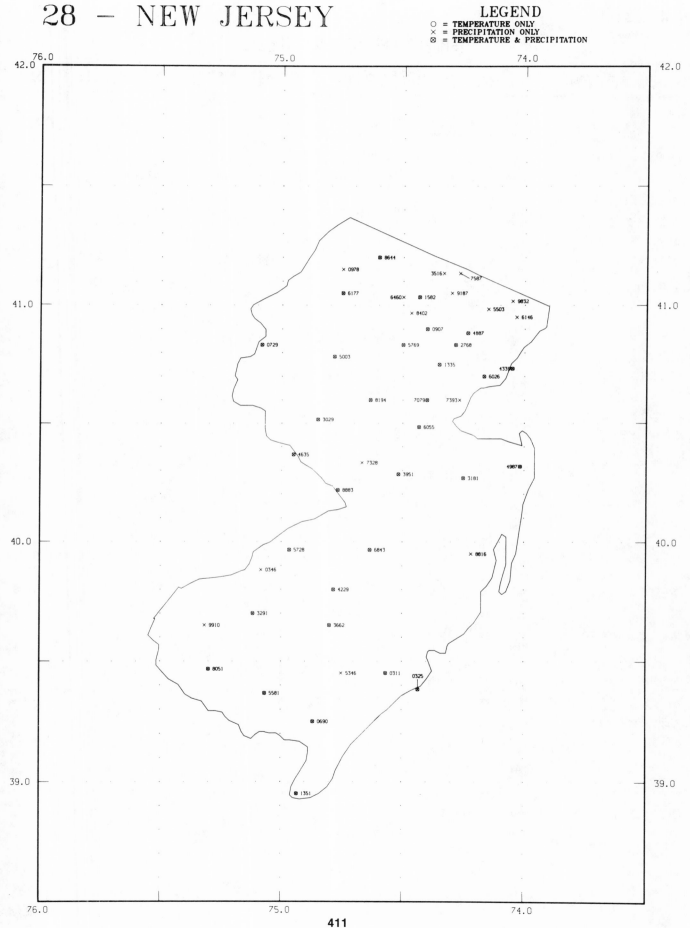

NEW MEXICO

TEMPERATURE NORMALS (DEG F)

STATION		JAN	FEB	MAR	APR	MAY	JUN	JUL	AUG	SEP	OCT	NOV	DEC	ANN
ALAMOGORDO	MAX	57.0	61.4	67.8	77.7	86.4	95.9	95.2	92.7	87.6	77.9	65.6	57.7	76.9
	MIN	27.9	31.0	36.7	44.1	52.3	61.8	65.3	63.7	57.5	46.1	34.1	28.1	45.7
	MEAN	42.5	46.2	52.3	60.9	69.4	78.9	80.3	78.2	72.6	62.1	49.9	42.9	61.4
ALBUQUERQUE WSO //R	MAX	47.2	52.9	60.7	70.6	79.9	90.6	92.8	89.4	83.0	71.7	57.2	48.0	70.3
	MIN	22.3	25.9	31.7	39.5	48.6	58.4	64.7	62.8	54.9	43.1	30.7	23.2	42.1
	MEAN	34.8	39.4	46.2	55.1	64.3	74.5	78.8	76.1	69.0	57.4	44.0	35.6	56.2
AMISTAD 1 SSW	MAX	50.7	55.2	62.2	71.9	80.0	89.5	92.4	90.1	83.4	73.8	59.8	52.8	71.8
	MIN	19.5	23.2	28.1	37.6	47.4	56.9	62.2	60.3	52.5	41.2	28.8	22.4	40.0
	MEAN	35.1	39.2	45.2	54.8	63.8	73.2	77.3	75.2	68.0	57.5	44.3	37.6	55.9
ARTESIA 6 S	MAX	57.1	61.7	68.7	78.1	86.0	94.4	95.1	93.3	87.0	78.0	65.4	58.7	77.0
	MIN	23.7	27.5	34.1	43.3	52.2	61.4	65.5	63.6	56.5	44.3	32.2	24.4	44.1
	MEAN	40.4	44.6	51.4	60.7	69.1	77.9	80.3	78.5	71.7	61.2	48.8	41.6	60.5
AZTEC RUINS NAT MON	MAX	42.3	49.7	57.9	67.8	77.5	87.9	92.6	89.5	82.9	71.0	55.1	43.7	68.2
	MIN	14.7	19.9	24.4	31.0	39.6	48.2	56.7	54.8	46.2	35.7	24.0	15.7	34.2
	MEAN	28.5	34.8	41.1	49.4	58.6	68.1	74.7	72.2	64.6	53.4	39.6	29.7	51.2
BELL RANCH	MAX	53.4	58.2	64.6	73.5	81.5	90.8	92.5	90.3	84.2	75.2	62.5	55.2	73.5
	MIN	18.0	22.4	27.6	37.9	48.0	57.6	63.8	61.9	53.6	40.5	27.8	19.7	39.9
	MEAN	35.7	40.4	46.2	55.7	64.8	74.2	78.2	76.1	68.9	57.9	45.2	37.5	56.7
BERNALILLO	MAX	49.2	54.9	62.6	72.1	81.2	91.2	94.0	90.8	84.6	73.8	59.1	49.8	71.9
	MIN	18.7	22.5	28.0	34.8	42.9	51.3	59.5	57.2	48.5	37.2	26.2	19.2	37.2
	MEAN	34.0	38.7	45.3	53.5	62.0	71.3	76.8	74.0	66.6	55.5	42.7	34.5	54.6
BITTER LAKES WL REF	MAX	57.7	62.6	70.3	79.5	87.4	95.8	96.0	93.9	87.4	78.7	65.6	59.1	77.8
	MIN	20.5	24.8	30.9	40.2	48.8	58.5	63.2	60.9	53.5	40.4	28.2	21.0	40.9
	MEAN	39.2	43.7	50.6	59.9	68.2	77.2	79.6	77.4	70.4	59.6	46.9	40.1	59.4
BLOOMFIELD 3 SE	MAX	40.7	48.1	56.3	66.6	77.0	88.5	92.7	89.3	82.3	70.0	53.8	42.7	67.3
	MIN	17.8	23.0	28.4	35.5	44.5	53.5	60.1	58.3	50.6	39.3	27.5	18.8	38.1
	MEAN	29.3	35.6	42.4	51.1	60.8	71.0	76.4	73.8	66.5	54.6	40.7	30.8	52.8
BOSQUE DEL APACHE	MAX	54.4	60.4	67.4	76.6	85.1	94.1	95.2	92.4	86.8	77.5	64.3	54.2	75.7
	MIN	20.7	24.2	30.3	37.4	45.5	54.2	61.3	59.3	50.9	39.1	27.1	21.0	39.3
	MEAN	37.6	42.3	48.9	57.0	65.3	74.2	78.3	75.9	68.9	58.3	45.7	37.6	57.5
CABALLO DAM	MAX	56.5	61.4	67.6	76.6	85.0	94.7	95.4	92.8	87.2	77.9	65.6	56.5	76.4
	MIN	25.7	29.0	34.6	41.9	49.8	59.5	65.9	63.7	56.7	44.5	33.2	26.3	44.2
	MEAN	41.1	45.2	51.1	59.3	67.4	77.1	80.7	78.3	72.0	61.2	49.4	41.4	60.4
CAMERON	MAX	50.6	54.6	62.1	71.2	79.5	88.6	90.5	88.5	82.0	72.4	58.9	52.0	70.9
	MIN	20.8	24.2	29.2	37.9	47.3	56.7	61.3	59.6	53.1	41.5	29.8	23.3	40.4
	MEAN	35.7	39.4	45.6	54.6	63.5	72.6	75.9	74.0	67.6	57.0	44.4	37.7	55.7
CARLSBAD FAA AIRPORT	MAX	57.1	61.9	69.4	79.1	87.4	95.5	95.6	93.6	86.9	77.5	64.8	58.3	77.3
	MIN	29.2	32.8	39.1	48.2	57.0	65.6	69.3	67.6	60.9	48.8	36.7	30.3	48.8
	MEAN	43.2	47.4	54.3	63.7	72.2	80.6	82.5	80.6	73.9	63.2	50.8	44.3	63.1
CARLSBAD CAVERNS	MAX	57.0	60.5	66.7	75.7	83.2	90.8	90.8	89.1	83.2	74.9	64.0	58.6	74.5
	MIN	33.7	35.9	41.8	49.8	57.4	64.6	66.6	65.9	60.8	52.6	41.6	36.0	50.6
	MEAN	45.4	48.3	54.3	62.8	70.3	77.7	78.7	77.5	72.1	63.8	52.8	47.3	62.6
CARRIZOZO	MAX	51.0	55.1	61.8	71.3	80.2	90.1	91.1	88.1	82.4	72.8	60.0	52.2	71.3
	MIN	22.0	24.8	30.9	38.3	46.5	55.8	60.9	58.6	51.9	40.3	28.7	22.2	40.1
	MEAN	36.6	40.0	46.4	54.8	63.4	73.0	76.0	73.4	67.2	56.6	44.4	37.2	55.8
CERRO 4 NE	MAX	36.4	40.9	48.5	58.9	68.5	78.8	82.8	80.1	74.6	64.3	48.8	38.8	60.1
	MIN	7.9	13.0	20.7	27.4	35.1	43.5	49.5	47.7	41.1	30.9	19.1	9.6	28.8
	MEAN	22.2	27.0	34.7	43.2	51.8	61.2	66.2	63.9	57.9	47.6	34.0	24.3	44.5
CHACO CANYON NAT MON	MAX	42.5	48.4	56.0	66.0	75.7	86.7	90.9	87.7	80.9	69.5	54.2	43.6	66.8
	MIN	12.3	17.0	21.6	27.9	36.6	45.9	54.4	52.6	43.6	31.5	20.4	11.8	31.3
	MEAN	27.4	32.7	38.8	47.0	56.2	66.3	72.7	70.2	62.3	50.6	37.4	27.8	49.1

NEW MEXICO

TEMPERATURE NORMALS (DEG F)

STATION		JAN	FEB	MAR	APR	MAY	JUN	JUL	AUG	SEP	OCT	NOV	DEC	ANN
CHAMA	MAX	36.8	40.3	45.9	56.2	66.0	77.0	80.8	77.7	72.2	62.6	47.8	39.1	58.5
	MIN	6.8	10.1	15.8	23.5	31.8	39.2	47.0	45.6	38.5	29.1	17.8	9.6	26.2
	MEAN	21.9	25.3	30.9	39.9	48.9	58.1	64.0	61.7	55.3	45.9	32.8	24.4	42.4
CIMARRON	MAX	46.2	49.7	54.6	63.1	71.7	81.4	83.7	81.1	76.3	68.2	54.8	48.5	64.9
	MIN	17.7	20.2	24.7	32.1	40.8	49.0	54.1	52.4	45.7	35.6	24.9	19.4	34.7
	MEAN	32.0	35.0	39.7	47.6	56.3	65.2	68.9	66.8	61.0	52.0	39.8	34.0	49.9
CLAYTON WSO R	MAX	47.4	50.6	56.3	65.9	74.5	84.4	87.8	85.6	78.4	69.2	55.8	49.6	67.1
	MIN	18.6	22.1	26.4	36.0	45.8	55.3	60.5	58.9	51.2	40.1	27.7	21.5	38.7
	MEAN	33.0	36.4	41.4	51.0	60.1	69.9	74.2	72.3	64.8	54.7	41.7	35.5	52.9
CLIFF 10 SE	MAX	55.5	60.0	65.3	74.0	82.8	92.9	93.6	90.5	86.3	76.9	64.8	56.4	74.9
	MIN	21.2	23.3	28.0	33.8	41.2	51.1	59.9	57.9	50.3	38.4	27.0	21.3	37.8
	MEAN	38.4	41.7	46.7	53.9	62.0	72.0	76.8	74.2	68.3	57.6	45.9	38.9	56.4
CLOVIS	MAX	51.3	55.3	62.2	72.1	80.2	89.1	91.1	89.1	82.6	73.2	60.4	53.5	71.7
	MIN	22.6	26.1	31.5	40.6	49.6	58.7	63.2	61.3	54.1	43.1	31.7	25.1	42.3
	MEAN	37.0	40.7	46.9	56.3	65.0	73.9	77.2	75.2	68.4	58.2	46.1	39.3	57.0
CLOVIS 13 N	MAX	52.9	57.2	64.3	73.4	81.3	90.0	91.3	89.3	83.5	74.4	61.0	54.4	72.8
	MIN	21.8	24.8	29.6	38.5	47.4	56.9	61.3	59.7	52.7	41.9	30.6	24.4	40.8
	MEAN	37.4	41.0	47.0	56.0	64.4	73.5	76.3	74.5	68.1	58.2	45.8	39.5	56.8
COLUMBUS	MAX	58.9	63.6	70.3	79.0	87.2	96.2	95.6	93.0	88.0	79.0	66.8	58.7	78.0
	MIN	27.4	30.3	36.3	43.5	51.7	61.9	66.5	64.3	57.6	45.7	33.8	27.4	45.5
	MEAN	43.2	47.0	53.3	61.3	69.5	79.1	81.0	78.7	72.9	62.4	50.3	43.1	61.8
CONCHAS DAM	MAX	53.4	58.0	63.9	72.8	81.3	91.1	93.4	91.4	85.0	75.7	62.3	55.1	73.6
	MIN	24.2	28.0	33.8	43.0	52.3	61.7	66.3	64.4	57.3	46.0	34.1	26.8	44.8
	MEAN	38.8	43.1	48.8	57.9	66.8	76.4	79.8	77.9	71.2	60.9	48.2	40.9	59.2
CUBA	MAX	40.9	44.5	50.7	60.8	70.8	82.1	86.2	82.8	77.3	67.2	52.8	43.9	63.3
	MIN	10.0	13.9	20.0	26.5	34.9	43.1	51.3	49.5	41.2	29.5	18.8	10.4	29.1
	MEAN	25.5	29.2	35.4	43.7	52.8	62.6	68.8	66.2	59.3	48.4	35.8	27.1	46.2
DES MOINES	MAX	45.3	47.9	53.3	62.3	70.8	80.7	84.0	82.0	76.1	66.7	53.3	47.1	64.1
	MIN	15.8	18.8	22.9	31.7	41.5	51.1	56.5	54.7	47.4	35.9	24.2	18.3	34.9
	MEAN	30.6	33.4	38.1	47.0	56.2	65.9	70.3	68.3	61.8	51.3	38.8	32.7	49.5
EAGLE NEST	MAX	37.3	40.9	46.2	55.1	64.3	74.7	77.8	75.1	70.3	62.0	48.0	39.2	57.6
	MIN	2.0	6.1	13.7	21.5	28.3	34.6	40.8	40.1	31.6	21.4	12.0	3.6	21.3
	MEAN	19.7	23.5	30.0	38.3	46.3	54.6	59.3	57.6	51.0	41.7	30.0	21.5	39.5
EL MORRO NAT MON	MAX	42.5	46.2	52.4	62.4	72.0	82.9	85.8	82.0	77.2	67.1	53.2	44.6	64.0
	MIN	13.0	16.5	22.1	27.6	34.6	43.1	51.6	49.7	42.5	31.7	21.3	13.8	30.6
	MEAN	27.8	31.4	37.3	45.0	53.3	63.0	68.7	65.8	59.8	49.4	37.3	29.2	47.3
EL VADO DAM	MAX	38.8	43.3	49.2	59.5	69.5	80.5	85.5	82.3	76.0	66.0	51.3	41.1	61.9
	MIN	3.3	10.2	18.7	25.2	32.8	40.0	48.7	47.3	38.0	27.5	17.2	7.9	26.4
	MEAN	21.1	26.8	34.0	42.4	51.2	60.3	67.1	64.8	57.0	46.8	34.2	24.5	44.2
ELEPHANT BUTTE DAM	MAX	55.1	60.3	66.8	75.3	83.2	92.8	93.4	90.7	85.6	76.4	63.9	55.0	74.9
	MIN	28.5	31.9	37.4	44.9	53.3	63.1	67.6	65.4	59.3	48.1	36.0	28.7	47.0
	MEAN	41.8	46.1	52.1	60.1	68.3	78.0	80.5	78.1	72.5	62.3	50.0	41.9	61.0
ELIDA	MAX	51.8	55.7	62.8	72.5	80.9	90.0	91.5	89.6	83.1	73.7	60.8	53.9	72.2
	MIN	22.8	25.8	31.6	40.7	49.8	59.1	63.0	61.5	54.7	43.3	31.4	24.7	42.4
	MEAN	37.3	40.8	47.2	56.6	65.4	74.6	77.3	75.6	68.9	58.5	46.1	39.3	57.3
ELK 3 E	MAX	55.5	57.6	62.8	70.3	77.6	84.9	84.3	82.6	78.0	71.6	62.4	57.0	70.4
	MIN	21.3	23.2	27.9	35.0	42.8	51.3	56.0	54.4	48.1	37.5	26.9	22.3	37.2
	MEAN	38.4	40.4	45.4	52.6	60.2	68.1	70.2	68.5	63.0	54.6	44.7	39.7	53.8
ESTANCIA	MAX	46.9	51.3	58.3	67.3	76.7	87.0	89.5	86.0	80.2	70.5	57.0	48.4	68.3
	MIN	16.5	19.8	24.5	30.7	38.6	46.6	53.4	51.6	43.8	32.6	22.3	16.0	33.0
	MEAN	31.7	35.5	41.4	49.0	57.7	66.9	71.5	68.8	62.0	51.6	39.7	32.2	50.7

413

NEW MEXICO

TEMPERATURE NORMALS (DEG F)

STATION		JAN	FEB	MAR	APR	MAY	JUN	JUL	AUG	SEP	OCT	NOV	DEC	ANN
FLORIDA	MAX	57.5	62.2	68.4	77.1	85.6	95.4	95.5	92.7	87.5	78.3	65.9	57.7	77.0
	MIN	23.8	25.6	30.2	37.1	45.1	55.6	63.2	60.9	53.6	42.0	30.4	23.7	40.9
	MEAN	40.6	43.9	49.3	57.2	65.4	75.5	79.4	76.8	70.6	60.2	48.2	40.7	59.0
FORT BAYARD	MAX	51.5	55.0	59.9	68.5	77.1	86.7	87.3	84.3	80.2	71.1	59.7	52.7	69.5
	MIN	24.3	26.0	29.6	36.0	43.4	53.3	58.5	56.2	50.6	41.1	30.3	24.8	39.5
	MEAN	37.9	40.5	44.8	52.3	60.3	70.0	72.9	70.2	65.4	56.1	45.0	38.8	54.5
FORT SUMNER	MAX	54.1	58.5	65.6	74.5	82.9	91.6	92.8	90.7	84.3	75.1	62.7	55.5	74.0
	MIN	22.3	26.1	31.5	40.6	49.5	59.0	63.7	61.7	53.8	41.5	30.7	24.3	42.1
	MEAN	38.2	42.3	48.5	57.6	66.2	75.3	78.3	76.2	69.1	58.4	46.7	39.9	58.1
FRUITLAND 2 E	MAX	41.8	50.2	58.3	68.0	77.9	88.8	93.3	90.1	83.4	72.0	55.5	43.7	68.6
	MIN	15.5	20.9	26.3	32.7	41.4	50.1	57.3	56.1	46.8	35.9	25.0	16.6	35.4
	MEAN	28.7	35.6	42.3	50.4	59.7	69.5	75.3	73.1	65.1	54.0	40 3	30.1	52.0
GALLUP 5 E	MAX	43.8	48.8	54.5	64.7	74.0	85.1	88.5	85.0	79.6	69.1	54.7	45.8	66.1
	MIN	13.7	18.2	23.3	29.2	36.9	45.0	53.3	52.3	44.4	33.2	21.9	14.4	32.2
	MEAN	28.8	33.6	38.9	46.9	55.5	65.1	70.9	68.7	62.0	51.2	38.3	30.2	49.2
GRAN QUIVIRA NAT MON	MAX	47.6	51.7	58.5	67.7	76.8	86.6	87.9	84.8	79.6	70.2	57.0	49.0	68.1
	MIN	21.6	23.6	27.9	34.7	42.8	52.1	56.6	55.0	48.8	38.9	27.8	22.0	37.7
	MEAN	34.6	37.7	43.2	51.3	59.8	69.4	72.2	69.9	64.2	54.6	42.4	35.5	52.9
GRENVILLE	MAX	45.9	48.9	54.1	63.5	71.9	81.7	85.9	84.2	77.4	68.0	55.1	48.4	65.4
	MIN	16.4	19.5	23.1	32.3	42.0	50.9	56.1	54.5	47.2	36.5	25.4	19.1	35.3
	MEAN	31.2	34.2	38.6	48.0	57.0	66.3	71.0	69.4	62.4	52.3	40.3	33.8	50.4
HATCH 2 W	MAX	58.6	63.4	69.4	77.9	86.1	95.2	95.4	92.6	87.7	79.2	67.2	58.8	77.6
	MIN	23.0	26.4	32.9	40.8	48.1	57.1	63.5	61.2	53.7	40.9	29.7	23.0	41.7
	MEAN	40.8	44.9	51.2	59.4	67.1	76.2	79.5	76.9	70.7	60.1	48.5	40.9	59.7
HOBBS	MAX	57.7	61.8	68.8	77.9	85.1	92.7	93.3	92.0	86.4	77.8	65.3	59.5	76.5
	MIN	28.0	31.4	37.7	47.0	55.4	63.7	66.9	65.5	59.1	48.5	36.8	30.3	47.5
	MEAN	42.9	46.6	53.3	62.5	70.3	78.2	80.1	78.8	72.8	63.2	51.1	44.9	62.1
JAL	MAX	60.2	65.5	72.8	82.1	89.6	96.8	97.5	96.0	89.6	80.9	67.9	61.5	80.0
	MIN	27.4	31.5	38.6	48.1	56.6	65.2	67.8	66.2	60.2	48.4	36.3	29.3	48.0
	MEAN	43.8	48.5	55.8	65.1	73.1	81.0	82.7	81.1	74.9	64.7	52.1	45.4	64.0
JEMEZ SPRINGS	MAX	46.0	50.1	56.2	65.2	74.4	84.7	87.5	84.4	79.5	69.6	55.9	47.4	66.7
	MIN	19.9	23.0	28.0	34.5	42.4	50.9	56.8	54.9	48.6	39.1	28.4	21.3	37.3
	MEAN	32.9	36.6	42.1	49.8	58.4	67.8	72.2	69.7	64.0	54.3	42.2	34.3	52.0
JORNADA EXP RANGE	MAX	57.7	61.9	68.8	77.3	85.6	95.4	96.2	93.2	87.8	78.5	66.3	57.6	77.2
	MIN	20.5	23.6	29.2	36.8	44.6	55.2	62.0	59.7	52.4	39.6	27.3	20.6	39.3
	MEAN	39.1	42.8	49.0	57.1	65.2	75.3	79.1	76.5	70.1	59.1	46.8	39.1	58.3
LAGUNA	MAX	47.7	52.4	58.9	68.5	77.6	88.2	90.6	87.3	81.5	71.4	57.7	48.9	69.2
	MIN	19.1	22.3	27.2	34.0	43.0	52.5	58.9	56.9	48.9	37.6	26.4	19.2	37.2
	MEAN	33.4	37.4	43.1	51.2	60.3	70.4	74.7	72.1	65.3	54.6	42.1	34.0	53.2
LAKE MALOYA	MAX	42.4	44.6	49.3	58.1	66.6	76.1	79.0	77.2	72.0	62.9	49.7	44.1	60.2
	MIN	9.5	12.0	17.1	26.0	35.0	42.8	48.2	46.8	40.1	30.8	20.2	13.1	28.5
	MEAN	26.0	28.3	33.3	42.1	50.8	59.5	63.6	62.0	56.1	46.9	35.0	28.6	44.4
LAS VEGAS FAA AIRPORT	MAX	45.1	48.1	53.3	61.9	70.9	80.8	83.0	80.3	74.8	65.8	53.6	46.7	63.7
	MIN	17.4	19.8	23.9	31.3	40.3	49.1	54.0	52.3	45.9	35.7	24.8	19.0	34.5
	MEAN	31.2	34.0	38.6	46.6	55.6	65.0	68.5	66.4	60.4	50.8	39.2	32.9	49.1
LORDSBURG 4 SE	MAX	58.3	63.3	69.2	78.6	86.9	96.3	97.0	94.1	89.4	79.7	66.9	59.0	78.2
	MIN	25.7	27.9	33.6	39.7	47.8	58.6	65.4	62.8	56.6	44.2	31.8	25.3	43.3
	MEAN	42.0	45.7	51.4	59.2	67.4	77.5	81.2	78.5	73.0	62.0	49.4	42.1	60.8
LOS ALAMOS	MAX	39.8	43.1	48.9	57.9	67.3	77.9	80.6	77.5	72.3	62.1	48.9	41.4	59.8
	MIN	18.5	21.5	26.4	33.6	42.7	52.2	56.0	54.2	48.2	38.5	27.1	20.1	36.6
	MEAN	29.2	32.3	37.7	45.8	55.0	65.1	68.3	65.9	60.3	50.4	38.0	30.8	48.2

NEW MEXICO

TEMPERATURE NORMALS (DEG F)

STATION		JAN	FEB	MAR	APR	MAY	JUN	JUL	AUG	SEP	OCT	NOV	DEC	ANN
MALJAMAR 4 SE	MAX	55.9	60.8	67.5	77.1	84.8	92.4	93.2	91.5	85.4	76.6	64.1	57.6	75.6
	MIN	25.9	28.9	35.0	43.1	51.8	60.2	63.9	62.5	55.9	45.0	33.2	27.2	44.4
	MEAN	40.9	44.9	51.3	60.1	68.3	76.3	78.6	77.0	70.7	60.8	48.7	42.4	60.0
MC GAFFEY 4 SE	MAX	39.2	42.0	46.9	55.8	66.2	77.7	81.3	78.1	73.1	63.2	49.7	41.9	59.6
	MIN	9.5	12.0	17.9	24.3	31.4	39.0	46.6	45.0	37.8	28.2	18.2	11.1	26.8
	MEAN	24.3	27.0	32.4	40.1	48.8	58.4	63.9	61.6	55.5	45.7	34.0	26.5	43.2
MELROSE	MAX	52.4	56.6	63.6	73.2	81.6	90.1	91.5	89.5	83.2	73.6	60.5	53.8	72.5
	MIN	21.7	24.9	30.6	39.2	48.5	57.9	62.6	60.8	53.6	42.3	30.7	24.1	41.4
	MEAN	37.1	40.8	47.1	56.2	65.1	74.0	77.1	75.2	68.4	58.0	45.6	39.0	57.0
MOUNTAIN PARK	MAX	49.3	51.7	57.5	65.8	73.7	82.5	81.4	78.4	75.0	68.0	57.5	51.2	66.0
	MIN	25.0	26.3	30.0	36.9	44.1	53.7	56.1	54.8	50.0	41.1	30.7	26.2	39.6
	MEAN	37.2	39.0	43.8	51.4	58.9	68.2	68.8	66.7	62.5	54.6	44.1	38.8	52.8
MOUNTAINAIR	MAX	45.9	50.5	57.3	66.5	75.4	85.5	87.5	84.4	79.0	69.5	55.6	47.0	67.0
	MIN	18.9	22.0	26.1	32.3	40.8	49.7	55.0	53.1	46.6	36.1	25.4	19.5	35.5
	MEAN	32.4	36.3	41.7	49.4	58.1	67.6	71.3	68.8	62.8	52.8	40.5	33.2	51.2
PASAMONTE	MAX	46.8	50.1	56.1	66.1	74.7	84.6	88.8	86.3	79.4	69.7	55.7	48.9	67.3
	MIN	15.4	18.7	23.5	32.3	42.1	51.8	56.8	54.8	47.1	35.9	24.2	17.6	35.0
	MEAN	31.1	34.4	39.8	49.2	58.5	68.2	72.8	70.6	63.3	52.8	40.0	33.3	51.2
PEARL	MAX	56.8	60.9	67.7	76.7	83.7	91.2	92.3	90.5	84.5	76.6	64.3	58.0	75.3
	MIN	26.1	29.5	35.7	44.9	53.7	61.9	65.4	63.9	57.4	46.9	34.6	28.0	45.7
	MEAN	41.5	45.2	51.8	60.8	68.7	76.6	78.9	77.2	71.0	61.8	49.5	43.0	60.5
PORTALES	MAX	55.0	59.3	66.7	75.6	83.6	91.3	92.7	91.1	85.1	75.8	62.9	56.2	74.6
	MIN	21.7	25.4	31.5	40.1	50.3	59.3	63.5	61.5	54.4	42.1	30.6	23.5	42.0
	MEAN	38.4	42.4	49.1	57.9	67.0	75.3	78.1	76.3	69.8	59.0	46.8	39.9	58.3
QUEMADO RANGER STATION	MAX	47.7	51.9	57.5	66.5	75.2	85.0	86.2	83.7	79.5	70.7	57.8	49.9	67.6
	MIN	12.5	15.8	20.2	25.0	32.4	40.6	50.6	48.5	40.5	28.8	18.4	12.5	28.8
	MEAN	30.2	33.9	38.9	45.7	53.8	62.8	68.4	66.1	60.0	49.8	38.1	31.2	48.2
RED RIVER	MAX	35.8	38.1	43.2	52.6	62.3	72.6	76.1	73.5	68.6	59.0	44.9	37.5	55.4
	MIN	3.7	5.9	13.4	20.7	28.2	34.3	40.6	39.6	33.1	24.4	13.2	5.5	21.9
	MEAN	19.8	22.0	28.4	36.7	45.2	53.5	58.4	56.6	50.8	41.7	29.1	21.5	38.6
ROSWELL WSO	MAX	55.4	60.4	67.7	76.9	85.0	93.1	93.7	91.3	84.9	75.8	63.1	56.7	75.3
	MIN	27.4	31.4	37.9	46.8	55.6	64.8	69.0	67.0	59.6	47.5	35.0	28.2	47.5
	MEAN	41.4	45.9	52.8	61.9	70.3	79.0	81.4	79.2	72.3	61.7	49.1	42.5	61.4
ROY	MAX	47.3	51.2	56.9	65.7	74.4	84.3	86.7	84.8	78.4	69.2	56.1	48.7	67.0
	MIN	18.1	20.9	25.1	33.7	43.8	52.8	57.7	56.1	48.9	37.9	26.3	19.9	36.8
	MEAN	32.8	36.1	41.0	49.7	59.1	68.6	72.2	70.5	63.6	53.6	41.2	34.3	51.9
RUIDOSO 2 NNE	MAX	49.5	51.6	56.8	65.2	73.3	82.0	81.3	79.3	75.3	67.5	57.2	50.7	65.8
	MIN	17.2	18.6	22.5	27.5	33.4	41.3	47.8	46.8	40.4	30.6	21.9	17.7	30.5
	MEAN	33.4	35.1	39.7	46.4	53.4	61.7	64.6	63.1	57.9	49.1	39.6	34.2	48.2
SAN JON	MAX	53.5	57.5	64.9	74.0	82.7	92.1	94.0	92.0	85.4	75.7	61.9	54.9	74.1
	MIN	22.2	25.9	32.0	41.4	51.2	61.1	65.5	63.8	56.1	44.2	31.6	24.6	43.3
	MEAN	37.8	41.7	48.5	57.7	67.0	76.6	79.8	78.0	70.7	59.9	46.8	39.8	58.7
SANTA ROSA	MAX	54.6	59.0	65.0	73.9	82.4	91.6	93.3	91.1	85.0	76.1	63.1	55.9	74.3
	MIN	24.2	26.7	31.8	39.8	48.8	57.9	63.0	60.8	53.5	41.9	31.5	25.7	42.1
	MEAN	39.4	42.9	48.5	56.9	65.7	74.8	78.2	76.0	69.3	59.0	47.3	40.9	58.2
SOCORRO	MAX	52.0	58.1	65.7	74.6	82.4	91.3	92.6	90.0	84.4	74.7	61.2	51.7	73.2
	MIN	21.9	25.4	31.5	39.0	47.0	55.8	62.2	60.0	52.4	40.5	28.5	22.2	40.5
	MEAN	37.0	41.8	48.6	56.8	64.7	73.6	77.4	75.0	68.4	57.7	44.8	37.0	56.9
SPRINGER	MAX	46.9	53.2	59.4	68.0	77.2	87.3	90.0	87.4	81.6	72.2	57.0	48.5	69.1
	MIN	12.2	16.5	22.6	30.6	40.0	48.5	54.3	52.7	44.6	32.8	21.4	13.7	32.5
	MEAN	29.6	34.9	41.1	49.3	58.6	68.0	72.2	70.1	63.1	52.6	39.2	31.1	50.8

NEW MEXICO

TEMPERATURE NORMALS (DEG F)

STATION		JAN	FEB	MAR	APR	MAY	JUN	JUL	AUG	SEP	OCT	NOV	DEC	ANN
STATE UNIVERSITY	MAX	58.4	62.9	69.3	77.8	85.7	94.6	94.6	92.1	87.1	78.5	66.4	58.6	77.2
	MIN	26.5	29.5	34.9	42.3	50.0	60.0	66.2	63.7	56.4	44.0	32.5	26.7	44.4
	MEAN	42.5	46.2	52.1	60.1	67.9	77.3	80.4	77.9	71.8	61.3	49.4	42.7	60.8
TAJIQUE	MAX	43.6	47.1	53.9	63.8	73.5	83.6	84.6	81.5	76.2	66.2	52.9	45.3	64.4
	MIN	19.1	20.9	24.9	30.5	38.6	47.0	52.5	51.1	44.6	35.0	25.7	19.8	34.1
	MEAN	31.4	34.1	39.4	47.2	56.1	65.3	68.6	66.3	60.4	50.6	39.3	32.6	49.3
TOHATCHI 1 ESE	MAX	42.4	48.0	53.8	63.3	72.8	84.5	89.4	85.6	79.5	68.9	53.7	44.1	65.5
	MIN	20.2	23.8	27.5	34.7	44.2	54.4	61.5	58.0	51.5	41.2	29.3	21.3	39.0
	MEAN	31.4	35.9	40.7	49.0	58.5	69.4	75.5	71.8	65.5	55.0	41.5	32.7	52.2
TRUTH OR CONSEQUENCES	MAX	53.9	58.8	65.3	74.3	82.6	92.3	92.6	89.8	84.4	74.8	62.3	53.9	73.8
	MIN	27.2	30.7	36.1	43.7	52.2	61.8	66.0	64.1	57.8	46.9	34.9	27.4	45.7
	MEAN	40.6	44.8	50.7	59.0	67.5	77.1	79.3	76.9	71.1	60.9	48.6	40.7	59.8
TUCUMCARI FAA AP	MAX	51.6	55.8	62.9	72.6	81.2	91.2	92.9	90.6	83.5	73.6	60.6	53.3	72.5
	MIN	22.6	26.6	32.5	42.0	51.5	61.2	65.8	63.8	56.0	44.5	32.3	25.2	43.7
	MEAN	37.1	41.2	47.7	57.3	66.4	76.2	79.4	77.2	69.8	59.1	46.5	39.3	58.1
TUCUMCARI 3 NE	MAX	54.3	58.6	65.5	74.6	82.5	91.8	93.6	91.5	85.5	76.5	62.9	56.0	74.4
	MIN	23.3	27.0	32.4	41.5	50.8	60.1	64.4	62.4	55.4	44.2	32.6	25.7	43.3
	MEAN	38.9	42.8	49.0	58.1	66.7	76.0	79.0	77.0	70.4	60.4	47.8	40.9	58.9
VALMORA	MAX	48.4	51.2	55.8	63.7	72.3	82.0	85.1	82.6	77.4	69.3	57.2	50.8	66.3
	MIN	13.4	17.0	21.9	29.3	38.2	46.2	52.6	50.5	43.0	32.0	21.3	15.0	31.7
	MEAN	30.9	34.1	38.9	46.5	55.3	64.1	68.9	66.6	60.3	50.6	39.3	32.9	49.0
WHITE SANDS NAT MON	MAX	57.4	62.7	70.2	79.4	88.0	97.0	97.3	94.8	89.5	79.3	65.9	57.1	78.2
	MIN	22.1	24.9	31.0	39.4	48.1	58.2	64.0	61.4	54.0	40.6	27.9	21.7	41.1
	MEAN	39.8	43.8	50.6	59.4	68.1	77.6	80.7	78.1	71.8	60.0	47.0	39.4	59.7
WINSTON	MAX	53.4	56.9	61.7	70.1	77.6	87.2	87.8	84.9	80.3	72.1	61.1	54.2	70.6
	MIN	18.2	20.2	24.4	31.0	38.0	47.8	54.9	52.5	45.3	34.8	23.8	18.4	34.1
	MEAN	35.8	38.5	43.1	50.6	57.8	67.5	71.4	68.7	62.8	53.5	42.5	36.3	52.4
WOLF CANYON	MAX	38.8	40.6	45.3	54.9	64.7	75.0	77.9	74.7	70.0	60.7	47.7	41.1	57.6
	MIN	7.7	10.0	16.2	22.4	28.6	34.7	42.3	41.4	33.9	25.0	16.0	9.3	24.0
	MEAN	23.2	25.3	30.8	38.7	46.6	54.9	60.1	58.1	52.0	42.9	31.9	25.2	40.8

NEW MEXICO

PRECIPITATION NORMALS (INCHES)

STATION	JAN	FEB	MAR	APR	MAY	JUN	JUL	AUG	SEP	OCT	NOV	DEC	ANN
AFTON 5 ESE	.35	.35	.27	.14	.27	.45	1.56	1.88	1.48	.86	.42	.44	8.47
ALAMOGORDO	.62	.53	.52	.24	.41	.78	2.18	2.15	1.63	1.13	.43	.56	11.18
ALBUQUERQUE WSO //R	.41	.40	.52	.40	.46	.51	1.30	1.51	.85	.86	.38	.52	8.12
ALEMAN RANCH	.35	.33	.26	.21	.29	.61	1.77	1.87	1.30	.90	.42	.53	8.84
AMISTAD 1 SSW	.29	.29	.66	.89	2.21	1.84	3.13	2.55	1.52	.95	.60	.40	15.33
ANIMAS	.60	.51	.54	.18	.13	.41	2.19	2.47	1.36	.98	.44	.80	10.61
ARTESIA 6 S	.34	.40	.38	.36	.92	1.18	1.54	1.78	1.80	1.26	.39	.32	10.67
AUGUSTINE	.31	.33	.33	.25	.29	.45	2.07	2.36	1.40	1.07	.33	.45	9.64
AZTEC RUINS NAT MON	.95	.63	.69	.64	.50	.29	.90	1.11	.84	1.28	.65	.83	9.31
BEAVERHEAD RANGER STA	.77	.72	.69	.39	.48	.59	2.57	2.84	1.84	1.43	.54	.95	13.81
BELL RANCH	.29	.21	.48	.70	1.31	1.37	2.89	2.58	1.33	1.01	.51	.40	13.08
BERNALILLO	.47	.46	.59	.43	.60	.58	1.52	1.66	.82	1.00	.50	.57	9.20
BITTER LAKES WL REF	.41	.43	.37	.41	.77	1.07	2.07	2.08	1.77	1.16	.39	.37	11.30
BLACK LAKE	.72	.73	1.16	1.14	1.86	1.61	3.50	3.50	1.37	1.22	.90	.67	18.38
BLOOMFIELD 3 SE	.59	.44	.66	.55	.41	.28	.93	1.27	.83	1.16	.64	.61	8.37
BOSQUE DEL APACHE	.25	.28	.34	.28	.46	.65	1.31	1.63	1.32	.99	.32	.45	8.28
BUCKHORN	1.06	.77	.70	.32	.21	.38	2.35	2.43	1.70	1.19	.52	1.07	12.70
CABALLO DAM	.30	.31	.26	.18	.28	.51	1.77	1.94	1.44	.89	.33	.50	8.71
CAMERON	.46	.70	.75	1.00	1.81	2.26	3.19	2.50	1.58	1.27	.65	.57	16.74
CANJILON RANGER STA	1.24	.81	1.05	.97	1.01	.75	2.27	2.46	1.17	1.20	.96	.89	14.78
CANTON	.34	.35	.40	.61	1.20	1.67	2.76	2.22	1.68	1.38	.36	.29	13.26
CARLSBAD	.34	.42	.41	.46	1.22	1.13	1.69	1.87	2.32	1.25	.51	.32	11.94
CARLSBAD FAA AIRPORT	.34	.35	.33	.40	.93	.71	1.70	1.88	2.16	1.16	.44	.26	10.66
CARLSBAD CAVERNS	.45	.41	.39	.65	1.19	1.18	1.94	2.43	2.94	1.43	.44	.41	13.86
CARRIZOZO	.60	.53	.64	.37	.63	.83	2.26	2.57	1.86	1.03	.62	.67	12.61
CERRO 4 NE	.62	.41	.59	.68	.96	.85	1.92	1.90	1.08	1.07	.75	.61	11.44
CHACO CANYON NAT MON	.42	.43	.48	.36	.57	.39	1.10	1.35	1.05	1.12	.58	.58	8.43
CHACON	.97	.79	1.18	1.11	1.67	1.36	3.18	3.85	1.35	1.21	1.02	.88	18.57
CHAMA	1.98	1.34	1.59	1.24	1.10	.79	2.02	2.62	1.70	1.66	1.37	1.60	19.01
CIMARRON	.33	.40	.69	.96	2.05	1.62	2.89	2.68	1.48	1.07	.67	.42	15.26
CLAYTON WSO R	.27	.28	.59	1.05	2.23	1.74	2.53	2.43	1.48	.75	.48	.29	14.12
CLIFF 10 SE	.98	.76	.86	.32	.19	.50	2.67	2.95	1.49	1.21	.56	1.00	13.49
CLOVIS	.44	.49	.59	.83	1.81	2.47	2.79	2.69	1.87	1.45	.58	.47	16.48
CLOVIS 13 N	.34	.42	.55	.75	1.77	2.28	2.94	2.53	1.77	1.30	.50	.42	15.57
COLUMBUS	.41	.44	.32	.14	.24	.25	2.17	1.75	1.47	.79	.48	.50	8.96
CONCHAS DAM	.31	.30	.56	.77	1.10	1.43	2.56	2.26	1.33	1.03	.41	.40	12.46
CROSSROADS 2 NE	.31	.41	.47	.73	1.55	2.02	2.62	2.42	1.98	1.58	.40	.26	14.75
CUBA	.91	.69	.87	.65	.80	.63	2.19	2.32	1.32	1.19	.75	.74	13.06
CURETON RANCH	1.05	.72	.79	.23	.24	.39	2.37	2.68	1.60	1.02	.58	.97	12.64
DES MOINES	.35	.44	.79	1.03	2.40	1.68	3.54	2.78	1.83	1.01	.59	.39	16.83
DILIA	.47	.48	.70	.78	1.15	1.03	2.59	2.72	1.64	1.12	.51	.54	13.73
DULCE	1.56	1.09	1.37	.98	.96	.67	1.72	2.65	1.42	1.49	1.13	1.51	16.55
EAGLE NEST	.72	.50	.77	.75	1.31	1.09	2.90	2.77	1.10	.80	.77	.68	14.16
EL MORRO NAT MON	1.02	.78	.95	.69	.55	.47	1.86	2.56	1.21	1.13	.73	.93	12.88
EL RITO	.75	.51	.74	.63	.96	.63	1.61	2.23	.99	1.07	.77	.56	11.45
EL VADO DAM	1.09	.79	.99	.78	1.00	.62	1.81	2.26	1.42	1.14	.91	.92	13.73
ELEPHANT BUTTE DAM	.31	.30	.27	.21	.33	.54	1.36	1.89	1.28	.92	.39	.49	8.29
ELIDA	.39	.37	.45	.65	1.32	1.59	2.33	2.31	1.89	1.18	.47	.28	13.23
ELK 3 E	.50	.55	.41	.57	.95	1.31	2.67	3.53	2.60	1.32	.57	.60	15.58
ESPANOLA	.48	.36	.48	.48	.83	.66	1.45	1.91	.91	.93	.51	.42	9.42
ESTANCIA	.46	.51	.56	.52	.84	.72	1.95	2.40	1.27	1.16	.50	.65	11.54
FARNSWORTH RANCH	.27	.47	.41	.36	.80	1.47	2.02	2.53	1.77	1.03	.35	.38	11.86
FAYWOOD	.63	.41	.40	.17	.17	.60	2.15	2.38	1.65	1.01	.42	.75	10.74
FLORIDA	.50	.48	.32	.21	.21	.41	2.16	2.12	1.53	.95	.39	.57	9.85
FLYING H	.44	.43	.36	.43	.94	1.17	2.28	2.69	2.47	1.32	.52	.49	13.54

417

NEW MEXICO

PRECIPITATION NORMALS (INCHES)

STATION	JAN	FEB	MAR	APR	MAY	JUN	JUL	AUG	SEP	OCT	NOV	DEC	ANN
FORT BAYARD	.80	.55	.58	.22	.35	.77	3.21	3.10	2.00	1.20	.53	.88	14.19
FORT SUMNER	.31	.33	.47	.51	1.00	1.40	2.56	2.41	1.70	1.30	.47	.33	12.79
FORT SUMNER 5 S	.32	.34	.42	.58	.99	1.36	2.59	2.47	1.82	1.43	.44	.38	13.14
FRUITLAND 2 E	.69	.42	.52	.49	.39	.22	.79	.92	.87	1.08	.58	.67	7.64
GAGE 4 ESE	.62	.47	.41	.18	.18	.39	2.05	1.81	1.28	.89	.37	.65	9.30
GALLUP 5 E	.63	.54	.63	.41	.38	.40	1.52	1.61	.95	1.30	.67	.62	9.66
GHOST RANCH	.66	.48	.66	.59	.84	.59	1.65	2.06	.95	.98	.58	.53	10.57
GLENWOOD	1.26	.90	.94	.40	.37	.65	2.63	2.27	1.56	1.55	.73	1.32	14.58
GLORIETA	.70	.78	.85	.69	1.01	1.20	3.04	3.01	1.55	1.07	.77	.69	15.36
GOLDEN	.61	.61	.86	.57	.95	.71	2.32	2.41	1.18	.93	.57	.71	12.43
GRAN QUIVIRA NAT MON	.62	.69	.72	.57	.71	.93	2.89	2.95	1.53	1.06	.65	.93	14.25
GRENVILLE	.22	.29	.61	1.06	2.37	2.23	3.20	2.47	1.66	.85	.48	.34	15.78
HACHITA 1 N	.63	.42	.48	.23	.13	.30	2.10	2.19	1.32	.89	.39	.63	9.71
HATCH 2 W	.43	.35	.27	.19	.27	.50	1.94	2.23	1.39	1.09	.38	.54	9.58
HICKMAN	.59	.72	.74	.49	.37	.51	2.23	2.75	1.48	.95	.51	.79	12.13
HILLSBORO	.51	.41	.36	.29	.36	.62	2.46	2.53	2.05	1.30	.45	.71	12.05
HOBBS	.33	.41	.51	.64	2.02	1.50	2.03	2.39	2.37	1.70	.54	.33	14.77
HOUSE	.24	.27	.45	.52	1.11	1.66	2.91	2.53	1.85	1.22	.42	.29	13.47
JAL	.28	.37	.40	.58	1.29	1.23	1.66	1.80	1.84	1.34	.49	.28	11.56
JEMEZ SPRINGS	.95	.86	.90	.77	.97	.89	2.56	3.16	1.20	1.56	.96	.89	15.67
JOHNSON RANCH	.64	.56	.62	.51	.64	.51	1.69	2.09	1.07	1.05	.60	.57	10.55
JORNADA EXP RANGE	.44	.34	.30	.17	.26	.49	1.81	1.96	1.30	.94	.43	.53	8.97
KELLY RANCH	.34	.51	.60	.47	.60	.82	2.80	3.21	1.92	1.38	.45	.64	13.74
LAGUNA	.33	.42	.40	.30	.58	.46	1.64	1.90	1.08	1.13	.30	.48	9.02
LAKE MALOYA	.87	1.27	1.51	1.64	3.24	2.04	3.50	3.03	1.55	1.38	1.39	1.08	22.50
LAS VEGAS FAA AIRPORT	.27	.26	.46	.70	1.57	1.35	3.22	3.50	1.53	1.11	.60	.42	14.99
LORDSBURG 4 SE	.84	.56	.63	.21	.19	.43	1.93	2.19	1.27	.89	.42	.84	10.40
LOS ALAMOS	.85	.68	.99	.86	1.14	1.12	3.18	3.93	1.63	1.52	.96	.95	17.81
LUNA RANGER STATION	.92	.68	.76	.46	.39	.65	2.75	3.00	1.48	1.80	.67	1.03	14.59
LYBROOK	.72	.65	.78	.46	.66	.47	1.33	2.23	1.10	1.16	.78	.73	11.07
MAGDALENA	.29	.45	.45	.32	.44	.45	2.35	2.70	1.63	1.05	.26	.53	10.92
MALJAMAR 4 SE	.33	.42	.47	.42	1.57	1.26	2.36	2.48	2.11	1.41	.52	.34	13.69
MAXWELL	.22	.21	.38	.72	1.67	1.53	2.60	3.13	1.50	1.02	.51	.26	13.75
MC GAFFEY 4 SE	1.74	1.28	1.54	.93	.60	.54	2.32	2.43	1.40	1.59	1.08	1.36	16.81
MELROSE	.32	.49	.53	.65	1.34	1.96	2.94	2.42	2.09	1.23	.46	.40	14.83
MIMBRES RANGER STATION	.99	.71	.93	.41	.32	.84	3.33	3.67	2.07	1.51	.66	1.15	16.59
MOSQUERO	.29	.28	.55	.89	1.98	1.72	3.00	3.05	1.68	1.02	.58	.41	15.45
MOUNTAIN PARK	1.29	1.04	1.09	.44	.67	1.22	3.89	3.72	2.24	1.47	.75	1.08	18.90
MOUNTAINAIR	.60	.77	.62	.57	.65	.76	2.58	2.28	1.66	1.16	.57	.76	12.98
OCHOA	.34	.35	.42	.54	1.09	1.14	1.29	1.67	1.89	1.12	.37	.20	10.42
OJO CALIENTE	.61	.41	.57	.49	.80	.53	1.49	2.03	.98	1.01	.70	.48	10.10
OROGRANDE	.43	.32	.27	.24	.36	.79	1.90	1.88	1.50	.93	.39	.38	9.39
PASAMONTE	.26	.24	.53	.82	2.13	1.89	3.07	2.65	1.41	.83	.46	.32	14.61
PEARL	.27	.39	.43	.55	1.73	1.80	2.10	2.09	2.13	1.31	.41	.32	13.53
PECOS RANGER STATION	.70	.69	.90	.73	1.00	1.04	2.90	3.32	1.61	1.06	.87	.60	15.42
PEDERNAL 4 E	.28	.35	.39	.40	.61	1.06	2.02	2.61	1.05	.84	.38	.44	10.43
PORTALES	.37	.42	.48	.64	1.69	2.23	2.72	2.65	1.84	1.27	.57	.46	15.34
PORTER	.37	.44	.61	.93	1.72	1.89	2.91	2.13	1.76	1.18	.61	.45	15.00
PROGRESSO	.44	.59	.57	.45	.70	1.02	2.74	2.61	1.29	.96	.53	.74	12.64
QUEMADO RANGER STATION	.45	.51	.62	.31	.34	.44	1.84	2.48	1.07	.91	.43	.49	9.89
RAGLAND	.37	.56	.60	.84	1.70	1.66	3.52	3.03	1.86	1.23	.65	.48	16.50
RED RIVER	1.08	.92	1.51	1.38	1.68	1.25	2.79	3.02	1.41	1.48	1.26	1.17	18.95
REDROCK 1 NE	.91	.74	.68	.22	.19	.46	2.63	2.38	1.77	1.01	.45	.84	12.28
RESERVE RANGER STATION	1.04	.82	1.08	.41	.42	.65	2.34	2.73	1.78	1.70	.79	1.28	15.04
RIENHARDT RANCH	.26	.22	.23	.27	.43	.64	1.74	2.13	1.47	1.08	.23	.35	9.05

NEW MEXICO

PRECIPITATION NORMALS (INCHES)

STATION	JAN	FEB	MAR	APR	MAY	JUN	JUL	AUG	SEP	OCT	NOV	DEC	ANN
ROSWELL WSO	.24	.28	.27	.37	.77	.91	1.38	2.17	1.72	.99	.33	.27	9.70
ROY	.30	.35	.57	.87	2.01	1.71	3.10	2.63	1.41	1.13	.47	.51	15.06
RUIDOSO 2 NNE	1.12	1.16	1.33	.63	.87	1.86	4.02	4.04	2.50	1.31	.88	1.63	21.35
SAN JON	.36	.50	.58	.88	1.68	1.94	3.06	2.59	1.63	1.17	.58	.47	15.44
SANDIA PARK	1.10	1.14	1.30	.85	1.10	.96	2.84	3.02	1.62	1.42	1.16	1.21	17.72
SANTA ROSA	.33	.39	.57	.59	1.12	1.37	2.56	2.95	1.49	1.14	.50	.55	13.56
SHIPROCK	.59	.33	.43	.35	.47	.28	.60	.78	.72	1.05	.55	.53	6.68
SKARDA	.54	.53	.74	.75	1.04	.85	2.43	2.56	1.20	1.07	.86	.66	13.23
SOCORRO	.27	.35	.34	.38	.36	.59	1.46	1.65	1.27	1.09	.32	.55	8.63
SPRINGER	.31	.28	.56	.90	1.97	1.45	2.86	3.28	1.15	1.20	.56	.36	14.88
STAR LAKE	.39	.33	.38	.32	.56	.43	1.34	1.76	.92	.85	.48	.41	8.17
STATE UNIVERSITY	.39	.45	.30	.14	.24	.63	1.50	1.84	1.15	.83	.40	.44	8.31
TAJIQUE	.87	1.08	1.05	.84	.90	.94	2.67	3.07	1.70	1.55	.77	1.20	16.64
TAOS	.80	.56	.78	.73	1.08	.87	1.70	2.02	1.17	1.02	.83	.67	12.23
TATUM	.39	.53	.58	.65	1.69	2.23	2.50	2.61	2.23	1.60	.49	.33	15.83
TIERRA AMARILLA 4 NNW	1.33	.92	1.12	.93	1.02	.64	2.10	2.64	1.50	1.27	1.02	1.03	15.52
TOHATCHI 1 ESE	.81	.49	.64	.42	.48	.34	1.41	1.60	1.00	.97	.63	.64	9.43
TRES PIEDRAS	.60	.58	.74	.71	.94	.79	2.08	2.49	1.09	1.18	.81	.67	12.68
TRUTH OR CONSEQUENCES	.30	.28	.26	.21	.41	.77	1.47	1.66	1.59	.99	.37	.47	8.78
TUCUMCARI FAA AP	.29	.38	.48	.87	1.46	1.40	2.93	2.31	1.32	1.06	.51	.39	13.39
TUCUMCARI 3 NE	.37	.42	.60	.87	1.48	1.60	3.09	2.61	1.34	1.04	.57	.50	14.49
TULAROSA	.57	.50	.53	.28	.52	.61	1.83	1.86	1.50	1.03	.46	.57	10.26
VALMORA	.32	.37	.57	.76	1.69	1.74	3.03	3.10	1.83	1.02	.63	.47	15.53
VILLANUEVA	.35	.51	.67	.56	.79	1.00	2.20	2.27	1.01	.86	.34	.52	11.08
WHITE SANDS NAT MON	.40	.34	.32	.20	.26	.61	1.55	1.45	1.01	.88	.35	.45	7.82
WHITE SIGNAL	1.20	.92	.77	.29	.25	.46	2.61	2.23	1.68	1.09	.60	1.13	13.23
WHITEWATER	.67	.44	.44	.16	.20	.37	2.00	1.89	1.42	.92	.32	.74	9.57
WINSTON	.44	.34	.33	.27	.50	.69	2.69	3.11	2.23	1.21	.34	.51	12.66
WOLF CANYON	1.63	1.55	1.68	1.17	1.22	1.01	3.41	3.47	1.60	1.72	1.33	1.37	21.16
YESO 2 S	.36	.38	.46	.56	1.02	1.33	2.42	2.56	1.74	1.16	.43	.47	12.89
ZUNI FAA AIRPORT	.93	.71	.87	.52	.42	.34	1.88	2.02	1.13	1.34	.71	.85	11.72

NEW MEXICO

HEATING DEGREE DAY NORMALS (BASE 65 DEG F)

STATION	JUL	AUG	SEP	OCT	NOV	DEC	JAN	FEB	MAR	APR	MAY	JUN	ANN
ALAMOGORDO	0	0	7	125	453	685	698	526	401	147	17	0	3059
ALBUQUERQUE WSO //R	0	0	12	242	630	911	936	717	583	302	81	0	4414
AMISTAD 1 SSW	0	0	30	246	621	849	927	722	614	313	109	11	4442
ARTESIA 6 S	0	0	25	168	486	725	763	571	427	169	35	0	3369
AZTEC RUINS NAT MON	0	0	78	360	762	1094	1132	846	741	468	216	35	5732
BELL RANCH	0	0	21	228	594	853	908	689	583	287	79	5	4247
BERNALILLO	0	0	36	295	669	946	961	736	611	345	124	10	4733
BITTER LAKES WL REF	0	0	18	186	543	772	800	596	451	177	27	0	3570
BLOOMFIELD 3 SE	0	0	39	326	729	1060	1107	823	701	417	164	11	5377
BOSQUE DEL APACHE	0	0	17	216	579	849	849	636	499	251	69	0	3965
CABALLO DAM	0	0	U	144	468	732	741	554	431	185	28	0	3283
CAMERON	5	5	50	256	618	846	908	717	601	319	104	14	4443
CARLSBAD FAA AIRPORT	0	0	8	116	426	642	676	493	346	118	8	0	2833
CARLSBAD CAVERNS	0	0	28	130	376	549	608	468	351	123	18	0	2651
CARRIZOZO	0	0	27	264	618	862	880	700	577	306	93	0	4327
CERRO 4 NE	38	76	218	539	930	1262	1327	1064	939	654	409	127	7583
CHACO CANYON NAT MON	0	11	110	446	828	1153	1166	904	812	540	279	62	6311
CHAMA	70	125	291	592	966	1259	1336	1112	1057	753	499	216	8276
CIMARRON	19	35	141	403	756	961	1023	840	784	522	277	66	5827
CLAYTON WSO R	0	0	80	327	699	915	992	801	732	420	177	25	5168
CLIFF 10 SE	0	0	12	235	573	809	825	652	567	333	116	6	4128
CLOVIS	0	0	32	225	567	797	868	680	561	272	69	5	4076
CLOVIS 13 N	0	0	20	222	576	791	856	672	558	282	86	6	4069
COLUMBUS	0	0	0	121	441	679	676	504	368	142	18	0	2949
CONCHAS DAM	0	0	13	167	504	747	812	613	507	233	52	0	3648
CUBA	16	45	186	515	876	1175	1225	1002	918	639	378	120	7095
DES MOINES	11	14	127	425	786	1001	1066	885	834	540	282	71	6042
EAGLE NEST	184	229	420	722	1050	1349	1404	1162	1085	801	580	312	9298
EL MORRO NAT MON	0	39	163	484	831	1110	1153	941	859	600	363	99	6642
EL VADO DAM	24	64	243	564	924	1256	1361	1070	961	678	428	156	7729
ELEPHANT BUTTE DAM	0	0	0	120	450	716	719	529	405	168	26	0	3133
ELIDA	0	0	31	220	567	797	859	678	552	263	61	6	4034
ELK 3 E	8	6	93	322	609	784	825	689	608	372	159	21	4496
ESTANCIA	0	10	103	415	759	1017	1032	826	732	480	231	32	5637
FLORIDA	0	0	0	168	504	753	756	591	487	242	55	0	3556
FORT BAYARD	0	13	64	282	600	812	840	686	626	381	164	18	4486
FORT SUMNER	0	0	23	219	549	778	831	636	512	238	49	5	3840
FRUITLAND 2 E	0	8	68	341	741	1082	1125	823	704	438	197	32	5559
GALLUP 5 E	5	10	116	428	801	1079	1122	879	809	543	302	67	6161
GRAN QUIVIRA NAT MON	0	8	75	322	678	915	942	764	676	411	174	14	4979
GRENVILLE	10	12	120	398	741	967	1048	862	818	510	261	68	5815
HATCH 2 W	0	0	0	160	495	747	750	563	428	181	32	0	3356
HOBBS	0	0	8	107	422	623	685	515	379	128	14	0	2881
JAL	0	0	0	91	393	608	657	462	304	87	0	0	2602
JEMEZ SPRINGS	0	9	87	338	684	952	995	795	710	456	214	41	5281
JORNADA EXP RANGE	0	0	10	199	546	803	803	622	496	248	65	0	3792
LAGUNA	0	0	52	326	687	961	980	773	679	414	162	12	5046
LAKE MALOYA	73	107	267	561	900	1128	1209	1028	983	687	440	174	7557
LAS VEGAS FAA AIRPORT	20	35	151	440	774	995	1048	868	818	552	295	67	6063
LORDSBURG 4 SE	0	0	7	138	468	710	713	540	422	194	61	0	3253
LOS ALAMOS	14	45	166	453	810	1060	1110	916	846	576	316	75	6387
MALJAMAR 4 SE	0	0	13	163	489	701	747	563	435	171	30	0	3312
MC GAFFEY 4 SE	64	120	285	598	930	1194	1262	1064	1011	747	502	205	7982
MELROSE	0	0	26	231	582	806	865	678	555	274	76	6	4099
MOUNTAIN PARK	31	41	99	328	627	812	862	728	657	408	204	44	4841

NEW MEXICO

HEATING DEGREE DAY NORMALS (BASE 65 DEG F)

STATION	JUL	AUG	SEP	OCT	NOV	DEC	JAN	FEB	MAR	APR	MAY	JUN	ANN
MOUNTAINAIR	0	14	107	378	735	986	1011	804	722	468	229	36	5490
PASAMONTE	6	9	101	378	750	983	1051	857	781	474	214	41	5645
PEARL	0	0	22	141	465	682	729	554	417	162	28	0	3200
PORTALES	0	0	11	196	546	778	825	633	493	228	38	0	3748
QUEMADO RANGER STATION	18	28	153	471	807	1048	1079	871	809	579	347	103	6313
RED RIVER	205	260	426	722	1077	1349	1401	1204	1135	849	614	345	9587
ROSWELL WSO	0	0	10	143	477	698	732	535	386	134	11	0	3126
ROY	9	7	91	353	714	952	998	809	744	459	200	38	5374
RUIDOSO 2 NNE	76	93	216	493	762	955	980	837	784	558	360	127	6241
SAN JON	0	0	12	186	546	781	843	652	516	237	58	6	3837
SANTA ROSA	0	0	10	199	531	747	794	619	512	253	56	0	3721
SOCORRO	0	0	28	232	606	868	868	650	508	254	84	6	4104
SPRINGER	0	6	86	384	774	1051	1097	843	741	471	208	33	5694
STATE UNIVERSITY	0	0	0	135	468	691	698	526	404	167	29	0	3118
TAJIQUE	11	37	145	446	771	1004	1042	865	794	534	279	67	5995
TOHATCHI 1 ESE	0	0	85	320	705	1001	1042	815	753	480	228	43	5472
TRUTH OR CONSEQUENCES	0	0	9	158	492	753	756	566	448	194	28	0	3404
TUCUMCARI FAA AP	0	0	12	199	555	797	865	666	536	245	55	0	3930
TUCUMCARI 3 NE	0	0	8	165	516	747	809	622	502	224	50	0	3643
VALMORA	19	35	154	446	771	995	1057	865	809	555	301	87	6094
WHITE SANDS NAT MON	0	0	0	169	540	794	781	594	446	186	29	0	3539
WINSTON	0	10	89	357	675	890	905	742	679	432	232	38	5049
WOLF CANYON	156	214	390	685	993	1234	1296	1112	1060	789	570	303	8802

NEW MEXICO

COOLING DEGREE DAY NORMALS (BASE 65 DEG F)

STATION	JAN	FEB	MAR	APR	MAY	JUN	JUL	AUG	SEP	OCT	NOV	DEC	ANN
ALAMOGORDO	0	0	8	24	153	417	474	409	235	36	0	0	1756
ALBUQUERQUE WSO //R	0	0	0	0	59	285	428	344	132	6	0	0	1254
AMISTAD 1 SSW	0	0	0	7	72	257	381	316	120	13	0	0	1166
ARTESIA 6 S	0	0	5	40	162	387	474	419	226	50	0	0	1763
AZTEC RUINS NAT MON	0	0	0	0	17	128	301	226	66	0	0	0	738
BELL RANCH	0	0	0	8	72	281	409	344	138	8	0	0	1260
BERNALILLO	0	0	0	0	31	199	366	279	84	0	0	0	959
BITTER LAKES WL REF	0	0	5	24	126	366	453	384	180	18	0	0	1556
BLOOMFIELD 3 SE	0	0	0	0	34	191	353	273	84	0	0	0	935
BOSQUE DEL APACHE	0	0	0	11	78	276	412	338	134	8	0	0	1257
CABALLO DAM	0	0	0	14	102	363	487	412	213	27	0	0	1618
CAMERON	0	0	0	7	57	242	343	284	128	8	0	0	1069
CARLSBAD FAA AIRPORT	0	0	15	79	231	468	543	484	275	60	0	0	2155
CARLSBAD CAVERNS	0	0	19	57	182	381	425	388	241	93	10	0	1796
CARRIZOZO	0	0	0	0	44	243	341	260	93	0	0	0	981
CERRO 4 NE	0	0	0	0	0	13	76	42	0	0	0	0	131
CHACO CANYON NAT MON	0	0	0	0	6	101	239	172	29	0	0	0	547
CHAMA	0	0	0	0	0	9	39	23	0	0	0	0	71
CIMARRON	0	0	0	0	7	72	140	90	21	0	0	0	330
CLAYTON WSO R	0	0	0	0	25	172	288	230	74	8	0	0	797
CLIFF 10 SE	0	0	0	0	23	216	366	285	111	5	0	0	1006
CLOVIS	0	0	0	11	69	272	378	316	134	14	0	0	1194
CLOVIS 13 N	0	0	0	12	67	261	350	295	113	11	0	0	1109
COLUMBUS	0	0	0	31	157	423	496	425	240	40	0	0	1812
CONCHAS DAM	0	0	0	20	108	346	459	400	199	40	0	0	1572
CUBA	0	0	0	0	0	48	134	82	15	0	0	0	279
DES MOINES	0	0	0	0	10	98	176	117	31	0	0	0	432
EAGLE NEST	0	0	0	0	0	0	8	0	0	0	0	0	8
EL MORRO NAT MON	0	0	0	0	0	39	120	64	7	0	0	0	230
EL VADO DAM	0	0	0	0	0	15	90	58	0	0	0	0	163
ELEPHANT BUTTE DAM	0	0	6	21	128	390	481	406	229	36	0	0	1697
ELIDA	0	0	0	11	74	294	386	333	148	18	0	0	1264
ELK 3 E	0	0	0	0	10	114	170	115	33	0	0	0	442
ESTANCIA	0	0	0	0	0	89	204	128	13	0	0	0	434
FLORIDA	0	0	0	8	68	315	446	366	173	19	0	0	1395
FORT BAYARD	0	0	0	0	18	168	249	174	76	6	0	0	691
FORT SUMNER	0	0	0	16	86	314	412	347	146	15	0	0	1336
FRUITLAND 2 E	0	0	0	0	32	167	323	259	71	0	0	0	852
GALLUP 5 E	0	0	0	0	7	70	188	125	26	0	0	0	416
GRAN QUIVIRA NAT MON	0	0	0	0	13	146	228	160	51	0	0	0	598
GRENVILLE	0	0	0	0	13	107	196	148	42	0	0	0	506
HATCH 2 W	0	0	0	13	97	336	450	369	174	8	0	0	1447
HOBBS	0	0	16	53	179	396	468	428	242	51	5	0	1838
JAL	0	0	19	90	255	480	549	499	301	82	6	0	2281
JEMEZ SPRINGS	0	0	0	0	9	125	228	155	57	6	0	0	580
JORNADA EXP RANGE	0	0	0	11	71	309	437	357	163	16	0	0	1364
LAGUNA	0	0	0	0	16	174	301	220	61	0	0	0	772
LAKE MALOYA	0	0	0	0	0	9	29	14	0	0	0	0	52
LAS VEGAS FAA AIRPORT	0	0	0	0	0	67	128	79	13	0	0	0	287
LORDSBURG 4 SE	0	0	0	20	136	375	502	419	247	45	0	0	1744
LOS ALAMOS	0	0	0	0	6	78	116	72	25	0	0	0	297
MALJAMAR 4 SE	0	0	10	24	133	339	422	372	184	33	0	0	1517
MC GAFFEY 4 SE	0	0	0	0	0	7	30	15	0	0	0	0	52
MELROSE	0	0	0	10	79	276	375	319	128	14	0	0	1201
MOUNTAIN PARK	0	0	0	0	15	140	148	93	24	5	0	0	425

NEW MEXICO

COOLING DEGREE DAY NORMALS (BASE 65 DEG F)

STATION	JAN	FEB	MAR	APR	MAY	JUN	JUL	AUG	SEP	OCT	NOV	DEC	ANN
MOUNTAINAIR	0	0	0	0	15	114	198	132	41	0	0	0	500
PASAMONTE	0	0	0	0	13	137	248	183	50	0	0	0	631
PEARL	0	0	8	36	143	348	431	378	202	42	0	0	1588
PORTALES	0	0	0	15	100	309	406	350	155	10	0	0	1345
QUEMADO RANGER STATION	0	0	0	0	0	37	123	62	0	0	0	0	222
RED RIVER	0	0	0	0	0	0	0	0	0	0	0	0	0
ROSWELL WSO	0	0	8	41	176	420	508	440	229	41	0	0	1863
ROY	0	0	0	0	17	146	233	178	49	0	0	0	623
RUIDOSO 2 NNE	0	0	0	0	0	28	64	34	0	0	0	0	126
SAN JON	0	0	0	18	120	354	459	403	183	28	0	0	1565
SANTA ROSA	0	0	0	10	77	294	409	341	139	13	0	0	1283
SOCORRO	0	0	0	8	74	264	384	310	130	5	0	0	1175
SPRINGER	0	0	0	0	10	123	226	164	29	0	0	0	552
STATE UNIVERSITY	0	0	0	20	119	369	477	400	204	21	0	0	1610
TAJIQUE	0	0	0	0	0	76	122	78	7	0	0	0	283
TOHATCHI 1 ESE	0	0	0	0	26	175	326	214	100	10	0	0	851
TRUTH OR CONSEQUENCES	0	0	5	14	106	363	443	369	192	31	0	0	1523
TUCUMCARI FAA AP	0	0	0	14	99	340	446	378	156	16	0	0	1449
TUCUMCARI 3 NE	0	0	6	17	103	333	434	372	170	22	0	0	1457
VALMORA	0	0	0	0	0	60	140	85	13	0	0	0	298
WHITE SANDS NAT MON	0	0	0	18	125	378	487	406	207	14	0	0	1635
WINSTON	0	0	0	0	9	113	202	125	23	0	0	0	472
WOLF CANYON	0	0	0	0	0	0	0	0	0	0	0	0	0

STATE-STATION NUMBER	STN TYP	NAME		LATITUDE DEG-MIN	LONGITUDE DEG-MIN	ELEVATION (FT)
29-0125	12	AFTON 5 ESE		N 3203	W 10652	4200
29-0199	13	ALAMOGORDO		N 3253	W 10557	4350
29-0234	13	ALBUQUERQUE WSO	//R	N 3503	W 10637	5311
29-0268	12	ALEMAN RANCH		N 3255	W 10656	4527
29-0377	13	AMISTAD 1 SSW		N 3554	W 10310	4500
29-0417	12	ANIMAS		N 3157	W 10849	4415
29-0600	13	ARTESIA 6 S		N 3246	W 10423	3320
29-0640	12	AUGUSTINE		N 3405	W 10741	7020
29-0692	13	AZTEC RUINS NAT MON		N 3650	W 10800	5640
29-0818	12	BEAVERHEAD RANGER STA		N 3325	W 10807	6670
29-0858	13	BELL RANCH		N 3532	W 10406	4500
29-0903	13	BERNALILLO		N 3519	W 10633	5045
29-0992	13	BITTER LAKES WL REF		N 3329	W 10424	3670
29-1000	12	BLACK LAKE		N 3618	W 10517	8358
29-1063	13	BLOOMFIELD 3 SE		N 3640	W 10758	5794
29-1138	13	BOSQUE DEL APACHE		N 3346	W 10654	4520
29-1252	12	BUCKHORN		N 3302	W 10843	4800
29-1286	13	CABALLO DAM		N 3254	W 10718	4190
29-1332	13	CAMERON		N 3454	W 10323	4600
29-1389	12	CANJILON RANGER STA		N 3629	W 10627	7828
29-1423	12	CANTON		N 3417	W 10410	4056
29-1469	12	CARLSBAD		N 3225	W 10414	3120
29-1475	13	CARLSBAD FAA AIRPORT		N 3220	W 10416	3232
29-1480	13	CARLSBAD CAVERNS		N 3211	W 10427	4435
29-1515	13	CARRIZOZO		N 3339	W 10553	5438
29-1630	13	CERRO 4 NE		N 3649	W 10535	7685
29-1647	13	CHACO CANYON NAT MON		N 3602	W 10754	6175
29-1653	12	CHACON		N 3610	W 10523	8500
29-1664	13	CHAMA		N 3655	W 10635	7850
29-1813	13	CIMARRON		N 3631	W 10455	6427
29-1887	13	CLAYTON WSO	R	N 3627	W 10309	4970
29-1910	13	CLIFF 10 SE		N 3252	W 10831	4800
29-1939	13	CLOVIS		N 3426	W 10312	4280
29-1963	13	CLOVIS 13 N		N 3436	W 10313	4435
29-2024	13	COLUMBUS		N 3150	W 10739	4010
29-2030	13	CONCHAS DAM		N 3524	W 10411	4244
29-2207	12	CROSSROADS 2 NE		N 3332	W 10319	4120
29-2241	13	CUBA		N 3602	W 10658	7045
29-2324	12	CURETON RANCH		N 3232	W 10834	5200
29-2453	13	DES MOINES		N 3645	W 10350	6632
29-2510	12	DILIA		N 3511	W 10504	5200
29-2608	12	DULCE		N 3657	W 10700	6950
29-2700	13	EAGLE NEST		N 3633	W 10516	8275
29-2785	13	EL MORRO NAT MON		N 3503	W 10821	7225
29-2820	12	EL RITO		N 3620	W 10611	6870
29-2837	13	EL VADO DAM		N 3636	W 10644	6750
29-2848	13	ELEPHANT BUTTE DAM		N 3309	W 10711	4576
29-2854	13	ELIDA		N 3357	W 10339	4345
29-2865	13	ELK 3 E		N 3256	W 10517	5700
29-3031	12	ESPANOLA		N 3601	W 10603	5685

STATE-STATION NUMBER	STN TYP	NAME	LATITUDE DEG-MIN	LONGITUDE DEG-MIN	ELEVATION (FT)
29-3060	13	ESTANCIA	N 3445	W 10604	6107
29-3145	12	FARNSWORTH RANCH	N 3354	W 10500	5400
29-3157	12	FAYWOOD	N 3237	W 10753	5180
29-3225	13	FLORIDA	N 3226	W 10729	4450
29-3237	12	FLYING H	N 3300	W 10506	5300
29-3265	13	FORT BAYARD	N 3248	W 10809	6142
29-3294	13	FORT SUMNER	N 3428	W 10415	4030
29-3296	12	FORT SUMNER 5 S	N 3422	W 10415	4050
29-3340	13	FRUITLAND 2 E	N 3644	W 10821	5145
29-3368	12	GAGE 4 ESE	N 3213	W 10801	4410
29-3420	13	GALLUP 5 E	N 3532	W 10839	6600
29-3511	12	GHOST RANCH	N 3620	W 10623	6460
29-3577	12	GLENWOOD	N 3320	W 10853	4725
29-3586	12	GLORIETA	N 3535	W 10546	7520
29-3592	12	GOLDEN	N 3516	W 10613	6700
29-3649	13	GRAN QUIVIRA NAT MON	N 3416	W 10605	6620
29-3706	13	GRENVILLE	N 3636	W 10337	5990
29-3775	12	HACHITA 1 N	N 3156	W 10820	4495
29-3855	13	HATCH 2 W	N 3240	W 10711	4052
29-3969	12	HICKMAN	N 3431	W 10756	7800
29-4009	12	HILLSBORO	N 3256	W 10734	5270
29-4026	13	HOBBS	N 3242	W 10308	3615
29-4175	12	HOUSE	N 3438	W 10354	4850
29-4346	13	JAL	N 3207	W 10312	3150
29-4369	13	JEMEZ SPRINGS	N 3547	W 10641	6250
29-4398	12	JOHNSON RANCH	N 3557	W 10705	7200
29-4426	13	JORNADA EXP RANGE	N 3237	W 10644	4275
29-4461	12	KELLY RANCH	N 3402	W 10708	6700
29-4719	13	LAGUNA	N 3502	W 10724	5800
29-4742	13	LAKE MALOYA	N 3659	W 10422	7400
29-4856	13	LAS VEGAS FAA AIRPORT	N 3539	W 10509	6857
29-5079	13	LORDSBURG 4 SE	N 3218	W 10839	4250
29-5084	13	LOS ALAMOS	N 3552	W 10619	7410
29-5273	12	LUNA RANGER STATION	N 3350	W 10856	7050
29-5290	12	LYBROOK	N 3614	W 10735	7160
29-5353	12	MAGDALENA	N 3407	W 10714	6540
29-5370	13	MALJAMAR 4 SE	N 3249	W 10342	4000
29-5490	12	MAXWELL	N 3633	W 10433	5909
29-5560	13	MC GAFFEY 4 SE	N 3521	W 10827	7800
29-5617	13	MELROSE	N 3426	W 10337	4599
29-5754	12	MIMBRES RANGER STATION	N 3256	W 10801	6247
29-5937	12	MOSQUERO	N 3547	W 10357	5485
29-5960	13	MOUNTAIN PARK	N 3257	W 10551	6780
29-5965	13	MOUNTAINAIR	N 3431	W 10615	6520
29-6281	12	OCHOA	N 3211	W 10326	3460
29-6321	12	OJO CALIENTE	N 3618	W 10603	6290
29-6435	12	OROGRANDE	N 3223	W 10606	4200
29-6619	13	PASAMONTE	N 3618	W 10344	5650
29-6659	13	PEARL	N 3239	W 10323	3800
29-6676	12	PECOS RANGER STATION	N 3535	W 10541	6900

STATE-STATION NUMBER	STN TYP	NAME	LATITUDE DEG-MIN	LONGITUDE DEG-MIN	ELEVATION (FT)
29-6687	12	PEDERNAL 4 E	N 3438	W 10534	6200
29-7008	13	PORTALES	N 3411	W 10321	4010
29-7026	12	PORTER	N 3513	W 10317	4100
29-7094	12	PROGRESSO	N 3425	W 10551	6300
29-7180	13	QUEMADO RANGER STATION	N 3421	W 10830	6879
29-7226	12	RAGLAND	N 3449	W 10344	5110
29-7323	13	RED RIVER	N 3642	W 10524	8676
29-7340	12	REDROCK 1 NE	N 3242	W 10844	4150
29-7386	12	RESERVE RANGER STATION	N 3343	W 10847	5847
29-7423	12	RIENHARDT RANCH	N 3345	W 10713	5450
29-7610	13	ROSWELL WSO	N 3318	W 10432	3649
29-7638	13	ROY	N 3557	W 10412	5890
29-7649	13	RUIDOSO 2 NNE	N 3322	W 10540	6838
29-7867	13	SAN JON	N 3507	W 10320	4230
29-8015	12	SANDIA PARK	N 3510	W 10622	7106
29-8107	13	SANTA ROSA	N 3457	W 10441	4620
29-8284	12	SHIPROCK	N 3647	W 10842	4870
29-8352	12	SKARDA	N 3646	W 10558	8280
29-8387	13	SOCORRO	N 3405	W 10653	4585
29-8501	13	SPRINGER	N 3623	W 10436	5857
29-8524	12	STAR LAKE	N 3556	W 10728	7100
29-8535	13	STATE UNIVERSITY	N 3217	W 10645	3881
29-8648	13	TAJIQUE	N 3445	W 10617	6698
29-8668	12	TAOS	N 3622	W 10537	6945
29-8713	12	TATUM	N 3316	W 10319	4100
29-8845	12	TIERRA AMARILLA 4 NNW	N 3645	W 10634	7425
29-8919	13	TOHATCHI 1 ESE	N 3551	W 10844	6420
29-9085	12	TRES PIEDRAS	N 3640	W 10559	8110
29-9129	13	TRUTH OR CONSEQUENCES	N 3314	W 10716	4820
29-9153	13	TUCUMCARI FAA AP	N 3511	W 10336	4050
29-9156	13	TUCUMCARI 3 NE	N 3512	W 10341	4096
29-9165	12	TULAROSA	N 3305	W 10600	4535
29-9330	13	VALMORA	N 3549	W 10456	6300
29-9496	12	VILLANUEVA	N 3516	W 10522	5790
29-9686	13	WHITE SANDS NAT MON	N 3247	W 10611	3995
29-9691	12	WHITE SIGNAL	N 3233	W 10822	6070
29-9720	12	WHITEWATER	N 3233	W 10808	5150
29-9806	13	WINSTON	N 3321	W 10739	6200
29-9820	13	WOLF CANYON	N 3558	W 10646	8135
29-9851	12	YESO 2 S	N 3424	W 10437	4850
29-9897	12	ZUNI FAA AIRPORT	N 3506	W 10847	6440

29 – NEW MEXICO

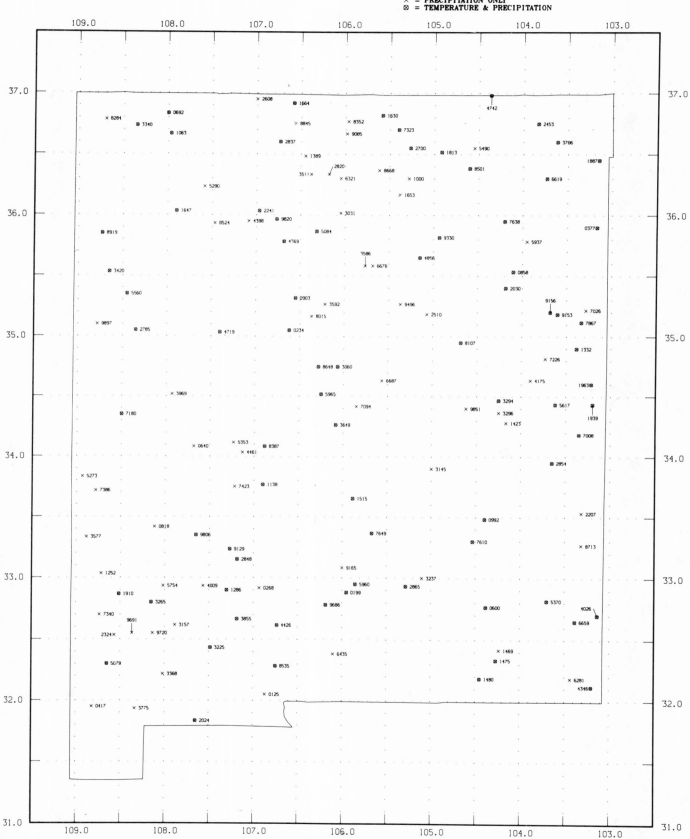

427

NEW YORK

TEMPERATURE NORMALS (DEG F)

STATION			JAN	FEB	MAR	APR	MAY	JUN	JUL	AUG	SEP	OCT	NOV	DEC	ANN
ALBANY WSO	R	MAX	30.2	32.7	42.5	57.6	69.5	78.3	83.2	80.7	72.8	61.5	47.8	34.6	57.6
		MIN	11.9	14.0	24.6	35.5	45.4	55.0	59.6	57.6	49.6	39.4	30.8	18.2	36.8
		MEAN	21.1	23.4	33.6	46.6	57.5	66.7	71.4	69.2	61.2	50.5	39.3	26.5	47.3
ALCOVE DAM		MAX	31.2	33.1	41.7	55.8	68.1	77.1	82.1	80.1	72.2	61.2	48.2	35.7	57.2
		MIN	9.5	10.8	21.6	32.8	42.4	51.9	56.5	54.4	46.8	35.9	28.1	15.9	33.9
		MEAN	20.4	22.0	31.7	44.3	55.3	64.5	69.3	67.3	59.5	48.6	38.2	25.8	45.6
ALFRED		MAX	30.2	32.4	41.4	55.6	67.0	75.1	78.6	76.8	70.2	59.5	45.9	34.1	55.6
		MIN	12.7	12.9	21.8	32.3	42.0	51.0	55.3	54.0	47.7	37.7	29.3	18.3	34.6
		MEAN	21.5	22.7	31.6	43.9	54.5	63.1	66.9	65.4	59.0	48.6	37.6	26.2	45.1
ALLEGANY STATE PARK		MAX	31.3	33.5	42.7	56.5	68.2	76.0	79.6	77.9	71.3	60.7	46.8	35.2	56.6
		MIN	13.3	13.3	21.7	31.9	40.5	49.8	53.4	52.5	46.5	36.7	29.2	18.9	34.0
		MEAN	22.3	23.4	32.2	44.2	54.4	62.9	66.5	65.2	58.9	48.7	38.1	27.1	45.3
ANGELICA		MAX	31.4	33.5	42.8	57.2	68.6	77.0	80.8	78.7	72.2	61.4	46.9	35.2	57.1
		MIN	12.6	12.8	21.9	31.6	40.7	49.6	53.6	52.5	46.2	36.2	29.0	18.3	33.8
		MEAN	22.0	23.2	32.3	44.4	54.7	63.3	67.2	65.7	59.2	48.8	38.0	26.7	45.5
BAINBRIDGE		MAX	31.7	34.4	44.1	58.9	70.9	79.4	83.4	81.1	73.5	62.0	47.5	35.3	58.5
		MIN	13.2	13.7	23.1	33.1	42.6	51.7	56.2	55.2	48.5	37.7	30.2	19.0	35.4
		MEAN	22.5	24.1	33.6	46.0	56.8	65.6	69.8	68.2	61.0	49.9	38.8	27.2	47.0
BATAVIA		MAX	31.0	33.0	42.2	56.3	68.1	77.5	81.3	79.4	72.7	61.8	47.9	35.7	57.2
		MIN	15.1	15.7	24.3	35.2	45.5	54.8	59.0	57.4	50.8	40.9	32.2	20.9	37.7
		MEAN	23.1	24.4	33.3	45.8	56.8	66.2	70.2	68.5	61.8	51.4	40.1	28.3	47.5
BINGHAMTON WSO	R	MAX	28.0	29.6	38.7	53.5	64.9	73.9	78.4	76.4	68.9	57.6	44.4	32.4	53.9
		MIN	14.3	15.1	24.0	35.1	45.5	54.6	59.4	57.9	50.6	40.5	31.3	19.9	37.4
		MEAN	21.2	22.4	31.4	44.3	55.2	64.3	68.9	67.2	59.8	49.1	37.9	26.2	45.7
BOONVILLE 2 SSW		MAX	24.5	26.5	35.4	50.4	63.3	72.1	76.5	74.5	66.8	55.5	41.6	28.8	51.3
		MIN	8.3	9.7	19.5	31.8	42.3	51.5	56.0	54.5	47.3	37.8	27.8	14.1	33.4
		MEAN	16.4	18.1	27.5	41.1	52.8	61.8	66.3	64.5	57.1	46.7	34.7	21.5	42.4
BRIDGEHAMPTON		MAX	37.7	38.2	44.8	55.0	64.2	73.6	79.4	78.9	72.5	62.8	52.6	42.4	58.5
		MIN	23.4	24.1	30.8	38.6	47.8	57.3	63.5	63.1	56.2	46.2	37.7	28.0	43.1
		MEAN	30.6	31.2	37.8	46.9	56.0	65.5	71.5	71.0	64.4	54.6	45.2	35.2	50.8
BROCKPORT 2 NW		MAX	30.9	32.5	41.2	55.4	67.1	76.8	81.4	79.3	72.4	61.3	47.9	35.7	56.8
		MIN	15.8	16.4	24.6	35.3	45.2	55.2	60.3	59.0	52.1	41.9	32.6	21.7	38.3
		MEAN	23.4	24.5	32.9	45.4	56.2	66.0	70.9	69.2	62.3	51.6	40.3	28.7	47.6
BUFFALO WSO	//R	MAX	30.0	31.4	40.4	54.4	65.9	75.6	80.2	78.2	71.4	60.2	47.0	35.0	55.8
		MIN	17.0	17.5	25.6	36.3	46.3	56.4	61.2	59.6	52.7	42.7	33.6	22.5	39.3
		MEAN	23.5	24.5	33.0	45.4	56.1	66.0	70.7	68.9	62.1	51.5	40.3	28.8	47.6
CANANDAIGUA 3 S		MAX	31.9	33.2	41.7	55.3	66.9	76.9	82.1	80.5	73.5	61.9	49.0	36.5	57.5
		MIN	15.7	15.7	24.3	34.9	44.6	54.9	59.8	58.5	52.0	41.8	32.8	21.8	38.1
		MEAN	23.8	24.5	33.0	45.1	55.8	65.9	71.0	69.5	62.8	51.9	40.9	29.2	47.8
CANTON 4 SE		MAX	25.7	27.7	38.3	53.2	66.2	75.0	79.7	77.6	69.8	58.3	44.9	30.5	53.9
		MIN	5.3	6.7	19.3	32.4	43.3	52.9	57.5	55.5	47.6	37.3	28.4	11.9	33.2
		MEAN	15.5	17.2	28.8	42.8	54.8	64.0	68.6	66.6	58.8	47.9	36.7	21.2	43.6
CARMEL 1 SW		MAX	34.6	37.0	45.8	59.0	69.9	78.4	83.4	81.8	74.6	63.9	51.3	38.5	59.9
		MIN	14.7	15.2	25.2	35.3	44.4	53.4	58.6	57.4	50.2	39.6	31.7	20.5	37.2
		MEAN	24.7	26.1	35.5	47.2	57.2	65.9	71.0	69.6	62.4	51.8	41.5	29.5	48.5
CHASM FALLS		MAX	26.6	29.1	39.0	52.9	67.1	75.8	79.3	76.8	69.4	58.5	44.0	30.4	54.1
		MIN	6.1	7.1	18.5	30.7	41.4	51.1	55.3	53.5	46.6	37.3	27.9	12.3	32.3
		MEAN	16.4	18.1	28.8	41.8	54.3	63.4	67.3	65.2	58.0	47.9	36.0	21.4	43.2
CHAZY		MAX	26.5	28.6	38.7	53.5	67.5	76.5	81.2	78.9	70.0	58.0	44.9	31.4	54.6
		MIN	6.0	7.5	19.6	32.4	43.3	53.4	58.0	55.8	47.6	37.9	28.6	13.5	33.6
		MEAN	16.3	18.0	29.2	43.0	55.4	65.0	69.6	67.3	58.8	48.0	36.8	22.5	44.2

NEW YORK

TEMPERATURE NORMALS (DEG F)

STATION		JAN	FEB	MAR	APR	MAY	JUN	JUL	AUG	SEP	OCT	NOV	DEC	ANN
CHERRY VALLEY 2 NNE	MAX	27.7	29.7	38.6	53.4	65.8	74.4	78.6	76.4	68.9	57.9	44.2	31.9	54.0
	MIN	10.7	12.0	21.6	32.9	43.1	52.7	57.3	55.6	48.8	38.8	29.1	16.5	34.9
	MEAN	19.2	20.9	30.1	43.1	54.5	63.6	68.0	66.0	58.9	48.4	36.7	24.2	44.5
COOPERSTOWN	MAX	30.3	32.6	41.5	55.9	67.9	76.2	80.3	78.3	71.1	60.5	46.6	34.4	56.3
	MIN	11.0	11.9	21.3	31.8	41.4	50.7	55.2	54.1	47.1	37.0	29.0	17.0	34.0
	MEAN	20.7	22.3	31.4	43.9	54.7	63.4	67.8	66.2	59.1	48.8	37.8	25.7	45.2
CORTLAND	MAX	29.6	30.8	40.1	54.3	66.4	75.9	80.5	78.6	71.3	59.9	46.3	33.6	55.6
	MIN	13.5	13.5	22.7	33.8	42.9	52.4	57.1	55.5	48.4	38.5	30.4	19.2	35.7
	MEAN	21.6	22.1	31.5	44.1	54.7	64.2	68.8	67.1	59.9	49.2	38.4	26.4	45.7
DANNEMORA	MAX	26.3	28.4	38.1	52.3	66.3	75.4	79.6	77.1	69.3	57.7	43.0	30.0	53.6
	MIN	7.7	9.5	20.2	32.5	44.0	53.5	58.1	56.2	48.6	38.6	28.1	13.5	34.2
	MEAN	17.0	19.0	29.2	42.4	55.2	64.5	68.9	66.7	59.0	48.2	35.6	21.8	44.0
DANSVILLE	MAX	33.0	34.3	43.7	57.3	69.5	79.0	83.5	82.0	74.9	63.3	49.9	37.5	59.0
	MIN	15.6	15.7	24.2	34.7	43.8	53.5	57.8	56.1	49.4	39.9	32.3	21.6	37.1
	MEAN	24.3	25.1	34.0	46.0	56.7	66.3	70.7	69.1	62.2	51.6	41.1	29.6	48.1
DOBBS FERRY	MAX	36.7	39.3	48.4	61.9	71.9	80.1	85.0	83.0	75.2	64.3	52.6	40.7	61.6
	MIN	23.1	24.1	31.6	41.2	50.9	60.1	65.5	64.4	57.6	47.4	38.3	27.7	44.3
	MEAN	29.9	31.7	40.0	51.6	61.4	70.2	75.2	73.7	66.4	55.8	45.5	34.2	53.0
ELIZABETHTOWN	MAX	27.5	30.3	40.2	54.3	68.1	77.1	81.6	78.9	70.7	58.7	44.9	31.6	55.3
	MIN	4.3	6.4	18.5	30.0	40.3	49.3	54.5	52.3	44.4	34.0	25.8	11.8	31.0
	MEAN	15.9	18.3	29.4	42.2	54.2	63.2	68.1	65.6	57.6	46.4	35.4	21.7	43.2
ELMIRA	MAX	32.7	34.6	43.9	57.7	69.3	78.6	83.5	81.4	73.8	62.1	49.0	36.8	58.6
	MIN	14.3	14.0	23.4	33.5	42.3	51.7	56.1	54.6	47.5	37.2	29.9	20.0	35.4
	MEAN	23.5	24.3	33.7	45.7	55.8	65.2	69.8	68.0	60.7	49.7	39.5	28.4	47.0
FRANKLINVILLE	MAX	29.7	31.6	40.5	54.4	65.9	74.8	78.5	76.9	70.3	59.6	45.5	33.6	55.1
	MIN	10.8	10.7	19.8	30.6	39.4	49.2	53.0	51.6	45.5	35.8	28.2	16.9	32.6
	MEAN	20.3	21.2	30.2	42.5	52.6	62.0	65.8	64.2	57.9	47.7	36.8	25.3	43.9
FREDONIA	MAX	32.5	33.9	43.1	56.3	67.6	77.2	81.1	79.3	73.4	62.7	49.2	37.5	57.8
	MIN	18.4	18.1	26.1	36.7	46.5	56.6	61.1	60.2	54.2	44.3	35.3	24.7	40.2
	MEAN	25.5	26.0	34.6	46.5	57.1	66.9	71.1	69.7	63.8	53.5	42.3	31.1	49.0
GENEVA RESEARCH FARM	MAX	30.6	31.8	40.9	54.9	66.5	76.3	81.2	79.5	72.2	60.6	47.9	35.3	56.5
	MIN	15.0	15.6	24.7	35.9	45.4	55.1	59.8	58.3	51.4	41.1	32.5	21.1	38.0
	MEAN	22.9	23.7	32.8	45.4	55.9	65.7	70.5	69.0	61.8	50.9	40.2	28.3	47.3
GLENHAM	MAX	35.7	38.0	47.1	60.4	72.0	80.8	85.8	84.0	76.5	65.1	52.2	39.6	61.4
	MIN	16.7	18.0	28.0	38.8	48.5	57.7	62.4	60.8	53.4	42.3	33.5	22.2	40.2
	MEAN	26.2	28.1	37.6	49.6	60.2	69.3	74.1	72.4	65.0	53.7	42.9	30.9	50.8
GLENS FALLS FAA AP	MAX	28.0	30.7	40.6	55.6	67.6	76.5	81.4	78.9	70.6	59.2	45.6	32.7	55.6
	MIN	8.0	10.6	22.7	34.0	44.2	53.5	58.0	56.0	48.0	37.4	29.2	15.7	34.8
	MEAN	18.0	20.7	31.7	44.8	55.9	65.0	69.7	67.5	59.4	48.3	37.4	24.2	45.2
GLOVERSVILLE	MAX	28.7	31.3	40.6	56.1	68.3	76.6	80.9	78.7	70.8	59.9	45.4	32.8	55.8
	MIN	11.5	13.0	23.1	34.0	44.3	53.5	58.2	56.5	49.3	38.9	30.4	17.7	35.9
	MEAN	20.1	22.2	31.8	45.1	56.3	65.1	69.5	67.7	60.1	49.4	37.9	25.3	45.9
GOUVERNEUR	MAX	27.1	29.7	39.9	54.9	67.5	76.1	80.3	78.2	70.3	59.5	45.5	31.5	55.0
	MIN	6.6	8.2	20.0	32.3	42.6	52.3	57.0	55.0	47.6	37.7	29.0	13.3	33.5
	MEAN	16.9	19.0	29.9	43.7	55.1	64.2	68.7	66.6	59.0	48.6	37.2	22.4	44.3
GRAFTON	MAX	28.4	30.6	39.7	54.0	66.3	74.2	78.4	76.3	68.5	58.3	44.6	32.6	54.3
	MIN	11.1	12.5	22.0	33.2	43.9	53.1	57.8	56.4	49.5	39.7	29.4	16.8	35.5
	MEAN	19.8	21.6	30.9	43.7	55.1	63.6	68.1	66.4	59.0	49.0	37.1	24.8	44.9
HEMLOCK	MAX	31.3	32.7	41.5	55.8	67.3	76.6	80.9	79.0	71.7	60.7	47.7	36.0	56.8
	MIN	14.0	13.7	22.7	33.7	44.0	53.7	58.8	57.2	50.7	40.5	31.8	21.0	36.8
	MEAN	22.7	23.2	32.1	44.7	55.7	65.2	69.9	68.1	61.2	50.6	39.8	28.5	46.8

429

NEW YORK

TEMPERATURE NORMALS (DEG F)

STATION		JAN	FEB	MAR	APR	MAY	JUN	JUL	AUG	SEP	OCT	NOV	DEC	ANN
INDIAN LAKE 2 SW	MAX	25.4	27.7	36.8	49.6	63.1	72.2	76.4	74.5	67.0	55.9	42.0	29.3	51.7
	MIN	1.9	2.4	13.6	26.1	36.5	46.0	50.4	48.6	42.1	32.1	23.7	9.2	27.7
	MEAN	13.6	15.1	25.2	37.8	49.8	59.1	63.4	61.6	54.6	44.0	32.9	19.3	39.7
ITHACA CORNELL UNIV. //	MAX	30.5	31.7	40.8	54.6	66.3	75.6	80.3	78.6	71.4	60.0	47.1	35.0	56.0
	MIN	13.8	13.7	23.6	34.4	43.3	52.9	57.2	55.6	48.9	39.1	31.4	20.1	36.2
	MEAN	22.2	22.7	32.2	44.5	54.8	64.3	68.8	67.1	60.2	49.6	39.3	27.6	46.1
LAKE PLACID CLUB	MAX	24.7	27.2	36.5	49.6	63.1	72.5	76.2	74.0	66.5	55.1	41.6	29.0	51.3
	MIN	3.2	3.8	15.1	27.7	38.5	48.3	52.2	50.2	43.0	33.4	24.2	9.5	29.1
	MEAN	13.9	15.5	25.8	38.7	50.8	60.4	64.3	62.1	54.7	44.3	32.9	19.3	40.2
LAWRENCEVILLE	MAX	25.5	27.8	38.3	53.7	67.2	75.7	79.9	77.6	69.9	58.4	44.3	29.9	54.0
	MIN	6.1	7.8	19.4	32.3	43.2	53.4	58.3	56.3	48.8	38.5	28.4	12.3	33.7
	MEAN	15.8	17.8	28.9	43.0	55.2	64.6	69.1	67.0	59.3	48.5	36.4	21.1	43.9
LIBERTY	MAX	29.8	31.5	40.3	54.3	65.4	73.9	78.3	76.9	69.6	59.2	46.2	33.5	54.9
	MIN	12.2	12.7	22.5	33.7	42.7	51.5	56.2	54.7	47.5	37.4	29.5	17.7	34.9
	MEAN	21.0	22.1	31.4	44.0	54.1	62.8	67.3	65.8	58.6	48.3	37.8	25.6	44.9
LITTLE FALLS CITY RES	MAX	27.7	30.1	39.6	55.1	67.8	76.7	80.9	78.7	71.0	59.6	45.2	32.3	55.4
	MIN	10.3	11.8	21.2	32.3	43.0	52.2	57.0	55.3	47.9	37.8	28.8	16.1	34.5
	MEAN	19.0	21.0	30.4	43.7	55.4	64.5	68.9	67.1	59.5	48.7	37.0	24.2	45.0
LITTLE VALLEY	MAX	30.0	31.5	40.6	54.0	65.9	75.0	79.3	77.4	70.3	59.1	45.9	34.1	55.3
	MIN	12.4	11.9	20.5	31.3	39.9	49.4	53.5	52.5	46.6	36.8	29.1	18.8	33.6
	MEAN	21.2	21.7	30.6	42.7	52.9	62.2	66.4	65.0	58.5	48.0	37.5	26.5	44.4
LOCKPORT 2 NE	MAX	31.0	33.0	41.9	56.2	67.7	77.3	81.7	79.7	72.6	61.2	47.9	35.7	57.2
	MIN	15.9	16.7	24.7	35.0	45.0	54.7	59.7	58.3	51.5	41.9	32.6	21.9	38.2
	MEAN	23.5	24.9	33.3	45.7	56.4	66.1	70.7	69.0	62.0	51.6	40.3	28.8	47.7
LOWVILLE	MAX	27.2	29.4	38.8	53.3	67.0	76.1	80.2	77.9	70.3	59.0	44.5	31.6	54.6
	MIN	7.8	9.0	20.2	32.2	42.7	51.5	55.7	54.0	47.0	37.5	28.6	14.1	33.4
	MEAN	17.5	19.2	29.5	42.8	54.9	63.8	68.0	66.0	58.7	48.3	36.6	22.9	44.0
MASSENA FAA AP	MAX	24.4	26.7	37.1	52.7	67.0	76.0	80.6	78.0	69.6	58.0	43.4	28.9	53.5
	MIN	4.9	6.9	19.4	32.8	43.7	53.0	57.6	55.6	47.6	37.4	27.9	11.4	33.2
	MEAN	14.7	16.8	28.3	42.8	55.4	64.5	69.1	66.8	58.6	47.7	35.7	20.2	43.4
MILLBROOK	MAX	33.0	35.6	44.4	58.2	69.4	76.9	81.4	79.8	72.6	62.4	49.6	36.9	58.4
	MIN	13.4	15.3	24.6	34.7	44.2	53.5	58.0	56.6	49.2	38.8	30.1	18.9	36.4
	MEAN	23.2	25.5	34.5	46.5	56.8	65.2	69.7	68.2	60.9	50.7	39.9	27.9	47.4
MINEOLA R	MAX	37.3	38.7	46.4	58.0	68.3	77.5	82.8	81.5	74.2	63.7	52.3	41.4	60.2
	MIN	25.5	26.1	33.1	41.8	51.2	60.5	66.4	65.5	58.7	48.5	39.8	29.8	45.6
	MEAN	31.4	32.4	39.8	49.9	59.7	69.1	74.6	73.5	66.5	56.1	46.1	35.7	52.9
MOHONK LAKE	MAX	31.0	33.7	42.6	56.8	67.2	74.4	78.6	77.0	69.9	59.7	47.0	34.8	56.1
	MIN	16.8	17.8	26.2	37.2	48.1	57.5	62.6	61.3	54.1	44.2	33.6	21.8	40.1
	MEAN	23.9	25.8	34.5	47.0	57.7	66.0	70.7	69.2	62.0	52.0	40.3	28.3	48.1
MOUNT MORRIS 2 W	MAX	31.1	32.3	41.5	55.5	67.4	77.1	81.8	79.9	72.9	61.5	48.0	35.6	57.1
	MIN	14.6	14.8	23.7	35.0	44.4	54.1	58.6	56.8	49.9	40.0	31.6	20.3	37.0
	MEAN	22.9	23.6	32.6	45.3	56.0	65.7	70.2	68.4	61.4	50.8	39.8	28.0	47.1
NEW YORK CNTRL PK WSO	MAX	38.0	40.1	48.6	61.1	71.5	80.1	85.3	83.7	76.4	65.6	53.6	42.1	62.2
	MIN	25.6	26.6	34.1	43.8	53.3	62.7	68.2	67.1	60.1	49.9	40.8	30.3	46.9
	MEAN	31.8	33.4	41.4	52.4	62.5	71.4	76.7	75.4	68.3	57.7	47.2	36.2	54.5
NEW YORK JFK INTL AP	MAX	37.5	39.1	46.6	58.3	67.7	76.9	82.7	81.7	75.2	64.7	53.2	41.8	60.5
	MIN	25.1	25.9	33.2	42.3	51.7	61.0	67.2	66.3	59.2	48.7	39.6	29.6	45.8
	MEAN	31.3	32.5	40.0	50.3	59.7	69.0	75.0	74.0	67.2	56.7	46.4	35.7	53.2
NEW YORK LA GUARDIA WSO	MAX	37.4	39.2	47.3	59.6	69.7	78.7	83.9	82.3	75.2	64.5	52.9	41.5	61.0
	MIN	26.1	27.3	34.6	44.2	53.7	63.2	68.9	68.2	61.2	50.5	41.2	30.8	47.5
	MEAN	31.8	33.3	41.0	51.9	61.7	71.0	76.4	75.3	68.2	57.5	47.1	36.2	54.3

430

NEW YORK

TEMPERATURE NORMALS (DEG F)

STATION		JAN	FEB	MAR	APR	MAY	JUN	JUL	AUG	SEP	OCT	NOV	DEC	ANN
NY WESTERLEIGH STAT IS	MAX	38.4	40.8	49.3	61.3	70.8	79.4	84.5	83.2	76.8	66.1	54.1	42.4	62.3
	MIN	23.5	24.5	32.0	41.0	50.4	59.7	65.1	64.0	57.0	46.2	37.5	27.8	44.1
	MEAN	31.0	32.7	40.7	51.2	60.6	69.6	74.8	73.7	66.9	56.2	45.9	35.2	53.2
NORWICH 1 NE	MAX	30.1	31.8	41.1	55.2	66.8	75.7	80.4	78.9	71.4	60.2	46.9	34.3	56.1
	MIN	9.1	9.6	20.3	31.5	40.5	49.8	54.1	52.4	45.1	34.9	27.7	15.9	32.6
	MEAN	19.6	20.7	30.7	43.3	53.7	62.8	67.3	65.6	58.3	47.5	37.3	25.1	44.3
OGDENSBURG 3 NE	MAX	26.2	28.4	38.8	53.8	67.0	76.1	80.9	79.0	70.7	59.5	45.3	31.2	54.7
	MIN	6.8	7.8	19.9	33.1	43.7	53.6	58.7	57.0	48.6	38.9	29.3	13.8	34.3
	MEAN	16.5	18.1	29.4	43.5	55.4	64.9	69.8	68.0	59.7	49.2	37.3	22.5	44.5
OSWEGO EAST //	MAX	30.1	31.1	39.4	51.8	63.5	73.3	78.4	77.1	70.4	59.5	46.9	34.8	54.7
	MIN	16.9	17.8	26.5	36.4	45.8	55.4	61.9	60.9	54.2	44.4	35.2	23.1	39.9
	MEAN	23.5	24.5	33.0	44.1	54.7	64.4	70.2	69.0	62.3	52.0	41.0	29.0	47.3
PATCHOGUE 2 N	MAX	38.3	39.6	47.1	57.8	67.6	76.5	82.1	81.3	74.7	64.8	53.6	42.8	60.5
	MIN	21.0	22.0	29.2	37.4	47.2	56.9	63.1	62.2	54.8	43.6	35.2	25.5	41.5
	MEAN	29.7	30.8	38.1	47.6	57.4	66.7	72.6	71.8	64.8	54.2	44.4	34.2	51.0
PENN YAN 2 SW	MAX	33.3	35.0	43.8	57.9	69.9	78.9	83.1	81.4	74.5	63.2	49.2	37.4	59.0
	MIN	16.6	16.3	25.0	35.7	45.7	55.5	60.6	59.1	52.5	42.7	33.7	22.7	38.8
	MEAN	25.0	25.7	34.4	46.9	57.8	67.3	71.9	70.3	63.5	53.0	41.5	30.1	49.0
PERU 2 WSW	MAX	28.1	30.1	40.1	54.4	67.7	76.9	81.2	78.7	70.3	59.0	45.3	31.9	55.3
	MIN	7.6	9.2	20.6	32.2	42.9	52.9	57.5	55.4	47.5	37.4	28.3	13.9	33.8
	MEAN	17.9	19.7	30.4	43.3	55.3	64.9	69.4	67.1	58.9	48.2	36.8	22.9	44.6
PORT JERVIS	MAX	34.6	37.9	47.5	62.3	72.9	81.1	85.2	82.6	74.1	63.2	50.4	38.1	60.8
	MIN	16.4	18.0	26.1	36.2	46.0	54.9	59.4	58.4	51.3	40.1	31.8	21.3	38.3
	MEAN	25.5	28.0	36.8	49.3	59.4	68.0	72.3	70.5	62.7	51.7	41.1	29.7	49.6
POUGHKEEPSIE FAA AP	MAX	34.0	36.9	46.1	59.7	70.5	79.2	84.1	82.1	74.3	63.4	50.6	38.2	59.9
	MIN	14.8	16.6	26.3	36.2	46.0	55.6	60.5	59.1	51.0	39.5	31.2	20.3	38.1
	MEAN	24.4	26.8	36.2	48.0	58.3	67.5	72.4	70.6	62.7	51.5	40.9	29.3	49.1
RIVERHEAD RESEARCH	MAX	37.9	39.0	46.7	58.5	69.5	78.1	82.8	81.5	74.8	64.6	53.4	42.4	60.8
	MIN	23.9	24.6	31.4	39.3	48.8	58.1	63.8	63.5	57.3	47.1	38.0	28.3	43.7
	MEAN	30.9	31.8	39.1	48.9	59.2	68.1	73.3	72.5	66.1	55.9	45.7	35.4	52.2
ROCHESTER WSO //R	MAX	30.8	32.2	41.2	56.0	67.7	77.7	82.3	80.1	72.8	61.5	48.0	35.5	57.2
	MIN	16.3	16.7	25.3	36.1	46.0	55.7	60.3	58.7	51.6	41.8	33.2	22.3	38.7
	MEAN	23.6	24.4	33.3	46.0	56.9	66.7	71.3	69.5	62.2	51.7	40.6	29.0	47.9
SALEM	MAX	30.9	32.7	42.4	56.4	68.1	77.1	82.1	80.2	72.4	61.2	48.2	35.3	57.3
	MIN	6.0	7.3	19.9	31.8	41.5	51.2	55.5	52.6	44.5	34.0	26.5	13.6	32.0
	MEAN	18.5	20.1	31.2	44.1	54.8	64.2	68.8	66.4	58.5	47.6	37.4	24.5	44.7
SCARSDALE	MAX	38.1	41.0	49.6	62.4	72.9	81.3	86.0	84.2	77.0	66.5	54.1	42.1	62.9
	MIN	20.9	22.3	29.4	38.5	47.5	56.7	62.1	61.0	54.0	43.1	34.8	25.4	41.3
	MEAN	29.6	31.7	39.5	50.5	60.3	69.0	74.1	72.7	65.5	54.8	44.5	33.8	52.2
SETAUKET	MAX	38.2	39.7	47.3	59.4	69.3	77.7	82.6	81.0	74.6	64.7	53.8	42.5	60.9
	MIN	24.1	24.6	31.4	39.9	49.2	58.9	64.8	64.6	58.4	48.2	39.0	28.8	44.3
	MEAN	31.2	32.2	39.4	49.7	59.3	68.3	73.7	72.8	66.6	56.5	46.4	35.7	52.7
SODUS CENTER //	MAX	31.7	33.5	42.0	55.8	68.0	77.7	81.9	80.0	73.0	61.9	48.4	36.1	57.5
	MIN	16.5	16.7	25.5	36.3	45.6	54.7	59.7	58.4	51.7	42.2	33.5	22.4	38.6
	MEAN	24.1	25.1	33.8	46.1	56.8	66.2	70.8	69.2	62.4	52.1	41.0	29.3	48.1
SPENCER 3 W	MAX	31.3	33.5	42.6	56.8	68.1	77.3	81.8	79.9	72.4	61.1	47.3	35.3	57.3
	MIN	10.9	11.5	21.2	31.3	40.6	49.7	54.0	52.7	45.7	35.2	27.8	17.0	33.1
	MEAN	21.1	22.5	31.9	44.1	54.4	63.5	67.9	66.3	59.1	48.2	37.6	26.2	45.2
STILLWATER RESERVOIR	MAX	24.6	26.6	36.1	49.1	63.0	71.9	76.1	74.2	66.6	55.5	41.7	28.7	51.2
	MIN	1.1	1.0	13.2	27.4	39.3	49.3	53.8	52.6	45.4	34.5	25.3	8.8	29.3
	MEAN	12.9	13.8	24.7	38.3	51.1	60.6	65.0	63.4	56.0	45.0	33.5	18.7	40.3

NEW YORK

TEMPERATURE NORMALS (DEG F)

STATION		JAN	FEB	MAR	APR	MAY	JUN	JUL	AUG	SEP	OCT	NOV	DEC	ANN
SYRACUSE WSO R	MAX	30.6	32.2	41.4	56.2	67.9	77.2	81.6	79.6	72.3	60.9	47.9	35.3	56.9
	MIN	15.0	15.8	25.2	36.0	46.0	55.4	60.3	58.9	51.8	41.7	33.3	21.3	38.4
	MEAN	22.8	24.0	33.3	46.1	57.0	66.3	70.9	69.3	62.1	51.3	40.6	28.3	47.7
TUPPER LAKE SUNMOUNT	MAX	24.7	26.9	36.6	49.6	63.6	72.6	76.6	74.2	66.8	55.7	41.7	28.7	51.5
	MIN	3.2	3.9	15.5	28.4	39.5	49.1	53.4	51.4	44.1	34.1	24.9	9.5	29.8
	MEAN	14.0	15.5	26.1	39.0	51.6	60.8	65.0	62.9	55.5	44.9	33.3	19.1	40.6
UTICA FAA AP	MAX	27.5	29.4	38.6	54.0	66.1	75.4	79.9	78.0	70.0	58.3	45.0	32.2	54.5
	MIN	12.7	14.1	23.7	35.3	45.3	54.7	59.5	58.2	50.7	40.5	31.8	18.9	37.1
	MEAN	20.1	21.8	31.2	44.6	55.7	65.1	69.8	68.1	60.4	49.4	38.4	25.6	45.9
WANAKENA RANGER SCHOOL//	MAX	26.5	28.7	38.2	52.0	65.9	74.2	77.7	75.4	67.9	56.9	42.7	30.2	53.0
	MIN	4.1	4.6	15.6	28.3	39.4	48.9	53.0	51.5	44.6	34.9	25.9	10.5	30.1
	MEAN	15.3	16.6	26.9	40.2	52.7	61.6	65.4	63.5	56.3	46.0	34.3	20.3	41.6
WATERTOWN	MAX	27.7	29.4	39.3	53.2	65.1	74.7	79.9	77.9	70.3	58.7	46.1	32.6	54.6
	MIN	8.9	10.5	22.2	34.9	45.8	55.7	60.6	59.0	51.3	40.6	31.5	16.1	36.4
	MEAN	18.4	20.0	30.8	44.1	55.5	65.2	70.2	68.5	60.8	49.7	38.8	24.4	45.5
WESTFIELD 3 SW	MAX	31.3	32.8	41.4	54.5	65.4	75.0	78.9	77.1	70.9	60.2	47.4	36.2	55.9
	MIN	18.3	18.1	26.3	37.3	47.4	57.6	62.2	61.3	55.2	45.1	35.2	24.6	40.7
	MEAN	24.9	25.5	33.9	45.9	56.4	66.3	70.6	69.2	63.1	52.7	41.3	30.4	48.4
WEST POINT	MAX	35.2	38.3	47.8	61.6	72.7	81.5	86.4	84.3	76.1	64.6	51.3	38.7	61.5
	MIN	19.1	20.2	29.0	39.2	49.3	58.4	63.4	62.1	55.3	44.8	35.4	24.1	41.7
	MEAN	27.2	29.3	38.4	50.4	61.0	70.0	74.9	73.2	65.7	54.7	43.3	31.4	51.6
WHITEHALL	MAX	29.3	32.4	42.6	58.0	71.8	80.9	85.5	82.5	73.4	61.1	47.1	33.7	58.2
	MIN	9.2	10.5	22.7	35.1	46.0	55.5	59.8	57.6	50.4	39.9	31.2	17.1	36.3
	MEAN	19.2	21.5	32.7	46.6	58.9	68.2	72.7	70.1	62.0	50.5	39.2	25.4	47.3

NEW YORK

PRECIPITATION NORMALS (INCHES)

STATION		JAN	FEB	MAR	APR	MAY	JUN	JUL	AUG	SEP	OCT	NOV	DEC	ANN
ADDISON		2.05	2.07	2.58	2.87	3.09	3.40	3.62	2.96	2.83	2.62	2.69	2.58	33.36
ALBANY WSO	R	2.39	2.26	3.01	2.94	3.31	3.29	3.00	3.34	3.23	2.93	3.04	3.00	35.74
ALCOVE DAM		2.33	2.18	3.32	3.20	3.66	3.40	3.71	3.24	3.55	3.43	3.40	3.16	38.58
ALFRED		2.42	2.17	3.01	3.11	3.21	4.09	3.37	3.24	3.31	2.95	3.03	2.86	36.77
ALLEGANY STATE PARK		2.81	2.58	3.20	3.57	3.94	4.25	4.15	3.98	4.34	3.58	4.07	3.53	44.00
ANGELICA		2.17	2.02	2.48	2.76	2.83	4.06	3.18	3.21	3.40	2.77	2.76	2.48	34.12
ARCADE		2.87	2.25	2.93	3.53	3.39	4.07	3.38	3.86	4.19	3.19	3.78	3.54	40.98
ARKVILLE 2 W		2.70	2.25	2.92	3.41	3.68	3.62	3.49	3.68	3.52	3.41	3.45	3.02	39.15
BAINBRIDGE		2.81	2.63	3.08	3.46	3.54	4.06	3.62	3.53	3.77	3.25	3.51	3.29	40.55
BALDWINSVILLE		3.00	3.04	3.13	3.27	3.15	3.54	3.42	3.55	3.53	3.43	3.75	3.48	40.29
BATAVIA		2.01	1.86	2.38	3.08	3.23	3.28	2.84	3.95	3.25	2.90	2.74	2.45	33.97
BEAVER FALLS		2.49	2.23	2.44	2.77	3.20	3.23	3.14	3.72	3.58	3.10	3.39	3.18	36.47
BENNETT BRIDGE		4.54	3.71	3.76	3.92	3.72	3.61	3.46	3.98	4.03	4.04	5.14	5.13	49.04
BIG MOOSE 3 E		4.04	3.52	4.04	4.13	4.10	4.16	4.13	4.75	4.61	3.98	4.87	4.81	51.14
BINGHAMTON WSO	R	2.54	2.33	2.94	3.07	3.19	3.60	3.48	3.35	3.32	3.00	3.04	2.92	36.79
BOONVILLE 2 SSW		5.31	4.83	5.04	4.58	4.25	4.35	4.25	4.43	4.85	4.20	5.64	5.85	57.58
BRADFORD 1 NW		1.79	1.84	2.33	2.71	2.96	3.39	3.33	2.99	2.95	2.60	2.63	2.25	31.77
BREWERTON LOCK 23		3.23	3.09	3.17	3.38	3.36	3.68	3.46	3.61	3.63	3.45	4.08	3.92	42.06
BRIDGEHAMPTON		4.21	3.82	4.64	3.66	3.63	2.64	2.75	4.25	3.53	3.69	4.42	4.73	45.97
BROADALBIN 5 NE		3.22	2.76	3.68	3.78	3.70	3.80	3.40	3.36	3.99	3.40	3.86	3.66	42.61
BROCKPORT 2 NW		1.53	1.54	2.13	2.73	2.55	2.65	2.28	3.00	2.75	2.53	2.54	2.10	28.33
BUFFALO WSO	//R	3.02	2.40	2.97	3.06	2.89	2.72	2.96	4.16	3.37	2.93	3.62	3.42	37.52
CANANDAIGUA 3 S		1.81	1.94	2.39	2.80	2.75	3.28	2.64	2.99	2.81	2.87	2.55	2.24	31.07
CANASTOTA		2.60	2.51	2.82	3.29	3.34	3.70	3.62	3.86	3.64	3.27	3.59	3.17	39.41
CANDOR		2.21	2.26	2.82	2.94	3.18	3.52	3.46	3.12	3.43	3.01	3.15	2.78	35.88
CANISTEO		1.89	1.79	2.39	2.77	3.01	3.72	3.00	3.12	3.15	2.65	2.62	2.18	32.29
CANTON 4 SE		2.18	2.06	2.29	2.92	3.18	3.09	3.37	3.88	3.65	3.13	3.08	2.79	35.62
CARMEL 1 SW		3.47	3.03	4.08	4.02	3.75	3.44	3.90	4.38	4.24	3.83	4.11	4.08	46.33
CAYUGA LOCK 1		2.05	2.14	2.55	2.88	3.06	3.29	3.33	3.30	3.04	3.05	2.99	2.76	34.44
CHASM FALLS		2.61	2.54	2.84	3.21	3.46	3.80	3.98	4.25	3.95	3.56	3.32	3.50	41.02
CHAZY		2.09	1.98	2.26	2.51	2.80	3.13	3.19	3.56	3.16	3.00	2.73	2.61	33.02
CHEMUNG		2.05	2.09	2.75	2.89	3.18	3.33	2.86	3.26	3.11	2.70	2.95	2.60	33.77
CHERRY VALLEY 2 NNE		2.58	2.34	3.29	3.54	3.75	3.78	3.89	3.77	3.61	3.33	3.72	3.33	40.93
CINCINNATUS		2.86	2.67	3.25	3.64	3.66	4.06	3.85	3.66	3.88	3.39	3.77	3.61	42.30
CLYDE LOCK 26		2.48	2.39	2.76	3.13	3.19	3.46	3.13	3.66	3.18	3.53	3.54	3.16	37.61
COLTON 2 N		2.58	2.49	2.83	3.22	3.25	3.39	3.66	4.07	3.83	3.35	3.28	3.34	39.29
CONKLINGVILLE DAM		3.28	3.01	3.98	3.74	3.54	3.44	3.17	3.39	3.62	3.22	3.89	3.83	42.11
COOPERSTOWN		2.57	2.29	3.24	3.37	3.50	3.89	3.65	3.39	3.66	3.13	3.38	3.21	39.28
CORTLAND		2.97	2.86	3.45	3.36	3.43	3.95	3.44	3.50	3.56	3.21	3.55	3.93	41.21
DANNEMORA		1.83	2.01	2.12	2.53	2.92	3.35	3.35	3.62	3.09	3.14	2.61	2.59	33.16
DANSVILLE		1.63	1.55	2.18	2.81	2.88	3.66	2.89	3.17	2.99	2.64	2.63	2.09	31.12
DOBBS FERRY		3.83	3.60	4.94	4.38	3.91	3.80	4.01	4.53	4.20	3.99	4.66	4.50	50.35
DOLGEVILLE		2.96	2.78	3.38	3.66	3.86	3.92	3.99	3.61	4.04	3.22	3.93	3.68	43.03
ELIZABETHTOWN		2.28	2.22	2.68	2.78	2.89	3.22	2.98	3.62	2.97	2.77	3.12	2.84	34.37
ELLENBURG DEPOT		1.71	1.72	2.07	2.29	2.59	3.15	3.24	3.61	2.82	2.92	2.33	2.07	30.52
ELLENVILLE		3.47	3.08	4.10	4.00	3.71	4.02	3.69	3.99	3.62	3.71	4.20	4.00	45.59
ELMIRA		1.99	1.96	2.59	2.79	3.18	3.34	3.23	3.03	3.07	2.82	2.72	2.38	33.10
FRANKFORT LOCK 19		2.45	2.41	2.84	3.39	3.51	4.24	4.01	3.61	4.01	3.32	3.89	2.97	40.65
FRANKLINVILLE		2.76	2.38	3.07	3.42	3.32	3.83	3.24	3.77	3.77	3.38	3.52	3.10	39.56
FREDONIA		2.54	1.85	2.58	3.19	2.92	3.36	2.97	3.80	4.39	3.67	3.71	3.01	37.99
FREEVILLE 2 NE		2.11	2.04	2.44	2.81	3.21	3.89	3.45	3.42	3.48	3.15	3.16	2.77	35.93
FROST VALLEY		4.09	3.58	4.58	4.57	4.54	4.55	4.26	4.77	4.70	4.64	4.96	4.88	54.12
GENEVA RESEARCH FARM		1.97	2.04	2.55	2.85	2.85	3.46	2.88	3.12	2.91	3.02	2.72	2.54	32.91
GLENHAM		3.24	2.87	3.58	3.70	3.49	3.77	3.84	4.12	3.93	3.40	3.69	3.78	43.41
GLENS FALLS FARM		2.74	2.56	3.50	3.68	3.86	3.97	3.60	3.60	3.74	3.34	3.76	3.48	41.83

NEW YORK

PRECIPITATION NORMALS (INCHES)

STATION	JAN	FEB	MAR	APR	MAY	JUN	JUL	AUG	SEP	OCT	NOV	DEC	ANN
GLENS FALLS FAA AP	2.51	2.36	3.09	3.10	3.05	3.14	3.00	3.14	3.03	2.86	2.97	2.96	35.21
GLOVERSVILLE	2.92	2.56	3.50	3.60	3.80	3.83	3.87	3.58	3.83	3.30	3.75	3.49	42.03
GOUVERNEUR	2.50	2.27	2.66	3.17	3.37	3.14	3.02	3.98	3.92	3.31	3.64	3.30	38.28
GRAFTON	2.77	2.39	3.29	3.71	4.17	4.68	4.00	4.19	4.03	3.54	3.92	3.29	43.98
GRAHAMSVILLE	3.24	2.87	3.69	3.96	3.85	4.06	4.37	4.24	3.86	3.86	3.99	3.78	45.77
HASKINVILLE	2.03	2.01	2.47	3.02	3.10	3.86	3.37	3.12	3.16	2.84	2.76	2.42	34.16
HEMLOCK	1.62	1.57	2.33	2.80	2.91	3.55	2.86	3.08	3.05	2.70	2.46	2.11	31.04
HIGHMARKET	3.97	3.35	3.59	4.24	4.71	4.40	4.21	4.79	5.37	4.72	5.15	4.37	52.87
HINCKLEY	3.64	3.20	4.03	4.08	4.47	4.29	4.26	4.39	4.66	3.89	4.47	4.25	49.63
HOOKER 4 N //	5.71	4.12	4.05	4.15	4.12	3.91	4.20	4.20	4.90	4.69	5.74	5.54	55.33
HOPE	3.32	3.07	4.12	3.89	3.66	3.66	3.58	3.74	3.95	3.40	4.24	4.01	44.64
INDIAN LAKE 2 SW	2.79	2.54	3.21	2.93	3.29	3.40	3.44	3.85	3.91	3.20	3.85	3.32	39.73
ITHACA CORNELL UNIV. //	2.08	2.08	2.57	2.80	3.05	3.73	3.54	3.38	3.38	3.09	2.97	2.60	35.27
LAKE DELAWARE	2.70	2.58	3.35	3.51	3.93	4.09	3.88	4.25	4.10	3.54	3.63	3.29	42.85
LAKE PLACID CLUB	2.69	2.37	2.86	2.93	3.18	3.54	3.79	4.16	3.61	2.99	3.23	3.04	38.39
LAWRENCEVILLE	1.69	1.87	2.01	2.65	3.03	3.32	3.47	4.02	3.45	3.05	2.56	2.21	33.33
LIBERTY	3.73	3.64	3.99	4.53	4.00	4.27	4.15	4.63	4.06	3.88	4.33	4.37	49.58
LITTLE FALLS CITY RES	2.72	2.48	2.90	3.69	3.79	4.05	4.01	3.79	4.19	3.23	3.79	3.30	41.94
LITTLE FALLS MILL ST	2.77	2.71	3.32	3.59	3.55	3.71	3.74	3.55	3.90	3.11	3.81	3.50	41.26
LITTLE VALLEY	3.94	3.40	3.92	4.13	3.75	4.36	3.65	4.21	4.39	3.79	4.76	4.74	49.04
LOCKE 2 W	2.03	2.09	2.72	3.10	3.21	4.12	3.36	3.45	3.53	3.30	3.26	2.95	37.12
LOCKPORT 2 NE	2.64	2.38	2.77	3.13	2.90	2.86	2.76	3.90	3.37	2.83	3.12	3.04	35.70
LOWVILLE	3.20	2.68	3.07	3.17	3.26	3.27	3.42	3.43	3.53	3.31	4.05	3.69	40.08
LYONS FALLS	3.67	3.12	3.35	3.54	3.67	3.69	3.69	3.95	4.20	3.52	4.24	4.18	44.82
MASSENA FAA AP	2.18	2.07	2.24	2.67	2.60	2.86	2.98	3.40	3.32	2.62	2.93	3.07	32.94
MAYS POINT LOCK 25	2.07	2.11	2.36	2.61	2.96	3.17	3.24	3.23	3.04	3.10	2.94	2.75	33.58
MILLBROOK	2.79	2.40	3.23	3.50	3.38	3.69	3.65	3.95	3.71	3.36	3.43	3.51	40.60
MINEOLA R	3.31	3.37	4.44	4.01	3.46	2.93	3.17	4.06	3.63	3.38	3.97	3.92	43.65
MOHONK LAKE	3.47	3.22	4.07	4.34	4.06	3.70	4.00	4.16	4.14	4.03	4.11	4.04	47.34
MOUNT MORRIS 2 W	1.50	1.50	1.80	2.69	2.48	3.16	2.62	2.79	2.76	2.57	2.21	1.93	28.01
NEWARK	2.09	2.18	2.45	2.89	3.13	3.26	2.73	3.10	3.18	3.31	3.00	2.86	34.18
NEW LONDON LOCK 22	2.88	2.66	2.94	3.66	3.71	4.04	3.69	3.79	3.80	3.45	3.93	3.33	41.88
NEW YORK AVE V BRKLYN	3.15	3.22	4.21	3.85	3.72	3.23	4.17	4.45	3.95	3.24	3.86	3.68	44.73
NEW YORK CNTRL PK WSO	3.21	3.13	4.22	3.75	3.76	3.23	3.77	4.03	3.66	3.41	4.14	3.81	44.12
NEW YORK JFK INTL AP	2.93	3.20	3.99	3.76	3.40	2.98	3.56	4.10	3.51	2.98	3.73	3.62	41.76
NEW YORK LA GUARDIA WSO	3.11	3.08	4.10	3.76	3.46	3.15	3.67	4.32	3.48	3.24	3.77	3.68	42.82
NY WESTERLEIGH STAT IS	3.31	3.35	4.39	3.89	3.73	3.23	4.56	4.96	3.99	3.51	3.99	3.83	46.74
NORWICH 1 NE	2.66	2.41	3.20	3.45	3.55	4.16	3.67	3.17	3.77	3.15	3.67	3.42	40.28
OGDENSBURG 3 NE	2.06	1.80	2.05	2.72	2.65	2.97	3.10	3.60	3.29	2.89	2.80	2.74	32.67
OLEAN	2.29	2.03	2.99	3.27	3.29	3.88	3.50	3.38	3.63	2.92	3.01	2.79	36.98
OSWEGO EAST //	3.24	3.02	2.91	3.28	3.07	3.19	2.65	3.09	3.51	3.51	4.02	3.81	39.30
PATCHOGUE 2 N	3.78	3.62	4.33	3.95	3.66	2.94	3.31	4.49	3.36	3.85	4.01	4.41	45.71
PENN YAN 2 SW	1.82	1.90	2.43	2.57	2.86	3.06	2.86	3.02	2.72	2.93	2.71	2.28	31.16
PERU 2 WSW	1.65	1.58	1.93	2.40	2.52	3.00	2.99	3.07	2.69	2.67	2.42	2.00	28.92
PORT JERVIS	3.17	2.69	3.75	3.78	3.64	3.58	4.08	4.06	3.54	3.39	3.85	3.48	43.01
POUGHKEEPSIE FAA AP	2.75	2.42	3.28	3.66	3.62	3.43	3.50	3.77	3.66	3.30	3.57	3.20	40.16
RIVERHEAD RESEARCH	4.07	3.63	4.28	3.74	3.53	2.90	3.20	4.17	3.60	3.56	4.18	4.46	45.32
ROCHESTER WSO //R	2.30	2.32	2.53	2.64	2.58	2.78	2.48	3.20	2.66	2.54	2.65	2.59	31.27
SABATTIS 3 NE	2.78	2.56	2.95	3.32	3.68	3.68	3.92	4.26	3.84	3.42	3.63	3.71	41.75
SALEM	2.79	2.29	2.98	3.32	3.58	3.84	3.63	3.56	3.78	3.13	3.24	3.08	39.22
SCARSDALE	3.40	3.27	4.63	4.13	3.80	3.39	4.02	4.55	3.96	3.71	4.46	4.10	47.42
SCHENECTADY	2.48	2.27	2.98	3.02	3.34	3.42	3.03	3.30	2.95	2.98	3.04	2.81	35.62
SETAUKET	3.62	3.35	4.35	3.92	3.52	3.02	3.26	4.02	3.77	3.62	4.14	4.30	44.89
SHERBURNE 2 S	2.29	2.13	2.81	3.02	3.11	3.59	3.42	3.33	3.50	3.05	3.14	2.76	36.15
SHOKAN BROWN STATION	3.52	3.17	4.20	4.33	4.06	3.76	4.05	3.85	4.16	4.26	4.51	4.30	48.17

NEW YORK

PRECIPITATION NORMALS (INCHES)

STATION	JAN	FEB	MAR	APR	MAY	JUN	JUL	AUG	SEP	OCT	NOV	DEC	ANN
SKANEATELES	2.32	2.38	2.94	3.04	3.21	3.76	3.71	3.60	3.37	3.45	3.31	3.14	38.23
SMITHS BASIN	2.46	2.16	2.82	3.07	3.25	3.47	3.50	3.47	3.47	2.94	3.35	2.97	36.93
SODUS CENTER //	2.47	2.37	2.38	3.14	3.08	3.24	2.77	3.38	3.44	3.79	3.70	2.81	36.57
SOUTH WALES EMERY PARK	3.64	2.72	3.09	3.70	3.71	3.65	3.29	4.06	4.09	3.32	3.98	3.82	43.07
SPENCER 3 W	2.35	2.34	2.96	3.24	3.26	3.53	3.49	3.36	3.42	3.14	3.34	3.01	37.44
STILLWATER RESERVOIR	3.23	2.91	3.42	3.72	3.89	4.00	4.06	4.37	4.25	3.78	4.22	3.96	45.81
SUFFERN WATER WORKS	3.30	3.12	4.18	4.37	4.01	3.73	4.08	4.85	4.56	3.80	4.39	3.84	48.23
SYRACUSE WSO R	2.61	2.65	3.11	3.34	3.16	3.63	3.76	3.77	3.29	3.14	3.45	3.20	39.11
TRENTON FALLS	3.70	3.17	3.77	3.94	4.26	4.49	4.55	4.39	4.37	3.90	4.50	4.34	49.38
TRIBES HILL	2.73	2.35	3.17	3.24	3.47	3.34	3.26	3.28	3.41	3.08	3.33	3.23	37.89
TUPPER LAKE SUNMOUNT	2.54	2.42	2.62	3.09	3.56	3.52	3.86	4.41	3.52	3.08	3.05	2.95	38.62
UTICA FAA AP	3.36	3.01	3.37	3.49	3.60	3.75	4.18	3.59	3.75	3.37	3.92	4.05	43.44
WANAKENA RANGER SCHOOL//	2.80	2.67	2.92	3.17	3.58	3.56	4.13	4.52	3.86	3.53	3.74	3.66	42.14
WATERLOO	1.85	1.81	2.36	2.65	2.87	3.12	3.29	2.96	2.95	2.89	2.78	2.40	31.93
WATERTOWN	3.29	2.54	2.75	3.17	3.30	3.17	3.02	3.78	3.93	3.46	4.06	4.06	40.53
WESTFIELD 3 SW	2.55	2.15	2.92	3.83	3.47	3.73	3.50	4.20	4.95	4.06	4.16	3.13	42.65
WEST POINT	3.66	3.33	4.36	4.21	4.15	3.79	3.69	4.23	4.37	4.22	4.53	4.22	48.76
WHITEHALL	2.71	2.35	2.95	3.10	3.15	3.21	3.11	3.63	3.40	2.96	3.17	2.99	36.73
WILSON 2 NE	1.95	1.59	2.07	2.68	2.75	2.71	2.43	3.30	2.80	2.44	2.48	2.26	29.46
WOLCOTT 3 NW	2.70	2.60	2.82	3.34	3.01	3.42	2.97	3.28	3.39	3.68	3.78	3.48	38.47

NEW YORK

HEATING DEGREE DAY NORMALS (BASE 65 DEG F)

STATION		JUL	AUG	SEP	OCT	NOV	DEC	JAN	FEB	MAR	APR	MAY	JUN	ANN
ALBANY WSO	R	7	15	149	450	771	1194	1361	1165	973	552	252	38	6927
ALCOVE DAM		17	48	188	508	804	1215	1383	1204	1032	621	317	69	7406
ALFRED		32	63	200	508	822	1203	1349	1184	1035	633	336	99	7464
ALLEGANY STATE PARK		44	66	204	505	807	1175	1324	1165	1017	624	342	106	7379
ANGELICA		33	60	194	502	810	1187	1333	1170	1014	618	332	95	7348
BAINBRIDGE		11	20	150	468	786	1172	1318	1145	973	570	275	55	6943
BATAVIA		11	32	136	426	747	1138	1299	1137	983	576	277	49	6811
BINGHAMTON WSO	R	16	35	178	493	813	1203	1358	1193	1042	621	313	79	7344
BOONVILLE 2 SSW		43	80	247	567	909	1349	1507	1313	1163	717	387	122	8404
BRIDGEHAMPTON		0	0	70	322	594	924	1066	946	843	543	282	37	5627
BROCKPORT 2 NW		8	22	126	420	741	1125	1290	1134	995	588	287	49	6785
BUFFALO WSO	//R	9	25	130	423	741	1122	1287	1134	992	588	294	53	6798
CANANDAIGUA 3 S		9	23	121	411	723	1110	1277	1134	992	597	301	48	6746
CANTON 4 SE		29	53	207	530	849	1358	1535	1338	1122	666	328	82	8097
CARMEL 1 SW		8	14	115	409	705	1101	1249	1089	915	534	255	43	6437
CHASM FALLS		40	77	226	530	870	1352	1507	1313	1122	696	343	92	8168
CHAZY		23	43	204	527	846	1318	1510	1316	1110	660	315	64	7936
CHERRY VALLEY 2 NNE		23	49	196	515	849	1265	1420	1235	1082	657	336	84	7711
COOPERSTOWN		23	46	193	502	816	1218	1373	1196	1042	633	327	85	7454
CORTLAND		19	39	181	490	798	1197	1345	1201	1039	627	333	85	7354
DANNEMORA		25	49	203	521	882	1339	1488	1288	1110	678	322	73	7978
DANSVILLE		14	24	131	423	717	1097	1262	1117	961	570	280	58	6654
DOBBS FERRY		0	0	50	292	585	955	1088	932	775	402	155	5	5239
ELIZABETHTOWN		35	80	239	577	888	1342	1522	1308	1104	684	348	104	8231
ELMIRA		18	28	159	474	765	1135	1287	1140	970	579	299	73	6927
FRANKLINVILLE		57	82	228	536	846	1231	1386	1226	1079	675	392	130	7868
FREDONIA		10	17	98	363	681	1051	1225	1092	942	555	264	44	6342
GENEVA RESEARCH FARM		9	20	132	437	744	1138	1305	1156	998	588	296	56	6879
GLENHAM		0	0	80	355	663	1057	1203	1033	849	462	179	17	5898
GLENS FALLS FAA AP		11	33	192	518	828	1265	1457	1240	1032	606	307	58	7547
GLOVERSVILLE		14	35	168	484	813	1231	1392	1198	1029	597	285	58	7304
GOUVERNEUR		20	49	200	508	834	1321	1491	1288	1088	639	319	68	7825
GRAFTON		18	38	198	496	837	1246	1401	1215	1057	639	320	85	7550
HEMLOCK		16	32	153	451	756	1132	1311	1170	1020	609	304	71	7025
INDIAN LAKE 2 SW		94	134	317	651	963	1417	1593	1397	1234	816	471	185	9272
ITHACA CORNELL UNIV.	//	19	36	169	477	771	1159	1327	1184	1017	615	326	77	7177
LAKE PLACID CLUB		81	121	313	642	963	1417	1584	1386	1215	789	446	154	9111
LAWRENCEVILLE		19	42	192	512	858	1361	1525	1322	1119	660	319	65	7994
LIBERTY		25	48	203	518	816	1221	1364	1201	1042	630	347	99	7514
LITTLE FALLS CITY RES		21	36	181	505	840	1265	1426	1232	1073	639	310	75	7603
LITTLE VALLEY		39	72	211	527	825	1194	1358	1212	1066	669	383	118	7674
LOCKPORT 2 NE		7	25	132	420	741	1122	1287	1123	983	579	287	47	6753
LOWVILLE		24	53	204	518	852	1305	1473	1282	1101	666	326	86	7890
MASSENA FAA AP		20	48	208	536	879	1389	1559	1350	1138	666	314	75	8182
MILLBROOK		16	18	151	443	753	1150	1296	1106	946	555	264	68	6766
MINEOLA	R	0	0	56	291	567	908	1042	913	781	453	193	24	5228
MOHONK LAKE		6	11	123	403	741	1138	1274	1098	946	540	249	45	6574
MOUNT MORRIS 2 W		13	30	144	446	756	1147	1305	1159	1004	591	296	61	6952
NEW YORK CNTRL PK WSO		0	0	36	240	534	893	1029	885	732	378	134	7	4868
NEW YORK JFK INTL AP		0	0	47	270	558	908	1045	910	775	441	192	23	5169
NEW YORK LA GUARDIA WSO		0	0	35	246	537	893	1029	888	744	393	149	8	4922
NY WESTERLEIGH STAT IS		0	0	44	278	573	924	1054	904	753	414	169	10	5123
NORWICH 1 NE		35	59	217	543	831	1237	1407	1240	1063	651	356	106	7745
OGDENSBURG 3 NE		17	48	178	490	831	1318	1504	1313	1104	645	316	68	7832
OSWEGO EAST	//	9	19	127	403	720	1116	1287	1134	992	627	327	80	6841

NEW YORK

HEATING DEGREE DAY NORMALS (BASE 65 DEG F)

STATION	JUL	AUG	SEP	OCT	NOV	DEC	JAN	FEB	MAR	APR	MAY	JUN	ANN
PATCHOGUE 2 N	0	0	65	335	618	955	1094	958	834	522	246	37	5664
PENN YAN 2 SW	12	12	105	379	705	1082	1240	1100	949	543	251	38	6416
PERU 2 WSW	20	42	198	521	846	1305	1460	1268	1073	651	315	55	7754
PORT JERVIS	0	7	109	412	717	1094	1225	1036	874	471	203	22	6170
POUGHKEEPSIE FAA AP	0	11	118	419	723	1107	1259	1070	893	510	228	28	6366
RIVERHEAD RESEARCH	0	0	51	287	579	918	1057	930	803	483	201	15	5324
ROCHESTER WSO //R	10	23	132	412	732	1116	1283	1137	983	570	274	41	6713
SALEM	19	47	218	539	828	1256	1442	1257	1048	627	326	75	7682
SCARSDALE	0	0	67	322	615	967	1097	932	791	435	178	15	5419
SETAUKET	0	0	38	269	558	908	1048	918	794	459	201	15	5208
SODUS CENTER //	13	21	119	404	720	1107	1268	1117	967	567	271	53	6627
SPENCER 3 W	22	47	196	521	822	1203	1361	1190	1026	627	337	97	7449
STILLWATER RESERVOIR	66	107	276	620	945	1435	1615	1434	1249	801	437	150	9135
SYRACUSE WSO R	12	25	133	425	732	1138	1308	1148	983	567	269	47	6787
TUPPER LAKE SUNMOUNT	68	114	292	623	951	1423	1581	1386	1206	780	420	148	8992
UTICA FAA AP	14	42	169	484	798	1221	1392	1210	1048	612	302	76	7368
WANAKENA RANGER SCHOOL//	60	95	268	589	921	1386	1541	1355	1181	744	388	121	8649
WATERTOWN	13	28	159	474	786	1259	1445	1260	1060	627	309	60	7480
WESTFIELD 3 SW	9	19	110	389	711	1073	1243	1106	964	573	286	51	6534
WEST POINT	0	0	65	326	651	1042	1172	1000	825	438	165	10	5694
WHITEHALL	0	11	130	450	774	1228	1420	1218	1001	552	220	21	7025

NEW YORK

COOLING DEGREE DAY NORMALS (BASE 65 DEG F)

STATION		JAN	FEB	MAR	APR	MAY	JUN	JUL	AUG	SEP	OCT	NOV	DEC	ANN
ALBANY WSO	R	0	0	0	0	19	89	206	145	35	0	0	0	494
ALCOVE DAM		0	0	0	0	16	54	151	119	23	0	0	0	363
ALFRED		0	0	0	0	11	42	91	75	20	0	0	0	239
ALLEGANY STATE PARK		0	0	0	0	13	43	90	72	21	0	0	0	239
ANGELICA		0	0	0	0	12	44	101	82	20	0	0	0	259
BAINBRIDGE		0	0	0	0	20	73	159	119	30	0	0	0	401
BATAVIA		0	0	0	0	22	85	172	141	40	0	0	0	460
BINGHAMTON WSO	R	0	0	0	0	9	58	137	104	22	0	0	0	330
BOONVILLE 2 SSW		0	0	0	0	9	26	83	65	10	0	0	0	193
BRIDGEHAMPTON		0	0	0	0	0	52	202	189	52	0	0	0	495
BROCKPORT 2 NW		0	0	0	0	14	79	191	152	45	0	0	0	481
BUFFALO WSO	//R	0	0	0	0	18	83	186	146	43	0	0	0	476
CANANDAIGUA 3 S		0	0	0	0	16	75	195	163	55	0	0	0	504
CANTON 4 SE		0	0	0	0	12	52	141	102	21	0	0	0	328
CARMEL 1 SW		0	0	0	0	13	70	194	156	37	0	0	0	470
CHASM FALLS		0	0	0	0	11	44	111	83	16	0	0	0	265
CHAZY		0	0	0	0	17	64	165	114	18	0	0	0	378
CHERRY VALLEY 2 NNE		0	0	0	0	10	42	116	80	13	0	0	0	261
COOPERSTOWN		0	0	0	0	8	37	110	83	16	0	0	0	254
CORTLAND		0	0	0	0	13	61	137	104	28	0	0	0	343
DANNEMORA		0	0	0	0	18	58	146	101	23	0	0	0	346
DANSVILLE		0	0	0	0	23	97	191	151	47	7	0	0	516
DOBBS FERRY		0	0	0	0	44	161	316	270	92	6	0	0	889
ELIZABETHTOWN		0	0	0	0	13	50	131	98	17	0	0	0	309
ELMIRA		0	0	0	0	14	79	167	121	30	0	0	0	411
FRANKLINVILLE		0	0	0	0	7	40	82	57	15	0	0	0	201
FREDONIA		0	0	0	0	19	101	199	163	62	7	0	0	551
GENEVA RESEARCH FARM		0	0	0	0	14	77	179	144	36	0	0	0	450
GLENHAM		0	0	0	0	31	146	282	233	80	0	0	0	772
GLENS FALLS FAA AP		0	0	0	0	25	58	157	110	24	0	0	0	374
GLOVERSVILLE		0	0	0	0	15	61	154	118	21	0	0	0	369
GOUVERNEUR		0	0	0	0	12	44	135	99	20	0	0	0	310
GRAFTON		0	0	0	0	13	43	114	81	18	0	0	0	269
HEMLOCK		0	0	0	0	15	77	168	128	39	0	0	0	427
INDIAN LAKE 2 SW		0	0	0	0	0	8	45	28	0	0	0	0	81
ITHACA CORNELL UNIV.	//	0	0	0	0	9	56	137	101	25	0	0	0	328
LAKE PLACID CLUB		0	0	0	0	6	16	59	31	0	0	0	0	112
LAWRENCEVILLE		0	0	0	0	15	53	146	104	21	0	0	0	339
LIBERTY		0	0	0	0	9	33	96	72	11	0	0	0	221
LITTLE FALLS CITY RES		0	0	0	0	12	60	142	101	16	0	0	0	331
LITTLE VALLEY		0	0	0	0	8	34	82	72	16	0	0	0	212
LOCKPORT 2 NE		0	0	0	0	20	80	184	149	42	0	0	0	475
LOWVILLE		0	0	0	0	13	50	117	84	15	0	0	0	279
MASSENA FAA AP		0	0	0	0	16	60	147	104	16	0	0	0	343
MILLBROOK		0	0	0	0	10	74	162	117	28	0	0	0	391
MINEOLA	R	0	0	0	0	29	147	301	266	101	15	0	0	859
MOHONK LAKE		0	0	0	0	23	75	183	141	33	0	0	0	455
MOUNT MORRIS 2 W		0	0	0	0	17	82	174	136	36	5	0	0	450
NEW YORK CNTRL PK WSO		0	0	0	0	56	199	363	322	135	14	0	0	1089
NEW YORK JFK INTL AP		0	0	0	0	28	143	310	279	113	13	0	0	886
NEW YORK LA GUARDIA WSO		0	0	0	0	47	188	353	319	131	13	0	0	1051
NY WESTERLEIGH STAT IS		0	0	0	0	32	148	304	272	101	6	0	0	863
NORWICH 1 NE		0	0	0	0	6	40	106	78	16	0	0	0	246
OGDENSBURG 3 NE		0	0	0	0	18	65	166	141	19	0	0	0	409
OSWEGO EAST	//	0	0	0	0	8	62	171	143	46	0	0	0	430

NEW YORK

COOLING DEGREE DAY NORMALS (BASE 65 DEG F)

STATION	JAN	FEB	MAR	APR	MAY	JUN	JUL	AUG	SEP	OCT	NOV	DEC	ANN
PATCHOGUE 2 N	0	0	0	0	10	88	236	215	59	0	0	0	608
PENN YAN 2 SW	0	0	0	0	28	107	226	176	60	7	0	0	604
PERU 2 WSW	0	0	0	0	15	52	157	107	15	0	0	0	346
PORT JERVIS	0	0	0	0	30	112	229	177	40	0	0	0	588
POUGHKEEPSIE FAA AP	0	0	0	0	21	103	233	185	49	0	0	0	591
RIVERHEAD RESEARCH	0	0	0	0	21	108	257	235	84	0	0	0	705
ROCHESTER WSO //R	0	0	0	0	23	92	205	163	48	0	0	0	531
SALEM	0	0	0	0	9	51	137	90	23	0	0	0	310
SCARSDALE	0	0	0	0	32	135	282	241	82	6	0	0	778
SETAUKET	0	0	0	0	24	114	270	242	86	5	0	0	741
SODUS CENTER //	0	0	0	0	17	89	193	151	41	0	0	0	491
SPENCER 3 W	0	0	0	0	8	52	112	88	19	0	0	0	279
STILLWATER RESERVOIR	0	0	0	0	6	18	66	58	6	0	0	0	154
SYRACUSE WSO R	0	0	0	0	21	86	195	158	46	0	0	0	506
TUPPER LAKE SUNMOUNT	0	0	0	0	0	22	68	48	7	0	0	0	145
UTICA FAA AP	0	0	0	0	14	79	163	138	31	0	0	0	425
WANAKENA RANGER SCHOOL//	0	0	0	0	7	19	73	48	7	0	0	0	154
WATERTOWN	0	0	0	0	15	66	174	136	33	0	0	0	424
WESTFIELD 3 SW	0	0	0	0	19	90	182	149	53	8	0	0	501
WEST POINT	0	0	0	0	41	160	307	257	86	7	0	0	858
WHITEHALL	0	0	0	0	30	117	242	169	40	0	0	0	598

30 — NEW YORK

STATE-STATION NUMBER	STN TYP	NAME		LATITUDE DEG-MIN	LONGITUDE DEG-MIN	ELEVATION (FT)
30-0023	12	ADDISON		N 4206	W 07714	1000
30-0042	13	ALBANY WSO	R	N 4245	W 07348	275
30-0063	13	ALCOVE DAM		N 4228	W 07356	600
30-0085	13	ALFRED		N 4215	W 07748	1740
30-0093	13	ALLEGANY STATE PARK		N 4206	W 07845	1500
30-0183	13	ANGELICA		N 4218	W 07801	1420
30-0220	12	ARCADE		N 4232	W 07825	1490
30-0254	12	ARKVILLE 2 W		N 4208	W 07439	1310
30-0360	13	BAINBRIDGE		N 4218	W 07529	1015
30-0379	12	BALDWINSVILLE		N 4309	W 07620	379
30-0443	13	BATAVIA		N 4300	W 07811	900
30-0500	12	BEAVER FALLS		N 4353	W 07526	740
30-0608	12	BENNETT BRIDGE		N 4332	W 07557	660
30-0668	12	BIG MOOSE 3 E		N 4349	W 07452	1850
30-0687	13	BINGHAMTON WSO	R	N 4213	W 07559	1590
30-0785	13	BOONVILLE 2 SSW		N 4327	W 07521	1580
30-0817	12	BRADFORD 1 NW		N 4223	W 07707	1360
30-0870	12	BREWERTON LOCK 23		N 4314	W 07612	377
30-0889	13	BRIDGEHAMPTON		N 4057	W 07218	60
30-0929	12	BROADALBIN 5 NE		N 4307	W 07409	800
30-0937	13	BROCKPORT 2 NW		N 4315	W 07758	413
30-1012	13	BUFFALO WSO	//R	N 4256	W 07844	705
30-1152	13	CANANDAIGUA 3 S		N 4251	W 07717	720
30-1160	12	CANASTOTA		N 4305	W 07546	410
30-1168	12	CANDOR		N 4214	W 07620	900
30-1173	12	CANISTEO		N 4217	W 07737	1130
30-1185	13	CANTON 4 SE		N 4434	W 07507	440
30-1207	13	CARMEL 1 SW		N 4125	W 07342	490
30-1265	12	CAYUGA LOCK 1		N 4257	W 07644	380
30-1387	13	CHASM FALLS		N 4445	W 07413	1060
30-1401	13	CHAZY		N 4453	W 07326	170
30-1413	12	CHEMUNG		N 4200	W 07638	810
30-1436	13	CHERRY VALLEY 2 NNE		N 4249	W 07444	1360
30-1492	12	CINCINNATUS		N 4232	W 07554	1450
30-1580	12	CLYDE LOCK 26		N 4304	W 07650	392
30-1664	12	COLTON 2 N		N 4435	W 07457	580
30-1708	12	CONKLINGVILLE DAM		N 4319	W 07356	808
30-1752	13	COOPERSTOWN		N 4242	W 07455	1240
30-1799	13	CORTLAND		N 4236	W 07611	1129
30-1966	13	DANNEMORA		N 4443	W 07343	1340
30-1974	13	DANSVILLE		N 4234	W 07742	685
30-2129	13	DOBBS FERRY		N 4101	W 07352	240
30-2137	12	DOLGEVILLE		N 4305	W 07446	685
30-2554	13	ELIZABETHTOWN		N 4413	W 07335	590
30-2574	12	ELLENBURG DEPOT		N 4454	W 07348	860
30-2582	12	ELLENVILLE		N 4143	W 07424	350
30-2610	13	ELMIRA		N 4206	W 07649	880
30-3010	12	FRANKFORT LOCK 19		N 4304	W 07507	410
30-3025	13	FRANKLINVILLE		N 4221	W 07827	1590
30-3033	13	FREDONIA		N 4225	W 07918	760

STATE-STATION NUMBER	STN TYP	NAME	LATITUDE DEG-MIN	LONGITUDE DEG-MIN	ELEVATION (FT)
30-3050	12	FREEVILLE 2 NE	N 4232	W 07619	1080
30-3076	12	FROST VALLEY	N 4158	W 07433	1840
30-3184	13	GENEVA RESEARCH FARM	N 4253	W 07702	718
30-3259	13	GLENHAM	N 4131	W 07356	275
30-3284	12	GLENS FALLS FARM	N 4320	W 07344	504
30-3294	13	GLENS FALLS FAA AP	N 4321	W 07337	321
30-3319	13	GLOVERSVILLE	N 4302	W 07421	760
30-3346	13	GOUVERNEUR	N 4420	W 07529	460
30-3360	13	GRAFTON	N 4247	W 07328	1560
30-3365	12	GRAHAMSVILLE	N 4151	W 07432	960
30-3722	12	HASKINVILLE	N 4225	W 07734	1640
30-3773	13	HEMLOCK	N 4247	W 07737	902
30-3851	12	HIGHMARKET	N 4335	W 07531	1790
30-3889	12	HINCKLEY	N 4319	W 07507	1190
30-3961	12	HOOKER 4 N //	N 4345	W 07544	1680
30-3970	12	HOPE	N 4321	W 07416	950
30-4102	13	INDIAN LAKE 2 SW	N 4345	W 07417	1660
30-4174	13	ITHACA CORNELL UNIV. //	N 4227	W 07627	960
30-4525	12	LAKE DELAWARE	N 4215	W 07454	1480
30-4555	13	LAKE PLACID CLUB	N 4417	W 07359	1880
30-4647	13	LAWRENCEVILLE	N 4445	W 07439	500
30-4731	13	LIBERTY	N 4148	W 07445	1610
30-4791	13	LITTLE FALLS CITY RES	N 4304	W 07452	900
30-4796	12	LITTLE FALLS MILL ST	N 4302	W 07452	360
30-4808	13	LITTLE VALLEY	N 4215	W 07848	1575
30-4836	12	LOCKE 2 W	N 4240	W 07628	1180
30-4844	13	LOCKPORT 2 NE	N 4311	W 07839	520
30-4912	13	LOWVILLE	N 4348	W 07530	960
30-4944	12	LYONS FALLS	N 4337	W 07522	800
30-5134	13	MASSENA FAA AP	N 4456	W 07451	214
30-5171	12	MAYS POINT LOCK 25	N 4300	W 07646	400
30-5334	13	MILLBROOK	N 4151	W 07337	815
30-5377	13	MINEOLA R	N 4044	W 07338	128
30-5426	13	MOHONK LAKE	N 4146	W 07409	1245
30-5597	13	MOUNT MORRIS 2 W	N 4244	W 07754	880
30-5679	12	NEWARK	N 4303	W 07705	430
30-5751	12	NEW LONDON LOCK 22	N 4313	W 07537	400
30-5796	12	NEW YORK AVE V BRKLYN	N 4036	W 07359	15
30-5801	13	NEW YORK CNTRL PK WSO	N 4047	W 07358	132
30-5803	13	NEW YORK JFK INTL AP	N 4039	W 07347	13
30-5811	13	NEW YORK LA GUARDIA WSO	N 4046	W 07354	11
30-5821	13	NY WESTERLEIGH STAT IS	N 4036	W 07410	80
30-6085	13	NORWICH 1 NE	N 4232	W 07530	1120
30-6164	13	OGDENSBURG 3 NE	N 4444	W 07527	285
30-6196	12	OLEAN	N 4205	W 07827	1420
30-6314	13	OSWEGO EAST //	N 4328	W 07630	350
30-6441	13	PATCHOGUE 2 N	N 4048	W 07301	55
30-6510	13	PENN YAN 2 SW	N 4239	W 07705	720
30-6538	13	PERU 2 WSW	N 4434	W 07334	510
30-6774	13	PORT JERVIS	N 4123	W 07441	470

STATE-STATION NUMBER	STN TYP	NAME	LATITUDE DEG-MIN	LONGITUDE DEG-MIN	ELEVATION (FT)
30-6820	13	POUGHKEEPSIE FAA AP	N 4138	W 07353	154
30-7134	13	RIVERHEAD RESEARCH	N 4058	W 07243	100
30-7167	13	ROCHESTER WSO //R	N 4307	W 07740	547
30-7348	12	SABATTIS 3 NE	N 4407	W 07440	1760
30-7405	13	SALEM	N 4310	W 07319	490
30-7497	13	SCARSDALE	N 4059	W 07348	199
30-7513	12	SCHENECTADY	N 4250	W 07355	225
30-7633	13	SETAUKET	N 4058	W 07306	40
30-7705	12	SHERBURNE 2 S	N 4239	W 07529	1080
30-7721	12	SHOKAN BROWN STATION	N 4157	W 07413	520
30-7780	12	SKANEATELES	N 4257	W 07626	875
30-7818	12	SMITHS BASIN	N 4321	W 07330	142
30-7842	13	SODUS CENTER //	N 4312	W 07701	420
30-8058	12	SOUTH WALES EMERY PARK	N 4243	W 07836	1090
30-8088	13	SPENCER 3 W	N 4212	W 07634	1020
30-8248	13	STILLWATER RESERVOIR	N 4353	W 07502	1690
30-8322	12	SUFFERN WATER WORKS	N 4107	W 07409	270
30-8383	13	SYRACUSE WSO R	N 4307	W 07607	410
30-8578	12	TRENTON FALLS	N 4316	W 07509	800
30-8586	12	TRIBES HILL	N 4257	W 07417	300
30-8631	13	TUPPER LAKE SUNMOUNT	N 4414	W 07426	1680
30-8737	13	UTICA FAA AP	N 4309	W 07523	718
30-8944	13	WANAKENA RANGER SCHOOL//	N 4409	W 07454	1510
30-8987	12	WATERLOO	N 4254	W 07652	452
30-9000	13	WATERTOWN	N 4358	W 07552	497
30-9189	13	WESTFIELD 3 SW	N 4217	W 07937	975
30-9292	13	WEST POINT	N 4123	W 07358	320
30-9389	13	WHITEHALL	N 4333	W 07324	119
30-9507	12	WILSON 2 NE	N 4319	W 07848	270
30-9544	12	WOLCOTT 3 NW	N 4315	W 07652	400

30 – NEW YORK

LEGEND

○ = TEMPERATURE ONLY
× = PRECIPITATION ONLY
⊗ = TEMPERATURE & PRECIPITATION

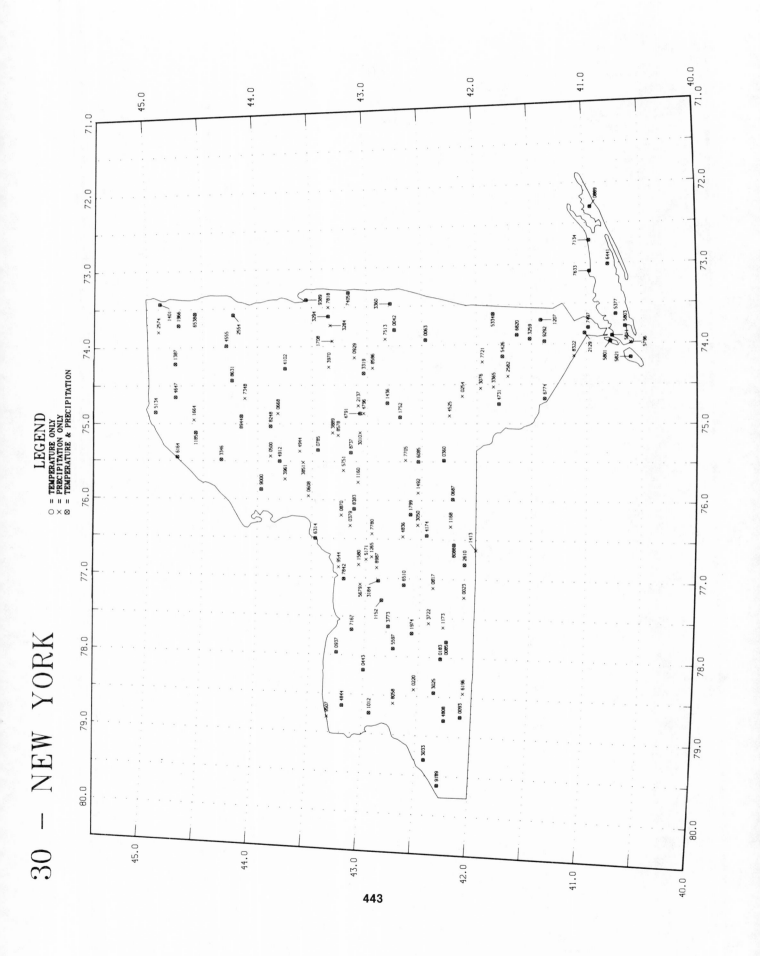

443

NORTH CAROLINA

TEMPERATURE NORMALS (DEG F)

STATION		JAN	FEB	MAR	APR	MAY	JUN	JUL	AUG	SEP	OCT	NOV	DEC	ANN
ALBEMARLE 4 N	MAX	52.3	55.9	64.0	74.1	79.9	85.6	88.7	87.7	82.3	72.7	63.8	54.6	71.8
	MIN	29.3	31.0	37.8	46.6	55.2	62.4	66.3	65.8	59.7	47.4	37.9	30.8	47.5
	MEAN	40.9	43.5	50.9	60.4	67.6	74.0	77.6	76.8	71.0	60.1	50.9	42.8	59.7
ANDREWS 2 E	MAX	48.3	51.7	59.3	69.8	76.8	82.9	85.6	85.4	80.7	71.1	60.3	51.4	68.6
	MIN	23.0	24.2	31.0	39.6	48.5	55.9	59.6	58.9	52.9	39.6	30.7	24.8	40.7
	MEAN	35.7	38.0	45.2	54.7	62.7	69.4	72.6	72.2	66.8	55.4	45.5	38.1	54.7
ASHEBORO 2 W	MAX	50.6	53.8	62.3	73.0	79.1	85.0	88.4	87.5	81.7	71.9	62.4	53.0	70.7
	MIN	30.8	32.1	39.3	48.3	56.0	62.9	67.0	66.4	60.6	48.7	40.1	33.0	48.8
	MEAN	40.7	43.0	50.8	60.7	67.6	74.0	77.7	77.0	71.2	60.4	51.3	43.0	59.8
ASHEVILLE WSO	MAX	47.5	50.6	58.4	68.6	75.6	81.4	84.0	83.5	77.9	68.7	58.6	50.3	67.1
	MIN	26.0	27.6	34.4	42.7	51.0	58.2	62.4	61.6	55.8	43.3	34.2	28.2	43.8
	MEAN	36.8	39.1	46.4	55.7	63.3	69.8	73.2	72.6	66.9	56.0	46.4	39.3	55.5
ASHEVILLE R	MAX	46.7	49.8	57.6	68.2	75.4	81.4	84.3	83.8	77.8	68.0	57.6	49.6	66.7
	MIN	27.6	29.2	36.1	45.1	52.7	59.4	63.3	62.6	57.1	45.3	36.5	30.2	45.4
	MEAN	37.2	39.6	46.9	56.7	64.1	70.4	73.8	73.2	67.4	56.7	47.0	39.9	56.1
BANNER ELK	MAX	41.8	43.5	50.8	60.8	67.7	73.6	76.4	76.3	71.4	62.3	52.2	44.7	60.1
	MIN	21.0	22.3	29.2	37.6	45.3	51.4	55.1	54.5	49.3	38.4	29.8	23.7	38.1
	MEAN	31.4	32.9	40.0	49.2	56.6	62.5	65.8	65.4	60.4	50.4	41.0	34.2	49.2
BENT CREEK	MAX	47.3	50.7	58.4	68.9	75.9	81.6	84.1	83.8	78.6	69.6	58.9	50.1	67.3
	MIN	24.5	26.1	32.9	40.5	48.6	55.8	60.1	59.4	53.8	41.6	32.9	26.8	41.9
	MEAN	35.9	38.4	45.7	54.8	62.3	68.8	72.2	71.6	66.2	55.7	45.9	38.5	54.7
BLACK MOUNTAIN	MAX	48.7	51.4	58.9	69.2	75.8	81.2	84.0	83.5	77.9	69.6	59.4	51.7	67.6
	MIN	26.3	27.3	34.6	42.7	50.0	56.6	60.6	60.0	54.5	42.7	34.3	28.3	43.2
	MEAN	37.5	39.4	46.8	56.0	62.9	69.0	72.3	71.8	66.2	56.2	46.9	40.0	55.4
BOONE	MAX	41.7	43.9	51.8	62.7	70.1	75.5	78.0	77.3	71.4	62.7	52.5	44.8	61.0
	MIN	22.8	24.2	30.8	39.6	48.1	54.7	58.5	57.6	51.6	40.7	32.0	25.6	40.5
	MEAN	32.2	34.1	41.3	51.2	59.1	65.1	68.3	67.5	61.6	51.7	42.3	35.2	50.8
BREVARD	MAX	49.5	53.0	60.6	70.2	76.5	81.9	84.5	83.7	78.4	70.2	60.2	52.0	68.4
	MIN	25.3	26.6	33.3	41.0	49.1	56.6	60.5	60.2	54.9	41.9	32.2	26.9	42.4
	MEAN	37.4	39.8	47.0	55.6	62.9	69.3	72.5	72.0	66.7	56.1	46.2	39.5	55.4
BURLINGTON FILTER PL	MAX	49.4	52.8	61.2	72.7	79.6	85.8	89.3	88.1	82.0	71.7	61.7	52.2	70.5
	MIN	29.0	30.5	37.7	47.3	56.1	63.5	67.4	66.5	60.1	47.8	38.5	31.1	48.0
	MEAN	39.2	41.7	49.4	60.0	67.9	74.7	78.3	77.4	71.0	59.8	50.1	41.7	59.3
CANTON 1 SW	MAX	47.2	49.9	57.3	67.5	74.3	80.0	82.7	82.3	76.8	67.5	57.7	50.1	66.1
	MIN	23.2	24.9	31.8	39.7	47.1	54.2	58.1	57.3	51.7	39.4	30.3	25.2	40.2
	MEAN	35.2	37.4	44.5	53.6	60.7	67.2	70.4	69.9	64.3	53.5	44.1	37.6	53.2
CAPE HATTERAS WSO	MAX	52.6	53.5	58.8	67.2	74.1	80.5	84.4	84.4	80.5	71.7	63.6	56.4	69.0
	MIN	37.6	37.7	43.3	51.1	59.7	67.5	71.9	72.0	67.9	58.1	48.3	40.9	54.7
	MEAN	45.1	45.6	51.1	59.2	66.9	74.0	78.2	78.2	74.2	64.9	56.0	48.7	61.9
CELO 2 S	MAX	45.6	48.0	55.3	65.7	72.6	77.8	80.6	79.9	74.4	65.7	56.3	48.4	64.2
	MIN	20.3	21.4	28.9	37.2	45.1	51.9	56.2	55.7	49.8	36.9	28.2	21.9	37.8
	MEAN	32.9	34.7	42.2	51.5	58.9	64.9	68.4	67.8	62.1	51.3	42.3	35.1	51.0
CHAPEL HILL 2 W	MAX	50.1	52.6	60.9	71.9	78.6	84.8	88.7	87.4	82.0	71.7	62.4	52.7	70.3
	MIN	27.3	28.8	36.5	45.7	54.1	61.6	65.7	64.8	58.4	45.9	36.7	29.2	46.2
	MEAN	38.7	40.8	48.7	58.8	66.4	73.2	77.2	76.1	70.2	58.8	49.6	41.0	58.3
CHARLOTTE WSO R	MAX	50.3	53.6	61.6	72.1	79.1	85.2	88.3	87.6	81.7	71.7	61.7	52.6	70.5
	MIN	30.7	32.1	39.1	48.4	57.2	64.7	68.7	68.2	62.3	49.6	39.7	32.6	49.4
	MEAN	40.5	42.9	50.4	60.3	68.2	75.0	78.5	77.9	72.0	60.7	50.7	42.6	50.0
CONCORD	MAX	50.7	54.0	62.3	73.4	80.6	86.7	90.1	89.3	83.6	73.3	63.2	53.3	71.7
	MIN	28.9	30.0	37.5	47.1	55.9	63.4	67.3	66.7	60.5	47.6	38.1	30.7	47.8
	MEAN	39.8	42.0	49.9	60.2	68.3	75.0	78.8	78.0	72.0	60.5	50.6	42.0	59.8

NORTH CAROLINA

TEMPERATURE NORMALS (DEG F)

STATION		JAN	FEB	MAR	APR	MAY	JUN	JUL	AUG	SEP	OCT	NOV	DEC	ANN
COWEETA EXP STATION	MAX	48.8	52.0	58.9	68.9	75.2	80.5	83.2	82.9	77.9	69.5	59.8	51.4	67.4
	MIN	24.5	25.9	32.6	40.5	47.7	54.4	58.2	57.6	52.7	40.2	31.8	25.9	41.0
	MEAN	36.7	39.0	45.7	54.8	61.5	67.5	70.7	70.3	65.3	54.9	45.8	38.7	54.2
CULLOWHEE	MAX	49.6	53.0	60.9	70.6	77.1	82.2	84.7	84.2	79.5	70.9	60.4	52.0	68.8
	MIN	25.6	27.1	33.4	41.0	49.1	56.6	61.0	60.5	55.1	42.1	32.3	27.2	42.6
	MEAN	37.6	40.1	47.2	55.8	63.1	69.4	72.9	72.4	67.3	56.5	46.4	39.6	55.7
DURHAM	MAX	50.9	53.5	62.0	73.1	80.0	86.3	89.6	88.5	82.9	72.6	62.8	53.2	71.3
	MIN	28.0	28.9	36.1	45.4	54.3	61.5	65.7	65.3	58.3	46.0	36.5	29.6	46.3
	MEAN	39.5	41.3	49.1	59.3	67.1	73.9	77.7	76.9	70.6	59.3	49.7	41.4	58.8
EDENTON	MAX	52.3	54.7	62.4	73.1	78.9	84.6	87.6	86.7	81.7	71.8	63.6	55.5	71.1
	MIN	32.7	33.8	40.5	49.7	57.9	65.7	70.0	69.2	63.6	52.1	42.9	35.0	51.1
	MEAN	42.5	44.3	51.5	61.4	68.4	75.2	78.9	78.0	72.7	62.0	53.3	45.2	61.1
ELIZABETH CITY	MAX	51.3	53.2	60.9	71.1	77.6	84.0	87.7	86.8	81.8	72.0	63.4	54.6	70.4
	MIN	31.0	32.4	39.3	48.0	56.7	64.2	68.9	68.5	62.4	50.8	41.2	33.6	49.8
	MEAN	41.2	42.8	50.2	59.6	67.2	74.1	78.3	77.7	72.1	61.5	52.3	44.2	60.1
ELIZABETH CITY FAA AP	MAX	50.5	52.6	59.8	69.9	76.8	83.6	87.3	86.4	81.2	71.4	62.6	53.9	69.7
	MIN	32.2	33.1	40.0	48.6	57.5	65.1	69.5	69.1	63.5	52.1	42.2	34.7	50.6
	MEAN	41.4	42.9	49.9	59.3	67.2	74.4	78.4	77.8	72.4	61.8	52.4	44.3	60.2
FAYETTEVILLE	MAX	53.1	55.6	63.6	74.2	81.1	86.8	89.8	89.0	84.0	74.4	65.2	55.8	72.7
	MIN	29.6	31.1	38.8	48.1	57.0	64.3	68.7	68.0	61.8	48.6	38.6	31.2	48.8
	MEAN	41.4	43.4	51.2	61.2	69.1	75.6	79.3	78.5	72.9	61.6	52.0	43.5	60.8
FLETCHER 2 NE	MAX	46.5	49.4	57.1	67.3	74.8	80.9	83.7	83.2	77.7	68.2	58.1	49.9	66.4
	MIN	22.1	23.4	30.2	38.4	47.0	55.0	58.9	58.2	51.6	38.9	29.8	23.8	39.8
	MEAN	34.4	36.4	43.7	52.9	60.9	68.0	71.4	70.7	64.6	53.6	44.0	36.8	53.1
FRANKLIN 1 SSW	MAX	50.1	53.3	60.7	70.7	77.6	82.9	85.5	85.3	80.4	71.7	61.0	52.7	69.3
	MIN	25.3	26.6	33.4	40.9	49.0	56.8	61.2	60.9	55.3	41.9	32.1	26.8	42.5
	MEAN	37.7	40.0	47.0	55.8	63.3	69.9	73.4	73.1	67.9	56.8	46.6	39.7	55.9
GASTONIA	MAX	52.1	55.7	63.9	74.4	81.1	86.7	89.6	88.9	83.1	73.3	63.4	54.3	72.2
	MIN	31.1	32.3	39.2	48.1	56.6	63.7	67.7	67.1	61.0	48.9	39.3	32.6	49.0
	MEAN	41.7	44.0	51.6	61.2	68.8	75.2	78.7	78.0	72.1	61.1	51.4	43.4	60.6
GOLDSBORO 1 SSW	MAX	52.5	54.7	63.0	73.9	80.9	86.9	90.2	89.1	84.3	74.2	64.9	55.4	72.5
	MIN	30.5	32.2	39.9	49.0	57.7	64.8	69.1	68.4	62.0	49.4	39.8	32.1	49.6
	MEAN	41.5	43.5	51.5	61.5	69.3	75.9	79.7	78.8	73.2	61.8	52.4	43.8	61.1
GREENSBORO WSO R	MAX	47.6	50.8	59.3	70.7	77.9	84.2	87.4	86.2	80.4	70.1	59.9	50.4	68.7
	MIN	27.3	29.0	36.5	45.9	55.0	62.6	66.9	66.3	59.3	46.7	37.1	29.9	46.9
	MEAN	37.5	39.9	48.0	58.3	66.5	73.5	77.2	76.3	69.9	58.4	48.5	40.2	57.9
HAMLET	MAX	53.5	57.1	65.2	75.6	82.5	88.3	91.1	90.1	84.8	75.0	65.1	55.6	73.7
	MIN	29.4	30.9	38.4	47.7	56.4	63.5	67.6	67.1	61.2	48.2	37.9	31.3	48.3
	MEAN	41.5	44.0	51.8	61.7	69.5	75.9	79.3	78.6	73.0	61.6	51.5	43.5	61.0
HATTERAS	MAX	52.9	53.7	59.4	67.8	74.7	81.0	84.9	85.1	81.3	72.8	64.5	56.9	69.6
	MIN	38.5	39.2	45.0	53.3	61.2	68.9	73.1	73.3	69.5	59.9	50.3	42.2	56.2
	MEAN	45.7	46.5	52.2	60.6	68.0	75.0	79.1	79.3	75.4	66.4	57.4	49.5	62.9
HENDERSONVILLE 1 NE	MAX	47.9	51.4	59.3	69.6	76.4	81.7	84.5	83.7	77.8	68.5	58.4	50.5	67.5
	MIN	25.5	26.9	33.7	41.7	49.8	56.9	60.9	60.0	54.0	41.7	33.0	27.2	42.6
	MEAN	36.7	39.2	46.5	55.7	63.1	69.3	72.7	71.8	65.9	55.1	45.7	38.9	55.1
HICKORY FAA AP	MAX	48.4	51.5	59.5	70.4	77.8	83.9	86.9	86.2	80.0	70.4	60.0	51.0	68.8
	MIN	28.4	29.9	37.4	46.6	55.3	62.6	66.5	65.7	59.6	47.4	37.7	30.8	47.3
	MEAN	38.4	40.7	48.5	58.5	66.6	73.3	76.7	76.0	69.8	58.9	48.9	40.9	58.1
HIGHLANDS 2 S	MAX	44.6	47.0	53.9	63.9	70.0	74.4	77.0	76.5	71.1	62.5	53.8	46.6	61.8
	MIN	27.1	28.2	34.3	42.6	50.6	56.7	60.4	60.1	55.3	44.1	35.1	29.4	43.7
	MEAN	35.9	37.6	44.1	53.3	60.3	65.6	68.7	68.3	63.2	53.3	44.5	38.0	52.7

NORTH CAROLINA

TEMPERATURE NORMALS (DEG F)

STATION		JAN	FEB	MAR	APR	MAY	JUN	JUL	AUG	SEP	OCT	NOV	DEC	ANN
HIGH POINT	MAX	49.7	53.2	61.4	72.4	79.5	85.2	88.3	87.1	81.4	71.5	61.3	52.1	70.3
	MIN	29.9	31.3	38.5	48.0	56.1	62.9	66.9	66.1	60.0	48.4	39.6	32.4	48.3
	MEAN	39.8	42.3	50.0	60.2	67.8	74.1	77.6	76.6	70.8	60.0	50.5	42.3	59.3
HOT SPRINGS 2	MAX	47.0	50.3	58.9	70.1	77.8	84.2	87.5	87.2	82.0	71.5	59.8	50.4	68.9
	MIN	27.8	29.5	36.3	44.5	52.6	59.5	63.3	62.7	57.7	45.8	36.4	30.2	45.5
	MEAN	37.5	39.9	47.6	57.3	65.2	71.9	75.5	75.0	69.9	58.7	48.1	40.3	57.2
JACKSON	MAX	50.5	53.1	61.9	73.2	79.7	86.0	89.4	88.2	82.7	72.3	63.0	53.4	71.1
	MIN	29.4	30.8	37.9	46.3	55.3	62.6	67.2	66.6	60.3	48.6	38.8	31.5	47.9
	MEAN	40.0	42.0	49.9	59.8	67.5	74.3	78.3	77.4	71.5	60.4	50.9	42.5	59.5
JACKSON SPRINGS 5 WNW	MAX	50.6	53.3	61.7	72.6	79.6	85.6	88.9	87.9	82.9	72.8	63.2	53.5	71.1
	MIN	30.2	31.1	38.8	48.3	56.7	63.7	67.5	66.7	61.0	49.3	40.4	32.4	48.8
	MEAN	40.4	42.2	50.3	60.5	68.1	74.7	78.2	77.3	72.0	61.1	51.8	43.0	60.0
KINSTON 5 SE	MAX	52.9	55.2	63.2	73.9	80.5	86.3	89.5	88.9	84.0	74.3	65.4	55.8	72.5
	MIN	30.7	31.9	39.6	48.1	56.3	63.5	67.8	67.2	61.6	49.2	39.6	32.1	49.0
	MEAN	41.8	43.6	51.4	61.0	68.4	74.9	78.7	78.1	72.8	61.8	52.5	44.0	60.8
LAURINBURG	MAX	55.4	58.8	66.6	76.8	83.5	88.8	91.2	90.6	85.6	76.1	66.6	57.8	74.8
	MIN	32.9	34.2	41.2	49.8	57.9	64.9	68.7	68.0	62.3	50.2	41.0	34.4	50.5
	MEAN	44.2	46.5	54.0	63.3	70.7	76.8	80.0	79.4	74.0	63.2	53.8	46.1	62.7
LENOIR	MAX	50.7	54.0	62.3	72.9	79.4	84.8	87.6	87.0	81.4	72.4	62.1	53.2	70.7
	MIN	27.6	29.1	35.9	44.7	52.8	59.8	63.5	62.9	57.3	45.3	35.9	29.4	45.4
	MEAN	39.2	41.6	49.1	58.8	66.1	72.3	75.6	75.0	69.4	58.9	49.0	41.3	58.0
LEXINGTON	MAX	50.8	54.4	63.3	74.4	81.7	87.7	90.9	89.7	83.7	73.5	62.7	53.2	72.2
	MIN	30.7	32.1	39.3	48.6	56.8	64.0	67.9	67.0	61.0	48.8	39.8	32.6	49.1
	MEAN	40.8	43.3	51.3	61.5	69.3	75.8	79.4	78.4	72.4	61.2	51.2	42.9	60.6
LOUISBURG	MAX	50.2	52.8	61.6	73.1	80.1	86.4	90.2	88.9	83.6	72.6	62.8	52.9	71.3
	MIN	24.7	26.0	33.4	42.2	51.7	59.7	64.6	63.9	56.4	43.2	33.0	26.2	43.8
	MEAN	37.5	39.4	47.5	57.7	65.9	73.0	77.4	76.5	70.0	57.9	47.9	39.5	57.5
LUMBERTON 6 NW	MAX	53.7	56.4	64.5	74.7	81.6	86.7	89.4	88.9	84.0	74.7	65.6	56.6	73.1
	MIN	31.2	32.7	40.2	48.8	57.0	64.0	68.0	67.3	61.3	48.4	39.2	32.4	49.2
	MEAN	42.5	44.6	52.4	61.8	69.3	75.4	78.8	78.1	72.7	61.6	52.4	44.5	61.2
MARION	MAX	50.2	53.8	62.4	72.7	79.1	84.8	87.5	86.5	80.7	71.3	61.3	52.4	70.2
	MIN	28.5	29.9	36.8	45.8	53.7	61.0	64.8	64.1	58.5	46.8	37.3	30.5	46.5
	MEAN	39.4	41.9	49.7	59.3	66.4	72.9	76.2	75.3	69.6	59.1	49.3	41.5	58.4
MARSHALL	MAX	46.1	49.5	57.9	68.2	75.8	81.6	84.4	83.9	79.2	69.1	57.8	49.4	66.9
	MIN	24.6	26.1	33.0	41.3	49.7	57.1	61.3	60.4	54.3	41.9	32.9	26.8	42.5
	MEAN	35.4	37.9	45.5	54.8	62.7	69.4	72.9	72.2	66.8	55.5	45.4	38.2	54.7
MAYSVILLE 6 SW	MAX	56.3	58.1	65.4	74.8	80.8	85.5	88.6	88.1	83.9	75.1	67.3	59.1	73.6
	MIN	31.1	32.1	38.4	46.1	54.6	61.8	66.5	65.8	60.1	48.9	39.4	32.8	48.1
	MEAN	43.7	45.2	51.9	60.5	67.7	73.7	77.6	77.0	72.1	62.1	53.4	46.0	60.9
MONCURE 3 SE	MAX	51.3	53.7	62.0	73.4	80.3	86.4	89.9	88.8	83.2	73.2	63.9	54.0	71.7
	MIN	25.1	26.5	34.6	43.5	52.6	60.2	64.7	64.4	57.7	44.1	33.9	26.8	44.5
	MEAN	38.2	40.1	48.3	58.5	66.5	73.3	77.3	76.7	70.5	58.7	48.9	40.5	58.1
MONROE 4 SE	MAX	52.6	56.3	63.9	73.9	80.6	86.5	89.3	88.6	83.4	73.6	64.1	55.1	72.3
	MIN	31.2	32.5	39.5	48.4	56.4	63.5	67.5	66.6	60.5	48.1	39.2	32.7	48.8
	MEAN	41.9	44.4	51.7	61.2	68.5	75.0	78.4	77.6	72.0	60.9	51.6	43.9	60.6
MOREHEAD CITY 2 WNW	MAX	54.5	55.9	62.1	70.5	77.8	83.4	86.5	86.7	83.1	74.8	66.2	57.9	71.6
	MIN	36.4	37.0	43.6	52.3	61.2	68.5	72.5	72.4	67.7	56.8	46.5	38.9	54.5
	MEAN	45.5	46.5	52.9	61.4	69.5	76.0	79.5	79.6	75.4	65.8	56.4	48.4	63.1
MORGANTON	MAX	51.3	54.5	62.5	73.1	79.9	85.4	88.2	87.6	81.6	72.5	62.2	53.4	71.0
	MIN	27.9	29.4	36.3	44.9	53.2	60.6	64.6	63.7	57.6	45.2	35.9	29.6	45.7
	MEAN	39.6	41.9	49.4	59.0	66.5	73.0	76.4	75.7	69.6	58.8	49.1	41.5	58.4

NORTH CAROLINA

TEMPERATURE NORMALS (DEG F)

STATION		JAN	FEB	MAR	APR	MAY	JUN	JUL	AUG	SEP	OCT	NOV	DEC	ANN
MOUNT AIRY	MAX	47.9	51.6	60.7	71.6	79.0	84.9	87.9	87.0	81.4	71.2	60.0	50.5	69.5
	MIN	25.9	27.2	34.5	42.7	51.8	59.2	63.1	62.5	56.0	43.9	34.6	27.9	44.1
	MEAN	36.9	39.5	47.6	57.2	65.4	72.1	75.6	74.8	68.7	57.6	47.3	39.2	56.8
NASHVILLE	MAX	51.2	53.6	61.7	72.9	79.3	85.4	89.0	87.7	83.0	72.6	63.8	54.0	71.2
	MIN	28.9	30.3	38.2	47.3	56.1	63.5	67.9	67.4	60.8	48.3	38.8	31.1	48.2
	MEAN	40.1	42.0	50.0	60.1	67.7	74.5	78.4	77.5	71.9	60.5	51.3	42.6	59.7
NEW BERN FAA AP	MAX	54.4	56.8	63.9	73.8	80.1	85.4	88.3	87.7	83.3	74.4	65.9	57.4	72.6
	MIN	33.6	34.8	41.5	50.0	58.8	65.9	70.2	69.8	64.6	52.9	42.6	35.5	51.7
	MEAN	44.0	45.9	52.7	61.9	69.5	75.7	79.2	78.8	74.0	63.7	54.3	46.5	62.2
NEW HOLLAND	MAX	54.8	56.8	63.7	72.5	78.7	84.2	87.4	86.9	82.6	74.1	65.9	57.8	72.1
	MIN	33.9	35.0	41.7	49.6	57.7	65.3	69.7	69.3	64.6	53.8	44.1	36.1	51.7
	MEAN	44.4	45.9	52.7	61.1	68.2	74.7	78.6	78.1	73.6	64.0	55.0	47.0	61.9
OXFORD 2 SW	MAX	50.3	53.1	61.7	72.6	79.5	85.6	88.9	87.7	82.4	72.2	62.4	52.9	70.8
	MIN	29.6	30.9	37.9	47.0	55.6	62.7	66.7	66.1	59.7	48.1	39.1	31.7	47.9
	MEAN	40.0	42.0	49.8	59.8	67.6	74.2	77.8	76.9	71.1	60.2	50.8	42.3	59.4
PISGAH FOREST 1 N	MAX	48.0	51.0	58.6	68.8	75.3	80.7	83.4	82.8	77.5	69.0	59.1	50.7	67.1
	MIN	24.1	25.2	31.8	40.3	48.3	55.7	59.7	59.2	53.9	40.9	31.5	25.6	41.4
	MEAN	36.1	38.1	45.2	54.6	61.8	68.2	71.6	71.0	65.7	55.0	45.3	38.2	54.2
PLYMOUTH 5 E	MAX	53.5	55.7	63.3	73.9	80.2	86.1	89.1	88.2	83.2	73.6	65.0	56.5	72.4
	MIN	30.7	32.1	38.5	46.6	55.2	62.5	66.9	66.4	60.8	49.4	39.7	32.7	48.5
	MEAN	42.1	43.9	50.9	60.3	67.7	74.3	78.0	77.3	72.0	61.5	52.4	44.6	60.4
RALEIGH DURHAM WSO R	MAX	50.1	52.8	61.0	72.3	79.0	85.2	88.2	87.1	81.6	71.6	61.8	52.7	70.3
	MIN	29.1	30.3	37.7	46.5	55.3	62.6	67.1	66.8	60.4	47.7	38.1	31.2	47.7
	MEAN	39.6	41.6	49.3	59.5	67.2	73.9	77.7	77.0	71.0	59.7	50.0	42.0	59.0
RALEIGH-N C STATE UNIV	MAX	50.3	52.8	61.0	72.2	79.0	85.1	88.8	87.6	82.1	71.7	62.6	53.1	70.5
	MIN	30.1	31.4	39.0	48.5	57.2	64.6	68.7	68.3	62.1	50.0	40.8	32.6	49.4
	MEAN	40.2	42.1	50.1	60.4	68.1	74.9	78.8	78.0	72.1	60.9	51.7	42.9	60.0
REIDSVILLE 2 NW	MAX	48.0	51.0	59.4	70.4	77.8	84.2	87.9	86.7	80.9	70.5	60.4	50.7	69.0
	MIN	27.9	29.3	36.9	46.4	54.9	62.0	65.8	65.0	58.6	46.9	38.3	30.5	46.9
	MEAN	38.0	40.2	48.2	58.4	66.4	73.1	76.9	75.9	69.8	58.7	49.4	40.6	58.0
SALISBURY	MAX	50.3	53.6	62.2	73.2	80.4	86.2	89.5	88.3	82.3	72.2	62.2	52.5	71.1
	MIN	29.9	31.1	38.4	47.4	55.7	63.0	67.0	66.2	59.9	47.6	38.4	31.5	48.0
	MEAN	40.1	42.4	50.4	60.3	68.1	74.6	78.3	77.2	71.1	59.9	50.3	42.0	59.6
SHELBY 2 NNE	MAX	50.8	54.0	62.3	72.7	79.6	85.8	88.8	87.9	82.0	72.1	62.5	53.1	71.0
	MIN	27.5	29.1	36.4	45.7	54.5	61.7	65.6	64.8	58.5	45.9	36.0	29.2	46.2
	MEAN	39.2	41.6	49.4	59.2	67.1	73.8	77.2	76.3	70.3	59.0	49.3	41.2	58.6
SILER CITY 2 NW	MAX	50.9	53.7	61.9	72.6	79.2	85.3	88.6	87.6	82.4	72.6	63.1	53.5	71.0
	MIN	24.9	26.2	33.8	42.8	51.9	59.3	63.6	62.9	56.1	42.9	33.5	26.3	43.7
	MEAN	37.9	40.0	47.9	57.7	65.6	72.3	76.1	75.3	69.3	57.8	48.4	39.9	57.4
SLOAN 3 S	MAX	56.5	59.1	66.6	76.1	81.9	86.7	89.6	88.9	84.2	75.2	67.1	59.0	74.2
	MIN	31.6	32.4	39.0	47.1	55.5	62.9	67.4	66.8	61.8	49.6	39.3	32.6	48.8
	MEAN	44.1	45.8	52.8	61.6	68.7	74.8	78.5	77.9	73.0	62.4	53.3	45.8	61.6
SMITHFIELD	MAX	53.2	56.1	64.4	75.4	82.0	88.0	90.8	89.8	84.5	74.3	65.0	55.9	73.3
	MIN	30.9	32.1	39.3	47.7	55.9	62.9	67.1	66.6	60.1	47.9	38.1	32.2	48.4
	MEAN	42.1	44.2	51.9	61.6	69.0	75.4	79.0	78.2	72.3	61.1	51.6	44.1	60.9
SOUTHPORT 5 N	MAX	55.8	57.2	63.4	72.0	79.0	84.3	87.1	87.1	83.3	75.1	66.8	58.8	72.5
	MIN	35.5	36.6	43.2	52.2	60.8	67.7	71.9	71.2	66.3	54.3	44.8	37.7	53.5
	MEAN	45.7	47.0	53.4	62.1	69.9	76.0	79.5	79.2	74.8	64.7	55.8	48.3	63.0
STATESVILLE 2 NNE	MAX	50.6	54.2	62.4	73.1	79.6	85.4	88.4	87.5	81.8	72.4	62.0	52.7	70.8
	MIN	26.5	28.1	34.8	44.0	53.1	60.6	64.6	63.8	57.2	44.7	35.0	28.1	45.0
	MEAN	38.6	41.2	48.7	58.6	66.4	73.0	76.5	75.7	69.5	58.6	48.6	40.4	58.0

NORTH CAROLINA

TEMPERATURE NORMALS (DEG F)

STATION		JAN	FEB	MAR	APR	MAY	JUN	JUL	AUG	SEP	OCT	NOV	DEC	ANN
TARBORO 1 S	MAX	52.0	54.2	62.8	74.2	80.9	86.8	89.9	89.2	83.9	73.6	64.1	54.9	72.2
	MIN	30.2	31.5	38.8	47.6	56.5	63.7	68.2	67.6	61.4	49.2	39.0	31.9	48.8
	MEAN	41.1	42.9	50.8	60.9	68.7	75.3	79.1	78.4	72.7	61.4	51.6	43.4	60.5
TRANSOU	MAX	42.7	45.4	53.5	64.0	70.9	76.5	79.6	78.8	73.2	63.7	53.5	45.5	62.3
	MIN	22.0	23.1	29.9	38.1	45.8	52.6	56.5	55.7	50.0	38.4	30.3	24.0	38.9
	MEAN	32.4	34.3	41.7	51.0	58.4	64.6	68.1	67.3	61.6	51.1	41.9	34.8	50.6
TRYON	MAX	53.0	56.9	65.0	74.7	80.9	86.1	88.4	87.9	82.3	73.5	63.9	55.1	72.3
	MIN	30.9	31.8	38.6	46.5	54.8	61.4	65.5	64.8	59.4	47.9	39.1	32.3	47.8
	MEAN	42.0	44.4	51.8	60.6	67.9	73.7	77.0	76.4	70.8	60.7	51.5	43.7	60.0
WADESBORO	MAX	51.6	55.0	62.8	73.5	80.5	86.4	89.7	89.0	83.6	73.6	64.0	54.4	72.0
	MIN	31.1	32.4	40.1	49.6	57.8	64.6	68.5	67.8	62.0	49.5	40.5	32.9	49.7
	MEAN	41.4	43.7	51.5	61.6	69.2	75.6	79.1	78.4	72.8	61.6	52.3	43.7	60.9
WAYNESVILLE 1 E	MAX	47.4	50.4	58.5	68.1	74.2	79.8	82.1	81.6	76.4	67.5	57.5	50.0	66.1
	MIN	24.6	26.1	33.0	40.9	47.9	54.7	58.7	58.0	52.5	40.3	31.9	26.7	41.3
	MEAN	36.0	38.3	45.8	54.5	61.1	67.3	70.4	69.8	64.5	53.9	44.7	38.4	53.7
WILLARD 4 SW	MAX	56.3	59.4	66.4	76.0	82.2	86.9	89.4	88.8	84.2	75.4	66.8	58.8	74.2
	MIN	33.1	34.3	40.9	48.6	57.3	64.0	68.0	67.6	62.4	50.8	41.2	34.6	50.2
	MEAN	44.7	46.9	53.7	62.3	69.8	75.5	78.7	78.2	73.3	63.1	54.0	46.7	62.2
WILMINGTON WSO R	MAX	55.9	58.1	64.8	74.3	80.9	86.1	89.3	88.6	83.9	75.2	66.8	59.1	73.6
	MIN	35.3	36.6	43.3	51.8	60.4	67.1	71.3	70.8	65.7	53.7	43.9	37.2	53.1
	MEAN	45.6	47.4	54.1	63.1	70.7	76.6	80.3	79.7	74.8	64.5	55.4	48.2	63.4
WILSON 2 W	MAX	52.1	54.8	62.7	74.0	80.4	86.7	89.6	88.8	83.6	73.5	64.0	54.8	72.1
	MIN	30.5	31.8	38.7	47.7	56.5	63.7	68.1	67.5	61.1	48.8	39.0	32.3	48.8
	MEAN	41.3	43.3	50.7	60.9	68.5	75.2	78.9	78.2	72.4	61.2	51.5	43.5	60.5

NORTH CAROLINA

PRECIPITATION NORMALS (INCHES)

STATION		JAN	FEB	MAR	APR	MAY	JUN	JUL	AUG	SEP	OCT	NOV	DEC	ANN
ALBEMARLE 4 N		3.74	3.85	4.74	3.53	4.07	4.12	4.96	4.56	4.36	3.06	2.86	3.49	47.34
ANDREWS 2 E		6.77	5.93	7.13	5.33	4.61	5.05	5.52	4.76	3.67	3.29	4.71	5.88	62.65
ASHEBORO 2 W		3.76	3.59	4.18	3.33	4.07	3.89	4.89	4.84	3.85	3.16	2.67	3.39	45.62
ASHEVILLE WSO		3.48	3.60	5.13	3.84	4.19	4.20	4.43	4.79	3.96	3.29	3.29	3.51	47.72
ASHEVILLE	R	2.73	2.93	4.34	3.36	3.31	3.27	2.91	3.76	3.53	2.72	2.71	2.79	38.36
BANNER ELK		3.89	3.92	5.06	4.36	4.35	4.38	4.81	4.56	4.20	3.55	3.63	3.55	50.26
BELHAVEN		3.98	3.54	3.92	3.04	4.54	4.93	6.31	5.69	5.56	3.54	3.17	3.36	51.58
BENT CREEK		3.51	3.72	5.38	3.98	4.07	3.78	4.11	4.58	4.13	3.56	3.52	3.63	47.97
BLACK MOUNTAIN		3.58	3.67	5.30	4.22	4.49	4.55	4.32	4.77	4.01	3.88	3.61	3.55	49.95
BOONE		4.09	4.48	5.86	4.64	4.59	4.63	4.67	4.33	4.73	4.21	4.04	3.86	54.13
BREVARD		5.44	5.50	7.15	5.39	5.75	5.76	5.86	6.53	5.37	4.91	4.71	5.52	67.89
BRIDGEWATER HYDRO		3.63	3.79	5.13	3.96	4.50	4.23	4.30	4.64	4.22	3.64	3.22	3.64	48.90
BURLINGTON FILTER PL		3.65	3.61	4.07	3.32	3.88	4.17	4.39	4.54	4.01	3.17	2.83	3.40	45.04
CANTON 1 SW		3.27	3.24	4.54	3.53	3.56	3.38	3.84	3.83	3.32	2.54	2.65	3.05	40.75
CAPE HATTERAS WSO		4.72	4.11	3.97	3.21	4.09	4.22	5.36	6.11	5.78	4.83	4.84	4.48	55.72
CARTHAGE 1 SSE		3.83	3.78	4.46	3.43	4.12	4.37	4.73	4.71	4.08	3.06	2.97	3.22	46.76
CELO 2 S		4.71	4.81	6.63	4.87	5.15	4.48	4.84	4.70	5.28	4.61	4.42	4.35	58.85
CHAPEL HILL 2 W		3.74	3.84	4.25	3.64	3.91	4.02	4.25	4.54	3.72	3.08	3.15	3.35	45.49
CHARLOTTE WSO	R	3.80	3.81	4.83	3.27	3.64	3.57	3.92	3.75	3.59	2.72	2.86	3.40	43.16
CONCORD		3.85	3.62	4.60	3.50	3.65	4.38	4.92	4.13	3.79	3.44	2.74	3.43	46.05
CONOVER OXFORD SHOALS		3.84	3.94	4.92	3.71	4.37	4.31	3.99	3.92	4.37	3.17	3.04	3.70	47.28
COWEETA EXP STATION		7.01	6.91	8.47	6.34	5.92	5.35	5.06	5.56	5.28	4.51	5.38	7.05	72.84
CULLOWHEE		4.51	4.39	5.72	3.97	4.38	4.24	4.42	4.10	3.51	3.11	3.40	4.51	50.26
DANBURY 1 NW		3.45	3.35	4.47	3.90	3.87	3.91	4.19	4.11	3.94	3.74	2.90	3.45	45.28
DURHAM		3.94	4.06	4.17	3.50	4.00	4.31	4.57	4.77	3.80	3.18	3.28	3.48	47.06
EDENTON		4.09	3.77	4.12	3.39	4.13	4.22	5.56	5.42	4.32	3.46	2.69	3.12	48.29
ELIZABETH CITY		4.03	3.77	4.15	2.96	3.92	4.02	5.70	5.71	4.72	3.91	2.96	3.10	48.95
ELIZABETH CITY FAA AP		4.10	3.58	4.02	3.04	4.14	3.78	5.27	5.48	4.56	3.41	2.94	3.19	47.51
ELIZABETHTOWN LOCK 2		3.49	3.29	4.23	2.90	3.79	4.61	6.07	5.66	4.09	2.88	2.82	2.95	46.78
ELKIN		3.67	3.65	4.77	4.05	4.28	4.34	3.62	4.68	4.58	3.50	2.95	3.48	47.57
ENFIELD		3.71	3.63	3.89	3.07	3.73	3.51	4.77	4.25	3.66	2.93	2.89	3.15	43.19
ENKA		3.12	3.37	4.68	3.57	3.87	3.48	3.93	4.50	3.61	3.08	3.06	2.90	43.17
FAYETTEVILLE		3.65	3.77	4.22	3.02	3.50	4.66	5.58	5.55	4.71	2.71	3.08	3.22	47.67
FLETCHER 2 NE		3.39	3.39	4.81	3.75	4.07	4.16	4.61	4.82	4.13	3.35	3.14	3.30	46.92
FT BRAGG WATER PLANT		3.62	3.94	4.29	3.51	3.67	4.62	5.81	5.04	4.14	3.23	3.22	3.32	48.41
FRANKLIN 1 SSW		4.70	4.67	6.08	4.41	4.10	4.35	4.27	4.34	4.06	3.29	3.57	4.55	52.39
GASTONIA		4.16	4.14	5.21	3.57	4.06	4.11	4.16	4.42	3.91	2.82	2.87	3.74	47.17
GLENDALE SPRINGS		3.81	4.16	5.72	4.59	4.78	5.26	5.02	4.92	5.48	4.94	4.24	3.78	56.70
GOLDSBORO 1 SSW		4.07	3.73	4.15	3.40	4.01	4.34	6.40	5.77	4.77	2.84	3.25	3.46	50.19
GRAHAM 2 ENE		3.67	3.46	4.01	3.25	3.77	3.64	4.22	4.46	3.84	3.04	2.67	3.24	43.27
GREENSBORO PUMP STA		3.46	3.45	3.82	3.25	3.60	4.33	4.30	4.02	3.99	3.33	2.56	3.31	43.42
GREENSBORO WSO	R	3.51	3.37	3.88	3.16	3.37	3.93	4.27	4.19	3.64	3.18	2.59	3.38	42.47
GREENVILLE		4.06	3.78	3.92	3.45	4.12	4.31	5.54	6.06	4.68	2.95	2.96	3.26	49.09
HAMLET		3.82	4.00	4.69	3.31	4.03	4.28	5.90	5.00	3.69	3.42	2.73	3.40	48.27
HATTERAS		4.82	3.96	4.07	3.03	4.09	4.30	5.27	5.78	5.45	4.61	4.79	4.29	54.46
HENDERSON 2 NNW		3.71	3.56	3.90	3.10	4.02	3.79	4.79	4.93	3.81	3.21	3.18	3.34	45.34
HENDERSONVILLE 1 NE		4.15	4.33	6.11	4.52	4.67	4.98	4.65	5.79	4.49	4.18	3.92	4.36	56.15
HICKORY FAA AP		3.74	4.21	5.18	3.81	4.09	4.85	4.30	4.19	4.52	3.41	3.09	3.82	49.21
HIGHLANDS 2 S		6.52	6.42	8.43	6.33	7.12	7.37	7.10	7.06	6.51	5.54	6.04	7.08	81.52
HIGH POINT		3.52	3.51	4.05	3.37	3.72	4.00	3.81	4.97	3.74	3.18	2.76	3.52	44.15
HOT SPRINGS 2		3.51	3.36	4.78	3.99	4.00	3.91	5.22	4.36	3.58	2.55	3.04	2.99	45.29
IDLEWILD		3.92	4.15	5.93	4.52	4.62	5.24	4.80	4.65	5.08	4.38	4.31	3.99	55.59
JACKSON		3.81	3.74	4.01	3.02	4.30	3.65	5.07	4.93	3.85	3.24	3.04	3.58	46.24
JACKSON SPRINGS 5 WNW		4.04	3.73	4.42	3.21	4.02	4.58	5.15	4.32	3.89	3.23	2.95	3.46	47.00
JEFFERSON		3.50	3.55	4.72	4.05	4.36	4.46	4.43	4.20	4.44	3.80	3.43	3.46	48.40

NORTH CAROLINA

PRECIPITATION NORMALS (INCHES)

STATION	JAN	FEB	MAR	APR	MAY	JUN	JUL	AUG	SEP	OCT	NOV	DEC	ANN
KINSTON 5 SE	4.11	3.66	3.96	3.39	4.42	5.39	6.20	5.76	5.30	3.06	2.92	3.42	51.59
LAURINBURG	3.86	3.94	4.40	3.12	3.72	5.34	5.64	4.73	4.38	2.99	2.86	3.19	48.17
LENOIR	3.65	4.01	5.02	4.20	4.11	4.56	4.87	4.26	4.59	3.79	3.25	3.63	49.94
LEXINGTON	3.95	3.68	4.42	3.27	3.80	4.14	4.43	4.22	3.57	3.15	2.87	3.71	45.21
LINCOLNTON 4 W	3.81	4.07	5.22	3.70	4.26	4.30	4.23	4.17	4.19	3.05	2.96	3.80	47.76
LOUISBURG	3.69	3.88	4.19	3.17	3.90	3.95	4.88	4.75	3.98	3.25	3.35	3.43	46.42
LUMBERTON 6 NW	3.47	3.46	4.28	2.94	3.73	4.94	5.26	4.91	4.35	2.87	2.76	2.99	45.96
MARION	4.12	4.55	6.01	4.65	4.55	5.34	4.43	5.41	4.50	4.69	4.12	4.23	56.60
MARSHALL	3.22	3.36	4.48	3.48	3.35	3.51	4.32	3.64	2.94	2.47	2.76	2.96	40.49
MAYSVILLE 6 SW	4.09	3.95	4.10	3.05	4.94	5.94	6.99	6.74	5.90	3.36	3.08	3.77	55.91
MC CULLERS 1 W	3.53	3.67	4.06	3.35	4.04	4.23	5.33	4.96	3.98	3.23	3.19	3.35	46.92
MONCURE 3 SE	3.88	3.74	4.29	3.26	4.00	4.41	5.56	5.15	4.00	3.28	3.28	3.51	48.36
MONROE 4 SE	4.11	3.91	4.68	3.21	3.61	3.96	5.10	4.74	4.19	3.22	2.78	3.42	46.93
MOREHEAD CITY 2 WNW	4.13	3.99	3.69	2.90	4.24	4.57	6.57	6.19	5.26	3.77	3.39	3.96	52.66
MORGANTON	3.99	4.22	5.11	4.08	4.57	4.60	4.13	3.90	4.99	3.81	3.22	3.73	50.35
MOUNT AIRY	3.37	3.50	4.46	3.95	3.96	3.87	4.02	4.36	4.22	3.35	2.91	3.47	45.44
MOUNT HOLLY 4 NE	3.82	4.10	5.01	3.50	3.76	3.95	3.94	3.91	3.57	2.93	2.87	3.69	45.05
MURPHY	5.36	5.30	6.43	5.02	4.24	4.33	5.23	4.67	3.75	2.84	3.98	4.77	55.92
NASHVILLE	3.49	3.70	3.96	3.11	3.93	3.86	4.70	4.46	3.74	2.95	3.15	3.19	44.24
NEUSE 2 NE	3.75	3.79	4.08	3.23	3.97	3.98	4.27	4.33	3.59	2.99	3.29	3.53	44.80
NEW BERN FAA AP	4.01	3.97	3.62	2.98	4.41	5.13	6.75	6.33	5.75	3.39	3.08	3.69	53.11
NEW HOLLAND	4.20	3.87	3.61	3.31	4.21	4.37	6.04	6.36	5.62	3.90	3.70	3.35	52.54
NORTH FORK 2	3.83	3.70	5.24	4.35	4.61	4.57	4.59	4.90	4.11	3.49	3.51	3.44	50.34
NORTH WILKESBORO	3.76	3.91	4.92	4.20	3.94	5.01	3.95	4.75	4.18	3.71	3.17	3.70	49.20
OXFORD 2 SW	3.51	3.59	3.92	3.09	3.94	3.93	4.64	4.70	3.54	2.97	3.13	3.28	44.24
PISGAH FOREST 1 N	5.50	5.26	7.05	5.19	5.73	5.46	5.57	6.18	5.36	4.81	4.71	5.35	66.17
PLYMOUTH 5 E	4.17	3.98	4.18	3.14	4.66	4.53	6.30	5.85	4.62	3.35	3.10	3.32	51.20
RALEIGH DURHAM WSO R	3.55	3.43	3.69	2.91	3.67	3.66	4.38	4.44	3.29	2.73	2.87	3.14	41.76
RALEIGH 4 SW	3.79	3.75	4.19	3.36	4.44	3.93	5.06	4.58	3.83	3.12	3.40	3.34	46.79
RALEIGH-N C STATE UNIV	3.84	3.75	4.09	3.31	4.22	3.73	4.84	4.52	3.85	3.28	3.24	3.33	46.00
RANDLEMAN	3.80	3.65	4.08	3.54	4.01	4.15	4.59	4.70	3.91	2.93	2.63	3.49	45.48
RED SPRINGS	3.60	3.71	4.14	3.09	3.66	4.76	5.26	4.50	4.31	2.63	2.88	3.26	45.80
REIDSVILLE 2 NW	3.57	3.29	4.03	3.34	3.72	3.92	4.14	3.89	3.75	3.58	2.78	3.25	43.26
RHODHISS HYDRO PLANT	3.76	4.08	4.93	3.92	3.85	4.79	4.10	4.26	4.57	3.24	2.97	3.62	48.09
ROCKY MOUNT POWER PL	3.73	3.57	3.95	3.08	3.60	4.41	5.25	4.97	4.08	2.96	3.12	3.23	45.95
ROCKY MOUNT 8 ESE	3.78	3.74	4.16	3.05	3.65	4.50	5.14	5.04	4.07	2.98	2.95	3.27	46.33
ROSMAN	6.82	6.63	8.66	6.56	7.07	6.91	6.89	7.52	6.71	5.95	6.19	7.53	83.44
ROUGEMONT	3.75	3.54	4.10	3.35	3.96	3.98	4.37	4.34	3.67	3.14	3.17	3.21	44.58
ROXBORO	3.65	3.36	3.75	3.25	3.74	3.95	4.34	4.50	3.78	3.21	3.08	3.20	43.81
SALISBURY	3.72	3.78	4.61	3.38	3.66	4.20	4.61	3.97	3.64	3.21	2.94	3.56	45.28
SCOTLAND NECK	3.93	3.90	3.99	3.01	3.90	4.50	5.21	4.77	4.29	3.42	2.89	3.37	47.18
SHELBY 2 NNE	4.14	4.03	5.36	3.86	4.12	4.51	4.35	4.57	4.01	3.35	3.15	3.98	49.43
SILER CITY 2 NW	4.05	3.85	4.34	3.45	3.96	3.97	4.87	5.19	3.89	3.34	2.91	3.44	47.26
SLOAN 3 S	4.20	3.78	4.45	3.29	4.69	6.19	7.26	5.88	5.22	2.69	3.17	3.41	54.23
SMITHFIELD	3.88	3.98	4.18	3.40	3.78	3.74	5.65	4.64	4.65	3.15	3.03	3.27	47.35
SOUTHPORT 5 N	4.32	3.84	4.52	2.70	3.93	4.85	6.29	6.84	6.96	3.53	3.15	3.93	54.86
SPARTA	3.47	3.77	5.21	4.16	4.28	3.82	3.90	4.85	4.31	4.14	3.73	3.64	49.28
STATESVILLE 2 NNE	3.80	3.89	4.81	3.52	3.85	4.52	3.62	4.37	4.12	3.15	2.98	3.84	46.47
SWANNANOA 2 E	3.21	3.19	4.56	3.73	4.13	4.18	3.88	4.04	3.90	3.30	3.26	2.99	44.37
TAPOCO	5.72	5.16	6.74	4.86	4.61	4.88	6.13	5.30	3.77	3.16	4.41	4.94	59.68
TARBORO 1 S	4.01	3.84	4.25	3.02	3.75	4.41	4.91	5.78	4.54	3.14	2.98	3.32	47.95
TRANSOU	3.71	4.11	5.85	4.69	4.77	4.62	4.66	4.88	4.89	4.49	4.07	3.94	54.68
TRYON	5.32	5.52	7.14	5.31	5.86	5.67	5.14	5.23	5.56	4.77	4.15	5.16	64.83
WADESBORO	4.19	3.83	4.66	3.17	3.98	4.38	4.97	4.84	4.20	3.09	2.70	3.38	47.39
WASHINGTON MAIN ST	4.00	3.73	3.78	3.45	4.28	4.56	6.49	5.81	4.57	3.40	3.13	3.43	50.63

NORTH CAROLINA

PRECIPITATION NORMALS (INCHES)

STATION		JAN	FEB	MAR	APR	MAY	JUN	JUL	AUG	SEP	OCT	NOV	DEC	ANN
WAYNESVILLE 1 E		4.21	4.02	5.47	3.96	3.87	3.56	4.06	3.88	3.48	2.79	3.24	4.07	46.61
WILKESBORO 2 W		3.82	3.79	5.04	4.19	3.94	4.88	4.09	4.92	4.41	3.62	3.20	3.66	49.56
WILLARD 4 SW		3.95	3.72	4.09	2.89	4.48	5.76	7.53	6.08	5.32	2.71	3.20	3.23	52.96
WILLIAMSTON 1 ENE		3.86	3.65	4.02	3.28	4.39	4.26	5.74	5.42	4.57	3.42	2.72	3.02	48.35
WILMINGTON WSO	R	3.64	3.44	4.04	2.98	4.22	5.65	7.44	6.64	5.71	2.97	3.19	3.43	53.35
WILSON 2 W		3.85	3.72	4.19	3.22	3.70	4.09	5.76	5.33	4.16	2.66	3.05	3.36	47.09
YANCEYVILLE 2 NNE		3.31	2.96	3.67	3.41	3.52	3.67	3.74	4.03	3.99	3.30	2.79	3.20	41.59

NORTH CAROLINA

HEATING DEGREE DAY NORMALS (BASE 65 DEG F)

STATION		JUL	AUG	SEP	OCT	NOV	DEC	JAN	FEB	MAR	APR	MAY	JUN	ANN
ALBEMARLE 4 N		0	0	6	180	423	688	747	602	444	163	58	0	3311
ANDREWS 2 E		0	0	48	302	585	834	908	756	614	312	130	16	4505
ASHEBORO 2 W		0	0	12	172	411	682	753	616	450	149	49	0	3294
ASHEVILLE WSO		0	0	57	286	558	797	874	725	577	283	114	23	4294
ASHEVILLE	R	0	0	44	268	540	778	862	711	568	255	100	13	4139
BANNER ELK		34	34	158	453	720	955	1042	899	775	474	270	109	5923
BENT CREEK		0	0	58	294	573	822	902	745	598	306	129	24	4451
BLACK MOUNTAIN		0	0	56	279	543	775	853	717	564	275	118	22	4202
BOONE		11	19	129	412	681	924	1017	865	735	414	201	66	5474
BREVARD		0	0	47	282	564	791	856	706	558	285	115	19	4223
BURLINGTON FILTER PL		0	0	13	190	447	722	800	652	491	171	51	0	3537
CANTON 1 SW		0	0	82	357	627	849	924	773	636	342	163	32	4785
CAPE HATTERAS WSO		0	0	0	76	276	510	617	543	437	186	37	0	2682
CELO 2 S		11	14	121	425	681	927	995	848	707	405	206	75	5415
CHAPEL HILL 2 W		0	0	16	220	462	744	815	678	511	201	72	5	3724
CHARLOTTE WSO	R	0	0	10	166	429	694	760	619	459	155	50	0	3342
CONCORD		0	0	12	176	432	713	781	644	474	160	48	0	3440
COWEETA EXP STATION		0	0	61	316	576	815	877	728	598	306	140	30	4447
CULLOWHEE		0	0	32	270	558	787	849	697	552	280	114	16	4155
DURHAM		0	0	14	207	459	732	791	664	498	189	58	0	3612
EDENTON		0	0	9	144	360	614	698	580	430	137	22	0	2994
ELIZABETH CITY		0	0	11	146	381	645	738	622	466	182	44	0	3235
ELIZABETH CITY FAA AP		0	0	6	139	378	642	732	619	473	183	40	0	3212
FAYETTEVILLE		0	0	6	153	390	667	732	605	436	134	32	0	3155
FLETCHER 2 NE		5	8	93	358	630	874	949	801	660	363	172	49	4962
FRANKLIN 1 SSW		0	0	29	262	552	784	846	700	558	280	113	14	4138
GASTONIA		0	0	10	157	408	670	722	588	423	134	46	0	3158
GOLDSBORO 1 SSW		0	0	5	148	378	657	729	602	426	131	26	0	3102
GREENSBORO WSO	R	0	0	12	221	495	769	853	703	533	215	73	0	3874
HAMLET		0	0	5	148	405	667	729	588	419	124	26	0	3111
HATTERAS		0	0	0	54	237	486	598	518	404	151	25	0	2473
HENDERSONVILLE 1 NE		0	0	64	311	579	809	877	722	574	283	113	16	4348
HICKORY FAA AP		0	0	18	205	483	747	825	680	517	205	68	6	3754
HIGHLANDS 2 S		14	11	89	363	615	837	902	767	648	351	171	59	4827
HIGH POINT		0	0	13	179	435	704	781	636	474	157	43	0	3422
HOT SPRINGS 2		0	0	20	218	507	766	853	703	545	240	83	9	3944
JACKSON		0	0	9	174	423	698	775	644	476	172	39	0	3410
JACKSON SPRINGS 5 WNW		0	0	9	156	396	682	763	638	464	153	42	0	3303
KINSTON 5 SE		0	0	6	154	375	651	719	599	430	146	44	0	3124
LAURINBURG		0	0	0	113	340	586	645	518	356	92	17	0	2667
LENOIR		0	0	19	202	480	735	800	655	498	196	69	6	3660
LEXINGTON		0	0	0	162	414	685	750	608	436	135	41	0	3231
LOUISBURG		0	0	26	244	513	791	853	717	543	229	77	7	4000
LUMBERTON 6 NW		0	0	10	160	378	636	704	578	406	135	36	0	3043
MARION		0	0	18	197	471	729	794	647	481	184	57	8	3586
MARSHALL		0	0	51	301	588	831	918	759	605	310	127	20	4510
MAYSVILLE 6 SW		0	0	8	147	353	589	660	554	417	153	34	0	2915
MONCURE 3 SE		0	0	15	219	483	760	831	697	518	208	69	7	3807
MONROE 4 SE		0	0	9	160	402	654	716	577	423	133	36	0	3110
MOREHEAD CITY 2 WNW		0	0	0	69	269	515	610	518	385	128	19	0	2513
MORGANTON		0	0	25	210	477	729	787	647	490	189	67	6	3627
MOUNT AIRY		0	0	22	241	531	800	871	714	539	241	78	5	4042
NASHVILLE		0	0	12	173	411	694	772	644	470	165	46	0	3387
NEW BERN FAA AP		0	0	5	114	327	580	651	535	392	122	31	0	2757
NEW HOLLAND		0	0	0	103	307	558	639	535	390	137	32	0	2701

NORTH CAROLINA

HEATING DEGREE DAY NORMALS (BASE 65 DEG F)

STATION	JUL	AUG	SEP	OCT	NOV	DEC	JAN	FEB	MAR	APR	MAY	JUN	ANN
OXFORD 2 SW	0	0	11	177	426	704	775	644	478	170	44	0	3429
PISGAH FOREST 1 N	0	0	64	316	591	831	896	753	614	312	145	32	4554
PLYMOUTH 5 E	0	0	7	147	378	632	710	591	445	156	34	0	3100
RALEIGH DURHAM WSO R	0	0	9	187	450	713	787	655	496	181	53	0	3531
RALEIGH-N C STATE UNIV	0	0	8	165	399	685	769	641	469	158	48	0	3342
REIDSVILLE 2 NW	0	0	25	229	468	756	837	694	521	215	81	10	3836
SALISBURY	0	0	13	181	441	713	772	633	462	158	45	0	3418
SHELBY 2 NNE	0	0	20	211	471	738	800	655	493	195	55	7	3645
SILER CITY 2 NW	0	0	32	242	498	778	840	700	530	234	95	11	3960
SLOAN 3 S	0	0	0	137	355	595	654	538	393	127	16	0	2815
SMITHFIELD	0	0	0	172	402	648	710	582	417	129	37	0	3097
SOUTHPORT 5 N	0	0	0	95	284	518	604	504	369	113	15	0	2502
STATESVILLE 2 NNE	0	0	28	228	492	763	818	666	512	206	70	9	3792
TARBORO 1 S	0	0	0	152	402	670	741	619	447	146	31	0	3208
TRANSOU	8	16	128	431	693	936	1011	860	722	420	218	70	5513
TRYON	0	0	9	155	405	660	713	577	419	149	51	0	3138
WADESBORO	0	0	8	148	381	660	732	596	428	126	32	0	3111
WAYNESVILLE 1 E	0	0	71	344	609	825	899	748	595	315	153	36	4595
WILLARD 4 SW	0	0	0	112	334	567	629	507	362	110	15	0	2636
WILMINGTON WSO R	0	0	0	94	295	521	607	498	350	94	10	0	2469
WILSON 2 W	0	0	11	168	405	667	735	608	451	149	39	0	3233

NORTH CAROLINA

COOLING DEGREE DAY NORMALS (BASE 65 DEG F)

STATION		JAN	FEB	MAR	APR	MAY	JUN	JUL	AUG	SEP	OCT	NOV	DEC	ANN
ALBEMARLE 4 N		0	0	7	25	139	273	391	366	186	29	0	0	1416
ANDREWS 2 E		0	0	0	0	59	148	238	223	102	0	0	0	770
ASHEBORO 2 W		0	0	10	20	129	270	394	372	198	29	0	0	1422
ASHEVILLE WSO		0	0	0	0	61	167	254	239	114	7	0	0	842
ASHEVILLE	R	0	0	7	6	72	175	273	254	116	11	0	0	914
BANNER ELK		0	0	0	0	9	34	58	46	20	0	0	0	167
BENT CREEK		0	0	0	0	45	138	223	205	94	5	0	0	710
BLACK MOUNTAIN		0	0	0	5	53	142	226	211	92	6	0	0	735
BOONE		0	0	0	0	18	69	114	96	27	0	0	0	324
BREVARD		0	0	0	0	50	148	233	217	98	6	0	0	752
BURLINGTON FILTER PL		0	0	8	21	141	291	412	384	193	29	0	0	1479
CANTON 1 SW		0	0	0	0	29	98	170	155	61	0	0	0	513
CAPE HATTERAS WSO		0	0	6	12	96	270	409	409	276	72	6	0	1556
CELO 2 S		0	0	0	0	16	72	116	101	34	0	0	0	339
CHAPEL HILL 2 W		0	0	6	15	116	251	378	344	172	28	0	0	1310
CHARLOTTE WSO	R	0	0	7	14	149	304	419	400	220	33	0	0	1546
CONCORD		0	0	6	16	150	300	428	403	222	36	0	0	1561
COWEETA EXP STATION		0	0	0	0	31	105	179	164	70	0	0	0	549
CULLOWHEE		0	0	0	0	55	148	245	229	101	6	0	0	784
DURHAM		0	0	0	18	123	271	394	369	182	30	0	0	1387
EDENTON		0	0	11	29	127	306	431	403	240	51	9	0	1607
ELIZABETH CITY		0	0	8	20	113	278	412	394	224	37	0	0	1486
ELIZABETH CITY FAA AP		0	0	5	12	108	282	415	397	228	40	0	0	1487
FAYETTEVILLE		0	0	8	20	159	318	443	419	243	47	0	0	1657
FLETCHER 2 NE		0	0	0	0	44	139	204	185	81	0	0	0	653
FRANKLIN 1 SSW		0	0	0	0	60	161	260	251	116	8	0	0	856
GASTONIA		0	0	8	20	164	310	425	403	223	36	0	0	1589
GOLDSBORO 1 SSW		0	0	8	26	160	327	456	428	251	48	0	0	1704
GREENSBORO WSO	R	0	0	6	14	120	259	378	350	159	17	0	0	1303
HAMLET		0	0	10	25	165	327	443	422	245	43	0	0	1680
HATTERAS		0	0	8	19	118	300	437	443	312	97	9	6	1749
HENDERSONVILLE 1 NE		0	0	0	0	54	145	239	214	91	0	0	0	743
HICKORY FAA AP		0	0	6	10	117	255	363	341	162	16	0	0	1270
HIGHLANDS 2 S		0	0	0	0	25	77	128	113	35	0	0	0	378
HIGH POINT		0	0	9	13	130	273	391	360	187	24	0	0	1387
HOT SPRINGS 2		0	0	5	9	89	216	326	310	167	23	0	0	1145
JACKSON		0	0	8	16	117	282	412	384	204	31	0	0	1454
JACKSON SPRINGS 5 WNW		0	0	8	18	138	291	409	381	219	35	0	0	1499
KINSTON 5 SE		0	0	8	26	150	301	425	406	240	55	0	0	1611
LAURINBURG		0	0	15	41	194	354	465	446	270	57	0	0	1842
LENOIR		0	0	5	10	103	225	329	310	151	13	0	0	1146
LEXINGTON		0	0	12	30	174	324	446	415	227	44	0	0	1672
LOUISBURG		0	0	0	10	105	247	384	357	176	24	0	0	1303
LUMBERTON 6 NW		7	7	15	39	169	315	428	406	241	54	0	0	1681
MARION		0	0	7	13	101	245	347	319	156	14	0	0	1202
MARSHALL		0	0	0	0	55	152	245	223	105	7	0	0	787
MAYSVILLE 6 SW		0	0	10	18	118	265	391	372	221	57	0	0	1452
MONCURE 3 SE		0	0	0	13	116	256	381	363	180	24	0	0	1333
MONROE 4 SE		0	0	11	19	145	300	415	391	219	33	0	0	1533
MOREHEAD CITY 2 WNW		6	0	10	20	158	330	450	453	312	94	11	0	1844
MORGANTON		0	0	7	9	114	246	353	332	163	18	0	0	1242
MOUNT AIRY		0	0	0	7	91	218	329	304	133	12	0	0	1094
NASHVILLE		0	0	5	18	130	289	415	388	219	34	0	0	1498
NEW BERN FAA AP		0	0	11	29	170	321	440	428	275	73	6	6	1759
NEW HOLLAND		0	0	9	20	131	291	422	406	258	72	7	0	1616

NORTH CAROLINA

COOLING DEGREE DAY NORMALS (BASE 65 DEG F)

STATION	JAN	FEB	MAR	APR	MAY	JUN	JUL	AUG	SEP	OCT	NOV	DEC	ANN
OXFORD 2 SW	0	0	7	14	125	276	397	369	194	29	0	0	1411
PISGAH FOREST 1 N	0	0	0	0	46	128	207	188	85	6	0	0	660
PLYMOUTH 5 E	0	0	8	15	118	279	403	381	217	39	0	0	1460
RALEIGH DURHAM WSO R	0	0	9	16	121	270	394	372	189	23	0	0	1394
RALEIGH-N C STATE UNIV	0	0	7	20	144	301	428	403	221	38	0	0	1562
REIDSVILLE 2 NW	0	0	0	17	125	253	369	338	169	34	0	0	1305
SALISBURY	0	0	10	17	141	288	412	378	196	23	0	0	1465
SHELBY 2 NNE	0	0	10	21	120	271	378	350	179	25	0	0	1354
SILER CITY 2 NW	0	0	0	15	113	230	344	319	161	19	0	0	1201
SLOAN 3 S	6	0	14	25	131	294	419	400	244	56	0	0	1589
SMITHFIELD	0	0	11	27	161	316	434	409	224	52	0	0	1634
SOUTHPORT 5 N	6	0	9	26	166	330	450	440	294	85	8	0	1814
STATESVILLE 2 NNE	0	0	7	14	114	249	357	332	163	29	0	0	1265
TARBORO 1 S	0	0	6	23	146	309	437	415	236	40	0	0	1612
TRANSOU	0	0	0	0	14	58	104	87	26	0	0	0	289
TRYON	0	0	10	17	141	265	372	353	183	21	0	0	1362
WADESBORO	0	0	10	24	163	318	437	415	242	43	0	0	1652
WAYNESVILLE 1 E	0	0	0	0	32	105	171	150	56	0	0	0	514
WILLARD 4 SW	0	0	12	29	164	315	425	409	252	53	0	0	1659
WILMINGTON WSO R	6	5	12	37	187	348	474	456	294	78	7	0	1904
WILSON 2 W	0	0	7	26	148	310	431	409	233	51	0	0	1615

31 — NORTH CAROLINA

STATE-STATION NUMBER	STN TYP	NAME		LATITUDE DEG-MIN	LONGITUDE DEG-MIN	ELEVATION (FT)
31-0090	13	ALBEMARLE 4 N		N 3525	W 08012	620
31-0184	13	ANDREWS 2 E		N 3512	W 08348	1827
31-0286	13	ASHEBORO 2 W		N 3542	W 07950	870
31-0300	13	ASHEVILLE WSO		N 3526	W 08233	2140
31-0301	13	ASHEVILLE	R	N 3536	W 08232	2242
31-0506	13	BANNER ELK		N 3610	W 08152	3750
31-0674	12	BELHAVEN		N 3533	W 07638	5
31-0724	13	BENT CREEK		N 3530	W 08236	2110
31-0843	13	BLACK MOUNTAIN		N 3537	W 08219	2395
31-0977	13	BOONE		N 3613	W 08141	3360
31-1055	13	BREVARD		N 3514	W 08244	2155
31-1081	12	BRIDGEWATER HYDRO		N 3544	W 08150	1100
31-1239	13	BURLINGTON FILTER PL		N 3605	W 07925	636
31-1441	13	CANTON 1 SW		N 3532	W 08251	2662
31-1458	13	CAPE HATTERAS WSO		N 3516	W 07533	7
31-1515	12	CARTHAGE 1 SSE		N 3520	W 07924	575
31-1624	13	CELO 2 S		N 3550	W 08211	2700
31-1677	13	CHAPEL HILL 2 W		N 3555	W 07906	500
31-1690	13	CHARLOTTE WSO	R	N 3513	W 08056	735
31-1975	13	CONCORD		N 3525	W 08036	690
31-1990	12	CONOVER OXFORD SHOALS		N 3550	W 08109	1100
31-2102	13	COWEETA EXP STATION		N 3502	W 08326	2240
31-2200	13	CULLOWHEE		N 3519	W 08311	2192
31-2238	12	DANBURY 1 NW		N 3625	W 08013	840
31-2515	13	DURHAM		N 3602	W 07858	406
31-2635	13	EDENTON		N 3603	W 07637	20
31-2719	13	ELIZABETH CITY		N 3619	W 07612	8
31-2724	13	ELIZABETH CITY FAA AP		N 3616	W 07611	10
31-2732	12	ELIZABETHTOWN LOCK 2		N 3438	W 07835	60
31-2740	12	ELKIN		N 3614	W 08051	890
31-2827	12	ENFIELD		N 3611	W 07741	111
31-2837	12	ENKA		N 3533	W 08239	2050
31-3017	13	FAYETTEVILLE		N 3504	W 07852	96
31-3101	13	FLETCHER 2 NE		N 3527	W 08229	2095
31-3168	12	FT BRAGG WATER PLANT		N 3511	W 07902	160
31-3228	13	FRANKLIN 1 SSW		N 3511	W 08323	2113
31-3356	13	GASTONIA		N 3517	W 08111	830
31-3455	12	GLENDALE SPRINGS		N 3621	W 08123	2910
31-3510	13	GOLDSBORO 1 SSW		N 3521	W 07801	82
31-3555	12	GRAHAM 2 ENE		N 3605	W 07922	656
31-3625	12	GREENSBORO PUMP STA		N 3605	W 07948	765
31-3630	13	GREENSBORO WSO	R	N 3605	W 07957	897
31-3638	12	GREENVILLE		N 3537	W 07723	30
31-3784	13	HAMLET		N 3454	W 07942	350
31-3897	13	HATTERAS		N 3513	W 07541	5
31-3969	12	HENDERSON 2 NNW		N 3622	W 07825	480
31-3976	13	HENDERSONVILLE 1 NE		N 3520	W 08227	2153
31-4020	13	HICKORY FAA AP		N 3545	W 08123	1142
31-4055	13	HIGHLANDS 2 S		N 3501	W 08312	3350
31-4063	13	HIGH POINT		N 3557	W 08000	912

STATE-STATION NUMBER	STN TYP	NAME	LATITUDE DEG-MIN	LONGITUDE DEG-MIN	ELEVATION (FT)
31-4265	13	HOT SPRINGS 2	N 3554	W 08250	1480
31-4385	12	IDLEWILD	N 3618	W 08127	2700
31-4456	13	JACKSON	N 3624	W 07725	125
31-4464	13	JACKSON SPRINGS 5 WNW	N 3513	W 07944	730
31-4496	12	JEFFERSON	N 3625	W 08129	2900
31-4684	13	KINSTON 5 SE	N 3513	W 07732	55
31-4860	13	LAURINBURG	N 3447	W 07927	226
31-4938	13	LENOIR	N 3555	W 08132	1294
31-4970	13	LEXINGTON	N 3549	W 08016	810
31-4996	12	LINCOLNTON 4 W	N 3528	W 08119	980
31-5123	13	LOUISBURG	N 3606	W 07819	260
31-5177	13	LUMBERTON 6 NW	N 3442	W 07904	132
31-5340	13	MARION	N 3541	W 08200	1425
31-5356	13	MARSHALL	N 3548	W 08240	2010
31-5420	13	MAYSVILLE 6 SW	N 3450	W 07718	44
31-5445	12	MC CULLERS 1 W	N 3540	W 07842	440
31-5763	13	MONCURE 3 SE	N 3535	W 07903	202
31-5771	13	MONROE 4 SE	N 3458	W 08030	586
31-5830	13	MOREHEAD CITY 2 WNW	N 3444	W 07644	10
31-5838	13	MORGANTON	N 3545	W 08141	1160
31-5890	13	MOUNT AIRY	N 3631	W 08037	1090
31-5913	12	MOUNT HOLLY 4 NE	N 3519	W 08059	700
31-6001	12	MURPHY	N 3505	W 08401	1550
31-6044	13	NASHVILLE	N 3558	W 07758	205
31-6091	12	NEUSE 2 NE	N 3555	W 07834	281
31-6108	13	NEW BERN FAA AP	N 3505	W 07702	18
31-6135	13	NEW HOLLAND	N 3527	W 07611	2
31-6236	12	NORTH FORK 2	N 3539	W 08221	2479
31-6256	12	NORTH WILKESBORO	N 3610	W 08109	1118
31-6507	13	OXFORD 2 SW	N 3617	W 07837	500
31-6805	13	PISGAH FOREST 1 N	N 3516	W 08242	2110
31-6853	13	PLYMOUTH 5 E	N 3552	W 07639	21
31-7069	13	RALEIGH DURHAM WSO R	N 3552	W 07847	434
31-7074	12	RALEIGH 4 SW	N 3544	W 07841	420
31-7079	13	RALEIGH-N C STATE UNIV	N 3547	W 07842	400
31-7097	12	RANDLEMAN	N 3549	W 07947	810
31-7165	12	RED SPRINGS	N 3449	W 07912	204
31-7202	13	REIDSVILLE 2 NW	N 3623	W 07942	890
31-7229	12	RHODHISS HYDRO PLANT	N 3546	W 08126	1050
31-7395	12	ROCKY MOUNT POWER PL	N 3557	W 07749	104
31-7400	12	ROCKY MOUNT 8 ESE	N 3554	W 07743	110
31-7486	12	ROSMAN	N 3508	W 08249	2180
31-7499	12	ROUGEMONT	N 3613	W 07855	540
31-7516	12	ROXBORO	N 3624	W 07858	690
31-7615	13	SALISBURY	N 3540	W 08029	764
31-7727	12	SCOTLAND NECK	N 3608	W 07725	102
31-7845	13	SHELBY 2 NNE	N 3519	W 08132	920
31-7924	13	SILER CITY 2 NW	N 3544	W 07930	625
31-7974	13	SLOAN 3 S	N 3447	W 07749	50
31-7994	13	SMITHFIELD	N 3531	W 07821	146

STATE-STATION NUMBER	STN TYP	NAME		LATITUDE DEG-MIN	LONGITUDE DEG-MIN	ELEVATION (FT)
31-8113	13	SOUTHPORT 5 N		N 3400	W 07801	20
31-8158	12	SPARTA		N 3630	W 08107	2940
31-8292	13	STATESVILLE 2 NNE		N 3549	W 08053	950
31-8442	12	SWANNANOA 2 E		N 3536	W 08222	2230
31-8492	12	TAPOCO		N 3527	W 08356	1110
31-8500	13	TARBORO 1 S		N 3553	W 07732	35
31-8694	13	TRANSOU		N 3625	W 08117	3300
31-8744	13	TRYON		N 3513	W 08214	1075
31-8964	13	WADESBORO		N 3457	W 08004	423
31-9100	12	WASHINGTON MAIN ST		N 3532	W 07703	10
31-9147	13	WAYNESVILLE 1 E		N 3529	W 08257	2658
31-9406	12	WILKESBORO 2 W		N 3609	W 08111	1040
31-9423	13	WILLARD 4 SW		N 3439	W 07802	55
31-9440	12	WILLIAMSTON 1 ENE		N 3551	W 07702	20
31-9457	13	WILMINGTON WSO	R	N 3416	W 07754	28
31-9476	13	WILSON 2 W		N 3543	W 07756	145
31-9700	12	YANCEYVILLE 2 NNE		N 3626	W 07920	620

31 — NORTH CAROLINA

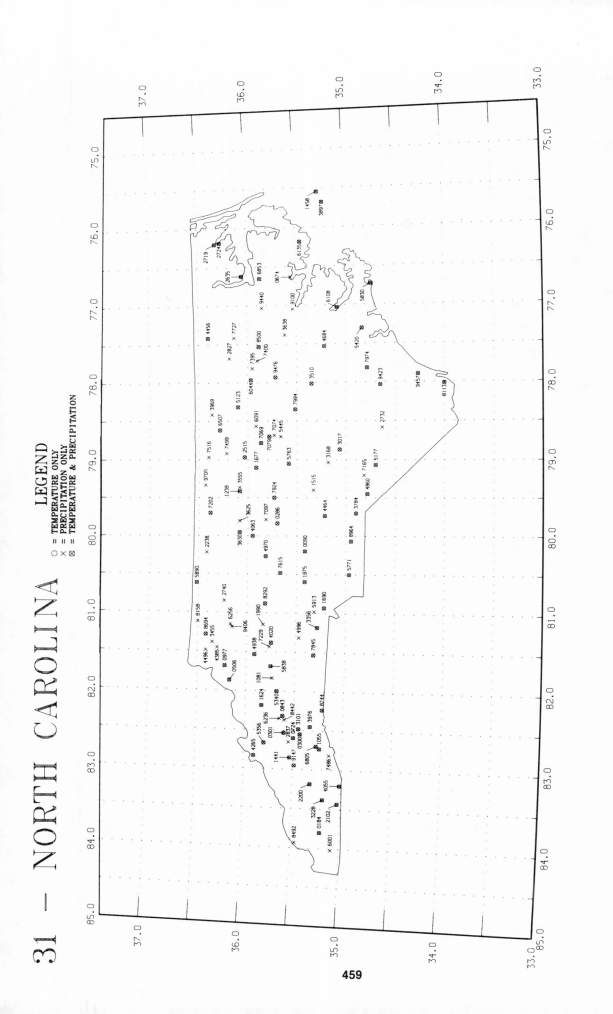

LEGEND

○ = TEMPERATURE ONLY
✕ = PRECIPITATION ONLY
⊗ = TEMPERATURE & PRECIPITATION

459

NORTH DAKOTA

TEMPERATURE NORMALS (DEG F)

STATION		JAN	FEB	MAR	APR	MAY	JUN	JUL	AUG	SEP	OCT	NOV	DEC	ANN
ALMONT 7 W	MAX	19.6	26.7	36.7	53.9	67.9	76.5	84.3	83.5	71.6	59.5	39.8	27.5	54.0
	MIN	.1	7.2	16.4	30.4	42.2	52.2	57.4	55.4	45.0	35.0	19.8	8.4	30.8
	MEAN	9.9	17.0	26.6	42.2	55.1	64.4	70.9	69.5	58.3	47.3	29.8	17.9	42.4
AMENIA	MAX	15.3	23.0	35.4	54.6	70.1	78.5	84.0	83.1	72.6	60.0	38.6	22.8	53.2
	MIN	-5.4	1.9	15.5	31.6	42.9	52.6	57.2	55.3	45.3	34.9	19.4	3.5	29.6
	MEAN	5.0	12.4	25.5	43.1	56.5	65.6	70.6	69.2	59.0	47.5	29.0	13.2	41.4
AMIDON	MAX	22.6	29.0	37.4	52.3	65.5	74.4	83.0	82.7	70.5	59.0	40.3	30.1	53.9
	MIN	1.5	8.6	16.6	29.6	40.5	50.3	55.8	53.9	43.5	33.1	19.6	9.0	30.2
	MEAN	12.1	18.8	27.0	41.0	53.0	62.4	69.4	68.4	57.0	46.1	30.0	19.6	42.1
ASHLEY	MAX	17.7	24.5	35.9	54.0	67.6	76.4	83.3	82.4	71.7	60.1	39.7	25.4	53.2
	MIN	-3.6	3.0	13.9	29.6	40.9	51.3	55.9	53.9	43.2	32.7	17.4	4.9	28.6
	MEAN	7.1	13.8	24.9	41.8	54.2	63.8	69.6	68.2	57.5	46.4	28.6	15.2	40.9
BELCOURT INDIAN RES	MAX	11.6	19.5	30.3	49.7	64.9	73.5	79.1	77.8	66.3	54.8	33.6	19.6	48.4
	MIN	-10.3	-3.0	7.9	25.7	37.5	47.4	52.1	49.7	39.5	29.6	14.4	-.9	24.1
	MEAN	.6	8.3	19.1	37.8	51.2	60.5	65.6	63.8	52.9	42.3	24.0	9.4	36.3
BISMARCK WSO //R	MAX	17.5	25.2	36.4	54.2	67.7	76.8	84.4	83.3	71.4	59.3	39.4	25.9	53.5
	MIN	-4.2	3.7	15.6	30.8	42.0	51.8	56.4	54.2	43.2	32.8	17.7	4.8	29.1
	MEAN	6.7	14.5	26.0	42.5	54.9	64.3	70.4	68.8	57.3	46.1	28.6	15.4	41.3
BOTTINEAU	MAX	10.2	17.9	29.4	49.8	65.4	74.5	80.4	79.4	67.2	55.3	33.7	18.9	48.5
	MIN	-10.6	-3.2	9.2	27.9	39.9	49.7	54.3	51.9	41.2	30.4	14.0	-1.0	25.3
	MEAN	-.2	7.4	19.3	38.9	52.7	62.1	67.4	65.7	54.2	42.9	23.9	9.0	36.9
BOWBELLS	MAX	13.2	20.8	31.2	50.4	65.7	74.8	81.6	80.5	68.0	56.4	35.4	22.1	50.0
	MIN	-6.8	.1	10.8	27.7	38.9	48.8	53.2	50.5	40.6	30.3	15.6	2.5	26.0
	MEAN	3.2	10.5	21.0	39.1	52.3	61.8	67.4	65.6	54.3	43.4	25.5	12.3	38.0
BOWMAN COURT HOUSE	MAX	23.4	29.8	38.6	53.6	66.2	75.3	84.0	83.2	71.3	59.7	41.0	30.3	54.7
	MIN	2.0	8.3	16.9	29.6	41.0	50.5	55.7	53.6	42.9	32.6	18.9	9.6	30.1
	MEAN	12.7	19.1	27.7	41.6	53.6	62.9	69.9	68.4	57.1	46.1	30.0	19.9	42.4
BUTTE	MAX	15.2	22.4	33.4	52.5	67.4	76.6	83.3	82.2	69.7	57.7	36.8	23.3	51.7
	MIN	-3.7	2.9	13.9	29.9	41.7	51.7	56.7	54.5	44.2	33.9	18.3	5.0	29.1
	MEAN	5.8	12.7	23.7	41.2	54.6	64.2	70.0	68.4	57.0	45.8	27.6	14.2	40.4
CARRINGTON	MAX	13.8	21.0	32.3	50.3	65.8	74.5	80.8	79.9	68.4	56.9	36.5	22.0	50.2
	MIN	-6.0	.0	11.6	28.4	40.0	50.6	55.4	52.5	42.0	32.0	16.6	2.7	27.2
	MEAN	3.9	10.6	21.9	39.4	52.9	62.6	68.1	66.2	55.2	44.5	26.6	12.4	38.7
CARSON	MAX	18.9	26.1	35.9	53.1	66.9	75.3	83.2	82.6	70.7	59.2	39.8	26.7	53.2
	MIN	-2.6	4.6	14.2	29.0	40.8	50.8	56.1	53.8	42.9	31.7	17.2	5.7	28.7
	MEAN	8.1	15.4	25.1	41.0	53.9	63.1	69.7	68.2	56.8	45.5	28.5	16.2	41.0
CAVALIER	MAX	11.7	19.7	31.4	51.5	68.3	77.0	81.6	80.3	69.1	56.8	35.3	19.5	50.2
	MIN	-7.7	-1.1	11.8	29.3	41.2	51.5	56.2	53.5	44.0	33.9	18.0	1.4	27.7
	MEAN	2.0	9.3	21.6	40.4	54.8	64.3	68.9	66.9	56.5	45.4	26.7	10.5	38.9
CENTER	MAX	18.5	25.8	36.6	54.2	68.5	76.7	83.7	83.1	71.2	59.7	39.6	26.5	53.7
	MIN	-3.7	3.7	13.7	28.7	40.0	50.1	54.6	52.2	41.6	31.3	17.1	5.0	27.9
	MEAN	7.4	14.7	25.2	41.4	54.2	63.4	69.2	67.7	56.4	45.6	28.4	15.8	40.8
COOPERSTOWN	MAX	14.2	21.8	34.5	53.4	68.9	77.5	83.5	82.3	70.9	58.6	37.1	22.1	52.1
	MIN	-6.2	.7	13.6	29.9	41.5	51.5	56.3	54.2	44.0	33.7	17.9	2.8	28.3
	MEAN	4.0	11.3	24.0	41.7	55.2	64.5	69.9	68.3	57.5	46.2	27.5	12.5	40.2
CROSBY	MAX	15.1	23.2	34.2	53.0	67.7	76.2	83.3	82.1	70.2	57.6	36.5	23.3	51.9
	MIN	-5.1	2.4	12.9	28.5	40.1	49.5	54.1	51.7	41.9	31.8	16.4	3.7	27.3
	MEAN	5.0	12.8	23.6	40.8	53.9	62.9	68.7	66.9	56.1	44.7	26.5	13.5	39.6
DEVILS LAKE KDLR	MAX	11.8	19.6	31.3	50.4	65.8	75.0	81.2	79.8	67.8	55.7	34.8	20.0	49.4
	MIN	-7.1	-.2	12.1	29.7	41.8	52.3	57.0	54.4	44.0	33.7	17.9	2.4	28.2
	MEAN	2.3	9.7	21.7	40.0	53.9	63.6	69.1	67.1	56.0	44.8	26.4	11.2	38.8

NORTH DAKOTA

TEMPERATURE NORMALS (DEG F)

STATION		JAN	FEB	MAR	APR	MAY	JUN	JUL	AUG	SEP	OCT	NOV	DEC	ANN
DICKINSON FAA AIRPORT //R	MAX	20.9	27.7	37.2	53.0	66.0	74.8	83.4	82.4	70.1	58.6	39.7	28.3	53.5
	MIN	1.0	7.9	16.1	29.5	40.6	50.5	55.5	53.9	43.3	33.4	18.9	8.9	30.0
	MEAN	11.0	17.8	26.6	41.2	53.3	62.6	69.5	68.2	56.7	46.0	29.3	18.6	41.7
DICKINSON EXP STATION	MAX	20.7	27.6	36.7	52.7	66.2	74.8	83.0	82.8	70.6	59.3	40.2	28.3	53.6
	MIN	-2.2	4.8	14.0	28.3	39.8	49.5	54.2	52.0	41.1	30.5	16.3	5.9	27.9
	MEAN	9.3	16.2	25.4	40.5	53.0	62.2	68.6	67.4	55.9	45.0	28.3	17.1	40.7
DRAKE	MAX	14.1	21.4	32.6	51.1	66.4	75.1	81.8	80.9	68.8	57.7	36.6	22.8	50.8
	MIN	-6.7	-.2	11.3	28.8	41.1	51.3	56.2	53.4	42.7	31.8	16.4	2.5	27.4
	MEAN	3.7	10.6	21.9	40.0	53.7	63.2	69.0	67.2	55.8	44.8	26.5	12.7	39.1
DUNN CENTER 2 SW	MAX	18.3	25.7	35.0	51.9	65.8	74.6	82.6	82.1	69.9	58.4	38.9	26.6	52.5
	MIN	-3.5	3.4	12.7	28.4	40.5	50.6	55.3	53.2	42.7	32.0	17.3	5.2	28.2
	MEAN	7.4	14.6	23.9	40.2	53.2	62.6	69.0	67.7	56.3	45.2	28.1	15.9	40.3
EDMORE 1 N	MAX	10.7	18.5	30.5	50.4	66.3	74.9	80.7	79.7	68.0	55.5	34.1	19.1	49.0
	MIN	-10.2	-3.2	9.4	27.6	38.9	49.1	53.3	50.9	41.0	30.6	14.7	-.7	25.1
	MEAN	.2	7.6	20.0	39.0	52.6	62.0	67.0	65.3	54.5	43.1	24.5	9.2	37.1
ELLENDALE	MAX	18.6	25.9	37.7	56.2	70.0	78.5	85.0	84.4	73.7	61.3	40.6	26.1	54.8
	MIN	-2.1	4.8	16.6	31.3	42.6	52.8	57.8	55.8	45.1	34.3	19.3	6.3	30.4
	MEAN	8.2	15.4	27.2	43.8	56.3	65.7	71.4	70.1	59.4	47.8	30.0	16.2	42.6
FARGO WSO //R	MAX	13.7	20.5	33.2	52.5	68.1	76.9	82.7	81.1	69.8	57.7	37.0	21.3	51.2
	MIN	-5.1	1.5	14.8	31.6	43.0	53.5	58.4	56.4	45.7	34.9	19.4	4.0	29.8
	MEAN	4.3	11.0	24.0	42.1	55.6	65.2	70.6	68.8	57.8	46.3	28.2	12.7	40.5
FESSENDEN	MAX	16.0	23.4	35.1	54.1	69.3	77.5	83.8	83.3	71.9	59.5	38.3	24.1	53.0
	MIN	-5.4	1.8	13.6	29.7	40.9	51.2	55.9	53.2	43.2	33.0	17.6	3.5	28.2
	MEAN	5.4	12.6	24.4	41.9	55.1	64.4	69.9	68.3	57.6	46.3	28.0	13.8	40.6
FORMAN 5 SSE	MAX	16.4	23.9	36.1	54.4	69.1	78.1	84.5	83.2	72.0	59.7	39.5	24.2	53.4
	MIN	-4.0	3.2	15.9	31.8	43.3	53.5	58.3	56.3	45.1	34.4	19.7	5.1	30.2
	MEAN	6.2	13.6	26.0	43.2	56.2	65.8	71.4	69.8	58.6	47.1	29.6	14.7	41.9
FORT YATES	MAX	21.2	28.5	38.7	55.5	68.8	78.0	85.6	84.3	73.5	61.9	42.2	28.9	55.6
	MIN	-.6	6.3	16.6	31.9	43.8	54.0	59.3	57.1	46.0	35.1	20.4	8.3	31.5
	MEAN	10.3	17.4	27.6	43.7	56.4	66.0	72.5	70.8	59.8	48.5	31.4	18.6	43.6
FOXHOLM 7 N	MAX	16.5	24.1	35.4	53.7	68.4	77.2	84.1	83.1	71.3	59.3	38.7	24.9	53.1
	MIN	-4.5	2.6	12.7	28.6	40.4	49.7	54.3	51.9	42.3	32.6	17.4	4.0	27.7
	MEAN	6.0	13.4	24.1	41.2	54.4	63.5	69.2	67.5	56.8	46.0	28.1	14.5	40.4
FULLERTON 1 ESE	MAX	17.5	25.3	37.5	56.1	70.3	78.6	85.2	84.4	73.5	61.2	40.1	25.4	54.6
	MIN	-3.1	4.1	15.8	31.2	42.3	52.6	57.3	55.2	44.3	33.7	19.0	5.4	29.8
	MEAN	7.2	14.7	26.7	43.6	56.3	65.7	71.3	69.8	58.9	47.5	29.6	15.4	42.2
GACKLE	MAX	16.4	23.7	35.2	53.7	68.5	77.6	84.2	83.1	71.6	59.2	38.4	24.3	53.0
	MIN	-3.9	3.0	14.5	30.4	42.4	52.5	57.4	55.5	44.7	34.5	18.6	5.0	29.6
	MEAN	6.3	13.4	24.9	42.1	55.5	65.1	70.9	69.3	58.2	46.9	28.5	14.6	41.3
GARRISON	MAX	15.2	22.8	33.5	51.3	65.9	74.9	82.2	81.6	69.4	57.5	37.5	24.0	51.3
	MIN	-4.7	1.6	12.5	29.2	40.6	50.5	55.6	53.3	42.6	32.0	17.1	4.7	27.9
	MEAN	5.3	12.2	23.0	40.3	53.2	62.7	68.9	67.5	56.0	44.8	27.3	14.4	39.6
GRAFTON	MAX	12.2	20.3	32.6	52.8	69.4	77.7	83.2	82.0	70.8	58.1	35.8	19.9	51.2
	MIN	-7.9	-1.3	11.8	29.5	41.2	51.3	55.4	53.5	44.0	33.9	18.0	1.5	27.6
	MEAN	2.2	9.5	22.3	41.2	55.4	64.5	69.3	67.8	57.4	46.0	26.9	10.7	39.4
GRAND FORKS FAA AP	MAX	11.6	19.1	31.4	50.6	67.2	76.2	81.6	80.1	68.4	56.0	34.8	19.7	49.7
	MIN	-7.4	-.6	12.7	30.8	41.5	51.6	56.0	53.7	44.0	33.7	18.2	2.3	28.0
	MEAN	2.2	9.3	22.1	40.7	54.4	63.9	68.8	66.9	56.2	44.9	26.5	11.1	38.9
GRAND FORKS UNIVERSITY	MAX	11.9	19.5	32.1	51.9	68.5	77.2	82.0	80.7	69.5	57.0	35.5	20.0	50.5
	MIN	-6.1	.5	13.7	31.5	42.9	53.1	57.5	55.3	45.5	35.3	19.5	3.4	29.3
	MEAN	2.9	10.0	22.9	41.7	55.7	65.1	69.8	68.0	57.5	46.2	27.5	11.8	39.9

NORTH DAKOTA

TEMPERATURE NORMALS (DEG F)

STATION		JAN	FEB	MAR	APR	MAY	JUN	JUL	AUG	SEP	OCT	NOV	DEC	ANN
GRANVILLE	MAX	15.2	23.2	34.5	53.7	68.2	76.5	83.4	82.4	70.6	58.8	37.9	23.4	52.3
	MIN	-6.0	1.3	12.6	28.9	40.4	50.5	55.3	52.9	42.8	32.4	17.2	3.1	27.6
	MEAN	4.6	12.3	23.6	41.3	54.3	63.5	69.4	67.7	56.7	45.6	27.6	13.3	40.0
GRENORA	MAX	16.7	25.1	36.0	54.1	68.0	76.5	83.5	82.4	70.6	58.8	37.8	24.9	52.9
	MIN	-4.3	3.5	13.5	28.3	39.8	49.2	54.1	51.7	41.7	31.6	16.7	4.6	27.5
	MEAN	6.2	14.3	24.8	41.2	53.9	62.9	68.8	67.1	56.2	45.3	27.3	14.7	40.2
HANKINSON R R STATION	MAX	17.0	23.1	34.7	53.0	67.9	77.1	83.3	81.6	70.3	59.3	40.0	24.9	52.7
	MIN	-4.2	1.8	14.2	31.0	43.2	53.8	59.0	56.3	45.2	34.2	19.3	4.9	29.9
	MEAN	6.4	12.5	24.5	42.0	55.6	65.5	71.1	69.0	57.8	46.8	29.7	14.9	41.3
HANNAH 2 N	MAX	9.3	16.8	28.8	48.6	65.4	74.2	79.0	78.0	66.9	54.6	32.8	17.5	47.7
	MIN	-11.4	-4.8	7.4	26.5	38.6	48.9	53.0	50.0	40.4	30.7	14.3	-2.0	24.3
	MEAN	-1.1	6.0	18.1	37.6	52.0	61.6	66.0	64.0	53.7	42.7	23.6	7.8	36.0
HANSBORO 3 W	MAX	11.5	19.8	30.5	49.9	65.7	74.6	80.1	78.4	67.1	55.3	34.1	19.9	48.9
	MIN	-9.1	-1.6	9.9	27.2	38.8	48.2	53.2	50.5	40.7	30.9	15.1	.9	25.4
	MEAN	1.2	9.1	20.2	38.6	52.3	61.4	66.7	64.5	53.9	43.1	24.6	10.4	37.2
HETTINGER	MAX	23.4	30.1	38.8	54.6	67.2	76.1	84.6	83.8	71.7	60.2	41.2	30.1	55.2
	MIN	1.7	8.7	16.9	30.1	40.8	50.8	55.8	53.7	42.6	32.5	18.7	8.9	30.1
	MEAN	12.6	19.4	27.8	42.3	54.0	63.5	70.2	68.8	57.2	46.4	30.0	19.5	42.6
HILLSBORO	MAX	14.0	21.5	34.0	53.4	69.6	78.1	83.7	82.5	71.5	58.9	37.3	21.7	52.2
	MIN	-4.7	1.9	14.7	31.6	42.9	53.3	58.1	55.6	45.8	35.3	19.8	4.3	29.9
	MEAN	4.7	11.7	24.3	42.5	56.3	65.7	70.9	69.0	58.7	47.2	28.6	13.1	41.1
JAMESTOWN FAA AIRPORT R	MAX	15.1	22.6	34.2	52.6	67.3	76.3	82.8	81.9	70.0	57.9	37.4	23.1	51.8
	MIN	-4.4	2.5	14.4	30.1	41.3	52.0	56.9	54.5	43.7	33.3	18.0	4.2	28.9
	MEAN	5.4	12.6	24.3	41.4	54.3	64.2	69.9	68.2	56.9	45.6	27.7	13.7	40.4
JAMESTOWN ST HOSPITAL	MAX	15.8	23.1	35.0	53.6	68.8	77.8	84.2	83.2	71.5	58.8	38.1	23.7	52.8
	MIN	-2.6	4.4	16.0	31.3	42.5	52.5	57.5	55.5	44.8	34.8	19.5	5.8	30.2
	MEAN	6.6	13.8	25.5	42.4	55.6	65.2	70.9	69.4	58.2	46.8	28.8	14.8	41.5
KENMARE 1 WSW	MAX	14.3	22.0	32.5	50.6	65.7	74.6	81.7	80.5	68.3	56.5	36.5	23.0	50.5
	MIN	-6.1	.9	11.0	27.5	39.7	49.5	54.2	51.0	40.8	30.8	16.1	3.4	26.6
	MEAN	4.1	11.4	21.8	39.1	52.7	62.1	68.0	65.8	54.6	43.6	26.3	13.2	38.6
LANGDON EXP FARM	MAX	9.6	17.9	29.9	49.7	66.2	74.7	79.7	78.6	67.6	55.2	33.2	17.6	48.3
	MIN	-10.5	-3.9	8.5	26.9	38.4	48.8	53.0	50.9	40.9	30.9	14.7	-1.1	24.8
	MEAN	-.5	7.0	19.2	38.3	52.4	61.7	66.4	64.8	54.3	43.1	24.0	8.3	36.6
LARIMORE	MAX	13.2	20.7	32.8	51.9	68.1	76.6	82.2	80.8	69.8	57.4	36.8	21.6	51.0
	MIN	-5.5	.5	13.0	29.7	41.4	51.9	56.2	53.8	44.2	34.2	18.8	2.4	28.4
	MEAN	3.9	10.6	22.9	40.8	54.8	64.3	69.3	67.4	57.0	45.9	27.9	12.0	39.7
LEEDS	MAX	12.0	19.7	31.8	51.5	67.0	75.9	81.8	80.6	68.6	56.3	35.1	20.3	50.1
	MIN	-7.8	-1.0	11.0	28.8	40.1	50.8	55.1	52.1	42.2	32.1	16.6	1.6	26.8
	MEAN	2.1	9.4	21.4	40.2	53.6	63.4	68.4	66.4	55.4	44.2	25.9	10.9	38.4
LINTON	MAX	19.8	27.9	39.2	57.4	71.1	79.9	87.1	86.5	74.9	62.7	41.9	27.5	56.3
	MIN	-4.2	3.6	15.3	30.5	42.0	51.9	57.2	55.0	43.6	32.2	17.0	4.5	29.1
	MEAN	7.9	15.7	27.3	44.0	56.6	65.9	72.2	70.8	59.3	47.5	29.5	16.0	42.7
LISBON	MAX	17.4	25.1	37.5	56.2	70.4	78.6	84.4	83.3	72.8	61.5	40.4	25.1	54.4
	MIN	-3.6	2.4	15.0	30.7	41.8	52.1	56.7	54.6	44.2	33.7	19.3	4.5	29.3
	MEAN	7.0	13.8	26.3	43.4	56.1	65.4	70.6	69.0	58.5	47.6	29.8	14.8	41.9
MANDAN EXP STATION	MAX	17.0	24.0	35.2	53.0	66.9	75.8	83.0	82.6	70.7	59.1	39.2	25.5	52.7
	MIN	-3.6	3.4	14.6	30.3	42.0	52.1	57.2	54.8	43.6	33.0	18.1	5.2	29.2
	MEAN	6.7	13.7	25.0	41.6	54.5	64.0	70.1	68.7	57.2	46.1	28.7	15.4	41.0
MAX	MAX	14.3	21.7	32.5	50.8	65.7	74.8	81.9	80.6	68.5	56.5	36.0	22.6	50.5
	MIN	-6.0	.4	11.6	28.4	40.1	50.2	54.8	52.2	41.6	30.9	15.9	2.8	26.9
	MEAN	4.2	11.1	22.0	39.7	52.9	62.5	68.4	66.4	55.0	43.7	26.0	12.8	38.7

NORTH DAKOTA

TEMPERATURE NORMALS (DEG F)

STATION		JAN	FEB	MAR	APR	MAY	JUN	JUL	AUG	SEP	OCT	NOV	DEC	ANN
MAYVILLE	MAX	13.6	21.0	33.5	52.9	68.8	77.2	82.7	81.5	70.2	58.0	36.6	21.5	51.5
	MIN	-5.0	1.7	14.6	31.4	42.7	53.1	57.7	55.6	45.9	35.5	19.7	4.4	29.8
	MEAN	4.3	11.3	24.1	42.2	55.8	65.2	70.2	68.6	58.0	46.8	28.2	13.0	40.6
MCCLUSKY	MAX	15.4	22.9	34.5	53.2	68.4	77.2	84.2	83.2	71.1	58.4	37.3	23.3	52.4
	MIN	-4.3	2.6	13.7	29.3	41.2	51.4	56.4	54.1	43.5	32.8	17.5	4.3	28.5
	MEAN	5.5	12.7	24.1	41.3	54.8	64.3	70.3	68.7	57.3	45.6	27.4	13.8	40.5
MCHENRY 5 NNW	MAX	12.3	19.9	31.9	51.3	66.8	75.2	81.3	80.0	68.4	56.5	35.7	20.6	50.0
	MIN	-8.2	-1.2	11.3	28.3	39.9	50.1	54.7	52.3	41.9	31.6	16.1	.8	26.5
	MEAN	2.0	9.3	21.6	39.8	53.4	62.6	68.0	66.2	55.2	44.1	25.9	10.7	38.2
MCLEOD 3 E	MAX	16.7	24.0	36.6	55.7	70.5	78.7	84.4	83.1	72.6	60.9	39.6	24.3	53.9
	MIN	-4.8	2.1	15.1	31.2	42.9	53.2	58.1	56.2	45.6	34.7	19.2	4.4	29.8
	MEAN	6.0	13.1	25.9	43.5	56.7	66.0	71.3	69.6	59.1	47.8	29.4	14.4	41.9
MEDORA	MAX	25.2	32.8	41.9	57.4	70.2	79.0	87.1	86.7	74.8	62.6	43.2	31.8	57.7
	MIN	-1.3	6.8	15.7	28.5	39.8	49.4	54.0	51.6	40.6	29.7	16.8	6.3	28.2
	MEAN	12.0	19.8	28.8	42.9	55.0	64.2	70.6	69.2	57.7	46.2	30.1	19.1	43.0
MINOT FAA AP	MAX	14.9	22.4	33.1	51.3	65.3	74.5	81.7	80.1	67.7	56.1	36.2	23.1	50.5
	MIN	-2.3	4.7	15.2	30.7	42.1	52.1	57.1	54.7	44.4	34.5	18.9	6.3	29.9
	MEAN	6.3	13.6	24.2	41.0	53.7	63.3	69.4	67.4	56.1	45.3	27.6	14.8	40.2
MINOT EXP STATION	MAX	14.3	21.5	32.6	51.3	65.9	74.9	81.5	80.5	68.3	56.3	36.1	22.7	50.5
	MIN	-5.4	1.6	12.4	29.1	41.0	51.0	55.7	53.1	43.0	32.6	17.1	3.8	27.9
	MEAN	4.5	11.6	22.5	40.2	53.5	63.0	68.6	66.9	55.7	44.5	26.6	13.3	39.2
MOHALL	MAX	13.0	21.0	32.4	51.8	66.5	75.4	81.7	80.7	69.0	56.7	36.3	21.7	50.5
	MIN	-7.5	-.4	11.2	28.2	39.7	49.7	54.2	51.4	40.9	30.6	15.9	2.2	26.3
	MEAN	2.7	10.3	21.8	40.1	53.1	62.6	68.0	66.1	55.0	43.7	26.1	12.0	38.5
MOTT	MAX	22.8	29.9	39.0	54.7	67.3	75.8	83.4	82.5	71.3	60.3	41.2	29.9	54.8
	MIN	-1.5	6.2	15.5	28.6	40.1	50.2	54.8	52.7	41.7	30.9	16.4	5.8	28.5
	MEAN	10.7	18.1	27.2	41.7	53.8	63.0	69.1	67.6	56.5	45.6	28.8	17.9	41.7
NAPOLEON	MAX	16.8	23.7	34.9	53.2	67.6	76.4	83.7	82.8	71.1	59.0	38.8	24.5	52.7
	MIN	-4.7	2.2	13.5	29.9	41.6	51.9	56.7	54.5	43.4	32.9	17.5	4.4	28.7
	MEAN	6.1	13.0	24.2	41.6	54.6	64.2	70.2	68.7	57.3	46.0	28.1	14.4	40.7
NEW ENGLAND	MAX	22.8	29.5	38.8	54.4	67.4	76.2	84.0	83.2	71.2	59.5	40.8	29.6	54.8
	MIN	.6	7.4	15.9	28.9	40.1	50.1	54.8	52.9	42.3	31.8	18.0	7.9	29.2
	MEAN	11.7	18.5	27.4	41.6	53.8	63.1	69.4	68.1	56.8	45.7	29.4	18.8	42.0
NEW SALEM 6 WNW	MAX	18.3	25.6	36.1	53.6	67.7	76.4	84.1	83.2	71.3	58.8	38.9	26.3	53.4
	MIN	-2.3	5.1	14.9	29.2	40.6	50.6	55.6	53.6	43.2	33.0	18.0	6.2	29.0
	MEAN	8.0	15.4	25.5	41.4	54.2	63.5	69.9	68.4	57.3	45.9	28.5	16.2	41.2
OAKES	MAX	16.3	23.5	35.6	53.4	67.8	76.6	82.9	82.1	71.4	59.7	39.4	24.8	52.8
	MIN	-5.7	1.0	13.9	30.5	42.7	53.3	58.4	56.0	44.3	32.5	17.7	3.4	29.0
	MEAN	5.3	12.3	24.8	42.0	55.3	65.0	70.7	69.1	57.8	46.1	28.6	14.1	40.9
PARK RIVER	MAX	13.1	20.9	32.9	52.8	69.1	77.5	83.1	82.0	70.9	58.5	36.4	20.8	51.5
	MIN	-5.4	1.1	13.3	30.0	41.9	52.5	57.2	54.8	45.2	35.1	19.2	3.5	29.0
	MEAN	3.8	11.0	23.2	41.4	55.5	65.0	70.2	68.4	58.1	46.8	27.8	12.2	40.3
PARSHALL	MAX	16.6	24.2	35.4	53.8	68.0	76.5	83.9	82.9	70.8	58.9	37.9	24.7	52.8
	MIN	-5.1	2.1	13.3	28.2	39.7	49.5	53.8	51.5	41.5	31.3	16.4	3.8	27.2
	MEAN	5.8	13.2	24.4	41.1	53.9	63.0	68.9	67.2	56.2	45.1	27.2	14.3	40.0
PEMBINA 1 S	MAX	9.4	17.4	29.8	50.8	67.3	76.3	81.3	79.8	69.0	56.6	34.7	18.5	49.2
	MIN	-11.2	-5.1	8.9	28.3	40.2	50.5	55.3	52.4	42.7	32.5	16.3	-.7	25.8
	MEAN	-.9	6.2	19.4	39.6	53.8	63.4	68.3	66.1	55.9	44.6	25.5	8.9	37.6
PETERSBURG 2 N	MAX	11.4	18.4	30.2	49.2	65.5	74.4	80.2	79.1	67.3	55.2	34.2	19.2	48.7
	MIN	-8.8	-2.2	10.4	28.3	40.2	50.3	54.5	51.9	41.7	31.5	15.9	.1	26.2
	MEAN	1.3	8.1	20.3	38.8	52.9	62.4	67.4	65.5	54.5	43.4	25.0	9.7	37.4

NORTH DAKOTA

TEMPERATURE NORMALS (DEG F)

STATION		JAN	FEB	MAR	APR	MAY	JUN	JUL	AUG	SEP	OCT	NOV	DEC	ANN
PETTIBONE	MAX	14.5	21.8	33.8	52.8	67.0	75.9	82.8	82.1	70.5	57.8	38.0	22.9	51.7
	MIN	-5.9	1.1	13.0	29.6	41.0	51.4	56.3	54.1	43.4	32.7	17.6	3.1	28.1
	MEAN	4.3	11.5	23.4	41.2	54.0	63.6	69.6	68.1	57.0	45.3	27.8	13.0	39.9
POWERS LAKE 1 N	MAX	13.4	21.5	31.6	50.0	65.3	73.9	80.9	79.9	67.9	56.4	35.6	22.4	49.9
	MIN	-8.6	-1.4	9.7	26.5	37.5	47.6	52.3	50.0	39.7	28.6	13.6	.9	24.7
	MEAN	2.4	10.1	20.7	38.3	51.4	60.8	66.6	65.0	53.8	42.5	24.6	11.7	37.3
RICHARDTON ABBEY	MAX	20.5	27.8	37.8	54.3	67.7	76.2	83.7	82.9	71.2	59.0	39.9	28.0	54.1
	MIN	.8	7.9	16.7	30.2	41.8	51.5	56.6	55.0	44.6	34.5	19.7	8.9	30.7
	MEAN	10.7	17.9	27.2	42.3	54.8	63.9	70.2	69.0	57.9	46.8	29.8	18.5	42.4
RIVERDALE	MAX	15.8	22.6	33.0	50.7	65.4	74.1	81.6	80.8	68.9	57.6	38.1	24.7	51.1
	MIN	-4.0	2.1	12.6	29.0	40.9	51.3	56.4	54.3	43.4	33.0	18.2	5.4	28.6
	MEAN	5.9	12.4	22.8	39.9	53.2	62.7	69.0	67.6	56.2	45.3	28.2	15.1	39.9
ROLLA	MAX	11.6	19.2	29.8	48.5	64.2	72.7	78.3	76.9	65.3	54.0	33.3	19.3	47.8
	MIN	-7.9	-.9	10.3	27.8	39.8	49.9	55.1	52.6	42.1	32.1	16.1	1.1	26.5
	MEAN	1.8	9.2	20.1	38.2	52.0	61.4	66.7	64.8	53.7	43.1	24.7	10.2	37.2
RUGBY	MAX	13.7	21.4	33.1	52.9	68.4	77.4	83.3	82.1	69.9	58.1	36.3	21.8	51.5
	MIN	-6.0	1.1	12.8	30.5	42.4	52.6	57.1	54.8	44.2	34.0	18.0	3.2	28.7
	MEAN	3.9	11.2	23.0	41.7	55.4	65.0	70.2	68.4	57.1	46.0	27.2	12.6	40.1
SAN HAVEN	MAX	11.2	19.5	30.9	50.0	65.2	73.4	79.3	77.8	66.3	54.9	34.0	19.4	48.5
	MIN	-7.7	-.3	10.7	28.1	40.2	50.1	54.9	52.6	41.8	32.3	16.0	1.4	26.7
	MEAN	1.8	9.6	20.8	39.1	52.7	61.7	67.1	65.2	54.1	43.6	25.0	10.4	37.6
SHARON	MAX	12.2	19.9	32.1	51.3	67.0	75.2	80.8	80.2	68.9	56.9	35.3	20.1	50.0
	MIN	-6.7	.5	12.9	29.6	41.2	51.4	55.8	53.9	43.8	33.7	17.3	2.2	28.0
	MEAN	2.8	10.2	22.5	40.5	54.1	63.3	68.3	67.1	56.4	45.3	26.3	11.2	39.0
STANLEY 3 NNW	MAX	14.1	21.7	32.6	50.8	65.4	74.3	81.4	80.5	67.9	56.2	35.8	22.5	50.3
	MIN	-5.8	1.4	11.4	26.8	38.6	48.5	53.5	50.1	39.8	29.9	15.5	2.8	26.0
	MEAN	4.1	11.5	22.0	38.8	52.0	61.4	67.5	65.3	53.9	43.1	25.6	12.7	38.2
STEELE	MAX	17.3	24.8	36.4	55.0	69.4	77.3	84.0	83.3	72.5	60.4	39.7	25.2	53.8
	MIN	-4.3	2.9	13.9	29.7	41.2	51.5	56.1	53.9	43.2	32.9	17.5	4.4	28.6
	MEAN	6.5	13.9	25.2	42.4	55.4	64.4	70.1	68.6	57.9	46.7	28.6	14.8	41.2
TOWNER	MAX	13.8	22.0	33.7	53.2	68.5	77.2	83.5	82.3	70.2	58.1	36.6	22.1	51.8
	MIN	-8.1	-.7	11.1	28.4	40.0	50.1	54.3	50.9	40.4	30.6	15.7	1.6	26.2
	MEAN	2.9	10.7	22.4	40.8	54.3	63.7	68.9	66.6	55.3	44.4	26.2	11.9	39.0
TURTLE LAKE	MAX	16.6	24.2	35.7	54.2	68.7	77.0	84.1	83.3	71.1	58.9	38.2	24.5	53.0
	MIN	-3.9	3.0	13.9	29.6	41.3	51.7	56.5	54.3	43.7	33.5	17.8	4.9	28.9
	MEAN	6.3	13.6	24.8	41.9	55.0	64.4	70.3	68.8	57.5	46.2	28.0	14.8	41.0
UPHAM 3 N	MAX	13.2	21.5	33.3	53.4	68.4	76.9	83.3	82.1	70.2	58.2	36.7	21.5	51.6
	MIN	-10.2	-3.3	8.7	27.3	39.3	49.4	53.4	50.6	40.2	29.5	14.3	-.7	24.9
	MEAN	1.5	9.1	21.0	40.4	53.9	63.2	68.4	66.4	55.2	43.8	25.5	10.4	38.2
VALLEY CITY	MAX	15.4	22.9	34.9	53.4	68.7	76.8	82.7	81.8	70.6	58.9	38.1	23.1	52.3
	MIN	-5.5	1.5	13.5	29.5	41.2	51.3	56.0	53.2	42.5	32.2	17.5	3.1	28.0
	MEAN	4.9	12.2	24.2	41.4	55.0	64.1	69.4	67.5	56.6	45.6	27.8	13.1	40.2
VELVA	MAX	16.8	24.8	36.0	54.6	69.2	77.5	83.6	82.3	70.6	59.0	38.2	24.4	53.1
	MIN	-4.4	3.0	14.0	29.6	41.2	51.2	55.9	53.3	43.2	33.1	17.9	4.4	28.5
	MEAN	6.2	13.9	25.1	42.1	55.2	64.4	69.8	67.8	56.9	46.1	28.1	14.4	40.8
WAHPETON	MAX	16.5	23.7	35.9	55.2	70.6	79.2	84.5	82.6	72.1	60.2	39.4	24.3	53.7
	MIN	-2.8	3.7	16.6	32.9	44.6	54.7	59.4	57.5	47.3	37.1	21.5	6.4	31.6
	MEAN	6.9	13.7	26.3	44.1	57.6	67.0	72.0	70.1	59.7	48.6	30.5	15.4	42.7
WASHBURN	MAX	18.2	26.1	37.2	55.3	69.2	78.0	85.0	84.1	72.5	60.0	39.4	26.2	54.3
	MIN	-2.7	5.1	15.4	30.5	41.8	51.9	56.7	54.7	43.9	34.0	19.5	6.3	29.8
	MEAN	7.8	15.6	26.3	42.9	55.5	65.0	70.9	69.4	58.2	47.0	29.5	16.2	42.0

NORTH DAKOTA

TEMPERATURE NORMALS (DEG F)

STATION		JAN	FEB	MAR	APR	MAY	JUN	JUL	AUG	SEP	OCT	NOV	DEC	ANN
WESTHOPE	MAX	11.6	19.6	31.4	52.0	67.2	75.5	81.4	80.2	68.5	56.6	34.8	19.8	49.9
	MIN	-8.6	-1.2	10.9	28.4	40.3	50.0	54.2	52.0	41.5	31.1	15.2	.9	26.2
	MEAN	1.5	9.2	21.2	40.2	53.8	62.8	67.8	66.1	55.0	43.9	25.0	10.4	38.1
WILLISTON WSO //R	MAX	17.5	25.6	36.3	54.0	67.5	76.5	84.0	82.5	70.2	58.2	38.1	25.5	53.0
	MIN	-4.3	3.5	14.3	29.6	41.5	51.1	56.0	53.7	43.0	32.0	17.0	4.3	28.5
	MEAN	6.6	14.6	25.3	41.8	54.5	63.8	70.0	68.1	56.6	45.1	27.6	14.9	40.8
WILLOW CITY	MAX	11.3	19.4	31.4	51.4	66.7	75.5	81.8	80.5	68.5	56.6	35.2	20.0	49.9
	MIN	-11.3	-4.5	7.9	27.1	38.7	49.0	53.2	49.9	39.5	28.8	13.6	-1.6	24.2
	MEAN	.0	7.5	19.7	39.3	52.7	62.3	67.5	65.2	54.0	42.8	24.4	9.2	37.1
WISHEK	MAX	16.2	22.9	34.3	52.3	66.9	76.2	83.2	82.3	71.0	58.9	38.8	24.6	52.3
	MIN	-6.4	-.2	11.2	27.6	39.1	49.3	53.9	51.2	40.5	30.1	15.0	2.7	26.2
	MEAN	4.9	11.4	22.8	40.0	53.0	62.7	68.6	66.8	55.8	44.5	27.0	13.7	39.3

NORTH DAKOTA

PRECIPITATION NORMALS (INCHES)

STATION	JAN	FEB	MAR	APR	MAY	JUN	JUL	AUG	SEP	OCT	NOV	DEC	ANN
ADAMS 7 SSW	.71	.39	.76	1.39	2.20	3.20	2.89	2.33	1.74	1.07	.62	.50	17.80
ALMONT 7 W	.42	.37	.60	1.69	2.28	3.48	2.14	2.25	1.39	.74	.44	.36	16.16
AMBROSE 3 N	.32	.31	.39	1.17	2.06	2.76	1.93	2.32	1.80	.76	.36	.36	14.54
AMENIA	.37	.37	.73	2.02	2.67	3.66	3.04	2.67	2.07	1.31	.66	.48	20.05
AMIDON	.37	.37	.54	1.53	2.54	3.85	2.23	1.49	1.41	.74	.42	.39	15.88
ASHLEY	.37	.38	.70	1.51	2.69	3.61	2.51	2.24	1.54	.95	.41	.36	17.27
BELCOURT INDIAN RES	.49	.43	.67	1.36	2.43	3.18	2.95	3.09	1.94	.87	.47	.49	18.37
BERTHOLD	.51	.48	.63	1.73	2.56	3.80	2.26	2.01	1.93	.84	.55	.54	17.84
BISMARCK WSO //R	.51	.45	.70	1.51	2.23	3.01	2.05	1.69	1.38	.81	.51	.51	15.36
BOTTINEAU	.47	.41	.57	1.31	2.15	3.31	2.80	2.87	1.96	.99	.48	.46	17.78
BOWBELLS	.60	.53	.62	1.57	2.34	3.32	2.14	2.18	2.03	.97	.48	.46	17.24
BOWMAN COURT HOUSE	.33	.42	.52	1.47	2.43	3.68	2.11	1.63	1.41	.76	.42	.35	15.53
BUTTE	.42	.39	.54	1.41	1.96	3.17	2.32	1.96	1.78	.78	.43	.39	15.55
CARRINGTON	.61	.50	.75	1.59	2.49	3.77	2.54	2.12	1.69	.98	.63	.51	18.18
CARSON	.33	.40	.60	1.85	2.68	3.62	2.39	1.89	1.38	.84	.43	.36	16.77
CAVALIER	.52	.39	.69	1.57	2.45	3.04	3.01	2.64	2.19	1.12	.63	.55	18.80
CENTER	.51	.54	.75	1.64	2.45	3.34	2.64	1.99	1.73	.91	.65	.59	17.74
COLGATE	.40	.27	.60	1.50	2.26	3.24	2.64	2.64	2.00	1.12	.51	.40	17.58
COOPERSTOWN	.51	.37	.70	1.48	2.37	3.57	3.05	2.60	2.23	1.15	.71	.54	19.28
COURTENAY 1 NW	.63	.47	.76	1.49	2.15	3.75	2.92	2.53	2.01	1.04	.54	.56	18.85
CROSBY	.40	.39	.43	1.30	2.00	2.71	2.04	1.99	1.78	.84	.40	.43	14.71
DEVILS LAKE KDLR	.55	.45	.76	1.00	2.18	3.32	2.21	2.11	1.88	.88	.63	.55	16.52
DICKINSON FAA AIRPORT //R	.41	.49	.68	1.71	2.36	3.34	2.06	1.58	1.53	.76	.47	.36	15.75
DICKINSON EXP STATION	.34	.40	.57	1.73	2.53	3.69	2.08	1.86	1.51	.85	.45	.34	16.35
DRAKE	.42	.35	.66	1.47	2.18	3.39	2.71	2.12	1.86	.93	.55	.45	17.09
DUNN CENTER 2 SW	.40	.48	.60	1.60	2.35	3.69	2.17	2.05	1.65	.87	.51	.39	16.76
EDMORE 1 N	.45	.30	.48	1.18	2.20	2.95	2.80	2.54	1.88	1.04	.41	.33	16.56
ELLENDALE	.45	.52	.69	2.14	3.04	4.01	2.50	2.28	1.89	1.14	.57	.43	19.66
EPPING	.54	.50	.60	1.38	2.14	2.90	1.80	1.65	1.51	.81	.48	.57	14.88
FAIRFIELD	.34	.35	.48	1.44	2.45	3.60	1.99	1.84	1.64	.77	.46	.31	15.67
FARGO WSO //R	.55	.42	.83	1.90	2.24	3.06	3.34	2.67	1.87	1.29	.79	.63	19.59
FESSENDEN	.61	.54	.74	1.48	2.49	3.49	2.50	2.24	1.76	1.00	.56	.57	17.98
FORBES 9 NNW	.63	.74	1.08	1.94	2.91	3.87	2.28	2.25	1.55	1.03	.64	.58	19.50
FORMAN 5 SSE	.55	.51	.87	2.03	2.76	3.78	2.77	2.59	1.82	1.06	.74	.49	19.97
FORT YATES	.33	.36	.58	1.55	2.32	2.80	1.73	1.61	1.33	.83	.34	.33	14.11
FOXHOLM 7 N	.46	.40	.47	1.54	2.12	3.49	1.99	2.04	1.79	.81	.50	.45	16.06
FULLERTON 1 ESE	.62	.68	.96	2.29	2.95	3.86	2.54	2.56	1.88	1.20	.76	.53	20.83
GACKLE	.34	.40	.76	1.77	2.69	3.69	2.81	2.23	1.77	1.02	.52	.35	18.35
GARRISON	.41	.48	.57	1.40	1.89	3.17	2.01	1.78	1.60	.68	.42	.46	14.87
GOLDEN VALLEY 10 S	.39	.35	.47	1.73	2.40	3.26	2.19	1.67	1.39	.72	.41	.34	15.32
GRAFTON	.59	.44	.74	1.52	2.19	3.25	2.54	2.56	2.06	1.08	.66	.53	18.16
GRAND FORKS FAA AP	.76	.51	.77	1.34	1.95	2.89	2.86	2.62	2.04	1.11	.80	.64	18.29
GRAND FORKS UNIVERSITY	.61	.47	.73	1.51	2.09	3.27	3.17	2.68	2.38	1.18	.84	.57	19.50
GRANVILLE	.45	.42	.61	1.55	2.17	3.30	1.95	2.42	1.76	.85	.44	.46	16.38
GRENORA	.43	.41	.46	1.28	2.06	2.93	2.12	1.79	1.50	.75	.43	.43	14.59
HANKINSON R R STATION	.55	.57	.78	2.05	2.62	4.03	2.74	2.59	1.96	1.19	.71	.57	20.36
HANNAH 2 N	.44	.43	.54	1.13	2.27	3.05	2.68	2.91	2.01	1.01	.56	.45	17.48
HANSBORO 3 W	.36	.31	.55	1.17	2.34	3.11	2.63	2.96	1.78	.83	.38	.38	16.80
HETTINGER	.31	.33	.49	1.68	2.76	3.64	2.04	1.77	1.43	.85	.42	.27	15.99
HILLSBORO	.43	.39	.75	1.71	2.21	3.73	3.22	2.68	2.16	1.36	.65	.53	19.82
JAMESTOWN FAA AIRPORT R	.60	.58	.79	1.52	2.32	3.59	3.02	2.12	1.61	.94	.52	.51	18.12
JAMESTOWN ST HOSPITAL	.37	.32	.56	1.51	2.29	3.58	3.28	2.26	1.68	.92	.39	.35	17.51
KENMARE 1 WSW	.53	.48	.60	1.46	2.11	3.10	1.85	1.82	1.83	.81	.41	.44	15.44
LANGDON EXP FARM	.59	.42	.72	1.33	2.44	2.97	2.99	3.01	1.97	1.12	.65	.51	18.72
LARIMORE	.59	.47	.88	1.44	2.19	3.16	2.77	2.75	2.33	1.23	.76	.60	19.17

466

NORTH DAKOTA

PRECIPITATION NORMALS (INCHES)

STATION	JAN	FEB	MAR	APR	MAY	JUN	JUL	AUG	SEP	OCT	NOV	DEC	ANN
LEEDS	.73	.46	.87	1.43	2.18	3.28	2.47	2.17	1.73	.88	.61	.60	17.41
LINTON	.51	.45	.79	1.85	2.57	3.56	2.33	1.67	1.53	.91	.55	.53	17.25
LISBON	.47	.47	.77	1.98	2.60	3.71	2.77	2.98	1.83	1.19	.69	.59	20.05
MANDAN EXP STATION	.38	.33	.51	1.54	2.26	3.51	2.24	1.97	1.56	.84	.44	.32	15.90
MAX	.44	.54	.65	1.70	2.17	3.77	2.44	2.18	1.74	.83	.57	.47	17.50
MAYVILLE	.65	.50	.82	1.54	2.24	3.46	2.48	2.57	2.12	1.25	.68	.65	18.96
MCCLUSKY	.55	.46	.67	1.61	2.16	3.99	2.35	2.03	1.64	.86	.52	.57	17.41
MCHENRY 5 NNW	.45	.37	.65	1.27	2.23	3.46	2.53	2.47	1.95	1.02	.58	.47	17.45
MCLEOD 3 E	.46	.42	.73	1.54	2.40	3.44	3.00	3.01	1.82	1.17	.77	.49	19.25
MEDINA	.35	.39	.74	1.50	2.29	3.33	2.84	2.14	1.61	.83	.39	.37	16.78
MEDORA	.43	.42	.57	1.41	2.40	3.53	1.96	1.45	1.42	.75	.50	.38	15.22
MINOT FAA AP	.70	.59	.78	1.79	2.34	3.28	2.27	2.03	1.91	.87	.61	.69	17.86
MINOT EXP STATION	.58	.53	.61	1.90	2.25	3.13	2.18	2.02	1.98	.87	.59	.54	17.18
MOHALL	.44	.40	.47	1.35	2.24	3.36	2.22	2.29	1.82	.96	.41	.40	16.36
MONTPELIER	.44	.49	.87	1.90	2.66	3.75	2.81	2.22	1.90	1.15	.61	.42	19.22
MOTT	.39	.40	.57	1.65	2.55	3.80	1.99	1.80	1.42	.70	.46	.36	16.09
MUNICH 4 SW	.50	.34	.61	1.14	2.07	3.07	2.81	2.53	1.74	.94	.48	.42	16.65
NAPOLEON	.45	.39	.76	1.55	2.57	3.19	2.83	2.06	1.73	.93	.53	.43	17.42
NEW ENGLAND	.39	.51	.65	1.75	2.70	3.71	2.13	1.65	1.52	.77	.55	.42	16.75
NEW SALEM 6 WNW	.47	.48	.72	1.70	2.34	3.62	2.20	2.18	1.48	.84	.52	.47	17.02
OAKES	.60	.62	.93	2.23	2.68	3.53	2.37	2.45	1.94	1.12	.69	.58	19.74
PARK RIVER	.52	.36	.66	1.70	2.32	3.40	2.66	2.45	2.02	1.09	.58	.49	18.25
PARSHALL	.34	.35	.42	1.39	2.24	3.66	2.16	1.98	2.01	.73	.39	.36	16.03
PEMBINA 1 S	.50	.41	.69	1.33	2.42	2.86	2.90	2.65	2.19	1.09	.55	.51	18.10
PETERSBURG 2 N	.59	.36	.78	1.34	2.19	3.34	2.73	2.54	2.06	1.08	.72	.45	18.18
PETTIBONE	.41	.35	.59	1.48	2.37	3.44	2.52	2.03	1.72	.96	.44	.38	16.69
POWERS LAKE 1 N	.35	.35	.49	1.31	2.07	3.14	2.14	1.96	1.85	.82	.37	.33	15.18
PRETTY ROCK	.29	.30	.63	1.65	2.85	3.59	2.22	2.08	1.38	.88	.39	.32	16.58
REEDER 13 N	.25	.33	.49	1.29	2.36	3.59	1.98	1.57	1.40	.74	.37	.26	14.63
RICHARDTON ABBEY	.47	.54	.76	1.81	2.77	3.78	2.35	1.81	1.55	.81	.56	.45	17.66
RIVERDALE	.38	.34	.42	1.52	2.28	3.69	2.32	2.09	1.80	.76	.38	.39	16.37
ROLLA	.51	.49	.66	1.20	2.25	3.44	2.82	2.92	1.97	.84	.53	.52	18.15
RUGBY	.59	.43	.64	1.29	1.93	3.14	2.56	2.61	1.79	.86	.50	.50	16.84
SAN HAVEN	.45	.40	.53	1.01	1.93	2.98	2.62	2.90	1.67	.92	.40	.41	16.22
SHARON	.55	.42	.83	1.64	2.54	3.43	2.93	2.55	2.40	1.17	.71	.52	19.69
SHERWOOD 3 N	.29	.30	.42	1.11	1.97	3.11	2.18	2.03	1.59	.73	.31	.31	14.35
SHIELDS	.33	.34	.66	1.70	2.39	3.16	2.05	1.94	1.32	.83	.43	.31	15.46
STANLEY 3 NNW	.47	.49	.62	1.64	2.44	3.83	2.32	2.02	2.06	.85	.51	.47	17.72
STEELE	.47	.41	.65	1.60	2.58	3.58	2.41	1.80	1.54	.92	.48	.45	16.89
TAGUS	.55	.56	.73	1.68	2.14	3.68	1.86	1.70	1.89	.85	.50	.61	16.75
TIOGA 1 E	.38	.42	.35	1.27	1.96	2.75	1.79	1.89	1.68	.69	.43	.38	13.99
TOWNER	.52	.40	.61	1.47	2.01	3.03	2.23	2.42	1.89	.87	.48	.50	16.43
TURTLE LAKE	.61	.50	.69	1.64	2.11	3.70	2.31	2.08	1.76	.86	.50	.56	17.32
TUTTLE	.46	.35	.62	1.48	2.65	3.70	2.30	1.91	1.77	.79	.42	.38	16.83
UPHAM 3 N	.46	.41	.49	1.36	2.11	3.08	2.15	2.41	1.77	.88	.48	.48	16.08
VALLEY CITY	.49	.44	.71	1.61	2.35	3.52	2.74	2.36	1.95	1.07	.64	.55	18.43
VELVA	.52	.49	.63	1.63	2.27	3.36	2.69	2.29	1.67	.90	.56	.56	17.57
WAHPETON	.62	.47	.79	2.14	2.88	4.09	3.21	2.88	2.00	1.25	.77	.62	21.72
WASHBURN	.41	.47	.70	1.83	2.48	3.79	2.71	2.06	1.90	1.08	.51	.41	18.35
WATAUGA S DAK 8 N	.31	.36	.61	1.66	2.61	3.31	2.14	1.93	1.35	.80	.41	.36	15.85
WESTHOPE	.47	.38	.52	1.19	1.96	3.04	2.13	2.31	1.86	.90	.43	.47	15.66
WILDROSE	.48	.48	.56	1.30	2.14	2.66	2.03	2.00	1.77	.83	.46	.49	15.20
WILLISTON WSO //R	.55	.50	.57	1.29	1.85	2.68	1.83	1.42	1.37	.74	.50	.55	13.85
WILLOW CITY	.46	.38	.53	1.20	1.84	2.88	2.28	2.50	1.77	.82	.40	.43	15.49
WILTON	.47	.39	.59	1.55	2.59	3.93	2.66	2.08	1.65	.93	.42	.41	17.67

NORTH DAKOTA

PRECIPITATION NORMALS (INCHES)

STATION	JAN	FEB	MAR	APR	MAY	JUN	JUL	AUG	SEP	OCT	NOV	DEC	ANN
WISHEK	.46	.46	.78	1.73	2.64	3.53	2.62	2.14	1.48	.88	.44	.45	17.61

NORTH DAKOTA

HEATING DEGREE DAY NORMALS (BASE 65 DEG F)

STATION	JUL	AUG	SEP	OCT	NOV	DEC	JAN	FEB	MAR	APR	MAY	JUN	ANN
ALMONT 7 W	23	49	239	549	1056	1460	1708	1344	1190	684	317	107	8726
AMENIA	16	33	208	543	1080	1606	1860	1473	1225	657	289	77	9067
AMIDON	32	55	275	586	1050	1407	1640	1294	1178	720	381	141	8759
ASHLEY	33	47	252	577	1092	1544	1795	1434	1243	696	345	115	9173
BELCOURT INDIAN RES	73	115	370	704	1230	1724	1996	1588	1423	816	438	166	10643
BISMARCK WSO //R	18	57	255	586	1092	1538	1807	1414	1209	675	324	100	9075
BOTTINEAU	51	94	335	685	1233	1736	2021	1613	1417	783	398	138	10504
BOWBELLS	58	93	336	670	1185	1634	1916	1526	1364	777	404	141	10104
BOWMAN COURT HOUSE	33	58	268	586	1050	1398	1621	1285	1156	702	361	134	8652
BUTTE	30	55	270	595	1122	1575	1835	1464	1280	714	343	105	9388
CARRINGTON	39	74	306	636	1152	1631	1894	1523	1336	768	386	127	9872
CARSON	33	60	266	605	1095	1513	1764	1389	1237	720	358	127	9167
CAVALIER	28	81	269	608	1149	1690	1953	1560	1345	738	340	101	9862
CENTER	30	62	276	601	1098	1525	1786	1408	1234	708	343	115	9186
COOPERSTOWN	20	52	249	583	1125	1628	1891	1504	1271	699	323	106	9451
CROSBY	39	70	288	629	1155	1597	1860	1462	1283	726	358	133	9600
DEVILS LAKE KDLR	27	65	285	626	1158	1668	1944	1548	1342	750	361	111	9885
DICKINSON FAA AIRPORT //R	38	57	277	589	1071	1438	1674	1322	1190	714	371	140	8881
DICKINSON EXP STATION	42	67	296	620	1101	1485	1727	1366	1228	735	379	134	9180
DRAKE	35	64	295	626	1155	1621	1900	1523	1336	750	368	116	9789
DUNN CENTER 2 SW	34	63	285	614	1107	1522	1786	1411	1274	744	375	131	9346
EDMORE 1 N	48	90	327	679	1215	1730	2009	1607	1395	780	396	136	10412
ELLENDALE	16	24	197	533	1050	1513	1761	1389	1172	636	283	77	8651
FARGO WSO //R	17	36	236	580	1104	1621	1882	1512	1271	687	311	86	9343
FESSENDEN	28	51	247	580	1110	1587	1848	1467	1259	693	324	98	9292
FORMAN 5 SSE	15	37	222	555	1062	1559	1823	1439	1209	654	293	81	8949
FORT YATES	15	26	208	512	1008	1438	1696	1333	1159	639	283	86	8403
FOXHOLM 7 N	31	73	267	589	1107	1566	1829	1445	1268	714	346	107	9342
FULLERTON 1 ESE	22	41	212	543	1062	1538	1792	1408	1187	642	285	73	8805
GACKLE	21	42	238	561	1095	1562	1820	1445	1243	687	310	94	9118
GARRISON	39	73	289	626	1131	1569	1851	1478	1302	741	376	129	9604
GRAFTON	28	52	254	589	1143	1683	1947	1554	1324	714	329	107	9724
GRAND FORKS FAA AP	42	69	278	623	1155	1671	1947	1560	1330	729	355	122	9881
GRAND FORKS UNIVERSITY	21	55	242	583	1125	1649	1925	1540	1305	699	316	93	9553
GRANVILLE	31	72	268	601	1122	1603	1872	1476	1283	711	344	110	9493
GRENORA	33	73	284	611	1131	1559	1823	1420	1246	714	352	128	9374
HANKINSON R R STATION	20	31	231	564	1059	1553	1817	1470	1256	690	313	76	9080
HANNAH 2 N	57	114	347	691	1242	1773	2049	1652	1454	822	417	146	10764
HANSBORO 3 W	50	122	346	679	1212	1693	1978	1565	1389	792	408	149	10383
HETTINGER	28	43	255	577	1050	1411	1624	1277	1153	681	349	122	8570
HILLSBORO	19	36	213	552	1092	1609	1869	1492	1262	675	296	80	9195
JAMESTOWN FAA AIRPORT R	23	50	263	601	1119	1590	1848	1467	1262	708	346	103	9380
JAMESTOWN ST HOSPITAL	17	36	232	564	1086	1556	1810	1434	1225	678	310	86	9034
KENMARE 1 WSW	47	104	329	663	1161	1606	1888	1501	1339	777	395	128	9938
LANGDON EXP FARM	57	102	332	679	1230	1758	2031	1624	1420	801	407	145	10586
LARIMORE	25	74	261	592	1113	1643	1894	1523	1305	726	338	95	9589
LEEDS	32	82	303	645	1173	1677	1950	1557	1352	744	365	109	9989
LINTON	22	35	212	543	1065	1519	1770	1380	1169	630	280	85	8710
LISBON	23	48	222	539	1056	1556	1798	1434	1200	648	297	77	8898
MANDAN EXP STATION	25	47	257	586	1089	1538	1807	1436	1240	702	339	105	9171
MAX	47	74	313	660	1170	1618	1885	1509	1333	759	386	121	9875
MAYVILLE	17	44	230	564	1104	1612	1882	1504	1268	684	310	84	9303
MCCLUSKY	24	45	260	601	1128	1587	1845	1464	1268	711	336	106	9375
MCHENRY 5 NNW	36	80	308	648	1173	1683	1953	1560	1345	756	371	123	10036
MCLEOD 3 E	15	30	204	533	1068	1569	1829	1453	1212	645	279	70	8907

NORTH DAKOTA

HEATING DEGREE DAY NORMALS (BASE 65 DEG F)

STATION	JUL	AUG	SEP	OCT	NOV	DEC	JAN	FEB	MAR	APR	MAY	JUN	ANN
MEDORA	25	38	251	583	1047	1423	1643	1266	1122	663	319	99	8479
MINOT FAA AP	29	79	288	611	1122	1556	1820	1439	1265	720	364	122	9415
MINOT EXP STATION	37	75	296	636	1152	1603	1876	1495	1318	744	371	121	9724
MOHALL	41	96	313	660	1167	1643	1931	1532	1339	747	382	123	9974
MOTT	36	62	277	601	1086	1460	1683	1313	1172	699	355	123	8867
NAPOLEON	27	56	260	589	1107	1569	1826	1456	1265	702	335	105	9297
NEW ENGLAND	31	52	272	598	1068	1432	1652	1302	1166	702	355	127	8757
NEW SALEM 6 WNW	29	55	260	592	1095	1513	1767	1389	1225	708	346	116	9095
OAKES	22	26	238	586	1092	1578	1851	1476	1246	690	321	93	9219
PARK RIVER	23	55	235	564	1116	1637	1897	1512	1296	708	322	88	9453
PARSHALL	39	69	284	617	1134	1572	1835	1450	1259	717	354	123	9453
PEMBINA 1 S	38	94	287	632	1185	1739	2043	1646	1414	762	371	124	10335
PETERSBURG 2 N	49	102	327	670	1200	1714	1975	1593	1386	786	395	133	10330
PETTIBONE	25	56	268	611	1116	1612	1882	1498	1290	714	356	116	9544
POWERS LAKE 1 N	70	103	349	698	1212	1652	1941	1537	1373	801	430	161	10327
RICHARDTON ABBEY	25	49	248	564	1056	1442	1683	1319	1172	681	329	109	8677
RIVERDALE	27	65	279	611	1104	1547	1832	1473	1308	753	375	117	9491
ROLLA	49	111	349	679	1209	1699	1959	1562	1392	804	417	144	10374
RUGBY	24	49	255	589	1134	1624	1894	1506	1302	699	324	91	9491
SAN HAVEN	57	112	338	663	1200	1693	1959	1551	1370	777	391	141	10252
SHARON	33	69	277	611	1161	1668	1928	1534	1318	735	354	112	9800
STANLEY 3 NNW	61	117	350	679	1182	1621	1888	1498	1333	786	409	144	10068
STEELE	23	63	243	567	1092	1556	1814	1431	1234	678	311	102	9114
TOWNER	30	97	300	639	1164	1646	1925	1520	1321	726	343	111	9822
TURTLE LAKE	20	52	250	583	1110	1556	1820	1439	1246	693	324	100	9193
UPHAM 3 N	43	79	307	657	1185	1693	1969	1565	1364	738	357	112	10069
VALLEY CITY	26	68	273	601	1116	1609	1863	1478	1265	708	327	99	9433
VELVA	31	49	265	586	1107	1569	1823	1431	1237	687	323	96	9204
WAHPETON	12	27	191	508	1035	1538	1801	1436	1200	627	258	54	8687
WASHBURN	19	47	238	558	1065	1513	1773	1383	1200	663	314	94	8867
WESTHOPE	43	83	314	654	1200	1693	1969	1562	1358	744	363	122	10105
WILLISTON WSO //R	23	58	277	617	1122	1553	1810	1411	1231	696	335	108	9241
WILLOW CITY	52	92	337	688	1218	1730	2015	1610	1404	771	391	129	10437
WISHEK	37	71	293	636	1140	1590	1863	1501	1308	750	381	137	9707

NORTH DAKOTA

COOLING DEGREE DAY NORMALS (BASE 65 DEG F)

STATION	JAN	FEB	MAR	APR	MAY	JUN	JUL	AUG	SEP	OCT	NOV	DEC	ANN
ALMONT 7 W	0	0	0	0	10	89	206	188	38	0	0	0	531
AMENIA	0	0	0	0	25	95	190	163	28	0	0	0	501
AMIDON	0	0	0	0	9	63	169	160	35	0	0	0	436
ASHLEY	0	0	0	0	10	79	175	147	27	0	0	0	438
BELCOURT INDIAN RES	0	0	0	0	10	31	92	78	7	0	0	0	218
BISMARCK WSO //R	0	0	0	0	10	79	186	174	24	0	0	0	473
BOTTINEAU	0	0	0	0	17	51	125	115	11	0	0	0	319
BOWBELLS	0	0	0	0	10	45	132	112	15	0	0	0	314
BOWMAN COURT HOUSE	0	0	0	0	8	71	185	163	31	0	0	0	458
BUTTE	0	0	0	0	21	81	185	160	30	0	0	0	477
CARRINGTON	0	0	0	0	11	55	135	112	12	0	0	0	325
CARSON	0	0	0	0	14	70	179	159	20	0	0	0	442
CAVALIER	0	0	0	0	24	80	149	140	14	0	0	0	407
CENTER	0	0	0	0	8	67	160	146	18	0	0	0	399
COOPERSTOWN	0	0	0	0	19	91	172	154	24	0	0	0	460
CROSBY	0	0	0	0	14	70	153	129	21	0	0	0	387
DEVILS LAKE KDLR	0	0	0	0	17	69	154	130	15	0	0	0	385
DICKINSON FAA AIRPORT //R	0	0	0	0	9	68	177	156	28	0	0	0	438
DICKINSON EXP STATION	0	0	0	0	7	50	154	142	23	0	0	0	376
DRAKE	0	0	0	0	18	62	159	132	19	0	0	0	390
DUNN CENTER 2 SW	0	0	0	0	10	59	158	147	24	0	0	0	398
EDMORE 1 N	0	0	0	0	11	46	110	99	12	0	0	0	278
ELLENDALE	0	0	0	0	13	98	214	182	29	0	0	0	536
FARGO WSO //R	0	0	0	0	19	92	191	154	20	0	0	0	476
FESSENDEN	0	0	0	0	17	80	180	153	25	0	0	0	455
FORMAN 5 SSE	0	0	0	0	20	105	213	185	30	0	0	0	553
FORT YATES	0	0	0	0	17	116	248	206	52	0	0	0	639
FOXHOLM 7 N	0	0	0	0	18	62	161	150	21	0	0	0	412
FULLERTON 1 ESE	0	0	0	0	15	94	217	190	29	0	0	0	545
GACKLE	0	0	0	0	16	97	204	175	34	0	0	0	526
GARRISON	0	0	0	0	10	60	160	150	19	0	0	0	399
GRAFTON	0	0	0	0	32	92	162	139	26	0	0	0	451
GRAND FORKS FAA AP	0	0	0	0	27	89	160	128	14	0	0	0	418
GRAND FORKS UNIVERSITY	0	0	0	0	28	96	170	148	17	0	0	0	459
GRANVILLE	0	0	0	0	13	65	167	155	19	0	0	0	419
GRENORA	0	0	0	0	8	65	151	138	20	0	0	0	382
HANKINSON R R STATION	0	0	0	0	21	91	210	155	15	0	0	0	492
HANNAH 2 N	0	0	0	0	14	44	88	83	8	0	0	0	237
HANSBORO 3 W	0	0	0	0	14	41	103	107	13	0	0	0	278
HETTINGER	0	0	0	0	8	77	190	161	21	0	0	0	457
HILLSBORO	0	0	0	0	26	101	202	160	24	0	0	0	513
JAMESTOWN FAA AIRPORT R	0	0	0	0	14	79	175	149	20	0	0	0	437
JAMESTOWN ST HOSPITAL	0	0	0	0	19	92	200	173	28	0	0	0	512
KENMARE 1 WSW	0	0	0	0	13	41	140	129	17	0	0	0	340
LANGDON EXP FARM	0	0	0	0	16	46	100	96	11	0	0	0	269
LARIMORE	0	0	0	0	22	74	158	148	21	0	0	0	423
LEEDS	0	0	0	0	11	61	138	125	15	0	0	0	350
LINTON	0	0	0	0	20	112	245	215	41	0	0	0	633
LISBON	0	0	0	0	21	89	196	172	27	0	0	0	505
MANDAN EXP STATION	0	0	0	0	13	75	183	162	23	0	0	0	456
MAX	0	0	0	0	10	46	152	118	13	0	0	0	339
MAYVILLE	0	0	0	0	25	90	179	155	20	0	0	0	469
MCCLUSKY	0	0	0	0	19	85	188	160	29	0	0	0	481
MCHENRY 5 NNW	0	0	0	0	12	51	129	117	14	0	0	0	323
MCLEOD 3 E	0	0	0	0	21	100	211	172	27	0	0	0	531

NORTH DAKOTA

COOLING DEGREE DAY NORMALS (BASE 65 DEG F)

STATION	JAN	FEB	MAR	APR	MAY	JUN	JUL	AUG	SEP	OCT	NOV	DEC	ANN
MEDORA	0	0	0	0	9	75	199	168	32	0	0	0	483
MINOT FAA AP	0	0	0	0	13	71	165	153	21	0	0	0	423
MINOT EXP STATION	0	0	0	0	15	61	148	134	17	0	0	0	375
MOHALL	0	0	0	0	13	51	134	130	13	0	0	0	341
MOTT	0	0	0	0	8	63	163	143	22	0	0	0	399
NAPOLEON	0	0	0	0	12	81	188	171	29	0	0	0	481
NEW ENGLAND	0	0	0	0	8	70	167	148	26	0	0	0	419
NEW SALEM 6 WNW	0	0	0	0	12	71	181	161	29	0	0	0	454
OAKES	0	0	0	0	20	93	199	153	22	0	0	0	487
PARK RIVER	0	0	0	0	27	88	184	160	28	0	0	0	487
PARSHALL	0	0	0	0	10	63	160	138	20	0	0	0	391
PEMBINA 1 S	0	0	0	0	23	76	140	128	14	0	0	0	381
PETERSBURG 2 N	0	0	0	0	20	55	124	117	12	0	0	0	328
PETTIBONE	0	0	0	0	15	74	168	152	28	0	0	0	437
POWERS LAKE 1 N	0	0	0	0	9	35	120	103	13	0	0	0	280
RICHARDTON ABBEY	0	0	0	0	13	76	186	173	35	0	0	0	483
RIVERDALE	0	0	0	0	10	48	151	145	15	0	0	0	369
ROLLA	0	0	0	0	14	36	102	105	10	0	0	0	267
RUGBY	0	0	0	0	26	91	185	154	18	0	0	0	474
SAN HAVEN	0	0	0	0	10	42	122	118	11	0	0	0	303
SHARON	0	0	0	0	16	61	135	134	19	0	0	0	365
STANLEY 3 NNW	0	0	0	0	6	36	139	126	17	0	0	0	324
STEELE	0	0	0	0	13	84	181	174	30	0	0	0	482
TOWNER	0	0	0	0	11	72	151	147	9	0	0	0	390
TURTLE LAKE	0	0	0	0	14	82	184	170	25	0	0	0	475
UPHAM 3 N	0	0	0	0	13	58	148	123	13	0	0	0	355
VALLEY CITY	0	0	0	0	17	72	163	146	21	0	0	0	419
VELVA	0	0	0	0	19	78	180	136	22	0	0	0	435
WAHPETON	0	0	0	0	28	114	229	185	32	0	0	0	588
WASHBURN	0	0	0	0	20	94	202	183	34	0	0	0	533
WESTHOPE	0	0	0	0	16	56	130	118	14	0	0	0	334
WILLISTON WSO //R	0	0	0	0	10	72	178	155	25	0	0	0	440
WILLOW CITY	0	0	0	0	10	48	129	98	7	0	0	0	292
WISHEK	0	0	0	0	9	68	149	127	17	0	0	0	370

STATE-STATION NUMBER	STN TYP	NAME	LATITUDE DEG-MIN	LONGITUDE DEG-MIN	ELEVATION (FT)
32-0022	12	ADAMS 7 SSW	N 4820	W 09807	1554
32-0136	13	ALMONT 7 W	N 4643	W 10139	2300
32-0189	12	AMBROSE 3 N	N 4900	W 10328	2027
32-0196	13	AMENIA	N 4700	W 09713	954
32-0209	13	AMIDON	N 4629	W 10319	2910
32-0382	13	ASHLEY	N 4602	W 09922	2001
32-0626	13	BELCOURT INDIAN RES	N 4850	W 09945	1960
32-0729	12	BERTHOLD	N 4819	W 10144	2080
32-0819	13	BISMARCK WSO //R	N 4646	W 10046	1647
32-0941	13	BOTTINEAU	N 4850	W 10027	1640
32-0961	13	BOWBELLS	N 4848	W 10215	1958
32-0995	13	BOWMAN COURT HOUSE	N 4611	W 10323	2980
32-1225	13	BUTTE	N 4750	W 10040	1740
32-1360	13	CARRINGTON	N 4727	W 09908	1586
32-1370	13	CARSON	N 4625	W 10134	2310
32-1435	13	CAVALIER	N 4848	W 09738	890
32-1456	13	CENTER	N 4707	W 10118	2100
32-1686	12	COLGATE	N 4714	W 09739	1180
32-1766	13	COOPERSTOWN	N 4726	W 09807	1428
32-1816	12	COURTENAY 1 NW	N 4714	W 09835	1515
32-1871	13	CROSBY	N 4854	W 10318	1952
32-2158	13	DEVILS LAKE KDLR	N 4807	W 09852	1464
32-2183	13	DICKINSON FAA AIRPORT //R	N 4647	W 10248	2585
32-2188	13	DICKINSON EXP STATION	N 4653	W 10248	2460
32-2298	13	DRAKE	N 4755	W 10022	1636
32-2365	13	DUNN CENTER 2 SW	N 4721	W 10239	2232
32-2525	13	EDMORE 1 N	N 4825	W 09828	1520
32-2605	13	ELLENDALE	N 4601	W 09832	1460
32-2735	12	EPPING	N 4817	W 10321	2224
32-2809	12	FAIRFIELD	N 4711	W 10313	2750
32-2859	13	FARGO WSO //R	N 4654	W 09648	896
32-2949	13	FESSENDEN	N 4739	W 09937	1620
32-3064	12	FORBES 9 NNW	N 4603	W 09853	2033
32-3117	13	FORMAN 5 SSE	N 4602	W 09736	1250
32-3207	13	FORT YATES	N 4606	W 10038	1653
32-3217	13	FOXHOLM 7 N	N 4827	W 10134	1609
32-3287	13	FULLERTON 1 ESE	N 4609	W 09824	1439
32-3309	13	GACKLE	N 4638	W 09908	1951
32-3376	13	GARRISON	N 4739	W 10125	1910
32-3529	12	GOLDEN VALLEY 10 S	N 4710	W 10203	1870
32-3594	13	GRAFTON	N 4825	W 09725	827
32-3616	13	GRAND FORKS FAA AP	N 4757	W 09711	839
32-3621	13	GRAND FORKS UNIVERSITY	N 4756	W 09705	830
32-3686	13	GRANVILLE	N 4816	W 10051	1504
32-3736	13	GRENORA	N 4837	W 10356	2131
32-3908	13	HANKINSON R R STATION	N 4604	W 09654	1068
32-3936	13	HANNAH 2 N	N 4900	W 09841	1575
32-3963	13	HANSBORO 3 W	N 4857	W 09927	1684
32-4178	13	HETTINGER	N 4559	W 10239	2680
32-4203	13	HILLSBORO	N 4724	W 09704	899

32 – NORTH DAKOTA

LEGEND
11 = TEMPERATURE ONLY
12 = PRECIPITATION ONLY
13 = TEMP. & PRECIP.

STATE-STATION NUMBER	STN TYP	NAME		LATITUDE DEG-MIN	LONGITUDE DEG-MIN	ELEVATION (FT)
32-4413	13	JAMESTOWN FAA AIRPORT	R	N 4655	W 09841	1492
32-4418	13	JAMESTOWN ST HOSPITAL		N 4653	W 09841	1457
32-4646	13	KENMARE 1 WSW		N 4840	W 10206	1810
32-4958	13	LANGDON EXP FARM		N 4845	W 09820	1615
32-5013	13	LARIMORE		N 4755	W 09738	1130
32-5078	13	LEEDS		N 4817	W 09926	1515
32-5210	13	LINTON		N 4616	W 10014	1685
32-5220	13	LISBON		N 4626	W 09740	1089
32-5479	13	MANDAN EXP STATION		N 4648	W 10054	1750
32-5638	13	MAX		N 4749	W 10118	2099
32-5660	13	MAYVILLE		N 4730	W 09719	975
32-5710	13	MCCLUSKY		N 4729	W 10028	1943
32-5730	13	MCHENRY 5 NNW		N 4738	W 09837	1545
32-5754	13	MCLEOD 3 E		N 4624	W 09714	1075
32-5798	12	MEDINA		N 4653	W 09919	1795
32-5813	13	MEDORA		N 4655	W 10331	2270
32-5988	13	MINOT FAA AP		N 4816	W 10117	1713
32-5993	13	MINOT EXP STATION		N 4811	W 10118	1769
32-6025	13	MOHALL		N 4846	W 10131	1645
32-6105	12	MONTPELIER		N 4642	W 09835	1380
32-6155	13	MOTT		N 4623	W 10220	2420
32-6195	12	MUNICH 4 SW		N 4838	W 09853	1598
32-6255	13	NAPOLEON		N 4630	W 09946	1955
32-6315	13	NEW ENGLAND		N 4633	W 10252	2621
32-6365	13	NEW SALEM 6 WNW		N 4652	W 10132	2115
32-6620	13	OAKES		N 4608	W 09805	1320
32-6857	13	PARK RIVER		N 4823	W 09745	998
32-6867	13	PARSHALL		N 4757	W 10208	1929
32-6947	13	PEMBINA 1 S		N 4857	W 09714	792
32-7027	13	PETERSBURG 2 N		N 4802	W 09800	1525
32-7047	13	PETTIBONE		N 4707	W 09931	1855
32-7281	13	POWERS LAKE 1 N		N 4834	W 10238	2205
32-7311	12	PRETTY ROCK		N 4610	W 10151	2480
32-7452	12	REEDER 13 N		N 4617	W 10257	2755
32-7530	13	RICHARDTON ABBEY		N 4653	W 10219	2470
32-7585	13	RIVERDALE		N 4730	W 10121	1950
32-7664	13	ROLLA		N 4852	W 09937	1819
32-7704	13	RUGBY		N 4822	W 10000	1550
32-7824	13	SAN HAVEN		N 4850	W 10002	1920
32-7986	13	SHARON		N 4736	W 09754	1516
32-8047	12	SHERWOOD 3 N		N 4900	W 10138	1647
32-8065	12	SHIELDS		N 4614	W 10107	1806
32-8276	13	STANLEY 3 NNW		N 4821	W 10225	2280
32-8366	13	STEELE		N 4651	W 09955	1857
32-8627	12	TAGUS		N 4820	W 10156	2165
32-8737	12	TIOGA 1 E		N 4824	W 10255	2245
32-8792	13	TOWNER		N 4820	W 10024	1482
32-8840	13	TURTLE LAKE		N 4731	W 10053	1889
32-8850	12	TUTTLE		N 4708	W 10000	1880
32-8913	13	UPHAM 3 N		N 4837	W 10044	1425

474

32 — NORTH DAKOTA

STATE-STATION NUMBER	STN TYP	NAME		LATITUDE DEG-MIN	LONGITUDE DEG-MIN	ELEVATION (FT)
32-8937	13	VALLEY CITY		N 4656	W 09800	1355
32-8990	13	VELVA		N 4803	W 10055	1511
32-9100	13	WAHPETON		N 4616	W 09636	960
32-9195	13	WASHBURN		N 4717	W 10102	1760
32-9219	12	WATAUGA S DAK 8 N		N 4601	W 10134	2070
32-9333	13	WESTHOPE		N 4855	W 10101	1504
32-9400	12	WILDROSE		N 4838	W 10310	2270
32-9425	13	WILLISTON WSO	//R	N 4811	W 10338	1899
32-9445	13	WILLOW CITY		N 4837	W 10018	1470
32-9455	12	WILTON		N 4709	W 10047	2171
32-9515	13	WISHEK		N 4616	W 09934	2015

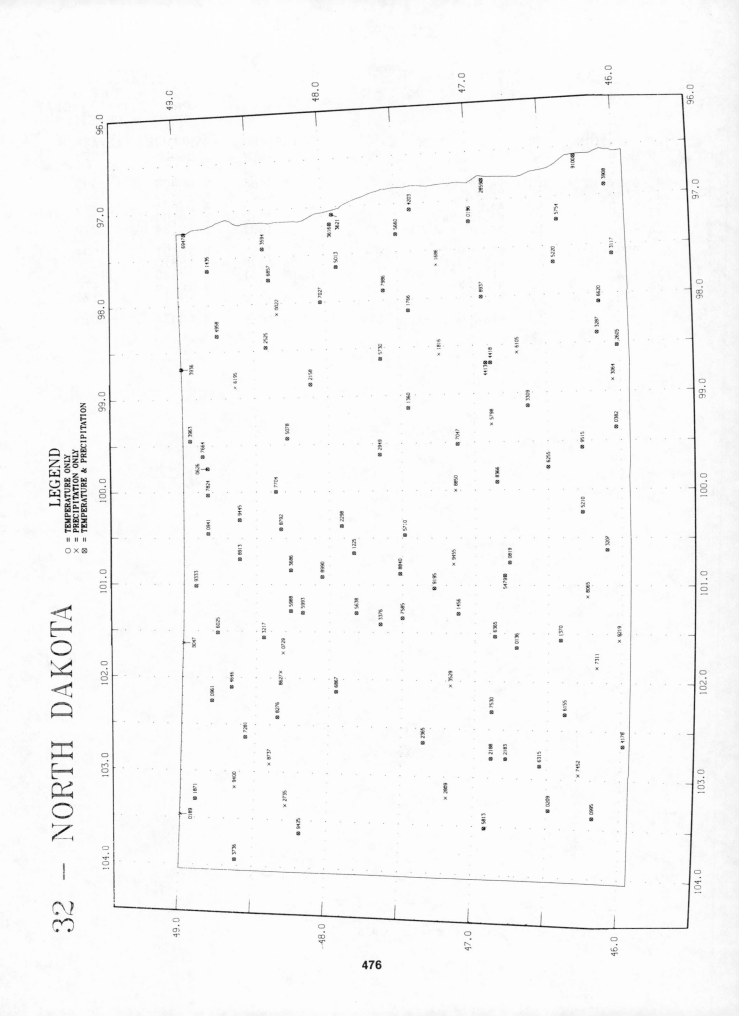

32 — NORTH DAKOTA

LEGEND
O = TEMPERATURE ONLY
X = PRECIPITATION ONLY
⊗ = TEMPERATURE & PRECIPITATION

OHIO

TEMPERATURE NORMALS (DEG F)

STATION		JAN	FEB	MAR	APR	MAY	JUN	JUL	AUG	SEP	OCT	NOV	DEC	ANN
AKRON CANTON WSO R	MAX	32.9	35.6	45.8	59.2	69.8	78.7	82.3	80.9	74.3	62.6	48.9	37.5	59.0
	MIN	17.2	18.8	27.5	38.0	47.7	56.8	61.0	59.9	53.2	42.4	33.0	23.0	39.9
	MEAN	25.1	27.2	36.7	48.6	58.8	67.8	71.6	70.4	63.8	52.5	41.0	30.3	49.5
ASHLAND	MAX	32.3	35.6	45.9	59.6	70.2	79.2	82.8	81.4	75.0	62.9	48.4	36.6	59.2
	MIN	16.8	18.7	27.3	37.7	47.4	56.4	60.2	58.8	52.4	41.7	32.0	22.2	39.3
	MEAN	24.6	27.2	36.6	48.7	58.8	67.9	71.5	70.1	63.7	52.3	40.3	29.4	49.3
ASHTABULA	MAX	32.7	34.3	43.3	56.3	67.6	77.5	81.3	79.8	73.8	62.5	49.5	37.8	58.0
	MIN	18.4	18.5	27.0	37.7	47.1	56.7	61.1	60.2	53.9	43.8	34.8	24.8	40.3
	MEAN	25.6	26.4	35.2	47.0	57.4	67.1	71.3	70.0	63.8	53.2	42.2	31.3	49.2
ATHENS 5 NW	MAX	39.5	43.2	54.4	66.7	76.3	83.1	85.7	85.0	79.1	68.1	54.3	43.4	64.9
	MIN	18.3	20.0	28.4	37.3	46.2	54.7	59.3	58.1	51.1	38.4	30.4	22.7	38.7
	MEAN	29.0	31.6	41.4	52.0	61.3	68.9	72.5	71.5	65.1	53.3	42.4	33.1	51.8
BARNESVILLE WATER WKS	MAX	35.5	38.2	48.7	60.8	71.2	79.4	82.9	81.9	76.4	64.3	51.0	39.6	60.8
	MIN	15.1	16.7	26.0	35.7	44.9	53.8	57.9	56.4	49.1	36.8	29.0	20.7	36.8
	MEAN	25.3	27.5	37.3	48.3	58.1	66.6	70.4	69.2	62.8	50.6	40.0	30.2	48.9
BELLEFONTAINE SEWAGE	MAX	33.2	37.0	47.8	61.0	71.5	80.3	83.4	81.9	76.3	64.4	49.7	37.9	60.4
	MIN	17.5	20.0	28.9	39.8	49.8	58.5	61.9	60.6	54.3	43.4	32.8	23.0	40.9
	MEAN	25.4	28.5	38.4	50.4	60.7	69.4	72.7	71.3	65.3	53.9	41.2	30.5	50.6
BOWLING GREEN SWG PL	MAX	32.8	36.1	46.9	61.4	72.6	82.1	85.4	83.4	77.4	65.5	50.2	37.8	61.0
	MIN	17.2	19.3	27.6	38.1	48.6	58.4	61.9	60.2	53.6	42.7	32.8	22.4	40.2
	MEAN	25.0	27.7	37.3	49.8	60.6	70.3	73.7	71.8	65.5	54.1	41.5	30.1	50.6
BUCYRUS	MAX	32.8	35.7	45.9	59.5	70.5	79.8	83.3	81.7	75.8	63.7	49.4	37.4	59.6
	MIN	15.9	17.3	26.3	37.0	47.0	56.6	60.3	58.3	51.3	40.0	31.0	21.2	38.5
	MEAN	24.4	26.5	36.1	48.3	58.8	68.2	71.8	70.0	63.5	51.9	40.2	29.3	49.1
CADIZ	MAX	35.9	39.1	50.0	62.7	72.4	79.9	83.0	82.1	76.6	65.3	51.6	40.1	61.6
	MIN	19.2	20.8	29.4	39.7	49.5	57.9	61.8	60.8	54.5	43.3	33.5	24.0	41.2
	MEAN	27.6	30.0	39.7	51.2	60.9	68.9	72.4	71.5	65.6	54.3	42.6	32.1	51.4
CALDWELL 6 NW	MAX	37.6	41.3	52.3	64.9	74.2	81.6	84.5	83.7	78.2	66.8	52.7	41.8	63.3
	MIN	19.0	20.7	29.4	39.5	48.1	56.1	60.2	59.1	52.4	41.1	32.6	23.7	40.2
	MEAN	28.3	31.0	40.8	52.2	61.2	68.9	72.4	71.4	65.3	54.0	42.7	32.8	51.8
CAMBRIDGE SEWAGE PLANT	MAX	37.8	41.3	52.0	64.7	74.7	82.1	85.1	84.2	78.5	66.9	53.0	41.9	63.5
	MIN	19.1	20.9	29.1	38.5	47.5	56.2	60.3	59.1	52.4	40.3	32.2	23.9	40.0
	MEAN	28.4	31.1	40.6	51.6	61.1	69.2	72.7	71.7	65.4	53.6	42.6	32.9	51.7
CANFIELD 1 S	MAX	34.1	37.2	47.5	61.0	71.4	79.7	83.0	81.8	76.1	64.2	50.1	38.3	60.4
	MIN	16.5	17.6	26.5	36.1	45.1	54.2	57.8	56.8	50.5	39.8	31.7	21.9	37.9
	MEAN	25.3	27.4	37.0	48.6	58.3	67.0	70.4	69.3	63.3	52.0	40.9	30.2	49.1
CHARDON	MAX	31.7	34.3	43.9	57.3	68.2	77.1	80.7	79.5	73.3	61.9	48.1	36.2	57.7
	MIN	15.6	15.9	25.0	35.6	45.4	54.6	58.8	57.6	51.1	41.1	31.9	21.5	37.8
	MEAN	23.7	25.1	34.5	46.5	56.8	65.9	69.8	68.6	62.3	51.5	40.1	28.9	47.8
CHARLES MILL DAM	MAX	33.0	35.7	45.7	58.7	69.6	78.6	82.3	81.1	75.3	63.4	49.4	37.3	59.2
	MIN	14.6	15.6	25.0	35.3	44.3	53.3	57.0	55.6	48.8	37.8	30.1	20.5	36.5
	MEAN	23.8	25.7	35.3	47.0	57.0	66.0	69.7	68.4	62.0	50.6	39.8	29.0	47.9
CHILO MELDAHL DAM	MAX	39.7	43.1	52.9	65.9	75.6	83.4	87.3	86.5	80.7	69.1	55.3	44.3	65.3
	MIN	19.9	22.1	30.6	40.4	49.7	58.5	62.9	61.6	54.5	42.5	32.9	24.9	41.7
	MEAN	29.8	32.7	41.8	53.2	62.7	71.0	75.1	74.0	67.6	55.8	44.1	34.6	53.5
CHIPPEWA LAKE	MAX	33.2	36.4	46.9	60.8	71.2	79.8	83.2	81.9	76.0	64.2	50.0	37.7	60.1
	MIN	16.6	17.9	26.6	36.5	45.7	54.6	58.1	56.8	50.8	40.5	32.1	22.2	38.2
	MEAN	24.9	27.2	36.8	48.7	58.5	67.2	70.7	69.3	63.4	52.4	41.1	30.0	49.2
CINCINNATI ABBE WSMO	MAX	38.0	42.0	52.3	65.2	75.0	82.9	86.3	85.7	79.4	67.4	52.8	42.2	64.1
	MIN	22.5	25.0	33.7	44.2	53.6	61.9	65.9	64.4	57.7	46.0	36.3	27.5	44.9
	MEAN	30.3	33.5	43.0	54.7	64.3	72.4	76.1	75.0	68.5	56.7	44.6	34.9	54.5

OHIO

TEMPERATURE NORMALS (DEG F)

STATION		JAN	FEB	MAR	APR	MAY	JUN	JUL	AUG	SEP	OCT	NOV	DEC	ANN
CIRCLEVILLE	MAX	38.3	42.1	53.0	65.8	75.7	83.6	86.6	85.6	80.2	68.6	54.0	42.6	64.7
	MIN	21.1	23.0	31.8	41.6	51.0	59.4	63.3	61.5	54.7	43.0	34.1	26.0	42.5
	MEAN	29.7	32.6	42.4	53.8	63.4	71.5	75.0	73.6	67.4	55.8	44.1	34.3	53.6
CLEVELAND WSO //R	MAX	32.5	34.8	44.8	57.9	68.5	78.0	81.7	80.3	74.2	62.7	49.3	37.5	58.5
	MIN	18.5	19.9	28.4	38.3	47.9	57.2	61.4	60.5	54.0	43.6	34.3	24.6	40.7
	MEAN	25.5	27.4	36.6	48.1	58.2	67.6	71.6	70.4	64.1	53.2	41.8	31.1	49.6
COLUMBUS SULLIVANT AVE	MAX	36.6	40.4	51.5	64.8	75.1	83.4	86.5	85.0	79.3	67.5	52.9	41.4	63.7
	MIN	19.1	21.4	30.0	40.2	50.1	58.7	63.2	61.5	54.5	42.7	33.2	23.8	41.5
	MEAN	27.8	30.9	40.8	52.5	62.7	71.1	74.9	73.3	66.9	55.1	43.1	32.6	52.6
COLUMBUS VALLEY CROSS	MAX	36.0	39.7	50.8	63.7	73.7	81.7	84.7	83.3	77.8	66.3	52.1	40.6	62.5
	MIN	19.5	22.1	30.9	40.7	50.5	59.2	63.0	61.4	54.5	42.7	33.4	24.5	41.9
	MEAN	27.8	31.0	40.8	52.3	62.1	70.5	73.9	72.4	66.2	54.5	42.8	32.6	52.2
COLUMBUS WSO //R	MAX	34.7	38.1	49.3	62.3	72.6	81.3	84.4	83.0	76.9	65.0	50.7	39.4	61.5
	MIN	19.4	21.5	30.6	40.5	50.2	59.0	63.2	61.7	54.6	42.8	33.5	24.7	41.8
	MEAN	27.1	29.8	40.0	51.4	61.4	70.2	73.8	72.4	65.8	53.9	42.1	32.1	51.7
COSHOCTON SEWAGE PL	MAX	36.3	39.9	50.6	63.4	73.8	82.0	84.9	83.6	77.7	66.0	52.1	40.7	62.6
	MIN	17.7	19.7	28.3	37.8	47.4	56.1	59.9	58.5	51.8	39.6	31.2	22.8	39.2
	MEAN	27.1	29.8	39.5	50.6	60.6	69.1	72.4	71.1	64.8	52.8	41.7	31.7	50.9
DAYTON	MAX	35.8	39.5	49.7	63.3	74.1	83.2	86.5	85.1	78.8	66.4	51.8	40.8	62.9
	MIN	21.1	23.4	32.5	43.4	53.7	62.8	66.8	64.9	57.6	45.9	36.1	26.5	44.6
	MEAN	28.5	31.5	41.1	53.4	64.0	73.0	76.7	75.0	68.3	56.1	44.0	33.6	53.8
DAYTON WSO R	MAX	34.5	38.0	48.6	62.0	72.4	81.6	84.9	83.4	77.1	65.1	50.5	39.3	61.5
	MIN	18.8	21.2	30.3	41.0	51.2	60.4	64.3	62.6	55.5	43.9	33.7	24.3	42.3
	MEAN	26.6	29.6	39.5	51.5	61.8	71.0	74.7	73.0	66.4	54.5	42.1	31.8	51.9
DEFIANCE	MAX	31.2	34.3	44.5	59.0	71.0	80.9	84.6	82.8	76.4	64.3	48.8	36.5	59.5
	MIN	13.9	15.5	24.8	36.3	46.5	56.5	60.3	58.3	51.0	39.3	29.9	19.6	37.7
	MEAN	22.6	24.9	34.7	47.6	58.8	68.7	72.5	70.6	63.7	51.8	39.4	28.1	48.6
DELAWARE	MAX	34.2	37.5	48.4	61.8	72.3	81.2	84.6	83.1	77.0	65.1	50.7	36.7	61.2
	MIN	16.4	18.5	27.4	37.8	47.4	56.7	60.5	58.8	51.6	39.6	31.0	21.9	39.0
	MEAN	25.4	28.0	37.9	49.8	59.9	69.0	72.6	71.0	64.4	52.4	40.9	30.3	50.1
ELYRIA 3 E	MAX	33.8	36.4	46.2	60.2	71.2	80.5	84.3	82.7	76.7	65.1	50.6	38.6	60.5
	MIN	18.0	19.2	27.8	37.8	47.4	56.8	60.8	59.6	53.3	42.6	33.5	23.8	40.1
	MEAN	25.9	27.8	37.0	49.0	59.3	68.7	72.6	71.2	65.1	53.9	42.1	31.2	50.3
FINDLAY FAA AP	MAX	31.4	34.6	45.1	59.1	70.3	79.9	83.3	81.6	75.3	63.2	48.3	36.3	59.0
	MIN	16.9	19.3	28.1	38.9	49.1	58.7	62.5	60.8	53.7	42.6	32.6	22.6	40.5
	MEAN	24.2	27.0	36.6	49.0	59.7	69.3	72.9	71.2	64.6	52.9	40.5	29.5	49.8
FINDLAY SEWAGE PLANT	MAX	31.6	34.9	45.6	59.4	70.9	80.3	83.8	81.9	75.5	63.4	48.2	36.3	59.3
	MIN	15.9	18.3	27.0	37.8	48.3	57.7	61.2	59.3	52.1	41.0	31.1	21.3	39.3
	MEAN	23.8	26.7	36.3	48.6	59.6	69.1	72.5	70.6	63.9	52.2	39.7	28.8	49.3
FREDERICKTOWN	MAX	33.3	36.4	47.1	60.1	71.1	80.2	83.9	82.8	76.7	64.6	49.9	37.8	60.3
	MIN	14.5	15.9	25.9	36.5	45.7	55.1	58.2	56.7	49.3	38.0	29.5	20.0	37.1
	MEAN	24.0	26.2	36.5	48.4	58.4	67.7	71.1	69.8	63.1	51.4	39.7	28.9	48.8
GALLIPOLIS 5 W	MAX	42.2	46.3	56.8	68.8	77.5	84.4	87.5	86.6	81.1	70.0	56.6	46.4	67.0
	MIN	22.3	24.1	32.1	41.2	50.1	58.3	62.7	61.5	55.0	42.6	33.6	26.4	42.5
	MEAN	32.3	35.2	44.5	55.1	63.8	71.4	75.1	74.1	68.0	56.4	45.1	36.4	54.8
GREENVILLE WATER PLT	MAX	33.4	36.6	47.3	60.7	71.6	80.7	84.1	82.8	77.0	64.8	49.9	38.1	60.6
	MIN	15.4	17.3	26.9	37.5	47.3	56.5	60.2	58.0	50.3	38.6	30.0	20.8	38.2
	MEAN	24.4	27.0	37.2	49.1	59.5	68.6	72.2	70.4	63.7	51.7	40.0	29.5	49.4
HAMILTON-FAIRFIELD	MAX	39.0	43.1	53.7	66.7	76.2	84.3	87.3	86.3	80.9	69.1	54.4	43.4	65.4
	MIN	20.7	23.2	31.7	41.8	51.1	59.7	63.4	61.4	54.6	42.7	33.9	25.6	42.5
	MEAN	29.9	33.2	42.7	54.2	63.7	72.0	75.4	73.9	67.8	55.9	44.2	34.5	54.0

OHIO

TEMPERATURE NORMALS (DEG F)

STATION		JAN	FEB	MAR	APR	MAY	JUN	JUL	AUG	SEP	OCT	NOV	DEC	ANN
HILLSBORO	MAX	36.9	40.1	50.8	63.3	72.7	80.3	83.9	83.0	77.6	66.1	52.2	41.1	62.3
	MIN	19.6	21.6	31.1	41.7	51.4	59.6	63.5	61.9	55.3	43.9	34.0	24.7	42.4
	MEAN	28.3	30.9	41.0	52.6	62.1	70.0	73.7	72.4	66.5	55.0	43.1	32.9	52.4
HIRAM	MAX	32.7	35.6	45.7	59.6	70.3	78.4	82.0	80.5	74.5	63.0	49.0	37.0	59.0
	MIN	16.8	18.1	26.7	37.5	47.4	55.9	60.3	59.0	53.0	42.3	32.5	22.2	39.3
	MEAN	24.8	26.9	36.2	48.6	58.9	67.2	71.2	69.8	63.7	52.7	40.8	29.6	49.2
IRONTON	MAX	42.7	46.9	57.7	69.8	78.7	85.4	88.2	87.1	81.3	70.2	56.8	46.7	67.6
	MIN	24.0	25.7	33.9	43.4	51.9	60.5	64.8	63.6	57.1	44.4	35.1	27.9	44.4
	MEAN	33.4	36.3	45.8	56.6	65.3	72.9	76.5	75.4	69.2	57.3	46.0	37.3	56.0
IRWIN	MAX	34.5	38.4	49.4	63.1	73.8	82.3	85.4	84.2	78.5	66.8	51.1	39.2	62.2
	MIN	16.7	19.3	28.1	38.3	48.2	56.7	59.9	58.1	51.1	40.0	31.1	22.0	39.1
	MEAN	25.7	28.9	38.8	50.8	61.0	69.5	72.7	71.2	64.9	53.4	41.1	30.6	50.7
JACKSON 2 NW	MAX	39.3	42.8	53.2	65.5	75.2	82.4	85.5	84.5	78.8	67.5	54.4	43.2	64.4
	MIN	17.6	19.7	28.5	37.8	47.0	55.6	60.1	58.5	51.2	38.0	29.8	22.2	38.8
	MEAN	28.5	31.3	40.9	51.7	61.1	69.0	72.9	71.5	65.0	52.8	42.1	32.7	51.6
KENTON	MAX	33.1	36.7	47.1	61.0	71.8	81.0	84.4	83.0	76.8	65.0	49.9	37.9	60.6
	MIN	16.2	18.3	27.1	37.9	47.9	56.7	60.5	58.6	51.9	40.9	31.4	21.5	39.1
	MEAN	24.7	27.5	37.1	49.4	59.9	68.9	72.4	70.8	64.4	53.0	40.7	29.7	49.9
LANCASTER 2 NW	MAX	36.5	39.4	50.2	62.8	73.4	81.8	85.4	84.2	78.6	66.5	52.3	40.8	62.7
	MIN	17.9	19.6	28.7	38.7	48.2	57.1	60.8	59.0	52.2	40.2	31.8	23.0	39.8
	MEAN	27.2	29.5	39.5	50.8	60.9	69.5	73.1	71.6	65.5	53.4	42.1	31.9	51.3
LIMA SEWAGE PLANT	MAX	33.4	37.1	47.5	61.4	72.5	81.7	85.1	83.2	77.3	65.3	50.3	38.5	61.1
	MIN	17.5	20.0	28.7	39.4	49.5	58.6	62.2	60.5	54.2	43.0	33.1	23.3	40.8
	MEAN	25.5	28.6	38.1	50.4	61.0	70.2	73.7	71.9	65.8	54.2	41.7	31.0	51.0
LONDON WATER WORKS	MAX	35.0	38.8	49.9	62.9	73.2	81.7	84.8	83.4	77.7	65.8	51.1	39.5	62.0
	MIN	18.5	20.8	29.7	39.3	48.9	57.5	61.1	59.5	53.1	41.8	32.6	23.6	40.5
	MEAN	26.7	29.8	39.8	51.1	61.1	69.6	73.0	71.5	65.4	53.8	41.9	31.6	51.3
MANSFIELD WSO	MAX	32.2	35.0	45.4	58.8	69.3	78.4	82.1	80.7	74.4	62.8	48.5	36.8	58.7
	MIN	17.4	19.2	27.9	38.4	48.2	57.5	61.8	60.5	53.8	43.0	33.0	22.9	40.3
	MEAN	24.8	27.2	36.7	48.6	58.8	68.0	72.0	70.6	64.1	52.9	40.8	29.9	49.5
MANSFIELD 6 W	MAX	32.3	35.4	45.9	59.2	69.6	78.2	81.8	80.3	74.2	62.7	48.5	36.8	58.7
	MIN	15.3	17.3	26.1	36.3	45.9	54.1	57.7	56.4	50.1	39.4	30.7	20.8	37.5
	MEAN	23.8	26.4	36.0	47.8	57.8	66.2	69.8	68.4	62.2	51.1	39.6	28.8	48.2
MARYSVILLE	MAX	33.9	37.3	48.1	61.5	72.1	80.8	84.1	82.8	76.6	64.6	49.9	38.2	60.8
	MIN	16.1	18.6	27.6	38.1	48.1	56.9	60.7	59.1	51.8	40.5	31.1	21.8	39.2
	MEAN	25.0	28.0	37.9	49.8	60.1	68.9	72.4	71.0	64.2	52.6	40.5	30.0	50.0
MC CONNELSVILLE LOCK 7	MAX	39.4	42.2	53.0	65.3	75.4	82.9	86.1	85.5	80.1	68.5	55.0	43.2	64.7
	MIN	17.1	18.5	27.5	37.7	47.3	56.4	60.8	59.6	52.0	39.1	30.2	21.9	39.0
	MEAN	28.3	30.4	40.3	51.6	61.4	69.7	73.5	72.6	66.1	53.8	42.6	32.5	51.9
MILFORD	MAX	38.5	42.4	52.9	66.4	75.9	84.1	87.5	86.5	80.7	68.9	54.1	42.8	65.1
	MIN	19.6	21.8	30.3	39.9	49.0	57.8	62.1	60.7	53.6	40.9	32.0	24.5	41.0
	MEAN	29.1	32.1	41.6	53.2	62.5	71.0	74.8	73.6	67.2	54.9	43.1	33.7	53.1
MILLERSBURG 1 W	MAX	34.8	38.3	48.9	61.5	71.3	79.4	82.5	81.5	75.7	64.4	50.7	39.1	60.7
	MIN	17.9	20.0	28.7	38.5	47.6	56.1	59.9	58.7	52.5	41.8	32.8	23.2	39.8
	MEAN	26.4	29.2	38.8	50.0	59.5	67.8	71.2	70.1	64.1	53.1	41.8	31.2	50.3
MILLPORT 2 NW	MAX	34.5	37.6	47.9	61.0	71.0	79.7	83.2	82.0	75.9	64.2	50.3	38.7	60.5
	MIN	16.1	17.6	26.5	35.6	44.7	53.2	57.3	56.0	49.2	38.0	30.3	21.5	37.2
	MEAN	25.3	27.6	37.2	48.3	57.9	66.5	70.3	69.1	62.5	51.1	40.3	30.1	48.9
MINERAL RIDGE WTR WKS	MAX	35.0	38.2	48.5	62.2	73.0	81.2	84.7	83.2	77.2	65.6	51.3	39.3	61.6
	MIN	18.1	18.9	27.2	36.9	46.2	55.2	59.1	58.3	51.9	41.3	33.3	23.8	39.2
	MEAN	26.6	28.6	37.9	49.6	59.6	68.3	71.9	70.8	64.6	53.4	42.3	31.6	50.4

OHIO

TEMPERATURE NORMALS (DEG F)

STATION		JAN	FEB	MAR	APR	MAY	JUN	JUL	AUG	SEP	OCT	NOV	DEC	ANN
MONTPELIER 1 WSW	MAX	31.4	35.1	45.4	60.2	71.8	81.0	84.4	82.9	76.7	65.0	49.1	36.4	60.0
	MIN	15.0	17.0	26.0	36.8	46.4	56.0	59.8	57.8	50.8	39.9	30.9	20.7	38.1
	MEAN	23.2	26.1	35.7	48.5	59.1	68.5	72.1	70.4	63.7	52.5	40.0	28.6	49.0
NAPOLEON WATER WORKS	MAX	32.6	36.2	47.0	61.7	73.1	82.5	85.5	83.8	77.6	65.8	50.0	37.6	61.1
	MIN	17.0	19.2	27.8	38.5	48.7	58.3	61.9	60.1	53.4	42.5	32.7	22.5	40.2
	MEAN	24.8	27.7	37.4	50.1	60.9	70.4	73.8	72.0	65.5	54.2	41.4	30.1	50.7
NEWARK WATER WORKS	MAX	36.1	39.5	50.6	63.5	74.1	82.3	85.4	83.9	77.7	65.8	51.7	40.2	62.6
	MIN	18.7	20.9	29.6	39.5	48.7	57.1	60.9	59.5	52.5	40.8	32.1	23.4	40.3
	MEAN	27.4	30.2	40.1	51.5	61.4	69.7	73.2	71.7	65.1	53.3	41.9	31.8	51.4
NEW LEXINGTON 2 NW	MAX	38.0	41.9	52.9	65.7	75.6	82.7	85.6	84.5	78.8	67.6	53.3	42.0	64.1
	MIN	17.4	19.5	28.1	37.6	46.9	55.4	59.1	57.7	50.9	38.8	30.4	22.3	38.7
	MEAN	27.7	30.7	40.5	51.7	61.2	69.1	72.4	71.1	64.9	53.2	41.9	32.2	51.4
NORWALK SEWAGE PLANT	MAX	33.3	36.0	45.7	59.5	70.5	79.8	83.4	81.9	75.9	64.1	50.1	37.9	59.8
	MIN	16.9	18.4	27.1	37.5	47.0	56.5	60.2	58.6	51.8	41.2	32.5	22.4	39.2
	MEAN	25.2	27.2	36.4	48.5	58.8	68.2	71.9	70.3	63.9	52.7	41.3	30.2	49.6
OBERLIN	MAX	34.0	37.2	47.4	61.5	72.6	81.3	84.8	83.1	76.8	65.4	50.8	38.5	61.1
	MIN	16.0	17.4	25.4	35.8	45.2	54.4	58.4	57.1	50.6	40.2	30.9	21.3	37.7
	MEAN	25.0	27.3	36.4	48.7	58.9	67.9	71.6	70.1	63.7	52.8	40.9	30.0	49.4
PAINESVILLE 4 NW	MAX	34.1	35.5	44.2	56.5	67.3	77.0	81.0	80.0	74.6	63.8	51.0	39.6	58.7
	MIN	19.6	20.0	27.9	38.2	48.3	57.8	61.9	61.4	55.5	45.7	36.1	26.0	41.5
	MEAN	26.9	27.8	36.0	47.3	57.8	67.4	71.5	70.7	65.1	54.8	43.6	32.8	50.1
PANDORA 2 NE	MAX	32.1	35.3	45.9	60.2	71.5	81.3	84.0	82.3	76.3	64.6	49.2	37.0	60.0
	MIN	16.3	18.5	27.4	38.2	48.4	57.9	61.0	58.9	52.2	41.3	32.1	21.9	39.5
	MEAN	24.2	26.9	36.7	49.2	60.0	69.6	72.5	70.6	64.3	53.0	40.6	29.5	49.8
PAULDING	MAX	32.1	35.5	46.2	60.7	72.1	81.6	85.1	83.3	77.1	65.1	49.6	37.3	60.5
	MIN	14.4	16.4	25.5	36.6	46.5	56.1	59.8	57.8	50.8	39.3	30.1	20.2	37.8
	MEAN	23.3	26.0	35.9	48.7	59.3	68.9	72.5	70.6	63.9	52.2	39.9	28.8	49.2
PEEBLES	MAX	40.3	44.2	54.4	66.6	75.6	82.8	86.0	85.1	79.6	68.4	54.8	44.3	65.2
	MIN	21.5	23.5	32.3	41.6	50.4	58.6	62.8	61.1	54.5	42.5	33.9	26.1	42.4
	MEAN	30.9	33.8	43.4	54.1	63.0	70.7	74.4	73.1	67.1	55.5	44.4	35.2	53.8
PHILO 3 SW	MAX	36.6	40.2	51.2	63.9	73.5	81.3	84.1	83.0	77.4	66.4	52.2	41.0	62.6
	MIN	20.1	22.1	30.7	41.1	50.0	58.0	61.5	60.1	53.8	43.0	33.8	24.8	41.6
	MEAN	28.4	31.1	41.0	52.5	61.8	69.7	72.9	71.6	65.6	54.7	43.1	32.9	52.1
PLYMOUTH	MAX	33.0	36.3	46.5	60.5	70.9	80.0	83.2	81.7	75.8	64.1	49.6	37.5	59.9
	MIN	16.1	17.5	26.6	37.5	47.0	56.1	59.8	58.3	51.9	41.5	31.6	21.2	38.8
	MEAN	24.6	26.9	36.5	49.0	58.9	68.1	71.5	70.1	63.9	52.8	40.6	29.4	49.4
PORTSMOUTH	MAX	40.6	43.8	54.2	67.1	76.3	83.2	86.4	85.4	79.8	68.7	55.9	45.1	65.5
	MIN	23.4	25.4	34.0	43.7	52.7	60.9	65.3	64.2	57.8	45.3	35.8	27.8	44.7
	MEAN	32.0	34.6	44.1	55.4	64.5	72.1	75.9	74.8	68.8	57.0	45.9	36.4	55.1
PUT IN BAY PERRY MON	MAX	32.3	34.9	43.9	57.4	68.6	78.7	82.9	81.4	75.4	63.6	49.5	37.2	58.8
	MIN	19.0	20.1	28.3	39.3	50.7	60.9	66.1	65.5	59.5	48.3	36.7	25.3	43.3
	MEAN	25.7	27.5	36.1	48.4	59.7	69.8	74.5	73.5	67.4	56.0	43.1	31.3	51.1
SANDUSKY R	MAX	32.7	34.8	44.0	57.5	68.6	78.6	82.7	81.3	74.8	63.1	49.5	37.8	58.8
	MIN	18.7	20.3	29.0	40.2	50.7	60.6	65.0	63.4	56.6	45.3	35.0	24.4	42.4
	MEAN	25.7	27.5	36.5	48.9	59.7	69.7	73.9	72.4	65.8	54.2	42.3	31.1	50.6
SENECAVILLE DAM	MAX	36.6	39.3	50.0	62.4	72.5	80.4	83.7	82.4	77.1	65.4	52.0	40.6	61.9
	MIN	17.2	18.2	27.5	37.8	46.8	55.9	59.8	57.9	51.0	38.9	30.7	22.7	38.7
	MEAN	26.9	28.8	38.8	50.2	59.7	68.1	71.7	70.2	64.1	52.2	41.4	31.7	50.3
STEUBENVILLE WATER WKS	MAX	36.7	39.8	50.1	62.8	72.7	80.7	84.0	82.8	76.9	65.4	52.2	40.9	62.1
	MIN	19.4	21.4	29.4	39.1	48.7	57.5	61.8	61.0	54.5	42.9	33.9	24.6	41.2
	MEAN	28.1	30.6	39.8	51.0	60.7	69.1	72.9	71.9	65.7	54.2	43.1	32.8	51.7

OHIO

TEMPERATURE NORMALS (DEG F)

STATION			JAN	FEB	MAR	APR	MAY	JUN	JUL	AUG	SEP	OCT	NOV	DEC	ANN
TIFFIN		MAX	33.1	36.4	47.1	61.3	72.3	81.2	84.3	82.4	76.4	64.6	49.7	37.9	60.6
		MIN	18.3	20.4	28.9	39.4	49.2	58.5	62.4	60.8	54.4	43.5	34.0	23.8	41.1
		MEAN	25.7	28.4	38.0	50.4	60.8	69.9	73.4	71.6	65.4	54.0	41.9	30.9	50.9
TOLEDO EXPRESS WSO	R	MAX	30.7	34.0	44.6	59.1	70.5	79.9	83.4	81.8	75.1	63.3	47.9	35.5	58.8
		MIN	15.5	17.5	26.1	36.5	46.6	56.0	60.2	58.4	51.2	40.1	30.6	20.6	38.3
		MEAN	23.1	25.8	35.4	47.8	58.6	68.0	71.8	70.1	63.2	51.7	39.3	28.1	48.6
TOLEDO BLADE		MAX	32.6	35.5	45.2	59.4	71.8	82.1	86.4	84.3	77.4	64.9	49.8	37.2	60.6
		MIN	18.5	20.3	28.9	40.5	51.1	61.3	65.5	63.6	56.4	44.9	34.4	23.7	42.4
		MEAN	25.5	27.9	37.0	50.0	61.5	71.7	76.0	74.0	66.9	54.9	42.1	30.5	51.5
UPPER SANDUSKY		MAX	33.4	37.0	47.8	61.7	72.7	81.9	85.1	83.6	77.4	65.4	50.2	38.2	61.2
		MIN	18.2	20.2	28.9	39.3	49.0	58.1	61.7	60.0	53.4	42.6	33.3	23.5	40.7
		MEAN	25.8	28.7	38.4	50.5	60.8	70.0	73.4	71.8	65.5	54.0	41.8	30.9	51.0
URBANA SEWAGE PLANT		MAX	33.9	37.1	47.7	61.0	71.8	80.8	84.5	83.0	77.2	64.7	50.2	38.6	60.9
		MIN	15.7	17.8	27.6	38.2	48.1	57.1	60.6	58.2	51.2	39.8	31.0	21.4	38.9
		MEAN	24.8	27.5	37.7	49.6	60.0	69.0	72.5	70.7	64.2	52.3	40.6	30.0	49.9
VAN WERT		MAX	33.4	37.4	47.9	62.4	73.3	82.2	85.4	83.8	77.6	65.9	50.4	38.1	61.5
		MIN	16.8	19.1	27.8	38.6	48.6	58.0	61.7	59.9	53.0	41.9	32.4	22.4	40.0
		MEAN	25.1	28.3	37.9	50.5	61.0	70.1	73.6	71.9	65.3	53.9	41.4	30.3	50.8
WARREN 3 S		MAX	34.3	37.1	47.2	61.0	71.6	80.1	83.7	82.4	75.9	64.1	50.4	38.6	60.5
		MIN	16.9	18.0	26.5	35.7	45.3	54.4	58.1	57.2	50.6	39.9	32.0	22.8	38.1
		MEAN	25.6	27.6	36.9	48.4	58.5	67.3	70.9	69.8	63.3	52.1	41.2	30.8	49.4
WASHINGTON COURT HOUSE		MAX	36.1	39.7	50.3	63.4	73.0	80.6	83.7	82.7	77.4	66.5	52.1	40.6	62.2
		MIN	19.8	22.1	30.7	40.8	50.9	59.6	63.5	61.9	55.0	43.4	33.2	24.6	42.1
		MEAN	27.9	30.9	40.5	52.1	62.0	70.1	73.6	72.4	66.2	55.0	42.7	32.7	52.2
WAUSEON WASTE WTR PLT		MAX	30.7	34.1	44.7	59.4	70.8	80.7	83.9	81.8	75.7	64.0	48.3	35.8	59.2
		MIN	14.4	16.6	25.8	36.3	46.5	56.2	59.5	57.6	50.5	39.7	30.2	20.1	37.8
		MEAN	22.6	25.4	35.2	47.9	58.7	68.5	71.7	69.7	63.1	51.9	39.3	28.0	48.5
WAVERLY		MAX	39.9	43.6	54.1	66.8	75.6	82.7	85.7	84.4	78.6	67.9	55.1	44.0	64.9
		MIN	19.6	21.7	30.6	40.1	49.4	57.8	62.4	60.8	53.9	41.0	32.1	24.2	41.1
		MEAN	29.8	32.7	42.4	53.4	62.5	70.2	74.1	72.6	66.3	54.5	43.7	34.1	53.0
WILMINGTON		MAX	36.4	40.0	50.6	63.3	73.5	82.0	85.1	84.0	78.1	66.1	51.6	40.6	62.6
		MIN	19.1	21.0	30.2	40.2	49.2	57.7	61.1	59.5	52.8	41.7	32.5	24.2	40.8
		MEAN	27.8	30.5	40.4	51.8	61.4	69.9	73.1	71.8	65.5	53.9	42.1	32.4	51.7
WOOSTER EXP STATION		MAX	32.7	35.7	46.2	59.0	69.5	78.3	81.8	80.6	74.3	62.6	48.7	37.0	58.9
		MIN	16.8	18.5	27.3	37.0	46.4	55.4	59.1	57.9	51.0	40.0	31.7	22.3	38.6
		MEAN	24.8	27.1	36.7	48.0	58.0	66.9	70.5	69.3	62.6	51.3	40.2	29.7	48.8
XENIA 5 SSE		MAX	36.3	40.0	50.9	64.1	73.7	81.0	84.1	82.8	77.6	66.2	51.9	40.7	62.4
		MIN	19.1	21.6	30.5	40.6	50.2	58.4	62.0	60.1	53.4	41.8	32.6	24.1	41.2
		MEAN	27.7	30.8	40.7	52.4	62.0	69.7	73.1	71.5	65.6	54.0	42.3	32.4	51.9
YOUNGSTOWN WSO	R	MAX	31.4	33.8	44.0	57.9	68.6	77.5	81.2	79.8	73.2	61.4	47.7	36.0	57.7
		MIN	16.9	18.0	26.5	36.9	46.0	55.1	59.0	58.2	51.5	41.5	32.7	22.6	38.7
		MEAN	24.2	25.9	35.3	47.4	57.3	66.3	70.1	69.0	62.4	51.5	40.2	29.4	48.3
ZANESVILLE FAA AP		MAX	36.0	39.2	49.8	62.5	72.7	80.7	83.7	82.6	76.7	65.1	51.4	40.2	61.7
		MIN	18.6	20.4	29.2	39.1	48.4	57.1	61.3	59.9	52.9	40.9	32.2	23.5	40.3
		MEAN	27.3	29.8	39.6	50.8	60.5	68.9	72.5	71.3	64.8	53.0	41.8	31.9	51.0

OHIO

PRECIPITATION NORMALS (INCHES)

STATION		JAN	FEB	MAR	APR	MAY	JUN	JUL	AUG	SEP	OCT	NOV	DEC	ANN
AKRON CANTON WSO	R	2.56	2.18	3.37	3.26	3.55	3.27	4.01	3.31	2.96	2.24	2.54	2.65	35.90
ALEXANDRIA 4 WSW		2.84	2.16	3.17	3.72	4.16	4.33	4.42	4.00	3.14	2.41	2.96	2.73	40.04
ASHLAND		2.68	2.15	3.26	3.63	3.84	3.32	4.11	3.35	3.27	2.14	2.78	2.66	37.19
ASHTABULA		2.35	1.78	2.63	3.49	3.41	3.52	4.07	3.81	3.63	3.29	3.48	2.85	38.31
ATHENS 5 NW		3.11	2.64	3.65	3.58	3.92	3.20	4.25	3.60	3.11	2.42	2.68	2.93	39.09
ATWOOD DAM		2.46	2.00	3.23	3.27	3.48	3.93	4.33	3.72	2.94	2.51	2.37	2.36	36.60
BARNESVILLE WATER WKS		2.98	2.65	3.77	3.97	4.07	4.22	4.50	3.79	3.18	2.66	2.74	2.94	41.47
BEACH CITY DAM		2.57	2.12	3.22	3.45	3.45	3.56	4.43	3.91	3.00	2.24	2.47	2.46	36.88
BELLEFONTAINE SEWAGE		2.40	1.92	2.99	3.63	3.78	3.68	3.78	3.07	2.74	2.19	2.49	2.62	35.29
BOLIVAR DAM		2.58	2.16	3.26	3.38	3.78	3.72	4.59	3.77	2.98	2.37	2.52	2.59	37.70
BOWLING GREEN SWG PL		1.91	1.53	2.51	3.25	3.26	3.59	3.99	2.79	2.60	2.07	2.47	2.23	32.20
BUCYRUS		2.43	1.94	2.90	3.70	3.57	3.81	3.98	3.09	3.08	2.05	2.66	2.49	35.70
CADIZ		2.77	2.29	3.36	3.51	3.85	3.81	4.29	3.82	2.72	2.63	2.59	2.59	38.23
CALDWELL 6 NW		2.68	2.24	3.27	3.33	3.79	4.20	4.42	3.80	3.00	2.43	2.51	2.45	38.12
CAMBRIDGE SEWAGE PLANT		2.77	2.54	3.69	3.51	3.83	4.40	4.07	3.60	2.74	2.45	2.66	2.65	38.91
CANFIELD 1 S		2.28	1.82	3.12	3.24	3.73	3.66	4.26	3.20	3.06	2.58	2.55	2.27	35.77
CARPENTER		2.84	2.44	3.69	3.41	4.27	3.61	4.57	3.93	3.41	2.36	2.53	2.67	39.73
CARROLLTON 2 NW		2.57	2.23	3.40	3.31	3.44	3.87	4.47	3.52	3.00	2.48	2.55	2.46	37.30
CENTERBURG		3.09	2.36	3.58	4.03	3.97	4.39	4.25	4.01	3.25	2.52	3.10	2.89	41.44
CHARDON		3.33	2.69	3.74	4.26	3.74	4.01	3.88	4.39	3.68	3.75	4.06	3.69	45.22
CHARLES MILL DAM		2.44	2.03	3.03	3.43	3.86	3.47	4.42	3.17	2.93	2.04	2.60	2.50	35.92
CHILO MELDAHL DAM		3.12	2.97	4.38	3.83	3.92	3.87	4.23	4.01	3.38	2.43	3.18	3.25	42.57
CHIPPEWA LAKE		2.38	2.00	3.10	3.44	3.67	3.73	3.96	3.06	3.17	2.30	2.75	2.56	36.12
CINCINNATI ABBE WSMO		3.09	2.58	3.83	3.71	4.07	3.83	4.04	3.45	3.11	2.47	3.03	2.89	40.10
CIRCLEVILLE		2.59	2.28	3.43	3.75	4.00	3.66	4.12	3.80	3.31	2.18	2.65	2.56	38.33
CLENDENING DAM		2.75	2.43	3.71	3.71	3.97	3.96	4.29	4.10	3.12	2.58	2.52	2.69	39.83
CLEVELAND WSO	//R	2.47	2.20	2.99	3.32	3.30	3.49	3.37	3.38	2.92	2.45	2.76	2.75	35.40
COLUMBUS SULLIVANT AVE		2.45	2.10	3.15	3.61	3.77	3.78	3.87	3.76	2.88	2.04	2.65	2.50	36.56
COLUMBUS VALLEY CROSS		2.70	1.96	3.28	3.64	3.98	3.85	3.95	4.17	2.97	2.14	2.79	2.51	37.94
COLUMBUS WSO	//R	2.75	2.18	3.23	3.41	3.76	4.01	4.01	3.70	2.76	1.91	2.64	2.61	36.97
COOPERDALE		2.86	2.19	3.54	3.72	3.85	3.93	4.52	3.50	3.03	2.45	2.80	2.64	39.03
COSHOCTON SEWAGE PL		2.76	2.39	3.57	3.98	3.65	3.88	4.57	3.64	2.98	2.48	2.86	2.80	39.56
DAYTON		2.77	2.38	3.31	3.92	3.75	4.35	3.74	3.15	2.80	2.20	2.86	2.66	37.89
DAYTON WSO	R	2.57	2.11	3.08	3.43	3.69	3.81	3.37	3.10	2.39	2.01	2.64	2.51	34.71
DEFIANCE		2.01	1.72	2.67	3.41	3.45	3.44	3.55	3.10	2.65	2.24	2.61	2.39	33.24
DELAWARE		2.50	2.04	3.11	3.72	3.60	3.77	3.99	3.76	2.73	2.04	2.66	2.60	36.52
DELAWARE DAM		2.64	2.11	3.08	3.53	3.42	3.78	4.02	3.46	2.66	2.10	2.63	2.60	36.03
DOVER DAM		2.73	2.29	3.59	3.60	3.59	4.21	4.56	3.71	3.07	2.50	2.67	2.65	39.17
EATON		2.91	2.52	3.53	3.96	4.26	3.95	3.84	3.36	2.79	2.41	2.87	3.01	39.41
ELYRIA 3 E		2.36	2.02	2.83	3.23	3.44	3.68	3.17	3.51	3.01	2.42	2.77	2.64	35.08
ENTERPRISE		3.02	2.43	3.65	3.76	3.85	3.75	4.25	3.68	3.12	2.48	2.77	2.71	39.47
FINDLAY FAA AP		2.02	1.59	2.85	3.46	3.55	3.41	3.90	3.18	2.80	1.93	2.40	2.22	33.31
FINDLAY SEWAGE PLANT		2.24	1.93	2.99	3.54	3.72	3.49	4.08	3.48	2.97	2.02	2.63	2.51	35.60
FRANKLIN		2.88	2.36	3.33	3.71	3.96	3.78	3.81	3.19	3.10	2.39	2.87	2.72	38.10
FREDERICKTOWN		2.79	2.11	3.40	3.77	3.71	4.25	3.88	3.78	3.06	2.33	2.77	2.71	38.56
FREMONT		2.10	1.65	2.74	3.07	3.25	3.54	3.75	3.37	2.68	1.92	2.42	2.29	32.78
GALION WATER WORKS		2.53	2.04	3.06	3.61	3.96	4.01	4.08	3.58	3.09	2.08	2.84	2.59	37.47
GALLIPOLIS 5 W		3.10	2.86	3.92	3.56	3.86	3.58	4.75	4.15	3.13	2.38	2.77	3.11	41.17
GREENVILLE WATER PLT		2.43	1.99	3.00	3.57	3.65	4.01	3.90	3.18	2.66	2.36	2.56	2.54	35.85
GREER		2.61	2.06	3.23	3.68	3.90	4.15	4.38	3.75	3.12	2.45	2.87	2.48	38.68
HAMILTON-FAIRFIELD		2.85	2.46	3.45	3.59	3.56	3.77	4.22	3.36	3.13	2.41	2.80	2.60	38.20
HIGGINSPORT		3.19	2.80	4.40	3.96	4.01	3.72	4.54	4.24	3.49	2.42	3.19	3.30	43.26
HILLSBORO		3.25	2.63	4.17	4.03	4.20	3.82	4.94	3.95	3.49	2.63	2.96	2.98	43.05
HIRAM		2.73	2.41	3.62	3.91	3.68	3.92	3.93	3.61	3.63	3.07	3.28	3.28	41.07
HOYTVILLE 2 NE		1.98	1.61	2.66	3.23	3.34	3.48	3.94	3.09	2.84	2.05	2.45	2.25	32.92

482

OHIO

PRECIPITATION NORMALS (INCHES)

STATION	JAN	FEB	MAR	APR	MAY	JUN	JUL	AUG	SEP	OCT	NOV	DEC	ANN
HUNTSVILLE 3 N	2.27	1.81	2.90	3.74	3.45	3.45	3.75	3.18	2.55	2.05	2.48	2.24	33.87
IRONTON	3.20	2.77	4.04	3.79	4.19	4.25	4.62	3.99	3.22	2.32	2.78	3.14	42.31
IRWIN	2.54	2.02	3.11	3.94	4.03	3.91	4.46	3.61	3.04	2.08	2.55	2.43	37.72
JACKSON 2 NW	3.10	2.72	4.21	3.91	3.94	3.46	4.85	3.91	3.46	2.34	2.66	3.08	41.64
KENTON	2.57	1.91	3.26	3.67	3.67	3.51	3.94	3.23	2.80	1.85	2.55	2.45	35.41
KINGS MILLS	3.35	2.63	3.71	4.14	4.14	4.16	4.32	3.47	3.43	2.67	3.10	2.81	41.93
LAKEVIEW 3 NE	2.61	2.03	3.32	3.66	3.48	3.67	3.78	3.03	2.63	2.07	2.56	2.64	35.48
LANCASTER 2 NW	2.48	2.27	3.11	3.15	3.40	3.74	3.84	3.63	2.72	2.06	2.44	2.51	35.35
LA RUE	2.40	1.92	3.30	3.73	3.40	3.20	3.66	3.26	2.80	1.98	2.46	2.44	34.55
LAURELVILLE	2.91	2.46	3.86	3.84	4.18	3.87	4.84	3.62	3.33	2.42	2.77	2.79	40.89
LEESVILLE DAM	2.78	2.36	3.52	3.39	3.65	4.15	4.34	3.76	3.14	2.46	2.52	2.60	38.67
LIMA SEWAGE PLANT	2.45	1.86	3.14	3.71	3.65	3.67	3.63	3.03	2.86	2.15	2.68	2.55	35.38
LITHOPOLIS 2 S	2.83	2.28	3.46	3.53	3.74	3.89	4.43	3.66	3.11	2.28	2.73	2.63	38.57
LONDON WATER WORKS	2.60	2.16	3.17	3.75	4.02	3.81	4.15	3.53	2.95	2.06	2.75	2.74	37.69
LOUISVILLE	2.56	2.06	3.24	3.26	3.64	3.73	4.32	3.42	3.14	2.47	2.49	2.52	36.85
MANSFIELD WSO	2.25	1.86	3.00	3.55	3.75	3.44	3.73	3.25	3.04	1.94	2.66	2.40	34.88
MANSFIELD 6 W	2.37	1.85	2.97	3.74	4.10	3.79	4.39	3.48	3.17	2.16	2.77	2.25	37.04
MARIETTA LOCK 1	2.77	2.42	3.51	3.38	3.31	3.17	3.98	3.79	2.94	2.46	2.39	2.61	36.73
MARION WATER WORKS	2.36	1.76	2.87	3.61	3.42	3.42	3.56	3.04	3.01	2.11	2.57	2.33	34.06
MARSHALLVILLE 1 SSW	2.40	1.94	3.27	3.45	3.73	3.50	4.17	3.63	3.17	2.50	2.63	2.43	36.82
MARYSVILLE	2.59	2.05	3.14	3.62	3.72	3.92	3.72	3.27	2.82	2.10	2.49	2.45	35.89
MC CONNELSVILLE LOCK 7	3.09	2.64	3.66	3.56	3.85	3.88	4.54	4.14	3.06	2.54	2.50	2.84	40.30
MIAMISBURG	2.83	2.27	3.26	3.80	4.12	4.12	3.98	3.34	3.11	2.42	2.75	2.60	38.60
MIDDLEBOURNE	2.48	2.16	3.05	3.24	3.41	3.83	4.05	3.28	2.55	2.16	2.50	2.39	35.10
MIDDLETOWN	2.93	2.41	3.33	3.68	4.09	3.74	4.07	3.24	3.07	2.43	2.87	2.81	38.67
MILFORD	3.31	2.76	4.09	3.84	4.03	3.56	4.29	3.52	3.11	2.69	3.19	3.10	41.49
MILLERSBURG 1 W	2.42	1.92	3.23	3.52	3.73	4.18	4.18	3.75	3.04	2.55	2.56	2.33	37.41
MILLPORT 2 NW	2.78	2.29	3.54	3.40	3.92	3.71	4.23	3.25	2.95	2.58	2.70	2.74	38.09
MINERAL RIDGE WTR WKS	2.41	1.93	3.01	3.23	3.15	3.58	3.81	3.20	3.12	2.53	2.44	2.38	34.79
MOHAWK DAM	2.99	2.38	3.68	3.84	3.75	3.97	4.31	3.84	3.07	2.64	2.93	2.75	40.15
MOHICANVILLE DAM	2.54	2.07	3.30	3.50	3.73	3.63	4.31	3.35	3.01	2.27	2.70	2.50	36.91
MONTPELIER 1 WSW	1.82	1.76	2.81	3.45	3.16	3.40	3.47	3.08	2.68	2.21	2.59	2.42	32.85
MOSQUITO CREEK DAM	2.32	1.87	2.95	3.53	3.24	3.57	3.91	3.26	3.09	2.80	2.60	2.40	35.54
MT GILEAD LAKES PARK	2.52	2.02	3.09	3.66	3.78	4.07	4.15	3.23	3.27	2.10	2.91	2.68	37.48
NAPOLEON WATER WORKS	2.22	1.86	2.86	3.47	3.42	3.39	4.06	3.30	2.50	2.24	2.64	2.53	34.49
NELSONVILLE 1 WNW	2.82	2.42	3.45	3.53	3.65	3.35	4.24	3.56	3.12	2.31	2.36	2.73	37.54
NEWARK WATER WORKS	3.00	2.30	3.58	3.88	4.03	4.19	4.20	3.92	3.07	2.46	2.84	2.84	40.31
NEW CARLISLE	2.63	2.14	3.11	3.62	4.10	4.17	3.77	3.93	2.70	2.16	2.76	2.54	37.63
NEW LEXINGTON 2 NW	2.81	2.41	3.60	3.84	3.77	4.23	4.62	3.81	2.74	2.33	2.68	2.82	39.66
NEW PHILADELPHIA 1 A	2.74	2.16	3.47	3.48	3.58	3.86	4.48	3.59	2.94	2.48	2.52	2.55	37.85
NEW STRAITSVILLE	2.91	2.45	3.69	3.65	3.81	3.80	3.92	3.76	2.84	2.33	2.52	2.74	38.42
NORWALK SEWAGE PLANT	2.26	1.83	2.92	3.48	3.51	3.95	4.32	3.30	2.94	2.07	2.60	2.42	35.60
NORWICH 1 E	2.60	2.20	3.44	3.38	3.58	4.58	4.37	3.94	2.86	2.38	2.61	2.43	38.37
OBERLIN	2.21	1.97	2.76	3.21	3.44	3.50	3.62	3.24	2.88	2.12	2.63	2.37	33.95
PAINESVILLE 4 NW	2.33	1.82	2.62	3.20	3.06	3.37	3.31	3.44	3.30	3.04	3.44	2.77	35.70
PANDORA 2 NE	2.28	1.95	3.04	3.46	3.47	3.28	3.56	2.88	2.89	2.03	2.53	2.57	33.94
PAULDING	2.03	1.71	2.82	3.54	3.60	3.85	3.54	2.75	2.72	2.26	2.61	2.43	33.86
PEEBLES	2.98	2.71	4.15	3.80	3.86	3.42	4.46	4.10	3.39	2.44	3.07	3.01	41.39
PHILO 3 SW	2.21	2.06	2.90	2.96	3.51	3.78	4.31	3.48	2.82	2.19	2.34	2.25	34.81
PIEDMONT DAM	2.55	2.29	3.47	3.48	3.76	3.56	4.14	3.68	2.78	2.39	2.48	2.52	37.10
PIKETON AEC PUMP STA	2.92	2.43	3.88	3.48	3.68	3.12	4.08	3.80	3.20	2.19	2.64	2.72	38.14
PIQUA	2.55	2.13	3.17	3.64	3.84	4.02	3.42	3.26	2.51	2.27	2.72	2.68	36.21
PLEASANT HILL	2.46	2.02	3.04	3.49	3.84	3.87	3.53	3.43	2.47	2.21	2.62	2.59	35.57
PLEASANT HILL DAM	2.84	2.17	3.43	3.70	4.16	3.98	4.23	3.51	2.95	2.32	2.78	2.76	38.83
PLYMOUTH	2.50	1.99	3.08	3.78	3.81	3.86	3.84	3.59	3.20	2.22	2.68	2.63	37.18

OHIO

PRECIPITATION NORMALS (INCHES)

STATION	JAN	FEB	MAR	APR	MAY	JUN	JUL	AUG	SEP	OCT	NOV	DEC	ANN
PORTSMOUTH	3.15	2.98	4.06	3.57	3.96	3.77	4.42	4.00	3.38	2.30	2.70	3.04	41.33
PORTSMOUTH US GRANT BR	3.17	2.88	4.11	3.62	3.99	3.88	4.26	4.04	3.49	2.36	2.79	3.03	41.62
PROSPECT	2.61	2.08	3.03	3.75	3.62	3.78	3.47	3.35	2.85	2.08	2.72	2.59	35.93
PUT IN BAY PERRY MON	2.04	1.56	2.53	3.10	2.97	3.23	3.06	3.41	2.55	2.00	2.26	2.25	30.96
RAVENNA 2 S	2.56	2.10	3.22	3.54	3.62	3.88	4.33	3.47	3.39	2.63	2.92	2.73	38.39
ST MARYS 3 W	2.01	1.65	2.68	3.50	3.43	3.44	3.42	3.02	2.81	2.11	2.37	2.16	32.60
SANDUSKY R	2.19	1.78	2.74	3.18	3.19	3.72	3.81	3.68	2.85	1.92	2.44	2.40	33.90
SAYRE	2.76	2.34	3.35	3.58	3.78	4.09	4.55	3.75	2.91	2.35	2.66	2.76	38.88
SEDALIA	2.87	2.39	3.36	3.87	4.10	3.90	4.20	3.86	3.26	2.12	2.75	2.61	39.29
SENECAVILLE DAM	2.80	2.38	3.46	3.52	3.75	4.57	4.50	3.82	2.88	2.55	2.66	2.70	39.59
SIDNEY 1 S	2.36	1.98	3.03	3.71	3.57	3.72	3.85	3.55	2.62	2.34	2.61	2.54	35.88
SPRINGFIELD SEWAGE PL	2.65	2.09	3.25	3.83	4.14	4.21	3.80	3.99	3.02	2.28	2.78	2.68	38.72
STEUBENVILLE WATER WKS	2.97	2.47	3.87	3.58	3.88	4.16	3.97	3.43	2.84	2.65	2.64	2.87	39.33
SUMMERFIELD 2 NE	2.79	2.49	3.25	3.69	3.68	4.25	4.39	3.77	2.89	2.61	2.42	2.69	38.92
TAPPAN DAM	2.78	2.42	3.68	3.47	3.73	4.02	4.25	3.98	2.82	2.49	2.62	2.62	38.88
TIFFIN	2.44	1.98	3.12	3.54	3.55	3.46	3.88	3.22	2.74	2.05	2.65	2.62	35.25
TIPP CITY	2.49	2.12	3.12	3.60	4.21	4.12	3.87	3.35	2.65	2.10	2.64	2.49	36.76
TOLEDO EXPRESS WSO R	1.99	1.80	2.64	3.04	2.90	3.49	3.26	3.19	2.53	1.94	2.41	2.59	31.77
TOLEDO BLADE	2.05	1.81	2.67	3.10	2.89	3.50	3.38	3.32	2.54	2.04	2.43	2.56	32.29
UPPER SANDUSKY	2.49	1.96	3.19	3.52	3.71	3.13	3.80	2.72	3.02	1.97	2.62	2.62	34.75
URBANA SEWAGE PLANT	2.58	2.14	3.17	3.73	4.08	3.76	4.05	3.32	2.93	2.24	2.71	2.56	37.27
UTICA	2.75	2.19	3.25	3.65	3.50	4.27	4.21	3.52	2.90	2.38	2.76	2.55	37.93
VAN WERT	2.36	2.07	3.27	3.82	3.74	3.94	3.44	2.83	2.98	2.41	2.74	2.60	36.20
VERSAILLES	2.58	2.25	3.41	3.80	3.49	4.20	3.18	3.27	2.70	2.28	2.70	2.75	36.61
WARREN 3 S	2.46	1.93	3.07	3.24	3.22	3.56	3.70	3.16	3.07	2.58	2.61	2.43	35.03
WASHINGTON COURT HOUSE	2.76	2.43	3.72	3.76	3.97	3.98	3.95	3.27	2.93	2.17	2.74	2.69	38.37
WAUSEON WASTE WTR PLT	2.15	1.93	2.88	3.53	3.35	3.40	3.75	3.01	2.64	2.18	2.70	2.68	34.20
WAVERLY	3.01	2.53	4.05	3.68	3.84	3.37	4.69	3.97	3.10	2.36	2.78	2.99	40.37
WESTERVILLE WATER PL	2.72	2.04	3.10	3.72	3.90	3.87	4.21	3.75	2.87	2.14	2.79	2.47	37.58
WEST MANCHESTER 3 SW	2.86	2.31	3.35	3.90	4.18	4.20	3.82	3.26	2.55	2.32	2.94	2.94	38.63
WILLS CREEK DAM	2.85	2.47	3.78	3.88	4.09	3.84	4.47	3.73	2.92	2.37	2.82	2.78	40.00
WILMINGTON	3.41	2.84	4.05	4.27	4.25	4.25	4.10	3.49	3.27	2.53	3.30	3.09	42.85
WOOSTER EXP STATION	2.48	1.93	3.07	3.25	3.72	3.35	4.18	3.64	3.19	2.06	2.45	2.39	35.71
XENIA 5 SSE	2.66	2.18	3.31	3.98	4.10	3.82	3.96	3.65	2.99	2.31	2.73	2.67	38.36
YOUNGSTOWN WSO R	2.69	2.23	3.29	3.46	3.29	3.53	4.04	3.47	3.10	2.65	2.82	2.76	37.33
ZANESVILLE FAA AP	2.59	2.39	3.49	3.55	3.63	4.15	4.42	3.77	2.81	2.25	2.58	2.73	38.36

OHIO

HEATING DEGREE DAY NORMALS (BASE 65 DEG F)

STATION		JUL	AUG	SEP	OCT	NOV	DEC	JAN	FEB	MAR	APR	MAY	JUN	ANN
AKRON CANTON WSO	R	0	14	108	394	720	1076	1237	1058	877	492	228	37	6241
ASHLAND		0	12	108	402	741	1104	1252	1058	880	489	229	36	6311
ASHTABULA		7	16	101	378	684	1045	1221	1081	924	540	257	47	6301
ATHENS 5 NW		0	8	83	371	678	989	1116	935	732	390	163	22	5487
BARNESVILLE WATER WKS		15	31	124	446	750	1079	1231	1050	859	501	247	53	6386
BELLEFONTAINE SEWAGE		0	7	82	357	714	1070	1228	1022	825	438	188	23	5954
BOWLING GREEN SWG PL		0	7	76	352	705	1082	1240	1044	859	456	186	16	6023
BUCYRUS		0	12	112	413	744	1107	1259	1078	896	501	235	34	6391
CADIZ		0	6	76	344	672	1020	1159	980	784	414	179	24	5658
CALDWELL 6 NW		6	9	86	350	669	998	1138	952	750	391	169	31	5549
CAMBRIDGE SEWAGE PLANT		0	6	80	359	672	995	1135	949	756	402	169	20	5543
CANFIELD 1 S		8	17	117	407	723	1079	1231	1053	868	492	237	45	6277
CHARDON		18	25	140	429	747	1119	1280	1117	946	555	276	72	6724
CHARLES MILL DAM		10	28	140	446	756	1116	1277	1100	921	540	271	61	6666
CHILO MELDAHL DAM		0	0	49	305	627	942	1091	904	719	358	149	12	5156
CHIPPEWA LAKE		7	17	108	395	717	1085	1243	1058	874	489	231	46	6270
CINCINNATI ABBE WSMO		0	0	39	279	612	933	1076	882	682	317	122	8	4950
CIRCLEVILLE		0	0	52	296	627	952	1094	907	701	343	132	10	5114
CLEVELAND WSO	//R	8	11	99	371	696	1051	1225	1053	880	507	244	33	6178
COLUMBUS SULLIVANT AVE		0	0	63	321	657	1004	1153	955	750	375	155	14	5447
COLUMBUS VALLEY CROSS		0	0	74	335	666	1004	1153	952	750	381	156	13	5484
COLUMBUS WSO	//R	0	5	78	355	687	1020	1175	986	775	408	178	19	5686
COSHOCTON SEWAGE PL		5	10	96	385	699	1032	1175	986	791	432	190	35	5836
DAYTON		0	0	44	296	630	973	1132	938	741	354	139	8	5255
DAYTON WSO	R	0	0	68	342	687	1029	1190	991	791	405	171	15	5689
DEFIANCE		0	15	101	418	768	1144	1314	1123	939	522	232	33	6609
DELAWARE		0	14	100	400	723	1076	1228	1036	840	456	210	30	6113
ELYRIA 3 E		0	8	84	351	687	1048	1212	1042	868	480	218	22	6020
FINDLAY FAA AP		0	14	96	386	735	1101	1265	1064	880	480	217	24	6262
FINDLAY SEWAGE PLANT		0	18	101	407	759	1122	1277	1072	890	492	217	21	6376
FREDERICKTOWN		0	16	126	430	759	1119	1271	1086	884	498	244	40	6473
GALLIPOLIS 5 W		0	0	47	282	597	887	1014	834	636	303	117	11	4728
GREENVILLE WATER PLT		0	12	106	418	750	1101	1259	1064	862	477	218	32	6299
HAMILTON-FAIRFIELD		0	0	43	296	624	946	1088	890	691	330	134	10	5052
HILLSBORO		0	0	68	327	657	995	1138	955	744	377	166	18	5445
HIRAM		0	11	101	388	726	1097	1246	1067	893	492	224	46	6291
IRONTON		0	0	32	270	570	859	980	804	601	271	102	6	4495
IRWIN		0	7	84	367	717	1066	1218	1011	812	426	181	21	5910
JACKSON 2 NW		0	7	91	387	687	1001	1132	944	747	399	175	26	5596
KENTON		0	12	93	385	729	1094	1249	1050	865	468	207	22	6174
LANCASTER 2 NW		0	14	92	371	687	1026	1172	994	791	431	195	30	5803
LIMA SEWAGE PLANT		0	8	79	347	699	1054	1225	1019	834	438	191	16	5910
LONDON WATER WORKS		0	7	82	355	693	1035	1187	986	781	417	172	16	5731
MANSFIELD WSO		0	10	105	381	726	1088	1246	1058	877	492	231	35	6249
MANSFIELD 6 W		9	23	138	431	762	1122	1277	1081	899	516	254	57	6569
MARYSVILLE		0	10	103	392	735	1085	1240	1036	840	456	204	22	6123
MC CONNELSVILLE LOCK 7		0	0	65	358	672	1008	1138	969	766	402	173	20	5571
MILFORD		0	6	71	339	657	970	1113	921	725	367	168	22	5359
MILLERSBURG 1 W		0	12	102	376	696	1048	1197	1002	812	450	209	36	5940
MILLPORT 2 NW		12	20	128	431	741	1082	1231	1047	862	501	247	50	6352
MINERAL RIDGE WTR WKS		0	7	92	363	681	1035	1190	1019	840	462	207	27	5923
MONTPELIER 1 WSW		6	17	109	406	750	1128	1296	1089	908	495	225	47	6476
NAPOLEON WATER WORKS		0	7	72	348	708	1082	1246	1044	856	447	183	16	6009
NEWARK WATER WORKS		0	5	90	369	693	1029	1166	974	772	405	166	17	5686
NEW LEXINGTON 2 NW		0	10	99	374	693	1017	1156	960	760	399	173	18	5659

OHIO

HEATING DEGREE DAY NORMALS (BASE 65 DEG F)

STATION		JUL	AUG	SEP	OCT	NOV	DEC	JAN	FEB	MAR	APR	MAY	JUN	ANN
NORWALK SEWAGE PLANT		0	13	101	391	711	1079	1234	1058	887	495	227	43	6239
OBERLIN		5	12	117	385	723	1085	1240	1056	887	489	231	43	6273
PAINESVILLE 4 NW		0	7	80	325	642	998	1181	1042	899	531	248	34	5987
PANDORA 2 NE		0	14	98	381	732	1101	1265	1067	877	474	209	19	6237
PAULDING		5	19	110	412	753	1122	1293	1092	902	489	228	32	6457
PEEBLES		0	0	52	308	618	924	1057	874	670	333	135	14	4985
PHILO 3 SW		0	8	85	336	657	995	1135	949	744	381	164	28	5482
PLYMOUTH		7	17	112	398	732	1104	1252	1067	884	480	230	46	6329
PORTSMOUTH		0	0	35	270	573	887	1023	851	648	294	112	9	4702
PUT IN BAY PERRY MON		0	0	45	296	657	1045	1218	1050	896	498	199	20	5924
SANDUSKY	R	0	5	72	349	681	1051	1218	1050	884	483	202	21	6016
SENECAVILLE DAM		9	24	111	409	708	1032	1181	1014	812	444	210	47	6001
STEUBENVILLE WATER WKS		0	0	76	343	657	998	1144	963	781	420	183	22	5587
TIFFIN		0	7	77	352	693	1057	1218	1025	837	438	185	17	5906
TOLEDO EXPRESS WSO	R	0	16	113	419	771	1144	1299	1098	918	516	237	39	6570
TOLEDO BLADE		0	0	59	331	687	1070	1225	1039	868	450	180	17	5926
UPPER SANDUSKY		0	6	77	353	696	1057	1215	1016	825	435	186	16	5882
URBANA SEWAGE PLANT		0	14	106	402	732	1085	1246	1050	846	462	207	22	6172
VAN WERT		0	0	72	358	708	1076	1237	1028	840	435	183	21	5958
WARREN 3 S		9	15	118	407	714	1060	1221	1047	871	498	232	49	6241
WASHINGTON COURT HOUSE		0	0	67	322	669	1001	1150	955	760	387	159	15	5485
WAUSEON WASTE WTR PLT		5	18	114	412	771	1147	1314	1109	924	513	231	29	6587
WAVERLY		0	6	74	332	639	958	1091	904	701	356	152	19	5232
WILMINGTON		0	11	103	369	687	1011	1153	966	763	402	182	21	5668
WOOSTER EXP STATION		6	18	127	429	744	1094	1246	1061	877	510	246	46	6404
XENIA 5 SSE		0	8	83	354	681	1011	1156	958	753	384	156	15	5559
YOUNGSTOWN WSO	R	7	19	130	423	744	1104	1265	1095	921	528	267	57	6560
ZANESVILLE FAA AP		0	7	91	378	696	1026	1169	986	787	426	186	25	5777

OHIO

COOLING DEGREE DAY NORMALS (BASE 65 DEG F)

STATION	JAN	FEB	MAR	APR	MAY	JUN	JUL	AUG	SEP	OCT	NOV	DEC	ANN
AKRON CANTON WSO R	0	0	0	0	36	121	208	181	72	7	0	0	625
ASHLAND	0	0	0	0	37	123	205	170	69	8	0	0	612
ASHTABULA	0	0	0	0	22	110	202	171	65	12	0	0	582
ATHENS 5 NW	0	0	0	0	48	139	236	209	86	8	0	0	726
BARNESVILLE WATER WKS	0	0	0	0	33	101	182	161	58	0	0	0	535
BELLEFONTAINE SEWAGE	0	0	0	0	55	155	243	203	91	13	0	0	760
BOWLING GREEN SWG PL	0	0	0	0	49	175	270	218	91	14	0	0	817
BUCYRUS	0	0	0	0	42	130	214	167	67	7	0	0	627
CADIZ	0	0	0	0	52	141	229	207	94	12	0	0	735
CALDWELL 6 NW	0	0	0	7	51	148	235	207	95	9	0	0	752
CAMBRIDGE SEWAGE PLANT	0	0	0	0	48	146	239	214	92	6	0	0	745
CANFIELD 1 S	0	0	0	0	30	105	176	151	66	0	0	0	528
CHARDON	0	0	0	0	22	99	167	137	59	11	0	0	495
CHARLES MILL DAM	0	0	0	0	23	91	156	133	50	0	0	0	453
CHILO MELDAHL DAM	0	0	0	0	78	192	313	279	127	20	0	0	1009
CHIPPEWA LAKE	0	0	0	0	30	112	184	150	60	0	0	0	536
CINCINNATI ABBE WSMO	0	0	0	8	101	230	344	310	144	22	0	0	1159
CIRCLEVILLE	0	0	0	7	82	205	310	270	124	11	0	0	1009
CLEVELAND WSO //R	0	0	0	0	33	111	213	178	72	5	0	0	612
COLUMBUS SULLIVANT AVE	0	0	0	0	84	197	307	261	120	15	0	0	984
COLUMBUS VALLEY CROSS	0	0	0	0	66	178	276	233	110	10	0	0	873
COLUMBUS WSO //R	0	0	0	0	66	175	273	235	102	11	0	0	862
COSHOCTON SEWAGE PL	0	0	0	0	53	158	235	199	90	7	0	0	742
DAYTON	0	0	0	6	108	248	363	310	143	20	0	0	1198
DAYTON WSO R	0	0	0	0	72	195	301	252	110	17	0	0	947
DEFIANCE	0	0	0	0	40	144	237	189	62	9	0	0	681
DELAWARE	0	0	0	0	52	150	238	200	82	10	0	0	732
ELYRIA 3 E	0	0	0	0	42	133	239	200	87	6	0	0	707
FINDLAY FAA AP	0	0	0	0	53	153	248	206	84	11	0	0	755
FINDLAY SEWAGE PLANT	0	0	0	0	49	144	236	191	68	11	0	0	699
FREDERICKTOWN	0	0	0	0	39	121	193	165	69	9	0	0	596
GALLIPOLIS 5 W	0	0	0	6	80	203	313	282	137	15	0	0	1036
GREENVILLE WATER PLT	0	0	0	0	48	140	227	179	67	6	0	0	667
HAMILTON-FAIRFIELD	0	0	0	6	94	220	322	276	127	14	0	0	1059
HILLSBORO	0	0	0	0	76	168	270	233	113	17	0	0	877
HIRAM	0	0	0	0	35	112	197	159	62	7	0	0	572
IRONTON	0	0	6	19	111	243	357	322	158	31	0	0	1247
IRWIN	0	0	0	0	57	156	241	199	81	7	0	0	741
JACKSON 2 NW	0	0	0	0	55	146	248	208	91	9	0	0	757
KENTON	0	0	0	0	49	139	233	192	75	13	0	0	701
LANCASTER 2 NW	0	0	0	5	68	165	256	219	107	12	0	0	832
LIMA SEWAGE PLANT	0	0	0	0	67	172	272	222	103	13	0	0	849
LONDON WATER WORKS	0	0	0	0	52	154	248	209	94	7	0	0	764
MANSFIELD WSO	0	0	0	0	39	125	220	184	78	6	0	0	652
MANSFIELD 6 W	0	0	0	0	31	93	157	128	54	0	0	0	463
MARYSVILLE	0	0	0	0	52	139	229	196	79	7	0	0	702
MC CONNELSVILLE LOCK 7	0	0	0	0	61	161	264	239	98	11	0	0	834
MILFORD	0	0	0	13	90	202	304	272	137	26	0	0	1044
MILLERSBURG 1 W	0	0	0	0	39	120	195	170	75	7	0	0	606
MILLPORT 2 NW	0	0	0	0	27	95	176	147	53	0	0	0	498
MINERAL RIDGE WTR WKS	0	0	0	0	40	126	216	187	80	0	0	0	649
MONTPELIER 1 WSW	0	0	0	0	42	152	226	185	70	18	0	0	693
NAPOLEON WATER WORKS	0	0	0	0	56	178	273	224	87	13	0	0	831
NEWARK WATER WORKS	0	0	0	0	55	158	254	213	93	7	0	0	780
NEW LEXINGTON 2 NW	0	0	0	0	55	141	229	199	96	8	0	0	728

OHIO

COOLING DEGREE DAY NORMALS (BASE 65 DEG F)

STATION		JAN	FEB	MAR	APR	MAY	JUN	JUL	AUG	SEP	OCT	NOV	DEC	ANN
NORWALK SEWAGE PLANT		0	0	0	0	35	139	219	177	68	10	0	0	648
OBERLIN		0	0	0	0	42	130	210	170	78	7	0	0	637
PAINESVILLE 4 NW		0	0	0	0	25	106	205	184	83	9	0	0	612
PANDORA 2 NE		0	0	0	0	54	157	236	187	77	9	0	0	720
PAULDING		0	0	0	0	51	149	238	192	77	15	0	0	722
PEEBLES		0	0	0	6	73	185	291	251	115	14	0	0	935
PHILO 3 SW		0	0	0	6	65	169	250	213	103	17	0	0	823
PLYMOUTH		0	0	0	0	41	139	208	175	79	19	0	0	661
PORTSMOUTH		0	0	0	6	96	222	338	304	149	22	0	0	1137
PUT IN BAY PERRY MON		0	0	0	0	35	164	295	266	117	17	0	0	894
SANDUSKY	R	0	0	0	0	38	162	276	235	96	14	0	0	821
SENECAVILLE DAM		0	0	0	0	46	140	217	185	84	12	0	0	684
STEUBENVILLE WATER WKS		0	0	0	0	49	145	245	218	97	8	0	0	762
TIFFIN		0	0	0	0	54	164	260	212	89	11	0	0	790
TOLEDO EXPRESS WSO	R	0	0	0	0	38	129	215	174	59	7	0	0	622
TOLEDO BLADE		0	0	0	0	72	218	341	284	116	18	0	0	1049
UPPER SANDUSKY		0	0	0	0	56	166	260	217	92	12	0	0	803
URBANA SEWAGE PLANT		0	0	0	0	52	142	235	191	82	8	0	0	710
VAN WERT		0	0	0	0	59	174	267	218	81	14	0	0	813
WARREN 3 S		0	0	0	0	31	118	192	164	67	7	0	0	579
WASHINGTON COURT HOUSE		0	0	0	0	66	168	267	234	103	12	0	0	850
WAUSEON WASTE WTR PLT		0	0	0	0	36	134	213	164	57	6	0	0	610
WAVERLY		0	0	0	8	75	175	282	242	113	7	0	0	902
WILMINGTON		0	0	0	6	70	168	255	222	118	25	0	0	864
WOOSTER EXP STATION		0	0	0	0	29	103	177	151	55	0	0	0	515
XENIA 5 SSE		0	0	0	6	63	156	251	210	101	13	0	0	800
YOUNGSTOWN WSO	R	0	0	0	0	28	96	166	143	52	0	0	0	485
ZANESVILLE FAA AP		0	0	0	0	47	142	233	203	85	6	0	0	716

33 — OHIO

STATE-STATION NUMBER	STN TYP	NAME		LATITUDE DEG-MIN	LONGITUDE DEG-MIN	ELEVATION (FT)
33-0058	13	AKRON CANTON WSO	R	N 4055	W 08126	1208
33-0083	12	ALEXANDRIA 4 WSW		N 4005	W 08241	1220
33-0256	13	ASHLAND		N 4052	W 08218	1050
33-0264	13	ASHTABULA		N 4151	W 08048	690
33-0274	13	ATHENS 5 NW		N 3923	W 08211	685
33-0298	12	ATWOOD DAM		N 4031	W 08117	950
33-0430	13	BARNESVILLE WATER WKS		N 3958	W 08110	1140
33-0493	12	BEACH CITY DAM		N 4038	W 08134	1000
33-0563	13	BELLEFONTAINE SEWAGE		N 4021	W 08346	1185
33-0823	12	BOLIVAR DAM		N 4039	W 08126	980
33-0862	13	BOWLING GREEN SWG PL		N 4123	W 08338	675
33-1072	13	BUCYRUS		N 4049	W 08259	1005
33-1152	13	CADIZ		N 4016	W 08100	1260
33-1178	13	CALDWELL 6 NW		N 3949	W 08136	980
33-1197	13	CAMBRIDGE SEWAGE PLANT		N 4002	W 08136	800
33-1245	13	CANFIELD 1 S		N 4100	W 08046	1140
33-1288	12	CARPENTER		N 3910	W 08213	710
33-1315	12	CARROLLTON 2 NW		N 4036	W 08106	1200
33-1399	12	CENTERBURG		N 4018	W 08242	1200
33-1458	13	CHARDON		N 4135	W 08113	1210
33-1466	13	CHARLES MILL DAM		N 4044	W 08222	1025
33-1536	13	CHILO MELDAHL DAM		N 3848	W 08410	500
33-1541	13	CHIPPEWA LAKE		N 4105	W 08154	1040
33-1561	13	CINCINNATI ABBE WSMO		N 3909	W 08431	760
33-1592	13	CIRCLEVILLE		N 3937	W 08257	670
33-1642	12	CLENDENING DAM		N 4016	W 08117	924
33-1657	13	CLEVELAND WSO	//R	N 4125	W 08152	777
33-1781	13	COLUMBUS SULLIVANT AVE		N 3957	W 08307	875
33-1783	13	COLUMBUS VALLEY CROSS		N 3954	W 08255	733
33-1786	13	COLUMBUS WSO	//R	N 4000	W 08253	812
33-1858	12	COOPERDALE		N 4013	W 08204	780
33-1890	13	COSHOCTON SEWAGE PL		N 4015	W 08152	760
33-2067	13	DAYTON		N 3946	W 08411	745
33-2075	13	DAYTON WSO	R	N 3954	W 08412	1002
33-2098	13	DEFIANCE		N 4117	W 08423	700
33-2119	13	DELAWARE		N 4017	W 08304	860
33-2124	12	DELAWARE DAM		N 4022	W 08304	930
33-2272	12	DOVER DAM		N 4034	W 08125	930
33-2485	12	EATON		N 3944	W 08438	1000
33-2599	13	ELYRIA 3 E		N 4123	W 08204	730
33-2626	12	ENTERPRISE		N 3934	W 08229	745
33-2786	13	FINDLAY FAA AP		N 4101	W 08340	797
33-2791	13	FINDLAY SEWAGE PLANT		N 4103	W 08340	768
33-2928	12	FRANKLIN		N 3933	W 08418	720
33-2956	13	FREDERICKTOWN		N 4029	W 08232	1145
33-2974	12	FREMONT		N 4120	W 08307	600
33-3021	12	GALION WATER WORKS		N 4043	W 08248	1170
33-3029	13	GALLIPOLIS 5 W		N 3850	W 08217	673
33-3375	13	GREENVILLE WATER PLT		N 4006	W 08439	1024
33-3393	12	GREER		N 4031	W 08212	900

STATE-STATION NUMBER	STN TYP	NAME	LATITUDE DEG-MIN	LONGITUDE DEG-MIN	ELEVATION (FT)
33-3483	13	HAMILTON-FAIRFIELD	N 3921	W 08435	575
33-3730	12	HIGGINSPORT	N 3847	W 08358	500
33-3758	13	HILLSBORO	N 3912	W 08337	1100
33-3780	13	HIRAM	N 4119	W 08109	1250
33-3874	12	HOYTVILLE 2 NE	N 4113	W 08346	700
33-3915	12	HUNTSVILLE 3 N	N 4029	W 08349	1030
33-3971	13	IRONTON	N 3832	W 08240	555
33-3987	13	IRWIN	N 4007	W 08329	1010
33-4004	13	JACKSON 2 NW	N 3904	W 08239	700
33-4189	13	KENTON	N 4038	W 08336	980
33-4238	12	KINGS MILLS	N 3921	W 08416	750
33-4363	12	LAKEVIEW 3 NE	N 4031	W 08353	1020
33-4383	13	LANCASTER 2 NW	N 3944	W 08238	860
33-4409	12	LA RUE	N 4035	W 08323	930
33-4434	12	LAURELVILLE	N 3928	W 08244	820
33-4473	12	LEESVILLE DAM	N 4028	W 08112	980
33-4551	13	LIMA SEWAGE PLANT	N 4043	W 08407	860
33-4616	12	LITHOPOLIS 2 S	N 3947	W 08249	990
33-4681	13	LONDON WATER WORKS	N 3953	W 08327	1020
33-4728	12	LOUISVILLE	N 4050	W 08115	1170
33-4865	13	MANSFIELD WSO	N 4049	W 08231	1295
33-4874	13	MANSFIELD 6 W	N 4045	W 08238	1310
33-4924	12	MARIETTA LOCK 1	N 3925	W 08127	610
33-4942	12	MARION WATER WORKS	N 4036	W 08310	920
33-4967	12	MARSHALLVILLE 1 SSW	N 4054	W 08144	1120
33-4979	13	MARYSVILLE	N 4014	W 08322	1000
33-5041	13	MC CONNELSVILLE LOCK 7	N 3939	W 08151	710
33-5185	12	MIAMISBURG	N 3939	W 08416	720
33-5199	12	MIDDLEBOURNE	N 4003	W 08120	880
33-5220	12	MIDDLETOWN	N 3931	W 08425	635
33-5268	13	MILFORD	N 3910	W 08418	570
33-5297	13	MILLERSBURG 1 W	N 4033	W 08156	930
33-5315	13	MILLPORT 2 NW	N 4043	W 08054	1145
33-5356	13	MINERAL RIDGE WTR WKS	N 4109	W 08047	890
33-5398	12	MOHAWK DAM	N 4021	W 08205	865
33-5406	12	MOHICANVILLE DAM	N 4044	W 08209	970
33-5438	13	MONTPELIER 1 WSW	N 4135	W 08434	840
33-5505	12	MOSQUITO CREEK DAM	N 4118	W 08046	910
33-5535	12	MT GILEAD LAKES PARK	N 4033	W 08248	1090
33-5669	13	NAPOLEON WATER WORKS	N 4123	W 08408	680
33-5718	12	NELSONVILLE 1 WNW	N 3927	W 08215	700
33-5747	13	NEWARK WATER WORKS	N 4005	W 08225	835
33-5786	12	NEW CARLISLE	N 3956	W 08402	870
33-5857	13	NEW LEXINGTON 2 NW	N 3944	W 08213	890
33-5904	12	NEW PHILADELPHIA 1 A	N 4030	W 08127	920
33-5947	12	NEW STRAITSVILLE	N 3935	W 08215	810
33-6118	13	NORWALK SEWAGE PLANT	N 4116	W 08237	670
33-6136	12	NORWICH 1 E	N 3959	W 08147	1080
33-6196	13	OBERLIN	N 4118	W 08213	817
33-6389	13	PAINESVILLE 4 NW	N 4145	W 08118	600

33 — OHIO

STATE-STATION NUMBER	STN TYP	NAME		LATITUDE DEG-MIN	LONGITUDE DEG-MIN	ELEVATION (FT)
33-6405	13	PANDORA 2 NE		N 4059	W 08357	760
33-6465	13	PAULDING		N 4107	W 08436	725
33-6493	13	PEEBLES		N 3857	W 08325	810
33-6600	13	PHILO 3 SW		N 3950	W 08155	1020
33-6616	12	PIEDMONT DAM		N 4011	W 08113	940
33-6630	12	PIKETON AEC PUMP STA		N 3904	W 08301	574
33-6645	12	PIQUA		N 4009	W 08415	870
33-6697	12	PLEASANT HILL		N 4002	W 08421	920
33-6702	12	PLEASANT HILL DAM		N 4037	W 08220	1125
33-6729	13	PLYMOUTH		N 4100	W 08240	1013
33-6781	13	PORTSMOUTH		N 3845	W 08255	540
33-6786	12	PORTSMOUTH US GRANT BR		N 3844	W 08300	565
33-6861	12	PROSPECT		N 4027	W 08311	900
33-6882	13	PUT IN BAY PERRY MON		N 4139	W 08248	580
33-6949	12	RAVENNA 2 S		N 4108	W 08114	1100
33-7383	12	ST MARYS 3 W		N 4032	W 08426	875
33-7447	13	SANDUSKY	R	N 4127	W 08243	606
33-7476	12	SAYRE		N 3941	W 08203	890
33-7538	12	SEDALIA		N 3945	W 08329	1070
33-7559	13	SENECAVILLE DAM		N 3955	W 08126	875
33-7693	12	SIDNEY 1 S		N 4016	W 08409	935
33-7932	12	SPRINGFIELD SEWAGE PL		N 3955	W 08352	910
33-8025	13	STEUBENVILLE WATER WKS		N 4023	W 08038	992
33-8148	12	SUMMERFIELD 2 NE		N 3949	W 08118	980
33-8240	12	TAPPAN DAM		N 4021	W 08114	950
33-8313	13	TIFFIN		N 4107	W 08310	760
33-8332	12	TIPP CITY		N 3958	W 08411	900
33-8357	13	TOLEDO EXPRESS WSO	R	N 4135	W 08348	669
33-8366	13	TOLEDO BLADE		N 4139	W 08332	595
33-8534	13	UPPER SANDUSKY		N 4050	W 08317	860
33-8552	13	URBANA SEWAGE PLANT		N 4006	W 08347	1000
33-8560	12	UTICA		N 4015	W 08227	1005
33-8609	13	VAN WERT		N 4052	W 08435	795
33-8642	12	VERSAILLES		N 4013	W 08429	975
33-8769	13	WARREN 3 S		N 4112	W 08049	900
33-8794	13	WASHINGTON COURT HOUSE		N 3931	W 08325	960
33-8822	13	WAUSEON WASTE WTR PLT		N 4133	W 08408	740
33-8830	13	WAVERLY		N 3908	W 08300	600
33-8951	12	WESTERVILLE WATER PL		N 4008	W 08257	810
33-8990	12	WEST MANCHESTER 3 SW		N 3952	W 08438	1090
33-9211	12	WILLS CREEK DAM		N 4009	W 08151	780
33-9219	13	WILMINGTON		N 3927	W 08350	1026
33-9312	13	WOOSTER EXP STATION		N 4047	W 08155	1020
33-9361	13	XENIA 5 SSE		N 3938	W 08354	915
33-9406	13	YOUNGSTOWN WSO	R	N 4116	W 08040	1178
33-9417	13	ZANESVILLE FAA AP		N 3957	W 08154	881

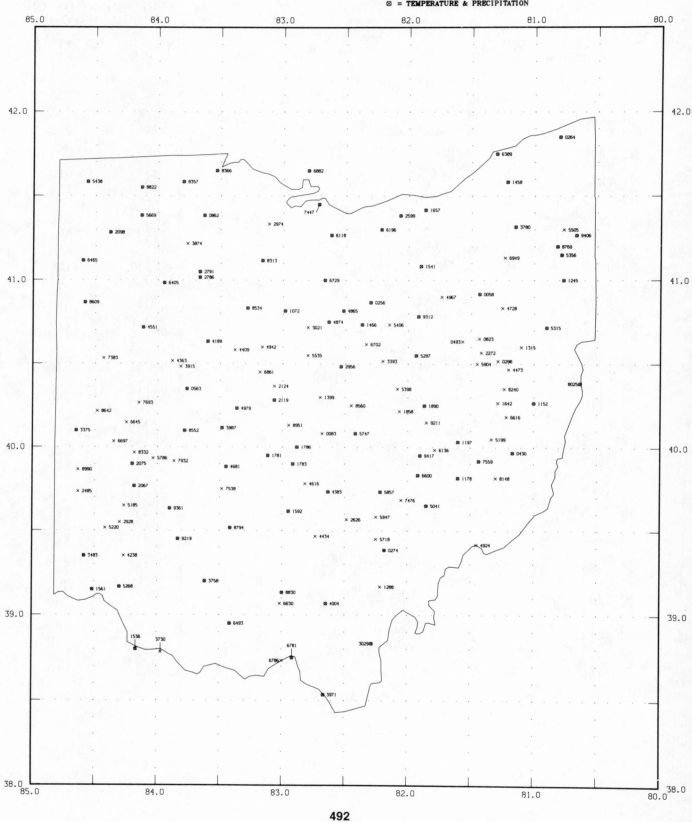

OKLAHOMA

TEMPERATURE NORMALS (DEG F)

STATION		JAN	FEB	MAR	APR	MAY	JUN	JUL	AUG	SEP	OCT	NOV	DEC	ANN
ADA	MAX	50.5	56.5	64.6	74.4	80.8	88.4	94.1	93.9	86.4	76.8	63.2	54.5	73.7
	MIN	28.6	32.8	40.1	50.6	58.6	66.9	71.2	69.5	62.7	52.0	40.3	32.4	50.5
	MEAN	39.6	44.7	52.4	62.5	69.7	77.7	82.7	81.7	74.6	64.4	51.8	43.5	62.1
ALTUS IRR. RESCH STN	MAX	52.5	58.4	67.4	77.9	85.1	93.8	98.2	96.8	88.6	78.7	64.4	55.7	76.5
	MIN	26.1	30.4	37.6	48.7	58.1	67.2	71.1	69.3	62.1	50.5	38.0	29.8	49.1
	MEAN	39.3	44.4	52.5	63.3	71.6	80.5	84.6	83.1	75.4	64.6	51.2	42.8	62.8
ALVA	MAX	46.2	52.0	60.8	72.1	80.6	91.1	96.5	95.4	86.1	75.8	59.7	49.9	72.2
	MIN	21.9	26.5	34.0	45.7	55.5	65.2	70.2	68.6	60.2	48.4	35.0	26.3	46.5
	MEAN	34.1	39.3	47.4	58.9	68.1	78.1	83.4	82.0	73.2	62.1	47.4	38.1	59.3
ANADARKO	MAX	49.9	56.4	64.9	75.3	82.5	90.9	96.2	95.4	87.8	77.2	63.0	54.1	74.5
	MIN	24.9	29.3	37.3	48.7	57.4	66.0	70.2	68.5	61.1	48.9	36.4	28.3	48.1
	MEAN	37.4	42.8	51.1	62.0	70.0	78.5	83.2	82.0	74.5	63.1	49.7	41.2	61.3
ANTLERS 2 ENE	MAX	52.3	57.7	65.7	75.1	81.7	88.9	94.3	94.2	87.1	77.6	64.7	56.0	74.6
	MIN	28.1	32.2	39.9	49.9	57.8	66.1	69.7	67.9	61.5	49.2	38.9	31.3	49.4
	MEAN	40.2	44.9	52.8	62.6	69.8	77.5	82.0	81.1	74.4	63.5	51.8	43.7	62.0
ARDMORE	MAX	53.8	59.6	67.5	76.9	83.4	91.1	96.1	95.8	88.6	78.9	65.6	57.4	76.2
	MIN	31.1	35.2	42.6	53.3	61.4	69.4	73.4	72.2	65.6	54.7	42.9	34.9	53.1
	MEAN	42.5	47.4	55.1	65.2	72.4	80.3	84.8	84.0	77.2	66.9	54.3	46.2	64.7
ARNETT	MAX	46.4	51.5	59.3	71.1	78.6	88.2	93.7	92.5	84.0	73.7	58.8	50.1	70.7
	MIN	20.2	24.8	31.5	43.6	53.9	63.2	67.9	66.1	57.8	45.9	32.6	24.3	44.3
	MEAN	33.3	38.2	45.4	57.4	66.3	75.7	80.8	79.3	70.9	59.8	45.7	37.2	57.5
BARTLESVILLE 2 W	MAX	46.6	53.1	62.2	74.1	81.0	88.9	94.7	94.1	86.0	75.8	60.9	50.7	72.3
	MIN	22.5	27.4	35.5	47.5	56.3	65.0	69.3	67.1	59.5	47.4	35.8	27.2	46.7
	MEAN	34.6	40.3	48.8	60.8	68.7	77.0	82.0	80.6	72.8	61.6	48.3	39.0	59.5
BEAVER 1 SW	MAX	47.3	53.1	60.6	72.4	80.6	90.6	95.7	94.1	85.6	75.5	59.5	50.9	72.2
	MIN	18.1	22.9	29.8	41.7	52.0	61.8	67.1	65.1	56.1	42.7	29.6	21.4	42.4
	MEAN	32.8	38.1	45.3	57.1	66.3	76.2	81.5	79.6	70.9	59.2	44.6	36.2	57.3
BIXBY 2 NE	MAX	46.8	52.8	61.5	72.9	79.9	88.0	93.7	93.0	85.4	75.7	61.4	51.9	71.9
	MIN	24.1	28.6	36.0	48.1	57.1	65.7	69.9	67.6	60.0	47.7	36.0	28.6	47.5
	MEAN	35.4	40.7	48.8	60.6	68.6	76.9	81.8	80.3	72.7	61.7	48.7	40.3	59.7
BOISE CITY 2 E	MAX	50.1	54.3	60.6	70.9	79.3	89.5	93.3	91.0	83.9	74.5	59.6	52.3	71.6
	MIN	18.0	22.3	27.6	37.9	47.1	57.5	62.7	60.4	52.2	40.4	28.1	21.3	39.6
	MEAN	34.1	38.3	44.1	54.4	63.2	73.5	78.0	75.7	68.1	57.4	43.9	36.8	55.6
BRISTOW	MAX	49.2	55.6	63.8	74.7	81.1	88.6	94.8	94.7	87.1	77.4	62.6	53.2	73.6
	MIN	24.2	29.2	37.1	49.0	57.0	65.7	69.6	67.6	60.2	49.2	36.7	28.4	47.8
	MEAN	36.7	42.4	50.5	61.9	69.1	77.2	82.2	81.2	73.7	63.3	49.7	40.8	60.7
BUFFALO	MAX	48.8	55.4	63.5	75.1	82.7	92.3	97.4	96.2	87.6	77.9	61.4	52.5	74.2
	MIN	20.6	25.3	32.5	44.4	54.4	64.4	69.3	67.4	58.7	46.2	32.7	24.5	45.0
	MEAN	34.7	40.4	48.0	59.7	68.5	78.4	83.4	81.8	73.2	62.1	47.0	38.5	59.6
CANTON DAM	MAX	48.5	54.3	63.0	73.5	81.0	90.1	95.8	95.0	86.4	76.2	61.1	51.7	73.1
	MIN	22.7	27.4	35.1	46.7	56.0	65.3	70.0	68.3	60.3	48.9	35.7	27.2	47.0
	MEAN	35.6	40.9	49.1	60.1	68.5	77.7	82.9	81.7	73.4	62.6	48.4	39.5	60.0
CARNEGIE 4 ENE	MAX	50.2	56.0	64.6	75.2	82.6	91.6	96.6	95.7	87.5	77.1	62.4	53.7	74.4
	MIN	24.5	29.1	37.0	48.4	57.3	66.7	70.6	68.8	61.3	49.2	36.6	28.3	48.2
	MEAN	37.3	42.6	50.8	61.8	70.0	79.2	83.7	82.3	74.4	63.2	49.5	41.1	61.3
CHANDLER NO 1	MAX	48.9	54.4	63.0	73.9	80.3	88.5	94.6	94.3	86.2	76.0	62.2	52.8	72.9
	MIN	25.9	30.4	38.4	50.0	58.0	66.5	70.9	69.3	62.0	50.6	38.6	30.2	49.2
	MEAN	37.4	42.4	50.7	62.0	69.2	77.5	82.8	81.8	74.1	63.3	50.4	41.5	61.1
CHATTANOOGA 3 NE	MAX	52.1	58.2	67.0	76.7	84.0	93.0	98.0	97.3	89.1	78.4	64.0	55.4	76.1
	MIN	25.7	30.3	37.6	48.8	57.6	66.7	70.5	69.2	62.3	50.1	37.7	29.3	48.8
	MEAN	39.0	44.3	52.3	62.8	70.8	79.9	84.3	83.3	75.7	64.3	50.9	42.4	62.5

OKLAHOMA

TEMPERATURE NORMALS (DEG F)

STATION		JAN	FEB	MAR	APR	MAY	JUN	JUL	AUG	SEP	OCT	NOV	DEC	ANN
CHEROKEE POWER PLANT	MAX	47.0	53.2	62.3	73.6	82.0	92.0	97.0	95.8	86.9	76.5	60.3	50.8	73.1
	MIN	22.0	26.5	34.0	45.9	55.4	65.6	70.4	68.5	60.0	47.8	34.4	25.7	46.4
	MEAN	34.5	39.9	48.2	59.8	68.7	78.9	83.7	82.2	73.5	62.2	47.3	38.3	59.8
CHICKASHA EXP STATION	MAX	50.2	56.3	65.3	75.6	82.2	90.6	95.3	93.8	86.6	76.9	62.8	54.0	74.1
	MIN	25.4	29.9	37.8	49.0	58.1	67.0	70.7	68.6	61.2	49.4	37.1	29.1	48.6
	MEAN	37.8	43.1	51.6	62.3	70.2	78.8	83.0	81.2	73.9	63.2	50.0	41.6	61.4
CLAREMORE 2 ENE	MAX	45.9	51.9	60.3	72.0	79.3	87.1	93.3	92.9	85.1	75.1	60.5	50.5	71.2
	MIN	23.1	27.6	35.9	47.7	56.6	65.5	69.8	67.6	60.3	47.7	36.1	27.4	47.1
	MEAN	34.5	39.8	48.1	59.9	67.9	76.3	81.6	80.3	72.7	61.4	48.4	39.0	59.2
CLEVELAND	MAX	48.5	54.3	63.2	74.7	81.3	89.0	94.7	94.4	86.4	76.6	62.0	52.4	73.1
	MIN	23.9	28.8	36.7	48.8	57.5	66.1	70.3	68.5	60.9	49.3	37.4	28.9	48.1
	MEAN	36.2	41.6	49.9	61.8	69.4	77.6	82.5	81.5	73.7	63.0	49.7	40.7	60.6
CLINTON	MAX	48.9	54.5	63.5	74.2	81.9	91.3	96.8	95.6	86.8	75.9	61.1	52.2	73.6
	MIN	23.7	28.2	35.5	46.9	56.1	65.5	69.8	68.4	60.3	48.4	35.8	27.5	47.2
	MEAN	36.4	41.4	49.5	60.6	69.0	78.4	83.3	82.0	73.6	62.2	48.5	39.9	60.4
CUSHING	MAX	46.3	52.1	60.9	72.6	79.7	88.1	94.3	93.8	85.8	75.3	60.6	51.0	71.7
	MIN	23.2	27.8	35.8	48.1	57.2	65.7	70.5	68.9	61.3	49.4	37.1	28.0	47.8
	MEAN	34.8	40.0	48.4	60.4	68.5	76.9	82.4	81.4	73.6	62.4	48.9	39.5	59.8
DUNCAN	MAX	51.6	57.4	66.1	76.1	82.8	90.5	96.0	95.7	87.9	77.7	64.0	55.4	75.1
	MIN	28.1	32.4	39.9	51.2	59.0	67.4	71.5	70.3	63.4	51.9	40.0	31.9	50.6
	MEAN	39.9	44.9	53.0	63.7	70.9	79.0	83.8	83.0	75.7	64.8	52.0	43.7	62.9
EL RENO 1 N	MAX	47.9	53.6	62.5	73.1	80.5	89.0	94.7	94.1	86.0	75.3	60.5	51.8	72.4
	MIN	24.4	28.9	36.4	47.9	56.8	65.8	70.2	68.5	60.8	49.5	36.5	28.3	47.8
	MEAN	36.2	41.3	49.5	60.5	68.7	77.4	82.5	81.3	73.4	62.4	48.5	40.1	60.2
ENID	MAX	45.9	52.1	61.4	72.4	80.6	90.3	95.6	94.4	85.7	75.3	59.3	49.7	71.9
	MIN	24.8	29.3	36.7	48.3	57.4	66.6	71.3	69.7	61.9	50.5	37.6	28.8	48.6
	MEAN	35.4	40.7	49.1	60.4	69.0	78.5	83.5	82.1	73.8	62.9	48.5	39.3	60.3
ERICK 4 E	MAX	51.2	56.6	64.8	75.2	82.5	91.4	96.1	95.0	87.3	77.0	62.5	54.2	74.5
	MIN	22.8	27.3	34.2	45.5	54.5	63.9	67.7	66.1	58.8	46.6	34.3	26.4	45.7
	MEAN	37.0	42.0	49.5	60.4	68.5	77.7	81.9	80.6	73.1	61.8	48.4	40.3	60.1
FORT SUPPLY DAM	MAX	48.4	54.0	62.1	73.1	79.8	89.0	94.3	93.2	85.3	75.8	60.6	51.8	72.3
	MIN	20.7	25.6	32.9	44.9	55.1	64.8	69.2	67.5	59.0	46.8	33.5	24.5	45.4
	MEAN	34.6	39.8	47.5	59.0	67.5	76.9	81.8	80.4	72.2	61.3	47.1	38.1	58.9
FREDERICK	MAX	53.5	59.6	68.5	78.7	85.8	94.4	99.5	98.6	90.0	79.5	65.0	56.6	77.5
	MIN	27.5	31.8	39.0	50.0	58.7	67.6	72.0	70.6	63.2	51.6	39.3	31.0	50.2
	MEAN	40.6	45.7	53.8	64.4	72.3	81.0	85.8	84.6	76.6	65.6	52.2	43.8	63.9
GAGE FAA AIRPORT	MAX	47.0	52.3	60.1	71.8	79.5	89.6	94.8	93.6	84.8	74.3	59.1	50.5	71.5
	MIN	19.4	24.2	31.5	43.2	53.5	63.5	68.2	66.6	57.7	45.0	31.5	23.0	43.9
	MEAN	33.3	38.3	45.9	57.5	66.5	76.6	81.5	80.1	71.3	59.7	45.3	36.8	57.7
GEARY	MAX	47.7	53.0	61.9	73.1	80.7	89.6	95.3	94.6	85.9	75.1	60.2	51.3	72.4
	MIN	24.9	29.2	36.7	48.2	57.0	66.0	70.6	69.1	61.6	50.2	37.4	29.0	48.3
	MEAN	36.3	41.1	49.3	60.7	68.9	77.8	83.0	81.9	73.8	62.6	48.8	40.2	60.4
GOODWELL	MAX	48.6	54.3	61.0	71.9	79.5	89.5	93.6	91.6	84.4	74.5	59.2	51.6	71.6
	MIN	18.3	22.9	28.6	39.9	49.8	59.7	65.1	63.2	54.6	42.3	29.3	21.8	41.3
	MEAN	33.5	38.6	44.8	55.9	64.7	74.6	79.4	77.4	69.5	58.4	44.3	36.7	56.5
GRAND RIVER DAM	MAX	46.2	52.0	60.9	72.9	80.1	88.0	93.3	92.6	84.8	74.4	60.3	50.9	71.4
	MIN	24.8	29.2	37.0	48.5	57.1	65.6	70.2	68.7	61.3	49.6	37.8	29.3	48.3
	MEAN	35.5	40.6	49.0	60.7	68.6	76.8	81.8	80.7	73.1	62.1	49.1	40.1	59.8
GUTHRIE 2 WNW	MAX	48.0	53.9	62.8	73.8	81.2	89.4	95.4	95.0	86.7	76.1	61.5	51.9	73.0
	MIN	24.3	28.7	36.7	48.5	57.3	66.3	70.8	69.2	61.6	49.7	37.0	28.1	48.2
	MEAN	36.2	41.3	49.8	61.2	69.3	77.9	83.1	82.1	74.1	63.0	49.3	40.0	60.6

OKLAHOMA

TEMPERATURE NORMALS (DEG F)

STATION		JAN	FEB	MAR	APR	MAY	JUN	JUL	AUG	SEP	OCT	NOV	DEC	ANN
HAMMON	MAX	49.6	55.3	63.9	74.8	81.9	91.5	96.9	95.4	86.5	76.4	61.3	52.7	73.9
	MIN	21.7	26.5	33.7	45.7	54.7	64.7	68.9	66.8	58.5	46.1	33.6	25.3	45.5
	MEAN	35.7	40.9	48.9	60.3	68.3	78.1	83.0	81.1	72.5	61.3	47.5	39.0	59.7
HENNESSEY 1 N	MAX	46.7	52.7	61.8	73.0	81.4	90.8	96.5	95.4	86.6	75.8	60.1	50.3	72.6
	MIN	24.2	28.5	35.9	47.4	56.9	66.1	70.8	69.1	61.1	49.7	36.8	28.2	47.9
	MEAN	35.5	40.6	48.9	60.2	69.2	78.5	83.7	82.3	73.9	62.8	48.5	39.3	60.3
HOBART FAA AIRPORT	MAX	47.8	53.5	62.3	73.1	81.0	90.8	95.6	94.2	85.6	74.7	60.2	51.5	72.5
	MIN	24.5	28.9	36.3	47.5	57.1	66.9	71.3	70.0	62.1	50.0	36.9	28.2	48.3
	MEAN	36.2	41.2	49.3	60.3	69.1	78.9	83.5	82.1	73.8	62.4	48.5	39.9	60.4
HOLDENVILLE	MAX	49.8	56.1	63.9	74.0	80.5	88.3	94.2	94.2	86.6	76.7	62.9	53.9	73.4
	MIN	27.8	32.1	39.5	50.4	58.8	66.7	70.9	69.5	62.5	51.5	39.8	31.9	50.1
	MEAN	38.8	44.1	51.7	62.2	69.7	77.5	82.6	81.9	74.6	64.1	51.4	42.9	61.8
HOLLIS	MAX	53.3	59.4	68.3	78.6	86.0	94.9	99.0	97.9	89.4	79.0	64.4	56.3	77.2
	MIN	24.5	29.1	36.3	47.8	57.5	66.9	70.8	68.8	61.3	48.9	36.3	28.1	48.0
	MEAN	38.9	44.3	52.3	63.2	71.8	81.0	84.9	83.4	75.3	64.0	50.4	42.2	62.6
HOOKER 1 N	MAX	48.0	53.7	61.2	72.1	80.4	90.7	95.0	92.9	85.1	75.0	58.9	51.0	72.0
	MIN	18.3	23.1	29.1	40.4	50.3	60.5	65.5	63.3	54.7	42.4	29.3	21.7	41.6
	MEAN	33.2	38.4	45.2	56.3	65.4	75.6	80.3	78.2	69.9	58.8	44.1	36.4	56.8
HUGO	MAX	53.3	58.6	66.7	76.0	82.7	89.9	95.0	94.8	88.0	78.3	65.3	56.7	75.4
	MIN	31.2	35.2	42.5	52.2	59.9	67.4	70.9	69.5	63.5	52.1	41.5	34.5	51.7
	MEAN	42.3	46.9	54.6	64.1	71.3	78.7	83.0	82.2	75.8	65.2	53.4	45.7	63.6
HULAH DAM	MAX	44.6	50.7	59.5	71.6	78.8	86.6	93.0	92.9	84.8	74.7	59.9	49.5	70.6
	MIN	20.2	24.5	33.8	46.8	55.7	64.5	69.1	67.0	59.2	46.4	34.3	24.8	45.5
	MEAN	32.4	37.6	46.7	59.2	67.3	75.6	81.1	80.0	72.0	60.6	47.1	37.2	58.1
IDABEL	MAX	53.5	58.6	66.6	75.6	82.3	89.5	94.0	94.0	87.5	77.9	65.1	56.8	75.1
	MIN	30.4	34.0	41.2	50.7	58.6	66.3	69.8	68.6	62.3	50.3	40.1	33.1	50.5
	MEAN	42.0	46.3	53.9	63.2	70.5	77.9	81.9	81.3	74.9	64.1	52.6	45.0	62.8
JAY	MAX	47.3	52.3	61.3	72.3	78.7	86.0	92.1	91.2	83.8	74.3	60.6	51.7	71.0
	MIN	23.5	28.0	35.8	47.0	54.8	63.3	67.6	65.6	57.8	46.8	35.6	27.9	46.1
	MEAN	35.4	40.1	48.6	59.7	66.8	74.7	79.8	78.4	70.8	60.5	48.1	39.8	58.6
JEFFERSON	MAX	46.3	52.3	61.4	72.8	81.5	91.7	96.7	95.3	86.5	76.1	59.9	50.0	72.5
	MIN	22.4	26.9	34.7	46.3	55.9	65.7	70.5	68.7	60.6	48.6	35.6	26.5	46.9
	MEAN	34.4	39.6	48.1	59.6	68.7	78.7	83.6	82.1	73.6	62.4	47.8	38.3	59.7
KENTON	MAX	51.3	54.9	60.7	70.5	79.1	89.7	93.7	91.6	84.7	75.0	60.3	53.2	72.1
	MIN	17.4	22.2	27.3	38.3	47.9	57.6	63.4	61.4	52.9	39.8	27.3	20.6	39.7
	MEAN	34.4	38.6	44.0	54.4	63.5	73.7	78.6	76.5	68.9	57.5	43.9	36.9	55.9
KINGFISHER	MAX	48.0	53.9	63.0	73.8	81.7	91.0	96.8	96.0	87.2	76.5	61.1	51.8	73.4
	MIN	24.0	28.6	36.2	47.7	57.0	66.3	70.6	68.8	61.1	49.3	36.6	27.9	47.8
	MEAN	36.0	41.2	49.6	60.8	69.4	78.6	83.7	82.4	74.2	62.9	48.9	39.9	60.6
LAWTON	MAX	51.2	56.8	65.5	75.5	82.7	90.9	96.4	95.9	87.9	77.5	63.5	54.7	74.9
	MIN	26.3	30.5	38.4	49.8	58.3	67.0	71.0	69.5	62.2	50.4	38.2	29.7	49.3
	MEAN	38.8	43.7	52.0	62.7	70.6	79.0	83.7	82.7	75.1	64.0	50.9	42.2	62.1
MADILL	MAX	51.8	57.4	65.4	75.0	81.9	89.7	95.5	95.6	87.8	77.6	64.2	55.5	74.8
	MIN	30.2	34.3	41.7	52.0	60.0	68.0	71.9	70.5	63.8	53.0	41.7	33.9	51.8
	MEAN	41.0	45.9	53.6	63.5	70.9	78.9	83.7	83.1	75.9	65.3	53.0	44.8	63.3
MANGUM RESEARCH STA	MAX	52.6	58.6	67.3	77.8	84.9	93.6	98.3	97.2	88.8	78.6	63.8	55.6	76.4
	MIN	24.6	29.2	36.4	47.6	57.0	65.9	69.4	67.9	60.8	49.0	36.4	28.3	47.7
	MEAN	38.6	43.9	51.9	62.7	71.0	79.8	83.9	82.6	74.9	63.8	50.2	41.9	62.1
MARIETTA 3 NW	MAX	52.8	58.5	66.3	75.5	82.0	90.0	95.7	95.5	88.1	78.1	64.8	56.5	75.3
	MIN	29.5	33.6	41.2	51.6	59.5	67.4	71.5	70.2	63.6	52.5	41.0	33.0	51.2
	MEAN	41.2	46.1	53.8	63.5	70.8	78.7	83.6	82.9	75.9	65.4	52.9	44.8	63.3

OKLAHOMA

TEMPERATURE NORMALS (DEG F)

STATION		JAN	FEB	MAR	APR	MAY	JUN	JUL	AUG	SEP	OCT	NOV	DEC	ANN
MCALESTER FAA AIRPORT	MAX	48.6	54.3	62.7	73.2	80.1	88.0	94.0	93.4	85.6	75.6	62.0	52.7	72.5
	MIN	27.5	31.8	39.8	50.5	58.9	67.6	71.4	70.1	62.8	50.9	39.4	31.3	50.2
	MEAN	38.1	43.1	51.3	61.9	69.5	77.8	82.7	81.7	74.2	63.2	50.8	42.0	61.4
MEEKER 1 E	MAX	48.3	54.6	62.9	73.8	80.4	88.5	94.6	94.4	86.4	76.1	61.8	52.6	72.9
	MIN	24.7	29.3	37.2	48.8	57.5	65.9	69.9	68.1	60.9	48.9	36.7	29.0	48.1
	MEAN	36.5	41.9	50.1	61.3	69.0	77.2	82.3	81.3	73.7	62.5	49.3	40.8	60.5
MIAMI	MAX	45.7	51.4	60.3	72.5	79.5	87.2	92.9	92.5	85.1	74.5	60.1	50.0	71.0
	MIN	23.7	28.2	36.1	47.7	56.3	65.1	69.3	67.2	60.1	48.3	36.7	28.2	47.2
	MEAN	34.7	39.8	48.2	60.1	67.9	76.2	81.1	79.9	72.6	61.4	48.4	39.2	59.1
MUSKOGEE	MAX	48.0	53.9	62.6	73.7	80.4	87.9	94.1	93.3	85.6	75.2	61.0	51.9	72.3
	MIN	27.3	31.8	39.6	50.3	58.5	66.9	71.2	69.7	62.7	50.7	39.2	31.5	50.0
	MEAN	37.7	42.9	51.1	62.0	69.5	77.5	82.6	81.5	74.2	62.9	50.1	41.7	61.1
MUTUAL	MAX	46.9	52.4	60.7	72.0	79.9	90.4	96.4	95.0	85.7	75.3	59.4	50.4	72.0
	MIN	21.2	25.8	32.9	44.4	54.1	64.1	68.6	67.0	58.9	46.6	33.8	25.2	45.2
	MEAN	34.1	39.2	46.8	58.2	67.1	77.2	82.6	81.0	72.3	60.9	46.6	37.8	58.7
NEWKIRK	MAX	43.8	50.4	59.6	71.8	79.8	88.8	94.4	93.3	84.7	74.2	58.4	48.1	70.6
	MIN	22.9	27.4	35.4	47.2	56.7	66.1	70.5	68.9	61.0	49.5	36.2	27.2	47.4
	MEAN	33.4	38.9	47.5	59.5	68.2	77.5	82.5	81.1	72.8	61.9	47.4	37.6	59.0
NOWATA	MAX	45.9	52.0	61.1	72.9	80.2	88.3	94.7	94.1	85.9	75.3	60.9	50.3	71.8
	MIN	23.4	27.8	35.6	46.9	56.1	64.7	69.5	67.5	59.9	48.1	36.2	27.6	46.9
	MEAN	34.7	40.0	48.4	59.9	68.2	76.5	82.1	80.8	72.9	61.8	48.6	39.0	59.4
OKEENE	MAX	49.4	55.4	64.3	74.9	82.7	92.2	97.4	96.8	88.1	78.0	62.1	52.8	74.5
	MIN	23.5	27.8	35.5	46.9	56.2	65.8	70.3	68.2	60.4	48.8	36.1	27.7	47.3
	MEAN	36.4	41.6	49.9	61.0	69.5	79.0	83.9	82.5	74.3	63.4	49.1	40.3	60.9
OKEMAH	MAX	48.4	54.4	62.8	73.2	79.8	87.7	93.9	93.5	85.6	75.5	61.6	52.4	72.4
	MIN	27.2	31.7	39.3	50.3	58.4	66.4	70.3	68.9	62.4	51.5	39.8	31.5	49.8
	MEAN	37.9	43.1	51.1	61.8	69.1	77.1	82.1	81.2	74.1	63.5	50.7	42.0	61.1
OKLAHOMA CITY WSFO //R	MAX	46.6	52.2	61.0	71.7	79.0	87.6	93.5	92.8	84.7	74.3	59.9	50.7	71.2
	MIN	25.2	29.4	37.1	48.6	57.7	66.3	70.6	69.4	61.9	50.2	37.6	29.1	48.6
	MEAN	35.9	40.8	49.1	60.2	68.4	77.0	82.1	81.1	73.3	62.3	48.8	39.9	59.9
OKMULGEE WATER WORKS	MAX	49.5	55.8	64.2	74.8	80.8	88.4	94.1	94.1	86.7	77.6	63.5	54.1	73.6
	MIN	25.3	30.2	38.3	49.7	57.7	66.1	69.3	67.5	60.2	48.5	37.7	29.7	48.4
	MEAN	37.4	43.0	51.3	62.3	69.3	77.3	81.7	80.8	73.5	63.1	50.6	41.9	61.0
PAULS VALLEY	MAX	51.1	57.4	65.8	76.0	83.1	91.2	96.6	96.2	88.2	77.9	63.9	54.7	75.2
	MIN	27.2	31.6	39.5	50.6	59.2	67.8	71.6	70.1	62.8	50.7	38.8	30.8	50.1
	MEAN	39.2	44.5	52.6	63.3	71.1	79.5	84.1	83.1	75.6	64.3	51.4	42.8	62.6
PAWHUSKA	MAX	46.7	52.8	61.6	73.4	80.1	87.8	93.8	93.4	85.3	75.3	60.5	50.7	71.8
	MIN	22.2	27.1	35.5	47.6	56.2	65.3	69.8	67.6	59.8	47.6	35.4	26.8	46.7
	MEAN	34.5	40.0	48.6	60.5	68.2	76.6	81.8	80.5	72.6	61.5	48.0	38.7	59.3
PERRY	MAX	47.9	53.8	62.9	74.3	81.3	89.7	95.2	94.6	86.5	76.6	61.3	51.7	73.0
	MIN	24.7	29.2	36.8	48.7	57.3	66.5	71.2	69.6	61.9	50.3	37.9	29.0	48.6
	MEAN	36.3	41.5	49.9	61.5	69.3	78.1	83.2	82.1	74.2	63.5	49.6	40.4	60.8
PONCA CITY FAA AIRPORT	MAX	42.8	48.8	58.2	70.3	78.8	88.0	93.8	92.5	83.4	73.0	57.5	47.0	69.5
	MIN	22.1	26.6	34.7	46.7	56.5	66.3	71.1	69.3	61.2	48.8	35.7	26.4	47.1
	MEAN	32.4	37.7	46.5	58.6	67.7	77.2	82.5	80.9	72.3	60.9	46.6	36.7	58.3
POTEAU	MAX	51.3	56.4	64.4	74.8	81.6	89.3	95.0	94.1	87.0	77.0	63.5	55.0	74.1
	MIN	28.5	32.8	40.6	50.8	58.3	66.1	70.5	68.8	62.0	50.5	39.9	32.4	50.1
	MEAN	39.9	44.6	52.5	62.8	70.0	77.7	82.7	81.5	74.5	63.8	51.7	43.7	62.1
PRYOR	MAX	46.2	52.1	60.9	72.3	79.7	87.6	93.5	93.0	85.4	75.1	60.7	50.8	71.4
	MIN	23.6	27.8	35.9	47.7	56.4	65.2	69.6	67.6	59.7	47.7	35.7	27.9	47.1
	MEAN	34.9	40.0	48.5	60.0	68.1	76.4	81.6	80.3	72.6	61.4	48.2	39.4	59.3

OKLAHOMA

TEMPERATURE NORMALS (DEG F)

STATION		JAN	FEB	MAR	APR	MAY	JUN	JUL	AUG	SEP	OCT	NOV	DEC	ANN
PURCELL	MAX	49.0	55.2	63.8	74.5	81.2	89.7	95.4	95.0	87.3	76.7	62.5	53.2	73.6
	MIN	24.8	29.3	37.0	48.9	57.7	66.3	70.2	68.7	61.2	49.0	36.6	28.7	48.2
	MEAN	36.9	42.2	50.4	61.7	69.5	78.0	82.8	81.9	74.2	62.9	49.6	41.0	60.9
SALLISAW	MAX	49.7	55.4	63.6	74.6	81.1	88.5	94.0	93.4	86.4	76.5	62.6	53.4	73.3
	MIN	27.1	31.4	39.0	49.8	58.3	66.2	70.1	68.6	62.1	50.3	38.7	30.9	49.4
	MEAN	38.4	43.4	51.3	62.2	69.7	77.4	82.1	81.0	74.2	63.4	50.7	42.2	61.3
SEMINOLE	MAX	50.7	57.0	65.1	75.6	81.9	89.7	95.6	95.3	87.7	77.9	63.7	54.6	74.6
	MIN	27.5	31.9	39.6	50.7	58.9	67.3	71.6	69.9	62.7	51.3	39.7	31.5	50.2
	MEAN	39.1	44.5	52.4	63.2	70.4	78.5	83.7	82.6	75.2	64.6	51.7	43.0	62.4
STILLWATER 2 W	MAX	47.1	52.9	61.6	73.0	79.9	88.0	93.9	93.4	85.3	75.4	61.0	51.5	71.9
	MIN	23.5	28.0	36.0	47.9	56.9	65.9	70.4	68.5	60.8	48.4	36.8	28.0	47.6
	MEAN	35.3	40.5	48.8	60.4	68.4	77.0	82.1	81.0	73.1	61.9	48.9	39.8	59.8
TAHLEQUAH 1 SE	MAX	48.5	54.0	62.4	73.5	79.9	87.4	93.2	92.9	85.8	75.5	61.6	52.2	72.2
	MIN	25.5	30.1	37.4	48.7	56.4	64.7	68.1	66.8	59.9	48.3	37.0	29.3	47.7
	MEAN	37.0	42.1	50.0	61.1	68.2	76.1	80.7	79.9	72.9	61.9	49.3	40.8	60.0
TALOGA	MAX	48.5	54.0	62.9	73.4	81.1	90.8	96.3	95.0	86.5	75.8	60.5	52.0	73.1
	MIN	21.7	26.1	33.6	45.2	54.6	64.2	68.1	66.4	58.7	46.2	33.5	25.3	45.3
	MEAN	35.1	40.1	48.3	59.3	67.9	77.5	82.2	80.7	72.6	61.0	47.0	38.7	59.2
TIPTON 4 S	MAX	52.6	58.5	67.5	77.5	84.8	93.8	97.8	95.9	88.3	78.4	64.2	55.8	76.3
	MIN	26.3	30.7	37.9	49.0	58.5	67.5	71.5	69.8	62.4	50.5	38.4	29.9	49.4
	MEAN	39.5	44.6	52.7	63.2	71.7	80.7	84.7	82.9	75.4	64.5	51.3	42.9	62.8
TULSA WSO //R	MAX	45.6	51.9	60.8	72.4	79.7	87.9	93.9	93.0	85.0	74.9	60.2	50.3	71.3
	MIN	24.8	29.5	37.7	49.5	58.5	67.5	72.4	70.3	62.5	50.3	38.1	29.3	49.2
	MEAN	35.2	40.7	49.3	60.9	69.1	77.7	83.2	81.7	73.8	62.6	49.2	39.8	60.3
VINITA 3 NNE	MAX	45.7	51.7	60.6	72.1	79.3	87.5	93.6	93.2	85.3	74.8	59.9	50.1	71.2
	MIN	23.2	27.8	35.6	47.2	56.0	64.4	68.5	66.4	59.3	47.6	36.0	27.6	46.6
	MEAN	34.5	39.8	48.1	59.7	67.6	76.0	81.1	79.8	72.3	61.2	47.9	38.9	58.9
WAGONER	MAX	48.0	54.0	62.7	73.8	80.5	88.3	94.6	93.8	86.4	76.3	62.0	52.4	72.7
	MIN	25.7	30.4	38.2	49.3	57.9	66.0	70.1	68.4	61.4	49.8	38.3	30.3	48.8
	MEAN	36.9	42.2	50.5	61.6	69.2	77.2	82.4	81.1	73.9	63.1	50.1	41.4	60.8
WALTERS	MAX	52.6	58.5	67.2	77.0	84.0	92.4	97.8	97.1	89.2	78.5	64.7	56.3	76.3
	MIN	27.2	31.4	39.1	50.3	58.9	67.5	71.2	70.2	63.2	51.0	38.6	30.6	49.9
	MEAN	39.9	45.0	53.1	63.6	71.5	80.0	84.5	83.7	76.2	64.8	51.7	43.5	63.1
WAURIKA	MAX	54.1	60.1	68.5	78.0	84.8	92.9	98.2	98.0	90.0	79.7	65.8	57.6	77.3
	MIN	27.9	32.3	39.9	50.9	59.0	67.3	71.1	69.7	63.0	51.3	39.3	31.4	50.3
	MEAN	41.0	46.2	54.2	64.5	71.9	80.1	84.7	83.8	76.5	65.5	52.6	44.6	63.8
WAYNOKA	MAX	48.2	54.3	63.1	74.3	81.8	91.1	96.4	95.4	86.9	76.9	61.3	51.6	73.4
	MIN	22.1	26.8	34.6	46.3	56.2	65.9	70.5	68.7	59.9	47.5	34.4	25.6	46.5
	MEAN	35.2	40.6	48.8	60.3	69.1	78.5	83.5	82.1	73.4	62.2	47.9	38.6	60.0
WEATHERFORD	MAX	48.2	54.0	63.0	73.7	81.6	90.7	95.8	94.6	86.1	75.5	60.5	51.6	72.9
	MIN	24.9	29.4	36.8	47.9	57.0	65.9	70.1	68.6	61.2	49.8	37.3	28.8	48.1
	MEAN	36.6	41.7	49.9	60.8	69.3	78.3	83.0	81.6	73.7	62.7	48.9	40.2	60.6
WEBBERS FALLS	MAX	47.5	53.0	61.5	72.9	80.2	88.1	94.0	93.5	86.2	75.9	62.2	52.3	72.3
	MIN	24.2	28.6	36.8	48.4	57.6	66.1	70.2	67.9	60.7	47.9	36.5	28.4	47.8
	MEAN	35.9	40.8	49.2	60.6	69.0	77.1	82.1	80.7	73.5	62.0	49.3	40.4	60.1
WICHITA MT WL REF	MAX	51.0	56.7	65.3	75.5	82.0	90.1	95.7	95.0	87.0	76.6	62.6	54.2	74.3
	MIN	24.5	28.9	36.5	48.0	56.3	65.3	69.4	68.0	60.3	48.7	36.4	28.2	47.5
	MEAN	37.8	42.8	50.9	61.8	69.2	77.7	82.6	81.5	73.7	62.7	49.5	41.2	61.0
WILBURTON 9 ENE	MAX	51.6	56.7	64.8	74.9	81.5	89.0	95.1	94.8	87.2	77.2	63.8	55.3	74.3
	MIN	27.0	31.2	38.7	49.3	56.8	65.0	68.8	66.9	60.2	48.4	37.8	30.4	48.4
	MEAN	39.3	44.0	51.8	62.1	69.2	77.0	82.0	80.9	73.7	62.8	50.9	42.9	61.4

OKLAHOMA

TEMPERATURE NORMALS (DEG F)

STATION		JAN	FEB	MAR	APR	MAY	JUN	JUL	AUG	SEP	OCT	NOV	DEC	ANN
WISTER DAM	MAX	52.0	57.2	65.1	75.9	82.2	89.7	95.1	94.5	87.2	78.0	65.0	56.0	74.8
	MIN	26.8	30.6	39.1	49.9	57.7	66.0	69.8	67.5	60.9	48.7	38.1	30.5	48.8
	MEAN	39.4	43.9	52.1	62.9	70.0	77.9	82.4	81.0	74.1	63.4	51.6	43.3	61.8
WOODWARD FIELD STATION	MAX	45.3	50.4	58.3	70.5	78.6	88.6	94.3	92.6	83.8	73.6	58.2	49.2	70.3
	MIN	20.6	25.2	32.9	45.1	55.1	64.7	69.8	67.9	59.1	46.8	34.1	25.0	45.5
	MEAN	33.0	37.8	45.6	57.8	66.9	76.6	82.0	80.3	71.5	60.3	46.2	37.1	57.9

OKLAHOMA

PRECIPITATION NORMALS (INCHES)

STATION	JAN	FEB	MAR	APR	MAY	JUN	JUL	AUG	SEP	OCT	NOV	DEC	ANN
ADA	1.36	1.88	2.90	3.77	5.63	3.73	2.69	3.09	4.01	3.92	2.55	1.94	37.47
ALTUS IRR. RESCH STN	.78	.92	1.28	2.03	4.65	2.94	1.92	2.24	2.85	2.55	1.02	.87	24.05
ALTUS DAM	.62	.94	1.30	1.98	4.78	3.48	2.60	2.13	2.74	2.70	1.02	.85	25.14
ALVA	.56	.87	1.62	2.43	4.06	3.80	2.59	2.89	2.47	1.57	1.20	.81	24.87
ANADARKO	.94	1.22	1.86	2.59	4.89	3.46	2.56	2.48	3.34	2.64	1.58	1.19	28.75
ANTLERS 2 ENE	2.20	2.75	3.57	5.11	5.94	3.97	3.17	3.23	5.27	3.91	3.18	3.02	45.32
ARDMORE	1.35	1.66	2.95	3.87	4.64	3.27	2.30	2.53	3.93	3.40	2.24	1.71	33.85
ARNETT	.43	.67	1.30	1.78	4.14	3.29	2.09	2.41	1.91	1.81	1.09	.63	21.75
BARNSDALL	1.20	1.43	3.11	3.29	5.27	4.54	3.20	3.17	4.72	3.07	2.32	1.62	36.94
BARTLESVILLE 2 W	1.16	1.45	2.72	3.32	4.67	4.09	2.99	3.02	4.13	3.21	2.25	1.48	34.49
BEAR MOUNTAIN TOWER	2.67	3.37	4.44	5.10	5.53	3.62	4.05	3.58	5.24	4.13	3.72	4.10	49.55
BEAVER 1 SW	.38	.58	1.17	1.25	3.26	2.84	2.87	2.81	1.52	1.22	.89	.45	19.24
BILLINGS	.91	1.22	2.07	2.92	4.60	4.11	3.52	2.89	4.22	2.47	1.89	1.22	32.04
BIXBY 2 NE	1.45	1.61	2.69	3.91	4.65	4.74	3.21	2.79	4.35	3.16	2.73	1.83	37.12
BOISE CITY 2 E	.36	.49	.82	1.35	2.43	1.99	2.60	2.38	1.56	.83	.63	.40	15.84
BOSWELL 5 NNW	2.10	2.78	3.32	4.57	4.95	3.62	2.65	2.68	4.91	3.70	3.02	2.64	40.94
BRISTOW	1.15	1.61	2.55	3.55	5.73	4.36	3.56	2.62	4.00	2.54	2.33	1.59	35.59
BROKEN BOW 1 N	3.03	3.29	4.47	5.33	5.69	3.81	3.87	2.96	4.72	3.83	4.02	3.82	48.84
BUFFALO	.53	.92	1.71	2.07	4.39	3.60	3.32	3.34	2.80	1.95	1.33	.69	26.65
CALVIN	1.40	1.90	3.38	4.43	5.82	4.53	3.55	2.58	4.32	3.71	2.69	1.96	40.27
CANTON DAM	.55	.96	1.67	2.29	4.95	3.70	2.41	2.25	3.13	2.08	1.56	.83	26.38
CARNASAW TOWER	3.17	3.33	4.65	5.48	6.34	4.03	4.14	3.10	4.99	4.14	4.28	3.90	51.55
CARNEGIE 4 ENE	.78	1.15	1.65	2.42	5.12	3.08	2.56	2.13	3.36	2.19	1.32	1.06	26.82
CARTER TOWER	2.69	3.30	4.57	5.26	5.94	3.85	4.39	3.67	4.96	4.58	3.82	3.91	50.94
CHANDLER NO 1	1.15	1.49	2.29	3.22	5.41	3.80	3.34	2.25	3.79	2.43	2.09	1.39	32.65
CHATTANOOGA 3 NE	.91	1.13	1.74	2.48	4.76	2.80	2.55	2.61	3.07	2.77	1.37	1.08	27.27
CHECOTAH	1.49	1.87	3.34	4.59	5.38	4.05	3.46	2.70	4.46	3.44	2.84	2.11	39.73
CHEROKEE POWER PLANT	.69	.92	1.93	2.55	3.85	3.99	2.76	2.58	2.67	1.82	1.28	.87	25.91
CHICKASHA EXP STATION	.90	1.21	1.94	2.84	5.12	3.09	2.52	2.52	3.48	2.71	1.55	1.08	28.96
CLAREMORE 2 ENE	1.38	1.62	3.16	3.76	4.67	4.63	3.08	2.91	3.88	3.42	2.49	1.85	36.85
CLEVELAND	1.13	1.50	2.70	3.18	4.63	3.90	3.60	3.30	4.50	3.02	2.17	1.35	34.98
CLINTON	.71	1.04	1.70	2.39	5.00	3.35	2.52	2.79	3.00	2.70	1.47	.91	27.58
COALGATE 2 S	1.63	2.16	3.89	4.96	5.19	3.84	2.96	2.74	5.04	3.93	2.89	2.19	41.42
CORDELL	.70	1.03	1.63	2.19	4.68	3.07	2.52	2.63	2.80	2.58	1.39	.91	26.13
CUSHING	1.04	1.31	2.47	3.18	5.35	4.29	3.72	2.69	3.89	2.68	2.01	1.31	33.94
DAISY 2 E	1.95	2.69	3.84	5.43	6.30	4.48	4.32	3.51	5.70	3.81	3.36	2.66	48.05
DEWAR 2 NE	1.41	1.81	3.12	4.28	5.11	4.07	3.54	2.61	4.31	3.26	2.69	1.87	38.08
DUNCAN	.98	1.21	2.14	2.71	5.62	3.46	2.33	2.35	3.65	2.95	1.90	1.35	30.65
DURANT SE STATE COL	1.74	2.25	3.27	4.54	5.00	3.72	2.54	2.47	5.61	3.47	2.80	2.18	39.59
EL RENO 1 N	.83	1.09	1.85	2.58	5.17	3.63	2.77	2.30	3.61	2.88	1.64	1.03	29.38
ELK CITY	.55	.95	1.52	2.21	4.93	3.32	2.41	2.33	2.58	1.99	1.36	.71	24.86
ENID	.91	1.16	1.89	2.78	5.01	4.12	3.18	3.36	3.21	2.80	1.78	1.03	31.23
ERICK 4 E	.48	.86	1.41	2.20	4.41	2.97	2.13	2.12	2.81	2.20	.99	.68	23.26
EUFAULA	1.53	2.07	3.97	4.68	5.48	4.12	3.65	2.73	4.20	3.41	2.96	2.44	41.24
FANSHAWE	1.88	2.79	4.42	5.00	5.91	4.21	4.03	3.07	4.68	3.08	3.95	2.94	45.96
FARGO	.46	.85	1.29	1.83	3.96	3.21	2.19	2.47	1.82	1.67	1.02	.65	21.42
FLASHMAN TOWER	2.73	3.23	4.56	5.70	6.08	4.32	4.13	4.19	5.12	4.14	3.99	3.74	51.93
FORAKER	1.02	1.22	2.39	3.13	4.82	4.19	3.47	3.51	4.08	3.10	2.37	1.34	34.64
FORT SUPPLY DAM	.50	.85	1.26	1.59	3.71	2.94	2.19	2.52	1.97	1.43	.92	.62	20.50
FREDERICK	.85	1.02	1.70	2.32	4.74	2.95	2.19	2.46	3.00	2.46	1.41	1.02	26.12
GAGE FAA AIRPORT	.45	.82	1.18	1.85	3.66	2.77	2.11	2.42	1.60	1.59	.84	.64	19.93
GEARY	.66	1.12	1.74	2.46	4.80	3.81	2.47	2.19	3.22	2.43	1.41	1.02	27.33
GOODWELL	.25	.31	.78	1.11	2.87	2.30	2.88	2.37	1.27	.95	.64	.27	16.00
GRAND RIVER DAM	1.56	1.92	3.12	3.91	4.81	4.94	3.56	3.38	4.25	3.43	2.83	2.00	39.71
GRANDFIELD	1.08	1.18	1.77	2.42	4.94	3.18	2.06	2.39	3.45	2.84	1.55	1.25	28.11

OKLAHOMA

PRECIPITATION NORMALS (INCHES)

STATION	JAN	FEB	MAR	APR	MAY	JUN	JUL	AUG	SEP	OCT	NOV	DEC	ANN
GREAT SALT PLAINS DAM	.62	.82	1.83	2.65	3.58	3.49	3.18	2.86	2.98	2.03	1.45	.79	26.28
GUTHRIE 2 WNW	.91	1.26	2.01	2.60	5.42	3.96	2.84	2.38	3.98	2.66	1.80	1.20	31.02
HAMMON	.51	.91	1.56	2.22	4.56	2.96	2.15	2.44	2.71	1.90	1.39	.71	24.02
HANNA	1.46	1.86	3.69	4.44	5.44	3.99	3.16	2.82	4.16	3.27	2.94	2.10	39.33
HASKELL 1 NW	1.63	1.92	3.17	4.11	4.97	4.82	3.18	2.33	3.97	3.07	2.89	1.97	38.03
HEALDTON 2 N	1.34	1.35	2.46	3.45	4.85	3.71	2.37	2.30	4.09	3.12	2.04	1.61	32.69
HEAVENER 1 SE	2.25	2.72	4.15	4.93	5.52	4.00	3.56	3.35	4.52	3.30	3.69	3.22	45.21
HELENA 1 SSE	.71	1.00	1.92	2.57	4.34	3.95	3.08	2.61	2.87	2.12	1.54	.94	27.65
HENNESSEY 1 N	.71	1.16	1.86	2.38	5.32	3.90	2.51	2.69	3.39	2.11	1.63	.99	28.65
HOBART FAA AIRPORT	.61	.91	1.27	2.24	4.98	2.90	2.49	1.88	2.87	2.52	1.08	.81	24.56
HOLDENVILLE	1.34	1.68	2.98	4.37	5.60	3.83	3.46	2.66	4.00	3.54	2.40	1.83	37.69
HOLLIS	.53	.77	1.05	2.20	4.07	2.98	1.87	2.03	2.68	2.25	.88	.73	22.04
HOLLOW 1 NE	1.35	1.55	3.14	3.72	4.86	4.58	3.75	3.30	4.84	3.53	2.99	1.89	39.50
HOMINY 4 N	1.07	1.39	2.82	3.12	4.64	4.15	3.42	3.03	4.48	2.95	2.06	1.28	34.41
HOOKER 1 N	.41	.46	1.23	1.19	3.43	2.95	2.93	2.78	1.62	1.11	.76	.39	19.26
HUGO	2.22	2.77	3.80	4.72	5.66	4.52	3.05	3.44	5.15	3.94	3.26	3.08	45.61
HULAH DAM	1.14	1.17	2.64	3.16	4.31	4.36	2.94	3.03	3.83	3.20	2.19	1.29	33.26
IDABEL	3.04	3.42	4.36	5.40	5.67	3.69	3.55	2.62	4.53	3.84	3.83	3.47	47.42
JAY	1.73	1.96	3.52	4.33	5.24	5.38	3.74	3.56	4.48	3.64	3.34	2.27	43.19
JEFFERSON	.70	.97	1.93	2.77	3.92	3.98	3.92	3.25	3.13	2.55	1.92	1.03	30.07
KENTON	.30	.28	.76	1.29	2.49	1.82	2.89	2.50	1.51	.90	.53	.30	15.57
KINGFISHER	.83	1.13	1.76	2.42	4.94	3.76	2.57	2.39	3.60	2.44	1.53	1.13	28.50
KINGSTON	1.71	2.26	3.16	4.10	5.04	3.62	2.33	2.49	4.67	3.64	2.53	2.01	37.56
KONAWA	1.33	1.65	2.89	4.12	6.10	3.72	2.53	2.46	4.12	3.59	2.34	1.86	36.71
LAVERNE	.63	.88	1.54	1.53	3.39	2.97	2.49	2.97	2.01	1.51	.99	.67	21.58
LAWTON	1.07	1.17	1.83	2.41	5.69	3.57	2.51	2.15	2.98	2.85	1.75	1.22	29.20
LEEDEY	.46	.90	1.34	2.50	4.78	3.24	1.97	2.58	2.23	1.76	1.33	.69	23.78
LINDSAY	1.13	1.42	2.26	3.31	6.28	3.41	2.61	2.31	3.80	3.07	2.08	1.47	33.15
LYONS 2 N	1.72	1.95	3.94	4.73	5.31	4.46	3.22	2.87	4.26	3.08	2.95	2.31	40.80
MADILL	1.69	2.12	3.01	4.51	5.10	3.85	2.28	2.43	4.60	3.57	2.46	1.97	37.59
MANGUM RESEARCH STA	.63	.86	1.18	1.89	4.72	2.85	2.69	2.05	2.78	2.62	.91	.76	23.94
MANNFORD 6 NW	1.12	1.47	2.57	3.29	4.79	3.88	3.20	3.08	4.23	2.64	2.20	1.42	33.89
MARAMEC	1.05	1.40	2.45	2.99	5.01	3.90	3.12	2.92	3.92	3.16	2.01	1.23	33.16
MARIETTA 3 NW	1.48	1.77	2.75	3.80	4.55	3.63	2.14	2.58	3.99	3.03	2.46	1.70	33.88
MARLOW 1 WSW	.90	1.20	2.00	2.68	6.01	3.82	2.57	2.42	3.66	2.95	1.95	1.36	31.52
MARSHALL	.76	1.16	1.99	2.38	5.25	4.00	2.59	2.75	3.51	2.60	1.63	1.14	29.76
MCALESTER FAA AIRPORT	1.62	2.26	3.85	4.54	5.62	3.66	3.41	3.25	4.96	3.90	3.07	2.38	42.52
MC CURTAIN 1 SE	1.88	2.53	3.91	4.77	5.67	4.28	3.81	3.01	4.46	3.31	3.58	2.64	43.85
MEEKER 1 E	1.07	1.47	2.43	3.56	5.64	3.70	3.01	2.53	3.86	2.78	2.05	1.43	33.53
MIAMI	1.53	1.88	3.44	3.72	5.03	4.88	3.93	3.51	4.60	3.74	2.95	2.15	41.36
MORAVIA 2 NNE	.50	.96	1.53	2.09	4.75	2.99	2.29	2.06	2.77	2.42	1.05	.80	24.21
MUSKOGEE	1.63	2.11	3.24	4.58	5.03	4.60	3.10	3.03	4.12	3.34	2.98	2.24	40.00
MUTUAL	.50	.93	1.58	2.45	4.32	3.17	2.56	2.20	2.48	1.52	1.15	.66	23.52
NEWKIRK	.86	1.10	1.98	2.95	4.72	4.59	3.55	3.50	3.54	2.77	1.94	1.22	32.72
NORMAN 3 S	1.13	1.33	2.33	3.30	5.89	3.62	3.23	2.56	3.73	2.63	2.04	1.35	33.14
NOWATA	1.28	1.64	3.27	3.50	4.62	4.78	2.94	3.39	4.31	3.30	2.55	1.80	37.38
OKEENE	.59	.94	1.82	2.33	4.99	3.97	2.34	2.57	2.93	2.12	1.60	.86	27.06
OKEMAH	1.38	1.45	2.70	4.18	5.02	4.47	3.38	2.60	3.80	2.87	2.44	1.83	36.12
OKLAHOMA CITY WSFO //R	.96	1.29	2.07	2.91	5.50	3.87	3.04	2.40	3.41	2.71	1.53	1.20	30.89
OKMULGEE WATER WORKS	1.63	1.79	3.03	4.52	5.08	4.71	3.05	2.63	3.80	2.89	2.63	2.05	37.81
PAULS VALLEY	1.31	1.47	2.30	3.50	5.46	3.37	2.33	2.32	3.67	3.57	2.17	1.71	33.18
PAWHUSKA	1.11	1.31	2.68	3.07	4.76	4.31	3.45	3.34	4.11	2.93	2.03	1.35	34.45
PAWNEE	1.01	1.31	2.48	2.97	4.84	4.02	3.13	3.01	4.37	2.72	1.88	1.25	32.99
PERKINS 1 SSE	1.12	1.26	2.41	2.65	5.20	4.17	3.53	2.61	4.22	3.15	2.07	1.35	33.74
PERRY	.87	1.32	2.36	2.70	5.28	4.13	3.53	3.33	3.74	2.63	1.80	1.20	32.89

OKLAHOMA

PRECIPITATION NORMALS (INCHES)

STATION	JAN	FEB	MAR	APR	MAY	JUN	JUL	AUG	SEP	OCT	NOV	DEC	ANN
PONCA CITY FAA AIRPORT	.91	1.22	2.10	2.90	4.49	4.17	4.10	3.36	3.84	2.60	2.05	1.27	33.01
PONTOTOC	1.33	1.93	3.29	4.09	5.73	3.55	2.89	2.71	4.12	3.78	2.59	1.87	37.88
POTEAU	1.84	2.68	4.12	4.70	5.92	3.47	3.68	3.30	4.25	3.35	4.02	2.93	44.26
PRAGUE	1.24	1.50	2.51	3.87	5.26	3.78	3.21	2.51	3.79	2.87	2.21	1.55	34.30
PRYOR	1.52	1.78	3.11	3.90	4.88	4.67	3.06	3.40	4.16	3.77	2.90	2.04	39.19
PURCELL	1.07	1.34	2.37	3.37	6.02	3.59	3.00	2.42	3.97	3.18	2.06	1.46	33.85
QUAPAW	1.55	1.82	3.32	3.98	5.18	4.77	3.77	3.44	4.80	3.66	2.88	2.01	41.18
QUINTON	1.62	2.10	3.69	4.33	5.57	4.03	3.80	3.10	4.41	3.61	3.24	2.36	41.86
RALSTON	1.00	1.30	2.52	2.97	4.72	4.39	3.49	2.92	3.86	2.69	1.95	1.36	33.17
REDROCK	.87	1.39	2.25	2.79	4.63	4.03	3.72	2.94	3.72	2.47	1.72	1.29	31.82
REGNIER	.27	.27	.68	1.11	1.92	1.80	2.50	1.91	1.41	.76	.51	.28	13.42
RENFROW	.71	1.01	1.91	2.56	3.83	3.93	3.51	2.88	3.21	2.32	1.70	.99	28.56
REYDON 4 W	.39	.79	1.40	2.27	4.29	3.31	2.09	2.23	2.34	1.68	.96	.62	22.37
ROOSEVELT	.68	.96	1.32	2.25	5.25	3.29	2.37	2.14	2.78	2.48	1.24	.97	25.73
SALLISAW	1.78	2.48	3.80	4.47	5.47	4.33	3.55	3.17	4.41	3.86	3.41	2.47	43.20
SAYRE 1 NE	.42	.72	1.28	2.05	4.41	3.17	2.07	2.05	2.44	2.13	1.09	.60	22.43
SEMINOLE	1.30	1.55	2.58	4.09	5.35	3.80	2.95	2.88	4.02	2.85	2.52	1.78	35.67
SHAWNEE	1.22	1.53	2.51	3.87	6.01	3.95	2.66	2.90	3.74	3.20	2.34	1.53	35.46
SKIATOOK	1.19	1.63	2.83	3.47	4.67	4.31	3.41	2.86	4.33	3.19	2.35	1.45	35.69
SNYDER	.84	1.05	1.43	2.05	5.00	2.88	2.47	2.21	2.81	2.36	1.23	1.02	25.35
SOBOL TOWER	2.34	2.85	4.14	5.10	5.90	3.96	3.72	3.36	5.42	4.06	3.41	3.52	47.78
SPAVINAW	1.53	1.79	3.13	4.08	5.06	4.78	3.73	3.60	4.38	3.65	3.21	2.03	40.97
SPIRO	1.82	2.70	4.14	4.62	5.36	3.55	3.79	2.60	4.03	3.31	3.85	2.79	42.56
STILLWATER 2 W	.90	1.20	2.19	2.58	5.08	3.92	3.79	2.83	3.93	2.90	1.78	1.22	32.32
STILWELL 1 NE	1.96	2.57	3.70	4.71	5.63	4.48	3.73	3.35	4.31	3.28	3.25	2.71	43.68
TAHLEQUAH 1 SE	1.78	2.42	3.64	4.56	5.47	4.63	3.39	3.06	4.34	3.39	3.20	2.46	42.34
TALOGA	.55	.94	1.62	2.44	5.13	3.27	2.62	2.44	2.63	1.86	1.47	.63	25.60
TIPTON 4 S	.82	.91	1.44	2.16	4.97	2.91	2.22	2.22	2.75	2.40	1.27	.93	25.00
TISHOMINGO NATIONAL WL	1.53	2.05	3.17	4.61	4.88	3.46	2.70	2.52	4.87	3.63	2.49	2.08	37.99
TULSA WSO //R	1.35	1.74	3.14	4.15	5.14	4.57	3.51	3.01	4.37	3.41	2.56	1.82	38.77
UNION CITY 4 ESE	1.09	1.41	2.37	3.33	5.90	4.21	2.84	2.55	3.76	3.09	2.06	1.34	33.95
VALLIANT	2.52	3.27	4.21	5.00	5.50	3.70	3.58	2.76	4.98	3.62	3.60	3.60	46.34
VINITA 3 NNE	1.53	1.81	3.54	4.07	5.35	4.87	3.38	3.61	4.75	3.72	2.96	2.14	41.73
VINSON 3 WNW	.47	.66	1.28	2.07	4.64	2.83	1.96	2.26	2.87	2.27	1.02	.78	23.11
WAGONER	1.72	1.89	3.39	4.67	4.83	5.09	3.50	2.85	4.09	3.10	3.20	2.06	40.39
WALTERS	1.20	1.27	2.13	2.83	5.31	3.59	2.96	2.57	3.26	2.92	1.83	1.42	31.29
WATONGA	.77	1.05	1.78	2.42	4.98	3.77	2.24	2.05	2.95	2.22	1.42	1.00	26.65
WAURIKA	1.13	1.30	1.94	2.96	4.85	3.25	2.28	2.55	3.40	2.69	1.93	1.48	29.76
WAYNOKA	.60	.98	1.63	2.18	4.44	3.75	2.55	2.70	2.50	1.71	1.28	.77	25.09
WEATHERFORD	.64	.99	1.59	2.23	4.72	3.63	2.49	2.69	3.28	2.73	1.36	.86	27.21
WEBBERS FALLS	1.63	2.31	3.59	4.60	5.31	4.09	3.15	2.88	4.34	3.75	2.99	2.29	40.93
WETUMKA 3 NE	1.42	1.60	3.12	4.37	5.42	4.32	3.19	2.42	4.02	3.13	2.77	1.89	37.67
WEWOKA 3 W	1.42	1.68	2.72	3.77	5.33	4.21	2.79	2.85	4.12	2.98	2.23	1.78	35.88
WICHITA MT WL REF	.90	1.17	1.89	2.45	5.24	3.46	2.49	2.04	3.11	2.73	1.54	1.12	28.14
WILBURTON 9 ENE	1.91	2.62	4.08	5.05	5.62	3.94	4.33	3.33	4.94	3.55	3.58	2.87	45.82
WISTER DAM	1.90	2.61	4.11	4.75	5.37	3.88	3.69	3.05	4.38	3.11	3.67	2.92	43.44
WOODWARD	.52	.96	1.50	2.00	4.07	3.14	2.82	2.82	2.03	1.82	1.09	.71	23.48
WOODWARD FIELD STATION	.55	.96	1.44	1.90	3.88	3.14	2.51	2.73	2.02	1.74	1.07	.69	22.63
ZOE 1 E	2.56	2.81	4.42	5.00	5.89	3.85	4.09	3.52	4.57	3.70	3.80	3.41	47.62

OKLAHOMA

HEATING DEGREE DAY NORMALS (BASE 65 DEG F)

STATION	JUL	AUG	SEP	OCT	NOV	DEC	JAN	FEB	MAR	APR	MAY	JUN	ANN
ADA	0	0	12	119	404	667	787	568	413	131	23	0	3124
ALTUS IRR. RESCH STN	0	0	7	107	414	688	797	577	407	124	18	0	3139
ALVA	0	0	20	154	528	834	958	720	558	214	57	5	4048
ANADARKO	0	0	13	127	459	738	856	622	444	142	26	0	3427
ANTLERS 2 ENE	0	0	9	124	400	660	769	563	398	121	26	0	3070
ARDMORE	0	0	0	69	330	583	698	501	340	81	7	0	2609
ARNETT	0	0	31	207	579	862	983	750	614	248	87	7	4368
BARTLESVILLE 2 W	0	0	18	160	501	806	942	692	516	171	36	0	3842
BEAVER 1 SW	0	0	31	219	612	893	998	753	617	254	94	10	4481
BIXBY 2 NE	0	0	21	172	489	766	918	680	514	168	43	0	3771
BOISE CITY 2 E	0	0	37	249	633	874	958	748	648	326	130	10	4613
BRISTOW	0	0	22	136	464	750	877	633	466	156	32	0	3536
BUFFALO	0	0	19	155	540	822	939	689	541	194	63	6	3968
CANTON DAM	0	0	15	145	498	791	911	675	508	192	50	0	3785
CARNEGIE 4 ENE	0	0	14	126	465	741	859	627	455	150	24	0	3461
CHANDLER NO 1	0	0	18	129	443	729	856	633	461	141	32	0	3442
CHATTANOOGA 3 NE	0	0	8	101	423	701	806	580	412	129	18	0	3178
CHEROKEE POWER PLANT	0	0	15	151	531	828	946	703	533	196	45	0	3948
CHICKASHA EXP STATION	0	0	13	128	450	725	843	613	435	142	24	0	3373
CLAREMORE 2 ENE	0	0	26	180	498	806	946	706	534	187	63	0	3946
CLEVELAND	0	0	20	127	459	753	893	655	484	148	23	0	3562
CLINTON	0	0	16	145	495	778	887	661	493	179	41	0	3695
CUSHING	0	0	20	150	483	791	936	700	532	169	49	0	3830
DUNCAN	0	0	8	105	397	660	778	570	396	112	17	0	3043
EL RENO 1 N	0	0	15	140	495	772	893	664	493	178	37	0	3687
ENID	0	0	15	134	495	797	918	680	507	178	40	0	3764
ERICK 4 E	0	0	13	145	498	766	868	644	491	178	46	0	3649
FORT SUPPLY DAM	0	0	25	164	537	834	942	706	558	216	69	6	4057
FREDERICK	0	0	7	93	389	657	756	550	377	105	15	0	2949
GAGE FAA AIRPORT	0	0	26	202	591	874	983	748	599	243	87	0	4353
GEARY	0	0	19	140	486	769	890	669	499	172	40	0	3684
GOODWELL	0	0	39	231	621	877	977	739	632	290	119	11	4536
GRAND RIVER DAM	0	0	18	158	482	772	915	683	514	180	42	0	3764
GUTHRIE 2 WNW	0	0	15	139	471	775	893	664	484	164	34	0	3639
HAMMON	0	0	22	170	525	806	908	675	515	183	63	6	3873
HENNESSEY 1 N	0	0	14	141	495	797	915	683	512	184	41	0	3782
HOBART FAA AIRPORT	0	0	16	142	495	778	893	666	497	180	39	0	3706
HOLDENVILLE	0	0	11	115	413	685	812	593	430	131	23	0	3213
HOLLIS	0	0	6	110	438	707	809	580	413	122	19	0	3204
HOOKER 1 N	0	0	29	226	627	887	986	745	621	274	97	10	4502
HUGO	0	0	0	94	357	598	704	513	347	94	9	0	2716
HULAH DAM	0	0	29	190	537	862	1011	767	577	203	65	0	4241
IDABEL	0	0	6	115	380	620	713	524	364	108	15	0	2845
JAY	0	0	34	192	507	781	918	697	521	195	62	9	3916
JEFFERSON	0	0	15	144	516	828	949	711	533	198	47	0	3941
KENTON	0	0	32	250	633	871	949	739	651	328	116	14	4583
KINGFISHER	0	0	14	129	483	778	899	666	490	174	35	0	3668
LAWTON	0	0	6	115	428	707	812	596	424	127	22	0	3237
MADILL	0	0	7	93	366	626	744	540	378	107	13	0	2874
MANGUM RESEARCH STA	0	0	6	118	444	716	818	591	426	140	24	0	3283
MARIETTA 3 NW	0	0	10	92	371	626	738	536	371	110	15	0	2869
MCALESTER FAA AIRPORT	0	0	16	133	433	713	834	613	441	144	34	0	3361
MEEKER 1 E	0	0	17	150	471	750	884	647	475	158	35	0	3587
MIAMI	0	0	27	174	498	800	939	706	531	186	58	7	3926
MUSKOGEE	0	0	17	140	447	722	846	619	448	138	32	0	3409

OKLAHOMA

HEATING DEGREE DAY NORMALS (BASE 65 DEG F)

STATION	JUL	AUG	SEP	OCT	NOV	DEC	JAN	FEB	MAR	APR	MAY	JUN	ANN
MUTUAL	0	0	18	175	552	843	958	722	574	230	79	6	4157
NEWKIRK	0	0	22	157	528	849	980	731	554	204	51	0	4076
NOWATA	0	0	21	157	492	806	939	700	524	194	46	0	3879
OKEENE	0	0	17	122	477	766	887	655	482	168	36	0	3610
OKEMAH	0	0	17	125	429	713	840	613	450	136	27	0	3350
OKLAHOMA CITY WSFO //R	0	0	15	145	486	778	902	678	506	184	41	0	3735
OKMULGEE WATER WORKS	0	0	16	138	432	716	856	616	448	131	30	0	3383
PAULS VALLEY	0	0	9	105	413	688	800	574	405	118	18	0	3130
PAWHUSKA	0	0	24	161	510	815	946	700	521	178	48	0	3903
PERRY	0	0	15	125	462	763	890	658	483	157	38	0	3591
PONCA CITY FAA AIRPORT	0	0	28	179	552	877	1011	764	580	223	65	0	4279
POTEAU	0	0	13	119	403	660	778	571	411	113	22	0	3090
PRYOR	0	0	24	167	504	794	933	700	527	196	55	0	3900
PURCELL	0	0	12	135	467	744	871	638	475	154	35	0	3531
SALLISAW	0	0	10	126	434	707	825	605	442	130	25	0	3304
SEMINOLE	0	0	10	100	404	682	803	574	411	132	23	0	3139
STILLWATER 2 W	0	0	18	158	483	781	921	686	515	183	48	0	3793
TAHLEQUAH 1 SE	0	0	24	169	476	750	868	641	480	163	56	0	3627
TALOGA	0	0	19	168	540	815	927	697	528	201	56	0	3951
TIPTON 4 S	0	0	6	99	411	685	791	571	402	123	15	0	3103
TULSA WSO //R	0	0	18	146	474	781	924	680	500	168	40	0	3731
VINITA 3 NNE	0	0	27	176	513	809	946	706	534	189	62	7	3969
WAGONER	0	0	17	141	447	732	871	638	466	150	30	0	3492
WALTERS	0	0	13	116	405	667	778	560	393	124	17	0	3073
WAURIKA	0	0	6	94	379	632	744	532	364	104	13	0	2868
WAYNOKA	0	0	16	158	513	818	924	683	518	177	48	0	3855
WEATHERFORD	0	0	15	134	483	769	880	652	482	168	33	0	3616
WEBBERS FALLS	0	0	15	152	471	763	902	678	501	171	36	0	3689
WICHITA MT WL REF	0	0	13	139	465	738	843	622	457	148	30	0	3455
WILBURTON 9 ENE	0	0	16	141	428	685	797	588	427	140	38	0	3260
WISTER DAM	0	0	15	130	410	673	794	591	417	126	41	0	3197
WOODWARD FIELD STATION	0	0	33	192	564	865	992	762	610	241	83	7	4349

OKLAHOMA

COOLING DEGREE DAY NORMALS (BASE 65 DEG F)

STATION	JAN	FEB	MAR	APR	MAY	JUN	JUL	AUG	SEP	OCT	NOV	DEC	ANN
ADA	0	0	23	56	168	381	549	518	300	100	8	0	2103
ALTUS IRR. RESCH STN	0	0	19	73	222	465	608	561	319	95	0	0	2362
ALVA	0	0	12	31	153	398	570	527	266	64	0	0	2021
ANADARKO	0	0	13	52	181	405	564	527	298	68	0	0	2108
ANTLERS 2 ENE	0	0	20	49	174	375	527	499	291	78	0	0	2013
ARDMORE	0	9	33	87	236	459	614	589	371	128	9	0	2535
ARNETT	0	0	6	20	128	328	490	443	208	46	0	0	1669
BARTLESVILLE 2 W	0	0	14	45	150	360	527	484	252	54	0	0	1886
BEAVER 1 SW	0	0	6	17	134	346	512	453	208	40	0	0	1716
BIXBY 2 NE	0	0	12	36	155	361	521	474	252	70	0	0	1881
BOISE CITY 2 E	0	0	0	8	74	265	403	332	130	14	0	0	1226
BRISTOW	0	0	17	63	159	366	533	502	283	83	5	0	2011
BUFFALO	0	0	14	35	171	408	570	521	265	65	0	0	2049
CANTON DAM	0	0	15	45	159	381	555	518	267	71	0	0	2011
CARNEGIE 4 ENE	0	0	14	54	179	426	580	536	296	71	0	0	2156
CHANDLER NO 1	0	0	17	51	163	375	552	521	291	77	0	0	2047
CHATTANOOGA 3 NE	0	0	19	63	198	447	598	567	329	79	0	0	2300
CHEROKEE POWER PLANT	0	0	12	40	160	417	580	533	270	64	0	0	2076
CHICKASHA EXP STATION	0	0	19	61	186	414	558	502	280	72	0	0	2092
CLAREMORE 2 ENE	0	0	10	34	152	339	515	474	257	68	0	0	1849
CLEVELAND	0	0	16	52	159	378	543	512	281	65	0	0	2006
CLINTON	0	0	13	47	165	402	567	527	274	58	0	0	2053
CUSHING	0	0	17	31	158	362	539	508	278	69	0	0	1962
DUNCAN	0	8	24	73	200	420	583	558	329	99	7	0	2301
EL RENO 1 N	0	0	13	43	152	372	543	505	267	59	0	0	1954
ENID	0	0	14	40	164	405	574	530	279	69	0	0	2075
ERICK 4 E	0	0	11	40	154	385	524	484	256	45	0	0	1899
FORT SUPPLY DAM	0	0	15	36	147	363	521	477	241	49	0	0	1849
FREDERICK	0	10	29	87	241	480	645	608	355	112	5	0	2572
GAGE FAA AIRPORT	0	0	7	18	134	353	512	468	215	38	0	0	1745
GEARY	0	0	12	43	160	384	558	524	283	66	0	0	2030
GOODWELL	0	0	6	17	109	299	446	384	174	27	0	0	1462
GRAND RIVER DAM	0	0	18	51	154	358	521	487	261	68	5	0	1923
GUTHRIE 2 WNW	0	0	13	50	167	387	561	530	288	77	0	0	2073
HAMMON	0	0	16	42	165	399	558	499	247	56	0	0	1982
HENNESSEY 1 N	0	0	13	40	172	405	580	536	281	73	0	0	2100
HOBART FAA AIRPORT	0	0	10	39	166	417	574	530	280	61	0	0	2077
HOLDENVILLE	0	8	18	47	169	375	546	524	299	87	0	0	2073
HOLLIS	0	0	20	68	230	480	617	570	315	79	0	0	2379
HOOKER 1 N	0	0	7	13	109	328	474	409	176	34	0	0	1550
HUGO	0	6	24	67	205	411	558	533	328	101	9	0	2242
HULAH DAM	0	0	10	29	136	323	499	465	239	53	0	0	1754
IDABEL	0	0	20	54	186	387	524	505	303	87	8	0	2074
JAY	0	0	12	36	118	300	459	415	208	52	0	0	1600
JEFFERSON	0	0	9	36	162	411	577	530	273	63	0	0	2061
KENTON	0	0	0	10	69	275	422	357	149	18	0	0	1300
KINGFISHER	0	0	12	48	171	408	580	539	290	64	0	0	2112
LAWTON	0	0	21	58	195	420	580	549	309	84	0	0	2216
MADILL	0	5	25	62	196	417	580	561	334	102	6	0	2288
MANGUM RESEARCH STA	0	0	20	71	210	444	586	546	303	81	0	0	2261
MARIETTA 3 NW	0	6	24	65	195	411	577	555	337	105	8	0	2283
MCALESTER FAA AIRPORT	0	0	17	51	174	384	549	518	292	78	7	0	2070
MEEKER 1 E	0	0	13	47	159	366	536	505	278	73	0	0	1977
MIAMI	0	0	10	39	148	343	499	462	255	63	0	0	1819
MUSKOGEE	0	0	17	48	172	375	546	512	293	75	0	0	2038

OKLAHOMA

COOLING DEGREE DAY NORMALS (BASE 65 DEG F)

STATION	JAN	FEB	MAR	APR	MAY	JUN	JUL	AUG	SEP	OCT	NOV	DEC	ANN
MUTUAL	0	0	9	26	145	372	546	496	237	48	0	0	1879
NEWKIRK	0	0	11	39	151	375	543	499	256	61	0	0	1935
NOWATA	0	0	10	41	145	349	530	490	258	58	0	0	1881
OKEENE	0	0	14	48	176	420	586	543	296	73	0	0	2156
OKEMAH	0	0	19	40	154	363	530	502	290	79	0	0	1977
OKLAHOMA CITY WSFO //R	0	0	13	40	147	360	530	499	264	61	0	0	1914
OKMULGEE WATER WORKS	0	0	23	50	163	369	518	490	271	79	0	0	1963
PAULS VALLEY	0	0	21	67	207	435	592	561	327	83	0	0	2293
PAWHUSKA	0	0	12	43	147	348	521	481	252	53	0	0	1857
PERRY	0	0	15	52	172	393	564	530	291	79	0	0	2096
PONCA CITY FAA AIRPORT	0	0	7	31	148	366	543	493	247	51	0	0	1886
POTEAU	0	0	23	47	177	381	549	512	298	82	0	0	2069
PRYOR	0	0	15	46	151	347	515	474	252	55	0	0	1855
PURCELL	0	0	23	55	175	390	552	524	288	70	0	0	2077
SALLISAW	0	0	18	46	170	372	530	496	286	76	5	0	1999
SEMINOLE	0	0	21	78	191	405	580	546	316	88	5	0	2230
STILLWATER 2 W	0	0	12	45	154	360	530	496	261	62	0	0	1920
TAHLEQUAH 1 SE	0	0	15	46	155	333	487	462	261	73	5	0	1837
TALOGA	0	0	10	30	146	375	533	487	247	44	0	0	1872
TIPTON 4 S	0	0	21	69	223	471	611	555	318	84	0	0	2352
TULSA WSO //R	0	0	14	45	167	381	564	518	282	72	0	0	2043
VINITA 3 NNE	0	0	10	30	143	337	499	459	246	58	0	0	1782
WAGONER	0	0	16	48	161	366	539	499	284	82	0	0	1995
WALTERS	0	0	24	82	219	450	605	580	349	110	6	0	2425
WAURIKA	0	5	30	89	227	453	611	583	351	109	7	0	2465
WAYNOKA	0	0	16	36	175	405	574	530	268	71	0	0	2075
WEATHERFORD	0	0	14	42	167	399	558	515	276	63	0	0	2034
WEBBERS FALLS	0	0	11	39	160	363	530	487	270	59	0	0	1919
WICHITA MT WL REF	0	0	20	52	160	381	546	512	274	68	0	0	2013
WILBURTON 9 ENE	0	0	18	53	168	360	527	493	277	73	0	0	1969
WISTER DAM	0	0	17	63	196	387	539	496	288	80	8	0	2074
WOODWARD FIELD STATION	0	0	9	25	142	355	527	474	228	46	0	0	1806

34 - OKLAHOMA

STATE-STATION NUMBER	STN TYP	NAME	LATITUDE DEG-MIN	LONGITUDE DEG-MIN	ELEVATION (FT)
34-0017	13	ADA	N 3447	W 09641	1015
34-0179	13	ALTUS IRR. RESCH STN	N 3435	W 09920	1380
34-0184	12	ALTUS DAM	N 3453	W 09918	1525
34-0193	13	ALVA	N 3648	W 09839	1374
34-0224	13	ANADARKO	N 3504	W 09814	1180
34-0256	13	ANTLERS 2 ENE	N 3415	W 09536	465
34-0292	13	ARDMORE	N 3410	W 09708	880
34-0332	13	ARNETT	N 3608	W 09946	2455
34-0535	12	BARNSDALL	N 3633	W 09610	740
34-0548	13	BARTLESVILLE 2 W	N 3645	W 09600	715
34-0584	12	BEAR MOUNTAIN TOWER	N 3408	W 09457	800
34-0593	13	BEAVER 1 SW	N 3648	W 10031	2500
34-0755	12	BILLINGS	N 3632	W 09727	1000
34-0782	13	BIXBY 2 NE	N 3558	W 09552	600
34-0908	13	BOISE CITY 2 E	N 3644	W 10229	4145
34-0980	12	BOSWELL 5 NNW	N 3405	W 09554	529
34-1144	13	BRISTOW	N 3550	W 09624	825
34-1162	12	BROKEN BOW 1 N	N 3403	W 09444	475
34-1243	13	BUFFALO	N 3650	W 09937	1795
34-1391	12	CALVIN	N 3458	W 09615	730
34-1445	13	CANTON DAM	N 3605	W 09836	1650
34-1499	12	CARNASAW TOWER	N 3409	W 09438	1000
34-1504	13	CARNEGIE 4 ENE	N 3508	W 09833	1340
34-1544	12	CARTER TOWER	N 3415	W 09447	1300
34-1684	13	CHANDLER NO 1	N 3542	W 09653	960
34-1706	13	CHATTANOOGA 3 NE	N 3427	W 09837	1154
34-1711	12	CHECOTAH	N 3528	W 09531	635
34-1724	13	CHEROKEE POWER PLANT	N 3646	W 09821	1180
34-1750	13	CHICKASHA EXP STATION	N 3503	W 09755	1085
34-1828	13	CLAREMORE 2 ENE	N 3619	W 09535	588
34-1900	13	CLEVELAND	N 3619	W 09628	815
34-1909	13	CLINTON	N 3531	W 09858	1610
34-1954	12	COALGATE 2 S	N 3431	W 09613	610
34-2125	12	CORDELL	N 3517	W 09859	1540
34-2318	13	CUSHING	N 3559	W 09646	950
34-2354	12	DAISY 2 E	N 3433	W 09543	740
34-2485	12	DEWAR 2 NE	N 3529	W 09554	615
34-2660	13	DUNCAN	N 3430	W 09757	1125
34-2678	12	DURANT SE STATE COL	N 3401	W 09623	675
34-2818	13	EL RENO 1 N	N 3533	W 09758	1315
34-2849	12	ELK CITY	N 3525	W 09925	1950
34-2912	13	ENID	N 3625	W 09752	1245
34-2944	13	ERICK 4 E	N 3512	W 09948	1985
34-2993	12	EUFAULA	N 3517	W 09535	635
34-3065	12	FANSHAWE	N 3457	W 09454	545
34-3070	12	FARGO	N 3623	W 09937	2110
34-3182	12	FLASHMAN TOWER	N 3429	W 09500	1749
34-3250	12	FORAKER	N 3652	W 09634	1275
34-3304	13	FORT SUPPLY DAM	N 3633	W 09935	2075
34-3353	13	FREDERICK	N 3424	W 09901	1300

STATE-STATION NUMBER	STN TYP	NAME	LATITUDE DEG-MIN	LONGITUDE DEG-MIN	ELEVATION (FT)
34-3407	13	GAGE FAA AIRPORT	N 3618	W 09946	2191
34-3497	13	GEARY	N 3537	W 09819	1575
34-3628	13	GOODWELL	N 3636	W 10139	3297
34-3700	13	GRAND RIVER DAM	N 3628	W 09503	766
34-3709	12	GRANDFIELD	N 3414	W 09841	1135
34-3740	12	GREAT SALT PLAINS DAM	N 3645	W 09808	1195
34-3821	13	GUTHRIE 2 WNW	N 3553	W 09727	980
34-3871	13	HAMMON	N 3538	W 09923	1780
34-3884	12	HANNA	N 3512	W 09553	680
34-3956	12	HASKELL 1 NW	N 3550	W 09541	600
34-4001	12	HEALDTON 2 N	N 3416	W 09730	896
34-4008	12	HEAVENER 1 SE	N 3453	W 09435	608
34-4019	12	HELENA 1 SSE	N 3632	W 09816	1350
34-4055	13	HENNESSEY 1 N	N 3607	W 09754	1170
34-4204	13	HOBART FAA AIRPORT	N 3500	W 09903	1552
34-4235	13	HOLDENVILLE	N 3505	W 09624	915
34-4249	13	HOLLIS	N 3441	W 09955	1615
34-4258	12	HOLLOW 1 NE	N 3654	W 09515	920
34-4289	12	HOMINY 4 N	N 3629	W 09623	770
34-4298	13	HOOKER 1 N	N 3653	W 10113	3015
34-4384	13	HUGO	N 3401	W 09530	570
34-4393	13	HULAH DAM	N 3655	W 09606	744
34-4451	13	IDABEL	N 3353	W 09449	460
34-4564	13	JAY	N 3625	W 09448	1040
34-4573	13	JEFFERSON	N 3643	W 09748	1070
34-4766	13	KENTON	N 3654	W 10258	4350
34-4861	13	KINGFISHER	N 3551	W 09756	1060
34-4865	12	KINGSTON	N 3400	W 09643	810
34-4915	12	KONAWA	N 3458	W 09645	962
34-5045	12	LAVERNE	N 3642	W 09954	2100
34-5063	13	LAWTON	N 3437	W 09827	1150
34-5090	12	LEEDEY	N 3552	W 09921	2055
34-5216	12	LINDSAY	N 3449	W 09736	978
34-5437	12	LYONS 2 N	N 3546	W 09444	1025
34-5468	13	MADILL	N 3405	W 09646	770
34-5509	13	MANGUM RESEARCH STA	N 3450	W 09926	1520
34-5522	12	MANNFORD 6 NW	N 3610	W 09626	830
34-5540	12	MARAMEC	N 3615	W 09641	955
34-5563	13	MARIETTA 3 NW	N 3359	W 09707	865
34-5581	12	MARLOW 1 WSW	N 3439	W 09759	1250
34-5589	12	MARSHALL	N 3609	W 09738	1045
34-5664	13	MCALESTER FAA AIRPORT	N 3453	W 09547	760
34-5693	12	MC CURTAIN 1 SE	N 3509	W 09458	655
34-5779	13	MEEKER 1 E	N 3530	W 09653	945
34-5855	13	MIAMI	N 3653	W 09452	795
34-6035	12	MORAVIA 2 NNE	N 3508	W 09930	1690
34-6130	13	MUSKOGEE	N 3545	W 09520	595
34-6139	13	MUTUAL	N 3614	W 09910	1865
34-6278	13	NEWKIRK	N 3653	W 09704	1135
34-6386	12	NORMAN 3 S	N 3511	W 09727	1109

STATE-STATION NUMBER	STN TYP	NAME	LATITUDE DEG-MIN	LONGITUDE DEG-MIN	ELEVATION (FT)
34-6485	13	NOWATA	N 3642	W 09538	720
34-6629	13	OKEENE	N 3607	W 09819	1200
34-6638	13	OKEMAH	N 3526	W 09618	935
34-6661	13	OKLAHOMA CITY WSFO //R	N 3524	W 09736	1285
34-6670	13	OKMULGEE WATER WORKS	N 3537	W 09601	645
34-6926	13	PAULS VALLEY	N 3445	W 09713	875
34-6935	13	PAWHUSKA	N 3640	W 09621	835
34-6940	12	PAWNEE	N 3620	W 09648	840
34-7003	12	PERKINS 1 SSE	N 3558	W 09702	840
34-7012	13	PERRY	N 3617	W 09718	1025
34-7201	13	PONCA CITY FAA AIRPORT	N 3644	W 09706	999
34-7214	12	PONTOTOC	N 3429	W 09637	1015
34-7246	13	POTEAU	N 3504	W 09438	572
34-7264	12	PRAGUE	N 3529	W 09642	985
34-7309	13	PRYOR	N 3619	W 09519	640
34-7327	13	PURCELL	N 3500	W 09722	1038
34-7358	12	QUAPAW	N 3658	W 09447	850
34-7372	12	QUINTON	N 3508	W 09522	654
34-7390	12	RALSTON	N 3630	W 09644	825
34-7505	12	REDROCK	N 3627	W 09711	905
34-7534	12	REGNIER	N 3656	W 10238	4020
34-7556	12	RENFROW	N 3656	W 09739	1214
34-7579	12	REYDON 4 W	N 3540	W 10000	2475
34-7727	12	ROOSEVELT	N 3451	W 09901	1464
34-7862	13	SALLISAW	N 3528	W 09447	531
34-7952	12	SAYRE 1 NE	N 3518	W 09937	1900
34-8042	13	SEMINOLE	N 3514	W 09640	865
34-8110	12	SHAWNEE	N 3519	W 09656	1025
34-8258	12	SKIATOOK	N 3622	W 09602	700
34-8299	12	SNYDER	N 3439	W 09857	1350
34-8305	12	SOBOL TOWER	N 3408	W 09514	750
34-8380	12	SPAVINAW	N 3623	W 09503	685
34-8416	12	SPIRO	N 3515	W 09437	490
34-8501	13	STILLWATER 2 W	N 3607	W 09705	895
34-8506	12	STILWELL 1 NE	N 3550	W 09437	1080
34-8677	13	TAHLEQUAH 1 SE	N 3554	W 09456	846
34-8708	13	TALOGA	N 3602	W 09858	1705
34-8879	13	TIPTON 4 S	N 3426	W 09908	1355
34-8884	12	TISHOMINGO NATIONAL WL	N 3411	W 09638	642
34-8992	13	TULSA WSO //R	N 3611	W 09554	668
34-9086	12	UNION CITY 4 ESE	N 3522	W 09753	1250
34-9118	12	VALLIANT	N 3400	W 09505	513
34-9203	13	VINITA 3 NNE	N 3641	W 09508	737
34-9212	12	VINSON 3 WNW	N 3455	W 09955	1945
34-9247	13	WAGONER	N 3558	W 09522	590
34-9278	13	WALTERS	N 3421	W 09819	995
34-9364	12	WATONGA	N 3551	W 09825	1550
34-9395	13	WAURIKA	N 3410	W 09800	875
34-9404	13	WAYNOKA	N 3635	W 09852	1510
34-9422	13	WEATHERFORD	N 3532	W 09842	1650

34 — OKLAHOMA

STATE-STATION NUMBER	STN TYP	NAME	LATITUDE DEG-MIN	LONGITUDE DEG-MIN	ELEVATION (FT)
34-9445	13	WEBBERS FALLS	N 3531	W 09508	480
34-9571	12	WETUMKA 3 NE	N 3516	W 09613	703
34-9575	12	WEWOKA 3 W	N 3510	W 09633	814
34-9629	13	WICHITA MT WL REF	N 3444	W 09843	1665
34-9634	13	WILBURTON 9 ENE	N 3456	W 09509	635
34-9724	13	WISTER DAM	N 3456	W 09443	498
34-9760	12	WOODWARD	N 3627	W 09923	1900
34-9762	13	WOODWARD FIELD STATION	N 3625	W 09924	1987
34-9985	12	ZOE 1 E	N 3446	W 09436	620

34 — OKLAHOMA

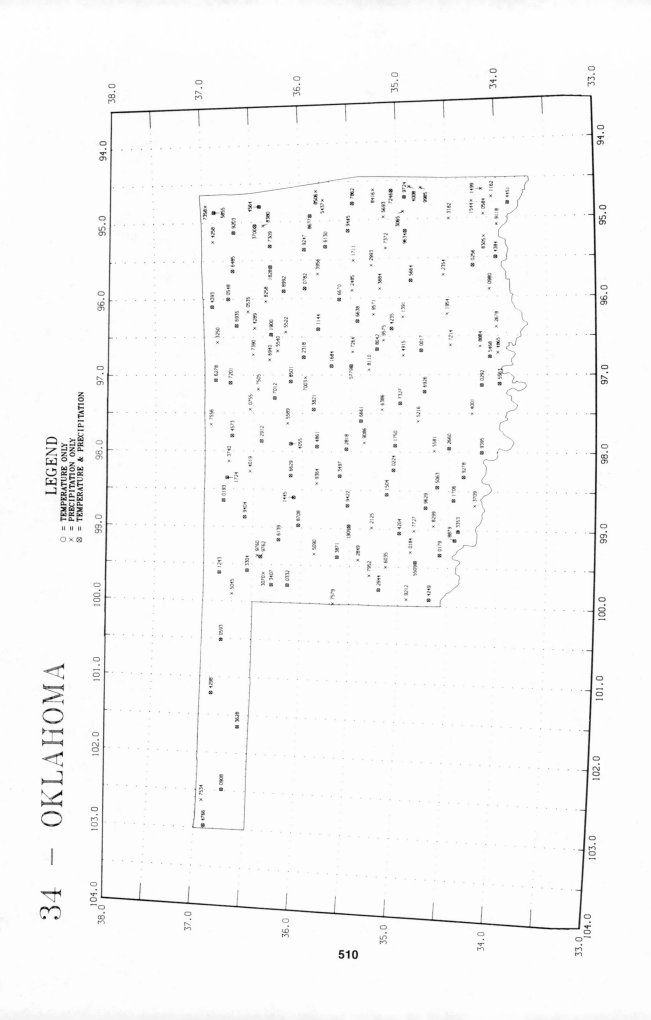

LEGEND

O = TEMPERATURE ONLY
X = PRECIPITATION ONLY
⊗ = TEMPERATURE & PRECIPITATION

OREGON

TEMPERATURE NORMALS (DEG F)

STATION		JAN	FEB	MAR	APR	MAY	JUN	JUL	AUG	SEP	OCT	NOV	DEC	ANN
ANTELOPE 1 N	MAX	40.2	46.6	51.4	58.8	67.8	76.3	86.0	83.9	76.5	64.2	49.5	42.6	62.0
	MIN	23.7	28.1	29.2	32.5	38.7	45.2	50.4	49.9	44.2	36.7	30.1	26.4	36.3
	MEAN	32.0	37.4	40.3	45.7	53.3	60.8	68.3	67.0	60.4	50.5	39.8	34.5	49.2
ARLINGTON	MAX	40.3	48.8	56.7	65.7	75.6	83.3	91.8	89.5	81.0	66.3	50.2	43.0	66.0
	MIN	27.1	32.1	35.1	40.4	47.6	55.0	60.7	59.8	51.5	41.8	33.8	30.4	42.9
	MEAN	33.7	40.5	45.9	53.1	61.6	69.1	76.3	74.7	66.3	54.1	42.0	36.7	54.5
ASHLAND 1 N	MAX	45.3	51.5	55.2	61.6	69.6	78.0	87.1	84.9	78.7	65.7	52.3	45.0	64.6
	MIN	29.7	32.0	33.1	35.7	41.5	47.6	51.8	51.1	46.1	39.3	34.1	30.3	39.4
	MEAN	37.5	41.8	44.2	48.7	55.6	62.8	69.5	68.0	62.4	52.5	43.2	37.7	52.0
ASTORIA WSO //R	MAX	46.8	50.6	51.9	55.5	60.2	63.9	67.9	68.6	67.8	61.4	53.5	48.8	58.1
	MIN	35.4	37.1	36.9	39.7	44.1	49.2	52.2	52.6	49.2	44.3	39.7	37.3	43.1
	MEAN	41.1	43.9	44.4	47.6	52.2	56.6	60.1	60.6	58.5	52.9	46.6	43.1	50.6
BAKER FAA AIRPORT	MAX	33.7	41.0	48.3	58.1	67.0	74.8	85.3	83.4	75.2	62.2	45.9	36.3	59.3
	MIN	16.7	22.4	25.7	30.5	37.9	44.2	47.9	45.8	38.3	30.0	24.3	18.9	31.9
	MEAN	25.2	31.7	37.0	44.3	52.5	59.5	66.6	64.6	56.8	46.1	35.1	27.6	45.6
BAKER KBKR	MAX	35.3	42.5	49.6	59.4	68.4	76.5	86.1	84.3	76.5	63.1	46.6	37.5	60.5
	MIN	18.4	23.6	26.4	31.4	38.5	45.1	50.4	48.8	41.6	33.9	26.8	21.0	33.8
	MEAN	26.9	33.1	38.0	45.4	53.5	60.8	68.3	66.6	59.1	48.5	36.8	29.3	47.2
BANDON 1 E-BATES BOG	MAX	52.5	54.5	54.4	56.3	59.6	62.9	64.6	65.8	65.9	62.6	57.3	53.6	59.2
	MIN	38.2	39.6	39.2	40.6	44.0	48.0	50.0	50.0	48.0	44.8	41.9	39.5	43.7
	MEAN	45.4	47.1	46.9	48.5	51.8	55.4	57.3	57.9	56.9	53.7	49.6	46.6	51.4
BEND	MAX	40.5	46.2	50.0	57.1	65.0	72.8	82.1	80.2	73.8	63.2	49.4	43.0	60.3
	MIN	21.0	24.2	24.4	28.0	34.2	40.2	44.3	43.2	37.0	30.6	26.0	23.1	31.4
	MEAN	30.8	35.2	37.2	42.6	49.6	56.6	63.2	61.7	55.4	46.9	37.7	33.1	45.8
BEULAH	MAX	36.6	44.4	51.9	61.8	71.5	80.4	90.9	88.6	80.3	67.1	49.6	39.2	63.5
	MIN	17.4	22.9	26.6	32.5	40.9	47.8	54.1	51.9	42.7	32.6	24.9	20.1	34.5
	MEAN	27.0	33.7	39.3	47.2	56.3	64.1	72.5	70.3	61.5	49.9	37.3	29.7	49.1
BONNEVILLE DAM	MAX	41.8	47.2	52.2	58.8	65.8	71.5	78.1	77.8	73.2	62.6	50.7	44.6	60.4
	MIN	31.8	35.2	36.7	41.1	46.5	52.1	56.2	56.0	52.7	46.3	39.3	34.8	44.1
	MEAN	36.8	41.3	44.5	50.0	56.2	61.8	67.2	66.9	62.9	54.5	45.0	39.7	52.2
BROOKINGS	MAX	53.8	55.7	56.2	58.8	62.5	65.8	66.4	66.8	68.1	64.1	58.5	54.9	61.0
	MIN	40.5	41.5	41.1	42.2	45.6	48.8	50.6	51.2	50.8	47.7	44.4	41.4	45.5
	MEAN	47.2	48.6	48.7	50.5	54.1	57.3	58.5	59.0	59.5	55.9	51.5	48.2	53.3
BURNS WSO AP //R	MAX	37.3	44.0	47.1	55.9	65.2	74.0	84.4	81.9	74.2	62.1	46.6	36.8	59.1
	MIN	17.7	23.0	25.8	30.2	38.9	46.9	54.4	52.5	43.6	33.7	24.6	18.2	34.1
	MEAN	27.5	33.5	36.5	43.1	52.1	60.5	69.4	67.2	58.9	47.9	35.6	27.5	46.6
CAPE BLANCO	MAX	50.7	51.8	51.2	52.7	54.7	57.1	58.2	59.5	59.4	57.6	54.8	52.2	55.0
	MIN	41.1	42.6	41.6	43.1	45.7	48.3	49.4	50.2	49.9	48.1	45.2	42.8	45.7
	MEAN	45.9	47.3	46.4	47.9	50.2	52.7	53.9	54.9	54.7	52.9	50.0	47.5	50.4
CHEMULT	MAX	36.5	41.9	45.3	53.6	63.9	72.7	82.2	79.5	73.4	61.0	44.8	37.6	57.7
	MIN	15.1	18.7	20.5	24.1	29.8	35.2	38.6	36.8	31.9	26.6	22.5	17.2	26.4
	MEAN	25.8	30.3	32.9	38.8	46.9	54.0	60.4	58.1	52.6	43.8	33.7	27.4	42.1
CHERRY GROVE 2 S	MAX	42.3	47.7	50.9	57.0	64.1	69.9	78.0	78.1	73.2	62.1	50.3	44.3	59.8
	MIN	31.3	34.4	34.9	38.0	42.6	47.2	50.6	51.0	49.4	43.6	37.1	33.4	41.1
	MEAN	36.9	41.1	43.0	47.5	53.3	58.6	64.3	64.6	61.3	52.9	43.7	38.9	50.5
CLATSKANIE 3 W	MAX	43.1	48.9	52.4	57.8	63.8	67.8	73.0	73.0	70.6	60.9	50.1	44.6	58.8
	MIN	33.5	35.7	36.4	39.4	44.4	49.7	53.1	53.8	50.5	44.5	38.5	35.6	42.9
	MEAN	38.4	42.3	44.4	48.6	54.1	58.7	63.1	63.4	60.6	52.7	44.3	40.1	50.9
CLOVERDALE 1 NW	MAX	48.9	52.8	53.7	57.1	62.3	66.2	70.4	70.8	70.3	64.0	55.6	50.6	60.2
	MIN	36.8	38.6	37.9	39.6	43.3	47.5	49.5	50.1	48.8	45.4	41.1	38.6	43.1
	MEAN	42.9	45.8	45.9	48.4	52.8	56.8	60.0	60.5	59.6	54.7	48.4	44.6	51.7

OREGON

TEMPERATURE NORMALS (DEG F)

STATION		JAN	FEB	MAR	APR	MAY	JUN	JUL	AUG	SEP	OCT	NOV	DEC	ANN
CONDON	MAX	37.8	44.0	49.4	56.5	65.2	73.9	83.4	81.4	73.5	61.5	46.6	40.8	59.5
	MIN	22.5	27.2	28.8	32.5	38.8	44.7	49.7	49.0	43.7	36.0	29.3	25.4	35.6
	MEAN	30.2	35.6	39.1	44.5	52.0	59.3	66.6	65.2	58.6	48.8	38.0	33.1	47.6
CORVALLIS ST COLLEGE	MAX	45.1	50.5	53.7	59.3	66.2	72.6	80.7	80.5	75.5	64.3	52.3	46.5	62.3
	MIN	32.9	35.0	36.1	38.8	43.2	48.3	50.6	50.7	47.7	41.7	37.2	34.6	41.4
	MEAN	39.0	42.8	44.9	49.1	54.7	60.5	65.6	65.6	61.6	53.0	44.7	40.6	51.8
COTTAGE GROVE 1 S	MAX	47.6	52.9	55.7	61.0	67.6	73.7	81.7	81.1	76.4	65.4	54.0	48.6	63.8
	MIN	32.5	34.5	34.8	37.0	41.2	45.9	47.6	47.6	44.5	41.0	37.1	34.2	39.8
	MEAN	40.1	43.7	45.3	49.0	54.4	59.8	64.7	64.4	60.5	53.2	45.6	41.5	51.9
COTTAGE GROVE DAM	MAX	46.2	51.2	53.7	59.0	65.3	71.7	79.8	79.6	75.0	64.2	52.7	47.6	62.2
	MIN	32.4	34.3	35.1	37.9	42.8	48.1	50.6	50.7	47.1	41.6	36.8	34.2	41.0
	MEAN	39.3	42.8	44.4	48.5	54.1	59.9	65.2	65.2	61.1	53.0	44.8	40.9	51.6
CRATER LAKE NP HQ	MAX	32.8	35.1	35.9	41.6	49.6	57.8	69.1	68.1	62.5	51.7	40.0	34.9	48.3
	MIN	17.0	18.3	17.6	21.3	27.7	33.9	40.9	40.4	36.6	30.5	23.6	19.3	27.3
	MEAN	24.9	26.7	26.8	31.5	38.7	45.9	55.0	54.3	49.5	41.1	31.8	27.1	37.8
DALLAS	MAX	45.6	51.2	54.8	61.2	68.9	75.0	83.0	82.1	77.5	65.7	52.8	46.9	63.7
	MIN	32.2	34.4	35.0	37.2	41.5	46.2	48.2	48.0	46.1	40.9	36.2	34.0	40.0
	MEAN	38.9	42.8	44.9	49.2	55.2	60.6	65.6	65.1	61.8	53.3	44.5	40.5	51.9
DORENA DAM	MAX	46.7	51.6	53.9	58.9	65.1	71.2	79.3	78.9	74.6	64.2	53.0	48.0	62.1
	MIN	31.8	33.8	34.4	37.4	41.9	47.2	49.5	49.4	46.0	41.0	36.6	33.7	40.2
	MEAN	39.3	42.7	44.2	48.2	53.5	59.2	64.4	64.2	60.3	52.6	44.8	40.8	51.2
DRAIN	MAX	47.8	53.3	56.7	62.7	69.8	76.1	84.1	83.3	78.5	66.7	54.1	48.4	65.1
	MIN	33.7	35.5	35.8	38.2	42.3	47.3	49.1	49.0	45.8	42.2	38.4	35.7	41.1
	MEAN	40.7	44.4	46.3	50.5	56.1	61.7	66.6	66.2	62.2	54.5	46.3	42.0	53.1
DUFUR	MAX	40.2	47.7	54.3	62.0	70.5	77.7	85.9	84.2	77.5	64.8	48.9	42.5	63.0
	MIN	23.1	28.0	29.7	33.3	38.8	44.4	48.1	47.6	43.1	35.6	29.9	26.0	35.6
	MEAN	31.7	37.9	42.0	47.6	54.7	61.1	67.0	65.9	60.3	50.2	39.4	34.3	49.3
ELGIN	MAX	37.5	44.3	50.5	59.9	69.0	77.2	88.4	86.5	78.6	65.0	48.2	40.1	62.1
	MIN	21.5	25.6	27.3	31.7	37.3	43.5	45.7	44.1	37.8	31.9	27.6	24.2	33.2
	MEAN	29.5	35.0	38.9	45.8	53.2	60.4	67.1	65.3	58.3	48.5	38.0	32.2	47.7
ELKTON 3 SW	MAX	48.8	53.8	57.7	63.3	70.2	76.6	84.0	83.7	79.3	67.4	54.6	49.3	65.7
	MIN	35.4	37.5	37.6	39.8	43.9	48.5	50.7	51.0	48.2	44.0	40.0	37.1	42.8
	MEAN	42.1	45.7	47.7	51.6	57.0	62.5	67.3	67.4	63.8	55.7	47.3	43.2	54.3
ENTERPRISE	MAX	34.1	40.7	46.6	55.6	65.2	72.5	82.9	81.1	73.1	61.1	45.0	36.1	57.8
	MIN	14.4	19.4	22.7	28.5	34.8	40.2	43.1	41.3	35.4	28.7	21.7	17.2	29.0
	MEAN	24.2	30.1	34.7	42.1	50.0	56.4	63.0	61.2	54.3	44.9	33.4	26.7	43.4
ESTACADA 2 SE	MAX	44.6	49.6	53.6	60.0	67.0	72.7	79.9	78.6	73.5	61.3	51.3	46.4	61.5
	MIN	33.1	35.6	36.2	39.1	43.5	48.2	50.9	50.9	47.8	43.0	38.0	34.9	41.8
	MEAN	38.9	42.6	44.9	49.6	55.3	60.5	65.5	64.8	60.7	52.2	44.6	40.6	51.7
EUGENE WSO R	MAX	46.3	51.4	55.0	60.5	67.2	74.2	82.6	81.3	76.4	64.6	52.8	47.3	63.3
	MIN	33.8	35.5	36.5	38.7	42.9	48.0	51.0	51.1	47.7	42.0	37.8	35.3	41.7
	MEAN	40.1	43.5	45.8	49.6	55.1	61.1	66.8	66.2	62.1	53.3	45.3	41.3	52.5
FERN RIDGE DAM	MAX	45.7	50.5	53.8	59.2	65.8	72.3	80.4	79.7	75.2	64.3	52.6	47.2	62.2
	MIN	32.9	34.8	35.8	38.7	43.4	48.7	51.6	52.0	48.9	43.0	37.6	34.8	41.9
	MEAN	39.4	42.7	44.8	49.0	54.6	60.5	66.0	65.9	62.1	53.7	45.1	41.0	52.1
FOREST GROVE	MAX	45.0	51.0	55.0	61.6	69.1	75.0	82.2	81.8	77.1	65.6	52.9	46.8	63.6
	MIN	32.4	34.8	35.9	38.7	43.6	48.8	51.8	51.2	47.7	41.6	36.9	34.5	41.5
	MEAN	38.7	43.0	45.5	50.2	56.4	61.9	67.1	66.6	62.4	53.7	44.9	40.7	52.6
FOSSIL	MAX	41.0	47.0	51.2	58.7	67.1	75.3	84.9	83.2	75.9	65.1	49.9	43.5	61.9
	MIN	22.8	27.0	27.5	30.4	35.1	41.3	44.5	44.5	39.7	33.5	28.8	26.1	33.4
	MEAN	31.9	37.0	39.4	44.5	51.1	58.3	64.7	63.9	57.8	49.3	39.4	34.8	47.7

OREGON

TEMPERATURE NORMALS (DEG F)

STATION		JAN	FEB	MAR	APR	MAY	JUN	JUL	AUG	SEP	OCT	NOV	DEC	ANN
GOLD BEACH RANGER STA	MAX	53.8	55.4	55.3	57.5	60.9	64.7	67.3	67.8	67.5	63.8	58.2	54.8	60.6
	MIN	40.2	41.2	40.4	42.0	45.4	49.0	50.8	51.4	50.7	47.2	43.8	41.2	45.3
	MEAN	47.0	48.3	47.9	49.8	53.2	56.9	59.0	59.6	59.1	55.5	51.0	48.0	52.9
GRANTS PASS	MAX	47.0	54.2	59.7	66.7	74.6	82.2	90.4	88.8	83.3	69.2	53.6	46.5	68.0
	MIN	32.8	34.3	35.2	37.9	43.4	49.1	52.7	51.8	46.6	40.9	37.4	34.0	41.3
	MEAN	39.9	44.3	47.5	52.3	59.1	65.7	71.6	70.3	65.0	55.1	45.5	40.3	54.7
GRIZZLY	MAX	39.7	44.9	48.6	55.7	63.7	72.5	82.2	79.9	72.9	61.7	48.3	42.4	59.4
	MIN	21.4	25.2	25.0	27.6	33.4	39.4	43.2	42.6	38.1	32.3	27.4	23.9	31.6
	MEAN	30.6	35.0	36.8	41.7	48.6	56.0	62.7	61.3	55.5	47.0	37.9	33.1	45.5
HALFWAY	MAX	33.3	40.7	49.9	61.3	70.8	78.8	89.0	86.9	78.4	64.7	47.0	35.7	61.4
	MIN	14.3	19.1	24.3	30.3	36.8	42.7	47.2	45.3	38.3	30.5	24.4	17.8	30.9
	MEAN	23.8	29.9	37.1	45.8	53.8	60.8	68.1	66.1	58.4	47.6	35.7	26.7	46.2
HART MOUNTAIN REFUGE	MAX	38.0	41.6	44.1	52.5	61.9	70.7	81.5	79.4	72.2	61.0	47.3	40.4	57.6
	MIN	17.6	21.5	21.6	25.7	32.3	38.7	44.1	43.1	37.6	30.7	24.2	19.5	29.7
	MEAN	27.8	31.6	32.8	39.1	47.1	54.7	62.8	61.3	55.0	45.9	35.8	30.0	43.7
HEADWORKS PTLD WATER	MAX	43.4	48.6	51.9	58.5	65.9	71.1	78.3	77.0	72.9	62.7	51.4	45.5	60.6
	MIN	32.8	35.3	35.6	38.6	43.4	48.4	51.4	51.9	49.2	44.1	38.1	34.9	42.0
	MEAN	38.1	42.0	43.8	48.6	54.7	59.8	64.9	64.5	61.1	53.4	44.8	40.2	51.3
HEPPNER	MAX	41.0	47.7	52.8	59.7	68.4	77.0	85.6	83.6	75.9	64.0	50.0	44.0	62.5
	MIN	25.1	30.0	31.9	35.6	41.7	47.9	52.3	52.0	46.0	38.5	31.6	27.9	38.4
	MEAN	33.1	38.9	42.4	47.7	55.1	62.5	69.0	67.8	61.0	51.3	40.9	36.0	50.5
HERMISTON 2 S	MAX	39.9	48.4	56.6	64.8	73.8	81.7	89.8	87.5	79.5	66.3	50.2	42.7	65.1
	MIN	24.1	29.5	32.7	38.4	45.8	53.0	57.8	56.3	48.0	37.9	31.1	27.2	40.2
	MEAN	32.0	38.9	44.7	51.6	59.8	67.4	73.8	71.9	63.7	52.1	40.6	35.0	52.6
HOOD RIVER EXP STATION	MAX	39.8	46.8	52.7	59.9	67.9	73.9	80.5	79.8	74.4	63.5	49.1	42.6	60.9
	MIN	27.3	31.6	33.5	38.1	43.6	49.7	52.9	52.4	46.1	38.7	33.6	30.5	39.8
	MEAN	33.6	39.2	43.1	49.0	55.8	61.8	66.7	66.1	60.3	51.2	41.4	36.6	50.4
HUNTINGTON	MAX	37.1	45.0	54.4	64.8	74.9	84.2	95.3	92.2	81.8	66.9	49.5	39.3	65.5
	MIN	21.0	26.3	31.0	37.9	47.0	55.2	64.0	61.5	50.5	38.8	29.3	23.1	40.5
	MEAN	29.1	35.7	42.8	51.4	61.0	69.7	79.6	76.8	66.2	52.9	39.4	31.2	53.0
KENT	MAX	37.2	43.5	48.8	56.2	64.9	73.8	83.5	81.7	73.9	62.0	46.7	40.4	59.4
	MIN	22.4	27.3	29.2	32.9	39.2	46.1	52.4	51.7	45.3	37.6	29.4	25.2	36.6
	MEAN	29.8	35.4	39.0	44.6	52.1	59.9	68.0	66.7	59.6	49.8	38.1	32.8	48.0
KLAMATH FALLS 2 SSW	MAX	38.7	44.8	49.6	57.7	67.0	75.3	84.7	82.5	76.2	63.8	48.2	39.8	60.7
	MIN	20.9	25.3	27.3	31.1	38.4	45.1	51.2	49.5	43.2	35.0	27.9	22.4	34.8
	MEAN	29.8	35.1	38.5	44.4	52.7	60.2	68.0	66.0	59.7	49.4	38.1	31.1	47.8
LA GRANDE	MAX	38.0	44.3	50.4	59.3	68.6	77.0	87.4	85.5	77.2	63.8	48.0	40.6	61.7
	MIN	23.7	27.9	29.8	34.8	41.5	48.1	53.3	51.9	44.6	36.6	30.8	26.5	37.5
	MEAN	30.9	36.1	40.1	47.0	55.1	62.6	70.4	68.7	60.9	50.2	39.4	33.6	49.6
LAKEVIEW	MAX	37.1	42.0	46.4	55.1	64.5	73.3	83.5	81.1	74.3	62.4	47.6	40.1	59.0
	MIN	19.6	23.8	25.7	30.8	37.9	44.6	50.5	47.8	41.6	33.7	26.9	21.6	33.7
	MEAN	28.4	32.9	36.1	43.0	51.2	59.0	67.0	64.5	58.0	48.1	37.3	30.9	46.4
LEABURG 1 SW	MAX	46.3	51.7	55.0	60.9	67.7	74.0	82.6	81.8	76.3	64.7	53.0	47.3	63.4
	MIN	33.4	35.5	35.8	38.9	43.3	48.2	50.5	50.3	47.8	42.9	38.0	35.1	41.6
	MEAN	39.8	43.6	45.4	49.9	55.5	61.1	66.6	66.1	62.1	53.8	45.5	41.2	52.6
MADRAS	MAX	42.1	49.0	54.3	61.5	69.9	78.0	87.1	85.5	78.1	65.3	50.6	44.4	63.8
	MIN	22.2	26.3	26.7	30.0	36.4	42.4	45.5	44.4	38.8	31.6	27.7	24.6	33.1
	MEAN	32.2	37.7	40.5	45.8	53.2	60.2	66.3	64.9	58.5	48.5	39.2	34.5	48.5
MALHEUR BRANCH EXP STA	MAX	35.0	43.5	54.0	63.9	73.8	82.5	92.2	89.3	79.3	65.0	47.9	37.3	63.6
	MIN	20.6	25.7	30.3	36.8	44.9	52.2	58.1	55.5	46.2	36.4	28.1	23.1	38.2
	MEAN	27.8	34.6	42.2	50.3	59.4	67.4	75.2	72.4	62.8	50.7	38.0	30.3	50.9

OREGON

TEMPERATURE NORMALS (DEG F)

STATION		JAN	FEB	MAR	APR	MAY	JUN	JUL	AUG	SEP	OCT	NOV	DEC	ANN
MALHEUR REFUGE HDQ	MAX	37.3	44.0	49.4	58.5	67.6	76.0	86.0	83.7	76.5	64.0	48.4	39.1	60.9
	MIN	17.7	23.0	25.0	29.6	37.3	43.8	49.0	46.8	38.6	29.9	24.1	19.3	32.0
	MEAN	27.5	33.5	37.2	44.1	52.5	59.9	67.5	65.3	57.5	47.0	36.3	29.3	46.5
MARION FRKS FISH HATCH	MAX	36.9	42.5	46.2	53.4	62.6	71.0	80.5	78.5	72.2	59.9	45.2	38.3	57.3
	MIN	25.4	27.9	28.6	31.9	37.2	43.6	46.3	45.4	40.4	35.1	30.5	27.8	35.0
	MEAN	31.2	35.2	37.4	42.7	49.9	57.3	63.5	62.0	56.3	47.5	37.9	33.1	46.2
MC MINNVILLE	MAX	45.6	50.9	55.1	61.5	68.8	74.8	82.8	81.9	76.8	64.9	52.7	46.9	63.6
	MIN	32.6	35.2	35.5	37.7	41.7	46.1	47.9	48.0	45.6	41.0	36.7	34.3	40.2
	MEAN	39.1	43.1	45.3	49.6	55.3	60.5	65.4	65.0	61.2	53.0	44.7	40.6	51.9
MEDFORD WSO R	MAX	45.0	52.9	57.1	63.8	72.2	81.0	90.7	88.8	82.8	68.7	52.6	44.2	66.7
	MIN	30.2	31.9	33.9	36.8	42.7	49.3	54.2	53.4	47.4	39.6	34.5	31.2	40.4
	MEAN	37.6	42.4	45.5	50.3	57.5	65.2	72.5	71.1	65.1	54.2	43.6	37.7	53.6
METOLIUS 1 W	MAX	40.3	47.1	52.4	59.6	67.8	75.4	84.1	82.4	75.7	63.3	48.8	42.8	61.6
	MIN	22.0	26.1	26.5	29.7	35.9	41.8	45.1	44.0	39.4	32.8	28.0	24.7	33.0
	MEAN	31.2	36.6	39.5	44.7	51.9	58.6	64.6	63.2	57.6	48.0	38.4	33.8	47.3
MIKKALO 6 W	MAX	39.0	46.5	52.7	60.4	69.8	78.2	86.7	85.0	76.7	64.0	48.4	42.3	62.5
	MIN	25.2	30.9	33.3	37.4	44.2	51.1	57.3	56.7	50.5	41.6	32.4	28.3	40.7
	MEAN	32.2	38.7	43.0	48.9	57.0	64.6	72.0	70.9	63.6	52.9	40.4	35.3	51.6
MILTON FREEWATER 4 NW	MAX	41.4	48.6	56.1	64.1	73.0	81.0	89.6	87.1	78.7	65.6	50.6	44.3	65.0
	MIN	27.1	32.7	36.3	41.2	47.7	53.9	59.0	57.5	50.3	41.4	33.9	30.2	42.6
	MEAN	34.3	40.7	46.2	52.7	60.4	67.5	74.3	72.3	64.5	53.5	42.3	37.3	53.8
MORO	MAX	37.1	44.1	50.2	57.5	66.1	74.1	83.1	81.3	74.1	61.7	46.7	40.2	59.7
	MIN	23.0	28.6	30.8	35.1	41.2	48.1	53.4	52.3	45.8	37.0	30.3	26.5	37.7
	MEAN	30.1	36.4	40.5	46.3	53.7	61.1	68.3	66.9	59.9	49.4	38.5	33.4	48.7
NEWPORT	MAX	49.6	52.4	52.6	54.8	58.3	61.9	64.2	64.9	65.2	61.4	55.3	51.0	57.6
	MIN	37.2	38.5	37.8	39.5	43.3	47.6	49.6	49.7	47.9	44.8	41.3	38.6	43.0
	MEAN	43.4	45.5	45.2	47.2	50.8	54.8	56.9	57.3	56.6	53.1	48.3	44.8	50.3
NORTH BEND FAA AP	MAX	51.2	53.5	53.7	55.9	59.8	63.4	65.8	66.8	66.4	62.9	57.0	52.7	59.1
	MIN	38.7	40.4	40.2	42.1	46.5	50.5	52.2	52.6	50.5	46.9	43.0	40.4	45.3
	MEAN	45.0	46.9	47.0	49.0	53.2	57.0	59.0	59.7	58.5	54.9	50.0	46.6	52.2
NYSSA	MAX	35.6	43.8	53.9	63.3	73.0	81.5	91.7	88.9	78.9	65.3	48.7	38.1	63.6
	MIN	21.2	25.8	30.0	37.4	45.6	53.2	58.6	56.1	46.2	36.1	28.4	23.7	38.5
	MEAN	28.4	34.8	42.0	50.4	59.3	67.3	75.2	72.5	62.5	50.7	38.6	30.9	51.1
OCHOCO RANGER STATION	MAX	35.5	41.6	46.5	55.0	63.4	71.9	81.8	80.6	73.5	61.0	44.3	37.2	57.7
	MIN	15.5	19.8	21.5	25.6	31.3	37.1	40.1	39.3	34.0	28.4	23.2	18.5	27.9
	MEAN	25.5	30.7	34.1	40.3	47.4	54.5	61.0	59.9	53.8	44.7	33.8	27.9	42.8
ONTARIO KSRV	MAX	36.7	45.2	55.7	66.1	76.4	85.8	96.4	93.2	82.8	67.6	49.8	38.8	66.2
	MIN	20.5	25.3	29.7	36.0	43.9	50.9	57.1	54.1	44.3	34.7	27.5	22.8	37.2
	MEAN	28.6	35.3	42.7	51.1	60.2	68.4	76.8	73.7	63.6	51.2	38.7	30.9	51.8
OREGON CITY	MAX	45.9	51.7	55.5	62.0	69.4	75.1	82.4	81.2	76.1	64.8	53.4	47.5	63.8
	MIN	34.4	36.9	38.0	41.1	46.1	51.3	54.4	54.3	51.1	45.0	39.2	36.1	44.0
	MEAN	40.2	44.3	46.8	51.5	57.8	63.3	68.4	67.8	63.6	54.9	46.3	41.8	53.9
OTIS 2 NE	MAX	46.6	51.2	53.1	56.3	61.3	65.3	69.3	70.1	68.9	61.5	52.8	48.0	58.7
	MIN	35.5	37.4	37.1	39.0	43.0	47.5	49.7	50.2	48.7	44.7	39.9	37.1	42.5
	MEAN	41.1	44.3	45.1	47.7	52.2	56.4	59.5	60.2	58.8	53.1	46.4	42.6	50.6
OWYHEE DAM	MAX	39.7	46.9	55.5	64.5	74.4	83.3	93.8	91.1	81.1	67.8	51.6	41.8	66.0
	MIN	22.7	28.0	31.7	37.0	43.5	49.0	53.1	52.0	46.1	38.3	29.7	24.9	38.0
	MEAN	31.2	37.5	43.7	50.8	59.0	66.2	73.5	71.5	63.6	53.0	40.7	33.4	52.0
PAISLEY	MAX	41.1	46.8	50.5	58.6	67.5	75.9	85.6	83.5	77.4	65.3	50.5	42.9	62.1
	MIN	21.6	25.4	26.4	30.9	38.1	44.9	50.1	48.3	41.7	34.2	26.7	22.6	34.2
	MEAN	31.4	36.2	38.5	44.8	52.8	60.4	67.9	65.9	59.6	49.8	38.6	32.8	48.2

OREGON

TEMPERATURE NORMALS (DEG F)

STATION			JAN	FEB	MAR	APR	MAY	JUN	JUL	AUG	SEP	OCT	NOV	DEC	ANN
PENDLETON WSO	//R	MAX	39.4	46.9	53.4	61.4	70.6	79.6	88.9	85.9	77.1	63.7	48.7	42.5	63.2
		MIN	26.3	31.8	34.4	39.2	46.1	52.9	58.6	57.5	50.5	41.3	33.4	29.5	41.8
		MEAN	32.8	39.4	43.9	50.3	58.4	66.2	73.8	71.7	63.8	52.5	41.1	36.0	52.5
PILOT ROCK 1 SE		MAX	41.4	48.2	54.1	61.6	70.4	79.9	89.9	87.8	79.1	65.9	51.1	44.6	64.5
		MIN	23.9	28.6	30.8	35.0	41.4	47.4	51.1	50.6	44.5	36.5	30.2	26.9	37.2
		MEAN	32.7	38.5	42.5	48.3	55.9	63.7	70.5	69.2	61.8	51.2	40.7	35.8	50.9
PORTLAND WSO	//R	MAX	44.3	50.4	54.5	60.2	66.9	72.7	79.5	78.6	74.2	63.9	52.3	46.4	62.0
		MIN	33.5	36.0	37.4	40.6	46.4	52.2	55.8	55.8	51.1	44.6	38.6	35.4	44.0
		MEAN	38.9	43.2	45.9	50.4	56.7	62.5	67.7	67.3	62.7	54.3	45.5	40.9	53.0
POWERS		MAX	51.9	55.7	57.2	61.0	66.0	71.2	76.7	77.6	75.9	68.1	58.3	53.1	64.4
		MIN	34.8	36.6	37.2	39.4	43.5	48.1	50.7	50.4	47.3	42.7	38.6	35.9	42.1
		MEAN	43.4	46.2	47.2	50.2	54.8	59.7	63.8	64.0	61.6	55.4	48.5	44.5	53.3
PRINEVILLE 4 NW		MAX	42.4	48.8	53.3	60.6	68.8	76.6	86.4	84.0	77.7	66.1	51.5	44.4	63.4
		MIN	21.2	25.3	25.1	27.7	34.6	40.3	42.5	41.0	35.2	29.2	25.9	22.7	30.9
		MEAN	31.8	37.1	39.2	44.2	51.8	58.5	64.5	62.5	56.5	47.7	38.7	33.6	47.2
PROSPECT 2 SW		MAX	45.2	51.3	54.4	61.4	70.0	77.8	87.7	86.3	81.1	68.5	52.9	45.6	65.2
		MIN	27.0	29.1	29.7	32.1	37.5	43.0	46.6	45.5	41.4	35.9	31.8	28.5	35.7
		MEAN	36.1	40.3	42.1	46.8	53.8	60.4	67.1	65.9	61.3	52.2	42.4	37.1	50.5
REDMOND 2 W		MAX	41.7	48.3	52.7	59.5	67.4	75.6	84.7	82.4	75.2	64.3	50.4	44.1	62.2
		MIN	22.1	26.2	26.5	29.6	35.9	42.1	46.0	44.9	39.4	33.1	27.8	24.6	33.2
		MEAN	31.9	37.3	39.6	44.6	51.7	58.9	65.4	63.7	57.3	48.7	39.1	34.4	47.7
REDMOND FAA AP		MAX	40.5	46.8	51.5	58.7	66.7	75.5	85.5	82.7	75.5	63.9	49.5	43.0	61.7
		MIN	20.8	25.1	25.4	28.5	35.0	41.9	46.6	45.5	39.3	32.2	26.9	23.1	32.5
		MEAN	30.7	36.0	38.5	43.6	50.9	58.7	66.1	64.1	57.5	48.1	38.2	33.1	47.1
REEDSPORT		MAX	50.3	53.8	55.3	58.1	62.3	66.3	69.5	70.3	70.2	64.9	56.3	51.5	60.7
		MIN	36.9	38.8	38.3	40.2	44.3	48.5	50.7	51.6	49.3	45.5	41.2	38.3	43.6
		MEAN	43.6	46.3	46.8	49.2	53.3	57.4	60.1	61.0	59.8	55.2	48.7	44.9	52.2
RICHLAND		MAX	39.2	47.9	56.4	66.9	76.9	85.4	94.6	92.7	84.0	71.0	53.7	41.9	67.6
		MIN	20.3	25.4	28.9	32.9	40.2	47.2	53.7	51.5	42.1	32.7	26.1	22.6	35.3
		MEAN	29.7	36.7	42.7	49.9	58.6	66.3	74.2	72.1	63.1	51.8	39.9	32.3	51.4
RIDDLE 2 NNE		MAX	48.1	54.2	57.7	63.3	69.7	76.4	84.0	83.0	78.8	67.5	54.6	48.7	65.5
		MIN	33.9	35.9	36.5	38.8	43.5	49.0	52.3	52.1	47.6	42.8	39.0	35.5	42.2
		MEAN	41.0	45.1	47.1	51.1	56.6	62.7	68.2	67.6	63.2	55.1	46.8	42.1	53.9
ROSEBURG KQEN		MAX	48.4	53.6	57.0	62.7	69.4	75.9	84.2	83.7	78.5	66.6	54.5	49.0	65.3
		MIN	33.9	35.6	36.2	38.5	43.6	49.0	52.5	52.8	48.5	42.8	38.6	35.4	42.3
		MEAN	41.2	44.6	46.6	50.6	56.5	62.4	68.4	68.2	63.5	54.7	46.5	42.2	53.8
ROUND GROVE		MAX	39.1	43.7	46.5	54.5	63.6	72.1	82.5	81.0	74.9	63.2	48.4	41.1	59.2
		MIN	18.2	21.8	22.5	25.9	31.5	37.2	42.1	40.5	34.9	29.1	24.1	19.4	28.9
		MEAN	28.7	32.8	34.5	40.3	47.6	54.7	62.3	60.8	54.9	46.2	36.2	30.3	44.1
SALEM WSO	R	MAX	45.7	51.1	54.6	60.3	67.3	73.9	82.2	81.2	76.2	64.5	52.6	47.0	63.1
		MIN	32.8	34.3	35.0	37.4	42.3	47.8	50.3	50.7	47.0	41.4	36.8	34.4	40.9
		MEAN	39.3	42.7	44.8	48.9	54.8	60.9	66.3	65.9	61.6	53.0	44.7	40.7	52.0
SEASIDE		MAX	49.7	53.2	53.8	56.7	61.0	64.1	67.3	67.7	69.1	63.9	56.2	51.5	59.5
		MIN	36.6	38.3	37.7	40.4	44.7	49.2	51.7	52.2	49.3	45.1	40.2	38.3	43.6
		MEAN	43.2	45.8	45.8	48.6	52.9	56.7	59.5	60.0	59.2	54.5	48.2	44.9	51.6
SENECA		MAX	33.6	39.2	43.9	53.0	62.3	71.0	81.6	80.1	71.9	60.1	44.8	36.1	56.5
		MIN	8.9	13.8	17.7	24.4	30.2	35.4	37.1	34.7	27.3	21.1	17.5	12.5	23.4
		MEAN	21.2	26.5	30.8	38.7	46.3	53.3	59.4	57.4	49.6	40.6	31.2	24.4	40.0
SEXTON SUMMIT WSO	//R	MAX	39.9	43.0	44.4	51.0	59.3	66.9	75.8	74.3	69.6	58.6	46.8	41.6	55.9
		MIN	30.1	31.8	30.8	33.1	38.9	45.1	51.9	51.8	49.6	43.2	36.0	31.8	39.5
		MEAN	35.0	37.4	37.6	42.1	49.1	56.1	63.9	63.1	59.6	50.9	41.4	36.8	47.8

OREGON

TEMPERATURE NORMALS (DEG F)

STATION		JAN	FEB	MAR	APR	MAY	JUN	JUL	AUG	SEP	OCT	NOV	DEC	ANN
SILVER CREEK FALLS	MAX	44.3	49.2	51.9	58.2	65.3	71.0	78.3	77.6	73.8	62.9	51.3	45.8	60.8
	MIN	29.9	31.7	32.1	34.3	39.1	44.1	45.6	45.6	42.0	37.8	33.5	31.7	37.3
	MEAN	37.1	40.5	42.0	46.3	52.2	57.6	62.0	61.6	57.9	50.4	42.4	38.8	49.1
SQUAW BUTTE EXP STA	MAX	35.9	42.0	46.7	55.4	64.3	73.6	84.2	82.3	74.5	62.4	47.0	38.2	58.9
	MIN	17.6	22.3	23.8	28.7	35.5	42.7	49.8	48.7	41.7	33.6	25.4	20.2	32.5
	MEAN	26.8	32.2	35.2	42.1	50.0	58.2	67.0	65.6	58.1	48.0	36.3	29.2	45.7
STAYTON	MAX	46.5	51.8	55.2	60.7	67.7	73.9	81.5	80.9	75.9	64.6	53.2	47.9	63.3
	MIN	32.8	35.2	36.2	38.9	43.1	48.1	50.2	50.2	47.1	42.4	37.2	34.5	41.3
	MEAN	39.7	43.5	45.7	49.8	55.4	61.0	65.9	65.6	61.5	53.5	45.2	41.2	52.3
THREE LYNX	MAX	40.7	46.7	50.3	57.2	64.2	70.2	77.9	77.3	73.2	61.6	48.9	42.9	59.3
	MIN	30.1	32.7	33.4	37.1	42.2	47.6	50.4	50.2	46.8	41.2	35.9	32.6	40.0
	MEAN	35.4	39.7	41.9	47.2	53.2	58.9	64.2	63.8	60.0	51.4	42.4	37.8	49.7
TILLAMOOK 1 W	MAX	48.8	52.4	53.4	56.7	60.9	64.5	67.3	68.2	68.5	63.0	55.0	50.4	59.1
	MIN	35.5	36.8	35.9	38.2	42.3	46.6	49.0	49.4	46.7	42.4	38.6	37.0	41.5
	MEAN	42.2	44.6	44.6	47.5	51.7	55.6	58.2	58.8	57.6	52.8	46.9	43.7	50.4
UNION	MAX	36.6	43.3	49.5	57.7	66.0	73.8	84.0	82.8	74.5	62.5	47.7	39.6	59.8
	MIN	23.5	27.6	29.0	33.4	39.3	45.3	49.1	47.9	41.3	34.6	30.1	26.2	35.6
	MEAN	30.1	35.5	39.3	45.6	52.7	59.6	66.6	65.4	57.9	48.6	38.9	32.9	47.8
VALE	MAX	37.5	45.9	55.8	66.1	75.7	84.5	94.9	91.5	81.1	67.2	50.5	40.1	65.9
	MIN	19.9	24.9	28.8	34.6	43.1	50.4	56.2	53.4	44.0	34.5	26.9	22.1	36.6
	MEAN	28.7	35.4	42.4	50.4	59.4	67.5	75.6	72.5	62.6	50.9	38.7	31.1	51.3
WICKIUP DAM //	MAX	36.8	41.8	45.0	52.9	62.3	70.5	80.5	78.8	72.3	60.5	45.9	38.9	57.2
	MIN	17.2	20.9	22.7	27.5	33.6	40.0	43.3	41.4	35.0	29.2	25.2	20.3	29.7
	MEAN	27.1	31.4	33.9	40.2	48.0	55.3	61.9	60.1	53.6	44.9	35.6	29.6	43.5

OREGON

PRECIPITATION NORMALS (INCHES)

STATION	JAN	FEB	MAR	APR	MAY	JUN	JUL	AUG	SEP	OCT	NOV	DEC	ANN
ANTELOPE 1 N	1.61	1.03	1.00	.78	1.29	1.01	.36	.67	.69	.91	1.66	1.68	12.69
ARLINGTON	1.57	.93	.70	.53	.55	.39	.22	.28	.37	.56	1.30	1.58	8.98
ASHLAND 1 N	2.75	1.77	1.82	1.35	1.35	1.01	.29	.50	.79	1.74	2.47	3.06	18.90
ASHWOOD	1.65	.96	1.03	.83	1.23	.88	.25	.59	.51	.80	1.49	1.57	11.79
ASTORIA WSO //R	11.29	7.81	7.26	4.60	2.84	2.43	1.04	1.56	3.11	6.21	9.88	11.57	69.59
AUSTIN 3 S	2.94	1.84	2.03	1.39	1.75	1.43	.58	.82	.98	1.42	2.35	3.06	20.59
BAKER FAA AIRPORT	1.02	.67	.80	.83	1.43	1.38	.47	.74	.69	.68	.88	1.04	10.63
BAKER KBKR	1.18	.73	1.01	.92	1.51	1.33	.68	.70	.67	.69	1.14	1.32	11.88
BANDON 1 E-BATES BOG	10.63	7.47	7.47	4.42	2.80	1.24	.35	.90	1.70	4.42	8.51	10.35	60.26
BEND	2.04	.97	.83	.53	.96	.95	.30	.54	.35	.67	1.48	1.91	11.53
BEULAH	1.63	.99	.87	.72	.97	1.01	.36	.42	.54	.83	1.31	1.59	11.24
BONNEVILLE DAM	12.59	8.56	8.29	5.29	3.69	2.44	.75	1.74	2.94	6.35	10.57	13.31	76.52
BROOKINGS	13.06	9.71	9.42	5.25	3.77	1.37	.46	1.22	2.31	5.72	11.36	12.57	76.22
BUENA VISTA STATION	.85	.51	.58	.69	1.07	1.01	.33	.54	.60	.69	.83	.79	8.49
BUNCOM 2 SE	3.96	2.44	2.27	1.43	1.28	.89	.25	.59	.83	1.93	3.01	4.22	23.10
BURNS WSO AP //R	.54	.33	.45	.51	.71	.80	.37	.84	.62	.86	.88	.88	7.80
BUTTE FALLS 1 SE	5.81	3.75	3.89	2.43	2.23	1.46	.26	.75	1.21	3.16	5.24	5.83	36.02
CAPE BLANCO	13.62	9.75	9.43	5.25	3.23	1.53	.52	1.29	2.11	5.45	11.18	13.34	76.70
CASCADIA STATE PARK	9.21	6.40	7.11	4.91	4.00	2.55	.53	1.37	2.38	5.17	8.36	10.12	62.11
CHEMULT	4.97	2.98	2.34	1.07	.99	1.00	.48	.71	.80	1.67	3.58	4.67	25.26
CHERRY GROVE 2 S	10.14	6.64	6.41	3.20	2.08	1.35	.46	1.07	1.84	4.26	8.08	10.35	55.88
CHILOQUIN 1 E	3.08	1.90	1.93	.96	1.05	.95	.29	.54	.67	1.40	2.56	3.18	18.51
CLATSKANIE 3 W	10.32	6.78	6.81	3.79	2.44	1.68	.65	1.35	2.48	4.72	8.50	10.18	59.70
CLOVERDALE 1 NW	14.08	9.31	9.92	5.77	3.89	3.09	1.19	1.79	3.52	7.15	11.22	14.10	85.03
CONDON	1.65	1.19	1.21	1.02	1.35	.99	.46	.68	.63	.96	1.81	1.85	13.80
CORVALLIS ST COLLEGE	7.55	4.86	4.63	2.46	1.92	1.20	.31	.81	1.48	3.39	6.17	7.77	42.55
CORVALLIS WATER BUREAU	12.83	8.87	8.30	3.86	2.40	1.05	.31	.84	1.54	4.30	9.37	12.33	66.00
COTTAGE GROVE 1 S	7.48	5.24	5.41	3.38	2.46	1.44	.31	.94	1.56	3.72	6.79	7.91	46.64
COTTAGE GROVE DAM	7.44	5.44	5.58	3.52	2.77	1.57	.34	.98	1.58	3.96	6.95	8.07	48.20
CRATER LAKE NP HQ	11.29	7.87	7.88	4.48	2.97	2.07	.53	1.27	2.19	5.17	9.60	11.55	66.87
DALLAS	9.65	6.40	5.76	2.92	2.00	1.06	.31	.74	1.57	3.57	7.23	9.37	50.58
DILLEY 1 S	8.40	5.41	5.18	2.60	1.82	1.30	.42	.86	1.60	3.70	6.71	8.75	46.75
DORENA DAM	6.99	5.05	5.37	3.68	2.95	1.83	.37	1.03	1.66	3.77	6.47	7.53	46.70
DRAIN	8.03	5.65	5.42	3.22	2.31	1.22	.25	.74	1.35	3.76	6.79	8.69	47.43
DUFUR	2.27	1.20	1.18	.68	.76	.61	.20	.41	.45	.80	1.58	2.02	12.16
ELGIN	3.22	2.37	2.22	1.71	1.79	1.39	.54	.78	1.14	1.94	3.02	3.66	23.78
ELKTON 3 SW	9.94	7.03	6.56	3.50	1.99	.96	.17	.68	1.39	4.15	7.60	10.55	54.52
ENTERPRISE	.95	.65	.92	1.09	1.75	1.98	.73	1.02	.99	.97	.97	1.06	13.08
ESTACADA 2 SE	9.37	6.10	6.68	4.59	3.54	2.45	.74	1.56	2.53	5.06	7.95	9.38	59.95
EUGENE WSO R	8.39	5.12	5.11	2.76	1.97	1.24	.27	.95	1.45	3.47	6.82	8.49	46.04
FERN RIDGE DAM	7.31	4.78	4.43	2.52	1.77	1.03	.20	.65	1.16	3.03	6.04	7.66	40.58
FOREST GROVE	8.20	5.21	4.87	2.46	1.70	1.14	.37	.90	1.58	3.54	6.54	8.34	44.85
FOSSIL	1.84	1.15	1.26	1.12	1.48	1.03	.45	.68	.73	1.11	1.70	1.76	14.31
FREMONT	1.94	.97	.95	.52	.90	.86	.37	.53	.38	.81	1.37	1.96	11.56
GOLD BEACH RANGER STA	14.54	10.47	10.59	6.07	4.10	1.45	.38	1.12	2.54	5.75	11.85	13.81	82.67
GRANTS PASS	6.46	4.21	3.46	1.73	1.34	.55	.21	.46	.89	2.47	4.59	5.94	32.31
GRIZZLY	1.60	1.06	1.07	.92	1.51	1.19	.28	.65	.69	.95	1.57	1.63	13.12
HALFWAY	3.45	2.20	1.84	1.45	1.53	1.37	.44	.64	.97	1.25	2.67	3.70	21.51
HART MOUNTAIN REFUGE	.98	.75	.99	1.13	1.60	1.43	.40	.50	.63	.99	.95	1.02	11.37
HASKINS DAM	13.81	9.59	9.40	4.34	2.56	1.28	.41	1.02	2.36	5.54	11.44	14.55	76.30
HEADWORKS PTLD WATER	12.37	8.18	8.91	6.39	5.17	3.90	1.20	2.30	3.82	6.76	10.26	12.41	81.67
HEPPNER	1.54	1.10	1.33	1.16	1.47	.98	.31	.69	.72	1.09	1.58	1.62	13.59
HERMISTON 2 S	1.35	.82	.71	.62	.68	.48	.19	.39	.39	.71	1.19	1.34	8.87
HOLLEY	8.70	5.93	5.99	3.58	2.88	1.78	.39	1.09	1.66	4.19	7.17	9.16	52.52
HOOD RIVER EXP STATION	5.86	3.54	3.21	1.63	1.02	.67	.18	.58	.92	2.47	4.79	5.98	30.85

OREGON

PRECIPITATION NORMALS (INCHES)

STATION		JAN	FEB	MAR	APR	MAY	JUN	JUL	AUG	SEP	OCT	NOV	DEC	ANN
HUNTINGTON		1.92	1.31	1.03	.84	1.02	.90	.25	.53	.61	.75	1.42	2.04	12.62
IONE 18 S		1.57	1.04	1.11	.96	1.25	.88	.31	.62	.54	.82	1.44	1.44	11.98
KENO		3.07	1.88	1.83	1.03	1.12	1.00	.29	.61	.66	1.67	2.60	3.34	19.10
KENT		1.50	1.06	.94	.86	.97	.72	.38	.53	.51	.74	1.55	1.62	11.38
KLAMATH FALLS 2 SSW		2.13	1.27	1.26	.67	.92	.80	.20	.56	.55	1.17	1.78	2.34	13.65
LA GRANDE		2.29	1.63	1.68	1.67	1.86	1.34	.53	.86	.98	1.43	2.12	2.40	18.79
LAKEVIEW		2.35	1.52	1.39	1.16	1.51	1.28	.31	.50	.59	1.27	1.76	2.16	15.80
LANGLOIS 2		13.66	10.02	9.80	5.46	3.79	1.53	.43	1.16	2.24	5.46	11.17	13.00	77.72
LEABURG 1 SW		9.58	6.95	7.26	4.83	3.79	2.25	.49	1.24	2.39	5.15	9.10	10.43	63.46
LOWER HAY CREEK		1.37	.82	.84	.74	1.01	.95	.26	.49	.47	.65	1.39	1.36	10.35
MADRAS		1.39	.89	.76	.63	.94	.92	.29	.46	.48	.63	1.32	1.38	10.09
MALHEUR BRANCH EXP STA		1.40	.98	.79	.77	.99	.80	.16	.48	.54	.71	1.12	1.27	10.01
MALHEUR REFUGE HDQ		1.06	.65	.80	.60	1.03	.96	.34	.59	.53	.75	.97	.94	9.22
MC KENZIE BRIDGE RS		11.35	7.70	7.73	4.70	3.77	2.33	.50	1.31	2.61	5.76	9.85	11.89	69.50
MC MINNVILLE		7.79	5.10	4.85	2.56	1.86	1.03	.36	.81	1.68	3.50	6.33	7.75	43.62
MEDFORD WSO	R	3.42	2.12	1.85	1.07	1.19	.67	.25	.46	.75	1.68	2.89	3.49	19.84
METOLIUS 1 W		1.42	.84	.75	.60	.97	.92	.24	.57	.43	.62	1.33	1.36	10.05
MIKKALO 6 W		1.57	1.00	.87	.77	.89	.68	.25	.41	.44	.65	1.46	1.54	10.53
MILTON FREEWATER 4 NW		1.79	1.12	1.27	1.12	1.21	.99	.37	.64	.65	1.12	1.53	1.81	13.62
MINAM 7 NE		3.48	2.47	2.31	1.92	2.26	1.81	.66	1.10	1.43	2.08	3.06	3.71	26.29
MORGAN 3 NE		1.44	.90	.75	.63	.71	.50	.17	.35	.34	.62	1.34	1.40	9.15
MORO		1.83	1.04	1.01	.73	.85	.65	.24	.42	.47	.77	1.63	1.74	11.38
NEWPORT		12.56	8.53	8.68	5.05	3.33	2.37	.79	1.49	2.82	5.92	10.17	12.91	74.62
NORTH BEND FAA AP		10.58	7.65	7.54	4.35	2.66	1.27	.33	.98	1.79	4.43	8.75	10.81	61.14
NYSSA		1.37	1.03	.79	.79	.91	.81	.18	.43	.58	.76	1.12	1.30	10.07
OAKRIDGE SALMON HATCH		7.36	5.01	5.17	3.26	2.84	1.73	.37	1.03	1.49	3.60	6.65	7.80	46.31
OCHOCO RANGER STATION		2.28	1.57	1.49	1.13	1.52	1.45	.51	.83	.78	1.39	2.45	2.65	18.05
ONTARIO KSRV		1.46	.93	.70	.72	.87	.70	.15	.41	.52	.68	1.14	1.32	9.60
OO RANCH		1.06	.62	.64	.59	1.05	.95	.31	.60	.49	.68	.97	.94	8.90
OREGON CITY		8.11	5.14	4.99	3.37	2.64	1.87	.56	1.24	2.08	3.79	6.59	8.02	48.40
OTIS 2 NE		16.31	11.07	11.34	6.94	4.31	3.44	1.32	2.09	4.05	8.21	13.36	16.64	99.08
OWYHEE DAM		1.04	.75	.69	.75	1.06	1.12	.23	.43	.50	.62	.82	.94	8.95
PAISLEY		1.45	.92	.89	.64	1.11	1.13	.40	.44	.42	.86	.96	1.53	10.75
PENDLETON WSO	//R	1.73	1.11	1.06	.99	1.09	.70	.30	.55	.58	.95	1.48	1.66	12.20
PILOT ROCK 1 SE		1.59	1.06	1.31	1.38	1.38	1.14	.32	.74	.76	1.07	1.48	1.65	13.88
PORTLAND WSO	//R	6.16	3.93	3.61	2.31	2.08	1.47	.46	1.13	1.61	3.05	5.17	6.41	37.39
POWERS		11.56	8.09	8.01	4.41	2.83	1.03	.26	.66	1.49	4.15	8.37	11.09	61.95
P-RANCH REFUGE		1.20	.83	.91	1.04	1.38	1.14	.29	.56	.62	.91	1.21	1.11	11.20
PRINEVILLE 4 NW		1.29	.83	.74	.63	1.10	1.04	.25	.57	.45	.77	1.41	1.42	10.50
PROSPECT 2 SW		7.02	4.59	4.54	2.64	2.26	1.21	.28	.89	1.23	3.67	6.33	7.14	41.80
REDMOND 2 W		1.07	.72	.61	.38	.82	.91	.30	.45	.31	.56	.98	1.14	8.25
REDMOND FAA AP		1.22	.70	.64	.46	.90	.83	.29	.48	.37	.55	1.04	1.06	8.54
REEDSPORT		13.28	9.62	9.61	5.51	3.33	1.74	.49	1.21	2.38	5.54	10.48	13.20	76.39
REX 1 S		7.22	4.46	4.28	2.58	2.11	1.50	.48	1.03	1.80	3.59	6.09	7.35	42.49
RICHLAND		1.42	.96	.85	1.01	1.43	1.14	.36	.55	.64	.69	1.30	1.51	11.86
RIDDLE 2 NNE		5.95	3.60	3.40	1.83	1.34	.73	.16	.49	.99	2.51	4.67	5.94	31.61
ROCK CREEK		2.88	1.93	1.79	1.41	1.92	1.73	.79	.93	.97	1.40	2.34	3.13	21.22
ROME 2 NW		.86	.52	.60	.67	1.01	1.18	.29	.49	.51	.60	.70	.77	8.20
ROSEBURG KQEN		5.96	3.85	3.51	2.07	1.52	.96	.19	.58	1.08	2.59	4.97	6.07	33.35
ROUND GROVE		2.25	1.70	1.75	1.25	1.65	1.35	.50	.72	.76	1.59	1.93	2.59	18.04
SALEM WSO	R	7.05	4.56	4.31	2.41	1.95	1.23	.35	.76	1.59	3.33	5.71	7.10	40.35
SEASIDE		12.48	8.95	8.19	5.14	3.02	2.70	1.20	1.65	3.23	6.39	10.44	12.53	75.92
SENECA		1.62	1.06	1.05	.95	1.27	.95	.43	.77	.63	.91	1.29	1.62	12.55
SEXTON SUMMIT WSO	//R	7.00	4.35	4.02	2.14	1.86	.96	.27	.70	1.31	3.15	5.96	6.42	38.14
SHEAVILLE		1.54	.96	.99	.98	1.08	1.07	.27	.44	.54	.88	1.20	1.28	11.23

OREGON

PRECIPITATION NORMALS (INCHES)

STATION	JAN	FEB	MAR	APR	MAY	JUN	JUL	AUG	SEP	OCT	NOV	DEC	ANN
SILVER CREEK FALLS	12.55	8.73	9.53	6.14	4.68	3.15	.78	1.75	3.29	6.53	10.75	12.76	80.64
SQUAW BUTTE EXP STA	1.40	.88	.90	.70	1.24	.98	.30	.65	.47	.83	1.16	1.36	10.87
STAYTON	8.08	5.58	5.76	3.56	3.00	2.11	.61	1.28	2.29	4.52	7.36	8.86	53.01
SUMMIT	12.15	8.54	8.31	4.78	3.04	1.71	.51	1.03	2.30	5.26	9.47	12.37	69.47
THREE LYNX	12.09	7.73	8.25	5.17	3.91	2.53	.56	1.37	2.58	5.82	9.83	12.50	72.34
TIDEWATER	16.39	11.81	11.65	6.68	3.96	2.20	.67	1.36	2.96	6.76	12.80	16.39	93.63
TILLAMOOK 1 W	14.87	10.11	10.43	6.13	4.07	3.04	1.29	2.00	3.73	7.64	12.52	15.07	90.90
UKIAH	2.07	1.31	1.47	1.30	1.74	1.31	.60	.84	.90	1.30	1.90	2.16	16.90
UNION	1.16	.82	1.14	1.28	1.92	1.53	.50	.91	1.01	1.09	1.26	1.39	14.01
UNITY	1.28	.77	.75	.65	1.25	1.20	.42	.69	.58	.75	1.16	1.33	10.83
VALE	1.29	.87	.73	.76	.98	.79	.23	.49	.55	.71	1.02	1.22	9.64
VALSETZ	22.62	15.85	14.85	8.15	4.58	2.75	1.09	2.13	4.59	10.41	18.71	22.89	128.62
WALLA WALLA 13 ESE	6.02	4.01	4.25	3.63	2.75	2.24	.59	1.28	1.93	3.29	5.22	5.73	40.94
WALLOWA	2.03	1.33	1.54	1.39	1.87	1.48	.67	.90	1.21	1.52	1.99	2.18	18.11
WASCO	1.90	1.08	1.02	.69	.67	.58	.25	.41	.44	.85	1.74	1.85	11.48
WATERLOO	7.25	4.98	5.09	3.10	2.75	1.75	.38	1.15	1.63	3.82	6.53	7.68	46.11
WICKIUP DAM //	3.65	2.12	1.81	.99	1.17	.99	.42	.71	.66	1.46	2.73	3.62	20.33
WILLAMINA 2 S	9.57	6.58	6.41	3.25	1.93	1.04	.35	.75	1.64	3.84	7.40	10.06	52.82

OREGON

HEATING DEGREE DAY NORMALS (BASE 65 DEG F)

STATION		JUL	AUG	SEP	OCT	NOV	DEC	JAN	FEB	MAR	APR	MAY	JUN	ANN
ANTELOPE 1 N		45	92	174	450	756	946	1023	773	766	579	370	181	6155
ARLINGTON		0	12	68	338	690	877	970	686	592	357	154	46	4790
ASHLAND 1 N		19	46	129	388	654	846	853	650	645	489	298	126	5143
ASTORIA WSO	//R	158	143	199	375	552	679	741	591	639	522	397	252	5248
BAKER FAA AIRPORT		40	107	263	586	897	1159	1234	932	868	621	388	191	7286
BAKER KBKR		17	72	210	512	846	1107	1181	893	837	588	357	163	6783
BANDON 1 E-BATES BOG		239	223	246	350	462	570	608	501	561	495	409	288	4952
BEND		96	147	297	561	819	989	1060	834	862	672	477	264	7078
BEULAH		0	22	155	468	831	1094	1178	876	797	534	283	110	6348
BONNEVILLE DAM		39	53	114	326	600	784	874	664	636	450	276	127	4943
BROOKINGS		202	186	176	282	405	521	552	459	505	435	338	231	4292
BURNS WSO AP	//R	18	48	215	530	882	1163	1163	882	884	657	405	177	7024
CAPE BLANCO		344	313	309	375	450	543	592	496	577	513	459	369	5340
CHEMULT		156	231	376	657	939	1166	1215	972	995	786	561	336	8390
CHERRY GROVE 2 S		90	96	150	375	639	809	871	669	682	525	363	208	5477
CLATSKANIE 3 W		89	89	146	381	621	772	825	636	639	492	338	198	5226
CLOVERDALE 1 NW		158	142	167	319	498	632	685	538	592	498	378	246	4853
CONDON		54	98	213	502	810	989	1079	823	803	615	403	197	6586
CORVALLIS ST COLLEGE		54	64	128	372	609	756	806	622	623	477	319	157	4987
COTTAGE GROVE 1 S		61	80	152	366	582	729	772	596	611	480	329	168	4926
COTTAGE GROVE DAM		57	73	142	372	606	747	797	622	639	495	338	168	5056
CRATER LAKE NP HQ		317	343	472	741	996	1175	1243	1072	1184	1005	815	573	9936
DALLAS		52	78	128	363	615	760	809	622	623	474	304	157	4985
DORENA DAM		69	88	159	384	606	750	797	624	645	504	357	185	5168
DRAIN		36	45	117	326	561	713	753	577	580	435	279	123	4545
DUFUR		48	78	170	459	768	952	1032	759	713	522	327	156	5984
ELGIN		31	88	226	512	810	1017	1101	840	809	576	366	163	6539
ELKTON 3 SW		31	36	99	288	531	676	710	540	536	402	252	114	4215
ENTERPRISE		92	161	328	623	948	1187	1265	977	939	687	465	269	7941
ESTACADA 2 SE		61	75	153	397	612	756	809	627	623	462	301	154	5030
EUGENE WSO	R	44	57	126	363	591	735	772	602	595	462	307	145	4799
FERN RIDGE DAM		47	60	118	350	597	744	794	624	626	480	322	154	4916
FOREST GROVE		49	67	116	350	603	753	815	616	605	444	273	136	4827
FOSSIL		84	112	231	487	768	936	1026	784	794	615	431	225	6493
GOLD BEACH RANGER STA		186	171	184	295	420	527	558	468	530	456	366	243	4404
GRANTS PASS		9	21	75	307	585	766	778	580	543	381	202	78	4325
GRIZZLY		114	167	297	558	813	989	1066	840	874	699	508	283	7208
HALFWAY		16	70	219	539	879	1187	1277	983	865	576	351	155	7117
HART MOUNTAIN REFUGE		105	156	311	592	876	1085	1153	935	998	777	555	317	7860
HEADWORKS PTLD WATER		68	94	146	360	606	769	834	644	657	492	319	174	5163
HEPPNER		27	61	154	425	723	899	989	731	701	519	312	136	5677
HERMISTON 2 S		0	18	103	400	732	930	1023	731	629	402	184	53	5205
HOOD RIVER EXP STATION		45	64	162	428	708	880	973	722	679	480	291	135	5567
HUNTINGTON		0	13	87	375	768	1048	1113	820	688	408	177	47	5544
KENT		42	81	197	471	807	998	1091	829	806	612	404	192	6530
KLAMATH FALLS 2 SSW		34	74	194	484	807	1051	1091	837	822	618	389	181	6582
LA GRANDE		11	51	174	459	768	973	1057	809	772	540	315	127	6056
LAKEVIEW		43	110	234	524	831	1057	1135	899	896	660	433	213	7035
LEABURG 1 SW		38	54	121	347	585	738	781	599	608	453	295	140	4759
MADRAS		49	103	213	512	774	946	1017	764	760	576	370	176	6260
MALHEUR BRANCH EXP STA		0	15	125	443	810	1076	1153	851	707	441	195	64	5880
MALHEUR REFUGE HDQ		29	76	245	558	861	1107	1163	882	862	627	388	183	6981
MARION FRKS FISH HATCH		98	139	272	543	813	989	1048	834	856	669	468	245	6974
MC MINNVILLE		56	94	146	372	609	756	803	613	611	462	301	154	4977
MEDFORD WSO	R	6	19	92	335	642	846	849	633	605	441	245	85	4798

OREGON

HEATING DEGREE DAY NORMALS (BASE 65 DEG F)

STATION	JUL	AUG	SEP	OCT	NOV	DEC	JAN	FEB	MAR	APR	MAY	JUN	ANN
METOLIUS 1 W	68	121	237	527	798	967	1048	795	791	609	406	207	6574
MIKKALO 6 W	13	40	119	375	738	921	1017	736	682	483	264	104	5492
MILTON FREEWATER 4 NW	0	17	95	357	681	859	961	680	583	369	168	52	4822
MORO	32	74	186	484	795	980	1082	801	760	561	355	156	6266
NEWPORT	251	239	255	369	501	626	670	546	614	534	440	306	5351
NORTH BEND FAA AP	186	168	201	313	450	570	620	507	558	480	366	242	4661
NYSSA	0	10	128	443	792	1057	1135	846	713	438	200	56	5818
OCHOCO RANGER STATION	141	185	344	629	936	1150	1225	960	958	741	546	319	8134
ONTARIO KSRV	0	15	110	428	789	1057	1128	832	691	417	182	58	5707
OREGON CITY	23	44	98	313	561	719	769	580	564	405	232	107	4415
OTIS 2 NE	175	153	192	369	558	694	741	580	617	519	397	258	5253
OWYHEE DAM	0	20	113	372	729	980	1048	770	660	426	207	72	5397
PAISLEY	28	80	197	471	792	998	1042	806	822	606	384	174	6400
PENDLETON WSO //R	7	27	120	388	717	899	998	717	654	441	220	75	5263
PILOT ROCK 1 SE	14	41	144	428	729	905	1001	742	698	501	289	111	5603
PORTLAND WSO //R	35	51	111	332	585	747	809	610	592	438	263	118	4691
POWERS	66	73	121	298	495	636	670	526	552	444	316	168	4365
PRINEVILLE 4 NW	74	138	267	536	789	973	1029	781	800	624	409	213	6633
PROSPECT 2 SW	42	78	154	397	678	865	896	692	710	546	352	168	5578
REDMOND 2 W	66	114	241	505	777	949	1026	776	787	612	412	206	6471
REDMOND FAA AP	69	121	246	524	804	989	1063	812	822	642	437	212	6741
REEDSPORT	154	131	164	304	489	623	663	524	564	474	363	228	4681
RICHLAND	0	19	133	417	753	1014	1094	792	691	453	212	72	5650
RIDDLE 2 NNE	23	40	103	307	546	710	744	557	555	417	264	115	4381
ROSEBURG KQEN	27	36	107	319	555	707	738	571	570	432	264	122	4448
ROUND GROVE	118	162	313	583	864	1076	1125	902	946	741	539	314	7683
SALEM WSO R	46	65	131	372	609	753	797	624	626	483	316	152	4974
SEASIDE	174	157	180	326	504	623	676	538	595	492	375	249	4889
SENECA	188	244	462	756	1014	1259	1358	1078	1060	789	580	351	9139
SEXTON SUMMIT WSO //R	112	140	217	445	708	874	930	773	849	687	493	283	6511
SILVER CREEK FALLS	120	144	226	453	678	812	865	686	713	561	397	233	5888
SQUAW BUTTE EXP STA	45	92	246	527	861	1110	1184	918	924	687	465	228	7287
STAYTON	54	74	134	357	594	738	784	602	598	456	298	147	4836
THREE LYNX	81	102	174	422	678	843	918	708	716	534	366	198	5740
TILLAMOOK 1 W	211	196	226	378	543	660	707	571	632	525	412	282	5343
UNION	29	86	234	508	783	995	1082	826	797	582	381	179	6482
VALE	0	15	132	437	789	1051	1125	829	701	438	196	72	5785
WICKIUP DAM //	134	180	347	623	882	1097	1175	941	964	744	527	301	7915

OREGON

COOLING DEGREE DAY NORMALS (BASE 65 DEG F)

STATION		JAN	FEB	MAR	APR	MAY	JUN	JUL	AUG	SEP	OCT	NOV	DEC	ANN
ANTELOPE 1 N		0	0	0	0	7	55	147	154	36	0	0	0	399
ARLINGTON		0	0	0	0	48	169	350	312	107	0	0	0	986
ASHLAND 1 N		0	0	0	0	7	60	159	139	51	0	0	0	416
ASTORIA WSO	//R	0	0	0	0	0	0	7	7	0	0	0	0	14
BAKER FAA AIRPORT		0	0	0	0	0	26	90	94	17	0	0	0	227
BAKER KBKR		0	0	0	0	0	37	119	122	33	0	0	0	311
BANDON 1 E-BATES BOG		0	0	0	0	0	0	0	0	0	0	0	0	0
BEND		0	0	0	0	0	12	40	45	9	0	0	0	106
BEULAH		0	0	0	0	13	83	235	186	50	0	0	0	567
BONNEVILLE DAM		0	0	0	0	0	31	107	112	51	0	0	0	301
BROOKINGS		0	0	0	0	0	0	0	0	11	0	0	0	11
BURNS WSO AP	//R	0	0	0	0	5	42	155	116	32	0	0	0	350
CAPE BLANCO		0	0	0	0	0	0	0	0	0	0	0	0	0
CHEMULT		0	0	0	0	0	6	14	17	0	0	0	0	37
CHERRY GROVE 2 S		0	0	0	0	0	16	68	83	39	0	0	0	206
CLATSKANIE 3 W		0	0	0	0	0	9	30	40	14	0	0	0	93
CLOVERDALE 1 NW		0	0	0	0	0	0	0	0	0	0	0	0	0
CONDON		0	0	0	0	0	26	104	104	21	0	0	0	255
CORVALLIS ST COLLEGE		0	0	0	0	0	22	72	82	26	0	0	0	202
COTTAGE GROVE 1 S		0	0	0	0	0	12	52	61	17	0	0	0	142
COTTAGE GROVE DAM		0	0	0	0	0	15	64	79	25	0	0	0	183
CRATER LAKE NP HQ		0	0	0	0	0	0	7	12	7	0	0	0	26
DALLAS		0	0	0	0	0	25	71	81	32	0	0	0	209
DORENA DAM		0	0	0	0	0	11	51	63	18	0	0	0	143
DRAIN		0	0	0	0	0	24	85	83	33	0	0	0	225
DUFUR		0	0	0	0	7	39	110	106	29	0	0	0	291
ELGIN		0	0	0	0	0	25	96	97	25	0	0	0	243
ELKTON 3 SW		0	0	0	0	0	39	102	111	63	0	0	0	315
ENTERPRISE		0	0	0	0	0	11	30	43	7	0	0	0	91
ESTACADA 2 SE		0	0	0	0	0	19	76	68	24	0	0	0	187
EUGENE WSO	R	0	0	0	0	0	28	100	94	39	0	0	0	261
FERN RIDGE DAM		0	0	0	0	0	19	78	88	31	0	0	0	216
FOREST GROVE		0	0	0	0	6	43	114	116	38	0	0	0	317
FOSSIL		0	0	0	0	0	24	75	78	15	0	0	0	192
GOLD BEACH RANGER STA		0	0	0	0	0	0	0	0	7	0	0	0	7
GRANTS PASS		0	0	0	0	20	99	213	186	75	0	0	0	593
GRIZZLY		0	0	0	0	0	13	43	52	12	0	0	0	120
HALFWAY		0	0	0	0	0	29	112	104	21	0	0	0	266
HART MOUNTAIN REFUGE		0	0	0	0	0	8	37	41	11	0	0	0	97
HEADWORKS PTLD WATER		0	0	0	0	0	18	65	78	29	0	0	0	190
HEPPNER		0	0	0	0	5	61	151	148	34	0	0	0	399
HERMISTON 2 S		0	0	0	0	22	125	278	232	64	0	0	0	721
HOOD RIVER EXP STATION		0	0	0	0	6	39	98	98	21	0	0	0	262
HUNTINGTON		0	0	0	0	53	188	453	379	123	0	0	0	1196
KENT		0	0	0	0	0	39	135	134	35	0	0	0	343
KLAMATH FALLS 2 SSW		0	0	0	0	8	37	127	105	35	0	0	0	312
LA GRANDE		0	0	0	0	8	55	178	166	51	0	0	0	458
LAKEVIEW		0	0	0	0	6	33	105	94	24	0	0	0	262
LEABURG 1 SW		0	0	0	0	0	23	87	88	34	0	0	0	232
MADRAS		0	0	0	0	0	32	90	100	18	0	0	0	240
MALHEUR BRANCH EXP STA		0	0	0	0	22	136	316	244	59	0	0	0	777
MALHEUR REFUGE HDQ		0	0	0	0	0	30	107	86	20	0	0	0	243
MARION FRKS FISH HATCH		0	0	0	0	0	14	51	46	11	0	0	0	122
MC MINNVILLE		0	0	0	0	0	19	68	94	32	0	0	0	213
MEDFORD WSO	R	0	0	0	0	12	91	239	208	95	0	0	0	645

OREGON

COOLING DEGREE DAY NORMALS (BASE 65 DEG F)

STATION		JAN	FEB	MAR	APR	MAY	JUN	JUL	AUG	SEP	OCT	NOV	DEC	ANN
METOLIUS 1 W		0	0	0	0	0	15	55	65	15	0	0	0	150
MIKKALO 6 W		0	0	0	0	16	92	230	223	77	0	0	0	638
MILTON FREEWATER 4 NW		10	0	0	0	25	127	288	243	80	0	0	0	773
MORO		0	0	0	0	0	39	134	133	33	0	0	0	339
NEWPORT		0	0	0	0	0	0	0	0	0	0	0	0	0
NORTH BEND FAA AP		0	0	0	0	0	0	0	0	6	0	0	0	6
NYSSA		0	0	0	0	23	125	316	243	53	0	0	0	760
OCHOCO RANGER STATION		0	0	0	0	0	0	17	27	8	0	0	0	52
ONTARIO KSRV		0	0	0	0	33	160	366	285	68	0	0	0	912
OREGON CITY		0	0	0	0	9	56	129	131	56	0	0	0	381
OTIS 2 NE		0	0	0	0	0	0	0	0	6	0	0	0	6
OWYHEE DAM		0	0	0	0	21	108	264	222	71	0	0	0	686
PAISLEY		0	0	0	0	6	36	118	108	35	0	0	0	303
PENDLETON WSO	//R	0	0	0	0	16	111	280	235	84	0	0	0	726
PILOT ROCK 1 SE		0	0	0	0	7	72	184	171	48	0	0	0	482
PORTLAND WSO	//R	0	0	0	0	6	43	119	122	42	0	0	0	332
POWERS		0	0	0	0	0	9	29	42	19	0	0	0	99
PRINEVILLE 4 NW		0	0	0	0	0	18	58	61	12	0	0	0	149
PROSPECT 2 SW		0	0	0	0	5	30	107	106	43	0	0	0	291
REDMOND 2 W		0	0	0	0	0	23	78	74	10	0	0	0	185
REDMOND FAA AP		0	0	0	0	0	23	103	93	21	0	0	0	240
REEDSPORT		0	0	0	0	0	0	0	7	8	0	0	0	15
RICHLAND		0	0	0	0	14	111	288	239	76	7	0	0	735
RIDDLE 2 NNE		0	0	0	0	0	46	123	121	49	0	0	0	339
ROSEBURG KQEN		0	0	0	0	0	44	133	136	62	0	0	0	375
ROUND GROVE		0	0	0	0	0	5	34	32	10	0	0	0	81
SALEM WSO	R	0	0	0	0	0	29	87	93	29	0	0	0	238
SEASIDE		0	0	0	0	0	0	0	0	6	0	0	0	6
SENECA		0	0	0	0	0	0	14	8	0	0	0	0	22
SEXTON SUMMIT WSO	//R	0	0	0	0	0	16	77	81	55	8	0	0	237
SILVER CREEK FALLS		0	0	0	0	0	11	27	38	13	0	0	0	89
SQUAW BUTTE EXP STA		0	0	0	0	0	24	107	111	39	0	0	0	281
STAYTON		0	0	0	0	0	27	82	93	29	0	0	0	231
THREE LYNX		0	0	0	0	0	15	56	65	24	0	0	0	160
TILLAMOOK 1 W		0	0	0	0	0	0	0	0	0	0	0	0	0
UNION		0	0	0	0	0	17	78	98	21	0	0	0	214
VALE		0	0	0	0	23	147	329	247	60	0	0	0	806
WICKIUP DAM	//	0	0	0	0	0	10	38	29	5	0	0	0	82

35 — OREGON

STATE-STATION NUMBER	STN TYP	NAME		LATITUDE DEG-MIN	LONGITUDE DEG-MIN	ELEVATION (FT)
35-0197	13	ANTELOPE 1 N		N 4455	W 12043	2775
35-0265	13	ARLINGTON		N 4543	W 12012	315
35-0304	13	ASHLAND 1 N		N 4213	W 12243	1780
35-0312	12	ASHWOOD		N 4444	W 12045	2484
35-0328	13	ASTORIA WSO	//R	N 4609	W 12353	8
35-0356	12	AUSTIN 3 S		N 4435	W 11830	4213
35-0412	13	BAKER FAA AIRPORT		N 4450	W 11749	3368
35-0417	13	BAKER KBKR		N 4447	W 11750	3444
35-0471	13	BANDON 1 E-BATES BOG		N 4307	W 12423	80
35-0694	13	BEND		N 4404	W 12119	3599
35-0723	13	BEULAH		N 4355	W 11810	3277
35-0897	13	BONNEVILLE DAM		N 4538	W 12157	60
35-1055	13	BROOKINGS		N 4203	W 12417	80
35-1124	12	BUENA VISTA STATION		N 4304	W 11852	4135
35-1149	12	BUNCOM 2 SE		N 4209	W 12258	1925
35-1176	13	BURNS WSO AP	//R	N 4335	W 11903	4140
35-1207	12	BUTTE FALLS 1 SE		N 4232	W 12233	2500
35-1360	13	CAPE BLANCO		N 4250	W 12434	211
35-1433	12	CASCADIA STATE PARK		N 4424	W 12229	850
35-1546	13	CHEMULT		N 4314	W 12147	4760
35-1552	13	CHERRY GROVE 2 S		N 4525	W 12315	780
35-1571	12	CHILOQUIN 1 E		N 4235	W 12151	4220
35-1643	13	CLATSKANIE 3 W		N 4606	W 12317	92
35-1682	13	CLOVERDALE 1 NW		N 4513	W 12354	20
35-1765	13	CONDON		N 4514	W 12011	2830
35-1862	13	CORVALLIS ST COLLEGE		N 4438	W 12312	225
35-1877	12	CORVALLIS WATER BUREAU		N 4431	W 12327	592
35-1897	13	COTTAGE GROVE 1 S		N 4347	W 12304	650
35-1902	13	COTTAGE GROVE DAM		N 4343	W 12303	831
35-1946	13	CRATER LAKE NP HQ		N 4254	W 12208	6475
35-2112	13	DALLAS		N 4456	W 12319	325
35-2325	12	DILLEY 1 S		N 4529	W 12307	165
35-2374	13	DORENA DAM		N 4347	W 12258	820
35-2406	13	DRAIN		N 4340	W 12319	292
35-2440	13	DUFUR		N 4527	W 12108	1330
35-2597	13	ELGIN		N 4534	W 11755	2655
35-2633	13	ELKTON 3 SW		N 4336	W 12335	114
35-2672	13	ENTERPRISE		N 4526	W 11716	3790
35-2693	13	ESTACADA 2 SE		N 4516	W 12219	410
35-2709	13	EUGENE WSO	R	N 4407	W 12313	359
35-2867	13	FERN RIDGE DAM		N 4407	W 12318	386
35-2997	13	FOREST GROVE		N 4532	W 12306	180
35-3038	13	FOSSIL		N 4500	W 12013	2650
35-3095	12	FREMONT		N 4320	W 12110	4512
35-3356	13	GOLD BEACH RANGER STA		N 4224	W 12425	50
35-3445	13	GRANTS PASS		N 4226	W 12319	925
35-3542	13	GRIZZLY		N 4431	W 12056	3639
35-3604	13	HALFWAY		N 4453	W 11707	2671
35-3692	13	HART MOUNTAIN REFUGE		N 4233	W 11939	5616
35-3705	12	HASKINS DAM		N 4519	W 12321	840

STATE-STATION NUMBER	STN TYP	NAME		LATITUDE DEG-MIN	LONGITUDE DEG-MIN	ELEVATION (FT)
35-3770	13	HEADWORKS PTLD WATER		N 4527	W 12209	748
35-3827	13	HEPPNER		N 4521	W 11933	1950
35-3847	13	HERMISTON 2 S		N 4549	W 11917	624
35-3971	12	HOLLEY		N 4421	W 12247	540
35-4003	13	HOOD RIVER EXP STATION		N 4541	W 12131	500
35-4098	13	HUNTINGTON		N 4421	W 11716	2120
35-4161	12	IONE 18 S		N 4519	W 11951	2130
35-4403	12	KENO		N 4207	W 12157	4116
35-4411	13	KENT		N 4512	W 12042	2707
35-4506	13	KLAMATH FALLS 2 SSW		N 4212	W 12147	4098
35-4622	13	LA GRANDE		N 4519	W 11805	2755
35-4670	13	LAKEVIEW		N 4211	W 12021	4774
35-4721	12	LANGLOIS 2		N 4256	W 12427	88
35-4811	13	LEABURG 1 SW		N 4406	W 12241	675
35-5080	12	LOWER HAY CREEK		N 4444	W 12059	1898
35-5139	13	MADRAS		N 4438	W 12108	2230
35-5160	13	MALHEUR BRANCH EXP STA		N 4359	W 11701	2240
35-5162	13	MALHEUR REFUGE HDQ		N 4317	W 11850	4109
35-5221	11	MARION FRKS FISH HATCH		N 4436	W 12157	2475
35-5362	12	MC KENZIE BRIDGE RS		N 4411	W 12207	1478
35-5384	13	MC MINNVILLE		N 4514	W 12311	148
35-5429	13	MEDFORD WSO	R	N 4222	W 12252	1312
35-5515	13	METOLIUS 1 W		N 4435	W 12111	2500
35-5545	13	MIKKALO 6 W		N 4528	W 12021	1550
35-5593	13	MILTON FREEWATER 4 NW		N 4558	W 11826	839
35-5610	12	MINAM 7 NE		N 4541	W 11736	3584
35-5726	12	MORGAN 3 NE		N 4535	W 11953	945
35-5734	13	MORO		N 4529	W 12043	1868
35-6032	13	NEWPORT		N 4438	W 12403	154
35-6073	13	NORTH BEND FAA AP		N 4325	W 12415	7
35-6179	13	NYSSA		N 4352	W 11700	2185
35-6213	12	OAKRIDGE SALMON HATCH		N 4345	W 12227	1275
35-6243	13	OCHOCO RANGER STATION		N 4424	W 12026	3975
35-6294	13	ONTARIO KSRV		N 4403	W 11658	2145
35-6302	12	OO RANCH		N 4317	W 11919	4136
35-6334	13	OREGON CITY		N 4521	W 12236	167
35-6366	13	OTIS 2 NE		N 4502	W 12356	150
35-6405	13	OWYHEE DAM		N 4339	W 11715	2400
35-6426	13	PAISLEY		N 4242	W 12032	4360
35-6546	13	PENDLETON WSO	//R	N 4541	W 11851	1482
35-6634	13	PILOT ROCK 1 SE		N 4529	W 11849	1697
35-6751	13	PORTLAND WSO	//R	N 4536	W 12236	21
35-6820	13	POWERS		N 4253	W 12404	230
35-6853	12	P-RANCH REFUGE		N 4249	W 11853	4205
35-6883	13	PRINEVILLE 4 NW		N 4421	W 12054	2840
35-6907	13	PROSPECT 2 SW		N 4244	W 12231	2482
35-7052	13	REDMOND 2 W		N 4416	W 12113	3010
35-7062	13	REDMOND FAA AP		N 4416	W 12109	3075
35-7082	13	REEDSPORT		N 4342	W 12407	60
35-7127	12	REX 1 S		N 4518	W 12255	490

35 — OREGON

STATE-STATION NUMBER	STN TYP	NAME		LATITUDE DEG-MIN	LONGITUDE DEG-MIN	ELEVATION (FT)
35-7160	13	RICHLAND		N 4446	W 11710	2215
35-7169	13	RIDDLE 2 NNE		N 4258	W 12321	663
35-7250	12	ROCK CREEK		N 4454	W 11805	4150
35-7310	12	ROME 2 NW		N 4252	W 11739	3410
35-7331	13	ROSEBURG KQEN		N 4312	W 12321	465
35-7354	13	ROUND GROVE		N 4220	W 12053	4888
35-7500	13	SALEM WSO	R	N 4455	W 12301	196
35-7641	13	SEASIDE		N 4559	W 12355	10
35-7675	13	SENECA		N 4409	W 11858	4666
35-7698	13	SEXTON SUMMIT WSO	//R	N 4237	W 12322	3836
35-7736	12	SHEAVILLE		N 4307	W 11703	4600
35-7809	13	SILVER CREEK FALLS		N 4452	W 12239	1350
35-8029	13	SQUAW BUTTE EXP STA		N 4329	W 11941	4675
35-8095	13	STAYTON		N 4448	W 12246	465
35-8182	12	SUMMIT		N 4438	W 12335	746
35-8466	13	THREE LYNX		N 4507	W 12204	1120
35-8481	12	TIDEWATER		N 4425	W 12354	50
35-8494	13	TILLAMOOK 1 W		N 4527	W 12352	10
35-8726	12	UKIAH		N 4508	W 11856	3355
35-8746	13	UNION		N 4513	W 11753	2765
35-8780	12	UNITY		N 4426	W 11814	4031
35-8797	13	VALE		N 4359	W 11715	2240
35-8833	12	VALSETZ		N 4450	W 12340	1135
35-8985	12	WALLA WALLA 13 ESE		N 4600	W 11803	2400
35-8997	12	WALLOWA		N 4534	W 11732	2923
35-9068	12	WASCO		N 4535	W 12042	1264
35-9083	12	WATERLOO		N 4430	W 12249	420
35-9316	13	WICKIUP DAM	//	N 4341	W 12141	4358
35-9372	12	WILLAMINA 2 S		N 4503	W 12330	236

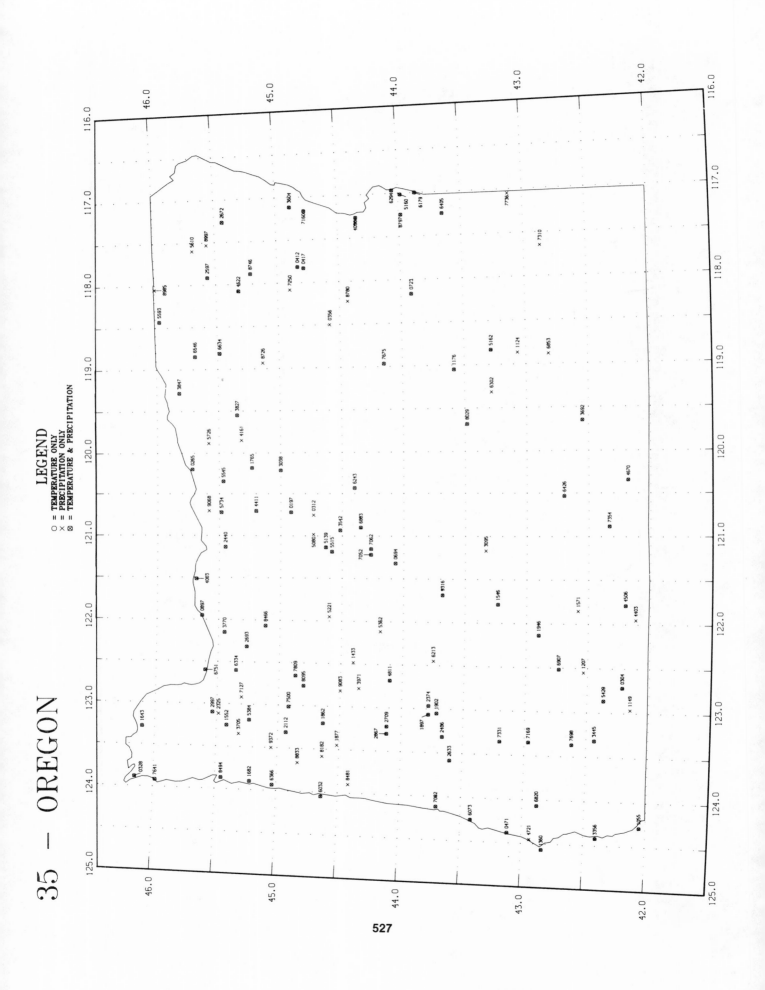

35 — OREGON

LEGEND

○ = TEMPERATURE ONLY
✕ = PRECIPITATION ONLY
⊗ = TEMPERATURE & PRECIPITATION

527

PENNSYLVANIA

TEMPERATURE NORMALS (DEG F)

STATION			JAN	FEB	MAR	APR	MAY	JUN	JUL	AUG	SEP	OCT	NOV	DEC	ANN
ALLENTOWN WSO	//R	MAX	34.9	37.8	47.6	61.0	71.1	80.1	84.6	82.4	75.3	64.2	51.3	39.2	60.8
		MIN	19.5	20.9	29.2	39.0	48.8	58.3	63.0	61.6	54.1	42.6	33.6	23.8	41.2
		MEAN	27.2	29.3	38.5	50.0	60.0	69.2	73.8	72.1	64.8	53.4	42.5	31.5	51.0
BAKERSTOWN 3 WNW		MAX	35.8	38.9	49.4	62.7	72.4	80.7	84.2	82.6	76.5	65.0	51.6	39.9	61.6
		MIN	18.9	20.4	28.9	39.2	48.3	56.3	60.5	59.7	53.1	42.0	33.5	24.0	40.4
		MEAN	27.4	29.7	39.2	51.0	60.4	68.5	72.3	71.2	64.8	53.5	42.6	32.0	51.1
BRADFORD FAA AP		MAX	27.5	29.9	39.2	53.1	64.4	72.8	76.5	75.1	68.4	57.1	43.2	31.7	53.2
		MIN	12.2	12.8	21.8	32.4	40.8	49.6	53.5	52.6	46.1	36.5	28.2	17.9	33.7
		MEAN	19.9	21.4	30.5	42.8	52.6	61.2	65.0	63.9	57.3	46.8	35.7	24.8	43.5
BRADFORD 4 W RES		MAX	30.8	33.4	42.5	56.6	67.8	75.7	78.7	77.0	70.7	60.5	46.3	34.2	56.2
		MIN	11.8	12.1	20.6	31.0	40.0	48.6	52.7	51.9	45.7	35.8	28.4	17.5	33.0
		MEAN	21.4	22.8	31.6	43.8	54.0	62.1	65.7	64.5	58.2	48.2	37.4	25.8	44.6
BURGETTSTOWN 2 W		MAX	35.1	37.7	48.3	60.9	71.4	79.2	82.7	81.3	75.4	63.7	51.2	39.3	60.5
		MIN	14.4	14.9	24.3	33.8	42.5	51.6	55.8	54.6	47.4	35.6	28.7	20.1	35.3
		MEAN	24.8	26.3	36.3	47.4	57.0	65.4	69.3	68.0	61.5	49.7	40.0	29.7	48.0
CARLISLE		MAX	37.2	40.2	50.8	64.1	74.2	82.9	87.2	85.4	*78.5	66.5	52.9	40.7	63.4
		MIN	21.4	22.8	30.9	41.2	50.1	59.5	63.9	62.8	55.9	43.9	35.0	25.5	42.7
		MEAN	29.3	31.5	40.9	52.7	62.2	71.2	75.6	74.2	67.2	55.2	43.9	33.1	53.1
CHAMBERSBURG 1 ESE		MAX	36.8	39.6	49.6	62.7	72.7	81.2	85.1	83.8	77.0	65.2	52.0	40.6	62.2
		MIN	20.4	21.7	29.7	39.5	48.7	57.5	61.6	60.4	53.2	41.6	33.0	24.3	41.0
		MEAN	28.6	30.7	39.7	51.1	60.7	69.3	73.4	72.1	65.1	53.4	42.5	32.5	51.6
CLARION 3 SW		MAX	33.4	36.3	46.7	60.6	71.4	79.0	82.3	80.4	74.4	62.9	49.1	37.4	59.5
		MIN	14.2	14.7	23.1	32.9	42.3	51.0	55.5	55.0	48.4	37.1	29.2	19.6	35.3
		MEAN	23.8	25.5	34.9	46.8	56.9	65.0	68.9	67.8	61.4	50.0	39.2	28.5	47.4
COATESVILLE 1 SW		MAX	37.9	40.1	49.8	62.4	72.2	80.8	85.3	83.8	77.6	66.2	53.9	41.9	62.7
		MIN	19.3	20.2	28.6	38.6	47.8	57.2	61.8	60.4	52.9	40.8	32.2	23.2	40.3
		MEAN	28.6	30.2	39.3	50.5	60.0	69.0	73.6	72.1	65.3	53.5	43.0	32.6	51.5
CONFLUENCE 1 SW DAM		MAX	35.4	37.9	47.6	60.2	70.8	79.0	82.4	81.2	75.5	63.8	50.9	39.2	60.3
		MIN	16.9	17.2	25.6	35.2	43.9	52.6	57.0	56.7	49.8	38.1	⁄30.2	21.5	37.1
		MEAN	26.2	27.5	36.6	47.7	57.4	65.8	69.7	69.0	62.7	51.0	40.6	30.4	48.7
CORRY		MAX	31.2	34.0	43.5	57.7	68.9	77.4	80.7	78.9	72.6	61.4	47.1	35.4	57.4
		MIN	15.4	15.5	23.9	34.1	43.3	52.4	56.2	55.2	49.2	39.3	31.4	21.3	36.4
		MEAN	23.3	24.8	33.8	46.0	56.1	64.9	68.4	67.1	60.9	50.4	39.3	28.4	47.0
DEVAULT 1 W		MAX	37.3	39.8	49.0	61.9	71.5	79.7	84.5	82.9	76.2	64.9	52.5	41.1	61.8
		MIN	19.5	21.0	28.9	39.3	48.7	58.0	62.3	61.4	53.9	43.1	33.4	23.7	41.1
		MEAN	28.4	30.4	39.0	50.6	60.1	68.9	73.4	72.2	65.0	54.1	42.9	32.4	51.5
DONEGAL		MAX	33.2	35.5	45.4	58.0	68.1	76.2	79.8	78.6	72.6	60.7	47.9	37.4	57.8
		MIN	14.6	15.7	24.0	33.9	42.5	51.3	55.5	54.5	47.7	36.9	28.4	19.4	35.4
		MEAN	23.9	25.6	34.8	45.9	55.4	63.8	67.6	66.6	60.1	48.8	38.2	28.4	46.6
DONORA 1 SW		MAX	39.4	42.6	53.2	65.8	75.0	82.3	85.5	84.3	78.8	67.6	54.8	43.5	64.4
		MIN	21.2	22.8	30.8	40.2	49.4	58.2	61.9	61.2	54.8	43.3	34.8	25.7	42.0
		MEAN	30.3	32.7	42.0	53.0	62.2	70.3	73.7	72.8	66.8	55.5	44.8	34.6	53.2
DREXEL UNIVERSITY		MAX	39.6	42.1	51.0	63.6	73.4	82.3	86.7	85.3	78.5	67.2	55.1	43.9	64.1
		MIN	26.2	27.5	35.1	44.7	54.4	63.6	68.8	68.2	61.5	50.2	40.8	31.1	47.7
		MEAN	33.0	34.8	43.1	54.2	63.9	72.9	77.8	76.8	70.0	58.8	48.0	37.5	55.9
EPHRATA		MAX	36.8	39.8	49.8	62.6	72.7	80.9	84.8	83.0	76.2	64.9	52.1	40.6	62.0
		MIN	21.2	22.6	30.6	40.6	50.0	58.8	63.2	62.2	55.5	44.6	35.2	25.7	42.5
		MEAN	29.0	31.2	40.2	51.6	61.4	69.9	74.0	72.6	65.9	54.8	43.6	33.2	52.3
ERIE WSO		MAX	30.9	32.2	41.1	53.7	64.6	74.0	78.2	77.0	71.0	60.1	47.1	35.7	55.5
		MIN	18.0	17.7	25.8	36.1	45.4	55.2	59.9	59.4	53.1	43.2	34.3	24.2	39.4
		MEAN	24.5	25.0	33.5	44.9	55.0	64.6	69.1	68.2	62.1	51.7	40.7	30.0	47.5

PENNSYLVANIA

TEMPERATURE NORMALS (DEG F)

STATION		JAN	FEB	MAR	APR	MAY	JUN	JUL	AUG	SEP	OCT	NOV	DEC	ANN
EVERETT 1 SW	MAX	36.8	38.7	48.1	61.5	72.1	80.4	84.4	83.1	76.7	65.0	51.9	40.1	61.6
	MIN	18.9	19.7	27.9	37.9	47.0	55.1	59.4	58.3	51.1	39.2	31.5	23.4	39.1
	MEAN	27.9	29.2	38.0	49.7	59.6	67.8	71.9	70.7	63.9	52.1	41.7	31.8	50.4
FARRELL SHARON	MAX	36.0	38.9	49.1	62.9	73.7	82.4	85.5	84.0	77.6	66.0	52.0	40.3	62.4
	MIN	20.1	21.2	29.3	38.7	48.1	56.8	60.7	59.9	53.1	42.5	34.8	25.2	40.9
	MEAN	28.1	30.1	39.2	50.8	60.9	69.6	73.2	72.0	65.4	54.3	43.4	32.8	51.7
FORD CITY 4 S DAM	MAX	34.6	37.0	47.2	60.0	70.7	78.9	82.8	81.5	75.5	63.9	50.8	39.1	60.2
	MIN	15.5	16.1	25.1	35.5	44.0	53.2	57.7	56.4	49.4	38.0	30.5	21.4	36.9
	MEAN	25.1	26.5	36.2	47.8	57.4	66.1	70.2	69.0	62.5	51.0	40.7	30.2	48.6
FRANKLIN	MAX	33.6	35.5	45.4	59.1	70.1	78.4	82.3	80.9	74.2	63.1	49.4	37.5	59.1
	MIN	16.1	16.2	25.0	35.0	44.1	53.4	57.5	56.7	50.0	39.4	31.6	22.0	37.3
	MEAN	24.9	25.9	35.2	47.1	57.1	65.9	69.9	68.8	62.1	51.3	40.6	29.8	48.2
FREELAND	MAX	31.1	33.5	42.4	56.3	67.3	75.3	79.1	77.5	70.3	60.0	46.4	34.8	56.2
	MIN	15.3	16.5	24.3	35.7	45.6	54.9	59.6	58.0	51.2	40.5	31.1	19.8	37.7
	MEAN	23.3	25.0	33.4	46.0	56.5	65.1	69.4	67.7	60.8	50.3	38.8	27.3	47.0
GETTYSBURG	MAX	37.6	40.0	49.7	62.5	72.7	81.4	86.0	84.6	77.7	65.8	53.2	41.5	62.7
	MIN	21.0	22.6	30.9	41.5	50.4	59.2	63.8	62.4	55.1	43.3	34.7	25.2	42.5
	MEAN	29.4	31.3	40.3	52.0	61.6	70.3	75.0	73.5	66.4	54.6	43.9	33.4	52.6
GREENVILLE	MAX	34.1	37.2	47.3	61.2	72.3	80.8	84.4	83.0	76.9	64.8	50.1	38.3	60.9
	MIN	16.9	18.0	26.4	36.0	45.1	54.1	57.8	56.7	50.3	39.8	32.0	22.8	38.0
	MEAN	25.5	27.6	36.9	48.6	58.7	67.5	71.1	69.9	63.6	52.3	41.1	30.6	49.5
HANOVER	MAX	38.4	40.7	50.7	63.4	73.5	81.9	86.4	85.0	78.4	66.6	54.0	42.3	63.4
	MIN	20.9	21.8	30.0	40.6	49.5	58.8	63.2	61.5	54.3	42.6	34.1	25.2	41.9
	MEAN	29.7	31.3	40.4	52.1	61.5	70.4	74.8	73.3	66.4	54.6	44.1	33.8	52.7
HARRISBURG WSO R	MAX	36.7	39.5	49.6	62.9	73.0	81.8	86.2	84.4	77.2	65.4	52.4	40.6	62.5
	MIN	22.1	23.5	31.5	41.5	51.0	60.5	65.3	64.2	56.6	44.6	35.4	26.2	43.5
	MEAN	29.4	31.5	40.6	52.2	62.0	71.2	75.8	74.3	66.9	55.0	43.9	33.4	53.0
HOLTWOOD	MAX	37.5	39.8	48.6	61.4	71.6	80.5	85.0	83.4	76.8	65.2	52.7	41.5	62.0
	MIN	23.4	24.5	32.6	42.5	52.5	62.0	67.1	66.2	59.1	47.1	36.8	27.5	45.1
	MEAN	30.5	32.1	40.6	52.0	62.1	71.3	76.1	74.8	67.9	56.2	44.8	34.6	53.6
INDIANA 3 SE	MAX	35.6	38.5	48.6	61.9	71.8	79.4	82.8	81.3	75.2	63.9	51.0	39.8	60.8
	MIN	17.1	18.1	26.7	35.9	45.1	53.5	57.8	57.1	50.6	39.1	31.1	22.1	37.9
	MEAN	26.4	28.3	37.7	48.9	58.5	66.5	70.3	69.2	62.9	51.5	41.1	31.0	49.4
JAMESTOWN 2 NW	MAX	32.0	33.9	43.7	57.3	68.4	77.3	81.4	80.1	74.0	62.4	48.6	36.5	58.0
	MIN	13.8	14.2	24.1	34.5	43.8	53.0	56.9	55.8	48.9	38.5	30.6	20.5	36.2
	MEAN	22.9	24.1	33.9	45.9	56.1	65.2	69.1	68.0	61.5	50.5	39.6	28.5	47.1
JOHNSTOWN	MAX	36.6	38.7	48.3	61.9	72.5	81.4	85.6	83.8	77.2	65.2	52.1	40.4	62.0
	MIN	19.2	20.4	29.0	39.2	48.2	56.1	60.3	58.9	52.0	40.7	33.2	24.0	40.1
	MEAN	27.9	29.6	38.7	50.6	60.4	68.8	72.9	71.4	64.7	53.0	42.7	32.2	51.1
KANE 1 NNE	MAX	30.1	32.1	41.7	55.1	67.0	75.1	79.1	77.2	70.6	59.3	45.5	33.6	55.5
	MIN	11.0	10.3	20.1	30.7	38.6	47.3	51.1	50.0	43.3	33.5	27.3	17.0	31.7
	MEAN	20.6	21.2	30.9	42.9	52.9	61.2	65.1	63.6	57.0	46.4	36.4	25.3	43.6
LANDISVILLE 2 NW	MAX	36.6	39.1	49.3	62.0	72.0	80.8	85.0	83.2	77.1	65.5	52.7	40.7	62.0
	MIN	19.2	20.3	29.6	38.7	48.1	57.5	61.8	60.4	53.0	41.2	32.7	23.5	40.5
	MEAN	27.9	29.8	39.5	50.4	60.1	69.2	73.5	71.9	65.1	53.4	42.7	32.1	51.3
LEWISTOWN	MAX	36.7	39.4	49.2	62.6	73.6	81.5	85.8	84.1	77.4	65.8	52.5	40.1	62.4
	MIN	19.4	20.4	28.6	38.3	47.1	56.0	60.6	59.7	52.8	41.2	33.0	23.8	40.1
	MEAN	28.1	29.9	38.9	50.5	60.4	68.8	73.2	71.9	65.1	53.5	42.8	32.0	51.3
MADERA	MAX	31.7	33.9	43.6	57.8	69.1	77.0	80.7	78.8	72.0	60.2	47.3	35.4	57.3
	MIN	12.5	12.4	21.4	31.3	39.9	48.8	53.2	52.4	45.3	34.0	27.2	17.5	33.0
	MEAN	22.1	23.2	32.5	44.5	54.5	63.0	67.0	65.6	58.7	47.1	37.3	26.4	45.2

PENNSYLVANIA

TEMPERATURE NORMALS (DEG F)

STATION		JAN	FEB	MAR	APR	MAY	JUN	JUL	AUG	SEP	OCT	NOV	DEC	ANN
MARCUS HOOK	MAX	39.3	42.1	50.9	63.1	73.4	82.5	86.7	85.1	77.9	66.4	54.4	43.4	63.8
	MIN	26.6	28.0	35.5	44.9	54.7	63.9	69.0	68.1	61.1	50.0	40.4	30.9	47.8
	MEAN	33.0	35.0	43.2	54.1	64.1	73.2	77.9	76.7	69.5	58.2	47.4	37.2	55.8
MEADVILLE 1 S	MAX	32.1	33.9	43.8	57.0	68.4	77.0	80.9	79.5	73.5	62.0	48.4	36.5	57.8
	MIN	14.8	14.6	24.1	34.4	43.2	52.6	56.4	55.3	48.6	38.4	30.9	21.0	36.2
	MEAN	23.5	24.3	33.9	45.7	55.8	64.8	68.7	67.4	61.1	50.3	39.7	28.8	47.0
MONTGOMERY L AND D	MAX	36.2	39.4	49.9	63.2	73.6	81.4	84.4	82.6	75.8	64.2	51.6	40.3	61.9
	MIN	20.1	21.5	29.4	38.9	48.2	57.0	61.4	60.6	54.3	43.0	34.4	25.1	41.2
	MEAN	28.2	30.5	39.6	51.1	60.9	69.2	72.9	71.6	65.1	53.6	43.0	32.7	51.5
MONTROSE	MAX	28.7	30.6	39.5	53.3	64.9	73.9	78.5	76.8	69.5	58.2	45.0	32.9	54.3
	MIN	12.0	12.4	21.5	32.8	42.4	51.7	56.0	54.3	47.5	37.3	28.6	17.6	34.5
	MEAN	20.4	21.5	30.5	43.1	53.7	62.8	67.3	65.6	58.5	47.8	36.8	25.3	44.4
MORGANTOWN	MAX	37.2	39.2	48.4	60.6	70.5	79.1	83.6	81.9	75.8	64.5	52.3	41.2	61.2
	MIN	18.8	20.3	28.6	38.4	47.7	57.0	62.0	60.6	53.5	41.9	33.2	23.2	40.4
	MEAN	28.0	29.8	38.5	49.5	59.1	68.0	72.8	71.3	64.7	53.2	42.8	32.2	50.8
NEW CASTLE 1 N	MAX	35.9	39.1	49.6	63.2	73.5	81.4	84.5	83.3	77.5	66.0	51.8	40.1	62.2
	MIN	18.2	19.0	27.0	36.5	45.6	54.8	58.6	57.8	51.2	40.0	32.7	23.6	38.8
	MEAN	27.1	29.1	38.4	49.9	59.6	68.1	71.6	70.6	64.4	53.1	42.3	31.9	50.5
NEWELL	MAX	38.0	40.5	51.4	64.2	74.0	82.1	85.9	85.0	79.3	67.0	54.0	42.5	63.7
	MIN	22.1	23.0	31.0	40.2	49.5	57.8	62.4	62.2	55.2	43.6	35.3	27.1	42.5
	MEAN	30.1	31.8	41.2	52.2	61.8	70.0	74.1	73.6	67.3	55.3	44.7	34.9	53.1
NEWPORT	MAX	36.4	39.1	49.5	63.0	73.1	81.7	85.9	84.1	76.8	64.9	52.1	39.9	62.2
	MIN	18.6	19.3	27.9	38.0	46.8	56.5	61.0	59.8	52.4	40.6	32.0	23.1	39.7
	MEAN	27.5	29.2	38.7	50.5	60.0	69.1	73.5	72.0	64.6	52.8	42.1	31.5	51.0
PALMERTON	MAX	35.3	37.8	47.2	60.7	71.2	79.7	84.3	82.5	74.7	63.9	50.7	38.9	60.6
	MIN	19.0	20.4	28.6	38.1	47.7	56.9	61.7	60.3	52.6	41.2	33.0	23.5	40.3
	MEAN	27.1	29.1	37.9	49.4	59.4	68.4	73.0	71.4	63.7	52.6	41.9	31.2	50.4
PHILADELPHIA WSO R	MAX	38.6	41.1	50.5	63.2	73.0	81.7	86.1	84.6	77.8	66.5	54.5	43.0	63.4
	MIN	23.8	25.0	33.1	42.6	52.5	61.5	66.8	66.0	58.6	46.5	37.1	28.0	45.1
	MEAN	31.2	33.1	41.8	52.9	62.8	71.6	76.5	75.3	68.2	56.5	45.8	35.5	54.3
PHILIPSBURG FAA AP	MAX	30.1	32.6	41.8	55.9	66.8	74.8	78.5	76.7	69.8	58.7	45.3	34.0	55.4
	MIN	14.1	14.9	23.7	33.9	42.8	50.8	54.6	53.6	46.6	37.1	29.3	19.2	35.1
	MEAN	22.1	23.8	32.8	44.9	54.8	62.8	66.6	65.2	58.2	47.9	37.3	26.6	45.3
PHOENIXVILLE 1 E	MAX	40.0	42.9	52.5	65.1	75.3	83.2	87.4	85.8	78.9	67.7	55.4	43.9	64.8
	MIN	20.1	21.6	30.1	39.1	48.9	57.8	62.3	60.6	53.9	42.3	33.7	24.7	41.3
	MEAN	30.1	32.2	41.3	52.1	62.1	70.5	74.9	73.2	66.5	55.0	44.6	34.3	53.1
PITTSBURGH AP WSO 2	MAX	34.1	36.8	47.6	60.7	70.8	79.1	82.7	81.1	74.8	62.9	49.8	38.4	59.9
	MIN	19.2	20.7	29.4	39.4	48.5	57.1	61.3	60.1	53.3	42.1	33.3	24.3	40.7
	MEAN	26.7	28.8	38.5	50.1	59.7	68.1	72.0	70.6	64.1	52.5	41.6	31.4	50.3
PLEASANT MOUNT 1 W	MAX	28.5	30.3	38.8	52.6	64.0	72.6	77.1	75.5	68.2	57.5	44.6	32.6	53.5
	MIN	9.7	9.9	19.3	31.1	40.3	49.3	53.5	52.0	45.0	34.7	26.5	15.2	32.2
	MEAN	19.1	20.1	29.1	41.9	52.2	61.0	65.3	63.8	56.6	46.1	35.6	23.9	42.9
PORT CLINTON	MAX	36.1	38.3	47.9	60.8	71.2	79.9	84.7	82.8	75.6	65.0	52.1	40.0	61.2
	MIN	15.3	16.3	25.1	34.8	43.4	52.6	56.8	55.6	48.0	36.1	28.9	20.0	36.1
	MEAN	25.8	27.3	36.5	47.8	57.3	66.3	70.8	69.3	61.8	50.6	40.5	30.0	48.7
PUTNEYVILLE 2 SE DAM	MAX	33.6	35.6	45.7	58.8	70.0	78.3	82.1	80.5	74.1	62.5	49.3	37.6	59.0
	MIN	13.9	14.3	23.6	34.0	42.9	51.8	55.9	54.7	47.8	37.0	29.3	19.6	35.4
	MEAN	23.8	25.0	34.7	46.4	56.5	65.1	69.1	67.6	61.0	49.8	39.3	28.6	47.2
RIDGWAY	MAX	32.6	34.9	44.6	57.9	69.5	77.3	80.9	79.3	72.9	61.7	48.0	36.4	58.0
	MIN	13.2	12.9	21.9	32.0	40.5	49.5	53.7	53.0	46.4	35.6	28.8	18.8	33.9
	MEAN	22.9	23.9	33.3	45.0	55.0	63.4	67.4	66.2	59.7	48.7	38.4	27.6	46.0

PENNSYLVANIA

TEMPERATURE NORMALS (DEG F)

STATION		JAN	FEB	MAR	APR	MAY	JUN	JUL	AUG	SEP	OCT	NOV	DEC	ANN
SHIPPENSBURG	MAX	36.8	40.0	50.6	63.6	73.6	81.5	85.5	84.3	78.0	66.2	52.0	40.5	62.7
	MIN	21.2	22.5	30.2	40.3	49.7	58.5	62.6	61.5	54.5	43.3	34.4	25.4	42.0
	MEAN	29.1	31.3	40.4	52.0	61.6	70.0	74.1	72.9	66.3	54.7	43.2	33.0	52.4
SLIPPERY ROCK 1 SSW	MAX	33.6	36.3	46.2	59.4	70.4	78.5	82.1	80.7	74.5	63.3	49.5	37.8	59.4
	MIN	16.1	17.1	25.9	35.9	45.1	53.7	57.5	56.2	49.9	39.5	31.5	22.0	37.5
	MEAN	24.9	26.8	36.1	47.6	57.8	66.1	69.8	68.5	62.3	51.4	40.5	30.0	48.5
STATE COLLEGE	MAX	33.4	35.6	45.2	58.7	69.9	78.1	82.2	80.5	73.5	61.9	48.8	37.0	58.7
	MIN	18.4	19.2	27.4	38.4	48.1	56.9	60.9	59.5	52.1	41.4	32.9	23.2	39.9
	MEAN	25.9	27.4	36.3	48.5	59.0	67.5	71.5	70.0	62.8	51.7	40.9	30.1	49.3
STROUDSBURG	MAX	35.2	38.1	47.9	61.5	72.8	80.8	85.5	83.4	75.9	64.6	50.9	38.8	61.3
	MIN	16.6	17.7	26.4	36.0	45.1	53.4	58.0	57.0	50.0	38.6	31.0	21.0	37.6
	MEAN	25.9	27.9	37.2	48.8	59.0	67.1	71.8	70.2	63.0	51.6	41.0	29.9	49.5
TIONESTA 2 SE DAM	MAX	32.1	33.9	44.0	57.3	69.0	77.0	80.8	79.3	73.1	61.7	48.2	35.9	57.7
	MIN	12.7	12.3	21.9	32.9	42.1	51.7	56.4	55.4	48.7	37.6	29.6	18.9	35.0
	MEAN	22.4	23.2	33.0	45.1	55.6	64.4	68.6	67.4	60.9	49.7	39.0	27.4	46.4
TOWANDA 1 ESE	MAX	34.4	36.8	45.9	60.0	70.9	79.3	83.2	81.4	74.3	63.2	49.8	38.3	59.8
	MIN	16.2	16.8	25.6	35.2	44.7	53.7	57.9	56.9	50.1	39.4	31.4	21.2	37.4
	MEAN	25.4	26.9	35.8	47.7	57.8	66.5	70.6	69.2	62.2	51.3	40.6	29.8	48.7
UNIONTOWN 1 NE	MAX	39.1	41.4	51.7	63.8	73.3	80.7	83.8	82.6	76.9	65.5	53.8	43.2	63.0
	MIN	21.0	22.2	30.2	39.8	49.0	57.8	61.7	60.6	53.7	41.8	33.7	25.6	41.4
	MEAN	30.1	31.8	41.0	51.8	61.2	69.3	72.8	71.6	65.3	53.7	43.8	34.4	52.2
WARREN	MAX	33.1	35.5	44.8	59.0	70.2	78.8	82.1	80.3	73.3	61.8	48.3	36.8	58.7
	MIN	16.5	16.3	24.5	34.3	43.9	53.0	57.3	56.6	50.1	39.3	31.5	21.9	37.1
	MEAN	24.8	26.0	34.7	46.7	57.1	65.9	69.7	68.4	61.7	50.6	39.9	29.4	47.9
WAYNESBURG 1 E	MAX	38.3	41.1	51.3	63.3	73.2	80.7	84.1	83.0	77.4	65.5	53.2	42.3	62.8
	MIN	17.5	18.7	27.0	36.2	45.1	54.0	58.1	57.0	49.8	37.3	30.0	22.3	37.8
	MEAN	27.9	29.9	39.2	49.8	59.1	67.4	71.1	70.0	63.6	51.4	41.7	32.3	50.3
WELLSBORO 3 S	MAX	28.9	30.7	39.7	53.4	64.6	73.0	77.8	76.1	69.2	58.1	45.1	32.8	54.1
	MIN	12.9	12.8	21.7	33.2	42.4	51.5	55.7	54.3	47.4	37.8	29.3	18.2	34.8
	MEAN	20.9	21.8	30.7	43.3	53.5	62.3	66.8	65.2	58.3	48.0	37.2	25.5	44.5
WEST CHESTER 1 W	MAX	38.5	40.8	50.4	62.7	72.7	81.4	85.9	84.2	77.7	66.6	54.4	42.5	63.2
	MIN	20.9	22.0	30.3	40.3	49.6	59.0	63.7	62.1	54.5	43.0	34.4	24.9	42.1
	MEAN	29.8	31.4	40.3	51.5	61.1	70.2	74.8	73.2	66.1	54.8	44.4	33.7	52.6
W-BARRE-SCRANTON WSO	MAX	32.1	34.4	44.1	58.2	69.1	77.8	82.1	80.0	72.7	61.4	48.2	36.3	58.0
	MIN	18.2	19.2	28.1	38.4	48.1	56.9	61.4	60.0	52.8	42.0	33.6	23.1	40.1
	MEAN	25.2	26.8	36.1	48.3	58.6	67.4	71.8	70.0	62.8	51.7	40.9	29.7	49.1
WILLIAMSPORT WSO	MAX	34.1	36.8	46.8	60.6	71.2	79.7	83.7	82.0	74.5	63.0	49.6	38.1	60.0
	MIN	18.3	19.5	28.4	38.5	48.0	56.8	61.3	60.3	53.2	41.6	33.2	23.3	40.2
	MEAN	26.2	28.2	37.6	49.6	59.6	68.3	72.5	71.1	63.9	52.3	41.4	30.7	50.1
YORK 3 SSW PUMP STA	MAX	39.3	42.6	52.8	65.3	75.6	83.4	87.2	85.9	79.6	68.4	54.6	43.0	64.8
	MIN	20.7	22.0	30.2	39.4	48.7	57.6	62.1	61.1	54.0	41.9	33.6	24.9	41.4
	MEAN	30.0	32.3	41.5	52.4	62.1	70.5	74.7	73.5	66.9	55.2	44.1	34.0	53.1

PENNSYLVANIA

PRECIPITATION NORMALS (INCHES)

STATION	JAN	FEB	MAR	APR	MAY	JUN	JUL	AUG	SEP	OCT	NOV	DEC	ANN
ACMETONIA LOCK 3	2.97	2.58	3.74	3.79	3.66	3.80	4.31	3.76	3.18	2.86	2.66	2.84	40.15
ALLENTOWN WSO //R	3.35	3.02	3.88	3.93	3.57	3.45	4.13	4.44	4.03	3.05	3.73	3.73	44.31
BAKERSTOWN 3 WNW	3.01	2.41	3.05	3.60	3.81	3.75	4.00	3.53	3.13	2.91	2.69	2.81	38.70
BARNES	3.30	2.72	3.54	3.79	3.94	4.57	4.55	4.03	3.92	3.27	3.71	3.52	44.86
BEAR GAP	2.55	2.30	3.16	3.51	4.02	4.05	4.06	3.68	4.29	3.38	3.78	2.88	41.66
BEAVER FALLS	2.55	2.26	3.25	3.20	3.57	3.68	3.85	3.37	3.08	2.57	2.40	2.54	36.32
BEAVERTOWN	2.73	2.47	3.43	3.59	3.89	3.81	3.94	3.61	3.96	3.25	3.56	3.11	41.35
BETHLEHEM	3.25	2.76	3.57	3.49	3.37	3.41	4.07	4.52	4.12	3.08	3.61	3.60	42.85
BLOSERVILLE 1 N	2.90	2.48	3.52	3.67	3.69	3.93	3.51	3.68	3.85	3.14	3.24	3.00	40.61
BRADDOCK LOCK 2	2.71	2.36	3.56	3.63	3.62	4.08	3.91	3.78	2.99	2.51	2.43	2.60	38.18
BRADFORD FAA AP	3.16	2.88	3.44	3.43	3.73	4.26	4.16	3.84	3.79	3.11	3.47	3.45	42.71
BRADFORD CENTRAL FS	2.75	2.27	3.22	3.64	3.57	4.16	3.91	3.82	4.10	3.17	3.54	3.17	41.32
BRADFORD 4 W RES	2.86	2.55	3.29	3.64	3.84	4.53	4.39	3.96	4.39	3.54	3.93	3.48	44.40
BRUCETON 1 S	2.78	2.32	3.41	3.51	3.58	4.35	4.18	3.57	3.08	2.42	2.37	2.66	38.23
BUFFALO MILLS	2.66	2.44	3.69	3.47	3.76	3.80	3.77	3.42	3.12	2.82	2.80	2.78	38.53
BURGETTSTOWN 2 W	3.00	2.50	3.69	3.62	4.20	3.92	4.34	3.70	3.15	2.65	2.65	2.77	40.19
CARLISLE	2.89	2.70	3.61	3.71	3.79	3.93	3.39	3.44	3.76	3.09	3.27	3.11	40.69
CARROLLTOWN 2 SSE	3.06	2.56	3.58	3.93	4.30	3.98	4.64	3.91	3.57	2.73	3.01	2.99	42.26
CEDAR RUN	2.67	2.42	3.50	3.39	4.07	4.03	3.96	3.58	3.97	3.14	3.39	3.09	41.21
CHADDS FORD	3.52	3.12	4.10	3.81	3.70	3.66	4.31	4.26	3.99	3.30	3.78	3.87	45.42
CHAMBERSBURG 1 ESE	2.96	2.59	3.86	3.57	3.58	4.34	3.41	3.71	3.42	2.84	3.13	3.20	40.61
CHARLEROI LOCK 4	2.94	2.42	3.67	3.66	3.73	3.57	3.71	3.58	3.00	2.47	2.48	2.68	37.91
CLARENCE	2.47	2.47	3.47	3.19	3.80	3.93	4.03	3.40	3.68	2.96	3.02	2.71	39.13
CLARION 3 SW	3.01	2.60	3.54	3.98	4.01	4.22	4.66	3.79	3.66	3.29	3.24	3.09	43.09
CLAUSSVILLE	3.54	3.02	3.99	3.96	3.83	3.99	4.20	4.75	4.38	3.22	3.91	3.78	46.57
CLEARFIELD	3.07	2.75	3.81	3.78	4.14	4.04	4.49	3.74	3.60	3.00	2.92	3.21	42.55
COATESVILLE 1 SW	3.29	2.88	3.84	3.83	3.88	4.62	4.38	4.27	3.94	3.28	3.55	3.83	45.59
CONFLUENCE 1 SW DAM	3.32	2.73	3.97	3.80	4.08	4.04	4.50	4.01	3.37	2.89	2.70	3.27	42.68
CONFLUENCE 1 NW	3.37	2.96	4.12	3.92	4.20	4.12	4.63	4.05	3.48	2.94	2.94	3.40	44.13
CONSHOHOCKEN	3.41	2.96	4.00	3.91	3.65	3.89	4.27	4.52	4.23	3.38	3.86	3.71	45.79
CORRY	3.42	2.79	3.62	4.00	3.73	4.21	4.46	4.32	3.70	3.69	4.35	4.11	46.40
CREEKSIDE	3.14	2.75	3.71	3.86	4.19	4.13	4.85	4.01	3.75	2.98	3.21	3.18	43.76
DANVILLE	2.66	2.41	3.15	3.34	3.81	3.83	4.13	3.75	3.81	3.00	3.20	2.95	40.04
DEVAULT 1 W	3.68	3.10	4.10	4.10	3.98	3.95	4.43	4.26	4.25	3.35	3.68	4.04	46.92
DONEGAL	3.53	3.06	4.09	4.24	4.12	4.74	4.42	4.43	3.74	3.19	3.11	3.54	46.21
DONORA 1 SW	2.65	2.20	3.40	3.30	3.64	3.65	3.55	3.77	2.93	2.54	2.26	2.44	36.33
DOYLESTOWN	3.22	2.73	3.69	3.69	3.86	3.67	4.17	4.42	4.06	3.51	3.62	3.55	44.19
DREXEL UNIVERSITY	3.47	3.08	4.08	3.80	3.58	3.89	4.62	4.42	3.70	3.14	3.67	3.62	45.07
ENGLISH CENTER	2.54	2.50	3.58	3.09	3.74	3.82	3.56	3.11	3.38	3.06	3.34	2.81	38.53
EPHRATA	3.11	2.53	3.54	3.88	3.53	4.19	4.20	4.47	4.05	3.16	3.42	3.42	43.50
ERIE WSO	2.49	2.12	2.91	3.49	3.28	3.72	3.28	3.85	3.89	3.37	3.74	3.25	39.39
EVERETT 1 SW	2.51	2.36	3.35	3.52	3.36	3.45	3.66	3.24	2.84	3.04	2.88	2.75	36.96
FARRELL SHARON	2.47	2.08	3.09	3.30	3.31	3.79	3.80	3.51	2.94	2.56	2.85	2.82	36.52
FORD CITY 4 S DAM	2.85	2.56	3.38	3.55	3.95	4.19	4.41	4.18	3.57	2.90	2.71	2.77	41.02
FRANKLIN	2.82	2.24	3.29	3.87	3.74	4.23	4.55	3.63	3.48	3.07	3.29	2.89	41.10
FREELAND	3.17	2.81	3.31	3.75	4.27	4.21	4.33	4.87	4.41	3.63	4.12	3.61	46.49
GALETON	2.63	2.38	3.47	3.30	3.69	3.91	4.01	3.58	3.38	2.98	3.34	3.00	39.67
GETTYSBURG	3.08	2.78	3.92	3.69	3.56	3.99	3.15	3.74	3.78	3.14	3.24	3.15	41.22
GLENWILLARD DASH DAM	2.69	2.21	3.40	3.43	3.56	3.38	3.68	3.40	2.81	2.56	2.33	2.40	35.85
GOULDSBORO	3.37	3.04	3.86	4.05	3.90	3.67	4.23	4.21	4.11	3.85	4.28	3.77	46.34
GREENSBORO LOCK 7	2.90	2.48	3.47	3.58	3.67	3.85	4.41	3.76	3.22	2.51	2.52	2.74	39.11
GREENVILLE	2.53	2.04	3.18	3.62	3.46	4.13	3.84	3.60	3.04	2.98	3.09	2.70	38.21
HANOVER	3.10	2.64	3.77	3.58	3.37	3.72	3.01	3.76	3.71	2.94	3.08	3.32	40.00
HARRISBURG WSO R	2.96	2.73	3.50	3.19	3.67	3.63	3.32	3.29	3.60	2.73	3.24	3.23	39.09
HAWLEY	2.87	2.60	3.24	3.46	3.44	3.33	3.24	3.93	3.48	3.23	3.44	3.27	39.53

PENNSYLVANIA

PRECIPITATION NORMALS (INCHES)

STATION		JAN	FEB	MAR	APR	MAY	JUN	JUL	AUG	SEP	OCT	NOV	DEC	ANN
HOLLISTERVILLE		3.04	2.82	3.55	3.73	3.79	3.89	3.55	4.00	3.69	3.49	3.83	3.50	42.88
HOLTWOOD		2.71	2.30	3.13	3.16	3.11	3.79	3.68	3.62	3.60	2.58	3.12	3.05	37.85
HONESDALE 4 NW		3.04	2.79	3.54	3.72	4.01	3.63	3.90	4.12	3.80	3.38	3.65	3.44	43.02
HUNTSDALE		2.93	2.62	3.72	3.57	3.66	3.95	3.26	3.31	3.64	2.91	3.20	3.14	39.91
HYNDMAN		2.45	2.41	3.53	3.38	3.56	3.65	3.81	3.43	3.00	2.64	2.63	2.71	37.20
INDIANA 3 SE		3.31	2.85	3.81	3.94	4.25	4.34	5.00	3.98	3.88	2.98	3.06	3.12	44.52
JAMESTOWN 2 NW		2.59	2.16	3.20	3.64	3.73	4.03	3.90	3.69	3.26	2.96	3.12	2.94	39.22
JOHNSTOWN		3.81	3.37	3.91	4.08	4.22	4.35	5.03	3.91	3.42	3.02	3.26	3.28	45.66
KANE 1 NNE		3.34	2.82	3.77	3.84	4.00	4.47	4.32	4.22	4.14	3.37	3.62	3.67	45.58
KITTANNING LOCK 7		2.92	2.55	3.52	3.77	3.77	4.03	4.06	4.15	3.56	2.87	2.77	2.99	40.96
KRESGEVILLE 3 W		3.51	2.96	4.00	4.06	3.98	3.86	4.57	4.83	4.32	3.66	4.29	4.05	48.09
LANDISVILLE 2 NW		2.80	2.17	3.20	3.45	3.58	4.10	4.37	3.61	3.63	2.91	3.26	3.11	40.19
LEHIGHTON		3.47	3.05	4.04	3.87	4.07	3.68	4.19	4.86	4.25	3.60	4.28	4.08	47.44
LE ROY		2.20	2.13	2.88	2.85	3.57	3.42	3.31	2.91	3.28	2.71	3.08	2.66	35.00
LEWISTOWN		2.39	2.25	3.24	3.07	3.74	4.05	3.71	3.40	3.24	2.96	3.05	2.76	37.86
LONG POND 2 W		3.82	3.32	4.28	4.28	4.33	4.15	4.52	4.53	4.65	4.53	4.64	4.18	51.23
MADERA		2.35	2.21	3.22	3.44	3.99	3.91	4.48	3.83	3.39	2.63	2.57	2.33	38.35
MAPLETON DEPOT		2.56	2.36	3.63	3.31	3.33	3.56	3.77	3.10	3.31	2.90	2.92	2.73	37.48
MARCUS HOOK		3.18	3.00	4.11	3.85	3.44	3.71	3.93	4.07	3.85	3.01	3.70	3.69	43.54
MARION CENTER 2 SE		3.72	3.21	4.21	4.24	4.44	4.48	4.99	4.28	4.05	3.22	3.36	3.77	47.97
MATAMORAS		3.20	2.87	3.77	4.01	3.59	3.80	3.99	4.11	3.90	3.42	3.99	3.56	44.21
MC KEESPORT		2.76	2.29	3.23	3.36	3.63	3.99	3.72	3.26	2.92	2.34	2.22	2.48	36.20
MEADVILLE 1 S		2.94	2.47	3.43	3.67	3.79	4.32	4.39	4.40	3.39	3.49	3.55	3.35	43.19
MERCER		3.00	2.55	3.62	3.93	3.89	4.47	4.09	3.74	3.41	3.00	3.27	3.02	41.99
MERCERSBURG		3.00	2.82	4.03	3.65	3.78	4.31	3.35	3.81	3.36	3.27	2.99	3.10	41.47
MILLHEIM		2.76	2.52	3.70	3.46	3.90	3.96	4.07	3.65	3.62	3.04	3.40	2.91	40.99
MILLVILLE 2 SW		2.71	2.47	3.25	3.34	3.86	3.96	3.73	3.66	3.75	3.17	3.24	3.00	40.14
MONTGOMERY L AND D		2.70	2.20	3.43	3.41	3.64	3.43	4.09	3.56	2.94	2.52	2.48	2.65	37.05
MONTROSE		2.95	2.69	3.44	3.51	3.72	3.99	3.71	3.86	3.97	3.43	3.58	3.29	42.14
MORGANTOWN		3.04	2.62	3.31	3.93	3.64	3.83	3.81	4.24	3.48	3.08	3.35	3.19	41.52
MYERSTOWN		3.31	2.66	3.55	3.61	3.86	4.09	3.70	3.99	4.18	3.04	3.51	3.41	42.91
NATRONA LOCK 4		2.93	2.49	3.59	3.80	3.62	3.81	4.13	3.83	3.15	2.74	2.52	2.79	39.40
NESHAMINY FALLS		3.37	3.10	4.38	3.80	3.86	3.65	4.92	5.19	4.10	3.36	3.88	3.74	47.35
NEW CASTLE 1 N		2.66	2.25	3.33	3.46	3.54	4.19	3.94	3.54	3.29	2.71	2.66	2.65	38.22
NEWELL		2.86	2.56	3.81	3.88	4.15	3.93	3.95	3.84	3.23	2.59	2.79	2.92	40.51
NEW PARK		2.86	2.67	3.86	3.80	3.95	4.16	4.09	4.44	4.09	3.04	3.56	3.41	43.93
NEWPORT		2.83	2.48	3.56	3.45	3.75	3.93	3.68	3.63	3.82	3.27	3.41	3.16	40.97
NEW TRIPOLI		3.46	3.19	4.10	4.30	4.08	4.23	3.93	4.46	4.14	3.50	4.20	3.97	47.56
NORRISTOWN		3.29	2.95	4.07	3.63	3.64	3.59	4.18	4.46	4.10	3.18	3.65	3.71	44.45
PALM 3 SE		3.52	2.99	4.04	4.34	3.71	3.82	4.02	4.42	3.90	3.32	3.71	3.97	45.76
PALMERTON		2.99	2.54	3.44	3.53	3.69	3.53	4.40	4.52	3.95	3.32	3.70	3.33	42.94
PARKER		2.77	2.28	3.21	3.55	3.61	4.28	4.03	3.48	3.17	2.91	2.83	2.85	38.97
PAUPACK 2 WNW		2.90	2.73	3.47	3.55	3.57	3.46	3.60	4.22	3.61	3.34	3.61	3.33	41.39
PHILADELPHIA WSO R		3.18	2.81	3.86	3.47	3.18	3.92	3.88	4.10	3.42	2.83	3.32	3.45	41.42
PHILIPSBURG FAA AP		2.52	2.65	3.54	3.32	3.81	4.27	4.02	3.63	3.57	3.08	3.04	2.65	40.10
PHOENIXVILLE 1 E		3.23	2.94	4.08	3.70	3.55	3.62	4.13	4.05	3.74	3.13	3.60	3.78	43.55
PITTSBURGH AP WSO 2		2.86	2.40	3.58	3.28	3.54	3.30	3.83	3.31	2.80	2.49	2.34	2.57	36.29
PLEASANT MOUNT 1 W		3.40	3.05	3.74	4.13	4.20	4.37	3.99	4.33	4.42	3.86	4.16	3.79	47.44
PORT CLINTON		3.73	3.10	4.07	4.14	4.19	4.30	4.55	4.44	4.53	3.60	4.08	4.03	48.76
PUTNEYVILLE 2 SE DAM		3.11	2.85	3.73	3.90	4.00	4.06	4.52	4.14	3.81	3.24	3.00	3.15	43.51
READING 3N R		3.32	2.83	3.65	3.83	3.63	3.91	3.91	3.75	3.78	2.83	3.62	3.62	42.68
RIDGWAY		2.78	2.44	3.52	3.67	4.09	4.12	4.30	3.91	3.43	2.96	3.16	2.99	41.37
RUSHVILLE		2.18	2.08	2.58	3.09	3.49	3.64	3.67	3.57	3.61	2.96	2.90	2.49	36.26
SAXTON		2.66	2.34	3.41	3.50	3.37	3.81	4.08	3.28	3.09	2.82	2.86	2.68	37.90
SCHENLEY LOCK 5		2.85	2.57	3.62	3.76	3.81	4.03	4.07	3.85	3.33	2.77	2.68	2.82	40.16

PENNSYLVANIA

PRECIPITATION NORMALS (INCHES)

STATION	JAN	FEB	MAR	APR	MAY	JUN	JUL	AUG	SEP	OCT	NOV	DEC	ANN
SHAMOKIN	3.12	2.76	3.52	3.43	3.87	3.85	3.77	3.68	4.20	3.17	3.64	3.43	42.44
SHIPPENSBURG	2.80	2.53	3.61	3.34	3.53	3.93	3.39	3.25	3.46	2.73	3.11	3.12	38.80
SINNEMAHONING	2.63	2.53	3.54	3.41	3.96	3.88	4.05	3.80	3.51	2.98	3.18	2.96	40.43
SLIPPERY ROCK 1 SSW	2.77	2.37	3.16	3.61	3.72	4.52	4.14	3.63	3.30	2.88	2.81	2.82	39.73
SOMERSET	3.40	2.86	4.01	4.12	4.39	4.38	4.09	3.98	3.29	2.82	3.20	3.24	43.78
SOUTH MOUNTAIN	3.53	3.16	4.20	4.02	4.04	4.81	3.63	4.01	4.25	3.59	3.66	3.48	46.38
SPRING GROVE	2.96	2.63	3.70	3.49	3.38	3.98	3.13	3.71	3.52	2.89	3.05	3.14	39.58
STATE COLLEGE	2.60	2.35	3.40	3.23	3.64	3.75	3.56	3.40	3.14	2.83	3.11	2.64	37.65
STRAUSSTOWN	3.41	2.84	3.85	3.90	3.96	4.12	4.31	4.42	4.36	3.42	3.98	3.74	46.31
STROUDSBURG	3.56	3.15	4.20	4.30	3.93	3.81	4.04	4.20	4.49	3.81	4.16	4.34	47.99
SUSQUEHANNA	2.78	2.48	3.19	3.46	3.35	3.67	3.34	3.87	3.58	3.18	3.39	2.93	39.22
TAMAQUA	3.75	3.24	4.40	4.23	4.26	4.23	4.44	4.94	4.54	3.94	4.58	4.16	50.71
TAMAQUA 4 N DAM	3.70	3.15	4.19	4.11	4.31	4.19	4.52	4.56	4.47	3.78	4.38	3.85	49.21
TIONESTA 2 SE DAM	2.93	2.53	3.54	4.01	4.07	4.08	4.47	3.83	3.43	3.28	3.34	3.20	42.71
TOWANDA 1 ESE	2.09	2.15	2.78	2.96	3.34	3.16	3.04	3.17	3.34	2.68	2.87	2.40	33.98
TROY	2.35	2.26	2.87	3.03	3.28	3.17	3.05	3.18	3.18	2.76	3.02	2.60	34.75
UNION CITY FILT PLANT	3.10	2.58	3.49	3.73	3.76	4.26	4.11	3.85	3.87	3.71	3.87	3.65	43.98
UNIONTOWN 1 NE	2.94	2.46	3.59	3.72	4.10	4.05	4.14	3.92	3.21	2.56	2.66	2.94	40.29
VANDERGRIFT	2.80	2.41	3.46	3.62	3.66	4.18	4.30	4.25	3.46	2.94	2.63	2.66	40.37
WARREN	2.83	2.37	3.28	3.61	3.91	4.51	4.08	4.10	3.88	3.43	3.61	3.28	42.89
WAYNESBURG 1 E	2.87	2.39	3.54	3.63	4.03	3.81	4.34	3.97	2.90	2.51	2.51	2.66	39.16
WELLSBORO 3 S	2.02	1.88	2.50	2.77	3.36	3.49	3.41	3.19	3.30	2.63	2.72	2.35	33.62
WEST CHESTER 1 W	3.48	2.96	4.01	3.79	3.70	4.00	4.48	4.36	4.12	3.25	3.75	3.83	45.73
WHITESBURG	3.01	2.56	3.24	3.59	3.76	4.01	4.22	3.91	3.62	2.93	2.81	2.80	40.46
WILKES-BARRE 4 NE	2.43	2.27	2.73	3.15	3.31	3.55	3.79	3.73	3.77	3.07	3.12	2.82	37.74
W-BARRE-SCRANTON WSO	2.27	2.05	2.63	3.01	3.16	3.42	3.39	3.47	3.36	2.78	2.98	2.54	35.08
WILLIAMSBURG	2.87	2.62	3.86	3.58	3.74	3.95	4.18	3.39	3.70	3.01	3.03	2.86	40.79
WILLIAMSPORT WSO	2.88	2.83	3.66	3.53	3.66	3.88	3.92	3.26	3.57	3.22	3.63	3.24	41.28
WOLFSBURG	2.58	2.34	3.50	3.34	3.62	3.55	4.12	3.33	3.16	2.89	2.70	2.69	37.82
YORK 3 SSW PUMP STA	2.92	2.52	3.52	3.70	3.47	4.29	3.39	3.81	3.64	2.96	3.22	3.19	40.63
YORK HAVEN	2.88	2.50	3.24	3.30	3.45	3.77	3.11	3.37	3.43	2.72	3.12	3.04	37.93
ZIONSVILLE 3 SE	3.21	2.73	3.67	4.04	3.79	3.90	4.01	4.48	4.10	3.51	3.82	3.66	44.92

PENNSYLVANIA

HEATING DEGREE DAY NORMALS (BASE 65 DEG F)

STATION		JUL	AUG	SEP	OCT	NOV	DEC	JAN	FEB	MAR	APR	MAY	JUN	ANN
ALLENTOWN WSO	//R	0	6	85	364	675	1039	1172	1000	822	450	190	12	5815
BAKERSTOWN 3 WNW		7	12	91	364	672	1023	1166	988	800	420	198	33	5774
BRADFORD FAA AP		59	82	243	564	879	1246	1398	1221	1070	666	390	136	7954
BRADFORD 4 W RES		51	67	220	521	828	1215	1352	1182	1035	636	349	117	7573
BURGETTSTOWN 2 W		13	30	146	474	750	1094	1246	1084	890	528	272	68	6595
CARLISLE		0	0	55	313	633	989	1107	938	747	374	154	13	5323
CHAMBERSBURG 1 ESE		0	0	71	364	675	1008	1128	960	784	417	172	15	5594
CLARION 3 SW		21	37	143	465	774	1132	1277	1106	933	546	273	80	6787
COATESVILLE 1 SW		0	0	71	362	660	1004	1128	974	797	435	181	18	5630
CONFLUENCE 1 SW DAM		7	16	119	434	732	1073	1203	1050	880	519	256	60	6349
CORRY		25	42	157	453	771	1135	1293	1126	967	570	293	73	6905
DEVAULT 1 W		0	8	96	342	663	1011	1135	969	806	432	185	19	5666
DONEGAL		32	58	185	502	804	1135	1274	1103	936	573	320	109	7031
DONORA 1 SW		0	0	59	306	606	942	1076	904	713	368	169	21	5164
DREXEL UNIVERSITY		0	0	19	211	510	853	992	846	679	328	112	0	4550
EPHRATA		0	0	59	322	642	986	1116	946	769	402	153	15	5410
ERIE WSO		17	28	130	420	729	1085	1256	1120	977	603	323	80	6768
EVERETT 1 SW		7	18	99	404	699	1029	1150	1002	837	459	213	51	5968
FARRELL SHARON		5	8	94	349	648	998	1144	977	800	431	187	36	5677
FORD CITY 4 S DAM		8	19	125	434	729	1079	1237	1078	893	516	258	55	6431
FRANKLIN		9	24	134	431	732	1091	1243	1095	924	537	273	62	6555
FREELAND		12	19	157	456	786	1169	1293	1120	980	570	282	67	6911
GETTYSBURG		0	0	57	331	633	980	1104	944	766	390	151	11	5367
GREENVILLE		0	16	116	403	717	1066	1225	1047	871	492	227	41	6221
HANOVER		0	0	55	332	627	967	1094	944	763	387	157	10	5336
HARRISBURG WSO	R	0	0	58	320	633	980	1104	938	756	384	150	12	5335
HOLTWOOD		0	0	36	281	606	942	1070	921	756	390	143	7	5152
INDIANA 3 SE		5	14	121	419	717	1054	1197	1028	846	483	229	44	6157
JAMESTOWN 2 NW		11	27	146	450	762	1132	1305	1145	964	573	294	65	6874
JOHNSTOWN		0	8	86	381	669	1017	1150	991	815	432	189	30	5768
KANE 1 NNE		56	88	250	577	858	1231	1376	1226	1057	663	381	137	7900
LANDISVILLE 2 NW		0	7	82	367	669	1020	1150	986	791	438	187	16	5713
LEWISTOWN		0	5	75	362	666	1023	1144	983	809	435	185	20	5707
MADERA		38	69	211	555	831	1197	1330	1170	1008	615	336	98	7458
MARCUS HOOK		0	0	21	225	528	862	992	840	676	331	106	0	4581
MEADVILLE 1 S		14	33	150	456	759	1122	1287	1140	964	579	301	75	6880
MONTGOMERY L AND D		0	9	87	358	660	1001	1141	966	787	417	174	19	5619
MONTROSE		33	56	212	533	846	1231	1383	1218	1070	657	358	100	7697
MORGANTOWN		0	0	77	370	666	1017	1147	986	822	465	206	28	5784
NEW CASTLE 1 N		0	10	98	374	681	1026	1175	1005	825	453	207	31	5885
NEWELL		0	0	56	315	609	933	1082	930	738	389	165	41	5258
NEWPORT		0	0	81	382	687	1039	1163	1002	815	435	189	16	5809
PALMERTON		0	10	100	389	693	1048	1175	1005	840	468	204	28	5960
PHILADELPHIA WSO	R	0	0	33	273	576	915	1048	893	719	363	127	0	4947
PHILIPSBURG FAA AP		54	86	223	530	831	1190	1330	1154	998	603	329	95	7423
PHOENIXVILLE 1 E		0	0	48	318	612	952	1082	918	735	387	136	7	5195
PITTSBURGH AP WSO 2		0	13	101	393	702	1042	1187	1014	822	447	201	28	5950
PLEASANT MOUNT 1 W		62	84	260	586	882	1274	1423	1257	1113	693	401	138	8173
PORT CLINTON		10	16	138	452	735	1085	1215	1056	884	516	254	45	6406
PUTNEYVILLE 2 SE DAM		15	28	150	471	771	1128	1277	1120	939	558	287	76	6820
RIDGWAY		35	61	187	505	798	1159	1305	1151	983	600	323	98	7205
SHIPPENSBURG		0	0	53	329	654	992	1113	944	763	390	151	14	5403
SLIPPERY ROCK 1 SSW		26	37	151	434	735	1085	1243	1070	896	522	259	73	6531
STATE COLLEGE		7	10	113	412	723	1082	1212	1053	890	495	216	34	6247
STROUDSBURG		11	20	118	415	720	1088	1212	1039	862	486	217	42	6230

PENNSYLVANIA

HEATING DEGREE DAY NORMALS (BASE 65 DEG F)

STATION	JUL	AUG	SEP	OCT	NOV	DEC	JAN	FEB	MAR	APR	MAY	JUN	ANN
TIONESTA 2 SE DAM	16	33	159	474	780	1166	1321	1170	992	597	307	80	7095
TOWANDA 1 ESE	7	16	123	425	732	1091	1228	1067	905	519	243	40	6396
UNIONTOWN 1 NE	0	6	74	360	636	949	1082	930	744	402	177	29	5389
WARREN	11	26	140	446	753	1104	1246	1092	939	549	265	58	6629
WAYNESBURG 1 E	6	16	106	426	699	1014	1150	983	800	456	218	42	5916
WELLSBORO 3 S	51	64	215	527	834	1225	1367	1210	1063	651	365	117	7689
WEST CHESTER 1 W	0	0	69	328	618	970	1091	941	766	405	165	17	5370
W-BARRE-SCRANTON WSO	7	10	117	417	723	1094	1234	1070	896	501	227	34	6330
WILLIAMSPORT WSO	0	7	101	398	708	1063	1203	1030	849	462	196	30	6047
YORK 3 SSW PUMP STA	0	0	49	311	627	961	1085	916	729	378	139	8	5203

PENNSYLVANIA

COOLING DEGREE DAY NORMALS (BASE 65 DEG F)

STATION		JAN	FEB	MAR	APR	MAY	JUN	JUL	AUG	SEP	OCT	NOV	DEC	ANN
ALLENTOWN WSO	//R	0	0	0	0	35	138	273	226	79	0	0	0	751
BAKERSTOWN 3 WNW		0	0	0	0	55	138	233	204	85	8	0	0	723
BRADFORD FAA AP		0	0	0	0	6	22	59	48	12	0	0	0	147
BRADFORD 4 W RES		0	0	0	0	8	30	72	52	16	0	0	0	178
BURGETTSTOWN 2 W		0	0	0	0	24	80	146	123	41	0	0	0	414
CARLISLE		0	0	0	0	67	199	329	288	121	9	0	0	1013
CHAMBERSBURG 1 ESE		0	0	0	0	38	144	260	225	74	0	0	0	741
CLARION 3 SW		0	0	0	0	22	80	142	123	35	0	0	0	402
COATESVILLE 1 SW		0	0	0	0	26	138	267	223	80	5	0	0	739
CONFLUENCE 1 SW DAM		0	0	0	0	21	84	153	140	50	0	0	0	448
CORRY		0	0	0	0	17	70	130	107	34	0	0	0	358
DEVAULT 1 W		0	0	0	0	33	136	264	231	96	0	0	0	760
DONEGAL		0	0	0	0	22	73	113	108	38	0	0	0	354
DONORA 1 SW		0	0	0	8	82	180	274	247	113	12	0	0	916
DREXEL UNIVERSITY		0	0	0	0	78	237	397	366	169	19	0	0	1266
EPHRATA		0	0	0	0	42	162	279	236	86	6	0	0	811
ERIE WSO		0	0	0	0	13	68	144	127	43	7	0	0	402
EVERETT 1 SW		0	0	0	0	45	135	221	195	66	0	0	0	662
FARRELL SHARON		0	0	0	0	60	174	259	225	106	18	0	0	842
FORD CITY 4 S DAM		0	0	0	0	23	88	169	143	50	0	0	0	473
FRANKLIN		0	0	0	0	28	89	161	142	47	6	0	0	473
FREELAND		0	0	0	0	18	70	149	103	31	0	0	0	371
GETTYSBURG		0	0	0	0	46	170	310	264	99	8	0	0	897
GREENVILLE		0	0	0	0	32	116	194	168	74	9	0	0	593
HANOVER		0	0	0	0	49	172	304	261	97	10	0	0	893
HARRISBURG WSO	R	0	0	0	0	57	198	335	291	115	10	0	0	1006
HOLTWOOD		0	0	0	0	53	196	344	304	123	9	0	0	1029
INDIANA 3 SE		0	0	0	0	28	89	169	144	58	0	0	0	488
JAMESTOWN 2 NW		0	0	0	0	18	71	138	120	41	0	0	0	388
JOHNSTOWN		0	0	0	0	46	144	245	207	77	9	0	0	728
KANE 1 NNE		0	0	0	0	6	23	59	45	10	0	0	0	143
LANDISVILLE 2 NW		0	0	0	0	35	142	264	221	85	8	0	0	755
LEWISTOWN		0	0	0	0	42	134	254	219	78	6	0	0	733
MADERA		0	0	0	0	11	38	100	88	22	0	0	0	259
MARCUS HOOK		0	0	0	0	78	248	400	363	156	14	0	0	1259
MEADVILLE 1 S		0	0	0	0	16	69	129	108	33	0	0	0	355
MONTGOMERY L AND D		0	0	0	0	47	145	248	213	90	0	0	0	743
MONTROSE		0	0	0	0	8	34	104	75	17	0	0	0	238
MORGANTOWN		0	0	0	0	23	118	242	198	68	0	0	0	649
NEW CASTLE 1 N		0	0	0	0	40	124	209	184	80	5	0	0	642
NEWELL		0	0	0	0	66	191	282	270	125	14	0	0	948
NEWPORT		0	0	0	0	34	139	264	221	69	0	0	0	727
PALMERTON		0	0	0	0	30	130	252	208	61	0	0	0	681
PHILADELPHIA WSO	R	0	0	0	0	59	202	357	319	129	9	0	0	1075
PHILIPSBURG FAA AP		0	0	0	0	13	29	104	92	19	0	0	0	257
PHOENIXVILLE 1 E		0	0	0	0	46	172	307	254	93	8	0	0	880
PITTSBURGH AP WSO 2		0	0	0	0	37	121	222	186	74	5	0	0	645
PLEASANT MOUNT 1 W		0	0	0	0	0	18	72	47	8	0	0	0	145
PORT CLINTON		0	0	0	0	16	84	190	149	42	5	0	0	486
PUTNEYVILLE 2 SE DAM		0	0	0	0	24	79	142	109	30	0	0	0	384
RIDGWAY		0	0	0	0	13	50	110	98	28	0	0	0	299
SHIPPENSBURG		0	0	0	0	46	164	282	248	92	10	0	0	842
SLIPPERY ROCK 1 SSW		0	0	0	0	36	106	174	146	70	13	0	0	545
STATE COLLEGE		0	0	0	0	30	109	208	165	47	0	0	0	559
STROUDSBURG		0	0	0	0	31	105	221	181	58	0	0	0	596

PENNSYLVANIA

COOLING DEGREE DAY NORMALS (BASE 65 DEG F)

STATION	JAN	FEB	MAR	APR	MAY	JUN	JUL	AUG	SEP	OCT	NOV	DEC	ANN
TIONESTA 2 SE DAM	0	0	0	0	16	62	127	107	36	0	0	0	348
TOWANDA 1 ESE	0	0	0	0	20	85	180	147	39	0	0	0	471
UNIONTOWN 1 NE	0	0	0	6	59	158	244	210	83	10	0	0	770
WARREN	0	0	0	0	20	85	156	132	41	0	0	0	434
WAYNESBURG 1 E	0	0	0	0	35	114	195	171	64	0	0	0	579
WELLSBORO 3 S	0	0	0	0	8	36	107	70	14	0	0	0	235
WEST CHESTER 1 W	0	0	0	0	44	173	308	257	102	11	0	0	895
W-BARRE-SCRANTON WSO	0	0	0	0	29	106	218	165	51	0	0	0	569
WILLIAMSPORT WSO	0	0	0	0	29	129	237	196	68	0	0	0	659
YORK 3 SSW PUMP STA	0	0	0	0	49	173	301	264	106	8	0	0	901

STATE-STATION NUMBER	STN TYP	NAME		LATITUDE DEG-MIN	LONGITUDE DEG-MIN	ELEVATION (FT)
36-0022	12	ACMETONIA LOCK 3		N 4032	W 07949	748
36-0106	13	ALLENTOWN WSO	//R	N 4039	W 07526	387
36-0355	13	BAKERSTOWN 3 WNW		N 4039	W 07959	1230
36-0409	12	BARNES		N 4140	W 07902	1310
36-0457	12	BEAR GAP		N 4050	W 07630	900
36-0475	12	BEAVER FALLS		N 4046	W 08019	760
36-0482	12	BEAVERTOWN		N 4045	W 07711	590
36-0629	12	BETHLEHEM		N 4037	W 07523	240
36-0763	12	BLOSERVILLE 1 N		N 4016	W 07722	650
36-0861	12	BRADDOCK LOCK 2		N 4024	W 07952	725
36-0865	13	BRADFORD FAA AP		N 4148	W 07838	2142
36-0867	12	BRADFORD CENTRAL FS		N 4157	W 07839	1500
36-0868	13	BRADFORD 4 W RES		N 4157	W 07844	1680
36-1033	12	BRUCETON 1 S		N 4018	W 07959	1040
36-1087	12	BUFFALO MILLS		N 3957	W 07839	1318
36-1105	13	BURGETTSTOWN 2 W		N 4023	W 08026	980
36-1234	13	CARLISLE		N 4013	W 07712	465
36-1255	12	CARROLLTOWN 2 SSE		N 4035	W 07842	2040
36-1301	12	CEDAR RUN		N 4131	W 07727	800
36-1342	12	CHADDS FORD		N 3952	W 07537	170
36-1354	13	CHAMBERSBURG 1 ESE		N 3956	W 07738	640
36-1377	12	CHARLEROI LOCK 4		N 4009	W 07954	749
36-1480	12	CLARENCE		N 4103	W 07756	1390
36-1485	13	CLARION 3 SW		N 4112	W 07926	1114
36-1505	12	CLAUSSVILLE		N 4037	W 07539	670
36-1519	12	CLEARFIELD		N 4102	W 07826	1100
36-1589	13	COATESVILLE 1 SW		N 3958	W 07550	342
36-1705	13	CONFLUENCE 1 SW DAM		N 3948	W 07922	1490
36-1710	12	CONFLUENCE 1 NW		N 3950	W 07922	1331
36-1737	12	CONSHOHOCKEN		N 4004	W 07519	70
36-1790	13	CORRY		N 4155	W 07938	1440
36-1881	12	CREEKSIDE		N 4041	W 07912	1060
36-2013	12	DANVILLE		N 4058	W 07637	460
36-2116	13	DEVAULT 1 W		N 4005	W 07533	360
36-2183	13	DONEGAL		N 4008	W 07924	1800
36-2190	13	DONORA 1 SW		N 4010	W 07952	762
36-2221	12	DOYLESTOWN		N 4018	W 07508	358
36-2236	13	DREXEL UNIVERSITY		N 3957	W 07511	30
36-2644	12	ENGLISH CENTER		N 4126	W 07717	880
36-2662	13	EPHRATA		N 4010	W 07610	485
36-2682	13	ERIE WSO		N 4205	W 08011	732
36-2721	13	EVERETT 1 SW		N 4000	W 07823	1010
36-2814	13	FARRELL SHARON		N 4114	W 08030	855
36-2942	13	FORD CITY 4 S DAM		N 4043	W 07930	950
36-3028	13	FRANKLIN		N 4123	W 07949	987
36-3056	13	FREELAND		N 4101	W 07554	1900
36-3130	12	GALETON		N 4144	W 07738	1365
36-3218	13	GETTYSBURG		N 3950	W 07714	500
36-3343	12	GLENWILLARD DASH DAM		N 4033	W 08013	705
36-3394	12	GOULDSBORO		N 4115	W 07527	1890

STATE-STATION NUMBER	STN TYP	NAME		LATITUDE DEG-MIN	LONGITUDE DEG-MIN	ELEVATION (FT)
36-3503	12	GREENSBORO LOCK 7		N 3947	W 07955	808
36-3526	13	GREENVILLE		N 4124	W 08023	980
36-3662	13	HANOVER		N 3948	W 07659	600
36-3699	13	HARRISBURG WSO	R	N 4013	W 07651	338
36-3758	12	HAWLEY		N 4129	W 07510	880
36-4008	12	HOLLISTERVILLE		N 4123	W 07526	1365
36-4019	13	HOLTWOOD		N 3950	W 07620	187
36-4043	12	HONESDALE 4 NW		N 4137	W 07519	1410
36-4166	12	HUNTSDALE		N 4006	W 07718	620
36-4190	12	HYNDMAN		N 3949	W 07844	960
36-4214	13	INDIANA 3 SE		N 4036	W 07907	1102
36-4325	13	JAMESTOWN 2 NW		N 4130	W 08028	1050
36-4385	13	JOHNSTOWN		N 4020	W 07855	1214
36-4432	13	KANE 1 NNE		N 4141	W 07848	1750
36-4611	12	KITTANNING LOCK 7		N 4049	W 07932	790
36-4672	12	KRESGEVILLE 3 W		N 4054	W 07534	720
36-4778	13	LANDISVILLE 2 NW		N 4007	W 07626	360
36-4934	12	LEHIGHTON		N 4050	W 07543	580
36-4972	12	LE ROY		N 4141	W 07643	1040
36-4992	13	LEWISTOWN		N 4036	W 07735	481
36-5160	12	LONG POND 2 W		N 4103	W 07530	1860
36-5336	13	MADERA		N 4050	W 07826	1460
36-5381	12	MAPLETON DEPOT		N 4024	W 07756	580
36-5390	13	MARCUS HOOK		N 3949	W 07525	12
36-5408	12	MARION CENTER 2 SE		N 4045	W 07902	1610
36-5470	12	MATAMORAS		N 4122	W 07442	439
36-5573	12	MC KEESPORT		N 4021	W 07952	720
36-5606	13	MEADVILLE 1 S		N 4138	W 08010	1065
36-5651	12	MERCER		N 4113	W 08014	1220
36-5662	12	MERCERSBURG		N 3950	W 07754	585
36-5790	12	MILLHEIM		N 4053	W 07729	1070
36-5817	12	MILLVILLE 2 SW		N 4106	W 07634	860
36-5902	13	MONTGOMERY L AND D		N 4039	W 08023	692
36-5915	13	MONTROSE		N 4150	W 07552	1560
36-5956	13	MORGANTOWN		N 4009	W 07554	595
36-6126	12	MYERSTOWN		N 4022	W 07618	480
36-6151	12	NATRONA LOCK 4		N 4037	W 07943	800
36-6194	12	NESHAMINY FALLS		N 4009	W 07457	60
36-6233	13	NEW CASTLE 1 N		N 4101	W 08022	825
36-6246	13	NEWELL		N 4005	W 07954	805
36-6289	12	NEW PARK		N 3944	W 07630	780
36-6297	13	NEWPORT		N 4029	W 07708	400
36-6326	12	NEW TRIPOLI		N 4041	W 07545	570
36-6370	12	NORRISTOWN		N 4007	W 07521	68
36-6681	12	PALM 3 SE		N 4023	W 07530	302
36-6689	13	PALMERTON		N 4048	W 07537	410
36-6721	12	PARKER		N 4105	W 07941	1140
36-6762	12	PAUPACK 2 WNW		N 4124	W 07514	1360
36-6889	13	PHILADELPHIA WSO	R	N 3953	W 07514	5
36-6916	13	PHILIPSBURG FAA AP		N 4054	W 07805	1942

STATE-STATION NUMBER	STN TYP	NAME	LATITUDE DEG-MIN	LONGITUDE DEG-MIN	ELEVATION (FT)
36-6927	13	PHOENIXVILLE 1 E	N 4007	W 07530	105
36-6993	13	PITTSBURGH AP WSO 2	N 4030	W 08013	1137
36-7029	13	PLEASANT MOUNT 1 W	N 4144	W 07527	1800
36-7116	13	PORT CLINTON	N 4035	W 07602	450
36-7229	13	PUTNEYVILLE 2 SE DAM	N 4055	W 07917	1270
36-7322	12	READING 3N R	N 4022	W 07556	270
36-7477	13	RIDGWAY	N 4125	W 07845	1360
36-7727	12	RUSHVILLE	N 4147	W 07607	870
36-7846	12	SAXTON	N 4012	W 07815	780
36-7863	12	SCHENLEY LOCK 5	N 4041	W 07940	783
36-7978	12	SHAMOKIN	N 4048	W 07633	770
36-8073	13	SHIPPENSBURG	N 4003	W 07731	680
36-8145	12	SINNEMAHONING	N 4119	W 07805	790
36-8184	13	SLIPPERY ROCK 1 SSW	N 4104	W 08004	1260
36-8244	12	SOMERSET	N 4000	W 07905	2100
36-8308	12	SOUTH MOUNTAIN	N 3951	W 07730	1520
36-8379	12	SPRING GROVE	N 3952	W 07652	470
36-8449	13	STATE COLLEGE	N 4048	W 07752	1175
36-8570	12	STRAUSSTOWN	N 4029	W 07611	600
36-8596	13	STROUDSBURG	N 4100	W 07511	480
36-8692	12	SUSQUEHANNA	N 4157	W 07536	1020
36-8758	12	TAMAQUA	N 4047	W 07559	925
36-8763	12	TAMAQUA 4 N DAM	N 4051	W 07559	1120
36-8873	13	TIONESTA 2 SE DAM	N 4129	W 07926	1200
36-8905	13	TOWANDA 1 ESE	N 4145	W 07625	745
36-8959	12	TROY	N 4147	W 07647	1100
36-9042	12	UNION CITY FILT PLANT	N 4154	W 07949	1400
36-9050	13	UNIONTOWN 1 NE	N 3955	W 07943	956
36-9128	12	VANDERGRIFT	N 4036	W 07933	800
36-9298	13	WARREN	N 4151	W 07908	1280
36-9367	13	WAYNESBURG 1 E	N 3954	W 08010	940
36-9408	13	WELLSBORO 3 S	N 4142	W 07716	1860
36-9464	13	WEST CHESTER 1 W	N 3958	W 07538	450
36-9655	12	WHITESBURG	N 4044	W 07924	1320
36-9702	12	WILKES-BARRE 4 NE	N 4117	W 07551	580
36-9705	13	W-BARRE-SCRANTON WSO	N 4120	W 07544	930
36-9714	12	WILLIAMSBURG	N 4027	W 07812	880
36-9728	13	WILLIAMSPORT WSO	N 4115	W 07655	524
36-9823	12	WOLFSBURG	N 4003	W 07832	1190
36-9933	13	YORK 3 SSW PUMP STA	N 3955	W 07645	390
36-9950	12	YORK HAVEN	N 4007	W 07643	310
36-9995	12	ZIONSVILLE 3 SE	N 4028	W 07527	680

36 – PENNSYLVANIA

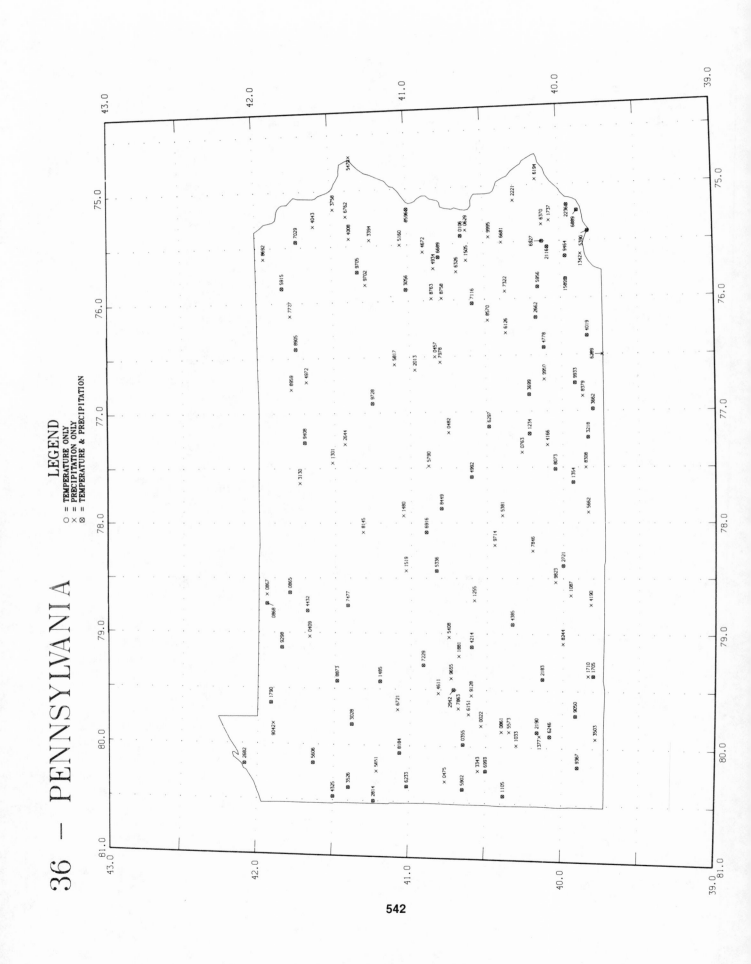

LEGEND

O = TEMPERATURE ONLY
X = PRECIPITATION ONLY
⊗ = TEMPERATURE & PRECIPITATION

542

RHODE ISLAND

TEMPERATURE NORMALS (DEG F)

STATION			JAN	FEB	MAR	APR	MAY	JUN	JUL	AUG	SEP	OCT	NOV	DEC	ANN
BLOCK ISLAND WSO	R	MAX	37.2	36.9	42.8	51.8	60.7	69.8	76.0	75.8	69.7	60.8	51.5	41.9	56.2
		MIN	25.0	25.1	31.4	38.9	47.6	56.9	63.6	63.8	57.9	48.9	40.2	29.6	44.1
		MEAN	31.1	31.0	37.1	45.4	54.2	63.4	69.8	69.8	63.8	54.8	45.9	35.8	50.2
KINGSTON		MAX	37.8	39.0	46.1	57.4	67.1	75.6	80.6	79.6	73.3	64.1	52.7	41.7	59.6
		MIN	18.6	19.5	27.5	35.6	44.4	53.7	59.6	58.7	51.2	40.5	33.0	22.8	38.8
		MEAN	28.2	29.3	36.8	46.5	55.8	64.7	70.1	69.2	62.2	52.3	42.8	32.3	49.2
PROVIDENCE WSO	R	MAX	36.4	37.7	45.5	57.5	67.6	76.6	81.7	80.3	73.1	63.2	51.9	40.5	59.3
		MIN	20.0	20.9	29.2	38.3	47.6	57.0	63.3	61.9	53.8	43.1	34.8	24.1	41.2
		MEAN	28.2	29.3	37.4	47.9	57.6	66.8	72.5	71.1	63.5	53.2	43.4	32.3	50.3

RHODE ISLAND

PRECIPITATION NORMALS (INCHES)

STATION		JAN	FEB	MAR	APR	MAY	JUN	JUL	AUG	SEP	OCT	NOV	DEC	ANN
BLOCK ISLAND WSO	R	3.53	3.38	3.98	3.55	3.37	2.28	2.71	4.06	3.51	3.21	3.99	4.34	41.91
KINGSTON		4.23	3.69	4.65	4.13	4.12	2.92	2.99	4.46	4.11	3.96	4.65	4.58	48.49
PROVIDENCE WSO	R	4.06	3.72	4.29	3.95	3.48	2.79	3.01	4.04	3.54	3.75	4.22	4.47	45.32

RHODE ISLAND

HEATING DEGREE DAY NORMALS (BASE 65 DEG F)

STATION		JUL	AUG	SEP	OCT	NOV	DEC	JAN	FEB	MAR	APR	MAY	JUN	ANN
BLOCK ISLAND WSO	R	7	5	75	316	573	905	1051	952	865	588	335	83	5755
KINGSTON		8	13	116	394	666	1014	1141	1000	874	555	290	50	6121
PROVIDENCE WSO	R	0	6	94	366	648	1014	1141	1000	856	513	239	31	5908

RHODE ISLAND

COOLING DEGREE DAY NORMALS (BASE 65 DEG F)

STATION		JAN	FEB	MAR	APR	MAY	JUN	JUL	AUG	SEP	OCT	NOV	DEC	ANN
BLOCK ISLAND WSO	R	0	0	0	0	0	35	155	154	39	0	0	0	383
KINGSTON		0	0	0	0	0	41	166	143	32	0	0	0	382
PROVIDENCE WSO	R	0	0	0	0	10	85	235	195	49	0	0	0	574

37 – RHODE ISLAND

STATE-STATION NUMBER	STN TYP	NAME		LATITUDE DEG-MIN	LONGITUDE DEG-MIN	ELEVATION (FT)
37-0896	13	BLOCK ISLAND WSO	R	N 4110	W 07135	110
37-4266	13	KINGSTON		N 4129	W 07132	100
37-6698	13	PROVIDENCE WSO	R	N 4144	W 07126	51

37 – RHODE ISLAND

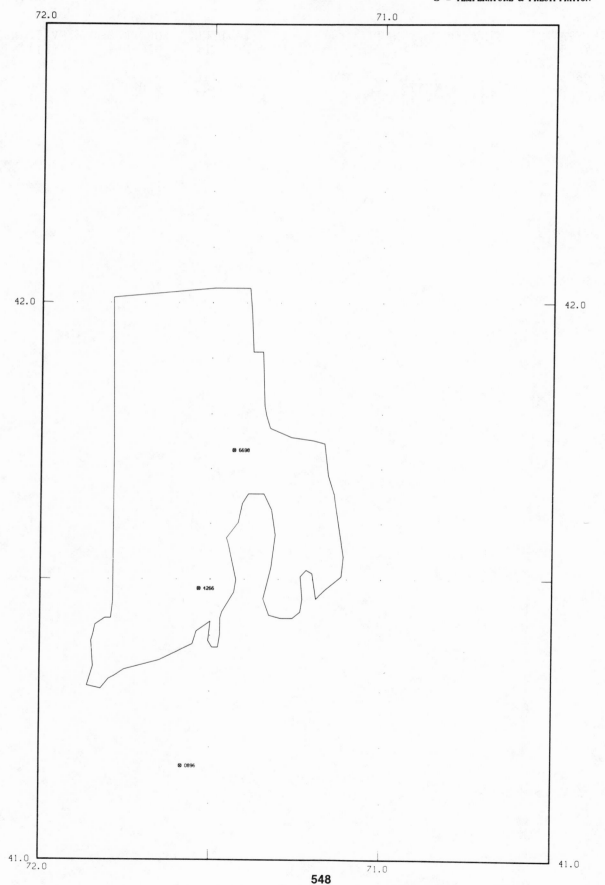

SOUTH CAROLINA

TEMPERATURE NORMALS (DEG F)

STATION			JAN	FEB	MAR	APR	MAY	JUN	JUL	AUG	SEP	OCT	NOV	DEC	ANN
AIKEN		MAX	57.4	60.8	68.3	77.0	83.5	88.9	91.2	90.3	84.9	75.9	67.1	59.5	75.4
		MIN	35.1	36.4	43.3	51.1	59.4	66.1	69.4	69.0	64.4	52.6	42.9	36.7	52.2
		MEAN	46.3	48.6	55.8	64.1	71.5	77.5	80.3	79.7	74.7	64.3	55.0	48.1	63.8
ANDERSON		MAX	53.3	57.1	64.8	74.4	81.1	86.9	89.8	89.3	83.4	74.4	64.5	55.3	72.9
		MIN	31.9	33.3	39.9	48.9	57.3	64.1	67.8	67.3	61.9	49.7	40.5	33.8	49.7
		MEAN	42.6	45.2	52.3	61.7	69.2	75.5	78.8	78.3	72.7	62.1	52.5	44.6	61.3
ANDERSON FAA AP		MAX	52.2	55.6	63.4	73.6	80.8	87.4	89.9	89.2	83.1	73.7	63.6	54.7	72.3
		MIN	32.2	33.9	40.8	49.8	58.4	65.9	69.5	68.9	63.3	50.3	40.7	34.5	50.7
		MEAN	42.2	44.8	52.1	61.7	69.7	76.7	79.8	79.1	73.2	62.0	52.2	44.6	61.5
BAMBERG		MAX	58.6	61.8	69.7	78.6	84.8	89.4	91.4	90.8	85.7	77.1	68.5	60.7	76.4
		MIN	35.2	36.6	43.2	50.9	58.9	65.7	69.3	68.8	64.2	51.9	42.4	36.6	52.0
		MEAN	46.9	49.2	56.5	64.8	71.9	77.6	80.4	79.8	75.0	64.5	55.5	48.7	64.2
BEAUFORT 7 SW		MAX	59.8	62.5	69.1	77.1	83.4	87.6	89.9	89.3	84.7	77.3	69.1	61.8	76.0
		MIN	38.3	40.0	46.5	54.3	62.6	68.6	71.7	71.4	67.3	56.6	47.0	40.6	55.4
		MEAN	49.1	51.3	57.8	65.7	73.0	78.1	80.8	80.4	76.0	67.0	58.1	51.2	65.7
BLACKVILLE 3 W		MAX	57.6	60.5	68.0	77.0	83.6	88.4	90.9	90.3	85.7	77.0	68.4	60.0	75.6
		MIN	34.4	35.9	43.2	51.2	59.3	65.8	68.9	68.4	63.5	51.6	42.2	36.2	51.7
		MEAN	46.0	48.2	55.7	64.1	71.5	77.1	79.9	79.4	74.6	64.4	55.3	48.1	63.7
CAESARS HEAD 1 NE		MAX	45.0	47.9	55.6	65.6	71.8	77.0	79.8	78.9	73.3	64.4	54.7	47.2	63.4
		MIN	27.6	29.4	35.8	45.4	53.4	59.4	62.9	62.2	57.1	46.5	37.5	30.2	45.6
		MEAN	36.3	38.7	45.7	55.5	62.6	68.2	71.3	70.5	65.2	55.4	46.1	38.7	54.5
CALHOUN FALLS		MAX	53.5	56.9	64.9	75.1	82.1	88.1	91.1	90.8	84.8	75.3	65.3	56.1	73.7
		MIN	30.5	31.7	38.5	47.3	56.4	63.9	67.5	67.0	61.4	48.1	38.1	31.8	48.5
		MEAN	42.0	44.3	51.7	61.2	69.3	76.0	79.3	78.9	73.1	61.8	51.7	44.0	61.1
CAMDEN 2 WSW		MAX	54.4	57.6	65.3	75.1	81.8	87.4	90.2	89.2	84.0	74.5	65.4	56.8	73.5
		MIN	29.7	30.5	37.5	46.2	55.6	63.3	67.7	66.9	60.9	47.6	37.2	30.6	47.8
		MEAN	42.0	44.0	51.4	60.7	68.8	75.4	78.9	78.1	72.5	61.1	51.3	43.7	60.7
CHARLESTON AP WSO		MAX	58.8	61.2	68.0	76.0	82.9	87.0	89.4	88.8	84.6	76.8	68.7	61.4	75.3
		MIN	36.9	38.4	45.3	52.5	61.4	68.0	71.6	71.2	66.7	54.7	44.6	38.5	54.2
		MEAN	47.9	49.8	56.7	64.3	72.2	77.6	80.5	80.0	75.7	65.8	56.7	50.0	64.8
CHARLESTON CITY WSO	R MAX		56.9	58.9	65.1	73.2	80.2	85.0	87.9	87.3	82.9	74.9	66.7	59.6	73.2
		MIN	41.4	43.0	49.7	58.3	66.3	72.2	75.2	74.9	70.9	60.4	50.9	44.1	58.9
		MEAN	49.2	51.0	57.4	65.8	73.2	78.7	81.6	81.1	76.9	67.7	58.8	51.8	66.1
CHERAW		MAX	53.1	56.1	64.2	74.9	82.2	87.9	90.9	89.6	84.5	74.8	65.2	55.8	73.3
		MIN	29.2	30.4	37.8	47.1	56.0	63.6	67.7	67.0	61.1	47.9	37.3	30.5	48.0
		MEAN	41.2	43.3	51.1	61.0	69.1	75.8	79.3	78.4	72.8	61.4	51.3	43.2	60.7
CHESTER		MAX	53.9	57.6	65.5	75.4	81.9	87.6	90.6	90.0	84.5	75.0	65.3	56.0	73.6
		MIN	30.4	32.1	38.8	47.7	56.2	63.3	67.4	66.8	61.0	48.4	38.7	31.9	48.6
		MEAN	42.2	44.9	52.2	61.6	69.1	75.5	79.0	78.4	72.8	61.7	52.0	44.0	61.1
CLEMSON UNIVERSITY		MAX	53.1	56.6	64.0	74.0	80.7	86.7	89.6	88.8	83.4	74.2	64.1	55.3	72.5
		MIN	30.7	32.2	38.8	47.6	56.0	63.3	67.2	66.5	61.1	48.2	38.9	32.1	48.6
		MEAN	41.9	44.4	51.4	60.8	68.4	75.0	78.4	77.7	72.3	61.3	51.5	43.8	60.6
COLUMBIA WSO	R MAX		56.2	59.5	67.1	77.0	83.8	89.2	91.9	91.0	85.5	76.5	67.1	58.8	75.3
		MIN	33.2	34.6	41.9	50.5	59.1	66.1	70.1	69.4	63.9	50.3	40.6	34.7	51.2
		MEAN	44.7	47.1	54.5	63.8	71.5	77.7	81.0	80.2	74.8	63.4	53.9	46.7	63.3
COLUMBIA UNI OF S C		MAX	56.5	60.2	68.0	77.8	84.1	89.6	92.0	90.9	85.4	76.0	66.8	58.7	75.5
		MIN	36.3	37.9	45.0	53.2	61.0	67.4	70.8	70.2	65.1	53.3	44.5	38.2	53.6
		MEAN	46.4	49.1	56.5	65.5	72.6	78.5	81.4	80.6	75.3	64.7	55.7	48.4	64.6
CONWAY		MAX	56.9	59.5	67.1	76.1	83.0	87.8	90.5	89.8	84.9	76.3	67.8	59.4	74.9
		MIN	33.7	35.1	42.6	51.0	59.4	66.2	70.2	69.7	64.8	52.5	42.1	35.4	51.9
		MEAN	45.3	47.3	54.9	63.6	71.2	77.0	80.4	79.8	74.9	64.4	55.0	47.4	63.4

SOUTH CAROLINA

TEMPERATURE NORMALS (DEG F)

STATION		JAN	FEB	MAR	APR	MAY	JUN	JUL	AUG	SEP	OCT	NOV	DEC	ANN
DARLINGTON	MAX	56.4	59.3	67.0	76.5	83.3	88.3	90.8	90.0	85.2	76.5	67.4	58.7	75.0
	MIN	33.6	35.0	42.0	50.1	58.3	65.2	68.9	68.0	62.7	50.3	40.9	35.0	50.8
	MEAN	45.0	47.2	54.5	63.3	70.8	76.8	79.9	79.0	74.0	63.4	54.2	46.9	62.9
DILLON 4 SW	MAX	54.7	57.7	65.6	75.5	82.5	87.8	90.8	90.2	85.2	75.7	66.8	57.6	74.2
	MIN	30.8	32.1	39.5	47.8	56.1	63.5	67.9	66.9	61.1	47.9	38.0	31.5	48.6
	MEAN	42.8	44.9	52.5	61.7	69.3	75.7	79.4	78.6	73.2	61.8	52.5	44.6	61.4
FLORENCE FAA AP	MAX	55.5	58.5	66.3	76.1	82.9	87.9	90.5	89.7	84.7	75.7	66.7	58.2	74.4
	MIN	34.4	35.9	43.0	51.2	59.8	66.7	70.6	69.9	64.7	52.2	42.8	36.2	52.3
	MEAN	45.0	47.2	54.7	63.7	71.4	77.3	80.6	79.8	74.7	63.9	54.8	47.2	63.4
FLORENCE 2 N	MAX	54.5	57.1	65.0	74.9	82.0	87.1	89.7	89.0	84.4	75.3	66.5	57.4	73.6
	MIN	33.0	34.2	42.2	51.1	59.3	66.1	69.8	68.8	63.1	50.5	41.4	34.5	51.2
	MEAN	43.8	45.7	53.6	63.0	70.7	76.7	79.8	78.9	73.8	62.9	54.0	46.0	62.4
GEORGETOWN 2 E	MAX	57.8	60.0	66.7	74.9	81.7	86.6	89.2	89.0	84.5	76.5	68.4	60.6	74.7
	MIN	36.6	37.7	45.0	53.3	62.0	68.3	71.7	71.2	66.7	55.2	45.5	38.3	54.3
	MEAN	47.2	48.9	55.9	64.1	71.8	77.5	80.5	80.1	75.6	65.9	57.0	49.4	64.5
GRNVLE-SPTNBRG WSO R	MAX	51.0	54.5	62.5	72.6	79.7	85.4	88.2	87.5	81.7	72.2	62.1	53.5	70.9
	MIN	31.2	32.6	39.4	48.3	56.9	64.2	68.2	67.4	61.7	49.1	39.6	33.2	49.3
	MEAN	41.1	43.6	51.0	60.5	68.3	74.8	78.2	77.5	71.7	60.7	50.9	43.4	60.1
GREENWOOD 3 ESE	MAX	52.6	56.3	64.3	74.5	81.5	87.8	90.5	89.9	83.8	74.0	64.3	55.0	72.9
	MIN	30.7	31.3	38.3	47.2	56.3	63.4	67.3	66.6	60.8	47.8	38.4	32.1	48.4
	MEAN	41.7	43.8	51.3	60.9	68.9	75.6	78.9	78.3	72.3	60.9	51.4	43.6	60.6
HAMPTON	MAX	60.6	63.8	71.2	79.3	85.4	89.7	92.0	91.3	86.3	78.5	70.0	62.6	77.6
	MIN	36.2	38.0	44.7	52.1	59.7	66.1	69.5	69.0	64.3	52.4	43.0	37.3	52.7
	MEAN	48.4	50.9	58.0	65.7	72.6	77.9	80.8	80.1	75.3	65.5	56.5	49.9	65.1
KINGSTREE 1 SE	MAX	57.0	59.4	67.2	76.1	83.4	88.4	91.2	90.2	85.3	76.6	68.1	59.6	75.2
	MIN	32.5	33.7	41.6	49.8	58.3	65.1	68.8	68.1	62.7	49.4	39.5	33.2	50.2
	MEAN	44.8	46.6	54.4	63.0	70.9	76.8	80.0	79.2	74.0	63.0	53.8	46.4	62.7
LAKE CITY	MAX	56.7	59.1	66.9	76.5	83.6	88.6	91.3	90.7	85.9	76.9	67.9	59.0	75.3
	MIN	34.1	35.7	42.7	50.6	59.3	66.1	69.4	69.0	63.7	51.5	42.0	35.6	51.6
	MEAN	45.5	47.5	54.9	63.6	71.4	77.4	80.3	79.9	74.8	64.2	55.0	47.3	63.5
LAURENS	MAX	52.9	56.8	64.8	75.1	81.8	87.9	90.8	90.1	83.8	73.9	64.5	55.2	73.1
	MIN	30.0	31.1	38.4	47.7	56.0	63.4	67.5	66.5	60.3	47.2	37.7	30.8	48.1
	MEAN	41.5	44.0	51.6	61.4	68.9	75.7	79.1	78.3	72.1	60.6	51.1	43.0	60.6
LITTLE MOUNTAIN	MAX	55.2	58.7	66.1	76.0	82.7	88.6	91.3	90.7	85.1	75.7	66.3	57.4	74.5
	MIN	34.4	35.8	42.8	51.3	59.4	65.9	69.5	68.8	63.6	51.6	42.9	36.3	51.9
	MEAN	44.8	47.3	54.5	63.7	71.1	77.3	80.4	79.8	74.4	63.7	54.6	46.9	63.2
LONGCREEK 1 N	MAX	51.0	54.3	61.7	71.7	77.7	83.6	86.5	85.4	80.1	70.8	61.6	53.5	69.8
	MIN	29.9	30.8	37.1	45.6	53.5	59.8	63.2	62.9	57.8	46.6	38.2	32.1	46.5
	MEAN	40.5	42.6	49.4	58.7	65.6	71.7	74.9	74.2	68.9	58.7	50.0	42.8	58.2
LORIS 1 S	MAX	56.5	58.8	66.3	75.4	82.1	87.1	89.9	89.3	84.6	76.1	67.8	59.1	74.4
	MIN	32.5	33.6	40.6	48.7	57.6	64.5	68.5	68.1	63.0	50.5	40.2	33.9	50.1
	MEAN	44.5	46.2	53.5	62.0	69.9	75.9	79.2	78.7	73.8	63.3	54.0	46.5	62.3
MARION	MAX	57.0	60.5	67.9	77.1	83.7	88.9	91.6	90.6	85.6	76.6	67.7	59.3	75.5
	MIN	33.9	35.3	42.4	50.0	58.3	65.0	68.9	68.1	62.5	50.0	41.1	35.0	50.9
	MEAN	45.5	47.9	55.2	63.6	71.0	77.0	80.3	79.4	74.1	63.3	54.4	47.2	63.2
MC COLL	MAX	55.1	58.4	66.1	76.1	83.0	88.5	91.0	90.2	85.5	75.8	66.5	57.5	74.5
	MIN	33.5	35.1	41.9	50.8	58.9	65.5	69.1	68.4	63.1	50.8	41.9	35.1	51.2
	MEAN	44.3	46.8	54.0	63.5	70.9	77.0	80.1	79.3	74.3	63.3	54.2	46.3	62.8
NEWBERRY	MAX	55.0	58.9	66.7	76.4	83.0	88.6	91.1	90.5	84.7	75.3	65.7	57.0	74.4
	MIN	32.2	33.6	40.3	48.8	57.1	64.4	68.2	67.4	61.8	49.2	39.7	33.6	49.7
	MEAN	43.6	46.3	53.6	62.6	70.1	76.5	79.7	79.0	73.3	62.3	52.7	45.3	62.1

SOUTH CAROLINA

TEMPERATURE NORMALS (DEG F)

STATION		JAN	FEB	MAR	APR	MAY	JUN	JUL	AUG	SEP	OCT	NOV	DEC	ANN
ORANGEBURG 2	MAX	56.2	59.2	67.0	76.7	83.4	88.1	90.7	90.4	85.3	76.3	67.5	58.7	75.0
	MIN	33.3	35.1	42.5	51.0	59.1	66.2	69.8	69.3	64.0	51.1	41.6	34.8	51.5
	MEAN	44.8	47.2	54.8	63.9	71.3	77.2	80.3	79.8	74.7	63.7	54.6	46.8	63.3
PARR	MAX	56.6	60.1	67.7	77.3	83.8	89.2	92.0	91.6	86.3	77.3	67.6	58.6	75.7
	MIN	31.1	32.4	39.1	47.6	55.8	63.2	67.6	66.8	60.9	47.8	38.3	32.2	48.6
	MEAN	43.9	46.2	53.4	62.5	69.9	76.2	79.8	79.2	73.6	62.6	53.0	45.4	62.1
PICKENS 5 SE	MAX	52.4	56.0	63.7	73.6	80.0	85.8	88.6	88.0	82.5	73.2	62.9	54.3	71.8
	MIN	32.1	33.3	40.2	49.0	56.6	63.3	66.8	66.4	61.7	50.4	41.3	34.3	49.6
	MEAN	42.3	44.7	52.0	61.3	68.4	74.6	77.7	77.2	72.1	61.8	52.1	44.3	60.7
PINOPOLIS DAM	MAX	56.7	59.1	66.7	75.4	82.4	87.4	90.2	89.7	85.1	76.5	67.9	59.4	74.7
	MIN	34.7	36.1	43.7	51.9	60.5	66.9	70.3	69.8	65.0	53.2	43.1	36.0	52.6
	MEAN	45.7	47.6	55.2	63.7	71.5	77.1	80.3	79.8	75.1	64.9	55.5	47.8	63.7
RAINBOW LAKE	MAX	52.8	56.3	64.3	74.4	81.3	86.9	89.8	89.5	83.4	74.2	63.7	54.9	72.6
	MIN	29.1	30.2	36.9	45.3	53.8	61.6	65.7	65.0	59.0	45.9	36.5	30.4	46.6
	MEAN	41.0	43.2	50.6	59.9	67.6	74.3	77.8	77.3	71.2	60.1	50.1	42.6	59.6
RIDGELAND 5 NE	MAX	60.4	63.9	71.2	79.2	84.9	89.0	91.4	90.6	85.8	77.9	69.3	62.4	77.2
	MIN	36.5	38.0	44.6	52.3	60.3	66.2	69.5	68.8	65.0	53.5	43.9	37.9	53.0
	MEAN	48.4	51.0	57.9	65.7	72.6	77.6	80.4	79.7	75.4	65.7	56.6	50.2	65.1
SALUDA	MAX	55.2	58.8	66.5	76.4	83.2	89.1	92.2	91.2	85.5	76.1	66.5	57.4	74.8
	MIN	30.7	32.2	39.3	48.3	56.3	63.7	67.4	66.6	61.1	47.7	38.2	32.1	48.6
	MEAN	42.9	45.5	52.9	62.4	69.8	76.4	79.9	78.9	73.3	61.9	52.4	44.8	61.8
SANTUCK	MAX	53.8	57.7	65.7	75.8	82.3	87.9	90.5	89.9	83.7	74.1	64.5	55.7	73.5
	MIN	32.1	33.4	40.3	49.1	57.5	64.2	68.0	67.4	61.8	49.7	40.5	33.7	49.8
	MEAN	43.0	45.6	53.0	62.4	69.9	76.1	79.3	78.7	72.8	61.9	52.5	44.8	61.7
SULLIVANS ISLAND	MAX	57.1	59.2	65.4	73.3	80.8	85.7	88.4	88.1	84.1	76.6	68.2	60.3	73.9
	MIN	39.6	40.7	47.7	55.9	64.0	70.4	73.7	73.6	69.2	58.7	49.3	42.2	57.1
	MEAN	48.4	50.0	56.6	64.7	72.4	78.1	81.1	80.9	76.7	67.7	58.8	51.3	65.6
SUMMERVILLE 2 WNW	MAX	58.0	60.7	68.0	76.3	82.9	87.4	89.9	89.6	85.1	77.0	68.7	60.6	75.4
	MIN	34.7	36.3	43.4	51.0	59.6	66.5	70.1	69.7	65.2	52.6	42.4	35.8	52.3
	MEAN	46.4	48.5	55.7	63.7	71.3	77.0	80.0	79.6	75.2	64.8	55.6	48.2	63.8
SUMTER	MAX	57.1	60.6	68.2	77.4	83.9	88.6	91.1	90.0	85.5	76.8	67.9	59.3	75.5
	MIN	34.0	35.8	42.9	50.9	59.1	65.6	69.1	68.4	63.1	50.8	41.7	35.4	51.4
	MEAN	45.6	48.2	55.6	64.2	71.5	77.1	80.2	79.3	74.3	63.8	54.8	47.4	63.5
TILGHMAN FOR NURSERY	MAX	58.2	61.7	69.1	79.1	85.7	90.3	92.8	91.7	86.7	78.2	68.8	60.7	76.9
	MIN	33.7	35.4	42.3	50.6	59.3	66.0	69.4	68.6	63.1	50.8	41.4	35.1	51.3
	MEAN	46.0	48.6	55.7	64.9	72.5	78.1	81.1	80.2	74.9	64.5	55.1	48.0	64.1
UNION 8 SW	MAX	52.9	56.2	64.1	74.8	81.6	87.6	90.7	90.2	84.3	74.8	64.8	55.4	73.1
	MIN	28.0	29.0	35.9	45.1	53.5	61.1	65.2	64.4	58.1	44.8	36.0	29.4	45.9
	MEAN	40.5	42.6	50.1	60.0	67.6	74.4	78.0	77.3	71.2	59.8	50.4	42.4	59.5
WALHALLA	MAX	52.6	56.4	63.7	73.3	80.1	85.6	88.4	87.7	82.4	73.4	63.1	54.6	71.8
	MIN	29.9	30.8	37.2	45.4	53.8	61.2	65.0	64.3	59.2	46.7	37.2	31.0	46.8
	MEAN	41.3	43.6	50.5	59.4	66.9	73.5	76.7	76.0	70.8	60.1	50.2	42.8	59.3
WALTERBORO 2 SW	MAX	59.3	62.6	70.1	78.3	84.5	89.0	91.3	90.6	85.7	77.6	69.2	61.7	76.7
	MIN	35.0	37.0	43.8	50.8	58.8	65.4	69.0	68.6	64.1	51.9	42.4	36.6	52.0
	MEAN	47.2	49.8	57.0	64.6	71.7	77.2	80.2	79.6	74.9	64.8	55.8	49.2	64.3
WINNSBORO 1 W	MAX	53.5	56.7	64.5	74.7	81.3	87.1	90.2	89.1	83.8	74.4	65.0	56.1	73.0
	MIN	30.8	32.0	39.9	49.1	57.5	64.4	68.4	67.6	62.1	49.4	40.0	32.6	49.5
	MEAN	42.2	44.4	52.2	61.9	69.4	75.8	79.3	78.4	73.0	61.9	52.5	44.4	61.3
WINTHROP COLLEGE	MAX	52.4	56.0	64.0	74.1	81.1	86.6	89.7	88.6	83.1	73.4	63.8	54.5	72.3
	MIN	32.0	33.3	40.4	49.4	57.7	64.7	68.4	67.7	62.0	50.2	41.0	33.9	50.1
	MEAN	42.2	44.7	52.2	61.8	69.4	75.7	79.1	78.2	72.6	61.8	52.4	44.3	61.2

SOUTH CAROLINA

TEMPERATURE NORMALS (DEG F)

STATION		JAN	FEB	MAR	APR	MAY	JUN	JUL	AUG	SEP	OCT	NOV	DEC	ANN
YEMASSEE 4 W	MAX	60.2	62.9	70.1	78.4	84.7	89.1	91.6	91.1	86.4	78.1	69.9	62.3	77.1
	MIN	34.5	36.2	43.3	50.7	58.5	65.0	68.7	68.6	63.9	51.9	41.7	35.5	51.5
	MEAN	47.4	49.6	56.7	64.6	71.6	77.1	80.2	79.9	75.2	65.0	55.8	48.9	64.3

SOUTH CAROLINA

PRECIPITATION NORMALS (INCHES)

STATION	JAN	FEB	MAR	APR	MAY	JUN	JUL	AUG	SEP	OCT	NOV	DEC	ANN
AIKEN	4.39	4.27	5.36	3.85	4.22	4.33	4.72	4.38	4.14	2.19	2.24	3.60	47.69
ANDERSON	4.91	4.18	6.14	4.24	4.03	4.13	4.49	3.84	4.17	3.32	3.61	4.34	51.40
ANDERSON FAA AP	4.38	3.98	5.67	3.94	3.93	3.47	4.00	3.72	3.96	2.74	3.09	3.91	46.79
ANTREVILLE	4.69	4.05	5.69	3.72	4.19	3.95	3.93	3.37	4.00	2.91	3.10	3.94	47.54
BAMBERG	3.90	4.19	4.55	3.18	4.38	5.07	5.22	5.26	4.06	2.58	2.17	3.48	48.04
BEAUFORT 7 SW	3.36	3.14	4.28	2.69	4.83	5.64	7.14	6.61	5.08	2.38	2.11	2.90	50.16
BISHOPVILLE 3 W	3.96	3.59	4.28	3.24	3.75	4.63	5.34	4.45	3.78	2.78	2.45	3.05	45.30
BLACKVILLE 3 W	4.07	4.22	4.95	3.55	4.19	5.27	5.00	4.48	3.87	2.39	2.24	3.60	47.83
BLAIR	4.40	3.82	5.26	3.66	3.74	3.87	4.96	4.39	4.21	2.71	2.45	3.55	47.02
BRANCHVILLE 6 S	3.57	3.98	4.45	3.01	5.06	4.77	5.18	5.66	4.43	2.80	1.96	3.17	48.04
CAESARS HEAD 1 NE	6.80	6.46	8.50	6.60	6.93	7.26	6.55	6.15	6.61	5.87	6.25	6.72	80.70
CALHOUN FALLS	4.92	3.99	5.56	3.98	4.22	4.03	4.63	3.32	3.94	2.68	2.99	3.81	48.07
CAMDEN 2 WSW	4.10	3.63	4.55	3.37	3.60	4.49	5.63	4.46	3.62	2.66	2.53	3.27	45.91
CATAWBA	4.25	3.86	4.98	3.51	3.72	4.55	5.05	4.21	3.62	3.07	2.93	3.55	47.30
CHAPPELLS	4.38	3.68	5.35	3.74	3.97	3.99	4.18	3.72	4.01	2.67	2.54	3.55	45.78
CHARLESTON AP WSO	3.33	3.37	4.38	2.58	4.41	6.54	7.33	6.50	4.94	2.92	2.18	3.11	51.59
CHARLESTON CITY WSO R	3.17	3.09	4.27	2.38	3.79	5.53	6.11	6.07	5.19	2.84	2.04	2.86	47.34
CHERAW	4.12	4.10	4.76	3.47	3.56	4.95	5.42	4.96	4.16	3.17	2.80	3.32	48.79
CHESTER	4.36	3.80	5.27	3.67	3.63	4.05	4.27	4.36	4.35	2.95	2.79	3.61	47.11
CLARK HILL DAM	4.82	4.06	5.20	3.94	4.12	3.64	4.56	3.66	3.53	2.66	2.47	3.78	46.44
CLEMSON UNIVERSITY	5.10	4.90	6.65	4.74	4.28	4.16	4.26	4.46	3.71	3.71	3.67	4.65	54.29
CLEVELAND 4 S	5.44	5.38	6.97	5.11	5.27	5.42	4.88	4.93	4.49	4.10	4.27	5.04	61.30
COLUMBIA WSO R	4.38	3.99	5.16	3.59	3.85	4.45	5.35	5.56	4.23	2.55	2.51	3.50	49.12
COLUMBIA UNI OF S C	4.11	3.82	4.82	3.47	3.66	4.48	5.46	4.83	4.23	2.56	2.40	3.17	47.01
CONWAY	3.76	3.53	4.27	2.96	4.59	5.62	6.15	5.69	5.76	3.08	2.41	3.27	51.09
DARLINGTON	3.83	3.92	4.44	3.12	3.69	4.88	5.64	5.49	4.14	2.66	2.47	3.26	47.54
DILLON 4 SW	3.63	3.51	4.30	3.15	3.56	4.69	5.11	4.85	3.86	3.00	2.58	3.17	45.41
EDGEFIELD 1 ENE	4.86	3.95	4.97	4.16	3.90	4.05	4.33	3.47	4.06	2.43	2.45	3.66	46.29
EFFINGHAM	3.66	3.42	4.22	2.99	3.75	4.60	5.68	4.98	3.76	2.71	2.21	3.13	45.11
FLORENCE FAA AP	3.53	3.37	4.17	2.92	3.30	4.65	5.69	4.73	4.05	2.38	2.20	2.99	43.98
FLORENCE 2 N	3.79	3.61	4.34	3.07	3.46	4.99	5.94	4.78	3.84	2.69	2.30	3.06	45.87
FORT MILL 4 NW	3.99	3.92	5.17	3.31	3.59	3.68	4.50	4.14	4.07	2.85	2.84	3.60	45.66
GAFFNEY 6 E	4.16	4.10	5.43	3.84	4.09	4.23	4.38	4.46	4.25	3.16	2.94	3.87	48.91
GASTON SHOALS	4.27	4.21	5.46	3.94	4.34	4.27	4.33	4.53	4.06	3.19	3.08	3.86	49.54
GEORGETOWN 2 E	3.57	3.40	4.23	2.24	4.19	5.11	6.82	6.14	5.90	3.71	2.52	3.35	51.18
GIVHANS FERRY STATE PK	3.56	3.63	4.41	2.98	4.44	5.32	6.42	6.00	4.30	2.63	2.28	3.34	49.31
GREAT FALLS	3.93	3.56	4.87	3.30	3.35	4.23	5.78	4.65	3.78	2.70	2.60	3.43	46.18
GRNVLE-SPTNBRG WSO R	4.21	4.39	5.87	4.35	4.22	4.77	4.08	3.66	4.35	3.49	3.21	3.93	50.52
GREENWOOD 3 ESE	4.69	3.95	5.53	3.88	3.97	3.53	4.16	3.62	4.12	2.87	2.93	3.61	46.86
HAMPTON	3.68	3.80	4.43	2.93	4.86	5.15	5.88	5.30	4.53	2.32	2.02	3.05	47.95
KERSHAW	4.14	3.64	4.88	3.50	3.64	4.07	5.16	4.58	4.00	2.72	2.96	3.37	46.66
KINGSTREE 1 SE	3.87	3.92	4.55	3.07	4.07	5.00	5.18	6.84	4.43	3.04	2.23	3.36	49.56
LAKE CITY	3.59	3.61	4.46	3.01	3.63	5.01	5.62	5.15	4.24	2.87	2.21	3.26	46.66
LAURENS	4.73	4.21	5.68	3.81	4.26	4.02	3.48	3.88	4.24	2.90	3.35	3.79	48.35
LITTLE MOUNTAIN	4.63	3.84	5.44	3.59	3.82	3.70	4.87	4.40	4.45	2.67	2.59	3.67	47.67
LONGCREEK 1 N	5.91	5.50	7.38	5.50	5.59	5.50	5.97	5.04	5.04	4.11	4.74	5.40	65.68
LORIS 1 S	3.57	3.58	4.33	2.82	4.15	5.33	5.86	5.72	5.07	2.63	2.40	3.24	48.70
MARION	3.51	3.76	4.72	3.08	3.89	4.64	5.96	5.65	4.24	3.15	2.72	3.15	48.47
MC COLL	4.09	3.87	4.54	3.06	3.86	4.65	4.96	4.49	4.01	2.96	2.62	3.20	46.31
MC CORMICK 9 E	4.88	4.21	5.07	4.11	3.99	4.00	4.73	3.93	3.80	2.60	2.54	3.64	47.50
NEWBERRY	4.39	3.66	5.46	3.48	3.86	3.85	4.73	3.93	4.51	2.44	2.60	3.52	46.43
NINETY NINE ISLANDS	4.33	4.14	5.48	3.60	4.25	3.99	4.24	4.10	4.27	2.99	2.97	3.93	48.29
OAKWAY	5.19	4.81	6.56	4.67	4.23	4.56	4.35	4.04	3.95	3.38	3.69	4.47	53.90
ORANGEBURG 2	4.00	3.99	4.58	3.32	4.22	4.56	5.46	5.32	4.16	2.42	2.28	3.28	47.59
PARR	4.37	3.65	5.08	3.38	3.79	3.85	4.34	3.71	3.82	2.59	2.40	3.35	44.33

SOUTH CAROLINA

PRECIPITATION NORMALS (INCHES)

STATION	JAN	FEB	MAR	APR	MAY	JUN	JUL	AUG	SEP	OCT	NOV	DEC	ANN
PEE DEE	3.70	3.51	4.35	3.19	3.49	4.88	5.53	5.16	4.27	3.06	2.50	3.11	46.75
PELION 4 NW	4.23	3.78	4.92	3.63	3.66	4.46	5.16	5.01	4.39	2.61	2.28	3.25	47.38
PICKENS 5 SE	5.25	5.03	6.68	4.91	4.49	5.01	4.50	4.16	3.87	3.92	3.94	4.87	56.63
PINOPOLIS DAM	3.55	3.55	4.22	2.78	4.22	5.77	7.05	5.90	4.36	2.97	2.00	3.20	49.57
RAINBOW LAKE	4.32	4.54	5.98	4.18	4.46	4.74	4.61	3.92	5.00	3.93	3.42	4.16	53.26
RIDGELAND 5 NE	3.42	3.58	4.38	3.01	4.96	4.92	7.02	6.44	5.29	2.39	2.22	3.14	50.77
RIMINI 2 SSW	3.76	3.89	4.43	2.96	3.63	5.19	4.94	4.80	3.92	2.35	2.21	3.04	45.12
SALEM	6.06	5.79	7.66	5.21	5.81	5.30	5.72	5.20	5.02	4.41	4.48	5.72	66.38
SALUDA	4.81	4.17	5.55	4.11	3.64	3.96	4.77	4.39	4.04	2.70	2.56	3.79	48.49
SANTUCK	4.44	4.15	5.63	3.51	3.64	4.27	4.89	4.18	4.30	2.78	2.87	3.63	48.29
SPARTANBURG 3E	4.17	4.13	5.58	4.21	4.08	4.79	4.38	3.94	4.37	3.27	3.14	4.01	50.07
SPRINGFIELD	4.13	4.23	4.95	3.69	4.01	4.84	5.33	4.49	4.23	2.36	2.19	3.47	47.92
SULLIVANS ISLAND	3.34	3.40	4.16	2.23	3.68	4.89	5.63	5.97	4.74	3.25	2.43	3.16	46.88
SUMMERVILLE 2 WNW	3.44	3.55	4.30	2.81	4.44	5.83	6.59	6.43	4.83	2.74	2.15	3.12	50.23
SUMTER	3.89	3.75	4.48	3.40	3.80	5.76	5.24	5.23	3.86	2.40	2.33	3.17	47.31
TILGHMAN FOR NURSERY	3.84	3.73	4.49	3.40	4.08	5.53	5.80	4.72	3.97	2.38	2.31	3.11	47.36
UNION 8 SW	4.91	4.21	5.61	3.58	3.92	4.22	4.29	4.05	4.37	3.05	3.08	3.82	49.11
WALHALLA	5.65	5.47	7.20	5.15	5.21	5.21	4.92	4.97	4.55	4.12	4.29	5.12	61.86
WALTERBORO 2 SW	3.54	3.82	4.63	2.92	4.90	5.82	6.88	5.54	5.37	2.93	2.23	3.47	52.05
WARE SHOALS	4.73	4.07	5.53	3.81	4.55	3.92	4.38	3.64	3.94	2.83	3.24	3.86	48.50
WATEREE DAM	3.32	3.11	4.17	2.81	3.16	3.51	4.37	3.86	3.09	2.32	2.32	2.85	38.89
WEST PELZER	4.88	4.43	6.15	4.37	4.33	4.20	4.21	3.87	4.57	3.56	3.35	4.15	52.07
WHITMIRE 2 NE	4.64	4.12	5.71	3.57	3.71	4.39	5.03	4.45	4.62	2.64	2.77	3.80	49.45
WINNSBORO 1 W	4.47	3.63	5.21	3.40	3.66	4.10	5.14	4.25	4.11	2.60	2.51	3.50	46.58
WINTHROP COLLEGE	4.40	4.13	5.24	3.46	3.78	4.02	5.03	4.34	4.41	3.05	2.97	3.74	48.57
WOODRUFF	4.60	4.35	5.79	4.10	4.01	4.01	3.63	3.30	3.88	3.14	3.21	3.88	47.90
YEMASSEE 4 W	3.42	3.78	4.29	2.91	4.92	5.17	6.64	6.10	4.94	2.50	2.02	3.17	49.86

SOUTH CAROLINA

HEATING DEGREE DAY NORMALS (BASE 65 DEG F)

STATION		JUL	AUG	SEP	OCT	NOV	DEC	JAN	FEB	MAR	APR	MAY	JUN	ANN
AIKEN		0	0	0	93	307	524	588	466	308	85	13	0	2384
ANDERSON		0	0	8	129	375	632	694	554	402	125	30	0	2949
ANDERSON FAA AP		0	0	7	143	389	632	707	566	410	131	36	0	3021
BAMBERG		0	0	0	95	294	511	572	450	290	61	9	0	2282
BEAUFORT 7 SW		0	0	0	52	228	434	508	394	257	46	0	0	1919
BLACKVILLE 3 W		0	0	0	96	297	530	598	470	309	80	11	0	2391
CAESARS HEAD 1 NE		10	8	73	304	567	815	890	736	598	291	128	33	4453
CALHOUN FALLS		0	0	0	139	399	651	713	580	421	141	42	0	3086
CAMDEN 2 WSW		0	0	10	170	411	660	713	588	429	148	42	0	3171
CHARLESTON AP WSO		0	0	0	76	262	471	543	434	286	69	6	0	2147
CHARLESTON CITY WSO	R	0	0	0	37	203	417	501	401	264	45	0	0	1868
CHERAW		0	0	12	165	415	676	738	608	439	142	47	0	3242
CHESTER		0	0	7	145	390	651	707	563	411	132	26	0	3032
CLEMSON UNIVERSITY		0	0	8	160	405	657	716	577	429	151	36	0	3139
COLUMBIA WSO	R	0	0	0	123	339	567	637	508	346	87	22	0	2629
COLUMBIA UNI OF S C		0	0	0	88	286	520	586	451	295	61	12	0	2299
CONWAY		0	0	0	98	306	546	618	496	330	83	13	0	2490
DARLINGTON		0	0	0	107	329	561	627	498	342	91	17	0	2572
DILLON 4 SW		0	0	8	148	375	632	688	563	397	127	30	0	2968
FLORENCE FAA AP		0	0	0	106	315	552	630	506	343	91	18	0	2561
FLORENCE 2 N		0	0	0	124	334	589	657	540	365	99	19	0	2727
GEORGETOWN 2 E		0	0	0	70	250	492	566	458	307	74	9	0	2226
GRNVLE-SPTNBRG WSO	R	0	0	7	162	423	670	741	599	442	154	41	0	3239
GREENWOOD 3 ESE		0	0	10	168	408	663	722	594	431	147	46	0	3189
HAMPTON		0	0	0	73	274	474	533	407	253	52	7	0	2073
KINGSTREE 1 SE		0	0	0	128	344	577	635	522	342	100	26	0	2674
LAKE CITY		0	0	0	108	307	549	616	498	333	99	18	0	2528
LAURENS		0	0	10	168	417	682	729	588	424	137	49	0	3204
LITTLE MOUNTAIN		0	0	0	105	317	561	635	502	344	85	23	0	2572
LONGCREEK 1 N		0	0	28	210	450	688	760	627	489	199	78	12	3541
LORIS 1 S		0	0	0	116	335	574	643	526	370	117	27	0	2708
MARION		0	0	0	109	325	552	614	486	324	82	15	0	2507
MC COLL		0	0	0	119	331	580	649	510	356	88	17	0	2650
NEWBERRY		0	0	0	129	369	611	663	524	367	105	23	0	2791
ORANGEBURG 2		0	0	0	119	318	564	635	505	337	83	19	0	2580
PARR		0	0	0	123	360	608	654	526	375	104	24	0	2774
PICKENS 5 SE		0	0	6	131	387	642	704	568	414	137	38	0	3027
PINOPOLIS DAM		0	0	0	93	291	533	609	493	324	88	15	0	2446
RAINBOW LAKE		0	0	8	180	447	694	744	610	453	165	46	0	3347
RIDGELAND 5 NE		0	0	0	70	266	467	530	403	248	54	6	0	2044
SALUDA		0	0	6	145	383	626	693	552	387	117	38	0	2947
SANTUCK		0	0	5	138	375	626	682	543	384	107	29	0	2889
SULLIVANS ISLAND		0	0	0	39	208	431	526	427	287	70	8	0	1996
SUMMERVILLE 2 WNW		0	0	0	94	290	526	588	470	310	82	13	0	2373
SUMTER		0	0	0	106	312	552	613	479	317	89	12	0	2480
TILGHMAN FOR NURSERY		0	0	0	99	303	533	598	464	311	74	14	0	2396
UNION 8 SW		0	0	13	190	438	701	760	627	468	165	65	5	3432
WALHALLA		0	0	9	176	444	688	735	599	456	180	60	0	3347
WALTERBORO 2 SW		0	0	0	85	291	498	566	437	282	70	13	0	2242
WINNSBORO 1 W		0	0	8	140	375	639	707	577	407	120	32	0	3005
WINTHROP COLLEGE		0	0	11	134	378	642	707	568	409	126	28	0	3003
YEMASSEE 4 W		0	0	0	99	288	505	563	443	287	69	13	0	2267

SOUTH CAROLINA

COOLING DEGREE DAY NORMALS (BASE 65 DEG F)

STATION		JAN	FEB	MAR	APR	MAY	JUN	JUL	AUG	SEP	OCT	NOV	DEC	ANN
AIKEN		8	7	23	58	215	375	474	456	291	71	7	0	1985
ANDERSON		0	0	8	26	161	315	428	412	239	39	0	0	1628
ANDERSON FAA AP		0	0	10	32	182	351	459	437	253	50	0	0	1774
BAMBERG		11	8	26	55	223	378	477	459	300	80	9	5	2031
BEAUFORT 7 SW		15	11	33	67	252	393	490	477	330	114	21	6	2209
BLACKVILLE 3 W		9	0	20	53	213	363	462	446	288	78	6	6	1944
CAESARS HEAD 1 NE		0	0	0	6	53	129	206	179	79	6	0	0	658
CALHOUN FALLS		0	0	8	27	175	330	443	431	248	40	0	0	1702
CAMDEN 2 WSW		0	0	7	19	160	315	431	406	235	49	0	0	1622
CHARLESTON AP WSO		13	9	29	48	229	378	481	465	321	101	13	6	2093
CHARLESTON CITY WSO	R	11	9	29	69	258	411	515	499	357	121	17	8	2304
CHERAW		0	0	8	22	174	324	443	415	246	54	0	0	1686
CHESTER		0	0	14	30	153	315	434	415	241	43	0	0	1645
CLEMSON UNIVERSITY		0	0	7	25	142	303	415	394	227	45	0	0	1558
COLUMBIA WSO	R	8	6	20	51	223	381	496	471	297	74	6	0	2033
COLUMBIA UNI OF S C		10	6	32	76	247	405	508	484	309	79	7	6	2169
CONWAY		7	0	17	41	205	360	477	459	300	80	6	0	1952
DARLINGTON		7	0	16	40	197	354	462	434	273	57	5	0	1845
DILLON 4 SW		0	0	10	28	163	321	446	422	254	49	0	0	1693
FLORENCE FAA AP		10	8	24	52	216	369	484	459	295	72	9	0	1998
FLORENCE 2 N		0	0	12	39	196	351	459	431	268	59	0	0	1815
GEORGETOWN 2 E		14	7	25	47	220	375	481	468	318	98	10	9	2072
GRNVLE-SPTNBRG WSO	R	0	0	8	19	143	297	409	388	208	29	0	0	1501
GREENWOOD 3 ESE		0	0	7	24	167	322	431	412	229	41	0	0	1633
HAMPTON		19	12	36	73	243	387	490	468	309	89	19	5	2150
KINGSTREE 1 SE		9	6	14	40	209	354	465	440	274	66	8	0	1885
LAKE CITY		11	8	20	57	216	372	474	462	294	83	7	0	2004
LAURENS		0	0	9	29	170	325	437	412	223	32	0	0	1637
LITTLE MOUNTAIN		8	7	18	46	212	369	477	459	287	65	0	0	1948
LONGCREEK 1 N		0	0	0	10	96	213	307	285	145	15	0	0	1071
LORIS 1 S		8	0	13	27	179	327	440	425	264	63	5	0	1751
MARION		10	7	20	40	201	360	474	446	273	56	7	0	1894
MC COLL		7	0	15	43	199	360	468	443	284	67	7	0	1893
NEWBERRY		0	0	14	33	181	345	456	434	253	46	0	0	1762
ORANGEBURG 2		9	7	21	50	215	366	474	459	291	79	6	0	1977
PARR		0	0	15	29	176	340	459	440	263	48	0	0	1770
PICKENS 5 SE		0	0	11	26	144	292	394	378	219	32	0	0	1496
PINOPOLIS DAM		10	6	20	49	216	363	474	459	303	90	6	0	1996
RAINBOW LAKE		0	0	7	12	127	282	397	381	194	28	0	0	1428
RIDGELAND 5 NE		16	11	28	75	241	378	477	456	312	92	14	8	2108
SALUDA		8	6	12	39	187	346	462	431	255	49	5	0	1800
SANTUCK		0	0	12	29	181	333	443	425	239	42	0	0	1704
SULLIVANS ISLAND		11	7	26	61	237	393	499	493	351	123	22	7	2230
SUMMERVILLE 2 WNW		12	8	22	43	208	360	465	453	306	88	8	5	1978
SUMTER		12	8	26	65	214	363	471	443	284	69	6	7	1968
TILGHMAN FOR NURSERY		9	5	23	71	246	393	499	471	300	84	6	6	2113
UNION 8 SW		0	0	6	15	145	287	403	381	199	29	0	0	1465
WALHALLA		0	0	6	12	119	260	363	341	183	24	0	0	1308
WALTERBORO 2 SW		14	12	34	58	221	366	471	453	297	79	15	8	2028
WINNSBORO 1 W		0	0	11	27	168	324	443	415	248	44	0	0	1680
WINTHROP COLLEGE		0	0	12	30	164	321	437	409	239	35	0	0	1647
YEMASSEE 4 W		17	11	29	57	218	363	471	462	306	99	12	6	2051

38 — SOUTH CAROLINA

STATE-STATION NUMBER	STN TYP	NAME	LATITUDE DEG-MIN	LONGITUDE DEG-MIN	ELEVATION (FT)
38-0074	13	AIKEN	N 3334	W 08144	527
38-0165	13	ANDERSON	N 3432	W 08240	800
38-0170	13	ANDERSON FAA AP	N 3430	W 08243	759
38-0204	12	ANTREVILLE	N 3420	W 08233	723
38-0448	13	BAMBERG	N 3317	W 08103	165
38-0559	13	BEAUFORT 7 SW	N 3223	W 08046	21
38-0736	12	BISHOPVILLE 3 W	N 3414	W 08018	249
38-0764	13	BLACKVILLE 3 W	N 3322	W 08119	324
38-0772	12	BLAIR	N 3425	W 08124	283
38-0972	12	BRANCHVILLE 6 S	N 3311	W 08048	95
38-1256	13	CAESARS HEAD 1 NE	N 3507	W 08236	3086
38-1277	13	CALHOUN FALLS	N 3405	W 08235	520
38-1310	13	CAMDEN 2 WSW	N 3415	W 08039	175
38-1462	12	CATAWBA	N 3451	W 08055	562
38-1530	12	CHAPPELLS	N 3411	W 08152	402
38-1544	13	CHARLESTON AP WSO	N 3254	W 08002	41
38-1549	13	CHARLESTON CITY WSO R	N 3247	W 07956	9
38-1588	13	CHERAW	N 3442	W 07953	155
38-1633	13	CHESTER	N 3442	W 08112	585
38-1726	12	CLARK HILL DAM	N 3340	W 08211	380
38-1770	13	CLEMSON UNIVERSITY	N 3441	W 08249	819
38-1804	12	CLEVELAND 4 S	N 3502	W 08229	1940
38-1939	13	COLUMBIA WSO R	N 3357	W 08107	213
38-1944	13	COLUMBIA UNI OF S C	N 3400	W 08101	310
38-1997	13	CONWAY	N 3350	W 07903	27
38-2260	13	DARLINGTON	N 3417	W 07952	175
38-2386	13	DILLON 4 SW	N 3422	W 07924	100
38-2712	12	EDGEFIELD 1 ENE	N 3347	W 08155	648
38-2757	12	EFFINGHAM	N 3404	W 07945	58
38-3106	13	FLORENCE FAA AP	N 3411	W 07943	146
38-3111	13	FLORENCE 2 N	N 3413	W 07946	144
38-3216	12	FORT MILL 4 NW	N 3500	W 08100	650
38-3356	12	GAFFNEY 6 E	N 3505	W 08134	650
38-3433	12	GASTON SHOALS	N 3508	W 08136	680
38-3468	13	GEORGETOWN 2 E	N 3321	W 07915	10
38-3525	12	GIVHANS FERRY STATE PK	N 3301	W 08023	60
38-3700	12	GREAT FALLS	N 3433	W 08053	356
38-3747	13	GRNVLE-SPTNBRG WSO R	N 3454	W 08213	957
38-3754	13	GREENWOOD 3 ESE	N 3410	W 08207	620
38-3906	13	HAMPTON	N 3252	W 08107	86
38-4690	12	KERSHAW	N 3433	W 08035	532
38-4753	13	KINGSTREE 1 SE	N 3339	W 07949	58
38-4886	13	LAKE CITY	N 3352	W 07945	70
38-5017	13	LAURENS	N 3430	W 08202	589
38-5200	13	LITTLE MOUNTAIN	N 3412	W 08125	711
38-5278	13	LONGCREEK 1 N	N 3447	W 08315	1610
38-5306	13	LORIS 1 S	N 3403	W 07853	90
38-5509	13	MARION	N 3411	W 07924	68
38-5633	13	MC COLL	N 3440	W 07933	190
38-5658	12	MC CORMICK 9 E	N 3355	W 08209	495

STATE-STATION NUMBER	STN TYP	NAME	LATITUDE DEG-MIN	LONGITUDE DEG-MIN	ELEVATION (FT)
38-6209	13	NEWBERRY	N 3417	W 08137	476
38-6293	12	NINETY NINE ISLANDS	N 3503	W 08130	500
38-6423	12	OAKWAY	N 3436	W 08302	990
38-6527	13	ORANGEBURG 2	N 3330	W 08052	260
38-6688	13	PARR	N 3416	W 08120	258
38-6749	12	PEE DEE	N 3412	W 07932	60
38-6775	12	PELION 4 NW	N 3348	W 08117	450
38-6831	13	PICKENS 5 SE	N 3451	W 08239	1040
38-6893	13	PINOPOLIS DAM	N 3315	W 07959	69
38-7113	13	RAINBOW LAKE	N 3507	W 08158	748
38-7281	13	RIDGELAND 5 NE	N 3232	W 08054	20
38-7313	12	RIMINI 2 SSW	N 3338	W 08031	80
38-7589	12	SALEM	N 3454	W 08259	1080
38-7631	13	SALUDA	N 3400	W 08146	540
38-7722	13	SANTUCK	N 3438	W 08130	512
38-8186	12	SPARTANBURG 3E	N 3459	W 81530	840
38-8219	12	SPRINGFIELD	N 3330	W 08117	300
38-8405	13	SULLIVANS ISLAND	N 3246	W 07952	9
38-8426	13	SUMMERVILLE 2 WNW	N 3302	W 08012	75
38-8440	13	SUMTER	N 3356	W 08019	169
38-8621	13	TILGHMAN FOR NURSERY	N 3354	W 08030	249
38-8786	13	UNION 8 SW	N 3439	W 08145	549
38-8887	13	WALHALLA	N 3445	W 08305	1061
38-8922	13	WALTERBORO 2 SW	N 3253	W 08041	56
38-8947	12	WARE SHOALS	N 3424	W 08214	642
38-8979	12	WATEREE DAM	N 3420	W 08040	226
38-9122	12	WEST PELZER	N 3439	W 08228	850
38-9218	12	WHITMIRE 2 NE	N 3432	W 08135	400
38-9327	13	WINNSBORO 1 W	N 3423	W 08106	555
38-9350	13	WINTHROP COLLEGE	N 3457	W 08103	690
38-9412	12	WOODRUFF	N 3444	W 08201	760
38-9469	13	YEMASSEE 4 W	N 3241	W 08055	82

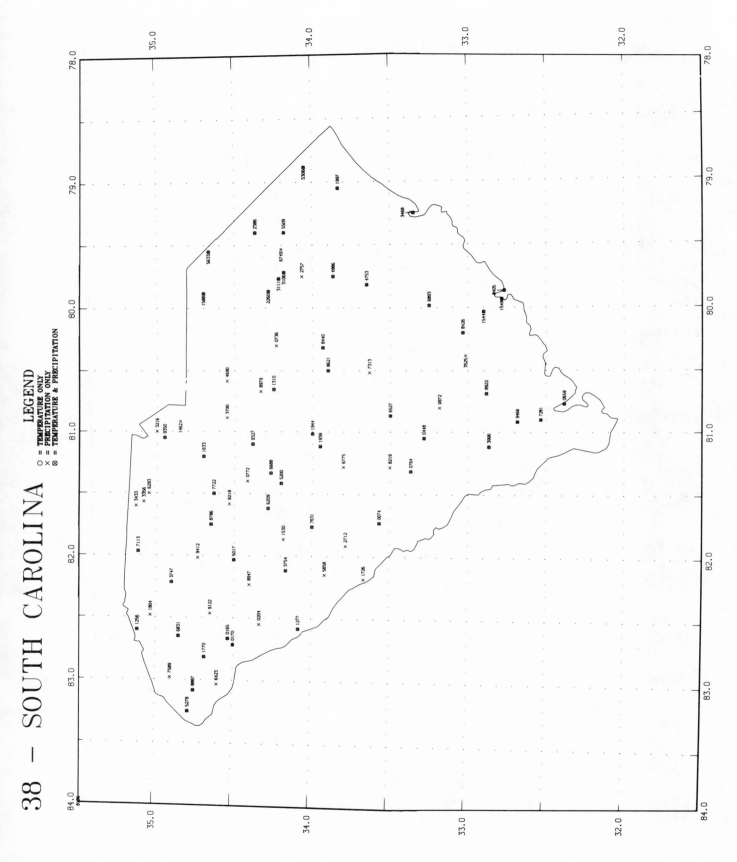

38 – SOUTH CAROLINA

LEGEND
O = TEMPERATURE ONLY
X = PRECIPITATION ONLY
⊗ = TEMPERATURE & PRECIPITATION

559

SOUTH DAKOTA

TEMPERATURE NORMALS (DEG F)

STATION			JAN	FEB	MAR	APR	MAY	JUN	JUL	AUG	SEP	OCT	NOV	DEC	ANN
ABERDEEN WSO	//R	MAX	18.9	26.1	37.9	56.5	69.6	78.6	85.3	84.4	73.4	61.3	41.0	26.5	55.0
		MIN	-2.2	5.3	17.5	32.6	43.9	54.0	58.9	56.5	45.5	34.1	19.4	6.4	31.0
		MEAN	8.3	15.7	27.7	44.6	56.8	66.3	72.1	70.5	59.5	47.7	30.2	16.5	43.0
ACADEMY		MAX	28.8	35.4	44.6	61.7	73.1	82.9	90.3	89.0	78.8	66.5	47.5	34.6	61.1
		MIN	6.0	12.6	21.4	35.0	45.9	56.1	61.4	59.6	49.3	38.1	23.9	13.0	35.2
		MEAN	17.5	24.0	33.0	48.4	59.6	69.5	75.9	74.3	64.1	52.3	35.7	23.8	48.2
ALEXANDRIA		MAX	25.0	32.0	42.6	60.8	72.6	81.8	87.8	86.0	76.4	64.6	45.8	31.5	58.9
		MIN	4.5	11.7	21.8	35.9	47.4	57.6	62.8	60.8	50.6	39.1	24.5	12.3	35.8
		MEAN	14.8	21.9	32.2	48.4	60.0	69.7	75.3	73.4	63.5	51.9	35.2	22.0	47.4
ARDMORE 2 N		MAX	33.9	40.5	48.0	60.1	70.7	81.4	90.5	88.9	79.0	66.2	48.1	37.8	62.1
		MIN	5.6	11.8	19.0	30.4	40.8	49.9	56.0	53.1	41.7	30.0	18.0	9.5	30.5
		MEAN	19.8	26.2	33.5	45.3	55.8	65.6	73.3	71.0	60.3	48.1	33.1	23.7	46.3
ARMOUR		MAX	27.3	34.5	43.8	61.0	73.0	82.4	89.0	87.2	77.0	65.2	46.8	33.8	60.1
		MIN	5.4	12.0	21.4	35.2	46.5	56.8	62.2	60.2	49.7	37.9	23.8	12.5	35.3
		MEAN	16.4	23.3	32.6	48.1	59.7	69.7	75.6	73.7	63.4	51.5	35.4	23.2	47.7
BELLE FOURCHE		MAX	33.3	39.6	46.2	59.9	71.1	80.3	89.0	87.9	77.7	66.0	47.6	38.4	61.4
		MIN	7.9	13.0	19.7	31.2	41.9	51.2	56.3	53.7	42.6	32.3	20.8	13.1	32.0
		MEAN	20.7	26.3	33.0	45.6	56.5	65.8	72.7	70.8	60.2	49.2	34.2	25.8	46.7
BONESTEEL		MAX	29.3	35.4	44.4	61.0	72.5	82.2	88.7	87.4	77.3	65.9	47.4	35.0	60.5
		MIN	6.1	12.2	21.2	34.7	45.7	55.8	62.0	59.8	49.1	37.4	23.3	12.6	35.0
		MEAN	17.7	23.8	32.8	47.9	59.1	69.0	75.4	73.6	63.2	51.7	35.4	23.8	47.8
BRIDGEWATER		MAX	24.5	32.0	42.6	61.1	73.5	82.8	88.6	86.6	76.6	64.9	45.2	31.2	59.1
		MIN	3.1	10.4	21.0	35.2	46.6	56.9	61.9	59.9	49.5	38.0	23.5	11.0	34.8
		MEAN	13.8	21.2	31.8	48.2	60.1	69.9	75.3	73.3	63.1	51.5	34.4	21.1	47.0
BRITTON		MAX	18.0	25.1	37.8	56.1	69.8	78.7	84.9	84.2	73.6	61.4	40.3	25.7	54.6
		MIN	-3.6	4.3	17.0	32.4	43.7	53.5	58.4	56.9	46.1	35.2	19.6	5.6	30.8
		MEAN	7.2	14.7	27.4	44.3	56.8	66.2	71.7	70.6	59.9	48.3	30.0	15.6	42.7
BROOKINGS 2 NE		MAX	20.1	26.3	37.4	55.5	68.9	77.9	83.7	81.9	71.8	60.4	41.5	27.0	54.4
		MIN	-1.7	4.7	17.0	31.8	43.0	53.2	57.8	55.3	44.7	33.7	20.0	7.1	30.6
		MEAN	9.2	15.5	27.2	43.7	56.0	65.6	70.7	68.6	58.3	47.1	30.8	17.1	42.5
CAMP CROOK		MAX	28.5	34.4	42.5	56.9	69.1	78.3	87.8	87.1	76.0	63.5	44.2	34.5	58.6
		MIN	3.5	9.4	17.2	29.1	39.9	49.5	54.6	52.2	41.2	30.6	18.1	9.7	29.6
		MEAN	16.1	21.9	29.9	43.0	54.6	64.0	71.2	69.7	58.6	47.1	31.2	22.1	44.1
CANTON		MAX	25.3	32.3	43.1	61.5	74.2	83.4	88.4	86.1	76.6	65.2	45.9	31.6	59.5
		MIN	3.0	10.1	21.2	35.1	46.6	56.7	61.4	59.3	48.8	37.3	23.1	10.9	34.5
		MEAN	14.2	21.2	32.2	48.3	60.4	70.1	74.9	72.7	62.7	51.2	34.5	21.3	47.0
CASTLEWOOD		MAX	20.7	28.2	39.4	57.5	70.9	79.8	85.7	84.4	74.0	62.1	41.8	27.8	56.0
		MIN	-2.1	5.2	17.1	31.3	42.5	53.1	57.5	55.9	45.2	34.0	19.4	6.6	30.5
		MEAN	9.4	16.7	28.2	44.5	56.7	66.5	71.6	70.1	59.6	48.1	30.6	17.2	43.3
CENTERVILLE 6 SE		MAX	25.3	32.1	42.7	61.1	73.4	82.9	87.9	85.7	76.3	65.2	46.4	31.6	59.2
		MIN	3.9	10.7	21.5	36.0	47.1	57.4	61.8	59.8	49.4	38.0	23.9	11.3	35.1
		MEAN	14.6	21.5	32.1	48.6	60.3	70.2	74.9	72.8	62.9	51.6	35.2	21.5	47.2
CLARK		MAX	20.2	27.3	38.4	56.6	69.6	78.3	84.5	83.4	73.0	61.1	41.0	27.0	55.0
		MIN	-1.1	6.3	17.6	32.4	44.2	54.8	59.7	58.0	47.4	36.0	20.7	7.6	32.0
		MEAN	9.6	16.8	28.0	44.5	56.9	66.6	72.1	70.7	60.2	48.6	30.9	17.3	43.5
COLUMBIA 8 N		MAX	18.8	26.0	38.2	56.5	70.0	79.2	85.8	84.7	74.3	61.5	40.9	26.3	55.2
		MIN	-3.6	3.7	16.3	32.0	43.6	53.9	58.9	57.0	45.9	34.5	19.5	5.7	30.6
		MEAN	7.6	14.9	27.3	44.3	56.8	66.6	72.3	70.8	60.1	48.0	30.3	16.0	42.9
CUSTER		MAX	34.3	38.7	42.0	51.4	62.1	72.2	80.3	79.1	69.7	59.6	44.7	38.5	56.1
		MIN	5.6	9.9	14.9	24.9	34.4	43.0	48.6	46.4	36.6	27.1	16.3	9.5	26.4
		MEAN	20.0	24.3	28.5	38.1	48.2	57.6	64.5	62.8	53.2	43.4	30.5	24.0	41.3

SOUTH DAKOTA

TEMPERATURE NORMALS (DEG F)

STATION		JAN	FEB	MAR	APR	MAY	JUN	JUL	AUG	SEP	OCT	NOV	DEC	ANN
DEADWOOD	MAX	32.7	37.5	41.7	52.2	63.5	73.6	81.6	80.2	69.6	58.4	43.1	36.5	55.9
	MIN	10.1	14.5	18.8	28.7	39.0	47.4	53.1	51.1	41.4	32.5	21.0	14.9	31.0
	MEAN	21.4	26.0	30.3	40.5	51.3	60.5	67.4	65.7	55.5	45.5	32.1	25.7	43.5
DE SMET	MAX	21.3	28.8	39.5	57.9	71.0	80.0	86.3	84.6	74.4	62.1	42.0	28.1	56.3
	MIN	1.0	8.5	19.2	33.4	45.1	55.3	60.3	58.3	48.0	36.7	21.9	9.4	33.1
	MEAN	11.2	18.7	29.4	45.7	58.1	67.7	73.3	71.5	61.2	49.4	32.0	18.8	44.8
DUPREE	MAX	25.5	32.1	41.4	57.8	70.3	79.8	89.0	88.1	76.5	63.8	44.0	31.5	58.3
	MIN	3.2	9.7	18.7	32.0	43.1	52.9	58.3	56.7	45.8	35.0	20.8	10.3	32.2
	MEAN	14.4	20.9	30.1	44.9	56.7	66.4	73.7	72.4	61.2	49.4	32.5	20.9	45.3
EUREKA	MAX	18.5	25.7	37.4	55.7	69.0	77.8	85.1	84.4	73.2	60.9	40.4	26.2	54.5
	MIN	-2.3	4.9	15.9	30.8	42.3	52.2	57.3	55.5	44.5	33.9	18.6	6.4	30.0
	MEAN	8.1	15.3	26.7	43.3	55.6	65.0	71.2	69.9	58.9	47.4	29.6	16.3	42.3
FAITH 2 W	MAX	25.6	31.8	40.7	57.0	69.4	79.3	88.2	86.9	75.4	62.8	43.7	31.6	57.7
	MIN	4.3	10.7	18.9	32.3	43.4	52.9	58.4	56.9	46.1	35.5	21.4	10.8	32.6
	MEAN	15.0	21.3	29.8	44.7	56.5	66.1	73.3	71.9	60.8	49.2	32.6	21.2	45.2
FAULKTON 1 NW	MAX	22.0	29.2	40.4	58.4	71.3	80.3	87.6	86.9	76.5	63.9	43.6	29.0	57.4
	MIN	.0	7.0	17.7	32.5	43.1	53.6	58.2	56.5	45.8	34.7	20.3	8.0	31.5
	MEAN	11.0	18.1	29.1	45.5	57.3	67.0	73.0	71.7	61.2	49.3	32.0	18.6	44.5
FLANDREAU	MAX	21.5	27.9	38.3	56.8	69.8	79.1	84.7	82.9	72.8	61.5	42.6	28.3	55.5
	MIN	-.5	6.0	18.3	33.2	44.6	55.0	59.8	57.3	46.8	35.2	21.2	8.2	32.1
	MEAN	10.5	17.0	28.3	45.1	57.3	67.1	72.3	70.1	59.8	48.3	31.9	18.3	43.8
FORESTBURG 3 NE //	MAX	24.0	31.1	41.6	59.4	71.4	80.7	87.2	85.9	76.1	63.9	44.9	30.7	58.1
	MIN	2.2	9.4	20.1	34.5	45.6	56.1	60.9	58.8	48.2	37.2	22.5	10.3	33.8
	MEAN	13.1	20.3	30.9	47.0	58.5	68.4	74.1	72.4	62.2	50.6	33.7	20.5	46.0
FORT MEADE	MAX	34.3	39.5	45.1	57.7	68.9	78.4	86.9	85.8	76.0	64.3	47.4	39.4	60.3
	MIN	11.2	16.3	22.0	33.3	43.9	53.2	59.1	57.4	47.2	37.3	24.4	16.7	35.2
	MEAN	22.8	27.9	33.6	45.5	56.4	65.8	73.0	71.7	61.6	50.9	35.9	28.1	47.8
GANN VALLEY	MAX	24.5	31.6	42.1	60.2	72.1	81.9	89.4	87.9	77.5	64.8	44.9	31.3	59.0
	MIN	1.7	8.9	19.1	33.1	44.2	54.6	60.0	58.1	47.3	35.7	20.9	9.3	32.7
	MEAN	13.1	20.2	30.6	46.7	58.2	68.3	74.7	73.0	62.4	50.3	33.0	20.3	45.9
GETTYSBURG	MAX	21.7	28.9	39.2	56.6	69.5	79.1	87.2	86.2	74.9	62.2	42.3	28.9	56.4
	MIN	1.0	7.6	17.3	31.9	42.9	53.2	58.4	56.4	45.6	34.9	20.5	8.8	31.5
	MEAN	11.4	18.3	28.3	44.3	56.2	66.2	72.8	71.3	60.3	48.6	31.4	18.9	44.0
GREGORY	MAX	31.0	37.4	45.5	61.7	73.0	82.2	89.3	88.1	78.7	67.3	48.6	36.6	61.6
	MIN	6.6	12.3	20.9	33.8	45.0	54.7	60.3	58.4	48.5	37.2	23.1	13.1	34.5
	MEAN	18.8	24.9	33.2	47.8	59.0	68.5	74.8	73.3	63.6	52.3	35.9	24.9	48.1
HIGHMORE 1 W	MAX	23.7	30.4	40.9	58.8	71.3	80.9	88.5	87.7	76.9	64.2	43.8	30.3	58.1
	MIN	1.5	8.4	18.5	32.5	43.6	53.7	59.4	57.7	47.0	35.9	21.0	9.4	32.4
	MEAN	12.6	19.5	29.7	45.7	57.5	67.3	74.0	72.7	62.0	50.1	32.4	19.9	45.3
HOT SPRINGS	MAX	36.5	42.1	48.4	60.0	70.5	81.1	89.2	88.0	78.4	66.5	49.0	40.6	62.5
	MIN	9.7	14.9	20.2	30.9	41.5	50.4	56.4	54.0	43.4	33.3	21.9	14.8	32.6
	MEAN	23.1	28.5	34.3	45.5	56.0	65.8	72.8	71.1	61.0	49.9	35.5	27.7	47.6
HOWARD	MAX	22.5	30.1	41.2	59.6	72.1	81.3	87.8	86.0	75.3	62.7	43.4	29.5	57.6
	MIN	1.7	8.8	19.7	34.1	45.4	55.4	60.7	58.6	48.1	37.1	22.5	9.9	33.5
	MEAN	12.1	19.5	30.5	46.9	58.7	68.4	74.2	72.3	61.7	49.9	33.0	19.7	45.6
HURON WSO //R	MAX	21.9	28.6	39.4	57.7	70.2	80.3	87.3	85.5	74.7	62.2	43.2	29.0	56.7
	MIN	.4	7.6	19.2	33.5	44.2	55.0	60.5	58.4	46.9	35.5	21.1	9.0	32.6
	MEAN	11.2	18.1	29.4	45.6	57.3	67.7	74.0	72.0	60.8	48.9	32.2	19.0	44.7
KENNEBEC	MAX	27.7	34.9	44.3	61.5	73.3	83.2	91.4	90.0	79.5	66.5	47.1	33.9	61.1
	MIN	3.5	10.0	19.6	33.1	44.4	55.0	60.2	58.7	47.8	35.7	21.3	10.2	33.3
	MEAN	15.6	22.5	32.0	47.3	58.9	69.1	75.8	74.4	63.7	51.1	34.2	22.1	47.2

SOUTH DAKOTA

TEMPERATURE NORMALS (DEG F)

STATION			JAN	FEB	MAR	APR	MAY	JUN	JUL	AUG	SEP	OCT	NOV	DEC	ANN
LEAD 1 SE	//	MAX	32.2	36.2	40.1	50.4	61.6	71.9	80.3	79.0	68.9	57.6	42.6	36.2	54.8
		MIN	13.1	17.2	20.5	29.6	39.8	48.8	55.5	54.3	44.8	36.2	24.0	18.3	33.5
		MEAN	22.7	26.7	30.3	40.0	50.7	60.4	67.9	66.7	56.9	47.0	33.3	27.3	44.2
LEMMON		MAX	22.3	29.2	38.2	54.0	66.8	76.0	83.9	82.8	71.1	59.3	40.7	29.3	54.5
		MIN	1.0	8.0	16.3	29.9	41.3	50.9	56.6	54.7	43.8	33.3	18.9	8.5	30.3
		MEAN	11.7	18.6	27.3	42.0	54.1	63.5	70.3	68.8	57.4	46.3	29.8	18.9	42.4
LONGVALLEY		MAX	34.3	39.5	45.8	59.8	71.1	81.5	90.2	88.8	78.9	66.3	49.1	39.2	62.0
		MIN	9.2	14.8	21.5	33.3	43.9	53.8	60.1	58.2	47.7	36.9	23.9	15.0	34.9
		MEAN	21.7	27.2	33.7	46.6	57.5	67.7	75.2	73.6	63.3	51.7	36.5	27.1	48.5
LUDLOW		MAX	25.6	31.9	39.7	55.0	67.4	76.4	85.4	84.8	72.9	60.8	42.1	32.0	56.2
		MIN	4.1	10.4	17.6	30.1	40.5	49.6	54.8	53.0	42.8	32.9	20.1	10.6	30.5
		MEAN	14.9	21.2	28.6	42.6	53.9	63.0	70.1	68.9	57.9	46.9	31.1	21.3	43.4
MARION		MAX	23.6	30.7	41.4	60.0	72.4	81.3	87.0	85.0	75.3	63.3	44.3	30.0	57.9
		MIN	3.0	10.0	20.5	35.0	46.6	56.8	61.7	59.5	49.2	37.7	23.4	10.8	34.5
		MEAN	13.4	20.4	31.0	47.5	59.6	69.1	74.4	72.3	62.3	50.5	33.9	20.5	46.2
MARTIN		MAX	32.8	38.7	44.9	58.6	70.0	80.2	88.3	86.7	76.9	64.6	47.0	37.1	60.5
		MIN	9.9	15.2	21.3	32.7	43.4	53.0	59.3	57.4	47.4	36.7	23.5	15.5	34.6
		MEAN	21.4	27.0	33.1	45.7	56.7	66.6	73.8	72.1	62.2	50.7	35.3	26.3	47.6
MCINTOSH		MAX	21.6	28.7	38.4	55.3	68.4	77.8	86.0	84.7	72.9	61.0	41.3	28.8	55.4
		MIN	.7	7.9	17.2	31.5	43.1	53.1	58.5	56.8	45.7	35.2	19.8	8.8	31.5
		MEAN	11.1	18.3	27.8	43.4	55.8	65.4	72.3	70.8	59.3	48.1	30.6	18.9	43.5
MENNO		MAX	27.1	34.3	44.4	62.5	74.3	83.6	89.1	87.3	77.9	66.1	47.2	33.1	60.6
		MIN	3.6	11.0	21.5	35.2	46.7	56.5	61.3	59.0	48.4	37.2	23.2	11.4	34.6
		MEAN	15.3	22.6	33.0	48.8	60.5	70.1	75.2	73.2	63.2	51.7	35.2	22.3	47.6
MIDLAND		MAX	31.5	38.4	46.7	62.4	74.0	83.6	92.4	91.0	80.3	67.6	49.3	37.0	62.9
		MIN	4.2	10.6	19.5	33.0	44.2	54.2	59.3	57.4	45.3	33.6	20.2	10.3	32.7
		MEAN	17.9	24.5	33.1	47.7	59.1	68.9	75.9	74.2	62.8	50.6	34.8	23.7	47.8
MILBANK		MAX	20.7	27.1	38.2	56.4	70.7	79.7	85.4	83.8	74.0	62.1	42.1	27.9	55.7
		MIN	.0	6.4	18.5	33.0	44.8	55.1	59.6	58.0	47.5	37.0	22.2	8.8	32.6
		MEAN	10.4	16.8	28.4	44.7	57.8	67.4	72.5	70.9	60.8	49.6	32.2	18.4	44.2
MILESVILLE 5 NE		MAX	27.5	34.3	43.3	59.1	71.0	80.8	89.5	88.7	77.5	64.5	45.8	33.2	59.6
		MIN	4.2	10.8	19.5	32.8	43.8	53.9	59.3	57.9	46.1	35.2	21.0	10.6	32.9
		MEAN	15.9	22.6	31.4	46.0	57.5	67.3	74.4	73.3	61.9	49.9	33.4	21.9	46.3
MILLER		MAX	23.6	30.7	41.0	59.1	71.6	80.8	87.8	86.7	76.2	63.6	43.8	30.3	57.9
		MIN	3.3	10.1	20.1	34.2	45.5	55.9	61.1	59.4	48.8	38.0	23.0	10.9	34.2
		MEAN	13.5	20.4	30.6	46.7	58.6	68.4	74.5	73.1	62.5	50.8	33.5	20.6	46.1
MITCHELL 2 SSE		MAX	24.8	31.9	42.3	60.6	73.1	82.4	89.0	87.1	76.9	64.7	45.8	31.4	59.2
		MIN	3.7	10.8	20.9	35.4	46.7	57.1	62.1	59.8	49.4	38.0	23.7	11.5	34.9
		MEAN	14.3	21.4	31.6	48.0	60.0	69.8	75.6	73.5	63.2	51.4	34.8	21.4	47.1
MOBRIDGE		MAX	21.3	28.9	39.5	56.3	69.0	78.1	86.0	84.5	73.1	61.4	42.2	28.5	55.7
		MIN	.4	7.2	17.9	32.6	44.6	54.9	60.6	58.6	47.4	36.0	21.1	8.8	32.5
		MEAN	10.9	18.1	28.7	44.5	56.8	66.5	73.3	71.6	60.3	48.7	31.7	18.7	44.2
MURDO		MAX	29.0	35.6	43.9	59.9	71.5	81.3	89.7	88.3	77.9	65.4	47.3	34.8	60.4
		MIN	6.2	11.9	20.0	33.3	44.6	54.7	60.7	58.8	48.2	37.2	23.0	12.6	34.3
		MEAN	17.6	23.8	32.0	46.6	58.1	68.0	75.2	73.6	63.1	51.3	35.2	23.7	47.4
NEWELL 2 NW		MAX	28.2	34.2	41.9	56.0	67.7	77.3	86.7	85.7	74.6	62.6	44.6	34.2	57.8
		MIN	4.6	10.3	18.6	31.1	42.3	51.9	57.6	55.4	44.3	32.9	20.0	10.5	31.6
		MEAN	16.4	22.3	30.3	43.6	55.0	64.6	72.2	70.6	59.5	47.7	32.3	22.4	44.7
OELRICHS		MAX	34.1	40.8	48.2	60.6	71.6	82.2	91.3	90.1	79.9	66.7	48.3	38.4	62.7
		MIN	9.5	14.7	20.8	31.7	42.0	51.3	57.5	55.6	44.7	34.0	21.9	14.3	33.2
		MEAN	21.9	27.8	34.5	46.2	56.8	66.8	74.5	72.9	62.3	50.4	35.1	26.4	48.0

SOUTH DAKOTA

TEMPERATURE NORMALS (DEG F)

STATION			JAN	FEB	MAR	APR	MAY	JUN	JUL	AUG	SEP	OCT	NOV	DEC	ANN
PHILIP		MAX	29.5	36.0	44.2	59.4	71.3	81.5	90.5	89.4	78.0	65.2	47.3	35.4	60.6
		MIN	4.8	10.8	19.0	32.1	43.6	53.8	59.4	57.4	45.9	34.3	20.8	10.8	32.7
		MEAN	17.2	23.4	31.6	45.8	57.5	67.7	75.0	73.4	62.0	49.8	34.0	23.1	46.7
PICKSTOWN	//	MAX	28.9	35.1	44.3	60.6	72.5	82.3	88.6	87.0	76.5	65.1	47.4	34.5	60.2
		MIN	7.7	13.9	22.8	36.5	47.9	58.1	64.0	62.2	51.6	40.4	26.0	14.7	37.2
		MEAN	18.3	24.5	33.6	48.6	60.2	70.2	76.3	74.6	64.1	52.8	36.7	24.7	48.7
PIERRE FAA AP		MAX	25.3	31.7	41.5	58.2	70.3	80.5	88.9	87.5	76.5	63.5	44.7	31.7	58.3
		MIN	4.9	11.4	20.8	34.1	45.1	55.5	61.2	59.6	48.4	37.3	23.4	12.3	34.5
		MEAN	15.1	21.6	31.2	46.2	57.7	68.0	75.1	73.6	62.5	50.4	34.1	22.0	46.4
POLLOCK		MAX	20.7	28.0	39.0	56.8	70.5	79.5	87.0	86.4	74.8	62.2	41.8	27.8	56.2
		MIN	-3.6	3.7	15.5	30.9	42.4	52.8	57.7	55.9	44.5	33.1	18.1	5.4	29.7
		MEAN	8.6	15.9	27.3	43.9	56.5	66.2	72.4	71.2	59.7	47.6	30.0	16.6	43.0
RALPH		MAX	26.2	32.8	41.2	56.3	68.4	77.1	86.0	85.2	73.8	62.2	43.6	32.7	57.1
		MIN	.4	7.2	15.6	28.0	39.1	48.4	53.2	50.5	39.0	28.6	16.0	6.8	27.7
		MEAN	13.3	20.0	28.4	42.2	53.8	62.7	69.6	67.9	56.5	45.4	29.8	19.7	42.4
RAPID CITY WSO	//R	MAX	32.4	37.4	44.2	57.0	68.1	77.9	86.5	85.7	75.4	63.2	46.7	37.4	59.3
		MIN	9.2	14.6	21.0	32.1	43.0	52.5	58.7	57.0	46.4	36.1	23.0	14.8	34.0
		MEAN	20.8	26.0	32.6	44.6	55.6	65.2	72.6	71.4	60.9	49.7	34.9	26.1	46.7
RAPID CITY		MAX	34.9	39.8	45.6	57.6	68.5	78.2	86.4	85.5	75.9	64.1	48.0	39.9	60.4
		MIN	11.8	16.9	22.9	33.8	44.6	53.8	60.1	58.2	47.9	38.1	25.2	17.4	35.9
		MEAN	23.4	28.4	34.3	45.7	56.5	66.0	73.3	71.9	61.9	51.1	36.6	28.7	48.2
REDFIELD 6 E		MAX	20.5	28.0	39.3	58.5	71.3	80.3	87.5	86.4	76.0	63.0	43.0	28.3	56.8
		MIN	-1.6	5.9	17.6	32.0	43.2	53.3	58.6	56.5	45.3	34.0	19.2	7.0	30.9
		MEAN	9.5	17.0	28.4	45.3	57.2	66.8	73.1	71.5	60.7	48.5	31.1	17.7	43.9
REDIG 11 NE		MAX	26.7	33.0	40.6	55.5	67.4	76.6	85.7	85.0	73.7	61.9	43.5	33.1	56.9
		MIN	3.3	9.6	17.1	29.5	40.3	49.8	55.0	53.1	42.1	31.9	18.9	9.7	30.0
		MEAN	15.0	21.3	28.9	42.5	53.9	63.2	70.4	69.1	57.9	46.9	31.2	21.4	43.5
SIOUX FALLS WSO	//R	MAX	22.9	29.3	40.1	58.1	70.5	80.3	86.2	83.9	73.5	62.1	43.7	29.3	56.7
		MIN	1.9	8.9	20.6	34.6	45.7	56.3	61.8	59.7	48.5	36.7	22.3	10.1	33.9
		MEAN	12.4	19.1	30.4	46.3	58.2	68.4	74.0	71.8	61.0	49.4	33.0	19.7	45.3
SISSETON		MAX	19.9	26.9	38.5	56.6	70.6	79.2	85.2	84.0	74.0	62.2	41.5	27.5	55.5
		MIN	-.4	6.5	18.4	32.9	44.6	54.8	59.9	58.0	47.4	36.9	22.1	8.3	32.5
		MEAN	9.8	16.7	28.5	44.7	57.6	67.0	72.6	71.0	60.7	49.6	31.8	17.9	44.0
SPEARFISH		MAX	34.5	39.5	44.3	56.4	67.5	77.0	85.6	84.6	74.6	63.3	47.3	39.6	59.5
		MIN	11.9	16.4	21.3	32.0	42.3	51.1	57.1	55.4	45.3	35.9	23.9	17.7	34.2
		MEAN	23.3	28.0	32.8	44.2	54.9	64.0	71.4	70.1	59.9	49.6	35.6	28.7	46.9
TIMBER LAKE		MAX	22.9	29.8	39.4	56.4	69.4	78.4	85.8	84.3	72.9	60.8	42.2	29.4	56.0
		MIN	1.7	8.6	17.8	32.1	43.4	53.4	58.6	56.8	45.9	35.3	20.7	9.2	32.0
		MEAN	12.3	19.2	28.7	44.3	56.4	65.9	72.2	70.6	59.4	48.1	31.5	19.3	44.0
TYNDALL		MAX	27.5	34.5	44.2	62.0	73.5	83.2	88.7	86.7	77.1	65.5	47.1	33.6	60.3
		MIN	5.8	12.6	22.8	36.3	47.9	57.7	62.8	60.8	50.8	39.1	24.8	12.9	36.2
		MEAN	16.7	23.6	33.5	49.1	60.7	70.5	75.8	73.8	64.0	52.3	35.9	23.3	48.3
VERMILLION 2 SE		MAX	28.2	35.4	45.8	63.7	75.0	84.4	89.0	86.6	77.9	67.0	48.4	34.5	61.3
		MIN	5.7	12.4	22.7	36.6	47.6	57.6	62.3	60.2	50.0	38.4	25.0	13.1	36.0
		MEAN	17.0	23.9	34.3	50.2	61.3	71.1	75.7	73.4	64.0	52.7	36.7	23.8	48.7
WAGNER		MAX	28.8	35.6	45.3	63.1	75.1	84.8	91.1	89.2	78.7	66.7	47.7	34.4	61.7
		MIN	6.8	13.2	22.7	36.1	47.6	57.6	63.2	61.2	50.8	39.0	24.7	13.5	36.4
		MEAN	17.8	24.4	34.0	49.6	61.4	71.3	77.2	75.2	64.8	52.9	36.2	24.0	49.1
WASTA		MAX	32.8	39.0	47.1	60.9	72.2	81.9	90.5	89.4	78.5	66.2	48.5	37.4	62.0
		MIN	6.8	13.0	21.1	33.1	43.9	53.6	59.3	57.2	45.9	34.5	21.7	12.0	33.5
		MEAN	19.8	26.0	34.1	47.0	58.1	67.7	74.9	73.3	62.2	50.4	35.1	24.7	47.8

SOUTH DAKOTA

TEMPERATURE NORMALS (DEG F)

STATION		JAN	FEB	MAR	APR	MAY	JUN	JUL	AUG	SEP	OCT	NOV	DEC	ANN
WATERTOWN FAA AIRPORT //R	MAX	18.4	25.2	36.3	53.9	67.6	77.2	83.4	81.8	70.8	58.8	39.5	25.4	53.2
	MIN	-2.2	4.6	16.6	31.7	43.2	54.0	58.9	56.8	46.0	35.0	20.0	6.8	31.0
	MEAN	8.1	14.9	26.5	42.8	55.4	65.6	71.2	69.3	58.4	46.9	29.8	16.2	42.1
WEBSTER	MAX	18.5	25.3	36.9	55.0	68.7	77.7	83.9	82.9	72.2	59.7	39.7	25.9	53.9
	MIN	-2.3	4.8	16.2	31.4	43.3	53.3	58.5	56.7	45.3	34.7	19.5	6.5	30.7
	MEAN	8.1	15.1	26.6	43.2	56.0	65.5	71.2	69.8	58.8	47.2	29.6	16.2	42.3
WENTWORTH 2 WNW	MAX	22.7	29.3	39.9	58.3	71.3	80.1	85.7	83.7	73.9	62.4	43.2	29.0	56.6
	MIN	1.6	8.4	19.3	33.5	45.0	55.3	60.0	57.7	47.7	37.0	22.4	9.8	33.1
	MEAN	12.2	18.9	29.6	45.9	58.2	67.7	72.9	70.7	60.8	49.7	32.8	19.4	44.9
WHITE LAKE	MAX	26.3	33.1	42.8	60.9	72.8	82.4	89.3	87.9	77.7	65.3	45.9	32.6	59.8
	MIN	3.5	10.6	20.1	34.1	45.4	55.8	61.1	59.3	48.7	37.0	22.8	11.2	34.1
	MEAN	14.9	21.8	31.5	47.5	59.1	69.2	75.2	73.6	63.3	51.2	34.4	21.9	47.0
WINNER	MAX	31.2	37.6	45.8	61.7	73.2	82.7	90.3	88.4	78.2	66.0	48.0	36.4	61.6
	MIN	9.0	14.7	22.3	35.2	46.7	56.9	62.7	60.7	50.4	39.3	25.2	15.2	36.5
	MEAN	20.1	26.1	34.0	48.5	60.0	69.9	76.5	74.6	64.3	52.7	36.6	25.8	49.1
WOOD	MAX	33.6	39.3	47.1	62.1	73.4	83.2	91.8	90.3	80.2	67.9	49.9	38.4	63.1
	MIN	8.4	14.0	21.8	34.5	45.5	55.5	61.2	59.6	49.3	38.2	24.5	14.5	35.6
	MEAN	21.0	26.7	34.5	48.3	59.4	69.3	76.5	75.0	64.7	53.1	37.2	26.5	49.4
YANKTON 3 N	MAX	25.6	32.2	41.9	59.4	71.7	81.4	87.3	85.2	75.4	64.4	46.3	32.4	58.6
	MIN	4.0	10.2	20.2	34.6	46.5	56.8	61.8	59.3	48.7	37.1	23.4	11.7	34.5
	MEAN	14.8	21.2	31.1	47.1	59.1	69.1	74.6	72.3	62.1	50.8	34.9	22.1	46.6

SOUTH DAKOTA

PRECIPITATION NORMALS (INCHES)

STATION		JAN	FEB	MAR	APR	MAY	JUN	JUL	AUG	SEP	OCT	NOV	DEC	ANN
ABERDEEN WSO	//R	.49	.64	.99	1.94	2.56	3.22	2.42	1.95	1.50	1.01	.60	.47	17.79
ACADEMY		.45	.76	1.25	2.47	3.36	3.72	2.54	2.17	1.80	1.30	.79	.45	21.06
ALEXANDRIA		.34	.58	1.17	2.24	2.94	3.61	2.87	2.60	2.14	1.37	.70	.52	21.08
ARDMORE 2 N		.39	.51	.87	1.76	2.72	2.54	2.35	1.32	1.11	.82	.45	.38	15.22
ARMOUR		.51	.89	1.37	2.27	3.01	3.72	2.97	2.56	2.08	1.32	.78	.74	22.22
BELLE FOURCHE		.33	.44	.63	1.64	2.56	3.23	1.60	1.46	1.13	.98	.47	.40	14.87
BISON		.36	.52	.83	1.85	2.56	3.28	2.20	1.75	1.21	.82	.51	.39	16.28
BONESTEEL		.39	.86	1.67	2.70	4.01	4.04	2.92	2.85	2.23	1.53	.86	.48	24.54
BRIDGEWATER		.52	.73	1.41	2.13	3.16	3.75	2.99	2.94	2.29	1.55	.88	.74	23.09
BRITTON		.37	.41	.57	1.79	2.64	3.75	2.42	2.32	1.57	1.14	.60	.42	18.00
BROOKINGS 2 NE		.34	.50	.97	2.02	3.06	4.41	2.85	3.15	1.95	1.33	.66	.45	21.69
BUSKALA RANCH		.78	.96	1.22	2.29	3.72	4.02	2.75	2.48	1.46	.97	.93	.91	22.49
CAMP CROOK		.25	.38	.48	1.26	2.51	2.91	1.79	1.36	1.12	.76	.40	.33	13.55
CANISTOTA 2 N		.43	.78	1.39	2.01	3.05	3.98	2.96	3.02	2.43	1.39	.83	.68	22.95
CANTON		.43	.93	1.42	2.39	3.05	3.85	2.97	3.46	2.41	1.47	.91	.76	24.05
CASTLEWOOD		.56	.70	1.07	2.04	3.26	4.10	3.10	2.78	1.80	1.54	.83	.65	22.43
CENTERVILLE 6 SE		.46	1.10	1.43	2.36	3.23	3.99	3.22	3.14	2.58	1.53	.98	.68	24.70
CHAMBERLAIN		.41	.69	.98	2.37	3.16	3.60	2.45	2.04	1.70	1.18	.69	.60	19.87
CLARK		.57	.68	1.00	2.08	2.88	3.75	2.82	2.57	1.50	1.43	.84	.65	20.77
CLEAR LAKE		.61	.72	1.40	2.35	3.09	3.97	3.29	3.16	1.97	1.66	.95	.69	23.86
COLUMBIA 8 N		.49	.55	.98	2.18	2.64	3.54	2.26	2.06	1.61	1.10	.63	.52	18.56
CONDE		.37	.60	.75	1.81	2.69	3.40	2.56	2.30	1.36	1.22	.65	.49	18.20
COTTONWOOD 2 E		.40	.51	.97	1.86	2.94	3.26	2.08	1.55	1.20	1.02	.46	.41	16.66
CUSTER		.39	.51	.88	1.86	3.14	3.51	2.95	2.05	1.15	.72	.53	.46	18.15
DEADWOOD		1.15	1.53	2.14	3.60	4.66	4.36	2.59	2.15	1.82	1.56	1.52	1.40	28.48
DE SMET		.60	.75	1.25	2.28	3.15	4.27	2.83	2.68	1.73	1.59	.93	.71	22.77
DUPREE		.26	.42	.74	1.69	2.84	3.15	1.96	1.59	1.08	.88	.36	.35	15.32
EAGLE BUTTE		.37	.66	1.06	2.11	2.94	3.18	2.39	1.74	1.20	1.03	.49	.48	17.65
ELM SPRINGS 3 ESE		.32	.48	.80	1.65	2.65	3.28	2.09	1.28	1.11	.93	.45	.40	15.44
EUREKA		.32	.39	.76	1.66	2.60	3.47	2.34	2.22	1.47	.97	.48	.34	17.02
FAITH 2 W		.32	.57	.76	1.67	2.71	3.30	2.08	1.67	1.17	.86	.43	.37	15.91
FARMINGDALE 4 N		.24	.38	.68	1.71	2.64	3.18	1.97	1.47	.97	.82	.36	.23	14.65
FAULKTON 1 NW		.33	.52	.90	2.08	2.79	3.53	2.26	2.34	1.18	1.16	.54	.36	17.99
FLANDREAU		.37	.64	1.12	2.04	2.97	3.97	2.73	2.93	2.35	1.58	.70	.52	21.92
FORESTBURG 3 NE	//	.51	.74	1.34	2.18	2.95	3.36	2.80	2.39	1.73	1.62	.85	.66	21.13
FORT MEADE		.47	.72	1.02	2.41	3.31	3.81	2.23	1.72	1.19	1.04	.70	.56	19.18
GANN VALLEY		.27	.43	.73	1.86	2.72	3.10	2.47	2.05	1.43	1.22	.49	.34	17.11
GETTYSBURG		.40	.59	.88	2.05	2.88	3.37	2.04	2.37	1.25	1.07	.61	.57	18.08
GLAD VALLEY 2 W		.38	.54	.91	1.85	2.84	3.46	2.11	1.83	1.28	.97	.48	.45	17.10
GREGORY		.52	.86	1.49	2.75	3.22	3.95	2.79	2.15	2.03	1.34	.93	.72	22.75
HIGHMORE 1 W		.34	.57	.91	2.08	2.69	3.28	2.57	2.33	1.32	1.24	.53	.47	18.33
HOT SPRINGS		.34	.44	.72	1.53	2.58	2.90	2.54	1.26	1.26	.81	.36	.34	15.08
HOWARD		.43	.69	1.21	2.17	2.99	3.71	2.54	2.60	1.99	1.41	.82	.60	21.16
HURON WSO	//R	.42	.77	1.22	1.99	2.72	3.31	2.26	2.01	1.36	1.39	.69	.52	18.66
IPSWICH		.41	.52	.88	2.17	2.83	3.58	2.36	2.00	1.28	1.05	.55	.46	18.09
IROQUOIS		.43	.61	1.04	1.99	3.00	3.55	2.52	2.22	1.60	1.43	.75	.48	19.62
KENNEBEC		.25	.48	.89	2.13	2.52	3.01	2.29	2.23	1.21	1.06	.51	.38	16.96
KIRLEY 6 N		.35	.58	1.01	1.72	2.62	3.00	1.97	1.85	1.36	.99	.43	.50	16.38
LEAD 1 SE	//	1.25	1.58	2.22	3.81	4.34	4.34	2.48	2.23	1.73	1.51	1.62	1.54	28.65
LEMMON		.52	.58	.87	1.85	2.68	3.70	2.41	1.81	1.39	.83	.56	.56	17.76
LEOLA		.45	.61	.90	2.20	3.02	3.70	2.53	2.08	1.49	1.02	.59	.45	19.04
LODGEPOLE 10 NW		.37	.45	.65	1.66	2.77	3.53	2.40	1.66	1.30	.77	.50	.36	16.42
LONGVALLEY		.24	.42	1.09	2.17	2.77	3.18	2.32	2.05	1.14	.93	.37	.32	17.00
LUDLOW		.31	.41	.52	1.60	2.64	3.89	1.99	1.73	1.32	.92	.48	.37	16.18
MANDERSON 1 SE		.34	.50	.99	2.05	2.96	3.43	2.27	1.64	1.28	1.02	.50	.47	17.45

SOUTH DAKOTA

PRECIPITATION NORMALS (INCHES)

STATION	JAN	FEB	MAR	APR	MAY	JUN	JUL	AUG	SEP	OCT	NOV	DEC	ANN
MARION	.49	.87	1.60	2.39	3.20	3.80	2.90	2.73	2.61	1.48	.95	.78	23.80
MARTIN	.26	.42	.93	1.96	2.95	3.37	2.36	2.09	1.26	.89	.38	.35	17.22
MCINTOSH	.36	.41	.74	1.69	2.85	3.34	2.19	1.91	1.41	.90	.43	.43	16.66
MCLAUGHLIN	.39	.49	.95	1.94	2.56	3.24	2.10	1.91	1.32	1.10	.52	.46	16.98
MELLETTE 2 SE	.38	.63	.95	2.20	2.95	3.66	2.47	2.26	1.52	1.19	.64	.43	19.28
MENNO	.44	.79	1.40	2.25	3.25	3.87	3.24	2.57	2.51	1.41	.94	.73	23.40
MIDLAND	.30	.46	.92	1.85	2.69	3.01	1.90	1.77	1.18	.95	.43	.36	15.82
MILBANK	.47	.58	1.07	2.31	2.91	4.00	2.65	2.65	1.64	1.55	.98	.54	21.35
MILESVILLE 5 NE	.40	.56	.99	1.82	3.18	3.17	1.92	1.83	1.32	1.11	.50	.54	17.34
MILLER	.43	.64	.99	2.21	2.75	3.14	2.30	2.24	1.30	1.25	.63	.49	18.37
MITCHELL 2 SSE	.37	.65	1.21	2.33	3.15	3.50	2.46	2.70	2.06	1.39	.75	.52	21.09
MOBRIDGE	.42	.46	.94	1.95	2.63	3.31	1.94	1.97	1.37	1.04	.56	.51	17.10
MURDO	.31	.46	1.08	2.20	2.67	3.28	2.09	1.85	1.11	1.15	.49	.43	17.12
NEWELL 2 NW	.34	.41	.60	1.49	2.51	3.13	1.85	1.39	1.06	.88	.43	.30	14.39
OELRICHS	.40	.54	1.03	1.95	2.92	2.83	2.19	1.54	1.16	.85	.52	.42	16.35
ONAKA	.34	.41	.75	1.68	2.43	3.81	2.23	2.10	1.19	1.04	.51	.43	16.92
PHILIP	.26	.39	.75	1.59	2.67	3.25	2.07	1.63	1.03	.87	.34	.33	15.18
PICKSTOWN //	.38	.68	1.21	2.29	2.92	3.85	2.65	2.51	2.21	1.27	.77	.63	21.37
PIERRE FAA AP	.46	.68	.89	1.94	2.71	3.76	2.15	2.00	1.32	1.11	.47	.59	18.08
PLATTE	.41	.73	1.34	2.46	3.32	4.11	2.59	2.41	1.99	1.45	.76	.63	22.20
POLLOCK	.37	.43	.76	1.77	2.67	3.24	2.20	1.84	1.32	.89	.49	.40	16.38
RALPH	.27	.32	.46	1.54	2.61	3.48	1.92	1.63	1.21	.83	.34	.28	14.89
RAPID CITY WSO //R	.42	.62	1.02	1.96	2.63	3.26	2.12	1.44	1.03	.81	.51	.45	16.27
RAPID CITY	.43	.63	1.01	2.30	3.06	3.71	2.33	1.79	1.07	.85	.52	.44	18.14
RAYMOND 3 NE	.35	.49	.79	1.88	2.62	3.46	2.91	2.19	1.34	1.37	.71	.42	18.53
REDFIELD 6 E	.33	.67	.90	2.05	2.94	3.68	2.48	1.97	1.25	1.26	.56	.42	18.51
REDIG 11 NE	.30	.41	.61	1.40	2.52	3.28	2.01	1.67	1.01	.79	.40	.31	14.71
REDOWL	.27	.41	.71	1.57	2.60	3.47	1.91	1.25	1.20	.81	.37	.33	14.90
SCENIC	.29	.48	.90	1.83	2.75	3.12	1.94	1.45	1.23	.90	.41	.30	15.60
SIOUX FALLS WSO //R	.50	.93	1.58	2.36	3.21	3.70	2.71	3.13	2.79	1.57	.92	.72	24.12
SISSETON	.61	.69	1.06	2.34	3.17	3.56	2.54	2.73	1.46	1.37	.93	.67	21.13
SPEARFISH	.56	.80	1.35	2.47	3.46	4.09	1.97	1.99	1.38	1.34	.93	.72	21.06
TIMBER LAKE	.41	.59	1.02	1.90	2.85	3.31	1.89	2.17	1.29	1.05	.53	.56	17.57
TYNDALL	.41	.75	1.34	2.36	3.40	3.86	3.40	2.65	2.50	1.37	.86	.65	23.55
VERMILLION 2 SE	.37	.74	1.39	2.22	3.66	3.83	3.37	3.22	2.21	1.49	.89	.68	24.07
VICTOR 1 ESE	.54	.55	.95	2.26	2.84	3.66	2.96	2.60	1.72	1.25	.86	.65	20.84
WAGNER	.53	.91	1.41	2.49	3.34	3.89	3.15	2.68	2.62	1.46	.98	.88	24.34
WASTA	.35	.44	.85	1.91	2.51	3.13	2.05	1.51	1.06	.92	.49	.38	15.60
WATERTOWN FAA AIRPORT //R	.56	.70	1.21	2.20	3.14	3.96	2.96	2.77	1.60	1.66	.89	.68	22.33
WEBSTER	.63	.63	.88	2.02	2.89	3.71	2.90	2.72	1.54	1.35	.80	.67	20.74
WENTWORTH 2 WNW	.46	.76	1.46	2.15	3.02	3.82	2.81	3.26	2.17	1.45	.83	.71	22.90
WESSINGTON SPRGS 8 SW	.58	.89	1.55	2.43	3.14	3.44	2.47	2.06	1.42	1.28	.85	.71	20.82
WHITE LAKE	.41	.70	1.18	2.33	3.37	3.55	2.58	2.55	1.81	1.27	.76	.53	21.04
WILMOT 1 ENE	.56	.72	1.15	2.37	3.25	3.72	2.81	2.70	1.57	1.63	.86	.61	21.95
WINNER	.46	.70	1.36	2.74	3.25	3.80	2.85	2.59	1.86	1.29	.74	.56	22.20
WOOD	.43	.71	1.19	2.33	3.22	3.06	2.50	2.15	1.34	1.18	.61	.56	19.28
YANKTON 3 N	.32	.64	1.38	2.19	3.49	3.92	3.17	3.08	2.33	1.35	.89	.52	23.28

SOUTH DAKOTA

HEATING DEGREE DAY NORMALS (BASE 65 DEG F)

STATION		JUL	AUG	SEP	OCT	NOV	DEC	JAN	FEB	MAR	APR	MAY	JUN	ANN
ABERDEEN WSO	//R	16	20	197	536	1044	1504	1758	1380	1156	612	274	73	8570
ACADEMY		8	8	116	404	879	1277	1473	1148	992	498	195	43	7041
ALEXANDRIA		8	8	116	414	894	1333	1556	1207	1017	498	190	36	7277
ARDMORE 2 N		19	17	190	524	957	1280	1401	1086	977	591	291	88	7421
ARMOUR		7	8	121	426	888	1296	1507	1168	1004	507	192	38	7162
BELLE FOURCHE		12	17	191	490	924	1215	1373	1084	992	582	271	89	7240
BONESTEEL		8	10	139	420	888	1277	1466	1154	998	513	211	48	7132
BRIDGEWATER		5	8	120	427	918	1361	1587	1226	1029	504	191	34	7410
BRITTON		22	27	196	518	1050	1531	1792	1408	1166	621	277	83	8691
BROOKINGS 2 NE		23	34	219	555	1026	1485	1730	1386	1172	639	298	79	8646
CAMP CROOK		23	33	230	555	1014	1330	1516	1207	1088	660	330	112	8098
CANTON		8	15	132	439	915	1355	1575	1226	1017	501	186	29	7398
CASTLEWOOD		21	20	191	524	1032	1482	1724	1352	1141	615	273	64	8439
CENTERVILLE 6 SE		9	23	137	430	894	1349	1562	1218	1020	492	198	39	7371
CLARK		18	21	186	508	1023	1479	1717	1350	1147	615	269	72	8405
COLUMBIA 8 N		13	17	182	527	1041	1519	1779	1403	1169	621	273	68	8612
CUSTER		93	102	363	670	1035	1271	1395	1140	1132	807	521	241	8770
DEADWOOD		54	84	307	605	987	1218	1352	1092	1076	735	425	173	8108
DE SMET		11	13	159	489	990	1432	1668	1296	1104	579	236	52	8029
DUPREE		14	18	173	484	975	1367	1569	1235	1082	603	274	87	7881
EUREKA		22	32	216	546	1062	1510	1764	1392	1187	651	305	90	8777
FAITH 2 W		13	23	187	495	972	1358	1550	1224	1091	609	279	87	7888
FAULKTON 1 NW		15	18	169	487	990	1438	1674	1313	1113	585	262	74	8138
FLANDREAU		16	22	181	524	993	1448	1690	1344	1138	597	265	60	8278
FORESTBURG 3 NE	//	10	11	141	453	939	1380	1609	1252	1057	540	226	52	7670
FORT MEADE		11	21	171	444	873	1144	1308	1039	973	585	276	91	6936
GANN VALLEY		11	12	153	461	960	1386	1609	1254	1066	549	233	57	7751
GETTYSBURG		13	30	186	514	1008	1429	1662	1308	1138	621	285	77	8271
GREGORY		12	10	129	405	873	1243	1432	1123	986	516	212	49	6990
HIGHMORE 1 W		16	17	157	469	978	1398	1624	1274	1094	579	254	75	7935
HOT SPRINGS		13	17	171	468	885	1156	1299	1022	952	585	285	87	6940
HOWARD		13	13	154	475	960	1404	1640	1274	1070	543	223	45	7814
HURON WSO	//R	15	14	171	505	984	1426	1668	1313	1104	582	259	62	8103
KENNEBEC		9	8	137	435	924	1330	1531	1190	1023	531	215	55	7388
LEAD 1 SE	//	56	72	279	558	951	1169	1311	1072	1076	750	443	186	7923
LEMMON		25	46	261	580	1056	1429	1652	1299	1169	690	347	118	8672
LONGVALLEY		10	12	141	421	855	1175	1342	1058	970	552	246	66	6848
LUDLOW		27	62	252	561	1017	1355	1553	1226	1128	672	352	132	8337
MARION		6	16	132	458	933	1380	1600	1249	1054	525	205	42	7600
MARTIN		14	18	163	449	891	1200	1352	1064	989	579	269	80	7068
MCINTOSH		16	31	211	524	1032	1429	1671	1308	1153	648	298	96	8417
MENNO		6	13	122	422	894	1324	1541	1187	992	486	179	33	7199
MIDLAND		9	7	144	446	906	1280	1460	1134	989	519	208	52	7154
MILBANK		13	25	159	487	984	1445	1693	1350	1135	609	249	62	8211
MILESVILLE 5 NE		13	14	169	473	948	1336	1522	1187	1042	570	251	74	7599
MILLER		11	11	141	447	945	1376	1597	1249	1066	549	224	53	7669
MITCHELL 2 SSE		5	11	126	429	906	1352	1572	1221	1035	510	192	36	7395
MOBRIDGE		12	21	185	505	999	1435	1677	1313	1125	615	270	75	8232
MURDO		9	13	144	433	894	1280	1469	1154	1023	552	233	55	7259
NEWELL 2 NW		16	22	210	536	981	1321	1507	1196	1076	642	316	108	7931
OELRICHS		11	15	151	453	897	1197	1336	1042	946	564	261	72	6945
PHILIP		12	14	158	477	930	1299	1482	1165	1035	576	248	66	7462
PICKSTOWN	//	0	5	119	390	849	1249	1448	1134	973	492	184	36	6879
PIERRE FAA AP		11	11	148	458	927	1333	1547	1215	1048	564	244	65	7571
POLLOCK		23	25	198	539	1050	1500	1748	1375	1169	633	281	87	8628

SOUTH DAKOTA

HEATING DEGREE DAY NORMALS (BASE 65 DEG F)

STATION		JUL	AUG	SEP	OCT	NOV	DEC	JAN	FEB	MAR	APR	MAY	JUN	ANN
RALPH		37	56	278	608	1056	1404	1603	1260	1135	684	354	140	8615
RAPID CITY WSO	//R	21	24	188	482	903	1206	1370	1092	1004	612	298	101	7301
RAPID CITY		11	20	164	440	852	1125	1290	1025	952	579	272	86	6816
REDFIELD 6 E		13	16	173	519	1017	1466	1721	1344	1135	591	262	65	8322
REDIG 11 NE		23	40	244	561	1014	1352	1550	1224	1119	675	349	122	8273
SIOUX FALLS WSO	//R	14	15	161	489	960	1404	1631	1285	1073	561	240	52	7885
SISSETON		13	25	174	484	996	1460	1711	1352	1132	609	254	66	8276
SPEARFISH		22	32	208	485	882	1125	1293	1036	998	624	321	122	7148
TIMBER LAKE		18	24	212	524	1005	1417	1634	1282	1125	621	281	86	8229
TYNDALL		0	8	112	406	873	1293	1497	1159	977	477	173	28	7003
VERMILLION 2 SE		7	11	119	398	849	1277	1488	1151	952	449	167	27	6895
WAGNER		0	5	107	383	864	1271	1463	1137	961	462	159	27	6839
WASTA		11	11	156	457	897	1249	1401	1092	958	540	228	60	7060
WATERTOWN FAA AIRPORT	//R	21	30	223	561	1056	1513	1764	1403	1194	666	312	79	8822
WEBSTER		21	29	213	552	1062	1513	1764	1397	1190	654	298	83	8776
WENTWORTH 2 WNW		11	19	166	481	966	1414	1637	1291	1097	573	234	55	7944
WHITE LAKE		9	11	121	436	918	1336	1553	1210	1039	525	212	38	7408
WINNER		7	9	116	392	852	1215	1392	1089	961	495	189	42	6759
WOOD		7	6	112	379	834	1194	1364	1072	946	501	199	45	6659
YANKTON 3 N		8	13	145	446	903	1330	1556	1226	1051	537	218	41	7474

SOUTH DAKOTA

COOLING DEGREE DAY NORMALS (BASE 65 DEG F)

STATION		JAN	FEB	MAR	APR	MAY	JUN	JUL	AUG	SEP	OCT	NOV	DEC	ANN
ABERDEEN WSO	//R	0	0	0	0	19	112	236	190	32	0	0	0	589
ACADEMY		0	0	0	0	27	178	346	296	89	10	0	0	946
ALEXANDRIA		0	0	0	0	35	177	327	269	71	8	0	0	887
ARDMORE 2 N		0	0	0	0	6	106	276	203	49	0	0	0	640
ARMOUR		0	0	0	0	28	179	335	277	73	8	0	0	900
BELLE FOURCHE		0	0	0	0	7	113	251	196	47	0	0	0	614
BONESTEEL		0	0	0	0	28	168	330	277	85	8	0	0	896
BRIDGEWATER		0	0	0	0	39	181	324	265	63	8	0	0	880
BRITTON		0	0	0	0	23	119	230	201	43	0	0	0	616
BROOKINGS 2 NE		0	0	0	0	19	97	200	145	18	0	0	0	479
CAMP CROOK		0	0	0	0	8	82	215	179	38	0	0	0	522
CANTON		0	0	0	0	43	182	315	254	63	12	0	0	869
CASTLEWOOD		0	0	0	0	15	109	225	178	29	0	0	0	556
CENTERVILLE 6 SE		0	0	0	0	53	195	316	265	74	15	0	0	918
CLARK		0	0	0	0	18	120	239	198	42	0	0	0	617
COLUMBIA 8 N		0	0	0	0	18	116	240	197	35	0	0	0	606
CUSTER		0	0	0	0	0	19	77	34	9	0	0	0	139
DEADWOOD		0	0	0	0	0	38	128	106	22	0	0	0	294
DE SMET		0	0	0	0	22	133	268	215	45	6	0	0	689
DUPREE		0	0	0	0	17	129	284	247	59	0	0	0	736
EUREKA		0	0	0	0	14	90	215	184	33	0	0	0	536
FAITH 2 W		0	0	0	0	15	120	271	237	61	5	0	0	709
FAULKTON 1 NW		0	0	0	0	23	134	263	226	55	0	0	0	701
FLANDREAU		0	0	0	0	26	123	243	180	25	7	0	0	604
FORESTBURG 3 NE	//	0	0	0	0	24	154	292	240	57	6	0	0	773
FORT MEADE		0	0	0	0	9	115	259	229	69	7	0	0	688
GANN VALLEY		0	0	0	0	22	156	312	260	75	6	0	0	831
GETTYSBURG		0	0	0	0	13	113	255	225	45	5	0	0	656
GREGORY		0	0	0	0	26	154	316	267	87	11	0	0	861
HIGHMORE 1 W		0	0	0	0	22	144	295	255	67	7	0	0	790
HOT SPRINGS		0	0	0	0	6	111	255	206	51	0	0	0	629
HOWARD		0	0	0	0	28	147	298	239	55	7	0	0	774
HURON WSO	//R	0	0	0	0	20	143	294	231	45	5	0	0	738
KENNEBEC		0	0	0	0	26	178	344	300	98	0	0	0	946
LEAD 1 SE	//	0	0	0	0	0	48	146	125	36	0	0	0	355
LEMMON		0	0	0	0	10	73	189	164	33	0	0	0	469
LONGVALLEY		0	0	0	0	13	147	327	278	90	9	0	0	864
LUDLOW		0	0	0	0	8	72	185	183	39	0	0	0	487
MARION		0	0	0	0	37	165	297	242	51	9	0	0	801
MARTIN		0	0	0	0	12	128	287	238	79	5	0	0	749
MCINTOSH		0	0	0	0	13	108	242	211	40	0	0	0	614
MENNO		0	0	0	0	39	186	322	267	68	10	0	0	892
MIDLAND		0	0	0	0	25	169	347	292	78	0	0	0	911
MILBANK		0	0	0	0	26	134	246	208	33	10	0	0	657
MILESVILLE 5 NE		0	0	0	0	18	143	305	271	76	0	0	0	813
MILLER		0	0	0	0	25	155	305	262	66	7	0	0	820
MITCHELL 2 SSE		0	0	0	0	37	180	334	274	72	8	0	0	905
MOBRIDGE		0	0	0	0	16	120	270	226	44	0	0	0	676
MURDO		0	0	0	0	19	145	325	279	87	8	0	0	863
NEWELL 2 NW		0	0	0	0	6	96	239	196	45	0	0	0	582
OELRICHS		0	0	0	0	7	126	306	260	70	0	0	0	769
PHILIP		0	0	0	0	16	147	322	274	68	5	0	0	832
PICKSTOWN	//	0	0	0	0	36	192	354	303	92	12	0	0	989
PIERRE FAA AP		0	0	0	0	17	155	324	278	73	6	0	0	853
POLLOCK		0	0	0	0	17	123	252	217	39	0	0	0	648

SOUTH DAKOTA

COOLING DEGREE DAY NORMALS (BASE 65 DEG F)

STATION	JAN	FEB	MAR	APR	MAY	JUN	JUL	AUG	SEP	OCT	NOV	DEC	ANN
RALPH	0	0	0	0	6	71	180	146	23	0	0	0	426
RAPID CITY WSO //R	0	0	0	0	7	107	257	223	65	8	0	0	667
RAPID CITY	0	0	0	0	8	116	268	234	71	9	0	0	706
REDFIELD 6 E	0	0	0	0	20	119	264	217	44	7	0	0	671
REDIG 11 NE	0	0	0	0	0	68	190	167	31	0	0	0	456
SIOUX FALLS WSO //R	0	0	0	0	29	154	293	226	41	6	0	0	749
SISSETON	0	0	0	0	25	126	249	211	45	6	0	0	662
SPEARFISH	0	0	0	0	7	92	220	191	55	7	0	0	572
TIMBER LAKE	0	0	0	0	15	113	241	198	44	0	0	0	611
TYNDALL	0	0	0	0	40	193	339	280	82	12	0	0	946
VERMILLION 2 SE	0	0	0	0	52	210	338	272	89	17	0	0	978
WAGNER	0	0	0	0	48	216	383	322	101	8	0	0	1078
WASTA	0	0	0	0	14	141	318	268	72	0	0	0	813
WATERTOWN FAA AIRPORT //R	0	0	0	0	15	97	213	163	25	0	0	0	513
WEBSTER	0	0	0	0	19	98	213	177	27	0	0	0	534
WENTWORTH 2 WNW	0	0	0	0	23	136	256	196	40	6	0	0	657
WHITE LAKE	0	0	0	0	29	164	325	277	70	8	0	0	873
WINNER	0	0	0	0	34	189	364	307	95	11	0	0	1000
WOOD	0	0	0	0	25	174	364	316	103	11	0	0	993
YANKTON 3 N	0	0	0	0	35	164	306	239	58	6	0	0	808

39 – SOUTH DAKOTA

LEGEND
11 = TEMPERATURE ONLY
12 = PRECIPITATION ONLY
13 = TEMP. & PRECIP.

STATE-STATION NUMBER	STN TYP	NAME		LATITUDE DEG-MIN	LONGITUDE DEG-MIN	ELEVATION (FT)
39-0020	13	ABERDEEN WSO	//R	N 4527	W 09826	1296
39-0043	13	ACADEMY		N 4328	W 09905	1675
39-0128	13	ALEXANDRIA		N 4339	W 09747	1350
39-0236	13	ARDMORE 2 N		N 4303	W 10339	3550
39-0296	13	ARMOUR		N 4319	W 09821	1510
39-0559	13	BELLE FOURCHE		N 4440	W 10351	3017
39-0701	12	BISON		N 4531	W 10228	2780
39-0778	13	BONESTEEL		N 4305	W 09857	1985
39-1032	13	BRIDGEWATER		N 4333	W 09730	1420
39-1049	13	BRITTON		N 4547	W 09745	1340
39-1076	13	BROOKINGS 2 NE		N 4419	W 09646	1642
39-1246	12	BUSKALA RANCH		N 4413	W 10349	6110
39-1294	13	CAMP CROOK		N 4533	W 10359	3120
39-1354	12	CANISTOTA 2 N		N 4338	W 09718	1555
39-1392	13	CANTON		N 4318	W 09635	1247
39-1519	13	CASTLEWOOD		N 4443	W 09702	1685
39-1579	13	CENTERVILLE 6 SE		N 4303	W 09654	1255
39-1609	12	CHAMBERLAIN		N 4348	W 09920	1400
39-1739	13	CLARK		N 4453	W 09744	1780
39-1777	12	CLEAR LAKE		N 4445	W 09641	1800
39-1873	13	COLUMBIA 8 N		N 4544	W 09818	1300
39-1917	12	CONDE		N 4509	W 09806	1330
39-1972	12	COTTONWOOD 2 E		N 4358	W 10152	2414
39-2087	13	CUSTER		N 4346	W 10336	5322
39-2207	13	DEADWOOD		N 4423	W 10344	4670
39-2302	13	DE SMET		N 4423	W 09733	1726
39-2429	13	DUPREE		N 4503	W 10136	2365
39-2468	12	EAGLE BUTTE		N 4500	W 10114	2412
39-2647	12	ELM SPRINGS 3 ESE		N 4419	W 10228	2645
39-2797	13	EUREKA		N 4546	W 09937	1884
39-2852	13	FAITH 2 W		N 4502	W 10205	2545
39-2888	12	FARMINGDALE 4 N		N 4402	W 10254	3150
39-2927	13	FAULKTON 1 NW		N 4502	W 09908	1565
39-2984	13	FLANDREAU		N 4403	W 09636	1550
39-3029	13	FORESTBURG 3 NE	//	N 4402	W 09804	1231
39-3069	13	FORT MEADE		N 4424	W 10328	3300
39-3217	13	GANN VALLEY		N 4402	W 09858	1750
39-3294	13	GETTYSBURG		N 4501	W 09957	2080
39-3316	12	GLAD VALLEY 2 W		N 4524	W 10149	2910
39-3452	13	GREGORY		N 4314	W 09926	2001
39-3832	13	HIGHMORE 1 W		N 4431	W 09928	1890
39-4007	13	HOT SPRINGS		N 4326	W 10328	3535
39-4037	13	HOWARD		N 4401	W 09731	1560
39-4127	13	HURON WSO	//R	N 4423	W 09813	1282
39-4206	12	IPSWICH		N 4527	W 09902	1530
39-4254	12	IROQUOIS		N 4422	W 09751	1395
39-4516	13	KENNEBEC		N 4355	W 09952	1700
39-4596	12	KIRLEY 6 N		N 4437	W 10120	2160
39-4834	13	LEAD 1 SE	//	N 4421	W 10346	5332
39-4864	13	LEMMON		N 4556	W 10210	2596

39 — SOUTH DAKOTA

STATE-STATION NUMBER	STN TYP	NAME	LATITUDE DEG-MIN	LONGITUDE DEG-MIN	ELEVATION (FT)
39-4891	12	LEOLA	N 4543	W 09856	1575
39-4960	12	LODGEPOLE 10 NW	N 4553	W 10251	2635
39-4983	13	LONGVALLEY	N 4328	W 10130	2470
39-5048	13	LUDLOW	N 4551	W 10323	3050
39-5154	12	MANDERSON 1 SE	N 4314	W 10227	3250
39-5228	13	MARION	N 4325	W 09715	1440
39-5281	13	MARTIN	N 4311	W 10144	3320
39-5381	13	MCINTOSH	N 4555	W 10121	2310
39-5406	12	MCLAUGHLIN	N 4549	W 10049	2000
39-5456	12	MELLETTE 2 SE	N 4509	W 09828	1290
39-5481	13	MENNO	N 4314	W 09735	1324
39-5506	13	MIDLAND	N 4404	W 10109	1890
39-5536	13	MILBANK	N 4513	W 09638	1145
39-5544	13	MILESVILLE 5 NE	N 4431	W 10137	2240
39-5561	13	MILLER	N 4431	W 09859	1587
39-5671	13	MITCHELL 2 SSE	N 4341	W 09801	1346
39-5691	13	MOBRIDGE	N 4532	W 10026	1668
39-5891	13	MURDO	N 4353	W 10043	2300
39-6054	13	NEWELL 2 NW	N 4444	W 10327	2870
39-6212	13	OELRICHS	N 4311	W 10314	3337
39-6282	12	ONAKA	N 4511	W 09927	1600
39-6552	13	PHILIP	N 4403	W 10136	2205
39-6574	13	PICKSTOWN //	N 4304	W 09832	1485
39-6597	13	PIERRE FAA AP	N 4423	W 10017	1734
39-6669	12	PLATTE	N 4323	W 09850	1610
39-6712	13	POLLOCK	N 4554	W 10017	1635
39-6907	13	RALPH	N 4547	W 10304	2800
39-6937	13	RAPID CITY WSO //R	N 4403	W 10304	3162
39-6947	13	RAPID CITY	N 4404	W 10312	3242
39-7007	12	RAYMOND 3 NE	N 4456	W 09754	1503
39-7052	13	REDFIELD 6 E	N 4453	W 09823	1296
39-7062	13	REDIG 11 NE	N 4523	W 10323	3070
39-7073	12	REDOWL	N 4442	W 10233	2765
39-7512	12	SCENIC	N 4346	W 10233	2808
39-7667	13	SIOUX FALLS WSO //R	N 4334	W 09644	1418
39-7742	13	SISSETON	N 4540	W 09703	1200
39-7882	13	SPEARFISH	N 4429	W 10353	3675
39-8307	13	TIMBER LAKE	N 4526	W 10104	2150
39-8472	13	TYNDALL	N 4259	W 09752	1420
39-8622	13	VERMILLION 2 SE	N 4245	W 09655	1190
39-8652	12	VICTOR 1 ESE	N 4552	W 09648	1080
39-8767	13	WAGNER	N 4305	W 09818	1430
39-8911	13	WASTA	N 4404	W 10226	2320
39-8932	13	WATERTOWN FAA AIRPORT //R	N 4455	W 09709	1746
39-9004	13	WEBSTER	N 4520	W 09732	1850
39-9042	13	WENTWORTH 2 WNW	N 4401	W 09700	1690
39-9077	12	WESSINGTON SPRGS 8 SW	N 4402	W 09844	1775
39-9232	13	WHITE LAKE	N 4344	W 09843	1655
39-9337	12	WILMOT 1 ENE	N 4525	W 09650	1160
39-9367	13	WINNER	N 4323	W 09952	1965

STATE-STATION NUMBER	STN TYP	NAME	LATITUDE DEG-MIN	LONGITUDE DEG-MIN	ELEVATION (FT)
39-9442	13	WOOD	N 4330	W 10029	2180
39-9502	13	YANKTON 3 N	N 4255	W 09723	1270

39 – SOUTH DAKOTA

LEGEND

○ = TEMPERATURE ONLY
✕ = PRECIPITATION ONLY
⊗ = TEMPERATURE & PRECIPITATION

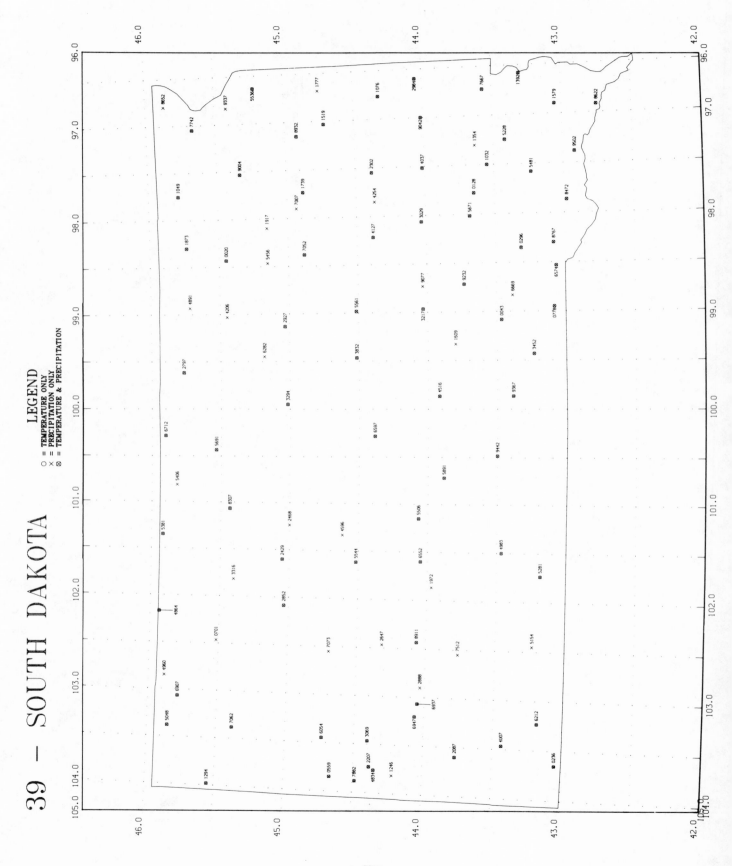

TENNESSEE

TEMPERATURE NORMALS (DEG F)

STATION		JAN	FEB	MAR	APR	MAY	JUN	JUL	AUG	SEP	OCT	NOV	DEC	ANN
ALLARDT	MAX	44.0	48.0	56.7	68.0	75.3	81.5	84.5	84.1	78.6	67.9	56.2	47.5	66.0
	MIN	24.7	26.7	34.7	44.1	51.7	58.8	62.6	61.6	56.1	44.0	35.1	28.4	44.0
	MEAN	34.4	37.4	45.7	56.1	63.6	70.2	73.6	72.9	67.4	56.0	45.7	38.0	55.1
BOLIVAR WTR WKS	MAX	48.2	52.8	61.4	73.0	80.4	87.7	91.0	90.2	84.4	74.2	61.5	52.3	71.4
	MIN	27.7	30.5	38.5	48.5	56.2	64.0	67.6	65.7	59.2	45.8	37.1	30.9	47.6
	MEAN	38.0	41.7	50.0	60.8	68.3	75.9	79.4	78.0	71.8	60.1	49.3	41.6	59.6
BRISTOL WSO R	MAX	44.5	48.4	57.6	68.3	76.3	82.9	85.5	85.2	80.4	69.5	57.3	48.1	67.0
	MIN	25.5	27.4	34.9	43.8	52.5	60.1	64.3	63.4	57.2	44.7	35.1	28.3	44.8
	MEAN	35.0	37.9	46.3	56.1	64.4	71.5	74.9	74.3	68.8	57.1	46.2	38.2	55.9
BROWNSVILLE	MAX	48.2	53.1	62.1	73.7	81.4	88.6	91.5	90.9	85.1	75.4	62.0	52.3	72.0
	MIN	29.1	32.1	40.3	50.4	58.3	65.9	69.5	67.7	61.1	48.6	39.4	32.6	49.6
	MEAN	38.7	42.7	51.2	62.1	69.9	77.3	80.6	79.3	73.1	62.0	50.7	42.5	60.8
CELINA	MAX	46.6	50.6	59.6	71.6	79.3	86.1	89.2	88.9	83.4	72.7	60.2	50.7	69.9
	MIN	24.7	26.5	34.2	43.3	51.6	60.0	64.2	63.0	56.6	43.6	34.5	28.0	44.2
	MEAN	35.7	38.6	46.9	57.5	65.5	73.1	76.7	76.0	70.0	58.2	47.4	39.4	57.1
CHATTANOOGA WSO R	MAX	48.2	52.6	60.9	72.6	79.8	86.4	89.3	88.8	83.0	72.3	60.3	51.4	70.5
	MIN	29.2	31.2	38.6	47.5	55.7	63.8	68.1	67.4	61.5	47.7	37.5	31.5	48.3
	MEAN	38.7	42.0	49.8	60.1	67.8	75.1	78.7	78.1	72.3	60.0	48.9	41.5	59.4
CLARKSVILLE SEW PLT	MAX	45.9	50.3	59.5	71.9	79.5	86.9	90.7	89.5	83.5	72.5	59.8	49.9	70.0
	MIN	25.4	27.7	36.2	46.1	54.3	62.7	66.6	65.2	58.5	45.2	35.9	29.1	46.1
	MEAN	35.7	39.0	47.9	59.0	66.9	74.8	78.7	77.4	71.0	58.9	47.8	39.5	58.1
COLUMBIA SEWAGE PLANT	MAX	47.7	52.2	60.5	72.4	79.6	86.5	89.8	89.1	83.2	73.0	60.6	51.8	70.5
	MIN	27.4	29.7	37.1	47.0	54.5	62.5	66.5	65.2	58.9	45.7	36.5	30.5	46.8
	MEAN	37.6	41.0	48.8	59.7	67.1	74.6	78.2	77.2	71.1	59.4	48.6	41.2	58.7
COPPERHILL	MAX	48.0	51.7	59.6	71.0	78.5	84.9	87.9	87.7	82.2	71.8	60.3	51.3	69.6
	MIN	26.2	27.6	34.3	42.7	50.9	58.6	63.0	62.1	56.0	42.8	34.1	28.3	43.9
	MEAN	37.1	39.7	47.0	56.9	64.8	71.8	75.5	74.9	69.2	57.3	47.2	39.9	56.8
COVINGTON 1 W	MAX	46.0	50.4	59.3	71.2	80.1	88.2	91.2	89.9	83.9	74.1	60.5	50.4	70.4
	MIN	27.4	30.5	38.7	49.4	57.8	65.6	69.1	66.9	60.2	47.4	38.0	31.1	48.5
	MEAN	36.7	40.4	49.0	60.3	69.0	76.9	80.2	78.5	72.0	60.8	49.3	40.8	59.5
CROSSVILLE EXP STA	MAX	42.2	46.0	54.3	65.8	73.7	80.2	83.7	83.5	78.2	67.6	55.6	46.3	64.8
	MIN	22.3	24.2	32.9	43.4	50.9	57.9	61.7	60.6	55.0	42.5	33.6	25.9	42.6
	MEAN	32.3	35.1	43.6	54.6	62.3	69.1	72.7	72.1	66.6	55.0	44.6	36.1	53.7
DICKSON	MAX	46.3	51.4	60.5	72.4	79.7	86.5	89.8	89.1	83.1	72.0	59.2	50.3	70.0
	MIN	27.5	30.0	37.8	47.7	55.4	62.9	66.9	65.6	59.5	47.3	37.8	30.9	47.4
	MEAN	36.9	40.7	49.2	60.1	67.6	74.8	78.4	77.4	71.3	59.7	48.6	40.6	58.8
DOVER 1 W	MAX	46.1	50.5	59.9	71.9	79.0	86.2	89.5	89.0	82.8	72.4	59.7	50.1	69.8
	MIN	24.4	26.8	35.7	45.6	53.3	61.6	65.4	63.7	57.0	43.5	35.0	28.2	45.0
	MEAN	35.2	38.7	47.8	58.8	66.2	74.0	77.4	76.3	69.9	58.0	47.4	39.2	57.4
DYERSBURG FAA AP	MAX	45.4	50.1	59.3	71.3	80.1	88.0	90.7	89.4	83.1	73.1	59.4	49.5	70.0
	MIN	28.9	32.4	40.6	51.2	59.6	67.4	70.8	68.8	62.1	49.8	39.9	33.0	50.4
	MEAN	37.2	41.3	50.0	61.3	69.8	77.7	80.8	79.1	72.6	61.5	49.7	41.3	60.2
FRANKLIN SEWAGE PLANT	MAX	48.4	53.1	61.9	73.2	80.4	87.4	90.5	89.8	84.3	74.1	61.2	52.3	71.4
	MIN	27.4	29.2	37.3	46.8	54.1	62.0	65.5	64.1	57.7	45.3	36.5	30.3	46.4
	MEAN	37.9	41.2	49.6	60.0	67.3	74.7	78.0	77.0	71.0	59.7	48.9	41.3	58.9
GATLINBURG 2 SW	MAX	48.2	51.9	61.0	71.5	78.0	83.5	86.3	85.5	80.8	70.8	60.0	51.7	69.1
	MIN	24.7	26.2	33.3	41.5	49.2	56.7	60.6	60.0	54.4	41.6	32.4	26.7	42.3
	MEAN	36.5	39.1	47.2	56.5	63.7	70.1	73.4	72.8	67.6	56.2	46.2	39.2	55.7
GREENEVILLE EXP STA	MAX	47.5	51.2	60.1	70.9	78.4	84.8	87.6	87.1	82.1	71.1	59.6	50.7	69.3
	MIN	26.1	27.8	34.9	43.4	52.3	60.0	64.1	63.0	57.0	44.0	34.6	28.1	44.6
	MEAN	36.8	39.5	47.6	57.2	65.4	72.4	75.9	75.1	69.6	57.6	47.1	39.5	57.0

TENNESSEE

TEMPERATURE NORMALS (DEG F)

STATION			JAN	FEB	MAR	APR	MAY	JUN	JUL	AUG	SEP	OCT	NOV	DEC	ANN
JACKSON FAA AP		MAX	47.1	51.8	60.6	72.4	80.7	88.2	91.3	90.6	84.3	74.1	60.4	51.3	71.1
		MIN	28.5	31.5	39.6	49.9	58.2	65.9	69.5	67.6	60.8	48.0	38.5	31.9	49.2
		MEAN	37.8	41.7	50.1	61.2	69.5	77.1	80.4	79.1	72.6	61.1	49.5	41.6	60.1
JACKSON EXP STA		MAX	46.3	50.7	59.5	71.6	79.3	86.8	89.9	89.4	83.5	73.4	60.2	50.6	70.1
		MIN	27.6	30.4	38.7	48.9	57.2	64.8	68.4	66.4	60.0	46.9	37.8	31.3	48.2
		MEAN	37.0	40.6	49.1	60.3	68.3	75.8	79.2	77.9	71.8	60.2	49.1	41.0	59.2
KINGSPORT		MAX	46.7	51.1	60.5	71.8	78.9	84.7	87.4	86.8	81.9	71.4	59.2	50.0	69.2
		MIN	27.3	28.8	35.9	44.6	52.9	59.9	63.9	63.2	57.3	45.3	36.3	29.9	45.4
		MEAN	37.0	40.0	48.2	58.2	65.9	72.3	75.7	75.0	69.6	58.4	47.8	40.0	57.3
KNOXVILLE WSO //R		MAX	46.9	51.2	60.1	71.0	78.3	84.6	87.2	86.9	81.7	70.9	59.1	50.3	69.0
		MIN	29.5	31.7	39.3	48.2	56.5	64.0	68.0	67.1	61.2	48.1	38.4	31.9	48.7
		MEAN	38.2	41.5	49.7	59.6	67.4	74.3	77.6	77.0	71.5	59.5	48.8	41.1	58.9
KNOXVILLE U OF TENN		MAX	46.7	50.6	59.8	71.2	78.6	85.1	88.2	87.8	82.6	71.6	59.6	50.1	69.3
		MIN	29.8	31.9	39.6	49.1	57.2	64.3	68.2	67.8	62.2	49.6	39.8	32.6	49.3
		MEAN	38.3	41.3	49.7	60.2	67.9	74.7	78.2	77.8	72.4	60.6	49.7	41.4	59.4
LENOIR CITY		MAX	47.0	51.0	60.1	71.2	78.7	85.2	88.4	88.1	82.9	72.1	59.8	50.3	69.6
		MIN	26.4	27.6	35.0	44.3	53.1	61.2	65.3	64.4	58.2	44.8	35.1	28.7	45.3
		MEAN	36.7	39.3	47.6	57.7	65.9	73.2	76.8	76.3	70.5	58.5	47.5	39.5	57.5
LEWISBURG EXP STA		MAX	46.3	50.3	58.7	70.1	78.1	85.4	88.7	88.6	82.8	72.0	59.4	50.3	69.2
		MIN	26.0	27.9	35.8	45.8	53.9	62.2	66.2	64.7	58.4	44.9	35.6	29.3	45.9
		MEAN	36.2	39.1	47.2	58.0	66.0	73.8	77.5	76.6	70.6	58.5	47.6	39.8	57.6
MARTIN U OF T BRANCH		MAX	45.9	50.6	60.0	72.1	80.1	88.0	90.6	90.1	84.0	73.9	60.2	49.8	70.4
		MIN	26.5	29.6	38.1	48.5	56.9	64.8	68.1	66.3	59.7	47.4	38.0	30.7	47.9
		MEAN	36.2	40.1	49.1	60.3	68.5	76.4	79.4	78.2	71.9	60.7	49.1	40.2	59.2
MC MINNVILLE		MAX	48.3	52.3	60.8	72.1	79.0	85.5	88.3	87.8	82.2	72.1	60.4	51.9	70.1
		MIN	28.7	30.8	38.2	47.0	54.5	62.1	65.8	64.9	59.3	46.3	37.1	31.5	47.2
		MEAN	38.5	41.6	49.5	59.6	66.7	73.8	77.1	76.4	70.8	59.2	48.8	41.7	58.6
MEMPHIS WSO R		MAX	48.3	53.0	61.4	72.9	81.0	88.4	91.5	90.3	84.3	74.5	61.4	52.3	71.6
		MIN	30.9	34.1	41.9	52.2	60.9	68.9	72.6	70.8	64.1	51.3	41.1	34.3	51.9
		MEAN	39.6	43.5	51.7	62.6	71.0	78.7	82.1	80.6	74.2	62.9	51.3	43.3	61.8
MILAN		MAX	45.6	50.0	59.2	71.0	79.1	86.6	90.1	89.4	83.4	73.1	59.8	50.0	69.8
		MIN	26.3	29.2	37.9	48.1	56.4	64.3	67.8	65.5	58.5	45.5	36.8	30.1	47.2
		MEAN	36.0	39.6	48.6	59.6	67.8	75.5	78.9	77.5	71.0	59.3	48.3	40.0	58.5
MONTEAGLE		MAX	45.2	49.5	57.7	69.3	75.7	82.0	84.5	84.2	78.9	68.9	57.2	48.4	66.8
		MIN	27.7	29.9	37.2	47.1	54.8	61.4	64.8	64.1	59.0	47.7	37.8	31.0	46.9
		MEAN	36.5	39.7	47.5	58.2	65.3	71.7	74.7	74.1	69.0	58.3	47.5	39.7	56.9
MURFREESBORO 5 N		MAX	47.7	52.0	60.8	72.3	79.7	87.0	90.2	89.8	84.1	73.3	60.7	51.6	70.8
		MIN	27.7	29.7	37.8	47.2	55.1	62.9	66.9	65.6	59.1	45.9	37.0	30.8	47.1
		MEAN	37.7	40.9	49.3	59.7	67.4	75.0	78.6	77.7	71.6	59.6	48.9	41.2	59.0
NASHVILLE WSO //R		MAX	46.3	50.7	59.6	71.2	79.2	86.7	89.8	89.0	83.2	72.3	59.2	50.4	69.8
		MIN	27.8	30.1	38.3	48.1	56.9	64.8	69.0	67.8	61.3	48.0	38.0	31.3	48.5
		MEAN	37.1	40.4	49.0	59.6	68.1	75.8	79.4	78.4	72.3	60.2	48.6	40.9	59.2
NEWBERN		MAX	45.7	50.3	59.8	71.7	80.2	87.8	90.8	89.7	83.8	73.9	60.1	50.0	70.3
		MIN	26.4	29.6	38.0	48.4	56.9	64.4	67.9	65.9	59.0	46.3	36.8	30.1	47.5
		MEAN	36.1	40.0	48.9	60.1	68.5	76.1	79.4	77.8	71.4	60.1	48.5	40.1	58.9
NEWPORT 1 NW		MAX	47.2	51.1	60.4	71.8	79.0	85.5	88.6	88.1	83.0	71.7	59.9	50.4	69.7
		MIN	25.9	27.5	34.8	44.0	52.6	60.6	64.5	63.6	57.5	44.0	34.5	27.7	44.8
		MEAN	36.5	39.3	47.6	57.9	65.8	73.1	76.6	75.9	70.3	57.9	47.2	39.1	57.3
OAK RIDGE WSO		MAX	45.7	50.2	59.0	70.5	78.1	84.6	87.2	86.7	81.3	70.4	58.1	48.9	68.4
		MIN	27.7	29.3	36.7	45.6	54.1	61.8	65.9	65.2	59.1	46.0	36.3	30.2	46.5
		MEAN	36.7	39.8	47.9	58.1	66.1	73.2	76.6	76.0	70.2	58.3	47.2	39.6	57.5

TENNESSEE

TEMPERATURE NORMALS (DEG F)

STATION		JAN	FEB	MAR	APR	MAY	JUN	JUL	AUG	SEP	OCT	NOV	DEC	ANN
PARIS 5 E	MAX	45.6	50.3	59.8	70.8	78.8	86.6	90.2	89.9	83.8	72.5	59.8	49.8	69.8
	MIN	25.1	27.9	35.9	45.1	53.4	61.7	65.7	63.9	56.9	43.7	35.8	29.1	45.4
	MEAN	35.4	39.1	47.9	58.0	66.2	74.2	78.0	77.0	70.4	58.1	47.8	39.5	57.6
ROGERSVILLE 1 NE	MAX	46.0	50.4	59.7	71.0	78.3	84.3	87.3	86.8	81.9	70.7	58.2	49.2	68.7
	MIN	25.7	27.5	34.6	43.5	51.4	58.3	62.1	61.3	55.4	43.0	34.2	28.0	43.8
	MEAN	35.9	38.9	47.2	57.3	64.9	71.3	74.7	74.0	68.7	56.9	46.3	38.7	56.2
SAMBURG WILDLIFE REF	MAX	44.0	48.3	57.8	70.5	79.9	88.3	91.6	90.1	83.6	73.0	59.3	48.3	69.6
	MIN	25.8	29.2	38.2	49.1	57.0	64.8	68.1	65.9	59.2	47.1	37.5	29.9	47.7
	MEAN	34.9	38.8	48.0	59.8	68.5	76.6	79.9	78.0	71.4	60.1	48.5	39.1	58.6
SAVANNAH	MAX	49.8	55.0	64.0	75.4	81.8	88.8	92.3	91.7	85.7	75.1	62.3	53.5	73.0
	MIN	28.8	31.3	39.2	48.6	56.0	63.4	67.0	65.9	59.7	47.0	38.3	32.1	48.1
	MEAN	39.3	43.2	51.6	62.0	68.9	76.1	79.7	78.8	72.7	61.0	50.3	42.8	60.5
SHELBYVILLE 3	MAX	48.6	53.0	61.5	72.7	80.0	87.1	89.8	89.6	84.0	73.5	61.0	52.3	71.1
	MIN	28.2	30.3	38.0	47.1	54.5	62.1	66.0	64.8	58.8	45.9	37.0	31.1	47.0
	MEAN	38.4	41.7	49.8	59.9	67.2	74.6	77.9	77.2	71.4	59.7	49.0	41.7	59.0
SPRINGFIELD EXP STA	MAX	43.7	48.0	57.1	68.9	77.2	85.0	88.6	87.6	82.0	70.7	57.6	48.0	67.9
	MIN	24.7	27.0	35.7	46.2	54.5	62.5	66.1	64.8	58.2	45.7	36.3	28.8	45.9
	MEAN	34.2	37.5	46.4	57.6	65.9	73.8	77.4	76.3	70.1	58.2	46.9	38.4	56.9
TULLAHOMA	MAX	47.9	52.4	60.9	72.2	79.0	85.5	88.3	88.1	82.5	72.4	60.3	51.7	70.1
	MIN	28.5	30.9	38.4	47.7	55.0	62.5	66.0	64.9	59.1	46.6	37.6	31.9	47.4
	MEAN	38.3	41.7	49.7	60.0	67.0	74.0	77.2	76.5	70.8	59.5	49.0	41.8	58.8
UNION CITY	MAX	43.7	48.1	57.4	69.5	78.0	86.0	89.1	88.3	82.6	72.3	59.0	48.4	68.5
	MIN	24.5	27.5	35.6	46.6	55.4	63.5	67.1	64.7	57.3	44.1	35.3	28.7	45.9
	MEAN	34.2	37.8	46.6	58.0	66.7	74.8	78.1	76.5	70.0	58.2	47.2	38.6	57.2
WATAUGA DAM	MAX	45.3	48.5	57.7	68.9	76.5	82.3	85.1	84.6	80.0	69.4	58.1	48.1	67.0
	MIN	25.1	26.2	33.0	41.6	50.6	58.1	61.9	61.4	56.0	43.9	34.4	27.6	43.3
	MEAN	35.2	37.3	45.4	55.2	63.6	70.2	73.5	73.0	68.0	56.7	46.3	37.9	55.2
WAYNESBORO	MAX	47.1	51.7	60.2	72.1	79.1	86.0	89.7	89.4	83.5	73.1	60.7	51.2	70.3
	MIN	24.0	25.6	33.9	43.9	51.9	60.0	64.0	62.1	55.7	41.7	32.7	26.9	43.5
	MEAN	35.6	38.7	47.0	58.0	65.5	73.0	76.9	75.7	69.6	57.5	46.7	39.1	56.9

TENNESSEE

PRECIPITATION NORMALS (INCHES)

STATION	JAN	FEB	MAR	APR	MAY	JUN	JUL	AUG	SEP	OCT	NOV	DEC	ANN
ALLARDT	4.93	4.43	5.88	4.64	4.34	4.79	5.51	3.88	4.00	3.00	4.06	4.83	54.29
BOLIVAR WTR WKS	4.93	4.47	5.56	5.18	4.63	3.47	4.02	3.45	4.07	2.43	4.23	4.89	51.33
BOLTON	4.35	4.54	5.35	5.52	4.84	3.49	3.99	3.33	4.05	2.49	4.23	4.77	50.95
BRISTOL WSO R	3.56	3.43	4.29	3.46	3.61	3.46	4.19	3.23	3.00	2.50	2.98	3.53	41.24
BROWNSVILLE	4.76	4.61	5.57	5.68	4.94	3.49	4.21	3.20	3.81	2.33	4.22	4.55	51.37
CELINA	4.98	4.64	5.92	4.46	4.14	4.41	4.77	3.22	4.17	2.75	4.12	4.98	52.56
CENTERVILLE WATER PL	5.01	4.39	5.71	4.97	4.60	3.78	4.59	3.46	3.41	2.76	4.24	4.98	51.90
CHATTANOOGA WSO R	5.20	4.69	6.31	4.57	4.00	3.32	4.55	3.41	4.30	2.92	4.19	5.14	52.60
CLARKSVILLE SEW PLT	4.52	4.26	5.92	4.46	4.14	3.74	3.81	3.79	3.31	2.83	4.22	4.64	49.64
COLUMBIA SEWAGE PLANT	5.41	4.83	6.21	5.15	4.97	3.74	4.25	3.43	3.57	3.47	3.91	4.99	53.93
CONASAUGA 1 NNW	4.87	4.44	6.03	4.61	4.33	3.55	4.64	4.44	4.45	2.82	4.08	4.41	52.67
COOKEVILLE	5.33	4.75	6.46	4.86	4.68	4.38	5.02	3.75	4.34	2.85	4.25	5.30	55.97
COPPERHILL	5.49	5.22	6.79	5.15	4.45	4.41	5.31	4.93	4.56	3.17	4.34	5.24	59.06
COVINGTON 1 W	4.64	4.54	5.55	5.61	4.85	3.32	3.98	3.13	3.89	2.49	4.55	4.61	51.16
CROSSVILLE EXP STA	5.67	4.98	6.80	5.17	4.97	4.59	5.10	3.40	4.20	3.18	4.59	5.81	58.46
DICKSON	4.91	4.50	5.85	4.98	4.54	3.87	4.35	3.66	3.67	2.90	4.23	4.78	52.24
DOVER 1 W	4.75	4.48	5.97	5.00	4.29	4.04	4.28	3.49	3.62	2.65	4.57	4.68	51.82
DYERSBURG FAA AP	4.13	4.40	5.40	4.54	4.61	3.57	3.95	3.27	3.51	2.27	4.33	4.42	48.40
FRANKLIN SEWAGE PLANT	4.91	4.49	6.00	4.66	4.64	3.85	4.32	3.90	3.66	3.18	3.83	4.87	52.31
GATLINBURG 2 SW	4.80	4.34	5.81	4.88	4.81	5.60	6.05	5.08	3.93	3.13	4.12	4.38	56.93
GREENEVILLE EXP STA	3.49	3.32	4.39	3.59	3.72	3.72	4.63	3.30	3.27	2.55	3.03	3.24	42.25
JACKSON FAA AP	4.77	4.45	5.24	5.44	4.99	3.81	4.44	2.98	3.52	2.63	4.12	4.60	50.99
JACKSON EXP STA	4.74	4.46	5.24	5.39	4.94	3.84	4.06	2.88	3.69	2.47	4.22	4.50	50.43
JAMESTOWN	5.16	4.64	6.05	4.81	4.43	4.73	5.50	3.86	4.37	2.97	4.24	4.89	55.65
JEFFERSON CITY EVAP	4.39	3.98	5.25	3.99	3.74	3.63	4.14	3.05	2.97	2.81	3.87	4.30	46.12
KINGSPORT	3.97	3.70	4.72	3.75	3.87	3.54	4.44	3.46	3.12	2.61	3.26	3.68	44.12
KINGSTON	5.12	4.57	6.23	4.36	4.48	3.88	4.83	3.36	3.87	2.92	4.40	5.53	53.55
KINGSTON SPRINGS 2 NNE	4.70	4.56	5.90	4.47	4.27	3.97	4.13	3.79	3.89	2.69	4.22	4.60	51.19
KNOXVILLE WSO //R	4.65	4.18	5.49	3.87	3.71	3.95	4.33	3.02	2.99	2.73	3.78	4.59	47.29
KNOXVILLE U OF TENN	4.57	4.00	5.40	3.78	3.85	3.94	4.32	3.31	3.20	2.78	3.85	4.36	47.36
LEBANON	4.92	4.34	5.90	4.56	4.27	3.95	4.37	3.89	3.82	2.94	3.91	4.80	51.67
LENOIR CITY	5.08	4.47	6.32	4.34	4.04	3.92	4.48	3.82	3.74	2.94	4.19	5.20	52.54
LEWISBURG EXP STA	5.26	4.59	6.41	4.95	5.01	3.71	4.56	3.13	4.02	3.36	4.24	4.95	54.19
LIVINGSTON	5.17	4.52	5.86	4.91	4.34	4.42	4.98	3.73	4.31	2.82	4.14	5.19	54.39
MARTIN U OF T BRANCH	4.32	4.21	5.84	5.02	4.72	4.42	4.87	3.29	3.72	2.66	4.50	4.44	52.01
MC MINNVILLE	5.21	4.57	6.25	4.60	4.80	4.15	4.73	3.48	4.16	2.87	4.13	5.29	54.24
MEMPHIS WSO R	4.61	4.33	5.44	5.77	5.06	3.58	4.03	3.74	3.62	2.37	4.17	4.85	51.57
MILAN	4.66	4.54	5.71	5.19	5.16	4.26	3.92	3.85	3.91	2.66	4.49	4.78	53.13
MONTEAGLE	6.31	5.57	7.13	5.44	5.32	4.28	5.15	3.83	4.59	3.61	5.21	5.90	62.34
MOSCOW	4.75	4.69	5.56	5.48	4.67	3.30	3.93	3.13	4.02	2.41	4.51	4.99	51.44
MURFREESBORO 5 N	4.81	4.48	5.85	4.87	5.04	3.70	4.28	3.59	3.99	2.85	3.83	4.66	51.95
NASHVILLE WSO //R	4.49	4.03	5.58	4.47	4.56	3.70	3.82	3.40	3.71	2.58	3.52	4.63	48.49
NEWBERN	4.12	4.06	5.51	4.92	4.81	4.08	4.05	3.29	3.70	2.41	4.49	4.51	49.95
NEWCOMB	4.20	3.88	5.32	4.18	4.18	4.15	4.95	3.86	3.31	3.01	4.03	4.24	49.31
NEWPORT 1 NW	3.98	3.61	4.97	3.86	4.22	3.81	4.37	3.63	3.20	2.57	3.18	3.48	44.88
OAK RIDGE WSO	5.25	4.60	6.21	4.41	4.23	4.26	5.21	3.75	3.80	2.89	4.50	5.65	54.76
ONEIDA	4.96	4.49	5.74	4.68	4.43	4.56	5.25	3.79	3.78	3.23	4.23	4.70	53.84
ORLINDA	4.56	4.28	5.62	4.31	4.53	3.87	3.88	3.29	3.41	2.61	4.15	4.65	49.16
PARIS 5 E	4.67	4.29	5.52	4.90	4.26	3.80	3.93	3.65	3.56	2.46	4.52	4.68	50.24
ROGERSVILLE 1 NE	4.33	3.89	5.08	4.17	4.14	3.31	4.37	3.61	3.24	2.50	3.62	4.27	46.53
SAMBURG WILDLIFE REF	4.04	3.91	5.05	4.74	4.89	3.75	3.95	3.22	3.08	2.55	4.36	4.36	47.90
SAVANNAH	5.26	4.83	6.15	5.28	4.91	4.05	4.03	3.13	3.57	2.87	4.71	5.56	54.35
SHELBYVILLE 3	5.27	4.63	6.32	4.88	4.85	3.73	4.63	3.64	4.06	3.29	4.16	4.87	54.33
SPRINGFIELD EXP STA	4.63	4.20	5.47	4.51	4.59	3.79	4.14	3.43	3.33	2.68	4.19	4.58	49.54
STATESVILLE	5.41	4.88	6.39	5.21	5.28	4.13	4.93	3.94	4.48	3.05	4.23	5.34	57.27

TENNESSEE

PRECIPITATION NORMALS (INCHES)

STATION	JAN	FEB	MAR	APR	MAY	JUN	JUL	AUG	SEP	OCT	NOV	DEC	ANN
TULLAHOMA	5.56	4.97	6.72	5.09	4.92	3.79	4.82	3.36	3.88	3.26	4.44	5.80	56.61
UNION CITY	4.23	4.05	5.63	4.95	4.80	4.32	4.13	3.36	3.60	2.73	4.54	4.49	50.83
WATAUGA DAM	4.17	3.96	4.77	4.29	4.31	3.94	5.41	4.36	3.49	3.15	3.38	3.79	49.02
WAYNESBORO	5.65	5.03	6.57	5.53	4.85	4.24	4.77	3.36	3.61	3.10	4.80	5.38	56.89

TENNESSEE

HEATING DEGREE DAY NORMALS (BASE 65 DEG F)

STATION		JUL	AUG	SEP	OCT	NOV	DEC	JAN	FEB	MAR	APR	MAY	JUN	ANN
ALLARDT		0	0	55	292	579	837	949	773	605	277	111	13	4491
BOLIVAR WTR WKS		0	0	19	197	471	725	837	659	481	164	55	0	3608
BRISTOL WSO	R	0	0	35	263	564	831	930	759	580	273	111	10	4356
BROWNSVILLE		0	0	11	163	429	698	815	624	449	134	33	0	3356
CELINA		0	0	25	240	528	794	908	739	569	236	92	7	4138
CHATTANOOGA WSO	R	0	0	11	190	483	729	815	644	478	171	56	6	3583
CLARKSVILLE SEW PLT		0	0	22	224	516	791	908	728	541	201	75	8	4014
COLUMBIA SEWAGE PLANT		0	0	25	208	492	738	849	672	513	193	64	7	3761
COPPERHILL		0	0	21	251	534	778	865	708	558	251	97	8	4071
COVINGTON 1 W		0	0	15	180	471	750	877	689	510	169	46	0	3707
CROSSVILLE EXP STA		7	5	61	320	612	896	1014	837	663	318	144	19	4896
DICKSON		0	0	16	203	492	756	871	680	508	173	55	5	3759
DOVER 1 W		0	0	28	248	528	800	924	736	548	210	87	6	4115
DYERSBURG FAA AP		0	0	11	160	459	735	862	664	483	152	33	0	3559
FRANKLIN SEWAGE PLANT		0	0	17	200	483	735	840	666	491	172	53	7	3664
GATLINBURG 2 SW		0	0	41	285	564	800	884	725	559	267	121	17	4263
GREENEVILLE EXP STA		0	0	24	251	537	791	874	714	539	244	83	6	4063
JACKSON FAA AP		0	0	11	169	465	725	843	652	481	154	40	0	3540
JACKSON EXP STA		0	0	15	190	477	744	868	683	506	173	51	0	3707
KINGSPORT		0	0	22	222	516	775	868	700	527	215	75	0	3920
KNOXVILLE WSO	//R	0	0	14	201	486	741	831	658	483	181	63	0	3658
KNOXVILLE U OF TENN		0	0	10	174	459	732	828	664	484	173	54	0	3578
LENOIR CITY		0	0	19	228	525	791	877	720	545	226	86	6	4023
LEWISBURG EXP STA		0	0	28	232	522	781	893	725	561	222	85	7	4056
MARTIN U OF T BRANCH		0	0	16	190	477	769	893	697	511	177	51	0	3781
MC MINNVILLE		0	0	21	210	486	722	822	655	492	181	69	5	3663
MEMPHIS WSO	R	0	0	9	137	415	673	787	602	433	126	25	0	3207
MILAN		0	0	27	216	501	775	899	711	519	189	66	6	3909
MONTEAGLE		0	0	27	233	525	784	884	708	551	219	89	10	4030
MURFREESBORO 5 N		0	0	17	209	483	738	846	682	501	190	68	0	3734
NASHVILLE WSO	//R	0	0	19	193	492	747	865	689	510	186	55	0	3756
NEWBERN		0	0	20	197	495	772	896	700	517	179	55	6	3837
NEWPORT 1 NW		0	0	24	243	534	803	884	720	539	228	97	9	4081
OAK RIDGE WSO		0	0	25	234	534	787	877	706	536	217	82	8	4006
PARIS 5 E		0	0	28	243	516	791	918	725	544	224	81	7	4077
ROGERSVILLE 1 NE		0	0	34	266	561	815	902	731	557	240	99	15	4220
SAMBURG WILDLIFE REF		0	0	17	195	495	803	933	734	541	183	52	0	3953
SAVANNAH		0	0	12	169	441	688	797	610	438	135	36	0	3326
SHELBYVILLE 3		0	0	17	197	480	722	825	652	485	175	57	0	3610
SPRINGFIELD EXP STA		0	0	34	247	543	825	955	770	589	238	96	8	4305
TULLAHOMA		0	0	19	201	480	719	828	652	487	172	60	0	3618
UNION CITY		0	0	29	239	534	818	955	762	583	229	75	0	4224
WATAUGA DAM		0	0	40	272	561	840	924	776	608	299	123	17	4460
WAYNESBORO		0	0	26	257	549	803	911	736	566	224	91	7	4170

TENNESSEE

COOLING DEGREE DAY NORMALS (BASE 65 DEG F)

STATION		JAN	FEB	MAR	APR	MAY	JUN	JUL	AUG	SEP	OCT	NOV	DEC	ANN
ALLARDT		0	0	7	10	67	169	267	245	127	13	0	0	905
BOLIVAR WTR WKS		0	6	16	38	157	327	446	403	223	45	0	0	1661
BRISTOL WSO	R	0	0	0	6	93	205	307	288	149	18	0	0	1066
BROWNSVILLE		0	0	21	47	185	369	484	443	254	70	0	0	1873
CELINA		0	0	8	11	107	250	363	341	175	29	0	0	1284
CHATTANOOGA WSO	R	0	0	7	24	142	309	425	406	230	35	0	0	1578
CLARKSVILLE SEW PLT		0	0	11	21	134	302	425	384	202	35	0	0	1514
COLUMBIA SEWAGE PLANT		0	0	10	34	129	295	409	378	208	34	0	0	1497
COPPERHILL		0	0	0	8	91	212	326	307	147	13	0	0	1104
COVINGTON 1 W		0	0	14	28	170	357	471	419	225	49	0	0	1733
CROSSVILLE EXP STA		0	0	0	6	60	142	246	225	109	10	0	0	798
DICKSON		0	0	18	26	136	299	415	384	205	39	0	0	1522
DOVER 1 W		0	0	15	24	124	276	384	350	175	31	0	0	1379
DYERSBURG FAA AP		0	0	18	41	182	381	490	437	239	52	0	0	1840
FRANKLIN SEWAGE PLANT		0	0	14	22	124	298	403	372	197	36	0	0	1466
GATLINBURG 2 SW		0	0	7	12	81	170	260	242	119	12	0	0	903
GREENEVILLE EXP STA		0	0	0	10	95	228	338	313	162	21	0	0	1167
JACKSON FAA AP		0	0	19	40	179	363	477	437	239	48	0	0	1802
JACKSON EXP STA		0	0	13	32	153	327	440	400	219	41	0	0	1625
KINGSPORT		0	0	7	11	103	223	332	310	160	18	0	0	1164
KNOXVILLE WSO	//R	0	0	8	19	137	283	391	372	209	30	0	0	1449
KNOXVILLE U OF TENN		0	0	9	29	144	295	409	397	232	37	0	0	1552
LENOIR CITY		0	0	6	7	114	252	366	350	184	27	0	0	1306
LEWISBURG EXP STA		0	0	9	12	116	271	388	360	196	31	0	0	1383
MARTIN U OF T BRANCH		0	0	18	36	159	342	446	409	223	56	0	0	1689
MC MINNVILLE		0	0	12	19	122	269	375	353	195	30	0	0	1375
MEMPHIS WSO	R	0	0	20	54	211	411	530	484	285	72	0	0	2067
MILAN		0	0	10	27	153	321	431	392	207	39	0	0	1580
MONTEAGLE		0	0	8	15	98	211	301	282	147	25	0	0	1087
MURFREESBORO 5 N		0	7	15	31	142	305	422	394	215	42	0	0	1573
NASHVILLE WSO	//R	0	0	14	24	151	328	446	415	238	45	0	0	1661
NEWBERN		0	0	18	32	163	339	446	397	212	46	0	0	1653
NEWPORT 1 NW		0	0	0	15	122	252	360	338	183	23	0	0	1293
OAK RIDGE WSO		0	0	5	10	117	254	360	341	181	26	0	0	1294
PARIS 5 E		0	0	13	14	118	283	403	372	190	29	0	0	1422
ROGERSVILLE 1 NE		0	0	6	9	96	204	301	279	145	14	0	0	1054
SAMBURG WILDLIFE REF		0	0	14	27	161	348	462	403	209	43	0	0	1667
SAVANNAH		0	0	23	45	157	337	456	428	243	45	0	0	1734
SHELBYVILLE 3		0	0	13	22	125	293	400	378	209	32	0	0	1472
SPRINGFIELD EXP STA		0	0	12	16	123	272	384	350	187	36	0	0	1380
TULLAHOMA		0	0	12	22	122	274	378	357	193	31	0	0	1389
UNION CITY		0	0	12	19	128	297	406	357	179	28	0	0	1426
WATAUGA DAM		0	0	0	0	80	173	267	248	130	15	0	0	913
WAYNESBORO		0	0	8	14	106	247	369	332	164	25	0	0	1265

40 — TENNESSEE

LEGEND
11 = TEMPERATURE ONLY
12 = PRECIPITATION ONLY
13 = TEMP. & PRECIP.

STATE-STATION NUMBER	STN TYP	NAME		LATITUDE DEG-MIN	LONGITUDE DEG-MIN	ELEVATION (FT)
40-0081	13	ALLARDT		N 3623	W 08453	1672
40-0876	13	BOLIVAR WTR WKS		N 3516	W 08859	455
40-0884	12	BOLTON		N 3519	W 08946	310
40-1094	13	BRISTOL WSO	R	N 3629	W 08224	1525
40-1145	13	BROWNSVILLE		N 3536	W 08916	360
40-1561	13	CELINA		N 3633	W 08530	550
40-1587	12	CENTERVILLE WATER PL		N 3545	W 08727	660
40-1656	13	CHATTANOOGA WSO	R	N 3502	W 08512	665
40-1790	13	CLARKSVILLE SEW PLT		N 3633	W 08722	382
40-1957	13	COLUMBIA SEWAGE PLANT		N 3538	W 08702	690
40-1978	12	CONASAUGA 1 NNW		N 3501	W 08444	775
40-2009	12	COOKEVILLE		N 3609	W 08531	1120
40-2024	13	COPPERHILL		N 3500	W 08423	1535
40-2108	13	COVINGTON 1 W		N 3534	W 08940	310
40-2202	13	CROSSVILLE EXP STA		N 3601	W 08508	1810
40-2489	13	DICKSON		N 3604	W 08724	814
40-2589	13	DOVER 1 W		N 3629	W 08751	475
40-2685	13	DYERSBURG FAA AP		N 3601	W 08924	334
40-3280	13	FRANKLIN SEWAGE PLANT		N 3556	W 08652	670
40-3420	13	GATLINBURG 2 SW		N 3541	W 08332	1454
40-3679	13	GREENEVILLE EXP STA		N 3606	W 08251	1320
40-4556	13	JACKSON FAA AP		N 3536	W 08855	413
40-4561	13	JACKSON EXP STA		N 3537	W 08851	400
40-4590	12	JAMESTOWN		N 3626	W 08456	1750
40-4609	12	JEFFERSON CITY EVAP		N 3607	W 08330	1200
40-4858	13	KINGSPORT		N 3631	W 08232	1284
40-4871	12	KINGSTON		N 3552	W 08432	755
40-4876	12	KINGSTON SPRINGS 2 NNE		N 3607	W 08706	480
40-4950	13	KNOXVILLE WSO	//R	N 3549	W 08359	980
40-4955	13	KNOXVILLE U OF TENN		N 3557	W 08355	895
40-5108	12	LEBANON		N 3613	W 08619	520
40-5158	13	LENOIR CITY		N 3548	W 08415	785
40-5187	13	LEWISBURG EXP STA		N 3527	W 08648	787
40-5327	12	LIVINGSTON		N 3623	W 08519	1040
40-5681	13	MARTIN U OF T BRANCH		N 3621	W 08852	387
40-5882	13	MC MINNVILLE		N 3541	W 08548	940
40-5954	13	MEMPHIS WSO	R	N 3503	W 09000	263
40-6012	13	MILAN		N 3556	W 08846	430
40-6162	13	MONTEAGLE		N 3515	W 08551	1930
40-6274	12	MOSCOW		N 3504	W 08924	345
40-6371	13	MURFREESBORO 5 N		N 3555	W 08622	550
40-6402	13	NASHVILLE WSO	//R	N 3607	W 08641	590
40-6471	13	NEWBERN		N 3607	W 08916	370
40-6493	12	NEWCOMB		N 3633	W 08411	1060
40-6534	13	NEWPORT 1 NW		N 3559	W 08312	1040
40-6750	13	OAK RIDGE WSO		N 3601	W 08414	905
40-6829	12	ONEIDA		N 3630	W 08431	1450
40-6861	12	ORLINDA		N 3636	W 08643	720
40-6977	13	PARIS 5 E		N 3619	W 08814	450
40-7884	13	ROGERSVILLE 1 NE		N 3625	W 08259	1355

40 — TENNESSEE

STATE-STATION NUMBER	STN TYP	NAME	LATITUDE DEG-MIN	LONGITUDE DEG-MIN	ELEVATION (FT)
40-8065	13	SAMBURG WILDLIFE REF	N 3623	W 08921	290
40-8108	13	SAVANNAH	N 3514	W 08815	440
40-8246	13	SHELBYVILLE 3	N 3529	W 08627	785
40-8562	13	SPRINGFIELD EXP STA	N 3628	W 08650	745
40-8609	12	STATESVILLE	N 3601	W 08608	723
40-9155	13	TULLAHOMA	N 3522	W 08612	1072
40-9219	13	UNION CITY	N 3625	W 08904	320
40-9455	13	WATAUGA DAM	N 3620	W 08207	1710
40-9502	13	WAYNESBORO	N 3518	W 08746	750

40 — TENNESSEE

LEGEND

O = TEMPERATURE ONLY
X = PRECIPITATION ONLY
⊗ = TEMPERATURE & PRECIPITATION

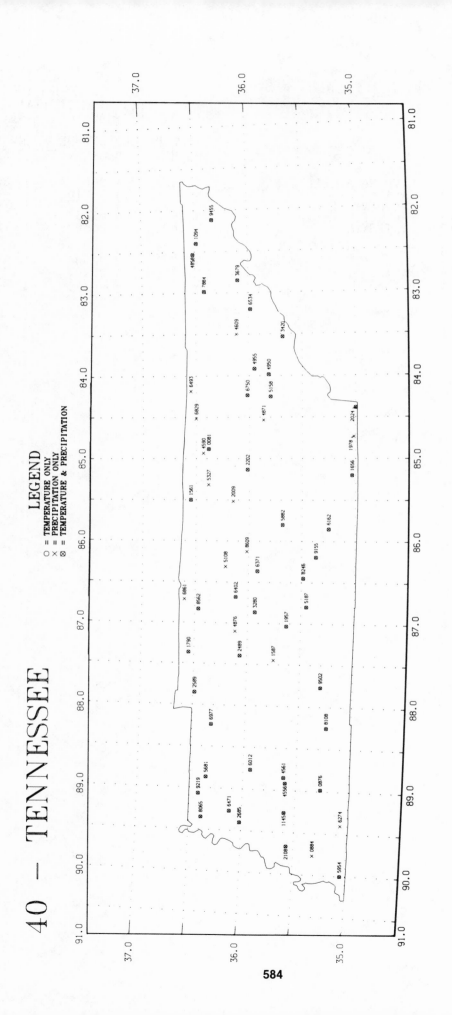

584

TEXAS

TEMPERATURE NORMALS (DEG F)

STATION			JAN	FEB	MAR	APR	MAY	JUN	JUL	AUG	SEP	OCT	NOV	DEC	ANN
ABILENE WSO	//R	MAX	55.5	60.3	68.6	77.6	84.1	91.8	95.4	94.5	87.1	77.6	64.8	58.4	76.3
		MIN	31.2	35.5	42.6	52.8	60.8	69.0	72.7	71.7	64.9	54.1	42.0	34.3	52.6
		MEAN	43.3	47.9	55.6	65.2	72.5	80.5	84.1	83.1	76.0	65.9	53.4	46.4	64.5
ALBANY		MAX	56.0	60.9	69.0	78.8	85.2	93.4	97.9	97.7	89.7	80.0	66.7	60.0	77.9
		MIN	28.6	32.3	38.8	50.1	58.0	66.6	70.1	68.3	61.5	50.1	38.3	31.7	49.5
		MEAN	42.3	46.7	53.9	64.5	71.7	80.0	84.0	83.0	75.6	65.1	52.5	45.9	63.8
ALICE		MAX	67.2	71.1	78.2	84.7	88.7	93.8	96.6	97.0	92.2	85.4	76.2	69.5	83.4
		MIN	43.0	46.3	53.1	61.9	67.8	72.4	73.8	73.4	70.2	61.2	52.4	45.8	60.1
		MEAN	55.1	58.7	65.7	73.3	78.3	83.1	85.2	85.2	81.2	73.3	64.3	57.7	71.8
ALPINE		MAX	60.6	63.6	70.4	78.4	85.0	90.8	89.5	88.3	83.5	77.3	67.3	61.5	76.4
		MIN	32.0	34.3	39.5	47.4	54.8	62.6	64.1	62.8	57.9	48.5	38.1	33.3	47.9
		MEAN	46.3	48.9	54.9	62.9	70.0	76.7	76.8	75.5	70.7	62.9	52.7	47.4	62.1
AMARILLO WSO	//R	MAX	49.1	53.1	60.8	71.0	79.1	88.2	91.4	89.6	82.4	72.7	58.7	51.8	70.7
		MIN	21.7	26.1	32.0	42.0	51.9	61.5	66.2	64.5	56.9	45.5	32.1	24.8	43.8
		MEAN	35.4	39.6	46.4	56.5	65.5	74.9	78.8	77.0	69.7	59.2	45.4	38.3	57.2
ANGLETON 2 W		MAX	63.2	65.7	71.7	77.8	83.6	89.1	91.9	91.5	88.1	81.7	72.6	66.6	78.6
		MIN	41.9	44.4	50.8	59.2	65.2	70.4	72.3	71.8	68.4	58.0	50.0	44.2	58.1
		MEAN	52.6	55.1	61.3	68.5	74.4	79.8	82.1	81.7	78.3	69.9	61.3	55.4	68.4
AUSTIN WSO	R	MAX	59.4	64.1	71.7	79.0	84.7	91.6	95.4	95.3	89.3	80.8	69.2	62.8	78.6
		MIN	38.8	42.2	49.3	58.3	65.1	71.5	73.9	73.7	69.1	58.7	48.1	41.4	57.5
		MEAN	49.1	53.2	60.5	68.7	74.9	81.6	84.7	84.5	79.2	69.8	58.7	52.1	68.1
BALLINGER 1 SW		MAX	59.0	63.9	72.1	80.8	86.2	93.3	96.4	95.7	88.8	80.0	67.4	61.6	78.8
		MIN	29.5	33.4	40.8	51.7	60.1	67.6	70.2	69.1	63.0	51.2	39.5	32.0	50.7
		MEAN	44.3	48.7	56.4	66.3	73.2	80.5	83.3	82.4	75.9	65.7	53.5	46.8	64.8
BAY CITY WATERWORKS		MAX	63.4	66.0	72.2	79.0	84.8	90.2	92.8	93.0	89.5	83.2	73.1	66.6	79.5
		MIN	42.4	45.0	51.6	60.5	66.8	71.9	74.0	73.1	69.3	59.8	51.1	44.7	59.2
		MEAN	52.9	55.5	61.9	69.8	75.8	81.1	83.5	83.1	79.4	71.5	62.2	55.7	69.4
BEEVILLE 5 NE		MAX	64.5	67.9	75.3	81.6	86.3	91.8	95.2	95.4	90.4	83.4	73.7	67.4	81.1
		MIN	41.4	44.3	51.5	60.1	66.0	70.9	72.5	72.4	68.9	59.5	50.5	44.0	58.5
		MEAN	53.0	56.2	63.4	70.9	76.2	81.4	83.9	83.9	79.6	71.5	62.1	55.7	69.8
BIG SPRING		MAX	57.1	61.3	69.8	79.0	85.7	93.0	94.6	93.6	86.9	78.3	65.8	59.7	77.1
		MIN	29.7	33.6	40.6	50.5	59.4	67.4	71.0	69.7	63.0	52.4	40.0	32.9	50.9
		MEAN	43.4	47.5	55.2	64.8	72.5	80.2	82.8	81.7	74.9	65.4	52.9	46.3	64.0
BLANCO		MAX	59.5	63.2	70.9	78.5	84.1	91.3	95.2	95.1	88.9	80.5	68.8	62.6	78.2
		MIN	33.3	36.7	43.4	53.2	60.5	67.6	69.8	69.0	64.0	53.2	42.4	35.8	52.4
		MEAN	46.4	49.9	57.1	65.9	72.3	79.5	82.6	82.1	76.5	66.9	55.6	49.2	65.3
BOERNE		MAX	60.2	64.2	71.8	78.5	83.4	89.9	93.3	93.3	87.8	79.7	68.9	62.9	77.8
		MIN	35.0	38.2	45.1	53.9	60.9	67.1	68.8	67.8	64.0	54.0	43.8	37.2	53.0
		MEAN	47.6	51.2	58.5	66.3	72.2	78.5	81.1	80.6	75.9	66.9	56.4	50.1	65.4
BONHAM		MAX	53.6	58.7	66.9	75.8	82.7	90.1	95.2	95.2	87.8	77.9	65.3	57.0	75.5
		MIN	31.4	35.4	42.5	52.4	60.3	68.0	71.9	70.6	63.8	52.7	41.8	34.6	52.1
		MEAN	42.5	47.1	54.7	64.1	71.5	79.1	83.6	82.9	75.8	65.4	53.6	45.9	63.9
BORGER 3 W		MAX	51.2	56.1	63.9	73.7	81.3	90.0	93.3	91.5	84.7	75.4	60.9	53.8	73.0
		MIN	23.6	27.9	33.8	43.9	53.2	62.4	67.1	65.4	57.9	46.8	34.0	27.0	45.3
		MEAN	37.4	42.0	48.9	58.8	67.3	76.3	80.2	78.4	71.4	61.1	47.4	40.4	59.1
BRADY 2 NNW		MAX	58.4	62.2	70.4	79.0	84.2	91.7	95.7	95.1	88.1	79.5	67.4	60.9	77.7
		MIN	30.4	34.1	40.9	51.2	59.1	66.4	69.6	68.5	63.0	51.9	40.1	33.2	50.7
		MEAN	44.4	48.2	55.6	65.1	71.7	79.0	82.7	81.8	75.6	65.7	53.8	47.0	64.2
BRENHAM		MAX	60.0	64.1	71.9	79.4	85.9	92.5	96.4	96.6	90.7	82.7	71.1	63.6	79.6
		MIN	38.5	41.4	48.1	57.2	64.2	70.1	72.6	72.2	67.7	57.3	47.5	41.3	56.5
		MEAN	49.3	52.8	60.0	68.3	75.0	81.3	84.5	84.5	79.3	70.0	59.3	52.4	68.1

TEXAS

TEMPERATURE NORMALS (DEG F)

STATION			JAN	FEB	MAR	APR	MAY	JUN	JUL	AUG	SEP	OCT	NOV	DEC	ANN
BRIDGEPORT		MAX	54.5	59.5	67.6	77.4	84.5	93.7	99.5	99.1	90.7	80.1	66.4	58.7	77.6
		MIN	28.6	32.9	40.3	51.0	59.2	67.4	70.9	69.2	62.5	50.9	39.7	32.1	50.4
		MEAN	41.6	46.2	54.0	64.3	71.9	80.5	85.2	84.2	76.6	65.5	53.1	45.4	64.0
BROWNSVILLE WSO	R	MAX	69.7	72.5	77.5	83.2	87.0	90.5	92.6	92.8	89.8	84.4	77.0	71.9	82.4
		MIN	50.8	53.0	59.5	66.6	71.3	74.7	75.6	75.4	73.1	66.1	58.3	52.6	64.8
		MEAN	60.3	62.8	68.6	74.9	79.2	82.6	84.1	84.1	81.4	75.3	67.7	62.3	73.6
BROWNWOOD		MAX	56.6	61.6	69.7	79.0	85.1	92.9	96.9	96.4	89.2	79.6	67.0	60.2	77.9
		MIN	31.4	35.2	42.9	53.6	61.5	69.1	72.5	71.3	65.1	53.9	41.8	34.1	52.7
		MEAN	44.0	48.4	56.3	66.3	73.3	81.0	84.7	83.8	77.2	66.8	54.4	47.2	65.3
CAMERON		MAX	60.6	65.2	72.9	79.9	85.3	92.0	96.2	96.4	90.1	81.8	70.5	64.0	79.6
		MIN	38.4	41.5	48.3	57.0	63.9	70.1	72.8	72.3	67.5	57.3	47.2	40.8	56.4
		MEAN	49.5	53.4	60.7	68.5	74.6	81.1	84.6	84.4	78.8	69.6	58.8	52.4	68.0
CANYON		MAX	53.7	57.8	65.1	74.2	81.6	89.8	92.0	90.2	84.3	75.7	62.4	55.9	73.6
		MIN	22.7	26.8	32.7	42.4	51.6	60.8	65.2	63.6	56.7	44.7	32.6	25.8	43.8
		MEAN	38.2	42.4	48.9	58.3	66.6	75.3	78.6	76.9	70.5	60.2	47.5	40.9	58.7
CENTER		MAX	57.2	61.9	69.3	77.4	83.9	90.3	94.4	94.4	88.9	80.6	68.4	60.9	77.3
		MIN	34.0	36.5	43.3	52.6	60.4	67.1	70.0	68.8	63.6	51.2	41.6	35.6	52.1
		MEAN	45.6	49.3	56.3	65.0	72.2	78.8	82.2	81.6	76.3	65.9	55.0	48.3	64.7
CENTERVILLE		MAX	57.7	62.2	70.1	77.8	84.3	91.3	95.6	95.9	89.7	81.2	69.2	62.0	78.1
		MIN	35.1	38.3	45.7	55.0	61.9	68.6	71.4	70.5	65.2	53.3	43.8	38.1	53.9
		MEAN	46.4	50.2	57.9	66.4	73.1	79.9	83.5	83.2	77.5	67.3	56.5	50.1	66.0
CHILDRESS FAA AIRPORT		MAX	51.8	56.9	65.0	75.5	82.9	91.7	95.6	94.3	86.2	76.3	62.7	54.9	74.5
		MIN	26.2	30.4	37.3	48.3	57.7	66.9	71.0	69.4	61.8	50.1	37.5	29.7	48.9
		MEAN	39.0	43.7	51.2	62.0	70.3	79.3	83.3	81.8	74.0	63.3	50.1	42.3	61.7
CHISOS BASIN		MAX	59.6	62.6	69.6	77.4	83.5	87.7	85.9	84.5	80.2	74.5	65.6	60.2	74.3
		MIN	37.0	38.7	43.9	51.6	57.6	62.7	63.3	62.1	58.2	51.2	42.4	37.9	50.6
		MEAN	48.3	50.7	56.8	64.5	70.6	75.2	74.6	73.3	69.2	62.9	54.0	49.1	62.4
CLARENDON		MAX	51.0	55.9	63.5	74.4	81.9	90.5	95.2	93.8	85.8	75.8	61.8	54.5	73.7
		MIN	21.9	25.9	32.1	43.1	52.7	62.0	66.0	64.1	56.8	44.8	32.5	25.5	44.0
		MEAN	36.4	40.9	47.8	58.8	67.3	76.3	80.6	79.0	71.3	60.3	47.2	40.0	58.8
CLARKSVILLE 2 NE		MAX	52.8	57.6	65.4	74.8	81.4	88.7	93.4	93.2	87.1	78.1	65.3	56.6	74.5
		MIN	30.1	34.0	41.2	51.0	59.5	67.0	70.5	69.1	62.9	50.9	40.4	32.9	50.8
		MEAN	41.5	45.8	53.4	62.9	70.5	77.9	82.0	81.2	75.0	64.5	52.9	44.8	62.7
CLEBURNE		MAX	57.6	63.0	70.7	78.8	84.8	92.9	97.6	97.6	90.2	80.7	67.9	61.0	78.6
		MIN	33.4	37.1	44.3	54.0	61.5	68.7	72.1	71.3	65.6	54.4	43.6	36.4	53.5
		MEAN	45.5	50.1	57.5	66.4	73.2	80.9	84.9	84.5	77.9	67.6	55.8	48.7	66.1
COLEMAN		MAX	59.4	64.1	72.0	80.4	85.6	92.6	96.4	95.8	88.7	79.5	67.3	61.5	78.6
		MIN	32.0	35.8	42.6	52.4	59.9	67.1	70.1	68.9	63.1	52.5	42.0	35.0	51.8
		MEAN	45.7	49.9	57.4	66.4	72.8	79.9	83.3	82.4	75.9	66.0	54.7	48.3	65.2
COLLEGE STATION FAA AP		MAX	59.0	63.4	70.7	78.1	84.5	91.1	94.8	95.0	89.0	80.6	69.1	62.1	78.1
		MIN	39.3	42.5	49.1	57.9	64.8	71.0	73.5	73.0	68.5	57.9	47.9	41.6	57.3
		MEAN	49.2	53.0	59.9	68.0	74.7	81.1	84.2	84.0	78.7	69.3	58.6	51.9	67.7
CONROE		MAX	60.5	64.5	72.0	79.1	85.3	91.3	94.6	94.5	89.5	81.9	70.9	63.7	79.0
		MIN	38.3	40.9	47.9	56.9	63.7	69.6	71.9	71.4	67.2	55.9	46.5	40.4	55.9
		MEAN	49.5	52.7	59.9	68.0	74.5	80.5	83.3	83.0	78.4	68.9	58.7	52.1	67.5
CORPUS CHRISTI		MAX	63.8	67.4	72.8	78.4	83.4	88.0	89.8	89.7	87.6	81.4	72.8	66.7	78.5
		MIN	48.5	52.3	59.6	67.3	72.9	77.7	79.1	78.9	75.7	68.0	58.3	51.7	65.8
		MEAN	56.2	59.9	66.2	72.9	78.2	82.9	84.4	84.3	81.7	74.7	65.6	59.2	72.2
CORPUS CHRISTI WSO	R	MAX	66.5	69.9	76.1	82.1	86.7	91.2	94.2	94.1	90.1	83.9	75.1	69.3	81.6
		MIN	46.1	48.7	55.7	63.9	69.5	74.1	75.6	75.8	72.8	64.1	54.9	48.8	62.5
		MEAN	56.3	59.3	65.9	73.0	78.1	82.7	84.9	85.0	81.5	74.0	65.0	59.1	72.1

TEXAS

TEMPERATURE NORMALS (DEG F)

STATION		JAN	FEB	MAR	APR	MAY	JUN	JUL	AUG	SEP	OCT	NOV	DEC	ANN
CORSICANA	MAX	55.3	59.9	67.8	76.5	83.2	91.1	96.3	96.5	89.6	80.1	67.6	59.4	76.9
	MIN	33.6	37.1	44.0	54.0	61.9	69.4	73.0	72.1	66.1	54.6	43.3	36.5	53.8
	MEAN	44.5	48.5	56.0	65.3	72.6	80.3	84.7	84.3	77.9	67.4	55.5	48.0	65.4
COTULLA FAA AIRPORT	MAX	64.9	69.9	77.9	85.4	90.4	95.9	98.6	98.1	92.2	84.1	73.6	67.3	83.2
	MIN	41.7	45.4	52.7	61.2	67.3	72.6	74.3	74.0	70.4	61.2	50.7	43.9	59.6
	MEAN	53.3	57.7	65.3	73.3	78.9	84.3	86.5	86.1	81.3	72.7	62.2	55.6	71.4
CROCKETT	MAX	57.6	61.8	69.6	77.5	84.1	90.5	94.5	94.8	89.0	80.7	68.8	61.3	77.5
	MIN	34.8	37.5	44.2	53.5	61.0	67.6	70.6	69.8	65.2	53.6	43.3	36.9	53.2
	MEAN	46.2	49.7	57.0	65.5	72.6	79.1	82.5	82.3	77.1	67.1	56.1	49.1	65.4
CROSBYTON	MAX	52.4	57.0	64.9	75.3	82.1	90.3	93.3	92.0	84.5	75.0	62.0	55.2	73.7
	MIN	23.5	26.8	33.4	44.4	53.8	63.2	66.6	64.7	57.9	46.7	34.4	27.0	45.2
	MEAN	38.0	41.9	49.2	59.8	68.0	76.8	80.0	78.4	71.3	60.9	48.2	41.1	59.5
CRYSTAL CITY	MAX	65.1	70.4	78.6	85.2	89.9	95.4	97.8	97.3	92.0	83.5	72.9	66.9	82.9
	MIN	41.1	45.2	52.8	61.2	67.4	72.7	74.3	74.0	70.1	61.0	50.1	43.1	59.4
	MEAN	53.1	57.8	65.7	73.3	78.7	84.1	86.1	85.6	81.1	72.3	61.6	55.1	71.2
DALHART FAA AIRPORT	MAX	48.8	53.3	59.8	69.8	78.2	88.2	91.4	89.2	81.7	72.1	58.2	51.5	70.2
	MIN	18.4	22.8	28.3	38.7	48.8	58.9	63.8	62.0	53.6	41.2	28.2	21.3	40.5
	MEAN	33.6	38.0	44.1	54.3	63.5	73.5	77.6	75.6	67.7	56.7	43.2	36.4	55.4
DALLAS-FORT WORTH REG WSO	MAX	54.0	59.1	67.2	76.8	84.4	93.2	97.8	97.3	89.7	79.5	66.2	58.1	76.9
	MIN	33.9	37.8	44.9	55.0	62.9	70.8	74.7	73.7	67.5	56.3	44.9	37.4	55.0
	MEAN	44.0	48.5	56.1	65.9	73.7	82.0	86.3	85.5	78.6	67.9	55.6	47.8	66.0
DALLAS FAA //R	MAX	55.0	59.9	68.0	76.9	83.8	91.8	96.3	96.0	88.7	79.0	66.2	58.7	76.7
	MIN	34.9	38.8	46.0	56.1	64.1	72.2	76.3	75.4	68.2	56.9	45.3	38.2	56.0
	MEAN	45.0	49.4	57.0	66.5	74.0	82.0	86.3	85.7	78.5	68.0	55.8	48.5	66.4
DANEVANG 2 SE	MAX	64.1	66.8	73.9	80.1	85.2	90.0	93.3	93.5	89.2	83.2	73.3	67.1	80.0
	MIN	44.3	46.4	52.9	60.9	66.6	71.5	73.1	73.1	69.9	60.5	52.0	46.3	59.8
	MEAN	54.2	56.6	63.4	70.5	75.9	80.8	83.2	83.3	79.5	71.9	62.7	56.7	69.9
DEL RIO WSO	MAX	63.2	68.6	76.5	84.2	89.1	95.1	97.7	97.0	91.7	82.4	71.2	64.8	81.8
	MIN	38.3	42.5	50.2	59.4	66.1	72.1	74.2	73.6	68.9	59.2	47.5	40.1	57.7
	MEAN	50.8	55.6	63.4	71.8	77.6	83.6	86.0	85.3	80.3	70.8	59.4	52.5	69.8
DENISON DAM	MAX	52.2	57.3	65.1	74.5	81.6	89.7	94.7	94.4	87.0	77.4	64.5	56.0	74.5
	MIN	30.5	34.8	41.9	52.4	60.3	68.4	72.2	70.8	64.0	53.3	42.4	34.4	52.1
	MEAN	41.4	46.1	53.6	63.5	71.0	79.1	83.5	82.6	75.6	65.3	53.5	45.2	63.4
DENTON 2 SE	MAX	56.0	61.1	69.0	77.4	83.6	91.3	95.9	95.8	89.0	79.6	66.9	59.3	77.1
	MIN	31.8	35.9	43.2	53.4	61.1	69.1	73.0	71.8	65.3	54.5	43.0	35.5	53.1
	MEAN	43.9	48.5	56.1	65.4	72.4	80.2	84.5	83.8	77.2	67.0	55.0	47.4	65.1
DILLEY	MAX	64.2	68.7	76.9	84.6	89.1	95.1	98.4	98.2	92.6	84.3	73.6	66.7	82.7
	MIN	38.5	42.1	49.5	58.6	65.4	70.8	72.1	72.0	68.5	58.6	48.1	41.2	57.1
	MEAN	51.4	55.4	63.2	71.6	77.2	83.0	85.3	85.1	80.6	71.5	60.9	54.0	69.9
DUBLIN	MAX	55.0	59.5	67.4	76.1	81.9	90.0	94.8	94.5	87.5	78.3	65.3	58.6	75.7
	MIN	31.1	34.7	41.4	52.1	59.9	67.5	71.0	70.2	64.3	53.5	42.1	34.5	51.9
	MEAN	43.1	47.1	54.5	64.1	70.9	78.8	82.9	82.4	75.9	65.9	53.7	46.6	63.8
EAGLE PASS	MAX	63.9	69.3	77.5	85.6	90.2	96.3	99.1	98.7	92.9	83.9	72.8	66.0	83.0
	MIN	38.1	42.5	50.1	59.8	66.5	72.3	74.4	73.7	69.3	59.2	47.4	39.8	57.8
	MEAN	51.0	55.9	63.9	72.8	78.4	84.4	86.8	86.2	81.1	71.6	60.1	52.9	70.4
EDEN 1	MAX	61.1	65.7	73.6	81.8	87.0	94.0	97.0	96.1	89.5	81.2	69.1	63.1	79.9
	MIN	33.0	36.7	43.5	52.7	59.8	66.5	69.1	68.1	62.9	53.2	42.1	35.2	51.9
	MEAN	47.1	51.2	58.6	67.3	73.5	80.2	83.1	82.1	76.2	67.2	55.7	49.2	66.0
EL PASO WSO R	MAX	57.9	62.7	69.6	78.7	87.1	95.9	95.3	93.0	87.5	78.5	65.7	58.2	77.5
	MIN	30.4	34.1	40.5	48.5	56.6	65.7	69.6	67.5	60.6	48.7	37.0	30.6	49.2
	MEAN	44.2	48.4	55.1	63.6	71.9	80.8	82.5	80.3	74.1	63.6	51.4	44.4	63.4

TEXAS

TEMPERATURE NORMALS (DEG F)

STATION		JAN	FEB	MAR	APR	MAY	JUN	JUL	AUG	SEP	OCT	NOV	DEC	ANN
ENCINAL 3 NW	MAX	66.2	70.7	79.4	87.4	91.9	97.3	100.1	99.7	94.2	86.0	75.5	68.5	84.7
	MIN	39.2	42.8	50.0	59.3	65.6	70.7	71.9	71.7	67.9	58.4	48.1	41.1	57.2
	MEAN	52.7	56.8	64.7	73.4	78.8	84.0	86.1	85.7	81.1	72.2	61.8	54.8	71.0
FALFURRIAS	MAX	68.0	72.3	79.3	86.1	90.2	94.8	97.6	97.4	92.4	85.6	76.2	70.1	84.2
	MIN	43.5	46.0	53.6	62.5	67.9	72.1	73.3	72.7	69.8	60.9	51.7	45.8	60.0
	MEAN	55.8	59.2	66.5	74.3	79.0	83.5	85.5	85.0	81.1	73.3	63.9	57.9	72.1
FLATONIA 2 W	MAX	62.3	66.4	74.0	80.4	85.7	91.7	95.4	95.7	90.1	82.8	72.0	65.5	80.2
	MIN	40.7	43.7	50.3	58.8	64.7	70.1	72.3	72.1	67.9	59.0	49.3	43.0	57.7
	MEAN	51.5	55.1	62.2	69.6	75.2	80.9	83.9	83.9	79.0	70.9	60.7	54.3	68.9
FOLLETT	MAX	46.1	51.3	58.6	70.2	78.1	87.7	93.0	92.2	83.8	73.5	58.1	49.8	70.2
	MIN	20.0	24.1	30.5	42.0	52.2	61.9	66.8	65.1	56.9	45.2	32.2	24.1	43.4
	MEAN	33.1	37.7	44.6	56.1	65.2	74.8	79.9	78.7	70.3	59.4	45.2	37.0	56.8
FT STOCKTON KFST RADIO	MAX	60.6	64.5	72.2	81.5	87.9	94.5	95.0	94.2	88.2	80.2	68.8	62.6	79.2
	MIN	30.2	33.5	39.0	48.2	57.0	65.2	67.9	66.2	60.5	49.7	38.2	32.1	49.0
	MEAN	45.5	49.0	55.6	64.9	72.5	79.9	81.5	80.2	74.4	65.0	53.6	47.3	64.1
FREDERICKSBURG	MAX	60.6	64.6	72.4	79.2	84.2	90.7	94.5	94.2	88.2	79.7	68.5	62.9	78.3
	MIN	35.5	38.8	45.7	54.6	61.1	67.2	69.1	68.4	64.1	54.6	44.2	37.9	53.4
	MEAN	48.1	51.7	59.1	66.9	72.7	79.0	81.8	81.3	76.2	67.2	56.4	50.5	65.9
GAINESVILLE	MAX	52.2	57.6	65.8	75.5	82.6	90.9	96.1	95.9	88.2	78.1	64.8	56.5	75.4
	MIN	28.2	32.3	39.9	50.7	59.2	68.0	71.9	70.6	63.4	51.5	39.6	31.7	50.6
	MEAN	40.3	45.0	52.9	63.1	70.9	79.4	84.0	83.2	75.8	64.8	52.2	44.1	63.0
GALVESTON WSO R	MAX	59.2	60.9	66.4	73.3	79.8	85.1	87.3	87.5	84.6	77.6	68.3	62.3	74.4
	MIN	47.9	50.2	56.5	64.9	71.6	77.2	79.1	78.8	75.4	67.7	57.6	51.2	64.8
	MEAN	53.6	55.6	61.4	69.1	75.7	81.2	83.2	83.2	80.0	72.7	63.0	56.8	69.6
GATESVILLE	MAX	58.7	62.9	71.0	79.1	84.8	92.4	97.2	97.3	90.3	81.3	69.3	62.7	78.9
	MIN	33.6	37.0	44.6	54.0	61.5	68.6	71.8	71.1	66.1	54.8	43.5	36.8	53.6
	MEAN	46.2	50.0	57.8	66.6	73.2	80.5	84.5	84.2	78.2	68.1	56.4	49.8	66.3
GILMER 2 W	MAX	54.2	59.2	66.8	75.5	82.0	89.2	94.0	94.0	87.5	78.3	66.4	58.3	75.5
	MIN	32.3	35.7	42.4	51.9	59.9	67.2	70.7	69.5	63.6	51.7	41.5	34.9	51.8
	MEAN	43.3	47.5	54.6	63.7	71.0	78.3	82.4	81.8	75.6	65.1	54.0	46.7	63.7
GOLIAD	MAX	66.4	69.8	76.5	81.8	86.9	92.3	95.6	96.0	91.3	84.8	75.1	68.8	82.1
	MIN	43.4	46.4	53.2	60.9	66.6	71.6	73.4	73.2	69.8	60.5	51.8	45.7	59.7
	MEAN	54.9	58.1	64.9	71.4	76.8	82.0	84.5	84.6	80.6	72.7	63.5	57.3	70.9
GRAHAM	MAX	54.8	60.0	68.3	77.8	83.8	92.0	97.2	97.0	89.6	79.8	66.6	59.1	77.2
	MIN	28.0	32.3	40.0	50.9	59.5	68.2	72.0	70.6	63.4	51.0	39.1	30.9	50.5
	MEAN	41.5	46.2	54.2	64.4	71.7	80.1	84.6	83.8	76.5	65.4	52.9	45.0	63.9
GREENVILLE 7 NW	MAX	52.1	56.9	64.8	74.2	81.4	89.7	94.5	94.7	87.8	78.1	64.7	56.2	74.6
	MIN	30.3	34.1	41.4	51.9	60.3	68.0	71.7	70.5	64.4	52.5	41.2	33.5	51.7
	MEAN	41.2	45.5	53.1	63.1	70.8	78.9	83.1	82.6	76.1	65.3	53.0	44.9	63.1
HALLETTSVILLE	MAX	64.4	68.0	75.2	81.2	86.6	92.5	96.1	96.7	91.5	84.4	73.7	67.1	81.5
	MIN	40.9	43.9	50.9	59.0	64.9	70.3	72.1	71.7	67.7	57.9	49.0	43.1	57.6
	MEAN	52.6	56.0	63.1	70.1	75.8	81.4	84.1	84.2	79.6	71.1	61.4	55.1	69.5
HARLINGEN	MAX	69.4	72.7	79.0	85.2	89.1	92.9	95.1	95.8	91.6	85.7	77.2	71.8	83.8
	MIN	48.0	50.4	57.1	64.6	69.3	72.5	73.4	73.3	70.9	63.3	55.8	50.2	62.4
	MEAN	58.7	61.6	68.1	75.0	79.2	82.7	84.2	84.6	81.2	74.6	66.5	61.0	73.1
HASKELL	MAX	54.5	59.2	67.8	78.3	84.9	93.0	96.9	96.3	88.3	78.4	65.1	57.9	76.7
	MIN	28.0	32.0	39.0	50.6	59.5	68.3	72.2	70.7	63.6	52.0	39.7	31.4	50.6
	MEAN	41.2	45.6	53.4	64.5	72.2	80.7	84.6	83.6	76.0	65.2	52.4	44.7	63.7
HENDERSON	MAX	56.5	60.8	68.3	76.6	82.6	89.3	93.1	92.8	86.6	78.4	67.3	60.0	76.0
	MIN	35.0	37.9	44.3	53.6	61.1	68.3	71.6	70.8	65.6	54.1	43.9	37.4	53.6
	MEAN	45.8	49.3	56.3	65.1	71.9	78.8	82.4	81.9	76.1	66.3	55.7	48.7	64.9

TEXAS

TEMPERATURE NORMALS (DEG F)

STATION		JAN	FEB	MAR	APR	MAY	JUN	JUL	AUG	SEP	OCT	NOV	DEC	ANN
HENRIETTA	MAX	53.1	58.4	66.5	76.5	83.2	91.9	97.4	97.0	89.2	78.9	65.4	57.3	76.2
	MIN	27.3	31.6	39.1	50.4	59.6	68.3	72.6	71.1	63.5	51.4	39.3	31.1	50.4
	MEAN	40.2	45.0	52.8	63.5	71.5	80.1	85.0	84.1	76.4	65.2	52.4	44.2	63.4
HEREFORD	MAX	50.3	54.0	61.3	71.2	79.1	88.1	90.4	88.6	82.2	73.4	59.8	52.9	70.9
	MIN	20.5	24.0	29.5	39.6	49.5	59.1	63.4	61.5	53.9	42.1	29.9	23.3	41.4
	MEAN	35.5	39.0	45.5	55.5	64.3	73.6	77.0	75.1	68.1	57.8	44.9	38.1	56.2
HICO	MAX	58.9	63.3	71.1	79.3	84.7	92.6	97.3	97.1	90.3	80.9	68.3	61.9	78.8
	MIN	32.6	36.2	43.4	53.5	60.8	68.1	71.6	70.5	64.4	53.3	42.5	35.5	52.7
	MEAN	45.8	49.8	57.3	66.4	72.8	80.4	84.5	83.8	77.4	67.1	55.4	48.7	65.8
HILLSBORO	MAX	55.6	60.5	68.8	77.2	83.7	91.6	96.0	96.2	89.5	80.0	67.6	60.3	77.3
	MIN	32.8	36.7	44.0	54.3	62.1	69.6	73.3	72.4	66.4	54.9	43.2	36.2	53.8
	MEAN	44.3	48.6	56.4	65.8	72.9	80.6	84.7	84.3	78.0	67.5	55.4	48.3	65.6
HOUSTON INCONT AP	MAX	61.9	65.7	72.1	79.0	85.1	90.9	93.6	93.1	88.7	81.9	71.6	65.2	79.1
	MIN	40.8	43.2	49.8	58.3	64.7	70.2	72.5	72.1	68.1	57.5	48.6	42.7	57.4
	MEAN	51.4	54.5	61.0	68.7	74.9	80.6	83.1	82.6	78.4	69.7	60.1	54.0	68.3
HUNTSVILLE	MAX	58.4	62.9	70.4	78.2	84.8	91.5	95.1	94.7	89.0	81.0	69.3	62.0	78.1
	MIN	38.3	41.1	48.1	57.2	63.9	69.8	72.1	71.4	66.6	56.8	47.2	40.8	56.1
	MEAN	48.4	52.0	59.3	67.7	74.4	80.6	83.6	83.1	77.8	68.9	58.3	51.4	67.1
JACKSBORO	MAX	57.2	62.3	70.2	79.0	84.9	92.4	96.9	96.5	89.7	80.3	67.6	60.7	78.1
	MIN	31.8	35.9	43.1	53.4	61.0	69.0	73.2	72.0	65.2	54.6	42.9	35.6	53.1
	MEAN	44.5	49.1	56.6	66.2	73.0	80.7	85.1	84.3	77.5	67.5	55.2	48.2	65.7
JUNCTION	MAX	62.1	66.4	74.2	81.3	86.9	93.3	96.7	95.9	90.0	81.3	70.2	64.1	80.2
	MIN	31.7	35.4	43.2	52.7	59.9	67.0	69.1	68.0	62.6	51.5	40.2	33.2	51.2
	MEAN	46.9	51.0	58.7	67.0	73.4	80.2	82.9	82.0	76.3	66.4	55.2	48.7	65.7
KAUFMAN 3 SE	MAX	54.3	58.8	67.0	76.0	83.1	91.2	96.3	96.5	89.5	79.9	66.8	58.0	76.5
	MIN	32.1	35.9	42.8	52.7	60.9	68.6	72.5	71.5	65.5	53.5	42.4	35.2	52.8
	MEAN	43.2	47.4	54.9	64.4	72.0	80.0	84.4	84.0	77.6	66.7	54.6	46.6	64.7
KINGSVILLE	MAX	69.2	72.7	79.4	84.5	88.5	92.9	95.4	95.4	91.5	85.6	77.0	71.8	83.7
	MIN	45.6	48.0	55.1	62.8	68.4	72.7	74.1	73.8	70.4	61.8	53.4	47.6	61.1
	MEAN	57.5	60.4	67.3	73.7	78.5	82.8	84.8	84.6	81.0	73.7	65.3	59.7	72.4
LA TUNA 1 S	MAX	58.9	63.8	70.1	78.8	87.7	96.3	96.4	93.7	88.8	79.8	67.2	59.3	78.4
	MIN	29.1	31.9	38.0	46.0	54.7	64.6	69.0	66.5	60.1	48.6	35.8	29.1	47.8
	MEAN	44.0	47.9	54.1	62.4	71.2	80.5	82.7	80.1	74.5	64.2	51.5	44.2	63.1
LAMESA 1 SSE	MAX	55.5	59.8	67.7	77.8	85.5	93.6	95.1	93.9	86.8	77.5	64.9	58.2	76.4
	MIN	25.1	28.6	35.1	45.9	55.0	64.1	66.9	65.3	59.3	47.4	35.3	28.0	46.3
	MEAN	40.3	44.2	51.4	61.9	70.3	78.9	81.0	79.6	73.1	62.4	50.1	43.2	61.4
LAMPASAS	MAX	58.0	62.2	70.0	78.1	83.6	91.3	95.9	95.7	88.9	80.1	67.9	61.4	77.8
	MIN	31.1	34.6	41.3	51.7	59.9	67.5	70.4	69.3	63.6	51.6	40.4	33.4	51.2
	MEAN	44.6	48.4	55.7	65.0	71.7	79.4	83.2	82.5	76.3	65.9	54.2	47.5	64.5
LAREDO NO 2	MAX	67.7	72.5	80.4	88.0	92.5	97.1	99.3	98.8	93.6	85.9	75.9	69.3	85.1
	MIN	44.5	48.3	56.1	64.2	69.6	73.9	75.5	75.0	71.6	63.4	53.1	46.6	61.8
	MEAN	56.2	60.4	68.3	76.1	81.1	85.5	87.4	86.9	82.6	74.7	64.5	58.0	73.5
LEVELLAND	MAX	54.6	58.4	66.4	76.0	84.0	91.8	92.2	90.6	83.9	75.7	63.4	56.8	74.5
	MIN	23.4	26.6	32.6	43.0	52.7	61.7	64.7	63.1	56.3	44.4	32.6	26.3	44.0
	MEAN	39.0	42.5	49.5	59.5	68.4	76.8	78.5	76.9	70.1	60.0	48.0	41.6	59.2
LIBERTY	MAX	61.5	65.1	71.7	78.7	85.5	91.1	93.9	94.0	89.9	82.6	72.0	64.9	79.2
	MIN	39.8	42.3	49.2	57.7	64.1	69.5	71.8	71.2	67.0	55.7	47.3	41.5	56.4
	MEAN	50.7	53.7	60.5	68.2	74.8	80.3	82.9	82.6	78.5	69.2	59.6	53.2	67.9
LIVINGSTON 2 NNE	MAX	59.5	63.4	70.6	78.0	84.4	90.6	94.0	93.9	89.1	81.2	70.0	62.9	78.1
	MIN	37.2	39.1	46.0	55.2	62.3	68.3	71.0	70.3	65.7	54.2	45.1	38.7	54.4
	MEAN	48.4	51.3	58.3	66.6	73.4	79.5	82.6	82.2	77.4	67.7	57.6	50.8	66.3

TEXAS

TEMPERATURE NORMALS (DEG F)

STATION			JAN	FEB	MAR	APR	MAY	JUN	JUL	AUG	SEP	OCT	NOV	DEC	ANN
LLANO		MAX	59.9	63.7	71.6	79.8	85.0	93.0	97.2	96.7	90.1	81.1	69.3	63.0	79.2
		MIN	31.9	35.8	43.8	54.6	62.2	69.6	72.1	70.8	65.1	53.4	41.8	33.9	52.9
		MEAN	46.0	49.8	57.7	67.2	73.6	81.3	84.7	83.8	77.6	67.3	55.6	48.4	66.1
LUBBOCK WSO	R	MAX	53.3	57.3	65.1	74.8	82.8	90.8	91.9	90.1	83.6	74.7	62.1	55.5	73.5
		MIN	24.3	27.9	35.2	45.8	55.2	64.3	67.6	65.7	58.7	47.3	34.8	27.4	46.2
		MEAN	38.8	42.6	50.2	60.3	69.0	77.6	79.8	77.9	71.2	61.0	48.5	41.5	59.9
LUFKIN FAA AP		MAX	59.2	63.5	71.1	78.6	84.8	90.6	93.8	93.8	88.8	81.0	69.4	62.4	78.1
		MIN	38.0	40.5	47.5	56.5	63.8	70.0	72.8	71.9	67.1	55.4	45.4	39.5	55.7
		MEAN	48.6	52.0	59.3	67.6	74.3	80.3	83.3	82.9	78.0	68.2	57.4	51.0	66.9
LULING		MAX	60.8	65.0	72.8	79.7	85.5	91.9	95.7	96.3	90.3	82.6	71.3	64.2	79.7
		MIN	37.0	40.6	47.6	57.0	64.0	70.0	72.0	71.5	67.0	56.2	46.2	39.6	55.7
		MEAN	48.9	52.9	60.2	68.3	74.7	81.0	83.9	83.9	78.7	69.4	58.8	52.0	67.7
MADISONVILLE		MAX	60.8	64.7	72.1	79.5	85.8	92.4	96.3	96.4	90.2	82.1	70.7	63.9	79.6
		MIN	38.8	41.5	48.2	56.7	63.5	69.6	72.3	71.7	66.8	56.2	46.6	40.9	56.1
		MEAN	49.8	53.2	60.2	68.1	74.7	81.0	84.3	84.0	78.5	69.2	58.7	52.4	67.8
MARLIN 3 NE		MAX	59.0	64.0	71.4	78.8	84.9	92.0	96.5	96.9	90.6	81.8	69.6	62.6	79.0
		MIN	35.8	39.5	46.3	55.4	62.7	69.3	72.3	71.7	66.7	55.7	45.0	38.6	54.9
		MEAN	47.4	51.8	58.9	67.1	73.8	80.7	84.5	84.3	78.7	68.7	57.3	50.6	67.0
MARSHALL		MAX	54.8	59.7	67.3	76.4	83.4	90.1	94.1	94.0	88.1	79.0	66.5	58.5	76.0
		MIN	33.3	36.7	43.4	53.0	60.8	67.9	71.5	70.4	64.8	52.5	42.3	35.6	52.7
		MEAN	44.1	48.2	55.4	64.7	72.1	79.0	82.8	82.2	76.5	65.7	54.4	47.1	64.4
MATADOR		MAX	54.0	58.4	66.2	76.5	83.4	91.4	95.5	94.2	86.5	77.3	64.1	57.5	75.4
		MIN	26.5	30.3	36.9	47.8	56.6	65.7	69.9	68.2	60.8	49.7	37.2	30.1	48.3
		MEAN	40.3	44.4	51.6	62.2	70.0	78.6	82.7	81.2	73.7	63.5	50.7	43.8	61.9
MATAGORDA NO 2		MAX	63.7	66.0	71.6	77.8	83.7	88.8	91.4	91.8	88.6	82.2	73.2	67.0	78.8
		MIN	45.8	48.3	54.8	62.8	69.4	74.9	76.9	75.8	71.9	62.7	53.9	48.1	62.1
		MEAN	54.8	57.2	63.2	70.3	76.6	81.8	84.1	83.8	80.2	72.5	63.6	57.6	70.5
MC ALLEN		MAX	69.9	73.5	80.1	86.2	89.6	93.2	95.1	96.2	92.6	86.3	77.6	71.9	84.4
		MIN	47.6	50.1	57.1	64.6	69.4	72.7	73.6	73.8	71.1	63.5	55.3	49.5	62.4
		MEAN	58.8	61.8	68.6	75.4	79.5	83.0	84.4	85.0	81.9	74.9	66.5	60.7	73.4
MC CAMEY		MAX	60.3	64.8	73.6	82.6	89.1	95.0	96.3	95.7	89.1	80.1	67.9	61.8	79.7
		MIN	31.2	35.1	42.9	52.7	60.8	69.2	71.8	70.6	64.6	53.7	41.1	33.2	52.2
		MEAN	45.8	50.0	58.3	67.7	75.0	82.1	84.1	83.2	76.9	67.0	54.5	47.5	66.0
MC COOK		MAX	69.5	73.6	80.8	87.4	90.1	94.0	96.4	97.5	92.7	86.0	77.2	71.5	84.7
		MIN	45.6	48.1	55.5	63.6	68.4	71.9	72.8	72.5	69.9	62.0	53.5	47.8	61.0
		MEAN	57.6	60.9	68.2	75.5	79.3	83.0	84.6	85.0	81.3	74.1	65.4	59.7	72.9
MCKINNEY 3 S		MAX	54.9	60.2	68.2	76.7	83.1	90.8	95.9	95.9	88.7	79.2	66.3	58.5	76.5
		MIN	32.6	36.6	43.8	53.6	61.4	69.0	72.6	71.3	65.2	54.3	43.2	35.8	53.3
		MEAN	43.8	48.4	56.0	65.2	72.3	79.9	84.3	83.6	77.0	66.8	54.8	47.2	64.9
MEMPHIS		MAX	52.4	57.7	65.1	75.8	83.2	91.8	96.5	95.3	87.2	77.1	63.1	56.0	75.1
		MIN	24.5	28.7	35.0	46.3	55.7	65.1	69.4	67.7	60.0	47.8	35.4	27.9	47.0
		MEAN	38.4	43.2	50.1	61.1	69.5	78.5	83.0	81.5	73.6	62.5	49.3	42.0	61.1
MEXIA		MAX	55.8	60.5	68.4	76.4	82.8	90.4	95.0	95.9	89.2	79.8	67.3	59.7	76.8
		MIN	34.4	38.1	45.2	55.0	62.6	69.6	73.1	72.3	67.0	55.9	44.8	37.4	54.6
		MEAN	45.1	49.3	56.8	65.7	72.7	80.0	84.1	84.1	78.1	67.9	56.1	48.6	65.7
MIAMI		MAX	48.3	52.9	60.2	71.6	79.4	88.8	93.8	92.4	84.6	74.2	59.6	52.0	71.5
		MIN	19.2	23.8	30.5	42.2	52.1	61.8	66.5	64.7	56.3	43.6	30.7	22.9	42.9
		MEAN	33.8	38.4	45.3	56.9	65.8	75.3	80.2	78.6	70.4	58.9	45.1	37.5	57.2
MIDLAND WSO	//R	MAX	57.6	62.1	69.8	78.8	86.0	93.0	94.2	93.1	86.4	77.7	65.5	59.7	77.0
		MIN	29.7	33.3	40.2	49.4	58.2	66.6	69.2	68.0	61.9	51.1	39.0	32.2	49.9
		MEAN	43.7	47.7	55.0	64.1	72.1	79.8	81.7	80.6	74.2	64.4	52.3	46.0	63.5

TEXAS

TEMPERATURE NORMALS (DEG F)

STATION		JAN	FEB	MAR	APR	MAY	JUN	JUL	AUG	SEP	OCT	NOV	DEC	ANN
MIDLAND 4 ENE	MAX	59.4	64.1	72.1	80.8	87.5	93.9	94.8	93.8	87.6	79.2	67.0	61.0	78.4
	MIN	29.2	32.9	39.7	49.8	58.2	66.3	68.7	67.6	61.3	50.1	38.3	31.4	49.5
	MEAN	44.3	48.5	55.9	65.3	72.9	80.1	81.7	80.7	74.5	64.7	52.7	46.2	64.0
MINERAL WELLS FAA AP	MAX	55.6	60.9	69.0	77.9	84.5	92.8	97.3	96.6	89.2	79.3	66.4	59.5	77.4
	MIN	32.1	35.9	43.5	53.6	61.4	69.2	72.9	71.9	65.5	54.4	42.6	35.1	53.2
	MEAN	43.9	48.5	56.3	65.8	73.0	81.0	85.1	84.3	77.4	66.9	54.5	47.3	65.3
MISSION 4 W	MAX	69.4	73.3	80.5	86.8	90.4	94.1	96.2	97.2	93.1	85.9	77.2	71.3	84.6
	MIN	46.3	48.7	56.2	64.5	69.4	72.9	74.0	73.8	71.3	63.2	54.6	48.2	61.9
	MEAN	57.9	61.0	68.4	75.7	79.9	83.5	85.1	85.5	82.2	74.5	65.9	59.8	73.3
MOUNT LOCKE	MAX	53.5	56.0	63.0	71.3	78.2	84.5	82.6	80.8	76.1	69.9	60.6	54.6	69.3
	MIN	32.0	33.0	37.5	45.1	51.7	57.8	58.6	57.7	54.2	47.3	37.7	33.2	45.5
	MEAN	42.7	44.5	50.3	58.2	65.0	71.2	70.6	69.2	65.2	58.7	49.2	43.9	57.4
MOUNT PLEASANT	MAX	53.6	58.7	66.4	75.8	83.0	90.5	94.8	94.6	88.1	79.2	66.1	57.7	75.7
	MIN	29.8	33.3	40.3	50.5	59.1	66.6	69.8	68.3	62.0	49.2	38.7	32.2	50.0
	MEAN	41.7	46.0	53.4	63.1	71.1	78.6	82.3	81.4	75.1	64.2	52.4	45.0	62.9
MULESHOE 1	MAX	52.2	56.1	63.5	73.4	81.3	89.9	91.4	89.4	83.0	73.9	60.9	54.5	72.5
	MIN	20.0	23.4	29.7	40.3	49.9	59.7	63.7	61.7	54.1	41.6	29.5	22.5	41.3
	MEAN	36.1	39.7	46.6	56.9	65.7	74.8	77.6	75.5	68.6	57.8	45.2	38.5	56.9
MUNDAY	MAX	56.7	62.1	70.5	80.3	86.7	94.3	98.3	97.3	89.5	79.9	66.9	59.6	78.5
	MIN	29.0	33.0	40.0	50.8	59.5	67.9	71.6	69.8	62.9	51.9	40.0	32.3	50.7
	MEAN	42.9	47.6	55.2	65.6	73.1	81.1	84.9	83.6	76.2	65.9	53.5	46.0	64.6
NEW BRAUNFELS	MAX	61.9	66.6	74.4	81.2	86.5	93.1	96.4	96.5	90.7	82.5	71.3	64.9	80.5
	MIN	37.8	41.5	48.1	57.1	64.1	70.7	72.7	72.2	68.0	57.6	47.1	40.2	56.4
	MEAN	49.9	54.0	61.3	69.2	75.3	81.9	84.6	84.4	79.4	70.1	59.2	52.6	68.5
NEW GULF	MAX	62.6	65.8	72.7	79.5	85.3	90.7	93.3	93.3	89.4	82.9	72.6	65.9	79.5
	MIN	41.3	44.5	51.4	60.1	65.8	71.0	72.9	72.5	68.9	58.9	50.2	43.7	58.4
	MEAN	52.0	55.2	62.1	69.8	75.6	80.9	83.1	82.9	79.2	70.9	61.5	54.8	69.0
NIXON	MAX	64.1	68.3	75.6	81.6	87.4	93.0	96.6	97.0	91.3	84.3	73.9	67.6	81.7
	MIN	40.6	43.5	50.4	58.8	64.8	70.3	72.4	72.0	67.8	58.1	49.0	42.7	57.5
	MEAN	52.3	55.9	63.1	70.3	76.1	81.7	84.5	84.5	79.6	71.2	61.5	55.2	69.7
PALACIOS FAA AIRPORT	MAX	62.5	65.2	71.5	77.7	83.3	88.4	90.6	90.8	87.9	81.5	71.9	65.6	78.1
	MIN	43.9	46.5	53.1	61.9	68.7	74.8	77.2	76.1	71.2	61.4	52.1	46.2	61.1
	MEAN	53.2	55.9	62.3	69.8	76.0	81.6	83.9	83.5	79.6	71.4	62.0	55.9	69.6
PALESTINE	MAX	55.8	60.6	68.2	76.5	83.2	90.1	94.6	94.9	88.6	80.0	67.4	59.7	76.6
	MIN	35.1	38.2	45.5	55.0	62.2	68.8	71.4	70.5	65.4	55.1	44.7	37.9	54.2
	MEAN	45.5	49.4	56.9	65.8	72.8	79.5	83.0	82.7	77.0	67.5	56.1	48.8	65.4
PARIS	MAX	51.0	56.2	64.3	74.1	81.6	89.3	94.3	94.1	86.9	77.5	64.2	55.3	74.1
	MIN	30.2	34.3	41.8	52.5	61.0	68.8	72.8	71.4	64.7	53.2	41.8	33.8	52.2
	MEAN	40.6	45.3	53.1	63.3	71.3	79.1	83.5	82.8	75.9	65.3	53.0	44.6	63.2
PECOS	MAX	60.9	66.2	74.5	84.4	91.9	99.3	99.7	98.5	91.7	82.5	69.6	63.0	81.9
	MIN	27.4	30.9	37.7	47.7	56.5	66.4	69.4	67.3	61.1	48.5	35.8	28.9	48.1
	MEAN	44.2	48.6	56.1	66.1	74.2	82.9	84.6	82.9	76.4	65.5	52.7	46.0	65.0
PIERCE 1 E	MAX	62.6	66.0	72.9	79.6	85.1	90.7	93.4	93.9	89.4	82.8	72.8	65.6	79.6
	MIN	40.1	42.6	49.5	58.2	64.5	70.0	71.7	71.4	68.0	57.9	49.0	42.2	57.1
	MEAN	51.4	54.4	61.2	68.9	74.8	80.4	82.6	82.7	78.7	70.4	60.9	53.9	68.4
PLAINS	MAX	54.4	58.2	65.7	75.4	83.1	91.5	92.4	91.2	84.7	75.9	62.6	56.5	74.3
	MIN	23.0	25.8	32.3	42.4	51.6	60.8	63.8	62.2	55.6	43.9	31.8	24.9	43.2
	MEAN	38.7	42.0	49.0	58.9	67.4	76.2	78.1	76.7	70.1	59.9	47.2	40.7	58.7
PLAINVIEW	MAX	51.9	56.2	63.6	73.7	81.6	90.1	92.4	90.5	83.9	74.9	61.6	54.8	72.9
	MIN	22.8	26.4	32.5	42.9	52.5	62.0	65.7	63.8	56.8	45.1	33.1	26.2	44.2
	MEAN	37.4	41.3	48.1	58.3	67.1	76.1	79.0	77.2	70.3	60.1	47.4	40.5	58.6

TEXAS

TEMPERATURE NORMALS (DEG F)

STATION			JAN	FEB	MAR	APR	MAY	JUN	JUL	AUG	SEP	OCT	NOV	DEC	ANN
PORT ARTHUR WSO	R	MAX	61.7	65.4	71.8	78.5	85.0	90.5	92.5	92.2	88.6	81.5	71.4	65.0	78.7
		MIN	42.1	44.3	51.0	59.4	66.1	71.8	73.7	73.3	69.8	58.9	49.8	44.4	58.7
		MEAN	51.9	54.9	61.4	69.0	75.6	81.2	83.1	82.8	79.2	70.2	60.6	54.7	68.7
PORT LAVACA NO 2		MAX	63.5	66.7	72.9	79.6	85.0	90.6	93.3	93.4	89.4	82.8	73.5	67.0	79.8
		MIN	44.1	47.6	54.4	62.7	68.9	74.1	75.9	75.5	71.9	63.2	53.9	47.0	61.6
		MEAN	53.8	57.2	63.6	71.2	76.9	82.4	84.6	84.5	80.7	73.1	63.7	57.0	70.7
PORT O CONNOR		MAX	61.6	64.2	70.2	76.7	82.4	87.7	90.3	90.5	87.9	81.4	72.1	65.2	77.5
		MIN	45.3	48.2	55.4	63.9	70.3	76.3	78.3	77.4	74.3	65.5	56.0	48.5	63.3
		MEAN	53.5	56.3	62.8	70.3	76.4	82.0	84.3	84.0	81.1	73.5	64.0	56.9	70.4
POTEET		MAX	64.2	68.5	76.6	82.9	87.3	93.6	96.7	96.7	91.4	83.4	73.0	66.6	81.7
		MIN	39.8	43.0	50.4	59.1	65.7	71.3	73.0	72.5	68.6	58.9	49.3	42.3	57.8
		MEAN	52.1	55.8	63.5	71.0	76.5	82.5	84.8	84.7	80.0	71.2	61.2	54.4	69.8
PRESIDIO		MAX	67.6	73.3	82.0	90.8	97.8	103.3	102.0	100.2	95.1	87.0	74.9	67.5	86.8
		MIN	33.1	36.7	43.8	53.0	61.6	70.9	72.8	70.9	65.6	54.1	41.0	33.8	53.1
		MEAN	50.4	55.0	62.9	71.9	79.7	87.2	87.4	85.6	80.4	70.5	58.0	50.6	70.0
QUANAH 5 SE		MAX	51.2	56.5	64.9	76.1	83.8	92.8	97.6	96.2	87.6	77.3	63.2	55.2	75.2
		MIN	23.8	28.4	36.3	48.3	57.9	67.5	71.7	69.8	61.7	48.8	36.2	27.2	48.1
		MEAN	37.6	42.5	50.6	62.3	70.8	80.2	84.7	83.0	74.7	63.1	49.7	41.2	61.7
RAYMONDVILLE		MAX	69.5	73.1	79.4	86.2	89.7	93.5	96.1	96.9	92.3	86.2	77.6	71.8	84.4
		MIN	47.0	49.2	56.5	64.4	68.9	72.2	72.9	73.0	70.7	62.8	54.9	49.1	61.8
		MEAN	58.3	61.2	68.0	75.3	79.3	82.9	84.6	84.9	81.5	74.5	66.3	60.5	73.1
RIO GRANDE CITY 3 W		MAX	69.5	73.9	81.9	89.4	92.5	97.0	99.3	99.8	94.3	86.7	77.4	71.0	86.1
		MIN	43.5	46.5	54.3	63.4	68.9	72.8	73.8	73.6	70.1	61.1	51.9	45.6	60.5
		MEAN	56.5	60.2	68.1	76.4	80.7	84.9	86.6	86.7	82.3	73.9	64.6	58.3	73.3
RISING STAR		MAX	54.8	59.2	67.5	76.5	82.7	90.5	95.2	94.5	87.3	77.6	65.0	58.1	75.7
		MIN	29.8	33.3	40.8	51.5	59.6	67.5	70.6	69.7	63.3	51.9	40.3	32.4	50.9
		MEAN	42.3	46.3	54.2	64.0	71.2	79.0	82.9	82.1	75.3	64.7	52.7	45.3	63.3
ROSCOE		MAX	56.9	61.8	70.1	79.2	85.5	92.4	95.0	93.9	86.6	77.9	65.4	59.3	77.0
		MIN	30.0	33.5	40.4	50.5	58.6	66.7	69.7	68.6	62.2	51.9	40.4	32.7	50.4
		MEAN	43.5	47.7	55.3	64.9	72.1	79.6	82.4	81.3	74.4	65.0	52.9	46.0	63.8
RUSK		MAX	56.0	60.4	68.1	76.1	82.4	89.5	93.5	93.7	87.6	79.5	67.3	60.0	76.2
		MIN	35.3	38.4	45.6	54.3	61.3	67.7	70.4	69.8	65.3	54.8	45.1	38.2	53.9
		MEAN	45.7	49.4	56.9	65.2	71.9	78.6	82.0	81.8	76.5	67.1	56.2	49.1	65.0
SAN ANGELO WSO	R	MAX	58.7	63.3	71.5	80.2	86.3	93.4	96.5	95.4	88.0	79.2	67.2	61.2	78.4
		MIN	32.2	36.1	43.4	53.4	61.5	69.3	72.0	71.2	65.0	54.0	42.0	34.7	52.9
		MEAN	45.5	49.7	57.5	66.8	73.9	81.4	84.3	83.3	76.5	66.6	54.6	48.0	65.7
SAN ANTONIO WSO	R	MAX	61.7	66.3	73.7	80.3	85.5	91.8	94.9	94.6	89.3	81.5	70.7	64.6	79.6
		MIN	39.0	42.4	49.8	58.8	65.5	72.0	74.3	73.7	69.4	58.9	48.2	41.4	57.8
		MEAN	50.4	54.3	61.8	69.6	75.5	81.9	84.6	84.2	79.4	70.2	59.5	53.0	68.7
SAN MARCOS		MAX	60.8	65.2	72.8	79.8	85.1	91.4	95.0	95.5	89.8	81.8	70.8	64.0	79.3
		MIN	35.9	39.3	46.0	55.6	62.9	69.3	71.3	70.8	66.4	55.7	44.9	38.0	54.7
		MEAN	48.4	52.3	59.4	67.7	74.0	80.3	83.2	83.2	78.1	68.8	57.9	51.0	67.0
SEMINOLE		MAX	56.9	61.4	69.0	78.7	85.8	93.6	94.5	93.2	86.9	78.1	65.6	59.1	76.9
		MIN	25.3	28.5	35.0	44.8	53.4	62.3	65.3	63.8	57.6	46.1	34.3	27.6	45.3
		MEAN	41.1	45.0	52.0	61.8	69.7	78.0	80.0	78.5	72.2	62.1	50.0	43.4	61.2
SEYMOUR		MAX	53.3	58.0	66.3	77.4	84.4	92.9	97.9	97.7	89.1	78.6	64.6	56.8	76.4
		MIN	25.8	30.2	37.5	49.0	58.2	67.2	71.1	69.6	61.9	49.8	37.1	29.1	48.9
		MEAN	39.6	44.1	51.9	63.2	71.3	80.1	84.5	83.6	75.5	64.2	50.9	43.0	62.7
SHERMAN PUMP STATION		MAX	51.6	56.7	64.7	74.6	81.8	90.1	95.3	95.2	87.7	77.6	64.3	55.7	74.6
		MIN	29.7	34.1	41.6	52.2	60.3	68.6	72.6	71.2	64.4	52.8	41.5	33.5	51.9
		MEAN	40.7	45.4	53.2	63.4	71.1	79.4	83.9	83.2	76.0	65.2	52.9	44.6	63.3

TEXAS

TEMPERATURE NORMALS (DEG F)

STATION		JAN	FEB	MAR	APR	MAY	JUN	JUL	AUG	SEP	OCT	NOV	DEC	ANN
SNYDER	MAX	55.0	59.5	67.9	77.8	84.7	92.6	95.2	94.2	86.9	77.7	64.7	58.1	76.2
	MIN	25.8	29.9	36.7	47.9	57.1	65.7	69.0	67.5	60.5	48.8	36.4	28.9	47.9
	MEAN	40.4	44.7	52.3	62.9	70.9	79.2	82.1	80.9	73.7	63.2	50.6	43.5	62.0
SONORA	MAX	61.7	66.4	74.4	81.8	87.1	92.6	95.2	94.6	88.9	80.0	69.1	63.2	79.6
	MIN	31.9	35.7	43.6	52.7	60.4	67.0	68.9	67.7	62.6	52.4	41.1	33.6	51.5
	MEAN	46.8	51.0	59.0	67.3	73.8	79.8	82.1	81.2	75.8	66.2	55.1	48.5	65.6
SPEARMAN	MAX	49.1	54.5	61.9	72.9	81.1	90.4	94.6	92.8	85.4	75.1	59.4	52.1	72.4
	MIN	19.5	24.2	30.7	41.4	51.2	60.9	65.5	63.6	55.8	43.4	30.4	23.0	42.5
	MEAN	34.3	39.4	46.3	57.2	66.2	75.7	80.1	78.3	70.6	59.3	44.9	37.6	57.5
STRATFORD	MAX	48.2	52.4	59.6	70.1	78.7	88.5	92.6	90.8	82.9	73.7	58.7	51.6	70.7
	MIN	17.7	21.9	27.4	37.9	48.1	58.5	63.6	61.7	53.1	40.9	28.0	20.9	40.0
	MEAN	33.0	37.2	43.5	54.0	63.4	73.5	78.1	76.2	68.1	57.3	43.4	36.3	55.3
SUGAR LAND	MAX	61.9	65.2	72.2	79.3	85.6	91.0	93.7	93.6	89.3	82.5	72.1	65.2	79.3
	MIN	41.4	44.0	50.9	59.7	66.0	71.5	73.6	73.0	69.1	59.0	50.2	43.6	58.5
	MEAN	51.7	54.7	61.6	69.5	75.8	81.3	83.7	83.4	79.2	70.8	61.2	54.4	68.9
SULPHUR SPRINGS	MAX	53.1	57.7	65.2	74.7	81.6	89.0	94.3	94.6	88.1	78.5	65.5	57.3	75.0
	MIN	30.7	34.6	41.3	50.9	59.5	67.1	70.7	69.3	62.9	51.3	40.7	34.1	51.1
	MEAN	41.9	46.2	53.3	62.9	70.6	78.1	82.5	82.0	75.5	65.0	53.1	45.7	63.1
TAYLOR	MAX	58.4	62.8	71.0	79.1	84.9	92.2	96.6	97.1	90.6	81.5	69.3	62.0	78.8
	MIN	35.4	38.7	45.9	55.6	63.0	69.5	72.1	71.5	66.3	55.6	44.9	38.0	54.7
	MEAN	46.9	50.8	58.5	67.4	74.0	80.9	84.4	84.3	78.5	68.6	57.1	50.0	66.8
TEMPLE	MAX	57.4	61.6	69.5	77.6	83.7	91.3	95.7	95.9	89.5	80.6	68.1	60.9	77.7
	MIN	35.7	39.0	45.9	55.7	63.3	70.2	73.5	73.0	67.8	57.2	45.7	38.4	55.5
	MEAN	46.6	50.3	57.7	66.7	73.5	80.8	84.6	84.5	78.7	69.0	56.9	49.7	66.6
THROCKMORTON	MAX	54.1	59.0	67.4	77.4	84.1	92.6	97.2	96.6	88.2	78.3	64.7	57.7	76.4
	MIN	28.1	32.3	39.9	51.1	59.6	68.2	72.2	71.1	63.7	51.8	39.6	31.5	50.8
	MEAN	41.1	45.7	53.6	64.2	71.9	80.4	84.8	83.9	76.0	65.1	52.2	44.6	63.6
TULIA	MAX	51.1	55.5	62.8	73.0	80.6	89.3	91.7	90.2	83.4	74.4	60.9	54.0	72.2
	MIN	21.5	24.8	30.6	40.7	50.7	60.2	64.5	62.7	55.5	44.0	31.6	24.8	42.6
	MEAN	36.3	40.2	46.7	56.9	65.7	74.8	78.1	76.5	69.5	59.2	46.2	39.4	57.5
UVALDE	MAX	64.4	69.0	76.9	83.9	88.2	94.2	97.1	96.8	91.8	83.6	72.9	66.3	82.1
	MIN	37.0	40.9	48.4	57.6	64.4	70.0	71.2	70.3	66.7	57.1	46.4	39.2	55.8
	MEAN	50.7	55.0	62.7	70.7	76.3	82.1	84.2	83.6	79.3	70.4	59.7	52.8	69.0
VEGA	MAX	49.5	53.4	60.4	70.2	78.7	87.8	90.9	88.8	82.1	72.6	58.8	52.0	70.4
	MIN	19.3	23.5	28.8	38.6	48.5	58.4	63.1	61.3	53.3	41.9	29.7	22.9	40.8
	MEAN	34.4	38.5	44.6	54.4	63.6	73.1	77.1	75.1	67.8	57.3	44.3	37.4	55.6
VERNON	MAX	55.0	60.4	69.1	79.1	86.0	94.4	98.8	97.8	89.9	79.7	66.0	58.4	77.9
	MIN	28.4	32.7	39.8	50.6	59.4	68.5	72.9	71.0	63.7	51.7	39.6	31.8	50.8
	MEAN	41.8	46.6	54.4	64.9	72.7	81.5	85.9	84.4	76.8	65.8	52.8	45.2	64.4
VICTORIA WSO R	MAX	63.6	67.1	73.8	80.2	85.6	90.8	93.7	93.7	89.3	82.8	73.0	66.7	80.0
	MIN	43.1	45.9	52.8	61.5	67.7	73.1	75.2	74.7	70.9	61.0	51.5	45.4	60.2
	MEAN	53.4	56.5	63.3	70.9	76.7	82.0	84.5	84.2	80.1	71.9	62.3	56.1	70.1
WACO WSO R	MAX	56.6	61.6	69.5	77.6	84.2	92.1	96.5	96.7	89.7	80.3	67.9	60.3	77.8
	MIN	35.7	39.4	46.6	56.5	64.2	71.5	75.2	74.5	68.6	57.2	46.1	38.5	56.2
	MEAN	46.2	50.5	58.1	67.1	74.2	81.9	85.9	85.6	79.2	68.8	57.0	49.5	67.0
WAXAHACHIE	MAX	56.3	61.3	69.4	77.8	84.5	92.4	96.9	96.8	89.8	80.7	67.7	60.1	77.8
	MIN	33.1	36.8	43.8	53.9	61.9	69.1	72.9	71.6	65.6	54.6	43.4	36.4	53.6
	MEAN	44.7	49.1	56.6	65.9	73.2	80.8	84.9	84.2	77.7	67.7	55.6	48.3	65.7
WEATHERFORD	MAX	53.2	57.9	66.2	75.6	82.5	91.6	96.7	96.2	88.3	77.9	64.8	57.2	75.7
	MIN	29.9	33.7	40.6	51.2	59.4	67.6	71.7	70.4	63.8	52.3	40.6	33.2	51.2
	MEAN	41.5	45.8	53.4	63.4	71.0	79.7	84.2	83.3	76.1	65.1	52.7	45.2	63.5

TEXAS

TEMPERATURE NORMALS (DEG F)

STATION		JAN	FEB	MAR	APR	MAY	JUN	JUL	AUG	SEP	OCT	NOV	DEC	ANN
WESLACO 2 E	MAX	70.6	74.2	80.6	86.1	89.3	92.5	94.9	95.7	92.0	86.1	77.7	72.6	84.4
	MIN	49.7	52.2	59.1	66.0	70.2	73.4	74.1	74.1	71.7	64.8	57.3	51.8	63.7
	MEAN	60.2	63.2	69.9	76.1	79.8	82.9	84.5	84.9	81.9	75.5	67.5	62.2	74.1
WHITNEY DAM	MAX	56.0	60.4	68.5	77.1	83.6	91.9	96.5	96.7	89.4	80.0	67.2	59.8	77.3
	MIN	32.1	36.0	43.6	54.0	61.6	69.0	71.9	71.3	65.8	54.3	43.1	35.2	53.2
	MEAN	44.1	48.2	56.1	65.6	72.6	80.5	84.2	84.0	77.7	67.2	55.2	47.6	65.3
WICHITA FALLS WSO R	MAX	52.3	58.0	66.7	76.8	84.1	93.2	98.5	97.3	88.7	78.2	64.4	56.2	76.2
	MIN	28.2	32.6	39.9	50.6	59.4	68.2	72.5	71.2	63.7	52.0	39.6	31.6	50.8
	MEAN	40.3	45.3	53.3	63.7	71.8	80.7	85.6	84.3	76.2	65.1	52.0	43.9	63.5
WILLS POINT	MAX	53.4	58.1	66.3	75.9	83.0	91.6	97.3	97.3	89.9	79.7	66.6	57.5	76.4
	MIN	32.1	35.9	42.9	52.3	60.4	67.9	71.9	70.9	65.1	53.8	42.9	35.6	52.6
	MEAN	42.8	47.0	54.6	64.1	71.7	79.8	84.6	84.2	77.5	66.7	54.8	46.6	64.5
WINK FAA AIRPORT	MAX	58.9	63.9	71.7	81.3	88.9	95.9	96.3	95.2	88.7	79.5	66.9	60.6	79.0
	MIN	28.0	32.2	39.6	49.5	58.5	67.5	70.5	69.0	62.4	50.2	37.2	29.5	49.5
	MEAN	43.5	48.1	55.7	65.4	73.7	81.7	83.4	82.1	75.6	64.9	52.1	45.1	64.3
YOAKUM	MAX	64.8	68.6	75.6	81.6	87.2	92.8	96.1	96.8	91.7	84.9	74.4	67.7	81.9
	MIN	41.9	44.9	51.9	60.1	65.8	71.4	73.1	72.7	68.8	59.3	50.2	44.3	58.7
	MEAN	53.4	56.8	63.8	70.9	76.5	82.1	84.6	84.8	80.3	72.1	62.3	56.0	70.3
YSLETA	MAX	59.3	63.9	70.9	79.8	88.1	97.0	97.0	94.5	89.2	80.3	67.9	59.8	79.0
	MIN	27.1	30.8	37.4	45.1	52.9	62.3	66.8	64.8	57.9	45.7	34.3	27.5	46.1
	MEAN	43.2	47.4	54.2	62.5	70.6	79.7	81.9	79.7	73.6	63.0	51.1	43.7	62.6

TEXAS

PRECIPITATION NORMALS (INCHES)

STATION		JAN	FEB	MAR	APR	MAY	JUN	JUL	AUG	SEP	OCT	NOV	DEC	ANN
ABERNATHY		.46	.68	.87	1.25	2.57	2.92	2.48	2.45	2.22	1.70	.60	.53	18.73
ABILENE WSO	//R	.97	.96	1.08	2.35	3.25	2.52	2.11	2.47	3.06	2.32	1.32	.85	23.26
ACKERLY 1 WSW		.46	.54	.80	1.10	2.19	1.69	2.18	1.69	2.80	1.61	.55	.50	16.11
ALBANY		1.30	1.35	1.27	2.87	3.74	2.34	2.36	3.05	3.87	2.44	1.50	1.01	27.10
ALICE		1.31	1.52	.75	1.66	2.95	3.33	1.96	2.81	6.35	3.14	1.59	1.14	28.51
ALPINE		.47	.43	.35	.35	1.03	1.95	2.91	2.58	2.61	1.16	.57	.42	14.83
ALTO 5 SW		3.39	3.11	3.12	4.60	3.95	3.56	2.92	2.24	4.37	2.90	3.55	3.66	41.37
ALVIN		3.49	3.23	2.77	3.15	3.98	5.02	4.97	3.80	6.51	3.59	3.88	3.60	47.99
AMARILLO WSO	//R	.46	.57	.87	1.08	2.79	3.50	2.70	2.95	1.72	1.39	.58	.49	19.10
ANAHUAC TBCD		3.89	3.35	2.71	4.17	4.28	4.23	4.72	5.09	5.82	3.50	4.07	4.23	50.06
ANGLETON 2 W		3.79	3.72	2.90	3.15	4.63	5.49	4.54	4.99	6.97	3.51	4.25	4.34	52.28
ANNA		1.68	2.34	2.83	4.43	5.21	3.29	2.16	2.11	4.44	3.18	2.69	2.29	36.65
ANTELOPE		1.18	1.29	1.68	2.83	4.21	2.93	2.00	2.11	3.70	2.69	1.78	1.27	27.67
ARCHER CITY		1.04	1.30	1.64	2.98	3.94	2.61	2.07	2.26	4.13	2.66	1.81	1.22	27.66
ARLINGTON		1.78	2.13	2.68	4.39	4.58	2.54	2.07	1.94	3.73	3.21	2.24	1.95	33.24
ASPERMONT 2 SSW		.60	.93	1.11	1.73	3.35	2.56	2.03	2.18	3.23	2.56	1.18	.73	22.19
ATHENS 3 SSE		2.57	3.01	2.97	4.35	5.11	3.01	1.57	2.14	4.21	3.81	3.36	3.32	39.43
ATLANTA 4 WSW		3.55	3.49	4.21	5.73	4.22	3.65	3.04	2.84	3.68	2.91	4.25	4.07	45.64
AUSTIN WSO	R	1.60	2.49	1.68	3.11	4.19	3.06	1.89	2.24	3.60	3.38	2.20	2.06	31.50
BAIRD		1.17	1.03	1.25	2.92	3.24	2.94	1.85	2.44	3.13	2.73	1.86	1.09	25.65
BAKERSFIELD		.51	.59	.44	.68	1.70	1.44	1.13	1.56	2.69	1.68	.64	.41	13.47
BALLINGER 1 SW		.97	1.06	.88	2.20	3.36	2.24	1.45	2.27	3.26	2.30	1.23	.89	22.11
BALMORHEA		.43	.47	.38	.47	1.49	1.29	1.72	1.70	2.39	1.24	.66	.41	12.65
BAY CITY WATERWORKS		2.83	2.90	2.02	2.97	4.33	4.43	3.77	4.33	6.62	3.41	3.18	3.10	43.89
BEAUMONT FILTER PLANT		4.47	3.80	3.01	4.36	4.79	4.30	5.40	4.64	6.23	4.37	4.16	4.97	54.50
BEDIAS		2.67	3.13	2.63	4.58	4.55	3.42	2.59	2.64	4.74	3.65	3.31	3.51	41.42
BEEVILLE 5 NE		1.78	2.11	1.05	2.33	3.61	3.25	1.94	2.95	5.47	2.97	1.95	1.70	31.11
BENBROOK DAM		1.64	1.81	2.06	4.03	4.35	2.51	2.46	1.98	3.16	2.99	1.84	1.78	30.61
BIG SPRING		.58	.64	.75	1.42	2.77	1.88	1.85	1.88	2.87	1.74	.73	.61	17.72
BIG WELLS		.71	1.05	.75	1.73	2.97	2.14	1.06	2.36	3.22	2.94	1.22	.88	21.03
BLANCO		1.88	2.54	2.05	3.37	4.24	3.14	1.98	2.61	4.75	3.93	2.19	1.97	34.65
BLEWETT 5 NW		.86	1.00	1.07	1.77	3.03	2.77	1.95	2.62	2.66	3.03	1.19	.80	22.75
BOERNE		1.60	2.17	1.79	3.28	3.88	2.90	1.59	3.06	4.55	3.33	2.34	1.75	32.24
BON WIER 2 E		4.61	4.12	4.03	4.90	5.48	4.07	5.12	3.96	4.63	3.12	4.55	5.76	54.35
BONHAM		2.06	2.67	3.58	4.77	5.07	4.08	2.90	2.33	4.83	3.50	2.99	2.77	41.55
BOOKER		.42	.65	1.25	1.34	3.75	2.91	2.98	2.63	1.78	1.19	.81	.48	20.19
BOQUILLAS RANGER STA		.36	.31	.21	.32	1.05	1.30	1.52	1.21	1.18	.86	.30	.30	8.92
BORGER 3 W		.47	.76	1.01	1.34	3.19	2.91	3.12	2.48	1.57	1.27	.68	.53	19.33
BOXELDER		3.00	3.44	4.06	5.48	4.82	3.91	3.84	2.65	4.49	3.56	4.13	3.68	47.06
BOYD		1.52	1.69	2.41	3.77	4.03	2.49	2.43	1.93	3.63	2.91	1.97	1.73	30.51
BRACKETTVILLE		.78	1.10	.93	2.04	2.62	2.99	1.42	2.12	2.85	2.61	.99	.81	21.26
BRADY 2 NNW		1.06	1.36	1.19	2.30	3.60	2.17	2.07	2.35	3.67	2.48	1.35	1.06	24.66
BRAVO		.31	.39	.60	1.04	2.07	2.15	2.85	2.73	1.72	.95	.63	.39	15.83
BRAZOS		1.65	1.48	1.79	3.23	4.43	2.33	2.45	1.75	3.37	2.95	1.90	1.43	28.76
BRENHAM		2.71	2.99	2.09	4.02	4.57	3.63	1.84	2.60	4.72	3.52	3.81	3.22	39.72
BRIDGEPORT		1.56	1.56	2.17	3.17	4.26	2.68	2.10	1.87	3.31	2.86	1.87	1.45	28.86
BROWNFIELD NO 2		.43	.55	.83	.95	2.34	2.45	2.34	2.10	2.31	1.77	.57	.33	16.97
BROWNSVILLE WSO	R	1.25	1.55	.50	1.57	2.15	2.70	1.51	2.83	5.24	3.54	1.44	1.16	25.44
BROWNWOOD		1.38	1.44	1.52	2.69	3.88	2.89	1.66	1.95	3.00	2.96	1.50	1.23	26.10
BUFFALO		3.38	3.30	3.35	4.68	4.47	3.35	2.28	2.17	4.17	3.95	3.70	3.43	42.23
BULVERDE		1.87	2.24	1.69	3.45	3.88	3.00	1.91	3.35	4.45	3.78	2.37	1.93	33.92
BUNKER HILL 5 N		.20	.27	.55	.81	2.11	1.70	2.86	2.43	1.39	.79	.51	.27	13.89
BURKETT		1.35	1.16	1.27	2.52	3.41	3.01	1.67	2.35	3.32	2.50	1.65	1.08	25.29
BURLESON 2 SSW		1.58	1.98	2.51	4.35	4.72	2.88	1.79	1.93	3.22	3.14	2.14	1.68	31.92
BURNET		1.58	2.11	1.78	3.26	4.47	2.49	1.91	2.06	3.59	3.45	1.99	1.68	30.37

TEXAS

PRECIPITATION NORMALS (INCHES)

STATION	JAN	FEB	MAR	APR	MAY	JUN	JUL	AUG	SEP	OCT	NOV	DEC	ANN
CAMERON	2.15	2.74	2.23	3.93	3.94	2.65	1.68	1.78	4.22	3.43	2.94	2.57	34.26
CAMP WOOD	.87	1.21	1.14	2.49	3.08	3.04	2.18	3.26	3.39	3.19	1.30	.90	26.05
CANADIAN 1 ENE	.38	.76	1.03	1.47	3.95	2.80	2.27	2.71	2.13	1.34	.71	.54	20.09
CANDELARIA	.28	.29	.21	.17	.53	1.28	2.05	2.25	1.99	.89	.37	.28	10.59
CANYON	.45	.56	.81	1.17	2.67	3.14	2.50	2.89	1.53	1.58	.64	.43	18.37
CARROLLTON	1.77	2.11	2.59	4.48	4.64	2.63	2.32	2.16	3.88	3.27	2.42	2.01	34.28
CARTHAGE	4.20	3.57	3.99	4.70	4.70	3.54	3.18	2.54	4.18	3.00	4.11	4.48	46.19
CASE RANCH 3 S	.60	.69	.88	1.52	2.24	1.69	1.42	2.02	2.73	2.02	.93	.76	17.50
CELINA	1.71	2.14	2.88	4.47	5.30	3.35	2.26	2.24	4.36	3.19	2.58	2.05	36.53
CENTER	4.38	3.80	4.26	4.89	5.01	3.94	3.60	3.96	4.47	2.93	4.04	4.44	49.72
CENTERVILLE	2.90	3.20	2.77	4.72	3.94	3.30	2.11	2.61	3.85	3.73	3.18	3.01	39.32
CHILDRESS FAA AIRPORT	.57	.71	1.07	1.76	3.50	2.60	1.95	1.91	2.32	2.04	.76	.70	19.89
CHISOS BASIN	.54	.46	.37	.53	1.47	1.85	3.16	3.04	2.74	1.45	.58	.52	16.71
CIBOLO CREEK	1.61	1.88	1.09	2.63	3.86	2.54	1.73	2.67	4.91	3.03	2.10	1.65	29.70
CLARENDON	.53	.67	1.09	1.60	3.96	3.42	2.41	2.58	2.09	1.73	.76	.61	21.45
CLARKSVILLE 2 NE	2.67	3.13	4.01	5.22	4.96	3.52	3.03	2.25	4.07	3.93	3.91	3.41	44.11
CLEBURNE	1.79	1.96	2.36	4.18	4.73	2.79	1.90	2.35	3.03	3.24	2.22	1.82	32.37
CLODINE	3.22	2.97	2.46	3.23	4.38	4.36	3.29	3.56	4.73	3.64	3.35	3.20	42.39
COLDWATER 3 W	.29	.32	.71	1.01	2.14	2.04	2.65	1.93	1.48	.88	.43	.26	14.14
COLEMAN	1.28	1.20	1.15	2.81	4.32	2.91	1.87	2.37	3.73	2.59	1.58	1.13	26.94
COLLEGE STATION FAA AP	2.48	2.97	2.39	4.34	4.35	3.21	2.39	2.30	4.93	3.42	3.33	2.97	39.08
COLUMBUS	2.86	3.05	2.29	4.02	5.29	4.10	2.48	2.74	5.09	3.35	3.17	3.01	41.45
COMANCHE	1.61	1.51	1.57	3.35	4.01	2.37	1.61	1.92	3.68	2.81	1.92	1.37	27.73
CONLEN	.34	.49	.86	1.33	2.80	2.37	2.53	2.42	1.49	1.07	.63	.42	16.75
CONROE	3.47	3.39	2.52	4.85	4.77	3.89	3.40	3.58	4.92	3.77	3.84	4.20	46.60
COOPER	2.66	2.92	3.58	4.79	5.04	3.93	2.85	2.24	4.47	3.63	3.31	3.41	42.83
COPE RANCH	.46	.67	.70	1.29	2.43	1.78	1.68	1.58	2.75	1.74	.88	.60	16.56
CORNUDAS SERVICE STA	.36	.33	.27	.22	.38	.89	1.58	1.93	1.43	.82	.33	.31	8.85
CORPUS CHRISTI	1.58	1.68	.90	1.93	3.23	3.13	1.54	3.24	6.06	3.33	1.61	1.45	29.68
CORPUS CHRISTI WSO R	1.63	1.55	.84	1.99	3.05	3.36	1.96	3.51	6.15	3.19	1.55	1.40	30.18
CORSICANA	2.32	2.47	2.55	4.59	5.09	2.83	1.61	2.22	3.49	3.73	2.79	2.94	36.63
COTTONWOOD	1.29	1.72	1.52	2.93	3.28	2.59	2.07	3.34	3.78	3.60	1.87	1.26	29.25
COTULLA	.88	1.20	.83	1.97	2.78	2.59	1.15	2.38	3.54	2.83	1.18	1.09	22.42
COTULLA FAA AIRPORT	.78	1.14	.76	1.93	2.92	2.22	1.13	2.66	3.19	2.80	1.13	.99	21.65
CRESSON	1.63	1.64	2.06	3.84	4.45	2.74	1.89	2.06	3.31	2.99	1.98	1.84	30.43
CROCKETT	3.43	3.05	2.79	4.72	4.24	3.69	2.45	2.83	4.35	3.33	3.84	3.51	42.23
CROSBYTON	.53	.78	1.02	1.42	3.04	2.58	2.47	2.42	2.69	2.03	.84	.66	20.48
CROWELL	.87	1.06	1.32	1.96	4.14	2.52	2.36	1.84	3.07	2.73	1.19	.83	23.89
CRYSTAL CITY	.77	1.13	.73	1.70	3.35	2.52	1.50	2.35	2.83	2.66	1.04	.76	21.34
CUERO 3 E	2.07	2.32	1.39	3.00	4.83	3.59	1.97	2.34	5.35	3.57	2.44	2.08	34.95
CYPRESS 1 SW	3.64	3.78	2.50	3.82	4.98	4.39	3.13	3.11	5.39	4.19	3.78	4.53	47.24
DAINGERFIELD 9 S	3.15	3.41	4.01	5.91	4.54	3.26	3.23	2.48	3.59	3.28	4.24	3.80	44.90
DALHART FAA AIRPORT	.38	.45	.73	1.18	2.61	2.11	2.98	2.60	1.33	1.12	.58	.38	16.45
DALLAS-FORT WORTH REG WSO	1.65	1.93	2.42	3.63	4.27	2.59	2.00	1.76	3.31	2.47	1.76	1.67	29.45
DALLAS FAA //R	1.82	2.11	2.93	4.45	4.40	2.77	2.02	2.16	3.59	3.36	2.38	2.17	34.16
DANEVANG 2 SE	2.49	2.82	1.99	2.89	4.44	4.74	2.66	4.28	5.75	3.48	2.89	2.87	41.30
DARROUZETT	.51	.77	1.38	1.37	3.60	2.83	2.76	2.66	1.63	1.29	.93	.60	20.33
DAVILLA	2.20	3.07	2.15	3.86	4.01	2.56	1.84	1.74	4.18	3.49	2.76	2.47	34.33
DEKALB	3.33	3.38	3.99	5.18	4.94	3.75	3.84	3.03	3.99	3.41	4.01	3.83	46.68
DEL RIO WSO	.51	.89	.63	1.85	1.99	1.72	1.69	1.60	2.73	2.24	.80	.55	17.19
DENISON DAM	1.60	2.17	3.07	4.18	4.55	3.37	2.36	2.51	5.08	3.06	2.51	2.01	36.47
DENTON 2 SE	1.55	2.00	2.63	4.13	4.71	3.04	2.12	2.04	4.14	3.03	2.21	1.93	33.53
DIALVILLE 2 W	3.33	3.15	3.63	4.97	4.85	3.13	2.49	2.48	4.02	3.24	3.34	3.62	42.25
DILLEY	1.08	1.42	.71	2.03	3.69	2.55	1.30	2.39	3.60	2.66	1.41	1.13	23.97
DIME BOX	2.47	2.95	2.28	3.84	4.52	3.83	2.01	2.06	4.97	3.10	2.94	2.84	37.81

TEXAS

PRECIPITATION NORMALS (INCHES)

STATION	JAN	FEB	MAR	APR	MAY	JUN	JUL	AUG	SEP	OCT	NOV	DEC	ANN
DIMMITT 6 E	.42	.55	.76	.90	2.08	2.92	2.44	2.17	1.67	1.56	.68	.47	16.62
DOOLE 6 NNE	1.15	1.29	1.34	2.44	4.09	2.60	1.87	2.26	3.67	2.51	1.45	1.03	25.70
DUBLIN	1.63	1.72	1.70	3.56	4.73	2.07	2.17	2.51	3.33	3.18	2.02	1.51	30.13
DUNDEE 6 NNW	.89	1.18	1.33	2.52	3.82	2.49	1.84	1.95	3.72	2.84	1.54	1.15	25.27
EAGLE PASS	.62	.81	.67	1.83	3.24	2.37	2.02	2.42	2.92	2.47	.88	.70	20.95
EASTLAND	1.46	1.35	1.51	2.83	3.92	2.43	1.87	2.16	2.83	2.84	1.81	1.21	26.22
EDEN 1	1.02	1.17	1.13	2.17	3.47	2.39	1.76	2.47	3.74	2.32	1.36	.88	23.88
EDNA HWY 59 BRIDGE	2.41	2.67	1.70	2.76	4.89	4.88	2.50	3.35	6.27	3.79	2.59	2.40	40.21
EDOM 2 NW	2.76	3.22	3.43	5.16	5.07	3.10	2.38	2.26	4.16	3.74	3.53	3.59	42.40
EL PASO WSO R	.38	.45	.32	.19	.24	.56	1.60	1.21	1.42	.73	.33	.39	7.82
ELDORADO 11 NW	.66	.85	.74	1.72	2.52	1.94	1.58	2.14	3.11	2.14	.99	.63	19.02
ELECTRA	1.00	1.26	1.77	2.47	4.56	2.87	1.92	2.10	3.32	2.78	1.55	1.20	26.80
EMORY	2.77	3.30	3.18	5.09	5.06	3.45	2.61	2.28	3.86	4.04	3.23	3.31	42.18
ENCINAL 3 NW	.96	1.18	.77	1.69	2.57	2.34	1.22	2.31	3.78	2.44	1.20	.95	21.41
ENNIS	2.19	2.47	2.45	4.72	5.38	3.31	2.10	2.17	3.89	3.65	2.77	2.56	37.66
EVADALE	4.98	4.11	3.59	5.23	4.63	4.55	4.93	4.20	5.58	3.71	4.64	5.22	55.37
FAIRFIELD 3 ESE	2.56	2.65	2.93	4.54	4.57	2.93	1.63	2.67	3.92	3.56	3.16	3.09	38.21
FALFURRIAS	1.40	1.35	.66	1.42	2.92	3.09	1.69	2.84	5.60	2.41	1.26	1.13	25.77
FALLS CITY 4 WSW	1.48	1.83	.91	2.45	4.01	2.81	1.65	2.55	4.45	2.61	2.15	1.38	28.28
FARMERSVILLE	2.33	2.62	3.45	4.89	5.53	3.19	2.99	2.09	4.48	3.27	2.94	2.77	40.55
FERRIS	1.98	2.62	2.59	4.77	4.75	2.80	2.03	2.05	4.02	3.38	2.66	2.37	36.02
FISCHERS STORE	1.58	2.26	1.74	3.23	3.79	3.25	1.99	2.66	4.47	3.60	2.38	1.84	32.79
FLATONIA 2 W	2.28	2.76	1.67	4.00	4.60	4.15	1.81	2.39	5.19	3.16	2.78	2.62	37.41
FLOMOT 1 N	.48	.68	.99	1.53	3.16	3.35	2.22	2.50	2.25	1.70	.70	.60	20.16
FLORESVILLE	1.74	2.02	1.20	2.77	3.96	2.88	1.48	2.65	4.12	2.97	2.13	1.66	29.58
FLOYDADA	.39	.63	.93	1.24	2.77	3.24	2.35	2.10	2.37	1.71	.71	.52	18.96
FLOYDADA 9 SE	.36	.56	.88	1.27	2.88	2.81	2.17	2.15	2.41	1.73	.64	.48	18.34
FOLLETT	.56	.86	1.52	1.53	3.63	2.94	2.71	2.81	1.84	1.31	.98	.57	21.26
FORESTBURG 4 S	1.48	1.46	2.17	3.50	4.03	2.97	2.15	1.97	4.19	3.23	2.17	1.51	30.83
FORSAN	.64	.67	.72	1.63	2.60	1.95	1.91	2.05	2.66	1.79	.84	.66	18.12
FT STOCKTON KFST RADIO	.42	.60	.43	.52	1.61	1.43	1.30	1.46	2.14	1.09	.77	.44	12.21
FOWLERTON	1.02	1.27	.64	1.60	3.17	2.39	1.26	2.49	3.74	2.57	1.34	1.08	22.57
FREDERICKSBURG	1.16	1.66	1.35	2.66	3.76	3.23	1.78	2.89	4.02	3.18	1.72	1.26	28.67
FREEPORT 2 NW	3.61	3.19	1.89	2.91	3.92	4.95	5.56	5.20	7.98	4.02	3.83	3.95	51.01
FREER	1.16	1.27	.82	1.59	3.19	2.71	1.17	2.40	4.76	2.84	1.42	1.10	24.43
FROST	2.04	2.36	2.22	4.23	5.18	2.65	1.66	2.34	3.54	3.59	2.69	2.36	34.86
FUNK RANCH	.66	.81	.73	1.58	2.46	1.98	1.55	2.41	2.86	1.88	1.01	.66	18.59
GAIL	.50	.56	.80	1.30	2.85	2.03	2.33	2.21	2.88	1.98	.68	.56	18.68
GAINESVILLE	1.65	1.95	2.74	3.40	4.34	3.24	2.15	2.23	4.25	3.18	2.19	1.67	32.99
GALVESTON WSO R	2.96	2.34	2.10	2.62	3.30	3.48	3.77	4.40	5.82	2.60	3.23	3.62	40.24
GARDEN CITY 1 E	.51	.55	.62	1.21	2.11	1.64	1.97	1.70	2.49	1.76	.72	.52	15.80
GATESVILLE	1.80	2.51	1.92	3.32	4.58	3.08	2.01	2.19	3.66	3.24	2.24	1.90	32.45
GEORGE WEST	1.68	1.63	.77	1.94	3.27	2.85	1.48	2.93	4.69	3.06	1.88	1.41	27.59
GIDDINGS	2.41	3.00	2.03	3.83	4.48	3.39	1.60	2.41	5.03	3.66	3.10	2.69	37.63
GILMER 2 W	3.08	3.35	3.93	6.00	4.78	3.60	2.95	2.42	4.24	2.96	4.03	3.78	45.12
GIRVIN	.56	.59	.56	.72	1.72	1.35	1.28	1.25	2.27	1.54	.77	.42	13.03
GOLD	1.35	1.89	1.59	2.86	4.12	3.44	1.85	2.51	4.51	3.20	1.95	1.49	30.76
GOLDTHWAITE 2 N	1.37	1.58	1.54	2.81	3.99	2.56	1.53	2.14	3.07	3.15	2.10	1.43	27.27
GOLIAD	2.02	2.46	1.53	3.02	4.31	3.81	2.19	3.42	5.75	3.77	2.33	2.16	36.77
GOLIAD 1 SE	1.81	2.34	1.38	2.94	4.14	3.35	1.87	2.96	5.13	3.43	2.13	2.06	33.54
GONZALES	1.86	2.38	1.61	3.68	4.27	3.58	1.35	2.10	4.54	3.26	2.43	2.09	33.15
GRAHAM	1.30	1.36	1.47	3.09	4.11	2.84	2.04	2.33	3.94	2.59	1.71	1.23	28.01
GRAPEVINE DAM	1.58	1.87	2.13	4.42	4.18	2.53	2.28	1.91	3.85	3.02	2.08	1.80	31.65
GREENVILLE 7 NW	2.28	2.67	3.40	4.73	5.30	3.24	2.73	2.14	4.47	3.68	3.04	2.75	40.43
GROVETON	3.48	3.31	3.23	4.56	4.46	3.96	3.10	3.07	4.10	3.08	3.57	4.18	44.10

TEXAS

PRECIPITATION NORMALS (INCHES)

STATION	JAN	FEB	MAR	APR	MAY	JUN	JUL	AUG	SEP	OCT	NOV	DEC	ANN
GRUVER	.43	.48	.92	1.08	2.94	2.71	3.13	2.48	1.48	1.16	.68	.43	17.92
GUNTER	1.57	2.05	2.83	4.56	5.23	3.51	2.17	2.25	4.32	3.10	2.42	2.09	36.10
GUTHRIE	.69	.91	1.03	1.67	3.47	2.62	1.88	2.34	3.07	2.17	1.01	.60	21.46
HAGANSPORT	2.64	2.80	3.47	5.20	4.68	3.43	3.30	2.48	3.88	3.33	3.77	3.29	42.27
HALLETTSVILLE	2.38	2.63	1.97	3.38	5.22	3.83	2.39	2.83	5.29	3.07	2.91	2.50	38.40
HAMILTON 2SSE	1.51	1.61	1.91	3.21	3.99	2.40	1.69	2.23	2.72	3.16	1.92	1.54	27.89
HAMLIN	.67	.87	.97	1.85	3.38	2.37	2.08	1.88	3.22	2.22	1.04	.84	21.39
HARLETON	3.81	3.38	4.04	5.76	4.63	3.78	2.81	2.24	4.02	3.22	3.77	4.26	45.72
HARLINGEN	1.38	1.46	.68	1.77	2.27	2.79	1.83	3.26	5.25	2.98	1.61	1.20	26.48
HARPER	1.09	1.35	1.33	2.49	3.33	2.37	1.78	2.87	3.38	3.03	1.41	1.13	25.56
HARTLEY	.34	.39	.77	1.15	2.38	2.62	2.93	2.61	1.47	1.04	.65	.37	16.72
HASKELL	.81	1.03	1.20	2.03	3.75	2.33	2.33	2.77	3.30	2.46	1.23	.89	24.13
HAWKINS 1 E	2.86	3.20	3.49	5.67	4.45	3.41	2.61	2.23	3.76	2.88	3.71	3.73	42.00
HENDERSON	3.70	3.50	3.75	4.85	4.98	3.83	2.91	2.74	3.92	3.30	3.49	3.77	44.74
HENRIETTA	1.30	1.39	2.05	2.90	4.60	3.49	2.10	2.45	3.73	2.94	1.70	1.44	30.09
HEREFORD	.38	.52	.70	.98	1.85	2.92	2.15	2.37	1.58	1.52	.64	.40	16.01
HEWITT 1 SE	1.88	2.28	2.35	3.55	4.59	2.58	1.63	2.20	3.44	3.27	2.66	2.11	32.54
HICO	1.76	1.86	1.93	3.61	3.84	2.29	2.23	1.98	3.07	3.47	2.17	1.55	29.76
HIGGINS	.43	.92	1.22	1.78	4.21	3.29	2.22	2.58	2.21	1.59	.98	.66	22.09
HILLSBORO	1.88	2.32	2.37	3.99	4.83	3.04	2.21	1.83	3.25	3.61	2.57	2.32	34.22
HONEY GROVE	2.27	2.54	3.61	4.70	5.05	4.51	2.98	2.30	4.76	3.63	3.39	2.90	42.64
HORGER	4.11	4.02	3.82	4.60	4.98	4.29	4.07	2.75	3.27	2.93	4.12	4.74	47.70
HOUSTON INCONT AP	3.21	3.25	2.68	4.24	4.69	4.06	3.33	3.66	4.93	3.67	3.38	3.66	44.77
HOUSTON-BARKER	3.17	3.07	2.12	3.30	4.07	3.42	3.19	3.65	4.77	3.67	3.52	3.23	41.18
HOUSTON-DEER PARK	3.72	3.16	2.56	3.67	4.24	5.21	4.81	4.73	6.00	3.69	4.16	3.65	49.60
HOUSTON-HEIGHTS	3.55	3.26	2.53	3.55	4.57	4.89	4.23	4.47	5.62	3.72	3.96	3.56	47.91
HOUSTON-INDEP HEIGHTS	3.58	3.36	2.68	3.48	4.25	4.45	4.42	3.85	5.22	3.81	3.93	3.44	46.47
HOUSTON-WESTBURY	3.33	3.36	2.29	3.52	4.20	5.17	3.74	3.83	5.05	3.45	3.67	3.35	44.96
HOUSTON-NORTH HOUSTON	3.23	3.20	2.47	3.50	4.36	3.90	3.89	3.73	4.56	3.81	3.67	3.55	43.87
HUNT	1.17	1.48	1.44	2.62	3.46	2.29	1.93	3.25	3.56	3.13	1.62	1.11	27.06
HUNTSVILLE	3.20	3.36	2.83	4.60	4.73	3.86	2.69	3.06	4.96	3.42	3.50	3.99	44.20
HYE	1.26	2.01	1.64	3.09	3.88	2.88	1.53	2.37	4.56	3.55	1.91	1.52	30.20
IMPERIAL	.42	.49	.33	.70	1.22	1.02	1.04	1.16	1.96	1.09	.56	.39	10.38
INDIAN GAP	1.36	1.58	1.57	2.80	3.98	2.40	1.69	1.94	3.03	3.38	1.95	1.29	26.97
JACKSBORO	1.42	1.30	1.78	3.24	3.99	2.43	2.46	2.09	3.35	2.87	1.83	1.23	27.99
JASPER 3 SW	4.39	4.12	3.80	4.59	5.05	4.30	4.28	3.50	4.10	3.22	4.35	4.89	50.59
JEDDO 2 NNE	2.20	2.46	1.79	3.87	4.59	3.59	1.74	2.10	4.93	3.75	2.98	2.26	36.26
JEFFERSON	3.86	3.52	3.92	5.28	4.60	3.43	3.13	2.48	3.59	3.04	3.76	4.08	44.69
JOURDANTON	1.21	1.82	.91	2.48	3.91	2.45	1.67	2.63	4.12	2.52	1.62	1.30	26.64
JUNCTION	.95	1.36	1.03	2.13	3.28	2.26	1.54	2.88	2.66	2.36	1.12	.95	22.52
KARNACK	4.00	3.71	4.09	4.91	4.74	4.14	3.39	2.56	3.83	2.94	4.06	4.21	46.58
KAUFMAN 3 SE	2.35	2.88	2.68	4.80	4.79	2.74	2.50	2.11	3.75	3.68	2.84	3.06	38.18
KENNEDALE 6 SSW	1.74	2.11	2.42	4.64	4.64	2.85	1.76	2.12	3.52	3.40	2.21	1.95	33.36
KINGSVILLE	1.51	1.57	.94	1.53	3.07	3.22	1.98	3.16	5.04	2.96	1.38	1.14	27.50
KNAPP 6 SW	.53	.59	.80	1.66	2.75	2.29	1.92	2.56	2.60	1.96	.73	.62	19.01
LA GRANGE	2.38	2.80	1.80	3.93	4.75	3.65	1.62	2.34	4.86	3.60	3.18	2.61	37.52
LA TUNA 1 S	.36	.38	.31	.14	.23	.70	1.53	1.92	1.29	.79	.32	.45	8.42
LAMESA 1 SSE	.52	.43	.74	.95	2.13	2.07	2.29	1.64	2.71	1.77	.58	.41	16.24
LAMPASAS	1.52	2.02	1.78	3.09	4.22	2.57	1.83	2.78	2.85	3.06	1.98	1.82	29.52
LAREDO NO 2	.73	1.05	.40	1.22	2.54	2.74	1.06	2.67	3.45	2.18	1.23	.87	20.14
LAVON DAM	2.06	2.31	3.03	4.32	4.94	2.69	2.46	1.85	4.32	3.16	2.50	2.42	36.06
LAWN	.98	.97	1.08	2.75	3.47	2.78	2.07	2.43	3.38	2.41	1.44	.85	24.61
LENORAH	.62	.76	.76	1.24	2.30	1.63	2.37	1.71	2.74	1.65	.80	.60	17.18
LEVELLAND	.43	.59	.71	1.03	2.20	2.46	2.66	2.59	2.53	1.98	.53	.43	18.14
LIBERTY	4.00	3.93	2.66	4.35	4.31	4.46	4.19	3.84	5.34	4.22	4.59	4.76	50.65

TEXAS

PRECIPITATION NORMALS (INCHES)

STATION	JAN	FEB	MAR	APR	MAY	JUN	JUL	AUG	SEP	OCT	NOV	DEC	ANN
LINDEN	3.61	3.62	4.26	5.85	4.57	3.91	3.05	2.46	3.53	3.02	4.34	4.05	46.27
LIPAN	1.55	1.55	1.95	3.20	4.46	2.36	2.04	2.30	2.98	3.09	1.85	1.43	28.76
LIPSCOMB	.37	.80	1.23	1.39	3.94	2.76	2.50	2.42	1.64	1.26	.80	.50	19.61
LITTLEFIELD NO 2	.46	.54	.60	1.15	2.65	3.18	2.43	2.28	2.12	1.62	.54	.37	17.94
LIVINGSTON 2 NNE	4.02	3.65	3.33	4.73	4.57	3.24	3.97	3.30	4.92	3.66	4.15	4.42	47.96
LLANO	1.13	1.72	1.35	3.06	3.92	2.34	1.54	2.35	3.49	2.77	1.61	1.28	26.56
LOCKHART 2 NW	1.85	2.35	1.61	3.57	3.76	3.63	1.20	2.26	4.60	3.47	2.77	1.93	33.00
LONG LAKE 5 SW	2.86	3.01	3.52	4.10	4.42	3.23	1.93	2.78	4.03	3.61	3.63	3.34	40.46
LONGVIEW	3.79	3.69	3.73	5.19	4.95	4.18	2.97	2.54	4.45	3.02	3.88	4.08	46.47
LOOP	.36	.41	.67	.76	2.06	1.76	2.22	2.09	1.91	1.77	.49	.27	14.77
LORENZO	.39	.59	.89	1.28	2.70	2.73	2.19	1.96	2.09	1.79	.65	.44	17.70
LUBBOCK WSO R	.38	.57	.90	1.08	2.59	2.81	2.34	2.20	2.06	1.81	.59	.43	17.76
LUFKIN FAA AP	3.55	3.05	3.38	4.27	4.31	3.39	2.81	2.46	3.72	2.98	3.59	3.97	41.48
LULING	1.98	2.43	1.62	3.60	4.41	4.27	1.58	2.17	4.54	3.27	2.85	1.94	34.66
MADISONVILLE	2.88	3.11	2.84	4.79	4.38	3.36	2.17	2.71	4.46	3.54	3.71	3.11	41.06
MARATHON	.40	.41	.31	.46	1.50	1.57	1.91	2.11	2.79	1.04	.57	.44	13.51
MARLIN 3 NE	2.12	2.54	2.63	4.22	5.06	2.81	1.89	2.03	3.57	3.52	2.96	2.65	36.00
MARSHALL	4.14	3.72	3.99	5.10	4.86	3.73	3.31	2.42	4.11	3.18	3.74	4.11	46.41
MASON	1.11	1.44	1.16	2.35	3.71	2.51	1.78	2.55	3.21	2.72	1.34	1.06	24.94
MATADOR	.51	.70	.92	1.61	3.12	3.13	2.24	2.30	2.39	2.02	.79	.70	20.43
MATAGORDA NO 2	2.92	2.54	1.70	2.91	4.79	4.03	3.23	4.05	7.30	3.37	3.50	2.86	43.20
MAUD 1 S	3.53	3.39	4.03	5.30	4.66	3.98	3.44	2.93	4.28	3.14	4.37	3.91	46.96
MC ALLEN	1.25	1.10	.64	1.56	2.08	2.79	1.59	2.38	4.29	3.20	1.15	1.01	23.04
MC CAMEY	.38	.55	.45	.71	1.75	1.34	1.28	1.46	2.25	1.48	.65	.43	12.73
MC COOK	1.26	1.08	.74	1.28	2.98	2.99	1.45	2.37	4.34	2.94	1.09	.96	23.48
MC GREGOR	2.05	2.34	2.18	3.76	4.72	3.13	2.34	2.04	3.47	3.26	2.42	2.06	33.77
MCKINNEY 3 S	1.88	2.33	3.03	4.46	5.02	3.20	2.61	2.19	4.52	2.88	2.64	2.12	36.88
MEMPHIS	.53	.72	1.07	1.96	3.99	3.25	1.93	1.82	2.19	1.64	.67	.57	20.34
MENARD	.87	1.09	1.14	2.05	3.06	2.36	1.89	2.20	3.14	2.45	1.21	.78	22.24
MERCEDES 6 SSE	1.26	1.25	.67	1.65	2.64	3.04	1.70	2.30	4.86	3.07	1.47	.99	24.90
MERTZON	.64	.90	.82	1.52	2.35	1.84	1.29	2.30	2.77	1.78	1.12	.71	18.04
MEXIA	2.66	2.78	2.93	4.37	4.72	3.15	1.61	2.11	4.57	3.54	3.01	2.83	38.28
MIAMI	.46	.86	1.07	1.73	3.71	2.82	2.61	2.45	2.11	1.46	.85	.57	20.70
MIDLAND WSO //R	.42	.58	.51	.84	2.05	1.44	1.72	1.60	2.08	1.41	.60	.45	13.70
MIDLAND 4 ENE	.52	.51	.49	.74	1.97	1.21	2.11	1.70	2.77	1.15	.55	.41	14.13
MINEOLA 7 SSW	2.75	3.02	3.09	5.17	4.65	3.15	2.56	2.25	3.76	3.31	3.40	3.24	40.35
MINERAL WELLS FAA AP	1.61	1.68	1.99	3.41	4.06	2.59	2.27	2.18	3.30	2.93	1.82	1.43	29.27
MISSION 4 W	1.24	1.07	.58	1.57	2.04	2.78	1.43	2.07	4.06	3.22	1.06	.85	21.97
MORGAN MILL	1.52	1.48	1.66	3.30	4.56	2.50	2.18	2.18	3.27	3.02	1.71	1.42	28.80
MORSE	.33	.49	.96	1.06	2.84	2.44	2.97	2.46	1.36	1.41	.62	.36	17.30
MORTON	.35	.48	.53	.90	1.78	2.39	2.27	2.92	2.58	1.56	.54	.32	16.62
MOUNT LOCKE	.60	.52	.43	.35	1.36	2.12	3.90	4.05	3.02	1.46	.65	.48	18.94
MOUNT PLEASANT	3.11	3.38	3.88	5.20	4.52	3.71	3.27	2.71	4.35	3.62	4.11	3.64	45.50
MUENSTER	1.66	1.80	2.62	3.68	4.65	3.00	2.19	2.18	4.32	3.22	2.34	1.71	33.37
MULESHOE 1	.50	.53	.55	.93	1.93	2.44	2.57	2.42	1.77	1.43	.65	.36	16.08
MULLIN	1.32	1.49	1.63	2.89	3.68	2.15	1.60	2.10	2.91	2.93	1.94	1.32	25.96
MUNDAY	.84	1.03	1.37	2.03	3.50	2.21	2.28	2.60	3.38	2.80	1.27	1.00	24.31
NAPLES 1 SW	3.40	3.65	4.05	5.31	4.31	3.55	3.37	2.51	3.93	3.11	4.03	3.65	44.87
NEGLEY 4 SSW	3.06	3.39	4.49	5.53	5.38	4.18	3.67	2.31	4.38	3.83	3.88	3.67	47.77
NELSON RANCH	1.22	1.36	1.41	2.69	3.40	2.79	2.13	3.18	3.26	3.36	1.69	1.11	27.60
NEW BRAUNFELS	1.86	2.33	1.60	3.08	4.50	3.20	1.69	2.77	4.27	3.50	2.80	2.01	33.61
NEW CANEY 4 NW	3.85	3.44	2.81	4.26	4.78	4.36	3.72	3.23	4.87	3.79	3.67	4.03	46.81
NEW GULF	2.68	2.74	2.01	2.94	4.95	4.93	3.65	4.32	5.77	3.53	3.13	2.97	43.62
NEWPORT	1.37	1.41	2.06	3.24	3.89	2.62	2.23	2.22	3.42	2.83	2.06	1.56	28.91
NIX STORE 1 W	1.20	2.02	1.50	3.39	4.84	2.86	1.72	2.27	2.97	3.14	1.89	1.63	29.43

TEXAS

PRECIPITATION NORMALS (INCHES)

STATION	JAN	FEB	MAR	APR	MAY	JUN	JUL	AUG	SEP	OCT	NOV	DEC	ANN
NIXON	1.89	2.17	1.45	3.30	3.90	3.33	1.58	2.36	4.93	3.50	2.39	1.76	32.56
NORTHFIELD	.59	.66	.85	1.59	3.32	2.78	2.10	2.11	2.37	2.08	.67	.63	19.75
NOTLA 3 SE	.40	.68	1.07	1.51	3.54	2.76	2.61	2.62	1.80	1.40	.90	.48	19.77
ODESSA	.38	.60	.42	.87	1.65	1.40	1.57	1.57	2.44	1.60	.58	.46	13.54
OLNEY 5 NNW	1.00	1.08	1.36	2.90	3.76	2.11	2.03	1.75	4.29	2.48	1.51	1.03	25.30
ORANGE 4 NW	4.84	4.49	3.29	4.27	5.09	4.96	6.32	5.66	6.44	3.75	4.50	5.59	59.20
OVERTON	3.85	3.59	3.69	4.99	4.83	3.65	3.14	1.81	4.13	3.26	3.77	4.00	44.71
OZONA	.63	.66	.73	1.63	2.29	2.05	1.75	2.01	2.78	2.01	.97	.52	18.23
PADUCAH	.65	.78	1.09	1.79	3.63	3.28	1.98	2.28	2.59	2.15	.93	.81	21.96
PAINT ROCK	1.03	1.10	1.09	2.29	3.38	2.37	1.67	1.95	3.78	2.47	1.21	.94	23.28
PALACIOS FAA AIRPORT	2.59	2.62	1.80	2.58	4.71	4.64	3.07	4.14	7.18	3.99	2.92	3.37	43.61
PALESTINE	3.13	2.98	3.62	4.52	4.74	3.75	2.07	2.76	3.76	3.56	3.45	3.38	41.72
PALO PINTO	1.77	1.51	1.84	3.78	3.95	2.72	2.71	2.03	3.10	2.95	1.88	1.49	29.73
PAMPA NO 2	.45	.84	1.05	1.31	3.09	2.92	2.62	2.47	2.02	1.47	.84	.49	19.57
PANDALE	.46	.68	.54	1.22	1.85	2.38	1.60	1.60	3.38	1.77	.67	.42	16.57
PANHANDLE	.51	.74	1.13	1.22	3.07	3.26	2.48	2.77	1.67	1.64	.72	.51	19.72
PARIS	2.31	2.96	3.71	4.80	5.14	3.87	3.54	2.94	5.01	3.95	3.39	3.35	44.97
PEARSALL	1.09	1.32	.79	2.15	3.72	2.59	1.47	2.77	3.21	2.79	1.52	1.12	24.54
PECOS	.33	.39	.28	.43	.95	1.02	1.34	1.19	1.79	1.08	.46	.31	9.57
PERRYTON 5 NNE	.45	.59	1.21	1.17	3.50	2.89	3.01	2.59	1.57	1.18	.97	.49	19.62
PERRYTON 10 WNW	.51	.62	1.23	1.10	3.18	2.73	2.83	2.37	1.70	1.10	.87	.53	18.77
PIERCE 1 E	2.82	2.89	2.05	3.05	4.26	4.63	3.04	3.53	5.81	4.13	2.75	2.76	41.72
PILOT POINT	1.91	2.26	3.03	4.52	5.10	4.17	2.46	2.28	4.23	3.22	2.48	1.92	37.58
PITTSBURG 5 S	2.90	3.33	3.83	5.36	4.81	3.36	2.70	2.22	4.05	3.24	3.97	3.52	43.29
PLAINS	.39	.60	.60	.98	1.92	1.94	2.46	2.41	2.17	1.85	.52	.39	16.23
PLAINVIEW	.49	.66	.79	1.24	3.27	3.07	2.57	2.03	2.03	1.68	.64	.50	18.97
PORT ARTHUR WSO R	4.18	3.71	2.93	4.05	4.50	3.96	5.37	5.45	6.13	3.63	4.33	4.55	52.79
PORT LAVACA NO 2	2.35	2.65	1.57	3.01	3.81	5.50	2.61	4.03	7.24	4.13	2.75	2.56	42.21
PORT O CONNOR	2.73	2.68	1.37	2.34	3.96	3.44	2.58	4.29	7.06	3.95	2.62	2.76	39.78
POST	.58	.65	.83	1.34	2.73	2.83	2.38	2.36	2.54	1.83	.80	.55	19.42
POTEET	1.30	1.93	1.06	2.94	4.05	2.57	1.49	2.42	4.00	2.75	1.88	1.42	27.81
PRESIDIO	.26	.31	.18	.26	.49	1.35	1.62	1.62	1.65	.78	.40	.28	9.20
PUTNAM	1.33	1.12	1.33	2.58	3.29	2.53	1.91	2.26	2.92	2.74	1.64	1.01	24.66
QUANAH 5 SE	.66	.85	1.22	1.93	3.90	2.91	2.33	2.18	3.07	2.50	1.02	.80	23.37
QUITMAN 2 S	2.86	3.42	3.46	4.98	5.03	3.31	2.20	2.84	3.82	3.21	3.56	3.34	42.03
RAINBOW	1.67	1.63	2.05	3.81	4.28	2.70	2.02	1.48	3.21	3.16	1.91	1.71	29.63
RAYMONDVILLE	1.36	1.29	.63	1.60	3.11	3.50	1.66	3.10	5.97	2.68	1.55	1.03	27.48
RED BLUFF CROSSING	1.22	1.73	1.41	3.05	3.75	2.27	1.75	2.31	3.27	2.49	1.64	1.27	26.16
RED BLUFF DAM	.32	.34	.29	.27	.86	1.16	1.45	1.55	2.07	1.15	.43	.25	10.14
REFUGIO	2.13	2.28	1.29	2.55	4.40	3.94	2.66	3.61	7.75	3.97	2.27	1.91	38.76
RICHARDSON	1.93	2.18	2.87	4.58	4.69	2.88	2.29	2.20	3.91	3.28	2.45	2.22	35.48
RICHLAND SPRINGS	1.21	1.49	1.47	2.67	3.72	2.16	1.68	2.30	3.22	2.61	1.52	1.32	25.37
RICHMOND	3.26	3.11	2.24	3.32	4.25	4.39	3.37	3.55	5.00	3.59	3.48	3.32	42.88
RINGGOLD	1.20	1.35	1.98	3.16	4.13	2.77	1.88	1.92	3.44	2.52	1.77	1.41	27.53
RIO GRANDE CITY 3 W	.81	.90	.57	1.50	2.28	2.02	1.30	2.16	5.11	2.28	.95	.69	20.57
RIOMEDINA 2 N	1.40	1.93	1.32	2.85	3.73	2.41	1.76	2.96	3.75	2.99	1.77	1.54	28.41
RISING STAR	1.43	1.29	1.53	2.97	4.16	2.78	1.97	2.04	3.25	2.97	1.70	1.15	27.24
ROANOKE	1.74	2.01	2.48	4.40	4.46	2.51	2.12	2.08	3.75	3.07	2.08	1.90	32.60
ROBERT LEE	.70	.92	.94	2.02	2.84	2.24	1.56	1.66	3.26	2.64	1.14	.75	20.67
ROBSTOWN	1.72	1.75	.96	1.90	3.21	3.37	2.26	3.69	6.12	3.07	1.68	1.33	31.06
ROCKLAND 1 W	4.29	3.73	3.63	4.78	4.78	3.76	4.06	2.81	4.01	3.42	3.94	4.35	47.56
ROCKWALL	2.15	2.39	3.08	4.66	5.21	3.09	2.40	2.04	3.79	3.26	2.49	2.38	36.94
ROSCOE	.85	.97	1.16	2.20	2.87	2.56	2.13	2.11	4.00	2.39	1.24	.87	23.35
ROSSER 1 SE	2.19	2.59	2.56	4.48	5.13	2.41	1.99	1.74	3.42	3.35	2.87	2.55	35.28
ROTAN	.74	.94	1.00	1.88	3.42	2.64	1.94	2.35	3.36	2.35	1.00	.84	22.46

TEXAS

PRECIPITATION NORMALS (INCHES)

STATION		JAN	FEB	MAR	APR	MAY	JUN	JUL	AUG	SEP	OCT	NOV	DEC	ANN
ROUND MOUNTAIN 4 WNW		1.31	1.84	1.52	2.97	3.78	2.35	1.54	2.06	4.59	3.13	1.81	1.52	28.42
RUNGE		1.91	1.98	1.18	2.74	4.10	3.08	1.39	2.61	5.09	2.86	2.19	1.94	31.07
RUSK		3.80	3.44	3.63	4.90	4.77	3.57	2.77	2.40	4.41	3.38	3.50	4.04	44.61
SABINAL 1 WSW		1.10	1.53	1.22	2.14	3.47	2.86	1.67	3.02	3.01	2.97	1.40	1.14	25.53
SAN ANGELO WSO	R	.64	.84	.79	1.75	2.52	1.88	1.22	1.85	3.04	2.05	.97	.64	18.19
SAN ANTONIO WSO	R	1.55	1.86	1.33	2.73	3.67	3.03	1.92	2.69	3.75	2.88	2.34	1.38	29.13
SAN MARCOS		1.86	2.69	1.60	3.37	4.42	3.38	1.81	2.39	3.46	2.81	2.13	34.31
SAN SABA		1.14	1.53	1.33	2.85	3.91	2.35	1.62	2.28	3.08	2.74	1.88	1.21	25.92
SANDERSON		.38	.52	.32	.92	1.37	1.59	1.24	1.58	2.56	1.37	.61	.37	12.83
SARITA 7 E		1.51	1.38	.63	1.69	2.60	2.65	1.49	2.93	5.18	2.91	1.50	1.02	25.49
SCHULENBURG		2.31	2.63	1.89	4.21	4.81	3.88	2.07	3.11	5.03	3.27	2.81	2.54	38.56
SEMINOLE		.47	.61	.66	1.02	2.04	1.81	2.14	2.18	2.28	1.58	.64	.37	15.80
SEYMOUR		.91	1.05	1.40	2.11	3.76	2.81	2.54	2.34	3.61	2.77	1.29	1.07	25.66
SHAMROCK		.46	.85	1.22	2.17	3.99	3.24	2.41	2.38	2.61	1.80	.85	.60	22.58
SHERMAN PUMP STATION		1.72	2.49	3.13	4.55	5.33	3.46	2.28	2.24	4.84	3.21	2.81	2.12	38.18
SILVERTON		.50	.61	.99	1.44	3.19	3.64	2.39	2.53	2.29	1.55	.69	.53	20.35
SINTON		1.80	2.01	1.01	2.16	3.80	3.17	2.42	3.79	6.67	3.99	1.78	1.73	34.33
SLATON 5 SE		.44	.58	.85	1.35	3.02	2.42	2.59	2.15	2.42	1.81	.79	.55	18.97
SLIDELL		1.59	1.76	2.49	3.61	4.58	3.36	2.26	2.22	3.66	2.97	2.15	1.67	32.32
SNYDER		.50	.64	.78	1.67	3.05	2.45	2.03	2.40	3.05	2.22	.91	.59	20.29
SONORA		.72	1.13	.75	2.04	2.67	2.23	1.92	2.26	2.82	2.44	1.07	.65	20.70
SPEARMAN		.52	.64	1.18	1.12	3.23	2.59	3.01	2.50	1.71	1.23	.92	.50	19.15
STAMFORD 1		.81	1.08	1.07	2.03	3.44	2.45	2.18	2.04	3.62	2.32	1.25	.94	23.23
STERLING CITY		.64	.73	.80	1.61	2.61	1.90	1.63	1.79	3.22	1.97	.96	.72	18.58
STRATFORD		.30	.40	.78	1.22	2.72	2.25	3.01	2.67	1.53	1.02	.64	.32	16.86
STRAWN 8 NNE		1.54	1.53	1.85	3.34	4.33	2.43	2.88	1.90	3.09	2.91	1.75	1.45	29.00
SUGAR LAND		3.27	3.28	2.34	3.52	4.29	4.28	3.11	3.88	5.15	3.77	3.69	3.29	43.87
SULPHUR SPRINGS		2.65	2.93	3.57	5.37	5.01	3.89	2.80	2.55	4.53	3.97	3.47	3.42	44.16
TAHOKA		.47	.62	.83	1.34	2.72	2.58	2.58	1.96	2.43	1.60	.71	.47	18.31
TAMPICO		.51	.53	.98	1.66	3.30	3.22	2.09	2.23	2.03	1.61	.67	.58	19.41
TASCOSA		.35	.41	.75	1.14	2.37	2.82	3.43	2.46	1.58	1.19	.57	.39	17.46
TAYLOR		1.97	2.83	2.06	3.56	3.95	3.17	1.54	2.14	4.41	3.54	2.66	2.38	34.21
TEAGUE RANCH		1.21	1.75	1.37	2.73	4.34	2.93	1.53	2.39	4.69	3.25	1.76	1.41	29.36
TEMPLE		1.94	2.57	1.97	3.56	4.65	2.94	1.75	2.50	3.51	3.30	2.69	2.37	33.75
TERRELL		2.36	2.78	2.99	4.84	4.54	3.14	2.12	1.94	4.15	4.08	2.87	2.55	38.36
THORNTON		2.40	2.55	2.52	4.48	4.82	2.54	1.90	2.14	3.38	3.91	2.97	2.82	36.43
THREE RIVERS		1.46	1.67	.92	1.89	3.30	2.46	1.15	2.54	4.33	2.60	1.58	1.34	25.24
THROCKMORTON		1.02	1.16	1.32	2.53	3.43	2.47	2.03	2.30	4.06	2.44	1.20	1.02	24.98
THURBER 5 NE		1.76	1.50	1.98	3.06	3.86	2.56	2.65	2.23	3.02	2.84	1.66	1.57	28.69
TOMBALL		3.31	3.18	2.26	4.02	4.24	3.67	3.09	2.99	4.39	3.66	3.30	3.74	41.85
TRENT 2 ENE		.83	.97	1.07	2.10	3.22	2.13	2.61	2.61	3.59	2.17	1.18	.85	23.33
TRENTON		1.89	2.58	3.26	4.58	5.04	3.47	2.68	2.12	4.77	3.11	2.83	2.62	38.95
TROY		1.99	2.51	2.21	3.75	4.88	2.83	1.90	2.25	3.60	3.34	2.77	2.31	34.34
TULIA		.41	.49	.79	1.09	2.44	3.59	2.11	2.02	1.88	1.52	.65	.48	17.47
UMBARGER		.34	.50	.81	.99	2.49	3.07	2.13	2.64	1.54	1.37	.51	.40	16.79
UVALDE		1.00	1.26	.95	2.17	3.26	2.81	1.74	2.94	2.73	3.08	1.19	.97	24.10
VALENTINE 10 WSW		.50	.41	.29	.19	.55	1.49	2.26	2.47	2.66	.98	.57	.39	12.76
VALLEY VIEW		1.63	1.97	2.75	3.89	4.56	3.40	2.23	2.16	4.34	3.07	2.44	1.75	34.19
VAN HORN		.48	.31	.20	.23	.60	.93	1.84	2.06	2.30	1.08	.62	.41	11.06
VEGA		.45	.61	.82	1.05	2.40	2.76	2.75	2.53	1.64	1.20	.70	.49	17.40
VERNON		.85	1.04	1.50	2.37	4.63	2.86	2.14	2.00	2.91	2.81	1.26	.88	25.25
VICTORIA WSO	R	1.87	2.24	1.34	2.61	4.47	4.53	2.58	3.33	6.24	3.31	2.24	2.14	36.90
VOSS 1 WSW		1.10	1.30	1.26	2.53	3.87	3.03	1.65	2.34	3.19	2.58	1.39	.96	25.20
WACO WSO	R	1.69	2.04	1.99	3.79	4.73	2.58	1.78	1.95	3.18	3.06	2.24	1.92	30.95
WALLER 3 SSW		2.66	3.03	2.26	3.85	4.60	3.75	2.31	2.63	5.04	3.59	3.65	3.37	40.74

TEXAS

PRECIPITATION NORMALS (INCHES)

STATION	JAN	FEB	MAR	APR	MAY	JUN	JUL	AUG	SEP	OCT	NOV	DEC	ANN
WARREN	4.43	4.36	3.35	4.92	5.12	4.35	4.13	3.41	4.48	3.90	4.38	5.20	52.03
WASHINGTON STATE PARK	2.80	3.08	2.28	4.18	4.13	3.16	2.05	2.87	4.62	3.74	3.26	3.15	39.32
WATER VALLEY	.66	.77	.73	1.66	2.54	1.83	1.28	2.47	2.86	2.06	.97	.71	18.54
WAXAHACHIE	1.94	2.43	2.68	4.56	5.02	2.77	1.78	2.26	4.21	3.52	2.57	2.51	36.25
WEATHERFORD	1.65	1.83	2.18	3.91	4.24	2.70	2.18	2.32	3.44	3.16	1.82	1.63	31.06
WELLINGTON 1 ESE	.49	.73	1.22	2.05	4.05	2.83	1.81	1.98	2.54	2.09	.82	.60	21.21
WESLACO 2 E	1.21	1.14	.64	1.39	2.22	2.85	1.88	2.33	4.58	2.65	1.41	.95	23.25
WHARTON	2.58	2.75	2.31	2.87	4.10	4.19	2.87	3.38	5.12	4.15	2.87	2.81	40.00
WHITNEY DAM	1.83	2.24	2.06	4.10	4.62	2.97	2.00	1.95	3.13	3.37	2.46	1.97	32.70
WICHITA FALLS WSO R	.93	1.00	1.82	2.99	4.34	2.85	2.00	2.14	3.41	2.61	1.42	1.22	26.73
WILLS POINT	2.92	3.14	3.10	5.75	5.32	3.26	2.29	2.06	4.04	3.92	3.25	3.34	42.39
WINK FAA AIRPORT	.36	.36	.35	.74	1.05	1.43	1.60	1.24	1.90	1.27	.43	.29	11.02
WINNSBORO 6 SW	2.73	3.13	3.74	4.97	4.79	3.44	2.42	2.69	3.68	3.18	3.38	3.43	41.58
WINTERS 1 NNE	.83	.98	1.12	2.45	3.77	2.73	2.11	2.16	3.64	2.49	1.24	.82	24.34
WOLFE CITY	2.21	2.54	3.38	4.69	5.15	3.49	2.48	2.08	4.61	3.50	3.12	2.81	40.06
YOAKUM	2.15	2.71	1.70	3.53	4.15	3.82	2.48	3.20	4.50	3.10	2.64	2.19	36.17
YORKTOWN	2.03	2.40	1.25	3.07	4.66	3.07	1.64	2.66	5.61	3.15	2.43	2.00	33.97
YSLETA	.34	.47	.26	.19	.23	.56	1.42	1.48	1.50	.74	.31	.37	7.87

TEXAS

HEATING DEGREE DAY NORMALS (BASE 65 DEG F)

STATION		JUL	AUG	SEP	OCT	NOV	DEC	JAN	FEB	MAR	APR	MAY	JUN	ANN
ABILENE WSO	//R	0	0	10	91	361	577	673	479	321	98	11	0	2621
ALBANY		0	0	13	104	385	592	704	518	369	116	22	0	2823
ALICE		0	0	0	11	124	257	341	223	106	11	0	0	1073
ALPINE		0	7	24	136	375	546	580	451	328	106	12	0	2565
AMARILLO WSO	//R	0	0	25	215	588	828	918	711	577	271	92	6	4231
ANGLETON 2 W		0	0	0	25	165	310	414	299	171	32	0	0	1416
AUSTIN WSO	R	0	0	0	37	221	406	505	347	203	41	0	0	1760
BALLINGER 1 SW		0	0	7	90	359	564	642	456	311	82	9	0	2520
BAY CITY WATERWORKS		0	0	0	23	153	303	396	289	153	22	0	0	1339
BEEVILLE 5 NE		0	0	0	19	157	308	407	278	156	19	0	0	1344
BIG SPRING		0	0	10	89	376	580	670	490	328	105	10	0	2658
BLANCO		0	0	0	74	298	490	582	432	280	84	8	0	2248
BOERNE		0	0	0	71	275	462	546	396	239	71	6	0	2066
BONHAM		0	0	6	88	354	592	698	506	348	95	9	0	2696
BORGER 3 W		0	0	11	166	528	763	856	644	509	210	67	0	3754
BRADY 2 NNW		0	0	10	100	343	558	639	477	324	95	10	0	2556
BRENHAM		0	0	0	31	207	397	502	358	210	39	0	0	1744
BRIDGEPORT		0	0	8	104	367	608	725	532	368	108	15	0	2835
BROWNSVILLE WSO	R	0	0	0	0	55	150	216	135	53	0	0	0	609
BROWNWOOD		0	0	0	74	329	552	651	469	304	84	6	0	2469
CAMERON		0	0	0	44	219	397	500	343	198	41	0	0	1742
CANYON		0	0	16	175	525	747	831	633	507	221	60	0	3715
CENTER		0	0	0	97	314	518	610	448	296	87	7	0	2377
CENTERVILLE		0	0	0	68	275	462	588	425	259	64	5	0	2146
CHILDRESS FAA AIRPORT		0	0	13	128	447	704	806	596	444	152	32	0	3322
CHISOS BASIN		0	5	26	131	339	493	518	404	275	89	14	0	2294
CLARENDON		0	0	24	187	534	775	887	675	542	217	58	9	3908
CLARKSVILLE 2 NE		0	0	9	121	372	626	729	544	381	116	15	0	2913
CLEBURNE		0	0	0	58	292	505	611	428	275	69	0	0	2238
COLEMAN		0	0	5	86	319	518	598	431	274	75	0	0	2306
COLLEGE STATION FAA AP		0	0	0	40	229	412	504	349	210	42	0	0	1786
CONROE		0	0	0	50	222	405	496	359	205	36	0	0	1773
CORPUS CHRISTI		0	0	0	9	109	219	311	196	94	7	0	0	945
CORPUS CHRISTI WSO	R	0	0	0	11	116	220	310	209	97	7	0	0	970
CORSICANA		0	0	0	66	302	527	642	468	309	82	6	0	2402
COTULLA FAA AIRPORT		0	0	0	19	147	305	388	244	121	12	0	0	1236
CROCKETT		0	0	0	69	285	493	589	436	279	82	0	0	2233
CROSBYTON		0	0	18	165	504	741	837	647	501	190	43	0	3646
CRYSTAL CITY		0	0	0	17	151	315	385	235	114	8	0	0	1225
DALHART FAA AIRPORT		0	0	42	269	654	887	973	756	648	327	111	10	4677
DALLAS-FORT WORTH REG WSO		0	0	0	56	300	533	651	469	313	85	0	0	2407
DALLAS FAA	//R	0	0	0	55	295	512	627	445	291	76	0	0	2301
DANEVANG 2 SE		0	0	0	17	138	274	363	260	133	16	0	0	1201
DEL RIO WSO		0	0	0	27	203	388	450	282	145	15	0	0	1510
DENISON DAM		0	0	6	98	358	614	732	535	376	116	11	0	2846
DENTON 2 SE		0	0	0	68	315	546	654	468	315	87	7	0	2460
DILLEY		0	0	0	27	167	349	434	293	156	20	0	0	1446
DUBLIN		0	0	7	89	348	570	679	508	353	108	10	0	2672
EAGLE PASS		0	0	0	26	189	379	444	278	138	11	0	0	1465
EDEN 1		0	0	5	70	298	490	565	396	252	67	0	0	2143
EL PASO WSO	R	0	0	0	96	408	639	645	465	318	93	0	0	2664
ENCINAL 3 NW		0	0	0	19	154	328	405	263	129	15	0	0	1313
FALFURRIAS		0	0	0	16	132	241	327	214	103	10	0	0	1043
FLATONIA 2 W		0	0	0	27	182	345	443	301	163	27	0	0	1488
FOLLETT		0	0	23	210	594	868	989	764	640	280	107	11	4486

TEXAS

HEATING DEGREE DAY NORMALS (BASE 65 DEG F)

STATION	JUL	AUG	SEP	OCT	NOV	DEC	JAN	FEB	MAR	APR	MAY	JUN	ANN
FT STOCKTON KFST RADIO	0	0	5	91	349	549	605	448	311	87	6	0	2451
FREDERICKSBURG	0	0	0	71	274	450	532	383	232	65	0	0	2007
GAINESVILLE	0	0	13	109	391	648	766	567	404	122	21	0	3041
GALVESTON WSO R	0	0	0	10	139	267	376	282	160	19	0	0	1253
GATESVILLE	0	0	6	79	285	479	594	430	268	82	10	0	2233
GILMER 2 W	0	0	9	107	342	567	673	498	346	110	16	0	2668
GOLIAD	0	0	0	15	130	259	346	229	113	14	0	0	1106
GRAHAM	0	0	7	96	371	620	729	532	367	111	20	0	2853
GREENVILLE 7 NW	0	0	7	102	368	623	738	546	388	115	17	0	2904
HALLETTSVILLE	0	0	0	26	166	319	408	279	144	24	0	0	1366
HARLINGEN	0	0	0	8	78	183	267	176	74	0	0	0	786
HASKELL	0	0	9	103	386	629	738	543	386	102	15	0	2911
HENDERSON	0	0	0	80	292	505	607	453	301	80	8	0	2326
HENRIETTA	0	0	15	106	386	645	769	560	399	120	23	0	3023
HEREFORD	0	0	35	239	603	834	915	728	605	295	102	6	4362
HICO	0	0	0	66	303	505	605	438	282	72	6	0	2277
HILLSBORO	0	0	0	77	305	518	642	465	301	78	9	0	2395
HOUSTON INCONT AP	0	0	0	36	201	349	442	314	175	32	0	0	1549
HUNTSVILLE	0	0	0	41	232	422	529	379	226	47	0	0	1876
JACKSBORO	0	0	5	61	308	521	636	457	309	76	6	0	2379
JUNCTION	0	0	7	77	303	505	567	403	249	69	0	0	2180
KAUFMAN 3 SE	0	0	0	78	326	570	676	493	338	104	8	0	2593
KINGSVILLE	0	0	0	7	100	206	286	184	76	8	0	0	867
LA TUNA 1 S	0	0	0	91	405	645	651	479	357	119	8	0	2755
LAMESA 1 SSE	0	0	0	131	447	676	766	582	432	143	13	0	3190
LAMPASAS	0	0	0	91	335	543	632	471	313	101	14	0	2500
LAREDO NO 2	0	0	0	10	99	245	311	188	73	0	0	0	926
LEVELLAND	0	0	30	183	510	725	806	630	487	196	45	0	3612
LIBERTY	0	0	0	43	213	372	461	333	183	43	0	0	1648
LIVINGSTON 2 NNE	0	0	0	68	248	445	532	399	246	62	0	0	2000
LLANO	0	0	0	65	298	515	595	434	266	70	6	0	2249
LUBBOCK WSO R	0	0	15	157	495	729	812	627	470	178	33	0	3516
LUFKIN FAA AP	0	0	0	60	253	438	523	380	225	51	0	0	1930
LULING	0	0	0	40	218	409	509	351	204	43	0	0	1774
MADISONVILLE	0	0	0	33	229	396	487	342	206	41	0	0	1734
MARLIN 3 NE	0	0	0	48	258	451	555	381	238	69	0	0	2000
MARSHALL	0	0	6	101	328	555	648	480	322	85	17	0	2542
MATADOR	0	0	12	130	436	657	766	577	437	155	33	0	3203
MATAGORDA NO 2	0	0	0	16	129	248	346	245	129	16	0	0	1129
MC ALLEN	0	0	0	7	75	187	260	162	64	0	0	0	755
MC CAMEY	0	0	0	69	327	543	595	424	248	50	0	0	2256
MC COOK	0	0	0	10	94	212	298	191	76	0	0	0	881
MCKINNEY 3 S	0	0	0	69	321	552	657	472	313	85	6	0	2475
MEMPHIS	0	0	12	145	471	713	825	610	474	173	43	0	3466
MEXIA	0	0	0	59	285	508	617	447	285	79	0	0	2280
MIAMI	0	0	24	220	597	853	967	745	617	261	91	8	4383
MIDLAND WSO //R	0	0	7	94	385	589	660	484	329	102	8	0	2658
MIDLAND 4 ENE	0	0	6	88	373	583	642	462	306	86	6	0	2552
MINERAL WELLS FAA AP	0	0	0	69	326	549	661	469	310	81	6	0	2471
MISSION 4 W	0	0	0	8	94	203	275	179	68	0	0	0	827
MOUNT LOCKE	18	34	72	219	474	654	691	574	462	213	65	9	3485
MOUNT PLEASANT	0	0	6	114	386	620	722	538	376	116	14	0	2892
MULESHOE 1	0	0	30	237	594	822	896	708	570	257	70	0	4184
MUNDAY	0	0	7	92	355	589	685	493	337	89	9	0	2656
NEW BRAUNFELS	0	0	0	39	207	392	488	337	193	39	0	0	1695

TEXAS

HEATING DEGREE DAY NORMALS (BASE 65 DEG F)

STATION		JUL	AUG	SEP	OCT	NOV	DEC	JAN	FEB	MAR	APR	MAY	JUN	ANN
NEW GULF		0	0	0	21	166	326	425	297	158	19	0	0	1412
NIXON		0	0	0	31	162	319	415	283	156	28	0	0	1394
PALACIOS FAA AIRPORT		0	0	0	19	159	295	389	276	151	19	0	0	1308
PALESTINE		0	0	0	59	284	502	616	447	286	76	0	0	2270
PARIS		0	0	6	97	370	632	756	552	393	116	13	0	2935
PECOS		0	0	0	79	375	589	645	459	294	71	0	0	2512
PIERCE 1 E		0	0	0	31	182	354	439	316	173	35	0	0	1530
PLAINS		0	0	27	183	534	753	815	644	496	210	47	0	3709
PLAINVIEW		0	0	15	181	528	760	856	664	531	224	50	0	3809
PORT ARTHUR WSO	R	0	0	0	33	190	327	431	306	167	23	0	0	1477
PORT LAVACA NO 2		0	0	0	14	130	269	372	250	138	13	0	0	1186
PORT O CONNOR		0	0	0	11	133	271	376	264	142	14	0	0	1211
POTEET		0	0	0	23	167	340	420	287	140	22	0	0	1399
PRESIDIO		0	0	0	21	226	446	453	286	123	9	0	0	1564
QUANAH 5 SE		0	0	9	132	459	738	849	630	462	143	28	0	3450
RAYMONDVILLE		0	0	0	8	84	197	283	183	79	0	0	0	834
RIO GRANDE CITY 3 W		0	0	0	15	103	237	308	197	81	0	0	0	941
RISING STAR		0	0	8	104	376	611	704	524	360	110	12	0	2809
ROSCOE		0	0	8	94	372	589	667	484	331	100	11	0	2656
RUSK		0	0	0	70	284	493	607	447	291	76	0	0	2268
SAN ANGELO WSO	R	0	0	0	75	326	527	605	428	274	73	5	0	2313
SAN ANTONIO WSO	R	0	0	0	41	199	378	463	319	178	28	0	0	1606
SAN MARCOS		0	0	0	48	240	434	523	369	218	55	0	0	1887
SEMINOLE		0	0	10	137	450	670	741	560	414	145	25	0	3152
SEYMOUR		0	0	10	119	430	682	787	585	431	127	23	0	3194
SHERMAN PUMP STATION		0	0	12	101	372	632	753	549	386	112	17	0	2934
SNYDER		0	0	13	122	437	667	763	568	410	133	18	0	3131
SONORA		0	0	0	82	306	512	564	398	229	55	0	0	2146
SPEARMAN		0	0	13	208	603	849	952	717	591	260	89	0	4282
STRATFORD		0	0	40	260	648	890	992	778	667	338	114	15	4742
SUGAR LAND		0	0	0	23	176	337	435	309	168	24	0	0	1472
SULPHUR SPRINGS		0	0	6	107	366	598	716	526	381	118	15	0	2833
TAYLOR		0	0	0	56	262	465	572	408	246	60	0	0	2069
TEMPLE		0	0	0	46	264	474	581	422	270	68	0	0	2125
THROCKMORTON		0	0	12	107	392	632	741	548	386	114	21	0	2953
TULIA		0	0	25	204	564	794	890	694	567	258	72	0	4068
UVALDE		0	0	0	36	188	386	453	305	169	20	0	0	1557
VEGA		0	5	47	260	621	856	949	742	632	328	124	16	4580
VERNON		0	0	5	92	372	614	719	521	362	95	11	0	2791
VICTORIA WSO	R	0	0	0	20	150	291	386	268	140	18	0	0	1273
WACO WSO	R	0	0	0	46	265	481	591	415	257	71	0	0	2126
WAXAHACHIE		0	0	0	55	305	518	636	454	306	90	0	0	2364
WEATHERFORD		0	0	8	101	379	614	729	545	384	124	16	0	2900
WESLACO 2 E		0	0	0	0	69	157	224	132	48	0	0	0	630
WHITNEY DAM		0	0	0	70	309	539	648	476	306	85	0	0	2433
WICHITA FALLS WSO	R	0	0	14	105	396	654	766	552	388	118	18	0	3011
WILLS POINT		0	0	6	80	320	570	688	509	344	97	9	0	2623
WINK FAA AIRPORT		0	0	6	86	392	617	667	473	305	81	0	0	2627
YOAKUM		0	0	0	21	151	295	391	267	131	22	0	0	1278
YSLETA		0	0	0	101	417	660	676	493	341	108	6	0	2802

TEXAS

COOLING DEGREE DAY NORMALS (BASE 65 DEG F)

STATION		JAN	FEB	MAR	APR	MAY	JUN	JUL	AUG	SEP	OCT	NOV	DEC	ANN
ABILENE WSO	//R	0	0	29	104	244	465	592	561	340	119	13	0	2467
ALBANY		0	5	25	101	229	450	589	558	331	107	10	0	2405
ALICE		34	46	127	260	412	543	626	626	486	269	103	30	3562
ALPINE		0	0	15	43	167	351	366	332	195	71	6	0	1546
AMARILLO WSO	//R	0	0	0	16	108	303	428	372	166	35	0	0	1428
ANGLETON 2 W		29	21	56	137	291	444	530	518	399	177	54	13	2669
AUSTIN WSO	R	12	16	63	152	307	498	611	605	426	186	32	6	2914
BALLINGER 1 SW		0	0	45	121	263	465	567	539	334	112	14	0	2460
BAY CITY WATERWORKS		21	23	57	166	335	483	574	561	432	224	69	15	2960
BEEVILLE 5 NE		35	31	107	196	347	492	586	586	438	220	70	20	3128
BIG SPRING		0	0	24	99	242	456	552	518	307	101	13	0	2312
BLANCO		6	10	35	111	235	435	546	530	345	133	16	0	2402
BOERNE		6	10	37	110	229	405	499	484	327	130	17	0	2254
BONHAM		0	5	29	68	210	423	577	555	330	100	12	0	2309
BORGER 3 W		0	0	10	24	138	339	471	415	203	45	0	0	1645
BRADY 2 NNW		0	7	33	98	218	420	549	521	328	122	7	0	2303
BRENHAM		15	16	55	138	310	489	605	605	429	186	36	6	2890
BRIDGEPORT		0	5	27	87	229	465	626	595	356	119	10	0	2519
BROWNSVILLE WSO	R	70	73	164	297	440	528	592	592	492	322	136	66	3772
BROWNWOOD		0	0	34	123	264	480	611	583	370	130	11	0	2606
CAMERON		20	18	64	146	298	483	608	601	414	187	33	7	2879
CANYON		0	0	8	20	110	312	422	369	181	27	0	0	1449
CENTER		9	9	27	87	230	414	533	515	343	124	14	0	2305
CENTERVILLE		11	10	39	106	256	447	574	564	375	140	20	0	2542
CHILDRESS FAA AIRPORT		0	0	17	62	196	429	567	521	283	75	0	0	2150
CHISOS BASIN		0	0	21	74	187	306	298	263	152	66	9	0	1376
CLARENDON		0	0	9	31	129	348	484	434	213	41	0	0	1689
CLARKSVILLE 2 NE		0	6	21	53	185	387	527	502	309	106	9	0	2105
CLEBURNE		6	11	42	111	257	477	617	605	387	138	16	0	2667
COLEMAN		0	8	38	117	246	447	567	539	332	117	10	0	2421
COLLEGE STATION FAA AP		14	13	52	132	301	483	595	589	411	173	37	6	2806
CONROE		16	15	47	126	295	465	567	558	402	171	33	5	2700
CORPUS CHRISTI		39	53	131	244	409	537	601	598	501	309	127	39	3588
CORPUS CHRISTI WSO	R	40	50	125	247	406	531	617	620	495	290	116	37	3574
CORSICANA		6	6	30	91	241	459	611	598	391	140	17	0	2590
COTULLA FAA AIRPORT		25	40	130	261	431	579	667	654	489	257	63	14	3610
CROCKETT		7	7	31	97	239	423	543	536	363	134	18	0	2398
CROSBYTON		0	0	11	34	136	354	465	415	207	38	0	0	1660
CRYSTAL CITY		16	33	136	257	425	573	654	639	483	244	49	8	3517
DALHART FAA AIRPORT		0	0	0	6	64	265	391	329	123	11	0	0	1189
DALLAS-FORT WORTH REG WSO		0	7	37	112	275	510	660	636	408	146	18	0	2809
DALLAS FAA	//R	7	8	43	121	283	510	660	642	405	148	19	0	2846
DANEVANG 2 SE		29	24	83	181	338	474	564	567	435	231	69	17	3012
DEL RIO WSO		10	18	95	219	391	558	651	629	459	207	35	0	3272
DENISON DAM		0	6	23	71	197	423	574	546	324	108	13	0	2285
DENTON 2 SE		0	6	39	99	237	456	605	583	370	130	15	0	2540
DILLEY		12	25	101	218	378	540	629	623	468	229	44	8	3275
DUBLIN		0	7	28	81	193	414	555	539	334	117	9	0	2277
EAGLE PASS		10	23	104	245	415	582	676	657	483	230	42	0	3467
EDEN 1		10	10	54	136	269	456	561	530	341	138	19	0	2524
EL PASO WSO	R	0	0	11	51	218	474	543	474	273	52	0	0	2096
ENCINAL 3 NW		23	33	119	267	428	570	654	642	483	242	58	12	3531
FALFURRIAS		42	52	150	289	434	555	636	620	483	273	99	21	3654
FLATONIA 2 W		25	24	77	165	316	477	586	586	420	210	53	14	2953
FOLLETT		0	0	7	13	113	305	462	425	182	37	0	0	1544

TEXAS

COOLING DEGREE DAY NORMALS (BASE 65 DEG F)

STATION	JAN	FEB	MAR	APR	MAY	JUN	JUL	AUG	SEP	OCT	NOV	DEC	ANN
FT STOCKTON KFST RADIO	0	0	19	84	239	447	512	471	287	91	7	0	2157
FREDERICKSBURG	8	10	49	122	243	420	521	505	336	139	16	0	2369
GAINESVILLE	0	7	29	65	204	432	589	564	337	103	7	0	2337
GALVESTON WSO R	23	18	48	142	332	486	564	564	450	248	79	13	2967
GATESVILLE	11	10	45	130	265	465	605	595	402	175	27	8	2738
GILMER 2 W	0	8	24	71	202	399	539	521	327	110	12	0	2213
GOLIAD	32	36	110	206	366	510	605	608	468	254	85	20	3300
GRAHAM	0	5	32	93	227	453	608	583	352	108	8	0	2469
GREENVILLE 7 NW	0	0	19	58	197	417	561	546	340	111	8	0	2257
HALLETTSVILLE	23	27	85	177	335	492	592	595	438	215	58	12	3049
HARLINGEN	72	80	170	300	440	531	595	608	486	306	123	59	3770
HASKELL	0	0	26	87	238	471	608	577	339	109	8	0	2463
HENDERSON	11	14	31	83	222	414	539	524	337	120	13	0	2308
HENRIETTA	0	0	21	75	225	453	620	592	357	112	8	0	2463
HEREFORD	0	0	0	10	81	264	372	313	128	16	0	0	1184
HICO	10	12	43	114	248	462	605	583	372	131	15	0	2595
HILLSBORO	0	6	34	102	254	468	611	598	390	154	17	0	2634
HOUSTON INCONT AP	20	20	51	143	307	468	561	546	402	181	54	8	2761
HUNTSVILLE	14	15	49	128	291	468	577	561	384	162	31	0	2680
JACKSBORO	0	12	49	112	254	471	623	598	380	139	14	0	2652
JUNCTION	5	11	54	129	264	456	555	527	346	121	9	0	2477
KAUFMAN 3 SE	0	0	25	86	225	450	601	589	378	130	14	0	2498
KINGSVILLE	53	55	148	269	419	534	614	608	480	277	109	42	3608
LA TUNA 1 S	0	0	19	41	200	465	549	468	289	66	0	0	2097
LAMESA 1 SSE	0	0	11	50	177	417	496	453	247	50	0	0	1901
LAMPASAS	0	6	25	101	222	432	564	543	342	119	11	0	2365
LAREDO NO 2	38	59	175	333	499	615	694	679	528	311	84	28	4043
LEVELLAND	0	0	7	31	150	354	419	369	183	28	0	0	1541
LIBERTY	18	17	44	139	304	459	555	546	405	173	51	6	2717
LIVINGSTON 2 NNE	17	15	38	110	263	435	546	533	372	152	26	0	2507
LLANO	6	8	40	136	273	489	611	583	382	137	16	0	2681
LUBBOCK WSO R	0	0	11	37	157	378	459	400	201	33	0	0	1676
LUFKIN FAA AP	14	16	48	129	288	459	567	555	390	160	25	0	2651
LULING	10	13	55	142	305	480	586	586	411	176	32	6	2802
MADISONVILLE	16	11	57	134	301	480	598	589	405	164	40	5	2800
MARLIN 3 NE	9	11	49	132	275	471	605	598	411	163	27	0	2751
MARSHALL	0	9	24	76	237	420	552	533	351	123	10	0	2335
MATADOR	0	0	21	71	188	408	549	502	273	83	7	0	2102
MATAGORDA NO 2	30	27	73	175	360	504	592	583	456	248	87	18	3153
MC ALLEN	68	72	176	312	450	540	601	620	507	314	120	53	3833
MC CAMEY	0	0	41	131	313	513	592	564	361	131	12	0	2658
MC COOK	68	76	175	315	443	540	608	620	489	292	106	48	3780
MCKINNEY 3 S	0	7	34	91	232	447	598	577	364	125	15	0	2490
MEMPHIS	0	0	12	56	182	405	558	512	270	67	0	0	2062
MEXIA	0	7	31	100	244	450	592	592	393	149	18	0	2576
MIAMI	0	0	6	18	116	317	471	422	186	31	0	0	1567
MIDLAND WSO //R	0	0	19	75	228	444	518	484	283	75	0	0	2126
MIDLAND 4 ENE	0	0	23	95	251	453	518	487	291	79	0	0	2197
MINERAL WELLS FAA AP	7	7	40	105	254	480	623	598	372	128	11	0	2625
MISSION 4 W	55	67	174	321	462	555	623	636	516	303	121	42	3875
MOUNT LOCKE	0	0	6	9	65	195	191	165	78	23	0	0	732
MOUNT PLEASANT	0	6	17	59	203	408	536	508	309	89	8	0	2143
MULESHOE 1	0	0	0	14	91	298	391	326	138	14	0	0	1272
MUNDAY	0	5	34	107	260	483	617	577	343	120	10	0	2556
NEW BRAUNFELS	20	29	78	165	319	507	608	601	432	197	33	8	2997

TEXAS

COOLING DEGREE DAY NORMALS (BASE 65 DEG F)

STATION		JAN	FEB	MAR	APR	MAY	JUN	JUL	AUG	SEP	OCT	NOV	DEC	ANN
NEW GULF		22	23	68	163	329	477	561	555	426	204	61	10	2899
NIXON		21	29	97	187	344	501	605	605	438	223	57	15	3122
PALACIOS FAA AIRPORT		23	21	67	163	341	498	586	574	438	217	69	13	3010
PALESTINE		11	10	35	100	246	435	558	549	360	137	17	0	2458
PARIS		0	0	24	65	208	423	574	552	333	107	10	0	2296
PECOS		0	0	18	104	289	537	608	555	346	95	6	0	2558
PIERCE 1 E		17	19	55	152	304	462	546	549	411	198	59	10	2782
PLAINS		0	0	0	27	122	336	406	363	180	24	0	0	1458
PLAINVIEW		0	0	7	23	115	333	434	378	174	29	0	0	1493
PORT ARTHUR WSO	R	25	23	56	143	329	486	561	552	426	194	58	8	2861
PORT LAVACA NO 2		24	31	95	199	369	522	608	605	471	265	91	21	3301
PORT O CONNOR		20	21	74	173	353	510	598	589	483	274	103	20	3218
POTEET		20	29	94	202	357	525	614	611	450	215	53	11	3181
PRESIDIO		0	6	58	216	456	666	694	639	462	192	16	0	3405
QUANAH 5 SE		0	0	15	62	208	456	611	558	300	73	0	0	2283
RAYMONDVILLE		75	77	172	309	443	537	608	617	495	303	123	58	3817
RIO GRANDE CITY 3 W		45	62	177	346	487	597	670	673	519	290	91	29	3986
RISING STAR		0	0	25	80	205	420	555	530	317	95	7	0	2234
ROSCOE		0	0	30	97	231	438	539	505	290	94	9	0	2233
RUSK		9	10	39	82	219	408	527	521	345	135	20	0	2315
SAN ANGELO WSO	R	0	0	42	127	281	492	598	567	350	125	14	0	2596
SAN ANTONIO WSO	R	10	19	78	166	326	507	608	595	432	202	34	6	2983
SAN MARCOS		8	13	45	136	283	459	564	564	393	166	27	0	2658
SEMINOLE		0	0	11	49	171	390	465	419	226	48	0	0	1779
SEYMOUR		0	0	25	73	218	453	605	577	325	94	7	0	2377
SHERMAN PUMP STATION		0	0	20	64	206	432	586	564	342	108	9	0	2331
SNYDER		0	0	16	70	201	426	530	493	274	66	0	0	2076
SONORA		0	6	43	124	273	444	530	502	324	119	9	0	2374
SPEARMAN		0	0	11	26	126	326	468	412	181	31	0	0	1581
STRATFORD		0	0	0	8	65	270	406	347	133	21	0	0	1250
SUGAR LAND		22	20	63	159	335	489	580	570	426	203	62	8	2937
SULPHUR SPRINGS		0	0	18	55	189	393	543	527	321	107	9	0	2162
TAYLOR		10	10	44	132	283	477	601	598	405	167	25	0	2752
TEMPLE		11	11	43	119	267	474	608	605	411	170	21	0	2740
THROCKMORTON		0	8	33	90	235	462	614	586	342	110	8	0	2488
TULIA		0	0	0	15	94	298	406	357	160	24	0	0	1354
UVALDE		10	25	98	191	350	513	595	577	429	203	29	8	3028
VEGA		0	0	0	10	80	259	379	318	131	21	0	0	1198
VERNON		0	6	34	92	250	495	648	601	359	117	6	0	2608
VICTORIA WSO	R	27	30	87	195	363	510	605	595	453	234	69	16	3184
WACO WSO	R	8	9	43	134	288	507	648	639	426	164	25	0	2891
WAXAHACHIE		7	9	45	117	259	474	617	595	381	139	23	0	2666
WEATHERFORD		0	7	24	76	202	441	595	567	341	104	10	0	2367
WESLACO 2 E		75	82	200	333	459	537	605	617	507	329	144	70	3958
WHITNEY DAM		0	5	31	103	240	465	595	589	381	138	15	0	2562
WICHITA FALLS WSO	R	0	0	26	79	229	471	639	598	350	108	6	0	2506
WILLS POINT		0	0	22	70	217	444	608	595	381	133	14	0	2484
WINK FAA AIRPORT		0	0	17	93	274	501	570	530	324	83	0	0	2392
YOAKUM		32	37	94	199	357	513	608	614	459	241	70	16	3240
YSLETA		0	0	6	33	179	441	524	456	258	39	0	0	1936

STATE-STATION NUMBER	STN TYP	NAME		LATITUDE DEG-MIN	LONGITUDE DEG-MIN	ELEVATION (FT)
41-0012	12	ABERNATHY		N 3350	W 10150	3359
41-0016	13	ABILENE WSO	//R	N 3225	W 09941	1762
41-0034	12	ACKERLY 1 WSW		N 3231	W 10144	2670
41-0120	13	ALBANY		N 3244	W 09918	1438
41-0144	13	ALICE		N 2744	W 09804	201
41-0174	13	ALPINE		N 3021	W 10340	4455
41-0190	12	ALTO 5 SW		N 3136	W 09508	280
41-0204	12	ALVIN		N 2925	W 09515	54
41-0211	13	AMARILLO WSO	//R	N 3514	W 10142	3607
41-0235	12	ANAHUAC TBCD		N 2947	W 09440	24
41-0257	13	ANGLETON 2 W		N 2909	W 09527	27
41-0262	12	ANNA		N 3321	W 09631	674
41-0271	12	ANTELOPE		N 3326	W 09822	1079
41-0313	12	ARCHER CITY		N 3335	W 09837	1060
41-0337	12	ARLINGTON		N 3244	W 09707	630
41-0394	12	ASPERMONT 2 SSW		N 3306	W 10014	1750
41-0404	12	ATHENS 3 SSE		N 3210	W 09550	455
41-0408	12	ATLANTA 4 WSW		N 3306	W 09414	310
41-0428	13	AUSTIN WSO	R	N 3018	W 09742	597
41-0478	12	BAIRD		N 3224	W 09925	1760
41-0482	12	BAKERSFIELD		N 3053	W 10218	2530
41-0493	13	BALLINGER 1 SW		N 3144	W 09958	1575
41-0498	12	BALMORHEA		N 3059	W 10345	3220
41-0569	13	BAY CITY WATERWORKS		N 2859	W 09558	48
41-0611	12	BEAUMONT FILTER PLANT		N 3006	W 09406	20
41-0635	12	BEDIAS		N 3046	W 09557	338
41-0639	13	BEEVILLE 5 NE		N 2827	W 09742	255
41-0691	12	BENBROOK DAM		N 3239	W 09727	770
41-0786	13	BIG SPRING		N 3215	W 10127	2500
41-0787	12	BIG WELLS		N 2835	W 09934	510
41-0832	13	BLANCO		N 3006	W 09825	1350
41-0852	12	BLEWETT 5 NW		N 2914	W 10006	970
41-0902	13	BOERNE		N 2947	W 09844	1422
41-0917	12	BON WIER 2 E		N 3045	W 09337	90
41-0923	13	BONHAM		N 3336	W 09611	566
41-0944	12	BOOKER		N 3627	W 10032	2760
41-0950	12	BOQUILLAS RANGER STA		N 2911	W 10258	1880
41-0958	13	BORGER 3 W		N 3539	W 10127	3140
41-0991	12	BOXELDER		N 3329	W 09453	350
41-0996	12	BOYD		N 3305	W 09734	705
41-1007	12	BRACKETTVILLE		N 2919	W 10024	1120
41-1017	13	BRADY 2 NNW		N 3109	W 09921	1722
41-1033	12	BRAVO		N 3539	W 10300	4160
41-1035	12	BRAZOS		N 3240	W 09807	812
41-1048	13	BRENHAM		N 3009	W 09624	353
41-1063	13	BRIDGEPORT		N 3313	W 09745	815
41-1128	12	BROWNFIELD NO 2		N 3311	W 10216	3290
41-1136	13	BROWNSVILLE WSO	R	N 2554	W 09726	19
41-1138	13	BROWNWOOD		N 3143	W 09859	1435
41-1188	12	BUFFALO		N 3128	W 09603	360

STATE-STATION NUMBER	STN TYP	NAME	LATITUDE DEG-MIN	LONGITUDE DEG-MIN	ELEVATION (FT)
41-1215	12	BULVERDE	N 2945	W 09827	1100
41-1224	12	BUNKER HILL 5 N	N 3612	W 10252	4370
41-1239	12	BURKETT	N 3200	W 09914	1545
41-1245	12	BURLESON 2 SSW	N 3231	W 09720	749
41-1250	12	BURNET	N 3046	W 09814	1320
41-1348	13	CAMERON	N 3051	W 09659	390
41-1398	12	CAMP WOOD	N 2940	W 10001	1470
41-1412	12	CANADIAN 1 ENE	N 3555	W 10022	2335
41-1416	12	CANDELARIA	N 3009	W 10441	2875
41-1430	13	CANYON	N 3459	W 10155	3577
41-1490	12	CARROLLTON	N 3257	W 09654	520
41-1500	12	CARTHAGE	N 3210	W 09421	317
41-1511	12	CASE RANCH 3 S	N 3138	W 10102	2190
41-1573	12	CELINA	N 3319	W 09647	688
41-1578	13	CENTER	N 3148	W 09410	325
41-1596	13	CENTERVILLE	N 3116	W 09559	330
41-1698	13	CHILDRESS FAA AIRPORT	N 3426	W 10017	1951
41-1715	13	CHISOS BASIN	N 2916	W 10318	5300
41-1741	12	CIBOLO CREEK	N 2901	W 09756	302
41-1761	13	CLARENDON	N 3456	W 10053	2700
41-1772	13	CLARKSVILLE 2 NE	N 3338	W 09502	445
41-1800	13	CLEBURNE	N 3220	W 09724	783
41-1838	12	CLODINE	N 2942	W 09541	87
41-1874	12	COLDWATER 3 W	N 3628	W 10238	4180
41-1875	13	COLEMAN	N 3150	W 09926	1711
41-1889	13	COLLEGE STATION FAA AP	N 3035	W 09621	314
41-1911	12	COLUMBUS	N 2943	W 09632	199
41-1914	12	COMANCHE	N 3154	W 09835	1345
41-1946	12	CONLEN	N 3614	W 10214	3820
41-1956	13	CONROE	N 3019	W 09527	245
41-1970	12	COOPER	N 3322	W 09542	495
41-1974	12	COPE RANCH	N 3134	W 10115	2570
41-2012	12	CORNUDAS SERVICE STA	N 3147	W 10528	4480
41-2014	13	CORPUS CHRISTI	N 2748	W 09724	14
41-2015	13	CORPUS CHRISTI WSO R	N 2746	W 09730	41
41-2019	13	CORSICANA	N 3205	W 09628	425
41-2040	12	COTTONWOOD	N 3010	W 09908	2250
41-2048	12	COTULLA	N 2826	W 09915	433
41-2050	13	COTULLA FAA AIRPORT	N 2827	W 09913	459
41-2096	12	CRESSON	N 3232	W 09737	1047
41-2114	13	CROCKETT	N 3118	W 09527	347
41-2121	13	CROSBYTON	N 3340	W 10114	3021
41-2142	12	CROWELL	N 3359	W 09943	1451
41-2160	13	CRYSTAL CITY	N 2841	W 09950	581
41-2173	12	CUERO 3 E	N 2905	W 09715	194
41-2206	12	CYPRESS 1 SW	N 2958	W 09543	150
41-2225	12	DAINGERFIELD 9 S	N 3255	W 09443	300
41-2240	13	DALHART FAA AIRPORT	N 3601	W 10233	3989
41-2242	13	DALLAS-FORT WORTH REG WSO	N 3254	W 09702	551
41-2244	13	DALLAS FAA //R	N 3251	W 09651	481

STATE-STATION NUMBER	STN TYP	NAME		LATITUDE DEG-MIN	LONGITUDE DEG-MIN	ELEVATION (FT)
41-2266	13	DANEVANG 2 SE		N 2903	W 09611	70
41-2282	12	DARROUZETT		N 3626	W 10019	2540
41-2295	12	DAVILLA		N 3047	W 09716	520
41-2352	12	DEKALB		N 3330	W 09436	400
41-2360	13	DEL RIO WSO		N 2922	W 10055	1026
41-2394	13	DENISON DAM		N 3349	W 09634	613
41-2404	13	DENTON 2 SE		N 3312	W 09706	630
41-2444	12	DIALVILLE 2 W		N 3152	W 09516	620
41-2458	13	DILLEY		N 2840	W 09910	569
41-2462	12	DIME BOX		N 3021	W 09649	330
41-2463	12	DIMMITT 6 E		N 3433	W 10213	3810
41-2527	12	DOOLE 6 NNE		N 3129	W 09934	1447
41-2598	13	DUBLIN		N 3206	W 09820	1502
41-2633	12	DUNDEE 6 NNW		N 3349	W 09856	1051
41-2679	13	EAGLE PASS		N 2842	W 10029	815
41-2715	12	EASTLAND		N 3224	W 09849	1450
41-2741	13	EDEN 1		N 3113	W 09951	2060
41-2768	12	EDNA HWY 59 BRIDGE		N 2858	W 09641	54
41-2772	12	EDOM 2 NW		N 3223	W 09538	500
41-2797	13	EL PASO WSO	R	N 3148	W 10624	3918
41-2812	12	ELDORADO 11 NW		N 3058	W 10042	2490
41-2818	12	ELECTRA		N 3402	W 09855	1216
41-2902	12	EMORY		N 3252	W 09545	463
41-2906	13	ENCINAL 3 NW		N 2805	W 09922	562
41-2925	12	ENNIS		N 3219	W 09637	517
41-3000	12	EVADALE		N 3020	W 09405	33
41-3047	12	FAIRFIELD 3 ESE		N 3143	W 09607	475
41-3063	13	FALFURRIAS		N 2713	W 09809	120
41-3065	12	FALLS CITY 4 WSW		N 2857	W 09804	300
41-3080	12	FARMERSVILLE		N 3310	W 09622	632
41-3133	12	FERRIS		N 3232	W 09640	475
41-3156	12	FISCHERS STORE		N 2959	W 09816	1080
41-3183	13	FLATONIA 2 W		N 2941	W 09708	425
41-3196	12	FLOMOT 1 N		N 3414	W 10059	2450
41-3201	12	FLORESVILLE		N 2908	W 09810	400
41-3214	12	FLOYDADA		N 3359	W 10120	3180
41-3215	12	FLOYDADA 9 SE		N 3352	W 10115	3125
41-3225	13	FOLLETT		N 3626	W 10008	2780
41-3247	12	FORESTBURG 4 S		N 3329	W 09734	1000
41-3253	12	FORSAN		N 3206	W 10122	2740
41-3280	13	FT STOCKTON KFST RADIO		N 3052	W 10254	3000
41-3299	12	FOWLERTON		N 2828	W 09849	335
41-3329	13	FREDERICKSBURG		N 3016	W 09852	1747
41-3340	12	FREEPORT 2 NW		N 2859	W 09523	8
41-3341	12	FREER		N 2753	W 09837	530
41-3379	12	FROST		N 3205	W 09648	520
41-3401	12	FUNK RANCH		N 3129	W 10048	2070
41-3411	12	GAIL		N 3246	W 10127	2530
41-3415	13	GAINESVILLE		N 3338	W 09708	758
41-3430	13	GALVESTON WSO	R	N 2918	W 09448	7

STATE-STATION NUMBER	STN TYP	NAME	LATITUDE DEG-MIN	LONGITUDE DEG-MIN	ELEVATION (FT)
41-3445	12	GARDEN CITY 1 E	N 3152	W 10128	2634
41-3485	13	GATESVILLE	N 3126	W 09746	850
41-3508	12	GEORGE WEST	N 2819	W 09808	185
41-3525	12	GIDDINGS	N 3011	W 09656	538
41-3546	13	GILMER 2 W	N 3244	W 09459	390
41-3557	12	GIRVIN	N 3105	W 10224	2302
41-3605	12	GOLD	N 3021	W 09842	1642
41-3614	12	GOLDTHWAITE 2 N	N 3029	W 09834	1570
41-3618	13	GOLIAD	N 2840	W 09724	160
41-3620	12	GOLIAD 1 SE	N 2838	W 09723	91
41-3622	12	GONZALES	N 2930	W 09727	312
41-3668	13	GRAHAM	N 3306	W 09835	1045
41-3691	12	GRAPEVINE DAM	N 3258	W 09703	585
41-3734	13	GREENVILLE 7 NW	N 3312	W 09613	610
41-3778	12	GROVETON	N 3104	W 09508	350
41-3787	12	GRUVER	N 3615	W 10124	3169
41-3822	12	GUNTER	N 3327	W 09645	723
41-3828	12	GUTHRIE	N 3337	W 10019	1740
41-3846	12	HAGANSPORT	N 3323	W 09514	315
41-3873	13	HALLETTSVILLE	N 2927	W 09656	235
41-3884	12	HAMILTON 2SSE	N 3141	W 09807	1268
41-3890	12	HAMLIN	N 3253	W 10008	1720
41-3941	12	HARLETON	N 3240	W 09434	355
41-3943	13	HARLINGEN	N 2613	W 09741	38
41-3954	12	HARPER	N 3018	W 09915	2080
41-3981	12	HARTLEY	N 3553	W 10224	3915
41-3992	13	HASKELL	N 3310	W 09944	1605
41-4020	12	HAWKINS 1 E	N 3235	W 09511	335
41-4081	13	HENDERSON	N 3211	W 09448	456
41-4093	13	HENRIETTA	N 3349	W 09812	900
41-4098	13	HEREFORD	N 3448	W 10228	3840
41-4122	12	HEWITT 1 SE	N 3127	W 09711	642
41-4137	13	HICO	N 3159	W 09802	1095
41-4140	12	HIGGINS	N 3607	W 10001	2570
41-4182	13	HILLSBORO	N 3201	W 09707	550
41-4257	12	HONEY GROVE	N 3335	W 09554	670
41-4280	12	HORGER	N 3100	W 09410	110
41-4300	13	HOUSTON INCONT AP	N 2958	W 09521	96
41-4313	12	HOUSTON-BARKER	N 2949	W 09544	130
41-4315	12	HOUSTON-DEER PARK	N 2943	W 09508	35
41-4321	12	HOUSTON-HEIGHTS	N 2947	W 09526	65
41-4323	12	HOUSTON-INDEP HEIGHTS	N 2952	W 09525	93
41-4325	12	HOUSTON-WESTBURY	N 2940	W 09528	65
41-4327	12	HOUSTON-NORTH HOUSTON	N 2953	W 09531	111
41-4374	12	HUNT	N 3004	W 09920	1763
41-4382	13	HUNTSVILLE	N 3043	W 09533	494
41-4402	12	HYE	N 3015	W 09834	1456
41-4425	12	IMPERIAL	N 3116	W 10242	2360
41-4440	12	INDIAN GAP	N 3140	W 09825	1575
41-4517	13	JACKSBORO	N 3314	W 09809	1098

STATE-STATION NUMBER	STN TYP	NAME		LATITUDE DEG-MIN	LONGITUDE DEG-MIN	ELEVATION (FT)
41-4563	12	JASPER 3 SW		N 3054	W 09402	290
41-4575	12	JEDDO 2 NNE		N 2950	W 09718	472
41-4577	12	JEFFERSON		N 3246	W 09420	199
41-4647	12	JOURDANTON		N 2854	W 09833	497
41-4670	13	JUNCTION		N 3030	W 09947	1760
41-4693	12	KARNACK		N 3241	W 09409	230
41-4705	13	KAUFMAN 3 SE		N 3233	W 09616	420
41-4761	12	KENNEDALE 6 SSW		N 3234	W 09714	752
41-4810	13	KINGSVILLE		N 2732	W 09753	65
41-4841	12	KNAPP 6 SW		N 3237	W 10113	2610
41-4903	12	LA GRANGE		N 2955	W 09652	307
41-4931	13	LA TUNA 1 S		N 3158	W 10636	3800
41-5013	13	LAMESA 1 SSE		N 3242	W 10156	2965
41-5018	13	LAMPASAS		N 3103	W 09811	1024
41-5060	13	LAREDO NO 2		N 2731	W 09928	491
41-5094	12	LAVON DAM		N 3302	W 09629	504
41-5097	12	LAWN		N 3209	W 09945	1970
41-5158	12	LENORAH		N 3218	W 10153	2800
41-5183	13	LEVELLAND		N 3334	W 10223	3550
41-5196	13	LIBERTY		N 3003	W 09449	35
41-5229	12	LINDEN		N 3301	W 09422	403
41-5243	12	LIPAN		N 3231	W 09803	931
41-5247	12	LIPSCOMB		N 3614	W 10016	2450
41-5265	12	LITTLEFIELD NO 2		N 3354	W 10220	3500
41-5271	13	LIVINGSTON 2 NNE		N 3044	W 09456	178
41-5272	13	LLANO		N 3045	W 09841	1040
41-5284	12	LOCKHART 2 NW		N 2954	W 09742	518
41-5327	12	LONG LAKE 5 SW		N 3137	W 09551	305
41-5341	12	LONGVIEW		N 3228	W 09444	330
41-5351	12	LOOP		N 3254	W 10225	3245
41-5363	12	LORENZO		N 3340	W 10132	3170
41-5411	13	LUBBOCK WSO	R	N 3339	W 10149	3254
41-5424	13	LUFKIN FAA AP		N 3114	W 09445	281
41-5429	13	LULING		N 2941	W 09740	405
41-5477	13	MADISONVILLE		N 3057	W 09555	230
41-5579	12	MARATHON		N 3012	W 10314	4080
41-5611	13	MARLIN 3 NE		N 3120	W 09651	372
41-5618	13	MARSHALL		N 3232	W 09421	352
41-5650	12	MASON		N 3044	W 09913	1586
41-5658	13	MATADOR		N 3401	W 10050	2280
41-5659	13	MATAGORDA NO 2		N 2842	W 09558	10
41-5667	12	MAUD 1 S		N 3319	W 09420	305
41-5701	13	MC ALLEN		N 2612	W 09813	122
41-5707	13	MC CAMEY		N 3108	W 10212	2454
41-5721	13	MC COOK		N 2630	W 09823	225
41-5757	12	MC GREGOR		N 3126	W 09725	718
41-5766	13	MCKINNEY 3 S		N 3310	W 09637	595
41-5821	13	MEMPHIS		N 3444	W 10033	2100
41-5822	12	MENARD		N 3055	W 09947	1951
41-5836	12	MERCEDES 6 SSE		N 2604	W 09754	75

41 – TEXAS

LEGEND
11 = TEMPERATURE ONLY
12 = PRECIPITATION ONLY
13 = TEMP. & PRECIP.

STATE-STATION NUMBER	STN TYP	NAME		LATITUDE DEG-MIN	LONGITUDE DEG-MIN	ELEVATION (FT)
41-5859	12	MERTZON		N 3116	W 10049	2180
41-5869	13	MEXIA		N 3141	W 09629	529
41-5875	13	MIAMI		N 3542	W 10038	2744
41-5890	13	MIDLAND WSO	//R	N 3157	W 10211	2851
41-5891	13	MIDLAND 4 ENE		N 3201	W 10201	2740
41-5954	12	MINEOLA 7 SSW		N 3235	W 09533	465
41-5958	13	MINERAL WELLS FAA AP		N 3247	W 09804	934
41-5972	13	MISSION 4 W		N 2613	W 09824	133
41-6060	12	MORGAN MILL		N 3223	W 09810	1050
41-6070	12	MORSE		N 3604	W 10129	3180
41-6074	12	MORTON		N 3343	W 10245	3740
41-6104	13	MOUNT LOCKE		N 3040	W 10400	6790
41-6108	13	MOUNT PLEASANT		N 3310	W 09500	416
41-6130	12	MUENSTER		N 3339	W 09722	1005
41-6135	13	MULESHOE 1		N 3414	W 10243	3780
41-6140	12	MULLIN		N 3133	W 09840	1430
41-6146	13	MUNDAY		N 3327	W 09938	1461
41-6190	12	NAPLES 1 SW		N 3311	W 09441	362
41-6247	12	NEGLEY 4 SSW		N 3342	W 09504	400
41-6257	12	NELSON RANCH		N 2957	W 09931	2120
41-6276	13	NEW BRAUNFELS		N 2942	W 09807	718
41-6280	12	NEW CANEY 4 NW		N 3012	W 09515	250
41-6286	13	NEW GULF		N 2916	W 09555	72
41-6331	12	NEWPORT		N 3328	W 09801	1050
41-6367	12	NIX STORE 1 W		N 3107	W 09822	1360
41-6368	13	NIXON		N 2916	W 09745	406
41-6433	12	NORTHFIELD		N 3417	W 10036	2070
41-6477	12	NOTLA 3 SE		N 3606	W 10036	2900
41-6502	12	ODESSA		N 3153	W 10224	2912
41-6641	12	OLNEY 5 NNW		N 3326	W 09847	1184
41-6664	12	ORANGE 4 NW		N 3007	W 09347	14
41-6722	12	OVERTON		N 3216	W 09459	507
41-6734	12	OZONA		N 3043	W 10112	2348
41-6740	12	PADUCAH		N 3401	W 10018	1890
41-6747	12	PAINT ROCK		N 3130	W 09955	1626
41-6750	13	PALACIOS FAA AIRPORT		N 2843	W 09615	12
41-6757	13	PALESTINE		N 3147	W 09539	600
41-6766	12	PALO PINTO		N 3246	W 09819	1045
41-6776	12	PAMPA NO 2		N 3532	W 10059	3250
41-6780	12	PANDALE		N 3012	W 10134	1720
41-6785	12	PANHANDLE		N 3521	W 10123	3440
41-6794	13	PARIS		N 3340	W 09534	542
41-6879	12	PEARSALL		N 2853	W 09905	635
41-6892	13	PECOS		N 3125	W 10330	2610
41-6950	12	PERRYTON 5 NNE		N 3628	W 10047	2930
41-6953	12	PERRYTON 10 WNW		N 3627	W 10058	3000
41-7020	13	PIERCE 1 E		N 2914	W 09611	105
41-7028	12	PILOT POINT		N 3324	W 09657	710
41-7066	12	PITTSBURG 5 S		N 3256	W 09458	350
41-7074	13	PLAINS		N 3311	W 10250	3640

STATE-STATION NUMBER	STN TYP	NAME		LATITUDE DEG-MIN	LONGITUDE DEG-MIN	ELEVATION (FT)
41-7079	13	PLAINVIEW		N 3411	W 10142	3370
41-7174	13	PORT ARTHUR WSO	R	N 2957	W 09401	16
41-7182	13	PORT LAVACA NO 2		N 2838	W 09638	20
41-7186	13	PORT O CONNOR		N 2826	W 09626	14
41-7206	12	POST		N 3312	W 10124	2620
41-7215	13	POTEET		N 2902	W 09835	480
41-7262	13	PRESIDIO		N 2933	W 10421	2590
41-7327	12	PUTNAM		N 3222	W 09911	1591
41-7336	13	QUANAH 5 SE		N 3415	W 09941	1495
41-7363	12	QUITMAN 2 S		N 3246	W 09528	371
41-7388	12	RAINBOW		N 3216	W 09742	648
41-7458	13	RAYMONDVILLE		N 2629	W 09746	29
41-7480	12	RED BLUFF CROSSING		N 3113	W 09835	1235
41-7481	12	RED BLUFF DAM		N 3154	W 10355	2600
41-7529	12	REFUGIO		N 2818	W 09717	45
41-7588	12	RICHARDSON		N 3257	W 09646	623
41-7593	12	RICHLAND SPRINGS		N 3116	W 09857	1375
41-7594	12	RICHMOND		N 2935	W 09545	101
41-7614	12	RINGGOLD		N 3349	W 09757	888
41-7622	13	RIO GRANDE CITY 3 W		N 2623	W 09852	176
41-7628	12	RIOMEDINA 2 N		N 2928	W 09853	950
41-7633	13	RISING STAR		N 3206	W 09858	1627
41-7659	12	ROANOKE		N 3300	W 09713	648
41-7669	12	ROBERT LEE		N 3154	W 10029	1780
41-7677	12	ROBSTOWN		N 2747	W 09740	85
41-7700	12	ROCKLAND 1 W		N 3101	W 09424	135
41-7707	12	ROCKWALL		N 3256	W 09628	543
41-7743	13	ROSCOE		N 3227	W 10032	2380
41-7773	12	ROSSER 1 SE		N 3227	W 09626	352
41-7782	12	ROTAN		N 3252	W 10028	1925
41-7787	12	ROUND MOUNTAIN 4 WNW		N 3028	W 09825	1365
41-7836	12	RUNGE		N 2852	W 09743	308
41-7841	13	RUSK		N 3148	W 09509	720
41-7873	12	SABINAL 1 WSW		N 2919	W 09929	951
41-7943	13	SAN ANGELO WSO	R	N 3122	W 10030	1903
41-7945	13	SAN ANTONIO WSO	R	N 2932	W 09828	788
41-7983	13	SAN MARCOS		N 2953	W 09757	600
41-7992	12	SAN SABA		N 3111	W 09843	1194
41-8022	12	SANDERSON		N 3009	W 10224	2800
41-8081	12	SARITA 7 E		N 2713	W 09741	38
41-8126	12	SCHULENBURG		N 2941	W 09654	374
41-8201	13	SEMINOLE		N 3243	W 10240	3340
41-8221	13	SEYMOUR		N 3336	W 09915	1294
41-8235	12	SHAMROCK		N 3512	W 10015	2320
41-8274	13	SHERMAN PUMP STATION		N 3339	W 09637	716
41-8323	12	SILVERTON		N 3429	W 10119	3280
41-8354	12	SINTON		N 2802	W 09731	59
41-8373	12	SLATON 5 SE		N 3322	W 10136	3050
41-8378	12	SLIDELL		N 3322	W 09724	980
41-8433	13	SNYDER		N 3243	W 10055	2325

STATE-STATION NUMBER	STN TYP	NAME		LATITUDE DEG-MIN	LONGITUDE DEG-MIN	ELEVATION (FT)
41-8449	13	SONORA		N 3034	W 10039	2140
41-8523	13	SPEARMAN		N 3611	W 10111	3090
41-8583	12	STAMFORD 1		N 3256	W 09948	1634
41-8630	12	STERLING CITY		N 3151	W 10059	2290
41-8692	13	STRATFORD		N 3621	W 10205	3693
41-8696	12	STRAWN 8 NNE		N 3240	W 09828	1177
41-8728	13	SUGAR LAND		N 2937	W 09538	85
41-8743	13	SULPHUR SPRINGS		N 3308	W 09536	495
41-8818	12	TAHOKA		N 3310	W 10149	3120
41-8833	12	TAMPICO		N 3428	W 10049	2253
41-8852	12	TASCOSA		N 3534	W 10218	3410
41-8861	13	TAYLOR		N 3035	W 09724	570
41-8877	12	TEAGUE RANCH		N 3026	W 09849	1630
41-8910	13	TEMPLE		N 3106	W 09721	700
41-8929	12	TERRELL		N 3245	W 09617	510
41-9004	12	THORNTON		N 3125	W 09635	489
41-9009	12	THREE RIVERS		N 2828	W 09811	150
41-9014	13	THROCKMORTON		N 3311	W 09911	1320
41-9015	12	THURBER 5 NE		N 3232	W 09820	964
41-9076	12	TOMBALL		N 3006	W 09537	200
41-9122	12	TRENT 2 ENE		N 3230	W 10006	1920
41-9125	12	TRENTON		N 3326	W 09620	764
41-9153	12	TROY		N 3112	W 09718	700
41-9175	13	TULIA		N 3432	W 10146	3500
41-9224	12	UMBARGER		N 3457	W 10207	3746
41-9265	13	UVALDE		N 2913	W 09946	910
41-9275	12	VALENTINE 10 WSW		N 3030	W 10438	4420
41-9286	12	VALLEY VIEW		N 3329	W 09710	720
41-9295	12	VAN HORN		N 3103	W 10450	4060
41-9330	13	VEGA		N 3515	W 10225	4010
41-9346	13	VERNON		N 3410	W 09918	1212
41-9364	13	VICTORIA WSO	R	N 2851	W 09655	104
41-9410	12	VOSS 1 WSW		N 3137	W 09935	1653
41-9419	13	WACO WSO	R	N 3137	W 09713	500
41-9448	12	WALLER 3 SSW		N 3001	W 09556	143
41-9480	12	WARREN		N 3037	W 09425	116
41-9491	12	WASHINGTON STATE PARK		N 3020	W 09609	210
41-9499	12	WATER VALLEY		N 3140	W 10043	2120
41-9522	13	WAXAHACHIE		N 3224	W 09651	628
41-9532	13	WEATHERFORD		N 3246	W 09749	1065
41-9565	12	WELLINGTON 1 ESE		N 3451	W 10012	2030
41-9588	13	WESLACO 2 E		N 2609	W 09758	75
41-9655	12	WHARTON		N 2919	W 09606	111
41-9715	13	WHITNEY DAM		N 3151	W 09722	574
41-9729	13	WICHITA FALLS WSO	R	N 3358	W 09829	994
41-9800	13	WILLS POINT		N 3242	W 09601	516
41-9830	13	WINK FAA AIRPORT		N 3147	W 10312	2807
41-9836	12	WINNSBORO 6 SW		N 3253	W 09520	429
41-9847	12	WINTERS 1 NNE		N 3158	W 09957	1862
41-9859	12	WOLFE CITY		N 3322	W 09604	670

41 — TEXAS

STATE-STATION NUMBER	STN TYP	NAME	LATITUDE DEG-MIN	LONGITUDE DEG-MIN	ELEVATION (FT)
41-9952	13	YOAKUM	N 2918	W 09709	377
41-9953	12	YORKTOWN	N 2859	W 09730	260
41-9966	13	YSLETA	N 3142	W 10619	3670

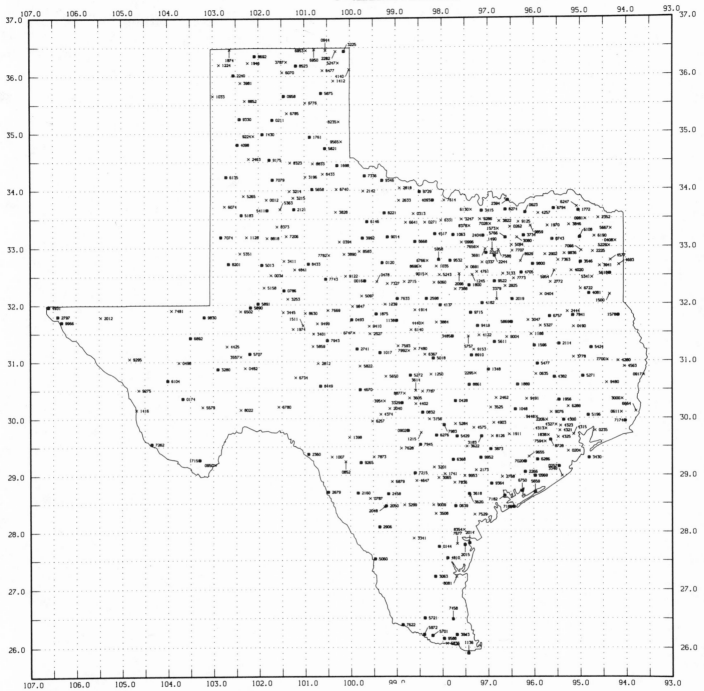

UTAH

TEMPERATURE NORMALS (DEG F)

STATION			JAN	FEB	MAR	APR	MAY	JUN	JUL	AUG	SEP	OCT	NOV	DEC	ANN
ALTON		MAX	40.4	43.7	47.6	57.0	66.9	77.3	83.8	81.3	74.5	64.3	50.7	42.8	60.9
		MIN	14.9	17.2	20.3	26.7	34.1	41.8	49.7	48.2	41.2	32.8	22.7	16.5	30.5
		MEAN	27.6	30.5	34.0	41.9	50.5	59.6	66.8	64.8	57.9	48.6	36.7	29.7	45.7
BEAR RIVER REFUGE		MAX	35.7	41.8	50.4	61.2	72.3	81.9	92.0	89.3	79.9	66.1	48.5	37.4	63.0
		MIN	14.9	19.5	28.3	36.9	46.3	53.5	59.7	56.9	47.7	38.0	28.2	19.8	37.5
		MEAN	25.3	30.7	39.4	49.1	59.3	67.7	75.8	73.1	63.8	52.1	38.4	28.6	50.3
BEAVER		MAX	42.8	46.1	52.1	60.6	71.1	81.9	88.6	85.9	78.9	67.5	52.9	44.2	64.4
		MIN	13.9	17.9	21.9	28.3	35.8	43.3	50.7	48.6	40.1	29.9	20.6	14.6	30.5
		MEAN	28.4	32.0	37.1	44.5	53.4	62.6	69.7	67.3	59.6	48.7	36.8	29.4	47.5
BLACK ROCK		MAX	41.6	48.0	55.7	65.4	75.3	85.5	93.0	90.0	81.9	69.7	53.6	43.2	66.9
		MIN	13.5	18.8	23.4	29.9	37.7	44.7	53.2	51.7	41.8	31.0	21.4	14.9	31.8
		MEAN	27.6	33.4	39.6	47.7	56.5	65.1	73.1	70.9	61.9	50.4	37.5	29.1	49.4
BLANDING		MAX	38.4	44.5	51.5	61.3	72.3	83.7	89.5	86.2	78.8	66.2	50.6	40.7	63.6
		MIN	16.2	21.5	26.3	32.9	41.4	50.0	57.5	55.3	47.3	37.3	26.2	18.3	35.9
		MEAN	27.3	33.0	38.9	47.1	56.9	66.9	73.5	70.8	63.1	51.8	38.4	29.5	49.8
BLUFF		MAX	42.5	51.8	60.7	70.5	79.9	90.5	95.9	92.8	85.6	72.7	56.7	44.4	70.3
		MIN	16.6	22.8	28.3	35.7	44.2	51.7	59.9	58.1	47.7	35.6	25.0	17.2	36.9
		MEAN	29.6	37.3	44.5	53.1	62.1	71.1	78.0	75.5	66.7	54.2	40.9	30.8	53.7
BRYCE CANYON FAA AP		MAX	35.5	38.6	43.3	52.4	62.8	73.8	80.1	77.0	70.6	59.7	45.2	37.4	56.4
		MIN	5.0	8.5	14.4	22.0	29.5	36.1	44.1	43.1	35.0	26.1	15.4	6.9	23.8
		MEAN	20.3	23.6	28.9	37.2	46.2	55.0	62.1	60.1	52.8	42.9	30.3	22.2	40.1
BRYCE CANYON NP HDQ		MAX	36.1	38.9	43.6	52.9	63.3	74.4	80.5	77.2	70.8	59.9	45.3	38.0	56.7
		MIN	8.4	11.0	15.6	22.7	30.1	37.8	45.4	43.9	35.9	26.9	17.1	10.1	25.4
		MEAN	22.2	25.0	29.6	37.8	46.7	56.1	63.0	60.6	53.4	43.4	31.2	24.1	41.1
CAPITOL REEF NATL MON		MAX	40.6	47.7	56.2	66.0	76.0	86.7	92.3	88.9	81.6	69.0	53.2	42.2	66.7
		MIN	17.5	24.0	29.8	37.5	46.3	55.1	62.5	60.4	52.7	42.9	30.0	20.3	39.9
		MEAN	29.1	35.9	43.0	51.8	61.2	70.9	77.4	74.7	67.2	56.0	41.6	31.2	53.3
CEDAR CITY FAA AP		MAX	42.1	46.6	52.4	61.2	71.6	83.3	90.1	87.3	79.9	67.8	52.7	43.9	64.9
		MIN	17.0	21.8	25.9	32.7	40.9	49.3	57.9	56.3	47.1	36.3	25.4	18.3	35.7
		MEAN	29.6	34.2	39.2	47.0	56.3	66.3	74.0	71.8	63.5	52.0	39.1	31.1	50.3
COTTONWOOD WEIR //		MAX	40.7	46.2	52.7	61.6	72.6	83.2	92.9	90.3	80.7	67.9	51.3	41.5	65.1
		MIN	21.9	26.0	30.8	38.6	47.6	56.5	66.1	63.8	54.9	43.4	31.4	23.1	42.0
		MEAN	31.3	36.1	41.8	50.1	60.1	69.9	79.6	77.1	67.8	55.7	41.4	32.3	53.6
COVE FORT //		MAX	41.8	44.9	50.2	59.5	70.1	82.0	90.5	87.7	79.6	67.0	52.1	43.4	64.1
		MIN	13.0	17.0	21.8	28.4	36.6	44.4	54.0	52.0	42.8	31.9	21.2	14.9	31.5
		MEAN	27.4	31.0	36.0	44.0	53.4	63.2	72.3	69.9	61.2	49.5	36.7	29.2	47.8
DEER CREEK DAM		MAX	33.2	37.9	45.8	56.6	67.8	77.8	87.5	84.8	76.2	64.6	47.6	37.3	59.8
		MIN	7.7	9.8	18.1	27.2	34.6	40.7	46.6	45.1	36.4	27.9	20.2	12.7	27.3
		MEAN	20.5	23.9	32.0	41.9	51.2	59.3	67.1	65.0	56.3	46.3	33.9	25.1	43.5
DESERET		MAX	38.6	46.1	54.5	64.0	74.5	85.1	93.6	90.1	80.8	67.9	51.2	40.2	65.6
		MIN	13.1	19.3	23.9	30.9	39.7	47.1	55.3	52.9	42.4	31.9	21.8	14.4	32.7
		MEAN	25.9	32.7	39.2	47.5	57.1	66.1	74.5	71.5	61.6	49.9	36.5	27.3	49.2
DESERT EXP RANGE		MAX	41.0	47.1	53.7	62.5	73.0	84.4	92.5	89.4	80.8	68.1	52.3	43.1	65.7
		MIN	11.8	18.3	23.3	30.3	38.8	47.0	55.1	53.3	43.4	32.3	21.1	13.5	32.4
		MEAN	26.4	32.7	38.5	46.4	55.9	65.7	73.8	71.4	62.2	50.3	36.7	28.4	49.0
DUGWAY		MAX	38.2	45.2	52.4	62.0	73.2	84.2	94.4	91.2	80.8	66.8	50.2	38.9	64.8
		MIN	17.2	23.7	27.9	35.7	44.6	53.5	62.5	60.0	48.7	36.8	26.1	18.7	38.0
		MEAN	27.7	34.5	40.2	48.9	59.0	68.9	78.5	75.6	64.7	51.8	38.2	28.8	51.4
ECHO DAM		MAX	35.5	40.4	47.1	57.1	68.4	78.7	88.8	86.4	77.8	65.6	48.0	37.9	61.0
		MIN	10.6	13.6	20.6	28.8	35.9	41.7	48.2	46.4	37.4	29.0	20.5	13.1	28.8
		MEAN	23.1	27.0	33.9	43.0	52.2	60.2	68.5	66.4	57.6	47.3	34.2	25.5	44.9

UTAH

TEMPERATURE NORMALS (DEG F)

STATION		JAN	FEB	MAR	APR	MAY	JUN	JUL	AUG	SEP	OCT	NOV	DEC	ANN
ELBERTA	MAX	38.2	44.2	53.1	62.8	73.5	83.9	92.0	89.0	80.2	67.1	51.0	39.7	64.6
	MIN	17.0	21.8	26.6	33.5	41.5	49.6	58.1	56.1	46.8	35.9	26.3	18.6	36.0
	MEAN	27.6	33.0	39.9	48.2	57.5	66.8	75.1	72.6	63.5	51.5	38.7	29.2	50.3
EPHRAIM SORENSENS FLD R	MAX	34.5	39.8	47.8	57.1	68.8	80.1	88.5	85.7	76.9	65.1	48.0	37.2	60.8
	MIN	13.2	17.9	24.1	31.3	39.3	47.1	54.7	52.7	44.0	34.5	23.6	15.3	33.1
	MEAN	23.9	28.9	36.0	44.2	54.1	63.6	71.7	69.2	60.5	49.8	35.8	26.3	47.0
ESCALANTE	MAX	40.7	46.6	54.2	63.3	73.2	83.8	89.6	86.2	79.5	67.9	52.8	42.7	65.0
	MIN	14.0	20.4	25.1	31.7	39.6	46.8	54.2	52.0	44.1	34.8	24.5	16.3	33.6
	MEAN	27.4	33.5	39.7	47.5	56.5	65.3	71.9	69.1	61.8	51.4	38.6	29.5	49.4
FAIRFIELD	MAX	38.1	43.6	51.6	61.1	71.8	81.4	89.6	87.2	79.7	67.9	50.9	39.8	63.6
	MIN	10.6	16.5	22.3	28.5	35.9	43.4	50.7	48.5	38.7	28.4	19.5	11.9	29.6
	MEAN	24.3	30.1	37.0	44.9	53.9	62.4	70.2	67.9	59.2	48.2	35.2	25.9	46.6
FARMINGTON USU FLD STA	MAX	38.7	44.9	52.4	61.9	72.8	82.6	92.1	89.6	80.0	67.2	50.8	40.0	64.4
	MIN	19.3	23.6	28.8	36.0	44.1	51.6	59.3	57.2	47.8	37.9	28.1	21.2	37.9
	MEAN	29.1	34.3	40.6	49.0	58.5	67.2	75.7	73.4	63.9	52.6	39.5	30.6	51.2
FERRON	MAX	34.9	41.3	49.3	59.5	69.9	80.2	87.3	84.3	77.0	65.9	49.5	38.1	61.4
	MIN	10.6	16.6	23.5	32.6	41.9	51.0	57.9	55.0	46.2	35.4	22.7	13.7	33.9
	MEAN	22.8	29.0	36.4	46.1	56.0	65.6	72.6	69.6	61.6	50.7	36.2	26.0	47.7
FILLMORE	MAX	40.9	46.5	53.7	62.5	72.8	83.9	92.1	89.5	81.3	68.6	52.1	41.9	65.5
	MIN	17.3	22.5	27.1	34.3	42.6	50.9	59.6	57.6	48.7	37.4	26.4	18.8	36.9
	MEAN	29.1	34.5	40.5	48.4	57.7	67.4	75.9	73.6	65.0	53.0	39.3	30.4	51.2
FORT DUCHESNE	MAX	28.3	36.5	49.5	62.0	73.2	83.3	91.3	87.9	78.9	65.5	46.8	32.7	61.3
	MIN	1.1	7.4	19.7	29.1	38.3	45.3	51.7	49.5	39.8	29.7	18.5	6.2	28.0
	MEAN	14.8	22.0	34.6	45.6	55.8	64.4	71.5	68.7	59.4	47.6	32.7	19.5	44.7
GARFIELD	MAX	36.7	41.9	49.4	58.9	70.1	80.7	91.0	88.1	77.5	64.0	48.6	38.7	62.1
	MIN	22.3	26.9	32.6	41.0	50.6	60.0	68.9	66.1	55.8	43.8	32.6	24.8	43.8
	MEAN	29.5	34.4	41.0	50.0	60.4	70.4	80.0	77.1	66.7	53.9	40.6	31.8	53.0
GARLAND	MAX	33.2	39.0	47.4	58.3	69.6	79.3	89.6	87.0	77.3	64.6	47.6	36.1	60.8
	MIN	14.3	19.0	25.5	33.7	42.3	49.9	57.3	55.3	45.6	35.4	25.3	17.1	35.1
	MEAN	23.8	29.0	36.5	46.1	56.0	64.6	73.5	71.2	61.5	50.0	36.5	26.6	47.9
GREEN RIVER AVN	MAX	36.8	47.7	58.1	68.6	79.0	89.9	96.4	93.3	84.9	71.2	53.9	40.9	68.4
	MIN	9.3	17.5	26.1	34.8	44.2	51.4	59.5	57.0	45.8	34.5	22.6	12.8	34.6
	MEAN	23.1	32.6	42.1	51.7	61.6	70.7	78.0	75.2	65.4	52.9	38.3	26.9	51.5
HANKSVILLE	MAX	39.8	48.6	58.4	68.7	80.0	91.4	98.0	94.4	85.6	71.8	54.0	42.1	69.4
	MIN	11.4	19.5	27.3	36.1	45.9	54.2	61.9	59.7	49.1	37.0	23.9	14.3	36.7
	MEAN	25.6	34.1	42.9	52.4	62.9	72.8	80.0	77.0	67.4	54.4	39.0	28.2	53.1
HEBER	MAX	34.7	39.7	47.3	57.6	68.8	77.8	87.0	84.7	76.8	65.7	48.3	37.6	60.5
	MIN	8.8	12.8	20.5	28.0	34.9	40.9	47.7	46.0	37.6	28.9	20.1	12.0	28.2
	MEAN	21.8	26.3	33.9	42.9	51.8	59.4	67.4	65.4	57.2	47.4	34.2	24.8	44.4
HIAWATHA	MAX	32.4	37.0	43.9	54.2	64.6	75.5	83.3	79.3	71.6	59.7	43.3	34.7	56.6
	MIN	13.7	17.6	22.1	30.3	39.2	49.1	56.1	54.2	46.4	36.4	23.8	16.1	33.8
	MEAN	23.1	27.3	33.1	42.2	51.9	62.3	69.7	66.8	59.0	48.1	33.5	25.4	45.2
JENSEN	MAX	29.2	37.7	50.8	63.4	74.8	84.5	92.4	89.2	80.5	67.0	48.2	33.6	62.6
	MIN	1.5	7.8	19.8	29.5	38.8	45.5	52.0	48.9	39.5	28.9	18.3	6.4	28.1
	MEAN	15.4	22.8	35.3	46.5	56.8	65.1	72.2	69.1	60.0	48.0	33.3	20.0	45.4
KANAB	MAX	47.8	53.5	58.6	67.4	76.8	87.5	93.1	90.2	84.0	73.6	59.1	50.0	70.1
	MIN	22.5	26.0	29.4	35.6	43.2	51.0	58.7	57.2	50.3	40.6	30.3	23.6	39.0
	MEAN	35.1	39.7	44.0	51.5	60.0	69.3	75.9	73.7	67.2	57.1	44.8	36.8	54.6
LAKETOWN	MAX	32.9	35.7	41.0	53.1	64.8	73.9	83.4	80.9	72.5	60.1	44.1	35.2	56.5
	MIN	10.6	10.7	16.4	26.1	34.2	40.3	46.7	44.9	37.1	28.7	20.9	14.2	27.6
	MEAN	21.7	23.2	28.7	39.7	49.5	57.1	65.1	62.9	54.8	44.4	32.5	24.7	42.0

UTAH

TEMPERATURE NORMALS (DEG F)

STATION		JAN	FEB	MAR	APR	MAY	JUN	JUL	AUG	SEP	OCT	NOV	DEC	ANN
LA VERKIN	MAX	52.1	58.4	63.9	72.2	81.9	91.9	97.6	94.9	88.7	77.6	62.8	53.5	74.6
	MIN	24.6	29.7	34.2	40.0	47.2	55.1	63.2	61.8	54.0	42.7	31.6	25.1	42.4
	MEAN	38.4	44.1	49.1	56.1	64.6	73.5	80.4	78.4	71.3	60.2	47.2	39.4	58.6
LEVAN	MAX	38.5	44.0	51.9	61.3	71.9	82.8	91.3	88.6	80.4	67.9	51.4	40.3	64.2
	MIN	14.0	19.2	24.5	31.7	39.9	47.6	55.8	53.8	44.8	34.7	24.3	16.3	33.9
	MEAN	26.3	31.6	38.3	46.5	55.9	65.2	73.6	71.2	62.6	51.4	37.9	28.3	49.1
LOA	MAX	39.4	43.1	48.3	57.0	67.0	77.0	82.5	79.5	73.3	63.4	49.1	40.8	60.0
	MIN	7.8	12.4	17.5	24.6	32.9	39.7	47.1	45.2	36.6	26.7	16.3	9.0	26.3
	MEAN	23.6	27.8	32.9	40.8	50.0	58.4	64.8	62.4	55.0	45.1	32.7	24.9	43.2
LOGAN UTAH STATE UNIV	MAX	33.1	38.0	46.0	56.8	68.0	77.1	87.2	85.1	75.3	62.7	45.7	35.3	59.2
	MIN	16.3	20.0	26.3	35.2	43.8	50.9	58.7	57.1	48.2	38.5	27.6	19.1	36.8
	MEAN	24.7	29.0	36.2	46.0	55.9	64.0	73.0	71.1	61.8	50.6	36.7	27.2	48.0
MANTI	MAX	37.4	42.4	50.7	59.9	70.0	80.0	87.4	85.0	77.0	65.5	49.4	39.1	62.0
	MIN	14.7	18.8	24.1	31.4	39.1	46.5	53.8	52.0	43.6	34.3	23.9	16.4	33.2
	MEAN	26.1	30.6	37.4	45.6	54.6	63.3	70.6	68.5	60.3	49.9	36.7	27.8	47.6
MEXICAN HAT	MAX	43.4	52.9	61.2	71.1	81.6	93.0	98.4	95.2	87.6	74.4	57.8	45.3	71.8
	MIN	18.9	24.8	30.1	38.4	48.1	57.2	65.2	63.1	52.8	40.1	28.7	20.0	40.6
	MEAN	31.2	38.9	45.7	54.8	64.9	75.1	81.8	79.2	70.2	57.3	43.3	32.7	56.3
MILFORD WSO //R	MAX	39.4	45.3	52.7	62.2	73.1	84.8	92.9	89.9	81.2	67.8	51.5	41.5	65.2
	MIN	13.4	18.8	23.6	30.4	38.6	46.8	55.6	54.2	43.9	32.6	22.0	14.8	32.9
	MEAN	26.4	32.1	38.2	46.3	55.9	65.8	74.3	72.1	62.6	50.3	36.8	28.2	49.1
MOAB 4 NW	MAX	41.8	50.9	60.8	71.2	81.9	92.7	99.4	96.1	88.0	74.9	57.4	44.5	71.6
	MIN	18.6	25.0	33.1	41.6	50.2	57.6	64.8	62.8	52.9	41.1	29.6	21.2	41.5
	MEAN	30.2	38.0	47.0	56.4	66.1	75.2	82.1	79.5	70.5	58.0	43.5	32.9	56.6
MODENA	MAX	42.7	48.1	54.6	63.7	73.7	85.0	91.9	88.7	81.8	70.0	54.1	44.8	66.6
	MIN	14.5	19.9	22.7	28.6	36.5	44.6	52.9	51.7	42.8	32.0	22.2	15.8	32.0
	MEAN	28.7	34.0	38.6	46.2	55.2	64.8	72.4	70.3	62.3	51.0	38.1	30.3	49.3
MONTICELLO	MAX	35.7	40.4	47.1	57.5	67.6	78.5	84.5	81.3	74.3	63.0	47.9	38.5	59.7
	MIN	14.2	17.6	22.6	29.6	37.7	45.4	52.8	50.9	43.4	34.1	23.3	15.8	32.3
	MEAN	25.0	29.0	34.9	43.6	52.7	62.0	68.6	66.1	58.9	48.6	35.6	27.2	46.0
MORGAN	MAX	35.3	40.7	48.4	59.0	70.4	80.8	89.5	86.9	78.1	66.3	48.0	38.1	61.8
	MIN	11.6	15.5	22.1	29.6	36.6	42.3	48.9	47.0	38.2	29.6	21.1	13.7	29.7
	MEAN	23.5	28.1	35.3	44.3	53.5	61.6	69.2	67.0	58.2	48.0	34.6	25.9	45.8
MORONI	MAX	36.5	42.3	50.8	60.7	71.5	81.9	89.6	86.7	79.4	67.3	50.0	38.7	63.0
	MIN	11.1	15.6	21.8	28.2	35.9	42.4	49.6	48.1	39.2	30.6	20.8	12.5	29.7
	MEAN	23.8	29.0	36.3	44.5	53.8	62.1	69.6	67.4	59.3	49.0	35.5	25.6	46.3
MOUNTAIN DELL DAM	MAX	38.5	43.0	48.8	58.2	69.1	78.7	88.3	86.0	77.3	64.9	48.7	39.6	61.8
	MIN	13.5	16.4	21.6	29.7	37.4	43.5	50.8	49.5	41.1	32.7	23.1	15.6	31.2
	MEAN	26.1	29.7	35.2	44.0	53.3	61.1	69.6	67.8	59.2	48.8	35.9	27.6	46.5
NEPHI	MAX	40.3	44.9	51.8	61.3	72.3	83.8	93.6	90.7	81.6	69.2	52.6	42.3	65.4
	MIN	17.4	21.9	26.9	34.0	42.1	50.1	58.3	56.2	47.1	36.4	26.5	19.0	36.3
	MEAN	28.9	33.4	39.4	47.7	57.2	67.0	76.0	73.5	64.4	52.9	39.5	30.7	50.9
OAK CITY	MAX	40.5	46.4	53.6	62.6	73.8	85.1	94.1	91.4	82.8	70.1	52.3	42.2	66.2
	MIN	18.8	23.8	28.5	35.5	44.6	53.6	62.6	60.0	50.6	39.4	27.6	20.2	38.8
	MEAN	29.7	35.1	41.1	49.1	59.2	69.4	78.4	75.7	66.7	54.8	40.0	31.2	52.5
OGDEN PIONEER PH	MAX	37.6	43.4	50.7	60.4	71.5	81.6	91.6	88.6	79.0	66.2	49.8	39.3	63.3
	MIN	19.5	23.8	29.4	37.6	46.4	54.3	62.4	59.9	50.5	39.9	29.1	21.7	39.5
	MEAN	28.6	33.6	40.0	49.0	59.0	68.0	77.0	74.3	64.8	53.1	39.4	30.5	51.4
OGDEN SUGAR FACTORY	MAX	37.2	42.9	50.9	61.1	72.2	82.2	92.2	89.3	79.7	67.0	50.2	39.5	63.7
	MIN	18.3	22.9	28.8	36.5	44.8	52.3	59.5	57.2	47.6	37.8	28.2	21.1	37.9
	MEAN	27.8	32.9	39.9	48.8	58.5	67.3	75.9	73.3	63.7	52.4	39.2	30.3	50.8

UTAH

TEMPERATURE NORMALS (DEG F)

STATION		JAN	FEB	MAR	APR	MAY	JUN	JUL	AUG	SEP	OCT	NOV	DEC	ANN
ORDERVILLE	MAX	45.9	50.5	55.0	64.0	73.8	84.5	90.8	88.2	81.4	71.2	57.1	48.4	67.6
	MIN	15.4	19.8	23.3	30.1	38.7	46.5	53.6	52.3	44.8	34.2	23.9	17.0	33.3
	MEAN	30.7	35.1	39.2	47.1	56.2	65.6	72.2	70.3	63.1	52.7	40.6	32.7	50.5
PAROWAN	MAX	42.5	46.6	52.1	60.4	70.6	81.5	87.7	85.1	78.6	68.0	53.1	44.4	64.2
	MIN	15.8	20.2	24.8	31.7	39.9	48.0	55.6	53.7	45.2	34.9	24.2	17.4	34.3
	MEAN	29.2	33.4	38.5	46.1	55.3	64.8	71.6	69.4	61.9	51.5	38.7	30.9	49.3
PARTOUN	MAX	41.0	47.3	54.7	63.8	74.5	85.4	94.9	91.9	82.8	69.0	52.8	42.0	66.7
	MIN	13.2	18.8	23.7	31.5	39.9	48.0	55.4	53.2	43.0	32.7	22.3	14.3	33.0
	MEAN	27.1	33.1	39.2	47.7	57.3	66.8	75.2	72.6	62.9	50.9	37.6	28.2	49.9
PINE VIEW DAM	MAX	31.0	36.5	44.6	56.3	67.8	77.5	87.7	85.1	75.7	63.0	44.8	33.7	58.6
	MIN	8.4	11.1	19.5	30.0	37.6	43.8	50.8	48.9	40.6	31.2	22.1	12.7	29.7
	MEAN	19.7	23.8	32.1	43.1	52.7	60.7	69.3	67.0	58.2	47.1	33.5	23.2	44.2
RICHFIELD RADIO KSVC	MAX	41.9	46.9	54.4	63.0	72.7	82.9	89.9	87.5	80.3	69.3	53.5	43.7	65.5
	MIN	14.1	18.9	23.3	29.5	37.2	44.0	51.6	50.0	40.4	30.5	21.5	15.1	31.3
	MEAN	28.0	32.9	38.9	46.3	55.0	63.5	70.8	68.8	60.4	49.9	37.5	29.4	48.5
RICHMOND	MAX	33.3	38.8	47.4	58.8	69.8	79.5	90.2	87.8	77.9	64.5	47.0	35.5	60.9
	MIN	14.3	17.8	23.5	31.4	39.2	45.9	52.7	51.5	42.8	33.5	24.6	16.7	32.8
	MEAN	23.8	28.4	35.5	45.1	54.5	62.7	71.5	69.7	60.4	49.0	35.9	26.1	46.9
RIVERDALE POWER HOUSE	MAX	37.0	42.5	50.3	60.1	71.1	81.0	91.1	87.9	78.5	65.4	49.1	38.4	62.7
	MIN	19.4	23.9	29.3	36.8	44.7	52.3	60.3	58.3	49.3	39.5	29.0	21.3	38.7
	MEAN	28.2	33.2	39.8	48.5	57.9	66.7	75.7	73.1	63.9	52.5	39.1	29.9	50.7
ROOSEVELT	MAX	29.8	38.1	51.4	63.1	74.4	84.1	91.7	88.4	79.7	66.5	47.9	33.8	62.4
	MIN	4.7	11.0	22.2	31.6	40.7	48.2	55.1	52.8	43.4	32.8	20.8	9.2	31.0
	MEAN	17.3	24.6	36.8	47.4	57.6	66.2	73.4	70.6	61.6	49.7	34.4	21.5	46.8
SAINT GEORGE	MAX	53.6	60.9	67.1	75.9	85.6	95.9	101.9	99.4	93.2	81.4	65.1	54.7	77.9
	MIN	26.9	31.5	36.6	43.7	52.2	60.5	67.9	66.3	56.8	45.1	33.9	27.1	45.7
	MEAN	40.3	46.2	51.9	59.8	68.9	78.3	84.9	82.8	75.0	63.3	49.5	40.9	61.8
SALINA	MAX	40.9	46.5	54.6	64.2	74.6	85.2	93.0	90.3	82.1	69.9	53.2	42.7	66.4
	MIN	14.2	19.1	24.1	30.7	38.8	46.3	54.4	51.8	42.3	31.6	22.3	15.0	32.6
	MEAN	27.6	32.9	39.4	47.5	56.7	65.8	73.7	71.1	62.2	50.8	37.7	28.9	49.5
SALT LAKE CITY WSFO R	MAX	37.4	43.7	51.5	61.1	72.4	83.3	93.2	90.0	80.0	66.7	50.2	38.9	64.0
	MIN	19.7	24.4	29.9	37.2	45.2	53.3	61.8	59.7	50.0	39.3	29.2	21.6	39.3
	MEAN	28.6	34.1	40.7	49.2	58.8	68.3	77.5	74.9	65.0	53.0	39.7	30.3	51.7
SANTAQUIN	MAX	39.3	44.3	51.2	60.8	71.8	82.4	90.9	88.0	79.0	66.2	50.2	40.4	63.7
	MIN	16.7	21.0	26.3	33.9	42.9	51.4	59.7	57.3	47.8	37.4	26.2	18.7	36.6
	MEAN	28.1	32.7	38.8	47.4	57.3	66.9	75.3	72.7	63.4	51.8	38.3	29.6	50.2
SCIPIO	MAX	38.9	44.2	52.2	61.6	72.0	81.6	89.6	87.2	79.4	67.4	51.3	40.5	63.8
	MIN	11.1	17.3	22.6	29.2	37.3	44.5	53.5	51.5	40.8	29.9	20.5	13.0	30.9
	MEAN	25.0	30.8	37.5	45.4	54.6	63.0	71.6	69.4	60.1	48.7	35.9	26.8	47.4
SCOFIELD DAM	MAX	28.0	32.2	37.9	47.9	59.8	70.1	78.0	75.2	67.6	56.5	40.1	31.0	52.0
	MIN	-.1	2.8	11.1	21.4	30.7	37.6	44.2	42.7	34.7	26.2	15.6	3.9	22.6
	MEAN	14.0	17.5	24.5	34.7	45.3	53.9	61.2	59.0	51.1	41.4	27.9	17.4	37.3
SILVER LAKE BRIGHTON	MAX	30.6	33.5	36.5	43.9	53.6	64.0	72.7	70.3	62.8	52.0	38.7	32.6	49.3
	MIN	7.3	8.5	11.3	19.3	28.3	36.1	43.7	42.0	34.5	26.1	15.1	9.0	23.4
	MEAN	19.0	21.0	24.0	31.6	40.9	50.1	58.2	56.2	48.7	39.1	27.0	20.8	36.4
SNAKE CREEK PH	MAX	33.9	38.4	45.2	56.1	67.4	76.9	85.0	82.2	74.1	63.0	45.9	36.0	58.7
	MIN	10.0	12.5	19.0	26.8	33.8	39.5	45.6	44.5	36.8	28.7	19.4	12.2	27.4
	MEAN	22.0	25.5	32.1	41.4	50.7	58.3	65.3	63.3	55.5	45.9	32.7	24.1	43.1
SPANISH FORK PWR HOUSE	MAX	38.1	43.8	52.2	62.1	73.4	84.0	92.9	89.7	80.5	67.4	50.3	39.5	64.5
	MIN	20.0	24.1	29.0	36.2	44.8	52.0	59.6	57.5	49.2	40.1	29.6	21.7	38.7
	MEAN	29.1	34.0	40.6	49.2	59.1	68.0	76.3	73.6	64.9	53.8	40.0	30.6	51.6

UTAH

TEMPERATURE NORMALS (DEG F)

STATION		JAN	FEB	MAR	APR	MAY	JUN	JUL	AUG	SEP	OCT	NOV	DEC	ANN
TIMPANOGOS CAVE	MAX	34.3	40.7	48.7	58.5	69.8	80.4	89.8	87.3	78.3	64.0	44.0	34.9	60.9
	MIN	20.2	22.9	26.8	33.9	42.1	49.4	57.6	55.7	48.0	39.2	28.6	21.8	37.2
	MEAN	27.3	31.9	37.8	46.2	56.0	64.9	73.7	71.6	63.2	51.6	36.3	28.3	49.1
TOOELE-	MAX	38.5	43.1	49.7	58.9	69.0	78.9	87.9	84.9	75.7	63.1	48.4	39.3	61.5
	MIN	20.4	24.6	29.4	37.1	46.4	55.1	63.6	61.1	52.0	40.5	29.2	22.0	40.1
	MEAN	29.5	33.9	39.6	48.0	57.7	67.0	75.8	73.0	63.9	51.8	38.8	30.7	50.8
TROPIC	MAX	41.0	45.7	51.5	60.3	69.7	79.7	85.6	82.4	75.6	66.3	51.9	43.6	62.8
	MIN	14.2	18.9	22.6	28.9	36.4	44.6	51.9	49.5	41.9	33.6	23.3	16.0	31.8
	MEAN	27.6	32.4	37.1	44.6	53.1	62.1	68.8	66.0	58.7	50.0	37.6	29.9	47.3
UTAH LAKE LEHI	MAX	36.7	42.1	50.3	60.3	71.6	81.5	90.0	87.1	77.9	65.2	48.5	38.4	62.5
	MIN	15.7	20.8	26.3	33.3	41.0	48.1	55.2	53.5	44.2	34.4	25.4	18.2	34.7
	MEAN	26.2	31.5	38.3	46.8	56.3	64.8	72.6	70.3	61.1	49.8	37.0	28.4	48.6
VERNAL AIRPORT	MAX	29.0	37.1	49.4	62.0	72.9	82.8	90.3	87.2	78.1	64.2	45.8	32.3	60.9
	MIN	4.6	10.2	20.6	29.3	38.3	45.9	52.1	49.7	40.5	30.6	19.6	8.9	29.2
	MEAN	16.8	23.7	35.0	45.6	55.6	64.4	71.2	68.5	59.3	47.4	32.7	20.6	45.1
WENDOVER WSO //R	MAX	36.5	43.5	51.5	61.3	71.9	82.0	92.1	89.0	78.4	63.3	47.1	37.0	62.8
	MIN	19.7	25.2	31.3	39.6	49.7	58.8	67.5	64.3	53.6	41.3	29.2	20.6	41.7
	MEAN	28.1	34.4	41.4	50.5	60.8	70.4	79.8	76.7	66.0	52.4	38.2	28.8	52.3
WOODRUFF	MAX	28.7	32.7	39.9	52.2	63.6	72.0	81.5	79.4	71.6	60.2	42.4	31.8	54.7
	MIN	2.7	5.1	13.9	23.8	31.3	38.8	43.7	41.2	32.0	22.8	13.9	5.6	22.9
	MEAN	15.8	18.9	26.9	38.1	47.5	55.4	62.6	60.3	51.8	41.5	28.2	18.8	38.8
ZION NATIONAL PARK	MAX	51.2	57.3	62.6	71.9	82.2	93.5	99.7	96.7	90.0	78.5	62.3	52.9	74.9
	MIN	28.9	32.7	35.9	42.8	51.7	61.1	68.6	66.9	60.1	49.6	37.4	30.1	47.2
	MEAN	40.1	45.0	49.3	57.4	67.0	77.3	84.2	81.8	75.1	64.1	49.9	41.5	61.1

UTAH

PRECIPITATION NORMALS (INCHES)

STATION	JAN	FEB	MAR	APR	MAY	JUN	JUL	AUG	SEP	OCT	NOV	DEC	ANN
ALPINE	1.68	1.45	1.51	1.96	1.48	1.00	.51	.92	.91	1.30	1.27	1.56	15.55
ALTON	2.04	1.68	1.57	1.13	.93	.51	1.49	1.77	1.32	1.12	1.39	1.61	16.56
BEAR RIVER REFUGE	1.16	.98	.89	1.40	1.31	1.16	.35	.64	.87	1.05	1.03	.96	11.80
BEAVER	.84	.91	.90	1.12	1.07	.57	1.10	1.36	1.00	.80	.72	.74	11.13
BEAVER CANYON PH	1.80	1.84	2.30	2.12	1.63	.84	1.47	1.66	1.27	1.17	1.32	1.50	18.92
BLACK ROCK	.61	.53	1.01	.97	.84	.53	.83	.74	.63	.69	.72	.50	8.60
BLANDING	1.34	.95	.80	.67	.59	.37	1.04	1.41	.89	1.46	.89	1.29	11.70
BLUFF	.78	.64	.55	.40	.37	.19	.76	.77	.60	1.15	.61	.79	7.61
BRYCE CANYON FAA AP	.84	1.04	.98	.74	.89	.53	1.09	1.77	1.27	1.04	.96	.96	12.11
BRYCE CANYON NP HDQ	1.25	1.35	1.37	.96	1.02	.58	1.22	2.15	1.52	1.22	1.18	1.16	14.98
CALLAO	.31	.34	.35	.43	.67	.72	.41	.53	.37	.45	.31	.25	5.14
CAPITOL REEF NATL MON	.28	.23	.38	.48	.68	.40	.92	1.07	.74	.81	.59	.28	6.86
CASTLE DALE	.53	.52	.44	.48	.74	.48	.66	.97	.72	.71	.55	.52	7.32
CEDAR CITY FAA AP	.64	.80	1.06	.98	.82	.45	1.10	1.17	.90	.78	.91	.65	10.26
CITY CREEK WATER PLANT	2.83	2.82	3.35	4.01	2.54	1.51	.97	1.12	1.48	2.17	2.32	2.79	27.91
COALVILLE	1.28	1.10	1.35	1.83	1.58	1.12	.83	.95	1.03	1.27	1.35	1.35	15.04
CORINNE //	1.78	1.52	1.36	1.73	1.66	1.42	.48	.80	1.04	1.18	1.39	1.50	15.86
COTTONWOOD WEIR //	1.98	1.96	2.61	3.03	2.24	1.20	.74	1.25	1.46	1.80	1.92	2.26	22.45
COVE FORT //	.97	1.35	1.54	1.54	1.20	.65	1.00	1.12	.95	.98	1.03	1.07	13.40
DEER CREEK DAM	3.09	2.43	2.02	1.78	1.49	1.06	.64	1.03	1.09	1.60	2.03	2.55	20.81
DESERET	.59	.52	.75	.80	.82	.43	.45	.63	.49	.67	.63	.55	7.33
DESERT EXP RANGE	.30	.31	.55	.60	.62	.42	.78	.84	.56	.50	.35	.29	6.12
DUGWAY	.51	.59	.63	.76	.83	.59	.44	.47	.48	.55	.52	.54	6.91
ECHO DAM	1.15	.94	1.20	1.58	1.54	1.17	.71	.92	.93	1.30	1.13	1.23	13.80
ELBERTA	.90	.80	.93	1.06	.98	.73	.65	1.04	.68	.85	.90	.94	10.46
EPHRAIM SORENSENS FLD R	.96	1.03	1.07	1.08	.99	.62	.62	.70	.94	.89	.85	.90	10.65
ESCALANTE	.91	.62	.76	.58	.72	.43	1.13	1.70	.97	1.05	.88	.80	10.55
EUREKA	1.68	1.53	1.74	1.69	1.46	1.13	1.21	1.39	.99	1.18	1.35	1.45	16.80
FAIRFIELD	1.00	.85	1.01	.94	1.08	.75	.91	.99	.75	.86	.87	.95	10.96
FARMINGTON USU FLD STA	2.11	1.89	2.03	2.94	2.22	1.36	.58	1.08	1.11	1.52	1.71	1.77	20.32
FERRON	.66	.60	.55	.47	.78	.51	.85	1.17	.78	.70	.58	.51	8.16
FILLMORE	1.45	1.52	1.79	1.75	1.26	.68	.63	.78	.93	1.07	1.31	1.34	14.51
FORT DUCHESNE	.44	.34	.50	.60	.62	.69	.52	.73	.61	.78	.47	.52	6.82
GARFIELD	1.21	1.18	1.64	2.30	1.71	1.22	.66	.76	1.13	1.38	1.36	1.30	15.85
GARLAND	1.56	1.35	1.26	1.55	1.62	1.55	.54	.98	.95	1.17	1.22	1.17	14.92
GARRISON	.46	.43	.82	.77	.81	.48	.53	.69	.53	.70	.59	.49	7.30
GREEN RIVER AVN	.40	.37	.46	.45	.61	.34	.38	.79	.61	.78	.46	.39	6.04
GUNLOCK POWER HOUSE	1.44	1.48	1.50	.76	.71	.31	.82	1.07	.73	.86	1.05	.91	11.64
HANKSVILLE	.30	.22	.35	.42	.49	.23	.44	.83	.60	.63	.43	.30	5.24
HANNA	1.10	.85	.73	.80	1.07	1.00	.87	1.40	1.06	1.08	.78	1.09	11.83
HATCH	.77	.72	.75	.55	.77	.52	1.63	1.52	1.08	.78	.87	.73	10.69
HEBER	2.09	1.52	1.27	1.32	1.18	.93	.65	.92	.92	1.29	1.50	1.73	15.32
HIAWATHA	1.05	1.03	1.04	1.01	1.19	.95	1.07	1.72	1.26	1.12	.89	1.18	13.51
IBAPAH	.49	.55	.80	.93	1.24	1.12	.58	.65	.51	.65	.46	.42	8.40
JENSEN	.51	.52	.61	.64	.75	.69	.43	.67	.71	.89	.53	.60	7.55
KAMAS RANGER STATION	1.80	1.88	1.53	1.79	1.56	1.15	.96	1.04	1.15	1.43	1.65	1.72	17.66
KANAB	1.75	1.25	1.41	.82	.68	.38	.87	1.37	.79	.90	1.11	1.24	12.57
KANOSH	1.21	1.36	1.62	1.67	1.34	.64	.74	.80	.87	1.01	1.19	1.17	13.62
KOOSHAREM	.64	.58	.56	.63	.88	.60	1.11	1.34	.95	.74	.50	.56	9.09
LAKETOWN	1.09	.88	.89	1.11	1.13	1.15	.53	.81	.84	.94	1.01	1.16	11.54
LA VERKIN	1.36	1.28	1.54	.78	.54	.30	.73	.86	.76	.69	.89	.86	10.59
LEVAN	1.31	1.32	1.52	1.66	1.33	.76	.68	.91	1.05	1.09	1.24	1.37	14.24
LOA	.39	.27	.34	.42	.69	.39	1.10	1.21	.87	.63	.42	.34	7.07
LOGAN UTAH STATE UNIV	1.68	1.57	1.75	2.06	1.71	1.53	.45	.96	1.06	1.43	1.53	1.63	17.36
MANTI	1.13	1.20	1.28	1.40	1.16	.69	.67	.89	1.08	.99	1.05	.99	12.53

UTAH

PRECIPITATION NORMALS (INCHES)

STATION	JAN	FEB	MAR	APR	MAY	JUN	JUL	AUG	SEP	OCT	NOV	DEC	ANN
MEXICAN HAT	.50	.43	.38	.31	.35	.19	.66	.65	.54	.96	.51	.61	6.09
MILFORD WSO //R	.69	.74	.99	.96	.73	.42	.61	.71	.69	.73	.69	.63	8.59
MINERSVILLE	.76	.82	1.05	1.00	.94	.42	1.00	1.25	.83	.80	.78	.74	10.39
MOAB 4 NW	.57	.52	.67	.91	.68	.37	.52	.83	.66	.94	.66	.67	8.00
MODENA	.69	.73	.80	.68	.70	.40	1.14	1.21	.80	.87	.73	.49	9.24
MONTICELLO	1.34	.97	.96	.86	1.00	.48	1.67	1.89	1.16	1.62	1.08	1.38	14.41
MORGAN	1.91	1.73	1.76	2.19	1.76	1.30	.52	.97	1.04	1.50	1.64	1.75	18.07
MORONI	.95	.66	.79	.73	.74	.46	.54	.84	.85	.77	.78	.95	9.26
MOUNTAIN DELL DAM	2.22	2.15	2.35	2.77	2.16	1.40	.85	1.10	1.46	1.94	1.90	2.40	22.70
MYTON	.42	.31	.51	.57	.68	.68	.48	.75	.60	.72	.44	.53	6.69
NEPHI	1.30	1.27	1.46	1.48	1.22	.76	.63	.95	.88	1.07	1.22	1.26	13.50
NEW HARMONY	2.15	2.26	2.18	1.20	.93	.52	1.07	1.60	1.24	1.15	1.59	1.64	17.53
OAK CITY	1.15	1.12	1.38	1.37	1.24	.63	.44	.85	.79	.95	1.13	1.10	12.15
OGDEN PIONEER PH	2.36	1.90	2.05	2.52	2.14	1.58	.65	.98	1.20	1.58	1.73	1.89	20.58
OGDEN SUGAR FACTORY	1.52	1.27	1.41	2.06	1.71	1.43	.50	.72	1.10	1.27	1.36	1.30	15.65
ORDERVILLE	1.98	1.63	1.47	.94	.74	.56	.93	1.43	1.10	1.10	1.23	1.44	14.55
PANGUITCH	.54	.65	.66	.60	.80	.58	1.46	1.56	1.10	.68	.74	.52	9.89
PARK VALLEY	1.01	.79	.73	.82	1.44	1.18	.92	1.00	.57	.71	.79	.74	10.70
PAROWAN	.87	1.07	1.29	1.24	.92	.48	1.22	1.38	.88	.88	1.02	.91	12.16
PARTOUN	.32	.42	.48	.69	.82	.71	.58	.45	.44	.52	.42	.34	6.19
PINE VIEW DAM	3.83	3.11	3.06	3.05	2.68	1.72	.71	1.09	1.46	2.11	2.76	3.21	28.79
RICHFIELD RADIO KSVC	.63	.62	.63	.71	.73	.41	.81	.69	.80	.64	.59	.56	7.82
RICHMOND	1.78	1.48	1.73	2.15	1.96	1.51	.55	1.00	1.11	1.52	1.44	1.52	17.75
RIVERDALE POWER HOUSE	1.87	1.56	1.84	2.47	2.03	1.42	.64	.97	1.20	1.49	1.55	1.50	18.54
ROOSEVELT	.54	.42	.56	.63	.63	.71	.40	.73	.66	.83	.50	.60	7.21
SAINT GEORGE	1.04	.90	.98	.47	.49	.21	.62	.65	.52	.56	.75	.72	7.91
SALINA	.90	.86	.92	1.09	.96	.50	.65	.80	.90	.80	.81	.76	9.95
SALT LAKE CITY WSFO R	1.35	1.33	1.72	2.21	1.47	.97	.72	.92	.89	1.14	1.22	1.37	15.31
SANTAQUIN	1.69	1.70	2.11	2.11	1.58	.97	.80	1.12	1.04	1.56	1.67	1.67	18.02
SCIPIO	1.27	1.24	1.40	1.25	1.10	.64	.70	.97	.88	.95	1.04	1.07	12.51
SCOFIELD DAM	2.00	1.64	1.33	1.09	1.07	.92	1.02	1.39	1.04	1.14	1.12	1.49	15.25
SILVER LAKE BRIGHTON	5.56	4.96	5.26	4.44	2.83	1.76	1.28	1.90	1.96	2.94	4.30	5.02	42.21
SNAKE CREEK PH	3.32	2.65	2.19	1.88	1.45	1.05	.69	1.19	1.07	1.65	2.21	2.82	22.17
SNOWVILLE	1.11	.88	.86	1.14	1.48	1.26	.54	.84	.70	.70	1.00	.94	11.45
SPANISH FORK PWR HOUSE	1.78	1.68	2.05	2.11	1.66	1.07	.74	1.06	1.20	1.52	1.71	1.82	18.40
SUMMIT	.77	.99	1.27	1.06	.83	.51	1.11	1.39	.86	.86	1.01	.85	11.51
THOMPSON	.82	.53	.74	.70	.87	.45	.61	1.08	.79	.95	.62	.53	8.69
TIMPANOGOS CAVE	2.74	2.35	2.45	2.77	2.33	1.54	1.02	1.42	1.30	1.95	1.87	2.31	24.05
TOOELE	1.22	1.32	1.94	2.38	1.58	1.06	.75	.86	.92	1.36	1.43	1.42	16.24
TROPIC	1.23	1.06	.95	.70	.77	.45	1.14	1.84	1.11	1.11	.96	.98	12.30
UTAH LAKE LEHI	.95	.76	1.09	1.25	.98	.71	.61	.88	.74	.92	.89	.88	10.66
VERNAL AIRPORT	.50	.40	.57	.69	.78	.73	.41	.67	.62	.82	.56	.63	7.38
WENDOVER WSO //R	.34	.36	.42	.43	.85	.61	.25	.42	.23	.47	.38	.30	5.06
WOODRUFF	.51	.48	.59	.88	.89	1.12	.72	.74	.79	.82	.62	.58	8.74
ZION NATIONAL PARK	1.76	1.71	1.78	1.12	.80	.60	.98	1.59	.88	.90	1.20	1.26	14.58

UTAH

HEATING DEGREE DAY NORMALS (BASE 65 DEG F)

STATION	JUL	AUG	SEP	OCT	NOV	DEC	JAN	FEB	MAR	APR	MAY	JUN	ANN
ALTON	24	65	223	508	849	1094	1159	966	961	693	450	182	7174
BEAR RIVER REFUGE	0	9	115	400	798	1128	1231	960	794	477	208	50	6170
BEAVER	11	26	184	505	846	1104	1135	924	865	615	365	119	6699
BLACK ROCK	0	7	137	453	825	1113	1159	885	787	519	277	71	6233
BLANDING	0	7	104	409	798	1101	1169	896	809	537	262	54	6146
BLUFF	0	0	37	335	723	1060	1097	776	636	357	130	14	5165
BRYCE CANYON FAA AP	99	160	366	685	1041	1327	1386	1159	1119	834	583	303	9062
BRYCE CANYON NP HDQ	81	147	348	670	1014	1268	1327	1120	1097	816	567	271	8726
CAPITOL REEF NATL MON	0	0	56	297	702	1048	1113	815	682	400	166	20	5299
CEDAR CITY FAA AP	0	0	114	403	777	1051	1097	862	800	540	285	62	5991
COTTONWOOD WEIR //	0	0	83	314	708	1014	1045	809	719	453	205	48	5398
COVE FORT //	0	15	155	481	849	1110	1166	952	899	630	368	118	6743
DEER CREEK DAM	14	69	267	580	933	1237	1380	1151	1023	693	428	181	7956
DESERET	0	10	153	468	855	1169	1212	904	800	525	259	76	6431
DESERT EXP RANGE	0	8	128	456	849	1135	1197	904	822	558	295	69	6421
DUGWAY	0	0	104	409	804	1122	1156	854	769	483	222	44	5967
ECHO DAM	9	48	236	549	924	1225	1299	1064	964	660	397	169	7544
ELBERTA	0	0	114	419	789	1110	1159	896	778	504	249	61	6079
EPHRAIM SORENSENS FLD R	0	15	168	471	876	1200	1274	1011	899	624	348	103	6989
ESCALANTE	0	22	119	422	792	1101	1166	882	784	525	273	75	6161
FAIRFIELD	0	30	193	521	894	1212	1262	977	868	603	344	111	7015
FARMINGTON USU FLD STA	0	8	120	388	765	1066	1113	860	756	480	228	61	5845
FERRON	0	14	147	443	864	1209	1308	1008	887	567	292	92	6831
FILLMORE	0	0	105	384	771	1073	1113	854	760	498	252	56	5866
FORT DUCHESNE	0	18	188	539	969	1411	1556	1204	942	582	289	84	7782
GARFIELD	0	0	77	351	732	1029	1101	857	744	450	192	40	5573
GARLAND	0	14	162	465	855	1190	1277	1008	884	567	291	94	6807
GREEN RIVER AVN	0	0	72	375	801	1181	1299	907	710	399	142	17	5903
HANKSVILLE	0	0	51	335	780	1141	1221	865	685	378	123	12	5591
HEBER	13	67	242	546	924	1246	1339	1084	964	663	409	183	7680
HIAWATHA	9	49	203	524	945	1228	1299	1056	989	684	413	149	7548
JENSEN	0	19	178	527	951	1395	1538	1182	921	555	263	71	7600
KANAB	0	0	50	259	606	874	927	708	651	412	184	32	4703
LAKETOWN	55	106	311	639	975	1249	1342	1170	1125	759	481	243	8455
LA VERKIN	0	0	17	175	534	794	825	585	493	279	95	7	3804
LEVAN	0	8	134	422	813	1138	1200	935	828	555	294	82	6409
LOA	38	102	300	617	969	1243	1283	1042	995	726	465	208	7988
LOGAN UTAH STATE UNIV	0	14	151	446	849	1172	1249	1008	893	570	294	105	6751
MANTI	0	22	169	468	849	1153	1206	963	856	582	330	105	6703
MEXICAN HAT	0	0	17	247	651	1001	1048	731	598	309	90	0	4692
MILFORD WSO //R	0	9	126	456	846	1141	1197	921	831	561	295	68	6451
MOAB 4 NW	0	0	15	230	645	995	1079	756	558	266	72	0	4616
MODENA	0	11	153	434	807	1076	1125	868	818	564	314	78	6248
MONTICELLO	10	54	194	508	882	1172	1240	1008	933	642	381	127	7151
MORGAN	7	52	218	527	912	1212	1287	1033	921	621	357	144	7291
MORONI	7	38	195	496	885	1221	1277	1008	890	615	353	125	7110
MOUNTAIN DELL DAM	7	32	198	502	873	1159	1206	988	924	630	367	152	7038
NEPHI	0	10	114	383	765	1063	1119	885	794	519	264	64	5980
OAK CITY	0	0	76	326	750	1048	1094	837	741	477	223	47	5619
OGDEN PIONEER PH	0	6	111	375	768	1070	1128	879	775	480	220	54	5866
OGDEN SUGAR FACTORY	0	5	123	395	774	1076	1153	899	778	486	225	59	5973
ORDERVILLE	0	16	110	381	732	1001	1063	837	800	537	287	82	5846
PAROWAN	0	16	141	419	789	1057	1110	885	822	567	317	83	6206
PARTOUN	0	8	114	437	822	1141	1175	893	800	519	258	67	6234
PINE VIEW DAM	11	50	229	555	945	1296	1404	1154	1020	657	387	156	7864

UTAH

HEATING DEGREE DAY NORMALS (BASE 65 DEG F)

STATION	JUL	AUG	SEP	OCT	NOV	DEC	JAN	FEB	MAR	APR	MAY	JUN	ANN
RICHFIELD RADIO KSVC	0	15	154	468	825	1104	1147	899	809	561	317	95	6394
RICHMOND	0	30	191	496	873	1206	1277	1025	915	597	332	123	7065
RIVERDALE POWER HOUSE	0	9	113	392	777	1088	1141	890	781	495	248	72	6006
ROOSEVELT	0	8	142	474	918	1349	1479	1131	874	528	241	65	7209
SAINT GEORGE	0	0	6	115	465	747	766	526	411	179	38	0	3253
SALINA	0	8	124	440	819	1119	1159	899	794	525	274	69	6230
SALT LAKE CITY WSFO R	0	0	97	377	759	1076	1128	865	753	474	220	53	5802
SANTAQUIN	0	7	140	414	801	1097	1144	904	812	528	264	88	6199
SCIPIO	0	23	177	505	873	1184	1240	958	853	588	329	124	6854
SCOFIELD DAM	122	194	417	732	1113	1476	1581	1330	1256	909	611	333	10074
SILVER LAKE BRIGHTON	211	273	489	803	1140	1370	1426	1232	1271	1002	747	447	10411
SNAKE CREEK PH	42	92	290	592	969	1268	1333	1106	1020	708	443	208	8071
SPANISH FORK PWR HOUSE	0	0	94	355	750	1066	1113	868	756	474	216	45	5737
TIMPANOGOS CAVE	0	12	130	422	861	1138	1169	927	843	564	295	98	6459
TOOELE	0	0	122	414	786	1063	1101	871	787	510	253	67	5974
TROPIC	26	70	201	465	822	1088	1159	913	865	612	376	127	6724
UTAH LAKE LEHI	0	14	148	471	840	1135	1203	938	828	546	280	75	6478
VERNAL AIRPORT	0	24	193	546	969	1376	1494	1156	930	582	298	99	7667
WENDOVER WSO //R	0	0	81	391	804	1122	1144	857	732	435	185	42	5793
WOODRUFF	85	163	396	729	1104	1432	1525	1291	1181	807	543	288	9544
ZION NATIONAL PARK	0	0	6	115	453	729	772	560	493	254	77	0	3459

UTAH

COOLING DEGREE DAY NORMALS (BASE 65 DEG F)

STATION	JAN	FEB	MAR	APR	MAY	JUN	JUL	AUG	SEP	OCT	NOV	DEC	ANN
ALTON	0	0	0	0	0	20	80	59	10	0	0	0	169
BEAR RIVER REFUGE	0	0	0	0	31	131	335	260	79	0	0	0	836
BEAVER	0	0	0	0	6	47	157	98	22	0	0	0	330
BLACK ROCK	0	0	0	0	13	74	251	190	44	0	0	0	572
BLANDING	0	0	0	0	10	111	264	187	47	0	0	0	619
BLUFF	0	0	0	0	40	197	403	326	88	0	0	0	1054
BRYCE CANYON FAA AP	0	0	0	0	0	0	9	8	0	0	0	0	17
BRYCE CANYON NP HDQ	0	0	0	0	0	0	19	11	0	0	0	0	30
CAPITOL REEF NATL MON	0	0	0	0	48	197	384	305	122	18	0	0	1074
CEDAR CITY FAA AP	0	0	0	0	15	101	279	215	69	0	0	0	679
COTTONWOOD WEIR //	0	0	0	6	53	195	453	375	167	25	0	0	1274
COVE FORT //	0	0	0	0	9	64	229	167	41	0	0	0	510
DEER CREEK DAM	0	0	0	0	0	10	79	69	6	0	0	0	164
DESERET	0	0	0	0	14	109	295	212	51	0	0	0	681
DESERT EXP RANGE	0	0	0	0	12	90	273	206	44	0	0	0	625
DUGWAY	0	0	0	0	36	161	419	332	95	0	0	0	1043
ECHO DAM	0	0	0	0	0	25	118	91	14	0	0	0	248
ELBERTA	0	0	0	0	16	115	313	236	69	0	0	0	749
EPHRAIM SORENSENS FLD R	0	0	0	0	10	61	208	145	33	0	0	0	457
ESCALANTE	0	0	0	0	10	84	214	149	23	0	0	0	480
FAIRFIELD	0	0	0	0	0	33	161	120	19	0	0	0	333
FARMINGTON USU FLD STA	0	0	0	0	27	127	332	268	87	0	0	0	841
FERRON	0	0	0	0	13	110	236	157	45	0	0	0	561
FILLMORE	0	0	0	0	25	128	338	270	105	12	0	0	878
FORT DUCHESNE	0	0	0	0	0	66	202	133	20	0	0	0	421
GARFIELD	0	0	0	0	49	202	465	375	128	7	0	0	1226
GARLAND	0	0	0	0	12	82	264	206	57	0	0	0	621
GREEN RIVER AVN	0	0	0	0	36	188	403	316	84	0	0	0	1027
HANKSVILLE	0	0	0	0	58	246	465	372	123	7	0	0	1271
HEBER	0	0	0	0	0	15	87	79	8	0	0	0	189
HIAWATHA	0	0	0	0	7	68	154	105	23	0	0	0	357
JENSEN	0	0	0	0	9	74	223	146	28	0	0	0	480
KANAB	0	0	0	7	29	161	338	273	116	14	0	0	938
LAKETOWN	0	0	0	0	0	6	58	41	0	0	0	0	105
LA VERKIN	0	0	0	12	83	262	477	415	206	26	0	0	1481
LEVAN	0	0	0	0	12	88	267	201	62	0	0	0	630
LOA	0	0	0	0	0	10	32	21	0	0	0	0	63
LOGAN UTAH STATE UNIV	0	0	0	0	12	75	248	203	55	0	0	0	593
MANTI	0	0	0	0	7	54	176	130	28	0	0	0	395
MEXICAN HAT	0	0	0	0	87	307	521	440	173	8	0	0	1536
MILFORD WSO //R	0	0	0	0	13	92	288	229	54	0	0	0	676
MOAB 4 NW	0	0	0	8	106	310	530	450	180	13	0	0	1597
MODENA	0	0	0	0	10	72	229	175	72	0	0	0	558
MONTICELLO	0	0	0	0	0	37	122	89	11	0	0	0	259
MORGAN	0	0	0	0	0	42	137	114	14	0	0	0	307
MORONI	0	0	0	0	5	38	149	112	24	0	0	0	328
MOUNTAIN DELL DAM	0	0	0	0	0	35	149	119	24	0	0	0	327
NEPHI	0	0	0	0	22	124	341	273	96	8	0	0	864
OAK CITY	0	0	0	0	43	179	415	332	127	10	0	0	1106
OGDEN PIONEER PH	0	0	0	0	34	144	372	294	105	6	0	0	955
OGDEN SUGAR FACTORY	0	0	0	0	24	128	338	263	84	0	0	0	837
ORDERVILLE	0	0	0	0	14	100	228	181	53	0	0	0	576
PAROWAN	0	0	0	0	17	77	208	152	48	0	0	0	502
PARTOUN	0	0	0	0	20	121	316	244	51	0	0	0	752
PINE VIEW DAM	0	0	0	0	6	27	144	112	25	0	0	0	314

UTAH

COOLING DEGREE DAY NORMALS (BASE 65 DEG F)

STATION	JAN	FEB	MAR	APR	MAY	JUN	JUL	AUG	SEP	OCT	NOV	DEC	ANN
RICHFIELD RADIO KSVC	0	0	0	0	7	50	182	133	16	0	0	0	388
RICHMOND	0	0	0	0	7	54	206	176	53	0	0	0	496
RIVERDALE POWER HOUSE	0	0	0	0	28	123	332	260	80	0	0	0	823
ROOSEVELT	0	0	0	0	12	101	260	181	40	0	0	0	594
SAINT GEORGE	0	0	0	23	159	399	617	552	306	63	0	0	2119
SALINA	0	0	0	0	16	93	270	197	40	0	0	0	616
SALT LAKE CITY WSFO R	0	0	0	0	28	152	388	311	97	5	0	0	981
SANTAQUIN	0	0	0	0	25	145	319	245	92	0	0	0	826
SCIPIO	0	0	0	0	7	64	207	159	30	0	0	0	467
SCOFIELD DAM	0	0	0	0	0	0	0	8	0	0	0	0	8
SILVER LAKE BRIGHTON	0	0	0	0	0	0	0	0	0	0	0	0	0
SNAKE CREEK PH	0	0	0	0	0	7	52	39	0	0	0	0	98
SPANISH FORK PWR HOUSE	0	0	0	0	33	135	350	269	91	8	0	0	886
TIMPANOGOS CAVE	0	0	0	0	16	95	270	217	76	7	0	0	681
TOOELE	0	0	0	0	26	127	335	253	89	0	0	0	830
TROPIC	0	0	0	0	7	40	144	101	12	0	0	0	304
UTAH LAKE LEHI	0	0	0	0	10	69	236	178	31	0	0	0	524
VERNAL AIRPORT	0	0	0	0	6	81	195	133	22	0	0	0	437
WENDOVER WSO //R	0	0	0	0	55	204	459	363	111	0	0	0	1192
WOODRUFF	0	0	0	0	0	0	11	18	0	0	0	0	29
ZION NATIONAL PARK	0	0	6	26	139	369	595	521	309	87	0	0	2052

STATE-STATION NUMBER	STN TYP	NAME	LATITUDE DEG-MIN	LONGITUDE DEG-MIN	ELEVATION (FT)
42-0061	12	ALPINE	N 4027	W 11147	4935
42-0086	13	ALTON	N 3726	W 11229	7040
42-0506	13	BEAR RIVER REFUGE	N 4128	W 11216	4208
42-0519	13	BEAVER	N 3817	W 11238	5920
42-0527	12	BEAVER CANYON PH	N 3816	W 11229	7275
42-0730	13	BLACK ROCK	N 3843	W 11257	4895
42-0738	13	BLANDING	N 3737	W 10928	6130
42-0788	13	BLUFF	N 3717	W 10933	4315
42-1002	13	BRYCE CANYON FAA AP	N 3742	W 11209	7595
42-1008	13	BRYCE CANYON NP HDQ	N 3739	W 11210	7915
42-1144	12	CALLAO	N 3954	W 11343	4330
42-1171	13	CAPITOL REEF NATL MON	N 3817	W 11116	5500
42-1214	12	CASTLE DALE	N 3913	W 11101	5660
42-1267	13	CEDAR CITY FAA AP	N 3742	W 11306	5620
42-1446	12	CITY CREEK WATER PLANT	N 4049	W 11150	5335
42-1588	12	COALVILLE	N 4055	W 11124	5550
42-1731	12	CORINNE	N 4133	W 11207	4230
42-1759	13	COTTONWOOD WEIR	N 4037	W 11147	4950
42-1792	13	COVE FORT	N 3836	W 11235	5980
42-2057	13	DEER CREEK DAM	N 4024	W 11132	5270
42-2101	13	DESERET	N 3917	W 11239	4585
42-2116	13	DESERT EXP RANGE	N 3836	W 11345	5252
42-2257	13	DUGWAY	N 4011	W 11256	4340
42-2385	13	ECHO DAM	N 4058	W 11126	5500
42-2418	13	ELBERTA	N 3957	W 11157	4690
42-2578	13	EPHRAIM SORENSENS FLD R	N 3921	W 11135	5580
42-2592	13	ESCALANTE	N 3746	W 11136	5810
42-2625	12	EUREKA	N 3957	W 11207	6460
42-2696	13	FAIRFIELD	N 4016	W 11205	4876
42-2726	13	FARMINGTON USU FLD STA	N 4101	W 11154	4340
42-2798	13	FERRON	N 3905	W 11108	5925
42-2828	13	FILLMORE	N 3857	W 11219	5160
42-2996	13	FORT DUCHESNE	N 4017	W 10952	4990
42-3097	13	GARFIELD	N 4043	W 11212	4310
42-3122	13	GARLAND	N 4144	W 11210	4350
42-3138	12	GARRISON	N 3856	W 11402	5275
42-3418	13	GREEN RIVER AVN	N 3900	W 11010	4070
42-3506	12	GUNLOCK POWER HOUSE	N 3717	W 11343	4060
42-3611	13	HANKSVILLE	N 3822	W 11043	4308
42-3624	12	HANNA	N 4024	W 11046	6745
42-3776	12	HATCH	N 3739	W 11226	6900
42-3809	13	HEBER	N 4031	W 11125	5580
42-3896	13	HIAWATHA	N 3929	W 11101	7230
42-4174	12	IBAPAH	N 4002	W 11359	5280
42-4342	13	JENSEN	N 4022	W 10921	4720
42-4467	12	KAMAS RANGER STATION	N 4039	W 11117	6495
42-4508	13	KANAB	N 3703	W 11232	4985
42-4527	12	KANOSH	N 3848	W 11226	5015
42-4764	12	KOOSHAREM	N 3831	W 11153	6950
42-4856	13	LAKETOWN	N 4149	W 11119	5988

42 — UTAH

STATE-STATION NUMBER	STN TYP	NAME	LATITUDE DEG-MIN	LONGITUDE DEG-MIN	ELEVATION (FT)
42-4968	13	LA VERKIN	N 3712	W 11316	3200
42-5065	13	LEVAN	N 3933	W 11152	5300
42-5148	13	LOA	N 3824	W 11139	7045
42-5186	13	LOGAN UTAH STATE UNIV	N 4145	W 11149	4785
42-5402	13	MANTI	N 3915	W 11138	5740
42-5582	13	MEXICAN HAT	N 3709	W 10952	4270
42-5654	13	MILFORD WSO //R	N 3826	W 11301	5028
42-5723	12	MINERSVILLE	N 3813	W 11255	5275
42-5733	13	MOAB 4 NW	N 3836	W 10936	3965
42-5752	13	MODENA	N 3748	W 11355	5460
42-5805	13	MONTICELLO	N 3752	W 10920	6980
42-5826	13	MORGAN	N 4102	W 11141	5070
42-5837	13	MORONI	N 3932	W 11135	5525
42-5892	13	MOUNTAIN DELL DAM	N 4045	W 11143	5420
42-5969	12	MYTON	N 4012	W 11004	5030
42-6135	13	NEPHI	N 3943	W 11150	5133
42-6181	12	NEW HARMONY	N 3729	W 11318	5290
42-6357	13	OAK CITY	N 3923	W 11220	5075
42-6404	13	OGDEN PIONEER PH	N 4115	W 11157	4350
42-6414	13	OGDEN SUGAR FACTORY	N 4114	W 11202	4280
42-6534	13	ORDERVILLE	N 3716	W 11238	5460
42-6601	12	PANGUITCH	N 3749	W 11227	6720
42-6658	12	PARK VALLEY	N 4149	W 11319	5520
42-6686	13	PAROWAN	N 3751	W 11250	5975
42-6708	13	PARTOUN	N 3939	W 11353	4750
42-6869	13	PINE VIEW DAM	N 4115	W 11150	4940
42-7260	13	RICHFIELD RADIO KSVC	N 3846	W 11205	5270
42-7271	13	RICHMOND	N 4154	W 11149	4680
42-7318	13	RIVERDALE POWER HOUSE	N 4109	W 11200	4390
42-7395	13	ROOSEVELT	N 4018	W 10959	5094
42-7516	13	SAINT GEORGE	N 3707	W 11334	2760
42-7557	13	SALINA	N 3857	W 11152	5190
42-7598	13	SALT LAKE CITY WSFO R	N 4047	W 11157	4222
42-7686	13	SANTAQUIN	N 3958	W 11147	5120
42-7714	13	SCIPIO	N 3915	W 11206	5306
42-7724	13	SCOFIELD DAM	N 3947	W 11107	7630
42-7846	13	SILVER LAKE BRIGHTON	N 4036	W 11135	8740
42-7909	13	SNAKE CREEK PH	N 4033	W 11130	5950
42-7931	12	SNOWVILLE	N 4158	W 11243	4560
42-8119	13	SPANISH FORK PWR HOUSE	N 4005	W 11136	4720
42-8456	12	SUMMIT	N 3748	W 11256	5946
42-8705	12	THOMPSON	N 3858	W 10943	5150
42-8733	13	TIMPANOGOS CAVE	N 4026	W 11143	5600
42-8771	13	TOOELE	N 4032	W 11218	4820
42-8847	13	TROPIC	N 3737	W 11205	6235
42-8973	13	UTAH LAKE LEHI	N 4022	W 11154	4497
42-9111	13	VERNAL AIRPORT	N 4027	W 10931	5280
42-9382	13	WENDOVER WSO //R	N 4044	W 11402	4237
42-9595	13	WOODRUFF	N 4132	W 11109	6343
42-9717	13	ZION NATIONAL PARK	N 3713	W 11259	4050

42 – UTAH

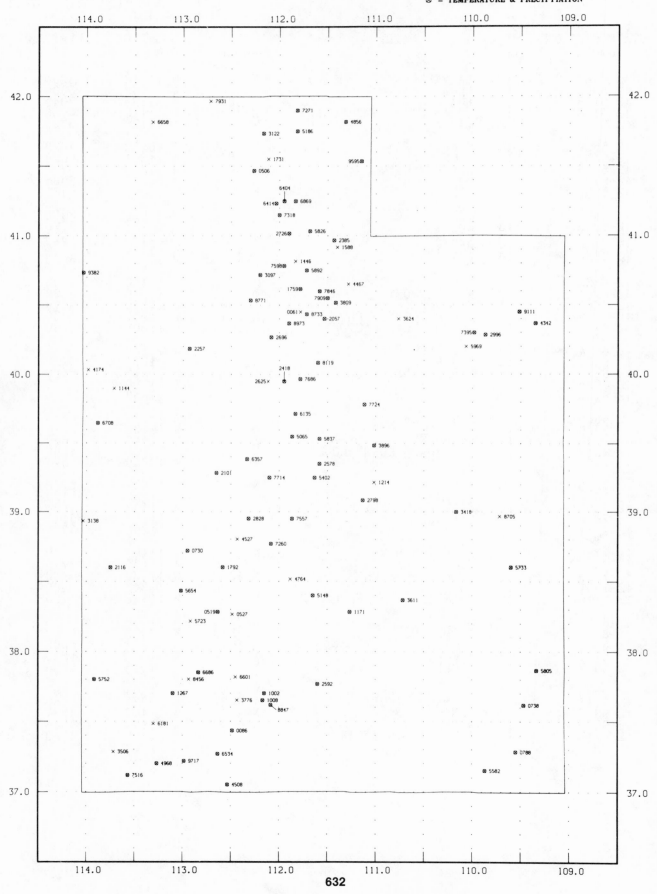

VERMONT

TEMPERATURE NORMALS (DEG F)

STATION			JAN	FEB	MAR	APR	MAY	JUN	JUL	AUG	SEP	OCT	NOV	DEC	ANN
BELLOWS FALLS		MAX	30.2	32.9	42.0	55.4	68.3	77.5	82.4	80.1	72.0	60.9	46.9	33.6	56.9
		MIN	9.2	9.9	22.2	33.2	42.8	53.0	57.6	55.5	47.8	37.0	29.1	15.6	34.4
		MEAN	19.7	21.5	32.1	44.4	55.6	65.3	70.0	67.8	59.9	49.0	38.1	24.6	45.7
BURLINGTON WSO	R MAX		25.4	27.3	37.7	52.6	66.4	75.9	80.5	77.6	68.8	57.0	43.6	30.3	53.6
		MIN	7.7	8.8	20.8	32.7	44.0	54.0	58.6	56.6	48.7	38.7	29.6	14.9	34.6
		MEAN	16.6	18.1	29.2	42.7	55.2	64.9	69.6	67.1	58.8	47.9	36.6	22.6	44.1
CAVENDISH		MAX	29.4	32.4	41.1	55.2	69.6	77.9	81.9	79.3	70.8	59.8	45.5	32.4	56.3
		MIN	5.8	7.2	18.6	29.2	39.6	49.0	53.3	51.6	44.4	33.9	26.0	12.7	30.9
		MEAN	17.7	19.8	29.9	42.2	54.6	63.5	67.6	65.5	57.6	46.9	35.8	22.6	43.6
CHELSEA		MAX	27.4	30.4	39.4	52.5	66.4	75.9	80.5	78.1	69.9	58.9	44.6	30.9	54.6
		MIN	.3	1.0	14.2	27.4	37.0	47.0	51.3	48.9	41.3	31.2	23.7	8.0	27.6
		MEAN	13.9	15.7	26.9	40.0	51.7	61.5	65.9	63.5	55.6	45.1	34.2	19.5	41.1
CORNWALL		MAX	28.1	31.2	41.2	55.9	69.2	78.0	82.6	80.1	71.7	59.8	46.0	32.6	56.4
		MIN	9.4	10.5	21.8	33.5	44.1	53.4	57.9	56.0	48.5	38.8	29.7	15.7	34.9
		MEAN	18.8	20.9	31.5	44.7	56.7	65.7	70.3	68.0	60.1	49.3	37.9	24.2	45.7
DORSET 1 S		MAX	30.7	33.1	42.0	55.7	68.0	76.6	80.3	78.2	70.7	60.1	46.9	34.5	56.4
		MIN	9.4	10.5	21.0	31.1	40.8	49.7	53.8	52.3	45.2	36.1	28.1	15.6	32.8
		MEAN	20.1	21.9	31.5	43.4	54.5	63.2	67.1	65.3	58.0	48.1	37.5	25.1	44.6
ENOSBURG FALLS		MAX	26.9	29.8	40.5	54.4	68.5	76.8	80.6	78.2	70.3	59.3	44.7	31.0	55.1
		MIN	4.0	4.9	18.2	30.3	41.2	50.7	55.2	53.3	46.0	36.3	27.0	11.4	31.5
		MEAN	15.5	17.4	29.4	42.4	54.9	63.7	68.0	65.8	58.2	47.8	35.9	21.2	43.4
MONTPELIER FAA AIRPORT		MAX	24.9	27.1	36.5	50.5	64.5	73.3	77.9	75.3	67.2	56.2	42.2	28.9	52.0
		MIN	6.3	7.7	19.1	31.1	41.6	50.9	55.2	53.1	45.6	36.1	27.1	12.4	32.2
		MEAN	15.6	17.4	27.8	40.8	53.1	62.1	66.6	64.2	56.4	46.2	34.6	20.6	42.1
NEWPORT		MAX	24.4	27.8	37.8	51.2	65.5	74.9	79.3	76.7	68.5	56.5	41.8	28.4	52.7
		MIN	3.3	4.3	16.3	29.2	40.6	50.4	54.9	52.6	44.9	35.7	26.2	10.4	30.7
		MEAN	13.8	16.1	27.1	40.2	53.1	62.7	67.1	64.7	56.7	46.1	34.0	19.4	41.8
READSBORO 1 SSE		MAX	29.6	31.3	39.4	53.2	66.0	75.0	79.7	77.7	69.9	59.1	46.0	33.2	55.0
		MIN	8.8	8.9	19.8	31.2	40.7	49.8	54.4	52.4	45.3	35.4	27.9	15.0	32.5
		MEAN	19.2	20.1	29.7	42.2	53.4	62.4	67.1	65.1	57.6	47.3	37.0	24.1	43.8
RUTLAND		MAX	30.2	32.9	42.5	56.3	69.1	77.3	81.5	79.1	71.4	60.8	47.3	34.3	56.9
		MIN	11.1	12.6	22.9	33.8	44.1	53.5	57.9	56.4	48.8	38.9	30.5	17.6	35.7
		MEAN	20.6	22.8	32.8	45.1	56.6	65.4	69.7	67.8	60.1	49.8	38.9	26.0	46.3
SAINT JOHNSBURY		MAX	27.4	31.0	40.9	54.9	69.4	78.4	82.2	79.6	71.4	59.8	44.7	30.9	55.9
		MIN	6.0	7.5	19.3	30.6	41.5	51.5	55.9	54.3	46.7	36.1	27.6	12.7	32.5
		MEAN	16.7	19.3	30.2	42.8	55.5	64.9	69.1	67.0	59.1	48.0	36.2	21.8	44.2
VERNON		MAX	32.1	34.6	43.8	57.2	69.3	78.1	83.1	81.1	73.4	62.4	48.7	35.5	58.3
		MIN	8.6	10.0	22.0	32.4	42.1	52.2	57.0	55.1	47.6	36.4	28.6	15.2	33.9
		MEAN	20.4	22.4	33.0	44.9	55.7	65.2	70.1	68.1	60.5	49.4	38.7	25.4	46.2
WEST BURKE		MAX	24.1	26.9	37.2	50.4	65.0	74.7	78.9	76.4	67.9	55.9	41.7	27.8	52.2
		MIN	-1.8	-1.9	11.3	25.4	35.5	45.3	49.8	47.7	40.0	30.5	22.2	5.7	25.8
		MEAN	11.1	12.5	24.3	37.9	50.3	60.0	64.4	62.1	54.0	43.2	31.9	16.8	39.0
WOODSTOCK 2 WSW		MAX	28.1	30.6	39.8	53.7	67.2	76.2	81.0	78.4	70.3	58.6	45.2	31.7	55.1
		MIN	3.3	4.6	18.1	29.9	40.1	49.6	54.1	52.2	44.5	33.4	25.8	11.1	30.6
		MEAN	15.7	17.6	29.0	41.8	53.6	62.9	67.6	65.3	57.4	46.0	35.5	21.4	42.8

VERMONT

PRECIPITATION NORMALS (INCHES)

STATION		JAN	FEB	MAR	APR	MAY	JUN	JUL	AUG	SEP	OCT	NOV	DEC	ANN
BELLOWS FALLS		3.08	2.84	3.36	3.51	3.36	3.04	3.06	3.77	3.49	3.39	3.77	3.73	40.40
BURLINGTON WSO	R	1.85	1.73	2.20	2.77	2.96	3.64	3.43	3.87	3.20	2.81	2.80	2.43	33.69
CAVENDISH		3.15	3.01	3.69	3.76	3.72	3.90	3.67	3.65	3.38	3.69	3.92	3.76	43.30
CHELSEA		2.62	2.27	2.78	2.99	3.33	3.10	3.46	3.33	3.33	3.12	3.27	3.06	36.66
CHITTENDEN		2.85	2.49	2.98	3.28	3.60	3.90	4.11	4.36	4.11	3.32	3.73	3.34	42.07
CORNWALL		2.20	1.94	2.45	2.56	2.96	3.06	3.34	3.79	3.15	2.80	2.96	2.79	34.00
DORSET 1 S		3.22	2.79	3.60	3.97	4.39	4.48	4.10	4.26	4.24	3.87	4.07	3.81	46.80
ENOSBURG FALLS		2.42	2.04	2.63	3.19	3.59	4.22	4.11	4.57	3.93	3.73	3.89	2.99	41.31
GILMAN		2.01	1.79	2.05	2.40	2.90	3.82	3.55	3.59	2.99	2.83	2.92	2.63	33.48
MONTPELIER FAA AIRPORT		2.35	2.39	2.48	2.59	2.97	3.20	3.03	3.33	2.90	2.88	2.96	2.86	33.94
NEWFANE		3.34	2.99	3.74	3.82	3.60	3.30	3.44	3.70	3.62	3.79	4.20	4.13	43.67
NEWPORT		2.64	2.49	2.69	3.14	3.22	4.16	3.95	4.32	3.46	3.23	3.35	3.29	39.94
PERU		4.03	3.32	4.29	4.07	4.56	4.34	4.38	4.51	4.21	4.26	4.45	4.41	50.83
READSBORO 1 SSE		3.80	3.28	4.18	4.35	4.24	4.14	3.95	4.25	4.13	3.85	4.66	4.42	49.25
ROCHESTER		3.17	2.86	3.46	3.56	3.76	3.44	3.87	3.91	3.77	3.51	4.18	3.88	43.37
RUTLAND		2.25	1.90	2.28	2.72	3.21	3.61	3.43	3.89	3.42	2.70	2.86	2.64	34.91
SAINT JOHNSBURY		2.44	2.22	2.55	2.84	3.08	3.58	3.67	3.54	3.06	3.07	3.21	3.15	36.41
SALISBURY		2.45	2.17	2.63	2.90	3.22	3.57	3.74	3.82	3.58	3.02	3.31	2.79	37.20
SEARSBURG STATION		4.35	3.73	4.69	4.62	4.53	3.93	4.02	4.41	4.37	4.35	5.18	5.11	53.29
SOUTH LONDONDERRY		3.36	3.01	3.69	3.53	3.54	3.51	3.61	3.94	3.53	3.49	3.97	3.71	42.89
SOUTH NEWBURY		2.42	1.94	2.21	2.65	2.99	3.22	3.50	3.57	3.21	2.85	3.02	2.93	34.51
UNION VILLAGE DAM	R	2.63	2.21	2.62	2.87	3.10	2.74	3.39	3.25	3.36	2.95	3.23	3.00	35.35
VERNON		3.34	2.92	3.75	3.84	3.79	3.40	3.40	3.70	3.64	3.49	4.03	3.98	43.53
WEST BURKE		2.71	2.40	2.81	2.98	3.47	4.16	4.20	4.18	3.44	3.37	3.50	3.52	40.74
WHITINGHAM 2 W		4.02	3.44	4.50	4.50	4.52	4.23	3.95	4.15	4.23	4.13	4.85	4.81	51.33
WOODSTOCK 2 WSW		3.03	2.64	3.22	3.38	3.53	3.05	3.31	3.51	3.39	3.30	3.50	3.57	39.43

VERMONT

HEATING DEGREE DAY NORMALS (BASE 65 DEG F)

STATION	JUL	AUG	SEP	OCT	NOV	DEC	JAN	FEB	MAR	APR	MAY	JUN	ANN
BELLOWS FALLS	13	31	177	496	807	1252	1404	1218	1020	618	304	60	7400
BURLINGTON WSO R	23	50	202	530	852	1314	1500	1313	1110	669	326	64	7953
CAVENDISH	29	62	231	561	876	1314	1466	1266	1088	684	329	83	7989
CHELSEA	56	109	291	617	924	1411	1584	1380	1181	750	419	129	8851
CORNWALL	14	31	172	487	813	1265	1432	1235	1039	609	280	46	7423
DORSET 1 S	27	68	219	524	825	1237	1392	1207	1039	648	334	85	7605
ENOSBURG FALLS	28	56	219	533	873	1358	1535	1333	1104	678	325	75	8117
MONTPELIER FAA AIRPORT	56	90	265	583	912	1376	1531	1333	1153	726	382	120	8527
NEWPORT	48	84	259	586	930	1414	1587	1369	1175	744	381	102	8679
READSBORO 1 SSE	31	63	232	549	840	1268	1420	1257	1094	684	365	108	7911
RUTLAND	12	33	166	471	783	1209	1376	1182	998	597	277	51	7155
SAINT JOHNSBURY	22	42	195	527	864	1339	1497	1280	1079	666	311	59	7881
VERNON	12	28	161	484	789	1228	1383	1193	992	603	301	55	7229
WEST BURKE	90	131	335	676	993	1494	1671	1470	1262	813	461	166	9562
WOODSTOCK 2 WSW	32	66	238	589	885	1352	1528	1327	1116	696	361	95	8285

VERMONT

COOLING DEGREE DAY NORMALS (BASE 65 DEG F)

STATION	JAN	FEB	MAR	APR	MAY	JUN	JUL	AUG	SEP	OCT	NOV	DEC	ANN
BELLOWS FALLS	0	0	0	0	13	69	168	118	24	0	0	0	392
BURLINGTON WSO R	0	0	0	0	22	61	165	115	16	0	0	0	379
CAVENDISH	0	0	0	0	6	38	110	77	9	0	0	0	240
CHELSEA	0	0	0	0	6	24	84	62	9	0	0	0	185
CORNWALL	0	0	0	0	22	67	179	124	25	0	0	0	417
DORSET 1 S	0	0	0	0	9	31	93	77	9	0	0	0	219
ENOSBURG FALLS	0	0	0	0	12	36	121	80	15	0	0	0	264
MONTPELIER FAA AIRPORT	0	0	0	0	13	33	106	65	7	0	0	0	224
NEWPORT	0	0	0	0	12	33	113	74	10	0	0	0	242
READSBORO 1 SSE	0	0	0	0	6	30	96	66	10	0	0	0	208
RUTLAND	0	0	0	0	17	63	158	120	19	0	0	0	377
SAINT JOHNSBURY	0	0	0	0	16	56	149	104	18	0	0	0	343
VERNON	0	0	0	0	13	61	170	124	26	0	0	0	394
WEST BURKE	0	0	0	0	5	16	71	41	0	0	0	0	133
WOODSTOCK 2 WSW	0	0	0	0	8	32	113	76	10	0	0	0	239

43 — VERMONT

STATE-STATION NUMBER	STN TYP	NAME		LATITUDE DEG-MIN	LONGITUDE DEG-MIN	ELEVATION (FT)
43-0499	13	BELLOWS FALLS		N 4308	W 07227	300
43-1081	13	BURLINGTON WSO	R	N 4428	W 07309	332
43-1243	13	CAVENDISH		N 4323	W 07236	800
43-1360	13	CHELSEA		N 4358	W 07228	750
43-1433	12	CHITTENDEN		N 4342	W 07257	1000
43-1580	13	CORNWALL		N 4357	W 07313	340
43-1786	13	DORSET 1 S		N 4315	W 07306	980
43-2769	13	ENOSBURG FALLS		N 4455	W 07249	422
43-3341	12	GILMAN		N 4425	W 07143	850
43-5278	13	MONTPELIER FAA AIRPORT		N 4412	W 07234	1126
43-5492	12	NEWFANE		N 4300	W 07238	470
43-5542	13	NEWPORT		N 4456	W 07212	766
43-6335	12	PERU		N 4315	W 07254	1670
43-6761	13	READSBORO 1 SSE		N 4245	W 07256	1120
43-6893	12	ROCHESTER		N 4351	W 07248	830
43-6995	13	RUTLAND		N 4336	W 07258	620
43-7054	13	SAINT JOHNSBURY		N 4425	W 07201	699
43-7098	12	SALISBURY		N 4354	W 07306	370
43-7152	12	SEARSBURG STATION		N 4252	W 07255	1560
43-7617	12	SOUTH LONDONDERRY		N 4311	W 07249	1060
43-7646	12	SOUTH NEWBURY		N 4403	W 07205	470
43-8556	12	UNION VILLAGE DAM	R	N 4348	W 07216	463
43-8600	13	VERNON		N 4246	W 07231	225
43-9099	13	WEST BURKE		N 4439	W 07159	900
43-9735	12	WHITINGHAM 2 W		N 4247	W 07255	1450
43-9984	13	WOODSTOCK 2 WSW		N 4337	W 07233	750

43 – VERMONT

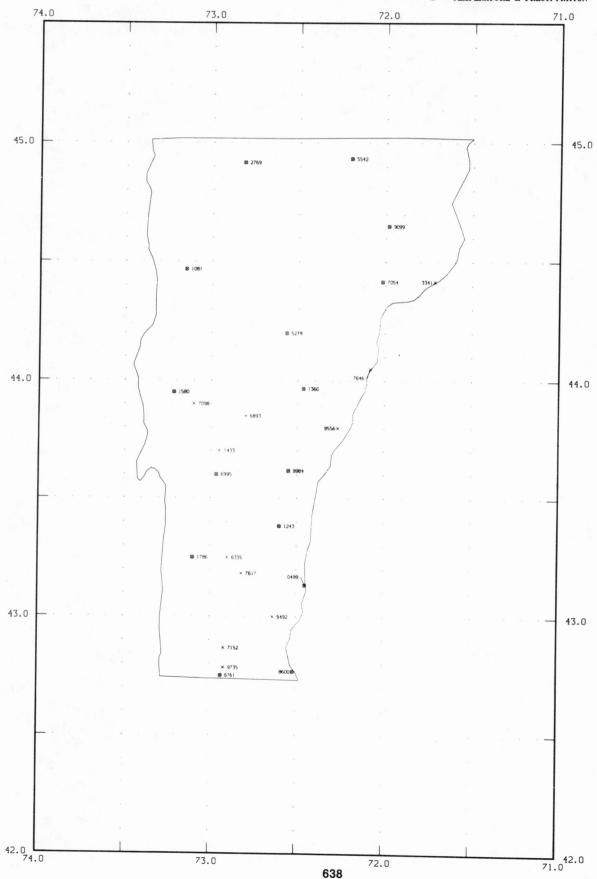

VIRGINIA

TEMPERATURE NORMALS (DEG F)

STATION		JAN	FEB	MAR	APR	MAY	JUN	JUL	AUG	SEP	OCT	NOV	DEC	ANN
BEDFORD	MAX	46.8	49.8	59.6	71.0	77.9	83.9	87.1	85.9	79.5	69.2	58.9	49.5	68.3
	MIN	26.5	27.9	35.1	43.9	52.7	59.7	63.9	63.0	56.7	45.7	36.9	29.5	45.1
	MEAN	36.7	38.9	47.4	57.4	65.3	71.8	75.5	74.5	68.1	57.5	47.9	39.6	56.7
BIG MEADOWS	MAX	35.9	37.3	45.2	57.1	65.3	71.9	75.3	74.1	68.3	58.2	47.8	38.7	56.3
	MIN	18.3	19.2	26.2	36.8	45.9	53.3	57.4	56.5	50.7	39.9	30.6	21.6	38.0
	MEAN	27.1	28.3	35.7	46.9	55.6	62.6	66.4	65.3	59.5	49.1	39.2	30.2	47.2
BLACKSBURG 3 SE	MAX	41.1	43.8	52.7	64.0	72.4	78.8	82.6	81.7	76.0	65.5	54.4	44.4	63.1
	MIN	20.6	21.9	29.4	38.7	47.5	54.7	59.3	58.5	51.6	39.5	30.7	23.6	39.7
	MEAN	30.9	32.9	41.1	51.3	60.0	66.8	71.0	70.1	63.8	52.5	42.6	34.0	51.4
BUCHANAN	MAX	46.5	50.1	59.6	70.9	78.4	84.5	87.7	86.7	80.6	70.4	58.9	49.1	68.6
	MIN	25.3	26.7	34.0	42.4	51.4	58.9	63.2	62.6	56.0	44.0	34.8	27.8	43.9
	MEAN	35.9	38.4	46.8	56.7	64.9	71.7	75.5	74.7	68.3	57.2	46.8	38.5	56.3
BURKES GARDEN	MAX	39.8	42.7	51.1	62.3	69.6	75.6	78.5	77.6	72.6	62.7	51.5	43.1	60.6
	MIN	20.1	21.6	28.9	36.6	45.1	51.8	56.1	55.0	48.8	37.4	29.6	23.1	37.8
	MEAN	30.0	32.1	40.0	49.5	57.4	63.7	67.3	66.3	60.7	50.1	40.6	33.2	49.2
CHARLOTTE COURT HOUSE	MAX	46.6	49.2	58.0	69.7	77.3	84.0	88.0	86.8	80.9	70.3	60.1	49.7	68.4
	MIN	24.0	25.3	33.8	43.3	52.0	59.7	64.3	63.4	56.5	44.1	34.6	26.9	44.0
	MEAN	35.3	37.3	45.9	56.5	64.7	71.9	76.1	75.2	68.8	57.2	47.4	38.3	56.2
CHARLOTTESVILLE 2 W	MAX	44.1	46.4	56.0	68.3	75.9	82.7	87.0	85.7	79.6	68.7	57.9	47.1	66.6
	MIN	26.4	27.6	35.7	45.9	54.8	62.3	66.6	65.6	59.6	48.7	39.1	29.8	46.8
	MEAN	35.3	37.0	45.9	57.1	65.4	72.5	76.8	75.7	69.6	58.7	48.5	38.5	56.8
CHATHAM 2 NE	MAX	46.5	49.0	57.7	69.0	77.3	83.7	87.6	86.5	80.8	70.4	59.7	49.7	68.2
	MIN	25.1	26.3	33.9	42.6	51.5	59.0	63.6	62.7	55.6	43.3	34.9	27.5	43.8
	MEAN	35.8	37.7	45.8	55.8	64.4	71.4	75.6	74.6	68.3	56.9	47.3	38.6	56.0
COLUMBIA	MAX	45.9	48.3	57.5	69.3	77.2	84.2	88.4	87.2	81.3	70.5	59.8	49.0	68.2
	MIN	23.0	24.3	32.9	42.6	51.5	59.8	64.3	63.3	56.0	43.2	33.7	26.1	43.4
	MEAN	34.5	36.3	45.2	56.0	64.4	72.0	76.4	75.3	68.7	56.9	46.7	37.6	55.8
CULPEPER	MAX	44.5	47.8	57.6	69.5	77.5	84.5	88.2	86.6	80.6	69.6	57.9	47.4	67.6
	MIN	23.7	25.3	33.3	42.7	52.1	60.2	64.6	63.4	56.7	44.4	35.2	27.0	44.1
	MEAN	34.1	36.6	45.5	56.1	64.9	72.4	76.4	75.1	68.7	57.0	46.6	37.2	55.9
DALE ENTERPRISE	MAX	41.8	44.8	54.4	65.7	74.5	81.9	85.6	84.2	77.9	66.5	54.8	45.1	64.8
	MIN	21.8	23.5	31.1	40.2	49.2	56.8	61.0	59.9	53.4	42.1	33.4	25.4	41.5
	MEAN	31.8	34.2	42.8	53.0	61.9	69.4	73.3	72.1	65.7	54.3	44.1	35.3	53.2
DANVILLE-BRIDGE ST	MAX	48.1	51.2	60.3	72.1	79.6	85.9	89.7	88.7	82.6	71.7	61.1	50.8	70.2
	MIN	27.0	28.3	35.5	44.9	54.0	61.7	65.9	65.5	58.5	45.6	36.2	29.1	46.0
	MEAN	37.6	39.8	48.0	58.6	66.8	73.8	77.8	77.1	70.5	58.7	48.7	40.0	58.1
DIAMOND SPRINGS	MAX	48.5	50.0	57.9	68.5	75.3	82.6	86.6	85.6	80.3	70.1	61.5	52.4	68.3
	MIN	31.1	31.9	38.9	48.0	56.6	64.8	69.4	69.0	63.7	52.6	42.9	34.3	50.3
	MEAN	39.9	41.0	48.4	58.3	66.0	73.7	78.0	77.3	72.0	61.4	52.2	43.4	59.3
ELKWOOD 6 SE	MAX	44.0	47.4	57.3	69.0	76.6	83.5	87.4	86.0	79.4	68.3	57.4	47.4	67.0
	MIN	23.1	24.6	32.6	41.7	50.6	58.3	63.1	62.3	55.2	43.1	34.5	26.5	43.0
	MEAN	33.6	36.0	44.9	55.4	63.6	70.9	75.2	74.2	67.3	55.7	46.0	37.0	55.0
FARMVILLE 2 N	MAX	49.1	51.9	61.1	72.4	79.1	85.7	89.1	88.0	81.7	71.7	61.3	51.4	70.2
	MIN	24.9	26.5	33.9	43.2	51.8	59.4	64.1	62.9	55.8	43.8	34.7	27.4	44.0
	MEAN	37.0	39.2	47.5	57.9	65.5	72.6	76.6	75.5	68.8	57.7	48.0	39.4	57.1
FLOYD 2 NE	MAX	43.2	46.3	54.6	65.5	73.0	79.1	82.3	81.8	76.1	66.3	54.8	46.4	64.1
	MIN	22.0	23.1	29.9	38.6	47.1	53.9	58.1	57.1	50.8	39.2	31.0	24.3	39.6
	MEAN	32.7	34.7	42.3	52.1	60.1	66.5	70.2	69.5	63.4	52.8	43.0	35.4	51.9
FREDERICKSBURG NATL PK	MAX	45.8	48.4	58.0	69.8	78.2	85.8	89.8	88.3	82.1	71.2	59.9	49.1	68.9
	MIN	23.6	24.4	32.8	42.1	51.7	59.8	64.4	63.0	56.3	43.9	34.4	26.7	43.6
	MEAN	34.7	36.4	45.4	55.9	65.0	72.8	77.1	75.7	69.2	57.6	47.1	37.9	56.2

VIRGINIA

TEMPERATURE NORMALS (DEG F)

STATION		JAN	FEB	MAR	APR	MAY	JUN	JUL	AUG	SEP	OCT	NOV	DEC	ANN
HOLLAND 1 E	MAX	49.1	50.9	59.3	70.6	77.7	84.5	88.0	86.9	81.9	71.4	62.3	52.5	69.6
	MIN	28.1	29.6	37.1	45.7	54.8	62.5	67.0	66.3	60.0	48.0	38.8	30.8	47.4
	MEAN	38.6	40.3	48.2	58.2	66.3	73.5	77.5	76.6	70.9	59.7	50.6	41.6	58.5
HOPEWELL	MAX	50.1	52.8	61.7	73.3	80.3	86.7	89.9	88.8	83.3	72.9	63.3	53.0	71.3
	MIN	29.2	31.0	37.9	47.1	55.8	63.5	68.2	67.6	61.1	49.2	39.8	32.2	48.6
	MEAN	39.7	41.9	49.9	60.2	68.1	75.1	79.0	78.2	72.2	61.1	51.6	42.7	60.0
HOT SPRINGS	MAX	40.7	43.5	52.5	64.5	73.3	79.6	82.9	81.8	75.5	64.9	53.2	43.6	63.0
	MIN	20.8	22.1	29.1	38.4	46.9	54.3	58.6	58.0	51.7	40.6	31.7	24.1	39.7
	MEAN	30.8	32.8	40.8	51.5	60.1	67.0	70.7	69.9	63.6	52.8	42.5	33.9	51.4
JOHN H KERR DAM	MAX	48.0	50.2	59.0	70.5	78.0	84.9	88.8	87.5	81.7	71.0	61.1	51.1	69.3
	MIN	26.1	27.5	35.4	45.0	54.4	62.0	66.5	65.4	58.1	45.8	36.3	29.1	46.0
	MEAN	37.1	38.9	47.2	57.8	66.3	73.5	77.6	76.5	69.9	58.4	48.8	40.1	57.7
LANGLEY AIR FORCE BASE	MAX	47.0	49.0	56.6	66.8	74.4	81.9	85.7	84.8	78.8	68.7	59.6	50.7	67.0
	MIN	31.4	32.8	40.0	48.7	58.1	66.3	71.1	70.7	64.9	53.2	42.9	34.7	51.2
	MEAN	39.2	40.9	48.3	57.8	66.2	74.1	78.4	77.8	71.9	61.0	51.3	42.7	59.1
LAWRENCEVILLE 5 W	MAX	49.7	52.3	61.1	72.3	79.1	85.5	89.2	88.3	82.3	71.9	62.3	52.2	70.5
	MIN	27.2	28.5	35.6	44.3	52.8	60.1	64.8	64.4	57.5	45.5	36.8	29.4	45.6
	MEAN	38.5	40.4	48.4	58.3	66.0	72.8	77.1	76.4	69.9	58.7	49.6	40.8	58.1
LEXINGTON	MAX	45.4	48.9	58.5	70.0	77.8	84.0	87.3	86.3	80.0	69.5	57.9	48.3	67.8
	MIN	23.8	25.4	32.8	41.6	50.6	58.1	62.6	61.8	55.2	43.1	33.9	26.5	43.0
	MEAN	34.6	37.1	45.7	55.9	64.2	71.1	74.9	74.0	67.6	56.3	46.0	37.4	55.4
LINCOLN	MAX	42.8	46.3	56.0	67.8	76.7	84.4	88.2	87.2	80.7	69.8	57.3	46.6	67.0
	MIN	23.7	25.2	32.8	42.6	51.6	60.1	64.5	63.3	56.6	44.9	35.9	27.4	44.1
	MEAN	33.3	35.8	44.4	55.2	64.2	72.3	76.3	75.3	68.7	57.4	46.6	37.0	55.5
LOUISA	MAX	45.3	48.7	58.2	69.7	77.0	83.7	87.3	85.9	79.9	69.5	58.5	48.0	67.6
	MIN	24.9	26.4	33.9	43.2	52.2	60.2	64.6	63.9	56.9	44.8	35.9	28.1	44.6
	MEAN	35.1	37.5	46.1	56.4	64.6	71.9	75.9	74.9	68.4	57.2	47.2	38.1	56.1
LURAY 5 E	MAX	43.9	46.9	55.9	67.6	76.3	83.1	86.8	85.5	79.4	68.6	57.2	46.8	66.5
	MIN	20.0	21.5	29.5	39.1	47.6	55.3	59.5	58.7	52.1	40.3	31.9	23.6	39.9
	MEAN	32.0	34.2	42.7	53.4	62.0	69.2	73.1	72.1	65.8	54.5	44.6	35.3	53.2
LYNCHBURG WSO R	MAX	44.4	47.1	56.2	68.1	75.8	82.5	86.1	85.0	78.8	68.1	57.2	47.5	66.4
	MIN	25.9	27.6	35.1	44.6	53.2	60.6	65.1	64.5	57.9	46.1	36.7	29.1	45.5
	MEAN	35.1	37.4	45.7	56.4	64.5	71.6	75.7	74.8	68.4	57.1	47.0	38.3	56.0
MARTINSVILLE FILTER PLT	MAX	47.6	50.7	59.9	71.4	78.4	84.5	87.9	86.7	80.7	70.6	59.9	50.2	69.0
	MIN	23.9	25.2	32.3	40.7	49.9	57.7	62.3	61.7	54.5	41.4	32.8	25.8	42.4
	MEAN	35.8	38.0	46.1	56.1	64.1	71.1	75.1	74.2	67.6	56.1	46.3	38.0	55.7
MOUNT WEATHER	MAX	36.7	38.9	48.1	60.4	69.8	77.4	81.3	80.2	73.7	62.6	50.8	40.3	60.0
	MIN	20.6	22.0	29.9	40.6	50.1	58.7	62.9	62.1	55.5	44.5	35.2	25.2	42.3
	MEAN	28.7	30.5	39.1	50.5	60.0	68.1	72.1	71.2	64.6	53.6	43.0	32.8	51.2
NEWPORT NEWS PRESS BLD	MAX	49.0	50.9	58.9	69.9	77.7	85.2	88.8	87.8	82.0	71.4	62.2	53.0	69.7
	MIN	31.7	32.7	39.4	48.5	57.4	65.7	70.5	70.3	64.8	53.4	43.4	35.5	51.1
	MEAN	40.4	41.8	49.2	59.2	67.6	75.5	79.7	79.1	73.4	62.4	52.8	44.3	60.5
NORFOLK WSO	MAX	48.1	49.9	57.5	68.2	75.7	83.2	86.9	85.7	80.2	69.8	60.8	51.9	68.2
	MIN	31.7	32.3	39.4	48.1	57.2	65.3	69.9	69.6	64.2	52.8	43.0	35.0	50.7
	MEAN	39.9	41.1	48.5	58.2	66.4	74.3	78.4	77.7	72.2	61.3	51.9	43.5	59.5
PAINTER 2 W	MAX	46.7	48.5	56.1	66.7	74.9	82.2	86.3	85.3	79.8	69.3	59.8	50.5	67.2
	MIN	28.6	29.7	36.6	45.2	54.3	62.6	67.6	66.6	60.4	49.5	40.3	32.1	47.8
	MEAN	37.7	39.1	46.4	56.0	64.6	72.5	77.0	76.0	70.1	59.4	50.0	41.3	57.5
PENNINGTON GAP	MAX	44.4	48.2	57.7	68.9	76.7	82.5	85.4	84.8	80.4	70.2	57.8	48.2	67.1
	MIN	23.0	24.7	32.3	40.6	49.1	56.5	60.9	60.0	53.5	40.6	31.5	25.4	41.5
	MEAN	33.7	36.5	45.0	54.8	62.9	69.5	73.2	72.4	67.0	55.4	44.7	36.8	54.3

VIRGINIA

TEMPERATURE NORMALS (DEG F)

STATION			JAN	FEB	MAR	APR	MAY	JUN	JUL	AUG	SEP	OCT	NOV	DEC	ANN
PIEDMONT RESEARCH STA		MAX	44.2	46.7	56.2	67.6	75.9	83.3	87.3	86.0	79.9	69.0	57.9	47.3	66.8
		MIN	23.1	24.7	32.6	43.0	52.1	60.3	64.7	63.7	57.0	44.7	35.1	26.5	44.0
		MEAN	33.7	35.7	44.4	55.3	64.0	71.9	76.1	74.9	68.5	56.9	46.6	36.9	55.4
RICHMOND WSO	R	MAX	46.7	49.6	58.5	70.6	77.9	84.8	88.4	87.1	81.0	70.5	60.5	50.2	68.8
		MIN	26.5	28.1	35.8	45.1	54.2	62.2	67.2	66.4	59.3	46.7	37.3	29.6	46.5
		MEAN	36.6	38.9	47.2	57.9	66.1	73.5	77.8	76.8	70.2	58.6	48.9	39.9	57.7
ROANOKE WSO	R	MAX	44.8	48.0	56.9	68.2	76.4	83.0	86.7	85.5	79.4	68.6	57.4	47.8	66.9
		MIN	26.2	27.8	35.3	44.3	53.0	60.1	64.6	63.8	57.0	44.9	36.3	28.7	45.2
		MEAN	35.5	37.9	46.1	56.3	64.7	71.6	75.7	74.7	68.2	56.8	46.9	38.3	56.1
STAUNTON SEWAGE PLANT		MAX	43.3	45.1	54.1	65.6	74.1	81.1	85.1	84.0	77.7	67.2	56.3	46.4	65.0
		MIN	21.5	22.7	30.9	40.6	49.4	57.4	61.6	60.2	53.2	41.5	32.6	24.8	41.4
		MEAN	32.4	34.0	42.5	53.1	61.8	69.3	73.4	72.1	65.5	54.4	44.5	35.6	53.2
SUFFOLK LAKE KILBY		MAX	48.7	51.2	59.1	70.2	77.4	84.4	87.9	86.8	81.0	70.4	61.4	52.2	69.2
		MIN	29.0	30.4	37.5	46.3	55.4	63.3	67.9	67.5	61.3	49.6	39.5	31.8	48.3
		MEAN	38.9	40.9	48.3	58.2	66.4	73.9	77.9	77.2	71.2	60.0	50.5	42.0	58.8
TIMBERVILLE 3 E		MAX	42.4	45.1	54.7	66.3	75.5	82.7	86.3	85.1	78.8	67.8	55.8	46.0	65.5
		MIN	21.3	22.9	30.8	40.0	49.2	57.1	61.3	60.3	53.7	41.9	33.2	25.0	41.4
		MEAN	31.9	34.0	42.8	53.2	62.4	69.9	73.8	72.7	66.3	54.9	44.5	35.5	53.5
WALKERTON 2 NW		MAX	47.4	50.4	59.4	71.0	78.3	85.1	88.6	87.3	81.3	70.8	60.7	50.5	69.2
		MIN	25.3	26.9	34.3	43.8	53.3	61.3	65.9	65.1	57.8	45.5	35.9	28.1	45.3
		MEAN	36.4	38.7	46.9	57.4	65.8	73.2	77.3	76.2	69.6	58.2	48.3	39.3	57.3
WARRENTON 3 SE		MAX	41.7	44.1	53.6	65.3	74.0	81.3	85.8	85.0	79.0	67.9	56.2	45.1	64.9
		MIN	22.8	24.1	32.3	42.6	51.6	60.1	64.5	63.4	57.0	45.2	36.0	26.8	43.9
		MEAN	32.3	34.1	42.9	54.0	62.8	70.7	75.2	74.2	68.0	56.6	46.2	36.0	54.4
WARSAW 2 NW		MAX	46.1	49.0	57.8	69.4	77.4	84.8	88.4	87.2	81.3	70.3	59.8	49.5	68.4
		MIN	26.8	28.2	35.4	44.5	53.7	61.8	66.3	65.5	58.8	47.2	38.3	30.2	46.4
		MEAN	36.5	38.6	46.6	57.0	65.6	73.4	77.4	76.4	70.0	58.8	49.0	39.9	57.4
WASH DC-DULLES WSO		MAX	40.9	43.9	53.5	65.7	74.6	82.6	87.0	85.8	79.3	68.0	55.9	44.9	65.2
		MIN	21.8	23.3	31.3	40.9	50.2	58.8	63.9	62.7	55.4	42.6	33.6	25.3	42.5
		MEAN	31.4	33.6	42.4	53.3	62.4	70.7	75.5	74.3	67.4	55.3	44.8	35.1	53.9
WASHINGTON NAT AP WSO	R	MAX	42.9	45.9	55.0	67.1	75.9	84.0	87.9	86.4	80.1	68.9	57.4	46.6	66.5
		MIN	27.5	29.0	36.6	46.2	56.1	65.0	69.9	68.7	62.0	49.7	39.9	31.2	48.5
		MEAN	35.2	37.5	45.8	56.7	66.0	74.5	78.9	77.6	71.1	59.3	48.7	38.9	57.5
WILLIAMSBURG 2 N		MAX	49.0	51.3	59.8	71.0	78.1	84.7	88.0	87.1	81.5	71.0	62.1	52.2	69.7
		MIN	27.7	29.0	36.0	45.2	54.4	62.0	66.9	66.3	60.3	48.4	38.8	30.9	47.2
		MEAN	38.4	40.2	47.9	58.1	66.3	73.4	77.5	76.7	70.9	59.8	50.4	41.6	58.4
WINCHESTER 1 N		MAX	40.9	44.0	54.0	66.3	75.4	82.8	86.5	85.5	78.8	67.2	55.0	44.2	65.1
		MIN	23.4	25.0	32.8	42.5	51.5	59.2	63.6	62.4	56.1	44.7	35.8	27.2	43.7
		MEAN	32.2	34.5	43.4	54.4	63.5	71.0	75.1	74.0	67.5	56.0	45.4	35.7	54.4
WOODSTOCK		MAX	43.8	46.8	56.3	67.9	76.7	83.9	87.6	86.6	80.3	69.4	57.4	47.1	67.0
		MIN	22.3	24.0	31.6	41.0	49.7	57.3	61.8	60.7	54.0	42.6	34.2	25.9	42.1
		MEAN	33.1	35.4	44.0	54.5	63.2	70.7	74.7	73.7	67.2	56.0	45.8	36.6	54.6
WYTHEVILLE 1 S		MAX	43.2	46.6	55.6	66.6	74.2	80.3	83.5	83.0	77.7	67.8	55.8	46.2	65.0
		MIN	22.4	23.7	30.4	38.6	47.5	54.3	58.6	57.7	51.7	39.5	31.2	24.9	40.0
		MEAN	32.8	35.2	43.0	52.6	60.9	67.3	71.1	70.3	64.7	53.7	43.5	35.6	52.6

VIRGINIA

PRECIPITATION NORMALS (INCHES)

STATION	JAN	FEB	MAR	APR	MAY	JUN	JUL	AUG	SEP	OCT	NOV	DEC	ANN
ALLISONIA	2.77	2.63	3.63	3.20	3.79	3.24	3.58	3.18	3.14	3.04	2.33	2.62	37.15
ALTAVISTA	3.21	3.07	3.60	2.92	3.81	3.25	3.61	3.71	3.14	3.29	2.80	3.04	39.45
AMISSVILLE	2.65	2.40	3.30	2.97	3.71	4.05	3.81	3.94	3.23	2.89	2.75	2.72	38.42
APPOMATTOX 3 W	3.00	3.08	3.87	3.33	3.79	3.40	4.03	3.78	3.61	3.44	3.06	3.15	41.54
BEDFORD	3.08	3.21	3.87	3.29	3.80	3.88	4.07	3.95	3.40	3.56	2.90	3.09	42.10
BERRYVILLE	2.58	2.22	3.10	3.28	3.62	3.90	3.65	3.99	3.11	3.12	2.86	2.52	37.95
BIG MEADOWS	3.50	3.06	4.20	3.83	4.48	4.32	4.05	5.27	4.99	5.18	4.41	3.36	50.65
BLACKSBURG 3 SE	2.95	2.94	3.91	3.55	3.62	3.61	3.65	3.53	3.48	3.14	2.68	2.92	39.98
BLAND	3.02	2.92	3.89	3.18	3.59	3.47	3.79	3.51	3.02	2.72	2.53	2.79	38.43
BREMO BLUFF	3.08	3.08	3.81	2.92	3.73	3.60	3.50	4.06	3.08	3.64	3.02	3.13	40.65
BROOKNEAL	3.42	3.13	3.93	3.35	3.55	3.22	3.80	4.09	3.47	3.27	2.85	3.12	41.20
BUCHANAN	2.94	2.91	3.72	3.04	3.73	3.57	4.13	4.04	3.41	3.70	2.85	2.91	40.95
BUCKINGHAM	3.09	2.94	3.72	3.15	3.80	3.32	3.76	4.02	3.72	3.55	2.95	3.13	41.15
BUENA VISTA	2.78	3.02	3.74	2.83	3.35	3.27	3.73	3.24	3.04	3.72	2.92	3.04	38.68
BURKES GARDEN	3.50	3.27	4.10	3.68	4.01	3.95	4.71	3.71	3.48	2.91	3.07	3.28	43.67
CHARLOTTE COURT HOUSE	3.25	3.16	3.80	3.26	3.47	3.73	3.71	3.99	3.66	3.55	2.90	3.16	41.64
CHARLOTTESVILLE 2 W	3.28	3.18	4.16	3.28	4.17	3.91	4.51	4.51	3.90	4.20	3.27	3.35	45.72
CHASE CITY	3.76	3.46	3.54	3.13	3.43	3.57	3.72	3.86	3.46	3.47	3.20	3.18	41.78
CHATHAM 2 NE	3.41	3.32	4.16	3.54	3.89	3.60	3.94	3.80	3.79	3.55	3.09	3.39	43.48
CLARKSVILLE	3.63	3.39	3.79	3.17	3.69	3.61	3.81	4.14	3.34	3.09	2.98	3.18	41.82
CLIFTON FORGE	2.88	2.93	4.09	3.17	3.52	3.54	3.49	3.50	3.05	3.38	2.82	2.86	39.23
COLUMBIA	2.96	2.77	3.51	2.99	3.62	3.18	3.63	4.01	3.37	3.92	3.00	3.02	39.98
CONCORD 5 S	3.33	3.06	3.89	3.22	3.60	3.54	3.77	3.51	3.48	3.69	3.14	3.14	41.37
COPPER HILL 1 NNE	2.97	3.15	4.00	3.86	4.06	4.15	3.70	3.86	4.57	4.46	3.20	2.83	44.81
COVINGTON	2.42	2.42	3.38	2.74	3.38	3.23	3.55	3.56	2.76	2.90	2.18	2.41	34.93
CULPEPER	2.98	2.81	3.79	3.22	3.60	3.96	3.83	4.10	3.68	3.45	3.22	3.00	41.64
DALE ENTERPRISE	1.99	2.07	2.87	2.57	3.36	3.27	3.68	3.90	3.23	2.86	2.41	2.20	34.41
DANVILLE-BRIDGE ST	3.43	3.32	4.01	3.45	3.73	3.86	4.06	4.04	3.68	3.23	2.82	3.31	42.94
DIAMOND SPRINGS	3.88	3.36	4.15	3.01	4.06	3.78	5.30	5.74	4.64	3.83	2.95	3.30	48.00
DRIVER 4 NE	3.75	3.62	3.96	2.87	3.78	3.76	4.86	5.39	4.23	3.60	3.05	3.40	46.27
ELKWOOD 6 SE	2.95	2.66	3.61	3.06	3.68	3.83	3.55	4.06	3.35	3.45	3.25	3.20	40.65
EMPORIA 1 WNW	3.24	3.29	3.46	2.90	3.89	3.48	4.91	4.42	3.65	2.96	2.83	3.39	42.42
FARMVILLE 2 N	3.46	3.30	3.89	3.16	3.82	3.63	4.24	4.07	3.51	3.17	3.30	3.42	42.93
FLOYD 2 NE	2.95	3.34	3.94	3.71	3.51	3.74	3.73	3.25	4.19	3.70	3.05	3.04	42.15
FREDERICKSBURG NATL PK	3.11	2.88	3.84	3.06	3.47	3.64	3.68	4.41	3.36	3.20	3.04	3.30	40.99
GALAX RADIO WBOB	2.73	3.10	3.92	3.43	3.93	4.02	4.08	3.57	3.89	3.74	2.94	3.08	42.43
GLEN LYN	2.66	2.65	3.49	3.15	3.52	3.16	3.97	3.25	2.95	2.59	2.45	2.61	36.45
GOSHEN	2.92	2.87	4.02	3.15	3.77	3.55	3.80	3.47	3.45	3.48	2.99	2.80	40.27
HILLSVILLE 1 S	2.86	3.09	4.15	3.54	4.20	4.02	4.14	3.63	3.83	3.57	2.83	2.99	42.85
HOLLAND 1 E	4.05	3.55	3.96	3.05	4.03	4.27	5.22	5.44	4.08	4.04	3.00	3.46	48.15
HOPEWELL	3.57	3.25	3.68	3.17	3.82	3.46	5.05	4.62	4.09	3.63	3.08	3.39	44.81
HOT SPRINGS	3.04	2.94	4.12	3.31	3.85	3.67	4.15	3.82	3.51	3.53	2.82	2.96	41.72
HUDDLESTON 2 SW	3.18	3.07	3.96	3.37	3.51	3.24	3.67	3.61	3.57	3.48	2.98	3.10	40.74
JOHN H KERR DAM	3.40	3.25	3.59	2.99	3.57	3.40	4.31	3.79	3.51	3.42	3.08	3.22	41.53
LAFAYETTE 1 NE	3.01	2.69	3.32	3.06	3.43	3.35	3.81	3.42	3.22	3.22	2.62	2.77	37.92
LANGLEY AIR FORCE BASE	3.72	3.49	4.26	2.84	3.77	3.74	4.80	4.77	4.60	3.39	2.91	3.26	45.55
LAWRENCEVILLE 5 W	3.62	3.68	3.99	3.08	3.95	4.01	4.06	4.17	3.84	3.57	3.13	3.65	44.75
LEXINGTON	2.75	2.89	3.79	2.87	3.33	3.36	3.66	3.09	3.28	3.25	2.67	2.92	37.86
LINCOLN	2.94	2.64	3.88	3.47	4.01	4.26	3.94	4.34	3.33	3.06	3.04	3.22	42.13
LOUISA	3.07	3.07	3.81	3.04	3.39	3.94	3.86	4.46	3.26	3.63	3.16	3.39	42.08
LURAY 5 E	2.99	2.39	3.25	2.89	3.44	3.59	3.57	4.12	3.59	3.59	3.01	2.82	39.25
LYNCHBURG WSO R	3.06	2.93	3.69	2.90	3.65	3.47	3.85	3.69	3.23	3.36	2.92	3.16	39.91
MANASSAS	2.75	2.32	3.14	2.95	3.49	3.44	3.38	3.61	3.59	2.88	2.88	3.03	37.46
MARTINSVILLE FILTER PLT	3.32	3.53	4.28	3.71	3.87	3.93	4.54	4.10	3.83	3.55	2.94	3.40	45.00
MEADOWS OF DAN 5 SW	4.02	3.97	5.24	4.90	4.68	5.11	5.35	5.03	5.23	4.56	3.71	3.95	55.75

VIRGINIA

PRECIPITATION NORMALS (INCHES)

STATION	JAN	FEB	MAR	APR	MAY	JUN	JUL	AUG	SEP	OCT	NOV	DEC	ANN
MONTEBELLO 3 NE	3.72	3.93	4.70	4.12	4.47	4.13	4.37	3.99	4.22	4.89	4.23	3.72	50.49
MONTEREY	2.96	2.90	3.85	3.47	3.65	3.88	3.75	3.32	3.36	3.39	2.68	3.20	40.41
MOUNT WEATHER	2.51	2.27	3.12	3.35	4.12	4.46	3.87	4.05	3.46	3.46	3.29	2.84	40.80
NEW CASTLE	2.89	2.91	3.91	3.17	3.56	3.59	3.43	3.57	3.49	3.19	2.69	2.91	39.31
NEWPORT 2 NW	2.65	2.57	3.36	2.99	3.50	3.40	3.22	3.07	3.42	3.24	2.56	2.64	36.62
NEWPORT NEWS PRESS BLD	3.73	3.55	4.05	2.93	3.68	3.79	4.64	4.82	4.60	3.40	2.86	3.16	45.21
NORFOLK WSO	3.72	3.28	3.86	2.87	3.75	3.45	5.15	5.33	4.35	3.41	2.88	3.17	45.22
NORTH RIVER DAM	2.76	2.73	3.65	3.04	3.99	3.70	4.04	3.78	3.94	3.33	3.00	2.87	40.83
PAINTER 2 W	3.41	3.31	4.13	2.92	3.47	3.51	4.10	4.28	3.41	3.57	2.96	3.37	42.44
PEDLAR DAM	3.24	3.40	4.39	3.55	4.06	3.76	4.32	4.02	3.95	3.91	3.39	3.56	45.55
PENNINGTON GAP	4.88	4.40	5.75	4.35	4.39	4.17	5.07	3.76	3.47	2.96	3.89	4.66	51.75
PIEDMONT RESEARCH STA	2.79	2.58	3.69	3.07	3.70	3.69	3.77	4.44	3.36	3.81	3.09	3.10	41.09
PILOT 1 ENE	2.70	2.62	3.47	3.49	3.57	3.48	3.73	3.62	3.38	3.42	2.59	2.62	38.69
RADFORD	2.73	2.61	3.49	3.18	3.10	3.17	3.84	3.27	2.94	3.03	2.47	2.64	36.47
RANDOLPH 5 NNE	3.50	3.34	4.06	3.48	3.68	3.50	4.06	3.90	4.14	3.54	3.00	3.23	43.43
RAPIDAN	3.30	2.93	3.91	3.07	3.72	3.96	3.37	4.03	3.35	3.83	3.14	3.25	41.86
RICHMOND WSO R	3.23	3.13	3.57	2.90	3.55	3.60	5.14	5.01	3.52	3.74	3.29	3.39	44.07
ROANOKE WSO R	2.83	3.19	3.69	3.09	3.51	3.34	3.45	3.91	3.14	3.48	2.59	2.93	39.15
ROCKFISH	3.42	3.33	4.24	3.55	3.98	4.55	4.14	4.24	4.26	4.37	3.36	3.61	47.05
ROCKY MOUNT	3.19	3.19	4.03	3.61	3.78	3.53	4.12	3.86	4.02	3.74	2.79	3.15	43.01
SPEEDWELL	2.64	2.79	3.36	2.96	3.57	3.28	4.47	3.24	2.85	2.99	2.55	2.71	37.41
STAFFORDSVILLE 3 ENE	2.74	2.70	3.61	3.36	3.43	3.30	3.77	3.53	3.28	2.77	2.48	2.68	37.65
STAUNTON SEWAGE PLANT	2.53	2.19	3.19	2.79	3.36	3.03	3.51	3.40	3.55	3.26	2.56	2.55	35.92
SUFFOLK LAKE KILBY	3.90	3.64	4.11	2.95	3.81	4.04	4.99	5.20	4.28	3.63	2.94	3.44	46.93
TIMBERVILLE 3 E	2.29	2.09	2.99	2.56	3.23	3.42	3.62	4.02	3.50	2.80	2.41	2.39	35.32
TROUT DALE	4.11	4.15	5.25	4.58	4.67	4.59	4.69	4.06	3.83	3.85	3.53	3.73	51.04
TYE RIVER 1 SE	3.31	3.12	4.04	3.34	3.90	3.66	4.12	3.77	3.55	3.80	3.11	3.31	43.03
VIENNA DUNN LORING	2.96	2.71	3.71	3.38	3.92	4.11	4.26	4.68	4.00	3.10	3.24	3.22	43.29
WALKERTON 2 NW	3.32	3.18	3.67	2.94	3.86	4.03	4.71	4.86	3.68	3.40	3.48	3.40	44.53
WALLACETON LK DRUMMOND	4.03	3.64	4.02	3.21	4.14	4.00	4.97	5.52	4.17	4.05	2.86	3.60	48.21
WARRENTON 3 SE	3.02	2.61	3.50	3.22	3.54	3.83	3.93	3.90	3.32	3.14	2.98	3.06	40.05
WARSAW 2 NW	3.17	2.74	3.51	2.96	4.01	3.69	4.38	4.33	4.19	3.53	3.30	3.20	43.01
WASHINGTON	2.42	2.20	3.09	3.29	3.85	3.64	4.04	4.47	3.42	3.47	3.09	2.60	39.58
WASH DC-DULLES WSO	2.83	2.64	3.43	3.14	3.62	4.23	3.75	4.16	3.26	3.01	2.99	3.29	40.35
WASHINGTON NAT AP WSO R	2.76	2.62	3.46	2.93	3.48	3.35	3.88	4.40	3.22	2.90	2.82	3.18	39.00
WILLIAMSBURG 2 N	3.74	3.49	4.22	2.98	4.40	4.21	5.23	4.68	4.42	3.56	3.21	3.43	47.57
WINCHESTER 1 N	2.50	2.36	3.28	3.24	3.60	4.06	4.01	3.74	3.15	2.97	2.73	2.63	38.27
WOODSTOCK	2.40	1.98	2.92	2.85	3.23	3.48	3.61	3.47	3.22	2.92	2.30	2.34	34.72
WOOLWINE 4 S	3.99	4.01	5.16	4.79	4.71	4.99	5.09	4.39	5.72	4.80	3.76	4.04	55.45
WYTHEVILLE 1 S	2.63	2.73	3.46	3.11	3.58	3.11	4.49	3.36	3.06	2.95	2.38	2.60	37.46

VIRGINIA

HEATING DEGREE DAY NORMALS (BASE 65 DEG F)

STATION	JUL	AUG	SEP	OCT	NOV	DEC	JAN	FEB	MAR	APR	MAY	JUN	ANN
BEDFORD	0	0	38	248	513	787	877	731	546	236	83	7	4066
BIG MEADOWS	40	50	181	493	774	1079	1175	1028	908	543	305	106	6682
BLACKSBURG 3 SE	8	10	101	388	672	961	1057	899	741	411	201	58	5507
BUCHANAN	0	0	27	253	546	822	902	745	564	255	91	7	4212
BURKES GARDEN	19	34	153	462	732	986	1085	921	775	465	250	88	5970
CHARLOTTE COURT HOUSE	0	0	37	264	528	828	921	776	592	261	97	8	4312
CHARLOTTESVILLE 2 W	0	0	24	220	495	822	921	784	592	245	86	0	4189
CHATHAM 2 NE	0	0	49	276	531	818	905	764	595	284	111	17	4350
COLUMBIA	0	0	28	268	549	849	946	804	614	274	100	5	4437
CULPEPER	0	0	29	258	552	862	958	795	605	274	84	0	4417
DALE ENTERPRISE	0	0	73	340	627	921	1029	862	688	360	140	16	5056
DANVILLE-BRIDGE ST	0	0	20	219	489	775	849	706	527	208	63	0	3856
DIAMOND SPRINGS	0	0	6	149	384	670	778	672	515	215	64	0	3453
ELKWOOD 6 SE	0	0	44	297	570	868	973	812	623	293	107	8	4595
FARMVILLE 2 N	0	0	27	240	510	794	868	722	543	220	75	0	3999
FLOYD 2 NE	13	15	110	384	660	918	1001	848	704	387	189	64	5293
FREDERICKSBURG NATL PK	0	0	30	247	537	840	939	801	608	280	78	0	4360
HOLLAND 1 E	0	0	12	190	432	725	818	692	521	216	55	0	3661
HOPEWELL	0	0	10	159	402	691	784	647	477	165	36	0	3371
HOT SPRINGS	0	8	104	382	675	964	1060	902	750	405	188	37	5475
JOHN H KERR DAM	0	0	27	224	486	772	865	731	552	224	73	6	3960
LANGLEY AIR FORCE BASE	0	0	16	168	411	691	800	675	518	231	64	0	3574
LAWRENCEVILLE 5 W	0	0	20	214	462	750	822	689	520	212	72	8	3769
LEXINGTON	0	0	39	281	570	856	942	781	598	278	101	5	4451
LINCOLN	0	0	25	248	552	868	983	818	639	300	107	0	4540
LOUISA	0	0	32	255	534	834	927	770	586	264	93	0	4295
LURAY 5 E	0	0	82	342	612	921	1023	862	691	352	157	20	5062
LYNCHBURG WSO R	0	0	32	258	540	828	927	773	598	263	97	7	4323
MARTINSVILLE FILTER PLT	0	0	37	285	561	837	905	756	586	273	98	9	4347
MOUNT WEATHER	0	0	79	362	660	998	1125	966	803	435	187	29	5644
NEWPORT NEWS PRESS BLD	0	0	0	136	366	642	763	650	490	199	51	0	3297
NORFOLK WSO	0	0	9	146	393	667	778	669	512	219	53	0	3446
PAINTER 2 W	0	0	13	200	450	735	846	725	577	276	76	0	3898
PENNINGTON GAP	0	0	48	307	609	874	970	798	626	313	128	15	4688
PIEDMONT RESEARCH STA	0	0	32	269	552	871	970	820	639	297	111	6	4567
RICHMOND WSO R	0	0	24	221	483	778	880	731	552	226	65	0	3960
ROANOKE WSO R	0	0	38	267	543	828	915	759	586	268	99	12	4315
STAUNTON SEWAGE PLANT	0	0	82	341	615	911	1011	868	698	361	160	22	5069
SUFFOLK LAKE KILBY	0	0	8	183	435	713	809	675	518	215	52	0	3608
TIMBERVILLE 3 E	0	0	63	318	615	915	1026	868	688	354	131	12	4990
WALKERTON 2 NW	0	0	22	229	501	797	887	736	561	238	66	0	4037
WARRENTON 3 SE	0	0	35	276	564	899	1014	865	685	334	127	14	4813
WARSAW 2 NW	0	0	18	211	480	778	884	739	570	249	70	0	3999
WASH DC-DULLES WSO	0	0	41	307	606	927	1042	879	701	355	138	8	5004
WASHINGTON NAT AP WSO R	0	0	13	197	489	809	924	770	595	257	68	0	4122
WILLIAMSBURG 2 N	0	0	10	191	438	725	825	694	530	216	55	0	3684
WINCHESTER 1 N	0	0	49	291	588	908	1017	854	670	322	115	9	4823
WOODSTOCK	0	0	47	285	576	880	989	829	651	315	120	11	4703
WYTHEVILLE 1 S	0	0	83	350	645	911	998	834	682	372	163	36	5074

VIRGINIA

COOLING DEGREE DAY NORMALS (BASE 65 DEG F)

STATION	JAN	FEB	MAR	APR	MAY	JUN	JUL	AUG	SEP	OCT	NOV	DEC	ANN
BEDFORD	0	0	0	8	92	211	326	295	131	16	0	0	1079
BIG MEADOWS	0	0	0	0	14	34	83	59	16	0	0	0	206
BLACKSBURG 3 SE	0	0	0	0	46	112	194	168	65	0	0	0	585
BUCHANAN	0	0	0	6	88	208	326	301	126	11	0	0	1066
BURKES GARDEN	0	0	0	0	15	49	90	74	24	0	0	0	252
CHARLOTTE COURT HOUSE	0	0	0	6	88	215	344	316	151	22	0	0	1142
CHARLOTTESVILLE 2 W	0	0	0	8	98	229	366	332	162	25	0	0	1220
CHATHAM 2 NE	0	0	0	8	92	209	329	298	148	24	0	0	1108
COLUMBIA	0	0	0	0	82	215	353	319	139	17	0	0	1125
CULPEPER	0	0	0	7	81	225	353	313	140	10	0	0	1129
DALE ENTERPRISE	0	0	0	0	43	148	257	223	94	8	0	0	773
DANVILLE-BRIDGE ST	0	0	0	16	119	268	397	375	185	24	0	0	1384
DIAMOND SPRINGS	0	0	0	14	95	265	403	381	216	37	0	0	1411
ELKWOOD 6 SE	0	0	0	0	64	185	316	285	113	9	0	0	972
FARMVILLE 2 N	0	0	0	7	91	232	360	326	141	13	0	0	1170
FLOYD 2 NE	0	0	0	0	37	109	175	155	62	6	0	0	544
FREDERICKSBURG NATL PK	0	0	0	7	78	237	375	332	156	18	0	0	1203
HOLLAND 1 E	0	0	0	12	95	259	388	360	189	26	0	0	1329
HOPEWELL	0	0	8	21	132	303	434	409	226	38	0	0	1571
HOT SPRINGS	0	0	0	0	36	97	180	160	62	0	0	0	535
JOHN H KERR DAM	0	0	0	8	113	261	391	357	174	19	0	0	1323
LANGLEY AIR FORCE BASE	0	0	0	15	101	277	415	397	223	44	0	0	1472
LAWRENCEVILLE 5 W	0	0	6	11	103	242	375	353	167	18	0	0	1275
LEXINGTON	0	0	0	0	77	188	307	279	117	11	0	0	979
LINCOLN	0	0	0	6	82	223	350	319	136	12	0	0	1128
LOUISA	0	0	0	6	80	212	338	307	134	13	0	0	1090
LURAY 5 E	0	0	0	0	64	146	254	224	106	17	0	0	811
LYNCHBURG WSO R	0	0	0	5	81	205	332	304	134	13	0	0	1074
MARTINSVILLE FILTER PLT	0	0	0	6	71	192	313	285	115	9	0	0	991
MOUNT WEATHER	0	0	0	0	32	122	224	196	67	9	0	0	650
NEWPORT NEWS PRESS BLD	0	0	0	25	132	315	456	437	255	56	0	0	1676
NORFOLK WSO	0	0	0	15	96	282	415	394	225	31	0	0	1458
PAINTER 2 W	0	0	0	6	64	229	372	341	166	27	0	0	1205
PENNINGTON GAP	0	0	6	7	63	150	254	233	108	9	0	0	830
PIEDMONT RESEARCH STA	0	0	0	6	80	213	344	307	137	18	0	0	1105
RICHMOND WSO R	0	0	0	13	99	258	397	366	180	23	0	0	1336
ROANOKE WSO R	0	0	0	7	89	210	332	301	134	12	0	0	1085
STAUNTON SEWAGE PLANT	0	0	0	0	61	151	263	224	97	13	0	0	809
SUFFOLK LAKE KILBY	0	0	0	11	96	270	400	378	194	28	0	0	1377
TIMBERVILLE 3 E	0	0	0	0	50	159	273	239	102	0	0	0	823
WALKERTON 2 NW	0	0	0	10	91	249	381	347	160	18	0	0	1256
WARRENTON 3 SE	0	0	0	0	58	185	316	285	125	16	0	0	985
WARSAW 2 NW	0	0	0	9	88	252	384	353	168	19	0	0	1273
WASH DC-DULLES WSO	0	0	0	0	57	179	326	288	113	7	0	0	970
WASHINGTON NAT AP WSO R	0	0	0	8	99	285	431	391	196	20	0	0	1430
WILLIAMSBURG 2 N	0	0	0	9	95	252	388	363	187	29	0	0	1323
WINCHESTER 1 N	0	0	0	0	68	189	313	279	124	12	0	0	985
WOODSTOCK	0	0	0	0	64	182	301	270	113	6	0	0	936
WYTHEVILLE 1 S	0	0	0	0	36	105	192	169	74	0	0	0	576

44 – VIRGINIA & D.C.

LEGEND
11 = TEMPERATURE ONLY
12 = PRECIPITATION ONLY
13 = TEMP. & PRECIP.

STATE-STATION NUMBER	STN TYP	NAME	LATITUDE DEG-MIN	LONGITUDE DEG-MIN	ELEVATION (FT)
44-0135	12	ALLISONIA	N 3656	W 08045	1870
44-0166	12	ALTAVISTA	N 3706	W 07918	510
44-0193	12	AMISSVILLE	N 3841	W 07801	550
44-0243	12	APPOMATTOX 3 W	N 3722	W 07853	925
44-0551	13	BEDFORD	N 3721	W 07931	975
44-0670	12	BERRYVILLE	N 3909	W 07759	580
44-0720	13	BIG MEADOWS	N 3831	W 07826	3535
44-0766	13	BLACKSBURG 3 SE	N 3711	W 08025	2000
44-0792	12	BLAND	N 3706	W 08107	2000
44-0993	12	BREMO BLUFF	N 3742	W 07818	225
44-1082	12	BROOKNEAL	N 3703	W 07857	500
44-1121	13	BUCHANAN	N 3732	W 07941	870
44-1136	12	BUCKINGHAM	N 3733	W 07833	485
44-1159	12	BUENA VISTA	N 3743	W 07922	840
44-1209	13	BURKES GARDEN	N 3705	W 08120	3300
44-1585	13	CHARLOTTE COURT HOUSE	N 3703	W 07838	560
44-1593	13	CHARLOTTESVILLE 2 W	N 3802	W 07831	870
44-1606	12	CHASE CITY	N 3650	W 07828	510
44-1614	13	CHATHAM 2 NE	N 3650	W 07922	700
44-1746	12	CLARKSVILLE	N 3638	W 07833	325
44-1801	12	CLIFTON FORGE	N 3749	W 07950	1050
44-1929	13	COLUMBIA	N 3746	W 07809	300
44-1955	12	CONCORD 5 S	N 3717	W 07858	650
44-1999	12	COPPER HILL 1 NNE	N 3706	W 08008	2720
44-2041	12	COVINGTON	N 3748	W 08000	1245
44-2155	13	CULPEPER	N 3828	W 07800	420
44-2208	13	DALE ENTERPRISE	N 3827	W 07856	1400
44-2245	13	DANVILLE-BRIDGE ST	N 3635	W 07923	410
44-2368	13	DIAMOND SPRINGS	N 3654	W 07612	25
44-2504	12	DRIVER 4 NE	N 3653	W 07629	17
44-2729	13	ELKWOOD 6 SE	N 3827	W 07746	300
44-2790	12	EMPORIA 1 WNW	N 3641	W 07733	100
44-2941	13	FARMVILLE 2 N	N 3720	W 07823	450
44-3071	13	FLOYD 2 NE	N 3656	W 08018	2600
44-3192	13	FREDERICKSBURG NATL PK	N 3818	W 07728	100
44-3267	12	GALAX RADIO WBOB	N 3640	W 08055	2385
44-3397	12	GLEN LYN	N 3722	W 08052	1524
44-3470	12	GOSHEN	N 3759	W 07930	1350
44-3991	12	HILLSVILLE 1 S	N 3644	W 08044	2585
44-4044	13	HOLLAND 1 E	N 3641	W 07647	80
44-4101	13	HOPEWELL	N 3718	W 07718	40
44-4128	13	HOT SPRINGS	N 3800	W 07950	2238
44-4148	12	HUDDLESTON 2 SW	N 3709	W 07930	1000
44-4414	13	JOHN H KERR DAM	N 3636	W 07817	323
44-4676	12	LAFAYETTE 1 NE	N 3714	W 08013	1380
44-4720	13	LANGLEY AIR FORCE BASE	N 3705	W 07621	10
44-4768	13	LAWRENCEVILLE 5 W	N 3646	W 07756	300
44-4876	13	LEXINGTON	N 3747	W 07926	1060
44-4909	13	LINCOLN	N 3907	W 07743	500
44-5050	13	LOUISA	N 3802	W 07800	420

LEGEND
11 = TEMPERATURE ONLY
12 = PRECIPITATION ONLY
13 = TEMP. & PRECIP.

STATE-STATION NUMBER	STN TYP	NAME		LATITUDE DEG-MIN	LONGITUDE DEG-MIN	ELEVATION (FT)
44-5096	13	LURAY 5 E		N 3840	W 07823	1200
44-5120	13	LYNCHBURG WSO	R	N 3720	W 07912	916
44-5213	12	MANASSAS		N 3847	W 07730	330
44-5300	13	MARTINSVILLE FILTER PLT		N 3642	W 07953	760
44-5453	12	MEADOWS OF DAN 5 SW		N 3640	W 08027	1500
44-5685	12	MONTEBELLO 3 NE		N 3753	W 07906	2450
44-5698	12	MONTEREY		N 3825	W 07935	2910
44-5851	13	MOUNT WEATHER		N 3904	W 07753	1720
44-6012	12	NEW CASTLE		N 3730	W 08006	1325
44-6046	12	NEWPORT 2 NW		N 3719	W 08031	1900
44-6054	13	NEWPORT NEWS PRESS BLD		N 3701	W 07627	50
44-6139	13	NORFOLK WSO		N 3654	W 07612	22
44-6199	12	NORTH RIVER DAM		N 3822	W 07916	2400
44-6475	13	PAINTER 2 W		N 3735	W 07549	30
44-6593	12	PEDLAR DAM		N 3740	W 07917	1013
44-6626	13	PENNINGTON GAP		N 3645	W 08303	1510
44-6712	13	PIEDMONT RESEARCH STA		N 3813	W 07807	515
44-6723	12	PILOT 1 ENF		N 3704	W 08021	2178
44-6999	12	RADFORD		N 3708	W 08033	1730
44-7025	12	RANDOLPH 5 NNE		N 3659	W 07842	350
44-7033	12	RAPIDAN		N 3818	W 07804	300
44-7201	13	RICHMOND WSO	R	N 3730	W 07720	164
44-7285	13	ROANOKE WSO	R	N 3719	W 07958	1149
44-7312	12	ROCKFISH		N 3748	W 07845	445
44-7338	12	ROCKY MOUNT		N 3700	W 07954	1232
44-7971	12	SPEEDWELL		N 3649	W 08110	2340
44-8022	12	STAFFORDSVILLE 3 ENE		N 3716	W 08043	1950
44-8062	13	STAUNTON SEWAGE PLANT		N 3809	W 07902	1385
44-8192	13	SUFFOLK LAKE KILBY		N 3644	W 07636	22
44-8448	13	TIMBERVILLE 3 E		N 3839	W 07843	1000
44-8547	12	TROUT DALE		N 3642	W 08126	3104
44-8600	12	TYE RIVER 1 SE		N 3738	W 07856	720
44-8737	12	VIENNA DUNN LORING		N 3854	W 07713	418
44-8829	13	WALKERTON 2 NW		N 3745	W 07703	50
44-8837	12	WALLACETON LK DRUMMOND		N 3636	W 07626	25
44-8888	13	WARRENTON 3 SE		N 3841	W 07746	500
44-8894	13	WARSAW 2 NW		N 3759	W 07646	140
44-8902	12	WASHINGTON		N 3843	W 07810	635
44-8903	13	WASH DC-DULLES WSO		N 3857	W 07727	291
44-8906	13	WASHINGTON NAT AP WSO	R	N 3851	W 07702	10
44-9151	13	WILLIAMSBURG 2 N		N 3718	W 07642	70
44-9186	13	WINCHESTER 1 N		N 3912	W 07810	760
44-9263	13	WOODSTOCK		N 3853	W 07831	887
44-9272	12	WOOLWINE 4 S		N 3643	W 08017	1490
44-9301	13	WYTHEVILLE 1 S		N 3656	W 08105	2450

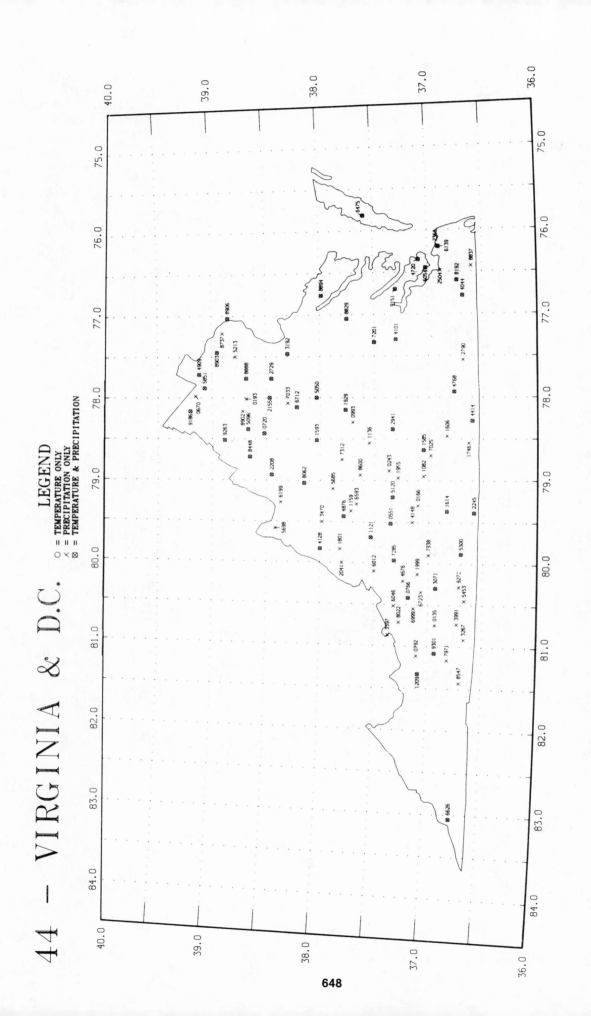

44 — VIRGINIA & D.C.

LEGEND

○ = TEMPERATURE ONLY
✕ = PRECIPITATION ONLY
⊗ = TEMPERATURE & PRECIPITATION

WASHINGTON

TEMPERATURE NORMALS (DEG F)

STATION		JAN	FEB	MAR	APR	MAY	JUN	JUL	AUG	SEP	OCT	NOV	DEC	ANN
ABERDEEN	MAX	45.4	49.8	51.8	56.3	61.4	65.1	69.0	69.4	68.9	61.5	52.2	47.1	58.2
	MIN	34.4	36.1	36.0	39.5	44.4	49.3	52.0	52.5	49.6	44.0	38.6	36.2	42.7
	MEAN	39.9	43.0	44.0	47.9	52.9	57.3	60.5	60.9	59.3	52.8	45.4	41.7	50.5
ANACORTES	MAX	44.2	48.4	51.0	56.8	63.0	67.3	71.6	71.1	66.9	58.8	50.7	46.5	58.0
	MIN	33.8	36.4	37.2	40.8	45.3	49.6	51.6	51.7	49.3	44.0	38.9	36.4	42.9
	MEAN	39.1	42.4	44.1	48.8	54.2	58.4	61.6	61.5	58.1	51.5	44.8	41.5	50.5
BATTLE GROUND	MAX	44.8	50.7	53.8	59.4	66.6	71.5	78.3	77.9	74.4	64.3	52.9	46.7	61.8
	MIN	31.7	34.0	34.8	37.8	42.1	47.5	49.6	49.6	46.2	40.8	36.0	33.8	40.3
	MEAN	38.3	42.4	44.3	48.6	54.4	59.5	64.0	63.8	60.3	52.6	44.5	40.3	51.1
BELLINGHAM 2 N	MAX	43.4	48.9	51.5	57.8	64.8	69.0	74.2	73.4	69.3	60.2	50.7	45.8	59.1
	MIN	30.6	33.5	34.2	37.7	42.7	48.1	50.1	50.0	46.1	40.8	35.8	33.2	40.2
	MEAN	37.0	41.2	42.9	47.7	53.8	58.6	62.2	61.7	57.7	50.5	43.2	39.6	49.7
BELLINGHAM FAA AIRPORT	MAX	42.3	47.5	50.1	56.0	62.4	66.7	71.5	71.1	66.9	58.6	49.8	44.8	57.3
	MIN	30.9	34.0	35.1	39.1	44.6	50.0	52.6	52.7	47.9	41.8	36.3	33.5	41.5
	MEAN	36.6	40.8	42.6	47.6	53.5	58.4	62.1	61.9	57.4	50.2	43.1	39.2	49.5
BICKLETON	MAX	35.1	41.0	46.0	54.7	63.8	72.3	81.8	80.3	72.1	59.4	44.0	38.0	57.4
	MIN	21.4	26.4	28.8	33.4	39.8	45.9	52.4	51.8	46.7	37.5	28.6	24.4	36.4
	MEAN	28.2	33.8	37.4	44.1	51.8	59.1	67.2	66.1	59.4	48.5	36.3	31.3	46.9
BLAINE 1 ENE	MAX	41.7	47.3	50.6	56.7	63.4	67.9	72.6	72.1	67.0	58.1	49.1	44.1	57.6
	MIN	30.7	33.8	34.7	38.5	43.7	48.8	50.7	50.8	47.1	41.7	36.1	33.4	40.8
	MEAN	36.2	40.6	42.7	47.6	53.6	58.4	61.7	61.5	57.1	49.9	42.6	38.8	49.2
BREMERTON	MAX	43.8	48.9	51.9	58.4	65.3	69.7	75.3	74.5	69.8	60.6	50.6	45.5	59.5
	MIN	33.0	35.0	36.0	39.7	44.9	49.7	52.8	53.0	49.7	44.0	38.3	35.1	42.6
	MEAN	38.4	42.0	44.0	49.1	55.1	59.7	64.1	63.8	59.8	52.3	44.5	40.3	51.1
BUCKLEY 1 NE	MAX	44.2	49.1	51.9	58.1	65.2	69.9	76.4	75.3	69.7	60.2	50.6	45.8	59.7
	MIN	31.8	34.3	34.6	37.6	42.7	47.6	49.8	50.0	47.0	41.6	36.0	33.5	40.5
	MEAN	38.0	41.7	43.2	47.9	54.0	58.8	63.1	62.7	58.4	50.9	43.4	39.7	50.2
CEDAR LAKE	MAX	39.3	44.3	46.5	52.8	60.0	64.9	71.8	70.9	66.3	57.8	46.8	41.4	55.2
	MIN	29.2	32.1	32.1	35.4	40.8	45.8	49.1	49.4	46.3	40.9	35.2	31.8	39.0
	MEAN	34.3	38.2	39.3	44.2	50.4	55.4	60.5	60.2	56.3	49.4	41.0	36.6	47.2
CENTRALIA	MAX	44.7	50.1	53.6	60.3	67.4	72.1	78.0	76.8	72.1	61.7	51.2	46.1	61.2
	MIN	33.1	35.1	35.5	38.5	43.4	48.6	51.5	51.6	48.3	42.7	37.6	35.3	41.8
	MEAN	39.0	42.6	44.6	49.4	55.4	60.4	64.8	64.3	60.2	52.3	44.4	40.7	51.5
CHELAN	MAX	31.9	40.2	49.7	60.7	70.5	77.6	85.6	84.4	75.0	61.1	44.3	35.1	59.7
	MIN	20.7	26.0	30.9	38.6	47.3	54.7	59.7	58.6	49.8	39.7	31.3	25.2	40.2
	MEAN	26.4	33.1	40.3	49.7	58.9	66.2	72.7	71.5	62.4	50.4	37.8	30.2	50.0
CHEWELAH 4 SSW	MAX	31.5	39.2	48.1	59.2	68.6	76.3	85.4	84.6	75.3	60.7	42.7	34.2	58.8
	MIN	15.8	21.6	26.1	32.9	39.5	44.9	46.6	45.1	38.3	30.8	26.1	21.2	32.4
	MEAN	23.7	30.4	37.1	46.0	54.1	60.6	66.0	64.9	56.8	45.8	34.4	27.7	45.6
CHIEF JOSEPH DAM	MAX	31.9	40.0	51.0	63.2	72.6	79.9	88.4	87.1	78.6	63.4	45.1	35.1	61.4
	MIN	18.5	24.5	30.4	37.7	45.4	52.4	57.3	56.3	48.0	37.8	29.9	23.5	38.5
	MEAN	25.2	32.3	40.7	50.5	59.0	66.2	72.8	71.7	63.3	50.6	37.5	29.3	49.9
CLEARBROOK	MAX	40.5	46.9	51.0	58.2	65.1	69.4	75.4	74.4	69.2	59.2	48.9	43.4	58.5
	MIN	29.3	33.3	34.2	37.6	42.9	47.7	49.5	48.9	45.8	40.6	35.1	32.1	39.8
	MEAN	34.9	40.1	42.7	47.9	54.0	58.6	62.4	61.7	57.5	49.9	42.1	37.8	49.1
CLEARWATER	MAX	45.8	50.5	52.0	56.4	62.1	65.3	69.7	69.9	69.0	61.9	52.5	47.3	58.5
	MIN	33.1	34.8	34.6	37.4	41.1	45.6	48.1	48.4	46.1	41.8	37.0	35.4	40.3
	MEAN	39.5	42.7	43.3	46.9	51.6	55.5	58.9	59.2	57.6	51.9	44.7	41.3	49.4
CLE ELUM	MAX	34.1	41.3	47.9	56.7	66.1	72.7	81.3	80.0	72.6	60.0	43.8	36.8	57.8
	MIN	18.6	24.3	26.7	32.4	38.9	45.8	50.6	49.2	41.1	33.0	27.5	23.0	34.3
	MEAN	26.3	32.9	37.3	44.6	52.5	59.3	66.0	64.6	56.9	46.6	35.7	29.9	46.1

649

WASHINGTON

TEMPERATURE NORMALS (DEG F)

STATION		JAN	FEB	MAR	APR	MAY	JUN	JUL	AUG	SEP	OCT	NOV	DEC	ANN
COLFAX 1 NW	MAX	36.6	43.9	49.6	58.5	67.4	74.9	84.5	83.3	75.2	62.7	46.8	39.5	60.2
	MIN	22.5	27.8	29.9	34.9	41.0	46.8	49.9	48.6	41.8	34.4	29.5	26.2	36.1
	MEAN	29.6	35.9	39.8	46.7	54.2	60.9	67.2	66.0	58.6	48.6	38.1	32.9	48.2
COLVILLE AP	MAX	31.4	39.4	47.9	59.1	68.4	75.7	85.4	83.6	73.9	58.6	41.0	33.2	58.1
	MIN	17.4	22.6	26.7	33.3	40.6	46.9	50.4	49.3	41.9	33.6	26.5	21.7	34.2
	MEAN	24.5	31.0	37.3	46.2	54.5	61.3	67.9	66.5	58.0	46.2	33.8	27.5	46.2
CONCRETE	MAX	41.4	47.5	51.9	59.7	67.3	71.3	77.1	76.3	71.4	61.4	49.4	43.0	59.8
	MIN	30.6	33.5	34.4	38.6	44.2	49.5	52.1	52.3	49.1	43.5	36.9	33.5	41.5
	MEAN	36.0	40.5	43.1	49.2	55.8	60.5	64.6	64.3	60.3	52.4	43.2	38.3	50.7
COULEE DAM 1 SW	MAX	31.5	39.5	48.9	60.3	69.9	78.1	86.7	84.9	75.6	60.6	43.2	34.9	59.5
	MIN	20.5	26.2	30.5	37.7	45.5	52.5	58.0	57.1	49.9	40.3	31.2	25.0	39.5
	MEAN	26.0	32.8	39.8	49.0	57.8	65.3	72.4	71.0	62.8	50.5	37.3	29.9	49.6
COUPEVILLE 1 S	MAX	43.8	48.4	51.2	56.7	62.9	67.5	72.1	72.0	67.9	58.8	50.1	45.6	58.1
	MIN	33.1	35.1	35.9	39.0	43.3	47.3	49.5	50.0	46.4	41.7	37.2	35.5	41.2
	MEAN	38.5	41.7	43.6	47.9	53.1	57.4	60.8	61.0	57.2	50.3	43.7	40.6	49.7
DALLESPORT FAA AIRPORT	MAX	39.6	48.0	55.5	63.8	72.8	79.4	87.3	85.9	79.7	66.2	49.7	42.5	64.2
	MIN	27.1	32.1	34.8	40.5	47.3	54.1	58.9	58.5	51.2	41.8	34.6	30.4	42.6
	MEAN	33.4	40.1	45.2	52.2	60.1	66.8	73.1	72.2	65.5	54.0	42.2	36.5	53.4
DAVENPORT	MAX	30.6	38.3	46.5	57.2	67.0	74.9	84.3	83.0	73.5	59.2	41.7	33.6	57.5
	MIN	18.2	24.0	27.7	33.0	39.5	45.4	50.3	49.3	43.1	34.6	27.4	22.1	34.6
	MEAN	24.4	31.2	37.1	45.2	53.3	60.2	67.4	66.1	58.3	46.9	34.5	27.9	46.0
DAYTON 1 WSW	MAX	39.6	46.4	52.2	60.3	69.1	77.3	87.5	85.4	76.9	64.4	49.4	42.5	62.6
	MIN	24.7	30.0	32.6	37.1	43.6	49.5	54.2	53.2	46.5	38.3	31.4	28.0	39.1
	MEAN	32.2	38.2	42.5	48.7	56.3	63.4	70.9	69.3	61.7	51.4	40.4	35.3	50.9
DIABLO DAM	MAX	37.1	43.6	47.9	56.6	65.5	70.6	77.7	77.0	70.6	58.3	45.6	39.4	57.5
	MIN	26.9	30.5	31.7	36.4	42.6	48.3	51.6	52.4	47.9	41.1	34.1	30.3	39.5
	MEAN	32.0	37.1	39.9	46.5	54.1	59.5	64.7	64.7	59.3	49.7	39.9	34.9	48.5
ELECTRON HEADWORKS	MAX	39.7	44.3	47.5	54.3	62.3	67.4	74.3	72.7	67.3	57.0	45.8	41.4	56.2
	MIN	28.6	30.6	30.7	33.5	38.3	43.6	45.9	45.6	41.9	36.9	32.3	30.6	36.5
	MEAN	34.2	37.5	39.1	43.9	50.3	55.5	60.1	59.2	54.6	47.0	39.0	36.0	46.4
ELMA	MAX	45.5	50.8	53.6	60.0	66.9	71.4	76.5	76.4	72.5	63.2	52.3	47.0	61.3
	MIN	33.0	35.0	34.9	37.9	42.5	47.4	50.2	50.5	47.4	42.0	37.1	35.2	41.1
	MEAN	39.3	42.9	44.3	49.0	54.7	59.5	63.4	63.5	60.0	52.6	44.7	41.1	51.3
ELWHA RANGER STATION	MAX	41.2	46.3	50.2	57.1	64.3	68.6	75.1	74.4	68.9	57.9	47.2	43.4	57.9
	MIN	30.5	32.7	33.3	36.6	41.4	46.2	49.4	50.4	47.2	41.0	35.2	32.8	39.7
	MEAN	35.9	39.6	41.8	46.9	52.9	57.4	62.3	62.5	58.0	49.5	41.2	38.1	48.8
EPHRATA FAA AIRPORT	MAX	32.4	41.7	51.8	62.5	72.6	80.6	89.0	86.7	77.8	63.0	44.6	35.6	61.5
	MIN	19.9	26.9	31.8	39.2	48.0	55.6	62.1	60.4	52.0	40.4	30.2	23.8	40.9
	MEAN	26.2	34.3	41.8	50.8	60.3	68.1	75.6	73.6	64.9	51.7	37.4	29.8	51.2
EVERETT	MAX	44.2	49.0	51.5	57.2	63.3	67.9	72.7	72.4	67.9	60.0	50.8	46.2	58.6
	MIN	32.6	35.0	36.0	40.0	45.2	50.5	53.1	52.7	48.5	42.7	37.1	35.1	42.4
	MEAN	38.4	42.0	43.8	48.6	54.3	59.2	62.9	62.5	58.2	51.4	44.0	40.7	50.5
FORKS 1 E	MAX	44.7	49.0	50.8	56.0	62.2	66.3	71.5	71.8	69.6	61.2	51.4	46.2	58.4
	MIN	32.9	34.6	34.0	36.8	41.2	46.1	48.8	49.2	46.7	41.9	36.7	34.9	40.3
	MEAN	38.8	41.8	42.4	46.4	51.7	56.2	60.2	60.5	58.2	51.6	44.1	40.6	49.4
GLENOMA 1 W	MAX	43.6	49.2	52.0	58.5	65.7	70.7	77.6	76.7	72.4	62.7	51.5	45.3	60.5
	MIN	29.8	32.4	33.0	36.1	41.3	46.7	49.3	48.6	44.9	39.0	34.4	32.1	39.0
	MEAN	36.7	40.9	42.5	47.3	53.5	58.7	63.5	62.7	58.7	50.8	43.0	38.7	49.8
GRAPEVIEW 3 SW	MAX	44.2	48.7	52.3	59.1	66.5	70.8	76.3	75.5	70.1	59.8	50.3	45.8	60.0
	MIN	34.5	36.3	36.5	39.7	44.6	49.7	52.5	53.2	49.6	44.4	39.4	36.6	43.1
	MEAN	39.4	42.6	44.5	49.4	55.6	60.3	64.4	64.4	59.9	52.1	44.9	41.2	51.6

WASHINGTON

TEMPERATURE NORMALS (DEG F)

STATION		JAN	FEB	MAR	APR	MAY	JUN	JUL	AUG	SEP	OCT	NOV	DEC	ANN
HATTON 9 ESE	MAX	35.5	44.5	53.0	63.1	72.6	81.3	90.6	87.7	78.8	64.1	46.5	38.1	63.0
	MIN	22.0	28.5	30.6	34.8	41.6	48.3	53.5	52.1	45.4	36.4	29.7	25.8	37.4
	MEAN	28.8	36.5	41.8	49.0	57.1	64.8	72.1	69.9	62.1	50.2	38.1	32.0	50.2
HOQUIAM FAA AIRPORT	MAX	45.3	49.4	51.3	55.6	60.8	64.3	68.0	68.6	67.9	60.7	51.9	47.0	57.6
	MIN	35.6	37.2	37.3	40.2	44.9	49.3	51.8	52.6	50.5	45.3	40.0	37.1	43.5
	MEAN	40.5	43.3	44.3	47.9	52.9	56.8	59.9	60.6	59.2	53.0	46.0	42.1	50.5
KENNEWICK	MAX	39.6	48.4	57.4	66.6	75.7	83.7	91.1	88.3	79.1	65.3	49.7	42.4	65.6
	MIN	26.4	31.5	34.7	40.4	47.6	54.2	58.8	57.6	50.0	41.2	33.8	29.6	42.2
	MEAN	33.1	40.0	46.0	53.5	61.7	68.9	75.0	73.0	64.6	53.3	41.8	36.0	53.9
KENT	MAX	45.2	50.3	53.7	60.3	67.8	72.6	78.4	77.4	71.8	61.6	51.3	46.5	61.4
	MIN	32.6	35.0	35.8	38.9	43.8	49.1	51.4	51.1	47.3	41.7	36.3	34.4	41.5
	MEAN	38.9	42.7	44.7	49.6	55.8	60.9	65.0	64.3	59.6	51.7	43.8	40.5	51.5
KID VALLEY	MAX	43.5	48.9	52.4	59.3	66.5	70.8	77.1	76.4	70.9	60.2	50.0	45.0	60.1
	MIN	30.4	32.6	33.0	36.1	40.8	45.9	48.3	48.5	45.3	39.6	34.8	32.6	39.0
	MEAN	37.0	40.8	42.7	47.7	53.7	58.3	62.7	62.5	58.1	49.9	42.4	38.8	49.6
LACROSSE 3 ESE	MAX	36.5	44.7	51.8	61.1	70.2	78.7	88.6	86.2	77.3	63.8	47.0	39.4	62.1
	MIN	23.2	29.3	30.9	35.2	40.9	47.2	51.7	50.6	44.2	36.1	30.1	26.7	37.2
	MEAN	29.9	37.0	41.3	48.2	55.6	63.0	70.2	68.4	60.8	50.0	38.5	33.1	49.7
LANDSBURG	MAX	43.0	48.5	51.4	57.2	64.4	69.0	75.4	74.2	69.7	60.3	49.8	44.5	59.0
	MIN	30.5	32.7	33.2	36.1	41.0	46.6	48.7	48.0	44.5	39.3	34.4	32.4	39.0
	MEAN	36.8	40.6	42.3	46.6	52.7	57.8	62.1	61.1	57.1	49.8	42.2	38.5	49.0
LAURIER	MAX	30.2	39.1	48.9	61.2	71.2	77.6	86.3	84.8	75.1	58.3	40.7	32.3	58.8
	MIN	15.5	20.9	24.5	31.4	39.4	46.2	49.4	48.6	41.0	32.9	25.6	20.1	33.0
	MEAN	22.9	30.0	36.7	46.4	55.3	62.0	67.9	66.7	58.1	45.6	33.2	26.3	45.9
LIND 3 NE	MAX	35.1	44.2	52.7	62.4	72.5	81.1	89.9	87.6	78.7	64.3	46.1	37.8	62.7
	MIN	21.7	27.9	30.1	34.3	41.0	47.3	52.6	51.5	45.0	36.4	29.4	25.2	36.9
	MEAN	28.4	36.0	41.4	48.4	56.8	64.2	71.3	69.5	61.9	50.4	37.8	31.6	49.8
LONGVIEW	MAX	44.7	50.5	53.9	59.9	66.5	71.2	77.2	77.2	73.0	63.2	52.2	46.4	61.3
	MIN	32.8	34.7	35.3	38.2	42.9	48.2	50.9	51.2	48.6	43.1	37.6	35.1	41.6
	MEAN	38.8	42.6	44.7	49.1	54.7	59.7	64.1	64.2	60.8	53.2	45.0	40.7	51.5
MC MILLIN RESERVOIR	MAX	43.7	48.8	51.4	57.4	64.2	68.9	75.1	74.7	69.8	60.5	50.3	45.6	59.2
	MIN	30.7	33.2	34.0	37.4	42.3	47.5	49.9	50.4	46.9	41.1	35.4	32.9	40.1
	MEAN	37.2	41.0	42.8	47.4	53.3	58.2	62.6	62.6	58.4	50.8	42.9	39.2	49.7
MONROE	MAX	44.3	50.1	53.0	59.7	66.5	71.4	76.9	76.1	71.0	61.6	51.1	45.9	60.6
	MIN	32.3	34.7	35.5	39.1	44.1	49.2	51.4	51.8	48.4	43.0	37.0	34.8	41.8
	MEAN	38.3	42.4	44.3	49.4	55.3	60.3	64.2	63.9	59.7	52.3	44.1	40.4	51.2
MOSES LAKE 3 E	MAX	34.1	43.1	52.6	62.3	71.1	78.8	86.8	85.1	76.5	63.0	45.9	37.3	61.4
	MIN	18.8	25.7	29.7	35.4	42.9	49.4	53.7	52.5	44.6	35.1	27.7	23.2	36.6
	MEAN	26.5	34.5	41.2	48.9	57.1	64.1	70.3	68.8	60.5	49.1	36.8	30.3	49.0
MOUNT ADAMS RANGER STA	MAX	36.4	42.5	47.9	56.8	66.8	73.8	82.7	80.5	73.4	60.8	44.9	38.2	58.7
	MIN	22.1	26.6	28.0	31.9	37.6	44.7	47.8	47.0	41.4	34.9	29.8	25.7	34.8
	MEAN	29.3	34.6	38.0	44.4	52.2	59.3	65.2	63.7	57.4	47.9	37.4	32.0	46.8
MOXEE CITY 10 E	MAX	35.5	44.0	51.8	60.5	69.4	77.4	85.2	83.5	75.3	62.2	45.9	37.7	60.7
	MIN	20.8	26.6	29.0	33.7	40.0	46.9	51.3	50.5	44.9	36.2	28.9	24.0	36.1
	MEAN	28.2	35.3	40.5	47.1	54.7	62.2	68.3	67.0	60.1	49.3	37.4	30.9	48.4
MUD MOUNTAIN DAM	MAX	42.4	46.6	48.2	53.5	60.3	65.1	71.8	71.2	66.9	58.7	49.4	44.7	56.6
	MIN	30.5	33.0	33.4	36.8	42.2	47.4	50.8	50.8	47.6	41.7	35.9	32.8	40.2
	MEAN	36.5	39.8	40.8	45.2	51.3	56.3	61.3	61.0	57.3	50.2	42.7	38.8	48.4
NEWHALEM	MAX	38.1	44.2	48.7	57.3	65.7	70.5	77.1	76.1	70.2	58.2	46.8	40.4	57.8
	MIN	29.6	32.6	33.6	38.2	44.3	49.1	52.4	52.8	49.7	43.2	36.4	32.4	41.2
	MEAN	33.9	38.4	41.2	47.7	55.0	59.8	64.8	64.5	60.0	50.7	41.7	36.4	49.5

WASHINGTON

TEMPERATURE NORMALS (DEG F)

STATION		JAN	FEB	MAR	APR	MAY	JUN	JUL	AUG	SEP	OCT	NOV	DEC	ANN
NEWPORT	MAX	31.3	39.1	46.9	58.5	68.4	76.0	85.5	83.9	73.8	58.2	40.6	33.3	58.0
	MIN	17.6	22.4	24.2	30.4	37.4	43.9	46.1	44.6	38.4	32.2	26.6	22.1	32.2
	MEAN	24.5	30.8	35.6	44.5	52.9	59.9	65.8	64.3	56.1	45.2	33.6	27.7	45.1
NORTHPORT	MAX	31.8	39.8	50.2	63.0	73.4	80.5	89.0	86.8	76.4	60.1	42.0	34.0	60.6
	MIN	19.4	24.2	27.5	34.2	41.5	47.5	50.5	49.3	42.8	36.0	28.9	23.7	35.5
	MEAN	25.6	32.0	38.9	48.6	57.5	64.0	69.8	68.1	59.7	48.1	35.5	28.9	48.1
OAKVILLE	MAX	45.3	50.1	53.3	59.5	65.8	70.0	76.1	75.8	72.3	62.9	52.5	46.8	60.9
	MIN	31.8	33.6	33.6	36.4	40.9	46.7	49.6	50.1	45.4	40.1	35.8	33.9	39.8
	MEAN	38.5	41.9	43.5	48.0	53.4	58.4	62.9	63.0	58.9	51.5	44.1	40.4	50.4
ODESSA	MAX	35.2	43.9	52.7	63.1	72.7	81.0	89.3	87.5	78.7	64.6	46.7	38.1	62.8
	MIN	19.6	25.9	28.7	33.9	40.7	47.7	52.4	50.8	43.5	34.1	27.5	23.1	35.7
	MEAN	27.4	34.9	40.7	48.5	56.7	64.3	70.9	69.2	61.1	49.4	37.1	30.7	49.2
OLGA 2 SE	MAX	43.3	47.4	50.1	56.0	62.1	66.1	69.9	69.2	65.4	57.3	49.4	45.3	56.8
	MIN	33.8	35.9	36.7	39.7	43.6	47.3	49.2	49.8	47.9	43.8	38.6	36.0	41.9
	MEAN	38.6	41.6	43.4	47.9	52.9	56.7	59.6	59.5	56.7	50.6	44.0	40.6	49.3
OLYMPIA WSO R	MAX	43.6	49.1	52.5	58.7	65.7	70.8	77.2	76.2	71.0	60.8	50.3	45.1	60.1
	MIN	30.8	32.5	32.8	35.8	40.5	46.0	48.7	48.8	45.1	39.4	34.8	32.8	39.0
	MEAN	37.2	40.8	42.7	47.3	53.1	58.4	63.0	62.5	58.1	50.1	42.6	39.0	49.6
OMAK 2 NW	MAX	29.6	38.5	49.6	61.7	70.8	78.6	86.1	83.9	75.2	60.3	42.3	32.8	59.1
	MIN	16.4	22.2	27.6	34.9	42.9	49.7	54.8	53.6	44.9	34.2	26.8	20.8	35.7
	MEAN	23.0	30.4	38.6	48.3	56.9	64.2	70.4	68.8	60.1	47.3	34.6	26.8	47.5
OTHELLO 6 ESE	MAX	35.6	44.7	53.9	63.0	72.3	80.2	87.9	85.9	77.8	63.9	46.8	38.4	62.5
	MIN	21.2	27.5	30.5	36.3	43.7	50.6	55.1	53.8	46.9	37.4	29.4	25.0	38.1
	MEAN	28.5	36.1	42.2	49.7	58.1	65.4	71.6	69.9	62.4	50.7	38.1	31.7	50.4
PALMER 3 ESE	MAX	40.9	46.3	49.0	55.3	62.6	67.1	73.9	73.0	68.7	59.6	48.6	42.9	57.3
	MIN	31.1	34.1	34.0	37.7	42.8	48.2	51.4	51.5	48.4	42.9	37.0	33.6	41.1
	MEAN	36.0	40.2	41.5	46.5	52.7	57.7	62.7	62.3	58.6	51.3	42.8	38.3	49.2
POMEROY	MAX	39.3	46.4	52.1	60.9	69.5	77.7	87.6	85.7	77.7	64.5	49.0	42.1	62.7
	MIN	24.5	29.9	31.6	37.0	43.3	49.7	54.1	52.9	46.4	38.5	31.1	27.5	38.9
	MEAN	32.0	38.2	41.9	49.0	56.4	63.7	70.9	69.4	62.1	51.5	40.1	34.8	50.8
PORT ANGELES	MAX	44.7	47.9	50.1	55.1	60.7	64.7	68.7	68.3	65.9	58.0	50.2	46.4	56.7
	MIN	33.2	35.6	36.1	39.4	44.1	48.3	50.9	51.0	48.5	43.1	37.7	35.3	41.9
	MEAN	39.0	41.8	43.1	47.3	52.4	56.5	59.8	59.7	57.3	50.6	44.0	40.9	49.4
PROSSER 4 NE	MAX	37.8	46.7	54.9	63.6	72.3	80.0	87.3	85.4	77.5	64.4	48.5	40.7	63.3
	MIN	23.3	28.6	31.4	37.0	43.8	49.7	53.2	52.1	46.5	38.0	30.8	26.3	38.4
	MEAN	30.6	37.7	43.2	50.3	58.1	64.9	70.3	68.8	62.0	51.3	39.7	33.5	50.9
PULLMAN 2 NW	MAX	34.2	40.8	46.0	55.1	64.1	71.6	81.8	81.1	72.4	59.8	43.9	37.3	57.3
	MIN	22.2	27.8	29.9	35.0	41.0	46.5	50.0	49.9	45.0	38.0	30.3	26.0	36.8
	MEAN	28.2	34.3	38.0	45.1	52.6	59.1	65.9	65.5	58.8	48.9	37.1	31.7	47.1
PUYALLUP 2 W EXP STA	MAX	45.7	50.8	54.0	60.4	68.0	72.8	78.6	77.4	72.0	62.3	51.7	47.1	61.7
	MIN	31.8	34.1	34.7	38.0	42.8	48.1	50.2	50.0	46.4	41.3	35.7	33.6	40.6
	MEAN	38.8	42.5	44.4	49.2	55.4	60.5	64.4	63.8	59.3	51.8	43.8	40.4	51.2
QUILCENE 2 SW	MAX	43.2	49.2	53.1	59.6	66.4	70.9	77.1	76.4	71.5	61.3	50.0	44.9	60.3
	MIN	30.1	32.4	33.3	36.8	42.1	47.3	50.0	49.6	45.6	39.6	34.5	32.2	39.5
	MEAN	36.7	40.9	43.2	48.2	54.3	59.1	63.5	63.0	58.5	50.5	42.3	38.6	49.9
QUILLAYUTE WSO	MAX	44.8	48.5	49.9	54.7	60.2	63.6	68.6	68.6	66.8	59.1	50.7	46.4	56.8
	MIN	33.2	34.8	34.4	37.2	41.7	46.5	49.4	49.7	46.7	41.6	36.9	35.1	40.6
	MEAN	39.0	41.7	42.2	46.0	51.0	55.1	59.0	59.2	56.8	50.4	43.8	40.8	48.7
QUINCY 1 S	MAX	33.7	43.1	53.2	63.4	72.7	80.0	87.5	85.7	77.6	63.7	45.4	36.5	61.9
	MIN	17.3	24.9	29.2	36.6	44.6	51.2	55.6	54.6	47.1	36.4	27.8	21.6	37.2
	MEAN	25.5	34.0	41.2	50.0	58.7	65.6	71.6	70.2	62.4	50.1	36.7	29.1	49.6

WASHINGTON

TEMPERATURE NORMALS (DEG F)

STATION			JAN	FEB	MAR	APR	MAY	JUN	JUL	AUG	SEP	OCT	NOV	DEC	ANN
RAINIER PARADISE RS	//	MAX	30.7	34.2	35.0	40.5	48.3	53.4	61.9	61.1	56.9	48.2	37.9	33.1	45.1
		MIN	19.2	21.6	20.9	24.8	31.0	36.6	43.1	43.2	39.9	33.6	25.8	21.6	30.1
		MEAN	25.0	27.9	28.0	32.7	39.7	45.0	52.5	52.1	48.4	40.9	31.9	27.4	37.6
REPUBLIC		MAX	28.4	37.8	45.2	56.5	65.7	73.0	81.7	80.3	71.4	56.8	39.0	30.6	55.5
		MIN	12.1	18.1	22.1	28.9	36.2	42.2	45.1	44.2	37.6	29.7	22.9	16.4	29.6
		MEAN	20.3	28.0	33.7	42.7	50.9	57.6	63.4	62.3	54.5	43.3	31.0	23.5	42.6
RICHLAND		MAX	41.1	49.5	58.0	66.9	75.9	83.6	91.6	89.5	81.4	67.9	51.7	43.8	66.7
		MIN	25.9	31.3	34.6	40.8	48.3	55.1	59.8	58.6	51.1	41.3	33.6	29.3	42.5
		MEAN	33.5	40.4	46.3	53.9	62.1	69.4	75.7	74.0	66.3	54.7	42.7	36.6	54.6
RITZVILLE 1 SSE		MAX	33.6	42.0	50.2	60.2	69.4	78.1	87.4	85.5	76.7	62.8	45.1	36.8	60.7
		MIN	20.1	25.9	28.8	33.4	40.2	46.7	52.3	51.4	45.0	36.0	28.0	23.8	36.0
		MEAN	26.9	34.0	39.5	46.8	54.8	62.5	69.9	68.5	60.9	49.4	36.5	30.3	48.3
ROSALIA		MAX	33.5	40.5	46.7	56.4	65.0	72.2	82.0	81.3	72.6	60.3	44.0	36.8	57.6
		MIN	21.1	26.6	29.1	34.0	40.6	46.3	50.5	50.0	43.3	35.2	28.6	24.4	35.8
		MEAN	27.3	33.6	37.9	45.2	52.8	59.3	66.3	65.7	58.0	47.8	36.3	30.6	46.7
SEATTLE EMSU WSO		MAX	45.3	50.1	52.6	58.3	64.8	69.0	74.6	73.6	69.2	60.4	51.3	46.9	59.7
		MIN	35.9	38.2	38.8	42.4	47.7	53.0	56.0	56.3	52.9	47.1	41.1	38.1	45.6
		MEAN	40.6	44.2	45.7	50.4	56.3	61.0	65.3	65.0	61.1	53.8	46.2	42.6	52.7
SEATTLE-TACOMA WSO	R	MAX	43.9	48.8	51.1	56.8	64.0	69.2	75.2	73.9	68.7	59.5	50.3	45.6	58.9
		MIN	34.3	36.8	37.2	40.5	46.0	51.1	54.3	54.3	51.2	45.3	39.3	36.3	43.9
		MEAN	39.1	42.8	44.2	48.7	55.0	60.2	64.8	64.1	60.0	52.5	44.8	41.0	51.4
SEATTLE U OF W		MAX	45.1	50.1	53.0	59.1	66.3	71.1	77.0	75.9	70.8	61.6	51.7	47.1	60.7
		MIN	34.9	37.2	38.1	41.7	47.2	51.9	55.3	55.7	52.3	46.4	40.3	37.5	44.9
		MEAN	40.0	43.7	45.5	50.4	56.8	61.5	66.2	65.8	61.6	54.0	46.0	42.3	52.8
SEDRO WOOLLEY		MAX	43.8	48.9	51.9	58.1	64.9	69.4	74.5	74.2	69.3	60.3	50.8	45.7	59.3
		MIN	31.6	34.4	35.4	39.1	43.8	48.5	50.1	50.4	47.5	42.1	36.7	34.4	41.2
		MEAN	37.7	41.7	43.7	48.6	54.4	59.0	62.3	62.4	58.5	51.2	43.8	40.1	50.3
SEQUIM		MAX	44.6	48.7	51.4	56.4	62.7	67.1	71.9	71.7	68.2	59.4	50.5	46.1	58.2
		MIN	32.1	34.5	35.2	38.8	43.8	48.2	50.9	51.3	48.5	42.4	36.5	34.1	41.4
		MEAN	38.3	41.6	43.3	47.6	53.3	57.7	61.4	61.5	58.4	50.9	43.5	40.2	49.8
SHELTON		MAX	44.4	49.3	52.7	59.2	66.6	71.4	76.9	76.1	71.2	61.2	51.0	45.7	60.5
		MIN	32.7	34.5	34.8	38.2	43.4	48.9	51.9	52.2	47.9	42.0	37.1	34.9	41.5
		MEAN	38.6	41.9	43.8	48.7	55.0	60.2	64.4	64.2	59.6	51.6	44.0	40.3	51.0
SNOQUALMIE FALLS		MAX	44.0	49.1	51.7	57.9	64.5	68.9	75.1	74.2	69.2	60.1	50.8	45.9	59.3
		MIN	32.2	34.7	34.6	37.8	42.8	48.1	50.7	50.8	46.9	41.5	36.6	34.4	40.9
		MEAN	38.1	41.9	43.2	47.9	53.7	58.6	62.9	62.5	58.1	50.9	43.8	40.2	50.2
SPOKANE WSO	//R	MAX	31.3	39.0	46.2	56.7	66.1	74.0	84.0	81.7	72.4	58.3	41.4	34.2	57.1
		MIN	20.0	25.7	29.0	34.9	42.5	49.3	55.3	54.3	46.5	36.7	28.5	23.7	37.2
		MEAN	25.7	32.4	37.6	45.8	54.3	61.7	69.7	68.1	59.4	47.6	34.9	29.0	47.2
SPRAGUE		MAX	34.1	42.3	50.0	59.9	69.5	77.6	86.6	84.5	75.4	61.8	45.0	37.0	60.3
		MIN	18.8	25.2	27.9	33.2	39.9	46.3	50.8	48.9	41.8	33.4	27.2	22.8	34.7
		MEAN	26.5	33.8	39.0	46.6	54.7	62.0	68.7	66.8	58.7	47.7	36.2	29.9	47.6
STAMPEDE PASS WSO	//R	MAX	28.1	32.1	34.7	41.0	49.5	56.8	65.1	64.0	58.6	47.8	34.8	30.2	45.2
		MIN	20.1	24.1	25.3	29.5	35.5	41.4	47.1	47.2	43.6	36.3	27.1	22.7	33.3
		MEAN	24.1	28.1	30.1	35.3	42.5	49.2	56.1	55.6	51.1	42.1	30.9	26.5	39.3
STARTUP 1 E		MAX	44.7	50.4	53.1	59.7	66.8	71.5	77.5	76.8	71.7	62.3	51.7	46.5	61.1
		MIN	32.4	34.6	34.9	38.3	43.1	48.1	50.0	50.0	46.8	41.6	36.8	34.7	40.9
		MEAN	38.6	42.5	44.0	49.0	55.0	59.8	63.8	63.4	59.3	52.0	44.3	40.6	51.0
STEHEKIN 3 NW		MAX	33.6	40.0	47.0	58.3	68.4	75.1	82.7	80.7	71.5	56.9	42.0	35.2	57.6
		MIN	23.6	27.2	29.3	34.9	41.7	48.1	51.9	51.6	44.5	36.2	30.2	27.0	37.2
		MEAN	28.6	33.6	38.1	46.7	55.1	61.6	67.3	66.2	58.0	46.6	36.1	31.1	47.4

WASHINGTON

TEMPERATURE NORMALS (DEG F)

STATION		JAN	FEB	MAR	APR	MAY	JUN	JUL	AUG	SEP	OCT	NOV	DEC	ANN
SUNNYSIDE	MAX	39.3	48.4	56.9	65.5	74.3	81.9	89.4	87.7	79.8	67.0	50.3	41.7	65.2
	MIN	23.1	28.3	31.8	37.9	45.4	51.9	55.0	53.1	46.5	37.2	30.4	26.2	38.9
	MEAN	31.2	38.4	44.4	51.7	59.8	66.9	72.2	70.4	63.1	52.1	40.4	34.0	52.1
TACOMA CITY HALL	MAX	45.6	50.3	52.8	58.8	65.9	70.4	76.0	74.9	70.1	61.3	51.8	47.3	60.4
	MIN	35.0	37.2	37.6	40.9	45.8	50.8	53.9	54.4	51.3	45.7	39.5	36.6	44.1
	MEAN	40.3	43.8	45.2	49.8	55.8	60.6	65.0	64.7	60.7	53.5	45.7	42.0	52.3
VANCOUVER 4 NNE	MAX	44.1	49.8	53.8	59.8	66.6	71.9	78.4	77.8	73.6	63.5	52.2	46.0	61.5
	MIN	32.3	34.8	36.3	39.7	44.9	50.2	53.0	52.7	49.2	42.9	37.3	34.5	42.3
	MEAN	38.2	42.4	45.1	49.7	55.8	61.0	65.7	65.3	61.5	53.2	44.8	40.3	51.9
WALLA-WALLA FAA AP	MAX	39.4	46.7	54.0	62.2	70.8	79.7	89.5	86.7	77.9	64.4	48.9	42.3	63.5
	MIN	27.1	32.7	35.6	40.5	47.3	54.0	60.5	60.0	52.6	43.3	34.7	30.3	43.2
	MEAN	33.3	39.7	44.8	51.4	59.1	66.9	75.1	73.4	65.2	53.8	41.8	36.3	53.4
WALLA WALLA WSO R	MAX	40.1	47.3	54.3	62.2	70.9	79.3	88.8	86.0	77.3	64.0	49.0	43.0	63.5
	MIN	28.4	33.9	37.3	42.4	49.3	55.8	62.2	61.2	53.6	44.5	35.9	31.4	44.7
	MEAN	34.3	40.7	45.8	52.3	60.1	67.6	75.5	73.6	65.5	54.3	42.5	37.2	54.1
WAPATO	MAX	38.9	48.2	56.8	65.9	75.2	82.4	89.8	87.7	79.7	66.4	49.7	41.2	65.2
	MIN	23.1	28.7	32.6	39.3	47.3	54.1	59.5	57.6	49.1	38.6	30.7	26.4	40.6
	MEAN	31.0	38.5	44.7	52.6	61.3	68.3	74.7	72.6	64.4	52.6	40.2	33.8	52.9
WATERVILLE	MAX	30.8	38.3	45.4	56.8	66.7	74.2	83.5	81.4	72.8	58.9	41.8	33.0	57.0
	MIN	13.8	20.3	25.7	32.7	40.1	46.6	51.5	51.1	43.8	34.0	24.8	17.4	33.5
	MEAN	22.3	29.3	35.6	44.8	53.4	60.4	67.5	66.3	58.3	46.4	33.3	25.2	45.2
WENATCHEE	MAX	34.2	43.2	53.5	64.0	72.8	80.0	87.9	86.3	77.5	63.0	46.3	37.4	62.2
	MIN	21.2	26.8	31.8	39.4	47.5	54.7	59.6	58.3	49.8	39.4	31.1	25.8	40.5
	MEAN	27.7	35.0	42.7	51.7	60.2	67.4	73.8	72.3	63.7	51.2	38.8	31.6	51.3
WILBUR	MAX	31.5	39.4	47.9	58.7	68.4	76.2	85.1	83.2	74.3	60.4	42.7	34.5	58.5
	MIN	18.4	24.1	27.8	33.3	40.0	45.9	50.5	49.5	43.3	34.8	27.8	22.5	34.8
	MEAN	25.0	31.8	37.8	46.0	54.2	61.1	67.8	66.4	58.8	47.6	35.3	28.5	46.7
WILLAPA HARBOR	MAX	46.4	50.8	53.1	58.1	64.2	67.6	72.3	72.4	70.4	62.6	53.0	47.8	59.9
	MIN	34.7	36.3	36.1	39.0	44.0	48.7	51.9	52.6	49.6	44.0	38.6	36.4	42.7
	MEAN	40.6	43.6	44.6	48.6	54.1	58.2	62.1	62.5	60.0	53.3	45.8	42.1	51.3
WILSON CREEK	MAX	34.5	43.5	52.9	63.5	73.7	82.0	90.1	88.0	79.0	64.4	46.0	37.0	62.9
	MIN	19.2	25.8	29.1	34.5	41.3	48.3	52.3	50.5	43.5	34.2	27.4	22.7	35.7
	MEAN	26.9	34.7	41.0	49.0	57.6	65.2	71.2	69.3	61.3	49.3	36.7	29.9	49.3
WINTHROP 1 WSW	MAX	27.7	38.0	48.0	61.2	71.2	78.5	86.7	85.2	76.8	62.1	41.5	30.3	58.9
	MIN	9.0	15.8	21.9	30.3	37.8	44.1	48.1	47.1	38.9	30.1	22.5	13.7	29.9
	MEAN	18.4	26.9	35.0	45.8	54.6	61.3	67.4	66.2	57.9	46.2	32.0	22.0	44.5
YAKIMA WSO //R	MAX	36.7	46.0	54.5	63.5	72.5	79.9	87.8	85.6	77.5	64.5	48.1	39.4	63.0
	MIN	19.7	26.1	29.2	34.7	42.1	49.1	53.0	51.5	44.3	35.1	28.2	23.6	36.4
	MEAN	28.2	36.1	41.9	49.2	57.3	64.5	70.4	68.6	60.9	49.9	38.2	31.5	49.7

WASHINGTON

PRECIPITATION NORMALS (INCHES)

STATION	JAN	FEB	MAR	APR	MAY	JUN	JUL	AUG	SEP	OCT	NOV	DEC	ANN
ABERDEEN	13.10	9.75	9.06	5.51	3.35	2.50	1.27	2.12	3.66	7.29	10.96	13.77	82.34
ABERDEEN 20 NNE //	21.12	16.34	14.44	8.84	4.95	3.41	2.33	3.18	6.37	11.97	17.86	22.81	133.62
ANACORTES	3.67	2.48	2.11	1.70	1.32	1.25	.96	1.00	1.44	2.44	3.22	4.04	25.63
ANATONE	2.43	1.58	1.82	1.72	2.08	1.98	.77	1.14	1.11	1.48	2.13	2.25	20.49
ARLINGTON	5.89	4.38	4.23	3.58	3.07	2.44	1.48	1.98	2.78	4.12	5.50	6.47	45.92
BATTLE GROUND	7.68	5.17	5.10	3.63	2.93	2.34	.72	1.62	2.34	4.56	7.10	8.52	51.71
BELLINGHAM 2 N	4.69	3.49	2.97	2.58	2.08	1.74	1.15	1.45	2.18	3.47	4.61	5.05	35.46
BELLINGHAM FAA AIRPORT	4.79	3.51	2.97	2.43	2.01	1.71	1.11	1.41	2.05	3.49	4.64	5.14	35.26
BICKLETON	2.60	1.56	1.15	.80	.80	.67	.32	.36	.45	.82	2.11	2.54	14.18
BLAINE 1 ENE	5.54	4.20	3.39	2.51	1.98	1.81	1.16	1.55	2.34	3.97	5.57	6.32	40.34
BREMERTON	8.25	5.94	5.34	2.91	1.78	1.50	.80	1.17	2.10	4.18	7.51	8.93	50.41
BUCKLEY 1 NE	6.52	4.79	4.33	3.91	3.13	2.92	1.20	2.03	2.94	4.20	6.15	6.85	48.97
CEDAR LAKE	14.47	10.73	10.24	8.16	5.69	5.15	2.35	3.17	5.53	8.77	12.92	15.23	102.41
CENTRALIA	7.38	4.98	4.73	3.01	2.03	1.78	.84	1.46	2.22	4.13	6.44	7.71	46.71
CHELAN	1.39	1.03	.85	.84	.62	.60	.30	.55	.52	.73	1.43	1.71	10.57
CHEWELAH 4 SSW	2.56	1.76	1.64	1.45	1.97	1.36	.80	1.12	1.04	1.41	2.46	3.12	20.69
CHIEF JOSEPH DAM	1.19	1.04	.75	.66	.68	.86	.21	.57	.40	.59	1.23	1.57	9.75
CHIMACUM 4 S	4.04	2.88	3.26	2.23	1.92	1.72	.89	1.24	1.27	2.28	3.38	4.76	29.87
CLEARBROOK	5.57	4.65	3.96	3.39	2.85	2.32	1.50	2.07	3.23	4.68	5.83	6.47	46.52
CLEARWATER	17.33	14.08	12.94	8.30	5.36	3.36	2.64	3.39	6.35	11.81	16.14	19.54	121.24
CLE ELUM	4.14	2.46	1.91	1.27	.77	.70	.27	.59	.81	1.63	3.51	4.59	22.65
COLFAX 1 NW	2.74	1.85	1.81	1.58	1.46	1.35	.50	.83	.94	1.45	2.25	3.04	19.80
COLVILLE AP	2.22	1.45	1.21	1.05	1.62	1.48	.77	1.16	.89	1.17	2.05	2.49	17.56
CONCONULLY	1.71	1.39	1.19	1.06	1.45	1.21	.60	1.20	.77	.93	1.38	1.89	14.78
CONCRETE	10.34	7.55	6.92	4.47	3.02	2.49	1.42	1.99	3.73	6.70	9.44	11.36	69.43
CONNELL 12 SE	1.22	.83	.79	.63	.79	.66	.24	.48	.41	.76	1.22	1.43	9.46
COULEE DAM 1 SW	1.17	.87	.71	.76	1.09	.72	.38	.58	.53	.65	1.26	1.51	10.23
COUPEVILLE 1 S	2.55	1.71	1.79	1.62	1.47	1.23	.76	1.00	1.30	1.66	2.35	3.00	20.44
DALLESPORT FAA AIRPORT	2.88	1.48	1.10	.50	.45	.31	.08	.32	.38	.93	2.03	2.71	13.17
DALLESPORT 9 N	4.31	2.64	2.03	1.11	.82	.58	.21	.57	.76	1.78	3.45	4.14	22.40
DARRINGTON RANGER STA	13.08	9.68	8.55	5.34	3.41	2.63	1.51	2.10	4.28	7.51	11.14	13.50	82.73
DAVENPORT	1.99	1.38	1.33	1.10	1.33	.97	.59	.77	.82	1.03	1.95	2.21	15.47
DAYTON 1 WSW	2.60	1.60	1.83	1.47	1.33	1.19	.45	.76	.89	1.50	2.22	2.79	18.63
DIABLO DAM	11.88	8.69	6.76	4.55	2.79	2.05	1.41	1.83	4.14	7.74	11.49	13.74	77.07
DOTY 3 E	9.13	6.35	6.29	3.57	2.21	1.80	.66	1.45	2.55	4.70	7.78	9.18	55.67
ELECTRON HEADWORKS	10.06	6.72	6.07	5.09	3.95	3.37	1.24	2.21	3.37	6.12	8.69	10.11	67.00
ELMA	11.05	7.79	7.30	4.47	2.52	2.00	1.02	1.66	2.91	6.31	9.43	11.14	67.60
ELWHA RANGER STATION	9.27	7.00	5.96	3.08	1.60	1.11	.71	1.21	2.05	5.10	8.39	10.35	55.83
EPHRATA FAA AIRPORT	.99	.67	.62	.53	.54	.53	.25	.28	.32	.49	.92	1.14	7.28
EVERETT	4.77	3.48	3.52	2.58	2.24	2.06	1.08	1.55	2.06	3.19	4.48	5.22	36.23
FORKS 1 E	17.99	14.42	12.85	7.95	4.84	3.05	2.29	2.78	5.46	11.47	15.92	20.04	119.06
GLACIER RANGER STATION	9.25	6.86	5.82	4.23	2.89	2.62	1.51	2.38	3.71	6.48	8.47	9.90	64.12
GLENOMA 1 W	9.82	6.57	6.04	4.80	3.37	2.78	.95	1.94	3.13	5.57	8.49	10.66	64.12
GRAPEVIEW 3 SW	8.63	6.30	5.20	3.26	1.88	1.43	.92	1.31	2.36	4.74	7.50	8.74	52.27
GREENWATER	8.79	5.62	4.91	4.07	2.88	2.55	1.07	1.89	2.99	5.05	7.45	8.96	56.23
HARRINGTON 5 S	1.53	1.14	.98	.91	1.08	.79	.43	.52	.61	.96	1.64	1.72	12.31
HARTLINE	1.24	.91	.82	.75	.93	.79	.41	.50	.57	.69	1.39	1.54	10.54
HATTON 9 ESE	1.27	.90	.81	.71	.84	.59	.26	.47	.43	.82	1.27	1.48	9.85
HOQUIAM FAA AIRPORT	10.95	8.22	7.71	4.53	2.96	2.09	1.22	1.78	3.57	6.66	9.93	11.60	71.22
IRENE MT WAUCONDA	1.19	.86	.81	1.08	1.81	1.91	1.01	1.38	.83	.94	1.09	1.39	14.30
KAHLOTUS 4 SW	1.35	.90	.86	.74	.86	.75	.27	.43	.47	.81	1.32	1.55	10.31
KENNEWICK	1.17	.66	.54	.46	.59	.44	.19	.40	.37	.56	.99	1.18	7.55
KENT	6.18	4.23	3.77	2.64	1.75	1.52	.81	1.34	2.05	3.47	5.68	6.48	39.92
KID VALLEY	8.81	6.03	6.47	4.98	3.54	2.84	1.09	2.11	2.75	5.20	7.29	9.07	60.18
LACROSSE 3 ESE	1.92	1.26	1.12	.90	.97	.90	.39	.60	.60	1.02	1.64	2.17	13.49

WASHINGTON

PRECIPITATION NORMALS (INCHES)

STATION	JAN	FEB	MAR	APR	MAY	JUN	JUL	AUG	SEP	OCT	NOV	DEC	ANN
LAKE WENATCHEE	8.30	4.60	3.09	1.63	1.10	.93	.43	.77	1.28	3.34	5.94	8.27	39.68
LANDSBURG	7.93	5.93	5.30	4.30	3.20	2.99	1.49	2.06	3.30	4.87	7.48	8.71	57.56
LAURIER	2.20	1.46	1.35	1.38	1.89	1.86	1.15	1.48	.95	1.33	1.84	2.38	19.27
LEAVENWORTH 3 S	5.15	2.95	2.15	1.14	.73	.65	.31	.69	.77	1.80	4.08	5.24	25.66
LIND 3 NE	1.11	.83	.71	.68	.81	.65	.27	.42	.49	.75	1.18	1.31	9.21
LONGVIEW	6.96	4.41	4.48	3.30	2.36	1.99	.85	1.58	2.21	4.07	6.28	7.65	46.14
MC MILLIN RESERVOIR	6.00	4.28	3.76	3.04	2.18	1.98	.89	1.64	2.21	3.56	5.66	6.20	41.40
MILL CREEK DAM	2.22	1.45	1.66	1.59	1.53	1.08	.44	.79	.98	1.50	2.17	2.32	17.73
MINERAL	13.73	9.31	8.64	5.49	3.51	2.82	1.29	1.95	3.16	6.58	10.97	13.21	80.66
MONROE	6.66	4.89	4.70	3.57	2.82	2.46	1.28	2.00	2.83	4.28	6.14	6.87	48.50
MOSES LAKE 3 E	.95	.69	.59	.53	.72	.57	.26	.35	.37	.55	.99	1.17	7.74
MOUNT ADAMS RANGER STA	9.14	5.24	4.74	2.25	1.29	.92	.30	.86	1.40	3.62	7.11	8.54	45.41
MOXEE CITY 10 E	1.00	.65	.61	.66	.61	.65	.26	.49	.43	.58	.96	.99	7.89
MUD MOUNTAIN DAM	7.08	4.89	4.66	4.58	3.96	3.75	1.58	2.54	3.46	4.73	6.64	7.45	55.32
NASELLE 1 ENE	18.52	13.77	13.20	7.26	4.56	3.31	1.71	2.68	4.74	9.92	15.96	19.49	115.12
NEAH BAY 1 E	15.57	12.23	10.27	6.91	3.73	2.80	2.35	2.65	4.80	10.22	14.33	17.50	103.36
NESPELEM 2 S	1.48	1.06	.89	.96	1.27	.94	.53	.65	.68	.94	1.54	1.94	12.88
NEWHALEM	12.51	9.42	7.17	4.79	3.05	2.59	1.67	2.24	4.30	8.02	14.37	14.37	81.58
NEWPORT	3.97	2.65	2.17	1.87	2.16	1.76	.93	1.29	1.32	2.07	3.55	4.20	27.94
NORTHPORT	2.33	1.64	1.35	1.28	1.87	1.92	.98	1.36	1.16	1.41	2.11	2.65	20.06
OAKVILLE	9.82	6.22	6.04	3.68	2.34	1.94	.77	1.61	2.73	5.21	8.02	9.89	58.27
ODESSA	1.29	.90	.83	.65	.90	.57	.31	.38	.51	.66	1.31	1.49	9.80
OLGA 2 SE	4.23	2.93	2.38	1.91	1.45	1.34	.96	1.20	1.75	2.91	3.85	4.53	29.44
OLYMPIA WSO R	8.50	5.77	4.85	3.13	1.85	1.44	.76	1.34	2.36	4.68	7.58	8.70	50.96
OMAK 2 NW	1.39	1.14	.89	.92	1.03	.90	.52	.68	.59	.82	1.38	1.77	12.03
OTHELLO 6 ESE	1.12	.79	.60	.51	.65	.57	.22	.33	.34	.54	.98	1.15	7.80
PALMER 3 ESE	12.24	9.41	9.01	7.71	5.57	5.24	2.38	3.28	5.30	7.59	11.00	13.47	92.20
PLAIN	5.45	3.29	2.27	1.14	.84	.72	.34	.68	.80	1.96	4.28	5.67	27.44
PLEASANT VIEW	1.75	1.18	1.16	.91	.96	.86	.28	.53	.62	.99	1.61	1.89	12.74
POINT GRENVILLE	12.82	10.02	9.67	6.00	3.89	2.86	1.98	2.54	4.21	8.75	12.60	13.80	89.14
POMEROY	2.14	1.36	1.45	1.18	1.27	1.07	.42	.76	.73	1.18	1.71	2.23	15.50
PORT ANGELES	4.39	2.74	2.12	1.32	.89	.86	.49	.85	1.26	2.44	3.77	4.25	25.38
PORT TOWNSEND	2.39	1.67	1.96	1.51	1.44	1.29	.82	.98	1.12	1.40	2.35	2.82	19.75
PROSSER 4 NE	1.05	.63	.50	.60	.68	.54	.24	.33	.36	.64	.98	1.15	7.70
PULLMAN 2 NW	2.89	2.09	1.96	1.58	1.52	1.49	.53	.95	.99	1.61	2.64	3.07	21.32
PUYALLUP 2 W EXP STA	6.23	4.51	3.81	2.82	1.82	1.63	.81	1.43	2.06	3.44	5.66	6.54	40.76
QUILCENE 2 SW	8.37	6.64	6.21	3.38	2.83	2.28	1.15	1.36	1.65	4.34	7.67	9.75	55.63
QUILLAYUTE WSO	15.07	12.10	11.27	7.10	4.70	3.06	2.32	2.85	5.27	10.51	13.94	16.31	104.49
QUINCY 1 S	.99	.75	.66	.62	.58	.51	.26	.36	.37	.52	1.04	1.19	7.85
RAINIER OHANAPECOSH	14.29	8.77	7.19	4.56	2.88	2.32	.87	1.89	3.30	6.62	11.27	14.15	78.11
RAINIER PARADISE RS //	18.44	13.16	11.46	7.55	4.62	3.87	1.71	3.22	5.50	9.70	15.43	19.31	113.97
RANDLE 1 E	9.44	5.94	5.58	4.12	2.91	2.55	.81	1.84	2.90	5.08	8.05	10.47	59.69
REPUBLIC	1.90	1.30	1.19	1.09	1.60	1.49	.84	1.31	.82	.96	1.52	2.11	16.13
RICHLAND	1.03	.69	.50	.42	.53	.44	.14	.32	.28	.46	.91	1.06	6.78
RITZVILLE 1 SSE	1.42	1.07	.91	.82	.88	.76	.34	.46	.58	.89	1.47	1.73	11.33
ROSALIA	2.35	1.56	1.43	1.30	1.48	1.32	.51	.85	.80	1.34	2.01	2.42	17.37
SAPPHO 8 E	14.97	11.66	10.05	6.50	3.89	2.67	1.86	2.47	4.62	9.29	13.80	15.86	97.64
SEATTLE EMSU WSO	5.94	4.20	3.70	2.46	1.66	1.53	.89	1.38	2.03	3.40	5.36	6.29	38.85
SEATTLE-TACOMA WSO R	6.04	4.22	3.59	2.40	1.58	1.38	.74	1.27	2.02	3.43	5.60	6.33	38.60
SEATTLE U OF W	5.24	3.79	3.46	2.39	1.52	1.59	.91	1.05	1.90	3.03	4.92	5.70	35.50
SEDRO WOOLLEY	6.03	4.64	3.99	3.56	2.63	2.29	1.52	1.98	3.09	4.35	5.77	6.59	46.44
SEQUIM	2.37	1.37	1.31	1.06	.94	1.03	.49	.78	.97	1.37	2.20	2.50	16.39
SHELTON	11.40	7.78	6.80	4.12	2.03	1.61	.92	1.38	2.61	5.78	9.72	11.45	65.60
SNOQUALMIE FALLS	9.07	6.49	5.97	4.50	3.21	2.81	1.43	1.86	3.17	5.19	8.13	9.54	61.37
SPOKANE WSO //R	2.47	1.61	1.36	1.08	1.38	1.23	.50	.74	.71	1.08	2.06	2.49	16.71

WASHINGTON

PRECIPITATION NORMALS (INCHES)

STATION		JAN	FEB	MAR	APR	MAY	JUN	JUL	AUG	SEP	OCT	NOV	DEC	ANN
SPRAGUE		2.07	1.39	1.19	1.04	1.14	.87	.48	.69	.66	1.06	1.93	2.24	14.76
SPRUCE		19.71	15.62	13.91	8.76	5.38	3.71	2.38	3.32	6.00	12.92	17.94	22.12	131.77
STAMPEDE PASS WSO	//R	14.59	10.19	8.88	6.28	3.97	3.84	1.56	2.85	4.65	7.74	12.14	15.88	92.57
STARTUP 1 E		8.68	6.20	5.93	5.31	4.31	3.64	1.99	2.49	3.84	5.86	8.02	9.18	65.45
STEHEKIN 3 NW		5.99	3.76	2.89	1.25	.87	.79	.47	.92	1.33	2.79	5.47	6.92	33.45
STEVENS PASS		14.18	9.66	7.89	4.70	3.29	2.68	1.25	2.10	3.64	6.63	11.16	14.57	81.75
SUNNYSIDE		1.03	.56	.42	.51	.53	.45	.20	.30	.37	.49	.83	.99	6.70
TACOMA CITY HALL		5.74	4.06	3.38	2.49	1.48	1.31	.75	1.25	1.95	3.27	5.47	6.02	37.17
TEKOA		2.70	1.67	1.66	1.44	1.75	1.62	.67	.89	.98	1.51	2.28	2.87	20.04
VANCOUVER 4 NNE		6.71	4.34	3.81	2.61	2.24	1.62	.57	1.22	1.88	3.46	5.61	7.00	41.07
WALLA-WALLA FAA AP		2.42	1.59	1.66	1.59	1.56	1.04	.42	.82	.95	1.54	2.29	2.48	18.36
WALLA WALLA WSO	R	2.12	1.40	1.41	1.35	1.40	.93	.35	.71	.83	1.40	1.87	2.19	15.96
WAPATO		1.20	.64	.56	.51	.45	.53	.19	.36	.34	.43	.93	1.10	7.24
WATERVILLE		1.50	.91	.73	.76	.82	.77	.30	.76	.51	.72	1.39	1.70	10.87
WAUNA 3 W		8.65	6.24	5.37	3.25	1.87	1.43	.92	1.31	2.21	4.50	7.47	8.82	52.04
WELLPINIT		2.50	1.75	1.52	1.48	1.74	1.09	.73	1.04	.89	1.23	2.47	2.98	19.42
WENATCHEE		1.37	.85	.60	.62	.55	.53	.15	.66	.35	.57	1.15	1.45	8.85
WHITE SWAN R. S.		1.74	.92	.68	.50	.36	.37	.24	.28	.28	.48	1.18	1.68	8.71
WILBUR		1.44	1.02	.94	.87	1.18	.88	.45	.70	.70	.87	1.58	1.68	12.31
WILLAPA HARBOR		13.71	10.03	9.38	5.96	3.53	2.86	1.36	2.17	3.76	7.54	11.20	13.83	85.33
WILSON CREEK		1.11	.78	.70	.60	.78	.59	.27	.38	.43	.61	1.15	1.36	8.76
WINTHROP 1 WSW		2.35	1.40	.89	.76	.83	.83	.58	.83	.64	.82	1.78	2.90	14.61
YAKIMA WSO	//R	1.44	.74	.65	.50	.48	.60	.14	.36	.33	.47	.97	1.30	7.98

WASHINGTON

HEATING DEGREE DAY NORMALS (BASE 65 DEG F)

STATION	JUL	AUG	SEP	OCT	NOV	DEC	JAN	FEB	MAR	APR	MAY	JUN	ANN
ABERDEEN	147	135	181	378	588	722	778	616	651	513	375	236	5320
ANACORTES	119	123	211	419	606	729	803	633	648	486	335	202	5314
BATTLE GROUND	88	107	161	384	615	766	828	633	642	492	329	179	5224
BELLINGHAM 2 N	114	124	223	450	654	787	868	666	685	519	347	201	5638
BELLINGHAM FAA AIRPORT	116	124	231	459	657	800	880	678	694	522	357	206	5724
BICKLETON	66	102	208	512	861	1045	1141	874	856	627	415	213	6920
BLAINE 1 ENE	118	120	237	468	672	812	893	683	691	522	353	204	5773
BREMERTON	82	89	169	394	615	766	825	644	651	477	307	174	5193
BUCKLEY 1 NE	96	119	206	437	648	784	837	652	676	513	341	201	5510
CEDAR LAKE	166	181	268	484	720	880	952	750	797	624	453	294	6569
CENTRALIA	66	93	162	394	618	753	806	627	632	468	301	161	5081
CHELAN	8	22	120	453	816	1079	1197	893	766	459	206	70	6089
CHEWELAH 4 SSW	48	89	259	595	918	1156	1280	969	865	570	343	156	7248
CHIEF JOSEPH DAM	8	21	121	446	825	1107	1234	916	753	435	205	72	6143
CLEARBROOK	107	128	229	468	687	843	933	697	691	513	341	202	5839
CLEARWATER	193	187	227	406	609	735	791	624	673	543	415	288	5691
CLE ELUM	65	102	255	570	879	1088	1200	899	859	612	388	194	7111
COLFAX 1 NW	25	73	214	508	807	995	1097	815	781	549	339	155	6358
COLVILLE AP	33	76	234	583	936	1163	1256	952	859	564	332	145	7133
CONCRETE	93	91	165	391	654	828	899	686	679	474	296	171	5427
COULEE DAM 1 SW	8	34	134	450	831	1088	1209	902	781	480	240	93	6250
COUPEVILLE 1 S	143	133	234	456	639	756	822	652	663	513	369	234	5614
DALLESPORT FAA AIRPORT	7	23	82	341	684	884	980	697	614	384	183	70	4949
DAVENPORT	42	88	233	561	915	1150	1259	946	865	594	368	183	7204
DAYTON 1 WSW	14	49	145	422	738	921	1017	750	698	489	277	124	5644
DIABLO DAM	92	98	193	474	753	933	1023	781	778	555	344	190	6214
ELECTRON HEADWORKS	167	194	312	558	780	899	955	770	803	633	456	290	6817
ELMA	91	96	165	384	609	741	797	619	642	480	319	184	5127
ELWHA RANGER STATION	130	121	216	481	714	834	902	711	719	543	375	241	5987
EPHRATA FAA AIRPORT	6	19	103	412	828	1091	1203	860	719	426	191	58	5916
EVERETT	93	108	209	422	630	753	825	644	657	492	332	187	5352
FORKS 1 E	162	156	214	415	627	756	812	650	701	558	412	269	5732
GLENOMA 1 W	94	112	201	440	660	815	877	675	698	531	357	204	5664
GRAPEVIEW 3 SW	74	79	167	400	603	738	794	627	636	468	291	161	5038
HATTON 9 ESE	8	41	145	459	807	1023	1122	798	719	480	255	98	5955
HOQUIAM FAA AIRPORT	164	146	187	372	570	710	760	608	642	513	375	251	5298
KENNEWICK	0	14	79	363	696	899	989	700	589	345	140	31	4845
KENT	56	81	172	412	636	760	809	624	629	462	288	144	5073
KID VALLEY	98	112	216	468	678	812	868	678	691	519	350	213	5703
LACROSSE 3 ESE	18	43	162	465	795	989	1088	784	735	504	297	123	6003
LANDSBURG	125	149	242	471	684	822	874	683	704	552	381	226	5913
LAURIER	27	71	228	601	954	1200	1305	980	877	558	310	122	7233
LIND 3 NE	11	45	152	453	816	1035	1135	812	732	498	267	105	6061
LONGVIEW	74	84	145	366	600	753	812	627	629	477	319	173	5059
MC MILLIN RESERVOIR	104	111	205	440	663	800	862	672	688	528	363	213	5649
MONROE	85	90	172	394	627	763	828	633	642	468	304	161	5167
MOSES LAKE 3 E	23	59	168	493	846	1076	1194	854	738	483	256	104	6294
MOUNT ADAMS RANGER STA	70	109	243	530	828	1023	1107	851	837	618	397	201	6814
MOXEE CITY 10 E	39	76	178	487	828	1057	1141	832	760	537	326	141	6402
MUD MOUNTAIN DAM	145	160	240	459	669	812	884	706	750	594	425	269	6113
NEWHALEM	87	96	181	443	699	887	964	745	738	519	317	183	5859
NEWPORT	47	109	276	614	942	1156	1256	958	911	615	375	175	7434
NORTHPORT	18	77	193	524	885	1119	1221	924	809	492	244	99	6605
OAKVILLE	103	107	191	419	627	763	822	647	667	510	360	207	5423
ODESSA	21	54	167	484	837	1063	1166	843	753	495	275	111	6269

WASHINGTON

HEATING DEGREE DAY NORMALS (BASE 65 DEG F)

STATION		JUL	AUG	SEP	OCT	NOV	DEC	JAN	FEB	MAR	APR	MAY	JUN	ANN
OLGA 2 SE		172	173	249	446	630	756	818	655	670	513	375	249	5706
OLYMPIA WSO	R	101	115	214	462	672	806	862	678	691	531	369	208	5709
OMAK 2 NW		14	32	178	549	912	1184	1302	969	818	501	262	95	6816
OTHELLO 6 ESE		33	47	143	443	807	1032	1132	809	707	459	238	95	5945
PALMER 3 ESE		118	129	207	425	666	828	899	694	729	555	381	233	5864
POMEROY		16	47	154	419	747	936	1023	750	716	480	277	121	5686
PORT ANGELES		170	170	235	446	630	747	806	650	679	531	391	259	5714
PROSSER 4 NE		10	37	137	425	759	977	1066	764	676	441	228	88	5608
PULLMAN 2 NW		56	96	217	499	837	1032	1141	860	837	597	384	197	6753
PUYALLUP 2 W EXP STA		67	84	181	409	636	763	812	630	639	474	298	151	5144
QUILCENE 2 SW		92	113	205	450	681	818	877	675	676	504	332	192	5615
QUILLAYUTE WSO		194	190	252	453	636	750	806	652	707	570	434	301	5945
QUINCY 1 S		9	34	135	462	849	1113	1225	868	738	450	219	85	6187
RAINIER PARADISE RS	//	394	415	507	747	993	1166	1240	1039	1147	969	784	600	10001
REPUBLIC		95	143	324	673	1020	1287	1386	1036	970	669	437	231	8271
RICHLAND		0	10	73	324	669	880	977	689	580	333	132	33	4700
RITZVILLE 1 SSE		22	59	184	484	855	1076	1181	868	791	546	325	133	6524
ROSALIA		46	93	238	533	861	1066	1169	879	840	594	378	194	6891
SEATTLE EMSU WSO		64	74	142	347	564	694	756	582	598	438	274	148	4681
SEATTLE-TACOMA WSO	R	76	97	169	388	606	744	803	622	645	489	313	169	5121
SEATTLE U OF W		51	62	132	341	570	704	775	596	605	438	260	138	4672
SEDRO WOOLLEY		108	104	202	428	636	772	846	652	660	492	329	189	5418
SEQUIM		131	126	203	437	645	769	828	655	673	522	363	228	5580
SHELTON		73	88	172	415	630	766	818	647	657	489	310	164	5229
SNOQUALMIE FALLS		106	119	213	437	636	769	834	647	676	513	350	205	5505
SPOKANE WSO	//R	17	63	209	539	903	1116	1218	913	849	576	339	140	6882
SPRAGUE		24	70	221	536	864	1088	1194	874	806	552	326	149	6704
STAMPEDE PASS WSO	//R	286	313	424	710	1023	1194	1268	1033	1082	891	698	474	9396
STARTUP 1 E		86	93	183	403	621	756	818	630	651	480	313	177	5211
STEHEKIN 3 NW		52	91	221	570	867	1051	1128	879	834	549	307	142	6691
SUNNYSIDE		7	23	114	400	738	961	1048	745	639	399	186	58	5318
TACOMA CITY HALL		56	73	146	357	579	713	766	594	614	456	289	153	4796
VANCOUVER 4 NNE		72	86	145	366	606	766	831	633	617	459	293	152	5026
WALLA-WALLA FAA AP		0	19	93	351	696	890	983	708	626	408	203	72	5049
WALLA WALLA WSO	R	0	16	85	332	675	862	952	680	595	381	175	54	4807
WAPATO		6	22	95	384	744	967	1054	742	629	376	160	54	5233
WATERVILLE		35	89	227	577	951	1234	1324	1000	911	606	366	183	7503
WENATCHEE		5	17	103	428	786	1035	1156	840	691	399	178	58	5696
WILBUR		38	81	216	539	891	1132	1240	930	843	570	342	170	6992
WILLAPA HARBOR		114	119	161	363	576	710	756	599	632	492	338	216	5076
WILSON CREEK		11	41	158	487	849	1088	1181	848	744	480	244	93	6224
WINTHROP 1 WSW		43	71	231	583	990	1333	1445	1067	930	576	326	151	7746
YAKIMA WSO	//R	18	46	161	468	804	1039	1141	809	716	474	254	101	6031

WASHINGTON

COOLING DEGREE DAY NORMALS (BASE 65 DEG F)

STATION	JAN	FEB	MAR	APR	MAY	JUN	JUL	AUG	SEP	OCT	NOV	DEC	ANN
ABERDEEN	0	0	0	0	0	0	8	8	10	0	0	0	26
ANACORTES	0	0	0	0	0	0	13	14	0	0	0	0	27
BATTLE GROUND	0	0	0	0	0	14	57	70	20	0	0	0	161
BELLINGHAM 2 N	0	0	0	0	0	9	27	21	0	0	0	0	57
BELLINGHAM FAA AIRPORT	0	0	0	0	0	8	26	28	0	0	0	0	62
BICKLETON	0	0	0	0	6	36	135	136	40	0	0	0	353
BLAINE 1 ENE	0	0	0	0	0	6	16	11	0	0	0	0	33
BREMERTON	0	0	0	0	0	15	54	51	13	0	0	0	133
BUCKLEY 1 NE	0	0	0	0	0	15	37	47	8	0	0	0	107
CEDAR LAKE	0	0	0	0	0	6	26	33	7	0	0	0	72
CENTRALIA	0	0	0	0	0	23	60	71	18	0	0	0	172
CHELAN	0	0	0	0	17	106	247	223	42	0	0	0	635
CHEWELAH 4 SSW	0	0	0	0	0	24	79	86	13	0	0	0	202
CHIEF JOSEPH DAM	0	0	0	0	19	108	250	229	70	0	0	0	676
CLEARBROOK	0	0	0	0	0	10	26	26	0	0	0	0	62
CLEARWATER	0	0	0	0	0	0	0	7	5	0	0	0	12
CLE ELUM	0	0	0	0	0	23	96	90	12	0	0	0	221
COLFAX 1 NW	0	0	0	0	0	32	93	104	22	0	0	0	251
COLVILLE AP	0	0	0	0	7	34	123	123	24	0	0	0	311
CONCRETE	0	0	0	0	10	36	80	70	24	0	0	0	220
COULEE DAM 1 SW	0	0	0	0	17	102	237	220	68	0	0	0	644
COUPEVILLE 1 S	0	0	0	0	0	6	13	9	0	0	0	0	28
DALLESPORT FAA AIRPORT	0	0	0	0	31	124	258	246	97	0	0	0	756
DAVENPORT	0	0	0	0	5	39	116	122	32	0	0	0	314
DAYTON 1 WSW	0	0	0	0	8	76	197	183	46	0	0	0	510
DIABLO DAM	0	0	0	0	6	25	83	88	22	0	0	0	224
ELECTRON HEADWORKS	0	0	0	0	0	5	15	14	0	0	0	0	34
ELMA	0	0	0	0	0	19	41	50	15	0	0	0	125
ELWHA RANGER STATION	0	0	0	0	0	13	46	44	6	0	0	0	109
EPHRATA FAA AIRPORT	0	0	0	0	45	151	335	285	100	0	0	0	916
EVERETT	0	0	0	0	0	13	28	30	5	0	0	0	76
FORKS 1 E	0	0	0	0	0	5	13	16	10	0	0	0	44
GLENOMA 1 W	0	0	0	0	0	15	48	40	12	0	0	0	115
GRAPEVIEW 3 SW	0	0	0	0	0	20	56	60	14	0	0	0	150
HATTON 9 ESE	0	0	0	0	11	92	229	193	58	0	0	0	583
HOQUIAM FAA AIRPORT	0	0	0	0	0	0	6	9	13	0	0	0	28
KENNEWICK	0	0	0	0	37	148	310	262	67	0	0	0	824
KENT	0	0	0	0	0	21	56	59	10	0	0	0	146
KID VALLEY	0	0	0	0	0	12	27	34	9	0	0	0	82
LACROSSE 3 ESE	0	0	0	0	5	63	179	148	36	0	0	0	431
LANDSBURG	0	0	0	0	0	10	35	28	5	0	0	0	78
LAURIER	0	0	0	0	9	32	117	124	21	0	0	0	303
LIND 3 NE	0	0	0	0	13	81	207	184	59	0	0	0	544
LONGVIEW	0	0	0	0	0	14	47	59	19	0	0	0	139
MC MILLIN RESERVOIR	0	0	0	0	0	9	29	37	7	0	0	0	82
MONROE	0	0	0	0	0	20	60	55	13	0	0	0	148
MOSES LAKE 3 E	0	0	0	0	11	77	187	177	33	0	0	0	485
MOUNT ADAMS RANGER STA	0	0	0	0	0	30	77	69	15	0	0	0	191
MOXEE CITY 10 E	0	0	0	0	6	57	142	138	31	0	0	0	374
MUD MOUNTAIN DAM	0	0	0	0	0	8	30	36	9	0	0	0	83
NEWHALEM	0	0	0	0	7	27	80	81	31	0	0	0	226
NEWPORT	0	0	0	0	0	22	71	88	9	0	0	0	190
NORTHPORT	0	0	0	0	11	69	166	173	34	0	0	0	453
OAKVILLE	0	0	0	0	0	9	37	45	8	0	0	0	99
ODESSA	0	0	0	0	18	90	204	184	50	0	0	0	546

WASHINGTON

COOLING DEGREE DAY NORMALS (BASE 65 DEG F)

STATION		JAN	FEB	MAR	APR	MAY	JUN	JUL	AUG	SEP	OCT	NOV	DEC	ANN
OLGA 2 SE		0	0	0	0	0	0	0	0	0	0	0	0	0
OLYMPIA WSO	R	0	0	0	0	0	10	39	38	7	0	0	0	94
OMAK 2 NW		0	0	0	0	11	71	182	149	31	0	0	0	444
OTHELLO 6 ESE		0	0	0	0	25	107	238	199	65	0	0	0	634
PALMER 3 ESE		0	0	0	0	0	14	47	45	15	0	0	0	121
POMEROY		0	0	0	0	10	82	198	183	67	0	0	0	540
PORT ANGELES		0	0	0	0	0	0	9	6	0	0	0	0	15
PROSSER 4 NE		0	0	0	0	14	85	175	155	47	0	0	0	476
PULLMAN 2 NW		0	0	0	0	0	20	84	111	31	0	0	0	246
PUYALLUP 2 W EXP STA		0	0	0	0	0	16	48	47	10	0	0	0	121
QUILCENE 2 SW		0	0	0	0	0	15	45	51	10	0	0	0	121
QUILLAYUTE WSO		0	0	0	0	0	0	8	10	6	0	0	0	24
QUINCY 1 S		0	0	0	0	24	103	214	195	57	0	0	0	593
RAINIER PARADISE RS	//	0	0	0	0	0	0	6	15	9	0	0	0	30
REPUBLIC		0	0	0	0	0	9	45	59	9	0	0	0	122
RICHLAND		0	0	0	0	42	165	332	289	112	0	0	0	940
RITZVILLE 1 SSE		0	0	0	0	9	58	174	168	61	0	0	0	470
ROSALIA		0	0	0	0	0	23	87	115	28	0	0	0	253
SEATTLE EMSU WSO		0	0	0	0	0	28	73	74	25	0	0	0	200
SEATTLE-TACOMA WSO	R	0	0	0	0	0	25	70	70	19	0	0	0	184
SEATTLE U OF W		0	0	0	0	5	33	88	87	30	0	0	0	243
SEDRO WOOLLEY		0	0	0	0	0	9	24	23	7	0	0	0	63
SEQUIM		0	0	0	0	0	9	20	18	0	0	0	0	47
SHELTON		0	0	0	0	0	20	55	63	10	0	0	0	148
SNOQUALMIE FALLS		0	0	0	0	0	13	41	41	6	0	0	0	101
SPOKANE WSO	//R	0	0	0	0	8	41	162	159	41	0	0	0	411
SPRAGUE		0	0	0	0	7	59	139	126	32	0	0	0	363
STAMPEDE PASS WSO	//R	0	0	0	0	0	0	10	22	7	0	0	0	39
STARTUP 1 E		0	0	0	0	0	21	49	44	12	0	0	0	126
STEHEKIN 3 NW		0	0	0	0	0	40	123	128	11	0	0	0	302
SUNNYSIDE		0	0	0	0	25	115	231	191	57	0	0	0	619
TACOMA CITY HALL		0	0	0	0	0	21	56	64	17	0	0	0	158
VANCOUVER 4 NNE		0	0	0	0	8	32	94	95	40	0	0	0	269
WALLA-WALLA FAA AP		0	0	0	0	20	129	316	279	99	0	0	0	843
WALLA WALLA WSO	R	0	0	0	0	23	132	326	282	100	0	0	0	863
WAPATO		0	0	0	0	45	153	306	258	77	0	0	0	839
WATERVILLE		0	0	0	0	6	45	113	129	26	0	0	0	319
WENATCHEE		0	0	0	0	29	130	278	243	64	0	0	0	744
WILBUR		0	0	0	0	7	53	125	125	30	0	0	0	340
WILLAPA HARBOR		0	0	0	0	0	12	24	41	11	0	0	0	88
WILSON CREEK		0	0	0	0	14	99	203	174	47	0	0	0	537
WINTHROP 1 WSW		0	0	0	0	0	40	118	108	18	0	0	0	284
YAKIMA WSO	//R	0	0	0	0	16	86	186	158	38	0	0	0	484

45 – WASHINGTON

LEGEND
11 = TEMPERATURE ONLY
12 = PRECIPITATION ONLY
13 = TEMP. & PRECIP.

STATE-STATION NUMBER	STN TYP	NAME	LATITUDE DEG-MIN	LONGITUDE DEG-MIN	ELEVATION (FT)
45-0008	13	ABERDEEN	N 4658	W 12349	10
45-0013	12	ABERDEEN 20 NNE //	N 4716	W 12342	435
45-0176	13	ANACORTES	N 4831	W 12237	30
45-0184	12	ANATONE	N 4608	W 11708	3570
45-0257	12	ARLINGTON	N 4812	W 12208	100
45-0482	13	BATTLE GROUND	N 4547	W 12232	295
45-0564	13	BELLINGHAM 2 N	N 4847	W 12229	140
45-0574	13	BELLINGHAM FAA AIRPORT	N 4848	W 12232	150
45-0668	13	BICKLETON	N 4600	W 12018	3000
45-0729	13	BLAINE 1 ENE	N 4900	W 12244	80
45-0872	13	BREMERTON	N 4734	W 12240	162
45-0945	13	BUCKLEY 1 NE	N 4710	W 12200	685
45-1233	13	CEDAR LAKE	N 4725	W 12144	1560
45-1276	13	CENTRALIA	N 4643	W 12257	185
45-1350	13	CHELAN	N 4750	W 12002	1120
45-1395	13	CHEWELAH 4 SSW	N 4813	W 11745	1675
45-1400	13	CHIEF JOSEPH DAM	N 4800	W 11939	810
45-1414	12	CHIMACUM 4 S	N 4757	W 12246	250
45-1484	13	CLEARBROOK	N 4858	W 12220	64
45-1496	13	CLEARWATER	N 4735	W 12418	75
45-1504	13	CLE ELUM	N 4711	W 12057	1930
45-1586	13	COLFAX 1 NW	N 4653	W 11723	1955
45-1650	13	COLVILLE AP	N 4833	W 11753	1885
45-1666	12	CONCONULLY	N 4834	W 11945	2275
45-1679	13	CONCRETE	N 4833	W 12146	195
45-1691	12	CONNELL 12 SE	N 4630	W 11846	1078
45-1767	13	COULEE DAM 1 SW	N 4757	W 11900	1630
45-1783	13	COUPEVILLE 1 S	N 4812	W 12242	50
45-1968	13	DALLESPORT FAA AIRPORT	N 4537	W 12109	222
45-1972	12	DALLESPORT 9 N	N 4545	W 12109	1919
45-1992	12	DARRINGTON RANGER STA	N 4815	W 12136	550
45-2007	13	DAVENPORT	N 4739	W 11809	2460
45-2030	13	DAYTON 1 WSW	N 4619	W 11800	1557
45-2157	13	DIABLO DAM	N 4843	W 12109	891
45-2220	12	DOTY 3 E	N 4638	W 12312	260
45-2493	13	ELECTRON HEADWORKS	N 4654	W 12202	1730
45-2531	13	ELMA	N 4700	W 12324	68
45-2548	13	ELWHA RANGER STATION	N 4802	W 12335	360
45-2614	13	EPHRATA FAA AIRPORT	N 4718	W 11932	1259
45-2675	13	EVERETT	N 4759	W 12211	60
45-2914	13	FORKS 1 E	N 4757	W 12422	350
45-3160	12	GLACIER RANGER STATION	N 4853	W 12157	935
45-3177	13	GLENOMA 1 W	N 4631	W 12210	870
45-3284	13	GRAPEVIEW 3 SW	N 4718	W 12252	30
45-3357	12	GREENWATER	N 4708	W 12138	1730
45-3502	12	HARRINGTON 5 S	N 4725	W 11815	2167
45-3529	12	HARTLINE	N 4741	W 11906	1910
45-3546	13	HATTON 9 ESE	N 4645	W 11839	1430
45-3807	13	HOQUIAM FAA AIRPORT	N 4658	W 12356	14
45-3975	12	IRENE MT WAUCONDA	N 4849	W 11854	2700

STATE-STATION NUMBER	STN TYP	NAME	LATITUDE DEG-MIN	LONGITUDE DEG-MIN	ELEVATION (FT)
45-4077	12	KAHLOTUS 4 SW	N 4636	W 11836	1340
45-4154	13	KENNEWICK	N 4613	W 11908	392
45-4169	13	KENT	N 4723	W 12216	32
45-4201	13	KID VALLEY	N 4622	W 12237	690
45-4338	13	LACROSSE 3 ESE	N 4648	W 11749	1546
45-4446	12	LAKE WENATCHEE	N 4750	W 12048	2005
45-4486	13	LANDSBURG	N 4723	W 12158	535
45-4549	13	LAURIER	N 4900	W 11814	1644
45-4572	12	LEAVENWORTH 3 S	N 4734	W 12040	1128
45-4679	13	LIND 3 NE	N 4700	W 11835	1630
45-4769	13	LONGVIEW	N 4610	W 12255	12
45-5224	13	MC MILLIN RESERVOIR	N 4708	W 12216	579
45-5387	12	MILL CREEK DAM	N 4605	W 11816	1175
45-5425	12	MINERAL	N 4643	W 12211	1480
45-5525	13	MONROE	N 4751	W 12159	120
45-5613	13	MOSES LAKE 3 E	N 4707	W 11912	1208
45-5659	13	MOUNT ADAMS RANGER STA	N 4600	W 12132	1960
45-5688	13	MOXEE CITY 10 E	N 4631	W 12010	1550
45-5704	13	MUD MOUNTAIN DAM	N 4709	W 12156	1308
45-5774	12	NASELLE 1 ENE	N 4622	W 12347	35
45-5801	12	NEAH BAY 1 E	N 4822	W 12437	15
45-5832	12	NESPELEM 2 S	N 4808	W 11859	1890
45-5840	13	NEWHALEM	N 4841	W 12115	525
45-5844	13	NEWPORT	N 4811	W 11703	2135
45-5946	13	NORTHPORT	N 4855	W 11747	1350
45-6011	13	OAKVILLE	N 4650	W 12313	85
45-6039	13	ODESSA	N 4720	W 11840	1540
45-6096	13	OLGA 2 SE	N 4837	W 12248	80
45-6114	13	OLYMPIA WSO R	N 4658	W 12254	195
45-6123	13	OMAK 2 NW	N 4826	W 11932	1228
45-6215	13	OTHELLO 6 ESE	N 4648	W 11903	1190
45-6295	13	PALMER 3 ESE	N 4718	W 12151	920
45-6534	12	PLAIN	N 4747	W 12039	1940
45-6553	12	PLEASANT VIEW	N 4631	W 11820	1665
45-6584	12	POINT GRENVILLE	N 4718	W 12417	100
45-6610	13	POMEROY	N 4628	W 11737	1810
45-6624	13	PORT ANGELES	N 4807	W 12326	99
45-6678	12	PORT TOWNSEND	N 4807	W 12245	100
45-6768	13	PROSSER 4 NE	N 4615	W 11945	903
45-6789	13	PULLMAN 2 NW	N 4646	W 11712	2545
45-6803	13	PUYALLUP 2 W EXP STA	N 4712	W 12220	50
45-6846	13	QUILCENE 2 SW	N 4749	W 12255	123
45-6858	13	QUILLAYUTE WSO	N 4757	W 12433	179
45-6880	13	QUINCY 1 S	N 4713	W 11951	1274
45-6896	12	RAINIER OHANAPECOSH	N 4644	W 12134	1950
45-6898	13	RAINIER PARADISE RS //	N 4647	W 12144	5427
45-6909	12	RANDLE 1 E	N 4632	W 12156	900
45-6974	13	REPUBLIC	N 4839	W 11844	2610
45-7015	13	RICHLAND	N 4617	W 11917	357
45-7059	13	RITZVILLE 1 SSE	N 4707	W 11822	1830

STATE-STATION NUMBER	STN TYP	NAME		LATITUDE DEG-MIN	LONGITUDE DEG-MIN	ELEVATION (FT)
45-7180	13	ROSALIA		N 4714	W 11722	2400
45-7319	12	SAPPHO 8 E		N 4804	W 12407	760
45-7458	13	SEATTLE EMSU WSO		N 4739	W 12218	20
45-7473	13	SEATTLE-TACOMA WSO	R	N 4727	W 12218	400
45-7478	13	SEATTLE U OF W		N 4739	W 12217	96
45-7507	13	SEDRO WOOLLEY		N 4830	W 12214	50
45-7538	13	SEQUIM		N 4805	W 12306	180
45-7584	13	SHELTON		N 4712	W 12306	22
45-7773	13	SNOQUALMIE FALLS		N 4733	W 12151	440
45-7938	13	SPOKANE WSO	//R	N 4738	W 11732	2349
45-7956	13	SPRAGUE		N 4718	W 11759	1930
45-7987	12	SPRUCE		N 4748	W 12404	365
45-8009	13	STAMPEDE PASS WSO	//R	N 4717	W 12120	3958
45-8034	13	STARTUP 1 E		N 4752	W 12143	170
45-8059	13	STEHEKIN 3 NW		N 4820	W 12042	1150
45-8089	12	STEVENS PASS		N 4744	W 12105	4070
45-8207	13	SUNNYSIDE		N 4619	W 12000	747
45-8286	13	TACOMA CITY HALL		N 4715	W 12226	267
45-8348	12	TEKOA		N 4713	W 11705	2610
45-8773	13	VANCOUVER 4 NNE		N 4541	W 12239	210
45-8928	13	WALLA-WALLA FAA AP		N 4606	W 11817	1170
45-8931	13	WALLA WALLA WSO	R	N 4602	W 11820	949
45-8959	13	WAPATO		N 4626	W 12025	850
45-9012	13	WATERVILLE		N 4739	W 12004	2620
45-9021	12	WAUNA 3 W		N 4722	W 12242	17
45-9058	12	WELLPINIT		N 4753	W 11759	2450
45-9074	13	WENATCHEE		N 4725	W 12019	634
45-9191	12	WHITE SWAN R. S.		N 4623	W 12043	970
45-9238	13	WILBUR		N 4745	W 11842	2160
45-9291	13	WILLAPA HARBOR		N 4641	W 12345	10
45-9327	13	WILSON CREEK		N 4725	W 11907	1276
45-9376	13	WINTHROP 1 WSW		N 4828	W 12011	1755
45-9465	13	YAKIMA WSO	//R	N 4634	W 12032	1064

45 — WASHINGTON

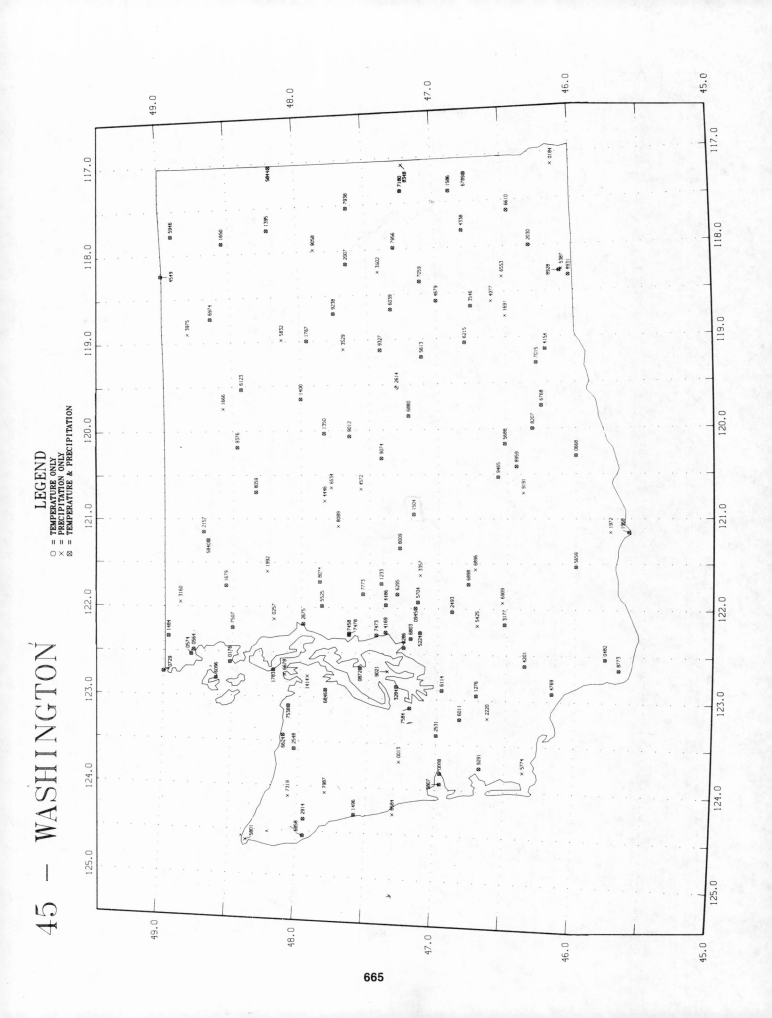

665

WEST VIRGINIA

TEMPERATURE NORMALS (DEG F)

STATION		JAN	FEB	MAR	APR	MAY	JUN	JUL	AUG	SEP	OCT	NOV	DEC	ANN
ATHENS CONCORD COLLEGE	MAX	41.8	44.9	54.0	65.2	72.9	78.5	81.6	80.8	75.6	65.7	54.1	45.2	63.4
	MIN	22.1	23.6	30.9	39.9	48.2	54.6	58.7	57.6	51.8	40.5	32.4	25.5	40.5
	MEAN	32.0	34.3	42.5	52.6	60.6	66.5	70.1	69.2	63.7	53.1	43.3	35.4	51.9
BAYARD	MAX	36.0	38.5	47.8	60.8	70.0	76.4	79.0	77.2	71.1	60.7	49.0	39.3	58.8
	MIN	16.1	17.3	24.6	33.9	42.7	49.9	54.3	53.1	46.8	35.4	27.9	19.9	35.2
	MEAN	26.1	28.0	36.2	47.3	56.4	63.2	66.7	65.1	58.9	48.1	38.5	29.6	47.0
BECKLEY V A HOSPITAL	MAX	40.9	44.1	53.5	65.5	73.3	77.7	80.6	79.6	74.4	64.3	53.2	44.2	62.6
	MIN	19.7	21.4	28.7	37.3	45.5	52.3	56.8	55.9	49.7	38.2	29.9	23.1	38.2
	MEAN	30.3	32.7	41.1	51.4	59.4	65.0	68.8	67.8	62.0	51.3	41.5	33.7	50.4
BECKLEY WSO //R	MAX	38.6	41.4	50.5	62.2	70.5	76.5	79.6	78.7	73.3	62.9	51.3	42.1	60.6
	MIN	21.6	23.2	31.4	40.8	49.2	55.9	59.8	59.0	52.9	41.7	33.0	25.3	41.1
	MEAN	30.1	32.3	41.0	51.5	59.9	66.2	69.7	68.9	63.1	52.3	42.2	33.7	50.9
BLUEFIELD FAA AP	MAX	39.2	42.0	51.0	62.4	70.0	75.6	78.4	77.6	72.2	62.5	51.4	42.7	60.4
	MIN	23.8	25.6	33.6	43.4	51.8	58.5	62.2	61.5	55.7	44.7	35.4	27.6	43.7
	MEAN	31.5	33.8	42.3	52.9	60.9	67.1	70.3	69.6	64.0	53.6	43.4	35.2	52.1
BLUESTONE DAM	MAX	41.0	44.5	53.8	65.8	74.8	81.2	84.8	83.8	78.3	67.1	54.8	44.0	64.5
	MIN	21.8	22.9	30.6	39.4	48.6	56.8	61.7	61.6	55.2	42.3	32.3	24.7	41.5
	MEAN	31.4	33.7	42.2	52.6	61.7	69.0	73.3	72.8	66.8	54.7	43.6	34.4	53.0
BRANDONVILLE	MAX	35.4	37.6	47.6	59.6	69.6	77.3	80.7	79.6	74.3	62.8	50.6	39.3	59.5
	MIN	16.1	17.4	26.2	36.2	44.4	52.2	56.7	55.3	48.2	37.0	29.5	21.0	36.7
	MEAN	25.8	27.5	36.9	47.9	57.0	64.8	68.7	67.5	61.3	49.9	40.1	30.2	48.1
BUCKHANNON 2 W	MAX	40.6	43.5	53.6	65.0	74.0	80.8	83.6	82.7	77.6	66.6	54.3	44.5	63.9
	MIN	20.1	21.3	29.3	38.1	46.7	54.3	58.9	57.8	51.3	39.3	31.4	24.4	39.4
	MEAN	30.4	32.4	41.5	51.6	60.4	67.5	71.3	70.3	64.5	52.9	42.9	34.5	51.7
CAIRO 3 S	MAX	41.7	45.6	56.6	68.6	77.9	84.0	86.5	85.4	80.0	69.6	56.4	45.4	66.5
	MIN	20.1	21.6	29.6	38.3	47.6	55.9	60.8	59.8	52.8	40.4	32.0	24.2	40.3
	MEAN	30.9	33.6	43.1	53.5	62.8	70.0	73.7	72.6	66.4	55.0	44.2	34.9	53.4
CANAAN VALLEY	MAX	34.8	37.1	46.0	58.6	67.7	74.5	77.4	76.5	71.0	60.1	48.0	38.4	57.5
	MIN	15.8	17.1	24.3	33.6	42.3	48.5	52.6	51.5	45.4	35.2	27.6	19.6	34.5
	MEAN	25.3	27.1	35.2	46.1	55.1	61.5	65.1	64.0	58.2	47.7	37.8	29.1	46.0
CHARLESTON WSO R	MAX	41.8	45.4	55.4	67.3	76.0	82.5	85.2	84.2	78.7	67.7	55.6	45.9	65.5
	MIN	23.9	25.8	34.1	43.3	51.8	59.4	63.8	63.1	56.4	44.0	35.0	27.8	44.0
	MEAN	32.9	35.6	44.8	55.3	63.9	71.0	74.5	73.7	67.6	55.9	45.3	36.9	54.8
CLARKSBURG 1	MAX	39.6	42.7	53.1	65.1	74.8	82.3	85.1	84.0	78.2	66.5	54.0	43.6	64.1
	MIN	19.9	21.0	28.8	37.8	47.2	55.9	60.4	59.6	52.7	40.0	31.3	24.3	39.9
	MEAN	29.8	31.8	41.0	51.5	61.0	69.1	72.8	71.8	65.4	53.2	42.6	34.0	52.0
ELKINS WSO //R	MAX	39.0	41.6	50.9	62.2	71.3	77.9	80.7	79.6	74.4	63.8	52.0	42.7	61.3
	MIN	17.6	19.1	27.3	36.0	44.8	52.4	57.0	56.3	49.7	37.1	29.1	21.6	37.3
	MEAN	28.3	30.4	39.1	49.1	58.1	65.2	68.9	68.0	62.1	50.5	40.6	32.2	49.4
FAIRMONT	MAX	38.3	41.5	52.0	64.2	73.5	80.7	83.7	82.6	76.8	65.2	52.7	42.2	62.8
	MIN	21.3	22.8	31.0	40.7	49.5	57.3	61.5	60.6	54.2	42.7	34.1	25.8	41.8
	MEAN	29.8	32.2	41.5	52.5	61.5	69.0	72.6	71.6	65.5	54.0	43.4	34.1	52.3
FLAT TOP	MAX	35.8	38.2	46.7	58.6	66.6	72.5	75.8	75.3	69.9	59.6	48.2	39.2	57.2
	MIN	18.8	20.1	27.5	37.5	46.7	53.5	58.1	57.3	51.1	39.7	30.4	22.5	38.6
	MEAN	27.3	29.2	37.1	48.1	56.7	63.0	67.0	66.3	60.5	49.7	39.3	30.8	47.9
FRANKLIN 2 NE	MAX	42.5	44.9	53.7	65.0	73.8	80.3	83.6	82.9	77.2	67.0	55.2	45.6	64.3
	MIN	20.2	21.6	29.0	37.9	46.6	53.4	57.6	56.7	50.4	39.1	31.2	23.7	39.0
	MEAN	31.3	33.3	41.4	51.5	60.2	66.9	70.6	69.8	63.8	53.1	43.2	34.7	51.7
GASSAWAY	MAX	43.3	47.0	57.5	69.3	77.5	83.2	85.8	84.4	79.2	68.6	57.1	47.2	66.7
	MIN	21.9	23.2	30.7	39.4	48.4	57.0	61.8	61.2	54.4	41.9	32.6	25.5	41.5
	MEAN	32.7	35.1	44.1	54.4	62.9	70.1	73.8	72.8	66.8	55.3	44.9	36.4	54.1

666

WEST VIRGINIA

TEMPERATURE NORMALS (DEG F)

STATION		JAN	FEB	MAR	APR	MAY	JUN	JUL	AUG	SEP	OCT	NOV	DEC	ANN
GLENVILLE 2 NE	MAX	41.3	44.5	54.7	66.7	76.0	82.3	85.2	84.3	79.0	67.8	55.6	45.2	65.2
	MIN	19.9	21.3	29.4	38.6	47.3	56.4	61.3	60.4	53.4	39.9	31.2	24.4	40.3
	MEAN	30.6	32.9	42.1	52.7	61.7	69.4	73.3	72.4	66.2	53.9	43.4	34.8	52.8
GRANTSVILLE 2 NW	MAX	41.5	44.5	54.9	66.9	76.4	83.1	86.0	85.3	79.9	68.7	56.5	45.6	65.8
	MIN	19.1	20.6	29.6	39.1	48.2	57.0	61.7	60.6	53.3	40.2	31.2	23.6	40.4
	MEAN	30.3	32.6	42.3	53.0	62.3	70.1	73.9	73.0	66.6	54.5	43.8	34.6	53.1
HAMLIN	MAX	42.5	45.7	55.6	67.5	76.4	83.0	86.5	85.5	80.2	69.3	57.0	46.6	66.3
	MIN	19.8	21.1	30.0	39.4	48.2	56.9	61.8	60.6	53.0	39.5	30.3	23.6	40.4
	MEAN	31.2	33.4	42.8	53.5	62.3	70.0	74.1	73.1	66.6	54.4	43.7	35.1	53.4
HOGSETT GALLIPOLIS DAM	MAX	40.8	43.8	54.0	66.1	75.4	82.4	85.9	84.9	79.5	68.1	55.8	45.1	65.2
	MIN	20.2	21.6	30.0	38.7	48.5	57.2	61.9	60.9	54.1	41.4	32.0	24.5	40.9
	MEAN	30.6	32.7	42.0	52.4	61.9	69.8	73.9	72.9	66.8	54.8	43.9	34.8	53.0
HUNTINGTON WSO	MAX	41.1	45.0	55.2	67.2	75.7	82.6	85.6	84.4	78.7	67.6	55.2	45.2	65.3
	MIN	24.5	26.6	35.0	44.4	52.8	60.7	65.1	64.0	57.2	44.9	35.9	28.5	45.0
	MEAN	32.8	35.8	45.1	55.8	64.3	71.7	75.4	74.2	68.0	56.3	45.6	36.9	55.2
KEARNEYSVILLE 1 NW WBASO	MAX	40.4	42.6	52.5	64.6	74.1	82.0	86.3	85.1	78.6	66.9	54.8	44.0	64.3
	MIN	20.6	22.4	30.2	40.0	48.4	56.9	61.5	60.1	53.0	40.7	32.8	24.4	40.9
	MEAN	30.5	32.5	41.4	52.3	61.3	69.5	73.9	72.6	65.8	53.8	43.8	34.2	52.6
LEWISBURG 2 SSW	MAX	41.1	44.7	54.7	66.0	74.3	80.1	83.3	82.4	77.2	66.8	54.2	44.5	64.1
	MIN	20.5	22.3	29.6	39.1	47.4	54.3	58.5	57.6	51.8	39.9	31.4	24.0	39.7
	MEAN	30.8	33.5	42.2	52.6	60.9	67.2	70.9	70.0	64.5	53.4	42.8	34.2	51.9
LONDON LOCKS	MAX	42.8	45.8	55.3	67.1	76.1	82.5	85.8	84.8	79.5	68.2	56.7	46.4	65.9
	MIN	22.8	24.0	31.7	40.1	49.6	58.3	63.1	63.2	56.9	44.2	33.9	26.4	42.9
	MEAN	32.8	34.9	43.5	53.7	62.9	70.4	74.5	74.1	68.2	56.2	45.3	36.4	54.4
MADISON	MAX	42.5	46.6	56.7	68.7	77.0	82.9	86.2	85.3	80.2	69.3	56.6	46.0	66.5
	MIN	22.2	23.3	31.4	40.4	49.6	58.2	63.1	62.4	55.4	42.0	31.8	25.2	42.1
	MEAN	32.4	35.0	44.1	54.6	63.3	70.6	74.6	73.9	67.8	55.7	44.3	35.6	54.3
MANNINGTON 1 N	MAX	40.0	43.0	54.2	66.4	76.0	82.7	85.6	84.4	78.8	67.5	54.4	43.7	64.7
	MIN	18.0	19.4	27.8	36.7	45.5	54.0	58.8	57.8	50.7	38.0	30.0	22.7	38.3
	MEAN	29.0	31.3	41.0	51.6	60.8	68.4	72.2	71.2	64.8	52.7	42.2	33.2	51.5
MARTINSBURG FAA AP	MAX	39.6	42.5	52.3	65.0	74.5	82.8	87.0	85.6	78.7	66.9	54.3	43.3	64.4
	MIN	21.3	23.1	31.1	40.8	49.9	58.3	63.0	61.8	54.3	42.2	33.5	25.1	42.0
	MEAN	30.5	32.8	41.7	52.9	62.2	70.6	75.0	73.7	66.5	54.6	43.9	34.2	53.2
MCROSS	MAX	39.9	43.3	53.1	64.9	72.7	77.8	80.5	79.8	75.1	65.4	53.4	43.4	62.4
	MIN	18.3	20.0	27.7	36.2	45.1	52.6	57.0	56.6	50.3	38.1	29.4	21.8	37.8
	MEAN	29.1	31.7	40.4	50.6	58.9	65.2	68.8	68.2	62.8	51.8	41.4	32.6	50.1
MIDDLEBOURNE 2 ESE	MAX	39.3	42.6	53.2	65.2	74.8	82.0	84.9	83.9	78.6	67.3	54.3	43.3	64.1
	MIN	19.5	21.0	29.0	37.9	47.3	56.3	61.0	59.8	53.1	40.0	31.2	23.8	40.0
	MEAN	29.4	31.8	41.1	51.6	61.0	69.2	73.0	71.9	65.8	53.7	42.8	33.6	52.1
MORGANTOWN FAA AIRPORT	MAX	38.0	40.7	51.2	63.5	73.1	80.5	83.6	82.4	76.6	65.0	52.5	42.2	62.4
	MIN	21.4	23.1	31.5	41.2	50.3	58.3	62.6	61.6	55.3	43.7	34.8	26.2	42.5
	MEAN	29.7	31.9	41.4	52.4	61.7	69.4	73.1	72.0	66.0	54.4	43.7	34.2	52.5
MORGANTOWN L AND D	MAX	39.8	42.8	53.0	65.2	74.6	81.6	84.5	83.3	77.8	66.5	54.1	43.9	63.9
	MIN	20.5	21.9	29.8	39.1	48.0	56.2	60.6	59.6	53.3	41.4	32.8	24.9	40.7
	MEAN	30.1	32.4	41.4	52.1	61.3	68.9	72.6	71.5	65.6	54.0	43.5	34.4	52.3
NEW CUMBERLAND	MAX	38.1	41.5	52.2	65.4	75.2	82.5	85.6	84.6	79.0	67.9	53.8	42.1	64.0
	MIN	20.6	22.1	29.9	39.1	48.1	57.0	61.2	60.2	54.1	42.6	34.3	25.4	41.2
	MEAN	29.4	31.8	41.1	52.2	61.7	69.8	73.4	72.5	66.6	55.3	44.1	33.8	52.6
OAK HILL	MAX	40.5	43.0	52.4	64.1	73.1	79.8	83.1	82.3	77.1	65.8	53.7	44.1	63.3
	MIN	19.6	20.9	29.0	38.7	46.9	54.3	58.7	58.0	51.7	39.7	30.8	23.5	39.3
	MEAN	30.1	32.0	40.7	51.4	60.1	67.0	70.9	70.2	64.4	52.8	42.3	33.8	51.3

WEST VIRGINIA

TEMPERATURE NORMALS (DEG F)

STATION		JAN	FEB	MAR	APR	MAY	JUN	JUL	AUG	SEP	OCT	NOV	DEC	ANN
PARKERSBURG FAA AP	MAX	38.7	42.0	52.6	64.8	74.3	81.5	84.3	83.3	77.5	66.1	53.4	43.0	63.5
	MIN	22.1	23.9	32.6	42.5	51.4	59.6	64.4	63.4	56.5	44.3	35.2	26.9	43.6
	MEAN	30.4	33.0	42.6	53.6	62.9	70.6	74.4	73.4	67.1	55.2	44.3	35.0	53.5
PARKERSBURG WSO R	MAX	39.5	42.7	53.1	65.3	74.8	82.0	84.9	84.0	78.0	66.7	53.9	43.5	64.0
	MIN	22.9	25.0	33.1	42.9	52.2	60.2	64.6	63.5	56.8	44.8	35.7	27.6	44.1
	MEAN	31.2	33.9	43.1	54.1	63.5	71.1	74.8	73.8	67.4	55.8	44.8	35.6	54.1
PICKENS 1	MAX	38.2	40.6	19.5	60.7	69.6	75.8	78.5	77.8	72.9	62.8	50.7	41.8	59.9
	MIN	17.2	18.9	26.7	36.1	44.3	51.3	55.3	54.6	48.7	37.8	28.9	21.2	36.8
	MEAN	27.7	29.8	38.1	48.4	57.0	63.6	66.9	66.2	60.8	50.3	39.8	31.5	48.3
PINEVILLE	MAX	42.2	46.0	55.6	68.0	76.6	82.9	85.7	85.1	79.8	68.8	56.4	45.8	66.1
	MIN	20.9	22.1	29.7	38.3	48.0	56.5	61.2	60.6	53.7	40.5	30.8	24.1	40.5
	MEAN	31.6	34.1	42.7	53.2	62.3	69.8	73.5	72.8	66.8	54.6	43.6	35.0	53.3
RAVENSWOOD LOCK PARK	MAX	42.7	46.2	56.6	68.5	77.2	83.7	86.5	85.8	80.6	70.0	57.2	47.0	66.8
	MIN	22.4	24.1	32.3	41.1	50.0	58.0	62.5	61.1	54.7	42.7	34.3	26.8	42.5
	MEAN	32.6	35.1	44.5	54.8	63.6	70.9	74.5	73.5	67.7	56.4	45.8	36.9	54.7
ROWLESBURG 1	MAX	39.3	42.2	52.1	64.3	73.9	80.9	83.5	82.7	77.4	66.0	53.7	42.7	63.2
	MIN	19.8	21.1	28.7	37.8	46.7	55.1	59.6	59.1	52.5	40.4	31.9	23.9	39.7
	MEAN	29.6	31.7	40.4	51.1	60.3	68.0	71.6	70.9	65.0	53.2	42.8	33.3	51.5
SPENCER	MAX	42.4	45.9	55.6	67.6	76.2	82.1	84.8	84.0	79.1	68.2	56.0	46.1	65.7
	MIN	22.1	23.5	31.4	40.9	49.2	57.1	61.3	60.5	53.9	42.0	33.4	26.3	41.8
	MEAN	32.3	34.7	43.5	54.3	62.7	69.6	73.1	72.3	66.5	55.1	44.7	36.2	53.8
SPRUCE KNOB	MAX	35.8	37.0	45.4	57.2	66.6	73.6	77.3	77.0	71.8	61.1	49.6	39.5	57.7
	MIN	16.4	17.5	25.0	36.1	45.0	52.5	56.3	55.3	48.8	38.8	29.1	20.1	36.7
	MEAN	26.1	27.3	35.3	46.7	55.8	63.1	66.8	66.2	60.3	50.0	39.4	29.8	47.2
UNION	MAX	40.6	43.6	52.7	63.6	72.7	79.1	82.7	82.0	76.6	66.1	54.2	44.2	63.2
	MIN	19.0	20.7	28.7	38.0	46.6	53.7	57.8	56.6	50.0	38.1	29.4	22.3	38.4
	MEAN	29.9	32.2	40.7	50.8	59.6	66.4	70.3	69.4	63.3	52.1	41.8	33.3	50.8
WARDENSVILLE R M FARM	MAX	41.0	42.9	52.1	64.0	73.6	81.3	85.4	84.4	78.1	67.2	55.5	44.5	64.2
	MIN	18.0	19.5	28.1	37.6	46.5	54.8	59.1	57.8	50.5	38.4	30.3	22.0	38.6
	MEAN	29.5	31.2	40.1	50.9	60.1	68.1	72.3	71.1	64.3	52.8	42.9	33.3	51.4
WEBSTER SPRINGS 1 E	MAX	43.1	46.5	56.2	68.0	76.4	82.3	84.9	83.5	78.2	68.0	55.9	46.3	65.8
	MIN	22.3	23.6	31.1	39.7	47.9	55.6	59.9	59.3	53.2	41.1	32.7	25.7	41.0
	MEAN	32.7	35.1	43.7	53.9	62.2	69.0	72.4	71.4	65.7	54.6	44.3	36.0	53.4
WELLSBURG 3 NE	MAX	38.0	41.4	52.1	65.2	75.0	82.6	85.7	84.5	78.6	66.6	53.4	42.2	63.8
	MIN	19.5	21.2	28.9	37.9	46.7	55.4	59.4	58.4	51.7	40.0	32.6	24.3	39.7
	MEAN	28.8	31.3	40.6	51.6	60.9	69.1	72.6	71.4	65.2	53.4	43.1	33.3	51.8
WESTON	MAX	40.4	43.0	53.4	64.9	75.2	82.4	85.6	84.6	79.1	67.4	55.0	44.2	64.6
	MIN	19.8	20.4	28.6	37.8	46.9	55.8	60.6	59.6	52.4	39.7	31.1	24.3	39.8
	MEAN	30.1	31.7	41.0	51.4	61.1	69.1	73.1	72.1	65.8	53.6	43.1	34.3	52.2
WHEELING FILT PLANT	MAX	37.5	39.8	50.4	62.7	73.3	81.5	84.8	83.6	78.1	66.4	53.1	41.8	62.8
	MIN	20.6	21.5	30.0	39.6	48.5	57.8	62.3	61.7	54.9	43.0	34.4	25.7	41.7
	MEAN	29.1	30.7	40.2	51.2	60.9	69.7	73.6	72.6	66.5	54.8	43.8	33.8	52.2
WHITE SULPHUR SPRINGS	MAX	43.0	46.7	56.6	68.4	76.9	82.6	85.8	84.7	78.9	68.7	55.8	46.0	66.2
	MIN	21.1	22.6	29.6	37.8	46.7	54.2	58.9	58.5	52.2	39.6	31.0	23.8	39.7
	MEAN	32.1	34.7	43.1	53.1	61.8	68.4	72.3	71.7	65.6	54.2	43.4	34.9	52.9
WINFIELD LOCKS	MAX	41.7	44.4	54.5	66.4	75.9	82.7	86.2	85.3	79.8	68.5	56.2	46.0	65.6
	MIN	22.4	23.4	31.3	40.3	50.0	59.0	63.8	63.1	56.6	44.1	34.2	26.5	42.9
	MEAN	32.1	34.0	42.9	53.4	62.9	70.8	75.0	74.2	68.2	56.3	45.2	36.3	54.3

WEST VIRGINIA

PRECIPITATION NORMALS (INCHES)

STATION	JAN	FEB	MAR	APR	MAY	JUN	JUL	AUG	SEP	OCT	NOV	DEC	ANN
ALDERSON	3.02	2.88	3.75	3.45	3.62	3.40	3.70	3.21	3.12	2.55	2.46	2.93	38.09
ATHENS CONCORD COLLEGE	2.98	2.80	3.66	3.52	3.80	3.63	3.89	2.90	3.24	2.53	2.46	2.77	38.18
BAYARD	4.04	3.45	4.56	4.42	4.38	4.18	4.48	4.87	3.16	3.30	3.14	3.84	47.82
BECKLEY V A HOSPITAL	3.49	3.00	3.80	3.54	3.57	3.99	4.65	4.28	3.80	2.63	2.80	2.99	42.54
BECKLEY WSO //R	3.44	3.19	4.13	3.59	3.86	3.82	4.46	3.68	3.38	2.54	2.81	3.23	42.12
BELINGTON	4.02	3.42	4.33	4.26	4.21	4.84	5.32	4.66	3.85	3.32	3.06	3.87	49.16
BELVA	3.67	3.01	4.04	3.56	4.06	3.90	5.29	4.47	3.44	2.94	2.93	3.24	44.55
BENS RUN 1 SSE	3.29	2.85	3.71	3.86	3.91	4.00	4.28	4.48	3.20	2.73	2.93	3.22	42.46
BLUEFIELD FAA AP	2.81	2.87	3.60	3.53	3.78	3.26	4.19	3.37	3.07	2.60	2.55	2.81	38.44
BLUESTONE DAM	2.61	2.56	3.59	3.17	3.59	3.15	4.04	3.11	2.91	2.35	2.24	2.53	35.85
BRANCHLAND	3.60	3.15	4.19	3.73	4.22	3.61	4.91	4.25	3.17	2.39	3.12	3.45	43.79
BRANDONVILLE	3.62	3.05	4.01	4.18	4.39	4.51	4.69	4.73	3.38	3.06	2.94	3.68	46.24
BRUSHY RUN	2.17	2.03	2.94	2.78	3.09	3.31	3.00	3.52	2.82	2.68	2.12	1.93	32.39
BUCKEYE	3.56	3.18	4.48	3.74	3.92	3.72	4.29	3.83	3.46	3.08	2.93	3.62	43.81
BUCKHANNON 2 W	4.00	3.25	4.13	4.20	4.10	4.60	4.93	4.24	3.69	3.16	3.17	3.93	47.40
BURNSVILLE	3.62	3.01	3.83	3.83	4.03	3.98	4.91	4.14	3.30	2.92	2.88	3.51	43.96
CAIRO 3 S	3.28	2.78	3.73	3.93	4.14	4.04	4.04	4.42	3.02	2.89	2.97	3.14	43.27
CAMDEN ON GAULEY	4.10	3.53	4.46	4.19	4.79	4.91	6.46	5.07	3.99	3.60	3.49	3.94	52.53
CANAAN VALLEY	4.62	4.09	5.08	4.30	4.63	4.90	4.81	5.10	3.52	3.55	3.59	4.28	52.47
CENTRALIA	4.12	3.33	4.17	3.98	4.19	4.63	5.71	4.87	3.77	3.26	3.24	3.61	48.88
CHARLESTON WSO R	3.48	3.11	4.00	3.52	3.68	3.32	5.36	4.15	3.01	2.63	2.90	3.27	42.43
CLARKSBURG 1	3.17	2.66	3.51	3.55	3.85	4.06	4.08	4.59	2.98	2.70	2.55	3.35	41.05
CLAY 1 SW	3.82	3.24	4.38	4.07	4.16	3.76	4.77	4.61	3.52	2.87	3.15	3.64	45.99
CLENDENIN 1 SW	3.78	3.23	4.29	3.86	3.97	3.93	5.43	4.34	3.38	2.81	3.18	3.47	45.67
CRESTON	3.57	2.95	4.00	3.84	4.03	3.83	4.55	4.40	3.25	2.94	2.91	3.58	43.85
DRY CREEK	3.94	3.53	4.63	4.08	4.32	4.16	5.36	4.67	3.56	2.88	3.35	3.74	48.22
ELKINS WSO //R	3.39	2.84	3.66	3.71	3.82	4.35	4.68	4.20	3.21	2.97	2.69	3.33	42.85
FAIRMONT	3.32	2.83	3.90	3.74	4.24	4.16	4.35	4.49	3.17	2.83	2.85	3.37	43.25
FLAT TOP	3.77	3.52	4.46	4.11	4.35	3.60	4.51	3.50	3.51	2.96	3.04	3.50	44.83
FRANKLIN 2 NE	2.00	1.97	2.80	2.66	3.10	3.31	3.29	3.34	2.86	2.97	2.09	2.07	32.46
GARY //R	3.20	2.91	3.97	3.70	3.83	3.44	4.49	3.87	3.23	2.65	2.73	2.93	40.95
GASSAWAY	3.56	3.12	3.93	3.90	3.88	4.37	5.16	4.49	3.62	3.08	3.01	3.56	45.68
GLENVILLE 2 NE	3.51	2.96	3.78	3.82	4.15	3.92	5.08	4.41	3.45	3.04	2.86	3.45	44.43
GRANTSVILLE 2 NW	3.42	3.02	3.89	3.62	3.73	3.53	4.62	4.22	3.28	2.71	2.86	3.37	42.27
HAMLIN	3.46	3.03	4.00	3.69	4.02	3.58	5.04	3.99	3.35	2.48	3.09	3.40	43.13
HOGSETT GALLIPOLIS DAM	3.11	2.81	4.00	3.51	3.85	3.46	4.48	3.89	3.12	2.39	2.77	3.03	40.42
HUNTINGTON WSO	3.24	2.83	4.08	3.48	3.94	3.56	4.47	3.73	3.07	2.40	2.82	3.12	40.72
KEARNEYSVILLE 1 NW WBASO	2.73	2.36	3.40	3.32	3.61	3.66	3.45	3.94	3.13	3.03	2.73	2.83	38.19
KERMIT	3.52	3.27	4.06	3.65	3.95	3.81	4.96	3.66	3.19	2.48	2.78	3.65	42.98
LAKE LYNN	2.96	2.39	3.48	3.65	3.81	4.08	4.23	4.16	3.38	2.58	2.50	2.83	40.05
LEWISBURG 2 SSW	2.72	2.85	3.62	3.36	3.48	3.63	4.47	3.73	2.99	2.74	2.46	2.82	38.87
LONDON LOCKS	3.41	3.13	4.31	3.86	4.13	3.70	5.55	4.30	3.42	2.83	3.14	3.48	45.26
MADISON	3.47	3.11	4.17	3.80	4.11	4.10	4.87	3.84	3.41	2.66	3.01	3.35	43.90
MANNINGTON 1 N	3.35	2.71	3.70	3.87	4.26	4.35	5.03	4.69	3.30	2.93	2.81	3.37	44.37
MANNINGTON 1 W	3.42	2.90	3.93	4.01	4.40	4.44	4.94	4.69	3.44	2.89	2.82	3.34	45.22
MARTINSBURG FAA AP	2.44	2.32	3.30	3.23	3.58	3.48	3.27	3.61	2.87	2.97	2.65	2.72	36.44
MATHIAS	2.14	2.10	2.96	2.88	3.45	3.56	3.50	3.71	3.18	2.84	2.47	2.17	34.96
MCROSS	4.15	3.83	4.86	4.51	4.53	4.70	5.56	4.61	3.70	3.41	3.70	3.95	51.51
MIDDLEBOURNE 2 ESE	3.18	2.74	3.56	3.79	4.37	3.95	4.69	4.81	3.47	2.87	2.74	3.04	43.21
MORGANTOWN FAA AIRPORT	2.95	2.52	3.61	3.51	3.69	3.91	3.99	4.57	3.36	2.64	2.70	3.14	40.59
MORGANTOWN L AND D	3.17	2.60	3.65	3.72	3.78	3.91	4.27	4.56	3.29	2.68	2.73	3.10	41.46
NEW CUMBERLAND	2.57	2.19	3.45	3.31	3.81	3.96	4.22	3.52	2.81	2.54	2.40	2.47	37.25
OAK HILL	3.63	3.08	4.17	3.72	4.07	3.86	5.20	4.33	3.57	2.98	3.09	3.33	45.03
PARKERSBURG FAA AP	3.19	2.85	3.84	3.56	3.72	3.76	4.44	3.99	2.94	2.71	2.75	2.92	40.67
PARKERSBURG WSO R	2.86	2.58	3.56	3.37	3.49	3.84	4.08	3.58	2.82	2.44	2.43	2.72	37.77

WEST VIRGINIA

PRECIPITATION NORMALS (INCHES)

STATION	JAN	FEB	MAR	APR	MAY	JUN	JUL	AUG	SEP	OCT	NOV	DEC	ANN
PHILIPPI	3.87	3.18	4.04	3.91	4.44	4.36	5.06	4.49	3.65	3.08	3.14	3.93	47.15
PICKENS 1	5.66	4.92	6.17	5.58	5.75	5.70	6.76	5.90	4.73	4.70	4.64	5.51	66.02
PINEVILLE	3.73	3.27	4.33	4.06	4.02	4.05	4.98	4.08	3.49	2.96	3.08	3.32	45.37
PRINCETON	2.77	2.65	3.57	3.35	3.44	3.52	3.80	3.08	3.03	2.59	2.44	2.60	36.84
RAVENSWOOD LOCK PARK	3.05	2.74	3.67	3.45	3.61	3.48	4.22	3.73	2.82	2.40	2.59	2.94	38.70
RIPLEY	3.79	3.28	4.27	4.07	4.36	3.82	5.02	4.13	3.38	2.97	3.17	3.68	45.94
ROMNEY 3 NNE	2.40	2.05	3.11	2.90	3.18	3.79	3.42	3.50	2.95	2.75	2.22	2.31	34.58
ROWLESBURG 1	4.57	3.72	5.06	4.96	4.97	5.05	5.26	5.40	3.69	3.62	3.82	4.84	54.96
SPENCER	3.41	2.75	3.74	3.55	3.60	3.53	4.93	4.40	3.36	2.81	2.69	3.16	41.93
SPRUCE KNOB	3.09	2.95	3.82	3.57	3.74	3.85	3.75	3.57	3.34	3.33	2.77	3.03	40.81
SUTTON RESERVOIR	3.54	2.87	3.76	3.85	4.06	4.42	5.24	4.50	3.75	3.11	2.87	3.39	45.36
THOMAS	4.65	3.89	5.24	5.02	5.06	5.33	5.40	5.51	3.67	3.77	3.67	4.68	55.89
UNION	2.51	2.58	3.48	3.07	3.41	3.17	3.84	3.02	3.14	2.58	2.27	2.45	35.52
VALLEY HEAD	4.08	3.53	4.53	4.09	4.29	4.81	5.43	4.33	3.60	3.64	3.23	3.87	49.43
VANDALIA	3.97	3.27	4.13	4.22	4.11	4.75	5.29	4.52	3.98	3.10	3.19	3.93	48.46
WARDENSVILLE R M FARM	2.22	1.95	2.95	2.94	3.46	3.60	3.39	3.65	3.05	2.91	2.42	2.17	34.71
WEBSTER SPRINGS 1 E	3.54	3.11	4.20	4.30	4.35	4.83	5.98	4.79	4.08	3.44	3.38	3.48	49.48
WELLSBURG 3 NE	2.71	2.30	3.52	3.63	3.82	3.81	4.23	3.78	2.79	2.68	2.48	2.56	38.31
WESTON	4.44	3.73	4.59	4.19	4.34	4.67	5.10	4.58	3.73	3.34	3.34	4.34	50.39
WHEELING FILT PLANT	2.79	2.37	3.65	3.57	3.71	3.62	4.14	3.55	3.30	2.43	2.37	2.56	38.06
WHITE SULPHUR SPRINGS	2.98	2.80	3.87	3.22	3.60	3.42	4.16	3.48	3.21	2.76	2.57	2.90	38.97
WILLIAMSON	3.60	3.54	4.41	3.72	3.99	4.06	4.69	3.58	3.42	2.51	3.22	3.56	44.30
WILLIAMSON 2	3.70	3.54	4.36	3.70	4.04	3.82	4.97	3.48	3.68	2.63	3.16	3.54	44.62
WINFIELD LOCKS	3.26	2.79	3.93	3.66	3.90	3.21	4.49	3.84	3.34	2.50	2.85	3.08	40.85

WEST VIRGINIA

HEATING DEGREE DAY NORMALS (BASE 65 DEG F)

STATION	JUL	AUG	SEP	OCT	NOV	DEC	JAN	FEB	MAR	APR	MAY	JUN	ANN
ATHENS CONCORD COLLEGE	6	8	101	369	651	918	1023	860	698	372	171	46	5223
BAYARD	31	66	199	524	795	1097	1206	1036	893	531	279	96	6753
BECKLEY V A HOSPITAL	16	22	130	425	705	970	1076	904	741	408	198	74	5669
BECKLEY WSO //R	11	14	110	398	684	970	1082	916	744	405	186	57	5577
BLUEFIELD FAA AP	0	10	93	357	648	924	1039	874	704	363	166	39	5217
BLUESTONE DAM	0	0	58	329	642	949	1042	876	707	372	158	28	5161
BRANDONVILLE	12	34	144	468	747	1079	1215	1050	871	513	267	78	6478
BUCKHANNON 2 W	0	9	88	380	663	946	1073	913	729	402	178	35	5416
CAIRO 3 S	0	0	62	319	624	933	1057	879	679	345	133	18	5049
CANAAN VALLEY	68	84	215	536	816	1113	1231	1061	924	567	316	141	7072
CHARLESTON WSO R	0	0	51	301	591	871	995	823	626	298	125	16	4697
CLARKSBURG 1	0	8	83	371	672	961	1091	930	744	405	167	27	5459
ELKINS WSO //R	11	19	127	450	732	1017	1138	969	803	477	233	69	6045
FAIRMONT	0	5	80	352	648	958	1091	918	729	379	162	32	5354
FLAT TOP	29	42	166	474	771	1060	1169	1002	865	507	272	108	6465
FRANKLIN 2 NE	7	15	100	373	654	939	1045	888	732	405	180	46	5384
GASSAWAY	0	0	52	313	603	887	1001	837	648	323	127	13	4804
GLENVILLE 2 NE	0	7	78	357	648	936	1066	899	710	375	165	27	5268
GRANTSVILLE 2 NW	0	0	65	335	636	942	1076	907	704	360	149	19	5193
HAMLIN	0	0	70	342	639	927	1048	885	688	349	150	19	5117
HOGSETT GALLIPOLIS DAM	0	0	60	329	633	936	1066	904	713	378	157	20	5196
HUNTINGTON WSO	0	0	62	293	582	871	998	818	617	293	125	17	4676
KEARNEYSVILLE 1 NW WBASO	0	0	66	354	636	955	1070	910	732	381	166	18	5288
LEWISBURG 2 SSW	8	9	87	363	666	955	1060	882	707	376	169	45	5327
LONDON LOCKS	0	0	44	296	591	887	998	843	667	345	142	25	4838
MADISON	0	0	47	308	621	911	1011	840	648	318	136	18	4858
MANNINGTON 1 N	0	8	89	387	684	986	1116	944	744	402	178	37	5575
MARTINSBURG FAA AP	0	6	62	327	633	955	1070	902	722	363	149	8	5197
MCROSS	13	21	113	414	708	1004	1113	932	763	432	211	72	5796
MIDDLEBOURNE 2 ESE	0	6	72	358	666	973	1104	930	741	402	172	22	5446
MORGANTOWN FAA AIRPORT	0	6	76	336	639	955	1094	927	732	384	169	30	5348
MORGANTOWN L AND D	0	8	69	355	645	949	1082	913	732	391	176	40	5360
NEW CUMBERLAND	0	0	57	310	627	967	1104	930	741	384	159	17	5296
OAK HILL	0	8	97	385	681	967	1082	924	753	408	193	46	5544
PARKERSBURG FAA AP	0	0	64	316	621	930	1073	896	694	347	150	16	5107
PARKERSBURG WSO R	0	0	59	300	606	911	1048	871	679	331	138	14	4957
PICKENS 1	29	38	154	456	756	1039	1156	986	834	498	263	97	6306
PINEVILLE	0	0	58	340	642	930	1035	865	691	358	150	25	5094
RAVENSWOOD LOCK PARK	0	0	63	283	576	871	1004	837	636	311	127	12	4720
ROWLESBURG 1	7	10	81	374	666	983	1097	932	763	422	195	43	5573
SPENCER	0	0	65	321	609	893	1014	848	667	327	136	22	4902
SPRUCE KNOB	45	50	179	471	768	1091	1206	1056	921	549	308	130	6774
UNION	8	14	113	400	696	983	1088	918	753	426	197	48	5644
WARDENSVILLE R M FARM	0	5	87	382	663	983	1101	946	772	423	186	24	5572
WEBSTER SPRINGS 1 E	0	0	66	330	621	899	1001	837	660	338	149	26	4927
WELLSBURG 3 NE	0	8	83	365	657	983	1122	944	756	402	173	20	5513
WESTON	0	0	66	362	657	952	1082	932	744	408	174	30	5407
WHEELING FILT PLANT	0	0	60	325	636	967	1113	960	769	414	187	19	5450
WHITE SULPHUR SPRINGS	0	0	77	341	648	933	1020	848	679	357	146	26	5075
WINFIELD LOCKS	0	0	43	286	594	890	1020	868	685	352	140	14	4892

WEST VIRGINIA

COOLING DEGREE DAY NORMALS (BASE 65 DEG F)

STATION	JAN	FEB	MAR	APR	MAY	JUN	JUL	AUG	SEP	OCT	NOV	DEC	ANN
ATHENS CONCORD COLLEGE	0	0	0	0	35	91	164	138	62	0	0	0	490
BAYARD	0	0	0	0	13	42	84	69	16	0	0	0	224
BECKLEY V A HOSPITAL	0	0	0	0	25	74	134	108	40	0	0	0	381
BECKLEY WSO //R	0	0	0	0	28	93	157	135	53	0	0	0	466
BLUEFIELD FAA AP	0	0	0	0	38	102	169	153	63	0	0	0	525
BLUESTONE DAM	0	0	0	0	56	148	257	244	112	10	0	0	827
BRANDONVILLE	0	0	0	0	19	72	127	111	33	0	0	0	362
BUCKHANNON 2 W	0	0	0	0	35	110	200	173	73	0	0	0	591
CAIRO 3 S	0	0	0	0	65	168	270	238	104	9	0	0	854
CANAAN VALLEY	0	0	0	0	9	36	71	53	11	0	0	0	180
CHARLESTON WSO R	0	0	0	7	91	196	295	270	129	19	0	0	1007
CLARKSBURG 1	0	0	0	0	43	150	246	218	95	6	0	0	758
ELKINS WSO //R	0	0	0	0	19	75	132	112	40	0	0	0	378
FAIRMONT	0	0	0	0	54	152	239	210	95	11	0	0	761
FLAT TOP	0	0	0	0	14	48	91	82	31	0	0	0	266
FRANKLIN 2 NE	0	0	0	0	31	103	181	164	64	0	0	0	543
GASSAWAY	0	0	0	0	62	166	273	242	106	12	0	0	861
GLENVILLE 2 NE	0	0	0	6	63	159	261	237	114	13	0	0	853
GRANTSVILLE 2 NW	0	0	0	0	65	172	276	252	113	9	0	0	887
HAMLIN	0	0	0	0	66	169	282	254	118	13	0	0	902
HOGSETT GALLIPOLIS DAM	0	0	0	0	61	164	276	249	114	12	0	0	876
HUNTINGTON WSO	0	0	0	17	103	218	322	285	152	24	0	0	1121
KEARNEYSVILLE 1 NW WBASO	0	0	0	0	51	153	276	239	90	7	0	0	816
LEWISBURG 2 SSW	0	0	0	0	42	111	191	164	72	0	0	0	580
LONDON LOCKS	0	0	0	6	77	187	299	282	140	24	0	0	1015
MADISON	0	0	0	6	83	186	298	276	131	20	0	0	1000
MANNINGTON 1 N	0	0	0	0	48	139	226	200	83	6	0	0	702
MARTINSBURG FAA AP	0	0	0	0	62	176	310	276	107	0	0	0	931
MCROSS	0	0	0	0	22	78	131	120	47	0	0	0	398
MIDDLEBOURNE 2 ESE	0	0	0	0	48	148	248	220	96	8	0	0	768
MORGANTOWN FAA AIRPORT	0	0	0	6	67	162	254	223	106	8	0	0	826
MORGANTOWN L AND D	0	0	0	0	61	157	240	209	87	14	0	0	768
NEW CUMBERLAND	0	0	0	0	56	161	260	237	105	10	0	0	829
OAK HILL	0	0	0	0	41	106	188	169	79	7	0	0	590
PARKERSBURG FAA AP	0	0	0	5	85	184	291	260	127	12	0	0	964
PARKERSBURG WSO R	0	0	0	0	92	197	304	273	131	14	0	0	1011
PICKENS 1	0	0	0	0	15	55	88	75	28	0	0	0	261
PINEVILLE	0	0	0	0	66	169	268	246	112	17	0	0	878
RAVENSWOOD LOCK PARK	0	0	0	0	84	189	295	267	144	17	0	0	996
ROWLESBURG 1	0	0	0	0	49	133	211	193	81	8	0	0	675
SPENCER	0	0	0	6	65	160	251	230	110	14	0	0	836
SPRUCE KNOB	0	0	0	0	23	73	101	88	38	6	0	0	329
UNION	0	0	0	0	29	90	172	150	62	0	0	0	503
WARDENSVILLE R M FARM	0	0	0	0	34	117	226	195	66	0	0	0	638
WEBSTER SPRINGS 1 E	0	0	0	0	62	146	233	202	87	8	0	0	738
WELLSBURG 3 NE	0	0	0	0	46	143	238	207	89	6	0	0	729
WESTON	0	0	0	0	53	153	251	225	90	9	0	0	781
WHEELING FILT PLANT	0	0	0	0	60	160	267	239	105	9	0	0	840
WHITE SULPHUR SPRINGS	0	0	0	0	47	128	226	212	95	6	0	0	714
WINFIELD LOCKS	0	0	0	0	75	188	310	285	139	16	0	0	1013

46 – WEST VIRGINIA

STATE-STATION NUMBER	STN TYP	NAME	LATITUDE DEG-MIN	LONGITUDE DEG-MIN	ELEVATION (FT)
46-0102	12	ALDERSON	N 3744	W 08038	1540
46-0355	13	ATHENS CONCORD COLLEGE	N 3726	W 08100	2590
46-0527	13	BAYARD	N 3916	W 07922	2375
46-0580	13	BECKLEY V A HOSPITAL	N 3747	W 08111	2330
46-0582	13	BECKLEY WSO //R	N 3747	W 08107	2504
46-0633	12	BELINGTON	N 3902	W 07956	1710
46-0661	12	BELVA	N 3814	W 08112	744
46-0687	12	BENS RUN 1 SSE	N 3928	W 08106	644
46-0921	13	BLUEFIELD FAA AP	N 3718	W 08113	2870
46-0939	13	BLUESTONE DAM	N 3739	W 08053	1388
46-1075	12	BRANCHLAND	N 3813	W 08212	600
46-1083	13	BRANDONVILLE	N 3940	W 07937	1798
46-1204	12	BRUSHY RUN	N 3850	W 07915	1375
46-1215	12	BUCKEYE	N 3811	W 08008	2100
46-1220	13	BUCKHANNON 2 W	N 3900	W 08016	1445
46-1282	12	BURNSVILLE	N 3852	W 08040	770
46-1328	13	CAIRO 3 S	N 3910	W 08110	680
46-1363	12	CAMDEN ON GAULEY	N 3822	W 08036	2030
46-1393	13	CANAAN VALLEY	N 3903	W 07926	3250
46-1526	12	CENTRALIA	N 3837	W 08035	1030
46-1570	13	CHARLESTON WSO R	N 3822	W 08136	939
46-1677	13	CLARKSBURG 1	N 3916	W 08021	977
46-1696	12	CLAY 1 SW	N 3827	W 08105	720
46-1723	12	CLENDENIN 1 SW	N 3829	W 08122	615
46-2054	12	CRESTON	N 3857	W 08117	660
46-2462	12	DRY CREEK	N 3752	W 08128	1264
46-2718	13	ELKINS WSO //R	N 3853	W 07951	1992
46-2920	13	FAIRMONT	N 3928	W 08008	1298
46-3072	13	FLAT TOP	N 3735	W 08106	3335
46-3215	13	FRANKLIN 2 NE	N 3840	W 07919	1900
46-3353	12	GARY //R	N 3722	W 08133	1426
46-3361	13	GASSAWAY	N 3840	W 08046	840
46-3544	13	GLENVILLE 2 NE	N 3857	W 08049	840
46-3648	13	GRANTSVILLE 2 NW	N 3856	W 08106	730
46-3846	13	HAMLIN	N 3817	W 08206	642
46-4200	13	HOGSETT GALLIPOLIS DAM	N 3841	W 08211	570
46-4393	13	HUNTINGTON WSO	N 3822	W 08233	827
46-4763	13	KEARNEYSVLE 1 NW WBASO	N 3923	W 07753	550
46-4816	12	KERMIT	N 3750	W 08224	629
46-5002	12	LAKE LYNN	N 3943	W 07951	900
46-5224	13	LEWISBURG 2 SSW	N 3746	W 08028	2185
46-5365	13	LONDON LOCKS	N 3812	W 08122	623
46-5563	13	MADISON	N 3803	W 08149	675
46-5621	13	MANNINGTON 1 N	N 3933	W 08021	975
46-5626	12	MANNINGTON 1 W	N 3932	W 08022	995
46-5707	13	MARTINSBURG FAA AP	N 3924	W 07759	537
46-5739	12	MATHIAS	N 3852	W 07852	1625
46-5871	13	MCROSS	N 3759	W 08045	2450
46-5963	13	MIDDLEBOURNE 2 ESE	N 3929	W 08052	750
46-6202	13	MORGANTOWN FAA AIRPORT	N 3939	W 07955	1240

46 — WEST VIRGINIA

STATE-STATION NUMBER	STN TYP	NAME		LATITUDE DEG-MIN	LONGITUDE DEG-MIN	ELEVATION (FT)
46-6212	13	MORGANTOWN L AND D		N 3937	W 07958	825
46-6442	13	NEW CUMBERLAND		N 4030	W 08036	750
46-6591	13	OAK HILL		N 3758	W 08109	1991
46-6849	13	PARKERSBURG FAA AP		N 3921	W 08126	831
46-6859	13	PARKERSBURG WSO	R	N 3916	W 08134	615
46-6982	12	PHILIPPI		N 3909	W 08002	1281
46-6991	13	PICKENS 1		N 3840	W 08013	2750
46-7029	13	PINEVILLE		N 3735	W 08132	1280
46-7207	12	PRINCETON		N 3722	W 08105	2410
46-7352	13	RAVENSWOOD LOCK PARK		N 3857	W 08146	584
46-7552	12	RIPLEY		N 3849	W 08143	610
46-7730	12	ROMNEY 3 NNE		N 3923	W 07844	640
46-7785	13	ROWLESBURG 1		N 3921	W 07940	1400
46-8384	13	SPENCER		N 3848	W 08121	964
46-8433	13	SPRUCE KNOB		N 3841	W 07931	3050
46-8662	12	SUTTON RESERVOIR		N 3839	W 08041	835
46-8807	12	THOMAS		N 3909	W 07930	3070
46-9011	13	UNION		N 3736	W 08032	1975
46-9086	12	VALLEY HEAD		N 3833	W 08002	2425
46-9104	12	VANDALIA		N 3856	W 08024	1120
46-9281	13	WARDENSVILLE R M FARM		N 3906	W 07835	960
46-9333	13	WEBSTER SPRINGS 1 E		N 3829	W 08025	1540
46-9368	13	WELLSBURG 3 NE		N 4018	W 08035	668
46-9436	13	WESTON		N 3902	W 08028	1026
46-9484	13	WHEELING FILT PLANT		N 4007	W 08042	665
46-9522	13	WHITE SULPHUR SPRINGS		N 3748	W 08018	1920
46-9605	12	WILLIAMSON		N 3740	W 08217	673
46-9610	12	WILLIAMSON 2		N 3741	W 08218	657
46-9683	13	WINFIELD LOCKS		N 3832	W 08155	571

46 - WEST VIRGINIA

LEGEND
○ = TEMPERATURE ONLY
× = PRECIPITATION ONLY
⊠ = TEMPERATURE & PRECIPITATION

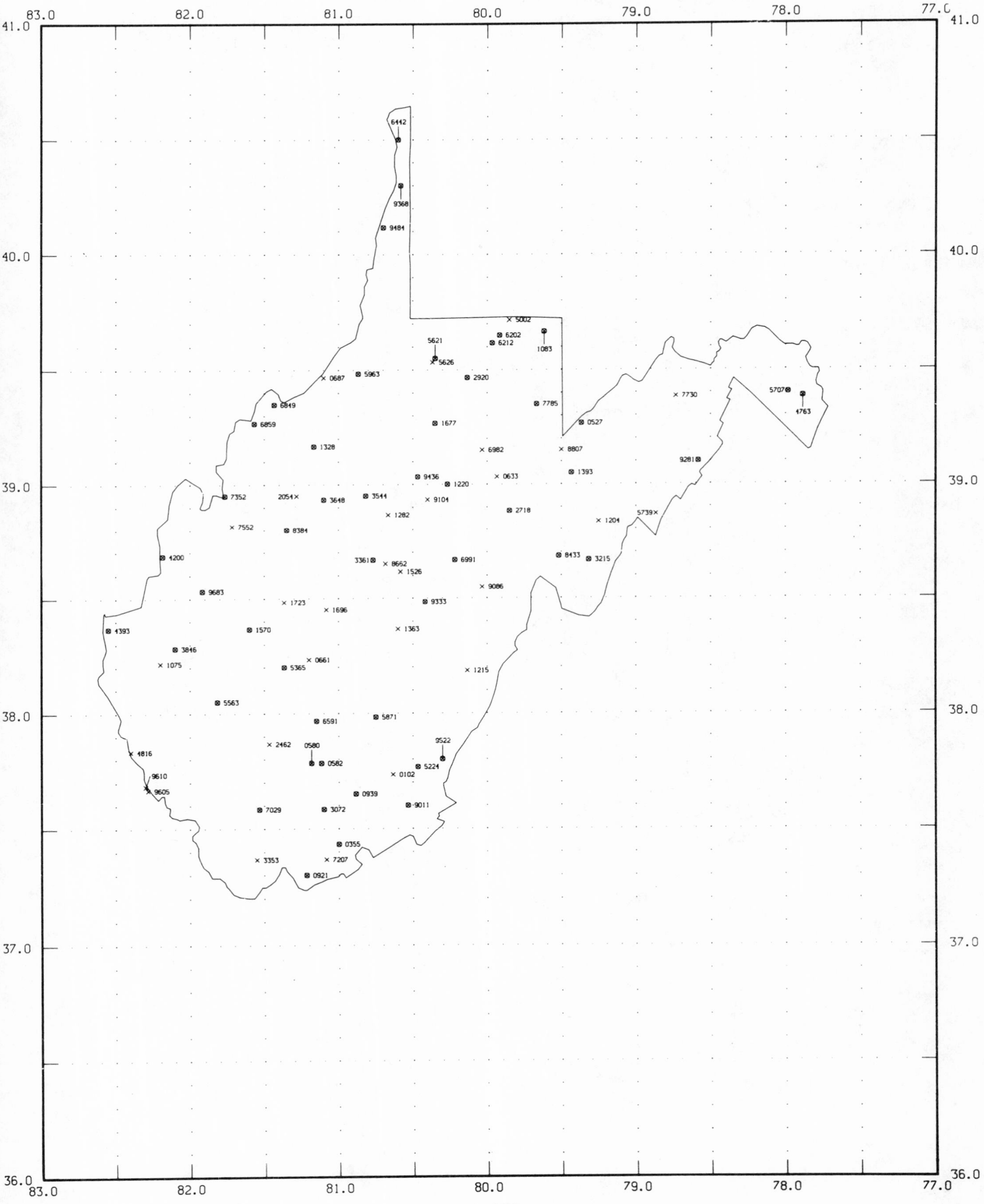

WISCONSIN

TEMPERATURE NORMALS (DEG F)

STATION		JAN	FEB	MAR	APR	MAY	JUN	JUL	AUG	SEP	OCT	NOV	DEC	ANN
AMERY	MAX	19.1	25.7	36.9	54.8	68.4	76.8	81.8	79.6	69.9	58.9	40.3	25.7	53.2
	MIN	-2.8	1.9	15.7	32.5	43.9	53.2	57.9	55.7	46.5	36.4	22.5	7.3	30.9
	MEAN	8.2	13.9	26.3	43.7	56.2	65.0	69.9	67.6	58.2	47.7	31.4	16.5	42.1
ANTIGO 1 SSW	MAX	21.5	27.0	37.2	53.9	67.8	75.8	80.3	77.8	68.6	57.9	40.0	26.3	52.8
	MIN	2.2	5.2	16.5	31.4	41.9	51.2	55.7	54.0	45.9	36.8	23.9	10.1	31.2
	MEAN	11.9	16.1	26.9	42.7	54.9	63.5	68.0	65.9	57.3	47.4	31.9	18.2	42.1
APPLETON	MAX	23.6	28.4	38.2	54.2	67.4	76.9	81.4	79.2	70.4	58.9	42.4	29.4	54.2
	MIN	7.0	11.0	21.5	34.7	45.9	55.9	61.3	59.5	51.0	41.0	27.9	14.8	36.0
	MEAN	15.3	19.7	29.9	44.5	56.7	66.4	71.4	69.4	60.7	49.9	35.2	22.1	45.1
ASHLAND EXP FARM	MAX	21.3	26.8	36.5	50.8	64.2	74.2	80.1	77.7	68.3	57.7	40.0	26.9	52.0
	MIN	-.4	2.5	14.4	28.0	38.2	47.9	54.5	53.3	45.3	35.5	23.2	8.5	29.2
	MEAN	10.5	14.7	25.4	39.5	51.2	61.1	67.3	65.5	56.8	46.6	31.6	17.7	40.7
BARABOO	MAX	24.8	30.4	40.5	56.6	69.1	78.0	82.2	80.3	71.7	60.8	44.1	30.9	55.8
	MIN	3.9	8.1	19.5	33.8	44.3	53.7	58.0	55.7	47.3	37.2	25.1	12.0	33.2
	MEAN	14.4	19.3	30.0	45.2	56.7	65.9	70.1	68.0	59.5	49.1	34.6	21.5	44.5
BAYFIELD 6 N	MAX	21.4	26.4	36.3	50.2	63.4	72.9	78.3	75.9	66.4	56.4	39.7	27.0	51.2
	MIN	4.1	6.7	17.2	30.2	39.4	48.6	55.7	55.0	47.4	38.3	25.5	12.1	31.7
	MEAN	12.8	16.5	26.8	40.3	51.5	60.8	67.0	65.5	56.9	47.4	32.6	19.5	41.5
BELOIT	MAX	27.9	33.2	44.0	60.8	72.8	81.5	84.9	82.6	75.7	64.6	47.3	33.5	59.1
	MIN	11.1	16.1	25.9	38.3	48.6	58.1	62.4	60.5	52.9	42.4	30.0	18.3	38.7
	MEAN	19.5	24.7	35.0	49.6	60.7	69.8	73.7	71.6	64.3	53.5	38.7	25.9	48.9
BLAIR	MAX	22.6	29.1	39.9	57.3	70.5	78.9	83.3	81.1	71.9	60.7	43.0	28.7	55.6
	MIN	1.0	5.9	18.3	33.4	44.1	53.7	58.0	55.9	46.7	36.7	24.2	10.5	32.4
	MEAN	11.8	17.5	29.1	45.4	57.4	66.3	70.7	68.5	59.3	48.7	33.7	19.6	44.0
BLOOMER	MAX	20.2	26.8	38.0	56.0	69.2	77.3	82.3	79.8	70.3	58.8	40.5	26.2	53.8
	MIN	1.0	5.9	18.1	33.1	44.2	53.5	58.3	56.1	47.2	37.3	23.7	10.0	32.4
	MEAN	10.6	16.4	28.1	44.6	56.8	65.4	70.3	68.0	58.8	48.1	32.1	18.1	43.1
BRODHEAD	MAX	25.8	30.9	41.7	58.3	70.7	80.0	84.3	82.1	74.6	63.0	46.0	32.0	57.5
	MIN	6.7	11.4	22.2	35.5	45.7	55.3	59.5	57.3	48.8	38.1	26.6	14.1	35.1
	MEAN	16.3	21.2	32.0	46.9	58.2	67.7	71.9	69.7	61.7	50.6	36.4	23.1	46.3
BRULE ISLAND	MAX	20.9	25.9	36.3	51.6	66.0	74.6	79.2	76.3	66.1	55.4	38.2	26.0	51.4
	MIN	-2.0	-1.0	10.6	26.1	36.8	46.0	50.8	49.3	41.5	32.8	21.2	6.7	26.6
	MEAN	9.5	12.5	23.5	38.9	51.4	60.3	65.0	62.8	53.8	44.1	29.7	16.4	39.0
BURLINGTON	MAX	26.1	30.5	40.6	56.0	68.4	78.0	82.6	80.7	73.5	61.9	45.9	32.2	56.4
	MIN	8.7	12.6	23.3	35.6	45.0	54.6	59.5	57.7	49.7	38.9	27.9	16.1	35.8
	MEAN	17.4	21.6	32.0	45.8	56.7	66.3	71.1	69.2	61.6	50.4	36.9	24.2	46.1
CHILTON	MAX	24.7	29.3	39.5	56.4	69.5	78.5	82.9	81.0	72.9	61.3	44.1	30.4	55.9
	MIN	7.5	10.8	21.1	34.5	44.3	53.9	59.0	57.6	49.8	40.3	28.2	15.2	35.2
	MEAN	16.2	20.1	30.3	45.5	57.0	66.2	71.0	69.4	61.4	50.8	36.2	22.9	45.6
CODDINGTON 1 E	MAX	22.1	27.8	38.1	55.6	68.6	76.1	80.1	78.1	69.5	59.0	41.9	27.9	53.7
	MIN	.7	4.8	16.2	31.1	41.3	50.8	55.0	52.9	44.7	35.1	22.5	9.1	30.4
	MEAN	11.4	16.3	27.2	43.4	55.0	63.5	67.6	65.5	57.1	47.1	32.2	18.6	42.1
CRIVITZ HIGH FALLS	MAX	24.7	29.2	38.8	54.4	68.3	77.0	81.6	79.0	69.8	59.4	42.2	29.5	54.5
	MIN	1.4	3.9	15.4	29.8	40.9	50.5	55.5	54.2	45.3	35.9	24.1	10.4	30.6
	MEAN	13.1	16.6	27.1	42.1	54.7	63.8	68.5	66.6	57.6	47.7	33.2	20.0	42.6
CUMBERLAND	MAX	19.4	26.6	37.7	55.7	69.4	77.7	82.2	79.8	69.7	58.6	39.6	25.4	53.5
	MIN	-.9	3.5	16.0	32.3	44.5	54.3	59.3	57.0	48.0	37.7	23.5	8.5	32.0
	MEAN	9.3	15.1	26.9	44.0	57.0	66.0	70.8	68.4	58.8	48.2	31.6	17.0	42.8
DALTON	MAX	24.7	29.9	40.4	57.2	70.1	78.8	82.9	81.2	72.5	61.4	44.3	30.5	56.2
	MIN	5.9	10.1	20.8	34.3	44.9	54.2	58.8	56.7	48.7	39.2	26.3	13.5	34.5
	MEAN	15.3	20.0	30.6	45.8	57.5	66.5	70.9	69.0	60.6	50.3	35.3	22.0	45.3

WISCONSIN

TEMPERATURE NORMALS (DEG F)

STATION		JAN	FEB	MAR	APR	MAY	JUN	JUL	AUG	SEP	OCT	NOV	DEC	ANN
DANBURY	MAX	19.7	27.0	38.1	55.4	69.0	76.6	81.2	78.7	68.7	58.2	39.5	25.5	53.1
	MIN	-4.1	.9	14.3	29.5	40.2	49.9	55.2	52.9	44.4	34.3	20.8	5.6	28.7
	MEAN	7.8	14.0	26.2	42.4	54.6	63.3	68.2	65.8	56.6	46.3	30.2	15.6	40.9
DARLINGTON	MAX	26.4	32.2	42.9	59.2	70.9	79.0	83.1	81.4	73.9	63.4	46.1	32.1	57.6
	MIN	6.7	11.7	22.3	35.3	45.9	55.3	59.7	57.8	49.2	38.4	26.6	14.4	35.3
	MEAN	16.6	22.0	32.6	47.3	58.4	67.2	71.4	69.6	61.6	50.9	36.4	23.3	46.4
DODGEVILLE 1 NE	MAX	24.4	29.9	40.5	57.3	69.2	77.8	82.0	80.2	71.9	60.7	44.0	30.2	55.7
	MIN	7.4	12.3	22.4	35.7	46.5	55.7	60.4	58.6	50.2	40.2	27.1	14.6	35.9
	MEAN	15.9	21.1	31.5	46.5	57.9	66.7	71.2	69.4	61.1	50.4	35.5	22.4	45.8
EAU CLAIRE FAA AP	MAX	19.8	26.4	37.6	55.7	68.9	77.7	82.3	79.7	69.9	58.9	40.7	26.4	53.7
	MIN	-.1	4.7	17.5	32.9	44.6	54.4	59.1	56.8	47.2	36.9	23.2	8.9	32.2
	MEAN	9.8	15.5	27.6	44.4	56.8	66.1	70.8	68.3	58.6	47.9	32.0	17.7	43.0
EL DORADO	MAX	24.3	29.1	39.4	56.1	69.2	78.0	82.6	80.7	72.3	60.7	43.6	30.0	55.5
	MIN	6.4	10.5	21.4	35.0	45.0	54.5	58.7	56.7	48.8	39.5	27.0	13.8	34.8
	MEAN	15.3	19.8	30.4	45.6	57.1	66.3	70.7	68.7	60.6	50.1	35.3	21.9	45.2
FOND DU LAC	MAX	24.4	29.0	39.2	55.3	68.4	77.6	81.9	79.8	71.5	60.3	43.7	30.3	55.1
	MIN	7.2	11.2	21.7	35.4	46.3	55.8	60.7	58.8	50.7	40.6	28.0	14.7	35.9
	MEAN	15.8	20.1	30.5	45.4	57.4	66.7	71.3	69.4	61.1	50.5	35.9	22.6	45.6
FORT ATKINSON	MAX	26.4	31.5	42.1	58.6	71.0	80.0	84.3	82.2	74.6	63.0	46.0	32.1	57.7
	MIN	8.2	12.8	23.4	36.6	46.8	56.2	60.7	58.9	51.1	40.9	28.8	16.1	36.7
	MEAN	17.3	22.2	32.8	47.6	58.9	68.2	72.5	70.6	62.9	52.0	37.4	24.1	47.2
GENOA DAM 8	MAX	24.7	30.7	41.3	57.7	70.8	79.4	83.6	81.4	72.8	61.8	44.5	30.7	56.6
	MIN	6.5	11.1	22.7	37.4	48.7	57.9	62.5	60.3	51.8	41.8	28.3	15.3	37.0
	MEAN	15.6	21.0	32.0	47.6	59.8	68.7	73.1	70.9	62.3	51.8	36.4	23.0	46.9
GERMANTOWN	MAX	26.0	30.6	40.2	55.4	67.8	77.3	82.0	80.3	72.6	61.3	45.4	32.0	55.9
	MIN	8.2	12.5	22.3	34.1	43.4	52.9	58.0	56.9	49.2	39.5	27.7	15.2	35.0
	MEAN	17.1	21.6	31.3	44.8	55.6	65.1	70.0	68.6	60.9	50.4	36.5	23.6	45.5
GORDON	MAX	19.9	27.3	38.2	55.0	68.7	77.5	82.2	79.2	68.8	58.0	39.2	25.5	53.3
	MIN	-5.5	-1.1	12.1	27.8	38.6	48.6	53.9	51.6	43.3	33.5	19.9	4.6	27.3
	MEAN	7.2	13.1	25.2	41.5	53.7	63.1	68.1	65.4	56.1	45.8	29.6	15.1	40.3
GRANTSBURG	MAX	18.6	25.8	37.1	55.4	68.5	76.7	81.6	79.1	68.6	58.1	40.0	25.3	52.9
	MIN	-4.5	.6	14.3	30.6	41.9	51.6	56.2	54.2	44.9	34.8	21.5	5.2	29.3
	MEAN	7.1	13.3	25.7	43.0	55.2	64.2	68.9	66.6	56.7	46.5	30.7	15.3	41.1
GREEN BAY WSO AP	MAX	22.5	26.9	37.0	53.7	66.6	76.2	80.9	78.7	69.8	58.5	42.0	28.5	53.4
	MIN	5.4	8.7	20.1	33.6	43.5	53.1	58.1	56.3	47.9	38.2	26.3	13.0	33.7
	MEAN	14.0	17.8	28.6	43.7	55.1	64.7	69.5	67.5	58.9	48.4	34.2	20.8	43.6
HANCOCK EXP FARM	MAX	23.5	29.0	39.4	57.2	70.6	79.1	82.8	80.9	72.1	60.8	43.2	29.0	55.6
	MIN	2.5	6.7	18.1	32.9	44.0	53.6	57.9	56.0	47.8	37.9	24.5	10.8	32.7
	MEAN	13.0	17.9	28.7	45.1	57.4	66.4	70.4	68.5	59.9	49.4	33.9	19.9	44.2
HATFIELD HYDRO PLANT	MAX	24.1	30.8	41.4	58.6	72.0	79.8	84.0	82.1	73.3	62.5	43.8	29.6	56.8
	MIN	-.1	4.4	16.4	31.1	41.7	51.0	55.7	53.6	44.9	35.3	22.7	9.4	30.5
	MEAN	12.0	17.6	28.9	44.9	56.9	65.4	69.9	67.9	59.1	48.9	33.3	19.5	43.7
HILLSBORO	MAX	25.2	31.5	41.9	58.1	71.1	79.8	84.3	82.1	73.6	62.7	45.0	31.1	57.2
	MIN	3.6	8.0	19.7	33.2	44.3	54.2	58.5	56.2	47.2	37.8	25.1	11.8	33.3
	MEAN	14.4	19.8	30.9	45.7	57.7	67.0	71.4	69.2	60.4	50.3	35.1	21.5	45.3
HOLCOMBE	MAX	21.4	28.0	38.9	56.2	69.8	77.8	82.4	79.9	70.7	59.9	41.3	27.2	54.5
	MIN	.0	4.0	16.9	32.8	43.7	52.9	57.6	55.5	47.1	37.5	23.9	9.6	31.8
	MEAN	10.7	16.0	27.9	44.5	56.8	65.4	70.0	67.7	58.9	48.7	32.6	18.4	43.1
JANESVILLE	MAX	27.5	32.7	43.3	59.9	72.5	81.6	85.5	83.4	75.6	63.8	46.7	33.2	58.8
	MIN	10.3	15.1	25.2	37.5	48.1	57.7	62.1	60.3	52.1	41.7	29.6	17.6	38.1
	MEAN	18.9	23.9	34.3	48.7	60.3	69.7	73.9	71.8	63.9	52.8	38.2	25.4	48.5

WISCONSIN

TEMPERATURE NORMALS (DEG F)

STATION		JAN	FEB	MAR	APR	MAY	JUN	JUL	AUG	SEP	OCT	NOV	DEC	ANN
KENOSHA	MAX	28.7	32.9	41.0	53.5	64.0	74.3	79.4	78.8	72.0	61.1	46.8	34.5	55.6
	MIN	12.1	16.7	25.7	36.0	44.7	54.2	60.6	60.5	52.9	42.5	30.3	19.0	37.9
	MEAN	20.4	24.8	33.4	44.8	54.4	64.3	70.1	69.6	62.5	51.8	38.6	26.8	46.8
KEWAUNEE	MAX	25.3	29.0	37.7	50.4	61.4	71.4	77.6	76.6	68.5	57.5	43.4	31.0	52.5
	MIN	10.0	12.6	22.9	33.8	42.3	51.2	58.0	58.2	50.4	40.8	29.5	17.3	35.6
	MEAN	17.7	20.8	30.3	42.1	51.9	61.4	67.8	67.5	59.5	49.2	36.5	24.1	44.1
LA CROSSE FAA AP	MAX	23.0	29.4	40.0	57.5	70.2	79.1	83.5	81.4	72.0	60.8	43.1	29.2	55.8
	MIN	5.0	9.9	21.8	36.9	48.5	57.8	62.4	60.3	51.2	41.0	27.4	13.8	36.3
	MEAN	14.0	19.7	30.9	47.2	59.4	68.5	73.0	70.9	61.6	50.9	35.3	21.5	46.1
LADYSMITH	MAX	20.9	27.6	38.3	55.6	68.8	76.6	80.6	77.9	68.4	58.3	40.0	26.5	53.3
	MIN	.3	4.7	17.1	31.7	42.5	51.7	56.5	54.5	46.3	36.8	23.5	9.5	31.3
	MEAN	10.6	16.1	27.7	43.6	55.7	64.1	68.6	66.2	57.4	47.6	31.8	18.0	42.3
LAKE GENEVA	MAX	27.2	32.6	42.8	58.4	71.0	80.5	84.5	82.4	74.5	62.5	46.1	32.7	57.9
	MIN	10.7	15.0	24.3	36.4	46.7	56.5	61.7	60.5	52.4	42.0	29.7	17.9	37.8
	MEAN	19.0	23.8	33.6	47.4	58.9	68.5	73.2	71.5	63.5	52.3	37.9	25.3	47.9
LAKE MILLS	MAX	26.4	31.8	42.0	58.3	70.8	79.2	83.3	81.6	74.0	62.8	45.8	32.0	57.3
	MIN	7.7	12.1	22.6	36.2	46.9	56.3	61.0	59.0	51.1	41.4	28.6	15.7	36.6
	MEAN	17.1	22.0	32.3	47.3	58.9	67.8	72.2	70.3	62.6	52.1	37.2	23.9	47.0
LANCASTER	MAX	24.4	30.7	41.2	58.1	70.3	79.2	83.3	81.4	73.1	62.0	44.4	30.6	56.6
	MIN	7.4	12.8	23.3	36.7	47.7	57.0	61.2	59.6	51.4	41.0	27.5	15.2	36.7
	MEAN	16.0	21.8	32.3	47.4	59.0	68.1	72.3	70.5	62.3	51.5	36.0	22.9	46.7
LONE ROCK FAA AIRPORT //R	MAX	24.7	30.7	41.5	58.3	70.8	79.6	83.4	81.3	73.0	61.8	44.7	30.9	56.7
	MIN	3.4	8.5	20.5	34.4	45.0	54.8	58.8	56.4	47.3	36.7	24.7	11.7	33.5
	MEAN	14.1	19.7	31.0	46.4	58.0	67.2	71.1	68.8	60.2	49.3	34.8	21.3	45.2
LONG LAKE DAM	MAX	20.9	26.8	37.4	52.7	66.7	74.6	79.0	76.4	66.7	56.2	38.4	25.7	51.8
	MIN	-2.3	-.7	9.9	25.8	37.1	47.0	51.8	49.8	42.0	33.0	20.3	5.9	26.6
	MEAN	9.3	13.0	23.7	39.3	51.9	60.8	65.4	63.1	54.4	44.6	29.4	15.8	39.2
LYNXVILLE DAM 9	MAX	25.6	31.5	42.0	58.5	71.4	80.2	84.3	82.0	73.4	62.2	45.0	31.4	57.3
	MIN	6.8	11.7	23.1	37.7	49.1	58.6	63.1	61.1	52.5	41.9	28.4	15.7	37.5
	MEAN	16.2	21.6	32.6	48.1	60.3	69.4	73.7	71.6	63.0	52.1	36.7	23.6	47.4
MADELINE ISLAND	MAX	21.6	26.0	35.2	47.9	59.4	69.4	76.5	75.1	66.3	56.2	40.3	28.0	50.2
	MIN	5.3	6.0	15.7	29.0	37.2	46.3	53.9	54.3	47.3	38.8	27.1	14.9	31.3
	MEAN	13.5	16.0	25.5	38.5	48.3	57.9	65.2	64.8	56.9	47.5	33.7	21.4	40.8
MADISON WSO //R	MAX	24.5	30.0	40.8	57.5	69.8	78.8	82.8	80.6	72.3	61.1	44.1	30.6	56.1
	MIN	6.7	11.0	21.5	34.1	44.2	53.8	58.3	56.3	47.8	37.8	26.0	14.1	34.3
	MEAN	15.6	20.5	31.2	45.8	57.0	66.3	70.6	68.5	60.1	49.5	35.1	22.4	45.2
MANITOWOC	MAX	26.0	29.7	38.2	52.2	64.3	74.5	79.6	78.2	70.1	58.6	44.0	31.8	53.9
	MIN	10.5	14.0	23.5	34.8	44.0	53.4	59.8	59.3	51.6	41.9	29.7	17.6	36.7
	MEAN	18.3	21.9	30.9	43.6	54.2	63.9	69.7	68.8	60.9	50.3	36.9	24.7	45.3
MARINETTE	MAX	26.7	30.9	40.6	55.5	68.4	78.5	83.5	81.1	72.1	60.5	44.7	32.0	56.2
	MIN	9.0	11.5	21.1	33.4	44.0	53.7	59.0	57.6	50.0	40.7	28.8	16.5	35.4
	MEAN	17.9	21.2	30.8	44.5	56.2	66.1	71.3	69.4	61.1	50.6	36.8	24.3	45.9
MARSHFIELD EXP FARM	MAX	21.0	26.9	37.2	54.7	68.0	76.2	80.7	78.5	69.7	58.6	40.7	26.9	53.3
	MIN	1.8	6.5	17.6	32.2	42.6	51.9	56.2	53.8	45.4	35.9	23.3	10.0	31.4
	MEAN	11.5	16.7	27.4	43.5	55.3	64.0	68.5	66.2	57.6	47.3	32.0	18.5	42.4
MATHER 5 NW	MAX	22.3	28.8	38.9	55.3	68.7	77.1	81.6	79.5	70.4	59.8	42.4	28.4	54.4
	MIN	1.5	5.6	17.7	33.0	43.7	52.9	57.0	54.5	45.6	35.9	23.6	10.4	31.8
	MEAN	11.9	17.2	28.3	44.2	56.3	65.0	69.3	67.0	58.0	47.9	33.0	19.4	43.1
MEDFORD	MAX	19.6	25.7	35.7	52.8	66.3	74.4	79.3	76.9	67.5	56.8	38.8	25.2	51.6
	MIN	.1	4.1	15.9	31.5	42.1	51.5	56.1	54.2	45.6	36.0	22.8	8.6	30.7
	MEAN	9.9	14.9	25.8	42.2	54.2	63.0	67.7	65.6	56.6	46.4	30.8	16.9	41.2

WISCONSIN

TEMPERATURE NORMALS (DEG F)

STATION		JAN	FEB	MAR	APR	MAY	JUN	JUL	AUG	SEP	OCT	NOV	DEC	ANN
MELLEN	MAX	20.4	26.3	36.8	52.1	66.1	74.4	79.4	76.4	67.1	56.7	38.7	25.5	51.7
	MIN	-1.7	.7	12.5	28.5	39.5	48.8	54.4	52.5	44.5	35.3	22.3	6.9	28.7
	MEAN	9.3	13.5	24.6	40.3	52.8	61.6	66.9	64.5	55.8	46.0	30.5	16.2	40.2
MENOMONIE	MAX	22.4	28.9	40.2	58.2	71.5	79.9	84.6	82.1	72.7	61.7	42.9	28.5	56.1
	MIN	1.8	6.9	18.9	33.8	44.9	54.6	58.9	56.8	48.1	38.1	24.7	11.1	33.2
	MEAN	12.1	17.9	29.5	46.0	58.3	67.3	71.8	69.5	60.5	49.9	33.8	19.8	44.7
MERRILL	MAX	21.5	27.4	37.7	54.7	68.5	76.6	81.3	78.5	68.8	57.9	40.2	26.7	53.3
	MIN	.0	3.3	15.1	30.8	41.5	50.9	55.3	53.3	44.9	35.5	22.8	8.6	30.2
	MEAN	10.8	15.4	26.4	42.8	55.0	63.8	68.4	66.0	56.9	46.7	31.5	17.7	41.8
MILWAUKEE MT MARY COL	MAX	27.2	31.6	41.4	56.7	69.3	78.6	83.3	81.7	74.1	61.9	46.1	33.1	57.1
	MIN	11.3	15.6	24.8	36.2	46.1	55.7	61.3	60.2	52.3	42.1	30.0	18.3	37.8
	MEAN	19.3	23.6	33.1	46.5	57.7	67.2	72.4	71.0	63.2	52.0	38.1	25.7	47.5
MILWAUKEE WSO R	MAX	26.0	30.1	39.2	53.5	64.8	75.0	79.8	78.4	71.2	59.9	44.7	32.0	54.6
	MIN	11.3	15.8	24.9	35.6	44.7	54.7	61.1	60.2	52.5	41.9	29.9	18.2	37.6
	MEAN	18.7	23.0	32.1	44.6	54.8	64.9	70.5	69.3	61.9	50.9	37.3	25.1	46.1
MINOCQUA DAM	MAX	20.6	26.6	36.9	52.1	66.5	74.1	78.4	75.8	66.4	56.1	38.2	25.1	51.4
	MIN	-1.9	.8	11.9	27.7	39.6	49.4	54.4	52.3	44.3	34.9	21.4	6.3	28.4
	MEAN	9.4	13.7	24.4	39.9	53.1	61.8	66.4	64.1	55.4	45.5	29.8	15.8	39.9
MONDOVI	MAX	22.8	29.5	40.1	57.9	70.8	79.1	83.3	81.0	71.9	61.3	43.0	28.7	55.8
	MIN	1.3	6.6	18.7	33.7	44.7	54.8	58.9	56.8	48.0	37.6	23.8	10.3	32.9
	MEAN	12.1	18.0	29.5	45.9	57.8	67.0	71.1	68.9	60.0	49.5	33.4	19.5	44.4
MONTELLO	MAX	23.9	29.1	39.7	56.6	69.4	78.7	82.8	80.5	71.8	60.5	43.5	30.1	55.6
	MIN	5.0	9.5	20.6	34.7	45.5	54.7	59.3	56.5	47.9	37.9	25.6	13.2	34.2
	MEAN	14.4	19.3	30.2	45.7	57.5	66.7	71.1	68.5	59.9	49.2	34.5	21.7	44.9
NEILLSVILLE 3 SW	MAX	21.9	28.3	38.6	55.7	69.5	77.6	81.9	79.9	70.7	59.6	41.4	27.8	54.4
	MIN	.1	5.1	17.0	32.3	43.0	52.0	56.3	54.5	45.9	36.0	22.9	9.5	31.2
	MEAN	11.1	16.7	27.8	44.0	56.3	64.8	69.1	67.2	58.3	47.8	32.2	18.7	42.8
NEW LONDON	MAX	24.4	29.5	39.8	56.9	70.4	79.1	83.5	81.1	72.6	61.3	43.6	29.8	56.0
	MIN	4.7	8.4	19.4	34.2	44.6	54.1	58.4	56.3	48.2	38.4	26.0	12.8	33.8
	MEAN	14.6	19.0	29.6	45.6	57.5	66.6	71.0	68.8	60.4	49.8	34.8	21.3	44.9
NORTH PELICAN	MAX	19.3	25.0	35.5	51.2	66.2	73.4	77.1	73.9	64.2	54.8	37.4	24.4	50.2
	MIN	-1.1	.7	12.1	28.5	39.9	49.9	54.7	53.0	44.7	35.6	22.0	7.1	28.9
	MEAN	9.1	12.9	23.8	39.8	53.1	61.7	65.9	63.5	54.5	45.2	29.7	15.8	39.6
OCONOMOWOC	MAX	25.4	30.3	40.7	57.2	70.0	78.8	83.3	81.1	73.0	61.3	44.9	31.4	56.5
	MIN	8.3	12.9	23.1	36.0	46.5	56.0	60.7	59.1	51.1	40.9	28.6	16.0	36.6
	MEAN	16.9	21.6	32.0	46.7	58.2	67.4	72.0	70.1	62.0	51.1	36.8	23.8	46.6
OCONTO 4 W	MAX	25.1	29.4	38.9	54.2	67.4	76.9	81.7	79.5	70.9	59.8	43.5	30.5	54.8
	MIN	5.6	8.0	18.5	32.0	42.6	52.6	57.6	55.7	47.3	37.9	26.1	13.4	33.1
	MEAN	15.4	18.7	28.7	43.1	55.0	64.7	69.7	67.6	59.1	48.9	34.9	22.0	44.0
OSHKOSH	MAX	24.2	29.1	39.0	54.9	68.1	77.5	82.2	80.2	71.6	59.9	43.2	30.0	55.0
	MIN	6.6	10.2	20.6	34.4	45.8	55.7	60.6	58.8	50.5	40.3	27.2	14.1	35.4
	MEAN	15.4	19.7	29.8	44.7	57.0	66.7	71.4	69.5	61.1	50.1	35.2	22.1	45.2
OWEN	MAX	18.1	23.9	34.6	51.9	65.9	74.8	79.5	77.4	68.0	57.0	39.2	24.7	51.3
	MIN	-2.5	1.4	14.6	31.6	41.9	51.2	55.4	53.0	44.3	34.5	21.5	6.6	29.5
	MEAN	7.8	12.7	24.6	41.7	53.9	63.0	67.5	65.2	56.2	45.8	30.4	15.7	40.4
PARK FALLS	MAX	18.6	25.0	35.7	51.7	64.9	73.3	77.5	75.0	65.5	55.3	37.2	23.7	50.3
	MIN	.3	4.0	15.5	30.7	42.0	51.4	56.1	54.0	45.5	36.4	22.7	8.4	30.6
	MEAN	9.5	14.5	25.7	41.3	53.4	62.4	66.9	64.5	55.5	45.9	30.0	16.0	40.5
PINE RIVER 3 NE	MAX	24.2	29.4	39.6	56.6	69.5	78.5	83.2	81.0	72.1	60.8	43.7	29.9	55.7
	MIN	5.2	9.2	20.0	34.0	44.0	53.4	58.2	56.3	48.0	38.3	26.0	12.9	33.8
	MEAN	14.8	19.4	29.8	45.3	56.8	65.9	70.7	68.7	60.0	49.6	34.9	21.4	44.8

WISCONSIN

TEMPERATURE NORMALS (DEG F)

STATION		JAN	FEB	MAR	APR	MAY	JUN	JUL	AUG	SEP	OCT	NOV	DEC	ANN
PITTSVILLE	MAX	23.4	29.3	39.7	56.8	70.2	78.1	82.3	80.1	71.0	60.2	42.5	28.8	55.2
	MIN	1.8	5.8	17.5	32.7	43.0	52.0	56.3	54.0	45.7	36.5	23.9	10.9	31.7
	MEAN	12.6	17.5	28.6	44.8	56.6	65.1	69.3	67.1	58.4	48.4	33.2	19.9	43.5
PLATTEVILLE	MAX	25.5	31.5	42.5	59.5	71.4	80.2	84.3	82.4	74.2	62.6	45.2	31.5	57.6
	MIN	7.9	13.1	23.4	36.3	47.0	56.2	60.3	58.4	50.0	39.9	27.4	15.4	36.3
	MEAN	16.7	22.4	32.9	47.9	59.2	68.2	72.4	70.4	62.1	51.3	36.3	23.5	46.9
PLYMOUTH	MAX	25.3	29.7	39.1	54.6	67.3	76.9	81.6	80.1	71.7	59.9	44.0	31.0	55.1
	MIN	9.0	13.0	22.4	34.4	44.1	53.8	59.3	58.1	50.1	40.6	28.4	16.2	35.8
	MEAN	17.2	21.4	30.8	44.6	55.7	65.4	70.5	69.1	60.9	50.3	36.2	23.6	45.5
PORTAGE	MAX	26.0	31.5	41.9	58.7	71.4	79.7	84.0	81.7	73.6	62.7	45.7	31.6	57.4
	MIN	6.5	11.0	21.9	35.3	45.5	54.8	59.2	57.1	48.7	39.2	26.9	14.6	35.1
	MEAN	16.3	21.3	31.9	47.0	58.5	67.3	71.6	69.4	61.2	50.9	36.3	23.1	46.2
PRAIRIE DU CHIEN	MAX	26.7	32.9	43.8	60.8	73.4	82.1	86.1	84.2	75.9	64.5	46.6	32.6	59.1
	MIN	7.1	12.1	23.4	37.2	48.5	58.0	62.0	60.1	51.5	41.1	28.1	15.5	37.1
	MEAN	16.9	22.6	33.6	49.0	61.0	70.1	74.1	72.2	63.8	52.9	37.3	24.1	48.1
PRAIRIE DU SAC 2 N //	MAX	24.6	30.0	40.0	56.1	69.0	78.4	82.7	80.5	71.9	60.5	43.7	30.3	55.6
	MIN	6.6	10.9	22.1	36.4	47.8	57.3	61.9	59.9	51.5	40.9	27.7	14.7	36.5
	MEAN	15.6	20.5	31.0	46.3	58.4	67.9	72.3	70.2	61.7	50.7	35.7	22.5	46.1
PRENTICE 1 N	MAX	19.8	26.0	36.9	53.1	66.7	74.4	78.7	76.4	67.2	56.8	38.5	24.9	51.6
	MIN	-2.6	.9	12.9	29.2	39.3	48.0	52.1	50.2	42.5	34.0	20.8	6.0	27.8
	MEAN	8.6	13.5	25.0	41.2	53.0	61.2	65.4	63.3	54.9	45.4	29.7	15.5	39.7
RACINE	MAX	28.3	32.4	41.3	54.9	66.6	77.1	81.6	80.5	73.3	61.7	46.5	33.9	56.5
	MIN	12.6	16.8	25.7	36.5	45.6	55.5	61.7	60.8	53.7	43.4	31.1	19.3	38.6
	MEAN	20.4	24.6	33.5	45.7	56.1	66.3	71.7	70.7	63.5	52.6	38.8	26.6	47.5
RAINBOW RESERVOIR	MAX	20.4	26.0	36.3	52.2	66.3	74.0	78.4	75.8	66.5	56.1	38.3	25.1	51.3
	MIN	-1.2	.9	12.0	27.6	39.4	49.0	53.7	52.0	44.0	34.7	21.7	7.0	28.4
	MEAN	9.6	13.5	24.2	39.9	52.9	61.5	66.1	63.9	55.3	45.4	30.0	16.1	39.9
REEDSBURG	MAX	26.1	31.7	42.0	58.7	70.8	79.0	83.2	81.3	73.0	62.2	45.0	31.4	57.0
	MIN	6.0	10.2	21.3	34.8	45.2	54.5	58.9	56.9	48.6	39.0	26.6	13.9	34.7
	MEAN	16.1	20.9	31.7	46.8	58.0	66.8	71.1	69.1	60.8	50.6	35.8	22.7	45.9
REST LAKE	MAX	19.2	25.0	35.7	52.0	66.3	73.5	77.4	74.7	65.3	55.4	37.3	24.3	50.5
	MIN	-.6	2.1	13.2	28.8	40.7	50.6	55.4	53.7	45.6	36.1	22.4	8.0	29.7
	MEAN	9.3	13.6	24.5	40.4	53.5	62.1	66.4	64.2	55.5	45.8	29.9	16.2	40.1
RHINELANDER	MAX	20.6	26.2	36.8	53.1	67.0	75.3	79.7	77.0	67.4	56.3	38.6	25.6	52.0
	MIN	.3	3.0	14.7	30.1	41.6	51.3	55.9	53.9	45.5	36.4	23.2	8.4	30.4
	MEAN	10.5	14.6	25.8	41.6	54.4	63.3	67.9	65.5	56.5	46.4	30.9	17.1	41.2
RICE LAKE	MAX	19.7	26.6	37.4	55.3	68.3	76.6	81.2	78.9	69.0	58.6	40.0	25.7	53.1
	MIN	-2.2	2.7	15.6	31.2	42.5	52.1	57.1	54.9	45.9	35.8	22.3	7.4	30.4
	MEAN	8.8	14.7	26.5	43.3	55.4	64.4	69.2	66.9	57.5	47.3	31.2	16.6	41.8
RICHLAND CENTER	MAX	26.3	32.6	43.1	59.8	72.1	80.4	84.9	82.8	74.3	63.2	45.6	31.8	58.1
	MIN	5.6	10.5	21.8	35.0	45.4	54.7	59.4	57.3	49.0	38.3	26.0	13.8	34.7
	MEAN	16.0	21.6	32.5	47.4	58.8	67.6	72.2	70.1	61.7	50.8	35.8	22.8	46.4
RIVER FALLS	MAX	20.9	27.4	38.5	56.8	70.0	78.5	83.0	80.5	70.9	60.2	41.7	27.4	54.7
	MIN	1.1	6.7	19.2	34.1	45.9	55.4	60.0	58.1	48.9	38.8	24.6	10.7	33.6
	MEAN	11.0	17.1	28.8	45.5	58.0	67.0	71.5	69.3	59.9	49.5	33.2	19.1	44.2
ROSHOLT	MAX	22.5	28.1	38.5	55.2	69.4	77.7	82.1	79.7	70.7	59.7	41.8	28.0	54.5
	MIN	2.0	5.9	17.0	31.8	42.7	52.1	56.9	54.9	46.6	37.0	24.1	10.4	31.8
	MEAN	12.3	17.0	27.8	43.5	56.1	64.9	69.5	67.3	58.6	48.4	33.0	19.2	43.1
ST CROIX FALLS	MAX	20.2	27.2	38.5	56.1	69.3	78.0	82.7	80.3	70.4	59.5	40.9	26.6	54.1
	MIN	-2.4	2.8	16.3	32.5	44.3	54.0	59.0	57.1	47.5	37.1	22.9	8.1	31.6
	MEAN	8.9	15.0	27.4	44.3	56.8	66.0	70.9	68.7	59.0	48.3	31.9	17.4	42.9

WISCONSIN

TEMPERATURE NORMALS (DEG F)

STATION		JAN	FEB	MAR	APR	MAY	JUN	JUL	AUG	SEP	OCT	NOV	DEC	ANN
SHAWANO	MAX	24.3	29.5	39.8	56.7	70.2	78.6	83.0	80.5	71.6	60.5	42.9	29.5	55.6
	MIN	4.8	7.9	19.1	32.7	43.1	52.7	57.4	55.2	46.8	37.6	25.7	12.7	33.0
	MEAN	14.6	18.8	29.4	44.7	56.7	65.7	70.2	67.8	59.2	49.1	34.3	21.1	44.3
SHEBOYGAN	MAX	26.9	30.9	39.4	52.2	63.5	74.0	80.1	79.1	71.4	59.6	44.7	32.4	54.5
	MIN	12.0	15.9	25.0	35.9	44.5	53.8	60.7	60.6	52.9	43.4	31.1	19.2	37.9
	MEAN	19.5	23.4	32.2	44.0	54.0	63.9	70.5	69.9	62.2	51.5	38.0	25.9	46.3
SOLON SPRINGS	MAX	20.2	27.2	38.2	55.0	69.3	77.8	82.5	79.8	69.4	58.3	39.0	25.4	53.5
	MIN	-3.1	1.8	14.1	28.8	39.6	49.0	54.7	52.7	44.3	34.7	21.1	6.0	28.6
	MEAN	8.5	14.5	26.2	41.9	54.5	63.4	68.6	66.3	56.9	46.6	30.1	15.8	41.1
SPARTA	MAX	23.3	29.9	40.7	58.2	71.3	79.5	83.9	81.7	72.7	61.5	43.3	29.3	56.3
	MIN	3.1	8.0	19.8	34.4	45.1	54.3	58.6	56.5	47.7	38.0	25.4	12.3	33.6
	MEAN	13.2	18.9	30.3	46.3	58.2	66.9	71.3	69.1	60.2	49.7	34.4	20.8	44.9
SPOONER EXP FARM	MAX	20.0	27.0	37.9	55.7	68.9	77.0	81.6	79.0	69.1	58.6	39.7	25.6	53.3
	MIN	-2.4	2.1	15.1	30.9	41.9	51.5	56.4	54.1	45.5	35.9	22.1	7.0	30.0
	MEAN	8.8	14.6	26.6	43.3	55.4	64.3	69.0	66.6	57.3	47.3	30.9	16.3	41.7
STANLEY	MAX	20.6	27.0	38.4	55.9	69.2	77.0	81.6	79.7	70.4	59.5	40.8	26.7	53.9
	MIN	.1	4.5	17.0	32.2	42.5	51.9	56.5	54.2	45.9	36.3	23.1	9.2	31.1
	MEAN	10.4	15.8	27.7	44.1	55.9	64.5	69.1	67.0	58.2	47.9	32.0	18.0	42.6
STEVENS POINT	MAX	22.4	27.8	38.2	55.2	68.3	77.1	81.5	79.2	70.3	59.5	42.4	28.2	54.2
	MIN	3.5	7.4	18.8	33.6	44.5	53.9	58.5	56.7	48.2	38.5	25.4	12.2	33.4
	MEAN	13.0	17.6	28.5	44.4	56.4	65.5	70.0	67.9	59.3	49.0	33.9	20.3	43.8
STOUGHTON	MAX	26.1	31.1	41.3	57.8	70.0	78.4	82.5	80.6	73.1	62.2	45.5	31.9	56.7
	MIN	8.4	13.1	23.5	36.2	47.0	56.5	61.2	59.2	50.7	40.5	28.5	16.0	36.7
	MEAN	17.3	22.1	32.4	47.1	58.5	67.5	71.8	69.9	61.9	51.4	37.0	23.9	46.7
STURGEON BAY EXP FARM	MAX	24.9	28.5	37.5	51.7	64.1	74.2	79.4	77.8	69.2	58.0	42.9	30.7	53.2
	MIN	8.8	10.9	20.6	32.8	41.7	51.2	57.5	56.9	49.7	40.8	29.6	17.3	34.8
	MEAN	16.9	19.7	29.0	42.3	52.9	62.7	68.5	67.4	59.5	49.4	36.3	24.0	44.1
SUPERIOR	MAX	20.0	25.9	35.3	49.2	61.0	70.3	78.0	75.8	66.7	56.6	39.4	25.9	50.3
	MIN	-1.6	3.5	14.8	28.9	37.5	46.1	53.5	53.2	45.3	35.8	23.0	8.1	29.0
	MEAN	9.2	14.7	25.1	39.1	49.3	58.2	65.8	64.5	56.1	46.2	31.2	17.1	39.7
TREMPEALEAU DAM 6	MAX	23.5	29.9	40.4	57.2	70.5	79.2	83.5	81.2	72.2	61.0	43.4	29.7	56.0
	MIN	4.1	9.0	21.0	36.5	48.2	57.4	61.8	59.5	50.8	40.9	27.0	13.2	35.8
	MEAN	13.8	19.5	30.7	46.9	59.4	68.3	72.7	70.4	61.5	51.0	35.2	21.5	45.9
TWO RIVERS	MAX	26.3	30.2	38.2	50.4	60.6	70.1	76.1	75.9	68.5	57.3	43.6	31.9	52.4
	MIN	10.7	14.2	23.3	34.1	42.2	50.6	57.2	57.8	50.6	41.3	29.7	17.8	35.8
	MEAN	18.5	22.2	30.8	42.3	51.5	60.3	66.7	66.9	59.6	49.3	36.7	24.9	44.1
VIROQUA	MAX	22.8	29.2	39.6	56.7	69.4	77.9	82.3	80.5	71.9	60.7	43.0	28.6	55.2
	MIN	3.9	8.7	20.1	34.7	45.6	55.1	59.3	57.0	48.6	38.6	25.2	12.3	34.1
	MEAN	13.4	19.0	29.9	45.7	57.6	66.5	70.8	68.8	60.3	49.7	34.2	20.5	44.7
WASHINGTON ISLAND	MAX	25.1	28.0	36.3	48.6	59.9	69.5	76.4	75.3	67.0	56.3	42.8	31.1	51.4
	MIN	10.2	10.5	19.6	30.6	39.4	49.3	56.4	56.3	49.6	40.9	30.2	18.9	34.3
	MEAN	17.7	19.3	27.9	39.6	49.7	59.4	66.4	65.8	58.3	48.6	36.5	25.0	42.9
WATERTOWN	MAX	26.0	30.9	41.3	57.8	70.8	80.0	84.0	82.0	73.8	62.3	45.2	31.7	57.2
	MIN	8.9	13.1	23.5	36.4	46.6	55.9	60.4	58.6	50.6	41.0	28.8	16.2	36.7
	MEAN	17.5	22.0	32.4	47.2	58.7	67.9	72.2	70.3	62.3	51.6	37.0	24.0	46.9
WAUKESHA	MAX	26.0	30.7	40.6	56.1	68.4	78.1	82.5	80.7	72.7	61.2	44.9	31.8	56.1
	MIN	10.1	14.6	24.1	36.0	46.2	55.9	61.1	59.7	51.6	41.2	29.2	17.3	37.3
	MEAN	18.1	22.7	32.4	46.1	57.3	67.0	71.8	70.2	62.2	51.2	37.1	24.6	46.7
WAUPACA	MAX	24.5	29.9	40.2	57.1	70.7	79.4	83.8	81.3	72.5	61.2	43.8	29.9	56.2
	MIN	5.2	9.0	20.0	33.8	44.4	53.8	58.7	56.6	48.2	38.5	26.1	13.1	34.0
	MEAN	14.9	19.5	30.2	45.5	57.6	66.6	71.3	69.0	60.4	49.9	35.0	21.5	45.1

WISCONSIN

TEMPERATURE NORMALS (DEG F)

STATION		JAN	FEB	MAR	APR	MAY	JUN	JUL	AUG	SEP	OCT	NOV	DEC	ANN
WAUSAU FAA AIRPORT	MAX	20.5	26.4	36.7	53.4	66.9	75.8	80.3	77.8	68.1	57.0	39.6	26.3	52.4
	MIN	1.8	5.3	17.4	32.8	43.8	53.3	58.4	56.3	47.3	37.5	24.4	10.3	32.4
	MEAN	11.2	15.9	27.0	43.1	55.4	64.6	69.4	67.0	57.7	47.3	32.0	18.3	42.4
WEST ALLIS	MAX	27.2	31.5	40.8	55.3	68.0	78.3	83.1	80.9	72.7	61.3	45.7	33.0	56.5
	MIN	12.1	16.9	26.3	37.9	47.7	57.6	63.6	62.6	54.6	44.2	31.3	19.4	39.5
	MEAN	19.7	24.2	33.6	46.6	57.9	68.0	73.4	71.8	63.7	52.8	38.5	26.2	48.0
WEST BEND	MAX	25.4	29.6	39.3	54.9	67.3	76.7	81.1	79.3	71.6	60.1	44.5	31.1	55.1
	MIN	8.8	13.0	22.7	34.7	44.5	54.1	59.6	58.4	50.7	41.0	28.5	16.0	36.0
	MEAN	17.1	21.3	31.0	44.8	55.9	65.4	70.3	68.9	61.2	50.5	36.5	23.6	45.5
WEYERHAUSER 2 SSE	MAX	19.5	26.0	36.9	54.5	67.6	75.0	80.0	77.5	67.8	57.2	39.1	25.1	52.2
	MIN	-1.0	4.0	16.2	31.3	41.8	50.7	55.6	53.2	44.9	34.8	22.0	7.4	30.1
	MEAN	9.3	15.0	26.6	42.9	54.8	62.9	67.9	65.4	56.4	46.1	30.6	16.2	41.2
WHITEWATER	MAX	27.6	32.6	43.1	59.7	72.0	80.7	84.6	82.6	75.4	64.1	47.2	33.4	58.6
	MIN	9.7	14.0	23.8	36.1	46.4	55.7	59.9	58.0	50.1	40.6	29.0	17.0	36.7
	MEAN	18.7	23.4	33.5	47.9	59.2	68.2	72.3	70.3	62.8	52.4	38.1	25.2	47.7
WILLOW RESERVOIR	MAX	19.6	25.3	35.7	51.9	66.3	74.6	78.9	76.2	66.7	56.0	38.1	24.6	51.2
	MIN	-1.6	.9	12.4	28.7	39.9	49.7	54.5	51.9	44.1	35.0	22.0	7.0	28.7
	MEAN	9.0	13.1	24.1	40.4	53.2	62.1	66.7	64.1	55.5	45.5	30.1	15.8	40.0
WINTER 6 NNW	MAX	17.7	23.9	34.8	51.2	64.7	73.2	77.8	75.2	65.7	55.7	37.8	23.9	50.1
	MIN	-5.5	-2.7	10.5	27.7	38.2	47.6	52.6	50.3	41.9	32.6	19.8	4.1	26.4
	MEAN	6.1	10.6	22.7	39.5	51.5	60.4	65.2	62.8	53.8	44.2	28.8	14.0	38.3
WISCONSIN DELLS	MAX	24.2	30.4	40.3	56.5	69.0	77.6	81.8	80.0	71.5	60.6	43.6	30.0	55.5
	MIN	5.3	9.2	20.4	34.2	45.4	55.0	59.8	57.7	49.0	38.5	26.0	13.3	34.5
	MEAN	14.8	19.8	30.4	45.4	57.2	66.3	70.9	68.9	60.2	49.6	34.8	21.7	45.0
WISCONSIN RAPIDS	MAX	22.5	28.5	38.8	55.6	69.0	77.9	82.3	80.0	70.5	59.3	41.9	28.1	54.5
	MIN	2.5	6.3	17.9	32.4	43.3	52.7	57.6	55.3	46.4	36.4	23.7	10.5	32.1
	MEAN	12.5	17.4	28.4	44.0	56.1	65.3	70.0	67.7	58.5	47.8	32.8	19.3	43.3

WISCONSIN

PRECIPITATION NORMALS (INCHES)

STATION	JAN	FEB	MAR	APR	MAY	JUN	JUL	AUG	SEP	OCT	NOV	DEC	ANN
ALMA DAM 4	.84	.64	1.85	2.76	3.65	4.47	4.44	3.89	3.36	2.42	1.57	.92	30.81
AMERY	.85	.71	1.51	2.49	3.53	4.81	3.79	4.76	3.30	2.09	1.43	1.05	30.32
ANTIGO 1 SSW	.90	.84	1.79	2.68	3.40	4.15	4.01	4.45	4.24	2.14	1.79	1.22	31.61
APPLETON	1.17	1.17	2.16	3.03	3.40	3.60	3.57	3.42	3.27	2.30	1.85	1.47	30.41
ASHLAND EXP FARM	1.05	.76	1.61	2.47	3.66	3.64	4.07	4.62	3.51	2.16	2.01	1.27	30.83
BALDWIN 1 SW	.88	.78	1.69	2.60	3.65	4.79	3.74	4.37	3.46	2.08	1.41	.98	30.43
BARABOO	.97	1.03	2.06	3.31	3.26	3.62	3.80	3.91	3.49	2.24	1.89	1.32	30.90
BAYFIELD 6 N	1.36	.83	1.95	2.10	3.67	3.48	3.86	4.39	3.75	2.47	2.19	1.62	31.67
BEAVER DAM	1.14	1.08	2.13	3.24	3.21	3.69	3.44	3.52	3.40	2.42	1.84	1.58	30.69
BELOIT	1.32	1.02	2.09	3.14	3.12	3.96	3.80	3.91	3.40	2.32	2.24	1.68	32.00
BLAIR	.87	.77	1.95	2.86	4.25	4.35	4.39	4.50	3.68	2.32	1.66	1.02	32.62
BLOOMER	.86	.65	1.57	2.71	3.71	4.72	3.74	4.55	3.63	2.12	1.46	1.07	30.79
BOWLER	.99	1.01	1.99	2.97	3.57	3.76	3.27	3.72	3.74	2.23	2.01	1.41	30.67
BREAKWATER	1.19	.93	1.93	2.61	3.44	3.73	3.60	4.01	3.63	2.26	1.85	1.50	30.68
BRILLION	1.05	.91	1.78	2.70	2.96	3.49	3.09	3.09	2.96	2.19	1.73	1.26	27.21
BRODHEAD	1.25	1.00	2.25	3.44	3.17	4.20	4.07	4.05	3.49	2.55	1.96	1.65	33.08
BRULE RANGER STATION	1.27	.90	1.72	2.20	3.59	4.09	4.31	4.51	3.66	2.42	2.06	1.20	31.93
BRULE ISLAND	1.19	.98	1.75	2.55	3.32	3.84	3.78	3.78	3.58	2.27	1.77	1.46	30.27
BUCKATABON	1.10	1.02	1.62	2.36	3.42	3.97	3.86	4.42	3.64	2.30	1.89	1.34	30.94
BURLINGTON	1.44	1.08	2.44	3.46	2.96	4.52	4.41	3.76	3.06	2.44	2.21	1.70	33.48
CEDAR FALLS HYDRO PL	1.00	.78	1.91	2.88	3.78	4.78	3.84	4.11	3.67	2.23	1.55	1.17	31.70
CHILTON	1.30	1.25	2.15	2.93	3.35	3.74	3.29	3.51	3.25	2.51	1.95	1.65	30.88
CLINTON 2 N	1.29	1.03	2.40	3.55	3.22	4.17	4.16	4.10	3.53	2.51	2.35	1.67	33.98
CLINTONVILLE	1.20	1.14	2.21	2.91	3.68	3.44	3.50	3.57	3.44	2.25	1.81	1.48	30.63
CODDINGTON 1 E	.74	.86	1.83	2.78	3.82	3.65	3.34	3.45	3.66	2.32	1.59	1.01	29.05
COUDERAY 8 W	.95	.79	1.55	2.43	3.61	4.45	4.52	4.88	3.97	2.48	1.71	1.26	32.60
CRIVITZ HIGH FALLS	1.04	.99	1.87	2.75	3.57	3.82	3.34	3.92	3.46	2.12	1.87	1.47	30.22
CUMBERLAND	1.04	.83	1.90	2.68	3.72	4.65	4.18	4.89	3.65	2.26	1.64	1.30	32.74
CURTISS	1.01	.93	1.85	2.61	3.79	4.70	4.34	4.11	3.94	2.17	1.76	1.27	32.48
DALTON	1.02	.99	2.04	2.98	3.34	3.56	3.40	3.57	3.42	2.40	1.76	1.32	29.80
DANBURY	1.02	.79	1.68	2.32	3.44	4.44	4.24	4.49	3.11	2.15	1.76	1.13	30.57
DARLINGTON	1.15	1.11	2.11	3.53	3.59	4.55	4.21	4.14	3.72	2.52	1.93	1.50	34.06
DODGE	.85	.75	2.01	2.84	4.00	4.22	4.15	4.06	3.69	2.15	1.73	1.04	31.49
DODGEVILLE 1 NE	1.19	1.12	2.47	3.45	3.73	4.21	3.92	4.05	3.35	2.39	2.01	1.53	33.42
EAGLE RIVER	1.17	.85	1.60	2.40	3.56	4.09	4.02	4.83	3.76	2.25	1.98	1.30	31.81
EAU CLAIRE FAA AP	.88	.67	1.57	2.94	3.85	4.37	3.74	4.28	3.47	2.16	1.37	1.01	30.31
EAU PLEINE RESERVOIR	.95	1.02	2.02	2.83	3.78	3.84	3.87	3.86	4.03	2.19	1.82	1.31	31.52
EL DORADO	.81	.80	1.64	2.79	3.16	3.47	3.49	3.44	3.25	2.43	1.74	1.21	28.23
FAIRCHILD RANGER STA	.96	.83	2.07	2.73	3.93	4.26	3.92	4.04	3.79	2.28	1.77	1.17	31.75
FLAMBEAU RESERVOIR	1.40	1.03	1.99	2.57	3.96	4.31	4.26	5.10	3.65	2.54	2.17	1.52	34.50
FOND DU LAC	.95	.93	1.78	2.80	3.18	3.41	3.74	3.49	3.16	2.33	1.66	1.39	28.82
FORT ATKINSON	1.38	1.05	2.25	3.17	3.04	3.92	4.35	3.79	3.31	2.43	2.05	1.69	32.43
GALESVILLE 3 ESE	.82	.69	1.78	2.75	4.15	4.33	4.09	4.41	3.59	2.09	1.48	1.03	31.21
GENOA DAM 8	.75	.85	1.85	3.22	3.68	4.48	4.31	3.88	3.49	2.18	1.61	.99	31.29
GERMANTOWN	1.04	.83	1.84	2.81	2.78	3.42	3.56	3.55	3.14	2.36	1.97	1.46	28.76
GOODRICH 1 E	1.05	.92	1.90	2.49	3.76	4.42	4.15	4.21	4.11	2.24	1.73	1.31	32.29
GORDON	1.13	.91	1.72	2.13	3.58	4.21	4.67	4.68	3.25	2.20	1.83	1.17	31.48
GRANTSBURG	.98	.80	1.52	2.14	3.37	4.53	4.40	4.79	3.35	2.21	1.49	1.08	30.66
GREEN BAY WSO AP	1.19	1.05	1.90	2.70	3.13	3.17	3.25	3.16	3.17	2.10	1.76	1.42	28.00
GURNEY	1.60	1.25	2.03	2.53	3.86	4.24	4.47	4.95	3.97	2.76	2.61	1.89	36.16
HANCOCK EXP FARM	.85	.94	1.92	2.87	3.52	3.47	3.43	3.77	3.76	2.28	1.72	1.10	29.63
HARTFORD	1.21	.98	2.09	2.96	3.15	3.92	3.82	3.45	3.66	2.48	1.91	1.58	31.21
HATFIELD HYDRO PLANT	.95	.86	2.07	3.00	4.04	4.19	3.89	4.20	3.69	2.28	1.69	1.27	32.13
HAYWARD RANGER STATION	1.13	.84	1.60	2.11	3.63	4.41	4.76	4.87	3.84	2.39	1.69	1.22	32.49
HILLSBORO	.80	.97	1.90	2.99	3.77	3.97	3.81	4.00	3.70	2.18	1.80	1.11	31.00

WISCONSIN

PRECIPITATION NORMALS (INCHES)

STATION	JAN	FEB	MAR	APR	MAY	JUN	JUL	AUG	SEP	OCT	NOV	DEC	ANN
HOLCOMBE	.95	.74	1.58	2.92	4.10	4.85	4.20	4.63	4.31	2.42	1.66	1.13	33.49
JANESVILLE	1.24	.95	2.15	3.31	3.21	4.02	4.11	3.81	3.44	2.49	1.97	1.59	32.29
JUMP RIVER 5 E	.86	.67	1.68	2.62	3.67	4.35	4.02	4.39	3.88	2.33	1.63	1.13	31.23
KENOSHA	1.44	1.01	2.27	3.55	2.98	3.76	4.02	3.53	3.31	2.46	2.09	1.82	32.24
KEWAUNEE	1.32	1.25	2.11	2.85	3.02	3.02	3.23	3.28	3.00	2.01	1.84	1.69	28.62
LA CROSSE FAA AP	.94	.89	1.96	3.05	3.61	4.15	3.83	3.70	3.47	2.08	1.50	1.07	30.25
LAC VIEUX DESERT	1.30	1.18	1.89	2.41	3.84	4.48	4.32	4.62	3.78	2.62	2.21	1.65	34.30
LADYSMITH	1.21	.89	1.86	2.71	3.59	4.66	4.51	4.34	3.89	2.26	1.71	1.36	32.99
LAKE GENEVA	1.82	1.35	2.72	3.77	3.16	4.25	4.42	4.05	3.41	2.54	2.31	2.10	35.90
LAKE MILLS	1.34	1.06	2.29	3.27	3.23	4.02	4.16	3.75	3.37	2.25	2.01	1.73	32.48
LANCASTER	1.00	.99	2.13	3.33	3.71	4.23	4.11	4.14	3.61	2.45	1.88	1.31	32.89
LAONA 6 SW	1.02	.85	1.80	2.67	3.41	3.59	3.77	4.19	3.95	2.15	1.77	1.35	30.52
LONE ROCK FAA AIRPORT //R	.91	.88	1.90	3.24	3.34	4.03	3.54	3.79	3.46	2.15	1.87	1.18	30.29
LONG LAKE DAM	1.19	.94	1.63	2.60	3.40	4.13	3.84	4.87	3.79	2.29	2.01	1.47	32.16
LYNXVILLE DAM 9	.84	.92	1.91	3.34	3.96	4.07	3.94	4.03	3.13	2.03	1.55	1.11	30.83
MADELINE ISLAND	1.45	1.09	1.85	2.30	3.62	3.38	3.54	4.27	3.03	2.10	2.05	1.52	30.20
MADISON WSO //R	1.11	1.02	2.15	3.10	3.34	3.89	3.75	3.82	3.06	2.24	1.83	1.53	30.84
MANITOWOC	1.30	1.25	2.21	2.82	2.92	3.14	3.20	3.11	2.79	2.24	2.05	1.80	28.83
MARINETTE	1.52	1.30	2.08	3.08	3.75	3.62	3.65	3.51	3.37	2.24	2.07	1.82	32.01
MARSHFIELD EXP FARM	.91	.87	1.82	3.02	4.33	4.11	3.79	4.07	4.09	2.67	1.76	1.27	32.71
MATHER 5 NW	1.12	.97	2.28	3.03	3.84	4.05	4.19	4.31	3.87	2.35	1.86	1.42	33.29
MAUSTON 1 SE	.98	1.09	2.18	3.17	3.74	3.89	3.64	4.34	3.98	2.42	1.84	1.30	32.57
MEDFORD	1.01	.96	1.75	2.55	3.67	4.54	4.15	4.29	3.96	2.03	1.62	1.28	31.81
MELLEN	1.29	.86	1.84	2.63	3.86	4.03	4.39	4.76	3.63	2.49	2.50	1.54	33.82
MENOMONIE	.80	.67	1.68	2.84	3.46	4.57	3.65	4.05	3.40	2.14	1.44	1.06	29.76
MERRILL	.97	.87	1.84	2.65	3.70	4.14	3.97	4.08	4.08	2.29	1.83	1.25	31.67
MILWAUKEE MT MARY COL	1.52	1.20	2.25	3.34	2.82	3.61	3.55	3.52	2.99	2.37	1.96	1.87	31.00
MILWAUKEE WSO R	1.64	1.33	2.58	3.37	2.66	3.59	3.54	3.09	2.88	2.25	1.98	2.03	30.94
MINOCQUA DAM	.99	.87	1.60	2.28	3.54	4.14	3.95	4.90	3.69	2.16	1.85	1.20	31.17
MONDOVI	.89	.75	1.76	2.94	3.85	4.14	4.06	4.12	3.55	2.22	1.53	1.10	30.91
MONROE 1 W	1.25	1.20	2.45	3.71	3.68	4.12	4.24	4.30	3.77	2.85	2.11	1.70	35.38
MONTELLO	1.12	1.10	2.14	3.30	3.47	3.56	3.32	3.70	3.36	2.34	1.87	1.55	30.83
MUSCODA	1.00	1.05	2.04	3.10	3.49	4.23	3.57	4.20	3.35	2.19	1.81	1.28	31.31
NEILLSVILLE 3 SW	.74	.77	1.71	2.73	3.58	4.11	4.09	3.92	3.65	2.14	1.52	1.00	29.96
NEW LONDON	1.26	1.23	2.06	2.96	3.62	3.60	3.32	3.48	3.45	2.26	1.83	1.49	30.56
NORTH PELICAN	1.03	.84	1.57	2.31	3.30	3.78	3.73	4.48	3.71	2.05	1.75	1.27	29.82
OCONOMOWOC	1.07	.84	1.71	2.91	2.88	3.78	4.10	3.96	3.37	2.26	1.83	1.43	30.14
OCONTO 4 W	1.42	1.13	1.91	2.74	3.41	3.45	3.48	3.34	3.18	2.18	1.96	1.52	29.72
OSHKOSH	1.18	1.12	2.08	2.93	3.35	3.22	3.35	3.34	3.29	2.26	1.80	1.56	29.48
OWEN	.84	.80	1.76	2.48	3.91	4.74	4.33	4.19	4.01	2.24	1.62	1.11	32.03
PARK FALLS	1.12	.84	1.78	2.59	3.78	4.37	4.02	5.03	3.66	2.44	1.90	1.32	32.85
PHELPS DEERSKIN DAM	1.17	.96	1.59	2.44	3.58	4.30	4.12	4.90	3.79	2.54	2.13	1.50	33.02
PINE RIVER 3 NE	1.11	1.04	2.19	3.07	3.64	3.30	3.03	3.44	3.41	2.35	1.74	1.42	29.74
PITTSVILLE	.89	.95	1.99	3.01	4.01	3.85	3.75	3.82	4.04	2.26	1.68	1.25	31.50
PLATTEVILLE	1.00	1.06	2.24	3.63	3.90	4.61	4.55	4.04	3.61	2.65	1.95	1.47	34.71
PLYMOUTH	1.29	1.25	2.33	3.37	3.25	3.59	3.66	3.47	3.79	2.68	2.07	2.00	32.75
PORTAGE	1.05	1.09	2.04	3.23	3.21	3.71	3.72	4.12	3.47	2.29	1.83	1.37	31.13
PORT WASHINGTON	1.32	1.04	1.97	3.07	2.86	3.33	3.25	3.28	3.19	2.36	1.98	1.72	29.37
PORT WING 5 SW	1.05	.74	1.71	2.23	3.53	3.82	3.81	4.00	3.44	2.13	1.87	1.28	29.61
PRAIRIE DU CHIEN	.95	1.03	2.11	3.30	3.74	3.95	3.78	4.07	2.88	2.24	1.66	1.22	30.93
PRAIRIE DU SAC 2 N //	1.03	1.00	1.93	2.95	3.25	3.58	3.83	3.63	3.34	2.10	1.84	1.42	29.90
PRENTICE 1 N	1.05	.84	1.79	2.71	3.66	4.37	3.83	4.82	3.84	2.37	1.74	1.19	32.21
RACINE	1.85	1.42	2.84	3.78	3.02	3.88	4.04	3.44	3.10	2.33	2.26	2.23	34.19
RAINBOW RESERVOIR	.99	.82	1.50	2.29	3.52	4.11	3.93	4.88	3.88	2.43	1.95	1.26	31.56
REEDSBURG	.94	1.11	2.14	3.31	3.37	3.95	3.75	4.05	3.53	2.05	1.75	1.32	31.27

WISCONSIN

PRECIPITATION NORMALS (INCHES)

STATION	JAN	FEB	MAR	APR	MAY	JUN	JUL	AUG	SEP	OCT	NOV	DEC	ANN
REST LAKE	1.15	.97	1.73	2.42	4.11	4.48	4.52	5.21	3.77	2.58	2.10	1.47	34.51
RHINELANDER	1.03	.81	1.50	2.39	3.45	3.97	3.64	4.74	3.87	2.26	1.83	1.23	30.72
RIB FALLS	.88	.83	1.73	2.65	3.82	4.00	3.59	4.06	3.82	2.22	1.75	1.15	30.50
RICE LAKE	.89	.81	1.67	2.70	3.59	4.77	4.11	4.66	3.64	2.16	1.52	1.09	31.61
RICE RESERVOIR	.96	.89	1.64	2.48	3.73	4.32	3.75	4.57	4.01	2.38	1.82	1.24	31.79
RICHLAND CENTER	1.05	1.12	2.30	3.56	3.68	4.40	3.90	4.05	3.74	2.31	1.91	1.39	33.41
RIDGELAND 1 NNE	.92	.83	1.80	2.80	3.78	4.92	3.65	4.56	3.59	2.10	1.46	1.12	31.53
RIPON 5 NE	1.31	1.21	2.22	2.75	3.28	3.64	3.43	3.32	3.24	2.32	1.72	1.62	30.06
RIVER FALLS	.81	.68	1.66	2.51	3.78	5.08	4.01	4.10	3.16	2.10	1.45	.98	30.32
ROSHOLT	1.05	1.01	1.98	3.11	3.86	3.87	3.93	3.83	3.85	2.35	1.95	1.35	32.14
ST CROIX FALLS	.88	.77	1.62	2.56	3.60	4.91	3.69	4.91	3.26	2.29	1.44	1.00	30.93
SHAWANO	1.30	1.20	1.95	3.04	3.69	3.67	3.42	3.77	3.55	2.37	1.87	1.49	31.32
SHEBOYGAN	1.48	1.34	2.34	2.96	2.76	3.43	3.35	3.15	3.24	2.58	1.99	1.92	30.54
SOLON SPRINGS	1.11	.86	1.70	2.32	3.89	4.49	4.60	4.67	3.20	2.32	1.96	1.19	32.31
SOUTH PELICAN	1.07	.95	1.65	2.55	3.56	4.18	4.06	4.55	4.37	2.41	1.86	1.30	32.51
SPARTA	.90	.89	1.93	2.91	3.68	4.12	3.65	3.77	3.81	2.19	1.56	1.12	30.53
SPIRIT FALLS	1.00	.88	1.81	2.52	3.63	4.11	3.57	4.29	3.83	2.31	1.92	1.32	31.19
SPOONER EXP FARM	.71	.63	1.36	2.07	3.53	4.14	4.18	4.50	3.44	2.10	1.39	.90	28.95
STANLEY	.95	.76	1.59	2.64	3.95	4.73	3.75	4.12	3.53	2.11	1.60	1.21	30.94
STEVENS POINT	.97	1.01	2.00	2.85	3.78	3.60	3.79	3.79	3.72	2.31	1.81	1.31	30.94
STOUGHTON	1.23	1.02	2.12	3.28	3.19	4.05	4.06	3.66	3.35	2.40	1.79	1.48	31.63
STRATFORD 2 NNW	1.04	.96	1.81	2.84	3.82	4.08	3.94	3.69	3.80	2.14	1.87	1.31	31.30
STURGEON BAY EXP FARM	1.34	1.07	1.93	2.90	3.28	3.26	3.37	3.32	3.54	2.33	1.99	1.73	30.06
SUGAR CAMP	.94	.77	1.44	2.32	3.43	3.93	3.93	4.58	3.76	2.31	1.73	1.18	30.32
SUMMIT LAKE RANGER STA	1.19	1.00	1.95	2.82	3.62	4.05	4.05	4.75	4.50	2.48	2.13	1.50	34.04
SUPERIOR	.95	.78	1.78	2.13	3.28	3.77	4.12	4.23	3.25	1.93	1.55	.96	28.73
TREMPEALEAU DAM 6	.85	.82	2.14	2.83	4.09	4.39	4.08	4.05	3.69	2.14	1.56	1.04	31.68
TWO RIVERS	1.36	1.27	2.31	2.97	2.83	2.92	3.10	3.15	2.80	2.27	1.91	1.81	28.70
UNION GROVE	1.31	1.08	2.29	3.36	3.00	4.11	3.93	3.95	2.98	2.34	2.02	1.84	32.21
VIROQUA	.86	.94	1.93	3.25	3.82	4.45	4.32	4.09	3.72	2.24	1.58	1.05	32.25
WASHINGTON ISLAND	1.37	.91	1.58	2.36	3.08	3.28	3.28	3.36	3.40	2.32	2.10	1.56	28.60
WATERTOWN	1.26	1.05	2.28	3.26	2.90	3.91	4.01	4.12	3.47	2.32	2.01	1.62	32.21
WAUKESHA	1.40	1.10	2.42	3.45	3.01	3.63	3.73	3.79	3.15	2.37	2.15	1.82	32.02
WAUPACA	1.16	1.06	2.12	3.00	3.90	3.66	3.55	3.55	3.90	2.38	1.79	1.40	31.47
WAUSAU FAA AIRPORT	.93	.95	1.90	2.86	3.77	3.93	3.97	4.14	3.88	2.26	1.78	1.25	31.62
WAUSAUKEE	1.19	1.05	1.84	2.87	3.60	3.60	3.53	3.74	3.52	2.23	1.90	1.51	30.58
WEST ALLIS	1.42	1.11	2.25	3.37	2.81	3.74	3.50	3.34	3.23	2.44	2.19	1.87	31.27
WEST BEND	1.31	.95	1.99	2.97	2.97	3.65	3.97	3.37	3.48	2.48	1.97	1.54	30.65
WEYERHAUSER 2 SSE	1.06	.83	1.78	2.77	3.81	4.52	4.37	4.48	3.95	2.35	1.57	1.22	32.71
WHITEWATER	1.23	.96	2.19	3.12	3.13	3.81	4.32	3.88	3.35	2.55	2.06	1.61	32.21
WILLOW RESERVOIR	.98	.84	1.55	2.21	3.56	4.05	3.84	4.67	3.60	2.27	1.70	1.20	30.47
WINTER 6 NNW	.90	.65	1.55	2.36	3.59	4.30	4.86	4.81	3.94	2.26	1.65	1.13	32.00
WISCONSIN DELLS	1.07	1.06	2.29	3.36	3.58	3.94	4.04	4.12	3.57	2.27	1.77	1.35	32.42
WISCONSIN RAPIDS	1.00	1.01	2.02	2.97	3.88	3.93	3.87	3.90	3.89	2.42	1.71	1.30	31.90
WIS RPDS GRAND AVE BR	.92	.90	1.94	2.88	3.98	3.92	3.84	3.77	3.98	2.46	1.75	1.24	31.58

WISCONSIN

HEATING DEGREE DAY NORMALS (BASE 65 DEG F)

STATION	JUL	AUG	SEP	OCT	NOV	DEC	JAN	FEB	MAR	APR	MAY	JUN	ANN
AMERY	21	52	218	544	1008	1504	1761	1431	1200	639	297	84	8759
ANTIGO 1 SSW	30	60	237	551	993	1451	1646	1369	1181	669	333	104	8624
APPLETON	8	21	143	476	894	1330	1541	1268	1088	615	284	60	7728
ASHLAND EXP FARM	44	67	251	570	1002	1466	1690	1408	1228	765	428	141	9060
BARABOO	19	38	177	502	912	1349	1569	1280	1085	594	283	69	7877
BAYFIELD 6 N	47	71	248	546	972	1411	1618	1358	1184	741	419	149	8764
BELOIT	0	9	79	375	789	1212	1411	1128	930	462	193	23	6611
BLAIR	17	29	185	514	939	1407	1649	1330	1113	588	271	69	8111
BLOOMER	21	34	195	529	987	1454	1686	1361	1144	612	277	76	8376
BRODHEAD	10	20	130	455	858	1299	1510	1226	1023	543	245	46	7365
BRULE ISLAND	73	109	336	648	1059	1507	1721	1470	1287	783	432	163	9588
BURLINGTON	8	30	130	460	843	1265	1476	1215	1023	576	279	60	7365
CHILTON	9	20	130	449	864	1305	1513	1257	1076	585	278	57	7543
CODDINGTON 1 E	39	64	241	555	984	1438	1662	1364	1172	648	327	99	8593
CRIVITZ HIGH FALLS	35	58	227	536	954	1395	1609	1355	1175	687	342	98	8471
CUMBERLAND	17	28	199	521	1002	1488	1727	1397	1181	630	279	69	8538
DALTON	12	25	150	464	891	1333	1541	1260	1066	576	262	56	7636
DANBURY	39	72	258	580	1044	1531	1773	1428	1203	678	340	111	9057
DARLINGTON	9	16	136	447	858	1293	1500	1204	1004	531	240	47	7285
DODGEVILLE 1 NE	11	21	141	460	885	1321	1522	1229	1039	555	249	51	7484
EAU CLAIRE FAA AP	17	28	203	535	990	1466	1711	1386	1159	618	284	66	8463
EL DORADO	9	29	151	469	891	1336	1541	1266	1073	582	270	57	7674
FOND DU LAC	9	20	135	458	873	1314	1525	1257	1070	588	265	54	7568
FORT ATKINSON	0	17	109	420	828	1268	1479	1198	998	522	229	38	7106
GENOA DAM 8	6	11	110	422	858	1302	1531	1232	1023	522	203	35	7255
GERMANTOWN	12	35	142	461	855	1283	1485	1215	1045	606	308	80	7527
GORDON	43	82	272	595	1062	1547	1792	1453	1234	705	368	115	9268
GRANTSBURG	26	59	257	574	1029	1541	1795	1448	1218	660	326	93	9026
GREEN BAY WSO AP	17	39	192	515	924	1370	1581	1322	1128	639	325	91	8143
HANCOCK EXP FARM	15	27	170	490	933	1398	1612	1319	1125	597	271	59	8016
HATFIELD HYDRO PLANT	22	38	188	504	951	1411	1643	1327	1119	603	281	76	8163
HILLSBORO	16	27	156	466	897	1349	1569	1266	1057	579	257	59	7698
HOLCOMBE	14	32	196	510	972	1445	1683	1372	1150	615	282	73	8344
JANESVILLE	0	7	89	391	804	1228	1429	1151	952	489	195	27	6762
KENOSHA	29	23	106	415	792	1184	1383	1126	980	606	337	109	7090
KEWAUNEE	25	43	174	490	855	1268	1466	1238	1076	687	406	141	7869
LA CROSSE FAA AP	10	14	135	448	891	1349	1581	1268	1057	534	216	37	7540
LADYSMITH	28	52	233	539	996	1457	1686	1369	1156	642	310	96	8564
LAKE GENEVA	0	11	93	402	813	1231	1426	1154	973	528	227	37	6895
LAKE MILLS	6	15	106	412	834	1274	1485	1204	1014	531	226	43	7150
LANCASTER	7	13	116	430	870	1305	1519	1210	1014	528	223	44	7279
LONE ROCK FAA AIRPORT //R	13	28	164	495	906	1355	1578	1268	1054	558	247	49	7715
LONG LAKE DAM	63	107	322	632	1068	1525	1727	1456	1280	771	420	156	9527
LYNXVILLE DAM 9	0	5	100	411	849	1283	1513	1215	1004	507	193	24	7104
MADELINE ISLAND	73	78	249	543	939	1352	1597	1372	1225	795	518	218	8959
MADISON WSO //R	12	29	161	490	897	1321	1531	1246	1048	576	273	58	7642
MANITOWOC	15	31	137	456	843	1249	1448	1207	1057	642	343	89	7517
MARINETTE	8	28	147	453	846	1262	1460	1226	1060	615	292	57	7454
MARSHFIELD EXP FARM	31	57	228	549	990	1442	1659	1352	1166	645	320	93	8532
MATHER 5 NW	26	38	216	536	960	1414	1646	1338	1138	624	296	80	8312
MEDFORD	31	57	256	577	1026	1491	1708	1403	1215	684	351	112	8911
MELLEN	49	85	279	589	1035	1513	1727	1442	1252	741	396	147	9255
MENOMONIE	13	17	154	476	936	1401	1640	1319	1101	570	242	44	7913
MERRILL	32	67	253	573	1005	1466	1680	1389	1197	666	332	105	8765
MILWAUKEE MT MARY COL	10	15	96	411	807	1218	1417	1159	989	555	259	49	6985

686

WISCONSIN

HEATING DEGREE DAY NORMALS (BASE 65 DEG F)

STATION	JUL	AUG	SEP	OCT	NOV	DEC	JAN	FEB	MAR	APR	MAY	JUN	ANN
MILWAUKEE WSO R	11	25	117	444	831	1237	1435	1176	1020	612	334	84	7326
MINOCQUA DAM	51	92	292	605	1056	1525	1724	1436	1259	753	388	144	9325
MONDOVI	16	20	171	489	948	1411	1640	1316	1101	573	252	53	7990
MONTELLO	16	34	168	498	915	1342	1569	1280	1079	579	267	64	7811
NEILLSVILLE 3 SW	19	39	210	533	984	1435	1671	1352	1153	630	292	77	8395
NEW LONDON	13	24	153	477	906	1355	1562	1288	1097	582	265	56	7778
NORTH PELICAN	61	107	318	614	1059	1525	1733	1459	1277	756	385	144	9438
OCONOMOWOC	9	19	125	441	846	1277	1491	1215	1023	549	246	44	7285
OCONTO 4 W	14	41	193	506	903	1333	1538	1296	1125	657	328	90	8024
OSHKOSH	9	17	137	470	894	1330	1538	1268	1091	609	278	51	7692
OWEN	37	82	269	595	1038	1528	1773	1464	1252	699	362	113	9212
PARK FALLS	45	86	289	592	1050	1519	1721	1414	1218	711	378	124	9147
PINE RIVER 3 NE	14	27	166	485	903	1352	1556	1277	1091	591	284	67	7813
PITTSVILLE	22	42	207	521	954	1398	1624	1330	1128	606	286	72	8190
PLATTEVILLE	8	14	121	434	861	1287	1497	1193	995	513	219	36	7178
PLYMOUTH	16	26	142	460	864	1283	1482	1221	1060	612	306	71	7543
PORTAGE	14	33	145	448	861	1299	1510	1224	1026	540	241	59	7400
PRAIRIE DU CHIEN	6	8	94	389	831	1268	1491	1187	973	480	180	20	6927
PRAIRIE DU SAC 2 N //	7	18	126	452	879	1318	1531	1246	1054	561	239	39	7470
PRENTICE 1 N	72	97	303	608	1059	1535	1748	1442	1240	714	387	145	9350
RACINE	6	21	91	393	786	1190	1383	1131	977	579	293	69	6919
RAINBOW RESERVOIR	58	94	296	608	1050	1516	1717	1442	1265	753	392	143	9334
REEDSBURG	11	22	148	454	876	1311	1516	1235	1032	546	247	51	7449
REST LAKE	55	89	285	595	1053	1513	1727	1439	1256	738	375	130	9255
RHINELANDER	38	76	262	582	1023	1485	1690	1411	1215	702	353	108	8945
RICE LAKE	25	48	235	549	1014	1500	1742	1408	1194	651	319	93	8778
RICHLAND CENTER	10	14	120	451	876	1308	1519	1215	1008	528	226	45	7320
RIVER FALLS	16	17	170	489	954	1423	1674	1341	1122	585	249	53	8093
ROSHOLT	20	47	200	521	960	1420	1634	1344	1153	645	303	81	8328
ST CROIX FALLS	21	32	196	524	993	1476	1739	1400	1166	621	282	69	8519
SHAWANO	19	44	184	499	921	1361	1562	1294	1104	609	285	71	7953
SHEBOYGAN	11	15	106	419	810	1212	1411	1165	1017	630	346	90	7232
SOLON SPRINGS	31	58	248	570	1047	1525	1752	1414	1203	693	345	98	8984
SPARTA	12	22	160	485	918	1370	1606	1291	1076	561	247	54	7802
SPOONER EXP FARM	27	56	240	549	1023	1510	1742	1411	1190	651	319	93	8811
STANLEY	32	41	214	530	990	1457	1693	1378	1156	627	304	85	8507
STEVENS POINT	16	35	182	503	933	1386	1612	1327	1132	618	293	70	8107
STOUGHTON	0	20	118	436	840	1274	1479	1201	1011	537	241	47	7204
STURGEON BAY EXP FARM	22	44	174	484	861	1271	1491	1268	1116	681	381	105	7898
SUPERIOR	76	98	271	583	1014	1485	1730	1408	1237	777	487	212	9378
TREMPEALEAU DAM 6	6	9	129	445	894	1349	1587	1274	1063	543	210	38	7547
TWO RIVERS	46	52	170	487	849	1243	1442	1198	1060	681	419	155	7802
VIROQUA	14	22	155	483	924	1380	1600	1288	1088	579	261	55	7849
WASHINGTON ISLAND	50	70	206	508	855	1240	1466	1280	1150	762	474	180	8241
WATERTOWN	7	13	113	425	840	1271	1473	1204	1011	534	235	40	7166
WAUKESHA	6	17	113	436	837	1252	1454	1184	1011	567	269	48	7194
WAUPACA	12	28	153	475	900	1349	1553	1274	1079	585	259	57	7724
WAUSAU FAA AIRPORT	21	45	226	549	990	1448	1668	1375	1178	657	320	88	8565
WEST ALLIS	0	11	83	389	795	1203	1404	1142	973	552	254	42	6848
WEST BEND	14	28	139	458	855	1283	1485	1224	1054	606	305	71	7522
WEYERHAUSER 2 SSE	38	61	261	586	1032	1513	1727	1400	1190	663	333	116	8920
WHITEWATER	6	18	108	401	807	1234	1435	1165	977	513	223	35	6922
WILLOW RESERVOIR	44	94	291	605	1047	1525	1736	1453	1268	738	382	127	9310
WINTER 6 NNW	72	113	336	645	1086	1581	1826	1523	1311	765	429	167	9854
WISCONSIN DELLS	14	31	162	485	906	1342	1556	1266	1073	588	265	62	7750

WISCONSIN

HEATING DEGREE DAY NORMALS (BASE 65 DEG F)

STATION	JUL	AUG	SEP	OCT	NOV	DEC	JAN	FEB	MAR	APR	MAY	JUN	ANN
WISCONSIN RAPIDS	22	37	201	538	966	1417	1628	1333	1135	630	305	75	8287

WISCONSIN

COOLING DEGREE DAY NORMALS (BASE 65 DEG F)

STATION	JAN	FEB	MAR	APR	MAY	JUN	JUL	AUG	SEP	OCT	NOV	DEC	ANN
AMERY	0	0	0	0	24	84	173	133	14	8	0	0	436
ANTIGO 1 SSW	0	0	0	0	20	59	123	88	6	5	0	0	301
APPLETON	0	0	0	0	27	102	207	158	14	7	0	0	515
ASHLAND EXP FARM	0	0	0	0	0	24	116	82	5	0	0	0	227
BARABOO	0	0	0	0	26	96	177	131	12	10	0	0	452
BAYFIELD 6 N	0	0	0	0	0	23	109	87	0	0	0	0	219
BELOIT	0	0	0	0	60	167	273	214	58	19	0	0	791
BLAIR	0	0	0	0	35	108	194	137	14	9	0	0	497
BLOOMER	0	0	0	0	23	88	185	127	9	6	0	0	438
BRODHEAD	0	0	0	0	35	127	224	166	31	9	0	0	592
BRULE ISLAND	0	0	0	0	10	22	73	41	0	0	0	0	146
BURLINGTON	0	0	0	0	22	99	197	160	28	8	0	0	514
CHILTON	0	0	0	0	30	93	195	156	22	9	0	0	505
CODDINGTON 1 E	0	0	0	0	17	54	120	80	0	0	0	0	271
CRIVITZ HIGH FALLS	0	0	0	0	22	62	143	108	5	0	0	0	340
CUMBERLAND	0	0	0	0	31	99	196	133	13	0	0	0	472
DALTON	0	0	0	0	30	101	195	149	18	8	0	0	501
DANBURY	0	0	0	0	18	60	139	96	6	0	0	0	319
DARLINGTON	0	0	0	0	35	113	207	159	34	9	0	0	557
DODGEVILLE 1 NE	0	0	0	0	29	102	204	157	24	7	0	0	523
EAU CLAIRE FAA AP	0	0	0	0	30	99	197	130	11	5	0	0	472
EL DORADO	0	0	0	0	25	96	185	144	19	7	0	0	476
FOND DU LAC	0	0	0	0	30	105	204	156	18	8	0	0	521
FORT ATKINSON	0	0	0	0	40	134	237	191	46	17	0	0	665
GENOA DAM 8	0	0	0	0	42	146	257	194	29	13	0	0	681
GERMANTOWN	0	0	0	0	16	83	167	146	19	9	0	0	440
GORDON	0	0	0	0	18	58	139	94	0	0	0	0	309
GRANTSBURG	0	0	0	0	22	69	147	108	8	0	0	0	354
GREEN BAY WSO AP	0	0	0	0	18	82	156	116	9	0	0	0	381
HANCOCK EXP FARM	0	0	0	0	36	101	182	135	17	6	0	0	477
HATFIELD HYDRO PLANT	0	0	0	0	30	88	174	128	11	5	0	0	436
HILLSBORO	0	0	0	0	31	119	214	157	18	10	0	0	549
HOLCOMBE	0	0	0	0	28	85	169	116	13	5	0	0	416
JANESVILLE	0	0	0	0	50	168	279	218	56	13	0	0	784
KENOSHA	0	0	0	0	8	88	187	165	31	6	0	0	485
KEWAUNEE	0	0	0	0	0	33	111	121	9	0	0	0	274
LA CROSSE FAA AP	0	0	0	0	42	142	258	197	33	11	0	0	683
LADYSMITH	0	0	0	0	22	69	139	90	5	0	0	0	325
LAKE GENEVA	0	0	0	0	38	142	258	213	48	8	0	0	707
LAKE MILLS	0	0	0	0	37	127	229	179	34	12	0	0	618
LANCASTER	0	0	0	0	37	137	233	183	35	11	0	0	636
LONE ROCK FAA AIRPORT //R	0	0	0	0	30	115	202	146	20	8	0	0	521
LONG LAKE DAM	0	0	0	0	14	30	76	49	0	0	0	0	169
LYNXVILLE DAM 9	0	0	0	0	47	156	274	210	40	11	0	0	738
MADELINE ISLAND	0	0	0	0	0	0	80	72	6	0	0	0	158
MADISON WSO //R	0	0	0	0	25	97	185	137	14	9	0	0	467
MANITOWOC	0	0	0	0	8	56	161	149	14	0	0	0	388
MARINETTE	0	0	0	0	19	90	203	164	30	7	0	0	513
MARSHFIELD EXP FARM	0	0	0	0	19	63	140	94	6	0	0	0	322
MATHER 5 NW	0	0	0	0	26	80	159	100	6	6	0	0	377
MEDFORD	0	0	0	0	16	52	115	75	0	0	0	0	258
MELLEN	0	0	0	0	18	45	108	70	0	0	0	0	241
MENOMONIE	0	0	0	0	34	113	224	156	19	8	0	0	554
MERRILL	0	0	0	0	22	69	137	98	10	6	0	0	342
MILWAUKEE MT MARY COL	0	0	0	0	32	115	239	201	42	8	0	0	637

WISCONSIN

COOLING DEGREE DAY NORMALS (BASE 65 DEG F)

STATION	JAN	FEB	MAR	APR	MAY	JUN	JUL	AUG	SEP	OCT	NOV	DEC	ANN
MILWAUKEE WSO R	0	0	0	0	18	81	182	158	24	7	0	0	470
MINOCQUA DAM	0	0	0	0	19	48	94	64	0	0	0	0	225
MONDOVI	0	0	0	0	29	113	205	141	21	9	0	0	518
MONTELLO	0	0	0	0	34	115	205	142	15	8	0	0	519
NEILLSVILLE 3 SW	0	0	0	0	22	71	146	108	9	0	0	0	356
NEW LONDON	0	0	0	0	32	104	199	142	15	6	0	0	498
NORTH PELICAN	0	0	0	0	16	45	89	61	0	0	0	0	211
OCONOMOWOC	0	0	0	0	35	116	226	177	35	10	0	0	599
OCONTO 4 W	0	0	0	0	18	81	160	121	16	7	0	0	403
OSHKOSH	0	0	0	0	30	102	207	156	20	8	0	0	523
OWEN	0	0	0	0	17	53	115	88	5	0	0	0	278
PARK FALLS	0	0	0	0	19	46	104	71	0	0	0	0	240
PINE RIVER 3 NE	0	0	0	0	30	94	190	142	16	7	0	0	479
PITTSVILLE	0	0	0	0	25	75	155	108	9	7	0	0	379
PLATTEVILLE	0	0	0	0	40	132	237	181	34	9	0	0	633
PLYMOUTH	0	0	0	0	17	83	186	153	19	0	0	0	458
PORTAGE	0	0	0	0	39	128	218	170	31	11	0	0	597
PRAIRIE DU CHIEN	0	0	0	0	56	173	288	231	58	14	0	0	820
PRAIRIE DU SAC 2 N //	0	0	0	0	34	126	233	179	27	9	0	0	608
PRENTICE 1 N	0	0	0	0	15	31	84	45	0	0	0	0	175
RACINE	0	0	0	0	17	108	214	198	46	8	0	0	591
RAINBOW RESERVOIR	0	0	0	0	17	38	93	60	0	0	0	0	208
REEDSBURG	0	0	0	0	30	105	200	149	22	8	0	0	514
REST LAKE	0	0	0	0	19	43	99	64	0	0	0	0	225
RHINELANDER	0	0	0	0	25	57	128	92	7	6	0	0	315
RICE LAKE	0	0	0	0	21	75	155	107	10	0	0	0	368
RICHLAND CENTER	0	0	0	0	34	123	234	173	21	11	0	0	596
RIVER FALLS	0	0	0	0	32	113	217	151	17	8	0	0	538
ROSHOLT	0	0	0	0	27	78	160	118	8	7	0	0	398
ST CROIX FALLS	0	0	0	0	28	99	204	147	16	6	0	0	500
SHAWANO	0	0	0	0	28	92	180	130	10	6	0	0	446
SHEBOYGAN	0	0	0	0	0	57	182	167	22	0	0	0	428
SOLON SPRINGS	0	0	0	0	19	50	143	99	5	0	0	0	316
SPARTA	0	0	0	0	36	111	208	149	16	11	0	0	531
SPOONER EXP FARM	0	0	0	0	21	72	151	105	9	0	0	0	358
STANLEY	0	0	0	0	22	70	159	103	10	0	0	0	364
STEVENS POINT	0	0	0	0	27	85	171	125	11	7	0	0	426
STOUGHTON	0	0	0	0	39	122	214	172	25	15	0	0	587
STURGEON BAY EXP FARM	0	0	0	0	6	36	130	118	9	0	0	0	299
SUPERIOR	0	0	0	0	0	8	101	83	0	0	0	0	192
TREMPEALEAU DAM 6	0	0	0	0	37	137	245	176	24	11	0	0	630
TWO RIVERS	0	0	0	0	0	14	98	111	8	0	0	0	231
VIROQUA	0	0	0	0	31	100	193	140	14	9	0	0	487
WASHINGTON ISLAND	0	0	0	0	0	12	93	95	0	0	0	0	200
WATERTOWN	0	0	0	0	40	127	230	178	32	9	0	0	616
WAUKESHA	0	0	0	0	30	108	217	178	29	8	0	0	570
WAUPACA	0	0	0	0	30	105	207	152	15	7	0	0	516
WAUSAU FAA AIRPORT	0	0	0	0	23	76	158	107	7	0	0	0	371
WEST ALLIS	0	0	0	0	34	132	265	222	44	10	0	0	707
WEST BEND	0	0	0	0	23	83	178	149	25	9	0	0	467
WEYERHAUSER 2 SSE	0	0	0	0	17	53	128	74	0	0	0	0	272
WHITEWATER	0	0	0	0	43	131	232	182	42	10	0	0	640
WILLOW RESERVOIR	0	0	0	0	16	40	97	66	6	0	0	0	225
WINTER 6 NNW	0	0	0	0	10	29	79	44	0	0	0	0	162
WISCONSIN DELLS	0	0	0	0	23	101	196	152	18	8	0	0	498

WISCONSIN

COOLING DEGREE DAY NORMALS (BASE 65 DEG F)

STATION	JAN	FEB	MAR	APR	MAY	JUN	JUL	AUG	SEP	OCT	NOV	DEC	ANN
WISCONSIN RAPIDS	0	0	0	0	29	84	177	120	6	5	0	0	421

STATE-STATION NUMBER	STN TYP	NAME	LATITUDE DEG-MIN	LONGITUDE DEG-MIN	ELEVATION (FT)
47-0124	12	ALMA DAM 4	N 4420	W 09156	670
47-0175	13	AMERY	N 4518	W 09222	1070
47-0239	13	ANTIGO 1 SSW	N 4508	W 08909	1450
47-0265	13	APPLETON	N 4415	W 08823	730
47-0349	13	ASHLAND EXP FARM	N 4634	W 09058	650
47-0486	12	BALDWIN 1 SW	N 4458	W 09223	1100
47-0516	13	BARABOO	N 4328	W 08944	823
47-0603	13	BAYFIELD 6 N	N 4653	W 09049	820
47-0645	12	BEAVER DAM	N 4327	W 08851	840
47-0696	13	BELOIT	N 4230	W 08902	780
47-0882	13	BLAIR	N 4418	W 09114	850
47-0904	13	BLOOMER	N 4506	W 09129	999
47-0991	12	BOWLER	N 4452	W 08859	1080
47-1039	12	BREAKWATER	N 4550	W 08815	1140
47-1064	12	BRILLION	N 4411	W 08804	815
47-1078	13	BRODHEAD	N 4237	W 08923	790
47-1131	12	BRULE RANGER STATION	N 4632	W 09135	994
47-1139	13	BRULE ISLAND	N 4557	W 08813	1250
47-1155	12	BUCKATABON	N 4601	W 08919	1650
47-1205	13	BURLINGTON	N 4240	W 08816	760
47-1308	12	CEDAR FALLS HYDRO PL	N 4456	W 09153	834
47-1568	13	CHILTON	N 4402	W 08809	840
47-1667	12	CLINTON 2 N	N 4237	W 08852	920
47-1676	12	CLINTONVILLE	N 4437	W 08845	800
47-1708	13	CODDINGTON 1 E	N 4422	W 08932	1060
47-1847	12	COUDERAY 8 W	N 4548	W 09128	1400
47-1897	13	CRIVITZ HIGH FALLS	N 4517	W 08812	827
47-1923	13	CUMBERLAND	N 4532	W 09201	1240
47-1931	12	CURTISS	N 4457	W 09026	1370
47-1970	13	DALTON	N 4339	W 08912	857
47-1978	13	DANBURY	N 4601	W 09222	950
47-2001	13	DARLINGTON	N 4241	W 09007	930
47-2165	12	DODGE	N 4408	W 09133	680
47-2173	13	DODGEVILLE 1 NE	N 4258	W 09007	1160
47-2314	12	EAGLE RIVER	N 4555	W 08915	1645
47-2428	13	EAU CLAIRE FAA AP	N 4452	W 09129	888
47-2447	12	EAU PLEINE RESERVOIR	N 4444	W 08945	1138
47-2507	13	EL DORADO	N 4348	W 08838	865
47-2678	12	FAIRCHILD RANGER STA	N 4436	W 09058	1080
47-2814	12	FLAMBEAU RESERVOIR	N 4604	W 09014	1564
47-2839	13	FOND DU LAC	N 4348	W 08827	760
47-2869	13	FORT ATKINSON	N 4254	W 08850	815
47-2996	12	GALESVILLE 3 ESE	N 4404	W 09117	730
47-3038	13	GENOA DAM 8	N 4334	W 09114	639
47-3058	13	GERMANTOWN	N 4313	W 08807	850
47-3182	12	GOODRICH 1 E	N 4509	W 09004	1390
47-3186	13	GORDON	N 4614	W 09145	1060
47-3244	13	GRANTSBURG	N 4547	W 09241	940
47-3269	13	GREEN BAY WSO AP	N 4429	W 08808	682
47-3332	12	GURNEY	N 4628	W 09030	980

47 — WISCONSIN

STATE-STATION NUMBER	STN TYP	NAME	LATITUDE DEG-MIN	LONGITUDE DEG-MIN	ELEVATION (FT)
47-3405	13	HANCOCK EXP FARM	N 4407	W 08932	1076
47-3453	12	HARTFORD	N 4319	W 08823	980
47-3471	13	HATFIELD HYDRO PLANT	N 4424	W 09044	953
47-3511	12	HAYWARD RANGER STATION	N 4600	W 09129	1196
47-3654	13	HILLSBORO	N 4339	W 09020	940
47-3698	13	HOLCOMBE	N 4513	W 09108	1030
47-3979	13	JANESVILLE	N 4240	W 08901	760
47-4080	12	JUMP RIVER 5 E	N 4521	W 09043	1245
47-4174	13	KENOSHA	N 4233	W 08748	600
47-4195	13	KEWAUNEE	N 4427	W 08731	689
47-4370	13	LA CROSSE FAA AP	N 4352	W 09115	651
47-4383	12	LAC VIEUX DESERT	N 4608	W 08908	1690
47-4391	13	LADYSMITH	N 4528	W 09105	1158
47-4457	13	LAKE GENEVA	N 4236	W 08826	880
47-4482	13	LAKE MILLS	N 4304	W 08855	840
47-4546	13	LANCASTER	N 4252	W 09042	1080
47-4582	12	LAONA 6 SW	N 4531	W 08846	1500
47-4821	13	LONE ROCK FAA AIRPORT //R	N 4312	W 09011	714
47-4829	13	LONG LAKE DAM	N 4554	W 08908	1630
47-4937	13	LYNXVILLE DAM 9	N 4313	W 09106	633
47-4953	13	MADELINE ISLAND	N 4650	W 09038	615
47-4961	13	MADISON WSO //R	N 4308	W 08920	858
47-5017	13	MANITOWOC	N 4406	W 08741	660
47-5091	13	MARINETTE	N 4506	W 08738	605
47-5120	13	MARSHFIELD EXP FARM	N 4439	W 09008	1250
47-5164	13	MATHER 5 NW	N 4411	W 09023	982
47-5178	12	MAUSTON 1 SE	N 4347	W 09004	865
47-5255	13	MEDFORD	N 4508	W 09021	1470
47-5286	13	MELLEN	N 4619	W 09039	1230
47-5335	13	MENOMONIE	N 4453	W 09156	780
47-5364	13	MERRILL	N 4511	W 08944	1270
47-5474	13	MILWAUKEE MT MARY COL	N 4304	W 08802	726
47-5479	13	MILWAUKEE WSO R	N 4257	W 08754	672
47-5516	13	MINOCQUA DAM	N 4553	W 08944	1589
47-5563	13	MONDOVI	N 4434	W 09140	815
47-5573	12	MONROE 1 W	N 4236	W 08940	990
47-5581	13	MONTELLO	N 4348	W 08919	790
47-5718	12	MUSCODA	N 4312	W 09026	700
47-5808	13	NEILLSVILLE 3 SW	N 4432	W 09038	1045
47-5932	13	NEW LONDON	N 4423	W 08844	780
47-6122	13	NORTH PELICAN	N 4538	W 08915	1610
47-6200	13	OCONOMOWOC	N 4306	W 08830	860
47-6208	13	OCONTO 4 W	N 4454	W 08757	660
47-6330	13	OSHKOSH	N 4403	W 08833	753
47-6357	13	OWEN	N 4457	W 09032	1240
47-6398	13	PARK FALLS	N 4556	W 09027	1517
47-6518	12	PHELPS DEERSKIN DAM	N 4603	W 08902	1683
47-6594	13	PINE RIVER 3 NE	N 4411	W 08902	875
47-6622	13	PITTSVILLE	N 4427	W 09007	1040
47-6646	13	PLATTEVILLE	N 4245	W 09029	1015

STATE-STATION NUMBER	STN TYP	NAME	LATITUDE DEG-MIN	LONGITUDE DEG-MIN	ELEVATION (FT)
47-6678	13	PLYMOUTH	N 4345	W 08759	865
47-6718	13	PORTAGE	N 4332	W 08926	800
47-6764	12	PORT WASHINGTON	N 4323	W 08752	600
47-6772	12	PORT WING 5 SW	N 4645	W 09129	670
47-6827	13	PRAIRIE DU CHIEN	N 4302	W 09109	658
47-6838	13	PRAIRIE DU SAC 2 N //	N 4319	W 08944	780
47-6859	13	PRENTICE 1 N	N 4533	W 09017	1568
47-6922	13	RACINE	N 4243	W 08751	730
47-6939	13	RAINBOW RESERVOIR	N 4550	W 08933	1600
47-7052	13	REEDSBURG	N 4332	W 09001	905
47-7092	13	REST LAKE	N 4608	W 08953	1600
47-7113	13	RHINELANDER	N 4538	W 08925	1580
47-7121	12	RIB FALLS	N 4458	W 08954	1260
47-7132	13	RICE LAKE	N 4530	W 09145	1115
47-7140	12	RICE RESERVOIR	N 4532	W 08945	1460
47-7158	13	RICHLAND CENTER	N 4320	W 09023	728
47-7174	12	RIDGELAND 1 NNE	N 4513	W 09153	960
47-7209	12	RIPON 5 NE	N 4352	W 08845	920
47-7226	13	RIVER FALLS	N 4452	W 09237	900
47-7349	13	ROSHOLT	N 4438	W 08918	1185
47-7464	13	ST CROIX FALLS	N 4525	W 09239	770
47-7708	13	SHAWANO	N 4446	W 08837	810
47-7725	13	SHEBOYGAN	N 4345	W 08743	648
47-7892	13	SOLON SPRINGS	N 4621	W 09149	1083
47-7980	12	SOUTH PELICAN	N 4532	W 08912	1600
47-7997	13	SPARTA	N 4357	W 09048	845
47-8018	12	SPIRIT FALLS	N 4527	W 08958	1470
47-8027	13	SPOONER EXP FARM	N 4549	W 09153	1100
47-8110	13	STANLEY	N 4458	W 09056	1120
47-8171	13	STEVENS POINT	N 4430	W 08934	1079
47-8229	13	STOUGHTON	N 4255	W 08913	840
47-8241	12	STRATFORD 2 NNW	N 4450	W 09005	1260
47-8267	13	STURGEON BAY EXP FARM	N 4452	W 08720	656
47-8288	12	SUGAR CAMP	N 4552	W 08924	1605
47-8324	12	SUMMIT LAKE RANGER STA	N 4523	W 08912	1730
47-8349	13	SUPERIOR	N 4642	W 09201	630
47-8589	13	TREMPEALEAU DAM 6	N 4400	W 09126	660
47-8672	13	TWO RIVERS	N 4409	W 08734	599
47-8723	12	UNION GROVE	N 4242	W 08803	760
47-8827	13	VIROQUA	N 4334	W 09054	1240
47-8905	13	WASHINGTON ISLAND	N 4522	W 08656	612
47-8919	13	WATERTOWN	N 4312	W 08843	837
47-8937	13	WAUKESHA	N 4301	W 08814	860
47-8951	13	WAUPACA	N 4421	W 08904	840
47-8968	13	WAUSAU FAA AIRPORT	N 4455	W 08937	1196
47-8978	12	WAUSAUKEE	N 4523	W 08757	745
47-9046	13	WEST ALLIS	N 4301	W 08759	723
47-9050	13	WEST BEND	N 4324	W 08811	940
47-9144	13	WEYERHAUSER 2 SSE	N 4524	W 09123	1225
47-9190	13	WHITEWATER	N 4251	W 08844	800

47 — WISCONSIN

STATE-STATION NUMBER	STN TYP	NAME	LATITUDE DEG-MIN	LONGITUDE DEG-MIN	ELEVATION (FT)
47-9236	13	WILLOW RESERVOIR	N 4543	W 08951	1560
47-9304	13	WINTER 6 NNW	N 4553	W 09104	1310
47-9319	13	WISCONSIN DELLS	N 4338	W 08947	880
47-9335	13	WISCONSIN RAPIDS	N 4423	W 08948	1035
47-9345	12	WIS RPDS GRAND AVE BR	N 4424	W 08949	1000

47 — WISCONSIN

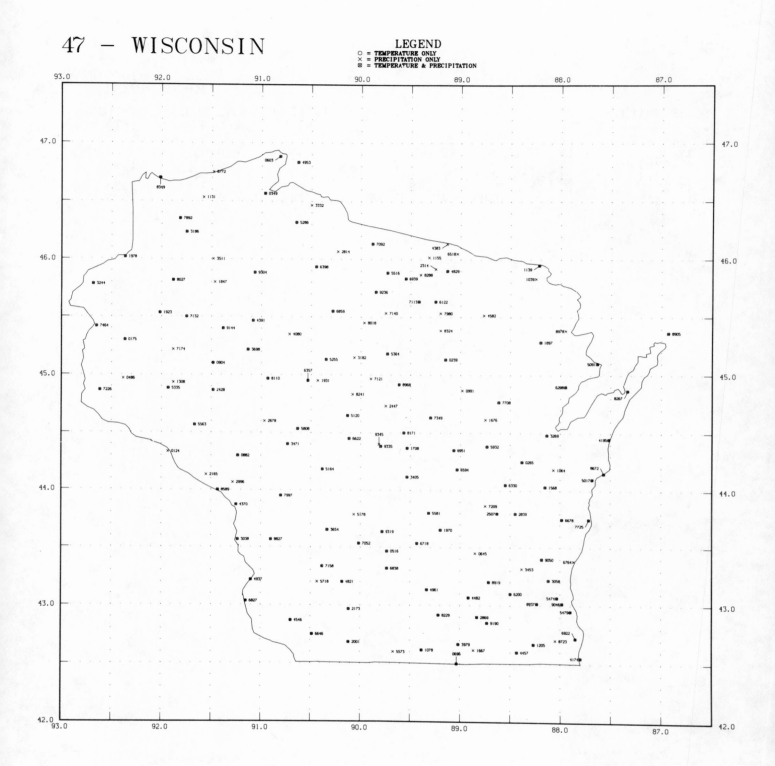

696

WYOMING

TEMPERATURE NORMALS (DEG F)

STATION			JAN	FEB	MAR	APR	MAY	JUN	JUL	AUG	SEP	OCT	NOV	DEC	ANN
AFTON		MAX	27.6	33.3	39.1	50.2	63.6	71.7	81.7	79.9	72.0	60.0	40.9	29.7	54.1
		MIN	4.9	7.6	12.8	23.4	31.5	36.9	41.2	39.8	32.6	24.5	15.6	6.5	23.1
		MEAN	16.3	20.5	26.0	36.8	47.6	54.3	61.5	59.9	52.3	42.3	28.3	18.1	38.7
ALBIN		MAX	37.4	42.0	46.4	57.1	67.5	78.8	86.4	84.7	76.0	63.9	47.7	40.5	60.7
		MIN	13.6	17.7	20.7	29.4	39.4	48.6	55.1	53.4	44.1	34.2	22.7	17.2	33.0
		MEAN	25.5	29.8	33.6	43.3	53.5	63.7	70.8	69.1	60.1	49.0	35.2	28.9	46.9
ALTA 1 NNW		MAX	29.0	34.9	39.2	48.3	61.1	70.1	80.2	77.6	69.0	57.1	40.3	31.7	53.2
		MIN	7.8	11.4	14.3	23.9	33.0	40.0	46.2	44.0	36.8	28.3	17.5	10.3	26.1
		MEAN	18.4	23.2	26.8	36.1	47.1	55.1	63.2	60.8	52.9	42.7	28.9	21.0	39.7
ALVA 5 SE		MAX	32.5	37.8	42.0	52.9	63.7	73.5	82.0	80.8	70.7	59.7	43.8	36.6	56.3
		MIN	5.4	10.3	15.1	25.9	35.6	44.1	48.8	46.9	37.5	28.2	17.2	10.4	27.1
		MEAN	19.0	24.1	28.6	39.4	49.7	58.8	65.5	63.9	54.1	43.9	30.5	23.5	41.8
ARCHER		MAX	38.8	42.3	45.4	56.0	66.5	77.3	85.7	83.6	75.2	63.8	47.9	41.7	60.4
		MIN	13.0	16.5	19.2	28.0	37.7	46.7	52.7	51.4	42.2	32.8	21.6	16.5	31.5
		MEAN	25.9	29.4	32.3	42.0	52.1	62.0	69.2	67.5	58.7	48.3	34.7	29.1	45.9
BASIN		MAX	29.0	39.0	49.9	62.0	72.9	82.7	91.9	89.2	77.9	65.0	45.4	33.7	61.6
		MIN	1.9	11.4	20.9	31.4	41.6	49.6	55.1	52.2	41.6	31.0	17.7	7.9	30.2
		MEAN	15.5	25.2	35.4	46.7	57.3	66.2	73.5	70.7	59.8	48.0	31.6	20.8	45.9
BONDURANT 3 NW		MAX	22.7	28.9	36.0	47.3	62.3	70.6	79.4	76.4	68.3	56.8	37.6	24.4	50.9
		MIN	-4.7	-2.7	3.1	17.8	27.5	32.2	34.3	32.7	25.1	16.9	8.0	-3.7	15.5
		MEAN	9.0	13.2	19.6	32.6	44.9	51.4	56.9	54.6	46.7	36.9	22.9	10.4	33.3
BORDER 3 N		MAX	26.5	31.7	38.9	52.0	64.7	73.3	83.6	81.7	73.4	61.4	41.9	29.7	54.9
		MIN	1.2	3.2	9.9	22.6	31.6	37.4	42.0	39.6	31.2	22.5	13.1	3.4	21.5
		MEAN	13.9	17.5	24.4	37.3	48.2	55.4	62.8	60.7	52.3	42.0	27.5	16.6	38.2
BOYSEN DAM		MAX	28.7	38.6	48.4	59.6	70.8	81.4	90.7	87.9	77.0	63.3	43.2	33.3	60.2
		MIN	7.3	14.4	23.3	33.8	44.3	53.3	60.4	58.7	48.3	37.5	23.6	13.3	34.9
		MEAN	18.0	26.5	35.9	46.7	57.5	67.4	75.6	73.3	62.7	50.4	33.4	23.3	47.6
BUFFALO BILL DAM		MAX	35.3	39.8	43.3	52.2	60.9	69.2	76.2	76.1	69.4	59.9	45.2	39.1	55.6
		MIN	19.0	23.9	26.3	34.4	43.2	50.5	57.0	56.5	49.8	41.9	30.4	24.2	38.1
		MEAN	27.2	31.9	34.8	43.3	52.1	59.9	66.6	66.3	59.6	50.9	37.8	31.7	46.8
CARPENTER 3 E		MAX	40.0	44.3	48.1	58.8	68.7	79.5	87.4	85.5	76.8	65.3	49.5	43.0	62.2
		MIN	12.4	16.5	19.7	29.0	38.5	47.6	53.7	51.7	42.8	32.5	21.2	15.8	31.8
		MEAN	26.2	30.4	33.9	43.9	53.6	63.6	70.6	68.6	59.8	48.9	35.4	29.4	47.0
CASPER 2 E		MAX	36.3	40.7	46.4	57.9	68.6	79.9	89.2	87.2	76.5	63.7	46.7	39.6	61.1
		MIN	15.6	19.5	23.0	30.9	40.0	48.3	54.8	52.7	42.5	34.1	24.1	19.4	33.7
		MEAN	26.0	30.1	34.7	44.4	54.3	64.1	72.0	70.0	59.5	48.9	35.4	29.5	47.4
CASPER WSO //R		MAX	32.5	37.4	43.4	54.9	66.2	78.1	87.1	84.8	74.2	61.0	43.9	35.6	58.3
		MIN	11.9	16.3	20.2	29.3	38.9	47.6	54.7	52.8	42.5	33.2	21.9	15.7	32.1
		MEAN	22.2	26.9	31.9	42.1	52.6	62.9	70.9	68.8	58.4	47.2	32.9	25.7	45.2
CHEYENNE WSO //R		MAX	37.3	40.7	43.6	54.0	64.6	75.4	83.1	80.8	72.1	61.0	46.5	40.4	58.3
		MIN	14.8	17.9	20.6	29.6	39.7	48.5	54.6	52.8	43.7	34.0	23.1	18.2	33.1
		MEAN	26.1	29.3	32.1	41.8	52.2	62.0	68.9	66.8	57.9	47.5	34.8	29.3	45.7
CHUGWATER		MAX	39.5	43.4	48.2	58.6	69.0	79.5	87.6	85.9	77.6	65.8	49.5	42.6	62.3
		MIN	14.9	17.8	20.2	28.0	37.4	45.7	51.1	49.0	39.1	30.8	21.8	17.3	31.1
		MEAN	27.2	30.6	34.2	43.3	53.2	62.6	69.4	67.5	58.3	48.3	35.7	29.9	46.7
CODY		MAX	35.3	41.1	46.2	55.9	66.5	76.1	85.3	82.7	72.6	62.1	46.0	39.0	59.1
		MIN	11.4	17.1	21.4	30.6	40.5	48.6	54.4	52.9	43.0	34.9	22.7	16.2	32.8
		MEAN	23.4	29.1	33.8	43.2	53.5	62.4	69.9	67.9	57.8	48.5	34.4	27.6	46.0
COLONY		MAX	30.3	36.4	42.6	56.3	67.8	77.7	87.3	86.2	74.7	62.4	44.8	35.5	58.5
		MIN	8.9	14.8	20.3	31.5	42.2	51.8	58.2	56.4	46.0	36.3	23.4	15.2	33.8
		MEAN	19.6	25.6	31.5	43.9	55.0	64.8	72.8	71.4	60.4	49.4	34.1	25.4	46.2

WYOMING

TEMPERATURE NORMALS (DEG F)

STATION		JAN	FEB	MAR	APR	MAY	JUN	JUL	AUG	SEP	OCT	NOV	DEC	ANN
CRANDALL CREEK	MAX	31.0	36.5	41.0	50.2	61.5	70.1	80.8	78.9	70.3	59.5	42.3	33.7	54.7
	MIN	4.0	6.7	10.6	21.1	29.2	34.9	39.6	38.1	31.1	24.3	14.7	6.9	21.8
	MEAN	17.5	21.7	25.8	35.7	45.4	52.5	60.2	58.5	50.7	41.9	28.6	20.3	38.2
DEAVER	MAX	29.9	39.3	48.8	60.6	70.9	80.2	89.4	86.8	75.7	63.6	44.7	34.1	60.3
	MIN	4.0	11.3	18.0	28.0	39.4	47.2	53.2	50.3	40.3	29.9	17.5	8.6	29.0
	MEAN	17.0	25.3	33.4	44.3	55.1	63.7	71.3	68.6	58.0	46.8	31.1	21.4	44.7
DILLINGER	MAX	31.8	37.8	43.6	55.5	66.3	77.4	87.5	85.7	74.3	61.7	44.9	36.1	58.6
	MIN	4.7	11.3	17.9	27.8	38.2	47.0	53.3	51.1	40.2	29.8	17.1	9.3	29.0
	MEAN	18.3	24.6	30.8	41.7	52.3	62.2	70.4	68.4	57.3	45.8	31.0	22.7	43.8
DIVERSION DAM	MAX	34.2	40.9	46.8	56.9	67.1	77.2	86.1	83.6	74.1	62.5	44.8	37.3	59.3
	MIN	7.2	13.4	19.3	28.5	38.0	45.5	50.6	48.5	39.9	31.1	17.9	10.8	29.2
	MEAN	20.7	27.2	33.1	42.7	52.6	61.4	68.4	66.1	57.0	46.8	31.4	24.1	44.3
DIXON	MAX	32.1	36.3	42.7	55.6	66.7	76.3	83.0	80.3	72.1	61.0	44.1	34.9	57.1
	MIN	3.9	8.0	16.5	26.4	34.4	41.3	47.5	45.2	36.0	25.9	15.6	5.9	25.6
	MEAN	18.0	22.2	29.6	41.0	50.6	58.8	65.3	62.7	54.1	43.5	29.9	20.4	41.3
DULL CENTER 1 SE	MAX	34.8	40.7	46.4	58.3	69.0	80.5	90.4	88.4	77.4	65.1	47.2	38.8	61.4
	MIN	9.3	15.1	20.0	30.0	40.3	49.0	55.0	53.1	42.5	32.4	21.1	14.0	31.8
	MEAN	22.1	27.9	33.2	44.2	54.7	64.8	72.7	70.8	59.9	48.8	34.2	26.4	46.6
ELK MOUNTAIN	MAX	32.1	35.2	39.4	50.6	62.4	72.6	80.6	78.3	70.2	58.5	42.9	35.4	54.9
	MIN	12.5	14.6	17.3	25.2	33.4	40.9	46.4	44.4	36.8	29.4	20.1	15.4	28.0
	MEAN	22.3	24.9	28.4	37.9	47.9	56.8	63.5	61.4	53.5	44.0	31.6	25.4	41.5
ENCAMPMENT 10 ESE	MAX	32.9	35.8	40.6	52.3	63.7	73.0	80.7	78.5	71.0	60.7	44.8	36.2	55.9
	MIN	10.9	12.7	16.3	25.2	34.3	40.8	47.3	45.6	37.0	28.3	18.4	12.5	27.4
	MEAN	21.9	24.3	28.5	38.8	49.0	56.9	64.0	62.1	54.1	44.6	31.6	24.4	41.7
EVANSTON 1 E	MAX	31.4	35.1	40.0	50.9	63.0	73.0	82.7	80.3	71.8	59.8	42.7	34.3	55.4
	MIN	6.2	8.4	13.7	23.1	31.6	38.3	43.9	42.0	33.8	25.4	15.3	8.1	24.2
	MEAN	18.8	21.8	26.9	37.0	47.3	55.7	63.3	61.2	52.8	42.6	29.0	21.2	39.8
FARSON	MAX	25.8	31.2	39.3	53.3	65.0	75.0	83.8	80.8	71.9	59.6	40.6	29.4	54.6
	MIN	-5.7	-1.1	9.9	21.7	30.7	38.1	43.9	41.3	31.7	21.7	8.0	-2.3	19.8
	MEAN	10.1	15.1	24.7	37.5	47.9	56.6	63.8	61.1	51.8	40.6	24.3	13.6	37.3
GILLETTE 2 E	MAX	31.5	37.3	43.4	55.2	65.9	76.3	86.6	85.3	74.3	62.2	44.2	35.9	58.2
	MIN	8.8	15.0	20.2	29.7	39.7	48.2	54.5	52.7	42.9	33.6	21.3	14.1	31.7
	MEAN	20.2	26.2	31.8	42.4	52.8	62.3	70.6	69.0	58.7	47.9	32.7	25.0	45.0
GILLETTE 18 SW	MAX	31.3	36.8	42.8	54.2	65.1	75.5	85.0	83.4	73.1	60.9	43.6	35.4	57.3
	MIN	9.7	15.1	19.4	28.8	38.5	47.1	53.5	51.8	42.0	32.8	20.8	14.2	31.1
	MEAN	20.5	26.0	31.1	41.5	51.8	61.3	69.3	67.6	57.6	46.9	32.2	24.8	44.2
GLENROCK 5 ESE	MAX	37.1	41.6	47.2	58.4	69.3	80.6	89.8	87.5	77.2	64.7	47.9	40.5	61.8
	MIN	14.0	18.9	22.4	31.5	40.6	49.3	55.8	53.6	43.2	34.0	23.9	17.9	33.8
	MEAN	25.6	30.3	34.8	45.0	55.0	65.0	72.8	70.6	60.2	49.4	35.9	29.2	47.8
GREEN RIVER	MAX	32.1	37.5	44.5	55.7	67.9	77.9	87.0	84.1	74.8	63.0	45.0	35.4	58.7
	MIN	4.7	8.9	16.8	26.9	36.0	43.4	49.4	47.2	36.8	26.8	15.8	7.3	26.7
	MEAN	18.4	23.2	30.7	41.3	52.0	60.7	68.2	65.7	55.9	44.9	30.4	21.4	42.7
HEART MOUNTAIN	MAX	31.5	39.8	46.6	57.1	67.8	77.2	85.1	82.8	72.9	61.6	44.0	35.8	58.5
	MIN	7.4	14.3	19.5	29.2	39.5	47.0	52.2	50.1	41.1	32.2	19.5	12.0	30.3
	MEAN	19.5	27.0	33.1	43.2	53.7	62.1	68.7	66.4	57.0	46.9	31.8	23.9	44.4
JACKSON	MAX	26.5	32.0	38.8	50.5	62.6	71.6	81.6	79.3	70.6	58.5	39.4	28.1	53.3
	MIN	5.4	8.4	13.8	23.9	30.6	36.5	39.9	38.0	30.8	22.7	15.5	6.9	22.7
	MEAN	16.0	20.2	26.3	37.2	46.6	54.0	60.8	58.7	50.7	40.6	27.5	17.6	38.0
KAYCEE	MAX	36.6	42.2	47.3	57.7	68.1	78.6	88.1	86.2	76.0	64.2	47.8	40.6	61.1
	MIN	5.4	12.5	18.5	27.8	37.5	45.1	51.0	48.6	38.9	29.0	17.1	10.0	28.5
	MEAN	21.0	27.4	32.9	42.8	52.8	61.9	69.6	67.4	57.4	46.6	32.4	25.3	44.8

WYOMING

TEMPERATURE NORMALS (DEG F)

STATION			JAN	FEB	MAR	APR	MAY	JUN	JUL	AUG	SEP	OCT	NOV	DEC	ANN
LAGRANGE		MAX	38.8	43.4	48.0	59.1	69.5	80.4	88.9	86.9	77.2	65.2	49.4	42.2	62.4
		MIN	12.1	16.8	20.8	29.6	39.6	48.7	54.7	52.2	42.0	31.8	20.9	15.6	32.1
		MEAN	25.4	30.1	34.4	44.4	54.6	64.6	71.8	69.6	59.6	48.5	35.2	29.0	47.3
LANDER WSO	//R	MAX	31.3	37.7	44.2	54.7	65.5	76.5	86.0	83.7	73.1	60.3	42.6	35.0	57.6
		MIN	7.7	13.5	19.9	29.8	39.6	48.0	55.4	53.4	43.6	33.3	18.9	11.3	31.2
		MEAN	19.6	25.7	32.1	42.3	52.6	62.3	70.8	68.6	58.3	46.8	30.8	23.2	44.4
LARAMIE FAA AIRPORT		MAX	32.1	35.4	39.6	50.3	61.8	73.0	80.6	78.2	69.6	57.9	42.1	34.9	54.6
		MIN	8.7	11.0	15.2	24.1	33.4	41.8	48.1	45.9	37.2	27.7	16.5	11.1	26.7
		MEAN	20.4	23.2	27.4	37.2	47.6	57.4	64.4	62.0	53.4	42.8	29.3	23.0	40.7
LOVELL		MAX	28.9	37.9	46.9	58.3	69.5	79.5	89.1	86.5	74.4	62.4	44.4	33.8	59.3
		MIN	4.1	12.3	19.8	30.4	41.5	49.4	54.5	51.4	40.5	30.0	18.3	9.0	30.1
		MEAN	16.5	25.1	33.4	44.4	55.6	64.5	71.8	69.0	57.5	46.2	31.4	21.4	44.7
MEDICINE BOW		MAX	31.6	35.4	41.2	53.6	65.3	75.8	83.3	80.8	71.8	59.7	42.4	34.2	56.3
		MIN	10.3	13.2	17.0	25.1	33.8	41.6	46.9	44.4	35.1	26.1	17.4	12.3	26.9
		MEAN	20.9	24.3	29.1	39.4	49.5	58.8	65.1	62.6	53.5	42.9	30.0	23.3	41.6
MIDWEST 1 SW		MAX	35.5	42.0	48.3	59.4	70.4	81.4	90.6	88.6	78.6	66.0	47.2	38.9	62.2
		MIN	10.9	16.3	21.1	30.0	39.5	47.9	54.3	52.2	42.1	32.9	21.7	15.3	32.0
		MEAN	23.2	29.2	34.7	44.7	55.0	64.7	72.5	70.4	60.4	49.5	34.5	27.1	47.2
MOORCROFT		MAX	29.8	35.2	41.9	54.3	65.4	76.0	85.7	84.2	73.1	60.6	43.0	34.0	56.9
		MIN	7.1	12.9	19.8	29.8	40.3	49.1	55.6	53.7	42.7	32.5	19.9	11.6	31.3
		MEAN	18.5	24.1	30.9	42.0	52.9	62.6	70.7	69.0	57.9	46.6	31.5	22.8	44.1
MORAN 5 WNW	R	MAX	24.6	30.9	36.5	46.0	57.5	67.2	77.4	75.0	65.9	54.0	36.9	27.0	49.9
		MIN	1.4	3.5	8.0	20.3	29.9	36.8	41.1	39.6	31.9	24.0	14.5	4.6	21.3
		MEAN	13.0	17.2	22.3	33.2	43.8	52.0	59.2	57.3	48.9	39.0	25.7	15.8	35.6
NEWCASTLE		MAX	32.5	38.3	44.8	57.1	68.6	79.4	87.5	84.9	73.8	61.6	44.6	36.3	59.1
		MIN	9.9	14.7	21.1	31.6	42.6	51.9	58.9	56.7	45.6	35.4	22.2	14.2	33.7
		MEAN	21.2	26.5	33.0	44.4	55.6	65.7	73.3	70.9	59.7	48.6	33.4	25.3	46.5
PATHFINDER DAM		MAX	31.6	35.8	41.7	53.8	65.4	76.6	86.1	83.7	73.8	61.8	44.3	35.1	57.5
		MIN	11.2	14.3	19.6	29.6	38.8	47.3	54.4	52.5	42.3	33.7	23.0	15.7	31.9
		MEAN	21.4	25.1	30.7	41.7	52.1	62.0	70.3	68.1	58.1	47.8	33.7	25.4	44.7
PAVILLION		MAX	32.2	39.8	47.1	57.6	67.8	77.4	85.5	82.8	73.1	60.9	43.6	35.4	58.6
		MIN	7.7	14.4	20.2	29.7	39.6	47.8	53.6	51.8	42.6	32.8	19.0	11.1	30.9
		MEAN	19.9	27.1	33.7	43.7	53.7	62.6	69.6	67.3	57.9	46.9	31.3	23.3	44.8
PHILLIPS		MAX	39.8	43.7	47.7	58.2	68.1	78.6	86.6	84.8	76.3	65.2	49.7	43.0	61.8
		MIN	15.2	18.9	22.1	30.5	40.0	48.4	54.6	52.5	43.0	33.9	23.3	18.7	33.4
		MEAN	27.5	31.3	34.9	44.4	54.1	63.5	70.6	68.7	59.7	49.6	36.5	30.9	47.6
POWELL		MAX	31.3	39.8	47.9	59.2	69.8	79.4	88.1	85.7	74.9	63.1	44.9	35.3	60.0
		MIN	9.7	16.6	22.0	31.7	42.7	50.8	57.0	54.8	44.4	34.5	22.1	14.0	33.4
		MEAN	20.5	28.2	35.0	45.4	56.3	65.1	72.5	70.3	59.7	48.8	33.5	24.7	46.7
REDBIRD 1 NW		MAX	34.1	41.0	48.0	59.9	70.6	81.5	90.5	88.6	78.1	65.5	47.4	38.2	62.0
		MIN	6.1	12.8	20.0	30.6	40.9	49.8	55.9	53.2	41.8	30.8	18.5	10.2	30.9
		MEAN	20.1	27.0	34.0	45.3	55.8	65.6	73.2	70.9	60.0	48.2	33.0	24.2	46.4
RIVERTON		MAX	29.2	37.5	46.7	57.8	68.8	79.7	89.3	86.5	75.3	62.1	42.6	32.7	59.0
		MIN	-.8	7.2	17.2	28.3	38.3	46.1	51.4	48.3	38.2	27.9	13.4	3.4	26.6
		MEAN	14.2	22.4	32.0	43.0	53.6	62.9	70.3	67.4	56.8	45.0	28.1	18.1	42.8
ROCHELLE 3 E		MAX	32.7	39.2	45.0	55.9	68.1	79.6	89.0	86.9	76.5	64.8	46.6	36.2	60.0
		MIN	6.2	13.1	19.0	28.2	38.2	48.1	54.6	52.6	39.7	29.0	19.3	10.6	29.9
		MEAN	19.5	26.2	32.0	42.0	53.2	63.9	71.9	69.8	58.1	46.9	32.9	23.5	45.0
ROCK SPRINGS		MAX	32.1	37.0	43.6	55.0	66.4	77.2	86.5	83.3	73.5	61.0	43.5	34.6	57.8
		MIN	11.2	15.5	20.7	28.7	37.6	45.2	51.8	49.5	39.9	30.6	20.1	13.7	30.4
		MEAN	21.6	26.3	32.1	41.9	52.0	61.2	69.1	66.4	56.7	45.8	31.8	24.2	44.1

WYOMING

TEMPERATURE NORMALS (DEG F)

STATION		JAN	FEB	MAR	APR	MAY	JUN	JUL	AUG	SEP	OCT	NOV	DEC	ANN
ROCK SPRINGS FAA AP	MAX	29.3	34.4	40.3	51.8	63.5	74.3	83.4	80.6	70.9	58.0	40.7	32.0	54.9
	MIN	10.6	14.5	19.2	27.7	37.1	45.5	52.9	50.9	41.3	31.5	19.5	13.0	30.3
	MEAN	20.0	24.5	29.8	39.8	50.3	59.9	68.2	65.8	56.1	44.8	30.1	22.5	42.6
SARATOGA	MAX	33.4	37.1	42.0	53.7	65.7	76.1	84.2	81.0	72.3	60.6	44.1	36.4	57.2
	MIN	10.0	13.1	17.6	26.3	35.1	42.8	48.6	46.5	37.5	28.3	18.3	12.0	28.0
	MEAN	21.7	25.1	29.8	40.0	50.4	59.5	66.5	63.8	54.9	44.5	31.2	24.2	42.6
SEMINOE DAM	MAX	29.3	33.6	39.8	52.1	63.8	74.7	83.4	81.0	71.4	58.7	41.2	32.8	55.2
	MIN	11.7	14.9	18.6	27.7	37.3	46.6	53.9	51.8	42.4	33.0	21.7	15.3	31.2
	MEAN	20.5	24.3	29.2	39.9	50.6	60.7	68.7	66.4	56.9	45.9	31.4	24.1	43.2
SHERIDAN WSO //R	MAX	31.8	38.0	44.1	55.6	66.2	75.9	86.0	84.5	73.3	61.9	45.3	36.9	58.3
	MIN	7.3	14.1	19.7	29.4	39.7	47.6	53.8	52.1	42.0	32.1	19.7	12.3	30.8
	MEAN	19.5	26.1	31.9	42.5	53.0	61.8	69.9	68.3	57.6	47.0	32.5	24.6	44.6
SHERIDAN FIELD STA	MAX	30.9	37.0	43.4	56.1	67.5	76.7	87.5	86.5	74.8	63.0	45.0	36.4	58.7
	MIN	4.1	11.0	17.9	29.3	39.4	47.5	53.0	50.6	40.5	30.1	18.0	9.8	29.3
	MEAN	17.5	24.0	30.7	42.7	53.5	62.2	70.3	68.6	57.7	46.6	31.6	23.1	44.0
SUNDANCE	MAX	29.9	35.5	41.8	53.5	64.5	74.4	82.8	81.7	71.9	60.1	42.3	33.4	56.0
	MIN	8.3	13.0	18.7	29.1	39.2	48.4	54.8	53.4	43.1	33.7	20.9	13.5	31.3
	MEAN	19.1	24.3	30.2	41.3	51.9	61.4	68.8	67.5	57.5	46.9	31.7	23.5	43.7
THERMOPOLIS	MAX	35.1	42.8	50.1	61.3	71.6	81.9	91.0	88.7	78.3	65.8	47.6	38.6	62.7
	MIN	5.8	13.9	20.6	30.5	39.8	47.2	53.3	51.0	41.1	31.1	18.7	10.2	30.3
	MEAN	20.5	28.4	35.4	45.9	55.7	64.6	72.2	69.9	59.7	48.5	33.2	24.4	46.5
TORRINGTON EXP FARM	MAX	40.4	46.3	51.7	62.5	72.2	82.8	90.0	87.8	79.2	67.9	51.9	43.6	64.7
	MIN	10.6	16.5	21.3	30.6	40.9	49.9	55.2	52.3	41.5	30.5	20.1	13.9	31.9
	MEAN	25.5	31.5	36.5	46.6	56.6	66.4	72.7	70.1	60.3	49.2	36.0	28.7	48.3
UPTON	MAX	29.0	35.2	43.1	56.3	67.5	78.2	87.1	85.2	74.5	61.3	42.5	32.6	57.7
	MIN	3.2	9.5	17.4	27.9	38.5	47.9	54.5	51.9	40.8	30.0	17.1	7.9	28.9
	MEAN	16.1	22.4	30.3	42.2	53.0	63.1	70.8	68.6	57.7	45.7	29.8	20.3	43.3
WHALEN DAM	MAX	38.5	43.9	48.7	59.3	69.9	81.1	89.6	87.8	78.1	66.8	50.3	42.4	63.0
	MIN	9.3	15.5	20.9	31.0	41.3	50.4	56.5	54.3	43.2	32.0	20.1	13.3	32.3
	MEAN	23.9	29.7	34.8	45.2	55.6	65.8	73.1	71.1	60.7	49.4	35.2	27.9	47.7
WHEATLAND 4 N	MAX	39.2	44.0	49.4	60.3	70.2	81.0	89.1	86.9	77.9	66.4	50.0	42.9	63.1
	MIN	15.7	19.7	23.1	31.5	41.4	49.6	55.7	53.1	43.4	34.4	24.8	19.7	34.3
	MEAN	27.5	31.9	36.3	45.9	55.8	65.3	72.4	70.0	60.7	50.4	37.4	31.3	48.7
WORLAND	MAX	28.0	37.8	47.2	58.7	69.8	80.0	89.4	86.6	74.9	62.8	44.4	32.7	59.4
	MIN	.8	9.8	19.5	30.5	41.6	49.6	54.5	51.0	40.4	30.1	17.0	7.0	29.3
	MEAN	14.4	23.8	33.3	44.6	55.7	64.8	72.0	68.8	57.7	46.5	30.7	19.9	44.4
YELLOWSTONE PARK //	MAX	28.4	34.2	38.9	49.0	60.8	70.2	81.0	78.1	68.3	56.4	39.4	31.1	53.0
	MIN	9.7	13.7	15.4	25.2	33.9	40.6	46.2	44.6	37.0	29.3	19.6	12.4	27.3
	MEAN	19.1	24.0	27.2	37.1	47.4	55.4	63.6	61.4	52.6	42.9	29.5	21.8	40.2
YODER	MAX	40.1	45.8	51.6	62.5	72.2	83.2	91.1	88.8	80.1	69.0	52.0	43.6	65.0
	MIN	11.8	16.6	21.4	29.8	39.8	48.6	54.3	51.5	40.9	31.0	20.9	15.4	31.8
	MEAN	26.0	31.2	36.5	46.2	56.0	65.9	72.7	70.2	60.5	50.0	36.5	29.5	48.4

WYOMING

PRECIPITATION NORMALS (INCHES)

STATION		JAN	FEB	MAR	APR	MAY	JUN	JUL	AUG	SEP	OCT	NOV	DEC	ANN
AFTON		1.71	1.35	1.30	1.67	1.95	1.90	1.05	1.26	1.38	1.34	1.44	1.68	18.03
ALBIN		.73	.57	1.49	1.96	3.14	2.57	1.96	1.40	1.13	.95	.77	.73	17.40
ALTA 1 NNW		1.94	1.47	1.53	1.96	2.43	2.44	1.16	1.68	1.66	1.65	1.59	1.81	21.32
ALVA 5 SE		.79	.94	1.42	2.84	3.87	3.91	2.07	1.89	1.59	1.43	1.12	.98	22.85
ARCHER		.39	.34	.99	1.34	2.57	2.43	1.86	1.52	1.25	.82	.56	.37	14.44
BASIN		.21	.17	.25	.74	1.13	1.16	.48	.58	.72	.49	.28	.22	6.43
BIG PINEY		.54	.47	.50	.83	1.21	.98	.76	.90	.80	.59	.51	.59	8.68
BONDURANT 3 NW		3.01	2.05	2.19	1.37	1.60	1.74	1.01	1.39	1.30	1.33	1.94	2.87	21.80
BORDER 3 N		1.37	1.14	.95	1.10	1.32	1.48	.76	.94	1.12	1.00	1.02	1.27	13.47
BOYSEN DAM		.30	.28	.56	1.61	1.89	1.48	.62	.60	.66	.86	.28	.25	9.39
BUFFALO BILL DAM		.41	.34	.69	1.64	2.15	1.85	.95	.80	1.07	.81	.52	.33	11.56
CARPENTER 3 E		.28	.21	.79	1.20	2.55	2.45	2.00	1.60	1.23	.67	.40	.34	13.72
CASPER 2 E		.50	.51	.90	1.62	2.38	1.61	1.06	.77	.87	1.04	.77	.50	12.53
CASPER WSO //R		.50	.56	.99	1.51	2.13	1.24	1.06	.63	.76	.88	.66	.51	11.43
CHEYENNE WSO //R		.41	.40	.97	1.24	2.39	2.00	1.87	1.39	1.06	.68	.53	.37	13.31
CHUGWATER		.67	.61	1.08	1.60	2.79	2.12	2.03	1.25	.94	.90	.70	.62	15.31
CLEARMONT 5 SW		.68	.56	.70	1.65	2.37	2.39	1.29	1.25	1.05	.99	.71	.59	14.23
CODY		.40	.29	.48	.99	1.69	1.62	.89	.81	.94	.61	.42	.26	9.40
COLONY		.33	.44	.66	1.43	2.49	2.91	1.65	1.45	1.13	.95	.53	.39	14.36
CRANDALL CREEK		1.67	1.17	1.26	1.16	1.68	1.67	1.18	1.29	1.33	1.13	.88	1.26	15.68
DEAVER		.20	.14	.13	.56	.91	1.21	.51	.67	.58	.28	.18	.11	5.48
DILLINGER		.39	.53	.58	1.51	2.60	2.47	1.49	1.19	1.03	.80	.44	.45	13.48
DIVERSION DAM		.17	.17	.44	1.14	2.11	1.54	.79	.60	.87	.64	.36	.20	9.03
DIXON		1.00	.70	.94	1.00	1.24	.87	.96	.99	1.02	1.13	.77	.99	11.61
DOUBLE FOUR RANCH		.44	.36	.96	1.62	2.85	2.34	1.79	1.19	1.08	.68	.59	.51	14.41
DUBOIS		.31	.20	.52	1.08	1.37	1.46	.75	.79	.88	.56	.38	.29	8.59
DULL CENTER 1 SE		.28	.33	.55	1.24	2.37	2.13	1.67	1.09	.91	.62	.38	.29	11.86
ELK MOUNTAIN		.85	.72	1.05	1.54	1.62	1.44	1.06	1.04	1.06	1.09	.85	.69	13.01
ENCAMPMENT 10 ESE		.85	.74	1.25	1.65	1.76	1.29	1.25	1.61	1.13	1.28	.98	.98	14.77
EVANSTON 1 E		.75	.63	.85	1.19	1.17	.98	.73	.91	.91	1.03	.77	.76	10.68
FARSON		.41	.31	.36	.54	1.19	1.07	.64	.69	.65	.65	.32	.39	7.22
FT LARAMIE 11 NNW		.28	.26	.52	1.27	2.37	2.44	1.55	1.04	1.00	.57	.33	.33	11.96
FORT WASHAKIE 2 S		.31	.39	.75	1.79	2.60	1.66	.74	.64	.94	.90	.67	.33	11.72
GILLETTE 2 E		.58	.63	.78	1.76	2.72	3.11	1.35	1.32	1.24	1.05	.71	.60	15.85
GILLETTE 18 SW		.56	.69	.95	1.94	2.74	3.06	1.22	1.31	1.03	1.03	.71	.60	15.84
GLENROCK 5 ESE		.49	.45	.85	1.69	2.58	1.94	1.20	.85	1.02	.98	.60	.38	13.03
GLENROCK 14 SSE		.67	.69	1.11	1.88	2.58	2.25	1.38	.84	1.09	1.05	.86	.65	15.05
GREEN RIVER		.39	.31	.50	.81	1.21	.98	.63	.76	.68	.73	.40	.34	7.74
GREYBULL		.40	.25	.35	.74	1.03	1.17	.53	.60	.70	.46	.34	.29	6.86
HEART MOUNTAIN		.23	.18	.47	.81	1.60	1.40	.78	.74	.81	.54	.28	.20	8.04
HECLA		.48	.50	1.17	1.69	2.49	2.03	2.19	1.44	.95	.96	.68	.46	15.04
JACKSON		1.63	1.04	1.05	1.08	1.72	1.72	.84	1.15	1.16	1.09	1.14	1.65	15.27
KAYCEE		.44	.37	.69	1.65	2.23	2.18	1.05	.83	.93	.81	.49	.43	12.10
KEELINE		.52	.53	.87	1.85	2.35	2.18	1.37	1.09	.99	.74	.58	.47	13.54
KEMMERER 4 SW		.69	.57	.51	.70	1.23	1.21	.62	.77	.81	.70	.62	.64	9.07
LAGRANGE		.55	.38	1.07	1.68	2.87	2.28	1.83	1.27	1.05	.80	.58	.52	14.88
LAKE YELLOWSTONE		2.20	1.45	1.62	1.33	1.69	2.25	1.28	1.72	1.53	1.32	1.47	1.93	19.79
LANDER WSO //R		.48	.63	1.13	2.22	2.69	1.45	.71	.49	.87	1.20	.76	.53	13.16
LARAMIE FAA AIRPORT		.46	.45	.74	.86	1.32	1.18	1.49	1.23	.82	.73	.51	.39	10.18
LEO 6 SW		.53	.66	.84	1.38	1.70	1.10	.80	.69	.70	.91	.66	.57	10.54
LOVELL		.23	.16	.23	.67	1.11	1.31	.56	.70	.69	.46	.27	.20	6.59
MEDICINE BOW		.53	.51	.78	1.18	1.35	1.13	1.13	.86	.88	.86	.56	.54	10.31
MIDWEST 1 SW		.71	.74	.86	1.84	2.47	2.01	1.23	.91	.97	.83	.73	.76	14.06
MOORCROFT		.37	.43	.44	1.11	2.57	2.59	1.42	1.24	1.01	.70	.43	.45	12.76
MORAN 5 WNW R		3.37	2.26	2.02	1.81	1.86	1.66	.98	1.23	1.41	1.44	2.23	2.99	23.26

WYOMING

PRECIPITATION NORMALS (INCHES)

STATION	JAN	FEB	MAR	APR	MAY	JUN	JUL	AUG	SEP	OCT	NOV	DEC	ANN
NEWCASTLE	.42	.49	.58	1.44	2.46	2.63	1.96	1.51	.93	.62	.47	.53	14.04
PARKMAN 5 WNW	.80	.83	1.23	2.88	3.29	2.82	1.07	1.29	1.80	1.42	1.08	.71	19.22
PATHFINDER DAM	.31	.41	.67	1.16	1.82	1.29	.77	.69	.67	.88	.46	.37	9.50
PAVILLION	.16	.12	.34	1.00	1.86	1.36	.67	.54	.66	.53	.33	.22	7.79
PHILLIPS	.45	.35	.86	1.48	2.87	2.25	1.95	1.31	1.02	.79	.55	.43	14.31
PINEDALE	.68	.52	.49	.78	1.44	1.32	.88	1.10	.93	.75	.61	.72	10.22
POWELL	.20	.17	.19	.56	1.18	1.36	.74	.69	.73	.40	.23	.15	6.60
RAIRDEN 2 WSW	.25	.19	.23	.74	1.08	.99	.43	.55	.64	.50	.28	.23	6.11
REDBIRD 1 NW	.29	.42	.66	1.56	2.44	2.61	1.95	1.38	1.02	.75	.37	.29	13.74
RIVERTON	.19	.22	.38	1.09	1.90	1.30	.65	.43	.60	.66	.41	.27	8.10
ROCHELLE 3 E	.27	.39	.63	1.45	2.58	2.16	1.67	1.29	.89	.73	.44	.37	12.87
ROCK SPRINGS	.45	.42	.60	.97	1.28	.99	.56	.72	.72	.74	.52	.46	8.43
ROCK SPRINGS FAA AP	.54	.53	.67	1.00	1.17	.95	.70	.69	.75	.76	.56	.53	8.85
SARATOGA	.47	.40	.70	.95	1.28	.94	.89	.95	.79	1.00	.50	.47	9.34
SEMINOE DAM	.56	.67	1.02	1.64	2.10	1.32	.90	.79	.83	1.18	.74	.64	12.39
SHERIDAN WSO //R	.74	.76	1.06	2.00	2.42	2.24	.94	.96	1.16	1.16	.81	.68	14.93
SHERIDAN FIELD STA	.56	.51	.78	1.75	2.66	2.88	1.06	1.09	1.28	1.26	.74	.53	15.10
SHOSHONI	.16	.12	.20	.91	1.58	1.08	.42	.52	.52	.55	.23	.17	6.46
SOUTH PASS CITY	1.56	.91	1.16	1.40	1.70	1.24	.69	.94	.99	1.03	.92	1.25	13.79
SUNDANCE	.58	.70	.79	1.77	2.80	3.50	1.72	1.61	1.30	.86	.69	.62	16.94
TENNYSON	.61	.51	1.28	1.75	2.73	2.37	2.00	1.41	1.22	.90	.70	.56	16.04
THERMOPOLIS	.43	.37	.79	1.61	2.36	1.64	.58	.74	.91	1.03	.59	.45	11.50
TORRINGTON EXP FARM	.31	.32	.68	1.57	2.53	2.48	1.72	1.01	.96	.80	.37	.38	13.13
TOWER FALLS	1.44	.97	1.13	1.25	1.93	2.25	1.56	1.57	1.46	1.15	1.14	1.40	17.25
UPTON	.41	.52	.52	1.45	2.58	2.76	1.75	1.47	.95	.77	.52	.56	14.26
UPTON 13 SW	.22	.44	.50	1.39	2.49	2.45	1.53	1.20	.93	.60	.30	.38	12.43
WHALEN DAM	.37	.33	.69	1.43	2.44	2.38	1.58	.99	1.02	.65	.37	.40	12.65
WHEATLAND 4 N	.27	.27	.64	1.34	2.54	2.12	1.38	1.02	.98	.61	.35	.28	11.80
WORLAND	.24	.16	.26	.87	1.34	1.10	.40	.58	.64	.59	.29	.19	6.66
YELLOWSTONE PARK //	1.32	.83	1.12	1.17	1.88	2.10	1.24	1.53	1.26	.92	1.06	1.19	15.62
YODER	.33	.25	.72	1.57	2.69	2.37	1.74	1.00	.99	.80	.37	.35	13.18

WYOMING

HEATING DEGREE DAY NORMALS (BASE 65 DEG F)

STATION		JUL	AUG	SEP	OCT	NOV	DEC	JAN	FEB	MAR	APR	MAY	JUN	ANN
AFTON		118	175	381	704	1101	1454	1510	1246	1209	846	539	321	9604
ALBIN		19	23	199	501	894	1119	1225	986	973	651	357	129	7076
ALTA 1 NNW		93	168	372	691	1083	1364	1445	1170	1184	867	555	305	9297
ALVA 5 SE		81	104	341	654	1035	1287	1426	1145	1128	768	474	202	8645
ARCHER		25	30	219	518	909	1113	1212	997	1014	690	400	142	7269
BASIN		5	17	195	527	1002	1370	1535	1114	918	549	249	73	7554
BONDURANT 3 NW		251	322	549	871	1263	1693	1736	1450	1407	972	623	408	11545
BORDER 3 N		89	156	381	713	1125	1500	1584	1330	1259	831	521	293	9782
BOYSEN DAM		0	9	154	453	948	1293	1457	1078	902	549	247	68	7158
BUFFALO BILL DAM		44	56	214	443	816	1032	1172	927	936	651	400	174	6865
CARPENTER 3 E		12	15	186	499	888	1104	1203	969	964	633	353	109	6935
CASPER 2 E		14	13	203	499	888	1101	1209	977	939	618	335	111	6907
CASPER WSO	//R	16	31	240	552	963	1218	1327	1067	1026	687	384	131	7642
CHEYENNE WSO	//R	24	37	235	543	906	1107	1206	1000	1020	696	397	139	7310
CHUGWATER		16	30	223	518	879	1088	1172	963	955	651	366	127	6988
CODY		20	55	252	512	918	1159	1290	1005	967	654	361	139	7332
COLONY		14	25	204	490	927	1228	1407	1103	1039	633	317	106	7493
CRANDALL CREEK		157	215	435	716	1092	1386	1473	1212	1215	879	608	375	9763
DEAVER		8	43	238	564	1017	1352	1488	1112	980	621	314	117	7854
DILLINGER		37	48	264	595	1020	1311	1448	1131	1060	699	394	147	8154
DIVERSION DAM		18	48	258	564	1008	1268	1373	1058	989	669	384	144	7781
DIXON		48	100	333	667	1053	1383	1457	1198	1097	720	446	201	8703
DULL CENTER 1 SE		18	21	205	502	924	1197	1330	1039	986	624	324	110	7280
ELK MOUNTAIN		86	122	349	651	1002	1228	1324	1123	1135	813	530	258	8621
ENCAMPMENT 10 ESE		75	122	333	632	1002	1259	1336	1140	1132	786	496	251	8564
EVANSTON 1 E		82	145	366	694	1080	1358	1432	1210	1181	840	549	285	9222
FARSON		69	141	396	756	1221	1593	1702	1397	1249	825	530	260	10139
GILLETTE 2 E		26	45	234	530	969	1240	1389	1086	1029	678	382	146	7754
GILLETTE 18 SW		35	52	252	561	984	1246	1380	1092	1051	705	409	162	7929
GLENROCK 5 ESE		9	11	193	484	873	1110	1221	972	936	600	315	84	6808
GREEN RIVER		17	69	284	623	1038	1352	1445	1170	1063	711	403	167	8342
HEART MOUNTAIN		31	73	272	561	996	1274	1411	1064	989	654	355	139	7819
JACKSON		133	203	429	756	1125	1469	1519	1254	1200	834	570	330	9822
KAYCEE		21	36	251	570	978	1231	1364	1053	995	666	378	145	7688
LAGRANGE		9	23	198	512	894	1116	1228	977	949	618	327	104	6955
LANDER WSO	//R	12	32	241	564	1026	1296	1407	1100	1020	681	384	142	7905
LARAMIE FAA AIRPORT		72	113	354	688	1071	1302	1383	1170	1166	834	539	239	8931
LOVELL		14	30	250	583	1008	1352	1504	1117	980	618	299	103	7858
MEDICINE BOW		52	98	345	685	1050	1293	1367	1140	1113	768	481	202	8594
MIDWEST 1 SW		10	14	188	481	915	1175	1296	1002	939	609	313	98	7040
MOORCROFT		33	45	253	570	1005	1308	1442	1145	1057	690	380	148	8076
MORAN 5 WNW	R	185	248	483	806	1179	1525	1612	1338	1324	954	657	390	10701
NEWCASTLE		13	20	213	508	948	1231	1358	1078	992	618	298	102	7379
PATHFINDER DAM		23	38	244	533	939	1228	1352	1117	1063	699	400	155	7791
PAVILLION		13	37	240	561	1011	1293	1398	1061	970	639	350	126	7699
PHILLIPS		16	19	198	477	855	1057	1163	944	933	618	338	119	6737
POWELL		8	27	217	502	945	1249	1380	1030	930	588	279	107	7262
REDBIRD 1 NW		17	25	199	521	960	1265	1392	1064	961	591	290	91	7376
RIVERTON		12	35	263	620	1107	1454	1575	1193	1023	660	353	120	8415
ROCHELLE 3 E		21	50	244	561	963	1287	1411	1086	1023	690	370	128	7834
ROCK SPRINGS		9	48	263	595	996	1265	1345	1084	1020	693	403	155	7876
ROCK SPRINGS FAA AP		13	56	279	626	1047	1318	1395	1134	1091	756	456	185	8356
SARATOGA		31	78	308	636	1014	1265	1342	1117	1091	750	453	187	8272
SEMINOE DAM		15	51	266	592	1008	1268	1380	1140	1110	753	446	172	8201
SHERIDAN WSO	//R	34	45	255	558	975	1252	1411	1089	1026	675	372	149	7841

WYOMING

HEATING DEGREE DAY NORMALS (BASE 65 DEG F)

STATION	JUL	AUG	SEP	OCT	NOV	DEC	JAN	FEB	MAR	APR	MAY	JUN	ANN
SHERIDAN FIELD STA	37	54	256	570	1002	1299	1473	1148	1063	669	362	140	8073
SUNDANCE	37	59	264	561	999	1287	1423	1140	1079	711	406	157	8123
THERMOPOLIS	9	35	190	512	954	1259	1380	1025	918	573	296	107	7258
TORRINGTON EXP FARM	5	14	174	490	870	1125	1225	938	884	552	268	69	6614
UPTON	26	40	258	598	1056	1386	1516	1193	1076	684	372	135	8340
WHALEN DAM	10	12	193	484	894	1150	1274	988	936	594	297	93	6925
WHEATLAND 4 N	12	19	182	458	828	1045	1163	927	890	573	295	96	6488
WORLAND	15	28	244	574	1029	1398	1569	1154	983	612	295	103	8004
YELLOWSTONE PARK //	83	151	379	685	1065	1339	1423	1148	1172	837	546	296	9124
YODER	8	12	174	465	855	1101	1209	946	884	564	286	85	6589

WYOMING

COOLING DEGREE DAY NORMALS (BASE 65 DEG F)

STATION		JAN	FEB	MAR	APR	MAY	JUN	JUL	AUG	SEP	OCT	NOV	DEC	ANN
AFTON		0	0	0	0	0	0	10	16	0	0	0	0	26
ALBIN		0	0	0	0	0	90	199	150	52	5	0	0	496
ALTA 1 NNW		0	0	0	0	0	8	37	38	9	0	0	0	92
ALVA 5 SE		0	0	0	0	0	16	96	70	14	0	0	0	196
ARCHER		0	0	0	0	0	52	156	108	30	0	0	0	346
BASIN		0	0	0	0	10	109	269	194	39	0	0	0	621
BONDURANT 3 NW		0	0	0	0	0	0	0	0	0	0	0	0	0
BORDER 3 N		0	0	0	0	0	0	21	23	0	0	0	0	44
BOYSEN DAM		0	0	0	0	15	140	332	267	85	0	0	0	839
BUFFALO BILL DAM		0	0	0	0	0	21	93	96	52	6	0	0	268
CARPENTER 3 E		0	0	0	0	0	67	186	127	30	0	0	0	410
CASPER 2 E		0	0	0	0	0	84	231	168	38	0	0	0	521
CASPER WSO //R		0	0	0	0	0	68	199	148	42	0	0	0	457
CHEYENNE WSO //R		0	0	0	0	0	49	145	93	22	0	0	0	309
CHUGWATER		0	0	0	0	0	55	153	107	22	0	0	0	337
CODY		0	0	0	0	0	61	172	145	36	0	0	0	414
COLONY		0	0	0	0	7	100	255	224	66	6	0	0	658
CRANDALL CREEK		0	0	0	0	0	0	8	14	6	0	0	0	28
DEAVER		0	0	0	0	7	78	204	154	28	0	0	0	471
DILLINGER		0	0	0	0	0	63	205	154	33	0	0	0	455
DIVERSION DAM		0	0	0	0	0	36	124	82	18	0	0	0	260
DIXON		0	0	0	0	0	15	57	29	6	0	0	0	107
DULL CENTER 1 SE		0	0	0	0	0	104	257	200	52	0	0	0	613
ELK MOUNTAIN		0	0	0	0	0	12	39	10	0	0	0	0	61
ENCAMPMENT 10 ESE		0	0	0	0	0	8	44	32	6	0	0	0	90
EVANSTON 1 E		0	0	0	0	0	6	29	27	0	0	0	0	62
FARSON		0	0	0	0	0	8	32	21	0	0	0	0	61
GILLETTE 2 E		0	0	0	0	0	65	199	169	45	0	0	0	478
GILLETTE 18 SW		0	0	0	0	0	51	168	132	30	0	0	0	381
GLENROCK 5 ESE		0	0	0	0	0	84	251	185	49	0	0	0	569
GREEN RIVER		0	0	0	0	0	38	116	91	11	0	0	0	256
HEART MOUNTAIN		0	0	0	0	5	52	145	117	32	0	0	0	351
JACKSON		0	0	0	0	0	0	0	8	0	0	0	0	8
KAYCEE		0	0	0	0	0	52	164	110	23	0	0	0	349
LAGRANGE		0	0	0	0	0	92	219	165	36	0	0	0	512
LANDER WSO //R		0	0	0	0	0	61	192	143	40	0	0	0	436
LARAMIE FAA AIRPORT		0	0	0	0	0	11	53	20	6	0	0	0	90
LOVELL		0	0	0	0	7	88	225	154	25	0	0	0	499
MEDICINE BOW		0	0	0	0	0	16	55	24	0	0	0	0	95
MIDWEST 1 SW		0	0	0	0	0	89	242	182	50	0	0	0	563
MOORCROFT		0	0	0	0	0	76	210	169	40	0	0	0	495
MORAN 5 WNW R		0	0	0	0	0	0	6	9	0	0	0	0	15
NEWCASTLE		0	0	0	0	7	123	270	203	54	0	0	0	657
PATHFINDER DAM		0	0	0	0	0	65	187	134	37	0	0	0	423
PAVILLION		0	0	0	0	0	54	155	108	27	0	0	0	344
PHILLIPS		0	0	0	0	0	74	190	133	39	0	0	0	436
POWELL		0	0	0	0	9	110	241	192	58	0	0	0	610
REDBIRD 1 NW		0	0	0	0	5	109	271	208	49	0	0	0	642
RIVERTON		0	0	0	0	0	57	176	109	17	0	0	0	359
ROCHELLE 3 E		0	0	0	0	0	95	235	199	37	0	0	0	566
ROCK SPRINGS		0	0	0	0	0	41	136	92	14	0	0	0	283
ROCK SPRINGS FAA AP		0	0	0	0	0	32	112	81	12	0	0	0	237
SARATOGA		0	0	0	0	0	22	78	41	0	0	0	0	141
SEMINOE DAM		0	0	0	0	0	43	130	94	23	0	0	0	290
SHERIDAN WSO //R		0	0	0	0	0	53	186	147	33	0	0	0	419

WYOMING

COOLING DEGREE DAY NORMALS (BASE 65 DEG F)

STATION	JAN	FEB	MAR	APR	MAY	JUN	JUL	AUG	SEP	OCT	NOV	DEC	ANN
SHERIDAN FIELD STA	0	0	0	0	6	56	201	166	37	0	0	0	466
SUNDANCE	0	0	0	0	0	49	154	137	39	0	0	0	379
THERMOPOLIS	0	0	0	0	8	95	232	187	31	0	0	0	553
TORRINGTON EXP FARM	0	0	0	0	7	111	244	172	33	0	0	0	567
UPTON	0	0	0	0	0	78	206	152	39	0	0	0	475
WHALEN DAM	0	0	0	0	6	117	261	201	64	0	0	0	649
WHEATLAND 4 N	0	0	0	0	9	105	242	174	53	5	0	0	588
WORLAND	0	0	0	0	7	97	232	146	25	0	0	0	507
YELLOWSTONE PARK //	0	0	0	0	0	8	40	39	7	0	0	0	94
YODER	0	0	0	0	7	112	247	173	39	0	0	0	578

STATE-STATION NUMBER	STN TYP	NAME		LATITUDE DEG-MIN	LONGITUDE DEG-MIN	ELEVATION (FT)
48-0027	13	AFTON		N 4244	W 11056	6210
48-0080	13	ALBIN		N 4125	W 10406	5345
48-0140	13	ALTA 1 NNW		N 4346	W 11102	6431
48-0200	13	ALVA 5 SE		N 4439	W 10421	4390
48-0270	13	ARCHER		N 4109	W 10439	6010
48-0540	13	BASIN		N 4423	W 10803	3837
48-0695	12	BIG PINEY		N 4232	W 11007	6821
48-0865	13	BONDURANT 3 NW		N 4314	W 11026	6504
48-0915	13	BORDER 3 N		N 4215	W 11102	6120
48-1000	13	BOYSEN DAM		N 4325	W 10811	4642
48-1175	13	BUFFALO BILL DAM		N 4430	W 10911	5156
48-1547	13	CARPENTER 3 E		N 4102	W 10418	5390
48-1565	13	CASPER 2 E		N 4251	W 10616	5195
48-1570	13	CASPER WSO	//R	N 4255	W 10628	5338
48-1675	13	CHEYENNE WSO	//R	N 4109	W 10449	6126
48-1730	13	CHUGWATER		N 4145	W 10449	5282
48-1816	12	CLEARMONT 5 SW		N 4435	W 10627	4055
48-1840	13	CODY		N 4433	W 10904	4990
48-1905	13	COLONY		N 4456	W 10412	3553
48-2135	13	CRANDALL CREEK		N 4454	W 10940	6600
48-2415	13	DEAVER		N 4453	W 10836	4105
48-2580	13	DILLINGER		N 4407	W 10507	4306
48-2595	13	DIVERSION DAM		N 4314	W 10856	5574
48-2610	13	DIXON		N 4102	W 10732	6360
48-2680	12	DOUBLE FOUR RANCH		N 4211	W 10524	6200
48-2715	12	DUBOIS		N 4333	W 10937	6917
48-2725	13	DULL CENTER 1 SE		N 4325	W 10457	4415
48-2995	13	ELK MOUNTAIN		N 4141	W 10625	7265
48-3045	13	ENCAMPMENT 10 ESE		N 4111	W 10637	7387
48-3100	13	EVANSTON 1 E		N 4116	W 11057	6780
48-3170	13	FARSON		N 4207	W 10927	6591
48-3490	12	FT LARAMIE 11 NNW		N 4223	W 10432	4760
48-3570	12	FORT WASHAKIE 2 S		N 4259	W 10853	5592
48-3855	13	GILLETTE 2 E		N 4417	W 10528	4556
48-3865	13	GILLETTE 18 SW		N 4405	W 10543	4903
48-3950	13	GLENROCK 5 ESE		N 4250	W 10547	4948
48-3960	12	GLENROCK 14 SSE		N 4240	W 10549	6427
48-4065	13	GREEN RIVER		N 4132	W 10929	6089
48-4080	12	GREYBULL		N 4430	W 10803	3810
48-4411	13	HEART MOUNTAIN		N 4441	W 10857	4790
48-4440	12	HECLA		N 4109	W 10511	6800
48-4910	13	JACKSON		N 4328	W 11046	6244
48-5055	13	KAYCEE		N 4343	W 10638	4660
48-5085	12	KEELINE		N 4245	W 10445	5280
48-5105	12	KEMMERER 4 SW		N 4146	W 11036	6936
48-5260	13	LAGRANGE		N 4138	W 10410	4582
48-5345	12	LAKE YELLOWSTONE		N 4433	W 11024	7762
48-5390	13	LANDER WSO	//R	N 4249	W 10844	5563
48-5415	13	LARAMIE FAA AIRPORT		N 4119	W 10541	7266
48-5525	12	LEO 6 SW		N 4211	W 10650	6000

48 — WYOMING

STATE-STATION NUMBER	STN TYP	NAME		LATITUDE DEG-MIN	LONGITUDE DEG-MIN	ELEVATION (FT)
48-5770	13	LOVELL		N 4450	W 10824	3837
48-6120	13	MEDICINE BOW		N 4154	W 10612	6570
48-6195	13	MIDWEST 1 SW		N 4325	W 10617	4840
48-6395	13	MOORCROFT		N 4416	W 10457	4275
48-6440	13	MORAN 5 WNW	R	N 4351	W 11035	6798
48-6660	13	NEWCASTLE		N 4351	W 10413	4265
48-7079	12	PARKMAN 5 WNW		N 4459	W 10726	4430
48-7105	13	PATHFINDER DAM		N 4228	W 10651	5930
48-7115	13	PAVILLION		N 4315	W 10841	5440
48-7200	13	PHILLIPS		N 4138	W 10429	4982
48-7260	12	PINEDALE		N 4252	W 10952	7175
48-7380	13	POWELL		N 4445	W 10846	4378
48-7473	12	RAIRDEN 2 WSW		N 4411	W 10757	4007
48-7555	13	REDBIRD 1 NW		N 4316	W 10418	3880
48-7760	13	RIVERTON		N 4301	W 10823	4954
48-7810	13	ROCHELLE 3 E		N 4336	W 10454	4496
48-7840	13	ROCK SPRINGS		N 4135	W 10913	6367
48-7845	13	ROCK SPRINGS FAA AP		N 4136	W 10904	6741
48-7990	13	SARATOGA		N 4127	W 10648	6790
48-8070	13	SEMINOE DAM		N 4208	W 10653	6838
48-8155	13	SHERIDAN WSO	//R	N 4446	W 10658	3964
48-8160	13	SHERIDAN FIELD STA		N 4451	W 10652	3800
48-8209	12	SHOSHONI		N 4314	W 10807	4830
48-8385	12	SOUTH PASS CITY		N 4228	W 10848	7875
48-8705	13	SUNDANCE		N 4424	W 10421	4750
48-8845	12	TENNYSON		N 4121	W 10423	5615
48-8880	13	THERMOPOLIS		N 4339	W 10812	4313
48-8995	13	TORRINGTON EXP FARM		N 4205	W 10413	4098
48-9025	12	TOWER FALLS		N 4455	W 11025	6266
48-9205	13	UPTON		N 4406	W 10437	4261
48-9207	12	UPTON 13 SW		N 4356	W 10446	4780
48-9604	13	WHALEN DAM		N 4215	W 10438	4294
48-9615	13	WHEATLAND 4 N		N 4207	W 10457	4638
48-9770	13	WORLAND		N 4401	W 10758	4061
48-9905	13	YELLOWSTONE PARK	//	N 4458	W 11042	6230
48-9925	13	YODER		N 4156	W 10418	4231

48 – WYOMING

LEGEND

O = TEMPERATURE ONLY
X = PRECIPITATION ONLY
⊗ = TEMPERATURE & PRECIPITATION

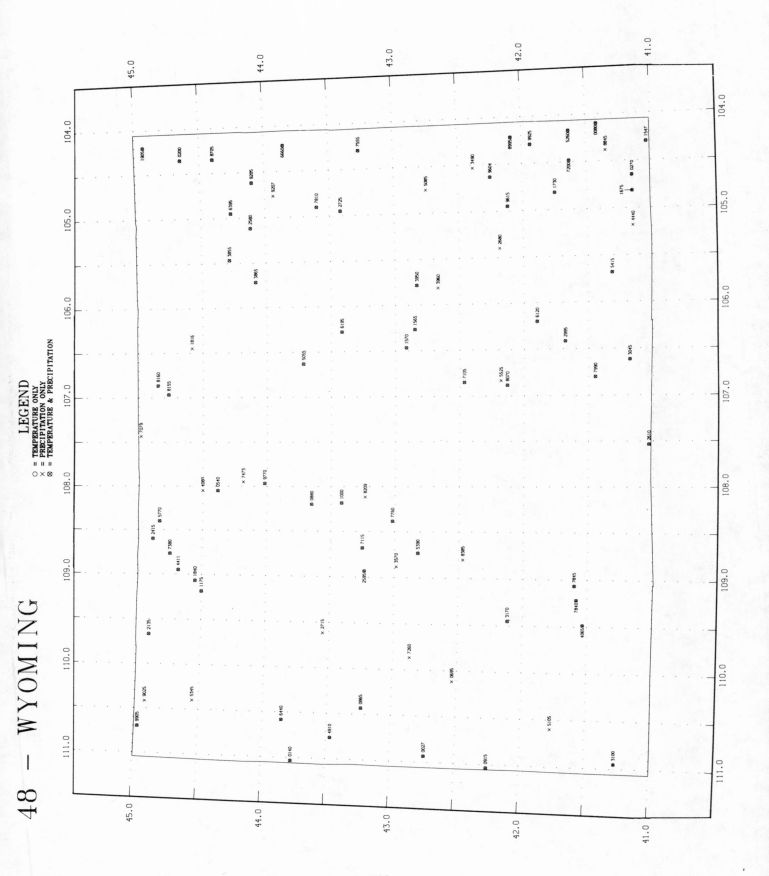

The concept of heating and cooling degree days

Heating Degree Days

Early this century heating engineers developed the concept of *heating degree days* as a useful index of heating fuel requirements. They found that when the daily mean temperature is lower than 65 degrees, most buildings require heat to maintain an inside temperature of 70 degrees.* The daily mean temperature is obtained by adding together the maximum and minimum temperatures reported for the day and dividing the total by two. Each degree of mean temperature below 65 is counted as one heating degree day. Thus, if the maximum temperature is 70 degrees and the minimum 52 degrees, four heating degree days would be produced. (i.e., 70 + 52 = 122; divided by 2 = 61; 65 - 61 = 4). If the daily mean temperature is 65 degrees or higher, the heating degree day total is zero.

For every additional heating degree day, more fuel is needed to maintain a comfortable 70 degrees indoors. A day with a mean temperature of 35 degrees—30 heating degree days—would require twice as as much fuel as a day with a mean temperature of 50—15 heating degree days, assuming, of course, similar meterological conditions such as wind speed and cloudiness.

So valuable has the heating degree concept become that daily, monthly and seasonal totals are routinely computed for all temperature observing stations in the National Weather Service's network. Daily figures are used by fuel companies for evaluation of fuel use rates and for efficient scheduling of deliveries. For example, if a heating system is known to use one gallon of fuel for every 5 heating degree days, oil deliveries will be scheduled to meet this burning rate. Gas and Electric Company dispatchers use the data to anticipate demand and to implement priority procedures when demand exceeds capacity.

The amount of heat required to maintain a certain temperature level is proportional to the heating degree days. A fuel bill usually will be twice as high for a month with 1,000 heating degree days as for a month with 500. For example, it can be estimated that about four times as much fuel will be required to heat a building in Chicago, where the annual average is about 6,100 heating degree days as it would to heat a building in New Orleans, where the average is about 1,500. All this is true only if building construction and living habits in these areas are similar. Since such factors are not constant, these ratios must be modified by actual experience. The use of heating degree days has the advantage that consumption rates are fairly constant, i.e., fuel consumed for 100 degree days is about the same whether the 100 heating degree days were accumulated on only three or four days or were spread over seven or eight days.

Accumulation of temperature data for a particular location has resulted in the establishment of "normal" values based on thirty years of record. Maps and tables of heating degree day normals, are published by the National Oceanic and Atmospheric Administration's Environmental Data Service (EDS). The maps are useful only for broad general comparisons, because temperatures, even in a small area, vary considerably depending on differences in altitude, exposure, wind, and other circumstances.

Heating degree day comparisons within a single area are the most accurate. For example, March heating degree day totals in the Midwest average about 70 percent of those for January. In Chicago, the coldest six months in order of decreasing coldness are January, December, February, March, November and April. Annual heating degree day data are published by heating season which runs from July of one year through June of the next year. This enables direct comparison of seasonal heating degree day data and seasonal heating fuel requirements.

* All temperature values in the above material are °F

Cooling Degree Days

The cooling degree day statistic—summer sister of the familiar heating degree day—serves as an index of air-conditioning requirements during the year's warmest months.

According to experts, the need for air-conditioning begins to be felt when the daily maximum temperature climbs to 80 degrees and higher. The cooling degree day is therefore a kind of mirror image of the heating degree day. After obtaining the daily mean temperature, by adding together the day's high and low temperatures and dividing the total by two, the base 65 is subtracted from the resulting figure to determine the cooling degree day total. For example, a day with a maximum temperature of 82 degrees and a minimum of 60 would produce six cooling degree days. ($82 + 60 = 142$; 142 divided by $2 = 71 - 65 = 6$). If the daily mean temperature is 65 degrees or lower, the cooling degree day total is zero.

The greater the number of cooling days, the more energy is required to maintain indoor temperatures at a comfortable level. However, the relationship between cooling degree days and energy use is less precise than that between heating degree days and fuel consumption. There is considerable controversy among meterologists, as well as air-conditioning engineers, as to what meteorological variables are most closely related to energy consumption by air-conditioning systems. Many experts argue that because high humidity levels make people feel more uncomfortable as temperatures rise, some measure of moisture should be included in calculating energy needs for air-conditioning. The Temperature-Humidity Index has been suggested as an alterative basis for calculating cooling degree days. In addition to humidity some experts feel there are other factors, such as cloudiness and wind speed, that should be included in computation of energy needs for air conditioning. All agree, however, that there is a need for a more effective measure of the influence of weather on air-conditioning loads.

Until a definitive study of the problem is conducted, NOAA's EDS is continuing to use and publish statistics based on simple cooling degree day calculations, employing air temperatures measured at National Weather Service Offices and cooperating stations throughout the country. As with heating degree days, normals of cooling degree days have been established, based on thirty years of record.

PLEASE NOTE THAT HEATING AND COOLING DEGREE DAYS DO NOT CANCEL EACH OTHER OUT. TOTALS FOR EACH ARE ACCUMUATED INDEPENDENTLY.